THE NEW PALGRAVE
DICTIONARY OF ECONOMICS
SECOND EDITION

THE NEW PALGRAVE
DICTIONARY OF ECONOMICS
SECOND EDITION

Edited by Steven N. Durlauf and Lawrence E. Blume

Volume 2   **command economy – epistemic game theory**

© Macmillan Publishers Ltd. 2008

This edition published 2008 by
PALGRAVE MACMILLAN
Houndmills, Basingstoke, Hampshire RG21 6XS and
175 Fifth Avenue, New York, N.Y. 10010
Companies and representatives throughout the world

PALGRAVE MACMILLAN is the global academic imprint of the Palgrave
Macmillan division of St. Martin's Press, LLC and of Palgrave Macmillan Ltd.
Macmillan® is a registered trademark in the United States, United Kingdom
and other countries. Palgrave is a registered trademark in the European
Union and other countries.

ISBN-10 0-230-22638-8   Volume 2
ISBN-13 978-0-230-22638-8   Volume 2

ISBN-10 0-333-78676-9   8-Volume set
ISBN-13 978-0-333-78676-5   8-Volume set

This book is printed on paper suitable for recycling and made from fully
managed and sustained forest sources. Logging, pulping and manufacturing processes
are expected to conform to the environmental regulations of the country of origin.

A catalogue record for this book is available from the British Library.

Library of Congress Cataloging-in-Publication Data

The new Palgrave dictionary of economics / edited by Steven N. Durlauf
and Lawrence E. Blume. – 2nd ed.
      v. cm.
   Rev. ed. of: The New Palgrave : a dictionary of economics. 1987.
   Includes bibliographical references.
   ISBN 978-0-333-78676-5 (alk. paper)

   1. Economics – Dictionaries.   I. Durlauf, Steven N.   II. Blume,
Lawrence.   III. New Palgrave.

HB61.N49 2008
330.03–dc22

                                                                    2007047205

10   9   8   7   6   5   4 · 3   2   1
17   16   15   14   13   12   11   10   09   08

Typesetting and XML coding by Macmillan Publishing Solutions, Bangalore, India
Printed in China

Contents

Publishing history

First edition of *Dictionary of Political Economy*,
edited by Robert Harry Inglis Palgrave, in three volumes:

Volume I, printed 1894.
Reprinted pages 1–256 with corrections, 1901, 1909.
Reprinted with corrections, 1915, 1919.

Volume II, printed 1896.
Reprinted 1900.
Reprinted with corrections, 1910, 1915.

Volume III, printed 1899.
Reprinted 1901.
Corrected with appendix, 1908.
Reprinted with corrections, 1910, 1913.
Reprinted, 1918.

New edition, retitled *Palgrave's Dictionary of Political Economy*,
edited by Henry Higgs, in three volumes:

Volume I, printed 1925.
Reprinted 1926.

Volume II, printed 1923.
Reprinted 1925, 1926.

Volume III, printed February 1926.
Reprinted May 1926.

The New Palgrave: A Dictionary of Economics,
edited by John Eatwell, Murray Milgate and Peter Newman.
Published in four volumes.

First published 1987.
Reprinted 1988 (twice).
Reprinted with corrections 1991.
Reprinted 1994, 1996.

First published in paperback 1998.
Reprinted 1999, 2003, 2004.

The New Palgrave Dictionary of Economics 2ⁿᵈ edition,
edited by Steven N. Durlauf and Lawrence E. Blume.
Published in eight volumes

List of entries A–Z

C
(CONTINUED)

command economy

A command economy is one in which the coordination of economic activity, essential to the viability and functioning of a complex social economy, is undertaken through administrative means – commands, directives, targets and regulations – rather than by a market mechanism. A complex social economy is one involving multiple significant interdependencies among economic agents, including significant division of labour and exchange among production units, rendering the viability of any unit dependent on proper coordination with, and functioning of, many others.

Economic agents in a command economy, and in particular production organizations, operate primarily by virtue of specific directives from higher authority in an administrative/political hierarchy, that is, under the 'command principle'. Thus the life cycle and activity of enterprises and firms, their production of output and employment of resources, adjustment to disturbances, and the coordination between them are primarily governed by decisions taken by superior organs responsible for managing those units' roles in the economic system. One of the most distinctive features of such an economy is the setting of the firm's production targets by higher directive, often in fine detail. The administrative means used include planning, material balances, quotas, rationing, technical coefficients, budgetary controls and limits, price and wage controls, and other techniques aimed at limiting the discretion of subordinate operational units/firms. The command principle strives to fully and effectively replace the operation of market forces in the key industrial and developmental sectors of the economy, and render the remaining (peripheral) markets manipulable and subordinate to political direction. Thus the command principle is likely to clash with the operation of market forces, yet a command economy may nonetheless contain and rely on the market mechanism in some of its sectors and areas, for example, influencing labour allocation, or stimulating small-scale private production of some consumables.

The term 'command economy' comes from the German *Befehlswirtschaft*, and was originally applied to the Nazi economy, which shared many formal similarities with that of the Soviet Union. It has received its fullest development in the analysis of the economic system of the Soviet Union, particularly under Stalin, although it has been applied to wartime administration of the US economy (1942–6; see Higgs, 1992), the Mormon economic system in mid-19th century Utah (Grossman, 2000), and the Inca production system in the 16th century Andes (La Lone and La Lone, 1987). Synonymous or near-synonymous terms include 'centrally planned economy', 'centrally administered economy',

'administrative command economy', 'Soviet-type economy', 'bureaucratic economy' and 'Stalinist economy'.

The command economy's conceptual origins go back to the Viennese economist Otto Neurath, who in the years before and after the First World War developed an extreme version (to the point of moneylessness) based chiefly on prior experience with wartime economies (Raupach, 1966). The concept of the command economy has since become a central conceptual framework in the analysis of economic systems, as it captures a logically coherent alternative to 'the market' as a way of organizing socially complex economic activity and interaction. The Soviet Union provided the most complete, and for a while successful, example of a command economy as a working alternative to a market system. Indeed, apart from the relatively short-lived Nazi case, and even briefer ones under emergency conditions in some other countries, especially in wartime, actual instances of command economies are virtually limited to Communist-ruled countries, with the USSR as the prototype and prime exemplar. Thus, what follows is mainly inspired by the Soviet example (Ericson, 1991) as it existed, essentially little altered since its appearance in the 1930s, until its collapse in the aftermath of President Gorbachev's *perestroika*, begun in 1987.

Nature of the command economy

The seminal analysis of the nature, characteristics, and problems of a command economy is Grossman (1963).

Any complex social economy must, for its very survival, maintain at least a 'tolerable' micro-balance, 'that minimal degree of coordination of the activities of the separate units (firms) which assures a tolerably good correspondence between the supply of individual producer and consumer goods and the effective demand for them' (Grossman, 1963, p. 101). In such an economy, appropriate balance can be achieved through decentralized, market-based (monetized, price-mediated) interaction of autonomous units, or by virtue of explicit specific coordinating directives (commands, targets) from some higher authorities. While the former is characteristic of a market economic system, the latter is defining of a 'command economy'. In the latter operational-level units (for example, firms) must merely 'implement' commands; they become 'executants' of plans and directives from above, plans which must insure balance through the coherence and consistency of the instructions they give. Thus the command mechanism requires relative centralization and severe restriction on the autonomy of subordinate operational units. It derives from the overwhelming priority of social goals, and requires the severe limitation, if not total destruction, of autonomous social and economic powers and the enforcement of strict obedience to directives.

A command economy is hence a creature of state authority, whose marks it bears and by whose hand it evolves, exists and survives. Command economies are imposed, whether through external duress or imitation, or indigenously in order to achieve specific purposes such as (*a*) maximum resource mobilization towards urgent and overriding national objectives, such as rapid industrialization or the prosecution of war; (*b*) radical transformation of the socio-economic system in a collectivist direction based on ideological tenets and power-political imperatives; and (*c*), not least, curing the disorganization of a market economy brought about by price control, possibly occasioned by inflationary pressure arising from (*a*) and/or (*b*).

The command economy therefore requires a formal, centralized, administrative hierarchy staffed by a bureaucracy, and it also needs to be embedded in (at least) an authoritarian, highly centralized polity if it is not to dissolve or degenerate into something else. And that bureaucracy, if it is to effectively implement the command principle, must exercise full control and discretion, if not necessarily formal ownership, with respect to the creation, use and disposal of all productive property and assets. At the same time, each office or firm and every economic actor within the command structure holds interests which, if only in part, do not coincide with those of superiors or of the overall leadership. This generates important problems of vested interests, principal–agent interaction, incentive provision, and general enforcement of the leadership's will, and calls for a variety of monitoring organizations (party, police, banks, and so on). The term 'command' must not be taken to preclude self-serving behaviour, bureaucratic politics, bargaining between superiors and subordinates, corruption, peculation and (dis)simulation. On the contrary, such behaviour tends to be widespread in a command economy; yet the concept of a 'command economy' remains valid so long as, in the main, authority relations and not a market mechanism govern the allocation of resources.

When not externally imposed, command economies typically arise from a millennialist elite, with unique access to 'the truth', achieving the political power to impose its will, while facing a crisis of apparently overwhelming proportions. The perception of a life-threatening crisis, driving the need for massive mobilization of all social resources and rendering potentially disastrous any hesitation or dissent, any questioning of ways and means, naturally leads, pushed by the 'logic of events', to the usurpation of all power of discretion, all legitimate authority, by the 'knowing' elite, which then becomes responsible for all that is done or not done in the society and the economy. The crisis may be artificial or real ('hostile encirclement'), externally or internally imposed (the need to industrialize, to 'catch up'), but it requires moving resources rapidly and massively, forcing new activities and interactions in the face of severe scarcities, of shortage of competent personnel, of massive uncertainties, and of strongly held, stark priorities. Indeed, a sense of overwhelming urgency and need for haste drove the elite of the Soviet Union in the 1930s to test and establish, through trial and error over several decades, the institutional structure of a 'command economy', albeit *less than absolute* from both necessity and choice (for example, due to the 'lessons' of 'War Communism') (Grossman, 1962; Zaleski, 1968).

Consequences of command

Rational application of the command principle calls for planning, which is basically of two types. Longer-term, developmental planning expresses the leadership's politico-economic strategy (for example, five-year and 'perspective' plans); shorter-term, coordinative planning (annual, quarterly, monthly, ten-day) ideally translates the strategy into resource allocation while aiming to match resource requirements and availabilities for individual inputs, goods, and so on, in a sufficiently disaggregated way for given time periods and locations. The task of elemental coordination, of micro-balance, so effortlessly accomplished by any functioning (however poorly) market system, is overwhelmingly large, and grows rapidly with industrialization and economic development, both of which lead to exponential growth of the complexity of the economy, and hence of the planning problem. With centralization and the abandonment of markets comes the need for massive, detailed coordinative planning, for 'making ends meet' in the expanding web of interconnections that must be maintained for economic life to continue. Coordinative planning serves, therefore, as the basis for specific operational directives to producers and users, thereby implementing the command principle to achieve the prime imperative of a social economy – 'balance'.

> It is *this* task that in fact consumes the largest part of the so-called planning in the command economy … Coordinative planning as it is conducted in the Soviet Union does little by way of consciously steering the economy's development or finding efficient patterns of resource allocation. Its overwhelming concern is simply to equate both sides of each 'material balance' by whatever procedure seems to be most expeditious. (Grossman, 1963, p. 108)

A major problem is that detailed planning and the corresponding directives are often late, are insufficiently detailed, may lack the requisite information, hence often cannot be effectively coordinated, and owing to their rigidity are peculiarly vulnerable to uncertainty (Ericson, 1983). Information in the command sector, by the logic of the system, tends to flow vertically up and down the administrative hierarchy rather than horizontally between buyer and seller, adding to difficulties of demand–supply coordination by informationally isolating operational units from their suppliers and users. In addition, problems of motivation, accountability (down as well as up), inappropriate decision-making parameters, and divergent interests complicate the procedure. Even at best, this manner of resource allocation can hope to attain only internal consistency (in the sense of effectively matching partially disaggregated requirements and availabilities) but not a higher order of economic efficiency. Economic calculation in pursuit of efficiency enters, if at all, at the project-planning stage, and not short-term resource allocation and use.

These problems are aggravated by the logic of haste that drove imposition of the command economy – 'the pressing contrast between urgent political goals and available resources'. (Grossman, 1963, p. 108) The necessary attention to the growing problem of balance further militates against any effort to consider developmental objectives or efficiency in making allocative decisions, so that a further bias against allocative efficiency is built into the command economy. Coupled with limited ability to gather, filter, process, and communicate information, and to compute solutions to planning problems, this creates a fundamental and growing inability to acceptably solve the underlying coordination problem, and hence further undermines any consideration of efficiency.

The logic of 'command' has a number of other consequences reflected in the institutions of such an economy. Planning in a command economy must be largely in *physical terms* due to the crucial importance of balance. The bottom line of the planning process must be available physical units of required inputs, in appropriate assortment, quantity and timing, necessitating physical targets for production and input utilization. Thus tens of thousands of materials and equipment balances must be drawn up and coordinated for each plan period, and then broken down and allocated in directives to specific implementers. And, to be directly usable, these must be in physical or crypto-physical (constant price) units that directly relate to the production processes being coordinated. Using economic-value units requires flexible and changing, marginal scarcity-based prices for valuation, as well as giving significant autonomy to subordinate units that inevitably then will make the trade-offs in assortment, quantity and timing within planned constraints on values (that is, 'budgets'). Hence, such valuations pose a fundamental challenge to the command economy.

Planning in physical terms, however, leads to 'enormous waste and inefficiency, to production for waste as much as for use' (Grossman, 1963, p. 110). There are at least three fundamental sources of this elemental waste: *grossness*, *aggregation*, and *unit of measure*. The need for these arises from the overwhelming complexity of the task of planning for, and directing the operation of, a complex social economy and the necessarily limited information gathering, processing, and dissemination capabilities of any economic agent or agency. However, the emphasis on *gross output* leads to 'input intensiveness', waste, and ignoring cost considerations. *Aggregation* leads to persistent subcategory imbalance in assortment, quality, type, timing, and so on, while *units of measurement* determine suboptimization objectives, distorting implementation decisions,

particularly when they are, for material balance reasons, input oriented. Thus each of these is essential for the feasibility of directive central planning, of the command mechanism, yet each loses, or destroys, essential information for the 'proper' (in the eyes of the system directors) implementation of plans, and opens space for creative interpretation of instructions/commands, and hence for 'suboptimization' by implementing units whose interests are not perfectly aligned with those of the centre (Nove, 1977). While the command mechanism logically requires unauthorized initiative to be forbidden, and strictly punished when exercised, the size of the task it faces inevitably opens the opportunity, indeed often the need, for such unauthorized initiative. Thus the physical quantity planning required by the command economy to maintain minimal functional 'balance' contains its own antithesis, unleashing forces that undermine the consistency of the plan and the coherence and balancedness of its realization. This fundamental contradiction lies behind most of the critical problems of the command economy in the Soviet Union and the myriad efforts to resolve them within the framework of the command mechanism that comprise the endless waves of reform following victory in the Great Fatherland War of 1941–5.

The 'logic of command' thus imposes a need to restrict autonomy, to restrict the capability of economic units to pursue any other than 'planned' or commanded purposes: economic agents must not have the capability to autonomously acquire and deploy resources for any purposes outside the plan. Comprehensive material balance planning and centralized materials and equipment allocation provide a necessary component, but one that is insufficient unless resources, including human, are denied the capability of autonomous movement and application. Severe restrictions on labour mobility, albeit not as severe as under Stalin, are required, as are comprehensive restrictions on the use of any 'generalized command over goods and services' – that is, money – that might be used to alter their patterns of allocation and use in the economy. The system must be substantially demonetized in order to '... constrict the ... range of choice in the face of the state's demands' (Grossman, 1966, p. 232).

Thus money must be deprived of 'moneyness' and prices must be kept 'passive', as mere accounting and measurement units. According to the logic of the command economy, the availability of money and the prices at which commodities and products are provided should have no essential impact on the allocation of goods and services, or on the nature and direction of economic/industrial development; all real activity is preordained in the plan and its subsequent implementing commands. The role of money is then to facilitate monitoring of commanded performance through the financial flows it generates. Thus monetary prices do not, and indeed should not, reflect to a substantial degree social goals and priorities; they merely reveal and measure the flow of commanded activity. Producer prices (and most retail prices), wages, prices of foreign currencies, and so on are generally centrally set and controlled, often remaining fixed for long periods of time. Micro-disequilibria naturally abound, while the widely perceived dubious meaningfulness of such prices and the administrative allocation of most producer goods in physical terms combine to sustain the system of detailed production plans and directives in terms of physical indicators – yet another bar to more efficient planning and management.

Finally, an absolutely essential, indeed defining, institution of the command economy is the *physical rationing* of resources and producers' goods. This is where the market is most fully and directly replaced, and where the central authorities have the ability to most directly influence and control the behaviour of subordinate operational units. It implements the centralized mobilization of resources to priorities, the most direct response to crises and challenges. And it most directly denies to subordinates the capability to produce, to develop, in ways outside those authorized in the plan. This makes the coexistence between the command principle and the market mechanism a source of continual conflict, as the market opens unauthorized opportunities to subordinates. In the Soviet Union the command principle, aided by the club of materials rationing, repeatedly pushed back and eliminated the market mechanism when (timidly) introduced in reforms, until the system collapsed in chaos, and the introduction of a full-fledged market economy was begun in 1992 (Schroeder, 1979; Aslund, 1995). Thus the nature of the command system makes it fundamentally incompatible with real markets, although some market institutions can, and indeed must, be allowed to function both within the non-state sectors and as the interface between them and state economic institutions/sectors.

Inherent challenges to the command economy

As Grossman notes in his seminal article (1963, p. 107), 'The chief persistent systemic problem of a command economy is the finding of the optimal degree of centralization (or decentralization) under given conditions and with reference to given social goals'. The fundamental dilemma is that full centralization poses an insoluble problem, while decentralization abandons the ability to direct, to control development, and to ensure the pursuit of social goals and priorities. With regard to the pure planning problem, a large body of theoretical literature arose in the late 1960s, and continued into the 1980s, on the problem of decentralizing the planning process to make its informational and computational burden manageable (Eckstein, 1971; Bornstein, 1973). But the problem is far greater, and less studied, with respect to implementation; rational planning is swamped by the struggle to maintain elemental coordination.

Decentralization versus priority

Looked at through the prism of relative advantages, operational decentralization shortens 'lines of communication', increasing flexibility, adaptation and responsiveness to a changing environment through local initiative and innovation, and vastly simplifying the decision problem of economic agents. But it does so at the cost of weakening or losing the 'advantages of centralization', including enforcement of regime values, capability for large-scale resource mobilization, concentration of scarce talent in central decision-making organs, and the maintenance of macro-balance. In particular, decentralization compromises the ability of the centre to directly manage the development and structure of the economy and to force the achievement of critical priorities regardless of cost. Furthermore, decentralization requires the introduction of the alternative coordination mechanism to insure tolerable micro-balance – the market – as decentralization undercuts the ability to directly coordinate, to balance from above. Thus, to prevent catastrophic imbalance, a more active money with economically flexible market prices must be allowed to function in a decentralized system.

The impossibility of planning and commanding the performance of all economic agents in full operational detail, however, forces some decentralization. This creates a chronic threat to balance which is thus a continuous argument for (re)centralization of planning and materials allocation. Furthermore, a partial decentralization of planning and management in a command economy may do more harm than good; it may impair balance without yielding sufficient benefit. Yet a complete decentralization, in the sense of a virtually full devolution of the major production decisions to the firm level, would be disastrous from the standpoint of balance, unless the price structure were properly altered to provide proper signals to firms *and* suitable behavioural rules were prescribed, that is, unless a market mechanism were introduced. Thus the logic of command predicts a 'treadmill of reforms' (Schroeder, 1979), an array of countervailing strengthenings of the oversight and control organs (in particular, the Party), and enhancements of their role in the economy, accompanying moves towards decentralization in the state sector. It also explains the Soviet institutional arrangement of inter-firm contracts as a decentralized implementation device. These are required to specify details of interaction within planned categories, and establish observable, and hence legally enforceable, commitments to planned implementation, constraining the autonomy necessarily granted through the minimal decentralization. And it explains the logic of the continuing restraints on the use of money and the continuing efforts at effective price control to keep the autonomy of agents restricted to the minimum necessary for the continued functioning of the less-than-absolute command economy.

Even limited decentralization requires that money be used in the command sector (as well as in the household sector), but its role as a bearer of options and as the means of pecuniary calculation for decision-making is necessarily limited and deliberately subordinated to the planners' will and the administrators' power. Banks and the treasury accommodate the money needs of production, ensuring a soft budget constraint for the individual firm. At the same time, the 'moneyness' of money at the firm level is low, hemmed in as it is by administrative constraints and impediments, including the rationing of nearly all producer goods, and by the widespread 'sellers' market' (shortages of goods and absolute lack of buyers' alternatives). This monetary ease, together with the sellers' market, plays an important role in ensuring individual workers' job security at the firm level and full employment in the large, while keeping the firm largely insensitive to money costs and/or benefits.

Within the command sector, money and prices have a necessary role in determining terms of alternate resource uses *only within* planned/commanded categories, and money has the role of limiting total claims to resources in areas, or at a level of detail, beyond the reach of plan directives. This requires 'businesslike management' within the firm – *khozraschet*, which is a 'set of behavioral rules that is supposed to govern the actions of Soviet managers beyond their primary responsibility, the fulfillment of output targets'. It pushes the firm toward 'technical efficiency' and limitation of 'claims on society's resources for productive use. ... *khozraschet* is a system that is well devised to control the behavior of managers in a command economy where a certain amount of devolution of power to them is inevitable, and where, further, managers' goals and values do not necessarily coincide with the official ones' (Grossman, 1963, p. 117). Thus money also has the role of facilitating the monitoring of performance in the command sector.

While administrative orders are the rule in a command economy, backed up by greater or lesser degree of state coercion (depending on country and period), any decentralization of implementation naturally relies on monetary ('material') incentives to elicit desired individual compliance and performance. Compounding the incentive problems arising from differences in information and interests between central authorities and implementing agents is the fact that the physical and other indicators to which the material incentives are linked may often be poor measures of social benefit (as seen by the leadership). Furthermore, resort to such rewards widens the distribution of official earnings and raises questions of permissible limits of income inequality. Yet there may be little choice in that the state must in effect compete with the much higher incomes from the *second economy*. Indeed, the Soviet Union during War Communism, Cuba in the 1960s, and the People's Republic of China during some periods before Mao's death in 1976 tended to downgrade material incentives in favour of normative controls, but never did quite abolish them.

The behaviour of the Soviet-type firm has been much studied (Granick, 1954; Berliner, 1957; Nove, 1977; Freris, 1984). Because its directives and the corresponding managerial incentives stress physical output, produced or shipped, and thanks to its low sensitivity to cost and the ambient sellers' market, the firm often sacrifices product cost, quality, variety, innovation and ancillary services to its customers to sheer product quantity. By the logic of command and the requirements of plan manageability, firms operate in an environment with sole suppliers and assigned users, reducing complexity by eliminating 'wasteful' redundancy in production and distribution. Thus firms in a command economy are largely insulated from any product competition, both from the outside world and from other domestic firms, thanks to the climate of administrative controls and the prevalent excess demand for their output. Difficulties with supply, frequent revision of its plans, interference by Party and other authorities, and other systemic problems also stand in the way of its more efficient and effective operation. Indeed, to function at all, the firm's management is frequently forced to break rules and even resort to criminally punishable acts.

This compounds a further critical challenge posed by necessary decentralization – the conflict between the will, purposes, incentives and priorities of the higher authorities and those at lower levels, particularly of the firms and their managements. Even the best-motivated managers, following all official rules and incentives, will sometimes fail to replicate the decisions that their superiors would have made had they been in a position to make them. This problem is aggravated by the inevitable ambiguity, incompleteness and inconsistency of those rules, incentives and the information available on the spot. Only binding physical constraints and observable outcomes can be systematically enforced, making 'centralized materials allocation the most powerful weapon at the disposal of the central authorities' (Grossman, 1963, p. 118). Thus, where material inputs are less determinate of a unit's activities, this information and incentive problem is greater, and the defiance of central will relatively more widespread and successful. This observation explains the non-viability of any reform that fails to fundamentally alter the materials allocation system.

Under-planned, ill-commanded sectors

A major challenge to the command economy also arises from the existence of sectors outside, or only partially affected by, the command principle. In the Soviet Union these included most of agriculture, much of housing, the household sector and some consumer goods and services. 'Markets' were allowed to function for the distribution of final consumer goods and services, including agricultural produce, for much of the activity of the 'collective' sector in agriculture and for household labour supply. For transactions with 'personal property' within the household and collective sectors, money was active and agents responded to market prices, while in the quasi-markets interfacing with the state sector – for example, labour and consumer goods – money was relatively active but prices remained largely controlled and non-market. These are sectors where information on needs/preferences and capabilities proved too difficult to acquire reliably in real time for acceptable allocation and balance to be commanded, and so at least one side of a market was allowed to function with an active money. Here, the command mechanism proves too crude and clumsy, and hence politically counterproductive, to be used outside of pressing emergencies. Indeed, this might be considered a lesson of War Communism, the first experience with a command economy in Soviet Russia, 1918–21.

In view of the theoretical incompatibility of command and market, how could these 'market' sectors be successfully grafted on to the command mechanism? An explanation (Grossman, 1963) rests on the trade-offs between the authorities' limited capabilities, the complexity of those sectors, and their centrality to regime priorities. A sector which provides significant inputs to physical planning and plan fulfilment, where the unpredictability in the flow of goods is unacceptable, cannot be left to the market without seriously undermining command. However, if a 'market' sector can be treated as a residual for purposes of materials planning and allocation, a buffer for planning, then its coexistence is acceptable. Further, if its operation is characterized by rapid change and complexities rather outside the core interests of the regime, if without disrupting the industrial core greater incentives and risk can be placed on those peripheral agents, and if non-market constraints can force the desired market response from it, then the centre will want to separate that sector from the command sphere, lowering its coordination burden by shifting it to the market.

These considerations were indeed active in the case of those sectors 'left to the market' in the Soviet Union: consumer goods retailing, the acquisition of labour services, the support of households in the countryside through a private agricultural sector, and a few peripheral and interstitial activities. Indeed, any attempt to truly 'marketize' any other sectors or activities in the command economy is doomed to fail unless the loss of fervour, of the sense of mission and urgency, leads to abandonment of the command mechanism. Yet even the existence of these limited market sectors, providing an outlet for incentive earnings and diverted resources, exerts a continuing corrosive pressure on the command economy and its control mechanisms.

The cancer of 'money'

A truly monumental challenge to the command economy lies in the role of money in any *less-than-absolute* command economy. As the complete centralization of decisions in the production sector (let alone in the household sector) is an impossibility, something must be left to local initiative and dispersed decision making.

Thus *khozraschet* is a logical necessity, '… an unfriendly bridgehead that threatens to seize ground whenever the planner fails or defaults' (Grossman, 1966, p. 228). With the inevitable devolution of some decision making to firms and households, money acquires a necessary and critical role in the command economy, going well beyond that consistent with the logic of command. That role arises from the need to economize in making decentralized decisions, and as a medium of exchange and store of value in the decentralized interactions that relate to all decisions. In acquiring this role, this 'moneyness', it allows accumulations of power outside the control of the regime. Money is a 'bearer of options' whose power and influence must be restrained if the command mechanism is to operate properly – to determine priorities and to insure maximal commitment to their achievement. As Grossman (1962, p. 214) noted, 'Money is a form of social power that may lead resources astray and is subject to only imperfect control by political authority.'

Thus the power of money has to be curbed in a command economy by limiting balances available to households and firms, by compartmentalizing money into cash and 'firm' circuits, and by erecting barriers and limits to the use of 'monies' in each category, although that undercuts the effectiveness of attempted decentralizations. Liquidity, 'moneyness', is constrained by the institutional structures and by all the characteristics and conditions of the 'sellers' market', rendering 'money' the only non-scarce commodity, in unusable excess to the extent the command mechanism is effective. Monetary policy in the properly functioning command economy reduces to limiting the volume of cash in the economy ('macro-monetary' control) through wage fund restrictions and cash control absorption plans of the retail sector, and the allocation of firm balances in restricted categories ('micro-financial' control) in just sufficient quantity to support the implementation of the plan, with confiscation of excess funds to prevent unauthorized activity by the firm (Garvy, 1977).

Similarly, the price system, expressed in terms of that money, must also be mobilized to the purpose of control. The inflexible, administratively segmented, average cost-based prices in command economies are a logical necessity of command- and haste-based shortage. For all the problems they cause, all the unintended consequences and distortions in the behaviour of subordinates, such prices help to keep money largely passive, at least in the core state sectors, and allow both money and prices to remain instruments, rather than disrupters, of command. More than being ideologically justified, such prices are a response to the pragmatic and pressing requirements of running a shortage economy with a rapidly developing system of centralized direction of enterprises and of materials allocation.

Money, however, is not so easily contained. Once in unobserved hands, it exercises its 'command over goods and services' without reference to plans, commands or regime priorities. Hence, given any discretion, in any sphere of activity not directly monitored agents will naturally use money in ways they find desirable, placing new demands on a physical system otherwise tautly planned and characterized by general scarcity. This is facilitated by the existence of agents and spheres of activity outside the command system, providing 'legitimate' sources and uses for monies, however acquired or disposed. And the possibility of acquiring money provides incentives for unauthorized activities, incentives to undertake unplanned interactions and reallocations. An active money vastly expands the sphere of discretion of 'subordinate' agents beyond any authorized by a decentralizing reform, and calls for severe administrative restrictions, a reduction to passivity, if it is not to disrupt the planned activities and discretion of the central authorities.

Yet attempts to administratively constrain the influence, the 'corruptive' power, of money become increasingly futile once the 'genie' has been 'let out of the bottle'. Even limited decentralizing reform, allowing money to influence some (subcategory) production and allocation decisions, inevitably lets loose more liquidity, more of a command over goods and services, than desired. This arises from a multitude of factors: errors in both physical and financial plans, inherent incompleteness of plans and commands due to limited information and time and to the necessity of aggregation, changing circumstances and shocks to the economy, mistakes in implementation and in responding to shocks, the irregularity and disruptions in the materials allocation system, the behavioural response of even the most enthusiastic and best-intentioned agents to these problems, and so on. All of these can lead to an unexpected lack of funds for doing what was commanded (if only implicitly), and hence disruption of commanded performance, unless additional liquidity is provided.

Thus monetary policy in a command economy, once money is allowed any room for activity, must be accommodating; a lack of funds can never be allowed to disrupt planned performance, just as an excess of funds cannot be allowed to facilitate unplanned or unauthorized activity. Thus the role, the influence, of money has a natural, inexorable tendency to grow: insufficient funds become an immediate problem generating new money through credits or additional allocations, while unused funds tend to stay hidden until ferreted out by inspection or accidental discovery. And as it grows, so does the challenge to the command principle. An increasing number of agents, in both the state and non-state sectors, has a growing ability to access resources, to divert them in the name, if not always the interest, of implementing decentralized plans, and thus to challenge the priorities of the political authorities. This growing challenge becomes a cancer in the system, a growth that undermines its health and feeds tendencies destructive of the priorities of the regime and its rulers.

The 'second economy'

As the command economy matures, as the messianic fervour with which it was imposed wanes and the use of extraordinary force diminishes in ensuring compliance with commands, these challenges to command metastasize into a competing yet symbiotically attached and dependent economic system: the *second economy*. This name highlights the distinction of this sphere of economic activity from the officially sanctioned, 'first', command economy. It is thus defined as 'all production and exchange activity that meets at least one of two criteria: (a) being directly for private gain; (b) being in some significant respect in knowing contravention of existing law' (Grossman, 1977, p. 25).

In the Soviet Union, attempts to strengthen 'material incentives' and activate 'the profit motive' in order to increase the effectiveness and technical efficiency of the implementation of central plans and directives and to stimulate technological progress and innovation, and the growing monetization of the agricultural sector, opened the door to massive expansion of money supply and eroded the barriers between the currency and the enterprise bank account monetary circuits. Collective farms and their subsidiary enterprises, owners of 'small means of transport', vodka manufacture and distribution, and the Caucasus republics (Georgia in particular) proved particularly rich sources of illicit (from the system's perspective) monetization and private 'entrepreneurial' activity. This both raised the spectre of inflation and opened the door to vastly increased opportunities for manipulation by self-interested subordinates in the command sector. Thus the use of 'economic levers' greatly increased the opportunity for and incidence of bribery, corruption, speculation, and even 'honest' private labour.

While the fundamental cause of the appearance and growth of the 'second economy' undoubtedly lies in the congenital institutional weaknesses of the command economy discussed above, there are a number of proximate sources that make it unsurprising. These include extensive price control, with consequent scarcity and misallocation, high taxes on non-state activities/incomes, prohibitions of private activity, unmet individual consumption needs, poorly protected impersonal (state) property, the personal power of bureaucrats and 'gatekeepers', and other historical factors, including the end of terror. These provide both motives and opportunities for officially illicit activity and for the authorities to overlook that activity. With the ageing of command and the decay in enthusiasm of its agents, the growth of such a second economy appears natural.

Thus growing 'monetization', the existence of ready and waiting market sectors, and the decline in the use of violent instruments of enforcement lead to a growing sphere and importance of activities outside the purview of 'planning' and 'command'. These market-mediated activities are at times supportive, helping to achieve tolerable micro-balance in the increasingly complex economy, but often are in violation of planned implementation and regime values. Private interests, necessarily allowed some leeway, grow in significance, increasingly seizing ground from command. In the Soviet Union, the private agricultural sector, initially permitted only to secure survival of the peasantry under the extractive pressure of rapid industrialization, and the consumers' personal services sectors provided the basis for a ubiquitous, if still systemically marginal, second economy.

But then even the core industrial sectors under the command mechanism find their managers and activities increasingly influenced by this illegitimate, shadow market, system, as managers are often forced to break rules and undertake illegal acts in order to do their job. Such acts, together with ubiquitous and protean illegal activity on private account, add up to a large underground economy characteristic of every command economy, which together with legal private activity (allowed in varying degree in different countries) both supports and supplements the 'first economy' and is inimical to it. While the second economy significantly adds to the supply of goods and services, especially for consumption, it also redistributes private income and wealth, contributes to the widespread official corruption, and generally criminalizes the population. Virtually every area of economic life is touched upon, and often entangled with, 'second economy' activities, while legal private activity naturally opens a loophole for illegal trading and entrepreneurship, generally below the purview of the authorities. And it goes hand in hand with the extension of corruption, ensuring that it remains outside of official notice.

Those 'violations' of legality within the command sector, a 'shadow economy', build informal inter-enterprise relations which are generally beneficial to the operation of state enterprises. They work to substantially correct the allocative failures of the command mechanism, improving firm performance and hence benefiting its management, and also provide lucrative opportunities for managers to directly benefit through the activization of barter, personal connections, and bribery. However, they also spawn further distortions in economic behaviour, as managers seek to generate access to cash, the life blood of the 'second economy', to extract rents, and to hide their activity from supervisory and statistical organs.

Thus the second economy plays a dual and contradictory role in the command economic system. First, it addresses a number of the problems of coordination and balance endemic to the command mechanism, reallocating both producers' and consumers' goods, facilitating plan fulfilment and the use of financial incentives, and generating new incomes and 'politically safe' outlets for private initiative. Hence it becomes important for enhancing consumer welfare, for production stability, and even for social stability. The 'second economy', and in particular its 'shadow' side, plays an essential role in

the first economy as a 'pressure valve', a release 'fixing command' by maintaining micro-balance and covering 'holes' in economic life left by the mistakes or oversight of the planners and central managers. And this role becomes increasingly important as the economy grows more complex and diversified, and hence becomes less susceptible to conscious oversight and direction.

As the central authorities struggle with their loss of control, searching for a solution through reform, decentralization and recentralization, monetization and administrative restriction, agents in the economy take advantage of gaps in control, of the autonomy and discretion offered by growing liquidity of the quasi-money in the system, to deal with problems of coordination and balance, inconsistency of plans and commands, and ubiquitous shortages and scarcities. Of course they operate in light of their own partial information, and in their own (private as well as official) interests, but in so doing save the system from collapsing under its own weight and rigidity (Powell, 1977). Thus the second/shadow economy provides a spontaneous surrogate economic reform that imparts a necessary modicum of flexibility, adaptability and responsiveness to a formal set-up that is too often paralyzing in its rigidity, slowness, and inefficiency. In doing so, the second economy also provides a valuable stabilizing influence on society and the polity, making life livable and the system humanly manipulable and responsive to private inducement. It makes everyone complicit in the way things work, equally 'guilty' before state and society, while providing an almost legitimate, and not politically dangerous or directly destructive, outlet for individual initiative and entrepreneurship. Finally, it relieves inflationary pressures (a 'monetary overhang') resulting from the command economy's necessary combination of monetary looseness and pricing rigidity.

Despite this positive functional role, the second economy also has a less positive *systemic* impact. It mocks the pretense of social direction and control, subverts its egalitarian impulse by accentuating differences in access and income, and gives the lie to the pretense of a 'new' ideologically correct ('Soviet') man. Its very existence and usefulness thus subvert the ideology of the regime, and it works against and undercuts regime priorities by exposing the incompetence and incapacity of the authorities. Its provision of alternatives weakens the 'plan, production, and labour discipline' so essential to the proper operation of the command mechanism. Indeed, it attacks the core of the command mechanism as it '… elevates the power of money in society to rival that of the dictatorship itself, rendering the regimes implements of rule less effective and less certain' (Grossman, 1977, p. 36). In particular, it corrupts officialdom and distorts prices, adding a (positive or negative) 'second economy margin', both 'in kind' and in money, breaking prices as an effective instrument of control. This weakens monetized incentives for state activities by providing competing, and often better, alternatives to them. Hence the second economy, and in particular the 'shadow economy' in the state sector, completes the cancerous development of agent autonomy, of the ability to work outside the plan and its subsequent commands, by providing viable alternatives to the plan.

Other dysfunctional impacts, undermining the operation of the command system, arise from its diverting of resources and products to unplanned sectors and activities, including diversion from development/investment priorities to consumers. This naturally generates undesirable (from a system perspective) redistribution of incomes, although recipients, including many high-placed officials, find it very desirable. Indeed, it is further disruptive of command by creating a 'two-tiered' system of prices and incomes, of consumer goods and labour markets. One tier is comprised of the low-priced, scarcity-ridden quasi-markets of the 'less-than-absolute' command economy, where the unenterprising, the overly scrupulous, and the 'slow' can survive. The other tier consists of real, albeit highly distorted, markets in the generally high-priced, risky but well-endowed second economy where the enterprising, entrepreneurial, and criminal can thrive. In this high tier, substantial incomes are generated and allocated, although they largely accrue to corrupt officials and 'gatekeepers' of scarce materials or permissions who can extract rather phenomenal 'rents'. The inequities this generates further undermine the legitimacy of the regime and generate potentially explosive social pressures, only partially relieved by the second economy's 'pressure valve' aspects.

Finally, it is worth noting that the second/shadow economy, through its activity outside of the officially measured sphere, seriously distorts statistical data and the information available to planners and allocators in the official economy, and, due to its illegality, also hides necessary information from other agents in the shadow economy. This aggravates the economic problems that spawn 'second/shadow economy' activities, deepening the contradictions between the centre and decentralized agents, and further corroding the institutional structures of the command economy.

Performance and fate

Command economies have been instrumental in radically transforming societies more or less according to their drafters' intents, in mobilizing resources for rapid industrialization and modernization, at times on a vast scale, and in rapidly amassing industrial power and military strength. Indeed, they have shown themselves highly effective in rapidly implementing large-scale projects and achieving overriding social goals, albeit at great cost. It is this effectiveness, when cost is no object, which explains why the command principle is resorted to in times of emergency and war. Hence in the Soviet Union command facilitated defence during, and rapid recovery and rebuilding of the Stalinist economy after, the massive trauma of the Great Fatherland War. Economic growth has been especially marked (though not unparalleled by market economies) where large amounts of unemployed and underemployed labour and rich natural resources could be mobilized and combined with existing (advanced, Western) technology, and where the public's material improvement could be restrained, or even seriously depressed, under strong political control. As these possibilities waned, and as the economies grew in size and complexity and thus became less amenable to centralized administrative management, rates of growth declined sharply. At the same time, the shortcomings of the command mechanism in adapting production to demand and its changes – providing consumer welfare, effecting innovation, serving export markets – became more apparent and less tolerable. This led to much discussion and repeated attempts at controlled institutional reform, at decentralizing and stimulating subordinate initiative without sacrificing ultimate control.

Some actual reforms in the externally imposed command economies of eastern Europe went so far as to introduce or extend the market mechanism to such a degree that one could no longer regard the system as a Soviet-type command economy, even if, before the 1990s, one could not speak of it as a full-fledged market economy either. Yugoslavia since the early 1950s, Hungary since 1968 and especially in the 1980s, and post-Mao China are the most important cases in point. Other actual reforms were of a minor or 'within-system' nature, aiming to decentralize certain types of decisions while eschewing the market mechanism and retaining the hierarchical form of organization and the command principle. In the hope of stimulating efficiency to revive growth rates, the decentralizing measures were accompanied by a number of other 'reforms' relating to organizational structure: prices (still controlled), incentives, indicators, materials rationing, and so on. The Soviet reforms of 1965, and those in the 1970s and 1980s prior to *perestroika*, were of that kind; many similar ones took place in other Communist countries after the mid-1950s and prior to the overthrow of Communism in 1989. On the whole, such reforms had little success in addressing the problems of the command economy. Bureaucratic and political obstacles apart, the attempt to decentralize economic decisions without bringing in a market mechanism almost inevitably leads to economic difficulties. The beneficiaries of devolution of decision-making lack the necessary information to produce just what the economy requires or to invest to meet prospective needs, and the coordination of plan-subsequent command is lost. Moreover, they may apply the additional power at their disposal to advance particularist causes or to divert resources into illegal channels. Microeconomic disequilibria mount, and soon superior authorities step in to recentralize on a case-by-case basis and the reform withers away (Grossman, 1963; Wiles, 1962, ch. 7; Kontorovich, 1988).

This failure of reform reflects the inherent contradictions of the command economy framed in the irreconcilable conflict between 'command' and 'money' discussed above (Ericson, 2005). The Soviet command economy, driven by the urgent need for and haste in industrialization and military development, initially relegated the influence of money and the market to the margins of the system, where they handled areas and activities in which command had been revealed as counterproductive during War Communism. That system, the 'less than absolute command economy', substantially industrialized, triumphed in the Great Fatherland War, and recovered to an almost perfect replica of its pre-war self by 1950. But by then the strains of its inherent inflexibility and the bounded rationality of the system's planners and managers began telling on continuing growth and the development of the economy. With economic growth came increasing complexity and growing intractability of the central planning and economic management problem. Some decentralization became essential, and increasingly so as time passed, opening the door once again to the rise of money as a significant influence on the operation and development of the economy. And that influence was only enhanced by the ageing and mellowing of the system. With the passing of 'terror' as an effective incentive mechanism, the stabilization of personnel and the regularization of procedures, it became ever harder to control agent behaviour, to contain the distractions of money and the self-interests it mobilized, and to uncover the rents that well-placed agents were able to extract, thus aggravating the inherent agency problems of the command economy.

The remaining years of the Soviet system thus witnessed an epic struggle, barely perceptible at first, but increasingly evident as reforms, decentralizations, reorganizations and recentralizations cycled around each other in the search for a solution to the increasingly evident and destructive malperformance and waste, and aggravating behavioural distortions in response thereto, generated by the struggle between the 'command principle' and the weak, but inexorably emerging, 'market'. Initially reflected in the dysfunctions of the marginal and quasi-markets of the command economy, and in the struggle to harness a 'passive' money to the purposes of command, the role of money grew along the 'treadmill of reforms' into the rival, if still largely subordinate and complementary, 'second economy', and in particular its 'shadow' component, on which the 'command principle' increasingly came to depend for its effectiveness. As long as the Soviet system remained a 'command economy', commands had to have last word, and money remained largely relegated to the sidelines, exercising its influence within the quasi-monetized instruments ('economic levers') of the command mechanism and the distorted markets of the second economy.

This inherent conflict, played out over Soviet history, revolves around a number of fundamental dualities, elemental oppositions which characterize these primary forces. The 'command principle' derives most basically from the urge, the will to control, to 'rationally' determine and direct the future, exercised by a 'gnostic' elite, immanent in the Party. It knows what needs to be done, by whom and how, and can tolerate no dissent or deviation. Juxtaposed to this 'Will of Society' stand the millions of independent 'wills', desires and objectives, anarchically coordinated through 'the market', whenever that set of institutions broke through the barriers and limits placed by 'command'. This provides the foundation for the eternal struggle between 'central priorities and control' and 'agent incentives and capabilities'.

This opposition is severely aggravated by urgency, by 'virtuous haste', in the pursuit of overriding social goals and central objectives. For the mobilization for, and focus of resources on, these priorities trample on the information, capabilities and goals of individual and organizational agents which must perforce implement that mobilization, implement those priorities. 'Effectiveness' in the pursuit of social objectives becomes opposed to 'efficiency' in the attainment of any objectives, denies trade-offs based on local information and incentives, and hence blocks flexibility in response to changing circumstances. Indeed, the single-minded pursuit of overriding objectives, of absolute priorities, naturally disrupts the fine coordination, the requirements of 'balance', necessary to consistently and efficiently pursue any objectives.

Throughout the history of the Soviet Union, the needs of centralization, given Soviet social goals, stood in fateful opposition to the necessity to decentralize in order to keep the system tolerably functioning. The latter necessity spawned repeated (partial) remonetizations and a 'second economy' that both shored up the operational foundations of the 'first economy' and undermined its long-term viability, corroding its ideological and systemic foundations. Money so unleashed intensified the dysfunctions and contradictions of the 'command economy', spurring further repeated 'reforms' and 'experiments' that merely further aggravated the inconsistencies, the 'oppositions' in the system, until the central leadership, largely unintentionally and out of ignorance, destroyed the 'command economy' in the radical systemic and economic 'restructurings' beginning with *perestroika* in 1987.

RICHARD E. ERICSON

See also **agency problems; decentralization; informal economy; second economy (unofficial economy); soft budget constraint; Soviet economic reform; Soviet growth record.**

Bibliography

Aslund, A. 1995. *How Russia Became a Market Economy.* Washington, DC: Brookings Institution.
Berliner, J. 1957. *Factory and Manager in the USSR.* Cambridge, MA: Harvard University Press.

Bornstein, M., ed. 1973. *Plan and Market*. New Haven, CT: Yale University Press.

Eckstein, A., ed. 1971. *Comparison of Economic Systems: Theoretical and Methodological Approaches*. Berkeley: University of California Press.

Ericson, R. 1983. A difficulty with the 'command' allocation mechanism. *Journal of Economic Theory* 31, 1–26.

Ericson, R. 1991. The classical Soviet-type economy: nature of the system and implications for reform. *Journal of Economic Perspectives* 5(4), 11–27.

Ericson, R. 2005. Command vs. shadow: the conflicted soul of the Soviet economy. *Comparative Economic Systems* 47(4), 1–27.

Freris, A. 1984. *The Soviet Industrial Enterprise*. New York: St Martin's Press.

Garvy, G. 1977. *Money, Financial Flows and Credit in the Soviet Union*. New York: Ballinger Publishing Company.

Granick, D. 1954. *Management of the Industrial Firm in the USSR*. New York: Columbia University Press.

Grossman, G. 1962. The structure and organization of the Soviet economy. *Slavic Review* 21, 203–22.

Grossman, G. 1963. Notes for a theory of the command economy. *Soviet Studies* 15(2), 101–23.

Grossman, G. 1966. Gold and the sword: money in the Soviet command economy. In *Industrialization in Two Systems*, ed. H. Rosovsky. New York: Wiley.

Grossman, G. 1977. The 'second economy' of the USSR. *Problems of Communism* 26(5), 25–40.

Grossman, G. 2000. Central planning and transition in the American desert: latter-day saints in present day sight. Online. Available as a preprocessed paper at http://www.econ.berkeley.edu/~grossman/mormons.pdf, accessed 23 June 2006.

Higgs, R. 1992. Wartime prosperity? A reassessment of the U.S. economy in the 1940s. *Journal of Economic History* 52, 41–60.

Kontorovich, V. 1988. Lessons of the 1965 Soviet economic reform. *Soviet Studies* 40, 308–16.

La Lone, M. and La Lone, D. 1987. The Inka state in the southern highlands: state administration and production enclaves. *Ethnohistory* 34, 47–62.

Nove, A. 1977. *The Soviet Economic System*. London: Allen and Unwin.

Powell, R. 1977. Plan execution and the workability of soviet planning. *Journal of Comparative Economics* 1, 57–76.

Raupach, H. 1966. Zur Entstehung des Begriffes Zentralverwaltungs-wirtschaft. *Jahrbuch für Sozialwissenschaft* 17(1), 86–101.

Schroeder, G. 1979. The Soviet economy on a treadmill of reforms. In *Soviet Economy in a Time of Change*, vol. 1, ed. Joint Economic Committee, US Congress. Washington, DC: Government Printing Office.

Wiles, P. 1962. *The Political Economy of Communism*. Cambridge, MA: Harvard University Press.

Zaleski, E. 1968. *Stalinist Planning for Economic Development*. Durham, NC: Duke University Press.

commodity fetishism

Since Plato, philosophy and then science have assumed first, that there is often a difference between appearance and reality; and, then, that it is sometimes possible to grasp what really is the case by investigating how things appear. Marx's account of commodity fetishism, a crucial step in his account of the capitalist mode of production, implements these assumptions explicitly. It describes how exchange relations appear to economic agents, where the appearance belies the reality at the same time that it provides cognitive access to it.

Market exchanges occur in all modes of production capable of sustaining an economic surplus. In capitalism, the process is generalized – not just in the sense that markets structure economic life but also, more importantly, because everything is commodified that can be. Universal commodification is the result of a protracted process that is definitively launched once labour – or, more precisely, labour power (labour time, adjusted for differences in intensity) – is commodified. The commodification of labour power is pivotal because this commodity is the sole source of value and therefore, ultimately, of wages, profits and rents. The generation and distribution of surplus value, of what is produced in excess of what is needed to reproduce the labour power expended in production processes, is the invisible underlying reality upon which perceptions of exchange relations depend. To persons engaged in buying and selling labour power, what appears is just that, as in any other exchange, individuals aim to do as well for themselves as they can, given their resources, their preferences, and the production technologies available to them. But what is really going on is a struggle over the distribution of the economic surplus at the point of production. That reality is opaque. Economic agents are therefore governed by the appearance of rational economic agents maximizing payoffs to themselves. In his account of commodity fetishism, Marx shows how this inevitable misperception helps to reproduce and sustain the underlying reality.

When Marx expressly addresses this phenomenon at the conclusion of the opening chapter of the first volume of *Capital* (1867), the economic agents he describes are property-holding individuals. Thus it is not exactly capitalism that he aims to model, but 'simple commodity production', an ahistorical idealization. However, the cogency of his account is unaffected as his analysis becomes more historical and concrete – to the point that the direct producers are, as in full-fledged capitalism, a propertyless proletariat with nothing to exchange except, of course, their own labour power. Commodity fetishism is therefore a general and pervasive fact wherever capitalist social relations hold sway. Thus the term denotes a systemic opacity at the level of appearance that helps to hold economic agents in thrall by masking the exploitation of labour. Because this misperception sustains the exploitation that engenders it, revolutionaries intent on overthrowing capitalism must tear away the veil of illusion by revealing the exploitation of workers that exchange relations conceal.

Marx does not directly address *how* commodity fetishism comes into being or how it is sustained. But he does provide fragments of an explanation when he focuses on the atomizing effects of market relations. All resource allocation mechanisms are social in the sense that they bring together a host of disparate and heterogeneous economic activities. However, where the commodity form prevails, the social character of market transactions is apparent only after goods and services are produced. The workers know that the corn they consume is produced by farmers, and the farmers know that the tools they use in growing corn are made by workers. Everyone also knows that, without food, workers would not be able to make the tools farmers use; and that, without tools, farmers would not be able to grow food for the workers. It is therefore evident in retrospect that workers and farmers are engaged in a collective endeavour. But it is not similarly evident prospectively. From that vantage point, it seems only that farmers and workers – and also the capitalists who provide them with means of production – are making individual choices aimed at bringing about the best outcomes for themselves, given the constraints they face. Even if they believe that these essentially egoistic activities are somehow socially beneficial, they can justify this belief by appealing to the workings of an 'invisible hand'. Because there is no visible hand that directs the process, the terms of interaction appear *as if* they are forces of nature to which individuals must accommodate. Thus market relations appear as infrangible constraints that human beings are obliged to operate within, not as social constructions that human beings can change. In terms that Kant introduced and that Marx, following Hegel, effectively assumed, freedom (autonomy) is then forfeit. Wills are heteronomously determined, governed by laws of an (apparently) impersonal *other* (the market system itself). To be free, we must therefore take control of the aggregation mechanism we have concocted. We do so by putting reason in command – not just at the individual level of the rational economic agent, but at the societal level as well.

What Marx says about commodity fetishism is concise and intriguing. For these reasons, and because it summarizes the very abstract analysis of the commodity form with which *Capital* begins, his account of the phenomenon has always been well known. 'Commodity fetishism' is one of those terms that everyone associates with Marx. But, even in what remains of Marxist circles, the basic tenets of Marx's account have faded from ongoing discussions. A number of factors have contributed to this turn of events: among them, the legacy of the so-called 'value controversy' of the 1970s; the efforts of mathematical economists in the 1970s and 1980s to put the categories of Marxist economic analysis on a sound, analytical footing; and attempts by analytical philosophers, working on Marxist themes, to reconstruct and, when possible, defend core Marxist positions. The conclusion that has emerged is that, *pace* Marx, there is nothing special about the commodification of labour power and therefore that the theory of surplus value cannot be sustained in the way that Marx believed. Nowadays, it is only the most doctrinaire Marxists who uphold the labour theory of value, the basis for Marx's account of commodity fetishism. This fact along with the decline of political movements that identify with the Marxist tradition and, its inevitable consequence, waning interest in Marx's work itself, has, for the time being, made commodity fetishism a matter of concern mainly to historians of economic thought.

Not long ago, the situation was quite the opposite. From roughly the 1950s through the 1970s, commodity fetishism played a central role in the two most important and innovative tendencies in Marxist theory: Marxist humanism and structuralist Marxism. These were opposing tendencies, politically and substantively. But they converged on according commodity fetishism centre stage.

Marxist humanists sought to de-Stalinize Marxism by recovering its Left Hegelian roots. This meant reading Marx's work through the prism of his early writings, before he broke with his 'erstwhile philosophical conscience', as he proclaimed in 1845 in *The German Ideology*. For the Left Hegelians, Ludwig Feuerbach's philosophical anthropology, elaborated in *The Essence of Christianity* (1841), was fundamental. There Feuerbach 'inverted' the theological dogma that 'God makes Man' by showing how the God idea is an 'objectification' of essential human traits. Lacking materiality, God is purely an objectification, an 'alienated' expression of the human essence. In taking consciousness of this fact, one recovers essential humanity and becomes emancipated from the thrall of its systemic misrepresentation. In the *Paris Manuscripts* (1844), Marx applied the Feuerbachian programme to objects of labour; 'objectifications' too of essential humanity, but also material things and therefore not objectifications only. Feuerbach arrived at his conclusions by 'interpreting' the theology of Right Hegelian theologians. His working hypothesis was that they had gotten the concept of God right, but that they radically misconstrued what the concept *means*. In the *Paris Manuscripts* Marx treated (Smithian) political economy the same way. He assumed that it correctly describes 'economic facts'. The task, then, was to interpret those facts – in order to reveal the alienation they express and, in so doing, to advance the emancipatory project of Left Hegelianism. How successful Marx was in implementing this programme is subject to debate. What is clear is that, as the focus of his theoretical work turned away from Hegelian philosophy towards political economy, history and politics, he became disabused of the idea that Adam Smith or any other classical economist had gotten political economy descriptively right. His life's project, thereafter, was to rework the conceptual apparatus of classical economics – more usually in its Ricardian, not Smithian, form – with a view to revealing the real 'laws of motion' of the capitalist mode of production. In this endeavour, Feuerbachian philosophical anthropology seemed to play no role. But, following the lead of Georg Lukacs (1923) several decades earlier, the Marxist humanists pointed out that there was, in *Capital*, an explicit point of connection – in the text on commodity fetishism. It was there that Marx brought his analysis of the commodity form to completion. But it was also there that, in modelling the commodity form, Marx identified the objectification of essential human traits in the process of capital accumulation. In consequence, capital, becomes a 'fetish', a *god* in Feuerbach's sense – one who controls economic behaviour by force of (illusory) power.

Structuralist Marxists, like Louis Althusser, were intent on reading Left Hegelianism out of the Marxist canon. They therefore treated Marx's references to fetishes and gods as ironic figures of speech, even as they attempted to enlist the text on commodity fetishism in the service of opposition to Marxist humanism. Borrowing a concept from the French philosopher Gaston Bachelard (1884–1962), Althusser (1965) disparaged Marx's early work by asserting the existence of an 'epistemological break' within the Marxist corpus. What he had in mind was roughly a 'paradigm shift' – not, however, within an ongoing scientific practice but between pre-scientific modes of thought and the inception of a new science. In Althusser's account, two previously monumental episte-mological breaks had occurred – one that established mathematics in ancient Greece, and one that established the sciences of nature in 17th-century Europe. Marx's achievement was supposedly on a par with these; he opened up a science of history. He did so by anticipating the structuralist turn the 'human sciences' (in France mainly) would later take – first in linguistics and psychology, later in anthropology and psychoanalysis. Specifically, in *Capital* and other writings of his maturity, Marx explained a range of diverse 'surface' phenomena by construing them as effects of the workings of a rel-atively small number of underlying, generally invariant 'deep' structures. The text on commodity fetishism lent itself to this construal of Marx's explanatory practice in as much as it depicted the perceptions of economic agents as effects of the unseen but causally efficacious process of surplus value extraction. Thus Marx's account can be seen as a theory of necessary (systemically induced) *misperception* – consonant with notions of explanation that contemporaneous structuralists endorsed. Perhaps the most innovative use Althusser made of commodity fetishism was in his theory of ideology, according to which modes of production constitute experiential subjectivity by 'interpellating' the human subjects who support or 'bear' them.

We now inhabit a different intellectual universe. In the past several decades, it has come to be widely believed, by erstwhile Marxists as much as by 'bourgeois economists', that Marx's focus on production rather than exchange inhibited the development of analytical economic tools. In so far as this belief is sound, the emphasis Marxists placed on commodity fetishism is partly to blame. The explanatory strategies of Marxist humanists and of struc-turalists have fallen into disrepute, too – largely because, in both cases, though for different reasons, the alleged connections between appearance and reality were never satisfactorily explained. No sustainable account was given either of how interpretation should proceed in the Marxist humanist case or, in the structuralist case, of how deep structures can be discerned in surface phenomena. Thus, commodity fetishism has fallen on hard times. However, we should not conclude that there is nothing viable in the concept or in the theoretical traditions that, until recently, magnified its importance. Hegelianism certainly, and structuralism possibly, still have much to teach us. The last word may not yet have been said on the theory of surplus value, either. If and when interest in Marx resumes, it will certainly be useful to revisit these issues. The notion of commodity fetishism played a key role in mid- and late 20th-century Marxism. The core idea it articulates – that necessary misperceptions sustain the capitalist order – can again provide useful insights. The concept may not be forever doomed to be of historical interest only.

ANDREW LEVINE

See also **capitalism; Marx's analysis of capitalist production.**

Bibliography

Althusser, L. 1965. *For Marx*. New York: Pantheon Press, 1969.
Althusser, L. 2002. Ideology and ideological state apparatus. In *Lenin and Philosophy and Other Essays*. New York: Monthly Review Press.
Althusser, L. and Baliber, E. 1968. *Reading Capital*. London: New Left Books, 1970.
Cohen, G. 1988. The labour theory of value and the concept of exploitation. In *History, Labour and Freedom: Themes from Marx*. Oxford: Clarendon Press.
Feuerbach, L. 1841. *The Essence of Christianity*. Amherst, NY: Prometheus, 1989.
Fromm, E. 1975. *Marx's Concept of Man*. New York: Continuum.
Lukacs, G. 1923. *History and Class Consciousness*. London: Merlin Press, 1967.
Marx, K. 1844. *Paris Manuscripts* (also called *Economic and Philosophical Manuscripts*). In Karl Marx and Frederick Engels, *Collected Works*, vol. 3. New York: International Publishers, 1975.
Marx, K. 1845. *The German Ideology*. In Karl Marx and Frederick Engels, *Collected Works*, vol. 5. New York: International Publishers, 1976.

Marx, K. 1867. *Capital*, vol. 1. Moscow: Progress Publishers, 1965.

Roemer, J. 1981. *Analytical Foundations of Marxist Economic Theory.* New York: Cambridge University Press.

Steedman, I. 1981. *Marx After Sraffa*, rev. edn. New York: Shocken.

commodity money

A commodity is an object that is intrinsically useful as an input to production or consumption. A medium of exchange is an object that is generally accepted as final payment during or after an exchange transaction, even though the agent accepting it (the seller) does not necessarily consume the object or any service flow from it. Money is the collection of objects that are used as media of exchange. Commodity money is a medium of exchange that may become (or be transformed into) a commodity, useful in production or consumption. This is in contrast to fiat money, which is intrinsically useless.

Commodity money can also be thought of as a medium of exchange that contains an option to consume a predetermined service flow at little or no cost. The option can be exercised in various ways, depending on the object. Coins can be melted down (at little cost) and the metal applied to non-monetary uses. In the case of paper or token money under a commodity money standard, the medium of exchange itself is intrinsically useless, but it is costlessly convertible into a specified quantity of the commodity on demand. Fiat money can also be converted into goods or services, but in quantities that will depend on market prices.

Commodity money is a thing of the past; countries worldwide now use fiat money standards. However, this is a relatively recent development. Commodity money, primarily in the form of coined metals, was the predominant medium of exchange for over two millennia. Although operating under a commodity money standard limits the scope for monetary policy, it does not eliminate it entirely. The history of commodity money is replete with numerous ways in which governments have altered the monetary system to achieve various goals.

From commodity money to fiat money

In early or primitive societies, it is often difficult to characterize the general patterns of trades and transactions, let alone determine how generally accepted a particular commodity might be. Nevertheless, a wide range of commodities have been reportedly used as money (cowry shells, wampum, salt, furs, cocoa beans, cigarettes and so on), perhaps the most exotic being the stone money of the island of Yap in Micronesia.

General acceptability of monetary objects is most clearly ascertained when the objects are standardized and exchanged repeatedly. With metallic commodities, the standardized objects are called coins. Coinage of metal

began in the eastern Mediterranean region or the Middle East, India and China between the sixth and fourth centuries BC. Coinage has developed in parallel and broadly similar ways in these areas.

The metals most commonly used have been gold, silver and copper (in decreasing order of scarcity), in varying degrees of fineness (silver mixed with substantial amounts of copper, called billon). Lead, tin and various copper alloys (bronze, brass, potin) have also been used, although less frequently than the more common metals. The metal is either mined or acquired through trade. The most common method of coinage is striking with a die, although cast coins are also found. In many legal traditions the right of coinage is a prerogative of the public or central authority, although it may be delegated or leased to regional authorities or private parties. This prerogative may also extend to mining. In other words, the rules governing the supply of commodity money vary from government monopoly to minimal regulation.

In Europe and the Mediterranean, coinage – an invention mythically linked to Croesus, King of Lydia – began near the Aegean Sea in the sixth century BC. The use of money developed considerably in Greek and Roman times, leading to a three-tiered system of gold, silver, and copper denominations. In the Roman empire, the provision of coinage was a government monopoly. The collapse of the empire in the West led, after a long transition, to a purely silver-based monetary system, with a largely decentralized provision of minting. Uniformity of coinage was restored under Charlemagne but quickly disappeared along with political fragmentation. Gold returned in common use from the mid-13th century. By the 14th century, most mints in western Europe operated along similar lines, with more or less unrestricted coinage on demand provided by profit-making mints. A great multiplicity of monetary systems persisted, giving rise to both foreign exchange markets (the earliest financial markets) and money changers (the first financial intermediaries).

The first instances of token coinage (coins that are intrinsically useless but are claims to fixed amounts of the commodity) appeared in the 15th century in Catalonia. Notes convertible on demand appeared in the 17th century, in Sweden and later in England. For a more complete discussion of medieval European coinage, see Spufford (1988).

Coins appear to have been used in India in the early fourth century BC and were probably used before then. The earliest coins were so-called punch-marked coins and were adaptations of Greek prototypes. Coins were first used in China and the Far East about the same time as in India. The distinctive bronze coinage with the square hole in the middle first appeared in the third century BC. Early coins in eastern Islamic lands were copies of Byzantine gold and bronze coins; those in the East were copies of Sassanian silver coins. For more on coinage in India and the Far East, see Williams (1997).

Until the 19th century, coins typically bore no indication of face value, and their market value could fluctuate even relative to one another. From the late Middle Ages, governments increasingly sought to regulate the value of coins in some manner, in particular assigning face value or legal tender value by decree. It became desirable to turn the collection of objects used as a medium of exchange into a stable system with fixed exchange rates between the objects. This was achieved to a large degree with bimetallism, a system in which gold and silver coins remained concurrently in circulation at a constant relative price. Its heyday was the mid-19th century, but beginning in 1873 the system was quickly abandoned, and by the First World War countries were using either gold only or (in Africa and eastern Asia) silver only. (Bimetallism is discussed in more detail in Redish, 2000, and Velde and Weber, 2000.) The development of banking in the 19th century also led to increased use of (convertible) notes and other monetary instruments.

The First World War brought about the suspension of convertibility of the notes in many countries. Most countries returned to convertibility between 1926 and 1931, but the onset of the Great Depression reversed the movement. After the Second World War the only major country whose currency was in any way directly tied to a commodity was the United States under the Bretton Woods system: dollars were convertible by non-residents of the United States into gold on demand, while other currencies of the system were convertible into dollars. The link between gold and the dollar was severed in 1971. Fiat money standards are now universal.

The nature of commodity money

The definitions of commodity and fiat monies given above make it seem as if there is a clear distinction between the two. It is more helpful, however, to think of media of exchange along a continuum. An object serving a purpose as a medium of exchange has value above its intrinsic content, reflecting the value of the service as a medium of exchange.

Because the value of a commodity qua commodity and the value as a medium of exchange can differ, the value of all commodity monies has a fiat component. A pure fiat money is one for which this fiat component makes up its entire value. A nice theoretical discussion of commodity and fiat monies is given by Sargent and Wallace (1983).

Price-level determination

It is natural that the medium of exchange in an economy is what becomes the unit of account, the unit in which debt contracts and the prices of goods and services are expressed. It is natural because the money appears on one side of virtually every transaction.

Because commodity money has an intrinsic value apart from that which it obtains by being a medium of exchange, its relative price will not be zero. Thus, in a commodity money economy, the value of money (the inverse of the price level) is bounded away from zero. Moreover, in a canonical commodity money system (see below) with unlimited minting at a set price, the value of money and its quantity tend to remain within a band. If the value of money falls far enough, it becomes preferable to exercise the option and convert some of it into other, non-monetary uses, thus reducing the quantity and preventing the value from falling further. Conversely, if the value of money rises high enough, it becomes worthwhile for agents to turn metal into coins at the mint at the set price, thus increasing the quantity of money. Such a self-regulating commodity money system provides an anchor to the price level. This has been touted as one of the advantages of a commodity money system, particularly in the case of the gold standard.

The question of price-level determination becomes more complicated when multiple commodity monies are made out of different commodities. An example is the circulation of full-bodied gold and silver coins. Should the unit of account be the gold coin or the silver coin? This matters because under a commodity money system a monetary authority does not have the ability to set the exchange rate between monies of different commodities forever. Thus, to the extent that the unit of account is used in contracts to determine the amount of future payments, the choice of the unit of account can affect the allocation of goods and services. This was one of the issues surrounding the possible adoption of a bimetallic standard mentioned above.

The inability of the monetary authority to set the exchange rate between different monies goes away under a pure fiat money system. Because fiat money is (virtually) costless to produce, the monetary authority can costlessly exchange one money for another to maintain whatever exchange rate is desired between different monies that it issues.

Monetary policy

The fact that a commodity is used as money alters its value. This is because part of the total quantity of the commodity – namely, the metal locked up in the form of coins, or the reserves held by the monetary authority – is not available for non-monetary uses. The allocation between monetary and non-monetary uses is determined in equilibrium. Restrictions on the ability to change this allocation, such as restrictions on melting or exporting coins, or limitations on the minting of metal, will have an effect on the equilibrium value of the money even if it has no immediate effect on the allocation itself. (Since money is an asset, its valuation is forward looking.) Thus, there is scope for monetary policy under a commodity money standard, although what constitutes monetary policy is different from and more limited in scope than what holds under a pure fiat money standard.

Monetary policy consists in actions that tend to alter the value of money. In a commodity money system, the value of money is the value of the option we have described. (The strike price of the option is zero, since the commodity is the money.) Most aspects of monetary policy with commodity money consist in modifying this option, typically by modifying the institutions governing the exercise of the option rather than by modifying the quantity of money, which the authority usually cannot control directly. When the monetary authority is directly involved in the provision of the money, it may directly profit from its actions. Potential profit is often an important consideration of monetary policy.

The canonical form of a commodity money standard comprises the following. One or more commodities are chosen to be the standard to which the monetary system will be anchored. The monetary authority defines the specifications of the monetary objects (weight, fineness) and defines the unit of account in terms of these monetary objects. The conversion of commodity into commodity money and vice versa is as costless as possible. In particular, the monetary authority provides for unlimited (and even costless) conversion of the commodity into monetary objects (coins or notes). Conversely, it places no hindrances on the conversion of monetary objects into commodities (coins can be melted, notes are convertible on demand), nor does it place limitations on the consumption of the commodity or its service flow (free possession, unrestricted import and export of the commodity). The monetary objects are unlimited legal tender.

One type of monetary policy modifies the specifications of monetary objects and units of account. An example is debasement, which is reducing the commodity content of a monetary object (and, frequently, of the corresponding unit of account). The result of debasement is inflation, since nominal prices will be adjusted to maintain the relative prices of goods and money. And, just as occurs with fiat money, inflation has the effect of transferring wealth from nominal creditors to nominal debtors. Since governments generally tended to be debtors, debasements were used to reduce the amount of their debts. Historically, debasements also had the secondary effect of increasing seigniorage revenue, since the quantity of coins minted tended to increase significantly after debasements that involved the introduction of new coins (see Rolnick, Velde and Weber, 1996; Sargent and Smith, 1997). Debasements were also used by governments to remedy malfunctions of a multiple-denomination commodity money system (see Sargent and Velde, 2002).

A second type of monetary policy adds or modifies restrictions on the conversion of commodity into money or money into commodity. For example, minting might be restricted by quantity, in which case the authority decides how much to mint. Minting might be unlimited but subject to a fee, called seigniorage. Governments typically charged such a fee, both to cover the actual costs of minting (called brassage) and as a tax (England was the first, in 1666, to provide minting at no cost). The rate of this tax or, equivalently, the price paid by the mint for bullion might be changed. These restrictions tended to alter the allocation of the commodity between monetary and non-monetary uses, and hence the value of the commodity and the money.

A third type of monetary policy sets limits to the legal tender quality of certain coins, or changes their legal tender value. Since coins did not have face values until the 19th century, it was up to monetary authorities to set, and from time to time alter, the legal tender values of coins. Frequently, foreign coins were authorized as legal tender at rates set for domestic coins. Countries attempting to maintain bimetallism in the face of fluctuations in the relative price of gold and silver often had to adjust the face value of either their gold or silver coins. Changes in the legal tender values could also be motivated by fiscal considerations or by attempts to target a particular price level or exchange rate.

The physical nature of the medium of exchange led to a particular set of concerns. Coins, like anything else, depreciate with use, through wear and tear. Since coins of different values have different usage rates, the depreciation rate varied by denomination. Also, being roughly constant over time, depreciation depended on the age of the coin. Finally, imperfect minting technology as well as actions by the public (clipping, sweating) aggravated the disparities between coins. This factor introduced heterogeneity among coins and hindered the achievement of a stable and uniform monetary system. Improvements in coin production partially remedied the problem, as did periodic recoinages.

When the monetary objects consist not only of coins but also of paper currency or tokens that are demand promises to the commodity, a fourth type of monetary policy is available: suspension of convertibility. The monetary authority can refuse to honour the promise of convertibility for some period of time. An example is the suspension of convertibility by the Bank of England between 1797 and 1819 during the wars with France. During the 19th century suspensions were not uncommon during financial or fiscal emergencies, with the understanding that the suspension would end after the emergency and convertibility would be restored at the pre-existing parity. This understanding has been described as a state-contingent gold standard (see Bordo and Kydland, 1996).

When there is a central bank, an additional monetary tool is to change the discount rate, the interest rate at which the central bank lends reserves to the banking system. During the gold standard period, this was the primary means by which central banks affected the exchange rate of their money against the monies of other countries.

Conclusion

Commodity money is a thing of the past; countries worldwide now use fiat money standards. This practice has led to an efficiency gain in the sense that resources that were once tied up in coins are now available for consumption and production (perhaps prompting John Maynard Keynes to refer to gold as the 'barbarous relic'). It has also led to a greater scope for monetary policy because the supply of money can be changed almost costlessly. However, along with this greater scope has come the greater potential for governments to use inflation to collect seigniorage revenue or to reduce the real value of their debts. How to use the freedom that commodity money restricted is still a matter of debate.

FRANÇOIS R. VELDE AND WARREN E. WEBER

See also **bimetallism; Bretton Woods system; fiat money; gold standard.**

Bibliography

Bordo, M. and Kydland, F. 1996. The gold standard as a commitment mechanism. In *Modern Perspectives on the Gold Standard*, ed. T. Bayoumi, B. Eichengreen and M. Taylor. Cambridge: Cambridge University Press.

Kiyotaki, N. and Wright, R. 1989. On money as a medium of exchange. *Journal of Political Economy* 97, 927–54.

Luschin von Ebengreuth, A. 1926. *Allgemeine Münzkunde und Geldgeschichte des Mittelalters und der neueren Zeit.* Munich: R. Oldenbourg.

Redish, A. 2000. *Bimetallism: An Economic and Historical Analysis.* Cambridge: Cambridge University Press.

Rolnick, A., Velde, F. and Weber, W. 1996. The debasement puzzle: an essay on medieval monetary history. *Journal of Economic History* 56, 789–808.

Sargent, T. and Smith, B. 1997. Coinage, debasements, and Gresham's laws. *Economic Theory* 10, 197–226.

Sargent, T. and Velde, F. 2002. *The Big Problem of Small Change.* Princeton, NJ: Princeton University Press.

Sargent, T. and Wallace, N. 1983. A model of commodity money. *Journal of Monetary Economics* 12, 163–87.

Spufford, P. 1988. *Money and Its Use in Medieval Europe.* Cambridge: Cambridge University Press.

Sussman, N. and Zeira, J. 2003. Commodity money inflation: theory and evidence from France in 1350–1436. *Journal of Monetary Economics* 50, 1769–93.

Velde, F. and Weber, W. 2000. A model of bimetallism. *Journal of Political Economy* 108, 1210–34.

Williams, J., ed. 1997. *Money: A History.* New York: St Martin's Press.

common factors

Economic analysis frequently involves the study of variables that exhibit similar behaviour, and it is often of interest to model this comovement. Well-known examples of comovement in multivariate data sets include business cycles in macroeconomic indicators and shifts in the entire term structure of interest rates, and researchers sometimes attribute this comovement to a small set of underlying forces or latent 'factors' that influence each variable in the system. It is then convenient to think of the variation in each variable in the system as the sum of two types of (unobserved) components, one of which captures variation that is due to 'common factors', while the other captures all other variation. Models that attribute comovement to common factors are called common factor models, and common factor analysis involves the identification and study of the common factors.

Common factor models are particularly popular in empirical settings because they offer parsimony, and simplify estimation by reducing the number of parameters that need to be estimated. Economists will typically be interested in interpreting common factors so that they can explain why comovement occurs. Economic theory sometimes predicts common factors. Perhaps the best-known example of this is the capital asset pricing model, in which the (excess) return for the market portfolio is the common factor in the (excess) return for each individual stock. Another well-known example arises when the term structure of interest rates is modelled, because the no arbitrage condition implies that the entire term structure is determined by a single factor, which is the instantaneous interest rate.

A simple model that captures the concept of common factors in a set of N time-series in the (demeaned) vector $Y_t = (Y_{1t}, Y_{2t}, \ldots, Y_{Nt})'$ is given by

$$Y_t = AF_t + \varepsilon_t, \qquad (1)$$

where F_t is an $r \times 1$ vector that contains r common factors, A is an $N \times r$ factor loading matrix (with $rank(A) = r < N$), and ε_t contains N idiosyncratic components. With the use of Σ_Y, Σ_F and Σ_ε to denote the variance covariance matrices of Y_t, F_t and ε_t, it is usual to assume that Σ_ε is diagonal, and it is also common to normalize the set of r factors in F_t by assuming that $\Sigma_F = I_r$.

Model (1) is similar to conventional factor models that are often used in cross-sectional settings, although the variables are specified here as time series, to facilitate discussion on dynamic factor models. If there is no serial correlation in Y_t or F_t, or if estimation is undertaken as if this is the case, then (1) is called a static factor model. It is usual to assume that F_t and ε_t are jointly stationary, that $E(\varepsilon_t) = 0$, $E(F_t \varepsilon_t') = 0$, and that ε_t contains no serial dependence, but these latter assumptions can be relaxed, depending on the type of factor model under consideration.

There are many ways to identify the factors in (1), and standard techniques include the use of principal component analysis, factor analysis and canonical correlations to estimate the parameters of various associated reduced rank regressions. More recently, researchers have focused on the time series properties of multivariate data-sets, and modern factor models include dynamic factor (or index) models, and models that incorporate common features.

These latter models incorporate various ways of allowing the factors to follow specific dynamic processes, or to contain specific time-series properties.

Principal component models

Principal component analysis involves the intuition that most of the variance in Y_t will be attributable to variance in the r components in F_t. The factors F_t are modelled as linear combinations of Y_t, and their identification is based on finding the r (orthogonal) linear combinations of Y_t that have the most variance. In practice this involves finding the eigen values $\lambda_1 > \lambda_2 > \ldots > \lambda_N$ and associated eigenvectors $f_1, \ldots, f_N$ of the form $f_i = \beta_i' Y_t$ that are associated with the roots of the equation $|\hat{\Sigma}_Y - \lambda I| = 0$, where $\hat{\Sigma}_Y$ is an estimate of Σ_Y. The β_i are picked so that $(\hat{\Sigma}_Y - \lambda_i I)\beta_i = 0$ and $\beta_i'\beta_i = 1$, and the factors F_t are then defined by $F_t = (f_1, \ldots, f_r)$. This decomposition ensures that $E(F_t \varepsilon_t') = 0$, but it implies that the ε_t (which are each linear combinations of $(f_{r+1}, \ldots, f_N)$) will be correlated with each other so that Σ_ε will not be diagonal. Principal components estimators of common factors and factor loadings are also the least squares estimators of the reduced rank regression given by

$$Y_t = \underset{(N \times r)}{A} \underset{(r \times N)}{B} Y_t + \varepsilon_t, \qquad (2)$$

where BY_t contains the r factors $F_t = (f_1, \ldots, f_r)$. Anderson (1984, ch. 11) provides a standard reference.

In practice, one needs to determine r before estimating common factor models, and this is often based on the ratio given by $\frac{\lambda_{r+1}^2 + \ldots + \lambda_N^2}{\lambda_1^2 + \ldots + \lambda_N^2}$. This ratio measures the loss of information in the reduced rank system relative to an unrestricted system, and typically investigators will choose r so that this ratio is kept small. Bai and Ng (2002) have developed model selection criteria that are consistent as $\{N, T\} \to \infty$.

Principal components are usually used for dimension reduction, and economic interpretation of the resulting factors is rarely straightforward. However, Stone (1947) has summarized a set of series from the US national accounts, associating the first three principal components with income, income growth and time and Chamberlain and Rothschild (1983) have promoted the use of principal components for estimating approximate factor models of asset prices. Stock and Watson (2002) have suggested the use of diffusion indexes (principal component factors associated with large macroeconomic data sets) for forecasting key macroeconomic variables, and the interest here centres on using information in the factors rather than interpreting the factors themselves.

Classical factor models

Classical factor models are closely related to principal components models, but the underlying intuition and the

assumptions are different. In this case the key assumption is that Σ_ε is diagonal so that the ε_t describe idiosyncratic effects that are unique to each variable in Y_t, while the factors describe joint effects in Y_t. The assumptions that $E(\varepsilon_t) = 0$ and $E(F_t \varepsilon_t') = 0$ still hold, and the ε_t are assumed to contain no serial dependence. Under these assumptions $\Sigma_Y = A\Sigma_F A' + \Sigma_\varepsilon$, and estimates for A, Σ_F and Σ_ε can be found by maximizing the function

$$L_T(A, \Sigma_\varepsilon) = -\frac{T}{2}\ln|\Sigma_Y| - \frac{1}{2}\sum_{t=1}^{t=T} Y_t'\Sigma_Y^{-1}Y_t$$

subject to the condition that $rank(A) = r$ and a set of normalization restrictions that will uniquely identify the $(r+1)(N + \frac{1}{2}r)$ parameters. Researchers often use the joint restrictions that $\Sigma_F = I_r$ and that $A'\Sigma_\varepsilon^{-1}A$ is diagonal for normalization, but other normalizations are common (see Anderson, 1984, ch. 14, for details). If Y_t and ε_t are normally distributed then $L_T(A, \Sigma_\varepsilon)$ is the log likelihood for Y_t (if we ignore the constant term), but, even when Y_t and ε_t are not normally distributed, the maximization of $L_T(A, \Sigma_\varepsilon)$ delivers quasi-maximum likelihood estimates. There are several ways of using the estimates of A and Σ_ε to obtain estimates of the factors in F_t, and perhaps the best-known of these is Bartlett's (1937; 1938) method based on generalized least squares given by

$$\hat{F}_t = (\hat{A}'\hat{\Sigma}_\varepsilon^{-1}\hat{A})^{-1}\hat{A}'\hat{\Sigma}_\varepsilon^{-1}Y_t.$$

As above, it is necessary to determine r prior to estimating the factors, and, on the assumption of normality, the likelihood ratio test statistic for testing $H_0 : r = s$ versus $H_A : r > s$ is given by

$$-T\left[\ln|\hat{\Sigma}_Y| - \ln\left|\hat{A}\hat{A}' + \hat{\Sigma}_\varepsilon\right|\right]$$
$$= -T\ \Sigma_{i=s+1}^{i=N}\ln(1 - \hat{\lambda}_i^2),$$

where the $\hat{\lambda}_i$ are the characteristic roots of $\hat{A}'\hat{\Sigma}_\varepsilon^{-1}\hat{A}$ (in decreasing order) and $\hat{A}$ is estimated under the null. The test statistic is asymptotically distributed as a χ_q^2 with $q = [(N - s)^2 - N - s]/2$ degrees of freedom under the null.

There are numerous applications of classical factor analysis to economic problems, and an early example includes Stone's (1945) factor analysis of the demand for N commodities. Another example includes a factor model of returns by Deistler and Hamann (2005).

Dynamic factor models

Classical factor models are not well suited to multivariate analysis of time series because they assume no serial correlation in ε_t, and, if there are any dynamics in F_t, then they are implicit and not explicitly modelled. Dynamic factor models address these concerns by treating the ε_t

and F_t as autoregressive moving average (ARMA) processes. The innovations that underlie the N processes for ε_t are assumed to be mutually uncorrelated, and uncorrelated with the innovations that underlie the F_t at all leads and lags, but the factors themselves can be mutually correlated. Different variables in Y_t can then move together because they are functions of the same factor(s), or because they are functions of different factors that are themselves correlated.

The identification and estimation of small-scale dynamic factor models is sometimes based on spectral techniques (see Geweke, 1977; or Sargent and Sims, 1977), and use of the Kalman filter in the time domain (as in Engle and Watson, 1981, or Harvey and Koopman, 1997) provides an alternative approach. Dynamic factor models have been particularly popular for estimating factor models of business cycles (as in Geweke and Singleton, 1981), but they have also been used for studying the term structure (Singleton, 1980) and fluctuations in employment across different industrial sectors (Quah and Sargent, 1993).

Recent work has shifted towards the identification and estimation of common factors in large-scale models, relying on the use of large N to obtain consistent estimates of the factors. One strand of this literature adopts a static framework and standard principal components to estimate the factors, and then builds dynamic models of the factors. The resulting models are sometimes called approximate dynamic factor models. Applications of this approach include Stock and Watson's diffusion index (2002), and Bernanke and Boivin's (2003) estimation of a monetary policy reaction function. Another strand of this literature allows different variables to depend on different lags of common factors. These 'generalized dynamic factor models' are estimated using 'dynamic principal components', which are the principal components of spectral density matrices at different frequencies. Applications of this latter approach include a study of business cycle dynamics in the United States (Forni and Reichlin, 1998) and the development of a coincident index for Europe (Forni et al., 2000).

Canonical correlation-based models

Principal component and factor models assume that the factors are linear combinations of the N variables in Y_t, but sometimes it is useful to assume that the factors are linear combinations of M variables contained in another multivariate time series denoted by X_t. The variables in X_t will often include lags of the variables in Y_t, but X_t can also include variables that would be classified as explanatory variables in a regression context. The factors in (1) can now be written in the form $F_t = B'X_t$ (where $rank(B) = r < \min\{N, M\}$). In what follows, we assume that the ε_t in (1) are white noise and uncorrelated with X_t.

The main idea behind a canonical correlations approach is to find linear combinations of X_t that are strongly correlated with linear combinations of Y_t, and, as for principal component models, the estimators of common factors and factor loadings are the least squares estimates of a reduced rank regression. In this case the regression is

$$Y_t = \underset{(N \times r)}{A} \underset{(r \times M)}{B} X_t + \varepsilon_t, \tag{3}$$

and the factors and factor loadings are related to the r largest roots of $R = \Sigma_Y^{-\frac{1}{2}} \Sigma_{YX} \Sigma_X^{-1} \Sigma_{XY} \Sigma_Y^{-\frac{1}{2}}$, which is the multivariate generalization of the (squared) correlation coefficient between two variables. If we order these roots (also called squared canonical correlations) so that $\lambda_1^2 > \lambda_2^2 > \ldots > \lambda_r^2$, and let $V_1, V_2, \ldots V_r$ be the r associated eigenvectors, then the factor loadings and factors are given by $A_i = \Sigma_Y^{-\frac{1}{2}} V_i$ and $B_i X_t = \Sigma_Y^{-\frac{1}{2}} \Sigma_{YX} \Sigma_X^{-1} V_i X_t$. Anderson (1984, ch. 12) provides a detailed discussion of canonical correlations, while Izenman (1980) discusses the associated reduced rank regressions. When the variables in X_t are simply lags of the variables in Y_t, then the first factor is the best predictor of Y_t based on past history, the second factor is the next best predictor, and so on, and the factors provide a set of leading indicators for Y_t. When X_t consists of explanatory variables for Y_t, then the factors are often called coincident indices. One can base a test of $H_0 : r = s$ versus $H_A : r > s$ on the test statistic $-T \sum_{i=s+1}^{i=N} \ln(1 - \hat{\lambda}_i^2)$, which has a χ^2 distribution with $(m - s)(n - s)$ degrees of freedom under the null.

Common feature models

Common feature models are a special class of factor models in which the common factors have a statistical characteristic of interest, while the idiosyncratic components fail to have this characteristic. Common features were first introduced by Engle and Kozicki (1993) when they discussed serial correlation features – a situation in which each of N variables is serially correlated, but there are linear combinations that are white noise. Here, the presence of $N - r$ white noise linear combinations implies a factor model in which there are r serially correlated factors (which are sometimes called common cycles). An earlier example of a common feature model is Stock and Watson's (1988) common trend model which is valid when variables are cointegrated (as in Engle and Granger, 1987). In this case the common factors are integrated of order one but the remaining components (often called error correction terms) are stationary. Other examples of common features include Vahid and Engle's (1993) common trend–common cycle representation, and common nonlinearity (Anderson and Vahid, 1998).

The identification of common features involves finding linear combinations of the data that do not have the feature, and this can be done using a canonical correlations approach in which the variables in X_t model the characteristic of interest. To illustrate, lags of Y_t are put into X_t when testing for serial correlation features in Y_t, and lagged levels are included in X_t when testing for common trends. Factors associated with the lowest eigen values define linear combinations that do not contain the feature, while factors associated with the highest eigen values are used to model the common features. Johansen's (1988) procedure provides a well-known example of this, although inference in this case is based on non-standard (rather than χ^2) distributions because the factors are non-stationary.

A well-known example of a common feature model is the real business cycle model of King, Plosser and Rebelo (1988), in which a common factor (productivity) generates the trend in output consumption and investment, and another factor (the deviation of capital stock from steady state) generates the common cycle.

HEATHER M. ANDERSON

See also **reduced rank regression; time series analysis.**

Bibliography

Anderson, H. and Vahid, F. 1998. Testing multiple equation systems for common nonlinear components. *Journal of Econometrics* 84, 1–36.

Anderson, T. 1984. *An Introduction to Multivariate Statistical Analysis*, 2nd edn. New York: Wiley.

Bai, J. and Ng, S. 2002. Determining the number of factors in approximate factor models. *Econometrica* 70, 191–221.

Bartlett, M. 1937. The statistical conception of mental factors. *British Journal of Psychology* 28, 97–104.

Bartlett, M. 1938. Methods of estimating mental factors. *Nature* 141, 609–10.

Bernanke, B. and Boivin, J. 2003. Monetary policy in a data rich environment. *Journal of Monetary Economics* 50, 525–46.

Chamberlain, G. and Rothschild, M. 1983. Arbitrage, factor structure and mean-variance analysis in large asset markets. *Econometrica* 51, 1305–24.

Deistler, M. and Hamann, E. 2005. Identification of factor models for forecasting returns. *Journal of Financial Econometrics* 3, 256–81.

Engle, R. and Granger, C. 1987. Cointegration and error correction representation, estimation and testing. *Econometrica* 55, 251–76.

Engle, R. and Kozicki, S. 1993. Testing for common features (with discussions). *Journal of Business and Economic Statistics* 11, 369–95.

Engle, R. and Watson, M. 1981. A one-factor multivariate time series model of metropolitan wage rates. *Journal of the American Statistical Association* 76, 774–81.

Forni, M. and Reichlin, L. 1998. Let's get real: a factor-analytic approach to disaggregated business cycle dynamics. *Review of Economic Studies* 65, 453–73.

Forni, M., Hallin, M., Lippi, M. and Reichlin, L. 2000. The generalized dynamic factor model: identification and estimation. *Review of Economics and Statistics* 82, 540–54.

Geweke, J. 1977. The dynamic factor analysis of economic times-series models. In *Latent Variables in Socioeconomic Models*, ed. D. Aigner and A. Goldberger. Amsterdam: North-Holland.

Geweke, J. and Singleton, K. 1981. Maximum likelihood 'confirmatory' factor analysis of economic time series. *International Economic Review* 22, 133–7.

Harvey, A. and Koopman, S. 1997. Multivariate structural time series models. In *Systematic Dynamics in Econometric and Financial Models*, ed. C. Heij, H. Schumacher and C. Praagman. Chichester: Wiley and Sons.

Izenman, A. 1980. Assessing dimensionality in multivariate regression. In *Handbook of Statistics*, vol. 1, ed. P. Krishnaiah. Amsterdam: North-Holland.

Johansen, S. 1988. Statistical analysis of cointegration vectors. *Journal of Economic Dynamics and Control* 12, 231–54.

King, R., Plosser, C. and Rebelo, S. 1988. Production, growth and business cycles II. New directions. *Journal of Monetary Economics* 21, 309–41.

Quah, D. and Sargent, T. 1993. A dynamic index model for large cross sections. In *Business Cycles, Indicators and Forecasting*, ed. J. Stock and M. Watson. Chicago: University of Chicago Press.

Sargent, T. and Sims, C. 1977. Business cycle modeling without pretending to have too much a-priori economic theory. In *New Methods in Business Cycle Research*, ed. C. Sims et al. Minneapolis: Federal Reserve Bank of Minneapolis.

Singleton, K. 1980. A latent time series model of the cyclical behavior of interest rates. *International Economic Review* 21, 559–75.

Stock, J. and Watson, M. 1988. Testing for common trends. *Journal of the American Statistical Association* 83, 1097–107.

Stock, J. and Watson, M. 2002. Macroeconomic forecasting using diffusion indexes. *Journal of Business and Economic Statistics* 20, 147–62.

Stone, J. 1945. The analysis of market demand. *Journal of the Royal Statistical Society, Series A*, 108, 286–382.

Stone, J. 1947. On the interdependence of blocks of transactions. *Journal of the Royal Statistical Society, Series B*, Supplement, 1–45.

Vahid, F. and Engle, R. 1993. Common trends and common cycles. *Journal of Applied Econometrics* 8, 341–60.

common knowledge. *See* **epistemic game theory.**

common property resources

The concept of common property has become famous in economics since Garett Hardin (1968) wrote his celebrated article on the 'The Tragedy of the Commons'. In this article, common property is taken to mean the absence of property rights in a resource, or what is equivalently known as a regime of 'open access'. Under such a regime, where a right of inclusion is granted to anyone who wants to use the resource, Hardin argued, inefficiency inevitably arises in the form of over-exploitation of the resource accompanied by an over-application of the variable inputs. Open access leads to efficiency losses because 'the *average product* of the variable input, not its *marginal product*, is equated to the input's rental rate when access is free and the number of exploiters is large' (Cornes and Sandler, 1983, p. 787). The root of the problem lies in the fact that the average product rule does not enable the users to internalize the external cost which their decisions impose on the users already operating in the resource domain. Of course, the efficiency losses are conceivable only in a world of resource scarcity, implying that the variable input is subject to decreasing returns. Such losses are considerable since they amount to the dissipation of the whole resource rent. Here is the crucial intuition behind the open access regime: when no property right is attached to a resource, the value of this resource is zero in spite of its scarcity.

Efficiency losses are to be measured not only in static but also in dynamic terms. Indeed, in an open access regime resource users are induced to compare average *instantaneous* returns with the input's rental price even though they may well be aware that they thereby contribute to reducing the future stock of the resource. The problem is simply that they are forced to follow a myopic rule because there is no way in which they can reap the future benefits of restraint in the present. Thus, for example, by refraining today from catching juvenile fish or from cutting down saplings in the forest, a villager can receive no assurance that he or she will be able in the next period to catch mature fish or to fell fully grown trees.

The main criticism levelled by numerous social scientists against the concept of open access is that the corresponding regime is rarely encountered on the ground. The typical regime, according to these critiques, is one under which a community possesses a collective ownership right over local natural resources. Under common property, therefore, a right of exclusion is assigned to a well-defined user group, and Hardin has created a lot of confusion by using the word 'commons' to refer to the alternative situation where no such right is granted to any agency. What is not always clear, however, is whether the ownership right involves only the ability to specify the rightful claimants to the resource, or whether it also involves the ability to define and enforce rules of use regarding that resource (for example, regulations about the harvesting season and production tools, allowed quotas of harvestable products of the resource, or taxes). Baland and Platteau (1996) have coined the term 'unregulated common property' to refer to the former situation, while the term 'regulated common property' is used for the latter.

Two polar situations can be considered on the basis of this analytically important distinction between two types of common property regimes. At one extreme, if common property is perfectly regulated, in the sense that the rules of use designed and enforced by the owner community allow a perfect internalization of the externalities, common property becomes equivalent to private property with a sole owner from an efficiency standpoint. This illustrates the general result that, absent transaction costs, institutions do not matter. At the other extreme, a strictly unregulated common property in the above sense implies that, as the number of users becomes quite large, over-exploitation of the resource becomes as important as under the open access regime: the rent attached to the resource is totally dissipated (see Platteau, 2000, ch. 3).

Between these two extremes we find the situations most typically observed on the ground and described in the numerous field studies devoted to this topic (see Ostrom, 1990; Baland and Platteau, 1996, for a review of such studies). In such instances, rules of use exist alongside membership rules, yet they tend to be imperfectly designed and imperfectly enforced by the village community. One key reason for these imperfections is the governance costs that unavoidably plague any collective decision-making process. Governance costs include all those costs incurred to reach a collective agreement and to organize a community of users. They are likely to be higher when the group is larger and when its membership is more heterogeneous (whether measured in terms of diversity of objectives or of wealth inequality). Moreover, governance costs are enhanced by the opportunistic tendencies of rights-holders not only to violate or circumvent collective rules but also to eschew efforts to create collective mechanisms of decision-making and enforcement. Costs arising from these proclivities are also dependent on the size of the user group: they are lower if the number of resource users is smaller and, at the limit, they are nil when there is a single user.

As a consequence of the aforementioned limitations, resources are less efficiently managed under a common property regime than they could be under a private ownership system. This is especially true if, owing to their scarcity, the resources carry high values which should be reflected in high rents. Population growth and market integration are thus two forces that tend to increase the monetary value of the efficiency losses arising from common property, that is, the forgone rents. This, at least, is the conclusion drawn by the so-called property rights school of Chicago economists (see, for example, Demsetz, 1967; Barzel, 1989). The advantages of private property appear all the more decisive as such a regime enables users to internalize externalities without incurring any governance costs. This is because it establishes a one-to-one relationship between individual actions and all their effects: 'A primary function of property rights is that of guiding incentives to achieve a greater internalization of externalities …' (Demsetz, 1967, p. 348).

Nevertheless, this ignores the costs of privatizing natural resources, which involve both directs costs and opportunity costs. Direct costs comprise transaction costs, such as the costs of negotiating, defining and enforcing private property rights. The usual argument is that such costs increase with the physical base of the resource. Thus, the wider the resource base (or the less concentrated the resource) the higher are the costs of delimiting and defending the resource 'territory' (Dasgupta, 1993, pp. 288–9). For many natural resources, the costs of dividing the resource domain appear prohibitive under the present state of technology. For example, the open sea – or, more exactly, the fish stock contained in it – presents insuperable difficulties for private appropriation. The enforcement of exclusive property rights to individual patches of the ocean would, indeed, be infinitely costly. This is especially evident when fish species are mobile and move within wide water spaces, since exclusive rights are too costly to establish and enforce whether over the resource or over the territory in which the resource moves.

The opportunity costs of privatization, for their part, correspond to the benefits that are lost when the common property regime is abandoned. Here, we can think of scale economies that may be present not only in the resource itself but also in complementary factors. The obvious advantage of coordinating the herding of animals so as to economize on shepherd labour in extensive grazing activities is probably the best illustration of the way scale economies in a complementary factor may prevent the division of a resource domain. Another important category of opportunity costs is the insurance benefits associated with common property. When returns to a resource are highly variable across time and space, the need to insure against such variability is yet another consideration that may militate against resource division. When a resource has a low predictability (that is, when the variance in its value per unit of time per unit area is high), users are generally reluctant to divide it into smaller portions because they would thereby lose the insurance benefits provided by keeping the resource whole.

For instance, herders (fishermen) may need to have access to a wide portfolio of pasture lands (fishing spots) in so far as, at any given time, wide spatial variations in yields result from climatic or other environmental factors. On the assumption that the probability distributions are not correlated too much across spatial groupings of land or water and that they are not overly correlated over time, a system offering access to a large area within which right-holding users can freely move appears highly desirable from a risk-reducing perspective.

The conclusion of the above discussion is, therefore, that the balance of the advantages and disadvantages of various property regimes is a priori undetermined. Economic theory, however, does provide useful guidance about which circumstances are more favourable to the persistence of common property or, conversely, to its

demise and replacement by private property. Furthermore, instead of being fixed once for all, the balance sheet is susceptible to evolution depending on the transformation of the parameters on which the benefits and costs of privatization depend. Thus, the direct costs of resource division may fall with technological progress. For example, the introduction of modern borehole drilling facilitates the privatization of common grazing areas (Peters, 1994). It is therefore not only the factors which enhance resource value but also those which reduce the direct costs of partitioning that may favour the private appropriation of natural resources.

JEAN-PHILIPPE PLATTEAU

See also **access to land and development; agriculture and economic development; land markets; population and agricultural growth; property law, economics and; tragedy of the commons.**

Bibliography

Baland, J. and Platteau, J.-P. 1996. *Halting Degradation of Natural Resources: Is There a Role for Rural Communities?* Oxford: Clarendon Press.

Barzel, Y. 1989. *Economic Analysis of Property Rights.* Cambridge: Cambridge University Press.

Cornes, R. and Sandler, T. 1983. On commons and tragedies. *American Economic Review* 83, 787–92.

Dasgupta, P. 1993. *An Inquiry into Well-Being and Destitution.* Oxford: Clarendon Press.

Demsetz, H. 1967. Toward a theory of property rights. *American Economic Review* 57, 347–59.

Hardin, G. 1968. The tragedy of the commons. *Science* 162, 1243–8.

Ostrom, E. 1990. *Governing the Commons: The Evolution of Institutions for Collective Action.* Cambridge: Cambridge University Press.

Peters, P. 1994. *Dividing the Commons: Politics, Policy, and Culture in Botswana.* Charlottesville and London: University Press of Virginia.

Platteau, J.-P. 2000. *Institutions, Social Norms and Economic Development.* London: Harwood Academic Publishers.

common rights in land

Common rights are rights to use land in common. The most important of these rights was the right to graze livestock on common grassland. But rights to gather fuel (wood, peat, gorse and turves), fertilizers, timber for building and other natural resources were also important. Common land is land used by a number of distinct individuals or households whose rights over the land are known as common rights.

Today we are accustomed to think of land as private property with a clear owner and possibly a tenant. Although in some countries there may be legal rights of public access to certain types of wild or agricultural land, it is generally the case that the owner or tenant of the land has exclusive rights to use the land and, within the limits of planning or zoning laws, may use it as he or she wishes. But in Europe, for at least a thousand years and ending only in the 19th century, a high proportion of land was 'common land' which many individuals were entitled to use for a variety of purposes.

It cannot be overemphasized that common land was generally not open-access land – land which anyone could use. There were regulations governing who could use the land, what they could use it for and how much they could use it. When economists think of common land and common rights they may have Garett Hardin's 'tragedy of the commons in mind' (Hardin, 1968). The principal subject of Hardin's article was in fact population growth, not historical common land or common rights. However, Hardin used a theoretical common land system as a model for the exploitation of open-access resources. In this system each herder could put as many animals as he wished on to the common pastures. Hardin argued that individual herders would choose to graze more and more animals on the common, thus inevitably leading to over-grazing and degradation of the resource. This model offers important insights into the destruction of, or damage to, unregulated open-access resources such as the atmosphere or fish stocks in the oceans. If common land and common rights had operated in this manner, it is unlikely that they would have remained a key part of European agriculture for so many centuries.

In the rest of this article the following questions are addressed: what were common rights? What was common land? Who had common rights? How was common land regulated? Was it efficient? How and why did it come to an end and with what consequences? The answers to these questions varied from one village to another across Europe and what follows is necessarily highly simplified (see de Moor, Shaw-Taylor and Warde, 2002, for a more detailed overview).

Common land

The types of common land and the terminology used to describe such land varied across Europe. Nevertheless, four major types of common land may be distinguished. First, the archetypical form of common land and the one with the widest geographical distribution is variously referred to as common waste, common pasture, waste, or common. This land was permanently common and most often grassland used for grazing animals. Usually such land was not suitable for arable cultivation typically because its natural fertility was low but sometimes for other reasons such a propensity to seasonal flooding. On some common wastes other resources were available, such as peat, turf, gorse or wood.

Second, in many parts of Europe much of the arable land (the land on which crops were grown) was also subject to common rights. Such land, known as open-fields, common

fields, or common arable, was privately owned and cultivated but subject to common grazing. In its classic form each farmer held a number of long thin strips of land scattered over an extensive area and intermixed with the strips of other farmers. Each farmer cultivated his own crops on the arable. But when the harvest was over, or in years when the land was being fallowed, all those with common rights could turn their livestock into the fields to graze. Thus the open fields alternated between private and common land over the course of the agricultural cycle.

Third, common woodland for the production of fuel and timber was widespread on the European continent but unusual in England. This was similar to common waste in that it was permanently in common use.

Fourth, common meadows, which were permanent grasslands for the production of hay, were divided into separate blocks in private use but after the hay had been harvested were open to common grazing. Thus, like the open fields, common meadows alternated between private land and common land over the agricultural cycle.

Common rights

As private property, the right to cultivate the common arable or to harvest the hay in common meadows lay with the owner of the land or the owner's tenant. Access to the common rights was considerably more complex and took different forms in different places; but it is possible to distinguish four main forms of access. First, in England and some parts of the Continent, the ownership or tenancy of particular buildings or landholdings was a prerequisite. Second, in many parts of the Continent citizenship of (as distinct from residence in) a commune or a municipality which itself owned the common resource was necessary, sometimes in combination with a property qualification. Third, in other parts of the Continent membership of a cooperative association which owned the common resources was necessary. Membership of these institutions was sometimes inherited, but sometimes it was attached to buildings or land (as in the first case). Fourth, there were cases were all residents in an area had common rights. But outside largely uninhabited areas, such as northern Sweden, this situation was unusual.

In consequence by no means all individuals or households enjoyed common rights. The proportion of the population that enjoyed common rights varied considerably from one region to another and changed over time. Where individuals or households did have common rights, the kinds and levels of the rights they enjoyed were determined by local regulations.

Regulation

Common land was almost invariably regulated by local institutions, often at the level of the individual village or manor. The institutions varied but were usually manor or village courts or village assemblies or committees of some kind, with the decisions made by a group of jurors. These institutions normally issued sets of rules, ordinances or by-laws which governed the usage of the commons and set fines for the infringement of rules. Officials or monitors were appointed to police the by-laws. The degree to which these institutions and their by-laws were subject to the influence of feudal overlords and the state varied considerably across Europe.

The by-laws provided the basic regulatory framework for managing the commons (for examples of by-laws see Ault, 1972). Their most critical function was to restrict the usage of common land and thus prevent a 'tragedy of the commons' developing. This was done in two ways. First, the by-laws would normally serve to restrict common rights to well-defined groups of users. For example, in much of England only those holding land in the open fields or with certain recognized dwellings, known as common-right houses, were allowed to pasture animals in the open fields or on the common pasture, while on much of the Continent pasture rights were restricted to citizens of communes or the holders of ancient farmsteads. Second, by-laws defined the amount of resources to which each commoner was entitled. Thus, by-laws might specify the amount of peat or wood each commoner was entitled to dig or cut each year or the number and type of animals which could be kept, and for which months of the year they might be kept on the common pastures, open fields and common meadows.

The number of animals each commoner could put on the common land was generally controlled by one of two types of rules. One, known as 'stinting', simply specified the number and type of animals (the stint) which each commoner might keep on the common. Often the stint was proportional to the area or the value of land held. The other form of access, known as 'levancy' and 'couchancy', stated merely that each commoner could keep as many animals on the common as he could over-winter (that is, feed when the common was closed) on his own holding. How this was policed in practice is a moot point, but it certainly served to limit numbers and may have differed little from stinting in practice.

One consequence of these types of rule is that some individuals had no common rights at all. Another is that different individuals who did have common rights could have very different levels of access. The situation varied too much to allow generalization, beyond the suggestion that the level of inequality in England was probably greater and had proceeded further at an early date than anywhere else.

Enclosure

The process by which common land and common rights were abolished and replaced by recognizably modern forms of private property was part and parcel of a broader reform of landholding known as 'enclosure' which could also entail the consolidation of scattered holdings and the wholesale reallocation of land to create ring-fenced farms. Enclosure in some form is probably as old as common land itself. In England significant enclosure took place in the medieval period and from the 17th to the early 19th centuries. In most of Europe the widespread attack on common land began in the late 18th century in the wake of Physiocratic critiques. The later Napoleonic reforms and a subsequent series of state-sponsored drives to modernize agriculture in the 19th century led to more sustained enclosure. Some common land survives to this day, generally in mountainous areas.

Efficiency

By the 18th century common rights and common land were being widely criticized by agricultural improvers and others for restricting agricultural productivity. Most agricultural writers have accepted this view of common land as inefficient, and associated enclosure with major increases in productivity (Ernle, 1936; Chambers and Mingay, 1966; Overton, 1996). Common rights and common land imposed two kinds of limitation on agricultural improvement. First, the communal regulation of common land made it more difficult to introduce new agricultural techniques and technologies or to respond to changes in market opportunities. Second, the sharing of the outputs from common land made individual investment less attractive. The spread of nitrogen fixing crops and new drainage technologies, which often allowed the cultivation of formerly uncultivable common land, together with better transport links made enclosure a steadily more pressing issue in the 18th and 19th centuries.

A number of economic historians have reconsidered the inefficiency of open fields in an English context. McCloskey (1976) has argued that the scattering of land in open fields in the medieval period was an efficient insurance against risk in a non-market economy. Allen (1992) has argued that enclosure did facilitate major technological changes obstructed by common land but that these innovations made only very marginal contributions to increased efficiency. Clark (1998) has argued that the inefficiencies imposed by common land were relatively modest and that, given the costs involved, enclosure was not economic until after 1750. However, the issue remains controversial essentially because it is inherently difficult to measure the agricultural productivity of farming in the 18th and 19th centuries with any degree of reliability. In other words, at present the data are too poor to allow an entirely plausible rebuttal of the views of 18th-century critics of the open fields. Moreover, much enclosure took place in the medieval period and in the 17th century (Wordie, 1983) and any fully satisfactory theory of the efficiency or otherwise of open fields would need to be able to account for the longer-term chronology of enclosure. The persistence of open-field farming in France has been investigated by Grantham (1980) and Hoffman (1989).

Another controversial issue is the importance of common land to the poor. Many historians have argued that the poor derived considerable benefits from common land and that enclosure was socially damaging; but this remains controversial (see Neeson, 1993; Shaw-Taylor, 2001. The extent to which the poor benefited from common land and common rights is hard to reconstruct, poorly understood, and varied considerably across Europe.

Common-pool resources

This article has been concerned exclusively with common land and common rights as they existed in Europe before the 20th century. However, it should be noted that while open fields and common meadow may be peculiarly European forms, common waste and institutions for its management can be found all over the world. Analytically, these systems are part of a larger family of 'common-pool-resource' systems (Ostrom, 1990) which have been adopted in many parts of the world to manage not just land but water resources and fish stocks as well.

LEIGH SHAW-TAYLOR

See also **access to land and development; common property resources; tragedy of the commons.**

Bibliography

Allen, R. 1992. *Enclosure and the Yeoman: The Agricultural Development of the South Midlands, 1450–1850.* Oxford: Oxford University Press.

Ault, W. 1972. *Open-field-farming in Medieval England: A Study of Village By-laws.* London: George Allen and Unwin.

Chambers, J. and Mingay, G. 1966. *The Agricultural Revolution 1750–1880.* London: Batsford.

Clark, G. 1998. Commons sense: common property rights, efficiency, and institutional change. *Journal of Economic History* 58, 73–102.

de Moor, M., Shaw-Taylor, L. and Warde, P., eds. 2002. *The Management of Common Land in North West Europe, c. 1500–1850.* Turnhout: Brepols.

Ernle, Lord. 1936. *English Farming Past and Present,* 5th edn. London: Longmans, Green and Co.

Grantham, G. 1980. The persistence of open-field farming in nineteenth century France. *Journal of Economic History* 40, 515–31.

Hardin, G. 1968. The tragedy of the commons. *Science* 162, 1243–8.

Hoffman, P. 1989. Institutions and agriculture in old-regime France. *Journal of Institutional and Theoretical Economics* 145, 166–81.

McCloskey, D. 1976. English open fields as behaviour towards risk. *Research in Economic History* 1, 124–70.

Neeson, J. 1993. *Commoners, Common-Right, Enclosure and Social Change in England 1700–1820.* Cambridge: Cambridge University Press.

Ostrom, E. 1990. *Governing the Commons: The Evolution of Institutions for Collective Action.* Cambridge: Cambridge University Press.

Overton, M. 1996. *Agricultural Revolution in England: The Transformation of the Agrarian Economy 1500–1850.* Cambridge: Cambridge University Press.

Shaw-Taylor, L. 2001. Parliamentary enclosure and the emergence of an English agricultural proletariat. *Journal of Economic History* 61, 640–62.

Wordie, J. 1983. The chronology of English enclosure, 1500–1914. *Economic History Review* 36, 483–505.

Commons, John Rogers (1862–1945)

Commons was born on 13 October 1862 in Hollandsburg, Ohio, and died on 11 May 1945 in Raleigh, North Carolina. He studied at Oberlin College (BA, 1888) and Johns Hopkins University (1888–90). He taught at Wesleyan, Oberlin, Indiana, Syracuse, and Wisconsin (1904–32).

The founder of the distinctive Wisconsin tradition of institutional economics, Commons derived his theoretical insights (generalized in his *Legal Foundations of Capitalism*, 1924, and *Institutional Economics*, 1934) from his practical, historical and empirical studies, particularly in the field of labour relations and in various areas of social reform. He drew insight not only from economics but also from the fields of political science, law, sociology and history. A principal adviser and architect of the Wisconsin progressive movement under Robert M. La Follette, Commons was active as an advisor to both state and federal governments. He was instrumental in drafting landmark legislation in the fields of industrial relations, civil service, public utility regulation, workmen's compensation and unemployment insurance. He served on federal and state industrial commissions, was a founder of the American Association for Labor Legislation, was active in the National Civic Federation, National Consumers' League (president, 1923–35), National Bureau of Economic Research (associate director, 1920–28), and the American Economic Association (president, 1917). He participated in antitrust litigation (especially the Pittsburgh Plus case) and in movements for reform of the monetary and banking system (often associated with Irving Fisher, who considered Commons one of the leading monetary economists of the period).

The critical thread uniting Commons's diverse writings was the development of institutions, especially within capitalism. He developed theories of the evolution of capitalism and of institutional change as a modifying force alleviating the major defects of capitalism. Commons came to recognize and stress that individual economic behaviour took place within institutions, which he defined as collective action in control, liberation, and expansion of individual action. The traditional methodologically individualist focus on individual buying and selling was not capable, in his view, of penetrating the forces, working rules and institutions governing the structural features of the economic system within which individuals operated. Crucial to the evolution and operation of the economic system was government, which was a principal means through which collective action and change were undertaken.

Commons rejected both classical harmonism and radical revolutionism in favour of a conflict and negotiational view of economic process. He accepted the reality of conflicting interests and sought realistic, evolutionary modes of their attenuation and resolution. These modes focused on a negotiational psychology in the context of a pluralist structure of power. He sought to enlist the open-minded and progressive leaders of business, labour and government in arrangements through which they could identify problems and design solutions acceptable to all parties.

In other contexts, he sought to use government as an agency for working out new arrangements to solve problems, such as worker insecurity and hardship, rather than promote systemic restructuring, although to many conservatives his ventures were radical enough. To these ends Commons and small armies of associates engaged in fact finding – his look-and-see methodology – in a spirit of bringing all scientific knowledge to bear on problem solving. From these experiences, indeed already manifest in the underlying strategy, Commons developed a theory of government as alternately a mediator of conflicting interests and an arena in which conflicting interests bargained over their differences; a theory of the complex organization – in terms of freedom, power and coercion – and evolution of the legal foundations of capitalism, which centred in part on the composing of major structural conflicts through the mutual accommodation of interests; and a theory of institutions with an affirmative view of their roles in organizing individual activity and resolving conflict.

The institutions Commons studied most closely were trade unions and government, particularly the judiciary. He developed his theory of the economic role of government in part on the basis of his study of the efforts of workers to improve their market position and in part on the use of government by both enemies and friends of labour. Commons's was an interpretation of trade unions as a non-revolutionary development, as collective action seeking to do for workers what the organizations of business attempted to do for their owners and managers. His study of the reception given unions and reform legislation led him to recognize the critical role of the United States Supreme Court (and the courts generally), and its conception of what was reasonable in the development and application of the working rules which governed the acquisition and use of power in the market. Accordingly, Commons developed a theory of property

which stressed its evolution and role in governing the structure of participation and relative withholding capacity in the market.

Commons also developed a theory of institutions which focused on their respective different mixtures of bargaining, rationing and managerial transactions, all taking place within a legal framework which was itself subject to change.

Although Commons's institutionalism had different emphases from that of Thorstein Veblen, for example, in that Commons stressed reform of the capitalist framework, they shared a view of economics as political economy and of the economy as comprising more than the market. Unlike Veblen, Commons was not antagonistic toward businessmen, and indeed accepted capitalism, though not necessarily on the terms given or preferred by the established power structure.

Commons was one of the few American economists to found a 'school', a tradition that was carried forward by a corps of students, especially Selig Perlman, Edwin E. Witte, Martin Glaeser and Kenneth Parsons. Much mid-20th-century American social reform, the New Deal for example, drew on or reflected the work of Commons and his fellow workers and students.

WARREN J. SAMUELS

Selected works

1893. *The Distribution of Wealth*. New York: Macmillan.
1905. *Trade Unionism and Labor Problems*. Boston: Ginn.
1910–11. (With others.) *Documentary History of American Industrial Society*, 10 vols. Cleveland: A.H. Clark.
1916. (With J.B. Andrews.) *Principles of Labor Legislation*. New York: Harper.
1919. *Industrial Goodwill*. New York: McGraw-Hill.
1919–1935. (With others.) *History of Labor in the United States*, 4 vols. New York: Macmillan.
1921. *Industrial Government*. New York: Macmillan.
1924. *The Legal Foundations of Capitalism*. New York: Macmillan.
1934. *Institutional Economics*. New York: Macmillan.
1934. *Myself*. New York: Macmillan.
1950. *The Economics of Collective Action*. Madison: University of Wisconsin Press.

community indifference curves

The idea of a community indifference curve, as the term is commonly used, is due to Scitovsky (1942). The genesis of the idea is the fact that comparative statics and welfare analysis in economic models is simplified considerably if there is a social preference ordering over aggregate commodity bundles which reflects the collective individual preferences of agents. Scitovsky's notion of a 'community indifference curve' essentially allows the analytical convenience of social indifference curves, in certain circumstances, without having to assume a specific Bergson–Samuelson social welfare function or having to assume the restrictive assumptions on agents' preferences needed to guarantee that agents act collectively as a single individual.

The definition of a community indifference curve is basically simple. Suppose there are m commodities and n agents. Let x denote a commodity vector (as m-vector with non-negative coordinates) and u_i a utility function representing agent i's preferences. We will assume that u_i is monotone increasing and quasi-concave. Given a vector $u' = (u'_1, \ldots, u'_n)$ of utility numbers, the community indifference curve at u', $CIC(u')$, is defined to be the set of all commodity vectors x such that there is a distribution $(x_1, \ldots, x_n)$ of commodity vectors satisfying $\sum_i x_i = x$ and $u_i(x_i) = u'_i$, $i = 1, \ldots, n$, and there is no $x' \leq x, x' \neq x$ which also has this property. Thus one can obtain any vector $x \in CIC(u')$ by fixing the quantities of all but one good and minimize the amount of the remaining good subject to achieving u'. As pointed out by Samuelson (1956), the community indifference curve can be interpreted as a 'dual' to the utility possibility frontier. The utility possibility frontier, for a given x, is the set of all vectors u' of utility numbers achievable by a Pareto efficient distribution of x to the agents. Let $U(x)$ denote the utility possibility frontier for the commodity vector x. Then it is easy to see that $CIC(u') = \{x : u' \in U(x)\}$ and that $U(x) = \{u' : x \in CIC(u')\}$.

We will now describe the most important properties of community indifference curves. First, each $CIC(u')$ looks like the indifference curve of a monotone quasi-concave utility function. That is, the set of vectors x such that $x \geq x^1$ for some $x^1 \in CIC(u')$ is a convex set. For example, when $m = 2$, $CIC(u')$ is a curve with a diminishing marginal rate of substitution. Second, unlike the utility possibility frontier, the community indifference curve is essentially an ordinal concept, that is it does not depend on the choice of utility functions representing agents' preferences, in the following sense. Suppose, for each i, u_i and v_i are two utility functions representing agent i's preferences, and let $(x_1, \ldots, x_n)$ be a Pareto efficient allocation to the agents. Define $u' = [u_1(x_1), \ldots, u_n(x_n)]$ and $v' = [v_1(x_1), \ldots, v_n(x_n)]$. Then $CIC(u') \equiv CIC(v')$. Clearly, community indifference curves can be parameterized by a given Pareto efficient allocation of goods rather than by a given vector of utilities. Third, assuming smooth utility functions, the marginal rate of substitution for any two commodities on a community indifference curve is equal to the common marginal rate of substitution of each agent. Specifically, pick an $x \in CIC(u')$, and let $(x_1, \ldots, x_n)$ be the Pareto efficient allocation of x such that $u_i(x_i) = u'_i$, $i = 1, \ldots, n$. Then for any two commodities h and h', the marginal rate of substitution of h and h' evaluated at $x \in CIC(u')$ is equal to the marginal rate of substitution of h for h' at x_i on agent i's indifference curve through x_i. Fourth, and very important, community indifference curves are not, in general, 'indifference' curves in the sense of being level curves of some function. Pick any x, and $u', u'' \in U(x)$, such that $u' \neq u''$. Then by

definition, $x \in CIC(u') \cap CIC(u'')$. Thus $CIC(u')$ must either coincide with $CIC(u'')$ or intersect properly. The condition for two community indifference curves never to intersect properly is then that $CIC(u') = CIC(u'')$ for all $u', u'' \in U(x)$, for all x. It turns out that this is true if and only if the agents have identical homothetic preferences, in which case the family of all community indifference curves will coincide with the family of indifference curves for the common preferences of the agents.

From the above definition and properties, the following observation constitutes the basic use of community indifference curves: if the economy is currently at a vector of utility numbers u', then x' is a commodity vector which lies above $CIC(u')$ if and only if there is some distribution of x' to the agents which will achieve a vector of utilities u'' such that $u'' > u'$. In this sense, x' is 'better' than any $x \in CIC(u')$. However, since from above community indifference curves can intersect properly, it may also be that there is a u''' such that $x' \in CIC(u''')$ and an $x' \in CIC(u')$ such that x lies above $CIC(u''')$, in which case x is also 'better' than x'. Thus it is important to realize that community indifference curves cannot be used to define a social ordering of aggregate output vectors. Nevertheless, community indifference curves can still be a useful analytical device. For example, consider a market economy with two produced goods. Consider an equilibrium in which all consumers face the same prices, in terms of the aggregate output vector x' and the vector of utilities u' obtained by the agents. Graphically this equilibrium can be represented by drawing the production possibility frontier and $CIC(u')$, noting they meet at x'. The slope of the production possibility frontier at x' represents the price ratio faced by firms, and the slope of the $CIC(u')$ at x' the common price ratio faced by consumers. If firms and consumers face the same price ratio, then the $CIC(u')$ must be tangent to the production possibility frontier at x. Thus no feasible x can be produced which can make all agents better off, so the situation is Pareto optimal. If, however, firms face different prices than the agents because of, for example, taxes or tariffs, then the slope of the $CIC(u')$ will be different from the slope of the production possibility frontier, and thus the two curves will intersect properly. In this case there must exist an x' on the production possibility frontier which lies above $CIC(u')$, so the original situation is Pareto inefficient.

WAYNE SHAFER

See also **Arrow's theorem; optimality and efficiency; social welfare function; welfare economics**.

Bibliography

Samuelson, P.A. 1956. Social indifference curves. *Quarterly Journal of Economics* 70(1), 1–22.
Scitovsky, T. 1942. A reconsideration of the theory of tariffs. *Review of Economic Studies* 9(2), 89–110.

comparative advantage

The modern economy, and the very world as we know it today, obviously depends fundamentally on specialization and the division of labour, between individuals, firms and nations. The principle of comparative advantage, first clearly stated and proved by David Ricardo in 1817, is the fundamental analytical explanation of the source of these enormous 'gains from trade'. Though an awareness of the benefits of specialization must go back to the dim mists of antiquity in all civilizations, it was not until Ricardo that this deepest and most beautiful result in all of economics was obtained. Though the logic applies equally to inter*personal*, inter*firm* and inter*regional* trade, it was in the context of inter*national* trade that the principle of comparative advantage was discovered and has been investigated ever since.

The basic Ricardian model

What constituted a 'nation' for Ricardo were two things – a 'factor endowment', of a specified number of units of labour in the simplest model, and a 'technology', the productivity of this labour in terms of different goods, such as cloth and wine in his example. Thus labour can move freely between the production of cloth and wine in England and in Portugal, but each labour force is trapped within its own borders. Suppose that a unit of labour in Portugal can produce one unit of cloth or one unit of wine, while in England a unit of labour can produce four units of cloth or two units of wine. Thus the opportunity cost of a unit of wine is one unit of cloth in Portugal while it is two units of cloth in England. On the assumption of competitive markets and free trade, it follows that *both* goods will never be produced in *both* countries since wine in England and cloth in Portugal could always be undermined by a simple arbitrage operation involving export of cloth from England and import of wine from Portugal. Thus wine in England or cloth in Portugal must contract until at least one of these industries produces zero output. If both goods are consumed in positive amounts, the 'terms of trade' in equilibrium must lie in the closed interval between one and two units of cloth per unit of wine. Which of the two countries specializes completely will depend upon the relative size of each country (as measured by the labour force *and* its productivity in each industry) and upon the extent to which each of the two goods is favoured by the pattern of world demand. Thus Portugal is more likely to specialize the smaller she is compared with England in the sense defined above and the more world demand is skewed towards the consumption of wine relative to the consumption of cloth.

The gains from trade

Viewed as a 'positive' theory, the principle of comparative advantage yields *predictions* about (*a*) the *direction* of trade: that each country exports the good in which it has the lower comparative opportunity cost ratio as defined by the technology in that country, and about (*b*) the *terms* of trade: that it is bounded above and below by these comparative cost ratios. From a 'normative' standpoint the principle implies that the citizens of each country become 'better off' as a result of trade, with the extent of the gains from trade depending upon the degree to which the terms of trade exceed the domestic comparative cost ratio. It is the 'normative' part of the doctrine that has always been the more controversial, and it is therefore necessary to evaluate it with the greatest care.

In Ricardo's example the total labour force in each country is presumably supplied by an aggregate of different households, each having the same *relative* productivity in the two sectors. Thus *all* households in *each* country must become better off as a result of trade if the terms of trade lie strictly in between the domestic comparative cost ratios. The import-competing sector in each country simply switches over instantaneously and costlessly to producing the export good (moving to the opposite corner of its linear production-possibilities frontier, in terms of the familiar geometry), obtaining the desired level of the other good by imports, raising utility in the process. When one country is incompletely specialized, then all households in that country remain at unchanged utility levels, all of the gain from trade going to the individuals in the 'small' country. Thus we have a situation in which *everybody gains*, in at least one country, while *nobody loses* in either country, as a result of trade.

This very strong result depends upon Ricardo's assumption of perfect occupational mobility in each country. Suppose we take the opposite extreme of completely *specific* labour in each sector, so that each country produces a fixed combination of cloth and wine, with no possibility of transformation. In this case, labour in the import-competing sector in each country must necessarily *lose*, as a result of trade, while labour in each country's export sector must gain. It can be shown, however, that trade will improve *potential* welfare in each country in the Samuelson (1950) sense that the utility-possibility frontier with trade will dominate the corresponding frontier without trade, so that no one need be worse off, and at least some one better off, if lump-sum taxes and transfers are possible (Samuelson, 1962).

International factor mobility and world welfare

Another very important normative issue is the question of the relationship between the free-trade equilibrium and *world* efficiency and welfare. In the Ricardian model world welfare in general will *not* be maximized by free trade alone. In the numerical example considered here Ricardo stresses the fact that England can still gain from trade even though she has an *absolute* advantage in the production of *both* goods, her productivity being greater in both cloth and wine, though comparatively greater in cloth. Suppose that labour in Portugal could produce at English levels, *if it moved to England*; that is, the English superiority is based on climatic or other 'environmental' factors and not on differences in aptitude or skill. Then, if labour was free to move, and in the absence of 'national' sentiment, all production would be located in England, and Portugal would cease to exist. The former Portuguese labour would be better off than under free trade, since their real wage in terms of wine will now be two units instead of one. The English labour would be worse off, if the terms of trade were originally better than 0.5 wine per unit of cloth, but it is easy to show that they could be sufficiently compensated since the utility-possibility frontier for the world economy as a whole is moved out by the integration of the labour forces.

The case when each country has an absolute advantage in one good is more interesting. As is easy to see, from Findlay (1982), this case will involve a movement of labour to the country with the higher real wage under free trade, increasing the production of this country's exportable and reducing that of the lower-wage country under free trade. The terms of trade turn against the higher-wage country until eventually the real wage is equalized. The terms of trade that achieve this equality of real wages will be equal to the ratio of labour product-ivities in each country's export sector; that is, the 'double factoral' terms of trade will be unity. This solution of free trade *combined* with perfect labour mobility will achieve not only efficiency for the world economy as a whole but equity as well. 'Unequal exchange' in the sense of Emmanuel (1972) would not exist, while liberal, utili-tarian and Rawlsian criteria of distributive justice would be satisfied as well, as pointed out in Findlay (1982). Despite all this, it still seems utopian to expect a policy of 'open borders', in *either* direction, for the contemporary world of nation-states.

Extensions of the basic Ricardian model

The two-country, two-good Ricardian model was extended to many goods and countries by a number of subsequent writers, whose efforts are described in detail by Haberler (1933) and Viner (1937). In the case of two countries and *n* goods the concept of a 'chain of com-parative advantage' has been put forward, with the goods listed in descending order in terms of the *relative* effi-ciency of the two countries in producing them. It is readily shown that with a uniform wage in each country all goods from 1 to some number *j* must be exported, while all goods from (*j* + 1) to *n* must be imported. The number *j* itself will depend upon the relative sizes of the two countries and the composition of world demand. Dornbusch, Fischer and Samuelson (1977) generalize this result to a continuum of goods in an extremely elegant and powerful model that has been widely used in sub-sequent literature. An analogous chain concept applies to the case of two goods and *n* countries, this time ranking the countries in terms of the ratio of their productivities in the two goods, with country 1 having the greatest *relative* efficiency in cloth and country *n* in wine. World demand and the sizes of the labour forces will determine the 'marginal' country *j*, with countries 1 to *j* exporting cloth and (*j* + 1) to *n* exporting wine.

The simultaneous consideration of comparative advantage with many goods and many countries presents severe analytical difficulties. Graham (1948) considered several elaborate numerical examples, his work inspiring the Rochester theorists McKenzie (1954) and Jones (1961) to apply the powerful tools of activity analysis to this particular case of a linear general equilibrium model. It is interesting to note in connection with mathematical programming and activity analysis that Kantorovich (1965) in his celebrated book on planning for the Soviet economy worked out an example of optimal specialization patterns for factories that corresponds *exactly* to the Ricardian model of trade between countries.

The three-factor Ricardian model

While most of the literature on the Ricardian trade model has concentrated on the model of Chapter 7 of the *Principles* in which it appears that labour is the sole scarce factor, his more extended model in the *Essay on Profits* has been curiously neglected, though the connections between trade, income distribution and growth which that analysis explores are quite fascinating. The formal structure of the model was laid out very thoroughly in Pasinetti (1960). The economy produces two goods, corn and manufactures, each of which has a one-period lag between the input of labour and the emergence of out-put. Labour thus has to be supported by a 'wage fund', an initially given stock that is accumulated over time by saving out of profits. Corn also requires land as an input, which is in fixed supply and yields diminishing returns to successive increments of labour. The wage-rate is given exogenously in terms of corn, and manufactures are a luxury good consumed only by the land-owning class, who obtain rents determined by the marginal product of land. Profits are the difference between the marginal product of labour and the given real wage, which is equal to the marginal product 'discounted' by the rate of interest, in this model equal to the rate of profit, defined as the ratio of profits to the real wage that has to be advanced a period before. Momentary equilibrium deter-mines the relative price of corn and manufactures, the rent per acre and the rate of profit, as well as the output levels and allocation of the labour force between sectors. The growth of the system is at a rate equal to the product of the rate of profit and the propensity to save of the capitalist class. It is shown that the system approaches a stationary state, with a monotonically falling rate of profit and rising rents per acre.

The opportunity to import corn more cheaply from abroad will have significant distributional and growth consequences. Just as Ricardo argued in his case for the repeal of the Corn Laws, cheaper foreign corn will reduce domestic rents and raise the domestic rate of profit, and thus the rate of growth. The approach to the stationary state is postponed, though of course it cannot be ultimately averted, while the growth consequences for the corn exporter are definitely adverse. The main doctrinal significance of this wider Ricardian model, however, is to reveal the extent to which the subsequent 'general equilibrium' or 'neoclassical' approach to international trade is already present within the Ricardian framework. For one thing, the pattern of comparative advantage itself depends upon the complex interaction of technology, factor proportions and tastes. In his Chapter 7 case the pattern of comparative advantage is *exogenous*, simply given by the four fixed technical coefficients indicating the productivity of labour in cloth and wine in England and Portugal. The production-possibility frontiers for each country are linear, and comparative advantage is simply determined by the relative magnitudes of the slopes. As demonstrated in Findlay (1974), however, the *Essay on Profits* model implies a concave production-possibilities frontier at any moment, since there are diminishing returns to labour in corn even though the marginal productivity of labour in manufactures is constant. With two countries the pattern of comparative advantage will depend upon the slopes of these curves at their autarky equilibria, which are *endogenous* variables depending upon the sizes of the 'wage fund' in relation to the supply of land and the consumption pattern of land-owners, as well as the technology for the two goods.

As Burgstaller (1986) points out, however, the steady-state solution of the model restores the linear structure of the pattern of comparative advantage. The zero profit rate in the steady state requires the marginal product of labour to be equal to the given real wage, and this implies a fixed land–labour ratio and hence output per unit of labour in corn. We thus once again have two fixed technical coefficients, so that the slope of the linear production-possibilities frontier is once again an exogenous indicator of comparative advantage.

The 'neo-Ricardian' approach of Steedman (1979a; 1979b) considers more general time-phased structures of production. Technology alone determines negatively sloped wage–profit or factor-price frontiers, any point on which generates a set of relative product prices and hence a pattern of comparative advantage relative to another such economy.

Factor proportions and the Heckscher–Ohlin model

While J.S. Mill, Marshall and Edgeworth all made major contributions to trade theory, the concept of comparative advantage did not undergo any evolution in their work beyond the stage at which Ricardo had left it. They essentially concentrated on the determination of the terms of trade and on various comparative static exercises.

The interwar years, however, brought fundamental advances, stemming in particular from the work of the Swedes Heckscher (1919) and Ohlin (1933). The development of a diagrammatic apparatus to handle general equilibrium interactions of tastes, technology and factor endowments by Haberler (1933), Leontief (1933), Lerner (1932) and others culminated in the rigorous establishment of trade theory and comparative advantage as a branch of neoclassical general equilibrium theory.

The essentials of this approach can be expounded in terms of the familiar two-country, two-good and two-factor model, on which see Jones (1965) for a detailed and lucid algebraic exposition. The given factor supplies and constant returns to scale technology define concave production-possibility frontiers, on the assumption that the goods differ in factor intensity. This determines the 'supply side' of the model, which is closed by the specification of consumer preferences. Economies that have identical technology, factor endowments and tastes will have the same autarky equilibrium price-ratio and so will have no incentive to engage in trade. Countries must therefore differ with respect to at least one of these characteristics for differences in comparative advantage to emerge. With identical technology and factor endowments, a country will have a comparative advantage in the good its citizens prefer *less* in comparison to the foreign country, since then this good will be cheaper at home. Similarly, if factor endowments and tastes are identical, differences in comparative advantage will be governed by relative technological efficiency; that is, a country will have a comparative advantage in the good in which its relative technological efficiency is greater, just as in the Ricardian model. These differences in technological efficiency could be represented, for example, by the magnitude of multiplicative constants in the production functions; that is, 'Hicks-neutral' differences.

In keeping with the ideas of Heckscher and Ohlin, however, it is differences in factor proportions that have dominated the explanation of comparative advantage in the neoclassical literature. The Heckscher–Ohlin theorem, that each country will export the commodity that uses its relatively abundant factor most intensively, has been rigorously established and the necessary qualifications carefully specified, as in Jones (1956). Among the more important of these is the requirement that factor-intensity 'reversals' do not take place; that is, that one good is always more capital-intensive than the other at all wage-rental ratios or at least within the relevant range defined by the factor proportions of the trading countries.

The Stolper–Samuelson theorem

Associated with the Heckscher–Ohlin theorem is the Stolper–Samuelson theorem (1941), that trade benefits the abundant and harms the scarce factor while protection does the opposite, and the celebrated factor price equalization theorem of Lerner (1952, though written in 1932) and Samuelson (1948; 1949; 1953), which states that under certain conditions free trade will lead to complete equalization of factor rewards even though factors are not mobile internationally. The normative significance of this theorem is that free trade alone can achieve world efficiency in production and resource allocation, unlike the case of the Ricardian model as pointed out earlier. The requirements for the theorem to hold, however, are very stringent, such as that the number of tradable goods produced be equal to the number of factors. It also requires factor proportions to be sufficiently close to each other in the trading partners so that the production patterns are fairly similar. Thus it would be far-fetched to expect the price of unskilled labour to be equalized between Bangladesh and the United States, for example.

The specific-factors model

An important and popular variant of the factor proportions approach is what Jones (1971) calls the 'specific factors' and Samuelson (1971) the Viner–Ricardo model. In this model each production sector has its own unique fixed factor, while labour is used in all sectors and is perfectly mobile internally between them. Trade patterns still reflect factor endowments but factor price equalization does not hold in this model since the number of factors is always one greater than the number of goods. Gruen and Corden (1970) present an ingenious three-by-three extension of this approach, in which one sector uses land and labour, while the two others use capital and labour, thus neatly integrating the 'specific factors' model with the regular two-by-two Heckscher–Ohlin model. Findlay (1995, chs 4 and 6) uses adaptations and extensions of the Gruen–Corden specification to introduce human capital formation and the concept of a natural resource 'frontier' into the Heckscher–Ohlin framework.

Long-run extensions of the factor proportions model

One limitation of the Heckscher–Ohlin model was that the stock of 'capital', however conceived, should be an endogenous variable determined by the propensity to save or time preference of each trading community, rather than being taken as exogenously fixed. Oniki and Uzawa (1965) extended the model to a situation where the labour force is growing in each country at an exogenous rate and capital is accumulated in response to given propensities to save in each country. One of the goods is taken to be the 'capital' good, conceived of as a malleable 'putty–putty' instrument. They demonstrated that the system converges in the long run to a particular capital–labour ratio for each country, which will be higher for the country with the larger saving propensity. In Findlay (1970; 1984), it is shown that as the capital–labour ratio evolves the pattern of comparative advantage for a given 'small' country in an open trading world will also shift over time towards more capital-intensive goods, thus formalizing the concept of a 'ladder of comparative advantage' that countries ascend in the process of economic development. Thus comparative advantage should not be conceived as given and immutable, but evolving with capital accumulation and technological change. Much of the loose talk about 'dynamic' comparative advantage in the development literature, however, is misconceived since it attempts to change the pattern of production by protection *before* the necessary changes in the capacity to produce efficiently have taken place. Other models which endogenize the capital stocks of the trading countries are Stiglitz (1970) and Findlay (1978) which utilizes a variable rate of time preference and an 'Austrian' point-input/point-output technology, which implies a continuum of capital goods as represented by the 'trees' of different ages, and Findlay (1995, ch. 2), which addresses the question posed by Samuelson (1965) of whether trade equalizes not only the marginal product or rental of capital but the rate of interest itself.

Empirical testing

Empirical testing of the positive side of the theory of comparative advantage begins in a systematic way only with the work of MacDougall (1951) on the Ricardian theory and the celebrated article of Leontief (1954) that uncovered the apparent paradox that US exports were more labour-intensive than her imports. Leontief's dramatic finding spurred considerable further empirical research motivated by the desire to find a satisfactory explanation. The increasing scarcity of natural resources in the USA, by causing capital to be substituted for it in import-competing production, was stressed as an explanation for the paradox by Vanek (1963). The role of 'human capital' as an explanation was pointed to by Kenen (1965) and a number of empirical investigators, who found that US exports were considerably more skill-intensive than her imports, even though physical capital-intensity was only weakly correlated with exports and imports. This pointed to the need to reinterpret the simple Heckscher–Ohlin model in terms of skilled and unskilled labour as the two factors, rather than labour of uniform quality and physical capital. Since the formation of skill through education is an endogenous variable, a function of a wage differential that is itself a function of trade, we need a general equilibrium model that can simultaneously handle both these aspects, as in Findlay and Kierzkowski (1983) and Findlay (1995, ch. 4).

Many other extensions of the Heckscher–Ohlin theory are surveyed in Jones and Neary (1984) and Ethier (1984), while Deardorf (1984) and Feenstra (2004) give very incisive accounts of the attempts at empirical testing of the theory of comparative advantage in its different manifestations. Further important progress in empirical testing of the Heckscher–Ohlin model has been achieved by the work of Leamer (1984), Trefler (1995), Harrigan (1997) and Davis and Weinstein (2001).

Increasing returns

Finally, the crucial role of increasing returns to scale in specialization and international trade has only recently been rigorously investigated, since it implies departures from perfect competition. Krugman (1979) and Lancaster (1980) introduced international trade into models of monopolistic competition with differentiated products, showing the possibility of gains from trade due to the provision of greater variety of similar goods rather than differences in comparative advantage, what is referred to as 'intra-industry' trade. Helpman and Krugman (1985) thoroughly examine and extend our knowledge in this area, while Grossman and Helpman (1991) expertly extend the monopolistic competition approach to deal with a number of issues involving endogenous technological progress and growth in the world economy.

RONALD FINDLAY

See also **Heckscher, Eli Filip; Heckscher–Ohlin trade theory; international trade theory; Leontief, Wassily; Ohlin, Bertil Gotthard; Ricardo, David; terms of trade.**

Bibliography

Burgstaller, A. 1986. Unifying Ricardo's theories of growth and comparative advantage. *Economica* 53, 467–81.

Davis, D.R. and Weinstein, D.E. 2001. An account of global factor trade. *American Economic Review* 91, 1423–53.

Deardorf, A. 1984. Testing trade theories. In *Handbook of International Economics*, vol. 1, ed. R.W. Jones and P.B. Kenen. Amsterdam: North-Holland.

Dornbusch, R., Fischer, S. and Samuelson, P.A. 1977. Comparative advantage, trade and payments in a Ricardian model with a continuum of goods. *American Economic Review* 67, 823–39.

Emmanuel, A. 1972. *Unequal Exchange*. New York: Monthly Review Press.

Ethier, W. 1984. Higher dimensional issues in trade theory. In *Handbook of International Economics*, vol. 1, ed. R.W. Jones and P.B. Kenen. Amsterdam: North-Holland.

Feenstra, R.C. 2004. *Advanced International Trade*. Princeton: Princeton University Press.

Findlay, R. 1970. Factor proportions and comparative advantage in the long run. *Journal of Political Economy* 78, 27–34.

Findlay, R. 1974. Relative prices, growth and trade in a simple Ricardian system. *Economica* 41, 1–13.

Findlay, R. 1978. An 'Austrian' model of international trade and interest equalization. *Journal of Political Economy* 86, 989–1007.

Findlay, R. 1982. International distributive justice. *Journal of International Economics* 13, 1–14.

Findlay, R. 1984. Growth and development in trade models. In *Handbook of International Economics*, vol. 1, ed. R.W. Jones and P.B. Kenen. Amsterdam: North-Holland.

Findlay, R. 1995. *Factor Proportions, Trade, and Growth*. Cambridge, MA: MIT Press.

Findlay, R. and Kierzkowski, H. 1983. International trade and human capital: a simple general equilibrium model. *Journal of Political Economy* 91, 957–78.

Graham, F. 1948. *The Theory of International Values*. Princeton: Princeton University Press.

Grossman, G.M. and Helpman, E. 1991. *Innovation and Growth in the Global Economy*. Cambridge, MA: MIT Press.

Gruen, F. and Corden, W.M. 1970. A tariff that worsens the terms of trade. In *Studies in International Economics*, ed. I.A. MacDougall and R. Snape. Amsterdam: North-Holland.

Haberler, G. 1933. *The Theory of International Trade*, Trans. by A. Stonier and F. Benham, London: W. Hodge, 1936; revised edn, 1937.

Harrigan, J. 1997. Technology, factor supplies and international specialization: estimating the neoclassical model. *American Economic Review* 87, 475–94.

Heckscher, E. 1919. The effects of foreign trade on the distribution of income. In *Ekonomisk Tidskrift*. English translation in *Readings in the Theory of International Trade*, ed. H.S. Ellis and L.A. Metzler. Philadelphia: Blakiston, 1949.

Helpman, E. and Krugman, P. 1985. *Market Structure and Foreign Trade*. Cambridge, MA: MIT Press.

Jones, R.W. 1956. Factor proportions and the Heckscher–Ohlin theorem. *Review of Economic Studies* 24, 1–10.

Jones, R.W. 1961. Comparative advantage and the theory of tariffs. *Review of Economic Studies* 28, 161–75.

Jones, R.W. 1965. The structure of simple general equilibrium models. *Journal of Political Economy* 73, 557–72.

Jones, R.W. 1971. A three-factor model in theory, trade and history. In *Trade, Balance of Payments, and Growth*, ed. J. Bhagwati et al. Amsterdam: North-Holland.

Jones, R.W. and Neary, P. 1984. Positive trade theory. In *Handbook of International Economics*, vol. 1, ed. R.W. Jones and P.B. Kenen. Amsterdam: North-Holland.

Kantorovich, L. 1965. *The Best Use of Economic Resources*. Cambridge, MA: Harvard University Press.

Kenen, P.B. 1965. Nature, capital and trade. *Journal of Political Economy* 73, 437–60; Erratum, December, 658.

Krugman, P.R. 1979. Increasing returns, monopolistic competition and international trade. *Journal of International Economics* 9, 469–79.

Lancaster, K.J. 1980. Intra-industry trade under perfect monopolistic competition. *Journal of International Economics* 10, 151–75.

Leamer, E.P. 1984. *Sources of International Comparative Advantage: Theory and Evidence*. Cambridge, MA: MIT Press.

Leontief, W.W. 1933. The use of indifference curves in the analysis of foreign trade. *Quarterly Journal of Economics* 47, 493–503.

Leontief, W.W. 1954. Domestic production and foreign trade: the American capital position re-examined. *Economia Internazionale* 7, 9–38.

Lerner, A.P. 1932. The diagrammatic representation of cost conditions in international trade. *Economica* 12, 345–56.

Lerner, A.P. 1952. Factor prices and international trade. *Economica* 19, 1–15.

MacDougall, G.D.A. 1951. British and American exports. *Economic Journal* 61, 697–724.

McKenzie, L.W. 1954. Specialization and efficiency in world production. *Review of Economic Studies* 21(3), 165–80.

Ohlin, B. 1933. *Inter-regional and International Trade.* Cambridge, MA: Harvard University Press.

Oniki, H. and Uzawa, H. 1965. Patterns of trade and investment in a dynamic model of international trade. *Review of Economic Studies* 32, 15–38.

Pasinetti, L. 1960. A mathematical formulation of the Ricardian system. *Review of Economic Studies* 27, 78–98.

Ricardo, D. 1951. *The Works and Correspondence of David Ricardo*, vols. 1 and 4, ed. P. Sraffa. Cambridge: Cambridge University Press.

Samuelson, P.A. 1948. International trade and the equalization of factor prices. *Economic Journal* 58, 163–84.

Samuelson, P.A. 1949. International factor price equalization once again. *Economic Journal* 59, 181–97.

Samuelson, P.A. 1950. Evaluation of real national income. *Oxford Economic Papers* 2, 1–29.

Samuelson, P.A. 1953. Prices of factors and goods in general equilibrium. *Review of Economic Studies* 21, 1–20.

Samuelson, P.A. 1962. The gains from international trade once again. *Economic Journal* 72, 820–29.

Samuelson, P.A. 1965. Equalization by trade of the interest rate along with the real wage. In *Trade, Growth and the Balance of Payments*, ed. R. Baldwin et al. Chicago: Rand McNally.

Samuelson, P.A. 1971. Ohlin was right. *Swedish Journal of Economics* 73, 365–84.

Steedman, I. 1979a. *Trade Amongst Growing Economies.* Cambridge: Cambridge University Press.

Steedman, I., ed. 1979b. *Fundamental Issues in Trade Theory.* London: Macmillan.

Stiglitz, J. 1970. Factor–price equalization in a dynamic economy. *Journal of Political Economy* 78, 456–88.

Stolper, W. and Samuelson, P.A. 1941. Protection and real wages. *Review of Economic Studies* 9, 58–73.

Trefler, D. 1995. The case of missing trade and other mysteries. *American Economic Review* 85, 1029–46.

Vanek, J. 1963. *The Natural Resource Content of U.S. Foreign Trade 1870–1955.* Cambridge, MA: MIT Press.

Viner, J. 1937. *Studies in the Theory of International Trade.* New York: Harper.

comparative statics

Comparative statics in competitive general equilibrium (GE) environments provide insight into the operation of GE models and a means, at least in principle, to confront GE models with data.

For concreteness, I focus most of this article on what is arguably the canonical GE comparative statics conjecture: in finite exchange economies (that is, no production), equilibrium price changes are negatively related to endowment changes. In particular, if the endowment of good 1 increases and the endowments of other goods remain the same, then the price of good 1 falls. At the end of this article, I briefly survey other GE comparative statics results.

I break the analysis into three cases of increasing complexity.

Case I

There is a single consumer and two commodities. In an equilibrium of this trivial economy, the consumer eats her own endowment. Equilibrium relative prices (which are well defined even though there is no trade) are given by the slope of the consumer's indifference curve through her consumption/endowment point, ω^*. Let E denote her wealth expansion path through her initial endowment; E is the set of points where the slope of her indifference curve is the same as at ω^*.

If the new endowment, $\hat{\omega}$, lies below E, as in Figure 1, then the equilibrium price ratio p_1/p_2 falls. If $\hat{\omega}$ lies above E, then p_1/p_2 rises. If $\hat{\omega}$ lies along E, then p_1/p_2 remains unchanged.

The differential version of Figure 1 is given by Figure 2. The vector μ, the tangent to E, is the derivative with respect to nominal wealth of each good's demand (μ is mnemonic for marginal propensity to consume vector). To first order, the rule is that p_1/p_2 falls if and only if

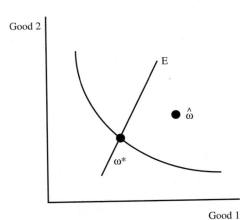

Figure 1 Comparative statics with one consumer and two commodities

Figure 2   First order comparative statics with one consumer and two commodities

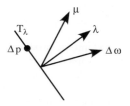

Figure 3   Condition CS with two goods

$\Delta\omega$, the vector of endowment changes, lies within $180°$ clockwise from μ. The vector μ incorporates second order information from the utility function and is, in particular, typically not collinear with the utility gradient.

If the endowment of good 1 increases while the endowment of good 2 remains unchanged, then $\Delta\omega$ lies along the positive good 1 axis. Figure 2 implies that, in this case, p_1/p_2 falls if and only if good 2 is normal ($\mu_2 > 0$); whether good 1 is normal or inferior (or even a Giffen good) is irrelevant.

Case II

There is again one consumer but L commodities. If $\Delta\omega$ lies along the positive good 1 axis, then a natural conjecture, to generalize the above observation for $L = 2$, is that p_1/p_ℓ falls for each good $\ell > 1$, provided each of these goods is normal. Hicks (1939) showed that this conjecture is false in general but that it is true if the gross substitute property holds (GS; the matrix of partial derivatives of excess demand with respect to price has positive off-diagonal entries). GS holds automatically in the one-consumer, $L = 2$, case because, at equilibrium, when $L = 2$, GS is equivalent to the weak axiom of revealed preference (WA).

Matters are more complicated if two or more endowments are shifting at the same time. For multivariate endowment shocks, there appears to be little one can say in general about changes in the price ratio of any specific pair of commodities. Instead, the conjecture is that $\Delta\omega$ is negatively related to Δp, the vector of equilibrium price changes. Formally,

$$\Delta p \cdot \Delta\omega \le 0.$$

Call this the comparative statics inequality, CS for short. Geometrically, CS says that the vectors Δp and $\Delta\omega$ lie at least $90°$ apart.

To interpret Δp as a change in relative prices, prices must be normalized. Consider linear price normalizations, in which all prices, in both the original economy and the perturbed economy, satisfy $p \cdot \lambda = 1$, where λ is an L vector. If all the coordinates of λ are positive, then a fall in the normalized price of good 1 means that the ratio

$$\frac{p_1}{p_{-1} \cdot \lambda_{-1}}$$

falls, where p_{-1} and λ_{-1} are subvectors corresponding to all goods other than the first: the price of good 1 falls relative to the value of a commodity bundle consisting of λ_ℓ units of each good $\ell > 1$. Standard choices of λ include $\lambda = (0, \ldots, 0, 1)$ (use the last commodity as numeraire) and $\lambda = \omega^*$ (normalize prices so that GNP remains constant; this is the Laspeyres normalization). Regardless of how, or whether, actual prices are normalized, one can re-normalize prices using whatever λ one chooses.

I can provide intuition for CS most easily by continuing to use figures for a two-good economy. Fix a normalizing vector λ. Since $p \cdot \lambda = 1$ for all p, $\Delta p \cdot \lambda = 0$. Therefore, Δp lies along the line that is at right angles to λ, labelled T_λ in Figure 3.

As drawn, $\Delta\omega$ lies within $180°$ clockwise from μ and hence p_1/p_2 falls. Therefore, Δp, normalized by λ, lies on the upper left-hand branch of T_λ. As illustrated, $\Delta\omega$ and Δp are more than $90°$ apart; hence CS holds.

On the other hand, suppose that $\Delta\omega$ lies in the cone spanned by λ and μ. Since $\Delta\omega$ again lies within $180°$ clockwise from μ, p_1/p_2 again falls and Δp again lies on the upper left-hand branch of T_λ. Now, however, $\Delta\omega$ and Δp are less than $90°$ apart. CS fails.

In general, in a one-consumer economy, for any number of commodities and for any preferences, CS fails whenever there is a gap between λ and μ and $\Delta\omega$ falls into this gap. Conversely, if $\lambda = \mu$ (or, more generally, if λ is a scalar multiple of μ) then CS holds for any endowment change: $\Delta p \cdot \Delta\omega \le 0$ with $\Delta p \cdot \Delta\omega = 0$ if and only if $\Delta\omega$ is collinear with μ (which is the differential analog of $\Delta\omega$ landing on the wealth expansion path E in Figure 1). In one consumer economies, $\lambda = \mu$ is thus the unique (up to scalar multiplication) linear price normalization for which CS holds for all endowment changes.

If preferences are quasi-linear in good L, and consumption is interior, then $\lambda = \mu$ implies $\lambda = (0, \ldots, 0, 1)$; the last good is used as numeraire. If the preferences are homothetic then μ is a scalar multiple of the reference endowment, ω^*, and so one can set $\lambda = \omega^*$. Typically, however, $\lambda = \mu$ is different from price normalizations commonly used in economics.

The $\lambda = \mu$ normalization, although non-standard, does have a sensible interpretation, provided μ is positive (all goods are weakly normal). If μ is positive, then a decrease in p_1 means that $p_1/(p_{-1} \cdot \mu_{-1})$ falls: the price of good 1 falls relative to the value of the consumer's *marginal* consumption of all other goods.

If $\Delta\omega$ lies along the positive good 1 axis and goods 2, ..., L are normal, then a minor variation on CS implies that $p_1/(p_{-1} \cdot \mu_{-1})$ falls, even if good 1 is inferior. This is a weaker conclusion than that of Hicks (1939) but it has a weaker hypothesis, since it does not assume the gross substitute property.

Case III

There are I consumers and L commodities. The generalization of CS is

$$\Delta p \cdot \Delta\bar{\omega} \leq 0,$$

where $\Delta\bar{\omega}$ denotes the change in the aggregate endowment. CS holds provided one uses an appropriate aggregate version of μ. Consider two alternatives. Each is a weighted sum of the individual marginal propensity to consume vectors, μ^i. In the first, $\bar{\mu}_{\Delta x}$, the weight on μ^i is consumer i's share in the change in the value of consumption, evaluated at the prices of the reference equilibrium. In the second, $\bar{\mu}_{\Delta\omega}$, the weight on μ^i is i's share in the change in the value of the endowment, again evaluated at reference equilibrium prices.

If one normalizes prices using $\lambda = \bar{\mu}_{\Delta x}$, then inequality CS holds provided *individual* excess demand satisfies the weak axiom (WA) at equilibrium. If $\lambda = \bar{\mu}_{\Delta\omega}$, then CS holds provided *aggregate* excess demand satisfies WA at equilibrium. See Nachbar (2002).

The hypothesis that aggregate excess demand satisfies WA is not implied by standard GE assumptions. One justification for nevertheless assuming WA is that it seems to be connected to the dynamic stability of the price adjustment process. WA holding at equilibrium is sufficient and almost necessary for local asymptotic stability under the Walrasian tâtonnement, for example. Comparative statics, by assuming that economies are at equilibrium, may therefore implicitly assume that aggregate excess demand satisfies WA. A second justification for assuming that aggregate excess demand satisfies WA is that this assumption, while strong, is not implausible in exchange economies. For some sufficient conditions for WA, see Becker (1962), Hildenbrand (1983), Grandmont (1992) and Quah (1997).

In the one-consumer case, the $\lambda = \mu$ price normalization was necessary as well as sufficient. There are analogous, but weaker, necessity results for $\bar{\mu}_{\Delta x}$ and $\bar{\mu}_{\Delta\omega}$. The important implication is that, because both $\bar{\mu}_{\Delta x}$ and $\bar{\mu}_{\Delta\omega}$ can vary with how the endowment changes are distributed across consumers, there may be *no* price normalization for which CS holds for all endowment changes. As the endowment distribution changes, the price normalization may have to change.

This illustrates an issue that has become a central theme in the recent literature on GE comparative statics. Given an arbitrary price normalization, standard GE assumptions impose no restrictions on the relationship between changes in equilibrium prices and changes in

the aggregate endowment (see Chiappori et al., 2004). This negative result, a cousin of the Debreu–Mantel–Sonnenschein theorem (DMS), has a loophole: standard GE assumptions do provide comparative statics restrictions if one works with micro-level information (for example, on the endowment distribution) rather than exclusively with aggregates. In the CS results, micro-level data is used in the price normalization. Note that CS requires micro data even if one assumes that aggregate excess demand satisfies WA.

Relative to the objectives laid out in the first paragraph of this article, the results on CS comparative statics fare reasonably well in providing insight into the operation of GE models. The $\bar{\mu}_{\Delta\omega}$ result is much the easier to interpret, since it is computed with the use of endowment changes, which are exogenous. The $\bar{\mu}_{\Delta x}$ result, on the other hand, extends easily to production economies. In contrast, I do not know whether the $\bar{\mu}_{\Delta\omega}$ result has a useful analog for production economies.

The CS inequality fares less well as a tool for empirical work, because it requires a large amount of data just to estimate the normalization vector. The necessity results imply that this difficulty is intrinsic to CS.

Other comparative statics results

Brown and Matzkin (1996), a path-breaking paper that has heavily influenced subsequent work in this area, exploits the DMS loophole noted above to give testable restrictions linking equilibrium prices with individual endowments. For related work, see Snyder (1999), Williams (2002), Kübler (2003) and Chiappori et al. (2004). Relative to CS, the Brown–Matzkin restrictions are easier to implement empirically because they do not require estimating normalization vectors, but they are harder to interpret.

As already noted, CS-type reasoning can be extended to production economies (see Quah, 2003; Nachbar, 2004). CS-type reasoning can also be extended to asset pricing environments (see Quah, 2003).

For shocks to preferences rather than endowments or technologies, the analog of CS is

$$\Delta p \cdot \Delta\bar{x} \geq 0,$$

where $\Delta\bar{x}$ is the change in equilibrium consumption. Profit maximization implies that this inequality holds for any price normalization. In this respect, the analysis of demand shocks is trivial compared with the analysis of supply shocks.

Interest in comparative statics has helped motivate research on the uniqueness, regularity, and stability of equilibria (see Kehoe, 1987). Note that some of the comparative statics results cited above (for example, the Brown–Matzkin results and the $\bar{\mu}_{\Delta x}$ CS result) do not assume uniqueness or stability.

Finally, perhaps the most famous comparative statics results are the Stolper-Samuelson theorem and its dual, the Rybcyznski theorem (for a recent treatment, see Echenique and Manelli, 2005). Stolper–Samuelson links changes in factor prices with factor intensities and changes in output prices. Rybcyznski links changes in final goods production with factor intensities and changes in factor supplies. Although it is possible to embed these results within a highly restricted GE model, they are partial equilibrium in spirit; wealth effects play no role.

JOHN NACHBAR

See also **general equilibrium; general equilibrium (new developments); international trade theory.**

Bibliography

Becker, G. 1962. Irrational behavior and economic theory. *Journal of Political Economy* 70, 1–13.
Brown, D. and Matzkin, R. 1996. Testable restrictions on the equilibrium manifold. *Econometrica* 64, 1249–62.
Chiappori, P., Ekelund, I., Kübler, F. and Polemarchakis, H. 2004. Testable implications of general equilibrium theory: a differentiable approach. *Journal of Mathematical Economics* 40, 105–19.
Echenique, F. and Manelli, A. 2005. Comparative statics, English auctions, and the Stolper–Samuelson theorem. Mimeo. Tempe: Arizona State University.
Grandmont, J. 1992. Transformations of the commodity space, behavioral heterogeneity and the aggregation problem. *Journal of Economic Theory* 57, 1–35.
Hicks, J. 1939. *Value and Capital*. Oxford: Clarendon Press.
Hildenbrand, W. 1983. On the law of demand. *Econometrica* 51, 997–1018.
Kehoe, T. 1987. Comparative statics. In *The New Palgrave: A Dictionary of Economics*, vol. 1, ed. J. Eatwell, M. Milgate and P. Newman. Basingstoke: Palgrave.
Kübler, F. 2003. Observable restrictions of general equilibrium with financial markets. *Journal of Economic Theory* 110, 137–53.
Nachbar, J. 2002. General equilibrium comparative statics. *Econometrica* 79, 2065–74.
Nachbar, J. 2004. General equilibrium comparative statics: the discrete case with production. *Journal of Mathematical Economics* 40, 153–63.
Quah, J. 1997. The law of demand when income is price dependent. *Econometrica* 65, 1421–42.
Quah, J. 2003. Market demand and comparative statics when goods are normal. *Journal of Mathematical Economics* 39, 317–33.
Snyder, S. 1999. Testable restrictions of Pareto optimal public good provision. *Journal of Public Economics* 71, 97–119.
Williams, S. 2002. Equations on the derivatives of an initial endowment-competitive equilibrium mapping for an exchange economy. Mimeo. Champaign-Urbana: University of Illinois.

compensating differentials

Compensating differentials represent a wage premium for unpleasant aspects of a job. Jobs differ along a number of dimensions. Some jobs offer generous health insurance benefits. Other jobs entail long hours or may expose workers to physical risks. Some jobs are only available in polluted cities. The theory of compensating differentials is based on the simple premise that there is 'no free lunch'. In market equilibrium, more unpleasant jobs will offer a wage premium relative to other jobs. Similarly, homes in nicer communities or high-quality-of-life cities will sell for a premium. To quote Sherwin Rosen (2002, p. 2), 'Markets accommodate diversity by establishing prices that tend to make different things relatively close substitutes at the margin. Adam Smith's insight that market prices tend to equalize their net advantages is fundamental to these problems. If one good has more desirable characteristics than another, the less preferred variety must compensate for its disadvantages by selling at a lower price.'

Defining compensating differentials

Jobs represent tied bundles of attributes. Suppose that a worker gains utility from earning a wage and from a job attribute. This attribute could represent job safety, or total days of vacation, or health insurance benefits. As shown in Figure 1, there are two jobs, A and B. Each job represents a different bundle of a wage and a non-market job-specific amenity level. The two jobs differ: job B is the more pleasant of the two. If all workers have the same utility function, then in equilibrium this representative worker must be indifferent between the two jobs. Thus, job A must pay a higher wage than job B to compensate this worker.

The econometrician can collect data on each job type's wage and amenities. In a more realistic economy where there are many types of jobs that differ with respect to the wage and their amenity level, the representative worker's indifference curve would be sketched out. The slope of the representative worker's indifference curve represents the compensating differential of how much lower a wage

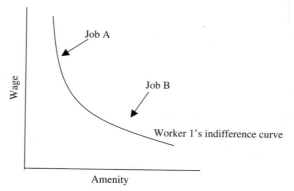

Figure 1

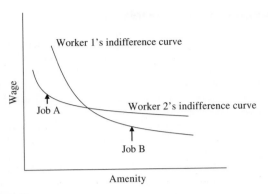

Figure 2

this worker would accept in return for a small increase in the job amenity.

To see how worker heterogeneity affects the interpretation about observed compensating differentials, consider the simple extension where we introduce two types of workers. These workers are equally productive but differ with respect to their demand for working in the more pleasant job. In Figure 2, worker 1 values the job amenity more than worker 2. In equilibrium, job A will pay a compensating differential to attract workers to be willing to work in this job. Worker 2 will choose to work in job A while worker 1 will choose to work in job B. Firm A will prefer to hire worker 2 rather than worker 1 because worker 1 requires a larger compensating differential for working in the more unpleasant job. The profit maximizing firm seeks to minimize its costs of production.

The econometrician will observe the equilibrium wage paid to workers in job A and B. As shown in Figure 2, this equilibrium wage–amenity relationship called the *hedonic wage function* does not represent either worker 1's or worker 2's indifference curves. Instead, this hedonic wage function represents the envelope of the minimum wage that heterogeneous workers are willing to accept to do a job. This simple example highlights how introducing worker heterogeneity affects inference from observed data (see Rosen, 2002). Figure 2 focuses on just one dimension of worker heterogeneity. The recent compensating differentials literature has explored the consequences of other dimensions of worker heterogeneity such as unobserved skill (IQ, for example) and a worker's ability to self-protect against injury on the job (Hwang, Reed and Hubbard, 1992; Shogren and Stamland, 2002).

Labour econometric applications of compensating differentials theory

An enormous applied econometrics literature has estimated various versions of hedonic wage functions to recover estimates of the marginal valuation of non-market job attributes. One major focus of this research has been to estimate the value of life by measuring how

much of a wage premium the marginal worker requires for working in a job with a higher probability of death (Viscusi and Aldy, 2003). Other studies have used hedonic methods to measure the compensating differential for mandated government health insurance benefits (Gruber, 1994).

The standard approach utilizes a large micro-data set. The dependent variable in such a study is a full-time worker's wage in a specific occupation, industry or city. For example, in equation (1) the dependent variable is the log of worker i's wage in industry j in year t. In an urban application, j would refer to a city rather than an industry. The researcher will include a large number of demographic controls, such as age, ethnicity, or education, to 'standardize' the worker. If one controls for these factors, the key variables of interest are the Z's in equation (1). In a labour economics application, the Z vector may represent a set of job specific attributes (length of day, job risk). In an urban economics application, the Z vector may represent attributes of the city where the job is located (climate, pollution, crime).

$$\text{Log}(\text{Wage}_{ijt}) = \gamma_0 + \gamma_1^* X_{it} + \gamma_2^* Z_{jt} + U_{ijt}$$

(1)

Ordinary least squares regression estimates of γ_2 are used to construct measures of the compensating differentials for job tasks and characteristics of employment locations. Estimates of such coefficients have been used to rank city quality of life (see Gyourko and Tracy, 1991) and represent the first stage of the hedonic two-step for recovering demand functions for non-market goods such as air quality or climate (Rosen, 1974; Ekeland, Heckman and Nesheim, 2004).

If the population differs with respect to its tastes for job attributes, then γ_2 can be used to construct a worker's budget constraint. For example, in a job-safety regression if γ_2 equals minus $100 then this means that a one-unit increase in job safety will cost the worker an extra $100 in wages. The rational worker facing this budget constraint will take this trade-off into account when choosing the job that maximizes her utility.

Hedonic estimates of compensating differentials can also be used to bound worker preferences. To return to Figure 2, a lower bound on worker 1's willingness to accept work in risky job A is the equilibrium wage paid to worker 2. Since we know that worker 1 chose the safe job and refused to work in job A at the wage that worker 2 accepted, worker 2's wage offer provides a lower bound (see Rosen, 2002).

The typical hedonic wage regression study estimates eq. (1) using ordinary least squares. This econometric approach will yield consistent estimates of γ_2 if the unobserved determinants of wages (that is, the error term) are uncorrelated with the explanatory variables. What is the error term in this hedonic pricing equation? While a researcher might hope that it represents measurement error in the dependent variable, it is more likely that the error term represents unobserved attributes of the worker and unobserved attributes of the geographical area where the worker lives and works.

Unfortunately for researchers, people self-select where to live and work. A researcher would like to know what wage the same worker would earn in every industry and in every city. In a cosmopolitan city such as New York, superstars of all fields, ranging from Don Trump in real estate to Derek Jeter in baseball, have all chosen to work there. A naive cross-city hedonic researcher would observe these stars living in New York City earning high wages *relative* to observationally identical people in Tulsa, and would conclude, based on the wage regression, that New York City's quality of life must be worse than Tulsa's. Clearly, the problem with this inference is the 'apples to oranges' comparison. New York City's amenities are a normal good. The high-skilled earn higher salaries and are attracted to living and working in this city.

Conclusion

A job's wage is not a sufficient statistic for its quality. Coal miners are paid a relatively high wage but the work is dangerous and unpleasant. A major research agenda in labour economics investigates how much people implicitly pay for non-market job attributes. Credible estimates of wage compensating differentials for living in less polluted cities or working in risky industries would greatly aid policy analysis that seeks to measure the benefits of environmental and safety regulation.

MATTHEW E. KAHN

See also **Roy model; wage inequality, changes in.**

Bibliography

Ekeland, I., Heckman, J. and Nesheim, L. 2004. Identification and estimation of hedonic models. *Journal of Political Economy* 112, S60–S109.

Gruber, J. 1994. The incidence of mandated maternity benefits. *American Economic Review* 84, 622–41.

Gyourko, J. and Tracy, J. 1991. The structure of local public finance and the quality of life. *Journal of Political Economy* 91, 774–806.

Hwang, H.-S., Reed, R. and Hubbard, C. 1992. Compensating wage differentials and unobserved productivity. *Journal of Political Economy* 100, 835–58.

Rosen, S. 1974. Hedonic prices and implicit markets: product differentiation in pure competition. *Journal of Political Economy* 82, 34–55.

Rosen, S. 2002. Markets and diversity. *American Economic Review* 92, 1–15.

Shogren, J. and Stamland, T. 2002. Skill and the value of life. *Journal of Political Economy* 110, 1168–73.

Viscusi, W. and Aldy, J. 2003. The value of statistical life: a critical review of market estimates throughout the world. *Journal of Risk and Uncertainty* 27, 5–76.

compensation principle

The term 'compensation principle' refers to the principle that, in comparing two alternative states in which a given community of persons might find itself, one of the states constitutes an improvement over the other (in the weak sense including equivalence) if it is possible for the gainers to compensate the losers for their losses and still be at least as well off as in the original state.

If the hypothetical compensation is actually carried out, the principle reduces to the Pareto criterion: all are at least as well off, in one state compared to the other. There is no need to invoke the compensation principle in such a case. On the other hand, if the principle is used to compare two unique alternative states in which a community might find itself, neither of which is Pareto-superior to the other, the principle seems quite arbitrary unless interpreted in a broader context. There is a sense in which one person might be said to be basically healthier than another even though, at the particular moment, such a person might have a cold and the other one not. The compensation principle is usually used to make comparisons in this sense; one state of the economy is sounder, healthier, more robust, or has greater productive potential, than another. What this implies is that states under comparison are usually not unique, singleton states but composite ones, or sets of states. Formally, the objects being compared are usually sets of commodity bundles that could be made available to the aggregate of consumers, described in the literature as 'situations' in contrast to single 'points' in such sets (cf. Baldwin, 1954).

Examples of comparisons in which the compensation principle is typically used are those between (*a*) a perfectly competitive system of industrial organization and an imperfectly competitive one; (*b*) free trade and no trade (or restricted trade); (*c*) the state of an economy before and after a war, or depression, or change in productive techniques. Most but not all of these types of comparisons are relevant to policy decisions; and the policy decisions are usually not of an ad hoc type (for which the compensation principle would hardly be appropriate) but of a fundamental nature concerning the underlying system of industrial organization and trade.

Inasmuch as the principle can be applied without the need to make interpersonal comparisons, some of its more ardent proponents have maintained that it is 'value-free'. However, there can be no doubt that it does require acceptance of some value judgements, since the Pareto criterion itself constitutes one – albeit a widely accepted one. Another value judgement implicit in the principle as it has usually been applied is that each individual is the best judge of his or her own well-being; while also quite widely accepted, this one is obviously controversial, and in fact government policy measures are often called for precisely in those instances where it is clearly an untenable assumption. But the most important and controversial way in which value judgements enter into the compensation principle is in the conflict between potentiality and actuality: one situation is judged better than another if everybody *could* be made better off in the new situation even though some in fact become worse off. This lacuna in the principle has led Little (1950) and Mishan (1969) to formulations in which compensation tests are combined with explicit distributional value judgements, and Samuelson (1947; 1956) into a full-fledged ethical system in which compensation is carried out to the extent that the ethical norms dictate.

In many applications the compensation principle is difficult to formulate in a precise manner unless one assumes absence of externalities in consumption, so it is usually formulated (but with some notable exceptions – for example, Coase, 1960) under the assumption that each person's welfare depends only on his or her own consumption of goods and services. In most applications, the data available for making comparisons are, almost inevitably, limited to aggregative information on the actual state of the economy in each situation; much of the work in applying the principle therefore consists in using economic theory to make inferences from the actual observations concerning underlying conditions in the economy. By its nature, the compensation principle is limited in its application to comparing alternative states (or sets of states) of a given community of individuals; thus, it cannot be applied (at least not literally) to historical comparisons of a country's condition over time (since the population has changed) or to comparisons of the living conditions of different countries (since the populations are different). However, extensions of the principle to cover such comparisons are possible provided suitable additional empirical assumptions and value judgements are accepted; for example, if all individuals are assumed to have identical preferences, one could ask whether there exists a redistribution of income in each period (or country) such that each individual in the one situation would be better off than each individual in the other. This would obviously entail additional value judgements along with the additional empirical assumptions.

1 Historical development: from Dupuit to Hotelling

The compensation principle may be traced back to Dupuit (1844, pp. 359–60; Arrow and Scitovsky, 1969, p. 272) and Marshall (1890, p. 447; 1920, p. 467) who used the concept of consumers' surplus to compare the losses of consumers (say from a bridge toll or an excise tax) with the gains to the government. The demonstration that the former exceed the latter, so that consumers cannot be compensated for their losses out of the government revenues, provided a convincing case for the superiority of income tax to an excise tax (or for the superiority of government subsidization of bridge construction to its financing of it by tolls), and at the same time provided scientific prestige and great intuitive appeal to a method that was able to reach such a definitive conclusion and furnish a measure of the 'deadweight loss'.

While Dupuit and Marshall used partial-equilibrium analysis, Pareto (1894, p. 58) was the first to introduce the concept into general-equilibrium theory, in the course of an article devoted to proving the optimality of competitive equilibrium. In the first part of this article (summarized by Sanger, 1895), Pareto used as his criterion of optimality the sum of individual utilities; in the second part, however – acknowledging the criticisms and suggestions of Pantaleoni and Barone (both admirers of Marshall, which Pareto was not) – he reformulated the problem so as to sum not the utilities of different consumers but the quantities they consume. His criterion of optimality (1894, p. 60) was that it should be impossible for one person to gain without another losing – 'Pareto optimality' – a criterion that had also been introduced by Marshall (1890, pp. 449–50; 1920, pp. 470–1). A more refined version of Pareto's argument later appeared in the *Cours* (Pareto, 1896–7, vol. 1, pp. 256–62; vol. 2, pp. 88–94).

The proposition formulated by Pareto (1894) anticipated what has now come to be known as the 'fundamental theorem of welfare economics', namely, that every competitive equilibrium is Pareto optimal and, conversely, every Pareto optimum can be sustained by a competitive equilibrium. Pareto considered the problem faced by a socialist state striving to attain an outcome in which it was impossible for one person to gain without another losing. The Ministry of Justice would concern itself with problems of income distribution, and the Ministry of Production with resource allocation and choice of production coefficients. A weakness of Pareto's argument was that he assumed a price system already to be established – perhaps our socialist state needs the prices of its capitalist neighbours to guide it. Pareto further assumed that each individual's budget constraint was adjusted by the addition of a parameter (a lump-sum subsidy or tax) controlled by the government. The government's objective was to maximize the sum of these parameters, which he showed was equal to aggregate profit – the value of commodities consumed less the value of factor services supplied, equal to the value of firms' output less the outlay on their factor inputs. If it were possible to increase all the parameters, the existing situation would not be Pareto optimal; if their sum were a maximum, it would not be possible to increase one of them without decreasing another, and the outcome would be Pareto optimal. Pareto showed that maximization of aggregate profit at the given prices, subject to the resource-allocation and production-function constraints, would lead to cost-minimization and zero profits. (For mathematical details of Pareto's arguments see Chipman, 1976, pp. 88–92.) Pareto summarized this result by stating (1896–7, vol. 2, p. 94):

Free competition of entrepreneurs yields the same values for the production coefficients as would be obtained by determining them by the condition that commodity outputs should be chosen in such a way that, for some appropriate distribution, maximum ophelimity would be achieved for each individual in society.

The last clause was Pareto's unfortunately awkward way of stating the criterion of Pareto optimality.

Barone (1908), who had originally spurred Pareto on to this line of argument, developed it further himself. He noted that a competitive equilibrium has the property that aggregate profit is at a maximum at the equilibrium prices, hence, for any feasible departure from this equilibrium, valuing consumption and factor services at the equilibrium prices, some individuals may gain and others will lose, the losses outweighing the gains so that, even if the gainers part with all their gains, the rest will still be worse off than originally. (Barone used what is now known as the criterion of revealed preference to make inferences concerning preferences from data on prices and incomes.) Such a state was described by Pareto and Barone as 'destruction of wealth', and its measure by aggregate income loss at the competitive-equilibrium prices provided an alternative to the deadweight loss considered by Dupuit and Marshall. Barone (1908) also related his arguments to those of Marshallian consumers'-surplus analysis.

Lerner (1934) invoked the compensation principle in his proposed method for measuring monopoly power, describing it as 'a loss to the consumer which is not balanced by any gain reaped by the monopolist'. In this paper Lerner also formulated, apparently independently, the concept of Pareto optimality.

Hotelling (1938) made a noteworthy contribution by providing an alternative demonstration of the inferiority of excise taxes to income taxes, using the compensation principle directly. He considered a single individual consuming n commodities in amounts q_j and facing market prices p_j. Prior to the imposition of the excise taxes (or tolls), the individual consumes a bundle q^0 at prices p^0 and income (or fixed component of income) m^0, which maximizes a utility function $U(q)$ subject to the budget constraint $p^0 \cdot q = m^0$. Subsequent to the introduction of taxes, market (tax-inclusive) prices and after-tax income are p^1 and m^1 respectively, and a bundle q^1 is chosen which maximizes $U(q)$ subject to $p^1 \cdot q = m^1$. The government collects $r = (p^1 - p^0) \cdot q^1 - (m^1 - m^0)$ in revenues. Since the government is assumed to collect $(p_j^1 - p_j^0) \cdot q_j^1$ in taxes on commodity j, p_j^0 must be identified with the production cost after the tax (as well as with the market price=production cost before the tax); this is a fairly restrictive assumption, since it implies that the tax does not affect production costs. (In this respect Hotelling's treatment is less general than Dupuit's and Marshall's, involving infinite elasticities of supply.) We may denote the ad valorem excise-tax rate on commodity j by $t_j = p_j^1/.p_j^0 - 1$, and a proportional income-tax rate by $t_0 = 1 - m^1/m^0$ (negative taxes are interpreted as subsidies). The government's revenues are

$$r = \sum_{j=1}^{n} t_j p_j^0 q_j^1 + t_0 m^0 = 0,$$

assumed zero since the government distributes the total proceeds of these excise taxes back to the consumer (or taxes the consumer if these are negative). The consumer's budget constraint after the imposition of the taxes is

$$\sum_{j=1}^{n} (1 + t_j) p_j^0 q_j^1 = (1 - t_0) m^0.$$

These two equations together imply that q^1 satisfies the budget constraint $p^0 \cdot q^1 = m^0$, hence q^1 was in the consumer's original budget set. Therefore, setting aside the 'infinitely improbable ...contingency' that q^0 and q^1 lie on the same indifference surface, Hotelling concluded (1938, p. 252) that 'if a person must pay a certain sum of money in taxes, his satisfaction will be greater if the levy is made directly on him as a fixed amount than if it is made through a system of excise taxes which he can to some extent avoid by rearranging his production and consumption'.

Unfortunately Hotelling overlooked the fact that if $t_j = t$ for all j then the government's budget constraint implies $p^0 \cdot q^1 = -m^0 t_0/t$, whence $t_0 = -t$ and $q^1 = q^0$. That is, a system of uniform ad valorem excise taxes is equivalent to a proportional income tax. This was pointed out by Frisch (1939) and accepted by Hotelling (1939). As Frisch made clear, what Hotelling really proved was the non-optimality of a system of non-proportional excise taxes or subsidies when selling prices are given. If these selling prices are equal to marginal costs, Hotelling's theorem shows that market prices should be proportional to marginal costs. Since incomes are fixed in Hotelling's formulation, income taxes may be regarded as lump-sum taxes. If institutional consideration make excise taxes impossible for one commodity (say leisure), then they must be zero for all commodities and optimality requires that prices be equal to marginal costs. (For a less charitable interpretation of Hotelling's contribution see Silberberg, 1980.)

Hotelling went on to assert that his proposition could be extended to many consumers (though no details or proof were provided), and he proceeded to examine the consumers'-surplus measure of loss $\frac{1}{2}(p^1 - p^0) \cdot (q^1 - q^0) = \frac{1}{2} Tp^0 \cdot (q^1 - q^0)$ (where T is a diagonal matrix of excise-tax rates t_j). He also made some general observations (1938, p. 267) that, to this day, constitute what is probably the best statement to be found of the philosophy underlying the compensation principle.

2 The years of the new welfare economics

In the cases to which the compensation principle was applied by Dupuit, Marshall, Lerner and Hotelling, compensation was made between the class of consumers on the one hand and a government or a monopolist on the other. While Pareto and Barone had discussed compensation between different classes of consumers (as had Hotelling in his general remarks) their work was unknown to English-speaking economists until the publication in 1935 of the English translation of Barone's 1908 work. Even this seems not to have struck home, however, since Kaldor (1939) cited passages from Harrod (1938) and Robbins (1938) to the effect that, since movement towards free trade would affect different classes differently, no scientific statement could be made concerning the beneficial effect of free trade without making interpersonal comparisons of utility.

Kaldor (1939) proceeded to sketch an argument to the effect that removal of an import duty (using the classical example of repeal of the Corn Laws) would result in a situation in which the losses incurred by the landlords could be compensated by the gains (through lower import prices) obtained by the other consumers. Such an argument cannot be correct, however, since, as Kaldor (1940) pointed out only a year later, it follows from Bickerdike's theory of optimal tariffs that a country can gain from the imposition of a sufficiently small duty, and, as Graaff (1949) and others later demonstrated, the compensation principle can be used to show that, with suitable compensation, all persons can gain. Unless the rate of corn duty was above the optimal tariff rate, the opposite conclusion would follow to that indicated by Kaldor (1939).

A previous attempt by Pareto (1895) to show by means of the compensation principle that a tariff would lead to 'destruction of wealth' was defective, since he assumed trade to be balanced in domestic prices and thus he failed to take account of the improvement in the terms of trade and the beneficial effect of the tariff revenues.

Other attempts prior to 1939 to make the case for free trade suffered from vagueness both in specifying the criterion of gain and in specifying the alternative with which free trade was being compared. Ricardo (1815, p. 25) stated: 'There are two ways in which a country may be benefited by trade – one by increase of the general rate of profits ... the other by the abundance of commodities, and by a fall in their exchangeable value, in which the whole community participate'. According to Cairnes (1874, p. 418), 'the true criterion of the gain on foreign trade [is] the degree in which it cheapens commodities, and renders them more abundant'. A hint of a compensation principle is found in Viner (1937, pp. 533–4):

> free trade ... necessarily makes *available* to the community *as a whole* a greater physical real income in the form of more of *all* commodities, and ... the state ... can, by appropriate supplementary legislation, make certain that removal of duties shall result in more of *every* commodity for *every* class of the community.

Like Kaldor's statement, this is formally incorrect; but it was sufficiently suggestive to stimulate Samuelson (1939) into providing a formal proof of a gains-from-trade theorem, albeit under very restrictive assumptions.

Samuelson (1939) assumed that an open economy had a locus $\phi(y, l) = 0$ of efficient combinations of outputs y and (variable) factor services l, and asserted that vectors of prices p and factor rentals w in competitive equilibrium would be such that aggregate profit $p \cdot y - w \cdot l$ is a maximum. This is the same as the proposition of Pareto (1894), and Barone (1908) referred to above. Letting x denote the bundle of commodities consumed, under both (balanced) free trade and autarky the budget equation $p \cdot x = p \cdot y$ holds. Letting superscripts 0 and 1 denote equilibrium values under autarky and free trade respectively, it follows that

$$p^1 \cdot x^1 - w^1 \cdot l^1 \geq p^1 \cdot x^0 - w^1 \cdot l^0.$$

Assuming all N individuals to be identical in their preferences and ownership of factors, and dividing this inequality through by the number of individuals, it states that each person chooses $(x^1/N, l^1/N)$ under free trade when $(x^0/N, l^0/N)$ is available, hence (if $p^1 \neq p^0$) each person prefers $(x^1/N, l^1/N)$ to $(x^0/N, l^0/N)$. Therefore free trade is Pareto-superior to autarky.

Samuelson went on to assert (1939, p. 204) that, if the assumption of identical individuals is dropped, then, although it could no longer be said that each individual was better off under free trade, 'it would always be possible for those who desired trade to buy off those opposed to trade, with the result that all could be made better off'. This argument went unchallenged until Olsen (1958) pointed out that, if compensation were paid from gainers to losers, a new equilibrium price constellation p^1 would result, and the argument no longer follows. For this reason Samuelson's 1939 results has come to be known as the gains-from-trade theorem for the 'small-country case', though this interpretation was not suggested by Samuelson at the time. But this description of Samuelson's result is inaccurate. Generalizing his argument we can say that if (x_i^t, l_i^t) are the allocations of (x^t, l^t) to individual i, where $\sum_{i=1}^{N} x_i^t = x^t$ and $\sum_{i=1}^{N} l_i^t = l^t$, the given the allocations (x_i^1, l_i^1) of (x^1, l^1) under free trade one can find Pareto-optimal allocations (x_i^0, l_i^0) of (y^0, l^0) under autarky such that

$$p^1 \cdot x_i^1 - w^1 \cdot l_i^1 \geq p^1 \cdot x_i^0 - w^1 \cdot l_i^0$$
$$\text{for} \quad i = 1, 2, \ldots, N.$$

This proves that for *any* free-trade equilibrium it is possible to find a weakly Pareto-inferior Pareto-optimal autarky equilibrium. It does not prove the obverse proposition that for any autarky equilibrium it is possible to find a weakly Pareto-superior free-trade equilibrium. A general gains-from-trade theorem was therefore yet to be established, but Samuelson had provided an important first step.

Hicks (1939) ushered in the 'new welfare economics' with a synthesis building on Hotelling (1938) and Kaldor (1939) and based on the compensation principle, making it possible, according to him, to make policy proposals in favour of economic efficiency which were free of value judgements. Hicks's most original contribution (Hicks, 1940) was his attempt to apply the compensation principle to data on a country's real national income. This was a natural thing to try to do, since Pigou's (1920) main work was devoted to evaluating a country's welfare by national-income comparison, and it was largely Pigou's resort to interpersonal comparisons in order to justify this that was the object of Robbins's (1938) criticism.

Hicks's (1940) basic tool was the 'revealed-preference' comparison which had been employed by Barone (1908) and Hotelling (1938). If observations are available at times 0 and 1 of a country's national income in period-1 prices, and it is recorded that $p^1 \cdot y^1 \geq p^1 \cdot y^0$ (where p^t, y^t are vectors of prices and outputs at time t), what can be inferred? In the first place, to make any headway one must assume that the observed situations are competitive equilibria. Let us define an allocation of a commodity bundle x as an $N \times n$ matrix X whose ith row, x_i, is the bundle of n commodities allocated to individual i, and whose row sum $\sigma(X) = \Sigma_{i=1}^N x_i$ is equal to x. As between two bundles x_i^0, x_i^1 consumed by individual i, let us define $x_i^1 R_i x_i^0$ to mean that x_i^1 is preferred or indifferent to x_i^0 by individual i, where R_i is a continuous, convex, monotonic total order, with P_i denoting strict preference and I_i indifference. (This relation assumes the absence of externalities in consumption.) Finally, let $X^1 R X^0$ (resp. $X^1 P X^0$) mean that X^1 is weakly (resp. strictly) Pareto-superior to X^0 (i.e. $x_i^1 R_i x_i^0$ for all i, resp. $x_i^1 R_i x_i^0$ for all i and $x_i^1 P_i x_i^0$ for some i). Then, from the real-income comparison $p^1 \cdot y^1 \geq p^1 \cdot y^0$, Hicks noted that there does not exist an allocation X of y^0 that is weakly Pareto-superior to the actual allocation X^1 of y^1. This follows from the same argument that establishes the Pareto optimality of the assumed competitive equilibrium in period 1. The non-existence of an allocation X of y^0 such that XRX^1, where $\sigma(X^1) = y^1$, constituted for Hicks the definition of an 'increase in real social income'.

Kuznets (1948) pointed out by an example that, in the case considered by Hicks, it could also be true that there is no allocation X of y^1 which is weakly Pareto superior to the actual allocation X^0 of y^0. Accordingly he suggested that Hicks's criterion be supplemented by the condition that there should exist an allocation $\bar{X}$ of y^1 that is weakly Pareto superior to the actual allocation X^0 of y^0. But while the latter criterion implies $p^0 \cdot y^1 \geq p^0 \cdot y^0$, it is not implied by it, so a national-income comparison using current and base prices would still not yield Kuznets's criterion.

Kuznets's criticism of Hicks was similar to the objection raised by Scitovsky (1941) to the criterion proposed by Kaldor (1939). According to Scitovsky's interpretation of Kaldor, an allocation X^1 of y^1 is better than an allocation X^0 of y^0, if there exists a reallocation $\bar{X}^1$ of y^1, which is Pareto superior to X^0. Scitovsky objected that this gave preference to the *status quo ante*, and besides, he pointed out that the criterion was internally inconsistent in the sense that it allowed two such pairs (X^t, y^t) to be superior to each other. He therefore proposed that Kaldor's test be supplemented by the criterion that there exist a reallocation $\bar{X}^0$ of y^0 that is Pareto inferior to X^1.

The literature on 'compensation tests' suffered from ambiguity as to the domain of definition of the relations and internal inconsistency of the relations. It was pointed out by Gorman (1955) that the relations were intransitive. It was shown in Chipman and Moore (1978) that the Hicks–Kuznets and Scitovsky double criteria, as well as the national-income comparisons in terms of base- and current-year prices, could lead to cycles of three competitive equilibria each superior to its successor.

The definitive contribution to the subject of national-income comparisons was that of Samuelson (1950) who introduced what Chipman and Moore (1971) described as the 'Kaldor–Hicks–Samuelson (KHS) ordering'. The objects under comparison in this approach are sets Y of commodity bundles y, e.g. production-possibility sets. Letting $A(Y)$ denote the set of allocation matrices X such that $\sigma(X) \in Y$, this ordering is defined by

$$Y^1 >_R Y^0 \Leftrightarrow \left[\forall X^0 \in A(Y^0)\right]\left[\exists X^1 \in A(Y^1)\right] X^1 R X^0.$$

In words, Y^1, is potentially superior to Y^0 if, for all allocations of commodity bundles in Y^0, there exists a (weakly) Pareto superior allocation of a commodity bundle in Y^1. This is a reflexive and transitive relation; it also satisfies the condition that $Y^0 \subseteq Y^1$ implies $Y^1 >_R Y^0$. Samuelson also introduced the important concept of a utility-possibility frontier, which is the relative boundary of a utility-possibility set $U(Y, R; f)$; this in turn is a set of N-tuples of individual utilities, $u = f(X)$, for some $X \in A(Y)$, where f is an N-tuple of positive-valued utility functions representing R. If the sets Y are 'disposable' (that is, containing for every $y \in Y$ the bundles y' with $0 \leq y' \leq y$), and the R_i continuous and monotonic, then the utility-possibility sets are also disposable. If Y is non-empty, compact disposable, and convex, and the R_i are continuous, monotonic, and convex, then, provided the f_i are continuous and concave, $U(Y, R; f)$ is non-empty, compact, and convex (cf. Chipman and Moore, 1971, p. 24). If the f_i are only quasi-concave not concave, $U(Y, R; f)$ need not be convex (cf. Kannai and Mantel, 1978). The KHS ordering among consumption-possibility sets translates into set-inclusion of the corresponding utility-possibility sets. Samuelson (1959, p. 10) gave an example of a case of crossing utility-possibility frontiers in which $X^2 \in A(Y^2)$ was Pareto superior to $X^1 \in A(Y^1)$ yet Y^1 would be ranked higher than Y^2 in terms of some value judgement. This established that the 'compensation tests' were not 'relatively *wertfrei*'.

Another approach was followed by Chipman and Moore (1973; 1976a), who asked the following question: if competitive equilibria (X^t, y^t, p^t) are observed satisfying $p^1 \cdot y^1 \geq p^1 \cdot y^0$ and $p^0 \cdot y^1 \geq p^0 \cdot y^0$, where $y^t \in Y_t$ for $t = 0, 1$, under what conditions on preferences must this imply that $Y^1 >_R Y^0$? For the case $Y^t = \{y^t\}$ they showed that the preference relations R_i must be identical and homothetic. This is a global result; with positive consumptions of all commodities the condition could no doubt be weakened to the aggregation criterion of Antonelli (1886), Gorman (1953), and Nataf (1953), namely, that consumer i's demand for commodity j have the form

$$x_{ij} = a_{ij}(p) + b_j(p)m_i$$

where m_i is consumer i's income.

Samuelson (1956) applied the compensation principle in a striking way in his proposed alternative to the new welfare economics. He discovered that, if a social-welfare function has the separable form $W[f(x)]$, then a social utility function $f_w(x) = \max\{W[f(x)] : X \in A(x)\}$ has the property that it can be achieved in a decentralized manner by means of an income-distribution policy assigning individual shares of aggregate income as functions of prices and aggregate income. The first complete proof of this result was presented in Chipman and Moore (1972) (see also Chipman and Moore, 1979; Chipman, 1982). The main tool of analysis used was the concept of a Scitovsky indifference surface (Scitovsky, 1942) which is defined as the boundary of the set $\Sigma_{i=1}^{N} R_i x_i$, where $R_i x_i$ is the set of all commodity bundles preferred or indifferent to x_i by individual i. This set is necessarily a subset of the set $R_w x$ of aggregate bundles preferred or indifferent to x by the Samuelson social ordering. In a competitive equilibrium the aggregate consumption bundle minimizes aggregate expenditure at the equilibrium prices over both sets, hence the bundle x_i minimizes each individual's expenditure over $R_i x_i$ (cf. Koopmans, 1957, pp. 12–13).

3 Gains from trade and optimal tariffs

The new tools developed by Scitovsky (1942) and Samuelson (1950; 1956) made possible a rigorous proof of a gains-from-trade theorem, as well as of the proposition that a country could gain by a tariff.

Kemp (1962) noted that Samuelson's 1939 theorem implied that for any point on the free-trade utility-possibility frontier, the autarky utility-possibility frontier must pass below it; he reasoned that, as a result, for any point on the autarky utility-possibility frontier, the free-trade utility-possibility frontier must pass above it. If this argument can be accepted, it follows that for every allocation $X^0 \in A(Y^0)$ where Y^0 is the autarkic production-possibility set, there exists a (weakly) Pareto-superior

allocation $X^1 \in A(Y^1)$ where Y^1 is the free-trade consumption-possibility set. Then free trade is superior to autarky by Samuelson's 1950 criterion.

The trouble with this argument, however, is that it requires that one can define a free-trade utility-possibility frontier (or consumption-possibility frontier) with the strong topological property of homeomorphism to the $(N-1)$-dimensional unit simplex (intuitively, absence of 'holes'). That this need not always be possible, was shown by Otani (1972, p. 149), and indeed admitted by Kemp and Wan (1972, p. 513). It is always possible if world prices are fixed, beyond our country's control. In that case the free-trade consumption-possibility set Y^1 is the budget set enclosing the production-possibility set Y^0 (cf. Samuelson, 1962, p. 821), and the gains-from-trade theorem follows immediately from the property $Y^1 \supseteq Y^0 \Rightarrow Y^1 >_R Y^0$. In similar fashion the famous 'Baldwin envelope' (Baldwin, 1948) defines a well-behaved consumption-possibility set containing the production-possibility set, from which one can prove the superiority of restricted trade (with an optimal tariff) to autarky (cf. Samuelson, 1962).

For the general case in which a country can influence world prices, a method was shown by Kenen (1957). If all but 1 of the N individuals are constrained to have the same level of satisfaction under trade as achieved under autarky, a net production-possibility set can be constructed which indicates the amount available for the Nth person. It remains only to show that the Nth person will gain from a movement from autarky to free trade. A similar approach was indicated by Vanek (1964).

Grandmont and McFadden (1972) and Chipman and Moore (1972) both used the concept of an income-distribution policy to establish the gains-from-trade theorem. In Chipman and Moore this policy was chosen to be one that maximizes a separable Bergson–Samuelson social-welfare function. A standard argument is used to show that social utility is at least as high under free trade as under autarky. It remains to show that a function $W(u)$ can be chosen so that the corresponding distribution policy ensures that an increase in social utility implies an increase in each individual's utility. This is achieved by choice of $W(u) = \min_i (u_i - u_i^0)/c_i$ where $c_i > 0$ and u_i^0 is the level of utility achieved by individual i under autarky.

4 General-equilibrium theory

The compensation principle is used in the proof of the theorem that every competitive equilibrium is Pareto-optimal (Arrow, 1952, pp. 516, 519; Koopmans, 1957, p. 49; Debreu, 1959, pp. 94–5), in the sense that arbitrary allocations of feasible output bundles among consumers are assumed possible, regardless of resource-ownership constraints. A pair (X^0, p^0) is a competitive equilibrium for the production-possibility set Y if $X^0 R X$ for all $X \in A(Y)$ satisfying $Xp^0 \leq X^0 p^0$ and $y^0 p^0 \geq y p^0$ for all $y \in Y$, where $y^0 = \sigma(X^0) \in Y$. Pareto-optimality means that one cannot find an $X \in A(Y)$ such that XPX^0. The proof is by contradiction: XPX^0 implies $Xp^0 \geq X^0p^0$ (the vector inequality being weak in all components and strict in at least one) hence taking column sums, $yp^0 > y^0p^0$.

The converse theorem, that every Pareto optimum can be sustained by a competitive equilibrium, requires stronger assumptions which are awkward to state (cf. Arrow, 1952, p. 518; Koopmans, 1957, p. 50; Debreu, 1959, p. 95). The basic idea of the proof (Koopmans, 1957, pp. 50–52; Debreu, 1959, p. 96) can be sketched in terms of the concept of a Scitovsky (1942) indifference surface. If X^0 is a Pareto-optimal allocation for a closed, convex production-possibility set Y, then the interior of the Scitovsky set of X^0 can be written $P_k x_k^0 + \sum_{i \neq k} R_i x_i^0$ for some k. Defining the allocation X^1 by $x_k^1 P_k x_k^0$ and $x_i^1 R_i x_i^0$ for $i \neq k$ we have $X^1 P X^0$ hence $X^1 \notin A(Y)$. Therefore the interior of the Scitovsky set does not intersect Y, and these convex sets can be separated by a hyperplane defining the equilibrium prices. It is then verified that at these prices the properties of a competitive equilibrium are satisfied.

Debreu (1954, p. 590) introduced an alternative equilibrium concept according to which the condition that consumer preferences be maximized subject to their budget constraints was replaced by the condition that consumer expenditures be minimized subject to the constraints that the bundles considered be at least as desirable as the equilibrium bundles. (The second of the above theorems follows more easily under this alternative definition.) For a given set of positive-valued utility functions representing consumer preferences, Arrow and Hahn (1971, p. 108) called this a 'compensated equilibrium'. As a means of proving existence of the latter they studied the utility-possibility frontier or 'Pareto frontier' (1971, p. 96), and obtained a new proof of the result of Chipman and Moore (1971) that the set of Pareto-optimal allocations X of Y (the 'contract curve') and the utility-possibility frontier are topologically homeomorphic to the unit simplex of dimension one less than the number of individuals. These results were further developed by Moore (1975).

5 Cost–benefit analysis

Hicks (1941, p. 112) made an interesting distinction between two tasks of welfare economics: (1) the study of (Pareto)-optimal organizations of the economy and (2) the study of deviations from such optima. More precisely, the first was concerned with when there was a deviation and the second with the size of the deviation. He also identified these two tasks with general- and partial-equilibrium analysis respectively, although there appears to be no justification for this other than the historical accident that consumers' surplus developed as a partial-equilibrium tool. He remarked that consumers' surplus is not needed for the first task, since lack of fulfillment of

the proportionality between marginal utilities and marginal costs provides the needed information immediately. For the second task, he was not content with a ranking of the non-optimal states, but with measuring the size of their deviations from optimality, which of course would provide such a ranking. Thus, the staunch ordinalist in consumer theory became an equally ardent cardinalist in consumer theory.

Hicks's concepts of compensating and equivalent variation (Hicks, 1942) may most conveniently be defined in terms of the minimum-income or income-compensation functions of McKenzie (1957) and Hurwicz and Uzawa (1971). Denoting the ith consumer's demand function by $x_i = h_i(p, m_i)$ (where x_i and p are n-vectors), and defining the indirect preference relation R_i^* by $(p^0, m_i^0) R_i^* (p^1, m_i^1)$ if and only if $h_i(p^0, m_i^0) R_i h_i(p^1, m_i^1)$, the income-compensation function is defined by

$$\mu_i(p; p^0, m_i^0) = \inf\{m_i : (p, m_i) R_i^* (p^0, m_i^0)\}.$$

Following Chipman and Moore (1980b), the generalized compensating variation in going from (p^0, m_i^0) to (p, m_i) is defined as

$$C_i(p, m_i; p^0, m_i^0) = m_i - \mu_i(p; p^0, m_i^0)$$

and the generalized equivalent variation by

$$E_i(p, m_i; p^0, m_i^0) = \mu_i(p^0; p, m_i) - m_i^0.$$

These reduce to Hicks's concepts when $m_i = m_i^0$.

The compensating variation expresses for each consumer the amount of money income he or she would be willing to give up (or the negative of the amount by which he or she would have to be compensated), at the new prices, to make up for the change in prices and income. One of the reasons for the great appeal of the concept is that these are amounts that can be added up over the set of consumers. In Hicks's words (1942, p. 127):

> the general test for a particular reform being an *improvement* is that the gainers should gain sufficiently for them to be able to compensate the losers and still remain gainers on balance. This test would be carried out by striking the balance of the Compensating Variations.

Denoting by m' the vector of N incomes in state t, and by M^t their sum, we can define a dual potential-improvement ordering between pairs of price-income pairs (p^t, M^t) as follows. Let $A^*(p, M)$ be the set of $(n+N)$-tuples (p, m) such that $\sum_{i=1}^{N} m_i = M$, and let R^* be the relation such that $(p^0, m^0) R^*(p^1, m^1)$ if and only if $(p^0, m_i^0) R_i^*(p^1, m_i^1)$ for $i = 1, 2, \ldots, N$. Then we define the dual KHS relation $>_{R^*}$ by

$$(p^0, M^0) >_{R^*} (p^1, M^1)$$
$$\Leftrightarrow [\forall (p', m') \in A^*(p^1, M^1)(\exists (p, m)$$
$$\in A^*(p^0, M^0)] : (p, m) R^*(p', m').$$

Choosing price-income pairs (p^0, m^0) and (p^1, m^1) satisfying this definition, since $\mu_i(p^t; p, m_i)$ is an indirect utility function representing R_i^* for $t = 0$ or 1, we have $\mu_i(p^t; p^0, m_i^0) \geq \mu_i(p^t; p^1, m_i^1)$ for all individuals i, hence

$$M^0 = \sum_{i=1}^{N} m_i^0 = \sum_{i=1}^{N} \mu_i(p^0; p^0, m_i^0)$$

$$\geqq \sum_{i=1}^{N} \mu_i(p^0; p^1, m_i^1)$$

so one obtains a multi-consumer analogue to the compensating variation from the formula

$$M^0 - \sum_{i=1}^{N} \mu_i(p^0; p^1, m_i^1) \geqq M^1$$

$$- \sum_{i=1}^{N} \mu_i(p^1; p^1, m_i^1) = 0$$

Likewise for the equivalent variation,

$$0 = \sum_{i=1}^{N} \mu_i(p^0; p^0, m_i^0)$$

$$- M^0 \geqq \sum_{i=1}^{N} \mu_i(p^0; p^1, m_i^1) - M^0.$$

In the latter case the same indirect utility functions are summed on both sides of the inequality sign; it is a case where Benthamites and compensationists can find common ground.

Boadway (1974) considered the relationship between the condition of positive summed compensating variations and the fulfillment of compensation tests and came to the negative conclusion that the former was neither necessary nor sufficient for satisfaction of the latter in general, but was sufficient in the case of identical and homothetic preferences. Foster (1976) showed that, if there are no price distortions (but not otherwise), satisfaction of the compensation tests implies satisfaction of the 'cost–benefit criterion' (positive summed compensating variations). This conclusion is in accord with the above inequalities.

What about the Hicksian tenet that the size of the compensating variation is important so that one can compare two suboptimal states? This would require one to be able to conclude that, if the compensating variation from state 0 to state 2 is positive and greater than the compensating variation from state 0 to state 1, then state 2 should be superior to state 1 in terms of the dual KHS ordering. But this is not true even in the case of the single consumer. It was shown in Chipman and Moore (1980) that the function $C_i(p, m_i; p^0, m_i^0)$ cannot be an indirect utility function for unrestricted domain

$(p, m_i) > 0$, and can be if m_i is held constant if and only if preferences are homothetic, and if p_1 is held constant if and only if preferences are 'parallel' with respect to commodity 1. If preferences are identical and homothetic, since $\mu_i = \mu$ is homogeneous of degree 1 in m_i, $\sum_{i=1}^{N} \mu_i(p^0; p, m_i) = \mu(p^0; p, M)$, so exact aggregative analogues are obtained to both the compensating and equivalent variations. If the equivalent variation, which is an indirect utility function, is used, restrictions on consumer preferences are not needed, and the problem of finding an adequate indicator of the size of the deviation from a given Pareto optimum is satisfactorily resolved.

6 Game theory

One of the striking aspects of von Neumann and Morgenstern's theory of games (1947) was not only its postulate of measurability of utility but also that of its transferability between players. Since this was introduced as a positive rather than a normative assumption, it has met with even greater resistance of the part of economists than the hedonist calculus. Indeed, it was not until Debreu and Scarf (1963) showed how game theory could be liberated from this restriction with their development of the concept of the core of an economy that game theory began to be taken really seriously by economists. The replacement of transferability of utilities by transferability of commodities bears a striking resemblance to the replacement in welfare economics of the calculus of utilities by the principle of compensation.

In some branches of game theory the assumption of transferable utility is still retained, but it has been made somewhat more plausible, or at least interpretable, by means of the postulate that the utility functions of all individuals are linear in one distinguished commodity used for making side payments (cf. Owen, 1982, p. 122). These utility functions have the form

$$U_i(x_{i1}, x_{i2}, \ldots, x_{in}) = c_i x_{i1} + V_i(x_{i2}, \ldots, x_{in}).$$

This form of the utility function goes back to Edgeworth (1891, p. 237n) and even earlier (though in garbled form) to Auspitz and Lieben (1889, p. 471). In Edgeworth it was used to illustrate the phenomenon of exchange when the marginal utility of one commodity serving as money was held constant, in accordance with one possible interpretation of Marshall's theory of consumers' surplus. (In the case $n = 2$ he showed that the exchange in commodity 2 would be constant, but in commodity 1 'indeterminate'; see the reply by Berry, 1891, on behalf of Marshall, and Marshall, 1891, p. 756; 1920, p. 845.) The above form for the utility function has been rediscovered many time, by Wilson (1939), Samuelson (1942), and others; cf. Chipman and Moore (1976b, p. 115). Barone (1894, p. 213n) gave the name 'ideal money (numéraire)' to a good with a constant marginal utility (commodity 1 in the above). For the case $c_i = c$ for all i, these 'parallel' preferences (cf. Boulding, 1945) yield a special case of the family of aggregable Antonelli–Gorman–Nataf demand functions referred to above.

7 Concluding observations

As Scitovsky (1941) pointed out, the compensation principle has been used in two quite different ways. Prior to Hicks (1940), it was used only to compare efficient with inefficient states of a given economy with a given technology or trading system. Starting with Hicks (1940), its use was extended to comparison of efficient states of an economy under different technologies. It has turned out that, in order for national-income comparisons to provide a correct indicator of potential-welfare improvement, very strong conditions are required concerning similarity of individual preferences: locally, the Antonelli–Gorman–Nataf conditions, and, globally, identical homothetic preferences. It is not even enough to assume that aggregate demand can be generated by an aggregate preference relation – for example, that preferences are homothetic and relative income-distribution constant (cf. Chipman and Moore, 1980a). Even in such cases, strong value judgements (such as acceptance of a particular Bergson–Samuelson social-welfare function) are required in order to draw welfare conclusions from national-income comparisons.

When attention is restricted to the efficient operation of an economy with a given technology, it turns out that, in most cases of interest, the ranking of consumption-possibility criterion sets according to the Kaldor–Hicks–Samuelson criterion follows from their ranking by set-inclusion. This does not mean, however, that the set-inclusion is always obvious or easy to prove.

The KHS ordering of consumption-possibility sets could be given simply a factual interpretation as indicating the 'productive potential' of an economy. But if it is given a normative interpretation then it obviously involves a value judgement, since a more efficient outcome, if it is not Pareto-superior, can obviously be judged worse in terms of some social-welfare function.

Samuelson's (1956) model of the 'good society', elegant though it is, is too sweeping for most economists to accept, and it begs the question of how the social-welfare function will be chosen. Little's (1950) and Mishan's (1969) attempts to link plausible distributional value judgements with compensation criteria have encountered unresolvable logical difficulties (cf. Chipman and Moore, 1978). The hope that the compensation principle would allow policy decisions to be made free of value judgements has not been fulfilled. Nevertheless, much has been learned about the interrelationships among values, facts, and policies, and it can certainly be said that the development of the compensation principle has led to clearer thinking about economic policy issues.

JOHN S. CHIPMAN

See also **social welfare function; welfare economics.**

Bibliography

Antonelli, G.B. 1886. *Sulla teoria matematica della economia politica*. Pisa: Folchetto. English translation: On the mathematical theory of political economy, in *Preferences, Utility, and Demand*, ed. J.S. Chipman, L. Hurwicz, M.K. Richter and H.F. Sonnenschein, New York: Harcourt Brace Jovanovich, 1971.

Arrow, K.J. 1952. An extension of the basic theorems of classical welfare economics. In *Proceedings of the Second Berkeley Symposium on Mathematical Statistics and Probability*, ed. J. Neyman. Berkeley and Los Angeles: University of California Press.

Arrow, K.J. and Hahn, F.H. 1971. *General Competitive Analysis*. San Francisco: Holden-Day.

Arrow, K.J. and Scitovsky, T., eds. 1969. *Readings in Welfare Economics*. Homewood, IL: Irwin.

Auspitz, R. and Lieben, R. 1889. *Untersuchungen über die Theorie des Preises*. Leipzig: Duncker & Humblot.

Baldwin, R.E. 1948. Equilibrium in international trade: a diagrammatic analysis. *Quarterly Journal of Economics* 62, 748–62.

Baldwin, R.E. 1954. A comparison of welfare criteria. *Review of Economic Studies* 21, 154–61.

Barone, E. 1894. Sulla 'consumers' rent'. *Giornale degli Economisti* Series 2, 9, September, 211–24.

Barone, E. 1908. Il Ministerio della produzione nello stato colletivista. *Giornale degli Economisti* Series 2, 37, August, 267–93; October, 391–414. English translation: The Ministry of Production in the collectivist state, in *Collectivist Economic Planning*, ed. F.A. Hayek. London: Routledge & Kegan Paul, 1935.

Berry, A. 1891. Alcune brevi parole sulla teoria del baratto di A. Marshall. *Giornale degli Economisti* Series 2,2, June, 549–53.

Boadway, R.W. 1974. The welfare foundations of cost-benefit analysis. *Economic Journal* 84, 926–39. A reply, *Economic Journal* 86 (1976), 358–61.

Boulding, K.E. 1945. The concept of economic surplus. *American Economic Review* 35, 851–69.

Cairnes, J.E. 1874. *Some Leading Principles of Political Economy Newly Expounded*. New York: Harper & Brothers.

Chipman, J.S. 1976. The Paretian heritage. *Revue européenne des sciences sociales et Cahiers Vilfredo Pareto* 14(37), 65–171.

Chipman, J.S. 1982. Samuelson and welfare economics. In *Samuelson and Neoclassical Economics*, ed. G.R. Feiwel. Boston: Kluwer-Nijhoff Publishing.

Chipman, J.S. and Moore, J.C. 1971. The compensation principle in welfare economics. In *Papers in Quantitative Economics*, vol. 2, ed. A.M. Zarley. Lawrence: University Press of Kansas.

Chipman, J.S. and Moore, J.C. 1972. Social utility and the gains from trade. *Journal of International Economics* 2(May), 157–72.

Chipman, J.S. and Moore, J.C. 1973. Aggregate demand, real national income, and the compensation principle. *International Economic Review* 14, 152–81.

Chipman, J.S. and Moore, J.C. 1976a. Why an increase in GNP need not imply an improvement in potential welfare. *Kyklos* 29(3), 391–418.

Chipman, J.S. and Moore, J.C. 1976b. The scope of consumer's surplus arguments. In *Evolution, Welfare, and Time in Economics*, ed. A. Tang, F.M. Westfield and J.S. Worley. Lexington, MA: Heath.

Chipman, J.S. and Moore, J.C. 1978. The New Welfare Economics, 1939–1974. *International Economic Review* 19, 547–84.

Chipman, J.S. and Moore, J.C. 1979. On social welfare functions and the aggregation of preferences. *Journal of Economic Theory* 21(August), 111–39.

Chipman, J.S. and Moore, J.C. 1980a. Real national income with homothetic preferences and a fixed distribution of income. *Econometrica* 48, 401–22.

Chipman, J.S. and Moore, J.C. 1980b. Compensating variation, consumer's surplus, and welfare. *American Economic Review* 70, 933–49.

Coase, R.H. 1960. The problem of social cost. *Journal of Law and Economics* 3(October), 1–44.

Debreu, G. 1951. The coefficient of resource utilization. *Econometrica* 19, 273–92.

Debreu, G. 1954. Valuation equilibrium and Pareto optimum. *Proceedings of the National Academy of Sciences* 40, 588–92.

Debreu, G. 1959. *Theory of Value*. New York: Wiley.

Debreu, G. and Scarf, H. 1963. A limit theorem on the core of an economy. *International Economic Review* 4, 235–46.

Dupuit, J. 1844. De la mesure de l'utilité des travaux publics. *Annales des Ponts et Chaussées, Mémoires et documents relatifs à l'art des constructions et au service de l'ingénieur* Series 2,2, 2e semestre, 332–75, Pl. 75. English translation: On the measurement of the utility of public works, in Arrow and Scitovsky (1969).

Edgeworth, F.Y. 1891. Osservazioni sulla teoria matematica dell'economia politica con riguardo speciale ai principi di economia di Alfredo Marshall. *Giornale degli Economisti* Series 2, 2, March, 233–45. Ancora a proposito della teoria del baratto, *Giornale degli Economisti* Series 2, 2, October, 316–18. Abridged English translation: On the determinateness of economic equilibrium, in F.Y. Edgeworth, *Papers Relating to Political Economy*, vol. 2, London: Macmillan, 1925.

Foster, E. 1976. The welfare foundations of cost-benefit analysis – a comment. *Economic Journal* 86, 353–8.

Frisch, R. 1939. The Dupuit taxation theorem. *Econometrica* 7, 145–50. A further note on the Dupuit taxation theorem, *Econometrica* 7, 156–7.

Gorman, W.M. 1953. Community preference fields. *Econometrica* 21, 63–80.

Gorman, W.M. 1955. The intransitivity of certain criteria used in welfare economics. *Oxford Economic Papers*, N.S. 7, February, 25–35.

Graaff, J.de V. 1949. On optimum tariff structures. *Review of Economic Studies* 17, 47–59. Reprinted in Arrow and Scitovsky (1969).

Graaff, J.de V. 1957. *Theoretical Welfare Economics*. Cambridge: Cambridge University Press.

Grandmont, J.M. and McFadden, D. 1972. A technical note on classical gains from trade. *Journal of International Economics* 2(May), 109–125.

Harrod, R.F. 1938. Scope and method of economics. *Economic Journal* 48, 383–412.

Hicks, J.R. 1939. The foundations of welfare economics. *Economic Journal* 49, 696–712.

Hicks, J.R. 1940. The valuation of social income. *Economica*, N.S. 7, 105–24.

Hicks, J.R. 1941. The rehabilitation of consumers' surplus. *Review of Economic Studies* 8(February), 108–16. Reprinted in Arrow and Scitovsky (1969).

Hicks, J.R. 1942. Consumers' surplus and index-numbers. *Review of Economic Studies* 9(Summer), 126–37.

Hicks, J.R. 1957. *A Revision of Demand Theory*. Oxford: Clarendon Press.

Hotelling, H. 1938. The general welfare in relation to problems of taxation and of railway and utility rates. *Econometrica* 6, 242–69. Reprinted in Arrow and Scitovsky (1969).

Hotelling, H. 1939. The relation of prices to marginal costs in an optimum system. *Econometrica* 7, 151–5. A final note, *Econometrica* 7, 158–9.

Hurwicz, L. and Uzawa, H. 1971. On the integrability of demand functions. In *Preferences, Utility, and Demand*, ed. J.S. Chipman, L. Hurwicz, M.K. Richter and H.F. Sonnenschein. New York: Harcourt Brace Jovanovich.

Kaldor, N. 1939. Welfare propositions in economics and interpersonal comparisons of utility. *Economic Journal* 49, 549–52. Reprinted in Arrow and Scitovsky (1969).

Kaldor, N. 1940. A note on tariffs and the terms of trade. *Economica*, N.S. 7, 377–80.

Kannai, Y. and Mantel, R. 1978. Non-convexifiable Pareto sets. *Econometrica* 46, 571–5.

Kemp, M.C. 1962. The gains from international trade. *Economic Journal* 72, 803–19.

Kemp, M.C. and Wan, H.Y., Jr. 1972. The gains from free trade. *International Economic Review* 13, 509–22.

Kenen, P.B. 1957. On the geometry of welfare economics. *Quarterly Journal of Economics* 71, 426–47.

Koopmans, T.C. 1957. *Three Essays on the State of Economic Science*. New York: McGraw-Hill.

Kuznets, S. 1948. On the valuation of social income – reflections on Professor Hicks' article. *Economica*, N.S. 15, February, 1–16, May, 116–31.

Lerner, A.P. 1934. The concept of monopoly and the measurement of monopoly power. *Review of Economic Studies* 1(June), 157–75.

Little, I.M.D. 1950. *A Critique of Welfare Economics*. 2nd edn, London: Oxford University Press, 1957.

Marshall, A. 1890. *Principles of Economics*. London: Macmillan. 2nd edn, 1891; 8th edn, 1920.

McKenzie, L.W. 1957. Demand theory without a utility index. *Review of Economic Studies* 24(June), 185–9.

Mishan, E.J. 1969. *Welfare Economics: An Assessment*. Amsterdam: North-Holland.

Moore, J.C. 1975. The existence of 'compensated equilibrium' and the structure of the Pareto efficiency frontier. *International Economic Review* 16, 267–300.

Nataf, A. 1953. Sur des questions d'agrégation en économétrie. *Publications de l'Institut de Statistique de l'Université de Paris* 2(4), 5–61.

von Neumann, J. and Morgenstern, O. 1947. *Theory of Games and Economic Behavior*, 2nd edn, Princeton: Princeton University Press.

Olsen, E. 1958. Udenrigshandelens gevinst [The gains of international trade]. *Nationaløkonomisk Tiddskrift* 98(1–2), 76–9.

Otani, Y. 1972. Gains from trade revisited. *Journal of International Economics* 2, 127–56.

Owen, G. 1982. *Game Theory*, 2nd edn. Orlando, FL: Academic Press.

Pareto, V. 1894. Il massimo di utilità dato dalla libera concorrenza. *Giornale degli Economisti* Series 2, 9 July, 48–66.

Pareto, V. 1895. Teoria matematica del commercio internazionale. *Giornale degli Economisti* Series 2, 10 April, 476–98.

Pareto, V. 1896, 1897. *Cours d'économie politique*, 2 vols. Lausanne: F. Rouge.

Pigou, A.C. 1920. *The Economics of Welfare*. London: Macmillan. 4th edn, 1932.

Ricardo, D. 1815. *An Essay on the Influence of a Low Price of Corn on the Profits of Stock*. London: John Murray. In *The Works and Correspondence of David Ricardo*, vol. 4, ed. P. Sraffa. Cambridge: Cambridge University Press, 1951.

Robbins, L. 1938. Interpersonal comparisons of utility: a comment. *Economic Journal* 48, 635–41.

Samuelson, P.A. 1939. The gains from international trade. *Canadian Journal of Economics and Political Science* 5(May), 195–205.

Samuelson, P.A. 1942. Constancy of the marginal utility of income. In *Studies in Mathematical Economics and Econometrics in Memory of Henry Schultz*, ed. O. Lange, F. McIntyre and T.O. Yntema. Chicago: University of Chicago Press.

Samuelson, P.A. 1947. *Foundations of Economic Analysis*. Cambridge, MA: Harvard University Press.

Samuelson, P.A. 1950. Evaluation of real national income. *Oxford Economic Papers*, N.S. 1, January, 1–29. Reprinted in Arrow and Scitovsky (1969).

Samuelson, P.A. 1956. Social indifference curves. *Quarterly Journal of Economics* 70(February), 1–22.

Samuelson, P.A. 1962. The gains from international trade once again. *Economic Journal* 72, 820–29.

Sanger, C.P. 1895. Recent contributions to mathematical economics. *Economic Journal* 5(March), 113–28.

Scitovsky, T. 1941. A note on welfare propositions in economics. *Review of Economic Studies* 9(November), 77–88. Reprinted in Arrow and Scitovsky (1969).

Scitovsky, T. 1942. A reconsideration of the theory of tariffs. *Review of Economic Studies* 9(Summer), 89–110.

Silberberg, E. 1980. Harold Hotelling and marginal cost pricing. *American Economic Review* 70, 1054–7.

Vanek, J. 1964. A rehabilitation of 'well-behaved' social indifference curves. *Review of Economic Studies* 31, 87–9.

Viner, J. 1937. *Studies in the Theory of International Trade*. New York: Harper & Brothers.

Wilson, E.B. 1939. Pareto versus Marshall. *Quarterly Journal of Economics* 53, 645–50.

competing risks model

A competing risks model is a model for multiple durations that start at the same point in time for a given subject, where the subject is observed until the first duration is completed and one also observes which of the multiple durations is completed first.

The term 'competing risks' originates in the interpretation that a subject faces different risks i of leaving the state it is in, each risk giving rise to its own exit destination, which can also be denoted by i. One may then define random variables T_i describing the duration until risk i is materialized. Only the smallest of all these durations $Y := \min_i T_i$ and the corresponding actual exit destination, which can be expressed as $Z := \arg\min_i T_i$, are observed. The other durations are censored in the sense that all is known is that their realizations exceed Y. Often those other durations are latent or counterfactual, for example if T_i denotes the time until death due to cause i.

In economics, the most common application concerns individual unemployment durations. One may envisage two durations for each individual: one until a transition into employment occurs, and one until a transition into non-participation occurs. We observe only one transition, namely, the one that occurs first. Other applications include the duration of treatments, where the exit destinations are relapse and recovery, and the duration of marriage, where one risk is divorce and the other is death of one of the spouses. More generally, the duration until an event of interest may be right-censored due to the occurrence of another event, or due to the data sampling design. The duration until the censoring is then one of the variables T_i.

Sometimes one is interested only in the distribution of Y. For example, an unemployment insurance (UI) agency may be concerned only about the expenses on UI and not in the exit destinations of recipients. In such cases one may employ standard statistical duration analysis for empirical inference with register data on the duration of

UI receipt. However, in studies on individual behaviour one is typically interested in one or more of the marginal distributions of the T_i. If these variables are known to be independent, then again one may employ standard duration analysis for each of the T_i separately, treating the other variables $T_j (j \neq i)$ as independent right-censoring variables. But often it is not clear whether the T_i are independent. Indeed, economic theory often predicts that they are dependent, in particular if they can be affected by the individual's behaviour and individuals are heterogeneous. It may even be sensible from the individual's point of view to use their privately observed exogenous exit rates into destinations j as inputs for the optimal strategy affecting the exit rate into destination $i (i \neq j)$ (see, for example, van den Berg, 1990). Erroneously assuming independence leads to incorrect inference, and in fact the issue of whether the durations T_i are related is often an important question in its own right.

Unfortunately, the joint distribution of all T_i is not identified from the joint distribution of Y, Z, a result that goes back to Cox (1959). In particular, given any specific joint distribution, there is a joint distribution with independent durations T_i that generates the same distribution of the observable variables Y, Z. In other words, without additional structure, each dependent competing risks model is observationally equivalent to an independent competing risks model. The marginal distributions in the latter can be very different from the true distributions.

Of course, some properties of the joint distribution are identified. To describe these it is useful to introduce the concept of the hazard rate of a continuous duration variable, say W. Formally, the hazard rate at time t is $\theta(t) := \lim_{dt \downarrow 0} \Pr(W \in [t, t + dt))/dt$. Informally, this is the rate at which the duration W is completed at t given that it has not been completed before t. The hazard rate is the basic building block of duration analysis in social sciences because it can be directly related to individual behaviour at t. The data on Y, Z allow for identification of the hazard rates of T_i at t given that $T \geq t$. These are called the 'crude' hazard rates. If the T_i are independent, then these equal the 'net' hazard rates of the marginal distributions of the T_i.

We now turn to a number of approaches that overcome the general non-identification result for competing risks models. In econometrics, one is typically interested in covariate or regressor effects. The main approach has therefore been to specify semi-parametric models that include observed regressors X and unobserved heterogeneity terms V. With a single risk, the most popular duration model is the mixed proportional hazard (MPH) model, which specifies that $\theta(t|X = x, V) = \psi(t)\exp(x'\beta)V$ for some function $\psi(.)$. V is unobserved, and the composition of the survivors changes selectively as time proceeds, so identification from the observable distributions of $T|X$ is non-trivial. However, it holds under the assumptions that $X \perp\!\!\!\perp V$ and $\mathrm{var}(X) > 0$ and

some regularity assumptions (see van den Berg, 2001, for an overview of results). With competing risks, the analogue of the MPH model is the multivariate MPH (MMPH) model. With two risks,

$$\theta_1(t|x, V) = \psi_1(t)\exp(x'\beta_1)V_1 \quad \text{and}$$
$$\theta_2(t|x, V) = \psi_2(t)\exp(x'\beta_2)V_2.$$

where $T_1, T_2|X, V$ are assumed independent, so that a dependence of the durations given X is modelled by way of their unobserved determinants V_1 and V_2 being dependent. Many empirical studies have estimated parametric versions of this model, using maximum likelihood estimation.

The semi-parametric model has been shown to be identified, under only slightly stronger conditions than those for the MPH model (Abbring and van den Berg, 2003). Specifically, $\mathrm{Var}(X) > 0$ is strengthened to the condition that the vector X includes two continuous variables with the properties that (a) their joint support contains a non-empty open set in $\mathbb{R}^2$, and (b) the vectors $\tilde{\beta}_1, \tilde{\beta}_2$ of the corresponding elements of β_1 and β_2 form a matrix $(\tilde{\beta}_1, \tilde{\beta}_2)$ of full rank. Somewhat loosely, X has two continuous variables that are not perfectly collinear and that act differently on θ_1 and θ_2. Note that, with such regressors, one can manipulate $\exp(x'\beta_1)$ while keeping $\exp(x'\beta_2)$ constant. The two terms $\exp(x'\beta_i)$ are identified from the observable crude hazards at $t = 0$ because at $t = 0$ no dynamic selection due to the unobserved heterogeneity has taken place yet. Now suppose one manipulates x in the way described above. If $T_1, T_2|X$ are independent, then the observable crude hazard rate of T_2 at $t > 0$, given that $T_1 \geq t$, does not vary along. But, if $T_1, T_2|X$ are dependent, then this crude hazard rate does vary along, for the following reason. First, changes in $\exp(x'\beta_1)$ affect the distribution of unobserved heterogeneity V_1 among the survivors at t, due to the well-known fact that V_1 and X are dependent conditional on survival (i.e. conditional on $T_1 \geq t > 0$) even though they are independent unconditionally. Second, if V_1 and V_2 are dependent, this affects the distribution of V_2 among the survivors at t, which in turn affects the observable crude hazard of T_2 at t given that $T_1 \geq t$. In sum, the variation in this crude hazard with $\exp(x'\beta_1)$ for given $\exp(x'\beta_2)$ is informative on the dependence of the durations. An analogous argument holds for the crude hazard rate corresponding to cause $i = 1$.

Note that identification is not based on exclusion restrictions of the sort encountered in instrumental variable analysis, which require a regressor that affects one endogenous variable but not the other. Here, all explanatory variables are allowed to affect both duration variables – they are just not allowed to affect the duration distributions in the same way. Identification with regressors was first established by Heckman and Honoré (1989), who considered a somewhat larger class of models than the MMPH model and accordingly imposed stronger conditions on the support of X.

Although the MPH model is identified from single-risk duration data where we observe a single spell per subject, there is substantial evidence that estimates are sensitive to misspecification of functional forms of model elements (see van den Berg, 2001, for an overview). This implies that estimates of MMPH models using competing-risks data should also be viewed with caution. It is advisable to include additional data. For example, longitudinal survey data on unemployment durations subject to right-censoring can be augmented with register data or retrospective data not subject to censoring (see for example van den Berg, Lindeboom and Ridder, 1994). More in general, one may resort to 'multiple-spell competing risks' data, meaning data with multiple observations of Y, Z for each subject. For a given subject, such observations can be viewed as multiple independent draws from the subject-specific distribution of Y, Z, on the assumption that the unobserved heterogeneity terms V_1, V_2 are identical across the spells of the subject. Here, a subject can denote a single physical unit, like an individual, for which we observe two spells in exactly the same state, or it can denote a set of physical units for which we observe one spell each. Multiple-spell data allow for identification under less stringent conditions than single-spell data. Abbring and van den Berg (2003) showed that such data identify models that allow for full interactions between the elapsed durations t and x in $\theta_i(t|x, V)$, and, indeed, allow the corresponding effects to differ between the first and the second spell. The assumptions on the support of X are similar to above. Fermanian (2003) developed a nonparametric kernel estimator of the Heckman and Honoré (1989) model.

Another approach to deal with non-identification of dependent competing risks models is to determine bounds on the sets of marginal and joint distributions that are compatible with the observable data. Peterson (1976) derived sharp bounds in terms of observable quantities. They are often wide. In case of the marginal distributions of two sub-populations distinguished by a variable X, the bounds associated with the different X may overlap, whether or not X (monotonically) affects (one of) the marginal distributions. With overlap, the causal effects of X cannot even be signed.

Bond and Shaw (2006) combined bounds with regressors. In the case of a single binary regressor, the only substantive assumption made is that there exist increasing functions g and h such that $T_1, T_2|X = 0$ equals $g(T_1), h(T_2)|X = 1$ in distribution. In words, the dependence structure is invariant to the values of the regressors, so the latter affect only the marginal distributions. Specifically, the copula (and therefore Kendall's τ) of the joint distribution is invariant to the value of X. The assumption is satisfied by the aforementioned competing risks models with regressors. Clearly, by itself the assumption is insufficient for point identification. The bounds concern the regressor effects on the marginal distributions. If it is assumed that X affects the marginal distributions of T_i in terms of first-order stochastic dominance, the bounds are sufficient to sign the effect of X on at least one of the marginal distributions (so, in case of MMPH models, also on at least one of the individual marginal distributions conditional on V).

We end this article by noting some connections between competing risks models and other models. First, they are related to switching regression models, or Roy models. For example, if $T_i|X, V$ in the MMPH model have Weibull distributions, then we can write log $T_i = x_i\alpha_i + \varepsilon_i (i = 1, 2)$ (for example, van den Berg, Lindeboom and Ridder, 1994), where we observe T_i iff $T_i < T_j (j \neq i)$. Second, competing risks models are building blocks of multivariate duration models, notably models where one of the durations is always observed (for example, T_1 captures the moment of a treatment and T_2 is the observed duration outcome of interest).

We have considered only continuous-time duration variables T_i that have different realizations with probability 1. Recently, semi-parametric and nonparametric results have been derived for discrete-time or interval-censored competing risks models and models where different risks can be realized simultaneously (see for example Bedford and Meilijson, 1997; van den Berg, van Lomwel and van Ours, 2003; Honoré and Lleras-Muney, 2006). The biostatistical literature contains many studies in which specific assumptions are made on the dependence structure of the two durations T_i, enabling inference on the marginal distributions from data on Y, Z (see for example Moeschberger and Klein, 1995, for a survey).

GERARD J. VAN DEN BERG

See also **partial identification in econometrics; proportional hazard model; selection bias and self-selection.**

Bibliography

Abbring, J. and van den Berg, G. 2003. The identifiability of the mixed proportional hazards competing risks model. *Journal of the Royal Statistical Society, Series B* 65, 701–10.

Bedford, T. and Meilijson, I. 1997. A characterization of marginal distributions of (possibly dependent) lifetime variables which right censor each other. *Annals of Statistics* 25, 1622–45.

Bond, S. and Shaw, J. 2006. Bounds on the covariate-time transformation for competing-risks survival analysis. *Life time Data Analysis* 12, 285–303.

Cox, D. 1959. The analysis of exponentially distributed lifetimes with two types of failure. *Journal of the Royal Statistical Society, Series B* 21, 411–21.

Fermanian, J. 2003. Nonparametric estimation of competing risks models with covariates. *Journal of Multivariate Analysis* 85, 156–91.

Heckman, J. and Honoré, B. 1989. The identifiability of the competing risks model. *Biometrika* 76, 325–30.

Honoré, B. and Lleras-Muney, A. 2006. Bounds in competing risks models and the war on cancer. *Econometrica*, 74, 1675–98.

Moeschberger, M. and Klein, J. 1995. Statistical methods for dependent competing risks. *Lifetime Data Analysis* 1, 195–204.

Peterson, A. 1976. Bounds for a joint distribution function with fixed sub-distribution functions: application to competing risks. *Proceedings of the National Academy of Sciences* 73, 11–13.

van den Berg, G. 1990. Search behaviour, transitions to nonparticipation and the duration of unemployment. *Economic Journal* 100, 842–65.

van den Berg, G. 2001. Duration models: specification, identification, and multiple durations. In *Handbook of Econometrics*, vol. 5, ed. J. Heckman and E. Leamer. Amsterdam: North-Holland.

van den Berg, G., Lindeboom, M. and Ridder, G. 1994. Attrition in longitudinal panel data, and the empirical analysis of dynamic labour market behaviour. *Journal of Applied Econometrics* 9, 421–35.

van den Berg, G., van Lomwel, A. and van Ours, J. 2003. Nonparametric estimation of a dependent competing risks model for unemployment durations. Discussion Paper No. 898. Bonn: IZA.

competition

Competition is a rivalry between individuals (or groups or nations), and it arises whenever two or more parties strive for something that all cannot obtain. Competition is therefore at least as old as man's history, and Darwin (who borrowed the concept from economist Malthus) applied it to species as economists had applied it to human behaviour.

A concept that is applicable to two cobblers or a thousand shipowners or to tribes and nations is necessarily loosely drawn. When Adam Smith launched economics as a comprehensive science in 1776, he followed this usage. He explained why a reduced supply of a good led to a higher price: the 'competition [which] will immediately begin' among buyers would bid up the price. Similarly if the supply become larger, the price would sink more, the greater 'the competition of the sellers' (Smith, [1776], 1976, pp. 73–4). Here competition was very much like a race: a race to obtain part of reduced supplies or to dispose of a part of increased supplies. Almost nothing except a number of buyers and sellers was necessary for competition to operate. And the greater the number of each, the greater the vigour of competition:

> If this capital [sufficient to trade in a town] is divided between two different grocers, their competition will tend to make both of them sell cheaper, than if it were in the hands of one only; and if it were divided among twenty, their competition would be just so much the

greater, and the chance of their combining together, in order to raise the price, just so much the less (ibid., pp. 361–2).

With such a loose concept, there was little occasion to speak of one market as being more or less competitive than another, although this very passage presented the commonsense idea that larger numbers of rivals increased the intensity of competition.

The competition of grocers in a town pertained to competition *within* a market or an industry. Smith made much of the competition of different markets or industries for resources, and he developed what has always remained the main theorem on the allocation of resources in an economy composed of private, competing individuals or enterprises. The argument may be stated: Each owner of a productive resource will seek to employ it where it will yield the largest return. As a result, under competition each resource will be so distributed that it yields the same rate of return in every use. For if a resource were earning more in one use than another, it would be possible for its return in the lower-yielding use to be increased by reallocating it to the higher-yielding use. And this theorem led to what John Stuart Mill called the most frequently encountered proposition in economics: 'There cannot be two prices in the same market' (Mill, 1848, Book II, ch. IV, s. 3).

The competition of different markets or industries for the use of the same resources called attention to some problems which are less important within a single market such as the grocery trade in a town. One must possess knowledge of the investment opportunities in these different employments, and that knowledge is less commonly possessed than knowledge within one market. It often requires a good deal of time to disengage resources from one field and instal them elsewhere. Both of these conditions were recognized by Smith, who spoke of the difficulty of keeping secret the existence of extraordinary profits, and of the long run sometimes required for the attainment of equality of rates of return.

For the next three-quarters of a century the prevailing treatment of competition followed the practice of Smith. One can find occasional hints of a more precise definition of competition, well illustrated by Nassau W. Senior:

> But though, under free competition, cost of production is the regulator of price, its influence is subject to much occasional interruption. Its operation can be supposed to be perfect only if we suppose that there are no disturbing causes, that capital and labour can be at once transferred, and without loss, from one employment to another, and that every producer has full information of the profit to be derived from every mode of production. But it is obvious that these suppositions have no resemblance to the truth. A large portion of the capital essential to production consists of buildings, machinery, and other implements, the results of much time and labour, and of little service for any except their

existing purposes ... few capitalists can estimate, except upon an average of some years, the amount of their own profits, and still fewer can estimate those of their neighbours (1836, p. 102).

Senior is hinting at a concept of perfect competition, but the hint is not pursued.

The classical economists felt no need for a precise definition because they viewed monopoly as highly exceptional: Harold Demsetz has counted only one page in 90 devoted to monopoly in *The Wealth of Nations* and only one in 500 in Mill's *Principles of Political Economy*. Indeed the word 'monopoly' was usually restricted to grants by the sovereign of exclusive rights to manufacture, import or sell a commodity; witness the entry in the *Penny Cyclopedia* (1839):

It seems then that the word monopoly was never used in English Law, except when there was a royal grant authorizing some one or more persons only to deal in or sell a certain commodity or article.

If a number of individuals were to unite for the purpose of producing any particular article or commodity, and if they should succeed in selling such article very extensively, and almost solely, such individuals in popular language would be said to have a monopoly. Now, as these individuals have no advantages given them by the law over other persons, it is clear they can only sell more of their commodity than other persons by producing the commodity cheaper and better (XV, p. 341).

The ability of rivals to seek out and compete away supernormal profits, unless prevented by legal obstacles, was believed to be the basic reason for the pervasiveness of competition.

In the last third of the 19th century the concept of competition became the subject of intense study. The most popular reason given for this attention is that the growth of large-scale enterprises, including railroads, public utilities, and finally great manufacturing enterprises, made obvious the fact that a simple concept of competition no longer fit the economy of an industrial nation such as England.

A second source of misgiving with the broad definition of competition is that it might not lead to the uniformity of returns to a resource predicted by the theory. The Irish economist Cliffe Leslie repeatedly made this charge:

Economists have been accustomed to assume that wages on the one hand and profits on the other are, allowing for differences in skill and so forth, equalized by competition, and that neither wages nor profits can anywhere rise above 'the average rate', without a consequent influx of labour or of capital bringing things to a level. Had economists, however, in place of reasoning from an assumption, examined the facts connected with the rate of wages, they would have found, from

authentic statistics, the actual differences so great, even in the same occupation, that they are double in one place what they are in another. Statistics of profits are not, indeed, obtainable like statistics of wages; and the fact that they are not so, that the actual profits are kept a profound secret in some of the most prominent trades, is itself enough to deprive the theory of equal profits of its base (1888, pp. 158–9).

The easiest way to combat such criticisms was not to confront them with data – that path was not chosen for many years – but to define competition in such a way as to ensure the desired results such as uniformity of price.

The complications possible with competition were raised also on the theoretical side. William T. Thornton, in his book *On Labour* (1869), denied the fact that prices were determined by the 'law of supply and demand', particularly within labour markets. He employed bizarre examples, such as supply and demand curves which coincided over a vertical range, to show that price could be indeterminate or unresponsive to changes in supply or demand. These objections naturally called forth responses, from both J.S. Mill (*Collected Works*, V) and Fleeming Jenkin, a famous engineer.

The most persuasive reason for the increasing attention to the concepts of economics was the gradual move of economic studies to the universities, which proceeded rapidly in the last decades of the century. The expanding use of mathematics was one major symptom of the development of the formal and abstract theory of economics by Walras, Pareto, Irving Fisher and others. That formalization would scarcely be possible without a more precise specification of the nature of competition, and the precise specification of the nature of competition, and the replies to Thornton's criticisms were a precursor to this literature.

The groundwork for the development of the concept of perfect competition was laid by Augustin Cournot in 1838 in his *Mathematical Principles of the Theory of Wealth*. He made the first systematic use of the differential calculus to study the implications of profit-maximizing behaviour. Starting with the definition, Profits = Revenue − Costs, Cournot sought to maximize profits under various market conditions. He faced the question: How does revenue (say, pq) vary with output (q)? The natural answer is to *define* competition as that situation in which p does not vary with q – in which the demand curve facing the firm is horizontal. This is precisely what Cournot did:

The effects of competition have reached their limit, when each of the partial productions D_k [the output of producer k] is *inappreciable*, not only with reference to the total production $D = F(p)$, but also with reference to the derivative $F'(p)$, so that the partial production D_k could be subtracted from D without any appreciable variation resulting in the price of the commodity (Cournot [1838] 1927, p. 90).

This definition of competition was especially appropriate in Cournot's system because, according to his theory of oligopoly, the excess of price over marginal cost approached zero as the number of like producers became large. The argument is as follows:

Let the revenue of the firm be q_ip, and let n identical firms have the same marginal costs, MC. Then the equation for maximum profits for one firm would be

$$p + q_i(\mathrm{d}p/\mathrm{d}q) = MC.$$

The sum of n such equations would be

$$np + q(\mathrm{d}p/\mathrm{d}q) = nMC,$$

for $nq_i = q$. This last equation may be written,

$$p = MC - p/nE,$$

where E is the elasticity of market demand (Cournot, 1838, p. 84).

Cournot believed that this condition of competition was fulfilled 'for a multitude of products, and, among them, for the most important products'.

Cournot's definition was enormously more precise and elegant than Smith's so far as the treatment of numbers was concerned. A market departed from unlimited competition to the extent that prices exceeded the marginal cost of the firm, and the difference approached zero as the number of rivals approached infinity. This definition, however, illuminated only the effect of number of rivals on the power of individual firms to influence the market price, on Cournot's special assumption that each rival believed that his output decisions did not affect the output decisions of his rivals. It therefore bore only on what we term market competition.

Cournot did not face the question of the role of information possessed by traders, and this question was taken up by William Stanley Jevons in 1871 in his *Theory of Political Economy*. He characterized a perfect *market* by two conditions:

> (1.) A market, then, is theoretically perfect only when all traders have perfect knowledge of the conditions of supply and demand, and the consequent ratio of exchange; ... (2.) ... there must be perfectly free competition, so that any one will exchange with any one else upon the slightest advantage appearing. There must be no conspiracies for absorbing and holding supplies to produce unnatural ratios of exchange (Jevons, 1871, pp. 86, 87).

By perfect knowledge Jevons meant only that each trader in a market knew the price bids of every other trader. The second condition ruled out any joint actions by two or more traders, without his noticing that with knowledge so perfect as to know the behaviour of rivals, there might appear the very conspiracies he ruled out. The two conditions dictated that 'there cannot be two prices for the

same kind of article' in a perfect market, which he called the 'law of indifference'.

The merging of the concepts of competition and the market was unfortunate, for each deserved a full and separate treatment. A market is an institution for the consummation of transactions. It performs this function efficiently when every buyer who will pay more than the minimum realized price for any class of commodities succeeds in buying the commodity, and every seller who will sell for less than the maximum realized price succeeds in selling the commodity. A market performs these tasks more efficiently if the commodities are well specified and if buyers and sellers are fully informed of their properties and prices. Also a complete, perfect market allows buyers and seller to act on differing expectations of future prices. A market may be perfect and monopolistic or imperfect and competitive. Jevons's mixture of the two has been widely imitated by successors, of course, so that even today a market is commonly treated as a concept subsidiary to competition.

Edgeworth was the first economist to attempt a systematic and rigorous definition of perfect competition. His exposition deserves the closest scrutiny in spite of the fact that few economists of his time or ours have attempted to disentangle and uncover the theorems and conjectures of the *Mathematical Psychics* (1881), probably the most elusively written book of importance in the history of economics. His exposition was the most influential in the entire literature.

The conditions of perfect competition are stated as follows:

> The *field of competition* with reference to a contract, or contracts, under consideration consists of all individuals who are willing and able to recontract about the articles under consideration ...
>
> There is free communication throughout a *normal* competitive field. You might suppose the constituent individuals collected at a point, or connected by telephones – an ideal supposition [1881], but sufficiently approximate to existence or tendency for the purposes of abstract science.
>
> A *perfect* field of competition professes in addition certain properties peculiarly favourable to mathematical calculation; ... The conditions of a *perfect* field are four; the first pair referable to the heading *multiplicity* or continuity, the second to *dividedness* or fluidity.
>
> I. An individual is free to *recontract* with any out of an indefinite number, ...
> II. Any individual is free to *contract* (at the same time) with an indefinite number; ... This condition combined with the first appears to involve the indefinite divisibility of each *article* of contract (if any X deal with an indefinite number of Ys he must give each an indefinitely small portion of x); which might be erected into a separate condition.

III. Any individual is free to *recontract* with another independently of, *without the consent* being required of, any third party, …

IV. Any individual is free to *contract* with another independently of a third party; …

The failure of the first [condition] involves the failure of the second, but not vice versa; and the third and fourth are similarly related (Edgeworth, 1881, pp. 17–19).

The essential elements of this formidable list of conditions are two:

(1) There are an indefinitely large number of independent traders on each side of a market (the Cournot condition).

(2) Each trader can costlessly make tentative contracts with everyone (hence the divisibility of commodities) and alter these contracts (recontract) so long as a more favourable contract can be made. The result is perfect knowledge (the Jevonian condition).

Edgeworth gave an intuitive argument for the need for an indefinitely large number of traders on both sides of a market. It proceeds as follows. Let there be one seller and two buyers, and let the seller gain all the benefits of the sale: each buyer is charged the maximum price he would pay rather than withdraw from the market. If now a second seller appears, he will find it advantageous to offer better terms to the two buyers: 'It will in general be possible for *one* of the [sellers] (without the consent of the other), to *recontract* with the two [buyers], so that for all those three parties the recontract is more advantageous than the previously existing contract' (ibid., p. 35). As the numbers of traders on each side increase, the price approaches the competitive equilibrium level where no individual trader can influence it.

A defect in this argument is that it ignores the fact that if the traders on one or both sides of the market, be they 2, or 2000 or 2,000,000, join together they can do better *individually* than by competing. If traders on each side join, however, there will be bilateral monopoly, not competition. Edgeworth gives no reason why the combination of traders fails to take place. Only in modern times has the reason for independent behaviour by rivals been established: the costs of reaching and enforcing agreements on joint action increase with both the number of rivals and the complexity of the transactions. At a certain level – quite possibly with only two traders under some conditions – the costs of joint action exceed the gain to at least some of the traders, and independent behaviour emerges.

Edgeworth's 'conjecture', as it is now often called, that a unique, competitive price would emerge when the number of traders became large, has given rise to a modern literature vast in scope and often highly advanced in its mathematical techniques (for references, see Hildenbrand, 1974). One result in this literature is that in the case of a large (infinite) number of traders, no coalition of a portion of the traders can exclude traders outside the coalition from trading at the price-taking equilibrium.

Edgeworth's introduction of the requirement that the commodity or service that is traded be highly divisible is a response to the following problem:

> Suppose a market, consisting of an equal number of masters and servants, offering respectively wages and service; subject to the condition that no man can serve two masters, no master employ more than one man; or suppose equilibrium already established between such parties to be disturbed by any sudden influx of wealth into the hands of the masters. Then there is no *determinate*, and very generally *unique*, arrangement towards which the system tends under the operation of, may we say, a law of Nature, and which would be predictable if we knew beforehand the real requirements of each, or of the average, dealer; … (Edgeworth, 1881, p. 46).

Consider the simple example: a thousand masters will each employ a man at any wage below 100; a thousand labourers will each work for any wage above 50. There will be a single wage rate: knowledge and numbers are sufficient to lead a worker to seek a master paying more than the going rate or a master to seek out a worker receiving less than the market rate. But any rate between 50 and 100 is a possible equilibrium. But if a single worker leaves the market, the wage will rise to 100, and if a single employer withdraws, the wage will fall to 50. This ability of a single trader to affect the price arises because of the lumpiness of the article traded (here a worker's labour for a given period). Once a worker can work for two masters, the withdrawal of one worker in a thousand will reduce the available hours of work per day to each employer by only 8/1000 hours or 4.8 minutes per day, with only negligible influence upon the wage rate. Alternatively, a distribution of wage offers and demands would also eliminate the indeterminacy and market power.

Edgeworth's analysis was limited to competition within a market, and it was left to John Bates Clark to emphasize the need for mobility of resources if the return on each resource was to be equalized in every use.

> …there is an ideal arrangement of the elements of society, to which the force of competition, acting on individual men, would make the society conform. The producing organism actually shapes itself about his model, and at no time does it vary greatly from it … We must use assumptions boldly and advisedly, make labour and capital absolutely mobile, and letting competition work in ideal perfection (Clark, 1899, pp. 68, 71).

Perfect and free mobility of resources is of course an even more extreme assumption than the other conditions required for perfect competition because there is less reason to believe that free movement of resources is even approached in the real economy. Nor is the assumption of perfect mobility necessary to eliminate monopoly power in a market: in the Victorian age, the price of wheat of Iowa was set in Liverpool even though transportation costs were substantial. The assumption is usually necessary to attain strict equality in the price of a good at every point (the law of one price), although even this is not strictly true (as in the factor price equalization theorem). Clark also demanded that the economy be stationary for perfect competition, a condition we shall return to later.

All the elements of a concept of perfect competition were in place by 1900, and this concept increasingly became the standard model of economic theory thereafter. The most influential statement of the conditions for perfect competition was made by Frank H. Knight in his doctoral dissertation, *Risk, Uncertainty and Profit* (1921). The conditions were stated in extreme form; for example, 'There must be perfect, continuous, costless intercommunication between all individual members of the society' (Knight, 1921, p. 78) – so Jones in Seattle would know the price of potatoes and be able costlessly to ship to Smith in Miami a bushel of potatoes at every moment of time.

Of course these conditions are not *necessary*, but only sufficient, to achieve the competitive equilibrium. For example, if even a considerable fraction of buyers knows that seller A is charging more than B for a given commodity, their patronage may be quite enough to force A to reduce his price to that of B. Nor are the various conditions independent of one another: for example, if it is very cheap for either a commodity or its buyers or sellers to move between two places, that will insure that the prices in the two places will be widely known.

Along with the development of the concept of competition as a standard component of the theory of prices and the allocation of resources, it acquired a growing role as the criterion by which to judge the efficiency of actual markets. Adam Smith had already advanced the proposition that output was maximized in a private enterprise economy with competition. If each owner of a resource maximized the return from his resources, then (in the absence of 'external' effects of one person's actions on others) aggregate output would be maximized. This theorem (labelled 'on maximum satisfaction') was developed and qualified by Léon Walras (1874), Alfred Marshall (1980), Pareto (1895–6, 1907), Pigou (1912) and a host of modern economists.

Competition is much too central a concept in economics to remain unaffected when economists change their interests or analytical methods. We may illustrate this fact by the problem of economic change.

In a regime of change, of growing population and capital or innovations or new consumer demands, the problem of defining competition is much more difficult than it is for the stationary economy. Unless the change is predictable with precision, knowledge must necessarily be incomplete and errors and lags in adaptation to new conditions can be large. For this reason, indeed, J.B. Clark believed that perfect competition was achievable only in the stationary economy.

Even short-run changes in market price raise the question: is the change in price initiated by a particular seller or buyer, and if so, is this trader not facing a negatively sloping demand curve or a positively sloping supply curve? The infinitely elastic supply and demand curves of perfectly competitive equilibrium seem inapplicable to periods of changing market conditions. Some economists nevertheless retain the condition that individual traders cannot influence price by introducing a hypothetical auctioneer who announces price changes.

A partial adaptation of the competitive concept to change is made by making it a long-run equilibrium concept. Even if resources are not costlessly mobile and even if entrepreneurs do not have perfect foresight, one can analyse the rate of approach of returns on resources to equality. If an industry experiences a once-for-all large change, it could be in competitive equilibrium before and after the change, and the equilibria could be studied by competitive theory (comparative statics).

This adaptation did not satisfy Joseph Schumpeter, who believed that incessant change in products and production methods was the very essence of competitive capitalism. He argued that the displacing of one product or method by another, a process which he called creative destruction, made the concept of perfect competition irrelevant to either positive analysis or welfare judgements. If the monopoly that reduced output, compared to competition, by 10 per cent in one year, increased output by 100 per cent over the next two decades, then monopoly might be preferred to stagnant competition.

It is crucial to this argument that monopoly provides large, though temporary, rewards to successful innovators but competition does not:

> But perfectly free entry into a *new* field may make it impossible to enter it at all. The introduction of new methods of production and new commodities is hardly conceivable with perfect – and perfectly prompt – competition from the start. And this means that the bulk of what we call economic progress is incompatible with it. As a matter of fact, perfect competition is and always has been temporarily suspended whenever anything new is being introduced – automatically or by measures devised for that purpose – even in otherwise perfectly competitive conditions. (Schumpeter, 1942, pp. 104–5)

Schumpeter relies on instantaneous rivalry to eliminate the incentives to innovation under competition, and the conclusion would not hold if competition is defined in terms of long-run equilibrium.

Nevertheless the issue is not disposed of so easily. If change is continuous rather than sporadic, long-run equilibria will never be fully achieved. Several economists have emphasized that alterations in the concept of competition are called for in periods of historical change. Kirzner has emphasized the role of entrepreneurial rivalry in competition, whereas such rivalry is nonexistent in a perfectly competitive equilibrium. Demsetz has proposed a concept of laissez-faire competition, in which freedom of resources to move into any use is the central element. Such realistic reversions to the competitive concept of the classical economists have not been systematically formalized into theoretical models.

The concept of perfect competition, or indeed any theoretically precise concept of competition, will not be met by the actual condition of competition in any industry. John Maurice Clark made the most influential effort to create a concept of 'workable competition' which would serve as a working rule for public policies which seek to preserve or increase competition.

Clark emphasized the fact that if one requisite of perfect competition is absent, it may be desirable that a second requisite also be unfulfilled. For example, with instantaneous mobility but imperfect knowledge, members of an occupation would keep shifting back and forth between two cities, always overshooting the amount of migration which would equalize wage rates. This propensity to overshoot equilibrium would be corrected with less mobility of labour. This problem was later formalized as the theory of the 'second best'.

The essence of the concept of workable competition was the belief that 'long-run curves, both of cost and of demand, are much flatter than short-run curves, and much flatter than the curves which are commonly used in the diagrams of theorists' (J.M. Clark, 1940, p. 460). This correct and sensible view led to a proliferation of studies, usually in doctoral dissertations, of individual industries, in which the workableness of competition in each industry was appraised. Unfortunately there were no objective criteria to guide these judgements, and there was no evidence that the studies were accepted by the governmental agencies which administered competitive policies.

The popularity of the concept of perfect competition in theoretical economics is as great today as it has ever been. The concept is equally popular as first approximation in the more concrete studies of markets and industries that comprise the field of 'industrial organization' (applied microeconomics). The limitations of the concept in dealing with conditions of persistent and imperfectly predicted change will not be removed until economics possesses a developed theory of change. Even within a stationary economic setting the concept is being deepened by mathematical economists (see Mas-Colell, 1982). Meanwhile the central elements of competition – the freedom of traders to use their resources where they will, and exchange them at any price they wish – will continue to play a major role in the economics of an enterprise economy.

GEORGE J. STIGLER

See also **exchange; large economies; perfect competition.**

Bibliography

Clark, J.B. 1899. *The Distribution of Wealth*. London: Macmillan.

Clark, J.M. 1940. Toward a concept of workable competition. *American Economic Review*, June; reprinted in *Readings in the Social Control of Industry*, Philadelphia: Blakiston, 1942.

Cournot, A. 1838. *Researches into the Mathematical Principles of the Theory of Wealth*. Reprinted New York: Macmillan, 1927.

Demsetz, H. 1982. *Economic, Legal and Political Dimensions of Competition*. Amsterdam: North-Holland.

Edgeworth, F.Y. 1881. *Mathematical Psychics*. London: Kegan Paul, 1932.

Hildenbrand, W. 1974. *Core and Equilibria of a Large Economy*. Princeton: Princeton University Press.

Jevons, W.S. 1871. *The Theory of Political Economy*. London: Macmillan.

Kirzner, I.M. 1973. *Competition and Entrepreneurship*. Chicago: University of Chicago Press.

Knight, F.H. 1921. *Risk, Uncertainty and Profit*, Part 2. Boston: Houghton Mifflin Co.

Cliffe Leslie, T.E. 1888. *Essays in Political Economy and Moral Philosophy*. London: Longmans Green.

McNulty, P.J. 1967. A note on the history of perfect competition. *Journal of Political Economy* 75, 395–9.

Marshall, A. 1890. *Principles of Economics*. London: Macmillan.

Mas-Colell, A., ed. 1982. *Noncooperative Approaches to the Theory of Perfect Competition*. New York: Academic Press.

Mill, J.S. 1848. *Principles of Political Economy*. In *Collected Works*, ed. J.M. Robson. Toronto: University of Toronto Press, 1965.

Nutter, G.W. 1951. *The Extent of Enterprise Monopoly in the United States, 1899–1939*. Chicago: University of Chicago Press.

Penny Cyclopedia of the Society for the Diffusion of Useful Knowledge. 1839.

Schumpeter, J.A. 1942. *Capitalism, Socialism and Democracy*. New York: Harper & Bros.

Senior, N.W. 1836. *Political Economy*. London: W. Clowes.

Shepherd, W.G. 1982. Causes of increased competition in the US economy, 1939–1980. *Review of Economics and Statistics* 64, 613–26.

Smith, A. 1776. *The Wealth of Nations*. Glasgow edn, Oxford University Press, 1976.

Stigler, G.J. 1957. Perfect competition, historically contemplated. *Journal of Political Economy* 65, 1–17.

Stigler, G.J. and Sherwin, R. 1985. The extent of the market. *Journal of Law and Economics* 28, 555–85.
Thornton, W.T. 1869. *On Labour*. London: Macmillan.

competition and selection

Under competitive conditions, a business firm must maximize profit if it is to survive – or so it is often claimed. This purported analogue of biological natural selection has had substantial influence in economic thinking, and the proposition remains influential today. In general, its role has been to serve as an informal auxiliary defence, or crutch, for standard theoretical approaches based on optimization and equilibrium. It appeared explicitly in this role in a provocative passage in Milton Friedman's famous essay on methodology (Friedman, 1953, ch. 1), and it seems that many economists are familiar with it in this context only.

There is, however, an alternative role that the proposition can and does play. It serves as an informal statement of the common conclusion of a class of theorems characterizing explicit models of economic selection processes. A model in this class posits, first, a range of possible behaviours for the firm. This range must obviously extend beyond the realm of profit maximization if the conclusion of the argument is to be non-trivial, and it must include behaviour that is appropriately termed 'profit maximizing' if the conclusion is to be logically attainable at all. The model must also characterize a particular dynamic process that in some way captures the general idea that profitable firms tend to survive and grow, while unprofitable ones tend to decline and fail. A stationary position of such a process is a 'selection equilibrium'.

Models of this type occupy an important but non-central position in evolutionary economic theory (Nelson and Winter, 1982). They establish that the equilibria of standard competitive theory can indeed be 'mimicked' (in several different senses) by the equilibria of selection models. More importantly, by making explicit the strong assumptions that apparently are required to generate this sort of result, they are the basis for a critique of its generality and an appraisal of the strength of the crutch on which standard theory leans. They also provide a helpful entry-way to the much broader class of evolutionary models in which mimicry results fail to hold. This entry-way has the convenient feature that the return path to standard theory is well marked; the sense in which evolutionary theory subsumes portions of standard theory becomes clear.

The concept of competition need not, of course, be considered only in the context of perfectly competitive equilibrium. In a broader sense of the term, any non-trivial selection model in which the 'fit' prosper and the 'unfit' do not is a model of a 'competitive' process. The process need not have a static equilibrium, or any

equilibrium, and it may easily lead to results that are clearly non-competitive by the standards of industrial organization economics.

The remainder of this essay first considers in more detail the theoretical links between selection processes and competitive equilibrium outcomes. It then examines a more interesting and less well-explored area that involves selection and, in a broad sense, competition; Schumpeterian competition.

Competitive equilibrium as a selection outcome

The intention here is to describe the heuristic basis of existing examples of this type of theorem, or, alternatively, to describe the basic recipe from which an obviously large class of broadly similar results could be produced. There may be other basic recipes, as yet unknown. There certainly are ways to ignore individual instructions of the recipe and yet preserve the result, though at the cost of delicately contrived adjustments in other assumptions.

(To avoid confusion, it should be noted at the outset that the word 'equilibrium' is used in two different senses in this discussion, the 'no incentives to change behaviour' sense employed in economic theory and the 'stationary position of a dynamic process' sense that is common outside of economics. The point of the discussion is, in fact, to relate these two equilibrium ideas in a particular way.)

(1) Constant returns to scale must prevail in the specific sense that the supply and demand functions of an individual firm at any particular time are expressible as the scale (or 'capacity') of that firm at that time multiplied by functions depending on prices, but not directly on scale or time. Increasing returns to scale must be excluded for familiar reasons. Decreasing returns must be excluded because they will in general give rise to equilibrium 'entrepreneurial rents' which could be partially dissipated by departures from maximization without threatening the survival of the firm. Thus, for example, the U-shaped long run average cost curve of textbook competitive theory does not provide a context in which selection necessarily mimics standard theory if competitive equilibrium would require some firms to be on the upward sloping portion of the curve.

(2) Firms must increase scale when profitable and decrease scale (or go out of business entirely) when unprofitable. Alternatively, profitability of a particular firm must lead to entry by perfect imitators of that firm's actions. In the absence of such assumptions, it is plain that there will in general be equilibria with non-zero profit levels, which under assumption (1) cannot mimic the competitive result. While the 'decline or fail' assumption is a plausible reflection of long-run breakeven constraints characteristic of actual capitalist institutions, no such realistic force attaches to the requirement that profitability lead to expansion. If firms do not pursue

profits in the long-run sense of expanding in response to positive profitability, stationary positions may involve positive profits. Such stationary positions fail to mimic competitive equilibria for that reason alone (given constant returns), but they also introduce once again the possibility that the short-run behavioural responses of surviving firms may dissipate some of the positive profit that is potentially achievable at selection equilibrium scale.

In standard theory, expansion in response to profitability may be seen as an aspect of the firm's profit-seeking on the assumption that it regards prices as unaffected by its capacity decisions. In turn, this ordinarily requires that the firm in question be but one of an indeterminately large number of firms that all have access to the same technological and organizational possibilities.

While the assumption that firms have identical production sets and behavioural rules is common and appears inoffensive in orthodox theorizing, it is very much at odds with evolutionary theory. The orthodox view comes down to the assertion that all productive knowledge is freely available to one and all – perhaps it is all in the public library. By contrast, evolutionary theory emphasizes the role of firms as highly individualized repositories of productive knowledge, not all of which is articulable. From the evolutionary perspective, the fact that mimicry theorems rely on assumptions of unimpaired access to a public knowledge pool is by itself sufficient to make it clear that the selection argument can provide only a weak and shaky crutch for standard competitive theory.

(3) A firm that is breaking even with a positive output at prevailing prices must not alter its behaviour; a potential entrant that would only break even at prevailing prices must not enter. This assumption is needed to assure that the competitive equilibrium position is in fact a stationary position of the selection process.

Models of natural selection in biology do not typically involve this sort of assumption, but neither do they conclude that only the fittest genotypes survive – the biological analogue of the proposition discussed here. Rather, they show how constant gene frequencies come to prevail as the selection forces that tend to eliminate diversity come into balance with mutation forces that constantly renew it. A strictly analogous treatment of economic selection would be much more appealing than the sort of result discussed here. It would admit that occasional disruptions may arise from random behavioural change, or from over-optimistic entrants. Thus, potentially at least, it could better serve the purpose of establishing the point that the results of standard competitive theory are in some sense robust with respect to its behavioural assumptions. Unfortunately, standard theory offers no clue as to what this sense might be. It is plain that the adjustment processes of the system are centrally involved, and there is no behaviourally plausible

theory of adjustment that is the dynamic counterpart in the disciplinary paradigm of static competitive equilibrium theory.

Within the limits defined by the requirement for a strictly static competitive outcome, the most plausible approach combines the idea of characterizing the firms in the selection process by their 'rules of behaviour' – an idea advanced in a seminal paper by Armen Alchian (1950) – with Herbert Simon's idea of satisficing (1955). In the simplest version, each firm simply adheres unswervingly to its own deterministic behavioural rule (or 'routine', in the language of Nelson and Winter, 1982). Such a rule subsumes or implies the firm's supply and demand functions, and given the conditions set forth in (1) and (2) above, a constant environment evokes a constant response. Satisficing may be introduced as a complication of this picture by an assumption that a firm that sustains losses over a period of time will search for a better behavioural rule; this adds behavioural plausibility to the adjustment process but does not introduce the possibility that random rule change might disrupt an otherwise stationary competitive equilibrium position.

(4) The final requirement can be succinctly but inadequately stated as 'some firms must actually be profit maximizers'. Although this formulation does adequately cover some simple cases, it does not suggest the depth and subtlety of the issues involved.

Two points deserve particular emphasis here. The first is the distinction between profit maximizing *rules of behaviour* (functions) and profit maximizing *actions*. In general, a selection equilibrium that mimics a particular competitive equilibrium must clearly be one in which some firms take actions that are profit maximizing in that competitive equilibrium, and in this sense are profit maximizers. But this observation does not imply that the survivors in the selection equilibrium possess maximizing *rules*, and in general it is not necessary that survivors be maximizers in this stronger sense. (Proof: Consider a competitive equilibrium with constant returns to scale. Restrict the firms' supply and demand functions to be constant up to a scale factor at the values taken in the given equilibrium. Embed this static equilibrium in a dynamic adjustment system in which firms' scales of output respond to profitability in accordance with assumption (2). Then the given competitive equilibrium becomes a selection equilibrium – since the only techniques in use make zero profit – but the firms are not profit maximizers in the stronger sense.)

The second point extends the first. The notion of profit maximizing behavioural rules itself rests on the conceptual foundation of a production set or function that is regarded as a given. In evolutionary theory, however, it is the rules themselves that are regarded as data and as logically antecedent to the values (actions) they yield in particular environments. Thus, in this context, a problem arises in interpreting the basic idea of a selection equilibrium mimicking a standard competitive one: there is no obvious set of 'possibilities' to which one should have reference.

The most helpful approach here emphasizes internal consistency. Assumptions about the structure of what is 'possible' can be invoked without the additional assumption that there is a given set of possibilities – for example, additivity and divisibility may be assumed without implying that the set of techniques to which these axioms apply is a given datum of the system. Such an approach provides a basis for discussing whether a particular selection equilibrium is legitimately *interpretable* as a competitive equilibrium given the other assumptions in force. Along this path one can explore a rich variety of selection equilibrium situations that may be thought of as competitive equilibria. Precisely because the variety is so rich, to know only that an outcome is interpretable in this fashion is to know very little about it.

In the light of formal analysis of selection models of the sort described above, how strong is the crutch that selection provides to standard theory? For many analytical purposes, it is a crucial weakness that the crutch relates only to equilibrium actions and not to behavioural rules; it is from the knowledge that the rules are maximizing that the results of comparative statics derive. A selection system disturbed by a parameter change from a 'mimicking' equilibrium does not necessarily go to a new 'mimicking' equilibrium, let alone to one that is consistent, in standard theoretical terms, with the information revealed in the original equilibrium. More fundamentally, selection considerations cannot compensate for the inadequacies of standard theory that arise from the basic assumption that production possibilities are given data of the system.

Schumpeterian competition

In two great works and in many other writings, Joseph Schumpeter proclaimed the central importance of innovative activity in the development of capitalism. His early book, *The Theory of Economic Development*, focused on the role and contribution of the individual entrepreneur. From today's perspective the work remains enormously insightful and provocative but may seem dated; the image of the late 19th-century captains of industry lurks implicitly in the abstract account of the entrepreneur. The late work, *Capitalism, Socialism and Democracy*, is likewise insightful, provocative and a bit anachronistic. In this case, the anachronism derives from the predictions of a future in which the innovative process is bureaucratized, the role of the individual entrepreneur is fully usurped by large organizations, and the sociopolitical foundations of capitalism are thereby undercut. Present reality does not correspond closely to Schumpeter's predictions, and it seems increasingly clear that he greatly underestimated the seriousness of the incentive problems that arise within large organizations, whether capitalist corporations or socialist states.

Substantial literatures have accumulated around a number of specific issues, hypotheses and predictions put forward in Schumpeter's various writings. Regardless of the verdicts ultimately rendered on particular points, everyday observation repeatedly confirms the appropriateness of his emphasis on the centrality of innovation in contemporary capitalism. It confirms, likewise, the inappropriateness of the continuing tendency of the economics discipline to sequester topics related to technological change in sub-sectors of various specialized fields, remote from the theoretical core.

The purpose of the present discussion is to assess the relationships of selection and competition from a Schumpeterian viewpoint, that is, to extend the discussion above by considering what difference it makes if firms are engaged in inventing, discovering and exploring new ways of doing things. Plainly, one difference it makes is that 'competition' must now be understood in the broad sense that admits a number of additional dimensions to the competitive process, along with price-guided output determination. In particular, costly efforts to innovate, to imitate the innovations of others, and to appropriate the gains from innovation are added to the firm's competitive repertoire.

Selection now operates at two related levels. The organizational routines governing the use made of existing products and processes in every firm interact through the market place, and the market distributes rewards and punishments to the contenders. These same rewards and punishments are also entries on the market's scorecard for the higher level routines from which new products and processes derive – routines involving, for example, expenditure levels on innovative and imitative R&D efforts. Over the longer term, selection forces favour the firms that achieve a favourable balance between the rents captured from successive rounds of innovation and the costs of the R&D efforts that yield these innovations.

In formal models constructed along these lines, it is easy to see how various extreme cases turn out. One class of cases formalizes the cautionary tale told by Schumpeter (1950, p. 105), in which competition that is 'perfect – and perfectly prompt' makes the innovative role non-viable. Sufficiently high costs of innovation and low costs of imitation (including costs of surmounting any institutional barriers such as patents) will lead to the eventual suppression of all firms that continue to attempt innovation, and the system will settle into a static equilibrium. (The character of this equilibrium may, however, depend on initial conditions and on random events along the evolutionary path; the production set ultimately arrived at is an endogenous feature of the process.) One can also construct model examples to illustrate the cautionary message 'innovate or die', the principal requirement being simply a reversal of the cost conditions stated above.

With the exception of some extreme or highly simplified cases, models of Schumpeterian competition describe complex stochastic processes that are not easily explored with analytical methods. Of course, the activity of writing down a specific formal model is often informative by itself in the sense that it illuminates basic conceptual issues and poses key questions about how complex features of economic reality can usefully be approximated by a model. Some additional insight can then be obtained using simulation methods to explore specific cases (Nelson and Winter, 1982, Part V; Winter, 1984). One of the most significant benefits from simulation is the occasional discovery of mechanisms at work that are retrospectively 'obvious' and general features of the model.

The discussion that follows pulls together a number of these different sorts of insights, emphasizing in particular some issues that do not arise in the related theoretical literature that explores various Schumpeterian themes using neoclassical techniques (For the most part these neoclassical studies explore stylized situations involving a single possible innovation, and thus do not address issues relating to the cumulative consequences of dynamic Schumpeterian competition. See Kamien and Schwartz (1981) and Dasgupta (1985) for references and perspectives on this literature.)

A fundamental constituent of any dynamic model of Schumpeterian competition is a model of technological opportunity. Such a model establishes the linkage between the resources that model firms apply to innovative effort and their innovative achievements. The long run behaviour of the model as a whole depends critically on the answers provided for a set of key questions relating to technological opportunity. Does the individual firm face diminishing returns in innovative achievement as it applies additional resources over a short period of time? If so, from what 'fixed factors' does the diminishing returns effect arise, and to what extent are these factors subject to change over time either by the firm's own efforts or by other mechanisms? Are selection forces to be studied in a context in which technological opportunity presents more or less the 'same problem' for R&D policy over an extended period, or is the evolutionary sorting out of different policies for the firm a process that proceeds concurrently with historical change in the criteria that govern the sorting?

Technological opportunity is said to be *constant* if R&D activity amounts to a search of an unchanging set of possibilities – in effect, there is a meta-production set or meta-production function that describes what is ultimately possible. *Increasing* technological opportunity means that possibilities are being expanded over time by causal factors exogenous to the R&D efforts in question – implying that, given a level of technological achievement and a level of R&D effort, the effort will be more productive of innovative results if applied later. With constant technological opportunity, returns to R&D effort must eventually be decreasing, approaching zero near the boundary of the fixed set of possibilities.

It is all too obvious that it may be very difficult to develop an empirical basis for modelling technological opportunity in an applied analysis of a particular firm, industry or national economy. There is no easy escape from the conundrum that observed innovative performance reflects both opportunity and endogenously determined effort, not to mention the fact that neither performance nor effort is itself easily measured or the even more basic question of whether analysis of the past can illuminate the future. These difficulties in operationalizing the concept of technological opportunity do not, unfortunately, in any way diminish its critical role in Schumpeterian competition.

The evolutionary analysis of Schumpeterian competition has not, thus far, produced any counterpart for the sorts of mimicry theorems that can be proved for static equilibria. That is, there is no model in which it can be shown that selection forces, alone or in conjunction with adaptive behavioural rules, drive the system asymptotically to a path on which surviving firms might be said to have solved the remaining portion of the dynamic optimization problem with which the model situation confronts them – except in the cases where the asymptotic situation is a static equilibrium with zero R&D. The list of identified obstacles to a non-trivial positive result is sufficiently long, and the obstacles are sufficiently formidable, so as to constitute something akin to an impossibility theorem. It seems extremely unlikely that a positive result can be established within the confines of an evolutionary approach – that is, without endowing the model firms with a great deal of correct information about the structure of the total system in which they are embedded.

The most formidable obstacle of all derives from the direct clash between the future-oriented character of a dynamic optimization and the fact that selection and adaptation processes reflect the experience of the past. If firms cannot 'see' the path that technological opportunity will follow in the future, if their decisions can only reflect past experience and inferences drawn therefrom, then in general they cannot position themselves optimally for the future. They might conceivably do so if the development of technological opportunity were simple enough to validate simple inference schemes. Such simplicity does not seem descriptively plausible; who is to say that it is implausible that in a particular case technological opportunity might be constant, or exponentially increasing, or following a logistic, or some stochastic variant of any of these? And without some restriction on the structural possibilities, how are model firms to make inferences to guide their R&D policies?

This obstacle is not featured prominently in the simulations reported by Nelson and Winter, which are largely confined to very tame and stylized technological regimes in which opportunity is summarized by a single exponentially increasing variable, called 'latent productivity'. Such an environment, reminiscent in some ways of neoclassical growth theory, seems at first glance to be a promising one for the derivation of a balanced growth outcome in which actual and latent productivity are rising at the same rate, the problem facing the firms is in a sense constant, and selection and adaptation might bring surviving firm R&D policies to optimal values.

In fact, such a result remains remote even under the very strong assumption just described. Demand conditions for the product of the industry (or the economy) affect the long run dynamics, and in this area also assumptions must be delicately contrived to avoid excluding a balanced growth outcome. For example, consider an industry model with constant demand in which demand is (plausibly) less than unit elastic at low prices. Then, cost reduction continued indefinitely would drive sales revenue to zero. Zero sales revenue will not cover the cost of continuing advance. What is involved here is a reflection of the basic economics of information; costs of discovery are independent of the size of the realm application, and on the assumption stated the economic significance of that realm is dwindling to nothing. The implication is that demand conditions may check progress even if technological opportunity is continually expanding. Indeed, this may well be the pattern that is typically realistic for any narrowly defined sector.

This difficulty too can be dispatched by an appropriately chosen assumption. Beyond it lie some further problems. A model that acknowledges the partially stochastic nature of innovative success will display gradually increasing concentration (Phillips, 1971), unless some opposing tendency is present. A good candidate for an opposing tendency is the actual exercise of market power that has been acquired by chance (Nelson and Winter, 1982, ch. 13). But this market power can, presumably, also shelter various departures from present value maximization, including departures from dynamically optimal R&D policy.

To reiterate, the quest for mimicry theorems in the context of Schumpeterian competition seems foredoomed to failure. Since models of Schumpeterian competition plainly provide a much better description of the world we live in than do models of static equilibrium, the overall conclusion with regard to the strength of the selection crutch is distinctly more negative than the conclusion for static models alone. Assumptions that firms maximize profit or present value will have to stand on their own, at least until somebody invents a better crutch for them. In the meantime, it will continue to be the case that predictions based on these assumptions are sometimes sound and sometimes silly, and standard theory does not offer a means of discriminating between the cases. More direct attention should be paid to the mechanisms of selection, adaptation and learning, which among them probably account for as much sense as economists have actually observed in economic reality, and also leave room for a lot of readily observable nonsense.

SIDNEY G. WINTER

Bibliography

Alchian, A.A. 1950. Uncertainty, evolution and economic theory. *Journal of Political Economy* 58, 211–21.

Dasgupta, P. 1985. The theory of technological competition. In *New Developments in the Analysis of Market Structure*, ed. J. Stiglitz and G.F. Mathewson. Cambridge, MA: MIT Press.

Friedman, M. 1953. *Essays in Positive Economics*. Chicago: University of Chicago Press.

Kamien, M. and Schwartz, N. 1981. *Market Structure and Innovation*. Cambridge: Cambridge University Press.

Nelson, R. and Winter, S. 1982. *An Evolutionary Theory of Economic Change*. Cambridge, Mass.: Belknap Press of The Harvard University Press.

Phillips, A. 1971. *Technology and Market Structure: A Study of the Aircraft Industry*. Lexington, Mass.: D.C. Heath.

Schumpeter, J.A. 1912. *The Theory of Economic Development*. Trans. Redvers Opie, Cambridge, MA: Harvard University Press, 1934.

Schumpeter, J.A. 1950. *Capitalism, Socialism and Democracy*. 3rd edn, New York: Harper.

Simon, H. 1955. A behavioral model of rational choice, *Quarterly Journal of Economics* 69, 99–118.

Winter, S.G. 1964. Economic 'natural selection' and the theory of the firm. *Yale Economic Essays* 4(1), 225–72.

Winter, S.G. 1971. Satisficing, selection and the innovating remnant. *Quarterly Journal of Economics* 85, 237–61.

Winter, S.G. 1984. Schumpeterian competition in alternative technological regimes. *Journal of Economic Behaviour and Organization* 5(3–4), 287–320.

competition, Austrian

The essence of Austrian economics is its emphasis on the ongoing economic process as opposed to the equilibrium analysis of neoclassical theory. Austrian concepts of competition reflect this emphasis. Indeed, one of the central challenges by Austrians to the neoclassical model, and a common denominator of virtually all Austrian economics, is the rejection of the concept of perfect competition. In this respect, a number of economists who cannot be considered Austrian in all aspects of their work, share, nonetheless, the Austrian emphasis on actual market activities and processes – for example, Joseph Schumpeter (1942), J.M. Clark (1961), Fritz Machlup (1942) and others.

When the concept of competition entered economics at the hands of Adam Smith and his predecessors, it was not clearly defined, but it generally meant entry by firms into profitable industries (or exit from unprofitable ones) and the raising or lowering of price by existing firms according to market conditions. There was little recognition, and virtually no analysis, of entrepreneurship as it might be reflected in these and other forms of competition, but there was a recognition that business firms do in most situations have some control over market prices, with the degree of control varying inversely with the number of firms in the industry. These basic ideas, expanded and supplemented, are generally compatible with most modern Austrian analysis.

What is objectionable to Austrian economists is the neoclassical concept of perfect competition, developed during the 19th and early 20th centuries. The development began with Cournot (1838), whose concern it was to specify as rigorously as possible the *effects* of competition, after the *process* of competition had reached its limits. His conceptualization of this situation was a market structure in which the output of any one firm could be subtracted from total industry output with no discernible effect on price. Later contributions by Jevons, Edgeworth, J.B. Clark and Frank Knight led to the model of perfect competition as we know it today (Stigler, 1957; McNulty, 1967).

The trouble with the concept from the Austrian point of view, as Hayek has emphasized, is that it describes an equilibrium situation but says nothing about the competitive process which led to that equilibrium. Indeed, it robs the firm of all business activities which might reasonably be associated with the verb 'to compete' (Hayek, 1948). Thus, firms in the perfectly competitive model do not raise or lower prices, differentiate their products, advertise, try to change their cost structures relative to their competitors, or do any of the other things done by business firms in a dynamic economic system. This was precisely the reason why Schumpeter insisted on the irrelevance of the concept of perfect competition to an understanding of the capitalist process.

For Schumpeter, any realistic analysis of competition would require a shift in analytical focus from the question of how the economy allocates resources efficiently to that of how it creates and destroys them. The entrepreneur, a neglected figure in classical and neoclassical economics, is the central figure in the Schumpeterian analytical framework. The entrepreneur plays a disequilibrating role in the market process by interrupting the 'circular flow' of economic life, that is, the ongoing production of existing goods and services under existing technologies and methods of production and organization. He does this by innovating – that is, by introducing the new product, the new market, the new technology, the new source of raw materials and other factor inputs, the new type of industrial organization, and so on. The result is a concept of competition grounded in cost and quality advantages which Schumpeter felt is much more important than the price competition of traditional theory and is the basis of the 'creative destruction' of the capitalist economic process. It produces an internal efficiency within the business firm, the importance of which for economic welfare is far greater, Schumpeter argued, than the allocative efficiency of traditional economic theory (Schumpeter, 1942).

His emphasis on the advantages of the firm's internal efficiency led Schumpeter to a greater tolerance for large-scale business organizations, even for those enjoying some degree of monopoly power, than was typical of many more traditional theorists of his time. This is a not uncommon characteristic of Austrian economics. Hayek, for example, makes the distinction between entrenched monopoly, with its probable higher-than-necessary costs, and a monopoly based on superior efficiency which does relatively little harm since in all probability it will disappear, or be forced to adjust to market conditions, as soon as another firm becomes more efficient in providing the same or a similar good or service (Hayek, 1948). And that is precisely Schumpeter's point. The ground under even large-scale enterprise is constantly shaking as a result of the competitive threat from the new firm, the new management, or the new idea. Schumpeter's competitive analysis was less a defence of monopoly power than of certain business activities which were judged to be monopolistic only from the comparative standpoint of the model of perfect competition. He insisted that the quality of a firm's entrepreneurship was of far greater significance than its mere size.

The leading contemporary Austrian theorist of competition is Israel Kirzner (1973). Kirzner's approach draws on the analysis of market processes and the concept of 'human action' developed earlier by Ludwig von Mises. For von Mises, entrepreneurship is human action in the market which successfully directs the flow of resources toward the fulfillment of consumer wants (Mises, 1949). Kirzner's more fully developed theory of competition is based on the idea that the means – end nexus of economic life is not given but is itself subject to creative human action. This creative role Kirzner defines as entrepreneurship, and it is essentially the ability to detect new but desired human wants, as well as new resources, techniques, or other ways through which to satisfy them. Whether he discovers new wants or new means of satisfying old ones, the Kirznerian entrepreneur is the one who sees and exploits what others fail to notice – the profit opportunities inherent in any situation in which the prices of factor inputs fall short of the price of the final product.

There is a difference between Kirzner's theory of entrepreneurship and that of Schumpeter. Schumpeter's entrepreneur is a disequilibrating force in the economic system; he initiates economic change. Kirzner's entrepreneur plays an equilibrating role; the changes he brings about are responses to the mistaken decisions and missed opportunities he detects in the market. Unlike Schumpeter's entrepreneur, he is not so much the creator of his own opportunities as a responder to the hitherto unnoticed opportunities that already exist in the market. Thus, in the competitive market process, the Schumpeterian and Kirznerian entrepreneurs may complement each other – the one creating change, the other responding to it.

Austrian dissatisfaction with the perfectly competitive model extends to the theories of imperfect and monopolistic competition. Hayek's and Kirzner's criticisms are the same as of perfect competition, namely, that the analysis is limited to an equilibrium situation in which the underlying data are assumed to be adjusted to each other, whereas the relevant problem is the process through which adjustment occurs. Schumpeter criticized monopolistic competition for its continued acceptance of an unvarying economic structure and forms of industrial organization. Nonetheless, the incorporation into economic theory of quality competition and sales efforts, complementing the traditional and limited focus on price competition, as well as the efforts on the part of some industrial organization specialists and institutional economists to analyse and explain actual market processes, are developments that are generally within the Austrian tradition.

PAUL J. MCNULTY

See also **Austrian economics; competition; creative destruction.**

Bibliography

Clark, J.M. 1961. *Competition as a Dynamic Process.* Washington, DC: Brookings Institution.
Cournot, A.A. 1838. *Recherches sur les principes mathématiques de la théorie des richesses.* Paris: Hachette.
Hayek, F.A. von. 1948. The meaning of competition. In *Individualism and Economic Order.* ed. F.A. Hayek. London: Routledge.
Kirzner, I. 1973. *Competition and Entrepreneurship.* Chicago: University of Chicago Press.
Machlup, F. 1942. Competition, pliopoly, and profit. Pts. I–II. *Economica*, N.S. 9, Pt. I, 1–23, Pt. II, 153–73.
McNulty, P. 1967. A note on the history of perfect competition. *Journal of Political Economy* 75, 395–9.
Mises, L. von. 1949. *Human Action.* New Haven: Yale University Press.
Schumpeter, J. 1942. *Capitalism, Socialism, and Democracy.* New York: Harper & Row, 1962.
Stigler, G. 1957. Perfect competition, historically contemplated. *Journal of Political Economy* 65, 1–17.

competition, classical

Only through the principle of competition has political economy any pretension to the character of a science. So far as rents, profits, wages, prices, are determined by competition, laws may be assigned for them. Assume competition to be their exclusive regulator, principles of broad generality and scientific precision may be laid down, according to which they will be regulated. (Mill, 1848, p. 242)

In all versions of economic theory 'competition', variously defined, is a central organizing concept. Yet the relationship between different definitions of competition and differences in the theory of value have not been fully appreciated. In particular, the characteristics of 'perfect' competition (notably the conditions which ensure price-taking) are often read back, illegitimately, into classical discussions of competition.

The mechanisms which determine the economic behaviour of industrial capitalism are not self-evident. As a form of economy in which production and distribution proceed by means of a generalized process of exchange (in particular by the sale and purchase of labour), it possesses no obvious direct mechanisms of economic and social coordination. Yet, in so far as these operations constitute a system, they must be endowed with some degree of regularity, the causal foundations of which may be revealed by analysis. The first steps in economic investigation which accompanied the beginnings of industrial capitalism consisted of a variety of attempts to identify such regularities, often by means of detailed description and enumeration, as in the works of Sir William Petty, and hence to establish the dominant causes underlying the behaviour of markets. But what was required was not simply the description and classification which precedes analysis, but abstraction, the transcendence of political arithmetic (Smith, 1776, p. 501).

The culmination of the search for a coherent abstract characterization of markets, and hence the foundation of modern economic analysis, is to be found in Chapter 7 of Book I of Adam Smith's *Wealth of Nations* – 'Of the Natural and Market Price of Commodities'. In this chapter Smith presented the first satisfactory formulation of the regularity inherent in price formation. The idea, partially developed earlier by Cantillon, and by Turgot in his discussion of the circulation of money, was that

There is in every society ... an ordinary or average rate of both wages and profits ... When the price of any commodity is neither more nor less than what is sufficient to pay the rent of land, the wages of labour, and the profits of stock employed ... according to their natural rates, the commodity is then sold for what may be called its natural price.

and that

The natural price ... is, as it were, the central price, to which the prices of all commodities are continually gravitating. Different accidents may sometimes keep them suspended a good deal above it, and sometimes force them down somewhat below it. But whatever may be the obstacles which hinder them from settling in this center of repose and continuance, they are continually tending towards it. (Smith, 1776, p. 65)

Thus the natural price encapsulates the persistent element in economic behaviour. And that persistence derives from the ubiquitous force of competition: or, as

Smith put it, the condition of 'perfect liberty' in which 'the whole of the advantages and disadvantages of the different employments of labour and stock must ... be either perfectly equal or continually tending to equality' (p. 111), for the natural price is 'the price of free competition' (p. 68).

The relationship between competition and the establishment of what Petty called 'intrinsic value' had been discussed in the works of Petty, Boisguillebert, Cantillon and Harris as the outcome of rival bargaining in price formation, competition being the greater when the number of bargainers was such that none has a direct influence on price. Quesnay expressed the formation of competitive prices as being 'independent of mens' will ... far from being an arbitrary value or a value which is established by agreement between the contracting parties' (in Meek, 1962, p. 90), but he did not relate the *organization of production* to the formation of prices in competitive markets. Consideration of that relationship required the development of a general conception of the role of capital, and with it the notion of a general rate of profit formed by the competitive disposition of capital between alternative investments (Vaggi, 1987).

A significant step in this direction was made by Turgot, who both conceived of the process of production as part of the circulation of money:

> We see ... how the cultivation of land, manufactures of all kinds, and all branches of commerce depend upon a mass of capitals, or movable accumulated wealth, which, having been first advanced by the entrepreneurs in each of these different classes of work, must return to them every year with a regular profit ... It is this continual advance and return of capitals which constitutes *what ought to be called the circulation of money*. (Turgot, 1973, p. 148)

and saw that the structure of investments would tend to be that which yielded a uniform rate of profit:

> It is obvious that the annual products which can be derived from capitals invested in these different employments are mutually limited by one another, and that all are relative to the existing rate of interest on money. (Turgot, 1973, p. 70)

However, Turgot neither related the determination of the rate of profit to production in general – he accepted the Physiocratic idea that the incomes of the industrial and commercial classes were 'paid' by agriculture – nor developed the conceptual framework which linked the formation of prices and of the rate of profit to the overall organization of the economy. These were to be Smith's achievements:

> If ... the quantity brought to market should at any time fall short of the effectual demand, some of the component parts of its price must rise above their

natural rate. If it is rent, the interest of all other landlords will naturally prompt them to prepare more land for the raising of this commodity; if it is wages or profit, the interest of all other labourers and dealers will soon prompt them to employ more labour and stock in preparing and bringing it to market. The quantity brought thither will soon be sufficient to supply the effectual demand. All the different parts of its price will soon sink to their natural rate, and the whole price to its natural price. (Smith, 1776, p. 65)

So in a competitive market there will be a tendency for the actual prices (or 'market prices' as Smith called them) to be relatively high when the quantity brought to market is less than the effectual demand (the quantity that would be bought at the natural price) and relatively low when the quantity brought to market exceeds the effectual demand. This working of competition was known as the 'Law of Supply and Demand'. The working of competition which constitutes the 'Law' do not identify the phenomena which *determine* natural prices. The 'Law' of supply and demand should not be confused with supply and demand 'theory', that is, the neoclassical theory of price determination which was to be developed one hundred years later. Nor should Smith's discussion of the tendencies of concrete market prices be confused with supply and demand function, which are loci of equilibrium prices.

Adam Smith's conception of 'perfect liberty' consists of the mobility of labour and stock between different uses – the mobility that is necessary for the establishment of 'an ordinary or average rate both of wages and profits' and hence for the gravitation of market prices toward natural prices. Smith identifies four reasons why market prices may deviate 'for a long time together' above natural price, creating differentials in the rate of profit, all of which involve restriction of mobility:

(a) extra demand can be 'concealed', though 'secrets of this kind ... can seldom be long kept';
(b) secret technical advantages;
(c) 'a monopoly granted either to an individual or a trading company';
(d) 'exclusive privileges of corporation, statutes of apprenticeship, and all those laws which restrain, in particular employments, the competition to a smaller number than might otherwise go into them'.

For Smith there is some similarity in the forces acting on wages and profits which derives from his conceiving of the capitalist as personally involved in the prosecution of a particular trade or business. So the rate of profit, like the rate of wages, may be differentiated between sectors by 'the agreeableness of disagreeableness of the business', even though 'the average and ordinary rates of profit in the different employments of stock should be more nearly upon a level than the pecuniary wages of the different sorts of labour' (1776, p. 124). Landlords, capitalists and workers are all active agents of mobility. In

Ricardo's discussion the emphasis shifted towards the distinctive role of capital:

> It is, then, the desire, which every capitalist has, of diverting his funds from a less to a more profitable employment, that prevents the market price of commodities from continuing for any length of time either much above, or much below their natural price. (Ricardo, 1817, p. 91)

Ricardo used the term 'monopoly price' to refer to commodities 'the value of which is determined by their scarcity alone', such as paintings, rare books and rare wines (1817, pp. 249–51) which have 'acquired a fanciful value', and he argued that for 'Commodities which are monopolised, either by an individual, or by a company ... their price has no necessary connexion with their natural value' (p. 385). His analysis of value and distribution is accordingly confined to 'By far the greatest part of those goods which are the object of desire ... such commodities only as can be increased in quantity by the exertion of human labour, and on the production of which competition operates without restraint' (p. 12).

For Marx competition is synonymous with the generalization of capitalist relations of production. Competition is thus related to the rise to dominance of the capitalist mode of production.

> While free competition has dissolved the barriers of earlier relations and modes of production, it is necessary to observe first of all that the things which were a barrier to it were the inherent limits of earlier modes of production, within which they spontaneously developed and moved. These limits became barriers only after the forces of production and the relations of intercourse had developed sufficiently to enable capital as such to emerge as the dominant principle of production. The limits which it tore down were barriers to its motion, its development and realization. It is by no means the case that it thereby suspended all limits, nor all barriers, but rather only the limits not corresponding to it ... Free competition is the real development of capital. (Marx, 1973, pp. 649–50)

And as capitalism itself develops so does competition:

> On the one hand... [capital] creates means by which to overcome obstacles that spring from the nature of production itself, and on the other hand, with the development of the mode of production peculiar to itself, it eliminates all the legal and extra-economic impediments to its freedom of movement in the different spheres of production. Above all it overturns all the legal or traditional barriers that would prevent it from buying this or that kind of labour-power as it sees fit, or from appropriating this or that kind of labour. (Marx, 1867, p. 1013)

The concentration of capital (increasing unit size of firms) and, in particular, the centralization of capital (cohesion of existing capitals) destroys and *recreates* competition. Competition is one of the most powerful 'levers of centralization', and

> The centralization of capitals, or the process of their attraction, becomes more intense in proportion as the specifically capitalist mode of production develops along with accumulation. In its turn centralization becomes one of the greatest levers of its development. (Marx, 1867, p. 778n)

Like Smith and Ricardo, Marx, relates the development of competition to the establishment of the general rate of profit:

> What competition, first in a single sphere, achieves is a single market value and market price derived from the individual values of commodities. And it is competition of capitals in various spheres, which first brings out the price of production equalising the rates of profit in the different spheres. The latter process requires a higher stage of capitalist production than the previous one. (Marx, 1894, p. 180)

It is in his conception of the circuit of capital that Marx best portrays capitalist competition. The image is one of capital as a homogeneous mass of value (money) seeking its maximum return. Profits are created by embodying capital in the process of production, the commodity outputs of which must be realized, that is, returned to the homogeneous money form to be reinvested. Competition is thus characteristic of the capitalist mode of accumulation; mobility and restructuring are two aspects of the same phenomenon.

Marx's general conception of capital as a system corroborates Quesnay's notion of an economy operating 'independent of men's will'. This does not mean that there may not be circumstances in which individual capitals exercise some control in particular markets – indeed, such limitations may be necessary for the accumulation process to proceed in certain lines. Capital removes only those barriers which *limit* its accumulation. The market control exercised in some lines of modern industry is not necessarily a limitation but may be a prerequisite of production on an extended scale. Aggregate capital flows discipline the actions of individual capitals, and hence endow the system with the regularity manifest in the perpetual tendency, successfully contradicted and recreated, towards a general rate of profit and associated prices.

Competition not only establishes the object of analysis, natural prices and the general rate of profit, but makes meaningful analysis possible, since it allows the operations of the capitalist economy to be characterized in a manner which permits theoretical statements of general validity to be made about them.

Theory proceeds by the extraction from reality of those forces which are believed to be dominant and persistent, and the formation of those elements into a formal system, the solution of which is to determine the magnitude or state of the variables under consideration.

It is obvious that the solution will not, except by a fluke, correspond to the actual magnitudes of the variables ruling at any one time, for these will be the outcome not solely of the elements grouped under the heading 'dominant and persistent', but also of the myriad of other forces excluded from the analysis as transitory, peculiar or specific (lacking general significance) which may, at any moment, exert a more or less powerful effect. Nonetheless, the practice of analysis embodies the assumption that the forces comprising the theory *are* dominant, and that the determined magnitudes will, on average, tend to be established. In any satisfactory analytical scheme these magnitudes must be centres of gravitation, capturing the essential character of the phenomena under consideration.

The importance of Smith's use of competition is now apparent. Theory cannot exist in a vacuum. Simply labelling forces dominant is not enough. These forces must operate through a process which establishes their dominance and through which the 'law-governed' nature of the system is manifest. That process is competition, which both enforces and expresses the attempt of individual capitals to maximize profits. Thus important aspects of the behaviour of a capitalist market economy may be captured at a sufficient level of generality to permit the formulation of general causal statements, that is, to permit analysis. Without this step, which constitutes the establishment of what was called above the *method* of analysis, it would have been impossible to develop any general form of economic *theory*.

The classical theory of value and distribution may be shown to provide a logically coherent explanation of the determination of the general rate of profit and hence of natural prices (prices of production) taking as data (see Sraffa, 1960):

(a) the size and composition of social output;
(b) the technique in use; and
(c) the real wage.

The classical achievement is thus composed of two independent elements: (*a*) the characterization of the object of the theory of value; and (*b*) the provision of a theory for the determination of that object. Underlying the former is the concept of gravitation imposed by competition, and underlying the latter the concept of gravitation inherent in theoretical abstraction. Any alternative system must not simply provide a different theory but also achieve a similar congruence with the traditional method.

The development in the final quarter of the 19th century of what was to become known as the neoclassical theory of value and distribution was an attempt to provide an alternative to a classical theory embroiled in the logical difficulties inherent in the labour theory of value and sullied by unsavoury associations with radicalism and Marxism. But despite the dramatic change in theory that was to be heralded by the works of Jevons, Menger

and Walras, the method of analysis which characterized the object the theory was to explain stayed fundamentally the same; the new theory was an alternative explanation of the same phenomena. Marshall labelled natural prices 'long-run normal prices', and declared that, as far as his discussion of value was concerned 'the present volume is chiefly concerned … with the normal relations of wages, profits, prices etc., for rather long periods' (1920, p. 315). The same continuity of method may be found in the work of Walras (1874–7, pp. 224, 380), Jevons (1871, pp. 86, 135–6), Böhm-Bawerk (1899, p. 380) and Wicksell (1934, p. 97).

Nonetheless, the structure of neoclassical theory is such that a different notion of competition is required. The classical emphasis on mobility must be supplemented by a precise definition of the relationships presumed to exist between individual agents. The fundamental concept of 'perfect' competition, for example, encompasses the idea that the influence of each individual participant in the economy is 'negligible', which in turn leads to the idea of an economy with infinitely many participants (Aumann, 1964). Such formulations are entirely absent from the classical conception of competition, since the classical theory is not constructed around individual constrained utility maximization.

JOHN EATWELL

See also **competition.**

Bibliography
Aumann, R.J. 1964. Markets with a continuum of traders. *Econometrica* 32, 39–50.
Böhm-Bawerk, E. von. 1899. *Capital and Interest*, vol. 2. South Holland, IL: Libertarian Press, 1959.
Jevons, W.S. 1871. *Theory of Political Economy.* Harmondsworth: Penguin, 1970.
Marshall, A. 1920. *Principles of Economics*, 8th edn. London: Macmillan.
Marx, K. 1867. *Capital*, vol. 1. Harmondsworth: Penguin, 1976.
Marx, K. 1894. *Capital*, vol. 3. New York: International Publishers, 1967.
Marx, K. 1973. *Grundrisse.* Harmondsworth: Penguin.
Meek, R.L. 1956. *Studies in the Labour Theory of Value.* London: Lawrence & Wishart.
Meek, R.L. 1962. *The Economics of Physiocracy.* London: Allen & Unwin.
Meek, R.L. 1973. Introduction to Turgot (1973).
Mill, J.S. 1848. *Principles of Political Economy.* London: Parker.
Ricardo, D. 1817. *Principles of Political Economy and Taxation*, ed. P. Sraffa, Cambridge: Cambridge University Press, 1951.
Smith, A. 1776. *An Inquiry into the Nature and Causes of the Wealth of Nations.* London: Methuen, 1961.
Sraffa, P. 1960. *Production of Commodities by Means of Commodities.* Cambridge: Cambridge University Press.

Turgot, A.J.R. 1973. *Turgot on Progress, Sociology and Economics*, ed. R.L. Meek. Cambridge: Cambridge University Press.

Vaggi, G. 1987. *The Economics of François Quesnay*. London: Macmillan.

Walras, M.E.L. 1874–7. *Elements of Pure Economics*. Homewood, IL: Irwin, 1954.

Wicksell, K. 1934. *Lectures on Political Economy*, vol. 1. London: Routledge & Kegan Paul.

computation of general equilibria

The general equilibrium model, as elaborated by Walras and his successors, is one of the most comprehensive and ambitious formulations in the current body of economic theory. The basic ingredients with which the Walrasian model is constructed are remarkably spare: a specification of the asset ownership and preferences for goods and services of the consuming units in the economy, and a description of the current state of productive knowledge possessed by each of the firms engaged in manufacturing or in the provision of services. The model then yields a complete determination of the course of prices and interest rates over time, levels of output and the choice of techniques by each firm, and the distribution of income and patterns of saving for each consumer.

The Walrasian model is essentially a generalization, to the entire economy and to all markets simultaneously, of the ancient and elementary notion that prices move to levels which equilibrate supply and demand. No intellectual construction of this scope, designed to address basic questions in a subject as complex and elusive as economics, can be described as simply true of false – in the sense in which these terms are used in mathematics or perhaps in the physical sciences. The assertions of economic theory are not susceptible to crisp and immediate experimental verification. Moreover, the Walrasian model disregards obvious aspects of human motivation which are of the greatest economic significance and which cannot be addressed in the language of our subject: economic theory is mute about our affective lives, about our opposing needs for community and individual assertion, and about the non-pecuniary determinants of entrepreneurial energy.

There are, in addition, aspects of economic reality which are capable of being described in the framework of the Walrasian model but which must be assumed away in order for the model to yield a determinate outcome. Uncertainty about the future is an ever-present fact of economic life, and yet the complete set of markets for contingent commodities required by the Arrow–Debreu treatment of uncertainty is not available in practice. Economies of scale in production are a central feature in the rise of the large manufacturing entities which dominate modern economic activity; their incorporation into the Walrasian model requires the introduction of non-convex production possibility sets for which the competitive equilibrium will typically fail to exist.

In spite of its many shortcomings, the Walrasian model – if used with tact and circumspection – is an important conceptual framework for evaluating the consequences of changes in economic policy or in the environment in which the economy finds itself. The effects of a major shock to the economy of the United States – such as the four-fold increase in the price of imported oil which occurred in late 1973 – can be studied by contrasting equilibrium prices, real wages and the choice of productive techniques both before and after the event in question. Generations of economists have used the Walrasian model to analyse the terms of trade, the impact of customs unions, changes in tariffs and a variety of other issues in the theory of International Trade. And much of the literature in the field of Public Finance is based on the assumption that the competitive model is an adequate description of economic reality.

In these discussions the analysis is frequently conducted in terms of simple geometrical diagrams whose use places a severe restriction on the number of consumers, commodities and productive sectors that can be considered. This is in contrast to formal mathematical treatments of the Walrasian model, which permit an extraordinary generality in the elaboration of the model at the expense of immediate geometrical visualization. Unfortunately, however, it is only under the most severe assumptions that mathematical analysis will be capable of providing unambiguous answers concerning the direction and magnitude of the changes in significant economic variables, when the system is perturbed in a substantial fashion. In order for a comparative analysis to be carried out in a multi-sector framework it is necessary to employ computational techniques for the explicit numerical solution of the highly non-linear system of equations and inequalities which represent the general Walrasian model.

The use of fixed-point theorems in equilibrium analysis

One of the triumphs of mathematical reasoning in economic theory has been the demonstration of the existence of a solution for the general equilibrium model of an economy, under relatively mild assumptions on the preferences of consumers and the nature of production possibility sets (see Debreu, 1982). The arguments for the existence of equilibrium prices inevitably make use of Brouwer's fixed-point theorem, or one of its many variants, and any effective numerical procedure for the computation of equilibrium prices must therefore be capable of computing the fixed points whose existence is asserted by this mathematical statement.

Brouwer's fixed-point theorem, enunciated by the distinguished Dutch mathematician L.E.J. Brouwer in 1912, is the generalization to higher dimensions of the elementary observation that a continuous function of a single variable which has two distinct signs at the two endpoints of the unit interval, must vanish at some intermediary point. In Brouwer's Theorem the unit interval is replaced by an arbitrary closed, bounded convex set S in R^n, and the continuous function is replaced by a continuous mapping of the set S into itself: $x \to g(x)$. Brouwer's Theorem then asserts the existence of at least one point x which is mapped into itself under the mapping; that is, a point x for which $x = g(x)$. To see how this conclusion is used in solving the existence problem let us begin by specifying, in mathematical form, the basic ingredients of the Walrasian model. (Figure 1)

The typical consumer is assumed to have a preference order for, say, the non-negative commodity bundles $x = (x_1, x_2, \ldots, x_n)$ in R^n; the preference ordering is described either by a specific utility function $u(x_1, x_2, \ldots, x_n)$ or by means of an abstract representation of preferences. The consumer will also possess, prior to production and trade, a vector of initial assets $w = (w_1, w_2, \ldots, w_n)$. When a non-negative price vector $p = (p_1, p_2, \ldots, p_n)$ is announced the consumer's income will be $I = p \cdot w$ and his demands will be obtained by maximizing preferences subject to the budget constraint $p \cdot x \le p \cdot w$. If the preferences satisfy sufficient regularity assumptions, the consumer's demand functions $x(p)$ will be single-valued functions of p, continuous (except possibly when some of the individual prices are zero), homogeneous of degree zero and will satisfy the budget constraint $p \cdot x(p) = p \cdot w$. (Figure 2)

The market demands are obtained by aggregating over individual demand functions and, as such, will inherit the properties described above. The market *excess*

demand functions, which I shall denote by $f(p)$, arise by subtracting the supply of assets owned by all consumers from the demand functions themselves. It is these functions which are required for a complete specification of the consumer side of the economy in the general equilibrium model: they may be obtained either by the aggregation of individual demand functions – as we have just described – or they may be directly estimated from econometric data. The following properties will hold, either as a logical conclusion or by assumption:

1. $f(p)$ is homogeneous of degree zero.
2. $f(p)$ is continuous in the interior of the positive orthant.
3. $f(p)$ satisfies the Walras Law $p \cdot f(p) = 0$.

The first of these properties permits us to normalize prices in any one of several ways; for example, $\Sigma p_j = 1$ or $\Sigma p_j^2 = 1$. Given either of these normalizations, I personally do not find it offensive to extend the property of continuity to the boundary, even though there are elementary examples of utility functions, such as the Cobb–Douglas function, for which this would not be correct.

The production side of the economy requires for its description a complete specification of the current state of technical knowledge about the methods of transforming inputs into outputs – with commodities differentiated according to their location and the time of their availability. This can be done by means of production functions, an input/output table with substitution possibilities and several scarce factors rather than labour alone, or by a

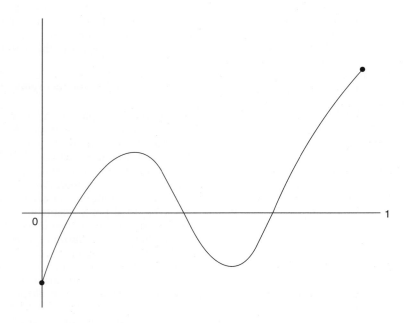

Figure 1

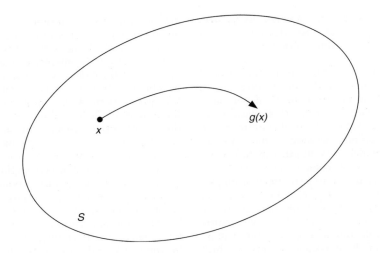

Figure 2

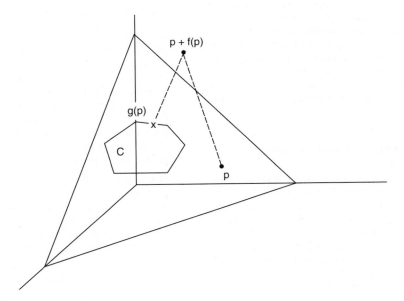

Figure 3

general activity analysis model:

$$
A = \begin{bmatrix}
-1 & 0 & \cdots & 0 & a_{1,n+1} & a_{1,k} \\
0 & -1 & & 0 & a_{2n+1} & a_{2,k} \\
\vdots & \vdots & & \vdots & \vdots & \vdots \\
0 & 0 & & -1 & a_{n,n+1} & a_{n,k}
\end{bmatrix}
$$

Each column of A describes a particular productive process, with inputs represented by non-negative entries and outputs by positive entries. The activities are assumed capable of being used simultaneously and at

arbitrary non-negative levels $x = (x_1, x_2, \ldots, x_k)$; the net production plan is then $y = Ax$. (Figure 3)

With this formulation, a competitive equilibrium is defined by a non-negative vector of prices $p = (p_1, p_2, \ldots, p_n)$ and a non-negative vector of activity levels $x = (x_1, x_2, \ldots, x_k)$ satisfying the following conditions:

1. $f(p) = Ax$,
2. $pA \leq 0$.

The first condition states that supply and demand are equal in all markets, and the second that there are not opportunities for positive profits when the profitability of each activity is evaluated at the equilibrium prices.

Taken in conjunction with the Walras's Law, these conditions imply that those activities which are used at a positive level in the equilibrium solution make a profit of zero.

Given the assumption of continuous and single-valued excess demand functions and the description of the production possibility set by means of an activity analysis model, the following rather direct application of Brouwer's Theorem is sufficient to demonstrate the existence of a equilibrium solution. Under weaker assumptions on the model, variants such as Kakutani's Fixed-Point Theorem may be required.

Let prices be normalized so as to lie on the unit simplex $S = \{p = (p_1, p_2, \ldots, p_n)|p_i \geq 0, \sum p_i = 1\}$. The set of prices p for which $pA \leq 0$ is termed the *dual* cone of the production possibility set generated by the activity matrix A. Its intersection with the unit simplex is a convex polyhedron C consisting of those normalized prices which yield a profit less than or equal to zero for all activities.

We construct a continuous mapping of S into itself as follows: for each p in S consider the point $p + f(p)$; a point which is generally not on the unit simplex itself. We then define $g(p)$ – the image of p under the mapping – to be that point in C which is closest, in the sense of Euclidean distance, to $p + f(p)$. It is then an elementary application of the Kuhn–Tucker Theorem to show that a fixed point of this mapping is, indeed, an equilibrium price vector.

The equilibrium model as a tool for policy evaluation

Brouwer's original proof of his theorem was not only difficult mathematically, but it was decidedly non-constructive; it offered no method for effectively computing a fixed point of the mapping. Brouwer did, in fact, reject his own argument during the later 'intuitionist' phase of his career, in which he proclaimed the acceptability of only those mathematical conclusions obtained by constructive procedures. In spite of the many simplifications in the proof of Brouwer's Theorem offered during the subsequent half-century, it was not until the mid-1960s that constructive methods for approximating fixed points of a continuous mapping finally made their appearance on the scene (Scarf, 1967) – aided by the development of the modern electronic computer and by the rapid methodological advances in the discipline of operations research.

In the early decades of this century, the question of the explicit numerical solution of the general equilibrium model was an active topic of discussion – not by numerical analysts – but rather by economists concerned with the techniques of economic planning in a socialist economy. The issue was raised in the remarkable paper published by Enrico Barone in 1908, entitled 'The Ministry of Production in a Socialist Economy'. Barone, and subsequently Oskar Lange (1936), accepted the Walrasian model – with suitable transfers of income – as an adequate description of ideal economic activity in an economy in which the means of production were collectively owned. In the absence of markets, prices, levels of output and the choice of productive techniques were to be obtained by an explicit numerical solution of the Walrasian system. A key feature of Barone's analysis was the concept of the 'technical coefficients of production' – the input/output coefficients associated with those activities in use at equilibrium. Barone's contention was that the equilibrium could be found – by an extremely laborious calculation which might indeed claim a significant share of the national product – only if the correct activities were known in advance. For Barone, rational economic calculation in a socialist economy was defeated by the many opportunities for substitution in production: the particular activities in use at equilibrium would be impossible to determine by a prior computation. It is instructive to quote Barone on this point.

> The determination of the coefficients economically most advantageous can only be done in an *experimental* way: and not on a *small scale*, as could be done in a laboratory; but with experiments on a *very large scale*, because often the advantage of the variation has its origin precisely in a new and greater dimension of the undertaking. Experiments may be successful in the sense that they may lead to a lower cost combination of factors; or they may be unsuccessful, in which case the particular organization may not be copied and repeated and others will be preferred, which *experimentally* have given a better result.

> The Ministry of Production could not do without these experiments for the determination of the *economically* most advantageous technical coefficients if it would realize the condition of the minimum cost of production which is *essential* for the attainment of the maximum collective welfare.

> It is on this account that the equations of the equilibrium with the maximum collective welfare are not soluble *a priori*, on paper.

An elementary algorithm

Barone's negative conclusion is certainly valid if the full production possibility set, including all of the possibilities for substitution in production, is not known to the central planner. In this event, numerical calculation is impossible, and Lange's suggestion, made some 20 years later, may be appropriate: the problem can be turned on its head and the market, itself, can be used as a mechanism of discovery as well as a giant analogue computer. But if the production possibility set can be explicitly constructed, substitution – in and of itself – does not seem to me to be a severe impediment to numerical computation.

At the present moment, some 20 years after the introduction and continued refinement of fixed-point computational techniques, I have in my possession a small floppy disk with a computer program which will routinely solve – on a personal computer – for equilibrium prices and activity levels in a Walrasian model in which the number of variables is on the order of 100. (The authors of the program suggest that examples with 300 variables can be accommodated on a mainframe computer.) Substantial possibilities of substitution, if known in advance, offer no difficulty to the successful functioning of this algorithm. In my opinion, the modern restatement of Barone's problem is rather that even 300 variables are extremely small in number in contrast to the millions of prices and activity levels implicit in his account. The computer, while expanding our capabilities immeasurably, has taught us a severe lesson about the role of mathematical reasoning in economic practice and forced us to shift our point of view dramatically from that held by our predecessors. We realize that our preoccupations are not with universal laws which describe economic phenomena with full and complete generality, but rather with intellectual formulations which are an imperfect representation of a complex and elusive reality. The application of general equilibrium theory to economic planning, and more generally to the evaluation of the consequences of changes in economic policy, must be based on highly aggregated models whose conclusions are at best tentative guides to action.

An exercise in comparative statics is begun by constructing a general equilibrium model whose solution reflects the economic situation existing prior to the proposed policy change. The number of parameters required to describe demand functions, initial endowments and the production possibility set is considerable, and in practice the constraint of reproducing the current equilibrium must be augumented by a variety of additional statistical estimates in order to specify the model. The limitations of data in the form required by the Walrasian model inevitably make this estimation procedure less than fully satisfactory.

The second step in the exercise is to calculate the solution after the proposed policy changes are explicitly introduced into the model. In some cases the policy variables being studied can be directly incorporated as parameters in the equations whose solution yields the equilibrium values; if the changes are small, their effects on the solution may be obtained by differentiating these equations and solving the resulting linear system for the corresponding changes in the equilibrium values themselves. This approach was adopted by Leif Johansen (1960) and by Arnold Harberger (1962) in his study of the incidence of a tax on corporate profits. The use of this method in policy analysis continues in Norway, and it forms the basis of the ambitious programme carried out by Peter Dixon and his collaborators in Australia (1982). If, on the other hand, the policy changes are large, the equilibrium position may be shifted substantially, and its determination may require the use of more sophisticated computational methods.

Fixed-point algorithms can be divided into two major classes: those based on the elements of differential topology, surveyed by Smale (1981), and those which are combinatorial in nature. The most elementary of the combinatorial algorithms for approximating a fixed point of a continuous mapping of the unit simplex $S = \{(x = (x_1, x_2, \ldots, x_n) | x_i \geq 0, \sum x_i = 1\}$ begins by dividing the simplex into a large number of small subsimplices as illustrated in Figure 4. In our notation the simplex is of dimension $n - 1$ and has faces of dimension $n - 2, \ldots, 1$. It is a requirement of the subdivision that the intersection of any two of the subsimplices is either empty or a full lower dimensional face of both of them.

Each vertex of the subdivision will have associated with it an integer label selected from the set $(1, 2, \ldots, n)$. When the method is applied to the determination of a

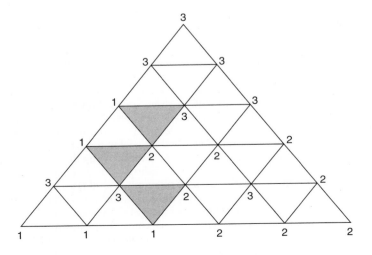

Figure 4

fixed point of a particular mapping, the labels associated with a vertex will depend on the mapping evaluated at that point. For the moment, however, the association will be arbitrary aside from the requirement that a vertex on the boundary of the simplex will have a label i only if the ith coordinate of that vertex is positive.

The remarkable combinatorial lemma demonstrated by Emanuel Sperner (1928) in his doctoral thesis is that at least one subsimplex must have all of its vertices differently labelled. Assuming this result to be correct, let us consider a mapping of the simplex in which the image of the vector $x = (x_1, \ldots, x_n)$ is $f(x) = [f_1(x), \ldots, f_n(x)]$. The requirement that the image be on the simplex implies that $f_i(x) \geq 0$ and that $\sum f_i(x) = 1$. It follows that for every vertex of the subdivision v, unless v is a fixed point of the mapping, there will be a least one index i for which $f_i(v) - v_i < 0$. If we select such an index to be the label associated with the vertex v, then the assumptions of Sperner's Lemma are clearly satisfied, and the conclusion asserts the existence of a simplex whose vertices are distinctly labelled.

If the simplicial subdivision is very fine, the vertices of this sub-simplex are all close together; at each vertex a different coordinate is decreasing under the mapping, and by continuity every point in the small subsimplex will have the property that each coordinate is not increasing very much under the mapping. Since the sum of the coordinate changes is by definition zero, the image of any point in the completely labelled subsimplex will be close to itself, and such a point will therefore serve as an approximate fixed point of the mapping. A formal proof of Brouwer's Theorem requires us to construct a sequence of finer and finer subdivisions, to find, for each subdivision, a completely labelled simplex, and to select a convergent sequence of these simplices tending to a fixed point of the mapping.

Sperner's Lemma may be applied to the equilibrium problem directly. For simplicity, consider the model of exchange in which the market excess demand functions are given by $g(p)$, with p on the unit price simplex. As before, we subdivide the simplex and associate an integer label from the set $(1, \ldots, n)$ with each vertex v of the subdivision, according to the following rule: the label i is to be selected from the set of those indices of which $g_i(p) \leq 0$. It is an elementary consequence of Walras's Law that a selection can be made which is consistent with the assumptions of Sperner's Lemma, and there will therefore be a subsimplex all of whose vertices bear distinct labels. By virtue of the particular labelling rule, any point in such a completely labelled simplex will be an approximate equilibrium price vector in the sense that all excess demands, at this price, will be either negative or, if positive, very small.

Sperner's original proof of his combinatorial lemma was not constructive; it was based on an inductive argument which required a complete enumeration of all completely labelled simplices for a series of lower

dimensional problems. In order to develop an effective numerical algorithm for the determination of such a simplex let us begin by embedding the unit simplex, and its subsimplices, in a larger simplex T, as in Figure 5. The larger simplex is subdivided by joining its n new vertices to those vertices of the original subdivision lying on the boundary of the unit simplex. The assumptions of Sperner's Lemma permit the new vertices to be given distinct labels from the set $(1, \ldots, n)$, in such a way that no additional completely labelled simplices are generated. For concreteness, let the new vertex receiving the label i be denoted by v^i.

We begin our search for a completely labelled simplex by considering the simplex with vertices $v^2, \ldots, v^n$ and one additional vertex, say v^*. If v^* has the label 1, this simplex is completely labelled and our search terminates; otherwise we move to an adjacent simplex by removing the vertex whose label agrees with that of v^* and replacing it with that unique other vertex yielding a simplex in the subdivision. As the process continues, we are, at each step, at a simplex whose vertices bear the labels $2, \ldots, n$, with a single one of these labels appearing on a pair of vertices. Precisely two $n - 2$ dimensional faces have a complete set of labels $2, \ldots, n$. The simplex has been entered through one of these faces; the algorithm proceeds by exiting through the other such face.

The argument first introduced by Lemke (1965) in his study of two person non-zero sum games was carried over by Scarf (1967) to show that the above algorithm never returns to a simplex previously visited and never requires a move outside of T. Since the number of simplices is finite, the algorithm must terminate, and termination can only occur when a completely labelled simplex is reached.

Improvements in the algorithm

The algorithm can easily be programmed for a computer, and it provides the most elementary numerical procedure for approximating fixed points of a continuous mapping and equilibrium prices for the Walrasian model. Since its introduction in 1967, the algorithm, in this particular form, has been applied to a great number of examples of moderate size, and it performs sufficiently well in practice to conclude that the numerical determination of equilibrium prices is a feasible undertaking. The algorithm does, however, have some obvious drawbacks which must be overcome to make it available for problems of significant size. For example, the information which yields the labelling of the vertices, and therefore the path taken by the algorithm, is simply the index of a coordinate which happens to be decreasing when the mapping is evaluated at the vertex. More recent algorithms make use of the full set of coordinates of the image of the vertex instead of a single summary statistic.

Second, this primitive algorithm is always initiated at the boundary of the simplex. If the approximation is not sufficiently good, the grid size must be refined, and a

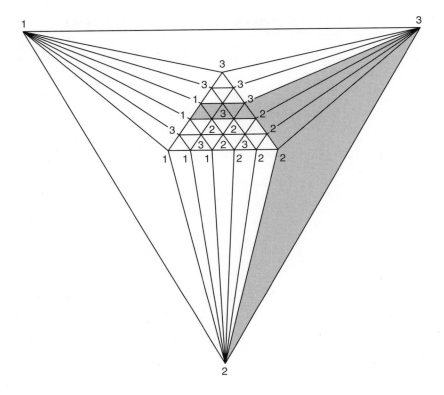

Figure 5

recalculation, which makes no use of previous information, must be performed. It is of the greatest importance to be able to initiate the algorithm at an arbitrary interior point of the simplex selected as our best a priori estimate of the answer.

The following geometrical setting (Eaves and Scarf, 1976) for the elementary algorithm suggests the form these improvements can take. Let us construct a piecewise linear mapping, $h(x)$, of T into itself as follows: for each vertex v in the subdivision let $h(v) = v^i$, where i is the label associated with v. We then complete the mapping by requiring h to be linear in each simplex of the subdivision. The mapping is clearly continuous on T and maps every boundary point of T into itself. Moreover, every subsimplex in the subdivision whose vertices are not completely labelled is mapped, by h, into the boundary of T. If none of the simplices were completely labelled, this construction would yield a most improbable conclusion: a continuous mapping of T into itself which is the identity on the boundary and which maps the entire simplex into the boundary. That such a mapping cannot exist is known as the Non-Retraction Theorem, an assertion which is, in fact, equivalent to Brouwer's Theorem. The impossibility of such a mapping reinforces our conclusion that a completely labelled simplex does exist.

Select a point c interior to one of the boundary faces of T and consider the set of points which map into c; that is,

the set of x for which $h(x) = c$. As Figure 6 indicates, this set contains a piecewise linear path beginning at the point c, and transversing precisely those simplices encountered in our elementary algorithm. There are however, other parts of the set $\{x|h(x) = c\}$: closed loops which do not touch the boundary of T and other piecewise linear paths connecting a pair of completely labelled simplices. Stated somewhat informally, the general conclusion, of which this is an example, is that the inverse image of a particular point, under a piecewise linear mapping from an n dimensional set to an $n-1$ dimensional set, consists of a finite union of interior loops, and paths which join two boundary points (see Milnor, 1965, for the differentiable version).

To see how this observation can be used, consider the product of the unit simplex S and the closed unit interval $[0, 1]$; that is, the set of points (x, t) with x in S and $0 \le t \le 1$, as in Figure 7. Extend the mapping from the unit simplex to this large set by defining $F(x, t) = (1 - t)f(x) + tx^*$, with x^* a preselected point on the simplex, taken to be an estimate of the true fixed point. The set of points for which $F(x, t) - x = 0$ is, by our general conclusion, a finite union of paths and loops. Precisely one of these paths intersects the upper boundary of the enlarged set. If the path is followed, its other endpoint must lie in the face $t = 0$ and yield a fixed point of the original mapping.

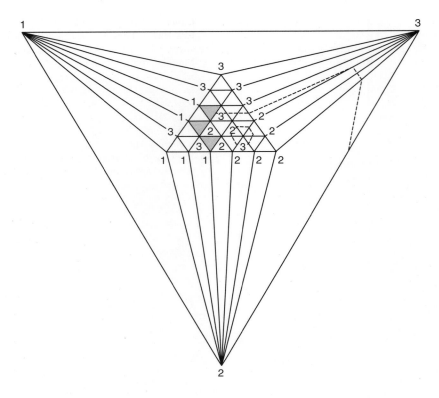

Figure 6

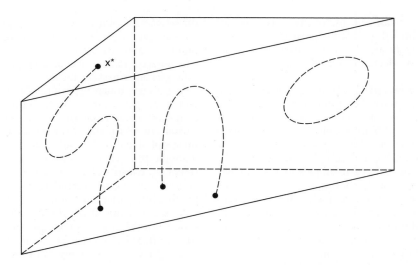

Figure 7

The path leading to the fixed point can be followed on the computer in several ways. We can, for example, introduce a simplicial decomposition of the set $S \times [0, 1]$ and approximate F by a piecewise linear mapping agreeing with F on the vertices of the subdivision. Following the path then involves the same type of calculation we have become accustomed to in carrying out linear programming pivot steps. There are a great many variations in the mode of simplicial subdivision leading to substantial improvements in the efficiency of our original fixed-point algorithm (Eaves, 1972; Merrill, 1971; van der Laan and Talman, 1979).

An alternative procedure, adopted by Kellogg, Li and Yorke (1976) and Smale (1976), is to impose sufficient regularity conditions on the underlying mapping so that differentiation of $F(x, t) - x = 0$ yields a set of differential equations for the path joining x^* to the fixed point on $t = 0$. This leads to a variant of Newton's method which is global in the sense that it need not be initiated in the vicinity of the correct answer. But, whichever of these alternatives we select, the numerical difficulties in computing equilibrium prices can be overcome for all problems of reasonable size.

Applied general equilibrium analysis

During the last 15 years, the field of Applied General Equilibrium Analysis has grown considerably; instead of the few tentative examples illustrating our ability to solve general equilibrium problems, we have seen the construction of a large number of models of substantial size designed to illuminate specific policy issues. The number of books and papers which have appeared in the field is far too large for a complete enumeration in this essay, and I shall mention only a few publications which may be consulted to obtain an indication of the diversity of this activity. The paper by Shoven and Whalley (1984) in the *Journal of Economic Literature* is a survey of applied general equilibrium models in the fields of taxation and international trade constructed by these authors and their colleagues. The volume by Adelman and Robinson (1978) is concerned with the application of general equilibrium analysis to problems of economic development. Whalley (1985) has written on trade liberalization, and Ballard, Fullerton, Shoven and Whalley (1985) on the evaluation of tax policy. Jorgenson (Hudson and Jorgenson, 1974), and Manne (1976) have made extensive applications of this methodology to energy policy, and Ginsburg and Waelbroeck (1981) provide a refreshing discussion of alternative computational procedures applied to a model of international trade involving over 200 commodities. The volume edited by Scarf and Shoven (1985) contains a collection of papers presented at one of an annual series of workshops in which both applied and theoretical topics of interest to researchers in the field of Applied General Equilibrium Analysis are discussed.

HERBERT E. SCARF

See also **general equilibrium; computation of general equilibria (new developments).**

Bibliography

Adelman, I. and Robinson, S. 1978. *Income Distribution Policy in Developing Countries: A Case Study of Korea*. Stanford: Stanford University Press.

Ballard, C.L., Fullerton, D., Shoven, J.B. and Whalley, J. 1985. *A General Equilibrium Model for Tax Policy Evaluation*. Chicago: University of Chicago Press.

Barone, E. 1908. Il Ministerio della Produzione nello stato colletivista. *Giornale degli Economisti e Revista di Statistica*. Trans. as 'The Ministry of Production in the Collectivist State', in *Collectivist Economic Planning*, ed. F.A. Hayek, London: G. Routledge & Sons, 1935.

Brouwer, L.E.J. 1912. Über Abbildungen von Mannigfaltigkeiten. *Mathematische Annalen* 71, 97–115.

Debreu, G. 1982. Existence of competitive equilibrium. In *Handbook of Mathematical Economics*, ed. K.J. Arrow and M. Intriligator, Amsterdam: North-Holland.

Dixon, P.B., Parmenter, B.R., Sutton, J. and Vincent, D.P. 1982. *ORANI: A Multisectoral Model of the Australian Economy*. Amsterdam: North-Holland.

Eaves, B.C. 1972. Homotopies for the computation of fixed points. *Mathematical Programming* 3, 1–22.

Eaves, B.C. and Scarf, H. 1976. The solution of systems of piecewise linear equations. *Mathematics of Operations Research* 1, 1–27.

Ginsburg, V.A. and Waelbroeck, J.L. 1981. *Activity Analysis and General Equilibrium Modelling*. Amsterdam: North-Holland.

Harberger, A. 1962. The incidence of the corporation income tax. *Journal of Political Economics* 70, 215–40.

Hudson, E.A. and Jorgenson, D.W. 1974. US Energy policy and economic growth. *Bell Journal of Economics and Management Science* 5, 461–514.

Johansen, L. 1960. *A Multi-Sectoral Study of Economic Growth*. Amsterdam: North-Holland.

Kellogg, R.B., Li, T.Y. and Yorke, J. 1976. A constructive proof of the Brouwer Fixed Point Theorem and computational results. *SIAM Journal of Numerical Analysis* 13, 473–83.

Kuhn, H.W. 1968. Simplicial approximation of fixed points. *Proceedings of the National Academy of Sciences* 61, 1238–42.

Lange, O. 1936. On the economic theory of socialism. *Review of Economic Studies* 4, 53–71, 123–42.

Lemke, C.E. 1965. Bimatrix equilibrium points and mathematical programming. *Management Science* 11, 681–9.

Manne, A.S. 1976. ETA: a model of energy technology assessment. *Bell Journal of Economics and Management Science* 7, 379–406.

Merrill, O.H. 1971. Applications and extensions of an algorithm that computers fixed points of certain non-empty convex upper semicontinuous point to set mappings. Technical Report 71–7, University of Michigan.

Milnor, J. 1965. *Topology from the Differentiable Viewpoint*. Charlottesville: University of Virginia Press.

Scarf, H.E. 1967. The approximation of fixed points of a continuous mapping. *SIAM Journal of Applied Mathematics* 15, 1328–43.

Scarf, H.E., with the collaboration of T. Hansen, 1973. *The Computation of Economic Equilibria*. London, New Haven: Yale University Press.

Scarf, H. and Shoven, J.B., eds. 1984. *Applied General Equilibrium Analysis*. Cambridge: Cambridge University Press.

Shoven, J.B. and Whalley, J. 1972. A general equilibrium calculation of the effects of differential taxation of income from capital in the U.S. *Journal of Public Economy* 1, 281–321.

Shoven, J.B. and Whalley, J. 1984. Applied general-equilibrium models of taxation and international trade. *Journal of Economic Literature* 22, 1007–51.

Smale, S. 1976. A convergent process of price adjustment and global Newton methods. *Journal of Mathematical Economics* 3, 107–20.

Smale, S. 1981. Global analysis and economics. In *Handbook of Mathematical Economics* Vol. I, ed. K.J. Arrow and M. Intriligator, Amsterdam: North-Holland.

Sperner, E. 1928. Neur Beweis für die Invarianz der Dimensionszahl und des Gebietes. *Abhandlungen an den mathematischen Seminar der Universität Hamburg* 6, 265–72.

van der Laan, G. and Talman, A.J.J. 1979. A restart algorithm for computing fixed points without an extra dimension. *Mathematical Programming* 17, 74–84.

Whalley, J. 1985. *Trade Liberalization among Major World Trading Areas.* Cambridge, Mass.: MIT Press.

computation of general equilibria (new developments)

1 Introduction

After Scarf (1967) showed that there exist globally convergent (and effectively applicable) algorithms to compute economic equilibria, there is now a class of computable applied models which are routinely used to evaluate the economic consequences of different taxes and tariff structures (see, for example, Shoven and Whalley, 1992). Research on efficient algorithms for the computation of general equilibria in these models largely took place outside of economics.

A large literature in numerical analysis has developed algorithms that are much faster than Scarf's original method and that can be used for large-scale applications. Efficient iterative schemes, mostly based on global Newton methods, now allow applied researchers to solve for competitive equilibria in models with hundreds of commodities and agents (see, for example, Ferris and Pang, 1997).

Recently, there has been substantial research in theoretical computer science on the development of polynomial time algorithms for the computation of general equilibria. For most existing methods, the number of operations needed to approximate equilibria within a fixed precision ε grows exponentially in $1/\varepsilon$. Under restrictive assumptions on preferences, in models without production, researchers have developed algorithms to approximate equilibria 'in polynomial time', that is, the running time of the algorithm increases polynomially in the input parameters and in the precision with which equilibria are computed. Codenotti,

Pemmaraju and Varadarajan (2004) give an overview on recent developments along this line.

In this article I will not discuss any of these practical aspects of the solution of large-scale models. I will instead focus on the following two unrelated developments in the computation of general equilibria in economics.

1. The computation of equilibria in models with time, uncertainty and missing asset markets.
2. The computation of all equilibria and the relationship between exact and approximate equilibria in the standard Arrow–Debreu model.

2 Models with asset markets

Due to their essential static nature, standard computable general equilibrium models suffer from an oversimplified treatment of uncertainty. Agents either solve a static problem or have myopic expectations, and the model can therefore not explicitly incorporate investment and saving decisions. The general equilibrium model with incomplete asset markets (GEI model) provides a basic framework with several agents and several commodities to incorporate uncertainty and financial markets. See, for example, Magill and Qunizii (1996) for an overview of the literature. The computation of equilibria in these models is challenging because in some specifications equilibria fail to exist while in others they are often numerically unstable.

Kehoe and Prescott (1995) argue that real business cycle models provide an alternative way to extend computable general equilibrium to models with time and uncertainty. There is now a large literature on the computation of equilibria in dynamic stochastic economies. This is reviewed elsewhere in this dictionary; see APPROXIMATE SOLUTIONS TO DYNAMIC MODELS (LINEAR METHODS); see also Judd (1998).

In the standard GEI model there are two time periods (Kubler and Schmedders, 2000, show how the problem of computation of equilibria in multi-period finance models can be essentially reduced to the two period case) and S possible states of the world in the second period. There are L perishable commodities available for trade at each state. There are H agents with endowments $e^h \in \mathbb{R}_+^{(S+1)L}$ and utility functions $u^h : \mathbb{R}_+^{(S+1)L} \to \mathbb{R}$. It is assumed throughout this article that utility functions are smooth in the sense of Debreu (1972) – that is, utility is C^2, strictly increasing, strictly quasi-concave, exhibits non-zero Gaussian curvature and indifference curves do not cut the axes.

There are J assets available for trade. In each state s, asset j pays a bundle of commodities $a_j(s) \in \mathbb{R}^L$. It is without loss of generality to assume that the $LS \times J$ matrix

$$A = \begin{pmatrix} a_1(1) & \cdots & a_J(1) \\ \vdots & \ddots & \vdots \\ a_1(S) & \cdots & a_J(S) \end{pmatrix}$$

has full rank J. Allowing assets to pay in different commodities is crucial when one wants to extend the model to several time periods and long-lived securities.

In the following, it will be useful to write commodity prices as

$$p = (p(0), p(1), \ldots, p(S)) \in \Delta^{(S+1)L-1}$$
$$= \left\{ p \in \mathbb{R}_+^{(S+1)L} : \sum_i p_i = 1 \right\},$$

and the $S \times J$ asset payoff matrix (as a function of spot prices $p(1) \ldots p(S)$), $R(p)$, as

$$R(p) = \begin{pmatrix} p(1) \cdot a_1(1) & \cdots & p(1) \cdot a_J(1) \\ \vdots & \ddots & \vdots \\ p(S) \cdot a_1(S) & \cdots & p(S) \cdot a_J(S) \end{pmatrix}.$$

In part of the discussion we assume an exogenous short-sale constraint, that is, there is a number $0 < K \leq \infty$ such that the two-norm of an agent's portfolio must always be less than or equal to K. One can then write an agent's aggregate excess demand function as the solution of his maximization problem in the GEI economy.

$$(z^h(p), \phi^h(p)) = \arg \max_{z \in \mathbb{R}^{L(S+1)}, \phi \in \mathbb{R}^J} u(e^h + z) \text{ s.t.}$$
$$p \cdot z = 0$$
$$(p(1) \cdot z(1), \ldots, p(S) \cdot z(S))^T = R(p) \cdot \phi$$
$$\|\phi\| \leq K.$$

A GEI equilibrium is a collection of prices, portfolios and a consumption allocation such that markets clear and each agent maximizes her utility, i.e. equilibrium prices p are characterized by $\sum_{h=1}^H z^h(p) = 0$.

In a slight idealization (see also the more precise definition in the next section), we assume that the maximization problem can be solved exactly and we define an ε-equilibrium as a price $\bar{p}$ such that

$$\left\| \sum_{h=1}^H z^h(\bar{p}) \right\| < \varepsilon.$$

2.1 A general algorithm

Although generally $R(p)$ will have full rank J, there will be so-called 'bad prices' at which the rank of $R(p)$ drops. When there are no short sale constraints, that is, $K = \infty$, this leads to a discontinuity of excess demand. Scarf's algorithm fails: no matter how fine the simplicial subdivision, if the algorithm terminates at some $\bar{p}$, one cannot necessarily infer a bound on $\|z(\bar{p})\|$ and hence cannot find an ε-equilibrium.

Homotopy continuation methods (see Garcia and Zangwill, 1981; Eaves, 1972) turn out to be ideally suited

for this numerical problem. In order to solve a system of equations $f(x) = 0$, $f : X \to Y$, the basic idea underlying homotopy methods is to find a smooth map $H : X \times [0, 1] \to Y$ with

$$H(x, 1) \equiv f(x) \text{ and } H(x, 0) \equiv g(x),$$

where $g : X \to Y$ has a known unique zero. The map H is called a smooth homotopy. In using homotopy methods it is crucial to set up the function, H, to ensure that there is a smooth path that connects $(x^s, 0)$ with $g(x^s) = 0$ to some $(\bar{x}, 1)$ with $f(\bar{x}) = 1$.

Brown, DeMarzo and Eaves (1996) develop a homotopy algorithm which can be shown to be globally convergent in that it finds an ε-equilibrium for any $\varepsilon > 0$ in a finite number of steps. Following the so-called Cass-trick, it is useful to introduce an unconstrained agent, that is, to define the first agent maximization problem as

$$z^u(p) = \arg \max_z u^1(e + z) \text{ s.t. } p \cdot z = 0,$$

and aggregate demand as $z(p) = z^u(p) + \sum_{h=2}^H z^h(p)$. Note that $\bar{p}$ is a GEI equilibrium (given that $K = \infty$) if and only if $z(p) = 0$. An ε-equilibrium is characterized by $\|z(p)\| < \varepsilon$.

Define the expenditure of the unconstrained agent y^u as

$$y^u = (p(1) \cdot z_1^u(p), \ldots, p(S) \cdot z_S^u(p)).$$

Define an extended payoff matrix $R^*(p)$ by

$$R^*(p) = [R(p), y^u(p)]$$

and let $R^*_{-i}(p)$ be $R^*(p)$ with the i'th column deleted. For the constrained agents $h = 2, \ldots, H$ define

$$z^h(p, R^*_{-i}(p)) = \arg \max_{z, \phi} u^h(e^h + z) \text{ s.t. } p \cdot z = 0$$
$$(p(1) \cdot z(1), \ldots, p(S) \cdot z(S))^T = R^*_{-i}(p) \cdot \phi.$$

Now consider a family of homotopies, indexed by i

$$H_i(p, t, \theta) = \begin{pmatrix} z^u(p) + t \sum_{h=2}^H z^h(p, R^*_{-i}(p)) \\ R^*(p)\theta \\ \theta \cdot \theta - 1 \end{pmatrix}.$$

To prove existence of a homotopy path, Brown, DeMarzo and Eaves (1996) show that $\cup_{i=1}^{J+1} H_i^{-1}(0)$ contains a smooth path connecting the starting point to a solution at $t = 1$.

While generically in endowments a homotopy path turns out to exist, the algorithm is hardly applicable in medium-sized problems, since the number of homotopies one has to consider can become quite large. An alternative is to focus on models with $K < \infty$ (or alternatively models with transaction costs) or to consider algorithms which might fail in a small class of problems but which are generally more efficient.

2.2 Short-sale constraints

In the presence of short-sale constraints, the excess demand function is continuous and equilibrium existence can be proven with Brouwer's theorem. Therefore, one could presumably use a version of Scarf's algorithm to compute equilibria in this case. However, while there are no new mathematical problems to be solved, the fact that the rank of the asset–payoff matrix can still collapse in equilibrium poses difficult numerical problems. Simple Newton method-based algorithms often do not work (see Kubler and Schmedders, 2000) unless one has a starting point very close to the actual solution. It turns out that, just as in the problem without short-sale constraints, homotopy continuation methods can provide a basis for reliable algorithms.

Schmedders (1998) develops a homotopy algorithm which can be used to solve models with a large number of heterogeneous households and goods. The basic idea of his algorithm is to modify the agents' problem by introducing a homotopy parameter $t \in [0, 1]$ as follows.

$$(z^h(p, t), \phi^h(p, t)) = \arg \max_{z \in \mathbb{R}^{L(S+1)}, \phi \in \mathbb{R}^J} u(e^h + z)$$

$$- (1 - t) \frac{1}{2} ||\phi||^2 \text{ s.t.}$$

$$p \cdot z = 0$$

$$(p(1) \cdot z(1), \ldots, p(S) \cdot z(S)) = R(p) \cdot \phi$$

$$||\phi|| \leq K.$$

Under the assumptions on utilities this is still a convex problem and the first order Kuhn–Tucker conditions are necessary and sufficient. Schmedders provides various examples that show that even for $K = \infty$ his algorithm, although not guaranteed to converge, performs well in practice.

For $K < \infty$, the Kuhn–Tucker inequalities can be converted into a system of equalities via a change of variables (see Garcia and Zangwill, 1981, ch. 4). Kubler (2001), Herings and Schmedders (2006) and others subsequently used this idea to solve models with transaction costs, trading constraints and other market imperfections.

Of course, it is an important practical problem how to trace out a homotopy path numerically. See Watson (1979) for a theoretical algorithm. For a practical description of numerical homotopy path-following methods see Schmedders (2004).

3 Equilibria in semi-algebraic economies

While it is clear that sufficient assumptions for the global uniqueness of competitive equilibria are too restrictive to be applicable to models used in practice, it remains an open problem how serious a challenge the non-uniqueness of competitive equilibrium poses to applied equilibrium modelling. In the presence of multiple equilibria, comparative statics exercises become meaningless. Furthermore, even when for a given specification of the economy equilibrium is globally unique, as Richter and Wong (1999) point out, the possibility of multiple equilibria for close-by economies implies that it is generally impossible to compute prices and allocations that are close-by exact equilibrium prices and allocations (as opposed to computing prices at which aggregate excess demand is close to zero). In this section I argue that one can solve these problems by focusing on so-called 'semi-algebraic' economies.

While the arguments are also applicable to the GEI model, for simplicity, consider a standard Arrow–Debreu exchange economy, $(u^h, e^h)_{h=1}^H$. There are H agents trading L commodities. Each agent h has individual endowments $e^h \in \mathbb{R}_+^L$ and 'smooth preferences' characterized by an utility function $u^h : \mathbb{R}_+^L \to \mathbb{R}$.

A Walrasian equilibrium is a collection of consumption vectors $(x^h)_{h=1}^H$ and prices $p \in \Delta^{L-1}$ such that

$$x^h \in \arg \max_{x \in \mathbb{R}_+^L} u^h(x) \text{ s.t. } p \cdot x \leq p \cdot e^h \quad (1)$$

$$\sum_{h=1}^H (x^h - e^h) = 0. \quad (2)$$

An approximate (ε-) equilibrium consists of an allocation an prices such that

$$||u^h(x^h) - [\max_{x \in \mathbb{R}_+^L} u^h(x) \text{ s.t. } p \cdot x \leq p \cdot e^h] || < \varepsilon \quad (3)$$

$$\left\| \sum_{h=1}^H (x^h - e^h) \right\| < \varepsilon. \quad (4)$$

Given any $\varepsilon > 0$, Scarf's algorithm (as well as the more efficient algorithms used in practice) finds a p, x^h which constitute an ε-equilibrium.

This leaves open two important theoretical questions.

1. Can one relate the approximate equilibrium prices and allocations, to exact equilibria, that is, given a computed ε-equilibrium $(\bar{p}, (\bar{x}^h))$, does there exist a Walrasian equilibrium $\tilde{p}, (\tilde{x}^h)$ with $||(\bar{p}, (\bar{x}^h)) - (\tilde{p}, (\tilde{x}^h))||$ small? Can one find good bounds on this distance which tend to zero as $\varepsilon \to 0$?

2. Given an economy $(u^h, e^h)_{h=1}^H$ with N Walrasian equilibria $(p^n, (x^h)^n)_{n=1}^N$ and any $\delta > 0$, is it possible to approximate all N equilibria, that is, to find N ε-equilibria $(\tilde{p}^n, (\tilde{x}^h)^n)_{n=1}^N$ with $||(p^n, (x^h)^n) - (\tilde{p}^n, (\tilde{x}^h)^n)|| < \delta$, for all $n = 1, \ldots, N$?

Clearly, the second problem is strictly more difficult to tackle than the first. Richter and Wong (1999) show that for general economies even the answer to the first question is negative. In order to obtain positive answers to both questions, one needs to restrict possible preferences. One approach is to assume that better sets are semi-algebraic sets. I will make the slightly more useful assumption that marginal utilities are semi-algebraic functions.

3.1 Semi-algebraic economies

We assume that for each h, $D_x u^h(x)$ is a semi-algebraic function, that is, its graph $\{(x, y) \in \mathbb{R}^{2L}_+ : y = D_x u^h(x)\}$ is a finite union and intersection of sets of the form

$$\{(x, y) \in \mathbb{R}^{2L} : g(x, y) > 0\}$$
$$\text{or } \{(x, y) \in \mathbb{R}^{2L} : f(x, y) = 0\}$$

for polynomials with real coefficients, f and g.

For practical purposes, the focus on semi-algebraic preferences is quite general. First note that Afriat's theorem implies that a finite set of observations on an individual's choices that can be rationalized by any utility function can also be rationalized by semi-algebraic preferences (in fact, Afriat's construction is piece-wise linear). Furthermore, note that the constant elasticity of substitution utility function which is often used in applied work is semi-algebraic if the elasticities of substitution are rational numbers.

It follows from the Tarski–Seidenberg theorem that for semi-algebraic economies the answers to both questions above are positive, since the relevant statements can be written as first order sentences (see Basu, Pollack and Roy, 2003). However, algorithmic quantifier elimination which needs to be used to answer general questions in this framework is so computationally inefficient that for practical purposes this does not help towards solving the above questions for interesting specifications of economies.

Nevertheless, given a semi-algebraic economy it is possible to find a system of polynomial equations $f(x) = 0$, $f : \mathbb{R}^{H(L+1)+L-1} \to \mathbb{R}^{H(L+1)+L-1}$, and finitely many inequalities $g^i(x) \geq 0, g^i : \mathbb{R}^{H(L+1)+L-1} \to \mathbb{R}^M$, $i = 1, \ldots, N < \infty$ such that $p, (x^h)$ is a Walrasian equilibrium for the economy (u^h, e^h) if and only if there exist $\lambda^h \in \mathbb{R}_{++}$, $h = 1, \ldots, H$ such that for some $i = 1, \ldots, N$,

$$f(p, (x^h, \lambda^h)) = 0, \quad g^i(p, (x^h, \lambda^h)) \geq 0.$$

Therefore, the problem of finding Walrasian equilibria reduces to finding the real roots of polynomial systems of equations and verifying polynomial inequalities (see Kubler and Schmedders, 2006).

Having reduced the problem of finding Walrasian equilibria to finding roots of a polynomial system of equations, one can then answer the two questions above affirmatively.

3.2 Question 1: Smale's alpha method

Smale's alpha method provides a simple sufficient conditions for approximate zeros to be close to exact zeros

and can be viewed as an extension of the Newton–Kantorovich conditions. The following results are from Blum et al. (1998, ch. 8).

Let $D \subset \mathbb{R}^n$ be open and let $f : D \to \mathbb{R}^n$ be analytic. For $z \in D$, define $f^{(k)}(z)$ to be the k'th derivative of f at z. This is a multi-linear operator which maps k-tuples of vectors in D into $\mathbb{R}^n$. Define the norm of an operator A to be

$$\|A\| = \sup_{x \neq 0} \frac{\|Ax\|}{\|x\|}.$$

Suppose that the Jacobian of f at z, $f^{(1)}(z)$ is invertible and define

$$\gamma(z) = \sup_{k \geq 2} \left\| \frac{(f^{(1)}(z))^{-1} f^{(k)}(z)}{k!} \right\|^{\frac{1}{(k-1)}}$$

and

$$\beta(z) = \|(f^{(1)}(z))^{-1} f(z)\|.$$

Theorem 1 Given a $\bar{z} \in D$, suppose the ball of radius $\left(1 - \frac{\sqrt{2}}{2}\right)/\gamma(\bar{z})$ around $\bar{z}$ is contained in D and that

$$\beta(\bar{z})\gamma(\bar{z}) < 0.157.$$

Then there exists a $\tilde{z} \in D$ with

$$f(\tilde{z}) = 0 \text{ and } \|\bar{z} - \tilde{z}\| \leq 2\beta(\bar{z}).$$

While the theorem applies to any locally analytic function, the bound $\gamma(z)$ can in general only be obtained if the system is in fact polynomial. For this case, the bound can be computed fairly easily. Given an ε-equilibrium the result gives an immediate bound on the distance between the approximation and an exact Walrasian equilibrium, hence answering Question 1 above.

3.3 Question 2: Polynomial system solving

In the following, I denote the collection of all polynomials in the variable $x_1, x_2, \ldots, x_n$ with coefficients in a field $\mathbb{K}$ by $\mathbb{K}[x_1, \ldots, x_n]$. The for this survey relevant examples of $\mathbb{K}$ are the field of rational numbers $\mathbb{Q}$, the field of real numbers $\mathbb{R}$, and the field of complex numbers $\mathbb{C}$. Polynomials over the field of rational numbers are computationally convenient since modern computer algebra systems perform exact computations over the field $\mathbb{Q}$. Economic parameters are typically real numbers, and equations characterizing equilibria lie in $\mathbb{R}[x]$. The algorithms to compute all solutions to polynomial systems always compute all solutions in an algebraically closed field, in this case $\mathbb{C}[x]$.

Given a polynomial system of equations $f : \mathbb{C}^M \to \mathbb{C}^M$ there is now a variety of algorithm to approximate numerically all complex and real zeros of f. Sturmfels's monograph (2002) provides an excellent overview. In this survey I briefly mention two possible approaches, homotopy continuation methods and solution methods based on Gröbner bases.

At the writing of this article, both approaches are too inefficient to be applicable to large economic models, but they can be used for models with four or five households and four or five commodities. To find all equilibria for a given economy, homotopy methods seem slightly more efficient, while Gröbner bases allow for statements about entire classes of economies.

3.3.1 All solution homotopies

Solving polynomial systems numerically means computing approximations to all isolated solutions. Homotopy continuation methods can provide paths to all approximate solutions. There are well-known bounds on the maximal number of complex solutions of a polynomial system. The basic idea is to start at a generic polynomial system $g(x)$ whose number of roots is at least as large as the maximal number of solutions to $f(x) = 0$ and whose roots are all known. Then one needs to trace out all paths (in complex space) of the homotopy $H(x,t) = tg(x) + (1 - t)f(x)$, which do not diverge to infinity. Smale's alpha method can be applied along the path to ensure that the approximate solutions are close to real exact solutions (see Blum et al., 1998). It can be shown that all solutions to $f(x) = 0$ can be found in this manner.

Sommese and Wampler (2005) provide a detailed overview. Applications of these methods in economics have so far been largely restricted to game theory, but the method is also applicable to Walrasian equilibria.

3.3.2 Gröbner basis

For given polynomials $f_1, \dots, f_k$ in $\mathbb{Q}[x]$ the set

$$I = \left\{ \sum_{i=1}^{k} h_i f_i : h_i \in \mathbb{Q}[x] \right\} = \langle f_1, \dots, f_k \rangle$$

is called the ideal generated by $f_1, \dots, f_k$. It turns out that under conditions which can often be shown to hold in practice, the so-called 'reduced Gröbner basis' of this ideal, I, in the lexicographic term order has the shape

$$\mathcal{G} = \{ x_1 - q_1(x_n), x_2 - q_2(x_n), \dots, x_{n-1} - q_{n-1}(x_n), r(x_n) \}$$

where r is a polynomial of degree d and the q_i are polynomials of degree $d - 1$.

This basis can be computed exactly, using Buchberger's algorithm (recently, much more efficient versions of the basic algorithm have been developed; see for example Faugère, 1999). The number of real solutions to the original system then equals the number of real solutions

of the univariate polynomial $r(.)$ which can be determined exactly by Sturm's method (see Sturmfels, 2002, for details). The roots of $r(.)$ can be approximated numerically with standard methods and the remaining solution to the original system is linear in these roots.

Kubler and Schmedders (2006) use the method to test for uniqueness of equilibria in semi-algebraic classes of economies.

FELIX KUBLER

See also **approximate solutions to dynamic models (linear methods); computation of general equilibria; general equilibrium.**

Bibliography

Basu, S., Pollack, R. and Roy, M.-F. 2003. *Algorithms in Real Algebraic Geometry.* Springer Verlag: New York.

Blum, L., Cucker, F., Shub, M. and Smale, S. 1998. *Complexity and Real Computation.* Springer Verlag: New York.

Brown, D.J., DeMarzo, P.M. and Eaves, B.C. 1996. Computing equilibria when asset markets are incomplete. *Econometrica* 64, 1–27.

Codenotti, B., Pemmaraju, S. and Varadarajan, K. 2004. Algorithms column: the computation of market equilibria. *ACM SIGACT News* 35(4), 23–37.

Debreu, G. 1972. Smooth preferences. *Econometrica* 40, 603–15.

Eaves, B.C. 1972. Homotopies for the computation of fixed points. *Mathematical Programming* 3, 1–22.

Faugère, J.C. 1999. A new efficient algorithm for computing Gröbner bases (f4). *Journal of Pure and Applied Algebra* 139, 61–88.

Ferris, M.C. and Pang, J.S. 1997. Engineering and economic applications of complementarity problems. *SIAM Review* 39, 669–713.

Garcia, C. and Zangwill, W. 1981. *Pathways to Solutions, Fixed Points, and Equilibria.* Englewood Cliffs, NJ: Prentice Hall.

Herings, P.J.J. and Schmedders, K. 2006. Computing equilibria in finance economies with incomplete markets and transaction costs. *Economic Theory* 27, 493–512.

Judd, K. 1998. *Numerical Methods in Economics.* Cambridge, MA: MIT Press.

Kehoe, T.J. and Prescott, E.C. 1995. Introduction to the symposium, the discipline of applied general equilibrium. *Economic Theory* 6, 1–11.

Kubler, F. 2001. Computable general equilibrium with financial markets. *Economic Theory* 18, 73–96.

Kubler, F. and Schmedders, K. 2000. Computing equilibria in stochastic finance economies. *Computational Economics* 15, 145–72.

Kubler, F. and Schmedders, K. 2006. Uniqueness of equilibria in semi-algebraic economies. Discussion paper, Northwestern University.

Magill, M.J.P. and Qunizii, M. 1996. *Theory of Incomplete Markets*. Cambridge, MA: MIT Press.

Richter, M.K. and Wong, K.-C. 1999. Non-computability of competitive equilibrium. *Economic Theory* 14, 1–27.

Scarf, H. 1967. On the computation of equilibrium prices. In *Ten Economic Studies in the Tradition of Irving Fisher*, ed. W.J. Fellner. New York: Wiley.

Schmedders, K. 1998. Computing equilibria in the general equilibrium model with incomplete asset markets. *Journal of Economic Dynamics and Control* 22, 1375–403.

Schmedders, K. 2004. Homotopy path-following with easyhomotopy: solving nonlinear equations for economic models. Working paper, Northwestern University.

Shoven, J.B. and Whalley, J. 1992. *Applying General Equilibrium*. Cambridge: Cambridge University Press.

Sommese, A.J. and Wampler, C.W. 2005. *The Numerical Solution of Systems of Polynomials Arising in Engineering and Science*. Singapore: World Scientific Press.

Sturmfels, B. 2002. *Solving Systems of Polynomial Equations*. CBMS Regional Conference Series in Mathematics No. 97. Providence, RI: American Mathematical Society.

Watson, L.T. 1979. A globally convergent algorithm for computing fixed points of C2 maps. *Applied Mathematics and Computation* 5, 297–311.

computational methods in econometrics

1 Introduction

In evaluating the importance and usefulness of particular econometric methods, it is customary to focus on the set of *statistical* properties that a method possesses – for example, unbiasedness, consistency, efficiency, asymptotic normality, and so on. It is crucial to stress, however, that meaningful comparisons cannot be completed without paying attention also to a method's *computational* properties. Indeed the practical value of an econometric method can be assessed only by examining the inevitable interplay between the two classes of properties, since a method with excellent statistical properties may be computationally infeasible and vice versa. Computational methods in econometrics are evolving over time to reflect the current technological boundaries as defined by available computer hardware and software capabilities at a particular period, and hence are inextricably linked with determining what the state of the art is in econometric methodology.

To give a brief illustration, roughly from the late 1950s until the early 1960s we had the 'Stone Age' of econometrics, when the most sophisticated computational instrument was the slide rule, which used two rulers on a logarithmic scale, one sliding into the other, to execute approximate multiplication and division. In this Stone Age, suitably named in honour of Sir Richard Stone,

winner of the 1984 Nobel Prize in Economics, the brightest Ph.D. students at the University of Cambridge were toiling for days and days in back rooms using slide rules to calculate ordinary linear regressions, a task which nowadays can be achieved in a split second on modern personal computers.

The classic linear regression problem serves to illustrate the crucial interaction between statistical and computational considerations in comparing competing econometric methods. Given data of size S, with observations on a dependent variable denoted by $S \times 1$ vector y and corresponding observations on k explanatory factors denoted by $S \times k$ matrix X ($k < X$), the linear plane fitting exercise is defined by Gauss's minimum quadratic distance problem:

$$\hat{\beta} = \arg\min_b (y - Xb)'(y - Xb)$$
$$\equiv \arg\min_b \sum_{s=1}^{S} (y_s - x_s'b)^2 \tag{1}$$

where x_s' is the sth row of matrix X and b is a $k \times 1$ vector of real numbers defining the regression plane Xb. Under the assumption that X has full column rank k, the solution to this *ordinary least squares* minimization problem is the linear-in-y expression $\hat{\beta} = (X'X)^{-1}X'y$, which only requires the matrix operations of multiplication and inversion.

Suppose, however, that Gauss had chosen instead as his measure of distance the sum of absolute value of the deviations, and defined instead:

$$\tilde{\beta} = \arg\min_b \sum_{s=1}^{S} |y_s - x_s'b|. \tag{2}$$

The vector $\tilde{\beta}$ that solves the second minimization is known as the *least absolute deviations* (LAD) estimator and has no closed-form matrix expression. In fact, calculation of $\tilde{\beta}$ requires highly nonlinear operations for which computationally efficient algorithms were developed only in the 1970s. To give a concrete example, consider the *intercept-only* linear regression model where X is the $S \times 1$ vector of ones. Then the single $\hat{\beta}$ coefficient that solves (1) is the sample mean of y, while $\tilde{\beta}$ that solves (2) is the sample *median* of y. The latter is orders of magnitude more difficult to compute than the former since it involves sorting y and finding the value in the middle, while the former simply adds all elements of y and divides by the sample size. Clearly, it could be quite misleading if $\hat{\beta}$ and $\tilde{\beta}$ where compared solely in terms of statistical properties without any consideration of their substantially different computational requirements.

A second example in a similar vein is the following parametric estimation problem. Suppose a sample of size S is observed on a single variable y. It is believed that each observation y_s is drawn independently from the same uniform distribution on the interval $[\theta,c]$ where the lower value of the support is the single unknown parameter that needs to be estimated, while c is known. Two parametric estimation methods with particularly attractive statistical properties are the generalized method of moments (GMM) and the method of maximum likelihood (MLE). Indeed, for relatively large sample sizes these two methods are comparably attractive in terms of statistical properties, while they differ *drastically* in terms of computational requirements: the GMM solution is $\hat{\theta}_{gmm} = \frac{2}{S}\sum_{s=1}^{S} y_s - c$, thus requiring only the simple calculation of the sample mean $\bar{y}$, while the MLE involves the highly nonlinear operation of finding the minimum of the data vector y, $\hat{\theta}_{mle} = min(y_1,\ldots,y_S)$.

In the following section we discuss in turn the leading classes of methods that are of particular importance in modern econometrics, while Section 3 introduces the concept of parallel processing and describes its current value and future promise in aiding dramatically econometric computation.

2 Computational methods important for econometrics

The advancement of computational methods for econometrics relies on understanding the interplay between the disciplines of econometric theory, computer science, numerical analysis, and applied mathematics. In the five subsections below we discuss the leading classes of computational methods that have proven of great value to modern econometrics.

2.1 Matrix computation and specialized languages

To start with the fundamental econometric framework of linear regression, the *sine qua non* of econometric computation is the ability to program and perform efficiently matrix operations. To this end, specialized matrix computer languages have been developed which include Gauss and Matlab. Fundamental estimators of the linear regression coefficient vector β, like the OLS $(X'X)^{-1}X'y$ and its generalized least squares (GLS) variant $(X'\Omega^{-1}X)^{-1}X'\Omega^{-1}y$, are leading examples of the usefulness of such matrix languages, where the $S \times S$ matrix Ω is a positive definite, symmetric variance-covariance matrix of the disturbance vector $\varepsilon \equiv y - X\beta$. Matrix operations are useful even for nonlinear econometric methods discussed below, since a generally useful approach is to apply linearization approximations through the use of differentiation and Taylor's expansions.

In implementing econometric methods that involve matrix operations, special attention needs to be paid to the dimensionality of the various matrices, as well as to any special properties a matrix may posses, which can affect very substantially the feasibility and performance of the computational method to be adopted. Looking at the OLS and GLS formulae, we see three different matrices that require inversion: $X'X$, Ω, and $X'\Omega^{-1}X$. The

first and the third are of dimension $k \times k$, while the second is $S \times S$. Since the number of regressors k is typically considerably smaller than the sample size S, the inversion of these matrices can involve vastly different burden in terms of total number of computer operations required as well as memory locations necessary for holding the information during those calculations. (For example, in panel data settings where multiple observations are observed in different time-periods for a cross-section of economic agents, it is not uncommon to have total sample sizes of 300,000 or more.) To this end, econometric analysts have focused on importing from numerical analysis matrix algorithms that are particularly efficient in handling sparse as opposed to dense matrices. By their very nature, sparse matrices exhibit a very high degree of compressibility and concomitantly lower memory requirements. See Drud (1977) for the use of sparse matrix techniques in econometrics. A matrix is called sparse if it is primarily populated by zeros, for example, the variance-covariance matrix of a disturbance vector following the moving-average-of-order-1 model:

$$\Omega_{ma1} = \sigma^2 \begin{pmatrix} 1 & \frac{\lambda}{1+\lambda^2} & 0 & \cdots & & 0 \\ \frac{\lambda}{1+\lambda^2} & 1 & \frac{\lambda}{1+\lambda^2} & \ddots & & \vdots \\ 0 & \frac{\lambda}{1+\lambda^2} & \ddots & \ddots & & 0 \\ \vdots & & \ddots & \ddots & 1 & \frac{\lambda}{1+\lambda^2} \\ 0 & \cdots & & 0 & \frac{\lambda}{1+\lambda^2} & 1 \end{pmatrix}.$$

In contrast, a stationary autoregressive disturbance of order 1 has a dense variance-covariance matrix:

$$\Omega_{ar1} = \sigma^2 \begin{pmatrix} 1 & \gamma & \gamma^2 & \cdots & \gamma^{S-1} \\ \gamma & 1 & \gamma & \ddots & \vdots \\ \gamma^2 & \gamma & \ddots & \ddots & \gamma^2 \\ \vdots & \ddots & \ddots & 1 & \gamma \\ \gamma^{S-1} & \cdots & \gamma^2 & \gamma & 1 \end{pmatrix}.$$

Other matrix algebra methods especially important in econometrics are the Cholesky factorization (see Golub, 1969) of a positive definite matrix A into the product $A=R'R$ where R is an upper-triangular matrix, and the singular value decomposition that allows the calculation of pseudo-inverse of any matrix B which may be non-square, and if square, not positive definite (see Belsley, 1974).

It is important to note that on occasion a brilliant theoretical development can simplify enormously the computational burden of econometric methods that, though possessing attractive statistical properties, were thought to be infeasible with existing computation technology in the absence of the theoretical development. A case in point is the GLS/MLE estimator for the one-factor random effects model proposed by Balestra and Nerlove (1966), which is of great importance in the analysis of linear panel data models. The standard formulation gives rise to the GLS formula requiring the inversion of an equi-correlated variance covariance matrix Ω of dimension $S \times S$, where S is of the order of the product of the number of available observations in the cross-section dimension times the number available in the time dimension. For modern panel data-sets, this can exceed 300,000, thus making the calculation of Ω^{-1} infeasible even on today's super-computers, let alone with the slide rules available in 1966. Fuller and Battese (1973), however, showed that the equi-correlated nature of the one-factor random effects model made calculation of the GLS estimator equivalent to an OLS problem, where the dependent variable $\tilde{y}$ and the regressors $\tilde{X}$ are simple linear combinations of the original data $y_{it}, x_{1it} \ldots, x_{kit}$ and its time averages $\bar{y}_{i.}, \bar{x}_{1i.}, \ldots, \bar{x}_{ki.}$ defined by $\bar{y}_{i.} \equiv \frac{1}{T}\sum_{t=1}^{T} y_{it}$ and $\tilde{y}_{it} \equiv y_{it} - \lambda\bar{y}_{i.}$, and analogously for the regressor variables. This realization allowed the calculation of the GLS estimator without the need for inverting the usually problematically large Ω matrix.

Another important case where a theoretical development in methodology led to a dramatic lowering of the computational burden and hence allowed the calculation of models that would otherwise have had to wait perhaps for decades for sufficient advancements in computer technology is the simulation-based inference for Limited Dependent Variable models, associated with the name of Daniel McFadden (1989). See Section 2.5 below, MCFADDEN, DANIEL and SIMULATION-BASED ESTIMATION.

2.2 Optimization

Many econometric estimators with attractive statistical properties require the optimization of a (generally) non-linear function of the form:

$$q \equiv \arg\max_{\theta} F(\theta; data) \qquad (3)$$

over a vector of unknown parameters θ of dimension p, typically considerably larger than 1. Examples are: the method of maximum likelihood, minimum-distance (OLS, LAD, GMM), and other extremum estimators. (The need to optimize functions numerically is also important for certain problems in computational economics, for example, the problem of optimal control.) Algorithms for optimizing functions of many variables are a key component in the collection of tools for econometric computation. The suitability of a certain algorithm to a specific optimization econometric problem depends on the following classification:

1. *Algorithms that require the calculation of first and possibly second derivatives* Versus *algorithms that do*

not. Clearly, if the function to be optimized is not twice continuously differentiable (as is the case with LAD) or even discontinuous (as is the case with the maximum score estimator for the semiparametric analysis of the binary response model – see Manski, 1975), algorithms that require differentiability will not be suitable. The leading example of an algorithm not relying on derivatives is the nonlinear simplex method of Nelder and Meade (1965).

2. *Local* Versus *global algorithms.* Optimization algorithms of the first type (for example, Gauss-Newton, Newton-Raphson, and Berndt et al. (1974)) search for an optimum in the vicinity of the starting values fed into the algorithm. This strategy may not necessarily lead to a global optimum over the full set of parameter space. This is of particular importance if the function to be optimized has multiple local optima, where typically the estimator with the desirable statistical properties corresponds to locating the overall optimum of the function. In such cases, global optimization algorithms (for example, simulated annealing and genetic optimization algorithm) should be employed instead.

Special methods are necessary for constrained optimization, where a function must be maximized or minimized subject to a set of equality or inequality constraints. These problems, in general considerably more demanding than unconstrained optimization, can be handled through three main alternative approaches: interior, exterior and re-parameterization methods.

Comprehensive reviews of optimization methods in econometrics can be found in Goldfeld and Quandt (1972), Quandt (1983), and Dennis and Schnabel (1984). These studies also discuss the related issue of the numerical approximation of derivatives and illustrate the fundamental link in terms of computation between optimization and the problem of solving linear and nonlinear equations. For similar methods used in economics, see NUMERICAL OPTIMIZATION METHODS IN ECONOMICS and NONLINEAR PROGRAMMING.

2.3 Sorting

Of special importance for computing the class of estimators known as robust or semiparametric methods is the ability to sort data rapidly and computationally efficiently. Such a need arises in the calculation of order statistics, for example, the sample median and sample minimum required by the first two estimation examples given above. The leading sorting algorithms, bubble-, heap- and quick-sort, have fundamentally different properties in terms of computation speed and memory requirements, in general depending on how close to being sorted the original data series happens to be. For a practical review of the leading sorting algorithms, see Press et al. (2001, ch. 8).

2.4. Numerical approximation and integration

Numerical approximation is necessary for any mathematical function that does not have a closed form solution, for example, exponential, natural logarithm and error functions. See Abramowitz and Stegun (1964) for an exhaustive study of mathematical functions and their efficient approximation. Judd (1996) focuses on numerical approximation methods particularly useful in economics and econometrics.

Numerical integration, also known as numerical quadrature, is a related approximation problem that is crucial to modern econometrics. There are two key fields of econometrics where integrals without a closed form must be evaluated numerically. The first is Bayesian inference where moments of posterior densities need to be evaluated, which take the form of high-dimensional integrals. See, inter alia, Zellner, Bauwens and VanDijk (1988). The second main class is classical inference in limited dependent variable (LDV) models; for example, Hajivassiliou and Ruud (1994). See Geweke (1996) for an exhaustive review of numerical integration methods in computational economics and econometrics, and Davis and Rabinowitz (1984) for earlier results.

It is important to highlight a crucial difference between the numerical integration problems in Bayesian inference and those in classical inference for LDV models, which makes various integration-by-simulation algorithms be useful to one field and not the other: in the Bayesian case, typically a single or a few high-dimensional integrals have to be evaluated accurately. In contrast, in the classical LDV inference case, quite frequently hundreds of thousands of such integrals need to be approximated.

2.5 Computer simulation

The need for efficient generation of pseudo-random numbers with good statistical properties on a computer appears very routinely in econometrics. Leading examples include:

- Statistical methods based on resampling, primarily the 'jackknife' and the 'bootstrap', as introduced by Efron (1982). These methods have proven of special value in improving the small sample properties of certain econometric estimators and test procedures, for example in reducing estimation bias. They are also used to approximate the small sample variance of estimators for which no closed form expressions can be derived.

- Evaluation of econometric estimators through Monte Carlo experiments, where hypothetical data-sets with certain characteristics are simulated repeatedly and the econometric estimators under study are calculated for each set. This allows the calculation of empirical (simulated) properties of the estimators, either to compare to theoretical mathematical calculations or because the latter are intractable.

- Calculation of frequency probabilities of possible outcomes in large-scale decision trees, for which the outcome probabilities are impossible to characterize theoretically.

- Sensitivity analyses and what-if studies, where an econometric model is 'run' on a computer under different scenarios of policy measures.

- Simulation-based Bayesian and classical inference, where integrals are approximated through computer simulation (known as Monte Carlo integration). Particularly important methods in this context are the following: frequency simulation; importance sampling; and Markov chain Monte Carlo methods (the leading exponents being Gibbs resampling and the Metropolis/Hastings algorithm). A related class of methods, known as variance-reduction simulation techniques, includes control variates and antithetics. See Geweke (1988) and Hajivassiliou, McFadden and Ruud (1996) for reviews. See also SIMULATION-BASED ESTIMATION.

3 Parallel computation

Parallel processing, where a computation task is broken up and distributed across different computers, is a technique that can afford huge savings in terms of total time required for solving particularly difficult econometric problems. For example, the simulation-based estimators mentioned in the previous section exhibit the potential of significant computational benefits by calculating them on computers with massively parallel architectures, because the necessary calculations can be organized in essentially an independent pattern. An example of such a computer is the Connection Machine CM-5 at the National Center for Supercomputing Applications in Illinois with 1,024 identical processors in a multiple-instruction/multiple-data (MIMDI) configuration. The benefits of such a parallel architecture on the problem of solving an econometric optimization classical estimator not involving simulation can also be substantial, since such estimators involve the evaluation of contributions to the criterion (for example, likelihood) function in the case of independently and identically distributed (i.i.d.) observations. Since typical applications in modern applied econometrics using cross-sectional and longitudinal data sets involve several thousands of i.i.d. observations, the potential benefits of parallel calculations of such estimators should be obvious. The benefits of a massively parallel computer architecture become even more pronounced in the case of simulation-based estimators. See Nagurney (1996) for a discussion of parallel computation in econometrics.

An alternative approach for parallel computation that does not involve a single computer with many processors has been developed recently and offers considerable promise for computational econometrics. Through the use of specialized computer languages, many separate computers are harnessed together over an organization's intranet or even over the internet, and an econometric computation task is distributed across them. The benefits of this approach depend critically on the relative burden of the overhead of communicating across the individual computers when organizing the splitting of the tasks and then collecting and processing the separate partial results. Such distributed parallel computation has the exciting potential of affording formidable super-computing powers to econometric researchers with only modest computer hardware.

VASSILIS A. HAJIVASSILIOU

See also **longitudinal data analysis; McFadden, Daniel; nonlinear programming; numerical optimization methods in economics; robust estimators in econometrics; simulation-based estimation.**

Bibliography

Abramowitz, M. and Stegun, I. 1964. *Handbook of Mathematical Functions*. Washington, DC: National Bureau of Standards.

Balestra, P. and Nerlove, M. 1966. Pooling cross-section and time-series data in the estimation of a dynamic model. *Econometrica* 34, 585–612.

Belsley, D. 1974. Estimation of system of simultaneous equations and computational specifications of GREMLIN. *Annals of Economic and Social Measurement* 3, 551–614.

Berndt, E.K., Hall, B.H., Hall, R.E. and Hausman, J.A. 1974. Estimation and inference in nonlinear structural models. *Annals of Economic and Social Measurement* 3, 653–66.

Davis, P.J. and Rabinovitz, P. 1984. *Methods of Numerical Integration*. New York: Academic Press.

Dennis, J.E. and Schnabel, R.B. 1984. *Unconstrained optimization and Nonlinear Equations*. Englewood Cliffs, NJ: Prentice-Hall.

Drud, A. 1977. An optimization code for nonlinear econometric models based on sparse matrix techniques and reduced grades. *Annals of Economic and Social Measurement* 6, 563–80.

Efron, B. 1982. *The Jackknife, the Bootstrap, and Other Resampling Plans*. CBMS-NSF Monographs No. 38. Philadelphia: SIAM.

Fuller, W.A. and Battese, G.E. 1973. Transformations for estimation of linear models with nested-error structure. *Journal of the American Statistical Association* 68, 626–32.

Geweke, J. 1988. Antithetic acceleration of Monte Carlo integration in Bayesian inference. *Journal of Econometrics* 38, 73–90.

Geweke, J. 1996. Monte Carlo simulation and numerical integration. In *Handbook of Computational Economics*, vol. 1, ed. H. Amman, D. Kendrik and J. Rust. Amsterdam: North-Holland.

Golub, G.H. 1969. Matrix decompositions and statistical calculations. In *Statistical Computation*, ed. R.C. Milton and J.A. Milder. New York: Academic Press.

Goldfeld, S. and Quandt, R. 1972. *Nonlinear Methods in Econometrics*. Amsterdam: North-Holland.

Hajivassiliou, V.A. and Ruud, P.A. 1994. Classical estimation methods using simulation. In *Handbook of Econometrics*, vol. 4, ed. R. Engle and D. McFadden. Amsterdam: North-Holland.

Hajivassiliou, V.A., McFadden, D.L. and Ruud, P.A. 1996. Simulation of multivariate normal rectangle probabilities and derivatives: theoretical and computational results. *Journal of Econometrics* 72(1, 2), 85–134.

Judd, K. 1996. Approximation, perturbation, and projection methods in economic analysis. In *Handbook of Computational Economics*, vol. 1, ed. H. Amman, D. Kendrik and J. Rust. Amsterdam: North-Holland.

Manski, C. 1975. Maximum score estimation of the stochastic utility model of choice. *Journal of Econometrics* 3, 205–28.

McFadden, D. 1989. A method of simulated moments for estimation of multinomial discrete response models. *Econometrica* 57, 995–1026.

Nagurney, A. 1996. Parallel computation. In *Handbook of Computational Economics*, vol.1, ed. H. Amman, D. Kendrik and J. Rust. Amsterdam: North-Holland.

Nelder, J.A. and Meade, R. 1965. A simplex method for function minimization. *Computer Journal* 7, 308–13.

Press, W.H., Flannery, B.P., Teukolsky, S.A. and Vetterling, W.T. 2001. *Numerical Recipes in Fortran 77: The Art of Scientific Computing*. Cambridge: Cambridge University Press.

Quandt, R. 1983. Computational problems and methods. In *Handbook of Econometrics*, vol. 1, ed. Z. Griliches and M. Intriligator. Amsterdam: North-Holland.

Zellner, A., Bauwens, L. and VanDijk, H. 1988. Bayesian specification analysis and estimation of simultaneous equation models using Monte Carlo methods. *Journal of Econometrics* 38, 73–90.

computer industry

The commercial computing industry accounts for a large fraction of economic activity. From its military and research origins in the late 1940s, it spread into the commercial realm and has since grown to include an extraordinary range of economic undertakings. Many economists believe this expansion of applications for computing has been a driver of economic growth.

Computing aids the automated tracking of transactions, a function that finds use, for example, in automating billing, managing the pricing of inventories of airline seating, and restocking retail outlets in a geographically dispersed organization. It also facilitates the coordination of information-intensive tasks, such as the dispatching of time-sensitive deliveries or emergency services. Computing also enables performance of advanced mathematical calculations, useful in such diverse activities as calculating interest on loans and generating estimates of underground geological deposits. Computer-aided precision also improves the efficiency of processes such as manufacturing metal shapes or the automation of communication switches, to name just two.

In any given era, computing markets are organized around platforms – a cluster of technically standardized components that buyers use together to make the aforementioned wide range of applications. Such platforms involve long-lived assets, both components sold in markets (that is, hardware and some software) and components made by buyers (that is, training and most software). Important computing platforms historically include the UNIVAC, the IBM 360 and its descendents, the Wang minicomputers, IBM AS/400, DEC VAX, Sun SPARC, Intel/Windows PC, Unix/Linux, and, after the mid-1990s, TCP/IP-based client-server platforms linked together.

Vendors tend to sell groups of compatible products under umbrella strategies aimed at the users of particular platforms. In the earliest eras of computing markets, the leading firms integrated all facets of computing and offered a supply of goods and services from a centralized source. In later eras, the largest and most popular platforms historically included many different computing, communications and peripheral equipment firms, software tool developers, application software writers, consultants, system integrators, distributors, user groups, news publications and service providers.

Until the early 1990s, most market segments were distinguished by the size of the tasks to be undertaken and by the technical sophistication of the typical user. Mainframes, minicomputers, workstations, and personal computers, in decreasing order, constituted different size-based market segments. Trained engineers or programmers made up the technical user base, while the commercial market was geared more towards administrators, secretaries and office assistants.

The most popular platform in the late 1980s and 1990s differed from the prominent platforms of earlier years. The personal computer (PC) began in the mid-1970s as an object of curiosity among technically skilled hobbyists, but became a common office tool after the entry of IBM's design. Unlike prior computing platforms, this one has diffused into both home and business use. From the beginning, this platform involved thousands of large and small software developers, third-party peripheral equipment and card developers, and a few major players. In more recent experience, control over the standard has completely passed from IBM to Microsoft and Intel. Microsoft produces the Windows operating system and Intel produces the most commonly used microprocessor. For this reason the platform is often called Wintel.

The networking and internet revolution in the late 1990s is responsible for blurring once-familiar distinctions. These new technologies have made it feasible to build client-server systems within large enterprises and across ownership boundaries. It employs internet-based computing systems networked across potentially vast geographic distances, supporting the emergence of a 'network of networks'.

Despite frequent and sometimes dramatic technical improvements in specific areas of technology, many features of the most common platforms in use tend to persist or change very slowly. Many durable components make up platforms. And, though they lose their market value as they become obsolete in comparison with frontier products, they do not as quickly lose their ability to provide a flow of services to users. Consequently, new technology tends to be most successful when new components enhance and preserve the value of previous investments, a factor that creates demand for 'backward compatible' upgrades or improvements. It also creates a demand for support and service activities to reduce the costs of making the transition from old to new.

Control over changes to design and other aspects of technical standards shapes the backward compatibility for key components. Control of these decisions is coincident with platform leadership – determining the rate and direction of change in technical features of components around which other firms build their businesses. In each platform, it is very rare to observe more than a small number of firms acquiring leadership positions. Since such positions have been historically associated with high firm profitability, firms compete fiercely for market dominance in component categories where standards are essential. Not surprisingly, competitive behaviour affiliated with obtaining and retaining market leadership does occasionally receive attention from antitrust authorities.

Though innovative change in computing began well prior to the invention of the integrated circuit, in popular discussion advances in computing have become almost synonymous with advances in microprocessors. This is due to an observation by Gordon Moore, who co-founded and became chairman at Intel. In 1965 he foresaw a doubling of circuits per chip every two years. This prediction about the rate of technical advance later became known as 'Moore's law'. In fact, microprocessors and DRAMS have been doubling in capability every 18 months since the mid-1970s.

Moore's prediction pertained narrowly to integrated circuits. However, a similar pattern of improvement – though with variation in the rate – characterizes other electronic components that go into producing a computer or that are complementary with computing in many standard uses. This holds for disk drives, display screens, routing equipment, and data-transmission capacity, to name a few. Such widespread innovation creates opportunities for new entry and rearrangements in the conditions of supply.

Accordingly, there has been an increasing secular trend in the number of firms that possess the necessary technical knowledge and commercial capabilities to bring to market some component or service of value to computing users. This factor alone explains the increasing complexity of supply chains for the supply of most computing hardware and software products. It is also coincident with their increasing geographic reach. In addition, as in other manufacturing processes, the increasing use of sophisticated information technology helps coordinate design and production involving firms from many countries and continents.

While the spawning of new information technology businesses in North America has tended to be concentrated in a small number of locations, such as the Boston area and Santa Clara Valley (popularly known as Silicon Valley), every other facet of the supply chain for computing involves firms headquartered and operating in a much wider set of locations. In North America, these range from Seattle, Austin, Los Angeles, the greater New York area, Denver–Boulder, Washington DC, the North Carolina Research Triangle, Chicago, and virtually all major cities in the United States. The supply chain for many complementary components has also been associated with many firms in Western Europe and as well as in India, Israel, South Korea, Singapore, Taiwan and China. Even more widespread are computing service firms, which follow business and home users dispersed across the globe.

Despite this geographic dispersion since the 1950s, US companies have retained leadership in generating new platforms and commercializing frontier technologies in forms that most users find valuable. Part of this results from the persistence of platform leadership for a time within a segment. In addition, US firms have historically been ascendant whenever platform leadership has changed. However, this pattern seems likely to change in the 21st century, as non-US firms already have found leadership positions in producing components of many platforms and in related areas of electronics, such as consumer electronics, communication equipment and specialized software.

While general improvements in technical capabilities are readily apparent, it is quite difficult to calculate the productivity improvements arising from increased investment in and use of computing. There is no question that existing computing activities have become less expensive, while new capabilities have been achieved. This has allowed economic actors to attain previously unobtainable outcomes. This shift in economic possibilities has generated a restructuring of organizational routines, market relationships, and other activities associated with the flow of goods, which inevitably improves the economy's ability to transform inputs into consumer welfare.

Yet altering the business use of computing can be slow. It often demands large adjustment costs and gradual learning about which organizational processes can best employ advances in computing. It can involve a reallocation of decision rights and discretion inside a large organization, especially when business units alter a wide array of intermediate routine processes (such as billing, account monitoring, and inventory management) or the coordination of services (such as the delivery of data for decision support). Moreover, the largest changes come from altering many complementary activities that respond to new and unanticipated opportunities, setting off new waves of invention. Each wave's productivity effect is interwoven with others.

Along with these improvements the boundaries of the 'computing market' have changed. A hardware-based definition for the computing market was barely adequate in the 1960s and is no longer adequate for economic analysis. However, there is no consensus about what alternative framing will be appropriate for understanding value creation, supplier behaviour, and user adoption in computing in the 21st century.

SHANE GREENSTEIN

See also **diffusion of technology; general purpose technologies; information technology and the world economy; Internet, economics of the; technical change.**

Bibliography

Bresnahan, T. and Greenstein, S. 1999. Technological competition and the structure of the computer industry. *Journal of Industrial Economics* 47(1), 1–40.

Brensnahan, T. and Malerba, F. 1999. Industrial dynamics and the evolution of firm's and nations' competitive capabilities in the world computer industry. In *Sources of Industrial Leadership*, ed. D. Mowery and R. Nelson. Cambridge, UK: Cambridge University Press.

Brynjolfsson, E. and Lorin, H. 2000. Beyond computation: information technology, organizational transformation and business performance. *Journal of Economic Perspectives* 14(4), 23–48.

Dedrick, J. and Kraemer, K. 2005. The impacts of IT on firm and industry structure: the personal computer industry. *California Management Review* 47(3), 122–42.

Flamm, K. 2003. The new economy in historical perspective: evolution of digital electronics technology. In *New Economy Handbook*, ed. D. Jones. San Diego: Academic Press/Elsevier.

Greenstein, S., ed. 2006. *The Industrial Economics of Computing*. Northampton, MA: Edward Elgar.

Jorgenson, D. and Wessner, C., eds. 2005. *Deconstructing the Computer: Report of a Symposium*. Washington, DC: National Academies Press.

McKinsey Global Institute. 2001. *U.S. Productivity Growth, 1995–2000: Understanding the Contribution of Information Technology Relative to Other Factors*. Washington, DC: McKinsey and Co.

computer science and game theory

1 Introduction

There has been a remarkable increase in work at the interface of computer science and game theory in the past decade. Game theory forms a significant component of some major computer science conferences (see, for example, Kearns and Reiter, 2005; Sandholm and Yokoo, 2003); leading computer scientists are often invited to speak at major game theory conferences, such as the World Congress on Game Theory 2000 and 2004. In this article I survey some of the main themes of work in the area, with a focus on the work in computer science. Given the length constraints, I make no attempt at being comprehensive, especially since other surveys are also available, including Halpern (2003), Linial (1994), Papadimitriou (2001), and a comprehensive survey book (Nisan et al., 2007).

The survey is organized as follows. I look at the various roles of computational complexity in game theory in Section 2, including its use in modelling bounded rationality, its role in mechanism design, and the problem of computing Nash equilibria. In Section 3, I consider a game-theoretic problem that originated in the computer science literature, but should be of interest to the game theory community: computing the *price of anarchy*, that is, the cost of using a decentralizing solution to a problem. In Section 4 I consider interactions between distributed computing and game theory. In Section 5, I consider the problem of implementing mediators, which has been studied extensively in both computer science and game theory. I conclude in Section 6 with a discussion of a few other topics of interest.

2 Complexity considerations

The influence of computer science in game theory has perhaps been most strongly felt through complexity theory. I consider some of the strands of this research here. There are a numerous basic texts on complexity theory that the reader can consult for more background on notions like NP-completeness and finite automata, including Hopcroft and Ullman (1979) and Papadimitriou (1994a).

2.1 Bounded rationality

One way of capturing bounded rationality is in terms of agents who have limited computational power. In economics, this line of research goes back to the work of Neyman (1985) and Rubinstein (1986), who focused on finitely repeated Prisoner's Dilemma. In n-round finitely repeated Prisoner's Dilemma, there are 2^{2^n-1} strategies (since a strategy is a function from histories to {cooperate, defect}, and there are clearly 2^n-1 histories of length $< n$). Finding a best response to a particular move can thus potentially be difficult. Clearly people do not find best responses by doing extensive computation. Rather, they typically rely on simple heuristics, such as 'tit for tat' (Axelrod, 1984). Such heuristics can often be captured by finite automata; both Neyman and Rubinstein thus focus on finite automata playing repeated Prisoner's Dilemma. Two computer scientists, Papadimitriou and Yannakakis (1994), showed that if both players in an n-round Prisoner's Dilemma are finite automata with at least 2^n-1 states, then the only equilibrium is the one where they defect in every round. This result says that a finite automaton with exponentially many states can compute best responses in Prisoner's Dilemma.

We can then model bounded rationality by restricting the number of states of the automaton. Neyman (1985) showed, roughly speaking, that if the two players in n-round Prisoner's Dilemma are modelled by finite automata with a number of states in the interval $[n^{1/k}, n^k]$ for some k, then collaboration can be approximated in equilibrium; more precisely, if the payoff for (cooperate, cooperate) is $(3, 3)$ there is an equilibrium in the repeated game where the average payoff per round is greater than $3 - \frac{1}{k}$ for each player. Papadimitriou and Yannakakis (1994) sharpen this result by showing that if at least one of the players has fewer than $2^{c_\varepsilon n}$ states, where $c_\varepsilon = \frac{\varepsilon}{12(1+\varepsilon)}$, then for sufficiently large n, there is an equilibrium where each player's average payoff per round is greater than $3-\varepsilon$. Thus, computational limitations can lead to cooperation in Prisoner's Dilemma.

There have been a number of other attempts to use complexity-theoretic ideas from computer science to model bounded rationality (see Rubinstein, 1998, for some examples). However, it seems that there is much more work to be done here.

2.2 Computing Nash equilibrium

Nash (1950) showed every finite game has a Nash equilibrium in mixed strategies. But how hard is it to actually find that equilibrium? On the positive side, there are well known algorithms for computing Nash equilibrium, going back to the classic Lemke–Howson (1964) algorithm, with a spate of recent improvements (see, for example, Govindan and Wilson, 2003; Blum, Shelton and Koller, 2003; Porter, Nudelman and Shoham, 2004). Moreover, for certain classes of games (for example, symmetric games, Papadimitriou and Roughgarden, 2005), there are known to be polynomial-time algorithms. On the negative side, many questions about Nash equilibrium are known to be NP-hard. For example, Gilboa and Zemel (1989) showed that, for a game presented in normal form, deciding whether there exists a Nash equilibrium where each player gets a payoff of at least r is NP-complete. Interestingly, Gilboa and Zemel also show that computing whether there exists a *correlated* equilibrium (Aumann, 1987) where each player gets a payoff of at least r is computable in polynomial time. In general, questions regarding correlated equilibrium seem easier than the analogous questions for Nash equilibrium; see Papadimitriou (2005) and Papadimitriou and Roughgarden (2005) for further examples. Chu and Halpern (2001) prove similar NP-completeness results if the game is represented in extensive form, even if all players have the same payoffs (a situation that arises frequently in computer science applications, where we can view the players as agents of some designer, and take the payoffs to be the designer's payoffs). Conitzer and Sandholm (2003) give a compendium of hardness results for various questions regarding Nash equilibria.

Nevertheless, there is a sense in which it seems that the problem of finding a Nash equilibrium is easier than typical NP-complete problems, because every game is guaranteed to have a Nash equilibrium. By way of contrast, for a typical NP-complete problem like propositional satisfiability, whether or not a propositional formula is satisfiable is not known. Using this observation, it can be shown that if finding a Nash equilibrium is NP-complete, then NP = coNP. Recent work has in a sense completely characterized the complexity of finding a Nash equilibrium in normal-form games: it is a *PPAD-complete* problem (Chen and Deng, 2006; Daskalis, Goldberg and Papadimitriou, 2006). PPAD stands for 'polynomial parity argument (directed case)'; see Papadimitriou (1994b) for a formal definition and examples of other PPAD problems. It is believed that PPAD-complete problems are not solvable in polynomial time, but are simpler than NP-complete problems, although this remains an open problem. See Papadimitriou (2007) for an overview of this work.

2.3 Algorithmic mechanism design

The problem of mechanism design is to design a game such that the agents playing the game, motivated only by self-interest, achieve the designer's goals. This problem has much in common with the standard computer science problem of designing protocols that satisfy certain specifications (for example, designing a distributed protocol that achieves Byzantine agreement; see Section 4). Work on mechanism design has traditionally ignored computational concerns. But Kfir-Dahav, Monderer and Tennenholtz (2000) show that, even in simple settings, optimizing social welfare is NP-hard, so that perhaps the most common approach to designing mechanisms, applying the Vickrey–Groves–Clarke (VCG) procedure (Clarke, 1971; Groves, 1973; Vickrey, 1961), is not going to work in large systems. We might hope that, even if we cannot compute an optimal mechanism, we might be able to compute a reasonable approximation to it. However, as Nisan and Ronen (2000; 2001) show, in general, replacing a VCG mechanism by an approximation does not preserve truthfulness. That is, even though truthfully revealing one's type is an optimal strategy in a VCG mechanism, it may no longer be optimal in an approximation. Following Nisan and Ronen's work, there has been a spate of papers either describing computationally

tractable mechanisms or showing that no computationally tractable mechanism exists for a number of problems, ranging from task allocation (Archer and Tardos, 2001; Nisan and Ronen, 2001) to cost-sharing for multicast trees (Feigenbaum, Papadimitriou and Shenker, 2000) (where the problem is to share the cost of sending, for example, a movie over a network among the agents who actually want the movie) to finding low-cost paths between nodes in a network (Archer and Tardos, 2002).

The problem that has attracted perhaps the most attention is *combinatorial auctions*, where bidders can bid on bundles of items. This becomes of particular interest in situations where the value to a bidder of a bundle of goods cannot be determined by simply summing the value of each good in isolation. To take a simple example, the value of a pair of shoes is much higher than that of the individual shoes; perhaps more interestingly, an owner of radio stations may value having a licence in two adjacent cities more than the sum of the individual licences. Combinatorial auctions are of great interest in a variety of settings including spectrum auctions, airport time slots (that is, take-off and landing slots), and industrial procurement. There are many complexity-theoretic issues related to combinatorial auctions. For a detailed discussion and references see Cramton, Shoham and Steinberg (2006); I briefly discuss a few of the issues involved here.

Suppose that there are n items being auctioned. Simply for a bidder to communicate her bids to the auctioneer can take, in general, exponential time, since there are 2^n bundles. In many cases, we can identify a bid on a bundle with the bidder's valuation of the bundle. Thus, we can try to carefully design a bidding language in which a bidder can communicate her valuations succinctly. Simple information-theoretic arguments can be used to show that, for every bidding language, there will be valuations that will require length at least 2^n to express in that language. Thus, the best we can hope for is to design a language that can represent the 'interesting' bids succinctly. See Nisan (2006) for an overview of various bidding languages and their expressive power.

Given bids from each of the bidders in a combinatorial auction, the auctioneer would like to then determine the winners. More precisely, the auctioneer would like to allocate the m items in an auction so as to maximize his revenue. This problem, called the *winner determination problem*, is NP-complete in general, even in relatively simple classes of combinatorial auctions with only two bidders making rather restricted bids. Moreover, it is not even polynomial-time approximable, in the sense that there is no constant d and polynomial-time algorithm such that the algorithm produces an allocation that gives revenue that is at least $1/d$ of optimal. On the other hand, there are algorithms that provably find a good solution, seem to work well in practice, and, if they seem to be taking too long, can be terminated early, usually with a good feasible solution in hand. See Lehmann, Müller and

Sandholm, (2006) for an overview of the results in this area.

In most mechanism design problems, computational complexity is seen as the enemy. There is one class of problems in which it may be a friend: voting. One problem with voting mechanisms is that of *manipulation* by voters. That is, voters may be tempted to vote strategically rather than ranking the candidates according to their true preferences, in the hope that the final outcome will be more favourable. This situation arises frequently in practice; in the 2000 US presidential election, American voters who preferred Nader to Gore to Bush were encouraged to vote for Gore, rather than 'wasting' a vote on Nader. The classic Gibbard–Satterthwaite theorem (Gibbard, 1973; Satterthwaite, 1975) shows that, if there are at least three alternatives, then in any nondictatorial voting scheme (that is, one where it is *not* the case that one particular voter dictates the final outcome, irrespective of how the others vote), there are preferences under which an agent is better off voting strategically. The hope is that, by constructing the voting mechanism appropriately, it may be computationally intractable to find a manipulation that will be beneficial. While finding manipulations for the plurality protocol (the candidate with the most votes wins) is easy, there are well-known voting protocols for which manipulation is hard in the presence of three or more candidates. See Conitzer, Sandholm and Lang (2007) for a summary of results and further pointers to the literature.

2.4 Communication complexity

Most mechanisms in the economics literature are designed so that agents truthfully reveal their preferences. However, in some settings, revealing one's full preferences can require a prohibitive amount of communication. For example, in a combinatorial auction of m items, revealing one's full preferences may require revealing what one would be willing to pay for each of the 2^m-1 possible bundles of items. Even if $m = 30$, this requires revealing more than one billion numbers. This leads to an obvious question: how much communication is required by various mechanisms? Formal work on this question in the economics community goes back to Hurwicz (1977) and Mount and Reiter (1974); their definitions focused on the dimension of the message space. Independently (and later), there was active work in computer science on *communication complexity*, the number of bits of communication needed for a set of n agents to compute the value of a function $f : x_{i=1}^n \Theta_i \rightarrow X$, where each agent i knows $\theta_i \in \Theta_i$. (Think of θ_i as representing agent i's type.) Recently there has been an explosion of work, leading to a better understanding of the communication complexity for many important economic allocation problems; see Segal (2006) for an overview. Two important themes in this work are understanding the role of price-based market mechanisms in solving allocation problems with minimal communication, and designing mechanisms that

provide agents with incentives to communicate truthfully while having low communication requirements.

3 The price of anarchy

In a computer system, there are situations where we may have a choice between a centralized and a decentralized solution to a problem. By 'centralized' here, I mean that each agent in the system is told exactly what to do and must do so; in the decentralized solution, each agent tries to optimize his own selfish interests. Of course, centralization comes at a cost. For one thing, there is a problem of enforcement. For another, centralized solutions tend to be more vulnerable to failure. On the other hand, a centralized solution may be more socially beneficial. How much more beneficial can it be?

Koutsoupias and Papadimitriou (1999) formalized this question by considering the ratio of the social welfare of the centralized solution to the social welfare of the Nash equilibrium with the worst social welfare (assuming that the social welfare function is always positive). They called this ratio the *price of anarchy*, and proved a number of results regarding the price of anarchy for a scheduling problem on parallel machines. Since the original paper, the price of anarchy has been studied in many settings, including traffic routing (Roughgarden and Tardos, 2002), facility location games (for example, where is the best place to put a factory) (Vetta, 2002), and spectrum sharing (how should channels in a WiFi network be assigned) (Halldórsson et al., 2004).

To give a sense of the results, consider the traffic-routing context of Roughgarden and Tardos (2002). Suppose that the travel time on a road increases in a known way with the congestion on the road. The goal is to minimize the average travel time for all drivers. Given a road network and a given traffic load, a centralized solution would tell each driver which road to take. For example, there could be a rule that cars with odd-numbered licence plates take road 1, while those with even-numbered plates take road 2, to minimize congestion on either road. Roughgarden and Tardos show that the price of anarchy is unbounded if the travel time can be a nonlinear function of the congestion. On the other hand, if it is linear, they show that the price of anarchy is at most 4/3.

The price of anarchy is but one way of computing the 'cost' of using a Nash equilibrium. Others have been considered in the computer science literature. For example, Tennenholtz (2002) compares the *safety level* of a game – the optimal amount that an agent can guarantee himself, independent of what the other agents do – to what the agent gets in a Nash equilibrium, and shows, for interesting classes of games, including load-balancing games and first-price auctions, that the ratio between the safety level and the Nash equilibrium is bounded. For example, in the case of first-price auctions, it is bounded by the constant e.

4 Game theory and distributed computing

Distributed computing and game theory are interested in much the same problems: dealing with systems where there are many agents, facing uncertainty and having possibly different goals. In practice, however, there has been a significant difference in emphasis between the two areas. In distributed computing, the focus has been on problems such as fault tolerance, asynchrony, scalability, and proving correctness of algorithms; in game theory, the focus has been on strategic concerns. I discuss here some issues of common interest. Most of the discussion in the remainder of this section is taken from Halpern (2003).

To understand the relevance of fault tolerance and asynchrony, consider the *Byzantine agreement* problem, a paradigmatic problem in the distributed systems literature. In this problem, there are assumed to be n soldiers, up to t of which may be faulty (the t stands for *traitor*); n and t are assumed to be common knowledge. Each soldier starts with an initial preference, to either attack or retreat. (More precisely, there are two types of nonfaulty agents – those that prefer to attack, and those that prefer to retreat.) We want a protocol that guarantees that (1) all *nonfaulty* soldiers reach the same decision, and (2) if all the soldiers are nonfaulty and their initial preferences are identical, then the final decision agrees with their initial preferences. (The condition simply prevents the obvious trivial solutions, where the soldiers attack no matter what, or retreat no matter what.)

The problem was introduced by Pease, Shostak and Lamport (1980), and has been studied in detail since then; Chor and Dwork (1989), Fischer (1983), and Linial (1994) provide overviews. Whether the Byzantine agreement problem is solvable depends in part on what types of failures are considered, on whether the system is *synchronous* or *asynchronous*, and on the ratio of n to t. Roughly speaking, a system is synchronous if there is a global clock and agents move in lockstep; a 'step' in the system corresponds to a tick of the clock. In an asynchronous system, there is no global clock. The agents in the system can run at arbitrary rates relative to each other. One step for agent 1 can correspond to an arbitrary number of steps for agent 2 and vice versa. Synchrony is an implicit assumption in essentially all games. Although it is certainly possible to model games where player 2 has no idea how many moves player 1 has taken when player 2 is called upon to move, it is not typical to focus on the effects of synchrony (and its lack) in games. On the other hand, in distributed systems, it is typically a major focus.

Suppose for now that we restrict to *crash failures*, where a faulty agent behaves according to the protocol, except that it might crash at some point, after which it sends no messages. In the round in which an agent fails, the agent may send only a subset of the messages that it is supposed to send according to its protocol. Further suppose that the system is synchronous. In this case, the following rather simple protocol achieves Byzantine agreement:

- In the first round, each agent tells every other agent its initial preference.
- For rounds 2 to $t+1$, each agent tells every other agent everything it has heard in the previous round. Thus, for example, in round 3, agent 1 may tell agent 2 that it heard from agent 3 that its initial preference was to attack, and that it (agent 3) heard from agent 2 that its initial preference was to attack, and it heard from agent 4 that its initial preferences was to retreat, and so on. This means that messages get exponentially long, but it is not difficult to represent this information in a compact way so that the total communication is polynomial in n, the number of agents.
- At the end of round $t+1$, if an agent has heard from any other agent (including itself) that its initial preference was to attack, it decides to attack; otherwise, it decides to retreat.

Why is this correct? Clearly, if all agents are correct and want to retreat (resp., attack), then the final decision will be to retreat (resp., attack), since that is the only preference that agents hear about (recall that for now we are considering only crash failures). It remains to show that if some agents prefer to attack and others to retreat, then all the nonfaulty agents reach the same final decision. So suppose that i and j are nonfaulty and i decides to attack. That means that i heard that some agent's initial preference was to attack. If it heard this first at some round $t' < t + 1$, then i will forward this message to j, who will receive it and thus also attack. On the other hand, suppose that i heard it first at round $t+1$ in a message from i_{t+1}. Thus, this message must be of the form 'i_t said at round t that … that i_2 said at round 2 that i_1 said at round 1 that its initial preference was to attack.' Moreover, the agents $i_1, \ldots, i_{t+1}$ must all be distinct. Indeed, it is easy to see that i_k must crash in round k before sending its message to i (but after sending its message to i_{k+1}), for $k = 1, \ldots, t$, for otherwise i must have gotten the message from i_k, contradicting the assumption that i first heard at round $t+1$ that some agent's initial preference was to attack. Since at most t agents can crash, it follows that i_{t+1}, the agent that sent the message to i, is not faulty, and thus sends the message to j. Thus, j also decides to attack. A symmetric argument shows that if j decides to attack, then so does i.

It should be clear that the correctness of this protocol depends on both the assumptions made: crash failures and synchrony. Suppose instead that *Byzantine* failures are allowed, so that faulty agents can deviate in arbitrary ways from the protocol; they may 'lie', send deceiving messages, and collude to fool the nonfaulty agents in the most malicious ways. In this case, the protocol will not work at all. In fact, it is known that agreement can be reached in the presence of Byzantine failures iff $t < n/3$, that is, iff fewer than a third of the agents can be faulty (Pease, Shostak and Lamport, 1980). The effect of asynchrony is even more devastating: in an asynchronous system, it is impossible to reach agreement using a deterministic protocol even if $t = 1$ (so that there is at most one failure) and only crash failures are allowed (Fischer, Lynch and Paterson, 1985). The problem in the asynchronous setting is that if none of the agents have heard from, say, agent 1, they have no way of knowing whether agent 1 is faulty or just slow. Interestingly, there are randomized algorithms (that is, behavioural strategies) that achieve agreement with arbitrarily high probability in an asynchronous setting [Ben-Or, 1983; Rabin, 1983].

Byzantine agreement can be viewed as a game where, at each step, an agent can either send a message or decide to attack or retreat. It is essentially a game between two teams, the nonfaulty agents and the faulty agents, whose composition is unknown (at least by the correct agents). To model it as a game in the more traditional sense, we could imagine that the nonfaulty agents are playing against a new player, the 'adversary'. One of the adversary's moves is that of 'corrupting' an agent: changing its type from 'nonfaulty' to 'faulty.' Once an agent is corrupted, what the adversary can do depends on the failure type being considered. In the case of crash failures, the adversary can decide which of a corrupted agent's messages will be delivered in the round in which the agent is corrupted; however, it cannot modify the messages themselves. In the case of Byzantine failures, the adversary essentially gets to make the moves for agents that have been corrupted; in particular, it can send arbitrary messages.

Why has the distributed systems literature not considered strategic behaviour in this game? Crash failures are used to model hardware and software failures; Byzantine failures are used to model random behaviour on the part of a system (for example, messages getting garbled in transit), software errors, and malicious adversaries (for example, hackers). With crash failures, it does not make sense to view the adversary's behaviour as strategic, since the adversary is not really viewed as having strategic interests. While it would certainly make sense, at least in principle, to consider the probability of failure (that is, the probability that the adversary corrupts an agent), this approach has by and large been avoided in the literature because it has proved difficult to characterize the probability distribution of failures over time. Computer components can perhaps be characterized as failing according to an exponential distribution (see Babaoglu, 1987, for an analysis of Byzantine agreement in such a setting), but crash failures can be caused by things other than component failures (faulty software, for example); these can be extremely difficult to characterize probabilistically. The problems are even worse when it comes to modelling random Byzantine behaviour.

With malicious Byzantine behaviour, it may well be reasonable to impute strategic behaviour to agents (or to an adversary controlling them). However, it is often difficult to characterize the payoffs of a malicious agent. The goals of the agents may vary from that of simply trying to delay a decision to that of causing disagreement. It is not clear what the appropriate payoffs should be for attaining these goals. Thus, the distributed systems literature has chosen to focus instead on algorithms that are guaranteed to satisfy the specification without making assumptions about the adversary's payoffs (or nature's probabilities, in the case of crash failures).

Recently, there has been some work on adding strategic concerns to standard problems in distributed computing; see, for example, Alvisi et al. (2005) and Halpern and Teague (2004). Moving in the other direction, there has also been some work on adding concerns of fault tolerance and asynchrony to standard problems in game theory; see, for example, Eliaz (2002), Monderer and Tennenholtz (1999a; 1999b) and the definitions in the next section. This seems to be an area that is ripe for further developments. One such development is the subject of the next section.

5 Implementing mediators

The question of whether a problem in a multiagent system that can be solved with a trusted mediator can be solved by just the agents in the system, without the mediator, has attracted a great deal of attention in both computer science (particularly in the cryptography community) and game theory. In cryptography, the focus on the problem has been on *secure multiparty computation*. Here it is assumed that each agent i has some private information x_i. Fix functions $f_1, \ldots, f_n$. The goal is to have agent i learn $f_i(x_1, \ldots, x_n)$ without learning anything about x_j for $j \neq i$ beyond what is revealed by the value of $f_i(x_1, \ldots, x_n)$. With a trusted mediator, this is trivial: each agent i just gives the mediator its private value x_i; the mediator then sends each agent i the value $f_i(x_1, \ldots, x_n)$. Work on multiparty computation (Goldreich, Micali and Wigderson, 1987; Shamir, Rivest and Adelman, 1981; Yao, 1982) provides conditions under which this can be done. In game theory, the focus has been on whether an equilibrium in a game with a mediator can be implemented using what is called *cheap talk* – that is, just by players communicating among themselves (cf. Barany, 1992; Ben-Porath, 2003; Forges, 1990; Gerardi, 2004; Heller, 2005; Urbano and Vila, 2004). As suggested in the previous section, the focus in the computer science literature has been in doing multiparty computation in the presence of possibly malicious adversaries, who do everything they can to subvert the computation, while in the game theory literature the focus has been on strategic agents. In recent work, Abraham et al. (2006) and Abraham, Dolev and Halpern (2007) considered deviations by both rational players, who have preferences and try to maximize them, and players who can viewed as malicious, although it is perhaps better to think of them as rational players whose utilities are not known by the other players or mechanism designer. I briefly sketch their results here; the following discussion is taken from Abraham, Dolev and Halpern (2007).

The idea of tolerating deviations by coalitions of players goes back to Aumann (1959); more recent refinements have been considered by Moreno and Wooders (1996). Aumann's definition is essentially the following.

Definition 1 $\vec{\sigma}$ is a k-resilient' equilibrium if, for all sets C of players with $|C| \leq k$, it is not the case that there exists a strategy $\vec{\tau}$ such that $u_i(\vec{\tau}_C, \vec{\sigma}_{-C}) > u_i(\vec{\sigma})$ for all $i \in C$.

As usual, the strategy $(\vec{\tau}_C, \vec{\sigma}_{-C})$ is the one where each player $i \in C$ plays τ_i and each player $i \notin C$ plays σ_i. As the prime notation suggests, this is not quite the definition we want to work with. The trouble with this definition is that it suggests that coalition members cannot communicate with each other during the game. Perhaps surprisingly, allowing communication can *prevent* certain equilibria (see Abraham, Dolev and Halpern, 2007, for an example). Since we should expect coalition members to communicate, the following definition seems to capture a more reasonable notion of resilient equilibrium. Let the cheap-talk extension of a game Γ be, roughly speaking, the game where players are allowed to communicate among themselves in addition to performing the actions of Γ and the payoffs are just as in Γ.

Definition 2 $\vec{\sigma}$ is a *k-resilient equilibrium* in a game Γ if $\vec{\sigma}$ is a k-resilient' equilibrium in the cheap-talk extension of Γ (where we identify the strategy σ_i in the game Γ with the strategy in the cheap-talk game where player i never sends any messages beyond those sent according to σ_i).

A standard assumption in game theory is that utilities are (commonly) known; when we are given a game we are also given each player's utility. When players make decisions, they can take other players' utilities into account. However, in large systems it seems almost invariably the case that there will be some fraction of users who do not respond to incentives the way we expect. For example, in a peer-to-peer network like Kazaa or Gnutella, it would seem that no rational agent should share files. Whether or not you can get a file depends only on whether other people share files. Moreover, there are disincentives for sharing (the possibility of lawsuits, use of bandwidth, and so on). Nevertheless, people do share files. However, studies of the Gnutella network have shown almost 70 per cent of users share no files and nearly 50 per cent of responses are from the top one per cent of sharing hosts (Adar and Huberman, 2000).

One reason that people might not respond as we expect is that they have utilities that are different from those we expect. Alternatively, the players may be irrational, or (if moves are made using a computer) they may

be playing using a faulty computer and thus not able to make the move they would like, or they may not understand how to get the computer to make the move they would like. Whatever the reason, it seems important to design strategies that tolerate such unanticipated behaviours, so that the payoffs of the users with 'standard' utilities do not get affected by the nonstandard players using different strategies. This can be viewed as a way of adding fault tolerance to equilibrium notions.

Definition 3 A joint strategy $\vec{\sigma}$ is *t-immune* if, for all $T \subseteq N$ with $|T| \leq t$, all joint strategies $\vec{\tau}$, and all $i \notin T$, we have $u_i(\vec{\sigma}_{-T}, \vec{\tau}_T) \geq u_i(\vec{\sigma})$.

The notion of t-immunity and k-resilience address different concerns. For t immunity, we consider the payoffs of the players not in T, and require that they are not worse due to deviation; for resilience, we consider the payoffs of players in C, and require that they are not better due to deviation. It is natural to combine both notions. Given a game Γ, let $\Gamma_{\vec{\tau}}^T$ be the game that is identical to Γ except that the players in T are fixed to playing strategy $\vec{\tau}$.

Definition 4 $\vec{\sigma}$ is a (k, t) *-robust* equilibrium if $\vec{\sigma}$ is t-immune and, for all $T \subseteq N$ such that $|T| \leq t$ and all joint strategies $\vec{\tau}$, $\vec{\sigma}_{-T}$ is a k-resilient strategy of $\Gamma_T^{\vec{\tau}}$.

To state the results of Abraham et al. (2006) and Abraham, Dolev and Halpern (2007) on implementing mediators, three games need to be considered: an *underlying game* Γ, an extension Γ_d of Γ with a mediator, and a cheap-talk extension Γ_{CT} of Γ. Assume that Γ is a *normal-form Bayesian game*: each player has a type from some type space with a known distribution over types, and the utilities of the agents depend on the types and actions taken. Roughly speaking, a cheap talk game *implements* a game with a mediator if it induces the same distribution over actions in the underlying game, for each type vector of the players. With this background, I can summarize the results of Abraham et al. (2006) and Abraham, Dolev and Halpern (2007).

- If $n > 3k + 3t$, a (k, t) -robust strategy $\vec{\sigma}$ with a mediator can be implemented using cheap talk (that is, there is a (k, t)-robust strategy $\vec{\sigma}'$ in a cheap talk game such that $\vec{\sigma}$ and $\vec{\sigma}'$ induce the same distribution over actions in the underlying game). Moreover, the implementation requires no knowledge of other agents' utilities, and the cheap talk protocol has bounded running time that does not depend on the utilities.
- If $n \leq 3k + 3t$, then, in general, mediators cannot be implemented using cheap talk without knowledge of other agents' utilities. Moreover, even if other agents' utilities are known, mediators cannot, in general, be implemented without having a $(k+t)$-punishment strategy (that is, a strategy that, if used by all but at

most $(k+t)$ players, guarantees that every player gets a worse outcome than they do with the equilibrium strategy) nor with bounded running time.

- If $n>2k+3t$, then mediators can be implemented using cheap talk if there is a punishment strategy (and utilities are known) in finite expected running time that does not depend on the utilities.
- If $n \leq 2k + 3t$ then mediators cannot, in general, be implemented, even if there is a punishment strategy and utilities are known.
- If $n>2k + 2t$ and there are broadcast channels then, for all ε, mediators can be ε-implemented (intuitively, there is an implementation where players get utility within ε of what they could get by deviating) using cheap talk, with bounded expected running time that does not depend on the utilities.
- If $n \leq 2k + 2t$, then mediators cannot, in general, be ε-implemented, even with broadcast channels. Moreover, even assuming cryptography and polynomially bounded players, the expected running time of an implementation depends on the utility functions of the players and ε.
- If $n>k + 3t$, then, assuming cryptography and polynomially bounded players, mediators can be ε-implemented using cheap talk, but if $n \leq 2k + 2t$, then the running time depends on the utilities in the game and ε.
- If $n \leq k + 3t$, then even assuming cryptography, polynomially bounded players, and a $(k + t)$-punishment strategy, mediators cannot, in general, be ε-implemented using cheap talk.
- If $n>k + t$, then, assuming cryptography, polynomially bounded players, and a public-key infrastructure (PKI), we can ε-implement a mediator.

The proof of these results makes heavy use of techniques from computer science. All the possibility results showing that mediators can be implemented use techniques from secure multiparty computation. The results showing that if $n \leq 3k + 3t$, then we cannot implement a mediator without knowing utilities, and that, even if utilities are known, a punishment strategy is required, use the fact that Byzantine agreement cannot be reached if $t<n/3$; the impossibility result for $n \leq 2k + 3t$ also uses a variant of Byzantine agreement.

A related line of work considers implementing mediators assuming stronger primitives (which cannot be implemented in computer networks); see Izmalkov, Micali and Lepinski (2005) and Lepinski et al. (2004) for details.

6 Other topics

There are many more areas of interaction between computer science than I have indicated in this brief survey. I briefly mention a few others here.

6.1 Interactive epistemology

Since the publication of Aumann's (1976) seminal paper, there has been a great deal of activity in trying to understand the role of knowledge in games, and providing epistemic analyses of solution concepts; see Battigalli and Bonanno (1999) for a survey. In computer science, there has been a parallel literature applying epistemic logic to reason about distributed computation. One focus of this work has been on characterizing the level of knowledge needed to solve certain problems. For example, to achieve Byzantine agreement common knowledge among the nonfaulty agents of an initial value is necessary and sufficient. More generally, in a precise sense, common knowledge is necessary and sufficient for coordination. Another focus has been on defining logics that capture the reasoning of resource-bounded agents. A number of approaches have been considered. Perhaps the most common considers logics for reasoning about *awareness*, where an agent may not be aware of certain concepts, and can know something only if he is aware of it. This topic has been explored in both computer science and game theory; see Dekel, Lipman and Rustichini (1998), Fagin and Halpern (1988), Halpern (2001), Halpern and Rêgo (2007), Heifetz, Meier and Schipper (2006), and Modica and Rustichini (1994; 1999) for some of the work in this active area. Another approach, so far considered only by computer scientists, involves *algorithmic knowledge*, which takes seriously the assumption that agents must explicitly compute what they know. See Fagin et al. (1995) for an overview of the work in epistemic logic in computer science.

6.2 Network growth

If we view networks as being built by selfish players (who decide whether or not to build links), what will the resulting network look like? How does the growth of the network affect its functionality? For example, how easily will influence spread through the network? How easy is it to route traffic? See Fabrikant et al. (2003) and Kempe, Kleinberg and Tardos (2003) for some recent computer science work in this burgeoning area.

6.3 Efficient representation of games

Game theory has typically focused on 'small' games, often two- or three-player games, that are easy to describe, such as Prisoner's Dilemma, in order to understand subtleties regarding basic issues such as rationality. To the extent that game theory is used to tackle larger, more practical problems, it will become important to find efficient techniques for describing and analysing games. By way of analogy, $2^n - 1$ numbers are needed to describe a probability distribution on a space characterized by n binary random variables. For $n = 100$ (not an unreasonable number in practical situations), it is impossible to write down the probability distribution in the obvious way, let alone do computations with it. The same issues will surely arise in large games. Computer scientists use graphical approaches, such as *Bayesian networks* and *Markov networks* (Pearl, 1988), for representing and manipulating probability measures on large spaces. Similar techniques seem applicable to games; see, for example, Kearns, Littman and Singh (2001), Koller and Milch (2001), and La Mura (2000) for specific approaches, and Kearns, (2007) for a recent overview. Note that representation is also an issue when we consider the complexity of problems such as computing Nash or correlated equilibria. The complexity of a problem is a function of the size of the input, and the size of the input (which in this case is a description of the game) depends on how the input is represented.

6.4 Learning in games

There has been a great deal of work in both computer science and game theory on learning to play well in different settings (see Fudenberg and Levine, 1998, for an overview of the work in game theory). One line of research in computer science has involved learning to play optimally in a reinforcement-learning setting, where an agent interacts with an unknown (but fixed) environment. The agent then faces a fundamental tradeoff between *exploration* and *exploitation*. The question is how long it takes to learn to play well (to get a reward within some fixed ε of optimal); see Brafman and Tennenholtz (2002) and Kearns and Singh (1998) for the current state of the art. A related question is efficiently finding a strategy minimizes *regret* – that is, finding a strategy that is guaranteed to do not much worse than the best strategy would have done in hindsight (that is, even knowing what the opponent would have done). See Blum and Mansour (2007) for a recent overview of work on this problem.

JOSEPH Y. HALPERN

See also **computation of general equilibria; computational methods in econometrics; computing in mechanism design; data mining; electronic commerce; epistemic game theory: an overview; epistemic game theory: beliefs and types; mathematics of networks; mechanism design (new developments); rationality, bounded; voting paradoxes.**

The work for this article was supported in part by NSF under grants CTC-0208535 and ITR-0325453, by ONR under grant N00014-02-1-0455, by the DoD Multidisciplinary University Research Initiative (MURI) program administered by the ONR under grants N00014-01-1-0795 and N00014-04-1-0725, and by AFOSR under grant F49620-02-1-0101. Thanks to Larry Blume, Christos Papadimitriou, Ilya Segal, Éva Tardos, and Moshe Tennenholtz for useful comments.

Bibliography

Abraham, I., Dolev, D., Gonen, R. and Halpern, J. 2006. Distributed computing meets game theory: robust mechanisms for rational secret sharing and multiparty computation. In *Proceedings of the 25th ACM Symposium on Principles of Distributed Computing*, 53–62.

Abraham, I., Dolev, D. and Halpern, J. 2007. Lower bounds on implementing robust and resilient mediators. Available at http://arxiv.org/abs/0704.3646, accessed 19 June 2007.

Adar, E. and Huberman, B. 2000. Free riding on Gnutella. *First Monday* 5(10).

Alvisi, L., Ayer, A.S., Clement, A., Dahlin, M., Martin, J.P. and Porth, C. 2005. BAR fault tolerance for cooperative services. In *Proceedings of the 20th ACM Symposium on Operating Systems Principles (SOSP 2005)*, 45–58.

Archer, A. and Tardos, É. 2001. Truthful mechanisms for one-parameter agents. In *Proceedings of the 42nd IEEE Symposium on Foundations of Computer Science*, 482–91.

Archer, A. and Tardos, É. 2002. Frugal path mechanisms. In *Proceedings of the 13th ACM-SIAM Symposium on Discrete Algorithms*, 991–9.

Aumann, R. 1959. Acceptable points in general cooperative *n*-person games. In *Contributions to the Theory of Games IV*, ed. A. Tucker and R. Luce. Princeton, NJ: Princeton University Press, pp. 287–324.

Aumann, R.J. 1976. Agreeing to disagree. *Annals of Statistics* 4, 1236–9.

Aumann, R.J. 1987. Correlated equilibrium as an expression of Bayesian rationality. *Econometrica* 55, 1–18.

Axelrod, R. 1984. *The Evolution of Cooperation*. New York: Basic Books.

Babaoglu, O. 1987. On the reliability of consensus-based fault-tolerant distributed computing systems. *ACM Translation on Computer Systems* 5, 394–416.

Barany, I. 1992. Fair distribution protocols or how the players replace fortune. *Mathematics of Operations Research* 17, 327–40.

Battigalli, P. and Bonanno, G. 1999. Recent results on belief, knowledge and the epistemic foundations of game theory. *Research in Economics* 53, 149–225.

Ben-Or, M. 1983. Another advantage of free choice: completely asynchronous agreement protocols. In *Proceedings of the Second ACM Symposium on Principles of Distributed Computing*, 27–30.

Ben-Porath, E. 2003. Cheap talk in games with incomplete information. *Journal of Economic Theory* 108, 45–71.

Blum, A. and Mansour, Y. 2007. Learning, regret minimization, and equilibria. In *Algorithmic Game Theory*, ed. N. Nisan, T. Roughgarden, É. Tardos, and V. Vazirani. Cambridge: Cambridge University Press.

Blum, B., Shelton, C.R. and Koller, D. 2003. A continuation method for Nash equilibria in structured games. In *Proceedings of the 18th International Joint Conference on Artificial Intelligence (IJCAI '03)*, 757–64.

Brafman, R.I. and Tennenholtz, M. 2002. R-MAX: a general polynomial time algorithm for near-optimal reinforcement learning. *Journal of Machine Learning Research* 3, 213–31.

Chen, X. and Deng, X. 2006. Settling the complexity of 2-player Nash equilibrium. In *Proceedings of the 47th IEEE Symposium on Foundations of Computer Science*, 261–72.

Chor, B. and Dwork, C. 1989. Randomization in Byzantine agreement. In *Advances in Computing Research 5: Randomness and Computation*, ed. S. Micali. Greenwich, CT: JAI Press, pp. 443–97.

Chu, F. and Halpern, J.Y. 2001. A decision-theoretic approach to reliable message delivery. *Distributed Computing* 14, 359–89.

Clarke, E.H. 1971. Multipart pricing of public goods. *Public Choice* 11, 17–33.

Conitzer, V. and Sandholm, T. 2003. Complexity results about Nash equilibria. In *Proceedings of the 18th International Joint Conference on Artificial Intelligence (IJCAI '03)*, 765–71.

Conitzer, V., Sandholm, T. and Lang, J. 2007. When are elections with few candidates hard to manipulate? *Journal of the ACM*. 54(3), Article 14.

Cramton, P., Shoham, Y. and Steinberg, R. 2006. *Combinatorial Auctions*. Cambridge, MA: MIT Press.

Daskalis, C., Goldberg, P.W. and Papadimitriou, C.H. 2006. The complexity of computing a Nash equilibrium. In *Proceedings of the 38th ACM Symposium on Theory of Computing*, 71–8.

Dekel, E., Lipman, B. and Rustichini, A. 1998. Standard state-space models preclude unawareness. *Econometrica* 66, 159–73.

Eliaz, K. 2002. Fault-tolerant implementation. *Review of Economic Studies* 69, 589–610.

Fabrikant, A., Luthra, A., Maneva, E., Papadimitriou, C.H. and Shenker, S. 2003. On a network creation game. In *Proceedings of the 22nd ACM Symposium on Principles of Distributed Computing*, 347–51.

Fagin, R. and Halpern, J.Y. 1988. Belief, awareness, and limited reasoning. *Artificial Intelligence* 34, 39–76.

Fagin, R., Halpern, J.Y., Moses, Y. and Vardi, M.Y. 1995. *Reasoning about Knowledge*. Cambridge, MA: MIT Press. A slightly revised paperback version was published in 2003.

Feigenbaum, J., Papadimitriou, C. and Shenker, S. 2000. Sharing the cost of multicast transmissions (preliminary version). In *Proceedings of the 32nd ACM Symposium on Theory of Computing*, 218–27.

Fischer, M.J. 1983. The consensus problem in unreliable distributed systems. In *Foundations of Computation Theory*, vol. 185, ed. M. Karpinski. Lecture Notes in Computer Science. Berlin/New York: Springer, pp. 127–40.

Fischer, M.J., Lynch, N.A. and Paterson, M.S. 1985. Impossibility of distributed consensus with one faulty processor. *Journal of the ACM* 32, 374–82.

Forges, F. 1990. Universal mechanisms. *Econometrica* 58, 1341–64.

Fudenberg, D. and Levine, D. 1998. *The Theory of Learning in Games*. Cambridge, MA: MIT Press.

Gerardi, D. 2004. Unmediated communication in games with complete and incomplete information. *Journal of Economic Theory* 114, 104–31.

Gibbard, A. 1973. Manipulation of voting schemes. *Econometrica* 41, 587–602.

Gilboa, I. and Zemel, E. 1989. Nash and correlated equilibrium: some complexity considerations. *Games and Economic Behavior* 1, 80–93.

Goldreich, O., Micali, S. and Wigderson, A. 1987. How to play any mental game. In *Proceedings of the 19th ACM Symposium on Theory of Computing*, 218–29.

Govindan, S. and Wilson, R. 2003. A global Newton method to compute Nash equilibria. *Journal of Economic Theory* 110, 65–86.

Groves, T. 1973. Incentives in teams. *Econometrica* 41, 617–31.

Halldórsson, M.M., Halpern, J.Y., Li, L. and Mirrokni, V. 2004. On spectrum sharing games. In *Proceedings of the 23rd ACM Symposium on Principles of Distributed Computing*, 107–14.

Halpern, J.Y. 2001. Alternative semantics for unawareness. *Games and Economic Behavior* 37, 321–39.

Halpern, J.Y. 2003. A computer scientist looks at game theory. *Games and Economic Behavior* 45, 114–32.

Halpern, J.Y. and Rêgo, L.C. 2007. Reasoning about knowledge of unawareness. *Games and Economic Behavior*. Also available at http://arxiv.org/abs/cs.LO/0603020, accessed 24 June 2007.

Halpern, J.Y. and Teague, V. 2004. Rational secret sharing and multiparty computation: extended abstract. In *Proceedings of the 36th ACM Symposium on Theory of Computing*, 623–32.

Heifetz, A., Meier, M. and Schipper, B. 2006. Interactive unawareness. *Journal of Economic Theory* 130, 78–94.

Heller, Y. 2005. A minority-proof cheap-talk protocol. Unpublished manuscript.

Hopcroft, J.E. and Ullman, J.D. 1979. *Introduction to Automata Theory, Languages and Computation*. New York: Addison-Wesley.

Hurwicz, L. 1977. On the dimensional requirements of informationally decentralized Pareto satisfactory processes. In *Studies in Resource AllocationProcesses*, ed. K.J. Arrow and L. Hurwicz. New York: Cambridge University Press, pp. 413–24.

Izmalkov, S., Micali, S. and Lepinski, M. 2005. Rational secure computation and ideal mechanism design. In *Proceedings of the 46th IEEE Symposium Foundations of Computer Science*, 585–95.

Kearns, M. 2007. Graphical games. In *Algorithmic Game Theory*, ed. N. Nisan, T. Roughgarden, É. Tardos and V. Vazirani. Cambridge: Cambridge University Press.

Kearns, M., Littman, M.L. and Singh, S.P. 2001. Graphical models for game theory. In *Proceedings of the 17th Conference on Uncertainty in Artificial Intelligence (UAI 2001)*, 253–60.

Kearns, M.J. and Reiter, M.K. 2005. *Proceedings of the Sixth ACM Conference on Electronic Commerce (EC '05)*. New York: ACM. Table of contents available at http://www.informatik.uni-trier.de/~ley/db/conf/sigecom/sigecom2005.html, accessed 19 June 2007.

Kearns, M. and Singh, S.P. 1998. Near-optimal reinforcement learning in polynomial time. In *Proceedings of the 15th International Conference on Machine Learning*, pp. 260–8.

Kempe, D., Kleinberg, J. and Tardos, É. 2003. Maximizing the spread of influence through a social network. In *Proceedings of the Ninth ACM SIGKDD International Conference on Knowledge Discovery and Data Mining*, 137–46.

Kfir-Dahav, N.E., Monderer, D. and Tennenholtz, M. 2000. Mechanism design for resource-bounded agents. In *International Conference on Multiagent Systems*, 309–16.

Koller, D. and Milch, B. 2001. Structured models for multiagent interactions. In *Theoretical Aspects of Rationality and Knowledge: Proceedings of the Eighth Conference (TARK 2001)*, 233–48.

Koutsoupias, E. and Papadimitriou, C.H. 1999. Worst-case equilibria. In *Proceedings of the 16th Conference on Theoretical Aspects of Computer Science*, vol. 1563, Lecture Notes in Computer Science. Berlin: Springer, pp. 404–13.

Kushilevitz, E. and Nisan, N. 1997. *Communication Complexity*. Cambridge: Cambridge University Press.

La Mura, P. 2000. Game networks. In *Proceedings of th 16th Conference on Uncertainty in Artificial Intelligence (UAI 2000)*, 335–42.

Lehmann, D., Müller, R. and Sandholm, T. 2006. The winner determination problem. In *Combinatorial Auctions*, ed. P. Cramton, Y. Shoham, and R. Steinberg. Cambridge, MA: MIT Press.

Lemke, C.E. and Howson, J.J.T. 1964. Equilibrium points of bimatrix games. *Journal of the Society for Industrial and Applied Mathematics* 12, 413–23.

Lepinski, M., Micali, S., Peikert, C. and Shelat, A. 2004. Completely fair SFE and coalition-safe cheap talk. In *Proceedings of the 23rd ACM Symposium Principles of Distributed Computing*, 1–10.

Linial, N. 1994. Games computers play: Game-theoretic aspects of computing. In *Handbook of Game Theory*, vol. 2, ed. R.J. Aumann and S. Hart. Amsterdam: North-Holland.

Modica, S. and Rustichini, A. 1994. Awareness and partitional information structures. *Theory and Decision* 37, 107–24.

Modica, S. and Rustichini, A. 1999. Unawareness and partitional information structures. *Games and Economic Behavior* 27, 265–98.

Monderer, D. and Tennenholtz, M. 1999a. Distributed games. *Games and Economic Behavior* 28, 55–72.

Monderer, D. and Tennenholtz, M. 1999b. Distributed Games: From Mechanisms to Protocols. In *Proceedings of the 16th National Conference on Artificial Intelligence (AAAI '99)*, 32–7.

Moreno, D. and Wooders, J. 1996. Coalition-proof equilibrium. *Games and Economic Behavior* 17, 80–112.

Mount, K. and Reiter, S. 1974. The informational size of message spaces. *Journal of Economic Theory* 8, 161–92.

Nash, J. 1950. Equilibrium points in *n*-person games. *Proceedings of the National Academy of Sciences* 36, 48–9.

Neyman, A. 1985. Bounded complexity justifies cooperation in finitely repeated Prisoner's Dilemma. *Economic Letters* 19, 227–9.

Nisan, N. 2006. Bidding languages for combinatorial auctions. In *Combinatorial Auctions*. Cambridge, MA: MIT Press.

Nisan, N. and Ronen, A. 2000. Computationally feasible VCG mechanisms. In *Proceedings of the Second ACM Conference on Electronic Commerce (EC '00)*, 242–52.

Nisan, N. and Ronen, A. 2001. Algorithmic mechanism design. *Games and Economic Behavior* 35, 166–96.

Nisan, N., Roughgarden, T., Tardos, É. and Vazirani, V. 2007. *Algorithmic Game Theory*. Cambridge, UK: Cambridge University Press.

Papadimitriou, C.H. 1994a. *Computational Complexity*. Reading, MA: Adison-Wesley.

Papadimitriou, C.H. 1994b. On the complexity of the parity argument and other inefficient proofs of existence. *Journal of Computer and System Sciences* 48, 498–532.

Papadimitriou, C.H. 2001. Algorithms, games, and the internet. In *Proceedings of the 33rd ACM Symposium on Theory of Computing*. 749–53.

Papadimitriou, C.H. 2005. Computing correlated equilibria in multiplayer games. In *Proceedings of the 37th ACM Symposium on Theory of Computing*, 49–56.

Papadimitriou, C.H. 2007. The complexity of finding Nash equilibria. In *Algorithmic Game Theory*, ed. N. Nisan, T. Roughgarden, É. Tardos and V. Vazirani. Cambridge: Cambridge University Press.

Papadimitriou, C.H. and Roughgarden, T. 2005. Computing equilibria in multi-player games. In *Proceedings of the 16th ACM–SIAM Symposium on Discrete Algorithms*, 82–91.

Papadimitriou, C.H. and Yannakakis, M. 1994. On complexity as bounded rationality. In *Proceedings of the 26th ACM Symposium on Theory of Computing*, 726–33.

Pearl, J. 1988. *Probabilistic Reasoning in Intelligent Systems*. San Francisco: Morgan Kaufmann.

Pease, M., Shostak, R. and Lamport, L. 1980. Reaching agreement in the presence of faults. *Journal of the ACM* 27, 228–34.

Porter, R., Nudelman, E. and Shoham, Y. 2004. Simple search methods for finding a Nash equilibrium. In *Proceedings of the 21st National Conference on Artificial Intelligence (AAAI '04)*, 664–9.

Rabin, M.O. 1983. Randomized Byzantine generals. In *Proceedings of the 24th IEEE Symposium on Foundations of Computer Science*, 403–9.

Roughgarden, T. and Tardos, É. 2002. How bad is selfish routing? *Journal of the ACM* 49, 236–59.

Rubinstein, A. 1986. Finite automata play the repeated prisoner's dilemma. *Journal of Economic Theory* 39, 83–96.

Rubinstein, A. 1998. *Modeling Bounded Rationality.* Cambridge, MA: MIT Press.

Sandholm, T. and Yokoo, M. 2003. *The Second International Joint Conference on Autonomous Agents and Multiagent Systems (AAMAS 2003).* Table of contents available at http://www.informatik.uni-trier.de/~ley/db/conf/atal/aamas2003.html, accessed 25 June 2007.

Satterthwaite, M. 1975. Strategy-proofness and Arrow's conditions: existence and correspondence theorems for voting procedures and social welfare functions. *Journal of Economic Theory* 10, 187–217.

Segal, I. 2006. Communication in economic mechanisms. In *Advances in Economics and Econometrics: Theory and Application, Ninth World Congress (Econometric Society Monographs,* ed. R. Blundell, W.K. Newey and T. Persson. Cambridge: Cambridge University Press, pp. 222–68.

Shamir, A., Rivest, R.L. and Adelman, L. 1981. Mental poker. In *The Mathematical Gardner,* ed. D.A. Klarner. Boston, MA: Prindle, Weber, and Schmidt, pp. 37–43.

Tennenholtz, M. 2002. Competitive safety analysis: robust decision-making in multi-agent systems. *Journal of A.I. Research* 17, 363–78.

Urbano, A. and Vila, J.E. 2002. Computational complexity and communication: coordination in two-player games. *Econometrica* 70, 1893–927.

Urbano, A. and Vila, J.E. 2004. Computationally restricted unmediated talk under incomplete information. *Economic Theory* 23, 283–320.

Vetta, A. 2002. Nash equilibria in competitive societies, with applications to facility location, traffic routing and auctions. In *Proceedings of the 43rd IEEE Symposium on Foundations of Computer Science,* 416–25.

Vickrey, W. 1961. Counterspeculation, auctions and competitive sealed tenders. *Journal of Finance* 16, 8–37.

Yao, A. 1982. Protocols for secure computation (extended abstract). In *Proceedings of the 23rd IEEE Symposium on Foundations of Computer Science,* 160–4.

computing in mechanism design

1 Introduction

Computational issues in mechanism design are important, but have received insufficient research interest until recently. Limited computing hinders mechanism design in several ways, and presents deep strategic interactions between computing and incentives. On the bright side, novel algorithms and increasing computing power have enabled better mechanisms. Perhaps most interestingly, limited computing of the agents can be used as a tool to implement mechanisms that would not be implementable among computationally unlimited agents. This article briefly reviews some of the key ideas, with the goal of alerting the reader to the importance of these issues and hopefully spurring future research.

I will discuss computing by the *centre,* such as an auction server or vote aggregator, in Section 2. Then, in Section 3, I will address the *agents'* computing, be they human or software.

2 Computing by the centre

Computing by the centre plays significant roles in mechanism design. In the following three subsections I will review three prominent directions.

2.1 Executing expressive mechanisms

As algorithms have advanced drastically and computing power has increased, it has become feasible to field mechanisms that were previously impractical. The most famous example is a *combinatorial auction (CA).* In a CA, there are multiple distinguishable items for sale, and the bidders can submit bids on self-selected packages of the items. (Sometimes each bidder is also allowed to submit exclusivity constraints of different forms among his bids.) This increase in the expressiveness of the bids drastically reduces the strategic complexity that bidders face. For one, it removes the exposure problems that bidders face when they have preferences over packages but in traditional auctions are allowed to submit bids on individual items only.

CAs shift the computational burden from the bidders to the centre. There is an associated gain because the centre has all the information in hand to optimize while in traditional auctions the bidders only have estimated projected (probabilistic) information about how others will bid. Thus CAs yield more efficient allocations.

On the downside, the centre's task of determining the winners in a CA (deciding which bids to accept so as to maximize the sum of the accepted bids' prices subject to not selling any item to more than one bid) is a complex combinatorial optimization problem, even without exclusivity constraints among bids. Three main approaches have been studied for solving it.

1. *Optimal winner determination using some form of tree search.* For a review, see Sandholm (2006). The advantage is that the bidding language is not restricted and the optimal solution is found. The downside is that no optimal winner determination algorithm can run in polynomial time in the size of the problem instance in the worst case, because the problem is $\mathcal{NP}$-complete (Rothkopf, Pekěc and Harstad, 1998). ($\mathcal{NP}$-complete problems are problems for which the fastest known algorithms take exponential time in the size of the problem instance in the worst case. $\mathcal{P}$ is the class of easy problems solvable in polynomial time. The statement of winner determination not being solvable in polynomial time in the worst case relies on the usual assumption $\mathcal{P} \neq \mathcal{NP}$. This is an open question in complexity theory, but is widely believed to be true. If false, that would have sweeping implications throughout computer science.)

2. *Approximate winner determination.* The advantage is that many approximation algorithms run in polynomial time in the size of the instance even in the worst case. For reviews of such algorithms, see Sandholm (2002a) and Lehmann, Müller and Sandholm (2006). (Other suboptimal algorithms do not have such time guarantees, such as local search, stochastic local search, simulated annealing, genetic algorithms and tabu search.) The downside is that the solution is sometimes far from optimal: no such algorithm can always find a solution that is within a factor

$$\min\left\{ \#bids^{1-\varepsilon}, \#items^{\frac{1}{2}-\varepsilon} \right\} \qquad (1)$$

of optimal (Sandholm, 2002a). (This assumes $\mathscr{LPP} \neq \mathscr{NP}$. It is widely believed that these two complexity classes are indeed unequal.) For example, with just nine items for sale, no such algorithm can extract even 33 per cent of the available revenue from the bids in the worst case. With 81 items, that drops to 11 per cent.

3. *Restricting the bidding language* so much that optimal (within the restricted language) winner determination can be conducted in worst-case polynomial time. For a review, see Müller (2006). For example, if each package bid is only allowed to include at most two items, then winners can be determined in worst-case polynomial time (Rothkopf, Pekěc and Harstad, 1998). The downside is that bidders have to shoehorn their preferences into a restricted bidding language; this gives rise to similar problems as in non-combinatorial mechanisms for multi-item auctions: exposure problems, need to speculate how others will bid, inefficient allocation, and so on.

Truthful bidding can be made a dominant strategy by applying the *Vickrey–Clarke–Groves* (*VCG*) *mechanism* to a CA. Such incentive compatibility removes strategic complexity of the bidders. The mechanism works as follows. The optimal allocation is used, but the bidders do not pay their winning bids. Instead each bidder pays the amount of value he takes away from the others by taking some of the items. This value is measured as the difference between the others' winning bids' prices and what the others' winning bids' prices would have been had the agent not submitted any bids. This mechanism can be executed by determining the winners once overall, and once for each agent removed in turn. (This may be accomplishable with less computing. For example, in certain network auctions it can be done in the same asymptotic complexity as one winner determination – Hershberger and Suri, 2001.)

Very few canonical CAs have found their way to practice. However, auctions with richer bid expressiveness forms (that are more natural in the given application and more concise) and that support expressiveness also by the bid taker have made a major breakthrough into practice (Sandholm, 2007; Bichler et al., 2006). This is sometimes called *expressive commerce* to distinguish it from vanilla CAs. The widest area of application is currently industrial sourcing. Tens of billions of dollars worth of materials, transportation and services are being sourced annually using such mechanisms, yielding billions of dollars in efficiency improvements. The bidders' expressiveness forms include different forms of flexible package bids, conditional discounts, discount schedules, side constraints (such as capacity constraints), and even hundreds of cost drivers (for example, fixed costs, variable costs, trans-shipment costs and costs associated with changes). The item specifications can also be left partially open, and the bidders can specify some of the item attributes (delivery date, insurance terms, and so on). in alternate ways. The bid taker also specifies preferences and constraints. Winner determination then not only decides who wins what, but also automatically configures the items. In some of these events it also configures the supply chain several levels deep as a side effect. On the high end, such an auction can have tens of thousands of items (multiple units of each), millions of bids, and hundreds of thousands of side constraints. Expressive mechanisms have also been designed for settings beyond auctions, such as combinatorial exchanges, charity donations and settings with externalities.

Basically all of the fielded expressive auctions use the simple pay-your-winning-bids pricing rule. There are numerous important reasons why few, if any, use the VCG mechanism. It can lead to low revenue. It is vulnerable to collusion. Bidders would not tell the truth because they do not want to reveal their cost structures, which the auctioneer could exploit the next time the auction is conducted, and so on (Sandholm, 2000; Rothkopf, 2007).

Basically all of the fielded expressive auctions use tree search for winner determination. In practice, modern tree search algorithms for the problem scale to the large and winners can be determined optimally. If winner determination were not done optimally in a CA, the VCG mechanism can lose its truth-dominance property (Sandholm, 2002b). In fact, any truthful suboptimal VCG-based mechanism for CAs is unreasonable in the sense that it sometimes does not allocate an item to a bidder even if he is the only bidder whose bids assign non-zero value to that item (Nisan and Ronen, 2000).

2.2 Algorithmic mechanism design
Motivated by the worry that some instances of *NP*-hard problems may not be solvable within reasonable time, a common research direction in theory of computing is approximation algorithms. They trade off solution quality for a guarantee that even in the worst case, the algorithm runs in polynomial time in the size of the input.

Analogously, Nisan and Ronen (2001) proposed *algorithmic mechanism design*: designing approximately optimal mechanisms that take the centre a polynomial number of computing steps even in the worst case. However, this is more difficult than designing approximately optimal algorithms because the mechanism has to motivate the agents to tell the truth.

Lehmann, O'Callaghan and Shoham (2002) studied this for CAs with single-minded bidders (each bidder being only interested in one specific package of items). They present a fast greedy algorithm that guarantees a solution within a factor $\sqrt{\#items}$ of optimal. They show that the algorithm is not incentive compatible with VCG pricing, but is with their custom pricing scheme. They also identify sufficient conditions for any (approximate) mechanism to be incentive compatible (see also Kfir-Dahav, Monderer and Tennenholtz, 2000). There has been substantial follow-on work on subclasses of single-minded CAs.

Lavi and Swamy (2005) developed a technique for a range of packing problems with which any k-approximation algorithm (that is, algorithm that guarantees that the solution is within a factor k of optimal) that also bounds the integrality gap of the linear programming (LP) relaxation of the problem by k can be used to construct a k-approximation mechanism. The LP solution, scaled down by k, can be represented as a convex combination of integer solutions, and viewing this convex combination as specifying a probability distribution over integer solutions begets a VCG-based randomized mechanism that is truthful in expectation. For CAs with general valuations, this yields an $O(\sqrt{\#items})$-approximate mechanism.

In a different direction, several mechanisms have been proposed where the agents can help the centre find better outcomes. This is done either by giving the agents the information to do the centre's computing (Banks, Ledyard and Porter, 1989; Land, Powell and Steinberg, 2006; Parkes and Shneidman, 2004), or by allowing the agents to change what they told the mechanism based on the mechanism's output and potentially also based on what other agents told the mechanism (Nisan and Ronen, 2000). In VCG-based mechanisms, an agent benefits from lying only if the lie causes the mechanism to find an outcome that is better overall.

2.3 Automated mechanism design
Conitzer and Sandholm (2002) proposed the idea of *automated mechanism design*: having a computer, rather than a human, design the mechanism. Because human effort is eliminated, this enables custom design of mechanisms for every setting. The setting can be described by the agents' (discretized) type spaces, the designer's prior over types, the desired notion of incentive compatibility (for example, dominant strategies vs. Bayes–Nash implementation), the desired notion of participation constraints (for example, *ex interim*, *ex post* or none), whether payments are allowed, and whether the mechanism is allowed to use randomization.) This can yield better mechanisms for previously studied settings because the mechanism is designed for the specific setting rather than a class of settings. It can also be used for settings not previously studied in mechanism design.

For almost all natural (linear) objectives, all variants of the design problem are $\mathcal{NP}$-complete if the mechanism is not allowed to use randomization, but randomized mechanisms can be constructed for all these settings in polynomial time using linear programming. Custom algorithms have been developed for some problems in each of these two categories. (Even the latter category warrants research. While the linear programme is polynomial in the size of the input, the input itself can be exponential in the number of agents.) Structured representations of the problem can also make the design process drastically faster.

Beyond the general setting, automated mechanism design has been applied to specific settings, such as creating revenue-maximizing CAs (without the need to discretize types) – Likhodedov and Sandholm, 2005 – (a recognized problem that eludes analytical characterization; even the two-item case is open), reputation systems (Jurca and Faltings, 2006), safe exchange mechanisms (Sandholm and Ferrandon, 2000), and supply chain settings (Vorobeychik, Kiekintveld and Wellman, 2006). Automated mechanism design software has recently also been adopted by several mechanism design theoreticians to speed up their research.

It turns out that even *multistage mechanisms* can be designed automatically (Sandholm, Conitzer and Boutilier, 2007). Furthermore, automated mechanism design has been applied to the design of *online mechanisms* (Hajiaghayi, Kleinberg and Sandholm, 2007), that is, mechanisms that execute while the world changes – for example, agents enter and exit the system.

3 Computing by the agents
I will now move to discussing computing by the agents.

3.1 Mechanisms that are hard to manipulate
This section demonstrates that one can use the fact that agents are computationally limited to achieve things that are not achievable via any mechanism among perfectly rational agents.

A seminal negative result, the *Gibbard–Satterthwaite theorem*, states that if there are three or more candidates, then in any non-dictatorial voting scheme there are candidate rankings of the other voters, and preferences of the agent, under which the agent is better off voting manipulatively than truthfully. One avenue around this impossibility is to construct desirable general non-dictatorial voting protocols under which *finding* a beneficial manipulation is prohibitively hard computationally.

There are two natural alternative goals of manipulation. In *constructive manipulation*, the manipulator tries to find an order of candidates that he can reveal so that his favourite candidate wins. In *destructive manipulation*, the manipulator tries to find an order of candidates that he can reveal so that his hated candidate does not win. These are special cases of the utility-theoretic notion of improving one's utility, so the hardness results, discussed below, carry over to the usual utility-theoretic setting.

Unfortunately, finding a constructive manipulation is easy (in $\mathcal{P}$) for the *plurality, Borda* and *maximin* voting rules (Bartholdi, Tovey and Trick, 1989), which are commonly used. On the bright side, constructive manipulation of the *single transferable vote* (STV) protocol is $\mathcal{NP}$-hard (Bartholdi and Orlin, 1991) (as is manipulation of the *second order Copeland* protocol (Bartholdi, Tovey and Trick, 1989), but that hardness is driven solely by the tie-breaking rule). Even better, there is a systematic methodology for slightly tweaking voting protocols that are easy to manipulate, so that they become hard to manipulate (Conitzer and Sandholm, 2003). Specifically, before the original protocol is executed, one pairwise elimination round is executed among the candidates, and only the winning candidates survive to the original protocol. This makes the protocols $\mathcal{NP}$-hard, #$\mathcal{P}$-hard (#$\mathcal{P}$-hard problems are at least as hard as counting the number of solutions to a problem in $\mathcal{P}$), or even $\mathcal{PSPACE}$-hard ($\mathcal{PSPACE}$-hard problems are at least as hard as any problem that can be solved using a polynomial amount of memory) to manipulate constructively, depending on whether the schedule of the pre-round is determined before the votes are collected, randomly after the votes are collected, or the scheduling and the vote collecting are carefully interleaved, respectively.

All of the hardness results of the previous paragraph rely on both the number of voters and the number of candidates growing. The number of candidates can be large in some domains, for example when voting over task or resource allocations. However, in other elections – such as presidential elections – the number of candidates is small. If the number of candidates is a constant, both constructive and destructive manipulation are easy (in $\mathcal{P}$), regardless of the number of voters (Conitzer, Sandholm and Lang, 2007). This holds even if the voters are weighted, or if a coalition of voters tries to manipulate. On the bright side, when a coalition of weighted voters tries to manipulate, complexity can arise even for a constant number of candidates; see Tables 1 and 2. Another lesson from that table is that randomizing over instantiations of the mechanism (such as schedules of a *cup*) can be used to make manipulation hard.

As usual in computer science, all the results mentioned above are worst-case hardness. Unfortunately, under weak assumptions on the preference distribution and voting rule, most instances of any voting rule are easy to manipulate (Conitzer and Sandholm, 2006).

All of the hardness results discussed above hold even if the manipulators know the non-manipulators' votes exactly. Under weak assumptions, if weighted coalitional manipulation with complete information about the others' votes is hard in some voting protocol, then individual unweighted manipulation is hard when there is uncertainty about the others' votes (Conitzer, Sandholm and Lang, 2007).

3.2 Non-truth-promoting mechanisms
A challenging issue is that even if it is prohibitively hard to find a beneficial manipulation, the agents might not tell the truth. For example, an agent might take a chance that he will do better with a lie. The following result

Table 2 *Complexity of destructive weighted coalitional manipulation*

Number of candidates:	2	≥3
STV	$\mathcal{P}$	$\mathcal{NP}$-complete
Plurality with runoff	$\mathcal{P}$	$\mathcal{NP}$-complete
Randomized cup	$\mathcal{P}$	?
Borda	$\mathcal{P}$	$\mathcal{P}$
Veto	$\mathcal{P}$	$\mathcal{P}$
Copeland	$\mathcal{P}$	$\mathcal{P}$
Maximin	$\mathcal{P}$	$\mathcal{P}$
Cup	$\mathcal{P}$	$\mathcal{P}$
Plurality	$\mathcal{P}$	$\mathcal{P}$

Source: Conitzer, Sandholm and Lang (2007).

Table 1 *Complexity of constructive weighted coalitional manipulation*

Number of candidates:	2	3	4, 5, 6	≥7
Borda	$\mathcal{P}$	$\mathcal{NP}$-complete	$\mathcal{NP}$-complete	$\mathcal{NP}$-complete
Veto	$\mathcal{P}$	$\mathcal{NP}$-complete	$\mathcal{NP}$-complete	$\mathcal{NP}$-complete
STV	$\mathcal{P}$	$\mathcal{NP}$-complete	$\mathcal{NP}$-complete	$\mathcal{NP}$-complete
Plurality with runoff	$\mathcal{P}$	$\mathcal{NP}$-complete	$\mathcal{NP}$-complete	$\mathcal{NP}$-complete
Copeland	$\mathcal{P}$	$\mathcal{P}$	$\mathcal{NP}$-complete	$\mathcal{NP}$-complete
Maximin	$\mathcal{P}$	$\mathcal{P}$	$\mathcal{NP}$-complete	$\mathcal{NP}$-complete
Randomized cup	$\mathcal{P}$	$\mathcal{P}$	$\mathcal{P}$	$\mathcal{NP}$-complete
Cup	$\mathcal{P}$	$\mathcal{P}$	$\mathcal{P}$	$\mathcal{P}$
Plurality	$\mathcal{P}$	$\mathcal{P}$	$\mathcal{P}$	$\mathcal{P}$

shows that, nevertheless, mechanism design can be improved by making the agents face complexity. (This is one reason why computational issues can render the *revelation principle* inapplicable. One of the things the principle says is that for any non-truth-promoting mechanism it is possible to construct an incentive-compatible mechanism that is at least as good. The theorem below challenges this.)

Theorem 1 (Conitzer and Sandholm, 2004) *Suppose the centre is trying to maximize social welfare, and neither payments nor randomization is allowed. Then, even with just two agents (one of whom does not even report a type, so dominant strategy implementation and Bayes–Nash implementation coincide), there exists a family of preference aggregation settings such that:*

- *the execution of any optimal incentive-compatible mechanism is $\mathcal{NP}$-complete for the centre, and*
- *there exists a non-incentive-compatible mechanism which (1) requires the centre to carry out only polynomial computation, and (2) makes finding any beneficial insincere revelation $\mathcal{NP}$-complete for the type-reporting agent. Additionally, if the type-reporting agent manages to find a beneficial insincere revelation, or no beneficial insincere revelation exists, the social welfare of the outcome is identical to the social welfare that would be produced by any optimal incentive-compatible mechanism. Finally, if the type-reporting agent does not manage to find a beneficial insincere revelation where one exists, the **social welfare of the outcome is strictly greater than the social welfare that would be produced by any optimal incentive-compatible mechanism**.*

An analogous theorem holds if, instead of counting computational steps, we count calls to a commonly accessible oracle which, when supplied with an agent, that agent's type, and an outcome, returns a utility value for that agent.

3.3 Preference (valuation) determination via computing or information acquisition

In many (auction) settings, even determining one's valuation for an item (or a bundle of items) is complex. For example, when bidding for trucking lanes (tasks), this involves solving two $\mathcal{NP}$-complete local planning problems: the vehicle routing problem with the new lanes of the bundle and the problem without them (Sandholm, 1993). The difference in the costs of those two local plans is the cost (valuation) of taking on the new lanes.

In these types of settings, the *revelation principle* applies only in a trivial way: the agents report their data and optimization models to the centre, and the centre does the computation for them. It stands to reason that in many applications the centre would not want to take on that burden, in which case such extreme direct mechanisms are not an option. Therefore, I will now focus on mechanisms where the agents report valuations to the centre, as in traditional auctions.

Bidders usually have limited computing and time, so they cannot exactly evaluate all (or even any) bundles – at least not without cost. This leads to a host of interesting issues where computing and incentives are intimately intertwined.

For example, in a one-object auction, should a bidder evaluate the object if there is a cost to doing so? It turns out that the Vickrey auction loses its dominant-strategy property: whether or not the bidder should pay the evaluation cost depends on the other bidders' valuations (Sandholm, 2000).

If a bidder has the opportunity to *approximate his valuation to different degrees*, how much computing time should the bidder spend on refining its valuation? If there are multiple items for sale, how much computing time should the bidder allocate on different bundles? A bidder may even allocate some computing time to evaluate other bidders' valuations so as to be able to bid more strategically; this is called *strategic computing*.

To answer these questions, Larson and Sandholm (2001) developed a deliberation control method called a *performance profile tree* for projecting how an anytime algorithm (that is, an algorithm that has an answer available at any time, but where the quality of the answer improves the more computing time is allocated to the algorithm) will change the valuation if additional computing is allocated toward refining (or improving) it. This deliberation control method applies to any anytime algorithm. Unlike earlier deliberation control methods for anytime algorithms, the performance profile tree is a *fully normative model of bounded rationality*: it takes into account all the information that an agent can use to make its deliberation control decisions. This is necessary in the game-theoretic context; otherwise a strategic agent could take into account some information that the model does not.

Using this deliberation control method, the auction can be modelled as a game where the agents' strategy spaces include computing actions. At every point, each agent can decide on which bundle to allocate its next step of computing as a function of the agent's computing results so far (and in open-cry auction format also the others' bids observed so far). At every point, the agent can also decide to submit bids. One can then solve this for equilibrium: each agent's (deliberation and bidding) strategy is a best-response to the others' strategies. This is called *deliberation equilibrium*.

This notion, and the performance profile tree, apply not only to computational actions but also to information gathering actions for determining valuations. (In contrast, most of the literature on information acquisition in auctions does not take into account that valuations can be determined to different degrees and that an agent may

Table 3 *Can strategic computing occur in deliberation equilibrium? The most interesting results are in bold. As a benchmark from classical auction theory, the table also shows whether or not perfectly rational agents, that can determine their valuations instantly without cost, would benefit from considering each others' valuations when deciding how to bid*

	Auction mechanism	Speculation by perfectly rational agents?	Strategic computing?	
			Limited computing	Costly computing
Single item	First price	yes	yes	yes
	Dutch	yes	yes	yes
	English	no	**no**	**yes**
	Vickrey	no	**no**	**yes**
Multiple items	First price	yes	yes	yes
	VCG	no	**yes**	**yes**

want to invest effort to determine others' valuations as well – even in private-value settings.)

Table 3 shows in which settings strategic computing can and cannot occur in deliberation equilibrium. This depends on the auction mechanism. Interestingly, it also depends on whether the agent has limited computing (for example, owning a desktop computer that the agent can use until the auction's deadline) or costly computing (for example, being able to buy any amount of supercomputer time where each cycle comes at a cost).

The notion of deliberation equilibrium can also be used as the basis for designing new mechanisms, which hopefully work well among agents whose computing is costly or limited. Unfortunately, there is an impossibility (Larson and Sandholm, 2005): there exists no mechanism that is *sensitive* (the outcome is affected by each agent's strategy), *preference formation independent* (does not do the computations for the agents; the agents report valuations), *non-misleading* (no agent acts in a way that causes others to believe his true type has zero probability), and *deliberation-proof* (no strategic computing occurs in equilibrium, that is, agents compute only on their own problems). Current work involves designing mechanisms that take part in preference formation in limited ways: for example, agents report their performance profile trees to the centre, which then coordinates the deliberations incrementally as agents report deliberation results. Current research also includes designing mechanisms where strategic computing occurs but its wastefulness is limited.

3.3.1 Preference elicitation by the centre
To reduce the agents' preference determination effort, Conen and Sandholm (2001) proposed a framework where the centre (also known as *elicitor*) explicitly elicits preference information from the agents incrementally on an as-needed basis by posing queries to the agents. The centre thereby builds a model of the agents' preferences, and decides what to ask, and from which agent, based on this model. Usually the process can be terminated with

the provably correct outcome while requiring only a small portion of the agents' preferences to be determined. Multistage mechanisms can yield up to exponential savings in preference determination and communication effort the agents need to go through compared to single-stage mechanisms (Conitzer and Sandholm, 2004).

The explicit preference elicitation framework was originally proposed for CAs (but the approach has since been used for other settings as well, such as voting). For general valuations, an exponential number of bits in the number of items for sale has to be communicated in the worst case no matter what queries are used (Nisan and Segal, 2006). However, experimentally only a small fraction of the preference information needs to be elicited before the provably optimal solution is found. Furthermore, for valuations that have certain types of structure, even the worst-case number of queries needed is small. Research has also been done on the relative power of different query types.

If enough information is elicited to also determine the VCG payments, and these are the payments charged to the bidders, answering the elicitor's queries truthfully is an *ex post* equilibrium (a strengthening of Nash equilibrium that does not rely on priors). (This assumes there is no explicit cost or limit to valuation determination; mechanisms have also been designed for settings where there is an explicit cost (Larson, 2006).) This holds even if the agents are allowed to answer queries that the elicitor did not ask (for example, queries that are easy for the agent to answer and which the agent thinks will significantly advance the elicitation process). We thus have a *pull–push mechanism* where both the centre and the agents guide the preference revelation (and thus also the preference determination/refinement by the agents). For a review, see Sandholm and Boutilier (2006). Ascending (combinatorial) auctions are an earlier special case, and have limited power compared to the general framework (Blumrosen and Nisan, 2005).

Preference elicitation can sometimes be computationally complex for the centre. It can be complex to intelligently decide what to ask next, and from whom. It can also be complex to determine whether enough information has been elicited to determine the optimal outcome. Even if the elicitor knows that enough has been elicited, it can be complex to determine the outcome – for example, allocation of items to bidders in some CAs.

3.4 Distributed (centre-free) mechanisms

Computer scientists often have a preference for distributed applications that do not have any centralized coordination point (centre). Depending on the application, the reasons for this preference may include avoiding a single vulnerable point of failure, distributing the computing effort (for computational efficiency or because the data is inherently distributed), and enhancing privacy. The preference carries over from traditional computer science applications to different forms of negotiation systems – for example, see Sandholm (1993) for an early distributed automated negotiation system for software agents.

Feigenbaum et al. (2005) have studied lowest-cost inter-domain routing on the Internet, modifying a distributed protocol so that the agents (routing domains) are motivated to report their true costs and the solution is found with minimal message passing. For a review of some other research topics in this space, see Feigenbaum and Shenker (2002).

One can go further by taking into account the fact that agents might not choose to follow the prescribed protocol. They may cheat not only on information-revelation actions, but also on message-passing and computational actions. Despite computation actions not being observable by others, an agent can be motivated to compute as prescribed by tasking at least one other agent with the same computation, and comparing the results (Sandholm et al., 1999). Careful problem partitioning can also be used to achieve the same outcome without redundancy by only requiring agents to perform computing and message passing tasks that are in their own interest (Parkes and Shneidman, 2004). Shneidman and Parkes (2004) propose a general proof technique and instantiate it to provide a non-manipulable protocol for inter-domain routing. Monderer and Tennenholtz (1999) develop protocols for one-item auctions executed among agents on a communication network. The protocols motivate the agents to correctly reveal preferences and communicate. For the setting where agents with private utility functions have to agree on variable assignments subject to side constraints (for example, meeting scheduling), Petcu, Faltings and Parkes (2006) developed a VCG-based distributed optimization protocol that finds the social welfare maximizing allocation and each agent is motivated to follow the protocol in terms of all three types of action. The only centralized party needed is a bank that can extract payments from the agents.

Cryptography is a powerful tool for achieving privacy when trying to execute a mechanism in a distributed way without a centre, using private communication channels among the agents. Consider first the setting with passive adversaries, that is, agents that faithfully execute the specified distributed communication protocol, but who try to infer (at least something about) some agents' private information.

- If agents are computationally limited – for example, they are assumed to be unable to factor large numbers – then arbitrary functions can be computed while guaranteeing that each agent maintains his privacy (except, of course, to the extent that the answer of the computation says something about the inputs) (Goldreich, Micali and Wigderson, 1987). Thus the desire for privacy does not constrain what social choice functions can be implemented.
- In contrast, only very limited social choice functions can be computed privately among computationally unlimited agents. For example, when there are just two alternatives, every monotonic, non-dictatorial social choice function that can be privately computed is constant (Brandt and Sandholm, 2005). With special structure in the preferences, this impossibility can sometimes be avoided. For example, with the standard model of quasi-linear utility, first-price auctions can be implemented privately; second-price (Vickrey) auctions with more than two bidders cannot (Brandt and Sandholm, 2004).

A more general model is that of active adversaries who can execute the distributed communication protocol unfaithfully in a coordinated way. A more game-theoretic model is that of rational adversaries that are not passive, but not malicious either. For a brief overview of such work, see COMPUTER SCIENCE AND GAME THEORY.

TUOMAS SANDHOLM

This work was funded by the National Science Foundation under ITR grant IIS-0427858, and a Sloan Foundation Fellowship. I thank Felix Brandt, Christina Fong, Joe Halpern, and David Parkes for helpful comments.

Bibliography

Banks, J.S., Ledyard, J. and Porter, D. 1989. Allocating uncertain and unresponsive resources: an experimental approach. *RAND Journal of Economics* 20, 1–25.

Bartholdi, J., III and Orlin, J. 1991. Single transferable vote resists strategic voting. *Social Choice and Welfare* 8, 341–54.

Bartholdi, J., III, Tovey, C. and Trick, M. 1989. The computational difficulty of manipulating an election. *Social Choice and Welfare* 6, 227–41.

Bichler, M., Davenport, A., Hohner, G. and Kalagnanam, J. 2006. Industrial procurement auctions. In Cramton, Shoham and Steinberg (2006).

Blumrosen, L. and Nisan, N. 2005. On the computational power of iterative auctions. *Proceedings of the ACM Conference on Electronic Commerce*, 29–43.

Brandt, F. and Sandholm, T. 2004. (Im)possibility of unconditionally privacy-preserving auctions. *Proceedings of the International Conference on Autonomous Agents and Multi-Agent Systems*, 810–17.

Brandt, F. and Sandholm, T. 2005. Unconditional privacy in social choice. *Proceedings of the Conference on Theoretical Aspects of Rationality and Knowledge*, 207–18.

Conen, W. and Sandholm, T. 2001. Preference elicitation in combinatorial auctions: extended abstract. *Proceedings of the ACM Conference on Electronic Commerce*, 256–9. More detailed description of algorithmic aspects in *Proceedings of the IJCAI-01 Workshop on Economic Agents, Models, and Mechanisms*, 71–80.

Conitzer, V. and Sandholm, T. 2002. Complexity of mechanism design. *Proceedings of the Conference on Uncertainty in Artificial Intelligence*, 103–10.

Conitzer, V. and Sandholm, T. 2003. Universal voting protocol tweaks to make manipulation hard. *Proceedings of the International Joint Conference on Artificial Intelligence*, 781–8.

Conitzer, V. and Sandholm, T. 2004. Computational criticisms of the revelation principle. *Conference on Logic and the Foundations of Game and Decision Theory*. Earlier versions: AMEC-03, EC-04.

Conitzer, V. and Sandholm, T. 2006. Nonexistence of voting rules that are usually hard to manipulate. *Proceedings of the National Conference on Artificial Intelligence*.

Conitzer, V., Sandholm, T. and Lang, J. 2007. When are elections with few candidates hard to manipulate? *Journal of the ACM* 54(3), Article 14.

Cramton, P., Shoham, Y. and Steinberg, R., eds. 2006. *Combinatorial Auctions*. Cambridge, MA: MIT Press.

Feigenbaum, J., Papadimitriou, C., Sami, R. and Shenker, S. 2005. A BGP-based mechanism for lowest cost routing. *Distributed Computing* 18, 61–72.

Feigenbaum, J. and Shenker, S. 2002. Distributed algorithmic mechanism design: recent results and future directions. *Proceedings of the International Workshop on Discrete Algorithms and Methods for Mobile Computing and Communications*, 1–13.

Goldreich, O., Micali, S. and Wigderson, A. 1987. How to play any mental game or a completeness theorem for protocols with honest majority. *Proceedings of the Symposium on Theory of Computing*, 218–29.

Hajiaghayi, M.T., Kleinberg, R. and Sandholm, T. 2007. Automated online mechanism design and prophet inequalities. *Proceedings of the National Conference on Artificial Intelligence*.

Hershberger, J. and Suri, S. 2001. Vickrey prices and shortest paths: what is an edge worth? *Proceedings of the Symposium on Foundations of Computer Science*, 252–9.

Jurca, R. and Faltings, B. 2006. Minimum payments that reward honest reputation feedback. *Proceedings of the ACM Conference on Electronic Commerce*, 190–9.

Kfir-Dahav, N., Monderer, D. and Tennenholtz, M. 2000. Mechanism design for resource bounded agents. *Proceedings of the International Conference on Multi-Agent Systems*, 309–15.

Land, A., Powell, S. and Steinberg, R. 2006. PAUSE: a computationally tractable combinatorial auction. In Cramton, Shoham and Steinberg (2006).

Larson, K. 2006. Reducing costly information acquisition in auctions. *Proceedings of the Autonomous Agents and Multi-Agent Systems*, 1167–74.

Larson, K. and Sandholm, T. 2001. Costly valuation computation in auctions. *Proceedings of the Theoretical Aspects of Rationality and Knowledge*, 169–182.

Larson, K. and Sandholm, T. 2005. Mechanism design and deliberative agents. *Proceedings of the International Conference on Autonomous Agents and Multi-Agent Systems*, 650–6.

Lavi, R. and Swamy, C. 2005. Truthful and near-optimal mechanism design via linear programming. *Proceedings of the Symposium on Foundations of Computer Science*, 595–604.

Lehmann, D., Müller, R. and Sandholm, T. 2006. The winner determination problem. In Cramton, Shoham and Steinberg (2006).

Lehmann, D., O'Callaghan, L.I. and Shoham, Y. 2002. Truth revelation in rapid, approximately efficient combinatorial auctions. *Journal of the ACM* 49, 577–602.

Likhodedov, A. and Sandholm, T. 2005. Approximating revenue-maximizing combinatorial auctions. *Proceedings of the National Conference on Artificial Intelligence*, 267–74.

Monderer, D. and Tennenholtz, M. 1999. Distributed games: from mechanisms to protocols. *Proceedings of the National Conference on Artificial Intelligence*, 32–7.

Müller, R. 2006. Tractable cases of the winner determination problem. In Cramton, Shoham and Steinberg (2006).

Nisan, N. and Ronen, A. 2000. Computationally feasible VCG mechanisms. *Proceedings of the ACM Conference on Electronic Commerce*, 242–52.

Nisan, N. and Ronen, A. 2001. Algorithmic mechanism design. *Games and Economic Behavior* 35, 166–96.

Nisan, N. and Segal, I. 2006. The communication requirements of efficient allocations and supporting prices. *Journal of Economic Theory* 129, 192–224.

Parkes, D. and Shneidman, J. 2004. Distributed implementations of generalized Vickrey–Clarke–Groves auctions. *Proceedings of the International Conference on Autonomous Agents and Multi-Agent Systems*, 261–8.

Petcu, A., Faltings, B. and Parkes, D. 2006. MDPOP: faithful distributed implementation of efficient social choice problems. *Proceedings of the International Conference on Autonomous Agents and Multi-Agent Systems*, 1397–404.

Rothkopf, M. 2007. Thirteen reasons why the Vickrey–Clarke–Groves process is not practical. *Operations Research* 55, 191–7.

Rothkopf, M., Pekěc, A. and Harstad, R. 1998. Computationally manageable combinatorial auctions. *Management Science* 44, 1131–47.

Sandholm, T. 1993. An implementation of the contract net protocol based on marginal cost calculations. *Proceedings of the National Conference on Artificial Intelligence*, 256–62.

Sandholm, T. 2000. Issues in computational Vickrey auctions. *International Journal of Electronic Commerce* 4, 107–29. Early version in ICMAS-96.

Sandholm, T. 2002a. Algorithm for optimal winner determination in combinatorial auctions. *Artificial Intelligence* 135, 1–54. Earlier versions: ICE-98 keynote, Washington U. tech report WUCS-99-01 Jan. 1999, IJCAI-99.

Sandholm, T. 2002b. eMediator: a next generation electronic commerce server. *Computational Intelligence* 18, 656–76. Earlier versions: Washington U. tech report WU-CS-99-02 Jan. 1999, AAAI-99 Workshop on AI in Ecommerce, AGENTS-00.

Sandholm, T. 2006. Optimal winner determination algorithms. In Cramton, Shoham and Steinberg (2006).

Sandholm, T. 2007. Expressive commerce and its application to sourcing: how we conducted $35 billion of generalized combinatorial auctions. *AI Magazine* 28(3), 45–58.

Sandholm, T. and Boutilier, C. 2006. Preference elicitation in combinatorial auctions. In Cramton, Shoham and Steinberg (2006).

Sandholm, T., Conitzer, V. and Boutilier, C. 2007. Automated design of multistage mechanisms. *Proceedings of the International Joint Conference on Artificial Intelligence*, 1500–6.

Sandholm, T. and Ferrandon, V. 2000. Safe exchange planner. *Proceedings of the International Conference on Multi-Agent Systems*, 255–62.

Sandholm, T., Larson, K., Andersson, M., Shehory, O. et al. 1999. Coalition structure generation with worst case guarantees. *Artificial Intelligence* 111, 209–38.

Shneidman, J. and Parkes, D.C. 2004. Specification faithfulness in networks with rational nodes. *Proceedings of the ACM Symposium on Principles of Distributed Computing*, 88–97.

Vorobeychik, Y., Kiekintveld, C. and Wellman, M. 2006. Empirical mechanism design: methods, with application to a supply chain scenario. *Proceedings of the ACM Conference on Electronic Commerce*, 306–15.

concentration measures

The term *concentration* (also *firm concentration, industry concentration* or *market concentration*) refers to aspects of the distribution of firm size within a specific market or industry that have traditionally been used to characterize the degree of competitiveness in the market. Even though the size of firms can be measured using many different variables, such as employment or assets, the sales level is the most commonly used size measure. Accordingly, if very few firms serve a very large portion of the market, it is said that the given market is highly 'concentrated', whereas if no single firm has a large share of sales it is said that the market is not 'concentrated'. Since concentration is an important reflection of the underlying market structure, its measurement is an important characterization of the interaction of firms within a specific market or industry.

The most common concentration measures are the 'n-firm concentration rate' and the 'Herfindahl index'. Let S_i be the market share of firm i; the 'n-firm concentration rate' is the sum of the market shares of the n biggest firms within the market:

$$C(n) = \sum_{i=1}^{n} S_i.$$

As indicated, the summation above is taken over the set of n biggest firms in the market. So, for example, the two-firm concentration rate of a given market is the sum of the market shares of the two biggest firms in the market where size is measured according to observed sales. In order to fully characterize the concentration of any given market, though, a number of these rates must be used, since there is no agreed-on value for n. This complicates its use for comparing concentration over time and across sectors, and for its use in statistical analysis.

The Herfindahl index, first devised by Albert Hirschman to measure the concentration of trade across sectors (so that the index is also known as 'Herfindahl–Hirschman index'; see Hirschman, 1964, for its history), is the sum of the squared market shares of all firms in the market:

$$H = \sum_{i=1}^{N} S_i^2.$$

The summation in this case is taken over the set of all N firms in the market. This index lies between zero and 1: if there is only one firm in the market, so that the market has the highest possible concentration, the index is 1. If, on the other hand, there are many equally sized firms in the market, the index will be close to zero. By squaring the individual market shares, this index gives relatively greater weight to the market shares of large firms. Conversely, the addition of one small firm to the market dilutes somewhat the market share of larger firms, and has a marginal negative effect on the index, which is consistent with any notion of market concentration. Any value of this index can correspond to multiple market configurations, being in that sense less illustrative of the actual concentration of a market than a set of n-firm concentration rates. On the other hand, this index can be easily correlated with other market characteristics and is therefore very useful for statistical analysis.

Other less commonly used concentration measures include entropy coefficients, the Gini coefficient and measures of the variance of market shares across firms within a market. The entropy coefficient is usually computed using the following formula:

$$E = \sum_{i=1}^{N} S_i \log_2 \left(\frac{1}{S_i} \right).$$

This index takes value zero if there is only one firm in the market and grows as market concentration decreases. The interpretation of this coefficient is complicated, because its formula weights both large and small firms less heavily than mid-size firms and grows unboundedly as market concentration decreases. It is therefore less commonly used than the Herfindahl index.

The Gini coefficient is commonly used to characterize the income or wealth inequality within a society. Its drawback as a measure of market concentration is that it is useful only to measure the concentration of firms' sizes within a market, given a number of firms. So according to the Gini coefficient a duopolistic market with two firms of equal size is as concentrated as a market with 100 firms with identical size. The same drawback applies for the use of measures of the variance of firm size – whose definition is simply the sample variance of firm sizes.

All the concentration measures mentioned above are very sensitive to the actual market definition that is used. In markets for differentiated products, for example, products may face a continuum of similar products, and determining which similar products exactly constitute a market is not always easy. Take the specific example of the market faced by US mobile phone services: with just a handful of national providers it is concentrated given the standard concentration measures. These national firms, nevertheless, are also competing with local companies in various segments of the market and even with long-distance phone companies and Internet companies as providers of communication services. The Internet, on the other hand, competes in some instances with cable and satellite companies, radio stations and even newspapers as sources of news and entertainment. What exactly the relevant market faced by mobile phone companies is will depend on the type of issue being addressed. Accordingly, the concentration measures will change depending on the adopted market definition.

On the other hand, even if the market is well defined, computed concentration measures may not reflect at all the real competitive structure of the market. For example, even in markets as highly concentrated as the market for computer processors, the dominant firms have to account for the invisible competition of potential entrants. The same happens in regional markets where outside firms are kept at bay by few local firms with a combination of low prices and high transportation costs. Computed concentration measures for specific markets cannot account for this unobserved competition and may therefore lead to wrong conclusions regarding the underlying behaviour of firms. In these instances, a behavioural measure, such as the Lerner index, which measures the relative size of firms' markups, may be a better indicator of the competitive structure of the market.

There is a body of empirical literature that uses market concentration measures across industries' to approximate the underlying differences in industries competitiveness. They were then used to infer statistically the relationship between market 'structure' and market 'performance'. For example, correlations of R&D expenditure and market concentration were computed to investigate whether firms in concentrated markets were more or less likely to innovate than firms in more competitive markets. The value of such correlations is limited because the observed concentration may be both a cause and an effect of individual firms' behaviour and the relationship is shaped by the specifics of the industry. In order to avoid the ambiguities of such an inter-industry approach, the more recent empirical microeconomic literature has generally focused instead on the understanding of firm behaviour within specific industries, for which the use of concentration measures is less relevant.

JUAN ESTEBAN CARRANZA

See also **competition; Gini ratio; market structure.**

Bibliography

Hirschman, A.O. 1964. The paternity of an index. *American Economic Review* 54, 761–2.
Ravenscraft, D. 1983. Structure-profit relationships at the line of business and industry level. *Review of Economics and Statistics* 65, 22–31.
Scherer, F.M. 1970. *Industrial Market Structure and Economic Performance*, 2nd edn. Boston: Houghton Mifflin.
Simon, H. and Bonini, C.P. 1958. The size distribution of business firms. *American Economic Review* 48, 607–17.

Condillac, Etienne Bonnot de, Abbé de Mureau (1714–1780)

Philosopher and economist. Born at Grenoble, the third son of a well-to-do aristocratic family, Condillac took his name from an estate purchased by his father in 1720. As a sickly child with poor eyesight he had little early education and was apparently still unable to read by the age of 12. After his father's death in 1727 he moved to Lyon to live with his oldest brother, continuing his education at its Jesuit college. Through this brother he may have first met Jean Jacques Rousseau, who was tutor to his nephews in 1740 and became a life-long friend. His second brother, l'Abbé de Mably, took Condillac to Paris in c.1733 to study theology at Saint Sulpice and the Sorbonne. He was ordained in 1740 and for the rest of his life 'ever faithful to the Christian church, would always wear his cassock, always remain l'Abbé' (Lefèvre, 1966, p. 11).

For the next 15 years he lived the life of a Paris intellectual, studying the philosophy of Descartes, Malebranche, Leibniz and Spinoza, 'to whose speculative systems he formed a life-long aversion, preferring the English philosophers Locke (who particularly influenced his thinking), Berkeley, Newton and rather belatedly, Bacon' (Knight, 1968, pp. 8–9). In this period he published the works which made his philosophical reputation: the *Essay on the Origin of Human Knowledge* (1746), the *Traité des Systèmes* (1749), his most famous philosophical work *Treatise on the Sensations* (1754) described as the 'most rigorous demonstration of the [18th-century] sensationalist psychology' (Knight, 1968, p. 12) and his *Traité des Animaux* (1755).

Apart from giving him entry to the Paris salons, where at Mlle de Lespinasse's salon he is reputed to have first met Turgot, another life-long friend (Le Roy, 1947, p. ix), his intellectual reputation gained him the position of tutor to Louis XV's grandson, the Duke of Parma. From 1758 to 1767 he resided in Parma. Because of its prime minister's economic development policies, inspired by a mixture of 'mercantilism, physiocracy and the ideas of Gournay', Condillac developed an interest in economic matters, an interest 'indirectly confirmed by his known contacts with the Italian political economists, Beccaria and Gherardo' (Knight, 1968, pp. 231–2). In 1768 he returned to Paris, but by 1773 had retired to his estate of Flux near Beaugency, where he died in 1780. During the last decade of his life he published his *Cours d'Etudes* (1775), his work on economics (1776), a text on logic (1780) for use in Polish Palatinate schools, and commenced the unfinished *La Langue des Calculs* (1798). In 1752, he became a member of the Royal Prussian Academy; in 1768 after his return from Parma he was elected to the French Academy. His works have been frequently collected, most recently by Le Roy (1947–51).

The impetus for Condillac's writing *Le Commerce et le Gouvernement* has been ascribed to a desire to assist his friend Turgot in the difficulties he faced in 1775 as finance minister over the grain riots induced by his restoration of the free trade in grain (Le Roy, 1947, p. xxv; Knight, 1968, p. 232). This fits with the work's unqualified support for free trade in general and the grain trade in particular (1776, esp. pp. 344–5, which seems directly inspired by the Paris events of 1775). Writing the book may also be explained as a return favour for Turgot's assistance in getting Condillac (1775) published (cf. Knight, 1968, pp. 13, 232). Despite Condillac's strong support for this major part of Physiocratic policy and his close adherence to other aspects of Physiocracy, his argument that manufacturing was productive brought critical replies from Baudeau and Le Trosne (1777). In this context it may be noted that his work bears little direct Physiocratic influence, the major influence being Cantillon (1755), the only work directly cited apart from Plumard de Dangeul (1754). It is, however, possible to detect some influence from the economics of Turgot, Galiani and Verri on the theory of value, price and competition (cf. Spengler, 1968, p. 212).

As published, the work is divided into two parts. The first provides the elements of the science. Its starting point is the foundation of value, which Condillac finds in the usefulness of an object relative to subjective needs making relative scarcity the key variable determining value. Value is distinguished from price because price can only originate in exchange. It is determined by the competition between buyers and sellers guided by their subjective estimation of value. Gains from exchange arise from differences in value; for Condillac, value cannot exchange for equal value. Although Condillac did discuss the costs of acquiring commodities, his emphasis is on exchange, trade and price. Exchange presumes surplus production and a need for consumption. Hence trade inspires and animates production and is essential to increasing wealth. Only simple pictures of production are presented: farm labourers producing prime necessities of food and materials; artisans transforming raw materials into essentials and luxuries; traders who circulate these products at home and abroad. By this circulation trade distributes the annual product and under competitive conditions settles its true prices. Condillac is more concerned with developing the institutions associated with trade: growth of towns and villages, money, banking, credit, interest and the foreign exchanges, the defence of property by government and hence the need for taxation, and the effects of restraints on trade, including the grain trade. The second part is almost completely devoted to examining effects of specific obstacles to trade ranging from war, tariffs, taxes, excessive government borrowing to luxury spending in the capital city and exclusive trading privileges. Moderate wants combined with complete freedom constitute his recipe for the best form of economic development.

Condillac's economic work received a mixed reception from later economists. J.B. Say (1805, p. xxxv) described it as an attempt 'to found a system of … a subject which [the author] did not understand'. Jevons (1871, p. xviii) praised Condillac's 'charming philosophic work [because] in the first few chapters … we meet perhaps the earliest distinct statement of the true connections between value and utility…'. Macleod (1896, p. 73) described it as a 'remarkable work … utterly neglected but in scientific spirit … infinitely superior to Smith'. Since then, it has remained neglected even though as 'a good if somewhat sketchy treatise on economic theory and policy [it was] much above the common run of its contemporaries' (Schumpeter, 1954, pp. 175–6).

PETER GROENEWEGEN

Selected works

1746. *An Essay on the Origin of Human Knowledge*, trans. Thomas Nugent. London: Nourse, 1756.

1749. *Traité des Systèmes, où l'on en démêle les inconvénine et les advantages*. Paris and Amsterdam.

1754. *Treatise on the Sensations*, trans. B.S. Geraldine Czar. London: Favill Press, 1930.

1755. *Traité des Animaux, où après avoir fain des observations critique sur le sentiment de Descartes et sur celui de M. Buffon on entreprend d'expliquer leurs principales facultés ....* Amsterdam.

1775. *Cours d'Étude pour l'instruction du Prince de Parma, Aujourd'hui Ferdinand, Duc de Parma.* Parma (and Paris).

1776. *Le Commerce et le gouvernement considerés relativement l'un à l'autre.* In *Oeuvres complètes de Condillac*, vol. 4. Paris: Brière, 1821.

1780. *The Logic*, trans. B.S. Joseph Neef. Philadelphia, 1809.

1798. *Le langue des Calculs, Ouvrage Posthume et élémentaire.* Paris.

Bibliography

Cantillon, R. 1755. *Essay on the Nature of Commerce in General*, trans. B.S.H. Higgs. London: Macmillan, 1931.

Jevons, W.S. 1871. *Theory of Political Economy*. 4th edn, London: Macmillan, 1911.

Knight, I.F. 1968. *The Geometric Spirit. The Abbé de Condillac and the French Enlightenment*. New Haven and London: Yale University Press.

Lefèvre, R. 1966. *Condillac ou la joie de vivre*. Paris: Editions Seghers.

Le Roy, G., ed. 1947–51. *Oeuvres philosophiques de Condillac*. Paris: Press Universitaires de France.

Le Trosne, G.F. 1777. *De L'intérêt Social, par rapport à la Valeur, à la circulation, à l'Industrie, & au commerce intérieur & extérieur: Ouvrage élémentaire dans lequel on discoute quelques Principes de M. l'Abbé de Condillac.* Paris.

Macleod, H.D. 1896. *The History of Economics*. London: Bliss, Sands & Co.

Plumard de Danguel, L.J. 1754. *Remarques sur les avantages et les désavantages de la France et de la Gr. Bretagne par rapport au commerce et aux autres sources de la puissance des états.* Leyden (and Paris).

Say, J.-B. 1805. *A Treatise on Political Economy of the Production, Distribution and Consumption of Wealth*, trans. C.R. Prinsep, New American Edition, Philadelphia, 1880; reissued New York: Kelley, 1963.

Schumpeter, J.A. 1954. *History of Economic Analysis*. London: Allen & Unwin, 1959.

Spengler, J.J. 1968. Condillac, Etienne Bonnot de. In *Encyclopedia of the Social Sciences*, vol. 3, 2nd edn. Chicago: Chicago University Press.

Condorcet, Marie Jean Antoine Nicolas Caritat, Marquis de (1743–1794)

Condorcet was a French mathematician and philosopher. With many of his fellow *encyclopédistes* he shared the conviction that social sciences are amenable to mathematical rigour. His pioneer work on elections, the *Essai sur l'application de l'analyse a la probabilité des decisions rendues a la pluralité des voix* (1785) is a major step in that direction.

The aim of the *Essai* is to 'inquire by mere reasoning, what degree of confidence the judgement of assemblies deserves, whether large or small, subject to a high or low plurality, split into several different bodies or gathered in one only, composed by men more or less wise' (*Discours préliminaire* to the *Essai*, p. iv).

In modern words, this is the jury problem: to decide whether the accused is guilty or not requires converting the opinions of several experts, with varying competence, into a single judgement. Systematic probabilistic computations for this problem occupy most of the *Essai*, often camouflaging the essential contributions. The opaqueness and technicality of the argument meant that a full recognition of its importance did not occur until more than 150 years later (Black, 1958). Since then Condorcet's findings have strongly influenced modern social choice theorists (for example, Arrow, Guilbaud and Black), and still play a central role in many of its recent developments.

The starting point is that majority voting is the unambiguously best voting rule when only two candidates are on stage. This fact, whose modern formulation is known as May's theorem (May, 1952) was clear enough to the encyclopedists, too. How, then, can we extend this rule to three candidates or more? The naive, yet widely used, answer is plurality voting (each voter casts a vote for one candidate; the candidate with most votes is elected). Both Condorcet and Borda (his colleague in the Academy of Sciences) raise the same objection against the plurality rule. Suppose, says Condorcet (*Discours préliminaire*, p. lviii) that 60 voters have the opinions shown in Table 1 about three candidates A, B, C.

In the illustration, candidate A wins by plurality. Yet if we oppose A against B only, A loses (25 to 35) and in A against C, A loses again (23 to 37). Thus the plurality rule does not convey accurately the opinion of the majority. From these identical premises, Borda proposes his well-known scoring method (each candidate receives 2 points from a voter who ranks him first, 1 point from one who ranks him second, and none from one who ranks him last; hence C is elected with score 78), whereas Condorcet opens a quite different route.

Condorcet posits a simple binomial model of voter error: in every binary comparison, each voter has a probability $1/2 < P < 1$ of ordering the candidates

Table 1

	23	19	16	2
Top	A	B	C	C
	C	C	B	A
Bottom	B	A	A	B

correctly. All voters are assumed to be equally able, and there is no correlation between judgements on different pairs. Thus for Condorcet the relevant data is contained in the 'majority tournament' that results from taking all pairwise votes:

B beats A, 35 to 25;   C beats A, 37 to 23;

C beats B, 41 to 19.

Condorcet proposes that the candidates be ranked according to 'the most probable combination of opinions' (*Essai*, p. 125). In modern statistical terminology this is a maximum likelihood criterion (see Young, 1986).

In the above example the most probable combination is given by the ranking: CBA since the three statements C over B, C over A, B over A agree with the greatest total number of votes. Condorcet's ranking criterion implies that an alternative (such as C) that obtains a majority over every other alternative must be ranked first. Such an alternative, if one exists, is known as a 'Condorcet winner'.

As Condorcet points out, some configurations of opinions may not possess such a winner, because the majority tournament contains a cycle (a situation known as 'Condorcet's paradox'). He exhibits the example shown in Table 2.

Here A beats B, 33 to 27; B beats C, 42 to 18; C beats A, 35 to 25. According to Condorcet's maximum likelihood criterion, this cycle should be broken at its weakest link (A over B), which yields the ranking B over C over A. Therefore in this case B is declared the winner.

Somewhat later in the *Essai* (pp. 125–6), Condorcet suggests that one may compute the maximum likelihood ranking of n candidates by, first, choosing the $n(n-1)/2$ binary propositions that have the majority in their favour; then, if there are cycles, *successively deleting* those with smallest majorities until a complete ordering of the candidates is obtained. Unfortunately, for $n > 3$ this heuristic algorithm does not necessarily yield the ranking that accords with the greatest number of votes. An axiomatic characterization of Condorcet's rule is given in Young and Levenglick (1978).

Condorcet's idea of reducing individual opinions to all pairwise comparisons between alternatives proved essential to the aggregation of preferences approach initiated by Arrow (1951). The key axiom independence of irrelevant alternatives (IIA) requires that voting on a pair of candidates be enough to determine the collective opinion on this pair: this generalizes majority tournaments by dropping the symmetry across voters and across candidates. In this sense Arrow's impossibility theorem means that the Condorcet paradox is inevitable in any non-dictatorial voting method satisfying IIA.

Many more useful insights can be discovered in the *Essai*. For instance the issue of strategic manipulations, which has played a central role in the theory of elections since the late 1960s, is suggested in places, although it is never systematically analysed. For example, on page clxxix of the *Discours Preliminaire*, Condorcet criticizes Borda's method as more vulnerable to a '*cabale*'. His argument is supported by the modern game theoretical approach: whenever the configurations of individual opinions guarantee existence of a Condorcet winner, it defines a strategy-proof voting rule. This is one of the principal arguments in favour of Condorcet consistent voting rules, namely, rules electing the Condorcet winner whenever it exists (see, for example, Moulin, 1983, ch. 4).

H. MOULIN AND H.P. YOUNG

See also **Borda, Jean-Charles de; social choice; voting paradoxes.**

Selected works

1785. Essai sur l'application de l'analyse à la probabilité des décisions rendues à la pluralité voix. Paris.

Bibliography

Arrow, K. 1951. *Social Choice and Individual Values*. New York: Wiley.
Black, D. 1958. *The Theory of Committees and Elections*. Cambridge: Cambridge University Press.
May, K. 1952. A set of independent necessary and sufficient conditions for simple majority decision. *Econometrica* 20, 680–84.
Moulin, H. 1983. *The Strategy of Social Choice*. Amsterdam: North-Holland.
Young, H.P. 1986. Optimal ranking and choice from pairwise comparisons. In *Information Pooling and Group Decisionmaking* ed. B. Grofman and G. Owen. Greenwich, CT: JAI Press.
Young, H.P. and Levenglick, A. 1978. A consistent extension of Condorcet's election principle. *SIAM Journal of Mathematics* 35, 285–300.

Table 2

23	17	2	10	8
A	B	B	C	C
B	C	A	A	B
C	A	C	B	A

congestion

'Congestion' is the phenomenon whereby the quality of service provided by a *congestible facility* degrades as its aggregate usage increases, when its capacity is held fixed. We shall develop the economic theory of congestion in the context of road traffic, but congestion is pervasive:

more telephone usage increases the probability of encountering a busy line; higher electricity demand may lead to voltage fluctuations, brownouts and eventually blackouts; more swimmers in a pool make comfortable swimming more difficult; more patients visiting a medical clinic results in longer waits and lower-quality care; in a more crowded classroom, students receive less individual attention, and more time is wasted on administration and discipline; and so on. The economic theory of congestion identifies how the capacity of a congestible facility and its usage fee should be chosen. Some degree of congestion is typically socially optimal.

The economic theory of congestion has much in common with the theory of clubs and local public goods (Scotchmer, 2002). The two literatures examine similar issues, but the economic theory of congestion has a policy perspective, while the theory of clubs and local public goods focuses on decentralized provision.

Formally, we may define congestion as follows. Consider a congestible facility in a steady state, that comprises I congestible elements. (Congestible elements for a sports stadium, for example, include nearby roads, parking facilities, the ticket office, washrooms, concessions, and seating.) Element i is characterized by a flow capacity, k_i, and a stock capacity, K_i; the flow capacity is the maximum throughput per unit time, the stock capacity the maximum number of users at a point in time. Similarly, the level of usage is described in terms of the throughput of congestible element i, n_i, and the number of users at a point in time, N_i. The congestible facility provides J dimensions of quality of service, with the level of dimension j indicated by s_j. Letting k, K, n, N and s denote the corresponding vectors,

$$s = S(k, K, n, N). \qquad (1)$$

Congestion occurs when there is at least one combination of j and i for which s_j is monotone decreasing in n_i (flow congestion) or N_i (stock congestion), that is, when some dimension of quality of service falls as the throughput or stock of users of some congestible element of capacity increases. This is the static or steady-state definition of congestion. The dynamic definition of congestion adds time subscripts to s, k, K, n and N, and appends equations of motion relating stocks and flows for the various elements of capacity.

For some congestible elements, such as a turnstile, the bottleneck in the Vickrey (1969) bottleneck model of traffic congestion, or a switching circuit, the flow capacity constraint is the more important; for others, such as a telephone line, an elevator, a swimming pool, or seating at a football stadium, the stock capacity constraint is the more important. It should also be noted that a congestible facility can take the form of a network of congestible elements of capacity; a natural distinction is then between link congestion (for example, highway links) and nodal congestion (for example, traffic intersections).

To develop the theory, we consider a particular congestible facility having a single element of capacity and identical users, that is in a steady state: a road of uniform width connecting a single entry point A and a single exit point B, for which an increase in traffic flow increases travel time and an increase in road width reduces it. In this context, the deterioration of quality of service with an increase in usage is the increase in travel time from an increase in traffic flow.

We start with the short-run problem of determining optimal flow and its decentralized attainment, holding road width fixed. Let f denote flow, w road width, $t = t(f, w)$ the travel time function with (functional subscripts denote partial derivatives) $t_f > 0$ and $t_w < 0$ and ρ the value of time. Then the cost to an individual driver of travelling from A to B, the *user cost*, is $\rho t(f, w)$. Total user costs per unit time equal flow times user cost: $\rho f t(f, w)$. The social cost per unit time from increasing flow by one unit, with capacity held fixed, the *short-run marginal social cost*, is $\rho t(f, w) + \rho f t_f(f, w)$. The first term is the user cost of the extra driver; the second, the *congestion externality cost*. A driver imposes a congestion externality by slowing other drivers down; increasing steady-state flow by one car increases each car's travel time by $t_f(f, w)$ and social cost by $\rho f t_f(f, w)$.

Figure 1 displays short-run equilibrium. p denotes trip price, $D(p)$ the aggregate trip demand function, and $uc(f)$ and $srmsc(f)$ the user cost and short-run marginal social cost as a function of f, holding w fixed. With no toll, a user's trip price equals his user cost, and equilibrium occurs where the demand and user cost functions intersect, with flow f^e. Assuming that the marginal social benefit from a trip equals the corresponding marginal willingness to pay, the optimum occurs where the demand and short-run marginal social cost curves intersect, with flow f^*. Thus, with no toll, equilibrium flow is excessive. Efficiency obtains when economic agents face the social costs of their decisions and derive the social benefits from them. In the no-toll case, the price of a trip falls short of its marginal social cost since a driver does not pay for slowing down other drivers. Following Pigou (1947), the standard remedy for internalizing the congestion externality is to impose a toll equal to the congestion externality cost, evaluated at the social optimum: τ^* in Figure 1. This causes the trip price function to shift up from $uc(f)$ to $uc(f) + \tau^*$ and equilibrium flow to fall to the optimal level.

The above argument illustrates the general principle that efficient utilization of a congestible facility requires that the price equal short-run marginal social cost and the toll the congestion externality cost. Different user types – for example, cars and trucks – may impose different congestion externality costs. Efficiency then requires that the toll be differentiated according to user type.

We now turn to the long-run planning problem in which both road width and flow are choice variables. We

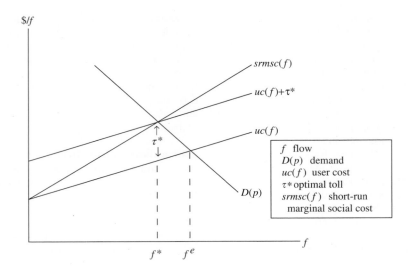

Figure 1

then consider decentralization of the optimum. Let $B(f)$ denote the social benefit per unit time from flow f, and $C(w)$ the amortized capital cost of road width w. (We ignore the complications that arise when the congestible facility is sufficiently large that its construction alters factor prices.) The social surplus generated by the road (per unit time) equals social benefit minus social cost, and social cost equals total user cost plus amortized capital cost:

$$SS(f, w) = B(f) - \rho f t(f, w) - C(w). \tag{2}$$

It is easily seen from (2) that the road width that maximizes social surplus is that which minimizes social cost. This means that, when the long-run planning problem is solved, production is carried out according to the long-run cost structure, and the short-run marginal cost pricing (which is again) required for optimal flow is equivalent to long-run marginal cost pricing:

$$p = LRMC. \tag{3}$$

Now, recall the basic result of production theory that $LRMC$ is equal to, less than or greater than $LRAC$ (long-run average cost) according to whether $LRAC$ is constant, decreasing or increasing. Combining this with (3), we have the result that, when $LRAC$ is constant, $p = LRAC$ holds at a long-run optimum. This is equivalent to equality between the total value of output and the total cost of output. Since total user cost is a component of both, this equality implies equality between toll receipts and amortized capital cost. Thus, *in the case of constant long-run average cost, the revenue raised from the optimal toll exactly covers the capital cost of providing a road of optimal width.* This is known as the 'self-financing' result. It was first derived by Mohring and Harwitz (1962) and subsequently generalized by Strotz (1965). (For a geometric derivation, see Arnott and Kraus, 2003.)

The self-financing result extends to congestible facilities with multiple elements of capacity, multiple dimensions of quality of service, and multiple user groups. If a congestible facility exhibits constant long-run average costs, provision of the facility can be decentralized via competing 'clubs'; competition will result in each club charging each user a fee for use of its congestible facility equal to the congestion externality cost he imposes, and choosing optimal capacity.

The above theory was developed on the assumption of a steady state. In the extension to treat nonstationary dynamics, which is conceptually straightforward, the distinction between flow externalities and stock externalities becomes sharper.

The theory relates to *first-best* pricing and capacity choice when congestion is the only externality. When usage entails other externalities, such as pollution, first-best pricing should take these into account. In any policy context, additional practical constraints that rule out attainment of the full first-best allocation need to be considered. These are treated by applying second-best theory (Diamond and Mirrlees, 1971). Consider, for example, the pricing problem facing a public transit authority. The underpricing of urban auto travel may call for the underpricing of mass transit (Lévy-Lambert, 1968; Marchand, 1968); since optimal lump-sum redistribution is infeasible for informational reasons, the authority may choose to sacrifice some efficiency to improve equity by charging lower fares to needy groups

(Atkinson and Stiglitz, 1980), rationing, or nonlinear pricing (Wilson, 1993); administrative costs may preclude fine-tuning the fare according to distance travelled or time of day, leading to variants of Ramsey pricing (Mohring, 1970); the authority may face a deficit constraint, requiring it to price above marginal social cost (Boiteux, 1956); with distortionary taxation, the social cost of financing an extra dollar of transit authority deficit may significantly exceed one dollar (Vickrey, 1959); and the government may choose to deviate from marginal social cost pricing to provide the public transit authority with higher-powered incentives (Laffont and Tirole, 1993) or to achieve political objectives. These considerations will also cause second-best capacity to deviate from first-best capacity.

RICHARD ARNOTT AND MARVIN KRAUS

See also **consumption externalities; France, economics in (before 1870); network goods (empirical studies); network goods (theory); Pigouvian taxes; urban economics; value of time.**

Bibliography

Arnott, R. and Kraus, M. 2003. Principles of transport economics. In *Handbook of Transportation Science*, 2nd edn, ed. R. Hall. Boston: Kluwer.

Atkinson, A. and Stiglitz, J. 1980. *Lectures on Public Economics*. New York: McGraw-Hill.

Boiteux, M. 1956. Sur la gestion des monopoles publics astreints à l'équilibre budgétaire. *Econometrica* 24, 22–40.

Diamond, P. and Mirrlees, J. 1971. Optimal taxation and public production I: production efficiency and II: tax rules. *American Economic Review* 61, 8–27, 261–78.

Laffont, J.-J. and Tirole, J. 1993. *A Theory of Incentives in Procurement and Regulation*. Cambridge, MA: MIT Press.

Lévy-Lambert, H. 1968. Tarification des services à qualité variable: Application aux péages de circulation. *Econometrica* 36, 564–74.

Marchand, M. 1968. A note on optimal tolls in an imperfect environment. *Econometrica* 36, 575–81.

Mohring, H. 1970. The peak-load problem with increasing returns and pricing constraints. *American Economic Review* 60, 693–705.

Mohring, H. and Harwitz, M. 1962. *Highway Benefits: An Analytical Framework*. Evanston, IL: Northwestern University Press.

Pigou, A. 1947. *A Study in Public Finance*, 3rd edn. London: Macmillan.

Scotchmer, S. 2002. Local public goods and clubs. In *Handbook of Public Economics*, vol. 4, ed. A. Auerbach and M. Feldstein. Amsterdam: North-Holland.

Strotz, R. 1965. Urban transportation parables. In *The Public Economy of Urban Communities*, ed. J. Margolis. Washington, DC: Resources for the Future.

Vickrey, W. 1959. Statement on the pricing of urban street use. *Hearings: US Congress, Joint Committee on Metropolitan Washington Problems*, 11 November, 454–77.

Vickrey, W. 1969. Congestion theory and transport investment. *American Economic Review Proceedings* 59, 251–60.

Wilson, R. 1993. *Nonlinear pricing*. Oxford: Oxford University Press.

conjectural equilibria

In an economy with very many agents the market environment of any one of these is independent of the market actions he decides upon. More generally one can characterize an economy as *perfectly competitive* if the removal of any one agent from the economy would leave the remaining agents just as well off as they were before his removal. (The economy is said to satisfy a 'no surplus' condition; see Makowski, 1980; and Ostroy, 1980.) When an economy is not perfectly competitive, an agent in making a decision must take note of its effect on his market environment, for example, the price at which he can sell. This effect may not be known (or known with certainty) and will therefore be the subject of *conjecture*. A conjecture differs from expectations concerning future market environments which may, say, be generated by some stochastic process. It is concerned with responses to the actions of the agent.

In the first instance then the topic of conjectural equilibria is that of an economy which is not perfectly competitive by virtue of satisfying a no surplus condition. But, as we shall see, an economy could fail to satisfy this condition and yet have a perfectly competitive equilibrium.

By an equilibrium in economics we usually mean an economic state which is a rest (critical) point of an (implicit) dynamic system. For instance, it is postulated in the textbooks that, when at going prices the amount agents wish to buy does not equal the amount they wish to sell, prices will change. Strictly this should mean that there would, in such a situation, be an incentive for some agent(s) to change prices. This causes difficulties when the economy is perfectly competitive (Arrow, 1959) since it implies that the agent can influence his market environment by his own actions. That is one reason why a fictitious auctioneer has been introduced to account for price changes.

When the economy is not perfectly competitive these difficulties are avoided. A price will be changed if some agent conjectures that such a change would be to his advantage. As a corollary then a conjectural equilibrium must be a state from which it is conjectured by each agent that it would be disadvantageous to depart by actions which are under the individual agent's control. (For a formal definition see below.)

But there are other difficulties. In particular, there is the question of the source of conjectures. If these are taken as given exogenously then there are many states which could be conjectural equilibria for *some* conjectures. It should be noted that a similar objection can be raised in conventional equilibrium analysis. There it is the preferences of agents which are taken as exogenous and there too there are many equilibria which are compatible with some (admissible) preferences. However, while conjectures may turn out to be false and this may occasion a change in conjectures, it is less easy to point to equally simple and convincing endogenous mechanisms of preference change. For that reason one may feel that conjectural equilibrium requires that conjectures are in some sense correct ('rational'). For if they are not they will change in the light of experience. This argument is considered below.

The reason why the idea of conjectural equilibria is of interest is that economies which are not intrinsically perfectly competitive (for example, because of the large number of agents) are of interest and because it allows one to study price formation without an auctioneer.

An illustration

Consider two agents each of whom can chose an action a_i from a set of action A_i. Let $A = A_1 \times A_2$ with elements $a = (a_1, a_2)$. Then a conjecture c_i is a map from $A \times A_i$ to A_j written as

$$C_i = \theta_i(a, a_i').$$

Its interpretation is this: given the actions of the two agents (a), C_i is the action of j conjectured by i to be result from his choice of a_i'. (In a more general formulation the conjecture can be a probability distribution but that is not considered here.) We require conjectures to be *consistent*:

$$\theta_i(a, a_i) = a_j \qquad (1)$$

This says that if agent i continues in his action a_i then he conjectures that j will do likewise. (This use of the word 'consistent' is *not* that of Bresnahan, 1981, and others who use it to mean 'correct'.)

Suppose now that there is a function v from A to R^2, written as $v(a) = [v_1(a), v_2(a)]$, which gives the payoffs to the agents as a function of their joint action a. Consider a^* to be one such joint action. One says that a^* is a *conjectural equilibrium* for the two agents if

$$v_i[a_i, \theta_i(a^*, a_i)] \leq V_i[a_i^*, \theta_i(a^*)] \qquad (2)$$
$$\text{all } a_i \in A_i, \quad i = 1, 2$$

That is, the joint action a^* is a conjectural equilibrium if no agent, given his conjecture, believes that he can improve his position by deviating to a different action.

It is not the case that conjectural equilibrium, as defined, always exists. For instance in the case of a duopoly in a homogeneous product where the action is 'setting the price', v may not be concave and a sensible conjecture may have discontinuities. One thus needs special assumptions to ensure existence or one must face the possibility that agents do not chose actions but probability distributions over actions (mixed strategies); for example, Kreps and Wilson (1982) in their work on sequential equilibrium employ conjectures which are probability distributions.

Supposing that a conjectural equilibrium exists, one may reasonably argue that until conjectures are less arbitrarily imposed on the theory not much has been gained – almost any pair of actions could be a conjectural equilibrium. A first attempt to remedy this is to ask that conjectures be correct (rational). If that is to succeed in any simple fashion it will be necessary to suppose that each agent has a unique best action under this conjecture. This is very limiting and it means that some of the classical duopoly problems cannot be resolved in this way.

Let the status quo again be a^*. Then if θ_1^* and θ_2^* are correct conjectures it must be that

$$v_2\{\theta_1^*(a^*, a_2), \theta_1^*[(a_1, a_1^*), \theta_2^*(a^*, a_2)]\}$$
$$> v_1\{a_s', \theta_1^*[(a_1^*, a_2), a_1']\} \quad \text{all } a_1' = A_2.$$
$$(3)$$

$$v_1\{\theta_2^*(a^*, a_2), \theta_1^*[(a_1^*, a_2), \theta_2^*(a^*, a_2)]\}$$
$$> v_1\{a_s', \theta_1^*[(a_1^*, a_2), a_1']\} \quad \text{all } a_1' = A_1.$$
$$(4)$$

A *rational conjectural equilibrium* is then a conjectural equilibrium a^* (with conjectures $\theta_1^*(\cdot), \theta_2^*(\cdot)$ which satisfy (3) and (4)). It must be re-emphasized that such an equilibrium may not exist for some A and v (see Gale, 1978; Hahn, 1978).

However, the idea is simple and, where applicable, coherent. It has however been criticized (in a somewhat intemperate and muddled paper) by Makowski (1983). This criticism appears to have had some appeal to some game theorists who like to think of games in extensive form (which they sometimes like to call dynamic). The criticism is this: when agent 1 deviates from a^* he is interested in the payoffs which he will get given this deviation and agent 2's response. This payoff Makowski thinks of as accruing in the 'period' after agent 1's deviation. But when agent 2 responds in that period he is interested in this payoff in the period following this response. So the agents expect 'the game to end' in different periods (Makowski, 1983, p. 8). Moreover, after agent 2 has responded, agent 1, in his turn, will again want to respond, that is, deviate from the deviation he started with. This criticism is then illustrated with an example in which one agent expects the other to return to the status quo *after* he has deviated from it.

All of this, however, is wrong. Firstly, if one wants to give a time interpretation to conjectures and so forth, then actions must be thought of as strategies. That is, the deviating agent deviates in one or more elements of his plan over the whole length of the game (perhaps infinite). Under correct conjectures responses and counter-responses are taken into account in evaluating the benefits of deviation. Hence, and secondly, a deviating agent is in this situation never surprised by the response of the other, which therefore does not lead him to further revise his deviation. On the definition, agent 1 expects the response to his deviation to be $\theta_1(a^*, a_1)$. Suppose this gives a_2 which is correct. Then that agent knows that the new status quo will be $(a_1, a_2) = a$ and if he has calculated benefits correctly he will not wish to deviate again.

However, there is the following to be said in favour of Makowski's criticism. Deviations in strategies may not be observable by the other agent. Therefore in traditional duopoly models with a sequential structure the re-interpretation of actions as strategies may be inappropriate. There is some evidence that in the duopoly literature with conjectures the consequent difficulties have not always been appreciated. It is also the case that too little attention has been paid to the assumption of a unique best response on which the above formulation depends.

An alternative to rational conjectures are *reasonable conjectures* (Hahn, 1978). A conjecture is reasonable if acting on any other conjecture would lower profits given the conjectures of other firms. Suppose that $\bar{\theta}$ is the set of all possible consistent conjectures. For any $\theta_i \in \bar{\theta}$, assume that there is a unique optimizing choice of output by firm i of $y_i(\theta_i)$. Then i's conjecture $\theta_i^0 \in \theta$ is reasonable if given jth conjecture θ_j:

$$v_i\left[y_i(\theta_i^0), y_j(\theta_j)\right] \geq \hat{v}_i\left[y_i(\theta_i'), y_j(\theta_j)\right]$$
$$\text{all } \theta_i' \in \theta. \tag{5}$$

But then a *reasonable conjectural equilibrium* is a pair (θ_1^0, θ_2^0) each in θ such that

$$v_i\left[y_i(\theta_i^0), y_j(\theta_j^0)\right] \geq \hat{v}_i\left[y_i(\theta_i'), y_j(\theta_j^0)\right],$$
$$i, j = 1, 2, \quad \theta_i' \in \bar{\theta}. \tag{6}$$

This is just a Nash equilibrium where conjectures are interpreted as strategies (Hart, 1982).

While this is still quite demanding, it is significantly weaker than (3). If equilibria exist they may be 'bootstrap equilibria', that is, they will depend on beliefs about the actions of others, which beliefs may be incorrect. There is certainly no ground for believing that they will be efficient.

One can go one step further in the direction of plausibility by requiring that conjectures be reasonable only for small, or infinitesimal, deviations from the status quo. After all, large experiments are likely to be costlier than small ones. This will allow a larger class of reasonable conjectures and equilibria.

General conjectural equilibrium

It is fair to say that at present general equilibrium theory is in some way complete only for a perfectly competitive economy, that is, one where the returns to an individual agent are just equal to the contribution which he makes (Makowski, 1980; Ostroy, 1980). In general (although there are exceptions) such an economy exists when it is large (for example, it consists of a non-atomic continuum of agents). But there is now another possibility: an economy can be perfectly competitive if agents conjecture that their market actions will have no effect on the prices at which they can trade.

The following assertion will be clear from what has already been discussed. Let us say that an economy is *intrinsically* perfectly competitive if it satisfies the *no-surplus condition*. Then perfectly competitive conjectures are rational if an economy is intrinsically perfectly competitive. But perfectly competitive conjectures can be reasonable even when the economy is not intrinsically perfectly competitive. That is, conjectures may be such that, if an agent acts on any conjecture other than the perfectly competitive one, his profits will be lower. For instance, this may even be the case for two duopolists with constant marginal costs whose conjectures refer to the price charged by the rival firm. It will also be clear that if we do not require conjectures to be either reasonable or rational then, in general, conjectures can be found to support a competitive equilibrium in an economy which is not intrinsically perfectly competitive.

In a general equilibrium context it is not clear what it is that firms are supposed to conjecture. In some sense the conjecture must refer to the reaction of the whole economy to the action of the conjecturing agent. In other words, it is not obvious how to define a game which adequately represents the economy. But in what sense?

Consider an economy with n produced goods and m non-produced goods. For simplicity suppose that all firms are single-product firms and that all firms producing the same good are alike, including their conjectures. There are very many households whose reasonable conjectures are always the competitive one. Households receive the profits of firms. Since the action of any one firm can affect the prices at which households can trade it is not at all clear what it is in the households' interest that the firms should maximize (Gabszewicz and Vial, 1972). If all households are alike it could be their common utility function, but that seems far removed from the world. I shall arbitrarily assume that firms maximize their profits in terms of one of the non-produced goods, say the first. This is arbitrary but it seems to me equally dubious to suppose that firms always choose in the 'best interests of shareholders', especially when that interest is often difficult and sometimes impossible to define.

Let $p \in R^n$, $w \in R_+^{m-1}$ be the price vectors in terms of good m of produced and non-produced goods respectively (so $w_m \equiv 1$). Let $y_j \in Y_j \subset R^{n+m}$ be the production of firm j where $y_{ij} > 0$ is its output of good j, $y_{ii} < 0$ is an input of good i, produced or non-produced. Let $y = \Sigma y_j$, where $y_j \in Y_j$ all j. Let $z \in R_+^m$ be the endowment of non-produced goods and

$$F = \{y \mid y \geq (0, -z)\}$$

so that F is the set of feasible net production vectors Y. Let θ_{hj} be the share of household h in firm j.

Given any $y \in F$ we think of each household as endowed with a certain strictly positive stock of non-produced goods and $\theta_{hj}Y_j$ of the production of firm j. To avoid unnecessary complications assume θ_{hj} ($j = 1, \ldots, n$) to be such that if z_h is the stock of non-produced goods owned by household h:

$$\text{For all } y \in F : z_h + \sum_j \theta_{hj} y_j \geq 0 \quad \text{all } h.$$

$$(7)$$

Households consume both types of goods. Hence for any $y \in F$ there is now an associated pure exchange economy where each household's endowment is given by (7). Making the usual assumptions there will exist at least one equilibrium $[p(y), w(y)]$. Suppose for the moment that there is only one for each $y \in F$.

Now firm j in this equilibrium observes $[p(y), w(y)]$ and will deviate from y_j (if it deviates at all) if it can thereby increase its conjectured profits. Let

$$\hat{\pi}_j \Big[p(y), w(y), y_j' \Big]$$

be the conjectural profit function of firm j. Then $y^0, p(y^0), w(y^0)$ is a conjectural equilibrium if for all $j = 1, \ldots, n$:

$$\hat{\pi}_j \Big[p(y^0), w(y^0), y_j^0 \Big] \geq \hat{\pi}_j \Big[p(y^0), w(y^0), y_j' \Big]$$
$$\text{all } y_j' \in Y.$$

$$(8)$$

Such a conjectural equilibrium will exist if all $\hat{\pi}_j(\cdot)$ are quasi-concave, an assumption for which there is scant justification (Hahn, 1978).

If we demand that conjectures be rational then conjectured and actual profit must coincide for all y_k' (the two coincide for $y_k' = y_k^0$ by the requirement that conjectures be consistent). One proceeds as follows. Let $y_k' = y_k^0$. Given the conjectures of the remaining firms find the conjectural equilibrium of the economy $p\{y^*(k), w[y^*(k), y^*(k)]\}$, where $y(k)$ is the vector y with y_k' in the kth place and condition (8) is not imposed for firm k. One then requires that for all $y_k' > 0$

$$\hat{\pi}_k \Big[p(y^0), w(y^0), y_k' \Big]$$
$$= \pi_k \{ p[y^*(k)], w[y^*(k)], y_k' \} \qquad (8a)$$

where $\pi_k(\cdot)$ is actual profit. For rational conjectures this should be true for all k.

It will be seen that rational conjectural equilibrium is very demanding. For a certain class of conjectures it will not even exist (Gale, 1978; Hahn, 1978). More importantly, the whole procedure breaks down if given a deviation by k, the conjectural equilibrium, is not unique. Lastly, even if by sufficient assumptions one overcomes these difficulties, it is not agreeable to common sense to suppose that firms can correctly calculate general equilibrium responses to their actions, nor is it obvious that they should always be concerned only with equilibrium states.

Reasonable conjectures do not fare much better, although a notable contribution to their study has recently been made by Hart (1982). Hart notices that conjectures of firms induce a supply correspondence (not generally convex) on their part. Here let us suppose that we can in fact speak of supply functions. These can be thought of as strategies in a manner already discussed. A reasonable conjectural equilibrium then satisfies the condition that, given the supply functions of other firms, no deviation by firm k to another supply response can increase its profits. In (8) one then substitutes on the right-hand side for $y_j', \eta_j'[p(y^0), w(y^0)]$, an admissible supply function (see Hart, 1982) of j and requires the inequality to hold for all such functions. Of course, one has

$$y_j^0 = \eta_j^0 \Big[p(y^0), w(y^0) \Big]$$

for a reasonable conjectural equilibrium.

To show existence of such an equilibrium will require strong assumptions. The technicalities will be found in Hart (1982). However, one of the assumptions which he makes is not only technically useful but economically sensible since it leads firms to face a simpler task in forming conjectures. Hart supposes the economy to consist of a number of islands each of which has many consumers and one firm of each type ($j = 1, \ldots, n$). The islands are small replicas of the whole economy. But households have shares in firms on all islands so that if there are enough islands their share in any firm on their own island is very small. That means that any firm can disregard the effect of a change in its own profits on the demand for the good it produces. To make this work one supposes that produced goods are totally immobile between islands while non-produced goods are totally mobile. By an appropriate assumption on consumers on each island one ensures that they all have the same demand. Lastly, since shares in a firm are held on many different islands the firm, in acting in the shareholder's interest is justified in neglecting the effect of its actions on relative prices on its own island and so is justified in maximizing profits.

From the point of view of conjectural equilibrium the island assumption allows firms (both reasonably and rationally) to ignore effects of their own actions on

w – the price vector of non-produced goods. These will be determined by demand and supply over all islands and in this determination any one firm can be regarded as playing a negligible role. This is some gain in realism. But after all allowances have been made it is still true that (*a*) the assumptions required for the existence of reasonable conjectural equilibrium are uncomfortably strong and (*b*) even when that is neglected such an equilibrium seems to have small descriptive power.

Simpler approaches

Negishi (1960) made the first, justifiably celebrated, attempt to incorporate imperfect competition in general equilibrium analysis. He did this by letting single product firms have consistent inverse demand conjectures (the case he studies most thoroughly makes these linear). Consistency is all he asked for of conjectures but he also needed the uncomfortable postulate that the resulting conjectural profit functions be quasi-concave. Later Hahn (1978), Silvestre (1977) and others added the requirement that, besides being consistent, the conjectured demand functions have, if differentiable, the correct slope at equilibrium (that is, that the conjecture be *infinitesimally* or 'first order' *rational*). It turns out that this extra requirement does not much restrict conjectures, nor thus the set of equilibria which can be generated by some conjectures. The reason roughly is this: in conjectural equilibrium, when conjectured profit functions are twice differentiable, the partial derivative of the conjectured profit function of firm j with respect to its own output much vanish. Suppose the economy to be in such an equilibrium and consider an infinitesimal output deviation by firm k. To find the equilibrium which ensues, differentiate all equilibrium relations, other than that for firm k, with respect to the output of firm k. Amongst these will be the condition that the marginal profit conjectured of every firm (other than k), be zero. Hence differentiation of that condition will yield second-order terms. But we can choose these arbitrarily since we are requiring only first-order rationality. One can show in fact that these second-order terms can be chosen so as to make the first-order conjectured change in profit of any firm k correspond to the actual change. (Details in Hahn, 1977.) Hence first-order rationality imposes few restrictions.

Both Hahn (1978) and Negishi (1979) have also considered kinked conjectures. The idea is this. If an agent can transact at the going price as much as he desires his conjectures are competitive. If he is quantity constrained (for example, if a firm cannot sell an amount determined by equality between marginal cost and price) his conjectures are non-competitive. That is, he considers that a price change is required to relax the quantity constraint. The fixprice methods of Drèze (1975) and others can be interpreted as an extreme form of such conjectures – for instance to relax a constraint on sales, price, it is conjectured, must be reduced to zero.

To such conjectures there have been two objections. Firstly, they assume that an agent's conjectures are not influenced by constraints on others. For instance, a firm which can hire as much labour as it wants at the going wage while workers cannot sell as much as they like does not conjecture that it could have the same amount of labour at a lower wage. To this one can answer that it is not easy for an agent to observe the quantity constraints on others. For instance, unemployment statistics do not tell us whether workers have chosen not to work or whether they are constrained in their sale of labour. None the less, this objection has some force and needs further study with proper attention to the information of agents.

The other objection is that these kinked conjectures are not explained. That is true if explanation turns on what an agent knows or can learn. None the less, the hypothesis seems to be to have psychological verisimilitude. If I can always sell my labour at the going wage there is little occasion for the difficult conjecturing of what would happen if I raised my wage. This is not so if I find that I cannot find employment at the going wage.

In any event these simpler approaches allow one to incorporate traditional monopolistic competition in a general equilibrium framework. Of course, some of the assumptions such as concave conjectured profit functions are strong. On the other hand, one can now allow for a certain amount of increasing returns (Silvestre, 1977).

Some conclusions

The conjectural approach has this merit: it takes proper and explicit note of the perceptions by individuals of their market environment. Economic theory perhaps too often neglects the possibility that what is the case may depend on what agents believe to be the case. Historians and others have long since studied the intimate mutual connection between beliefs and events but economists have not made much headway here. The conjectural approach is perhaps a small beginning. For it deals with the theories agents hold and this must plainly enter into our theory of agents.

In particular one should not pay too much attention to the objection that conjectures may not be derivable from some first principles of rationality. It seems to me quite proper to find their description in history. Nor, as has been argued, will an appeal to learning render conjectures in some sense objectively justifiable. This is clear from the discussion of reasonable conjectures and from the costs of experimentation. For hundreds of years witches were burned in the light of a reasonable theory which few would now regard as having proper objective correlatives. There is no reason to suppose that it is possible for businesses or governments now to do better than some of the best minds of the past.

From a more immediately relevant standpoint, conjectural theories are of interest because they attempt a general equilibrium analysis of non-perfect competition. It is good to know that in a proper sense perfectly competitive economies can be viewed as limiting Cournot conjectural equilibrium economies (Novshek and Sonnenschein, 1978). But this knowledge does not contribute to the study of properly imperfectly competitive economies. Again the study of fixprice equilibria has borne some fruits, but not those which were first sought by Triffin (1940) when he proposed a framework for general equilibrium with monopolistic competition. If it is the case that actual economies are not perfectly competitive nor that they behave 'as if' they were, then the task set by Triffin requires serious attention, and it is likely that conjectural theories will have a role to play.

Recent developments in game theory (for example, Kreps and Wilson, 1982) suggest that these two conjectures will have to play a part. Indeed, quite generally in that theory players conjecture that their opponent is 'rational' in an appropriate sense. It is not the case that the conjectural equilibrium approach is an alternative to the game theoretic one.

F. HAHN

See also **auctioneer.**

Bibliography

Arrow, K.J. 1959. Toward a theory of price adjustment. In M. Abramovitz et al., *The Allocation of Economic Resources*. Stanford: Stanford University Press.

Bresnahan, T.F. 1981. Duopoly models with consistent conjectures. *American Economic Review* 71, 934–45.

Drèze, J. 1975. Existence of equilibrium under price rigidity and quantity rationing. *International Economical Review* 16, 301–20.

Gabszewicz, J.J. and Vial, J.D. 1972. Oligopoly 'à la Cournot' in general equilibrium analysis. *Journal of Economy Theory* 4, 381–400.

Gale, D. 1978. A note on conjectural equilibria. *Review of Economic Studies* 45(1), 33–8.

Hahn, F.H. 1977. Exercise in conjectural equilibria. *Scandinavian Journal of Economics* 79, 210–26.

Hahn, F.H. 1978. On non-Walrasian equilibria. *Review of Economic Studies* 45, 1–17.

Hart, O. 1982. Reasonable conjectures. Theoretical Economics Paper No. 61. STICERD, London School of Economics.

Kreps, D.M. and Wilson, R.B. 1982. Sequential equilibria. *Econometrica* 50, 863–94.

Makowski, L. 1980. A characterisation of perfectly competitive economies with production. *Journal of Economic Theory* 22, 208–21.

Makowski, L. 1983. 'Rational conjectures' aren't rational and 'reasonable conjectures' aren't reasonable. Discussion Paper No. 60. SSRC Project on Risk, Information and Quantity Signals, Cambridge University.

Negishi, T. 1960. Monopolistic competition and general equilibrium. *Review of Economic Studies* 28, 196–202.

Negishi, T. 1979. *Micro-Economic Foundations of Keynesian Macro-Economics*. Amsterdam: North-Holland.

Novshek, W. and Sonnenschein, H. 1978. Cournot and Walras equilibrium. *Journal of Economic Theory* 19, 223–66.

Ostroy, J. 1980. The no-surplus condition as a characterisation of perfectly competitive equilibrium. *Journal of Economic Theory* 22, 183–207.

Silvestre, J. 1977. A model of general equilibrium with monopolistic behavior. *Journal of Economic Theory* 16, 425–42.

Triffin, R. 1940. *Monopolistic Competition and General Equilibrium Theory*. Cambridge, MA: Harvard University Press.

Ulph, D. 1983. Rational conjectures in the theory of oligopoly. *International Journal of Industrial Organization* 1(2), 131–54.

conspicuous consumption

Conspicuous consumption means the use of consumer goods in such a way as to create a display for the purpose of impressing others rather than for the satisfaction of normal consumer demand. It is consumption intended chiefly as an ostentatious display of wealth. The concept of conspicuous consumption was introduced into economic theory by Thorstein Veblen (1899) in the context of his analysis of the latent functions of 'conspicuous consumption' and 'conspicuous waste' as symbols of upper-class status and as competitive methods of enhancing individual prestige.

Veblen argued that the leisure class is chiefly interested in this type of consumption, but that, to a certain degree, it exists in all classes. The leisure class undoubtedly has much more opportunity for this kind of consumption. The criterion as to whether a particular outlay fell under the heading of conspicuous consumption was whether, aside from acquired tastes and from the canons of usage and conventional decency, its result was a net gain in comfort or in fullness of life.

It is widely though that Veblen introduced the concept of conspicuous consumption into economic literature, but it was known much earlier. Adam Smith (1776, Book I, ch.11) wrote about people who like to possess those distinguishing marks of opulence that nobody but themselves can possess. In the eyes of such people the merit of an object that is in any degree either useful or beautiful is greatly enhanced by its scarcity, or by the great amount of labour required to accumulate any considerable quantity of it. This is the labour for which nobody but themselves can afford to pay. Smith concluded that this domain was ruled by fashion. J.-B. Say and McCulloch wrote about this issue in a similar way. But the author who first used the term 'conspicuous consumption' was the Canadian economist John Rae (1796–1872). His explanation of the nature and effects of luxury was based on the meaning of vanity in human life. He understood vanity to be the mere desire for superiority over others without any reference to merit. The aim is to have what others cannot have, whereas the stimulus to productivity in economic life is the passion for effective accumulation: 'Articles of which consumption is conspicuous, are incapable of gratifying this passion' (Rae, 1834).

However, it was Veblen who introduced the concept of conspicuous consumption as a phenomenon important for the understanding of consumption as a whole. He gave Rae no reference at all.

Veblen's historical and socio-economic explanation of this institution gave as a result the so-called 'Veblen effect'. This is the phenomenon whereby as the price of an article falls some consumers construe this as a reduction in the quality of the good or loss of its 'exclusiveness' and cease to buy it.

F. STANKOVIĆ

See also **Rae, John; Veblen, Thorstein Bunde.**

Bibliography

Mason, R.S. 1981. *Conspicuous Consumption: A Study of Exceptional Consumption Behaviour*. New York: St Martin's Press.

Rae, J. 1834. *The Sociological Theory of Capital*, ed. C. Mixter. New York: Macmillan.

Smith, A. 1776. *An Inquiry into the Nature and Causes of the Wealth of Nations*, ed. E. Cannan, London: Methuen, 1981.

Sweezy, P. 1952. Veblen and Marx. In *Socialism and American Life*, 2 vols. ed. D.D. Egbert and S. Persons. Princeton: Princeton University Press.

Veblen, T. 1899. *The Theory of the Leisure Class*. London: George Allen & Unwin.

constant and variable capital. *See* **Marx's analysis of capitalist production.**

constitutions, economic approach to

The economic approach to constitutions applies the methodology of economics to the study of constitutions, just as public choice applies this methodology to the full range of topics of political science.

The economic approach to constitutions began with *The Calculus of Consent* by James Buchanan and Gordon Tullock (1962, hereafter B&T). Theirs was largely a *normative* analysis of what ought to go into a constitution. Their main findings and the literature that grew out of their work are reviewed first, after which the *positive* stream of the literature is discussed.

Normative research on constitutions

Arguably the most important contribution of *The Calculus* was to view democracy as a two-stage process. In stage one, institutions to make future collective decisions are placed into the constitution. In stage two, collective decisions are made using these rules. The long-run nature of the choices at the first stage creates considerable uncertainty about the consequences of different voting rules. This uncertainty makes unanimous agreement on the rules of the political game likely, even though individuals would disagree in stage two about the outcomes of the game. This unanimity at the constitutional stage provides the normative underpinning for the constitution (B&T, p. 7). Harsanyi (1955) also used uncertainty over future positions to produce unanimity and to provide a normative argument for a Benthamite social welfare function (SWF), as did Rawls (1971) in his ethical theory of a social contract. Mueller (1973) discussed conditions under which a B&T constitution maximizes a Harsanyian SWF.

Another innovation in *The Calculus* was to introduce the *external costs of collective decisions* (B&T, pp. 63–8). When a collective choice is made without the consent of all members of the community, the decision can make some members worse-off. The votes of those favouring the decision thus impose a negative externality on those opposing it. The smaller the majority required to pass an issue, the more likely it is that an individual is on the losing side. However, the amount of time required to make a collective decision is also likely to increase with the required majority. The optimal majority minimizes the sum of collective decisions' external and decision-making costs.

There is nothing in B&T's costs-minimization-approach that implies that the optimal majority is likely to be a simple majority, and thus their approach does not account for this rule's ubiquitous use. The approach does imply the widespread use of the simple majority rule, if one of the two cost curves – most plausibly decision-making costs – has a sharp discontinuity at 50 per cent (Mueller, 2003, pp. 76–8).

Rae (1969) used the two-stage approach to provide a completely different normative justification for the simple majority rule. At the constitutional stage, each individual is uncertain of whether he will favour x or $\sim x$ in future votes on these binary issues. The expected gain if an individual favours x and x wins equals the expected loss if x wins and the individual favours $\sim x$. Rae further assumed that the probability of favouring x equals the probability of favouring $\sim x$. An egoist chooses the voting rule that minimizes the probability that she favours x in the future and $\sim x$ is imposed, or that she favours $\sim x$ and x is imposed. The simple majority is the only rule satisfying this condition. (For additional discussion and references see, Rae and Schickler, 1997.)

Mueller (2001) generalized the two-stage approach to show that the optimal majority for binary choices depends on the relative payoffs from the two issues. (Riley, 2001, presents a game theoretic analysis of a two-stage constitutional process.) As the loss to those favouring x rises relative to the gain to those favouring $\sim x$, higher required majorities become optimal to implement $\sim x$, with unanimity being optimal when the asymmetry in payoffs is very large. Mueller (1991; 1996, ch. 14) employed this analysis to explain why placing rights to act into a constitution would maximize the expected utilities of those writing it.

Positive research on constitutions

The positive literature of constitutions falls into two categories: studies of constitutional conventions and of the consequences of constitutions. The second category is obviously very large, and so I provide only the flavour of this type of work.

Charles Beard's work (1913) might well be regarded as the first *economic* analysis of the Philadelphia Convention. Beard stressed the self-interest of the participants, and claimed that the final product reflected the interests of the landowning aristocracy. In an equally cynical analysis, Landes and Posner (1975, p. 893) claimed that the First Amendment was a result of pressure from 'publishers, journalists, pamphleteers, and others who derive pecuniary and nonpecuniary income from publication and advocacy of various sorts'. Case studies of constitutional conventions confirm the importance of the self-interest of the participants in determining the constitution's content. For example, representatives from small parties favour rules that produce proportional representation and low percentage thresholds for taking seats in the parliament. Representatives from large parties favour the reverse. If delegates are selected geographically, the constitution protects geographic interests. (For further discussion and references to the literature, see Elster, 1991, and Mueller, 1996, ch. 21). Econometric analyses confirm these findings. McGuire and Ohlsfeldt (1986) and McGuire (1988) concluded that the votes of delegates to the Philadelphia convention reflected both their personal interests and those of their constituencies. Eavey and Miller (1989) reached the same conclusion from the voting patterns of those who ratified the Pennsylvania and Maryland constitutions.

A key decision facing any constitutional convention is whether to design institutions that will produce a two-party system or a multiparty system. In practice, this choice appears to rest upon the number of representatives elected from each electoral district (Taagepera and Shugart, 1989; Lijphart, 1990; Mueller, 1996, chs. 8–10). Recent theoretical and empirical work by Persson and Tabellini (1999; 2000; 2003; 2004a; 2004b) and Persson, Roland and Tabellini (2000) demonstrates the economic importance of this choice. They find more rent seeking, more corruption, more redistribution and larger deficits in multiparty systems. Presidential systems lead to smaller governmental sectors because they generally contain stronger checks and balances than parliamentary systems. (For a review and references to other contributions, see Persson and Tabellini, 2004a.)

Conclusions

There are two kinds of people in the world: those who believe that constitutions matter and those who do not. The contributors to the literature reviewed here fall into the former category. Their work helps illustrate *why and in what way* constitutions matter, and further illustrates the fruitfulness of undertaking an economic approach to the study of constitutions.

DENNIS C. MUELLER

See also **Buchanan, James M.; collective rationality.**

Bibliography

Beard, C. 1913. *An Economic Interpretation of the Constitution of the United States.* New York: Macmillan, 1941.

Buchanan, J. and Tullock, G. 1962. *The Calculus of Consent.* Ann Arbor: University of Michigan Press.

Eavey, C. and Miller, G. 1989. Constitutional conflict in state and nation. In *The Federalist Papers and the New Institutionalism,* ed. B. Grofman and D. Wittman. New York: Agathon Press.

Elster, J. 1991. Arguing and bargaining in two constituent assemblies. Mimeo. Storrs Lectures, Yale Law School.

Harsanyi, J. 1955. Cardinal welfare, individualistic ethics, and interpersonal comparisons of utility. *Journal of Political Economics* 63, 309–21.

Landes, W. and Posner, R. 1975. The independent judiciary in an interest-group perspective. *Journal of Law and Economics* 18, 875–901.

Lijphart, A. 1990. The political consequences of electoral laws, 1945–85. *American Political Science Review* 84, 481–96.

McGuire, R. 1988. Constitution making: a rational choice model of the Federal Convention of 1787. *American Journal of Political Science* 32, 483–522.

McGuire, R. and Ohlsfeldt, R. 1986. An economic model of voting behavior over specific issues at the Constitutional Convention of 1787. *Journal of Economic History* 46, 79–111.

Mueller, D. 1973. Constitutional democracy and social welfare. *Quarterly Journal of Economics* 87, 60–80.

Mueller, D. 1991. Constitutional rights. *Journal of Law, Economics, and Organization* 7, 313–33.

Mueller, D. 1996. *Constitutional Democracy.* Oxford: Oxford University Press.

Mueller, D. 2001. The importance of uncertainty in a two-stage theory of constitutions. *Public Choice* 108, 223–58.

Mueller, D. 2003. *Public Choice III.* Cambridge: Cambridge University Press.

Persson, T., Roland, G. and Tabellini, G. 2000. Comparative politics and public finance. *Journal of Political Economy* 108, 1121–61.

Persson, T. and Tabellini, G. 1999. The size and scope of government: comparative politics with rational politicians. *European Economic Review* 43, 699–735.

Persson, T. and Tabellini, G. 2000. *Political Economics – Explaining Economic Policy*. Cambridge, MA: MIT Press.

Persson, T. and Tabellini, G. 2003. *Economic Effects of Constitutions*. Cambridge, MA: MIT Press.

Persson, T. and Tabellini, G. 2004a. Constitutions and economic policy. *Journal of Economic Perspectives* 18, 75–98.

Persson, T. and Tabellini, G. 2004b. Constitutional rules and fiscal policy outcomes. *American Economic Review* 94, 25–45.

Rae, D. 1969. Decision-rules and individual values in constitutional choice. *American Political Science Review* 63, 40–56.

Rae, D. and Schickler, E. 1997. Majority rule. In *Perspectives on Public Choice*, ed. D. C. Mueller. Cambridge, MA: Cambridge University Press.

Rawls, J. 1971. *A Theory of Justice*. Cambridge, MA: Belknap Press.

Riley, J. 2001. Constitutional democracy as a two-stage game. In *Constitutional Culture and Democratic Rule*, ed. J. Ferejohn, J. Rakove and J. Riley. Cambridge: Cambridge University Press.

Taagepera, R. and Shugart, M. 1989. *Seats and Votes*. New Haven: Yale University Press.

consumer expenditure

The study of consumers' expenditure, both in total and in composition, has always been of major concern to economists. Neoclassical economics sees the delivery of individual consumption as the main object of the economic system, so that the efficiency with which the economy achieves this goal is the criterion by which alternative systems, institutions and policies are to be judged. Within a capitalist economy, such considerations lead to an examination of the relationship between *prices* and consumption behaviour, and theoretical development and empirical analysis have been a major continuous activity since the middle of the last century. Even older is the tradition of using individual household budgets to dramatize poverty, and the relationship between household incomes and household expenditure patterns has occupied social reformers, statisticians and econometricians since at least the 18th century. In more modern times, it has been recognized that the study of public finance and of taxation depends on a knowledge of how price changes affect the welfare and behaviour of individuals, and the recent development of optimal tax theory and of tax reform analysis has placed additional demands on our understanding of the links between prices, expenditures and welfare.

In the last fifty years, aggregate consumption has become as much of an object of attention as has its composition, and in spite of a common theoretical structure, there has been a considerable division of labour between macro economists, interested in aggregate consumption and saving, and micro economists whose main concern has been with composition, and with the study of the effects of relative prices on demand. The interest of macroeconomics reflects both long-term and short-term interests. What is not consumed is saved, saving is thrift and the basis for capital formation, so that the determinants of saving are the determinants of future growth and prosperity. More immediately, aggregate consumption accounts for a large share of national income, typically more than three-quarters, so that fluctuations in behaviour or 'consumption shocks' have important consequences for output, employment, and the business cycle. Since Keynes's *General Theory*, the consumption function, the relationship between consumption and income, has played a central role in the study of the macroeconomy. Since the 1930s, there has been a continuous flow of theoretical and empirical developments in consumption function research, and some of the outstanding scientific achievements in economics have been in this field.

In this essay, the major themes will be the interplay between theory and evidence in the study of consumers' expenditure and its composition. If economists have any serious claim to being scientists, it should be clearly visible here. The best minds in the profession have worked on the theory of consumption and on its empirical implementation, and there have always been more data available than could possibly be examined. I hope to show that there have been some stunning successes, where elegant models have yielded far from obvious predictions that have been well vindicated by the evidence. But there is much that remains to be done, and much that needs to be put right. Many of the standard presumptions of economics remain just that, assumptions unsupported by evidence, and while modern price theory is logically consistent and theoretically well developed, it is far from having that solid body of empirical support and proven usefulness that characterizes similar central theories in the natural sciences.

1 A simple theoretical framework

Almost all discussions of consumer behaviour begin with a theory of *individual* behaviour. I follow neoclassical tradition by supposing that such behaviour can be described by the maximization of a utility function subject to suitable constraints. The axioms that justify utility maximization are mild, see any microeconomic text such as Varian (1978/1984) or Deaton and Muellbauer (1980b), so that utility maximization should be seen as no more than a convenient framework that rules out the grossest kind of behavioural inconsistencies. The assumptions that have real force are those that detail the constraints facing individuals or else put specific structure on utility functions. Perhaps the most general specification of preferences that could be considered is one that is written

$$u_t = E_t\{f(q_1, q_2, \ldots, q_t, \ldots, q_T)\} \tag{1}$$

where u_t is utility at time t, E_t is the expectation operator for expectations formed at time t, q_1 to q_T are vectors of consumption in periods 1 to T, and $f(\cdot)$ is a quasi-concave function that is non-decreasing in each of its arguments. Several things about this formulation are worth brief discussion. The function $f(\cdot)$ yields the utility that would be obtained from the consumption vector under certainty, and it represents the utility from a *life-time* of consumption; the indices 1 to T therefore represent *age* with 1 the date of birth and T that of death. The expectation operator is required because choice is made subject to uncertainty, not about the choices themselves, which are under the consumer's control, but about the consequences of current choices for future opportunities. It is not possible to travel backward through time, so that choices once made cannot be undone, and yet the cost of current consumption in terms of future consumption foregone is uncertain, as is the amount of resources that may become available at future dates. The consumer must therefore travel through life, filling in the slots in (1) from left to right as best as he or she can, and at time (or age) t, everything to the left will be fixed and unchangeable, whether now seen to be optimal or not, while everything ahead of t is subject to the random buffeting of unexpected changes in interest rates, prices, and incomes. The solution to this sort of maximization problem has been elegantly characterized by Epstein (1975); here I shall work with something that is more restrictive but more useful and note in Section 3 below some phenomena that are better handled by the more general model.

Intertemporal utility functions are frequently assumed to be *intertemporally additive*, so that the preference rankings between consumption bundles in any two periods or ages are taken to be independent of consumption levels in any third period. If so, the utility function (1) takes the more mathematically convenient form

$$u_t = E_t \sum_{r=1}^{T} v_r(q_r). \tag{2}$$

Note that by writing utility in the form (2), since the expectation operator is additive over states of the world preferences are in effect assumed to be *simultaneously* additive over both states and periods, an assumption that can be formally defended, see Gorman (1982) and Browning, Deaton and Irish (1985). It has the consequence that risk aversion and intertemporal substitutability become two aspects of the same phenomenon. Individuals that dislike risk, and will pay to avoid it, will also attempt to smooth

their consumption over time and will require large incentives to alter their preferred consumption and saving profiles. Note also that the additive structure of (2) means that, unlike the case of (1), previous decisions are irrelevant for current ones. For decision-making at time t, bygones are bygones, and conditional on asset and income positions, future choices are unaffected by what has happened in the past. There can therefore be no attempt to make up for lost opportunities, nor can such phenomena as habit formation be easily modelled.

Because utility in (2) is intertemporally separable, maximization of life-time utility implies that, within each period, the period subutility function $v_t(\cdot)$ must be maximized subject to whatever total it is optimal to spend in that period. The period by period allocation of consumption expenditure to individual commodities need not, therefore, be planned in advance, but can be left to be determined when that period or age is reached, and period t allocation will follow according to the rule

$$\text{maximize } v_t(q_t) \quad \text{subject to } p_t \cdot q_t = x_t, \tag{3}$$

where p_t is the price vector corresponding to q_t and x_t is the total amount to be spent in t. Problem (3) is one of standard (static) utility maximization, though note that x_t is not given to the consumer, but is determined by the wider intertemporal choice problem. Nevertheless, not the least advantage of the intertemporally additive formulation is its implication that the composition of expenditure follows the standard utility maximization rule. It allows separate attention to be given to demand analysis on the one hand, i.e. to the problem (3), and to the consumption function on the other hand, this being understood to be the intertemporal allocation of resources, i.e. the determination of x_t.

Write the maximized value of utility from the period t problem as $\psi_t(x_t, p_t)$, where $\psi(\cdot)$ is a standard indirect utility function. The original intertemporal utility function then takes the form

$$u_t = E_t \sum_{r=0}^{T-t} \psi_r(x_{t+r}, p_{t+r}). \tag{4}$$

The constraints under which this function is maximized are most conveniently analysed through the conditions governing the evolution of wealth from period to period. If A_t is the (ex-dividend) value of assets at the start of period t, N_{it} is the nominal holdings of asset i with price P_{it}, d_{it} is the dividend on i paid immediately before the beginning of t, and y_t is income in period t, then

$$A_{t+1} = \sum_i N_{it}(P_{it+1} + d_{it+1}) \tag{5}$$

$$\sum_i N_{it} P_{it} = A_t + y_t - x_t. \tag{6}$$

Conditions (5) and (6) determine how wealth evolves from period to period, and the picture is completed by requiring that the consumer's terminal assets be positive, i.e.

$$A_{T+1} \geq 0 \tag{7}$$

To solve this problem, the technique of backward recursion is used. This rests on the observation that it is impossible to know what to do in period t without taking into account the problem in period $(t + 1)$, nor that in $(t + 1)$ without thinking about $(t + 2)$, and so on. However, in period T there is no future, so that looking ahead from date t, we can write subutility in period T in terms of that period's price and inherited assets, and we write this as v_T, i.e.

$$v_T = v_T(A_T) = \psi_T(A_T + y_T, p_T). \tag{8}$$

Given this, the consumer can look ahead from period t to period $(T - 1)$ and foresee that the problem then will be to choose the composition of assets N so as to maximize v_{T-1}, where

$$v_{t-1}(A_{T-1}) = \max_N [\![\psi_{T-1}(A_{T-1} + y_{T-1} - N \cdot P_{T-1}, p_{T-1})$$
$$+ E_{T-1}\{v_T[N \cdot (P_T + d_T)]\}]\!]. \tag{9}$$

At the next stage, assets in $(T - 2)$ will be allocated so as to trade off the benefits of consumption in $(T - 2)$ versus the benefits of A_{T-1} in v_{T-1} in (9) above and again yielding a maximized value v_{T-2}. As we follow this back through time, the consumer finally reaches the current period t, where he or she faces an only slightly complicated version of the usual 'today tomorrow' trade-off; the asset vector N must be chosen to solve the problem,

$$u_t = \max_N [\![\psi_t(A_t + y_t - N \cdot P_t, p_t)$$
$$+ E_t\{v_{t+1}[N \cdot (P_{t+1} + d_{t+1})]\}]\!]. \tag{10}$$

From this sequence of problems, several important results readily follow. First, consider the derivatives of each of the functions $v_r(A_r)$ which represent the marginal value of an extra unit of currency for the remaining segment of life time utility from r through to T. By the envelope theorem (see for example Dixit (1976) for a good exposition), it is legitimate to differentiate through the maximization problem, from which

$$v'_r(A_r) = \partial\psi_r/\partial x_r = \lambda_r, \text{ say,} \tag{11}$$

so that λ_r is the marginal utility of money in period r. Secondly, the maximization of (10) with respect to portfolio choice gives the relationship, for each asset i,

$$P_{it}\partial\psi_t/\partial x_t = E_t\{(P_{it+1} + d_{it+1})\partial\psi_{t+1}/\partial x_{t+1}\} \tag{12}$$

which, defining the asset *return* R_{it+1} as $(P_{it+1} + d_{it+1})/P_{it}$, and using (11) can be rewritten in the simple form

$$\lambda_t = E_t(\lambda_{t+1}R_{it+1}). \tag{13}$$

This equation, in current parlance often referred to as the 'Euler equation', can be used to derive many of the implications of the theory of consumption. Note first that it is little more than the standard result that the marginal rate of substitution between today's and tomorrow's consumption should be equal to the relative price. However, the equation is set in a multiperiod framework, not a two-period one, and it explicitly recognizes the uncertainty in both asset returns and in the value of money in subsequent periods. The equation also holds for all i, i.e. for all assets, so that the result also has implications for asset pricing as well as for consumption and saving, and for this reason the model is often referred to as the consumption-asset pricing model. I shall return to these implications below.

The theory as presented above is the modern equivalent of the life-cycle theory of consumption that dates back to Irving Fisher (1930) and Frank Ramsey (1928), and that had its modern genesis in the papers by Modigliani and Brumberg (1954) and (1954, published 1979). Modigliani and Brumberg's treatment differs from the above only in not explicitly modelling uncertainty, and by including only a single asset. The modern version appears first in Breeden (1979) and in Hall (1978), see also Grossman and Shiller (1981).

2 Predictions and evidence

One of the most important implications of the theory above, and of equation (13) in particular, is that the evolution of consumption over the life-cycle is independent of the pattern of income over the life-cycle. The asset evolution equations (5) and (6) allow consumers to borrow and lend at will, so that the only ultimate constraint on their consumption is one of life-time solvency. In consequence, consumption patterns are free to follow tastes, the evolution of family structure, or the different needs that come with ageing, provided that in the end total life-time expenditure lies within (total) life-time resources, whether from inherited wealth or from labour income. It is often assumed that tastes are such that consumers prefer to have a relatively smooth consumption stream, and this can be illustrated from a special case of equation (13). Assume that the within-period utility function is homothetic so that $\psi(x, p)$ is $\phi(x/a(p))$ for some linearly homogeneous function $a(\cdot)$, and that $\phi(\cdot)$ has the isoelastic from with elasticity $(1 - \sigma)$. Life-time utility takes the form

$$u_t = \sum_{r=0}^{T-t}(1 + \delta)^{-r}[x_{t+r}/a(p_{t+r})]^{1-\sigma} \tag{14}$$

where δ is the rate of pure time preference, and $\sigma \geq 0$ is the coefficient of relative risk aversion and the reciprocal of the intertemporal elasticity of substitution. Equation (14) can be used to evaluate (13), and gives immediately

$$E[\{(1 + r_{t+1})/(1 + \delta)\}\{c_t/c_{t+1}\}^\sigma] = 1$$

(15)

where r_{t+1} is the real after tax rate of interest from t to $t+1$ on any asset, and c_t is real consumption, x_t/a (p_t). Equation (15) shows that, if expectations are fulfilled, consumption will grow over the life-cycle if the real rate of interest is greater than the rate of pure time preference, and vice versa, while with $r_t = \delta$, consumption is constant with age. These results are of course an artefact of the specific assumptions about utility, and for any real household consumption can be expected to vary predictably with age according to patterns of family formation, growth, and ageing; Modigliani and Ando (1957) have suggested that consumption per 'equivalent adult' might be constant over the life-cycle. But whatever the shape of preferences, there need be no relationship between the profiles of consumption and of income; income can be saved until it is needed, or borrowed against if it is not yet available.

Independent of the life-time *pattern* of consumption is its level, which under the life-cycle model is determined by the level of total life-time resources, so that individuals with the same tastes but with higher incomes or higher inherited assets will have higher levels of consumption throughout their lives. If the future were entirely predictable, the consumption plan at any point in time could be decided with reference to the level of total wealth, this being the value of financial assets and the discounted present value of current and future incomes. In this sense, the life-cycle model is a permanent income theory of consumption, where permanent income is the annuity value of lifetime wealth, though the lifetime interpretation is only one of the many that are offered in Friedman's (1957) original statement. Whether life-cycle or not, linking consumption to *future* incomes has important consequences. First, consumption will respond only to 'surprises' or 'shocks' in income; changes in income that have been foreseen are already discounted in previous behaviour and should not induce any changes in plans. Of course, this does *not* mean that consumption will not change along with changes in income; a change may have been planned in any case, and some proportion of any actual change may well have been unforeseen. However, if a substantial fraction of the regular changes in income over the business cycle are foreseen by consumers, or if unanticipated fluctuations in income are regarded as only temporary with limited consequences for total life time resources, then consumption will not respond very much to cyclical fluctuations in income. Aggregate consumption is indeed much smoother than is aggregate income, and this has been traditionally accepted as an important piece of confirmatory evidence. I shall take up the matter again below when I deal with the recent econometric evidence.

The distinction between measured income and permanent income is also important for the interpretation of cross-sectional evidence. Since measured income can be regarded as an error-ridden proxy for permanent income, the regression of consumption on measured income will be biased downward (rotated clockwise) compared with the true regression of consumption on permanent income. Cross-sectional regressions, or time-series regressions of simple Keynesian consumption functions will therefore tend to understate the long-run marginal propensity to consume. Well before the work on life-cycle models, Kuznets (1946) showed that the long-run saving ratio in the United States had been roughly constant in spite of repeated cross-sectional analyses showing that the saving ratio rose with income, and the life-cycle theory could also readily account for these findings. It is interesting to note that the constancy of the saving ratio is far from being well established as an empirical fact; the evidence for other countries with long-run data is very mixed, and even the United States saving ratio is clearly influenced in the long-run by technical change, migration patterns, and demographic shifts, see Kuznets (1962) and Deaton (1975). Life-cycle and permanent income theories also predict that households with atypically high income will tend to save a great deal of it, a prediction which explained the apparently anomalous finding that black households tend to save more than white households at the same level of measured income; since blacks typically have lower household income than whites, those with the same measured income can be expected to have a higher transitory component.

The Modigliani and Brumberg life-cycle story was also important because it offered a story of capital accumulation in society as a whole that relied on the way in which people made preparation for their own futures, particularly for their future retirement. In a stationary life-cycle economy, in which there is neither economic nor population growth, aggregate saving is zero, and the old, as they dissave, pass on the ownership of the capital stock to the next generation who are, in turn, saving for their own retirement. With either population or income growth, the aggregate scale of saving by the young would be greater than that of dissaving by the old, so that, to a first approximation, the aggregate saving ratio, while in the long run independent of the *level* of national income, would depend on the sum of its population and per capita real income growth rates. Modigliani (1986), in his Nobel address, has given an account of how very simple stylized models of saving and refinement yield quite accurate predictions of the saving ratio and of the ratio of wealth to national income, and the predictions about the growth effects have been repeatedly borne out in international comparisons of saving rates, see Modigliani (1970), Houthakker (1961, 1965), Leff (1969) and Surrey (1974).

Perhaps the only problem with these interpretations is that there is little evidence that the old actually dissave, except by running down state social security or pension schemes; see for example Mirer (1979). Partly, this may be a rational response to uncertainty about the date of death and about possible medical expenses near the end of life (Davies, 1980), partly there may be statistical problems of measurement (Shorrocks, 1975), and partly consumers may wish to leave bequests. However, most countries' tax systems penalize donors who do not pass on assets prior to death, so the reason for the size of actual bequests remains something of a mystery. Bernheim, Schleiffer and Summers (1985) have gone so far as to suggest that parents retain their wealth until death in order to control their heirs and to solicit attention from them. They claim empirical support for a positive relationship between visits by children to their parents and parents' bequeathable assets; visits are apparently especially frequent to rich sick parents, but not at all frequent to poor sick parents. Related to the dispute about the reason for bequests is a parallel dispute on their importance in the transmission of the capital stock, see the original contribution by Kotlikoff and Summers (1981) and Modigliani's reply, summarized in his (1986) Nobel lecture.

The life-cycle and permanent income models also provided the econometric specifications for a generation of macroeconometric models. Ando and Modigliani (1963) suggested a simple form for the aggregate consumption function in which real aggregate consumption was a linear function of expected real labour income, YL, and of the real value of financial assets, i.e.

$$c_t = \alpha E_t(YL) + \delta W_t. \tag{16}$$

In practical econometric work, the expectation was typically replaced by a linear function of current and past values of labour income, a procedure that can be formally justified by modelling labour income as a linear ARIMA process, a topic to which I shall return below. Wealth or a subset of wealth was included as data allowed, although sometimes the return to wealth was included with labour income which could then be replaced by total income, so that, with smoothing, (16) becomes a permanent (total) income model of consumption. A favourite variant, suggested in Friedman (1957), was to model permanent income as an infinite moving average of current income with geometrically declining weights,

$$y_t^p = (1 - \lambda) \sum_{r=0}^{\infty} \lambda^r y_{t-r}, \tag{17}$$

so that if current consumption is proportional to permanent income, substitution yields

$$c_t = kc_{t-1} + k(1 - \lambda)y_t, \tag{18}$$

a formulation that is also easy to defend if consumers 'partially adjust' to changes in current income. Models like (18), possibly with additional lags, and with the occasional appearance of more or less 'exotic' regressors, such as wealth, interest rates, inflation rates, money supply, as well as various dummy variables for 'problem' observations, were the standard fare of macroeconometric models in their heyday, from the early sixties for about a decade and a half. They fit the data well, they accounted for the smoothness of consumption relative to income, and they accorded at least roughly with the general features of the life-cycle and permanent income formulations which provided them with pedigree and general theoretical legitimacy. Dozens of papers could be cited within this tradition; those by Stone (1964, 1966), Evans (1967), and Davidson et al. (1978) will perhaps stand as good examples.

3 Recent econometric experience

In the mid-1970s, the general state of complacency of macroeconomic modelling was rapidly eroded, largely by the apparent inability of the standard models to explain, let alone to predict, the coexistence of unemployment and inflation. The relationship between consumption and income did not escape some of the blame, although the main focus of attack was elsewhere. Standard consumption functions, which had worked well into the early seventies, seriously under-predicted aggregate saving during the period of (at least relatively) rapid inflation that characterized most Western economies in the middle of the decade. The implementation of the theory of the consumption function was also singled out for discussion in Lucas's famous (1976) essay that became known as the Lucas 'critique'. As Lucas forcefully argued, if consumption is determined by the discounted present value of *expected* future incomes, the response of consumption to a change in income is not well-defined until we know how expectations of income are formed. Each observed realization will cause a re-evaluation of future prospects in accordance with formulae that depend on the nature of the stochastic process governing income. If the nature of the stochastic process is changed, for example by a fundamental change in the tax code, then the way in which information is processed will change, and new information about incomes will have different implications for future expectations and for future consumption. This insight is of great importance, although its implications for econometric modelling were initially taken much too negatively; if the rules keep changing, econometric models will be inherently unstable (as evidenced by their performance in the mid-seventies) and we should give up trying to find stable relationships. Instead, as events have shown, the introduction of rational expectations has given a whole new lease of life to the study of consumption, with developments as positive as anything that has happened since the life-cycle and permanent income models were the 'new' theories in the mid-fifties. Lucas's critique suggested at least two lines for research. First, could the failure of consumption functions, or indeed of macroeconometric models in general, really be traced to a change in the way expectations were formed? If so, it ought to be possible to detect changes in the stochastic process generating real income. Second, and more generally, if expectations are important, there ought to be high returns to the simultaneous modelling of consumption and income, so that knowledge of the structure of the latter can be used either to estimate the consumption function or to test for the validity of the expectations mechanism. My own reading of the evidence is that the Lucas critique is *not* capable of explaining the failure of the empirical consumption function, but that the under-prediction of saving resulted from ignorance of the fact that saving appears to respond positively to inflation, or at least to unanticipated inflation. There is overwhelming evidence from a large number of countries, see in particular Koskela and Viren (1982a, 1982b), that saving increased with inflation in the 1970s, even when we allow for real income and its various lags. Such a finding is also consistent with the life-cycle theory since unanticipated inflation imparts a negative shock to real assets, so that risk-averse, low inter-temporal elasticity consumers will save to replace the lost assets so as to avoid the chance of low consumption later. It is also possible to explain the relationship through the confusion between relative and absolute price changes that is engendered by unanticipated inflation in an environment in which goods are bought sequentially, see Deaton (1977), but it would be hard to devise a test that would separate this from the life-cycle explanation. But if inflation was indeed the cause of the failure of the empirical consumption functions, then it is a standard enough story. An important variable was omitted from the analysis, it had not been very variable in the past so that its omission was hard to detect, and economists had not been imaginative enough to perceive its importance in advance. The Lucas critique is only one of the many problems that can beset an econometric equation, and it does not seem to have been the fatal one in this case.

The second research direction, the joint examination of income and consumption, has proved more productive. The first important step was taken by Hall (1978), who pointed out that equation (15) implies that, as an approximation consumption should follow a random walk with drift. To see why, assume that the real interest rate r is constant and known, and write (15) in the form

$$c_{t+1}^{-\sigma} = \{(1 + \delta)/(1 + r)\}c_t^{-\sigma} + \varepsilon_{t+1} \quad (19)$$

where the expectation at t of ε_{t+1} is zero. Equation (19) is exact, but a convenient expression can be reached by factoring c_t out of the right hand side, taking logarithms, and approximating. This gives

$$\ln c_{t+1} = \ln c_t + g + v_{t+1} \quad (20)$$

where g is positive or negative as r is greater than or less than δ, and the 'innovation' v_{t+1}, like ε_{t+}, has expectation zero at time t. Equation (20) shows that, in the absence of 'news', consumption will grow or decline at a steady rate g, so that nothing that is known by the consumer at time t or earlier should have any value for predicting the deviation of the rate of change of consumption from its constant mean. The result is often referred to as the 'random walk' property of consumption, though the theory does not predict that v_{t+1} has constant variance, so that, strictly speaking, the stochastic process is not a random walk.

For someone used to thinking about the consumption function as the relationship between consumption and income, Equation (2) is notable for the apparent absence of any reference to income. But of course income can appear through the stochastic term v_{t+1} if current income contains new information about its own value or about future values of income, and this will generally be the case. The random walk model does not predict that consumption should not respond to current income. It does however predict that, conditional on lagged consumption, past income or changes in income should not be correlated with the current change in consumption, and a considerable amount of effort has recently gone into testing this proposition. In Hall's (1978) original paper, to the surprise of the author and of much of the profession, the model worked well for an aggregate of United States consumption of non-durables and services. The level of consumption certainly depends on its own lagged value, but the addition of one or more lagged values of income or of further lagged values of consumption did not significantly add to the explanatory power of the model. Hall examined the role of the number of other lagged variables and discovered that lagged stockmarket prices had predictive power for the change in consumption, so that he concluded by formally rejecting the model. However, the overwhelming impression was favourable, at least relative to expectations.

Hall's test procedures are attractive because they do not depend on the properties of the income process, and focus only on consumption and its lags. But robustness comes at the price of power, and later work has devoted considerable attention to the joint properties of consumption and real income. Perhaps the natural route to modelling is to find a representation of real income as a stochastic process, typically as some sort of ARIMA. Once this is known, changes in income can be decomposed into anticipated and unanticipated components using the standard forecasting formulae from statistical time series analysis, so that it becomes possible to test whether consumption responds to one but not to the other. The random walk model seemed not to survive these tests so well. Papers by Flavin (1981) and by Hayashi (1982) showed that, for United States data, consumption is sensitive to *anticipated* changes in income, something that should not be the case in a thoroughgoing life-cycle model in which consumers are efficiently looking into the future. The phenomenon became known as the 'excess sensitivity' result, and was typically ascribed to the existence of a substantial number of consumers who wish to borrow against future income but are unable to do so. Such liquidity constrained consumers can be expected to consume all their available income, so that their consumption will increase one for one with all income changes, whether anticipated or not.

However, it is not clear that the excess sensitivity finding is itself robust. First, it is becoming increasingly recognized that the problems of econometric testing in the time-series models are more severe than had been generally supposed. The time series of both consumption and income are non-stationary, and it sometimes seems as if hypothesis testing in models involving non-stationary variables is like building on shifting sands; see Mankiw and Shapiro (1985, 1986) and Durlauf and Phillips (1986) for some of the problems. Second, there are a large number of variables other than income which can affect consumption, so that, according to (20), surprises in wealth and in inflation should affect consumption, as should the level of real interest rates. Adding even a few of these variables reduces degrees of freedom and diminishes the probability of being able to reject the basic model. Both Bean (1985) and Blinder and Deaton (1985) find that time-series models of consumption with several variables are more easily reconciled with the theory than are the simple two variable models. Not all of this should be ascribed to lack of degrees of freedom; for example Blinder and Deaton consistently find that unanticipated changes in wealth affect consumption and that anticipated changes do not. Third, even in a bivariate income-consumption model, Campbell (1987) has found that the model is largely consistent with the time-series evidence. Campbell recognizes the possibility of time-series feedback from lagged consumption to income, and models saving and the change in income as a bivariate vector-autoregressive system in which each series is regressed on lagged values of both. The structure of this representation then turns out to be very close to what it would have to be if the life-cycle rational expectations model were correct. The conflict between Campbell's results and the excess sensitivity findings are presumably accounted for by the feedback from saving to changes in labour income, since his model is otherwise compatible with the earlier ones.

Similarly mixed findings are also being uncovered from longitudinal panels that follow individual households over time. In contrast to the situation with labour supply, there are few panel data in the United States that cover household consumption, and most work has used the data on expenditure on food that is contained in the Michigan Panel Study of Income Dynamics (PSID). In an elegant paper, Hall and Mishkin (1982) found results that were in accord with the excess sensitivity results; there is a strong negative correlation in their data between

changes in consumption and changes in lagged income that is inconsistent with the view that only surprises in income should matter. However, since in their data changes in income are negatively correlated over time, a negative correlation between the lagged income change and the change in consumption can be interpreted as a positive correlation between consumption changes and changes in actual income, as predicted by the model of liquidity constraints. Hall and Mishkin conclude that these results would be consistent with a model in which about one fifth of consumers were unable to borrow as much as they wished. Once again, these results were supported by other similar evidence, see in particular Zeldes (1985) and Bernanke (1984), also using the PSID, Runkle (1983), using data from the Denver Income Maintenance Experiment, and Hayashi (1985a) using panel data from Japan. However, one potential problem with the use of panels is the importance of errors of measurement in such data. There is a considerable body of evidence that PSID income changes are subject to very substantial reporting errors, see in particular Altonji (1986), Duncan and Hill (1985), and Abowd and Card (1985). Altonji and Siow (1985) have recently estimated a model similar to Hall and Mishkin's using the PSID but with allowance for measurement error, and they find little conflict with the view that consumption responds only to news. However, it is unclear, at least to this reader, whether the acceptance of the model represents low power once errors of measurement are allowed for, or whether such errors really offer a plausible explanation for Hall and Mishkin's findings.

A more formal line of research has attempted to estimate the Euler condition (15) directly, thus avoiding the approximations made by Hall and by others. Rewrite (15) once more, this time as

$$(1 + r_{t+1})(c_{t+1})^{-\sigma} - (1 + \delta)(c_t)^{-\sigma} = \varepsilon_{t+1} \tag{21}$$

where, as before ε_{t+1} is orthogonal to any variable known in period t or earlier. Hansen and Singleton (1982) proposed that the parameters in (21) be estimated by a generalized methods of moments scheme. Suppose that we have two variables or instruments z_{1t} and z_{2t}, each known at time t, so that we have $E_t(z_{it}\,\varepsilon_{t+1}) = 0$ for $i = 1, 2$. We can then estimate the two unknown parameters, σ and Δ, by equating sample and theoretical moments, and solving the two equations, $i = 1, 2$

$$T^{-1} \sum_{t=0}^{T-1} [z_{it}\{(1 + r_{t+1})(c_{t+1})^{-\sigma} \tag{22}$$
$$- (1 + \delta)(c_t)^{-\sigma}\}] = 0.$$

If, as is typically the case, we have more than two z-variables, then it will not generally be possible to choose the two parameters so that (22) is exactly zero.

Instead, the vector can be made as small as possible, or more specifically, the parameters can be estimated by minimizing a quadratic form that can be thought of as a weighted sum of squares of the left-hand side of (22); see Hansen and Singleton for details. If the model were true, this minimized value ought to be small, so that with more instruments than parameters, the generalized method of moments procedure yields a test-statistic that is diagnostic for model adequacy.

Test procedures based directly on the Euler conditions have several notable advantages. As was the case for Hall's procedures, few assumptions have to be made about the structure of the income process, and the model satisfies the best professional standards of seeking a direct confrontation between theory and data with as few approximations and supplementary assumptions as possible. The model can also be readily extended to test the implications of the consumption asset pricing model by repeating the tests using the returns on a range of alternative assets, see (13) above. Hansen and Singleton's study, as well as several others, find that the test statistics are much too large to be consistent with the theory and so reject the intertemporal model implied by the Euler conditions. Given the apparent superiority of the tests, these results have been accorded a great deal of weight in the literature. However, while I believe that Hansen and Singleton's work represents a very important methodological advance, I think that there are good reasons for not treating their results as a definitive rejection of life-cycle theory. The high level of technique that is embodied in deriving the Euler equation, not to mention the complexity of generalized methods of moments estimation, should not blind us to the very simple, even simple-minded, economic story that underlies these models. Fundamentally, the Euler equation says that the marginal rate of substitution between today's and tomorrow's consumption should be equal to the rate of return on assets between today and tomorrow, so that estimation of the Euler equation, unlike the Hall or excess-sensitivity tests, focuses very directly on the relationship between real interest rates and changes in real consumption, and the model will not fit the data if there is no close association between the two. And it only takes a very cursory inspection of United States time-series data to see that there is no such association. Real consumption grew in all but one year between 1954 and 1984, while real after-tax interest rates were as often negative as positive, so that consistency with the theory would require that the pure rate of time preference be negative. Nor is there any association between the rate of growth of consumption and the level of real after-tax interest rates, see Deaton (1986b) for some data. But this in no way reflects badly on the life-cycle theory. As was made perfectly clear in the original Modigliani and Brumberg papers, and it is the *essence* of the life-cycle model, aggregate consumption cannot be expected to behave like individual consumption. Imagine a stationary economy with neither population nor real income growth, in which there is an excess of real interest rates over the rate of pure time preference, and in which all consumers have identical additive life-time preferences with isoelastic subutility functions. In such an economy, each individual has a consumption path that is growing over time, but aggregate consumption is constant, a result that is achieved by old people dying and being replaced by young people who have much lower consumption levels relative to their incomes. Unless we believe that there is some automatic and immediate relationship between real interest rates, time preference and growth, as would obtain for example along a 'golden age' growth path, or unless we believe that consumers have infinite lives, then there is no reason at all to suppose that aggregate consumption should look at all like the life-cycle path of a representative consumer. Representative agent models are frequently useful, and it is not very constructive to dismiss macro-economics because it requires implausible aggregation assumptions. However, the life-cycle model provides a well-worked-out account of individual and aggregate saving, an account that is consistent with a good deal of other evidence and theory, and it *does not* predict that aggregate consumption should be consistent with the intertemporal optimization conditions for a single individual. The general question of the effects of interest rates on consumption is something that has remained in dispute for a long time, and in spite of repeated attempts to isolate the effect, careful studies have tended to be unable to do so, or at least to find effects that are at all robust, or that can be replicated on even slightly different data sets or data periods. Economic theories or policy prescriptions that rely on intertemporal substitution of consumption in response to changes in real interest rates are not well-buttressed by any solid body of empirical evidence.

Another useful approach to testing the life-cycle model is to consider the stylized facts of the income and consumption processes, and to see whether consumption behaves in the way that is to be expected given the stochastic process of income. Most people who have studied the time series for quarterly real disposable income in the United States agree that, like GDP, the series can be parsimoniously described by a model that is linear in its first two lags, i.e. an autoregression of the form

$$y_t = \alpha_1 + \alpha_2 y_{t-1} + \alpha_3 y_{t-2} + u_t \qquad (23)$$

where u_t is the income innovation, that part of current income that cannot be anticipated from previous observation of the series. Of course, real income is not a stationary series, but has a strong upward trend, and there is considerable disagreement about the nature of this trend, what is the economic story behind it, and how it should be modelled. One possibility is that real income contains a *deterministic* time trend, so that there is some sort of equilibrium growth path that cannot be altered by shocks to the economy. Shocks certainly exist, but they cause only short term temporary deviations from the path and have little or no long-term temporary deviations from the path and have little or no long-term significance. In this view, equation (23) applies to the *deviations* of income from trend, not to income itself; equivalently, (23) can be modified by including a linear or quadratic time trend. The alternative view is that there is no deterministic trend, but that the rate of change of income is a stationary stochastic series with constant mean. In practice, this can look very like the previous model, but there is the vital conceptual difference that in the second, non-deterministic model, there is nothing that will ever bring income back to any deterministic path. In consequence, shocks to current income have permanent and long-lasting effects. The version of (23) that corresponds to this view can be written.

$$\{(y_t - y_{t-1}) - \gamma\} = \rho\{(y_{t-1} - y_{t-2}) - \gamma\} + u_t$$
(24)

which can readily be seen to be a special case of (23), though note that it is the case where the time series possesses a unit root, or is stationary in first differences. For (24) to be a valid specialization of (23), the quadratic equation with the α's of (23) as coefficients must have a unit root, hence the term. Equation (24) appears to fit the data well and the parameter ρ turns out to be around 0.4, so that (24) says that if the increase in real income in one quarter is greater than its long term mean, then the next quarter's increase is also likely to be above the mean, though by less. While the long-term mean of the rate of change of income is constant and equal to y, good fortune (positive u's) and bad fortune (negative u's) never have to be paid for (or made up), since shocks are immediately consolidated into the income level, and growth goes on in the same way as before, but from the new base. As Campbell and Mankiw (1986) have emphasized, the unit root model exhibits shock *persistence*, while the deterministic trend model does not; they suggest that shock persistence is what we should expect if supply shocks predominate over demand shocks, with the reverse in standard Keynesian models where shocks are typically attributed to fluctuations in aggregate demand.

It turns out that it is almost impossible to tell these two processes apart on United States time-series data. Processes with unit roots are inherently difficult to tell apart from processes that are stationary around deterministic trends, and the tests that are available, Dickey and Fuller (1981), Phillips and Perron (1986), certainly cannot reject the hypothesis that (24) is a valid specialization of (23). Nor would the tests convince a believer in the deterministic model that income does not have a deterministic trend, even though it will readily be recognized that the deviations from trend are themselves close to non-stationarity. Since both process are special

cases of (23) with the inclusion of a trend, and since each assumes parameter values that are very close to one another, one might think (and hope) that the two models would have very similar implications. But it is easy to see this is not true. If permanent income is taken as the annuity value of discounted future incomes, then (24) implies that any innovation u_t to current income, because it will persist forever, and because it can be expected to be followed by another infinitely persistent innovation of the same sign, will change permanent income by more than the amount of the innovation. Equation (25) below gives the formula for the change in permanent income, if the real interest rate is r, and if real income follows (24), see Flavin (1981) or Deaton (1986b),

$$\Delta y_t^p = \frac{(1+r)^2}{r+1-\rho} u_t$$
(25)

so that the change in permanent income is between one and a half and twice as large as the innovation in current income. By contrast, fitting the deterministic model yields a much smaller effect, with the change in permanent income about one fifth of the shock in measured income. Since consumption should change by about the same amount as does permanent income, the life-cycle model, together with the unit root formulation, yields the uncomfortable prediction that consumption should be *more* variable than income over the business-cycle, not less. If the unit root model is correct, then the life-cycle and permanent income models can be rejected because they predict what they were designed to predict, that consumption is smooth relative to real income! The deterministic model gives no such problems, but as yet we have no way of being sure that it is correct, unless, of course we assume from the start that the life-cycle story is true.

There is insufficient space in this essay to follow these issues further, or to discuss in detail the evidence for and against the two formulations of the stochastic process governing real income; the interested reader can refer to Deaton (1986b) and to the evidence on persistence in GDP presented by Campbell and Mankiw (1986) and by Cochrane (1986). There are a number of possible solutions to these puzzles, and a great deal of empirical work remains to be done, though I suspect that the time-series data on income are insufficiently long to allow the isolation of the very long-run properties on which the permanent income theory rests, see in particular the interesting paper by Watson (1986).

4 Variations on the basic theme

There exist many interesting developments of the basic life-cycle model, and I have space to discuss only a few. I have already mentioned the role of liquidity constraints, and many people would take it as transparent that many consumers do not have access to unlimited credit, or else face borrowing rates that are higher than the rates at which they can lend. Of course, many consumers may be able to smooth their consumption without recourse to borrowing, and the borrowing needs of many others may be met by the typically rather good markets in home mortgages. For consumers who nevertheless wish to borrow but cannot, their spending will be closely tied to their actual income. For some of the theoretical and empirical literature on this point see Flemming (1973), Dolde and Tobin (1971), and Hayashi (1985b). The theoretical consequences of uncertainty about the date of death have been worked out by Yaari (1965), and as argued above, play a possibly important part in the explanation of the saving behaviour of the elderly.

Another line of research is the possible relaxation of the assumption that preferences are intertemporally additive. Allowing all periods (or ages) to interact with all other periods in an unrestricted way, as in equation (1), would be much too general to be useful, and the search has been for simple models that break the restriction in a natural and straightforward way. One useful analogy is with the theory of durable good purchases, where utility depends on the *stock* of assets possessed, the stock in turn being the integral of past purchases less depreciation. Purchases in one period therefore have consequences for utility in subsequent periods, something that will be taken into account by a forward looking consumer. In the case of durable goods, the assumption of perfect capital markets effectively converts durable into non-durable goods, with the price of a unit of stock for one period being the implicit rental or user cost, the latter being defined as the sum of interest cost, depreciation, and expected capital loss, see for example Diewert (1974) or Deaton and Muellbauer (1980b, ch. 13).

However, various authors, Houthakker and Taylor (1970) perhaps being the first, have extended the durable model to encompass 'psychic' stocks which, like physical stocks, are augmented by purchases and diminished by depreciation, but unlike physical stocks, can either increase or decrease utility. The latter case covers habit formation; consumption of an addictive good generates pleasure now, but engenders a hungry habit that is pleasureless but costly in the future. The model has been given an elegant formulation in two papers by Spinnewyn (1979a, 1979b). As an example, see also Muellbauer (1985), take the utility function

$$u_t = \sum_{k=0}^{T-t} (1+\delta)^{-k} v(c_{t+k} - \alpha c_{t+k-1}) \qquad (26)$$

where α is a measure of habit formation. Spinnewyn maximizes this function with respect not to c_t, but with respect to the 'net' quantities $z_t = c_t - \alpha c_{t-1}$, and shows how to rewrite the budget constraint so as to define corresponding prices of the z's that reflect not only market prices of the goods, but also the costs of consumption

now in terms of pleasure foregone later. Under certainty, and looking ahead from time t, the full shadow price of an additional unit of consumption now is

$$p_z = \sum_{k=0}^{T-t} [\alpha/(1+r)]^k p_{t+k} \qquad (27)$$

because the habits that are built up now have to be paid for later. Note that this sort of formulation also predicts that it is $c_t - \alpha c_{t-1}$ not c_t, that is proportional to permanent income, so that consumption itself will adjust only sluggishly to changes in permanent income with habits causing a drag. Other formulations of nonseparable preferences can be found in the papers by Kydland and Prescott (1982), and by Eichenbaum, Hansen, and Singleton (1984), both of which are concerned to reconcile fluctuations in the aggregate economy with the behaviour of a single representative agent.

Many of the models discussed so far assume that the consumption function actually exists, hence taking for granted the essentially Keynesian assumption that income is given to the consumer, and is not chosen together with consumption. A considerable body of work has grown up in the last ten years that is concerned with the simultaneous choice of labour supply and consumption in a life-cycle setting. Heckman (1971) and Ghez and Becker (1975) are among the pioneers of this approach. Unlike the price of goods, the price of leisure tends to show a systematic pattern over the life-cycle, so that, if consumers are free to choose their hours, and if they can freely borrow and lend so as to transfer resources between periods, it will pay them to work hardest during those periods in their life-cycles when the rewards for doing so are highest, and to take their life-time leisure when wage rates are low and leisure is cheap. There is superficial evidence in favour of this story, and Ghez and Becker, followed by Smith (1977) and Browning, Deaton and Irish (1985), all find that workers tend to work longest hours in middle age when wage rates are high and the lowest number of hours at the beginning and end of the economically active life, when wage rates are relatively low. Consumption also tends to peak in the middle age, and this can be brought into the story by assuming that consumption and leisure are complements, so that the lack of leisure in middle age is partially compensated by high levels of expenditure. This elegant fable has also been made much of in equilibrium theories of the business cycle, which accounts 'unemployment' as a voluntary vacation taken when the real wage is low and leisure is on sale, see in particular Lucas and Rapping (1969) and Lucas (1981).

There now exists a growing volume of literature that shows just how much violence to the facts is done by this story. All the evidence quoted above looks across different individuals at different points in their life-cycles, while the theory says that the same individual will change his or her hours of work along with changes in the real wage over the life-cycle. Time-series and panel data from the United States and time-series of cross-sections from the United Kingdom suggest that this is simply not the case, see for example Mankiw, Rotemberg and Summers (1985), Ashenfelter and Ham (1979), Ashenfelter (1984), and Browning, Deaton and Irish (1985). Even MaCurdy's (1981) more postive study provides only very weak evidence, see in particular Altonji (1986). The joint consumption and labour supply story fares even less well than the labour supply model alone, and there is clear evidence that the way in which consumption and hours fluctuate over the cycle (sometimes together and sometimes in opposite directions) is not consistent with the way in which they move together over the life-cycle. The attempt to provide a unified theory of business and life-cycles has been an interesting and important one, but it cannot be said to have been successful.

I have been somewhat cavalier in my treatment of aggregation issues, choosing to emphasize them when I believe them to be important, for example in the fitting of Euler conditions, and ignoring them when it has been convenient to do so. Attempts to do better than this have not been notably successful. Formal conditions that allow aggregation in consumption function models are typically too restrictive to be useful, so that, in theory, changes in the distribution of income should have detectable effects on aggregate consumption. However, attempts such as that by Blinder (1975) to link the distribution of income to consumption have not been notably successful, perhaps because the income distribution is not variable, or because it changes smoothly enough over time to preserve a stable relationship between average income and average consumption. There is also an issue of aggregation over goods in order to define real consumption at all, even at the level of the individual agent. In the derivation in section 1 above, I made the convenient assumption that within-period preferences were homothetic, so that an index number of real consumption could be formed. But homotheticity, although very convenient for studying the consumption function, is very inconvenient for studying the allocation of expenditure among goods since it implies that the within-period total expenditure elasticities of each good are all equal to unity. Fortunately, there are aggregation results of Gorman's (1959), see also Deaton and Muellbauer (1980b, ch. 5) for an exposition that allows us to have the best of both worlds, at least if we remain with intertemporally additive preferences. If the single-period indirect utility function $\psi(x, p)$ takes the form known as the 'generalized Gorman polar form'

$$\psi(x,p) = F[x/a(p)] + b(p) \qquad (28)$$

where $a(p)$ and $b(p)$ are linearly homogeneous functions of prices and $F(\cdot)$ is monotone increasing, then the real expenditure index $x/a(p)$ can serve as an indicator of real

consumption just as in the homothetic case. This happens because when the consumer chooses the allocation of life-time expenditure over periods so as to maximize the intertemporal sum of terms like (28), the $b(p)$ terms are irrelevant. However, the intra-period demand functions that correspond to (28) do not display unitary elasticities unless the $b(p)$ is identically equal to zero, and quite general functional forms are permitted. There is therefore no real conflict between the analysis of the consumption function on the one hand, and the analysis of demand on the other. It is to the latter that I now turn.

5 Theoretical and empirical demand functions
Demand functions are the relationships between the purchase of individual goods, income or total expenditure, prices, and a variety of other factors depending on the context. Economists have attempted to make empirical links between demand and price since Gregory King's famous demand curve for wheat, see Davenant (1699), and since the middle of the 19th century, there has been a great development in the theory of consumer behaviour. Much practical work continues in the tradition of King, paying little attention to formal theory, concerning itself instead with finding empirical regularities. For a firm studying the demand for its product, or for anyone interested in establishing a single price elasticity, this probably remains the best approach; the major developments in econometric technique and empirical formulation have not been much concerned with, or relevant to, these very practical questions. The pragmatic approach (the term comes from Goldberger's famous but unpublished (1967) study), probably reached its peak with the publication of Richard Stone's great monograph, (Stone, 1954a), and much is still to be learned by a careful study of Stone's procedures for measuring income and price elasticities. However, in this essay, I shall follow the literature, and follow its more methodological approach.

The theory outlined in Section 1 above suggests that the demand functions of an individual consumer can be derived by maximizing a utility function $v(q)$ subject to a budget constraint $p.q = x$, where x is total expenditure. In the analysis here, x is chosen at some previous level of decision making, but traditionally it is treated as if it were a datum by the consumer, the utility maximization yields a vector q that is some function $g(x,p)$, say, of total expenditure and prices. These demand functions cannot simply be any functions, but must have certain properties as a result of their origins in utility maximization. Obviously, the total value of the demands should be equal to total outlay x, the 'adding-up' property, and it must be true that proportional changes in x and in p do not have any effect on quantities demanded, the 'homogeneity' or 'absence of money illusion' property. Somewhat less obvious are the famous symmetry and negativity properties. These apply to the Slutsky (1915) matrix, S, the typical element of which is defined as

$$s_{ij} = \partial q_i/\partial p_j + q_i \partial q_j/\partial x. \qquad (29)$$

As any intermediate text shows, see for example Deaton and Muellbauer (1980b, ch. 2), the Slutsky matrix must be symmetric and negative semi-definite. The symmetry property is not readily turned into simple intuition; negativity implies that the diagonal elements of the matrix are non-positive, a proposition often referred to as 'the law of demand'. The four properties, adding-up, homogeneity, symmetry and negativity, essentially exhaust the implications of utility maximization, so that any empirical demand functions that satisfy them can be regarded as having been generated by utility maximization, or by rational choice, with 'rational' defined, following Gorman (1981), as 'having smooth strictly quasi-concave preferences, and being greedy'.

Stone (1954b) was the first to attempt to use this theory directly to confront the data. He started from a (general) linear expenditure system of the form

$$p_i q_i = \sum_j a_{ij} p_j + b_i x \qquad (30)$$

where a_{ij} and b_i are unknown parameters. Stone showed that, in general, the system (30) does not satisfy the four requirements, but will do so if, and only if, the parameters are restricted so that the model can be written in the form

$$p_i q_i = p_i \gamma_i + \beta_i (x - p \cdot \gamma) \qquad (31)$$

with the β-parameters summing to unity. In this form the model is known as the linear expenditure system. As Samuelson (1947–8) and Geary (1949–50) had earlier shown, the utility function corresponding to (31) has the form

$$u = \sum_i \sum \beta_i \ln (q_i - \gamma_i), \qquad (32)$$

sometimes referred to (somewhat inappropriately) as the Stone–Geary utility function. It can be thought of as a sum of Bernoulli utility functions of the quantity of each good above the minimal γ's.

Stone's achievement lay not in deriving the demand functions, but in thinking to estimate them. The demand functions (30), even if fitted to the data by least-squares, require non-linear optimization, and Stone invented a simple and not very efficient scheme, but one that allowed him to obtain parameter estimates and a good fit to interwar British data for a six commodity disaggregation of expenditures. This was a major breakthrough, not only in demand analysis, but also in applied econometrics in general. Indeed, much of demand analysis for a decade or so after Stone's paper consisted of applying

better algorithms and faster computers to the fitting of Stone's model to different data sets.

The linear expenditure system offers a demand model for a system of, say n goods, and requires only $2n - 1$ parameters, a degree of parsimony that was very important in allowing the model to be estimated on very short time-series data. However, such economy brings its own price, and the linear expenditure system is very restrictive in the sort of behaviour that it can allow. In particular, and pathological cases apart, the model cannot allow inferior goods (goods the demand for which falls as total outlay increases), nor can it allow goods to be complements rather than substitutes. (As defined by Hicks (1939) goods i and j are complements if the (i,j)th term in the Slutsky matrix is negative, so that the utility compensated cross-price response of i to an increase in the price of j is positive.) Normal (non-inferior) goods that are substitutes for one another may be the most important case, but they do not encompass everything that we might want to study. The linear expenditure system also implies that the marginal propensity to consume each good is the same no matter what is the total to be spent, and many cross-section studies of household budgets have suggested that this is not in fact the case.

Unfortunately, it is quite difficult to write down utility functions that will lead to more general demand functions than those of the linear expenditure system, nor is there any obvious way of generalizing Stone's procedure of writing down functions and making them consistent with the theory. Progress was only really made once applied demand analysis started using 'dual' formulations of preferences to specify demands. In the demand context, duality refers to a switch of variables, from quantities to prices, so that utility becomes a function, not directly of quantities consumed, but indirectly of prices and total expenditure. This indirect utility formulation is given by the function $\psi(x, p)$, already used above, and this is simply the maximum attainable utility from total outlay x at prices p. Since $\psi(x,p) = u$, and the function is monotone increasing in x, it can be inverted to give $x = c(u, p)$, known as the 'cost function', since it gives the minimum necessary cost that is required to reach the utility level u. By a theorem usually attributed to Shephard (1953) and to Uzawa (1964), these two functions contain a complete representation of preferences; provided preferences are convex, and provided the functions satisfy homogeneity and convexity (or concavity) conditions, preferences can be reconstructed from knowledge of either of the two functions. It is also very easy to move from either cost or indirect utility functions to the demand functions. For the indirect utility function, we have Roy's identity (Roy, 1943).

$$q = -\nabla_p \psi(x, p)/\psi_x(x, p) \equiv g(x, p) \qquad (33)$$

which immediately yields demand functions from preferences in a form that are suitable for estimation, while for the cost function, we have Shepard's Lemma (1953),

$$q = \nabla_p c(u,p) = \nabla_p c[\psi(x,p),p] \equiv g(x,p) \tag{34}$$

where, as in (33), the operator ∇ denotes a vector of partial derivatives.

Demand analysis now had a high road to specification. Think of some quasi-convex decreasing function of the ratios of price to total outlay and call it an indirect utility function, or think of some function of utility and prices that is increasing in its arguments and linearly homogeneous and concave in prices and call it a cost function. Either way, and with only simple differentiation, new (and sometimes) interesting demand functions will be generated. Alternatively, and even more importantly, it is possible to use theory to aid and check out empirical knowledge. If it is known that the marginal propensity to spend on food is a declining function of total expenditure, or if it is thought likely that some goods do not depend very directly on the prices of other goods, it is relatively straightforward to find out what preferences (if any) will yield the result. It becomes possible, not just to generate demand functions serendipitously, but to generate good and useful ones deliberately.

There are many examples that could be cited from the literature. One of the most widely used in the *translog* model which was first proposed in 1970 by Jorgenson and Lau, see Christensen, Jorgenson and Lau (1973) for a convenient reference. To derive the translog, write the indirect utility function in terms of the ratios of prices to outlay, $r = p/x$, and approximate the indirect utility function as a second order polynomial in the logarithms of r. Application of Roy's identity yields demand functions in which the budget share of each good is the ratio of two functions, each of which is linear in the logarithms of the price to outlay ratios. Estimation of these rational functions, like estimation of the linear expenditure system, requires the use of non-linear maximization techniques. A related model, the 'almost ideal demand system' (AIDS) has been proposed by Deaton and Muellbauer (1980a), and I use this to illustrate some of the issues that arise with the current generation of demand models. The AIDS is specified by the logarithm of its cost function which takes the form

$$\ln c(u,p) = \alpha_0 + \sum_k \alpha_k \ln p_k$$
$$+ 0.5 \sum_k \sum_m \gamma_{km} \ln p_k \ln p_m$$
$$+ u \exp \left\{ \sum_k \beta_k \ln p_k \right\}, \tag{35}$$

so that, applying Shephard's lemma and rearranging, we have demand functions

$$p_i q_i / x \equiv w_i = \alpha_i + \beta_i \ln(x/P) + \sum_j \gamma_{ij} \ln p_j \tag{36}$$

where P is a linearly homogeneous price index, the form of which can readily be inferred from (35). The parameters of the model must satisfy certain restrictions if (35) is to be a proper (log) cost function, and (36) a proper system of demand functions. The matrix of γ-parameters can be taken to be symmetric in (35), but must be so in (36), and its rows and columns must add to zero for the homogeneity and adding-up properties to be satisfied. The β-parameters can be positive or negative, with positive values indicating luxury goods, and negative values necessities. The main advantage of the AIDS model in time-series applications is that the price index P can typically be approximated by some known price index selected before estimation, so that the demand system is linear in its parameters. In consequence, it can be estimated by ordinary least squares on an equation by equation basis, at least if the symmetry of the γ-matrix is ignored. The homogeneity restrictions can be tested equation by equation using a t- or F-test, and while imposing or testing symmetry requires an iterative procedure, estimation can be done by straightforward iterated restricted generalized least-squares, see Barten (1979) or Deaton (1974a) for further discussion.

The results of estimating the AIDS model are sufficiently similar to those from other models and other studies, see e.g. Barten (1969), Deaton (1974a), Christensen, Jorgenson and Lau (1973), and many others, that perhaps they can be taken as representative. What typically seems to happen is that the homogeneity restrictions appear *not* to be satisfied, so that in the application of AIDS to British data, Deaton and Muellbauer found, for example, that the F-test for transport had a value of 172 compared with the 5 per cent critical value of 4.8. Results on symmetry from AIDS and other systems are more mixed, and it now seems clear that testing symmetry is not usually possible given the amount of data typically available in time series, or put more positively, that there is no convincing evidence against symmetry. The difficulty is that symmetry involves a set of restrictions *across* different equations, so that unlike homogeneity, which involves tests *within* each equation, exact, small sample tests are not available. Researchers have therefore fallen back on asymptotically valid tests, and it turns out that these work very badly for the usual sort of samples, especially when there are more than a very small number of goods in the demand system. The papers by Laitinen (1978) and Meisner (1979) first established the problem, see also Evans and Savin (1982) and Bera, Byron and Jarque (1981) for further evidence.

The AIDS model, like the translog and several others, e.g. Diewert's (1973) 'generalized Leontief' system, fall into the class of 'flexible functional forms'. This criterion of flexibility, first proposed by Diewert (1971), is an important guarantee that the model is sufficiently richly parametrized so as to allow estimation of what are thought to be the main parameters of interest, typically the total expenditure elasticities, and the matrix of own and cross-price elasticities. A 'second order' flexible functional form is one that has sufficient parameters, so configured, that it is possible to set the value of the function, and of its first and second partial derivatives to any arbitrary set of (theoretically permissible) values. By applying Roy's identity or Shephard's lemma, it is clear that a cost or indirect utility function that is a second order flexible functional form will yield demand functions that are first-order flexible, so that it is possible for estimation to yield any set of price and expenditure elasticities that are consistent with utility theory. For empirical work, such a guarantee is important, because it ensures that the elasticities are being measured, not assumed. Contrast, for example, the linear expenditure system (31) with the AIDS model (36). Both could be fitted to the same set of data, and the parameter estimates of each could be used to generate a complete set of expenditure and price elasticities. But the linear expenditure system is *not* a flexible functional form, and so its estimated elasticities are not independent of one another, as is apparent from the fact that there are $2n - 1$ parameters compared with the total number of potentially independent elasticities, which is $(n - 1)(1 + n/2)$. (There are $n - 1$ independent demand equations, each of which has an expenditure elasticity, and n price elasticities; however, one price elasticity per equation is lost to homogeneity, and symmetry imposes a further $(n - 1)(n - 2)/2$ constraints.) The linear expenditure system does not therefore *measure* all the price and income elasticities, but determines them by a mixture of measurement and assumption, the main assumption being that of additive preferences, see Deaton (1974b) for further details. The AIDS, by contrast, has exactly the right number of parameters to allow for intercepts and a full set of elasticities, so that when it (or the translog, or the generalized Leontief) is estimated, so is the full set of elasticities.

Being able to do this is a great step forward in methodology, but just as the linear expenditure system probably asks too little of modern data, (although not of the data available to Stone and the early pioneers of the systems approach), the second-order flexible functional forms probably ask too much, or equivalently, put too little structure on the problem. The consequences show up in large standard errors, a high frequency of apparently chance correlations, and a lack of robustness to functional form changes within the class of flexible functional forms, in other words, in all the standard symptoms of over-parametrization. These problems are particularly acute for

the measurement of *price* elasticities, because in most time-series data, commodity prices tend to move together with relatively little variation in relative prices. And although the focus of most research on demand analysis over the last thirty years has been on the estimation and testing of price responses, there is certainly no consensus on what numbers, if any, are correct. Estimates obtained from the linear expenditure system are not credible because they are forced to satisfy an implausibly restrictive structure, while those from flexible functional forms are not credible because the data are not informative enough to supplement the lack of prior structure. Some intermediate forms are clearly required.

One of the attractions of flexible functional forms is their ability to approximate quite general forms for preferences. However, the models so far considered offer only approximations, and there is no guarantee that they have satisfactory *global* properties. Partly this is the standard problem that a fitted model will be forced to give a reasonable account of the data over the sample used for estimation, but may predict very badly elsewhere. But there are other deeper issues. Taking the AIDS as an example, estimation of (36) subject to symmetry and homogeneity will produce a system of estimated demand functions that will satisfy adding-up, homogeneity and symmetry for *all* values of x and p. However, there are two other important properties that are not assured. First, there is no guarantee that the predicted budget shares will necessarily lie between zero and one, so that there may be regions of price space in which the estimated model yields nonsensical predictions. Second, there is no way that the AIDS can be guaranteed to have a negative semi-definite Slutsky matrix for all prices, at least not without restricting parameters to the point where the model ceases to be a flexible functional form. The parameters could be chosen so as to satisfy negativity for some particular combination of prices and outlay, but there will be no guarantee that the law of demand will be satisfied elsewhere. In the translog model, it is possible to impose a restriction that guarantees negativity everywhere, but the model with the restriction has the property that all estimated own price elasticities must be less than minus one, independently of whether this is in fact true, and it almost certainly is not, see Diewert and Wales (1987). A demand system is described as 'regular' if it has a negative definite Slutsky matrix and predicts positive demands, and several empirical studies, see e.g. Wales (1977) for one of the first, found that estimated flexible functional forms were not regular over disturbingly large regions of even the parameter space used to estimate them. Caves and Christensen (1980), and later Barnett and Lee (1985) and Barnett, Lee and Wolfe (1985), investigated the same problem theoretically by taking a known utility function, choosing the parameters of flexible functional forms to match its level and derivatives at a point, and then mapping out the regions of price space in which the systems remained regular. The results at least for the translog and the generalized Leontief model, were not good.

These regularity issues may seem of limited importance in practice, but this is far from being the case. One of the major reasons for being interested in complete empirical demand systems is to be able to examine the consequences of price changes, particularly of price changes that follow changes in government policy. The United States relies relatively little on indirect taxation as a source of public finance, but such is not the case in most of Europe, and the vast majority of developing countries maintain complex systems of price wedges, particularly for foods and for agricultural production. The effects of such systems cannot be predicted without good information on how demands respond to price changes, nor can reforms be intelligently discussed. However, estimated demand systems that are not regular are not a great deal of help. All of the theory of welfare economics, of consumer surplus, of optimal taxation and of tax reform, *assumes* that demand behaviour is generated by utility maximization at the individual level, and implementation without regularity risks internal contradiction. For example, if compensated demand functions slope *upwards*, the government can generate a dead-weight gain by imposing a distortionary tax. Of course, it may not be the empirical work that is wrong, but the theory that we used to try to model behaviour. If so, the estimated demand functions are still not useful, since we now have no idea what to do with them. But I doubt that evidence goes so far; it is not that behaviour itself is irregular, but that we have not yet found a good modelling strategy that contains a reasonable amount of prior information to supplement the paucity of data, and at the same time can deliver global regularity if it is warranted by the evidence.

A number of interesting experiments are currently under way that involve new modelling techniques. One possibility is that the Taylor series expansions that motivate most flexible functional forms are themselves inadequate to the task. In particular, Taylor approximations lose their ability to approximate if they are also asked to possess other properties of the functions that they are approximating. For example, we might want to test whether or not preferences are additively separable, as in the linear expenditure system. One strategy would be to write down some second-order approximation to preferences, estimate the resulting demand model, and then test whether or not the conditions imposed on the demands by additivity are satisfied. But this will not work in general, because there may be no additive system of demand equations that has the precise functional form demanded by the approximation. The same phenomenon is well illustrated by Stone's derivation of the linear expenditure system itself. The original general linear expenditure equations (29) can clearly be justified as a Taylor approximation to any set of homogeneous demand functions, and yet the imposition of only *symmetry* generates the demand system (30) which comes from the *additive* utility function (31). Additivity is not imposed, but linear expenditure systems are only symmetric if they are additive. Similarly many flexible function forms are only globally regular if they are homothetic, see for example, Blackorby, Primont and Russell (1977). Several recent studies have proposed alternative ways of making functional approximations. Gallant (1982) has proposed using Fourier series approximations while Barnett (1983) has suggested that Laurent series can be used to generate demand models with good properties. Gallant's models are even more heavily parametrized than standard flexible functional forms, and there must be some question as to the suitability of trigonometrical functions for demand functions. Barnett's 'miniflex Laurent' model does not use the full flexibility of the Laurent series, but appears to have quite good approximation and regularity properties in practice, see Barnett and Lee (1985) and Barnett, Lee and Wolfe (1985); even so, its estimation is complex, and many of the parameters have to be estimated subject to inequality constraints.

A second line of current research has abandoned the standard approach of econometric analysis, taking instead a completely non-parametric approach. Since many of the difficulties discussed above arise from choice of functional form, it is useful to ask how far it is possible to go without assuming any functional form at all. We know from standard revealed preference theory that two observed vectors of prices and quantities can be inconsistent with utility maximization; if bundle one is chosen when bundle two is available, so that bundle one is revealed preferred to bundle two, then no subsequent choice should reveal bundle two to be preferred to bundle one. Before embarking on the exercise of fitting some specific utility function to any finite collection of price and quantity pairs, one might then ask whether the collection is conceivably consistent with any set of preferences. If it is, then contradictions between an estimated system and the theory must be a matter of inappropriate functional form. The conditions for utility consistency of a finite set of data were originally derived by Afriat (1967), who proposed a condition called cyclical consistency. Much later Varian (1982) not only provided an accessible and clear account of Afriat's results, but also recast the cyclical consistency condition into a 'generalized axiom of revealed preference (GARP)' that runs as follows. A bundle q^i is strictly directly revealed preferred to a bundle q if $p^i q^i > p^i q$, while q^i is revealed preferred to q, if there exists a sequence, $j, k, \ldots, m$ such that $p^i q^i \geq p^i q^j$, $p^j q^j \geq p^j q^k$, $\ldots$, $p^m q^m \geq p^m q$, so that q^i is directly or indirectly (weakly) revealed preferred to q. GARP is satisfied if for all q^i revealed preferred to q^j, it is not true that q^j is strictly directly revealed preferred to q^i, and given GARP the data can be rationalized by a continuous, strictly concave, and non-satiated utility function. Differentiability can also be ensured by a sight strengthening of GARP, see Chiappori and Rochet (1987). GARP is readily tested for any given set of data by checking the pairwise inequalities and using a simple algorithm provided by Varian to map out the patterns of indirect revealed preference. Repeated applications of the method to time-series data have nearly always confirmed the consistency of the data with the theory. In retrospect, it is clear that violations of GARP cannot occur unless some budget lines intersect, so that if, over time, economic growth has resulted in the aggregate budget line moving steadily outward with little change in slopes, GARP is bound to be satisfied. (However, post-war United States data budget planes do occasionally intersect, and Bronars (1987) has recently shown that hypothetical demands generated by selecting random points on the actual budget lines would more often than not fail GARP.)

The contradictions between the parametric and non-parametric approaches can perhaps be resolved by thinking of the latter as a modelling technique that uses a very large number of parameters, so that the failure of the parametric models to fit theory to data can be thought of as failure to parametrize the models sufficiently richly. But I have already argued that these models already have too many parameters, and adding more would only exacerbate the already serious problems of measurement. For many purposes, the theory is only useful if it is capable of delivering a description of the data that is reasonably parsimonious. There is also something rather simple minded about non-parametric techniques that tends to be disguised by the sophisticated and elegant expositions that have been given them by Varian and others. Consider a very simple theory that says variable x should move directly with variable y as, for example, in the Euler equation (15) above which says that, under certainty consumption should grow from period t to $t+1$ if and only if the real interest rate from t to $t+1$ is greater than some fixed constant. A non-parametric test on a finite set of data would accept the theory if, in fact, x, and y always did move together, and reject it if x and y ever moved in opposite directions. That such testing procedures are widely employed in the press and by the uninformed public is no reason for treating them seriously in economics.

I have so far discussed the formulation and estimation of demand functions, meaning the relationships between quantities, outlay, and prices, and this has been the topic of most applied demand analysis over the last 30 years. However, there is an older tradition of demand analysis, in which the object of attention is household budget data, and this literature has recently been enjoying something of a revival. Since household budget data typically come from a cross-section of households over a short period of time, usually within a single year, prices are treated as common to all sample points, so that the focus of attention becomes the relationship between demand and outlay and the influence of household composition on the pattern of household expenditures. The oldest, and perhaps only law of economics, Engel's Law

that the share of food in the budget declines as total outlay increases, comes from Engel's (1857, published 1895) study of Belgian working-class families, and early empirical studies of demand were almost inevitably based on household surveys (see Stigler (1954) for a masterly review). The modern study of Engel curves, the relationships between expenditure and total outlay, begins (and almost ended) with Prais and Houthakker (1955). Prais and Houthakker studied the shapes of Engel curves, the relationship between demand and households, particularly in relation to the choice of quality, a topic that has subsequently been unjustly neglected. The functional forms for Engel curves that Prais and Houthakker examined became the staple menu for most subsequent studies, even though only one of their forms, the linear Engel curve, is capable of satisfying adding-up, and the linear form typically performs very badly on the data. Since 1955 a number of other Engel curves have been proposed, notably the lognormal Engel curve of Aitchison and Brown (1957), and Leser's (1963) revival of the form suggested much earlier by Holbrook Working (1943). Working's form, which apparently escaped the attention of Prais and Houthakker, makes the budget share of each commodity a linear function of the logarithm of total outlay. The formulation is particularly useful, for not only is it capable of accounting for most of the curvature that is discovered in empirical Engel curves, but it is also consistent with utility theory, and corresponds to the case where the welfare elasticity of the cost of living is independent of income. Gorman (1981) has provided a general characterization theorem for Engel curves of the form

$$p_i q_i = \sum_k a_{ik}(p) \xi_k(x) \tag{37}$$

and has shown that the $\xi_k(\cdot)$ functions can be powers of x (polynomial Engel curves), or x multiplied by powers of $\log x$ (Engel curves relating budget shares to powers of the logarithm of outlay), or have trigonometric forms. This last form includes Fourier representations of Engel curves, while the first two allow Taylor or Laurent expansions for the expenditure/outlay and for the share/log-outlay forms. The Working–Engel curve is the first member of Gorman's 'share to log' class, and the theorem tells us that we may add quadratic or higher order terms to improve the fit. However, Gorman's paper contains a remarkable result; the matrix of the a-coefficients in (37) has rank at most equal to three. In consequence, the share to log and log-squared Engel curves are as general as any, as are the Engel curves of the quadratic expenditure system, see Howe, Pollak and Wales (1979). Given Gorman's results, and the empirical success of the Working form, it and its quadratic generalization deserve wide use in the analysis of budget studies. There is also accumulating evidence that such forms are indeed necessary. Thomas (1986), in a wide-ranging examination of household survey data from developing countries, has shown that Engel's Law itself does not appear to hold among the very poor, so that, in many cases, the share of the budget devoted to food at first rises with total outlay before falling in conformity with the Law.

Prais and Houthakker also proposed a much-used formulation for the effects of household composition on behaviour. It can be written

$$p_i q_i / m_i(a) = f_i \{x/m_0(a)\} \qquad (38)$$

where a is a vector of household demographic characteristics (perhaps a list of numbers of people in each age and sex category) and m_i and m_0 are scalar valued functions known as the 'specific' and 'general scales' respectively. In this literature, scales are devices that convert family structure into numbers of equivalent adults, so that a family of two adults and two children might be two equivalent adults for theatre entertainment, three equivalent adults for food, and six equivalent adults for milk. The general scale is supposed to reflect the overall number of equivalent adults, so that the Prais and Houthakker model is a simple generalization of the idea that *per capita* demand should be a function of *per capita* outlay. Barten (1964), in a very important paper, took up the Prais–Houthakker idea of specific scales, but assumed that the arguments of the household utility function were the household consumption levels each deflated by the corresponding specific scale. The consequences of Barten's formulation are similar to those of Prais and Houthakker, but embody the additional insight that changes in family composition affect the effective shadow prices of goods, so that demographic changes will exercise, not only income, but also substitution effects on the pattern of demand. The story is often summarized by the phrase, 'if you have a wife and child, a penny bun costs three-pence', quoted in Gorman (1976), but the really far-reaching substitution effects of children are probably on time use and labour supply, particularly of women.

Since household surveys typically contain large samples of households, there is less need for theory to save degrees of freedom, and it is possible to estimate quite general functional forms that link expenditures to household composition patterns and then to interpret the results in terms of the various models. In addition, neither the Prais–Houthakker nor the Barten model seem to yield easily implemented functional forms, e.g. linear ones, nor is it clear that either model is even identified on a single cross-sectional household survey in which all prices are constant, see for example Muellbauer (1980) and Deaton (1986a). However, some empirical results for the two models can be found in Muellbauer (1977, 1980) and in Pollak and Wales (1980, 1981) who also examine Gorman's (1976) extension of Barten's model in which additional people are supposed to bring with them fixed needs for particular commodities. The fixed needs model

is close to the formulation proposed by Rothbarth (1943) for measuring the costs of children. Rothbarth pointed out that there are certain commodities, adult goods, that are not consumed by children, so that when children are added to a household, the only effects on the household's consumption of adult goods will be the income effects that reflect the fact that, with unchanged total resources, the household is now poorer. Deaton, Ruiz-Castillo and Thomas (1985) have recently attempted to test Rothbarth's contention, and in their Spanish data it seems possible to identify a sensible group of adult goods, the expenditure on each of which changes with additional children in the same way as they change in response to changes in outlay.

Studies of the effects of family composition on household expenditure patterns have frequently been concerned, not only with estimating demands, but also with attempts to measure the 'cost' of children. It would take me too far afield to do justice to this topic here. Readers interested in this controversial area should perhaps start with Rothbarth (1943), who in a few pages makes a very simple and quite convincing case, and look also at Nicholson (1976). Pollak and Wales (1979) weigh in on the opposite side, and claim that it is impossible to measure child costs from expenditure data. My own position is argued in Deaton and Muellbauer (1986); there are certainly grave problems to be overcome in moving from the analysis of household survey data to the measurement of the costs of children, and it is clear that identifying assumptions must be made that are more severe and more controversial than those required, for example, to go from demand functions to consumer surplus. But that does not mean that it is not possible for such assumptions to be proposed and to be sensibly discussed.

ANGUS DEATON

See also **bequests and the life cycle model; consumer expenditure (new developments and the state of research); demand theory; Euler equations; Friedman, Milton; rational expectations.**

Bibliography

Abowd, J.M. and Card, D. 1985. The covariance structure of earnings and hours changes in three panel data sets. Princeton University, mimeo.

Afriat, S.N. 1967. The construction of a utility function from expenditure data. *International Economic Review* 8, 67–77.

Aitchison, J. and Brown, J.A.C. 1957. *The Lognormal Distribution.* Cambridge: Cambridge University Press.

Altonji, J. 1986. Intertemporal substitution in labor supply: evidence from micro data. *Journal of Political Economy* 94, S176–215.

Altonji, J. and Siow, A. 1985. Testing the response of consumption to income changes with (noisy) panel data.

Industrial Relations Section Working Paper No.186, Princeton University, mimeo.

Ando, A. and Modigliani, F. 1963. The life-cycle hypothesis of saving: aggregate implications and tests. *American Economic Review* 53, 55–84.

Ashenfelter, O. 1984. Macroeconomic analyses and microeconomic analyses of labor supply. Presented to Carnegie-Rochester Conference, Bal Harbor, Florida, November 1983.

Ashenfelter, O. and Ham, J. 1979. Education, unemployment, and earnings. *Journal of Political Economy* 87, S99–116.

Barnett, W.A. 1983. New indices of money supply and the flexible Laurent demand system. *Journal of Business and Economic Statistics* 1, 7–23.

Barnett, W.A. and Lee, Y.W. 1985. The global properties of the miniflex Laurent, generalized Leontief, and translog flexible functional forms. *Econometrica* 53, 1421–37.

Barnett, W.A., Lee, Y.W. and Wolfe, M. 1985. The three dimensional global properties of the miniflex Laurent, generalized Leontief, and translog flexible functional forms. *Journal of Econometrics*, 3–31.

Barten, A.P. 1964. Family composition, prices, and expenditure patterns. In *Econometric Analysis for National Economic Planning*, ed. P.E. Hart, G. Mills, and J.K. Whitaker. London: Butterworth.

Barten, A.P. 1969. Maximum likelihood estimation of a complete system of demand equations. *European Economic Review* 1, 7–23.

Bean, C.R. 1985. The estimation of surprise models and the surprise consumption function. Centre for Economic Policy Research (London), Discussion Paper No.54, mimeo.

Bera, A.K., Byron, R. and Jarque, C.M. 1981. Further evidence on asymptotic tests for homogeneity in large demand systems. *Economics Letters* 8, 101–5.

Bernanke, B.S. 1984. Permanent income, liquidity, and expenditure on automobiles: evidence from panel data. *Quarterly Journal of Economics* 99, 587–614.

Bernheim, B.D., Schleiffer, A. and Summers, L.H. 1985. Bequests as a means of payment. *Journal of Political Economy*, 1045–76.

Blackorby, C., Primont, D. and Russell, R.R. 1977. On testing separability restrictions with flexible functional forms. *Journal of Econometrics* 5, 195–209.

Blinder, A.S. 1975. Distribution effects and the aggregate consumption function. *Journal of Political Economy* 83, 447–75.

Blinder, A.S. and Deaton, A.S. 1985. The time series consumption function revisited. *Brookings Papers on Economic Activity* 2, 465–511.

Breeden, D. 1979. An intertemporal asset pricing model with stochastic consumption and investment opportunities. *Journal of Financial Economics* 7, 265–96.

Bronars, S.G. 1987. The power of non-parametric tests of preference maximization. *Econometrica* 55, 693–8.

Browning, M.J., Deaton, A.S. and Irish, M.J. 1985. A profitable approach to labor supply and commodity demands over the life cycle. *Econometrica* 53, 503–43.

Campbell, J.Y. 1987. Does saving anticipate declining labor income? An alternative test of the permanent income hypothesis. *Econometrica* 55, 1249–73.

Campbell, J.Y. and Mankiw, N.G. 1986. Are output fluctuations transitory? National Bureau of Economic Research Working Paper 1916, processed.

Caves, D.W. and Christensen, L.R. 1980. Global properties of flexible functional forms. *American Economic Review* 70, 422–32.

Chiappori, P.-A. and Rochet, J.-C. 1987. Revealed preferences and differentiable demand. *Econometrica* 55, 687–91.

Christensen, L.R., Jorgenson, D.W. and Lau, L.J. 1973. Transcendental logarithmic production frontiers. *Review of Economics and Statistics* 55, 28–45.

Cochrane, J.H. 1986. How big is the random walk in GNP? Department of Economics, University of Chicago, processed.

Davenant, C. 1699. *Essay upon the Probable Methods of Making a People Gainers in the Balance of Trade.* London.

Davidson, J.E.H. et al. 1978. Econometric modelling of the aggregate time-series relationship between consumers expenditure and income in the United Kingdom. *Economic Journal* 88, 661–92.

Davies, J.B. 1980. Uncertain lifetime, consumption and dissaving in retirement. *Journal of Political Economy* 89, 561–77.

Deaton, A.S. 1974a. The analysis of consumer demand in the United Kingdom, 1900–1970. *Econometrica* 42, 341–67.

Deaton, A.S. 1974b. A reconsideration of the empirical implications of additive preferences. *Economic Journal* 84, 338–48.

Deaton, A.S. 1975. The structure of demand in Europe 1920–1970. In *The Fontana Economic History of Europe*, ed. C.M. Cippola. London: Collins-Fontana, Vol. 5.

Deaton, A.S. 1977. Involuntary saving through unanticipated inflation. *American Economic Review* 67, 899–910.

Deaton, A.S. 1986a. Demand analysis. In *Handbook of Econometrics*, ed. Z. Griliches and M. Intriligator. Amsterdam: North-Holland, Vol. 3.

Deaton, A.S. 1986b. Life-cycle models of consumption: is the evidence consistent with the theory? NBER Working Paper No.1910, processed.

Deaton, A.S. and Muellbauer, J. 1980a. An almost ideal demand system. *American Economic Review* 70, 312–26.

Deaton, A.S. and Muellbauer, J. 1980b. *Economics and Consumer Behaviour.* New York: Cambridge University Press.

Deaton, A.S. and Muellbauer, J. 1986. On measuring child costs, with applications to poor countries. *Journal of Political Economy* 94, 720–44.

Deaton, A.S., Ruiz-Castillo, J. and Thomas, D. 1985. The influence of household composition on household expenditure patterns: theory and Spanish evidence. Woodrow Wilson School, Princeton University, processed.

Dickey, D.A. and Fuller, W.A. 1981. Likelihood ratio statistics for autoregressive time series with a unit root. *Econometrica* 49, 1057–72.

Diewert, W.E. 1971. An application of the Shephard duality theorem: a generalized Leontief production function. *Journal of Political Economy* 79, 481–507.

Diewert, W.E. 1973. Functional forms for profit and transformation functions. *Journal of Economic Theory* 6, 284–316.

Diewert, W.E. 1974. Intertemporal consumer theory and the demand for durables. *Econometrica* 42, 497–516.

Diewert, W.E. and Wales, T.J. 1987. Flexible functional forms and global curvature conditions. *Econometrica* 55, 43–68.

Dixit, A.K. 1976. *Optimization in Economic Theory*. Oxford: Oxford University Press.

Dolde, W. and Tobin, J. 1971. Monetary and Fiscal Effects on Consumption in Consumer Spending and Monetary Policy: The Linkages. Boston: Federal Reserve Bank of Boston, Conference Series no. 5.

Duncan, G.J. and Hill, D.H. 1985. An investigation of the extent and consequences of measurement error in labor economic survey data. *Journal of Labor Economics*, 508–32.

Durlauf, S.N. and Phillips, P.C.B. 1986. Trends versus random walks in time-series analysis. Cowles Foundation Discussion Paper No.788. Yale University, New Haven, processed.

Eichenbaum, M.S., Hansen, L.P. and Singleton, K. 1984. A time series analysis of representative agent models of consumption and leisure choice under uncertainty. Graduate School of Industrial Administration, Carnegie-Mellon University, Pittsburgh, mimeo.

Engel, E. 1895. Die lebenkosten Belgischer Arbeiter-Familien früher und jetzt. *International Statistical Institute Bulletin* 9, 1–74.

Epstein, L. 1975. A disaggregate analysis of consumer choice under uncertainty. *Econometrica* 43, 877–92.

Evans, G.B.A. and Savin, N.E. 1982. Conflict among the criteria revisited; the W, LR, and LM tests. *Econometrica* 50, 737–48.

Evans, M.K. 1967. The importance of wealth in the consumption function. *Journal of Political Economy* 75, 335–51.

Fisher, I. 1930. *The Theory of Interest*. New York: The Macmillan Company.

Flavin, M. 1981. The adjustment of consumption to changing expectations about future income. *Journal of Political Economy* 89, 974–1009.

Flemming, J.S. 1973. The consumption function when capital markets are imperfect: the permanent income hypothesis reconsidered. *Oxford Economic Papers* 25, 160–72.

Friedman, M. 1957. *A Theory of the Consumption Function*. Princeton: Princeton University Press.

Gallant, A.R. 1982. Unbiased determination of production technologies. *Journal of Econometrics* 20, 285–323.

Geary, R.C. 1949–50. A note on 'A constant utility index of the cost of living'. *Review of Economic Studies* 18, 65–6.

Ghez, G. and Becker, G.S. 1975. *The Allocation of Time and Goods over the Life-cycle*. New York: Columbia University Press.

Goldberger, A.S. 1967. Functional form and utility: a review of consumer demand theory. Social Systems Research Institute, University of Wisconsin, processed.

Gorman, W.M. 1959. Separable utility and aggregation. *Econometrica* 27, 469–81.

Gorman, W.M. 1976. Tricks with utility functions. In *Essays in economic analysis*, ed. M. Artis and A.R. Nobay. Cambridge: Cambridge University Press.

Gorman, W.M. 1981. Some Engel curves. In *Essays in the Theory and Measurement of Consumer Behaviour in honour of Sir Richard Stone*, ed. A.S. Deaton. Cambridge: Cambridge University Press.

Gorman, W.M. 1982. Facing an uncertain future. IMSS Technical Report No.359, Stanford University, processed.

Grossman, S.J. and Shiller, R.J. 1981. The determinants of the variability of stock market prices. *American Economic Review, Papers and Proceedings* 71, 222–7.

Hall, R.E. 1978. Stochastic implications of the life cycle-permanent income hypothesis: theory and evidence. *Journal of Political Economy* 86, 971–87.

Hall, R.E. and Mishkin, F.S. 1982. The sensitivity of consumption to transitory income: estimates from panel data on households. *Econometrica* 50, 461–81.

Hansen, L.P. and Singleton, K.J. 1982. Generalized instrumental variables estimation of non-linear rational expectations models. *Econometrica* 50, 1269–86.

Hayashi, F. 1982. The permanent income hypothesis: estimation and testing by instrumental variables. *Journal of Political Economy* 90, 895–916.

Hayashi, F. 1985a. Permanent income hypothesis and consumption durability: analysis based on Japanese panel data. *Quarterly Journal of Economics*, 183–206c.

Hayashi, F. 1985b. Tests for liquidity constraints: a critical survey. Osaka University and NBER, processed. Presented at the Fifth World Congress of the Econometric Society, Cambridge, MA, August 1985.

Heckman, J.J. 1971. Three essays on the supply of labor and the demand for goods. Unpublished PhD thesis, Princeton University.

Hicks, J.R. 1939. *Value and Capital*. Oxford: Oxford University Press.

Houthakker, H.S. 1961. An international comparison of personal saving. *Bulletin of the International Statistical Institute* 38, 55–70.

Houthakker, H.S. 1965. On some determinants of saving in developed and underdeveloped countries. In *Problems in*

Economic Development, ed. A.G. Robinson. London: Macmillan.

Houthakker, H.S. and Taylor, L.D. 1970. *Consumer Demand in the United States: Analysis and Projections*. 2nd edn, Cambridge, MA: Harvard University Press.

Howe, H., Pollak, R.A. and Wales, T.J. 1979. Theory and time series estimation of the quadratic expenditure system. *Econometrica* 47, 1231–47.

Koskela, E. and Viren, M. 1982a. Saving and inflation: some international evidence. *Economics Letters* 9, 337–44.

Koskela, E. and Viren, M. 1982b. Inflation and savings: testing Deaton's hypothesis. *Applied Economics* 14, 579–90.

Kotlikoff, L.J. and Summers, L.H. 1981. The role of intergenerational transfers in aggregate capital accumulation. *Journal of Political Economy* 89, 706–32.

Kuznets, S. 1946. *National Income: A summary of findings*. National Bureau of Economic Research. New York: Arno Press.

Kuznets, S. 1962. Quantitative aspects of the economic growth of nations: VII: the share and structure of consumption. *Economic Development and Cultural Change* 10, 1–92.

Kydland, F.E. and Prescott, E.C. 1982. Time to build and aggregate fluctuations. *Econometrica* 50, 1345–70.

Laitinen, K. 1978. Why is demand homogeneity so often rejected? *Economics Letters* 1, 187–91.

Leff, N. 1969. Dependency rates and saving rates. *Economic Journal* 59, 886–96.

Leser, C.E.V. 1963. Forms of Engel functions. *Econometrica* 31, 694–703.

Lucas, R.E. 1976. Econometric policy evaluation: a critique. In *The Phillips Curve and Labor Markets*, ed. K. Brunner and A. Meltzer, Carnegie-Rochester Conference Series on Public Policy 1, Amsterdam: North-Holland.

Lucas, R.E. 1981. Introduction. In *Studies in Business Cycle Theory*, ed. R.E. Lucas. Cambridge, MA: MIT Press.

Lucas, R.E. and Rapping, L. 1969. Real wages, employment, and inflation. *Journal of Political Economy* 77, 721–54.

MaCurdy, T.E. 1981. An empirical model of labor supply in a life-cycle setting. *Journal of Political Economy* 89, 1059–85.

Mankiw, N.G. and Shapiro, M. 1985. Trends, random walks, and tests of the permanent income hypothesis. *Journal of Monetary Economics* 16, 165–74.

Mankiw, N.G. and Shapiro, M. 1986. Do we reject too often? Small sample properties of tests of rational expectations models. *Economics Letters* 20, 139–45.

Mankiw, N.G., Rotemberg, J.J. and Summers, L.H. 1985. Intertemporal substitution in macroeconomics. *Quarterly Journal of Economics* 100, 225–51.

Meisner, J.F. 1979. The sad fate of the asymptotic Slutsky symmetry test. *Economics Letters* 2, 231–3.

Mirer, T.W. 1979. The wealth–age relationship among the aged. *American Economic Review* 69, 435–43.

Modigliani, F. 1970. The life-cycle hypothesis of saving and inter-country differences in the saving ratio. In *Induction,*

Growth and Trade: Essays in Honour of Sir Roy Harrod, ed. W.A. Eltis et al., Oxford: Clarendon Press.

Modigliani, F. 1986. Life cycle, individual thrift, and the wealth of nations. *American Economic Review* 76, 297–313.

Modigliani, F. and Ando, A. 1957. Tests of the life-cycle hypothesis of savings. *Bulletin of the Oxford Institute of Economics and Statistics* 19, 99–124.

Modigliani, F. and Brumberg, R. 1954. Utility analysis and the consumption function: an interpretation of cross-section data. In *Post-Keynesian Economics*, ed. K.K. Kurihara. New Brunswick: Rutgers University Press.

Modigliani, F. and Brumberg 1979. Utility analysis and aggregate consumption functions: an attempt at integration. In *The Collected Papers of Franco Modigliani*, ed. A. Abel, Vol. 2, Cambridge, MA: MIT Press.

Muellbauer, J. 1977. Testing the Barten model of household composition effects and the cost of children. *Economic Journal* 87, 460–87.

Muellbauer, J. 1980. The estimation of the Prais–Houthakker model of equivalence scales. *Econometrica* 48, 153–76.

Muellbauer, J. 1985. Habits, rationality and the life-cycle consumption function. Nuffield College, Oxford, mimeo.

Nicholson, J.L. 1976. Appraisal of different methods of estimating equivalence scales and their results. *Review of Income and Wealth* 22, 1–11.

Phillips, P.C.B. and Perron, P. 1986. Testing for a unit root in a time series regression. Cowles Foundation Discussion Paper No.795, Yale University, New Haven, processed.

Pollak, R.A. and Wales, T.J. 1979. Welfare comparisons and equivalent scales. *American Economic Review* 69, 216–21.

Pollak, R.A. and Wales, T.J. 1980. Comparisons of the quadratic expenditure system and translog demand system with alternative specifications of demographic effects. *Econometrica* 48, 595–612.

Pollak, R.A. and Wales, T.J. 1981. Demographic variables in demand analysis, *Econometrica* 49, 1533–51.

Prais, S.J. and Houthakker, H.S. 1955. *The Analysis of Family Budgets*. Cambridge: Cambridge University Press.

Ramsey, F.P. 1928. A mathematical theory of saving. *Economic Journal* 38, 543–59.

Rothbarth, E. 1943. Note on a method of determining equivalent income for families of different composition. Appendix 4 in C. Madge ed., *War-time Patterns of Saving and Spending*, Occasional Paper 4, National Institute of Economic and Social Research, London.

Roy, R. 1943. *De l'utileé: contribution à la théorie des choix*. Paris: Herman.

Runkle, D.E. 1983. Liquidity constraints and the permanent income hypothesis: evidence from panel data. MIT, processed.

Samuelson, P.A. 1947–8. Some implications of linearity. *Review of Economic Studies* 15, 88–90.

Shephard, R. 1953. *Cost and Production* Functions. Princeton: Princeton University Press.

Shorrocks, A.F. 1975. The age-wealth relationship: a cross-section and cohort analysis. *Review of Economics and Statistics* 57, 155–63.

Slutsky, E. 1915. Sulla teoria del bilancio del consumatore. *Giornale degli Economisti* 15, 1–26. English translation in *Readings in Price Theory*, ed. G.J. Stigler and K. Boulding, Chicago: Chicago University Press, 1952.

Smith, J.P. 1977. Family labor supply over the life cycle. *Explorations in Economic Research* 4, 205–76.

Spinnewyn, F. 1979a. Rational habit formation. *European Economic Review* 15, 91–109.

Spinnewyn, F. 1979b. The cost of consumption and wealth in a model with habit formation. *Economics Letters* 2, 145–8.

Stigler, G.J. 1954. The early history of empirical studies of consumer behavior. *Journal of Political Economy* 62, 95–113.

Stone, J.R.N. 1954a. *The Measurement of Consumers' Expenditure and Behaviour in the United Kingdom, 1920– 1938*, Volume 1. Cambridge: Cambridge University Press.

Stone, J.R.N. 1954b. Linear expenditure systems and demand analysis: an application to the pattern of British demand. *Economic Journal* 64, 511–27.

Stone, J.R.N. 1964. Private saving in Britain, past, present, and future. *The Manchester School* 32, 79–112.

Stone, J.R.N. 1966. Spending and saving in relation to income and wealth. *L'Industria*, 471–99.

Surrey, M.J.C. 1974. Saving, growth, and the consumption function. *Bulletin of the Oxford Institute of Statistics* 36, 125–42.

Thomas, D. 1986. Essays on the analysis of Engel curves in developing countries. PhD thesis, Princeton University.

Uzawa, H. 1964. Duality principles in the theory of cost and production. *International Economic Review* 5, 216–20.

Varian, H.R. 1978. *Microeconomic Analysis*. 2nd edn, New York: Norton, 1984.

Varian, H.R. 1982. The non-parametric approach to demand analysis. *Econometrica* 50, 945–73.

Wales, T.J. 1977. On the flexibility of flexible functional forms: an empirical approach. *Journal of Econometrics* 5, 183–93.

Watson, M.W. 1986. Univariate detrending method with stochastic trends. *Journal of Monetary Economics* 18, 49–75.

Working, H. 1943. Statistical laws of family expenditure. *Journal of the American Statistical Association* 38, 43–56.

Yaari, M.E. 1965. Uncertain lifetime, life insurance, and the theory of the consumer. *Review of Economic Studies* 32, 137–50.

Zeldes, S. 1985. Consumption and liquidity constraints: an empirical investigation. The Wharton School, University of Pennsylvania, processed.

consumer expenditure (new developments and the state of research)

The state of research on consumer expenditure up to the mid-1980s is described in CONSUMER EXPENDITURE. Here, we provide an overview of recent developments on the intertemporal model of consumer behaviour under uncertainty. We organize our discussion around what has been the workhorse model for the analysis of dynamic consumption behaviour – the life-cycle permanent income model. Although our discussion of the intertemporal model is self-contained, it is not meant to be an exhaustive survey of this large literature. We do not cover demand analysis, despite the many exciting developments that have occurred in recent years.

The permanent income life-cycle (PILC) model, introduced during the 1950s by Modigliani and Brumberg (1954) and Friedman (1957), still plays an important role in the consumption literature. The PILC model can be loosely defined as a framework where individuals maximize utility over time given a set of intertemporal trading opportunities. Consumption at different points in time is treated as different commodities, so that, given intertemporal trading opportunities, consumption in a given period depends on total (life-cycle) resources and (intertemporal) prices. Optimal consumption choices are such that the ratio of (expected) marginal utilities of consumption at different times equals the ratio of inter-temporal prices. Therefore, the relationship between consumption and total resources is likely to depend on preferences (and in particular on the elasticity of inter-temporal substitution and the rate at which the future is discounted) and on interest rates (as they represent intertemporal prices). If we allow for uncertainty, as we discuss below, risk will also enter as a potentially important determinant of consumption.

This model can generate implications and insights for many important questions not only in macroeconomics but also in public finance, and has therefore attracted much attention, both theoretically and empirically. Recent research has stressed the need to look at preferences on the one hand and markets on the other, as the policy implications are the result of both.

The permanent income life-cycle model

In its simplest incarnation the PILC model considers a finite horizon, no uncertainty and very simple preferences. In such a situation, it is simple to translate the basic intuition of the model, to which we referred above, into a closed form solution for consumption that depends not just on current income but on the total amount of resources available to an individual and inter-temporal prices. The problem of this specification, of course, is its lack of realism. Not only do consumers in reality face much more complicated intertemporal environments, but it is likely that these complications have a first-order effect on consumption choices. Therefore, the simplest version of the model is a useful way to convey the main ideas behind PILC, but it needs to be complicated considerably to be of use for policy analysis.

The introduction of uncertainty in the model, which makes it much more realistic, complicates the problem enormously. The first formalizations of the life-cycle model under uncertainty date back to the 1970s (Bewley, 1977). Typically, one assumes that consumers maximize expected life-cycle utility choosing consumption and, in more general settings, leisure and financial asset holdings. Consumers are assumed to know the stochastic nature of their environment. Even with many simplifications on the nature of preferences, the model does not yield closed form solutions for consumption, except in the most special cases.

MaCurdy (1981; see also 1999) uses dynamic optimization techniques to derive necessary conditions for the optimal solution of the intertemporal optimization problem faced by consumers. The attractiveness of this approach lies in the fact that it cuts through the necessity of solving the model completely, which is a very hard task indeed, to focus on some useful implications of the model. In particular, these contributions focus on the basic first order condition, the so-called Euler equation, that equates the ratio of marginal utilities to intertemporal prices.

The first macro paper to take this approach is Hall (1978): under strong assumptions on preferences and returns, (non-durable) consumption is a random walk, that is:

$$E(C_{t+1}|I_t) = C_t \qquad (1)$$

where I_t denotes information available at time t. This remarkable proposition requires that utility be quadratic in consumption (and additively separable over time, states of nature and in its other arguments, notably male and female leisure and durable goods). It also requires that there is at least one financial asset with fixed real return, and that this equals the time-preference parameter. If consumers have rational expectations, then:

$$C_{t+1} = C_t + \varepsilon_{t+1} \qquad E(\varepsilon_{t+1}|Z_t) = 0 \qquad (2)$$

for all variables Z known at time t. A notable feature of Hall's model is that the Euler equation for consumption aggregates perfectly, because it involves linear transformations of the data. Hall used the Euler equation to test for the prediction implied by (2): no variable known to the consumer at time t should help predict the change in consumption between t and $(t+1)$.

Hall's paper was the first of many contributions that exploited the Euler equation and the fact that such an approach does not require the complete specification of the environment in which the consumer lives, or even the complete budget constraint. Moreover, the approach is robust to the presence of various imperfections in some intertemporal markets. And while the specification with

quadratic utility yields a linear equation for consumption, alternative specifications, with more plausible preferences, are easily introduced. For instance, in the case of power utility, an expression similar to (2) can be obtained for the log of consumption.

The price that one pays in using the Euler equations approach, which we discuss below, is that one does not obtain a closed form solution for consumption. An approach that goes beyond the consideration of the Euler equations is taken up in an important paper by Flavin (1981).

Flavin (1981) adopts the same theoretical framework as Hall (1978), and assumes that no other asset is available to the consumer (as in Bewley, 1977). However, Flavin develops a solution for consumption. To do so, she has to specify completely the stochastic environment in which the consumer lives and use particularly simple preferences. In particular, Flavin (1981) assumes that the only stochastic variable is labour income, that preferences are quadratic and that the consumer can save or borrow in a single asset with a fixed rate of interest. Under these conditions, Flavin shows that consumption is set equal to permanent income, and this is in turn defined as the present value of current and expected future incomes:

$$C_t = \frac{r}{1+r} A_t + \frac{r}{1+r} \sum_{k=0}^{\infty} E(y_{t+k}|I_t) \qquad (3)$$

where A denotes financial wealth and y is labour income. In this model, the first difference in consumption equals the present value of income revisions, due to the accrual of new information between periods t and $(t+1)$:

$$\Delta C_t = \frac{r}{1+r} \sum_{k=0}^{\infty} \frac{1}{(1+r)^k} \\ \times \left[E(y_{t+k}|I_t) - E(y_{t+k}|I_{t-1}) \right]. \qquad (4)$$

Equation (3) makes clear the main implications of the model: consumption depends on the present discounted value of future expected income. The interest rate plays the important role of converting future resources to present ones and therefore constitutes an important determinant of consumption. Flavin (1981) noticed that eq. (3) imposes cross-equation restrictions on the joint time series process for income and consumption. A similar approach had been followed by Sargent (1978) and, subsequently, by Campbell (1987) who noticed that an implication of (4) is that saving predicts future changes in income, the so-called 'saving for a rainy day' motive.

One of the main implications of the PILC model, particularly evident in eq. (3), is that, in appraising the effects of a given policy, for instance a tax reform that affects disposable income, a distinction must be drawn between permanent and temporary changes (Blinder and Deaton, 1985; Poterba, 1988).

Another feature of Flavin's model is that the closed form solution for consumption is the same under certainty and uncertainty, as long as expected values of future incomes are taken. This is a direct consequence of the assumption of quadratic utility that makes the marginal utility linear in consumption. For this reason, it is often referred to as the certainty equivalent model.

Extensions of the simple certainty equivalent model

The certainty equivalent model is appealing for its simplicity, but its implications are typically rejected by the data: Hall and Mishkin (1982) were particularly influential in suggesting that some of the model implications were rejected in micro data. At the same time, the model with quadratic preferences was perceived to be too restrictive in its treatment of financial decisions: quadratic preferences imply increasing absolute risk aversion in consumption (or wealth), something that is unappealing on theoretical grounds and strongly counterfactual (riskier portfolios are normally held by wealthier households). Quadratic preferences also imply that the willingness to substitute over time is a decreasing function of consumption: poor consumers should react much more to interest rate changes than rich consumers, after allowance has been made for the wealth/income effect.

The alternative adopted in much of the literature has been to assume power utility and to allow for the existence of a number of risky financial assets. Once one deviates from quadratic utility, however, and/or allows for stochastic interest rates, one loses the ability to obtain a closed form solution for consumption. Many of the studies that made this choice, therefore, focused on the study of the Euler equations derived from the maximization problem faced by the consumer. The basic first-order conditions used in this literature are two:

$$U_{ct} = \lambda_t, \qquad (5)$$

$$\lambda_t = E\left(\lambda_{t+1} \frac{1 + r^k_{t+1}}{1 + \delta}\middle| I_t\right). \qquad (6)$$

Equation (5) says that, at each point in time, the marginal utility of consumption equals the Lagrange multiplier associated with the budget constraint relevant for that period, which is sometimes referred to as the marginal utility of wealth. The second condition, eq. (6), that is derived from intertemporal optimality, dictates the evolution of the marginal utility of wealth (δ is a subjective discount rate). An equation of this type has to hold for each asset k for which the consumer is not at a corner. This is because the consumer is exploiting that particular intertemporal margin.

The attractiveness of Euler equations is that one can be completely agnostic about the stochastic environment faced by the consumer, about the time horizon,

about the presence of imperfections in financial markets (as long as there is at least one asset that the consumer can freely trade), about the presence of transaction costs in some component of consumption or labour supply. All relevant information is summarized in the level of the marginal utility of wealth. The approach is conceptually similar to the use of an (unobservable) fixed effect in econometrics. By taking first differences, one eliminates the unobservable marginal utility of wealth and is left only with the innovations to eq. (6).

Early papers along these lines were Hansen and Singleton (1982; 1983), who used power utility (also known as isoelastic, isocurvature or CRRA) as it has more appealing theoretical properties (relative risk aversion is constant in wealth or consumption, the elasticity of intertemporal substitution is also a constant). If we substitute eq. (5) into (6) and using the properties of the CRRA utility function, the Euler equations for consumption corresponding to each asset (k) will be:

$$E\left\{\left(\frac{C_{t+1}}{C_t}\right)^{-\gamma} \frac{1 + r^k_{t+1}}{1 + \delta}\right\} = 1 \qquad (7)$$

where γ is a curvature parameter (equal to the relative risk aversion parameter and to the reciprocal of the elasticity of intertemporal substitution) and δ, the subjective discount rate, measures impatience.

An equation such as (7) can be log-linearized to obtain (see Hansen and Singleton, 1983):

$$\Delta \log C_{t+1} = k + \frac{1}{\gamma} \log(1 + r^k_{t+1}) + \varepsilon_{t+1}. \qquad (8)$$

Although consumption appears on the left-hand side of eq. (8), this equation is not a consumption function, but an equilibrium condition. It cannot explain or predict consumption levels: consumption is crucially determined by the residual term ε_{t+1} and there is nothing that tells us what determines such a term or how this term changes with news about income, interest rates or any other relevant variable.

The Euler equation for a single asset can identify the elasticity of intertemporal substitution, a key parameter for the evaluation of the welfare costs of interest taxation (Boskin, 1978; Summers, 1981) and for the analysis of real business cycles (King and Plosser, 1984). The joint estimation of several Euler equations can help identify the pure discount rate parameter (governing patience), but also shed light on risk aversion, given that different assets typically have different risk characteristics.

The derivation of a closed form solution for consumption when certainty equivalence does not hold was first successfully tackled by Caballero (1990; 1991). Caballero (1991) took the Flavin model with known finite life, and constant absolute risk aversion (CARA) preferences, and showed that, when the optimal consumption age profile is flat with no uncertainty, it is increasing with income uncertainty. This change in the slope of the consumption profile was described as precautionary saving, because early in life consumers save more if labour income is more uncertain. Later work by Gollier (1995) and Carroll and Kimball (1996) established that a similar result holds whenever the third derivative of the utility function is positive, a feature of preferences labelled prudence. Both CARA and power utility exhibit prudence. The presence and size of precautionary savings is a matter of great relevance for public policy, in so far as public insurance schemes covering such risks as unemployment, health and longevity should reduce the need for consumers to accumulate assets.

The great merit of the model with prudence is that it highlights the need to save for rainy days even if sunny days are equally likely. An increased variance in the shocks to income reduces consumption even if expected income does not change. In the case of discrete variables, such as unemployment or illness, changes in first and second moments occur simultaneously, but this is not the case for continuous variables. The ability to distinguish between first and second moment effects is of crucial importance in the analysis of public policy, because of its social insurance characteristics.

The solution of the Bewley model with more general utility functions has to be computed numerically or rely on approximations. Several studies in the early 1990s took up the challenge of characterizing such solutions. Deaton (1991) studied a model with power utility and infinite life. Deaton considered the existence of liquidity or no-borrowing constraints, and showed that impatient consumers would hold limited assets to insure against low income draws. Carroll (1992) instead covered the case of finite lives, and showed that, if consumers are sufficiently impatient and their labour income is subject to both permanent and temporary shocks, they set consumption close to income. The model with impatient consumers under labour income uncertainty has been labelled 'the buffer stock model', because saving is kept to the lowest level compatible with the need to buffer negative income shocks. Later work by Attanasio et al. (1999) and Gourinchas and Parker (2002) clarifies the role played by age-related changes in demographics and the hump-shape age profile of labour income in generating income tracking for relatively young consumers (micro data show that financial asset accumulation starts around age 40). Hubbard, Skinner and Zeldes (1994; 1995) show instead how precautionary motives interact with the insurance properties of Social Security in the United States.

Many of the papers cited in the preceding paragraph consider relatively simple versions of the life-cycle model. In particular, a single non-durable commodity is assumed and preferences are assumed to be additively separable with leisure and over time. While this greatly simplifies the solution, the construction of a more realistic and complex model has become an important area of research. This development follows from the recognition that, for many purposes, and in particular for policy analysis, a model that delivers consumption as a function of exogenous variable is a very useful tool indeed.

This area of research has to deal with two important issues. First of all, the model can become very quickly, from a numerical point of view, very difficult to solve. The large number of state variables that characterize the solution of reasonably realistic models and the consideration of discrete choices and non-convexities linked to transaction costs can push the numerical capabilities of even very powerful computers. Second and even more importantly, if one wants to obtain solutions for consumption in a dynamic context, one has to characterize completely the stochastic environment in which the consumer lives. This contrast sharply with the Euler equation approach that allowed the researcher to be agnostic about most aspects of the environment and, under certain conditions, avoid solving difficult problems, such as labour supply, housing and other durable choices and so on. The Euler equation would hold regardless of the presence of non-convexities and other type of difficulties connected with these choices. These, instead have to be fully specified if one wants to work with a model that delivers a solution for consumption. These two difficulties constitute limits for the research in this area that, in all likelihood will not be overcome in the near future.

The empirical evidence on the PILC model
Since its introduction in the 1950s, there is no consensus about the empirical relevance of the PILC model. While the model it is one of the main tools in modern macroeconomics and public finance, its empirical performance is mixed. In this section, we discuss two branches of the literature.

The life-cycle model with various sources of uncertainty and generic preferences generates decision rules and behaviour of great complexity. Consumption and saving choices depend in an unknown fashion on every single aspect of the stochastic environment faced by the consumer, for instance on the entire distribution of future wages and earnings opportunities, on pension arrangements, on the asset markets the consumer can access, on mortality risks and so on and so forth. The Euler equation approach allows researchers to deal in a rigorous fashion with extremely rich models and yet derive relatively simple implications to test some aspects of the model and, with the help of additional assumptions, to identify some of the structural parameters that inform individual behaviour. We now understand that Euler equations can be used to determine what type of preferences fits the available data and can therefore provide one of the building blocks (preferences) in the study of the questions above. We also know that the presence of liquidity constraints does not necessarily produce violations of Euler equations because, even when liquidity constraints are present, they might be rarely binding.

The Euler equation is robust to a number of market imperfections, but is silent about how consumption or its growth reacts to specific news about shocks, changes in interest rates, taxation and so on. It is therefore useless for specific policy analysis. In other words, while the parameters of an Euler equation can be estimated in a wide set of circumstances, and one can use the equation to test the specification of the model, none of these results will provide an answer to questions like what is the effect of a change in taxation or interest rates on the level of consumption and saving?

This important shortcoming of the Euler equation approach explains why such an approach, which has informed and dominated the large empirical literature on the validity of the life-cycle permanent income model is virtually absent in the public economics literature on, say, the effect of pension reforms on saving or on the effect of changes in the taxation of interest on saving. And yet the conceptual framework that is behind the study of these issues is the same as that used to study consumption behaviour.

Policy analysis requires instead the availability of a consumption function, that is, a relation that explains consumption as a function of those variables that the consumer can take as exogenous at any given moment. Only in the simplest versions of the life-cycle model is it possible to derive an analytical expression for the consumption function. In general, given a set of assumptions on preference parameters and market and non-market opportunities, one has to rely on numerical solutions and/or approximations.

A less ambitious but potentially profitable approach that does not require numerical methods or incredibly rich data-sets is the estimation of reduced form equations, whose specification is informed by the life-cycle model. These are particularly useful in situations in which one analyses large (and possibly exogenous) changes to some of the likely determinants of consumption or saving. Such studies can address substantive issues and even test some aspects of the life-cycle model. Examples of studies of this kind include the reaction of consumption (and saving) to changes in pension entitlements (Attanasio and Brugiavini, 2003; Attanasio and Rohwedder, 2003; Miniaci and Weber, 1999), to swings in the value of important wealth components (such as housing, Attanasio and Weber, 1994) and to changes in specific taxes (Parker, 1999; Souleles, 1999; Shapiro and Slemrod, 2003).

Below we review the empirical evidence on the PILC model, organizing it in two subsections. First we start with the empirical evidence derived from Euler equations. We then move on to evidence that considers the *levels* of consumptions, rather than its changes.

Evidence from Euler equations

Two important empirical issues can be addressed with the study of Euler equations:

- What is the empirical relevance of the model? Is there a sensible specification of preferences that fits the observed data?
- What is the magnitude of the relevant preference parameters?

Tests of the model

As mentioned above, a prediction of the model is that changes in consumption cannot be predicted by expected changes in income or any other variable known to the consumer at time $t - 1$. This is the essence of the Hall (1978) test and of many others. Evidence that consumption can be predicted by lagged variables has been interpreted as indicative of liquidity constraints, myopic behaviour, misspecification of preferences and so on. The relationship between consumption and income has received considerable attention. The first to observe that the life-cycle model predicts no relation between the life-cycle profile of income and consumption was Thurow (1969). Thurow argued that the fact that consumption tracked income over the life-cycle was a rejection of the main implications of the PILC model. To this argument, essentially identical to many others proposed subsequently, Heckman (1974) replied that non-separability between consumption and leisure could explain such a relationship.

Despite this early exchange, after Hall (1978) a large fraction of the literature based on consumption Euler equations focused on the relationship between predictable changes in income and expected consumption growth. Hall and Mishkin (1982), as well as Campbell and Mankiw (1990; 1991) all report violations of this prediction, and label this finding 'excess sensitivity'. Excess sensitivity can be explained by the presence of liquidity constrained consumers, or of rule-of-thumb consumers, that is, consumers who let their expenditure track their income as a way to avoiding the complexities of choosing the optimal consumption path. However, consistently with Heckman's (1974) argument, excess sensitivity can be reconciled with the intertemporal optimization model if more general, and sensible, utility functions are used. In particular, if one assumes that leisure affects utility in a non-additive way, consumption changes respond to predictable labour income changes, whether or not leisure is a freely chosen variable. Finally, and importantly, the aggregation issue proves to be important. Attanasio and Weber (1993) show that results obtained with improperly aggregated micro data are consistent with results obtained with aggregate data and indicate rejections of the model that instead disappear with properly aggregated data and rich enough preference structures.

To summarize the discussion so far, it seems that while simple tests of the life-cycle model seem to reject the implications from the model and in particular those derived from Euler equations, it is possible to find specification of preferences that do a good job at fitting the available data, especially for households that are headed by prime-aged individuals. Aspects that are crucial for fitting the data are the use of household level data, allowing for changes in consumption needs induced by changes in family composition and the use of preferences specifications that allow for the marginal utility of consumption to depend on labour supply.

Estimation of preference parameters
Recent research on consumption and saving has singled out three preference parameters for attention: the elasticity of intertemporal substitution, the relative risk aversion parameter and the subjective discount rate. The size of these parameters has important implications in many applications of the model, ranging from macroeconomics to public finance to financial economics.

Perhaps surprisingly, not much evidence has been accumulated on the discount factor from the estimation of Euler equations. This can be explained by the fact that in log-linearized versions of the Euler equation, the parameter is not identified, while non-linear versions of the model are ridden by a number of econometric problems, particularly in relatively small samples of the type used in Euler equation estimation (see Attanasio and Low, 2004).

As for the distinction between the elasticity of intertemporal substitution (EIS) and the coefficient of risk aversion, it is absent in the most popular specifications used in the literature: a model where consumers maximize expected utility and preferences are iso-elastic and additively separable over time. In such a situation, the EIS is the reciprocal of the coefficient of relative risk aversion. Not many empirical papers have worked with preferences that allow for these two parameters to be disjoint.

An influential paper by Hall (1988) claimed that this parameter is close to zero. This finding has been challenged on various grounds. Attanasio and Weber (1993; 1995) point out that aggregation bias could be responsible for such a low estimate: they estimated a much higher elasticity (around 0.8) using UK and US cohort data (that is, data from repeated cross-sections, consistently aggregated over individuals born in the same years).

In the macro literature little attention has been paid to the possibility that the EIS may differ across consumers, particularly as a function of their consumption. A simple way to capture the notion that poor consumers may be less able to smooth consumption across periods and states of nature is to assume that the utility function does not depend on total (non-durable) consumption, but rather on the difference between consumption and needs. Thus we could retain the analytical attraction of power utility, but have $(C - C^*)$ as its argument, where C^* is an

absolute minimum that the consumer must reach in each and every period. This functional form is known as Stone–Geary utility in demand analysis, and is the simplest way to introduce non-homotheticity in a demand system. One could interpret 'external habits' (Abel, 1990; Campbell and Cochrane, 1999) as a special way to parameterize C^* (by making it a fraction of past consumption of other consumers). Attanasio and Browning (1995), Blundell, Browning and Meghir (1994) and Atkenson and Ogaki (1996) are among the few examples of papers that explicitly allow for wealth-dependent EIS.

Demographics might also affect preferences, and might explain consumption changes and the shape of the consumption age profile, as argued by Attanasio et al. (1999) as well as Browning and Ejrnaes (2002).

Evidence from the levels of consumption
As stressed above, the Euler equation imposes some restrictions on the dynamics of consumption but, on its own, does not determine the level of consumption. If one neglects numerical complications, a solution for consumption can be obtained by considering jointly the Euler equation and the sequence of budget constraints faced by the consumer as well as his or her initial wealth and a terminal condition. As noted by Sargent (1978), Flavin (1981) and later by Campbell (1987), the Euler equation and the intertemporal budget constraint imply a number of cross-equation restrictions for the joint time series processes of consumption and income. When one is able to obtain a closed form solution for consumption, as is the case with quadratic utility, these restrictions can be easily expressed in terms of a linear time series model, and tested.

Some of these restrictions are also implied by the Euler equation, while others are not. In particular, the restrictions on the contemporaneous correlation between income and consumption are not implied: as we stressed above, the Euler equation is silent about how news about income is translated into news about consumption.

Campbell and Deaton (1989) and West (1988) proposed a test that links the innovation to permanent income to consumption and presented evidence that aggregate consumption seems to be 'excessively smooth' in that it does not react enough to news about income. Campbell and Deaton make a connection between excess sensitivity and excess smoothness. Within the certainty equivalent model, they jointly model the consumption and income processes as a vector autoregression, assuming that income has a unit root plus some persistence. In this context, consumption changes reflect the permanent income innovation more than one-to-one: not only is the income shock permanent, but it also predicts future, smaller shocks of the same sign. This implies that over the business cycle consumption should be more volatile than income. But in actual aggregate data consumption is smoother than income: this is labelled 'excess smoothness', and is shown to be exactly equivalent to excess sensitivity.

Clearly the implications of a given set of intertemporal preferences for policy relevant questions depend crucially on the markets individuals have access to, on their imperfections and on the nature of the equilibrium they give rise to. The implications of complete markets would be very different from those one would derive if liquidity constraints or other markets imperfections were prevalent.

Insurance and credit markets

So far we have taken the assets the consumer can use to move resources over time as given and, in the simplest versions of the model, we have made very strong assumptions on this crucial aspect. For instance, we have assumed that consumers can borrow and lend at a fixed interest rate. The reality is, obviously, much more complex and, from a theoretical point of view, very many different environments have been studied. In particular, the possibilities open to a consumer depend on the market arrangements available. Below we discuss several of these market arrangements and briefly mention their implications for the determination of consumption.

Perfect insurance markets

If markets are complete and consumers can trade a full set of contingent claims without cost, individual risk will be completely diversified. In such a situation, a number of results deliver very useful predictions. In particular, it can be shown that a competitive equilibrium is symmetric and it is therefore possible to characterize the properties of competitive equilibria by considering the problem of a fictitious social planner, which, given a set of Pareto weights, maximizes social welfare. A strong implication of perfect markets is that the marginal utility of different consumers will move proportionally over time. The implication is very intuitive: the social planner faces a unique resource constraint, and marginal utility of all individuals, multiplied by the appropriate (and arbitrary) Pareto weight, will be equal to the multiplier associated to this unique constraint. As a consequence, marginal utility will move proportionally. If utility is isoelastic, consumption moves proportionally. These implications, stressed by Townsend (1994), have been tested in several papers (Cochrane, 1991; Attanasio and Davis, 1996; Hayashi, Altonji and Kotlikoff, 1996).

Many assets

When there are many assets, one can derive an Euler equation such as (7) for each of the assets for which the consumer is not at a corner. The Euler equations for consumption with different assets naturally ties up with asset pricing equations. This approach to asset pricing was developed by Breeden (1979) and Lucas (1978), and

extended to the case of non-additive separability of consumption and leisure in an incomplete markets setting by Bodie, Merton and Samuelson (1992). The model we sketched above is quite restrictive: the relative risk aversion parameter is inversely related to the elasticity of intertemporal substitution: Epstein and Zin (1989) show how this restriction can be relaxed in a more general model with power utility where the timing of uncertainty resolution matters (see also Epstein and Zin, 1991; Attanasio and Weber, 1989).

Interestingly, an Euler equation for an asset holds even if there are important imperfections in some other assets. As long as the consumer is exploiting a given margin to move resources over time, an equation such as (7) will apply. If the interest rate for a given asset changes with the level of the asset, then the Euler equation (7) will have to be augmented with a term reflecting this effect (Pissarides, 1978).

Liquidity constraints

The Euler equation will be violated when the consumer is able, for some reason, to borrow against future income. In such a situation, eq. (7) will hold as an inequality and the marginal utility of current consumption will be higher than the present discounted value of future consumption. Consumers who are liquidity constrained will be very sensitive to changes in current income. This case has received a considerable amount of attention in the literature. Many of the tests of violation of the Euler equation, such as Zeldes (1989), have focused on the so-called 'excess sensitivity' of consumption changes to predictable changes in income. It should be mentioned that, in a model with finite lives and a non-zero probability that income would be zero in each time period, standard regularity conditions on the utility function imply that a consumer will never want to borrow. If income is bounded away from zero, then the maximum the consumer will want to borrow is the present discounted value of the minimum value of income repeated in the future. This type of constraint has been sometimes referred to as a 'natural' liquidity constraint. Notice that such a constraint does not imply a violation of the Euler equation. If the restriction to borrowing is tighter, the Euler equation will instead be occasionally violated. And, even in periods in which it is not violated, the level of consumption will be affected by the possibility that the constraint will be binding in the future. As Hayashi (1987) explains, the presence of an operative, albeit not binding, liquidity constraint is equivalent to a shortening of the planning horizon or an increase in the discount rate. Evidence can be obtained by noting that consumers who are liquidity constrained will not be sensitive to changes in the level of the interest rate. As they will be at a kink of an intertemporal budget constraint, the demand for loans will be inelastic to changes in the slope of such an intertemporal budget constraint: the interest rate.

Endogenous liquidity constraint

In recent years, several studies have tried to model the shortcomings of credit and insurance markets by allowing for specific imperfections and frictions explicitly. The two main causes of imperfections that have been considered are: (*a*) private and asymmetric information and (*b*) the inability to perfectly enforce contracts. Models of this type can be seen as ways to endogenize specific market structures (such as one where consumers have access to a single asset in which they cannot borrow). In an influential paper, for instance, Cole and Kocherlakota (2001) show that an economy where individuals have a single bond in which they can borrow can be derived as a constrained equilibrium outcome where individuals have private information both on their income and on their savings.

Further extensions and alternative models

While the evidence on the relevance of the life-cycle model is still inconclusive, a number of empirical puzzles have directed attention to more complex preference structures. In particular, the equity premium puzzle and the evolution of aggregate saving rates in high-growth economies (South East Asia) has led macroeconomists to incorporate habits into the model. However, there is still little formal evidence on the empirical relevance of habits in micro data. The widely documented retirement consumption puzzle (that is, a sudden drop of consumption at retirement) as well as a number of more or less anecdotal pieces of evidence on the inadequacy of saving for retirement and other forms of 'irrational' behaviour, have been interpreted as potentially supportive of time-inconsistent preferences. The most elegant way to introduce time-inconsistent preferences is provided by the hyperbolic discounting assumption (Laibson, 1997).

Habits

Habits cause consumers to adjust slowly to shocks to permanent income, thus potentially explaining the excess smoothness of aggregate consumption, but also increase the utility loss associated with consumption drops, and may therefore help explain the equity premium puzzle.

Habits can take various forms: today's marginal utility may depend on the consumer's own past consumption level (internal habits) or the past consumption level of other consumers (external habits). This latter model seems to work better on aggregate data (Campbell and Cochrane, 1999), even though a recent survey by Chen and Ludvigson (2004) challenges this conclusion.

Empirical macro-evidence on the presence of habits is mixed, and this may be due to the very nature of aggregate consumption data, as stressed in Dynan (2000). The serial correlation of aggregate consumption growth is affected by time aggregation (Heaton, 1993), by aggregation over consumers, and by data construction methods (particularly for the services from durable goods). For this reason micro data seem preferable.

The simplest way to introduce habits (or durability) of consumption is to write the utility function as follows:

$$\Sigma_t u(x_t - \gamma' x_{t-1}; z_t) \qquad (11)$$

where x is a vector of goods or services and z is any other variable that affects marginal utility (demographics, leisure, other goods that are not explicitly modelled). The γ parameters are positive for goods that provide services across periods (durability), negative for goods that are addictive (habit formation) or zero for goods that are fully non-durable, non-habit forming (Hayashi, 1985).

The Euler equations corresponding to (11) involves x at four different periods of time, and their estimation typically requires panel data. High-quality consumption panel data are rare, and this has limited the scope for empirical analysis. Meghir and Weber (1996) have used Consumer Expenditure Survey (CEX) quarterly data on food, transport and services (and a more flexible specification of intertemporal non-separabilities than is implied by eq. 11), and found no evidence of either durability or habits once leisure, stock of durables and cars as well as other conditioning variables are taken into consideration.

Similarly negative evidence on habits has been reported by Dynan (2000), using Panel Study of Income Dynamics (PSID) annual food at home data. Carrasco, Labeaga and López-Salido (2005) use Spanish panel data and find some evidence for habits.

The few studies that have used micro data on non-durable consumption items to investigate the issue find little or no evidence of habits, at least once preferences capture the presence of non-separabilities between goods and leisure.

Durable goods

The presence of durable goods has received less attention in the micro-based literature than in the macro-literature, which has stressed the importance of their high volatility to explain business cycle fluctuations (Mankiw, 1982; Chah, Ramey and Starr, 1995).

The simplest way to introduce durable goods into the analysis is to let the stock of durables affect utility (on the assumption that services are proportional to the stock), and to posit a relation between current stock, S_t, previous stock, S_{t-1}, and current purchases q_t (or maintenance and repairs) in physical terms like:

$$S_t = (1 - \rho)S_{t-1} + q_t \qquad (12)$$

where ρ is a constant depreciation rate. This leads to the standard first-order condition for the durable good, according to which the relevant price is the user cost.

Typically, durable goods are costly to adjust, because of transaction costs (resale markets are dominated by information problems, known as the 'lemon' problem, and search costs are non-negligible). Sometimes these costs are modelled as a convex, differentiable function (Bernanke, 1985), but the recent literature has stressed the need to take into account their non-differentiable nature (Grossman and Laroque, 1990; Eberly, 1994; Attanasio, 2000; Bertola, Guiso and Pistaferri, 2005). This generates infrequent adjustment: consumers do not adjust continuously in response to depreciation, or income and price shocks, but wait until the actual stock hits either a lower limit, s, or an upper limit, S, and then adjust it to a target level. An interesting feature of this literature is that aggregate behaviour reflects changes in both the number of consumers that adjust and in the target level.

Durable goods might also play an insurance role, because they can be used to sustain consumption when times are bad. Postponing the purchase of food, or clothing, is certainly harder than failing to replace an old refrigerator or car, and housing maintenance can be put off for very long periods before structural damage occurs (Browning and Crossley, 2000). Durable goods also play a more specific insurance role, against changes in the price of the corresponding services. This is particularly relevant in the case of housing, where owning your home may be the best way to hedge the risk of future increases in the market price of housing services (Sinai and Souleles, 2005). Durable goods can also play a liquidity role, if they can be used as collateral to obtain a loan that pays for current consumption (Alessie, Devereux and Weber, 1997). A typical example could be the ability to remortgage a house, or to borrow 100 per cent of the value of a newly purchased car.

Even if one is not interested in modelling durable goods, the existence of a stock of durables should not be neglected when estimating preference parameters if utility is not additive in non-durable goods and durable services. Significant effects of durable goods (cars) on the Euler equation for non-durables have been found in UK data (Alessie, Devereux and Weber, 1997), and US data (Padula, 1999).

Quasi-hyperbolic discounting
The widely documented consumption puzzle (that is the sudden drop of consumption at retirement, see Hamermesh, 1984; Banks, Blundell and Tanner, 1998; and Bernheim, Skinner and Weinberg, 2001), as well as a number of more or less anecdotal pieces of evidence on the inadequacy of saving for retirement and other forms of 'irrational' behaviour, have been interpreted as potentially supportive of time-inconsistent preferences. The most elegant way to introduce time-inconsistent preferences is provided by the quasi-hyperbolic discounting assumption (Laibson, 1997). Consumers maximize the expected value of the following life-time utility index:

$$u(c_t) + \beta \sum_{\tau=1}^{T-t} \delta^t u(c_{t+\tau}) \qquad (13)$$

This implies that a different, lower discount factor is used to choose between this period and the next (the product of β and δ) and between any two other periods (δ). This generates time-inconsistent plans, with too little saving for retirement. For this reason, consumers may choose to enter long-term commitment plans, such as 401(k)s in the United States.

The quasi-hyperbolic discounting model lends itself to estimation and testing, but requires solving for the consumption function numerically. Even though an Euler equation for this model has been derived, its empirical use is limited, because it involves the marginal propensity to consume out of wealth (Harris and Laibson, 2001). It also suffers from some potential difficulties related to the definition of the time period, which crucially affects the properties of the solution, the length of which is arbitrarily set by the researcher.

A more tractable specification of preferences that may be used to model quasi-rational impatience has been put forward by Gul and Pesendorfer (2001; 2004), who stress the importance of self-control problems leading to the postponement of saving.

Where do we stand?
Since the 1970s we have learned much about the empirical implications of the life-cycle model and about the details of the model that need to be modified to fit the available evidence. Much work, however, remains to be done. In particular, there is scope to develop more complex numerical models that incorporate several realistic features. The areas of labour supply and housing are, in our opinion, particularly important. We also need to develop our understanding of the empirical implications of alternative models, such as hyperbolic discounting and check the extent to which they are empirically distinguishable from more standard models with complex preferences. Finally, it is important to stress the need for more and better data. One of the lessons learned from the development of new surveys that have been used to measure household wealth is that with enough ingenuity and creativity one can measure several of the variables that are relevant for our understanding of consumption and saving behaviour.

Our analysis of consumption and saving requires that more comprehensive measures of consumption are included in existing surveys, and that we learn to make systematic use of records on expectations, perceived uncertainty and so on.

ORAZIO P. ATTANASIO AND GUGLIELMO WEBER

See also **consumer expenditure; consumption-based asset pricing models (empirical performance); consumption-based asset pricing models (theory); elasticity of intertemporal substitution; Engel curve; Friedman, Milton; Modigliani, Franco; precautionary saving and precautionary wealth; revealed preference theory.**

Bibliography

Abel, A.B. 1990. Asset prices under habit formation and catching up with the Joneses. *American Economic Review* 80, 38–42.

Alessie, R., Devereux, M.P. and Weber, G. 1997. Intertemporal consumption, durables and liquidity constraints: a cohort analysis. *European Economic Review* 41, 37–59.

Atkenson, A. and Ogaki, M. 1996. Wealth varying intertemporal elasticities of substitution: evidence from panel and aggregate data. *Journal of Monetary Economics* 38, 507–34.

Attanasio, O.P. 2000. Consumer durables and inertial behaviour: estimation and aggregation of (S, s) rules for automobile purchases. *Review of Economic Studies* 67, 667–96.

Attanasio, O.P., Banks, J., Meghir, C. and Weber, G. 1999. Humps and bumps in life-time consumption. *Journal of Business and Economic Statistics* 17, 22–35.

Attanasio, O.P. and Browning, M. 1995. Consumption over the life cycle and over the business cycle. *American Economic Review* 85, 1118–37.

Attanasio, O.P. and Brugiavini, A. 2003. Social security and households' saving. *Quarterly Journal of Economics* 118, 1075–119.

Attanasio, O.P. and Davis, S.J. 1996. Relative wage movements and the distribution of consumption. *Journal of Political Economy* 104, 1227–62.

Attanasio, O.P. and Low, H. 2004. Estimating Euler equations. *Review of Economic Dynamics* 7, 405–35.

Attanasio, O.P. and Rohwedder, S. 2003. Pension wealth and household saving: evidence from pension reforms in the United Kingdom. *American Economic Review* 93, 1499–521.

Attanasio, O.P. and Weber, G. 1989. Intertemporal substitution, risk aversion and the Euler equation for consumption. *Economic Journal* 99, 59–73.

Attanasio, O.P. and Weber, G. 1993. Consumption growth, the interest rate and aggregation. *Review of Economic Studies* 60, 631–49.

Attanasio, O.P. and Weber, G. 1994. The UK consumption boom of the late 1980s: aggregate implications of microeconomic evidence. *Economic Journal* 104, 1269–1302.

Attanasio, O.P. and Weber, G. 1995. Is consumption growth consistent with intertemporal optimization? Evidence from the consumer expenditure survey. *Journal of Political Economy* 103, 1121–57.

Banks, J., Blundell, R. and Tanner, S. 1998. Is there a retirement-savings puzzle? *American Economic Review* 88, 769–88.

Bernanke, B. 1985. Adjustment costs, durables and aggregate consumption. *Journal of Monetary Economics* 15, 41–68.

Bernheim, B.D., Skinner, J. and Weinberg, S. 2001. What accounts for the variation in retirement wealth among U.S. households? *American Economic Review* 91, 832–57.

Bertola, G., Guiso, L. and Pistaferri, L. 2005. Uncertainty and consumer durables adjustment. *Review of Economic Studies* 72, 973–1007.

Bewley, T.F. 1977. The permanent income hypothesis: a theoretical formulation. *Journal of Economic Theory* 16, 252–59.

Blinder, A. and Deaton, A. 1985. The time series consumption function revisited. *Brookings Papers on Economic Activity* 1985(2), 465–521.

Blundell, R., Browning, M. and Meghir, C. 1994. Consumer demand and the life-cycle allocation of household expenditures. *Review of Economic Studies* 61, 57–80.

Bodie, Z., Merton, R.C. and Samuelson, W.F. 1992. Labor supply flexibility and portfolio choice in a life-cycle model. *Journal of Economic Dynamics and Control* 16, 427–49.

Boskin, M.J. 1978. Taxation, saving and the rate of interest. *Journal of Political Economy* S3–S28.

Breeden, D.T. 1979. An intertemporal asset pricing model with stochastic consumption and investment opportunities. *Journal of Financial Economics* 7, 265–96.

Browning, M. and Crossley, T.F. 2000. Luxuries are easier to postpone: a proof. *Journal of Political Economy* 108, 1022–6.

Browning, M. and Ejrnaes, M. 2002. Consumption and children. Working Paper No. 2002–6, Centre for Applied Microeconometrics, University of Copenhagen.

Caballero, R.J. 1990. Consumption puzzles and precautionary savings. *Journal of Monetary Economics* 25, 113–36.

Caballero, R.J. 1991. Earnings uncertainty and aggregate wealth accumulation. *American Economic Review* 81, 859–71.

Campbell, J.Y. 1987. Does saving anticipate declining labor income? An alternative test of the permanent income hypothesis. *Econometrica* 55, 1249–73.

Campbell, J.Y. and Cochrane, J. 1999. Force of habit: a consumption-based explanation of aggregate stock market behavior. *Journal of Political Economy* 107, 205–51.

Campbell, J.Y. and Deaton, A. 1989. Why is consumption so smooth? *Review of Economic Studies* 56, 357–73.

Campbell, J.Y. and Mankiw, N.G. 1990. Permanent income, current income, and consumption. *Journal of Business and Economic Statistics* 8, 265–79.

Campbell, J.Y. and Mankiw, N.G. 1991. The response of consumption to income: a cross-country investigation. *European Economic Review* 35, 723–56.

Carrasco, R., Labeaga, J.M. and López-Salido, J.D. 2005. Consumption and habits: evidence from panel data. *Economic Journal* 115, 144–65.

Carroll, C.D. 1992. The buffer-stock theory of saving: some macroeconomic evidence. *Brookings Papers on Economic Activity* 1992(2), 61–156.

Carroll, C.D. 1997. Buffer-stock saving and the life cycle/permanent income hypothesis. *Quarterly Journal of Economics* 112, 1–55.

Carroll, C.D. and Kimball, M.S. 1996. On the concavity of the consumption function. *Econometrica* 64, 981–92.

Chah, E.Y., Ramey, V.A. and Starr, R.M. 1995. Liquidity constraints and intertemporal consumer optimization: theory and evidence from durable goods. *Journal of Money, Credit and Banking* 27, 272–87.

Chen, X. and Ludvigson, S.C. 2004. *A Land of Addicts? An Empirical Investigation of Habits-Based Asset Pricing Models*. New York: New York University Press.

Choi, J.J., Laibson, D., Madrian, B.C. and Metrick, A. 2006. Saving for retirement on the path of least resistence. In *Behavioral Public Finance: Toward a New Agenda*, ed. E. McCaffrey and J. Slemrod. New York: Russell Sage Foundation.

Cochrane, J.H. 1991. A simple test of consumption insurance. *Journal of Political Economy* 99, 957–76.

Cole, H.L. and Kocherlakota, N.R. 2001. Efficient allocations with hidden income and hidden storage. *Review of Economic Studies* 68, 523–42.

Deaton, A. 1991. Saving and liquidity constraints. *Econometrica* 59, 1221–48.

Dynan, K. 2000. Habit formation in consumer preferences: evidence from panel data. *American Economic Review* 90, 391–406.

Eberly, J.C. 1994. Adjustment of consumers' durables stocks: evidence from automobile purchases. *Journal of Political Economy* 102, 403–36.

Epstein, L.G. and Zin, S.E. 1989. Substitution, risk aversion, and the temporal behavior of consumption and asset returns: a theoretical framework. *Econometrica* 57, 937–69.

Epstein, L.G. and Zin, S.E. 1991. Substitution, risk aversion, and the temporal behavior of consumption and asset returns: an empirical analysis. *Journal of Political Economy* 99, 263–86.

Flavin, M.A. 1981. The adjustment of consumption to changing expectations about future income. *Journal of Political Economy* 89, 974–1009.

Friedman, M. 1957. *A Theory of the Consumption. Function.* Princeton, NJ: Princeton University Press.

Gollier, C. 1995. The comparative statics of changes in risk revisited. *Journal of Economic Theory* 66, 522–36.

Gourinchas, P.-O. and Parker, J.A. 2002. Consumption over the life cycle. *Econometrica* 70, 47–89.

Grossman, S.J. and Laroque, G. 1990. Asset pricing and optimal portfolio choice in the presence of illiquid durable consumption goods. *Econometrica* 58, 25–51.

Gul, F. and Pesendorfer, W. 2001. Temptation and self-control. *Econometrica* 9, 1403–35.

Gul, F. and Pesendorfer, W. 2004. Self-control and the theory of consumption. *Econometrica* 72, 119–58.

Hall, R.E. 1978. Stochastic implications of the life cycle-permanent income hypothesis: theory and evidence. *Journal of Political Economy* 86, 971–87.

Hall, R.E. 1988. Intertemporal substitution in consumption. *Journal of Political Economy* 96, 339–57.

Hall, R.E. and Mishkin, F.S. 1982. The sensitivity of consumption to transitory income: estimates from panel data on households. *Econometrica* 50, 461–81.

Hamermesh, D.S. 1984. Consumption during retirement: the missing link in the life cycle. *Review of Economics and Statistics* 66, 1–7.

Hansen, L.P. and Singleton, K.J. 1982. Generalized instrumental variables estimation of nonlinear rational expectations models. *Econometrica* 50, 1269–86.

Hansen, L.P. and Singleton, K.J. 1983. Stochastic consumption, risk aversion, and the temporal behavior of asset returns. *Journal of Political Economy* 91, 249–65.

Harris, C. and Laibson, D. 2001. Dynamic choices of hyperbolic consumers. *Econometrica* 69, 935–57.

Hayashi, F. 1985. The permanent income hypothesis and consumption durability: analysis based on Japanese panel data. *Quarterly Journal of Economics* 100, 1083–113.

Hayashi, F. 1987. Tests for liquidity constraints: a critical survey. In *Advances in Econometrics II: Fifth World Congress*, ed. T. Bewley. Cambridge: Cambridge University Press.

Hayashi, F., Altonji, J. and Kotlikoff, L. 1996. Risk-sharing between and within Families. *Econometrica* 64, 261–94.

Heaton, J. 1993. The interaction between time-nonseparable preferences and time aggregation. *Econometrica* 61, 353–85.

Heckman, J.J. 1974. Life cycle consumption and labor supply: an explanation of the relationship between income and consumption over the life cycle. *American Economic Review* 64, 188–94.

Hubbard, R.G., Skinner, J. and Zeldes, S.P. 1994. The importance of precautionary motives in explaining individual and aggregate saving. *Carnegie-Rochester Conference Series on Public Policy* 40, 59–125.

Hubbard, R.G., Skinner, J. and Zeldes, S.P. 1995. Precautionary saving and social insurance. *Journal of Political Economy* 103, 360–99.

King, R.G. and Plosser, C.I. 1984. Money, credit, and prices in a real business cycle. *American Economic Review* 74, 363–80.

Laibson, D. 1997. Golden eggs and hyperbolic discounting. *Quarterly Journal of Economics* 62, 443–77.

Lucas, R.E., Jr. 1978. Asset prices in an exchange economy. *Econometrica* 46, 1429–45.

MaCurdy, T.E. 1981. An empirical model of labor supply in a life-cycle setting. *Journal of Political Economy* 89, 1345–70.

MaCurdy, T.E. 1999. An essay on the life cycle: characterizing intertemporal behavior with uncertainty, taxes, human capital, durables, imperfect capital markets, and nonseparable preferences. *Research in Economics* 53, 5–46.

Mankiw, G.N. 1982. Hall's consumption hypothesis and durable goods. *Journal of Monetary Economics* 10, 417–25.

Meghir, C. and Weber, G. 1996. Intertemporal nonseparability or borrowing restrictions? A disaggregate analysis using a U.S. consumption panel. *Econometrica* 64, 1151–81.

Miniaci, R. and Weber, G. 1999. The Italian recession of 1993: aggregate implications of microeconomic evidence. *Review of Economics and Statistics* 81, 237–49.

Modigliani, F. and Brumberg, R. 1954. Utility analysis and the consumption function: an interpretation of cross-section data. In *Post Keynesian Economics*, ed. K. Kurihara. New Brunswick, NJ: Rutgers University Press.

Padula, M. 1999. Euler equations and durable goods. Working Paper No. 30, CSEF.

Parker, J.A. 1999. The reaction of household consumption to predictable changes in social security taxes. *American Economic Review* 89, 959–73.

Pissarides, C.A. 1978. Liquidity considerations in the theory of consumption. *Quarterly Journal of Economics* 92, 279–96.

Poterba, J. 1988. Are consumers forward looking? Evidence from fiscal experiments. *American Economic Review* 78, 413–18.

Sargent, T.J. 1978. Rational expectations, econometric exogeneity, and consumption. *Journal of Political Economy* 86, 673–700.

Shapiro, M.D. and Slemrod, J. 2003. Consumer response to tax rebates. *American Economic Review* 93, 381–96.

Sinai, T. and Souleles, N.S. 2005. Owner-occupied housing as a hedge against rent risk. *Quarterly Journal of Economics* 120, 763–89.

Souleles, N.S. 1999. The response of household consumption to income tax refunds. *American Economic Review* 89, 947–58.

Summers, L. 1981. Capital taxation and capital accumulation in a life-cycle growth model. *American Economic Review* 71, 533–44.

Thurow, L.C. 1969. The optimum lifetime distribution of consumption expenditures. *American Economic Review* 59, 324–30.

Townsend, R.M. 1994. Risk and insurance in village India. *Econometrica* 62, 539–91.

West, K.D. 1988. The insensitivity of consumption to news about income. *Journal of Monetary Economics* 21, 17–33.

Zeldes, S.P. 1989. Consumption and liquidity constraints: an empirical investigation. *Journal of Political Economy* 97, 305–46.

consumer surplus

How does the market power exercised by firms influence consumer welfare? What is the effect of excise taxes on households with different levels of income? Does governmental regulation increase the welfare of consumers? Topical issues such as these indicate that the measurement of welfare is a fundamental element of public policy analysis. Indeed, a full consideration of taxes, subsidies, transfer programmes, health care reform, regulation, environmental policy, the social security system, and educational reform must ultimately address the question of how these policies affect individual well-being.

While centrally important to many problems of economic analysis, confusion persists concerning the relationship between commonly used indicators of welfare and well-established theoretical formulations. For more than 150 years, consumer surplus has been used to measure the welfare effects of changes in prices and incomes. Its popularity can be ascribed to its intuitive appeal, the ease with which it is implemented, and its modest data requirements. Although it is generally accepted that Dupuit (1844) was the originator of the concept of consumer surplus, it is largely attributed to Marshall (1890). (Chipman and Moore, 1976, provide a brief survey of the history of the debate related to consumer surplus.) We begin with the following notation:

$\mathbf{p} = (p_1, p_2, \ldots, p_n)$ – a vector of commodity prices.
Y_k – the income of individual k.
A_k – a vector of demographic characteristics of individual k.
$x_{ik} = x_i(\mathbf{p}, Y_k, \mathbf{A_k})$ is the demand for good i by individual k.

Suppose we are interested in the welfare impact of a change in the price of a single commodity from p_1^0 to p_1^1. The change in consumer surplus is given by:

$$\Delta CS_k = -\int_{p_1^0}^{p_1^1} x_1(t, p_2, \ldots p_n, Y_k, \mathbf{A_k}) \, dt.$$

$$(1)$$

If ΔCS_k is positive (negative), the price change is judged to have increased (decreased) the welfare of individual k. Is it ordinally equivalent to the change in utility? A necessary condition is that the demand function is generated by a rational consumer who maximizes utility subject to a budget constraint. Unless consumers have optimized and are at the boundaries of their budget sets, it is impossible to assess the welfare effects of changes in prices and incomes. (That is, demands must be 'integrable' and consistent with a well-behaved utility function. Hurwicz and Uzawa, 1971, provide a formal statement of the integrability conditions.)

If demands are consistent with rational consumer behaviour, an indirect utility function $V(\mathbf{p}, Y_k, \mathbf{A_k})$ represents the maximum utility attained at prices $\mathbf{p}$ and income Y_k, and Roy's Identity provides the link between demands and utility:

$$x_1(\mathbf{p}, Y_k, \mathbf{A_k}) = -\frac{\partial V(\mathbf{p}, Y_k, \mathbf{A_k})/\partial p_1}{\partial V(\mathbf{p}, Y_k, \mathbf{A_k})/\partial Y_k}.$$

$$(2)$$

If the marginal utility of income is constant, substitution of (2) into (1) yields an explicit expression for the change in consumer surplus that is ordinally equivalent to the change in utility:

$$\Delta CS_k = \frac{V(\mathbf{p}^1, Y_k, \mathbf{A_k}) - V(\mathbf{p}^0, Y_k, \mathbf{A_k})}{\partial V / \partial Y_k}.$$

While constancy of the marginal utility of income is restrictive, Chipman and Moore (1976; 1980) have shown that application of consumer surplus is more problematical if there are changes in more than one price. In such circumstances, the change in consumer surplus must be evaluated using a line integral defined over the path of price changes from $\mathbf{p}^0$ to $\mathbf{p}^1$:

$$\Delta CS_k = \int_{\mathbf{p}^0}^{\mathbf{p}^1} \sum_i x_i(\mathbf{p}, Y_k, \mathbf{A_k}) d\mathbf{p_i}. \quad (3)$$

Price paths are not observed so it is essential that (3) be path independent. This holds if the uncompensated price effects are symmetric (see, for example, Angus Taylor and Robert Mann, 1972, pp. 500–4):

$$\frac{\partial x_i}{\partial p_j} = \frac{\partial x_j}{\partial p_i} \quad \text{for all } i \neq j.$$

This form of symmetry requires preferences to be homothetic, which is a restriction that is inconsistent with well-established empirical regularities.

In the most general circumstance of changes in prices and income, consumer surplus is defined as:

$$\Delta CS_k = -\int_Z \sum_i x_i(\mathbf{p}, Y_k, \mathbf{A_k}) dp_i \quad (4)$$
$$+ (Y_k^1 - Y_k^0),$$

where $\mathbf{Z}$ is a path between $(\mathbf{p}^0, Y_k^0)$ and $(\mathbf{p}^1, Y_k^1)$. Chipman and Moore (1976) have demonstrated that there are no circumstances under which (4) is path independent and ordinally equivalent to the change in utility of a rational consumer.

Hicksian surplus measures

Given the problems with consumer surplus, how should the welfare effects of price and income changes be measured? Hicks (1942) developed an approach that is exactly analogous to (4) once we substitute compensated for uncompensated demand functions:

$$\Delta HS_k = -\int_Z \sum_i x_i^c(\mathbf{p}, V, \mathbf{A_k}) dp_i \quad (5)$$
$$+ (Y_k^1 - Y_k^0),$$

where $x_i^c(\mathbf{p}, V, \mathbf{A_k})$ is the compensated demand for the ith good evaluated at utility level V. Compensated price effects are symmetric, so the line integral in (5) is path independent and the surplus measure is single-valued.

For simple binary comparisons of policies, the utility level at which ΔHS_k is evaluated is often treated as a matter of little consequence. If it is calculated at the utility attained at prices $\mathbf{p}^1$ and income Y_k^1 (denoted V^1), a generalized version of the equivalent variation is obtained:

$$EV_k = E(\mathbf{p}^0, V^1, \mathbf{A_k}) - E(\mathbf{p}^1, V^1, \mathbf{A_k}) \quad (6)$$
$$+ (Y_k^1 - Y_k^0)$$
$$= E(\mathbf{p}^0, V^1, \mathbf{A_k}) - E(\mathbf{p}^0, V^0, \mathbf{A_k})$$

where $E(\mathbf{p}, V, \mathbf{A_k})$ is the expenditure function, defined as the minimum income needed for individual k to attain utility V at prices $\mathbf{p}$. Not only is the generalized equivalent variation single-valued, but it is ordinally equivalent to the change in utility. That is, EV_k is positive if and only if $V^1 > V^0$.

The utility level at which (5) is evaluated is important for multiple comparisons of price and income changes. The generalized equivalent variation will give an ordering of outcomes that is identical to that based on utility levels. If (5) is evaluated at $V^0 = V(\mathbf{p}^0, Y_k^0, \mathbf{A_k})$, we obtain the generalized compensating variation:

$$CV_k = E(\mathbf{p}^0, V^0, \mathbf{A_k}) - E(\mathbf{p}^1, V^0, \mathbf{A_k}) \quad (7)$$
$$+ (Y_k^1 - Y_k^0)$$
$$= E(\mathbf{p}^1, V^1, \mathbf{A_k}) - E(\mathbf{p}^1, V^0, \mathbf{A_k}).$$

Because the utility levels are 'cardinalized' using different prices for each set of binary comparisons, the ordering of multiple outcomes based on (7) need not match the ordering based on utility levels. Chipman and Moore (1980) have shown that consistent rankings of outcomes require restrictions on preferences that are the same as for consumer surplus.

While the simple static formulation of consumer surplus is the most frequent application, the conceptual framework can be extended to analyse the effects of changes in utility in more general settings. For example, intertemporal welfare effects are often represented as the discounted sum of the within-period equivalent or compensating variations.

Keen (1990) has shown that this will differ from the lifetime equivalent variation to the extent that individuals are able to substitute intertemporally. As an alternative approach, he defines V_L to be the maximum level of lifetime utility of an individual who lives T periods when the profiles of prices and interest rates are $\{\mathbf{p_t}\}$ and $\{r_t\}$ respectively. If the (optimal) time path of utility corresponding to V_L at these prices and interest rates is $\{V_{kt}\}$, the lifetime expenditure function can be represented as:

$$\Omega_L(\{\mathbf{p_t}\}, \{r_t\}, V_L) = \sum_t g_t \, E(\mathbf{p_t}, V_{kt}, \mathbf{A_{kt}}),$$

where $g_t = \prod_{s=0}^{t} (1 + r_s)^{-1}$.

As in the static framework, the lifetime expenditure function can be used to represent an exact measure of the change in lifetime welfare. Define V_L^1 to be the maximum level of lifetime welfare when the profile of prices and interest rates are $\{\mathbf{p_t^1}\}$ and $\{r_t^1\}$ and denote the corresponding time path of utility as $\{V_{kt}^1\}$. The reference prices and interest rates, $\{\mathbf{p_t^0}\}$ and $\{r_t^0\}$, yield a lifetime utility level of V_L^0 and within-period utilities $\{V_{kt}^0\}$. Keen's exact measure of the change in lifetime welfare, evaluated at the reference prices, is exactly analogous to the generalized equivalent variation:

$$\Delta W_L = \Omega_L(\{\mathbf{p_t^0}\}, \{r_t^0\}, V_L^1) \\ - \Omega_L(\{\mathbf{p_t^0}\}, \{r_t^0\}, V_L^0).$$

The concepts of the equivalent and compensating variation can also be extended to cases in which the choices made by consumers are discrete rather than continuous. Dagsvik and Karlstrom (2005) describe the compensating variation in the context of a random utility model defined as:

$$U_{jk} = V(\mathbf{p_j}, Y_k, \mathbf{A_k}) + \varepsilon_{jk} \quad (j = 1, 2, \ldots, J),$$

where U_{jk} is the utility of individual k in alternative j, $V(.)$ is a deterministic indirect utility function, and ε_{jk} are random variables. There are a total of J choices available to the consumer and, for simplicity, it is assumed that only prices vary across alternatives.

Consider the welfare effect of a change in the set of prices and income facing individual k from $(\mathbf{p_1^0}, \mathbf{p_2^0}, \ldots, \mathbf{p_J^0}, Y_k^0)$ to $(\mathbf{p_1^1}, \mathbf{p_2^1}, \ldots, \mathbf{p_J^1}, Y_k^1)$. If the consumer chooses the alternative that maximizes U_{jk}, the compensating variation is defined implicitly as that value CV_k that satisfies the following equality:

$$\max_j \; V(\mathbf{p_j^0}, Y_k^0, \mathbf{A_k}) + \varepsilon_{jk} \\ = \max_j \; V(\mathbf{p_j^1}, Y_k^1 - CV_k, \mathbf{A_k}) + \varepsilon_{jk}.$$

Although conceptually analogous to the equivalent and compensating variation described previously, CV_k is now random and cannot, in general, be represented in closed form.

From demand functions to welfare measurement

While it was understood that the equivalent variation resolved the conceptual problem of welfare measurement, it had little influence on applied welfare economics because compensated demand functions were presumed to be unobservable. Willig (1976) made the first attempt to bridge the gap between theory and application by showing that, for a single price change, consumer surplus can provide an approximation to the equivalent or compensating variation. However, with multiple price and income changes, consumer surplus is not single-valued and is of no use in approximating changes in economic welfare (McKenzie, 1979).

Shortly after the publication of Willig's paper, however, empirical procedures were developed to estimate the equivalent or compensating variation. Each method begins with the specification of a demand function and, under the assumption of integrability, is used to recover the utility or expenditure functions. The complexity of this procedure diminishes if demand functions are linear, and consideration is restricted to changes in the price of a single good.

Hausman (1981) provided an analytic solution to this problem for a demand function given by:

$$x_1 = \gamma_p p_1 + \gamma_Y Y_k + \gamma_\mathbf{A} \mathbf{A_k},$$

where γ_p, γ_Y, and $\gamma_\mathbf{A}$ are unknown parameters to be estimated econometrically. Roy's Identity provides a partial differential equation that can be solved to obtain an expenditure function of the form:

$$E(p_1, V, \mathbf{A_k}) = V e^{\gamma_Y p_1} - (1/\gamma_Y)[\gamma_p p_1 \\ + (\gamma_p/\gamma_Y) + \gamma_\mathbf{A} \mathbf{A_k}]. \tag{8}$$

The expenditure function allows the equivalent variation to be computed exactly as in (6) and Willig-type approximations are unnecessary. Hausman's method has the same data requirements as consumer surplus, and only linear regression methods are needed to estimate the unknown parameters.

Closed form solutions to the partial differential equation implied by Roy's Identity can be obtained for only a limited class of demand functions. An alternative approach is to begin with an assumed form of the indirect utility function and use Roy's Identity to obtain a system of demand equations. Since the form of the utility function is assumed from the outset, it is unnecessary to solve a complex system of partial differential equations.

Muellbauer (1974) provided an early example of this approach. He assumed that demands were consistent with a Stone–Geary utility function given by:

$$V(\mathbf{p}, Y_k) = \frac{(Y_k - \Sigma p_i \delta_i)}{\Pi p_i^{\alpha_i}} \tag{9}$$

where $\delta = (\delta_1, \delta_2, \ldots, \delta_n)$ and $\alpha = (\alpha_1, \alpha_2, \ldots, \alpha_n)$ are unknown parameters. The corresponding expenditure function is:

$$E(\mathbf{p}, V) = \Sigma p_i \delta_i + V(\Pi p_i^{\alpha_i}).$$

The unknown parameters can be estimated by fitting the linear expenditure system to household budget data:

$$p_i x_i = p_i \delta_i + \alpha_i (Y_k - \Sigma p_i \delta_i) \\ (i = 1, 2, \ldots, n). \tag{10}$$

Given estimates of $\boldsymbol{\alpha}$ and $\boldsymbol{\delta}$, the expenditure function can be used to compute the equivalent or compensating variation as in (6) and (7).

While this is more general than Hausman's approach, it has its own disadvantages. For an assumed form of the utility function, the functional forms of the demands are the same for every good, which may hinder the ability of the model to fit the data. Is it possible to start with an arbitrary demand system (rather than a utility function) and measure the welfare effects of multiple price changes? Two elegant procedures were proposed that required more complicated calculations to recover the expenditure function, but did not impose restrictions on the form of the demand functions other than the standard integrability conditions.

The first method is based on an approximation to McKenzie's (1957) indirect money metric utility function defined as:

$$\mu(\mathbf{p}, Y_k, \mathbf{A_k}; \mathbf{p}^0) = E(\mathbf{p}^0, V(\mathbf{p}, Y_k, \mathbf{A_k}), \mathbf{A_k}).$$

McKenzie and Pearce (1976) showed that $\Delta\mu$ can be approximated by a Taylor's series expansion about the initial equilibrium:

$$\Delta\mu = \frac{\partial\mu}{\partial\mathbf{p}'}\Delta\mathbf{p}(1/2) + \Delta\mathbf{p}'\frac{\partial^2\mu}{\partial\mathbf{p}\partial\mathbf{p}'}\Delta\mathbf{p}$$
$$+ \left(\frac{\partial\mu}{\partial Y} + \frac{\partial^2\mu}{\partial\mathbf{p}\partial Y'}\Delta\mathbf{p} + 1/2\frac{\partial^2\mu}{\partial Y^2}\Delta Y\right)$$
$$\times \Delta Y + R$$

$$(11)$$

where R represents higher order terms in the series.

The expression in (11) can be represented as a function of uncompensated demand functions when μ is evaluated at the reference prices (this follows from Roy's Identity and from the fact that at these prices the marginal utility of income is equal to one and all higher income derivatives are zero – see McKenzie and Pearce, 1976, for details):

$$\Delta\mu = -\mathbf{x}'\Delta\mathbf{p} - (1/2)\Delta\mathbf{p}'\left(\frac{\partial\mathbf{x}}{\partial\mathbf{p}} - \mathbf{x}\frac{\partial\mathbf{x}}{\partial Y'}\right)\Delta\mathbf{p}$$
$$+ \left(1 - \frac{\partial\mathbf{x}}{\partial\mathbf{Y}'}\Delta\mathbf{p}\right)\Delta Y + R.$$

$$(12)$$

Given knowledge of the demand functions and the magnitudes of the price and income effects, one has all of the information necessary to get as accurate an estimate of the change in utility as desired.

Vartia (1983) developed an algorithm that recovers the expenditure function numerically to any desired level of accuracy. Let $\mathbf{p}(t)$ and $Y_k(t)$ be the paths of price and income changes for $0 \leq t \leq 1$. As prices and income change, the movements of demands along an indifference curve can be represented implicitly by the differential equation:

$$\frac{dY_k(t)}{dt} = \sum_{i=1}^{n} x_i(\mathbf{p}(t), Y_k(t), \mathbf{A_k})\frac{dp_i(t)}{dt}.$$

Integrating over t yields an expression that can, in principle, be solved to obtain $E(\mathbf{p}(t), V^0, \mathbf{A_k})$ which is the centrepiece of the welfare calculations:

$$E(\mathbf{p}(t), V^0, \mathbf{A_k}) - E(\mathbf{p}^0, V^0, \mathbf{A_k})$$
$$= \sum_{i=1}^{n}\int_0^t x_i(\mathbf{p}(t), E(\mathbf{p}(t), V^0, \mathbf{A_k}), \mathbf{A_k})$$
$$\times \frac{dp_i(t)}{dt}dt.$$

$$(13)$$

Vartia described several algorithms that can be used to solve this equation numerically over the price path $\mathbf{p}(t)$ so that, when evaluated at $t=1$, we obtain $E(\mathbf{p}^1, V^0, \mathbf{A_k})$. As long as the demands satisfy the integrability conditions, the solution to (13) will be independent of the price path used in the algorithm. This method is valid for multiple price and expenditure changes and, because a closed-form solution is unnecessary, facilitates flexibility in estimating demand patterns.

Aggregation

The methods described to this point provide estimates of the change in welfare for individuals. In practice, analysts are more concerned about the impact of policies on groups. Micro-level estimates are an essential first step, but, for welfare economics to be useful to practitioners, a method of aggregation is essential. The easiest approach is to assume that market demands are generated by a representative consumer. Under this condition, the methods described previously can be applied to aggregate demands and the utility function of the representative agent can be recovered.

While frequently applied, this is unsatisfactory for a number of reasons. Market demands need not be consistent with a rational representative consumer. Even if every individual has demands that are consistent with utility maximization, aggregate demands need not satisfy any of the integrability conditions other than homogeneity of degree zero in prices and income (Sonnenschein, 1972). Moreover, it is unclear what this utility function actually represents. Kirman (1992) presents an example in which the representative agent prefers (aggregate) market basket A to B even though all individuals prefer the reverse. This violation of the most basic principle of social choice suggests that the utility of the representative agent should not be used for policy analysis even in the unlikely event that aggregate demands are integrable.

An alternative approach is to define aggregate welfare to be a function of the individual surplus measures. Such an approach was advocated by Harberger (1971) in his effort to make consumer surplus the standard tool for applied welfare analysis. At a conceptual level, such an indicator of aggregate welfare appears to be a natural extension of the positive analysis of welfare measurement at the micro level. This is obviously not the case because aggregation necessitates normative judgements in which the gains to some must be weighed against the losses to others. Simply summing the surplus measures, for example, embodies a version of utilitarianism and ignores distributional concerns.

Since any method of measuring welfare for groups of individuals necessarily involves subjective judgements, it seems reasonable to state explicitly the underlying ethical basis for the method of ordering outcomes in the aggregate. The social choice theoretic framework used by Sen (1970) provides a reasonable way of presenting the normative assumptions related to the measurability and comparability of individual welfare levels that facilitate well-behaved social orderings of outcomes. Under conditions described by Sen and others, these orderings can be represented by a social welfare function:

$$W = W(V_1, V_2, \ldots, V_K)$$

where V_k is a welfare indicator of individual k.

A monetary measure of social welfare can be obtained using Pollak's (1981) concept of a social expenditure function:

$$M(\mathbf{p}, W) = \min \ \{Y : W(V_1, \ldots, V_K) \geq W, \Sigma Y_k = Y\}.$$

This function is exactly analogous to its micro-level counterpart and is the minimum level of aggregate income required to attain a specified social welfare contour. If W^0 is the social welfare under policy 0 and W^1 is the welfare under policy 1, the monetary measure of the change is social welfare is exactly analogous to the generalized equivalent variation:

$$\Delta W = M(\mathbf{p}, W^1) - M(\mathbf{p}, W^0).$$

ΔW is clearly ordinally equivalent to the changes in social welfare, and normative judgements are represented explicitly through the specification of the social welfare function.

DANIEL T. SLESNICK

See also **cost–benefit analysis; cost minimization and utility maximization; Hicksian and Marshallian demands; indirect utility function; social welfare function.**

Bibliography

Chipman, J.S. and Moore, J. 1976. The scope of consumer's surplus arguments. In *Evolution, Welfare and Time in Economics: Essays in Honor of Nicholas Georgescu-Roegen*, ed. A.M. Tang, F.M. Westfield, and J.S. Worley. Lexington, MA: Heath-Lexington Books.

Chipman, J.S. and Moore, J. 1980. Compensating variation, consumer's surplus, and welfare. *American Economic Review* 70, 933–49.

Dagsvik, J. and Karlstrom, A. 2005. Compensating variation and Hicksian choice probabilities in random utility models that are nonlinear in income. *Review of Economic Studies* 72, 57–76.

Dupuit, J. 1844. On the measurement of the utility of public works. Trans R.H. Barback. In *Readings in Welfare Economics*, ed. K.J. Arrow and T. Scitovsky. Homewood, IL: Richard D. Irwin.

Harberger, A.C. 1971. Three basic postulates for applied welfare economics. *Journal of Economic Literature* 9, 785–97.

Hausman, J.A. 1981. Exact consumer's surplus and deadweight loss. *American Economic Review* 71, 662–76.

Hicks, J.R. 1942. Consumer's surplus and index numbers. *Review of Economic Studies* 9(2), 126–37.

Hurwicz, L. and Uzawa, H. 1971. On the integrability of demand functions. In. *Preferences, Utility and Demand*, ed. J. Chipman et al. New York: Harcourt, Brace and Jovanovich.

Keen, M. 1990. Welfare analysis and intertemporal substitution. *Journal of Public Economics* 42, 47–66.

Kirman, A.P. 1992. Whom or what does the representative individual represent? *Journal of Economic Perspectives* 6(2), 117–36.

Marshall, A. 1890. *Principles of Economics*. London: Macmillan.

McKenzie, L. 1957. Demand theory without a utility index. *Review of Economic Studies* 24(65), 185–9.

McKenzie, G.W. 1979. Consumer's surplus without apology: comment. *American Economic Review* 69, 465–8.

McKenzie, G.W. and Pearce, I. 1976. Exact measures of welfare and the cost of living. *Review of Economic Studies* 43, 465–8.

Muellbauer, J. 1974. Prices and inequality: the U.K. experience. *Economic Journal* 84(333), 32–55.

Pollak, R.A. 1981. The social cost of living index. *Journal of Public Economics* 15, 311–36.

Sen, A.K. 1970. *Collective Choice and Social Welfare*. San Francisco: Holden Day.

Sonnenschein, H. 1972. Market excess demand functions. *Econometrica* 40, 549–63.

Taylor, A.E. and Mann, R. 1972. *Advanced Calculus*, 2nd edn. Lexington: Xerox College Publishing.

Vartia, Y.O. 1983. Efficient methods of measuring welfare change and compensated income in terms of ordinary demand functions. *Econometrica* 51, 79–98.

Willig, R.E. 1976. Consumer's surplus without apology. *American Economic Review* 66, 589–97.

consumption-based asset pricing models (empirical performance)

The aim of consumption-based asset pricing models is to explain a number of important and puzzling features of asset returns using standard economic theory. Perhaps the best-known challenge for these models is the *equity premium puzzle*. Let us start from the Euler equations for stock and bond choice, and let us assume that both of these Euler equations hold with equality. If agents have constant relative risk aversion (CRRA) preferences and if returns and consumption growth are jointly log-normal, then the Sharpe ratio (that is, the equity premium per unit of risk) can be decomposed as:

$$\frac{E(R^e)}{std(R^e)} \approx \alpha \times std(\Delta c) \times corr(\Delta c, R^e), \quad (1)$$

where R^e is the excess return on stocks over bonds, α is the relative risk aversion (RRA) parameter, and Δc denotes log consumption growth. The equity premium is about six per cent per year in the US data with a standard deviation of 15 per cent, producing a Sharpe ratio ($E(R^e)/std(R^e)$) of 0.4. Mehra and Prescott (1985) used the construct of a representative agent who consumes the aggregate endowment stream. Constantinides (1982), Rubinstein (1974) and Wilson (1968) derived aggregation results that rely on either complete markets or the absence of idiosyncratic income risk. By appealing to these aggregation results, Mehra and Prescott could substitute *per-capita* consumption growth into (1). This series has a standard deviation of less than two per cent in the post-war US data, and a low correlation with stock returns – less than 0.25 by most estimates. Substituting these values into the expression above implies a lower bound for the relative risk aversion coefficient of 80, which is implausibly high judging by its implications for an individual's choices in other settings. In other words, we need extremely high risk aversion to rationalize the observed equity premium, and that is the puzzle. Furthermore, even if one is willing to accept such a high coefficient of risk aversion, this choice creates different puzzles itself – a point first noted by Weil (1989).

To understand Weil's 'risk-free rate puzzle', first note that the Euler equation for the risk-free asset choice can be linearized to obtain:

$$E[R^f] \approx -\ln \beta + \alpha E(\Delta c) - \frac{\alpha^2}{2} var(\Delta c).$$

$$(2)$$

Let us assume a positive time discount rate ($\beta < 1$), and an average consumption growth rate of 1.5 per cent per year. Let us also abstract from uncertainty for the moment. Then a risk aversion of 40 would imply an implausibly high interest rate of nearly 60 per cent per year simply because these households are extremely unwilling to substitute consumption over time. As a result, they desire a flat consumption profile and, therefore, would like to transfer resources from the future to today. But since this is not feasible in an endowment economy, the equilibrium risk-free rate needs to be very high to discourage this type of consumption smoothing and make individuals willing to consume their endowment every period.

The last term in (2) captures the precautionary savings motive, which becomes active in the presence of uncertainty. For very high levels of risk aversion, this effect dominates the intertemporal substitution effect, and an increase in the RRA coefficient *reduces* the risk-free rate. Epstein and Zin (1989) developed a class of recursive preferences that disentangles the inverse of the elasticity of intertemporal substitution from the coefficient of risk aversion. As discussed below, these preferences allow one to make progress on the equity premium puzzle without running into the risk-free rate puzzle.

Against the backdrop of Mehra and Prescott's benchmark model, subsequent papers that attempt to resolve these puzzles can be categorized according to whether they modify (i) the preferences, (ii) the endowment process, or (iii) the market and asset structure. We discuss each of these approaches in turn.

The utility function

Recursive preferences

In the case of CRRA utility, the stochastic discount factor (SDF) has the following form: $M_{t,t+1} = \beta(C_{t+1}/C_t)^{-\alpha}$, where C denotes the level of consumption. A drawback of this specification is that it restricts the elasticity of intertemporal substitution (EIS) to be the reciprocal of the RRA parameter when in fact these two parameters capture conceptually distinct aspects of individuals' preferences. Building on work by Kreps and Porteus (1978), Epstein and Zin (1989) and Weil (1989) introduced 'recursive preferences' (also called 'non-expected utility'):

$$U_t = \left[(1-\beta)C_t^\rho + \beta E_t(U_{t+1}^{1-\alpha})^{\rho/1-\alpha} \right]^{1/\rho},$$

$$(3)$$

where α is still the RRA parameter, but now the EIS is captured by a separate parameter: $1/(1-\rho)$. In this case, the SDF is given by:

$$M_{t,t+1} = \left[\beta \left(\frac{C_{t+1}}{C_t} \right)^{\rho-1} \right]^\gamma \left(\frac{1}{R_t^M} \right)^{1-\gamma},$$

where $\gamma = \alpha/\rho$, and R_t^M is the total return on the investors' wealth portfolio (including human capital which must be tradable for this representation to be derived; see Epstein and Zin, 1989, and Weil, 1989). An appealing feature of this SDF is that it combines two components that are each central to separate asset pricing theories: in particular, the SDF is a geometric average of consumption growth and the market return, where the latter is the relevant SDF in the standard capital asset pricing model (CAPM). Moreover, when $\alpha = 0$ (logarithmic risk preferences), then the CAPM emerges as a special case whereas $\alpha = \rho$ reduces it to the standard case of expected utility (see Epstein and Zin, 1989; Campbell, 2000).

In addition, this preference specification is flexible enough to allow a choice of a coefficient of relative risk aversion that is high enough to match the equity premium without being forced to accept a very low EIS. The low EIS is responsible for the risk-free rate puzzle, as explained above. Bansal and Yaron (2004) exploit this agent's concern for long-run consumption risk by introducing a small predictable component in consumption growth.

Habit formation and catching-up with the Joneses

Another approach, pioneered by Sundaresan (1989), Abel (1990) and Constantinides (1990), starts from the following specification of the investor's preferences over consumption streams C_t:

$$U_t = \frac{(C_t - X_t)^{1-\alpha}}{1-\alpha}$$

where X_t is some function of either (i) the individual's own past consumption or (ii) the past consumption of a reference group, such as an individual's peers, neighbours, or the population as a whole. Abel's specification features the ratio of C_t to X_t instead of the level difference. The first approach allows an individual's marginal utility to depend on her own past consumption history. This is commonly referred to as habit formation, endogenous habit, or internal habit. The second interpretation allows an individual's utility to depend on her status *relative* to her peers, neighbours or the population as a whole. This is referred to as catching-up with the Joneses or as external habit. These preference specifications amplify the effect of consumption growth shocks on the marginal utility growth of investors, in turn generating a high equity premium.

A particularly successful version of the catching-up-with-the-Joneses specification was developed by Campbell and Cochrane (1999) (henceforth CC) who choose the sensitivity of X to consumption growth shocks to match the conditional and unconditional moments of returns. In the baseline CC model, aggregate consumption and dividend growth are i.i.d. over time. Menzly, Santos and Veronesi (2004) introduce additional cash flow dynamics to explain the time series and cross-section of stock returns, while Santos and Veronesi (2005) emphasize the importance of labour income share variation to understand time variation in risk premia. Wachter (2002) applies a version of the CC model to the term structure, while Verdelhan (2004) uses the same model to explain the forward premium puzzle.

Looks like habit

Several recent papers have proposed models with standard preferences (such as CRRA) but consider economic environments that give rise to SDFs similar to those resulting from external habit preferences (such as the one used in CC). Examples include work by Piazzesi, Schneider and Tuzel (2007) who introduce housing services consumption into this framework, and by Yogo (2006) who considers durable consumption broadly defined, building on earlier work by Dunn and Singleton (1986) and Eichenbaum and Hansen (1990). Finally, Guvenen (2005) studies a model with limited stock market participation and shows that while the asset pricing implications of his model are similar to those in CC, the implications for macroeconomic questions (such as policy analysis, and so on) are quite different.

Additional arguments in the utility function

The models discussed so far assume that investors only derive utility from non-durable consumption. In exchange economy models (in which the consumption process is exogenous) this is equivalent to assuming that non-durable consumption enters the utility function in a separable manner. Some recent papers explicitly model the utility flow from housing consumption (in a non-separable manner), and find that such an extension improves the asset pricing performance (see Grossman and Laroque, 1990; Piazzesi, Schneider and Tuzel, 2007; Flavin and Yamashita, 2002). Similarly, a labour–leisure choice was introduced by Boldrin, Christiano and Fisher (2001) and Danthine and Donaldson (2002), in a representative agent framework, and by Uhlig (2006) in an incomplete markets framework. However, these authors find that this extension negatively affects the performance of asset pricing models, because it allows households to smooth their marginal utility by adjusting on the labour–leisure margin. As a result, one needs to introduce additional – typically labour market – frictions to counteract this new smoothing opportunity.

Consumption dynamics

In consumption-based asset pricing models, it is common to assume that aggregate consumption growth is i.i.d. over time, because the evidence for consumption growth predictability in the data is weak. In the i.i.d. case, the conditional market price of risk, which can be approximated by the conditional standard deviation of the log SDF, $\sigma_t(\log M_{t,t+1}) = \alpha \times \sigma_t(\Delta c)$, is constant. Therefore, these models cannot generate any time variation in risk premia on equity or any other asset.

In the context of a standard representative agent model, Kandel and Stambaugh (1990) generate time-variation in risk premia by introducing heteroskedasticity in aggregate consumption growth. Bansal and Yaron (2004) deviate from the i.i.d. assumption by introducing a small predictable component in consumption growth that is statistically hard to detect. This long-run component increases the market price of consumption risk. In addition, they add some time variation in the size of the long-run risk component. Colacito and Croce (2005) show these long-run risk models can reconcile the low volatility of exchange rate changes with the large market price of risk. Finally, Longstaff and Piazzesi (2002) argue that corporate earnings are much more risky than aggregate consumption growth, and that this can account for a large share of the equity premium puzzle.

Production economy models

These asset pricing puzzles have also attracted a lot of attention from macroeconomists because the same basic framework used in Mehra and Prescott (1985) also forms the backbone of the Kydland and Prescott (1982) model and the subsequent real business cycle literature. Therefore, understanding why individuals dislike risk in financial markets could help shed light on individuals' perceptions of macro risk and consumption fluctuations, which are key issues for macroeconomic policy. However, macroeconomists are also interested in the determination of quantities, such as output, investment and consumption, making the exchange economy framework unsuitable for their purposes. Therefore, macroeconomists replace the exogenous endowment stream with the endogenous equilibrium consumption process generated by a standard neoclassical production economy that faces technology shocks. One of the first findings of this approach, summarized in Rouwenhorst (1995), is that resolving the equity premium puzzle in a production economy is far more challenging than in an exchange economy, because this endogenous consumption process becomes too smooth if one increases risk aversion. As a result, one needs to resort to real frictions such as large adjustment costs in Jermann's (1998) model. Furthermore, and as noted above, allowing for an endogenous labour supply choice, as is common in macroeconomic analysis, gives consumers another margin to smooth marginal utility and further reduces the equity premium. Boldrin, Christiano and Fisher (2001) and Uhlig (2006) have successfully introduced labour market frictions to effectively shut down this channel.

Market and asset structure

The aggregation results we appeal to in order to use a representative agent in asset pricing depend on market completeness. A natural question is to ask what happens if some of these markets are shut down.

Incomplete markets

In an attempt to resolve the equity premium puzzle, *uninsurable* idiosyncratic income risk has been introduced into consumption-based asset pricing models by Aiyagari and Gertler (1991), Telmer (1993), Lucas (1994), Heaton and Lucas (1996), Krusell and Smith (1997) and Marcet

and Singleton (1999), among others. Their main results, obtained numerically for a range of parameter values, suggest that the impact of uninsurable labour income risk on the equity premium is small, because agents manage to smooth consumption quite well by trading a risk-free bond. In fact, Levine and Zame (2002) show that under general conditions the equilibrium allocations and prices in incomplete market economies converge to the complete market counterparts as households become more patient, rendering the incompleteness moot.

So when does imperfect risk sharing matter? Mankiw (1986) derives a sufficient condition for imperfect risk sharing to increase the equity risk premium: the cross-sectional variance of consumption growth needs to increase when returns are low (that is, in recessions). Constantinides and Duffie (1996) embed this counter-cyclical cross-sectional variance mechanism in a general equilibrium model. Grossman and Shiller (1982) show that the Mankiw–Constantinides–Duffie (MCD) mechanism breaks down in continuous-time diffusion models, because the cross-sectional variance of consumption growth is deterministic.

Discussion of other models

Rietz (1988) was the first to argue that countries like the United States may simply have been very lucky. Hence, the observed history of the US economy may understate the actual probability of economic disasters, such as the Great Depression (at least as perceived by investors). In this case, the volatility of the SDF may be significantly higher than the one estimated from historical time series. As a result, investors will shun stocks and demand a much higher equity premium to hold them. One difficulty with this explanation is that many economic disasters also result in governments reneging on their debt obligations. Barro (2006) extends Rietz's framework by distinguishing between two types of disasters – those that only affect the stock market and those that affect all asset markets – and explores the empirical implications of this mechanism in recent work.

FATIH GUVENEN AND HANNO LUSTIG

See also **capital asset pricing model; consumption-based asset pricing models (theory); elasticity of intertemporal substitution; incomplete markets; recursive preferences.**

Bibliography

Abel, A.B. 1990. Asset prices under habit formation and catching up with the Jones. *American Economic Review* 80, 38–42.

Aiyagari, S.R. and Gertler, M. 1991. Asset returns with transaction costs and uninsured individual risk. *Journal of Monetary Economics* 27, 311–31.

Bansal, R. and Yaron, A. 2004. Risks for the long run: a potential resolution of asset prizing puzzles. *Journal of Finance* 59, 1481–509.

Barro, R. 2006. Rare disasters and asset markets in the twentieth century. *Quarterly Journal. of Economics* 121, 823–66.

Boldrin, M., Christiano, L. and Fisher, J. 2001. Habit persistence, asset returns, and the business cycle. *American Economic Review* 91, 149–66.

Campbell, J.Y. 2000. Asset pricing at the millennium. *Journal of Finance* 55, 1515–67.

Campbell, J.Y. and Cochrane, J.H. 1999. By force of habit: a consumption-based explanation of aggregate stock market behavior. *Journal of Political Economy* 107, 205–51.

Colacito, R. and Croce, M. 2005. Risks for the long-run and the real exchange rate. Working paper, New York University.

Constantinides, G.M. 1982. Intertemporal asset pricing with heterogeneous consumers and without demand aggregation. *Journal of Business* 55, 253–67.

Constantinides, G.M. 1990. Habit-formation: a resolution of the equity premium puzzle. *Journal of Political Economy* 98, 519–43.

Constantinides, G.M. and Duffie, D. 1996. Asset pricing with heterogeneous consumers. *Journal of Political Economy* 104, 219–40.

Danthine, J.-P. and Donaldson, J.B. 2002. Labour relations and asset returns. *Review of Economic Studies* 69, 41–64.

Dunn, K. and Singleton, K. 1986. Modeling the term structure of interest rates under nonseparable utility and durability of goods. *Journal of Financial Economics* 17, 769–99.

Eichenbaum, M. and Hansen, L.P. 1990. Estimating models with intertemporal substitution using aggregate time series data. *Journal of Business and Economic Statistics* 8, 53–69.

Epstein, L.G. and Zin, S. 1989. Substitution, risk aversion and the temporal behavior of consumption and asset returns: a theoretical framework. *Econometrica* 57, 937–69.

Flavin, M. and Yamashita, T. 2002. Owner-occupied housing and the composition of the house-hold portfolio. *American Economic Review* 79, 345–62.

Grossman, S. and Laroque, G. 1990. Asset pricing and optimal portfolio choice in the presence of illiquid durable consumption goods. *Econometrica* 58, 25–51.

Grossman, S. and Shiller, R. 1982. Consumption correlatedness and risk measurement in economies with non-traded assets and heterogeneous information. *Journal of Financial Economics* 10, 195–210.

Guvenen, F. 2005. A parsimonious macroeconomic model for asset pricing: habit formation or cross-sectional heterogeneity? Working paper, University of Texas at Austin.

Hansen, L.P. and Singleton, K. 1983. Stochastic consumption, risk aversion, and the temporal behavior of asset returns. *Journal of Political Economy* 91, 249–65.

Heaton, J. and Lucas, D. 1996. Evaluating the effects of incomplete markets on risk sharing and asset pricing. *Journal of Political Economy* 104, 668–712.

Jermann, U. 1998. Asset pricing in production economies. *Journal of Monetary Economics* 257–75.

Kandel, S. and Stambaugh, R.F. 1990. Expectations and volatility of consumption and asset returns. *Review of Financial Studies* 3, 207–32.

Kreps, D. and Porteus, E.L. 1978. Temporal resolution of uncertainty and dynamic choice theory. *Econometrica* 46, 185–200.

Krusell, P. and Smith, J.A. 1997. Income and wealth heterogeneity, portfolio selection, and equilibrium asset returns. *Macroeconomic Dynamics* 1, 387–422.

Kydland, Finn, E. and Prescott, Edward, C. 1982. Time to build and aggregate fluctuations. *Econometrica* 50, 1350–70.

Levine, D. and Zame, W. 2002. Does market incompleteness matter? *Econometrica* 70, 1805–39.

Longstaff, F. and Piazzesi, M. 2002. Corporate earnings and the equity premium. Working paper, UCLA Anderson School.

Lucas, D. 1994. Asset pricing with unidiversifiable income risk and short sales constraints: deepening the equity premium puzzle. *Journal of Monetary Economics* 34, 325–41.

Mankiw, G.N. 1986. The equity premium and the concentration of aggregate shocks. *Journal of Financial Economics* 17, 211–9.

Marcet, A. and Singleton, K. 1999. Equilibrium asset prices and savings of heterogeneous agents in the presence of incomplete markets and portfolio constraints. *Macroeconomic Dynamics* 3, 243–77.

Mehra, R. and Prescott, E. 1985. The equity premium: a puzzle. *Journal of Monetary Economics* 15, 145–61.

Menzly, L., Santos, T. and Veronesi, P. 2004. Understanding predictability. *Journal of Political Economy* 112, 1–47.

Piazzesi, M., Schneider, M. and Tuzel, S. 2007. Housing, consumption, and asset pricing. *Journal of Financial Economics* 83, 531–69.

Rietz, T.A. 1988. The equity risk premium: a solution? *Journal of Monetary Economics* 22, 117–31.

Rouwenhorst, G. 1995. Asset pricing implications of equilibrium business cycle models. In *Frontiers of Business Cycle Research*, ed. T.F. Cooley. Princeton, NJ: Princeton University Press.

Rubinstein, M. 1974. An aggregation theorem for security markets. *Journal of Financial Economics* 1, 225–44.

Santos, J. and Veronesi, P. 2005. Labor income and predictable stock returns. *Review of Financial Studies* 19, 1–43.

Sundaresan, S. 1989. Intertemporally dependent preferences and the volatility of consumption and wealth. *Review of Financial Studies* 2, 73–89.

Telmer, C. 1993. Asset-pricing puzzles and incomplete markets. *Journal of Finance* 48, 1803–32.

Uhlig, H. 2006. Macroeconomics and asset prices: some mutual implications. Working paper, Humboldt University.

Verdelhan, A. 2004. Habit-based explanation of the exchange rate risk premium. Working paper, Boston University.

Wachter, J. 2002. Habit formation and returns on bonds and stocks. Unpublished paper, Stern School of Business, New York University.

Weil, P. 1989. The equity premium puzzle and the risk-free rate puzzle. *Journal of Monetary Economics* 24, 401–24.

Wilson, R. 1968. The theory of syndicates. *Econometrica* 36, 119–32.

Yogo, M. 2006. A consumption-based explanation of the cross-section of expected stock returns. *Journal of Finance* 61, 539–80.

consumption-based asset pricing models (theory)

Consumption-based asset pricing models study the pricing of payoff streams using the covariance of these payoffs with the marginal utility growth of investors.

The central component of a consumption-based asset pricing model is the Euler equation, which imposes restrictions on the covariance between asset returns and the marginal utility growth of investors. An easy and intuitive way to derive this equation is by using a variational argument. Suppose that the optimal consumption path of investor i is given by $\{C_t^i\}_{t=0}^T$ where T is possibly infinite. Suppose further that an asset j is available with a return $R_{t,t+1}^j$ between periods t and $t+1$, and the investor is not facing a binding portfolio constraint with respect to this asset. Then a feasible strategy is to reduce consumption at time t by a small amount ε, invest it in asset j, and consume the proceeds, $C_{t+1}^i + \varepsilon R_{t+1}^j$, in the next period. Assuming a time-separable utility function, with the one-period felicity function denoted by U and a time discount factor of β, this strategy changes the investors' expected lifetime utility by $-U_c(C_t^i, X_t)\varepsilon + E_t[\beta U_c(C_{t+1}^i, X_{t+1})\varepsilon R_{t+1}^j]$, where E_t is the mathematical conditional expectation operator; X represents the arguments of the utility function other than consumption; and U_c denotes the partial derivative with respect to consumption. The optimality of the original sequence implies that this strategy cannot be profitable for any amount ε and any asset available. Setting this gain to zero and rearranging yields the Euler equation:

$$E_t\left[M_{t,t+1}R_{t,t+1}^j\right] = 1$$

$$\text{where} \quad M_{t,t+1} = \beta \frac{U_c(C_{t+1}^i, X_{t+1})}{U_c(C_t^i, X_t)}. \tag{1}$$

This Euler equation was first derived by Rubinstein (1976) and Lucas (1978) in discrete time, and by Breeden (1979) in continuous time. While this class of models can in principle be used to study a broad variety of assets, this article will focus on stocks and short-term bonds, which have received the greatest attention in the consumption-based asset pricing literature.

In the case of a one-period discount bond with gross return $R_{t,t+1}^f = 1/P_t^f$ – a bond that costs P_t^f dollars today and pays off 1 dollar tomorrow – the Euler equation can be rewritten as

$$P_t^f = E_t[M_{t,t+1}]. \tag{2}$$

Similarly, when the asset is a stock with ex-dividend price P_t^s and dividend payment D_t, the Euler equation can be rearranged to read $P_t^s = E_t[M_{t,t+1}(P_{t+1}^s + D_{t+1})]$. By forward substitution this equation yields:

$$P_t^s = E_t\left[\sum_{s=1}^{\infty} M_{t,t+s}D_{t+s}\right], \tag{3}$$

which determines the price of a share of equity as the value of all future dividends it entitles discounted by the SDF.

Lucas (1978) and Mehra and Prescott (1985) used a representative-agent endowment economy structure in which the dividend stream, $\{D_t\}_{t=1}^{\infty}$, is exogenously produced by a 'tree'. Furthermore, these dividends are assumed to be perishable ('fruit'), so in equilibrium the price of equity (in the tree) adjusts to the point where the representative agent is *willing* to consume all available dividends: $C_t = D_t$. Substituting this condition into the expression for M in eq. (1), and then using M in eqs (2) and (3) shows that the price of this stock and that of the one-period bond are entirely determined by the stochastic process for D_t together with the functional form for U (we ignore X_t for now).

Hansen and Singleton (1983) tested the representative agent's Euler equation on US consumption data, and found that the model was rejected. In a famous paper, Mehra and Prescott (1985) showed that when one chooses the properties of C_t to match the moments of aggregate consumption in the data ('calibrate the model to data'), the equity premium $E(R_{t+1}^s - R_t^f)$ generated by the model was about 60 *times* smaller than that observed in the historical US data. This 'equity premium puzzle' has generated enormous interest and led to the development of a wide range of consumption-based asset pricing models in an attempt to resolve it. For further discussion of the empirical performance of these models, see CONSUMPTION-BASED ASSET PRICING MODELS (EMPIRICAL PERFORMANCE).

An alternative way to explain the hurdles these models face is by deriving an empirical lower bound on the volatility of the stochastic discount factor (SDF). Subtracting the Euler equation for bond returns from the one for stock returns yields: $E[M_{t,t+1}(R^s_{t+1} - R^f_t)] = 0$. Noting that the left-hand side of this condition can be rewritten as $Cov(M_{t,t+1}(R^s_{t+1} - R^f_t)) + E(M_{t,t+1})E(R^s_{t+1} - R^f_t)$, some simple manipulations yield the following key decomposition:

$$\frac{E\left(R^s_{t+1} - R^f_t\right)}{\sigma\left(R^s_{t+1} - R^f_t\right)} = -\frac{\sigma\left(M_{t,t+1}\right)}{E\left(M_{t,t+1}\right)} \qquad (4)$$
$$\times corr\left(M_{t,t+1}, R^s_{t+1} - R^f_t\right),$$

where $\sigma(\cdot)$ denotes the standard deviation. Observing that the correlation term is bounded from above in absolute value by 1, we get

$$\frac{E\left(R^s_{t+1} - R^f_t\right)}{\sigma\left(R^s_{t+1} - R^f_t\right)} \leq \frac{\sigma\left(M_{t,t+1}\right)}{E\left(M_{t,t+1}\right)}. \qquad (5)$$

The left-hand side of this inequality is the 'Sharpe ratio' – the (expected) excess return demanded by investors per unit (standard deviation) of risk they bear – which averages about 0.40 in annual US data. The right-hand side is called the 'market price of risk' or the 'maximum Sharpe ratio'. This inequality bound implies that a consumption-based model must be able to generate an SDF with a coefficient of variation (standard deviation normalized by mean) of at least 40 per cent to be consistent with the Sharpe ratio observed in the data. Hansen and Jagannathan (1991) developed and generalized this observation to provide a 'volatility bounds' test for potential candidate models. As discussed in CONSUMPTION-BASED ASSET PRICING MODELS (EMPIRICAL PERFORMANCE), the majority of plausibly calibrated asset pricing models fail this test.

When the investor faces a binding borrowing constraint, she cannot increase her consumption today by reducing the holdings of asset j. As a result, her marginal utility today will remain higher than the value implied by the equality condition in (1), and the Euler condition for that asset will instead be an inequality: $E_t[M_{t,t+1}R^j_{t,t+1}] < 1$. This relaxes the lower bound on the volatility of the SDF derived in equation (5) (cf. Luttmer, 1996).

To develop further implications of consumption-based models it is necessary to impose additional structure on $M_{t,t+1}$, which requires being more specific about (i) the functional form and the arguments of the utility function; (ii) the stochastic properties of variables affecting marginal utility (that is, consumption, leisure, and so on); and (iii) the market structure. The latter determines whether an appropriate aggregation theorem holds (which happens for example when markets are complete), in which case C^i can be replaced with aggregate consumption. Therefore, consumption-based models can be broadly categorized based on the assumptions they

make along these three dimensions. These different models are discussed in CONSUMPTION-BASED ASSET PRICING MODELS (EMPIRICAL PERFORMANCE).

Another feature of asset markets that has received much attention in the literature concerns the high volatility of stock prices. For example, the standard deviation of the log price/dividend (P/D) ratio of stocks is about 40 per cent per annum in the US data. In a world with a constant SDF (as would be the case with risk-neutral investors), it is impossible to rationalize this high volatility with the relatively low variability of the underlying dividend stream (LeRoy and Porter, 1981; Shiller, 1981). Let p_t denote the log price, d_t denote the log dividend, and r_t denote the log stock return. Using a first-order approximation, Campbell and Shiller (1988) show that the log P/D ratio can be decomposed as follows:

$$p_t - d_t = constant + E_t \sum_{j=1}^{\infty} \rho^{j-1}[\Delta d_{t+j} - r_{t+j}]$$

with $\rho = \exp(\overline{pd})/(1 + \exp(\overline{pd}))$ and $\overline{pd}$ denotes the average log P/D ratio. The first term in the square brackets is referred to as the cash flow component, and the second part is referred to as the discount rate component. This decomposition implies that the variance of the log P/D ratio can be stated as:

$$var(p_t - d_t) = cov\left(p_t - d_t, \sum_{j=1}^{\infty} \rho^{j-1}\Delta d_{t+j}\right)$$
$$- cov\left(p_t - d_t, \sum_{j=1}^{\infty} \rho^{j-1}r_{t+j}\right).$$

This expression shows that the P/D ratio moves only because it predicts future returns on stocks or because it predicts future dividend growth. In the data, most of the volatility in P/D ratio is due to news about future expected returns ('discount rates'), not due to future dividend growth ('cash flows') (Campbell, 1991; Cochrane, 1991). There is a large literature that documents the predictability of stock returns over longer holding periods, starting with work by Campbell and Shiller (1988; 1998), Poterba and Summers (1986) and Fama and French (1988; 1989). Other variables that predict returns include the spread between long and short bonds (Fama and French, 1989) and the T-bill rate (Lamont, 1998). More recently, more attention has been paid to macroeconomic variables that predict returns, most notably in the work by Lettau and Ludvigson (2001a) who document that the consumption/wealth ratio is a powerful predictor of stock returns.

So, the volatility of P/D ratio implies that excess returns on stocks are highly predictable. In other words, expected excess returns change a lot over time, even per unit of risk. We use the conditional version of the expression in (4) to understand the implications of this finding:

$$\frac{E_t\left(R_{t+1}^s - R_t^f\right)}{\sigma_t\left(R_{t+1}^s - R_t^f\right)} = -\frac{\sigma_t\left(M_{t,t+1}\right)}{E_t\left(M_{t,t+1}\right)} \tag{6}$$
$$\times corr_t\left(M_{t,t+1}, R_{t+1}^s - R_t^f\right),$$

where σ_t denotes the conditional standard deviation. Good models need to produce a lot of time variation in the right-hand side of (6) and this happens mostly through variation in the conditional market price of risk (first term). This is an upper bound on the conditional Sharpe ratio. (See also Lettau and Ludvigson, 2001b, on how to measure variation in the conditional Sharpe ratio.) Another test of consumption-based asset pricing models is whether they are able to generate as much predictability as found in the data. Examples of early models that match the variation in the conditional market price of risk include Kandel and Stambaugh (1990), Campbell and Cochrane (2000) and Barberis, Huang, and Santos (2001). More recent work includes the work by Santos and Veronesi (2005), Menzly, Santos and Veronesi (2004), Piazzesi, Schneider and Tuzel (2007), Guvenen (2005), Lustig and Van Nieuwerburgh (2005; 2006) and Bansal and Yaron (2004). These models are discussed in detail in CONSUMPTION-BASED ASSET PRICING MODELS (EMPIRICAL PERFORMANCE).

FATIH GUVENEN AND HANNO LUSTIG

See also **consumption-based asset pricing models (empirical performance); Euler equations.**

Bibliography

Bansal, R. and Yaron, A. 2004. Risks for the long-run: a potential resolution of asset pricing puzzles. *Journal of Finance* 59, 1481–509.

Barberis, N., Huang, M. and Santos, T. 2001. Prospect theory and asset prices. *Quarterly Journal of Economics* 116, 1–53.

Breeden, D.T. 1979. An intertemporal asset pricing model with stochastic consumption and investment opportunities. *Journal of Financial Economics* 7, 265–96.

Campbell, J.Y. 1991. A variance decomposition for stock returns. *Economic Journal* 101, 157–79.

Campbell, J.Y. and Cochrane, J.H. 2000. Explaining the poor performance of consumption-based asset pricing models. *Journal of Finance* 55, 2863–78.

Campbell, J.Y. and Shiller, R.J. 1988. The dividend-price ratio and expectations of future dividends and discount factors. *Review of Financial Studies* 1, 195–227.

Campbell, J.Y. and Shiller, R.J. 1998. Stock prices, earnings and expected dividends. *Journal of Finance* 43, 661–76.

Cochrane, J.H. 1991. Explaining the variance of price-dividend ratios. *Review of Financial Studies* 5, 243–80.

Fama, E.F. and French, K.R. 1988. Dividend yields and expected stock returns. *Journal of Financial Economics* 22, 3–27.

Fama, E.F. and French, K.R. 1989. Business conditions and expected returns on stocks and bonds. *Journal of Financial Economics* 25, 23–49.

Guvenen, F. 2005. A parsimonious macroeconomic model for asset pricing: habit formation or cross-sectional heterogeneity? Working paper, University of Texas at Austin.

Hansen, L.P. and Jagannathan, R. 1991. Implications of security markets data for models of dynamic economies. *Journal of Political Economy* 99, 252–62.

Hansen, L.P. and Singleton, K. 1983. Stochastic consumption, risk aversion, and the temporal behavior of asset returns. *Journal of Political Economy* 91, 249–65.

Kandel, S. and Stambaugh, R.F. 1990. Expectations and volatility of consumption and asset returns. *Review of Financial Studies* 3, 207–32.

Lamont, O. 1998. Earnings and expected returns. *Journal of Finance* 53, 1563–87.

LeRoy, S.F. and Porter, R.D. 1981. The present-value relation: tests based on implied variance bounds. *Econometrica* 49, 555–74.

Lettau, M. and Ludvigson, S.C. 2001a. Consumption, aggregate wealth and expected stock returns. *Journal of Finance* 56, 815–49.

Lettau, M. and Ludvigson, S.C. 2001b. Measuring and modeling variation in the risk-return tradeoff. In *Handbook of Financial Econometrics*, ed. Y. Ait-Sahalia and L.P. Hansen. Amsterdam: North-Holland.

Lucas, R. 1978. Asset prices in an exchange economy. *Econometrica* 46, 1429–54.

Lustig, H. and van Nieuwerburgh, S. 2005. Housing collateral, consumption insurance and risk premia: an empirical perspective. *Journal of Finance* 60, 1167–219.

Lustig, H. and Nieuwerburgh, S.V. 2006. Can housing collateral explain long-run swings in asset returns? Working Paper NYU Stern and UCLA.

Luttmer, E. 1996. Asset pricing in economies with frictions. *Econometrica* 64, 1439–67.

Mehra, R. and Prescott, E. 1985. The equity premium: a puzzle. *Journal of Monetary Economics* 15, 145–61.

Menzly, L., Santos, T. and Veronesi, P. 2004. Understanding predictability. *Journal of Political Economy* 112, 1–47.

Piazzesi, M., Schneider, M. and Tuzel, S. 2007. Housing, consumption, and asset pricing. *Journal of Financial Economics* 83, 531–69.

Poterba, J.M. and Summers, L.H. 1986. The persistence of volatility and stock market fluctuations. *American Economic Review* 76, 1142–51.

Rubinstein, M. 1976. The valuation of uncertain income streams and the pricing of options. *Bell Journal of Economics* 7, 407–25.

Santos, J. and Veronesi, P. 2005. Labor income and predictable stock returns. *Review of Financial Studies* 19, 1–43.

Shiller, R.J. 1981. The use of volatility measures in assessing market efficiency. *Journal of Finance* 36, 291–304.

consumption externalities

Consumption externalities occur when consumption by some creates external costs or benefits for others. Their recognition by economists dates at least as far back as Adam Smith's discussion of how local consumption standards influence the goods that people consider essential (or 'necessaries', as Smith called them). In the following passage, for example, he described the factors that influence the amount someone must spend on clothing in order to be able appear in public 'without shame':

> By necessaries I understand, not only the commodities which are indispensably necessary for the support of life, but whatever the custom of the country renders it indecent for creditable people, even of the lowest order, to be without. A linen shirt, for example, is, strictly speaking, not a necessary of life. The Greeks and Romans lived, I suppose, very comfortably, though they had no linen. But in the present times, through the greater part of Europe, a creditable day-labourer would be ashamed to appear in publick without a linen shirt, the want of which would be supposed to denote that disgraceful degree of poverty which, it is presumed, no body can well fall into without extreme bad conduct. (Smith, 1776, pp. 869–70)

Consumption externalities received only limited attention in Smith's *Wealth of Nations* and only occasional mention by economists during the century that followed its publication. Karl Marx (1847), for example, noted that 'A house may be large or small; as long as the neighboring houses are likewise small, it satisfies all social requirement for a residence. But let there arise next to the little house a palace, and the little house shrinks to a hut.'

It was not until Thorstein Veblen's *The Theory of the Leisure Class* appeared in 1899 that consumption externalities received their first serious, book-length treatment in economics. Veblen's thesis was that much of consumption is undertaken to signal social position. But although his book is still widely read and cited by scholars in numerous disciplines, its general theme was largely ignored by economists during the 50 years following its publication.

Duesenberry's relative income hypothesis

Interest in this theme was rekindled with the publication of James Duesenberry's *Income, Saving, and the Theory of Consumer Behavior* in 1949. In this volume, Duesenberry offered his 'relative income hypothesis', in which he argued that an individual's spending behaviour is influenced by two important frames of reference – the individual's own standard of living in the recent past and the living standards of others in the present. Thus, in Duesenberry's account, people are subject to both intrapersonal and interpersonal consumption externalities.

His theory attempted to explain three important empirical regularities: (*a*) long-run aggregate savings rates remain roughly constant over time, even in the face of substantial income growth; (*b*) aggregate consumption is much more stable than aggregate income in the short run; and (*c*) individual savings rates rise substantially with income in cross-section data. When Duesenberry's book was first published, individual consumption was generally modelled by economists as a linear function of income with a positive intercept term. This model could accommodate rising savings rates in cross-section data and the stability of consumption over the business cycle, but not the long-run stability of aggregate savings rates.

Duesenberry's hypothesis was hailed as an advance because of its ability to track all three stylized fact patterns. The poor save at lower rates, he argued, because they are more likely to encounter others with desirable goods that are difficult to afford. Moreover, since this will be true no matter how much national income grows, unfavorable comparisons will always occur more frequently for the poor – and hence the absence of any tendency for savings rates to rise with income in the long run.

To explain why consumption is more stable than income in the short run, Duesenberry argued that families compare their living standards not only to those of others around them but also to their own standards from the past. The high consumption level once enjoyed by a formerly prosperous family thus constitutes a frame of reference that makes cutbacks difficult when income falls.

Despite Duesenberry's success in tracking the data, many economists felt uncomfortable with his relative income hypothesis, which to them seemed more like sociology or psychology than economics. The profession was therefore immediately receptive to alternative theories that purported to explain the data without reference to softer disciplines. The most important among these theories was Milton Friedman's permanent income hypothesis, variants of which still dominate today's research on spending.

In hindsight, however, there remain grounds for scepticism about whether Friedman's theory was a real step forward. For example, its fundamental premise – that savings rates are independent of permanent income – has been refuted by numerous careful studies (see, for example, Carroll, 1998). Some modern consumption theorists have responded by positing a bequest motive for rich consumers, a move that begs the question of why leaving bequests should entail greater satisfaction for the rich than for the poor.

Another problem is that, contrary to Friedman's assertion that the marginal propensity to consume out of windfall income should be nearly zero, people actually consume such income at almost the same rate as permanent income (Bodkin, 1959). To this observation, Friedman (1963) himself responded that consumers appear to have unexpectedly short planning horizons. But if so, then consumption does not really depend primarily on permanent income.

Abundant evidence suggests that context influences evaluations of living standards (see, for example, Veenhoven, 1993; Easterlin, 1995; Luttmer, 2005). In the light of this evidence, it seems fair to say that Duesenberry's hypothesis not only has been more successful than Friedman's in tracking how people actually spend but also rests on a more realistic model of human nature. And yet the relative income hypothesis is no longer even mentioned in most leading economics textbooks. Its absence appears to signal the profession's continuing reluctance to acknowledge concerns about relative consumption.

Welfare implications

In traditional economic models, individual utility depends only on absolute consumption. These models lie at the heart of claims that pursuit of individual self-interest promotes aggregate welfare. In contrast, models that include concerns about relative consumption identify a fundamental conflict between individual and social welfare. This conflict stems from the fact that concerns about relative consumption are stronger in some domains than in others. The disparity gives rise to expenditure arms races focused on 'positional goods' – those for which relative position matters most. The result is to divert resources from 'non-positional goods', causing welfare losses. (The late Fred Hirsch, 1976, coined these terms.)

The nature of the misallocation can be made clear with the help of two simple thought experiments. In each, you must choose between two worlds that are identical in every respect except one. The first choice is between world A, in which you will live in a 4,000-square-foot house and others will live in 6,000-square-foot houses; and world B, in which you will live in a 3,000-square-foot house, others in 2,000-square-foot houses. Once you choose, your position on the local housing scale will persist.

If only absolute consumption mattered, A would be clearly better. Yet most people say they would pick B, where their absolute house size is smaller but their relative house size is larger. Even those who say they would pick A seem to recognize why someone might be more satisfied with a 3,000-square-foot house in B than with a substantially larger house in A.

In the second thought experiment, your choice is between world C, in which you would have four weeks a year of vacation time and others would have six weeks; and world D, in which you would have two weeks of vacation, others one week. This time most people pick C, choosing greater absolute vacation time at the expense of lower relative vacation time.

The modal responses in these two thought experiments suggest that housing is a positional good and vacation time a non-positional good. The point is not that absolute house size and relative vacation time are of no concern. Rather, it is that positional concerns weigh more heavily in the first domain than in the second.

When the strength of positional concerns differs across domains, the resulting conflict between individual and social welfare is structurally identical to the one inherent in a military arms race. When deciding how to apportion available resources between domestic consumption and military armaments, each country's valuations are typically more context-dependent in the armaments domain than in the domain of domestic consumption. After all, being less well armed than a rival nation could spell the end of political independence. The familiar result is a mutual escalation of expenditure on armaments that does not enhance security for either nation. Because the extra spending comes at the expense of domestic consumption, its overall effect is to reduce welfare. Note, however, that if each country's valuations were equally context-sensitive in the two domains, there would be no arms race, for in that case the attraction of having more arms than one's rival would be exactly offset by the penalties of having lower relative consumption.

For parallel reasons, the modal responses to the two thought experiments suggest an equilibrium in which people consume too much housing and too little leisure (for a formal demonstration of this result, see Frank, 1985a). In contrast, conventional welfare theorems, which assume that individual valuations depend only on absolute consumption, imply optimal allocations of housing and leisure.

In addition to leisure, goods that have been classified as non-positional by various authors include workplace safety, workplace democracy, savings and insurance. And since public goods are, by definition, available in equal quantities to all consumers, they, too, are inherently non-positional. The general claim is that unregulated market exchange will tend to emphasize the production of positional goods at the expense of these and other non-positional goods (Frank, 1985b). Among the policies suggested as remedies for this imbalance have been income and consumption taxes, overtime laws, hours laws for commercial establishments, legal holidays, workplace safety and health regulation, non-waivable workers' rights, and tax-financed savings accounts.

Consumption externalities also have implications for the theory of revealed preference, which says that, if a well-informed individual chooses a risky job that pays $600 a week rather than a safer one that pays only $500, he reveals that the safety increment is worth less than $100 to him. If safety is a non-positional good, however, this inference does not follow, for it ignores the fact that, if all workers exchange safety for increased income, the anticipated increase in relative consumption does not occur. The value that workers assign to safety may thus be revealed as much in the patterns of safety regulation they favour as in the nature of the jobs they choose.

ROBERT H. FRANK

See also **leisure; time use; Veblen, Thorstein Bunde.**

Bibliography

Bodkin, R. 1959. Windfall income and consumption. *American Economic Review* 49, 602–14.

Carroll, C. 1998. Why do the rich save so much? In *Does Atlas Shrug: The Economic Consequences of Taxing the Rich*, ed. J. Slemrod. New York: Oxford University Press.

Duesenberry, J. 1949. *Income, Saving, and the Theory of Consumer Behavior*. Cambridge, MA: Harvard University Press.

Easterlin, R. 1995. Will raising the incomes of all increase the happiness of all? *Journal of Economic Behavior and Organization* 27, 35–47.

Frank, R. 1985a. The demand for unobservable and other nonpositional goods. *American Economic Review* 75, 101–16.

Frank, R. 1985b. *Choosing the Right Pond*. New York: Oxford University Press.

Friedman, M. 1957. *A Theory of the Consumption Function*. Princeton, NJ: Princeton University Press.

Friedman, M. 1963. Windfalls, the horizon, and related concepts in the permanent income hypothesis. In *Measurement in Economics*, ed. C. Christ. Stanford: Stanford University Press.

Hemenway, D. and Solnick, S. 1998. Is more always better? *Journal of Economic Behavior and Organization* 37, 373–83.

Hirsch, F. 1976. *Social Limits to Growth*. Cambridge, MA: Harvard University Press.

Luttmer, E. 2005. Neighbors as negatives: relative earnings and well-being. *Quarterly Journal of Economics* 120, 936–1002.

Marx, K. 1847. Relation of wage-labour to capital. In *Wage Labour and Capital*. Marx/Engels Internet Archive, 1999. Online. Available at http://www.marxists.org/archive/marx/works/1847/wage-labour/ch06.htm, accessed 16 August 2005.

Smith, A. 1776. *An Inquiry into the Nature and Causes of the Wealth of Nations*, ed. R. Campbell and A. Skinner. Oxford: Oxford University Press, 1976.

Veblen, T. 1899. *The Theory of the Leisure Class*. New York: Modern Library.

Veenhoven, R. 1993. *Happiness in Nations: Subjective Appreciation of Life in 56 Nations*. Rotterdam: Erasmus University.

consumption sets

The idea of consumption sets was introduced into general equilibrium theory in July 1954 in Arrow and Debreu (1954, pp. 268–9) and Debreu (1954, p. 588), the name itself appearing only in the latter paper. Later expositions were given by Debreu (1959) and Arrow and Hahn (1971) and a more general discussion by Koopmans (1957, Essay 1). Although there have been several articles concerned with non-convex consumption sets (e.g. Yamazaki, 1978), in more recent years their role in general equilibrium theory has been muted, especially in approaches that use global analysis (see for example, Mas-Colell, 1985, p. 69). Such sets play no role in partial equilibrium theories of consumer's demand, even in such modern treatments as Deaton and Muellbauer (1980). Since general equilibrium theory prides itself on precision and rigour (e.g. Debreu, 1959, p. x), it is odd that on close examination the meaning of consumption sets becomes unclear. Indeed, three quite different meanings can be distinguished within the various definitions presented in the literature. These are given below (in each case the containing set is the commodity space, usually R^n): M1 The consumption set C1 is that subset on which the individual's preferences are defined. M2 The consumption set C2 is that subset delimited by a natural bound on the individual's supply of labour services, i.e. 24 hours a day. M3 The consumption set C3 is the subset of all those bundles, the consumption of any one of which would permit the individual to survive. Each definition in the literature can (but here will not) be classified according to which of these meanings it includes. In probably the best known of them (Debreu, 1959, ch. 4), the consumption set appears to be the intersection of all three subsets C1–C3. M1 is plain. After all, preferences have to be defined on *some* proper subset of the commodity space, since the whole space includes bundles with some inadmissibly negative coordinates. M2 is also reasonable, although a full treatment of heterogeneous labour services does raise problems for what is meant by an Arrow–Debreu 'commodity' (see for example that of Arrow–Hahn, 1971, pp. 75–6). It is M3 that gives real difficulty, both in itself and in relation to the others.

First, there is little reason to expect either C1 or C3 to be a subset of the other, and so still less to expect M1 and M3 to define the same set. No individual would have any problem in preferring one bundle, the consumption of which would ensure her survival, to a second bundle, the consumption of which would result in her death by starvation. However, she might well prefer the second bundle to a third, whose consumption would cause her to die from thirst (the representation of such preferences by a real-valued utility function might pose problems, but that is another matter). On the other hand, the same individual might not be able to rank in order of preference two bundles each of which contains exotic food and drink, even though fully assured that the consumption of either bundle would allow her to survive.

More importantly, M3 implicitly introduces *consumption* activities, the actual eating and drinking and sheltering that are essential to survival. Such activities constitute what are sometimes called, by analogy with production, the consumption technology. Some partial equilibrium models, such as 'the new home economics' and the theory of characteristics, have treated aspects of such technologies but so far general equilibrium theory has not. In particular, Arrow–Debreu theory has not done so. As a consequence (and unlike some forms of the classical 'corn model') it does not give a coherent account of the birth and death of individual persons, any more than it does of the birth and death of individual firms (see GENERAL EQUILIBRIUM). Hence the third meaning M3, which in effect presumes that the model contains such an account when it does not, is hard to interpret. One major difficulty of interpretation arises with the Slater-like condition that each individual's endowment of goods and services, valued at the competitive prices p^*, should be strictly greater than $\inf \{ \langle x, p^* \rangle : x \in C\}$, where $\langle .,. \rangle$ denotes inner product and C is 'the' consumption set (see COST MINIMIZATION AND UTILITY MAXIMIZATION). This condition is important in proofs of existence of competitive equilibrium, to ensure for example that the budget correspondence is continuous, or that a compensated equilibrium is a competitive equilibrium. It is itself guaranteed by assumptions (discussed by McKenzie, 1981, pp. 821–5) on the relations between 'individual' consumption sets and the aggregate production set.

If C is taken to contain C3 then the assumptions just referred to imply that every consumer survives in every competitive equilibrium, not merely for one period but over the whole (finite) Arrow–Debreu span. This is a breathtaking assertion of fact which recalls irresistibly Hicks's wry observation: 'Pure economics has a remarkable way of producing rabbits out of a hat – apparently *a priori* propositions which apparently refer to reality. It is fascinating to try to discover how the rabbits got in' (1939, p. 23).

On the other hand if C is taken to be C1, then the assumptions take on a purely technical (and so less objectionable) aspect, whose role is essentially to ensure that the system stays within the (relative) interior of the sets concerned and so displays appropriate continuity. But then there is no presumption that individual agents survive in a competitive equilibrium, even for one period (cf. Robinson, 1962, p. 3). The multi-period versions of the Arrow–Debreu model are then at risk, since individuals disappear and take their labour service endowments with them. This should not come as a surprise – the problems of time in economics are really too complicated to be overcome simply by adding more dimensions to the one-period model.

Some models that include C3 in C attempt to justify Slater-like conditions directly, on the grounds that 'Not many economies in the present day are so extremely laissez faire as to permit people to starve' (Gale and Mas-Colell, 1975, p. 12). This justification clearly fails as long as the behaviour of the public agency whose actions allegedly prevent such starvation is not modelled *explicitly*, like that of the private agents.

It is usually assumed that consumption sets are bounded below, closed and convex. The first two assumptions are innocuous but the third poses issues of a conceptual kind, which spring from difficulties in interpreting the idea of a convex combination $x^t = tx^1 + (1-t)x^2$ of two bundles x^1 and x^2, where $t \varepsilon [0,1]$. Consider the example, sometimes used, in which x^1 is a house in London and x^2 a house in Paris. We cannot take seriously the claim that x^t is a house in the Channel, so t cannot refer to distance. An alternative claim that t refers to the proportion of the period that is spent in London could arise from many different finite partitions of the time interval, not all of which need to be ranked equally by the individual. In effect, convexity of the consumption set comes down to the divisibility of consumer goods, an assumption which in the past has proved not such a bad approximation if one is interested mainly in general equilibrium aspects of market demand, and representative rather than actual consumers. Indivisibilities of producer goods are of course much more serious.

PETER NEWMAN

See also **Arrow–Debreu model of general equilibrium; cost minimization and utility maximization; general equilibrium; indivisibilities.**

Bibliography

Arrow, K.J. and Debreu, G. 1954. Existence of an equilibrium for a competitive economy. *Econometrica* 22, 265–90.

Arrow, K.J. and Hahn, F.H. 1971. *General Competitive Analysis.* San Francisco: Holden-Day.

Deaton, A. and Muellbauer, J. 1980. *Economics and Consumer Behaviour.* Cambridge: Cambridge University Press.

Debreu, G. 1954. Valuation equilibrium and Pareto optimum. *Proceedings of the National Academy of Sciences* 40(7), 588–92.

Debreu, G. 1959. *Theory of Value.* Cowles Commission Monograph No.17. New York: Wiley.

Gale, D. and Mas-Colell, A. 1975. An equilibrium existence theorem for a general model without ordered preferences. *Journal of Mathematical Economics* 2, 9–15.

Hicks, J.R. 1939. *Value and Capital.* Oxford: Clarendon Press.

Koopmans, T.C. 1957. *Three Essays on the State of Economic Science.* New York: McGraw-Hill.

McKenzie, L.W. 1981. The classical theorem on existence of competitive equilibrium. *Econometrica* 49, 819–41.

Mas-Colell, A. 1985. *The Theory of General Economic Equilibrium. A Differentiable Approach.* Cambridge: Cambridge University Press.

Robinson, J.V. 1962. The basic theory of normal prices. *Quarterly Journal of Economics* 76(1), 1–20.

Yamazaki, A. 1978. An equilibrium existence theorem without convexity assumptions. *Econometrica* 46, 541–55.

consumption taxation

Whether household income or household consumption constitutes a better measure of a household's ability to pay taxes, and whether there are substantial efficiency gains to choosing one tax base rather than the other, are two of the central questions of public finance. The debate between advocates of income taxes and advocates of consumption taxes has spanned several centuries. While income has often been viewed as the basis for taxation, and Adam Smith discusses taxation relative to household incomes, Thomas Hobbes, John Stuart Mill and Irving Fisher were all strong proponents of taxing consumption. Consumption tax supporters argue that the amount that an individual draws from the economy's resource pool should determine his or her tax burden. They also point out that an income tax levies a 'double tax' on saving, since saved income is taxed both when it is earned and when the savings yield a return to capital. Kaldor (1955) offers a broad review of the case for consumption taxation. Two notable reports in the late 1970s, one by the Meade Commission (Meade, 1978) in the United Kingdom and the other by the staff of the US Treasury Department (1977), outlined the modern cases for consumption taxation and developed specific proposals.

Proponents of income taxation argue that the change in an individual's command over resources between one period and the next is an appropriate measure of 'ability to pay', even if those resources are not immediately consumed. This is the measure of taxable capacity suggested by Robert Murray Haig and Henry Simons: 'Haig–Simons' income. Moreover, they argue that changes in resources should be taxed regardless of whether they arise from labour income or from the returns to past saving.

Income taxes and consumption taxes exhibit different time profiles over the course of a lifetime. When individuals experience a period of retirement before they die, the time profile of tax payments under a consumption tax will fall later in the lifetime than the corresponding payments under an income tax. This is because individuals continue to consume after they stop earning labour income. Retirees under an income tax pay tax only on their capital income, while retirees under a consumption tax pay tax on their total outlays, which are likely to exceed their capital income.

The debate between proponents of consumption taxation and proponents of income taxation concerns whether or not capital income should be taxed. The foregoing philosophical issues notwithstanding, the efficiency cost of taxing capital income has been an active subject of economic research. Chamley (1986) and Judd (1985) argue that the effective distortions from capital taxes cumulate over time as the difference between discounting the future at before-tax and after-tax interest rates increases with the compounding horizon. They claim that the optimal steady-state capital income tax rate should be zero. However, they also point out that a one-time capital levy is an efficient device for raising revenue. A number of recent studies, described in Auerbach (2006), have examined the robustness of the theoretical claim that the optimal capital tax rate is zero.

Consumption tax proponents, such as Bradford (1980), claim not only that taxing consumption rather than income avoids intertemporal distortions, but also that it solves many of the most difficult measurement and accounting problems associated with income taxation. Under a consumption tax, for example, there would be no distinction between the tax burden on investment projects financed with debt and those financed with equity, or between realized and unrealized capital gains. There would be no need to measure the rate at which long-lived physical assets depreciate, as one must do under an income tax. Income tax proponents respond that some components of consumption may be difficult to measure, and that it is more difficult to tailor consumption taxes than income taxes to achieve redistributive goals.

Formalizing consumption taxation vs. income taxation

The essential difference between a consumption tax and an income tax can be illustrated by comparing the lifetime budget constraints that consumers would face under each tax system. An income tax is levied on both labour and capital income. When a household has assets of A_{t-1} at the beginning of period t, these assets earn a pre-tax return r and the household earns labour income of wL where w equals the real wage and L denotes labour supply, the income tax base is $wL+rA_{t-1}$. The income tax not only reduces the after-tax real wage but also lowers the after-tax return to saving. In a life-cycle model in which a household lives for T periods and in which there is no inflation, the life-cycle budget constraint with an income tax is

$$\sum_{t=1}^{T} C_t/(1 + r(1 - \tau))^t$$
$$= \sum_{t=1}^{T}(1 - \tau)w_t L_t/(1 + r(1 - \tau))^t + A_0$$

$$(1)$$

In this expression, C denotes real consumption spending, and A_0 is the household's initial wealth endowment.

In contrast, the life-cycle budget constraint with a consumption tax levied at rate θ is

$$\sum_{t=1}^{T}(1+\theta)C_t/(1+r)^t \qquad (2)$$
$$= \sum_{t=1}^{T}w_t L_t/(1+r)^t + A_0$$

The discount rate in this case is the pre-tax return. The consumption tax levied on outlays in each period is equivalent to a tax on labour income *and* the household's initial endowment. If $(1-v)=1/(1+\theta)$, then eq. (2) can be rewritten as

$$\sum_{t=1}^{T}C_t/(1+r)^t$$
$$= \sum_{t=1}^{T}(1-v)w_t L_t/(1+r)^t + (1-v)A_0$$
$$(3)$$

The timing of tax payments under the 'wage-and-endowment tax' in (3) is different from that under the consumption outlays tax in (2), but the present value of taxes and the effects on economic incentives are the same under the two systems. The tax on initial endowment is an essential component of this equivalence: a wage tax alone is *not* equivalent to a consumption tax because initial assets escape taxation when only wages are taxed.

The current tax system in most developed nations is a hybrid structure, reflecting some elements of income taxation but also embodying components of a consumption tax. This is most apparent in nations that rely on both an income tax and a consumption tax, such as a value added tax, for a substantial share of government revenue. Even within many income tax systems, however, there are provisions that move toward an income tax-consumption tax hybrid. In the United States, for example, capital income that accrues in employer-provided pension plans and in a variety of taxpayer-directed retirement saving accounts, such as Individual Retirement Accounts (IRAs), is excluded from income taxation. Some types of capital income are taxed at rates below the top statutory tax rates on wage income. Realized capital gains have often been taxed at preferential rates, and in some cases dividend income to households is also subject to reduced rates of tax. There is substantial variation in tax structures across nations, but the principle of allowing some tax reduction on capital income is widespread. This makes it difficult to assess where any particular nation's tax system falls on the spectrum between an income tax and a consumption tax.

Types of consumption taxes

In practice, there are many ways to implement a consumption tax. Two, the retail sales tax and the value added tax, are widely used in practice. Both are examples of indirect consumption taxes, because they are levied without any reference to the consumer's identity. Direct consumption taxes, in contrast, are levied on households by computing their total consumption. In contrast to indirect consumption taxes, direct consumption taxes can be levied at progressive rates. While direct consumption taxes have never been used as the primary revenue source in any nation, they have been actively debated in the policy reform literature. Tax structures that closely resemble direct consumption taxes have been adopted as components of existing tax systems. The two most widely discussed direct consumption tax options are the savings-exempt income tax and the 'X-tax,' a combination of a cash-flow tax on business income and a household wage tax.

A *retail sales tax* (RST) is the simplest consumption tax. It is collected by retailers at the point of final sale, and it corresponds directly to the tax on consumption spending described in eq. (2) above. In 2006, 44 of the 50 US states levied some form of sales tax, with rates typically between four and seven per cent. There is little experience with RSTs above ten per cent. One unresolved question with regard to proposals that call for significantly higher RSTs is whether the difficulty of monitoring all points of purchase would lead to substantial problems of tax evasion.

A *value added tax* (VAT) is a very common form of consumption tax. Virtually all developed nations with the exception of the United States levy some form of VAT, with rates ranging up to 25 per cent in Denmark, Norway and Sweden. The VAT is collected from businesses on the difference between the gross value of their sales and the cost of any inputs that they purchase from other entities that have already paid VAT.

To illustrate the operation of VAT, consider a bakery that produces and sells bread for $100. The baker's input costs are $30 for flour and $65 for an employee. The bakery earns a $5 profit. If flour is purchased from another firm that has already paid VAT, then the bakery's VAT liability equals $70 times the VAT tax rate, since its value added equals its sales of $100 minus input purchases that have already paid VAT, or $30. Wages are *not* deducted from sales when computing value added. Although the VAT is collected in stages from all firms in a production chain, it is equivalent to an RST at the same rate. One attractive feature of the VAT is that downstream firms, such as the baker in this example, help ensure VAT compliance by upstream firms that supply intermediate goods. In this example if the flour seller cannot provide documentation for its VAT payment, the baker will face tax on value added of $100. Thus the baker has an incentive, all else equal, to purchase inputs from suppliers who pay VAT.

Ebrill et al. (2001) offer a comprehensive discussion of VAT implementation issues and summarize experience with the VAT in both developed and developing nations. The VAT accounts for a substantial share of revenue in most industrialized nations. The treatment of international transactions has proven a source of difficulty in some nations, since exporting firms are typically granted a rebate for their VAT payments. Some tax evasion schemes involve exporting goods to qualify for the rebate and re-importing the same goods without paying VAT on the import. The taxation of financial services also proves challenging under the VAT.

A *savings-exempt income tax* (SEIT) is a consumption tax that is built on an income tax model. For those who are familiar with an income tax system, it provides a way of shifting to a consumption tax without drastic administrative changes in the tax system. The Nunn-Domenici 'USA Tax', introduced in the US Senate in the mid-1990s and analysed in Ginsburg (1995), was a based on this type of consumption tax.

Under the SEIT, the tax base is income less saving. To prevent taxpayers from simply claiming high levels of saving and thereby avoiding tax liability, saving must be documented in the form of a contribution to a 'qualified account'. Income earned on assets held in the qualified account is not taxed, but withdrawals from the qualified account are included in the tax base. Thus a taxpayer who earns $50,000 and contributes $5,000 would be taxed on $45,000 in the contribution period. If, some years later, when earnings equal $25,000, the taxpayer withdraws $10,000 from the qualified account, she would be taxed on $35,000.

Even though the SEIT taxes the earnings that have accrued on the contributions to the qualified account when the funds are withdrawn from this account, the return on capital is untaxed in this setting. Taxing accumulated capital income when the proceeds are withdrawn is *not* equivalent to taxing capital income as it accrues: this is the reason Individual Retirement Accounts, 401(k) plans and other tax-deferred saving programmes provide an incentive for personal saving. When capital income is taxed as it accrues, the value of earning one dollar, paying tax on it at rate τ, and then investing it for T periods at a pretax rate of return r but with an accrual tax rate τ, is $(1 - \tau)(1 + (1 - \tau)r)^T$. In contrast, if the initial earnings are excluded from taxation, there is no taxation of accruing capital income, and withdrawals are taxed at 100τ per cent, then the value after T periods is $(1 - \tau)(1 + r)^T$. The qualified account approach eliminates the tax burden on the 'inside build up' of capital assets.

One of the key challenges in implementing a SEIT is avoiding the wholesale reallocation of existing wealth into 'qualified accounts' at the time the SEIT is adopted. Such transfers could sharply reduce tax collections, but, since they involve previously accumulated assets, they would not translate into marginal incentives for new saving. If it were possible to inventory the assets of each

taxpayer when the SEIT was implemented, this would make it possible to design regulations to limit the transfer problem. Absent such information on previously accumulated wealth, however, transfers of pre-existing wealth into qualified accounts are likely to prove a difficult implementation issue for the savings-exempt income tax.

An *X-tax* combines a cash flow tax on businesses, much like a VAT with a deduction for wages, with a household-level tax on wage income. The X-tax and its relatives are descended from proposals in the US Treasury Department's (1977) report on fundamental tax reform. Bradford (1986) discusses several plans of this type, and one widely discussed variant was developed by Hall and Rabushka (1995). The X-tax has greater flexibility than a VAT for achieving distributional goals, since the household level tax can include progressive rates or transfers to low-earning households. This illustrates the distributional flexibility of direct rather than indirect consumption taxes. If the household tax is a flat rate tax on wages at the same rate as the corporate cash flow tax, then the X-tax is equivalent to a VAT or an RST. When the rates are different, then the X-tax becomes a combination of a VAT and an additional tax or subsidy on labour income. The cash flow nature of the business tax eliminates the need to measure depreciation, since firms can claim an immediate deduction – expensing – for purchases of capital goods.

In practice, neither the RST nor the VAT is implemented strictly along the principles described above. Proposals for both the SEIT and the X-tax also include additional features that often introduce efficiency costs that would not arise in 'textbook' versions of these taxes. The RST, for example, typically exempts some goods and services. Expenditures on food, medical care and clothing are often excluded from the tax base, thereby achieving a more progressive distribution of tax burdens while creating distortions between various classes of consumption goods. The VAT is often implemented at different rates on different goods, with exemptions for some goods, creating the same distortionary effects. Because both the savings-exempt income tax and the X-tax require households to file tax returns, they are prone to modification to allow deductions for some expenditure categories, such as mortgage interest or health insurance premiums. While neither of these consumption tax plans has been tried in practice, they probably would be influenced by the same political pressures that have generated a wide array of tax expenditures in the current income tax code.

Efficiency gains from replacing an income tax with a consumption tax

Income taxes create two distortions: one between the before-tax and the after-tax real product wage, which distorts the labour–leisure margin, and one between the before-tax and the after-tax real rate of return to saving. The latter distorts the lifetime allocation of consumption relative to the pattern that would be chosen if the return to delaying consumption equalled the economy's pre-tax marginal product of capital. Shifting from an income tax to a consumption tax eliminates the second distortion. The key analytical issue in evaluating the welfare consequences of replacing an income tax with a consumption tax is therefore measuring the efficiency costs associated with the taxation of saving and investment. This efficiency cost depends on the underlying structure of consumer preferences. The interest elasticity of saving is often invoked as a summary measure of the key preference parameters. When changes in after-tax returns induce only modest changes in household saving, the efficiency gain from switching from an income tax to a consumption tax will be smaller than when the interest elasticity of saving is large.

Auerbach and Kotlikoff (1987) use a dynamic general equilibrium model, including a realistic treatment of household life-cycle income and consumption streams, to evaluate the efficiency gains from replacing an income tax with a consumption tax. Their results suggest that for a given revenue requirement, the steady-state capital stock is larger with a consumption tax than with an income tax. This translates into higher steady-state per capita utility under the consumption tax than the income tax.

The steady-state comparison is not the only consideration when evaluating two alternative tax systems, however. It is possible to design tax reforms that raise steady-state welfare but cause welfare losses in the transition from an initial equilibrium to the new steady state. The trade-off between short-run and long-run policy effects depends on the policymaker's discount rate and in calibrated general equilibrium models it is possible to compute the present discounted value of the gains and losses to the cohorts alive at different dates.

Transition from one tax regime to another

Focusing on the present value of welfare gains and losses draws attention to the transitional rules that govern the switch from one tax system, say an income tax, to another, such as a consumption tax. These transition rules can determine whether a policy reform represents a net gain or a net loss relative to continuation of the initial income tax regime. Altig et al. (2001) illustrate this important point using a more elaborate version of the model developed in Auerbach and Kotlikoff (1987). They find that if the tax basis of existing assets is extinguished when the income tax is replaced by a consumption tax, so that depreciation allowances are no longer claimed after the reform, and if investors who accumulated savings under the income tax regime do not receive any relief from the consumption tax burden they will face when they draw down their assets, then the efficiency gains from adopting a consumption tax may be as large as five per cent of national income.

'Grandfathering' existing assets sharply reduces these efficiency gains, because it reduces the base of the consumption tax and requires higher tax rates to satisfy a given revenue constraint. This results in greater distortions on the labour–leisure margin. Designing transition relief that participants in the political process will view as fair, without forgoing most of the efficiency gains from a stark consumption tax transition, is likely to be one of the greatest challenges in any consumption-oriented tax reform.

JAMES M. POTERBA

See also **tax expenditures; taxation of income; value-added tax.**

Bibliography

Altig, D., Auerbach, A.J., Kotlikoff, L.J., Smetters, K.A. and Walliser, J. 2001. Simulating fundamental tax reform in the United States. *American Economic Review* 91, 574–95.

Auerbach, A.J. 2006. The choice between income and consumption taxes: a primer. Working Paper No. 12307. Cambridge, MA: NBER.

Auerbach, A.J. and Kotlikoff, L.J. 1987. *Dynamic Fiscal Policy*. Cambridge: Cambridge University Press.

Bradford, D. 1980. The case for a personal consumption tax. In *What Should be Taxed? Income or Expenditure*, ed. J. Pechman. Washington, DC: Brookings Institution.

Bradford, D. 1986. *Untangling the Income Tax*. Cambridge, MA: Harvard University Press.

Chamley, C. 1986. Optimal taxation of capital income in general equilibrium with infinite lives. *Econometrica* 54, 607–22.

Ebrill, L., Keen, M., Bodin, J.-P. and Summers, V. 2001. *The Modern VAT*. Washington, DC: International Monetary Fund.

Ginsburg, M. 1995. Some thoughts on working, saving, and consuming in Nunn-Domenici's tax world. *National Tax Journal* 48, 585–602.

Hall, R. and Rabushka, A. 1995. *The Flat Tax*, 2nd edn. Stanford, CA: Hoover Institution Press.

Judd, K.L. 1985. Redistributive taxation in a simple perfect foresight model. *Journal of Public Economics* 29, 59–83.

Kaldor, N. 1955. *An Expenditure Tax*. London: George Allen & Unwin.

Meade, J.E. 1978. *The Structure and Reform of Direct Taxation*. London: Allen & Unwin.

US Treasury Department. 1977. *Blueprints for Basic Tax Reform*. Washington, DC: US Treasury Department.

contemporary capitalism

1 What is 'capitalism'?

At the beginning of the 21st century, 'capitalism' has triumphed as the dominant system for allocating a society's economic resources. The last time in history in which the persistence of capitalism in the world's most advanced economies was seriously called into question was the Great Depression of the 1930s – a decade during which the unemployment rate in the United States remained at 15 per cent or higher, notwithstanding unprecedented state intervention under the New Deal. It took the Second World War to pull the United States and the world economy out of depression, and in the subsequent decades it took substantial and sustained government spending in the rich economies of North America and western Europe to hold unemployment to acceptable levels.

In the post-war era, the Soviet Union's highly planned economy posed as a possible alternative to capitalism. The purported strength of the Soviet challenge, however, turned out to be based at least as much on Cold War ideology emanating from the United States as on the actual productive power of the Soviet Union and its satellites. By the 1990s the Soviet model had virtually vanished, as Russia itself sought to make the transition to a 'market economy', guided, tragically, by a mythical ideology of how capitalism is supposed to operate, imported from the United States.

Over the same period capitalism entrenched itself in East Asia. During the 1970s and 1980s Japan became a rich economy on the basis of a distinctive model of 'collective capitalism', and in the 1980s and 1990s the East Asian 'Tigers' – Hong Kong, Singapore, South Korea and Taiwan – closed the gap, each with its own variant of the Japanese model. More recently China and India, with one-third of the world's population, have experienced rapid economic growth, driven by what many would call 'capitalist' institutions. Yet, even as firms cross the globe to access Indian software engineers, and vice versa, India remains a nation with one-third of the world's illiterates. Meanwhile the fact that China, the world's second largest economy since the early 1990s, continues to be guided by an avowedly Communist government raises the question of what 'capitalism' really is.

Defining contemporary capitalism is not merely a question of semantics. If, as has been demonstrated since the mid-20th century, 'capitalism' is a powerful engine of economic growth, we want to know how it functions as a mode of resource allocation and the social conditions under which capitalist growth is not only strong but also stable, and equitable. We also want to know how the institutions of contemporary capitalism that generate growth might be transferred to those parts of the world – first and foremost Africa but also eastern Europe and Latin America as well as parts of the Middle East – that have economically been left behind. Given its pervasiveness and dominance, a depiction of the institutions that define contemporary capitalism is tantamount to a description of the economic world in which perhaps one-half of the world's population now lives and to which much of the other half now aspires.

There is no consensus among economists on the definition of contemporary capitalism. The dominant approach to analysing resource allocation and the economic performance of an advanced economy rests on the notion that a capitalist economy is essentially a market economy that allocates resources to their most productive uses. But what at any time and in any place, the student of economic development asks, explains how those most productive uses come to exist? And why in certain times and places? Fundamental to capitalist growth is 'innovation', the process that generates goods and services that, even with factor prices held constant, are of higher quality and lower cost than those previously available (Lazonick, 2006c). Can a theory of capitalism as a market economy comprehend the innovation process?

In the early 20th century a young Joseph Schumpeter asked this question. As a Viennese economics student, Schumpeter was versed in the relatively recent, and increasingly influential, Austrian and Walrasian theories of how, through the equilibrating mechanism of the market, the economy could achieve an 'optimal' allocation of resources across productive uses. Schumpeter's insight was to recognize that such a view of the economic world could not explain economic development. In 1911 Schumpeter wrote *The Theory of Economic Development* (first translated into English in 1934) to argue that entrepreneurial activity that results in innovation – what he called the 'Fundamental Phenomenon of Economic Development' – can disrupt the 'Circular Flow of Economic Life as Conditioned by Given Circumstances' to change the ways in which the economy operates and performs. Without such disruption of equilibrium conditions, the economy would not develop. Over the next four decades Schumpeter sought to elaborate a theory of economic development informed by his own, evolving, understanding of the changing reality of the most advanced capitalist economies.

In particular, Schumpeter sought to understand the role of the business enterprise in advanced capitalist development. By the 1940s he had taken definitive leave of his youthful conceptions of the innovative entrepreneur as an individual actor and innovation as simply 'new combinations' of existing resources. Rather, he saw that powerful business organizations both developed and utilized productive resources to create new technologies and access new markets. The creation of new technologies, moreover, destroyed the commercial viability of old technologies. In *Capitalism, Socialism, and Democracy*, first published in 1942, Schumpeter argued that the process of 'creative destruction' had become embodied in established corporations as 'technological "progress" tends, through systematization and rationalization of research and of management, to become more effective and sure-footed', being 'the business of teams of trained specialists who turn out what is required and make it work in predictable ways' (Schumpeter, 1950, pp. 118, 132).

This article takes as its point of departure the proposition, suggested by Schumpeter, that the key to understanding 'capitalism' as a mode of resource allocation that generates economic growth is the organization and performance of its most innovative business enterprises. That is not to say that markets and states are unimportant to the operation, and hence definition, of capitalism. Historically, however, well-functioning markets are outcomes of successful capitalist development. For the individual, markets create the possibility of choosing what to consume and for whom to work, including the prospect of working for oneself. But markets cannot explain the development of the new products and processes that drive the growth of the capitalist economy. The innovation process is uncertain, collective and cumulative (see O'Sullivan, 2000). The uncertain character of innovation means that investments in innovation require *strategic control* over resource allocation by individuals who have intimate knowledge of the technologies, markets and competitors that an innovative strategy must confront. The collective character of innovation means that the implementation of an innovation strategy requires the *organizational integration* of a hierarchical and functional division of labour into a process of organizational learning. The cumulative character of innovation means that the process requires *financial commitment* until it can generate financial returns. Enterprises, not markets, engage in strategic control, organizational integration and financial commitment (Lazonick, 2003).

Nor can one explain innovation by appealing to the notion of the developmental state as its driving force, as has often been done for the East Asian economies. Implicit, and at times explicit, in this view is an acceptance of the ideology that the economic development of the United States is an exemplar of the workings of the market economy. Yet from gun manufacture and interchangeable parts in the first half of the 19th century to the computer revolution and Internet in the late 20th century, as well as railroads, aviation and the life sciences in between, the history of US capitalism is replete with examples of the critical role of the developmental state in allocating resources to the processes of knowledge creation that then provided the foundations for US industrial leadership. Yet, as important as the developmental state has been even in a so-called 'market economy' such as the United States, the allocation of resources to knowledge creation would have been wasted, and would probably never have been made, had it not been for the presence and influence of innovative enterprises that have made use of this knowledge to generate higher-quality, lower-cost products than had previously been available.

In this article I focus on the changing role of innovative enterprise in determining resource allocation and economic performance in contemporary capitalism. Space constraints dictate that I confine the analysis of contemporary capitalism to the case of the United States, with the caveat that, even in a highly globalized economy in which one might expect convergence to a common business model, there are almost as many distinctive 'varieties of capitalism' in terms of governance, employment

and investment institutions, as there are advanced capitalist nations. The US economy is, however, the world's largest and richest economy. It is also the one in which market ideology is most virulent and the actual mode of resource allocation most misunderstood. Section 2 of this article provides historical background to understanding contemporary US capitalism by describing the key characteristics of the 'Old Economy business model' (OEBM) that made the United States the world's most powerful nation in the decades after the Second World War. Section 3 analyses the challenges that confronted OEBM in the 1970s and 1980s, and how the ideology of 'maximizing shareholder value' arose to legitimize a redistribution of income from labour interests to financial interests. Section 4 shows how the 'New Economy business model' (NEBM) emerged in the 1980s and 1990s to drive the innovation process, but in ways that have contributed to unstable and inequitable economic growth. Section 5 concludes with some questions about the future of the US model in a global economy in which many distinctive business models still compete.

2 The Old Economy business model

The United States emerged from the Second World War as the undisputed world leader in GDP per capita, a position that it still retains. With western Europe and Japan still in recovery from the war, the United States was at its peak of dominance in the 1950s on the basis of a highly collective model of capitalism embodied in the managerial corporation, and personified in the concept of the 'organization man' (Whyte, 1956). The stereotypical 'organization man' was white, Anglo-Saxon and Protestant, obtained a college education, got a well-paying job with an established company early in his career, and then worked his way up and around the corporate hierarchy over decades of employment, with a substantial 'defined benefit' pension, complete with highly subsidized medical coverage, awaiting him on retirement. The employment stability offered by an established corporation was highly valued, while inter-firm labour mobility was shunned.

'Organization men' rose to top executive positions where, as salaried managers rather than owners, they exercised strategic control. This separation of share ownership and managerial control, which continues to characterize the US industrial corporation, resulted from the widespread distribution among shareholders of the corporation's publicly traded stock. In principle, boards of directors representing the interests of shareholders monitor the decisions of these managers. In practice, incumbent top executives choose the outside directors and are themselves members of the board. Shareholders can challenge management through proposals to the annual general meeting, but over the course of the 20th century a body of law evolved that enables management to exclude shareholder proposals that deal with normal business matters (for example, downsizings) as distinct from social issues (for example, sex discrimination).

The separation of ownership from control has worked effectively to generate innovation when the interests of salaried executives who exercise strategic control have been aligned with those of employees who engage in the development and ensure the utilization of the company's productive resources. In the post-Second World War decades the organizational integration of the capabilities of administrative and technical specialists enabled US firms to develop the world's most competitive systems of mass production. These personnel were products of the US system of higher education, which since the early decades of the century had prepared the labour force to enter employment in bureaucratic organizations.

A distinctive feature of the US business model was the organizational segmentation between these salaried managers, in whose training and experience the corporation made substantial investments, and so-called 'hourly' workers. (Non-salaried employees were classified as 'hourly', or 'non-exempt', workers because of the stipulation of the National Labor Relations Act that emerged from the New Deal era that required employees who were paid an hourly wage receive 150 per cent of that wage if they worked longer than the normal working hours. The overtime work of salaried personnel is exempt from this provision.) The corporation viewed these operatives, who were typically high-school graduates, as interchangeable commodities in whose capabilities the company had no need to invest, notwithstanding the fact that they often spent their entire working lives with one company. At the same time, these industrial corporations needed reliable even if low-skill workers to tend mass production processes. The combination of dominant product-market positions and union power, which advanced the pay and protected the employment of senior workers, enabled the hourly worker to receive good pay and benefits, including a defined-benefit pension that assumed long-term employment with a single company.

The developmental state played an indispensable role in the innovation process by partially funding the system of higher education as well as, in the forms of research labs, subsidies and contracts, programmes for technology development in sectors such as aerospace, computers and life sciences. The development of the productive potential of these government investments relied in turn on corporate research capabilities. Retained earnings formed the foundation of committed finance for new corporate investments in innovation. When corporations needed additional investment financing, they issued corporate bonds at favourable rates that reflected the established position of the company as well as its conservative debt–equity ratios. Companies used bank loans almost exclusively for working capital, and made only limited use of the stock market as a source of investment funds.

These social conditions enabled US corporations to grow very large in the post-war decades. The 50 largest US industrial corporations by revenues on the Fortune 500 list averaged 87,070 employees in 1957, 117,393 in 1967, and 119,093 in 1977. These figures do not include employment at AT&T, the regulated telephone monopoly, which in 1971 employed 1,015,000 people, of whom 700,000 were union members with good wages, stable employment and excellent benefits. By the late 1960s and early 1970s increasing numbers of blacks were moving into union jobs in the steel, automobile, electrical equipment, consumer durable and telecommunications industries. The growth of established corporations in these industries in the three decades after the Second World War contributed to a more equal distribution of family income in the US economy.

3 'Maximizing shareholder value'

During the 1970s the US model faltered in the face of Japanese competition. Building on innovative capabilities developed for their home markets during the 1950s and 1960s, Japanese companies gained competitive advantage over US companies in industries such as steel, memory chips, machine tools, electrical machinery, consumer electronics and automobiles. US companies had entered the 1970s as world leaders in these industries. Many US observers attributed the rapid increase in Japanese exports to the United States in the 1970s to Japan's lower wages and longer working hours. By the early 1980s, however, with real wages in Japan continuing to rise, it became clear that Japanese advantage was based on the superior organization of their enterprises, and in particular on a more thoroughgoing integration of participants in the functional and hierarchical divisions of labour for the dual purposes of transforming technologies and accessing new markets. Indeed, during the 1980s Japan exported management practices as well as material goods to the West. From the second half of the 1980s, with the yen strengthening and trade surpluses generating political backlash, Japanese companies made a transition to direct investment in the United States and other advanced economies.

A growing financial orientation of US business that had surfaced in the conglomerate movement of the 1960s undermined the abilities and incentives of established US corporations to respond to the Japanese challenge. To some extent the growth of the US industrial corporation in the post-war decades had been based on strategic investments in new product lines and geographic areas that built on the corporation's existing productive capabilities, and yielded economies of scale and scope. The conglomerate movement, however, saw major corporations invest in scores of *unrelated* businesses, often through mergers and acquisitions, based on the prevailing, but erroneous, ideology that a good corporate executive could manage any type of business, and that conglomeration offered the synergies of superior corporate management. The conglomerate movement failed because it segmented top executives, in positions of strategic control, from the rest of the managerial organization that had to develop and utilize productive resources to sustain the firm's competitive advantage (Lazonick, 2004).

In the late 1970s and early 1980s the conglomerates unraveled. In the mid-1970s Michael Milken, a Drexel Burnham investment banker, had created the junk bond market by convincing institutional investors, in search of higher yields in an inflationary era, to hold downgraded corporate securities, many of them 'fallen angels' from unsuccessful conglomeration. By the late 1970s, with the junk-bond market well developed, it became possible to issue new junk bonds to finance leveraged buyouts (LBOs) in which the top managers of a conglomerate division turned it into an independent company to recapture strategic control over resource allocation. By the late 1980s, however, the junk bond had become an instrument for the hostile takeover of entire companies, with KKR's 1989 LBO of RJR Nabisco for $24.5 billion marking the height of what became known as 'the deal decade'.

The ideology that justified hostile takeovers was that the corporation should be run to 'maximize shareholder value' (see Lazonick and O'Sullivan, 2000). Proponents of shareholder value charged that, either because of opportunism or incompetence, many incumbent corporate managers were making poor allocative decisions. By exercising their influence through the market for corporate control, shareholders could force incumbents to alter their allocative decisions, replace them with those who would maximize shareholder value, or distribute cash to shareholders in the forms of dividends and stock repurchases so that shareholders themselves could, so the argument goes, reallocate the economy's resources to their best alternative uses.

While the hostile takeover movement did not directly threaten high-tech companies (in which the most valuable assets could walk out the door), by the end of the 1980s the top executives of virtually all US industrial corporations had embraced the ideology of maximizing shareholder value and made it their own. By the 1980s executive stock option compensation was a well-established practice. Since in the United States option awards did not require that the company's stock price outperform the stock market or even the stock prices of a group of competitors, those who received these awards could only gain from what, from July 1982 to August 2000, turned out to be the longest stock market boom in US history, with the Dow Jones Industrial Average and the S&P500 Index both rising about 1,300 per cent.

As Table 1 shows, stock-price appreciation drove the extraordinary real stock yields that were sustained over the 1980s and 1990s. The relatively low dividend yields in the 1990s did not reflect stinginess on the part of US corporations; the US corporate payout ratio – the amount of dividends as a percentage of after-tax

Table 1 *US corporate stock and bond yields, 1960–2005. Average annual per cent change*

	1960–9	1970–9	1980–9	1990–9	2000–5
Real stock yield	6.63	−1.66	11.67	15.01	−1.87
Price yield	5.80	1.35	12.91	15.54	−0.76
Dividend yield	3.19	4.08	4.32	2.47	1.58
Change in CPI	2.36	7.09	5.55	3.00	2.67
Real bond yield	2.65	1.14	5.79	4.72	3.60

Notes: Stock yields are for Standard and Poor's composite index of 500 US corporate stocks (424 of which are, as of 28 March 2006, NYSE). Bond yields are for Moody's Aaa-rated US corporate bonds. *Source*: Council of Economic Advisers (2006, Tables B-62, B-73, B-95 and B-96).

corporate profits (with inventory evaluation and capital consumption adjustments) – averaged 48 per cent in the 1980s and 57 per cent in the 1990s compared with 39 per cent in the 1960s and 41 per cent in the 1970s. It was just that the rate of increase of stock prices outstripped the rate of increase of dividend payments, thus depressing the dividend yield. The form that the stock yield takes is of significance because investors can capture the dividend yield by holding stocks, whereas they can capture the price yield only by buying and selling stocks. Inherent in high-price yields, therefore, is a volatile stock market.

A volatile stock market benefits those who are compensated in stock options on an annual basis, especially when, as is the case in the United States, options vest as quickly as one year from the date of grant and can be exercised for up to ten years. It has been estimated that, largely because of the gains from exercising stock options, on average the ratio of CEO pay of an S&P500 company to that of a production worker was 42 in 1985, 107 in 1990, 525 in 2000, and 411 in 2005. Top executives took a keen interest in their company's stock price, and in the 1980s and 1990s, in the name of 'maximizing shareholder value', they found ways in which they could use their positions of strategic control over corporate resource allocation to influence it. They could cook the corporate books to boost current earnings, a practice that became widespread in these decades and one for which a few executives have been fined or even jailed. The American Competitiveness and Corporate Accountability Act of 2002, better known as Sarbanes–Oxley, has sought to stem this practice. But quite apart from artificially inflating corporate earnings, top corporate executives also found that downsizing the labour force and repurchasing corporate stock helped to boost a company's stock price, even though these resource allocations did not necessarily improve the company's competitive performance.

The era of corporate downsizing took hold in the recession of 1980–2 when hundreds of thousands of stable, well-paid blue-collar jobs were lost that were never subsequently restored (see Lazonick, 2004). It would appear that the blacks who had relatively recently moved into these types of jobs were particularly hard hit; last hired, they tended to be the first fired. The subsequent 'boom' years of the mid-1980s witnessed hundreds of plant closings. In the 'white-collar' recession of the early 1990s tens of thousands of professional, administrative and technical employees found that their jobs had been eliminated, although once again it was blue-collar workers who bore the brunt of the downturn. In 1980 manufacturing employment was 22 per cent of the labour force; by 1990 it had fallen to 17 per cent and by 2001 to 14 per cent. While the employment picture generally became much better during the Internet boom of the last half of the 1990s, job cutting remained a way of life for many major US corporations. According to data on layoff announcements by companies in the United States collected by the recruitment firm, Challenger, Gray and Christmas, announced job cuts averaged just under 550,000 per year for the period 1991–4, 450,000 per year in 1995–7, and 656,000 per year during the boom years 1998–2000.

Meanwhile, from the mid-1980s US corporations began to actively support their stock prices through large-scale stock repurchases. Companies included in the S&P500 in March 2006 distributed more cash to shareholders in repurchases than in dividends in 1997 through 2000 and again in 2004, and just slightly less in 2001 through 2003. Since 1978 net equity issues by US nonfinancial corporations has been positive in only six of 28 years (1980, 1982, 1983, 1991, 1992, 1993); since the early 1980s US industrial corporations have in aggregate been supplying capital to the stock market rather than vice versa. In 2005 the net flow of cash from non-financial corporations to the stock market was a record $366 billion, 1.42 times in real dollars the previous high in 1998 (Lazonick, 2006d).

4 The New Economy business model

On 29 December 1995, AT&T announced that, as part of the process of breaking itself up into three separate companies, it would be cutting 40,000 jobs. AT&T was a company that could trace its origins back to the 1870s, had created the world's most advanced telephone system, was the home of the famous Bell Labs that among many other accomplishments invented the transistor in 1947, and, despite having lost its status as a regulated monopoly in 1984, still employed 308,700 people. Now, however, AT&T became emblematic of the failure of US Old Economy corporations to continue to provide employment opportunities. With campaigning for the 1996 presidential election picking up steam, Patrick Buchanan, a right-wing politician, caught the attention of the media by denouncing the highly paid executives of AT&T and other downsizing corporations as 'corporate hit men'. Fuel was added to the fire by the revelation that, in the name of 'creating shareholder value', Al Dunlap, whom the

American public came to know as 'Chainsaw Al', had in 20 months as CEO of Scott Paper devastated the 115-year old company while putting an estimated $100 million in his own pocket. In March 1996, the *New York Times* ran a seven-part series, later released as a paperback, on 'the downsizing of America' (Lazonick, 2004).

By the spring of 1996, however, the furor over corporate downsizing had disappeared. In its place, Americans became enthralled by the prosperity promised by what in the second half of the 1990s came to be called the 'New Economy'. In the United States the previous half-century had seen a massive accumulation of information and communications technology (ICT) capabilities. The development of computer chips from the late 1950s had provided the technological foundation for the microcomputer revolution from the late 1970s, which in turn had provided the technological infrastructure for the Internet boom of the second half of the 1990s. The research funding for this accumulation of ICT capabilities had come mainly from the US government and the research laboratories of established Old Economy hightechnology corporations. Each wave of technological innovation, however, created opportunities for the emergence of start-up companies that were to become central to the commercialization of the new technologies.

Although by the mid-1980s the Japanese had outcompeted even the leading US semiconductor firms in the memory chip market, US companies such as Intel, Motorola and Texas Instruments continued to dominate the microprocessor and logic chip markets that drove product innovation in the microelectronics industry (Lazonick, 2006a). While Silicon Valley was not the only US location for innovation in this industry, the concentration of semiconductor start-ups in the region from the late 1950s resulted in the emergence by the 1980s of a distinctive mode of combining strategy, finance and organization: the 'New Economy business model' (NEBM) (see Table 2). During the 1990s NEBM spread beyond Silicon Valley start-ups and was adopted successfully by leading Old Economy ICT companies such as Hewlett-Packard and IBM. In the Internet boom of the late 1990s elements of NEBM diffused to other ICT companies, including an Old Economy company such as Lucent Technologies, spun off from AT&T in its 1996 trivestiture, which almost destroyed itself in attempting to adopt the business model. In the 2000s NEBM characterizes the most innovative sectors of the US economy (for the case of biotechnology, see Pisano, 2006).

The founders of New Economy firms have typically been scientists and engineers who have gained specialized experience in existing firms, although in some cases they have been university faculty members intent on commercializing their academic knowledge. Some of these entrepreneurs have come from existing Old Economy companies, where it was often difficult for their new ideas to get internal backing. But New Economy companies themselves have become increasingly important as

Table 2 *Comparing business models in ICT*

	Old Economy business model (OEBM)	New Economy business model (NEBM)
Strategy, product	Firm growth based on multidivisional structure: multi-product firm	New firm entry into specialized ICT markets; accumulate new capabilities by acquiring (other) young technology firms
Strategy, process	Vertical integration of the value chain; in-house standards and proprietary R&D	Vertical specialization of the value chain; industry technology standards; R&D for cross-licensing and alliances; outsourcing routine work to specialist contract manufacturers and/or offshoring routine work to low-wage nations
Finance	Venture finance from savings, family and business associates; NYSE listing, growth finance from retentions, after dividends, and bond issues	Organized venture capital; early IPO on NASDAQ; retentions with zero dividends; use of own stock as a compensation and combination currency; systematic stock repurchases to support stock price
Organization	Secure employment; 'organization man' (career with one company), industrial union; defined-benefit pension, good medical coverage in employment and retirement	Insecure employment; interfirm mobility of labour, broad-based stock options, non-union, defined contribution pension, employees bear more burden of medical insurance

sources of new entrepreneurs who left their current employers to start new firms. Large numbers of high-tech entrepreneurs in the United States have been foreign-born, coming mainly from India and China (Saxenian, 2006).

Typically, the founding entrepreneurs of a New Economy start-up seek committed finance from venture capitalists with whom they share not only ownership of the company but also strategic control. In the 2000s Silicon Valley remains by far the leading location in the United States and the world for venture-backed high-tech start-ups. The region acquired this position from the 1960s as a distinctive venture capital industry emerged out of the opportunities for start-ups created by the microelectronics revolution. Besides sitting on the board of directors of the new company, the venture capitalist generally recruits professional managers, who are given company stock along with stock options, to lead the transformation of the firm from a new venture to a going concern. This stock-based compensation gives these managers a powerful financial incentive to develop the innovative capabilities of the company to the point where it can do an initial public offering (IPO) or private sale to a listed company, thus enabling the start-up's privately held shares to be transformed into publicly traded shares. Both before and after making this transition, their tenure with, and value to, the company depends on their managerial capabilities, not their fractional ownership stakes (Lazonick, 2006a).

The stock market speculation of the 'dotcom' era made it all too easy to cash out of a start-up, as many high-tech firms that had not engaged in innovation did IPOs or were sold to established companies. When start-ups do innovate, the key to making the transition from new venture to going concern has been the organizational

integration of an expanding body of technical and administrative 'talent'. As Silicon Valley developed from the 1960s, this educated and experienced labour had to be induced to trade secure employment with an Old Economy company for insecure employment with a start-up. To attract these highly mobile people and retain their services, Silicon Valley firms increasingly adopted 'broad-based' employee stock option plans that extended this form of compensation to a large proportion, sometimes all, of the firm's non-executive employees rather than just to top executives. In start-ups, stock options usually served as a partial substitute for cash salaries, and the eventual gains from exercising options were viewed as a substitute for a company-funded pension (Lazonick, 2006a).

Again, the underlying stock would become valuable if and when the start-up did an IPO or a private sale to a publicly listed company. Shortening the expected period between the launch of a company and an IPO was the practice of most venture-backed high-tech start-ups of going public on NASDAQ, created in 1971 as an electronic exchange for the over-the-counter markets with less stringent listing requirements than the 'Old Economy' New York Stock Exchange (NYSE). The 1978 cut in the capital gains tax rate to 28 per cent, after it had been raised to 49 per cent just two years before, provided further encouragement to entrepreneurs and venture capitalists to found new companies, and for employees of these companies, rewarded with stock options, to provide the skills and efforts needed to transform new ventures into going concerns. In 1979 the clarification of the 'prudent man' rule as applied to the Employee Retirement Income Security Act (ERISA) of 1974 gave asset managers the green light to allocate a portion of their portfolios to riskier stocks and venture capital funds, and resulted in a flood of new money, especially from pension funds, into the venture capital industry. The American Electronics Association and the National Venture Capital Association, with their strongest and deepest roots in Silicon Valley, were the frontline Washington lobbyists for these regulatory changes.

While institutional money provided capital to NEBM, high-tech labour became more mobile from one firm to another than it had been in the Old Economy. Employee stock options induced this mobility, but what made it possible in terms of the knowledge bases that managers and engineers possessed were *industry* standards, as distinct from *in-house* standards, that emerged in the various sectors of ICT. In the Old Economy in-house standards promoted the growth of large vertically integrated firms on the basis of proprietary technologies, whereas in the New Economy industry standards encouraged new entry. Nevertheless, as demonstrated by the important cases of Intel and Microsoft in the development of the microcomputer industry, those New Economy firms that dominated in the setting of the industry standards could also grow very large (at the end of fiscal 2005 Intel employed 99,900 and Microsoft 61,000). By establishing industry standards, their growth encouraged rather than discouraged start-ups, which in turn depended on the availability of not only venture capital (which came from many sources besides the formal venture capital industry) but also mobile labour whose knowledge and experience could be easily integrated into the start-up's learning processes.

Of critical importance in setting industry standards in microelectronics was IBM's decision in 1980 to enter what became known as the personal computer (PC) industry with Intel supplying the microprocessor and Microsoft the operating system. At the time IBM controlled about 80 per cent of the computer market, had over 341,000 employees, and, with an explicit system of 'lifelong employment', trumpeted the fact that since 1921 it had not terminated an employee involuntarily. Yet between 1990 and 1994 IBM slashed its employment from 374,000 to 220,000. In 1991–3, the company had losses of $16 billion (including more than $8 billion in 1993, at the time the largest annual loss in US corporate history) on total revenues of $192 billion, and encouraged the media to believe that the mass layoffs were necessary to avoid bankruptcy. Yet virtually all of the losses came from 'restructuring' charges, that is, the cost of terminating employees (Lazonick, 2006a).

In retrospect, it is clear that these charges were the cost of ridding the company of its 70-year-old system of lifelong employment. The industry standards in ICT, which IBM had played a leading role in establishing, served to reduce the value to the company of older employees with experience accumulated at IBM over the course of their careers and to increase the value of younger employees who may have had experience working for other ICT companies. Explicitly reflecting this change in employment policy, in 1999 IBM announced that it would replace its traditional defined-benefit pension plan, which favoured long-term employees, with a portable 'cash-balances' plan that would be much more attractive to younger employees who did not envisage a lifelong career with IBM. In December 2004, as its employment reached 329,000, IBM announced that new employees would no longer be eligible for the cash-balances pension fund. Instead the company would offer them a defined-contribution pension, with the company matching the employee contribution up to six per cent of his or her compensation.

From the mid-1990s, with the Old Economy commitment to its employees out of the way, IBM adopted all of the elements of NEBM. It shifted out of hardware into services, and outsourced its manufacturing. It became by far the leading patenter in the United States, even as it cut R&D from the ten per cent of sales that prevailed in the 1980s to six per cent of sales since the mid-1990s, this change reflecting an expressed shift to product development and away from basic research. Since the early 1990s IBM has engaged in patenting much less to control proprietary technologies, as had been the case in the past, and much more to gain access through cross-licensing to technologies controlled by other companies and to generate intellectual property revenue ($1.3 billion per year in the 2000s).

As it rid itself of lifelong employment in the early 1990s IBM began to extend stock options, previously reserved for top executives, to a broad base of employees. In 1990 options outstanding were only four per cent of all shares outstanding; in 2005, 15.2 per cent. As for distributions to shareholders, in New Economy fashion, subsequent to its early 1990s restructuring IBM has favoured repurchases over dividends. In 1981–90 IBM's dividends were 48 per cent and repurchases 12 per cent of net income; in 1993–2005 dividends were 15 per cent and repurchases 91 per cent. In an effort to offset dilution of shareholdings as employees exercise stock options, and more generally to boost its stock price, in 1995–2005 IBM has spent $62.6 billion on stock repurchases. Over the same period the company has spent $56.6 billion on R&D.

As for a New Economy company that, unlike IBM, started out that way, Cisco Systems, which since the late 1990s has controlled about 75 per cent of the Internet router market, is a prime example of the importance, and implications, of broad-based stock options in NEBM compensation. Founded in Silicon Valley in 1984, Cisco grew from about 200 employees at the time of its IPO in 1990 to 40,000 employees during 2000. Throughout its history Cisco has awarded stock options to virtually all of its employees. By the end of fiscal 2000 stock options outstanding accounted for 14 per cent of the company's total stock outstanding; by 2005 that number was 23 per cent. In March 2000, at the peak of the Internet boom, Cisco had the highest market capitalization of any company in the world. Under such conditions its stock options were very lucrative. I have estimated that over the 11 years 1995–2005 (all years for which data are reported refer to fiscal year's end, the last week in July), Cisco employees, totaling about 256,000 employee-years, shared $21.5 billion in gains from exercising stock options, for an average of $84,000 per employee-year. The annual averages per employee ranged from less then $9,000 in 2003 to more than $281,000 in 2000. Of the total amount, the highest paid executives, totaling 57 executive-years, shared $893 million, for an average of $15.7 million per executive-year, with annual averages ranging from $1.3 million in 2003 to $51.3 million in 2000. The annual ratios of average top-executive to average employee gains from exercising stock options ranged from 36:1 in 1997 to 594:1 in 2005 (Lazonick, 2006d). Cisco employees have a clear financial interest in the company's stock price, and the company's top executives even more so.

Besides using their own stock as a compensation currency, during the 1990s some New Economy companies grew large by using their stock, instead of cash, to acquire other, smaller and typically younger, New Economy firms in order to gain access to new technologies and markets. Cisco mastered this growth-through-acquisition strategy. From 1993 through 2005 Cisco made 106 acquisitions valued in nominal terms at $46.9 billion, over 80 per cent of which was paid in the company's stock rather than cash. In 1999 and 2000 alone, Cisco did 41 acquisitions at a cost of $26.7 billion with over 99 per cent paid in stock (Carpenter, Lazonick and O'Sullivan, 2003).

At the same time, like many if not most New Economy companies, Cisco conserved cash by paying no dividends. Along with its use of stock as a combination currency, this payout policy enabled Cisco to become a giant company in the 1990s without taking on any long-term debt. Since the bursting of the Internet bubble from mid-2000 through 2005, however, Cisco has spent $27.2 billion repurchasing its own stock to support its sagging stock price. In 2004–5, as it spent $19.3 billion on stock repurchases, Cisco used $8.3 billion in cash – including $6.5 billion of it raised through its first-ever bond issue – to do 24 acquisitions rather than continue to use its stock as an acquisition currency that it would then feel compelled to offset with repurchases. (Cisco's decision to use cash rather than stock for acquisitions was helped by the Financial Accounting Standards Board's 2001 closing of the 'pooling-of-interests' loophole that enabled companies like Cisco that did all-stock acquisitions to record them on their balance sheets at book values, which were generally a small fraction of market values, and thus inflated future earnings. Nevertheless, in 2002 and 2003, with pooling-of-interests accounting outlawed, Cisco still used stock for payment of over 97 per cent of the price of its nine acquisitions.)

The corporate obsession with supporting its stock price through massive stock repurchases has therefore taken hold of companies in the most innovative sectors of the US economy. As further notable examples, for the years 1995–2005 Intel distributed $51.3 billion in repurchases along with $6.0 billion in dividends compared with R&D spending of $38.0 billion, while Microsoft distributed $45.4 billion in repurchases and $38.7 billion in dividends compared with R&D spending of $40.8 billion. Microsoft's dividends included a one-time payment of $36.1 billion in November 2004.

These companies would argue that R&D spending and stock repurchases are both working toward the same end: to enhance the company's innovative capabilities by, in the case of R&D, generating new knowledge, and in the case of repurchases, competing for high-tech labour capable of transforming that knowledge into innovative products and processes. By boosting stock prices, it is argued, repurchases help to attract, retain and motivate people who choose to work for companies in which they are partially compensated with stock options. In the case of Microsoft the argument has had less weight since July 2003 when the company ended its option programme (although many Microsoft employees still have unexercised options awarded prior to that date). In the 2000s, moreover, the extent and location of the talented labour supplies for which companies like Cisco, IBM, Intel and Microsoft compete have changed dramatically with the rise of India and China (Lazonick, 2006b). These dramatically changed labour market conditions for high-tech labour raise serious questions concerning which employees benefit from a company's stock price performance and for how long, and indeed whether established high-tech companies even need to use employee stock options to compete successfully for high-tech labour.

The offshoring to India and China in the 2000s of high value-added jobs of software engineers and computer programmers that it was previously thought could not go abroad represents the latest stage in four decades of the globalization of NEBM. Beginning in the early 1960s Silicon Valley semiconductor companies were among the first to offshore assembly to East Asia, and by the early 1970s virtually every US semiconductor manufacturer had followed suit. When these companies set up plants in places like South Korea, Taiwan, Hong Kong, Singapore and Malaysia, they employed, alongside unskilled and predominantly female assembly labour, indigenous university graduates as managers and engineers. Over time the US companies upgraded their facilities in these locations, and offered more and better employment opportunities for the indigenous well-educated labour force. As a striking example, in 1984 Intel claimed that, of its 8,500 employees outside of the United States (of 26,000 employees worldwide), only 60 were US citizens. This indigenous employment through foreign direct investment encouraged the national governments to increase the level of investment in their already well-developed systems of higher education, thus augmenting the future high-tech labour supply (Lazonick, 2006b).

In the 1990s established US ICT companies, led by IBM and Hewlett-Packard, dramatically reduced their employment of production workers by outsourcing manufacturing operations to electronic manufacturing service providers, also known as contract manufacturers (Lazonick, 2006c). Indeed, younger companies like Cisco grew rapidly without doing any in-house manufacturing. Initially the contract manufacturers would set up operations or take over existing plants of their customers in the United States. But a key capability of the leading contract manufacturers is to shift production that has become more routine and cost-sensitive to lower wage areas of the world. In the late 1990s and early 2000s the leading contract manufacturers grew at a rapid pace; at the end of 2005 employment at the five largest – Flextronics, Solectron, Sanmina-SCI, Celestica and Jabil Circuit – totalled 260,000. While we do not know the global distribution of this labour force, North America accounts for only an estimated 25 per cent of the sales of these five companies.

Meanwhile, in the 1990s and 2000s hundreds of thousands of foreigners, especially Indians, with college degrees in science and engineering have migrated to the United States for graduate education and work experience (Lazonick, 2006b). Many acquired permanent resident (immigrant) status in the United States, as the US government expanded employment-based preferences in the issuance of immigrant visas. For access to US work experience, however, the most important mode of entry for high-tech employees has been on non-immigrant H-1B and L-1 visas. The H-1B programme enables non-immigrants, the vast majority of whom have at least a bachelor's degree and whose skills are purportedly unavailable in the United States, to work in the United States for up to six years. In the first half of the 2000s about 70 per cent of H-1B visa holders had science or technology degrees, and 40–50 per cent came from India (the next largest national group is from China, at less than ten per cent). The L-1 visa programme permits a company with operations in the United States to transfer foreign employees to the United States to acquire work experience, with no limitation of time. In 2001, there were an estimated 810,000 people on H-1B visas in the United States, and possibly as many on L-1 visas.

Many of these non-immigrant visa holders have continued to work in the United States by obtaining permanent resident status. But most have returned to their native countries with valuable industrial experience that can be used to start new firms and, more typically, to work as technical specialists for indigenous or foreign companies. As a result of both the migration of US companies abroad in search of high-tech labour as well as the migration of foreign high-tech labour to the United States, and then back to their home countries, in the 2000s, to an extent never before imagined, even the best-educated US high-tech employees compete with a truly global labour supply for jobs.

5 Stable and equitable growth?

On 16 March 2005 the Semiconductor Industry Association (SIA) organized a Washington, DC press conference in which it exhorted the US government to step up support for research in the physical sciences, including nanotechnology, to assure the continued technological leadership of the United States. Intel CEO Craig Barrett was there as a SIA spokesperson to warn: 'U.S. leadership in technology is under assault' (*Electronic News*, 2005):

> The challenge we face is global in nature and broader in scope than any we have faced in the past. The initial step in responding to this challenge is that America must decide to compete. If we don't compete and win, there will be very serious consequences for our standard of living and national security in the future…U.S. leadership in the nanoelectronics era is not guaranteed. It will take a massive, coordinated U.S. research effort involving academia, industry, and state and federal governments to ensure that America continues to be the world leader in information technology.

At the time Barrett was a member of the US National Academy of Sciences Committee on Prospering in the Global Economy in the 21st Century, which delved into deficiencies in the development of science and engineering capabilities in the United States. Notwithstanding his obvious concern about these problems from a public policy perspective, on a radio talk show in February 2006 Barrett (by this time Chairman of Intel) remarked: 'Companies like Intel can do perfectly well in the global marketplace without hiring a single US employee' (wbur.org, 2006).

The problem with this statement is not that US workers should have privileged access to jobs at a US-based company like Intel (which still employs half of its almost 100,000 employees in the United States). The problem is that, if a powerful company like Intel is not dependent on US high-tech employees for its future labour force, why should it be concerned about supporting the mass educational infrastructure in the United States needed to develop this future labour force? And what does it mean to say that 'America must decide to compete' if, as I would argue is the case (Lazonick, 2006b), the most innovative US corporations have more of an interest in the Malaysian or Indian system of mass education than in the US system?

Since the mid-1970s the US mass education system has been performing poorly in science and mathematics by the standards of both the advanced and many developing economies. Such was much less the case in the three decades or so after the Second World War, when the Old Economy corporation was more dependent upon a labour force that was well-educated at the primary and secondary levels in the United States. This shift in the performance of the mass education parallels the reversal of post-war progress towards a more equal distribution of income that began about three decades ago. The much less secure employment of most US corporate employees in the shift from OEBM to NEBM would seem to have contributed to this reversal.

Meanwhile in the 2000s the compensation of the CEOs of US corporations has long since passed levels that are at a minimum unseemly and some would say obscene. The 'explosion in CEO pay', which has been discussed in the United States since the mid-1980s, seems to have no limits, especially if, when the corporate stock price falls, it can be once again pumped up or boards of directors can replace the 'lost' stock option income by other forms of remuneration such as salaries, bonuses or restricted stock. The seemingly endless explosions in top-executive pay reflect the obsession of US corporate executives with 'maximizing shareholder value' and, cash flow permitting, disgorging billions upon billions of corporate cash to shareholders in the forms of repurchases and dividends to try to make it happen.

In terms of public policy initiatives, virtually nothing has been done to control top executive pay in the United States. One well-known attempt was misguided. In 1993 President Clinton carried out a campaign promise to control CEO pay by legislating a cap of $1 million on the amount of 'non-performance-related' top executive compensation – salary and bonus – that a corporation could claim as a tax deduction. One perverse result of this law was that companies that were paying CEOs less than $1 million in salary and bonus *raised* these components of CEO pay towards $1 million, which executives now viewed as the government-approved CEO 'minimum wage'. The other perverse result was that companies increased CEO option compensation, for which tax deductions were not in any case being claimed, as an alternative to exceeding the $1 million salary-and-bonus cap.

That having been said, the limits to the gains from stock options, not just for top executives but also for broad bases of the employees of US high-tech corporations, would long ago have been reached if not for the fact that many of these corporations have been in the forefront of innovation. Given the unchallenged sway that the ideology of 'maximizing shareholder value' has over the governance of these corporations, I have no doubt that instability, as reflected in the boom and bust of the stock market in the late 1990s and early 2000s, and inequity, as reflected in the worsening of the distribution of income, will continue to beset the US economy.

Whether US corporations will remain in the forefront of innovation that, by necessity, must underpin long-term economic growth is another matter. Notwithstanding globalization, the US model of contemporary capitalism is not a global model. No other contemporary capitalist economy has made the commitment to 'shareholder value' that is the most distinctive feature of the US model. Japan has come through the stagnation of the 1990s as a highly innovative economy, while eschewing shareholder value ideology and practices (Lazonick, 2005). In western European nations the ideology has been tempered by a commitment to 'social inclusion'; the question is whether the equity and stability that social inclusion brings can be harnessed to support innovative enterprise. In the emerging giants, India and China, the stock market has come to play a more important, and possibly dangerous, role. In all of these economies, the success of innovative companies has been based, however, not on the stock market, but on the principles of strategic control, organizational integration, and financial commitment. Historically these principles also underpinned innovative enterprise in the United States. Many corporate executives who exercise control over resource allocation in the US economy may, however, have forgotten these principles, or worse yet, while they have been busy enriching themselves, they may have never bothered to learn them.

WILLIAM LAZONICK

Bibliography

To conserve both the word-count and flow of this essay, I have kept bibliographic references to a minimum, indicating instead works of mine in which these references can be found.

Carpenter, M., Lazonick, W. and O'Sullivan, M. 2003. The stock market and innovative capability in the New Economy: the optical networking industry. *Industrial and Corporate Change* 12, 963–1034.

Council of Economic Advisers. 2006. *Economic Report of the President, 2006*. Washington, DC: Executive Office of the President.

Electronic News. 2005. US could lose race for nanotech leadership, SIA panel says. 16 March. Online. Available at http://www.reed-electronics.com/electronicnews/article/CA511197?nid=2019&rid=1344283927, accessed 5 September 2006.

Lazonick, W. 2003. The theory of the market economy and the social foundations of innovative enterprise. *Economic and Industrial Democracy* 24, 9–44.

Lazonick, W. 2004. Corporate restructuring. In *The Oxford Handbook of Work and Organization*, ed. S. Ackroyd et al. Oxford: Oxford University Press.

Lazonick, W. 2005. The institutional triad and Japanese development [translated into Japanese]. In *The Contemporary Japanese Enterprise*, vol. 1, ed. G. Hook and A. Kudo. Tokyo: Yukikaku Publishing.

Lazonick, W. 2006a. Evolution of the New Economy business model. In *Internet and Digital Economics*, ed. E. Brousseau and N. Curien. Cambridge: Cambridge University Press.

Lazonick, W. 2006b. Globalization of the ICT labor force. In *The Oxford Handbook on ICTs*, ed. R. Mansell et al. Oxford: Oxford University Press.

Lazonick, W. 2006c. Innovative enterprise and economic development. In *Business Performance in Twentieth Century: A Comparative Perspective*, ed. Y. Cassis and A. Colli. Cambridge: Cambridge University Press.

Lazonick, W. 2006d. The US stock market and the governance of innovative enterprise. Working paper, INSEAD.

Lazonick, W. and O'Sullivan, M. 2000. Maximizing shareholder value: a new ideology for corporate governance. *Economy and Society* 29, 13–35.

New York Times. 1996. *The Downsizing of America.* New York: Times Books.

O'Sullivan, M. 2000. The innovative enterprise and corporate governance. *Cambridge Journal of Economics* 24, 393–416.

Pisano, G. 2006. *The Science Business: Strategy, Organization, and Leadership in Biotechnology.* Cambridge, MA: Harvard Business School Press.

Saxenian, A. 2006. *The New Argonauts: Regional Advantage in a Global Economy.* Cambridge, MA: Harvard University Press.

Schumpeter, J. 1934. *The Theory of Economic Development.* Oxford: Oxford University Press.

Schumpeter, J. 1950. *Capitalism, Socialism and Democracy,* 3rd edn. New York: Harper.

Wbur.org. 2006. Sharpening the cutting edge. *On Point.* 9 February. Online. Available at http://www.onpointradio.org/shows/2006/02/20060209_b_main.asp, accessed 5 September 2006.

Whyte, W. 1956. *The Organization Man.* New York: Simon and Schuster.

contestable markets

Contestable markets are those in which competitive pressures from potential entrants exercise strong constraints on the behaviour of incumbent suppliers. For a market to be contestable, there must be no significant entry barriers. Then, in order to offer no profitable opportunities for additional entry, an equilibrium configuration of the industry must entail no significant excess profits, and must be efficient in its pricing and in its allocation of production among incumbent suppliers. This is so of a contestable market whether it is populated with only a monopolist or with a large number of actively competing firms, because it is potential competition from potential entrants rather than competition among active suppliers that effectively constrains the equilibrium behaviour of the incumbents.

Perfectly contestable markets (PCMs) are a benchmark for the analysis of industry structure – a benchmark based on an idealized limiting case. Perfectly contestable markets are open to entry by entrepreneurs who face no disadvantages vis-à-vis incumbent firms and who can exit without loss of any costs that entry required to be sunk. The potential entrants have available the same best-practice production technology, the same input markets and the same input prices as those available to the incumbents. There are no legal restrictions on market entry and exit, and there are no special costs that must be borne by an entrant that do not fall on incumbent firms as well. Consumers have no preferences among firms except those arising directly from price or quality differences in firms' offerings.

Potential entrants into perfectly contestable markets are profit-seekers who respond with production to profitable opportunities for entry. They assess the profitability of their marketing plans by making use of the current prices of incumbent firms. Thus, for example, an entrepreneur will enter a market if he anticipates positive profit from undercutting the incumbent's price and serving the entire market demand at the new lower price. Potential entrants are undeterred by prospects of retaliatory price cuts by incumbents and, instead, are deterred only when the existing market prices leave them no room for profitable entry.

These features of the behaviour of potential entrants are key to the workings of perfectly contestable markets, and they are fully rational only where entry faces no disadvantages and is costlessly reversible. Hence, the benchmark case of perfect contestability excludes the sunk costs, precommitments, asymmetric information and strategic behaviour that characterize many real markets and that are the focus of much current research attention in the field of industrial organization. With irreversibilities and the inducements for strategic behaviour absent, industry structure in PCMs is determined by the fundamental forces of demand and production technology.

Of course, this is also true of perfectly competitive markets. However, this most familiar idealized limiting case is not a satisfactory benchmark for the study of industry structure in general, because it is intrinsically inapplicable to a variety of significant cases. In particular, where increasing returns to scale are present, perfectly competitive behaviour is logically inconsistent with the long-run financial viability of unsubsidized firms.

Perfectly contestable markets can serve in place of perfectly competitive markets as the general standard of comparison for the organization of industry whether or not scale economies are prevalent. Where they are not, perfectly competitive behaviour is necessary for equilibrium in PCMs, and, where scale economies do prevail, equilibrium in PCMs entails behaviour different from that found in perfectly competitive markets but which none the less tends to exhibit desirable welfare properties. In other words, perfect contestability is a generalization of perfect competition that has strong implications in significant circumstances where the latter is inapplicable.

In order to clarify and expand on these ideas, subsequent sections offer analytic outlines of the theory of perfectly contestable markets and applications of the theory to single-product and multi-product industries. Finally, observations are offered on the implications of this theory for the formulation of government policy towards industry.

Perfectly contestable markets: definitions and basic properties

The theory presented here lies in the realm of partial equilibrium. It deals with the provision of the set of products $N = \{1, \ldots, n\}$, some of which may not actually be produced, and which is a proper subset of all the goods in the economy. The prices of these products are represented by vectors $p \in R^n_{++}$, and other prices are assumed to be exogenous and are suppressed in the notation. $Q(p) \in R^n_+$ is the vector-valued market demand function for the products in N, and it suppresses consumers' incomes which are assumed to be exogenous. For any output vector $y \in R^n_+$, $C(y)$ is the cost at exogenously fixed factor prices when production is efficient. The underlying technology is assumed to be freely available to all incumbent firms and all potential entrants. Where necessary, $C(y)$ and $Q(p)$ will be assumed to be differentiable.

Definition 1 A *feasible industry configuration* is composed of m firms producing output vectors $y^1, \ldots, y^m \in R^n_+$, at prices $p \in R^n_{++}$ such that the markets clear, $\sum_{i=1}^m y^i = Q(p)$, and that each firm at least breaks even, $p \cdot y^i - C(y^i) \geq 0, i = 1, \ldots, m$.

Thus, the industry configuration is taken to be comprised of m firms, where m can be any positive integer, so that the industry structure is monopolistic if $m = 1$, competitive if m is sufficiently large, or oligopolistic for intermediate values of m. The term 'feasibility' refers to the requirements that each of the firms involved selects a non-negative output vector that permits its production costs, $C(y^i)$ to be covered at the market prices, p, and that the sum of the outputs of the m firms satisfies market demands at those prices.

Definition 2 A feasible industry configuration over N, with prices p and firms' outputs $y^1, \ldots, y^m$, is *sustainable* if $p^e \cdot y^e \leq C(y^e)$, for all $p^e \in R^n_{++}$, $y^e \in R^n_+$, $p^e \leq p$, and $y^e \leq Q(p^e)$.

The interpretation of this definition is that a sustainable configuration affords no profitable opportunities for entry by potential entrants who regard incumbents' prices as fixed (for a period sufficiently long to make $C(\cdot)$ the relevant flow cost function for an entrant). Here, a feasible marketing plan of a potential entrant is comprised of prices, p^e, that do not exceed the incumbents' quoted prices, p, and a quantity vector, y^e, that does not exceed market demand at the entrant's prices, $Q(p^e)$. The configuration is sustainable if no such marketing plan for an entrant offers a flow of profit, $p^e \cdot y^e - C(y^e)$, that is positive.

Definition 3 A *perfectly contestable market* (PCM) is one in which a necessary condition for an industry configuration to be in equilibrium is that it be sustainable.

A PCM so defined may be interpreted, heuristically, as a market subject to potential entry by firms that have no disadvantage relative to incumbents, and that assess the profitability of entry on the supposition that incumbents' prices are fixed for a sufficiently long period of time. Then, since one requirement for equilibrium is the absence of new entry, an equilibrium configuration in a PCM must offer no inducement for entry; that is, it must be sustainable.

Definition 4 A feasible industry configuration over $N, p; y^1, \ldots, y^m$, is a *long-run competitive equilibrium* if $p \cdot y \leq C(y) \forall y \in R^n_+$.

So defined, a long-run competitive equilibrium has precisely the characteristics usually ascribed to it. Together, $p \cdot y^i \geq C(y^i)$ and $p \cdot y \leq C(y), \forall y \in R^n_+$, imply that $p \cdot y^i = C(y^i)$ and that the $y^i \in \arg \max_y [p \cdot y - C(y)]$. Thus, each firm in the configuration takes prices as parametric, chooses output to maximize profits, earns zero profit, and equates marginal costs to prices of produced outputs. It is now easy to show

Proposition 1 A long-run competitive equilibrium is a sustainable configuration, so that a perfectly competitive market is a PCM.

Proposition 2 Sustainable configurations need not be long-run competitive equilibria, and a PCM need not be perfectly competitive.

The simplest example sufficient to prove this second proposition is an industry producing a single product with increasing returns to scale over the relevant range of output. Here, the only feasible configuration that is sustainable entails one firm producing the maximal output level y^* given by the intersection of the declining average cost curve with the industry demand curve, and selling at the price p^* given by the corresponding level of average cost. This configuration is sustainable because, at a price equal to or less than p^*, sale of any quantity on or inside the demand curve yields revenue no greater than production cost; in this range, price does not exceed average cost. Yet this configuration is not a long-run competitive equilibrium, as defined above, because sale of a quantity greater than y^* would earn positive profit if the price could remain at p^*, and because at y^* price exceeds marginal cost which is less than average cost. In fact, in this example there is no possible long-run competitive equilibrium since marginal cost lies below average cost throughout the relevant range of output levels given by demand. In contrast, there is a sustainable configuration.

Hence, Propositions 1 and 2 show that the sustainable industry configuration is a substantive generalization of the long-run competitive equilibrium, and that the PCM is a substantive generalization of the perfectly competitive market. The following propositions summarize some characteristics of equilibria in PCMs.

Proposition 3 Let $p; y^1, \ldots, y^m$ be a sustainable industry configuration. Then each firm must (i) earn zero profit by operating efficiently, $p \cdot y^i - C(y^i) = 0$; (ii) avoid cross-subsidization, $p_s \cdot y_s^i \geq C(y^i) - C(y_{N-s}^i)$, $\forall SCN$ (where the vector x_T agrees with the vector x in components $j \in T$ and has zeros for its other components); (iii) price at or above marginal cost, $p_j \geq \partial C(y^i)/\partial y_j$.

The interpretation of condition (ii) is that the revenues earned from the sales of any subset of the goods must not fall short of the incremental costs of producing that subset. Otherwise, in view of the equality of total revenues and costs, the revenues collected from the sales of the other goods must exceed their total stand-alone production cost. In PCMs, such pricing invites entry into the markets for the goods providing the subsidy.

Proposition 4 Let $p; y^1, \ldots, y^m$ be a sustainable configuration with $y_j^k < \sum_{h=1}^m y_j^h$. Then $p_j = \partial C(y^k)/\partial y_j$. That is, if two or more firms produce a given good in a PCM, they must select input–output vectors at which their marginal costs of producing it are equal to the good's market price.

The implications of this result are surprisingly strong. The discipline of sustainability in perfectly contestable markets forces firms to adopt prices just equal to marginal costs, provided only that they are not monopolists of the products in question. Conventional wisdom implies that, generally, only perfect competition involving a multitude of firms, each small in its output markets, can be relied upon to provide marginal-cost prices. Here we see that potential competition by prospective entrants, rather than rivalry among incumbent firms, suffices to make marginal-cost pricing a requirement of equilibrium in PCMs, even those containing as few as two active producers of each product. The conventional view holds that the enforcement mechanism of full competitive equilibrium requires the smallness of each active firm in its product market, in addition to freedom of entry. We see that the smallness requirement can be dispensed with, almost entirely, with exclusive reliance on the freedom of entry that characterizes PCMs.

Proposition 5 Let $p; y^1, \ldots, y^m$ be a sustainable configuration. Then, for any $\hat{y}^1, \ldots, \hat{y}^k$ with

$$\sum_{j=1}^k \hat{y}^j = \sum_{j=1}^m y^j, \quad \sum_{j=1}^k C(\hat{y}^j) \geq \sum_{j=1}^m C(y^j).$$

That is, a sustainable configuration minimizes the total cost to the industry of producing the total industry output.

This proposition is a generalization to PCMs of a well-known result for perfect competition. It can be interpreted as a manifestation of the power of unimpeded potential entry to impose efficiency upon the industry. For example, the proposition implies that if a monopoly occupies a PCM it must be a *natural* monopoly – production by a single firm must minimize industry cost for the given output vector. Thus, Propositions 3, 4 and 5 are powerful tools for the analysis of industry structure in PCMs. Proposition 5 permits information on the properties of production costs to be used to assess the scale and scope of firms' activities in PCMs. Then, Propositions 3 and 4 permit inferences to be drawn about the corresponding equilibrium prices.

PCMS with a single product

This analytic approach leads to very strong results in the single-product case. Propositions 3–5 show that there are only two possible types of sustainable configurations in single-product industries. The first type involves a single firm which charges the lowest price that is consistent with non-negative profit. The firm must be a natural monopoly when it produces the quantity that is demanded at this price. And, in this circumstance, the result maximizes welfare subject to the constraint that all firms in question be viable financially without subsidies. Such a second-best maximum is referred to as a 'Ramsey optimum'.

The second type of sustainable configuration involves production by one or more firms of outputs at which both marginal cost and average cost are equal to price. Here, in the long run, all active firms exhibit the behaviour that characterizes perfectly competitive equilibrium. And, of course, the result involves both (first-best) welfare optimality and financial viability. Hence, in this case, Ramsey optimality and the first-best coincide. This establishes the result that in a single-product industry any sustainable configuration is Ramsey optimal.

However, in general, because of the 'integer problem', sustainable configurations may generally not exist. This problem arises, for example, where there is only one output at which a firm's marginal and average costs coincide, and where the quantity of output demanded by the market at the competitive price is greater than this, but is not an integer multiple of that amount. Then, no sustainable configurations exist.

There is, however, a plausible assumption, supported by empirical evidence, at least to some degree, that eliminates the integer problem. Suppose that a firm's average cost curve has a flat-bottom rather than being 'U'-shaped. In particular, suppose that the minimum level of average cost is attained not only at one output, but (at least) at all outputs between the minimum efficient scale, y_m, and twice the minimum efficient scale. Then any industry output, y^I, that is at least equal to y_m can be apportioned among an integer number of firms, each of which achieves minimum average cost. Specifically, y^1 can be divided evenly among $\lfloor y^1/y_m \rfloor$ firms (where $\lfloor x \rfloor$ is the largest integer not greater than x) and each firm's output, $y^1/\lfloor y^1/y_m \rfloor$, must lie in the (half-open) interval between y_m and $2y_m$. Hence, in this case, the Ramsey optimum can either be a sustainable configuration of two or more firms performing competitively, or a sustainable natural monopoly. Such a monopoly may either produce an output at which there are increasing returns to scale and it will then price at average cost, or it may produce an output between y_m and $2y_m$ with locally constant returns to scale and adopt a price equal both to average and marginal cost. This, together with the preceding argument, establishes the following result.

Proposition 6 In a single-product industry in which the firm's average cost curve has a flat-bottom between minimum efficient scale and twice minimum efficient scale, a configuration is sustainable if and only if it is Ramsey optimal.

This result shows that, under the conditions described, there is equivalence between welfare optimality and equilibrium in PCMs. This extends the corresponding result for perfectly competitive equilibria to cases of increasing returns to scale. Moreover, since the behavioural assumptions required for a PCM are weaker than those underlying perfectly competitive markets, the equivalence result is more sweeping. In particular, Proposition 6 implies that PCMs can be expected to perform well, whatever the number of firms participating in equilibrium. It is the potential competition of potential entrants, rather than the active competition of existing rivals, that drives equilibrium in PCMs with a single product to welfare optimality.

Multi-product perfectly contestable markets

In industries that produce two or more goods, a rich variety of industry structures become possible, even in PCMs. Here, while the constraints imposed upon incumbents by perfect contestability are not nearly as effective in limiting the range of possible outcomes as they are in single product industries, they nevertheless provide a helpful basis for analysis. In particular, Propositions 3–5 indicate connections among various qualitative properties of multi-product cost functions and various elements of industry structure in PCMs. These connections constitute one theme of this section. The other theme is the normative evaluation of the industry structures that arise in multi-product PCMs.

Before proceeding, it may be useful to provide definitions of some of the multi-product cost properties that are used in the analysis.

Definition 5 Let $P = \{T_1, \ldots, T_k\}$ be a non-trivial partition of $S \subseteq N$. There are (weak) *economies of scope* at y_s

with respect to the partition P if $\sum_{i=1}^{k} C(y_{Ti}) > (\geq) C(y_s)$. If no partition is mentioned explicitly, then it is presumed that $T_i = \{i\}$.

Definition 6 The *degree of scale economies* defined over the entire product set, $N = \{1, \ldots, n\}$, at y, is given by $S_N(y) = C(y)/y \cdot \nabla C(y)$.

Returns to scale are said to be increasing, constant or decreasing as S_N is greater than, equal to or less than unity. This occurs as the elasticity of ray average cost with respect to t is negative, positive or zero; where *ray average cost* is $\mathrm{RAC}(ty^0) \equiv C(ty^0)/t$.

Definition 7 The *incremental cost* of the product set $T \subseteq N$ at y is given by $\mathrm{IC}_T(y) \equiv C(y) - C(y_{N-T})$. The *average incremental cost* of T is $\mathrm{AIC}_T(y) \equiv \mathrm{IC}_T(y)/\sum_{j \in T} y_j$.

The average incremental cost of T is decreasing, increasing or constant at y if $\mathrm{AIC}_T(ty_T + y_{N-T})$ is a decreasing, increasing or locally constant function of t at $t = 1$. These cases are labelled respectively, increasing, decreasing or constant *returns to the scale of the product line T*. The degree of scale economies specific to T is

$$\mathrm{IC}_T(y)/\sum_{i \in T} y_i \frac{\partial C(y)}{\partial y_i}.$$

Definition 8 A cost function $C(y)$ is *trans-ray convex* through some point $y^* = (y_1^*, \ldots, y_n^*)$ if there exists at least one vector of positive constants $w_1, \ldots, w_n$ such that for every two output vectors $y^a = (y_1^a, \ldots, y_n^a)$ and $y^b = (y_1^b, \ldots, y_n^b)$ that lie on the hyperplane $\sum w_i y_i = w_0$ through point y^*, $C[ky^a + (1-k)y^b] \leq kC(y^a) + (1-k)C(y^b)$ for $k \in (0, 1)$.

In view of the general result that sustainable configurations minimize industry-wide costs (Proposition 5), these cost properties permit inferences to be drawn about industry structure in multi-product PCMs. The first issue that arises is when multi-commodity production is characteristic of equilibrium in a PCM.

Proposition 7 A multi-product firm in a PCM must enjoy (at least weak) economies of scope over the set of goods it produces. When strict economies of scope are present, there must be at least one multi-product firm in any PCM that supplies more than one good.

The second basic question that arises is whether there can be two or more firms actively producing a particular good in a PCM. If there are, then, by Proposition 4, marginal cost pricing must result. The answer depends upon the availability of product-specific scale economies.

Proposition 8 Any product with average incremental costs that decline throughout the relevant range (that is, that offers product-specific increasing returns to scale) must be produced by only a single firm (if it is produced at all) in a PCM. Further, such a product must be priced above marginal cost, unless the degree of product-specific scale economies is exactly one.

Thus, regardless of the presence or absence of economies of scope, globally declining average incremental costs imply that a product must be monopolized in a PCM. It is an immediate corollary that if all goods in the set N exhibit product-specific scale economies, and if there are economies of scope among them all, then the industry is a natural monopoly that must be monopolized in a PCM.

Another route to this result is provided by the 'weak invisible hand theorem of natural monopoly'.

Proposition 9 Trans-ray convexity of costs together with global economies of scale imply natural monopoly. If, in addition certain other technical conditions are met, a monopoly charging Ramsey-optimal prices is a sustainable configuration.

In general, there may exist natural monopoly situations in which no sustainable prices are possible for the Ramsey optimal product set. Further, even where sustainable prices exist, the Ramsey optimal prices may not be among them. However, under the conditions of the weak invisible hand theorem, the Ramsey optimal prices for the Ramsey optimal product set are guaranteed to be sustainable, so that PCMs are consistent with (second-best) welfare optimal performance by a natural monopoly.

PCMs will yield first-best welfare optimality if there exist sustainable configurations with at least two firms actively producing each good. For in this case Propositions 4 and 5 guarantee industry-wide cost efficiency and marginal-cost pricing of all products. Here, two issues must be resolved: Does industry-wide cost minimization require at least two producers of each good? And if so, do sustainable configurations exist?

The existence problem can be solved in a manner analogous to its solution in the case of single-product industries: by assuming that ray average costs remain at their minimum levels for output vectors that lie (on each ray) between minimum efficient scale and twice minimum efficient scale. And the presence of at least two producers (or one operating in the region where constant returns prevail) of each good is assured if the quantities demanded by the market at the relevant marginal-cost prices are no smaller than minimum efficient scale (along the relevant ray) and if the cost function exhibits trans-ray convexity.

Policy implications of PCMS

One of the principal lessons of the analysis of PCMs is that monopoly does not necessarily entail welfare losses.

Rather, the 'weak invisible hand theorem' shows that under certain conditions sustainability and Ramsey optimality are consistent, so that the total of consumers' and producers' surpluses may well be maximized (subject to the constraint that firms be self-supporting) in the equilibrium of a monopoly which operates in contestable markets.

Even stronger results follow from the discussed results that under certain conditions sustainability and a first-best solution are consistent in an oligopoly with a small number of firms. When minimization of industry cost requires that each good be produced by at least two firms, sustainability requires any equilibrium to satisfy the necessary conditions for a first-best allocation of resources. Thus, in these cases, the invisible hand has the same power over oligopoly in perfectly contestable markets that it exercises over a perfectly competitive industry.

This theory suggests that in a market that approximates perfect contestability, the general public interest is well-served by a policy of laissez-faire rather than active regulation by administrative or antitrust means. Small numbers of large firms, vertical and even horizontal mergers and other arrangements which have traditionally been objects of suspicion of monopolistic power, are rendered harmless and perhaps even beneficial by the presence of contestability.

On the other hand, contestability theory does not lend support to the proposition that the unrestrained market automatically solves all economic problems and that virtually all regulation and antitrust activity entails unwarranted and costly intervention. The economy of reality is composed of industries which vary widely in the degree to which they approximate the attributes of perfect contestability. Before the theory of contestability can be legitimately applied to reach a conclusion that intervention is unwarranted in a specific sector, it must first be shown that the sector lies unprotected by entry barriers and that the force of potential entry therefore actively constrains the behaviour of incumbent firms. This then becomes the appropriate first stage in an analysis of efficient government policy towards an industry. Only where the conditions of contestability are found to characterize the reality of an industry can there be validity in applying the normative conclusions of contestability theory concerning the power of potential competition actually to enforce efficient behaviour on incumbents.

Even where contestability is absent in reality, the formulation on efficient regulation can be usefully guided by the theory of contestable markets instead of the theory of perfectly competitive markets. The first-best lesson of the perfect competition model, calling for prices to be set equal to marginal costs, has no doubt contributed to the common regulatory ethos which *equates* price to *some* measure of cost. This doctrine has been used frequently where it is completely inappropriate and without logical foundation, that is, in cases where prices should be based on demand as well as cost considerations, because of the presence of economies of scale and scope. Such arbitrary measures as fully distributed costs cannot substitute for marginal cost measures as decision rules for proper pricing, and the search for a substitute is a remnant of inappropriate reliance on the model of perfect competition for guidance in regulation.

In contrast, contestability theory suggests cost measures that are appropriate guideposts for regulated pricing – incremental and stand-alone costs. The incremental cost of a given service is, of course, the increment in the total costs of the supplying firm when that service is added to its product line. In perfectly contestable markets, the price of a product will lie somewhere between its incremental and its stand-alone cost, just where it falls in that range depending on the state of demand. One cannot legitimately infer that monopoly power is exercised from data showing that prices do not exceed stand-alone costs, and stand-alone costs constitute the proper cost-based ceilings upon prices, preventing both cross-subsidization and the exercise of monopoly power. A simple example will show why this is so.

First, suppose that a firm supplies two services, A and B, which *share no costs* and that each costs 10 units a year to supply. The availability of effective potential competition would force revenues from each service to equal 10 units a year. For higher earnings would attract (profitable) entry, and lower revenues would drive the supplier out of business. In this case, in which common costs are absent, incremental and stand-alone costs are equal to each other and to revenues, and the competitive and contestable benchmarks yield the same results.

Next, suppose instead that of the 20 unit total cost 4 are fixed and common to A and B, while 16 are variable, 8 of the 16 being attributable to A and 8 to B. If, because of demand conditions, at most only a bit more than 8 can be generated from consumers of A, then a firm operating and surviving in contestable markets will earn a bit less than 12 from B. These prices lie between incremental costs (8) and stand-alone costs (12), are mutually advantageous to consumers of both services, and will attract no entrants, even in the absence of any entry barriers. In contrast, should the firm attempt to raise the revenues obtained from B above the 12 unit stand-alone cost, it would lose its business to competitors willing to charge less. Similarly, the same fate would befall it in contestable markets if it priced B in a way that earned more than 8 plus the contribution towards the common cost of 4, less the contribution towards that common cost from service A.

Thus, the forces of idealized potential competition in perfectly contestable markets enforce cost constraints on prices, but prices remain sensitive to demands as well. Actual competition and potential competition are *effective* if they constrain rates in this way, and in such circumstances regulatory intervention is completely unwarranted. But if, in fact, market forces are not sufficiently strong, then there may be a proper role for regulation of natural monopoly, and the theoretical guidelines derived from the workings of contestable markets are the appropriate ones to apply. That is, prices must be constrained to lie between incremental and stand-alone costs. (This is the approach recently adopted by the Interstate Commerce Commission to determine maximum rates for US railroad services, and the method has already withstood appeals to the federal courts.)

ROBERT D. WILLIG

See also **barriers to entry.**

Bibliography

Baumol, W.J., Panzar, J.C. and Willig, R.D. 1982. *Contestable Markets and the Theory of Industry Structure.* New York: Harcourt Brace Jovanovich.

Baumol, W.J., Panzar, J.C. and Willig, R.D. 1985. On the theory of perfectly contestable markets. In *New Developments in the Analysis of Market Structure*, ed. J. Stiglitz and F. Mathewson. New York: Harcourt Brace.

Baumol, W.J. and Willig, R.D. 1986. Contestability: developments since the book. *Oxford Economic Papers* 38, 9–36.

contingent commodities

The theory of general competitive equilibrium was originally developed for environments where no uncertainty prevailed. Everything was certain and phrases like 'it might rain' or 'the weather might be hot' were outside the scope of the theory. The idea of *contingent commodity*, that was introduced by Arrow (1953) and further developed by Debreu (1953), was an ingenious device that enabled the theory to be interpreted to cover the case of uncertainty about the availability of resources and about consumption and production possibilities. Basically, the idea of contingent commodity is to add the environmental event in which the commodity is made available to the other specifications of the commodity. With no uncertainty every commodity is specified by its physical characteristics and by the location and date of its availability. It is fairly clear, however, that such a commodity can be considered to be quite different where two different environmental events have been realized. The following examples clarify this: an umbrella at a particular location and at a given date in case of rain is clearly different from the same umbrella at the same location and date when there is no rain; some ice cream when the weather is hot is clearly different from the same ice cream (and at the same location and date) when the weather is cold; finally, the economic role of wheat with specified physical characteristics available at some location and date clearly depends on the precipitation during its growing season. Thus, specifying commodities by both the standard characteristics and the environmental events seems very natural, whereas the role of the adjective in 'contingent commodities' is simply to make it clear that one is dealing with commodities the availability of which is contingent on the occurrence of some environmental event. With this specification the model with contingent commodities is very similar to the classical model of general competitive equilibrium and thus questions like the existence of equilibrium and its optimality (with the additional aspect of efficient allocation of risk bearing) are answered in a similar way. Note that, although this model deals with uncertainty, no concept of probabilities is needed for its formal description.

To make things more explicit we look at a simple model with contingent commodities. Assume that, without referring to uncertain events, there are $k \geq 1$ commodities, indexed by i, and that there are $n > 1$ mutually exclusive and jointly exhaustive events (or states of nature), indexed by s, where k and n are finite. Thus a contingent commodity is denoted by x_{is} and the total number of these commodities is kn, which is greater than k but still finite. Consumption and production sets are thus defined as subsets of the kn-dimensional Euclidean space, and the economic behaviour of firms and consumers naturally follows from profit maximization (by firms) and utility maximization (by consumers). The price p_{is} of the contingent commodity x_{is} is the number of units of account that have to be paid in order to have the ith commodity being delivered at the sth event. It is assumed that the market is organized before the realization of the possible events. Thus payment for the contingent commodity x_{is} is done at the beginning while delivery takes place after the realization of events and only in case event s has occurred. Note that the price of the (certain) ith commodity, that is, the number of units of account that have to be paid in order to have the ith commodity *for sure*, is the sum over s of the prices p_{is}. For example, assume that the price of one quart of ice cream if the weather is hot is \$2.00, the price of one quart of the same ice cream if the weather is cold is \$1.00 and that $n = 2$ (either it is hot or cold). Thus the price of having one quart of that ice cream for sure is \$2.00 + \$1.00 = \$3.00.

It should be noted that, although the probabilities of the possible events do not explicitly enter the model, the attitude towards risk of both consumers and producers is of interest and does play a significant role in this framework. The preference relations of consumers defined on subsets of the kn-dimensional Euclidean space reflect not only their 'tastes' but also their subjective beliefs about the likelihoods of different events as well as their attitude towards risk. Convexity of consumers' preferences, for example, is interpreted as risk aversion while, in the same spirit, profit maximization of firms is interpreted as risk neutrality. It should be mentioned that both Arrow and Debreu basically assume expected utility maximizing behaviour, in the sense of the Savage (1954) framework. A more general approach to such preference relations can be found in Yaari (1969), where, again, convexity is taken to mean risk aversion.

A unified and more formal treatment of time and uncertainty using contingent commodities can be found in Debreu (1959, ch. 7). Radner (1968) presents an extension of the above model to the case in which different economic agents have different information.

ZVI SAFRA

See also **Arrow–Debreu model of general equilibrium; uncertainty; uncertainty and general equilibrium.**

Bibliography

Arrow, K.J. 1953. Le rôle de valeurs boursières pour la répartition la meilleure des risques. *Econométrie*. Paris: CNRS. English translation 'The role of securities in the optimal allocation of risk-bearing' in *Review of Economic Studies* (1964); reprinted in K.J. Arrow, *Essays in the Theory of Risk-Bearing*. Chicago: Markham, 1971.
Debreu, G. 1953. Une économie de l'incertain. Mimeo, Paris: Electricité de France.
Debreu, G. 1959. *Theory of Value*. New York: Wiley.
Radner, R. 1968. Competitive equilibrium under uncertainty. *Econometrica* 36, 31–58.
Savage, L.J. 1954. *Foundations of Statistics*. New York: Wiley.
Yaari, M.E. 1969. Some remarks on measures of risk aversion and their uses. *Journal of Economic Theory* 1, 315–29.

contingent valuation

Most economists would agree that no researcher should prefer demand data from hypothetical markets if data concerning the identical goods or services, based on real markets, are readily available. However, there are many situations when even the cleverest empirical economist cannot come up with revealed preference data from actual markets that can be relied upon for information about household demands for some types of goods. Environmental goods are one class of goods where real-market demands sometimes cannot be measured adequately. In the 1980s, environmental economists began in earnest to exploit stated-preference demand information, usually collected using household surveys. This demand information is used primarily to produce utility-theoretic measures of the social benefits of environmental protection measures for benefit–cost analyses.

Environmental economists called these methods 'contingent valuation methods' (CVM) because the valuations were elicited 'contingent upon the conditions described in the survey'. Research that focused on the development and assessment of CVM in environmental economics was well under way by the mid-1980s. However, two events in 1989 thrust the method to the forefront of the field. First, the *Exxon Valdez*, an ocean tanker, ran aground in Prince William Sound in Alaska, spilling 11 million gallons of oil in an environmental disaster that attracted a huge amount of media attention worldwide. Second, just a few months later, the US Court of Appeals held that the economic damages assessed for spills of oil or other hazardous substances could include 'lost passive use values', and that these values could legally be measured via CVM.

Plaintiffs and defendants in the Exxon Valdez case thus had a big incentive to advocate and derogate CVM, respectively. For at least a time, the discussion in the literature teetered on the brink of losing its polite academic tone. Given the escalation of the controversy over CVM, the US National Oceanic and Atmospheric Administration (NOAA) convened a panel of experts (untainted by any active role in the Exxon Valdez litigation) to assess CVM. This exercise, by Arrow et al. (1993), produced a set of pronouncements concerning best practices for the conduct of CVM studies. While the 1993 NOAA Panel report cannot be considered the last word on CVM, it was very influential, and there has since been strong pressure on researchers either to conform to the NOAA best practices or to fully justify any departures from them.

As a result of the Exxon Valdez case, much doubt about the reliability of stated preference data led to numerous comparisons of the implications of stated and revealed preference data (for example, Carson et al., 1996). CVM works best when respondents have a clear sense of the consequences of their choices – in terms of both their own budgets and the exact nature of the good that they are being asked to consider paying for – and when they are reasonably familiar with market transactions involving that good. This means that CVM is, unfortunately, most successful when it is least needed. The challenge for researchers is to ensure that demand information gathered using CVM, in less-than-ideal contexts, is as valid and useful as possible.

Myriad biases and qualifications may afflict poorly executed CVM studies. A partial list includes incentive compatibility, hypothetical bias (if the choice is perceived to have absolutely no real consequences), strategic bias (when people try to manipulate the outcome by misrepresenting their preferences), non-response bias (since people cannot be compelled to participate), starting-point bias (for surveys with iterative bids), interviewer bias (for in-person surveys), and information bias (when some portion of the value is constructed during the survey where it did not exist before). Other problems include yea-saying, part–whole bias or embedding, scenario rejection, and the potential for respondents not to pay sufficient attention to their real budget constraints.

Choice formats have been an important issue in the development of CVM. For example, in some early applications of CVM survey respondents were asked directly to identify the single highest dollar amount that they would pay to obtain some change in conditions. These

were called open-ended CV questions. Researchers quickly realized that such a task was difficult for consumers who were unfamiliar with naming their own price, especially for goods they may never before have thought much about having to pay for. CVM elicitation techniques evolved fairly quickly to a dichotomous-choice format, where respondents are given a choice between two states of the world. One state is typically the status quo, while the other involves a specified change or set of changes (such as an improvement in environmental quality, or some other rationed public good) that come at a price (typically a lump-sum payment). This binary choice format was found to fit naturally into a random utility model (RUM) framework that had also become a popular approach to consumer choice problems, both real and hypothetical, in the transportation mode-choice literature and elsewhere in economics.

Respondents' preferences, based on their answers to dichotomous-choice CV questions, can be characterized either in an ad hoc fashion or in a more formal utility-theoretic framework. One standard approach is to specify an indirect utility function shared by all respondents. In its simplest form, the level of indirect utility is assumed to depend on the individual's net income under each of the two alternatives, and upon a discrete indicator of whether there is a change in the rationed public good, or no change, under each alternative. Respondents can choose the environmental improvement programme along with its associated cost (implying lower net income), or decline the environmental improvement programme in order to avoid the cost (preserving their net income). If a respondent prefers the programme with its associated cost, the researcher assumes that the respondent's utility level is higher under that alternative. Equivalently, this means that the *net* indirect utility associated with the programme alternative is positive.

A discrete-choice econometric model, typically involving a binary logit estimator, is used to estimate the sample average marginal utilities of (a) net income and (b) the discrete bundle of changes represented by the programme in question. It is of course possible to allow for heterogeneity across the sample in these marginal utilities. Most often, heterogeneity is introduced by allowing the otherwise scalar marginal utility associated with going from 'no programme' to 'programme' to become a systematically varying parameter. Of course, if the identical programme is offered to all respondents, it is not possible to allow this marginal utility to vary with attributes of the programme. However, it can easily be allowed to vary with characteristics of the respondent.

Less commonly, the marginal utility of income is also allowed to vary across respondents, either with the respondent's income (to allow for diminishing marginal utility of income) or with other respondent characteristics. However, there is a premium on simplicity for the marginal utility of income, stemming from the need to use the estimated marginal utility of income parameter(s) to recover demand information. For this reason, many researchers will, if it is justified by the data, prefer a choice model that is linear and additively separable in net income under each alternative.

Linearity and additive separability in income is convenient (when warranted) because the willingness to pay (WTP) function associated with the fitted model is given by the marginal rate of substitution (MRS) between the programme and income. This MRS is given by the ratio of the marginal utility of the programme to the marginal utility of income, producing a result that can be expressed in dollars per 'unit' of the programme, where the program indicator is either zero or 1 in the simple binary case. In the non-stochastic case, for a simple dichotomous choice CVM model, this is a single number − 'WTP for the program' − if the researcher has assumed homogeneous preferences throughout the sample.

Some extra empirical housekeeping is necessary when it is acknowledged that this point estimate is constructed as the ratio of two estimated quantities, each of which (due to the use of maximum likelihood estimation methods for the logit or probit model) is an asymptotically normally distributed random variable. In theoretical terms, the ratio of two normally distributed random variables has an undefined mean, because zero is a possible value for the denominator. As a practical matter, some researchers use simulation methods to build up a sampling distribution for the estimated WTP. It is possible to use packaged software to make a large number of random draws from the joint distribution of the logit or probit parameters (based on the estimated parameter point estimates and the parameter variance–covariance matrix). One can then build up a sampling distribution for the needed ratio. Other strategies for dealing with this inconvenience involve estimating the model in 'WTP-space' rather than 'utility-space' or employing the newer mixed logit (random-parameters logit models) and stipulating that the marginal utility of income parameter be distributed lognormal (since it should be strictly positive, on average), rather than normal, so that the potential divide-by-zero problem goes away.

Over the 1990s contingent valuation researchers in environmental economics gradually made better contact with their counterparts working in other literatures who were confronted by similar problems where there is a lack of market data for products or public goods that need to be valued. In the transportation literature, researchers had grappled early on with the problem of forecasting demand for public transportation projects, or new types of vehicles, that did not yet exist. Researchers began to introduce hypothetical new transportation options which could be characterized in the same terms as existing options (in 'attribute space') but which had some attributes that lay well outside the set of existing options on some dimensions, or which involved attributes that were not relevant for existing options (such as travel range or recharge time for prospective electric vehicles).

One key difference from contingent valuation was the practice of asking survey respondents to consider more than just 'the status quo versus a single alternative'. Furthermore, the alternatives were more richly specified. Instead of using simply a dummy variable to indicate whether the policy, programme or public good was present or absent, each alternative was characterized in terms of an array of attributes.

Similar problems were also being addressed in the marketing literature, particularly in the context of 'pre-test' marketing. Companies considering whether to develop and introduce new products needed to know in advance about the likely demand for these products, perhaps as a function of alternative possible product configurations. Market researchers developed a set of techniques they called 'conjoint analysis'. In the marketing literature, the specifications used for the choice models were initially very ad hoc. Little attention was paid to the interpretation of the estimated coefficients as marginal utilities, and simple linear and additively separable specifications were very common. The slope coefficients were known as 'part-worths' rather than marginal utilities. However, much was learned about the degree of consistency between planned purchase behaviour and actual purchase behaviour.

CVM has also recently grown in popularity in other sub-disciplines, notably health economics. However, Smith (2003) surveys that literature and suggests that researchers in that field have not yet developed a set of best practices for the use of CVM with the types of choices that are most common in health economics contexts.

In the transportation and marketing literatures, the desired demand information often spanned a number of possible alternative products or services. Stated preference studies were often conducted not just to determine respondents' willingness to pay for a single well-defined good but to understand how willingness to pay might be affected by variations in the mix of attributes making up a prospective good. It was often necessary to anticipate demands for differentiated products where each product could be characterized as a bundle of attributes and the levels of these attributes differed across alternatives.

In contrast, more of the impetus for CVM non-market valuation research in environmental economics stemmed from a number of significant lawsuits. In the legal context, there is a premium on simplicity in economic modelling so as not to confuse the jury or the judge. It is often best to produce one value for one clearly defined commodity. (Providing a judge or jury with a function that describes demand, where WTP depends upon a wide array of attributes, conditions or respondent attributes, can actually be a liability when attorneys are trying to make a simple, clear and persuasive case. In a legal context, it is most incisive to value one thing, and to value it as precisely as possible.) Eventually, however, environmental economists began to acknowledge the value of understanding the heterogeneity in demands for environmental goods, since this knowledge can be very helpful to policymakers who wish to consider how different versions of a policy might affect different constituencies.

There are many commonalities between the tasks faced by environmental economists and those faced by transportation economists and market researchers, but there is one key difference. In transportation economics and market research, it is often the case that the public transit system in question will actually be built, or the new product will actually be developed and put on the market. There is an opportunity to go back and see whether the level of demand predicted by the stated preference study actually materializes when there is a real market. In the environmental economics literature, there are typically fewer opportunities to 'validate' the stated preference demand information with revealed preferences for the same product.

One common expectation for a good CVM study is now a demonstration that the demand function that has been estimated should 'walk and talk' like a demand function. For example, is willingness to pay to preserve big-game hunting opportunities lower, on average, for elderly women than for middle-aged males? Is willingness to pay to preserve air quality higher for people with lung disease or asthma, or for people who have family members with these illnesses? These tests are commonly called 'construct validity' assessments. Contingent valuation studies that pass a battery of plausibility tests such as these can generally be viewed as more reliable.

Another common test of contingent valuation estimates that these stated preference demand functions are typically expected to satisfy is something called a 'scope test'. This means that, on average, respondents' willingness to pay for an alternative that involves more of the 'good' in question should be greater than that for an alternative that involves less of the 'good'. It is of course possible that marginal utility may be positive (as the scope test implies) at low levels of the good, but also that it may go to zero if the quantity of the good is high enough for satiation to set in, and there is no theoretical basis for expecting willingness to pay to be proportional to the amount of the good in question.

CVM data can also sometimes be pooled with actual choice data. This can allow portions of the underlying indirect utility function to be anchored upon real choices, even though the variability in attributes in the real alternatives may not span the full domain relevant to pending policy decisions. The stated choice questions can be used to extend the domain of the estimated demand function.

While economists will remain uncomfortable about reliance upon stated preference information, many now acknowledge that there are circumstances where stated preference data are all that can be collected. In fact, the need for economists to rely upon survey data (what people say as opposed to what they actually do in markets) is now being acknowledged in the other contexts in economics. For example, expectations about future income or life expectancy figure prominently in a number of economic theories. These expectations typically cannot be measured directly, but can sometimes be elicited using surveys and put to good use empirically (see Manski, 2004).

It is worth noting that not just stated preference data but also revealed preference data can be highly variable in its quality. Much revealed preference demand data is also drawn from consumer surveys. It is not always clear that the individual respondent sees the need for accuracy and completeness to be as critical as researchers using the data might hope. In consumer expenditure surveys, for example, interviewers prompt subjects for different types of expenditures, but the enthusiasm and engagement of the survey subject often determines the accuracy and completeness of the data. Rather than viewing revealed preference data as of unambiguously high quality and stated preference data (such as that produced by CVM) as of unambiguously low quality, it may be prudent simply to acknowledge that both types of data can have their problems.

A partial list of current frontiers in CVM-related research is possible. These frontiers include continuing assessment of (a) alternative elicitation formats (there are many candidates beyond the simple NOAA-recommended binary choice format), (b) the choice contexts presented to subjects, (c) the effects of allowing subjects to express uncertainty about their choices, (d) the effects of practice and fatigue when several CVM questions are presented to each respondent, (e) integrating stated choices with additional types of real market information, (f) how the degree of complexity of the CVM choice scenarios interacts with the cognitive capacity of the subject and/or the subject's inclination to pay attention, and (g) the neuroeconomics of real as opposed to stated choices.

Two of the classic books on CVM are Cummings, Brookshire and Schulze (1986) and Mitchell and Carson (1989). Following the Exxon Valdez case, a provocative debate was featured in the *Journal of Economic Perspectives* (Diamond and Hausman, 1994; Hanemann, 1994; Portney, 1994). McFadden (1994) raised some specific concerns about the reliability of CVM data in the context of an empirical application to the existence value of wilderness areas in the western United States. In the intervening years, however, research concerning CVM has continued apace. Helpful expositions and discussions of recent innovations have made their way into textbook form, with one particularly useful summary being provided in Chapter 6 of Freeman (2003). A brief, accessible and very helpful introduction to CVM for non-specialists is contained in Carson (2000). Louviere, Hensher and Swait (2000) offer a comprehensive discussion of stated choice methods broadly defined, including experimental design, modelling, estimation and combining revealed and stated preference data, with illustrations in marketing, transportation and environmental economics. An inventory of the wide range of practical issues to consider in actually implementing a CVM study is provided by Boyle (2003), while Holmes and Adamowicz (2003) update the state of the art for attribute-based (conjoint choice) methods.

There is still considerable variation in individual researchers' levels of comfort with CVM and stated preference data more generally. We might reconsider the question posed by Diamond and Hausman (1994): 'Contingent valuation – is some number better than no number?' There are now many economists who would agree that a value based on stated preferences – from a study that is carefully conceived and executed, based on a sufficiently large sample that is representative of its intended population, that has been put through a battery of consistency and validity assessments, and that produces an implied demand function that behaves the way we would expect a 'real' demand function to behave – is almost certainly better than no number. This is especially true when 'no number' creates the risk that a value of zero would otherwise be imputed, by default, for use in policy decisions.

TRUDY ANN CAMERON

See also **environmental economics**.

Bibliography

Arrow, K., Solow, R., Portney, P.R., Leamer, E.E., Radner, R. and Schuman, H. 1993. Report of the NOAA panel on contingent valuation. *Federal Register* 58, 4601–14.

Boyle, K.J. 2003. Contingent valuation in practice. In *A Primer on Nonmarket Valuation*, ed. P.A. Champ, K.J. Boyle and T.C. Brown. Boston, MA: Kluwer Academic Publishers.

Carson, R.T. 2000. Contingent valuation: a user's guide. *Environmental Science & Technology* 34, 1413–18.

Carson, R.T., Flores, N.E., Martin, K.M. and Wright, J.L. 1996. Contingent valuation and revealed preference methodologies: comparing the estimates for quasi-public goods. *Land Economics* 72, 80–99.

Cummings, R.G., Brookshire, D.S. and Schulze, W.D., eds. 1986. *Valuing Environmental Goods: An Assessment of the Contingent Valuation Method*. Totowa, NJ: Rowman and Allanheld.

Diamond, P.A. and Hausman, J.A. 1994. Contingent valuation – is some number better than no number? *Journal of Economic Perspectives* 8(4), 45–64.

Freeman, A.M.I. 2003. *The Measurement of Environmental and Resource Values: Theory and Methods*. Washington, DC: Resources for the Future.

Hanemann, W.M. 1994. Valuing the environment through contingent valuation. *Journal of Economic Perspectives* 8(4), 19–43.

Holmes, T.P. and Adamowicz, W.L. 2003. Attribute-based methods. In *A Primer on Nonmarket Valuation*, ed. P.A. Champ, K.J. Boyle and T.C. Brown. Boston, MA: Kluwer Academic Publishers.

Louviere, J.J., Hensher, D.A. and Swait, J.D. 2000. *Stated Choice Methods*. New York: Cambridge University Press.

Manski, C.F. 2004. Measuring expectations. *Econometrica* 72, 1329–76.

McFadden, D. 1994. Contingent valuation and social choice. *American Journal of Agricultural Economics* 76, 689–708.

Mitchell, R.C. and Carson, R.T. 1989. *Using Surveys to Value Public Goods: The Contingent Valuation Method*. Washington, DC: Resources for the Future.

Portney, P.R. 1994. The contingent valuation debate – why economists should care. *Journal of Economic Perspectives* 8(4), 3–17.

Smith, R.D. 2003. Construction of the contingent valuation market in health care: a critical assessment. *Health Economics* 12, 609–28.

continuous and discrete time models

Discrete time models are generally only an approximation, and the error induced by this approximation can under some conditions be important.

Most economists recognize that the use of discrete time is only as an approximation, but assume (usually implicitly) that the error of approximation involved is trivially small relative to the other sorts of simplification and approximation inherent in economic theorizing. We consider below first the conditions under which this convenient assumption may be seriously misleading. We discuss briefly how to proceed when the assumption fails and the state of continuous time economic theory.

Approximation theory

Some economic behaviour does involve discrete delays, and most calculated adjustments in individual patterns of behaviour seem to occur following isolated periods of reflection, rather than continually. These notions are sometimes invoked to justify economic theories built on a discrete time scale. But to say that there are elements of discrete delay or time discontinuity in behaviour does not imply that discrete time models are appropriate. A model built in continuous time can include discrete delays and discontinuities. Only if all delays were discrete multiples of a single underlying time unit, and synchronized across agents in the economy, would modelling with a discrete time unit be appropriate.

Nonetheless, sometimes discrete models can avoid extraneous mathematical complexity at little cost in approximation error. It is easy enough to argue that time is in fact continuous and to show that there are in principle cases where use of discrete time models can lead to error. But it is also true in practice that more often than not discrete time models, translated intuitively and informally to give implications for the real continuous time world, are not seriously misleading. The analytical task, still not fully executed in the literature, is to understand why discrete modelling usually is adequate and thereby to understand the special circumstances under which it can be misleading.

The basis for the usual presumption is that, when the time unit is small relative to the rate at which variables in a model vary, discrete time models can ordinarily provide good approximations to continuous time models. Consider the case, examined in detail in Geweke (1978), of a dynamic multivariate distributed leg regression model, in discrete time.

$$Y(t) = A^*X(t) + U(t), \qquad (1)$$

where * stands for convolution, so that

$$A^*X(t) = \sum_{s=-\infty}^{\infty} A(s)X(t-s). \qquad (2)$$

We specify that the disturbances are uncorrelated with the independent variable vector X, that is, $cov[X(t), U(s)] = 0$, all t, s. The natural assumption is that, if approximation error from use of discrete time is to be small, $A(s)$ must be smooth as a function of s, and that in this case (1) is a good approximation to a model of the form

$$y(t) = a^*x(t) + u(t) \qquad (3)$$

where

$$a^*x(t) = \int_{-\infty}^{\infty} a(s)x(t-s)ds \qquad (4)$$

and y, a and x are functions of a continuous time parameter and satisfy $y(t) = Y(t)$, $x(t) = X(t)$ and $a(t) = A(t)$ at integer t. In this continuous time model we specify, paralleling the stochastic identifying assumption in discrete time, $cov[x(t), u(s)] = 0$, all t, s. If the discrete model (2) corresponds in this way to a continuous time model, the distributed lag coefficient matrices $A(s)$ are uniquely determined by a and the serial correlation properties of x.

We should note here that, though this framework seems to apply only to the case where X is a simple discrete sampling of x, not to the time-averaged case where $X(t)$ is the integral of $x(s)$ from $t-1$ to t, in fact both cases are covered. We can simply redefine the x process to be the continuously unit-averaged version of the original x process. This redefinition does have some effect on the nature of limiting results as the time unit goes to zero (since the unit-averaging transformation is different at each time unit) but turns out to be qualitatively of minor importance. Roughly speaking, sampling a unit-averaged process is like sampling a

process whose paths have derivatives of one higher order than the unaveraged process.

Geweke shows that under rather general conditions

$$\sum_{s=-\infty}^{\infty} ||A(s) - \tau a(s\tau)||^2 \to 0 \qquad (5)$$

as the time unit τ goes to zero, where $|| \, ||$ is the usual root-sum-of-squared-elements norm. In this result, the continuous time process x and lag distribution a are held fixed while the time interval corresponding to the unit in the discrete time model shrinks.

This is the precise sense in which the intuition that discrete approximation does not matter much is correct. But there are important limitations on the result. Most obviously, the result depends on a in (3) being an ordinary function. In continuous time, well-behaved distributed lag relations like (3) are not the only possible dynamic relation between two series. For example, if one replaces (3) by

$$y(t) = \alpha(d/dt)x(t) + u(t), \qquad (6)$$

then the limit of A in (2) is different for different continuous x processes. In a univariate model with second-order Markov x (for example, one with $cov[x(t), x(t-s)] = (1 + \theta|s|)e^{-\theta|s|} \, var[x(t)]$, the limiting discrete time model, as τ goes to zero, is

$$y(t) = a\{-0.02X(t+4) + 0.06X(t+3) \\ - 0.22X(t+2) + 0.80X(t+1) \\ - 0.80X(t-1) + 0.22X(t-2) \\ - 0.06X(t-3) + 0.02X(t-4)\} \\ + U(t) \qquad (7)$$

(see Sims, 1971).

This result is not as strange as it may look. The coefficients on X sum to zero and are anti-symmetric about zero. Nonetheless, (7) is far from the naive approximation which simply replaces the derivative operator with the first difference operator. In fact, if the estimation equation were constrained to involve only positive lags of X, the limiting form would be

$$Y(t) = \alpha\{1.27X(t) - 1.161X(t-1) \\ + 0.43X(t-2)\} - 0.12X(t-3) \\ + 0.03X(t-4) - 0.01X(t-5) \\ + U(t). \qquad (8)$$

The naive approximation of (3) by $Y(t) = \alpha[X(t) - X(t) - 1] + U(t)$ is valid only in the sense that, if this form is imposed on the discrete model a priori, the least squares estimate of α will converge to its true value. If the resulting estimated model is tested for fit against (8) or (7), it will be rejected.

Although the underlying model involves only the contemporaneous derivative of x, (8) and (7) both involve fairly long lags in X. If x paths have higher than first-order derivatives (for example, if they are generated by a third-order stochastic differential equation) the lag distributions in (8) and (7) are replaced by still higher-order limiting forms. Thus, different continuous time processes for x which all imply differentiable time paths produce different limiting discrete A. Here the fact that the time unit becomes small relative to the rate of variation in x does not justify the assumption that approximation of continuous by discrete models is innocuous. In particular, the notion that discrete differencing can approximate derivatives is potentially misleading.

It should not be surprising that the discrete time models may not do well in approximating a continuous time model in which derivatives appear. Nonetheless, empirical and theoretical work which ignores this point is surprisingly common.

If a is an ordinary function, there is still chance for error despite Geweke's result. His result implies only that the mean square deviation of a from A is small. This does not require that individual $A(t/\tau)'$s converge to the corresponding $a(t)$ values. For example, in a model where x is univariate and $a(t) = 0$, $t < 0$, $a(0) = 1$, $a(s)$ continuous on $[0, \infty]$, the limiting value for $A(0)$ is 0.5, not 1.0. Thus, if $a(t) = e^{-\theta t}$ on $[0, \infty)$, making a monotone decreasing over that range, $A(t)$ will not be monotone decreasing. It will instead rise between $t = 0$ and $t = 1$. This is not unreasonable on reflection: the discrete lag distribution gives a value at $t = 0$ which averages the continuous time distribution's behaviour on either side of $t = 0$. It should therefore not be surprising that monotonicity of a does not necessarily imply monotonicity of A, but the point is ignored in some economic research.

Another example of possible confusion arises from the fact that, if the x process has differentiable paths, $a(t) = 0$ for $t < 0$ does not imply $A(t) = 0$ for $t < 0$. The mean-square approximation result implies that when the time unit is small the sum of squares of coefficients on $X(t - s)$ for negative s must be small relative to the sum of squares on $X(t - s)$ for positive s, but the first few lead coefficients will generally be non-zero and will not go to zero as the time interval goes to zero. This would lead to mistaken conclusions about Granger causal priority in large samples, if significance tests were applied naively.

Geweke's exploration of multivariate models shows that the possibilities for confusing results are more numerous and subtle in that case. In particular, there are ways by which poor approximation of $\alpha_j(s)$ by $A_j(s/\tau)$ in some s interval (for example, around $s = 0$) can lead to contamination of the estimates of other elements of the A matrix, even though they correspond to x_j's and a_j's that in a univariate model would not raise difficulties.

In estimation of a dynamic prediction model for a single vector y, such as a vector autoregression (VAR) or dynamic stochastic general equilibrium model (DSGE), the question for approximation theory becomes whether the continuous time dynamics for y, summarized in a Wold moving average representation

$$y(t) = a^* u(t) \qquad (9)$$

has an intuitively transpareted connection to the corresponding discrete time Wold representation

$$Y(t) = A^* U(t). \qquad (10)$$

In discrete time the $U(t)$ of the Wold representation is the one-step-ahead prediction error, and in continuous time $u(t)$ also represents new information about y arriving at t. There are two related sub-questions. Is the A function the same shape as the a function; and is the U vector related in a natural way to the u vector? The u vector is a continuous time white noise, so that U cannot possibly be a simple discrete sampling of u.

If y is stationary and has an autoregressive representation, then $U(t) = A^{-1*}a^* u_t$, with the expression interpreted as convolution in continuous time, but with A^{-1} putting discrete weight on integers. The operator connecting U and u is then $A^{-1*}a$. There are cases where the connection between continuous and discrete time representations is intuitive. For example, if $a(s) = \exp(-Bs)$ (with the exponentiation interpreted as a matrix exponential in a multivariate case), then

$$U(t) = \int_0^1 e^{-Bs} u(t - s) ds \qquad (11)$$

and $A(s) = a(s)$ at integers. This is a more intuitive and precise matching than in any case we examined above for projection of one variable on another. If $a(0)$ is full rank and right-continuous at zero and if $a(s)$ is differentiable at all $s > 0$, then a similar intuitively simple matching of A to a arises when the time unit is small enough.

However, non-singularity of $a(0)$ rules out differentiability of time paths for y. When time paths for y, or some elements of it, are differentiable, no simple intuitive matching between A and a arises as the time unit shrinks.

There is one clear pattern in the difference in shape between A and a that stands in contrast to the case of distributed lag projection considered above. If both the continuous time and the discrete time moving average representations are fundamental, then by definition the one-step-ahead prediction error in $y(t)$ based on $y(t - s)$, $s \geq 1$ is

$$\int_0^1 a(s) u(t - s) ds, \qquad (12)$$

while the one-step-ahead prediction error in $Y(t)$ based on $Y(t - s)$, $s = 1, 2, \ldots$ is $A_0 Y(t)$. Now the information set we use in forecasting based on the past of Y at integer values alone is smaller than the information set based on all past values of y, so the one-step-ahead error based on the discrete data alone must be larger. If we normalize in the natural way to give U an identity covariance matrix and to make $\text{var}(g^*u(t)) = \int g(s)g'(s)$ (so u is a unit white noise vector), then it must emerge that

$$A_0 A_0 \geq \int_0^1 a(s)a(s)' ds, \tag{13}$$

where the inequality is interpreted as meaning that the left-hand-side matrix minus the right-hand-side matrix is positive semi-definite. In other words, the initial coefficient in the discrete MAR will always be as big or bigger than the average over $(0,1)$ of the coefficients in the continuous MAR. This tendency of the discrete MAR to seem to have a bigger instant response to innovations is proportionally larger the smoother a is near zero.

More detailed discussion of these points, together with numerous examples, appears in Marcet (1991).

Estimation and continuous time modelling

How can one proceed if one has a model like, say, (6), to which a discrete time model is clearly not a good approximation? The only possibility is to introduce explicitly a model for how x behaves between discrete time intervals, estimating this jointly with (6) from the available data. Doing so converts (6) from a single-equation to a multiple-equation model. That is, the device of treating x as 'given' and non-stochastic cannot work because an important part of the error term in the discrete model arises from the error in approximating a^*x by A^*X. Furthermore, because separating the approximation error component of U from the component due to u is essential, one would have to model serial correlation in u explicitly. The model could take the form

$$\begin{bmatrix} y(t) \\ x(t) \end{bmatrix} = \begin{bmatrix} c(s) & a^*b(s) \\ 0 & b(s) \end{bmatrix} * \begin{bmatrix} w(t) \\ v(t) \end{bmatrix}, \tag{14}$$

where w and v are white noise processes fundamental (in the terminology of Rosanov, 1967), for y and x. To give b and c a convenient parametric form, one might suppose them rational, so that (14) can be written as a differential equation, that is,

$$P(D) y(t) = P(D)a^*x(t) + w(t) \tag{15}$$

$$Q(D)x(t) = v(t), \tag{16}$$

where P and Q are finite-order polynomials in the derivative operator, $Q^{-1}(D)v = b^*v$, and $P^{-1}(D)w = c^*w$.

A discrete time model derived explicitly from a continuous time model is likely to be nonlinear at least in parameters and therefore to be more difficult to handle than a more naive discrete model. However with modern computing power, such models are usable. Bergstrom (1983) provides a discussion of estimating continuous time constant coefficient linear stochastic differential equation systems from discrete data, the papers in the book (1976) he edited provide related discussions, and Hansen and Sargent (1991), in some of their own chapters of that book, discuss estimation of continuous time rational expectations models from discrete data.

Estimating stochastic differential equation models from discrete data has recently become easier with the development of Bayesian Markov chain Monte Carlo (MCMC) methods. Though implementation details vary across models, the basic idea is to approximate the diffusion equation

$$dy_t = a(y_t)dt + b(y_t)dW_t, \tag{17}$$

where W_t is a Wiener process, by

$$y_t = e^{-a(y_{t-\delta})}y_{t-\delta} + b(y_{t-\delta})\varepsilon_t. \tag{18}$$

Such an approximation can be quite inaccurate unless δ is very small. But one can in fact choose δ very small, much smaller than the time interval at which data are observed. The values of y_t at times between observations are of course unknown, but if they are simply treated as unknown 'parameters it may be straightforward to sample from the joint posterior distribution of the y's at non-observation times and the unknown parameters of the model. The Gibbs sampling version of MCMC samples alternately from conditional posterior distributions of blocks of parameters. Here, sampling from the distribution of y at non-observation dates conditioning on the values of model parameters is likely to be easy. If the model has a tractable form, it will also be easy to sample from the posterior distribution of the parameters conditional on all the y values, both observed and unobserved. Application of these general ideas to a variety of financial models is discussed in Johannes and Polson (2006).

Another approach that has become feasible with increased computing power is to develop numerical approximations to the distribution of $y_{t+\delta}$ conditional on data through time t. Aït-Sahalia (2007) surveys methods based on this approach.

Modelling in continuous time does not avoid the complexities of connecting discrete time data to continuous time reality – it only allows us to confront them directly. One reason this is so seldom done despite its technical feasibility is that it forces us to confront the weakness of economic theory in continuous time. A model like (15)–(16) makes an assertion about how many times y and x are differentiable, and a mistake in that assertion can result in error as bad as the mistake of ignoring the time aggregation problem. Economic theory does not have much to say about the degree of differentiability of most aggregate macroeconomic time series.

When the theory underlying the model has no believable restrictions to place on fine-grained dynamics, it may be better to begin the modelling effort in discrete time. As is often true when models are in some respect under-identified, it is likely to be easier to begin from a normalized reduced form (in the case the discrete time model) in exploring the range of possible interpretations generated by different potential identifying assumptions.

Recent developments in financial economics have produced one area where there are continuous time economic theories with a solid foundation. Stochastic differential equations (SDEs) provide a convenient and practically useful framework for modelling asset prices. These SDE models imply non-differentiable time paths for prices, and it is known (Harrison, Pitbladdo and Schaefer, 1984) that differentiable time paths for asset prices would imply arbitrage opportunities, if there were no transactions costs or bounds on the frequency of transactions.

However, there are in fact transactions costs and bounds on transactions frequencies, and no-arbitrage models for asset prices break down at very fine, minute-by-minute, time scales. Successful behavioural modelling of these fine time scales requires a good theory of micro-market structure, which is still work in progress.

It is worthwhile noting that a process can have non-differentiable paths without producing white noise residuals at any integer order of differentiation: for example, a model satisfying (3) with $a(s) = s^{0.5}e^{-s}$. Such a process has continuous paths with unbounded variation and is not a semimartingale. That is, it is not the sum of a martingale and a process with bounded variation, and therefore cannot be generated from an integer-order SDE. Similarly, if $a(s) = s^{-0.5}e^{-s}$, the process has non-differentiable paths but is nonetheless not a semimartingale. The existence of such non-semimartingale processes and their possible applications to financial modelling is discussed in Sims and Maheswaran (1993).

CHRISTOPHER A. SIMS

See also **time series analysis.**

Bibliography

Aït-Sahalia, Y. 2007. Estimating continuous-time models using discretely sampled data. In *Advances in Economics and Econometrics, Theory and Applications. Ninth World Congress*, vol. 3, ed. R. Blundell, T. Persson and W.K. Newey. Cambridge: Cambridge University Press.

Bergstrom, A.R., ed. 1976. *Statistical Inference in Continuous Time Economic Models*. Amsterdam: North-Holland.

Bergstrom, A.R. 1983. Gaussian estimation of structural parameters in higher order continuous time dynamic models. *Econometrica* 51, 117–52.

Geweke, J. 1978. Temporal aggregation in the multiple regression model. *Econometrica* 46, 643–62.

Hansen, L.P. and Sargent, T.J., eds. 1991. *Rational Expectations Econometrics*. Boulder and Oxford: Westview Press.

Harrison, J.M., Pitbladdo, R. and Schaefer, S.M. 1984. Continuous price processes in frictionless markets have infinite variation. *Journal of Business* 57, 353–65.

Johannes, M. and Polson, N. 2006. MCMC methods for continuous-time financial econometrics. In *Handbook of Financial Econometrics*, ed. Y. Aït-Sahalia and L.P. Hansen. Amsterdam: North-Holland.

Marcet, A. 1991. Temporal aggregation of economic time series. In *Rational Expectations Econometrics*, ed. L.P. Hansen and T.J. Sargent. Boulder and Oxford: Westview Press.

Rozanov, Yu.A. 1967. *Stationary Random Processes*, trans A. Feinstein. San Francisco, Cambridge, London, Amsterdam: Holden-Day.

Sims, C.A. 1971. Approximate specifications in distributed lag models. In Proceedings of the 38th Session, *Bulletin of the International Statistical Institute* 44, Book 1.

Sims, C.A. and Maheswaran, S. 1993. Empirical implications of arbitrage-free asset markets. In *Models, Methods and Applications of Econometrics*, ed. P.C.B. Phillips. Oxford: Blackwell.

contract theory

As with so many major concepts in economics, contract theory was introduced by Adam Smith who, in his monumental *Wealth of Nations* (1776, book III, ch. 2), considered the relationship between peasants and farmers through this lens. For instance, he pointed out the perverse incentives provided by sharecropping contracts, widespread in 18th-century Europe. However, it is fair to say that the issues of incentives and contract theory were largely ignored by economists until the end of the 20th century. By then, the focus of economic theory was on the working of markets and price formation. Firms were viewed only as production technologies, and the issue of the separation between ownership and control was most often put aside. This black-box approach was, of course, quite unsatisfactory. At the turn of the 1970s, with the methodological revolution of game theory, more emphasis was placed on strategic interactions between a small number of players in a world where informational problems matter. From this new perspective, the allocation of resources is no longer ruled by the price system but by *contracts* between asymmetrically informed partners. Contract theory has deeply changed our view of the functioning of organizations and markets.

This article aims to provide a brief overview of contract theory, stressing a few major insights and illustrating them with useful applications. Due to space constraints, it does not do justice to several aspects of contract theory, and will mostly reflect my own tastes in the field. In particular, I focus on the so-called *theory of* complete contracts, leaving aside the burgeoning theory of incomplete contracts which is covered elsewhere in this dictionary. Successive sections deal respectively, with adverse selection, moral hazard and non-verifiability: the three different paradigms which have been used in the field of complete contract theory. Since the distinction between complete and incomplete contracts is easier to draw once these notions have already been explained, I will postpone such discussion to the end of the article.

Adverse selection

Consider the following buyer–seller relationship as the archetypical example of contractual relationship between a principal (the buyer) and his agent (the seller) who produces some good or service on his behalf. The mere delegation of this task to the agent gives the agent access to private information about the technology. This *adverse selection* environment is captured by assuming that a technological parameter θ is known only by the agent. It is drawn from a distribution in an exogenous type space Θ which is common knowledge. Neither the principal nor a court of law observes this parameter. Contracts cannot specify outputs and prices as a function of the realized state of nature.

The buyer enjoys a net benefit $S(\theta, q) - t$ when buying q units of output at a price t. The seller enjoys a profit $t - C(\theta, q)$ from producing that good. We will assume that these functions are concave in q. Notice that the state of nature θ might affect both the agent's and the principal's utility functions. This can, for instance, be the case if this parameter also determines the quality of the good to be traded.

Under complete information, efficiency requires that the buyer and the seller trade the first-best quantity $q^*(\theta)$ such that the buyer's marginal benefit from consumption equals the seller's marginal cost of production:

$$\frac{\partial S}{\partial q}(\theta, q^*(\theta)) = \frac{\partial C}{\partial q}(\theta, q^*(\theta)). \qquad (1)$$

Many mechanisms or institutions lead to this outcome. Both the price mechanism and a take-it-or-leave-it offer by one party to the other would achieve the same allocation, although with different distributions of the surplus between the traders. If the principal retains all bargaining power (for instance, because there is a competitive fringe of potential sellers), he could offer a forcing contract stipulating an output $q^*(\theta)$ and a transfer $t^*(\theta)$ which just covers the seller's cost. This forcing contract maximizes the buyer's net gains from trade and leaves the seller just indifferent between participating or not.

In what follows, we mostly focus on the case where the uninformed principal has full bargaining power in contracting. In this framework, the *contract* between the buyer and the seller does not only have the allocative and

distributive roles it has under complete information. It also has the role of *communicating information* from the informed party to the uninformed party. This communication role suggests that the informed party should be given a choice among different options and that this choice should reveal information about the adverse selection parameter.

A first step in the analysis consists of describing the set of allocations which are feasible under asymmetric information. The basic tool for doing so is the revelation principle (see Gibbard, 1973; Green and Laffont, 1977; Dasgupta, Hammond and Maskin, 1979; and Myerson, 1979, among others), which states that there is no loss of generality in restricting the analysis to *revelation mechanisms* that are *direct*, that is, of the form $\{t(\hat{\theta}), q(\hat{\theta})\}_{\hat{\theta} \in \Theta}$ with $\hat{\theta}$ a message ('report') sent by the informed seller to the uninformed buyer, and *truthful*, that is, such that the agent finds it optimal to report his true type.

Therefore, incentive feasible contracts satisfy the following *incentive* constraints

$$t(\theta) - C(\theta, q(\theta)) \geq t(\hat{\theta}) - C(\theta, q(\hat{\theta}))$$
$$\forall (\theta, \hat{\theta}) \in \Theta^2. \qquad (2)$$

To be acceptable, a contract must also satisfy the seller's participation constraints

$$t(\theta) - C(\theta, q(\theta)) \geq 0 \qquad \forall \theta \in \Theta \qquad (3)$$

which ensure that, irrespective of his type, the agent by contracting gets at least his reservation payoff (exogenously normalized to zero).

Once the set of incentive feasible allocations is described, the analysis may proceed further. Keeping in mind that the uninformed buyer designs his offer under asymmetric information, we might characterize an *optimal contract*. Such a contract maximizes the uninformed buyer's expected net surplus subject to the feasibility constraints (2) and (3).

Much of the theoretical literature developed over the 1980s and early 1990s has investigated the structure of the set of incentive feasible allocations and its consequences for optimal contracting. A key property is the so-called Spence–Mirrlees condition (see Spence, 1973; 1974; and Mirrlees, 1971) for early contributions which put forward that condition). This condition is satisfied when the slope of the agent's indifference curves can be ranked with respect to his type. In our example, this condition holds when $\frac{\partial^2 C}{\partial \theta \partial q} > 0$, that is, when higher types also have higher marginal costs and should thus produce less. Therefore, the monotonicity condition

$$q(\theta) \geq q(\theta') \quad \text{for } \theta \leq \theta' \qquad (4)$$

is a direct consequence of the incentive constraints. The Spence–Mirrlees condition can be viewed as a regularity assumption making the incentive problem well-behaved.

It ensures that only incentive constraints between 'nearby' types matter in the optimization. Intuitively, this means that the seller with a given marginal cost may be tempted to overstate slightly its costs, receiving the higher transfer targeted to less efficient types but producing at a lower marginal cost. By so doing, this more efficient type receives an information rent. Once these local constraints are taken into account and when the Spence–Mirrlees condition holds, the incentives to mimic more distant types are no longer relevant. With this reduction of the set of relevant incentive constraints, the principal's optimization problem is significantly simplified.

The result of this optimization is straightforward. Inducing information revelation by the most efficient types requires giving up an *information rent* to those types. The basic intuition of most adverse-selection models is that reducing this rent requires production to be distorted. For instance, when efficient types want to mimic less efficient ones, the latter's allocation should be made less attractive. This is obtained by distorting their production downward and modifying transfers accordingly.

To see more formally the nature of the output distortion, consider the case where types are distributed over a compact set $[\underline{\theta}, \bar{\theta}]$ according to the cumulative distribution function $F(\cdot)$ (with a positive density $f(\cdot)$). The second-best optimal output $q^{SB}(\theta)$ under adverse selection is the solution to:

$$\frac{\partial S}{\partial q}(\theta, q^{SB}(\theta)) = \frac{\partial C}{\partial q}(\theta, q^{SB}(\theta))$$
$$+ \frac{F(\theta)}{f(\theta)} \frac{\partial^2 C}{\partial q \partial \theta}(\theta, q^{SB}(\theta)).$$
$$(5)$$

Condition (5) states that, for any type θ, the buyer's marginal benefit must equal the seller's *marginal virtual cost* (see Laffont and Martimort, 2002, chs 2 and 3, for details). The virtual cost of a given type takes into account not only its cost of production but also the cost of deterring other types (here more efficient types) from mimicking that type. The allocation is no longer efficient, as under complete information, but *interim efficient* in the sense of Holmström and Myerson (1983).

Condition (5) is crucial, and is found in various forms in any adverse-selection model. It states that, under asymmetric information, there is a fundamental trade-off between implementing allocations close to efficiency and giving information rents to the most efficient types to induce information revelation. This trade-off calls for distortions away from efficiency.

Provided that the output schedule defined by (5) satisfies the monotonicity condition (4), this is the exact solution of our problem. To guarantee monotonicity, on top of assumptions on the concavity of $S(\cdot)$ and $\frac{\partial S}{\partial \theta}(\cdot)$,

convexity of $C(\cdot)$ and $\frac{\partial C}{\partial \theta}(\cdot)$, $\frac{\partial^2 C}{\partial \theta \partial q}(\cdot) > 0$, $\frac{\partial^3 C}{\partial \theta^2 \partial q}(\cdot) > 0$ and $\frac{\partial^2 S}{\partial \theta \partial q}(\cdot) < 0$, one needs also to impose a property on the type distribution, the so-called *monotonicity of the hazard rate* $\frac{F(\theta)}{f(\theta)}$ (see Bagnoli and Bergstrom, 2005). Otherwise, the optimal contract may entail some area of pooling such that all types belonging to a set with positive measure produce the same amount and are paid the same price. The optimal solution may then be obtained using 'ironing techniques' (see for instance Guesnerie and Laffont, 1984).

Direct extensions

Adverse-selection methodology has been successfully extended in various directions allowing for multidimensional types (Armstrong and Rochet, 1999), and/or multiple outputs (Laffont and Tirole, 1993, ch. 3), and type-dependent reservation utilities (Lewis and Sappington, 1989; Jullien, 2000). There, the analysis is substantially more complex as types can no longer be ranked as easily as in the model sketched above. The Spence–Mirrlees condition might fail to hold and global incentive constraints may bind, leading to pooling allocations being optimal. Another interesting extension is the case of hidden knowledge, in which contracting takes place before the agent becomes informed. The logic of such models is very close to that we discuss below in the section on moral hazard. In a nutshell, the trade-off between allocative efficiency and rent extraction is now replaced by the trade-off between insuring the agent against shocks on costs and inducing him to reveal his cost once it is known. Output distortions still arise (see Laffont and Martimort, 2002, ch. 2, for details). Others have endogenized the asymmetric information structure and examined the incentives to learn about the unknown parameter (see, for instance, Crémer, Khalil and Rochet, 1998). Finally, there exists a literature that considers the case where the principal is the informed party (Maskin and Tirole, 1990; 1992). New difficulties arise from the fact that the mere offer of the contract may signal information.

Multi-agent organizations

The most important extensions of the adverse selection paradigm certainly concern multi-agent organizations. Such complex organizations emerge because of the need to share common resources, produce public goods, internalize production externalities or enjoy information economies of scale. Although any such reason calls for a specific analysis, a few common themes of the literature can be highlighted by remaining at a rather general level.

Regarding the implementation concept, different notions of incentive compatibility may be used depending on the context. First, agents may know each other's types and play a Nash equilibrium of the direct revelation mechanism offered by the principal (see Maskin, 1999, and the discussion of the non-verifiability paradigm below). Second, agents may only know their own type, form beliefs on each others' types and play a Bayesian–Nash equilibrium (see D'Aspremont and Gérard-Varet, 1979). Third, one may also insist on dominant strategy implementation because it does not depend on the specification of beliefs (see Gibbard, 1973; Groves, 1973; Green and Laffont, 1977). To each implementation concept corresponds a notion of incentive feasibility. Once the set of incentive feasible contracts is defined, one can proceed to optimization. It is a trivial observation that, the more restrictive the implementation concept, the lower is the principal's payoff at the optimum.

In some cases, such as the provision of public goods within a society of privately informed agents or in bargaining models between a buyer and a seller with equal bargaining power, the goal is no longer to design a multilateral contract which would extract the rents of all agents but, instead, to maximize some *ex ante* efficiency criterion under incentive constraints. Groves (1973) showed that dominant strategy mechanisms suffice to implement the first-best decision in a public good context. One caveat is that the budget generally fails to be balanced. D'Aspremont and Gérard-Varet (1979) proposed a Bayesian incentive-compatible mechanism which implements the first-best and still satisfies budget balance. As argued by Laffont and Maskin (1979), such a mechanism may conflict with the agents' participation constraint. In a bargaining environment, Myerson and Satterthwaite (1983) showed in a similar vein that there exists no Bayesian bargaining mechanism that is efficient, budget-balance and individually rational.

The optimal multilateral contract can be very sensitive to the information structure. In environments where risk-neutral agents have correlated types but know only their own type, the principal can condition one agent's compensation on another's report. By doing so, the principal can fully extract the rent from both agents in a Bayesian-Nash equilibrium. One may view this result as a strong rationale for relative performance evaluation, yardstick competition, benchmarking and internalization of similar activities within the same organization. This puzzling insight of Crémer and McLean (1988) no longer holds when one introduces risk-aversion, *ex post* participation constraints or limited liability constraints. These assumptions reintroduce information rents in the multi-agent organization, and the standard trade-off between efficiency and rent extraction reappears.

When the agents' types are independently distributed, yardstick competition is ineffective and the agents derive information rents. However, the externality that one agent's task may exert on another can shape the distribution of these rents. In competitive environments, such as procurement auctions among sellers, it is no longer the distribution of the agents' marginal costs but the distribution of their virtual marginal costs (see Myerson, 1981) which determines who should produce and how much. Because virtual costs may be ranked differently from true costs, inefficiencies arise under asymmetric information. Moreover, competition may help reduce rents by putting each agent under the threat of being excluded from production if he overstates his cost too much. There is then a positive externality among competing agents.

Instead, more cooperative environments, such as public good problems or procurement of complementary inputs by several suppliers, involve negative externalities between agents. Given that each agent has a limited impact on the organization's overall production, the incentives to overstate costs and thereby receive greater transfers are exacerbated. 'Free riding' arises in such organizations (see Mailath and Postlewaite, 1990).

When competition between agents or between agents and the supervisors supposed to monitor them would benefit the principal, one must consider the possibility of collusion aimed at securing more rent. Reducing the scope for collusion requires using mechanisms that are less sensitive to information and reducing supervisory discretion. Incentive contracts look more like inflexible bureaucratic rules (see Tirole, 1986; Laffont and Martimort, 2000). The optimal response to collusion may also entail more delegation to lower levels of the hierarchy, as in Laffont and Martimort (1998) and Faure-Grimaud, Laffont and Martimort (2003).

Dynamics

Different extensions of the static framework correspond to different abilities of the contractual partners to commit themselves inter-temporally and/or different ways for the cost parameters to vary over time. Under full commitment, the lessons of the static rent–efficiency trade-off can be easily extended, although the precise features of the optimal contract depend on how types evolve over time (see, for instance, Baron and Besanko, 1984, for the case of persistent types). The case of limited commitment is more interesting. Long-term contracts may either be renegotiated (Dewatripont, 1989; Hart and Tirole, 1988; Laffont and Tirole, 1990) or even are not feasible, in which cases the parties resort to spot contracts (Laffont and Tirole, 1988). The rent–efficiency trade-off must be adapted to take into account how information is revealed progressively over time. However, the basic idea still holds. As past performances reveal information about the agent's type, the optimal contract trades off *ex post* efficiency gains in contracting against the agent's desire to hide information in the earlier periods of the relationship so as to secure more rent in the later periods.

Applications

Since the mid-1980s, models of optimal contracting under adverse selection have spanned the economic literature. Let us quote only a few major applications. Mirrlees (1971) analysed optimal taxation schemes when the agent's productivity is privately observed. He introduced the Spence–Mirrlees condition and derived the implementability conditions. He also used optimal control techniques (Pontryagyn Principle) to compute the optimal taxation scheme. (The taxation problem differs from our buyer–seller example because participation in the mechanism is mandatory and the state's budget constraint must be added to the characterization of feasible allocations.)

Mussa and Rosen (1978) studied the problem of a monopolist selling one unit of a good to a continuum of consumers vertically differentiated with respect to their willingness to pay for the quality of this good. This was the first model using adverse selection techniques in a framework without income effect. Maskin and Riley (1984) were interested in characterizing the optimal nonlinear price used by a monopolist in a second-degree price discrimination context.

Baron and Myerson (1982) applied the methodology to the regulation of natural monopolies privately informed about their marginal costs of production. Laffont and Tirole (1986) extended this analysis to allow for cost observability but also introduced moral hazard elements (the possibility for the regulated firm to reduce its costs by undertaking some non-observable effort). They derived cost-reimbursement rules and pricing policies. They showed that menus of linear contracts might implement the optimal contract.

Green and Kahn (1983) and Hart (1983) studied labour market contracts and discussed distortions towards overemployment or underemployment that may arise depending on the contractual environment considered.

Finally, Townsend (1979) and Gale and Hellwig (1985) analysed optimal financial contracts in a framework where the borrower's income is observable only *ex post* and at a cost. Optimal contracts may look like debt in such environments.

Moral hazard

To return to our buyer–seller example, we now assume that there is only one unit of a good to be traded whose quality q is random and which yields a surplus $S(q)$ to the buyer. The distribution of quality is affected by an effort e undertaken by the agent at a cost $\psi(e)$ (where $\psi' > 0$ and $\psi'' > 0$). The cumulative distribution is $F(q|e)$ (with density $f(q|e)$) on a support $Q = [\underline{q}, \bar{q}]$ independent of the agent's effort. To simplify, the agent's preferences are separable in money and effort: $U = u(t) - \psi(e)$ where $u(\cdot)$ is increasing and concave ($u' > 0, u'' \leq 0$). The agent's outside option is not to produce, which gives him a payoff normalized to zero.

The agent's effort is observable neither by the principal nor by a court of law. This is a *moral hazard* setting. Contracts stipulate the agent's payment as a function of the realized quality assumed to be observable and verifiable (contractible) by a court of law. Therefore, contracts are of the form $\{t(\tilde{q})\}_{\tilde{q} \in Q}$.

If the effort were observable, its value could also be specified by contract. Therefore, the seller can at the same time be forced to exert the first-best level of effort and be fully insured against uncertainty on realized quality with a flat payment independent of his performance:

$$u(t^*) = \psi(e^*).$$

This is no longer the case when the agent's effort is non-verifiable. The first step of the analysis is to describe the set of feasible incentive contracts implementing a given level of effort e.

In a moral hazard setting, incentive constraints write as:

$$\int_{\underline{q}}^{\bar{q}} u(t(q))f(q|e)dq - \psi(e)$$

$$\geq \int_{\underline{q}}^{\bar{q}} u(t(q))f(q|e')dq - \psi(e')$$

$$\forall(e, e'). \tag{6}$$

The agent's participation constraint is:

$$\int_{\underline{q}}^{\bar{q}} u(t(q))f(q|e)dq - \psi(e) \geq 0. \tag{7}$$

Risk neutrality

A first case of interest is when the agent is risk-neutral ($u(t) \equiv t$). The simple 'sell-out' contract, $t(q) = S(q) - C$ where C is a constant, implements the first-best level of effort e^*. Provided that $C = \int_{\underline{q}}^{\bar{q}} S(q)f(q|e^*) - \psi(e^*)$, this scheme also extracts all the surplus from the agent who is just indifferent between producing or not.

Intuitively, with such a 'sell-out' contract, the agent's private incentives to exert effort are aligned with the social incentives. This efficient outcome is obtained by, first, having the agent pay a bond worth C for the right to serve the principal, and second, having the principal pay an amount $S(q)$ contingent on the quality realized.

Such a 'sell-out' contract requires that the agent bears the full consequences of a bad performance. It might not be feasible when the agent has limited liability and cannot be punished for bad performances. (For details, see Laffont and Martimort, 2002, ch. 4). The conjunction of moral hazard and limited liability allows the agent to derive a limited liability rent. Intuitively, only rewards, not punishments, can be used to provide incentives, and this restriction on instruments is costly for the principal. This rent creates a trade-off between efficiency and rent extraction, as in the adverse selection framework. Effort is distorted below the first-best level.

Risk aversion

Let us turn to the more complex case of risk aversion. A first concern of the literature has been to 'simplify' the set of incentive constraints (2) by replacing it with a first-order condition:

$$\int_{\underline{q}}^{\bar{q}} u(t(q))f_e(q|e)dq = \psi'(e). \qquad (8)$$

Denoting by λ (resp. μ) the positive multiplier of the incentive (resp. participation) constraint (8) (resp. (7)), the optimal second-best schedule $t^{SB}(q)$ satisfies

$$\frac{1}{u'(t^{SB}(q))} = \mu + \lambda \frac{f_e(q|e)}{f(q|e)}. \qquad (9)$$

This condition yields two important insights. First, the contract must simultaneously provide the risk-averse agent with insurance, which requires a fixed payment, and with incentives to exert effort, which requires that payments be linked to performance. There is now a trade-off between *insurance* and *incentives*.

Second, the monotonicity of the agent's compensation with respect to the quality level (a priori quite intuitive property) is obtained only when the *monotone likelihood ratio property* holds, namely, when $\frac{\partial}{\partial q}\left(\frac{f_e(q|e)}{f(q|e)}\right) > 0$. This property means that higher levels of performance are more informative about the agent's effort.

Finally, the optimal contract must use all signals which are informative about the agent's effort but no uninformative signals. Using them would only let the agent bear more risk without any beneficial impact on incentives. This is the so-called informativeness principle of Holmström (1979).

Extensions

In a model with a finite number of quality and effort levels, Grossman and Hart (1983) offered a careful study of the set of incentive constraints and its consequences for the shape of optimal contracts. There is no general result on the ranking between the first-best and the second-best effort levels in such environments. The discrete version of the first-order approach requires that only nearby constraints matter in the agent's problem. This concavity of the agent's problem is ensured when $F(q|e)$ is itself convex in q. In models with a continuum of effort levels and outcomes, this first-order approach was suggested in Mirrlees (1999), more rigorously justified in Rogerson (1985) and Jewitt (1988) and applied in Holmström (1979) and Shavell (1979).

The moral hazard methodology has been used to justify the optimality of linear incentive schemes in well-structured environments (Holmström and Milgrom, 1987); an often found feature of real world contracts. Equipped with this tool, Holmström and Milgrom (1991; 1994) investigated how multiple tasks and jobs should be arranged in an organization.

To avoid the complexity of models with a continuum of effort levels, modellers have found it useful to focus on simplified environments with two levels of effort. This approach was instrumental in the work on corporate finance of Holmström and Tirole (1997).

Multi-agent organizations

When applied to multi-agent organizations, the 'informativeness principle' suggests that an agent's compensation should be linked to another's performance if it is informative about his own effort (see Mookherjee, 1984). Relative performance evaluation and benchmarking can help eliminate common shocks affecting all agents' performances. Of particular importance in this respect are tournaments which use only the ranking of the agents' performances to determinate their compensations. Tournaments provide agents with insurance against common shocks, which has a positive incentive effect. More generally, the properties of tournaments and how they compare with (a priori suboptimal) linear schemes have been investigated in Nalebuff and Stiglitz (1983) and Green and Stokey (1983).

In more cooperative environments where different agents contribute to a joint project, the fundamental difficulty is how to share the proceeds of production among agents of the team and still provide some incentives. Since each agent enjoys only a fraction of those proceeds but bears the full cost of his effort, he reduces his effort supply. This leads to a free-rider problem within teams, which is analysed in Holmström (1982).

If we remain in cooperative environments but allow now for a principal acting as a budget breaker, this principal may find it worthwhile to reduce the agency cost of implementing a given effort profile by having agents behave cooperatively (Itoh, 1993). Even when agents do not cooperate, mutual observability of effort levels can also help to eliminate agency cost, as in Ma (1988). This last argument relies on the logic of non-verifiability models, developed below.

Dynamics

The basic issue investigated by dynamic models of moral hazard is the extent to which repeated relationships alleviate the moral hazard problem. The intuition is that the principal should filter out the agent's effort by looking at the whole history of his performances. This may eliminate any agency problem, at least when parties do not discount too much the future (see Laffont and Martimort, 2002, ch. 8, for an example). More generally, the insurance–incentives trade-off may be relaxed when the risk-averse agent's rewards and punishments can be smoothed over the whole relationship, as shown in Spear and Srivastava (1987). A direct consequence of intertemporal smoothing is that the optimal dynamic contract exhibits *memory*; good (resp. bad) performance today will also affect positively (resp. negatively) future compensations. This insight has been used to formalize a theory of the wage dynamics inside the firm (Harris and Holmström, 1982).

Fama (1980) argued that reputation in the labour market exerts enough discipline on managers to alleviate moral hazard even in the absence of explicit contracts. Holmström (1999) built a model of career concerns where the manager's interest in influencing the labour market's beliefs concerning his or her quality provides incentives to exert effort. Career concerns are nevertheless in general not enough to induce first-best effort levels, and some inefficiencies remain.

Non-verifiability

Let us return to the buyer–seller model above. Although we now assume that it is observable by both the principal and the agent, the state of nature θ may still not be verifiable by a court of law, in which case it cannot be part of the contract. This shared knowledge stands in sharp contrast with the asymmetric information structures examined in previous sections.

The first difficulty consists of building a mechanism based only on verifiable variables (namely, the quantities traded and corresponding payments) which implements the first-best quantity $q^*(\theta)$ and transfers $t^*(\theta)$. This problem was addressed by Maskin (1999). He demonstrated that the first-best quantities and transfers can easily be implemented with a direct revelation mechanism $\{t(\hat{\theta}_a, \hat{\theta}_b), \; q(\hat{\theta}_a, \hat{\theta}_b)\}_{(\hat{\theta}_a, \hat{\theta}_b) \in \Theta^2}$ where both the buyer and the seller report simultaneously the state of nature they commonly know. Truth-telling is obviously a Nash equilibrium of this mechanism provided that both traders are severely punished when making different reports, since such cases would be inconsistent with the underlying information structure.

A more subtle issue is how to design a mechanism such that this truthful Nash equilibrium is unique. Maskin (1999) proposed a condition for players' preferences such that this is the case. Moore and Repullo (1988) significantly extended the domain of preferences by hardening the implementation concept, replacing Nash behaviour by subgame-perfection in a sequential moves mechanism (see Laffont and Martimort, 2002, ch. 6, for an example, and Moore, 1992, for an exhaustive survey of the literature).

The basic thrust of the non-verifiability paradigm is that a court of law can get around non-verifiability by building such revelation mechanisms, at least as long as the non-verifiable state is payoff-relevant. If one sticks to that interpretation, non-verifiability does not present a significant limit on contracting.

A second issue of the literature is the impact of non-verifiability on the incentives of traders to perform specific and non-verifiable investments. Given our previous claim that non-verifiability is generally not a constraint, the model resembles the standard moral hazard model. Providing incentives for investments meets the same difficulties as in the previous section.

Extensions

In practice, revelation mechanisms have been criticized as overly complex, as relying on threats which may either be non-credible or violate limited liability constraints. The so-called incomplete contracts literature has thus focused on cases where such revelation mechanisms are not feasible. In such environments, either no contract at all or only a very rough one can be written *ex ante*. For instance, parties can agree *ex ante* on a simple fixed-price/fixed quantity contract which serves as a threat point for the bargaining which takes place *ex post* when the state of nature is realized (see Edlin and Reichelstein, 1996, among others).

Alternatively, this threat point may be determined by the allocation of ownership rights where such a right gives the owner the opportunity to use assets as he prefers in case bargaining fails (see Grossman and Hart, 1986; Hart and Moore, 1988). The issue is then to derive from those exogenous constraints distortions of investments and optimal organizations which may mitigate those distortions.

The incomplete contracts paradigm is similar to the complete contracts one (adverse selection, moral hazard and non-verifiability) in the sense that it also imposes limits on what a court may verify. It differs from it because it also imposes exogenous restrictions on the set of mechanisms available to the parties. The justification for these restrictions is found either in the bounded rationality of players or the difficulties in describing or foreseeing contingencies, all theoretical issues which remain high on the agenda of economic theorists and are still unsettled. The relevant literature on incomplete contracts is too large to be summarized in this short article. The interested reader may refer to Tirole (1999) for an overview or to the entry for this term in this Dictionary.

DAVID MARTIMORT

See also **adverse selection; agency problems; incomplete contracts; mechanism design; mechanism design (new developments); moral hazard.**

I thank D. Gromb and J. Pouyet for helpful comments on an earlier version.

Bibliography

Armstrong, M. and Rochet, J. 1999. Multidimensional screening: a user's guide. *European Economic Review* 43, 959–79.

Bagnoli, M. and Bergstrom, T. 2005. Log-concave probability and its applications. *Economic Theory* 26, 445–69.

Baron, D. and Besanko, D. 1984. Regulation and information in a continuing relationship. *Information Economics and Policy* 1, 447–70.

Baron, D. and Myerson, R. 1982. Regulating a monopolist with unknown costs. *Econometrica* 50, 911–30.

Crémer, J., Khalil, F. and Rochet, J. 1998. Strategic information gathering before a contract is offered. *Journal of Economic Theory* 81, 163–200.

Crémer, J. and McLean, R. 1988. Full extraction of surplus in Bayesian and dominant strategy auctions. *Econometrica* 56, 1247–57.

D'Aspremont, C. and Gérard-Varet, L. 1979. Incentives and incomplete information. *Journal of Public Economics* 11, 25–45.

Dasgupta, P., Hammond, P. and Maskin, E. 1979. The implementation of social choice rules. *Review of Economic Studies* 46, 185–216.

Dewatripont, M. 1989. Renegotiation and revelation information over time: the case of optimal labour contracts. *Quarterly Journal of Economics* 104, 489–520.

Edlin, A. and Reichelstein, S. 1996. Hold-ups, standard breach remedies, and optimal investments. *American Economic Review* 86, 478–501.

Fama, E. 1980. Agency problem and the theory of the firm. *Journal of Political Economy* 88, 288–307.

Faure-Grimaud, A., Laffont, J.-J. and Martimort, D. 2003. Collusion, delegation and supervision with soft information. *Review of Economic Studies* 70, 253–80.

Gale, D. and Hellwig, M. 1985. Incentive-compatible debt contracts: the one-period problem. *Review of Economic Studies* 52, 647–63.

Gibbard, A. 1973. Manipulations of voting schemes: a generalized result. *Econometrica* 41, 587–601.

Green, J. and Kahn, C. 1983. Wage-employment contracts. *Quarterly Journal of Economics* 98, 173–88.

Green, J. and Laffont, J.-J. 1977. Characterization of satisfactory mechanisms for the revelation of preferences for public goods. *Econometrica* 45, 427–38.

Green, J. and Stokey, N. 1983. A comparison of tournaments and contracts. *Journal of Political Economy* 91, 349–64.

Grossman, S. and Hart, O. 1983. An analysis of the principal–agent problem. *Econometrica* 51, 7–45.

Grossman, S. and Hart, O. 1986. The costs and benefits of ownership: a theory of lateral and vertical integration. *Journal of Political Economy* 94, 691–719.

Groves, T. 1973. Incentives in teams. *Econometrica* 41, 617–31.

Guesnerie, R. and Laffont, J.-J. 1984. A complete solution to a class of principal–agent problems with an application to the control of a self–managed firm. *Journal of Public Economics* 25, 329–69.

Harris, M. and Holmström, B. 1982. A theory of wage dynamics. *Review of Economic Studies* 49, 315–33.

Hart, O. 1983. Optimal labour contracts under asymmetric information: an introduction. *Review of Economic Studies* 50, 3–35.

Hart, O. and Moore, J. 1988. Property rights and the nature of the firm. *Journal of Political Economy* 98, 1119–58.

Hart, O. and Tirole, J. 1988. Contract renegotiation and Coasian dynamics. *Review of Economic Studies* 55, 509–40.

Holmström, B. 1979. Moral hazard and observability. *Bell Journal of Economics* 10, 74–91.

Holmström, B. 1982. Moral hazard in teams. *Bell Journal of Economics* 13, 324–40.

Holmström, B. 1999. Managerial incentive problems: a dynamic perspective. *Review of Economic Studies* 66, 169–82.

Holmström, B. and Milgrom, P. 1987. Aggregation and linearity in the provision of intertemporal incentives. *Econometrica* 55, 303–28.

Holmström, B. and Milgrom, P. 1991. Multi-task principal agent analysis. *Journal of Law, Economics, and Organization* 7, 24–52.

Holmström, B. and Milgrom, P. 1994. The firm as an incentive system. *American Economic Review* 84, 972–91.

Holmström, B. and Myerson, R. 1983. Efficient and durable decision rules with incomplete information. *Econometrica* 51, 1799–819.

Holmström, B. and Tirole, J. 1997. Financial intermediation, loanable funds, and the real sector. *Quarterly Journal of Economics* 112, 663–91.

Itoh, H. 1993. Coalition incentives and risk-sharing. *Journal of Economic Theory* 60, 416–27.

Jewitt, I. 1988. Justifying the first-order approach to principal–agent problems. *Econometrica* 56, 1177–90.

Jullien, B. 2000. Participation constraints in adverse-selection models. *Journal of Economic Theory* 93, 1–47.

Laffont, J.-J. and Martimort, D. 1998. Collusion and delegation. *Rand Journal of Economics* 29, 280–305.

Laffont, J.-J. and Martimort, D. 2000. Mechanism design with collusion and correlation. *Econometrica* 68, 309–42.

Laffont, J.-J. and Martimort, D. 2002. *The Theory of Incentives: The Principal–Agent Model*. Princeton: Princeton University Press.

Laffont, J.-J. and Maskin, E. 1979. A differentiable approach to expected utility maximizing mechanisms. In *Aggregation and Revelation of Preferences*, ed. J.-J. Laffont. Amsterdam: North-Holland.

Laffont, J.-J. and Tirole, J. 1986. Using cost observation to regulate firms. *Journal of Political Economy* 94, 614–41.

Laffont, J.-J. and Tirole, J. 1988. The dynamics of incentive contracts. *Econometrica* 56, 1153–75.

Laffont, J.-J. and Tirole, J. 1990. Adverse selection and renegotiation in procurement. *Review of Economic Studies* 57, 597–626.

Laffont, J.-J. and Tirole, J. 1993. *A Theory of Incentives in Procurement and Regulation*. Cambridge, MA: MIT Press.

Lewis, T. and Sappington, D. 1989. Countervailing incentives in agency problems. *Journal of Economic Theory* 49, 294–313.

Ma, C. 1988. Unique implementation of incentive contracts with many agents. *Review of Economic Studies* 55, 555–72.

Mailath, G. and Postlewaite, A. 1990. Asymmetric information bargaining problems with many agents. *Review of Economic Studies* 57, 351–638.

Maskin, E. 1999. Nash equilibrium and welfare optimality, *Review of Economic Studies* 66, 23–38.

Maskin, E. and Riley, J. 1984. Monopoly with incomplete information. *Rand Journal of Economics* 15, 171–96.

Maskin, E. and Tirole, J. 1990. The principal–agent relationship with an informed principal. I: Private values. *Econometrica* 58, 379–410.

Maskin, E. and Tirole, J. 1992. The principal–agent relationship with an informed principal. II: Common values. *Econometrica* 60, 1–42.

Mirrlees, J. 1971. An exploration in the theory of optimum income taxation. *Review of Economic Studies* 38, 175–208.

Mirrlees, J. 1999. The theory of moral hazard with unobservable behaviour. Part I. *Review of Economic Studies* 66, 3–22.

Mookherjee, D. 1984. Optimal incentive schemes with many agents. *Review of Economic Studies* 51, 433–46.

Moore, J. 1992. Implementation in environments with complete information. In *Advances in Economic Theory*, ed. J.-J. Laffont. Cambridge: Cambridge University Press.

Moore, J. and Repullo, R. 1988. Subgame-perfect implementation. *Econometrica* 56, 1191–20.

Mussa, M. and Rosen, S. 1978. Monopoly and product quality. *Journal of Economic Theory* 18, 301–17.

Myerson, R. 1979. Incentive compatibility and the bargaining problem. *Econometrica* 47, 61–73.

Myerson, R. 1981. Optimal auction design. *Mathematics of Operations Research* 6, 58–63.

Myerson, R. and Satterthwaite, M. 1983. Efficient mechanisms for bilateral trading. *Journal of Economic Theory* 28, 61–73.

Nalebuff, B. and Stiglitz, J. 1983. Prizes and incentives: towards a general theory of compensation. *Bell Journal of Economics* 14, 21–43.

Rogerson, W. 1985. The first-order approach to principal–agent problems. *Econometrica* 53, 1357–68.

Shavell, S. 1979. On moral hazard and insurance. *Quarterly Journal of Economics* 93, 541–62.

Smith, A. 1776. *The Wealth of Nations*. New York: Prometheus Books, 1991.

Spear, S. and Srivastava, S. 1987. On repeated moral hazard with discounting. *Review of Economic Studies* 54, 599–617.

Spence, M. 1973. Job market signaling. *Quarterly Journal of Economics* 87, 355–74.

Spence, M. 1974. *Market Signalling: Informational Transfer in Hiring and Related Processes*. Cambridge, MA: Harvard University Press.

Tirole, J. 1986. Hierarchies and bureaucracies: on the role of collusion in organizations, *Journal of Law, Economics and Organization* 2, 181–214.

Tirole, J. 1999. Incomplete contracts: where do we stand? *Econometrica* 67, 741–82.

Townsend, R. 1979. Optimal contracts and competitive markets with costly state verification. *Journal of Economic Theory* 21, 417–25.

contracting in firms

In many realms of economic life, the actions of individuals affect the welfare of others. Nowhere is this more relevant than in firms, where employees act on behalf of owners or shareholders to provide services for customers and clients. This separation of the interests of employees from those whose actions they benefit has generated a large literature on incentive contracting, where the overarching objective is the alignment of such interests. The early literature on agency theory, described in the first edition of this volume by Lazear (1987), conceptually mimics that on externalities – the other area of economics that deals with welfare consequences of actions on others – by showing a variety of ways in which the compensation of agents can be constructed to internalize the effects on one's actions on others. There are two ways of doing this. First, one could simply tell employees what to do and to penalize them if they fail to do so. In the literature, this is referred to as input monitoring. While this can sometimes help, it is often hard to monitor either what workers do, or the intensity with which they do so – a salesman on the road would be a good example. Similarly, while overseers can sometimes identify what it is that agents are doing, they may not know what they *should* be doing – a board of directors monitoring a CEO would be apposite here. Accordingly, the second solution to misaligned incentives is to design compensation plans such that the agent's pay depends on her contribution – 'output' – so that the concerns of other parties are internalized.

A simple model can illustrate this point, and is useful to describe other complications that can arise. The agent is assumed to take some action ('effort') $e \geq 0$, which is unobserved by the principal. She is averse to exerting effort. Consider a simple parameterization of the agent's utility function, where the agent cares about wages w and effort; assume that the agent has exponential utility $V = -exp[-r(w - C(e))]$, where w is the worker's wage, $r \geq 0$ is the constant rate of absolute risk aversion, the worker's cost of supplying effort is $C(e) = \frac{ce^2}{2}$, and her reservation utility is U^*. To focus attention on the role of output contacting, assume that the principal cannot observe effort e (so monitoring of inputs is not possible), but instead only observes a signal on effort $y = e + \varepsilon$, where $\varepsilon \sim \mathcal{N}(0, \sigma^2)$, with σ^2 representing measurement error. Assume also that the principal chooses to reward the agent in a linear fashion on output – a piece rate: $w = \beta_0 + \beta_y y$. (There is a large literature on the optimal shape of compensation contracts – see Prendergast, 1999; Gibbons, 1996, for an overview.) Then there is a simple solution to attaining efficient effort: choose the contact to internalize the benefit to others by setting $\beta_y = 1$. In words, efficiency arises when the agent is residual claimant on the benefit of others.

This solution, providing a simple prescription for how compensation contacts should be designed, is both simple and intuitive. And empirically false. There are, of course, some occupations where one can find evidence of such 'high-powered incentives', where agents are essentially residual claimants on output. Indeed, the literature on agency theory is replete with references to such occupations – taxi cab drivers, franchisees, share-cropper farmers and the self-employed. Yet these are exceptions; instead, 'low-powered incentives' in firms are more the norm (see Prendergast, 1999, for details). Consequently, one of the quandaries of the literature has become why so few workers seem to have contracts where their pay is strongly linked to their performance, and much of the subsequent literature to that outlined in the first edition of the *New Palgrave* has identified relevant constraints on incentive contracting.

The earliest candidate to explain why high-powered incentives are rare is that high-powered contracts impose *risk* on workers (Holmstrom, 1979). Consider the contract that induces efficient effort above: $\beta_y = 1$. The objective of the firm is to maximize profits subject to the worker's willingness to take the position. This implies that the fixed component, β_0, is changed to guarantee that agents earn their reservation utility, so the principal's objective becomes a surplus maximization exercise. When the worker is risk neutral, the fixed component is reduced sufficiently such that the total compensation cost is $U^* + \frac{c}{2}$. In words, the only cost that the employer incurs in addition to U^* is the effort cost. This is not true when the worker is risk averse. In the context of the preferences V above, compensation costs increase when incentive contracts are used for two reasons – the cost of increased effort as above, but also a risk cost imposed on workers. Both costs are increasing in β_y. With exponential preferences and linear contracts, this trade-off results in the optimal contract being $\beta_y^* = \frac{1}{1+rc\sigma_y^2}$. This approach to studying incentive contracting has become knows as the 'trade-off of risk and incentives', where firms trade off the benefits of great effort with higher compensation costs induced by a risk premium, such that the chosen level of effort falls below the level that internalizes benefits to others. Only in the case where there is either no measurement error ($\sigma_y^2 = 0$) or risk neutrality ($r = 0$) does efficient effort arise.

At its most general, this costliness of exposing a worker to large degrees of risk (or its analogue, liquidity constraints) surely explains some part of the absence of high-powered incentives. In much the same way as financial assets with higher undiversifiable risk require higher expected returns, so also are risky jobs likely to demand higher compensation. Despite this, the empirical literature on how compensation contracts trade off such risk issues against higher effort has shown little evidence in its favour. There are two principal empirical implications of the theory. First, riskier environments should have lower incentives – β_y^* declines with σ_y^2. There have been many studies of the relationship between risk and the strength of

incentives in a variety of occupations. If anything, this literature suggests that the relationship between risk and the provision of incentives is positive rather than the negative relationship posited by this theory. See Prendergast (2002) for details and an explanation as to why this may be. Second, the trade-off of risk and incentives implies that compensation should not depend on measures that workers cannot control. Again, this has found little support in the data. For example, Bertrand and Mullainathan (2001), have examined executive contracts in the United States, and found little evidence that contracts reward executives any less for measures that they cannot control (say, where an oil company's profits change simply because the price of crude changed) than for those that they can (such as a merger). More evidence on this failure to filter out uncontrollable factors concerns the infrequency of relative performance evaluation. Consider two sales-force workers (or executives) who carry out a similar job. If demand for the products that they sell varies for common reasons beyond their control, an efficient way of limiting risk exposure is to (at least partially) reward the workers on how well they do relative to each other. Yet empirically there is relatively little evidence of such benchmarking (for example, see Janakiraman, Lambert and Larker, 1992).

A second limitation on incentive contracting arises when measures do not reflect the objectives of the principal. Workers often carry out a host of activities in their jobs, yet measures of performance may not reflect all these aspects. A good example of this would be measuring the performance of a teacher. While measures may be available on some component of what they do – such as test scores for a teacher – many important aspects may remain unmeasured. When contracts are designed on the subset of things that can be measured, there is a danger that they ignore the unmeasured aspects. For instance, there is evidence of teachers 'teaching for the test' or cheating to achieve higher test scores (Jacob and Levitt, 2003). This phenomenon has become known as *multi-tasking* (Holmstrom and Milgrom, 1992), which becomes potentially important when there is no single measure that reflects the contribution of an agent. Accordingly, it is not surprising that a consistent empirical finding is that jobs which are described by firms as complex tend not to offer significant incentive pay (see Prendergast, 1999, for details).

Another limitation on the ability of firms to provide incentives to workers comes from team production. Measures of performance for most workers reflect not only what they do but also the contributions of others. In itself, this does not change the calculus above in any conceptual sense, other than that the measurement error now includes the actions of others. As an example, assume that two agents (1 and 2) work on a team and that output measures the true contributions of both plus an error term $y = e_1 + e_2 + \varepsilon$. Efficient effort arises as before by setting $\beta_y = 1$ for each worker. However, there is now a potential problem of budget breaking, where marginal payments exceed marginal output. In this example, when total output rises by one dollar, compensation costs increase by two dollars. In many firms – for instance, partnerships – such budget breaking is not possible. If instead the principal can pay out no more than one dollar for every dollar extra on output, this naturally places an upper bound of $\beta_y = \frac{1}{2}$ on average for the agents. Hence, budget balancing places a natural limitation on firm incentives. This also leads to a free rider problem in teams, where maximum incentive compensation in an N member team mechanically declines as N increases. (This is known as the '$\frac{1}{N}$ problem'.) There is also considerable empirical evidence (such as Gaynor and Pauly, 1990) on such free riding – mostly from legal and medical partnerships – illustrating how various measures of performance disimprove as the size of the team being rewarded increases.

Many measures of output are not denominated in dollar terms, but instead come in the form of evaluations by others. For instance, it would be difficult to measure the contribution of a social worker or a customer service representative without using feedback from supervisors or clients. Another limitation on contracts arises when such subjective measures can be corrupted by evaluators with vested interests. Two particular sources of such vested interests have been considered in the literature. First, information on performance often originates with clients as they are the only ones with first-hand experience of the agent's efforts. For instance, compensation for many customer service representatives depends on client evaluations. When clients have relatively similar preferences to the principal – such as that the agent should be courteous and efficient – contracts based on evaluations can mimic the objective contracts above. Yet in other instances, the vested interest of clients can render incentives difficult to implement. A good example of this arises in occupations such as police or immigration control, whose objective is not necessarily to make their clients happy. Making pay depend on evaluations in these instances can be harmful as it gives agents incentives to keep clients happy when they should not, such as a police officer not arresting a suspect. In these cases, incentive contracting on evaluations typically needs to be curtailed to avoid such incentives (see Prendergast, 2003, for details).

The second example of vested interests with subjective evaluations is where the principal has an incentive not to implement the (*ex ante*) efficient contract by reneging on a promised payment to save costs. Thus, even though an agent exerts effort and performs well, the supervisor claims otherwise to keep costs down. This can arise either by outright lying or perhaps by manipulating whatever measures are available. A relevant example here is the movie industry, where actors are sometimes paid on the 'net profits' of a film. As a result, there have been numerous court cases regarding firms using creative transfer pricing arrangements to reduce profits for very successful movies. See Cheatham, Davis and Cheatham (1996) for more details on this. Such incentives to renege are likely even worse when there are no objective measures of performance. Because the desire to renege is greater when discretionary incentive payments are higher, it follows that the only credible contracts often involve few incentives. (Clive Bull, 1987, considers a role for repeated interaction between the principal and agent as a means of reducing incentives to renege. While repeating the relationship can result in sufficient incentives for complete honesty by the principal, it remains the case that, if the relationship's value is not sufficiently great, incentives must be muted to reduce incentives for cheating.)

It is incorrect to assume that the ability to manipulate measures of performance always mute incentives – sometimes it can result in incentive pay being inefficiently high. Consider again two occupations where agents are typically residual claimants – taxi drivers and sharecroppers. At first blush, it would seem odd that they have such extreme incentives. Aren't these as likely candidates for trading off risk against incentives as any? However, one characteristic of each of these occupations is that they have opportunities for hiding output, either by taking fares without using the metre (in the case of cab drivers) or selling crops privately (in the case of farmers). In both cases, the only outcome that makes this incentive irrelevant is to render them residual claimants, even if risk considerations would suggest otherwise.

Another issue which can constrain efficient incentives, yet which has received almost no attention in the empirical literature, is where agents hold *private information*. Take a specific instance – real estate agents. In Chicago, real estate contracts take a simple form – agents make three per cent of the sales price of the house. Assume that my home is worth $500. This linear contract not only offers only three per cent on the relevant margin for improving the selling price of the house, but predominantly rewards the agent for selling the house for say $450. Yet anyone could sell the house for $450 and it seems highly inefficient to reward in this way. So why not renegotiate to something better? An example of such an improvement (subject to risk issues) would be to offer nothing on the first $450, but to pay a piece rate of 30 per cent on anything over $450. In this way, the agent has more incentives on margin, yet breaks even relative to the original contract if the house sells for its original price.

One reason why such renegotiation does not arise is that the agent may privately know the true value of the home, while the owner believes it to be worth $500 on average. Consider a homeowner who offers the new contract above to the agent. It is clear that the agent rejects the new contract if it is truly worth less than expected, and accepts it if worth more. But this implies that the agent earns *information rents* on average. As a result, on average the homeowner loses money from the renegotiation unless effort increases enough. This option available to the agent limits the ability of contracts to attain efficiency. Instead, in the usual monopoly fashion, the homeowner would offer a contract to trade off the efficiency gains of increased effort with infra-marginal losses of the type described above, resulting in lower-powered incentives. (There is a large mechanism design literature on this topic that has largely been ignored by the empirical literature on incentives; see Laffont and Tirole, 1986, for example. This is surely partly because of the empirical conundrum as to why mechanisms are so rare in reality.)

Much of the recent literature has been focused on how incentive contracts can cause adverse behavioural responses. Another possible mechanism for such responses is where *intrinsic motivation* can be crowded out by the use of incentive contracts. The premise of this literature has been that in many occupations agents enjoy carrying out the activity or care about the outcomes of their actions. As a result, they will exert effort beyond that which they can get away with even in the absence of incentive contracts. This, in itself, is not enough to limit incentive contracting. However, there is some psychological evidence that agents enjoy their jobs less when incentive contracting is used. In effect, they feel that they are only doing it 'for the money' and hence lose interest. A commonly cited example is the willingness of people to donate blood, where the warm feeling from donating declines when payments are made. In some instances, this can imply that incentive contracting can *reduce* effort if these crowding out effects are strong enough. As a result, it can be optimal to provide no incentives even when effort is one-dimensional. This area of research, whose empirical testing has largely been restricted to the laboratory, is still in its early stages and is likely to see much refining over the coming years. See Frey and Jegen (2001) for a survey.

Another likely fruitful area of future research concerns non-monetary ways of motivating workers. This literature largely began as an exercise in how workers could be motivated to internalize the benefits of others, yet has almost exclusively become an exercise in how to motivate through monetary contracting. Yet it is clear that there are a myriad of means of motivating workers – sense of achievement, 'doing good', status, and so on – that firms tap into. How such mechanisms operate, and the way in which they interact with monetary contracts, remain an unstudied topic of research, though see Besley and Ghatak (2005), for some theoretical work on this issue.

It is worthwhile to note a caveat before concluding. The discussion above concerns the absence of *observed* incentive contracts. Yet workers often have unobserved carrots and sticks that can motivate them. For instance, many workers exert effort in the hope of attaining a promotion (Lazear and Rosen, 1981), or a better job offer (Holmstrom, 1999). Many of these mechanisms for inducing desired behaviour are dynamic, where good performance today results in a greater likelihood of promotion, or better job offers in future. Such incentives are clearly important for workers. However, it remains the case that explicit incentive payments remain limited even in those cases where the above types of career concern are negligible. (For example, it is well known that promotion prospects become very limited for workers who remain in a job grade for a long period. Yet explicit incentives are no more common for those workers than for any other.) The interaction of unobserved (typically career) incentives with the more explicit set of piece rates and bonuses that have been considered above is surely of first-order importance to firms, though it remains surprisingly unexplored in the literature (see Baker, Gibbons and Murphy, 1994, for an exception).

To conclude, perhaps the central foundation of modern economics is the idea that appropriate prices guide behaviour in efficient ways. Despite this, one of the defining characteristics of the employment relationship in many firms is the absence of the kind of explicit prices whereby wages depend in a clear way on observed outcomes. The early incarnations of agency theory were concerned with designing prices in a way that could serve to fully internalize the effects of agents' actions on the welfare of their employers. Yet this initial optimism has now been tempered with a somewhat more nuanced view that shows trade-offs that will ultimately help in defining more precisely the nature of labour market relationships.

CANICE PRENDERGAST

Bibliography

Baker, G., Gibbons, R. and Murphy, K.J. 1994. Subjective performance measures in optimal incentive contracts. *Quarterly Journal of Economics* 109, 1125–56.

Bertrand, M. and Mullainathan, S. 2001. Are CEOs rewarded for luck? The ones without principals are. *Quarterly Journal of Economics* 116, 901–32.

Besley, T. and Ghatak, M. 2005. Competition and incentives with motivated agents. *American Economic Review* 95, 616–36.

Bull, C. 1987. The existence of self-enforcing wage contracts. *Quarterly Journal of Economics* 102, 147–59.

Cheatham, C., Davis, D. and Cheatham, L. 1996. Hollywood profits: gone with the wind? *CPA Journal* 12, 32–4.

Frey, B. and Jegen, R. 2001. Motivation crowding theory: a survey of empirical evidence. *Journal of Economic Surveys* 15, 589–611.

Gaynor, M. and Pauly, M. 1990. Compensation and productive efficiency in partnerships. Evidence from medical group practice. *Journal of Political Economy* 98, 544–74.

Gibbons, R. 1996. Incentives and careers in organizations. In *Advances in Economics and Econometrics: Theory and Applications*, ed. D. Kreps and K. Wallis. Cambridge: Cambridge University Press.

Holmstrom, B. 1979. Moral hazard and observability. *Bell Journal of Economics* 10, 74–91.

Holmstrom, B. 1999. Managerial incentive problems: a dynamic perspective. *Review of Economic Studies* 66, 169–82.

Holmstrom, B. and Milgrom, P. 1992. Multi-task principal agent analyses: linear contracts, asset ownership and job design. *Journal of Law, Economics, and Organization* 7, 24–52.

Jacob, B.A. and Levitt, S.D. 2003. Rotten apples: an investigation of the prevalence and predictors of teacher cheating. *Quarterly Journal of Economics* 118, 843–77.

Janakiraman, S.N., Lambert, R.A. and Larker, D.F. 1992. An empirical investigation of the relative performance evaluation hypothesis. *Journal of Accounting Research* 30, 53–69.

Laffont, J.-J. and Tirole, J. 1986. Using cost observation to regulate firms. *Journal of Political Economy* 94, 614–41.

Lazear, E.P. 1987. Incentive contracts. In *The New Palgrave: A Dictionary of Economics*, vol. 2., ed. J. Eatwell, M. Milgate and P. Newman. London: Macmillan.

Lazear, E. and Rosen, S. 1981. Rank order tournaments as optimal labor contracts. *Journal of Political Economy* 89, 841–64.

Prendergast, C. 1999. The provision of incentives in firms. *Journal of Economic Literature* 37, 7–63.

Prendergast, C. 2002. The tenuous trade-off between risk and incentives? *Journal of Political Economy* 110, 1071–102.

Prendergast, C. 2003. The limits of bureaucratic efficiency. *Journal of Political Economy* 111, 929–59.

control functions

The control function approach is an econometric method used to correct for biases that arise as a consequence of selection and/or endogeneity. It is the leading approach for dealing with selection bias in the correlated random coefficients model (see Heckman and Robb, 1985; 1986; Heckman and Vytlacil, 1998; Wooldridge, 1997; 2003; Heckman and Navarro, 2004), but it can be applied in more general semiparametric settings (see Newey, Powell and Vella, 1999; Altonji and Matzkin, 2005; Chesher, 2003; Imbens and Newey, 2006; Florens et al., 2007).

The basic idea behind the control function methodology is to model the dependence between the variables not observed by the analyst on the observables in a way that allows us to construct a function K such that, conditional on the function, the endogeneity problem (relative to the object of interest) disappears.

In this article I deal exclusively with the problem of identification. That is, I assume access to data on an arbitrarily large population. As a consequence, I do not discuss estimation, standard errors or inference. In the examples, I analyse how to recover parameters in a way that, I hope, shows directly how to perform estimation via sample analogues.

The Set-up

The general set-up I consider is the following two-equation structural model; an outcome equation:

$$Y = g(X, D, \varepsilon), \qquad (1)$$

and an equation describing the mechanism assigning values of D to individuals:

$$D = h(X, Z, v), \qquad (2)$$

where X and Z are vectors of observed random variables, D is a (possibly vector valued) observed random variable, and ε and v are general disturbance vectors not independent of each other but satisfying some form of independence of X and Z.

The problem of endogeneity arises because D is correlated with ε via the dependence between ε and v. Because eq. (2) represents an assignment mechanism in many economic models, it is generically called the 'selection' or 'choice' equation. This set-up has been applied to problems like earnings and schooling (Willis and Rosen, 1979; Cunha, Heckman and Navarro, 2005), wages and sectoral choice (Heckman and Sedlacek, 1985) and production functions and productivity (Olley and Pakes, 1996), among others.

The goal of the analysis is to recover some functional of $g(X, D, \varepsilon)$ of interest

$$a(X, D) \qquad (3)$$

that cannot be recovered in a straightforward way because of the endogeneity/selection problem. As an example, when D is binary interest sometimes centres on the effect of going from $D = 0$ to $D = 1$ for an individual chosen at random from the population, the so-called average treatment effect: $a(X, D) = E(g(X, 1, \varepsilon) - g(X, 0, \varepsilon))$.

The key behind the control function approach is to notice that (conditional on X, Z) the only source of dependence is given by the relation between ε and v. If v was known, we could condition on it and analyse eq. (1) without having to worry about endogeneity. The main idea behind the control function approach is to recover some function of v via its relationship with the model observables so that we can now condition on it and solve the endogeneity problem.

Definition The control function approach proposes a function K (the control function) that allows us to recover $a(X, D)$ such that K satisfies

A-1. K is a function of X, Z, D.
A-2. ε satisfies some form of independence of D conditional on $\rho(X, K)$, with ρ a knowable function.
A-3. K is identified.

Assumption **A-2** is the key assumption of the approach. It states that, once we condition on K, the dependence between ε and D (that is, the endogeneity) is no longer a problem. To help fix ideas, consider the following example of a simple linear in parameters additively separable version of the model of eqs (1) and (2).

Example 1 *Linear regression with constant effects.* Write the outcome eq. (1) as

$$Y = X\beta + D\alpha + \varepsilon$$

and assume that our object of interest (3) is α. Assume that we can write eq. (2) as

$$D = X\rho + Z\pi + v \tag{4}$$

with $v, \varepsilon \perp\!\!\!\perp X, Z$ where $\perp\!\!\!\perp$ denotes statistical independence. Such a model arises, for example, if Y is logearnings and D is years of schooling as in Heckman, Lochner and Todd (2003). If ability is unobservable since high ability is associated with higher earnings but also with higher schooling, then ε and v would be correlated.

If we let $K = v$ be the residual of the regression in (4), then we can recover α from the following regression

$$Y = X\beta + D\alpha + K\psi + \eta,$$

where it follows that $E(\eta|X, K) = 0$. It is easy to show that in this case the control function estimator and the two-stage least squares estimator are equivalent. (To my knowledge, although in a different context – a SUR model – Telser, 1964, was the first to use the residuals from other equations as regressors in the equation of interest.)

The previous case is a simple example of a control function where $K = D - E(D|X, Z)$. In this case, because of the constant effects assumption (that is, α is not random), standard instrumental variables methods and the control function approach coincide. In general, this is not the case.

In the next section I describe in detail the control function methodology for the binary choice case (Roy, 1951). This case is interesting both because it is the workhorse of the policy evaluation literature and because, by virtue of its nonlinearity, it highlights the implications of a nonlinear structure in a relatively simple context. I then briefly describe extensions to more general cases. For simplicity, I focus on the additively separable in unobservables case, but recent research provides generalizations to non-additive functions (see Blundell and Powell, 2003; Imbens and Newey, 2006, among others).

The case of a binary endogenous variable

In this section I describe how the control function approach solves the selection/endogeneity problem when the endogenous variable is binary. This problem has a long tradition in economics going back (at least) to Roy (1951). In Roy's original version of the model (see ROY

MODEL) an individual is deciding whether to become a fisherman ($D = 0$) or a hunter ($D = 1$).

Associated with each occupation is a payoff $Y_D = g_D(X) + \varepsilon_D$. Since we can only observe individuals in one sector at a time, the *observed* outcome for an individual is given by Y_1 if he becomes a hunter ($D = 1$) and by Y_0 if he becomes a fisherman ($D = 0$). That is, the observed outcome (Y) can be written as:

$$\begin{aligned} Y &= DY_1 + (1 - D)Y_0 \\ &= g_0(X) + D\big(g_1(X) - g_0(X)\big) \\ &\quad + \varepsilon_0 + D(\varepsilon_1 - \varepsilon_0). \end{aligned} \tag{5}$$

The model is closed by assuming that individuals choose the occupation with the highest payoff. That is,

$$\begin{aligned} D &= \mathbf{1}(Y_1 - Y_0 > 0) \\ &= \mathbf{1}\big(g_1(X) - g_0(X) + \varepsilon_1 - \varepsilon_0 > 0\big), \end{aligned} \tag{6}$$

where $\mathbf{1}(a)$ is an indicator function that takes value 1 if a is true and 0 if it is false. Endogeneity arises because the error term in choice eq. (6) contains the same random variables as the outcome eq. (5). A generalized version of the model replaces the simple income maximization rule in (6) with a more general decision rule

$$D = \mathbf{1}(h(X, Z) - v > 0). \tag{7}$$

The model described by eqs (5) and (7) is general enough to be used in many different cases. Many questions of interest in economics fit this framework if, instead of thinking of two sectors, fishing and hunting, we think of two generic potential states, the treated state ($D = 1$) and the untreated state ($D = 0$) with their associated potential outcomes. The decision rule in (7) is general enough to capture not only income maximization but also utility maximization and even a deciding actor different from the agent directly affected by the outcomes (parents deciding for their children, for example). The simple income maximization rule in (6) shows why, *in general* if $\varepsilon_1 \neq \varepsilon_0$, then $\varepsilon_1 - \varepsilon_0$ is likely to be correlated with D.

The correlated random coefficients model is a special case of the model described by (5) and (7) when $\varepsilon_1 - \varepsilon_0$ is not independent of D and $g_j(X) = \alpha_j + X\beta$ for $j = 0, 1$. (For simplicity I assume $\beta_1 = \beta_0 = \beta$. The case where $\beta_1 \neq \beta_0$ follows directly.) To see why simply rewrite (5) as

$$Y = \alpha_0 + X\beta + D(\alpha_1 - \alpha_0 + \varepsilon_1 - \varepsilon_0) + \varepsilon_0 \tag{8}$$

so that now the coefficient on D is (a) random and (b) correlated with D. In this case we have that the gains from treatment ($\alpha_1 - \alpha_0 + \varepsilon_1 - \varepsilon_0$) are heterogeneous (that is, they are not constant even after controlling for X) and they are correlated with D. I come back to this special linear in parameters case in example 2.

Though other parameters of interest can be defined, I consider the case in which we are interested in the two particular functionals that receive the most attention in the evaluation literature – the average treatment effect and the average effect of treatment on the treated. I impose that $\varepsilon_1, \varepsilon_0, v$ are absolutely continuous with finite means, and that $\varepsilon_1, \varepsilon_0, v \perp\!\!\!\perp X, Z$. (One could weaken the assumption to be $\varepsilon_1, \varepsilon_0 \perp\!\!\!\perp X|Z$ and $v \perp\!\!\!\perp X, Z$.)

Under these assumptions the average treatment effect is given by

$$ATE(x) = E(Y_1 - Y_0|X = x)$$
$$= g_1(x) - g_0(x) = x(\beta_1 - \beta_0)$$

where the last equality follows if eq. (8) applies. $ATE(X)$ is of interest to answer questions like the average effect of a policy that is mandatory, for example. When receipt of treatment is not mandatory or randomly assigned, the average effect of treatment among those individuals who are selected into treatment is commonly the functional of interest (see Heckman, 1997; Heckman and Smith, 1998). This effect is measured by the average effect of treatment on the treated:

$$TT(x) = E(Y_1 - Y_0|X = x, D = 1)$$
$$= g_1(x) - g_0(x) + E(\varepsilon_1 - \varepsilon_0|X$$
$$= x, D = 1) = \alpha_1 - \alpha_0 + E(\varepsilon_1 - \varepsilon_0|X$$
$$= x, D = 1),$$

where the last equality follows for the linear in parameters case of eq. (8).

Now, suppose we ignored the endogeneity problem and attempted to recover either of these objects from the data on outcomes at hand. In particular, if we used the (observed) conditional means of the outcome

$$E(Y|X = x, D = 1) - E(Y|X = x, D = 0) = g_1(x)$$
$$- g_0(x) + E(\varepsilon_1|X = x, D = 1)$$
$$- E(\varepsilon_0|X = x, D = 0)$$

we would not recover either $ATE(X)$ or $TT(x)$. Notice too that, since the endogenous variable D is binary, we cannot directly recover v and use it as a control as we did in the linear case of example 1 above. Instead, we can recover a function of v that satisfies the definition of a control function.

Let $F_v()$ denote the cumulative distribution function of v. To form the control function in this case, first take eq. (7) and write the choice probability

$$P(x, z) \equiv \Pr(D = 1|X = x, Z = z)$$
$$= \Pr(v < h(x, z)) = F_v(h(x, z)),$$

which under our assumptions implies

$$h(x, z) = F_v^{-1}(P(x, z)).$$

Following the analysis in Matzkin (1992), we can recover both $h(x, z)$ and $F_v()$ nonparametrically up to normalization.

Next, take the conditional (on X, Z) expectation of the outcome for the treated group

$$E(Y|X = x, Z = z, D = 1)$$
$$= g_1(x) + E(\varepsilon_1|X = x, Z = z, D = 1).$$

We can write the last term as

$$E(\varepsilon_1|X = x, Z = z, D = 1)$$
$$= E(\varepsilon_1|v < h(x, z)) = E(\varepsilon_1|v < F_v^{-1}(P(x, z))).$$

That is, we can write it as a function of the known $h(x, z)$ or, equivalently, as a function of the probability of selection $P(x, z)$,

$$E(Y|X = x, Z = z, D = 1) = g_1(x) + K_1(P(x, z)),$$

where $K_1(P(X, Z))$ satisfies our definition of a control function. So, provided that we can vary $K_1(P(X, Z))$ independently of $g_1(X)$, we can recover $g_1(X)$ up to a constant. We can identify the constant in a limit set such that $P \to 1$ since $\lim_{P \to 1} K_1(P) = 0$. Provided that we have enough support in the probability of treatment – that is, provided that some people choose treatment with probability arbitrarily close to (1) –we can recover the constant. (See example 2.) Using the same argument we can form

$$E(Y|X = x, Z = z, D = 0) = g_0(x) + K_0(P(x, z))$$

and identify $g_0(X)$ (up to a constant) and the control function $K_0(P(X, Z))$. As before, we can recover the constant in $g_0(X)$ by noting that $\lim_{P \to 0} K_0(P) = 0$.

Intuitively, we need to be able to vary the $K_1(P(X, Z))$ function relative to the $g_1(X)$ function so that we can identify them from the observed variation in Y_1. One possibility is to impose that g_1 and K_1 are measurably separated functions. (That is, provided that, if $g_1(X) = K_1(P(X, Z))$ almost surely then $g_1(X)$ is a constant almost surely; see Florens, Mouchart and Rolin, 1990.) The simplest way to satisfy this restriction is by exclusion. That is, if $K_1(P(X, Z))$ is a nontrivial function of Z conditional on X and Z shows enough variation, we can vary the K_1 function by varying Z while keeping $g_1(X)$ constant. Another related possibility is to assume that g_1 and K_1 live in different function spaces. For example, g_1 a linear function and K_1 the nonlinear mills ratio term that results from assuming that $(\varepsilon_0, \varepsilon_1, v)$ are jointly normal as in the original Heckman (1979) selection correction model.

Once we have recovered $g_0(X), g_1(X), K_0(P(X, Z))$, $K_1(P(X, Z))$ we can now form our parameters of interest. Given $g_0(X)$ and $g_1(X)$, $ATE(X) = g_1(X) - g_0(X)$

immediately follows. To recover $TT(X)$, first notice that, by the law of iterated expectations

$$E(\varepsilon_0|X = x, Z = z)$$
$$= E(\varepsilon_0|X = x, Z = z, D = 1)P(x, z)$$
$$+ E(\varepsilon_0|X = x, Z = z, D = 0)$$
$$(1 - P(x, z)) = 0,$$

where $P(X, Z)$ is known from our analysis above and $E(\varepsilon_0|X = x, Z = z, D = 0) = K_0(P(x, z))$. Rewriting the expression above we get $E(\varepsilon_0|X = x, Z = z, D = 1) = -\frac{K_0(P(x,z))(1-P(x,z))}{P(x,z)}$. With this expectation in hand we can recover $TT(X, Z) = g_1(X) - g_0(X) + K_1(P(X, Z)) + \frac{K_0(P(X,Z))(1-P(X,Z))}{P(X,Z)}$. By integrating against the appropriate distribution, we can recover $TT(X) = \int TT(X, z)dF_{Z|X,D} = 1(z)$.

The following example shows how the control function methodology can be applied to recover average effects of treatment in a linear in parameters model with correlated random coefficients. This model arises when there are unobservable gains that vary over individuals and these gains are correlated with the choice of treatment (that is, when there is essential heterogeneity. See Heckman, Urzua and Vytlacil, 2006; Basu et al., 2006). The Roy model of eqs (5) and (6) in which the unobservable individual gains $(\varepsilon_1 - \varepsilon_0)$ are correlated with the choice of sector is an example of this case.

Example 2 *Correlated random coefficients with binary treatment.* Assume we can write the outcome equations in linear in parameters form,

$$Y_j = \alpha_j + X\beta_j + \varepsilon_j \quad j = 0, 1.$$

Let D be an indicator of whether an individual receives treatment ($D=1$) or not ($D=0$). We also write a linear in parameters decision rule:

$$D = 1(X\delta + Z\gamma - v > 0).$$

From the analysis in Manski (1988) we can recover δ, γ and F_v (up to scale). With $P(x, z) = \Pr(D = 1|X = x, Z = z)$ in hand, we then form

$$Y_j = \alpha_j + X\beta_j + K_j(P(X, Z)) + \eta_j$$

where $E(\eta_j|X = x, K_j(P(X, Z)) = k_j) = 0$. To emphasize the problem of identification of the constant α_j we can rewrite the outcome as

$$Y_j = \tau_j + X\beta_j + \tilde{K}_j(P(X, Z)) + \eta_j$$

where $K_j(P(X, Z)) = \kappa_j + \tilde{K}_j(P(X, Z))$ and $\tau_j = \alpha_j + \kappa_j$.

The elements of the outcome equations can be recovered by various methods. One could, for example, use Robinson (1988) and use residualized nonparametric regressions to recover β_j, τ_j and $K_j(P(X, Z))$. Alternatively, one could approximate $K(P(X, Z))$ with a polynomial on $P(X, Z)$. In this case we would have

$$Y_j = \tau_j + X\beta_j + \pi_1 P(X, Z) + \pi_2 P(X, Z)^2$$
$$+ \cdots + \pi_n P(X, Z)^n + \eta_j$$

where $\tilde{K}_j(P(X, Z)) = \sum_{i=1}^{n} \pi_{j1} P(X, Z)^i$. When $j = 0$ then $\lim_{P \to 0} K_0(P) = 0$ and it follows that $\tilde{K}_0(P) = K_0(P)$ and $\tau_0 = \alpha_0$. For the treated case ($j = 1$) we have that $\lim_{P \to 1} K_1(P(X, Z)) = 0$. Since $\tilde{K}_1(1) = \sum_{i=1}^{n} \pi_{1i}$ it follows that $\kappa_1 = -\sum_{i=1}^{n} \pi_{1i}$ and $\alpha_1 = \tau_1 - \sum_{i=1}^{n} \pi_{1i}$.

Extensions for a continuous endogenous variable
In this section I briefly review the use of the control function approach for the case in which the endogenous variable D is continuous and we assume that $X, Z \perp\!\!\!\perp \varepsilon, v$. Following Blundell and Powell (2003) I assume that the object of interest is the average structural function

$$a(X, D) = \int g(X, D, \varepsilon)dF_\varepsilon(\varepsilon),$$

which, in the additively separable case $g(X, D, \varepsilon) = \mu(X, D) + \varepsilon$ is simply the regression function $\mu(X, D)$.

If we assume that the choice equation

$$D = h(X, Z, v)$$

is strictly monotonic in v (which would follow automatically if it were additively separable in v), we can recover $h()$ and F_v from the analysis of Matzkin (2003) up to normalization. A convenient normalization is to assume that $v \sim \text{Uniform}(0, 1)$ in which case we can directly recover v from the quantiles of F_v, but other normalizations are possible. From the independence assumption it follows that $E(\varepsilon|X, D, Z) = E(\varepsilon|v)$, so we can write the outcome equation as

$$Y = \mu(X, D) + E(\varepsilon|v)$$

$$= \mu(X, D) + K(v)$$

which allows us to recover $\mu(X, D)$ directly (up to normalization). In the additively separable case we analyse, we can relax the full independence assumption and instead assume directly that the weaker mean independence assumption $E(\varepsilon|X, D, Z) = E(\varepsilon|v)$ holds.

SALVADOR NAVARRO

See also **endogeneity and exogeneity; identification; Roy model; selection bias and self-selection.**

Bibliography

Altonji, J.G. and Matzkin, R.L. 2005. Cross section and panel data estimators for nonseparable models with endogenous regressors. *Econometrica* 73, 1053–102.

Basu, A., Heckman, J.J., Navarro, S. and Urzua, S. 2006. Use of instrumental variables in the presence of heterogeneity and self-selection: an application in breast cancer patients. Unpublished manuscript, Department of Medicine, University of Chicago.

Blundell, R. and Powell, J. 2003. Endogeneity in nonparametric and semiparametric regression models. In *Advances in Economics and Econometrics: Theory and Applications, Eighth World Congress*, vol. 2, ed. L.P. Hansen, M. Dewatripont and S.J. Turnovsky. Cambridge: Cambridge University Press.

Chesher, A. 2003. Identification in nonseparable models. *Econometrica* 71, 1405–41.

Cunha, F., Heckman, J.J. and Navarro, S. 2005. Separating uncertainty from heterogeneity in life cycle earnings. *Oxford Economic Papers* 57, 191–261.

Florens, J.-P., Heckman, J.J., Meghir, C. and Vytlacil, E.J. 2007. Identification of treatment effects using control functions in models with continuous, endogenous treatment and heterogeneous effects. Unpublished manuscript, Columbia University.

Florens, J.-P., Mouchart, M. and Rolin, J.M. 1990. *Elements of Bayesian Statistics*. New York: M. Dekker.

Heckman, J.J. 1979. Sample selection bias as a specification error. *Econometrica* 47, 153–62.

Heckman, J.J. 1997. Instrumental variables: a study of implicit behavioral assumptions used in making program evaluations. *Journal of Human Resources* 32, 441–62. Addendum published in 33(1) (1998).

Heckman, J.J., Lochner, L.J. and Todd, P.E. 2003. Fifty years of Mincer earnings regressions. Technical Report No. 9732. Cambridge, MA: NBER.

Heckman, J.J. and Navarro, S. 2004. Using matching, instrumental variables, and control functions to estimate economic choice models. *Review of Economics and Statistics* 86, 30–57.

Heckman, J.J. and Robb, R. 1985. Alternative methods for evaluating the impact of interventions: an overview. *Journal of Econometrics* 30, 239–67.

Heckman, J.J. and Robb, R. 1986. Alternative methods for solving the problem of selection bias in evaluating the impact of treatments on outcomes. In *Drawing Inferences from Self-Selected Samples*, ed. H. Wainer. New York: Springer. Repr. Mahwah, NJ: Lawrence Erlbaum Associates, 2000.

Heckman, J.J. and Sedlacek, G.L. 1985. Heterogeneity, aggregation, and market wage functions: an empirical model of self-selection in the labor market. *Journal of Political Economy* 93, 1077–125.

Heckman, J.J. and Smith, J.A. 1998. Evaluating the welfare state. In *Econometrics and Economic Theory in the Twentieth Century: The Ragnar Frisch Centennial Symposium*, ed. S. Strom. New York: Cambridge University Press.

Heckman, J.J., Urzua, S. and Vytlacil, E.J. 2006. Understanding instrumental variables in models with essential heterogeneity. *Review of Economics and Statistics* 88, 389–432.

Heckman, J.J. and Vytlacil, E.J. 1998. Instrumental variables methods for the correlated random coefficient model: estimating the average rate of return to schooling when the return is correlated with schooling. *Journal of Human Resources* 33, 974–87.

Imbens, G.W. and Newey, W.K. 2006. Identification and estimation of triangular simultaneous equations models without additivity. Unpublished manuscript, Department of Economics, MIT.

Manski, C.F. 1988. Identification of binary response models. *Journal of the American Statistical Association* 83, 729–38.

Matzkin, R.L. 1992. Nonparametric and distribution-free estimation of the binary threshold crossing and the binary choice models. *Econometrica* 60, 239–70.

Matzkin, R.L. 2003. Nonparametric estimation of nonadditive random functions. *Econometrica* 71, 1393–75.

Newey, W.K., Powell, J.L. and Vella, F. 1999. Nonparametric estimation of triangular simultaneous equations models. *Econometrica* 67, 565–603.

Olley, G.S. and Pakes, A. 1996. The dynamics of productivity in the telecommunications equipment industry. *Econometrica* 64, 1263–97.

Robinson, P.M. 1988. Root-n-consistent semiparametric regression. *Econometrica* 56, 931–54.

Roy, A.D. 1951. Some thoughts on the distribution of earnings. *Oxford Economic Papers* 3, 135–46.

Telser, L.G. 1964. Iterative estimation of a set of linear regression equations. *Journal of the American Statistical Association* 59, 845–62.

Willis, R.J. and Rosen, S. 1979. Education and self-selection. *Journal of Political Economy* 87(5, Par 2), S7–S36.

Wooldridge, J.M. 1997. On two stage least squares estimation of the average treatment effect in a random coefficient model. *Economics Letters* 56, 129–33.

Wooldridge, J.M. 2003. Further results on instrumental variables estimation of average treatment effects in the correlated random coefficient model. *Economics Letters* 79, 185–91.

conventionalism

Conventionalism is the methodological doctrine that asserts that explanatory ideas should not be considered true or false but merely better or worse. At the beginning of the 20th century the status of the laws of physics was the burning issue. It was the famous philosopher Henri Poincaré who in 1902 asked whether the laws of physics were 'only arbitrary conventions'. He answered 'Conventions, yes; arbitrary, no'. Obviously, languages and measurement units are arbitrary conventions but nobody would seriously claim they were explanatory ideas. In Poincaré's day, the question bothering physicists who were dealing with Albert Einstein's new theory (namely, relativity) was whether the choice between Euclidian and non-Euclidian geometry was a matter of convention – that is, a matter of convenience. For everyday questions Euclidian geometry is convenient but perhaps for Einstein's physics non-Euclidian is the better choice. For some matters, such as the choice of language to express an idea or of units to measure a distance, most people would allow that such a choice may be completely arbitrary.

Although few of them have ever heard of Poincaré, most economists will say almost the same thing whenever they make a methodological pronouncement concerning the truth status of economic theories, models or assumptions. Rarely, however, have economists been concerned with the questions raised about non-Euclidian geometry (except for John Maynard Keynes's metaphorical suggestion at the beginning of his *General Theory*). Of course, hardly any economist questions language being a matter of convenience; moreover, economists often justify the use of mathematics by claiming that its use is like that of language and thus should be judged by its convenience, not its truth status (Samuelson, 1952; 1954). But in the 1940s critics of Marshallian and Walrasian (that is, neoclassical) economics argued that the truth status of a theory's assumptions should matter. In his 1953 response to the critics of the realism of assuming perfect competition when explaining the economy, Milton Friedman advocated an alternative methodology: instrumentalism. Instrumentalism, unlike conventionalism, claims merely that the truth status of assumptions does not matter so long as the theory is useful. For those economists who still think the truth status of their theories matters, but realize that one can never prove a theory's truth status by induction, the most common response is something like Poincaré's conventionalism.

There are many examples of economists making methodological pronouncements that exhibit adherence to conventionalism. Paul Samuelson denied that any economic explanation was true, writing that 'An explanation … is a better kind of description' (1965, p. 1165). Obviously, some descriptions are better than others, and thus he claimed that we give the honorific title of 'explanation' to the best description. If one were to agree with Samuelson then one certainly could never claim that one's explanation was true. Herbert Simon chose to express this differently; he said all explanations are approximations. Specifically, he said (1963, p. 231) 'Unreality of premises is not a virtue in scientific theory; it is a necessary evil – a concession to the finite computing capacity of the scientist that is made tolerable by the principle of continuity of approximation'. Robert Lucas agreed with that when he said 'Any model that is well enough articulated to give clear answers to the questions we put to it will

necessarily be artificial, abstract, patently "unreal"' (1980, p. 696). Robert Aumann, the game theorist, has advocated an even more limited view for explanatory theories. As he put it 'scientific theories are not to be considered "true" or "false"'. Going further, he said, 'In constructing such a theory, we are not trying to get at the truth, or even to approximate to it: rather, we are trying to organize our thoughts and observations in a useful manner.' In this regard, he argued that a theory is like 'a filing system in an office operation, or to some kind of complex computer program' (1985, pp. 31–2). Lucas and Aumann were merely restating Samuelson's 1965 position on methodology.

The philosophy of conventionalism

For followers of philosophers Willard Quine and Karl Popper, the truth status of explanations or theories cannot be so easily dismissed or limited. While any choice of language or units of measurement may be conventional, the truth status of theories is not a matter of choice, convenient or otherwise.

Unfortunately, the methodological doctrine of conventionalism is often confused with instrumentalism. As the philosopher Joseph Agassi (1966a) points out, they are responses to two different questions. One concerns the role of theories and the other the truth status of theories. Specifically, if we ask 'What is the *role* of a theory?', instrumentalism's answer is that theories are tools and should not be judged by epistemological standards of truth status or by conventionalist criteria of approximate truth or relative merit (except, perhaps, by simplicity or economy). Conventionalism's different answer is the one stated by Aumann: theories are filing systems or catalogues of observed data. Of course, every description is also an appeal to a filing system in that one depicts or locates it within a system by referring to other defined dimensions and concepts. If, instead, we chose the question, 'What is the *status* of a theory?', conventionalism's answer is that, of course, theories are approximations and thus should not be considered true or false but better or worse. Instrumentalism's position is simply that truth status does not matter. With this in mind, it is easy to find economists advocating both methodological positions depending on which question is asked. For example, after saying that a theory is like a filing system, Aumann goes on to say that 'We do not refer to such a system as being "true" or "untrue"; rather, we talk about whether it "works" or not, or, better yet, how well it works' (1985, p. 32).

From the perspective of the philosopher Karl Popper, the main question is: what problem is solved by the doctrine of conventionalism? Since the time when Adam Smith's friend David Hume observed that there was no logical justification for the common belief that much of our empirical knowledge was based on inductive proofs (see Russell, 1945), methodologists and philosophers have been plagued with what they call the 'problem of induction'. The paradigmatic instance of the problem of induction is the realization that we cannot provide an inductive proof that 'the sun will rise tomorrow'. This leads many of us to ask, 'So *how* do we know the sun will rise tomorrow?' If it is impossible to provide a proof, then presumably we would have to admit we do not know the answer to this burning question! Several writers have claimed to have solved this famous problem (for a discussion of such claims, see Miller, 2002). Such a claim is quite surprising since it is a problem that is impossible to solve. Nevertheless, what it is and how it is either 'solved' or circumvented is fundamental to understanding all contemporary methodological discussions.

Up to the time of Popper's entry into the discussion in the mid-1930s, most philosophers took it for granted that all claims to knowledge must be justified. Inductive arguments were seen to be the obvious method. But Popper acknowledged the problem that as a matter of simple logic an inductive argument is impossible. A logical argument is one in which, whenever all the premises are true, any logically derived statements must also be true. An inductive argument is one in which one would argue logically from the truth of particular statements (for example, observation statements such as 'the sun rose today at 7 a.m.') alone to prove the truth of a general statement (for example, the sun always rises). The 'problem of induction' would be solved if one could demonstrate the existence of such an inductive logic. The importance of this problem arises once one realizes that, without some premise of a general nature (such as we find in physics concerning the movement of the earth around the sun and earth's rotation), no finite set of observations could ever prove the non-existence of a counter-example (a refuting instance that would be denied by the general statement in question) somewhere or sometime in the future. For example, to prove that the statement 'All ravens are black' is true requires a proof that there does not exist anywhere in the universe a 'non-black raven'. Everyone agrees that one cannot provide such a literal negative proof. So, it has been argued (Boland, 1982; 2003), most discussions of methodology in economics are concerned with the problem *with* induction rather than the problem *of* induction.

Conventionalism can be seen as a solution to the problem *with* induction. Conventionalism presumes that this problem can be solved even though the problem *of* induction cannot. That is, if there were an inductive logic, then the truth status of a true theory or model could in principle be provable since all assumptions of a universal form could be inductively proven. Without such a logic, many think – still insisting that any claim to knowledge must be justified – that some other means must be found to sort through competing theories. That is, how can we choose the best from a set of competing theories? More specifically, by what criteria do we choose between competing theories? Obvious examples of such criteria are simplicity, generality, robustness, testability, falsifiability, verifiability, confirmability, operational meaningfulness, plausibility, probability, and so on. None of these criteria are considered substitutes for truth status (truth or falsity); they are only choice criteria (truthlikeness). If a criterion can be quantified, one could even see the choice as a matter of applying economics (see Boland, 1971). For example, one might choose the theory that is most confirmed – but it still must be remembered that today's most confirmed theory could be a false theory even today.

For many philosophers, such theory-choice criteria are just short-run solutions to the problem *with* induction. That is, in the short run we might be satisfied with invoking such criteria, so that we can choose between theories and thereby be able to push on, but it is hoped that in the long run someone can come up with a solution to the problem *of* induction.

Conventionalism as employed by economists

Among economists who openly practise conventionalism, it is a doctrine with many variants and relatives. The most common practice is the avoidance of using the words 'true' (or 'false') when discussing theories, models and assumptions. Instead, we see 'best' being invoked with the use of some conventionalist theory-choice truthlikeness criterion. Also common is the use of the word 'valid' to avoid saying 'true'. Sometimes it is used to mean that a theory is valid if it is logically consistent with available data or evidence. The difficulty is that 'valid' is a question of the logicality of an argument (do the conclusions necessarily follow from the assumptions made?) A logically valid argument can still be false, so it is not always clear what is meant by a valid statement or a valid theory.

One weak form conventionalism is old-fashioned relativism. Another weak form is what the followers of McCloskey (1983) call modernism. In yet another weak form it can be seen to be the rationale for so-called methodological pluralism. The most common form is stronger in that it involves the notion that theories are to be evaluated or compared by means of some form of probability calculus.

Those adherents to conventionalism who advocate the objective form of probability calculus seem unaware of the logical difficulties involved. One might wish to use probability as the measure of confirmation of a theory so that one could use such a measure as the criterion for theory choice. The difficulty arises when one asks what constitutes positive evidence – namely, evidence to be used to calculate the probability measure that would serve as, say, the 'degree of confirmation'. Of course, if one requires all observational evidence to be exactly true, then to be an actual confirmation the objective probability measure would have to be 1.00. That is, just one true observation that contradicts the theory in question requires the rejection of the theory. So it would seem that objective probability measures are inappropriate. But econometrics-based hypothesis testing is not as strict since it allows for errors in the observations of the variables. Hence, the objective probability measure can be of some value less that 1.00. Theory choice in this case would seem to be a simple matter of choosing the theory with the highest probability, that is, the highest degree of confirmation.

Among those who openly advocate a subjective form of probabilities, the most common view is based on Bayesian probabilities which provide a compromise by allowing for explicit roles for both subjectivism in the form of prior probability assessments and objectivism in the form of adjustments based on new objective evidence. Again, the main question for using probabilities concerns what would count as confirming evidence or evidence that increases the subjective probability. Like all confirmation criteria, even if everyone attaches a high subjective probability to the theory in question being true it could still be false and perhaps refuted by the next observation report.

The common element underlying all probability measures to be used for theory choice is the notion that the number of confirming observations should somehow matter. Of course, such an expectation does not require the questionable use of probabilities as a measure of confirmation. But avoiding any reliance on probability will not circumvent the more well-known logical problems of confirmation. All conceptions of a logical connection between positive evidence and degrees of confirmation suffer from a profound logical problem called, by some philosophers, the 'paradox of confirmation' or the 'paradox of the ravens' (cf. Sainsbury, 1995; Agassi, 1966b).

The philosopher's paradox of confirmation merely points out that *any* evidence which does not refute a simple universal statement, say, 'All ravens are black' must increase the degree of confirmation. The paradox is based on the observation that, *in terms of what observable evidence would count*, this example of a simple universal statement is logically equivalent to its 'contra-positive' statement 'All non-black things are non-ravens'. Any true observation that is consistent with one of the statements is consistent with the other (equivalent) statement. But in these terms it must be recognized that positive evidence consistent with the contra-positive statement includes red shoes as well as white swans – since in both cases we have non-black things which are not ravens. That is, the set of all confirming instances must include all things which are not non-black ravens. In other words, the more red shoes we observe, the more evidence there is in favour of the contra-positive statement – that is, a red shoe increases the universal statement's degree of confirmation – and, since the contra-positive statement is logically equivalent to the universal statement in question, the latter's degree of confirmation also increases.

Obviously, this consideration merely divides the contents of the universe into non-black ravens and everything else (Hempel, 1966). This consideration calls into question all claims of confirmation.

Few economists who make pronouncements concerning the appropriate methodology to use in economics are aware of the philosophical problems involved. Almost all think we must have some criterion to choose between competing theories or models. All of them take for granted the necessity of justifying their choice. No recognition seems to be given to the simple fact that one's favourite theory can be true even though it cannot be proven true. That is, whether one's theory is true is a separate question from how one knows it to be true.

The notion of a conventionalist theory-choice criterion presumes that there is a philosophical necessity to choose one theory from among its competitors. But there is no such necessity, even though it will always be difficult to convince economists of this whenever they are naive concerning the philosophy of science. But, given that there are so many different criteria to use, one would think any theory that is best by all criteria should be the chosen theory. But it is doubtful that any theory could satisfy all criteria; so the question is begged as to which criterion is the best criterion. This question seems to put us on the road of an infinite regress: by what criterion do we choose the best criterion to choose between theories? Not a promising journey.

LAWRENCE A. BOLAND

See also **assumptions controversy; instrumentalism and operationalism; pluralism in economics.**

Bibliography

Agassi, J. 1966a. Sensationalism. *Mind* 75, 1–24.
Agassi, J. 1966b. The mystery of the ravens: discussion. *Philosophy of Science* 33, 395–402.
Aumann, R. 1985. What is game theory trying to accomplish? In *Frontiers of Economics*, ed. K. Arrow and S. Honkapohja. Oxford: Basil Blackwell.
Boland, L. 1971. Methodology as an exercise in economic analysis. *Philosophy of Science* 38, 105–17.
Boland, L. 1982. *The Foundations of Economic Method*. London: Allen & Unwin.
Boland, L. 2003. *The Foundations of Economic Method: A Popperian Perspective*. London: Routledge.
Friedman, M. 1953. Methodology of positive economics. In *Essays in Positive Economics*. Chicago: Univ. of Chicago Press.
Hempel, C. 1966. *Foundations of Natural Science*. Englewood Cliffs: Prentice-Hall.
Keynes, J.M. 1936. *General Theory of Employment, Interest and Money*. New York: Harcourt, Brace & World.
Lucas, R. 1980. Methods and problems in business cycle theory. *Journal of Money, Credit and Banking* 12, 696–715.

McCloskey, D. 1983. The rhetoric of economics. *Journal of Economic Literature* 21, 481–517.

Miller, D. 2002. Induction: a problem solved. In *Karl Poppers kritischer Rationalismus heute*, ed. J. Böhm, H. Holweg and C. Hoock. Tübingen: Mohr Siebeck.

Poincaré, H. 1902. *La science et l'hypothèse*. Paris: Flammarion.

Poincaré, H. 1905. *Science and Hypothesis*. London: Walter Scott Publishing Company.

Russell, B. 1945. *A History of Western Philosophy*. New York: Simon and Schuster.

Sainsbury, R. 1995. *Paradoxes*. Cambridge: Cambridge University Press.

Samuelson, P. 1952. Economic theory and mathematics: an appraisal. *American Economic Review* 42, 56–66.

Samuelson, P. 1954. Some psychological aspects of mathematics and economics. *Review of Economics and Statistics* 36, 380–2.

Samuelson, P. 1965. Professor Samuelson on theory and realism: reply. *American Economic Review* 55, 1164–72.

Simon, H. 1963. Problems of methodology: discussion. *American Economic Review, Proceedings* 53, 229–31.

convergence

The general question of convergence, understood as the tendency of differences between countries to disappear over time, is of long-standing interest to social scientists. In the 1950s and early 1960s, many analysts discussed whether capitalist and socialist economies would converge over time, in the sense that market institutions would begin to shape socialist economies just as government regulation and a range of social welfare policies grew in capitalist ones.

In modern economic parlance, convergence usually refers specifically to issues related to the persistence or transience of differences in per capita output between economic units, be they countries, regions or states. Most research has focused on convergence across countries, since the large contemporaneous differences between countries generally dwarf intra-country differences. In the context of economic growth, the convergence hypothesis arguably represents the most commonly studied aspect of growth, although the effort to identify growth determinants is arguably the main area of contemporary growth research.

In this overview of convergence, our primary emphasis will be on the development of precise statistical definitions of convergence. This reflects an important virtue of the current literature, namely, the introduction of statistical methods to adjudicate whether convergence is present. At the same time, there is no single definition of convergence in the literature, which is one reason why empirical evidence on convergence is indecisive. Our discussion focuses on convergence across countries,

which has dominated empirical studies, although there is reference to studies that focus on other units.

β-convergence

The primary definition of convergence used in the modern growth literature is based on the relationship between initial income and subsequent growth. The basic idea is that two countries exhibit convergence if the one with lower initial income grows faster than the other. The local (relative to steady state) dynamics of the neoclassical growth model in both its Solow and Cass–Koopmans variants imply that lower-income economies will grow faster than higher-income ones.

As a statistical question, this notion of convergence can be operationalized in the context of a cross-country regression. Let g_i denote real per capita growth of country i across some fixed time interval and $y_{i,0}$ denote the initial per capita income for country i. Then, unconditional β-convergence is said to hold if, in the regression

$$g_i = k + \log y_{i,0}\beta + \varepsilon_i, \beta < 0. \tag{1}$$

For cross-country regression analysis, one typically does not find unconditional β-convergence unless the sample is restricted to very similar countries, for example, members of the OECD. This finding is in some ways not surprising, since unconditional β-convergence is typically not a prediction of the existing body of growth theories. The reason for this is that growth theories universally imply that growth is determined by factors other than initial income. While different theories may propose different factors, they collectively imply that (1) is misspecified. As a result, most empirical work focuses on conditional β-convergence. Conditional β-convergence holds if $\beta < 0$ for the regression

$$g_i = k + \log y_{i,0}\beta + Z_i\gamma + \varepsilon_i \tag{2}$$

where Z_i is a set of those growth determinants that are assumed to affect growth in addition to a country's initial income. While many differences exist in the choice of controls, it is nearly universal to include those determinants predicted by the Solow growth model, that is, population growth and human and physical capital accumulation rates.

Unlike unconditional β-convergence, evidence of conditional β-convergence has been found in many contexts. For the cross-country case, the basic finding is generally attributed to Barro (1991), Barro and Sala-i-Martin (1992) and Mankiw, Romer and Weil (1992). The Mankiw, Romer and Weil analysis is of particular interest as it is based on a regression suggested by the dynamics of the Solow growth model. Hence, their findings have been widely interpreted as evidence in favour of decreasing returns to scale in capital (the source of $\beta < 0$ in the Solow model), and therefore as evidence against the Lucas–Romer endogenous growth approach, which emphasizes increasing returns in capital accumulation (either human or physical) as a source of perpetual growth.

From the perspective of the neoclassical growth model, the term $-\beta$ also measures the rate at which an economy's convergence towards its steady-state growth rate, that is, the growth rate determined exclusively by the exogenous rate of technical change. The many findings in the cross-country literature are often summarized by the claim countries converge towards their steady-state growth rates at a rate of about two per cent per year, although individual studies produce different results. The convergence rate has received inadequate attention in the sense that a finding of convergence may have little consequence for questions such as policy interventions if it is sufficiently slow.

As is clear from (2), any claims about conditional convergence necessarily depend on the choice of control variables Z_i. This is a serious concern given the lack of consensus in growth economics on which growth determinants are empirically important. Doppelhofer, Miller and Sala-i-Martin (2004) and Fernandez, Ley and Steel (2001) use model averaging methods to show that the cross-country findings that have appeared for conditional β-convergence are robust to the choice of controls. A number of additional statistical issues such as the role of measurement error and endogeneity of regressors are surveyed and evaluated in Durlauf, Johnson and Temple (2005).

The assumption in cross-section growth regressions that the unobserved growth terms ε_i are uncorrelated with $\log y_{i,0}$ rules out the possibility that there are country-specific differences in output levels; if such effects were present, they would imply a link between the two. For this reason, a number of researchers have investigated convergence using panel data. This leads to models of the form

$$g_{i,t} = c_i + \log y_{i,t-1}\beta + Z_{i,t}\gamma + \varepsilon_{i,t} \qquad (3)$$

where growth is now measured between $t-1$ and t. This approach not only can handle fixed effects, but can allow for instrumental variables to be used to address endogeneity issues. Panel analyses have been conducted by Caselli, Esquivel and Lefort (1996), Islam (1995) and Lee, Pesaran and Smith (1997). These studies have generally found convergence with rather higher rates than appear in the cross-section studies; for example, Caselli, Esquivel and Lefort (1996) report annual convergence rate estimates of ten per cent.

As discussed in Durlauf and Quah (1999) and Durlauf, Johnson and Temple (2005), panel data approaches to convergence suffer from the problem that, once country specific effects are allowed, it becomes more difficult to interpret results in terms of the underlying economics. The problem is that, once one allows for fixed effects, then the question of convergence is changed, at least if

the goal is to understand whether initial conditions matter; simply put, the country-specific effects are themselves a form of initial conditions. When studies such as Lee, Pesaran and Smith (1997) allow for rich forms of parameter heterogeneity across countries, β-convergence become equivalent to the question of whether there is some mean reversion in a country's output process, not whether certain types of contemporaneous inequalities diminish. This does not diminish the interest of these studies as statistical analyses, but means their economic import can be unclear.

σ-convergence and the cross-section distribution of income

A second common statistical measure of convergence focuses on the whether or not the cross-section variance of per capita output across countries is or is not shrinking. A reduction in this variance is interpreted as convergence. Letting $\sigma^2_{\log y,t}$ denote the variance across i of $\log y_{i,t}$, σ-convergence occurs between t and $t+T$ if

$$\sigma^2_{\log y,t} - \sigma^2_{\log y,t+T} > 0. \qquad (4)$$

There is no necessary relationship between β- and σ-convergence. For example, if the first difference of output in each country obey $\log y_{i,t} - \log y_{i,t-1} = \beta \log y_{i,t-1} + \varepsilon_{i,t}$, then $\beta < 0$ is compatible with a constant cross-sectional variance (which in this example will equal the variance of $\log y_{i,t}$). The incorrect idea that mean reversion in time series implies that its variance is declining is known as Galton's fallacy; its relevance to understanding the relationship between convergence concepts in the growth literature was identified by Friedman (1992) and Quah (1993a). While it is possible to construct a cross-section regression to test for σ-convergence (cf. Cannon and Duck, 2000), they do not test β-convergence per se.

Work on β-convergence has led to general interest in the evolution of the cross-country income distribution. Quah (1993b; 1996) has been very influential in his modelling of a stochastic process for the distribution itself, with the conclusion that it is converging towards a bimodal steady-state distribution. Other studies of the evolution of the cross-section distribution include Anderson (2004) who uses nonparametric density methods to identify increasing polarization between rich and poor economies across time. Increasing divergence between OECD and non-OECD economies is shown in Maasoumi, Racine and Stengos (2007), working with residuals from linear growth regressions.

One difficulty with convergence approaches that emphasize changes in the shape of the cross-section distribution is that they may fail to address the original question of the persistence of contemporaneous inequality. The reason for this is that it is possible, because of movements in relative position within the distribution, for the cross-section distribution to flatten out while at the same time differences at one point in time are reversed; similarly, the cross-section distribution can become less diffuse while gaps between rich and poor widen. That being said, an examination of the locations of individual countries in various distribution studies typically indicates that the increasing polarization of the world income distribution is mirrored by increasing gaps between rich and poor. A useful extension of this type of research would be to employ the dynamics of individual countries to provide additional information on how the cross-section distribution evolves.

Time series approaches to convergence

An alternative approach to convergence is focused on direct evaluation of the persistence of transitivity of per capita output differences between economies. This approach originates in Bernard and Durlauf (1995), who equate convergence with the statement that

$$\lim_{T \Rightarrow \infty} E(\log y_{i,t+T} - \log y_{j,t+T} | F_t) = 0$$

$$(5)$$

where F_t denotes the history of the two output series up to time t. They find that convergence does not hold for OECD economies, although there is some cointegration in the individual output series. Hobijn and Franses (2000) find similar results for a large international data-set. Evans (1996) employs a clever analysis of the evolution of the cross-section variance to evaluate the presence of a common trend in OECD output, and finds one is present; his analysis allows for different deterministic trends in output and so in this sense is compatible with Bernard and Durlauf (1995).

The relationship between cross-section and time series convergence tests is complicated. Bernard and Durlauf (1996) argue that the two classes of tests are based on different assumptions about the data under study. Cross-section tests assume that countries are in transition to a steady state, so that the data for a given country at time t is drawn from a different stochastic process from the data at some future $t + T$. In contrast, time series tests assume that the underlying stochastic processes are time-invariant parameters, that is, that countries have transited to an invariant output process. They further indicate how convergence under a cross-section test can in fact imply a failure of convergence under a time series test, because of these different assumptions. For these reasons, time series tests of convergence seem appropriate for economies that are at similar stages and advanced stages of development.

From statistics to economics

The various concepts of convergence we have described are all purely statistical definitions. The economic questions that motivated these definitions are not, however, equivalent to these questions, so it is important to consider convergence as an economic concept in order to assess what is learned in the statistical studies. As argued in Durlauf, Johnson and Temple (2005), the economic questions that underlie convergence study revolve around the respective roles of initial conditions versus structural heterogeneity in explaining differences in per capita output levels or growth rates. It is the permanent effect of initial conditions, not structural features that matters for convergence. If we define initial conditions as $\rho_{i,0}$ and the structural characteristics as $\theta_{i,0}$, convergence can be defined via

$$\lim_{t \to \infty} E(\log y_{i,t} - \log y_{j,t} | \rho_{i,0}, \theta_{i,0}, \rho_{j,0}, \theta_{j,0})$$
$$= 0 \text{ if } \theta_{i,0} = \theta_{j,0}.$$

$$(6)$$

The gap between the definition (6) and the statistical tests that have been employed is evident when one considers whether the statistical tests can differentiate between economically interesting growth models, some of which fulfil (6) and others of which do not. One such contrast is between the Solow growth model and the Azariadis and Drazen (1990) model of threshold externalities, in which countries will converge to one of several possible steady states, with initial conditions determining which one emerges. By definition (6), the Solow model produces convergence whereas the Azariadis–Drazen model does not. However, as shown by Bernard and Durlauf (1996) it is possible for data from the Azariadis–Drazen model to produce estimates that are consistent with a finding of β-convergence.

There is in fact a range of empirical findings of growth nonlinearities that are inconsistent with convergence in the sense of (6). Durlauf and Johnson (1995) is an early study of this type, which explicitly estimated a version of the Azariadis–Drazen model in which the Solow model, under the assumption of a Cobb–Douglas aggregate production function, is a special case. Durlauf and Johnson rejected the Solow model specification and found multiple growth regimes indexed by initial conditions. Their findings are consistent with the presence of convergence clubs in which different groups of countries are associated with one of several possible steady states. These results are confirmed by Papageorgiou and Masanjala (2004) using a CES production function specification.

The Durlauf and Johnson analysis uses a particular classification procedure, known as a regression tree, to identify groups of countries obeying a common linear model. Other statistical approaches have also identified convergence clubs. For example, Bloom, Canning and Sevilla (2003) use mixture distribution methods to model countries as associated with one of two possible output processes, and conclude that individual countries may be classified into high-output manufacturing- and service-based economies and low-output agriculture-based economies. Canova (2004) uses Bayesian methods to identify convergence clubs for European regions.

As discussed in Durlauf and Johnson (1995) and Durlauf, Johnson and Temple (2005), studies of nonlinearity also suffer from identification problems with respect to questions of convergence. One problem is that a given data-set cannot fully uncover the full nature of growth nonlinearities without strong additional assumptions. As a result, it becomes difficult to extrapolate those relationships between predetermined variables and growth to infer steady-state behaviour. Durlauf and Johnson give an example of a data pattern that is compatible with both a single steady and multiple steady states. A second problem concerns the interpretation of the conditioning variables in these exercises. Suppose one finds, as do Durlauf and Johnson, that high- and low-literacy economies are associated with different aggregate production functions. One interpretation of this finding is that the literacy rate proxies for unobserved fixed factors, for example culture, so that these two sets of economies will never obey a common production function, and so will never exhibit convergence in the sense of (6). Alternatively, the aggregate production function could structurally depend on the literacy rate, so that, as literacy increases, the aggregate production functions of currently low-literacy economies will converge to those of the high-literacy ones. Data analyses of the type that have appeared cannot distinguish between these possibilities.

Conclusions

While the empirical convergence literature contains many interesting findings and has helped identify a number of important generalizations about cross-country growth behaviour, it has yet to reach any sort of consensus on the deep economic questions for which the statistical analyses were designed. The fundamentally nonlinear nature of endogenous growth theories renders the conventional cross-section convergence tests inadequate as ways to discriminate between the main classes of theories. Evidence of convergence clubs may simply be evidence of deep nonlinearities in the transitional dynamics towards a unique steady state. Cross-section and time series approaches to convergence not only yield different results but are predicated on different views of the nature of transitory versus steady-state behaviour of economies, differences that themselves have yet to be tested.

None of this is to say that convergence is an empirically meaningless question. Rather, progress requires continued attention to the appropriate statistical definition of convergence and the use of statistical procedures consistent with the definition. Further, it seems important to move beyond current ways of assessing convergence both in terms of better use of economic theory and by a broader view of appropriate data sources. Graham and Temple (2006) illustrate the potential for empirical analyses of convergence that employ well-delineated structural models. The research programme developed in Acemoglu, Johnson and Robinson (2001; 2002) provides a perspective on the micro-foundations of country-specific heterogeneity that speaks directly to the convergence question and which shows the power of empirical analysis based on careful attention to economic history. For these reasons, research on convergence should continue to be productive and important.

STEVEN N. DURLAUF AND PAUL A. JOHNSON

See also **economic growth, empirical regularities in; endogenous growth theory; neoclassical growth theory; neoclassical growth theory (new perspectives).**

Bibliography

Acemoglu, D., Johnson, S. and Robinson, J. 2001. The Colonial origins of comparative development: an empirical investigation. *American Economic Review* 91, 1369–401.

Acemoglu, D., Johnson, S. and Robinson, J. 2002. Reversal of fortune: geography and institutions in the making of the modern world income distribution. *Quarterly Journal of Economics* 117, 1231–94.

Anderson, G. 2004. Making inferences about the polarization, welfare, and poverty of nations: a study of 101 countries 1970–1995. *Journal of Applied Econometrics* 19, 530–50.

Azariadis, C. and Drazen, A. 1990. Threshold externalities in economic development. *Quarterly Journal of Economics* 105, 501–26.

Barro, R. 1991. Economic growth in a cross-section of countries. *Quarterly Journal of Economics* 106, 407–43.

Barro, R. and Sala-i-Martin, X. 1992. Convergence. *Journal of Political Economy* 100, 223–51.

Bernard, A. and Durlauf, S. 1995. Convergence in international output. *Journal of Applied Econometrics* 10(2), 97–108.

Bernard, A. and Durlauf, S. 1996. Interpreting tests of the convergence hypothesis. *Journal of Econometrics* 71, 1–2, 161–73.

Bloom, D., Canning, D. and Sevilla, J. 2003. Geography and poverty traps. *Journal of Economic Growth* 8, 355–78.

Canova, F. 2004. Testing for convergence clubs in income per capita: a predictive density approach. *International Economic Review* 45, 49–77.

Cannon, E. and Duck, N. 2000. Galton's fallacy and economic convergence. *Oxford Economic Papers* 53, 415–19.

Caselli, F., Esquivel, G. and Lefort, F. 1996. Reopening the convergence debate: a new look at cross country growth empirics. *Journal of Economic Growth* 1, 363–89.

Doppelhofer, G., Miller, R. and Sala-i-Martin, X. 2004. Determinants of long-term growth: a Bayesian averaging of classical estimates (BACE) approach. *American Economic Review* 94, 813–35.

Durlauf, S. and Johnson, P. 1995. Multiple regimes and cross-country growth behaviour. *Journal of Applied Econometrics* 10, 365–84.

Durlauf, S., Johnson, P. and Temple, J. 2005. Growth econometrics. In *Handbook of Economic Growth*, ed. P. Aghion and S. Durlauf. Amsterdam: North-Holland.

Durlauf, S. and Quah, D. 1999. The new empirics of economic growth. In *Handbook of Macroeconomics*, ed. J. Taylor and M. Woodford. Amsterdam: North-Holland.

Evans, P. 1996. Using cross-country variances to evaluate growth theories. *Journal of Economic Dynamics and Control* 20, 1027–49.

Fernandez, C., Ley, E. and Steel, M. 2001. Model uncertainty in cross-country growth regressions. *Journal of Applied Econometrics* 16, 563–76.

Friedman, M. 1992. Do old fallacies ever die? *Journal of Economic Literature* 30, 2129–32.

Graham, B. and Temple, J. 2006. Rich nations, poor nations: how much can multiple equilibria explain? *Journal of Economic Growth* 11, 5–41.

Hobijn, B. and Franses, P. 2000. Asymptotically perfect and relative convergence of productivity. *Journal of Applied Econometrics* 15, 59–81.

Islam, N. 1995. Growth empirics: a panel data approach. *Quarterly Journal of Economics* 110, 1127–70.

Lee, K., Pesaran, M. and Smith, R. 1997. Growth and Convergence in multi country empirical stochastic Solow model. *Journal of Applied Econometrics* 12, 357–92.

Maasoumi, E., Racine, J. and Stengos, T. 2007. Growth and convergence: a profile of distribution dynamics and mobility. *Jounal of Econometrics* 136(2) 483–508.

Mankiw, N., Romer, D. and Weil, D. 1992. A contribution to the empirics of economic growth. *Quarterly Journal of Economics* 107, 407–37.

Papageorgiou, C. and Masanjala, W. 2004. The Solow model with CES technology: nonlinearities with parameter heterogeneity. *Journal of Applied Econometrics* 19, 171–201.

Quah, D. 1993a. Galton's fallacy and tests of the convergence hypothesis. *Scandinavian Journal of Economics* 95, 427–43.

Quah, D. 1993b. Empirical cross-section dynamics in economic growth. *European Economic Review* 37, 426–34.

Quah, D. 1996. Convergence empirics across economies with (some) capital mobility. *Journal of Economic Growth* 1, 95–124.

convex programming

1 Introduction

Firms maximize profits and consumers maximize preferences. This is the core of microeconomics, and under conventional assumptions about decreasing returns it is an application of convex programming. The paradigm of convex optimization, however, runs even deeper through economic analysis. The idea that competitive markets perform well, which dates back at least to Adam Smith, has been interpreted since the neoclassical revolution as a variety of conjugate duality for the primal optimization problem of finding Pareto-optimal allocations. The purpose of this article and the companion article DUALITY is (in part) to explain this sentence. This article surveys without proof the basic mathematics of convex sets and convex optimization with an eye towards their application to microeconomic and general equilibrium theory, some of which can be found under DUALITY.

Unfortunately there is no accessible discussion of concave and convex optimization outside textbooks and monographs of convex analysis such as Rockafellar (1970; 1974). Rather than just listing theorems, then, this article attempts to provide a sketch of the main ideas. It is certainly no substitute for the sources. This article covers only convex optimization in finite-dimensional vector spaces. While many of these ideas carry over to infinite-dimensional vector spaces and to important applications in infinite horizon economies and economies with non-trivial uncertainty, the mathematical subtleties of infinite-dimensional topological vector spaces raise issues which cannot reasonably be treated here. The reader looking only for a statement of the Kuhn–Tucker theorem is advised to read backwards from the end, to find the theorem and notation.

A word of warning. This article is written from the perspective of constrained maximization of concave functions because this is the canonical problem in microeconomics. Mathematics texts typically discuss the constrained minimization of convex functions, so textbook treatments will look slightly different.

2 Convex sets

A subset C of a Euclidean vector space V is convex if it contains the line segment connecting any two of its members. That is, if x and y are vectors in C and t is a number between 0 and 1, the vector $tx + (1 - t) y$ is also in C. A linear combination with non-negative weights which sum to 1 is a *convex combination* of elements of C; a set C is convex if it contains all convex combinations of its elements.

The key fact about convex sets is the famous *separation theorem*. A linear function p from the vector space V to **R** and a real number a define a *hyperplane*, the solutions to the equation $p \cdot x = a$. Every hyperplane divides V into two *half-spaces*; the upper (closed) half-space, containing those vectors x for which $p \cdot x \geq a$, and the lower (closed) half-space, containing those vectors x for which $p \cdot x \leq a$. The separation theorem uses linear functionals to describe closed convex sets. If a given vector is not in a closed convex set, then there is a hyperplane such that the set lies strictly inside the upper half-space while the vector lies strictly inside the lower half-space:

Separation theorem If C is a closed convex set and x is not in C, then there is a linear functional p and a real number a such that $p \cdot y > a$ for all $y \in C$, and $p \cdot x < a$.

This theorem implies that every closed convex set is the intersection of the half-spaces containing it. This half-space description is a *dual* description of closed convex sets, since it describes them with linear functionals. From the separation theorem the existence of a *supporting hyperplane* can also be deduced. If x is on the boundary of a closed convex set C, then there is a (non-zero) linear functional p such that $p \cdot y \geq p\, x$ for all $y \in C$; p is the hyperplane that supports C at x.

The origin of the term 'duality' lies in the mathematical construct of the dual to a vector space. The *dual space* of a vector space V is the collection of all *linear functionals*, that is, real-valued linear functions, defined on V. The distinction between vector spaces and their duals is obscured in finite dimensional spaces because each such space is its own dual. If an n-dimensional Euclidean vector space is represented by column vectors of length n, the linear functionals are $1 \times n$ matrices; that is the dual to $\mathbf{R}^n$ is $\mathbf{R}^n$. (This justifies the notation used above.) Self-duality (called reflexivity in the literature) is not generally true in infinite-dimensional spaces, which is reason enough to avoid discussing them here. Nonetheless, although V will be $\mathbf{R}^n$ throughout this article, the usual notation V^* will be used to refer to the dual space of V simply because it is important to know when we are discussing a vector in V and when we are discussing a member of its dual, a linear functional on V.

If the weights in a linear combination sum to 1 but are not constrained to be non-negative, then the linear combination is called an *affine combination*. Just as a convex set is a set which contains all convex combinations of its elements, an affine set in a vector space V is a set which contains all affine combinations of its elements. The set containing all affine combinations of elements in a given set C is an affine set, $A(C)$. The purpose of all this is to define the *relative interior* of a convex set C, ri C. The relative interior of a convex set C is the interior of C relative to $A(C)$. A line segment in $\mathbf{R}^2$ has no interior, but its relative interior is everything on the segment but its endpoints.

3 Concave functions

The neoclassical assumptions of producer theory imply that production functions are concave and cost functions are convex. The quasi-concave functions which arise in consumer theory share much in common with concave functions, and quasi-concave programming has a rich duality theory.

In convex programming it is convenient to allow concave functions to take on the value $-\infty$ and convex functions to take on the value $+\infty$. A function f defined on $\mathbf{R}^n$ with range $[-\infty, \infty)$ is concave if the set $\{(x,a): a \in, a \leq f(x)\}$ is convex. This set, a subset of $\mathbf{R}^{n+1}$, is called the *hypograph* of f and is denoted hypo f. Geometrically, it is the set of points in $\mathbf{R}^{n+1}$ that lie on or below the graph of f. Similarly, the *epigraph* of f is the set of points in $\mathbf{R}^{n+1}$ that lie on or above the graph of f: epi $f = \{(x,a): a \in, a \geq f(x)\}$. A function f with range $(-\infty, \infty]$ is convex $-f$ is concave, and convexity of f is equivalent to convexity of the set epi f. Finally, the *effective domain* of a concave function is the set dom $f = \{x \in \mathbf{R}^n : f(x) > -\infty\}$, and similarly for a convex function. Those familiar with the literature will note that attention here is restricted to *proper* concave and convex functions. Functions that are everywhere $+\infty$ will also be considered concave, and those everywhere $-\infty$ will be assumed convex when Lagrangeans are discussed below.

Convex optimization does not require that functions be differentiable or even continuous. Our main tool is the separation theorem, and for that closed convex sets are needed. A concave function f is *upper semi-continuous* (usc) if its hypograph is closed; a convex function is *lower semi-continuous* (lsc) if its epigraph is closed. Upper and lower semi-continuity apply to any functions, but these concepts interact nicely conveniently with convex and concave functions. In particular, usc concave and lsc convex functions are continuous on the relative interiors of their domain. A famous example of a usc concave function that fails to be continuous is $f(x,y) = -y^2/2x$ for $x > 0$, 0 at the origin and $-\infty$ otherwise. Along the curve $y = \sqrt{\alpha x}$, $y \to 0$ as $x \to 0$, but f is constant at $-\alpha/2$, so f is not continuous at $(0,0)$, but it is usc because the supremum of the limits at the origin is 0.

It is useful to know that, if f is concave and usc, then $f(x) = \inf q(x) = a \cdot x + b$ where the infimum is taken over all a and b such that $a \cdot x + b$ is everywhere at least as big as f. This is another way of saying that, since hypo f is closed, it is the intersection of all half-spaces containing it.

4 The Fenchel transform

The concave Fenchel transform associates with each usc function on a Euclidean space V, not necessarily concave, a usc concave function on its dual space V^* (which, we recall, happens to be V since its dimension is finite). The adjective 'concave' is applied because a similar transform is defined slightly differently for convex functions. The concave Fenchel transform of f is

$$f^*(p) = \inf_{x \in V} \{p \cdot x - f(x)\},$$

which is often called the *conjugate* of f. (From here on out we will drop the braces.) The conjugate f^* of f is concave because, for fixed x, $p \cdot x - f(x)$ is linear, hence concave, in p, and the pointwise infimum of concave functions is concave. The textbooks all prove that, if hypo f is closed, so is hypo f^*, that is, upper semi-continuity is preserved by conjugation. So what is this transformation doing, and why is it interesting?

The conjugate f^* of a concave function f describes all the non-vertical half-spaces containing hypo f. This should be checked. A half-space in $\mathbf{R}^{n+1}$ can be represented by the inequality $(p, q) \cdot (x, y) \geq a$ where q is a real number (as is a) and $p \in V^*$. The half-space is non-vertical if $p^* \neq 0$. In $\mathbf{R}^2$ this means geometrically that the line defining the boundary of the half-space is not vertical. So choose a linear functional $p \neq 0$ in V^*. For any $(x, z) \in$ hypo f, and any $p \in V^*$,

$$p \cdot x - z \geq p \cdot x - f(x)$$
$$\geq \inf_{x \in V} p \cdot x - f(x)$$
$$= f^*(p).$$

In other words, the upper half-space $(p, -1) \cdot (x, z) \geq f^*(p)$ contains hypo f. It actually supports hypo f because of the infimum operation: If $a > f^*(p)$, there is an $(x, z) \in$ hypo f such that $p \cdot x - z < a$, so the upper half-space fails to contain hypo f.

Before seeing what the Fenchel transform is good for, we must answer an obvious question. If it is good to transform once, why not do it again? Define

$$f^{**}(x) = \inf_{p \in V^*} p \cdot x - f^*(p),$$

the *double dual* of f. The fundamental fact about the Fenchel transform is the following theorem, which is the function version of the dual descriptions of closed convex sets.

Conjugate duality theorem If f is usc and concave, then $f^{**} = f$.

This is important enough to explain. Notice that just as p is a linear functional acting on x, so x is a linear functional acting on p. Suppose that f is concave and usc. For all x and p, $p \cdot x - f(x) \geq f^*(p)$, and so $p \cdot x - f^*(p) \geq f(x)$. Taking the infimum on the left, $f^{**}(x) \geq f(x)$.

On the other hand, take a $p \in V^*$ and a real number b such that the half-space $(p, -1) \cdot (x, z) \geq b$ in $\mathbf{R}^{n+1}$ contains the hypograph of f. This is true if and only if $p \cdot x - b \geq f(x)$ for all x and because f is usc, $f(x)$ is the infimum of $p \cdot x - b$ over all such p and b. Since $p \cdot x - f(x) \geq b$ for all x, take the infimum on the left to conclude that $f^*(p) \geq b$. Thus $p \cdot x - b \geq p \cdot x - f^*(p)$, and taking the infimum now on the right, $p \cdot x - b \geq f^{**}(x)$. Taking the infimum on the left over all the p and b such that the half-space contains hypo f, $f(x) \geq f^{**}(x)$.

It is worthwhile to compute an example to get the feel of the concave Fenchel transform. If C is a closed convex set, the *concave indicator function* of C is $\phi(x)$ which is 0 for x in C, and $-\infty$ otherwise. This is a good example to see the value of allowing infinite values. The Fenchel

transform of ϕ is $\phi^*(p) = \inf_{x \in \mathbf{R}^n} p \cdot x - \phi(x)$. Clearly the infimum cannot be reached at any $x \notin C$, for the value of ϕ at such an x is $-\infty$, and so the value of $p \cdot x - \phi(x)$ is $+\infty$. Consequently $\phi^*(p) = \inf_{x \in C} p \cdot x$. This function has the enticing property of positive homogeneity: If t is a positive scalar, then $\phi^*(tp) = t\phi^*(p)$.

Compute the double dual, first for $x \notin C$. The separating hyperplane theorem claims the existence of some p in V^* and a real number a such that $p \cdot x < a \leq p \cdot y$ for all $y \in C$. Take the infimum on the right to conclude that $p \cdot x < \phi^*(p)$, which is to say that $p \cdot x - \phi^*(p) < 0$. Then, multiply both sides by an arbitrary positive scalar t to conclude that $tp \cdot x - \phi^*(tp)$ can be made arbitrarily negative. Hence $\phi^{**}(x) = -\infty$ if $x \notin C$. And if x is in C? Then $p \cdot x - 0 \geq \phi^*(p)$ for all p (recall $\phi(x) = 0$). So $p \cdot x - \phi^*(p) \geq 0$. But $\phi^*(0) = 0$, so $\phi^{**}(x)$, the infimum of the left-hand side over all possible p functionals, is 0. Thus the Fenchel transform of ϕ^* recovers ϕ.

A particularly interesting version of this problem is to suppose that C is an 'at least as good as' set for level u of some upper semi-continuous and quasi-concave utility function (or, more generally, a convex preference relation with closed weak upper contour sets). Then $\phi^*(p)$ is just the minimum expenditure necessary to achieve utility u at price p. See DUALITY for more discussion. Another interesting exercise is to apply the Fenchel transform to concave functions which are not usc, and to non-concave functions. These constructions have important applications in optimization theory which we will not pursue.

The theory of convex functions is exactly the same if, rather than the *concave Fenchel transform*, the *convex Fenchel transform* is employed: $\sup_{x \in \mathbf{R}^n} p \cdot x - f(x)$. This transform maps convex lsc functions on V into convex lsc functions on V^*. Both the concave and convex Fenchel transforms will be important in what follows.

5 The subdifferential

The separation theorem applied to hypo f implies that usc concave functions have tangent lines: For every $x \in$ ri dom f there is a linear functional p_x such that $f(y) \leq f(x) + p_x(y - x)$. This inequality is called the *subgradient inequality*, and p_x is a *subgradient* of f; p_x defines a tangent line for the graph of f, and the graph lies on or underneath it. The set of subgradients of f at $x \in$ dom f is denoted $\partial f(x)$, and is called the *subdifferential* of f at x. Subdifferentials share many of the derivative's properties. For instance, if $0 \in \partial f(x)$, then x is a global maximum of f. In fact, if $\partial f(x)$ contains only one subgradient p_x, then f is differentiable at x and $Df(x) = p_x$. The set $\partial f(x)$ need not be single-value, however, because f may have kinks. The graph of the function f defined on the real line such that $f(x) = -\infty$ for $x < 0$ and $f(x) = \sqrt{x}$ for $x \geq 0$ illustrates why the subdifferential may be empty at the boundary of the effective domain. At 0, a subgradient would be infinitely steep.

There is a corresponding *subgradient inequality* for convex $f : f(y) \geq f(x) + p_x \cdot (y - x)$. With these definitions, $\partial(-f)(x) = -\partial f(x)$. Note that some texts refer to superdifferentials for concave functions and subdifferentials for convex functions. Others do not multiply the required terminology, and we follow them.

The multivalued map $x \mapsto \partial f(x)$, is called the *subdifferential correspondence* of f. An important property of subdifferential correspondences is monotonicity. From the subgradient inequality, if $p \in \partial f(x)$ and $q \in \partial f(y)$, then $f(y) \leq f(x) + p \cdot (y - x)$ and $f(x) \leq f(y) + q \cdot (x - y)$, and it follows that $(p-q) \cdot (x-y) \leq 0$. For convex f the inequality is reversed.

The Fenchel transforms establish a clear relationship between the subdifferential correspondences of concave functions and their duals. If f is concave, then the subdifferential inequality says that $p \in \partial f(x)$ if and only if for all $z \in X$, $p \cdot x - f(x) \leq p \cdot z - f(z)$. The map $z \mapsto p \cdot z - f(z)$ is minimized at $z=x$, and so p is in $\partial f(x)$ if and only if $f^*(p) = p \cdot x - f(x)$. If f is usc, then $f^{**} \equiv f$, and so $f^{**}(x) = f(x) = p \cdot x - f^*(x)$. That is, $p \in \partial f(x)$ if and only if $x \in \partial f^*(p)$.

6 Optimization and duality

Economics most often presents us with constrained maximization problems. Within the class of problems with concave objective functions, there is no formal difference between constrained and unconstrained maximization. The constrained problem of maximizing concave and usc f on a closed convex set C is the same as the unconstrained problem of maximizing $f(x) + I_C(x)$ on $\mathbf{R}^n$, where $I_C(x)$ is the concave indicator function of C.

The general idea of duality schemes in optimization theory is to represent maximization (or minimization) problems as half of a minimax problem which has a saddle value. There are several reasons why such a seemingly odd construction can be useful. In economics it often turns out that the other half of the minimax problem, the dual problem, sheds additional light on properties and interpretations of the primal problem. This is the source of the 'shadow price' concept: The shadow price is the value of relaxing a constraint. Perhaps the most famous example of this is the Second Theorem of Welfare Economics.

6.1 Lagrangeans

The *primal problem* (problem P) is to maximize $f(x)$ on a Euclidean space V. Suppose there is a function $L : V \times V^* \to \mathbf{R}$ such that $f(x) = \inf_{p \in V^*} L(x, p)$. Define $g(p) = \sup_{x \in V} L(x, p)$, and consider the problems of maximizing $f(x)$ on V and minimizing $g(p)$ on V^*. The first problem is the primal problem, and the second is called the *dual problem*. For all x and p it is clear that $f(x) \leq L(x, p) \leq g(p)$, and thus that

$$\sup_x \inf_p L(x, p) = \sup_x f(x) \leq \inf_p g(p) = \inf_p \sup_x L(x, p).$$

If the inequality is tight, that is, it holds with equality, then the common value is called a *saddle value* of L. In particular, a saddle value exists if there is a *saddle point* of L, a pair (x^*, p^*) such that for all $x \in V$ and $p \in V^*$, $L(x, p^*) \leq L(x^*, p^*) \leq L(x^*, p)$. A pair (x^*, p^*) is a saddlepoint if and only if x^* solves the primal, p^* solves the dual, and a saddle value exists. The function L is the *Lagrangean*, which is familiar from the analysis of smooth constrained optimization problems. Here it receives a different foundation.

The art of duality schemes is to identify an interesting L, and here is where the Fenchel transforms come in. Interesting Lagrangeans can be generated by embedding the problem max f in a parametric class of concave maximization problems. Suppose that there is a (Euclidean) parameter space P, and a usc and concave function $F: V \times Y \to$ such that $f(x)=F(x, 0)$, and consider all the problems $\max_{x \in V} F(x, y)$. A particularly interesting object of study is the *value function* $\phi(y) = \sup_x F(x, y)$, which is the indirect utility function in consumer theory, and the cost function in the theory of the firm (with concave replaced by convex and max by min). The map $y \mapsto -F(x, y)$ is closed and convex for each x, so define on $V \times V^*$

$$L(x, p) = \sup_y p \cdot y + F(x, y),$$

its (convex) Fenchel transform. The map $p \mapsto L(x, p)$ is closed and convex on V^*. Transform again to see that $F(x, y) = \inf_{p \in V^*} L(x, p) - p \cdot y$. In particular, $f(x) = \inf_p L(x, p)$.

An example of this scheme is provided by the usual concave optimization problem given by a concave objective function f, K concave constraints $g_k(x) \geq 0$, and an implicit constraint $x \in C$: $\max_x f(x)$ subject to the constraints $g_k(x) \geq 0$ for all k and $x \in C$. Introduce parameters y so that $g_k(x) \geq y_k$, and define $F(x, y)$ to be $f(x)$ if all the constraints are satisfied and $-\infty$ otherwise. The supremum defining the Lagrangean cannot be realized for y such that x is infeasible, and so

$$L(x, p) = \begin{cases} f(x) + \sum_k p_k g_k(x) & \text{if } x \in C \text{ and } p \in \mathbf{R}_+^K, \\ +\infty & x \in C \text{ and } p \notin \mathbf{R}_+^K, \\ -\infty & \text{if } x \notin C \end{cases} \tag{1}$$

if there are feasible x, then $F(x,y)$ is everywhere $-\infty$, and so $L(x, p) \equiv -\infty$.

Here, in summary, are the properties of the Lagrangean for the problems discussed here:

Lagrangean theorem If $F(x, y)$ is lsc and concave then (1) the Lagrangean L is lsc and convex in p for each $x \in V$, (2) L is concave in x for each $p \in V^*$, and (3) $f(x) = \inf_p L(x, p)$.

Following the original scheme, the objective for the dual problem is $g(p) = \sup_x L(x,p)$, and the *dual problem* (problem D) is to maximize g on V^*. Perhaps the central fact of this dual scheme is the relationship between the dual objective function g and the value function ϕ. The function ϕ is easily seen to be concave, and simply by writing out the definitions, one sees that $g(p) = \sup_y p \cdot y + \phi(y)$, the convex Fenchel transform of the convex function $-\phi$. So $g(p)$ is lsc and convex, $g(p) = (-\phi)^*(p)$ and whenever ϕ is usc, $\inf_p g(p) = \phi(0)$.

To make the duality scheme complete, the min problem should be embedded in a parametric class of problems in a complementary way. Take $G(p,q) = \sup_{x \in V} L(x,p) - q \cdot x$, so that $g(p) = G(p,0)$. With this definition, $-G(p,q) = \inf_{x \in V} q \cdot x - L(x,p)$, the concave Fenchel transform of $x \mapsto L(x,p)$. The value function for the parametric class of minimization problems is $\gamma(q) = \inf_p G(p,q)$. The relationship between F and G is computed by combining the definitions:

$$G(p,q) = \sup_{x,y} F(x,y) - q \cdot x + p \cdot y$$

$$= -\inf_{x,y} q \cdot x - p \cdot y - F(x,y)$$

$$= -F^*(q,-y) \text{ and so}$$

$$F(x,y) = \inf_{p,q} G(p,q) + q \cdot x - p \cdot y$$

where the F^* is the concave Fenchel transform of the map $(x,y) \mapsto F(x,y)$. Computing from the definitions, $f(x) = \inf_{p,q} q \cdot x + G(p,q) = \inf_q q \cdot x + \gamma(q)$, so $f = (-\gamma)^*$, and whenever γ is lsc, $\sup_x f(x) = \gamma(0)$.

In summary, if $F(x,y)$ is concave in its arguments, and usc, then we have constructed a Lagrangean and a dual problem of minimizing a concave and lsc $G(p,q)$ over p. If the value functions $\phi(y)$ and $\gamma(q)$ are usc and lsc, respectively, then $\sup_x F(x,0) = \gamma(0)$ and $\inf_p G(p,0) = \phi(0)$, so a saddle value exists. Upper and lower semi-continuity of the value functions can be an issue. The hypograph of ϕ is the set of all pairs (y,a) such that $\sup_x F(x,y) \geq a$, and this is the projection onto y and a of the set of all triples (x,y,a) such that $F(x,y) \geq a$, that is, hypo F. Unfortunately, even if hypo F is closed, its projection may not be, so upper semi-continuity of ϕ does not follow from the upper semi-continuity of F.

In the constrained optimization problem with Lagrangean (1), the parametric class of dual minimization problems is to minimize $G(p,q) = \sup_{x \in C} f(x) + \sum_k p_k g_k(x) - q \cdot x$ if $y \in \mathbf{R}_+^K$ and $+\infty$ otherwise. Specialize this still further by considering linear programming. The canonical linear program is to *max* $a \cdot x$ subject to the explicit constraints $b_k \cdot x \leq c_k$ and the implicit constraint $x \geq 0$. Rewrite the constraints as $-b_k \cdot x + c_k \geq 0$ to be consistent with the formulation of (1).

Then

$$G(p,q) = \sup_{x \geq 0} a \cdot x - \sum_k p_k(b_k \cdot x - c_k) - q \cdot x$$

$$= \sum_k c_k p_k + \sup_{x \geq 0} \left(a - \sum_k p_k b_k - q \right) \cdot x$$

for $p \in \mathbf{R}_+^K$ and $+\infty$ otherwise. The dual problem is to minimize this over p. The sup term in G will be $+\infty$ unless the vector in parentheses is non-positive, in which case the sup will be 0. So the dual problem, taking $q = 0$, is to minimize $\sum_k c_k p_k$ over p subject to the constraints that $\sum_k p_k b_k \geq a$ and $p \in \mathbf{R}_+^K$. If the primal constraints are infeasible, $\phi(0) = -\infty$. If the dual is infeasible, $\gamma(0) = +\infty$, and this serves as an example of how the dual scheme can fail over lack of continuity. For linear programs there is no problem with the hypographs of ϕ and γ, because these are *polyhedral convex sets*, the intersection of a finite number of closed half-spaces, and projections of closed polyhedral convex sets are closed.

6.2 Solutions

Subdifferentials act like partial derivatives, particularly with respect to identifying maxima and minima: x^* in V solves problem P if and only if $0 \in \partial f(x^*)$. When f is identically $-\infty$, there are no solutions which satisfy the constraints. Thus dom f is the set of *feasible solutions* to the primal problem P. Similarly, $p^* \in V^*$ solves the dual problem D if and only if $\partial g(p^*) = 0$, and here dom g is the set of *dual-feasible solutions*. Saddlepoints of the Lagrangean also have a subdifferential characterization. Adapting the obvious partial differential notion and notation, (x^*, p^*) is a saddle point for L if and only if $0 \in \partial_x L(x^*, p^*)$ and $0 \in \partial_p L(x^*, p^*)$ (these are different 0's since they live in different spaces), which we write $(0,0) \in \partial L(x^*, p^*)$. This condition is often called the *Kuhn–Tucker* condition. The discussion so far can be summarized in the following theorem, which is less general than can be found in the sources:

Kuhn–Tucker theorem Suppose that $F(x, y)$ is concave and usc. Then the following are equivalent:

1. $\sup f = \inf g$,
2. ϕ is usc and concave,
3. the saddle value of the Lagrangean L exists,
4. γ is lsc and convex. In addition, the following are equivalent:
5. x^* solves P, p^* solves D, and the saddle value of the Lagrangean exists.
6. (x^*, p^*) satisfy the Kuhn–Tucker condition. For economists, the most interesting feature of the dual is that it often describes how the value of the primal problem will vary with parameters. This follows from properties of the subdifferential and the key relation between the primal value function and the dual objective function, $g = (-\phi)^*: -\partial \phi(0) = \partial(-\phi)(0)$, and this

equals the set $\{p : (-\phi)^*(p) = p \cdot 0 - (-\phi)(0)\}$, and this is precisely the set $\{p : g(p) = \sup_x f(x)\}$. In words, if p is a solution to the dual problem D, then $-p$ is in the subgradient of the primal value function. When $\partial\phi(0)$ is a singleton, there is a unique solution to the dual, and it is the derivative of the value function with respect to the parameters. More generally, from the subdifferential of a convex function one can construct directional derivatives for particular changes in parameter values. Similarly, $-\partial\gamma(0) = \{x : f(x) = \inf_p g(p)\}$, with an identical interpretation. In summary, add to the Kuhn–Tucker theorem the following equivalence:

7. $-p^* \in \partial\phi(0)$ and $-x^* \in \partial\gamma(0)$

The remaining question is, when is any one of these conditions satisfied? A condition guaranteeing that the subdifferentials are non-empty is that $0 \in$ ri dom ϕ, since concave functions always have subdifferentials on the relative interior of their effective domain. In the constrained optimization problem whose Lagrangean is described in (1), an old condition guaranteeing the existence of saddlepoints is the *Slater condition*, that there is an $x \in$ ri C such that for all k, $g_k(x) > 0$. This condition implies that $0 \in$ ri dom ϕ, because there is an open neighbourhood around 0 such that for y in the neighbourhood and for all k, $g_k(x) > y_k$. Thus $\phi(y) \geq F(x,y) > -\infty$ for all y in the neighbourhood. Conditions like this are called *constraint qualifications*. In the standard calculus approach to constrained optimization, they give conditions under which derivatives sufficiently characterize the constraint set for calculus approximations to work (see Arrow, Hurwicz and Uzawa, 1961).

Finally, it is worth noting that infinite dimensional constrained optimization problems, such as those arising in dynamic economic models and the study of uncertainty, can be addressed with extensions of the methods discussed here. The main difficulty is that most infinite dimensional vector spaces are not like $\mathbf{R}^n$. There is no 'natural' vector space topology, and which topology one chooses has implications for demonstrating the existence of optima. The existence of separating hyperplanes is also a difficulty in infinite dimensional spaces. These and other problems are discussed in Mas-Colell and Zame (1991). Nonetheless, much of the preceding development does go through. See Rockafellar (1974).

LAWRENCE E. BLUME

See also **convexity; duality; Lagrange multipliers; quasi-concavity.**

Bibliography

Arrow, K., Hurwicz, L. and Uzawa, H. 1961. Constraint qualifications in maximization problems. *Naval Logistics Research Quarterly* 8, 175–91.

Mas-Colell, A. and Zame, W. 1991. Equilibrium theory in infinite dimensional spaces. In *Handbook of Mathematical Economics*, vol. 4, ed. W. Hildenbrand and H. Sonnenschein. Amsterdam: North-Holland.

Rockafellar, R.T. 1970. *Convex Analysis*. Princeton, NJ: Princeton University Press.

Rockafellar, R.T. 1974. *Conjugate Duality and Opttimization*. CBMS Regional Conference Series No. 16. Philadelphia: SIAM.

convexity

Convexity is the modern expression of the classical law of diminishing returns, which was prominent in political economy from Malthus and Ricardo through the neo-classical revolution. Its importance today rests less on any utilitarian or behavioural psychological rationale or physical principle than on its utility as a tool of mathematical analysis. In general equilibrium and game theory, proofs of the existence of equilibrium, competitive and Nash, respectively, rely on the application of a fixed-point theorem to a set-valued, convex-valued map from a convex set to itself. Welfare economics provides another example: The second theorem of welfare economics, which asserts that optimal allocations can be supported by competitive prices, relies on an application of the supporting hyperplane theorem to an appropriate convex set.

Convexity is a property of real vector spaces, and its domain of application in economic analysis is not just Euclidean spaces but also the infinite dimensional vector spaces which arise in the study of uncertainty and dynamics, where infinite numbers of goods are required. Nonetheless, this brief exposition will be confined to Euclidean spaces.

Definitions

A set $C \subset \mathbf{R}^n$ is *convex* if the line segment connecting any two points in C lies wholly within C. Formally put, C is convex if and only if for all points x and y in C and all scalars t in the unit interval $[0, 1]$, the point $tx + (1-t)y$ is also in C. A ball is convex; a boomerang is not. An extended real-valued function f defined on a convex set $C \subset \mathbf{R}^n$ is *convex* if its epigraph or supergraph, $\{(x, \mu) : x \in C, \mu \in \mathbf{R}, f(x) \leq \mu\}$ is convex. For real-valued functions, this is equivalent to the more familiar definition that for all x and y in C and $t \in [0,1]$, $f(tx + (1-t)y) \geq tf(x) + (1-t)f(y)$. A function f is *concave* if $-f$ is convex.

Optimization

Students of economics first encounter convexity in the study of optimization. If $x^* \in \mathbf{R}^n$ is a critical point of a smooth function, and if x^* a local maximum, then the

Hessian matrix at x^*, the matrix of second-order partial derivatives, must be negative semi-definite; that is, it is locally concave. Any critical point with a negative definite Hessian must be a local maximum. Negative definiteness of the Hessian implies but is not implied by strict (local) concavity. For Jevons, utility was additively separable, and so the principle of diminishing marginal utility itself was enough to derive concavity. Edgeworth, the first economist to consider non-separable utility functions, realized that diminishing marginal utility was not, in general, enough to guarantee convexity. His development of demand theory relied on a differential condition that can be shown to imply *quasi-concavity*. A real-valued function f with a convex domain C is *quasi-concave* if for each real number α, the set $\{x \in C : f(x) \geq \alpha\}$ is convex. To appreciate the difference between concavity and quasi-concavity, note that any strictly increasing function on the real line is quasi-concave. The differential description of convexity and its variants (quasi-convexity, pseudo-convexity) and the associated necessary and sufficient second-order conditions for constrained optimization problems has produced a volume of analysis, most of which is of second-order importance to contemporary economic theory. Exhaustive coverage can be found in Simon and Blume (1994).

Duality

The representation of consumers by expenditure functions and firms by profit functions is said to be 'dual' to the 'primal' representations by preferences and production sets, respectively. These representations rely on alternative ways of representing closed convex sets: The 'primal' description is a list of its elements, and the 'dual' description is the list of closed half-spaces containing it. The dual representation for closed convex sets is equivalent to the separating hyperplane theorem: If x in $\mathbf{R}^n$ is not in a closed convex set C, then there is a hyperplane $H \subset \mathbf{R}^n$ with x on one side and C on the other. That is, there is a $p \in \mathbf{R}^n$ and a number α such that $p \cdot x < \alpha$ and, $p \cdot y > \alpha$ for all $y \in C$. (See CONVEX PROGRAMMING and DUALITY.)

Large numbers and convexity

Convexity is sometimes an inappropriate assumption. Half a box of two left shoes and half a box of two right shoes is surely preferred to either box, but the 50:50 mixture of a good burgundy and a good stout is only a headache. Fortunately, the analysis of perfectly competitive markets rests not on the preferences of any individual consumer, but on the average behaviour of a large number of consumers. A central insight behind much research of the 1970s and 1980s (and which was anticipated by Edgeworth, 1881, a century before) is that averaging is a convexifying operation. This is the content of the Shapley–Folkman theorem as applied to large finite economies, and Lyapunov's theorem in the analysis of economies with a continuum of agents. (See CORES, LARGE ECONOMIES and PERFECT COMPETITION.) For economies with large numbers of small consumers and small firms, the important analytical constructs are approximately convex. With respect to the existence of equilibrium and its welfare properties, large economies look like convex economies. Hildenbrand (1974) is an entry point to this important body of research.

LAWRENCE E. BLUME

See also **convex programming; cores; duality; large economies; perfect competition.**

Bibliography

Edgeworth, F.Y. 1881. *Mathematical Psychics*. London: C. Kegan Paul & Co.
Hildenbrand, W. 1974. *Core and Equilibria of a Large Economy*. Princeton: Princeton University Press.
Simon, C. and Blume, L. 1994. *Mathematics for Economists*. New York: W.W. Norton & Co.

convict labour

Some European countries banished convicts to labour in overseas colonies – sometimes using private markets to transport and employ this labour.

Punishing felons who did not warrant execution and were too poor to pay monetary fines posed a dilemma for early modern societies. The long-standing punishments of one-off physical chastisements, such as whippings, increasingly seemed too barbaric and returned malefactors to society too quickly. While long-term incarceration was more civilized and removed malefactors from society, penitentiaries were expensive to build and operate, and the criminal's labour was lost to society. Sentencing felons to labour in overseas colonies thus became an attractive solution.

Between 1854 and 1920 France sent between 20,000 and 30,000 convicts to French Guiana and New Caledonia. Spain sent convicts to North Africa, Cuba, and Puerto Rico. Britain, however, was the largest participant, sending 6,000–10,000 convicts to its colonies between 1614 and 1718 and another 50,000 mostly to its American colonies Virginia and Maryland between 1718 and 1775 (Coldham, 1992; Ekirch, 1987). After the United States closed its shores to British convicts, convict transportation was shifted to Australia where approximately 160,000 were landed between 1787 and 1868 (Nicholas, 1988). Another 18,000 were shipped to Bermuda and Gibraltar.

The Transportation Act of 1718 shifted the overseas banishment of British felons from a case-by-case petitioning of the Crown to a routine sentence imposed by courts. The sentences allowed were seven years, 14 years, or a lifetime of banishment – 74 per cent, 24 per cent and two per cent, respectively, of those transported – which became the length of the convict's overseas labour contract. Most transported convicts were guilty of property crimes and were Englishmen. Between 13 per cent and 23 per cent were Irish, and between 10 per cent and 15 per cent were female (Ekirch, 1987). Sentences were not rigidly tied to crimes; for example, highway robbers received 7-year, 14-year, and lifetime sentences – 38 per cent, 50 per cent, and 12 per cent, respectively (Grubb, 2000). Not until convicts had completed their sentences could they return to Britain without facing being hanged if caught.

The privatization of overseas convict disposal reached its zenith after the Transportation Act. The government minimized its cost of overseas convict disposal by channelling convicts through the existing competitive markets for voluntary servant labour, where emigrants traded forward-labour contracts to shippers for passage to America. Shippers recouped their cost by selling these contracts (emigrants) to private employers in America. Potential shipping profits related to labour heterogeneity were arbitraged away by bargaining over contract length. The typical voluntary servant negotiated a four-year labour contract and sold in America for eight and half pounds sterling. By contrast, courts fixed the length of convict sentences (labour contracts) independently of labour heterogeneity. Convicts were then transferred to private shippers for transportation overseas. Shippers sold their convicts as servant labour to private employers in America to recoup their shipping expense. The average convict sold for 11 pounds sterling (Grubb, 2000).

By fixing contract lengths – the parameter used to arbitraged shipping profits in the voluntary servant market – the courts altered the convict auction price distribution and profit arbitrage process from that which existed in the voluntary servant market. The distribution of convict contract prices had a higher mean, higher standard deviation, and lower kurtosis than that of voluntary servant contract prices. Shippers did not earn excess profits on convicts. The higher sale price was matched by the higher cost of chaining convicts during shipment and paying variable fees charged by county jailers. Jailers played shippers off against each other for access to convict cargo. The government subsidized one shipper in the London market who earned, net of political bribes, excess profits (Grubb, 2000).

Shippers carried both voluntary and convict servants concurrently. Potential employers were shown the conviction papers that stated each convict's sentence and crime. Post-auction convicts were largely indistinguishable from voluntary servants. Most were employed in agriculture and at iron forges alongside slaves and voluntary servants. They lived in their employer's house and ate at their employer's table. Criminal conviction, however, carried a stigma that led to price discounts. A year's worth of convict labour sold for a 21 per cent discount on average over that of comparable voluntary servant labour. Convicts guilty of more serious and professional crimes, such as arsonists and receivers of stolen goods, sold for even greater discounts. Convicts also ran away more often than did voluntary servants: 16 per cent versus six per cent, respectively (Grubb, 2001).

Per given crime, a 14-year versus a 7-year sentence signalled the courts' perception of the severity of harm inflicted by, and incorrigibility of, the convict. American employers responded to this information by demanding additional price discounts of 48 per cent and 68 per cent per year of labour for convicts sentenced to 14 years and to life, respectively, as opposed to seven years for the same crime. Employers also paid premiums and received discounts for certain convict attributes, other things equal. For example, taller convicts sold for a substantial premium, and female convicts with venereal disease sold for 19 per cent less than females without the disease (Grubb, 2001).

For underpopulated colonies lacking competitive labour markets, such as Australia, European governments typically had to transport convicts to the colonies themselves, directly employing them on government projects there (Nicholas, 1988). During the 19th century, European governments also became increasingly reluctant to use existing competitive markets to auction convict labour for fear that it would look like government-sanctioned slavery. Instead, convicts were transferred via bureaucratic petition or assignment systems. Under these conditions, the system struggled to employ convict labour efficiently and to be a cost-effective punishment. Convict transportation waned as social reformers succeeded in replacing it with incarceration in newly built penitentiaries and as maturing colonies increasingly resisted being convict dumping-grounds.

FARLEY GRUBB

See also **auctions (empirics); compensating differentials; human capital, fertility and growth; indentured servitude; international migration; labour market institutions.**

Bibliography

Coldham, P. 1992. *Emigrants in Chains*. Baltimore, MD: Genealogical Publishing.

Ekirch, A. 1987. *Bound for America: The Transportation of British Convicts to the Colonies, 1718–1775*. New York: Oxford University Press.

Grubb, F. 2000. The transatlantic market for British convict labor. *Journal of Economic History* 60, 94–122.

Grubb, F. 2001. The market evaluation of criminality: evidence from the auction of British convict labor in America, 1767–1775. *American Economic Review* 91, 295–304.

Nicholas, S., ed. 1988. *Convict Workers: Reinterpreting Australia's Past*. New York: Cambridge University Press.

cooperation

Cooperation is said to occur when two or more individuals engage in joint actions that result in mutual benefits. Examples include the mutually beneficial exchange of goods, the payment of taxes to finance public goods, team production, common pool resource management, collusion among firms, voting for income redistribution to others, participating in collective actions such as demonstrations, and adhering to socially beneficial norms.

A major goal of economic theory has been to explain how wide-scale cooperation among self-regarding individuals occurs in a decentralized setting. The first thrust of this endeavour involved Walras's general equilibrium model, culminating in the celebrated 'invisible hand' theorem of Arrow and Debreu (Arrow and Debreu, 1954; Debreu, 1959; Arrow and Hahn, 1971). But, the assumption that contracts could completely specify all relevant aspects of all exchanges and could be enforced at zero cost to the exchanging parties is not applicable to many important forms of cooperation. Indeed, such economic institutions as firms, financial institutions, and state agencies depend on incentive mechanisms involving strategic interaction in addition to explicit contracts (Blau, 1964; Gintis, 1976; Stiglitz, 1987; Tirole, 1988; Laffont, 2000).

The second major thrust in explaining cooperation eschewed complete contracting and developed sophisticated repeated game-theoretic models of strategic interaction. These models are based on the insights of Shubik (1959), Taylor (1976), Axelrod and Hamilton (1981) and others that repetition of social interactions plus retaliation against defectors by withdrawal of cooperation may enforce cooperation among self-regarding individuals. A statement of this line of thinking, applied towards understanding the broad historical and anthropological sweep of human experience is the work of Ken Binmore (1993; 1998; 2005). For Binmore, a society's moral rules are instructions for behaviour in conformity with one of the myriad of Nash equilibria of a repeated n-player social interaction. Because the interactions are repeated, and these rules form a Nash equilibrium, the self-regarding individuals who comprise the social order will conform to the moral rules.

We begin by reviewing models of repeated dyadic interaction in which cooperation may occur among players who initially cooperate and in the next round adopt the action of the other player in the previous round, called *tit for tat*. These models show that as long as the probability of game repetition is sufficiently great and individuals are sufficiently patient, a cooperative equilibrium can be sustained once it is implemented. This reasoning applies to a wide range of similar strategies. We then analyse *reputation maintenance* models of dyadic interaction, which are relevant when individuals interact with many different individuals, and hence the number of periods before a repeat encounter with any given individual may be too great to support the tit-for-tat strategy.

We then turn to models of cooperation in larger groups, arguably the most relevant case, given the scale on which cooperation frequently takes place. The folk theorem (Fudenberg and Maskin, 1986) shows that, in groups of any size, cooperation can be maintained on the assumption that the players are sufficiently future-oriented and termination of the interaction is sufficiently unlikely. We will see, however, that these models do not successfully extend the intuitions of the dyadic models to many-person interactions. The reason is that the level of cooperation that may be supported in this way deteriorates as group size increases and the probability of either behavioural or perceptual error rises, and because the theory lacks a plausible account of how individuals would discover and coordinate on the complicated strategies necessary for cooperation to be sustained in these models. This difficulty bids us investigate how other-regarding preferences, *strong reciprocity* in particular, may sustain a high level of cooperation, even with substantial errors and in large groups.

Repetition allows cooperation in groups of size two

Consider a pair of individuals who play the following *stage game* repeatedly: each can *cooperate* (that is, help the other) at a cost $c > 0$ to himself, providing a benefit to the other of $b > c$. Alternatively, each player can *defect*, incurring no cost and providing no benefit. Clearly, both would gain by cooperating in the stage game, each receiving a net gain of $b - c > 0$. However, the structure of the game is that of a *Prisoner's Dilemma*, in which a self-regarding player earns higher payoff by defecting, no matter what his partner does.

The behaviour whereby each individual provides aid as long as this aid has been reciprocated by the other in the previous encounter, is called *tit for tat*. Although termed 'reciprocal altruism' by biologists, this behaviour is self-regarding, because each individual's decisions depend only on the expected net benefit the individual enjoys from the long-term relationship.

On the assumption that after each round of play the interaction will be continued with probability δ, and that players have discount factor d (so $d = 1/(1 + r)$, where r is the rate of time preference), then provided

$$\delta db > c, \tag{1}$$

each individual paired with a tit-for-tat player does better by cooperating (that is, playing tit for tat) rather than by defecting. Thus tit for tat is a best response to itself. To see this, let **v** be the present value of cooperating when paired with a tit-for-tat player. Then

$$\mathbf{v} = b - c + \delta d \, \mathbf{v}, \tag{2}$$

which gives

$$\mathbf{v} = \frac{b - c}{1 - \delta d}. \tag{3}$$

The present value of defecting for ever on a tit-for-tat playing partner is b (the single period gain of b being followed by zero gains in every subsequent period as a result of the tit-for-tat player's defection), so playing tit-for-tat is a best response to itself if and only if $(b-c)(1-\delta d) > b$, which reduces to (1). Under these conditions unconditional defect is also a best response to itself, so either cooperation or defection can be sustained.

But suppose that, instead of defection for ever, the alternative to tit for tat is for a player to defect for a certain number of rounds, before returning to cooperation on round $k > 0$. The payoff to this strategy against tit for tat is $b - (\delta d)^k c + (\delta d)^{k+1} \mathbf{v}$. This payoff must not be greater than $\mathbf{v}$ if tit for tat is to be a best response to itself. It is an easy exercise in algebra to show that the inequality

$$\mathbf{v} \geq b - (\delta d)^k c + (\delta d)^{k+1} \mathbf{v}$$

simplifies to (1), no matter what the value of k. A similar argument shows that when (1) holds, defecting for ever (that is, $k = \infty$) does not have a higher payoff than cooperating.

Cooperation through reputation maintenance

Tit for tat takes the form of frequent repetition of the Prisoner's Dilemma stage game inducing a pair of self-regarding individuals to cooperate. In a sizable group, an individual may interact frequently with a large number of partners but infrequently with any single one, say on the average of once every k periods. Players then discount future gains so that a payoff of $\mathbf{v}$ in k periods from now is worth $d^k \mathbf{v}$ now. Then, an argument parallel to that of the previous section shows that cooperating is a best response if and only if

$$\frac{b-c}{1-\delta d^k} > b$$

which reduces to

$$\delta d^k b > c. \qquad (4)$$

Note that this is the same equation as (1) except that the effective discount factor falls from d to d^k. For sufficiently large k, it will not pay to cooperate. Therefore, the conditions for tit-for-tat reciprocity will not obtain.

But cooperation may be sustained in this situation if each individual keeps a mental model of exactly which group members cooperated in the previous period and which did not. In this case, players may cooperate in order to cultivate a *reputation for cooperation*. When individuals tend to cooperate with others who have a reputation for cooperation, a process called *indirect reciprocity* can sustain cooperation. Let us say that an individual who cooperated in the previous period *in good standing*, and specify that the only way an individual can fall into *bad standing* is by defecting on a partner who is in good standing. Note that an individual can always defect when his partner is in bad standing without losing his good standing status. In this more general setting the tit-for-tat strategy is replaced by the following *standing strategy*: cooperate if and only if your current partner is in good standing, except that, if you accidentally defected the previous period, cooperate this period unconditionally, thereby restoring your status as a member in good standing. This *standing model* is due to Sugden (1986).

Panchanathan and Boyd (2004) have proposed an ingenious deployment of indirect reciprocity, assuming that there is an ongoing dyadic helping game in society based on the indirect reciprocity information and incentive structure, and there is also an n-player public goods game, played relatively infrequently by the same individuals. In the dyadic helping game, two individuals are paired and each member of the pair may confer a benefit b upon his partner at a cost c to himself, an individual remaining in good standing so long as he does not defect on a partner who is in good standing. This random pairing is repeated with probability δ and with discount factor d. In the public goods game, an individual produces a benefit b_g that is shared equally by all the other members, at a cost c_g to himself. The two games are linked by defectors in the public goods game being considered in bad standing at the start of the helping game that directly follows. Then, cooperation can be sustained in both the public goods game and in the dyadic helping game so long as

$$c_g \leq \frac{b(1-\varepsilon)-c}{1-\delta d}, \qquad (5)$$

where ε is the rate at which cooperators unintentionally fail to produce the benefit. Parameters favouring this solution are that the cost c_g of cooperating in the public goods game be low, the factor δd be close to unity, and the net benefit $b(1-\varepsilon)-c$ of cooperating in the reputation-building reciprocity game be large.

The major weakness of the standing model is its demanding informational requirements. Each individual must know the current standing of each member of the group, the identity of each member's current partner, and whether each individual cooperated or defected against his current partner. Since dyadic interactions are generally private, and hence are unlikely to be observed by more than a small number of others, errors in determining the standing of individuals may be frequent. This contrasts sharply with the repeated game models of the previous section, which require only that an individual know how many of his current partners defected in the previous period. Especially serious is that warranted non-cooperation (because in one's own mental accounting one's partner is in bad standing) may be perceived to be unwarranted defection by some third parties but not by others. This will occur with high frequency if information partially private rather than public (not everyone has the same information). It has been proposed that gossip and other forms of communication can transform private into public information, but how this might occur among self-regarding individuals has not been (and probably cannot be) shown, because in any practical setting individuals may benefit by reporting dishonestly on what they have observed, and self-regarding individuals do not care about the harm to others induced by false information. Under such conditions, disagreements among individuals about who ought to be punished can reach extremely high levels, with the unravelling of cooperation as a result.

In response to this weakness of the standing model, Nowak and Sigmund (1998) developed an indirect reciprocity model which they term *image scoring*. Players in the image scoring need not know the standing of recipients of aid, so the informational requirements of indirect reciprocity are considerably reduced. Nowak and Sigmund show that the strategy of cooperating with others who have cooperated in the past, *independent of the reputation of the cooperator's partner*, is stable against invasion by defectors, and weakly stable against invasion by unconditional cooperators once defectors are eliminated from the population. Leimar and Hammerstein (2001), Panchanathan and Boyd (2003), and Brandt and Sigmund (2004; 2005), explore the applicability of image scoring.

Cooperation in large groups of self-regarding individuals

Repeated game theory has extended the above two-player results to a general n-player stage game, the so-called *public goods game*. In this game each player cooperates at cost $c > 0$, contributing an amount $b > c$ that is shared equally among the other $n-1$ players. We define the *feasible payoff set* as the set of possible payoffs to the various players, assuming each cooperates with a certain probability, and each player does at least as well as the payoffs obtaining under mutual defection. The set of feasible payoffs for a two-player public goods game is given in Figure 1 by the four-sided figure ABCD. For the n-player game, the figure ABCD is replaced by a similar n-dimensional polytope.

Repeated game models have demonstrated the so-called folk theorem, which asserts that any distribution of payoffs to the n players that lies in the feasible payoff set can be supported by an equilibrium in the repeated public goods game, provided the discount factor times the probability of continuation, δd, is sufficiently close to unity. The equilibrium concept employed is a refinement of subgame perfect equilibrium. Significant contributions to this literature include Fudenberg and Maskin (1986), assuming perfect information, Fudenberg, Levine and Maskin (1994), assuming imperfect information, so that cooperation is sometimes inaccurately reported as defection, and Sekiguchi (1997), Piccione (2002), Ely and

Välimäki (2002), Bhaskar and Obara (2002) and Mailath and Morris (2006), who assume that different players receive different, possibly inaccurate, information concerning the behaviour of the other players.

The folk theorem is an *existence theorem* affirming that any outcome that is a Pareto improvement over universal defection may be supported by a Nash equilibrium, including point C (full cooperation) in the figure and outcomes barely superior to A (universal defection). The theorem is silent on which of this vast number of equilibria is more likely to be observed or how they might be attained. When these issues are addressed two problems are immediately apparent: first, equilibria in the public goods game supported in this manner exhibit very little cooperation if large numbers of individuals are involved or errors in execution and perception are large, and second, the equilibria are not robust because they require some mechanism allowing coordination on highly complex strategies. While such a mechanism could be provided by centralized authority, decentralized mechanisms, as we will see, are not sustainable in a plausible dynamic.

The dynamics of cooperation

The first difficulty, the inability to support high levels of cooperation in large groups or with significant behavioural or perceptual noise, stems from the fact that the only way players may punish defectors is *to withdraw their own cooperation*. In the two-person case, defectors are thus targeted for punishment. But for large n, withdrawal of cooperation to punish a single defector punishes all group members equally, most of whom, in the neighbourhood of a cooperative equilibrium, will be cooperators. Moreover, in large groups, the rate at which erroneous signals are propagated will generally increase with group size, and the larger the group, the larger the fraction of time group members will spend punishing (miscreants and fellow cooperators alike). For instance, suppose the rate at which cooperators accidentally fail to produce b, and hence signal defection, is five per cent. Then, in a group of size two, a perceived defection will occur in about ten per cent of all periods, while in a group of size 20, at least one perceived defection will occur in about 64 per cent of all periods.

As a result of these difficulties, the folk theorem assertion that we can approximate the per-period expected payoff as close to the efficient level (point C in Figure 1) as desired as long as the discount factor δ is sufficiently close to unity is of little practical relevance. The reason is that as $\delta \to 1$, the current payoff approximates zero, and the expected payoff is deferred to future periods at very little cost, since future returns are discounted at a very low rate. Indeed, with the discount factor δ held constant, the efficiency of cooperation in the Fudenberg, Levine and Maskin model declines at an exponential rate with increasing group size (Bowles and Gintis, 2007,

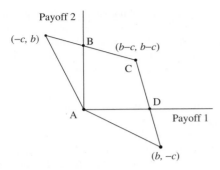

Figure 1   Two-player public goods game

ch. 13). Moreover, in an agent-based simulation of the public goods with punishment model, on the assumption of a benefit/cost ratio of $b/c = 2$ (that is, contributing to the public good costs half of the benefit conferred on members of the group) and a discount factor times probability of repetition of $d\delta = 0.96$, even for an error rate as low as $\varepsilon = 0.04$, fewer than half of the members contribute to the public good in groups of size $n = 4$, and less that 20 per cent contribute in groups of size $n = 6$ (Bowles and Gintis, 2007, ch. 5).

The second limitation of the folk theorem analysis is that it has not been shown (and probably cannot be shown) that the equilibria supporting cooperation are dynamically robust, that is, asymptotically stable with a large basin of attraction in the relevant dynamic. Equilibria for which this is not the case will seldom be observed because they are unlikely to be attained and if attained unlikely to persist for long.

The Nash equilibrium concept applies when each individual expects all others to play their parts in the equilibrium. But, when there are multiple equilibria, as in the case of the folk theorem, where there are many possible patterns of response to given pattern of defection, each imposing distinct costs and requiring distinct, possibly stochastic, behaviours on the part of players, there is no single set of beliefs and expectations that group members can settle upon to coordinate their actions (Aumann and Brandenburger, 1995).

While game theory does not provide an analysis of how beliefs and expectations are aligned in a manner allowing cooperation to occur, sociologists (Durkheim, 1902; Parsons and Shils, 1951) and anthropologists (Benedict, 1934; Boyd and Richerson, 1985; Brown, 1991) have found that virtually every society has such processes, and that they are key to understanding strategic interaction. Borrowing a page from sociological theory, we posit that groups may have *focal rules* that are common knowledge among group members. Focal rules could suggest which of a countless number of strategies that could constitute a Nash equilibrium should all individuals adopt them, thereby providing the coordination necessary to support cooperation. These focal rules do not ensure equilibrium, because error, mutation,

migration, and other dynamical forces ensure that on average not all individuals conform to the focal rules of the groups to which they belong. Moreover, a group's focal rules are themselves subject to dynamical forces, those producing better outcomes for their members displacing less effective focal rules.

In the case of the repeated public goods game, which is the appropriate model for many forms of large-scale cooperation, Gintis (2007) shows that focal rules capable of supporting the kinds of cooperative equilibria identified by the folk theorem are not evolutionarily stable, meaning that groups whose focal rules support highly cooperative equilibria do worse than groups with less stringent focal rules, and as a result the focal rules necessary for cooperation are eventually eliminated.

The mechanism behind this result can be easily explained. Suppose a large population consists of many smaller groups playing n-person public goods games, with considerable migration across groups, and with the focal rules of successful groups being copied by less successful groups. To maintain a high level of cooperation in a group, focal rules should foster punishing defectors by withdrawing cooperation. However, such punishment is both costly and provides an external benefit to other groups by reducing the frequency of defection-prone individuals who might migrate elsewhere. Hence, groups that 'free ride' by not punishing defectors harshly will support higher payoffs for its members than groups that punish assiduously. Such groups will then be copied by other groups, leading to a secular decline in the frequency of punishment suggested by focal rules in all groups. Thus, suppose that the groups in question were competitive firms whose profits depend on the degree of cooperation among firm members. If all adopted a zero-tolerance rule (all would defect if even a single defection was perceived), then a firm adopting a rule that tolerated a single defection would sustain higher profits and replace the zero-tolerance firms. But this firm would in turn be replaced by a firm adopting a rule that tolerates two defections.

These two problems – the inability to support efficient levels of cooperation in large groups with noisy information, and dynamic instability – have been shown for the case where information is public. Private information, in general the more relevant case, considerably exacerbates these problems.

Cooperation with other-regarding individuals

The models reviewed thus far have assumed that individuals are entirely self-regarding. But cooperation in sizable groups is possible if there exist other-regarding individuals in the form of *strong reciprocators*, who cooperate with one another and punish defectors, even if they sustain net costs. Strong reciprocators are altruistic in the standard sense that they confer benefits on other members of their group (in this case, because their altruistic punishment of defectors sustains cooperation) but would increase their own payoffs by adopting self-regarding behaviours. A model with *social preferences* of this type can explain large-scale decentralized cooperation with noisy information as long as the information structure is such that defectors expect a level of punishment greater than costs of cooperating.

Cooperation is not a puzzle if a sufficient number of individuals with social preferences are involved. The puzzle that arises is how such altruistic behaviour could have become common, given that bearing costs to support the benefits of others reduces payoffs, and both cultural and genetic updating of behaviours is likely to favour traits with higher payoffs. This evolutionary puzzle applies to strong reciprocity. Since punishment is costly to the individual, and an individual could escape punishment by cooperating, while avoiding the costs of punishment by not punishing, we are obliged to exhibit a mechanism whereby strong reciprocators could proliferate when rare and be sustained in equilibrium, despite their altruistic behaviour.

This is carried out in Sethi and Somanathan (2001), Gintis (2000), Boyd et al. (2003), Gintis (2003) and Bowles and Gintis (2004). The evolutionary viability of other types of altruistic cooperation is demonstrated in Bowles, Jung-Kyoo and Hopfensitz (2003), Boyd et al. (2003), Bergstrom (1995) and Salomonsson and Weibull (2006). The critical condition allowing the evolution of strong reciprocity and other forms of altruistic social preferences is that individuals with social preferences are more likely than random to interact with others with social preferences. Positive assortment arises in these models due to deliberate exclusion of those who have defected in the past (by ostracism, for example), random differences in the composition of groups (due to small group size and limited between-group mobility), limited dispersion of close kin who share common genetic and cultural inheritance, and processes of social learning such as conformism or group level socialization contributing to homogeneity within groups. As in the repeated game models, smaller groups favour cooperation, but in this case for a different reason: positive assortment tends to decline with group size. But the group sizes that sustain the altruistic preferences that support cooperative outcomes in these models are at least an order of magnitude larger than those indicated for the repeated game models studied above.

In sum, we think that other-regarding preferences provide a compelling account of many forms of human cooperation that are not well explained by repeated game models with self-regarding preferences. Moreover, a number of studies have shown that strong reciprocity and other social preferences are a common human behaviour (Fehr and Gächter, 2000; Henrich et al., 2005) and could have emerged and been sustained in a gene-culture co-evolutionary dynamic under conditions experienced by ancestral humans (Bowles, 2006). The above models also show that strong reciprocity and other social preferences that support cooperation can evolve and persist even when there are many self-regarding players, where group sizes are substantial, and when behavioural or perception errors are significant.

Conclusion: economics and the missing choreographer

The shortcomings of the economic theory of cooperation based on repeated games strikingly replicate those of economists' other main contribution to the study of decentralized cooperation, namely, general equilibrium theory. Both prove the existence of equilibria with socially desirable properties, while leaving the question of how such equilibria are achieved as an afterthought, thereby exhibiting a curious lack of attention to dynamics and out-of equilibrium behaviour. Both purport to model decentralized interactions but on close inspection require a level of coordination that is not explained, but rather posited as a *deus ex machina*. To ensure that only equilibrium trades are executed, general equilibrium theory resorts to a fictive 'auctioneer'. No counterpart to the auctioneer has been made explicit in the repeated-game approach to cooperation. Highly choreographed coordination on complex strategies capable of deterring defection are supposed to materialize quite without the need for a choreographer.

Humans are unique among living organisms in the degree and range of cooperation among large numbers of substantially unrelated individuals. The global division of labour and exchange, the modern democratic welfare state, and contemporary warfare alike evidence our distinctiveness. These forms of cooperation emerged historically and are today sustained as a result of the interplay of self-regarding and social preferences operating under the influence of group-level institutions of governance and socialization that favour cooperators, in part by protecting them from exploitation by defectors.

The norms and institutions that have accomplished this evolved over millennia through trial and error. Consider how real-world institutions addressed two of the shoals on which the economic models foundered. First, the private nature of information, as we have seen, makes it virtually impossible to coordinate the targeted punishment of miscreants. Converting private information about transgressions into public information that can provide the basis of punishment often involves civil or criminal trials, elaborate processes that rely on commonly agreed upon rules of evidence and ethical norms of appropriate behaviour. Even these complex institutions frequently fail to transform the private protestations of innocence and guilt into common knowledge. Second, as in the standing models with private information, cooperation often unravels when the withdrawal of cooperation by the civic-minded intending to punish a defector is interpreted by others as a violation of a cooperative norm, inviting further defections. In all successful modern societies, this problem was eventually addressed by the creation of a corps of specialists entrusted with carrying out the more severe of society's punishments, whose uniforms conveyed the civic purpose of the punishments they meted out, and whose professional norms, it was hoped, would ensure that the power to punish was not used for personal gain. Like court proceedings, this institution works imperfectly. It is hardly surprising then that economists have encountered difficulty in devising simple models of how large numbers of self-regarding individuals might sustain cooperation in a truly decentralized setting.

Modelling this complex process is a major challenge of contemporary science. Economic theory, favouring parsimony over realism, has instead sought to explain cooperation without reference to other-regarding preferences and with minimalist or fictive descriptions of social institutions.

SAMUEL BOWLES AND HERBERT GINTIS

See also **agent-based models; behavioural economics and game theory; behavioural game theory; evolutionary economics; game theory; group selection; public goods experiments; repeated games; social preferences.**

Bibliography

Arrow, K.J. and Debreu, G. 1954. Existence of an equilibrium for a competitive economy. *Econometrica* 22, 265–90.

Arrow, K.J. and Hahn, F. 1971. *General Competitive Analysis*. San Francisco: Holden-Day.

Aumann, R.J. and Brandenburger, A. 1995. Epistemic conditions for Nash equilibrium. *Econometrica* 65, 1161–80.

Axelrod, R. and Hamilton, W.D. 1981. The evolution of cooperation. *Science* 211, 1390–6.

Benedict, R. 1934. *Patterns of Culture*. Boston: Houghton Mifflin.

Bergstrom, T.C. 1995. On the evolution of altruistic ethical rules for siblings. *American Economic Review* 85, 58–81.

Bhaskar, V. and Obara, I. 2002. Belief-based equilibria the repeated Prisoner's Dilemma with private monitoring. *Journal of Economic Theory* 102, 40–69.

Binmore, K. 1993. *Game Theory and the Social Contract: Playing Fair*. Cambridge, MA: MIT Press.

Binmore, K. 1998. *Game Theory and the Social Contract: Just Playing*. Cambridge, MA: MIT Press.

Binmore, K.G. 2005. *Natural Justice*. Oxford: Oxford University Press.

Blau, P. 1964. *Exchange and Power in Social Life*. New York: Wiley.

Bowles, S. 2006. Group competition, reproductive leveling, and the evolution of human altruism. *Science* 314, 1669–72.

Bowles, S. and Gintis, H. 2004. The evolution of strong reciprocity: cooperation in heterogeneous populations. *Theoretical Population Biology* 65, 17–28.

Bowles, S. and Gintis, H. 2007. A cooperative species: human reciprocity and its evolution, in preparation.

Bowles, S., Jung-Kyoo, C. and Hopfensitz, A. 2003. The co-evolution of individual behaviors and social institutions. *Journal of Theoretical Biology* 223, 135–47.

Boyd, R., Gintis, H., Bowles, S. and Richerson, P.J. 2003. Evolution of altruistic punishment. *Proceedings of the National Academy of Sciences* 100, 3531–5.

Boyd, R. and Richerson, P.J. 1985. *Culture and the Evolutionary Process*. Chicago: University of Chicago Press.

Brandt, H. and Sigmund, K. 2004. The logic of reprobation: assessment and action rules for indirect reciprocation. *Journal of Theoretical Biology* 231, 475–86.

Brandt, H. and Sigmund, K. 2005. Indirect reciprocity, image scoring, and moral hazard. *Proceeding of the National Academy of Sciences* 102, 2666–70.

Brown, D.E. 1991. *Human Universals*. New York: McGraw-Hill.

Debreu, G. 1959. *Theory of Value*. New York: Wiley.

Durkheim, E. 1902. *De La Division du Travail Social*. Paris: Presses Universitaires de France, 1967.

Ely, J.C. and Välimäki, J. 2002. A robust folk theorem for the Prisoner's Dilemma. *Journal of Economic Theory* 102, 84–105.

Fehr, E. and Gächter, S. 2000. Cooperation and punishment. *American Economic Review* 90, 980–94.

Fudenberg, D., Levine, D.K. and Maskin, E. 1994. The Folk Theorem with imperfect public information. *Econometrica* 62, 997–1039.

Fudenberg, D. and Maskin, E. 1986. The Folk Theorem in repeated games with discounting or with incomplete information. *Econometrica* 54, 533–54.

Gintis, H. 1976. The nature of the labor exchange and the theory of capitalist production. *Review of Radical Political Economics* 8(2), 36–54.

Gintis, H. 2000. Strong reciprocity and human sociality. *Journal of Theoretical Biology* 206, 169–79.

Gintis, H. 2003. The hitchhiker's guide to altruism: genes, culture, and the internalization of norms. *Journal of Theoretical Biology* 220, 407–18.

Gintis, H. 2007. *Modeling Cooperation with Self-regarding Agents*. Santa Fe: Santa Fe Institute.

Henrich, J., Boyd, R., Bowles, S., Camerer, C. et al. 2005. Economic man in cross-cultural perspective: behavioral experiments in 15 small-scale societies. *Behavioral and Brain Sciences* 28, 795–815.

Laffont, J.J. 2000. *Incentives and Political Economy*. Oxford: Oxford University Press.

Leimar, O. and Hammerstein, P. 2001. Evolution of cooperation through indirect reciprocity. *Proceedings of the Royal Society of London, Series B* 268, 745–53.

Mailath, G.J. and Morris, S. 2006. Coordination failure in repeated games with almost-public monitoring. *Theoretical Economics* 1, 311–40.

Nowak, M.A. and Sigmund, K. 1998. Evolution of indirect reciprocity by image scoring. *Nature* 393, 573–7.

Panchanathan, K. and Boyd, R. 2003. A tale of two defectors: the importance of standing for evolution of indirect reciprocity. *Journal of Theoretical Biology* 224, 115–26.

Panchanathan, K. and Boyd, R. 2004. Indirect reciprocity can stabilize cooperation without the second-order free rider problem. *Nature* 432, 499–502.

Parsons, T. and Shils, E. 1951. *Toward a General Theory of Action*. Cambridge, MA: Harvard University Press.

Piccione, M. 2002. The repeated Prisoner's Dilemma with imperfect private monitoring. *Journal of Economic Theory* 102, 70–83.

Salomonsson, M. and Weibull, J. 2006. Natural selection and social preferences. *Journal of Theoretical Biology* 239, 79–92.

Sekiguchi, T. 1997. Efficiency in repeated Prisoner's Dilemma with private monitoring. *Journal of Economic Theory* 76, 345–61.

Sethi, R. and Somanathan, E. 2001. Preference evolution and reciprocity. *Journal of Economic Theory* 97, 273–97.

Shubik, M. 1959. *Strategy and Market Structure: Competition, Oligopoly, and the Theory of Games*. New York: Wiley.

Stiglitz, J. 1987. The causes and consequences of the dependence of quality on price. *Journal of Economic Literature* 25, 1–48.

Sugden, R. 1986. *The Economics of Rights, Co-operation and Welfare*. Oxford: Basil Blackwell.

Taylor, M. 1976. *Anarchy and Cooperation*. London: Wiley.

Tirole, J. 1988. *The Theory of Industrial Organization*. Cambridge, MA: MIT Press.

cooperative game theory.   *See* **cores; games in coalitional form; Shapley value.**

cooperative game theory (core).   *See* **core convergence; cores.**

coordination problems and communication

Lewis (1969) defined a *coordination equilibrium* as a Nash equilibrium in which no agent would be better off if any other agent had chosen a different action. When there are multiple coordination equilibria, agents face an obvious coordination problem. The resolution of coordination problems rests upon individuals coming to understand the intentions of one another. The most explicit way of developing this understanding is for the individuals to communicate with one another. Common knowledge of a language must precede communication. Even with common knowledge of a language, individuals may not be bound to do what they say they will do. In such circumstances, talk is 'cheap'.

When will the receiver, having received a message from a sender, behave differently from how the receiver would have behaved if no message had been sent? According to Farrell and Rabin (1996) *highly credible* messages will not be ignored. A message that signals an intention to take action X is highly credible if it satisfies two conditions: it is (*a*) *self-signalling* and (*b*) *self-committing*. A message that the sender is taking action X is self-signalling if, and only if, it is both true and it is in the sender's interest to have it believed to be true. A message is self-committing if a belief by the receiver that the message is true creates an incentive for the sender to do what the sender said he or she would do. A message that is self-committing, if believed, will lead to an outcome that is a Nash equilibrium. A message can be self-committing without being self-signalling. For example, in the classic game of Chicken, if one player announces that he will be Passive, that message is self-committing since, if it is believed by the receiver then the receiver's best response is to be Aggressive, and the best response of the sender to the receiver's aggression is to be Passive. However, the sender would prefer to have the receiver believe that the sender will play Aggression. So the message, 'I intend to play Passive', is not self-signalling because it is not in the interest of the sender to have the receiver believe it is true.

A message is *cheap talk* if the sender is not bound to do what the message says. Crawford (1998) provides a survey of a number of cheap talk experiments. In experiments with structured communication, either only one player may send a message (one-sided communication) or more than one player can send a message. When the payoff functions of the players are symmetrical, one-sided communication breaks the symmetry of the game without communication. This is sufficient to allow a very high level of coordination. Indeed, in such games one-sided communication is much more effective in promoting coordination that is simultaneous, two-sided communication. This suggests that, when payoff functions are symmetric but players have different preference orderings over equilibria, as in the Battle of the Sexes, the principal impact of one-sided communication is to create an extensive form game in which the symmetry is broken by designating one player as the first mover. In games with Pareto-ordered equilibria communication is not needed to break symmetry, but may be effective in reducing uncertainty about the intentions or, in Crawford's terms, to give 'reassurance'. Empirically this 'reassurance' appears to be most effective in achieving coordination on the Pareto-dominant equilibrium when communication is two-sided, but even one-sided communication has a positive effect on the likelihood of achieving the Pareto-optimal outcome. Furthermore, this effect has been found to be greater when a message was self-signalling than when such a message was only self-committing.

When there are multiple players each player must be interested in, and possibly condition his actions on, the entire message profile. Therefore, the concepts of self-signalling and self-committing messages may not have much meaning in this context. Nevertheless, there is some evidence that costless pre-play communication can help groups whose members repeatedly interact to achieve more efficient outcomes than is attainable without such communication (Blume and Ortmann, 2007).

A signal that is commonly observed may be used to coordinate actions even if the signal does not emanate from any of the players. Traffic signals play this role. We do as these signals say we should do because we believe that others will also do what the signals say they should do. This belief is reinforced by experience, so doing as the signals suggest has simply become a convention that is adopted by drivers. While this convention is backed by law, there is good reason to believe that it is so ingrained in people's expectations that they would continue to act as the signals suggest even in the absence of any law. Can signals be effective in coordinating actions when the signals are not sent by any of the players and do not themselves have any payoff consequences? Van Huyck, Gillette and Battalio (1992) found that, when a game has multiple coordination equilibria, all of which yield the same payoff, a signal from an outside 'moderator' that specifically says 'play a particular equilibrium' produces a very high degree of coordination on the suggested equilibrium, even though absent any signal there is a high frequency of coordination failure. However, in games where the equilibria are Pareto ordered the introduction of a recommendation to play any equilibrium other than the payoff-dominant equilibrium significantly reduces the degree of coordination that is achieved. The authors also found that when there was an equilibrium that provided equal payoff a recommendation to play an equilibrium with unequal payoffs had little influence on how the game was played. Evidently some features, such as symmetry, may be sufficiently strong focal points that the introduction of extrinsic signals may have little influence. Similarly, some features of a game may make some coordination equilibria, once achieved by repeated interaction, exceedingly difficult to displace through the introduction of communication, even if everyone would gain by moving to another coordination equilibrium (Cooper, 2006).

A 'sunspot' is a commonly observable event that may have been correlated in the past with different outcomes. For example, published forecasts may have this property. When agents coordinate their actions on a 'sunspot' the resulting equilibrium is called a 'sunspot equilibrium'. Marimon, Spear and Sunder (1993) devised an experiment to see whether they could generate a sunspot equilibrium where prices fluctuate with an extrinsic signal even though the fundamental parameter values remained fixed. During a 'training interval', the colour of a blinking light on a screen was perfectly correlated with a change in a parameter that induced changes in equilibrium prices. After this 'training period' the parameter value was fixed, but the signal continued to vary according to the same process. Prices continued to be volatile but there was little evidence that the variation in the sunspot variable had any effect on the observed price volatility. Duffy and Fisher (2005), using a quite different design, were able to induce sunspot equilibria under restricted conditions. They found that the semantics of the sunspot variable mattered. There were two fundamental equilibria in their design. One equilibrium had a high price, the other a low price. When the sunspot message was either 'high' or 'low' the outcomes of the actions were sometimes correlated with the message. But when the message was either 'sunshine' or 'rain' this correlation was never observed. Evidently, correlation of expectations with the signal depends upon how confident people are that everyone is interpreting the signal in the same way. They also found that information that is generated by observable actions subsequent to the observation of the signal itself tends to diminish the focal power of the signal.

Sometimes actions might 'speak' louder than words. In a Prisoners' Dilemma game the cooperative outcome is not a Nash equilibrium, but it does Pareto-dominate the Nash equilibrium. Since non-cooperation is a dominant strategy a message that one intends to play 'Cooperate' is neither self-committing nor self-signalling. Nevertheless, Duffy and Feltovich (2002) found that when this message was sent it tended to be truthful and also tended to induce a cooperative response. Similarly, when their past actions with other players were observable, subjects were more likely to cooperate than if neither communication nor observability was possible. Furthermore, observation increased the frequency of cooperative choices by more than cheap talk. This suggests that observability of past actions may sometimes be more effective than mere words in helping people achieve a good outcome.

JACK OCHS

See also **cheap talk; experimental economics; game theory.**

Bibliography

Blume, A. and Ortmann, A. 2007. The effects of costless pre-play communication: experimental evidence from games with Pareto-ranked equilibria. *Journal of Economic Theory* 132, 274–90.

Cooper, D. 2006. Are experienced managers experts at overcoming coordination failure? *Advances in Economic Analysis & Policy* 6(2), Article 6.

Crawford, V. 1998. A survey of experiments on communication via cheap talk. *Journal of Economic Theory* 78, 286–98.

Duffy, J. and Feltovich, N. 2002. Do actions speak louder than words? An experimental comparison of observation and cheap talk. *Games and Economic Behavior* 39, 1–27.

Duffy, J. and Fisher, E. 2005. Sunspots in the laboratory. *American Economic Review* 95, 510–29.

Farrell, J. and Rabin, M. 1996. Cheap talk. *Journal of Economic Perspectives* 10(3), 103–18.

Lewis, D. 1969. *Convention: A Philosophical Study.* Cambridge, MA: Harvard University Press.

Marimon, R., Spear, S. and Sunder, S. 1993. Expectationally driven market volatility: an experimental study. *Journal of Economic Theory* 61, 74–103.

Van Huyck, J., Gillette, A. and Battalio, R. 1992. Credible assignments in coordination games. *Games and Economic Behavior* 4, 606–26.

copulas

Sklar introduced copulas in 1959 (Sklar, 1973; 1996). Concisely stated, copulas are functions that connect multivariate distributions to their one-dimensional margins. If F is an m-dimensional continuous cumulative distribution function (CDF) with one-dimensional margins $F_1, \ldots, F_m$, then there exists an m-dimensional unique copula C such that $F(x_1, \ldots, x_m) = C(F_1(x_1), \ldots, F_m(x_m))$. In general, marginal distributions alone cannot determine the joint distributions.

Copulas are useful because, first, they represent a method for deriving joint distributions given the fixed marginals, even when marginals belong to different parametric families of distributions; second, in a bivariate context copulas can be used to define nonparametric measures of dependence for pairs of random variables that can capture asymmetric (tail) dependence as well as correlation or linear association.

Copulas and dependence

We begin with Sklar's theorem. An m-copula is an m-dimensional CDF whose support is contained in $[0,1]^m$ and whose one-dimensional margins are uniform on $[0,1]$. In other words, an m-copula is an m-dimensional distribution function with all m univariate margins being $U(0,1)$. To see the relationship between distribution functions and copulas, consider a continuous m-variate distribution function $F(y_1, \ldots y_m)$ with univariate marginal distributions $F_1(y_1), \ldots, F_m(y_m)$ and inverse probability transforms (quantile functions) $F_1^{-1}, \ldots, F_m^{-1}$. Then $y_1 = F_1^{-1}(u_1) \sim F_1, \ldots, y_m = F_m^{-1}(u_m) \sim F_m$ where $u_1, \ldots, u_m$ are uniformly distributed variates. Copulas are expressed in terms of marginal CDFs. The transforms of uniform variates are distributed as $F_i (i = 1, \ldots, m)$. Hence

$$F(y_1, \ldots, y_m) = F(F_1^{-1}(u_1), \ldots, F_m^{-1}(u_m))$$
$$= C(u_1, \ldots, u_m),$$

$$(1)$$

is the unique copula associated with the distribution function. The copula parameterizes a multivariate

distribution in terms of its marginals. For an m-variate distribution F, the copula satisfies

$$F(y_1, \ldots, y_m) = C(F_1(y_1), \ldots, F_m(y_m); \theta),$$

$$(2)$$

where θ is usually a scalar-valued dependence parameter. For many empirical applications, the dependence parameter is the main focus of estimation. Because the marginal distributions may come from different families, copulas are a 'recipe' for generating joint distributions by combining given marginal distributions using a known copula. This construction allows researchers to consider marginal distributions and dependence as two separate, but related, issues.

The functional form of a copula places restrictions on the dependence structure; for example, it may support only positive dependence. Therefore, a pivotal modelling problem is to choose a copula that adequately captures dependence structures of the data without sacrificing attractive properties of the marginals. Copulas are multivariate distribution functions, hence Fréchet bounds apply. A copula may impose restrictions such that the full coverage between the bounds is not attained.

An important advantage of copulas is that they generate more general measures of dependence than the correlation coefficient. Correlation is a symmetric measure of linear dependence, bounded between $+1$ and -1 and invariant with respect to only linear transformations of the variables. By contrast, copulas have an attractive invariance property: the dependence captured by a copula is invariant with respect to increasing and continuous transformations of the marginal distributions. The same copula may be used for, say, the joint distribution of (Y_1, Y_2) as $(ln\, Y_1, ln\, Y_2)$.

Measures of dependence based on concordance, such as Spearmans's rank correlation (ρ) and Kendall's τ, overcome limitations of the correlation coefficient. In some cases the concordance between extreme (tail) values of random variables is of interest. For example, one may be interested in the probability of the event that stock indexes in two countries exceed (or fall below) given levels. This requires a dependence measure for upper and lower tails of the distribution, rather than a linear correlation measure. Measures of lower and upper tail dependence can be readily derived for a stated copula. The copula dependence parameter θ can be converted to measures of concordance such as Spearman's ρ and Kendall's τ (Nelsen, 1999).

Examples

Nelsen (1999) and Joe (1997) catalogue many functional forms for copulas. Particularly important is the Archimedean class. Bivariate Archimedean copulas take the general symmetric form

$$C(u_1, u_2; \theta) = \varphi^{-1}(\varphi(u_1) + \varphi(u_2)), \quad (3)$$

where the generator function $\varphi(\cdot)$ is a convex decreasing function; for example, $\varphi(t) = -ln(t)$. The dependence parameter θ in imbedded in the functional form of the generator.

The *Clayton copula*, a member of the Archimedean class, takes the form

$$C(u_1, u_2; \theta) = (u_1^{-\theta} + u_2^{-\theta} - 1)^{-1/\theta} \quad (4)$$

with the dependence parameter θ restricted on the region $(0, \infty)$. As θ approaches zero, the marginals become independent. The Clayton copula cannot account for negative dependence. It has been used to study correlated risks because it exhibits strong left tail dependence and relatively weak right tail dependence.

The *Gumbel copula* is another member of the Archimedean class and takes the form

$$C(u_1, u_2; \theta) = \exp(-(\tilde{u}_1^{\theta} + \tilde{u}_2^{\theta})^{1/\theta}) \quad (5)$$

where $\tilde{u}_j = -\log u_j$. The dependence parameter is restricted to the interval $[1, \infty)$. Like the Clayton copula, Gumbel does not allow negative dependence, but in contrast it exhibits strong right tail dependence and relatively weak left tail dependence. If outcomes are strongly correlated at high values but less correlated at low values, then the Gumbel copula is an appropriate choice.

The (non-Archimedean) *Gaussian copula* takes the form

$$C(u_1, u_2; \theta) = \Phi_G(\Phi^{-1}(u_1), \Phi^{-1}(u_2); \theta), \quad (6)$$

where Φ is the CDF of the standard normal distribution, and $\Phi_G(u_1, u_2)$ is the standard bivariate normal distribution with correlation parameter θ restricted to the interval $(-1, 1)$. This copula allows equal degrees of positive and negative dependence.

Figures 1, 2, and 3 illustrate lower and upper tail dependence using three samples generated using Monte Carlo draws from the above three copulas. The samples have comparable degrees of linear dependence but different tail dependence properties.

Estimation and applications

In some applications it would natural to parameterize the marginals in terms of a regression function with covariates z, that is, $u_j = F(y_j | z_j; \beta_j)$, where z_j is a vector of covariates. Then the bivariate copula takes the form $C(y_1, y_2 | z_1, z_2; \beta_1, \beta_2, \theta) = C(F(y_1 | z_1; \beta_1), F(y_2 | z_2; \beta_2); \theta)$. The copula density, defined as

$$\frac{d}{dy_2 dy_1} C(F_1(\cdot), F_2(\cdot); \theta)$$
$$= C_{12}(F_1(\cdot), F_2(\cdot)) f_1(\cdot) f_2(\cdot), \quad (7)$$

$f_j(\cdot) = \partial F_j(\cdot)/\partial y_j$, can be used to build the likelihood, which can be maximized simultaneously with respect to all unknown parameters. Alternatively, the marginal densities can be estimated first, either parametrically or nonparametrically, and then the likelihood can be maximized with respect to θ only at the second stage.

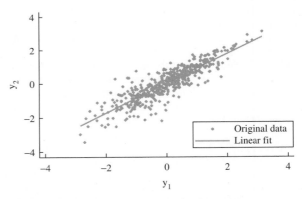

Figure 2   Sample from Gumbel copula, theta = 3.3

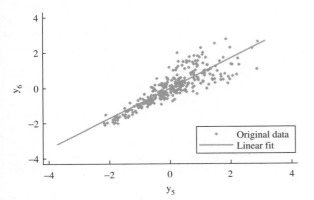

Figure 1   Sample from Clayton copula, theta = 4.67

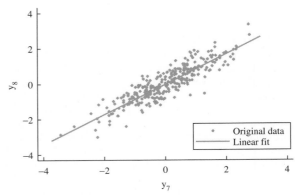

Figure 3   Sample from Gaussian copula, theta = .89

Multivariate models of survival data pioneered the application of copulas. Econometric applications are more recent, but growing rapidly. There are numerous time series and financial market applications of copulas (Cherubini, Luciano and Vecchiato, 2004). Few models in this literature include regressors. Other areas of applications include volatility and exchange rate modelling where GARCH effects and tail dependence are expected (Patton, 2006). Selection models provide leading examples of microeconometric applications of copulas (Smith, 2003; Zimmer and Trivedi, 2006).

PRAVIN K. TRIVEDI

See also **seemingly unrelated regressions; simultaneous equations models.**

Bibliography

Cherubini, U., Luciano, E. and Vecchiato, W. 2004. *Copula Methods in Finance.* New York: John Wiley.

Joe, H. 1997. *Multivariate Models and Dependence Concepts.* London: Chapman and Hall.

Nelsen, R. 1999. *An Introduction to Copulas.* New York: Springer.

Patton, A. 2006. Estimation of multivariate models for time series of possibly different lengths. *Journal of Applied Econometrics* 21, 147–73.

Sklar, A. 1973. Random variables, joint distributions, and copulas. *Kybernetica* 9, 449–60.

Sklar, A. 1996. Random variables, distribution functions, and copulas – a personal look backward and forward. In *Distributions with Fixed Marginals and Related Topics,* ed. L. Ruschendorf, B. Schweizer and M. Taylor. Hayward, CA: Institute of Mathematic Statistics.

Smith, M. 2003. Modeling selectivity using Archimedean copulas. *Econometrics Journal* 6, 99–123.

Zimmer, D. and Trivedi, P. 2006. Using trivariate copulas to model sample selection and treatment effects: application to family health care demand. *Journal of Business and Economic Statistics* 24, 63–76.

core convergence

The core of an economy, first defined by Edgeworth (1881), is the set of all economic outcomes such that no group of individuals ('coalition') can make each of its members better off ('improve on' or 'block' the outcome), using only the resources available to the group. (A common mistake is to ask, in reference to a particular core allocation, 'what coalition(s) have formed?' An allocation is in the core precisely when no coalition can improve on it, and a core allocation does not identify an associated coalition or coalitions. It is when an allocation is *not* in the core that one can identify one or more coalitions that are associated with it, because they can improve on it and thus demonstrate that the coalition is not in the core.)

The most important reason for studying the core is the light it sheds on Walrasian equilibrium, introduced by Walras (1874). While the notion of Walrasian equilibrium is based entirely on the institution of trading via prices, and assumes that individuals take prices as given, the definition of the core is completely institution-free; this is one of its major virtues.

The core has both normative and positive significance apart from its relationship to Walrasian equilibrium. Normatively, if one accepts the distribution of the economy's initial resources as equitable, then any allocation outside the core is unfair to at least one coalition. Regardless of whether the distribution of initial resources is equitable, it would be surprising to find the economy settling on an allocation outside the core, since that would indicate there is a coalition which *could* have made each of its members better off, using only its own resources, but for some reason has failed to coalesce and do so; this is the positive significance.

While there has been much work on the cores of production economies, the bulk of the work on the core has been carried out in exchange economies, in which trading and consuming are the only economic activities. In part, this is because there are a number of competing definitions of the core in production economies, based on how the ownership of the production technology is assigned to individuals and groups. For simplicity, we shall focus our attention on exchange economies.

Walrasian equilibrium is an economic equilibrium notion based on market clearing, mediated by prices. Consumers choose the consumption vector which maximizes utility over their budget sets; firms choose production plans which maximize profit. Critically, it is assumed that individuals and firms take prices as given, without taking into account any ability they may have to influence those prices through their actions. A price vector is a Walrasian equilibrium price if the choices made by individuals and firms, taking prices as given, are consistent in the sense that market supply equals market demand. A Walrasian allocation is the vector of individual consumptions and firm productions generated by a Walrasian equilibrium price. A Walrasian equilibrium is a pair consisting of a Walrasian equilibrium price and its associated Walrasian allocation.

An income transfer is a vector which assigns to each agent a real number, and which satisfies budget balance: the sum of the numbers is zero. An allocation is a Walrasian equilibrium with transfers if there is an income transfer and a price vector such that the demand of each agent, given the prices and the budget of the agent, taking into account the agent's endowment of goods and income transfers, just equals the individual's consumption at the allocation.

The First and Second Welfare Theorems are two of the most important results concerning Walrasian equilibrium. Recall that, in an exchange economy, an allocation is Pareto optimal if there is no reallocation of consumption which makes every agent better off. In other words, the coalition consisting of all agents (coalition of the whole) cannot improve upon the allocation. Thus, it is clear that every core allocation is Pareto optimal.

The First Welfare Theorem asserts that every Walrasian allocation with transfers is Pareto optimal. A slight modification of the proof suffices to show that every Walrasian allocation lies in the core. (Note that it is *not* true that every Walrasian allocation with transfers lies in the core. The income transfers allow us to move consumption among agents. For example, consider the allocation which gives the entire social consumption to a single agent. If we choose a price vector which supports that agent's preference at the social consumption, then there is an income transfer that makes this allocation a Walrasian allocation with transfers. But this allocation will rarely lie in the core, since the coalition consisting of all the other agents will generally be able to improve on it.) This is an important strengthening of the First Welfare Theorem, which has both positive and normative significance. On the positive side, it is a strong stability property of Walrasian equilibrium, since it asserts that no group of individuals would choose to upset the equilibrium by recontracting among themselves, making it more plausible that we will see Walrasian equilibrium arise in real economies. On the normative side, if we accept the distribution of initial endowments as equitable, it tells us that Walrasian allocations are fair to all groups in the economy.

The Second Welfare Theorem asserts that, in an exchange economy with standard assumptions on preferences (convexity is the crucial assumption), every Pareto optimal allocation is a Walrasian equilibrium with transfers. Note that while the definition of Pareto optimality makes no mention of prices, the Second Welfare Theorem asserts that every Pareto optimal allocation is closely associated to a price vector. The price vector appears magically; mathematically, this is a consequence of the separating hyperplane theorem, for which convexity is a critical assumption. As noted above, the most important use of the core is as a test of the price-taking assumption inherent in the definition of Walrasian equilibrium; a number of other tests have been proposed, but *core convergence* is the most commonly used. Core convergence is closely analogous to the conclusion of the Second Welfare Theorem. The definition of the core makes no mention of prices. However, if an exchange economy is sufficiently large, it is a remarkable fact that every core allocation is closely associated with a price vector that 'approximately decentralizes' it; in other words, every core allocation approximately satisfies the definition of Walrasian equilibrium, *without transfers*. This is an important strengthening of the Second Welfare Theorem. The notion of approximate decentralization depends to a considerable

extent on the assumptions one is willing to make on the preferences and endowments of the individuals in the economy. (One version states that core allocations can be realized as *exact* Walrasian equilibrium with *small* income transfers.)

Core convergence has a number of implications, both normative and positive. The extent to which each of these implications is justified in a particular setting depends a great deal on the form of convergence, and thus on the assumptions one is willing to make on the economy. For an extensive survey focusing on the relationship between assumptions and the form of convergence, see Anderson (1992).

On the normative side, core convergence is a strong 'unbiasedness' property of Walrasian equilibrium, since it asserts that restricting attention to Walrasian allocations does not narrow the set of outcomes much beyond the narrowing that occurs in the core. Thus, Walrasian equilibrium has no hidden implications for the welfare of different groups, beyond whatever equity concerns one might have over the initial endowments. If one accepts the distribution of initial endowments as equitable, then any allocation that is far from Walrasian will not be in the core, and hence will treat some group of agents unfairly. On the positive side, if one accepts the core as a positive description of the allocations one is likely to see in practice in any economy, then core convergence tells us that the allocations we see will be nearly Walrasian.

However, the greatest significance of core convergence is as a test of the reasonableness of the price-taking assumption that is hidden in plain sight in the definition of Walrasian equilibrium. In real markets, we see prices used to equate supply and demand, but this does not guarantee Walrasian outcomes. Agents possessing market power may choose to demand quantities different from their price-taking demands at the prevailing price, thereby altering that price and leading to a non-Walrasian outcome. If the outcome is not at least approximately Walrasian, then the welfare theorems and the results on existence and generic determinacy of Walrasian allocations would have limited implications for real economies.

Core convergence and non-convergence allows us to identify situations in which price-taking is more or less reasonable. Core convergence implies that all trade takes place at almost a single price. An agent who tries to bargain cannot influence the prices much, so there is little incentive to be anything other than a price-taker. On the other hand, core non-convergence makes price-taking an implausible assumption.

Edgeworth (1881) doubted the positive significance of Walrasian equilibrium, and argued that the core, not the set of Walrasian equilibria, was the best positive description of the outcomes of a market mechanism. Moreover, while Edgeworth's name is closely associated with core convergence, and he did prove a core convergence theorem, he argued that in real economies, the presence of firms and syndicates which possess market power ensures that the core does *not* converge.

Edgeworth's argument about the effects of market power applies most strongly to the production side of the economy, where we do in fact see large firms, syndicates and labour unions. However, on the consumption side, the wealthiest individual in the world consumes a small part of the world's annual consumption. In exchange economies in which each consumer is small, core convergence holds. So core convergence provides a justification for the price-taking assumption on the consumption side, provided one views the world as an exchange economy in which the production decisions have been previously made by some exogenous process, outside the scope of the model, endowments include the income obtained by selling one's labour in the exogenous production process, and the only economic activity is trade and consumption of what has been produced.

The proof of the most basic core convergence theorem, which assumes very little about preferences and endowments, and establishes approximate decentralization in a relatively weak sense, is closely analogous to the proof of the Second Welfare Theorem. The approximately decentralizing price vector appears magically, as a consequence of the separating hyperplane theorem and the Shapley–Folkman theorem, which asserts that the sum of a large number of sets is approximately convex. Convexity of preferences plays no role. Indeed, the definition of the core, because it allows for individuals to be included or excluded from potential coalitions, introduces a non-convexity which is not present in the Second Welfare Theorem, and the Shapley–Folkman theorem controls that non-convexity, whether or not preferences themselves are convex.

The definitions and results just described verbally are presented more formally below.

Many people have made important contributions to the study of core convergence. A survey of these contributions is given in Anderson (1992), and a list of some of the more important contributions is included in the bibliography.

Now, we turn to a more formal presentation

Definition 1 In an exchange economy with agents $i = 1, \ldots, I$ having strict preferences $\succ_i$ and endowments $\omega_i \in \mathbf{R}_+^L$, a *coalition* is a set $S \subseteq \{1, \ldots, I\}$. An *exact allocation* is $x \in \left(\mathbf{R}_+^L\right)^I$ such that $\sum_{i=1}^I x_i = \sum_{i=1}^I \omega_i$. An exact allocation is *weakly Pareto optimal* if there is no other exact allocation x' satisfying $x_i' \succ_i x_i (i = 1, \ldots, I)$. A coalition S *blocks* or *improves on* an exact allocation x by x' if $\sum_{i \in S} x_i' = \sum_{i \in S} \omega_i$ and $\forall_{i \in S} \; x_i' \succ_i x_i$. The *core* is the set of all exact allocations which cannot be improved on by any nonempty coalition. The *price simplex* is $\Delta = \{p \in \mathbf{R}_+^k : \sum_{\ell=1}^L p_\ell = 1\}$.

Theorem 2 *In an exchange economy, every core allocation is weakly Pareto optimal.*

Proof If x is not weakly Pareto optimal, then there exists x' such that $\sum_{i=1}^I x_i' = \sum_{i=1}^I x_i$, $x_i' \succ_i x_i$. Then $S = \{1, \ldots, I\}$ improves on x by x', so x is not in the core.

Theorem 3 (Strong First Welfare Theorem) *In an exchange economy, every Walrasian Equilibrium lies in the core.*

Proof Suppose (p^*, x^*) is a Walrasian Equilibrium. If x^* is not in the core, there exists $S \subseteq I$, $S \neq \emptyset$ and $x_i'(i \in S)$ such that $\sum_{i \in S} x_i' = \sum_{i \in S} \omega_i$ and $x_i' \succ_i x_i^*$ for each $i \in S$. Since x_i^* lies in i's demand set at the price p^*, $p^* \cdot x_i > p^* \omega_i$, so $p^* \cdot \sum_{i \in S} x_i' = \sum_{i \in S} p^* \cdot x_i' > \sum_{i \in S} p^* \cdot \omega_i = p^* \cdot \sum_{i \in S} \omega_i$ but $\sum_{i \in S} x_i' = \sum_{i \in S} \omega_i$, a contradiction. Therefore, x^* is in the core.$\Diamond$

Theorem 4 (Core convergence, E. Dierker, 1975, and Anderson, 1978) Suppose we are given an exchange economy with L commodities, I agents and preferences $\succ_1, \ldots, \succ_I$ satisfying weak monotonicity (if $x \gg y$, then $x \succ_i y$) and the following free disposal condition: $x \gg y$, $y \succ_i z \Rightarrow x \succ_i z$. If x is in the core, then there exists $p \in \Delta$ such that

$$\frac{1}{I} \sum_{i=1}^I |p \cdot (x_i - \omega_i)| \tag{1}$$
$$\leq \frac{2L}{I} \max\{\|\omega_1\|_\infty, \ldots \|\omega_I\|_\infty\}$$

$$\frac{1}{I} \sum_{i=1}^I |\inf\{p \cdot (y - x_i) : y \succ_i x_i\}| \tag{2}$$
$$\leq \frac{4L}{I} \max\{\|\omega_1\|_\infty, \ldots \|\omega_I\|_\infty\}$$

where $\|x\|_\infty = \max\{|x_1|, \ldots, |x_L|\}$.

If there are many more agents than goods and the endowments are not too large, the bounds on the right-hand sides of eqs (1) and (2) will be small. In that case, eq. (1) says that trade occurs almost at the price p, and that each x_i is almost in the budget set, while eq. (2) says that the price p almost supports $\succ_i$ at x_i, in the sense that everything preferred to x_i costs almost as much as x_i. Taken together, eqs (1) and (2) say that the pair (p, x) satisfies a slightly perturbed version of the *definition* of Walrasian equilibrium. Indeed, if we knew the left sides of eqs (1) and (2) were zero, then $p \cdot (x_i - w_i) = 0$, so x_i lies in i's budget set, and $y \succ_i x_i \Rightarrow p \cdot y \geq p \cdot \omega_i$, so x would be a Walrasian quasi-equilibrium! (A pair (p^*, x^*) is said to be a Walrasian quasi-equilibrium if it satisfies the definition of a Walrasian equilibrium except that instead of requiring that x_i^* lie in i's demand set, we only require that x_i^* lie in i's quasi-demand set, that is $p^* \cdot x_i^* \le p^* \cdot \omega_i$ and every $y \succ_i x_i^*$ satisfies $p^* \cdot y \ge p^* \cdot \omega_i$.)

Outline of Proof: Follow the proof of the Second Welfare Theorem.

- Suppose x is in the core. Define $B_i = \{y - \omega_i : y \succ_i x_i\} \cup \{0\} = (\{y : y \succ_i x_i\} \cup \{\omega_i\}) - \omega_i$ and $B = \sum_{i=1}^I B_i$. The first term in the definition of B_i corresponds to members of a potential improving coalition; for accounting purposes, we assign members outside the coalition their endowments. Note that B_i is *not* convex, even if $\succ_i$ is a convex preference. *Claim:* If x is in the core, then $B \cap \mathbf{R}_{--}^L = \emptyset$. Suppose $z \in B \cap \mathbf{R}_{--}^L$. Then there exists $z_i \in B_i$ such that $z = \sum_{i=1}^I z_i$. Let $S = \{i : z_i \ne 0\}$; since $z \ll 0$, $S \ne \emptyset$. For $i \in S$, let $x_i' = \omega_i + z_i - \frac{z}{|S|}$. Then $x_i' \gg \omega_i + z_i \succ_i x_i$ by the definition of B_i, $x_i' \succ_i x_i$ by free disposal, and $\sum_{i \in S} x_i' = \sum_{i \in S} \omega_i$, so S can improve on x by x', so x is not in the core.
- Let $v = -L(\max_{i=1,\&,I} \|\omega_i\|_\infty, \ldots, \max_{i=1,\&,I} \|\omega_i\|_\infty)$. *Claim:* $(\text{con } B) \cap (v + \mathbf{R}_{--}^L) = \emptyset$. If $z \in \text{con} B$, by the Shapley–Folkman theorem, and relabelling the agents, we may write

$$z = \sum_{i=1}^I z_i, \quad z_i \in \text{con } B_i \ (i = 1, \ldots, I),$$

$$z_i \in B_i \ (i \ne \{1, \ldots, L\})$$

Choose

$$\hat{z}_i = \begin{cases} 0 & \text{if } i = 1, \ldots, L \\ z_i & \text{if } i = L+1, \ldots, I \end{cases}$$

Then $\sum_{i=1}^I \hat{z}_i \in B$ so $\sum_{i=1}^I \hat{z}_i < / < 0$. If $z \ll v$, then $\sum_{i=1}^I \hat{z}_i = \sum_{i=1}^L 0 + \sum_{i=L+1}^I z_i \le \sum_{i=1}^L (\omega_i + z_i) + \sum_{i=L+1}^I z_i = \sum_{i=1}^L \omega_i + \sum_{i=1}^I z_i = \sum_{i=1}^L \omega_i + z \ll \sum_{i=1}^L \omega_i + v \le 0$, so $B \cap \mathbf{R}_{--}^L \ne \emptyset$, a contradiction which proves the claim.
- By the separating hyperplane theorem, there exists $p \ne 0$ such that $\sup p \cdot (v + \mathbf{R}_{--}^L) \le \inf p \cdot (\text{con } B)$. If $p_\ell < 0$ for some ℓ, then $\sup p \cdot (v + \mathbf{R}_{--}^L) = +\infty$, while $\inf p \cdot (\text{con } B) \le 0$, a contradiction, so $p > 0$ and we can normalize $p \in \Delta$. Then $\inf p \cdot B \ge \inf p \cdot (\text{con } B) \ge p \cdot v = -L\max\{\|\omega_1\|_\infty, \ldots, \|\omega_I\|_\infty\}$.
- Adapt the remainder of the proof of the Second Welfare Theorem; this requires a few tricks.

ROBERT M. ANDERSON

See also **Arrow–Debreu model of general equilibrium; cores; Edgeworth, Francis Ysidro; existence of general equilibrium; general equilibrium; general equilibrium (new developments).**

Bibliography

Anderson, R.M. 1978. An elementary core equivalence theorem. *Econometrica* 46, 1483–7.

Anderson, R.M. 1981. Core theory with strongly convex preferences. *Econometrica* 49, 1457–68.

Anderson, R.M. 1985. Strong core theorems with nonconvex preferences. *Econometrica* 53, 1283–94.

Anderson, R.M. 1986. Core allocations and small income transfers. Working Paper No. 8621, Department of Economics, University of California at Berkeley.

Anderson, R.M. 1987. Gap-minimizing prices and quadratic core convergence. *Journal of Mathematical Economics* 16, 1–15. Correction, *Journal of Mathematical Economics* 20 (1991), 599–601.

Anderson, R.M. 1992. The core in perfectly competitive economies. In *Handbook of Game Theory with Economic Applications*, vol. I, eds. R.J. Aumann and S. Hart. Amsterdam: North-Holland.

Aumann, R.J. 1964. Markets with a continuum of traders. *Econometrica* 32, 39–50.

Bewley, T.F. 1973a. Edgeworth's conjecture. *Econometrica* 41, 425–54.

Brown, D.J. and Robinson, A. 1974. The cores of large standard exchange economies. *Journal of Economic Theory* 9, 245–54.

Brown, D.J. and Robinson, A. 1975. Nonstandard exchange economies. *Econometrica* 43, 41–55.

Debreu, G. 1975. The rate of convergence of the core of an economy. *Journal of Mathematical Economics* 2, 1–7.

Debreu, G. and Scarf, H. 1963. A limit theorem on the core of an economy. *International Economic Review* 4, 236–46.

Dierker, E. 1975. Gains and losses at core allocations. *Journal of Mathematical Economics* 2, 119–28.

Edgeworth, F.Y. 1881. *Mathematical Psychics*. London: Kegan Paul.

Grodal, B. 1975. The rate of convergence of the core for a purely competitive sequence of economies. *Journal of Mathematical Economics* 2, 171–86.

Grodal, B. and Hildenbrand, W. 1974. Limit theorems for approximate cores. Working Paper IP-208, Center for Research in Management, University of California, Berkeley.

Hildenbrand, W. 1974. *Core and Equilibria of a Large Economy*. Princeton: Princeton University Press.

Kannai, Y. 1970. Continuity properties of the core of a market. *Econometrica* 38, 791–815.

Khan, M.A. 1974. Some equivalence theorems. *Review of Economic Studies* 41, 549–65.

Vind, K. 1964. Edgeworth allocations in an exchange economy with many traders. *International Economic Review* 5, 165–77.

Vind, K. 1965. A theorem on the core of an economy. *Review of Economic Studies* 32, 47–8.

Walras, L. 1874. *Eléments d'économie politique pure*. Lausanne: L. Corbaz.

cores

The *core* of an economy consists of those states of the economy which no group of agents can 'improve upon'. A group of agents can improve upon a state of the economy if, by using the means available to that group, each member can be made better off. Nothing is said in this definition of how a state in the core actually is reached. The actual process of economic transactions is not considered explicitly.

To keep the presentation as simple as possible, we shall consider only the core for exchange economies with an arbitrary number l of commodities, even though the core concept applies to more general situations.

Consider a finite set A of economic agents; each agent a in A is described by his *preference relation* $\precsim_a$ (defined on the positive orthant R^l_+) and his *initial endowments* e_a (a vector in R^l_+). The outcome of any exchange, that is to say, a state (x_a) of the exchange economy $\mathcal{E} = \{\precsim_a, e_a\}_{a \in A}$, is a *redistribution* of the total endowments, i.e.

$$\sum_{a \in A} x_a = \sum_{a \in A} e_a.$$

A *coalition* of agents, say $S \subset A$, can *improve upon* a redistribution (x_a), if that coalition S, by using the endowments available to it, can make each member of that coalition better off, that is to say, there is a redistribution, say $(y_a)_{a \in S}$, such that

$$y_a \succ_a x_a \text{ for every } a \in S \text{ and } \sum_{a \in S} y_a = \sum_{a \in S} e_a.$$

The set of redistributions for the exchange economy $\mathcal{E}$ that no coalition can improve upon is called the *core* of the economy $\mathcal{E}$, and is denoted by $C(\mathcal{E})$.

The core is a rather theoretical, however, fundamental equilibrium concept. Indeed, the core provides a theoretical foundation of a more operational equilibrium concept, the *competitive equilibrium* which, in fact, is a very different notion of equilibrium. The allocation process is organized through markets; there is a price for every commodity. All economic agents take the price system as given and make their decisions independently of each other. The equilibrium price system coordinates these independent decisions in such a way that all markets are simultaneously balanced.

More formally, an allocation (x^*_a) for the exchange economy $\mathcal{E} = \{\precsim_a, e_a\}_{a \in A}$ is a *competitive equilibrium* (or a *Walras allocation*) if there exists a price vector $p^* \in R^l_+$ such that for every $a \in A, x^*_a \in \phi_a(p^*)$ and

$$\sum_{a \in A} x^*_a = \sum_{a \in A} e_a.$$

Here $\phi_a(p^*)$ or more explicitly, $\phi(p^*, e_a, \precsim_a)$ denotes the demand of agent a with preferences $\precsim_a$ and endowment e_a, i.e. the set of most desired commodity vectors (with respect to $\precsim_a$) in the budget-set $\{x \in R^l_+ | p^* \cdot x \leq p^* \cdot e_a\}$.

The set of all competitive equilibria for the economy $\mathcal{E}$ is denoted by $W(\mathcal{E})$.

The core and the set of competitive equilibria for an economy with two agents and two commodities can be represented geometrically by the well-known Edgeworth–Box (see Figure 1). The size of the box is determined by the total endowments $e_1 + e_2$. Every point P in the box represents a redistribution; the first agent receives $x_1 = P$ and the second receives $x_2 = (e_1 + e_2) - P$.

It is easy to show that for every exchange economy $\mathcal{E}$ a competitive equilibrium belongs to the core,

$$W(\mathcal{E}) \subset C(\mathcal{E}).$$

Thus, a state of the economy $\mathcal{E}$ which is decentralized by a price system cannot be improved upon by cooperation. This proposition strengthens a well-known result of Welfare Economics—every competitive equilibrium is Pareto-efficient.

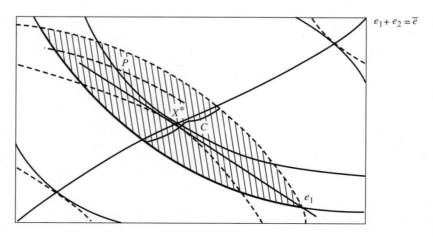

Figure 1

The inclusion $W(\mathscr{E}) \subset C(\mathscr{E})$ is typically strict. Indeed, if the initial allocation of endowments is not Pareto-efficient, which is the typical case, then, if there are any allocations in the core at all, there are core-allocations which are not competitive equilibria.

This leads us to the *basic problem* in the theory of the core:

> For which kind of economies is the 'difference' between the core and the set of competitive equilibria small? Or in other words, under which circumstances do cooperative barter and competition through decentralized markets lead essentially to the same result?

Naturally, the answer depends on the way one measures the 'difference' between the two equilibrium concepts. However this is done one expects that the economy must have a large number of participants.

In answering the basic question we try to be comprehensible (for example by avoiding the use of measure-theoretic concepts) but not comprehensive. Therefore, if we refer in the remainder of this entry to an economy $\mathscr{E} = \{\precsim_a, e_a\}_{a \in A}$ we shall always assume that preference relations are continuous, complete, transitive, monotone and strictly convex. The total endowments $\Sigma_{a \in A} e_a$ of an economy are always assumed to be strictly positive. We shall not repeat these assumptions. Furthermore, if we call an economy smooth, then we assume in addition that preferences are smooth (hence representable by sufficiently differentiable utility functions) and individual endowments are strictly positive.

These assumptions simplify the presentation tremendously. For generalizations we refer to the extensive literature.

We remark that under the above assumptions there always exists a competitive equilibrium, and hence, the core is not empty.

Large economies

The simplest and most stringent measure of difference between the two equilibrium sets, $C(\mathscr{E})$, and $W(\mathscr{E})$, which we shall denote by $\delta(\mathscr{E})$, can be defined as follows.

Let $\delta(\mathscr{E})$ be the smallest number δ with the property: for every allocation $(x_a) \in C(\mathscr{E})$ there exists an allocation $(x_a^*) \in W(\mathscr{E})$ such that

$$|x_a - x_a^*| \le \delta$$

for every agent a in the economy $\mathscr{E}$.

Thus, if $\delta(\mathscr{E})$ is small, then from every agent's view a core allocation is like a competitive equilibrium.

Unfortunately for this measure of difference, it is not true that $\delta(\mathscr{E})$ can be made arbitrarily small provided the number of agents in the economy $\mathscr{E}$ is sufficiently large (even if one restricts the agents' characteristics $(\precsim_a, e_a)$ to an a priori given finite set).

Consequently one considers also weaker measures for the 'difference' between the two equilibrium concepts $C(\mathscr{E})$ and $W(\mathscr{E})$. For example, define $\delta_1(\mathscr{E})$ and $\delta_2(\mathscr{E})$, respectively, as the smallest number δ with the property: for every $(x_a) \in C(\mathscr{E})$ there exists a price vector $p \in R_+^l$ such that

$$(\delta_1) \quad |x_a - \phi_a(p)| \le \delta \quad \text{for every agent } a \text{ in } \mathscr{E}$$

or

$$(\delta_2) \quad \frac{1}{\#A} \sum_{a \in A} |x_a - \phi_a(p)| \le \delta.$$

Clearly, the measures δ_1 and δ_2 are weaker than δ since the price vector p is not required to be an equilibrium price vector for the economy $\mathscr{E}$. The number $\delta_1(\mathscr{E})$ (and, *a fortiori*, $\delta_2(\mathscr{E})$) does not measure the distance between the sets $C(\mathscr{E})$ and $W(\mathscr{E})$. but the degree by which an allocation in the core can be decentralized via a price system. Obviously one has $\delta_2(\mathscr{E}) \le \delta_1(\mathscr{E}) \le \delta(\mathscr{E})$.

One can show that $\delta_2(\mathscr{E})$ becomes arbitrarily small for sufficiently large economies. More precisely,

Theorem 1 Let T be a finite set of agents' characteristics $(\precsim, e)$ and let b be a strictly positive vector in R^l. Then for every $\varepsilon > 0$ there exists an integer N such that for every economy $\mathscr{E} = \{\precsim_a, e_a\}_{a \in A}$ with $\#A \ge N$,

$$\frac{1}{\#A} \sum_{a \in A} e_a \ge b$$

and $(\succsim_a, e_a) \in T$ one has

$$\delta_2(\mathscr{E}) \le \varepsilon.$$

(The finite set T in Theorem 1 can be replaced by a compact set with respect to a suitably chosen topology: see Hildenbrand, 1974.) We emphasize that this result does not imply that in large economies core-allocations are near to competitive equilibria. In fact, Theorem 1 does not hold if δ_2 is replaced by the measure of difference δ or even δ_1. Theorem 1 does imply, however, that for sufficiently large economies one can associate to every core-allocation a price vector which 'approximately decentralizes' the core-allocation. Some readers might consider this conclusion as a perfectly satisfactory answer to our basic problem. If one holds this view, then the rest of the paper is a superfluous intellectual pastime. We would like to emphasize, however, that the meaning of 'approximate decentralization' is not very strong. First, the demand $\phi_a(p)$ is not necessarily near to x_a for every agent a in the economy; only the mean deviation

$$\frac{1}{\#A} \sum_{a \in A} |x_a - \phi_a(p)|$$

becomes small. Second, total demand is not equal to total supply; only the mean excess demand

$$\frac{1}{\#A}\sum_{a\in A}[\phi_a(p)-e_a]$$

becomes small.

There are alternative proofs in the literature, e.g. Bewley (1973), Hildenbrand (1974), Anderson (1981) or Hildenbrand (1982). These proofs are based either on a result by Vind (1965) or Anderson (1978).

Sharper conclusions than the one in Theorem 1 will be stated in the following sections. There we consider a sequence $(\mathcal{E}_n)_{n=1,\dots}$ of economies and then study the asymptotic behaviour of $\delta(\mathcal{E}_n)$.

Before we present these limit theorems we should mention another approach of analysing the inclusion $W(\mathcal{E})\subset C(\mathcal{E})$. Instead of analysing the asymptotic behaviour of the difference $\delta(\mathcal{E}_n)$ for a sequence of finite economies one can define a large economy where every agent has strictly no influence on collective actions. This leads to a *measure space without atoms* of economic agents (also called a *continuum of agents*). For such economies the two equilibrium concepts coincide. See Aumann (1964).

Replica economies

Let $\mathcal{E}=\{\preceq_i,e_i\}$ be an exchange economy with m agents. For every integer n we define the n-fold *replica economy* $\mathcal{E}_n$ of $\mathcal{E}$ as an economy with $n\cdot m$ agents; there are exactly n agents with characteristics $(\preceq_i,e_i)$ for every $i=1,\dots,m$.

More formally,

$$\mathcal{E}_n=\left\{\preceq_{(i,j)},e_{(i,j)}\right\}_{\substack{1\le i\le m\\1\le j\le x}}$$

where $\preceq_{(i,j)}=\preceq_i$ and $e_{(i,j)}=e_i, 1\le i\le m$ and $1\le j\le n$. Thus, an agent a in the economy $\mathcal{E}_n$ is denoted by a double index $a=(i,j)$. We shall refer to agent (i,j) sometimes as the jth agent of type i.

Replica economies were first analysed by F. Edgeworth (1881) who proved a limit theorem for such sequences in the case of two commodities and two types of agents. A precise formulation of Edgeworth's analysis and the generalization to an arbitrary finite number of commodities and types of agents is due to Debreu and Scarf (1963).

Here is the basic result for replica economies.

Theorem 2 For every sequence $(\mathcal{E}_n)$ of replica economies the difference between the core and the set of competitive equilibria tends to zero, i.e.,

$$\lim_{n\to\infty}\delta(\mathcal{E}_n)=0.$$

Furthermore, if $\mathcal{E}$ is a smooth and regular economy then $\delta(\mathcal{E}_n)$ converges to zero at least as fast as the inverse

of the number of participants, i.e., there is a constant K such that

$$\delta(\mathcal{E}_n)\le\frac{K}{n}.$$

The proof of this remarkably neat result is based on the fact that a core–allocation (x_{ij}) assigns to every agent of the same type the same commodity bundle, i.e., $x_{ij}=x_{ik}$, This 'equal treatment' property simplifies the analysis of $\delta(\mathcal{E}_n)$ tremendously. Indeed, an allocation (x_{ij}) in $C(\mathcal{E}_n)$, which can be considered as a vector in $R^{l\cdot m\cdot n}$, is completely described by the commodity bundle of one agent in each type, thus by a vector $(x_{11},x_{21},\dots,x_{m1})$ in $R^{l\cdot m}$, a space whose dimension is independent of n.

Thus, let

$$C_n=\left\{(x_{11},x_{21},\dots,x_{m1})\in R^{l\cdot m}\,|\,(x_{ij})\in C(\mathcal{E}_n)\right\}.$$

One easily shows that $C_{n+1}\subset C_n$. It is not hard to see that Theorem 1 follows if

$$\cap_{n=1}^{\infty}C_n=W(\mathcal{E}_1).$$

But this is the well-known theorem of Debreu and Scarf (1963). The essential arguments in the proof go as follows. Let $(x_1,\dots,x_m)\in\cap_{n=1}^{\infty}C_n$. One has to show that there is a price vector p^* such that $x\succ_i x_i$ implies $p^*\cdot x>p^*\cdot e_i$. For this it suffices to show that there is a p^* such that

$$p^*\cdot z\ge 0\quad\text{for every}$$
$$\in\cup_{i=1}^m\left(\{x\in R_+^l\,|\,x\succ_i x_i\}-e_i\right)=Z,$$

i.e., there is a hyperplane (whose normal is p^*) which supports the set z. One shows that the assumption $(x_1,\dots,x_m)\in\cap_{n=1}^{\infty}C_n$ implies that 0 does not belong to the convex hull of z. Minkowski's Separation Theorem for convex sets then implies the existence of the desired vector p^*.

The second part of the conclusion of Theorem 2 is due to Debreu (1975).

Type economies

The limit theorem on the core for replica economies is not fully satisfactory since replication is a very rigid way of enlarging an economy. The conclusion '$\delta(\mathcal{E}_n)\to 0$' in Theorem 2, to be of general relevance, should be robust to small deviations from the strict replication procedure.

Consider a sequence $(\mathcal{E}_n)$ of economies where the characteristics of every agent belong to a given finite set of types $T=\{(\preceq_1,e_1),\dots,(\preceq_m,e_m)\}$. We do not consider this as a restrictive assumption (considered as an approximation, one can always group agents' characteristics into a finite set of types). Let the economy $\mathcal{E}_n$ have N_n agents; $N_n(1)$ agents of the first type, $N_n(i)$ agents of type i. Of course the idea is that N_n tends to ∞ with increasing n. Consider the fraction $\nu_n(i)$ of agents in the economy $\mathscr{E}_n$ which are of type i, i.e.,

$$\nu_n(i) = \frac{N_n(i)}{N_n}.$$

The sequence $(\mathscr{E}_n)$ is a replica sequence of an economy $\mathscr{E}$ (not necessarily of $\mathscr{E}_1$) if and only if the fractions $\nu_n(i)$ are all independent of n. It is this rigidity which we want to weaken now.

A sequence $(\mathscr{E}_n)$ of economies with characteristics in a finite set T is called a *sequence of type economies* (over T) if

(i) the number N_n of agents in $\mathscr{E}_n$ tends to infinity and
(ii) $\nu_n(i) = \frac{N_n(i)}{N_n} \xrightarrow[(n \to \infty)]{} \nu(i) > 0.$

EXAMPLE (random sampling of agents' characteristics):
Let π be a probability distribution over the finite set T. Define the economy $\mathscr{E}_n$ as a random sample of size n from this distribution $\pi(\cdot)$. The law of large numbers them implies property (ii): $\nu_n(i) \to \pi(i)$.

The step from replica economies to type economies – as small as it might appear to the reader – is conceptually very important. Yet with this 'small' generalization the analysis of the limit behaviour of $\delta(\mathscr{E}_n)$ or $\delta_1(\mathscr{E}_n)$ is made more difficult. Even worse, it is no longer true that for *every* sequence $(\mathscr{E}_n)$ of type economies one obtains $\delta(\mathscr{E}_n) \to 0$ — even if the preferences of all types are assumed to be very nice, say smooth. There are some 'exceptional cases' where the conclusion $\delta(\mathscr{E}_n) \to 0$ does not hold. But these are 'exceptional' cases and the whole difficulty in the remainder of this section is to explain in which precise sense these cases are 'exceptional' and can therefore be ignored. We shall first exhibit the 'cases' where the conclusion fails to hold. Then we shall show that these cases are exceptional.

We denote by $\Pi(\mathscr{E})$ the set of normalized *equilibrium price vectors* for the economy $\mathscr{E} = \{\precsim_a, e_a\}_{a \in A}$. Thus, for $p^* \in \Pi(\mathscr{E})$ the excess demand is zero, i.e.,

$$\sum_{a \in A} [\phi_a(p^*) - e_a] = 0.$$

To every sequence $(\mathscr{E}_n)$ of type economies we associate a 'limit economy' $\mathscr{E}_\infty$. This economy has an 'indefinitely large' number of agents of every type; the fraction of agents of type i is given by $\nu(i)$. The mean (per capita) excess demand of that limit economy $\mathscr{E}_\infty$ is defined by

$$z_\nu(p) = \sum_{i=1}^{m} \nu(i)[\phi(p, e_i, \precsim_i) - e_i].$$

An equilibrium price vector p^* of the limit economy $\mathscr{E}_\infty$ is defined by $z_\nu(p^*) = 0$. Let $\Pi(\nu)$ denote the set of normalized equilibrium price vectors for $\mathscr{E}_\infty$. Obviously for a replica sequence $(\mathscr{E}_n)$ we have $\Pi(\mathscr{E}_n) = \Pi(\nu)$ for all n. However, for a sequence of type economies the set $\Pi(\mathscr{E}_n)$

of equilibrium prices of the economy $\mathscr{E}_n$ depends on n, and it might happen that the set $\Pi(\nu)$ is not similar to $\Pi(\mathscr{E}_n)$ even for arbitrarily large n. To fix ideas, it might happen that $\Pi(\mathscr{E}_n) = \{p_n\}$ and $\Pi(\nu)$ contains not only $p = \lim p_n$ but also another equilibrium price vector. Such a situation has to be excluded.

We call a sequence of type economies *sleek* if $\Pi(\mathscr{E}_n)$ converges (in the Hausdorff-distance) to $\Pi(\nu)$.

It is known (Hildenbrand, 1974) that the sequence $(\Pi(\mathscr{E}_n))$ converges to $\Pi(\nu)$ if $\Pi(\nu)$ is a singleton (i.e., the limit economy has a unique equilibrium) or, in general, if (and only if) for every open set O in R^ℓ with $O \cap \Pi(\nu) \neq \emptyset$ it follows that $O \cap \Pi(\mathscr{E}_n) \neq \emptyset$ for all n sufficiently large.

We now have exhibited the cases where a limit theorem on the core holds true.

Theorem 3. For every sleek sequence $(\mathscr{E}_n)$ of type economies

$$\lim_{n \to \infty} \delta(\mathscr{E}_n) = 0.$$

Unfortunately there seems to be no short and easy proof. The main difficulty arises from the fact that for allocations in the core of a type economy the 'equal treatment' property, which made the replica case so manageable is no longer true. For a proof see Hildenbrand and Kirman (1976) or Hildenbrand (1982) and the references given there. The main step in the proof is based on a result of Bewley (1973).

It remains to show that non-sleek sequences of type economies are 'exceptional cases'.

The strongest form of 'exceptional' is, of course, 'never'. We mentioned already that a sequence $(\mathscr{E}_n)$ is sleek if its limit economy has a unique equilibrium. Unfortunately, however, only under very restrictive assumptions on the set T of agents' characteristics does uniqueness prevail; for example,

(1) if every preference relation leads to a demand function which satisfies gross-substitution (Cobb–Douglas utility functions are typical examples),
(2) if every preference relation is homothetic and the endowment vectors $e_i(i = 1, \ldots, m)$ are collinear.

Since there is no reasonable justification for restricting the set T to such special types of agents we have to formulate a model in which we allow non-sleek sequences to occur provided, of course, this can be shown to be 'exceptional cases'. Let S^{m-1} denote the open simplex in R^m, i.e.

$$S^{m-1} = \left\{ x \in R^m \,\Big|\, x_i > 0, \sum_{i=1}^{m} x_i = 1 \right\}.$$

The limit distribution $\nu(i)$ of a sequence of type economies with m types is a point in S^{m-1}.

A closed subset C in S^{m-1} which has $(m-1$ dimensional Lebesgue) measure zero is called *negligible*. Thus, if a distribution v is not in C then a sufficiently small change will not lead to C. Furthermore, given any arbitrary small positive number $\mathscr{E}$ one can find a countable collection of balls in S^{m-1} such that their union covers C, and that the sum of the diameters of these balls is smaller than $\mathscr{E}$. Thus, in particular, if $v \in C$ then one can approximate v by points which do not belong to C. Clearly, a negligible set is a small set in S^{m-1}.

Theorem 4. Given a finite set T of m smooth types of agents, there exists a negligible subset C in S^{m-1} and a constant K such that for every sequence $(\mathscr{E}_n)$ of type economies over T whose limit distribution v does not belong to C one has $\delta(\mathscr{E}_n) \leq K/\#A_n$, thus in particular, $\lim\limits_{n \to \infty} \delta(\mathscr{E}_n) = 0$.

The convergence of $\delta(\mathscr{E}_n)$ follows from Theorem 3 and Theorems 5.4.3 and 5.8.15 in Mas–Colell (1985). For the rate of convergence see Grodal (1975).

WERNER HILDENBRAND

See also **Edgeworth, Francis Ysidro; existence of general equilibrium.**

Bibliography

There is an extensive literature on limit theorems on the core which contains important generalizations of the results given here. For a general reference we refer to Hildenbrand (1974) or (1982), Mas-Colell (1985), Anderson (1981) and the references given there.

Anderson, R.M. 1978. An elementary core equivalence theorem. *Econometrica* 46, 1483–7.
Anderson, R.M. 1981. Core theory with strongly convex preferences. *Econometrica* 49, 1457–68.
Aumann, R.J. 1964. Markets with a continuum of traders. *Econometrica* 32, 39–50.
Bewley, T.F. 1973. Edgeworth's conjecture. *Econometrica* 41, 425–54.
Debreu, G. 1975. The rate of convergence of the core of an economy. *Journal of Mathematical Economics* 2, 1–8.
Debreu, G. and Scarf, H. 1963. A limit theorem on the core of an economy. *International Economic Review* 4, 235–46.
Edgeworth, F.Y. 1881. *Mathematical Psychics*. London: Kegan Paul.
Grodal, B. 1975. The rate of convergence of the core for a purely competitive sequence of economies. *Journal of Mathematical Economics* 2, 171–86.
Hildenbrand, W. 1974. *Core and Equilibria of a Large Economy*. Princeton: Princeton University Press.
Hildenbrand, W. 1982. Core of an economy. In *Handbook of Mathematical Economics*, ed. K.J. Arrow and M.D. Intriligator. Vol. 2 , Amsterdam: North-Holland.
Hildenbrand, W. and Kirman, A.P. 1976. *Introduction to Equilibrium Analysis*. Amsterdam: North-Holland.
Mas-Colell, A. 1985. *The Theory of General Economic Equilibrium, A Differentiable Approach*. Cambridge: Cambridge University Press.
Vind, K. 1965. A theorem on the core of an economy. *Review of Economic Studies* 32, 47–8.

Corn Laws, free trade and protectionism

The Corn Laws were the parliamentary statutes that regulated the import and export of grain for the benefit of British producers in the early 19th century. Though these laws derived from legislation in the period 1804–15, they were but the extension or modification of a system that had been introduced in 1773 to prohibit exports of wheat when prices rose above a given level and that limited imports through a variety of duties based on a sliding scale. The goal of these laws was ostensibly the desire to stabilize the price of grain, which had been a regular goal of parliament since the late 17th century.

The debates about the abolition of the Corn Laws in the early to mid-1800s hold a special place in the economic history of Great Britain on account of their central role in shifting commercial policy to nearly free trade. Because of Britain's dominance of industrial trade in the 19th century and the leadership she exerted in international commerce, the struggles over the Corn Laws have been seen as emblematic of all debates about the advisability of free trade or protectionism. Despite the symbolic importance of these events, it is easy to overlook the facts that Britain after the repeal of the Corn Laws did not immediately move to perfectly free trade and that the political struggle over their abolition had at least as much to do with domestic concerns over the importance of agriculture in a modern economy as with ideological questions about the advisability of free trade.

Mercantilism and the rise of British liberalism

The regulation or promotion of international trade has been perhaps the oldest policy issue in the political economy of international relations.

It is a common belief that trade is a primary source of a nation's wealth. But this has often been misunderstood to mean that exports enhance wealth while imports detract from it. This view, a central component of what is called mercantilism, stems from the mistaken belief that the benefits of trade flow only one way. One view was that a nation's wealth derived from the quantity of specie or gold and silver coin in the country. Therefore, exports contributed to this while imports detracted from it.

Some of this reasoning was theoretical, but more commonly mercantile theory was simply the evolution of a set of policies deriving from the fiscal needs of the newly emerging nation-states. Unsurprisingly, many states viewed the success of the state as synonymous with the success of the nation itself. Revenue was essential to the maintenance of the large armies that were a prerequisite for the nation-state. So trade was viewed as an essentially zero-sum game with both losers and winners. Moreover, this concern about revenue often translated into a concern for specie. Whereas modern economics treats specie as virtually irrelevant to the supply of money, contemporaries viewed coin itself as a necessary prerequisite of sound financial policy. Hence trade surpluses were preferred because they brought more precious metals in than they took out of the kingdom.

One of the earliest theoretical discussions of this view comes from Thomas Mun, who wrote 'The ordinary means therefore to increase our wealth and treasure is by foreign trade, wherein we must ever observe this rule; to sell more to strangers yearly than we consume of theirs in value' (1664, p. 11).

Adam Smith, the founder of modern economics, was the most prominent critic of this view. Starting from the observation that voluntary trade was mutually beneficial, and noting that the wealth of a nation's inhabitants, not its quantity of coin, made for true wealth, Smith argued in the *Wealth of Nations* against what he labelled the 'mercantile system'. He articulated the virtues of free and open trade, both in international and in home commerce. Indeed, the term 'free trade' was employed throughout the 18th and 19th centuries to refer to unregulated domestic trade as least as often as it referred to the free flow of goods from abroad.

These ideas were later modelled more systematically by the English economist David Ricardo, who formalized the analysis and showed that nations could maximize their welfare by specializing in the production of goods with the lowest opportunity cost and trading with other nations. This is the central idea behind the law of comparative advantage, usually attributed to David Ricardo, and developed more thoroughly by Paul Samuelson and others in the 20th century. Most important for this claim was the idea that a nation did not even have to be the 'best' producer of any product for there to be gains from trade. A nation that was more productive that another in all industries would still do better by specializing in some areas and trading for the other goods with another country. Thus, any claim that a nation could not benefit if it had no comparative advantage would be false. Every nation has a comparative advantage in producing some product, even if it has an absolute advantage in none.

Smith's ideas and those of his successors provided the philosophical basis for the classical liberal movements of the late 18th and early 19th centuries. By the early 1800s, the idea of a limited state that minimized regulation and promoted welfare through the encouragement of open trade at home and abroad had emerged as an important ideological view, promoted by prominent intellectuals and supported by an influential subset of the British political class. Nonetheless, the strong interest in the liberal ideas derived from the Scottish Enlightenment persuaded states not to fully adopt a policy of free trade. This was often not so much the result of any ideological predisposition as a response to the state's desire for greater revenue. Taxing trade – both at home and from abroad – was one of the most common means of generating the income that supported the expanding bureaucracy of the modern state. Furthermore, special interests often worked to distort policy to favour of specific producers or economic sectors.

Since the late 17th century, Britain had been especially dependent on customs and excise taxes of various sorts. The rise of British liberalism had come in the same century (the 18th) that had seen the British state grow to an unprecedented size. Growth of government revenue had vastly outstripped the rate of overall economic growth and served to fund a professional bureaucracy at home and an expanding imperialist policy abroad. This enabled the British to either defeat or stalemate their traditional rival, France, in a series of military struggles that extended from the late 1600s to the era of Napoleon a century later. Moreover, this expansion of the central government came with little change in the revenues from land, the traditional source of income. Most of the gains came from steep increases in revenue from trade; and rising excises were some of the abuses cited by the American colonists as the basis for the independence movement.

However, changes in the landscape of the British economy – most notably the urban and industrial expansion that began in the late 1700s and is known as the Industrial Revolution – made Britain the premier industrial producer of the early 19th century and put pressure on the government to transform legislation that had kept agricultural prices high and had limited imports for the benefit of the farmers who were an increasingly small share of the economy.

The 19th-century Corn Law repeal: free trade rhetoric vs. protectionist reality

The interests of industrial producers who felt that workers would be better served by cheap bread and the ideas of liberal elites saw concrete expression in the creation of the Anti-Corn Law League beginning in the 1830s. Statesmen such as Richard Cobden explicitly saw the movement as the first step in an attempt to push the British government to adopt a general policy of free trade.

However, it is not clear that theoretical ideas played a large role in the actual dismantling of the Corn Laws. Furthermore, Smith had always held up the staple industries and national defence as areas that might be exceptions to the doctrine of pure laissez-faire. However, the end of the Napoleonic Wars in 1815 removed the basis for wartime support of the Corn Laws and pushed the government to consider modifying or abolishing the restrictions in a transition to a peacetime economy.

As early as 1821 the government of Lord Liverpool had begun to consider reforming a system that it regarded as temporary and motivated by a desire to secure stable prices during wartime with a mix of regulation and protection. The Corn Laws did not seem to be fulfilling that function and, in the absence of war, their maintenance seemed unnecessary for the public good. Of course, the farm interests that gained from these rules would have fought for the continuation of these protections. Nonetheless, the increased voting power of urban workers empowered by the 1832 Reform Act reinforced Prime Minister Peel's conviction that support for industry was vital to the future development of Britain and led him to push for the abolition of all Corn Laws in the 1840s. The onset of the Irish potato famine in 1845 gave a special impetus to the desire to promote lower prices for basic staples and allowed Peel to push for the full abolition of the Corn Laws in 1846.

This legislation repealing the Corn Laws is often cited as the pivotal moment in the rise of free trade in Britain and in Europe because it was followed over the next decade with the reduction or removal of duties on hundreds of imports in Britain – hence the claim that henceforth Britain moved swiftly to full free trade. However, this accomplishment has been somewhat exaggerated in conventional history. Partly because of the need for continued revenue and partly because of pressure from special interests, a few large and important tariffs on coffee, tea, wine, spirits, sugar and tobacco continued up to the 1860s, tariffs which had a disproportionate impact on the trade of Britain.

The 1860 Anglo-French Trade Treaty and the true coming of free trade

The wine and spirit tariffs were especially important and had been mentioned prominently in Smith's criticism of the mercantile system in the *Wealth of Nations*. These tariffs had arisen from Britain's desire to punish her rival France and had developed as a means of protecting domestic beverage interests such as beer and gin at home, and colonial imports such as rum. Lacking an equivalent slogan to that of the cry for 'cheap bread', there was no great movement to reform these substantial duties.

Consequently, despite the British reputation as the leading free trader in the 19th century, Britain in fact had higher average tariffs than the more openly interventionist nation of France for the first three quarters of that century. The burden on the working classes from the combination of high tariffs on imported wine and liquor and the regulation and taxation of domestic production meant that consumption of alcohol was repressed throughout the 18th and early 19th centuries, despite all the income gains during the Industrial Revolution.

Where basic alcoholic beverages had been seen as a necessary staple in the 17th century, they were more likely to be treated as luxuries in the 19th.

Full reform had to wait until 1860, when Britain and France concluded the Anglo-French Treaty of Commerce. This landmark treaty can be said to have truly ushered in the age of free trade in Europe. Brokered by Cobden in Britain and Michel Chevalier in France, the treaty had come after many years of negotiation. Early overtures to the French to sign such a treaty had been rebuffed in the 1840s because Britain had been unwilling to compromise their duties on wine – which had been the category of greatest concern to the French. However, changes in British fiscal structure arising from the imposition of an income tax in the 1850s made it easier for the British government to contemplate tariff cuts that might have compromised the budget in the short run. (British Liberals believed that given enough time, lower rates on imported wine would be offset by increased trade, a belief that proved accurate.) Moreover, the political considerations that led to wine duties being designed from the early 1700s to favour the products of friendly nations such as Portugal and Spain over that of France grew less important in the decades of peace following the defeat of Napoleon Bonaparte in 1815.

Thus, it became possible to conclude a treaty in 1860 in which Britain lowered and modified all its wine and spirit tariffs to remove any anti-French bias and caused France to lower tariffs and remove all prohibitions on goods – primarily textiles – imported from Britain. The 1860 Treaty was also significant for being a Most Favoured Nation agreement in which any subsequent treaties with third countries negotiated by either party would cause concessions to be applied equally to the original signatories. Concern by other Western nations that they would be left out of a trading arrangement between the two leading European powers led to almost the whole of Europe concluding equivalent treaties with either Britain or France over the next decade. By the 1870s virtually the whole of Europe was an extremely open trading area with free movement of goods, capital, and labour that in some ways has never been matched even by today's European Union. And by the end of the 19th century Britain could be said to have genuinely become a free trader with few or modest tariffs on most items, and possibly the lowest average tariffs in all Europe.

It is also interesting to note that Britain provides something of a counter-example to the tendency of modern-day protectionists to fret about the trade balance. Britain was the undoubted leader in world trade throughout the 19th century yet she also ran a merchandise trade deficit for virtually the whole of that period up to the First World War.

The one major counter-example to the tendency in the West to move towards freer international commerce had been the United States. Whereas Europe was busy lowering or abolishing tariffs and trade restrictions after 1860, the USA raised tariffs substantially from the 1860s onwards. Tariffs were the major source of revenue for the federal government before the constitutional amendment that permitted an income tax. Furthermore, the civil war gave control of the government to the Republicans under Lincoln, who had made protection an important plank in the party's platform. To some extent the United States was fortunate in that many of the negative potential effects of the tariffs were somewhat offset by the free movement of capital, the large size of the internal US market, and the benefits of an extremely open immigration policy. Thus, while goods trade was restricted, capital and labour remained mostly mobile.

By the end of the 19th century, however, the free trade regime brought on by the 1860 Anglo-French Treaty began to unravel. As early as 1878 Germany began to modify her agricultural tariffs in response to pressure from farmers due to increased competition from Russia and the United States. French textile manufacturers pushed the government to abandon the treaty in 1882 and a new set of tariffs were put into place at the beginning of 1892. However, it is worth noting that in both cases the resulting tariff regimes were still relatively moderate and not comparable to the high protection of early Britain or mid-19th-century USA, and Europe still enjoyed vigorous exchange up to 1914, when the European system of open trade was effectively destroyed, first by the war and then by the high tariff walls that nations began to enact during the Great Depression.

The 19th-century trade debates have remained an important touchstone for both scholars and political elites. The same general issues persist to this day. How vigorously should a nation pursue free trade? Is it best to liberalize unilaterally or bilaterally with treaties or collectively through groups like the World Trade Organization? Today we continue to hear concerns about the importance of the trade deficit in hampering or restraining economic growth. Large and small nations often invoke the need to protect infant industries as a justification for tariffs, although it is interesting that in most cases throughout the world it is ageing and decaying industries that are likely to receive protection rather than the newer, more innovative sectors of the economy. And, as with Great Britain in the 19th century, the USA today is seen as the leader in world trade, with some of the same questions being asked about the extent to which trade is manipulated to improve world welfare or merely to enhance the narrow interests of the leading nations. And with the rise of treaties such as the North American Free Trade Agreement and the Central American Free Trade Agreement, as well as the Eurozone, there remain questions as to the virtues of piecemeal reform or the extent to which these agreements are merely mechanisms for obstructing trade by parcelling out the world into separate trading blocs.

JOHN NYE

See also **globalization; historical economics, British; Smith, Adam.**

Bibliography

Hilton, B. 1977. *Corn, Cash and Commerce: The Economic Policies of the Tory Government 1815– 1830.* Oxford: Oxford University Press.

Irwin, D.A. 1996. *Against the Tide: An Intellectual History of Free Trade.* Princeton: Princeton University Press.

Mun, T. 1664. *England's Treasure by Forraign Trade.* London: Printed by J.G. for Thomas Clark.

Nye, J.V.C. 1991. The myth of free trade in Britain and Fortress France: tariffs and trade in the nineteenth century. *Journal of Economic History* 51, 23– 46.

Schonhardt-Bailey, C., ed. 1997. *The Rise of Free Trade. Vol. 1, Protectionism and its Critics, 1815– 1837.* London: Routledge.

Smith, A. 1776. *An Inquiry into the Nature and Causes of the Wealth of Nations.* Oxford: Clarendon Press, 1976.

corporate governance

While some of the questions have been around since Berle and Means (1932), the term 'corporate governance' did not exist in the English language until the mid-1980s. Since then, however, corporate governance issues have become important not only in the academic literature but also in public polity debates. During this period, corporate governance has been identified with takeovers, financial restructuring and institutional investors' activism. But what exactly is corporate governance? Why is there a corporate governance 'problem'? Why does Adam Smith's invisible hand not automatically provide a solution? What role do takeovers, financial restructuring and institutional investors play in a corporate governance system?

In this article I will try to provide a systematic answer to these questions, making explicit the essential link between corporate governance and the theory of the firm. My goal is to provide a common framework that helps to analyse the results obtained in these two fields and identify the questions left unanswered. This is not a survey, so I make no attempt to be comprehensive. For an excellent survey on the topic the reader is referred to Shleifer and Vishny (1997).

1 When do we need a governance system?

The word 'governance' is synonymous with the exercise of authority, direction, and control. These words, however, seem strange when used in the context of a free-market economy. Why do we need any form of authority? Isn't the market responsible for allocating all resources efficiently without the intervention of authority? The basic (neoclassical) undergraduate microeconomics courses rarely mention the words 'authority' and 'control'.

In fact, neoclassical microeconomics describes well only one set of transactions, which Williamson (1985) calls 'standardized'. Consider, for instance, the purchase of a commodity, like wheat. There are many producers of the same quality of wheat and many potential customers. In this context, Adam Smith's invisible hand ensures that the good is provided efficiently without the need of any form of authority.

Many daily transactions, however, do not fit this simple example. Consider, for instance, the purchase of a customized machine. The buyer must contact a manufacturer and agree upon the specifications and the final price. Unlike the case of wheat, the signing of the agreement does not represent the end of the relationship between the buyer and the seller. Producing the machine requires some time. During this time many events can occur, which alter the cost of producing the machine as well as the buyer's willingness to pay for it. More importantly, before the agreement was signed, the market for manufacturers was competitive. Once production has begun, though, the buyer and the seller are trapped in a situation of bilateral monopoly. The customized machine probably has a higher value to the buyer than to the market. On the other hand, the contracted manufacturer has probably the lowest cost, to finish the machine. The difference between what the two parties generate together and what they can obtain in the marketplace represents a quasi-rent, which needs to be divided *ex post*. In dividing this surplus Adam Smith's invisible hand is of no help, while authority does play a role.

In the spirit of Williamson (1985), I define a *governance system* as the complex set of constraints that shape the *ex post* bargaining over the quasi-rents generated in the course of a relationship. A main role in this system is certainly played by the initial contract. But the contract, most likely, will be incomplete, in the sense that it will not fully specify the division of surplus in every possible contingency (this might be too costly to do or outright impossible because the contingency was unanticipated). This creates an interesting distinction between decisions made *ex ante* (when the two parties entered a relationship and irreversible investments were sunk) and *ex post* (when the quasi-rents are divided). This contract incompleteness also creates room for bargaining.

The outcome of the bargaining will be affected by several factors besides the initial contract. First, which party has ownership of the machine while it is being produced. Second, the availability of alternatives: how costly is it for the buyer to delay receiving the new machine; how costly is it for the manufacturer to delay the receipt of the final payment; how much more costly is it to have the job finished by another manufacturer, and so on. Finally, a major role in shaping the bargaining outcome is played by the institutional environment. For example: how effective and rapid is law enforcement; what are the professional norms; how quickly and reliably does information about the manufacturer's performance travel across potential clients, and so on. All these conditions constitute a governance system.

As illustrated by the machine example, there are two necessary conditions for a governance system to be needed. First, the relationship must generate some quasi-rents. In the absence of quasi-rents, the competitive nature of the market will eliminate any scope for bargaining. Second, the quasi-rents are not perfectly allocated *ex ante*. If they were, then there would be no scope for bargaining either.

2 Corporate governance

The above definition of governance is quite general. One can talk about the governance of a transaction, of a club, and, in general, of any economic organization. In a narrow sense, corporate governance is simply the governance of a particular organizational form – a corporation.

Yet the bargaining over the *ex post* rents, which I defined as the essence of governance, is influenced, but not uniquely affected, by the legal structure used. A corporation, in principle, is just an empty legal shell. What makes a corporation valuable is the claims the legal shell has on an underlying economic entity, which I shall refer to as the firm. While often the legal and the economic entity coincide, this is not always the case. For this reason, I define *corporate governance* as the complex set of constraints that shape the *ex post* bargaining over the quasi-rents generated by a firm.

To be sure, many problems that fall within the realm of corporate governance can be (and have been) profitably analysed without necessarily appealing to such a broad definition. Nevertheless, all the governance mechanisms discussed in the literature can be reinterpreted in light of this definition. Allocation of ownership, capital structure, managerial incentive schemes, takeovers, boards of directors, pressure from institutional investors, product market competition, labour market competition, organizational structure, and so on can all be thought of as institutions that affect the process through which quasi-rents are distributed. The contribution of this definition in simply to highlight the link between the way quasi-rents are distributed and the way they are generated. Only by focusing on this link can one answer fundamental questions such as who should control the firm.

Of course, this definition of corporate governance raises the age-old question of what a firm is. But this question should be central to corporate governance. Before we can discuss how a firm should be governed, we need to define the firm. This question is also important because it helps us identify to what extent, if any, corporate governance is different from the governance of a simple contractual relationship (such as in the machine example).

There are two main definitions of the firm available in the literature. The first, introduced by Alchian and

Demsetz (1972), is that the firm is a nexus of contracts. According to this definition, there is nothing unique in corporate governance, which is simply a more complex version of standard contractual governance.

The second definition, due to Grossman and Hart (1986) and Hart and Moore (1990) (henceforth GMH), is that the firm is a collection of physical assets that are jointly owned. Ownership matters because it confers the right to make decisions in all the contingencies unspecified by the initial contract. On the one hand, this definition has the merit of differentiating between a simple contractual relationship and a firm. Since the firm is defined by the non-contractual element (that is, the allocation of ownership), corporate governance (as opposed to contractual governance) is defined by the effect of this non-contractual element. Not surprisingly, the focus of the corporate governance literature since the mid-1990s has been the allocation of ownership (hence this literature is called the property rights view of the firm). On the other hand, this definition has the drawback of excluding any stakeholder other than the owner of physical assets from being important to our understanding of the firm.

More recently, Rajan and Zingales (2001; 1998) have proposed a broader definition. They define the firm as a nexus of specific investments: a combination of mutually specialized assets and people. Unlike the nexus of contracts approach, this definition explicit recognizes that a firm is a complex structure that cannot be instantaneously replicated. Unlike the property rights view, this definition recognizes that all the parties who are mutually specialized belong to the firm, be they workers, suppliers or customers. While this definition does not necessarily coincide with the legal definition, it does coincide with the economic essence of a firm: a network of specific investments that cannot be replicated by the market.

3 Incomplete contracts and governance
In an Arrow–Debreu economy it is assumed that agents can costlessly write all state-contingent contracts. As a result, all decisions are made *ex ante* and all quasi-rents are allocated *ex ante*. Thus, there is no room for governance. More surprisingly, even if we relax the assumption that every state-contingent contract can be written and admit that certain future contingencies are not observable (and thus not contractible), we still find no room for governance as long as one can costlessly write contracts on all future observable variables.

Recall the example of the customized machine, and assume that the manufacturer's effort is unobservable to others and is, therefore, not contractible. The neoclassical approach to this problem is to design a mechanism (hence the term 'mechanism design'), contingent on all publicly observable variables, which provides the manufacturer with the best possible incentives to exert effort. Myerson (1979) shows that all optimal mechanisms are equivalent to a revelation (direct) mechanism in which

the agent (manufacturer) publicly announces his information and receives compensation contingent on his announcement. An important consequence of this result is that, in the mechanism design approach, delegation (giving an agent discretion over certain decisions) is always weakly dominated by a fully centralized mechanism, where all decisions are made *ex ante* by the designer. The mechanism design approach reproduces several distinguishing features of an Arrow–Debreu economy: all decisions are made *ex ante* and executed only *ex post*; as a result, all conflicts are resolved and all rents are allocated *ex ante*. This leaves no room for *ex post* bargaining. All these features are incompatible not only with my definition of governance, but also with any meaningful (that is, related to the sense in which this term has been used) definition of governance. This is best illustrated with two examples.

One of the crucial questions in corporate governance concerns the party in whose interest corporate directors should act. In the mechanism design approach this question cannot even be raised. All possible future conflicts are resolved *ex ante* and the initial contract specifics how directors will behave in any observable state of the world. However, since this question is raised all the time, it must be that all possible conflicts are not resolved *ex ante*.

Second, the mechanism design approach avoids renegotiation: the initial contract is so designed that the agents do not want to renegotiate. As a result, the designer wants to make renegotiation as inefficient as possible: this reduces the costs of providing incentives to the agents with no efficiency costs, since renegotiation never occurs in equilibrium (Aghion, Bolton and Felli, 1997). If this result were to be taken seriously, the optimal public policy approach would be to preserve any inefficiency in the system in order to avoid destroying its beneficial incentives *ex ante*. In reality, though, the jurisprudential approach is completely different. For example, courts do not support punitive damages that are considered excessive with respect to the issue at stake.

Only in a world where some contracts contingent on future observable variables are costly (or impossible) to write *ex ante* is there room for governance *ex post*. Only in such a world are there quasi-rents that must be divided *ex post* and real decisions that must be made *ex post*. Finally, only in a world of incomplete contracts can we define what a firm is and discuss corporate governance as being different from contractual governance. Not surprisingly, the theory of governance is intimately related to the emergence and evolution of the incomplete contracts paradigm.

A fundamental milestone in this evolution is the residual rights of control concept introduced by Grossman and Hart (1986). In a world of incomplete contracts, it is necessary to allocate the right to make *ex post* decisions in unspecified contingencies. This residual right is both meaningful and valuable. It is meaningful because it confers the discretion to make decisions *ex post*. It is valuable because this discretion can be used strategically in bargaining over the surplus.

4 Why does corporate governance matter?

By definition, corporate governance matters for distribution of rents, but to what extent does it matter for economic efficiency? There are three main channels through which the conditions that affect the division of quasi-rents also affect the total surplus produced. In presenting these channels I make a sharp distinction between *ex ante* (when specific investments need to be sunk) and *ex post* (when quasi-rents are divided) effects, as though the firm lasted just one period. Of course, this is not true in reality because *ex post* considerations of one period are mixed with *ex ante* considerations for the next period.

4.1 Ex ante incentive effects

The process through which surplus is divided *ex post* affects the *ex ante* incentives to undertake some actions, which can create or destroy some value, in two main ways.

First, rational agents will not spend the optimal amount of resources in value-enhancing activities that are not properly rewarded by the governance system. In fact, one goal in designing a governance system is to motivate those investments that are not properly rewarded in the marketplace. The canonical example of how a change in the governance structure can change the incentives to make a value-enhancing relationship-specific investment is the Fisher Body case. In the early 1920s, Fisher Body (an auto body manufacturer) refused to locate its plants close to General Motors' plants in spite of the obvious efficiency improvement generated by such a move. Locating close to GM would have reduced Fisher Body's ability to supply other car manufacturers, which would have weakened its bargaining position *ex post* and possibly reduced its share of the quasi-rents generated by the relationship with GM (see Klein, Crawford and Alchian, 1978). A change in the governance system (the acquisition of Fisher Body by GM) eventually led to the efficient plant location decision. Another famous illustration of the same phenomenon is managerial shirking. A manager will shirk if her *ex post* bargaining payoff does not increase sufficiently with her effort and, therefore, fails to compensate her for the cost of this effort.

Second, rational agents will spend resources in inefficient activities whose only (or main) purpose is to alter the outcome of the *ex post* bargaining in their favour. For example, a manager may specialize the firm in activities she is best at running because this increases her marginal contribution *ex post* and, thus, her share of the *ex post* rents (Shleifer and Vishny, 1989). Interestingly, this problem is not limited to the top of the hierarchy, but is present throughout. Subordinates, who do not have

much decision power, will waste resources trying to capture the benevolence of their powerful superiors (Milgrom, 1988). Even the well-known tendency of managers to overinvest in growth can be interpreted as a manifestation of this problem. Managers like to expand the size of their business because this makes them more important to the value of the firm and, thus, increases the payoff they can extract in the *ex pest* bargaining.

Of course, a governance system might promote or discourage these activities. For example, Chandler (1966) reports that, under the Durand reign, GM's capital allocation was highly politicized ('a sort of horse trading'). The move to a multi-divisional structure, with the resulting increase in divisional managers' autonomy, reduced the managers' payoff from rent-seeking. Similarly, Milgrom and Roberts (1990) explain many organizational rules as a way to minimize influence costs. Finally, Rajan, Servaes and Zingales (2000) argue that inefficient 'power-seeking' is more severe the more investment opportunities a firm's divisions have. Consistent with this claim, they find that the value of a diversified firm is negatively related to the diversity of the investment opportunities of its divisions.

Thus, a governance system affects the incentives to invest or seek power, altering the marginal payoffs that these actions have in *ex post* bargaining. For instance, for an independent Fisher Body, the marginal effect on the bargaining payoff of localizing its plants close to GM is negative (it reduces the value of its outside options), but is positive for Fisher Body as a unit of GM, which does not have the authority to supply other manufacturers without GM's consent (see Rajan and Zingales, 1998). Thus, a different ownership structure alters the incentives to make specific investments.

4.2 Inefficient bargaining

A second channel through which a governance system affects total value is by altering *ex post* bargaining efficiency. This is tantamount to saying that the governance system affects the degree to which the assumptions of the Coase theorem are violated. A governance system, therefore, can affect the degree of information asymmetry between the parties, the level of coordination costs, or the extent to which a party is liquidity constrained.

For example, if control rights are assigned to a large and dispersed set of claimants (like the shareholders of most publicly traded companies), free-rider problems may prevent an efficient action from being undertaken even if property rights are well defined and perfectly tradeable (Grossman and Hart, 1980). Alternatively, the allocation of control rights can affect efficiency by determining the direction in which a compensating transfer must be made. The direction of the transfer matters when one of the parties to the *ex post* bargaining is liquidity constrained (Aghion and Bolton, 1992) or when it faces a different opportunity to invest these resources productively rather than in power-seeking activities (Rajan and

Zingales, 1996). In both cases an efficient transaction may not be agreed upon – in the first case because the party that should compensate does not have the resources, in the second case because the transaction (while efficient *per se*) may generate such an increase in wasteful power-seeking as to more than offset its benefits.

To this standard list of imperfections, Hansmann (1996) adds the divergence of interests among the parties who have control rights. Citing the political economy literature, Hansmann argues that *ex post* inefficiency is increasing in the divergence of interests among control holders. For example, he argues that allocating control to workers is more costly when they differ in their professional skills, hierarchical position and tenure. While Hansmann does not provide a formal model of why this relation occurs, he does provide very compelling evidence that in practice control rights are rarely allocated to parties with conflicting interests. His conjecture is intriguing because there is no well-established general theory of how different governance systems lead to different levels of *ex post* inefficiency. There is little doubt, however, that these inefficiencies exist and are important. For example, Wiggins and Libecap (1985) document that an excessively dispersed initial allocation of drilling rights leads to an inefficient method of extracting oil, with estimated losses as big as 50 per cent of the total value of the reservoir.

4.3 Risk aversion

Finally, a governance system might affect the *ex ante* value of the total surplus by determining the level and the distribution of risk. If the different parties have different degrees of risk aversion (or different opportunities to diversify or hedge risk), then the efficiency of a governance system is also measured by how effectively it allocates risk to the most risk-tolerant party. This idea is the cornerstone of Fama and Jensen's (1983a; 1983b) analysis of organizational structure and corporate governance.

Different governance systems can also *generate* a different amount of risk. Suppose, for instance, that the total amount of surplus generated is constant. It is still possible that the payoff of each party is stochastic, if the governance structure generates a stochastic bargaining outcome. For example, a life insurance contract written in nominal dollars creates a pure gamble between the policy holders and the insurance company with respect to the future rate of inflation. This additional 'governance' risk (in this case created by the contract, in general created by the governance structure) reduces the value of the total surplus, if the parties are risk averse and cannot diversify away the risk.

In summary, the objectives of a corporate governance system should be: (*a*) to maximize the incentives for value-enhancing investments, while minimizing inefficient power seeking; (*b*) to minimize inefficiency in *ex post* bargaining; (*c*) to minimize any 'governance' risk and allocate the residual risk to the least risk averse parties.

5 Who should control the firm?

To show the utility of the framework developed thus far, I will use it to address one of the most controversial issues in corporate governance: who should control the firm? In particular, I will analyse it with regard to the first of the three above objectives of a corporate governance system. For an analysis focused on the second objective the reader is referred to Hansmann (1996), and for an analysis focused on the third to Fama and Jensen (1983a; 1983b).

As far as the first objective is concerned, the allocation of control is important because it affects the division of surplus. By controlling a firm's decisions, a party can ensure for itself of more and more valuable options without the collaboration of the other parties. This guarantees the controlling party a larger share of the surplus within the relationship. Thus, in the framework outlined above, the question of who should control the firm can be rephrased as: whose investments need more protection in the *ex post* bargaining? Again, the answer to this question is indissolubly linked to the underlying theory of the firm.

In the nexus of contracts view, the firm 'is just a legal fiction which serves as a focus for the complex process in which the conflicting objectives of individuals … are brought in equilibrium within a framework of contractual relationship' (Jensen and Meckling, 1976, p. 312). Thus, according to this view each party is fully protected by its contract with the exception of the shareholders, who accept a residual payoff because they possess a comparative advantage in diversifying risk. As a result, shareholders need the protection insured by control.

While widely popular, this explanation is unsatisfactory. The contractual protection provided to the parties involved in the nexus of contracts is complete only if contracts are complete. But if contracts are complete, then the statement that shareholders are in control is meaningless. In fact, in a world of complete contracts all the decisions are made *ex ante*, and thus shareholders are no more in control than are the workers: everything is contained in the initial grand contract. Furthermore, as I have already argued in Section 3, this conclusion is inconsistent with the existence of a debate on what a company should do.

Alternatively, if contracts are incomplete, then the argument that all other parties are fully protected by their contractual relationships does not automatically follow. In fact, in this context one should ask why shareholders need more protection than other parties to the nexus of contracts. I return to this issue below.

In the property rights view of the firm, the reason why shareholders should be in control is straightforward. Control is allocated so as to maximize the incentives to make human capital-specific investments. The owner of the firm will generally be the worker with the most expropriable investment. In other words, the property rights approach does not deal with outside shareholders and, thus, it applies only to entrepreneurial firms.

Outside of the GHM framework, the typical justification for why shareholders (or more generally the providers of finance) are in control is based on a combination of three arguments. Shareholders need more protection because: (*a*) their investment is more valuable; (*b*) other stakeholders can protect their investments better through contracts; (*c*) other stakeholders have other sources of power *ex post* that protect their investments. Of the three arguments, the first is clearly unfounded. Reviewing the empirical evidence on the return on specialized human capital, Blair (1995) estimates that the quasi-rents generated by specialized human capital are as big as accounting profits, which are likely to overestimate the quasi-rent generated by physical capital. Hence, there is no ground for dismissing human capital investments as second order to physical capital investments.

The second argument is harder to dismiss. Since we lack a fully satisfactory theory of why contracts are incomplete, we cannot easily argue which contracts are more incomplete. Nevertheless, it is hard to argue that human capital investments are easier to contract than physical capital investments. If there is one contingency that is easily verifiable, it is the provision of funds. Thus, it is not obvious why providers of funds are at a comparative disadvantage.

The most convincing argument is probably the third. As Williamson (1985) puts it,

> the suppliers of finance bear a unique relation to the firm: The whole of their investment in the firm is potentially placed at hazard. By contrast, the productive assets (plant and equipment; human capital) of suppliers of raw material, labor, intermediate product, electric power, and the like normally remains in the suppliers' possession.

Thus, the other stakeholders have a better outside option in the *ex post* bargaining, and they do not need the protection ensured via the residual rights of control.

Even this argument, however, is not fully satisfactory. In fact, it suggests only that the suppliers of finance should have some form of contractual protection – it does not necessarily imply that they should be protected via the residual rights of control.

A satisfactory explanation of why the residual right belongs to the shareholders can be obtained only in a theory of the firm that explicitly accounts for the existence of different stakeholders, and models the interaction between contractual (for example, ownership) and non-contractual (for example, unique human capital investments) sources of power. An attempt in this direction is made by Rajan and Zingales (1998).

To understand the argument, note that the residual right of control over an asset always increases the share of surplus captured by its owner (who has the opportunity to walk away with the asset), but it does not necessarily increase her marginal incentive to specialize. If, as is likely, a more specialized asset has less value outside the

relationship for which it has been specialized, then specialization decreases the owner's outside opportunity and, thus, her share of the quasi-rents. Owning a physical asset, then, makes an agent more reluctant to specialize it. As a result, the residual right of control is best allocated to a group of agents who need to protect their investment against *ex post* expropriation, but who have little control over how much the asset is specialized.

Consider now the different members of the specific investments nexus that makes up the firm. Most of the specific investments which form this nexus are in human capital and, therefore, can neither be contracted nor delegated *ex ante*. Granting the residual right to any of these members will have a negative effect on their incentive to specialize. By contrast, since the provision of funds is easily contractible, funds will be provided in the optimal amount as long as their providers receive sufficient surplus *ex post*. Thus, allocating the residual rights of control to the providers of funds has the positive effect of granting them enough surplus *ex post*, while avoiding the negative effect of reducing their marginal incentives.

Once they have provided funds, however, financial investors might be reluctant to use these funds for very specialized projects, for fear of seeing their share of the return fall. Thus, it is optimal that, while retaining a residual right of control over the assets, the providers of funds delegate the right to specialize the assets to a third party, who does not internalize the opportunity loss generated by this specialization. This third party, thus, should not be in the position of a mere agent, who owes a duty of obedience to the principal, but should be granted the independence to act in the interest of the firm (that is, the whole body of members of the nexus of specific investments), and not only of the shareholders. Blair and Stout (1997) claim that this is the role American corporate law attributes to the board of directors.

In sum, a broader definition of the firm allows us to understand why the residual right of control is allocated to the providers of capital and why its use is mostly delegated to a board of directors.

6 Normative analysis

An interesting, and largely unexplored, application of the incomplete contract approach to corporate governance is the analysis of its normative implications. In a world of complete contracts, such analysis has limited scope. A benevolent social planner would be unable to improve the *ex ante* allocation reached by private contracting, because this will achieve the constrained-efficient outcome. *Ex post*, the outcome might be inefficient, but that inefficiency is always part of the written contract and needs to be preserved to maintain *ex ante* future efficiency. By contrast, in a world of incomplete contracts, there is ample scope to analyse both *ex ante* and *ex post* efficiency.

First, a privately optimal governance system may not be socially efficient. In fact, a world of incomplete contracts generates some incentives to 'arbitrage power' through time. Consider an entrepreneur, who has immense bargaining power today, but anticipates losing it in the near future. If she could write all the contracts she could succeed in extracting all the present and future surplus arising from a relationship without any distortion. But, if some contracts cannot be written, then the entrepreneur has an incentive to distort her choices so as to transfer some of her bargaining power today into the future, enabling her to capture some of the future surplus as well. This is the idea underlying the choice of ownership in Zingales (1995a) and Bebchuk and Zingales (1996), and of the hierarchical structure in Rajan and Zingales (2001). It can also be used to provide a rationale for the existing mandatory rules (see Bebchuk and Zingales, 1996).

Second, in a world of incomplete contracts one can discuss the welfare effects of different institutions. For example, in a world of complete contracts the type of legal system a country adopts is irrelevant, as long as private contracts are enforced. By contrast, it is at least conceivable that in an incomplete contract world it may have a significant effect. This is intriguing because empirically it has been shown that legal institutions have an effect on the appropriability of quasi-rents by outside investors (Zingales, 1995b), on the way corporate governance is structured (La Porta et al., 1996), and on the amount of external finance raised (La Porta et al., 1997).

Finally, the incomplete contract approach generates a potential role for government intervention *ex post*. Unlike in the mechanism design approach, in an incomplete contract world *ex post* inefficiency is not necessarily desirable *ex ante*. Thus, a selective intervention that eliminates *ex post* inefficiency, while preserving the distributional consequences sought *ex ante*, will improve welfare.

7 Limitations of the incomplete contract approach

While the incomplete contracts approach to corporate governance has brought tremendous insights to the corporate governance debate, it has two weaknesses.

First, its predictions for the optimal allocation of ownership are extremely sensitive to what contracts can be written. Consider, for instance, the plant localization problem discussed above. If no contracts can be written, then – according; to the property rights approach – Fisher Body (who makes the bigger specific investment) should own the asset. However, if General Motors could credibly commit through a contract to buy all its car bodies from Fisher Body (as it did), then giving ownership to Fisher Body will confer too much power on it, and, thus, it is optimal for General Motors to own the asset (Hart, 1989), Thus, who should have the residual right of control depends crucially upon what the contractable rights are. But this is very difficult to argue on a priori grounds without a general theory of why contracts are incomplete (see Maskin and Tirole, 1997).

Second, this approach relies heavily (as does the complete contract approach) on the agents anticipating all future possible contingencies (Hellwig, 1997). This requirement can be reasonable when the subject of analysis is a small entrepreneurial firm, but it loses credibility when it is applied to large publicly held companies formed decades ago. Can we really interpret the capital structure of IBM today as the outcome of the design by Charles Flint (its founder) in 1911 attempting to allocate control optimally? Hart (1995) argues that the 'founding father' interpretation is simply a metaphor for the capital structure that a manager will choose under the pressure of the corporate control market. Yet Novaes and Zingales (1995) show that the two approaches lead to different predictions, not only about the level of debt but also about its sensitivity to the cost of financial distress and times. Thus, in the current state of knowledge, the *ex ante* approach to the capital structure of non-entrepreneurial companies lacks theoretical foundations.

8 Summary and conclusions

In this article I have tried to summarize the results obtained by applying the incomplete contracts approach to corporate governance. In a world where all future observable contingencies can be costlessly contracted upon *ex ante*, there is no room for governance. By contrast, in an incomplete contracts world, corporate governance can be defined as the set of conditions that shapes the *ex post* bargaining over the quasi-rents generated by a firm. A governance system has efficiency effects both *ex ante*, through its impact on the incentive to make relationship-specific investments, and *ex post*, by altering the conditions under which bargaining takes place. A governance system also affects the level and the distribution of risk.

The incomplete contracts approach has been very successful in explaining the corporate governance of entrepreneurial firms. It can explain how ownership is allocated and how capital structure is chosen. By contrast, it is difficult for this approach to cope with the complexity of large publicly traded complies. Nevertheless, recent contributions in the area are able to account for some important features of large corporations: allocation of ownership to the providers of capital who are dispersed, and the importance of internal organization.

Many aspects, however, remain to be investigated. First and foremost is the role of the board of directors. The second is the interaction between the different mechanisms of corporate governance. While we have many models that describe how each mechanism works in isolation, we know very little about how they interact. The effects are not obvious. For example, debt and takeovers are generally thought, in isolation, to be two instruments that reduce the amount of quasi-rents appropriated by management. But the use of debt may crowd out the effectiveness of takeovers, increasing rather than decreasing managerial rents (Novaes and Zingales, 1995). Third, the normative implications of this approach deserve more attention. In a world of incomplete contracts, privately optimal governance can be inefficient *ex ante* and *ex post*. Of course, this is only a theoretical possibility, whose relevance needs to be assessed in the data. The most important contribution, however, will arise from a development of the underlying theory. Without a better understanding of why contracts are incomplete, all the results are merely provisional.

LUIGI ZINGALES

See also **hold-up problem; incomplete contracts.**

Bibliography

Aghion, P. and Bolton, P. 1992. An incomplete contract approach to financial contracting. *Review of Economic Studies* 59, 473–94.

Aghion, P., Bolton, P. and Felli, L. 1997. Some issues on contract incompleteness. Working paper, London School of Economics.

Alchian, A. and Demsetz, H. 1972. Production, information costs and economic organization. *American Economic Review* 62, 777–95.

Bebchuk, L. and Zingales, L. 1996. Corporate ownership structures: private versus social optimality. Working Paper No. 5584. Cambridge, MA: NBER.

Berle, A. and Means, G. 1932. *The Modern Corporation and Private Property*. New York: Macmillan.

Blair, M.M. 1995. *Ownership and Control*. Washington, DC: Brookings Institution.

Blair, M.M. and Stout, L. 1997. A theory of corporation law as a response to contracting problems in team production. Working Paper, Brookings Institution.

Chandler, A. 1966. *Strategy and Structure*. Garden City, NY: Anchor Books.

Fama, F. and Jensen, M.C. 1983a. Separation of ownership and control. *Journal of Law and Economics* 26, 301–25.

Fama, E. and Jensen, M.C. 1983b. Agency problems and residual claims. *Journal of Law and Economics* 26, 327–49.

Grossman, S. and Hart, O. 1980. Takeover bids, the free rider problem and the theory of the corporation. *Bell Journal of Economics* 11, 42–69.

Grossman, S. and Hart, O. 1986. The costs and the benefits of ownership: a theory of vertical and lateral integration. *Journal of Political Economy* 94, 691–719.

Hansmann, H. 1996. *The Ownership of Enterprise*. Cambridge, MA: Belknap Press of Harvard University Press.

Hart, O. 1989. An economist's perspective on the theory of the firm. *Columbia Law Review* 89, 1757–74.

Hart, O. 1995. *Firms, Contracts, and Financial Structure*. Oxford: Oxford University Press.

Hart, O. and Moore, J. 1990. Property rights and the nature of the firm. *Journal of Political Economy* 98, 1119–58.

Hellwig, M. 1997. Unternchmensfinanzierung, Unterichmenskontrolle und Ressourcenallokation: Was leister das Finanzsystem. Arbeitspapier Nr. 97/02, University of Mannheim.

Jensen, M.C. and Meckling, W. 1976. Theory of the firm; managerial behavior, agency costs and ownership structure. *Journal of Financial Economics* 3, 305–60.

Klein, H., Crawford, R. and Alchian, A. 1978. Vertical integration, appropriable rents and the competitive contracting process. *Journal of Law and Economics* 21, 297–326.

La Porta, R., Lopez de Silancs, F., Shleifer, A. and Vishny, R. 1996. Law and finance. Working Paper No. 5661. Cambridge, MA: NBER.

La Porta, R., Lopez de Silancs, F., Shleifer, A. and Vishny, R. 1997. Legal determinants of external finance. *Journal of Finance* 52, 1131–50.

Maskin, F. and Tirole, J. 1997. Unforeseen contingencies, property rights and incomplete contracts. Working Paper No. 1796, Institute of Economic Research, Harvard University.

Milgrom, P. 1988. Employment contracts, influence activities, and efficient organization design. *Journal of Political Economy* 96, 42–60.

Milgrom, P. and Roberts, J. 1990. Bargaining costs, influence costs and the organization of economics activity. In *Perspectives on Positive Political Economy*, ed. J. Alt and K. Shepsle. Cambridge: Cambridge University Press.

Myerson, R. 1979. Incentive compatibility and the bargaining problem. *Econometrica* 47, 61–73.

Novaes, W. and Zingales, L. 1995. Capital structure choice when managers are in control: entrenchment versus efficiency. Working Paper No. 5384. Cambridge, MA: NBER.

Rajan, R., Servaes, H. and Zingales, L. 2000. The cost of diversity: diversification discount and inefficient investment. *Journal of Finance* 55, 35–80.

Rajan, R. and Zingales, L. 1996. The tyranny of the inefficient: an enquiry into the adverse consequences of power struggles. Working Paper No. 5396. Cambridge, MA: NBER.

Rajan, R. and Zingales, L. 1998. Power in a theory of the firm. *Quarterly Journal of Economics* 113, 387–432.

Rajan, R. and Zingales, L. 2001. The firm as a dedicated hierarchy. *Quarterly Journal of Economics* 116, 805–51.

Shleifer, A. and Vishny, R. 1989. Management entrenchment: the case of manager-specific investments. *Journal of Financial Economics* 25, 123–40.

Shleifer, A. and Vishny, R. 1997. A survey of corporate governance. *Journal of Finance* 52, 737–83.

Wiggins, S.N. and Libecap, G.D. 1985. Oil field unitization: contractual failure in the presence of imperfect information. *American Economic Review* 75, 368–85.

Williamson, O. 1985. *The Economic Institutions of Capitalism*. New York: The Free Press.

Zingales, L. 1995a. Insider ownership and the decision to go public. *Review of Economic Studies* 62, 425–48.

Zingales, L. 1995b. What determines the value of corporate votes? *Quarterly Journal of Economics* 110, 1047–73.

corporate law, economic analysis of

The economic analysis of corporate law focuses primarily on publicly held corporations. Following Coase (1937), the corporation is conceptualized as a nexus of contracts. Because corporate law focuses primarily on the authority of management and its obligations to shareholders, the primary 'contract' of interest is that between management and shareholders. The content of the manager–shareholder contract is conceptualized in terms of the agency-cost model of Jensen and Meckling (1976), with management viewed collectively as agent, and shareholders viewed collectively as principal. Ideally, the terms of the manager–shareholder contract minimize agency costs and thereby maximize the value of the firm.

Most of the economics-oriented corporate law literature can be divided into three areas, all of which focus on the United States. First, there are papers that analyse the economic forces by which corporate law is created by states and adopted by firms, and the proper role of corporate law in light of those forces. Second, there are papers that analyse particular monitoring mechanisms that law creates – shareholder voting, shareholder lawsuits, takeovers. A third set of papers analyses the basic features of a corporation, focusing on limited liability.

This review will discuss these three sets of papers. We do not address the substantial literature on law and finance that suggests that a country's corporate law rules may affect its financial markets and economic growth (see La Porta et al., 1997; 1998; and Rajan and Zingales, 1998). Nor do we address corporate governance strategies, such as CEO pay, that are largely independent of corporate law.

The role of corporate law

Economics-oriented scholarship on the role of corporate law can be roughly divided into two generations. The first generation, which spanned the period from the late 1970s to the mid-1990s, tended to reach the conclusion that market forces would yield socially optimal corporate governance outcomes. The second generation spans the period from the mid-1990s to the present. This generation, which includes more empirical work than the first, has painted a less perfect picture of the relationship between market forces and socially optimal corporate governance (see Klausner, 2006).

First-generation scholarship

The central insight of the first generation of economics-oriented corporate law scholarship was the conceptualization of the public corporation as a contractual arrangement between managers and shareholders. This insight has its origin in Coase (1937). It was developed within the agency cost framework in Jensen and Meckling (1976), and extended to the analysis of corporate law by Easterbrook and Fischel (1989; 1991). Although managers and shareholders do not negotiate governance arrangements, the price mechanism for a company's shares in an initial public offering (IPO) is expected to serve the same function, just as it does in other markets where buyers and sellers do not explicitly negotiate contracts. Consequently, the legally enforceable elements of the corporation's governance structure are viewed as the product of a market-mediated contracting process. Scholars writing in this framework therefore argue that firms' governance structures tend to minimize the agency costs associated with the separation of ownership and control, and thereby maximize the value of the corporation.

Legally enforceable governance commitments can take either of two forms. First, firms select the corporate law rules that govern the rights of shareholders and the obligations of management. Each of the 50 US states has enacted corporate law rules. Firms are free to elect to be governed by any of these rules, regardless of where they do business. To be governed by any state's legal rules, a firm need only incorporate in that state at the time of its IPO. Subsequent disputes between managers and shareholders will then by the decided according to the corporate law of that state. A firm cannot change its state of incorporation unless its board of directors and shareholders holding at least a majority of its shares agree. Second, pre-IPO manager/shareholders must draft a charter that will govern the corporation once it goes public. A charter begins as a blank slate and can include any governance arrangements that a firm's pre-IPO shareholders choose to adopt. To a substantial degree, the law allows a firm's charter to override provisions of corporate law. Thus, corporate law rules are often simply default rules that can be superseded by a corporation's charter terms.

Thus, one insight of this first generation was that corporate law was a product that states produce and firms consume. Winter (1977) was the first to argue that states are engaged in a 'race to the top' to produce corporate law that would tend to minimize agency costs. In order to obtain revenues from franchise fees and to create business for their local lawyers, states were expected to offer corporate law (that is, default rules) that would maximize the value of many firms and thereby save firms the trouble of customizing their own charter terms. Romano (1985) provided empirical evidence consistent with the proposition that a race to the top was occurring. She found, however, that Delaware had already achieved a substantial lead and questioned whether the race would actually make it to the top. The argument that market forces would produce legally enforceable governance commitments that would minimize agency costs stood in contrast to an earlier claim by Cary (1974) that states were engaged in a 'race to the bottom' to create legal rules that appeal to management at the expense of shareholders.

Second-generation scholarship

The second generation of scholarship has cast both empirical and theoretical doubt on the contractarian claims described above.

Empirical findings

A central claim of the contractarian conception of the corporation and corporate law is that corporations are heterogeneous in their corporate governance needs – hence the value of atomistic contracting. Empirical studies have now shown, however, that there is a high degree of uniformity in firms' governance commitments at the time they go public.

Daines (2002) found that, between 1978 and 2002, 50 per cent of firms incorporated in Delaware, and that during the second half of this period over 70 per cent of firms incorporated in Delaware. More importantly, however, Daines found that, among firms that did not incorporate in Delaware, nearly all incorporated in the state in which they were headquartered – whatever that state happened to be. Bebchuk and Cohen (2003) and Kahan (2006) confirmed Daines's findings.

These findings regarding incorporation decisions have three implications. First, the decision to incorporate in one's home state (when no out-of-state firms incorporate there) cannot be motivated primarily by the content of a state's laws. Something else must be at work. Daines's findings suggest that this choice may be made by the firm's local lawyer, hoping to keep the firm's business following the IPO, or by management wanting access to the state legislature if it needs a law passed. Romano (1987) found that most state anti-takeover legislation enacted in the 1980s was initiated by in-state management seeking protection from hostile bids. Bebchuk and Cohen (2003) found that states seem to retain more home-state incorporations if they already have state anti-takeover statutes on their books, but Kahan (2006) refuted this finding. Kahan did find, however, that states with very low-quality corporate law retained fewer home-state incorporations than did other states.

Second, these findings imply a high degree of uniformity in the governance commitments reflected in a firm's incorporation decision. Firms that focus on the quality of corporate law choose Delaware law. This uniformity casts some doubt on the contractarian assumption that firms are heterogeneous in their governance needs. Alternatively, the findings may suggest that there is value in uniformity itself, a point addressed below. Either way, there is evidence that the choice of Delaware as a state of incorporation enhances firm value. Romano (1987) and Daines (2001) found evidence consistent with this conclusion. Subramanian (2004) argues that this is a small-firm effect.

Third, the findings on incorporation choices cast doubt on the proposition that states compete to attract incorporations – whether racing to the top or to the bottom, Delaware seems to be the only state competing.

This is what Kahan and Kamar (2002) find in a study of states' efforts, or lack thereof, to attract incorporations and to earn revenues from them.

Empirical research has also revealed a high degree of uniformity in corporate charters. These supposed vehicles of customized contracting and innovation turn out to be fairly empty vessels. The only dimension on which they vary is in that of takeover defences (Klausner, 2006), and variability in that respect sits uneasily with the proposition that IPO charters maximize firm value. Three studies by Daines and Klausner (2001), Field and Karpoff (2002) and Coates (2001) have shown that firms commonly go public with charters providing for staggered boards, which are an effective anti-takeover defence that tends to reduce share value.

Theoretical challenges to the contractarian framework

It is possible that the contractarians overstated their premise that firms are heterogeneous in their governance needs. When it comes to legally enforceable governance commitments, perhaps one size fits all.

There are theoretical reasons, however, to doubt that homogeneous governance needs explain the uniformity described above. The essentially complete absence of customization or innovation in corporate charters suggests there are market imperfections in the contracting process. There has been no lack of innovation in corporate governance since the mid-1980s. None, however, originated in a corporate charter. Innovation at the firm level has taken the form of unilateral adoption of governance structures – for instance, an independent board or separation of CEO and chair – with no legally binding commitment to maintain those structures. The absence of legally binding commitments suggests that the cost of legal enforcement plays some role in the relative emptiness of corporate charters. While there have been innovations in legally enforceable governance mechanisms, they have not occurred at the level of the individual firm charter or even state law, as the contractarian thesis predicts. Instead, they have occurred, for better or worse, through Securities and Exchange Commission (SEC) regulation and federal statute (Sarbanes–Oxley Act, described below).

Klausner (1995; 2006) and Kahan and Klausner (1996) posit that there are learning and network externalities associated with state corporate law and corporate charter terms. As a result of these externalities, commonly used governance mechanisms have value independent of their intrinsic content; they tend to be better understood and less uncertain in their application than customized mechanisms. These externalities may thus explain the attraction of Delaware and the lack of customization or innovation in corporate charters.

In this context learning externalities take the form of judicial precedents interpreting and applying legal rules, and lawyers' familiarity with these precedents. Because many firms have been incorporated in Delaware, there is a large body of Delaware precedent. As a result, there is less uncertainty regarding how a legal rule will be applied. This may make Delaware valuable because firms have adopted it in the past.

Future judicial interpretations are valuable as well. The larger the number of firms that use the same legal rule or charter term over time, the more the rule or term will be litigated in the future, and the more frequently it will be interpreted. As Hansmann (2006) explains, the alternative would be periodic charter amendments, which could be difficult to accomplish because of the need to have a majority of shareholders and the company's board agree. Consequently, the market dynamic by which firms choose a state of incorporation can be expected to mirror that of product markets in network industries. The equilibrium in those industries can be socially suboptimal uniformity, which may be what is reflected in the attraction of Delaware incorporation and the 'plain vanilla' charter – that is, a charter with no customization that adopts essentially all default rules.

Kahan and Klausner (1996) offer two additional explanations of uniformity in charter terms and incorporation choices. One is that lawyers who draft charters on behalf of their corporate clients may be exhibit the same sort of individually rational herd behaviour that Scharfstein and Stein (1990) and Zwiebel (1995) model for agents such as money managers. These models are based on reputational payoffs to winning or losing with or without the herd. The second explanation relies on results in psychological experiments that reveal a 'status quo' bias, an 'anchoring' bias and a 'conformity' bias in other settings.

Law-intensive monitoring mechanisms

Corporate law creates three monitoring mechanisms and influences a fourth. First, corporate law gives shareholders the right to vote for the board of directors and to approve certain major changes, such as a change to the firm's charter or a merger or sale of the firm. Second, corporate law specifies managers' duties to shareholders and provides a way for shareholders to collectively sue management for its failure to fulfil these obligations. Third, corporate law regulates the takeover process, which allows a poorly run firm to be acquired by a third party. Finally, US federal securities law imposes mandatory disclosure obligations on publicly held firms, which facilitates each of these monitoring mechanisms and enables non-legal monitoring mechanisms (such as the press).

Shareholder voting

Corporate law gives control of the firm to the board of directors. Shareholder influence over managers comes from their right to elect the board and their implicit (or explicit) threat to vote out incumbent directors. Board elections are held annually and shareholders frequently have the ability to call interim elections. Today, voting is also the means by which control over firms changes hands in a takeover (Gilson and Schwartz, 2001).

Shareholders' ability to oust directors is thus an important check on managerial misbehaviour. The primary limitation on the effectiveness of the shareholder vote is economic rather than legal: shareholders' collective action problems. Individual shareholders with small stakes may not find it worthwhile to become informed and therefore typically either fail to vote or simply vote with management.

An important question is whether institutional investors, by virtue of their larger stakes, will solve the collective action problem and monitor managers more effectively. Money managers, pension funds, mutual funds, banks, insurance companies, and hedge funds all aggregate large pools of equity capital and may be more effective monitors. Rock (1991) and Romano (1993) give some reason to be cautious about their impact, however. They point out that the interests of money managers sometimes conflict with those of other shareholders. Banks, pension funds and insurance companies may side with incumbent managers if doing so gives them other opportunities to profit by managing the firm's pension funds, making loans or selling other services. Index funds have different disincentives to monitor. They compete on cost, and activism would increase their costs. Public pension funds are frequently active in pressuring managers, but these funds are run by political appointees and may favour politically popular proposals unrelated to firm value. Thus, the empirical evidence suggests that institutional shareholder activism has had only weak effects on firm performance (see, for example, Romano, 2001).

Others, focusing on the rules that govern shareholder ownership and voting, are also cautious about the potential impact about institutional investors. Roe (1994) and Black (1992) argue that shareholder passivity and collective action problems are created not solely by economic forces but also by politically motivated legal constraints that limit the institutional shareholder's incentives and ability to check incumbent managers. In this political view of shareholder passivity, a variety of banking, insurance and financial regulations prevent institutional investors from owning larger stakes or from monitoring managers more closely.

More recently, hedge funds have begun to aggregate large blocks of stock and to use their voting power to influence firm policies. Some investigate whether hedge funds have interests that conflict with other shareholders, which would suggest that hedge fund activism should be regulated (see Kahan and Rock, 2007; and Hu and Black, 2006). The alternative view is that hedge funds' large stakes and relative freedom from regulatory restrictions allows them to overcome collective action problems and to monitor managers.

Shareholder suits

The law provides mechanisms by which shareholders can collectively sue managers for mismanagement. As a means of controlling agency costs, however, shareholder suits are flawed. Because most shareholders will gain little from a successful lawsuit, shareholders often have no incentive to initiate or monitor these suits. Unless a major institutional shareholder is involved as lead plaintiff, lawyers initiate the suits, pay all costs, make litigation decisions, including settlement decisions, and collect a fee if the plaintiff class collects. To the extent the lawyer's interests diverge from the interests of the shareholders, agency costs are present on the plaintiffs' side of these lawsuits.

On the defendants' side, the familiar agency costs are present. Managers can use corporate funds to protect themselves – appropriately in some cases and inappropriately in others. They use corporate funds to purchase directors' and officers' liability insurance, which covers their personal liability and defence costs, unless they are proven to have engaged in deliberate fraud or the equivalent. Management can also use corporate funds to settle suits. Alexander (1991), Macey and Miller (1991), Coffee (1985; 1986), Romano (1991), Bohn and Choi (1996), among others, have argued that meritorious suits against management settle too easily, and that the prospect of settlement encourages frivolous suits.

The result of this battle of agents is nearly always a settlement in which the corporation and/or its directors' and officers' liability (D&O) insurer are the sole sources of payments. Consequently, payments go from shareholder to shareholder either directly or via insurance companies through premiums. Unless these suits deter mismanagement, the net winners are the lawyers on both sides. The shareholders in the aggregate are net losers (see Arlen and Carney, 1992; Langevoort, 1996; Mahoney, 1992; Easterbrook and Fischel, 1985).

Without commenting on the merits of these suits, Black, Cheffins and Klausner (2006) found only 13 cases, out of several thousand filed since 1980, in which outside directors have made personal payments. Inside managers bear personal liability more often, but settlement dynamics leave their assets untouched in all but a handful of cases per year (Alexander, 1991). Consequently, there is a question whether these suits have a significant deterrent effect.

The Public Securities Litigation Reform Act of 1995 (PSLRA) created several mechanisms designed to deter the filing of non-meritorious suits and to deter early settlement of meritorious suits. For instance, the law empowered the courts to select a lead plaintiff to monitor the shareholders' lawyer, with a presumption favouring institutional shareholders with substantial shareholdings. The law also requires a court to dismiss a suit unless the plaintiffs have alleged particular facts that support a 'strong inference' that a violation of the securities laws was committed with the legally required intent. This requirement was directed at the reported practice by which lawyers would file suits simply because a company's shares took a sharp drop in price, and then force the company into an expensive discovery process.

Ever since its enactment, scholars have tried to assess the impact of the PSLRA on securities class actions. Event studies, on the whole, have indicated that the law had a positive impact on share prices (Spiess and Tkac, 1997; Johnson, Kasznik and Nelson, 2000; Johnson, Nelson and Pritchard, 2000.) However, Ali and Kallapur (2001) found that the legislation had a negative impact on share prices. Studies have also tended to show that the PSLRA reduced the filing of non-meritorious suits (Johnson, Nelson and Pritchard, 2000; Bajaj, Muzumdar and Sarin, 2003). Others suggest, however, that some meritorious suits are deterred as well (Choi, 2007; Sale, 1998).

Choi, Fisch and Pritchard (2005), Thomas and Cox (2006) and Perino (2006) have shown that, while private institutional shareholders have not assumed the role of lead plaintiff, public pension plans have assumed that role to some extent. Perino (2006) found evidence consistent with monitoring by public pension plans.

Market for corporate control
The market for corporate control in the United States is regulated by state and federal law and is an important check on agency costs. If a firm is run poorly, either because managers are inattentive, consume too many perks or miss profitable merger opportunities, it may become the target of a takeover and its managers replaced. An active market in corporate control thus gives managers incentive to increase firm value (Manne, 1965).

In a 'hostile takeover' a buyer attempts to purchase a large block of stock, use its voting power to oust incumbent managers, purchase the remaining shares, and replace management. Alternatively, in the shadow of a hostile takeover, managers can agree to a 'friendly merger'. Both are associated with large gains to target shareholders. The evidence generally suggests that the premium comes from improvements in firm performance (see Andrade, Mitchell and Stafford, 2001; Romano, 1992).

The law and economics literature has focused on three questions. First, what should managers do when the firm becomes the target of a hostile takeover? Easterbrook and Fischel (1982; 1991) argue that target management should remain passive and that the law should prohibit them from resisting a takeover. They argue that resistance will reduce bidder returns and thus bidders' incentive to engage in takeovers. This will in turn reduce the disciplinary threat of takeovers and increase agency costs generally. Gilson (1982) and Bebchuk (1982) argue that managers should resist a takeover attempt to the extent necessary to hold an auction, which will assure that the assets of the firm end up in their highest valued use.

A second question involves whether managers' negotiating over a potential merger should be allowed to grant termination fees or 'lock-ups' to favoured bidders. Such measures may discourage competition and affect the outcome of an auction, raising the risk that managers will favour particular bidders in exchange for private benefits, such as job security. Ayres (1990) and Fraidin and

Hanson (1994) argue that termination fees and lock-ups will often not change the outcome of the auction and should therefore not be disfavoured. Kahan and Klausner (1996) examine how termination fees and lock-ups affect bidders' incentives to make a bid in the first place and their impact on agency costs generally. They explain that there is no reason for a target to grant a termination fee greater than a bidder's cost of making a bid.

Finally, a large literature examines whether, on average, takeover defences help or harm shareholder wealth. The typical research strategy examines how a firm's stock price reacts to the adoption of a takeover defence (see, for example, Comment and Schwert, 1995). This strategy usually suffers from a fatal flaw: it ignores the fact that the most potent defence, the 'poison pill', is freely available to all firms even after a hostile bid is received. Therefore, in effect, all firms have a poison pill and most other takeover defences are relatively unimportant. To disable a poison pill, a hostile bidder must first wage a proxy fight to unseat incumbent managers, install new managers who can remove the poison pill, and then go forward with the merger. The only takeover defences that are relevant other than a poison pill are those that either prevent an acquirer from replacing a target board or delay an acquirer's effort to do so. The most common defence is a classified (or staggered) board, which prevents an acquirer from taking control of a target board for two annual election cycles (see, for example, Daines, 2006; Faleye, 2007; Coates and Subramanian, 2002). Dual class stock, which is rarely used, allows management to control the election of the board and can therefore prevent an acquisition altogether.

A related literature examines whether firm takeover defences and shareholder rights predict stock returns (see, for example, Gompers, Ishii and Metrick, 2003; Cremers and Nair, 2005).

Mandatory disclosure
The monitoring mechanisms described above all depend, in part, on informed shareholders. Shareholder monitoring (of the kind contemplated by voting, law suits and the market for corporate control) is more effective when investors are informed. Thus, in many ways, the central regulatory event in US financial history was probably the 1933 and 1934 Acts, which created the Securities and Exchange Commission and required that publicly traded firms disclose detailed information about their historical performance and financial condition. These rules force firms both to disclose what they would otherwise prefer to keep private and to keep private information they might otherwise wish to disclose.

It is easy to see why disclosure might be valuable to investors. Accurate information allows investors to price securities and to monitor managers' performance. It is less easy to see why disclosure rules must be mandatory. Firms that fail to disclose will find it hard to raise money, as investors may take silence for bad news and refuse to

invest (Ross, 1979; Grossman, 1981; Milgrom, 1981). Therefore, firms and entrepreneurs may find it in their own interest to disclose information, whether or not it is required by law.

However, firms would not always find full disclosure to be in their interest. Disclosure imposes direct costs as firms produce and verify the information, as well as indirect costs if competitors, customers and others can use the information to the firm's disadvantage. Moreover, the costs and benefits of disclosure are likely to vary between firms. Left to their own devices, therefore, firms will commit to varying levels of disclosure. Some therefore argue that markets can sort out the costs and benefits of disclosure and believe that uniform and mandatory disclosure requirements are unnecessary and even harmful (Romano, 1998; 2005 Mahoney, 1997; Choi and Guzman, 1997). Others believe that there are externalities from a firm's disclosures and that a mandatory rule may therefore be socially beneficial (Easterbrook and Fischel, 1991; Coffee, 1984; Dye, 1990; Admati and Pfleiderer, 2000.)

Empirical evidence has not conclusively resolved this debate. Stigler (1964), Benston (1969; 1973) and Simon (1989), report evidence that mandatory disclosure did not improve investor welfare, but may have changed the characteristics of firms that go public. Recent evidence examines the effect of mandatory disclosure on firm returns and on asymmetric information (see Greenstone, Oyer and Vissing-Jorgenson, 2006; Daines and Jones, 2007).

A related debate involves whether managers should be allowed to trade on non-public information. Some hold that trading by informed insiders reveals valuable information and reduces agency costs (Manne, 1966; Carlton and Fischel, 1983). Others argue that insider trading is inefficient (Cox, 1986; Kraakman, 1991) or reduces stock market liquidity (Goshen and Parchomovsky, 2000). Beny (2007) reviews international evidence.

Creditors and the corporation
Because the corporation is a legal entity, distinct from its shareholders and managers, shareholders in the firm have 'limited liability' in that they are generally not personally liable for the debts of the corporation. At worst, public shareholders can lose their equity in the firm if the firm becomes insolvent.

This separate legal status gives rise to two issues. First, because shareholders will reap the upside of the firm's successes but will not bear the full downside of its failures, managers may promote the interests of shareholders at the expense of creditors (Jensen and Meckling, 1976). The legal rule of 'veil piercing' developed to respond to this problem, though to a very limited extent. Under extreme circumstances in which a corporation is undercapitalized and other conditions are met, a court may impose liability on the corporation's shareholders. As a practical matter, however, this rule is not applied to public companies' shareholders, and in the private company context the courts' application of this rule is notoriously unpredictable (Thompson, 1991).

The rule of limited liability makes sense for contract creditors, who can negotiate their own protection from default or charge and interest rate that compensates for the risk. Tort creditors, however, are different. Those owed compensation for, say, a firm's pollution emissions, will not have had the opportunity to negotiate with the firm *ex ante* to address the possibility that it will not have sufficient net assets to pay them. Thus, to deter corporate management from externalizing costs in the form of accidents and other torts and to prevent excess investment in risky activities, Hansmann and Kraakman (1991) argue that it may be desirable, and practical, to hold public shareholders personally liable for a corporation's torts. Grundfest (1992) and Alexander (1992) disagree as to the practicality of this proposal.

A second issue involving limited liability is the use of the corporate form to 'partition' assets to create separate pools of assets to bond separate debts and other contractual commitments. Hansmann and Kraakman (2000) explain how the partitioning of assets to separately bond the commitments of the corporate entity, individual shareholders and corporate entities within a group of affiliated corporations can promote efficiencies in creditor monitoring.

Sarbanes–Oxley Act of 2002

The Sarbanes–Oxley Act of 2002 (SOX) introduced sweeping corporate governance mandates on firms whose shares trade on US securities exchanges. Until this legislation, legal rules regarding substantive corporate governance were the province of US state law, and federal law was limited primarily to disclosure requirements. SOX imposed a series of federal requirements on the board operation and structure. Event studies of various legislative events leading to the enactment of SOX yielded mixed results. Li, Pincus and Rego (2004), Jain and Rezaee (2006), and Chhaochhaira and Grinstein (2004) show a positive reaction, but Zhang (2005) shows a negative reaction. Litvak (2007) finds a negative reaction by comparing foreign cross-listed firms subject to SOX with cross-listed firm not subject to SOX. Aggarwal and Williamson (2006) found that six of the governance structures mandated by SOX (all related to board independence) had a positive impact on share value when adopted by firms voluntarily prior to SOX. Romano (2005), on the other hand, looked at other SOX requirements (loans to officers, executive certification of financials, auditors' provision of non-audit services, and audit committee independence) and reports that there is no evidence to support their value to shareholders. Linck, Netter and Yang (2006) find that whatever the benefit of SOX, it increased the cost of boards, especially for small firms.

ROBERT DAINES AND MICHAEL KLAUSNER

See also **corporate governance.**

Bibliography

Admati, A.R. and Pfleiderer, P. 2000. Forcing firms to talk: disclosure regulation and externalities. *Review of Financial Studies* 13, 479–519.

Aggarwal, R. and Williamson, R. 2006. Did new regulations target the relevant corporate governance attributes? Working paper, McDonough School of Business, Georgetown University.

Alexander, J.C. 1991. Do the merits matter? A study of settlements in securities class actions. *Stanford Law Review* 43, 497–528.

Alexander, J.C. 1992. Unlimited shareholder liability through a procedural lens. *Harvard Law Review* 106, 387–445.

Ali, A. and Kallapur, S. 2001. Securities price consequences of the Private Securities Litigation Reform Act of 1995 and related events. *Accounting Review* 76, 431–60.

Andrade, G., Mitchell, M. and Stafford, E. 2001. New evidence and perspectives on mergers. *Journal of Economic Perspectives* 15(2), 103–20.

Arlen, J. and Carney, W. 1992. Vicarious liability for fraud on securities markets. *University of Illinois Law Review* 1992, 691–745.

Ayres, I. 1990. Analyzing stock lock-ups: do target treasury sales foreclose or facilitate takeover options? *Columbia Law Review* 90, 682–718.

Bajaj, M., Muzumdar, S. and Sarin, A. 2003. Securities class action settlements: an empirical analysis. *Santa Clara Law Review* 43, 1001–33.

Bebchuk, L. 1982. The case for facilitating competing tender offers. *Stanford Law Review* 35, 24–47.

Bebchuk, L.A. and Cohen, A. 2003. Firms' decisions where to incorporate. *Journal of Law and Economics* 46, 383–425.

Benston, G.J. 1969. The value of the SEC's accounting disclosure requirements. *Accounting Review* 44, 515–32.

Benston, G.J. 1973. Required disclosure and the stock market: an evaluation of the Securities Exchange Act of 1934. *American Economic Review* 63, 132–55.

Beny, L. 2007. Insider trading laws and stock markets around the world. *Journal of Corporate Law* 32, 237–300.

Black, B. 1992. Next steps in proxy reform. *Journal of Corporation Law* 18, 1–55.

Black, B., Cheffins, B. and Klausner, M. 2006. Outside director liability. *Stanford Law Review* 58, 1055–60.

Bohn, J. and Choi, S.J. 1996. Fraud in the new-issues market: empirical evidence on securities class actions. *University of Pennsylvania Law Review* 144, 903–64.

Carlton, D. and Fischel, D. 1983. The regulation of insider trading. *Stanford Law Review* 35, 857–95.

Cary, W. 1974. Federalism and corporate law: reflections upon Delaware. *Yale Law Journal* 83, 663–705.

Chhaochhaira, V. and Grinstein, Y. 2004. Corporate governance and firm value: the impact of the 2002 governance rules. Working paper, Cornell University.

Choi, S.J. 2004. The evidence on securities class actions. *Vanderbilt Law Review* 57, 1465–525.

Choi, S. 2007. Do the merits matter less after the Private Securities Litigation Reform Act? *Journal of Law, Economics & Organization.*

Choi, S.J., Fisch, J.E. and Pritchard, A.C. 2005. Do institutions matter? The impact of the Lead Plaintiff Provision of the Private Securities Litigation Reform Act. *Washington University Law Quarterly* 83, 869–905.

Choi, S. and Guzman, A. 1997. National laws, international money: regulation in a global capital market. *Fordham Law Review* 65, 1855–908.

Coase, R. 1937. The nature of the firm. *Economica* 4, 386–405.

Coates, J.C., IV. 2001. Explaining variation in takeover defenses: blame the lawyers. *California Law Review* 89, 1301–415.

Coates, J.C., IV and Subramanian, G. 2002. The powerful antitakeover force of staggered boards. *Stanford Law Review* 54, 887–951.

Coffee, J.C., Jr. 1984. Market failure and the economic case for a mandatory disclosure system. *Virginia Law Review* 70, 717–53.

Coffee, J.C., Jr. 1985. The unfaithful champion: the plaintiff as monitor in shareholder litigation. *Law and Contemporary Problems* 48, 5–81.

Coffee, J.C., Jr. 1986. Understanding the plaintiff's attorney: the implications of economic theory for private enforcement of law through class and derivative actions. *Columbia Law Review* 86, 669–727.

Comment, R. and Schwert, G.W. 1995. Poison or placebo? Evidence on the deterrence and wealth effects of modern antitakeover measures. *Journal of Financial Economics* 39, 3–43.

Cox, J.D. 1986. Insider trading regulation and the production of information: theory and evidence. *Washington University Law Quarterly* 64, 475–505.

Cremers, M. and Nair, V.B. 2005. Governance mechanisms and equity prices. *Journal of Finance* 60, 2859–94.

Daines, R. 2001. Does Delaware law improve firm value? *Journal of Finance & Economics* 62, 525–58.

Daines, R. 2002. The incorporation choices of IPO firms. *New York University Law Review* 77, 1559–605.

Daines, R. 2006. Do classified boards affect firm value? Takeover defenses after the poison pill. Working paper, Stanford Law School.

Daines, R. and Jones, C. 2007. Mandatory disclosure, asymmetric information and liquidity: the impact of the 1934 Act. Working paper, Law School, Stanford University.

Daines, R. and Klausner, M. 2001. Do IPO charters maximize firm value? Antitakeover protection in IPOs. *Journal of Law, Economics, & Organization* 17, 83–120.

Dye, R.A. 1990. Mandatory v. voluntary disclosures: the cases of financial and real externalities. *Accounting Review* 65, 1–24.

Easterbrook, F. and Fischel, D. 1982. Auctions and sunk costs in tender offers. *Stanford Law Review* 35, 1–19.

Easterbrook, F. and Fischel, D. 1985. Optimal damages in securities cases. *University of Chicago Law Review* 52, 611–42.

Easterbrook, F. and Fischel, D. 1989. The corporate contract. *Columbia Law Review* 89, 1416–48.

Easterbrook, F. and Fischel, D. 1991. *The Economic Structure of Corporate Law.* Cambridge, MA: Harvard University Press.

Faleye, O. 2007. Classified boards, firm value, managerial entrenchment. *Journal of Financial Economics* 83, 501–29.

Field, L.C. and Karpoff, J.M. 2002. Takeover defenses of IPO firms. *Journal of Finance* 57, 1857–89.

Fraidin, S. and Hanson, J. 1994. Toward unlocking lockups. *Yale Law Journal* 103, 1739–834.

Gilson, R. 1982. Seeking competitive bids versus pure passivity in tender offer defense. *Stanford Law Review* 35, 51–67.

Gilson, R.J. and Schwartz, A. 2001. Sales and elections as methods for transferring corporate control. *Theoretical Inquiries in Law* 2, 783–814.

Gompers, P.A., Ishii, J.L. and Metrick, A. 2003. Corporate governance and equity prices. *Quarterly Journal of Economics* 118, 107–55.

Goshen, G. and Parchomovsky, G. 2000. On insider trading, markets, and 'negative' property rights in information. Fordham Law & Economics Research Paper No. 06.

Greenstone, M., Oyer, P. and Vissing-Jorgensen, A. 2006. Mandated disclosure, stock returns, and the 1964 Securities Acts Amendments. *Quarterly Journal of Economics* 121, 399–460.

Grossman, S.J. 1981. The information role of warranties and private disclosure about product quality. *Journal of Law and Economics* 24, 461–83.

Grundfest, J.A. 1992. The limited future of unlimited liability: a capital markets perspective. *Yale Law Review* 102, 387–425.

Hansmann, H. 2006. Corporation and contract. *American Law and Economics Review* 8, 1–19.

Hansmann, H. and Kraakman, R. 1991. Toward unlimited shareholder liability for corporate torts. *Yale Law Review* 100, 1879–934.

Hansmann, H. and Kraakman, R. 2000. The essential role of organizational law. *Yale Law Journal* 110, 387–440.

Hu, H. and Black, B. 2006. The new vote buying: empty voting and hidden ownership. *Southern California Law Review* 79, 811–908.

Jain, P. and Rezaee, Z.J. 2006. The Sarbanes–Oxley Act of 2002 and capital-market behavior: early evidence. *Contemporary Accounting Research* 23, 629–54.

Jensen, M.C. and Meckling, W.H. 1976. Theory of the firm: managerial behavior, agency costs and ownership structure. *Journal of Financial Economics* 3, 303–60.

Johnson, M., Kasanik, R. and Nelson, K. 2000. Shareholder wealth effects of the Private Securities Litigation Reform Act of 1995. *Review of Accounting Studies* 5, 217–33.

Johnson, M., Kasznik, R. and Nelson, K. 2001. The impact of securities litigation reform on the disclosure of forward-looking information. *Journal of Accounting Research* 39, 297–328.

Johnson, M., Nelson, K. and Pritchard, A. 2000. In re Silicon Graphics Securities Litigation: shareholder wealth effects of the interpretation of the Private Securities Litigation Reform Act's pleading standard. *Southern California Law Review* 73, 773–809.

Kahan, M. 2006. The demand for corporate law: statutory flexibility, judicial quality, or takeover protection. *Journal of Law, Economics, & Organization* 22, 340–65.

Kahan, M. and Kamar, E. 2002. The myth of state competition in corporate law. *Stanford Law Review* 55, 679–760.

Kahan, M. and Klausner, M. 1996. Path dependence in corporate contracting: increasing returns, herd behavior and cognitive biases. *Washington University Law Quarterly* 74, 347–66.

Kahan, M. and Klausner, M. 1997. Standardization and innovation in corporate contracting (or 'The economics of boilerplate'). *Virginia Law Review* 83, 713–55.

Kahan, M. and Rock, E. 2007. Hedge funds in corporate governance and corporate control. *University of Pennsylvania Law Review* (forthcoming).

Klausner, M. 1995. Corporations, corporate law, and networks of contracts. *Virginia Law Review* 81, 757–833.

Klausner, M. 2006. The contractarian theory of corporate law: a generation later. *Journal of Corporate Law* 31, 779–97.

Kraakman, R. 1991. The legal theory of insider trading regulation in the United States. In *European Insider Dealing*, ed. K. Hopt and E. Wymeersch. London: Butterworths.

Langevoort, D.C. 1996. Capping damages for open-market securities fraud. *Arizona Law Review* 38, 639–68.

La Porta, R., Lopez-de-Silanes, F., Shleifer, A. and Vishny, R.W. 1997. Legal determinants of external finance. *Journal of Finance* 52, 1131–50.

La Porta, R., Lopez-de-Silanes, F., Shleifer, A. and Vishny, R.W. 1998. Law and finance. *Journal of Political Economy* 106, 1113–55.

Li, H., Pincus, M. and Rego, S. 2004. Market reaction to events surrounding the Sarbanes–Oxley Act of 2002 and earnings management. Working paper, University of Iowa.

Linck, J., Netter, J. and Yang, T. 2006. Effects and unintended consequences of the Sarbanes–Oxley Act on corporate boards. Working paper, University of Georgia.

Litvak, K. 2007. The effect of the Sarbanes–Oxley Act on non-US companies cross-listed in the US. *Journal of Corporate Finance*.

Macey, J. and Miller, G. 1991. The plaintiffs' attorney's role in class action and derivative litigation: economic analysis and recommendations for reform. *Chicago Law Review* 58, 1–94.

Mahoney, P. 1992. Precaution costs and the law of fraud in impersonal markets. *Virginia Law Review* 78, 623–60.

Mahoney, P.G. 1997. The exchange as regulator. *Virginia Law Review* 83, 1453–500.

Manne, H. 1965. Mergers and the market for corporate control. *Journal of Political Economy* 73, 110–20.

Manne, H. 1966. *Insider Trading and the Stock Market*. New York: Free Press.

Milgrom, P. 1981. Good news and bad news: representation theorems and application. *Bell Journal of Economics* 12, 380–91.

Perino, M. 2006. Institutional activism through litigation: an empirical analysis of public pension fund participation in securities class actions. Legal Studies Research Paper No. 06-0055, St. John's University.

Pritchard, A.C. and Sale, H. 2005. What counts as fraud? An empirical study of motions to dismiss under the Private Securities Litigation Reform Act. *Journal of Empirical Legal Studies* 2, 125–49.

Rajan, R.G. and Zingales, L. 1998. Financial dependence and growth. *American Economic Review* 88, 559–86.

Rock, E. 1991. The logic and (uncertain) significance of institutional shareholder activism. *Georgetown Law Review* 79, 445–506.

Roe, M. 1994. *Strong Managers, Weak Owners*. Princeton: Princeton University Press.

Romano, R. 1985. Law as a product: some pieces of the incorporation puzzle. *Journal of Law, Economics, & Organization* 1, 225–83.

Romano, R. 1987. The political economy of takeover statutes. *Virginia Law Review* 73, 111–98.

Romano, R. 1991. The shareholder suit: litigation without foundation? *Journal of Law, Economics, & Organization* 7, 55–87.

Romano, R. 1992. A guide to takeovers: theory, evidence and regulation. *Yale Journal on Regulation* 9, 119–80.

Romano, R. 1993. Public pension fund activism in corporate governance reconsidered. *Columbia Law Review* 93, 795–852.

Romano, R. 1998. Empowering investors: a market approach to securities regulation. *Yale Law Journal* 107, 2359–430.

Romano, R. 2001. Less is more: making institutional investor activism a valuable mechanism of corporate governance. *Yale Journal on Regulation* 18, 174–252.

Romano, R. 2005. The Sarbanes–Oxley Act and the making of quack corporate Governance. *Yale Law Journal* 114, 1521–611.

Ross, S.A. 1979. Disclosure regulation in financial markets: implications of modern finance theory and signaling theory. In *Issues in Financial Regulation: Regulation of American Business and Industry*, ed. F.R. Edwards. New York: McGraw Hill.

Sale, H. 1998. Heightened pleading and discovery stays: an analysis of the effect of the PSLRA's Internal-Information Standard on '33 and '34 Act claims. *Washington University Law Quarterly* 76, 537–95.

Scharfstein, D.S. and Stein, J. 1990. Herd behavior and investment. *American Economic Review* 80, 465–89.

Schwert, G.W. 1995. Comment: Poison or placebo: evidence on the deterrence and wealth effects of modern antitakeover measures. *Journal of Financial Economics* 39, 3–43.

Simon, C. 1989. The effect of the 1933 Securities Act on invertor information and the peformance of new issues. *American Economic Review* 79, 295–318.

Spiess, D.K. and Tkac, P. 1997. The Private Securities Litigation Reform Act of 1955: the stock market casts its vote. *Managerial and Decision Economics* 18, 545–61.

Subramanian, G. 2004. The disappearing Delaware effect. *Journal of Law, Economics, & Organization* 20, 32–59.

Stigler, G. 1964. Public regulation of the securities markets. *Journal of Business* 2, 117–42.

Thomas, R. and Cox, J. 2006. Does the plaintiff matter? An empirical analysis of lead plaintiffs in securities class actions. *Columbia Law Review* 100, 101–55.

Thompson, R. 1991. Piercing the corporate veil: an empirical study. *Cornell Law Review* 76, 1036–74.

Thompson, R. and Sale, H. 2003. Securities fraud as corporate governance: reflections upon federalism. *Vanderbilt Law Review* 56, 859–915.

Thompson, R. and Thomas, R. 2004. The public and private faces of derivative lawsuits. *Vanderbilt Law Review* 58, 1747–823.

Winter, R. 1977. State law, shareholder protection and the theory of the corporation. *Journal of Legal Studies* 6, 251–92.

Zhang, I. 2005. Economic consequences of the Sarbanes–Oxley Act of 2002. Related Publication 05–07. AEI–Brookings Joint Center for Regulatory Studies.

Zwiebel, J. 1995. Corporate conservatism, herd behavior and relative compensation. *Journal of Political Economy* 103, 1–25.

corporations

A *corporation* is an artificial person, with many of the legal rights of a biological one. This modern legal and economic usage arose in the 16th century from the term's now archaic meaning of 'a group acting as one body' – encompassing municipal governments, businesses and other groups of individuals united towards a common goal. In that century and the next, trade with the Orient and the New World promised immense returns, but only after vast capital outlays on fleets of ships, networks of forts and private armies to defend them. The first business corporations, such as the Dutch East Indies Company, the British East India Company and the Hudson's Bay Company, were formed to pool the savings of many individuals and permit ventures on a scale none could afford individually. Each owner of a *share* of the corporation was periodically entitled to a *dividend* – a pro rata division of the corporation's free cash flow.

Polling all a corporation's shareholders for each business decision was impractical in an age of sailing ships and horse-drawn carriages. Instead, the shareholders elected *boards of governors* (later *directors*) – reputable men trusted by the majority of shareholders to direct the corporation's affairs.

This did not prevent all dispute. The Dutch East Indies Company (*Vereenigde Oostindische Compagnie* in Dutch) was formed as a limited time venture. When that limit drew near, the board boldly announced that the corporation would persist indefinitely. The shareholders sued to force a liquidating dividend – and lost! Fortunately, they found they could sell their shares to other investors for the value of a liquidating dividend – or even more (Frentrop, 2002/3). Thus was born the first modern stock market, and the *alienability*, or unhindered sale, of shares became a defining characteristic of a corporation. Letting shareholders realize their investments by selling their shares, rather than liquidating the business, gave corporations a second defining characteristic: *indefinitely long lives*.

Boards occasionally betrayed their shareholders' trust and caused a corporation to contravene the law. Since individual shareholders were not consulted, holding them fully to account for the corporation's misdeeds seemed wrong. Since the corporation is a legal person, plaintiffs could sue it directly, and need not sue its shareholders personally. Thus, *limited liability* statutes came to shield individual shareholders from personal lawsuits for wrongs by corporations whose shares they own. Limited liability, a third defining characteristic of the modern business corporation, is an important innovation because it frees individuals to invest their savings in corporations run by strangers, undertaking risky ventures, or doing business in far off places. Vulnerability to personal lawsuits would otherwise make such investments seem indefensibly reckless to most savers.

Early corporations, like the Hudson's Bay Company, assigned one vote to each share in board elections. This essentially let the wealthiest shareholders appoint the board and, if they wished, run the corporation in their narrow interest rather than in the interests of all shareholders equally. For example, a large shareholder might force the corporation to do business with another corporation she controlled on disadvantageous terms. This sort of self-dealing, which Johnson et al. (2000) dub 'tunneling', remains a widespread corporate governance concern where firms typically have dominant shareholders. Or a dominant shareholder might simply relish the perks, power and prestige of running the corporation, and refuse to make way for more qualified managers – a corporate governance problem called 'entrenchment' (Morck, Shleifer and Vishny, 1988). Entrenchment and tunnelling provide controlling shareholders with *private benefits of control* – returns not shared with small shareholders (Dyck and Zingales, 2004). Distorted corporate governance associated with private benefits of control remains a first-order governance concern wherever corporations typically have a controlling shareholder. According to La Porta, Lopez-de-Silanes and Shleifer (1999), this includes the large corporate sectors of virtually all countries except Germany, Japan, the United Kingdom and the United

States. Small and middle-sized corporations everywhere tend to have controlling shareholders.

In the 19th century, *democratic* corporate governance became associated with *one vote per shareholder*, rather than one vote per share (Dunlavy, 2004). Echoes of this remain in the *voting caps* of modern Canadian and European corporations, which limit any single shareholder's voting power regardless of shares owned. However, large shareholders in many countries later turned deviations from one vote per share to their advantage by granting themselves special classes of common stock with many votes per share. In most countries, such *dual class shares* now virtually always magnify, rather than limit, the voting power of large shareholders, and so amplify, rather than dampen, problems associated with private benefits of control (Nenova, 2003).

In the United States and the United Kingdom, however, one vote per share is the norm in shareholder meetings. Disclosure rules, regulatory oversight, officer and director liability, and other restraints on private benefits of control also seem more effective in America and Britain than elsewhere in curtailing private benefits of control (LaPorta, Lopez-de-Silanes and Shleifer, 1999; Dyck and Zingales, 2004). This makes being a large shareholder less attractive, especially if holding a diversified portfolio of small stakes in many firms reduces risk (Burkart, Panunzi and Shleifer, 2003). Unsurprisingly, most large American and British corporations now lack controlling shareholders (LaPorta, Lopez-de-Silanes and Shleifer, 1999). They are run by professional managers who often own few shares (Morck, Shleifer and Vishny, 1988).

A small shareholder who monitored and controlled these corporate top managers would bear all the investigative, legal and administrative costs involved, but the benefits of better governance would be spread across all shareholders. The cost therefore typically exceeds the benefit for any small shareholder acting alone (Grossman and Hart, 1988). The consequent general lack of monitoring and control in corporations with no large shareholder gives rise to *other people's money* corporate governance problems. Adam Smith (1776) famously explains that, since corporate managers who own few or no shares are more

> the managers of other people's money than of their own, it cannot well be expected that they should watch over it with the same anxious vigilance with which partners in a private copartnery frequently watch over their own. Like the stewards of a rich man, they... consider attention to small matters as not for their master's honour and very easily give themselves a dispensation from having it.

Unmonitored professional managers can thus enjoy the perks and privileges of running large corporations without any real concern for the returns they generate. Berle and Means (1932) argue that this sort of governance problem occurs in many large American corporations.

But in other countries, other people's money governance problems probably also afflict many corporations that, on first inspection, seem to have a controlling shareholder. This is because large corporations in most countries are not freestanding entities, but belong to *corporate groups* (LaPorta, Lopez-de-Silanes and Shleifer, 1999). These are typically pyramidal structures, in which an apex shareholder, usually an extremely wealthy family, controls one or more listed corporations, which each control more listed corporations, which each control yet more listed corporations, *ad valorem et infinitum*. A family that controls 51 per cent of a listed corporation that controls 51 per cent of another that controls 51 per cent of yet another and so on actually owns only 0.51^n of the corporation n tiers down the in pyramid, with the remainder of each corporation financed by public or minority shareholders. Pyramids with a dozen or more layers are not uncommon, rendering the controlling shareholder's actual ownership of corporations at the pyramid's base negligible. Pyramidal business groups thus permit controlling shareholders to extract private benefits of control from corporate empires financed largely from other people's money (Morck, Stangeland and Yeung, 2000; Bebchuk, Kraakman and Triantis, 2000). Pyramids were common in the United States until the 1930s (Berle and Means, 1932; Bonbright and Means, 1932), but were eliminated by various New Deal initiatives, including the double and multiple taxation of inter-corporate dividends (Morck, 2005). British pyramids apparently withered under sustained attacks from institutional investors (Franks, Mayer and Rossi, 2005). However, the relevant unit of economic analysis for many purposes elsewhere in the world should often be the *business group*, not the corporation.

Jensen and Meckling (1976) show that *agency costs*, the present value of the costs of expected future governance shortfalls of any sort, are born by the corporation's initial shareholders. A corporation's founders receive less per share when they first sell shares to outside investors if worse corporate governance problems seem likely.

This gives rise to a *time inconsistency* problem in securities and corporations law. Investors and entrepreneurs selling shares to the public benefit from credible guarantees of good governance because these limit agency costs and so raise share prices. But top corporate decision makers in firms that have already issued shares, who foresee issuing no more, wish to maximize their utility (Baumol, 1959; 1962; Williamson, 1964) and understandably value the freedom to spend public shareholders' money as they like and to capture such private benefits of control as they can. Actual public policy probably reflects these groups' relative political lobbying power, which can change over time (Morck, Stangeland and Yeung, 2000; Morck, Wolfenzon and Yeung, 2005).

The normative view that a corporation *should* be run to maximize shareholder value derives from economists' assumption that firms maximize profits. In neoclassical economic theory, a firm that maximizes the present value of all its expected future economic profits necessarily maximizes the market value of its shares. This follows from modelling the corporation as a *nexus* of contracts, with the shareholders the *residual claimants* to the firm's cash flows (Fama and Jensen, 1983a; 1983b). Neoclassical theory further allows that profit maximization (value maximization in a multi-period setting) accords with economic efficiency under certain idealized conditions; see, for example, Varian (1992) and Malliaris and Brock (1983).

This normative view conflicts with the actual legal duties of corporate officers, directors and controlling shareholders in many countries. For example, many northern European countries and some US states impose a duty to balance shareholders' interests with those of *stakeholders*, especially employees. This is formalized in the German legal principle of *Mitbestimmung* (co-determination), which requires members of the *Aufsichtsrat* (supervisory board) of a large corporation to balance the interests of shareholders, employees and the state (Fohlin, 2005). Common law legal systems assign officers and directors a duty to act *for the corporation*. In Britain and the United States, this is often interpreted as a duty to act for the corporation's owners, its shareholders. A duty to maximize share value seems implicit (Jensen and Meckling, 1976; Black and Coffee, 1997). However, the Canadian Supreme Court holds in *Peoples v. Wise* that the duty of the officers and directors of a corporation is not to shareholders, nor to any other stakeholders, but to the corporation per se. The social welfare implications of assigning different legal duties to corporate top decision makers are incompletely understood. Giving labour a voice in corporate decision making seems to impede risk taking and hamper growth (Faleye, Mehrotra and Morck, 2006). Moreover, regardless of their assigned objective, if those entrusted to govern great corporations occasionally put their own interests ahead of their legal duties, agency costs must arise in some form.

The view that a corporation's top managers ought to maximize shareholder value also collides with evidence that stock prices are sometimes set by investors with incomplete information (Myers and Majluf, 1984) or behavioural biases (Shleifer, 2000). Coase (1937) argues that firms come about to alleviate information asymmetries and other market imperfections, collectively denoted *transactions costs*, and that the boundaries of the firm correspond to an efficient solution to these problems. Alchian and Demsetz (1972) argue that the critical market imperfections arise from people working in teams. Williamson (1975) argues that interdependent assets are more generally important. Jensen (2004) calls for more research on normative theories about the boundaries of the corporation and the objective function of its top decision makers if stock prices are set by *noise traders*, that is, investors with behavioural biases. One

approach holds that corporations actually exist primarily to lock the economy's capital into productive uses by isolating capital allocation decisions from maniac or panicked investors (Stout, 2004). This view long dominated discussions of corporate management in Japan (for example, Aoki and Dore, 1994) but appears to give rise to its own set of inefficiencies (see, for example, Morck and Nakamura, 1999).

RANDALL MORCK

See also **corporate governance; firm, theory of the.**

Bibliography

Aoki, M. and Dore, R.P. 1994. *The Japanese Firm: The Sources of Competitive Strength*. New York: Oxford University Press.

Alchian, A. and Demsetz, H. 1972. Production, information costs and economic organization. *American Economic Review* 62, 777–95.

Baumol, W. 1959. *Business Behavior, Value and Growth*. New York: Macmillan.

Baumol, W. 1962. On the theory of expansion of the firm. *American Economic Review* 52, 1078–87.

Bebchuk, L., Kraakman, R. and Triantis, G. 2000. Stock pyramids, cross ownership and dual class equity: the mechanisms and agency costs of separating control from cash flow rights. In *Concentrated Corporate Ownership*, ed. R. Morck. Chicago: University of Chicago Press.

Berle, A. and Means, G. 1932. *The Modern Corporation and Private Property*. New York: Macmillan.

Black, B.S. and Coffee, J.C., Jr. 1997. Hail Britannia? Institutional investor behavior under limited regulation. *Michigan Law Review* 92, 1997–2087.

Bonbright, J. and Means, G. 1932. *The Holding Company – Its Public Significance and Its Regulation*. New York: McGraw-Hill.

Burkart, M., Panunzi, F. and Shleifer, A. 2003. Family firms. *Journal of Finance* 58, 2173–207.

Coase, R. 1937. The nature of the firm. *Economica* 4, 386–405.

Dunlavy, C. 2004. The unnatural origins of one vote per share – a chapter in the history of corporate governance. Working paper, Department of History, University of Wisconsin–Madison.

Dyck, A. and Zingales, L. 2004. Private benefits of control: an international comparison. *Journal of Finance* 59, 537–601.

Faleye, O., Mehrotra, V. and Morck, R. 2006. When labor has a voice in corporate governance. *Journal of Financial and Quantitative Analysis* 41, 489–510.

Fama, E. and Jensen, M. 1983a. Agency problems and residual claims. *Journal of Law and Economics* 26, 327–49.

Fama, E. and Jensen, M. 1983b. Separation of ownership and control. *Journal of Law and Economics* 26, 301–25.

Fohlin, C. 2005. The history of corporate ownership and control in Germany. In *The History of Corporate*

Governance around the World: Family Business Groups to Professional Managers, ed. R. Morck. Chicago: University of Chicago Press.

Franks, J., Mayer, C. and Rossi, S. 2005. Spending less time with the family: the decline of family ownership in the UK. In *The History of Corporate Governance around the World: Family Business Groups to Professional Managers*, ed. R. Morck. Chicago: University of Chicago Press.

Frentrop, P. 2002/3. *A History of Corporate Governance*. Amsterdam: Deminor Press.

Grossman, S. and Hart, O. 1988. One share one vote and the market for corporate control. *Journal of Financial Economics* 20, 175–202.

Jensen, M. 2004. Agency costs of overvalued equity. Harvard NOM Research Paper No. 04-26, Harvard Business School.

Jensen, M. and Meckling, W. 1976. Theory of the firm: managerial behavior, agency costs and ownership structure. *Journal of Financial Economics* 3, 305–60.

Johnson, S., La Porta, R., Lopez-de-Silanes, F. and Shleifer, A. 2000. Tunneling. *American Economic Review* 90, 22–7.

La Porta, R., Lopez-de-Silanes, F. and Shleifer, A. 1999. Corporate ownership around the world. *Journal of Finance* 54, 471–517.

Malliaris, A.G. and Brock, W. 1983. *Stochastic Methods in Economics and Finance*. Amsterdam: North-Holland.

Morck, R. 2005. How to eliminate pyramidal business groups: the double-taxation of intercorporate dividends and other incisive uses of tax policy. *Tax Policy and the Economy* 19, 135–79.

Morck, R. and Nakamura, M. 1999. Banks and corporate control in Japan. *Journal of Finance* 54, 319–40.

Morck, R., Shleifer, A. and Vishny, R. 1988. Management ownership and market valuation: an empirical analysis. *Journal of Financial Economics* 20, 293–315.

Morck, R., Stangeland, D.A. and Yeung, B. 2000. Inherited wealth, corporate control, and economic growth: the Canadian disease. In *Concentrated Corporate Ownership*, ed. R. Morck. Chicago: University of Chicago Press.

Morck, R., Wolfenzon, D. and Yeung, B. 2005. Corporate governance, economic entrenchment, and growth. *Journal of Economic Literature* 43, 655–720.

Myers, S. and Majluf, N. 1984. Corporate financing and investment decisions when firms have information that investors do not have. *Journal of Financial Economics* 13, 187–222.

Nenova, Tatiana. 2003. The value of corporate voting rights and control: a cross-country analysis. *Journal of Financial Economics* 68, 325–51.

Shleifer, A. 2000. *Inefficient Markets: An Introduction to Behavioral Finance*. Oxford: Oxford University Press.

Smith, Adam. 1776. *An Inquiry into the Nature and Causes of the Wealth of Nations*. London: Ward, Lock and Tyler.

Stout, Lynn. 2004. On the nature of corporations. Law & Economics Research Paper No. 04-13, School of Law, UCLA.

Varian, H. 1992. *Microeconomic Analysis*. 3rd edn. New York: W. W. Norton.

Williamson, O. 1964. *The Economics of Discretionary Behavior: Managerial Objectives in a Theory of the Firm*. Englewood Cliffs, NJ: Prentice Hall.

Williamson, O. 1975. *Markets and Hierarchies: Analysis and Antitrust Implications*. New York: Free Press.

correspondence principle

The correspondence principle is the relation, which exists in certain economic models, between comparative statics of equilibria and the properties of out-of-equilibrium dynamics.

The correspondence principle (CP) implies that one obtains unambiguous comparative statics by selecting equilibria with desirable dynamic properties. Generally, the CP determines comparative statics in models with a one-dimensional endogenous variable, and in monotone multidimensional models. It does not determine comparative statics in general multidimensional models, such as Walrasian general equilibrium models with more than two goods.

One-dimensional models

The CP holds quite generally in one-dimensional models. Consider, for example, a two-good economy with excess-demand function for good 1 given by z_1, shown in Figure 1. We fix the price of good 2; by Walras's Law the equilibrium prices are the zeroes of z_1: there are three equilibria, p_1^1, p_1^2 and p_1^3.

Now consider the comparative-statics exercise of shifting excess demand up to $\hat{z}_1$. What is the effect on equilibrium price? Locally, the price increases if the equilibrium is p_1^1 or p_1^3, but it decreases if it is p_1^2. The different comparative statics at p_1^1 and p_1^2 corresponds exactly to the different behavior of tâtonnement dynamics after a small perturbation: p_1^1 is stable while p_1^2 is unstable.

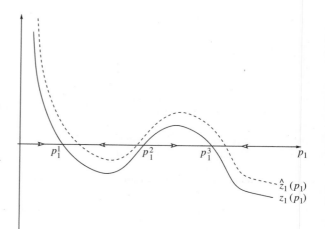

Figure 1   Two-good economy.

The difference between comparative statics at p_1^1 and at p_1^2 is easy to explain. The comparative statics at p_1^1 says: slightly larger prices than p_1^1 are reached by increasing excess demand, and smaller prices are reached by decreasing excess demand. Since excess demand is zero at p_1^1, there must be *positive* excess demand at slightly larger prices and *negative* excess demand at slightly smaller prices. Hence, tâtonnement dynamics, which respond to the sign of excess demand, converges to p_1^1 after a small perturbation from p_1^1. On the other hand, at p_1^2 larger prices result from a decrease in excess demand; hence excess demand is positive at larger prices. Similarly, excess demand is negative at smaller prices. As a result, tâtonnement dynamics will not approach p_1^2 after a small perturbation from p_1^2.

If the economy is subject to sporadic shocks, one should not observe p_1^2, the unstable equilibrium. Hence, as a consequence of the correspondence between comparative statics and dynamics, one should expect an increase in excess demand to produce an increase in equilibrium price.

I shall give a general statement of the correspondence principle for the one-dimensional case. Consider a model where the endogenous variable takes values in $[0,1]$ and equilibria are determined as the fixed points of $f(\cdot, t) : [0, 1] \to [0, 1]$; $t \in T \subseteq \mathbb{R}$ is an exogenous parameter. Assume that T is convex and that f is C^1.

A *selection of equilibria* is a function $e : T \to [0, 1]$ such that $e(t) = f(e(t), t)$ for all $t \in T$. Say that a fixed point $x \in [0, 1]$ is *stable* if there is a neighbourhood V of x such that any sequence x_n satisfying $x_0 \in V$ and $x_{n+1} = f(x_n)$ for $n \geq 1$, converges to x. Say that $x \in [0, 1]$ is *unstable* if, for any neighbourhood V of x, there is a neighbourhood W of x such that all sequences defined as above eventually lie in the complement of W.

Proposition 1 *Let f be monotone increasing in t. If e is a continuous selection of equilibria that is strictly decreasing over some interval $[\underline{t}, \bar{t}]$, then for all $t \in (\underline{t}, \bar{t})$, $e(t)$ is unstable.*

Multidimensional models

The one-dimensional CP is a relation between the sign of the comparative-statics change in prices, and the sign of excess demand for smaller and larger prices. When more than one price is determined, this relation does not need to exist. Still, the CP holds for monotone models – models where the different dimensions of the endogenous variables are in some sense complements. Monotone economic models stem mainly from game theoretic models with strategic complementarities.

I proceed to give a statement of the CP. Consider a model where the endogenous variable takes values in a compact rectangle $X \subseteq \mathbb{R}^n$, and equilibria are determined as the fixed points of $f(\cdot, t) : X \to X$; $t \in T \subseteq \mathbb{R}$ is a parameter and T is convex.

Proposition 2 *Let f be monotone increasing in (x, t) and let e be a continuous selection of equilibria.*

- *If e is strictly decreasing over $[\underline{t}, \bar{t}] \subseteq T$, then for all $t \in (\underline{t}, \bar{t})$, $e(t)$ is unstable.*
- *If e is strictly increasing over $[\underline{t}, \bar{t}]$, then for all $t \in (\underline{t}, \bar{t})$, if $e(t)$ is locally isolated, it is stable.*

Literature

The CP was formulated by Paul Samuelson (1941; 1942; 1947), who also coined the term (though Hicks, 1939, stated the CP informally). Samuelson formulated the one-dimensional CP. The version in Proposition 1 is taken from Echenique (2000). Basset, Maybee and Quirk (1968) study the scope of the CP. Arrow and Hahn (1971) present a critical discussion of the CP, and, because it fails in economies with more than two goods, conclude that 'very few useful propositions are derivable from this principle'. The monotone multidimensional CP is from Echenique (2002), who presents a general version of Proposition 2. Echenique (2004) presents a CP that does not rely on continuous selections of equilibria. The CP is also effective in dynamic optimization models (Brock, 1983; Burmeister and Long, 1977; Magill and Sheinkman, 1979) and in models of international trade (Bhagwati, Brecher and Hatta, 1987).

FEDERICO ECHENIQUE

See also **comparative statics.**

Bibliography

Arrow, K.J. and Hahn, F.H. 1971. *General Competitive Analysis.* San Francisco: Holden-Day.

Bassett, L., Maybee, J. and Quirk, J. 1968. Qualitative economics and the scope of the correspondence principle. *Econometrica* 36, 544–63.

Bhagwati, J.N., Brecher, R.A. and Hatta, T. 1987. The global correspondence principle: a generalization. *American Economic Review* 77, 124–32.

Brock, W.A. 1983. A revised version of Samuelson's correspondence principle. In *Models of Economic Dynamics*, ed. H. Sonnenschein. New York: Springer-Verlag.

Burmeister, E. and Long, N.V. 1977. On some unresolved questions in capital theory: an application of Samuelson's correspondence principle. *Quarterly Journal of Economics* 91, 289–314.

Echenique, F. 2000. Comparative statics by adaptive dynamics and the correspondence principle. Working Paper No. E00-273, Department of Economics, University of California, Berkeley.

Echenique, F. 2002. Comparative statics by adaptive dynamics and the correspondence principle. *Econometrica* 70, 833–44.

Echenique, F. 2004. A weak correspondence principle for models with complementarities. *Journal of Mathematical Economics* 40, 145–52.

Hicks, J.R. 1939. *Value and Capital.* Oxford: Clarendon Press.

Magill, M.J.P. and Sheinkman, J.A. 1979. Stability of regular equilibria and the correspondence principle for symmetric variational problems. *International Economic Review* 20, 297–315.

Samuelson, P.A. 1941. The stability of equilibrium: comparative statics and dynamics. *Econometrica* 9, 97–120.

Samuelson, P.A. 1942. The stability of equilibrium: linear and nonlinear systems. *Econometrica* 10, 1–25.

Samuelson, P.A. 1947. *Foundations of Economic Analysis*. Cambridge, MA: Harvard University Press.

correspondences

A *correspondence* Q from a domain set X to a range set Y associates with each element x in X, a non-empty subset of Y, $Q(x)$. A *function* is a correspondence such that $Q(x)$ is a singleton for each x in X. It is for this reason that a correspondence is also termed a *multi-valued function* or, more simply, a *multi-function*. Another name for a correspondence is a *set-valued mapping*.

Correspondences arise naturally in economic theory. One may think of an individual consumer's demand correspondence, which associates with each price system the set of utility maximizing consumption plans; see, for example, Hildenbrand (1974, p. 92). An equally pervasive example is an individual producer's supply correspondence which associates with each price system the set of profit-maximizing production plans (see, for example, Arrow and Hahn, 1971, pp. 54–5). The fact that these individual responses are correspondences rather than functions is simply a consequence of 'flats' in the underlying indifference surfaces and isoquants or, more precisely, of the constancy of marginal rates of substitution in consumption and in production over a range of commodity bundles. Indeed, the association of these marginal rates with the point at which they are evaluated is another example of a correspondence that arises naturally in economic theory, particularly in the study of marginal cost pricing equilibria in economies with increasing returns to scale (for example, Brown et al., 1986). The fact that there is no unique rate of substitution is simply a consequence of 'kinks' in the underlying function. In the case of a convex function, such a correspondence is termed the *subdifferential* correspondence, and, for more general functions, it is *Clarke's generalized derivative*.

If the domain and range of a correspondence are *topological spaces*, one can formulate various notions of continuity of a correspondence. Recall that (X, τ_X) is a topological space if X is a set and τ_X is a collection of subsets of X that contains X and the empty set ø and is closed under finite intersection and arbitrary union. We can now present one formalization of the intuitive idea of continuity of a correspondence. A correspondence $Q : X \to Y, X, Y$, both topological spaces, is said to be upper *semicontinuous* (u.s.c.) if for any V in τ_Y, the set $\{x \in X : Q(x) \subset V\}$ is in τ_X. Q is said to be lower *semicontinuous* (l.s.c.) if for any V in τ_Y, the set $\{x \in X : Q(x) \cap V \neq ø\}$ is in τ_X. It is easy to convince oneself that

a correspondence may be u.s.c. without being l.s.c. and vice versa. It is also easy to show that, if Y is a compact space, a correspondence Q is u.s.c. if and only if its graph, $\mathrm{Gr}Q$, $\mathrm{Gr}Q = \{(x, y) \in X \times Y : y \in Q(x)\}$, is such that its complement belongs to $\tau_X \times \tau_Y$. A correspondence is said to be *continuous* if it is both u.s.c. and l.s.c.

A very useful result for establishing u.s.c. of correspondences arising from maximization is Berge's *maximum theorem*. This states, in particular, that for any continuous correspondence Q from a topological space X to a topological space Y and any continuous function f from $X \times Y$ into the reals, the associated correspondence $\mu : X \to Y$ given by $\mu(x) = \{y \in Q(x) : f(y, x) \geq f(y', x) \text{ for all } y' \in Q(x)\}$ is u.s.c. This theorem is used to show u.s.c. of the demand and supply correspondences in the theory of the consumer and of the producer.

A result which plays a significant role in the proof of the existence of a competitive equilibrium is Kakutani's *fixed point theorem* for convex valued, u.s.c. correspondences which take a non-empty convex compact subset of an Euclidean space to itself. The theorem states that such correspondences Q have a fixed point, that is, an element x such that $x \in Q(x)$. Kakutani's theorem yields as an immediate corollary Brouwer's fixed point theorem and generalizes, word for word, to locally convex spaces as has been shown by Glicksberg and Ky Fan (see, for example, Berge, 1963, p. 251).

It is of interest to know of conditions under which a correspondence $Q : X \to Y$ yields a *continuous selection*, that is, a continuous function $f : X \to Y$ such that $f(x) \in Q(x)$ for all x in X. The celebrated selection theorems of Michael (see, for example, Bessaga and Pelczynski, 1975, ch. II.7) give a variety of sufficient conditions for this. One of these requires X to be a paracompact topological space, Y to be a separable Banach space and Q to be convex valued and l.s.c. This theorem has been used by Gale and Mas-Colell (1974) to show the existence of competitive equilibrium for economies in which consumer preferences need neither be complete nor transitive. If Q is u.s.c. rather than l.s.c., recent work of Cellina gives sufficient conditions under which one may obtain an *approximate continuous selection*.

So far in this exposition we have been considering results on correspondences whose domain and range are both topological spaces. An alternative setting is one where the range is a topological space but the domain is a *measurable space*. (T, Σ) is a measurable space if T is a set and Σ is a family of subsets that includes T and is closed under complementation and countable unions, that is, Σ is a *σ-algebra*. Such correspondences arise naturally in the study of economies in which the set of agents is modelled as a measurable space. An obvious example of such a correspondence is one which associates with every agent his/her set of utility maximizing consumption plans under a given price system.

One can develop concepts analogous to continuity for correspondences from a measurable space to a topological space. A correspondence $Q : T \to Y$ is said to be *measurable* if, for any set V in τ_Y, the set $\{t \in T : Q(t) \cap V \neq \varnothing\}$ is an element of Σ. Variants of this definition have been presented in the literature along with conditions under which these variants are all equivalent. One particularly fruitful variant requires the measure space to be *complete* and the correspondence to have a *measurable graph*, that is, GrQ is a subset of $\Sigma \otimes \mathcal{B}(Y)$, the smallest σ-algebra generated by the sets in $\Sigma \times \mathcal{B}(Y)$ and where $\mathcal{B}(Y)$ is the smallest σ-algebra generated by sets in τ_Y.

We can now state a measure-theoretic analogue of Berge's theorem. Let Q be a correspondence with a measurable graph and f a $\Sigma \otimes \mathcal{B}(Y)$ measurable function from $T \times Y$ into the reals. Then a result due to the collective efforts of Debreu and Castaing–Valadier states that under a mild restriction on Y, namely Souslin, the correspondence $\mu : T \to X, \mu(t) = \{x \in Q(t) : f(t, x) \geq f(t, x')$ for all x' in $Q(t)\}$, has a measurable graph.

We have developed enough terminology to state a fundamental theorem due to the collective efforts of von Neumann, Aumann and St. Beuve. This states that under a restriction on the range space Y, namely, Souslin, every correspondence Q with a measurable graph yields a *measurable selection*, that is, a measurable function $f : T \to Y$ such that $f(t) \in Q(t)$ for all t in T.

Once we have a measurable selection theorem, we are in a position to formulate a satisfactory notion of an integral of a correspondence, a notion which may also be seen as a formalization of a sum of an infinite number of sets. However, one preliminary notion that still needs to be stated is that of a *measure* μ on (T, Σ). A measure μ is a set-valued function from Σ into (say) Euclidean space R^n such that

$$\mu(A) \geq 0, \mu\left(\bigcup_{i=1}^{\infty} A_i\right) = \sum_{i=1}^{\infty} \mu(A_i)$$

for all A, A_i in Σ and such that A_i are mutually disjoint. Now let us assume we know how to integrate a function with respect to μ and can therefore specify a function $f : T \to R^n$ to be an *integrable function* if its integral (Lebesgue integral) is finite. Following Aumann, we can define the integral of a correspondence Q, $\int_T Q(t) \mathrm{d}\mu$ to be the set $\{\int_T f(t) \mathrm{d}t : f$, an integrable function which is a measurable selection from $Q\}$. It is now clear that $\int_T Q(t) \mathrm{d}\mu$ is non-empty if Q has a measurable graph and if there exists an integrable function g with non-negative values and such that $|x| \leq g(t)$ for all $x \in Q(t)$ and for all $t \in T$.

Finally, we can state a consequence of Lyapunov's theorem on the range of an *atomless measure* that has played a fundamental role in the development of the theory of economies with a continuum of agents. A measure μ on a measurable space (T, Σ) is atomless if (T, Σ, μ) has no *atoms*, that is $A \in \Sigma$ such that $\mu(A) > 0$ and $B \in \Sigma, B \subset A$

implies $\mu(B) = \mu(A)$ or $\mu(B) = 0$. The Lyapunov–Richter theorem states that the integral of a correspondence $Q : T \to R^n$ is convex if μ is an atomless measure on (T, Σ).

In summary, a correspondence is a versatile mathematical object for which a deep and rich theory can be developed and which arises naturally in many diverse areas of applied mathematics, including economic theory. For an introduction to this theory and to its applications, the reader is referred to the following references which also contain all the concepts and results not referenced in this entry.

M. ALI KHAN

See also **fixed point theorems; Lyapunov functions.**

Bibliography

Arrow, K.J. and Hahn, F.H. 1971. *General Competitive Analysis*. San Francisco: Holden-Day.

Aubin, J.P. and Cellina, A. 1984. *Differential Inclusions*. New York: Springer-Verlag.

Berge, C. 1963. *Topological Spaces*. New York: Macmillan.

Bessaga, C. and Pelczynski, A. 1975. *Selected Topics in Infinite-Dimensional Topology*. Warsaw: Polish Scientific Publishers.

Brown, D., Heal, G., Ali Khan, M. and Vohra, R. 1986. On a general existence theorem for marginal cost pricing equilibria. *Journal of Economic Theory* 38, 111–19.

Castaing, C. and Valadier, M. 1977. *Convex Analysis and Measurable Multifunctions*. Lecture Notes in Mathematics No. 580. New York: Springer-Verlag.

Clarke, F.H. 1983. *Optimization and Nonsmooth Analysis*. New York: John Wiley.

Gale, D. and Mas-Colell, A. 1974. An equilibrium existence theorem for a general model without ordered preferences. *Journal of Mathematical Economics* 2, 9–15. Erratum in *Journal of Mathematical Economics* 6, 297–8.

Hildenbrand, W. 1974. *Core and Equilibria of a Large Economy*. Princeton, NJ: Princeton University Press.

Klein, E. and Thompson, E.A. 1985. *Introduction to the Theory of Correspondences*. New York: John Wiley.

Rockafellar, R.T. 1970. *Convex Analysis*. Princeton, NJ: Princeton University Press.

cost–benefit analysis

Public policies, such as infrastructure projects, social welfare programmes, tax laws and regulations, typically have diverse effects in the sense that people would be willing to pay something to obtain effects they view as desirable and would require compensation to accept voluntarily effects they view as undesirable. If, across all members of society, the total amount willing to be paid by those who enjoy desirable effects (benefits) exceeds the total amount needed to compensate those who suffer undesirable effects (costs), then adopting the policy would make it potentially possible to achieve a Pareto improvement on the status quo. If the benefits do not exceed the costs, then adopting the policy does not offer a potential Pareto improvement. How should such costs and benefits be determined? Cost–benefit analysis (CBA) is the collection of generally accepted methods and rules for assessing the social costs and benefits of alternative public policies.

The US Flood Control Act of 1936 appears to be the first call for CBA to be systematically used to inform public policy (Steiner, 1974); it became embedded within modern welfare economics with articles by John R. Hicks (1939) and Nicholas Kaldor (1940) that set out the efficiency rationale for requiring policies to have positive net benefits. Two forces have contributed to the increased use of CBA since the 1960s. First, budget pressures and the desire to avoid inefficient regulations have led many governments to promote, or even require, the subjection of certain types of policies to CBA. Its use in the United States, particularly in the area of economic regulation, has been mandated by a series of Executive Orders (Hahn and Sunstein, 2002). Her Majesty's Treasury in the United Kingdom publishes the *Green Book* to help public sector organizations apply CBA to ensure that 'public funds are spent on activities that provide the greatest benefits to society, and that they are spent in the most efficient way' (HM Treasury, 2002: v). Second, economists have shown ingenuity in finding ways to value goods not traded in efficient markets, thereby expanding the range of policies to which CBA can be reasonably applied. For example, the travel-cost method provides a way to value recreational facilities that charge an administratively determined entry fee (Clawson and Knetsch, 1966); hedonic pricing models facilitate valuation of spatially varying local public goods (Smith and Huang, 1995); and the development of the contingent valuation survey method, propelled by environmental damage assessment suits in US courts, permits the valuation of public goods, such as existence value, that lack readily observable behavioural traces needed for revealed preference estimation (David, 1963; Bateman and Willis, 2000).

CBA promotes efficiency by identifying the set of feasible projects that would yield the largest positive net benefits. Three conceptual criticisms can be made against this proposition. First, because those who suffer costs from a policy are almost never fully compensated, CBA in any particular application generally will not guarantee a Pareto improvement. The counter-argument is that, if CBA is consistently used to select policies offering the largest net benefits and there are no consistent losers, then it is likely that overall everyone will actually be made better off. Second, the CBA techniques for measuring net benefits cannot guarantee a coherent social ordering of policy alternatives. For example, it is possible to

identify situations in which moving from one policy to another offers positive net benefits as does moving back to the original policy (Scitovsky, 1941; Blackorby and Donaldson, 1990). As no fair social choice rule can guarantee a transitive social ordering (Arrow, 1963), this result is not surprising and is of minor consequence compared with the practical difficulties encountered in applying CBA. Third, and most important, only a few economists argue that public policies should be selected solely to promote the goal of efficiency. Other goals, such as equity and preservation of human dignity, are often legitimately viewed as relevant to policy choice, so that CBA is inappropriate as a decision rule. Nonetheless, as efficiency is almost always one of the relevant goals of public policy, CBA remains useful as a method for assessing efficiency in the context of a broader multi-goal analysis.

Social perspective

CBA assesses social costs and benefits, which distinguishes it from the self-regarding calculus of individual economic actors. The meaning of 'social' in this context is twofold. First, it involves the definition of the relevant society; that is, it requires a determination of whose costs and benefits have standing (Whittington and MacRae, 1986). Economists generally argue for national standing, recognizing that those in a particular country live under the same political contract, or constitution, and share a common economy with its own fiscal and monetary policy. In practice, however, sub-national governments often base their decisions only on their own costs and benefits and therefore demand CBA with standing restricted to those under their jurisdictions. Even when geographic standing is resolved, issues remain as to whether the costs and benefits of all residents – citizens, legal aliens, illegal aliens, those with legally proscribed preferences – should count (Zerbe, 1998).

Second, it requires comprehensive assessment of the valued effects of policies on those with standing. The effects are commonly divided into the categories of active and passive use. Policies affect active use by changing the observable quantities of goods consumed, such as day care or fishing. Passive use includes all those effects that cannot be readily identified with observable changes in behaviour: existence value, or the willingness to pay for some good, such as wilderness, that one never expects to consume actively (Krutilla, 1967); option value, or the willingness to pay for some good that one may wish to consume actively in the future (Weisbrod, 1964); donor value, or the willingness to pay for redistributions of goods to others (Hochman and Rogers, 1969). The absence of observable behaviour precludes valuation of passive use through the revealed preference methods most favoured by economists. Stated preference methods, such as contingent value surveys, are thus necessary for undertaking comprehensive assessments of policies with effects on passive use.

Social benefits: willingness to pay

A common metric for policy effects is required if these effects are to be aggregated across individuals within the relevant society. If more than one policy alternative is to be compared with the status quo, then this metric must have ordinal properties. Further, if it is to be compared with the resource costs of implementing the policy, then it must be measured in the monetary unit of the society. Equivalent variation (EV) satisfies these conditions (McKenzie, 1983). Consider the expenditure, or cost-utility, function C(U,P), where C is the amount of money needed to achieve utility U with price vector P. If U_1 is the person's utility under the price vector P_1 that would result from the policy change and P_0 is the price vector that would result under the status quo, then the equivalent variation of the policy change is given by

$$EV = C(U_1, P_0) - C(U_1, P_1)$$

the difference between the expenditure needed to achieve U_1 without the policy and with it. The EV is the amount of money that one would have to give to the person instead of implementing the policy so that the person is as well off as he or she would have been had the policy been implemented. A negative EV indicates that the person finds the net effects of the policy undesirable.

In its actual use, CBA almost always evaluates policy effects with willingness to pay, which differs conceptually from EV. Willingness to pay answers the question: how much money could be taken away from a person in conjunction with the policy so that he or she has the same utility with the policy as without it? Rather than corresponding to EV, which holds utility constant at a level with the policy, willingness to pay corresponds to compensating variation, which holds utility constant at the pre-policy level. Although compensating variation is more intuitively appealing, it does not provide a fully satisfactory money metric like EV.

The equivalent or compensating variation of a price change in a single market can be calculated as the change in social surplus as measured under the appropriate Hicksian, or utility-compensated, demand schedule. In practice, however, analysts typically work with econo-metrically derived demand curves that do not hold utility constant. Changes in consumer surplus measured with these Marshallian demand curves only approximate the compensating variation, with differences driven by income effects that can be large for either large income elasticities or large price changes. Some progress has been made to put bounds on the differences between the Marshallian and Hicksian measures (Willig, 1976; Seade, 1978), but these bounds are rarely applied in practice.

The interpretation of Marshallian consumer surplus as willingness to pay becomes even more complicated when policies have secondary effects in the markets of complements and substitutes of the goods primarily affected by policies. Although a general equilibrium model would be most appropriate for taking account of these secondary market effects, common practice is to approximate the combined effect of the primary and secondary markets by measuring surplus changes with the use of an estimated demand schedule for the primary market that does not hold the prices of substitutes and complements constant (Sugden and Williams, 1978; Gramlich, 1990; Boardman et al., 2006). In such cases, analysts need not account for price changes in undistorted secondary markets. Indeed, doing so would likely result in double counting of benefits.

Social costs: opportunity costs

Public policies generally require the use of real resources to produce their effects. The guiding principle for monetizing the forgone value of these resources is opportunity cost: what is the value of the resources in their next-best use? That is, what is the value forgone by using the resources for the project? When factor markets are undistorted and the additional demand created by the project does not increase price, the opportunity cost of the resource just equals its market value, which, if the resource is obtained by purchase, just equals the expenditure on the resource. When factor markets are undistorted but the additional demand induced by the policy drives up price, then the opportunity cost of the resource equals the sum of expenditure and the change in social surplus, the algebraic sum of the change in consumer surplus and the change in rents usually measured as change in producer surplus based on the short-run supply schedule (Mishan, 1968). For example, if supply and demand curves are linear, then the opportunity cost equals the average of the pre- and post-purchase prices of the resource times the quantity purchased.

If markets are distorted, then even if price does not change the opportunity cost does not necessarily equal the expenditure required to secure supply. For example, a common factor-market distortion is involuntary unemployment resulting from minimum wages imposed by either law or custom. The expenditures needed to hire workers from a market with involuntary unemployment for a project clearly overestimate the opportunity cost of this labour. Nonetheless, the opportunity cost is almost certainly not zero, as sometimes argued by policy advocates, because the time of the workers hired by the project has an opportunity cost in terms of forgone leisure and household production.

Accommodating uncertainty

CBA requires prediction of the effects of adopting a policy. Predictions are inherently uncertain. In addition to uncertainty about such parameters as price elasticities required for predictions of changes in social surplus, CBA often requires analysts to confront fundamental uncertainty about future states of the world. For example, preparing a vaccine to guard against a potential pandemic is costly but offers large benefits in the event that a pandemic actually materializes. CBA requires analysts to convert these uncertainties into risks by specifying representative states of the world and assigning probabilities to these states. Common practice is to model the policy choice as a decision analysis problem, or game against nature, and to choose the policy that maximizes the expected value of social surplus.

A more conceptually valid measure of the benefits of a project with certain costs in the face of risk about the future state of the world is option price (Graham, 1981). Option price answers that question: what is the maximum certain payment that an individual would be willing to make to obtain the project? The sum of these certain payments for all those with standing can then be compared with the certain cost of implementing the policy. In general, however, option price does not equal the expected value of an individual's surplus over the possible states of the world; it differs from expected surplus by the option value of the policy for the individual. Although contingent valuation surveys seek to elicit individuals' option prices directly, more commonly analysts estimate benefits as expected surpluses, and consider option value as an excluded value. Some progress has been made in signing option value (Larson and Flacco, 1992), but analysts rarely have enough information for confidently including it as a monetized correction to expected surplus.

Discounting for time
Policies typically have effects that extend far into the future. Infrastructure projects in particular are usually characterized by large initial investments followed by beneficial use over years or even scores of years. CBA requires that costs and benefits accruing in the future be converted into their present value equivalents. On the assumption that future costs and benefits are predicted in real dollars, then a dollar of cost or benefit occurring t periods beyond the present equals in present value terms

$$1/(1 + d)^t$$

where d is the real discount rate for the period length. In practice, discounting is usually done on an annual basis. As valid comparison of projects requires that they be assessed over the same time horizon, it is often necessary to convert present values to equivalent perpetual streams of constant values through the use of an annuity factor.

The appropriate value for the real discount rate remains controversial. One approach is to set the discount rate equal to the marginal rate of pure time preference, the rate at which consumers are indifferent between exchanging current for future consumption. Another approach is to set the discount rate equal to the opportunity cost of capital, the marginal rate of return

on private investment. In an ideal capital market these two rates would be equal. In the presence of transaction costs and taxes, however, these rates differ substantially. For example, an estimate of the marginal rate of pure time preference based on the after-tax real rate of return on US treasury bonds is 1.5 per cent, while an estimate of the opportunity cost of capital based on the expected real yield on AAA corporate bonds is 4.5 per cent (Moore et al., 2004).

If all costs and benefits correspond to changes in consumption, then the marginal rate of pure time preference is the appropriate discount rate. Instead, if all costs and benefits correspond to changes in private investment, then the marginal rate of return on private investment is the appropriate discount rate. However, most projects involve changes in both consumption and investment. The shadow price of capital approach involves expressing all costs and benefits in terms of consumption changes so that the marginal rate of pure time preference can be applied (Bradford, 1975). In application, this means applying a shadow price to changes in private investment so that they are converted to the present values of their associated streams of consumption changes.

Shadow prices
Much of the challenge of CBA lies in deriving appropriate shadow prices when policies have effects beyond those that can be taken into account as changes of prices or quantities in undistorted markets. In developing countries, for example, import and export controls and the presence of subsistence agriculture often distort virtually all prices, necessitating the determination of a complete set of shadow prices based on prices in international markets (Little and Mirrlees, 1974; Squire and van der Tak, 1975; Dinwiddy and Teal, 1996). Economic research provides a number of shadow price estimates that can be used in conducting CBA. Indeed, were these shadow prices not readily available, the plausible range of application of CBA would be much narrower.

One of the most commonly needed shadow prices is the value of a statistical life. That is, what is the willingness of a representative member of a population to pay for reductions in mortality risk? Economists have used a variety of methods to estimate the value of a statistical life, most commonly taking advantage of differences in risks and wages across occupations or the purchases of safety devices. The number of studies is sufficiently large that a number of meta-analyses have been conducted to develop estimates of the value of a statistical life for the United States in the range of roughly $4 million to $6 million in 2002 dollars (Miller, 2000; Viscusi and Aldy, 2003). Tied to any estimate of the value of a statistical life is the value of a life year. Health economists have developed a number of methods for estimating the quality of life in various health states, so that, in conjunction with the value of a life year, they can monetize a quality-adjusted life year (QALY) for use in CBAs of health care interventions (Dolan, 2000). Estimates of shadow prices for injuries, noise, recreational activities, air pollutants, commuting time, and the marginal excess burden of taxation (for application to changes in government revenue) are also readily available (Boardman et al., 2006).

DAVID L. WEIMER

See also **consumer surplus; contingent valuation; hedonic prices; Pareto principle and competing principles; rent; social discount rate; value of life; value of time.**

Bibliography

Arrow, K. 1963. *Social Choice and Individual Values*, 2nd edn. NewHaven: Yale University Press.

Bateman, I. and Willis, K., eds. 2000. *Valuing Environmental Preferences Theory and Practice of the Contingent Valuation Method in the US EC and Developing Countries*. Oxford: Oxford University Press.

Blackorby, C. and Donaldson, D. 1990. A review article: the case against the use of the sum of compensating variation in cost–benefit analysis. *Canadian Journal of Economics* 23, 471–94.

Boardman, A., Greenberg, D., Vining, A. and Weimer, D. 2006. *Cost–benefit Analysis: Concepts and Practice*. 3rd edn. Upper Saddle River, NJ: Prentice Hall.

Bradford, D. 1975. Constraints on government investment opportunities and the choice of discount rate. *American Economic Review* 65, 887–99.

Clawson, M. and Knetsch, J. 1966. *Economics of Outdoor Recreation*. Baltimore, MD: Johns Hopkins University Press.

David, R. 1963. Recreation planning as an economic problem. *Natural Resources Journal* 3, 239–49.

Dinwiddy, C. and Teal, F. 1996. *Principles of Cost-benefit Analysis for Developing Countries*. Cambridge: Cambridge University Press.

Dolan, P. 2000. The Measurement of health-related quality of life for use in resource allocation in health care. In *Handbook of Health Economics* 1B, ed. A. Culyer and J. Newhouse. Amsterdam: Elsevier.

Graham, D. 1981. Cost–benefit analysis under uncertainty. *American Economic Review* 71, 715–25.

Gramlich, E. 1990. *A Guide to Benefit-Cost Analysis*. 2nd edn. Englewood Cliffs, NJ: Prentice-Hall.

Hahn, R. and Sunstein, C. 2002. A new executive order for improving federal regulation? Deeper and wider cost-benefit analysis. *University of Pennsylvania Law Review* 150, 1389–552.

Hicks, J.R. 1939. The Valuation of Social Income. *Economica* 7, 105–24.

HM Treasury. 2002. *The Green Book: Appraisal and Evaluation in Central Government*. London: The Stationary Office.

Hochman, H. and Rogers, J. 1969. Pareto optimal redistribution. *American Economic Review* 59, 542–57.

Kaldor, N. 1940. Welfare propositionsof economics and interpersonal comparisons of utility. *Economic Journal* 49, 549–52.

Krutilla, J. 1967. Conservation reconsidered. *American Economic Review* 57, 777–86.

Larson, D. and Flacco P. 1992. Measuring option prices from market behavior. *Journal of Environmental Economics and Management* 22, 178–98.

Little, I. and Mirlees, J. 1974. *Project Appraisal and Planning for Developing Countries*. London: Heinemann Educational.

McKenzie, G. 1983. *Measuring Economic Welfare*. Cambridge: Cambridge University Press.

Miller, T. 2000. Variations between countries in the values of statistical life. *Transport Economics and Policy* 34, 169–88.

Mishan, E. 1968. What is producer's surplus? *American Economic Review* 58, 1269–82.

Moore, M., Boardman, A., Vining, A., Weimer, D. and Greenberg, D. 2004. 'Just give me a number!' Practical values for the social discount rate. *Journal of Policy Analysis and Management* 23, 789–812.

Scitovsky, T. 1941. A note onwelfare propositions in economics. *Review of Economic Studies* 41, 77–88.

Seade, J. 1978. Consumer's surplus and linearity of Engel curves. *Review of Economic Studies* 9, 77–88.

Smith, V. and Huang, J. 1995. Can markets value air quality? A meta analysis of hedonic property value models. *Journal of Political Economy* 103, 209–77.

Squire, L. and van der Tak, H. 1975. *Economic Analysis of Projects*. Baltimore MD: Johns Hopkins University Press.

Steiner, P. 1974. Public expenditure budgeting. In *The Economics of Public Finance*, ed. A. Blinder et al. Washington, DC: The Brookings Institution.

Sugden, R. and Williams, A. 1978. *Principles of Practical Cost–benefit Analysis*. Oxford: Oxford University Press.

Viscusi, W. and Aldy, J. 2003. The value of statistical life: a critical review of market estimates throughout the world. *Journal of Risk and Uncertainty* 27, 5–76.

Weisbrod, B. 1964. Collective consumption services of individual consumption goods. *Quarterly Journal of Economics* 78, 71–7.

Whittington, D. and MacRae, D. 1986. The issue of standing in cost–benefit analysis. *Journal of Policy Analysis and Management* 5, 665–82.

Willig, R. 1976. Consumer's surplus without apology. *American Economic Review* 66, 589–97.

Zerbe, R., Jr., 1998. Is cost-benefit analysis legal? Three rules. *Journal of Policy Analysis and Management* 17, 419–56.

cost functions

1 Introduction

Cost and expenditure functions are widely used in both theoretical and applied economics. Cost functions are often used in econometric studies which describe the technology of firms or industries while their consumer theory counterparts, expenditure functions, are frequently used to describe the preferences of consumers.

Cost and expenditure functions also play an important role in many theoretical investigations. This is due to the fact that a cost function embodies the consequences of cost minimizing behaviour on the part of a consumer or producer and so it is not necessary to spell out the details of the primal minimization problem that defined the cost function. This may seem like a very minor advantage, but when one is dealing with, say, the comparative statics of a general equilibrium problem, the use of cost functions leads to the analysis of a much smaller system of equations and hence the structure of the problem can be more easily understood.

Sections 2–5 below develop the theoretical properties of cost functions while Sections 6–8 are devoted to empirical applications of cost functions in the producer and consumer contexts.

2 Properties of cost functions

One of the fundamental paradigms in economics is the one which has a producer competitively minimizing costs subject to his technological constraints. Competitive means that the producer takes input prices as fixed during the given period of time irrespective of the producer's demand for those inputs.

We assume that only one output can be produced using N inputs and that the producer's technology can be summarized by a *production function* F: $y = F(x)$ where $y \geq 0$ is the maximal amount of output that can be produced during a period, given the non-negative vector of inputs $x \equiv (x_1, \ldots, x_N) \geq 0_N$. We further assume that the cost of purchasing one unit of input i is $p_i > 0, i = 1, \ldots, N$ and that the positive vector of input prices that the producer faces is $p \equiv (p_1, \ldots, p_N) \gg 0_N$.

For $y \geq 0, p \gg 0_N$, the producer's *cost function* C is defined as the solution to the following constrained minimization problem:

$$C(y,p) \equiv \min_x \{p \cdot x : F(x) \geq y, x \geq 0_N\}$$

$$(1)$$

where $p \cdot x \equiv \Sigma_{n=1}^N p_n x_n$. Thus $C(y, p)$ is the minimum input cost of producing at least the output level y, given that the producer faces the input price vector p.

The minimization problem (1) can also be given a consumer theory interpretation: let F be a consumer's preference or *utility function*, let y be a utility or welfare level, let x be a vector of commodity purchases (rentals in the case of consumer durables), and let p be a vector of commodity (rental) prices. In this case, the consumer attempts to minimize the cost of achieving at least the target welfare level indexed by y, and the solution to (1) defines the consumer's *expenditure function*.

Unfortunately, the minimum (1) may not exist in general. However, if we impose the following very weak regularity condition on F, it can be shown that C will be well defined as a minimum: *Assumption 1 on F: F is continuous from above.*

Assumption 1 means that for every y in the range of F, the *upper level set* $L(y) \equiv \{x : F(x) \geq y, x \geq 0_N\}$ is a closed set. The assumption is a technical one of minimal economic interest. It is also a very weak condition from an empirical point of view, since it cannot be contradicted by a finite set of data on the inputs and output of a producer.

If we assume that the production function F satisfies Assumption 1, it turns out that the cost function C has the following properties: *Property 1: C* is a non-negative function; that is, $C(y,p) \geq 0$; *Property 2: C* is *linearly homogeneous* in input prices p for each fixed output level y; that is, $C(y^1,p) \geq C(y^2,p)$ for $y^1 \geq y^2 \geq 0$ and $p \gg 0_N$; *Property 3: C* is *nondecreasing* in p for fixed y; that is, $C(y,p^1) \geq C(y,p^2)$ for $y \geq 0, p^1 \geq p^2 \gg 0_N$; *Property 4: C* is *concave* in p for fixed y; that is, $C(y, \lambda p^1 + (1-\lambda)p^2) \geq \lambda C(y,p^1) + (1-\lambda)C(y,p^2)$ for $y \geq 0, 0 \leq \lambda \leq 1, p^1 \gg 0_N$ and $p^2 \gg 0_N$; *Property 5:C* is *nondecreasing* in y for fixed p; that is, *Property 6: C* is *continuous from below* in y for fixed p; that is $\{y : C(y,p) \leq \alpha\}$ is a closed set for every α and $p \gg 0_N$.

Properties 1–4 for C were derived by Shephard (1953) under stronger regularity conditions on F and Properties 4, 5, and 6 were obtained by McKenzie (1957), Uzawa (1964) and Shephard (1970) respectively.

From the viewpoint of economies, all of the properties of C are intuitively obvious except Properties 4 and 6. Property 6 on C is the technical counterpart to Assumption 1 on F and is of minimal economic interest. However, Property 4 has some significant economic implications as we shall see in Section 5 below.

We can already draw some useful empirical implications from the fact that a cost function must satisfy Properties 1–6 above. For example, in industrial organization and applied econometrics, it is quite common to assume that the true functional form for a firm's or industry's cost function has the following functional form:

$$C(y,p) \equiv \alpha + \beta \cdot p + \gamma y \qquad (2)$$

where $\alpha, \beta \equiv (\beta_1, \ldots, \beta_N)$ and γ are unknown parameters. However, Property 2 implies that α and γ must be zero in order for the cost function to be linearly homogeneous in input prices. But then $C(y,p) = \beta \cdot p$ does not depend on the output level y, which is very implausible.

3 Duality between cost and production functions

It is easy to see that the family of upper level sets, $L(y) \equiv \{x : F(x) \geq y, x \geq 0_N\}$, completely determines the production function F. Furthermore, the cost function C may be defined in terms of the production function by (1) or equivalently, in terms of the family of upper level sets as follows:

$$C(y,p) = \min_x \{p \cdot x : x \text{ belongs to } L(y)\}. \quad (3)$$

Thus given the production function F or the family of level sets $L(y)$, the cost function C is determined.

We now ask the following question: given a cost function C which has Properties 1 to 6, can we use C to define the underlying production function F?

For a given output level y and input price vector $p \gg 0_N$, define the corresponding isocost plane by $\{x : p \cdot x = C(y,p)\}$. From the definitions of $C(y, p)$ and $L(y)$, it is obvious that the set $L(y)$ must lie above this isocost plane and be tangent to it; that is, $L(y)$ must be a subset of the set $\{x : p \cdot x \geq C(y,p)\}$ and this conclusion must be true for every positive input price vector p. Thus $L(y)$ must be a subset of the intersection of all these sets which we denote by $M(y)$:

$$M(y) \equiv \bigcap_{p \gg 0}^{N} \{x : p \cdot x \geq C(y,p)\}. \quad (4)$$

The set $M(y)$ is called the *disposal, convex hull of $L(y)$*; see McFadden (1966).

Each set $\{x : p \cdot x \geq C(y,p)\}$ is a halfspace and is a convex set. A set S is *convex* if and only if x^1 and x^2 belong to S and $0 \leq \lambda \leq 1$ implies $\lambda x^1 + (1 - \lambda)x^2$ also belongs to S. Since $M(y)$ is the intersection of a family of convex sets, $M(y)$ is also a convex set. $M(y)$ also has the following *free disposal* property:

$$\begin{aligned} &x^1 \text{ belongs to } M(y), x^1 \leq x^2, \\ &\text{then } x^2 \text{ belongs to } M(y). \end{aligned} \quad (5)$$

We know $L(y)$ must be a subset of $M(y)$. If we want $L(y)$ to coincide with $M(y)$, then $L(y)$ must also be a convex set with the free disposal property. It can be shown that $L(y)$ will have these last two properties for every output level y if and only if the production function F has the following two properties: *Assumption 2 on F: F is quasiconcave* function; that is, for every y belonging to the range of F, $L(y) \equiv \{x : F(x) \geq y\}$ is a convex set. *Assumption 3 on F: F is nondecreasing*; that is, if $x^2 \geq x^1 \geq 0_N$, then $F(x^2) \geq F(x^1)$.

We may now answer our earlier question about whether a cost function C can completely determine the production function F: the answer is yes if the production satisfies Assumptions 1–3.

More precisely, we have the following result: given a cost function C which satisfies Properties 1–6, then the production function F defined by

$$\begin{aligned} F(x) \equiv \max_y \{y : C(y,p) \leq p \cdot x \\ \text{for every } p \gg 0_N\}, \quad x \geq 0_N \end{aligned} \quad (6)$$

satisfies Assumptions 1–3. Moreover, if we define the cost function C^* which corresponds to the F defined by (6) in the usual way [recall (1)], then $C^* = C$; that is, this derived cost function C^* coincides with the original cost function C. Thus there is a *duality between production functions F satisfying Assumptions 1–3 and cost functions C having Properties 1–6: each function completely determines the other under these regularity conditions.*

Duality theorems similar to the above results have been established under various regularity conditions by Shephard (1953; 1970), Uzawa (1964), McFadden (1966; 1978a) and Diewert (1971; 1982).

4 The derivative property of the cost function

The following result is the basis for most of the theoretical and empirical applications of cost functions.

Suppose the cost function C satisfies Properties 1–6 listed in Section 2 and in addition, C is differentiable with respect to the components of p at the point (y^*, p^*). Then the solution $x^* \equiv (x_1^*, \ldots, x_N^*)$ to the cost minimization problem $\min_y \{p^* \cdot x : F(x) \geq y^*, x \geq 0_N\}$ is unique and

$$x_i^* = \partial C(y^*, p^*)/\partial p_i, \quad i = 1, \ldots, N; \quad (7)$$

that is, the cost minimizing demand for the ith input is equal to the partial derivative of the cost function with respect to the ith input price.

The result (7) is known as the derivative property of the cost function (see McFadden, 1978a) or Shephard's Lemma, since Shephard (1953) was the first to obtain the result. It should be noted that Hicks (1946) and Samuelson (1947) obtained the result (7) earlier, but under different hypotheses: they assumed the existence of a utility or production function F and deduced (7) by analysing the comparative statics properties of the cost minimization problem (1). On the other hand, Shephard (1953; 1970) assumed only the existence of a cost function satisfying the appropriate regularity conditions.

A very elegant proof of (7) using the hypotheses of Hicks and Samuelson is due to Karlin (1959) and Gorman (1976). Their proof proceeds as follows.

Let x^* be a solution to $\min_x \{p^* \cdot x : F(x) \geq y^*, x \geq 0_N\} = C(y^*, p^*)$. Then for every $p \gg 0_N$, x^* is feasible for the cost minimization problem defined by $C(y^*, p) = \min_x \{p \cdot x : F(x) \geq y^*, x \geq 0_N\}$ but it is not necessarily optimal. Thus for every $p \gg 0_N$, we have the following inequality:

$$p \cdot x^* \geq C(y^*, p). \quad (8)$$

We also have

$$p^* \cdot x^* = C(y^*, p^*). \quad (9)$$

For $p \gg 0_N$, define the function $g(p) \equiv p \cdot x^* - C(y^*, p)$. From (8), $g(p) \geq 0$ for all $p \gg 0_N$, and from (9), $g(p^*) = 0$. Thus $g(p)$ attains a global minimum at

$p = p^*$. Since g is differentiable, the first-order necessary conditions for a minimum must be satisfied at p^*:

$$\nabla_p g(p^*) = x^* - \nabla_p C(y^*, p^*) = 0_N \quad (10)$$

where $\nabla_p g(p^*) \equiv [\partial g(p^*)/\partial p_1, \ldots, \partial g(p^*)/\partial p_N]$ denotes the vector of first-order partial derivatives of g with respect to the components of p evaluated at p^* and $\nabla_p C(y^*, p^*)$ denotes the vector of first-order partial derivatives of C with respect to the components of p evaluated at (y^*, p^*). The second set of equalities in (10) can be rearranged to yield (7).

From an econometric point of view, Shephard's Lemma is a very useful result. In order to obtain a valid system of cost minimizing input demand functions, $x(y, p) \equiv [x_1(y, p), \ldots, x_N(y, p)]$ all we have to do is postulate a functional form for C which satisfies Properties 1–6 and then differentiate C with respect to the components of the input price vector p; that is, $x(y, p) = \nabla_p C(y, p)$. It is not necessary to compute the production function F that corresponds to C via the Shephard Duality Theorem nor is it necessary to undertake the often complex algebra involved in deriving the input demand functions using the production function and Lagrangian techniques. In Section 6 below, we shall consider several functional forms for C that have been suggested for their econometric convenience.

5 The comparative statics properties of cost functions

Suppose that we are given a cost function C satisfying Properties 1–6 that is also twice continuously differentiable at (y^*, p^*) where $y^* > 0$ and $p^* \gg 0_N$. Applying Shephard's Lemma (7), the above differentiability assumption ensures that the cost minimizing input demand functions $x_i(y, p)$ exist and are once continuously differentiable at (y^*, p^*).

Define $[\partial x_i/\partial p_j] \equiv [\partial x_i(y^*, p^*)/\partial p_j]$ to be the N by N matrix of partial derivatives of the N demand functions $x_i(y^*, p^*)$ with respect to the N prices p_j, $i, j = 1, \ldots, N$. From (7), it follows that

$$\begin{aligned}[\partial x_i/\partial p_j] &= [\partial^2 C(y^*, p^*)/\partial p_i \partial p_j] \\ &\equiv \nabla_{pp}^2 C(y^*, p^*)\end{aligned} \quad (11)$$

where $\nabla_{pp}^2 C(y^*, p^*)$ is the matrix of second-order partial derivatives of the cost function with respect to the components of the input price vector evaluated at (y^*, p^*). The twice continuous differentiability property of C implies by Young's Theorem in calculus that $\nabla_{pp}^2 C(y^*, p^*)$ is a symmetric N by N matrix. Thus using (11), we have

$$[\partial x_i/\partial p_j] = [\partial x_i/\partial p_j]^T = [\partial x_j/\partial p_i] \quad (12)$$

where A^T denotes the transpose of the Matrix A. Thus we have established the Hicks (1946) and Samuelson (1947)

symmetry restrictions on input-demand functions, $\partial x_i(y^*, p^*)/\partial p_j = \partial x_j(y^*, p^*)/\partial p_i$ for all i and j.

Since C is concave in p and is twice continuously differentiable with respect to the components of p at the point (y^*, p^*), it follows from a characterization of concave functions that $\nabla_{pp}^2 C(y^*, p^*)$ is a negative semidefinite matrix. Thus by (11),

$$z^T [\partial x_i/\partial p_j] z \leq 0 \quad \text{for all vectors } z. \quad (13)$$

In particular, letting $z = e_i$, the ith unit vector, (13) implies

$$\partial x_i(y^*, p^*)/\partial p_i \leq 0 \quad \text{for } i = 1, \ldots, N; \quad (14)$$

that is, the ith cost minimizing input demand function cannot slope upwards with respect to the ith input price for $i = 1, \ldots, N$.

Since C is linearly homogeneous in p, we have $C(y^*, \lambda p^*) = \lambda C(y^*, p^*)$ for all $\lambda > 0$. Partially differentiating this equation with respect to p_i for λ close to 1 yields the equation $C_i(y^*, \lambda p^*)\lambda = \lambda C_i(y^*, p^*)$ where $C_i(y^*, p^*) \equiv \partial C(y^*, p^*)/\partial p_i$. Thus $C_i(y^*, \lambda p^*) = C_i(y^*, p^*)$ and differentiation of this last equation with respect to λ yields when $\lambda = 1$:

$$\sum_{j=1}^{N} p_j^* \partial^2 C(y^*, p^*)/\partial p_i \partial p_j = 0 \quad (15)$$

$$\text{for} \quad i = 1, \ldots, N.$$

Equations (11) and (15) imply that the input-demand functions $x_i(y^*, p^*)$ satisfy the following N restrictions:

$$\sum_{j=1}^{N} p_j^* \partial x_i(y^*, p^*)/\partial p_j = 0 \quad (16)$$

$$\text{for} \quad i = 1, \ldots, N.$$

A final general restriction on the derivatives of the input-demand functions may be obtained as follows: for λ near 1 differentiate both sides of $C(y^*, \lambda p^*) = \lambda C(y^*, p^*)$ with respect to y and then differentiate the resulting equation with respect to λ. When $\lambda = 1$, the last equation becomes:

$$\sum_{j=1}^{N} p_j^* \partial^2 C(y^*, p^*)/\partial y \partial p_j = \partial C(y^*, p^*)/\partial y. \quad (17)$$

The twice continuous differentiability of C at (y^*, p^*) and (7) imply:

$$\begin{aligned}\partial^2 C(y^*, p^*)/\partial y \partial p_j &= \partial^2 C(y^*, p^*)/\partial p_j \partial y \\ &= \partial x_j(y^*, p^*)/\partial y.\end{aligned} \quad (18)$$

Property 5 for cost functions implies that

$$\partial C(y^*, p^*)/\partial y^* \geq 0. \tag{19}$$

Using (18) and (19), (17) is equivalent to:

$$\sum_{j=1}^{N} p_j^* \partial x_j(y^*, p^*)/\partial y \geq 0. \tag{20}$$

Thus for at least one j, we must have $\partial x_j(y^*, p^*)/\partial y \geq 0$; that is, as output increases, not every input demand can decrease.

We have shown that the assumption of cost minimizing behaviour implies a number of restrictions on input demand functions that are potentially testable. Hicks (1946) and Samuelson (1947) obtained the restrictions (12), (13), and (16) using the first-order conditions for the primal cost minimization problem (1) and the properties of determinants of bordered Hessian matrices; Samuelson also obtained (20). Our derivation of the restrictions on input-demand functions using the dual approach is due to McKenzie (1957), Karlin (1959) and McFadden (1978a).

Hicks (1946) also showed that when $N = 2$, so that there are only two inputs, then (12), (13), and (16) imply that

$$\begin{aligned}\partial x_1(y^*, p_1^*, p_2^*)/\partial p_2 \\ = \partial x_2(y^*, p_1^*, p_2^*)/\partial p_1 \geq 0.\end{aligned} \tag{21}$$

Hicks (1946) called two distinct goods i and j *substitutes* if and only if $\partial x_i(y, p)/\partial p_j \geq 0$ and *complements* if and only if $\partial x_i(y, p)/\partial p_j < 0$. Thus in the two input case, the two goods must be substitutes. Hicks also showed that in the three input case, at least two of the three pairs of goods must be substitutes.

We turn now to empirical applications of cost functions.

6 Functional forms for cost functions

Shephard's Lemma (7) provides a convenient method for generating systems of cost minimizing input demand functions: simply postulate a functional form for $C(y, p)$ and then partially differentiate C with respect to each input price. Below, we present three examples to illustrate the technique.

Our first example is the *translog cost function due to Christensen, Jorgenson and Lau (1971; 1973). The logarithm of the cost function is defined as follows:*

$$\ln C(y, p) \equiv \alpha_0 + \sum_{i=1}^{N} a_i \ln p_i$$

$$= (1/2) \sum_{i=1}^{N} \sum_{j=1}^{N} a_{ij} \ln p_i \ln p_j$$

$$+ \sum_{i=1}^{N} a_{ij} \ln p_i \ln y + a_y \ln y$$

$$+ (1/2) a_{yy} \ln y \ln y \tag{22}$$

where the $a_i, a_{ij}, = a_{ij}, a_i, a_y$ and a_{yy} are $1 + N + (1/2) N(N+1) + N + 2 = 3 + 2N + (1/2)N(N+1)$ parameters determined by the technology of the firm or industry. Differentiating both sides of (22) with respect to the logarithm of the ith input price, $\ln p_i$, for $i = 1, \ldots, N$ yields the following system of equations:

$$s_i = a_i + \sum_{j=1}^{N} a_{ij} \ln p_j + a_{iy} \ln y, \tag{23}$$
$$i = 1, \ldots, N$$

where the ith input cost share is defined as $s_i \equiv [p_i \partial C(y, p)/\partial p_i]/C(y, p) = p_i x_i(y, p)/C(y, p)$ where the last equality follows using (7).

By Property 2 for cost functions, $C(y, p)$ must be linearly homogeneous in input prices. This property will be satisfied by the translog cost function cost function if and only if the following $N+2$ linear restrictions on the parameters hold:

$$\sum_{i=1}^{N} a_i = 1, \sum_{i=1}^{N} a_{iy} = 0 \text{ and }$$

$$\sum_{j=1}^{N} a_{ij} = 0 \text{ for } i = 1, \ldots, N. \tag{24}$$

It is possible to append errors to equations (22) and $N - 1$ of the equations (23) and econometrically estimate the unknown parameters, given data on inputs, input prices and output. The symmetry restrictions $a_{ij} = a_{ij}$ and the restrictions (24) may be imposed or one can test for their validity. If these restrictions are imposed, then the resulting translog cost function will have $1 + N + (1/2)N(N+1)$ free parameters.

What considerations are relevant in choosing a functional form for a cost function? The following four properties are desirable: (1) *flexibility*; that is, the functional form for C should have a sufficient number of free parameters to be able to provide a second-order approximation to an arbitrary twice continuously differentiable function with the appropriate theoretical properties, (ii) *parsimony*; that is, the functional form for C should have the minimal number of free parameters required to have the flexibility property, (iii) *linearity*; that is, the unknown parameters of C should appear in the system of estimating equations in a linear fashion in order to facilitate econometric estimation, and (iv) *consistency*; that is, the functional form for C should be consistent with Properties 1–6 for cost functions. These considerations were first suggested by Diewert (1971) in an informal manner; the term parsimony is due to Fuss, McFadden and Mundlak (1978) and the term flexible is due to Diewert (1974). The equivalence of various definitions of the flexibility property is discussed by Barnett (1983).

How satisfactory is the translog cost function in the light of the above considerations? We consider the flexibility property first. In order to be able to approximate a function of $1+N$ variables to the second order, we require $1 + (1 + N) + (1 + N)^2$ free parameters. However, if we assume that the cost functions are twice continuously differentiable, then we can reduce the number by $N(N + 1)/2$ due to the symmetry property of the second-order partial derivatives. The linear homogeneity property of the cost function, Property 2, yields an additional $N+1$ restriction on the first and second derivatives of C, (15) and (17), plus the following restriction (which follows using Euler's Theorem on homogeneous functions):

$$C(y,p) = \sum_{i=1}^{N} p_i \partial C(y,p)/\partial p_i. \qquad (25)$$

Thus a flexible functional form for a cost function should have $1 + (1 + N) + (1 + N)^2 - [(1/2)N(N + 1) + N + 1 + 1] = 1 + N + (1/2)N(N + 1)$ free parameters, which is precisely the number the translog cost function has when the restrictions (24) are imposed. It can be shown that the translog cost function is indeed flexible and we have just shown that it is also parsimonious.

As can be seen by inspecting (22) and (23), the estimating equations are linear in the unknown parameters, so the linearity property is also satisfied.

If the restrictions (24) are imposed, Property 2 will be satisfied. In practice, the other properties that a cost function must have will be satisfied with the exception of Property 4, the concavity in prices property. If all of the a_{ij} and a_{iy} parameters are zero, then the translog cost function reduces to a Cobb–Douglas cost function which satisfies the concavity property globally. However, in the general case, the best we can hope for is that the concavity property is satisfied locally for a range of input prices.

If a production function is linearly homogeneous (that is, $F(\lambda x) = \lambda F(x)$ for $\lambda \geq 0$ and $x \geq 0_N$) so that the technology is subject to constant returns to scale, then the corresponding cost function has the following property:

$$C(y,p) = yC(1,p); \qquad (26)$$

that is, total cost is equal to the output level y times the cost of producing one unit of output, $C(1,p) \equiv c(p)$, the *unit cost function.*

If C is twice continuously differentiable and satisfies (26), then one can show that the following $2+N$ restrictions on the first and second derivatives of C must hold:

$$C(y,p) = y\partial C(y,p)\partial y; \qquad (27)$$

$$\partial^2 C(y,p)/\partial y^2 = 0; \qquad (28)$$

$$\partial C(y,p)/\partial p_i = y\partial^2 C(y,p)/\partial y\partial p_i, \qquad (29)$$
$$i = 1,\ldots,N.$$

However, in view of (25), it can be seen that only $N-1$ of the restrictions (29) are new. Thus the assumption of a constant returns to scale technology imposes new restrictions on the derivatives of the cost function C.

It can be shown that necessary and sufficient conditions for the translog cost function defined by (22) and (24) to satisfy (26) are the following $N+1$ restrictions:

$$a_y = 1, a_{yy} = 1 \text{ and } a_{iy} = 0 \qquad (30)$$
$$\text{for } i = 1,\ldots,N-1.$$

Of course (30) and (24) imply that $a_{Ny} = 0$ as well.

It can be shown that if the restrictions (24) and (30) are imposed on the parameters of the translog cost function defined by (22), then the resulting functional form is flexible in the class of cost functions that satisfy the constant returns to scale property (26). Note that we can test for the validity of the constant returns to scale property by testing whether the $N+1$ linear restrictions (30) hold.

For our second example, consider the following functional form for a cost function:

$$C(y,p) \equiv c(p)y + \sum_{i=1}^{N} b_i p_i$$
$$+ b_{yy}\left(\sum_{i=1}^{N} \beta_i p_i\right)y^2; \qquad (31)$$

$$c(p) \equiv \sum_{i=1}^{N}\sum_{j=1}^{N} b_{ij} p_i^{1/2} p_j^{1/2} \qquad (32)$$

where the $b_{yy}, b_i, b_{ij} = b_{ji}$ and β_i are parameters which characterize the technology. If $b_i = 0$ for $i = 1,\ldots,N$ and $b_{yy} = 0$, then (31) reduces to the *Generalized Leontief cost function* defined by Diewert (1971). If in addition, $b_{ij} = 0$ for all $i \neq j$, then (31) reduces to the cost function $\sum_{i=1}^{N} b_{ii} p_i y$, which is dual to the Leontief (no substitution) production function, $F(x_1,\ldots,x_N) \equiv \min\{x_i/b_{ii} : i = 1, \ldots, N\}$.

In order for the cost function defined by (31) and (32) to satisfy the parsimony property, it is necessary for the empirical investigator to prespecify the β_i parameters; for example, one could set β_i equal to 1 or to the average input quantity x_i observed in the sample of data. Under these conditions, the Generalized Leontief cost function has $(1/2)N(N + 1) + N + 1$ free parameters, which is just the required number for the flexibility property. In fact, Diewert and Wales (1987) show that this cost function is flexible and parsimonious when the β_i are predetermined.

Applying (7), the input-demand functions that correspond to (31) and (32) are:

$$x_i(y, p) = \sum_{j=1}^{N} b_{ij} p_i^{-1/2} p_j^{1/2} y + b_i$$

$$+ b_{yy} \beta_i y^2. \quad i = 1, \ldots, N.$$

(33)

For the purpose of econometric estimation, errors can be appended to the N equations (33). If the β_i are predetermined, it can be seen that the system of estimating equations is linear in the unknown parameters.

If we wish to test for a constant returns to scale technology, then the following $1 + N$ linear restrictions on the parameters are necessary and sufficient for this property:

$$b_{yy} = 0 \quad \text{and} \quad b_i = 0 \quad \text{for} \quad i = 1, \ldots, N.$$

(34)

Note that the linear homogeneity in prices property is satisfied by the Generalized Leontief cost function. The other properties for cost functions will also be satisfied in practice with the exception of Property 4, the concavity in prices property. If all $b_{ij} \geq 0$ for $i \neq j$, then the concavity property will be globally satisfied, but this assumption rules out complementary pairs of inputs (recall the discussion about substitutes and complements at the end of the previous section). Thus in general, one can only hope that the concavity property will be satisfied locally, as was the case with the translog cost function.

For our third and final example, consider the following *normalized quadratic cost function defined by (31) but now c(p) is defined as follows:*

$$c(p) \equiv \sum_{i=1}^{N} b_{ii} p_i + (1/2)$$

$$\sum_{i=1}^{N} \sum_{j=1}^{N} a_{ij} p_i p_j \left/ \left(\sum_{n=1}^{N} \alpha_n p_n \right) \right.$$

(35)

where the N by N matrix $A \equiv [a_{ij}]$ *is symmetric and satisfies the following restriction for some input price vector* $p^* \gg 0_N$;

$$Ap^* = 0_N.$$

(36)

This functional form is due to Diewert and Wales (1987); it generalizes some functional forms due to Fuss (1977) and McFadden (1978b). The functional form has Nb_{ii} parameters, $(1/2)N(N-1)$ free a_{ij} parameters taking into consideration (36), N b_i parameters, 1 b_{yy}, N β_i and N α_n parameters or $1 + 3N + (1/2)N(N+1)$ free parameters in all. In order for this functional form to have the parsimony property, it is necessary for the empirical investigator to prespecify the β_i and α_n parameters; we assume that this has been done and these parameters are

non-negative and not identically equal to zero. Under these conditions, Diewert and Wales (1987) show that this cost function is parsimonious and flexible at the point (y^*, p^*) where p^* is the price vector which appears in (36).

Applying (7), the system of input-demand functions divided by the output y is:

$$x_i(y, p)/y = b_{ii} + \sum_{j=1}^{N} a_{ij} p_j \left(\sum_{n=1}^{N} \alpha_n p_n \right)^{-1}$$

$$- \left(\sum_{j=1}^{N} \sum_{k=1}^{N} a_{jk} p_j p_k \right)$$

$$\times \left(\sum_{n=1}^{N} \alpha_n p_n \right)^{-2} \alpha_i + b_i y^{-1}$$

$$+ b_{yy} \beta_i y, \quad i = 1, \ldots, N.$$

(37)

Errors can be appended to (37) and we obtain a system of estimating equations which is linear in the unknown parameters, provided that the α_n and β_i are prespecified.

If we wish to test for a constant returns to scale technology, then again the $1 + N$ linear restrictions (34) are necessary and sufficient for this property.

The normalized quadratic cost function with prespecified α_n and β_i is flexible, parsimonious and has linear estimating equations. As was the case with our first two examples, our third example has no problem in satisfying Properties 1, 2, 3, 5 and 6 for cost functions. It also turns out that our third example has no problem in satisfying Property 4: Diewert and Wales (1986) using some results due to Lau (1978), show that the normalized quadratic cost function is globally concave if and only if the A matrix is negative semidefinite. They also indicate how this negative semidefiniteness property can be imposed if necessary without destroying the flexibility of the functional form; simply set $A = -SS^T$ where S is a lower triangular N by N matrix and S^T is its transpose. However, in this latter case, nonlinear regression techniques must be used in order to estimate the unknown parameters.

The extensive empirical literature on estimating cost functions is nicely reviewed by Jorgenson (1984).

7 Applications to the estimation of consumer preferences

The cost function techniques described in the previous section can be used to obtain empirical descriptions of technologies. Those techniques can also be adapted to obtain empirical descriptions of consumer preferences.

As was noted in Section 2, y may be interpreted as a household's welfare level, F as a utility or preference function, p as a vector of commodity prices and $C(y, p)$ as the minimum cost of achieving at least the welfare level y.

However, the econometric techniques described in the previous section cannot be utilized immediately in the consumer context because utility cannot be observed whereas output can. We acknowledge this difference by using u, the consumer's utility or welfare level, in place of y in what follows.

The theory outlined in the previous sections is still valid: given a differentiable functional form for the cost function $C(u, p)$ that satisfies Properties 1 to 6, we may form the consumer's system of *constant real income* or *Hicksian demand functions* $x(u, p) \equiv [x_1(u, p), \ldots, x_N(u, p)]$ by differentiating the cost function with respect to each commodity price p_i [recall (7)]:

$$x_i(u, p) = \partial C(u, p)/\partial p_i, \quad i = 1, \ldots, N.$$
(38)

We determine u as a function of the prices p and the consumer's observed expenditure on commodities during the period Y, say, by equating the minimum cost of achieving the welfare level u to the observed expenditure; that is, we solve the following equation for u:

$$C(u, p) = Y.$$
(39)

The solution function g where $u = g(Y, p)$ is known as the consumer's *indirect utility function*. Now replace u in the right-hand side of (38) by $g(Y, p)$ and obtain the consumer's system of *market demand functions*:

$$x_i = \partial C(g(Y, p), p)/\partial p_i, \quad i = 1, \ldots, N.$$
(40)

If we multiply equation i in (40) by p_i, sum the resulting equations and use (7), (25) and (39), then we obtain the identity $\Sigma_{i=1}^{N} p_i x_i = Y$, so only $N - 1$ of the N equations in (40) are independent. Thus for econometric estimation purposes, we may add errors to $N - 1$ of the equations in (40), and given a functional form for C, we may use these equations to estimate the unknown parameters in C. We shall discuss this technique in more detail shortly, but first, we must discuss the problems involved in cardinalizing utility.

The scaling of utility is irrelevant in describing a consumer's preferences. However, when we postulate a functional form for a cost function, we are implicitly imposing a cardinalization of the consumer's utility. Hence, we might as well impose a convenient cardinalization: *money metric scaling of utility* (the term is due to Samuelson, 1974). This involves setting utility u equal to 'income' Y, holding prices constant at some specified price vector p^*, that is, we have

$$Y = g(Y, p^*) \quad \text{for all } Y > 0.$$
(41)

In terms of the cost function, (41) may be rewritten as

$$u = C(u, p^*) \quad \text{for all } u > 0.$$
(42)

In examples 2 and 3 in the previous section, the cost function had the following form:

$$C(u, p) = c(p)u + \sum_{i=1}^{N} b_i p_i$$
$$+ b_{yy} \left(\sum_{i=1}^{N} \beta_i p_i \right) u^2.$$
(43)

In order to make (43) consistent with money metric scaling, (42), the following three restrictions on the parameters of C must be satisfied:

$$c(p^*) = 1, \sum_{i=1}^{N} b_i p_i^* = 0 \text{ and } b_{yy} = 0.$$
(44)

Using $b_{yy} = 0$ we find that the indirect utility function that corresponds to the C defined by (43) is

$$g(Y, p) = \left(Y - \sum_{i=1}^{N} b_i p_i \right) \Big/ c(p).$$
(45)

Substitution of (45) into (40) yields the following system of consumer demand functions:

$$x_i = b_i + [\partial c(p)/\partial p_i] \left(Y - \sum_{i=1}^{N} b_i p_i \right) \Big/ c(p).$$
$$i = 1, \ldots, N.$$
(46)

Now add errors to $N - 1$ of the equations (46), calculate the partial derivatives of the $c(p)$ defined by (32) or (35), impose the normalizations (44) and we have a system of nonlinear estimating equations. An empirical example of this technique for estimating consumer preferences may be found in Diewert and Wales (1986).

Finally, we note that cost functions of the type defined by (43) with $b_{yy} = 0$ have very convenient aggregation over consumers' properties; see Gorman (1953) and Deaton and Muellbauer (1980).

8 Cost functions and measures of welfare gain

Consider a consumer whose preferences can be represented by the differentiable cost function, $C(u, p)$. Suppose we can observe the consumer's choices x^1 and x^2 during periods 1 and 2 when prices p^1 and p^2 prevail. Let u^1 and u^2 be the welfare levels attained during those two periods. Then by (7),

$$x^i = \nabla_p C(u^i, p^i), \quad i = 1, 2.$$
(47)

For many purposes in applied welfare economics, it is useful to evaluate the *ex post* welfare change of the consumer. Two natural measures, suggested originally by

Hicks (1942), are his *equivalent* and *compensating varia-tions* which we denote by $V(p^1)$ and $V(p^2)$:

$$V(p^1) \equiv C(u^2, p^1) - C(u^1, p^1);$$
$$V(p^2) \equiv C(u^2, p^2) - C(u^1, p^2). \quad (48)$$

From (47) and (25), $C(u^1, p^1) = p^1 \cdot x^1$ and $C(u^2, p^2) = p^2 \cdot x^2$. However, the costs $C(u^2, p^1)$ and $C(u^1, p^2)$ are not observable. Hence the following question arises: can we form approximations to $V(p^i)$ that use only observable data?

Linear approximations to $C(u^i, p^j)$ may be obtained using Taylor's Theorem. Thus we have:

$$V(p^1) \cong [C(u^2, p^2) + \nabla_p C(u^2, p^2) \cdot (p^1 - p^2)]$$
$$- C(u^1, p^1) = [p^2 \cdot x^2 + x^2 \cdot (p^1 - p^2)]$$
$$- p^1 \cdot x^1 \text{ using } (47)$$
$$= p^1 \cdot (x^2 - x^1) \quad (49)$$

and

$$V(p^2) \cong C(u^2, p^2)$$
$$- [C(u^1, p^1) + \nabla_p C(u^1, p^1) \cdot (p^2 - p^1)]$$
$$= p^2 \cdot x^2 - [p^1 \cdot x^1 + x^1 \cdot (p^2 - p^1)]$$
$$\text{using } (47) = p^2 \cdot (x^2 - x^1). \quad (50)$$

The first-order approximations (49) and (50) are essentially due to Hicks (1942; 1946).

To obtain a second-order approximation result, we proceed indirectly. Suppose the consumer's cost function is defined by

$$C(u, p) \equiv c(p) + \sum_{i=1}^{N} b_i p_i u \quad (51)$$

where $c(p)$ is the normalized quadrative unit cost function defined by (35) for some prespecified $\alpha \equiv (\alpha_1, \ldots, \alpha_n) > 0_N$.

It can be shown that the cost function defined by (51) can provide a second-order approximation to an arbitrary twice continuously differentiable cost function that satisfies the money metric scaling of utility property (42).

Now use the parameters vector α which occurred in the definition of $c(p)$ in order to define the *normalized prices* v^i

$$v^i \equiv p^i / (p^i \cdot \alpha), \quad i = 1, 2. \quad (52)$$

Straightforward calculations show that if C is defined by (51), then the following identity holds *exactly*:

$$(1/2)V(v^1) + (1/2)V(v^2)$$
$$= (1/2)(v^1 + v^2) \cdot (x^2 - x^1) \quad (53)$$

where $V(v^1)$ and $V(v^2)$ are equivalent and compensating variations evaluated using the normalized prices v^i in place of the commodity price vectors p^i. Thus (53) says that an average of the Hicksian variations using normalized prices is exactly equal to the average of the normalized prices inner producted with the vector of quantity differences, $x^2 - x^1$, provided that preferences are defined by the cost function (51). Note that the right-hand side of (53) can be evaluated using observable price and quantity data. Since the formula on the right-hand side is exact for preferences which have a second-order approximation property, we could call it a *superlative welfare gain measure* in analogy to the terminology used in index number theory. The term gain measure is due to King (1983).

W.E. DIEWERT

See also **duality; production functions.**

Bibliography

Barnett, W.A. 1983. New indices of money supply and the flexible Laurent demand system. *Journal of Business and Economic Statistics* 1, 7–23.

Christensen, L.R., Jorgenson, D.W. and Lau, L.J. 1971. Conjugate duality and the transcendental logarithmic production function. *Econometrica* 39, 255–6.

Christensen, L.R., Jorgenson, D.W. and Lau, L.J. 1973. Transcendental logarithmic production frontiers. *Review of Economics and Statistics* 55, 28–45.

Deaton, A. and Muellbauer, J. 1980. *Economics and Consumer Behaviour.* Cambridge: Cambridge University Press.

Diewert, W.E. 1971. An application of the Shephard duality theorem: a generalized Leontief production function. *Journal of Political Economy* 79, 481–507.

Diewert, W.E. 1974. Applications of duality theory. In *Frontiers of Quantitative Economics*, vol. 2, ed. M.D. Intriligator and D.A. Kendrick. Amsterdam: North-Holland, 106–171.

Diewert, W.E. 1982. Duality approaches to microeconomic theory. In *Handbook of Mathematical Economics*, vol. 2, ed. K.J. Arrow and M.D. Intriligator. Amsterdam: North-Holland.

Diewert, W.E. and Wales, T.J. 1986. Normalized quadratic systems of consumer demand functions. Discussion Paper No. 86–16, Department of Economics, University of British Columbia, Vancouver, Canada, May.

Diewert, W.E. and Wales, T.J. 1987. Flexible functional forms and global curvature conditions. *Econometrica.* 55, 43–68.

Fuss, M.A. 1977. Dynamic factor demand systems with explicit costs of adjustment. In *Dynamic Models of the Industrial Demand for Energy*, ed. E.R. Berndt, M. Fuss and L. Waverman. Palo Alto, CA: Electric Power Research Institute.

Fuss, M., McFadden, D. and Mundlak, Y. 1978. A survey of functional forms in the economic analysis of production. In *Production Economics: A Dual Approach to Theory and Applications*, vol. 1, ed. M. Fuss and D. McFadden. Amsterdam: North-Holland.

Gorman, W.M. 1953. Community preference fields. *Econometrica* 21, 63–80.

Gorman, W.M. 1976. Tricks with utility functions. In *Essays in Economic Analysis*, ed. M. Artis and R. Nobay. Cambridge: Cambridge University Press.

Hicks, J.R. 1942. Consumers' surplus and index-numbers. *Review of Economic Studies* 9, 126–37.

Hicks, J.R. 1946. *Value and Capital*. 2nd edn. Oxford: Clarendon Press.

Jorgenson, D.W. 1984. Econometric methods for modeling producer behaviour. Discussion Paper No. 1086, Harvard Institute for Economic Research, Harvard University, Cambridge, MA. In *Handbook of Econometrics*, vol. 3, ed. Z. Griliches and M.D. Intriligator. Amsterdam: North-Holland.

Karlin, S. 1959. *Mathematical Methods and Theory in Games, Programming and Economics*, vol. 1. Palo Alto, CA: Addison-Wesley.

King, M.A. 1983. Welfare analysis of tax returns using household data. *Journal of Public Economics* 21, 183–214.

Lau, L.J. 1978. Testing and imposing monotonicity, convexity and quasi-convexity constraints. In *Production Economics: A Dual Approach to Theory and Applications*, vol. 1, ed. M. Fuss and D. McFadden. Amsterdam: North-Holland.

McFadden, D. 1966. Cost, revenue and profit functions: a cursory review. Working Paper No. 86, IBER, University of California at Berkeley, March.

McFadden, D. 1978a. Cost, revenue and profit functions. In *Production Economics: A Dual Approach to Theory and Applications*, vol. 1, ed. M. Fuss and D. McFadden. Amsterdam: North-Holland.

McFadden, D. 1978b. The general linear profit function. In *Production Economics: A Dual Approach to Theory and Applications*, vol. 1, ed. M. Fuss and D. McFadden. Amsterdam: North-Holland.

McKenzie, L. 1957. Demand theory without a utility index. *Review of Economic Studies* 24, 185–9.

Samuelson, P.A. 1947. *Foundations of Economic Analysis*. Cambridge, MA.: Harvard University Press.

Samuelson, P.A. 1974. Complementarity – an essay on the 40th anniversary of the Hicks–Allen revolution in demand theory. *Journal of Economic Literature* 12, 1255–89.

Shephard, R.W. 1953. *Cost and Production Functions*. Princeton: Princeton University Press.

Shephard, R.W. 1970. *Theory of Cost and Production Functions*. Princeton: Princeton University Press.

Uzawa, H. 1964. Duality principles in the theory of cost and production. *International Economic Review* 5, 216–20.

cost minimization and utility maximization

Consider the following standard problem in the theory of demand: Find $x^* \geq 0$ so as to max $u(x)$ subject to $\langle x, p \rangle \leq \omega$ where $\langle x, p \rangle$ is the inner product of the n-dimensional commodity and price vectors, and $\omega > 0$ and u are the consumer's income and utility function respectively; this problem is here labelled max(p, ω).

The functional dependence of the *value* $v^*[\equiv u(x^*)]$ of this nonlinear programming problem on its parameters (p, ω) is denoted by $v(p, \omega)$, where v is the *indirect* utility function. The similar dependence of the *solution* x^* of max (p, ω) is written $f(p, \omega)$, where f is the *ordinary* (or *Marshallian*) demand function (or correspondence). If v^* does not exist then neither do v, x^* or f. Important though they are such non-existence problems are irrelevant here, so without further ado assume that every optimization problem has a solution.

Consider next a problem whose form is similar to that of max (p, ω) but whose objective is different, that is, cost minimization rather than utility maximization. Specifically, find $x^{**} \geq 0$ so as to minmin $\langle x, p \rangle$ subject to $u(x) \geq \tau$ where x, p and u are as before and τ is a *target* level of utility; this new problem is labelled min(p, τ). The functional dependence of the value $\mu^{**}(\equiv \langle x^{**}, p \rangle)$ of min(p, τ) on its parameters (p, τ) is denoted $c(p, \tau)$, where c is the *cost* (or *expenditure*) function. The similar dependence of x^{**} on (p, τ) is written $h(p, \tau)$, where h is the *compensated* (or *Hicksian*) demand function (or correspondence).

Suppose now that max(p, ω) is solved and its value v^* is inserted into the second optimization problem, thus creating the problem min(p, v^*). Is each solution x^* of max(p, ω) necessarily also a solution of min(p, v^*)? Call this Question I. A similar question can be asked of the reverse situation, which is: For arbitrary (p, τ) solve min(p, τ), obtain it value μ^{**} and then solve the resulting max problem, (p, μ^{**}). Question II is then: Is each solution x^{**} of min(p, τ) necessarily also a solution of max(p, μ^{**})?

Problem min(p, v^*) has often been called the *dual* of max(p, ω), from as far back as Arrow and Debreu (1954, pp. 285–6) to Deaton and Muellbauer (1980, pp. 37 ff.) and beyond. Indeed, this usage is now so common that for most economists min(p, v^*) seems to be *the* leading species of the genus *dual problem*.

One can see why. It appears to be quite analogous to dual problems in linear programming (lp), with max becoming min, and objective and constraint functions becoming constraint and objective functions, respectively. However, the analogy with duals in lp is misleading, for each solution x^{**} of the alleged 'dual' min(p, v^*) is located in the *same* space as each solution x^* of its 'primal' max(p, ω), whereas in lp the solutions to the dual all lie in the *dual* space. As Deaton and Muellbauer justly remark: 'The essential feature of the duality approach is a *change of variables*' (1980, p. 47, their italics). So a new term for the relation that $\min(p, v^*)$ bears to $\max(p, \omega)$ is needed in order to distinguish it from genuine duality; the 'mirrored' (or 'reflected') problem is suggested in Newman (1982).

In demand theory it is sometimes recognized explicitly that Question I needs an answer (for example, Samuelson, 1947, p. 103; McKenzie, 1957, p. 186) but more often not, probably because the usual assumptions on preferences are quite sufficient for coincidence of x^{**} with x^*. An explicit treatment appears unnecessary: '... clearly, the vector of commodities must in both cases be the same' (Deaton and Muellbauer, 1980, p. 37). In welfare economics, however, it has long been recognized that a suitably generalized form of Question I, simple as it is, has importance for the first fundamental theorem of welfare economics, namely that every competitive allocation is (strongly) Pareto-optimal.

Question II has always been considered more delicate than Question I. Indeed, it was not even put until Arrow (1951, pp. 527–8) exhibited his famous 'exceptional case' (now often known as the Arrow Corner) in which it receives a negative answer. Its relevance for proofs of existence of competitive equilibrium was fully grasped by Arrow and Debreu (1954, sections 4 and 5), and later Debreu (1959, pp. 67–71), for essentially this reason, devoted four pages of his terse classic to a detailed examination of both Questions.

It is interesting that although the second Question is economically more subtle than the first, from a sufficiently abstract point of view the two are logically isomorphic (see Newman, 1982, where in both Theorem (c) and Theorem (c') the assertion 'iff' is wrong and should be replaced by 'if'). While such extreme abstraction is irrelevant here, both $\max(p, \omega)$ and $\min(p, \tau)$ do need to be put into a form suitable for general equilibrium theory.

The setting

The consumer is now endowed, not with an exogenous positive income, but with a nonzero bundle x^0, whose worth $\langle x^0, p \rangle$ may be zero. For simplicity (and only that), free disposal is assumed.

Assumptions about preferences

The consumer has two disjoint binary relations $\succ$ ('preference') and $\sim$ ('indifference') each defined on some non-empty $S \subset R^n$; the union of $\succ$ and $\sim$ is denoted $\succsim$. Indifference is reflexive and symmetric (so that preference is irreflexive) and the statements $x^1 \succ x^3$ and $x^2 \sim x^3$ together imply $x^1 \succ x^3$. Neither completeness nor transitivity of preference is assumed, so a utility function need not exist.

The generalized version of $\max(p, \omega)$ is then: Find $x^* \in S$ for which $\langle x^*, p \rangle \leq \langle x^0, p \rangle$ and such that $x \succ x^*$ implies $\langle x, p \rangle > \langle x^0, p \rangle$. In words, '$x^*$ is feasible and

anything preferred to it is unaffordable'. This problem is labelled $\max(p, x^0)$.

The generalized version of $\min(p, \tau)$ is: Find $x^{**} \in S$ for which $x^{**} \succsim t$ and such that $x \succsim t$ implies $\langle x, p \rangle \geq \langle x^{**}, p \rangle$. In words, 'anything at least as good as the target bundle $t \in S$ costs at least as much as x^{**}'. This problem is labelled $\min(p, t)$.

Note that in the absence of a utility function $\max(p, x^0)$ can have a solution but not a value, while $\min(p, t)$ can have both value and solution, just as before.

Some definitions

Any bundle $x^\sharp$ to which no $x \in S$ is preferred is called *bliss*, while a bundle $x_\sharp$ for which at prices p there is no cheaper $x \in S$ is called *p-minimal*. Preferences are *locally nonsatiated* at x^1 if *any* neighbourhood $N(x^1)$ contains $x \succ x^1$, while $x^2 \in S$ has *locally cheaper points* at p (a term apparently due to McKenzie, 1957) if *any* neighbourhood $N(x^2)$ contains a bundle $x \in S$ which at prices p is cheaper than x^2. If $x^\sharp$ is bliss it cannot be locally nonsatiated, and if $x_\sharp$ is p-minimal it has no locally cheaper points at p.

Following Bergstrom, Parks and Rader (1976), preferences are said to have *open upper sections* if $x' \succ x^2$ implies the existence of a neighbourhood $N(x^1) \subset S$ for which $x \succ x^2$ for every x in it.

The following simple result answers both Questions satisfactorily and generalizes easily to a wide class of infinite-dimensional commodity spaces.

Theorem (i) Assume (a) that if $x \in S$ is not bliss it is locally nonsatiated, and (b) that the solution x^* of $\max(p, x^0)$ is not bliss. Then x^* also solves $\min(p, x^*)$. Moreover, the value μ^{**} of $\min(p, x^{**})$ equals $\langle x^0, p \rangle$.

(ii) Assume (c) that preferences have open upper sections, (d) that if $x \in S$ is not p-minimal it has locally cheaper points at p, and (e) that the solution x^{**} of $\min(p, t)$ is not p-minimal. Then x^{**} also solves $\max(p, \mu^{**})$, where $\mu^{**} = \langle x^{**}, p \rangle$.

Proof (i) Suppose the result false, so there exists $x^1 \succsim x^*$ such that $\langle x^1, p \rangle < \langle x^*, p \rangle$. Now $x^1 \succ x^*$ cannot occur because if it did $\langle x^1, p \rangle < \langle x^*, p \rangle \leq \langle x^0, p \rangle$ would imply that x^* does not solve $\max(p, x^0)$, contrary to hypothesis. So $x^1 \sim x^*$.

Since the vector p represents a continuous linear function (a1) there is a neighbourhood $N(x^1)$ all of whose points are cheaper at prices p than x^*. From (b) there exists $x \succ x^*$, and this and the symmetry of $\sim$ imply that $x \succ x^1$ as well, so that x^1, is not bliss either. Hence from (a) at least one member of $N(x^1)$, say x^2, is such that $x^2 \succ x^1$. Because $x^1 \succ x^*$ this leads to $x^2 \succ x^*$, which again contradicts the hypothesis. Thus x^* solves $\min(p, x^*)$, which implies $\mu^{**} = \langle x^*, p \rangle$.

Suppose $\langle x^*, p \rangle < \langle x^0, p \rangle$. By the continuity of p there exists $N(x^*)$ all of whose points are cheaper at prices p than is x^0, while from (b) and (a) at least one of them, say x^3, is such that $x^3 \succ x^*$. Yet again, this contradicts the hypothesis. So $\langle x^*, p \rangle = \langle x^0, p \rangle$.

(ii) By assumption $x^{**} \succsim t$. Suppose $x^{**} \succ t$. From (c) there exists $N(x^{**})$ such that $x \succ t$ for every x in it. From (e) and (d) x^{**} has locally cheaper points at p, so at least one x in $N(x^{**})$, say x^1, is cheaper than x^{**} at p. Since $x^1 \succ t$, this contradicts the hypothesis that x^{**} solves $\min(p, t)$. Hence $x^{**} \sim t$.

Suppose now that x^{**} does not solve $\max(p, \mu^{**})$, so there is an x^2 such that $x^2 \succ x^{**}$ and $\langle x^2, p \rangle \leq \langle x^{**}, p \rangle$. Hence $x^2 \succ t$. If x^2 were cheaper at p than x^{**} that would again contradict the hypothesis. So $\langle x^2, p \rangle \leq \langle x^{**}, p \rangle$.

From (e) there is an $x \in S$ cheaper at p than x^{**}, hence cheaper than x^2, so x^2 is not p-minimal either. Since $x^2 \succ t$, from (c) there exists $N(x^2)$ such that $x \succ t$ for every x in it and from (d) at least one of these must be cheaper than x^2 at p, and so cheaper than x^{**}, which again contradicts the hypothesis. Q.E.D.

One sees just how few and how weak are the assumptions on preferences that enable Questions I and II to be answered, as distinct (for example) from those needed to guarantee the *existence* of solutions x^* and x^{**}. Note that two assumptions are used for Question I and three for II, an inequality which occurs because the constraint in $\max(p, x^0)$ is linear and hence continuous, whereas in the problem $\min(p, t)$ some continuity in the (nonlinear) constraint has to be *imposed* by means of the 'extra' assumption (c). This asymmetry disappears in a more abstract treatment, with more general constraints.

The intuitions behind the proof help to see why Question II is a serious problem for general equilibrium theory. In the proof of (i) the bundle x^1 that is cheaper than x^* is made a little bigger, in effect increasing satisfaction by increasing expenditure, until a bundle is reached that is still affordable at income $\langle x^0, p \rangle$ but which is better than x^*; that expenditure can always be thus 'traded' for satisfaction is assured by local nonsatiation. In the proof of (ii) the bundle x^1 that is better than x^{**} is made a little smaller, lessening satisfaction in return for less cost, until a bundle is reached that is still as good as t but which costs less than x^{**}; such 'trading' of satisfaction for expenditure is guaranteed by the existence of locally cheaper points. However, if the expenditure on x^{**} at prices p is already least possible (that is, if x^{**} is p-minimal) then 'trading' in *that* direction cannot occur – one cannot go below least cost.

Of the five assumptions of the Theorem the only one whose meaning is not transparent and whose restriction is not 'reasonable' is (e), so that it comes as no surprise that the main thing wrong at the Arrow Corner is that (e) does not hold there. For further discussion of this Slater-like assumption and its role in general equilibrium theory, see CONSUMPTION SETS.

PETER NEWMAN

See also **duality**.

Bibliography

Arrow, K.J. 1951. An extension of the basic theorems of classical welfare economics. In *Proceedings of the Second Berkeley Symposium on Mathematical Statistics and Probability*, ed. J. Neyman. Berkeley: University of California Press.

Arrow, K.J. and Debreu, G. 1954. Existence of an equilibrium for a competitive economy. *Econometrica* 22, 265–90.

Bergstrom, T.C., Parks, R.P. and Rader, T. 1976. Preferences which have open graphs. *Journal of Mathematical Economics* 3, 265–8.

Deaton, A. and Muellbauer, J. 1980. *Economics and Consumer Behaviour*. Cambridge: Cambridge University Press.

Debreu, G. 1959. *Theory of Value*. Cowles Commission Monograph No. 17. New York: John Wiley.

McKenzie, L. 1957. Demand theory without a utility index. *Review of Economic Studies* 24, 185–9.

Newman, P. 1982. Mirrored pairs of optimization problems. *Economica* 49, 109–19.

Samuelson, P.A. 1947. *Foundations of Economic Analysis*. Cambridge, MA.: Harvard University Press.

cost-push inflation

The concept of cost-push inflation emerged in the period after the Second World War. The Keynesian model of that time emphasized that the economy could operate with inefficiently low utilization of its capital and labour resources, and that expanding demand would employ those resources. Once full employment was achieved, further expansion of demand would only pull up nominal wages and prices. In contrast to this demand-pull inflation, cost-push described the price increases that came from labour unions pushing up wages despite the existence of excessive unemployment. Since the 1970s, when oil prices rose by many times in two abrupt steps, the idea of cost push has been extended to describe price increases arising from any important shift up in supply schedules at given levels of aggregate demand.

The key distinction between price increases arising from monopoly power in wage settings or from any other supply shock and price increases arising from an increase in aggregate demand along an unchanged supply curve is important both for empirical modelling of inflation and for stabilization policy. By the 1960s, the short-run Phillips curve had emerged as a description of the relation between inflation and unemployment over the business cycle. It described an empirical regularity according to which wages rose gradually faster as unemployment declined, with the relation becoming steeper the lower the unemployment rate. Subsequent amendments to this model took explicit account of learning and expectations and of the interrelation between wages and prices. In the dominant model that emerged, inflation will accelerate (decelerate) indefinitely if the economy operates persistently below (above) a natural rate of unemployment. And in models that stress the importance of expectations, the anticipation of faster or slower price increases speeds up this process of acceleration or deceleration.

While inflation is responsive to aggregate demand in all these models, its responsiveness to supply shocks is more nuanced. In models that stress inflationary expectations, shocks that are widely perceived as one-time shifts up nominal supply curves will lead only to one-time shocks to the price level. In models with institutions that partially or fully index wages to prices, or models with adaptive expectations of inflation, such shocks will have larger and more protracted effects.

Empirically, the distinction between cost-push and excess-demand effects is not always easily drawn. The inflation identified with unemployment below the natural rate or with the steep portion of the short-run Phillips curve is attributable to excess demand. The more modest variations in inflation that may occur as unemployment varies above the natural rate are not characterized so readily. A useful interpretation of these systematic cyclical tendencies is that they represent the normal operation of heterogeneous labour markets in response to cyclical variations in aggregate demand, with wages and prices in some sectors rising faster as their markets tighten while slack is still present in other sectors. On this interpretation, they neither signal that the economy is at a natural rate nor indicate the presence of exogenous cost-push effects on prices. However, these modest variations in inflation may also indicate cost-push effects in wage settings that interfere with achieving full employment, and at times past policymakers have interpreted them in this way and tried to suppress them.

The interdependence between prices and wages presents another difficulty in distinguishing endogenous from exogenous changes in wages. If labour supply depends on real wages – that is, wages relative to the average price level – then labour supply will not change if nominal wages change proportionally in response to disturbances to the cost of living. The narrowest concept of cost-push would, therefore, include only shifts up in labour supply schedules that raise wages relative both to their normal response to cyclical demand conditions and to their normal response to consumer prices.

Such complications obscure the possible presence of cost push from wages in typical circumstances. However, when the exercise of market power in wage setting is extreme, it becomes more apparent. In the United States, large wage increases in the early post-war years are examples of cost push. Coming after wartime controls, these did not raise the concerns that the abrupt acceleration of wages in many industrialized economies did in the late 1960s and early 1970s. For example, in Germany annual increases in hourly compensation jumped from 7.5 per cent in 1968 to 17.5 per cent in 1970, and in the United Kingdom the acceleration over the same period was from seven per cent to 15.5 per cent.

During the 1970s, supply shocks to important raw materials prices dominated world price developments in the decade, producing the second main type of cost-push inflation. These supply shocks included the historic increases in oil prices in 1973–4 and again in 1979, and the food price explosion of 1973. Although world aggregate demand was relatively strong in both 1973 and 1979, the magnitude of the price increases that resulted would not be expected, and is better seen as a consequence of major shifts in world supplies. A succession of poor crops provoked the food price rise, while the successful organization of the OPEC oil cartel, aided by a levelling off in United States oil production, caused the oil price explosion.

Coincident with the 1973–5 supply shocks, further large jumps in wage inflation occurred in several countries. In both the United Kingdom and Japan, annual increases in hourly compensation rose to more than 30 per cent from less than half that rate in 1972. Most other major industrial countries experienced similar, though less dramatic, accelerations in wages. Although these wage developments were doubtless fuelled by the effects of the supply shocks on consumer prices, the differences across countries indicate another round of wage push in many, even when one allows for a normal response to price changes. The rapid wage increases in turn further boosted consumer prices. The eventual changes in real wages, as well as the eventual increase in price levels, varied significantly among the industrial countries during the mid-1970s. In the United States, the speed-ups and slowdowns in wage increases and in prices were far less dramatic than in Europe and Japan. However, over the entire decade of the 1970s wages in the highly unionized sectors of the US economy outpaced economy-wide wages substantially, indicating a moderate but persistent wage push from important major industries.

While this post-war record shows that both wage push and supply shocks have at times been important in pushing up price levels, several difficulties remain with the idea of cost push as a distinct source of inflation, and some analysts reject the idea altogether. First of all, inflation refers to an ongoing rate of increase in prices. A one-time rise in the average price level will translate into some rate of increase in prices over a period spanning the rise. Without quibbles over how long a time period is needed before a measurement qualifies as an 'inflation rate', the distinction between a one-time rise in the price level and an ongoing inflation rate is important. Second, inflation refers to the general price level, not to a subset of prices. A rise in oil prices is, first of all, a rise in the relative price of oil. If wages and prices were fully flexible and responded instantly to changes in the balance between demand and supply, then, in the presence of non-accommodating macroeconomic policies, cost-push shocks would indeed create only relative price changes; inflation, in the aggregate, would be impossible. Those who see monetary policy as able to control the overall price level, if not instantly at least over a relatively short period of time, see a cost push from some sectors as a relative price change that becomes a change in the overall price level only if accommodated by monetary policy. On this view, the accommodation rather than the cost-push causes the inflation.

However, such reasoning ignores the considerable downward rigidity in wages and stickiness in many prices as well as the interactions between prices and wages in modern economies, and thus loses the important role that cost-push shocks played in shaping economic performance in these inflationary periods. There are positive correlations among most prices and wages in the economy. In part these reflect common reactions to aggregate developments and in part they represent causal links among wages and prices throughout the economy.

When the links are strong, as they were in the inflationary periods of the late 1960s and 1970s, a cost-push supply shock will not only add directly to the average price level but will set in motion increases in other prices and wages strong enough to persist for some time, even in the face of slowing demand and increasing underutilization of resources. Consequently, an attempt by monetary policy to hold the overall price level unchanged in the face of such a cost-push shock will result mainly in reducing output and employment. Only gradually will the upward movement of prices originating from a supply shock yield to restrictive monetary policies. On the other hand, because the initial shock induces positively correlated responses in wages and other prices, an accommodative policy that aims to maintain output and employment in the face of the shock will result in a rise in the overall price level that is substantially greater than the direct effect of the shock itself. The question confronting stabilization policy is thus how much to accommodate. And the best answer will differ with different institutions and at different times.

The idea that cost-push inflation originating in excessive union wage demands would interfere with the attainment of full employment prompted attempts in several countries to design incomes policies as part of the stabilization policy arsenal. The idea was that demand management by government would aim at keeping the economy around full employment, while understandings among government, labour and business would aim at heading off wage-push inflation that might otherwise arise before full employment was achieved. There was some evidence of success from incomes policies, known as wage-price guideposts in the United States, in the mid-1960s (Perry, 1967). But whatever chance such policies may have had in the longer run in a relatively benign environment, they were overwhelmed once economies

were driven into the excess demand region during the Vietnam War, and the oil and food supply shocks of the early 1970s sharply raised average price levels everywhere. There has been little interest in incomes policies since that time.

By the 1990s, conventional stabilization policies had achieved low inflation rates throughout the industrial world, and the power of unions to originate more inflationary wage increases was very sharply reduced in almost all countries. Both these developments have lessened the problems of stabilization policy. There is evidence that the low-inflation environment has sharply reduced the links that formerly caused price shocks to spark a wage-price inflationary spiral, as they did in the 1970s (Brainard and Perry, 2000). Wages did not accelerate in response to the world oil price shocks of the mid-2000s, and monetary policymakers were able to focus largely on the core inflation rate – the aggregate inflation rate excluding food and energy prices – in setting policy. At least for now, inflation originating from cost push poses a much smaller risk for stabilization policies today than it has at times in the past.

GEORGE L. PERRY

See also **demand-pull inflation; inflation; inflation expectations.**

Bibliography

Akerlof, G., Dickens, W. and Perry, G. 1996. The macroeconomics of low inflation. *Brookings Papers on Economic Activity* 1996(1), 1–59.

Brainard, W. and Perry, G. 2000. Making policy in a changing world. In *Economic Events, Ideas, and Policies*, ed. and J. Tobin. Washington, DC: Brookings Institution.

Okun, A. 1981. *Prices and Quantities: A Macroeconomic Analysis*. Washington, DC: Brookings Institution.

Okun, A. and Perry, G. 1978. *Curing Chronic Inflation*. Washington, DC: Brookings Institution. Also available as G. Perry, Slowing the wage-price spiral: the macroeconomic view. *Brookings Papers on Economic Activity* 1978(2), 259–99.

Perry, G. 1967. Wages and the guideposts. *American Economic Review* 57, 897–904.

Schultze, C. 1985. Microeconomic efficiency and nominal wage stickiness. *American Economic Review* 75, 1–15.

Tobin, J. 1972. Inflation and unemployment. *American Economic Review* 62, 1–18.

countertrade

Countertrade is a commercial transaction in which a seller, typically from an industrialized country, supplies goods, services or technology to a buyer in a developing country or a formerly planned economy, and in which, in return, the seller purchases from the buyer an agreed amount of goods, services or technology. A distinctive feature of countertrade is the existence of a link between the two transactions, the original import in the developing country and the subsequent export.

Countertrade takes a variety of forms. The three most commonly distinguished are 'barter', 'counterpurchase' and 'buyback'. Barter in the strict sense of the word refers to an import that is paid entirely or partly with an export from the importing country without using foreign exchange. Counterpurchase refers to a transaction in which the import is paid with foreign exchange, but the industrialized country commits to buy export goods from the developing country in return. Buyback is a transaction in which the seller supplies a production facility and the parties agree that the supplier of the facility will buy goods produced with that production facility. All three forms of countertrade are frequently observed in international trade.

Under central planning, countertrade was especially observed in international trade among countries belonging to the Council for Mutual Economic Assistance (Comecon, an economic organization of communist states) as well as in East–West trade. In particular in the 1980s, in the aftermath of the international debt crisis, countertrade became prevalent in international trade with developing countries and Eastern Europe. Before 1989 countertrade accounted for up to 40 per cent of total trade between East and West. After 1989, with the domestic debt crisis in transition countries, barter became dominant in domestic trade in these countries. While countertrade continues to be significant in North–South trade, reliable estimates are not available.

Explanations for countertrade

One of the most frequently cited explanations of countertrade is that it allows countries to overcome a shortage of hard currency. The observation that countertrading countries are highly indebted is taken as evidence that these countries face a shortage of foreign exchange and that their low creditworthiness makes it impossible to finance imports with a simple loan from an international bank (for example, OECD, 1981; 1985). This interpretation is not fully plausible because countertrade uses export goods which otherwise could have been used to generate foreign exchange to pay for future imports. Furthermore, if the foreign-exchange shortage were the main explanation of countertrade we would expect barter to be the prevalent form of countertrade since only barter does in fact avoid the use of hard currency. However, barter accounts for only a small portion of total international countertrade (Marin and Schnitzer, 2002a). Mirus and Yeung (1987) find that countertrade in the form of simple barter or counterpurchase does not improve a country's foreign exchange position unless it improves economic efficiency in the sense that it leads to an increase in national income.

Empirical evidence points to another explanation, starting from the observation that in international trade contract enforcement is problematic and hence conventional contracts cannot be relied on as the main mechanism to sustain economic exchange. International countertrade (as well as domestic barter, as pointed out below) can be explained as an institutional response to such contractual problems arising in imperfect capital and goods markets. Difficulties in contract enforcement are an important impediment to international transactions in the world economy. In international trade, national sovereignty interferes with contract enforcement because national borders demarcate national jurisdictions. Such demarcations segment markets and impose severe transaction costs on exchanges across national jurisdictions. The hazards involved in international transactions are often disregarded, but they make headlines each time a sovereign debtor threatens to stop servicing its debt, as it happened in the international debt crisis in the 1980s or in the Russian financial crisis in 1998.

If contract enforcement is weak, problems may arise on both ends of a business transaction: the seller may fail to deliver the goods, and the buyer may fail to pay for them. If buyers have no cash to pay, and thus face liquidity constraints at the time of delivery, the business transaction can take place only if the seller can trust the buyer to pay in due course. On the other hand, the buyer is willing to engage in a business transaction only if she can trust the seller to deliver the right goods. Both problems are prevalent in international trade. Enforcing the payment of goods can pose serious problems. In the aftermath of the debt crisis, highly indebted countries were liquidity constrained and could not finance necessary imports. Given their level of indebtedness, debt repayment could not be relied on. The debtor country could create more liquidity by not repaying its debt rather than by receiving a new loan. There are also important problems arising on the seller's side. In international trade, the most conspicuous example is the technology transfer problem. It is often reported that explicit contracts cannot be relied on to make sure that developing countries receive the advanced technology promised (Parsons, 1987; Kogut, 1986). These countries often complain that firms from industrialized countries sell inferior technology to them, technology that is outdated and cannot be sold on Western markets.

Solving the creditworthiness problem

Countertrade can be interpreted as the institutional response to the lack of creditworthiness of countries and firms. Countertrade introduces a deal-specific collateral for the credit granted for the original import. This collateral protects the interests of the creditor for one particular business transaction and thus mitigates the contractual hazards associated with indebtedness that would otherwise prevent the transaction from taking place.

The argument that payments in kind may have advantages over payments in cash contradicts the conventional wisdom in the theory of money. The common view is that barter is inefficient because it does not overcome the 'double coincidence of wants problem' (where each trading partner wants to buy exactly what the other partner wants to sell and vice versa) as money does. A seller may need to accept goods for which she has no use herself. The point, however, is that goods have superior credit enforcement properties to those of money. Money is an anonymous medium of exchange. This anonymity can prove disadvantageous in trade with countries which lack creditworthiness, since the debtor in the developing country or eastern Europe can use it for purposes other than repaying debt. Goods, by contrast, can more easily be earmarked as the property of the creditor and can thus serve as collateral. However, payment in goods is problematic if it is difficult to judge the quality of the goods offered as means of payment. Thus, it is important to choose goods that are very liquid and hardly anonymous, making it both easy to determine their value and easy to earmark. Goods can be ranked with respect to their liquidity and anonymity properties, providing an explanation for the export pattern of countertrade and barter (Marin and Schnitzer, 2002b).

Solving the technology transfer problem

Buyback contracts have been interpreted as incentive contracts that ensure the transfer of desirable quality technology and post-installation service performance if standard forms of internalization, like joint ventures or foreign direct investment, are not possible due to political and ownership constraints (Hennart, 1989; Chan and Hoy, 1991; Mirus and Yeung, 1993). But for the argument to work, it is essential that there be a technological relation between the two goods to be traded. However, buyback accounts for a surprisingly small fraction of all countertrade transactions. Thus, even though this explanation is theoretically appealing, it cannot explain the great majority of technology imports, which take the form of counterpurchase.

Interestingly, the technology transfer problem can be solved with a simple counterpurchase transaction as well, despite the lack of a technological link (Marin and Schnitzer, 1995). Although the lack of liquidity makes it difficult to finance imports, it is this very lack of liquidity that can actually help when it comes to dealing with problems on the supplier's side. The idea is that the export from the developing country serves as a hostage that deters cheating on technology quality and defaulting on the payment of the original import from the industrialized country. For this mechanism to work, the export has to be profitable to both the industrialized country firm and developing country, and the contract is so designed that the export becomes sufficiently less profitable for either party that does not fulfil its obligations in the original import, be it technology transfer or payment of the import. The technology seller offers high-quality technology because otherwise she loses her collateral for the credit as the developing country firm would lack the revenues that are generated with the technology and that are necessary to produce the export goods. This contractual arrangement makes the technology supplier internalize the externality her technology imposes on the developing country. The developing country party will deliver the export goods because the terms of the contract are designed such that this is more profitable than selling them otherwise. So although the import and the export are not technologically related, the countertrade contract establishes a financial link that improves the incentives of the parties involved. Thus, countertrade is a first-best substitute for foreign direct investment when these countries are reluctant to give access to foreign ownership in their markets. This goes to prove that, in an imperfect world in which contract enforcement is weak, as in developing countries or imperfect capital markets, something that seems to be worse – that contractual problems arise on both sides of the business transaction rather than on only one side – can improve contract enforcement. In international trade, the liquidity constraint helps to solve the technology transfer problem.

Other explanations

Some other possible explanations of countertrade are that developing countries use countertrade transactions to promote the export of 'new' goods – goods they have not previously exported to industrialized countries – in order to gain access to new markets and to diversify their exports (OECD, 1981; 1985). The empirical evidence gives some support for the view that countertrade has helped to stimulate and diversify exports. Other studies confirm that the goods exported by developing countries through countertrade arrangements are often goods for which export markets have yet to be established. Readily marketable products, like raw materials, are usually not available for countertrade. It can also be observed that a country removes goods from the countertrade shopping list once it has gained some experience with exporting these particular goods (Banks, 1983). Furthermore, it has been argued that countertrade corrects distortions in non-competitive markets (Caves, 1974). Using barter may allow competing more aggressively without openly violating collusive agreements. It may also allow more effective price discrimination. There is indeed some evidence that barter is used as a vehicle to change the terms of trade to allow price discrimination by Western monopolists (Caves and Marin, 1992). Mandated countertrade has also been discussed as a policy response to contracting failures arising from asymmetric information about goods valuations (Ellingson and Stole, 1996).

Barter trade in transition economies

Barter trade has received renewed attention in the 1990s, when it became a dominant phenomenon in domestic trade in a number of transition countries, most notably in the successor states of the former Soviet Union. After 1989, domestic barter in Russia increased manifold after macroeconomic stabilization in 1994, from five per cent of GDP to 60 per cent in 1998. In Ukraine, the share of barter in industrial sales is estimated to have been more than 50 per cent in 1997. Only since the financial crisis in August 1998 have barter and the use of other money surrogates started to decline again.

A number of different explanations have been put forward for this phenomenon. Some experts have viewed it as a tax-avoidance mechanism because it allows a distortion of the true value of profits, and thus reduces tax liabilities. Furthermore, since the banking sector acts as a tax collection agency that transfers firms' cash income in bank accounts to the state to pay for outstanding tax arrears, barter allows tax avoidance because it avoids payments in cash. While there may be some truth in this kind of argument, few firms report tax advantages as a major reason for using barter (Marin and Schnitzer, 2002a).

A more popular explanation refers to soft budget constraints and the lack of market discipline. The absence of hard budget constraints, so the argument goes, leads managers and workers to avoid the costs arising from restructuring by maintaining production in inefficient activities. Barter would allow concealing the true market value of output. But the empirical evidence suggests that barter is not a phenomenon of state-owned enterprises. Newly established private firms display an exposure to barter that is similar to or greater than that of state-owned firms or cooperatives (Marin and Schnitzer, 2002a).

The 'virtual economy' argument of Gaddy and Ickes (1998) has been one of the most influential explanations of barter in Russia. The virtual economy argument claims that barter helps to create the image that the manufacturing sector in Russia is producing value while in fact it is not. This argument rests on the assumption that the manufacturing sector is value-subtracting, and most participants in the economy have an interest to pretend that it is not. Barter allows the parties to keep up this illusion by allowing the manufacturing sector to sell its output at a higher price than its market value and the value-adding natural resource sector (Gazprom) to accept this high price because of a lack of other sources. This way the manufacturing sector survives by drawing resources from the natural resource sector. According to the argument, keeping up the illusion of a value-adding manufacturing sector is highly costly for the Russian economy at large because this cross-subsidizing from the value-adding natural resource sector to the value-subtracting manufacturing sector prevents the manufacturing sector from moving into valuable activity.

This argument appeals to experts of central planning and policy observers in transition economies, because the practice of cross-subsidizing across different activities in the economy was a widespread feature of central planning. But it raises a number of questions. If the natural resource sector is producing valuable output, why does the sector not have other opportunities than to subsidize the manufacturing sector? In fact, the natural resource sector is supposed to have significant bargaining power in the interaction with other sectors when it is producing goods which the market values highly. Why then does the sector end up subsidizing the rest of the economy? And in fact, evidence from barter transactions in the Ukraine suggests that, in contrast to the assertions of the virtual economy proponents, the electricity and gas industries in the natural resource sector gained from barter transactions, instead of losing (Marin, 2002).

A more plausible explanation refers to the similarities between barter in international trade and barter in transition economies, and links the surge of domestic barter to a 'lack of trust' problem (Marin and Schnitzer, 2005). In transition countries, poorly developed legal and financial institutions made contract enforcement unreliable and imposed severe transaction costs on any economic activity. These costs became prohibitively large in times of historic change and revolution. Unstable business partner relationships and rapidly changing social norms limited the extent to which economic exchanges could be sustained by reputation, by repeated interactions or by embedding them in social networks. This led to a lack of trust, meaning that reliable input supplies on the one hand and credit enforcement on the other hand were difficult to sustain, resulting in economic disorganization and a tremendous output fall. In such an environment, barter can be used as a commitment device to overcome the problems of unreliable input supplies and credit enforcement, by linking transactions and specifying terms of trade that give the right incentives to adhere to the terms of the barter contract.

DALIA MARIN AND MONIKA SCHNITZER

See also **barter; international trade theory; planning; third world debt; transfer of technology.**

Bibliography

Banks, G. 1983. The economics and politics of countertrade. *World Economy* 6, 159–82.

Caves, R. 1974. The economics of reciprocity: theory and evidence on bilateral trading arrangements. In *International Trade and Finance: Essays in Honour of Jan Tinbergen*, ed. W. Sellekaerts. London: Macmillan.

Caves, R. and Marin, D. 1992. Countertrade transactions: theory and evidence. *Economic Journal* 102, 1171–83.

Chan, R. and Hoy, M. 1991. East–West joint ventures and buyback contracts. *Journal of International Economics* 30, 331–43.

Ellingson, T. and Stole, L.A. 1996. Mandated countertrade as a strategic commitment. *Journal of International Economics* 40, 67–84.

Gaddy, C.G. and Ickes, B.W. 1998. Russia's virtual economy. *Foreign Affairs* 77, 53–67.

Hennart, J.-F. 1989. The transaction-cost rationale for countertrade. *Journal of Law, Economics and Organization* 5, 127–53.

Kogut, B. 1986. On designing contracts to enforce contractibility: theory and evidence from East–West trade. *Journal of International Business Studies* 17, 47–61.

Marin, D. 2002. Trust versus illusion: what is driving demonetization in Russia? *Economics of Transition* 10, 173–200.

Marin, D. and Schnitzer, M. 1995. Tying trade flows: a theory of countertrade with evidence. *American Economic Review* 85, 1047–64.

Marin, D. and Schnitzer, M. 2002a. *Contracts in Trade and Transition: The Resurgence of Barter*. Cambridge: MIT Press.

Marin, D. and Schnitzer, M. 2002b. The economic institution of international barter. *Economic Journal* 112, 293–316.

Marin, D. and Schnitzer, M. 2005. Disorganization and financial collapse. *European Economic Review* 47, 387–408.

Mirus, R. and Yeung, B. 1987. Countertrade and foreign exchange shortages: a preliminary assessment. *Weltwirtschaftliches Archiv* 123, 535–44.

Mirus, R. and Yeung, B. 1993. Why countertrade? An economic perspective. *International Trade Journal* 7, 409–33.

OECD (Organization for Economic Cooperation and Development). 1981. *East–West Trade: Recent Developments in Countertrade*. Paris: OECD.

OECD. 1985. *Countertrade: Developing Countries Practices*. Paris: OECD.

Parsons, J.E. 1987. Forms of GDR economic cooperation with the nonsocialist countries. *Comparative Economic Studies* 29, 7–18.

countervailing power

'Countervailing power' is a term coined by J.K. Galbraith (1952) to describe the ability of large buyers in concentrated downstream markets to extract price concessions from suppliers. Galbraith saw countervailing power as an important force offsetting suppliers' increased market power arising from the general trend of increased concentration in US industries. He provided examples such as a nationwide grocery chain extracting wholesale price discounts from food producers, and large auto manufacturers extracting price discounts from steel producers.

The concept of countervailing power was controversial in Galbraith's day (see Stigler's, 1954, criticism), and continues to be so today. Formalizing the concept is difficult because it is difficult to model bilateral monopoly or oligopoly, and there exists no single canonical model. Whether and how wholesale discounts to large downstream firms are passed through to final-good consumers is unclear. The concept has the controversial antitrust implication that horizontal mergers between downstream firms may be pro-competitive.

There are a number of theories explaining why large buyers obtain price discounts from sellers. A simple theory is that the cost of serving large buyers is lower per unit than that of serving small buyers. Serving large buyers may involve lower distribution costs. For example, the supplier may be able to ship its product to a large buyer's central warehouse rather than having to ship it to the individual retail outlets owned by small buyers. Serving large buyers may also involve lower production costs. For example, if the supplier's production function exhibits increasing returns to scale and the supplier serves one buyer at a time each production period, per-unit production costs will be lower when serving a large buyer.

Other theories involve more subtle strategic effects. A literature including Horn and Wolinsky (1986), Stole and Zwiebel (1996), Chipty and Snyder (1999), Inderst and Wey (2003) and Raskovich (2003) considers a model in which a monopoly supplier bargains under symmetric information separately and simultaneously with each of a number of buyers. Each buyer regards itself as marginal, conjecturing that all other buyers consummate their negotiations with the supplier efficiently. If aggregate surplus across all negotiations is concave in quantity, the marginal surplus from a transaction involving a large quantity is higher per unit than that from one involving a small quantity. This higher per-unit marginal surplus for large buyers translates into a lower per-unit price. The aggregate surplus function would be concave, for example, if the supplier has increasing marginal production costs. Even if the supplier's cost function were linear, the total surplus function effectively becomes concave if the supplier is assumed to be risk averse, as in Chae and Heidhues (2004) and DeGraba (2005).

Size discounts also emerge if large buyers' outside options are better. In Katz (1987) and Sheffman and Spiller (1992), for example, the larger the buyer, the more credible are its threat of integrating backward and producing the good itself. Size discounts also emerge if the supplier's outside option is worse when facing a large buyer. In Inderst and Wey (2007), for example, if bargaining with a large buyer breaks down, it is difficult for the supplier to unload this large quantity on the other buyers since this involves marching down these other buyers' declining marginal surplus functions.

Size discounts also emerge if one departs from the bargaining model with a monopoly supplier and instead considers competing suppliers. In Snyder (1998), collusion is difficult to sustain in the presence of a larger buyer because the benefit from undercutting and supplying the buyer is greater. To prevent undercutting in equilibrium, suppliers collude on a lower price for large buyers. In Dana (2004) and Inderst and Shaffer (2007), by pooling their demands and buying as a group from one supplier, buyers can increase the intensity of competition among suppliers of differentiated products.

Several papers have begun to examine the question of whether a downstream firm's countervailing power translates into lower final-good prices, using a model with competing downstream firms (Dobson and Waterson, 1997; von Ungern-Sternberg, 1996; Chen, 2003). This work suggests that an increase in countervailing power can have the opposite effect, raising consumer prices and/or lowering social welfare.

Early empirical studies of countervailing power (see Scherer and Ross, 1990, for a survey) took the standard structure–conduct–performance regressions (regressions of supplier profits or markups on supplier concentration using cross-sectional observations at the industry level) and added a buyer-concentration variable, often finding a significantly negative coefficient. Later intra-industry studies found more nuanced circumstances under which buyer-size discounts emerge. Ellison and Snyder (2002) and Sorensen (2003) observed size discounts in pharmaceutical and hospital-services markets only if there were competing, not monopoly, suppliers. In an experimental study, Normann, Ruffle and Snyder (2007) observed buyer-size discounts only when the total surplus function exhibited a certain curvature, consistent with theory.

CHRISTOPHER M. SNYDER

See also **bargaining; Galbraith, John Kenneth; monopsony; price discrimination (theory).**

Bibliography

Chae, S. and Heidhues, P. 2004. Buyers' alliances for bargaining power. *Journal of Economics and Management Strategy* 13, 731–54.

Chen, Z. 2003. Dominant retailers and the countervailing power hypothesis. *RAND Journal of Economics* 34, 612–25.

Chipty, T. and Snyder, C.M. 1999. The role of firm size in bilateral bargaining: a study of the cable television industry. *Review of Economics and Statistics* 81, 326–40.

Dana, J. 2004. Buyer groups as strategic commitments. Mimeo, Northwestern University.

DeGraba, P. 2005. Quantity discounts from risk averse sellers. Working Paper No. 276, U.S. Federal Trade Commission.

Dobson, P.W. and Waterson, M. 1997. Countervailing power and consumer prices. *Economic Journal* 107, 418–30.

Ellison, S.F. and Snyder, C.M. 2002. Countervailing power in wholesale pharmaceuticals. Mimeo, MIT.

Galbraith, J.K. 1952. *American Capitalism: The Concept of Countervailing Power.* Boston: Houghton Mifflin.

Horn, H. and Wolinsky, A. 1986. Bilateral monopolies and incentive for merger. *RAND Journal of Economics* 19, 408–19.

Inderst, R. and Shaffer, G. 2007. Retail mergers, buyer power, and product variety. *Economic Journal* 117, 45–67.

Inderst, R. and Wey, C. 2003. Bargaining, mergers, and technology choice in bilaterally oligopolistic industries. *RAND Journal of Economics* 34, 1–19.

Inderst, R. and Wey, C. 2007. Buyer power and supplier incentives. *European Economic Review* 51, 647–67.

Katz, M.L. 1987. The welfare effects of third degree price discrimination in intermediate goods markets. *American Economic Review* 77, 154–67.

Normann, H.-T., Ruffle, B.J. and Snyder, C.M. 2007. Do buyer-size discounts depend on the curvature of the surplus function? Experimental tests of bargaining models. *RAND Journal of Economics*.

Raskovich, A. 2003. Pivotal buyers and bargaining position. *Journal of Industrial Economics* 51, 405–26.

Scherer, F.M. and Ross, D. 1990. *Industrial Market Structure and Economic Performance.* Boston: Houghton Mifflin.

Sheffman, D.T. and Spiller, P.T. 1992. Buyers' strategies, entry barriers, and competition. *Economic Inquiry* 30, 418–36.

Snyder, C.M. 1998. Why do large buyers pay lower prices? Intense supplier competition. *Economics Letters* 58, 205–9.

Sorensen, A. 2003. Insurer-hospital bargaining: negotiated discounts in post-deregulation Connecticut. *Journal of Industrial Economics* 51, 471–92.

Stigler, G.J. 1954. The economist plays with blocs. *American Economic Review* 44, 7–14.

Stole, L.A. and Zwiebel, J. 1996. Organizational design and technology choice under intrafirm bargaining. *American Economic Review* 86, 88–102.

von Ungern-Sternberg, T. 1996. Countervailing power revisited. *International Journal of Industrial Organization* 14, 507–20.

Courcelle-Seneuil, Jean Gustave (1813–1892)

French economist and economic adviser. Born in the Dordogne, he studied law in Paris, then returned to his native region to manage an industrial firm. At the same time, during the July monarchy, he wrote for Republican newspapers and economic periodicals. After the 1848 revolution, he held briefly a high position in the Ministry of Finance. In the following years he became a frequent contributor to the *Journal des économistes*, and published a successful textbook on banking in 1852. In 1853, the Chilean government contracted him to teach economics at the University of Chile in Santiago, and to be available as official economic adviser; he stayed for ten years, until 1863, when he returned to France. While in Chile he published his most ambitious work in economics, the *Traité théorique et pratique d'économie politique* (1858), which the Chilean government arranged to bring out in a Spanish translation. After his return to France, he resumed his activity as prolific writer of books and articles on economic affairs. He also published several works on political and historical topics and translated into French John Stuart Mill's *Principles of Political Economy*, Summer Maine's *Ancient Law* and William Graham Sumner's *What Social Classes Owe to Each Other*. He was appointed councillor of state in 1879, and three years later was elected member of the Académie des Sciences Morales et Politiques.

Throughout his life, Courcelle-Seneuil was a stalwart defender of free trade and laissez-faire. Charles Gide, the co-author (with Charles Rist) of a well-known history of economic doctrines, wrote about him in rather sarcastic terms:

> He was virtually the *pontifex maximus* of the classical school; the holy doctrines were entrusted to him and it was his vocation to denounce and exterminate the heretics. During many years he fulfilled this mission through book reviews in the *Journal des économistes* with priestly dignity. Argus-eyed, he knew how to detect the slightest deviations from the liberal school. (Gide, 1895, p. 710)

Courcelle-Seneuil's special interest, starting with the publication of a small book on bank reform in 1840, was the introduction of more freedom into banking or, to use a modern term, the 'deregulation' of this industry. Above all, he advocated the abolition of the Bank of France's exclusive right of issue. According to Gide, Courcelle-Seneuil was more esteemed in England and the United States than in France. In any event, adoption of his monetary and banking proposals was never seriously considered in his own country.

Once in Chile, Courceile-Seneuil became a powerful policymaker and influential teacher. He arrived at a time when the international prestige of the laissez-faire doctrine was at its height and when gold booms and subsequent busts in California and Australia caused considerable fluctuations in Chile's agricultural exports to these areas, creating a need for flexible short- and long-term credit facilities. This combination of events, joined with the prestige emanating from the foreign savant, permitted him to obtain in Chile what he had failed to achieve in his own country: under his guidance, the administration of Manuel Montt (1851–61) promulgated a banking law that established total freedom for any solvent person to found a bank and permitted all banks to issue currency subject only to one limitation: the banknotes in circulation were not to exceed 150 per cent of the issuing bank's capital.

Courcelle-Seneuil's advice was also sought in connection with a new customs tariff and here again he achieved substantial change: the level of protection was severely cut back, although some tariffs were retained for revenue purposes.

But the principal influence exercised by Courcelle-Seneuil resided in his forceful teaching: as the University of Chile's first professor of economics, he was apparently successful in instilling doctrinaire zeal in his students, some of whom later became influential policymakers. Thus, Chilean historians have not only traced the abandonment of convertibility in 1878 to the permissiveness of the 1860 Banking Law and the lack of industrial development to the 1864 tariff; they also see Courcelle-Seneuil's indirect influence in the acquisition of the nitrate mines of Tarapacá by private foreign interests after Chile's victory over Peru in the War of the Pacific (1882) had given it title to the mines. Alienation of the mines was indeed recommended by a government committee dominated by Courcelle-Seneuil's disciples, who felt, like their teacher, that state ownership and management of business enterprises was to be strictly shunned. Secular inflation, industrial backwardness, domination of the country's principal natural resources by foreigners – all of these protracted ills of the Chilean economy have been attributed to the French expert.

Since the economically advanced countries were also those where economic science first flourished, they soon produced a peculiar export product: the foreign economic expert or adviser. Courcelle-Seneuil is probably the earliest prototype of the genre and his ironic career in Chile exhibits characteristics that were to remain typical of numerous later representatives. First, the adviser is deeply convinced that, thanks to the advances of economic science, he knows the correct solutions to economic problems no matter they may arise. Secondly, the country which invites the expert looks forward to his advice as to some magic medicine which will work even when (perhaps especially when) it hurts. Some countries seem particularly prone to this attitude. In Chile foreign or foreign-trained experts have played key roles at crisis junctures, from Courcelle-Seneuil in the mid-19th century to Edwin Kemmerer in the 1920s, the Klein–Saks Mission in the 1950s, and finally to the 'Chicago boys' in the 1970s. Thirdly, the influence of the adviser derives not only from the intrinsic value and persuasiveness of his message, but from the fact that he usually has good connections in his home country and can therefore facilitate access to its capital market. Courcelle-Seneuil, for example, suspended his university courses in 1858–9 to accompany a Chilean financial mission that travelled to France in search of a railroad construction loan. Fourthly, the foreign adviser is often criticized for wishing to transplant the institutions of his own country to the country he advises, but his real ambition is more extravagant: it is to endow the country with those ideal institutions which exist in his mind only, for he has been unable to persuade his own countrymen to adopt them. Fifthly, history in general, and nationalist historiography in particular, is likely to be unkind to the foreign adviser. In retrospect he can easily become a universal scapegoat: whatever went wrong is attributed to his nefarious influence. This demonization is more damaging than the adviser himself could possible have been: it forestalls authentic learning from past experience.

ALBERT O. HIRSCHMAN

Selected works

1840. *Le crédit et la Banque*. Paris.
1858. *Traité théorique et pratique d'économie politique*, 2 vols. Paris: Amyot.
1867. *La Banque libre*. Paris: Guillaumin.

Bibliography

Encina, F. 1951. *Historia de Chile*, vol. 18, ch. 58. Santiago: Nascimiento.
Fuentealba, H.L. 1946. *Courcelle-Seneuil en Chile: errores del liberalismo económico*. Santiago: Prensas de la Universidad de Chile.
Gide, C. 1895. Die neuere volkswirtschaftliche Litteratur Frankreichs. *Schmollers Jahrbuch*.
Hirschman, A.O. 1963. *Journeys toward Progress*. New York: Twentieth Century Fund, 163–8.
Journal des économistes. 1892. Obituary [of M.J.G. Courcelle-Seneuil]. July.
Juglar, C. 1895. Notice sur la vie et les travaux de M.J.G. Courcelle-Seneuil. Académie des Sciences Morales et Politiques, *Compte Rendu*, 850–82.
Pinto, S.C. 1959. *Chile, un caso de desarrollo frustrado*. Santiago: Edit. Universitaria.
Will, R.M. 1964. The introduction of classical economics into Chile. *Hispanic-American Historical Review* 44(1), 1–21.

Cournot, Antoine Augustin (1801–1877)

Cournot was born at Gray (Haute-Saône) on 28 August 1801 and died in Paris on 30 March 1877. Until the age of 15 his education was at Gray. After studying at Besançon he was admitted to the Ecole Normale Supérieure in Paris in 1821. In 1823 he obtained his licentiate in sciences and in October of that year was employed by Marshal Gouvion-Saint-Cyr as literary adviser to the Marshal and tutor to his son. In 1829 he obtained his doctorate in science with a main thesis in mechanics and a secondary one in astronomy. Through the sponsorship of Poisson in 1834 he obtained the professorship in analysis and mechanics at Lyon.

After a year of teaching he became primarily involved in university administration. In 1835 he became rector of the Académie de Grenoble and subsequently became inspector general of education and from 1854 to 1862 was rector of the Académie de Dijon. He became a Knight of the Legion of Honour in 1838 and an Officer in 1845. He was afflicted with failing eyesight and in the last part of his life was nearly blind. In 1862 he retired from public life but continued his own researches in Paris until his death.

Cournot was a prolific writer. His writings can be broadly divided into three categories: (1) mathematics;

(2) economics and (3) the philosophy of science and philosophy of history.

In considering Cournot as an economist it is necessary to place his major economic work, *Recherches sur les principes mathématiques de la théorie des richesses* (1838) in the context not only of *Principes de la théorie des richesses* (1863), which can be regarded as a literary version of his work of a quarter of a century earlier, and his *Revue sommaire des doctrines économiques* (1877) which appeared in the last year of his life, but also of his writings on probability and the philosophy of science, in particular *Exposition de la théorie des chances et des probabilités* (1843) and *Matérialisme, vitalisme, rationalisme: Etudes des données de la science en philosophie* (1875).

It is possible to weave a broad cloth of interpretation taking into account not merely Cournot's other works but what appears to be known of his personality and the considerable social and political flux in France during the times in which he lived. Guitton (1968) has suggested that Cournot had a rather melancholic and solitary temperament and 'did nothing to make his books attractive'. He notes that: 'Cournot was a pioneer. He did nothing to court his contemporaries, and they, in turn, not only failed to appreciate him but ignored him.' Palomba ([1981] 1984) provides a sketch of the historical background of his time, noting the growth of socialist ideas in Europe, the political actions and reactions to the French Revolution and the challenges to the concept of ownership. Rather than challenge or repeat the broad contextual interpretation of Cournot provided by Palomba, this article is confined primarily to the direct interpretation of his works in economics and supporting texts in the light of many of the developments in economics which are consistent with and may be indebted to his original ideas.

The texts followed here include the French given in the complete works of Cournot (1973) and the Nathaniel T. Bacon translation (1899) entitled *Researches into the Mathematical Principles of the Theory of Wealth*, which also contains an essay by Irving Fisher on Cournot and Mathematical Economics as well as a bibliography on Mathematical Economics from 1711 to 1897. The 1929 reprint of the 1897 edition was used.

The preface sets forth with great clarity Cournot's fundamental approach to political economy. He states:

> But the title of this work sets forth not only theoretical researches; it shows also that I intend to apply to them the forms and symbols of mathematical analysis. Most authors who have devoted themselves to political economy seem also to have had a wrong idea of the nature of the applications of mathematical analysis to the theory of wealth.

> But those skilled in mathematical analysis know that its object is not simply to calculate numbers, but that it is also employed to find the relations between magnitudes which cannot be expressed in numbers and between functions whose law is not capable of algebraic

expression. Thus the theory of probabilities furnishes a demonstration of very important propositions, although without the help of experience it is impossible to give numerical values for contingent events, except in questions of mere curiosity, such as arise from certain games of chance. (p. 3)

Cournot continues in the preface to note that only the first principles of differential and integral calculus are required for his treatise. Professional mathematicians could be interested in it for the questions raised rather than the level of mathematics presented. He ends the preface with the caveat:

I am far from having thought of writing in support of any system, and from joining the banners of any party; I believe that there is an immense step in passing from theory to governmental applications; I believe that theory loses none of its value in thus remaining preserved from contact with impassioned polemics; and I believe, if this essay is of any practical value, it will be chiefly in making clear how far we are from being able to solve, with full knowledge of the case, a multitude of questions which are boldly decided every day. (p. 5)

The first chapter, 'Of Value in Exchange or of Wealth in General', provides insight into the breadth of Cournot's concern for the social and historical context of wealth.

Property, power, the distinctions between masters, servants and slaves, abundance, and poverty, rights and privileges, all these are found among the most savage tribes, and seem to flow necessarily from the natural laws which preside over aggregations of individuals and of families; but such an idea of wealth as we draw from our advanced state of civilization, and such as is necessary to give rise to a theory, can only be slowly developed as a consequence of the progress of commercial relations, and of the gradual reaction of those relations on civil institutions. (pp. 7–8)

He notes that: 'it is a long step to the abstract idea of *value in exchange* which supposes that the objects to which such value is attributed *are in commercial circulation*.'

In order to illustrate the distinction between the word *wealth* in ordinary speech and value in exchange, he presents an example of a publisher who destroys two-thirds of his stock expecting to derive more profit from the remainder than the entire edition. The economics of elasticity is developed more formally in Chapter 4 on demand, but the concept is clear.

Chapter 2, 'On Changes in Value, Absolute and Relative', begins by noting that 'we can only assign value to a commodity by reference to other commodities'. This leads to a discussion of the use of a corrected money which would serve as 'the equivalent of the mean sun of the astronomers'.

Chapter 3, 'Of the Exchanges', is the first in which formal mathematical manipulation is employed. He considers a silver standard in which all currencies are fixed in ratio to a gram of fine silver. He observes that the ratios of exchange for the same weight of fine silver cannot differ by more than transportation and smuggling costs. Given the volume of trade measured in silver he considers the arbitrage conditions for the $m(m-1)/2$ ratios among m centres. Fisher (1892) notes, however, that Cournot did not appear to be acquainted with determinants as he did not attempt a general solution of the exchange equations he proposed, but limited his calculations to three centres of exchange.

It is in Chapter 4, 'On the Law of Demand', that the modernity of his approach stands out. He is interested in demand as it is revealed in sales at a given price. He represents the relationship between sales and price by the continuous function $D = F(p)$ and observes that this function generally increases in size with a fall in price and that the empirical problem is to determine the form of $F(p)$. He indicates an appreciation of the concept of elasticity of demand although he did not develop the formal measure.

Chapters 5 and 6 deal with monopoly without and with taxation; Chapter 7 is on the competition of producers and Chapter 8 on unlimited competition. The ninth chapter is on the mutual relations of producers and the tenth on the communication of markets. The final two chapters are somewhat macroeconomic in scope. Chapter 11 is entitled 'Of the Social Income' and 12 'Of Variations in the Social Income, Resulting from the Communication of Markets'.

As our commentary is primarily on Chapters 5–8, the order is reversed and 11 and 12 are dealt with first. Cournot explicitly avoids setting up the whole closed microeconomic system.

It seems, therefore, as if, for a complete and rigorous solution of the problems relative to some parts of the economic system, it were indispensable to take the entire system into consideration. But this would surpass the powers of mathematical analysis and of our practical methods of calculation, even if the values of all the constants could be assigned to them numerically. The object of this chapter and of the following one is to show how far it is possible to avoid this difficulty, while maintaining a certain kind of approximation, and to carry on, by the aid of mathematical symbols, a useful analysis of the most general questions which this subject brings up.

We will denote by *social income* the sum, not only of incomes properly so called, which belong to members of society in their quality of real estate owners or capitalists, but also the wages and annual profits which come to them in their capacity of workers and industrial agents. We will also include in it the annual amount of the stipends by means of which individuals or the state sustain those classes of men which economic writers have characterized as unproductive, because the product of their labour is not anything material or saleable. (pp. 127–8)

But, using a first order approximation, he studies the effect of a change in price and consumption of a good on social income as a whole under competition, under monopoly and when a new product is introduced.

Finally, although we make continuous and almost exclusive use of the word *commodity*, it must not be lost sight of (Article 8) that in this work we assimilate to commodities the rendering of services which have for their object the satisfaction of wants or the procuring of enjoyment. Thus when we say that funds are diverted from the demand for commodity A to be applied to the demand for commodity B, it may be meant by this expression that the funds diverted from the demand for a commodity properly so called, are employed to pay for services or vice versa. When the population of a great city loses its taste for taverns and takes up that for theatrical representations, the funds which were used in the demand for alcoholic beverages go to pay actors, authors, and musicians, whose annual income, according to our definition, appears on the balance sheet of the social income, as well as the rent of the vineyard owner, the vine-dresser's wages, and the tavern-keeper's profits. (p. 149)

The last chapter considers international trade and national income and uses a first order approximation rather than a closed equilibrium system to study the benefits of opening up trade.

Moreover (and this is the favourite argument of writers of the school of Adam Smith), it should be inferred from the asserted advantage assigned to the exporting market, and the asserted disadvantage suffered by the importing market, that a nation should so arrange as always to export and never to import, which is evidently absurd, as it can only export on condition of importing, and even the sum of the values exported, calculated at the moment of leaving the national market, must necessarily be equal to the sum of the values imported, calculated at the moment of arrival on the national market. (p. 161)

Cournot also notes the problem of analysing a tariff war:

The question would no longer be the same if establishment of a barrier for the benefit of A producers might provoke, by way of retaliation, the establishment of another barrier for the benefit of B producers, against whom the first barrier was raised. The government of A would then have to weigh the advantage resulting from the first measure to the citizens of A against the drawbacks caused by the retaliation. The two markets A and B would thus again be placed in symmetrical conditions, and each should be considered as acting the double part of an exporting and importing market. (p. 164)

He closes his comments with:

We have just laid a finger on the question which is at the bottom of all discussions on measures which prohibit or restrict freedom of trade. It is not enough to accurately analyse the influence of such measures on the national income; their tendency as to the distribution of the wealth of society should also be looked into. We have no intention of taking up here this delicate question, which would carry us too far away from the purely abstract discussions with which this essay has to do. If we have tried to overthrow the doctrine of Smith's school as to barriers, it was only from theoretical considerations, and not in the least to make ourselves the advocates of prohibitory and restrictive laws. Moreover, it must be recognized that such questions as that of commercial liberty are not settled either by the arguments of scientific men or even by the wisdom of statesmen. (p. 171)

He closes his work with the observation about theory that:

By giving more light on a debated point, it soothes the passions which are aroused. Systems have their fanatics, but the science which succeeds to systems never has them. Finally, even if theories relating to social organization do not guide the doings of the day, they at least throw light on the history of accomplished facts. (p. 171)

Although the contribution of these last chapters is not as great as those to which we now turn, the spirit and style is that of a major theorist concerned deeply and objectively with application to practical affairs.

In Chapters 5–9 Cournot develops his theory of monopoly, oligopoly and unlimited competition. This can be contrasted with Ricardo (1817) before and Walras (1874) after, who concentrated on unlimited competition with no aim at producing a unified theory involving numbers.

In Chapter 5 Cournot deals with monopoly, considering increasing, decreasing and constant returns and in Chapter 6 the influence of taxation on a monopoly is considered. He notes direct taxes and indirect taxes as well as bounties and their influences on both producers and consumers; and closes with an examination of two variations of taxation in kind.

Chapter 7 provides a smooth transformation from single person maximization to non-cooperative optimization where agents who mutually influence each other act without explicit cooperation.

We say *each independently*, and this restriction is very important, as will soon appear; for if they should come to an agreement so as to obtain for each the greatest possible income, the results would be entirely different, and would not differ, so far as consumers are concerned, from those obtained in treating of a monopoly.

Instead of adopting $D = F(p)$ as before, in this case it will be convenient to adopt the inverse notation

$p = f(D)$; and then the profits of proprietors (1) and (2) will be respectively expressed by

$$D_1 f(D_1 + D_2), \text{ and } D_2 f(D_1 + D_2),$$

i.e. by functions into each of which enter two variables, D_1 and D_2. (p. 80)

It is at this point that Cournot switches from price to quantity of a homogeneous product as the strategic variable used by the competitors. His words and the mathematics do not quite match. He says, 'This he will be able to accomplish by properly adjusting his price.' The first order condition for the existence of a non-cooperative equilibrium with quantity as the strategic variable is given. A diagram showing a stable equilibrium and another with a non-stable equilibrium are presented. The analysis is generalized to n producers including the possibility of an extra group of producers beyond n, all of whom produce at capacity. He obtains n symmetric equations for the firms with interior production levels and sets the others at capacity.

When he introduces n different general cost functions for the n firms he handles the situation with all having an equilibrium defined by the simultaneous satisfaction of the equations arising from the first order conditions. But he does not deal with the possibility that costs could be such that different subsets of firms could be active in different equilibria.

The criticism levelled by Bertrand (1883) in his review written well after Cournot's death concerns the modelling rather than the mathematics. As Cournot considered competition without entry among firms selling an identical product it was fairly natural to avoid the discontinuity in the payoff function caused by selecting price as an independent variable. But the observation of Bertrand matters for markets with a finite number of firms. The choice of strategic variable causes not only mathematical difficulties but raises questions concerning economic realism and relevance. Quantity, price, quality, product differentiation and scope can all be considered as playing dominant roles in different markets. But the general explanation of price and quantity as strategic variables was and is critical to the development of economic theory. Cournot provided the foundations for the understanding of quantity. Bertrand, whose review of the books of Cournot and Walras was somewhat tangential to his professional interests offered only an example rather than a developed theory of price competition. It remained for Edgeworth (1925, pp. 111–42) to explore the underlying difficulties with the payoff functions for duopoly with increasing marginal costs; and it has only been since the 1950s with the advent of the theory of games that there has been an adequate study of the properties of non-cooperative equilibria in games with price and quantity as strategic variables, without or with product differentiation.

The thesis of Nash (1951) on the existence of non-cooperative equilibria for a class of games in strategic form provided a broad general underpinning for the concept of non-cooperative equilibrium. It was then immediately observable that, although Cournot's work with equilibria of games with a continuum of strategies was not strictly covered by Nash's work, conceptually Cournot's solution could be viewed as an application of non-cooperative equilibrium theory to oligopoly (see Mayberry, Nash and Shubik, 1953). The broader investigation of the price model and the interpretation of the instability of the Edgeworth cycle in terms of mixed strategy equilibria has only taken place recently. This also includes a growing literature on how to embed both the Cournot and Bertrand–Edgeworth models into a closed economic system or Walrasian framework. A summary of much of this work is presented by Shubik (1984).

It is important to appreciate that the developments in the theory of monopolistic competition such as those of Hotelling (1929) and Chamberlin (1933) and J. Robinson (1933) were based upon the Cournot non-cooperative game model. Although it may be argued that Chamberlin's and Mrs Robinson's works possibly contained broader and richer models of competition among the few than that of Cournot, they represented a step backwards in their lack of mathematical sophistication and analysis. The Chamberlin discussion of large group equilibrium does have price as the strategic variable along with product differentiation and entry, but the solution concept is the non-cooperative equilibrium à la Cournot with the caveat that an attempt to produce a strict formal mathematical model of Chamberlin's large group equilibrium leads one to conclude that the game having price as a strategic variable is closer to Edgeworth's analysis than that of Cournot and a price strategy non-cooperative equilibrium may not exist.

In Chapter 8 Cournot shows his basic grasp of the important strategic difference between pure competition and oligopolistic competition. Using his own words, he states:

> The effects of competition have reached their limit, when each of the partial productions D_2 is *inappreciable*, not only with reference to the total production $D = F(p)$, but also with reference to the derivative $F'(p)$, so that the partial production D_k could be subtracted from D without any appreciable variation resulting in the price of the commodity. This hypothesis is the one which is realized, in social economy, for a multitude of products, and, among them, for the most important products. It introduces a great simplification into the calculations, and this chapter is meant to develop the consequences of it. (p. 90)

In modern mathematical economics, in the linking of competition among the few and the Walrasian system into a logically consistent whole, two approaches to the study of large numbers have been adopted. The first is replication and has its roots in Cournot and, more formally, Edgeworth (1881). Following Edgeworth this method was used in cooperative core theory by Shubik

(1959). The second involves considering a continuum of economic agents where each agent can be regarded as a set of measure zero. Cournot clearly saw the need to consider a market in which each individual firm is too small to influence price. But it remained for Aumann (1964) to fully formalize the concept of an economic game with a continuum of agents.

After 25 years during which his seminal work in mathematical economics was essentially ignored, Cournot demonstrated his concern for his ideas by publishing *Principes de la théorie des richesses* (1863), where he offered a non-mathematical rendition of his early work. This book is of considerably greater length than its predecessor and is divided into four books: Book 1, Les Richesses (eight chapters); Book 2, Les Monnaies (seven chapters); Book 3, Le Système économique (ten chapters) and Book 4, L'Optimisme économique (seven chapters).

This book met with no more immediate success than his original work and is not as deep. For example the chapters on money, although they contain discursive and historical material of interest, have little material of analytic depth.

In spite of the indifference of the environment to his writings in economics, Cournot regarded his contribution as sufficiently important that some 14 years later, in the year of his death, he published his *Revue sommaire des doctrines économiques* (1877). This book was also longer, non-mathematical and of less significance than the work of almost 40 years earlier. But Cournot's own sense of having been at least partially vindicated after 40-odd years is indicated in his *avant-propos*:

I was at that point in 1863, when I had the desire to find out whether I had sinned in the substance of ideas or only in their form. To that end, I went back to my work of 1838, expanding it where needed, and, most of all, removing entirely the algebraic apparatus which intimidates so much in these subjects. Whence the book entitled: 'Principes de la théorie des richesses'. 'Since it took me,' I said in the preface, 'twenty-five years to lodge an appeal of the first sentence, it goes without saying that I do not intend, whatever happens, to resort to any other means. If I lose my case a second time, I will be left only with the consolation which never abandons disgraced authors: that of thinking that the sentence that condemns them will one day be quashed in the interest of the law, that is of the truth.'

When I took this engagement in 1863, I did not think that I would live long enough to see my 1838 case reviewed as a matter of course. Nevertheless, more than thirty years later, another generation of economists, to put it like Mr. the commander Boccardo, discovered that I opened up back then, though too timidly and too partially, a good path to be followed, on which I was even somewhat preceded by a man of merit, the doctor Whewell. While another Englishman, Mr. Jevons, was undertaking to enlarge this path, a young Frenchman,

Mr. Leon Walras, professor of Political Economy at Lausanne, dared to maintain right in the Institute that it was wrong to pay so little attention to my method and my algorithm, which he used rightfully to expose a new theory, more amply developed.

Now, look at my bad luck. If I won a little late, without any involvement, my 1838 case, I lost my 1863 case. If one wanted in retrospective to make a case for my algebra, my prose (I am ashamed of saying it) did not get better success from the publisher. The *Journal des Economistes* (August 1864) criticized me mainly 'for not having moved on from Ricardo,' for not having taken into account the discoveries that so many men of merit have made in twenty-five years in the field of political economy; thus the poor author that no one of the official world of French economists wanted to quote incurred the reproach of not having quoted others enough.

Cournot was central to the founding of modern mathematical economics. The average reader tends not to be aware that the textbook presentations of the 'marginal cost equals marginal revenue' optimizing condition for monopoly and 'marginal cost equals price' for the firm in pure competition come directly from the work of Cournot (including an investigation of the second order conditions).

He had to wait many years for recognition, but when it came in the works of Jevons, Marshall, Edgeworth, Walras and others, it moved the course of economic theory. Marshall notes (*Memorials of Alfred Marshall*, pp. 412–13, letter 2, July 1900) 'I fancy I read Cournot in 1868', this was when Marshall was 26, some 30 years after the book appeared. He acknowledges him both as a great master and as his source 'as regards the form of thought' for Marshall's theory of distribution. Jevons, in his preface to the second edition of *The Theory of Political Economy* records 'I procured a copy of the work as far back as 1872' and that it 'contains a wonderful analysis of the laws of supply and demand, and of the relations of prices, production, consumption, expenses and profits'. He excuses himself for his lateness in coming to Cournot observing: 'English economists can hardly be blamed for their ignorance of Cournot's economic works when we find French writers equally bad.' Walras in the preface to the fourth edition of *Elements of Pure Economic* (Jaffé translation, p. 37) acknowledges his 'father Auguste Walras, for the fundamental principles of my economic doctrine'; and 'Augustin Cournot for the idea of using the calculus of functions in the elaboration of this doctrine'. His liberal references to Cournot include his discussion of monopoly and the description of supply and demand.

The art of formal modelling is different from but related to the use of mathematical analysis in economics. The clarity and parsimony of Cournot's modelling stand out and have served as beacons guiding the development of mathematical economics.

An important feature missing from Cournot's seminal work is the discussion of the role of chance and uncertainty in the economy. He stressed the importance of chance in both his book *Exposition de la théorie des chances et des probabilités* (1843) and in *Matérialisme, vitalisme, rationalisme* (1875).

Although economics was the only social science he attempted to mathematize, he was well aware of the simplifications being made in cutting economic analysis from the context of history and society.

> The economist considers the body social in a state of division and so to say of extreme pulverization, where all the particularities of organization and of individual life offset each other and vanish. The laws that he discovers or believes to discover are those of a mechanism, not those of a living organism. For him, it is no longer a question of social physiology, but of what is rightfully called social physics (p. 56). We mention that these cases of regression which imply abstractions of the same kind, if not of the same type and of the same value, reappear in various stages of scientific construction.

Cournot's work on chance and probability does not appear to have provided any new mathematical analysis, but he made three distinctions concerning the nature of probability. His book of 1843 was a text with the dual purpose of teaching the non-mathematician the rules of the calculus of probability and of dissipating the obscurities on the delicate subject of probability. He stressed the distinction between objective and subjective probability. His opening chapters provide a discussion of the appropriate combinatorics and frequency of occurrence interpretation of probability.

Cournot stressed the distinction between objective probability where frequencies are known and subjective probability. He noted:

> We could, since then, relying on the theorems of Jacques Bernoulli, who was already aware of their meaning and scope, pass immediately to the applications those theorems had in the sciences of facts and observations. However, a principle, first stated by the Englishman Bayes, and on which Condorcet, Laplace and their successors wanted to build the doctrine of 'a posteriori' probabilities, became the source of much ambiguity which must first be clarified, of serious mistakes which must be corrected and which are corrected as soon as one has in mind the fundamental distinction between probabilities which have an objective existence, which give a measure of the possibility of things, and subjective probabilities, relating partly to one's knowledge, partly to one's ignorance, depending on one's intelligence level and on the available data. (p. 155)

Subjective probability rests on the consideration of events which our ignorance calls for us to treat as equiprobable due to insufficient cause.

He added a third category which he entitled 'philosophical probability' (Chapter 17) 'where probabilities are not reducible to an enumeration of chances' but

'which depend mainly on the idea that we have of the simplicity of the laws of nature' (p. 440).

Cournot's views on probability appear to be intimately related to his concern for social statistics and economic modelling. Although he did not establish formal links between his mathematical economics models and chance he regarded history and the development of institutions as dependent on chance and economics as set in the context of institutions.

Cournot was at best an indifferent mathematician. Bertrand clearly dominated him in that profession. But from his own writings it is clear that Cournot was well aware of both his purpose in applying mathematics to economics and his limitations as a mathematician. At the age of 58 he wrote his *Souvenirs* which he finished in Dijon in October 1859. They were published many years later with an introduction by Botinelli (1913). In these writings Cournot provides his self-assessment as a mathematician.

> I was starting to be a little known in the academic world through a fairly large number of scientific articles. This was the basis of my fortune. Some of these articles ended up with Mr. Poisson, who was then the leader in Mathematics at the Institute, and mainly at the University, and he liked them particularly. He found in them philosophical insight, which I think was not all that wrong. Furthermore, he foresaw that I would go a long way in the field of pure mathematical speculation, which was (I always thought it and never hesitated to say it) one of his mistakes.

The general tenor of his *Souvenirs* is of a moderately conservative, quietly humourous, self-effacing man with considerable understanding of his environment and a broad belief in science and its value to society.

Regarding his work as a whole, his dedication and power as the founder of mathematical economics and the promoter of empirical numerical investigations emerges. He strove for around 40 years to have his ideas accepted. He did so with persistence and humour (referring to his major work as 'mon opuscule'). He understood the need to wait for a generation to die. And before his death with the work and words of Jevons and Walras he saw the vindication of his approach.

MARTIN SHUBIK

See also **Bertrand, Joseph Louis François.**

Selected works

1838. *Researches into the Mathematical Principles of the Theory of Wealth.* Trans. N.T. Bacon, New York: Macmillan, 1929.

1841. *Traité élémentaire de la théorie des fonctions et du calcul infinitésimal.* 2nd edn, Paris: Hachette, 1857.

1843. *Exposition de la théorie des chances et des probabilités.* Paris: Hachette.

1861. *Traité de l'enchaînement des idées fondamentales dans les sciences et dans l'histoire.* New edn, Paris: Hachette, 1911.

1863. *Principes de la théorie des richesses*. Paris: Hachette.

1872. *Considérations sur la marche des idées et des évènements dans les temps modernes*. 2 vols, Paris: Boivin, 1934.

1875. *Matérialisme, vitalisme, rationalisme: Études des données de la science en philosophie*. Paris: Hachette, 1923.

1877. *Revue sommaire des doctrines économiques*. Paris: Hachette.

1913. *Souvenirs 1760–1860*. With an introduction by E.P. Bottinelli. Paris: Hachette. Published posthumously.

1973. *A.A. Cournot Oeuvres Complètes*, 5 vols, ed. André Robinet. Paris: Librairie Philosophique J. Vrin.

Bibliography

Aumann, R.J. 1964. Markets with a continuum of traders. *Econometrica* 32, 39–50.

Bertrand, J.L.F. 1883. (Book reviews of) *Théories Mathematique de la richesse sociale* par Léon Walras; *Recherches sur les principes mathématiques de la théorie de la richesse* par Augustin Cournot. *Journal des Savants* 67, 499–508.

Chamberlin, E.H. 1933. *The Theory of Monopolistic Competition*. Cambridge, MA: Harvard University Press.

Edgeworth, F.Y. 1881. *Mathematical Psychics*. London: Kegan Paul.

Edgeworth, F.Y. 1925. *Papers Relating to Political Economy*, I. London: Macmillan.

Fisher, I. 1892. *Mathematical Investigations in the Theory of Value and Prices*. New Haven: Connecticut Academy of Arts and Sciences, *Transactions 9*. Reprinted, New York: Augustin M. Kerlley, 1961.

Guillebaud, C.W., ed. 1961. *Marshall's Principles of Economics*. Vol. 2, *Notes*. London: Macmillan.

Guitton, H. 1968. Antoine Augustin Cournot. In *The International Encyclopedia of the Social Sciences*, vol. 3. New York: Macmillan and Free Press.

Hotelling, H. 1929. Stability in competition. *Economic Journal* 34, 41–57.

Jevons, W.S. 1911. *The Theory of Political Economy*. 4th edn, London: Macmillan, 1931.

Mayberry, J., Nash, J.F. and Shubik, M. 1953. A comparison of treatments of a duopoly situation. *Econometrica* 21, 141–55.

Nash, J.F., Jr. 1951. Noncooperative games. *Annals of Mathematics* 54, 289–95.

Palomba, G. 1984. Introduction à l'oeuvre de Cournot. *Economie Appliquée* 37, 7–97. Trans. from Italian, extracted from *Cournot Opere*, Turin: UTET (1981).

Ricardo, D. 1817. *The Principles of Political Economy and Taxation*. London: J.M. Dent, 1965.

Robinson, J. 1933. *The Economics of Imperfect Competition*. London: Macmillan.

Shubik, M. 1959. Edgeworth market games. In *Contributions to the Theory of Games IV*, ed. A.W. Tucker and R.D. Luce. Princeton, NJ: Princeton University Press.

Shubik, M. 1984. *A Game Theoretic Approach to Political Economy*. Cambridge, MA: MIT Press.

Walras, L. 1874–7. *Elements of Pure Economics*. Trans. W. Jaffée, London: George Allen & Unwin, 1954.

Cournot competition

The classic Cournot model is static in nature, with each (single-product) firm's strategy being the quantity of output it will produce in the market for a specific homogeneous good; as Kreps (1987) observed, Cournot's model was an early progenitor of Nash's famous paper. Many recent applications have involved multi-stage games; for example, each of n firms might first simultaneously choose investment levels (say, in cost-reducing R&D) and then simultaneously choose output levels in the second stage. Often now used in such a manner, we will see that the Cournot model is doing well, contributing to a range of new research, as it moves towards the two-century mark.

1 The basic one-stage model and associated concepts

Consider an industry comprised of n firms, each firm choosing an amount of output to produce. Firm i's *output level* is denoted as q_i, $i = 1, \ldots, n$; let the vector of firm outputs be denoted $\boldsymbol{q} \equiv (q_1, q_2, \ldots, q_n)$. The firms' products are assumed to be perfect substitutes (the *homogeneous-goods* case); let Q denote the aggregate industry output level (that is, $Q \equiv \sum_{i=1}^{n} q_i$). We will refer to the $(n-1)$ vector of output levels chosen by firm i's rivals as $\boldsymbol{q}_{-i}$; so, let $(\boldsymbol{q}_{-i}, q_i)$ also be the n-vector $\boldsymbol{q}$. Market demand for the perfect-substitutes case is a function of aggregate output and its inverse is denoted as $p(Q)$; furthermore, let firm i's cost of producing q_i be denoted as $c_i(q_i)$. Thus, firm i's *profit* function is written as $\pi^i(\boldsymbol{q}) \equiv p(Q)q_i - c_i(q_i)$. All elements of the model are assumed to be commonly known by the firms, though extensions allowing incomplete information are not uncommon.

A *Cournot equilibrium* consists of a vector of output levels, $\boldsymbol{q}^{CE}$, such that no firm wishes to unilaterally change its output level when the other firms produce the output levels assigned to them in the (purported) equilibrium. Alternatively put (and reversing history), it is a Nash equilibrium of the normal-form game with quantities as strategies chosen from a compact space (for example, q_i in $[0, Q^*]$, for some appropriate Q^*, such as $p(Q^*) = 0$) and with the $\pi^i(\boldsymbol{q})$ as the payoff functions. Thus, $\boldsymbol{q}^{CE}$ is a Cournot equilibrium if the following n equations are satisfied: $\pi^i(\boldsymbol{q}^{CE}) \geq \pi^i(\boldsymbol{q}_{-i}^{CE}, q_i)$ for all values of q_i, for $i = 1, \ldots, n$.

In analysing his model applied to a duopoly (he also considered the n-firm version), Cournot provided the notion of *best-response functions*. In the duopoly case, this is a pair of functions, $\psi^1(q_2)$ and $\psi^2(q_1)$, which provide the profit-maximizing choice of output for firm 1 and 2 (respectively), given conjectures about the output level chosen by the rival firm (that is, each firm's choice of its output level reflects a *best-response property*). Hence, $\psi^i(q_j) = \arg\max_q \pi^i(q, q_j)$, $i, j = 1, 2$, $i \neq j$. That is, we want $\psi^i(q_j)$ to be the solution to firm i's first-order condition: $p(\psi^i(q_j) + q_j) + p'(\psi^i(q_j) + q_j)\psi^i(q_j) - c'_i(\psi^i(q_j)) = 0$, $i, j = 1, 2, i \neq j$. We'll assume for now that the problem has a nice solution and that some sort of sufficiency condition holds (for example, strict quasi-concavity of profits), but the discussion below on existence and uniqueness of equilibrium shows that such classical assumptions are overly strong and are overly restrictive for some modern applications, such as those involving multi-stage games or discontinuous cost functions. More generally, $\psi^i(q_j)$ could be a correspondence (a point-to-set map); we generally restrict the discussion below to functions, and assume as much differentiability as needed.

If output-level choices are best responses to conjectures about each firm's rival's choice of output, *and* if these conjectures are correct in equilibrium, then the resulting vector of output levels provides a Cournot equilibrium: $q_i^{CE} = \psi^i(q_j^{CE})$ for $i, j = 1, 2$, and $j \neq i$. In other words, the equilibrium occurs where the best-response functions cross when graphed in the space of output levels. Generalizing to n firms, this condition can be written as $q_i^{CE} = \psi^i(\boldsymbol{q}_{-i}^{CE})$ for $i = 1, \ldots, n$: $\boldsymbol{q}^{CE}$ is a Cournot equilibrium if it consists of mutual best-responses for all the firms.

Some variations on the basic model are worth mentioning. If the cost function for a firm has both fixed and variable components, and if the fixed component is avoidable (that is, is zero at zero output), then the best-response function for the firm will be discontinuous at the positive output level where variable profits just cover the avoidable cost. This is important for two reasons. First, avoidable fixed costs are not unusual in many entry scenarios: think of an airline entering a market where there are already some competitors, with the avoidable cost being advertising. Second, this discontinuity could mean that the only equilibrium might involve some or all firms choosing to not enter (or to exit) the market, even if absent these avoidable costs $\boldsymbol{q}^{CE}$ would be strictly positive.

Another avenue for interaction would consider imperfect factor markets, so that instead of $c_i(q_i)$ the cost function for firm i would be written as $c_i(\boldsymbol{q}_{-i}, q_i)$; then strategic interaction occurs not only through revenue but also via factor markets. Finally, if the model is one of short-run competition, then the output level of the firm may be restricted to be less than some predetermined capacity level; a simple version is that there are parameters k_i, $i = 1, \ldots, n$, such that a constraint on firm i's quantity choice is $q_i \leq k_i$, $i = 1, \ldots, n$; this induces a vertical segment (at the capacity level) in a firm's best-response function. Such capacity levels might be choices made in an earlier stage.

Finally, a number of papers develop 'non-Cournot' models which generate Cournot-model results. Kreps and Scheinkman (1983) provide a two-stage model of capacity choice followed by price setting in a homogeneous-goods duopoly; the result is a unique subgame-perfect equilibrium with Cournot capacities and a market-clearing price consistent with the standard Cournot model (however, Davidson and Deneckere, 1986, show that this result is especially sensitive to the basis for rationing consumers over firms when out-of-equilibrium firm-level demand exceeds capacity). Klemperer and Meyer (1986) analyse a one-stage game wherein duopolists producing heterogeneous goods non-cooperatively choose either a price or a quantity as the firm's strategy; under either multiplicative or additive error in the demand function, if marginal costs are upward sloping, the outcome is that predicted by the Cournot model (applied to the heterogenous-goods case; see the discussion of this case in Section 2 below). The classic embedding of the Cournot model is that of Bowley (1924), the best-known developer of models with 'conjectural variations' (CV). This is a static story wherein the first-order conditions in the analysis include firm i's conjecture of each rival's reaction to a small change in firm i's quantity (for example, $\partial q_j/\partial q_i$ need not be zero for each $j \neq i$); different values of the CV generate competitive, collusive, or Cournot outcomes (among others). Such a handy static embedding of alternative degrees of competition has been employed in a number of theoretical applications, and in a variety of empirical analyses trying to estimate market power. However, Daughety (1985) shows that a basic rationality requirement (that each firm's CV be the same as the actual slope of the best-response function) leads to the Cournot outcome, so that alternative CV values violate this form of rational expectations. Furthermore, Korts (1999) shows that empirical analyses using the CV approach to assess market power will generally mis-measure the degree of competitiveness of the industry.

2 Properties of the Cournot equilibrium

For most of this section we emphasize results for an n-firm, homogeneous-goods, complete-information model, where a firm's cost function depends only on that firm's output level. As suggested earlier, possibly one of the most important reasons for the continuing interest in the properties of the Cournot equilibrium is that Cournot competition is frequently used as a final stage in a variety of models; analysis employing such refinements as subgame perfection rely on a well-behaved subgame.

Existence, uniqueness and stability

Novshek (1985) provides an existence theorem that has quite practical uses (for expository purposes we consider a slightly less general version). Besides continuity and twice differentiability of the inverse demand function,

$p(Q)$, Novshek's existence theorem requires that: (1) $p(Q)$ crosses the quantity axis at a finite value and is strictly decreasing for quantities below that cut point; (2) the marginal revenue for each firm is decreasing in the aggregate output of its rivals; and (3) each firm's cost function is non-decreasing and lower semi-continuous. Requirement (2) is written formally as $p'(Q_{-i} + q_i) + p''(Q_{-i} + q_i)q_i < 0$, where $Q_{-i} \equiv Q - q_i$, for all i. This is equivalent to the assumption that $\partial^2 \pi^i(\boldsymbol{q})/\partial Q_{-i}\partial q_i < 0$ for all i, that is, that Q_{-i} and q_i are *strategic substitutes*, which means that an expansion in Q_{-i} implies that the optimal q_i falls. The third requirement means that costs cannot fall as the output level is increased and that cost functions can have jumps (discontinuities), as long as the functions are continuous from the left. This was a substantial improvement over previous existence theorems and it allows for an important case: avoidable fixed costs, such as those in the airline-entry example mentioned earlier. Amir (1996) applies an ordinal version of the theory of *supermodular games* to the existence issue (see Vives, 2005, for a recent survey of supermodular games; see also Amir, 2005, for a comparison of *ordinal* and *cardinal complementarity* in this context); this change of techniques allows for weaker demand conditions (primarily that log $p(Q)$ is concave) but requires a slightly stronger condition on each firm's cost function (marginal costs are positive, so models wherein marginal costs might be zero – as might occur with capacity competition – are left out) in order to guarantee that a Cournot equilibrium exists. As an example of the advantages concerning demand analysis, let $p(Q) = (Q - \overline{Q})^2$ for $Q \le \overline{Q}$, and zero otherwise. Such a function satisfies (1) above, is log-concave (actually, convex), but is excluded from consideration by Novshek's second condition.

Gaudet and Salant (1991) provide conditions for a Cournot equilibrium to be *unique* which address an important consideration when Cournot models are used in a subgame of a larger game: their theorem allows for degeneracy (one or more firms produce zero output but have marginal cost equal to the equilibrium price); thus, such firms are just at the shutdown point in the equilibrium. In a one-stage application this could be eliminated via a small perturbation in the parameters, but in a multi-stage application such an outcome need not be pathological, as some of the second-stage 'parameters' are strategic variables in the first-stage model (the authors provide a simple, full-information entry game to illustrate this). The sufficient conditions for uniqueness are (not surprisingly) more restrictive than those for existence (on the assumption that Novshek's conditions hold as well): (1) each firm's cost function must be twice continuously differentiable and strictly increasing; and (2) the slope of the marginal cost function is strictly bounded above the slope of the demand function. Thus, concave costs are allowed, to some degree, but the cost function cannot be 'too concave', even on subsets of its domain.

Cournot provided an explicit dynamic stability argument for his model by imagining sequential play by each agent (myopically best-responding in the current period to the existing output levels of all rivals); this is referred to as *best-reply dynamics* and when this process converges the solution is termed *stable*. Using best-reply dynamics to rationalize a static solution has, historically, been a source of substantial criticism, but nonetheless some papers use the requirement of Cournot stability to select an equilibrium when there are multiple equilibria (dynamic stability should not be confused with equilibrium refinement criteria in game theory such as strategic stability). A sufficient condition in the duopoly case is that $|\partial \psi^1(q_2)/\partial q_2||\partial \psi^2(q_1)/\partial q_1| < 1$ (see Fudenberg and Tirole, 1991); see Seade (1980) for more general conditions (and problems) for best-reply dynamics in the n-firm case. For an approach employing an explicit evolutionary process via replicator dynamics with noise, with firms able to choose 'behavioural' strategies (including, but not limited to, best-reply), see Droste, Hommes and Tunistra (2002).

Welfare

Two types of inefficiency can occur in a Cournot equilibrium: the equilibrium price exceeds the marginal cost of production, and aggregate output is inefficiently distributed over the firms. Compare the first-order conditions for firms in a duopoly, each producing under conditions of non-decreasing marginal costs (that is, $p(Q) + p'(Q)q_i = c_i'(q_i)$, $i = 1, 2$) with those for a central planner choosing q_1 and q_2 so as to maximize total surplus: $p(Q) = c_i'(q_i)$, $i = 1, 2$. Clearly, if demand is downward-sloping at the equilibrium, aggregate output in the Cournot equilibrium will be less than what the social planner would choose. However, a second distortion can be seen in this comparison: under the social planner, each firm's marginal costs are equalized with the others'. This will hold only in a symmetric Cournot equilibrium (where $q_1 = q_2$): production is, in general, inefficiently allocated across the firms.

The maldistribution of production implies that strategic interaction readily may yield counter-intuitive welfare results. As a simple example, consider a duopoly wherein (inverse) industry demand is $p = a - Q$ and firm i's cost function is $c_i(q) = C_i q$, $i = 1, 2$, with $a > C_1 > C_2 > 0$; that is, the linear demand, constant-but-unequal-marginal-cost case. It is straightforward to find the equilibrium and show that it is interior and unique. Let W be the sum of producers' and consumers' surplus. Then a little work shows that $dW/dC_1 > 0$ if $11C_1 - 7C_2 - 4a > 0$; to see that these conditions are non-empty, consider the parameter specification ($a = 20$, $C_1 = 13$, $C_2 = 8$), which satisfies all the foregoing requirements. The point of the example is that a *reduction* of firm 1's marginal cost leads to a *decrease* in equilibrium welfare. Thus, strategic interaction by the firms in the marketplace can lead to reversals of the usual welfare intuition that cost-improving technological change is beneficial. The reason this occurs is that the cost reduction results in an increase in the high-cost firm's equilibrium output level and a (smaller) decrease in the low-cost firm's output level; this increased inefficiency in aggregate production can be sufficient to overwhelm other efficiency improvements (such as the increase in industry output). This is similarly true if in the above model firm 2 is an incumbent monopolist (using simple monopoly pricing) and firm 1 an entrant: welfare will fall due to entry.

In the n-firm version of the constant-marginal-cost model, changes in the distribution of production costs (holding the mean fixed) do not affect industry output; this is seen by summing over the first-order conditions, whence $np(Q) + p'(Q)Q = \sum_{i=1}^{n} C_i$. Bergstrom and Varian (1985) showed that (on the assumption that the pre- and post-change equilibria are interior) such mean-preserving changes in the marginal costs strictly improve welfare if and only if the variance of the marginal costs strictly increases; the reason is that the aggregate cost of production has decreased if the variance increases. Salant and Shaffer (1999) extend this idea to consider the effects of changes in first-stage parameters (for example, cost-reducing R&D investments) on second-stage costs in models wherein Cournot competition is employed in the second stage. They argue that, since aggregate production costs are *maximized* when all firms have the same costs, it is the asymmetric equilibria in such games (which are often assumed away) which may yield the most important outcomes to examine, from both a social and a private perspective.

Does entry necessarily reduce the equilibrium price? A recent contribution provides a clean result if we restrict attention to the symmetric case wherein all firms have the same twice continuously differentiable and non-decreasing cost function, and demand is continuously differentiable and downward-sloping. Amir and Lambson (2000) show that the equilibrium price falls with an increase in the number of competitors if, for all Q, $p'(Q) < c''(q)$ for all q in $[0, Q]$. Thus, even with some degree of returns to scale (for example, as might occur with U-shaped average costs), entry will reduce price, at least with identical firms. However, Hoernig (2003) shows that, even if the equilibria are stable and there are no returns to scale, price can rise with entry if products are differentiated.

If the products of the firms are imperfect substitutes (that is, products are differentiated), then (in general) there is no aggregate demand function $p(Q)$; rather firm i's inverse demand function would be written as $p_i(q)$ and profits would be written as $\pi^i(q) = p_i(q)q_i - c_i(q_i)$, $i = 1, \ldots, n$. Welfare in this model can be contrasted with a reformulation of the model so that each firm chooses a price for its product; standard parlance is to call the price-strategy model the *(differentiated products) Bertrand model* (even though Bertrand's famous review

of Cournot did not envision heterogeneity in products; see Friedman's 1988 translation of Bertrand's review). Without going into detail on the (differentiated products) Bertrand model, Singh and Vives (1984) have shown (for linear, symmetric demand and constant marginal costs in a duopoly setting) that, while profits under Cournot competition exceed those under Bertrand competition, total surplus is higher under Bertrand competition than under Cournot competition. Note that this result holds in the one-stage game. However, these results may be reversed in a two-stage application. For example, Symeonidis (2003) considers R&D investment with spillovers in a two-stage game, and shows that (at least for a portion of the parameter space) Cournot competition leads to higher welfare than Bertrand competition. The basic intuition is that, if profits are higher for second-stage Cournot competition than for second-stage Bertrand competition, and first-stage investment is inefficiently low in either case, then the increased second-stage profits may partly correct the inefficiently low first-stage investment, leading to an overall welfare gain for competition in quantities rather than prices.

Finally, convergence of a Cournot equilibrium to a competitive equilibrium, as the number of firms grows, was considered by Cournot in Chapter 8 of his book, and has been the subject of a number of papers; see Novshek and Sonnenschein (1978; 1987) for a general equilibrium treatment where appropriate replication of Cournot economies yields equilibria arbitrarily close to the Walrasian equilibrium; see Alos-Ferrer (2004) for an evolutionary model (which allows for memory) at the level of an industry.

3 Applications

The literature exploring and applying the Cournot model is vast; an earlier extended bibliography can be found in Daughety (1988/2005). The more recent literature employing the Cournot model is already becoming significant in size: a survey of articles in 16 top mainline and field journals, for the period 2001–5, netted approximately 125 articles exploring or applying the Cournot model in one of its various common forms. An online Excel file of (abbreviated) citations and some characteristics of each article (number of firms, number of stages, welfare considerations, informational regime, and topic classification), as accessed on 21 November 2006, is available at http://www.vanderbilt.edu/Econ/faculty/Daughety/ExtendedCournotBib2001-2005.xls

However, some excellent papers have undoubtedly been missed (not to mention papers from the 1990s), and space limitations preclude anything beyond the briefest of tours and just a taste of the literature, so only a very few can be discussed below. This section addresses five topics which account for a significant portion of the literature, three areas that overlap other fields, and two (comparatively) new areas of research.

Delegation

Vickers (1985) uses an n-firm, two-stage model to examine performance measures for managers. Restricting the manager's performance measure to be a weighted average of profits and output, with the weights determined by the owner of each firm in the first stage, he shows that the weight on output is non-zero. This makes each manager more aggressive (each chooses to produce a higher output level), thereby leading to lower profits per firm. Sklivas (1987) considers the differentiated-products Bertrand version and shows that owners choose weights on revenue and profits so as to make managers more passive (they post higher prices), leading to increased profits. Miller and Pazgal (2001) have unified this literature, showing that incentive schemes based on own and rival's profits result in an equilibrium which is insensitive to whether the firm chooses price or quantity as its strategic variable.

Information transfer

Vives (1984), Gal-Or (1985), and Li (1985) all consider variants of 'information transfer' models to examine the possibility of information sharing, whereby firms may choose to pool information on either demand or cost parameters. These models are analysed as *Bayesian–Nash games*, so that, before seeing a private signal about the parameter of interest (for example, the demand intercept), each firm chooses whether or not to share the information with the other firms; then information is received and production (or pricing) occurs in the second stage. The nature of the good (substitutes or complements), the type of information (common or individual), and the strategy space (quantities or prices) all affect whether firms will share information. Ziv (1993) relaxes the verifiability of information and finds that firms will send misleading information if they can; he then considers mechanisms for eliciting truthful messages.

Intellectual property

Katz and Shapiro (1985) and Kamien and Tauman (1986) consider the licensing of innovations in an oligopoly. Katz and Shapiro employ a three-stage duopoly game in which the innovation is developed, then a single license is auctioned, and then the firms compete. Kamien and Tauman use a two-stage, n-firm game with a posted price for the innovation (a fee or a royalty), followed by competition. More recently, Fauli-Oller and Sandonis (2003) consider optimal competition policy when considering licences as an alternative to merger. Anton and Yao (2004) allow for weak patent protection and consider how disclosure of information about an innovation (for example, through the patent application) can be a signalling device to influence competitors, but those same competitors may be able to employ the information to successfully use (infringe on) the patent; here small innovations are patented and substantial innovations are protected through secrecy.

Mergers

Salant, Switzer and Reynolds (1983) show that exogenously determined mergers of a subset of firms in the constant-marginal-cost set-up yields a problematical result: a sufficient condition for a merger to be unprofitable is that it involve less than 80 per cent of the industry, hardly a resounding endorsement of using such a model to analyse mergers. This result, however, is partly driven by the assumptions of homogeneous products, constant unit costs, and industry structure. Perry and Porter (1985) show that various mergers can be profitable if firms have sufficiently increasing marginal costs. Daughety (1990), using a two-tiered-industry, n-firm model, with m firms choosing output in the first stage (tier) and $n - m$ firms choosing output in the second stage, shows that if $1 < m < n$, then, when m is comparatively small ($m < n/3$), mergers of two second-tier firms to make a first-tier firm can be both profitable and social-welfare-enhancing, even though such mergers increase concentration and have no cost synergies (all firms have identical constant unit costs). Recently, Pesendorfer (2005), using a repeated game model with entry, has found that merger to monopoly may not be profitable, but merger in a non-concentrated industry can be; these differences from the previous literature partly reflect long-run versus short-run profitability computations.

R&D

D'Aspremont and Jacquemin (1988) considered cost-reducing R&D in the presence of spillovers, and considered both non-cooperative and cooperative R&D decision-making; there have been a number of recent papers on cost-reducing spillovers (see, for example, Zhao, 2001, for more on the negative welfare effects of cost-reducing innovation, and Symeonidis, 2003, cited in Section 2 above, as well as the work discussed below under the subject of auctions with competition). Toshimitsu (2003) considers the incentive and welfare properties of quality-based R&D subsidies for firms in a model of endogenously determined product quality (and thus product differentiation); subsidizing high quality is welfare-enhancing (independent of whether the Cournot or Bertrand model is employed).

Other fields

Areas of ongoing effort which extend into other fields include *experimental economics*, *the financial structure of the firm* (see, for example, Brander and Lewis, 1986, on determinate debt-equity due to imperfect competition, and see Povel and Raith, 2004, extending Brander and Lewis via endogenously determined debt contracts); and *international trade* (see, for example, Brander and Spencer, 1985, analysing the strategic use of subsidies in international competition; Mezzetti and Dinopoulos, 1991, discussing domestic firm–union bargaining and import competition; and Spencer and Qiu, 2001, concerning relationship-specific investments and trade).

New topics

Finally, a few examples of comparatively new topics. While auctions with private information has long been an area of interest, the developing literature on *auctions with competition* has started to take seriously the combination of incomplete information and post-auction competition. For example, see Das Varma (2003) or Goeree (2003), who find that signalling by winners of an auction causes bids to be biased when post-auction interaction between the auction's winner and losers can be influenced by the size of the bid. A nice example is when firms have private information about how acquiring a cost-reducing innovation might affect the firm's production costs, and bidding for a licence for the innovation precedes Cournot oligopoly interaction; here signalling with a high bid suggests that the winner will have low costs and will produce a high level of output.

A second new area is *networks*; one recent example is Goyal and Moraga-Gonzalez (2001), who model bilateral agreements to share knowledge, and allow for the possibility of partial collaboration, via considering possible networks of relationships. They examine how the nature of the firms' interaction in markets can contribute to the instability of certain types of strategic alliances and the stability of other ones.

4 A broader perspective on Cournot competition

If alive to critique this essay, Cournot might view the interpretation of the term 'Cournot competition' being limited merely to the legacy of his oligopoly analysis to be an overly restrictive interpretation of the assignment. And well he should. Hicks (1935; 1939) argues that Cournot was the first to present a modern model of monopoly as well as the precise conditions for perfect competition; furthermore, as noted earlier, Cournot's eighth chapter concerned 'unlimited competition'. In the 1937 Cournot Memorial session of the Econometric Society, A. J. Nichol (1938) observed that, if ever there was an apt illustration of Carnegie's dictum that 'It does not pay to pioneer', then Cournot's life and work would be it. Cournot's oligopoly model was essentially ignored for many years, or was relegated to dusty corners of microeconomics texts, but over recent decades it has come to be an essential tool in many an economist's toolbox, and is likely to continue as such.

ANDREW F. DAUGHETY

See also **Bertrand competition; experimental economics.**

Bibliography

Alos-Ferrer, C. 2004. Cournot versus Walras in dynamic oligopolies with memory. *International Journal of Industrial Organization* 22, 193–217.

Amir, R. 1996. Cournot oligopoly and the theory of supermodular games. *Games and Economic Behavior* 15, 132–48.

Amir, R. 2005. Ordinal versus cardinal complementarity: the case of Cournot oligopoly. *Games and Economic Behavior* 53, 1–14.

Amir, R. and Lambson, V.E. 2000. On the effects of entry in Cournot markets. *Review of Economic Studies* 67, 235–54.

Anton, J.A. and Yao, D.A. 2004. Little patents and big secrets: managing intellectual property. *RAND Journal of Economics* 35, 1–22.

Bergstrom, T.C. and Varian, H.R. 1985. Two remarks on Cournot equilibria. *Economics Letters* 19, 5–8.

Bertrand, J. 1883. Review of Walras's *Théorie mathématique de la richesse social* and Cournot's *Recherches sur les principes mathématiques de la théorie des richesses*. Trans. J.W. Friedman, in A.F. Daughety (1988).

Bowley, A.L. 1924. *The Mathematical Groundwork of Economics*. Oxford: Oxford University Press.

Brander, J.A. and Lewis, T.R. 1986. Oligopoly and financial structure: the limited liability effect. *American Economic Review* 76, 956–70.

Brander, J.A. and Spencer, B. 1985. Export subsidies and international market share rivalry. *Journal of International Economics* 18, 83–100.

Cournot, A. 1838. *Researches into the Mathematical Principles of the Theory of Wealth*. Trans. N.T. Bacon, New York: Macmillan, 1929.

d'Aspremont, C. and Jacquemin, A. 1988. Cooperative and noncooperative R&D in duopoly with spillovers. *American Economic Review* 78, 1133–7.

Das Varma, G. 2003. Bidding for a process innovation under alternative modes of competition. *International Journal of Industrial Organization* 21, 15–37.

Daughety, A.F. 1985. Reconsidering Cournot: the Cournot equilibrium is consistent. *RAND Journal of Economics* 16, 368–79.

Daughety, A.F. 1988. *Cournot Oligopoly – Characterization and Applications*. New York: Cambridge University Press (reprinted 2005).

Daughety, A.F. 1990. Beneficial concentration. *American Economic Review* 80, 1231–37.

Davidson, C. and Deneckere, R. 1986. Long-run competition in capacity, short-run competition in price, and the Cournot model. *RAND Journal of Economics* 17, 404–15.

Droste, E., Hommes, C. and Tunistra, J. 2002. Endogenous fluctuations under evolutionary pressure in Cournot competition. *Games and Economic Behavior* 40, 232–69.

Fauli-Oller, R. and Sandonis, J. 2003. To merge or to license: implications for competition policy. *International Journal of Industrial Organization* 21, 655–72.

Fudenberg, D. and Tirole, J. 1991. *Game Theory*. Cambridge, MA: MIT Press.

Gal-Or, E. 1985. Information transmission – Cournot and Bertrand. *Review of Economic Studies* 53, 85–92.

Gaudet, G. and Salant, S. 1991. Uniqueness of Cournot equilibrium: new results from old methods. *Review of Economic Studies* 58, 399–404.

Goeree, J.K. 2003. Bidding for the future: signaling in auctions with an aftermarket. *Journal of Economic Theory* 108, 345–64.

Goyal, S. and Moraga-Gonzalez, J.L. 2001. R&D networks. *RAND Journal of Economics* 32, 686–707.

Hicks, J.R. 1935. Annual survey of economic theory: the theory of monopoly. *Econometrica* 3, 1–12.

Hicks, J.R. 1939. *Value and Capital*, 2nd edn. London: Oxford University Press.

Hoernig, S.H. 2003. Existence of equilibrium and comparative statics in differentiated goods Cournot oligopolies. *International Journal of Industrial Organization* 21, 989–1019.

Kamien, M.I. and Tauman, Y. 1986. Fees versus royalties and the private value of a patent. *Quarterly Journal of Economics* 101, 471–92.

Katz, M.L. and Shapiro, C. 1985. On the licensing of innovations. *RAND Journal of Economics* 16, 504–20.

Klemperer, P. and Meyer, M. 1986. Price competition vs. quantity competition: the role of uncertainty. *RAND Journal of Economics* 17, 618–38.

Korts, K.S. 1999. Conduct parameters and the measurement of market power. *Journal of Econometrics* 88, 227–50.

Kreps, D.M. 1987. Nash equilibrium. In *The New Palgrave: A Dictionary of Economics*, vol. 3, ed. J. Eatwell, M. Milgate and P. Newman. London: Macmillan.

Kreps, D.M. and Scheinkman, J.A. 1983. Quantity precommitment and Bertrand competition yield Cournot outcomes. *Bell Journal of Economics* 14, 326–37.

Li, L. 1985. Cournot oligopoly with information sharing. *RAND Journal of Economics* 16, 521–36.

Mezzetti, C. and Dinopoulos, D. 1991. Domestic unionization and import competition. *Journal of International Economics* 31, 79–100.

Miller, N.H. and Pazgal, A.I. 2001. The equivalence of price and quantity competition with delegation. *RAND Journal of Economics* 32, 284–301.

Nichol, A.J. 1938. Tragedies in the life of Cournot. *Econometrica* 3, 193–7.

Novshek, W. 1985. On the existence of Cournot equilibrium. *Review of Economic Studies* 52, 85–98.

Novshek, W. and Sonnenschein, H. 1978. Cournot and Walras equilibrium. *Journal of Economic Theory* 19, 223–66.

Novshek, W. and Sonnenschein, H. 1987. General equilibrium with free entry. *Journal of Economic Literature* 25, 1281–306.

Perry, M.K. and Porter, R.H. 1985. Oligopoly and the incentive for horizontal merger. *American Economic Review* 75, 219–27.

Pesendorfer, M. 2005. Mergers under entry. *RAND Journal of Economics* 36, 661–79.

Povel, P. and Raith, M. 2004. Financial constraints and product market competition: ex ante vs. ex post incentives. *International Journal of Industrial Organization* 22, 917–49.

Salant, S.W. and Shaffer, G. 1999. Unequal treatment of identical agents in Cournot equilibrium. *American Economic Review* 89, 585–604.

Salant, S.W., Switzer, S. and Reynolds, R.J. 1983. Losses from horizontal merger: the effects of an exogenous change in industry structure on Cournot–Nash equilibrium. *Quarterly Journal of Economics* 98, 185–99.

Seade, J. 1980. The stability of Cournot revisited. *Journal of Economic Theory* 23, 15–27.

Singh, N. and Vives, X. 1984. Price and quantity competition in a differentiated duopoly. *RAND Journal of Economics* 15, 546–54.

Sklivas, S.D. 1987. The strategic choice of managerial incentives. *RAND Journal of Economics* 18, 452–58.

Spencer, B.J. and Qiu, L.D. 2001. Keiretsu and relationship-specific investment: a barrier to trade? *International Economic Review* 42, 871–901.

Symeonidis, G. 2003. Comparing Cournot and Bertrand equilibria in a differentiated duopoly with product R&D. *International Journal of Industrial Organization* 21, 39–53.

Toshimitsu, T. 2003. Optimal R&D policy and endogenous quality choice. *International Journal of Industrial Organization* 21, 1159–78.

Vickers, J. 1985. Delegation and the theory of the firm. *Economic Journal Supplement* 95, 138–47.

Vives, X. 1984. Duopoly information equilibrium: Cournot and Bertrand. *Journal of Economic Theory* 34, 71–94.

Vives, X. 2005. Complementarities and games: new developments. *Journal of Economic Literature* 43, 437–79.

Zhao, J. 2001. A characterization for the negative welfare effects of cost reduction in a Cournot oligopoly. *International Journal of Industrial Organization* 19, 455–69.

Ziv, A. 1993. Information-sharing in oligopoly: the truth-telling problem. *RAND Journal of Economics* 24, 455–65.

creative destruction

Creative destruction refers to the incessant product and process innovation mechanism by which new production units replace outdated ones. It was coined by Joseph Schumpeter (1942), who considered it 'the essential fact about capitalism'.

The process of Schumpeterian creative destruction (restructuring) permeates major aspects of macroeconomic performance, not only long-run growth but also economic fluctuations, structural adjustment and the functioning of factor markets.

At the microeconomic level, restructuring is characterized by countless decisions to create and destroy production arrangements. These decisions are often complex, involving multiple parties as well as strategic and technological considerations. The efficiency of those decisions not only depends on managerial talent but also hinges on the existence of sound institutions that provide a proper transactional framework. Failure along this dimension can have severe macroeconomic consequences once it interacts with the process of creative destruction (see Caballero and Hammour, 1994; 1996a; 1996b; 1996c; 1998a; 1998b; 2005). Some of these limitations are natural, as they derive from the sheer complexity of these transactions. Others are man-made, with their origins ranging from ill-conceived economic ideas to the achievement of higher human goals, such as the inalienability of human capital. In moderate amounts, these institutional limitations give rise to business cycle patterns such as those observed in the most developed and flexible economies. They can help explain perennial macroeconomic issues such as the cyclical behaviour of unemployment, investment and wages. In higher doses, by limiting the economy's ability to tap new technological opportunities and adapt to a changing environment, institutional failure can result in dysfunctional factor markets, resource misallocation, economic stagnation, and exposure to deep crises.

Given the nature of this short piece, I will skip any discussion of models, and refer the reader to Caballero (2006) for a review of the models behind the previous paragraph, and to Aghion and Howitt (1998) for an exhaustive survey of Schumpeterian growth models. Instead, I focus on reviewing recent empirical evidence on different aspects of the process of creative destruction.

Recent evidence on the pace of creative destruction

There is abundant recent empirical evidence supporting the Schumpeterian view that the process of creative destruction is a major phenomenon at the core of economic growth in market economies.

The most commonly used empirical proxies for the intensity of the process of creative destruction are those of factor reallocation and, in particular, job flows. Davis, Haltiwanger and Schuh (1996) (henceforth DHS) offered the clearest peek into this process by documenting and characterizing the large magnitude of job flows within US manufacturing. They defined job creation (destruction) as the positive (negative) net employment change at the establishment level from one period to the next. Using these definitions, they concluded that over ten per cent of the jobs that exist at any point in time did not exist a year before or will not exist a year later. That is, over ten per cent of existing jobs are destroyed each year and about the same amount is created within the same year. Following the work by DHS for the United States, many authors have constructed more or less comparable measures of job flows for a variety of countries and episodes. Although there are important differences across them, there are some common findings. In particular, job creation and destruction flows are large, ongoing and persistent. Moreover, most job flows take place within rather than between narrowly defined sectors of the economy.

Given the magnitude of these flows and that they take place mostly within narrowly defined sectors, the presumption is strong that they are an integral part of the process by which an economy upgrades its technology. Foster, Haltiwanger and Krizan (2001) provide empirical support for this presumption. They decompose changes in industry-level productivity into within-plant and reallocation (between-plant) components, and conclude that the latter – the most closely related to the creative destruction component – accounts for over 50 per cent of the ten-year productivity growth in the US manufacturing sector between 1977 and 1987. Moreover, in further decompositions they document that entry and exit account for half of this contribution: exiting plants have lower productivity than continuing plants. New plants, on the other hand, experience a learning and selection period through which they gradually catch up with incumbents. Other studies of US manufacturing based on somewhat different methodologies (see Baily, Hulten and Campbell, 1992; Bartelsman and Dhrymes, 1994) concur with the conclusion that reallocation accounts for a major component of within-industry productivity growth. Bartelsman, Haltwanger and Scarpetta (2004) provide further evidence along these lines for a sample of 24 countries and two-digit industries over the 1990s.

Recent evidence on the cyclical features of creative destruction

At the business cycle frequency, sharp liquidations (rises in job destruction) constitute the most noted impact of contractions on creative destruction. In contrast, job creation is substantially less volatile and mildly procyclical. There is an extensive literature that, extrapolating from the spikes in liquidations (recently measured in job flows but long noticed in other contexts), finds that recessions are times of increased reallocation. In fact, this has been a source of controversy among economists at least since the pre-Keynesian 'liquidationist' theses of such economists as Hayek, Schumpeter and Robbins. These economists saw in the process of liquidation and reallocation of factors of production the main function of recessions. In the words of Schumpeter (1934, p. 16): 'depressions are not simply evils, which we might attempt to suppress, but ... forms of something which has to be done, namely, adjustment to ... change.'

In Caballero and Hammour (2005) we turned the liquidationist view upside down. While we sided with Schumpeter and others on the view that increasing the pace of restructuring of the economy is likely to be beneficial, we provided evidence that, contrary to conventional wisdom, restructuring *falls* rather than rises during contractions.

Since the rise in liquidations during recessions is not accompanied by a contemporaneous increase in creation, implicit in the increased-reallocation view is the idea that increased destruction is followed by a surge in creation during the recovery phase of the cyclical downturn. This presumption is the only possible outcome in a representative firm economy, as the representative firm must replace each job it destroys during a recession by creating a new job during the ensuing recovery. However, once one considers a heterogeneous productive structure that experiences ongoing creative destruction, other scenarios are possible. The cumulative effect of a recession on overall restructuring may be positive, zero, or even negative, depending not only on how the economy contracts but also on how it recovers. Thus, the relation between recessions and economic restructuring requires one to examine the effect of a recession on aggregate separations not only at impact, but cumulatively throughout the recession-recovery episode. We explored this issue using quarterly US manufacturing gross job flows and employment data for the 1972–93 period, and found that, along the recovery path, job destruction declines and falls below average for a significant amount of time, more than offsetting its initial peak. On the other hand, job creation recovers, but it does not exceed its average level by any significant extent to offset its initial decline. As a result, our evidence indicates that, on average, recessions *depress* restructuring.

Similarly, in Caballero and Hammour (2001) we approached the question of the pace of restructuring over the cycle from the perspective of corporate assets. Studying the aggregate patterns of merger and acquisition (M&A) activity and its institutional underpinnings, we reached a conclusion that also amounts to a rejection of the liquidationist perspective. Essentially, a liquidationist perspective in this context would consider fire sales during sharp liquidity contractions as the occasion for intense restructuring of corporate assets. The evidence points, on the contrary, to briskly expansionary periods characterized by high stock market valuations and abundant liquidity as the occasion for intense M&A activity.

Recent evidence on institutional impediments to creative destruction and their cost

For all practical purposes, some product or process innovation is taking place at every instant in time. Absent obstacles to adjustment, continuous innovation would entail infinite rates of restructuring. What are these obstacles to adjustment? The bulk of it is technological – adjustment consumes resources – but (over-?) regulation and other man-made institutional impediments are also a source of depressed restructuring.

While few economists would object to the hypothesis that labour market regulation hinders the process of creative destruction, its empirical support is limited. In Caballero et al. (2004) we revisited this hypothesis using a sectoral panel for 60 countries. We found that job security provisions – measured by variables such as grounds for dismissal protection, protection regarding

dismissal procedures, notice and severance payments, and protection of employment in the constitution – hamper the creative destruction process, especially in countries where regulations are likely to be enforced. Moving from the 20th to the 80th percentile in job security cuts the annual speed of adjustment to shocks by a third. By impairing worker movements from less to more productive units, effective labour protection reduces aggregate output and slows down economic growth. We estimated that moving from the 20th to the 80th percentile of job security lowers annual productivity growth by as much as 1.7 per cent.

Similarly, the idea that well-functioning financial institutions and markets are important factors behind economic growth is an old one. The process of creative destruction is likely to be a chief factor behind this link. In Caballero, Hoshi and Kashyap (2006) we analysed the decade-long Japanese slowdown of the 1990s and early 2000s. The starting point of our analysis is the well-known observation that many large Japanese banks would have been out of business had regulators forced them to recognize all their loan losses. Because of this, the banks kept many zombie firms alive by rolling over loans that they knew would not be collected (evergreening). Thus, the normal competitive outcome whereby the zombies would shed workers and lose market share was thwarted. Using an extensive data-set, we documented that roughly 30 per cent of firms were on life support from the banks in 2002 and about 15 per cent of assets resided in these firms. The main idea in our article is that the counterpart to the congestion created by the zombies is a reduction in profits for potential and more productive entrants, which discourages their entry. We found clear evidence of such a pattern in firm-level data and of the corresponding reduced restructuring in sectoral data.

Bertrand, Schoar and Thesmar (2004) further drive home the point that problems in the banking sector can have grave consequences for the health of the restructuring process. They use a differences-in-differences approach on firm-level data for the period 1977–99 to analyse the impact of the banking reforms of the mid-1980s on firm and bank behaviour. These reforms eliminated government interference in bank lending decisions, eliminated subsidized bank loans, and allowed French banks to compete more freely in the credit market. They find that, after the reforms, firms' exit rates and asset reallocation rise, and are more correlated with performances.

International competition is an important source of creative destruction. Trefler (2004) concludes that there are significant productivity and reallocation effects from trade openness, even in industrialized economies. To reach this conclusion, Trefler takes advantage of the Canada–US Free Trade Agreement (FTA) to study the effects of a reciprocal trade agreement on Canada. He finds that, for industries that experienced the deepest Canadian tariff reductions, the contraction of low-productivity plants reduced employment by 12 per cent while raising industry-level labour productivity by 15 per cent. Moreover, he finds that at least half of this increase is related to exit and/or contraction of low-productivity plants. Finally, for industries that experienced the largest US tariff reductions, plant-level labour productivity soared by 14 per cent. Consistent with this evidence, Bernard, Jensen and Schott (2006) find that in the United States productivity growth is fastest in industries where trade costs (barriers) have declined the most.

Domestic deregulation of goods markets can have similar effects. For example, Olley and Pakes (1996) find that deregulation in the US telecommunications industry increased productivity predominantly through factor reallocation towards more productive plants rather than through intra-plant productivity gains. More broadly, Klapper, Laeven and Rajan (2004) study the effect of entry regulation on firm behaviour in a sample including firm-level data from countries of western and eastern Europe. Their findings support the notion that regulation affects entry: 'naturally high-entry' industries have relatively lower entry in countries that have higher entry regulations. Moreover, both the growth rate and share of high-entry industries are depressed in countries with more stringent barriers to entry. Finally, Fishman and Sarria-Allende (2004) extend the Klapper, Laeven and Rajan study to countries outside Europe and include both industry- and firm-level data from the UNIDO and WorldScope databases, and reach similar conclusions.

Final remarks

Evidence and models coincide in their conclusion that the process of creative destruction is an integral part of economic growth and fluctuations. Obstacles to this process can have severe short- and long-run macroeconomic consequences.

RICARDO J. CABALLERO

See also **Schumpeter, Joseph Alois; structural change.**

Bibliography

Aghion, P. and Howitt, P. 1998. *Endogenous Growth Theory.* Cambridge, MA: MIT Press.
Baily, N., Hulten, C. and Campbell, D. 1992. Productivity dynamics in manufacturing establishments. In *Brookings Papers on Economic Activity: Microeconomics*, ed. M. Baily and C. Winston. Washington, DC: Brookings Institution.
Bartelsman, E. and Dhrymes, P. 1994. Productivity dynamics: US manufacturing plants, 1972–1986. Finance and Economics Discussion Series No. 94-1. Washington, DC: Board of Governors, Federal Reserve System.
Bartelsman, E., Haltiwanger, J. and Scarpetta, S. 2004. Microeconomic evidence of creative destruction in industrial and developing countries. Mimeo, University of Maryland.
Bernard, A., Jensen, J. and Schott, P. 2006. Survival of the best fit: exposure to low-wage countries and the (uneven) growth of US manufacturing plants. *Journal of International Economics* 68, 219–37.
Bertrand, M., Schoar, A. and Thesmar, D. 2004. Banking deregulation and industry structure: evidence from the French banking reforms of 1985. Discussion Paper No. 4488. London: Centre for Economic Policy Research.
Caballero, R. 2006. *Specificity and the Macroeconomics of Restructuring.* Yrjo Jahnsson Lectures. Cambridge, MA: MIT Press.
Caballero, R., Cowan, K., Engel, E. and Micco, A. 2004. Effective labor regulation and microeconomic flexibility. Mimeo, MIT.
Caballero, R. and Hammour, M. 1994. The cleansing effect of recessions. *American Economic Review* 84, 1350–68.
Caballero, R. and Hammour, M. 1996a. The fundamental transformation in macroeconomics. *American Economic Review* 86(2), 181–6.
Caballero, R. and Hammour, M. 1996b. On the timing and efficiency of creative destruction. *Quarterly Journal of Economics* 111, 805–52.
Caballero, R. and Hammour, M. 1996c. On the ills of adjustment. *Journal of Development. Economics* 51, 161–92.
Caballero, R. and Hammour, M. 1998a. The macroeconomics of specificity. *Journal of Political Economy* 106, 724–67.
Caballero, R. and Hammour, M. 1998b. Jobless growth: appropriability, factor substitution and unemployment. *Carnegie-Rochester Conference Series on Public Policy* 48, 51–94.
Caballero, R. and Hammour, M. 2001. Institutions, restructuring, and macroeconomic performance. In *Advances in Macroeconomic Theory*, ed. J. Dreze. New York: Palgrave Macmillan.
Caballero, R. and Hammour, M. 2005. The cost of recessions revisited: a reverse-liquidationist view. *Review of Economic Studies* 72, 313–41.
Caballero, R., Hoshi, T. and Kashyap, A. 2006. Zombie lending and depressed restructuring in Japan. Working Paper No. 12129. Cambridge, MA: NBER.
Davis, S., Haltiwanger, J. and Schuh, S. 1996. *Job Creation and Destruction.* Cambridge, MA: MIT Press.
Fishman, R. and Sarria-Allende, V. 2004. Regulation of entry and the distortion of industrial organization. Working Paper No. 10929. Cambridge, MA: NBER.
Foster, L., Haltiwanger, J. and Krizan, C. 2001. Aggregate productivity growth: lessons from microeconomic evidence. In *New Developments in Productivity Analysis*, ed. E. Dean, M. Harper and C. Hulten. Chicago: University of Chicago Press.
Klapper, L., Laeven, L. and Rajan, R. 2004. Business environment and firm entry: evidence from international data. Working Paper No. 10380. Cambridge, MA: NBER.
Olley, S. and Pakes, A. 1996. The dynamics of productivity in the telecommunications equipment industry. *Econometrica* 64, 1263–98.

Schumpeter, J. 1934. Depressions. In *Economics of the Recovery Program*, ed. D. Brown et al. New York: McGraw-Hill.

Schumpeter, J. 1942. *Capitalism, Socialism, and Democracy*. New York: Harper & Bros.

Trefler, D. 2004. The long and short of the Canada–US Free Trade Agreement. *American Economic Review* 94, 870–95.

creative destruction (Schumpeterian conception)

Schumpeter invented the phrase 'creative destruction' in his famous book on the development of capitalism into socialism (Schumpeter, 1942). In his view the process of creative destruction is the essential fact about capitalism and refers to the incessant mutation of the economic structure from within, destroying the old and creating a new.

In the footsteps of Karl Marx, Schumpeter argues that in dealing with capitalism we are dealing with an evolutionary process. It is by nature a form or method of economic change and not only never is, but never can be, stationary. The fundamental impulse that sets and keeps the capitalist engine in motion comes from new goods and new methods of production and transportation, created by the Schumpeterian entrepreneur, who is always on the outlook for new combinations of the factors of production (Heertje, 2006).

The process of creative destruction takes time. For that reason there is no point in appraising its performance within a static framework. A system may produce an optimal allocation of resources at every point of time and may yet in the long run be inferior to a system without such optimal allocation, because the non-optimality may be a condition for the level and speed of long-run performances; in other words, for dynamic efficiency. Furthermore, the process of creative destruction in Schumpeter's vision must be seen as the background for individual decisions and strategies. Economic theory has a tendency to concentrate on decisions about prices by firms, which are assumed to maximize profits, within a given structure. Schumpeter argues that the relevant problem is how capitalism creates and destroys these structures (Metcalfe, 1998).

Schumpeter's conception of creative destruction overturns the idea that price competition is the only component of the market behaviour of entrepreneurs. In fact, it is not that kind of competition which counts, but the competition from the new commodity, the new technology, the new source of supply and the new type of organization. Instead of marginal changes, fundamental upheavals are brought about by process and product innovations of existing firms and potential competitors.

Restrictive practices of monopolists and large firms are to be judged against the background of the perennial gale of creative destruction, rather than in the context of stationary development. The potential threat of process and product innovation reduces the scope and importance of restrictive practices that aim to guarantee the monopolist or big firm a quiet life. If however the profits are used to counterattack, restrictive practices may help to deepen the process of creative destruction and, therefore, the dynamic effects of capitalism (Reisman, 2004).

The process of creative destruction as described by Schumpeter has been experienced again since the 1980s in the United States, Japan and Western Europe and since the 1990s in China and India as well. On the basis of new technologies many old firms, structures and professions have been swept away and new industrial organizations and labour relations have emerged. In particular, the application of information technology and the Internet with the dramatic decrease in transaction costs of communication is leading to major changes of a quantitative and qualitative nature in both the private and public sector of the economy. On the one hand, 'external' growth of already large firms which take over others is a feature of modern capitalism; on the other hand, every day new small firms are established, often created by former executives of existing (and long-lived) companies.

This extensive discussion of the process of creative destruction illustrates Schumpeter's strong emphasis on the supply side of the economy. It would be an interesting question to study the impact of the process of creative destruction on employment. My guess would be that, on balance, the process of creative destruction is more creative than destructive, not only with regard to employment but also concerning broader perspectives of growth and welfare. This may be one of the reasons why Schumpeter's work has had a lasting and ever-increasing influence on economic theory.

ARNOLD HEERTJE

See also **creative destruction; market structure; Schumpeter, Joseph Alois.**

Bibliography

Heertje, A. 2006. *Schumpeter on the Economics of Innovation and the Development of Capitalism*. Cheltenham: Edward Elgar.

Metcalfe, J.S. 1998. *Evolutionary Economics and Creative Destruction*. London: Routledge.

Reisman, D. 2004. *Schumpeter's Market*. Cheltenham: Edward Elgar.

Schumpeter, J. 1942. *Capitalism, Socialism, and Democracy*. New York: Harper.

credit card industry

The concept of a general purpose credit card originated in 1949, when Frank McNamara dined in a New York restaurant and discovered that he could not pay for his meal (Evans and Schmalensee, 1999). By the 1980s credit cards had become ubiquitous, and they remain a popular form of payment in most economies. Banks offer cards, setting terms such as interest rates and annual fees. Transactions are handled by networks such as Visa and MasterCard, which emerged in the 1970s as joint member associations. Early research examining the market typically focused on the retail level, while more recent work has tended to focus on the network level, mirroring a shift in policy concerns in the 1980s.

In its early years the US retail credit card market was characterized by extreme interest rate 'stickiness' – credit card rates remained virtually constant over time, regardless of economy-wide changes in interest rates. Credit card issuers also appear to have earned super-normal profits during the same period. This presents a puzzle in an industry displaying many classic characteristics of a perfectly competitive market (Ausubel, 1991). Ausubel suggests a variety of explanations for this puzzle, including the possibility that credit card borrowers do not fully anticipate the degree to which they will use the cards.

Ausubel's research spurred a wave of subsequent work proposing explanations for interest rate stickiness. Mester (1994) and Brito and Hartley (1995) provide theoretical explanations for interest rate stickiness based on asymmetric information or consumer transaction costs. Calem and Mester (1995) provide empirical evidence that consumer search and switching costs might explain interest rate stickiness. A complementary explanation for interest rate stickiness is that state-level interest rate ceilings during the 1980s facilitated tacit collusion among card issuers, leading to greater-than-normal interest rate stability (Knittel and Stango, 2003).

By the early 1990s interest rates had become much more flexible as credit card issuers switched to variable interest rates. By most accounts, the market also became more competitive during this time. One explanation for the change is technological progress that allowed more efficient credit scoring by large nationally marketed card issuers, creating a truly national market that fostered aggressive competition. Other explanations include the threat of interest rate regulation and the entry of new issuers.

At the network level, the key economic issue is that payment card systems like MasterCard and Visa are two-sided markets: they have to attract cardholders to get merchants and merchants to get cardholders. Diners Club did this in 1950 by initially giving away cards to consumers and charging merchants seven per cent of their bill. These days, consumers obtain rewards for using their cards. This structure of pricing has raised the concern of some policymakers. In their view, retailers pay too much to accept credit cards, costs that end up being covered by consumers who do not use credit cards (by way of higher retail prices). Card associations sustain such a price structure through the setting of an interchange fee, which determines how much the merchant's bank must pay the cardholder's bank for each card transaction. A high interchange fee results in a high merchant fee and a low (or negative) fee for cardholders.

The issue of how much to charge each type of user is a common one in other two-sided markets. Magazines and newspapers decide how much to charge readers versus advertisers, and shopping malls decide how much to charge shoppers versus shops. The interest of policymakers in credit cards has spurred research in two-sided markets more generally.

Baxter (1983) provides an early analysis of interchange fees (see Rochet, 2003, for a survey). His key insight is that efficiency calls for card transactions whenever the *joint* benefits to the consumer and merchant of using the card exceed the joint costs of doing so. In the absence of an interchange fee, each type of user will face only the private costs and benefits of cards. A payment from the merchant's bank (acquirer) to the cardholder's bank (issuer) via the interchange fee can align the private incentive to use cards with the social incentive. This provides a justification for setting an interchange fee, but does not imply that card associations will set it at the right level.

One reason a card association might set the interchange fee too high is that acquirers may pass through a larger proportion of interchange fees into merchant fees than issuers pass back to cardholders (in the form of lower fees or higher rewards). Then associations will want to pass revenues to the issuing side, via high interchange fees, where they are competed away less aggressively. A second possible reason is that, if merchants accept cards to attract customers from each other, their private willingness to accept cards includes the surplus their customers get from using cards. As a result, cardholder surplus is over-represented, and card associations tend to charge merchants too much and cardholders too little. Although these theoretical possibilities highlight possible divergences between privately and socially optimal interchange fees, they provide no basis for the cost-based regulation of interchange fees.

VICTOR STANGO AND JULIAN WRIGHT

See also **two-sided markets.**

Bibliography

Ausubel, L. 1991. The failure of competition in the credit card market. *American Economic Review* 81, 50–81.

Baxter, W. 1983. Bank interchange of transactional paper: legal perspectives. *Journal of Law and Economics* 26, 541–88.

Brito, D. and Hartley, P. 1995. Consumer rationality and credit cards. *Journal of Political Economy* 103, 400–33.

Calem, P. and Mester, L. 1995. Consumer behavior and the stickiness of credit-card interest rates. *American Economic Review* 85, 1327–36.

Evans, D. and Schmalensee, R. 1999. *Paying with Plastic*. Cambridge, MA: MIT Press.

Knittel, C. and Stango, V. 2003. Price ceilings as focal points for tacit collusion: evidence from credit cards. *American Economic Review* 93, 1703–29.

Mester, L. 1994. Why are credit card rates sticky? *Economic Theory* 4, 505–30.

Rochet, J.-C. 2003. The theory of interchange fees: a synthesis of recent contributions. *Review of Network Economics* 2(2), 97–124.

credit cycle

Prior to Keynes's *General Theory*, the resolution of the question why, in capitalist economies, aggregate variables undergo repeated fluctuations about the trend was regarded by economists as a main challenge for the profession. What was then called business (or trade) cycle theory grew quite independently from the classical and subsequently neoclassical corpus of price theory. In fact, for all economists, a clear-cut distinction existed between the long-run forces at work in an economy – the subject of a rigorous value and distribution theory – and the more or less ad hoc explanations of the short-run oscillations around such an (equilibrium) centre of gravity. Of course, from Ricardo and Thornton down the 19th century to Overstone and Mill, money and credit played a substantial, but independent, part in these exogenous explanations of the business cycle. Along the same line, the founding fathers of marginalism (in particular Walras, Marshall and Jevons) failed to coordinate, even in a remotely satisfactory way, money and trade cycle with their then novel price theory.

Following Wicksell's and Mises's lead, it is only with the post-First World War attempts to integrate marginalist value and monetary theory that theorists started pondering the possible 'incorporation of cyclical phenomena into the system of economic equilibrium theory' (Hayek, 1929, p. 33n.). The rediscovery of Tooke's (1844) income approach to the quantity theory of money is probably one of the earliest stepping-stones in the development of credit-cycle theories. This line of thought suggests that the explanation of money prices should start not from the quantity of money but from nominal income. Though another way of writing a Marshallian cash balance equation, Wicksell's (1898, p. 44) or Hawtrey's (1913, p. 6) emphasis on the 'aggregate of money income', on how it varies, is expanded or held, is a crucial turning-point on the road towards an analysis in terms of income, saving and investment. This shift of emphasis, together with the simultaneous progress in monetary theory proper (notably the development of a comprehensive and integrated monetary theory of interest), the 1914–18 inflationary episode and the post-war cyclical upheavals provided in the 1920s and 1930s the right intellectual stimulus for credit-cycle theories to grow and multiply.

Explicitly or implicitly, to tackle this issue, Continental economists (for example, Mises, Cassel, Hayek, Schumpeter and Aftalion), members of the Cambridge School then dominating in England (Keynes, Robertson, Pigou, Hawtrey), Fisher and Mitchell in the United States all used the common analytical framework established jointly by Walras, Menger, Marshall and Jevons. This is made up of two basic (though familiar) propositions: on the one hand, there is an inverse relation between the volume of investment and the rate of interest (that is, a downward-sloping investment demand curve) and, on the other, despite short-run 'frictions', the interest rate is assumed to be sensitive enough to divergences between investment decisions and full employment saving.

The central theme of this argument (first expressed with great clarity in Wicksell's cumulative process) is that the market rate of interest oscillates in the short run around a natural rate of interest determined in the long run by the supply of and the demand for capital as a stock, which, in turn, guarantees the equality between planned investment and full employment saving. Once this logic is understood, it then emerges that the entire development of interwar trade-cycle theories took place within the second proposition outlined above; namely, that, in the long run, the interest rate is assumed to be sensitive enough to divergences between investment decisions and full employment saving. Hence, since the twin concepts of an interest-elastic demand curve for investment and natural rate of interest were never called into question, the orgy of debates that took place in the 1920s and 1930s was conducted in terms of an analysis of various short-run forces which temporarily keep at bay the long-run forces of saving and investment.

These forces are, of course, of multiple nature. Of particular interest to interwar economists, and one of the essential features of business cycle, with its recurrence of upswings and downswings, is a *credit cycle*, an alternation of credit expansion and credit contraction. But it was assumed neither that an alternation of prosperity and depression would not exist in a barter economy (or in a purely specie system) nor that cycles could be viewed as functions of monetary factors only.

In fact, and thanks to their common capital theory, none of the leading interwar credit cycle theorists fell into either of these traps. Even Hawtrey who, with remarkable consistency kept claiming that business cycles are a purely monetary phenomenon, had clearly in mind a Wicksell-like cumulative process derived from Marshall's oral tradition in monetary theory. This common theoretical background and a deep interest in a then fast-developing monetary theory make similarities between credit cycle theorists

sufficiently pronounced to entitle us to speak of a single monetary theory [of the cycle], the votaries of which disagree on one issue only: whether bank-loan rates act primarily on 'durable capital' [Keynes, Robertson,

Hayek] or via the stocks of wholesalers [Hawtrey]. (Schumpeter, 1954, p. 1121)

In 1913, Hawtrey was amongst the first to provide a detailed analysis of the financial working of the cumulative process in an Anglo-Saxon environment. However, even if his theory usefully describes the ways in which money and credit behave in the cycle, the main weakness of his contribution is, of course, its almost exclusive emphasis on dealers' stocks in the course of a credit cycle. If Hawtrey does not deny altogether that a credit expansion/contraction has an influence on the volume of investment, he holds it however to be unimportant when compared with the direct influence on the wholesalers' stocks. He then logically disputes the existence of forced saving on the very ground of this availability of stocks and fails completely to link his credit cycle theory with the dominant Marshallian capital theory. Such a model led Hawtrey not only to give Bank Rate the crucial part to play in any counter-cyclical policy but also to consider its fluctuations as the only explanation of cyclical fluctuations. To sketch British interwar depressions as almost exclusively functions of Bank Rate (itself a function of Britain's absorption of gold) is a rather bold simplification Hawtrey was never quite ready to abandon.

If the theoretical apparatus underlying the *Treatise on Money* proceeds from the same logic, Keynes's fundamental equations introduce, however, a number of very sophisticated and new variations on the basic credit-cycle theme. In particular, causes of credit cycles are of non-monetary nature (they result from fluctuations in the rate of investment relative to the rate of saving), the influence of Bank Rate on investment is not limited 'to one particular kind of investments, namely, investments by dealers in liquid goods [stocks]' (Keynes, 1930, vol. 1, p. 173), the cumulative process includes a theory of the demand for money beyond the traditional income motive (that is, an early version of liquidity preference), and, in the short run, there is no longer a direct relation between the quantity of money/credit and the price level: monetary or credit changes do not foster ipso facto a forced/abortive saving process. Despite the higher degree of sophistication shown in the *Treatise*, in a classic chapter on the modus operandi of the Bank Rate, Keynes displays bold confidence in this mechanism to smooth any credit cycle, to fill the gap between saving and investment and to correct all temporary monetary divergences from the long-run full employment equilibrium. However, Keynes's disaffection with the forced saving doctrine and the purely static nature of his fundamental equations drew sharp criticisms from Robertson and Hayek. Though from different standpoints, they both considered Keynes's credit cycle analysis as no more than an attempt to spell out the appropriate banking policy which could maintain a monetary equilibrium. In particular, Keynes's version of the credit cycle lacked, for the former, a proper sequential stability analysis and, for the latter, an explicit integration with capital theory.

Along lines very similar to Keynes's and, up to the late 1920s, in close cooperation with him, Robertson worked out a detailed sequential analysis of the interdependence of real and monetary magnitudes during the cycle. But clearly, for him, the cycle results from over-investment, this tendency to over-invest being a typical feature of decentralized economies stemming from the repercussions on the volume of investment of its gestation period. However, the largest part of Robertson's professional output was devoted to studying the monetary or credit symptoms of such economic fluctuations, that is, how banks may respond to an increased demand for credit during expansion.

This led Robertson to a redefinition of the concept of saving in a monetary economy and to the rôle of this new concept in the cycle. This approach was linked with a sequential analysis of the lagged adjustments of output to monetary flows. In the 'forced saving' debate, central to all credit cycle theories, and contrary to Hayek who considered it as the villain of the piece, Robertson saw that phenomenon as only a relatively minor component of his theory, the factors at the root to his 'credit inflation' being the *real* cause of this expansion. Dragged among others by Keynes into endless discussions in the realm of monetary and interest theory, Robertson never managed however to offer an articulate and full-blown version of his theory of industrial fluctuations. In particular, the problem of the alteration in the structure of production, a question forming the core of Hayek's cycle theory, never received more than passing comment.

Grounded of course in the Austrian tradition and Wicksell's cumulative process (first extended by Mises, 1912, and Cassel, 1918), the distortion of the production time structure is absolutely central to Hayek's monetary cycle theory. The divergence between 'natural' and market rates of interest is linked by Hayek to the variability in forced saving and considered as the cause of cyclical fluctuations. Hayek's 'additional credit' theory places the cause of this gap between these two rates upon newly created money. The increase in loan capital resulting from a 'trailing market rate' makes investment surpass voluntary saving: a cumulative expansion results. Such an increase in investment alters the relative prices of capital and consumer goods in favour of the former. The increased output of capital goods distorts the production time structure. At a later stage, higher factor incomes drive up the demand for consumption goods, which through increased withdrawals from bank accounts will raise the market rate of interest and, finally, make some investment unprofitable. Then, the turnabout that takes place in the cycle brings a change in the other direction in the production structure, this time in favour of consumer goods. Clearly, crises are caused by over-investment, that is, by a decline in the desire to purchase the flow of capital goods coming on the market. The reversal of the process initiated by credit inflation does take place (as in most credit cycle theories) whenever the market rate catches up with prices; and since, sooner or later, banks run up against the limits set to their lending by their reserves, this process cannot be explosive (Fisher, 1911, also noticed, at least in his earliest writings, this stabilizing influence of the banking system).

Hayek's credit cycle theory thus marks a real break with what had come before. The theory of money is no longer a theory of the value of money 'in general' because relative prices may be changed by monetary influences and the Wicksellian full-employment assumption is dropped. The specific task of the trade cycle theorist is, for Hayek, to analyse short-period positions of the economy 'in successive moments of time' (1941, p. 23). The adoption of such an 'intertemporal equilibrium' approach to cycles (conceptually not different from modern temporary equilibrium) marks not only a crucial methodological turning point, but also the swan song of credit cycle theories.

On the one hand, this new method of 'intertemporal equilibrium' heralds the abandonment of the traditional framework in which cycles (defined as short-run disequilibria) are seen as temporary deviations from long-period equilibrium conditions determined by systematic and persistent forces at work in decentralized economies. In the present case, the 'natural' rate of interest determined in the long run by the supply of and the demand for capital is no longer the norm towards which the system is tending. It is in fact a property of such an 'intertemporal equilibrium' that not only will the price of the same commodity be different at different points in time but also that the stock of capital will not yield a uniform 'natural' rate of interest on its supply-price.

On the other, the publication of Keynes's *General Theory* redirected research efforts away from this question into the problem of the determination of output at a point in time. It is only since the late 1960s, with the search for 'microfoundations for macroeconomics', and the subsequent advent of rational expectations and non-Walrasian equilibria, that this line of thought has been back on the theoretical agenda. However, given the extreme complexity of the problem and the relative crudeness of models still in their infancy, progress has so far been very modest.

P. BRIDEL

See also **Hawtrey, Ralph George.**

Bibliography

Fisher, I. 1911. *The Purchasing Power of Money*. New York: Macmillan.
Hawtrey, R.G. 1913. *Good and Bad Trade*. London: Constable.
Hayek, F.A. von. 1929. *Monetary Theory and the Trade Cycle*. Trans. N. Kaldor and H.M. Crome, London: Jonathan Cape, 1933.
Hayek, F.A. von. 1941. *The Pure Theory of Capital*. London: Routledge.
Keynes, J.M. 1930. *A Treatise on Money*. Vol. 1, *The Pure Theory of Money*. As in *Collected Writings*, vol. 5, London: Macmillan, 1971.
Mises, L. von. 1912. *The Theory of Money and Credit*. Trans. H.E. Batson, London: Jonathan Cape, 1934.
Schumpeter, J.A. 1954. *History of Economic Analysis*. London: Oxford University Press.
Tooke, T. 1844. *An Inquiry into the Currency Principle*. London: Longman, Brown, Green & Longmans.
Wicksell, K. 1898. *Interest and Prices*. Trans. R.F. Kahn, London: Macmillan, 1936.

credit rationing

Broadly speaking, 'credit rationing' refers to any situation in which lenders are unwilling to advance additional funds to a borrower even at a higher interest rate. In the words of Jaffee and Modigliani (1969, pp. 850–1), 'credit rationing [is] a situation in which the demand for commercial loans exceeds the supply of these loans at the commercial loan rate quoted by the banks'. Key to this definition is that changes in the interest rate cannot be used to clear excess demand for loans in the market. In essence, this definition treats credit rationing as a supply side phenomenon, with the lender's supply function becoming perfectly price inelastic at some point.

If the projects that are being funded by the loan are not scalable, however, then a distinction must be made between a situation in which a lender eventually restricts the size of loan it will provide to any individual borrower and one in which 'rationed' borrowers are denied credit altogether. This phenomenon arises in circumstances in which lending is not scalable. Stiglitz and Weiss (1981, pp. 394–5) therefore define credit rationing as follows:

> We reserve the term credit rationing for circumstances in which either (a) among loan applicants who appear to be identical some receive a loan and others do not, and the rejected applicants would not receive a loan even if they offered to pay a higher interest rate; or (b) there are identifiable groups of individuals in the population who, with a given supply of credit, are unable to obtain loans at any interest rate, even though with a larger supply of credit, they would.

According to this definition, lenders fully fund some borrowers but deny loans to others despite the fact that the latter are identical in the lender's eyes to those who receive loans.

Thus, there are two working definitions of credit rationing in the literature. The first focuses on situations in which increases in the interest rate cannot clear excess demand in the loan market, whether this excess demand reflects a single borrower (who would like a larger loan amount) or many. Under this definition, rationing would exist if every potential borrower received a loan but a smaller one than that desired at the equilibrium interest rate. The second definition — the Stiglitz–Weiss definition — restricts its attention to situations in which some borrowers are completely rationed out of the market, even though they would be willing to pay an interest rate higher than that prevailing in the market.

Both of these definitions focus on the supply side of the market. One could argue, however, that it is useful to think of non-price rationing as any phenomenon that limits the amount of funding used by firms such that firms are not able to use the price mechanism to successfully bid for additional funds, whether this is caused by supply-side constraints (as under the narrow definitions of credit rationing described above) or by other distortions in credit markets (related, for example, to regulation). This would allow a broader definition of 'credit rationing' in which regulatory constraints, rather than just informational problems, lead to non-price allocations of credit.

Why care about credit rationing?

Early interest in credit rationing was driven in part by questions about the role that credit rationing might play in transmitting the macroeconomic effects of monetary policy, which was related to research on the so-called 'availability doctrine' in the 1950s and 60s (Scott, 1957). To the extent that monetary policy operates through a 'credit channel' (in which contractionary policy affects the economy through a decline in the supply of funds available for banks to lend), and to the extent that changes in the terms of lending include not only changes in loan pricing but also changes in the quantities of credit available to borrowers, credit rationing may play an important role in the transmission of monetary policy's effects on the economy (Blinder and Stiglitz, 1983).

In addition to the cyclical effects of rationing in credit markets related to monetary policy, development economists, especially Ronald McKinnon (1973), argued that a different credit rationing problem is more relevant for the long-term growth prospects of developing countries. High inflation, high zero-interest reserve requirements, government-mandated loan allocations to favoured borrowers, and interest rate ceilings on loans or deposits in developing economies (a combination which McKinnon termed 'financial repression') subjected many developing countries' banking systems to an extreme form of regulation-induced credit rationing. High reserves, high inflation, and interest ceilings on deposits meant that banks were rationed in the deposit market, and thus had few funds to lend, while lending mandates and loan interest-rate ceilings meant that what funds were available to lend were often rationed by restrictions on who could bid for those funds.

Additionally, George Akerlof (1970), in his path-breaking article on the role of adverse selection in preventing market development, drew attention at an early date to the possible effects of information problems in retarding the development of lending markets, particularly in developing countries. In an ideal world, in the absence of any government policies limiting beneficial lending, all borrowers with positive net present value projects would be able to obtain outside funding (whether through debt or equity instruments, or bank or non-bank sources of funds). But Akerlof showed that, if markets were unable to distinguish good risks from bad ones, lending might not be feasible. The failure to develop institutions capable of producing credible information about borrowers and using that information to screen applicants could, according to Akerlof, play an important role in financial underdevelopment.

Many development economists have come to recognize that the failure to properly allocate funds in the loan market – a broad phenomenon, within which credit rationing is a special and extreme case – can be an especially important potential impediment to growth in developing countries because of the relative absence of institutions in those countries that allow effective screening of borrowers (to mitigate adverse selection) or ongoing monitoring of borrowers' actions (to mitigate moral hazard).

An additional motivation for an interest in credit rationing comes from the literature on bank fragility. Credit rationing can also apply to the market in which financial intermediaries raise their funds. Financial institutions go to great pains to attract and maintain deposits through (a) the structure of their contracts (which typically afford withdrawal options to depositors), (b) their long-term relationships with market monitors who track their progress, and (c) their established reputations for good management. But sometimes the market suddenly decides to ration credit to a particular bank or to the whole banking system; and when this happens the affected banks find it hard to attract and maintain deposits at any price. Thus, the literature on 'bank runs' as an historical phenomenon can be thought of as a literature on credit rationing in the markets in which financial institutions raise their funds. Depositors that decide to participate in a bank run ration credit to their bank in the sense that the decision to withdraw is a quantity, not a price, decision. They are simply unwilling to leave their money in the bank.

Finally, much of the current research on discrimination in credit markets is driven by evidence that black and Hispanic minority loan applicants are denied more frequently than comparable whites (for example, Munnell et al., 1996; Cavalluzzo and Cavalluzzo, 1998; Cavalluzzo and Wolken, 2005). Of course, this begs the question of why borrowers are denied loans in the first place, rather than simply priced according to their risk. In other words, understanding why there are differences in denial rates across groups necessarily entails exploring why rationing (loan denial) occurs.

The development of credit rationing theory

Early views on credit rationing

The earliest discussions of credit rationing viewed it as a non-equilibrium phenomenon, arising either because of exogenous interest rate rigidities (for example, interest rate ceilings or usury laws) or because of a lack of competition in the loan market (Scott, 1957). Soon authors made a distinction between temporary credit rationing, in which market interest rates are slow to adjust to exogenous shocks such as changes in the lender's cost of funds or borrower demand, and 'equilibrium' credit rationing, which persists after the market has fully adjusted to these shocks. Clearly the more interesting and difficult to explain phenomenon is equilibrium credit rationing.

Hodgman (1960) was the first to try to explain how credit rationing can persist in a rational, equilibrium framework. In this model, lenders evaluate potential borrowers on the basis of the loan's expected return–expected loss ratio. In addition, it is assumes that there is a maximum repayment that the borrower can credibly promise, which effectively limits how much the lender will offer the borrower regardless of the interest rate: eventually the expected losses become too great relative to the expected return. This model was much debated in the ensuing years. In particular, Miller (1962) argued that Hodgman's analysis could be made consistent with rational expectations between the borrower and lender by incorporating bankruptcy costs that would be incurred by the lender upon the borrower's default. The real significance of the Hodgman article, however, was that it established as an important theoretical goal the objective of explaining how credit rationing could persist as an equilibrium phenomenon.

Freimer and Gordon (1965) resolved many of the issues regarding the structure of the Hodgman and Miller models by showing that credit rationing can occur with a risk-neutral lender if the borrower has a fixed-sized funding need. But this was done assuming an exogenous interest rate. Jaffee and Modigliani (1969) completed the picture by endogenizing the equilibrium interest rate by modelling both the supply and demand sides of the market. Credit rationing in their model, however, is the direct result of an exogenous assumption that borrowers within a given group must be charged the same interest rate, even though the lender can distinguish differences among them.

This early work was important in that it firmly established the idea that credit rationing could be a persistent equilibrium phenomenon. Ultimately, however, the solutions proposed relied on very restrictive assumptions about agent preferences or the contracts they could employ. More satisfactory explanations of credit rationing had to wait for the information economics revolution of the 1970s.

Modern credit rationing theory

Akerlof's (1970) pioneering article on adverse selection was motivated in part by the desire to explain extreme cases of credit rationing (the absence of a credit market), but Jaffee and Russell (1976) provide the first explicit asymmetric information rationale for credit rationing in the general sense. In their model, lenders cannot distinguish *ex ante* between high- and low-quality borrowers (that is, those who will repay their loans and those who will default). Contracts are written to determine the size of the loan offered and the interest rate. As in the Rothschild–Stiglitz (1976) insurance framework, low-quality borrowers must accept the contract that is preferred by the high-quality borrowers, lest they be identified as the deadbeats they are. Although a market-clearing interest rate/loan amount combination does exist, high-quality borrowers prefer a contract that entails a slightly lower interest rate with a reduced loan amount. As a result, the pooling outcome entails credit rationing. The primary problem with this model is that the 'equilibrium' is not stable, in that unsustainable separating contracts dominate the pooling outcome.

In 1981, Joseph Stiglitz and Andrew Weiss published what has become the canonical model of credit rationing, because it was the first model that fully endogenized contract choices with a stable, rationing equilibrium. In the Stiglitz–Weiss framework, credit rationing occurs because the lender's expected return is not monotonically increasing in the interest rate. Instead, adverse selection or moral hazard problems eventually cause the lender's expected return to decline as the interest rate rises.

In the adverse selection version of the model, borrowers and lenders are both risk neutral. Borrowers are characterized by their projects, which are assumed to have the same expected returns but differ from one another in their risk. Specifically, borrower projects differ on the basis of mean-preserving spreads (Rothschild and Stiglitz, 1970). These projects are also assumed to require a fixed investment (that is, they are indivisible) and borrowers have a fixed amount of internal equity that they can invest in the project. Limited liability upon default means that the lender's payoff is a concave function of the project's return, while the borrower's profit function is convex.

These assumptions imply that, at any given interest rate, a subset of the least risky borrowers will drop out of the market, choosing instead to forgo their projects. In essence, the borrower's limited liability means that he reaps all of the project's gain (beyond the cost of debt service) when its return is high, but loses his collateral (his paid-in capital invested in the project, if any) only when the project's return is low. For low-risk projects, however, the potential upside gains are small. If those low-risk borrowers are pooled with high-risk borrowers, they will face higher than warranted interest rates.

Low-risk borrowers will increasingly withdraw from the market as interest rates rise; as rates rise, borrowers with low-risk projects are better off withdrawing from the market and simply consuming their endowments rather than agreeing to invest and pay a high interest rate. As a result, increases in the interest rate cause more and more good borrowers to drop out of the market, lowering the average creditworthiness of the lender's remaining applicant pool. The size of the adverse selection premium faced by low-risk borrowers (the amount of interest low-risk borrowers have to pay in excess of what their project risks warrant) becomes larger with each interest rate rise because the interest rate must compensate for the default risk of an ever-worsening pool of borrowers.

Thus, increases in the interest rate affect lender returns in two ways. The first is the direct effect that a higher interest rate raises the lender's return (for a given pool of borrowers). Rising interest rates, however, also have the indirect effect of lowering the average quality of the lender's applicant pool, thereby lowering the lender's expected return from any given loan. Eventually, this secondary, adverse selection effect may outweigh the first interest rate effect, causing lender profits to decline as the interest rate rises.

Once the non-monotonicity of the lender's return in the interest rate is established, the possibility of credit rationing follows immediately. Profit-maximizing lenders will never voluntarily choose to raise the interest rate beyond where the adverse selection effect dominates. If excess demand exists in the market at this rate, credit rationing will be the equilibrium.

Paradoxically, in this model the very best credit risks do not seek funding because they do not find it worthwhile. This may seem odd, but it is important to remember that these borrowers are not rationed. Instead, they voluntarily drop out of the market because the cost of being pooled with higher-risk borrowers is too great. The rationed borrowers are the higher-risk borrowers who stay in the market and request funding.

Alternatively, Stiglitz and Weiss show how changes in the interest rate may also affect the borrower's choice of project, so that moral hazard in project choice (sometimes referred to as 'asset substitution' in the finance literature) can be another reason that the lender's expected return is non-monotonic in the interest rate. Suppose that the borrower is able to choose among projects with different risk profiles. If, at a given interest rate, the borrower is indifferent between two projects, Stiglitz and Weiss show that an increase in the interest rate will cause the borrower to prefer the project that has the higher probability of default. Of course, the lender prefers the safer project. Thus (with slightly more restrictive distributional assumptions than in the adverse selection case), increases in the interest rate once again can eventually lower the lender's expected return, leading to credit rationing.

Models of credit rationing need not posit rationing for all borrowers. Realistically, some borrowers (certain firms for which information control problems are particularly acute) may be subject to rationing while other borrowers are not. Borrowers not subject to rationing may be able to avoid rationing because their prospects are more observable, or because their behaviour is more controllable.

Bank runs as credit rationing

The theoretical literature on credit rationing in the deposit market (bank runs) has some features that distinguish it from the literature on credit rationing in the loan market. The ultimate causes of deposit market rationing can be similar to, or very different from, the causes of loan market rationing. As discussed above, loan market rationing can reflect either information and incentive problems in the loan market or exogenous regulations. In the case of the deposit market, rationing can result either from incentive and information problems relating to the depositor–bank relationship or from exogenous liquidity needs of depositors.

With respect to the former, under some circumstances a bank run may reflect a loss of confidence in the market value of the bank's asset portfolio and changes in bank behaviour that attend such a loss. If the value of the portfolio falls sufficiently, and if the information and incentive problems are sufficiently severe, the perceived risk of losses in the bank can prompt depositors to ask for their money back because depositors have reason to be risk-intolerant (that is, to be unwilling to leave their money in a bank that has too high a level of risk). An example of such a model is Calomiris and Kahn (1991). Here the depositor withdraws funds in bad states of the world because doing so is necessary to prevent the banker from abusing his control over the bank's portfolio.

An alternative cause of credit rationing in the deposit market is a shock to the liquidity needs of depositors, which forces depositors to demand their funds from their banks irrespective of the portfolio performance of the banks. Diamond and Dybvig (1983) is an example of a model of this phenomenon.

Bank depositor runs are but one specific example of how financial intermediaries may be credit rationed due to creditor risk intolerance and/or liquidity shocks. During the 1998 Russian financial crisis, for example, it was widely reported that many emerging market hedge funds dumped their holdings of risky securities of all kinds in a scramble to reduce their risks and thus re-establish the high-quality credit ratings needed to retain their debtors. Intermediaries were also scrambling to accumulate liquidity, as many of their claimants needed to withdraw funds to meet other obligations related to the financial market upheaval.

The limits of credit rationing

Credit rationing as a problem of information and control (as it was modelled by Jaffee and Russell, 1976, and

Stiglitz and Weiss, 1981) is properly seen as an extreme case of the more general phenomenon of capital market misallocation, which includes cases where capital is misallocated (due to adverse selection and moral hazard) without any rationing occurring. It is important to recognize that, from the standpoint of either cyclical concerns about the transmission of monetary policy or developmental concerns about the efficiency of the allocation of capital, the important phenomenon is not rationing per se but rather the extent to which the market fails to allocate resources efficiently. Even a market that never suffers from credit rationing can be highly inefficient in its allocation of capital. In that sense, credit rationing may be somewhat beside the point. Indeed, the corporate finance literature is full of examples of models of market imperfections involving moral hazard and adverse selection in which credit is misallocated, and in which positive net present-value projects are not funded or negative net present-value projects *are* funded.

In some cases, firms may even be priced out of the market for funds entirely, so that they avoid funding profitable investments. For example, Jensen and Meckling (1976) show that the potential for asset substitution at the expense of creditors can make it much more costly for firms to access debt markets. Indeed, asset substitution can make it prohibitively expensive to issue debt. Note that this is not a case of credit rationing as defined by Stiglitz and Weiss, since suppliers are not refusing credit. Rather, the high asset substitution premium that firms would be charged if they sought credit can result in a decision by the firm not to fund a positive net present-value investment. Similarly, Myers and Majluf (1984) show that because of adverse selection problems – which are particularly acute in the public equity market – some firms may decide to avoid issuing equity to fund a positive net present-value investment. Here, again, a firm is not being rationed by suppliers, but is unwilling to seek financing because of its prohibitive pricing.

As the literature on capital market misallocations and credit rationing developed in the late 1970s and early 1980s, critics pointed out some limiting circumstances in which capital markets did not have a tendency to underfund positive net present-value projects. For example, both adverse selection and moral hazard problems can be overcome by sufficient collateral. By placing collateral at risk a firm could signal its high quality, or commit itself not to abuse creditors by undertaking excessive risk (see Bester, 1985). Of course, collateral is not always available, nor is it costless to place collateral at risk. In the case of a limited liability enterprise, the firm's net worth limits its available collateral. Firms that can finance themselves from internal funds and limited amounts of low-risk debt can avoid the adverse selection and moral hazard costs associated with external finance, but young, growing firms tend to be in need of substantial amounts of external finance, far in excess of their accumulated net worth. If borrowers use all of their available 'collateral', then, on

the margin, collateral cannot mitigate adverse selection or moral hazard problems.

In the consumer context, it is also important to recognize that the moral hazard and adverse selection problems that arise in corporate lending may differ in importance across the various areas of consumer lending. For example, moral hazard may be limited in the context of mortgage lending where actions destructive to the lender's interest are likely to harm the homeowner as well (consider inadequate protection against the risk of fire, for example). Furthermore, the modern use of credit scores and loan-to-value ratios may make mortgage lenders more knowledgeable about an applicant's true credit risk than the applicant himself, particularly if that applicant has significant equity invested in the house and lacks experience in the credit market (Calomiris, Kahn and Longhofer, 1994). Under such circumstances, the implications of adverse selection models (which depend on the superiority of the information of the borrower about his type) may be irrelevant, or even reversed. On the other hand, in the context of uncollateralized credit card borrowing based only on past credit records, unobservably high-risk borrowers (those who know that they are about to have major medical costs, lose their job, or become divorced) may have strong incentives to borrow, implying the possibility for severe adverse selection.

How is credit rationing measured empirically?
Although credit rationing is a widely discussed phenomenon, there is a surprising paucity of evidence confirming its existence. The key problem is that, while the concept of a credit-rationed borrower is easy to understand in theory, under each of the various models of credit rationing discussed above it is extremely difficult to measure 'excess demand' of individual borrowers or the similitude of borrowers' creditworthiness.

Indirect methods
Jaffee and Modigliani (1969) attempt to infer the presence of credit rationing by measuring the proportion of new commercial loans originated at the prevailing prime rate and/or with very large loan sizes. The intuition they use is that prime and/or large borrowers have the lowest risk and are therefore the least likely to be rationed. As a result, a larger proportion of loans will go to these low-risk borrowers when credit rationing is severe. Jaffee and Modigliani use this proxy to see how market factors affect the prevalence of credit rationing. Of particular interest is their result that increases in the average commercial loan rate are associated with higher levels of rationing, which seems to confirm the appropriateness of their proxy for credit rationing.

Other authors have attempted to measure whether commercial loan rates are 'sticky' in response to changes in open-market interest rates. The idea here is that in most credit rationing models there is an implicit cap above which lenders will ration credit. As open-market rates rise, this cap is more likely to become binding, meaning that commercial loan rates will not fully respond to changes in open-market rates. Following this approach, a number of authors, including Goldfeld (1966) and Jaffee (1971), have found that commercial loan rates are, in fact, slow to adjust to changes in open-market rates, and offer this as evidence in support of credit rationing.

Berger and Udell (1992), however, provide convincing evidence that, although commercial-loan rate stickiness does occur, it does so in a fashion that is inconsistent with information-based credit rationing models. In particular, they find that nearly half of the observed loan rate stickiness occurs for loans made to borrowers who are exploiting a previously contracted bank loan commitment. Such borrowers are precluded from rationing by contract. Furthermore, they show that the fraction of loans made under commitment actually decreases during times of credit market tightness, exactly the opposite of what one would expect should credit rationing be an important phenomenon.

Direct methods

Other authors have attempted to directly measure credit rationing using survey data to identify 'rationed' borrowers. For example, Cox and Jappelli (1990) and Chakravarty and Scott (1999) use data from the Survey of Consumer Finances (SCF) in which households are directly asked whether they recently have been denied credit or been unable to obtain as much credit as they requested. Although these articles purport to measure how some outside factor affects the likelihood of being rationed, it is not clear that borrowers who self-report being denied credit have, in fact, been 'rationed' in the Stiglitz–Weiss meaning of the term. After all, their denial of credit could simply reflect a failure to properly select into the right risk class in order to be approved, or the fact that the borrower was simply uncreditworthy at any interest rate.

With regard to business lending, Cressy (1996) uses a sample of new businesses that opened accounts with a major British bank to ascertain whether credit rationing affects the likelihood of business survival. He concludes that firms self-select for finance based on the entrepreneur's human capital, implying that no credit rationing is occurring.

One strand of the empirical literature on credit rationing, broadly defined, focuses on whether differential mortgage loan denial rates between white and minority borrowers constitutes evidence of discrimination (a much cited reference is Munnell et al., 1996; Ross and Yinger, 2002, provide an excellent review of this literature). Although the discrimination literature does not specifically focus on the question of whether borrowers are credit rationed, any conclusion that one group is denied loans at a greater rate than others after creditworthiness is controlled for would imply that a form of credit rationing is occurring. This 'rationing', however, is distinct from that in Stiglitz–Weiss because the borrowers are not observably identical, and the underlying cause of 'rationing' is either lender preferences (Becker, 1971) or some form of statistical discrimination (Calomiris, Kahn and Longhofer, 1994; Longhofer and Peters, 2005).

Evidence on 'intermediary rationing'

In contrast to the limited evidence of traditional borrower credit rationing, there is a significant body of evidence supporting the idea that financial institutions are rationed by their depositors. In recent years, a large literature has developed examining the determinants of deposit withdrawal from individual banks, and a parallel literature has developed on systemic banking panics. These articles find that in circumstances where the condition of banks is perceived to have deteriorated, depositors withdraw funds rather than simply demand a higher interest rate on deposits (Calomiris and Mason, 2003; Calomiris and Wilson, 2004). The links between bank characteristics and deposit withdrawals observed in these and other similar studies suggest that deposit rationing is related to information and incentive problems, rather than just liquidity shocks to depositors, although such shocks may still play a role.

Final thoughts

It is worth noting that improvements in underwriting processes may have dramatically altered the practical impact of credit rationing in recent years. The use of risk-based pricing in consumer lending, including credit card loans and mortgages, has become widespread, reflecting the increased ability of lenders to distinguish between borrowers with different risk profiles (see, for example, Edelberg, 2003; Chomsisengphet and Pennington-Cross, 2006). The same is true for commercial credit markets, in which instruments such as junk bonds, senior-subordinated securitization issues, and the like serve to provide financial market access to broader classes of instruments, borrowers and risks. As a result, 'sorting' among borrowers overall has increased, and today there is likely much less diversity in pools of 'observably identical' borrowers than there was when Stiglitz and Weiss first developed their model. While this suggests that in some markets credit rationing is a very different and perhaps less important phenomenon today than it once was, an important potential role remains for credit rationing, particularly as it pertains to financial allocations in emerging markets, the pricing of particularly opaque segments of the lending markets of developed economies, and the ways in which financial institutions may be rationed in response to shocks to their portfolios.

CHARLES W. CALOMIRIS AND STANLEY D. LONGHOFER

See also **Akerlof, George Arthur; banking crises; micro-credit; Stiglitz, Joseph E.**

Bibliography

Akerlof, G.A. 1970. The market for 'lemons': quality uncertainty and the market mechanism. *Quarterly Journal of Economics* 84, 488–500.

Becker, G.S. 1971. *The Economics of Discrimination*, 2nd edn. Chicago: University of Chicago Press.

Berger, A.N. and Udell, G.F. 1992. Some evidence on the empirical significance of credit rationing. *Journal of Political Economy* 100, 1047–77.

Bester, H. 1985. Screening vs. rationing in credit markets with imperfect information. *American Economic Review* 75, 850–5.

Blinder, A.S. and Stiglitz, J.E. 1983. Money, credit constraints, and economic activity. *American Economic Review* 73, 297–302.

Calomiris, C.W. and Kahn, C.M. 1991. The role of demandable debt in structuring optimal banking arrangements. *American Economic Review* 81, 497–513.

Calomiris, C.W., Kahn, C.M. and Longhofer, S.D. 1994. Housing finance intervention and private incentives: helping minorities and the poor. *Journal of Money, Credit and Banking* 26, 634–74.

Calomiris, C.W. and Mason, J.R. 2003. Fundamentals, panics, and bank distress during the depression. *American Economic Review* 93, 1615–47.

Calomiris, C.W. and Wilson, B. 2004. Bank capital and portfolio management: the 1930s 'capital crunch' and the scramble to shed risk. *Journal of Business* 77, 421–55.

Cavalluzzo, K.S. and Cavalluzzo, L.C. 1998. Market structure and discrimination: the case of small businesses. *Journal of Money, Credit, and Banking* 30, 771–92.

Cavalluzzo, K. and Wolken, J. 2005. Small business loan turndowns, personal wealth, and discrimination. *Journal of Business* 78, 2153–77.

Chakravarty, S. and Scott, J.S. 1999. Relationships and rationing in consumer loans. *Journal of Business* 72, 523–44.

Chomsisengphet, S. and Pennington-Cross, A. 2006. The evolution of the subprime mortgage market. *Federal Reserve Bank of St. Louis Review* 88(1), 31–56.

Cox, D. and Jappelli, T. 1990. Credit rationing and private transfers: evidence from survey data. *Review of Economic Statistics* 72, 445–54.

Cressy, R. 1996. Are business startups debt-rationed? *Economic Journal* 106, 1253–70.

Diamond, D.W. and Dybvig, P.H. 1983. Bank runs, deposit insurance, and liquidity. *Journal of Political Economy* 91, 401–19.

Edelberg, W. 2003. Risk-based pricing of interest rates in household loan markets. FEDS Working Paper No. 2003–62.

Ferguson, M.F. and Peters, S.R. 2000. Is lending discrimination always costly? *Journal of Real Estate Finance and Economics* 21, 23–44.

Freimer, M. and Gordon, M.J. 1965. Why bankers ration credit. *Quarterly Journal of Economics* 79, 397–416.

Goldfeld, S.M. 1966. *Commercial Bank Behavior and Economic Activity: A Structural Study of Monetary Policy in the Postwar United States*. Amsterdam: North-Holland.

Hodgman, D.R. 1960. Credit risk and credit rationing. *Quarterly Journal of Economics* 74, 258–78.

Jaffee, D.M. 1971. *Credit Rationing and the Commercial Loan Market*. New York: Wiley.

Jaffee, D.M. and Modigliani, F. 1969. A theory and test of credit rationing. *American Economic Review* 59, 850–72.

Jaffee, D.M. and Russell, T. 1976. Imperfect information, uncertainty, and credit rationing. *Quarterly Journal of Economics* 90, 651–66.

Jensen, M.C. and Meckling, W.H. 1976. Theory of the firm: managerial behavior, agency costs and ownership structure. *Journal of Financial Economics* 3, 305–60.

Longhofer, S.D. and Peters, S.R. 2005. Self-selection and discrimination in credit markets. *Real Estate Economics* 33, 237–68.

McKinnon, R.I. 1973. *Money and Capital in Economic Development*. Washington, DC: Brookings Institution.

Miller, M.H. 1962. Credit risk and credit rationing: further comments. *Quarterly Journal of Economics* 76, 480–8.

Munnell, A.H., Tootell, G.M., Browne, L.E. and McEneaney, J. 1996. Mortgage lending in Boston: interpreting HMDA data. *American Economic Review* 86, 25–53.

Myers, S.C. and Majluf, N.S. 1984. Corporate financing and investment decisions when firms have information that investors do not have. *Journal of Financial Economics* 13, 187–221.

Ross, S.L. and Yinger, J. 2002. *The Color of Credit: Mortgage Discrimination, Research Methodology, and Fair-Lending Enforcement*. Cambridge, MA: MIT Press.

Rothschild, M. and Stiglitz, J.E. 1970. Increasing risk I: a definition. *Journal of Economic Theory* 2, 225–43.

Rothschild, M. and Stiglitz, J.E. 1976. Equilibrium in competitive insurance markets: an essay on the economics of imperfect information. *Quarterly Journal of Economics* 90, 630–49.

Scott, I.O. 1957. The availability doctrine: theoretical underpinnings. *Review of Economic Studies* 25, 41–8.

Stiglitz, J.E. and Weiss, A. 1981. Credit rationing in markets with imperfect information. *American Economic Review* 71, 393–410.

crime and the city

Crime is defined as an act committed in violation of a law forbidding it and for which punishment is imposed upon conviction. Crime is, however, not evenly distributed across space as it tends to be concentrated in specific areas where people are generally poor and uneducated. In both the United States and Europe, the typical urban pattern is that large cities have higher crime rates than smaller cities, and poor, largely minority neighbourhoods experience higher crime rates than more affluent white neighbourhoods (Raphael and Sills, 2005). According to the United Nations Interregional Crime and Justice Research Institute, (see Alvazzi del Frate, 1997), the percentage of population who are victims of burglary in urban areas with more than 100,000 inhabitants over a five-year period (between 1992 and 1996) is: 16 for Western Europe, 24 for North America, 20 for South America, 18 for Eastern Europe, 13 for Asia and 38 for Africa. Another typical pattern common to both the United States and Europe is that ethnic minorities are overrepresented in criminal activities. In the United States, the proportion of 20–29-year-old black men directly in trouble with the law (in jail or prison or on probation or parole) reached 23 per cent in 1989 (Freeman, 1999). There is, however, one notable difference. Since the mid-1980s, crime has declined in the United States but increased in Europe, especially in large urban areas (Blumstein and Wallman, 2000).

Theories

In the standard crime model (Becker, 1968), each individual has to implement a cost–benefit analysis in order to choose between becoming a criminal and participating in the labour market. The cost is the severity of punishment, which obviously depends on the probability of being arrested. The benefit consists in the proceeds from crime. If crime is localized, then criminals will trade off a lower probability of being arrested (since, in some areas, a host of criminals are active and the number of policemen is not sufficient) against lower proceeds from crime (more criminals also imply less booty). In this context, Sah (1991) examines the influence of the social environment on individuals' perceptions of the probability of arrest. Indeed, people develop their ideas about the relative benefits and costs of crime based on the observations they make every day. If a person lives in an area with a high crime rate, and particularly if the criminals are seen to be relatively successful, then that person is more likely to engage in criminal activity. The main result of this paper is that individuals in some areas tend to commit more crime than the Beckerian model would predict because of the gap between the perceived and the real cost of committing crime, which leads to a lower sense of impunity based on the information provided by their criminal friends.

Another approach (Verdier and Zenou, 2004) proposes that distance to jobs plays a role in crime behaviour and provides a unified explanation for why blacks commit more crime, are located in poorer neighbourhoods and receive lower wages than whites. The mechanism is as follows. If everybody believes that blacks are more prone to crime than whites, even if there is no basis for this, then blacks are offered lower wages and, as a result, locate further away from jobs. Because distant residence implies more tiredness and higher commuting costs, the

black–white wage gap is widened further. Blacks have thus a lower opportunity cost of committing crime (lower outside option) and become indeed more criminal than whites. The loop is closed and the beliefs are self-fulfilling.

Whereas the standard Beckerian approach focuses on individual behaviour, Glaeser, Sacerdote, and Scheinkman (1996) stress the role of peers and social interactions in criminal activities, especially in urban areas because of the high variance in crime rates. Two types of individuals are assumed: those who, as in the standard model, base their crime decision on a cost–benefit analysis, and those who only imitate their neighbours. Because of these social interactions, the benefits from crime are greater than in the Beckerian model. Moreover, if these interactions are localized (as is usually the case), then it becomes easy to explain very high levels of crime in some areas of the city. Indeed, if there are already a lot of criminals in a particular location, then crime becomes 'contagious' by spreading like a virus and amplifies the number of criminals in this location. There are social *multiplier* effects through a feedback loop: negative social behaviour such as crime leads to more negative social behaviour.

Calvó-Armengol and Zenou (2004), and Ballester, Calvó-Armengol and Zenou (2004) propose a model along these lines but represent social interactions in terms of a social network of criminal friends. People in a network not only imitate but also influence each other. Here, the cost of committing crime is reduced thanks to the network of friends. Indeed, delinquents learn from other criminals belonging to the same network how to commit crime in a more efficient way by sharing the know-how about the 'technology' of crime. They show that the influence of peers on the individual's criminal activity depends on his or her position in the network, and each agent's criminal effort is proportional to his or her Bonacich centrality measure (see Bonacich, 1987). For a given network, the Bonacich network centrality counts, for each agent, the total number of direct and indirect paths of any length in the network stemming from this agent. Such paths are weighted by a geometrically decaying factor (with path length). In other words, the 'location' of each individual in a network of friends, as measured by the Bonacich centrality measure, is a key determinant of his or her criminal activity.

As a result, in a spatial or social context, an efficient policy aiming at reducing crime would not be, as in the Beckerian model, to increase at random the cost of committing crime, but rather to target criminals according to their location in the urban or social space. Ballester, Calvó-Armengol and Zenou (2004) propose a policy that consists in finding and getting rid of the key player, that is, the criminal who, once removed, leads to the highest aggregate crime reduction. They show that the key player is not necessarily the most active criminal (that is, the one with the highest Bonacich centrality). Indeed, removing a criminal from a network has both a direct and an indirect effect. The direct effect is that fewer criminals contribute to the aggregate crime level. The indirect effect is that the network topology is modified, and the remaining criminals adopt different crime efforts. The key player is the one with the highest overall effect.

Empirical studies

One of the first tests of the Becker model was undertaken by Ehrlich (1973), who used as explanatory variables the imprisonment rate and the average sentence for the crime in question. More recently, the focus has been on urban or social problems because this is particularly fruitful for understanding personal and property crime as opposed to white-collar crime. Cullen and Levitt (1999), using data for 137 US cities from 1976 to 1993, explore the relationship between crime and urban flight (that is, the flight of the white population from city centres to suburbs). They find that each additional reported crime in city centre is associated with a net decline of about one resident. Causality runs from rising crime rates to city depopulation. Pursuing this area of research, Glaeser and Sacerdote (1999) provide three reasons for higher crime rates in big cities. They report that 27 per cent of the difference between urban and rural crime rates in the United States is due to higher pecuniary benefits for crime in cities, 20 per cent to a lower probability of arrest and recognition in cities, and the remaining 45–60 per cent to the observable characteristics of individuals. This last number can be explained by a positive covariance across agents' decisions about crime, so that the variance of crime rate is higher than the variance predicted by local conditions. This implies that social interactions should matter, especially in cities.

Case and Katz (1991) were among the first to investigate this last issue. Using data from the 1989 NBER survey of youths living in low-income Boston neighbourhoods, they find that the behaviours of neighbourhood peers appear to substantially affect youth behaviours in a manner suggestive of contagion models of neighbourhood effects. The direct effect of moving a youth with given family and personal characteristics to a neighbourhood where 10 per cent more of the youths are involved in crime than in his or her initial neighbourhood is to raise the probability the youth will become involved in crime by 2.3 per cent.

Glaeser, Sacerdote and Scheinkman (1996) find that, across crimes, crime committed by younger people has higher degrees of social interaction, while, across cities, for serious crimes in general and for larceny and auto theft in particular, the degree of social interactions is larger in those communities where families are less intact, that is, have more female-headed households. Ludwig, Duncan and Hirschfield (2001) and Kling, Ludwig and Katz (2005) explore this last result by using data from the Moving to Opportunity (MTO) experiment that assigned

a total of 638 families from high-poverty Baltimore neighbourhoods into three 'treatment groups': (*a*) Experimental group families receive housing subsidies, counselling and search assistance to move to private-market housing in low-poverty census tracts; (*b*) Section 8-only comparison group families receive private-market housing subsidies with no programme constraints on relocation choices; and (*c*) a Control group receives no special assistance under MTO. They show that relocating families from high- to low-poverty neighbourhoods reduces juvenile arrests for violent offences by 30 to 50 per cent of the arrest rate for control groups. This also suggests very strong social interactions in crime behaviours.

Using a very detailed data-set of friendship networks in the United States from the National Longitudinal Survey of Adolescent Health (AddHealth), Calvó-Armengol, Patacchini and Zenou (2005) test the main results of Ballester, Calvó-Armengol and Zenou (2004). Contrary to the standard approach, here peer effects are conceived not as an average intra-group externality that affects identically all the members of a given group, but as a collection of dyadic bilateral relationships, which constitutes a social network. The position and thus the centrality of each individual are thus crucial to understand criminal behaviour. Calvó-Armengol, Patacchini and Zenou (2005) show that, after observable individual characteristics and unobservable network specific factors are controlled for, the individual's position in a network (as measured by his or her Bonacich centrality) is a key determinant of his or her level of criminal activity. A standard deviation increase in the Bonacich centrality increases the level of individual delinquency by 45 per cent of one standard deviation.

YVES ZENOU

See also **labour market discrimination; law, economic analysis of; neighbours and neighbourhoods; racial profiling; residential segregation; social interactions (theory); social multipliers; social networks in labour markets; spatial mismatch hypothesis; urban economics.**

Bibliography

Alvazzi del Frate, A. 1997. Preventing crime: citizens' experience across the world. Issues and Reports No. 9. Rome: United Nations Interregional Crime and Justice Research Institute.

Ballester, C., Calvó-Armengol, A. and Zenou, Y. 2004. Who's who in crime networks. Wanted: the key player. Discussion Paper No. 4421. London: CEPR.

Becker, G. 1968. Crime and punishment: an economic approach. *Journal of Political Economy* 76, 169–217.

Blumstein, A. and Wallman, J. 2000. *The Crime Drop in America*. New York: Cambridge University Press.

Bonacich, P. 1987. Power and centrality: a family of measures. *American Journal of Sociology* 92, 1170–82.

Calvó-Armengol, A., Patacchini, E. and Zenou, Y. 2005. Peer effects and social networks in education and crime. Discussion Paper No. 5244. London: CEPR.

Calvó-Armengol, A. and Zenou, Y. 2004. Social networks and crime decisions: the role of social structure in facilitating delinquent behavior. *International Economic Review* 45, 939–58.

Case, A. and Katz, L. 1991. The company you keep: the effects of family and neighborhood on disadvantaged youths. Working Paper No. 3705. Cambridge, MA: NBER.

Cullen, J. and Levitt, S. 1999. Crime, urban flight, and the consequences for cities. *Review of Economics and Statistics* 81, 159–69.

Ehrlich, I. 1973. Participation in illegitimate activities: a theoretical and empirical investigation. *Journal of Political Economy* 81, 521–65.

Freeman, R. 1999. The economics of crime. In *Handbook of Labor Economics*, ed. O. Ashenfelter and D. Card. Amsterdam: North-Holland.

Glaeser, E. and Sacerdote, B. 1999. Why is there more crime in cities? *Journal of Political Economy* 107, S225–S258.

Glaeser, E., Sacerdote, B. and Scheinkman, J. 1996. Crime and social interactions. *Quarterly Journal of Economics* 111, 508–48.

Kling, J., Ludwig, J. and Katz, L. 2005. Neighborhood effects on crime for female and male youth: evidence from a randomized housing voucher experiment. *Quarterly Journal of Economics* 120, 87–130.

Ludwig, J., Duncan, G. and Hirschfield, P. 2001. Urban poverty and juvenile crime: evidence from a randomized housing-mobility experiment. *Quarterly Journal of Economics* 116, 655–79.

Raphael, S. and Sills, M. 2005. Urban crime in the United States. In *A Companion to Urban Economics*, ed. R. Arnott and D. McMillen. Boston: Blackwell.

Sah, R. 1991. Social osmosis and patternsof crime. *Journal of Political Economy* 99, 1272–95.

Verdier, T. and Zenou, Y. 2004. Racial beliefs, location and the causes of crime. *International Economic Review* 45, 731–60.

cross-cultural experiments

A large number of well-replicated results using a wide variety of experimental games are inconsistent with the assumption that people are money maximizers. Instead, people's behaviour is consistent with choices based on social preferences in which people place a positive value on fairness, reciprocity, or equity (see Camerer, 2003, for a review). For example, subjects typically make significant positive contributions in the public goods games, reject positive offers in the ultimatum game, and impose costly punishment in the third-party punishment game (see, Camerer, ch. 2, for descriptions of these games.) In some games these results are insensitive to framing and whether behaviour is anonymous to the experimenter ('double blind').

These experiments are open to two qualitatively different interpretations: It could be that pro-social behaviours like cooperation in the public goods game and punishment in the third-party punishment game reflect human nature. Cooperation in the public goods game could result from universal cognitive systems that cause people everywhere to behave as if all acts have reputational consequences, even when facts suggest no one will know what they have done. Punishment in the third-party punishment game could result from a pan-human motivational system that causes people to prefer outcomes that are fair or mutually beneficial, and to derive satisfaction from punishing unfair behaviour. However, with few exceptions experimental subjects have been university students in urbanized, industrial societies. Thus, it also could be that observed pro-social behaviour results from culturally evolved beliefs and values that are specific to such social environments. It is obviously of great importance to determine which of these two interpretations is correct.

To answer this question, a team of anthropologists and economists performed two rounds of experimental games in a wide range of cultural environments. The first round (Henrich et al., 2004; 2005) comprised a diverse group of 15 societies including peoples like the Aché and Hadza who live in nomadic foraging bands, the Achuar and Au who live in small villages and mix hunting and horticulture, Mongol and Sangu pastoralists, and sedentary Shona farmers in Zimbabwe. The ultimatum game was performed in all 15 societies, and the public goods game and the dictator game were performed in different subsets. The second round (Henrich et al., 2006) included a similar and overlapping range of 15 societies. Based on experience in the first round, experimental protocols were improved and standardized, and a greater effort was made to collect standardized data on individual characteristics. During the second round the ultimatum, dictator, and third-party punishment games were performed in all 15 societies. In addition complete strategies for second players in the ultimatum game and punishers in the third-party punishment game were elicited using the strategy method.

These experiments reveal a number of interesting results.

1. *Behaviour in non-Western populations can be quite different from that of Western university subjects.* Figure 1 shows the distribution of ultimatum game offers in the first round of experiments. The Pittsburgh data taken from Roth et al. (1991) are typical for university populations – the modal offer is 50 per cent but many subjects make somewhat lower offers. Behaviour in other populations can be very different. For example, modal offers are much lower among two lowland tropical forest groups; the Achuar and the Machiguenga are quite low. Interestingly, these very low offers were usually accepted,

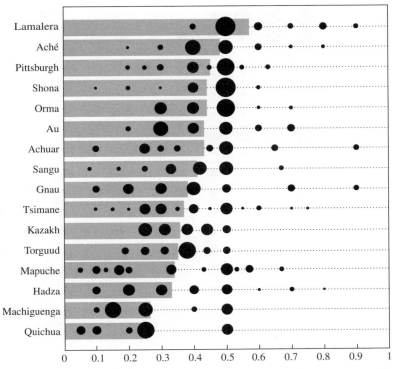

Figure 1   Ultimatum game offer. *Note*: A bubble plot showing the distribution of ultimatum game offers for each group. The diameter of the circle at each location along each row represents the proportion of the sample that made a particular offer. The right edge of the lightly shaded horizontal grey bar is the mean offer for that group. In the Machiguenga row, for example, the mode is 0.15, the secondary mode is 0.25, and the mean is 0.26. *Source*: Henrich et al. (2005).

behaviour much closer to the predictions of money maximization than the behaviour of Western university subjects. Non-western populations also exhibited novel behaviours not seen in university populations. Figure 2 shows the rejection probabilities for different ultimatum game offers. Notice that in several populations increasing offer level above 50 per cent *increased* the rate of rejections, a phenomenon not observed among student subjects.

2. *Behavioural differences are correlated with group characteristics but not individual characteristics.* The ethnographers who performed most of these experiments have studied these groups for many years and have detailed data on subjects about income, wealth, education, market contact, and a variety of other factors. None of these factors was significantly correlated with ultimatum game offers within social groups in first round, or offers or rejections in the second round. Because measures of wealth, income, and so on are not comparable across groups, these measures could not be aggregated to derive group characteristics. However, during the first round, ethnographers who were blind to the results ranked each of the groups along five dimensions: extent of cooperation in subsistence, degree of market contact,

amount of privacy, amount of anonymity, and social complexity. We also had comparable data on settlement size. It turned out that market contact, settlement size, and social complexity were all highly correlated, so these were collapsed into a single variable labelled 'aggregate market contact'. Multiple linear regression showed that increasing aggregate market contact and cooperation in subsistence significantly predicted increased ultimatum game offers, and together the two variables accounted for more than half of the variance among groups in average offers.

3. *Variation in punishment predicts variation in altruism across societies.* In the third-party punishment game, an individual, the 'punisher' observes a dictator game and can punish the dictator at a cost to him or herself. The average minimum offer acceptable to the punisher in this game provides a measure of the level of punishment in that society. As is shown in Figure 3, this measure of punishment also predicts the level of altruism measured by dictator offers in the ordinary dictator game.

Taken together these results indicate that pro-social behaviour in economic experiments does not result from an invariant property of our species, and instead suggest that there are significant cultural differences between

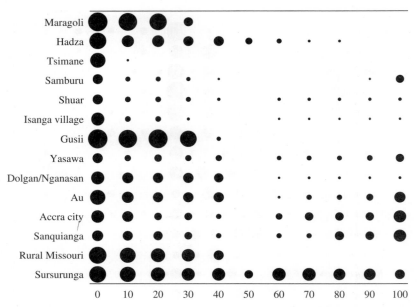

Figure 2   Ultimatum game rejection rates. *Note*: The diameter of the black circles is proportional to the fraction of offers that would have been rejected in the ultimatum game during the second round of experiments plotted as a function of the offer as a percentage of the maximum offer. For scale, note that the Gusii and Maragoli rejected all offers of zero. Notice that in all societies offering 50% of the stake minimizes the probability of rejection, but that in a number of societies increasing offers above 50% increases the rate of rejection. *Source*: Henrich et al. (2006).

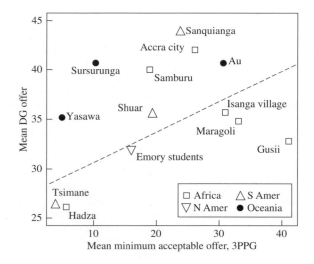

Figure 3   Mean minimum acceptable offer, third-party punishment game. *Note*: The mean offer in the dictator game for a society plotted against the mean value of the minimum acceptable offer in the third-party punishment game. The different symbols indicate continents. The size of each symbol is proportional to the number of DG pairs at each site. The dotted line gives the weighted regression line, with continental controls of mean dictator game offers against mean minimum acceptable offer in the third-party punishment game. *Source*: Henrich et al. (2006).

societies. The fact that ultimatum game behaviour is predicted by the average level of cooperation and average level of market contact further indicates that these cultural differences are not arbitrary, but may reflect economic, ecological and social differences between societies. However, the lack of correlation between individual characteristics and individual behaviour indicates that the differences between societies are not likely to be explained as the simple aggregation of individual experiences. Instead, it is more plausible that cultures evolve over time in response to the average conditions which they face, and that individual behaviour is, in turn, shaped by these cultural differences.

ROB BOYD

See also **experimental economics.**

Bibliography

Camerer, C. 2003. *Behavioral Game Theory: Experiments on Strategic Interaction.* Princeton: Princeton University Press.

Henrich, J.R.B., Bowles, S., Camerer, C., Fehr, E. and Gintis, H. 2004. *The Foundations of Human Sociality: Economic Experiments and Ethnographic Evidence from Fifteen Small-Scale Societies.* New York: Oxford University Press.

Henrich, J., Boyd, R., Bowles, S., Camerer, C., Fehr, E., Gintis, H., McElreath, R., Alvard, M., Barr, A., Ensminger, J., Hill, K., Gil-White, F., Gurven, M., Marlowe, F., Patton, J.Q., Smith, N. and Tracer, D. 2005. 'Economic man' in cross-cultural perspective: behavioral experiments in 15 small-scale societies. *Behavioral and Brain Sciences* 28, 795–855.

Henrich, J., McElreath, R., Barr, A., Ensminger, J., Barrett, C., Bolyanatz, A., Cardenas, J.C., Gurven, M., Gwako, E., Henrich, N., Lesorogol, C., Marlowe, F., Tracer, D. and Ziker, J. 2006. Costly punishment across human societies. *Science* 312, 1767–70.

Roth, A.E., Prasnikar, V., Okuno-Fujiwara, M. and Zamir, S. 1991. Bargaining and market behavior in Jerusalem, Ljubljana, Pittsburgh, and Tokyo: an experimental study. *American Economic Review* 81, 1068–95.

crowding out

'Crowding out' refers to all the things which can go wrong when debt-financed fiscal policy is used to affect output.

A first line of argument questions whether fiscal policy has any effect at all on spending. Changes in the pattern of taxation which keep the pattern of spending unaffected do not affect the intertemporal budget constraint of the private economy and thus may have little effect on private spending. This argument, known as the 'Ricardian equivalence' of debt and taxation, holds only if taxes are lump sum (Barro, 1974). Some taxes which

induce strong intertemporal substitution, such as an investment tax credit for firms, will have stronger effects if they are temporary; for most others, such as income taxes, changes in the intertemporal pattern may have only a small effect on the pattern of spending.

The Ricardian equivalence argument is not settled empirically and its validity surely depends on the circumstances. A change in the intertemporal taxation of assets such as land or housing, leaving the present value of taxes the same, will have little effect on their market value, thus on private spending. An explicitly temporary income tax increase may have little effect on spending while the anticipation of prolonged deficits may lead taxpayers to ignore the eventual increase in tax liabilities. Evidence from specific episodes, such as the 1968 temporary tax surcharge in the United States, suggests partial offset at best.

Changes in the pattern of government spending obviously have real effects. But here again, various forms of direct crowding out may be at work. Public spending may substitute perfectly or imperfectly for private spending, so that changes in public spending may be directly offset, fully or partially, by consumers or firms. Even if public spending is on public goods, the effect will depend on whether the change in spending is thought to be permanent or transitory. Permanent changes, financed by a permanent increase in taxes, will, as a first approximation, lead to a proportional decrease in private spending, with no effect on total spending. Temporary changes in spending, associated with a temporary increase in taxes, lead to a smaller reduction in private spending and thus to an increase in total spending.

In summary, one should not expect any change in taxation or government spending to have a one-for-one effect on aggregate demand. An eclectic reading of the discussion above may be that only sustained decreases in income taxation, or the use of taxes that induce strong intertemporal substitution, or temporary increases in spending, can reliably be used to boost aggregate demand. The focus in what follows will be on these forms of fiscal expansion.

Crowding out at full employment

Not every increase in aggregate demand translates into an increase in output.

This is clearly the case if the economy is already at full employment (I use 'full employment' to mean employment when unemployment is equal to its natural rate). While tracing the effects of fiscal expansion at full employment is of limited empirical interest, except perhaps as a description of war efforts, it is useful for what follows. If labour supply is inelastic, output is fixed and any increase in aggregate demand must be offset by an increase in interest rates, leaving output unchanged. In the case of an increase in public spending, private spending will decrease; in the case of a decrease in income taxation, private spending will in the end be the same, but its composition will change as the share of interest sensitive components decreases. (If labour supply can vary, the story is more complicated. See, for example, Baxter and King, 1993, for an analysis of changes in government spending in an otherwise standard RBC model.)

This is just the beginning of the story, however. Over time, changes in capital and debt lead to further effects on output. The decrease in investment in response to higher interest rates leads to a decline in capital accumulation and output, reducing the supply of goods. If fiscal expansion is associated with sustained deficits, the increase in debt further increases private wealth and private spending at given interest rates, further increasing interest rates and accelerating the decline in capital accumulation (see, for example, Blanchard, 1985, for a characterization of these dynamic effects in an economy with finite horizon consumers). How strong is this negative effect of debt on capital accumulation likely to be? One of the crucial links in this mechanism is the effect of government debt on interest rates; empirical evidence, both across countries and from the last two centuries, shows surprisingly little relation between the two. This probably reflects, however, more the difficulty of identifying and controlling for other factors than the absence of an effect of debt and deficits on interest rates.

Worse can happen. It may be that the fiscal programme becomes unsustainable. There is no reason to worry about a fiscal programme in which debt grows temporarily faster than the interest rate. But there is reason to worry when there is a positive probability that, even under the most optimistic assumptions, debt will have to grow for ever faster than the interest rate. When this is the case, it implies that the government can meet its interest payments on existing debt only by borrowing more and more. What happens then may depend on the circumstances. Bond holders may start anticipating repudiation of government debt and require a risk premium on the debt, further accelerating deficits and the growth of the debt. If they instead anticipate repudiation through inflation, they will require a higher nominal rate and compensation for inflation risk in the form of a premium on all nominal debt, private and public. What is sure is that there will be increased uncertainty in financial markets and that this will further contribute to decreases in output and in welfare. The historical record suggests that it takes very large deficits and debt levels before the market perceives them as potentially unsustainable. England was able in the 19th century to build debt-to-GDP ratios close to 200 per cent without apparent trouble. Some European countries are currently running high deficits while already having debt-to-GDP ratios in excess of 100 per cent, without any evidence of a risk premium on government debt. The threshold seems lower for Latin American economies. But even if one excludes this worst-case scenario, fiscal expansion can clearly have adverse effects on output at full employment. The relevant issue, however, is whether the same dangers are present when fiscal expansion is implemented to reduce unemployment, which is presumably when it is most likely to be used.

Crowding out at less than full employment

The historical starting point of the crowding out discussion is the fixed price IS–LM model. In that model, a fiscal expansion raises aggregate demand and output. The pressure on interest rates does not come from the full employment constraint as before but from the increased demand for money from increased output. Thus the fiscal multiplier is smaller the lower the elasticity of money demand to interest rates, or the larger the elasticity of private spending to interest rates. Fiscal expansion crowds out the interest-sensitive components of private spending, but the multiplier effect on output is positive. As output and interest rates increase, it is quite possible for both investment and consumption to increase. But what happens when the model is extended to take into account dynamics, expectations and so on? Can one overturn the initial result and get full crowding out or even negative multipliers?

Even within the static IS–LM, one can in fact get zero or negative multipliers. This is the case, for example, if money demand from agents is higher than that from the government and the change in policy redistributes income from the government to agents. While this case is rather exotic, a much stronger case can be made if the economy is small, open, and with capital mobility and flexible exchange rates, as in the 'Mundell–Fleming' model. In this case, with the interest rate given from outside, and fixed money supply, money demand determines output; fiscal policy leads only to exchange rate appreciation. Exchange rate-sensitive components are now crowded out by fiscal expansion. The multiplier is equal to zero.

When dynamic effects are taken into account, other channels arise for crowding out. The analysis of these dynamic effects, with the dynamics of debt accumulation taken into account, was initially conducted under the maintained assumption of fixed prices and demand determination of output (Tobin and Buiter, 1976). Then, as debt was accumulating, private wealth and spending increased, leading to even larger effects of fiscal policy on output in the long run than in the short run. But the assumption of fixed prices, while debt and capital accumulation are allowed to proceed, is surely misleading; when prices are also allowed to adjust, the effects of fiscal policy become more complex, and crowding out more likely. This is because some of the full employment effects come back into prominence: if fiscal expansion is maintained even after the economy has reached full employment, then the perverse effects of higher interest rates on capital accumulation and full employment output come again into play. This is true even if deficits disappear before the economy returns to full employment; the economy inherits a larger level of debt, and thus must have higher interest rates and lower capital accumulation than it would otherwise have had. The fiscal expansion trades off a faster return to full employment for lower full-employment output.

Anticipations of these full employment effects are likely to feed back and modify the effects of fiscal policy at the start, when the economy is still at less than full employment. Anticipations of higher interest rates, perhaps also of higher distortions due to the higher taxes needed to service the debt, may dominate the direct effects of higher government spending on demand, and lead to an initial decrease rather than an initial increase in demand and output. Symmetrically, fiscal consolidation, to the extent that it implies lower interest rates and lower distortions in the future, may be expansionary. This is even more likely to be the case if fiscal consolidation decreases the risk of default on government debt, and thus decreases the risk of major economic disruptions. There is indeed some evidence that, when initial fiscal conditions are very bad, and the fiscal consolidation is large and credible, the net effect of consolidation may be expansionary (Giavazzi and Pagano, 1990).

Crowding out: an assessment

Should one conclude from this that fiscal policy is an unreliable macroeconomic tool, with small and sometimes negative effects on output? The answer is 'no'. Fiscal policy is likely to partly crowd out some components of private spending, even in the best circumstances, but there is little reason to doubt that it can help the economy return to full employment. Ricardian equivalence and direct crowding out warn us that not any tax cut or spending increase will increase aggregate demand. But there is little question that temporary spending or sustained income tax cuts will do so. Results of full crowding out at less than full employment, such as the Mundell–Fleming result, are simply a reminder that the monetary-fiscal policy mix is important.

In all cases, monetary accommodation of the increased demand for money removes the negative or the zero multipliers. That fiscal expansion affects capital accumulation, and output adversely at full employment, and that unsustainable fiscal programmes may lead to crises of confidence, is a reminder that fiscal expansion should not be synonymous with steady increases in the debt-to-GDP ratio even after the economy has returned to full employment. This shows one of the difficulties associated with fiscal expansion: if done through tax cuts, it has to be expected to last long enough to affect private spending, but not so long as to lead to expectations of runaway deficits in the long run. The room for manoeuvre is, however, substantial. Some taxes, such as the investment tax credit, work best when temporary. These can be used,

as they work in the short run and have few adverse implications for the long run.

OLIVIER J. BLANCHARD

See also **budget deficits; real business cycles; Ricardian equivalence theorem; Tobin, James.**

Bibliography

Barro, R. 1974. Are government bonds net wealth? *Journal of Political Economy* 82, 1095–117.

Baxter, M. and King, R. 1993. Fiscal policy in general equilibrium. *American Economic Review* 83, 315–34.

Blanchard, O. 1985. Debt, deficits, and finite horizons. *Journal of Political Economy* 93, 223–47.

Giavazzi, F. and Pagano, M. 1990. Can severe fiscal contractions be expansionary? *NBER Macroeconomics Annual* 5, 75–122.

Tobin, J. and Buiter, W. 1976. Long-run effects of fiscal and monetary policy on aggregate demand. In *Monetarism*, ed. J. Stein. Amsterdam: North-Holland.

cultural transmission

Preferences, beliefs, and norms that govern human behaviour are partly formed as the result of genetic evolution, and partly transmitted through generations and acquired by learning and other forms of social interaction. The transmission of preferences, beliefs and norms of behaviour which is the result of social interactions across and within generations is called *cultural transmission*. Cultural transmission is therefore distinct from, but interacts with, genetic evolution.

Cultural transmission is an object of study of several social sciences, such as evolutionary anthropology, sociology, social psychology and economics, as well as of evolutionary biology. The theoretical contributions of Cavalli-Sforza and Feldman (1981) and Boyd and Richerson (1985), who apply models of evolutionary biology to the transmission of cultural traits, as well as the empirical study of cultural socialization in American schools by Coleman (1988), had a great multidisciplinary impact. Recently, economists have also studied the determination and the dynamics of preferences, beliefs, norms and, more generally, cultural and cognitive attitudes.

Cultural transmission arguably plays an important role in the determination of many fundamental preference traits, like discounting, risk aversion and altruism. It plays a central role in the formation of cultural traits and norms, like attitudes towards the family and fertility practices, and in the job market. It is, however, the pervasive evidence of the resilience of ethnic and religious traits across generations that motivates a large fraction of the theoretical and empirical literature on cultural transmission. For instance, the fast assimilation of immigrants into a 'melting pot', which many social scientists predicted until the 1960s (see, for example, Gleason, 1980, for a survey), simply did not materialize. Moreover, the persistence of 'ethnic capital' in second- and third-generation immigrants has been documented by Borjas (1992), and recently also by Fernandez and Fogli (2005) and Giuliano (2007) for norms of behaviour regarding, respectively, work and fertility practices and living arrangements. Orthodox Jewish communities in the United States constitute another example of the strong resilience of culture (see Mayer, 1979, and the discussion of a 'cultural renaissance' rather than the complete assimilation of Jewish communities in New York in the 1970s). Outside the United States, Basques, Catalans, Corsicans, and Irish Catholics in Europe, Quebecois in Canada, and Jews of the diaspora have all remained strongly attached to their languages and cultural traits even through the formation of political states which did not recognize their ethnic and religious diversity.

Models of cultural transmission have implications regarding the determinants of the persistence of cultural traits and more generally regarding the population dynamics of cultural traits. In the economic literature in particular, cultural transmission is modelled as the result of purposeful socialization decisions inside the family ('direct vertical socialization') as well as of indirect socialization processes like social imitation and learning ('oblique and horizontal socialization'). Therefore, the persistence of cultural traits or, conversely, the cultural assimilation of minorities is determined by the costs and benefits of various family decisions pertaining to the socialization of children in specific socio-economic environments, which in turn determine the children's opportunities for social imitation and learning.

Evolutionary biology models

L. Cavalli-Sforza and M. Feldman are the first to formally study the transmission of cultural traits. Their formal models are adopted from evolutionary biology. In a baseline version of these models, they obtain a simple differential equation which describes the population dynamics of cultural traits. Consider the dynamics of a dichotomous cultural trait in the population; formally, a fraction q^i of the population has trait i, and a fraction $q^j = 1 - q^i$ has trait j. Families are composed of one parent and a child, and hence reproduction is asexual. All children are born without defined preferences or cultural traits, and are each first exposed to their parent's trait, which they adopt with probability d^i. If a child from a family with trait i is not directly socialized, which occurs with probability $1 - d^i$, he or she picks the trait of a role model chosen randomly in the population (that is, he or she picks trait i with probability q^i and trait j with probability $1 - q^i$). Therefore, the probability that the child of parents of trait i will also have trait i is $\Pi^{ii} = d^i + (1 - d^i)q^i$; while the probability that he or she will have trait j is $\Pi^{ij} = (1 - d^i)(1 - q^i)$. It follows

that the dynamics of the fraction of the population with trait i, in the continuous time limit, are characterized by:

$$\dot{q}^i = (d^i - d^j)q^i(1 - q^i) \qquad (1)$$

The dynamics that equation (1) describes imply that the distribution of cultural traits in the population converges to a degenerate distribution concentrated on trait i whenever $d^i > d^j$ (and on trait j when $d^i < d^j$), while any initial distribution is stationary in the knife-edge case in which $d^i = d^j$. This model therefore predicts the complete assimilation of the trait with weaker direct vertical socialization. Moreover, it predicts faster assimilation for smaller minorities. Both predictions are at odds with the documented strong resilience of cultural traits discussed above. Cavalli-Sforza and Feldman show how these extreme predictions can be relaxed by considering other effects like mutations, migrations and horizontal cultural transmission among peers. Boyd and Richerson (1985) in turn extend the analysis of Cavalli-Sforza and Feldman (1981) by considering forms of direct vertical socialization called *frequency dependent* biased transmission, which depend on the distribution of the population by cultural trait. Formally, they allow d^i to be a function of q^i.

Bisin and Verdier (2001a) study the same differential equation for the population dynamics of cultural traits, with the objective of characterizing the conditions which give rise to culturally heterogeneous stationary distributions, that is, limit population with a positive fraction of either cultural trait, $0 < q^i < 1$. They show that the crucial determinant of the composition of the stationary distribution consists in whether the socio-economic environment (oblique socialization) acts as a substitute or as a complement to direct vertical socialization. More precisely, when direct vertical socialization and oblique transmission are *cultural substitutes*, parents by definition socialize their children less the more widely dominant are their cultural traits in the population. In such a case, $d^i(q^i)$ is a strictly decreasing function in q^i, and in the long run a non-degenerate stable stationary distribution exists. It is characterized by a q^i such that the direct vertical socialization of the two cultural types are equalized (that is, $d^i(q^i) = d^j(1 - q^i)$). Intuitively, when family and society are substitutes in the transmission mechanism, in fact families socialize children more intensely whenever the set of cultural traits they wish to transmit is common only to a minority of the population. Conversely, families which belong to a cultural majority spend fewer resources directly socializing their children, since their children adopt or imitate with high probability the predominant cultural trait in society at large, which is the one their parents desire for them. *Cultural substitutability* tends to preserve cultural heterogeneity in the population because in this case minorities directly socialize their children more than majorities. The other typical situation is the opposite one in which direct vertical transmission is a *cultural complement* to oblique transmission;

that is, when parents socialize their children more intensely the more widely dominant their cultural trait is the population. In such a case, $d^i(q^i)$ is a strictly increasing function in q^i and in the long run the dynamics converges to a culturally homogeneous cultural population (with either $q^i = 0$ or $q^i = 1$ depending on the initial distribution).

Economic models of cultural transmission

Economic models of cultural transmission induce testable restrictions on the form of the function $d^i(q^i)$. In their baseline specification, for instance, Bisin and Verdier (2001a) assume that parents are altruistic towards their children and hence might want to socialize them to a specific cultural model if they think this will increase their children's welfare. If we let V^{ij} denote the utility to a type i parent of a type j child, $i, j \in \{a, b\}$, the formal assumption is

for all i, j with $i \neq j$, $V^{ii} > V^{ij}$

This assumption, called *imperfect empathy*, can be interpreted as a form of myopic or paternalistic altruism. Parents are aware of the different traits children can adopt and are able to anticipate the socio-economic choices a child with trait i will make in his or her lifetime. However, parents can evaluate these choices only through the filter of their own subjective evaluations and cannot 'perfectly empathize' with their children. As a consequence of imperfect empathy, parents, while altruistic, tend to prefer children with their own cultural trait and hence attempt to socialize them to this trait. (Some justifications of imperfect empathy from an evolutionary perspective are provided by Bisin and Verdier, 2001b. The assumption can be relaxed, as for example in Sáez-Martí and Sjogren, 2005.) Assume socialization is costly and let costs be denoted by $C(d^i)$. Parents of type i then choose d^i to maximize:

$$-C(d^i) + (\Pi^{ii} V^{ii} + \Pi^{ij} V^{ij}) \qquad (2)$$

$$s.t \Pi^{ii} = d^i + (1 - d^i)q^i,$$
$$\Pi^{ij} = (1 - d^i)(1 - q^i) \qquad (3)$$

Under standard assumptions, the solution to this problem provides a continuous map $d^i = d(q^i, \Delta V^i)$, where $\Delta V^i = V^{ii} - V^{ij}$ is the subjective utility gain of having a child with trait i. It reflects the degree of 'cultural intolerance' of type i's parents with respect to cultural deviations from their own trait. Given imperfect empathy on the part of parents, $\Delta V^i > 0$. The dynamics of the fraction of the population with cultural trait i is then determined by equation (1) evaluated at $d^i(q^i) = d(q^i, \Delta V^i)$. It is straightforward to demonstrate that this class of socialization mechanisms generates cultural substitutability and therefore the preservation of

cultural heterogeneity. Other micro-founded specifications and examples are provided in Bisin and Verdier (2001a), some of which illustrate the contrary possibility of cultural complementarity and the tendency of cultural homogenization over time.

Direct socialization mechanisms and socio-economic interactions

Several specific choices contribute to direct family socialization and hence to cultural transmission. Prominent examples are education decision, family location decisions, and marriage choices. While education choices have been studied by Cohen-Zada (2004), and marriage choices by Bisin and Verdier (2000), the literature has to date shown little interest in the socialization effects of location choices, for instance, the socialization effects of urban agglomeration by ethnic or religious trait.

The simple analysis of the economic model of cultural transmission of Bisin and Verdier depends crucially on the assumption that the utility to a type i parent of a type j child, V^{ij} is independent of the distribution of the population by cultural trait, that is, independent of q^i. Many interesting analyses of cultural transmission require this assumption to be relaxed. In many instances the adoption of the cultural trait of the majority in fact favours children, for example in the labour market; a typical example is language adoption. In this case altruistic parents, even if paternalistic, might favour (or discourage less intensely) the cultural assimilation of their children. If we allow for interesting socio-economic effects interacting with the socialization choices of parents, the basic cultural transmission model of Bisin and Verdier has been applied to several different environments and cultural traits and social norms of behaviour, from preferences for social status (Bisin and Verdier, 1998) to corruption (Hauk and Sáez-Martí, 2002), hold-up problems (Olcina and Penarrubia, 2004), development and social capital (François, 2002), inter-generational altruism (Jellal and Wolff, 2002), labour market discrimination (Sáez-Martí and Zenou, 2005), globalization and cultural identities (Olivier, Thoenig and Verdier, 2005), and work ethics (Bisin and Verdier, 2005).

Empirical analysis of cultural transmission models

While an interesting literature has documented the relevance of cultural factors in several socio-economic choices, much less is known about cultural transmission per se. Nonetheless, several important questions are beginning to be answered. First of all, several important correlations have been documented in sociology, in particular with regard to the role of marriage in socialization (see, for instance, Hayes and Pittelkow, 1993; Ozorak, 1989; Heaton, 1986). The literature in economics has instead concentrated more specifically on the direct empirical validation of the economic approach to cultural transmission surveyed above, thereby estimating the relative importance of direct and oblique socialization for different specific traits and the prevalence of cultural substitution or complementarity in specific socio-economic environments. Patacchini and Zenou (2004) find evidence of cultural complementarity in education in the United Kingdom. Cohen-Zada (2004) finds instead for the United States that the demand for private religious schooling decreases with the share of the religious minority in the population, in accord with cultural substitution. Fernandez, Fogli and Olivetti (2004) find evidence of an important role for mothers in the transmission to their sons of attitudes favouring the participation of women in the labour force and acquisition of higher education. Finally, Bisin, Topa and Verdier (2004a), using the General Social Survey data for the United States over the period 1972–96, estimate for religious traits the structural parameters of the model of marriage and child socialization in Bisin and Verdier (2000). They find that observed intermarriage and socialization rates are consistent with Protestants, Catholics and Jews having a strong preference for children who identify with their own religious beliefs, and taking costly decisions to influence their children's religious beliefs. The estimated 'relative intolerance' parameters are high and asymmetric across religious traits, suggesting an interestingly rich representation of 'cultural distance'.

Genetic and cultural evolution

Cultural transmission possibly has a role also in the determination of fundamental preference parameters, such as time discounting, risk aversion, altruism, and interdependent preferences. Purely evolutionary models have been complemented by alternative models of cultural transmission and genetic and cultural co-evolution. The wealth of different approaches proposed is best exemplified by the study of preferences for cooperation. The observation that humans often adhere to collectively beneficial actions which are not in their private interest (or which are not rationalizable as strategic equilibria) has led to a theoretical literature explaining how psychological 'preferences for cooperation' can be sustained in the context of genetic and/or cultural evolution (this is called the *puzzle of pro-sociality* by Gintis, 2003a). For instance, in the context of the Prisoner's Dilemma, Becker and Madrigal (1995) exploit the ability of habits to induce preferences; Guttman (2003), Stark (1995), and Bisin, Topa and Verdier (2004b) show how cooperation can be sustained by different modes of cultural evolution; Gintis (2003b) shows that a general capacity to internalize fitness-enhancing norms of behaviour can be genetically adaptive, and hence that cooperation can also be internalized by 'hitchhiking' on this general capacity.

The empirical evidence on the nature–nurture debate (see Ceci and Williams, 1999, for a review) has not yet been systematically taken to the point of distinguishing the genetic from the cultural factors in the determination of fundamental preference parameters. Similarly, the empirical evidence distinguishing the different cultural transmission models of fundamental preference traits is almost non-existent. The only exception is by Jellal and Wolff (2002), who study the implication of the pattern of *inter vivos* transfers within the family in France for the transmission of inter-generational altruism. They argue that the evidence is more consistent with a cultural transmission model such as that of Bisin and Verdier (2001a) rather than with a 'demonstration effect' model, as in Stark (1995), where parents take care of their elders in order to elicit similar behaviour in their children.

ALBERTO BISIN AND THIERRY VERDIER

See also **culture and economics; identity; social interactions (empirics).**

Bibliography

Becker, G. and Madrigal, V. 1995. On cooperation and addiction. Mimeo, University of Chicago.

Bisin, A., Topa, G. and Verdier, T. 2004a. Religious intermarriage and socialization in the United States. *Journal of Political Economy* 112, 615–64.

Bisin, A., Topa, G. and Verdier, T. 2004b. Cooperation as a transmitted cultural trait. *Rationality and Society* 16, 477–507.

Bisin, A. and Verdier, T. 1998. On the cultural transmission of preferences for social status. *Journal of Public Economics* 70, 75–97.

Bisin, A. and Verdier, T. 2000. Beyond the melting pot: cultural transmission, marriage and the evolution of ethnic and religious traits. *Quarterly Journal of Economics* 115, 955–88.

Bisin, A. and Verdier, T. 2001a. The economics of cultural transmission and the dynamics of preferences. *Journal of Economic Theory* 97, 298–319.

Bisin, A. and Verdier, T. 2001b. Agents with imperfect empathy might survive natural selection. *Economics Letters* 2, 277–85.

Bisin, A. and Verdier, T. 2005. Work ethic and redistribution: a cultural transmission model of the welfare state. Mimeo, New York University.

Borjas, G. 1992. Ethnic capital and intergenerational income mobility. *Quarterly Journal of Economics* 57, 123–50.

Boyd, R. and Richerson, P. 1985. *Culture and the Evolutionary Process.* Chicago: University of Chicago Press.

Cavalli-Sforza, L. and Feldman, M. 1981. *Cultural Transmission and Evolution: A Quantitative Approach.* Princeton, NJ: Princeton University Press.

Ceci, S. and Williams, W. 1999. *The Nature–Nurture Debate: The Essential Readings.* Oxford: Blackwell.

Cohen-Zada, D. 2004. Preserving religious identity through education: economic analysis and evidence from the US. Mimeo, Ben-Gurion University.

Coleman, J. 1988. Social capital in the creation of human capital. *American Journal of Sociology* 94, S95–S120.

Fernandez, R. and Fogli, A. 2005. Culture: an empirical investigation of beliefs, work, and fertility. Working paper No. 11268. Cambridge, MA: NBER.

Fernandez, R., Fogli, A. and Olivetti, C. 2004. Mothers and sons: preference formation and female labor force dynamics. *Quarterly Journal of Economics* 119, 1249–99.

François, P. 2002. *Social Capital and Economic Development.* New York: Routledge.

Gintis, H. 2003a. Solving the puzzle of prosociality. *Rationality and Society* 15, 155–87.

Gintis, H. 2003b. The hitchhikers guide to altruism: genes, culture and the internalization of norms. *Journal of Theoretical Biology* 220, 407–18.

Giuliano, P. 2007. Living arrangements in Western Europe: does cultural origin matter? *Journal of the European Economic Association* 5, 927–52.

Gleason, P. 1980. American identity and Americanization. In *Harvard Encyclopedia of American Ethnic Groups*, ed. T. Stephan, O. Ann and H. Oscar. Cambridge, MA: Harvard University Press.

Guttman, J. 2003. Repeated interaction and the evolution of preferences for reciprocity. *Economic Journal* 113, 631–56.

Hauk, E. and Sáez-Martí, M. 2002. On the cultural transmission of corruption. *Journal of Economic Theory* 107, 311–35.

Hayes, B. and Pittelkow, Y. 1993. Religious belief, transmission, and the family: an Australian study. *Journal of Marriage and the Family* 55, 755–66.

Heaton, T. 1986. How does religion influence fertility? The case of Mormons. *Journal for the Scientific Study of Religion* 28, 283–99.

Jellal, M. and Wolff, F. 2002. Cultural evolutionary altruism: theory and evidence. *European Journal of Political Economy* 18, 241–62.

Mayer, E. 1979. *From Suburb to Shetl: The Jews of Boro Park.* Philadelphia: Temple University Press.

Olcina, G. and Penarrubia, C. 2004. Hold-up and intergenerational transmission of preferences. *Journal of Economic Behavior and Organization* 54, 111–32.

Olivier, J., Thoenig, M. and Verdier, T. 2005. Globalization and the dynamics of cultural identity. Mimeo. Paris: Paris-Jourdan Sciences Économiques.

Ozorak, E. 1989. Social and cognitive influences on the development of religious beliefs and commitment in adolescence. *Journal for the Scientific Study of Religion* 28, 448–63.

Patacchini, E. and Zenou, Y. 2004. Intergenerational education transmission: neighborhood quality and/or parents' involvement? Working Paper No. 631. Stockholm: Research Institute of Industrial Economics.

Sáez-Martí, M. and Sjogren, A. 2005. Peers and culture. Working Paper No. 642. Stockholm: Research Institute of Industrial Economics.

Sáez-Martí, M. and Zenou, Y. 2005. Cultural transmission and discrimination. Mimeo. Stockholm: Research Institute of Industrial Economics.

Stark, O. 1995. *Altruism and Beyond: An Economic Analysis of Transfers and Exchanges within Families and Groups.* Cambridge: Cambridge University Press.

culture and economics

Economic decisions are made within a social context; as Aristotle reminds us, man is a social animal. The relevance of this statement to economics, however, is far from clear. In what ways, if any, do we need to consider the social nature of man in order to study economic questions? This article attempts to provide a partial answer to this question.

Traditionally, economists seek to explain differences in economic outcomes by studying how agents, with given preferences and beliefs, react to changes in the policy environment, institutions and technology. At a deeper level than the taste for apples versus oranges, however, few would deny that preferences and beliefs must be, to some extent, endogenous. Our level of trust in others, the determinants of status in society, our beliefs about the correct trade-off between efficiency and equity, or the 'proper' roles for men and women, are all examples of beliefs or preferences that have differed across societies and over time. These beliefs and preferences impact on individual behaviour and how society allocates scarce resources. At the individual level they help determine whether a woman participates in the formal labour market and the career she follows, the extent to which racism is tolerated, or the degree of assortative matching on wealth in marriages. At a collective level, they help determine, for example, the range and depth of the welfare state, the legality of slavery, or the proportion of the budget that is dedicated to foreign aid.

Although at some general level few may disagree that preferences, beliefs, or values of the type discussed above are endogenous (and may therefore differ across societies), whether they have a quantitatively significant impact on economic outcomes is another matter. Do differences in beliefs and preferences that vary systematically across groups of individuals separated by space (either geographic or social) or time – what I shall henceforth term *culture* – play an important role in explaining differences in outcomes? (For the purposes of this article, I will not give a more rigorous definition of culture than the abbreviated one here. See Elster, 1989, for a discussion of social norms and culture and Manski, 2000, for a discussion of peer effects and social interactions.) Modern economics (as opposed to sociology or anthropology) has largely been, until recently, reluctant to investigate this question. Although in principle there is nothing non-standard about positing preference/belief heterogeneity among individuals to explain differences in outcomes, the Stigler–Becker dictum *de gustibus non est disputandum* (Stigler and Becker, 1977) and its assertion that 'no scientific behavior has been illuminated by assumptions of differences in taste' has cast a long shadow in economics. Thus, the main challenge faced by those who believe that culture might matter has been to find a convincing way to show that culture can be studied rigorously and, in particular, that it is possible to separate the influence of culture from institutions and standard economic variables. In this sense, running, say, cross-country regressions on variables that one suspects reflect cultural attitudes (for example, different savings patterns may reflect attitudes towards thrift) to study the effect of culture has long (and correctly) been considered unsatisfactory. Despite one's best efforts to control for differences in countries' economic environments, identifying the residual with culture is ultimately unconvincing. It is difficult, if not impossible, to summarize the economic environment faced by agents with a few aggregate variables. Thus, there are bound to be omitted variables and problems of endogeneity, which are all further confounded by mismeasurement.

Hence, despite a long history of writers on the relationship between culture and economics (which includes Marx, Weber, Gramsci, Polanyi, Banfield and, more recently, Putnam and Landes, among others), modern neoclassical economics has been by and large silent on the topic of culture and only in recent years have economists started to think seriously again about how culture may help explain economic phenomena. In this article I will selectively review some recent attempts to empirically identify the effects of culture on important economic outcomes and to answer the question, 'does culture matter?' Answering this question affirmatively naturally leads one to explore the propagation mechanisms of culture, to theorize about the relationship between institutions and culture, and to investigate the dynamic of culture – all topics that I will briefly touch upon at the end.

Empirical evidence on culture

In this section I examine some of the recent evidence on the importance of culture for economic outcomes. For expository ease, I have divided the empirical evidence into that which uses survey data, evidence based on immigrants or their descendants (what I call the 'epidemiological approach'), and historical case studies. There is also a small body of experimental work that, by showing that across societies there exist marked differences in how individuals play games such as the ultimatum, public good or dictator game, has also shed light on the relationship between culture and economics (see, for example, Henrich et al., 2001).

Survey-based evidence

Perhaps the most natural approach to doing empirical work on culture consists in using the beliefs expressed by individuals in surveys (for instance, the World Value Surveys) on a variety of issues as expressions of culture and correlating them with economic outcomes. This approach, however, must overcome the problem of reverse causality. That is, differences in beliefs may be solely a consequence of different economic and institutional environments. Hence, the use of instrumental variables is required in order to identify causality. Overall, this has been difficult to achieve.

As shown by Guiso, Sapienza and Zingales (2003), the intensity of religious beliefs and religious denomination are correlated with a variety of individual attitudes such as trust in others, government's role, views of working women and the importance of thrift. Guiso, Sapienza and Zingales (2006) show that these attitudes, aggregated at the country level, are correlated with cross-country aggregate outcomes (for example, savings, redistributive versus regressive taxation, and trade). In order to ensure that the reverse causality is not at play, the attitudes are instrumented, usually by the religious composition in the country. This work is suggestive but there are several concerns associated with it. In addition to questions about omitted variables, it is not clear that religious composition is a valid instrument since it may also help explain the aggregate outcome through other channels. (Indeed, the coefficients on the instrumental variable results tend to look very high relative to the ones obtained by ordinary least squares. Running regressions at the individual outcome level would be more convincing, but opinion surveys unfortunately tend not to have high-quality economic data (the World Value Survey, for example, classifies income levels into ten categories). Recent work by Guiso, Sapienza and Zingales (2005) on the relationship between trust and trade, instead instruments trust with the genetic distance between indigenous populations. This seems a promising avenue of research.

Tabellini (2005) takes a significant step towards overcoming some of the weaknesses discussed above. To study whether culture affects economic development across European regions, he also aggregates (at the regional level) individual responses from the World Value Surveys to questions about trust, respect and the link between individual effort and economic success. The scope for omitted variables is reduced by focusing on within-country variation in Europe (by including country fixed effects). The attitudes are then instrumented with historical variables, such as regional literacy rates at the end of the 19th century and indicators of political institutions in the period from 1600 to 1850. The author finds that the proxies for culture are quantitatively significant determinants of per capita GDP levels and growth rates across regions. It is possible of course that the instruments are not valid. For example, they could affect output directly via sectoral composition or public investment. The paper contains a good discussion of these and other alternative hypotheses.

The epidemiological approach

A very different approach to relying on opinion data is to examine the economic outcomes of immigrants or their descendants. This is reminiscent of the epidemiology literature that, in order to attempt to identify the contribution of the environment broadly defined (namely, physical and cultural) relative to genes in disease, studies various health outcomes for immigrants and compares them to outcomes for natives (see, for example, the classic study by Marmot et al., 1975).

To understand the strengths and weaknesses of such an approach, suppose that the level of, say, heart disease differs markedly between two countries (the source and host countries). If heart disease in immigrants converges to that of natives in the host country, the difference between the two countries is unlikely to be driven by genetics and instead results from the environment. Failure to find convergence, on the other hand, does not imply the opposite. There are many reasons why the environment may be solely responsible and still sustain differential levels of heart disease. For example, cultural assimilation may occur slowly (for instance, if immigrants maintain the same dietary patterns as in the source country), or living in the source country at a young age may confer some degree of immunity, or selection into immigration may be correlated with a particular health outcome.

The epidemiological strategy in economics has its own set of problems. In particular, it is important to recognize that immigrants may be subject to many shocks (language difficulties, worse employment opportunities, greater uncertainty and so forth) which cause them to deviate from their traditional behaviour. Culture, furthermore, is socially constructed: to be replicated, the behaviour may require the incentives – rewards and punishments – provided by a larger social body such as a neighbourhood, school, or ethnic network. Furthermore, immigrants are unlikely to be a representative sample of their home-country's population. Their beliefs, preferences, and unobserved differences in their economic circumstances may differ significantly from the country average. Lastly, the exposure of immigrants (or their descendants) to a different culture from the one prevalent in their country of heritage presumably weakens the latter's impact on their behaviour. Note that all the factors mentioned above introduce a bias towards finding culture to be insignificant. Thus, on the whole, comparisons of behaviour or outcomes across different immigrant groups are a very demanding test of the importance of culture. In epidemiology, when differences across groups remain, one must be careful not to conclude that genetics is determinative when the underlying cause may be cultural; in economics, when significant differences are not observed, one must be careful not to rule out cultural forces.

In economics, the paper by Carroll, Rhee and Rhee (1994) is the first that, to my knowledge, follows an approach similar to the one described above. The authors are interested in exploring whether cross-country differences in savings rates may be culturally driven. Using individual-level data on immigrants to Canada, they estimate individual consumption levels as a function of permanent income (as captured by labour and asset income), the interaction of this variable with demographic variables, some measures of wealth, and finally the interaction of a region of origin dummy (and years since arrival to Canada) with their measure of permanent income. If there exist different cultural attitudes towards savings, and if this attitude is maintained in immigrants, then one should observe different propensities across immigrants, by region of origin, to consume out of permanent income (that is, the regional dummies should be significantly different from one another). The authors find that the saving patterns of immigrants do not vary significantly by region of origin. Recent immigrants as a whole save less than native-born Canadians, but there is no statistically significant difference in behaviour across immigrant groups.

There are several weaknesses in the data-set used in the study above that may bias it against finding results that show a significant impact of culture. Wealth, for example, is not well measured. In particular, as only South East Asia's saving rate differed markedly from those of other regions in the immigrant population (31 per cent relative to 18–20 per cent across the remaining regions), the small number of immigrants from this group in the sample limits the power of the test. Note also that, if the motivation to save more stems from the desire to provide one's child with greater status via a larger bequest, the incentive to do this may be much less marked in a society in which savings are generally low or in which status stems from consumption behaviour.

Fernández and Fogli (2005; 2006) use a similar, but arguably less problematic, methodology by studying second-generation Americans in order to investigate the quantitative importance of culture. Their research focuses on the fertility and work behaviour of married second-generation American women (that is, women who were born in the United States but whose parents were born elsewhere). The use of second-generation immigrants attenuates the problems associated with the first generation's adjustment to a foreign setting (for example, language difficulties) and even some selection problems are less likely to play a role for the second generation. On the other hand, second-generation individuals have been more exposed to the new culture, and that will tend to diminish the role of culture from the country of heritage. Our hypothesis is that attitudes towards woman's 'proper' role in society and towards ideal family size are culturally different across countries and that this culture is likely to be transmitted intergenerationally and show up in systematic differences in female labour force participation (LFP) and fertility, even if individuals were raised in the United States.

In our 2005 paper, the challenge was how to best capture the attitudes towards women and family size in the parents' country of origin. We chose not to use country dummies (as in Carroll, Rhee and Rhee, 1994) but to instead examine whether past values of economic variables in the country of origin that should reflect this culture – in particular, past values of female LFP and total fertility rates (TFR) – are able to play a quantitatively significant role in explaining differences in outcomes across second-generation women in the United States. Our argument is that these economic variables reflect the institutions (for example, markets, legal framework, minimum wages and so on), the strictly economic environment (demand and supply, transportation costs, access to day care, for example), as well as the preferences and beliefs (that is, the culture) of individuals in the country making decisions at that time. If these variables are able to explain the behaviour of women who, by virtue of living in the USA and in a different time period, face different institutions and economic variables, then solely the cultural component of these variables should affect their choices. This is a more demanding test that is superior to the 'black box' approach of using country dummies which leaves open the question of what it is about the country that matters to outcomes.

In individual level regressions, we find that our cultural proxies – past values of female LFP and TFR – help explain both how much second-generation American women work and their fertility. As our data-set – the 1970 US Census – does not allow us to control for family factors such as parental wealth, income, and education, we include the woman's education, her spouse's education, and total personal income (as well as location, age, and so on) in our regressions. By including these variables, the coefficient on the cultural proxy only captures the direct effect of culture rather than its full direct and indirect effects (for example, a woman who wants engage in market work is more likely to invest in education and hence, by controlling for education, we are eliminating the effect of culture on this variable), but this is preferable to not controlling for differences in parental background, other than culture, that may affect women's work and fertility outcomes. We find that the cultural proxies still matter even after including these additional variables. Furthermore, the cultural proxies are quantitatively significant: a one standard-deviation increase in the corresponding cultural proxy is associated with approximately an eight per cent increase in hours worked per week and about a 14 per cent increase in the number of children. The forces of assimilation means that these numbers should be taken, if anything, as a downward biased estimate of the true power of culture in the original setting (that is, in the country of ancestry).

We also examine the most compelling alternative economic explanation for our results, namely, the hypothesis that these are driven by unobserved human capital. We do this by showing that the results are robust to the inclusion of the country of ancestry's level of per capita GDP in various years and to the years of education of immigrants (by country of ancestry) in 1940 (this remains the case when Hanushek and Kimko's (2000) measures of education quality in the parents' country of origin are included). We also demonstrate that the work cultural proxy does not have explanatory power in a Mincer wage regression which it would be expected to have if it captured unobserved human capital. Lastly, we show that the work cultural proxy is insignificant in explaining how much married second-generation American men work whereas the fertility cultural proxy retains its explanatory power. (If the work cultural proxy had a negative effect on how much these men work, that might indicate a substitution effect. In our regressions, the coefficient is basically zero and insignificant.) This is important because it implies that there does not exist some omitted economic variable at the parental country-of-origin level that affects the productivity of both men and women and that helps explain how much they work.

The methods used in Fernández and Fogli (2005) could be profitably extended to examine other issues, such as entrepreneurship or savings behaviour. It might also be interesting to elaborate upon the recent approach by Algan and Cahuc (2006) that attempts to combine survey evidence with the epidemiological approach in order to study the effects of culture on cross-country labour market outcomes. Although this work is too preliminary to discuss in depth, using the attitudes of, say, second-generation Americans to instrument for the attitudes of individuals in the home country seems cleaner than relying on variation in religious denominations. As usual, the question will be whether there is some omitted background economic variable correlated with the country of origin (particularly given the quality of the survey data-sets) that could be driving the results, but it seems a promising avenue of research (see also the interesting work on culture and migrants within regions in Italy by Ichino and Maggi, 2000). As shown recently in Fernández (2007a) using the World Value Survey, the attitudes of individuals in the country of ancestry towards women's market work and housework have explanatory power for the work outcomes of second-generation American women in 1970.

Historical case studies

The analysis of historical episodes in which changes in either culture or environment yield 'natural experiments' is likely to add richness and depth to our understanding of culture and the economy. Greif's 1994 paper is probably the best-known work in economics that makes the link between culture and institutional development. In brief, Greif argues that cultural beliefs (collectivist versus individualist) are reflected in the different ways in which in the 11th century Genoese traders and Maghrebi traders set up their trading institutions. Both groups of merchants required agents to conduct their business overseas, and in both cases there was an agency problem as the overseas agent might be tempted to cheat the merchant. Maghrebi traders set up 'horizontal' relations in which merchants served as agents for traders and vice versa. Information was shared among merchants/traders and an agent who was dishonest with one merchant could expect to be shunned by other merchants. The Genoese, on the other hand, set up 'vertical' relationships in which individuals specialized as merchants or agents. Information was not shared among merchants. This led the Genoese to set up more formal enforcement institutions. The two different responses, argues Greif, then had important consequences once trading opportunities were expanded in previously inaccessible areas. The Maghrebi expanded trade using other Maghrebi agents whereas the Genoese were able to establish agency relations with non-Genoese, leading to very different economic development paths thereafter (see also Greif's, 2005, recent book on the topic).

Another compelling example is provided by Botticini and Eckstein (2005) who present the thesis that an 'exogenous' cultural change gave rise to the pattern of Jewish occupational selection that we see to this day. They argue that with the destruction of the Temple in Jerusalem in 70 CE, the Pharisees became the dominant religious group and transformed Judaism from a religion based on sacrifices to one whose main rule required each male to read and to teach his sons the Torah. This reform was implemented in places where most Jews were farmers who would not gain anything from investing in education. When urbanization expanded many centuries later, Jews had a comparative advantage in the skilled occupations demanded in the new urban centres. Thus, culture – the religious requirement of reading skills for other than human capital reasons – gave rise to the pattern of Jewish occupational selection seen since the ninth century.

Theories of culture

Is it necessary to modify the standard economic model in order to incorporate culture? The answer definitely is 'no'. What appear to be societal differences in preferences may only be choice of equilibrium strategies in a game with multiple equilibria and standard preferences. This is in fact the most common way to think about the role of culture in economics, and is fully in keeping with our working definition of culture as systematic differences (across groups) in preferences or beliefs. Here the heterogeneity lies in the expectations (beliefs) over the strategies that will be played in equilibrium. Hence differences in culture can be identified with, for example, which equilibrium we play in a static game (for example, do we drive on the right- or left-hand side of the road) or the degree of cooperation ('trust') sustained in a repeated Prisoner's Dilemma game.

Within the 'culture as multiple equilibria' literature, I find particularly interesting the research that attempts to

generate behaviour that looks like social norms (such as determinants of status). Take, for example, a dynamic matching model in which individuals who differ in wealth choose a partner with whom to match and obtain utility from joint consumption and the utility of their child. As shown in Mailath and Postlewaite (2003), in addition to an equilibrium in which there is assortative matching on wealth, there may also be an equilibrium with imperfectly assortative matching that depends also on non-economic characteristics such as whether one has blue eyes. In this equilibrium, blue eyes matter not because of their intrinsic value, but simply because the matching rule allocates, for the same wealth level, a wealthier partner to individuals with blue eyes. Thus, a woman would be willing to match with a man with blue eyes and slightly lower wealth than another man without blue eyes, because although she obtains lower joint consumption, there is a 50 per cent chance that her child would inherit blue eyes and hence a better match and higher consumption in the future. To an outside observer, it might therefore appear that in this society people had an intrinsic preference for blue eyes, although this inference would be incorrect.

Although the example above is interesting, its explanation for a particular social norm seems incomplete and intuitively less than compelling. The preference for blue eyes or light skin may perhaps initially come about as a choice among many equilibria and involve solely a calculation about the trade-off between one's own consumption and that of one's child (though that too seems doubtful and is more likely the result of a history in which these traits are correlated with higher status). Over the longer run, however, one may conjecture that what sustains these equilibria – what makes these cultural traits less fragile to perturbations – is that these calculations are embodied in the individual and in society as preferences and beliefs about the inherent superiority/desirability of such features. People come to prefer blue eyes; people become racist. Thus, what is missing more generally in the theory of culture is an analysis of how preferences and beliefs (about things other than equilibrium strategies) themselves evolve.

The hypothesis that certain features of culture (those that have greater depth than driving on the left or the right side of the street) become part of preferences and beliefs implies that they cannot be discarded easily simply because they are no longer useful or beneficial, though over time this will certainly lessen their appeal. In this way, the operation of culture may be clearest to perceive when it no longer serves any useful societal purpose or particular group interest but nonetheless, at least for some time, persists – for example, religious prohibition on eating pork. (One reason speculated for this prohibition is that consumption of undercooked pork is linked to trichinosis. It is now known that this problem can be eliminated, however, by thoroughly cooking the meat.) In the context of the matching example above, individuals may eventually be willing to match with lower wealth people with blue eyes because this matching rule is incorporated into preferences/beliefs over what type of mate is intrinsically better even if the benefit derived by passing this trait on to their offspring is no longer substantial (say, because family size falls and decreases the payoff from the inheritable trait relative to the decrease in immediate joint consumption).

So far, we have discussed differences in culture as systematic differences in preferences and beliefs without distinguishing much between the two. This is not accidental, since, in general, the distinction between preferences and beliefs for our purposes is rather fuzzy. Even for simple preferences such as the trade-off between apples and oranges, what one knows (or believes) about the nutritional contents of the two may affect how one 'feels' about them, as may any other mental associations (for example, whether one is considered more exotic, how they were grown and so forth). In general, there are few pure (or naive) preferences – what one thinks or believes influences how one feels (and the same may be true vice versa. See Damasio, 1995, for an interesting exposition of evidence in favour of the hypothesis that emotions affect – and in fact are necessary for – the ability to think well). This is not to deny that people have some inherent tastes (for example, it is believed that human beings have a taste for fat, probably because of the evolutionary advantage associated with an inclination to eat meat in an environment in which protein and iron were not easily obtained).

For more complex questions the above is even more likely to be true. Consider, for example, the large increase in female labour force participation in the 20th century. Is it that woman's disutility from market work decreased or that her beliefs about the meaning or consequences of her working that changed over time? The dichotomy between the two alternatives does not seem very useful in this case. If the focus is on understanding why actions change over time, then using standard preferences and modelling the evolution of beliefs as giving rise to changes in expected payoffs may be the more useful strategy (the latter is the approach taken by Fernández, 2007b, who shows that a model of the evolution of female LFP as an intergenerational learning process does a good job of replicating a century of US female LFP data). If instead one wished to understand the utility from a given action, particularly one in which identity is concerned, then incorporating cultural beliefs into preferences may be a better route (see, for instance, Akerlof and Kranton, 2000). For example, wearing a dress or having a woman as a boss may decrease a man's utility, independently of any expectations of future consequences, simply because it makes him feel (culturally) less masculine.

Culture and institutions
As seen previously, the main challenge faced by most empirical work on culture is to convincingly isolate its effects from the incentives provided by traditional economic variables and institutions. This should not be taken to mean that culture and institutions are independent variables. Indeed, one way to think about institutions is as congealed culture: that is, which institutions are set up and how these evolve depends not only on the problems faced by society (or by a particular group in society) at a particular moment in time but also the beliefs/preferences – the culture – that are prevalent. As elaborated on in our earlier discussion of Greif (1994), cultural beliefs (collectivist versus individualist), for example, were reflected in the different ways in which in the 11th century Genoese traders and Maghrebi traders set up their trading institutions, leading to very different economic development paths thereafter. My hypothesis is that the reverse causality is also likely to hold: that is, not only does culture affect institutions but also institutions affect the dynamic evolution of culture. In this sense, work that attempts to establish whether institutions or culture are the most important determinants of economic development seems misconceived (see Fernández, 2007c, for a theoretical analysis of the dynamic dependency of culture and institutions; also Bowles, 1998, for a review of some of the theoretical and empirical evidence on the effect of markets on culture).

Concluding remarks
The rigorous study of culture and economics is in its infancy. We would like to understand, for example, how culture propagates and evolves. In particular, what is the relative importance of family versus other institutions as cultural transmission mechanisms for different beliefs or in different environments? To what extent is cultural transmission purposeful, that is, optimizing on the part of an individual or her parents (as in Bisin and Verdier, 2000) or for a social group, and to what extent is it involuntary? (Fernández, Fogli and Olivetti, 2004, show that whether a man's mother worked while he was growing up is correlated with whether his wife works, even after controlling for a whole series of socioeconomic variables. They interpret this as preference transmission, but whether it is voluntary – optimizing – or simply by example is an open question.) When and why does culture change abruptly whereas at other times it proceeds glacially?

The relationship between technology and culture also needs to be investigated. How does technology influence culture and how does culture shape technological change? Some papers (for instance, Greenwood and Guner, 2005; Greenwood, Seshadri and Yorukoglu, 2002) argue that sexual norms and female LFP changed because of changes in technology. These papers ignore, among other things, the endogeneity of demand for new technology. Despite the convenient simplification of treating technology as a primitive, it too is endogenous. The extent to which societies put resources into developing technology that 'liberates' individuals from household work, for example, depends on things such as whether slavery is available or whether women expect to work in the market or at home. Put differently, both the relative price of market versus household labour and the elasticity of labour supply depend on the institutions (for example, slavery) and expected division of labour (for example, clearly differentiated gender roles) that are in place. The opposite is also true – the extent to which one can substitute capital for labour, whether at work or at home, helps determine which institutions are viable and may determine the pace and ease with which beliefs or preferences change.

From a theoretical perspective, the endogeneity of preferences and beliefs raises difficult questions for welfare. How should we evaluate policies once we recognize that preferences can change? While this is indeed a vexing and problematic question for welfare economics, recognizing that man is a social animal that is (perhaps uniquely) capable of reflecting upon, and hence changing, his preferences and beliefs greatly enriches our view of ourselves and the world and within it the potential role of economic discourse. In the words of A.O. Hirschman, 'de valoribus est disputandum'.

RAQUEL FERNÁNDEZ

See also **cultural transmission; social norms.**

Bibliography

Algan, Y. and Cahuc, P. 2006. Minimum wage: the price of distrust. Mimeo, CREST-INSEE.

Akerlof, G. and Kranton, R.E. 2000. Economics and identity. *Quarterly Journal of Economics* 115, 715–33.

Bisin, A. and Verdier, T. 2000. Beyond the melting pot: cultural transmission, marriage, and the evolution of ethnic and religious traits. *Quarterly Journal of Economics* 115, 955–88.

Botticini, M. and Eckstein, Z. 2005. Jewish occupational selection: education, restrictions, or minorities? *Journal of Economic History* 65, 922–48.

Bowles, S. 1998. Endogenous preferences: the cultural consequences of markets and other economic institutions. *Journal of Economic Literature* 36, 75–111.

Carroll, C., Rhee, B. and Rhee, C. 1994. Are there cultural effects on saving? Some cross-sectional evidence. *Quarterly Journal of Economics* 109, 685–99.

Damasio, A. 1995. *Descartes' Error: Emotion, Reason, and the Human Brain.* New York: Harper Perennial.

Elster, J. 1989. Social norms and economic theory. *Journal of Economic Perspectives* 3(4), 99–117.

Fernández, R. 2007a. Women, work, and culture. *Journal of the European Economic Association* 5, 305–32.

Fernández, R. 2007b. Culture as learning: the evolution of female labor force participation over a century. Mimeo, New York University.

Fernández, R. 2007c. The co-evolution of culture and institutions. Mimeo, New York University.

Fernández, R. and Fogli, A. 2005. Culture: an empirical investigation of beliefs, work, and fertility. Working Paper No. 11268. Cambridge, MA: NBER.

Fernández, R. and Fogli, A. 2006. Fertility: the role of culture and family experience. *Journal of the European Economic Association* 4, 552–61.

Fernández, R., Fogli, A. and Olivetti, C. 2004. Mothers and sons: preference formation and female labor force dynamics. *Quarterly Journal of Economics* 119, 1249–99.

Greenwood, J. and Guner, N. 2005. *Social Change.* Economie d'Avant Garde Research Reports No. 9, Economie d'Avant Garde.

Greenwood, J., Seshadri, A. and Yorukoglu, M. 2005. Engines of liberation. *Review of Economic Studies* 72, 109–33.

Greif, A. 1994. Cultural beliefs and the organization of society: a historical and theoretical reflection on collectivist and individualist societies. *Journal of Political Economy* 102, 912–50.

Greif, A. 2005. *Institutions: Theory and History. Comparative and Historical Institutional Analysis.* Cambridge: Cambridge University Press.

Guiso, L., Sapienza, P. and Zingales, L. 2003. People's opium? Religion and economic attitudes. *Journal of Monetary Economics* 50, 225–82.

Guiso, L., Sapienza, P. and Zingales, L. 2005. Cultural biases in economic exchange. Working Paper No. 11005. Cambridge, MA: NBER.

Guiso, L., Sapienza, P. and Zingales, L. 2006. Does culture affect economic outcomes? *Journal of Economic Perspectives* 20(2), 23–48.

Hanushek, E. and Kimko, D. 2000. Schooling, labor-force quality, and the growth of nations. *American Economic Review* 90, 1184–208.

Henrich, J., Boyd, R., Bowles, S., Camerer, C., Fehr, E., Gintis, H. and McElreath, R. 2001. In search of homo economicus: behavioral experiments in 15 small-scale societies. *American Economic Review* 91(2), 73–8.

Ichino, A. and Maggi, G. 2000. Work environment and individual background: explaining regional shirking differentials in a large Italian firm. *Quarterly Journal of Economics* 115, 1057–90.

Mailath, G. and Postlewaite, A. 2003. The social context of economic decisions. *Journal of the European Economic Association* 1, 354–62.

Manski, C. 2000. Economic analysis of social interactions. *Journal of Economic Perspectives* 14(3), 115–36.

Marmot, M.G., Syme, S.L., Kagan, A., Kato, H., Cohen, J.B. and Belsky, J. 1975. Epidemiologic studies of coronary heart disease and stroke in Japanese men living in Japan, Hawaii and California: prevalence of coronary and hypertensive heart disease and associated risk factors. *American Journal of Epidemiology* 102, 514–25.

Stigler, G. and Becker, G. 1977. De gustibus non est disputandum. *American Economic Review* 67, 76–90.

Tabellini, G. 2005. Culture and institutions: economic development in the regions of Europe. Working Paper No. 1492, CESifo.

Cunningham, William (1849–1919)

A member of the English Historical School, Cunningham was educated at the Universities of Edinburgh and Cambridge. He held various posts as lecturer at Cambridge and was elected Fellow of Trinity College in 1891. From 1891 to 1897 he was Tooke Professor of Statistics of King's College London. In addition, he pursued a religious career. He was ordained in 1874 and rose to be Archdeacon of Ely (1907–19).

Cunningham was one of the most important pioneers in economic history. His *Growth of English Industry and Commerce* (1882) was the first textbook in the field, widely used for several decades and an important foundation on which English economic history was to be constructed, and he relentlessly fought for the recognition and establishment of economic history as an independent discipline.

Cunningham became increasingly hostile towards economic theory. He felt that its assumptions about human behaviour and the institutional framework were leading to insufficiently complete analyses and were blatantly unrealistic for most periods in history. In 1892 he started the English Methodenstreit by attacking Marshall for constructing economic history from general principles instead of empirical data. The debate was partly the result of his personal and professional antagonism towards Marshall and his wish to apply economics to politics.

Cunningham shifted from an internationalist and free trader to a nationalist and protectionist, making the preservation and strengthening of the nation-state his most weighty political and economic objective. By the time of the fiscal controversy in 1903 he fully endorsed the tariff reform movement and subscribed to imperialism, with the great empire securing peace and order.

O. KURER

Selected works

1882. *The Growth of English Industry and Commerce*. Cambridge: Cambridge University Press.

1892a. The perversion of economic history. *Economic Journal* 2, 491–506.

1892b. The perversion of economic history. A reply to Professor Marshall. *Pall Mall Gazette*, 29 September, and *Academy*, 1 October, 288.

1904. *The Rise and Decline of the Free Trade Movement*. Cambridge: Cambridge University Press.

1911. *The Case Against Free Trade*. Preface by Joseph Chamberlain. London: John Murray.

Bibliography

Cunningham, A. 1950. *William Cunningham: Teacher and Priest*. Preface by F.R. Salter. London: Society for Promoting Christian Knowledge.

Foxwell, H.S. 1919. Archdeacon Cunningham (obituary). *Economic Journal* 29, 382–90.

Maloney, J. 1976. Marshall, Cunningham, and the emerging economics profession. *Economic History Review* 29, 440–51.

Maloney, J. 1985. *Marshall, Orthodoxy and the Professionalisation of Economics*. Cambridge: Cambridge University Press.

Scott, W.R. 1920. William Cunningham, 1849–1919. *British Academy, Proceedings* 9, 465–74.

Semmel, B. 1960. *Imperialism and Social Reform: English Imperial Thought 1895–1914*. London: George Allen & Unwin.

Wood, J.C. 1983. *British Economists and the Empire, 1860–1914*. Beckenham, Kent: Croom Helm.

currency boards

A currency board is defined as an exchange rate arrangement in which the exchange rate is fixed to an anchor currency and the central bank operates with a simple rule that precludes the monetary authorities from issuing money unless they obtain an equivalent amount of international assets to back it. From a practical point of view this means that the central bank has no independent monetary policy and that it creates or contracts the money supply only as the result of its interventions in the foreign exchange market. If there is excess demand for domestic currency capital will flow in (probably in response to an increase in interest rates) and the central bank, by acquiring these flows, will expand the money supply. If there is excess supply of domestic currency, the central bank will take in this excess supply by giving away international assets, thus contracting the money supply. In some cases this rule is implemented by forcing the central bank to have full backing of domestic base money with international reserves. In some cases a currency board does not require a one-to-one backing of the monetary base, but it still precludes the conduct on an independent monetary policy beyond very strict limits. In fact, a currency board also differs from a typical peg in its commitment to the system, which is usually enshrined in law and in the Central Bank charter.

As of July 2006 the exchange rate arrangement classification published by the International Monetary Fund (IMF) identifies 13 countries with currency boards. Of these, six correspond to countries in the Eastern Caribbean Currency Union (Antigua and Barbuda, Dominica, Grenada, St Kitts and Nevis, St Lucia and St Vincent and the Grenadines), plus seven others: Bosnia and Herzegovina, Brunei Darussalam, Bulgaria,

China-Hong Kong SAR, Djibouti, Estonia and Lithuania. Because all these countries are relatively small, currency boards are placed in a relatively unpopular category amongst potential exchange rate regimes.

There are two main reasons why countries have typically used currency boards. In some cases the currency board is more attractive than a common currency. For example, for the Eastern Caribbean countries mentioned above it seems relatively obvious they should use the US dollar as currency to maximize the benefits from a stable exchange rate arrangement with their almost sole trading partner. However, the currency board allows them to keep the exchange rate credibly fixed without giving up the seigniorage revenue of domestic currency. In other cases countries have resorted to a currency board as a way out of monetary and inflation chaos. Argentina's currency board experience in the 1990s and Bulgaria's currency board are appropriate examples. Even though, as we will see below, the evidence points to large trade benefits of currency boards, it is typically assumed that the main benefit of currency boards is as a tool to fight inflation.

The interest in and excitement about currency boards reflects both the need that countries have faced to solve either of the two problems mentioned above – currency integration without seigniorage cost and exiting from a high inflation situation – and the assessment made at the time of whether a currency board is the most efficient way to reach those objectives. Recent years have been unkind to currency boards on both counts. While the use of a currency board as a replacement for a common currency remains a valid motive, its effect as an anti-inflation device has become less relevant as inflation rates fell throughout the 1990s. In 2007 most countries exhibit single-digit inflation rates, and only a handful of exotic cases appear to have a monetary policy that is out of control. The high-inflation history of yesteryear has been critical to this improvement by fostering much stronger fiscal policies and monetary policies that are much freer from political pressures (both when central banks are independent and when they are not) and increasingly within an inflation targeting framework. As inflation has decreased, so have the benefits of a currency board, thus making it a relatively less attractive proposition. Furthermore, while before the demise of Argentina's currency board in early 2002 no currency board had been forced to end, the fact that Argentina's currency board came to an end in the midst of a major crisis (after enduring a long period of high interest rates) raised some questions as to how much credibility the regime actually bought. As a result, many countries have opted to jump directly all the way to dollarization (for example, El Salvador and Ecuador) or to pursue integration into a currency union (Slovenia) thus making currency boards lose ground even to alternative 'harder' exchange-rate commitments.

In spite of the recent drop in interest in this specific regime, nothing precludes a rise in interest again in the future, so a discussion of the specifics of currency boards remains useful. The best way to organize the discussion is to present the advantages of a currency board, then move to the disadvantages, and then attempt a synthesis.

Advantages of a currency board

The main advantage that is ascribed to a currency board is the credibility gains that it allows, helping deliver lower inflation and better fiscal results. The argument is simple: a currency board represents a strong commitment that if broken can have a large and costly effect on expectations. Because politicians fear this loss of credibility, the currency board, while in place, lowers inflation expectations and inflation itself and should provide the incentives for an improvement in fiscal behaviour.

These predictions have been broadly borne out. On the inflation front Ghosh, Gulde and Wolf (1998), drawing on a data-set for all IMF countries between 1970 and 1996, found that countries with currency boards delivered an inflation rate that was about four per cent lower, a sizable effect. This result has held up in later work (see for example Levy-Yeyati and Sturzenegger, 2001; and Kuttner and Posen, 2001).

The record on fiscal discipline is also relatively favourable. Ghosh, Gulde and Wolf (1998) and Culp, Hanke and Miller (1999) find that countries on currency boards tend to run tighter fiscal policies. Fatas and Rose (2000) also find that currency boards are associated with fiscal restraint (though, somewhat surprisingly, this restraint does not carry on to dollarized economies or those operating within the context of a common currency). Anecdotal evidence also seems to point in the same direction. In 2001, as Argentina's currency board was under fire, fiscal authorities implemented large budget adjustments in an attempt to strengthen the system.

Currency boards may also have an effect on trade as a result of the stability it induces on the exchange rate, an effect similar to the one that has been identified for countries that adopt a common currency with other countries. This exercise is specifically undertaken in Frankel and Rose (2002), who find that the effect of a currency board is a more than tripling of trade (in fact they find that the trade effects for currency boards and common currencies are statistically indistinguishable). Thus the trade motive for a currency board seems to be important. Added to the benefits of saving on seigniorage, it explains why currency boards may remain an attractive option for some small countries.

Disadvantages of a currency board

Four main arguments have been advanced against currency boards. First, the fact that it precludes monetary authorities from running an independent monetary policy and that the exchange rate cannot adjust in response to real shocks; second, that it may 'hide' underlying problems, leading to larger crises down the road; third, that it stimulates large currency mismatches in the portfolio structures of government and the private sector; and fourth, that it limits the ability of the central bank to act as a lender of last resort, thus hindering the possibility of developing a locally based financial sector.

The debate has focused mostly on whether alternative mechanisms and policies within the context of the currency board are available to deal with these problems. Let us review each of them briefly.

On the loss of monetary/exchange rate policy, the question is how relevant a loss this it. It can be argued that the idea of a currency board is indeed to limit the scope for an independent monetary policy, which had otherwise proven unable to contain high inflation. To the extent that inflation and fiscal policy improve, not much may be lost relative to the situation in which monetary policy merely induced inflation without any particular benefit in terms of macroeconomic stabilization. Thus, assessing whether this is a cost requires us to evaluate what the counterfactual is. Proponents of currency boards could argue that only countries where monetary policy serves no purpose choose currency boards as a commitment device.

Of course, if monetary policy *were* possible, the costs of doing without it may turn out to be particularly costly for currency boards. The case of Argentina helps illustrate why this should be so. Argentina had established a currency board with the dollar to quell inflation expectations in the early 1990s. Like any other emerging country, it was hurt by the rush out of emerging markets following Russia's default in 1998. This rush strongly appreciated the dollar, making Argentina's currency stronger exactly when the country needed it to weaken. The fact that currency boards require a strong anchor currency and that capital flows may strengthen these currencies when there is turmoil in emerging countries – thus moving the exchange rate exactly in the opposite direction to the one the country would have otherwise chosen – poses a problem for currency boards during periods of high turbulence in international financial markets. Of course, as much as in the optimal currency area debate, how costly the loss of the monetary instrument is depends on the availability of alternative adjustment mechanisms: fiscal transfers, remittances, labour market mobility, or internal price flexibility, which may all operate as substitutes for the loss of monetary policy (the effectiveness of these alternative mechanisms may explain the different fates of Hong Kong's and Argentina's currency boards). Fiscal policy can also be used as a stabilizer that may substitute for the lack of exchange or monetary policy, though the ability of countries to use it seems relatively limited, particularly for those countries that opted for a currency board as a result of their poor fiscal policies. Some evidence for the fact that the lack of monetary policy may hurt is provided by Levy-Yeyati and Sturzenegger (2001), who compare the growth performance of hard pegs generally (including currency boards) with other regimes. They find that hard pegs trail floating regimes in growth performance (though not by more than pegs or intermediate regimes). However, this allows us to conclude that, in the end, the lack of policy responses may have a detrimental effect on overall economic performance.

The fact that currency boards may delay an adjustment has also been a cause of concern. Aizenman and Glick (2005) and Kuttner and Posen (2001) have both found that the harder and longer the peg, the larger are the depreciations upon exiting. This is to be expected, because the stronger the commitment, the fixed exchange rate spell will be typically longer, and only under more unfavourable conditions will the peg be abandoned, suggesting that an earlier adjustment may have been beneficial. This conclusion, however, should be treated with care because it fails to take into account the fact that this stringency also helps avoid many exits that later on would have turned out to be unnecessary.

The same caution should be used when evaluating the tendency of currency boards to foster the evolution of mismatches in government and private sector debt structures. The basic idea is that as long as the currency board holds countries develop a tendency to 'dollarize' their financial sectors (see Catao and Terrones, 2000), with banks piling foreign currency deposits on their liability side, firms borrowing in dollars abroad and governments issuing debt in dollars. This is a problem because the asset side of these borrowers is in most cases linked mostly to the local economy, and thus, whether denominated in foreign currency or not, subject to currency risk in the event of a devaluation. This mismatch, however, is a double-edged sword. On the one hand it increases the commitment of the authorities to the peg (and this is why sometimes it is encouraged by the authorities as an additional credibility booster), but on the other it may also trigger large capital outflows in anticipation of a crisis. In the presence of large mismatches, agents would correctly anticipate a devaluation to produce a costly crisis, thus accelerating the run and the likelihood that the currency will sink. How these two factors play out during a crisis depends on the specifics of each individual country.

Finally, a currency board limits the ability of the central bank to operate as lender of last resort, particularly in the event of a bank run. This has been suggested as an explanation of why countries with currency boards quickly develop an international based banking system (typically with local institutions bought by foreign banks) which is better insured against runs at any specific location. Proponents of currency boards have suggested several alternatives to replace the central bank's function as lender of last resort with other mechanisms. Among these are the possibility of the government operating as lender of last resort, potentially by borrowing in dollars in times of need; the setting up of insurance schemes by which financial institutions buy in advance the access to funds in the context of a systemic liquidity run (these schemes were implemented by Mexico and Argentina); tighter capital and liquidity requirements on the banking sector; and the piling up of 'extra reserves' as far as possible. The first of these mechanisms is doubtful, as the government may have limited access to financing when it faces a crisis, and the others entail a cost. However, it may be said that some of these schemes have been implemented and used successfully. Specifically, Argentina used its contingent credit line with private banks during its 2001 crisis and banks honoured their pledge at the time.

Where does this leave us?

The conclusion is then that, as much as with currency unions, there seems to be a strong trade motive to set up a currency board. In fact, for a fiscally sound small country with the ability to conduct fiscal policy with some flexibility a currency board may be superior to a common currency as it allows the country to retain the seigniorage on its money stock. For larger middle-income countries a currency board has been pursued more as a way of improving credibility than anything else. While currency boards seem to have delivered, the Argentina case also suggests that their role in improving credibility cannot be taken fully for granted. If a currency board is implemented in times of easy access to international financial markets, fiscal discipline may be side-stepped and a fiscal and currency crisis may still occur at the end of the day. Additionally, policymakers should ask themselves if it makes sense to buy the credibility through a peg, or to buy it the hard way, day by day, implementing reasonable fiscal policies while maintaining some degree of flexibility in monetary policy. The successful experience since the mid-1990s of many countries with managed floating regimes and inflation targeting seems to point to this direction. If this trend continues, currency boards may become even rarer in the future.

FEDERICO STURZENEGGER

See also **currency unions; dollarization.**

Bibliography

Aizenman, J. and Glick, R. 2005. Pegged exchange rate regimes – a trap? Working Paper No. 2006–07, Federal Reserve Bank of San Francisco.

Catao, L. and Terrones, M. 2000. Determinants of dollarization: the banking side. Working Paper No. 00/146, International Monetary Fund.

Culp, C., Hanke, S. and Miller, M. 1999. The case for an Indonesian currency board. *Journal of Applied Corporate Finance* 11, 57–65.

Fatas, A. and Rose, A.K. 2000. Do monetary handcuffs restrain Leviathan? Fiscal policy in extreme exchange rate regimes. Discussion Paper No. 2692, CEPR.

Frankel, J. and Rose, A. 2002. An estimate of the effect of common currencies on trade and income. *Quarterly Journal of Economics* 117, 437–66.

Ghosh A., Gulde, A.-M. and Wolf, H.C. 1998. Currency boards: the ultimate fix? Working Paper No. 98/8, International Monetary Fund.

Kuttner, K. and Posen, A. 2001. Beyond bipolar: a three-dimensional assessment of monetary frameworks. Working Paper No. 52, Oesterreichische Nationalbank.

Levy-Yeyati, E. and Sturzenegger, F. 2001. Exchange rate regimes and economic performance. *IMF Staff Papers* 47, 62–98.

currency competition

'Currency competition' refers to the free, or virtually free, entry of private-sector firms into the issuance of a circulating medium of exchange in lieu of a government monopoly on currency issue. Although there is little analytical basis for focusing on the private issuance of securities that circulate at the expense of those that do not, that is exactly the approach of the literature on currency competition and thus of this article.

The best real-world examples of currency competition come from periods, some lasting more than a century, in which countries allowed banks to operate relatively free from regulation. This freedom allowed, among other things, banks to issue paper notes. Shuler (1992) identified 66 countries as having free banking for some period in the 19th and 20th centuries, and all of them reportedly had multiple private-sector note issuers.

Today, there is no true private note issuance. Any privately issued notes are issued by banks that operate as agents of their respective central banks. Shuler (1992) attributed the demise of privately issued notes to several factors. One factor was a shift in attitudes about the need for and proper role of central banks. The view took hold that currency issuance could be destabilizing if left to the private sector, and governments nationalized currency issuance in their central banks. This was the case in England, for example, where the Currency School came to dominate and the Bank Act of 1844 eliminated private note issuance. Another major factor leading to government monopolies over currency issuance was the First World War and governments' need for additional sources of revenue. The ability to issue currency directly became very appealing.

By the 1970s, governments' monopoly on currency creation was raising its own concerns. These government issuers had an incentive to overissue to generate additional seigniorage revenue. When inflation began rising in the 1970s, some blamed this incentive to overproduce and called for denationalization of currency issuance. Friedrich Hayek (1990) was perhaps the most prominent proponent of a return to currency competition. Hayek argued, in the terms of today, that an equilibrium could exist with competitive issuance and that it would likely dominate the equilibrium arising when the government monopolizes currency issuance. The logic was that the demand for a privately issued currency depends in part on the currency's quality because such currencies are distinguishable. The more units of a currency supplied, the lower is the currency's value in exchange and thus its perceived quality and the public's demand for it. Competition would thus give issuers an incentive to protect the value of their currencies by not overissuing.

In considering what currency competition might look like, economists rediscovered the free banking periods, and a literature arose studying them. The first wave of that literature consisted of historical studies of free banking and private note issuance, although there were also a few theoretical models. Later, in the 1990s, the potential for new electronic means of payment, such as stored value and digital currencies for the Internet, led to another generation of research on currency competition, this time primarily theoretical.

Most discussions of currency competition, whether from a theoretical or an historical perspective, failed to distinguish inside money from outside money. Hellwig's work (1985) was an exception. Inside money is a claim that obligates its issuer to redeem or exchange the money for some specified monetary or non-monetary object. Failure to do so, perhaps because of insufficient reserves held against the money, can result in a failure to fulfil that obligation and ultimately bankruptcy. The value of a privately issued inside money depends in part, then, on the likelihood of the issuer fulfilling its claim, and only in part on the value of using the money in exchange. Outside money is not a claim against the issuer or anyone else. The issuer makes no promise to redeem its currency at any time for anything of value. The value of a privately issued outside money derives solely from its value in exchange.

The experience in the US free banking era (1837–63) is an example of the importance of the claim that backs an inside money. Bank notes issued in the free banking era were supposed to be fully backed to guarantee the issuer's ability to redeem them, but often they were not. In some cases, no backing was held. Bank note reporters kept track of the financial condition of issuers and of the prices at which notes were trading. Weber (2002) found that notes traded for one another at flexible exchange rates that often depended in part on the extent to which the notes were backed. When the public became aware that an issuer's notes lacked backing, the notes stopped circulating.

The distinction between inside and outside money is important for studying currency competition. Competition in outside note issuance is likely to divert fewer resources from consumption and production than competition in inside note issuance because there is no need to hold reserves against the outside notes. However, without reserves to back outside money, the money is likely to be overissued because of its near-zero marginal

cost of production. Thus, the welfare gain from avoiding overissuing with an inside money must be balanced against the welfare loss from holding full or fractional reserves against such money.

Historical experience with outside money has almost always involved a single, government-issued fiat currency. The existing theoretical literature suggests why privately issued outside money is virtually never observed. In many different economic environments, economists have shown that there can be no equilibrium with competitive issuance of outside money if issuers cannot make binding commitments about the volume of notes they will issue. Taub (1985) and Bryant (1981) showed this in an overlapping-generations model. Ritter (1995) did so in a search model of money. In all cases, the argument is as follows, and similar to Hayek's. If issuing new money is costless, issuers cannot make binding commitments, and money has some positive value, then any private agent that issues notes will issue an unlimited quantity, driving the inflation rate to infinity and the real value of the money to zero. Rational agents would anticipate this ultimate outcome and be unwilling to hold the money at any earlier date. The inability to make binding commitments, coupled with a time inconsistency problem, is a key feature of this argument because issuers always want to believe they will constrain their note issuance, but when they need to they never have the incentive to do so.

A few models have gotten around this result. Klein (1974), for example, provided an early argument based on reputation formation for the existence of equilibria with free entry into private issuance. He argues that the monies of different issuers can be distinguishable by quality, so they can circulate at flexible exchange rates with one another. His discussion, however, blurs the distinction between inside and outside money.

In another example, Martin and Schreft (2006) showed that privately issued outside money can be valued if agents believe that all notes issued up to some threshold will be valued, but additional notes will be worthless. These beliefs create a discontinuity in the value of the marginal unit of currency. Because the value of a marginal unit of currency reaches zero for some finite supply, the limit argument no longer applies. Martin and Schreft derived their existence result in both an overlapping generations and a search-theoretic environment, though it should hold in any environment in which fiat currency could be valued. Interestingly, welfare is not necessarily greater with competitive issuance than with monopoly issuance and depends on the environment considered. In the search environment, neither competitive issuers nor a monopolist achieve the efficient quantity of money in the long run. In the overlapping-generations environment, the efficient allocation is achieved in finitely many periods if agents incur a cost of becoming money issuers. A monopoly issuer might achieve as desirable an allocation, but only if its actions are sufficiently constrained by agents' beliefs.

In contrast, the historical experience with inside money has involved multiple inside monies that are all convertible into some single dominant outside money. A modern literature on privately issued inside notes, largely attributable to Wallace and others, has considered this case. Cavalcanti and Wallace (1999a; 1999b) studied a search-theoretic model with an exogenously given and indivisible outside money and inside money issued by private agents known as banks. To get the private money to be valued, they assumed that issuers who do not accept a note when presented with one face a stiff punishment: they lose the ability to issue notes and revert to autarky. This assumption is reminiscent of the redemption requirements of successful systems for private inside currency issuance, like the Suffolk Banking System that operated in New England in the early 1800s. The authors found that, if the stock of outside money is sufficiently small, then the optimal mechanism has private notes issued and also redeemed on demand. Additionally, expected utility is greater in economies with inside money than only outside money because the set of implementable allocations is larger.

In the United States, at least, it is claimed that little currently prohibits private-sector issuance of outside currency in either paper or digital form. The laws prohibiting it have either expired or been repealed. It will be interesting to see if a resurgence of private issuance occurs.

STACEY L. SCHREFT

See also **fiat money; free banking era; Hayek, Friedrich August von; inflation; inside and outside money.**

Bibliography

Bryant, J. 1981. The competitive provision of fiat money. *Journal of Banking and Finance* 5, 587–93.
Cavalcanti, R. de O. and Wallace, N. 1999a. A model of private bank-note issue. *Review of Economic Dynamics* 2, 104–36.
Cavalcanti, R. de O. and Wallace, N. 1999b. Inside and outside money as alternative media of exchange. *Journal of Money, Credit, and Banking* 31, 443–57.
Hayek, F. 1990. *Denationalisation of Money: The Argument Refined*, 3rd edn. London: Institute of Economic Affairs.
Hellwig, M. 1985. What do we know about currency competition? *Zeitschrift für Wirtschafts- und sozialwissenchaften* 105, 565–88.
Klein, B. 1974. The competitive supply of money. *Journal of Money, Credit, and Banking* 6, 423–53.
Martin, A. and Schreft, S. 2006. Currency competition: a partial vindication of Hayek. *Journal of Monetary Economics* 53, 2085–111.
Ritter, J. 1995. The transition from barter to fiat money. *American Economic Review* 85, 134–49.
Schreft, S. 1997. Looking forward: the role for government in regulating electronic cash. *Federal Reserve Bank of Kansas City Economic Review* (Fourth Quarter), 59–84.
Shuler, K. 1992. The world history of free banking: an overview. In *The Experience of Free Banking*, ed. K. Dowd. New York: Routledge.
Taub, B. 1985. Private money with many suppliers. *Journal of Monetary Economics* 16, 195–208.
Weber, W. 2002. Banknote exchange rates in the antebellum United States. Working Paper No. 623. Research Department, Federal Reserve Bank of Minneapolis.

currency crises

A currency crisis occurs when investors flee from a currency en masse out of fear that it might be devalued. Currency crises are episodes characterized by sudden depreciations of the domestic currency, large losses of foreign exchange reserves of the central bank, and (or) sharp hikes in domestic interest rates.

There have been numerous currency crises since 1980. The so-called debt crisis erupted in 1982 following Mexico's default and devaluation in August. This crisis spread rapidly to all Latin American countries, and by the time it was over, most Latin American countries had devalued their currencies and defaulted on their foreign debts. The debt crisis was followed by a decade of negative growth and isolation from international capital markets. The output costs of this crisis were so large that the 1980s became known as the 'lost decade' for Latin America.

Crises are not just emerging-market phenomena. The 1990s opened with crises in industrial Europe – the European Monetary System (EMS) crises of 1992 and 1993. By the end of these crises, in the summer of 1993, the lira and the sterling had been driven from the Exchange Rate Mechanism (ERM); Finland, Norway, and Sweden had abandoned their unofficial peg to the European Currency Unit (ECU); the Spanish peseta, the Portuguese escudo and the Irish punt had devalued; and Europe's central bank governors and finance ministers had widened the ERM's intervention margins to ± 15 per cent from ± 2.25 per cent. Only then did the currency market stabilize.

Crises are hardy perennials. Within one year of the EMS crises, a currency crisis exploded in Mexico, with currency jitters spreading around the Latin American region. In 1997, it was Asia's turn. A new episode of currency turbulences started in July of that year with the depreciation of the Thai baht. Within a few days the crisis had spread to Indonesia, Korea, Malaysia and the Philippines. Turmoil in the foreign exchange market heightened in 1998 with the Russian default and devaluation in August. The Russian crisis spread around the world with speculative attacks in economies as far apart as South Africa, Brazil and Hong Kong. Currency crises have continued to erupt in the new millennium, with Argentina's crisis in December 2001 including the largest foreign-debt default in history.

The numerous financial crises that have ravaged emerging markets as well as mature economies have fuelled a continuous interest in developing models to explain why speculative attacks occur. Models are even catalogued into three generations. The first-generation models focus on the fiscal and monetary causes of crises. These models were mostly developed to explain the crises in Latin America in the 1960s and 1970s. In these models, unsustainable money-financed fiscal deficits lead to a persistent loss of international reserves and ultimately to a currency crash (see, for example, Krugman, 1979).

The second-generation models aim at explaining the EMS crises of the early 1990s. These models focus on explaining why currency crises tend to happen in the midst of unemployment and loss of competitiveness. To explain these links, governments are modelled facing two targets: reducing inflation and keeping economic activity close to a given target. Fixed exchange rates may help in achieving the first goal but at the cost of a loss of competitiveness and a recession. With sticky prices, devaluations restore competitiveness and help in the elimination of unemployment, thus prompting the authorities to abandon the peg during recessions. Importantly, in this setting of counter-cyclical policies, the possibility of self-fulfilling crises becomes important, with even sustainable pegs being attacked and frequently broken (see, for example, Obstfeld, 1994).

The next wave of currency crises, the Mexican crisis in 1994 and the Asian crisis in 1997, fuelled a new variety of models – also known as third-generation models – which focus on moral hazard and imperfect information. The emphasis here has been on 'excessive' booms and busts in international lending and asset price bubbles. These models also link currency and banking crises, sometimes known as the 'twin crises' (Kaminsky and Reinhart, 1999). For example, Diaz-Alejandro (1985) and Velasco (1987) model difficulties in the banking sector as giving rise to a balance of payments crisis, arguing that, if central banks finance the bail-out of troubled financial institutions by printing money, we have the classical story of a currency crash prompted by excessive money creation. Within the same theme, McKinnon and Pill (1995) examines the role of capital flows in an economy with an unregulated banking sector with deposit insurance and moral hazard problems of the banks. Capital inflows in such an environment can lead to over-lending cycles with consumption booms, real exchange rate appreciations, exaggerated current account deficits, and booms (and later busts) in stocks and property markets. Importantly, the excess lending during the boom makes banks more prone to a crisis when a recession unfolds. In turn, the fragile banking sector makes the task of defending the peg by hiking domestic interest rates more difficult and may lead to the eventual collapse of the domestic currency. Following the crisis in Argentina in 2001, the links between debt sustainability, sovereign defaults, and currency crises again attracted the attention of the

economics profession. Finally, currency crises have also been linked to the erratic behaviour of international capital markets. For example, Calvo (1998) has brought to general attention the possibility of liquidity crises in emerging markets due to sudden reversals in capital flows, in large part triggered by developments in the world financial centres.

To summarize, all models suggest that currency crises erupt in fragile economies. Importantly, the three generations of models conclude that vulnerabilities come in different varieties. Still, the first attempts to study the vulnerabilities that precede crises have adopted 'the one size fits all' approach (see, for example, Frankel and Rose, 1996; and Kaminsky, 1998). That is, the regressions estimated to predict crises include all possible indicators of vulnerability. These indicators include those related to sovereign defaults, such as high foreign debt levels, or indicators related to fiscal crises, such as government deficits, or even indicators related to crises of financial excesses, such as stock and real estate market booms and busts. In all cases, researchers impose the same functional form on all observations. When some indicators are not robustly linked to all crises, they tend to be discarded even when they may be of key importance for a subgroup of crises. Naturally, these methods leave many crises unpredicted and, furthermore, cannot capture the evolving nature of currency crises.

The next step in the empirical analysis of crises should be centred on whether crises are of different varieties. The first attempt in this direction is in Kaminsky (2006). In this article, a different methodology is used to allow for *ex ante* unknown varieties of currency crises. To identify the possible multiple varieties of crises, regression tree analysis is applied. This technique allows us to search for an unknown number of varieties of crises and of tranquil times using multiple indicators. This technique was also applied to growth by Durlauf and Johnson (1995).

Interestingly, this method catalogues crises into six classes:

1. *Crises with current account problems*. This variety is characterized by just one type of vulnerability, that of loss of competitiveness, that is, real exchange rate appreciations.
2. *Crises of financial excesses*. The fragilities are associated with booms in financial markets. In particular, they are identified as crises that are preceded by the acceleration in the growth rate of domestic credit and other monetary aggregates.
3. *Crises of sovereign debt problems*. These crises are characterized by fragilities associated with 'unsustainable' foreign debt.
4. *Crises with fiscal deficits*. This variety is just related to expansionary fiscal policy.
5. *Sudden-stop crises*. This type of crisis is only associated with reversals in capital flows triggered by sharp

hikes in world interest rates, with no domestic vulnerabilities.

6. *Self-fulfilling crises.* This class of crises is not associated with any evident vulnerability, domestic or external.

These estimations allow us to answer four important questions about crises.

1. *Do crises occur in countries with sound fundamentals?* Even though this estimation allows for the identification of self-fulfilling crises (crises in economies with sound fundamentals), the results indicate that basically all crises are preceded by domestic or external vulnerabilities. Only four per cent of the crises are unrelated to economic fragilities.

2. *How important are sudden reversals in capital flows in triggering crises?* While many have stressed that the erratic behaviour of international capital markets is the main culprit in emerging market currency crises, only two per cent of the crises in developing countries are just triggered by sudden-stop problems. While sudden-stop problems do occur, the reversals in capital flows mostly occur in the midst of multiple domestic vulnerabilities (see, Calvo, Izquierdo and Talvi, 2004).

3. *Are crises different in emerging economies?* Crises in emerging markets are preceded by far more domestic vulnerabilities than those in industrial countries. Overall, 86 per cent of the crises in emerging economies are crises with multiple domestic vulnerabilities, while economic fragility characterizes only 50 per cent of the crises in mature markets.

4. *Are some crises more costly than others?* It is a well-established fact that financial crises impose substantial costs on society. Many economists have emphasized the output losses associated with crises. But these are not the only costs of crises. In the aftermath of crises, most countries lose access to international capital markets, losing the ability to reduce the effect of adverse income shocks by borrowing in international capital markets. In most cases, countries have to run current account surpluses to pay back their debt. Finally, the magnitude of the speculative attack is itself important. For example, large depreciations may cause adverse balance sheet effects on firms and governments when their liabilities are denominated in foreign currencies. *Crises of financial excesses,* those also associated with banking crises – twin crisis episodes – are the costliest. Not only does the domestic currency depreciate the most, but also output losses are higher and the reversal of the current account deficit is attained via a dramatic fall in imports. In the aftermath of these crises, exports fail to grow even though the depreciations in this type of crises are massive. This evidence suggests that countries are even unable to attract trade credits to finance exports when their economies are mired in financial problems. In contrast, *self-fulfilling crises* and *sudden-stop crises* (but

with no domestic vulnerabilities) have no adverse effects on the economies. Output (relative to trend) is unchanged or continues to grow in the aftermath of crises with no observed domestic fragility. In these crises, booming exports are at the heart of the recovery of the current account.

GRACIELA LAURA KAMINSKY

See also **currency crises models.**

Bibliography

Calvo, G. 1998. Capital flows and capital-market crises: the simple economics of sudden stops. *Journal of Applied Economics* 1, 35–54.

Calvo, G., Izquierdo, A. and Mejia, L. 2004. *On the empirics of sudden stops: the relevance of balance-sheet effects.* Working paper No. 10520. Cambridge, MA: NBER.

Diaz-Alejandro, C. 1985. Good-bye financial repression, hello financial crash. *Journal of Development Economics* 19, 1–24.

Durlauf, S. and Johnson, P. 1995. Multiple regimes and cross-country growth behavior. *Journal of Applied Econometrics* 10, 365–84.

Frankel, J. and Rose, A. 1996. Currency crises in emerging markets: an empirical treatment. *Journal of International Economics* 41, 351–66.

Kaminsky, G.L. 1998. Currency and banking crises: the early warnings of distress. International Finance Discussion Papers No. 629. Board of Governors of the Federal Reserve System.

Kaminsky, G.L. and Reinhart, C. 1999. The twin crises: the causes of banking and balance-of-payments problems. *American Economic Review* 89, 473–500.

Kaminsky, G.L. 2006. Currency crises: are they all the same? *Journal of International Money and Finance* 25, 503–27.

Krugman, P. 1979. A model of balance-of-payments crises. *Journal of Money, Credit, and Banking* 11, 311–25.

McKinnon, R.I. and Pill, H. 1995. Credible liberalizations and international capital flows: the 'overborrowing syndrome'. In *Financial Deregulation and Integration in East Asia,* ed. T. Ito and A.O. Krueger. Chicago: University of Chicago Press.

Obstfeld, M. 1994. The logic of currency crises. *Cahiers Economiques et Monétaires* 43, 189–213.

Obstfeld, M. 1996. Models of currency crises with self-fulfilling features. *European Economic Review* 40, 1037–47.

Velasco, A. 1987. Financial and balance-of-payments crises. *Journal of Development Economics* 27, 263–83.

currency crises models

There have been many currency crises during the post-war era (see Kaminsky and Reinhart, 1999). A currency crisis is an episode in which the exchange rate depreciates substantially during a short period of time. There is an extensive literature on the causes and consequences of a currency crisis in a country with a fixed or heavily managed exchange rate. The models in this literature are often categorized as first-, second- or third-generation.

In first-generation models the collapse of a fixed exchange rate regime is caused by unsustainable fiscal policy. The classic first-generation models are those of Krugman (1979) and Flood and Garber (1984). These models are related to earlier work by Henderson and Salant (1978) on speculative attacks in the gold market. Important extensions of these early models incorporate consumer optimization and the government's intertemporal budget constraint into the analysis (see Obstfeld, 1986; Calvo, 1987; Drazen and Helpman, 1987; Wijnbergen, 1991). Flood and Marion (1999) provide a detailed review of first-generation models.

In a fixed exchange rate regime a government must fix the money supply in accordance with the fixed exchange rate. This requirement severely limits the government's ability to raise seigniorage revenue. A hallmark of first-generation models is that the government runs a persistent primary deficit. This deficit implies that the government must either deplete assets, such as foreign reserves, or borrow to finance the deficit. It is infeasible for the government to borrow or deplete reserves indefinitely. Therefore, in the absence of fiscal reforms, the government must eventually finance the deficit by printing money to raise seigniorage revenue. Since printing money is inconsistent with keeping the exchange rate fixed, first-generation models predict that the regime must collapse. The precise timing of its collapse depends on the details of the model.

The key ingredients of a first-generation model are its assumptions regarding purchasing power parity (PPP), the government budget constraint, the timing of deficits, the money demand function, the government's rule for abandoning the fixed exchange rate, and the post-crisis monetary policy. In the simplest first-generation models there is a single good whose domestic currency price is P_t and whose foreign currency price is 1. Let S_t denote the nominal exchange rate. PPP implies $P_t = S_t$. Suppose for simplicity that the government has a constant ongoing primary deficit, δ. It finances this deficit by reducing its stock of foreign reserves, f_t, which can either evolve as a smooth function of time or jump discontinuously. In the former case, f_t evolves according to $\dot{f}_t = rf_t - \delta + \dot{M}_t/S_t$, where r is the real interest rate, M_t is the monetary base, and a dot over a variable denotes its derivative with respect to time. When foreign reserves change discontinuously, $\Delta f_t = \Delta(M_t/S_t)$. When $\delta > rf_0$ interest income from foreign assets will not be sufficient to finance the deficit.

To illustrate the key properties of first-generation models, we make three simplifying assumptions. First, money demand takes the Cagan (1956) form, $M_t = \theta P_t \exp[-\eta(r + \pi_t)]$, where $\theta > 0$ and $\pi_t = \dot{P}_t/P_t$

is the inflation rate. Second, the government abandons the fixed exchange rate regime when its foreign reserves are exhausted. Third, as soon as foreign reserves are exhausted, the government prints money at a constant rate μ to fully finance its deficit.

These assumptions imply that after the crisis the level of real balances, $m_t = M_t/P_t$, is constant and equal to $\overline{m} = \theta \exp[-\eta(r + \mu)]$. The post-crisis government budget constraint reduces to $\delta = \mu\overline{m}$. This equation determines μ. Let t^* denote the date at which foreign reserves are exhausted and the government abandons the fixed exchange rate regime. PPP implies $S_{t^*} = P_{t^*} = \overline{M}/\overline{m}$, where $\overline{M}$ is the monetary base the instant after date t^*. Under perfect foresight the exchange rate cannot jump discontinuously at t^* since such a jump would imply the presence of arbitrage opportunities. Given that the exchange rate must be a continuous function of time at t^*, $S_{t^*} = S$ and $\overline{M} = \overline{m}S$.

Prior to the crisis real balances are given by $m = \theta \exp(-\eta r)$. Therefore, at date t^* there is a sudden drop in real money demand from m to $\overline{m}$ implying that reserves drop discontinuously to zero at time t^*: $\Delta f_{t^*} = \overline{m} - m$. This is why the literature refers to t^* as the date of the speculative attack. Prior to the crisis the government's reserves fall at the rate $\dot{f}_t = rf_t - \delta$. The budget constraint implies that $t^* = \ln\{[\delta - r(m - \overline{m})]/(\delta - rf_0)\}/r$. While the collapse of the fixed exchange rate regime is inevitable, it does not generally occur at time zero unless $m - \overline{m} > f_0$.

A shortcoming of this type of first-generation model is that the timing of the speculative attack is deterministic and the exchange rate does not depreciate at the time of the attack. These shortcomings can be remedied by introducing shocks into the model, as in Flood and Garber (1984).

Early first-generation models predict that ongoing fiscal deficits, rising debt levels, or falling reserves precede the collapse of a fixed exchange rate regime. This prediction is inconsistent with the 1997 Asian currency crisis. This inconsistency led many observers to dismiss fiscal explanations of this crisis. However, Corsetti, Pesenti and Roubini (1999), Burnside, Eichenbaum and Rebelo (2001a), and Lahiri and Végh (2003) show that bad news about prospective deficits can trigger a currency crisis. Under these circumstances a currency crisis will not be preceded by persistent fiscal deficits, rising debt levels, or falling reserves. These models assume that agents receive news that the banking sector is failing and that banks will be bailed out by the government. The government plans to finance, at least in part, the bank bailout by printing money beginning at some time in future. Burnside, Eichenbaum and Rebelo (2001a) show that a currency crisis will occur before the government actually starts to print money. Therefore, in their model, a currency crisis is not preceded by movements in standard macroeconomic fundamentals, such as fiscal deficits and money growth. Burnside, Eichenbaum and Rebelo argue that their model accounts for the main characteristics of the Asian currency crisis.

This explanation of the Asian currency crisis stresses the link between future deficits and current movements in the exchange rate. This link is also stressed by Corsetti and Mackowiak (2006), Daniel (2001), and Dupor (2000), who use the fiscal theory of the price level to argue that prices and exchange rates jump in response to news about future deficits.

In first-generation models the government follows an exogenous rule to decide when to abandon the fixed exchange rate regime. In second-generation models the government maximizes an explicit objective function (see, for example, Obstfeld, 1994; 1996). This maximization problem dictates if and when the government will abandon the fixed exchange rate regime. Second-generation models generally exhibit multiple equilibria so that speculative attacks can occur because of self-fulfilling expectations. In Obstfeld's models (1994; 1996) the central bank minimizes a quadratic loss function that depends on inflation and on the deviation of output from its natural rate (see Barro and Gordon, 1983, for a discussion of this type of loss function). The level of output is determined by an expectations-augmented Phillips curve. The government decides whether to keep the exchange rate fixed or not. Suppose agents expect the currency to devalue and inflation to ensue. If the government does not devalue then inflation will be unexpectedly low. As a consequence output will be below its natural rate. Therefore the government pays a high price, in terms of lost output, in order to defend the currency. If the costs associated with devaluing (lost reputation or inflation volatility) are sufficiently low, the government will rationalize agents' expectations. In contrast, if agents expect the exchange rate to remain fixed, it can be optimal for the government to validate agents' expectations if the output gains from an unexpected devaluation are not too large. Depending on the costs and benefits of the government's actions, and on agents' expectations, there can be more than one equilibrium. See Jeanne (2000) for a detailed survey of second-generation models.

Morris and Shin (1998) provide an important critique of models with self-fulfilling speculative attacks. They emphasize that standard second-generation models assume that fundamentals are common knowledge. Morris and Shin demonstrate that introducing a small amount of noise into agents' signals about fundamentals will lead to a unique equilibrium.

Many currency crises coincide with crises in the financial sector (Diaz-Alejandro, 1985; Kaminsky and Reinhart, 1999). This observation has motivated a literature that emphasizes the role of the financial sector in causing currency crises and propagating their effects. These third-generation models emphasize the balance-sheet effects associated with devaluations. The basic idea is that banks and firms in emerging market countries have explicit currency mismatches on their balance sheets because they borrow in foreign currency and lend in local currency. Banks and firms face credit risk because their income is related to the production of non-traded goods whose price, evaluated in foreign currency, falls after devaluations. Banks and firms are also exposed to liquidity shocks because they finance long-term projects with short-term borrowing. Eichengreen and Hausmann (1999) argue that currency mismatches are an inherent feature of emerging markets. In contrast, authors such as McKinnon and Pill (1996) and Burnside, Eichenbaum and Rebelo (2001b) argue that, in the presence of government guarantees, it is optimal for banks and firms to expose themselves to currency risk.

Different third-generation models explore various mechanisms through which balance-sheet exposures may lead to a currency and banking crisis. In Burnside, Eichenbaum and Rebelo (2004) government guarantees lead to the possibility of self-fulfilling speculative attacks. In Chang and Velasco (2001) liquidity exposure leads to the possibility of a Diamond and Dybvig (1983) style bank run. In Caballero and Krishnamurthy (2001) firms face a liquidity problem because they finance risky long-term projects with foreign loans but have access to limited amounts of internationally accepted collateral.

An important policy question is: what is the optimal nature of interest rate policy during and after a currency crisis? There has been relatively little formal work on this topic. Christiano, Braggion and Roldos (2006) take an important first step in this direction. They argue that it is optimal to raise interest rates during a currency crisis and to lower them immediately thereafter. Studying optimal monetary policy in different models of currency crises remains an important area for future research.

CRAIG BURNSIDE, MARTIN EICHENBAUM AND SERGIO REBELO

See also **currency crises; fiscal theory of the price level.**

Bibliography

Barro, R. and Gordon, D. 1983. A positive theory of monetary policy in a natural rate model. *Journal of Political Economy* 91, 589–610.

Burnside, C., Eichenbaum, M. and Rebelo, S. 2001a. Prospective deficits and the Asian currency crisis. *Journal of Political Economy* 109, 1155–98.

Burnside, C., Eichenbaum, M. and Rebelo, S. 2001b. Hedging and financial fragility in fixed exchange rate regimes. *European Economic Review* 45, 1151–93.

Burnside, C., Eichenbaum, M. and Rebelo, S. 2004. Government guarantees and self-fulfilling speculative attacks. *Journal of Economic Theory* 119, 31–63.

Caballero, R. and Krishnamurthy, A. 2001. International and domestic collateral constraints in a model of emerging market crises. *Journal of Monetary Economics* 48, 513–48.

Cagan, P. 1956. Monetary dynamics of hyperinflation. In *Studies in the Quantity Theory of Money*, ed. M. Friedman. Chicago: University of Chicago Press.

Calvo, G. 1987. Balance of payments crises in a cash-in-advance economy. *Journal of Money Credit and Banking* 19, 19–32.

Chang, R. and Velasco, A. 2001. A model of financial crises in emerging markets. *Quarterly Journal of Economics* 116, 489–517.

Christiano, L., Braggion, F. and Roldos, J. 2006. The optimal monetary response to a financial crisis. Mimeo, Northwestern University.

Corsetti, G. and Mackowiak, B. 2006. Fiscal imbalances and the dynamics of currency crises. *European Economic Review* 50, 1317–38.

Corsetti, G., Pesenti, P. and Roubini, N. 1999. What caused the Asian currency and financial crisis? *Japan and the World Economy* 11, 305–73.

Daniel, B. 2001. The fiscal theory of the price level in an open economy. *Journal of Monetary Economics* 48, 293–308.

Diamond, D. and Dybvig, P. 1983. Bank runs, deposit insurance and liquidity. *Journal of Political Economy* 91, 401–19.

Diaz-Alejandro, C. 1985. Good-bye financial repression, hello financial crash. *Journal of Development Economics* 19, 1–24.

Drazen, A. and Helpman, E. 1987. Stabilization with exchange rate management. *Quarterly Journal of Economics* 102, 835–55.

Dupor, W. 2000. Exchange rates and the fiscal theory of the price level. *Journal of Monetary Economics* 45, 613–30.

Eichengreen, B. and Hausmann, R. 1999. Exchange rates and financial fragility. In *New Challenges for Monetary Policy: A Symposium Sponsored by the Federal Reserve Bank of Kansas City*. Kansas City: Federal Reserve Bank of Kansas City.

Flood, R. and Garber, P. 1984. Collapsing exchange rate regimes: some linear examples. *Journal of International Economics* 17, 1–13.

Flood, R. and Marion, N. 1999. Perspectives on the recent currency crisis literature. *International Journal of Finance and Economics* 4, 1–26.

Henderson, D. and Salant, S. 1978. Market anticipations of government policies and the price of gold. *Journal of Political Economy* 86, 627–48.

Jeanne, O. 2000. Currency crises: a perspective on recent theoretical developments. Special Papers in International Economics, No. 20, International Finance Section, Princeton University.

Kaminsky, G. and Reinhart, C. 1999. The twin crises: the causes of banking and balance-of-payments problems. *American Economic Review* 89, 473–500.

Krugman, P. 1979. A model of balance of payments crises. *Journal of Money, Credit and Banking* 11, 311–25.

Lahiri, A. and Végh, C. 2003. Delaying the inevitable: interest rate defense and BOP crises. *Journal of Political Economy* 111, 404–24.

McKinnon, R. and Pill, H. 1996. Credible liberalizations and international capital flows: the overborrowing syndrome.
In *Financial Deregulation and Integration in East Asia*, ed. T. Ito and A. Krueger. Chicago: University of Chicago Press.

Morris, S. and Shin, H. 1998. Unique equilibrium in a model of self-fulfilling currency attacks. *American Economic Review* 88, 587–97.

Obstfeld, M. 1986. Speculative attack and the external constraint in a maximizing model of the balance of payments. *Canadian Journal of Economics* 29, 1–20.

Obstfeld, M. 1994. The logic of currency crises. *Cahiers Economiques et Monétaires* 43, 189–213.

Obstfeld, M. 1996. Models of currency crises with self-fulfilling features. *European Economic Review* 40, 1037–47.

Wijnbergen, S. van. 1991. Fiscal deficits, exchange rate crises and inflation. *Review of Economic Studies* 58, 81–92.

currency unions

Currency unions (also known as monetary unions) are groups of countries that share a single money. Currency unions are unusual, since most countries have their own currency. For instance, the United States, Japan and the United Kingdom all have their own monies. But a reasonable number of countries participate in currency unions, and their importance is growing. In May 2005, 52 of the 184 IMF members participated in currency unions.

Currency unions present and past

Currency unions commonly come about when a small or poor country unilaterally adopts the money of a larger, richer 'anchor' country. For instance, a number of countries currently use the US dollar, including Panama, El Salvador, Ecuador, and a number of smaller countries and dependencies in the Caribbean and Pacific. Swaziland, Lesotho and Namibia all use the South African rand. Both the Australian and New Zealand dollars are used by a number of countries in the Pacific; Liechtenstein uses the Swiss franc; and so forth. In the past, a number of countries have used the currency of their colonizer; over 50 countries and dependencies have used the British pound sterling at one time or another. Cases like this are known as official dollarization (unofficial dollarization occurs when the currency of a foreign country circulates widely but is not formally the national currency). In such cases, the small country essentially relinquishes its right to sovereign monetary policy. It loses its ability to independently influence its exchange and interest rates; these are determined by the anchor country, typically on the basis of the interests of the anchor.

There are also a number of multilateral currency unions between countries of more or less equal size and wealth. For instance, the East Caribbean dollar circulates in Anguilla, Antigua and Barbuda, Dominica, Grenada, Montserrat, Saint Kitts and Nevis, Saint Lucia, and Saint

Vincent and the Grenadines. The Central Bank of the West African States circulates the Communauté française d'Afrique (CFA) franc in Benin, Burkina Faso, Côte d'Ivoire, Guinea-Bissau, Mali, Niger, Senegal, and Togo. The Bank of the Central African States circulates a slightly different CFA franc in Cameroon, the Central African Republic, Chad, Republic of Congo, Equatorial Guinea, and Gabon.

The largest and most important currency union is the Economic and Monetary Union of the European Union (EMU). EMU technically began on 1 January 1999, although the euro was physically introduced only three years later. Twelve countries are formally members of EMU: Austria, Belgium, Finland, France, Germany, Greece, Ireland, Italy, Luxembourg, the Netherlands, Portugal, and Spain. (A number of smaller European territories and French dependencies also use the euro.) These countries jointly determine monetary policy for EMU through the international European Central Bank. The number of members in EMU is expected to grow with time, especially as countries that acceded to the European Union in 2004 become eligible for EMU entry. However, both Sweden and Denmark have rejected membership in referenda, and the euro remains unpopular in the UK.

While a number of currency unions currently exist, many have not survived. The Latin Monetary Union began in 1865 when France, Belgium, Italy and Switzerland (later joined by Greece, Romania, and others) adopted common regulations for their individual currencies to encourage the free international flow of money. This essentially amounted to a commitment to mint silver and gold coins to uniform specifications, but without other restrictions on monetary policy. The union effectively ended with the onset of the First World War. The war also ended the Scandinavian Monetary Union which Denmark, Norway, and Sweden began in 1873. The economic union between Belgium and Luxembourg that began in 1921 has been absorbed into EMU. Multilateral currency unions in East Africa, Central Africa, West Africa, South Asia, South-East Asia, and the Caribbean have also disappeared.

Theory: why should countries enter currency union?

Historically, most countries have had their own moneys. There seems to be a tight connection between national identity and national money; a country's money is a potent symbol of sovereignty. Still, some countries have entered into currency union. Why? Economists have theorized about the potential economic benefits of currency union which can, in certain circumstances, overwhelm the perceived political costs.

Like all other monetary regimes, currency unions are fully compatible with Robert Mundell's (1968) celebrated 'Trilemma' or 'Incompatible Trinity'. A country would like its monetary regime to deliver three desirable goals that turn out to be mutually exclusive: domestic monetary sovereignty, capital mobility, and exchange rate stability. Currently, large rich countries like the United States, Japan and the UK have domestic monetary sovereignty and open capital markets but have floating exchange rates. By way of contrast, members of a currency union essentially relinquish the first objective (monetary independence) in exchange for the latter benefits (capital mobility and stable exchange rates). Indeed, some economists think of currency unions as simply extreme forms of fixed exchange rates, with all the associated pros and cons. Countries inside currency union receive more microeconomic benefits than they would from a fixed exchange rate, since sharing a single money leads to deeper integration of real and financial markets. On the other hand, a country can devalue or float the exchange rate more easily than it can leave a currency union. Still, this is an unsatisfying theoretical approach to the issue of currency unions. It does not address the vital question: what is the optimal size of a currency union? If the right size for a currency union is not necessarily the country, how should we tackle the problem?

The theoretical analysis of currency unions began with a seminal paper by Mundell (1961). Mundell's analysis answered the question: what is the appropriate domain for a currency? Mundell briefly argued there are advantages to regions that use a common money. In particular, currency union facilitates international trade; a single medium of exchange reduces transactions costs, as does a common unit of account. However, a common currency can also cause problems in the dual presence of asymmetric shocks and nominal rigidities (in prices and wages). Suppose demand shifts from Western to Eastern goods. The increase in demand for Western output results in inflationary pressures there, while East goes into recession. Mundell argued that, if unemployed labour could move freely from East to relieve inflationary pressures in West, the two problems could be resolved simultaneously. However, in the absence of labour mobility, the asymmetric shock could be better handled by allowing the Western currency to appreciate. But in order for this to happen, both East and West must have their own monies! Mundell concluded that the optimal currency area was the area within which labour is mobile; regions of labour mobility should have their own currencies.

Two other classic contributions to the theory of optimal currency areas are worthy of note. McKinnon (1963) examined the effects of country size on currency unions; he concluded that smaller countries tend to be more open and have fewer nominal rigidities, making them better candidates for currency union. Kenen (1969) considered the effects of the economy's degree of diversification, and argued that more diversification resulted in fewer asymmetric shocks, and accordingly fewer benefits from national monetary policy.

The key focus of Mundell's theoretical optimum currency area framework – the adjustment to asymmetric shocks – has stood the test of time well. The ability of a region to respond to such shocks is viewed as a critical part of a sustainable and desirable currency union. Still, hardly anyone now takes the narrow specifics of Mundell's original article seriously. In particular, Mundell's conclusion that the optimum currency area is a region of labour mobility is no longer widely believed. The problem of asymmetric business cycles that Mundell described is intrinsically a problem of … business cycles. The costs of shifting labour are high almost everywhere in the world, which is why labour moves only slowly, even within countries with relatively flexible labour markets like the United States. Accordingly, most economists are uncomfortable thinking that labour could or should shift in response to the shocks and propagation mechanisms that cause business cycles. After all, the nominal rigidities that are responsible for business cycles do not last for ever. Thus, Mundell's idea of labour mobility is no longer viewed as a viable adjustment mechanism. (This conclusion is tempered if one believes that real shocks cause business cycles without nominal rigidities.)

Still, there are other ways to share the risks of, or adjust to, asymmetric shocks, and much of the relevant work has incorporated these other mechanisms. Mundell originally ignored capital mobility. But private capital markets can, in principle, spread shocks internationally if investors diversify across regions or sectors. However, more attention has been paid to the public sector, since a federal system of taxes and transfers may be an efficient way to spread risks across regions. To continue with the East and West example, a progressive federal tax structure reduces inflationary Western pressures, and allows benefits to be paid to the unemployed in the East. Both regions suffer less macroeconomic volatility with such automatic stabilizers in place. The most controversial adjustment mechanism is counter-cyclical fiscal policy. In response to an asymmetric shock, regions that are free and capable of deploying discretionary fiscal policy can uses changes in taxes and government spending to respond to asymmetric shocks, even within the monetary confines of a currency union. More generally, mechanisms to handle asymmetric shocks are still an integral part of the theory of currency unions.

Mundell originally thought the great benefit of currency union was the facilitation of trade since money is a convenience that lowers transactions costs. But suppose that countries produce moneys of different qualities. Argentina has gone through five currencies since 1970; high Argentine inflation results in a low convenience value for Argentine money. Suppose Argentina decides to give up on a national money altogether and enter into a currency union with a foreign producer of higher-quality money: the United States, say. Argentina will surely experience different shocks from the United States, and these shocks have to be handled. Perhaps then Argentina should enter a currency union with a country with more similar shocks? The problem is that the most obvious contender, Brazil, also has a history of monetary incompetence. The larger point is that a low-quality domestic monetary authority increases a country's willingness to enter currency union, as does the availability of high-quality foreign money. Alesina and Barro (2002) provide an elegant model that incorporates such features. In their model, countries enter currency unions with neighbours in order to facilitate trade, so long as the neighbours possess monetary institutions of quality. Lower inflation and reduced transactions costs of trade provide gains, while the inability to respond to idiosyncratic asymmetric shocks generates losses.

Empirics: what do we know in practice about currency unions?

During the run-up to EMU, a considerable empirical literature developed that quantified different aspects of optimal currency areas. Much attention was paid to estimating the synchronization of business cycles for potential EMU candidates; Bayoumi and Eichengreen (1992) was the first important paper. The tradition has since been generalized to more countries by Alesina, Barro and Tenreyro (2002), who characterized co-movements in prices as well as output. Frankel and Rose (1998) showed that the intensity of trade had a strong positive effect on business cycle synchronization; that is, the optimum currency area criteria are jointly endogenous. If currency union lowers the transactions costs of trade and thus leads to an increase in trade, it may also thereby reduce the asymmetries in business cycles; areas that do not look like currency unions *ex ante* may do so *ex post*. Bayoumi and Eichengreen (1998) successfully link optimum currency area criteria (principally the asymmetry of business cycle shocks) to exchange rate volatility and intervention, and show that a number of features of the optimum currency area theory appear in practice, even for countries not in currency unions.

Somewhat curiously, little work was done to analyse actual currency unions until around 2000. This is probably because the currency unions that preceded EMU consisted mostly of small or poor countries, which were viewed as irrelevant for EMU. But this gap in the literature implicitly allowed economists to focus their attention on the costs of currency union, which tend to be macroeconomic in nature (resulting from the absence of national monetary policy as a tool to stabilize business cycles). As Mundell clearly pointed out, there are also benefits from a currency union, mostly microeconomic in nature. Fewer monies mean lower transactions costs for trade, and thus higher welfare. An unresolved issue of importance is the size of the benefits that stem from currency union. There is evidence that currency unions have been associated with increased trade in goods, though its size is much disputed. Using data on pre-EMU currency unions (such as the CFA franc zone), Rose (2000) first estimated the effect of currency union on trade, and found it to result in an implausibly high tripling of trade. This finding and the intrinsic interest of EMU have resulted in a literature that has almost universally found smaller estimates, which are yet of considerable economic size. Rose and Stanley (2005) provide a quantitative survey that concludes that currency union increases trade by between 30 and 90 per cent. Engel and Rose (2002) examine other macroeconomic aspects of pre-EMU currency unions, and find that currency union members are more integrated than countries with their own monies, but less integrated than regions within a single country. Edwards and Magendzo (2003) compare inflation, output growth and output volatility in countries inside currency unions and those outside them, and find that currency unions have lower inflation and higher output volatility than countries with their own currencies.

Areas of ignorance

The impact of currency union on financial markets is not something that is currently well understood. Yet this is an area of great interest, since currency union might result in deeper financial integration – or it might not. It is clearly of concern to the British government, which has made the financial effects one of its five tests for EMU entry (see HM Treasury, 2003).

More generally, Europe's experiment with currency union is still young. It is simply too early to know whether EMU has resulted in substantial changes in the real economy, financial or labour markets, or political economy. As the data trickles in, most expect a continuing reassessment of currency unions in theory and especially practice.

ANDREW K. ROSE

See also **Mundell, Robert.**

I thank Steven Durlauf for helpful comments.

Bibliography

Alesina, A. and Barro, R. 2002. Currency unions. *Quarterly Journal of Economics* 117, 409–36.

Alesina, A., Barro, R. and Tenreyro, S. 2002. Optimal currency areas. In *NBER Macroeconomics Annual 2002*, ed. M. Gertler and K. Rogoff. Cambridge, MA: MIT Press.

Bayoumi, T. and Eichengreen, B. 1992. Shocking aspects of European monetary unification. In *The Transition to Economic and Monetary Union in Europe*, ed. F. Torres and F. Giavazzi. New York: Cambridge University Press.

Bayoumi, T. and Eichengreen, B. 1998. Exchange rate volatility and intervention: implications of the theory of optimum currency areas. *Journal of International Economics* 45, 191–209.

Edwards, S. and Magendzo, I. 2003. A currency of one's own? An empirical investigation on dollarization and independent currency unions. Working Paper No. 9514.Cambridge, MA: NBER.

Engel, C. and Rose, A. 2002. Currency unions and international integration. *Journal of Money Credit and Banking* 34, 1067–89.

Frankel, J. and Rose, A. 1998. The endogeneity of the optimum currency area criteria. *Economic Journal* 108, 1009–25.

HM Treasury. 2003. *UK Membership of the Single Currency: EMU Studies*. Online. available at http://www.hm-treasury.gov.uk./documents/international_issues/the_euro/assessment/studies/euro_assess03_studindex.cfm, accessed 25 March 2006.

Kenen, P. 1969. The theory of optimum currency areas: an eclectic view. In *Monetary Problems of the International Economy*, ed. R. Mundell and A. Swoboda. Chicago: University of Chicago Press.

McKinnon, R. 1963. Optimum currency areas. *American Economic Review* 53, 717–24.

Mundell, R. 1961. A theory of optimum currency areas. *American Economic Review* 51, 657–65.

Mundell, R. 1968. *International Economics*. New York: Macmillan.

Rose, A. 2000. One money, one market: estimating the effect of common currencies on trade' *Economic Policy* 30, 7–46.

Rose, A. and Stanley, T. 2005. A meta-analysis of the effect of common currencies on international trade. *Journal of Economic Surveys* 19, 347–65.

Currie, Lauchlin (1902–1993)

Lauchlin Currie was born on 8 October 1902 in West Dublin, Nova Scotia, and died in Bogotá, Colombia, on 23 December 1993 after an unusually long and varied career as an academic economist and top-level policy adviser. After two years at St Francis Xavier University, Nova Scotia, 1920–2, he moved to the London School of Economics (LSE), where his teachers included Edwin Cannan, Hugh Dalton, A. L. Bowley, R. H. Tawney and Harold Laski. In 1925 he obtained his BsC. and moved to Harvard, where the chief inspiration for his Ph.D. thesis, 'Bank Assets and Banking Theory' (January 1931), was Allyn Abbott Young. However, when Young moved to the LSE in 1927 his formal supervisor was John H. Williams.

He remained at Harvard until 1934 as teaching assistant to Williams, Ralph Hawtrey and Joseph Schumpeter. His Ph.D. thesis attacked the 'commercial loan' or 'needs of trade' theory of banking by showing that it was not only unsound in theory but had been more honoured in the breach than the observance – until its disastrous influence on monetary policy in the late 1920s and early 1930s.

In a January 1932 memorandum, Currie, Harry Dexter White and Paul Theodore Ellsworth presented a radical anti-depression programme (see Laidler and Sandilands, 2002). In keeping with their explanation of the contraction as due to a collapsing money supply, they urged vigorous open-market operations and deficit spending financed by money creation. This memorandum was part of an early Harvard influence (through Young, Hawtrey, Williams and Currie; see Laidler, 1999) on what had been claimed as a unique Chicago monetary tradition.

In Currie (1933a) he showed the hopeless confusion that resulted from the ambiguity of the word 'credit'. He stressed control over the quantity of money (defined as cash plus demand deposits, for which there had been no estimates until Currie published a series in 1934) rather than the quantity or quality of credit or loans. He also computed the first estimate of the income velocity of money in the United States (Currie, 1933b), with an explanation of its cyclical variations.

His 'The Failure of Monetary Policy to Prevent the Depression of 1929–32' (1934a) fully anticipated Milton Friedman and Anna Schwartz's (1963) diagnosis of this period. He argued that apart from the stock market there were none of the traditional signs of a boom in the 1920s. Tight monetary policies had been ineffectual in checking the rise in stock prices but only too effective in contributing to the decline in building activity and the pressure on foreign countries that preceded the Depression.

He also demonstrated the perverse elasticity of money in the business cycle due to differences in reserve requirements for different classes of bank and bank deposit (1934b). In the face of the banks' reserve losses in 1929–32 and their abhorrence of heavy indebtedness to the reserve banks, the administration's policy was 'one of almost complete passivity and quiescence', so the self-generating forces of the Depression continued unchecked.

In 1934 Jacob Viner recruited him to the 'freshman brain trust' at the US Treasury where he developed a blueprint for a system of 100 per cent reserves against demand deposits, to break the link between the lending and the creation of money and to strengthen central bank control (see Phillips, 1995). Later that year Marriner Eccles, the new governor of the Federal Reserve Board, hired Currie as his top adviser, from 1934 to 1939. (Many of his memoranda to Eccles are published in Sandilands, 2004.)

At the Fed Currie drafted what became the 1935 Banking Act that gave the Fed increased powers to raise reserve requirements. In 1936–7 these powers were used, 'as a precautionary measure', to reduce the huge build-up of banks' excess reserves. This has been widely blamed for the sharp recession of 1937–8, a view Currie consistently rejected (1938). Instead, he invoked his newly constructed 'net federal income-creating expenditure series' (1935; and see Sweezy, 1972) to show the strategic role of fiscal policy in complementing monetary policy to revive

an acutely depressed economy. In November 1937 he had a four-hour meeting with President Roosevelt to explain that the recession was due to sharp fiscal contraction and that balancing the budget was not the way to restore business confidence. He insisted on the need for better coordination of monetary and fiscal policy. In May 1939 the rationale for this was explained in theoretical and statistical detail by Currie and Alvin Hansen (respectively 'Mr Inside' and 'Mr Outside', according to Tobin, 1976), in joint testimony before the Temporary National Economic Committee.

From 1939 to 1945, Currie was President Roosevelt's special adviser on economic affairs in the White House. He was also in charge of lend-lease to China, 1941–3, and ran the Foreign Economic Administration, 1943–4. In early 1945 he headed a tripartite (United States, British and French) mission to Bern to persuade the Swiss to freeze Nazi bank balances and stop shipments of German supplies through Switzerland to the Italian front. He was also closely involved in loan negotiations with British and Soviet allies and in preparations for the 1944 Bretton Woods conference (staged primarily by his friend Harry White).

After the war it was alleged by Elizabeth Bentley, an ex-Soviet agent, that Currie and White had participated in Soviet espionage. Though she had never met them herself, she claimed they had passed information to other Washington economists who were abetting her own espionage, and that they probably knew this. White and Currie were heavily involved in official wartime cooperation with the Soviets, but Bentley put a sinister interpretation on these activities. They appeared together before the House Committee on Un-American Activities in August 1948 to rebut Bentley's charges. Their testimony satisfied the Committee at that time, though the strain contributed to the fatal heart attack that White suffered three days after the hearing.

No charges were laid against Currie, and in 1949 he headed a major World Bank survey of Colombia. In 1950 the Colombians invited him to return to Bogotá, where he remained for most of the next 40 years as a top presidential adviser. He has been falsely accused of fleeing the United States to avoid charges of disloyalty. In fact in December 1952 he was a witness before a grand jury in New York investigating Owen Lattimore's role in the famous *Amerasia* case that involved the publication of secret State Department documents by that magazine, though his next visit to the United States was not until 1961 when he had a meeting in the White House with Walt Rostow, then President Kennedy's National Security Adviser, to discuss a development plan for Colombia.

By that time Currie had assumed Colombian citizenship (personally conferred on him by President Alberto Lleras in 1958), partly because in 1954 the US government had refused to renew his passport, ostensibly because he was only a naturalized US citizen and was now residing abroad. However, the reality was probably connected with the then secret 'Venona' project that had deciphered wartime Soviet cables that mentioned Currie. The related cases of Currie and White are discussed in Sandilands (2000) and Boughton and Sandilands (2003), where it is shown that the evidence against them is far from conclusive. After reading the latter paper, Major-General Julius Kobyakov, deputy director of the KGB's American desk in the late 1980s, wrote to the present writer on 22 December 2003 to confirm our conclusions. After extensive archival research on Soviet intelligence in the 1930s and 1940s he found that

> there was nothing in [Currie's] file to suggest that he had ever wittingly collaborated with the Soviet intelligence... However, in the spirit of machismo, many people claimed that we had an 'agent' in the White House. Among the members of my profession there is a sacramental question: 'Does he know that he is our agent?' There is very strong indication that neither Currie nor White knew that.

There were two breaks to Currie's advisory and academic work in Colombia: during a military dictatorship, 1953–8, he retired to develop a prize-winning herd of Holstein cattle; and from 1966 to 1971 he was a professor at Michigan State (1966), Simon Fraser (1967–8 and 1969–71), Glasgow (1968–9), and Oxford (1969) universities. He returned permanently to Colombia in 1971 at the behest of President Misael Pastrana to prepare a national plan of development known as the Plan of the Four Strategies, with a focus on urban housing and export diversification. The plan was implemented and the institutions that were established in support of the plan played a major role in accelerating Colombia's urbanization.

He remained as chief economist at the National Planning Department for ten years, 1971–81, followed by 12 years at the Colombian Institute of Savings and Housing until his death in 1993. There he defended the unique index-linked housing finance system (based on 'units of constant purchasing power' for both savers and borrowers) that he had established in 1972. The system thus continued to boost Colombia's growth rate and urban employment opportunities year by year. Currie was also a top adviser on urban planning, and played a major part in the first United Nations Habitat conference in Vancouver in 1976. His 'cities-within-the-city' urban design and financing proposals (including the public recapture of land's socially created 'valorización', or 'unearned land value increments', as cities grow) were elaborated in *Taming the Megalopolis* (1976). To the time of his death he was a regular teacher at the National University of Colombia, Javeriana University, and the University of the Andes, and continued to publish widely (a comprehensive bibliography is in Sandilands, 1990, reviewed by Charles Kindleberger, 1991). His writings and policy advice were heavily influenced by his old Harvard mentor, Allyn Young. Notable is his posthumous

(1997) paper that offers a unique macroeconomic interpretation of Youngian increasing returns and the endogenous nature of self-sustaining growth.

ROGER SANDILANDS

See also **development economics; endogenous growth theory; Federal Reserve System; Great Depression, monetary and financial forces in; Hansen, Alvin; housing supply; monetary policy, history of; urbanization; Viner, Jacob; Young, Allyn Abbott.**

Selected works

1931. Bank assets and banking theory. Ph.D. thesis, Harvard University.
1932. (With P. Ellsworth and H. White.) Memorandum on anti-depression policy. *History of Political Economy* 34(2002), 533–52.
1933a. The treatment of credit in contemporary monetary theory. *Journal of Political Economy* 41, 509–25.
1933b. Money, gold and incomes in the United States, 1921–32. *Quarterly Journal of Economics* 48, 77–95.
1934a. The failure of monetary policy to prevent the Depression of 1929–32. *Journal of Political Economy* 42, 145–77. Reprinted in *Landmarks in Political Economy*, ed. E. Hamilton, H. Johnson and A. Rees. Chicago: University of Chicago Press, 1962.
1934b. *The Supply and Control of Money in the United States.* Cambridge, MA: Harvard University Press.
1935. Comments and observations on 'Federal Income-Increasing Expenditures, 1933–35'. *History of Political Economy* 10(1978), 507–48.
1938. Causes of the recession. *History of Political Economy* 12(1980), 303–35.
1976. *Taming the Megalopolis: A Design for Urban Growth.* Oxford: Pergamon Press.
1997. Implications of an endogenous theory of growth in Allyn Young's macroeconomic concept of increasing returns. *History of Political Economy* 29, 414–43.

Bibliography

Boughton, J. and Sandilands, R. 2003. Politics and the attack on FDR's economists: from Grand Alliance to Cold War. *Intelligence and National Security* 18(3), 73–99.
Friedman, M. and Schwartz, A. 1963. *A Monetary History of the United States, 1867–1960.* Princeton: Princeton University Press.
Kindleberger, C. 1991. Review of Roger J. Sandilands, *The Life and Political Economy of Lauchlin Currie. Journal of Political Economy* 99, 1119–22.
Laidler, D. 1999. *Fabricating the Keynesian Revolution.* Cambridge: Cambridge University Press.
Laidler, D. and Sandilands, R. 2002. An early Harvard *Memorandum* on anti-depression policies. *History of Political Economy* 34, 515–32.

Phillips, J. 1995. *The Chicago Plan and New Deal Banking Reform.* Armonk, NY: M.E. Sharpe.

Sandilands, R. 1990. *The Life and Political Economy of Lauchlin Currie: New Dealer, Presidential Adviser, and Development Economist.* Durham, NC: Duke University Press.

Sandilands, R. 2000. Guilt by association? Lauchlin Currie's alleged involvement with Washington economists in Soviet espionage. *History of Political Economy* 32, 473–515.

Sandilands, R., ed. 2004. New light on Lauchlin Currie's monetary economics in the New Deal and beyond. *Special issue of Journal of Economic Studies* 31(3/4).

Sweezy, A. 1972. The Keynesians and government policy, 1933–1939. *American Economic Review* 62, 116–24.

Tobin, J. 1976. Hansen and public policy. *Quarterly Journal of Economics* 90, 32–7.

cyclical markups

Firms that have increasing returns to scale, that produce differentiated products, or that are part of a small oligopoly can generally be expected to set a price above marginal cost. In so far as a firm's ratio of price to marginal cost is larger than one, there is no particular reason to suppose that this ratio, or markup, will stay constant when overall economic conditions change. Indeed, different models of imperfect competition have different predictions concerning how this markup should vary as aggregate income and activity expands and contracts. Thus, an analysis of whether markups rise when aggregate activity rises or whether they rise when aggregate activity declines provides a useful lens for determining which theories of firm behaviour have more validity.

Markup variations are also of central importance for macroeconomics. One of the central questions for macroeconomics is why the economy expands and contracts at cyclical frequencies in the first place, and cyclical movements in markups are potentially an important nexus that allows such fluctuations to occur. When a single firm (or industry) raises the ratio of its price to its marginal cost, one expects its relative price to rise so that the quantity it sells falls. However, when every firm in the economy tries to raise its price relative to its marginal cost, relative prices need not be affected.

When every firm raises its markup two important consequences follow. The first is that real marginal cost, which can be defined as nominal marginal cost divided by the typical price charged by firms, must fall. Thus, the question of whether markups are countercyclical is the same as the question of whether real marginal costs are procyclical. The second consequence of all firms varying their markups at the same time is that the aggregate demand for labour changes. To see this, notice that nominal marginal cost is equal to the nominal wage divided by the marginal product of labour. Thus, a generalized increase in markups means that prices must rise relative to nominal wages if employment is to remain at a level that keeps the marginal product of labour constant. Alternatively, firm are willing to pay the same real wage only if the marginal product of labour rises, and this requires that employment fall if labour is subject to diminishing returns. In either way of seeing this change, the demand for labour at any given wage falls.

The role of markup changes in economic fluctuations

The capacity of markup changes to generate changes in aggregate labour demand is important because several pieces of evidence suggest that short-run business fluctuations are the result of changes in the demand for labour. That the willingness of firms to hire labour at any given real wage increases in economic expansions is suggested first of all by the tendency of real wages to increase when the economy expands. As shown by Bils (1985), this tendency is particularly strong when one looks at the wages of individuals (as opposed to looking at average wages paid to all workers). Moreover, as emphasized by Bils (1987), firms tend to use more overtime hours in economic booms, and firms are legally obliged to pay higher hourly wages for these overtime hours. When combined with the pro-cyclicality of real wages, other pieces of evidence also suggest that labour demand is higher in booms. In booms, both the unemployment rate and the fraction of the unemployed who have been unemployed for longer than five weeks tend to be lower (both of which suggest that finding jobs is easier) and that the number of help-wanted advertisements is larger (suggesting that it is more difficult for firms to find workers even as they pay them higher wages).

The real business cycle literature stresses a different source of labour demand movements: namely, exogenous changes in the productivity of the typical firm. This hypothesis has the advantage that it explains in a straightforward fashion why labour productivity is somewhat pro-cyclical. However, as discussed below, movements in markups lead to pro-cyclical productivity under a variety of plausible assumptions. In this regard, a clear advantage of the view that markup movements are responsible for important labour demand movements is that labour productivity and real wages rise together with output also when output increases appear to be due to non-technological factors such as increases in military spending, expansionary monetary policy or reductions in the price of oil. (Evidence of these conditional correlations of productivity and output can be found in Hall, 1988.)

Relative to markup variations, exogenous short-run changes in technical progress have another disadvantage as sources of cyclical fluctuations. This is that technical progress not only increases the willingness of firms to hire workers but also reduces the willingness of workers to work at any given wage. These contractionary movements in labour supply are the result of 'wealth effects': technical progress makes people richer and thus induces them to consume both more goods and more leisure. These effects are particularly large if technical progress is somewhat permanent, as tends to be true with actual examples of such progress. These reductions in labour supply imply that shocks to technical progress have only small expansionary effects on employment. By contrast, reductions in markups induce only modest wealth effects, so employment responds more strongly to the resulting increases in labour demand.

These conceptual benefits of countercyclical markups raise the question of whether markups do indeed rise in economic contractions and fall in booms. To discuss this, it is worth starting with the case where the value added production function takes the Cobb–Douglas form. With capital essentially fixed in the short run, this implies that aggregate value added Y is equal to the labour input H to the power α. The marginal product of labour is then equal to α times the average product of labor Y/H The ratio of marginal cost to price is then the wage divided by both the marginal product of labour and the price, so that it is proportional to the labour share in value added (or unit labour cost) WH/PY.

Measuring markup variations

If aggregate data are used, the labour share in value added is not a very cyclical variable. Labour productivity Y/H tends to rise mildly in expansions, as does the average real wage – though the size of these effects depends on how one measures economic expansions. Because cyclical productivity changes are slightly larger than the corresponding average changes in real wages, the labour share has a modest tendency to fall in expansions. If the labour share were seen as equal to the inverse of the markup (as implied by the Cobb–Douglas assumptions), markups would be pro-cyclical and actually dampen cyclical fluctuations.

As suggested in the survey by Rotemberg and Woodford (1999), this Cobb–Douglas case is a good baseline, but a number of corrections to the resulting measure of the markup immediately suggest themselves, and these tend to make measured markups more counter-cyclical. The first of these is that, as already alluded to above, what matters for marginal cost is not the average wage but the marginal wage for an additional hour of work. The average wage is dragged down in booms by the absorption into employment of many relatively low-wage workers who are not employed in recessions. If these workers are less productive, their wage per effective unit of labour input may actually be relatively large. Whatever the case, individual workers who remain employed do see their wages rise more substantially, as emphasized by Bils (1987). Admittedly, these wage increases are concentrated among workers who change jobs, and the increases in the 'straight-time' wages of people who stay in the same job

are more modest. The marginal hour of work, on the other hand, is more likely to be an overtime hour in booms, and this is probably the most important reason for believing that the marginal hour of labour is more expensive then.

It also seems important to correct the way the Cobb–Douglas approach measures the marginal product of labour. According to this functional form. the marginal product of labour is simply proportional to the average product of labour. Given that the average product of labour actually rises slightly in booms, this functional form essentially requires that the economy become 'more productive' in booms, perhaps as a result of increased technical progress.

The tendency of labour productivity to be pro-cyclical can be interpreted in two rather different ways, both of which have a direct bearing on calculations of the cyclical properties of the marginal product of labour. The first is that firms are subject to increasing returns to scale. The simplest functional form that captures this supposes that there are fixed costs, that is, that some of their inputs are 'overhead' inputs that are required to produce even a minuscule positive quantity of output for sale. Suppose for example, that $\bar{H}$ units of labour are overhead units so that output continues to be given by the Cobb–Douglas form but is now proportional to $(H - \bar{H})$ to the power α. The marginal product of labour is then proportional to the ratio $Y/(H - \bar{H})$. In booms, the percentage increase in $H - \bar{H}$ obviously exceeds the percentage by which H rises so that $Y/(H - \bar{H})$ falls by more than Y/H. This means that for $\bar{H}$ sufficiently large, the marginal product of labour falls, marginal costs rise and measured markups fall. Assuming that some of the labour input takes this overhead form can thus easily lead to the inference that markups are indeed counter-cyclical.

A second possible reason for the observation that the average product of labour is pro-cyclical is that firms do not fully utilize all their labour in recessions. They 'hoard' labour to avoid having to incur hiring and training costs when economic activity recovers. This raises two important questions. The first is whether workers produce something else other than measured output when they are being hoarded. The second is whether the firm needs to pay them less when their GDP-producing effort is lower. Given that real wages are only slightly pro-cyclical, it is probably more realistic to suppose that the cost of an hour of labour services to the firm is the same whether the worker incurs effort (and produces) or not. Particularly if the workers are not producing much unmeasured output in recessions, this implies that marginal cost in recessions is considerable smaller than is implied by H/Y. Real marginal cost is more pro-cyclical than WH/PY and markups are more counter-cyclical. One attractive feature of this explanation for pro-cyclical labour productivity is that it is very compatible with the idea that markups are counter-cyclical. Firms are willing to keep workers idle in recessions even though marginal cost is extremely low precisely because they are keeping their prices high relative to marginal cost.

There are two additional types of evidence suggesting that markups are relatively low in booms and high in recessions. The first comes from the behaviour of intermediate inputs relative to final goods. A crude view of materials is that these are used in fixed proportions relative to the gross output of final goods. However, Basu (1995) shows that the ratio of materials to final goods tends to rise when the economy expands. If the material intensity of output is a choice variable, the ratio of marginal cost to price must also equal the real price of materials divided by the marginal product of materials. It is reasonable to suppose with Basu (1995) that the marginal product of materials diminishes as the level of materials inputs rises. With constant returns, the increase in the ratio of materials to output in booms thus implies that real marginal cost is pro-cyclical even if the price of materials relative to final output were constant. In fact, Murphy, Shleifer and Vishny (1989) show that that prices of more processed goods tend to fall relative to prices of less processed goods in economic expansions, and this too indicates a tendency of price to fall relative to marginal cost during booms.

The second additional source of evidence comes from the behaviour of inventories. Inventories rise in booms but, as stressed by Bils and Kahn (2000), they rise by less in percentage terms than sales. At the same time, long-run growth in sales does tend to be associated with equiproportonate increases in inventories in the industries they consider. In addition, they discuss cross-sectional evidence that shows that, within industries, the inventory–sales ratios of products with large sales are not smaller than the inventory-sales ratios for products with low sales. This suggests that there is something special about the decline in inventory–sales ratios that is observed in booms. It suggests, in particular, that conditions in booms lead firms to economize on inventory holding. As Bils and Kahn (2000) argue, the evidence seems most consistent with the idea that firms keep their inventories relatively low in booms because real marginal cost is relatively high.

Theories of cyclical markup variations

A considerable body of evidence, then, seems consistent with counter-cyclical markups, and suggests that countercyclical markups might be central to aggregate fluctuations because they rationalize the changes in employment that characterize such fluctuations. The question that remains is why markups should vary cyclically. There are basically five types of models that explain these movements in markups. These are: models of variable demand elasticity, models of variable entry, models of sticky prices, models of investment in market share and models of implicit collusion.

In a monopolistically competitive setting, markups are equal to the elasticity of demand over the the elasticity of

demand minus 1. Increases in the elasticity of demand thus lower markups (towards the competitive level of 1) and could thus be a source of business expansions. This still leaves the question of why the elasticity of demand facing the typical firm should vary over time. One possibility is that the proportion of demand that comes from highly elastic customers rises in booms. Gali (1994) obtains such composition effects under the supposition that investment is more price sensitive than consumption. Ravn, Schmidt-Rohe and Uribe (2004) obtain a related effect by supposing that people have formed a 'habit' for at least a fraction of past purchases, and the elasticity of demand for these habitual purchases is negligible relative to the elasticity of demand for non-habitual ones. As consumption rises in economic expansions, more of the purchases are non-habitual so that the elasticity of demand is higher and markups have to be correspondingly lower.

Devereux, Head and Lapham (1996) show that changes in demand induced, for example, by changes in government purchases lead new firms to enter existing industries. Entry of new firms is indeed quite pro-cyclical. Such entry can, in turn, make each firm's perceived elasticity of demand higher (because they fear more competitors). Thus variable entry can be seen as a reason for changes in elasticities that lead to counter-cyclical markups. Even if the expansion in the number of firms that takes place in booms is seen as too small for this effect to be large, the potential for increases in entry may lead incumbents to keep their prices low to avert the creation of an even larger number of new firms. This limit pricing might also be able to rationalize counter-cyclical markups.

Sticky prices, which are widely assumed in new Keynesian macroeconomics, probably provide the most straightforward model of counter-cyclical markups. Firms that keep their prices constant when demand increases (as a result of expansionary government policy, for example) will generally see their marginal costs rise both because of diminishing returns and because of increases in the costs of factor inputs. Thus, keeping their prices relatively constant will lead them to have lower markups. The argument that sticky prices derive their influence on the economy from their consequences for variable markups is presented in more detail in Kimball (1995).

If customers who have already purchased a good have relatively inelastic demand, keeping price low is like an investment activity for the firm. It encourages new customers (those whose demand is elastic) to become addicted. Changes in economic conditions can lead firms to desire to either increase or decrease these investments. Increases in interest rates in particular might lead firms to wish to reduce these investments, at least temporarily. Chevalier and Scharfstein (1996) provide evidence that the cash condition of firms plays a large role in these investments as well. They show that recessions have a disproportionate effect on the pricing of cash-strapped firms, who turn out to be more eager to raise prices and thereby reduce their investment in market share.

Lastly, Rotemberg and Saloner (1986) have emphasized that high prices may be more difficult to sustain for implicitly collusive oligopolists in economic expansions. When current sales are high, each firm perceives a greater benefit from undercutting the implicit agreement because it can thereby secure even higher sales. To prevent this, the oligopolists must lower their markups of price relative to marginal cost. Some cross-sectional evidence suggests that markups are indeed more counter-cyclical in more concentrated sectors, as a theory that applies only to implicitly collusive oligopolists suggests. As shown by Rotemberg and Woodford (1992), the model can be embedded in a general equilibrium structure so that increases in government purchases raise output together with real wages. The increased rate of interest induced by additional government purchases lowers the present value of the future benefits from cooperation. It thus forces oligopolies to be less ambitious in the profits that they seek from current prices, so that markups fall and labour demand rises.

JULIO J. ROTEMBERG

See also **microfoundations; new Keynesian macroeconomics.**

Bibliography

Bils, M.J. 1985. Real wages over the business cycle: evidence from panel data. *Journal of Political Economy* 93, 666–89.

Bils, M.J. 1987. The cyclical behavior of marginal cost and price. *American Economic Review* 77, 838–57.

Bils, M. and Kahn, J.A. 2000. What inventory behavior tells us about business cycles. *American Economic Review* 90, 458–81.

Basu, S. 1995. Intermediate inputs and business cycles: implications for productivity and welfare. *American Economic Review* 85, 512–31.

Chevalier, J.A. and Scharfstein, D.S. 1996. Capital-market imperfections and countercyclical markups: theory and evidence. *American Economic Review* 86, 703–25.

Devereux, M.B., Head, A.C. and Lapham, B.J. 1996. Monopolistic competition, increasing returns, and the effects of government spending. *Journal of Money, Credit, and Banking* 28, 233–54.

Gali, J. 1994. Monopolistic competition, business cycles, and the competition of aggregate demand. *Journal of Economic Theory* 63, 73–96.

Hall, R.E. 1988. The relation between price and marginal cost in U.S. industry. *Journal of Political Economy* 96, 921–47.

Kimball, M.S. 1995. The quantitative analytics of the basic neomonetarist model. *Journal of Money, Credit, and Banking* 27, 1241–77.

Murphy, K.M., Shleifer, A. and Vishny, R.W. 1989. Building blocks of market clearing business cycle models. In *NBER Macroeconomics Annual*, ed. O.J. Blanchard and S. Fischer. Cambridge, MA: MIT Press.

Ravn, M., Schmitt-Grohe, S. and Uribe, M. 2004. Deep habits. Working Paper No. 10261. Cambridge, MA: NBER.

Rotemberg, J.J. and Saloner, G. 1986. A supergame-theoretic model of price wars during booms. *American Economic Review* 76, 390–407.

Rotemberg, J.J. and Woodford, M. 1992. Oligopolistic pricing and the effects of aggregate demand on economic activity. *Journal of Political Economy* 100, 1153–207.

Rotemberg, J.J. and Woodford, M. 1999. The cyclical behavior of prices and costs. In *Handbook of Macroeconomics*, vol. 1B, ed. J.B. Taylor and M. Woodford. Amsterdam; New York and Oxford: North-Holland.

D

Dalton, Edward Hugh John Neale (1887–1962)

British fiscal economist and prominent Labour politician, Hugh Dalton was a student of A.C. Pigou and J.M. Keynes. His main professional interest was in the use of taxation as an instrument for the redistribution of income and wealth, an interest inspired by Pigou's teaching and by his revulsion at the contrast between the sufferings inflicted on younger generations by the First World War and the material gains of those who financed or profited from the war itself. (Dalton spent four years on military service in France and Italy and lost several close friends, including the poet Rupert Brooke.) His main contribution was to investigate the properties of a modification of Bernoulli's formula $dw = dw/x$ where $w =$ economic welfare and $x =$ income but in which equal increases in welfare should correspond to more than proportionate increases in income, a condition satisfied by Dalton's formula $dw = dx/x^2$ so that $w = c - 1/x$ where c is a constant. Using this formula he concluded that economic welfare would be improved by transfers from rich to poor (Dalton, 1935), a proposition that has excited the interest of 'modern' public finance theorists of the neo-utilitarian school (see Fishburn and Willig, 1984). He elaborated his ideas in several works including his highly successful standard text *Principles of Public Finance* and in his lectures as Reader in Economics at the London School of Economics (1923–36). There he was responsible for teaching and for recommending Lionel Robbins to be Professor of Economics, a typical example of his desire not only to 'corrupt the young' (as he termed it) but also to promote the interests even of those with whom he disagreed.

Dalton combined teaching with a political career throughout the 1920s and 1930s, rising to political eminence as a member of Churchill's coalition government during the Second World War. As Minister of Economic Warfare he was responsible for setting up the famous sabotage team, the Special Operations Executive (SOE). Later as President of the Board of Trade he formulated plans for post-war distribution of industry designed to prevent mass unemployment. In the Attlee Labour government of 1945 he reached the pinnacle of his political career as Chancellor of the Exchequer, one of his first acts being to nationalize the Bank of England. His famous attempt to drive down interest rates through a cheap money policy in order to float off an issue of Treasury stock at 2.5 per cent is a classic example of the failure of even an experienced and able economist to understand that, other than in the short run, governments can control either the price or the supply of bonds but not both.

ALAN PEACOCK

Selected works

1923. *Principles of Public Finance*. London: George Routledge & Sons.
1935. *The Inequality of Incomes*. 4th Impression, London: George Routledge & Sons, especially the Appendix.

Bibliography

Davenport, N. 1961–70. Hugh Dalton. In *Dictionary of National Biography*, Oxford: Oxford University Press.
Fishburn, P.C. and Willig, R.D. 1984. Transfer principles in income redistribution. *Journal of Public Economics* 25, 323–8.
Pimlott, B. 1985. *Hugh Dalton*. London: Jonathan Cape.

Dantzig, George B. (1914–2005)

George Dantzig is known as the 'father of linear programming' and the 'inventor of the simplex method'. Employed at the Pentagon (the US government's defence establishment) in 1947 and motivated to 'mechanize' programming in large time-staged planning problems, George Dantzig gave a general statement of what is now known as a linear program, and invented an algorithm, the simplex method, for solving such optimization problems. By the force of Dantzig's theory, algorithms, practice, and professional interaction, linear programming flourished. Linear programming has had an impact on economics, engineering, statistics, finance, transportation, manufacturing, management, and mathematics and computer science, among other fields. The list of industrial activities whose practice is affected by linear programming is very long.

Over the subsequent half century, Dantzig remained a major contributor to the subject of linear programming as researcher, practitioner, teacher, mentor, and leader. The impact of linear programming and extensions on theory, business, medicine, government, the military, all in the broadest sense, is now hard to overstate. In the words of the editors of the Society for Industrial and Applied Mathematics: 'In terms of widespread application, Dantzig's algorithm is one of the most successful of all time: linear programming dominates the world of industry, where economic survival depends on the ability to optimize within budgetary and other constraints …' (quoted in Dongarra and Sullivan, 2000).

There were some significant contributions to what became linear programming prior to Dantzig's work. In their time, however, these results were not applied, linked together, or continued. In fact, they were nearly lost, perhaps because the prevailing historical setting was not favourable. As these contributions have been recognized, they have been drawn into the history of linear programming.

A linear program defined

In mathematical terms, a linear program is most simply stated as the problem of minimizing a multivariate linear function constrained by linear inequalities. Dantzig's first formulation of a linear program was the equivalent problem of minimizing a linear function over non-negative variables constrained by linear (material balance) equations. In matrix notation such a linear program is:

$$\text{Minimize}_{x,z} \quad c^T x = z$$

$$LP(A, b, c):$$

$$\text{subject to } Ax = b, \quad x \geq 0.$$

Here the given data are the $m \times n$ matrix A and vectors b and c; the unknowns to be determined are the objective scalar value z and the decision-variable vector x. The simplex method solves a linear program in a comprehensive sense; in particular, no conditions are imposed on the data (A, b, c). Dantzig assessed a linear program as the simplest optimization model with broad applicability.

The study, solution, and application of linear programs constitute the subject of linear programming. The use of the words 'programming' and 'program' has changed somewhat over time. The original idea was that 'programming' is the activity of deciding now upon a plan, called a program, for some system that would be executed later in time. The same meaning was subsequently adopted in computer programming where the system is a computer. (See LINEAR PROGRAMMING.)

Early life and education

George Bernard Dantzig was born to Tobias Dantzig and his wife, Anja Ourisson, in Portland Oregon, on 8 November 1914. Tobias, a housepainter and pedlar in his early years in the United States, later held professional positions at John Hopkins University (1919–20) and the University of Maryland (1927–46) where he chaired the mathematics department from 1930 to 1941. He is best known for his book *Number, The Language of Science*, which is still in print (T. Dantzig, 1930).

In 1936, George Dantzig both received an A.B. in Mathematics and Physics at the University of Maryland and married Anne Shmuner (1917–2006), who at age 19 received an A.B. in French at Maryland. In 1938 Dantzig received an MA in Mathematics at the University of Michigan; he was a Horace Rackham Scholar. In 1937–39 Dantzig worked at the U.S. Bureau of Labor Statistics as a junior statistician. Inspired by a paper of J. Neyman on which he had been assigned to report, Dantzig wrote to Neyman, then at University College London, asking if he could study under his supervision. Neyman relocated to the University of California at Berkeley, and Dantzig became his student in 1939. As is now folklore, one day Dantzig arrived late for one of Neyman's theoretical statistics classes and proceeded to copy two problems from the blackboard. In a few weeks time, with some effort, Dantzig solved the problems and submitted his

homework, whereupon it was tossed onto a large pile of papers on Neyman's desk. Early one Sunday morning, about two weeks later, George and Anne were awakened by a pounding on their apartment door. There was Neyman waving George's homework. As it turned out, the assumed homework problems were, in fact, important unsolved problems. Furthermore, Neyman continued, these solutions, suitably presented, would suffice for George's Ph.D. dissertation. A. Wald independently obtained one of the same results, and the work was eventually published jointly in Dantzig and Wald (1951). Before Dantzig could complete his degree, Pearl Harbor was attacked, and he took leave of absence to work at the U.S. Air Force Comptroller's Office.

Dantzig in Washington, DC, 1941–52

At the outbreak of the Second World War, Dantzig began working at the War Department, again as a junior statistician. By the war's end, he was in charge of the Combat Analysis Branch of the Statistical Control Division of the United States Air Force. His office collected and consolidated data with hand-operated mechanical desk calculators about sorties flown, tons of bombs dropped, planes lost, personnel attrition rates, and so on. By end of the war, Dantzig had a personnel force of 300 reporting to him.

In 1946 Dantzig returned to Berkeley for one semester to defend his thesis and complete his minor thesis in dimension theory. Throughout his life, Dantzig acknowledged a great debt to J. Neyman, his mentor. Dantzig nonetheless turned down a position in mathematics at UC Berkeley for the greater financial security of a position at the Pentagon. There he undertook the challenge to 'mechanize' the planning process. War planning required coordination of an entire nation and yet was executed with desk calculators; the need for mechanization was clear. To this end, a group in the Air Force was organized under the name Project SCOOP (Scientific Computation of Optimum Programs) and headed by M.K. Wood. Dantzig was a principal. Two movements suggested that progress was possible: Leontief's (1936) work and the emergence of the computer; indeed, Project SCOOP arranged for Pentagon support of computer development (see Dantzig, 1947).

In early 1947, Dantzig formulated the general statement of a linear program. In June of that year he learned from T.C. Koopmans, who had been studying transportation problems (Koopmans, 1947) that economists had no algorithm for solving a linear program. By July Dantzig had designed the simplex method, a name suggested by Leo Hurwicz (see SIMPLEX METHOD FOR SOLVING LINEAR PROGRAMS). Experiments with the simplex method in the following year at the Pentagon were encouraging. Linear programs were also solved with the simplex method at the National Bureau of Standards (NBS) in coordination with SCOOP. At NBS a 'large one', the diet problem, was undertaken by J. Laderman. It had been studied earlier by Stigler (1945). The question was: what selection of 77 foods produces a diet meeting nine nutritional criteria at the least cost? The problem was solved by the simplex method with five statistical clerks using desk calculators. According to (Dantzig 1963), 'approximately 120 man-days were required to obtain a solution'. The simplex method was gaining acceptance. Air Force applications of linear programming in years following included contract bidding, crew training, deployment scheduling, maintenance cycles, personnel assignments, and airlift logistics.

From special cases as a triangular model to the general algorithm, the simplex method was first implemented on a computer in 1949 by M. Mantalbano (NBS) on an IBM 602-A, in 1950 by C. Diehm on the SEAC, in 1951 by A. Orden (Air Force) and A. Hoffman (NBS) on the SEAC, and in 1952 by the Air Force for the Univac. The next generation of codes, circa 1952–56, which achieved commercial quality, was developed by W. Orchard-Hays at the RAND Corporation on a sequence of IBM machines. For the matrix A of $LP(A, b, c)$ of size 200 by 1000, linear programs could be solved in five hours (Orchard-Hays, 1954). In years following, there was a flood of computer implementations, both by commercial vendors and in research institutions. As of 2006, linear programs where both m and n exceed hundreds of thousands are routinely solved in hours by the simplex method on personal computers.

After describing and testing the simplex method, Dantzig had an audience with J. von Neumann at Princeton in 1947. Among world-class mathematicians, von Neumann had the broadest interests. Dantzig began his explanation of linear programming with the 30-minute version when von Neumann snapped 'Get to the point'. Dantzig began again, this time with his one-minute version. Von Neumann responded, 'Oh, that!' He envisioned an analogy with matrix games as developed in von Neumann and Morgenstern (1944). Extrapolating from what he knew about duality in matrix games, von Neumann expounded on what was to become known as duality in linear programming. As a by-product of the meeting, it was evident that any matrix game problem could be solved by a linear program. Volume VI of John von Neumann: Collected Works contains his previously uncirculated manuscript dated 15–16 November 1947 on duality in linear programming (von Neumann, 1947). The following January, Dantzig (1948a) wrote 'A Theorem on Linear Inequalities'. This memorandum clarified his understanding of von Neumann's duality monologue. Von Neumann's (1947) paper is regarded as the earliest on this subject; Dantzig's memorandum is the second. A.W. Tucker, also at Princeton, took an interest in the relationship of linear programming and game theory and involved his students, D. Gale and H.W. Kuhn. These three subsequently wrote the definitive account of duality in linear programming (Gale, Kuhn and Tucker, 1951).

First linear programming conference, 1949

Koopmans organized a conference on 'linear programming' and economics in Chicago at the Cowles Commission for Research in Economics in 1949. Koopmans and others (including Dantzig) edited the conference proceedings volume *Activity Analysis of Production and Allocation* (1951). Dantzig's work was the focus of the proceedings; of the 25 papers, Dantzig co-authored a paper with M.K. Wood and authored four others, including the two leading papers which developed linear programming for time-staged planning. Earlier versions of these two papers appeared in *Econometrica* (1949). Four of the 20 contributors to these proceedings – K.J. Arrow, T.C. Koopmans, P.A. Samuelson, and H.A. Simon – were later to win Nobel Prizes. Hundreds of books on, or inspired by, linear programing followed over the years. Four of note are Dorfman, Samuelson and Solow (1958), Arrow, Hurwicz and Uzawa (1958), Dantzig (1963), and Schrijver (1986). The terminology 'linear programming' was not in regular use at the time of this conference; Koopmans had suggested it to Dantzig (1948b) in lieu of expressions like 'programming in a linear structure'. Even so, Koopmans (1951) observed, 'To many economists the term linearity is associated with narrowness, restrictiveness, and inflexibility of hypotheses'. R. Dorfman, at the Pentagon with Dantzig, had suggested the broader expression of 'mathematical programming'.

Nonlinear programming, 1950

Following the early successes of linear programming, there was a natural inclination to generalize the model, the algorithm, and duality to results beyond linear functions to a next layer of difficulty such as differentiable, convex, quadratic, or polynomial functions. This body of research has become known as 'nonlinear programming'. As for optimality conditions and duality, the paper 'Nonlinear Programming' of Kuhn and Tucker (1951) was pivotal at the time: their investigation proceeded through the Lagrangian function and saddle points thereof with the duality in linear programming as a target. The Lagrangian had been used in equality-constrained optimization, and results obtained there were less general. Kuhn and Tucker cited the fundamental paper of John (1948), which includes inequality constraints. Some 25 years later, the master's thesis of Karush (1939) came to light in the mathematical programming community; Karush, as far as is known, was the first to lay down optimality conditions for a nonlinear (inequality constrained) program. Rockafellar (1970) carried the convex duality analysis to a new level. As for nonlinear programming algorithms, tens, and eventually hundreds, were forthcoming, many using ideas from the simplex method in one way or another.

Dantzig at RAND, 1952–60

Reorganization of the Air Force preceded Dantzig's taking a position at the RAND Corporation in Santa

Monica, California, as a research mathematician. Awareness of the power of linear programming set the scene for a second growth. For the next few years most theoretical development of linear programming took place at RAND and Princeton. Dantzig's book *Linear Programming and Extensions* (1963) records his own (and collaborative) contributions during this period.

Transportation and network optimization problems
The war years has seen interest in optimal transportation research. Historically significant papers from this period include Hitchcock (1941), Kantorovich (1942), Koopmans (1947; 1949), Kantorovich and Gavurin (1949), and Flood (1956). Flood, through M. Shiffman, had come upon the Kantorovich papers on translocation and transportation; however, linear programming launched the general analysis of optimal transportation. Dantzig made several contributions here, starting with Dantzig (1951). Dantzig, Fulkerson and Johnson (1954) is a seminal work on the travelling salesman problem. Others are Dantzig and Fulkerson (1954) on tanker routing, and Dantzig and Fulkerson (1955) on maximizing flow through a network. For networks with non-negative arc distances, Dantzig (1960a) stated an algorithm for shortest distances. Dijkstra (1959) produced similar results at about the same time. *Flows in Networks* by the RAND Corporation's Ford and Fulkerson (1962) was then the definitive work on the subject.

Large-scale methods and decomposition
Dantzig and Orchard-Hays (1954) described the 'revised simplex method' as a more efficient version of the simplex method. As linear programming was applied to more applications and with a broader scope, including time and alternate scenarios, the size of linear programs that needed to be solved continued to grow. Dantzig was among the first to observe that large linear programs typically had two convenient features: sparsity and structure. Sparsity refers to the fact that a very small percentage, often less than one hundredth of one per cent, of the *A* data matrix is non-zero. Structure refers to the fact that the non-zeros typically occur an orderly pattern of submatrices of *A*. Dantzig (1955a) wrote the first paper on methods for large-scale linear programs addressing upper bounds, block triangular systems, and secondary constraints. Building on the Dantzig, Orden and Wolfe (1955) paper on generalized linear program, Dantzig and Wolfe (1960) devised a generalization of the simplex method, called the *decomposition principle*, for certain structured large-scale linear programs, wherein the problem is decomposed allowing for use for what is now called distributed computation.

Quadratic programming
A most natural first extension of a linear program is a quadratic program, that is, a linear program except that the objective is a quadratic function such as $x^T Q x + q^T x$.

A convex quadratic program is one with a convex objective function to be minimized. Following the success of linear programming, there was a proliferation of studies on convex quadratic programming and associated algorithms.

Convex programming
Convex programming is also a natural extension of linear programming. Here a convex function is minimized over a convex region; the latter is specified by convex inequality constraints. If the feasible region is bounded, the convex program can be approximated as close as desired by a linear program, and one can improve the approximation as the simplex method runs. A special case of a convex program is one having linear inequality constraints and a *separable* objective function, that is, a function that is the sum of univariate convex functions. Charnes and Lemke (1954) and Dantzig (1956) solved such problems with linear programming approximations.

Stochastic programming with recourse
Linear programming offered a breakthrough for mathematical approximation and solution of planning problems. Dantzig knew that to move to the next level of approximation of planning, an accommodation of uncertainty and of discrete variables was needed; he made inroads on each. Linear programming has been extended in a number of directions to incorporate uncertainty. An elementary example is a linear program where the costs $c = (c_1, c_2, \ldots, c_n)$ are random variables and the desire is to minimize the expected value. In this case the problem is solved as the linear program where the costs are simply taken as their expected value. More interesting is the Markowitz (1956) portfolio selection where quadratic programming is used to obtain at least a desired level of expected return while minimizing risk.

Dantzig's early work on stochastic programming was stimulated by his work with A.R. Ferguson on the assignment of aircraft to routes, where a deterministic formulation proved insufficient, and so uncertain demand needed to be considered (Ferguson and Dantzig, 1955). Subsequently, Dantzig (1955b) applied linear programming to solve multistage decision problems sequenced amidst uncertainty; this topic is often referred to as stochastic programming with recourse. Such a multistage problem concerns the optimization of a sequence of decisions in time where each decision depends on random events which in turn are dependent on previous decisions. The vision in this paper was truly extraordinary, and has been reprinted as one of the ten most influential papers in management science since the mid-1950s in Hopp (2004).

Integer programming and cuts
An integer program is a linear program except that some, or all, of the variables $x_1, x_2, \ldots, x_n$ are required to take on integer values, as in $x_i = 0, 1, 2, \ldots$. Dantzig,

Fulkerson and Johnson (1954) took the first steps towards obtaining integer solutions for a large problem with the simplex method. They addressed an instance of the travelling salesman problem, find the shortest route, by car, through major cities of the 48 states and Washington, DC. Let a directed network represent the available roads and let costs represent distances. The variables are flows on each link of the network. Constrain for one unit of flow into each capital, constrain for one unit of flow out of each capital, constrain for conservation of flow at other nodes, and find the minimum cost flow. The linear programming solutions here, which yield flows of 0 or 1, are deficient as a solution for the travelling salesman problem in that isolated loops of flow may occur. To combat such loops, Dantzig et al. sequentially and dynamically (as the simplex method was stopped and continued) introduced additional constraints, called cuts, which would prohibit those loops which had occurred in a solution of the expanding linear program, without constraining out desired solutions. The concept of a cut or cutting planes was so conceived. In addition, this study revealed the inherent difficulty of the travelling salesman problem. Over the following decades, aspects of this matter would grow to become a major issue in applied mathematics. There is a vast difference between linear constraints and linear inequality constraints (both with unconstrained variables); there is an even larger difference between real variables and integer variables. Subsequently, Gomory (1958), at Princeton, began the design of several general purpose cutting plane algorithms for solving integer programs, and gave proofs for finite convergence. These algorithms did not work well for a reason not understood at the time: namely, that general integer programs are inherently hard to solve.

Other edge path descent algorithms

By 1955 the simplex method was regarded as *the* algorithm for solving linear programs. Indeed, the simplex method inspired dozens of related fundamental ideas for algorithms, and hundreds of variations. In particular, there was steady research on variations of edge path descent algorithms, that is, those which accept the simplex method strategy but strive to improve upon it. One target was to reduce computation time by reducing the number of pivots and the work per pivot. Example contributions include: the dual simplex method of Lemke (1954), the parametric method of Orchard-Hays (1954), the primal-dual method of Dantzig, Ford and Fulkerson (1956) and the parametric objective method of Gass and Saaty (1955). In a slightly different direction were the column generation and the decomposition method of Dantzig and Wolfe (1960; 1961). Essentially all of these variants of the simplex method have proved valuable for various specialized tasks related to linear programs, and sometimes nonlinear programs. For nonlinear programs the main ideas of the simplex method have been adopted; here one can think of solving linear or quadratic programs that are approaching the nonlinear program. It is interesting to note that as late as the early 1970s an eminent speaker of a plenary session of a national mathematical programming conference said that the simplex method was the best algorithm for linear programming and that it always would be; the statement was accepted, without objection.

Problem reduction

The mathematical subject of computational complexity aims to categorize problems by their solution difficulty. Several of Dantzig's papers (1957; 1960b; 1968) contributed to the foundation of this subject. A basic technique of computational complexity is the reduction of one class of problems to another. For the reduction of discrete problems, Dantzig focused on problems in mixed binary form, *MBP*, and the related relaxed form *RMBP* obtained by replacing binary constraints with corresponding interval constraints. $MBP(A, b, c)$ is a linear program $LP(A, b, c)$ plus the discrete constraints $x_i = 0, 1$ for $i = 1, \ldots, k$ for some $k \leq n$. $RMBP(A, b, c)$ is the linear program $LP(A, b, c)$ plus the linear inequalities $0 \leq x_i \leq 1$ for $i = 1, \ldots, k$. For emphasis, $MBP(A, b, c)$ is not a linear program whereas $RMBP(A, b, c)$ is.

A few problem classes of form *MBP* can be solved as the corresponding linear program *RMBP*; that is if (x, z) is an extreme point solution, as the simplex method would generate, of *RMBP*, then (x, z) is a solution to *MBP*. Problem classes *MBP* which can be so solved by *RMBP* include the assignment problem, shortest route problems with non-negative distances, and the tanker scheduling problem. Other problems, such as the empty container problem, most scheduling problems, fixed charge problems, and travelling salesman problems, do not permit such solution; nevertheless, the corresponding *RMBP* can be most helpful in solving or approximately solving *MBP*. As time and theory have revealed, general problems of type *MBP* are difficult to solve.

Let C^* be the convex hull of all feasible solutions of *MBP* and let C be the set of all solutions of *RMBP*. Then C^* is a subset of C, and all extreme points of C^* are extreme points of C; the issue is, however, that there are extreme points of C that are not in C^*. Note that, if there is but one binary variable, then *MBP* can be solved as two linear programs, one with $x_1 = 0$ and one with $x_1 = 1$; but for general k, this scheme requires the solution of an exponential number 2^k of linear programs. For reducing problems to the *MBP* form, Dantzig (1960b) illustrated a number of examples such as: (*a*) dichotomies, (*b*) discrete variables, (*c*) piecewise linear objective functions, (*d*) conditional constraints, and (*e*) the fixed charge problem.

Recognition of earlier work, 1958–60

Towards the end of the 1950s, the mathematical programming community became aware of three relevant works from the past. The first two are pertinent to the simplex method and the third relevant to the formulation of real problems as linear programs. Fourier (1826) had also written on the idea of descending from vertex to adjacent vertex in the polyhedron defined by linear inequalities for minimizing a linear error over linear inequalities. De la Vallée Poussin (1911), independently of Fourier's work, made a similar suggestion and gave two examples. There appears to have been no follow-up on their suggestions. Also, neither Fourier nor de la Vallée Poussin described his ideas fully enough to reveal any awareness of degeneracy considerations and corresponding non-convergence possibilities, much less any procedures for coping with the matter. Made aware of Kantorovich's transportation papers by Flood (1956), Koopmans (1960) corresponded with Kantorovich. In due course, an English translation of Kantorovich's remarkable 1939 paper was made available to the West as 'Mathematical Methods of Organizing and Planning Production' (Kantorovich, 1960). Therein Kantorovich had formulated a collection of problems as what we now call linear programs. These problems were: machine utilization, production planning, scrap management, refinery scheduling, fuel utilization, construction planning, and arable land distribution. Using the Minkowski separation theorem, Kantorovich proved in this work that optimal multipliers exist. He suggested some ideas based on 'resolving multipliers' (essentially dual variables, or marginal costs) towards an algorithm, but none has emerged following this line of thought. According to Dantzig (1963), 'Kantorovich should be credited with being the first to recognize that certain important broad classes of production problems had well-defined mathematical structures which, he believed, were amenable to practical numerical evaluation and could be numerically solved'. But although Koopmans (1960) argued that, with a suitable transformation, one of Kantorovich's problems had the generality of Dantzig's linear program, Koopmans's conclusion was not justified as the argument did not and could not cover the possibilities of infeasibility and an unbounded objective, a point made by Charnes and Cooper (1962). Koopmans's argument notwithstanding, the statement of a general linear program belongs to Dantzig.

Dantzig returns to UC Berkeley, 1960–66

Dantzig left RAND to become a professor in the industrial engineering department at the University of California at Berkeley. There, that year, he established the Operations Research Center. Operations research (OR) was a term that emerged in the Second World War to describe the activity of studying an operation (process, system, and so on) with mathematical methods with the intent of improving performance. In 1963 Dantzig completed his classic *Linear Programming and Extensions*. The book was based on his research which began at the Pentagon and

continued through RAND and UC Berkeley. By the time Dantzig left UC Berkeley in 1966, he had produced 11 Ph.D. students, and written about 25 research papers on the theory and practice of linear programming and extensions (integer, nonlinear, stochastic, and so on). As a mentor of Ph.D. students, Dantzig was among the very best. Within a course or two he could bring students to the frontier on some aspect of linear programming. His new book offered a full perspective of linear programming right up to 1963. Dantzig supplied the time, inspiration, guidance, knowledge, and example that students needed. He lived and breathed research.

Interest in the study of linear and nonlinear complementarity problems, as such, began in the early 1960s. Dantzig's second student, Cottle (1964), wrote on this topic, and his work was extended in Cottle and Dantzig (1968). Problems in this category can be viewed as abstractions of optimality conditions or of (economic or physical) equilibrium conditions. In a complementarity problem, one has a mapping W of R^N into itself and seeks a solution z of the conditions $W(z) \geq 0$, $z \geq 0$, $z^T W(z) = 0$. In the linear complementarity problem, the mapping would be of the form $W(z) = Mz + q$. The linear complementarity problem is related to the minimization of $z^T(Mz + q)$ subject to the constraints $Mz + q \geq 0$ and $z \geq 0$. This would be easy enough to solve as a quadratic program, if the objective function were convex. However, the excitement arose from the fact that the problem could be solved, effectively, in the absence of convexity. From the classic paper of Lemke (1965) followed the computation of points in the core of a balanced game and the computation of economic equilibria (Scarf, 1967; 1973), the computation of fixed points with piecewise linear homotopies (Eaves, 1972), and the computation with differentiable functions (Smale, 1976).

Dantzig at Stanford University, 1966–96

Dantzig joined the Stanford faculty in 1966, half-time in the inter-departmental Operations Research Program and half-time in Computer Science. In 1967 the OR Program became the Department of Operations Research in the School of Engineering; this is where Dantzig conducted his work. He was away for two years: in 1973–74 at the International Institute for Applied Systems Analysis in Austria, and in 1978–79 at the Center for Advanced Study in the Behavioral Sciences on the Stanford campus. In 1973 he was appointed to the C.A. Criley Professorship in Transportation Science. While at Stanford, Dantzig produced 41 Ph.D. students and published about 115 research papers on the theory and applications of mathematical programming. Dantzig's Ph.D. progeny, if Berkeley and Stanford graduates and subsequent generations are counted, as of 2006 exceeded 200. Dantzig had long felt that the development of good software was key to widespread usage of linear programming in industry. This vision led him to create the Systems Optimization Laboratory (SOL) at Stanford for research and development of numerical algorithms for mathematical programming. Under the SOL banner were the PILOT and planning under uncertainty programs (see Dantzig et al., 1973; Gill et al., 2007).

Stochastic programming with recourse, continued, 1989–2005

Cognizant of the potentially enormous size of multi-stage stochastic linear programs, Dantzig and Madansky (1961) suggested the incorporation of statistical sampling of uncertainties together with approximating time-staged models to solve the full problem. Following this avenue some 30 years later, Dantzig and Glynn (1990) brought together decomposition, Monte Carlo sampling, and multiprocessing to solve time-staged linear programs (see also Infanger, 1991; Dantzig and Infanger, 1992). In a series of papers, importance sampling was used to estimate second-stage costs and Benders cuts. Portfolio optimization and electric power planning were among the applications envisioned; the latter problems, with 39 uncertain parameters leading to 15 million scenarios, were solved to high accuracy with a confidence level of 95 per cent; in equivalent deterministic form, such problems would have more than four billion constraints. However, Dantzig, to the end, regarded stochastic linear programming as a major unresolved problem.

Computational complexity, 1972–2006

Since its inception, the question of the number of steps required by the simplex method for a given linear program has been of interest. In the 1970s the field of 'computational complexity' emerged; a theory of problem difficulty which draws a sharp distinction between categories of problems that could be solved in polynomial time (number of steps) in the size of their data, and those which could not. How did linear programming fit into this scheme? Klee and Minty (1972) produced a worst case example of a simple linear program on which the simplex method takes an exponential number of iterations. But the expected number of pivots of the simplex method over a random selection of problems was shown to be polynomial in (m, n) (Smale, 1983). This raised the question: could a linear program be solved in polynomial time? Khachiyan (1979) defined a polynomial time algorithm for linear programs based on a sequence of convergent ellipsoids; however, unexpectedly according to computational complexity, the algorithm was very slow, and certainly no competitor of the simplex method. Later, Karmarkar (1984) gave a polynomial time interior point algorithm for linear programs which was claimed to be superior to the simplex method in the sense of solving linear programs much faster on a computer; the method required the linear program to be expressed in a special form with an optimal objective value of zero and viewed each iterate as being at the centre of a polyhedron

in a different coordinate system. The method typically required considerably fewer iterations than the simplex method, but each iteration required significantly more computations. The method was patented by AT&T and published as a theoretical result. There was considerable secrecy associated with the particulars of its implementation; and, thus, no independent verification was possible regarding its claimed superiority in computational speed over the simplex method. It was later shown to be equivalent under the same special form to the logarithmic barrier method, a method traceable back to Frisch (1955) and Fiacco and McCormick (1968). The logarithmic barrier method, however, could be applied to a linear program in standard form. The logarithmic barrier method was in the public domain and so allowed researchers to focus on computational improvements. Today, it is known that there are problems for which the logarithmic barrier method is superior to the simplex method; notable are those very large problems for which AA^T is sparse. For a survey of interior point methods, see Todd (1996). It is also interesting to note that most practical interior-point algorithms include an option to move the ε-optimal interior point solution to the nearest extreme point, a procedure requiring a significant number of simplex-type pivots. A technique to do this was proposed by Dantzig (1963, ch. 6, exercise 11). As of 2006, the simplex method (and various realizations thereof) remains the algorithm of choice for the majority of linear programs.

Dantzig in retirement, 1996–2005

Dantzig was retired from Stanford in stages, each firmly resisted. He was formally retired from the regular faculty at age 65 in 1980, but was recalled until age 82 in 1996. Until that year he remained as active as formal members of the faculty. After that he met at home with all who wished to consult him: students, colleagues, and strangers. Whenever presented with an idea, Dantzig would respond, as always, with something of value. Until around 2001 he continued to travel and present papers. At his 90th birthday celebration, he attended a full day of presentations followed by a banquet and additional talks. He was full of energy, enthusiasm, keen observations, and wit. Dantzig's mind was razor-sharp up to the end.

In retirement, Dantzig's principal project was the writing of a multi-volume book on linear programming and extensions. Dantzig had always felt that software was a key element that would contribute to the success of linear programming usage. He wanted to write another book on linear programming that incorporated software to aid students in learning both the theory and the practice of linear programming, and in particular in learning how to implement the simplex method and other algorithms for commercial use. In 1985 he invited M.N. Thapa to coauthor such a book. As work on the book progressed, it became apparent to the authors that the amount of material required a really huge book. One volume became two, and two became four. In the end only two volumes were completed (Dantzig and Thapa, (1997; 2003). Dantzig continued to be fascinated by interior point methods; von Neumann's and Karmarkar's algorithms were reanalysed and included in the second volume. According to M. Thapa, Dantzig never tired of editing and re-editing to improve proofs and readability. He would say: 'it is like polishing a stone; the more you polish it, the more it will shine.' Dantzig also continued his work with G. Infanger on planning under uncertainty. In addition to their research together, Dantzig and Infanger consulted on financial portfolio design. They intended to edit a collection of papers (including work of their own) on planning under uncertainty. Dantzig was convinced that the way to get further exposure for, and research into, planning under uncertainty was to set up an institute; to no avail, he tried at Stanford, tried at EPRI, and finally tried to create a stand-alone non-profit organization. In addition to these projects, Dantzig reworked the text of a science fiction novel he had begun in 1980.

Dantzig's honours

In 1975 L.V. Kantorovich and T.C. Koopmans received the Nobel Prize in Economics for 'their contributions to the theory of optimum allocation of resources'. Both mentioned Dantzig in their Nobel Lectures. That Dantzig did not participate in this prize came as a great shock and disappointment to those familiar with his contributions. Himself aside, Dantzig regarded Leontief, Kantorovich, von Neumann, and Koopmans as the principal early contributors to linear programming.

Dantzig, the man, and his contributions have nevertheless been honoured extensively. His honours include distinguished memberships, prizes, honorary doctorates, and dedications. He was elected to membership in the National Academy of Sciences, the National Academy of Arts and Sciences, and the National Academy of Engineering. He was a fellow of the Econometric Society, the Institute of Mathematical Statistics, the Association for the Advancement of Science, the Operations Research Society, IEEE, and the Omega Rho Society. He was awarded the War Department Exceptional Civilian Service Medal, the National Medal of Science, the John von Neumann Theory Prize, the NAS Award in Applied Mathematics and Numerical Analysis, the Harvey Prize (Technion), the Silver Medal of Operational Research Society (England), the Adolph Coors American Ingenuity Award, the Special Recognition Award of Mathematical Programming Society (MPS), the Harold Pender Award, and the Harold Lardner Memorial Prize (Canada). He received honorary doctorates from the Israel Institute of Technology (Technion), University of Linköping (Sweden), University of Maryland, Yale University, Université Catholique de Louvain (Belgium), Columbia University, the University of

Zurich, and Carnegie-Mellon University. Dantzig was also honoured as the dedicatee of a symposium of MPS, in two volumes of *Mathematical Programming*, in the first issue of the *Journal of Optimization of the Society for Industrial and Applied Mathematics* (SIAM), with the joint MPS-SIAM Dantzig Prize, and with the INFORMS Dantzig Prize for students. In 2006, a fellowship in his name was established in the Department of Management Science and Engineering at Stanford University.

Acknowledgements
The authors are grateful to David Dantzig, Jessica Dantzig Klass, and many of Dantzig's friends and colleagues who have contributed to this biographical article. These include A.J. Hoffman, G. Infanger, E. Klotz, J.C. Stone, M.J. Todd, J.A. Tomlin and M.H. Wright. This article has also benefited from other writings on G.B. Dantzig's life, namely: Albers and Reid (1986), Albers, Alexanderson and Reid (1990), Cottle (2003; 2005; 2006), Cottle and Wright (2006), Dantzig (1982; 1991), Dorfman (1984), Gill et al. (2007), Kersey (1989), Lustig (2001), Gass (1989; 2002; 2005).

RICHARD W. COTTLE, B. CURTIS EAVES AND MUKUND N. THAPA

See also **Kantorovich, Leonid Vitalievich; Koopmans, Tjalling Charles; Leontief, Wassily; linear programming; simplex method for solving linear programs; von Neumann, John.**

Selected works

1947. Prospectus for the AAF electronic computer. Unpublished manuscript.

1948a. A theorem on linear inequalities. Unpublished manuscript, 5 January.

1948b. Programming in a linear structure. Washington, DC: Comptroller, USAF. February. Abstract in *Econometrica* 17(1949), 73–4.

1951. Application of the simplex method to a transportation problem. In *Activity Analysis of Production and Allocation*, ed. T.C. Koopmans. New York: John Wiley & Sons.

1951. (With A. Wald.) On the fundamental lemma of Neyman and Pearson. *Annals of Mathematical Statistics* 22, 87–93.

1954. (With D.R. Fulkerson.) Minimizing the number of tankers to meet a fixed schedule. *Naval Research Logistics Quarterly* 1, 217–22.

1954. (With D.R. Fulkerson and S.M. Johnson.) Solution of a large scale traveling-salesman problem. *Journal of Operations Research Society of America* 2, 393–410.

1954. (With W. Orchard-Hays.) The product form for the inverse in the simplex method. *Mathematical Tables and Other Aids to Computation* 8, 64–7.

1955a. Upper bounds, secondary constraints, and block triangularity in linear programming. *Econometrica* 23, 174–83.

1955b. Linear programming under uncertainty. *Management Science* 1, 197–206.

1955. (With A.R. Fergusson.) The problem of routing aircraft. *Aeronautical Engineering Review* 14(4), 51–5. RAND Research Memorandum RM1369, 1954.

1955. (With D.R. Fulkerson.) Computation of maximal flows in networks. *Naval Research Logistics Quarterly* 2, 277–83.

1955. (With A. Orden, A. and P. Wolfe.) The generalized simplex method for minimizing a linear form under linear inequality restraints. *Pacific Journal of Mathematics* 5(2), 183–195. RAND Research Memorandum RM-1264, 1954.

1956. Recent advances in linear programming. *Management Science* 2, 131–44.

1956. (With L.R. Ford, Jr. and D.R. Fulkerson.) A primal-dual algorithm for linear programs. In *Linear Inequalities and Related Systems*, Annals of Mathematics Study No. 38, ed. H.W. Kuhn and A.W. Tucker. Princeton, NJ: Princeton University Press.

1957. Discrete variable extremum problems. *Operations Research* 5, 226–77.

1960a. On the shortest route through a network. *Management Science* 6, 187–90. RAND memorandum P-1345, 1959.

1960b. On the significance of solving linear programming problems with some integer variables. *Econometrica* 28, 30–44.

1960. (With P. Wolfe.) Decomposition principle for linear programs. *Operations Research* 8, 101–11.

1961. (With A. Madansky.) On the solution of two-stage linear programs under uncertainty. In *Proceedings, Fourth Berkeley Symposium on Mathematical Statistics and Probability*, I, ed. J. Neyman. Berkeley University of California Press. RAND memorandum P-2039, 1960.

1961. (With P. Wolfe.) The decomposition algorithm for linear programming. *Econometrica* 29, 767–78.

1963. *Linear Programming and Extensions*. Princeton, NJ: Princeton University Press.

1968. Large-scale linear planning. In *Mathematics of the Decision Sciences*, vol. 1, ed. G.B. Dantzig and A.F. Veinott, Jr. Providence, RI: American Mathematical Society.

1968. (With R.W. Cottle.) Complementary pivot theory of mathematical programming. *Linear Algebra and its Applications* 1, 103–25.

1973. (With others.) On the need for a System Optimization Laboratory. In *Mathematical Programming*, ed. T.C. Hu and S.M. Robinson. New York: Academic Press.

1982. Reminiscences about the origins of linear programming. *Operations Research Letters* 1(2), 43–48.

1990. (With P.W. Glynn.) Parallel processors for planning under uncertainty. *Annals of Operations Research* 22, 1–21.

1991. Linear programming. In *History of Mathematical Programming: A Collection of Personal Reminiscences*, ed. J.K. Lenstra, A.H.G. Rinnooy Kan and A. Schrijver. Amsterdam: North-Holland.

1992. (With G. Infanger.) Large-scale stochastic linear programs: importance sampling and Benders decomposition. In *Computational and Applied Mathematics I – Algorithms and Theory, Proceedings of the 13th IMACS World Congress*, ed. C. Brezinski and U. Kulisch. Amsterdam: North-Holland.

1997. (With M.N. Thapa.) *Linear Programming, 1. Introduction*. New York: Springer.

2003. (With M.N. Thapa.) *Linear Programming 2: Theory and Extensions*. New York: Springer.

Bibliography

Albers, D.J., Alexanderson, G.L. and Reid, C. 1990. *More Mathematical People*. New York: Harcourt Brace Jovanovich.

Albers, D.J. and Reid, C. 1986. An interview with George B. Dantzig: the father of linear programming. *College Mathematics Journal* 17, 292–314.

Arrow, K.J., Hurwicz, L. and Uzawa, H. 1958. *Studies in Linear and Non-Linear Programming*. Stanford, CA: Stanford University Press.

Benders, J.K. 1962. Partitioning procedures for solving mixed-variables programming problems. *Numerische Mathematik* 4, 238–52.

Charnes, A. and Cooper, W.W. 1962. On some works of Kantorovich, Koopmans and others. *Management Science* 8, 246–63.

Charnes, A. and Lemke, C.E. 1954. Minimization of non-linear separable convex functionals. *Naval Research Logistics Quarterly* 1, 301–12.

Cottle, R.W. 1964. Nonlinear programs with positively bounded Jacobians. Ph.D. thesis. Department of Mathematics. University of California at Berkeley.

Cottle, R.W. 2003. *The Basic George B. Dantzig*. Stanford, CA: Stanford University Press.

Cottle, R.W. 2005. George B. Dantzig: operations research icon. *Operations Research* 53, 892–8.

Cottle, R.W. 2006. George B. Dantzig: a life in mathematical programming. *Mathematical Programming* 105, 1–8.

Cottle, R.W. and Wright, M.H. 2006. Remembering George Dantzig. *SIAM News* 39(3), 2–3.

Dantzig, T. 1930. *Number, The Language of Science*. New York: Macmillan. 4th edition, ed. J. Mazur, republished New York: Pi Press, 2005.

de la Vallée Poussin, M.C.J. 1911. Sur la méthode de l'approximation minimum. *Annales de la Société Scientifique de Bruxelles* 35, 1–16.

Dijkstra, E. 1959. A note on two problems in connection with graphs. *Numerische Mathematik* 1, 269–71.

Dongarra, J. and Sullivan, F. 2000. The top 10 algorithms. *Computing in Science and Engineering* 2(1), 22–3.

Dorfman, R. 1984. The discovery of linear programming. *Annals of the History of Computing* 6, 283–95.

Dorfman, R., Samuelson, P.A. and Solow, R.M. 1958. *Linear Programming and Economic Analysis*. New York: McGraw-Hill.

Eaves, B.C. 1972. Homotopies for the computation of fixed points. *Mathematical Programming* 3, 1–22.

Fiacco, A.V. and McCormick, G.P. 1968. *Nonlinear Programming: Sequential Unconstrained Minimization Techniques*. New York: John Wiley & Sons.

Flood, M.M. 1956. The traveling-salesman problem. *Operations Research* 4, 61–75.

Ford, L.R., Jr. and Fulkerson, D.R. 1962. *Flows in Networks*. Princeton, NJ: Princeton University Press.

Fourier, J.B.J. 1826. Solution d'une question particulière du calcul des inégalités. *Nouveau Bulletin des Sciences par la Société Philomatique de Paris*, 99–100. Reprinted in *Oeuvres de Fourier, Tome II*, ed. G. Darboux. Paris: Gauthier, 1890.

Frisch, R.A.K. 1955. *The Logarithmic Potential Method of Convex Programs*. Oslo: University Institute of Economics.

Gale, D., Kuhn, H.W. and Tucker, A.W. 1951. Linear programming and the theory of games. In *Activity Analysis of Production and Allocation*, ed. T.C. Koopmans. New York: John Wiley & Sons.

Gass, S.I. 1989. Comments on the history of linear programming. *IEEE Annals of the History of Computing* 11(2), 147–51.

Gass, S.I. 2002. The first linear-programming shoppe. *Operations Research* 50, 61–8.

Gass, S.I. 2005. In Memoriam, George B. Dantzig. 2005. Online. Available at http://www.lionhrtpub.com/orms/orms-8-05/dantzig.html, accessed 12 January 2007.

Gass, S.I. and Saaty, T.L. 1955. The computational algorithm for the parametric objective function. *Naval Research Logistics Quarterly* 2, 39–45.

Gill, P.E., Murray, W., Saunders, M.A., Tomlin, J.A. and Wright, M.H. 2007. George B. Dantzig and systems optimization. Online. Available at http://www.stanford.edu/group/SOL/GBDandSOL.pdf, accessed 2 February 2007.

Gomory, R.E. 1958. Essentials of an algorithm for integer solutions to linear programs. *Bulletin of the American Mathematical Society* 64, 256.

Hitchcock, F.L. 1941. The distribution of a product from several sources to numerous localities. *Journal of Mathematics and Physics* 20, 224–30.

Hoffman, A. 1953. *Cycling in the Simplex Algorithm*. Report No. 2974. Washington, DC: National Bureau of Standards.

Hopp, W.J., ed. 2004. Ten most influential papers of *Management Science*'s first fifty years. *Management Science* 50(12 Supplement), 1764–9.

Infanger, G. 1991. Monte Carlo (importance) sampling within a Benders decomposition algorithm for stochastic linear programs. *Annals of Operations Research* 39, 41–67.

John, F. 1948. Extremum problems with inequalities as side conditions. In *Studies and Essays, Courant Anniversary Volume*, ed. K.O. Friedrichs, O.E. Neugebauer and J.J. Stoker. New York: Wiley-Interscience.

Kantorovich, L.V. 1942. Translocation of masses. *Doklady Akademii Nauk SSSR* 37, 199–201. Reprinted in *Management Science* 5 (1958–59), 1–4.

Kantorovich, L.V. 1960. Mathematical methods of organizing and planning production *Management Science* 6, 363–422. English translation of original monograph published in 1939.

Kantorovich, L.V. and Gavurin, M.K. 1949. The application of mathematical methods to freight flow analysis. *Problems of Raising the Efficiency of Transport Performance*. Akademiia Nauk, USSR. (Kantorovich confirmed the paper was completed and submitted in 1940, but publication delayed by the Second World War.)

Karmarkar, N. 1984. A new polynomial-time algorithm for linear programming. *Combinatorica* 4, 373–95.

Karush, W. 1939. Minima of functions of several variables with inequalities as side constraints. MsC. dissertation, Department of Mathematics, University of Chicago.

Kersey, C. 1989. *Unstoppable*. Naperville, IL: Sourcebooks, Inc.

Khachiyan, L.G. 1979. A polynomial algorithm in linear programming in Russian. *Doklady Akademii Nauk SSSR* 244, 1093–6. English translation: *Soviet Mathematics Doklady* 20 (1979), 191–4.

Klee, V. and Minty, G.J. 1972. How good is the simplex algorithm? In *Inequalities III*, ed. O. Shisha. New York: Academic Press.

Koopmans, T.C. 1947. Optimum utilization of the transportation system. *Proceedings of the International Statistical Conferences, 1947*, vol. 5. Washington D.C. Also in *Econometrica* 16 (1948), 66–8.

Koopmans, T.C. 1949. Optimum utilization of the transportation system. *Econometrica* 17(Supplement), 136–46.

Koopmans, T.C., ed. 1951. *Activity Analysis of Production and Allocation*. New York: John Wiley & Sons.

Koopmans, T.C. 1960. A note about Kantorovich's paper, 'Mathematical Methods of Organizing and Planning Production.' *Management Science* 6, 363–5.

Kuhn, H.W. and Tucker, A.W. 1951. Nonlinear Programming. In *Proceedings of the Second Berkeley Symposium on Mathematical Statistics and Probability*, ed. J. Neyman. Berkeley: University of California Press.

Lemke, C.E. 1954. The dual method of solving the linear programming problem. *Naval Research Logistics Quarterly* 1, 36–47.

Lemke, C.E. 1965. Bimatrix equilibrium points and mathematical programming. *Management Sciences* 11, 681–9.

Leontief, W.W. 1936. Quantitative input and output relations in the economic system of the United States. *Review of Economic Statistics* 18, 105–25.

Lustig, I. 2001. e-optimization.com. Interview with G.B. Dantzig. Online. Available at http://e-optimization.com/directory/trailblazers/dantzig, accessed 29 December 2006.

Markowitz, H.M. 1956. The optimization of a quadratic function subject to linear constraints. *Naval Research Logistics Quarterly* 3, 111–33. RAND Research Memorandum RM-1438, 1955.

Orchard-Hays, W. 1954. A composite simplex algorithm-II. RAND Research Memorandum RM-1275. Santa Monica, CA: RAND Corporation.

Rockafellar, R.T. 1970. *Convex Analysis*. Princeton, NJ: Princeton Press.

Scarf, H. 1967. The core of an N-person game. *Econometrica* 35, 50–69.

Scarf, H. 1973. *The Computation of Economic Equilibria*. New Haven, CT: Yale University Press.

Schrijver, A. 1986. *Theory of Linear and Integer Programming*. Chichester: John Wiley & Sons.

Smale, S. 1976. A convergent process of price adjustment and global Newton methods. *Journal of Mathematical Economics* 3, 1–14.

Smale, S. 1983. The problem of the average speed of the simplex method. In *Mathematical Programming: The State of the Art*, ed. A. Bachem, M. Grötschel and B. Korte. Berlin: Springer-Verlag.

Stigler, G.J. 1945. The cost of subsistence. *Journal of Farm Economics* 27, 303–14.

Todd, M.J. 1996. Potential-reduction methods in mathematical programming. *Mathematical Programming* 76, 3–45.

von Neumann, J. 1947. Discussion of a maximum problem. Unpublished working paper dated 15–16 November. In *John von Neumann: Collected Works*, vol. 6, ed. A.H. Taub. Oxford: Pergamon Press, 1963.

von Neumann, J. and Morgenstern, O. 1944. *Theory of Games and Economic Behavior*. Princeton, NJ: Princeton University Press.

data filters

Economic models are by definition incomplete representations of reality. Modellers typically abstract from many features of the data in order to focus on one or more components of interest. Similarly, when confronting data, empirical economists must somehow isolate eatures of interest and eliminate elements that are a nuisance from the point of view of the theoretical models they are studying. Data filters are sometimes used to do that.

For example, Figure 1 portrays the natural logarithm of US GDP. Its dominant feature is sustained growth, but business cycle modellers often abstract from this feature in order to concentrate on the transient ups and downs. To relate business cycle models to data, empirical macroeconomists frequently filter the data prior to analysis to remove the growth component. Until the 1980s, the most common way to do that was to estimate and subtract a deterministic linear trend. Linear de-trending is conceptually unattractive, however, because it presupposes that all shocks are neutral in the long run. While some disturbances – such as those to monetary policy – probably are neutral in the long run, others probably are not. For instance, a technical innovation is likely to remain

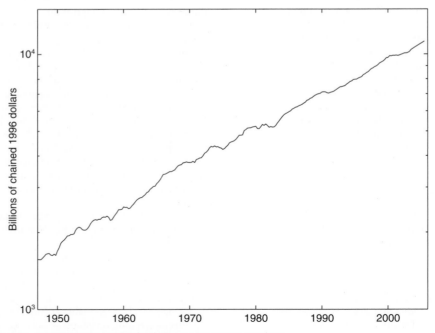

Figure 1   Real US GDP, 1947–2006. *Source*: Federal Reserve Economic Database.

relevant for production until it is superseded by another, later technical innovation.

The desire to model permanent shocks in macroeconomic time series led to the development of a variety of stochastic de-trending methods. For example, Beveridge and Nelson (1981) define a stochastic trend in terms of the level to which a time series is expected to converge in the long run. Blanchard and Quah (1989) adopt a more structural approach, enforcing identifying restrictions in a vector autoregression that separate permanent shocks that drive long-run movements from the transitory disturbances which account for cyclical fluctuations.

Another popular way to measure business cycles involves application of band-pass and high-pass filters. Engle (1974) was one of the first to introduce band-pass filters to economics. In the business cycle literature, the work of Hodrick and Prescott (1997) and Baxter and King (1999) has been especially influential. Figure 2 illustrates measures of the business cycle that emerge from the Baxter–King and Hodrick–Prescott filters.

In this article, I describe how data filters work and explain the theoretical rationale that lies behind them. I focus on the problem of measuring business cycles because that is one of the principal areas of application. Many of the issues that arise in this context are also relevant for discussions of seasonal adjustment. For a review of that literature, see Fok, Franses and Paap (2006).

How data filters work

The starting point is the Cramer representation theorem. Cramer's theorem states that a covariance stationary random variable x_t can be expressed as

$$x_t - \mu_x = \int_{-\pi}^{\pi} \exp(i\omega t) dZ_x(\omega), \qquad (1)$$

where μ_x is the mean, t indexes time, $i = \sqrt{-1}$, ω represents frequency, and $dZ_x(\omega)$ is a mean zero, complex-valued random variable that is continuous in ω. The complex variate $dZ_x(\omega)$ is uncorrelated across frequencies, and at a given frequency its variance is proportional to the spectral density $f_{xx}(\omega)$. If we integrate the spectrum across frequencies, we get the variance of x_t,

$$\sigma_x^2 = \int_{-\pi}^{\pi} f_{xx}(\omega) d\omega. \qquad (2)$$

This theorem provides a basis for decomposing x_t and its variance by frequency. It is perfectly sensible to speak of long- and short-run variation by identifying the long run with low-frequency components and the short run with high-frequency oscillations. High frequency means that many complete cycles occur within a given time span, while low frequency means the opposite.

Baxter and King (1999) define a business cycle in terms of the periodic components $dZ_x(\omega)$. They partition x_t into three pieces: a trend, a cycle, and irregular fluctuations. Inspired by the NBER business cycle chronology, they say

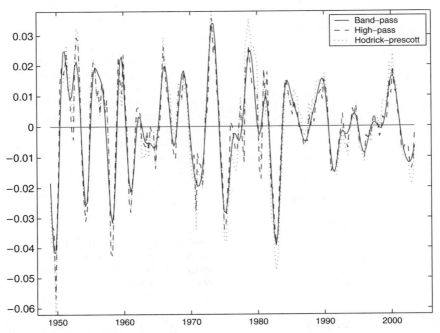

Figure 2   Filtered GDP, 1949–2003. *Source*: Federal Reserve Economic Database and author's calculations

the business cycle consists of periodic components whose frequencies lie between 1.5 and 8 years per cycle. Those whose cycle length is longer than 8 years are identified with the trend, and the remainder are consigned to the irregular component.

The units for ω are radians per unit time. A more intuitive measure of frequency is units of time per cycle, which is given by the transformation $\lambda = 2\pi/\omega$. Often we work with quarterly data. To find the ω corresponding to a cycle length of 1.5 years, just set $\lambda_h = 6$ quarters per cycle and solve for $\omega_h = 2\pi/6 = \pi/3$. Similarly, the frequency corresponding to a cycle length of 8 years is $\omega_l = 2\pi/32 = \pi/16$. Baxter and King define the interval $[\pi/16, \pi/3]$ as 'business cycle frequencies'. The interval $[0,\pi/16)$ corresponds to the trend, and $(\pi/3, \pi]$ defines irregular fluctuations. One nice feature of the Baxter–King filter is that it can be easily adjusted to accommodate data sampled monthly or annually, just be resetting ω_l and ω_h.

To extract the business cycle component, we need to weigh the components $dZ_x(\omega)$ in accordance with Baxter and King's definition and integrate across frequencies,

$$x_t^B = \int_{-\pi}^{\pi} B(\omega) \exp(i\omega t) dZ_x(\omega), \qquad (3)$$

where

$$B(\omega) = 1 \quad \text{for } \omega \in [\pi/16, \pi/3] \text{ or } [-\pi/3, -\pi/16],$$
$$= 0 \quad \text{otherwise.} \qquad (4)$$

In technical jargon, $B(\omega)$ is an example of a 'band-pass' filter: the filter passes periodic components that lie within a pre-specified frequency band and eliminates everything else. The Baxter–King filter suppresses all fluctuations that are too long or short to be classified as part of the business cycle and allows the remaining elements to pass through without alteration.

Many economists are more comfortable working in time domain, and for that purpose it is helpful to express the cyclical component as a two-sided moving average,

$$x_t^B = \sum_{j=-\infty}^{\infty} \beta_j(x_{t+j} - \mu_x). \qquad (5)$$

The lag coefficients can be found by solving

$$\beta_j = \frac{1}{2\pi} \int_{-\pi}^{\pi} B(\omega) \exp(i\omega j) d\omega. \qquad (6)$$

The solution is

$$\beta_0 = \frac{\omega_h - \omega_l}{\pi},$$
$$\beta_j = \frac{\sin(\omega_h j) - \sin(\omega_l j)}{\pi j} \quad \text{for } j \neq 0. \qquad (7)$$

Notice that an ideal band-pass filter cannot be implemented in actual data samples because it involves infinitely many leads and lags. In practice, economists approximate x_t^B with finite-order moving averages,

$$x_t^B \doteq \sum_{j=-n}^{n} \tilde{\beta}_j(x_{t+j} - \mu_x). \qquad (8)$$

Baxter and King (1999) and Christiano and Fitzgerald (2003) analyse how to choose the lag weights $\tilde{\beta}_j$ in order to best approximate the ideal measure for a given n.

For real-time applications, the two-sided nature of the filter is a drawback because the current output of the filter depends on future values of x_{t+j}, which are not yet available. Kaiser and Maravall (2001) address this problem by supplementing the filter with an auxiliary forecasting model such as a vector autoregression or univariate ARIMA model, replacing future x_{t+j} with forecasted values. This substantially reduces the approximation error near the end of samples.

That the filter is two-sided is also relevant for models that require careful attention to the timing of information. Economic hypotheses can often be formulated as a statement that some variable z_t should be uncorrelated with any variable known in period $t-1$ or earlier. These hypotheses can be examined by testing for absence of Granger causation from a collection of potential predictors to z_t. The output of a two-sided filter should never be included among those predictors, however, for that would put information about present and future conditions on the right-hand side of the regression and bias the test towards a false finding of Granger causation. Similar comments apply to the choice of instruments in generalized-method-of-moments problems. For applications like these, one-sided filters are needed in order to respect the integrity of the information flow.

While Baxter and King favour a three-part decomposition, other economists prefer a two-part classification in which the highest frequencies also count as part of the business cycle. The trend component is still defined in terms of fluctuations lasting more than eight years, but the cyclical component now consists of all oscillations lasting eight years or less. To construct this measure, we define a new filter $H(\omega)$ such that

$$H(\omega) = 1 \quad \text{for } \omega \in [\pi/16, \pi] \text{ or } [-\pi, -\pi/16],$$
$$= 0 \quad \text{otherwise.} \qquad (9)$$

This is known as a 'high-pass' filter because it passes all components at frequencies higher than some pre-specified value and eliminates everything else. If we use this filter in the Cramer representation, we can extract a new measure of the business cycle by computing

$$x_t^H = \int_{-\pi}^{\pi} H(\omega) \exp(i\omega t) dZ_x(\omega). \qquad (10)$$

Once again, this corresponds to a two-sided, infinite-order moving average of the original series x_t,

$$x_t^H = \int_{j=-\infty}^{\infty} \gamma_j(x_{t+j} - \mu_x), \quad (11)$$

with lag coefficients $\gamma_0 = 1 - \omega_l/\pi$ and $\gamma_j = -\sin(\omega_l j)/\pi j$. As before, this involves infinitely many leads and lags, so an approximation is needed to make it work. The approximation results of Baxter and King (1999) and Christiano and Fitzgerald (2003) apply here as well.

Hodrick and Prescott (1997) also seek a two-part decomposition of x_t. They proceed heuristically, identifying the trend τ_t and the cycle c_t by minimizing the variance of the cycle subject to a penalty for variation in the second difference of the trend,

$$\min_{\{\tau_t\}} \left\{ \sum_{t=-\infty}^{\infty} [(x_t - \tau_t)^2 + \varphi(\tau_{t+1} - 2\tau_t + \tau_{t-1})^2] \right\}. \quad (12)$$

The Lagrange multiplier φ controls the smoothness of the trend component. After experimenting with US data, Hodrick and Prescott set $\varphi = 1600$, a choice still used in most macroeconomic applications involving quarterly data. After differentiating (12) with respect to τ_t and rearranging the first-order condition, one finds that c_t can be expressed as an infinite-order, two-sided moving average of x_t,

$$c_t = HP(L)x_t = \frac{\varphi(1-L)^2(1-L^{-1})^2}{1 + \varphi(1-L)^2(1-L^{-1})^2} x_t, \quad (13)$$

where L is the lag operator. Although Hodrick and Prescott's derivation is heuristic, King and Rebelo (1993) demonstrate that $HP(L)$ can be interpreted rigorously as an approximation to a high-pass filter with a cut-off frequency of eight years per cycle. The close connection between the two filters is also apparent in Figure 2, which shows that high-pass and Hodrick–Prescott filtered GDP are highly correlated.

Data filters for measuring of business cycles?

While data filters are very popular, there is some controversy about whether they represent appealing definitions of the business cycle. For one, there is a disconnect between the theory and macroeconomic applications, for the theory applies to stationary random processes and applications involve non-stationary variables. This is not critical, however, because the time-domain filters $\beta(L)$, $\gamma(L)$, and $HP(L)$ all embed difference operators, so business cycle components are stationary even if x_t has a unit root.

A more fundamental criticism concerns the fact that the Baxter–King definition represents a deterministic

vision of the business cycle. According to a theorem of Szego, Kolmogorov, and Krein, the prediction error variance can be expressed as

$$\sigma_\varepsilon^2 = 2\pi \, \exp\left[\frac{1}{2\pi}\int_{-\pi}^{\pi} \log f_{BC}(\omega)d\omega\right], \quad (14)$$

where $f_{BC}(\omega)$ is the spectrum for the business-cycle component (see Granger and Newbold 1986, pp. 135–6). For an ideal band-pass filter, the spectrum of x_t^B is

$$f_{BC}(\omega) = |B(\omega)|^2 f_{xx}(\omega). \quad (15)$$

Since $B(\omega) = 0$ outside of business cycle frequencies, it follows that $f_{BC}(\omega) = 0$ on a measurable interval of frequencies. But then eq. (14) implies $\sigma_\varepsilon^2 = 0$, which means that x_t^B is perfectly predictable from its own past. The same is true of measures based on ideal high-pass filters. A variable that is perfectly predictable based on its own history is said to be 'linearly deterministic'. Thus, according to the Baxter–King definition, the business cycle is linearly deterministic.

In practice, of course, measured cycles are not perfectly predictable because actual filters only approximate the ideal. But this means that innovations in measured cycles are due solely to approximation errors in the filter, not to something intrinsic in the concept. The better the approximation, the closer the measures are to determinism.

How to square this deterministic vision with stochastic general equilibrium models is not obvious. Engle (1974), Sims (1993) and Hansen and Sargent (1993) suggest one rationale. They were interested in estimating models that are well specified at some frequencies but mis-specified at others. Engle studied linear regressions and showed how to estimate parameters by band-spectrum regression. This essentially amounts to running regressions involving band-pass filtered data, but band-pass filtering induces serial correlation in the residuals, and Engle showed how to adjust for this when calculating standard errors and other test statistics. He also developed methods for diagnosing mis-specification on particular frequency bands.

Sims (1993) and Hansen and Sargent (1993) are interested in fitting a rational-expectations model of the business cycle to data that contain seasonal fluctuations. They imagine that the model abstracted from seasonal features, as is common in practice, and they wonder whether estimates could be improved by filtering the data with a narrow band-pass filter centred on seasonal frequencies. They find that seasonal filtering does help, because otherwise parameters governing business cycle features would be distorted to fit unmodelled seasonal fluctuations. Filtering out the seasonals lets the business cycle parameters fit business cycle features.

Business cycle modellers also frequently abstract from trends, and Hansen and Sargent conjectured that the same rationale would apply to trend filtering. Cogley

(2001) studies this conjecture but finds disappointing results. The double-filtering strategy common in business cycle research (which applies the filter to both the data and the model) has no effect on periodic terms in a Gaussian log likelihood, so it is irrelevant for estimation. The seasonal analogy (which filters the data but not the model) also fails, but for a different reason. The key assumption underlying the work of Engle, Sims, and Hansen and Sargent is that specification errors are confined to a narrow frequency band whose location is known a priori. That is true of the seasonal problem but not of the trend problem. Contrary to intuition, trend-specification errors spread throughout the frequency domain and are not quarantined to low frequencies. That difference explains why the promising results on seasonality do not carry over to trend filtering.

Finally, some economists question whether filter-based measures capture an important feature of business cycles. Beveridge and Nelson (1981) believe that trend reversion is a defining characteristic of the business cycle. They say that expected growth should be higher than average at the trough of a recession because agents can look forward to a period of catching up to compensate for past output losses. By the same token, expected growth should be lower than average at the peak of an expansion. Cochrane (1994) confirms that this is a feature of US business cycles by studying a vector autoregression for consumption and GDP.

Cogley and Nason (1995) consider what would happen if x_t were a random walk with drift. For a random walk, expected growth is constant regardless of whether the level is a local maximum or minimum. Because it lacks the catching-up feature, many economists would say that a random walk is acyclical. Nevertheless, when the Hodrick–Prescott filter is applied to a random walk, a large and persistent cycle emerges. Thus the Hodrick–Prescott filter can create a business cycle even if no trend reversion is present in the original data. Cogley and Nason call this a spurious cycle. Furthermore, the problem is not unique to the Hodrick–Prescott filter; Benati (2001), Murray (2003) and Osborn (1995) document similar results for band-pass filters and for other approximations to high-pass filters.

Conclusion

Christiano and Fitzgerald remark that data filters are not for everyone. They are certainly convenient for constructing rough and ready measures of the business cycle, and they produce nice pictures when applied to US data. But some economists worry about the spurious cycle problem, especially in applications to business cycle models where the existence and properties of business cycles are points to be established. In much of that literature, attention has shifted away from replicating properties of filtered data to matching the shape of impulse response functions.

TIMOTHY COGLEY

See also **business cycle measurement; seasonal adjustment; spectral analysis; structural vector autoregressions; trend/cycle decomposition.**

Bibliography

Baxter, M. and King, R. 1999. Measuring business cycles: approximate band-pass filters for economic time series. *Review of Economics and Statistics* 81, 575–93.

Benati, L. 2001. Band-pass filtering, cointegration, and business cycle analysis. Working Paper No. 142, Bank of England.

Beveridge, S. and Nelson, C. 1981. A new approach to decomposition of economic time series into permanent and transitory components with particular attention to measurement of the business cycle. *Journal of Monetary Economics* 7, 151–74.

Blanchard, O. and Quah, D. 1989. The dynamic effects of aggregate demand and supply disturbances. *American Economic Review* 79, 655–73.

Christiano, L. and Fitzgerald, T. 2003. The band pass filter. *International Economic Review* 44, 435–65.

Cochrane, J. 1994. Permanent and transitory components of GNP and stock prices. *Quarterly Journal of Economics* 109, 241–65.

Cogley, T. 2001. Estimating and testing rational expectations models when the trend specification is uncertain. *Journal of Economic Dynamics and Control* 25, 1485–525.

Cogley, T. and Nason, J. 1995. Effects of the Hodrick–Prescott filter on trend and difference stationary time series: implications for business cycle research. *Journal of Economic Dynamics and Control* 19, 253–78.

Engle, R. 1974. Band-spectrum regression. *International Economic Review* 15, 1–11.

Fok, D., Franses, P. and Paap, R. 2006. Comparing seasonal adjustment methods. In *Palgrave Handbook of Econometrics: Volume1; Econometric Theory*, ed. T. Mills and K. Patterson. Basingstoke: Palgrave Macmillan.

Granger, C. and Newbold, P. 1986. *Forecasting Economic Time Series*. New York: Academic Press.

Hansen, L. and Sargent, T. 1993. Seasonality and approximation errors in rational expectations models. *Journal of Econometrics* 55, 21–55.

Hodrick, R. and Prescott, E. 1997. Postwar U.S. business cycles: an empirical investigation. *Journal of Money Credit and Banking* 29, 1–16.

Kaiser, R. and Maravall, A. 2001. *Measuring Business Cycles in Economic Time Series*. New York: Springer-Verlag.

King, R. and Rebelo, S. 1993. Low-frequency filtering and real business cycles. *Journal of Economic Dynamics and Control* 17, 207–31.

Murray, C. 2003. Cyclical properties of Baxter–King filtered time series. *Review of Economics and Statistics* 85, 472–76.

Osborn, D. 1995. Moving average detrending and the analysis of business cycles. *Oxford Bulletin of Economics and Statistics* 57, 547–58.

Sims, C. 1993. Rational expectations modeling with seasonally adjusted data. *Journal of Econometrics* 55, 9–19.

data mining

'Data mining' and the older word 'fishing' are pejorative terms for illusory or distorted statistical inference from an empirical regression model, where the distortion results from explorations of various models in a single sample of data. This process usually involves adding or dropping variables, but may involve exploring a variety of alternative nonlinear functional forms or data sub-samples. Data mining properly applies as a derogatory term only when exploratory results are used for inference within the sample used in exploration. But the term is sometimes used to refer to the exploratory process itself, as economists emphasize inference over data exploration, and even use inference to discuss exploratory activities. Some take data mining to be a more serious offence when there is conscious effort to manipulate, although data mining will distort results regardless of intent.

Importance and history

Some economists consider data mining to be pervasive in applied work. But the portion subscribing to this view is unclear, since those who do so understandably retreat from applied work into economic or econometric theory. Leamer and Leonard (1983, p. 306) give voice to the view that collective data mining renders standard inference meaningless, and hence in general 'statistical analyses are either greatly discounted or completely ignored'. This stance may have reached a peak in the late 1970s, fuelled by an explosion in the volume of regression studies. But contemporary suspicion is still quite common. Kennedy (2003, pp. 82–3) characterizes the 'average economic regression' as perpetrating some of the worst data mining practices.

The issue was known to the originators of econometrics. Ragnar Frisch (1934) advocated methods to deal with the data mining issue which were applied into the 1950s, then neglected for two decades and reincarnated in modern form by Leamer (1983). Because Frisch found that differing but reasonable specifications could yield disparate results, he came to believe attempts at formal inference were illegitimate. Malinvaud (1966, chs. 1 and 2) provides a wonderful exposition of Frisch's methods and of why Frisch's stance was replaced by contemporary textbook assumptions. Even Haavelmo's (1944, ch. 7, sect. 17) founding statement of the contemporary inferential approach discusses data mining.

Econometrics textbooks quite properly warn against data mining, yet it is difficult to avoid and is pervasive in published work. This places the new practitioner in a difficult position. It is helpful to be armed with an understanding of the consequences of data mining and why data mining is difficult to avoid. Econometrics in the contemporary sense began when we decided that economic data could be treated as equivalent to sampling from an uncontrolled experiment (Haavelmo, 1944), borrowing from R.A. Fisher's methods for experimental data. The following illustration clarifies these issues.

An illustration

Suppose two students of the economy live in parallel universes. Both are interested in a variable y, believing the most important determinant of this variable y to be another variable x_1, but also supposing that variables x_2 and x_3 may be relevant. Their initial data-sets are identical, and they propose to model y via a linear regression model. Both start out assuming that the errors of the model (ε) are independent and normally distributed with constant variance. Thus they propose the model $y = b_1x_1 + b_2x_2 + b_3x_3 + \varepsilon$, where the coefficients 'b_i' are to be estimated.

The first student lives in a universe in which he can generate more data via experiments. The second student must wait passively for the passage of time before she can see more data; data generated by events she does not control. Thus, the first student is confident of his science, while the second student is in the actual universe of economics.

Now suppose that in their initial regression results for the coefficient on x_1 they find the sign is the opposite of what they expected. As in standard practice they take this to imply that they have omitted an important variable. After fiddling with their specifications they find that adding a variable x_4 yields a more sensible coefficient estimate for the variable x_1. Suppose also they find that, for the coefficients on x_2 and x_3, the null hypothesis for coefficients of zero would be accepted individually (leaving the other variable coefficient unrestricted, as in a t-test). But suppose they find the joint hypothesis ($b_2 = b_3 = 0$) would be rejected. They find the fit of the regression is penalized least by dropping the variable x_3 and do so. They have used a process of specification search to arrive at a model for y as a function of x_1, x_2 and x_4.

The first student takes the results to his professor. The professor commends the effort to learn from the world, but corrects the student on one point. He notes that, although the estimated standard error for the coefficient on x_3 included zero, it also included (we will suppose) five, and if this coefficient is truly so far from zero then (given expected variation in x_3) the variable x_3 would have appreciable effects. So the professor tells him to run another experiment designed so that the resulting data-set is large enough (and so standard errors of coefficient estimates are small enough) to usefully distinguish between large and small values of b_3. The student does so, and publishes the results with the statistics and standard critical values treated as valid 'tests'. This is not data mining.

Now the second student takes her results to her professor. This professor says the first regression result (employing x_1, x_2 and x_3) can be treated as possibly generating test statistics drawn from standard distributions. However, in the final model (x_1, x_2 and x_4) some of the t-statistics were created by design. Since one 'fished' or fiddled with variables included in the model until the coefficient on x_1 had the correct sign, the t-statistic was drawn from a distribution such that there was 100 per cent probability it would have the 'correct' sign. Likewise the student explored specifications until the t-statistic for the coefficient on x_2 appeared to be significant. This implies for the final specification that within the interval bounded by the standard critical values (approximately plus or minus 2) the probability of the t-statistic for b_2 falling within this standard range must actually be zero, hardly a standard t-distribution. This process of modifying the model and re-estimating it using the same sample used to suggest those modifications will also affect in an unknown manner the distribution of other test statistics, even those that were not direct objects of exploration and design. These are data mined results.

Note that the two professors agree that something was potentially learned in the exploratory stage. Both students could use data exploration to reveal aspects of the first sample, but the results of exploration over this same sample could not then provide a formal test. As in any legitimate science, the first professor views taking inspiration from observation to be a process separate from confirmation or testing. The second student also hopes to have learned something from the sample, but her professor objects to treating the statistics resulting from this exploration as providing a test. The second student treated each regression as though it was a separate experiment, but regressions and their associated statistics are mere calculations that organize the data. Also note that, when these students took the initial estimate of b_1 as having the 'wrong sign', they were applying strong prior beliefs which led them to place little weight upon this empirical result. Bayesian inference provides a formal treatment of such priors.

The second student continues the consultation with her professor. The professor says these first results are not publishable because economists are interested in inference, and all she has shown is that the first model did not make sense. The professor may advise that she should first have chosen a successful regression model from the empirical literature, modifying it only slightly if at all. If the student is alert, she will notice the data available to her is identical to that in the literature, except for a few more recent observations.

So this alert student will go back to her professor and tell him she already knows the regression results will be the same as those already published, except to the extent the new data observations have some effect when averaged with the old. The test statistics will not have the usual distributions; instead, the distributions are a function of the previous results and the portion of new observations relative to those used in the previously published results. The student has discovered that, to the extent data-sets overlap, taking guidance from the regressions of other researchers is collective data mining, even if one runs only one regression oneself. Thus collective data mining is pervasive, and the meaning of published test statistics is unclear. Only if each data-set is entirely distinct can one learn from the work of others while preserving known statistic distributions.

Contemporary practice and remedies

Three partial remedies for data mining are practised in the current literature. One is to insist upon seeing all the possible regression results a reasonable researcher might propose, supplementing imperfect 'tests' with a range of results. This is most associated with Leamer (1983), but we have already mentioned the earlier work of Frisch. Current practice is moving towards this approach, more often presenting multiple specifications.

A second remedy is inspired by noting that it is possible to calculate probabilities for statistics resulting from specification search, if the process begins with a model including a set of variables large enough that the true model is reasonably assumed nested within, and respecification deletes and does not add variables. An example is the general-to-specific approach. This approach is now common when specifying lag-lengths of time-series models, but in other contexts is controversial. The statistical consequences of such an approach fall under the heading of 'pretest' estimators discussed in most econometrics textbooks, but the best introductory discussion is found in Campos, Ericsson and Hendry (2005, Introduction, sects. 3.3–3.4). Interestingly, Hoover and Perez (1999) show that when pretest distributions are not accounted for this second remedy leads to an acceptable level of distortion.

A third remedy reserves some of the available data for 'out-of-sample' tests. Here one engages in specification search in one portion of the data and then tests in the reserved portion. We place 'out-of-sample' in quotes because this is not confirmation in a new sample. This response cannot avoid collective data mining because it is likely that among many projects the more satisfactory reserved-sample results will be selected for publication, if not by individual authors then through the collective filter of journal referees. But this remedy is useful to the individual researcher.

The first two remedies focus on data exploration, and only the third remedy adds the key scientific step of confirmation in separate data. Followers of the second remedy such as David Hendry and others of the 'London School' are often accused of data mining. Yet they have been the strongest proponents and practitioners of the third remedy, which provides the legitimate test in separate data, even inventing new out-of-sample tests such as for forecast encompassing. A good introduction to the second and third remedies is found in Charemza and Deadman (1997).

As noted in our discussion of the third remedy, universal adoption of these remedies cannot avoid collective data mining. Collective data mining would be avoided if upon accepting a paper the journal offered an explicit or implicit contract to accept a follow-up study. Formal and precise testing would be performed in the subsequent study employing only data not available for the initial paper. This is yet to be practised by any journal, so as a result the methodological issues remain troublesome, leaving room for vague and inconsistent norms across referees and journals. New practitioners must develop their own approaches to navigating these norms and practices, while deciding how to preserve their own sense of integrity.

CLINTON A. GREENE

See also **Bayesian statistics; extreme bounds analysis; model selection; specification problems in econometrics; spurious regressions.**

Bibliography

Campos, J., Ericsson, N. and Hendry, D., eds. 2005. *General-To-Specific Modeling*. Cheltenham, UK: Edward Elgar.

Charemza, W. and Deadman, D. 1997. *New Directions in Econometric Practice*. Cheltenham, UK: Edward Elgar.

Fisher, R. 1925. *Statistical Methods for Research Workers*. Edinburgh: Oliver and Boyd.

Frisch, R. 1934. *Statistical Confluence Analysis by Means of Complete Regression Systems*. Oslo: University Economics Institute.

Haavelmo, T. 1944. The probability approach in econometrics. *Econometrica* 12(supplement), iii–115.

Hoover, K. and Perez, S. 1999. Data mining reconsidered: encompassing and the general-to-specific approach to specification search. *Econometrics Journal* 2, 167–91.

Kennedy, P. 2003. *A Guide to Econometrics*. Cambridge, MA: MIT Press.

Leamer, E. 1983. Let's take the con out of econometrics. *American Economic Review* 73, 31–43.

Leamer, E. and Leonard, H. 1983. Reporting the fragility of regression estimates. *Review of Economics and Statistics* 65, 306–17.

Malinvaud, E. 1966. *Statistical Methods of Econometrics*. Chicago: Rand McNally and Company.

Davanzati, Bernardo (1529–1606)

Merchant, classical scholar, translator and economist, Davanzati was born in Florence where, apart from a period of residence in Lyon as a merchant, he worked until his death. His contributions to economics are contained in *Notizia dei cambi* (1582) which explains the operation of the foreign exchanges, and *Lezione delle Monete* (1588), translated into English in 1696 as *A Discourse Upon Coin* presumably because of its relevance to the recoinage controversies. Besides these economic writings, Davanzati produced a history of the English Reformation (1602) and a translation of Tacitus (1637) frequently described as a masterpiece of Italian literature.

Davanzati's observations on the foreign exchanges present a detailed discussion of the origins and practice of this art classified by him as the third type of mercantile transaction, the others being barter (goods for goods) and trade (goods for money). The analysis demonstrates how exchange rates fluctuate between gold points according to the supply and demand of bills, the gold points being determined by a risk premium, transport costs and interest lost while the funds are in transit. His illustration of a foreign exchange transaction by bills of exchange involving six parties residing in Lyon and Florence (1582, pp. 62–8) has been argued by De Roover (1963, p. 113) to be so instructive that, had it been more thoroughly studied by historians and economists, 'fewer blunders in the history of banking' would have been made.

Davanzati's lecture on coin is one of the earliest presentations of the metallist view of the origin and nature of money. He stresses the advantages of money over barter in facilitating both the division of labour and trade of 'superfluities' between cities and nations. In the metallist tradition, money is defined as 'Gold, Silver, or Copper, coin'd by Publick Authority at pleasure, and by the consent of Nations, made the Price and Measure of Things' (1588, p. 12). Non-metallic and non-convertible money can only be made acceptable to the public through coercion. Money is therefore a human convention and its intrinsic value is small relative to its value as means of exchange. To explain this value, Davanzati presents an early quantity theory which relates the value of stocks of commodities to the world's money stock. Although he is aware of the importance of monetary circulation (he compares it to the importance of the circulation of blood in the animal body), he does not develop a concept of its velocity. The lecture on money concludes with a forceful critique of the practice of debasing the coinage, based on analysing its consequences and illustrated with many examples of the practice. Davanzati argues that this 'evil' can be avoided only by making 'Money pass according to its Intrinsick Value' (1588, p. 24). Davanzati's lecture has also been noted because of its hints at the so-called 'paradox of value' and its references to elements of scarcity and usefulness in the determination of commodity prices. This and other aspects of his work were noted by Galiani (1750). Earlier his views appear to have been well received by Locke who owned, annotated and may even have inspired the Toland translation (Harrison and Laslett, 1965, p. 120).

PETER GROENEWEGEN

Selected works

1582. *Notizia de'Cambi a M. Giulio del Caccia*. In *Scrittori classici Italiani di economia politica*. Parte Antica vol. 2, ed. Pietro Custodi, Milan: G.G. Destefanis, 1804.

1588. *Lezione delle Monete*. Trans. John Toland as *A Discourse Upon Coin*, London: Awnsham and Churchill, 1969.

1602. *Scisma d'Inghilterra sino alla morta della reina Maria ristretto in lingua propria Fiorentina*. Milan.

1637. *Gli Annali di C. Cornelio Tacito … con la traduzione in volgar Fiorentino*. Florence: Landini.

Bibliography

De Roover, R. 1963. *The Rise and Decline of the Medici Bank*. Cambridge, MA: Harvard University Press.

Galiani, F., ed. 1750. *Della Moneta*. In *Della Moneta e scritti inedite*, ed. A. Merola, Milan: Feltrinelli, 1963.

Harrison, J. and Laslett, P. 1965. *The Library of John Locke*. Oxford: Clarendon Press, 1971.

Davenant, Charles (1656–1714)

Economist and administrator. Born in London, eldest son of William Davenant, the playwright and Poet Laureate, he was educated at Cheam School, Surrey, and entered Balliol College, Oxford, in 1671, going down in 1673 without a degree to take over the management of his father's theatre. In 1675 he wrote a tragedy, *Circe* (Davenant, 1677), but the theatre gained him little financial success. He also obtained an LL.D from Cambridge in 1675 and practised law for a short period. From 1678 to 1689 he was Commissioner of Excise. He sat as MP for St Ives from 1685 to 1688 and represented Great Bedwin in the Tory interest following the elections of 1698 and 1700. The financial consequences of his loss of office as Excise Commissioner in 1689 and unsuccessful attempts in 1692 and 1694 to obtain other positions in the revenue service appear to have inspired a career as pamphleteer, starting in 1695. Until 1702, when he again obtained preferment by being appointed Secretary to the Commission for negotiating the union between England and Scotland, he produced a steady flow of political and economic writings dealing with aspects of taxation, public debt, monetary and trade questions, foreign policy and criticisms of Whig policy in general. In June 1703 he obtained the post of Inspector-General of Exports and Imports in the Customs Office, a position he retained till his death in 1714. Most of his political and commercial writings were collected by C.E. Whitworth (1771) but two manuscript works on money and credit (Davenant, 1695b and 1696) were not published till 1942 (Evans, 1942).

Davenant's position in the history of economics rests on a variety of contributions. Initially, his work was largely depicted as typically that of an 'adherent of the mercantile theory' (Hughes, 1894, p. 483), but 'Tory free trader' (Ashley, 1900, p. 269) better describes his pronouncements on foreign trade policy as he particularly advocated the removal of trade restrictions, such as those affecting woollen exports, which benefited the landed interest by raising land values (Davenant, 1695a, pp. 16–17; 1697, pp. 98–104). His free trade position is not unambiguous. Although Davenant's remark that 'Trade is by its nature free, finds its own channel, and best directeth its own course.' (1697, p. 98) is often quoted, the contradictory view that 'it is the prudence of a state to see that [its] industry, and stock, be not diverted from things profitable to the whole, and turned upon objects unprofitable, and perhaps dangerous to the public' (1697, p. 107) is less frequently noticed. Schumpeter's (1954, p. 196, n.4, and p. 242) depiction of Davenant's work as 'comprehensive quasi-system' emphasizing the interdependence of economic activity is also rather difficult to sustain, though it is possible to quote isolated remarks from Davenant's works in support. For example, Davenant's statement that 'all trades have a mutual dependence one upon the other, and one begets another, and the loss of one frequently loses half the rest' (1697, p. 97) cannot really be described as the general theoretical proposition it appears to be. Its only use is to provide a basis for some special pleading on behalf of the East India trade. Waddell's conclusion (1958, p. 288) that Davenant was a person neither of 'exceptional ability, nor of any great strength of character' and 'a competent publicist' rather than 'an original thinker' or 'practical man of affairs' seems a more appropriate assessment from an examination of his economic writings.

Davenant's plea for the importance of 'political arithmetic' or 'the art of reasoning by figures, upon things relating to government' (1698, p. 128) provides a further claim to fame, partly because it made more readily available the fairly sophisticated national income and expenditure estimates of his friend Gregory King (1696). Most of Davenant's political arithmetic application relates to taxation and estimating the gains from trade in terms of bullion, but he himself also made a useful contribution to the collection of international trade data as part of his duties as Inspector-General of Exports and Imports.

The precise details of Davenant's association with Gregory King are not fully known, but their names are also linked in another famous 'statistical' exercise, the so-called King–Davenant law of demand, first noted by Thornton (1802) and Lauderdale (1804), and later extensively discussed by Jevons (1871, pp. 154–8), who on the evidence available to him cautiously attributed to Davenant the data on which the law is based (but see Barnett, 1936, pp. 6–7). However, apart from providing these data, Davenant himself characteristically drew no such analytical conclusions from this information (1698, Part II, pp. 224–5; see Creedy, 1986, for a detailed discussion).

Davenant's contributions to the recoinage debates (1695b; 1696) are less well known because they were not included in Whitworth (1771). Full recoinage was not necessary in Davenant's view when the inferior (because clipped or worn) coins were still usefully employed in small retail transactions. In addition, the detrimental effects on the exchange rate and commodity prices of the deteriorating currency were greatly exaggerated. The rise in prices, Davenant argues, could be attributed to a great many other causes; the depreciated exchange rate was more easily explained by the substantial overseas remittances induced by the European war and was therefore better remedied by floating a public loan in Holland. Although in these essays, Davenant's exposition is not always complete, Evans (1942, p. vi) regards them as containing 'all the essential elements of the analysis of money and credit' and integrating 'the entire problem of currency and public finance'. Finally, Davenant's contributions to tax administration need to be recognized. They have been described as 'translating into principles, and trying to provide a reasonable justification for the practices that the more methodical and innovating officials (such as Pepys at the Navy Office and Admiralty, and Downing and Lowndes at the Treasury …) were adopting and enforcing' and that in these matters of administrative thinking, unlike his economics, 'Davenant's viewpoint steadily became [dominant] in the course of the next century or so' (Hume, 1974, p.477). His writings also remain a useful source for much information on trade and finance over the final decades of the Stuart monarchy.

PETER GROENEWEGEN

See also **King, Gregory.**

Selected works

1677. Circe, *A Tragedy. As it is acted at His Royal Highness the Duke of York's Theatre.* London: Richard Tonson.
1695a. *An Essay on Ways and Means of Supplying the War.* In Whitworth (1771, vol. 1).
1695b. *A Memorial Concerning the Coyn of England.* Reprinted in Evans (1942).
1696. *A Memoriall Concerning Creditt.* Reprinted in Evans (1942).
1697. *An Essay on the East-India Trade.* Reprinted in Whitworth (1771, vol. 1).
1698. *Discourses on the Public Revenues, and on the Trade of England in Two Parts.* Reprinted in Whitworth (1771, Part I in vol. 1; Part II in vol. 2).

Bibliography

Ashley, W.J. 1900. The Tory origin of free trade policy. In *Surveys Historic and Economic,* ed. W.J. Ashley. London: Longmans.
Barnett, G.E. 1936. *Two Tracts by Gregory King.* Baltimore: Johns Hopkins Press.
Creedy, J. 1986. On the King–Davenant 'law' of demand. *Scottish Journal of Political Economy* 33(3), 193–212.
Evans, G.H. 1942. *Two Manuscripts by Charles Davenant.* Baltimore: Johns Hopkins Reprints of Economic Tracts.
Hughes, D. 1894. Charles D'Avenant (1686–1714). In *Dictionary of Political Economy,* vol. 1, ed. R.H.I. Palgrave. London: Macmillan.
Hume, L.J. 1974. Charles Davenant on financial administration. *History of Political Economy* 6, 463–77.
Jevons, W.S. 1871. *Theory of Political Economy.* 4th edn, London: Macmillan, 1911.
King, G. 1696. *Natural and Political Observations and Conclusions upon the State and Condition of England.* Reprinted in Barnett (1936).
Lauderdale, J.M., Eighth Earl of. 1804. *An Inquiry into the Nature and Origin or Public Wealth.* Edinburgh. Repr. with an introduction and revisions from the 2nd edn, New York: Kelley, 1966.
Schumpeter, J.A. 1954. *History of Economic Analysis.* London: Allen & Unwin, 1959.
Thornton, H. 1802. *An Inquiry into the Nature and Effects of the Paper Credit of Great Britain.* London.
Waddell, D. 1958. Charles Davenant (1656–1714), a biographical sketch. *Economic History Review* 11, 279–88.
Whitworth, Sir C.E. 1771. *The Political and Commercial Works of that Celebrated Writer Charles D'Avenant LL.D.* London.

Davenport, Herbert Joseph (1861–1931)

Davenport was born on 10 August 1861, in Wilmington, Vermont, and died on 16 June 1931, in New York City. He commenced a professorial career at the age of 41 after having been a land speculator (initially successful, but wiped out in the Panic of 1893) and high school teacher and principal. His academic work was at the University of South Dakota, Harvard Law School, Leipzig, Paris and Chicago (Ph.D., 1898). He taught at Chicago (1902–8), Missouri (1908–16) and Cornell (1916–29). He was President of the American Economic Association in 1920.

A leading, albeit somewhat iconoclastic, economic theorist of his day, he contributed to the reformulation of microeconomics from absolutist value theory to relativistic price theory. He stressed that, while there were real forces at work in the economy, identifying them as human desires and productive capacities, price itself reflected nothing more fundamental than a temporary equation of demand and supply. Prices are not determined *by* the margins but *at* the margins. Recognizing the limits imposed by a resultant superficiality and simultaneity of determination, he felt that economists qua economists need not inquire into the formation of desires or institutions but should study the pecuniary logic of phenomena from the standpoint of price in a society dominated by the private and acquisitive point of view. His economics focused on entrepreneurial opportunity-cost adjustments and encompassed a non-normative distribution theory based directly on price theory.

While differing from his close friend Thorstein Veblen on certain substantive issues, Davenport's work nonetheless reflected the impact of Veblen's critiques of traditional theory and of the actual market economy. Emphasizing positive economics and rejecting apologetics (economic theory was not to be the monopoly of reactionaries), Davenport was willing to recognize that the search for private gain did not always conduce to social welfare, but this conclusion was not to be considered a part of economic science per se.

WARREN J. SAMUELS

Selected works

1896. *Outlines of Economic Theory*. New York: Macmillan.
1897. *Outlines of Elementary Economics*. New York: Macmillan.
1908. *Value and Distribution*. Chicago: University of Chicago Press.
1913. *Economics of Enterprise*. New York: Macmillan.
1935. *The Economics of Alfred Marshall*. Ithaca: Cornell University Press.

Davidson, David (1854–1942)

Born into a Jewish merchant family in Stockholm, Davidson studied law and economics at Uppsala University from 1871, became a docent in 1878, professor extraordinarius from 1880 to 1889, and then professor ordinarius for 30 years until he retired in 1919. Frequently called on to serve on parliamentary committees from 1891 to 1931, Davidson's influence was strongly felt on Sweden's monetary and tax policies, for instance the 'gold exclusion policy' of 1916–24.

In 1899 Davidson launched Sweden's first economic journal, *Ekonomisk Tidskrift*, to which he contributed almost all his work over 40 years as its owner and editor (in 1965 it was renamed *The Swedish Journal of Economics* and issued in English). This journal greatly stimulated economic research in Sweden with numerous contributions from, among others, Wicksell, Cassel, Lindahl, Myrdal and Ohlin.

Unlike Wicksell and Cassel, who published their works in German (later translated into English), all of Davidson's writings are in Swedish, none of them translated. This, and the fact that his work – five tracts 1878–89, over 200 articles in his journal on a variety of subjects, plus chapters in several government reports – was never systematized in treatise form, accounts for his contributions to economics having been known, until recently, only to Scandinavian academics.

In his dissertation, *Bidrag till läran om de ekonomiska lagarna för kapitalbildningen* (A Contribution to the Theory of Capital Formation), Davidson anticipated Böhm-Bawerk's *Positive Theory of Capital* (1884). To Davidson, capital was generated in the main by the unequal distribution of income. To the wealthy, increases in present goods have small and declining utility relative to that of future goods. The latter are obtained in greater quantity, variety and value by investing savings for a return – interest – in production of capital goods which, indirectly, increase productivity. This perspective inverts the first of Böhm-Bawerk's famous 'three grounds' for interest, and transforms the third to a marginal productivity theory of waiting. In his later work Davidson adopted the substance of Wicksell's amendments and reconstruction of Böhm-Bawerk's capital theory.

Davidson's monetary theory is best understood from his response in articles of 1908–25 to his friend Wicksell's path-breaking work in this area. *Inter alia*, Davidson criticized Wicksell's monetary norm of price level stability as inappropriate in conditions of 'commodity shortage'. Eventually, by 1925 Wicksell was moved to amend his norm to accommodate Davidson's critique (Uhr, 1960, chs 10 and 11).

In his early tract *Om beskattningsnormen vid inkomstskatten* (A Taxation Norm for the Income Tax, 1889), Davidson urged the replacement of Sweden's several property taxes and most of its excises by a progressive income tax with a uniquely broad base. It base was to include 'the citizen's potential consumption power' by levying the tax (*a*) on any increment in his net worth accrued (*whether realized or not*) between the end and the beginning of the tax year; and (*b*) also on his actual consumption spending during the year. Net worth increments accrue to a person as the value of his assets increases over that of his liabilities, due to savings, capital gains, bequests, and so on. Such gains confer potential consumption power, which should be taxed along with actual consumption spending out of income.

Over the years, aware of difficulties his proposed tax base would encounter as it called for annual balance sheet and income–consumption statements, Davidson conceded some simplifications on the tax declarations, and to taxing capital gains only when realized by the sale of value-appreciated assets. He also agreed that the tax rates levied on net worth increments would have to be lower than the rates levied on consumption expenditures.

These concessions notwithstanding, Sweden's parliament in its first comprehensive income tax of 1910 adopted only one part of Davidson's proposal. It passed a progressive tax on income as usually defined (rather than on consumption spending as such), and added to it a second title, a tax on net worth increments at rates substantially lower than on income. Largely due to Davidson, this combination of an income and a net worth increments tax has remained a standard feature in Sweden's tax system since 1910.

CARL G. UHR

Selected works

1878. *Bidrag till läran om de ekonomiska lagarna för kapitalbildningen* [A Contribution to the Theory of Capital Formation]. Uppsala.

1889. *Om beskattningsnormen vid inkomstskatten*
[A Taxation Norm for the Income Tax]. Uppsala.

Bibliography

Böhm-Bawerk, E. von. 1884. *The Positive Theory of Capital.*
Trans. W. Smart, New York: G.E. Stechert & Co., 1930.

Heckscher, E.F. 1952. David Davidson. *International Economic Papers* 2, 111–35.

Uhr, C.G. 1960. *Economic Doctrines of Knut Wicksell.*
Berkeley: University of California Press.

Uhr, C.G. 1975. *Economic Doctrines of David Davidson.*
Uppsala: Studia Oeconomica Upsaliensis.

Debreu, Gerard (1921–2004)

Life

Gerard Debreu, the son of a Calais lace manufacturer, was born on 4 July 1921. He took his baccalauréat in 1939, just before the outbreak of the Second World War. Instead of entering university, he then began an improvised mathematics curriculum in Ambert and, later, in Grenoble. In 1941 he was admitted to the École normale supérieure, where he studied with Henri Cartan and the Bourbaki group. After D-Day he enlisted in the French Army, and served in Algeria and Germany. Returning to his studies, he completed the agrégation de mathématiques in early 1946. While pursuing his mathematical studies in Paris, he was captivated by Maurice Allais's (1943) exposition of the Walrasian general equilibrium analysis, which became the central pillar of his research programme. It was the flip of a coin which determined that he, rather than Edmond Malinvaud, would receive a travelling fellowship from the Rockefeller Foundation. This funded a year at Harvard, Berkeley and the Cowles Commission at Chicago, followed by studies at Uppsala and, with Ragnar Frisch, in Oslo. Debreu returned to Chicago and the Cowles Commission, and moved with it to Yale in 1955 with his wife of ten years and his nine- and five-year-old daughters. A year at the Center for Advanced Study in the Behavioral Sciences at Stanford gave the Debreu family a taste for California, and in 1962 Debreu accepted a position at the University of California at Berkeley. There he remained until his retirement. Debreu became a US citizen in 1975, having been deeply moved by America's response to the Watergate affair.

Gerard Debreu received numerous honours and awards. He was a Fellow of the American Academy of Arts and Sciences (1970), vice president and president of the Econometric Society (1970, 1971), a Chevalier de la Légion d'honneur (1976), a member of the National Academy of Sciences (1977), a Distinguished Fellow of the American Economic Association (1982) and its president in 1990, a Foreign Associate of the French Académie des sciences (1984) and a Fellow of the American Association

for the Advancement of Science (1984). He was awarded honorary degrees from, among many, the University of Bonn, Université de Lausanne, Northwestern University, Université des sciences sociales de Toulouse, and Yale University. Most prominent of all, in 1983 he was the recipient of the Bank of Sweden Prize in Economic Sciences in Memory of Alfred Nobel.

The elegance of Gerard Debreu's work was reflected in his personal style. He was also a competitive bridge player, and perhaps his first publication was a monograph on the game. In contrast to his revealed preference for the spare prose and clean, elegant arguments of the *Theory of Value* (1959) was his love of *A La Recherche du Temps Perdu*. 'My appreciation of Proust', he said in a 1983 *New York Times* interview, 'is in his style, subtlety and taste. I prize conciseness very much, and that is certainly something that you cannot accuse Proust of. His compulsion, as you know, eventually killed him. I'll try to escape that fate.' Debreu was reserved in person, but displayed a quick and subtle wit. I remember his beginning a lecture on the computation of economic equilibrium with the observation that the existence of equilibrium had been established and that now Herbert Scarf has taught us how to compute the zeros of the excess demand function. It only remains, he said, for the econometricians to estimate it, and we would be done. Gerard Debreu died in Paris on New Year's Eve 2004. His ashes were placed in a niche in the Père Lachaise cemetery, the final resting place of many of Frances's most eminent artists and intellectuals, including Marcel Proust.

Work

The influence of Gerard Debreu's work can be seen throughout contemporary economics, but his research output was largely confined to general equilibrium theory and its requirements.

The existence of competitive equilibrium

Gerard Debreu's broad fame in the economics community is due to his work on the existence of competitive equilibrium. The complexity of simultaneous price and quantity determination in multiple markets of related and unrelated goods stands in stark contrast to the cutting power of the simple Marshallian scissors of supply and demand in a market with a single good. It is certainly not obvious that a multi-market equilibrium should exist. The existence problem, open since the publication of Léon Walras's *Eléments d'économie politique pure* (1874), was first given a broad and general treatment by Arrow and Debreu (1954a). As Arrow tells the story, in earlier work on the problem, he and Debreu had each made a mistake for which the other had a solution. It was suggested that they collude, and the outcome was displayed at the remarkable 1952 Winter Meeting of the Econometric Society in Chicago where both the Arrow and Debreu's paper (1954a) and McKenzie's (1954) paper were presented. The Arrow and Debreu 'private ownership economy' is today the standard reference for a general competitive model. McKenzie's treatment of technology is somewhat more special, although the two models are not directly comparable. The method of proof is to introduce a fictitious agent, a Walrasian auctioneer, whose role is to choose prices. Then the entire problem sets up like a non-cooperative game, with the added wrinkle that feasible strategies for one player may depend upon the choices of the others. Fortunately, Debreu (1952) had already established the existence of a kind of Nash equilibrium for these games, which he called a 'social equilibrium'. This approach to the existence of equilibrium is quite different from the approach through the excess demand correspondence, which was already developed in 1954 and appears in Debreu's (1959) essential masterwork, the *Theory of Value*. The social equilibrium approach is particularly well-suited to economies in which it is difficult to get one's hands on excess demand directly, such as economies with externalities, public sector decision-making, non-convexities, and incomplete and intransitive preferences.

Welfare economics

The central question of economic analysis, the workings of the invisible hand, is formulated today as the achievement (or not) of an optimal allocation of resources. The characterization of optimality by means of marginal rates of substitution was first completed by Oscar Lange (1942). This characterization, however, is unsatisfactory for several reasons, including the facts that marginal rates of substitution may fail to exist for otherwise unremarkable preference orders, the treatment of corners is complicated, and the corresponding second-order conditions are sufficient only for local optimality. At about the same time on two different American coasts, Kenneth Arrow (1952) and Gerard Debreu (1951) proposed an alternative analysis of the relationship between equilibrium and optimality, making use of convexity assumptions and, in particular, the separating hyperplane theorem instead of the calculus. Debreu (1954b) extended his geometric analysis from finite dimensional vector spaces to linear topological vector spaces, that is, from finite to an infinite number of commodities. This advance is important for such diverse topics as financial markets, uncertainty, dynamic modelling and commodity differentiation. The first half of Debreu (1951) establishes the classical welfare theorems, relying only on convexity and topological assumptions on preferences. The second half of the paper introduces the coefficient of resource utilization, a measure of deadweight loss. Debreu (1954a) applied this measure to the deadweight loss associated with tax-subsidy schemes, a measure that has been implemented empirically by Farrell (1957) and Whalley (1976) to study productive efficiency and the deadweight loss of alternative tax schemes. A comparison of the Debreu coefficient with other measures of deadweight loss, including that of his contemporary M. Boiteux at the École normale supérieure, can be found in Diewert (1981).

The theory of value

Debreu's *Theory of Value* (1959) is not simply about the existence and optimality of equilibrium. It is a statement of method that has profoundly changed the way economics is practised. For this alone it is among the most original books of 20th-century economic thought. Most economists identify Debreu with mathematics, manipulating formulas and proving theorems. But for Debreu this, although pleasurable, was the easy part of economic theory. He once told me that it was harder to be an economist than a mathematician. A mathematician had to be correct and elegant; but an economist had to be all that and also interesting. The power of a model lies in the economist's ability to interpret with it, and this is the point of all the 'elegance' and clarity in Debreu's exposition. In the preface, he writes (1959, p. x), 'Allegiance to rigor dictates the axiomatic form of the analysis where the theory, in the strict sense, is logically entirely disconnected from its interpretations. ... Such a dichotomy reveals all the assumptions and the logical structure of the analysis.' Debreu taught that the separation of logical analysis from interpretation is crucial to good theory. The logic of market equilibrium is independent of what commodities actually are, except in so far as what they are may suggest additional structure on the primitives of the equilibrium model. This is most clearly demonstrated in Chapter 7. Here Debreu reinterprets the model by appending to the description of commodities the state of nature in which it is available. The use of Arrow's (1953) contingent commodities 'allows one to obtain a theory of uncertainty free from any probability concept and formally identical with the theory of certainty developed in the preceding chapters' (1959, p. 98). Three pages later, Debreu observes that the convexity assumptions required by the theoretical analysis could be understood as risk aversion. And although Debreu stops here, it is not a big step to observe that natural preference models, like Savage's subjective expected utility model, lead to an additive structure for preferences that may have implications for the nature of equilibrium.

Large economies and the core

Competitive equilibrium requires prices, and prices in turn already require a sophisticated set of market institutions. Nonetheless, 'general' is a key word in the phrase general competitive equilibrium. The principle behind the abstract treatment of market equilibrium is that the workings of supply and demand are more or less the same whether the market under discussion is a modern financial market in London or New York or a village market of farmers and petty traders in India or East Africa. This is quite a claim. Support for this idea comes from the fact that the Walrasian outcome from markets with quoted prices can also be supported by a seemingly more fundamental equilibrium concept that makes no mention of prices at all: the core.

The core comes from F. Y. Edgeworth's *Mathematical Psychics* (1881), in which the contract curve is first introduced, and which, remarkably, undertakes a limit analysis of the economy with two types of traders and two goods. Edgeworth showed that the set of core allocations shrinks to the set of competitive equilibria as the number of agents becomes large. Debreu and Scarf (1963) pick up this question and quickly dispatch it for replica economies, which are generalizations of the large population structures Edgeworth studied. Immediately thereafter came Aumann's (1964) equivalence theorem for the core and equilibrium set of an economy with a continuum of agents, which, among other things, launched the subject of economies described by a measure space of agents. These developments are important because perfect competition is most naturally expressed as a large economy (large number of agents) phenomenon, and because empirical descriptions of large markets may be best described by distributions on the space of agent characteristics.

Smooth economies

It is often said that Gerard Debreu took the calculus out of economics with his topological equilibrium analysis of the 1950s and early 1960s. If so, it returned with a vengeance in his 1970 and 1972 papers on economies with differentiable excess demand. It has been clear since the Edgeworth box that economies with multiple equilibria are inescapable, a fundamental indeterminacy of the analysis. One can easily construct exchange economies with a continuum of equilibria. But how far does it extend? Is this the norm or are these economies pathological? In a path-breaking series of papers Debreu drew the line between normal and bizarre. He demonstrated that if individual demand is differentiable, then the 'generic' case is one in which there are only a finite number of isolated equilibria; that is, equilibria are locally unique. 'Economies with a Finite Set of Equilibria', his 1970 paper, is particularly striking in its simplicity. Once it is determined that an economy is regular, the main result follows from the inverse function theorem – surely a result known to anyone who has taken a multivariate calculus course. Only the deeper fact that regularity is generic requires more advanced tools such as Sard's theorem. Again, Debreu's intuition was geometric. In lectures this was explained with a simple diagram. Subsequent work has used the tools of differential topology to uncover the deeper structure of the equilibrium manifold, the graph of the equilibrium correspondence. These tools are also of fundamental importance for economies with incomplete markets. With incomplete markets and financial assets rather than real assets, indeterminacy is no longer unusual, and this is of critical importance for applications to macroeconomics and finance. Some of this work is surveyed in the monographs of Balasko (1988) and Mas-Colell (1985).

Excess demand

It is important to ask of any theory, 'what can it say?' That is, what kinds of predictions will the theory make, and what patterns in data will contradict the theory? In general equilibrium theory this question was first asked by Sonnenschein (1972) in the following way: in exchange economies, the market excess demand function satisfies the restrictions of continuity, homogeneity and Walras's Law. This and a boundary condition is enough to prove the existence of equilibrium prices. Sonnenschein asked if excess demand functions had any additional structure beyond these three requirements. Sonnenschein (1972), Mantel (1974) and Debreu (1974), with an important extension by (Mas-Colell, 1977), showed that the answer is 'no'. Any function defined for strictly positive prices and satisfying these three conditions is identical up to boundary behaviour with an excess demand function for an exchange economy containing no more agents than goods, each agent with continuous, strictly convex and monotonic preferences. Thus the hypothesis of utility maximization in exchange economies, with no additional assumptions about agents' characteristics, will place few restrictions on comparative static results or on the nature of the equilibrium price set.

These results are often incorrectly interpreted to mean that general equilibrium theory is empty, that it predicts nothing. This is entirely incorrect. General equilibrium theory is not so much a theory as a theoretical framework within which theories can be built by making explicit assumptions about the nature of tastes, technologies and endowments. To say that the framework does not limit market behaviour without any assumptions about its primitive objects is to say that the framework is maximally expressive. Its power to predict market behaviour comes from assumptions about the population of agents participating in the market. The so-called 'anything goes' theorems simply imply that more results will require more assumptions about the preferences and endowments of agents. It had been Debreu's hope that restrictions on the distributions of agents' characteristics would lead to interesting conclusions, but progress has been slow.

Other contributions

Debreu has produced seminal papers in areas of economic theory other than general equilibrium analysis. Which preference orders have a continuous utility representation? This question is answered by (1954c). Which preferences have additive separable representations? Debreu's (1958) answer to this very difficult question is topological in nature, and quite distinct from the algebraic answers found in the mathematical psychology literature.

Debreu was exceptional in the classroom and in seminar. His lectures were crystalline, elegantly shaped, and parsimonious. Often they were too clear; we students left the class convinced we understood, only to discover on problem sets how subtle were the arguments that had seemed so obvious on the blackboard. Debreu's expository writings, especially his Nobel Address (1984), are required for everyone with a serious interest in contemporary economics.

Conclusion

It is impossible to imagine modern economics without the scholarship of Gerard Debreu. Debreu, Kenneth Arrow and a few others who solved the big open questions of general equilibrium theory in the 1950s had an impact that reached far beyond the confines of formal competitive analysis. They were responsible for making formal modelling a requirement for serious economic analysis of any kind. Formal modelling is not merely a theoretical discourse; the availability of formal models requires a means for the models to confront data. Modern econometrics is inconceivable without the idea of formal modelling as a strategy of enquiry. It is not by accident that, just as the general equilibrium theory was taking off at the Cowles Commission in the 1950s, so too was modern econometrics. The contributions of the 'mathematical economists' launched a revolution that has touched on every area of economic practice.

LAWRENCE E. BLUME

See also **core convergence; cores; general equilibrium; welfare economics.**

Selected works

1951. The coefficient of resource utilization. *Econometrica* 19, 273–92.
1952. A social equilibrium existence theorem. *Proceedings of the National Academy of Sciences* 38, 597–607.
1954a. (With K.J. Arrow.) Existence of an equilibrium for a competitive economy. *Econometrica* 22, 265–90.
1954b. Valuation equilibrium and Pareto optimum. *Proceedings of the National Academy of Sciences* 40, 584–92.
1954c. A classical tax-subsidy problem. *Econometrica* 22, 14–22.
1954d. Representation of a preference ordering by a numerical function. In *Decision Processes*, ed. R.M. Thrall, C.H. Coombs and R.L. Davis. New York: Wiley.
1958. Stochastic choice and cardinal utility. *Econometrica* 26, 440–4.
1959. *Theory of Value*. New York: Wiley. Repr. New Haven: Yale University Press, 1971.
1963. (With and H. Scarf.) A limit theorem on the core of an economy. *International Economic Review* 4, 235–46.
1970. Economies with a finite set of equilibria. *Econometrica* 38, 387–92.
1972. Smooth preferences. *Econometrica* 40, 603–15.
1974. Excess demand functions. *Journal of Mathematical Economics* 1, 15–21.
1984. Economic theory in the mathematical mode. *American Economic Review* 74, 267–78.

Bibliography

Arrow, K.J. 1952. An extension of the basic theorems of classical welfare economics. In *Proceedings of the Second Berkeley Symposium on Mathematical Statistics and Probability*, ed. J. Neyman. Berkeley: University of California Press.
Arrow, K.J. 1953. Le rôle des valeurs boursières pour la répartition la meilleure des risques. *Econométrie* 40, 41–8, Cahiers du CNRS.
Aumann, R.J. 1964. Markets with a continuum of traders. *Econometrica* 32, 39–50.
Balasko, Y. 1988. *Foundations of the Theory of General Equilibrium*. Boston: Academic.
Diewert, W.E. 1981. The measurement of deadweight loss revisited. *Econometrica* 49, 1225–44.
Edgeworth, F.Y. 1881. *Mathematical Psychics: An Essay on the Application of Mathematics to the Moral Sciences*. London: C. Kegan Paul.
Farrell, M.J. 1957. The measurement of productive efficiency. *Journal of the Royal Statistical Society* 120, 253–91.
Lange, O. 1942. The foundations of welfare economics. *Econometrica* 10, 215–28.
Mantel, R. 1974. On the characterization of aggregate excess demand. *Journal of Economic Theory* 7, 348–53.
Mas-Colell, A. 1977. On the equilibrium price set of an exchange economy. *Journal of Mathematical Economics* 4, 117–26.
Mas-Colell, A. 1985. *The Theory of General Economic Equilibrium: A Differentiable Approach*. Cambridge: Cambridge University Press.
McKenzie, L. 1954. On equilibrium in Graham's model of world trade and other competitive systems. *Econometrica* 22, 147–61.
Sonnenschein, H. 1972. Market excess demand functions. *Econometrica* 40, 549–63.
Walras, M.E.L. 1974. *Éléments d'économie politique pure, ou théorie de la richesse sociale*. Lausanne: Corbaz.
Whalley, J. 1976. Some general equilibrium analysis applied to fiscal harmonization in the European community. *European Economic Review* 8, 290–312.

decentralization

The main question to be answered by the theory of resource allocation, or by the theory of economic organization, concerns the performances of alternative systems characterized by different degrees of centralization of decision taking. A fully centralized system runs the risk of being inefficient because it does not create proper economic incentives and the centre is poorly informed. A pure market system with its high degree of decentralization runs the risk of bringing inequitable results and being inefficient because markets can never be complete, externalities exist and public wants tend to be neglected. Can these risks be avoided within the two opposite extremes of pure centralization or full decentralization? Can intermediate systems better resolve the difficulties? And if so, how?

Basic to the discussion are two features: the nature of the *information* held by various agents, and the *incentives* that should lead them to behave in conformity with collective requirements. These features and the issue of decentralization do not only appear for full economic systems, which this entry will consider, but also for the internal organization of firms or communities. They are stylized in the principal–agent problem: which rules should determine how to share the proceeds of an activity between the principal owner and his better-informed agent? (Ross, 1973; Grossman and Hart, 1983).

For the clarification of the complex issues involved, theory starts from a model of the conditions of economic activity. It makes assumptions such that, independently of economic organization, there exists a best outcome, or at least a set of 'optimal' outcomes. It then asks how well alternative forms of organization succeed in finding, implementing or at least approaching this best outcome or set of optimal outcomes.

By so doing, the theory discussed here neglects two related questions: how to determine what should be considered as 'the best' outcome in a society with many individuals, and which non-economic considerations interfere with the issue of decentralization? The theory of social choice shows the fundamental difficulty of the first question (Arrow, 1951), which is avoided when optimality is identified with Pareto efficiency. As for the second, philosophers may find in human nature or in the aims pursued by human societies reasons that favour some organization, beyond its economic performance; in particular, the right of individuals to autonomy appears fundamental in Western culture and is an important justification of decentralization, and even of the market system for such economists as Hayek (1944).

Formal concepts and preliminaries

The following conceptual apparatus, although not yet common, is well suited to the purpose (see Hurwicz, 1960; Mount and Reiter, 1974).

An *economic environment* is defined by a set of commodities and their possible uses, by a list of agents and their characteristics (technology, endowments, preferences, and so on), and by an initial information structure (what each agent knows). The feasible set of economic environments defines 'the economy'.

An important property of an economy is its higher or lower degree of *decomposability*, which concerns agents' characteristics and the information structure. The highest decomposability is assumed in competitive equilibrium theory, where all consumption is private, no external effect exists and a *private information structure* prevails (each agent perfectly knows its own characteristics and the situation on all markets, but nothing else). But models with public goods, for instance, usually admit some decomposability, which matters for the validity of the results.

An *optimality correspondence* $P: E \to A$ defines which vectors of actions simultaneously taken by the various agents are optimal when the economic environment is e, i.e. optimal vectors belong to $P(e)$ (clearly, E is the set of feasible e, that is 'the economy', while A is the set of feasible vectors a, each one of them defining the actions taken by all the agents). For instance $P(e)$ may be the set of Pareto efficient vectors. But in the theory discussed here, it is often more narrowly defined so as to take equity considerations into account: a social utility function may have to be maximized or a rule on the consumers 'income distribution' satisfied.

A *resource allocation mechanism* $f: E \to A$ should select one $a = f(e)$ for each environment e (in some cases f may be multivalued, i.e. become a correspondence). The best formalized mechanism is the competitive equilibrium of a 'private ownership economy'. A study of decentralization requires a careful specification of the mechanism, which is typically viewed as operating in two stages: first, an iterative exchange of messages, usually between the agents and a centre, resulting in a message correspondence $g : E \to M$ (the message $m = g(e)$ specifies what information about e has been collected at the centre), second an outcome function $h : M \to A$. For instance, the competitive mechanism is often specified as resulting from the tâtonnement process, in which an auctioneer learns which demands and supplies are announced at various proposed vectors of prices, and searches for the equilibrium prices; once these prices are found, the outcome function gives the equilibrium exchanges, hence productions and consumptions.

The performances of alternative mechanisms of course concern the final result: one must know whether the outcome $f(e)$ belongs to the optimal set $P(e)$ for all environments in E, or at least for a precise subset of E, and how close it is to $P(e)$ otherwise. But interesting performances also concern intermediate features of the mechanism, which usually is iterative. At step t the previously collected message m_{t-1} is enriched according to $m_t = g_t(m_{t-1}, e)$ and, if necessary, the process could end by $a = h_t(m_t)$. In a *finite* procedure it does end at T with $m = m_T$ and $h(m) = h_T(m_T)$; but most mechanisms assume an infinite sequence of m_t for $t = 1, 2 \ldots$ ad infinitum. One must then know whether and how $h_t(m_t)$ approaches $P(e)$, monotonically or otherwise. Since the transmission of information is costly, the nature and size of the message space M_t to which m_t belongs are also important characteristics (Mount and Reiter, 1974).

The planning problem

Early in this century many economists objected to socialist planning programmes that could not be implemented, because they unrealistically assumed that a central administration could have the knowledge and computing power required for an efficient control of economic activity. The leading figure was L. von Mises (1920 in particular); but Hayek (1935) was first to emphasize the problems raised by the decentralization of information. Socialist economists answered that decentralized mechanisms could operate, either mimicking the market system while being free of its deficiencies (Lange, 1936) or using different well conceived modes of information gathering (Taylor, 1929). The debate was, in the interwar years, the subject of the '*economic theory of socialism*'. (For a well-documented survey, see Bergson, 1948.)

The problem was again taken up during the 1960s, in particular because the logic of efficient planning was discussed in Eastern and Western Europe (Arrow and Hurwicz, 1960; Kornai, 1967; Malinvaud, 1967; Heal, 1973). Many planning procedures were rigorously studied as resource allocation mechanisms. Their definition implied an iterative exchange of information between a Central Planning Board and firms, sometimes also representative consumers. The additional messages provided by the function g_t at step t then consisted of *prospective indices* announced by the Board, for instance prices for the various commodities, and replies called *proposals* sent to the Board by firms and other agents, for instance preferred techniques of production and their input requirements, or supplies and demands.

In this discussion it is common to distinguish between price-guided procedures, in which the Board announces price vectors, and other procedures, in which quantity indices or targets worked out at the centre play a more or less important role. The nature and properties of the environment are then found to be crucial for the determination of the relative performances of alternative procedures, in particular of price-guided against quantity-guided procedures (Weitzman, 1974).

The analytical study of various procedures usually assumes that decentralized agents exactly follow specified rules for the determination of their proposals and so faithfully reveal part of their private information. Some procedures are then found to be efficient and to permit achievement of distributive objectives. But efficiency is typically easier precisely in those environments that are also favourable to the efficiency of free competition. Besides the possibility of incorrect reporting, the main difficulty concerning the relevance of this literature is to know whether its models provide an approximate representation of procedures that are actually used, or at least administratively feasible. Manove (1976) has made this claim for his representation of Soviet planning.

The public good problem

The most relevant field of application may very well be the theory of public goods. Decisions concerning the provision of public services and their financing cannot be fully decentralized; but the knowledge required is dispersed and must be gathered in a proper way. Hence even the positive theory of public goods was often formulated along lines that look like those of planning procedures (Malinvaud, 1971). The same remark applies to decisions concerning public projects with large fixed costs, even if their output is privately consumed.

Considered as a planning procedure, the search for the best decision is often viewed as involving 'prospective indices' that define amounts of service to be provided, ask for corresponding individual marginal utilities and look whether the sum of the latter would cover the cost of additional service. This is compatible with the dual arrangement for private goods, prices being announced, supplies and demands being the replies. The procedure is then quantity-guided for public goods and price-guided for private goods (Drèze and Vallée Poussin, 1971).

The collective consumption of many types of public goods is not really national but limited to local communities (primary education, city transports, and so on). Administrative science sees the decentralization issue as being to know at which level should decisions be taken: at the national level, so as to distribute fairly these services among communities, or at the local level, so as to permit better adaptation to local needs and wishes. Economists do not seem to have contributed to this issue; their discussion of local public goods assumes full administrative decentralization (Tiebout, 1956).

Incentive compatibility

The study of a decentralized system has to consider whether the actual reports and behaviour of individual agents do not deviate from what they are supposed to report and do; in case of deviations, how are the performances of the system affected? The problem is serious: once the rules of organization and decisions are known, individual agents may benefit from misreporting their private information or from behaving in a way that, although deviant, does not clearly appear to be so. In other words, they may act as players in a game, rather than as members of a team, and this may be more or less detrimental for the optimality of the final result.

The problem has long been known for organizations in which some agents do not individually benefit from what is achieved and therefore lack the incentive to do their best. Monopolistic or other non-competitive behaviour is often interpreted as a breach of the normal rules of resource allocation. In the theory of public good the 'free rider problem' occurs as soon as some individuals, having a high marginal utility for the public good, would benefit from hiding this fact so as to contribute little to the financing of the good.

Study of the problem has been active during the past two decades (Green and Laffont, 1979). The fundamental difficulty has been exhibited by such results as the following one: in the classical model of an exchange economy with a finite number of consumers, no procedure can be found that would necessarily lead to a Pareto efficient result in which individuals, acting as players in a noncooperative game, would faithfully report (Hurwicz, 1972). However, misreporting may not prevent a procedure from eventually leading to an optimum, as was proved in a number of cases.

Experiments moreover show that the game-theoretic approach to the incentive problem may be misleading because it neglects non-economic motivations that individuals may find for accepting a team-like behaviour and therefore for faithfully reporting (Smith, 1980).

E. MALINVAUD

Bibliography

Arrow, K. 1951. *Social Choice and Individual Values*. New York: Wiley.

Arrow, K. and Hurwicz, L. 1960. Decentralization and computation in resource allocation. In *Essays in Economics and Econometrics in Honour of Harold Hotelling*, ed. R. Pfouts. Chapel Hill: University of North Carolina Press.

Bergson, A. 1948. Socialist economics. In *A Survey of Contemporary Economics*, ed. H. Ellis. Philadelphia: Blakiston.

Drèze, J. and Vallée Poussin, D. de la. 1971. A tâtonnement process for public goods. *Review of Economic Studies* 38, 133–50.

Green, J. and Laffont, J.-J. 1979. *Incentives in Public Decision-Making*. Amsterdam: North-Holland.

Grossman, S. and Hart, O. 1983. An analysis of the principal-agent problem. *Econometrica* 51(1), 7–45.

Hayek, F. 1935. Socialist calculation: the state of the debate. In *Collectivist Economic Planning*, ed. F. Hayek. London: G. Routledge & Sons.

Hayek, F. 1944. *The Road to Serfdom*. Chicago: University of Chicago Press.

Heal, G. 1973. *The Theory of Economic Planning*. Amsterdam: North-Holland.

Hurwicz, L. 1960. Optimality and information efficiency in resource allocation processes. In *Mathematical Methods in the Social Sciences*, ed. K.J. Arrow, S. Karlin and P. Suppes. Stanford: Stanford University Press.

Hurwicz, L. 1972. On informationally decentralized systems. In *Decision and Organization*, ed. R. Radner and C. McGuire. Amsterdam: North-Holland.

Kornai, J. 1967. *Mathematical Planning of Structural Decisions*. Amsterdam: North-Holland.

Lange, O. 1936. On the economic theory of socialism. *Review of Economic Studies* 4, 53–71, 123–42.

Malinvaud, E. 1967. Decentralized procedures for planning. In Malinvaud, E. and Bacharach, M., *Activity Analysis in the Theory of Growth and Planning*, Macmillan: London.

Malinvaud, E. 1971. A planning approach to the public good problem. *Swedish Journal of Economics* 11, 96–112.

Manove, M. 1976. Soviet pricing, profit and technological choice. *Review of Economic Studies* 43, 413–21.

Mises, L. von. 1920. Economic calculation in the socialist commonwealth. First published in German in *Archiv für Sozialwissenshaft*, April; English translation in *Collectivist Economic Planning*, ed. F. Hayek, London: G. Routledge & Sons, 1935.

Mount, K. and Reiter, S. 1974. The informational size of message spaces. *Journal of Economic Theory* 8(2), 161–92.

Ross, S. 1973. The economic theory of agency: the principal's problem. *American Economic Review* 63(2), 134–9.

Smith, V. 1980. Experiments with a decentralized mechanism for public good decisions. *American Economic Review* 70, 584–99.

Taylor, F.M. 1929. The guidance of production in a socialist state. *American Economic Review* 19, 1–8.

Tiebout, C.M. 1956. A pure theory of local expenditures. *Journal of Political Economy* 64, 416–24.

Weitzman, M. 1974. Prices versus quantities. *Review of Economic Studies* 41, 477–91.

decision theory. *See* **rationality; statistical decision theory.**

decision theory in econometrics

The decision-theoretic approach to statistics and econometrics explicitly specifies a set of models under consideration, a set of actions available to the analyst, and a loss function (or, equivalently, a utility function) that quantifies the value to the decision-maker of applying a particular action when a particular model holds. Decision rules, or procedures, map data into actions, and can be evaluated on the basis of their expected loss.

Abraham Wald, in a series of papers beginning with Wald (1939) and culminating in the monograph (Wald, 1950), developed statistical decision theory as an extension of the Neyman–Pearson theory of testing. It has since played a major role in statistical theory for point estimation, hypothesis testing, and forecasting, especially in the construction of 'optimal' procedures. Some textbooks such as Ferguson (1967) and Berger (1985) emphasize statistical decision theory as a foundation for statistics. But the decision theory framework is sufficiently flexible that it can be used for many empirical applications that do not fit neatly into the usual statistical set-ups. Some examples are discussed below.

Like the Neyman–Pearson theory, Wald's approach emphasizes evaluating the performance of a decision rule under various possible parameter values. There does not always exist a single rule that dominates all others uniformly over the parameter space, just as there does not always exist a uniformly most powerful test in the special case of hypothesis testing. Wald, who also made contributions to game theory, proposed to evaluate a procedure by its minmax risk – the worst-case expected loss over the parameter space. Savage (1951) discusses the minmax principle and suggests an alternative, the minmax-regret principle. Alternatively, one can place a probability measure on the parameter space, and evaluate rules by their weighted average (Bayes) risk.

Basic framework

In Wald's basic framework, we start with a set of actions $\mathscr{A}$, and a parameter space Θ, which characterizes the set of models under consideration. A loss function $L(\theta, a)$ gives the loss or disutility suffered from taking action $a \in \mathscr{A}$ when the parameter is $\theta \in \Theta$. The decision maker observes some random variable Z, distributed according to a probability measure P_θ when θ is the 'true' parameter. Here, the parameter space Θ could be finite-dimensional (corresponding to a parametric family of distributions) or infinite-dimensional (corresponding to semiparametric and nonparametric models). The observed random variable Z could be a vector, as for example in the situation of observing a random sample of size n from some distribution. Often, the set of possible probability measures $\{P_\theta : \theta \in \Theta\}$ is called a *statistical experiment*.

A decision rule or procedure $d(z)$ maps observations on Z into actions. In some cases, it is useful to allow for randomization over the actions. A randomized decision rule is a mapping from observations into probability measures over the action space. A simpler, usually equivalent formulation is to consider rules $\delta(z, u)$ which are allowed to depend on the observed value z and the value u of a random variable U, distributed standard uniform independently of Z. The *risk*, or expected loss, of a decision rule δ under θ is defined as

$$R(\theta, \delta) = E_\theta[L(\theta, \delta(Z, U))]$$
$$= \int_0^1 \int L(\theta, \delta(z, u)) dP_\theta(z) du.$$

A rule δ is *admissible* if there exists no other rule δ' with

$$R(\theta, \delta') \leq R(\theta, \delta), \quad \forall \theta \in \Theta,$$

and

$$R(\theta, \delta') < (\theta, \delta) \quad \text{for some } \theta.$$

Ordering decision rules

In general, there are many admissible decision rules, which may do well in different parts of the parameter space. Thus, while the admissibility criterion eliminates obviously inferior rules, it may not provide concrete guidance on how to 'solve' the decision problem. Additional criteria can help by providing a sharper partial ordering of decision rules.

One way to rank decision rules is to average their risk over the parameter space. Let Π be a probability measure on Θ. The *Bayes risk* of a decision rule δ is

$$r(\Pi, \delta) = \int R(\theta, \delta) d\Pi(\theta).$$

A rule is a *Bayes rule* if it minimizes this weighted average risk. Let the probabilities P_θ have densities p_θ with respect to some dominating measure, and let the prior Π have density π. Typically, a Bayes rule can be implemented by choosing, for any given observed data z, the action that minimizes the posterior expected loss

$$\int L(\theta, a) d\Pi(\theta|z),$$

where $\Pi(\theta|z)$ is the posterior distribution with density

$$\pi(\theta|z) = \frac{\pi(\theta) p_\theta(z)}{\int p_\theta(z) d\Pi(\theta)}.$$

There is a close connection between the admissible rules and the Bayes rules. If the parameter set is finite, a Bayes rule for a prior that places positive probability on every element of Θ is admissible. Furthermore, 'complete class theorems' give results in the opposite direction. In particular, if the parameter set is finite, any admissible rule is Bayes for some prior distribution. If Θ is not finite, some care needs to be taken to make a precise statement of the relationship between the admissible and Bayes rules; see for example Ferguson (1967).

An alternative ordering is based on the worst-case risk $\sup_{\theta \in \Theta} R(\theta, \delta)$. A *minmax* rule δ_m satisfies

$$\sup_{\theta \in \Theta} R(\theta, \delta_m) = \inf_\delta \sup_{\theta \in \Theta} R(\theta, \delta).$$

In general, a minmax rule need not be admissible.

A closely related criterion is the *minmax regret* criterion. The regret loss of a rule is the difference between its loss and the loss of the best possible action under θ:

$$L_r(\theta, a) = L(\theta, a) - \inf_{a \in \mathscr{A}} L(\theta, a).$$

We can then define regret risk as $R_r(\theta, \delta) = E_\theta(L_r(\theta, \delta(Z, U))$. The *minmax regret rule* minimizes the worst-case regret risk. This rule was suggested by Savage (1951) as an alternative to the minmax criterion. He argued that in cases where the minmax criterion is unduly conservative, minmax regret rules can be reasonable.

Savage (1954) showed that a decision-maker who satisfied certain axioms of coherent behaviour would act as if she placed a prior on the parameter space and minimized posterior expected loss. Gilboa and Schmeidler (1989) showed that, under a different set of axioms, a decision-maker would follow the minmax principle.

Calculation of Bayes and minmax rules can be difficult in many applications. Bayesian posterior distributions can be calculated directly when the prior and likelihood have a conjugate form. One way to solve for a minmax rule is to guess the form of a 'least favourable' prior and solve for the associated Bayes rule. If the risk function of the Bayes rule is everywhere less than the Bayes risk, then the rule is minmax. A related method is to construct a least favourable sequence of prior distributions, and calculate the limit of the Bayes risks. If a particular rule has worst-case risk lower than the limit of Bayes risks, then the rule is minmax. Another useful technique for obtaining minmax rules makes use of invariance properties of the decision problem. If the model and loss are invariant with respect to a group of transformations, and that group satisfies a condition called amenability, then the best equivariant procedure is minmax by the Hunt–Stein theorem. These techniques are discussed in Ferguson (1967) and Berger (1985).

If Bayes and minmax rules cannot be obtained analytically, computational methods can sometimes be useful. Recently developed simulation methods such as Markov chain Monte Carlo have greatly expanded the range of settings where Bayes rules can be numerically computed. Chamberlain (2000) develops algorithms for computing minmax rules, and applies them to an estimation problem for a dynamic panel data model.

Asymptotic statistical decision theory

Despite advances in computational methods, many statistical decision problems remain intractable. In such cases, large-sample approximations may be used to show that certain rules are approximately optimal. Le Cam (1972; 1986) proposed to approximate complex statistical decision problems by simpler ones, in which optimal decision rules can be calculated relatively easily. One then finds sequences of rules in the original problem that approach the optimal rule in the limiting version of the problem.

As an example, suppose we observe n i.i.d. draws from a distribution P_θ where $\theta \in \Theta \subset \mathbb{R}^k$ and the probability measures $\{P_\theta\}$ satisfy conventional regularity conditions with non-singular Fisher information I_θ. We can think of this as defining a sequence of experiments, where the nth experiment consists of observing an n-dimensional random vector distributed according to P_θ^n, the n-fold product of P_θ. Since, in the limit, θ can be determined exactly, we fix a centring value θ_0, and reparametrize the model in terms of local alternatives $\theta_0 + h/\sqrt{n}$, for $h \in \mathbb{R}^k$. This sequence of experiments has as its 'limit experiment' the experiment consisting of observing a single draw $Z \sim N(h, I_{\theta_0}^{-1})$, and we say that the original sequence of experiments satisfies local asymptotic normality (LAN). More precisely, according to an asymptotic representation theorem (see van der Vaart, 1991), for any sequence of procedures δ_n in the original experiments that converge in distribution under every local parameter h, these limit distributions are matched by the distributions associated with some randomized procedure $\delta(Z)$ in the limit experiment. Thus, the limit experiment characterizes the set of attainable limit distributions of procedures in the original sequence of experiments. Solving the decision problem in the limit experiment leads to bounds on the best possible asymptotic behaviour of procedures in the original problem, and often suggests the form of asymptotically optimal procedures.

Le Cam's theory underlies the classic result that in regular parametric models, Bayes and maximum likelihood point estimators of θ are 'asymptotically efficient'. In the LAN limit experiment $Z \sim N(h, I_{\theta_0}^{-1})$, a natural estimator for the parameter h is $\delta(Z) = Z$. This can be shown to be minmax and best equivariant for 'bowl-shaped' loss functions. Both the Bayes and MLE estimators in the original problem are matched asymptotically by this optimal estimator, so they are locally asymptotically minmax and best equivariant. The ideas have been extended to models with an infinite-dimensional parameter space (see Bickel et al. (1993) and van der Vaart, 1991, among others), to obtain semiparametric efficiency bounds for finite-dimensional sub-parameters. More recently, a body of work has developed limit experiment theory for nonparametric problems such as nonparametric regression and nonparametric density estimation (see Brown and Low, 1996, and Nussbaum, 1996, among others). These results show that nonparametric regression and density estimation are asymptotically equivalent to a white-noise model with drift, for which a number of optimality results are available.

Applications in economics

Portfolio choice
A number of authors have used statistical decision theory to study portfolio allocation when the distribution of returns is uncertain. Some examples include Klein and Bawa (1976), Kandel and Stambaugh (1996), and Barberis (2000), who develop Bayes rules for portfolio choice problems.

Treatment choice
Another econometric application of statistical decision theory is to treatment assignment problems, in which a social planner wishes to assign individuals to different treatments (for example, different job training programmes) to maximize some measure of social welfare. Manski (2004) develops minmax-regret results for the treatment assignment problem, Dehejia (2005) develops Bayesian rules, and Hirano and Porter (2005) obtain asymptotic minmax regret-risk bounds and show that certain simple rules are optimal according to this criterion.

Model uncertainty and macroeconomic policy
Brainard (1967) studied a macroeconomic policy problem, in which a parameter describing the effect of a policy instrument on a macroeconomic outcome is not known with certainty but is given a distribution. The policymaker has a utility function over outcomes and chooses the policy that makes expected utility. More recently, a number of authors have continued this line of work, extending the analysis to more general forms of model uncertainty and developing both Bayesian and minmax solutions. Some examples include Hansen and Sargent (2001), Rudebusch (2001), Onatski and Stock (2002), Giannoni (2002), and Brock, Durlauf and West (2003).

Instrumental variables models

Decision-theoretic ideas underlie recent work on the linear instrumental variables model in econometrics. Chamberlain (2005) develops minmax optimal point estimators in the IV model using invariance arguments. Andrews, Moreira and Stock (2004) have developed tests in the IV model that are optimal under an invariance restriction, and Chioda and Jansson (2004) have developed optimal conditional tests.

Time series models

Asymptotic statistical decision theory has been useful in studying certain time series models which do not satisfy standard regularity conditions. Jeganathan (1995) shows that a number of models for econometric time series have limit experiments that are not of the standard LAN form, but are locally asymptotically mixed normal (LAMN) or locally asymptotically quadratic (LAQ). Ploberger (2004) obtains a complete class theorem for hypothesis tests in the LAQ case, which nests the LAMN and LAN cases.

Auction and search models

Some parametric auction and search models, in which the support of the data depends on some of the model parameters, do not satisfy the LAN regularity conditions. For such models, Hirano and Porter (2003) showed that the maximum likelihood point estimator is not generally optimal in the local asymptotic minmax sense, but that Bayes estimators are asymptotically efficient.

KEISUKE HIRANO

See also **Bayesian econometrics; Markov chain Monte Carlo methods; maximum likelihood; Savage, Leonard J. (Jimmie); Wald, Abraham.**

Bibliography

Andrews, D.W.K., Moreira, M.M. and Stock, J.H. 2004. Optimal invariant similar tests for instrumental variables regression. Discussion Paper No. 1476. Cowles Foundation, Yale University.

Barberis, N.C. 2000. Investing for the long run when returns are predictable. *Journal of Finance* 55, 225–64.

Berger, J.O. 1985. *Statistical Decision Theory and Bayesian Analysis.* New York: Springer-Verlag.

Bickel, P.J., Klaasen, C.A., Ritov, Y. and Wellner, J.A. 1993. *Efficient and Adaptive Estimation for Semiparametric Models.* New York: Springer-Verlag.

Brainard, W.C. 1967. Uncertainty and the effectiveness of policy. *American Economic Review* 57, 411–25.

Brock, W.A., Durlauf, S.N. and West, K.D. 2003. Policy evaluation in uncertain economic environments. *Brookings Papers on Economic Activity* 2003(1), 235–322.

Brown, L.D. and Low, M.G. 1996. Asymptotic equivalence of nonparametric regression and white noise. *Annals of Statistics* 24, 2384–98.

Chamberlain, G. 2000. Econometric applications of maxmin expected utility. *Journal of Applied Econometrics* 15, 625–44.

Chamberlain, G. 2005. Decision theory applied to an instrumental variables model. Working paper, Harvard University.

Chioda, L. and Jansson, M. 2004. Optimal conditional inference for instrumental variables regression. Working paper, UC Berkeley.

Dehejia, R.H. 2005. Program evaluation as a decision problem. *Journal of Econometrics* 125, 141–73.

Ferguson, T.S. 1967. *Mathematical Statistics: A Decision Theoretic Approach.* New York: Academic Press.

Giannoni, M.P. 2002. Does model uncertainty justify caution? Robust optimal monetary policy in a forward-looking model. *Macroeconomic Dynamics* 6(1), 111–44.

Gilboa, I. and Schmeidler, D. 1989. Maxmin expected utility with non-unique prior. *Journal of Mathematical Economics* 18, 141–53.

Hansen, L.P. and Sargent, T.J. 2001. Acknowledging misspecification in macroeconomic theory. *Review of Economic Dynamics* 4, 519–35.

Hirano, K. and Porter, J. 2003. Asymptotic efficiency in parametric structural models with parameter-dependent support. *Econometrica* 71, 1307–38.

Hirano, K. and Porter, J. 2005. Asymptotics for statistical treatment rules. Working paper, University of Arizona.

Jeganathan, P. 1995. Some aspects of asymptotic theory with applications to time series models. *Econometric Theory* 11, 818–87.

Kandel, S. and Stambaugh, R.F. 1996. On the predictability of stock returns: an asset-allocation perspective. *Journal of Finance* 51, 385–424.

Klein, R.W. and Bawa, V.S. 1976. The effect of estimation risk on optimal portfolio choice. *Journal of Financial Economics* 3, 215–31.

Le Cam, L. 1972. Limits of experiments. *Proceedings of the Sixth Berkeley Symposium of Mathematical Statistics* 1, 245–61.

Le Cam, L. 1986. *Asymptotic Methods in Statistical Decision Theory.* New York: Springer-Verlag.

Manski, C.F. 2004. Statistical treatment rules for heterogeneous populations. *Econometrica* 72, 1221–46.

Nussbaum, M. 1996. Asymptotic equivalence of density estimation and Gaussian white noise. *Annals of Statistics* 24, 2399–430.

Onatski, A. and Stock, J.H. 2002. Robust monetary policy under model uncertainty in a small model of the U.S. economy. *Macroeconomic Dynamics* 6(1), 85–110.

Ploberger, W. 2004. A complete class of tests when the likelihood is locally asymptotically quadratic. *Journal of Econometrics* 118, 67–94.

Rudebusch, G.D. 2001. Is the fed too timid? Monetary policy in an uncertain world. *Review of Economics and Statistics* 83, 203–17.

Savage, L.J. 1951. The theory of statistical decision. *Journal of the American Statistical Association* 46, 55–67.

Savage, L.J. 1954. *The Foundations of Statistics*. New York: Wiley.

van der Vaart, A.W. 1991. An asymptotic representation theorem. *International Statistical Review* 59, 97–121.

Wald, A. 1939. Contributions to the theory of statistical estimation and testing hypotheses. *Annals of Mathematical Statistics* 10, 299–326.

Wald, A. 1950. *Statistical Decision Functions*. New York: Wiley.

default and enforcement constraints

Intertemporal exchange, that is the exchange of resources today for a promise of resources at a later date in a given state, is key for promoting economic efficiency. For example, to finance an investment, a government borrows capital abroad in exchange for a promise of repayment once the investment has paid off. Or, to finance consumption, an individual who loses her job borrows resources in exchange for the promise of repayment once she gets a new job. If the enforcement of promises is limited, the extent of intertemporal exchange can be reduced by so-called *enforcement constraints* and, under some conditions, *default*, that is, the breaking of promises, can arise. This article presents a simple general equilibrium set-up to analyse these issues and provide some direction for the design of enforcement policies. Key references for the theory of limited enforcement without default are Kehoe and Levine (1993), Kocherlakota (1996) and Alvarez and Jermann (2000), while for limited enforcement with default see Zame (1993) and Dubey, Geanakoplos and Shubik (2005).

The set-up

The goal of this set-up is to capture the need for intertemporal exchange, as described in the examples above. There are two agents which live for two periods and consume a single good. Agent 1, the borrower, owns a technology such that, if k units of the good are invested in period 1, AK^α, $0 < \alpha < 1$, units are produced in period 2, where A is a random variable realized in period 2, with positive support and distribution $F(A)$ known to both agents. Agent 2, the lender, is endowed with e units of the consumption good in period 1. Consumption allocations of agent i are consumption at date 1, c_{i1} and the function $c_{i2}(A)$ which assigns period 2 consumption for each possible realization of A. Borrower's utility is given by $u(c_{11}) + \int u(c_{12}(A))dF(A)$ where u is a concave utility function satisfying Inada conditions. The lender has linear utility given by $c_{21} + \int c_{22}(A)dF(A)$. Linear utility implies that lender's equilibrium utility is constant across different market structures so that borrower's utility is the only statistic needed to Pareto-rank equilibria. In all the economies described below the following resource constraints hold

$$c_{11} + c_{21} + k = e$$
$$c_{12}(A) + c_{22}(A) = Ak^\alpha \quad \text{for every } A$$

A frictionless benchmark

Assume agents can trade a complete set of Arrow–Debreu promises which are fully and costlessly enforceable. The budget constraints of the borrower are

$$c_{11} + k = \int p(A)dF(A) \qquad (1)$$

$$c_{12}(A) = Ak^\alpha - p(A) \quad \text{for every } A \qquad (2)$$

where $p(A)$ denotes the amount that the borrower promises to repay in state A. Equilibrium allocations display complete risk sharing, that is, the ratio of marginal value of consumption of the two agents is constant across dates and states of the world. We denote with c^{AD} the constant, across dates and states, level of consumption of the borrower in this economy.

Limited enforcement

This section describes an economy denoted as ADLE (Arrow–Debreu Limited Enforcement) and shows that limited enforcement prevents full risk sharing, reduces investment and welfare. Assume that in period 2 the borrower can walk away from any promise made to the lender by suffering a default deadweight cost proportional to its output and equal to δAk^α where $\delta > 0$ is a parameter that measures the strength of enforcement. This implies that any Arrow–Debreu promise $p(A) > \delta Ak^\alpha$ will not be honoured by the borrower and thus will not be purchased by the lender. Also, promises satisfying $p(A) \leq \delta Ak^\alpha$ will be fully honoured and priced as in the frictionless economy. So limited enforcement limits the use of state-contingent promises but does not induce default. A convenient way of capturing this, following Alvarez and Jermann (2000), is to assume that the borrower faces constraints on the sales of each promise so as to guarantee no default. These *enforcement* constraints have the form

$$p(A) \leq \delta Ak^\alpha \quad \text{for every } A \qquad (3)$$

as the borrower can sell each promise only up to the point where the cost of keeping it is equal to the cost of defaulting on it. Equilibrium allocations can be characterized by substituting budget constraints (1) and (2) into the borrower's utility and taking first-order conditions with respect to k and $p(A)$ subject to constraints (3). This yields

$$u'(c_{11}) = \int [A\alpha k^{\alpha-1}u'(c_{12}(A)) + A\alpha k^{\alpha-1}\delta\mu(A)]\, dF(A) \qquad (4)$$

where

$$\mu(A) = u'(c_{11}) - u'(c_{12}(A))$$

are the Lagrange multipliers on the enforcement constraints. If the cost of default δ is sufficiently small and the distribution of A is sufficiently spread out, $c_{11} = c(A) = c^{AD}$ is not a solution of (4) as enforcement constraints on the high A promises would be violated. The solution is then characterized by a level of productivity A^* such that for all $A > A^*$ enforcement constraints are binding and $c(A) = (1 - \delta)Ak^{\alpha} > c_{11}$. For $A \leq A^*$ enforcement constraints are not binding and $c(A) = c_{11} < c^{AD}$. Complete risk sharing involves the borrower selling promises to repay in states with high A, in order to finance consumption today (when she has no output) and consumption tomorrow in states with low A. But if the distribution of A is spread out, complete risk sharing calls for promises of a large transfer of resources from the borrower to the lender in the states with high A. When enforcement is limited (δ is low) the lender, in period 1, correctly anticipates that these transfers will not be made and buys a smaller amount of the promises. So, relative to complete risk sharing, the borrower has fewer resources in period 1 and in the period 2 states with low A, but consumes more in period 2 states high A. This allocation of consumption increases the marginal value of resources in period 1 relative to the expected marginal value of resources in period 2 and thus reduces k relative to the full enforcement case. Finally, equilibria in economies with strong enforcement (high δ) Pareto-dominate equilibria with weak enforcement (low δ). To see this, note that, for the borrower, the equilibrium allocation in the weak enforcement economy is budget-feasible in the strong enforcement economy, so, if it is not chosen, it must yield her lower utility.

ADLE economies have been used extensively in a variety of applications such as asset pricing (Alvarez and Jermann, 2000), international business cycles (Kehoe and Perri, 2002) and consumption inequality (Krueger and Perri, 2006). All these studies show that limited enforcement prevents complete risk sharing, and for this reason it provides a much better fit with the data than standard Arrow–Debreu economies. This environment, though, cannot be used to understand equilibrium default (that is, the actual break of a promise and the suffering of the associated cost) as the trade in contingent promises makes incurring the default cost unnecessary. In order to understand when default arises and what its consequences are, the next section considers an economy in which contingent promises cannot be traded, either because markets are exogenously missing or because the borrower has private information about realizations of A.

Limited enforcement and non-contingent promises

The borrower finances consumption and investment only by selling a non-contingent promise p which can be defaulted on in state A by suffering the default cost δAk^{α}. Since the cost of repaying the promise does not vary with the state while the default cost is increasing with A, if there is equilibrium default it will happen in the low A states. In particular, if the borrower invests k and sells a promise p, she will default in all the states such that $A \leq \frac{p}{\delta k^{\alpha}}$.

As a consequence, the equilibrium price of the promise is given by

$$q(p,k) = 1 - F\left(\frac{p}{\delta k^{\alpha}}\right). \tag{5}$$

The problem of the borrower is then

$$\max_{p,k} u(q(p,k)p - k) + \int_0^{\frac{p}{\delta k^{\alpha}}} u((1-\delta)Ak^{\alpha})dF(A)$$
$$+ \int_{\frac{p}{\delta k^{\alpha}}}^{\infty} u(Ak^{\alpha} - p)dF(A). \tag{6}$$

The equilibrium is characterized by a couple p, k which solve (5) and (6). It can be immediately shown that equilibria in this economy are, generically, Pareto-inferior to equilibria in the corresponding ADLE economy. Also, for many parameter values equilibria in this set-up differ from those in the ADLE economy along two important dimensions: (a) there is a positive measure of states for which default occurs and (b) there is a positive measure of values for δ for which welfare is *decreasing* in the strength of enforcement. As a simple example, consider the case in which A can take only two values: a high value A_h with probability π and a low value A_l with probability $1 - \pi$, with $\pi > A_l/A_h$. In this case there is a range of values for δ for which the equilibrium promise and capital satisfy

$$\delta A_l k^{\alpha} < p < \delta A_h k^{\alpha}, \tag{7}$$

so that default happens only when state A_l is realized and consequently $q(p,k) = \pi$. Now consider the effect of a marginal reduction in δ. Equation (7) shows that, if the borrower kept k and p unchanged in response to the change in δ, default patterns, and hence $q(p, k)$, would not change; however reducing δ increases the returns of borrower in the default state so its utility would increase relative to the initial equilibrium. Here weakening enforcement allows the borrower to implicitly transfer, through default, more resources to the low A state and thus to achieve a better allocation of risk across states. In the ADLE economy this transfer was achieved through the Arrow–Debreu promises so default was not necessary. When promises cannot be made state-contingent, increasing payoffs in the default states is the only way of obtaining this transfer.

In this simple example weakening enforcement does not affect default frequency, but in more general set-ups it does and as a consequence increases equilibrium interest rates and hampers intertemporal exchange. This effect is detrimental for welfare. But the example above suggests that the detrimental effect can be offset by the positive effect of the better risk allocation across states. Note that this result does not rely on the two-state assumption, and it can be shown to hold, for example, also when A is log-normally distributed.

Summary

Limiting contract enforcement in otherwise frictionless environments constrains intertemporal exchange and hampers risk sharing, investment and welfare, but does not induce default. When additional frictions, such as incomplete markets or private information, limit the span of tradable promises, then limited enforcement can play a positive role by inducing equilibrium default, which can be used as a (costly) way of providing better allocation of risk across states. The analysis sheds light on how enforcement policies should be related to the observed frequency of default.

When limited enforcement is the only friction, default is never observed, yet tightening enforcement is socially beneficial. When limited enforcement coexists with other frictions, default happens in equilibrium but this does not necessarily mean that enforcement should be tightened. Indeed, tightening enforcement without ameliorating the additional friction might reduce default but also risk sharing and welfare.

FABRIZIO PERRI

See also **risk sharing; sovereign debt.**

Bibliography

Alvarez, F. and Jermann, U. 2000. Efficiency, equilibrium, and asset pricing with risk of default. *Econometrica* 68, 775–97.

Dubey, P., Geanakoplos, J. and Shubik, M. 2005. Default and punishment in general equilibrium. *Econometrica* 73, 1–37.

Kehoe, P. and Perri, F. 2002. International business cycles with endogenous incomplete markets. *Econometrica* 70, 907–28.

Kehoe, T. and Levine, D. 1993. Debt-constrained asset markets. *Review of Economic Studies* 60, 865–88.

Kocherlakota, N. 1996. Implications of efficient risk sharing without commitment. *Review of Economic Studies* 63, 595–609.

Krueger, D. and Perri, F. 2006. Does income inequality lead to consumption inequality? Evidence and theory. *Review of Economic Studies* 73, 163–93.

Zame, W. 1993. Efficiency and the role of default when securities markets are incomplete. *American Economic Review* 83, 1142–64.

defence economics

Defence economics is a relatively new part of the discipline of economics. One of the first specialist contributions in the field was by C. Hitch and R. McKean, *The Economics of Defense in the Nuclear Age* (1960). This book applied basic economic principles of scarcity and choice to national security. It focused on the quantity of resources available for defence and the efficiency with which such resources were used by the military. For example, defence consumes scarce resources that are therefore not available for social welfare spending (for example, missiles versus education and health trade-offs). Once resources are allocated to defence, military commanders have to use them efficiently, combining their limited quantities of arms, personnel and bases to 'produce' security and protection. Within such a military production function, there are opportunities for substitution. For example, capital (weapons) can replace (and have replaced) military personnel; imported arms can replace nationally produced weapons; and nuclear forces have replaced large standing armies. Defence economics is about the application of economic theory to defence-related issues.

The development of defence economics and its research agenda reflected current events. For example, during the cold war there was a focus on the superpower arms races, alliances (NATO and the Warsaw Pact), nuclear weapons and 'mutually assured destruction'. The end of the cold war resulted in research into disarmament, the challenges of conversion and the availability of a peace dividend. Since the end of the cold war, the world remains a dangerous place with regional and ethnic conflicts (for example, Bosnia, Kosovo, Iraq), threats from international terrorism (for example, terrorist attacks on USA on 11 September 2001), rogue states and weapons of mass destruction (that is, biological, chemical and nuclear weapons). NATO has accepted new members (for example, former Warsaw Pact states) and has developed new missions, and the European Union has developed a European Security and Defence Policy. Changing threats and new technology require the armed forces and defence industries to adjust to change and new challenges. Peace-keeping has become a major mission for armed forces and is an example of the trend towards globalization.

The modern era of globalization involves more international transactions in goods, services, technology and factors of production, which brings new security challenges for both nation states and the international community. Defence firms have become international companies with international supply networks. Globalization also highlights the importance of international collective action to respond to new threats such as international terrorism and to maintain world peace (for example, through international peacekeeping missions under UN, NATO or EU control). But international collective action experiences the standard problems of burden-sharing and free riding.

This article outlines the development of defence economics; it defines the field and describes the 'stylized facts' of world military expenditure; the defence economics problem is considered; and a case study of conflict and terrorism illustrates some of the new developments in the field.

A brief history

Defence issues have existed throughout history as nations have been involved in armed conflict of various forms and durations (for example, the Hundred Years War). Great powers have used military force to dominate regions and parts of the world (for example, Alexander the Great; Roman legions; Genghis Khan; Ottoman Turks; Nazi Germany), with such powers rising and falling (Kennedy, 1988). Conflict has also been characterized by major technical changes ranging from bows and arrows to cannons and machine guns, from sailing ships to iron and steel warships and nuclear-powered vessels, from horse cavalry to tanks, from flag communications to radios and satellite communications and from balloons to aircraft, missiles, nuclear weapons and space systems. Historically, the economic base for conflict was first an agricultural society, then an industrial society followed by a knowledge economy.

Some of the classical economists studied war and conflict (for example, Smith, Ricardo, Malthus, J. S. Mill: see Goodwin, 1991, ch. 2). For these economists, war departed from much of their conventional thinking: it involved chaos and disorder rather than market equilibrium, and it required government action rather than private market behaviour. Yet it remains surprising that, with a long history of wars, including two world wars and the superpower arms race of the Cold War, relatively few economists have been attracted to the field. A review of the economics literature on conflict concludes that 'We were surprised at the relative absence of applied economics studies of actual conflicts' (Sandler and Hartley, 2003, p. xl). There are various possible explanations for the relative absence of economists studying war and conflict. These include data and security problems, the difficulty of applying conventional market analysis to the chaos and disequilibrium of conflict, a traditional reluctance to analyse the public sector (with defence assumed to be exogenous), and the feeling that defence and security issues are not as important as other social welfare issues, with war viewed as an immoral and unethical subject. Furthermore, security issues have not been as an attractive career path for economists (compared with issues such as inflation, unemployment, growth and developing countries), and conflicts are usually of short duration so that they offer only limited research prospects before peace returns to remove war-related problems (Goodwin, 1991, pp. 1–2).

Definitions

Defence economics studies all aspects of war and peace and embraces defence, disarmament and conversion. This definition includes studies of both conventional and non-conventional conflict such as civil wars, revolutions and terrorism. It involves studies of the armed forces and defence industries and the efficiency with which these sectors use scarce resources in providing defence output in the form of peace, protection and security. Reductions in defence spending (such as those following the end of the cold war) result in disarmament, which involves reallocating resources from the defence to the civilian

sector. This raises questions about the impact of disarmament on the employment and unemployment of both military personnel and defence industry workers; the possibilities for converting military bases and arms industries to civil uses (the Biblical swords to ploughshares); and the role of public policies in assisting the transition and reallocation of resources.

The coverage of the subject is extensive and involves economic theory, empirical testing and policy-related issues, including applications of public choice analysis. Both defence and peace have distinctive economic characteristics in that they are public goods which are non-rival and non-excludable. There are large literatures dealing with the determinants of military expenditure, including economic theories of military alliances and arms races (that is, threats) and the impact of defence spending on economic growth and development. Armed forces are major buyers of both equipment (arms/weapons) and military personnel, and such procurement choices affect defence industries and both local and national labour markets. For example, government procurement of weapons involves choices between competition and preferential purchasing and between various types of contracts (for example, fixed-price, cost-plus), each with different implications for contractor efficiency and profitability. There is a related literature on industrial and alliance policies comparing the economics of supporting a national defence industrial base with alternative industrial policies such as international collaboration, licensed production or importing foreign equipment. Imports also involve the international arms trade, its economic impacts on both buyers and suppliers, and policy initiatives to regulate such trade. More generally, there is an extensive literature on arms control and disarmament, the adjustment costs of disarmament, the economics of conversion and the contribution of public policy to minimizing such adjustment costs. Finally, there have been some new developments involving the application of economics to the study of conflict, civil wars, revolutions and terrorism (Brauer, 2003; Hegre and Sandler, 2002; Sandler and Hartley, 1995; 2007).

Defence economics became established in the 1960s with the publication of a number of pioneering contributions, mostly by US economists. These contributions applied economics to some novel areas and included economic models of alliances (Olson and Zeckhauser, 1966), the economics of arms races (Richardson, 1960; Schelling, 1966), the procurement of weapons and military personnel (Peck and Scherer, 1962; Oi, 1967), and the impact of military spending on economic development (Benoit, 1973). A further development confirming the emergence of defence economics as an accepted part of the discipline of economics was the launch in 1990 of a field journal, *Defence Economics*, later renamed *Defence and Peace Economics* (initially it was published four times per year, but in 2000 it was expanded to six issues per year).

Inevitably, defence economics generates controversy reflected in myths and emotion. Critics point to the 'wastes' of defence spending and its 'crowding-out' of 'valuable' civil expenditure. Classic examples include the sacrifice of schools and hospitals associated with major weapons projects such as modern combat aircraft and aircraft carriers (for example, the US F-22 aircraft and the European Typhoon). Peace economists are similarly critical of defence economics and military spending: they focus on peace topics such as disarmament and the maintenance of peace, arms control and international security, conflict analysis and management, and crises and war studies. Defence economists are not, however, 'warmongers': they are instead interested in understanding the economics of the military–industrial–political complex and all aspects of defence whereby a proper understanding of these issues will contribute to a more peaceful world. A starting point in showing how economists analyse defence is to review the 'stylized facts' of world military spending.

The stylized facts of world military spending

What is known about military spending, and where are the gaps in the data? Good quality data exist on world military spending, the world's armed forces and the arms trade. Cross-section and time-series data are available at the country level; some examples are shown in Table 1. The data on world military expenditure show aggregate spending by the USA accounting for 45 per cent of total world military spending and NATO accounting for some 70 per cent. Similarly, in 2004 the USA dominated defence R&D spending, accounting for some 75 per cent of the world total and 31 per cent of world arms exports.

Table 1 shows examples of defence shares of GDP to illustrate the burdens of defence spending, especially for developing nations such as Eritrea, India and Pakistan (an arms race situation) and for the Middle East (a conflict region). Burundi and Sudan have defence burdens similar to or greater than those of the UK and Germany. Table 1 also shows other measures of the economic burdens of defence for the world's poorer nations (that is, nations which cannot feed, house or educate their populations and which have poor health records). Developing nations accounted for 70 per cent of the world total of 21.3 million military personnel, and such totals further show the importance of military manpower economics. Similarly, developing nations are major importers of arms, while the developed nations are the major arms exporters. Such data provide an introduction to some of the major themes of defence economics, namely, the determinants of military expenditure, arms races, alliances, the relationship between defence spending and economic development, the arms trade and the economics of military personnel.

Micro-level data are more limited but there are some useful sources especially on defence contractors and defence industries. Table 2 provides examples of such micro-level data based on the 100 largest arms-producing

Table 1   *World military spending and armed forces, various years*

World military expenditure	US$ billion, 2004
NATO	722
USA	467
France	52
Germany	38
UK	54
China	37
Russia	23
World total	**1,035**

Defence share of GDP	%[a]
USA	3.9
France	2.6
Germany	1.4
UK	2.3
Eritrea	19.4
Burundi	5.9
Sudan	2.4
India	2.1
Pakistan	4.4
Israel	9.1
Jordan	8.9
Oman	12.2

Defence research and development[b]	US$ billion, 2004 (2001 prices and PPP rates)
USA	67.5
Russia	6.1
UK	4.7
USA and EU total	80.9
Estimated world total of defence R&D	**90.0 +**

World armed forces	Number of military personnel, 1999 ('000s)
Developed nations	6,550
Developing nations	14,700
NATO	4,580
USA	1,490
UK	218
Eritrea	215
China	2,400
World	**21,300**

World arms trade	US$ million, 2000–04 (1990 prices)
Major importers	
China	11,677
India	8,526
Greece	5,263
UK	3,395
Turkey	3,298
World total	**84,490**

(Continued)

Table 1   *(Continued)*

Major exporters	
Russia	26,925
USA	25,930
France	6,358
Germany	4,878
UK	4,450
World total	**84,490**

Notes: [a]Defence share data for USA, France, Germany and UK are for 2004; all other data are for 2003.
[b]Defence R&D data are for government-funded defence R&D.
PPP=purchasing power parity.
Sources: US DoS (2002); NATO (2005); OECD (2004); SIPRI (2005).

Table 2   *Defence companies and industries*

Major defence companies	Arms sales, 2003 (US$ million)
Lockheed Martin (USA)	24,910
Boeing (USA)	24,370
Northrop Grumman (USA)	22,720
BAE Systems (UK)	15,760
Raytheon (USA)	15,450
General Dynamics (USA)	13,100
Thales (France)	8,350
EADS (Europe)	8,010
United Technologies (USA)	6,210
Finmeccanica (Italy)	5,290

Major defence industries	Employment numbers, 2003 ('000s)
Industrialized countries	4,710
Developing countries	2,769
NATO	3,452
EU	645
USA	2,700
China	2,100
Russia	780
France	240
UK	200
World total	**7,479**

Sources: BICC (2005); SIPRI (2005).

companies (SIPRI, 2005) and employment in national defence industries (BICC, 2005). Again, these data are available on a cross-section and time-series basis, and the company data include total sales, total profits and aggregate employment. From Table 2 it can be seen that the USA has six of the world's top ten arms companies and that the American firms have a substantial scale advantage over their European rivals: the average size of a US firm from the top ten is almost twice the corresponding average of the European companies. These data are the basis for research questions about the determinants of firm size, the impact of economies of scale, scope and

learning, and the determinants of performance in terms of labour productivity and profitability.

Table 2 also shows data on defence industry employment. The industrialized nations accounted for 63 per cent of total employment in the world's defence industries, with the developing countries accounting for the remaining 37 per cent. The USA, China and Russia have the largest defence industries by employment, accounting for 75 per cent of the world total. Overall, the world military–industrial complex employed almost 29 million personnel in the armed forces and defence industries, reinforcing its role as a major employer of labour, including some highly qualified R&D staff and other highly skilled workers. Such scarce labour has alternative uses in the civilian sector, raising questions as to whether defence spending 'crowds out' valuable civil investment and diverts scientific manpower from civil research projects.

Despite the available data, there remain significant gaps in our knowledge of the world's military sector. Typically, new defence projects are surrounded by secrecy; there are problems in identifying some defence goods (for example, dual use goods, such as civil airliners which can be used as military transport aircraft); there is a lack of good-quality data on defence R&D, including employment in defence R&D; and little is known about China, especially its defence R&D programmes (Hartley, 2006a). International comparisons of military expenditure data are also sensitive to the choice of exchange rate adjustments, with country rankings sensitive to the use of market exchange rates or purchasing power parity rates (SIPRI, 2005). At the firm and industry levels, analysis of the military business in terms of defence output, employment and profitability is complicated because the typical output comprises a mix of military and civil components, making it difficult to compare the performance of defence contractors and civil firms. Further gaps exist in our knowledge of the world regional distribution of military bases and defence plants, so that it is difficult to assess the economic dependence of various regions on defence spending. Little is known about defence industry supply chains both within countries and within the global economy. Finally, there is a need for more reliable data on the international trade (including illegal transactions) in small arms (these are often the main weapons used in many regional conflicts, such as in Bosnia).

The defence economics problem

This is the standard choice problem of economics, but applied to defence. Typically, following the end of the cold war defence budgets have been either constant or falling in real terms; and these limited budgets are faced with rising input costs of both capital and labour. Equipment costs have been rising by some ten per cent per annum in real terms, which means a long-run reduction in the numbers of weapons acquired for the armed forces (for example, the US Air Force's original requirement for

F-22 combat aircraft for 750 units was later reduced to some 180 aircraft). Similarly, with an all-volunteer force, the costs of military personnel have to rise faster than wage increases in the civil sector. This wage differential is required to attract and retain military personnel by compensating them for the net disadvantages of military life. Here, the military employment contract is unique in that armed forces personnel are subject to military discipline; they are required to deploy to any part of the world at short notice; they could remain overseas indefinitely; and some might never return (that is, death and injury are a feature of this contract). This combination of constant or falling defence budgets and rising input costs means that governments and defence policymakers cannot avoid the need for difficult choices in a world of uncertainty (that is, where the future is unknown and unknowable, and no one can accurately predict the future).

Faced with this defence choice problem, governments have adopted various solutions. They can adopt a policy of 'equal misery' whereby each of the services is subject to budget cuts (for example, reduced training, cancelling some new equipment projects and delaying others); or they can undertake a major revision of a nation's defence commitments (for example, a defence review such as the UK's 1998 Strategic Defence Review); or they can seek to improve efficiency in the armed forces and defence industries (for example, via a competitive equipment procurement policy and military outsourcing). Other policy options include joining a military alliance (such as NATO; EU) or avoiding the defence choice problem by increasing the defence budget (as in the USA since 11 September 2001); but then choices are needed between defence and social welfare spending.

Economics offers three broad policy principles for formulating an efficient defence policy, namely, final outputs, substitution and competition. Take first the principle of *final outputs*. Measuring defence output is notoriously difficult, but it can be expressed in such general terms as peace, security and threat reduction. The UK has solved the problem by committing (and funding) its armed forces to having the capacity to fight simultaneously three small to medium conflicts (for example, Bosnia, Kosovo, Sierra Leone) or one large-scale conflict as part of an international coalition (for example, the Gulf War, Iraq). This approach is a departure from the traditional focus on measuring inputs in terms of the numbers of infantry regiments, warships, tanks and combat aircraft. Such a focus fails to address the key issue of the contribution of these inputs to final defence output in the form of peace and protection. A focus on inputs also fails to address the marginal contribution of each of the armed forces: what would be the implications for defence output if, say, the air force were expanded by five to ten per cent, or the navy was reduced by five to ten per cent?

The second economic principle is that of *substitution*. There are alternative methods of achieving protection, each with different cost implications. Possible examples of partial substitutes include reserves replacing regular personnel, civilians replacing regulars (for example, police in Northern Ireland replacing army personnel), attack helicopters replacing tanks, ballistic and cruise missiles and unmanned combat air vehicles replacing manned strike and bomber aircraft, air power replacing land forces, and imported equipment replacing nationally produced equipment. Some of these substitutions might alter the traditional monopoly property rights of each of the armed forces. For example, surface-to-air missiles operated by the army might replace manned fighter aircraft operated by the air force, and maritime anti-submarine aircraft operated by the air force might replace frigates supplied by the navy.

The third economic principle is that of *competition* as a means of achieving efficiency. Standard economic theory predicts that, compared with monopoly, competition results in lower prices, higher efficiency, and competitively determined profits and innovation in both products and industrial structure. For equipment procurement, competition means allowing foreign firms to bid for national defence contracts and awarding fixed-price contracts rather than cost-plus contracts; it also means ending any 'cosy' relationship between the defence ministry and its national champions and any preferential purchasing and guaranteed home markets.

Competition can be extended to activities undertaken by the armed forces. Here, there is a public sector monopoly problem whereby the armed forces have traditionally undertaken a range of activities 'in house' without being subject to any rivalry. Military outsourcing allows private contractors to bid for and undertake such activities. Examples include accommodation, catering, maintenance, repair, training, transport and management tasks (for example, managing stores or depots and firing ranges). In some cases, outsourcing involves private finance initiatives whereby the private sector finances the activity (for example, new buildings or an aircrew simulator training facility) and then enters into a long-term contract with the defence ministry to provide services to the armed forces in return for rental payments. Another variant is a public–private partnership whereby the private sector finances an activity or asset in return for rental payments from the defence ministry, but the contractor is allowed to sell any peacetime spare capacity to other users (for example, tanker aircraft capacity which when not needed in peacetime can be rented to other users).

Application of the policy guidelines to an efficient defence policy requires that individuals and groups in the military–industrial–political complex are provided with sufficient incentives to behave efficiently. There are the inevitable principal–agent problems where agents have considerable opportunities to pursue their own interests which may conflict with those of their principals (for example, leading a quiet life rather than bearing the costs of change). Individuals and groups in the armed forces and defence ministries will be reluctant to apply the substitution principle if there are no personal or group incentives and rewards for achieving efficient substitution (that is, interest groups can be barriers to change). Compare the private sector, where there are market and institutional arrangements promoting efficiency in the form of rivalry between suppliers, the profit motive and the capital market as a 'policing and monitoring' mechanism through the threats of takeover and bankruptcy. Such market arrangements are absent in the armed forces (and elsewhere in the public sector).

There is also the challenge of achieving 'top level' efficiency in defence provision. Economic theory solves this challenge as a standard optimization problem involving the maximization of a social welfare function subject to resource or budget constraints (where welfare is dependent on civil goods and security, with security provided by defence). Operationalizing this apparently simple optimization rule is much more difficult. Individual preferences for defence are subject to its public good characteristics and free riding problems and the continued difficulty of defining defence output. In democracies, society's preferences are usually expressed through voting at elections. However, elections are limited as a means of obtaining an accurate indication of society's preferences for defence and its willingness to pay. Elections occur infrequently; they are usually for a range of policies of which defence is only one element in the package (which includes policies on, for example, education, health, transport, the environment, foreign policy and taxation); and the 'voting paradox' shows the difficulty of deriving a society's preferences using the voting system. Nor do voters have reliable information on the output of defence spending.

Defence economics explains military spending using a demand model of the form:

$$ME = M(P, Y, T, A, Pol, S, Z)$$

where ME = real military spending; P = relative prices of military and civil goods and services; Y = real national income; T = threats in the form of the military expenditure of a rival nation (arms race models); A = membership of a military alliance and the real military expenditure of the allies (such as NATO); Pol = variable for the political composition of the government (for example, left- or right-wing, with the latter favouring 'strong defences'); S = a variable representing the security and strategic environment (such as the end of the cold war; conflicts such as Korea, Vietnam, the Gulf War and Iraq); and Z = other relevant influences (for example, land mass to be protected). Estimation of the demand model usually proceeds without a price variable, mainly because most nations do not provide relative price data. This omission can be justified if the price of military goods and services has inflated at the same rate as civil goods and services; but such an assumption is not always realistic. A survey of empirical results is presented in Sandler and Hartley (1995; 2007).

Conflict and terrorism

The demand model for military expenditure recognized the relevance of threats such as terrorism and conflict as determinants of defence spending. Traditionally, conflict and terrorism have been the preserve of disciplines other than economics. For example, debates and decisions about war involve political, military, moral and legal judgements. But conflict has an economic dimension, namely, its costs. Wars are not costless: they can involve massive costs (for example, the Second World War). Economics has also made further contributions in analysing the causes of conflict and in identifying potential targets during conflict (for example, the Second World War selection of aircraft factories, dams, submarine yards and oil fields as targets for Allied bombing raids on Germany).

Economic models start by analysing conflict as the use of military force to achieve a reallocation of resources within and between nations (that is, civil wars and international conflict). Nations invade to capture or steal another nation's property rights over its resources (such as land, minerals, oil, population, water). Conflict has a distinctive feature: it destroys goods and factors of production, and it is easier to destroy than to create. In peacetime, civilian economies aim to create more goods and services through growth and expanding a nation's production possibility frontier. Conflict uses military force and destructive power to enable a nation to acquire resources from another state, so expanding its production boundary through military force (Vahabi, 2004).

Conflict and terrorism provide opportunities for applying game theory. They involve strategic behaviour, interactions and interdependence between adversaries ranging from small groups of terrorists, rebels and guerrillas to nation states. Strategic interaction means that conflict can be analysed as games of bluff, chicken and 'tit-for-tat' with first-mover advantage and possibilities of one-shot or repeated games. For example, first-mover advantage might indicate a pre-emptive strike (for example, Pearl Harbour in 1941; Kuwait in 1990). However, there are other, non-economic explanations of conflict. These include religion, ethnicity and grievance (for example, Germany after the First World War); the desire for a nation state (such as Palestine); the absence of democracy; and mistakes and misjudgement.

The costs of war are a relatively neglected dimension of conflict. War involves both one-off and continuing costs. One-off costs are those of the actual conflict, while continuing costs are any post-conflict costs including those of occupation and peacekeeping. A further distinction is necessary between military and civilian costs. In principle, the military costs of conflict are the marginal resource costs arising from the conflict (that is, those costs which would not otherwise have been incurred). Examples include the costs of preparation and deployment prior to a conflict; the costs of the conflict, including the costs of basing forces overseas and the use of ammunition, missiles and equipment, including human capital and equipment losses in combat; the post-conflict occupation and peacekeeping missions and the costs of returning armed forces to their home nation.

There are further costs of conflict in the form of impacts on the civilian economies of the nations involved in the war. For example, the US and UK involvement in the Iraq war that began in 2003 had possible short- and long-term impacts for both economies. There were possible impacts on oil prices, share prices, the airline business, tourism, defence industries, private contractors, aggregate demand and future public spending plans. Further substantial costs were imposed on the Iraq economy in the form of deaths and injuries of military and civilian personnel, together with the damage and destruction of physical assets. Table 3 shows some examples of the costs of various conflicts for the UK and USA. The

Table 3   *Costs of conflict*

UK: Conflict	Military costs to UK (US$ billion, 2005 prices)
World War I	357
World War II	1,175
Gulf War	6.0
Bosnia	0.7
Kosovo	1.7
Iraq	6.0 +

USA: Conflict	Military costs to USA (US$ billions, 2005 prices)
World War I	208
World War II	3,148
Korea	365
Vietnam	537
Gulf War	83
Iraq	440

Estimated civilian costs:	Civilian costs (US$ billion, 2005 prices)
Iraq war	
Costs to US economy[a] from Iraq war	557
Costs to world economy[b] from Iraq war	1,183

Iraq war: costs to Iraq	US$ billion (2005 prices)
Reconstruction costs	20–60

Notes: [a]US civilian costs are of lost GDP for the period 2003–10.
[b]Cost to world economy is lost GDP for the period 2003–10.
Source: Hartley (2006b).

general point remains that wars are costly and require scarce resources which have alternative uses (that is, wars involve the sacrifice of hospitals, schools and social welfare programmes). Questions also arise as to whether the benefits of conflict exceed its costs.

Defence economists have also contributed to the analysis of terrorism using both choice-theoretic and game-theoretic models. Terrorism shows that nonconventional conflict is also costly. The attacks of 11 September 2001 on the USA resulted in almost 3,000 deaths and economic losses of $80 billion–90 billion (Barros, Kollisa and Sandler, 2005). Other terrorist-related costs include nations spending on homeland security measures, on terrorist-related intelligence, on security measures in airports, the lost time waiting at airports to clear security, the losses of liberty and freedoms and the war on terror (for example, in Afghanistan and Iraq).

Choice-theoretic models of terrorism apply standard consumer choice theory with terrorists maximizing a utility function subject to budget constraints. The utility function can be specific, such as a choice between attack modes, say, skyjackings and bombings, or more generally involve a choice between terrorist and peaceful activities. The approach offers some valuable insights into terrorist behaviour and possible policy solutions. The model shows that terrorist behaviour and activities can be influenced by governments acting to reduce terrorist funds (that is, an income effect), by changing relative prices (that is, promoting a substitution effect), and by efforts to change terrorist preferences towards more peaceful activities (for example, Northern Ireland). The substitution effect is an especially powerful insight showing that policies which increase the relative price of one attack mode, such as skyjackings, will encourage terrorists to substitute an alternative and lower-cost method of attack (for example, assassinations, bombings, or kidnappings: Frey and Luechinger, 2003; Anderton and Carter, 2005).

Conclusion

Defence economics is now established as a reputable subdiscipline of economics. It shows how economic theory and methods can be applied to the defence sector embracing the armed forces, defence industries and the political–institutional arrangements for making defence choices. But this is only the beginning. Massive opportunities remain for further research in the field. Changes in threats, new technology and continued budget constraints will require further adjustments in the armed forces and defence industries, and will generate a new set of research problems. Examples include space warfare, the economics of nuclear weapons policy, assessing the efficiency of armed forces, improving the efficiency of military alliances and developing more efficient approaches to international governance and international collective action.

KEITH HARTLEY

See also **arms races; arms trade; terrorism, economics of; war and economics; World Wars, economics of.**

Bibliography

Anderton, C. and Carter, J. 2005. On rational choice theory and the study of terrorism. *Defence and Peace Economics* 16, 275–82.

Barros, C., Kollisa, C. and Sandler, T. 2005. Security challenges and threats in a post-9/11 world. *Defence and Peace Economics* 16, 327–9.

Benoit, E. 1973. *Defense and Economic Growth in Developing Countries*. Boston: DC Heath.

BICC (Bonn International Centre for Conversion). 2005. *Conversion Survey 2005*. Baden-Baden: Nomos Verlagsgellschaft.

Brauer, J. 2003. Economics of conflict, war and peace in historical perspective, Special Issue, *Defence and Peace Economics* 14, 151–236.

Frey, B. and Luechinger, S. 2003. How to fight terrorism: alternatives to deterrence. *Defence and Peace Economics* 14, 237–49.

Goodwin, C., ed. 1991. *The Economics of National Security*. Durham and London: Duke University Press.

Hartley, K. 2006a. Defence R&D: data issues. *Defence and Peace Economics* 17(3), 1–10.

Hartley, K. 2006b. The economics of conflict. In T*he Elgar Companion to Public Economics: Empirical Public Economics*, ed. A. Ott and R. Cebula. Cheltenham: Edward Elgar.

Hegre, H. and Sandler, T. 2002. Economic analysis of civil wars. Special Issue of *Defence and Peace Economics* 13, 429–96.

Hitch, C. and McKean, R. 1960. *The Economics of Defense in the Nuclear Age*. Cambridge, MA: Harvard University Press.

Kennedy, P. 1988. *The Rise and Fall of the Great Powers*. London: Fontana Press.

NATO. 2005. NATO – *Russia Compendium of Financial and Economic Data Relating to Defence*. Brussels: NATO.

OECD. 2004. *Main Science and Technology Indicators*. Paris: OECD.

Oi, W. 1967. The economic cost of the draft. *American Economic Review* 57(2), 39–62.

Olson, M. and Zeckhauser, R. 1966. An economic theory of alliances. *Review of Economics and Statistics* 48, 266–79.

Peck, M. and Scherer, F. 1962. *The Weapons Acquisition Process*. Boston: Harvard University Press.

Richardson, L. 1960. *Arms and Insecurity: A Mathematical Study of the Causes and Origins of War*. Pittsburgh, PA: Homewood.

Sandler, T. and Hartley, K. 1995. *The Economics of Defense*. Cambridge Surveys of Economic Literature. Cambridge: Cambridge University Press.

Sandler, T. and Hartley, K., eds. 2003. *The Economics of Conflict*, 3 vols. Cheltenham: Edward Elgar.

Sandler, T. and Hartley, K., eds. 2007. *Handbook of Defense Economics*, vol. 2, Amsterdam: North-Holland.

Schelling, T. 1966. *Arms and Influence*. New Haven, CT: Yale University Press.

SIPRI (Stockholm International Peace Research Institute). 2005. *SIPRI Yearbook 2005*. Oxford: Oxford University Press.

US DoS (US Department of State). 2002. *World Military Expenditures and Arms Transfers, 1999–2000*. Washington, DC: Bureau of Verification and Compliance, US Department of State.

Vahabi, M. 2004. *The Political Economy of Destructive Power*. Cheltenham: Edward Elgar.

de Finetti, Bruno (1906–1985)

De Finetti was born in Innsbruck, Austria, and died in Rome. After a degree in mathematics at Milan University, he chose practical activities rather than an academic career, and worked at the Istituto Centrale di Statistica (1927–31) and then at the Assicurazioni Generali (1931–46). Only later did he turn to an academic career and win a chair in financial mathematics at Trieste University (1939); from 1954 to 1961 he held the chair in the same subject at the University of Rome and from 1961 to 1976 the chair of calculus of probabilities at the same university. He was a member of the Accademia Nazionale dei Lincei and Fellow of the International Institute of Mathematical Statistics.

De Finetti's fame rests on his contributions to probability and to decision theory, but he also worked in descriptive statistics, mathematics and economics.

Together with Ramsey and Savage, de Finetti is one of the founders of the subjectivist approach to probability theory. The first illustrations (in non-technical terms) of his conception are in (1930a) and (1931b). He considers probability as a purely subjective entity 'as it is conceived by all of us in everyday life'. The probability that a person attributes to the occurrence of an event is nothing more or less than the measure of the person's degree of confidence (hope, fear, …) in this event actually taking place. This can be interpreted as the amount (say, 0.72) that the person deems it fair to pay (or receive) in order to receive (or pay) the amount 1 if the event in question occurs. The mathematical theory was presented in his 1935 lectures at the Institut Poincaré (1937); see also (1970) and (1972).

De Finetti also introduced the important concept of *exchangeability* in probability (1929; 1930b; 1937; 1938) and proved the theorem on exchangeable variables named after him. Exchangeability is a weaker concept than independence and has been receiving increasing attention in probability theory (in fact, the natural assumption for a Bayesian is not independence, but exchangeability). In his 1935 Poincaré lectures (1937) he also treated the relations between the subjectivist point of view and the concept of exchangeability, which in his vision are at the basis of sound inductive reasoning and behaviour and, hence, of (statistical) decision theory (1959; 1961). It goes without saying that his position on the subject of statistical inference is fundamentally Bayesian.

In descriptive statistics he adhered to the functional concept according to which a statistic is an index selected on the basis of the single case (the aspects that one wants to stress, the aim of the statistical investigation, and so on); in (1931a) he stressed the importance of means which have the property of being associative.

Among his mathematical contributions the (1949) paper is especially interesting for economists. Here de Finetti investigates the conditions under which a concave function can be associated with a given 'convex stratification' (that is, a one-parameter family of convex sets, one interior to the other as the parameter varies). The author also discusses the conditions for a quasi-concave function to be transformed into a concave one by means of an increasing function. This paper started the literature on the 'concavification' of quasi-concave functions. As the author pointed out, these investigations also bear on consumer theory – where the convex stratification is the indifference map and the associated function is the utility function.

De Finetti also wrote on economic problems, where he stressed the importance of rigorous reasoning and verification, and emphasized the idea that the scope of economics, freed from the tangle of individual and corporative interests, should always and only be that of realizing a collective optimum (in Pareto's sense) inspired by criteria of equity (1969). An important initiative of his for the diffusion and correct application of mathematical and econometric methods in economics was the annual CIME (Centro Internazionale Matematico Estivo) seminar that he organized from 1965 to 1975; this enabled young Italian economists to benefit from courses given by Frisch, Koopmans, Malinvaud, Morishima, Zellner, to mention only a few of the lecturers.

GIANCARLO GANDOLFO

See also **Bayesian statistics; convexity; Savage, Leonard J. (Jimmie).**

Selected works

A full bibliography of de Finetti's works up to 1980 is contained in B. de Finetti, *Scritti (1926– 1930)*, ed. L. Daboni et al., Padua: Cedam, 1981, with an autobiographical note.

1929. Funzione caratteristica di un fenomeno aleatorio. In *Atti del Congresso Internazionale dei Matematici* (1928), Bologna: Zanichelli, 179–190.
1930a. Fondamenti logici del ragionamento probabilistico. *Bollettino dell'Unione Matematica Italiana* 9, December, 258–61.
1930b. Funzione caratteristica di un fenomeno aleatorio. *Memorie della Reale Accademia dei Lincei*, Classe di scienze fisiche, matematiche e naturali, vol. 4, fasc. 5.
1931a. Sul concetto di media. *Giornale dell'Istituto Italiano degli Attuari* 2, 369–96.
1931b. *Probabilismo. Saggio critico sulla teoria delle probabilità e sul valore della scienza*. Naples: Perrella; also in *Logos* (1931), 163–219.
1937. La prévision: ses lois logiques, ses sources subjectives. *Annales de l'Institut Henri Poincaré*, vol. 7, fasc. I. Trans. as 'Foresight: its logical laws, its subjective sources', in *Studies in Subjective Probability*, ed. H.E. Kyburg Jr. and H.E. Smokler, New York: Wiley, 1964.
1938. Sur la condition de 'équivalence partielle'. (Conférence au Colloque consacré à la théorie des probabilités, University of Geneva, 1937.) In *Actualités Scientifiques et Industrielles*, No. 739. Paris: Herman.
1949. Sulle stratificazioni convesse. *Annali di matematica pura e applicata*, series 4, vol. 30, 173–83.
1959. La probabilità e la statistica nei rapporti con l'induzione, secondo i diversi punti di vista. In *Atti corso CIME su Induzione e Statistica* (Varenna), Rome: Cremonese.
1961. Dans quel sens la théorie de la décision est-elle et doit-elle être 'normative'. In *Colloques internationaux du Centre National de la Récherche Scientifique*. Paris: CNRS.
1969. *Un matematico e l'economia*. Milan: F. Angeli (anthology of previously published papers).
1970. *Teoria delle probabilità. Sintesi introducttiva con appendice critica*, 2 vols. Turin: Einaudi. Trans. as *Theory of Probability*, 2 vols. New York: Wiley, 1974–5.
1972. *Probability, Induction and Statistics*. New York: Wiley (anthology of writings).

de-industrialization, 'premature' de-industrialization and the Dutch Disease

One of the most notable stylized facts of the world economy since the late 1960s is the rapid decline in manufacturing employment in industrialized countries (a fall of about 25 million jobs). Although the structure of employment has changed substantially over the long-term course of economic development, changes of the scale and speed during this period constitute an unprecedented phenomenon – manufacturing employment in the European Union, for example, has fallen by more than a third. Manufacturing is an activity considered by many as the most effective engine of growth – either because it is a crucial driver of outward shifts of the production frontier, or due to its capacity to set in motion processes of cumulative causation based on increasing returns (for example, in Post Keynesian, structuralist and Schumpeterian thought). It has therefore been argued that a process of de-industrialization on this scale is likely to have significant negative long-term effects on growth (on both its rate and its sustainability), investment, and employment.

This concern has been particularly pronounced in countries that experienced drastic de-industrialization following the discovery of mineral resources – a phenomenon that became known as the 'Dutch Disease'. The key issue is the double-edged effect of a mineral discovery. On the one hand, it allows for an expansion of expenditure and employment; but on the other, it could easily lead to a contraction of the non-commodity tradable sector. This phenomenon was first analysed in the 1950s in relation to the mixed effects of sudden increases in mineral exports or in the price of wool for the Australian economy.

OECD countries began de-industrializing in the late 1960s, while some high-income developing countries in East Asia entered this phase in the late 1980s. Soon afterwards, some middle-income Latin American countries and South Africa also began to de-industrialize after radical economic reforms, despite their level of income per capita being far lower than other countries which began to de-industrialize earlier. This latter process has been labelled 'premature' de-industrialization (Palma, 2005), and should not be confused with the so-called 'resource curse' hypothesis, which refers to the poor macroeconomic performance of many mineral-exporting economies.

The following tables (Tables 1 and 2) show the above-mentioned regional trends in the share of manufacturing in total employment and the share of manufacturing value added in GDP.

The four sources of de-industrialization

The first source of de-industrialization: an 'inverted-U' relationship between manufacturing employment and income per capita.
The most commonly used concept of de-industrialization emerges from an understanding of the relationship

Table 1   *Manufacturing employment (% of total), 1960–2003*

Region	1960	1970	1980	1990	2003
Sub-Saharan Africa	4.4	4.8	6.2	5.5	5.5
South Africa	11.3	12.8	18.2	15.7	14.1
Latin America	15.4	16.3	16.5	16.8	14.2
Southern Cone and Brazil	17.2	17.5	17.3	17.9	13.1
Middle East and North Africa	7.9	10.7	12.9	15.1	15.3
South Asia	8.7	9.2	10.7	13.0	13.9
East Asia (excluding China)	10.0	10.4	15.8	16.6	14.9
NICs	14.6	19.2	27.5	28.7	19.4
China	10.9	11.5	10.3	13.5	12.3
Third World	10.2	10.8	11.5	13.6	12.5
OECD	26.5	26.8	24.1	20.1	17.3

Notes: Averages are weighted by economically active population. Southern Cone: Argentina, Chile and Uruguay. NICs: Korea, Taiwan, Singapore and Hong Kong.
Sources: Calculations using International Labour Organization (ILO) databank; for Taiwan, *The Republic of China Yearbook of Statistics*.

Table 2   *Manufacturing value added (% of GDP), 1960–2003*

Region	1960	1970	1980	1990	2003
Sub-Saharan Africa	15.3	17.8	17.4	14.9	13.8
South Africa	21.0	23.9	22.5	25.5	18.1
Latin America	28.1	26.8	28.2	25.0	16.7
Southern Cone and Brazil	32.2	29.8	31.7	27.7	16.9
Middle East and North Africa	10.9	12.2	10.1	15.6	14.2
South Asia	13.8	14.5	17.4	18.0	16.2
East Asia (excluding China)	14.0	19.2	23.3	25.5	27.6
NICs	15.4	22.5	27.1	26.5	24.9
China	23.7	30.1	40.6	33.0	31.3
Third World	21.6	22.1	24.3	23.9	22.7
OECD	28.9	28.3	24.5	22.1	17.3

Note: NICs does not include Taiwan.
Source: Calculations using data (in real terms) from World Bank (1984; 2006).

between manufacturing employment and income per capita as an 'inverted-U'. De-industrialization is simply the drop in manufacturing employment occurring when countries reach a certain level of income per capita – that is, mature economies switching employment to specialized services as part of their 'normal' process of development. As such, de-industrialization could well have positive long-term growth effects. According to Rowthorn (1994), using data for 1990, this drop begins at US$12,000 (Figure 1).

Although many other analyses are consistent with this hypothesis, Palma (2005) has suggested that de-industrialization has been a more complex phenomenon. He argues that, in addition to the 'inverted-U', there are three further processes at work.

The second source of de-industrialization: a declining relationship over time between income per capita and manufacturing employment
One additional source of de-industrialization has been the remarkable collapse of the 'inverted-U' relationship over time.

In essence, for high- and middle-income countries, the level of manufacturing employment associated with a given level of income per capita has been falling over time. In fact, the four better-known hypotheses originally developed to explain the 'inverted-U' relationship mentioned above are more relevant to this 'second source' of de-industrialization, as until the mid-1980s no country had reached the level of income corresponding to the turning point of the respective curve. These hypotheses are as follows:

- The fall in manufacturing employment is merely a statistical illusion caused primarily by the re-allocation of labour from manufacturing to services through

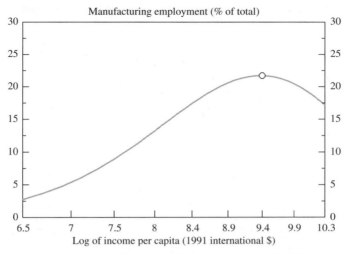

Figure 1   Rowthorn's regression: manufacturing employment and income per capita, 1990. *Source*: Rowthorn (1994).

contracting-out of activities such as transport, cleaning, design, security, catering, recruitment and data processing. This process could be the result of a further movement in a long line of progressive transformations aiming at enlarging the scope for specialization and the division of labour, or just a cost-cutting operation aimed, for example, at bypassing labour legislation.

- The fall in manufacturing employment results from a reduction in the income elasticity of demand for manufactures, particularly in high income countries.
- It is the consequence of higher productivity growth in manufacturing than in other sectors of the economy.
- It is the result of a new international division of labour (including 'outsourcing'), which has a negative impact on manufacturing employment in industrialized countries, especially for non-skilled labour.

Although a detailed analysis of the role of each of these factors in de-industrialization is outside the scope of this article (see Rowthorn and Ramaswamy, 1999), it is important at least to add that the 1980s switch in 'policy regime' in OECD countries (broadly speaking, from post-war Keynesianism to demand-constraining monetarism) did also contribute to the huge 1980s drop in manufacturing employment. (For the 1980s debate on de-industrialization, see Singh, 1987.) The technological revolution that took off in the 1980s also played a major role (Pérez, 2002).

The third source of de-industrialization: changing income per capita corresponding to the turning point of the regression

This additional source of de-industrialization is also evident in Figure 2 (see also Figure 3). This concerns the

remarkable leftwards movement in the turning point of the regressions during the 1980s. (Rowthorn and Wells, 1987, had discussed the possibility of the 'inverted-U' relationship peaking at a lower level of income per capita over time.) During the 1980s the income per capita at which the curve peaked fell by about half – from approximately $21,000 in 1980 to just over $10,000 in 1990 (in 1985 international US$). Until 1980 no country was located to the right of the turning point of the corresponding cross-section curve, but in the 1990 regression there were more than 30 countries beyond that critical point. However, during the 1990s this process was reversed, and by 2000 again no country was beyond that critical point.

The changing shape of the curves is crucial to the understanding of the dynamic of the interrelationship between the three sources of de-industrialization discussed so far. Basically, as the arrows of Figure 3 indicate, during the 1980s there was a remarkable degree of de-industrialization in high-income countries; during the 1990s, by contrast, de-industrialization affected mainly middle-income countries.

The fourth source of de-industrialization: the Dutch Disease

Finally, in several countries we can observe a further degree of de-industrialization. These countries experienced a fall in their manufacturing employment that was clearly greater than would have been expected, given the three sources of de-industrialization discussed above (Figure 4).

Rather than simple cases of 'overshooting', Palma (2005) identifies this phenomenon with a specific conceptualization of the Dutch Disease: in countries that have an export surge of commodities or services, or a major shift in economic policy, a unique *additional* degree of de-industrialization is typical (that is,

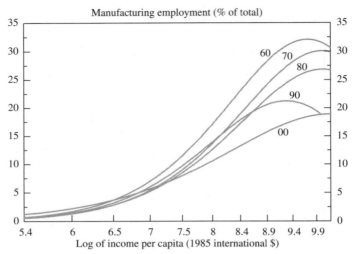

Figure 2   Second source of de-industrialization: a declining relationship, 1960–2000. *Notes*: The range of the horizontal axis is the actual income range of the sample for 2000. The regressions are based on a sample of 105 countries. In all regressions in this and the following figures, all parameters are significant at the 1% level, and the adjusted R^2 are between 66% and 77%. All regressions pass the relevant diagnostic tests. Note that these regressions are simply a cross-sectional description of cross-country differences in manufacturing employment, when categorized by income per capita; hence they should not be interpreted in a 'predicting' way, because there are a number of difficulties with a curve estimated from a single cross-section – especially regarding the homogeneity restrictions that are required to hold. *Source*: Palma (2005), using ILO Databank and the Penn World Tables; this is the source for all figures other than Figure 8.

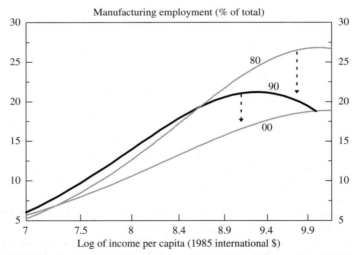

Figure 3   The changing nature of de-industrialization between the 1980s and the 1990s. *Note*: The range of the horizontal axis is the actual income range of the sample for 2000.

additional to the three de-industrialization forces already discussed).

Originally, Dutch Disease had a narrow meaning – the appreciation of the real exchange rate resulting from a boom in commodity exports. (For an analysis of the macro-processes at work, see Corden and Neary, 1982;

Ros, 2000.) Elsewhere, mostly in neoclassical models, it simply referred to the adverse terms of trade effect for tradables following a sudden shift in their production frontier. However, with time the meaning has widened to include all possible negative macroeconomic effects associated with the 'resource curse' hypothesis—for

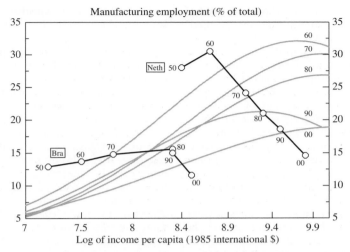

Figure 4   Fourth source of de-industrialization: cases of 'overshooting'? *Notes*: Neth: The Netherlands; Bra: Brazil.

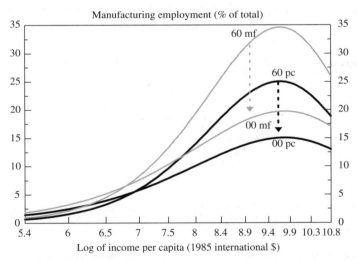

Figure 5   Changes in manufacturing employment and income per capita, 1960–2000. *Note*: An intercept dummy differentiates the two groups of countries.

Woolcock, Pritchett and Isham (2001), for example, resource-rich countries are not very good at accumulating social capital. (See also Mehlum, Moene and Torvik, 2006; for a critical analysis of the 'resource curse' hypothesis, see DiJohn, 2007.)

The origins of this 'disease' lie in the fact that the relationship between manufacturing employment and per capita income tends to differ between those countries that generate a trade surplus in manufacturing and those that do not. Note that the 'trade surplus in manufacturing' group includes economies that find themselves in this position out of *necessity* as well as others due to *growth policy*. In the first case, given resource

endowments force some countries to aim for a manufacturing surplus to finance inevitable trade deficits in commodities and/or services (for example, Japan and India). In the second, some resource-rich countries still try to achieve a trade surplus in manufacturing by implementing a growth policy based on a strong 'industrialization agenda' (for example, Finland, Malaysia and Vietnam).

Figure 5 shows the long-term changes between manufacturing employment and income per capita in the 'trade surplus in manufacturing' (mf) and in the 'trade surplus in primary commodities or services' (pc) groups of countries.

Although the 'pc' countries tend to reach a lower level of industrialization at any given point in time, the 'pc effect' per se has not led to a higher degree of de-industrialization. In fact, if we take the highest point of the curves, in these four decades the share of manufacturing employment in both 'mf' and 'pc' countries dropped by about half.

After this introduction, it is now possible to explain the concept of Dutch Disease. There is a group of countries – both industrialized and developing – that exhibit a specific *additional* degree of de-industrialization. The Netherlands rightly gives its name to this phenomenon.

From this perspective, what happened in the Netherlands was a discovery of a natural resource (gas) leading manufacturing employment to switch from an 'mf' structure to a 'pc' one. When this occurs, as Figure 6A shows, the country experiencing this 'disease' moves along *two* different paths of de-industrialization. The first path consists of the three processes of de-industrialization discussed above (from '60-mf' to '00-mf'). The second corresponds to a further component of de-industrialization resulting from the change in the reference group (from '00-mf' to '00-pc'). In this context, the Dutch Disease should be regarded only as the

additional level of de-industrialization associated with the latter movement. In the case of the Netherlands, then, it is the (five percentage points) difference between manufacturing employment falling by 10.9 percentage points between 1960 and 2000 (hypothetical non-Dutch Disease scenario), or by 15.9 percentage points (actual Dutch-Disease situation) – that is, manufacturing employment falling from 30.5 per cent of total employment in 1960, to 19.6 per cent in 2000 in the former scenario, or to 14.6 per cent in the latter one.

Dutch Disease is thus clearly not a phenomenon limited to the Netherlands or to the discovery of mineral resources; it has also occurred in other countries and for other reasons. One case is the United Kingdom, which had a boom in both oil and financial-services exports (see Figure 6B). As a result of this (and of Prime Minister Thatcher) the trade balance in manufacturing switched from a surplus of four per cent of GDP (late 1970s) to a deficit of four per cent (mid-2000s). Figure 6C shows that, by contrast, the share of manufacturing employment in other EU countries fell only according to the changes in the 'mf' scenario. In turn, Figure 6D shows that, although four other industrialized countries (major

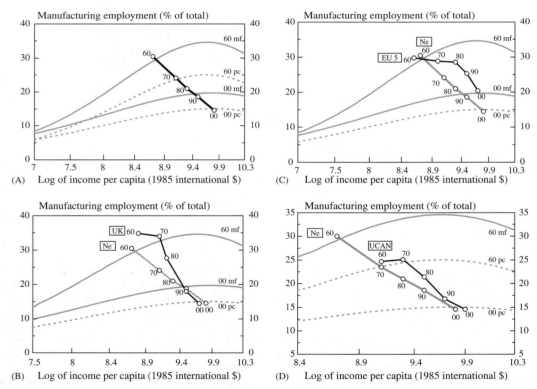

Figure 6   (A) The Netherlands: unravelling the Dutch Disease, 1960–2000. (B) The United Kingdom; catching the Dutch Disease, 1960–2000. *Notes*: Ne: The Netherlands; UK: United Kingdom. (C) Five countries of the European Union, 1960–2000. *Notes*: EU5: Germany, France, Italy, Belgium and Austria. (D) Four traditional primary commodity exporters, 1960–2000. *Notes*: UCAN: United States, Canada, Australia and New Zealand.

commodity exporters throughout the period) also found themselves in the 'pc' category in 2000, they did not suffer from the Dutch Disease simply because they were in the 'pc' category from the start. Although both the 'EU-5' and the 'UCAN' countries experienced a *similar* drop in the share of manufacturing employment (9.7 and 10.5 percentage points, respectively), neither switched from one reference group to another.

The phenomenon of the Dutch Disease also occurred in countries that developed flourishing service-exporting sectors, such as tourism (for example, Greece, Cyprus and Malta) and financial services (for example, Switzerland, Luxembourg and Hong Kong); see Palma (2005).

Finally, this 'disease' was also experienced after 1980s in some middle-income Latin American countries (and to some extent in South Africa) where state-led import-substituting industrialization (ISI) had achieved industrialization levels characteristic of the 'mf' group (despite the fact that these countries generated large trade surpluses in commodities). In these cases, radical change of the economic policy regime (from ISI to comprehensive trade and financial liberalization) resulted in the Dutch Disease; that is, the transformation of their employment structure from a policy-induced 'mf' to a more 'Ricardian' resource-rich 'pc'.

Brazil and the three Southern Cone countries experienced the greatest de-industrialization following their economic reforms, while also being among the countries of the region that had previously industrialized the most and that had subsequently implemented the most drastic reforms (Figure 7).

These four Latin American countries began this period – as did the Netherlands – with a level of manufacturing employment typical of countries aiming at a trade surplus in manufacturing ('60-mf'), even though these resulted from different causes. The case of the Netherlands is due to its (pre-natural gas) poor resource endowment, whereas in the four Latin American countries their position was the result of a 'structuralist' industrialization agenda (see STRUCTURALISM). And if both reached 2000 with levels of manufacturing employment typical of the 'pc' group, this was once again for different reasons: in the Netherlands, the discovery of a natural resource at a 'mature' stage of industrialization was decisive, whereas in Latin America the sharp reversal of the ISI policies was responsible. Note that in the latter 'extra' degree of de-industrialization ('mf' to 'pc') took place over and above the already mentioned huge collapse of the 'mf' path for middle-income countries during the 1990s (Figure 3).

From this perspective, the key difference between developing Asia and 'premature de-industrializers' in Latin America in terms of the implementation of economic reforms is that in the latter these reforms seem to have obstructed their transition towards a more mature – that is, self-sustaining – form of industrialization. (For the concept of 'self-sustaining industrialization', see Kaldor, 1967.) Resource-poor *and* resource-rich developing Asia, instead, succeeded in combining these reforms with a dynamic 'mf' path (Figure 8).

Perhaps Latin America is in desperate need of a touch of the so-called East Asian Confucianism; that is, once a new development path has been chosen, a significant degree of pragmatism, self-confidence, a progressive capitalist elite and an avant-garde political leadership can be of great assistance in policymaking success.

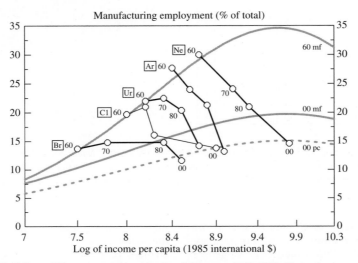

Figure 7   Argentina, Brazil, Chile and Uruguay: catching the Dutch Disease, 1960–2000. *Notes*: Ar: Argentina; Br: Brazil; Cl: Chile; Ur: Uruguay; Ne: The Netherlands. The year 1990 has been omitted not to 'congest' the figure. South Africa's share of manufacturing employment also fell from an 'mf' level in 1980 to close to a 'pc' one in 2000.

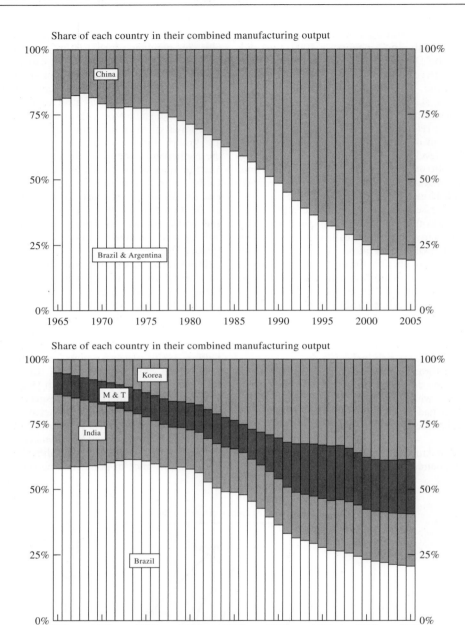

Figure 8   Brazil, Argentina and China: manufacturing production, 1965–2005. *Notes*: M&T: Malaysia and Thailand. Three-year moving averages. The relative decline of South Africa's manufacturing sector is even greater than Brazil's (though not as extreme as Argentina's). Manufacturing output measured in US$, 2000. *Source*: World Bank (2006).

In sum, the Dutch Disease should not be seen as simple 'overshooting' of de-industrialization, but as a specific type of 'additional' de-industrialization. In general, this has taken place for one of three different reasons: the discovery of natural resources (for example, the Netherlands); the development of export-service activities,

mainly tourism and finance (for example, Greece in the former, and Hong Kong in the latter); and finally, changes in economic policy (for example, Brazil and South Africa).

All the above types of de-industrialization should also be distinguished from those of the late 1980s and 1990s in many sub-Saharan economies and countries of the former Soviet Union and eastern Europe, which experienced a process of de-industrialization associated with a *fall* in income per capita: a case of de-industrialization 'in reverse'.

Finally, Finland, Sweden, Denmark, Malaysia, Vietnam and, to a lesser extent, other south-east Asian resource-rich countries (such as Thailand and Indonesia) prove both that economic policies do exist to avoid the Dutch Disease in commodities- and export-services-booming economies (see Pesaran, 1984; Palma, 2000), and that there is no such thing as an unavoidable 'curse of natural resources'. Countries with high potential for developing commodities and export-services activities have sufficient degrees of policy freedom to follow ambitious and successful 'industrialization agendas' (not least of the commodities themselves, as in the Nordic case and Malaysia). Also, export rents could be used effectively in that direction. However, as the Latin American experience in particular shows, it seems that, as globalization progresses, fewer and fewer countries are willing to take advantage of such degrees of policy freedom. This is not only because forces in the new international institutional and financial order are rapidly working to reduce these degrees of policy freedom, but also because of domestic changes in economic ideologies and the structure of property rights.

However, whether a process of structural change that includes 'premature' de-industrialization can deliver rapid and *sustainable* economic growth is another matter altogether; so is the issue of whether the current 'premature' de-industrialization occurring in Latin America and South Africa contains important components of policy-induced 'uncreative destruction'.

De-industrialization: does it matter?

Rapid de-industrialization has reopened an age-old debate in economic theory: is a unit of value added in manufacturing equal to one in commodities or services, especially in terms of its growth-enhancing properties?

Although a detailed discussion of this debate is beyond the scope of this article, from the perspective of de-industrialization we may classify growth theories into three groups (in doing this, of course, we have to acknowledge the necessary degree of simplification which every classification of intellectual tendencies entails). This requires a distinction between two concepts: 'activity' and 'sector'. Examples of the former include R&D and education, and of the latter manufacturing. The first camp of growth theories includes those (mainly neoclassical models) that treat growth as both 'sector-*indifferent*' and 'activity-*indifferent*'. Examples are Solow-Swam-type models (both traditional and 'augmented' ones), and the branch of 'endogenous' theories that associates growth with increasing returns which are activity-indifferent. Examples are early 'AK' models and more recent ones in which changes in the rate of growth are the result of the

cumulative effect of market imperfections arising in the process of technical change. However, these imperfections, and the associated increasing returns, are somehow seen as stemming directly from within the production function (rather than being based on the use of R&D or the production of human capital).

The second camp still regards growth as 'sector-indifferent', but models it as 'activity-*specific*' (for example, Romer's work and neo-Schumpeterian models). In these models, increasing returns, though generated by research-intensive activities, are explicitly not associated with manufacturing activities as such or with investment in manufacturing; nor do they allow for specific effects from manufacturing on R&D activities (except that investment in *any* sector could be 'complementary' to R&D through its effect on the profitability of research; see Aghion and Howitt, 1998). Therefore, in these models there is no room for Kaldorian-style effects concerning investment embedding or embodying technical change.

Finally, in the third camp are those (mainly Post Keynesian, Schumpeterian and structuralist theories) that argue that growth is both 'sector-*specific*' and 'activity-*specific*' (but the latter only in the sense that it is specific to the nature of the sector involved). For instance, the approaches to growth found in Hirschman, Kaldor, Kalecki, Prebisch, Furtado, Thirlwall and (arguably) Schumpeter follow this line of argument. What is common to these 'sector-specific' growth theories is that the pattern, the dynamic and the sustainability of growth are crucially dependent on the activities being developed. In particular, there are specific growth enhancing effects associated with manufacturing due to its capacity to set in motion processes of cumulative causation. This is because 'learning by doing', dynamic economies of scale, increasing returns, externalities and spillover effects are more prevalent in manufacturing than elsewhere in the economy. Therefore, the crucial feature distinguishing this camp from the previous two is that issues such as technological change, synergies, balance-of-payments sustainability and the capacity of developing countries to 'catch up', are directly linked to the size, strength and depth of the manufacturing sector.

Then, in terms of the possible growth consequences of de-industrialization, the first growth camp does not regard de-industrialization as a particularly relevant growth issue per se. Even when it becomes a major growth or employment issue, this is only due to market imperfections. For example, Sachs and Warner (1997) argue that if neoclassical competitive conditions prevail, a declining manufacturing sector implies no hindrance to growth or full employment. Furthermore, for these growth theories, even if the discovery of natural gas did produce some structural changes in output and employment in the Dutch economy, labelling these transformations a 'disease' would be a misleading dramatization. Also, from this perspective, if 'premature' de-industrialization in resource-rich countries consists of the transformation of employment structures from an artificially policy-induced 'mf' to a more Ricardian 'pc' path, that can hardly be bad for growth!

From the point of view of the second camp, de-industrialization in 'mature' economies may or may not have an impact on growth per se; this would all depend on the specific form that the de-industrialization takes. For instance, it could actually result in a stimulus for growth if the 'upward' de-industrialization in mature economies is associated with the reallocation of resources within manufacturing into more R&D-intensive products. However, in the case of 'premature' de-industrialization in middle-income countries it is more difficult to argue from this approach that such a phenomenon could be positive for long-term growth.

Finally, except for normal (or 'upward') de-industrialization in properly mature economies, the third approach to economic growth understands de-industrialization and the Dutch Disease as unequivocally negative for growth and employment – especially if it involves 'premature' de-industrialization in developing countries. The same is true of the current narrowing-down of the policy space to fight them. For example, an interpretation from this perspective of the industrialized countries' remarkable slowdown in productivity growth since the mid-1970s could be that this may well be the result of 'wrong' policies (such as monetarism) and 'wrong' structural change (such as 'financialization') excessively intensifying de-industrialization in the 1980s. ('Financialization' is the rise in size and dominance of the financial sector relative to the non-financial sector, as well as the diversification towards financial activities in non-financial corporations.) And one interpretation of the remarkably poor growth performance of most Latin American economies and South Africa since economic reform, especially Brazil, would be that this is the likely consequence of 'premature' de-industrialization – affecting not just the pace of their economic growth but its sustainability.

JOSÉ GABRIEL PALMA

See also **Kaldor, Nicholas; Kalecki, Michal; Post Keynesian economics; Prebisch, Raúl; Robinson, Joan Violet; Schumpeter, Joseph Alois; structuralism; unemployment; Young, Allyn Abbott.**

I am extremely grateful to Fiona Tregenna for many constructive comments.

Bibliography

Aghion, P. and Howitt, P. 1998. *Endogenous Growth Theory*. Cambridge, MA: MIT Press.
Auty, R.M., ed. 2001. *Resource Abundance and Economic Development*. New York: Oxford University Press.
Corden, W.M. and Neary, J.P. 1982. Booming sector and Dutch disease economics. *Economic Journal* 92, 825–48.
DiJohn, J. 2007. *The Political Economy of Late Industrialization in Oil-Exporting Countries*. Pennsylvania: Penn State University Press.

Kaldor, N. 1967. *Problems of Industrialization in Underdeveloped Countries*. Ithaca: Cornell University Press.
Mehlum, H., Moene, K. and Torvik, R. 2006. Institutions and the resource curse. *Economic Journal* 116, 1–20.
Palma, J.G. 2000. Trying to 'tax and spend' oneself out of the Dutch-Disease: the Chilean economy from the war of the Pacific to the Great Depression. In *An Economic History of Latin America*, ed. E. Cárdenas, J.A. Ocampo and R. Thorp. Basingstoke: Palgrave.
Palma, J.G. 2005. Four sources of 'de-industrialisation' and a new concept of the 'Dutch-disease'. In *Beyond Reforms: Structural Dynamic and Macroeconomic Vulnerability*, ed. J.A. Ocampo. Palo Alto: Stanford University Press and the World Bank.
Pérez, C. 2002. *Technological Revolutions and Financial Capital*. Cheltenham: Elgar.
Pesaran, H. 1984. Macroeconomic policy in an oil-exporting economy with foreign exchange controls. *Economica* 49, 253–70.
Pieper, U. 2000. De-industrialisation and the social and economic sustainability nexus in developing countries. *Journal of Development Studies* 36, 66–99.
Ros, J. 2000. *Development Theory and Economic Growth*. Ann Arbor: University of Michigan Press.
Rowthorn, R. 1994. Korea at the cross-roads. Working Paper No. 11, Centre for Business Research, Cambridge.
Rowthorn, R. and Ramaswamy, R. 1999. Growth, trade and deindustrialization. *IMF Staff Papers* 46(1), 18–41.
Rowthorn, R. and Wells, J. 1987. *De-Industrialisation and Foreign Trade*. Cambridge: Cambridge University Press.
Sachs, J.D. and Warner, A.M. 1997. Natural resource abundance and economic growth. Working paper, HIID, Harvard University.
Singh, A. 1987. Manufacturing and de-industrialization. In *The New Palgrave: A Dictionary of Economics*, vol. 3, ed. J. Eatwell, M. Milgate and P. Newman. London: Macmillan.
Thirlwall, A. 2002. *The Nature of Economic Growth*. Cheltenham: Elgar.
Woolcock, M., Pritchett, L. and Isham, J. 2001. The social foundations of poor economic growth in resource-rich countries. In *Resource Abundance and Economic Development*, ed. R.M. Auty. New York: Oxford University Press.
World Bank. 1984. *World Development Indicators 1984*. New York: Oxford University Press.
World Bank. 2006. *World Development Indicators 2006*. Washington, DC: World Bank.

demand price

Earlier economic literature doubtless contains casual usages of the phrase 'demand price', but its appropriation as a technical term appears to date from Alfred Marshall's *Principles of Economics* (Marshall, 1890: see Marshall, 1920, pp. 95–101). Marshall applied the term in the contexts of both individual and market demand. Starting with a commodity (tea) purchasable in integral units of a pound's weight, an individual's demand price for the xth pound is the price he is just willing to pay for it given that he has already acquired $x - 1$ pounds. The basic assumption is that this demand price is lower the larger is x. A schedule of demand prices for all possible quantities (values of x) defines the consumer's demand schedule. Its graph is naturally drawn with quantity on the horizontal axis. In the case of a perfectly divisible commodity, the demand price of quantity x must be redefined as the price *per unit* which the consumer would be willing to pay for a tiny increment, given that he already possesses amount x. The demand schedule then graphs as a continuous negatively sloped demand curve showing demand price in this sense as a function of x.

If the individual is free to buy any quantity at a fixed price, his 'marginal demand price' is the demand price for that quantity 'which lies at the margin or terminus or end of his purchases' (Marshall, 1920, p. 95). For a perfectly divisible commodity, marginal demand price must equal market price. For a commodity purchasable in integral units only, market price may lie anywhere below marginal demand price, but not so low as to make the next unit marginal.

Marshall's discussion of consumer behaviour is based on two general assumptions, although these are informally relaxed at various points. The first is that the utility obtained from consuming a commodity depends only on the amount of that commodity. The second is that the marginal utility of 'money', or expenditure on all other goods, remains approximately constant with respect to variation in the expenditure on any particular commodity – the presumption being that the latter expenditure is only a small fraction of total expenditure. These assumptions have convenient consequences for the concept of demand price. If $u(x)$ denotes the utility a consumer obtains from consuming quantity x of a given good in a specified period, while λ is the constant marginal utility of money to him, then demand price for quantity x is $(\mathrm{d}u/\mathrm{d}x)/\lambda$ in the case of divisible quantity and $[u(x) - u(x - 1)]/\lambda$ in the case when only integral quantities are feasible. In either case, given the value of λ, demand price depends on x alone and is proportional to marginal utility. The hypothesis of diminishing demand price is tantamount to that of diminishing marginal utility. A further advantage is that the demand price for quantity x is independent of the pecuniary terms on which the earlier units were, or are to be, acquired, as these terms will not change the marginal utility of money.

Although demand price is, on the above assumptions, proportional to marginal utility it has the great advantage of being measured in operational money units. This permits a monetary measure of the net benefit or consumer surplus obtained from the option of buying the commodity in question on specified monetary terms, rather than having to divert the expenditure to other goods. The

distinction between demand price and market price is an operational version of the classical distinction between value in use and value in exchange.

The concept of demand price features prominently in Marshall's analysis of the market for a single commodity sold at a fixed price which is uniform to all buyers. Demand price is now interpreted as the maximum *uniform* price at which any specified aggregate quantity of the commodity can be sold on the market during a given period. The negatively sloped market demand curve is simply a lateral addition of the individual demand curves and expresses the common demand price as a function of the aggregated quantity. Marshall recognized (1920, p. 457n) that it would be more natural when dealing with market demand to view quantity as a function of price, as Cournot (1838, pp. 44–55) had done, but chose the converse approach to maintain symmetry with his treatment of supply. Believing in the importance of scale economies in production, he deemed it generally impossible to treat quantity supplied per unit of time as a single-valued function of market price. Instead, adopting what he took to be the businessman's perspective, he introduced the concept of 'supply price'; the minimum uniform price at which any given quantity will be supplied to the market.

Market equilibrium occurs at any quantity whose demand price and supply price are equal, so that the market demand curve intersects the market supply curve – the latter the graph of supply price as a function of aggregate quantity supplied, a lateral sum of individual supply curves. Equilibrium is locally stable if the demand curve cuts the supply curve from above at the equilibrium quantity. This result is justified by the argument that the rate of supply will increase if the current market price (always determined by demand price) exceeds supply price at the current quantity, so that additional production offers excess profit, decreasing in the opposite case (Marshall, 1920, pp. 345–7). The resulting dynamic process is usually referred to as the Marshallian adjustment process.

It is probably due to Marshall's influence that English-speaking economists still graph demand and supply curves with quantity on the horizontal axis even though adopting a more Walrasian perspective which treats quantities demanded and supplied as functions of market price.

Marshall's conception of the demand price of a lone commodity, segregated from other commodities by an assumed constancy of the marginal utility of money, does not feature prominently in modern theoretical work. Instead, a multi-commodity formulation of utility and demand is typically adopted. Consider a consumer maximizing the utility function $u(x_1, x_2, \ldots, x_n)$ subject to the budget constraint $\Sigma_{p_i x_i} = M$. (Here the x_i are quantities and the p_i prices of the n commodities and M is a preset total expenditure level. The utility function, u, is assumed strictly increasing, strictly quasi-concave, and differentiable.) This maximization implies the consumer's direct demand functions $x_i = d_i(p_1/M, p_2/M, \ldots, p_n/M)$, $i = 1, 2, \ldots, n$, sometimes (but with dubious justification) referred to as Marshallian demand functions to distinguish them from Hicksian compensated or constant-utility demand functions. These demand functions can usually be inverted to yield the indirect or inverse demand functions $p_i/M = g_i(x_1, x_2, \ldots, x_n)$, $i = 1, 2, \ldots, n$. However, these can be obtained more immediately from the budget constraint and the first-order conditions $\partial u/\partial x_i = \lambda p_i, i = 1, 2, \ldots, n$ (where λ is the Lagrange multiplier associated with the budget constraint). We have, for $i = 1, 2, \ldots, n$,

$$\frac{p_i}{M} = \frac{\partial u/\partial x_i}{\lambda M} = \frac{\partial u/\partial x_i}{\sum \lambda p_j x_j} = \frac{\partial u/\partial x_i}{\sum x_j(\partial u/\partial x_j)}$$
$$\equiv g_i(x_1, x_2 \ldots, x_n) \qquad (1)$$

(The g_i are clearly unaffected by a monotone increasing transformation of u and reduce to $(\partial u/\partial x_i)/u$ if u is homogeneous of degree one.) The indirect demand functions (1) are the natural generalization of Marshall's demand-price concept at the individual level, defining an n-vector of normalized prices at which a given n-vector of commodities will be demanded.

Indirect demand functions may be useful in the contexts of central planning or rationing, where they can indicate the prices planners should choose to clear markets given the quantities available, or the notional prices at which ration allotments would just be freely purchases (see Pearce, 1964, pp. 57–64). But unfortunately, although indirect demand functions are readily obtained for the individual, they are not as easily aggregated to the market level as are direct demand functions. The asymmetry arises from the fact that individuals face identical prices but do not make identical quantity choices. Thus, market-level indirect demand functions must generally be obtained by first aggregating the individual direct demand functions and then inverting the resulting market functions.

The modern duality approach to consumer behaviour has revealed fundamental symmetries in the roles of prices and quantities. The alternatives of viewing quantity demanded as a function of price or demand price as a function of quantity can now be seen as only one of a variety of dual alternatives which considerably enrich theoretical and econometric analysis. (See Gorman, 1976, for a simple treatment.)

J.K. WHITAKER

See also **Marshall, Alfred.**

Bibliography

Cournot, A.A. 1838. *Mathematical Principles of the Theory of Wealth*. New York: Macmillan, 1897.
Gorman, W.M. 1976. Tricks with utility functions. In *Essays in Economic Analysis*, ed. J.J. Artis and A.R. Nobay, London: Cambridge University Press.
Marshall, A. 1890. *Principles of Economics*, Vol. I. London: Macmillan.
Marshall, A. 1920. *Principles of Economics: An Introductory Volume*. London: Macmillan. Eighth edition of Marshall (1890).
Pearce, I.F. 1964. *A Contribution to Demand Analysis*. Oxford: Clarendon Press.

demand-pull inflation

The term 'demand-pull' inflation originated with the simple Keynesian model of the macroeconomy and was used as a contrast to price increases arising from shocks to aggregate supply. In the Keynesian model, there is a well-defined level of potential GDP corresponding to full employment levels of employment and unemployment. Nominal wages are downwardly rigid, so that below full employment aggregate supply increases with prices while aggregate demand decreases. The difference between potential and actual GDP is the output gap, and there is an asymmetry in the economy's response to shifts in demand when output gaps are positive and when they are not. With a positive gap – that is, in the operating region below full employment – an expansion of aggregate demand mainly raises employment and output and only moderately raises prices. But at full employment the aggregate supply curve is vertical, and an expansion of demand only pulls up wages and prices. Hence the term 'demand-pull inflation'.

Macro models of fluctuations have evolved in important ways from this simple Keynesian case. The early empirical Phillips curves described an empirical relation between the level of unemployment and rates of change, rather than levels, of prices. Such relations were estimated from periods characterized by frequent cycles in activity. They did not control for expected or ongoing rates of inflation, so did not directly address the consequences of maintaining real aggregate demand at levels that raised prices. James Tobin (1972), among others, reasoned that the average wage and price increases associated with approaching full employment in the empirical Phillips curves came from the operation of a heterogeneous labour market in which demand constantly shifted among sectors. In his model, the short-run inflation that was observed in the typical cyclical episode reflected wage and price changes that reduced wasteful search unemployment, rather than a misguided attempt to sustain employment above the full employment level.

The first important departure came from theoretical models based on representative agents and firms that examined the consequences of permanently maintaining demand at levels that raised wages and prices in the short run. In the late 1960s Milton Friedman (1968) and Edmund Phelps (1969) independently formulated models of a natural rate of unemployment in which inflation fed back fully into wages and hence prices, so that an unemployment rate below the natural rate could be sustained only by ever-higher inflation rates. In effect, these accelerationist price models resurrected the vertical Keynesian supply curve at full employment for the long run, but allowed demand policies that raised the inflation rate in the short run to achieve lower levels of unemployment, but only temporarily. Since the higher employment associated with price increases could not be sustained, a corollary was that zero inflation was the appropriate target for policy. Tobin's model, with its heterogenous economy, denied that a natural rate identified by prices rising faster corresponded to full employment. However, the natural rate model became widely accepted as a theoretical construct, especially after the introduction of rational expectations models in which anticipation of faster or slower price increases would speed up the process of price acceleration or deceleration. Some theoretical models also assumed price and wage flexibility rather than stickiness. And some even rejected the idea that aggregate demand could leave the economy below full employment, modelling all cyclical variations in output and employment as shocks to aggregate supply. Modern neo-Keynesian models retain both the assumption of price and wage stickiness, which is supported by empirical research, and the implication that output can depart from its potential level. But they attach a more central role to expectations than do early Keynesian models.

All these models share the original idea of demand-pull inflation in that inflation arises when aggregate demand is excessive. They differ in their description of how the process works out over different time horizons and empirically in how the region of excess demand can be identified for informing forecasters and policymakers. Empirical implementation of rational expectations models continues to be elusive, and most empirical work has used adaptive expectations with accelerationist models to estimate the natural rate and the level of potential output. These estimates proved to be unreliable in the 1990s when economic expansion steadily reduced unemployment rates well below those predicted to cause accelerating inflation in those models. Some recent research has supported the idea that a modest rate of inflation, rather than complete price stability, is necessary to maintain the fullest utilization of resources. This can be so for a variety of reasons. With downward wage rigidity, price stability will keep real wages above their efficient level in a noticeable fraction of firms. Moderate inflation will minimize this problem, permitting the economy to achieve optimal employment (Akerlof, Dickens and Perry, 1996). Furthermore, very low inflation rates will be ignored by many economic agents, leading firms to sustain output and employment at levels above those of a full expectational equilibrium (Akerlof, Dickens and Perry, 2000). And on the demand side, with very low or zero inflation, the zero floor on nominal interest rates may prevent monetary policy from getting real interest rates low enough to achieve full employment. The experience of Japan after its financial bubble burst is an example (Krugman, 1998).

Originally, the explicit modelling of demand-pull inflation was important because of the distinction it drew between price increases arising from excess demand and price increases originating in shifts up in the aggregate supply schedule, also referred to as cost-push. The sharp increases in wage costs that occurred in the heyday of union strength in industrialized economies are important historical examples of shifts in aggregate supply schedules. In the 1960s and 1970s, the experience with such cost-push shocks motivated the attempts to impose wage-price guideposts in the United States, and similar incomes policies in the United Kingdom and elsewhere. Such incomes policies were seen as a way to contain excessive wage and price increases that arose when the economy was operating below its full employment level.

Although there has been no recent interest in incomes policies, the distinction between price increases originating in excess aggregate demand and those originating from shifts in important supply schedules continues to be a feature of policy deliberations and of empirical work today. Core inflation rates, which omit the impact effect of energy and food prices on aggregate price indices, are routinely reported in monthly statistical releases, reflecting a distinction most analysts find useful. Core inflation rates are seen as more likely to feed back into wage increases, and are a better indicator of demand-pull effects on prices. And policymakers regularly make allowances for the effect of supply shocks in considering their stabilization response to changes in reported inflation rates.

History provides examples of significant inflation in which excess demand or major supply shocks or both were important. In the United States, during the Second World War and the Korean War maximizing output was the paramount goal of government even though it meant expanding demand well beyond the normal full-employment point. The potential inflation generated by operating in this excess-demand region was moderated, if not completely suppressed, by rationing and price controls. Demand-pull inflation was also a feature of the industrial economies in the late 1960s, when US military spending was greatly enlarged and labour and product markets became tight for an extended period throughout the industrial world. An abrupt explosion of wage increases at the end of the 1960s and in the early 1970s in most industrialized countries suggests that cost-push contributed importantly to the inflation of that period. The rise in food prices in 1973 and the oil supply shocks of 1973 and 1979 added further to the ongoing inflation of that decade and doubtless contributed to an increase in inflationary expectations and to the response of unions and firms to those expectations.

It was particularly striking that inflation was so little affected by the very deep recessions of the mid-1970s in the advanced economies. That episode convinced most economists of the shortcomings of the simple short-run Phillips curve model, which predicted that inflation would slow cyclically in the mid-1970s. But it was also not consistent with flexible price accelerationist models which predict that prices and wages will fall when the economy is operating below its natural rate. It did support the pessimistic verdict that a well-established inflation can persist long after the initiating shocks have disappeared and long after a reduction of demand has eliminated any excess demand from the economy.

The stabilization challenge confronting policymakers in that period was seen not merely as avoiding excess aggregate demand, but also as choosing how much to accommodate inflation in order to maintain real growth and how much to give up in output and employment in order to suppress inflation. After the second OPEC oil price shock in 1979, Paul Volcker was appointed Chairman of the US Federal Reserve and, under his leadership, the Fed chose to strongly suppress demand until inflation receded sharply. The lower inflation that ensued is consistent with the predictions of some conventional cyclical models. The severity of the policy used, as reflected in the record high interest rates it produced and the very deep recession that policymakers tolerated, can also be interpreted as evidence that policymakers can shape expectations and that doing so affects how promptly the inflation rate changes.

In the United States, the period that began in the 1990s was a sharp contrast to the 1970s in that inflation had been moderate for many years. As noted above, by the end of the decade the unemployment rate had fallen well below existing empirical estimates from natural-rate models. Yet inflation remained very low, both before the modest recession of the early 2000s and in the several years after it, even after a new oil price shock. Most European economies experienced similarly low inflation in this period. However, several suffered from chronically high rates of unemployment. While considerable controversy surrounds the reasons for this persistence of unemployment, some analysts believe inadequate aggregate demand over an extended period is partly to blame. There are several implications for stabilization policies aimed at avoiding inflation from all this experience: While empirical estimates from the 1970s suggested inflation was prone to quicken through a wage-price spiral, the recent period suggests no such tendency so long as inflation rates are modest (Brainard and Perry, 2000). Furthermore, the economy's potential output and the attainable unemployment rate – the thresholds of the demand-pull region of resource utilization – cannot be adequately estimated using typical accelerationist models. Finally, the contrasting experiences across the United States and European economies show that policies targeting inflation alone are not sufficient to assure full employment.

GEORGE L. PERRY

See also **cost-push inflation; inflation; monetary business cycle models (sticky prices and wages); monetary and fiscal policy overview.**

Bibliography

Akerlof, G., Dickens, W. and Perry, G. 1996. The macroeconomics of low inflation. *Brookings Papers on Economic Activity* 1996(1), 1–59.

Akerlof, G., Dickens, W. and Perry, G. 2000. Near-rational wage and price setting and the long-run Phillips curve. *Brookings Papers on Economic Activity* 2000(1), 1–44.

Blanchard, O. and Summers, L. 1986. Hysteresis and the European unemployment problem. In *NBER Macroeconomics Annual*, ed. S. Fischer. Cambridge, MA: MIT Press.

Brainard, W. and Perry, G. 2000. Making policy in a changing world. In *Economic Events, Ideas, and Policies*, ed. G. Perry and J. Tobin. Washington, DC: Brookings Institution.

Calvo, G. 1983. Staggered prices in a utility maximizing framework. *Journal of Monetary Economics* 12, 383–98.

Friedman, M. 1968. The role of monetary policy. *American Economic Review* 58, 1–17.

Gordon, R. 1981. Output fluctuations and gradual price adjustment. *Journal of Economic Literature* 19, 493–530.

Hall, R., ed. 1982. *Inflation: Causes and Effects*. Chicago: University of Chicago Press for the NBER.

Krugman, P. 1998. It's baaack: Japan's slump and the return of the liquidity trap. *Brookings Papers on Economic Activity* 1998(2), 137–87.

Lucas, R. 1972. Econometric testing of the natural rate hypothesis. In *The Econometrics of Price Determination*, ed. O. Eckstein. Washington, DC: Board of Governors of the Federal Reserve System.

Nickell, S. 1997. Unemployment and labor market rigidities: Europe versus North America. *Journal of Economic Perspectives* 11(3), 55–74.

Nordhaus, W. 1981. Macroconfusion: the dilemma of economic policy. In *Macroeconomics, Prices, and Quantities*, ed. J. Tobin. Washington, DC: Brookings Institution.

Okun, A. 1975. Inflation: its mechanics and welfare costs. *Brookings Papers on Economic Activity* 1975(2), 351–90.

Okun, A. 1981. *Prices and Quantities: A Macroeconomic Analysis*. Washington, DC: Brookings Institution.

Perry, G. 1980. Inflation in theory and practice. *Brookings Papers on Economic Activity* 1980(1), 207–41.

Phelps, E. 1969. The new microeconomics in inflation and employment theory. *American Economic Review* 59, 147–60.

Sargent, T. 1982. The ends of four big inflations. In *Inflation: Causes and Consequences*, ed. R. Hall. Chicago: University of Chicago Press.

Schultze, C. 1981. Some macro foundations for micro theory. *Brookings Papers on Economic Activity* 1981(2), 521–76.

Taylor, J. 1980. Aggregate dynamics and staggered contracts. *Journal of Political Economy* 88, 1–24.

Tobin, J. 1972. Inflation and unemployment. *American Economic Review* 62, 1–18.

demand theory

The main purpose of demand theory is to describe and explain observed consumer choices of commodity bundles. Market parameters, typically prices and income, determine constraints on commodity bundles. Given a combination of market parameters, a commodity bundle or a non-empty set of commodity bundles, which satisfies the corresponding constraints, is called a demand vector or a demand set. The mapping which assigns to every admissible combination of market parameters a unique demand vector (or a non-empty demand set) is called a demand function (or a demand correspondence, respectively). Traditional demand theory considers the demand function (or correspondence) as the outcome of some optimizing behaviour of the consumer. Its primary goal is to determine how alternative assumptions on the constraints, objectives and behavioural rules of the consumer affect his observed demands for commodities. The traditional model of the consumer postulates preferences over alternative commodity bundles to describe the objectives of the consumer. Its behavioural rule consists in maximizing these preferences on the set of feasible commodity bundles which satisfy the budget constraint imposed by the market parameters. If there is a unique preference maximizer under each budget constraint, then preference maximization determines a demand function. If there is at least one preference maximizer under each budget constraint, then preference maximization determines a demand correspondence.

Once the traditional view is adopted, the occurrence of demand correspondences cannot be avoided. Compatibility of observed demand, which is always unique, with some demand correspondence poses a minor problem in general. However, the correspondence should be obtained through preference maximization. The last requirement leads to the main issues of modern demand theory: Which demand correspondences are compatible with preference maximization? Given any conditions necessary for demand correspondences to be compatible with preference maximization, are they sufficient? Which demand correspondences are compatible with a special class of preferences? What type of preferences yields a particular class of demand correspondences? When addressing these issues, modern demand theory attempts to link two concepts: preferences and demand.

Historically, the important concept was utility rather than preference. Before Fisher (1892) and Pareto (1896), utility was conceived as cardinal: that is, it was assumed to be a measurable scale for the degree of satisfaction of the consumer. Fisher and Pareto were the first to observe that an arbitrary increasing transformation of the utility function has no effect on demand. Edgeworth (1881) had already written utility as a general function of quantities of all commodities and had employed indifference curves. It is now widely accepted in demand theory that only ordinal utility matters. That is, a utility function serves merely as a convenient device to represent a preference relation, and any increasing transformation of the utility function will serve this purpose as well.

Representability by utility functions imposes some restrictions on preferences. The problem of representability of a preference relation by a numerical function was solved by Debreu (1954; 1959; 1964) based on work by Eilenberg (1941), and by Rader (1963) and Bowen (1968). While still assuming cardinal utility, Walras (1874) developed the first 'theory of demand'. His demand was a function of all prices and the endowment bundle, obtained through utility maximization. Slutsky (1915) finally assumed an ordinal utility function with enough restrictions to yield a maximum under any budget constraint and testable properties of the resulting demand functions. In particular, he obtained negativity of diagonal elements and symmetry of the 'Slutsky matrix'.

Antonelli (1886) was the first to go the opposite way: construct indifference curves and a utility function from the so-called inverse demand function. Pareto (1906) took the same route. Katzner (1970) reports on recent results in this direction. The construction of preference relations from demand functions was achieved in two ways:

1. Samuelson (1947) and Houthakker (1950) introduced the concept of revealed preference into demand theory. Considerable progress in relating utility and demand in terms of revealed preference was achieved by Uzawa (1960), further refinements being due to Richter (1966).
2. Hurwicz and Uzawa (1971) contributed to the following so-called integrability problem: construct a twice continuously differentiable utility representation from a continuously differentiable demand function which satisfies certain integrability conditions (including symmetry and negative semi-definiteness of the Slutsky matrix).

Kihlstrom, Mas-Colell and Sonnenschein (1976) unified the two approaches by relating the axioms of revealed preference to properties of the Slutsky matrix.

Since there exists a sizable literature on demand theory, many of the concepts and results are well established and well-known. These have become so much part of standard knowledge in economic theory that they are included in any contemporary microeconomic textbook and other surveys. It would substantially reduce the space available for a presentation of the new results of recent decades if an extended introductory account of demand theory were to be included here as well.

Commodities and prices

Consumers purchase or sell commodities, which can be divided into goods and services. Each commodity is specified by its physical quality, its location, and the date of its availability. In the case of uncertainty, the state of nature in which the commodity is available may be added to the specification of a commodity. This leads to the notion of a contingent commodity (see Arrow, 1953; Debreu, 1959). We assume as in traditional theory that there exists a finite number l of such commodities. Quantities of each commodity are measured in real numbers. A *commodity bundle* is an l-dimensional vector $x = (x_1, \ldots, x_l)$. The set of all l-dimensional vectors $x = (x_1, \ldots, x_l)$ is the l-dimensional Euclidean space $\mathbb{R}^l$ which we interpret as the *commodity space*. $|x_h|$ indicates the quantity of commodity $h = 1, \ldots, l$. Commodities are assumed to be perfectly divisible, so that their quantity may be expressed as any (non-negative) real number. The standard sign convention for consumers assigns positive numbers for commodities made available to the consumer (inputs) and negative numbers for commodities made available by the consumer (outputs). Hence, a priori any commodity bundle $x \in \mathbb{R}^l$ is conceivable.

The price p_h of a commodity h, $h = 1, \ldots, l$, is a real number which is the amount of account that has to be paid in exchange for one unit of the commodity. For the consumer, p_h is given and has to be paid now for the delivery of commodity h under the circumstances (location, date, state) specified for commodity h. A *price system* or *price vector* is a vector $p = (p_1, \ldots, p_l)$ in $\mathbb{R}^l$ and contains the prices for all commodities. The value of a commodity bundle x given the price vector p is $px = \sum_{h=1}^{l} p_h x_h$. This means that commodity bundles are *priced linearly*.

Consumption sets and budget sets

Typically, some commodity bundles cannot be consumed by a consumer for physical reasons. Those consumption bundles which can be consumed form the consumer's *consumption set*. This is a non-empty subset X of the commodity space $\mathbb{R}^l$. A consumer must choose a bundle x from his consumption set X in order to subsist. Traditionally, inputs in consumption are described by positive quantities and outputs by negative quantities. So in particular, the labour components of a consumption bundle x are all non-positive, unless labour is hired for a service. One usually assumes that the consumption set X is closed, convex, and bounded below. Vectors $x \in X$ are sometimes called *consumption plans*.

Given the sign convention on inputs and outputs and a price vector p, the value px of a consumption plan x defines the net outlay of x, that is the value of all purchases (inputs) minus the value of all sales (outputs) for the bundle x. Trading the bundle x in a market at prices p implies payments and receipts for that bundle. Therefore, the value of the consumption plan should not exceed the initial wealth (or income) of the consumer which is a given real number w. If the consumer owns a vector of initial resources ω and the price vector p is given, then w may be determined by $w = p\omega$. The consumer may have other sources of wealth: savings and pensions, bequests, profit shares, taxes, or other liabilities. Given p and w, the

set of possible consumption bundles whose value does not exceed the initial wealth of the consumer is called the *budget set* and is defined formally by

$$\beta(p, w) = \{x \in X | px \leq w\}.$$

The ultimate decision of a consumer is to choose a consumption plan from his budget set. Those vectors in $\beta(p, w)$ which the consumer eventually chooses form his *demand set* $\varphi(p, w)$.

Preferences and demand

The choice of the consumer depends on his tastes and desires. These are represented by his *preference relation* $\succsim$ which is a binary relation on X. For any two bundles $x, y \in X$, $x \succsim y$ means that x is at least as good as y. If the consumer always chooses a most preferred bundle in his budget set, then his demand set is defined by

$$\varphi(p, w) = \{x \in \beta(p, w) | x' \in \beta(p, w)$$
$$\text{implies } x \succsim x' \text{ or not } x' \succsim x\}.$$

Three basic axioms are usually imposed on the preference relation $\succsim$ which are taken as a definition of a rational consumer:

Axiom 1 (reflexivity). If $x \in X$, then $x \succsim x$, that is, any bundle is as good as itself.
Axiom 2 (transitivity). If $x, y, z \in X$ such that $x \succsim y$ and $y \succsim z$, then $x \succsim z$.
Axiom 3 (completeness). If $x, y \in X$, then $x \succsim y$ or $y \succsim x$.

A preference relation $\succsim$ which satisfies these three axioms is a complete preordering or weak order on X and will be called a *preference order*. Already Axioms 2 and 3 define a preference order, since Axiom 3 implies Axiom 1. A preference relation $\succsim$ on X induces two other relations on X, the relation of strict preference, $\succ$, and the relation of indifference, $\sim$.

Definition Let $\succsim$ be a preference relation on the consumption set X. A bundle x is said to be *strictly preferred* to a bundle y, that is $x \succ y$, if and only if $x \succsim y$ and not $y \succsim x$. A bundle x is said to be *indifferent to* a bundle y, that is $x \sim y$, if and only if $x \succsim y$ and $y \succsim x$.

Lemma *Suppose $\succsim$ is reflexive and transitive. Then*

(i) *$\succ$ is irreflexive, that is, not $x \succ x$, and transitive;*
(ii) *$\sim$ is an equivalence relation on X, which means that $\sim$ is reflexive, transitive, and symmetric: that is, $x \sim y$ if and only if $y \sim x$.*

For $Z \subseteq X, x \in Z$, x is called *maximal in Z*, if for all $z \in Z$: not $z \succ x$. x is called a *best element of Z* or *most preferred in Z*, if for all $z \in Z : x \succsim z$. Best elements are maximal; maximal elements are not necessarily best elements. If $\succsim$ is complete, then best and maximal elements coincide. Obviously for any price vector p and initial wealth w,

$$\varphi(p,w) = \{x \in \beta(p,w) | x \text{ is maximal in } \beta(p,w)\}.$$

Axioms 1–3 are not questioned in most of consumer theory. However, transitivity and completeness may be violated by observed behaviour. Recent developments in the theory of consumer demand indicate that some weaker axioms suffice to describe and derive consistent demand behaviour (see, for example, Sonnenschein, 1971; Katzner, 1971; Shafer, 1974; Kihlstrom, Mas-Colell and Sonnenschein, 1976; Kim and Richter, 1986). In an alternative approach, one could start from a strict preference relation as the primitive concept. This may sometimes be convenient. However, the weak relation $\succsim$ seems to be the more natural concept. If the consumer chooses x, although y was a possible choice as well, then his choice can only be interpreted in the sense of $x \succsim y$, but not as $x \succ y$.

For the remainder of this section, let us fix a preference order $\succsim$ on X and a non-empty subset B of $\mathbb{R}^{l+1}$ such that for every $(p,w) \in B$, there is a unique $\succsim$-best element in $\beta(p,w)$: that is, maximization of $\succsim$ defines a *demand function* $f{:}B \to X$ such that $\varphi(p,w)=\{f(p,w)\}$ for all $(p,w) \in B$.

Let x, $x' \in X$, $x \neq x'$. We call x *revealed preferred* to x' and write xRx', if there is $(p,w) \in B$ such that $x=f(p,w)$ and $px' \leq px$. xRx' implies that both x and x' belong to the budget set $\beta(p,w)$ and x is chosen. Since f is derived from $\succsim$-maximization, xRx' implies $x \succ x'$. We call x *indirectly revealed preferred* to x' and write xR^*x', if there exists a finite sequence $x_0 = x, x_1, \ldots, x_n = x'$ in X such that $x_0 R x_1, \ldots, x_{n-1} R x'$. Obviously, R^* is transitive. Since $\succ$ is transitive, xR^*x' implies $x \succ x'$. Consequently, the following must hold (otherwise $x \succ x'$!):

(SARP) $xR^*x' \Rightarrow$ not $(x'R^*x)$.

(SARP) implies

(WARP) $xRx' \Rightarrow$ not $(x'Rx)$.

(SARP) is the *strong axiom of revealed preference*; (WARP) is the *weak axiom*. Hence $\succsim$-maximization implies the strong axiom and a fortiori the weak axiom. For the inverse implication, see Chipman et al. (1971, chs. 1, 2, 3 and 5). For $l \geq 3$, there exist demand functions which satisfy (WARP) but not (SARP), whereas for $l = 2$, (WARP) and (SARP) are equivalent; see Section 3.J of Mas-Colell, Whinston and Green (1995) and Kihlstrom, Mas-Colell and Sonnenschein (1976, p. 977).

Continuous preference orders and utility functions

Axioms 1–3 have intuitive appeal. This is less so with the topological requirements of the following Axiom 4.

Axiom 4 (continuity). For every $x \in X$, the sets $\{y \in X | y \succsim x\}$ and $\{y \in X | x \succsim y\}$ are closed relative to X.

If $\succsim$ is a preference order, then Axiom 4 is equivalent to: For every $x \in X$, the sets $\{y \in X | y \succ x\}$ and $\{y \in X | x \succ y\}$ are open in X.

Closedness of $\{y \in X | y \succsim x\}$ requires that for any sequence y^n, $n \in \mathbb{N}$, in X such that y^n converges to $y \in X$ and $y^n \succsim x$ for all n, the limit y also satisfies $y \succsim x$. Openness of $\{y \in X | y \succ x\}$ means that if $y \succ x$, then $y' \succ x$ for any y' close enough to y.

The sets $\{y \in X | y \succsim x\}$ are called *upper contour sets* of the relation $\succsim$ and the sets $\{y \in X | x \succsim y\}$ are called *lower contour sets* of $\succsim$. For $x \in X$, the set $I(x) := \{y \in X | y \sim x\}$ is called the *indifference class* of x with respect to $\succsim$ or the $\succsim$-*indifference surface* through x or the $\succsim$-*indifference curve* through x. In the case $\succsim$ is reflexive and transitive, $I(x)$ is the equivalence class of x with respect to the equivalence relation $\sim$.

There is a preference order $\succsim$ on $\mathbb{R}^l$, $l \geq 2$, which does not satisfy Axiom 4, namely the *lexicographic order* defined by $(x_1, \ldots, x_l) \succsim (y_1, \ldots, y_l)$ if and only if $x=y$ or there exists $k \in \{1, \ldots, l\}$ such that: $x_j = y_j$ for $j < k$ and $x_k > y_k$. Few studies of the relationship between the order properties of Axioms 1–3 and the topological property of Axiom 4 have been made. We emphasize the following result.

Theorem (Schmeidler, 1971). *Let $\succsim$ denote a transitive binary relation on a connected topological space X. Assume that there exists at least one pair $\bar{x}, \bar{y} \in X$ such that $\bar{x} \succ \bar{y}$. If for every $x \in X$, (i) $\{y \in X | y \succsim x\}$ and $\{y \in X | x \succsim y\}$ are closed and (ii) $\{y \in X | y \succ x\}$ and $\{y \in X | x \succ y\}$ are open, then $\succsim$ is complete.*

Definition Let X be a set and $\succsim$ be a preference relation on X. Then a function u from X into the real line $\mathbb{R}$ is a *(utility) representation* or a *utility function for* $\succsim$, if for all $x, y \in X : u(x) \geq u(y)$ if and only if $x \succsim y$. Clearly, if u is a utility representation for $\succsim$ and $f : \mathbb{R} \to \mathbb{R}$ is an increasing transformation, then the composition $f \to u$ is also a representation of $\succsim$. If $u : X \to \mathbb{R}$ is any function, then $\succsim$, defined by $x \succsim y$ if and only if $u(x) \geq u(y)$ for $x, y \in X$, is a preference order on X and u is a utility representation for $\succsim$.

Most utility functions used in consumer theory are continuous. If u is continuous and $\succsim$ is represented by u, then by necessity $\succsim$ is a continuous preference order. In our case where $X \subseteq \mathbb{R}^l$, the opposite implication also holds: If $\succsim$ is a continuous preference order, then it has a continuous utility representation.

Theorem (Debreu, Eilenberg, Rader). *Let X be a topological space with a countable base of open sets (or a connected, separable topological space) and $\succsim$ be a continuous preference order on X. Then $\succsim$ has a continuous utility representation.*

In our context of Euclidean commodity spaces, explicit constructions of continuous utility representations for continuous and monotonic preference orders are available. See Arrow and Hahn (1971) for the 'Euclidean distance approach' and Neuefeind (1972) for the 'Lebesgue measure approach'. For topological spaces X with a countable base of open sets, it has further been shown by Rader (1963) and Bosi and Mehta (2002) that an upper semi-continuous preference order on X has an upper semi-continuous utility representation.

As an immediate consequence of the representation theorem for preference relations, one obtains one of the standard results on the non-emptiness of the demand set $\varphi(p,w)$, since any continuous function attains its maximum on a compact set (Weierstrass's theorem), though a direct proof is also possible.

Corollary *Let $X \subseteq \mathbb{R}^l$ be bounded below and closed, $\succsim$ be a continuous preference order on X, $p \in \mathbb{R}^l_{++}$ (that is, $p \gg 0$) and $w \in \mathbb{R}$. Then $\beta(p, w) \neq \emptyset$ implies $\varphi(p, w) \neq \emptyset$.*

There has been a recent shift from proving existence to a more systematic study of the non-existence of utility representations. Needless to say that there are many preference orders on $\mathbb{R}^l$ or on subsets thereof with continuous utility representations. There are also total orders $\succsim$ (that is, preference orders $\succsim$ with $x \sim y \Leftrightarrow x = y$) on $\mathbb{R}^l$, $l \geq 2$, which admit utility representations, since there exist bijections $u : \mathbb{R}^l \to \mathbb{R}$. However, for $l \geq 2$, there is no total order on $\mathbb{R}^l$, $\mathbb{R}^l_+$ or $[0,1]^l$ which has a continuous utility representation; see Candeal and Induráin (1993). Moreover, a preference order $\succsim$ on X, which is not continuous, need not have a utility representation. For instance, the lexicographic order on $\mathbb{R}^l$, $l \geq 2$, a total order first discussed by Debreu (1954), does not have a utility representation, nor even a discontinuous one. Beardon et al. (2002) provide a classification of total orders which do not admit a utility representation. Estévez Toranzo and Hervés Beloso (1995) show that, if X is a non-separable metric space, then there exists a continuous preference order on X which cannot be represented by a utility function.

Some properties of preferences and utility functions Some of the frequent assumptions on preference relations correspond almost by definition to analogous properties of utility functions, while other analogies need demonstration. We discuss the assumptions most commonly used.

Monotonicity A preference order $\succsim$ on $X \subseteq \mathbb{R}^l$ is *monotonic*, if $x, y \in X, x \geq y, x \neq y$ implies $x \succ y$.

This property means desirability of all commodities. If a monotonic preference order has a utility representation u, then u is an increasing function (in all arguments). Inversely, if $\succsim$ is represented by an increasing function, then $\succsim$ is monotonic.

Non-satiation Let $\succsim$ be the preference relation of a consumer over consumption bundles in X and let $x \in X$.

(i) x is a *satiation point for* $\succsim$ if $x \succsim y$ for all $y \in X$: that is, x is a best element in X.

(ii) The preference relation is *locally not satiated at x*, if for every neighbourhood U of x there exists $z \in U$ such that $z \succ x$.

Consider a utility representation u for $\succsim$. Then $x \in X$ is a satiation point if and only if u has a global maximum at x. $\succsim$ is locally not satiated at x if and only if u does not attain a local maximum at x. Local non-satiation rules out that u is constant in a neighbourhood of x. If $\succsim$ is locally not satiated at all x, then $\succsim$ cannot have thick indifference classes or satiation points.

Convexity A preference relation $\succsim$ on $X \subseteq \mathbb{R}^l$ is called

(i) *convex*, if the set $\{y \in X | y \succsim x\}$ is convex for all $x \in X$;
(ii) *strictly convex*, if X is convex and $\lambda x + (1 - \lambda)x' \succ x'$ for any two bundles x, $x' \in X$ such that $x \neq x', x \succsim x'$ and for any λ such that $0 < \lambda < 1$;
(iii) *strongly convex*, if X is convex and $\lambda x + (1 - \lambda)x' \succ x''$ for any three bundles $x, x', x'' \in X$ such that $x \neq x', x \succsim x'', x' \succsim x''$ and for any λ such that $0 < \lambda < 1$.

Quasi-concavity A function $u : X \to \mathbb{R}$ is called

(i) *quasi-concave*, if $u(\lambda x + (1 - \lambda)y) \geq \min\{u(x), u(y)\}$ for all $x, y \in X$ and any $\lambda \in [0,1]$;
(ii) *strictly quasi-concave*, if $u(\lambda x + (1 - \lambda)y) > \min\{u(x), u(y)\}$ for all $x, y \in X$ with $x \neq y$ and any $\lambda \in (0,1)$.

Let u be a representation of the preference order $\succsim$. Then u is (strictly) quasi-concave if and only if $\succsim$ is (strictly) convex. Quasi-concavity is preserved under increasing transformations: that is, it is an ordinal property. In contrast, concavity is a cardinal property which can be lost under increasing transformations. With respect to the difficult problem to characterize those preference orders which have a concave representation, we refer to Kannai (1977). Clearly, if $\succsim$ is locally not satiated at all x, then $\succsim$ does not have a satiation point. In general, the inverse implication is false. If, however, $\succsim$ is strictly convex and does not have a satiation point, then $\succsim$ is locally not satiated at all x. Moreover, if $\succsim$ is strictly convex, then it has at most one satiation point. An immediate implication is the following lemma.

Lemma *Let $X \subseteq \mathbb{R}^l$ be bounded below, convex, and closed. Let $\succsim$ be a strictly convex, continuous preference order on X, $p \in \mathbb{R}^l_{++}$, and $w \in \mathbb{R}$. Then $\beta(p, w) \neq \emptyset$ implies that $\varphi(p,w)$ is a singleton.*

Separability Separable utility functions were used in classical consumer theory long before associated properties of preferences had been defined. All early contributions to utility theory assumed without much discussion an additive form of the utility function over different commodities. It was not until Edgeworth (1881) that utility was written as a general function of a vector of commodities. The particular consequences of separability for demand theory were discussed well after the general non-separable case in demand theory had been treated and generally accepted. Among the many contributors are Sono (1945), Leontief (1947), Samuelson (1947), Houthakker (1960), Debreu (1960), and Koopmans (1972). We follow Katzner (1970) in our presentation.

Let $N = \{N_j\}_{j=1}^k$ be a partition of the set $\{1,\ldots,l\}$ and assume that $X = S_1 \times \cdots \times S_k$. Let $J=\{1,\ldots,k\}$ and for any $j \in J, y \in X, y = (y_1,\ldots,y_k) \in \prod_{i \in J} S_i$ write $y_{-j} = (y_1,\ldots,y_{j-1},y_{j+1},\ldots,y_k)$ for the vector of components different from j. For any y_{-j}, a preference order $\succsim$ on X induces a preference order $\succsim_{y_{-j}}$ on S_j which is defined by $x_j \succsim_{y_{-j}} x_j'$ if and only if $(y_{-j}, x_j) \succsim (y_{-j}, x_j')$ for $x_j, x_j' \in S_j$. In general, the induced ordering $\succsim_{y_{-j}}$ will depend on y_{-j}. The first notion of separability states that for any j, the preference orders $\succsim_{y_{-j}}$ are independent of $y_{-j} \in \prod_{i \neq j} S_i$. The second notion of separability states that for any proper subset I of J, the induced preference orders $\succsim_{y_{J \setminus I}}$ on $\prod_{i \in I}$ are independent of $y_{J \setminus I} \in \prod_{i \notin I} S_i$.

Definition Let $\succsim$ be a preference order on $X = \prod_{j \in J} S_j$.

(i) $\succsim$ is called *weakly separable* with respect to N if $\succsim_{y_{-j}} = \succsim_{z_{-j}}$ for each $j \in J$ and any $y_{-j}, z_{-j} \in \prod_{i \neq j} S_i$.
(ii) $\succsim$ is called *strongly separable* with respect to N if $\succsim_{y_{J \setminus I}} = \succsim_{z_{J \setminus I}}$ for each $I \subseteq J, I \neq \emptyset, I \neq J$ and any $y_{J \setminus I}, z_{J \setminus I} \in \prod_{i \notin I} S_i$.

Definition Let $u : \prod_{j \in J} S_j \to \mathbb{R}$. u is called

(i) *weakly separable* with respect to N, if there exist continuous functions $v_j : S_j \to \mathbb{R}, j \in J$, and $V : \mathbb{R}^k \to \mathbb{R}$ such that $u(x) = V(v_1(x_1),\ldots,v_k(x_k))$;
(ii) *strongly separable* with respect to N, if there exist continuous functions $v_j : S_j \to \mathbb{R}, j \in J$, and $V : \mathbb{R} \to \mathbb{R}$ such that $u(x) = V(\sum_{j \in J} v_j(x_j))$.

The two important equivalence results on separability are due to Debreu and Katzner. The version of Debreu's theorem given here is slightly weaker than his original result.

Theorem (Katzner, 1970). *Let $\succsim$ be a continuous, monotonic preference order on $X = \prod_{j \in J} S_j$ with $S_j = \mathbb{R}^{N_j}$ for all $j \in J$. Then $\succsim$ is weakly separable if and only if every continuous representation of $\succsim$ it is weakly separable.*

Theorem (Debreu, 1960). *Let $\succsim$ be a continuous, monotonic preference order on $X = \prod_{j \in J} S_j$ with $S_j = \mathbb{R}^{N_j}$ for all $j \in J = \{1,\ldots,k\}$ and $k \geq 3$. Then $\succsim$ is strongly separable if and only if every continuous representation is strongly separable.*

Under the assumptions of this theorem, if $\succsim$ is strongly separable with representation $u(x) = V(\sum_{j \in J} v_j(x_j))$, then

V must be increasing or decreasing. Therefore,

$$v(x) = \begin{cases} \sum_{j \in J} v_j(x_j) & \text{for } V \text{ increasing} \\ -\sum_{j \in J} v_j(x_j) & \text{for } V \text{ decreasing} \end{cases}$$

is also a representation of $\succsim$. This is the additive form of separable utility used by early economists who thought that each commodity h had its own intrinsic utility representable by a scalar function u_h. The overall utility was then simply obtained as the sum of these functions, $u(x) = \sum_h u_h(x_h)$. Such a formulation is given by Jevons (1871) and Walras (1874) and implicitly contained in Gossen (1854).

In the case of uncertainty, with finitely many states of nature $j \in J = \{1,\ldots,k\}$, respective probabilities $\pi_j > 0$ and consumption $x_j \in S_j$ in state $j \in J$, an additively separable utility representation $u(x) = \sum_{j \in J} v_j(x_j)$ is tantamount to an expected utility representation $u(x) = \sum_{j \in J} \pi_j u_j(x_j)$ with $u_j = v_j/\pi_j$. Hence, an expected utility representation in the tradition of Savage (1954) implies separability with respect to states of nature. In contrast, the novel concept of Choquet expected utility à la Schmeidler (1986; 1989) typically violates separability with respect to states of nature.

For $k=2$, weak and strong separability of preferences coincide. But there are separable preferences which do not admit a strongly separable utility representation, for instance $X = \mathbb{R}^2_+$, $N_j = \{j\}$ for $j=1, 2$, $\succsim$ given by $u(x_1, x_2) = \sqrt{x_1} + \sqrt{x_1 + x_2}$. Separability of preferences imposes restrictions on demand correspondences and on demand functions (for details see Barten and Böhm, 1982, Sections 9, 14, and 15).

Continuous demand

Given any price–wealth pair $(p, w) \in \mathbb{R}^{l+1}$, the budget set of the consumer was defined as $\beta(p, w) = \{x \in X \mid px \leq w\}$. Let $S \subseteq \mathbb{R}^{l+1}$ denote the set of price–wealth pairs for which the budget set is non-empty. Then β describes a correspondence from S into X: that is, β associates to any $(p, w) \in S$ the non-empty subset $\beta(p, w)$ of X. There are two standard notions of continuity of correspondences, upper hemi-continuity and lower hemi-continuity (see Hildenbrand, 1974).

Definition A compact-valued correspondence Ψ from S into an arbitrary subset T of $\mathbb{R}^l$ is *upper hemi-continuous* (u.h.c.) at a point $y \in S$, if for all sequences $(y^n, z^n) \in S \times T$ such that $y^n \to y$ and $z^n \in \Psi(y^n)$ for all n, there exist $z \in \Psi(y)$ and a subsequence z^{n_k} of z^n such that $z^{n_k} \to z$.

Definition A correspondence Ψ from S into an arbitrary subset T of $\mathbb{R}^l$ is *lower hemi-continuous* (l.h.c.) at a point $y \in S$, if for any $z \in \Psi(y)$ and any sequence y^n in S with $y^n \to y$ there exists a sequence z^n in T such that $z^n \to z$ and $z^n \in \Psi(y^n)$ for all n.

Definition A correspondence is *continuous* if it is both lower and upper hemi-continuous.

For single-valued correspondences, the notions of lower and upper hemi-continuity coincide with the usual notion of continuity for functions. For proofs of the following lemmas, see Debreu (1959) or Hildenbrand (1974).

Lemma *Let $X \subseteq \mathbb{R}^l$ be a convex set. Then the budget correspondence $\beta : S \to X$ has a closed graph and is lower hemi-continuous at every point $(p,w) \in S$ for which $w > \min\{px \mid x \in X\}$ holds.*

Combining a previous corollary on the non-emptiness of the demand set and a fundamental theorem of Berge (1966) yields the next result.

Lemma *Let $X \subseteq \mathbb{R}^l$ be a convex set. If the preference relation has a continuous utility representation, then the demand correspondence is defined (that is, non-empty valued), compact-valued, and upper hemi-continuous at each $(p,w) \in S$ such that $\beta(p,w)$ is compact and $w > \min\{px \mid x \in X\}$.*

It follows immediately from the definitions that $\varphi(\lambda p, \lambda w) = \varphi(p,w)$ for any $\lambda > 0$ and any price–wealth pair (p,w): that is, demand is homogeneous of degree zero in prices and wealth. For convex preference orders, the demand correspondence is convex-valued. For strictly convex preference orders, the demand correspondence is single-valued: that is, one obtains a demand function. The results of this section and of the section on continuous preference orders and utility functions are summarized in the following lemma, which uses the weakest assumptions of traditional demand theory to generate a continuous demand function.

Lemma *Let $S' := \{(p,w) \in S \mid \beta(p,w) \text{ is compact and } w > \min\{px \mid x \in X\}\}$. If $\succsim$ denotes a strictly convex and continuous preference order, then $\varphi(p,w)$ defines a continuous demand function which satisfies: (i) homogeneity of degree zero in prices and wealth and (ii) the strong axiom of revealed preference.*

Continuous demand without transitivity
Transitivity is often violated in empirical studies. This excludes utility maximization, but not necessarily preference maximization. However, as the next theorem indicates, existence and continuity of demand do not depend on transitivity as crucially as one may expect. The theorem follows from a result by Sonnenschein (1971).

Theorem *Let $S^* = \{(p,w) \in S \mid \varphi(p,w) \neq \emptyset\}$. Suppose that X is compact and $\succsim$ is complete and has a closed graph.*

(i) *If $\{x' \in X \mid x' \succ x\}$ is convex for all $x \in X$, then $\varphi(p,w) \neq \emptyset$ whenever $\beta(p,w) \neq \emptyset$ (that is, $S^* = S$).*

(ii) *If $S^* = S$ and $(p^0, w^0) \in S$ such that β is continuous at (p^0, w^0), then φ is u.h.c. at (p^0, w^0).*

The assumption that X is compact is not necessary. For case (i) it suffices that all budget sets $\beta(p,w)$ under consideration be compact. For case (ii) it is sufficient that there exist a compact subset X^0 of X and a neighbourhood S^0 of (p^0, w^0) such that $\varphi(S^0) \subseteq X^0$.

To complete this section we state a lemma on the properties of a demand function obtained under preference maximization without transitivity. This contrasts with the lemma at the end of the previous section. Intransitivity essentially implies that the strong axiom of revealed preference need not hold. The lemma follows from the theorem by Sonnenschein and from the result by Shafer (1974).

Lemma *Let $X = \mathbb{R}^l_+, B = \mathbb{R}^{l+1}_{++}$. Suppose continuity and strong convexity of $\succsim$ (in addition to completeness). Then preference maximization yields a continuous demand function $f : B \to X$ which satisfies (i) homogeneity of degree zero in prices and wealth and (ii) the weak axiom of revealed preference.*

The converse statement of the lemma does not hold. For $l = 2, X = \mathbb{R}^2_+, B = \mathbb{R}^3_{++}$, there is a C^1-function $f : B \to X$ which fulfils (i), (ii), and (iii) $pf(p,w) = w$ for all $(p,w) \in B$, but which cannot be obtained as the demand function for a continuous, complete and strictly convex preference relation (John, 1984; Kim and Richter, 1986). In addition, John (1995) has shown that continuity of f, (ii) and (iii) imply (i).

Smooth preferences and differentiable utility functions
Owing to the representation theorem of Debreu, Eilenberg and Rader, continuity of a utility function and continuity of the represented preference order are identical under the perspective of demand theory. When continuous differentiability of demand is required, continuity of the preference relation will not suffice in general. The first rigorous attempt to study 'differentiable preference orders' goes back to Antonelli (1886). We follow the more direct approach of Debreu (1972) to characterize 'smooth preference orders'. Smoothness of preferences is closely related to sufficient differentiability of utility representations and the solution of the integrability problem (see Debreu, 1972; also Debreu, 1976; Hurwicz, 1971; and the section below on integrability). For the purpose of this and subsequent sections, let $P = \mathbb{R}^l_{++}$ denote the (relative) interior of $\mathbb{R}^l_+$ and assume that $X = P$. Let $\succsim$ be a continuous and monotonic preference order on P which we may consider as a subset of $P \times P$: that is, $(x,y) \in \succsim \Leftrightarrow x \succsim y$ for $(x,y) \in P \times P$. Also, the associated indifference relation $\sim$ will be considered as a subset of $P \times P$. To describe a smooth preference order, differentiability assumptions

will be made on the (graph of the) indifference relation in $P \times P$.

For $k \geq 1$, let C^k denote the class of functions which have continuous partial derivatives up to order k, and consider two open sets X and Y in an Euclidean space $\mathbb{R}^n$. A bijection $h:X \to Y$ is a C^k-diffeomorphism if both h and h^{-1} are of class C^k. $M \subseteq \mathbb{R}^n$ is a C^k-hypersurface, if for every $z \in M$, there exist an open neighbourhood U of z, an open subset V of $\mathbb{R}^n$, a hyperplane $H \subset \mathbb{R}^n$ and a C^k-diffeomorphism $h:U \to V$ such that $h(M \cap U) = V \cap H$. A C^k-hypersurface has locally the structure of a hyperplane up to a C^k-diffeomorphism. Considering the indifference relation $\sim$ as a subset of $P \times P$, the set $\tilde{I} = \{(x,y) \in P \times P | x \sim y\} \subset \mathbb{R}^{2l}$ constitutes the 'indifference surface' of the preference relation. Then $\succsim$ is called a C^2-preference order (or smooth preference order), if $\tilde{I}$ is a C^2-hypersurface.

Theorem (Debreu, 1972). *Let $\succsim$ be a continuous and monotonic preference order on P and $\tilde{I}$ be its indifference surface. Then $\succsim$ is a C^2-preference order if and only if it has a monotonic utility representation of class C^2 with no critical point.*

Properties of differentiable utility functions

Utility functions of class C^2 provide the truly classical approach to demand theory (see, for example, Slutsky, 1915; Hicks, 1939; Samuelson, 1947).

Let $\succsim$ be a monotonic, strictly convex C^2-preference order on P and $u : P \to \mathbb{R}$ be a C^2-utility representation of $\succsim$ with no critical point. Then u is continuous, increasing in all arguments, and strictly quasi-concave. Moreover, all second-order partial derivatives $u_{ij}(x) = (\partial^2 u/\partial x_i \partial x_j)(x)$, $i,j = 1, \ldots, l$, $x \in P$, exist, all u_{ij} are continuous functions of x and $u_{ij} = u_{ji}$ for $i,j = 1, \ldots, l$. Let $D^2 u = (u_{ij})$ denote the Hessian matrix of u. Then $D^2 u$ is symmetric. The first-order derivatives $u_i(x) = (\partial u/\partial x_i)(x)$, $i = 1, \ldots, l$, are continuous functions of x. Assume that $u_i(x) > 0$ for $i = 1, \ldots, l$, $x \in P$ and define

$$Du(x) = \begin{bmatrix} u_1(x) \\ \vdots \\ u_l(x) \end{bmatrix}$$

as the gradient of u at x. For any $m \times n$-matrix M, let M' denote the transpose of M.

Theorem *If $u : P \to \mathbb{R}$ is a strictly quasi-concave utility function of class C^2, then $z'D^2u(x)z \leq 0$ for all $x \in P$ and $z \in \{\tilde{z} \in \mathbb{R}^l | \tilde{z}Du(x) = 0\}$.* (For a proof, see Barten and Böhm, 1982.)

It will be shown in the next section that the conclusion of this theorem does not guarantee the existence of a differentiable demand function. The following definition strengthens the property of strict quasi-concavity.

Definition u is called *strongly quasi-concave* if

$$z'D^2u(x)z < 0 \quad \text{for all} \quad x \in P, \quad z \neq 0$$
$$\text{and} \quad z \in \{\tilde{z} \in \mathbb{R}^l | \tilde{z}Du(x) = 0\}.$$

Consider the bordered Hessian matrix

$$H(x) = \begin{bmatrix} D^2u(x) & Du(x) \\ [Du(x)]' & 0 \end{bmatrix}.$$

Then u is strongly quasi-concave whenever u is strictly quasi-concave and $H(x)$ is non-singular. (For a proof, see Barten and Böhm, 1982).

The properties of strict and strong quasi-concavity are invariant under increasing C^2-transformations. For other results and consequences of differentiable utility functions the reader may consult Barten and Böhm (1982) and the references listed there, or Debreu (1972), Mas-Colell (1974).

Differentiable demand

The earlier section on continuous demand without transitivity provides sufficient conditions on preferences for the existence of a continuous demand function which is homogeneous of degree zero in prices and wealth and satisfies the strong axiom of revealed preference. In this section, the implications of smooth preferences for differentiability of demand will be studied. Consider an assumption (D), consisting of the following three parts:

(D1) $X = P$.
(D2) $\succsim$ is a monotonic, strictly convex C^2-preference order on X and the closure relative to $\mathbb{R}_+^l \times \mathbb{R}_+^l$ of its indifference surface $\tilde{I}$ is contained in $P \times P$.
(D3) The price-wealth space is $B = \mathbb{R}_{++}^{l+1}$.

Given (D), there exists a demand function $f:B \to X$ with $p \cdot f(p,w) = w$ for all $(p,w) \in B$. Let u be an increasing strictly quasi-concave C^2-utility representation for $\succsim$. The following key result on the differentiability of demand was first given by Katzner (1968). For a detailed proof see Barten and Böhm (1982).

Theorem *Let $(\bar{p}, \bar{w}) \in B$ and $\bar{x} = f(\bar{p}, \bar{w})$. Then the following assertions are equivalent:*

(i) *f is C^1 in a neighbourhood of $(\bar{p}, \bar{w})$.*

(ii) $\begin{bmatrix} D^2u(\bar{x}) & \bar{p}' \\ \bar{p} & 0 \end{bmatrix}$ *is non-singular.*

(iii) *$H(\bar{x})$ is non-singular.*

Once the demand function f is continuously differentiable, it is straightforward to derive all of the well-known comparative statics properties, for the proof of which we refer again to Barten and Böhm (1982). Let $f = (f^1, \ldots, f^l)$ be a demand function of class C^1 and define the respective partial derivatives

$$f_w = (f^1_w, \ldots, f^l_w) = \left(\frac{\partial f^1}{\partial w}, \ldots, \frac{\partial f^l}{\partial w} \right),$$

$$f^i_j = \frac{\partial f^i}{\partial p_j}, \qquad i,j = 1, \cdot, l;$$

$$s^i_j = f^i_j + f^i_w f^j, \qquad i,j = 1, \ldots, l.$$

From these we obtain the Jacobian matrix of f with respect to prices, $J = (f^i_j)$, and the so-called Slutsky matrix $S = (s^i_j)$.

Theorem

(i) $p f_w = 1$, $pJ = -f$,
(ii) $Sp' = 0$,
(iii) S is symmetric,
(iv) $ySy' < 0$, if $y \in \mathbb{R}^l, y \neq \alpha p$ for all $\alpha \in \mathbb{R}$,
(v) rank $S = l - 1$.

Property (iv) implies that all diagonal elements of S are strictly negative: that is, $s^i_i = f^i_i + f^i_w f^i < 0$. If $f^i_w > 0$, commodity i is called a *normal good* which implies that $f^i_i < 0$: that is, demand is downward sloping in its own price. On the other hand, a negative income effect $f^i_w < 0$, that is, when commodity i is an *inferior good*, is a necessary, but not a sufficient condition for a positive own price effect $f^i_i > 0$, that is, for commodity i to be a *Giffen good*.

Duality approach to demand theory

With the notion of an expenditure function, an alternative approach to demand analysis is possible which was suggested by Samuelson (1947). For the further development and details, we refer to Diewert (1974; 1982).

As a matter of convenience and for ease of presentation, assumption (D) will be imposed on the preference relation $\succsim$. Let u denote a strictly quasi-concave increasing C^2-utility representation for $\succsim$ and let $f : B \to X$ be the demand function derived from preference maximization. Let us further assume that $u(X) = \mathbb{R}$. (This requirement can always be fulfilled by means of an increasing transformation.) Define the *indirect utility function* $v : B \to \mathbb{R}$ associated with u by $v(p,w) = u(f(p,w))$ for $(p,w) \in B$.

Given a price system $p \in \mathbb{R}^l_{++}$ and a utility level $c \in \mathbb{R}$, let $e(p,c) = \min\{p \cdot x | x \in X, u(x) \geq c\}$. Since u is strictly quasi-concave and increasing, there exists a unique minimizer $h(p,c)$ of this problem such that $e(p,c) = ph(p,c)$. $h : \mathbb{R}^l_{++} \times \mathbb{R} \to \mathbb{R}^l_{++}$ is called the *Hicksian (income-compensated) demand function* and $e : \mathbb{R}^l_{++} \times \mathbb{R} \to \mathbb{R}_{++}$ is called the *expenditure function* for u. Since assumption (D) holds, preference maximization and expenditure minimization imply the following properties and relationships:

(i) $c = v[p, e(p,c)]$ for all (p,c).
(ii) $w = e[p, v(p,w)]$ for all (p,w).

(iii) $v(p, \cdot)$ and $e(p, \cdot)$ are inverse functions for any p.
(iv) $h(p,c) = f[p, e(p,c)]$ for all (p,c).
(v) $f(p,w) = h[p, v(p,w)]$ for all (p,w).
(vi) e is strictly increasing and continuous in c.
(vii) e is non-decreasing, positive linear homogeneous, and concave in prices.
(viii) v is strictly increasing in w, and continuous.
(ix) v is non-increasing in prices and homogeneous of degree zero in income and prices.

Moreover, some interesting and important consequences of these properties can be obtained if the functions are sufficiently differentiable.

Theorem

(i) e is C^k if and only if v is C^k. ($k = 1, 2$).
(ii) If e is C^1, then $\partial e / \partial p = h$.
(iii) If f is C^1, then: v is C^2.
(iv) $f = -(\partial v / \partial p) / (\partial v / \partial w)$ (*Roy's identity*).
(v) h is C^1 and e is C^2.
(vi) $\partial h / \partial p = S$ (*Slutsky equation*) with $\partial h / \partial p$ evaluated at $[p, v(p,w)]$ and S evaluated at (p,w).

Integrability

A review of the previous discussions and analytical results involving the concepts of

$\succsim$ preference
u utility
h income-compensated demand function
e expenditure function
v indirect utility
f (direct) demand function

makes apparent their relationships which can be characterized schematically by the following diagram:

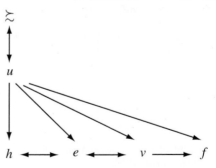

where $a \to b$ indicates that concept b can be derived from concept a under certain conditions. The integrability problem is to establish $f \to u$: that is, to recover the utility function from the demand function f.

Two recent developments

Advanced microeconomic theory assumes a distribution of consumer characteristics to determine mean demand of a consumption sector. In accordance with traditional demand theory, the primitive characteristics of a consumer are his preference relation $\succsim$ and his wealth w, and possibly his consumption set X. If we disregard the latter, the corresponding distribution of consumer characteristics is a preference–wealth distribution (see Hildenbrand, 1974). This approach lends itself to both positive and normative analysis. In contrast, Hildenbrand (1994) and others adopt a purely positive point of view and take pairs (f,w) as the primitive concepts, where f is a demand function not necessarily derived from preference maximization of 'rational' consumers.

Like traditional demand theory, most of theoretical and empirical economics has not distinguished between households and individual consumers. Chiappori (1988; 1992) and others have developed models of collective rationality of multi-person households where each member has his or her own preferences.

VOLKER BÖHM AND HANS HALLER

See also **aggregation (theory); collective rationality; correspondences; Hicksian and Marshallian demands; integrability of demand; revealed preference theory; separability.**

Bibliography

Antonelli, G.B. 1886. *Sulla Teoria Matematica della Economia Politica*. Pisa: Nella Tipografia del Folchetto. Trans. as 'On the mathematical theory of political economy', in Chipman et al. (1971).

Arrow, K.J. 1953. Le rôle des valeurs boursières pour la répartition la meilleure des risques. *Econométrie*, Colloques Internationaux du Centre National de la Recherche Scientifique, Paris, vol. 11, pp. 41–7.

Arrow, K.J. and Hahn, F. 1971. *General Competitive Analysis*. San Francisco: Holden-Day and Edinburgh: Oliver and Boyd.

Arrow, K.J. and Intriligator, M.D., eds. 1982. *Handbook of Mathematical Economics*, vol. 2. Amsterdam: North-Holland.

Arrow, K.J., Karlin, S. and Suppes, P. 1960. *Mathematical Methods in the Social Sciences*. Stanford: Stanford University Press.

Barten, A.P. and Böhm, V. 1982. Consumer theory. In Arrow and Intriligator (1982).

Beardon, A.F., Candeal, J.C., Herden, G., Induráin, E. and Mehta, G.B. 2002. The non-existence of a utility function and the structure of non-representable preference relations. *Journal of Mathematical Economics* 37, 17–38.

Berge, C. 1966. *Espaces Topologiques. Fonctions Multivoques*. Paris: Dunod. Trans. as 'Topological Spaces', Edinburgh: Oliver and Boyd, 1973.

Bosi, G. and Mehta, G.B. 2002. Existence of a semicontinuous or continuous utility function: a unified approach and an elementary proof. *Journal of Mathematical Economics* 38, 311–28.

Bowen, R. 1968. A new proof of a theorem in utility theory. *International Economic Review* 9, 374.

Candeal, J.C. and Induráin, E. 1993. Utility functions on chains. *Journal of Mathematical Economics* 22, 161–8.

Chipman, J.S., Hurwicz, L., Richter, M.K. and Sonnenschein, H.F. 1971. *Preferences, Utility, and Demand*. New York: Harcourt Brace Jovanovich.

Chiappori, P.-A. 1988. Rational household labor supply. *Econometrica* 56, 63–89.

Chiappori, P.-A. 1992. Collective labor supply and welfare. *Journal of Political Economy* 100, 437–67.

Debreu, G. 1954. Representation of a preference ordering by a numerical function. In *Decision Processes*, ed. R.M. Thrall et al. New York: Wiley.

Debreu, G. 1959. *Theory of Value*. New York: Wiley.

Debreu, G. 1960. Topological methods in cardinal utility theory. In *Mathematical Methods in the Social Sciences*, ed. K.J. Arrow et al. Stanford: Stanford University Press.

Debreu, G. 1964. Continuity properties of Paretian utility. *International Economic Review* 5, 285–93.

Debreu, G. 1972. Smooth preferences. *Econometrica* 40, 603–15.

Debreu, G. 1976. Smooth preferences. A corrigendum. *Econometrica* 44, 831–2.

Diewert, W.E. 1974. Applications of duality theory. In *Frontiers of Quantitative Economics*, vol. 2, ed. M.D. Intriligator and D.A. Kendrick. Amsterdam: North-Holland.

Diewert, W.E. 1982. Duality approaches to microeconomic analysis. In Arrow and Intriligator (1982).

Edgeworth, F.Y. 1881. *Mathematical Psychics*. London: Kegan Paul.

Eilenberg, S. 1941. Ordered topological spaces. *American Journal of Mathematics* 63, 39–45.

Estévez Toranzo, M. and Hervés Beloso, C. 1995. On the existence of continuous preference orderings without utility representations. *Journal of Mathematical Economics* 24, 305–9.

Fisher, I. 1892. Mathematical investigations in the theory of value and prices. *Transactions of the Connecticut Academy of Arts and Sciences* 9, 1–124. Repr. in *The Works of Irving Fisher*, vol. 1, ed. W.J. Barber. London: Pickering and Chatto, 1997.

Gossen, H.H. 1854. *Entwicklung der Gesetze des menschlichen Verkehrs und der daraus fließenden Regeln für menschliches Handeln*. Braunschweig, 2nd edn. Berlin: Prager, 1889.

Hicks, J.R. 1939. *Value and Capital*. Oxford: Clarendon.

Hildenbrand, W. 1974. *Core and Equilibria of a Large Economy*. Princeton: Princeton University Press.

Hildenbrand, W. 1994. *Market Demand: Theory and Empirical Evidence*. Princeton: Princeton University Press.

Houthakker, H.S. 1950. Revealed preference and the utility function. *Economica* N.S. 17, 159–74.

Houthakker, H.S. 1960. Additive preferences. *Econometrica* 28, 244–57; Errata: *Econometrica* 30 (1962), 633.

Hurwicz, L. 1971. On the problem of integrability of demand functions. In Chipman et al. (1971).

Hurwicz, L. and Uzawa, H. 1971. On the integrability of demand functions. In Chipman et al. (1971).

Jevons, W.S. 1871. *Theory of Political Economy*. London: Macmillan.

John, R. 1984. A counterexample to a conjecture concerning the nontransitive consumer. Discussion Paper No. 151. Sonderforschungsbereich 21, University of Bonn.

John, R. 1995. The weak axiom of revealed preference and homogeneity of demand functions. *Economics Letters* 47, 11–16.

Kannai, Y. 1977. Concavifiability and construction of concave utility functions. *Journal of Mathematical Economics* 4, 1–56.

Katzner, D.W. 1968. A note on the differentiability of consumer demand functions. *Econometrica* 36, 415–8.

Katzner, D.W. 1970. *Static Demand Theory*. New York: Macmillan.

Katzner, D.W. 1971. Demand and exchange analysis in the absence of integrability conditions. In Chipman et al. (1971).

Kihlstrom, R., Mas-Colell, A. and Sonnenschein, H. 1976. The demand theory of the weak axiom of revealed preference. *Econometrica* 44, 971–8.

Kim, T. and Richter, M.K. 1986. Nontransitive-nontotal consumer theory. *Journal of Economic Theory* 38, 324–63.

Koopmans, T. 1972. Representation of preference orderings with independent components of consumption. In *Decision and Organization*, ed. C.B. McGuire and R. Radner. Amsterdam: North-Holland.

Leontief, W. 1947. Introduction to a theory of the internal structure of functional relationships. *Econometrica* 15, 361–73. Repr. in *Selected Readings in Economic Theory*, ed. K.J. Arrow. Cambridge, MA: MIT Press, 1971.

Mas-Colell, A. 1974. Continuous and smooth consumers: approximation theorems. *Journal of Economic Theory* 8, 305–36.

Mas-Colell, A., Whinston, M.D. and Green, J. 1995. *Microeconomic Theory*. Oxford: Oxford University Press.

Neuefeind, W. 1972. On continuous utility. *Journal of Economic Theory* 5, 174–6.

Pareto, V. 1896. *Cours d'Économie Politique*. Lausanne: Rouge.

Pareto, V. 1906. L'ofelimità nei cicli non chiusi. *Giornale degli economisti* 33, 15–30. Trans. as 'Ophelimity in non-closed cycles', in Chipman et al. (1971).

Rader, T. 1963. The existence of a utility function to represent preferences. *Review of Economic Studies* 30, 229–32.

Richter, M.K. 1966. Revealed preference theory. *Econometrica* 34, 635–45.

Samuelson, P.A. 1947. *Foundations of Economic Analysis*. Cambridge, MA: Harvard University Press.

Samuelson, P.A. 1950. The problem of integrability in utility theory. *Economica* 17, 355–85.

Savage, L. 1954. *Foundations of Statistics*. New York: Wiley.

Schmeidler, D. 1971. A condition for the completeness of partial preference relations. *Econometrica* 39, 403–4.

Schmeidler, D. 1986. Integral representation without additivity. *Proceedings of the American Mathematical Society* 97, 255–61.

Schmeidler, D. 1989. Subjective probability and expected utility without additivity. *Econometrica* 57, 571–87.

Shafer, W. 1974. The nontransitive consumer. *Econometrica* 42, 913–9.

Slutsky, E. 1915. Sulla teoria del bilancio del consumatore. *Giornale degli Economisti e Rivista di Statistica* 51, 1–26. Trans. as 'On the theory of the budget of the consumer', in *Readings in Price Theory*, ed. G.J. Stigler and K.E. Boulding. Homewood, IL: Irwin, 1953.

Sonnenschein, H. 1971. Demand theory without transitive preferences, with applications to the theory of competitive equilibrium. In Chipman et al. (1971).

Sono, M. 1945. The effect of price changes on the demand and supply of separable goods. *Kokumni Keizai Zasski* 74, 1–51 [in Japanese]. English translation: *International Economic Review* 2 (1960), 239–71.

Uzawa, H. 1960. Preference and rational choice in the theory of consumption. In Chipman et al. (1971).

Walras, L. 1874. *Elements d'économie politique pure*. Lausanne: Corbaz. Trans. W. Jaffé as *Elements of Pure Economics*. London: Allen and Unwin, 1954.

democratic paradoxes

Models of elections tend to give two quite contradictory predictions about the result of political competition. In two-party competition, if the 'policy space' involves two or more independent issues, then 'pure strategy Nash equilibria' generally do not exist and instability or *chaos* may occur (see Plott, 1967; McKelvey, 1976; 1979; Schofield, 1978; 1983; 1985; McKelvey and Schofield, 1986; 1987; Saari, 1997; Austen-Smith and Banks, 1999). That is to say, whatever position is picked by one party, there always exists another policy point which will give the second party a majority over the other. Moreover, vote maximizing strategies could lead political candidates to wander all over the policy space.

On the other hand, the earlier electoral models based on the work of Hotelling (1929) and Downs (1957) suggest that parties will converge to an electoral centre (at the electoral *median*) when the policy space has a single dimension. (An equilibrium can also be guaranteed as long as the decision rule requires a sufficiently large majority – Schofield, 1984; Strnad, 1985; Caplin and Nalebuff, 1988 – or when the electoral distribution has a certain concavity property – Caplin and Nalebuff, 1991.) Although a pure strategy Nash equilibrium generically fails to exist in competition between two agents under majority rule in high enough dimension, there will exist mixed strategy Nash equilibria (Kramer, 1978) whose support lies within a subset of the policy space known as the 'uncovered set' (see McKelvey, 1986; Banks, Duggan and Le Breton, 2002). These various and contrasting theoretical results can be seen as a paradox: will democracy tend to generate centrist compromises, or can it lead to chaos? This question is of fundamental importance in a world in which many countries are experimenting with democracy for the first time.

Partly as a result of these theoretical difficulties with the 'deterministic' electoral model, and also because of the need to develop empirical models of voter choice (Poole and Rosenthal, 1984), attention has focused on 'stochastic' vote models. A formal basis for such models is provided by the notion of 'quantal response equilibria' (McKelvey and Palfrey, 1995). In such models, the behaviour of each voter is modelled by a vector of choice probabilities (Lin, Enelow and Dorussen, 1999). A standard result in this class of models is that all parties converge to the electoral origin when the parties are motivated to maximize vote share or plurality (in the two-party case) (see McKelvey and Patty, 2006; Banks and Duggan, 2005). The predictions concerning convergence are at odds with empirical evidence that parties appear to diverge from the electoral centre (Merrill and Grofman, 1999; Adams, 2001; Schofield and Sened, 2006).

The *paradox* that actual political systems display neither *chaos* nor *convergence* is the subject of this article. The key idea is that the convergence result need not hold if there is an asymmetry in the electoral perception of the 'quality' of party leaders (Stokes, 1992). The average weight given to the perceived quality of the leader of the jth party is called the party's 'valence'. In empirical models this valence is independent of the party's position, and adds to the statistical significance of the model. In general, valence reflects the overall degree to which the party is perceived to have shown itself able to govern effectively in the past, or is likely to be able to govern well in the future (Penn, 2003). The early empirical model of US presidential elections by Poole and Rosenthal (1984) included these valence terms. The authors noted that there was no evidence of candidate convergence.

Formal models of elections incorporating valence have been developed (Ansolabehere and Snyder, 2000; Groseclose, 2001; Aragones and Palfrey, 2002), but the theoretical results to date have been somewhat inconclusive. Extension to the multiparty case is of interest because of recent empirical models of voting in the Netherlands and Germany (Schofield et al., 1998; Quinn, Martin, and Whitford, 1999; Quinn, and Martin, 2002), Britain (Schofield, 2005a; 2005b), Israel (Schofield, Sened and Nixon, 1998; Schofield and Sened, 2002; 2005; 2006) and Italy (Giannetti and Sened, 2004). All these empirical models have suggested that divergence is generic. Most of these empirical models have been based on the 'multinomial logit' assumption that the stochastic errors had a 'Type I extreme value distribution' (Dow and Endersby, 2004).

Schofield (2007) provides a 'classification theorem' for the formal vote model based on the same stochastic distribution assumption. The 'policy space' is assumed to be of dimension w, and there is an arbitrary number, p, of parties. The party leaders exhibit differing valence. A 'convergence coefficient' incorporating all the parameters of the model can be defined. Instead of using the notion of a Nash equilibrium, the result is given in terms of the existence of a 'local Nash equilibrium'. It is shown that there are necessary and sufficient conditions for the existence of a 'pure strategy vote maximizing local Nash equilibrium' (LNE) at the mean of the voter distribution. When the necessary condition fails, then parties, in equilibrium, will adopt divergent positions. In general, parties whose leaders have the lowest valence will take up positions furthest from the electoral mean. Moreover, because a pure strategy Nash equilibrium (PNE) must be a local equilibrium, the failure of existence of the LNE at the electoral mean implies non-existence of such a centrist PNE. The failure of the necessary condition for convergence has a simple explanation: if the variance of the electoral distribution is sufficiently large in contrast to the expected vote share of the lowest-valence party at the electoral mean, then this party has an incentive to move away from the origin towards the electoral periphery. Other low-valence parties will follow suit, and the local equilibrium will result with parties distributed along a 'principal electoral axis'.

An empirical study of voter behaviour for Israel for the election of 1996 (based on Schofield and Sened, 2005) is used to show that the necessary condition for party convergence failed for this election. The equilibrium positions obtained from the formal result, under vote maximization, are in general comparable with, but not identical to, the estimated positions. The two highest-valence parties (Labour and Likud) were symmetrically located on either side of the electoral origin, while the lowest-valence parties were located far from the origin. In such a polity, based on a proportional electoral system, it is generally necessary to form coalition governments. The existence of small, low-valence, radical parties on the electoral periphery may create serious difficulties in the formation of majority government. It is possibly for this reason that Ariel Sharon, formerly leader of Likud, and Shimon Peres, formerly leader of Labour, in 2005 formed Kadima, a new centrist party.

This article also presents results from analysis of the 1997 election in Britain (Schofield, 2005a; 2005b). In this case the empirical estimates of the parameters of the model, taken together with the formal analysis, suggest that convergence should have occurred. Instead the Conservative Party was estimated to be at a position far from the electoral centre. It is suggested that the discrepancy between the formal and the empirical models can be accommodated by considering the effect of activists on the optimal party position. Since concerned activists will raise funds for the party, but only if the party adopts a policy position that accords with activists' concerns, there is a tension between activist demands and the electoral concerns of the party leadership. The model based on activist support estimates the marginal trade-off generated by opposed activist groups within a party. It is suggested that the low valence of recent Conservative leaders obliged them to seek support from activists supporting British sovereignty against the European Union, and thus to take up radical positions on the second, 'European' axis.

In contrast, the apparent move by the Labour Party towards the electoral centre between 1992 and 1997 was a consequence of the increase of the electoral valence of Tony Blair, the leader of the party, rather than a cause of this increase.

Recent work by Miller and Schofield (2003) using this model suggests that, in the United States, the movement of presidential candidates in a two-dimensional policy space generated by economic and social dimensions is the result of contending and opposed activist groups.

The underlying premise of the notion of the *local Nash equilibrium*, used in these models, is that party leaders will not consider 'global' changes in party policies, but will instead propose small changes in the party position in response to changes in beliefs about electoral response.

These models regard elections as the aggregation of both electoral evaluation or 'valence' and electoral preferences. Valence can be regarded as that element of a voter's choice which is determined by judgement rather than preference. This accords well with the arguments of James Madison in *Federalist 10* of 1787 (Rakove, 1999) and of Condorcet (1785) in his treatise on social choice theory. Schofield (2005c; 2006) provides a discussion of the relevance of these valence models for the constitutional basis of the US polity.

Empirical analysis for Israel

Figure 1 shows the estimated positions of the parties at the time of the 1996 Israeli election, Figure 1 also gives the estimated distribution of voter ideal points for the 1996 election, based on a factor analysis of the survey responses derived from the survey of Arian and Shamir (1999). The two dimensions of policy deal with attitudes to the Palestine Liberation Organization (PLO) (the horizontal axis) and religion (the vertical). The party positions were obtained from analysis of party manifestos (Schofield, Sened and Nixon, 1998; Schofield and Sened

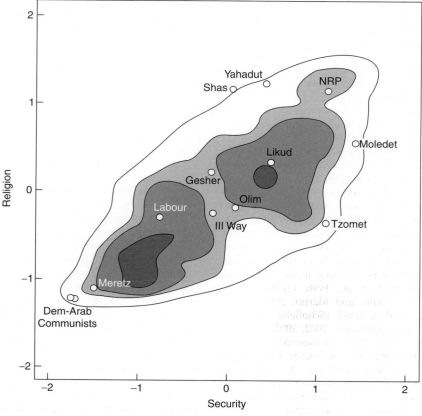

Figure 1   Voter distribution and estimated party positions in the Knesset at the 1996 election

2005; 2006). With the use of information on the individual voter intentions, it is possible to construct a
multinomial logit model (based on the Type I extreme
value distribution).

The model assumes that the voter utility vector has the
form $\mathbf{u}_i(xi, \mathbf{z}) = (u_{i1}(x_i, z_1), \ldots, u_{ip}(x_i, z_p))$, where

$$u_{ij}(x_i, z_j) = u_{ij}^*(x_i, z_j) + \varepsilon_j \quad and$$
$$u_{ij}^*(x_i, z_j) = \lambda_j - \beta \|x_i - z_j\|^2.$$

Here the position of voter i is x_i while the position of
party j is z_j. The term $\|x_i - z_j\|$ is the distance between
these two points. The *valences* of the p parties are given
by the vector $\lambda = (\lambda_p, \lambda_{p-1}, \ldots, \lambda_2, \lambda_1)$ and are ranked

$$\lambda_p \geq \lambda_{p-1} \geq \ldots \geq \lambda_2 \geq \lambda_1.$$

The error terms $\{\varepsilon_j\}$ have the *Type I extreme value dis-
tribution*, Ψ.
(The cumulative distribution, Ψ, takes the closed form
$\Psi(h) = \exp[-\exp[-h]]$.)
The probability that a voter i chooses party j is

$$\rho_{ij}(\mathbf{z}) = \Pr[[u_{ij}(x_i, z_j) > u_{il}(x_i, z_l)], \quad \text{for all } l \neq j.$$

Here Pr stands for the probability operator associated
with Ψ. The expected vote share of agent j is

$$V_j(\mathbf{z}) = \frac{1}{n} \sum_{i \in N} \rho_{ij}(\mathbf{z}).$$

This model is denoted $M(\lambda, \beta; \Psi)$. A *local pure strategy
Nash equilibrium* (LNE) is simply a vector $\mathbf{z} =
(z_1, \ldots, z_p)$ of party positions with the property that
each z_j locally maximizes $V_j(\mathbf{z})$, taking the other party
positions A necessary condition for $\mathbf{z}^* = (0, \ldots, 0)$ to be
pure strategy Nash equilibrium (PNE) is that it be a LNE
and thus that all Hessians have eigenvalues at $\mathbf{z}^*$ that are
non-positive. This can be expressed as a single necessary
condition on a 'convergence coefficient' defined in terms
of the Hessian of the vote share function of the party
with the lowest valence (Schofield, 2006b). Since the
lowest-valence party is the National Religious Party
(NRP) (for the 1996 model for Israel), a *necessary* condition for the NRP vote share to be maximized at the
origin is that *both* eigenvalues of this Hessian be non-
positive. However, the calculation given below shows that
one of the eigenvalues was positive. It follows that
the NRP position that maximizes its vote share is *not* at
the origin. Thus the convergent position $(0, \ldots, 0)$ cannot
be a Nash equilibrium to the vote maximizing game.

Indeed it is obvious that there is a principal component
of the electoral distribution, and this axis is the eigenspace
of the positive eigenvalue. It follows that low-valence
parties should then position themselves on this eigenspace, as illustrated in the simulation given in Figure 2.

To present the calculation, we use the fact that the
valence of the NRP was -4.52. The spatial coefficient is
$\beta = 1.12$. Because the valences of the major parties are
4.15 and 3.14, the formal analysis implies that, when all
parties are at the origin, the vote share, ρ_{NRP} can be
computed to be

$$\rho_{NRP} \simeq \frac{1}{1 + e^{4.15+4.52} + e^{3.14+4.52}} \simeq 0.$$

Moreover, the Hessian of the NRP at the origin
depends on the electoral variance and this is

$$C_{NRP} = 2(1.12) \begin{pmatrix} 1.0 & 0.591 \\ 0.591 & 0.732 \end{pmatrix} - I$$
$$= \begin{pmatrix} 1.24 & 1.32 \\ 1.34 & 0.64 \end{pmatrix}.$$

The eigenvalues of the NRP Hessian at the origin are
2.28 and -0.40, giving a saddle point. Thus, the origin
cannot be a Nash equilibrium. The 'convergence
coefficient' can be calculated to be 3.88, larger than the
necessary upper bound of 2.0. The major eigenvector for
the NRP is (1.0,0.8), and along this axis the NRP vote
share function increases as the party moves away from
the origin. The minor, perpendicular axis is given by the
vector $(1,-1.25)$ and on this axis the NRP vote share
decreases. Figure 2 gives one of the local equilibria in
1996, obtained by simulation of the model. The figure
makes it clear that the vote maximizing positions lie
on the principal axis through the origin and the point
(1.0, 0.8). In all, five different LNE were located. However, in all the equilibria the two high-valence parties,
Labour and Likud, were located at precisely the same
positions, as shown in Figure 2. The only difference
between the various equilibria was that the positions of
the low-valence parties were perturbations of each other.

Figure 2 suggests that the simulation was compatible
with the predictions of the formal model based on the
extreme value distribution. All parties were able to
increase vote shares by moving away from the origin,
along the principal axis, as determined by the large, positive principal eigenvalue. In particular, the simulation
confirms the logic of the above analysis. Low-valence
parties, such as NRP and Shas, in order to maximize
vote shares must move far from the electoral centre.
Their optimal positions will lie in either the north-east
quadrant or the south-west quadrant. The vote
maximizing model, without any additional information,
cannot determine which way the low-valence parties
should move. As noted above, the simulations of the
empirical models found multiple LNE essentially differing
only in permutations of the low-valence party positions.

In contrast, since the valence difference between Labour
and Likud was relatively low, their optimal positions
would be relatively close to, but not identical to, the

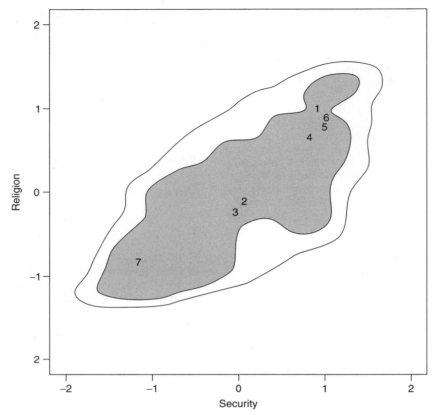

Figure 2   A simulated local Nash equilibrium in the vote maximizing game in Israel in 1996. *Note*: 1: Shas; 2: Likud; 3: Labour; 4: NRP; 5: Molodet; 6: III Way; 7: Meretz.

electoral mean. The simulation for the elections of 1988 and 1992 are also compatible with this theoretical inference. Figure 2 also suggests that every party, in local equilibrium, should adopt a position that maintains a minimum distance from every other party. The formal analysis, as well as the simulation exercise, suggests that this minimum distance depends on the valences of the neighbouring parties. Intuitively it is clear that, once the low-valence parties vacate the origin, then high-valence parties like Likud and Labour will position themselves almost symmetrically about the origin, and along the major axis.

Comparison between Figure 1, of the estimated party positions, and Figure 2, of simulated equilibrium positions, reveals a notable disparity particularly in the position of Shas. In 1996 Shas was pivotal between Labour and Likud, in the sense that, to form a winning coalition government, either of the two larger parties required the support of Shas. It is obvious that the location of Shas in Figure 1 suggests that it was able to bargain effectively over policy and, presumably, perquisites. Indeed, it is plausible that the leader of Shas was aware of this situation, and incorporated this awareness in the utility function of the party.

The close correspondence between the simulated LNE based on the empirical analysis and the estimated actual political consuggests that the true utility function for each party j has the form $U_j(\mathbf{z}) = V_j(\mathbf{z}) + \delta_j(\mathbf{z})$, where $\delta_j(\mathbf{z})$ may depend on the beliefs of party leaders about the post-election coalition possibilities, as well as the effect of activist support for the party.

This hypothesis leads to the further hypothesis that, for the set of feasible strategy profiles in the Israel polity, $\delta_j(\mathbf{z})$ is 'small' relative to $V_j(\mathbf{z})$. A formal model to this effect could indicate that the LNE for $\{U_j\}$ would be close to the LNE for $\{V_j\}$. Note, however, that this perturbation of the party utility function causes parties to leave the main electoral axis. It is possibly for this reason that coalition politics in Israel has been very complex.

The Likud Party, under Ariel Sharon, was constrained by the religious parties in its governing coalition. This apparently caused Sharon to leave Likud to set up a new centrist party, Kadima ('Forward') with Shimon Peres, previously leader of Labour. The reason for this reconfiguration was the victory on 10 November 2005 of Amir Peretz over Peres for leadership of the Labour Party, and Peretz's move to the left along the principal electoral axis.

Consistent with the model presented here, Sharon's intention was to position Kadima very near the electoral centre on both dimensions, to take advantage of his high valence among the electorate. Sharon's subsequent hospitalization had an adverse effect on the valence of Kadima, under its new leader, Ehud Olmert. Even so, in the election of 28 March 2006 Kadima took 29 seats, against 19 seats for Labour, and only 12 for Likud. One surprise was a new centrist pensioners' party with 7 seats. Because Kadima with Labour and the other parties of the left had 70 seats, Olmert was able to put together a majority coalition on 28 April, including the Orthodox party Shas. As Figure 1 illustrates, Shas is centrist on the security dimension, indicating that this was the key issue of the election.

Empirical analysis for Britain

This section analyses the general election in Britain in 1997 in order to suggest how activists for the parties may influence party positioning. The analysis shows that the valence model as presented above cannot always explain divergence of party positions. For example, Figure 3 shows the estimated positions of the party leaders, based on a survey of party MPs in 1997 (Schofield, 2005a; 2005b).

In addition to the Conservative Party, Labour Party, and Liberal Democrats, responses were obtained from Ulster Unionists, Scottish Nationalists and Plaid Cymru (Welsh Nationalists). The axis is economic, the second pro or anti the European Union. The electoral model was estimated for the election in 1997, using only the economic dimension.

For this election, we $(\lambda_{con}, \lambda_{lab}, \lambda_{lib}, \beta)_{1997} = (+1.24, 0.97, 0.0, 0.5)$ so the probability ρ_{lib}, that a voter chooses the Liberal Democrats is

$$\rho_{lib} = \frac{e^0}{e^0 + e^{1.24} + e^{0.97}} = \frac{1}{7.08} = 0.14.$$

The Hessian for this party at the origin is $C_{lib} = -0.28$, which is compatible with a Nash equilibrium at the origin. Extending the model to two dimensions gives a Hessian

$$C_{lib} = (0.72)\begin{pmatrix} 1.0 & 0 \\ 0 & 1.5 \end{pmatrix} - I$$
$$= \begin{pmatrix} -0.28 & 0 \\ 0 & +0.8 \end{pmatrix}.$$

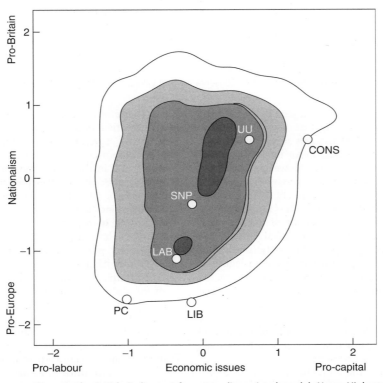

Figure 3 Estimated party positions in the British Parliament for a two-dimensional model. *Notes*: Highest-density contours of the voter sample distribution at the 95%, 75%, 50% and 10% levels. CONS: Conservative Party; LAB: Labour Party; LIB: Liberal Democrats; PC: Plaid Cymru (Welsh Nationalists); SNP: Scottish National Party; UU: Ulster Unionist Party. *Source*: MP survey data and a National Election Survey.

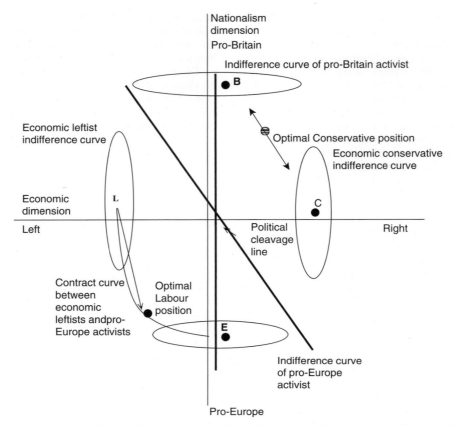

Figure 4   Illustration of vote maximizing positions of Conservative and Labour Party leaders in a two-dimensional policy space

According to the formal model, all parties should have converged to the origin on the first axis. Because the eigenvalue for the Liberal Democrats is positive on the second axis, we have an explanation for its position away from the origin on the Europe axis. However, there is no explanation for the location of the Conservative Party so far from the origin on both axes. Schofield (2005a; 2005b) adapts the activist model of Aldrich (1983a; 1983b) wherein the falling exogenous valence of the Conservative Party leader increases the marginal importance of two opposed activist groups in the party: one group 'pro-capital' and one group 'anti-Europe', as in Figure 4.

The optimal Conservative position will be determined by balancing the electoral effects of these two groups. The optimal position for this party will be one which is 'closer' to the locus of points that generates the greatest activist support. This locus is where the joint marginal activist pull is zero. This locus of points can be called the 'activist contract curve' for the Conservative Party.

Note that in Figure 4 the indifference curves of representative activists for the parties are described by ellipses. This is meant to indicate that preferences of different activists on the two dimensions may accord different saliences to the policy axes. The 'activist contract curve' given

in the figure, for Labour, say, is the locus of points satisfying the first order condition for vote maximization. This curve represents the balance of power between Labour supporters more interested in economic issues (centred at L in the figure and those more interested in Europe (centred at E). The optimal positions for the two parties will be at appropriate positions that satisfy the optimality condition.

According to this model, a party's optimal position will tend to be nearer to the electoral origin when the valence of the party leader is higher. In contrast, a party whose leader has low valence will be more influenced by activist groups, and will tend to adopt a position further from the electoral centre and nearer to the position preferred by the dominant activist group. This model has been applied to the US polity by Miller and Schofield (2003) and Schofield, Miller and Martin (2003).

Proportional representation and plurality rule

Most of the early work in formal political theory focused on two-party competition, and generally concluded that there would be strong centripetal electoral forces causing parties to converge to the electoral centre. The extension of this theory to the multiparty context, common in

European polities, has proved very difficult because of the necessity of dealing with coalition governments (Riker, 1962). However, the symmetry conditions developed by McKelvey and Schofield (1987) showed that a large, centrally located party could dominate policy if it occupied what is known as a 'core position'. Thus, in situations where there is a stable policy core there would be certainty over the post-election policy outcome of coalition negotiation (Laver and Schofield, 1998). Absent a policy core, the post-election outcome will be a lottery across various possible coalitions, all of which are associated with differing policy outcomes and cabinet allocations. Modelling this post-election 'committee game' can be done with cooperative game theoretical concepts (Banks and Duggan, 2000).

Although the non-cooperative stochastic electoral model presented here can give insight into the relationship between electoral preferences and beliefs (regarding the valences of party leaders), it is still incomplete. The evidence suggests that party leaders pay attention not only to electoral responses but also to the post-election coalition consequences of their choices of policy positions. Nonetheless, the combination of the electoral model and post-election bargaining theory (Schofield and Sened, 2002) suggests the following.

In a polity based on a proportional electoral rule, the high-valence parties will be attracted towards the electoral centre. However, if there are two such competing parties of similar valence neither will locate quite at the centre. There may be many low-valence parties, whose equilibrium, vote maximizing positions will be far from the electoral centre. In order to construct winning coalitions, one or other of the high-valence parties must bargain with more 'radical' low-valence parties, and this could induce a degree of coalitional instability. However, it is possible that a charismatic leader, such as Sharon in Israel, can adopt a centrist position and dominate politics by controlling the policy core.

In a polity based on a plurality electoral rule, the disproportionality between votes and seats may increase the importance of activist groups. A party with a relatively low-valence leader will be forced to depend on activist support. Consequently, the party will be obliged to move to a more radical position so to attract activist support.

This may provide a reason why Britain's Labour Party appeared to acquiesce to the demands of its left-wing supporters during the leadership of Michael Foot in 1980–3 and of Neil Kinnock in 1983–92. This led to Labour defeats in the elections between 1983 and 1992. Tony Blair became Labour leader following the death of John Smith in 1994 and his high valence allowed him to overcome union opposition and to craft the centrist 'New Labour' policies that led to Labour victories in the elections of 1997, 2001 and 2005.

Concluding remarks

To sum up, these models suggest how the democrat paradox can be resolved: convergence to an electoral centre is not a generic phenomenon, but can occur when a party leader is generally regarded by the electorate to be of superior quality or valence. Chaos does not occur in these models, though a degree of coalitional instability is possible under proportional electoral rule when there is no highly regarded political leader at the policy core.

NORMAN SCHOFIELD

See also **political competition; rational behaviour; rational choice and political science.**

This article is based on research supported by NSF Grant SES 024173. The table and figures are reproduced from Schofield and Sened (2006) by permission of Cambridge University Press.

Bibliography

Adams, J. 2001. *Party Competition and Responsible Party Government*. Ann Arbor: University of Michigan Press.

Adams, J. and Merrill III, S. 1999. Modeling party strategies and policy representation in multiparty elections: why are strategies so extreme? *American Journal of Political Science* 43, 765–91.

Aldrich, J. 1983a. A spatial model with party activists: implications for electoral dynamics. *Public Choice* 41, 63–100.

Aldrich, J. 1983b. A Downsian spatial model with party activists. *American Political Science Review* 77, 974–90.

Ansolabehere, S. and Snyder, J. 2000. Valence politics and equilibrium in spatial election models. *Public Choice* 103, 327–36.

Aragones, E. and Palfrey, T. 2002. Mixed equilibrium in a Downsian model with a favored candidate. *Journal of Economic Theory* 103, 131–61.

Aragones, E. and Palfrey, T. 2005. Spatial competition between two candidates of different quality: the effects of candidate ideology and private information. In *Social Choice and Strategic Decisions*, ed. D. Austen-Smith and J. Duggan. Heidelberg: Springer.

Arian, A. and Shamir, M. 1999. *The Election in Israel: 1996*. Albany: SUNY Press.

Austen-Smith, D. and Banks, J. 1999. *Positive Political Theory I*. Ann Arbor: University of Michigan Press.

Banks, J. and Duggan, J. 2000. A bargaining model of collective choice. *American Political Science Review* 94, 73–88.

Banks, J. and Duggan, J. 2005. *The theory of probabilistic voting in the spatial model of elections*. In *Social Choice and Strategic Decisions*, ed. D. Austen-Smith and J. Duggan. Heidelberg: Springer.

Banks, J., Duggan, J. and Le Breton, M. 2002. Bounds for mixed strategy equilibria and the spatial model of elections. *Journal of Economic Theory* 103, 88–105.

Caplin, A. and Nalebuff, B. 1988. On 64% majority rule. *Econometrica* 56, 787–814.

Caplin, A. and Nalebuff, B. 1991. Aggregation and social choice: a mean voter theorem. *Econometrica* 59, 1–23.

Condorcet, N. 1785. *Essai sur l'application de l'analyse à la probabilité des décisions rendues à la pluralité des voix*. Paris: Imprimerie Royale. Translated in part in I. McLean and F. Hewitt, *Condorcet: Foundations of Social Choice and Political Theory*. Aldershot: Edward Elgar, 1994.

Coughlin, P. 1992. *Probabilistic Voting Theory*. Cambridge: Cambridge University Press.

Dow, J. and Endersby, J. 2004. Multinomial probit and multinomial logit: a comparison of choice models for voting research. *Electoral Studies* 23, 107–22.

Downs, A. 1957. *An Economic Theory of Democracy*. New York: Harper and Row.

Enelow, J. and Hinich, M. 1984. *The Spatial Theory of Voting*. Cambridge: Cambridge University Press.

Giannetti, D. and Sened, I. 2004. Party competition and coalition formation: Italy 1994–1996. *Journal of Theoretical Politics* 16, 483–515.

Groseclose, T. 2001. A model of candidate location when one candidate has a valance advantage. *American Journal of Political Science* 45, 862–86.

Hinich, M. 1977. Equilibrium in spatial voting: the median voter result is an artifact. *Journal of Economic Theory* 16, 208–19.

Hotelling, H. 1929. Stability in competition. *Economic Journal* 39, 41–57.

Kramer, G. 1978. Existence of electoral equilibrium. In *Game Theory and Political Science*, ed. P. Ordeshook. New York: New York University Press.

Laver, M. and Schofield, N. 1998. *Multiparty Government: The Politics of Coalition in Europe*. Ann Arbor: Michigan University Press.

Lin, T.-M., Enelow, J. and Dorussen, H. 1999. Equilibrium in multicandidate probabilistic spatial voting. *Public Choice* 98, 59–82.

McKelvey, R. 1976. Intransitivities in multidimensional voting models and some implications for agenda control. *Journal of Economic Theory* 12, 472–82.

McKelvey, R. 1979. General conditions for global intransitivities in formal voting models. *Econometrica* 47, 1085–111.

McKelvey, R. 1986. Covering, dominance and institution-free properties of social choice. *American Journal of Political Science* 30, 283–314.

McKelvey, R. and Palfrey, T. 1995. Quantal response equilibria for normal form games. *Games and Economic Behavior* 10, 6–38.

McKelvey, R. and Patty, J. 2006. A theory of voting in large elections. *Games and Economic Behavior* 57, 155–80.

McKelvey, R. and Schofield, N. 1986. Structural instability of the core. *Journal of Mathematical Economics* 15, 179–98.

McKelvey, R. and Schofield, N. 1987. Generalized symmetry conditions at a core point. *Econometrica* 55, 923–33.

Merrill III, S. and Grofman, B. 1999. *A Unified Theory of Voting*. Cambridge: Cambridge University Press.

Miller, G. and Schofield, N. 2003. Activists and partisan realignment in the U.S. *American Political Science Review* 97, 245–60.

Penn, E. 2003. A model of far-sighted voting. Working paper, Institute of Quantitative Social Science, Harvard University.

Plott, C. 1967. A notion of equilibrium and its possibility under majority rule. *American Economic Review* 57, 787–806.

Poole, K. and Rosenthal, H. 1984. U.S. presidential elections 1968–1980: a spatial analysis. *American Journal of Political Science* 28, 283–312.

Quinn, K. and Martin, A. 2002. An integrated computational model of multiparty electoral competition. *Statistical Science* 17, 405–19.

Quinn, K., Martin, A. and Whitford, A. 1999. Voter choice in multiparty democracies. *American Journal of Political Science* 43, 1231–47.

Rakove, J., ed. 1999. *James Madison: Writings*. New York: Library of America.

Riker, W. 1962. *The Theory of Political Coalitions*. New Haven, CT: Yale University Press.

Saari, D. 1997. The generic existence of a core for q-rules. *Economic Theory* 9, 219–60.

Schofield, N. 1978. Instability of simple dynamic games. *Review of Economic Studies* 45, 575–94.

Schofield, N. 1983. Generic instability of majority rule. *Review of Economic Studies* 50, 695–705.

Schofield, N. 1984. Social equilibrium and cycles on compact sets. *Journal of Economic Theory* 33, 59–71.

Schofield, N. 1985. *Social Choice and Democracy*. Heidelberg: Springer.

Schofield, N. 2005a. A valence model of political competition in Britain: 1992–1997. *Electoral Studies* 24, 347–70.

Schofield, N. 2005b. Local political equilibria. In *Social Choice and Strategic Decisions: Essays in Honor of Jeffrey S. Banks*, ed. D. Austen-Smith and J. Duggan. Heidelberg: Springer.

Schofield, N. 2005c. The intellectual contribution of Condorcet to the founding of the US republic. *Social Choice and Welfare* 25, 303–18.

Schofield, N. 2006. *Architects of Political Change: Constitutional Quandaries and Social Choice Theory*. Cambridge: Cambridge University Press.

Schofield, N. 2007. The mean voter theorem: necessary and sufficient conditions for convergent equilibrium, *Review of Economic Studies* 74, 965–80.

Schofield, N., Martin, A., Quinn, K. and Whitford, A. 1998. Multiparty electoral competition in the Netherlands and Germany: a model based on multinomial probit. *Public Choice* 97, 257–93.

Schofield, N., Miller, G. and Martin, A. 2003. Critical elections and political realignment in the U.S.: 1860–2000. *Political Studies* 51, 217–40.

Schofield, N. and Sened, I. 2002. Local Nash equilibrium in multiparty politics. *Annals of Operations Research* 109, 193–211.

Schofield, N. and Sened, I. 2005. Multiparty competition in Israel: 1988–1996. *British Journal of Political Science* 35, 635–63.

Schofield, N. and Sened, I. 2006. *Multiparty Government: Elections and Legislative Politics*. Cambridge: Cambridge University Press.

Schofield, N., Sened, I. and Nixon, D. 1998. Nash equilibrium in multiparty competition with stochastic voters. *Annals of Operations Research* 84, 3–27.

Stokes, D. 1963. Spatial models and party competition. *American Political Science Review* 57, 368–77.

Stokes, D. 1992. Valence politics. In *Electoral Politics*, ed. D. Kavanagh. Oxford: Clarendon Press.

Strnad, J. 1985. The structure of continuous-valued neutral monotonic social functions. *Social Choice and Welfare* 2, 181–95.

Train, K. 2003. *Discrete Choice Methods for Simulation*. Cambridge: Cambridge University Press.

demographic transition

The demographic transition is the process whereby fertility and mortality move from initially high levels to subsequent low levels, with accompanying changes in the size, growth rate and age distribution of the population.

Before the start of the demographic transition, life was short, fertility was high, growth was slow, and the population was young. Declining mortality starts the typical transition, followed after a considerable lag by fertility decline (France and the United States were important exceptions to this ordering). This pattern of change causes growth rates first to accelerate and then to slow again, as population moves towards low fertility, long life and an old age structure.

The transition began around 1800 with declining mortality in Europe. It has now spread to all parts of the world and is projected to be completed by 2100. This global demographic transition has brought momentous changes, reshaping the economic and demographic life cycles of individuals and restructuring populations. Global population size increased by a factor of 6.5 between 1800 and 2000, and by 2100 will have risen by a factor of ten. There will then be 50 times as many elderly but only five times as many children: the ratio of elders to children will have risen by a factor of ten. The length of life, which has already more than doubled, will have tripled, while births per woman will have dropped from six to two. In 1800, women spent about 70 per cent of their adult years bearing and rearing young children, but that fraction has decreased in many parts of the world to only about 14 per cent due to lower fertility and longer life (Lee, 2003). These changes are sketched in Table 1.

Before the demographic transition

According to Thomas Malthus (1798), slow population growth in the pre-industrial past was no accident. Faster population growth would depress wages, causing fertility to fall and mortality to rise due to famine, war or disease. Thus, population size was held in equilibrium with the slowly growing economy. The need to establish a separate household at marriage kept mean age at first marriage high, averaging around 25 years for women, and overall fertility low, at four to five births per woman. Mortality was moderately high, with life expectancy between 25 and 35 years. Outside of Europe and its offshoots, fertility and mortality were higher in the pre-transitional period. In India in the late 19th century, life expectancy was in the low twenties, while fertility was six or seven births per woman (Bhat, 1989). In Taiwan, the picture was similar around 1900. In the 1950s and 1960s, fertility in the less developed countries (LDCs, see UNPD, 2005, for definition) was typically six or higher.

Declining mortality

The demographic transition began first in north-west Europe, where mortality started its secular decline around 1800. In many low-income countries, the decline in mortality began in the early 20th century and then accelerated dramatically after the Second World War. The first stage of mortality decline is due to reductions in contagious and infectious diseases. Starting with the development of smallpox vaccine in the late 18th century, preventive medicine played a role in mortality decline in Europe. Public health measures were important from the

Table 1   *Global population trends over the transition: estimates, guesstimates and forecasts, 1700–2100*

Year	Life expectancy (years at birth)	Total fertility rate (births per woman)	Pop. Size (billions)	Pop. growth rate (%/year)	Pop. <15 (% total pop.)	Pop >65 (% total pop)
1700	27	6.0	.68	0.50	36	4
1800	27	6.0	.98	0.51	36	4
1900	30	5.2	1.65	0.56	35	4
1950	47	5.0	2.52	1.80	34	5
2000	65	2.7	6.07	1.22	30	7
2050	74	2.0	9.08	0.33	20	16
2100	81	2.0	9.46	0.04	18	21?

Sources: United Nations estimates and projections, 1900–2100; other sources for earlier years (see Lee, 2003, for details).

late 19th century, and some quarantine measures may have been effective in earlier centuries. Improved personal hygiene also helped as the germ theory of disease became more widely known and accepted. Improving nutrition was also important in the early phases of mortality decline. Famine mortality was reduced by improvements in storage and transportation that permitted integration of regional and international food markets. Secular increases in incomes led to improved nutrition in childhood and throughout life. Better-nourished populations with stronger organ systems were better able to resist disease.

Today, the high-income countries have already largely achieved the potential mortality reductions through control of contagious disease and improved nutrition. For them, further reduction in mortality must continue to come from reductions in chronic and degenerative diseases, notably heart disease and cancer (Riley, 2001).

Most LDCs did not begin the mortality transition until the 20th century but then made rapid gains. Between 1950–4 and 2000–4, life expectancy in LDCs has increased from 41.1 years to 63.4, with average gains of 0.45 years of life per calendar year. Such rapid rates of increase in low-income countries will surely taper off as mortality levels approach those of the more developed countries (MDCs), whose gains have been less than half as rapid at 0.19 years per year. There are notable exceptions to this generally favourable picture. In sub-Saharan Africa, life expectancy has been declining since the early 1980s, largely due to HIV/AIDS. In the southern African region, life expectancy dropped from 62 to 48 between the early 1990s and the early 2000s. On average, eastern European (including the former USSR) life expectancy is lower now than it was in the late 1960s (UNPD, 2005).

How far and how fast will mortality fall and life expectancy rise in the 21st century? Methods that extrapolate historical trends in mortality by age suggest greater longevity gains than MDC government actuaries typically project, but past official projections have under-predicted subsequent gains, particularly at the older ages. Some experts argue that we are approaching biological limits and that these historical trends cannot be expected to continue; they foresee an upper limit of 85 years for life expectancy. Others, impressed by advances in genetic and stem cell research, foresee much more rapid gains for the future than occurred in the past.

Fertility transition

Most economic theories of fertility start with the idea that couples wish to have some number of surviving children rather than a number of births per se. On this assumption, once potential parents recognize an exogenous increase in child survival, fertility should decline. However, mortality and fertility interact in complicated ways. For example, increased survival raises the return on post-birth investments in children, while some of the

improvement in child survival is itself a response to parental decisions to invest more in the health and welfare of a smaller number of children. Nonetheless, it is very likely that mortality decline has exerted an important independent influence on fertility decline.

Economic change also influences fertility by altering the costs and benefits of childbearing and rearing, which are time-intensive. Technological progress and increasing physical and human capital make labour more productive, raising the value of time in all activities and making children increasingly costly relative to consumption goods. Since women have had primary responsibility for childbearing and rearing, variations in the productivity of women have been particularly important. For example, physical capital may substitute for human strength, reducing or eliminating the productivity differential between male and female labour, and thus raising the opportunity cost of children. Rising incomes have shifted consumption demand towards non-agricultural goods and services, for which educated labour is a more important input. A rise in the rate of return to education then leads to increased investments in education. Overall, these patterns have several effects: children become more expensive, their economic contributions are diminished by school time, and educated parents have higher value of time, which raises the opportunity costs of childrearing. Furthermore, parents with higher incomes choose to devote more resources to each child, and, since this raises the cost of each child, it also leads to fewer children. Developing markets and governments replace many economic functions of the traditional family and household, to which children contributed, further weakening the value of children.

The importance of contraceptive technology for fertility decline in the past and future is hotly debated, with many economists viewing it as of relatively little importance (Pritchett, 1994). The European fertility transition, for example, was achieved using *coitus interruptus*, a widely known traditional method requiring no modern technology.

Between 1890 and 1920, fertility within marriage began to decline in most European provinces, with a median decline of about 40 per cent from 1870 to 1930 (Coale and Watkins, 1986). The fertility transition in the MDCs largely occurred before the Second World War. After the war, many of these countries experienced baby booms and busts, followed by the 'second fertility transition' as fertility fell far below replacement level, marriage rates fell, and increasing proportions of births occurred outside marriage. Many LDCs began the fertility transition in the mid-1960s, and these later transitions have typically been more rapid than earlier ones, with fertility reaching replacement level (around 2.1 births per woman) within 20 to 30 years after onset. Fertility transitions in East Asia have been particularly early and rapid, while those in South Asia and Latin America have been slower (Bulatao and Casterline, 2001). The transition in sub-Saharan Africa started from a higher initial level of fertility and began later. By now, almost all countries have begun the fertility transition (UNPD, 2005; Bulatao and Casterline, 2001).

Currently, 66 countries with 44 per cent of the world's population have fertility at or below replacement level. Of these, 43 are MDCs, but 23 are LDCs. Average fertility in the MDCs is 1.56 births per woman, and in many it has fallen below 1.3. Many LDCs, particularly in East Asia, also have fertility far below replacement. It is not yet clear whether fertility will fall farther, rebound towards replacement, or stay at current levels.

Age at first marriage and first birth are generally moving to older ages throughout the industrial world and much of the developing world as well. This depresses the total fertility rate, which is a synthetic cohort measure, by 10–40 per cent below the underlying completed fertilities of generations. When the average age of childbearing stops rising, the total fertility rate should increase to this underlying level.

Population growth

A steady state population growth rate for a hypothetical zero migration population can be associated with each level of fertility and life expectancy, as depicted in Figure 1.

Figure 1 plots growth rate contours or isoquants. These differ from actual growth rates due to net migration and the transitory influence of age distribution. In this figure, a demographic transition begins as a move to the right, representing a gain in life expectancy with little change in fertility and therefore movement to a higher population growth contour. Next, a diagonal downward movement to the right reflects a simultaneous decline in fertility and mortality, recrossing contours towards lower rates of growth and perhaps going negative, as do the MDCs. Historical data are extended using UNPD (2005) data and projections, by development status.

India, shown separately, had higher initial fertility and mortality than Europe, as did the least developed countries relative to the LDCs in 1950, which in turn had far higher mortality and fertility than the MDCs in that year. In all cases, the initial path is horizontally to the right, indicating that mortality decline preceded fertility decline, causing accelerating population growth approaching three per cent for the LDCs and least developed countries. Europe briefly attains 1.5 per cent steady state population growth, but then fertility plunges, a decline picked up after 1950 by the group of LDCs, ending with population decline at 1 per cent annually (the actual European population growth rate is slightly higher than this hypothetical steady state one due to age distribution and immigration).

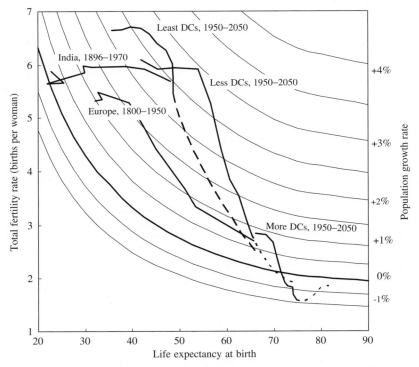

Figure 1 Life expectancy and total fertility rate with population growth isoquants: past and projected trajectories for more, less, and least developed countries. *Source*: Bhat (1989); UNPD (2005); see Lee (2003) for further details.

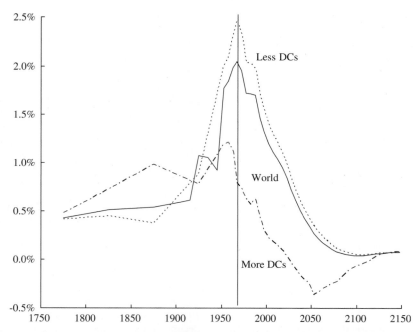

Figure 2   Population growth rates, 1750–2150. *Source*: UNPD (2005); see Lee (2003) for further details.

All three groups are projected by the UN to approach the zero-growth contour by 2050.

Historical and projected population growth rates, as opposed to hypothetical steady state ones, can be seen over a longer time period in Figure 2. Growth rates in the MDCs rose about a half of one per cent above those in the LDCs in the century before 1950. But after the Second World War, population growth surged in the LDCs, with the growth rate peaking at 2.5 per cent in the mid-1960s, then dropping rapidly. The population share of the MDCs is projected to drop from its current 20 per cent to only 13 per cent in 2050. Long-term United Nations projections suggest that global population growth will be close to zero by about 2100. The projection for the MDC population is nearly flat, with population decrease in Europe and Japan offset by population increase in the United States and other areas.

Changing age distribution over the demographic transition

Figure 3 plots the changes in age distribution that accompany a classic demographic transition, using historical data from India from 1896 to 2000 (stars) and United Nations projections through 2050 (hollow circles). These data are superimposed on a stylized transition that was simulated with the use of mathematical functions for the trajectories of fertility and life expectancy. Simulated fertility starts close to six births per woman and ends at 2.1. Life expectancy starts at 24 years and ends at 80. Mortality decline starts in 1900, 50 years

before the fertility decline begins in 1950. The Indian fertility transition is slower than that of East Asia but similar to that of Latin America.

The distinctive changes in the age distribution can be seen in the 'dependency ratios', which take either the younger or the older population and divide it by the working-age population. The initial mortality decline, while fertility remains high, raises the proportion of surviving children in the population, as reflected in the increasing child dependency ratios. Counter-intuitively, mortality decline initially makes populations younger rather than older in a phase which here lasts 70 years. Families find themselves with increasing numbers of surviving children, and both families and governments may struggle to achieve human capital investment goals for the unexpectedly high number of children.

Next, as fertility declines, child dependency ratios decline and soon fall below their pre-transition levels. The working age population grows faster than the population as a whole, so the total dependency ratio declines. This second phase may last 40 or 50 years. Some analysts have worried that the rapidly growing labour force in this phase might cause rising unemployment and falling capital–labour ratios, while others have stressed the advantages of this phase, calling these a demographic gift or bonus. Figure 3 shows that in India the bonus occurs between 1970 and 2015, when the total dependency rate is declining. The decline in dependents per worker would by itself raise per capita income by 22 per cent, other things equal, adding 0.5 per cent per year to per capita income growth over the 45-year span.

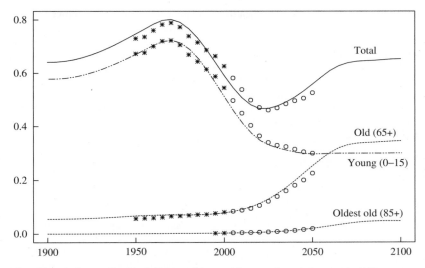

Note: Lines indicate a simulated demographic transition superimposed over actual (*)
and projected (o) dependency ratios for India.

Figure 3   Changing age distribution over a classic demographic transition: actual and projected dependency ratios for India and simulations, 1900–2100. *Source*: Actual India data for the period 1891–1901 to 1941–51 are taken from Bhat (1989). Other actual and projected data are taken from UNPD (2003).

In a third phase, increasing longevity leads to a rapid increase in the elderly population while low fertility slows the growth of the working age population. The old-age dependency ratio rises rapidly, as does the total dependency ratio. In India, population ageing will occur between 2015 and 2060. If the elderly are supported by transfers, either from their adult children or a public-sector pension system supported by current tax revenues, then the higher total dependency ratio means a greater burden on the working-age population. However, to the extent that the elderly prepare for their retirement by saving and accumulating assets earlier in their lives and then dissave in retirement, population ageing may cause lower aggregate saving rates, as life-cycle savings models and some empirical analyses suggest. But even with lower savings rates the capital–labour ratio may rise, since the labour force is growing more slowly. The net effect would then be to stimulate growth in labour productivity due to capital deepening.

At the end of the full transitional process, the total dependency ratio is back near its level before the transition began, but now child dependency is low and old-age dependency is high. Presumably mortality will continue to decline in the 21st century, and the process of individual and population ageing will continue. No country in the world has yet completed this phase of population ageing, since even the industrial countries are projected to age rapidly over the next three or four decades. In this sense, no country has yet completed its demographic transition.

Population ageing is due both to low fertility and to long life. Low fertility raises the ratio of elderly to working-age people, with no corresponding improvement in health to facilitate a prolongation of working years. For this reason, it imposes important resource costs on the population, regardless of institutional arrangements for old-age support. Lower total expenditures on children and increased capital per worker will offset these costs.

By contrast, population ageing due to declining mortality is generally associated with increasing health and vitality of the elderly. Such ageing may put pressure on pension programmes that have rigid retirement ages, but this problem is a curable institutional one, since the ratio of healthy, vigorous years over the life cycle to frail or disabled years has not changed, and individuals can adjust by keeping the fractions of their adult life spent working and retired constant, for example.

Some consequences of the demographic transition

The three centuries of demographic transition from 1800 to 2100 will reshape the world's population in a number of ways. Population will rise from 1 billion in 1800 to 9.5 billion in 2100. The average length of life will increase by a factor of two or three, fertility will have declined by two-thirds, and the median age of the population will double from the low twenties to the low forties. The population of Europe will decline by ten per cent between 2005 and 2050, and its share of world population will have declined by two-thirds since 1950. But many other changes will also have been set in train in family structure, health, institutions for saving and supporting retirement, and even in international flows of people and capital.

At the level of families, as the number of children born declines sharply, childbearing becomes concentrated into only a few years of a woman's life; combined with greater longevity, this means that many more adult years are available for other activities. Parents with fewer children are able to invest more in each child, reflecting the quality–quantity trade-off, which may also be one of the reasons parents reduced their fertility.

The processes which lead to longer life also alter the health status of the surviving population. For the United States, it appears that years of healthy life are growing roughly as fast as total life expectancy. In other industrial populations the story is more mixed. Trends in health, vitality and disability are of enormous importance for human welfare.

The economic pressures on pension programmes caused by the increasing proportion of elderly are exacerbated in the MDCs by dramatic declines in the age at retirement, which for US men fell from 74 in 1910 to 63 in 2000. Population ageing will also generate intense financial pressures on publicly funded systems for health care and for long-term care.

At the international level, the flow of people and capital across borders may offset these demographic pressures. As population growth has slowed or even turned negative in the MDCs, it is not surprising that international migration from Third World countries has accelerated. Net international migration to the MDCs experienced a roughly linear increase from near-zero in the early 1950s to around 2.6 million per year in the 1990s. Of course, these net numbers for large population aggregates conceal a great deal of offsetting international gross migration flows within and between regions (UNPD, 2005). For example, prior to 1970 Europe was a net sending region, but from 1970 to 2000 it received 18 million net immigrants. During the 1990s, repatriation of African refugees reversed the net flows from the least developed countries. But overall, while MDCs may seek to alleviate their population ageing through immigration, United Nations simulations indicate that the effect will be only modest, since immigrants also grow old, and their fertility converges to levels in the receiving country.

Might international flows of capital cushion the financial effects of population ageing? Population ageing may cause declining aggregate saving rates, but, with slowing labour force growth, capital–labour ratios will probably rise and profit rates fall, particularly if there is a move towards funded pensions. Capital flows from the MDCs into the LDCs might help keep the rate of return on investments from falling, but the possibilities are limited by the much smaller size of Third World economies.

Dramatic population ageing is the inevitable final stage of the global demographic transition, and it will bring serious economic and political challenges. Meeting these challenges will require flexible institutional structures, adjustments in life-cycle planning, and a willingness to pay for rising costs of health care and retirement.

RONALD D. LEE

See also **fertility in developed countries; historical demography; international migration; mortality; population ageing.**

Bibliography

Bhat, P. 1989. Mortality and fertility in India, 1881–1961: a reassessment. In *India's Historical Demography: Studies in Famine, Disease and Society*, ed. T. Dyson. London: Curzon Press.

Bulatao, R. and Casterline, J., eds. 2001. *Global Fertility Transition*, a supplement to *Population and Development Review* 27.

Coale, A. and Watkins, S., eds. 1986. *The Decline of Fertility in Europe*. Princeton: Princeton University Press.

Lee, R. 2003. The demographic transition: three centuries of fundamental change. *Journal of Economic Perspectives* 17(4), 167–90.

Malthus, T. 1798. *An Essay on the Principle of Population*, ed. D. Winch and P. James. Cambridge: Cambridge University Press, 1992.

Pritchett, L. 1994. Desired fertility and the impact of population policies. *Population and Development Review* 20, 1–55.

Riley, J. 2001. *Rising Life Expectancy: A Global History*. Cambridge: Cambridge University Press.

UNPD (United Nations Population Division). 2003. *World Population Prospects: The 2002 Revision*. New York: United Nations.

UNPD. 2005. *World Population Prospects: The 2004 Revision*. New York: United Nations.

Denison, Edward (1915–1992)

Edward Denison was a major contributor to the development of the US national income accounts and one of the originators of growth accounting. He received a Ph.D. in economics from Brown University in 1941. Denison's early career (1941–56) was spent in the national income division of the US Commerce Department where he worked with Milton Gilbert, George Jaszi, and Charles Schwartz to develop the national accounts of the United States. The United States had published estimates of national income and its components in 1934; and Richard Stone and others developed both expenditure and income-side estimates of GNP for the United Kingdom that were published in 1941. The US expenditure-side estimates were first published in 1942.

Denison participated in a 1944 tripartite meeting with Canada, the United Kingdom, and the United States that worked to establish consensus on a set of concepts and methods for the national accounts. That meeting and subsequent work provided much of the basis for the standardized system of national accounts (SNA) that was adopted and expanded by the United Nations and the OECD. The United States did not initially adopt the SNA; but by 2000 it was following the SNA in all of its important respects.

Denison moved to the Committee on Economic Development (CED) in 1956 where his research focused on identifying the sources of economic growth. In expanding the framework of growth accounting, Denison sought to go beyond a simple partitioning of economic growth into the contributions of the factor inputs and a residual of total factor production. He incorporated changes in the quality of the inputs, such as job skills, economies of scale, and other contributors to the residual, such as research and development. His initial analysis was published by the CED in 1962 as *The Sources of Economic Growth in the United States and the Alternatives Before Us*. A distinctive feature of his approach was the extent to which he anchored it in the basic accounting framework of the national accounts rather than the concepts of neoclassical production theory employed a few years later by Jorgenson and Griliches (1967). This aspect made it easy for other researchers to duplicate his methodology within their own countries.

Denison moved to the Brookings Institution in 1963 and extended his analysis to international comparisons with publication of *Why Growth Rates Differ* (1967). Two important later contributions were *How Japan's Economy Grew So Fast* (with W.K. Chung, 1976) and *Accounting for Slower Economic Growth: The United States in the 1970s* (1979). In *Accounting for Slower Growth*, he explored a wide range of popular explanations for the productivity slowdown, including higher energy prices, government regulation, and reduced R&D expenditures, and argued that their effects were too small to account for the magnitude and persistence of the slowdown. He received the Distinguished Fellow Award of the American Economic Association in 1981.

Denison's exchanges with Jorgenson and Griliches (1967; 1972a), while centred around differences in their approaches to measuring the contributions to growth, served to highlight an ongoing debate about the relative importance of capital accumulation and total factor productivity gains. Denison's approach, by minimizing several aspects of the measurement of capital's contribution, tended to support the conventional wisdom of the time that TFP accounted for a substantial portion of growth. Jorgenson and Griliches were attempting to argue that careful measurement of the factor inputs could drastically shrink the residual contribution of TFP. Denison won out on the issue of the relative importance of TFP by pointing to some problems with Jorgenson–Griliches adjustment for variations in capacity utilization; but the longer-term value of the debate was in showing that their approaches were quite similar. In subsequent years, the

Jorgenson–Griliches approach, with its anchor in production theory, has dominated the conceptual discussion. However, many of the empirical studies continue to follow Denison's careful use of national income accounts data.

BARRY BOSWORTH

See also **economic growth; growth accounting; total factor productivity.**

Selected works

1962. *The Sources of Economic Growth in the United States and the Alternatives Before Us*. New York: Committee for Economic Development.
1967. *Why Growth Rates Differ?* Washington, DC: Brookings Institution.
1969. Some major issues in productivity analysis: an examination of estimates by Jorgenson and Griliches. *Survey of Current Business* 49(5, Part 2), 1–27.
1972. Final comments. *The Measurement of Productivity*. Washington, DC: Brookings Institution.
1979. *Accounting for Slower Economic Growth: The United States in the 1970s*. Washington, DC: Brookings Institution.
1976. (With W. K. Chung.) *How Japan's Economy Grew So Fast: The Sources of Postwar Expansion*. Washington, DC: Brookings Institution.

Bibliography

Jorgenson, D. and Griliches, Z. 1967. The explanation of productivity change. *Review of Economic Studies* 34, 249–83.
Jorgenson, D. and Griliches, Z. 1972a. Issues in growth accounting: a reply to Edward F. Denison. *The Measurement of Productivity*. Washington, DC: Brookings Institution.
Jorgenson, D. and Griliches, Z. 1972b. Final reply. *The Measurement of Productivity*. Washington, DC: Brookings Institution.

dependency

Dependency theories emerged in Latin America in the early 1960s as a challenge to traditional Marxist and structuralist thinking regarding whether capitalist development in the periphery was both still *viable* (given the transformations of the world economy after the Second World War), and still *necessary* (as an unavoidable transition step towards socialism).

There can be little doubt that the Cuban Revolution was a turning point in Marxist analysis of capitalist development in the periphery. The events in Cuba gave rise to a new approach, of which most of the 'dependency analyses' form part. This argued that capitalism had totally lost its historical 'progressive' role in the periphery (if it ever had one); that is, it was both no longer capable of developing the productive forces of backward societies, and (thus) no longer able to bring them closer towards socialism. Consequently, this approach also argued against the politics of the popular fronts in the periphery and in favour of an immediate transition towards socialism.

Following traditional Marxist analysis, the pre-dependency, pre-Cuban Revolution approach saw capitalism as still historically progressive in the periphery; however, it argued that its key historical task – the 'bourgeois-democratic' revolution – was being inhibited by a new alliance between imperialist forces and the traditional oligarchies. The bourgeois-democratic revolution was the revolt of the emerging capitalist forces of production against the old pre-capitalist order. This revolution would be based on an alliance between the rising bourgeoisie and other progressive forces of society; the principal battle line would be between the new capitalist elites and the traditional oligarchies – between industry and land, capitalism versus pre-capitalist forms of monopoly and privilege. Because it would be the result of the pressure of a rising class whose path was being blocked in political, economic and social terms, this revolution would bring to the periphery (as it had done in the centre) not only political emancipation but economic progress as well.

One of the main analytical challenges facing the pre-dependency Marxist analysis was to explain why the 'bourgeois–democratic' revolution in the periphery was not really happening as expected (a phenomenon that was seriously hindering the process of capitalist development there). Since Lenin, this analysis had identified imperialism as the unmistakable main obstacle facing this revolution. The traditional oligarchies could not be the reason for this as on their own, they were not expected to prove any match for the new emerging capitalist classes. Therefore, the principal target in this struggle was unmistakable: North American imperialism. The allied camp for this fight, by the same reasoning, was also clear: everyone, except those (pre-capitalist) internal groups allied with imperialism. Thus, the anti-imperialist struggle was at the same time a struggle for domestic capitalist development and industrialization. The state and the 'national' bourgeoisie appeared as the potential leading agents for the development of the new capitalist economy, which in turn was viewed as a necessary stage towards socialism.

The Cuban Revolution questioned the very essence of this approach, insisting that the domestic bourgeoisies in the periphery no longer existed as a progressive social force but had become 'lumpen', 'rent seekers', incapable of rational accumulation and rational political activity, dilapidated by their consumerism, and blind to their 'real' long-term interests. It is within this framework, and with the explicit motive of developing theoretically and documenting historically this new approach that

dependency analysis appeared on the scene. At the same time, both inside and outside the Economic Commission for Latin America (ECLAC), two other major Dependency Schools began to develop (see STRUCTURALISM).

The general focus of all 'dependency' analyses is the development of peripheral capitalism (or, rather, the lack of it). More specifically, these studies attempted to analyse the obstacles to capitalist development in the periphery from the point of view of the new interplay between 'internal' and 'external' structures that had emerged after the Second World War. However, this interplay was analysed in several different ways.

With the necessary degree of simplification that every classification of intellectual tendencies entails, I distinguish between three major approaches – not mutually exclusive from the point of view of intellectual history – in 'dependency' analysis. First is the approach begun by Paul Baran, Paul Sweezy and Andre Gunder Frank; its essential characteristic is that it attempted to construct a comprehensive theory of the practical impossibility of capitalist development in the periphery. In these theories the 'dependent' character of peripheral economies is the crux on which the whole analysis of underdevelopment turns; that is, dependency is seen as causally linked to permanent capitalist underdevelopment.

The second approach is associated with the ECLAC Structuralist School, especially Celso Furtado, Aníbal Pinto and Osvaldo Sunkel. These writers sought to reformulate the classical ECLAC analysis of Latin American development from the perspective of a critique of the obstacles to 'national' development. This attempt at reformulation was not just process of adding new elements (mainly political and social) that were lacking in the original Prebisch–ECLAC analysis (see PREBISCH, RAÚL), but a thoroughgoing attempt to proceed beyond that analysis, adopting an increasingly different perspective.

Finally, the third approach, started by Fernando Henrique Cardoso and Enso Faletto, attempted to distance itself from the first by deliberately avoiding the formulation of a mechanico-formal theory of dependency and underdevelopment – specifically, by trying to avoid a mechanico-formal theory of the inevitability of underdevelopment in the capitalist periphery based on its dependent character. In turn, it concentrated on the study of what have been called 'concrete situations of dependency'; that is to say, the precise forms in which the different economies and polities of the periphery have been articulated with those of the advanced nations at different times, and how their specific dynamics have thus been generated.

The first approach: dependency as a formal theory of the inevitability of capitalist underdevelopment: on cutting a knot that could not be unravelled

There is no doubt that the 'father' of this approach was Baran. His principal contribution (1957) took up the approach of the Sixth Congress of the COMINTERN regarding the supposedly irresolvable nature of the contradictions between the economic and political needs of imperialism and those of the processes of political transformation, economic development and industrialization of the periphery.

To defend its interests, international monopoly capital would not only form alliances with pre-capitalist domestic oligarchies intended to block progressive capitalist transformations in the periphery, but its activities would also have the effect of distorting the process of capitalist development in these countries. As a result, international monopoly capital would have easy access to peripheral resources and finance, and the traditional élites in the periphery would be able to maintain their monopoly on power and their traditional (mostly predatory and rent-seeking) modes of surplus extraction. Within this context the possibilities for any form of dynamic economic growth in dependent countries were extremely limited or non-existent; the surplus they were able to generate (mainly from primary commodity export activities) was largely appropriated by foreign capital, or otherwise squandered by traditional elites. Therefore, long-term economic stagnation and underdevelopment was inevitable. The only way out was political. At a very premature stage, capitalism had become a fetter on the development of the productive forces in the periphery and, consequently, its historical role had already come to an early end.

Baran developed his ideas influenced both by the Frankfurt School's general pessimism regarding the nature of capitalist development (see Jay, 1996) and by Sweezy's (1946) proposition that the rise of monopolies imparts to capitalism a tendency towards stagnation and decay (see MONOPOLY CAPITALISM). He also followed the main growth paradigm of his time, the Harrod–Domar model, which held that the size of the investable surplus was the crucial determinant of growth (together with the efficiency with which it was used: the incremental capital–output ratio).

Starting out with Baran's analysis, Frank (1967) attempted to prove the thesis that the only political and economic solution to capitalist underdevelopment was a radical transformation of an immediately socialist character. For our purposes we may identify three levels of analysis in Frank's model of the 'development of underdevelopment'. In the first (arguing against 'dualistic' analyses), he attempted to demonstrate that the periphery had been incorporated and fully integrated into the world capitalist economy since the very early stages of colonial rule. In the second, he tried to show that such incorporation into the world capitalist economy had transformed the countries in question immediately into capitalist economies. Finally, in the third level, Frank attempted to prove that the integration of these supposedly capitalist economies into the world capitalist system was achieved through an interminable metropolis–satellite chain, through which the surplus generated at each stage was successfully siphoned off towards the centre. Therefore, for Frank the choice was clear: continue to endlessly underdevelop within capitalism, or socialist revolution.

In my opinion, the real value of Frank's analysis is his critique of the supposedly dual structure of peripheral societies. Frank argues convincingly that the different sectors of the economies in question are and have been, since very early in their colonial history, well integrated to the world economy. Moreover, he has correctly emphasised that this integration has not automatically brought about capitalistic economic development, such as 'optimistic' models (derived from Adam Smith) would have predicted, in which increased international trade and the division of labour would inevitably bring about economic growth and prosperity. Nevertheless, Frank's error lies in his attempt to explain this phenomenon by using the same economic deterministic framework of the model he purports to transcend. In fact, he merely turns it upside-down: integration into the world economy cannot possibly bring about capitalism development in the periphery because the development of the industrialised centre necessarily requires the underdevelopment of the periphery. Frank's error is characteristic of the whole tradition of which he is part, including Baran (1957), Sweezy (1946), Amin (1970) and Wallerstein (1974; 1980) among the better known. In their analysis, there is always a priority of external over internal structures; in order to do this, they have to separate almost metaphysically the two sides of the opposition (the internal and the external), losing in the process the notion of movement through the dynamic of the contradictions between these two structures. The analysis which emerges is one typified by 'antecedent causation and inert consequences'.

It is not surprising that this type of analysis leads Frank to develop a circular concept of capitalism. Although it is evident that capitalism is a system where production for profits via exchange predominates, the opposite is not necessarily true: the existence of production for profits in the market is not necessarily an indication of capitalist relationship of production. For Frank, this is a sufficient condition for the existence of capitalist forms of surplus extraction (and for the periphery to have been 'capitalist' since the beginning of colonial rule).

Although Frank did not go very far in his analysis of the world capitalist system as a whole, of its origins and its development, Amin (1970) and Wallerstein (1974; 1980) tackled this tremendous challenge. The central concerns of Frank's theory of the 'development of underdevelopment' are also addressed by dos Santos (1970), Marini, Caputo, Pizarro, Hinkelammert, and continued later on by many non-Latin American social scientists. The most thoroughgoing critiques of these theories of underdevelopment have come from Brenner (1977), Cardoso (1972), Kay (1989), Laclau (1971), Lall (1975), Palma (1978), and Warren (1980).

I would argue that the theories of dependency examined here are mistaken not only because they do not 'fit the facts', but also – and equally important – because their mechanico-formal nature renders them both static and ahistorical. Their analytical focus has not been directed to the understanding of how new forms of capitalist development in the periphery have been marked by a series of specific economic, political, and social contradictions, instead only to assert the claim that capitalism had lost, or never had, a historically progressive role in the periphery.

Now, if the argument is that the progressiveness of capitalism has manifested itself in the periphery differently from in advanced capitalist countries, or in diverse ways in the different branches of the peripheral economies, or that it has generated inequality at regional levels and in the distribution of income, and has been accompanied by such phenomena as unemployment, and has benefited the elite almost exclusively, or again that it has taken on a cyclical nature, then this argument does no more than affirm that the development of capitalism in the periphery has been characterized by its contradictory and exploitative nature. The specificity of capitalist development in the Third World stems precisely from the particular ways in which these contradictions have been manifested, the different ways in which many of these countries have faced and temporarily overcome them, the ways in which this process has created further contradictions, and so on. It is through this process that the specific dynamic of capitalist development in different peripheral countries has been generated.

Reading their political analysis, one is left with the impression that the whole question of what course the revolution should take in the periphery revolves solely around the problem of whether or not capitalist development is viable. In other words, their conclusion seems to be that, if one accepts that capitalist development is feasible on its own terms, one is automatically bound to adopt the political strategy of waiting for and/or facilitating such development until its full productive powers have been exhausted, and only then to seek to move towards socialism. As it is precisely this option that these writers wish to reject, they have been obliged to make in their work a forced march back towards a pure ideological position to deny any possibility of capitalist development in the periphery.

The second approach: dependency as a reformulation of the ECLAC analysis of Latin American development

Towards the end of the 1960s the analysis of ECLAC regarding Latin American development suffered a gradual decline due to several key factors (see FURTADO, CELSO). The apparently gloomy panorama of capitalist development in Latin America in the 1960s led to substantial ideological changes in many influential ECLAC thinkers, and it strengthened the convictions of the Marxist 'dependency' writers reviewed earlier. The former were faced with the problem of trying to explain the apparent failure of their structuralist policies, particularly concerning import-substituting industrialization (see STRUCTURALISM). The latter felt vindicated in their view of the unfeasibility of any form of 'dependent capitalist development'.

Finally, by making a basically ethical distinction between 'economic growth' and 'economic development', most of the research done within the perspective of this second approach followed two separate lines, one concerned with the obstacles to economic growth (and in particular to manufacturing), the other with the apparently perverse character taken by capitalist development. The fragility of this formulation lies in its inability to distinguish between a socialist critique of capitalism and the analysis of the actual obstacles to capitalist development in the periphery.

The third approach: dependency as a methodology for the analysis of 'concrete situations of development'

In my critique of the dependency studies reviewed so far, I have described the fundamental elements of what I understand to be the third of the three approaches within the dependency school. This approach is primarily associated with the work of Cardoso and Faletto, dating from the completion of their 1967 book.

Briefly, this third approach to the analysis of dependency can be summarized as follows.

1. In common with the two other approaches to 'dependency' discussed already, this third approach sees the Latin American economies as an integral part of the world capitalist system, in the context of increasing internationalization of the system as a whole. It also argues that the central dynamic of that system lies outside the peripheral economies and that, therefore, the options which are open to them are limited (but not determined) by the development of the system at the centre. In this way the 'particular' is in some way conditioned by the 'general'. Therefore, a basic element for the analysis of these societies is given by the understanding of the general determinants of the world capitalist system, which is itself rapidly changing. However, the theory of imperialism, which was originally developed to provide an understanding of the dynamics of that system, has had enormous difficulty in keeping up with the significant and decisive changes in the capitalist system since the death of Lenin. During this period, capitalism underwent substantial changes, and the theory failed to keep up with them properly.

One widely recognized characteristic of the third approach to dependency has been its effort to incorporate these transformations. For example, this approach was quick to grasp that the rise of the multinational corporations after the Second World War progressively transformed centre–periphery relationships, as well as relationships between the countries of the centre. As foreign capital became increasingly directed towards manufacturing industry in the periphery, the struggle for industrialization, which was previously seen as an anti-imperialist struggle, in some cases increasingly become the goal of foreign capital itself. Thus dependency and industrialization ceased to be necessarily contradictory processes, and a path of 'dependent development' for important parts of the periphery became possible.

2. The third approach has not only accepted but has also tried to enrich the analysis of how developing societies are structured through unequal and antagonistic patterns of social organization, showing the social asymmetries, the exploitative character of social organization and its relationship with the socio-economic base. This approach has also given considerable importance to the particular aspects of each economy like the effect of the diversity of natural resources, geographic location and so on, thus also extending the analysis of the 'internal determinants' of the development of peripheral economies.

3. However, while these improvements are important, the most significant feature of this approach is that it attempts to go beyond the analysis these internal and external elements, and insists that from the premises so far outlined one arrives at only a partial, abstract and inde-terminate characterization of the historical process in the periphery, which can only be overcome by understanding how the 'general' and the 'specific' determinants interact in particular and concrete situations. It is only by under-standing the specificity of 'movement' in the peripheral societies as a dialectical unity of both these internal and external factors that one can explain the particularity of social, political and economic processes in these societies.

Only in this way can one explain how, for example, the same process of mercantile expansion could simultane-ously produce systems of slave labour, systems based on other forms of exploitation of indigenous populations, and incipient forms of wage labour. What is important is not simply to show that mercantile expansion was the basis of the transformation of most of the periphery, and even less to deduce mechanically that that process made these countries immediately capitalist. Rather, this approach emphasizes the specificity of history and seeks to avoid vague, abstract concepts by demonstrating how, throughout the history of backward nations, different sectors of local classes allied or clashed with foreign interests, organized different forms of the state, sustained distinct ideologies or tried to implement various policies or defined alternative strategies to cope with a constantly changing imperialist challenge.

The study of the dynamic of dependent societies as a dialectical unity of internal and external factors implies that the conditioning effect of each on the development of these societies can be separated only by undertaking a static (and metaphysical) analysis. Equally, if the internal dynamic of the dependent society is a particular aspect of the general dynamic of the capitalist system, it does not imply that the latter produces concrete effects in the former, but only that it finds concrete expression in that internal dynamic.

The system of 'external domination' reappears as an internal phenomenon through the social practices of local groups and classes, who share the interests and val-ues of external forces. Other internal groups and forces oppose this domination, and in the concrete develop-ment of these contradictions the specific dynamic of the society is generated. It is not a case of seeing one part of the world capitalist system as 'developing' and another as 'underdeveloping', or of seeing imperialism and dependency as two sides of the same coin, with the underdeveloped or dependent world reduced to a passive role determined by the other.

There are, of course, elements within the capitalist sys-tem that affect all developing economies, but it is precisely *the diversity within this unity* that characterizes historical processes. Thus the analytical focus should be oriented towards the elaboration of concepts capable of explaining how the general trends in capitalist expansion are trans-formed into specific relationships between individuals, classes and states, how these specific relations in turn react upon the general trends of the capitalist system, how internal and external processes of political domination reflect one another, both in their compatibilities and their contradictions, how the economies and polities of periph-eral countries are articulated with those of the centre, and how their specific dynamics are thus generated.

However, as is obvious, this third approach to the anal-ysis of peripheral capitalism is not unique to 'dependency' studies and as such, in time, has superseded them.

JOSÉ GABRIEL PALMA

See also **Baran, Paul Alexander; Engels, Friedrich; Furtado, Celso; Lenin, Vladimir Ilyich [Ulyanov]; Marx, Karl Heinrich; Marx's analysis of capitalist production; monopoly capitalism; structuralism; Sweezy, Paul Marlor.**

I am extremely grateful to Fiona Tregenna for many constructive comments.

Bibliography

Amin, S. 1970. *L'accumulation à l'échelle mondiale: critique de la théorie du sous-développement*. Paris: Anthropos; New York: Monthly Review Press, 1975.

Baran, P. 1957. *The Political Economy of Growth*. New York: Monthly Review Press.

Brenner, R. 1977. The origins of capitalist development: a critique of neo–Smithian Marxism. *New Left Review* 104, 25–93.

Cardoso, F.H. 1972. Dependency and development in Latin America. *New Left Review* 74, 83–95.

Cardoso, F.H. and Faletto, E. 1967. *Dependencia y Desarrollo en América Latina*. Mexico, Siglo XXI; Berkeley: University of California Press, 1977.

Dos Santos, T. 1970. The structure of dependence. *American Economic Review* 60, 231–6.

Frank, A.G. 1967. *Capitalism and Underdevelopment in Latin America: Historical Studies of Chile and Brazil.* New York: Monthly Review Press.

Jay, M. 1996. *The Dialectical Imagination: A History of the Frankfurt School and the Institute for Social Research 1923–1950.* Berkeley: University of California Press.

Kay, C. 1989. *Latin American Theories of Development and Underdevelopment.* London: Routledge.

Laclau, E. 1971. Feudalism and capitalism in Latin America. *New Left Review* 67, 19–38.

Lall, S. 1975. Is dependence a useful concept in analysing underdevelopment? *World Development* 11, 799–810.

Owen, R. and Sutcliffe, B., eds. 1972. *Studies in the Theory of Imperialism.* London: Longman.

Palma, J.G. 1978. Dependency: a formal theory of underdevelopment or a methodology for the analysis of concrete situations of underdevelopment? *World Development* 6, 881–924.

Sweezy, P.M. 1946. *The Theory of Capitalist Development,* London: D. Dobson.

Wallerstein, I. 1974. *The Modern World System: Capitalist Agriculture and the Origins of the European World–Economy in the Sixteenth Century.* New York: Academic Press.

Wallerstein, I. 1980. *The Modern World System II: Mercantilism and the Consolidation of the European World-Economy, 1600–1750.* New York: Academic Press.

Warren, B. 1980. *Imperialism: Pioneer of Capitalism.* ed. J. Sender. London: Verso.

depletion. *See* **exhaustible resources.**

deposit insurance

People living in countries where bank deposits are insured would never question the wisdom of an explicit insurance scheme. The idea that their savings are protected by a government-backed guarantee is something they simply take for granted. Only some crazy economist would ask whether deposit insurance makes sense. Well, does it? Surprisingly, the evidence is that it may not. Deposit insurance, which is supposed to stabilize the financial system, may do more harm than good.

This article examines the nature of deposit insurance by answering the following series of questions: (*a*) What do financial intermediaries do that warrants government intervention? (*b*) What is the history of deposit insurance? (*c*) Does deposit insurance do what it is designed to do? And (*d*), are there any alternatives?

Financial intermediaries, banks, and bank runs

The term 'financial intermediaries' encompasses a large set of institutions that include depository institutions as well as insurance companies, securities firms and pension funds. The first of these – what we all call 'banks' – are both the most commonly known to individuals and provide the broadest array of services. They pool savings, accepting resources from a large number of small savers in order to provide large loans to borrowers; provide access to the payments system, so that individuals can make and receive payments; provide liquidity, allowing depositors to transform their financial assets into money quickly and easily at low cost; and diversify risk, giving even the smallest saver a mechanism for diversification.

To appreciate the importance of financial intermediaries, consider what it would be like without them. If banks didn't exist, all finance would be direct, with borrowers obtaining funds straight from the lenders. Such a system would be costly and ultimately ineffective. It would be so difficult and expensive for borrowers and lenders to find each other, and then to come to agreement over the terms of a loan, that it is unlikely there would be any transactions at all. And without a financial system to transfer funds from savers to investors, there would be no economic development. The world would be a very different place.

Because of the services they provide, banks face a risk that other financial institutions (and industrial firms) do not. They are vulnerable to runs. Here's why. Banks issue liquid liabilities in the form of short-term demand deposits, and hold illiquid long-term assets, structured as securities and loans. The bank promises all its depositors that, if they want the entire balance of their checking account, they just have to come and ask. If a bank has insufficient funds to meet requests for withdrawal on demand, it will fail.

Banks not only guarantee their depositors immediate cash on demand; they promise to satisfy depositors' withdrawal requests on a first-come, first-served basis – what is called a 'sequential service constraint'. This commitment has important implications. Suppose depositors begin to lose confidence in a bank's ability to meet their withdrawal requests. True or not, reports that a bank has become *insolvent* can spread fear that it will run out of cash and close its doors. Mindful of the bank's first-come, first-served policy, panicked depositors rush to convert their account balances into cash before other customers arrive. Such a *bank run* can cause a bank to fail. Importantly, if people believe that a bank is in trouble, that belief alone can make it so.

While banking system panics and financial crises can result from false rumours, they can also come about for more concrete reasons. Widespread downturns in economic activity drive down the value of loans and securities, so bank capital (the difference between assets and liabilities) falls. If things get bad enough, banks become insolvent and fail. A big economic downturn can put the entire financial system at risk. Gorton (1988) reports that significant contractions are associated with all seven of the severe financial panics in the United States that occurred between 1871 and 1914.

In a market-based economy, the opportunity to succeed is also an opportunity to fail. It would be natural to dismiss bank failures as analogous to the closing of an unpopular restaurant. But, while individual banks should be, and are, allowed to fail, the fact that banks are dependent on one another (in a way that restaurants are not) means that when one bank fails it puts others at risk.

Banks are linked both on their balance sheets and in their customers' minds. In recent years in the United States, inter-bank loans make up roughly four per cent of bank assets – an amount that represents almost half of bank capital. If one bank fails, it could put the system at risk. Information asymmetries are the reason that a depositor run on a single bank can turn into a bank panic that threatens the entire financial system. Most of us are not in a position to assess the quality of a bank's balance sheet. So, when rumours spread that a certain bank is in trouble, depositors everywhere begin to worry about their own banks' financial condition. Concern about even one bank can create a panic that causes profitable banks to fail, leading to a complete collapse of a country's banking system. Bank failure is contagious.

All of this leads to the following conclusions. Not only are individual banks fragile and vulnerable to runs, but the entire banking system is prone to panics. Contagion creates an externality that provides the economic justification for government intervention in the system.

Deposit insurance and the government safety net

Government officials intervene in the financial system both to protect small investors and to ensure financial stability. They do it with two related tools: the lender of last resort, where a central bank that can issue liabilities without limit provides loans to banks that are illiquid but not insolvent; and deposit insurance.

History reveals that the presence of a lender of last resort significantly reduces, but does not eliminate, bank panics. The series of three bank panics in the United States during the Great Depression of the 1930s, described in Friedman and Schwartz (1963), is one example of a failure of this sort. The Federal Reserve System was in place and had the capacity to operate as a lender, but did not.

The first national deposit insurance scheme was enacted by the US Congress in 1935 as a direct response to the bank panics in the 1930s. White (1995) sets out the history, noting that the debate was contentious, and that the stated purpose of deposit insurance was to stabilize the banking system. As surprising at it may seem from a modern perspective, investor protection per se was not the point.

When one thinks about deposit insurance, it is important to keep in mind that no private fund can be large enough to withstand a system-wide panic. Only the fiscal authority (possibly combined with the central bank) has the necessary resources.

For decades the US system was nearly unique. In 1974 only 12 countries had explicit national deposit insurance systems. Explicit deposit insurance is a phenomenon of the last quarter of the 20th century, when it became a part of the generally accepted best-practice advice international organizations gave to developing countries. Demirgüç-Kunt and Kane (2002) report that by 1999 the number of countries with deposit insurance had risen to 71 (with the insurable limits ranging up to more than eight times a country's per capita GDP). Prior to this, most systems were implicit, whereby depositors would exert their substantial political influence to force fiscal authorities to supply unlimited deposit guarantees in the event of a bank failure. This is all somewhat surprising, given the obvious political appeal of any system that has no immediate budgetary outlay associated with it. What politician wouldn't want to make an apparently costless promise to protect the bank deposits of his or her constituents?

Does deposit insurance work?

In their classic theoretical treatment of deposit insurance, Diamond and Dybvig (1983) show that, if self-fulfilling depositor runs result from information asymmetries, then government-supplied insurance can improve social welfare. But at what cost?

Insurance changes people's behaviour. Protected depositors have no incentive to monitor their bankers' behaviour. Knowing this, a bank's managers take on more risk than they would otherwise, since they get the benefit of risky bets that pay off while the government assumes the costs of the ones that don't. In protecting depositors, then, the government creates moral hazard. This is not just a theory. In 1980, the deposit insurance limit in the United States was raised to $100,000, four times its earlier level. Over the following ten years, several thousand depository institutions (banks and savings and loans) failed. That was more than four times the number that failed in the first 46 years of explicit deposit insurance. While a vast majority of the institutions that failed in the 1980s were small, the cost of reimbursing depositors exceeded 3 percent of one year's GDP. The bill was ultimately paid by US taxpayers.

The problem of excessive risk taking did not stop with the resolution of the 1980s crisis. Today, the US banking system's assets are worth between 10 and 12 times their equity. In the 1920s, this same leverage ratio was closer to four. Industrial firms typically have leverage that is half that lower number. In other words, deposit insurance has driven up leverage in banking. And with the increase in leverage comes an equal increase in risk (as measured by the standard deviation of returns).

So, in an attempt to solve one problem, deposit insurance created another. And to combat bankers' excessive risk taking, governments were forced to set up regulatory and supervisory structures. Among other things, there are now constraints on the assets banks can hold, rules governing the minimum levels of capital that banks must maintain, and requirements that banks make public information about their balance sheets. Supervisors have to enforce the detailed web of regulations.

Does this complex mechanism actually work to stabilize the financial system? The evidence is not encouraging. Demirgüç-Kunt and Kane (2002) summarize international research and conclude that explicit deposit insurance actually makes financial crises more likely. When countries have either implemented a new scheme or expanded an existing one, the probability of crises has increased.

To make matters worse, the creation of deposit insurance retards the evolution of non-bank financing mechanisms. Cecchetti and Krause (2005) find that countries with more extensive deposit insurance schemes tend to have both smaller financial markets and a fewer publicly traded firms per capita. To put it bluntly, deposit insurance is bad for financial development, and may be bad for real economic growth.

Are there alternatives?

So, if deposit insurance schemes do more harm than good, what should we do to stabilize the financial system? The natural response of an economist is to use the price system. Measure how risky a bank's balance sheet is, and set its deposit insurance premiums accordingly. Beginning in 1991, the US Federal Deposit Insurance Corporation did implement a risk-based premium structure. But this is extremely difficult to do well. Banks can always find ways to evade detailed rules, exploiting the system to reduce the prices they pay. In the end, this is not a solution.

There are three other options. We could implement changes that further restrict the assets held by banks, eliminating their asset transformation function. We could increase our reliance on the central bank's lender-of-last-resort function. Or it may be possible to design a scheme to ensure that large depositors will impose discipline on the risk taking of bank managers.

Proposals for narrow banking are in the first category. A narrow bank is an institution that holds only a very limited set of very low-risk, highly liquid assets, such as short-term government securities. Since insolvency is impossible for such an institution, liability holders would not have to worry about the quality of the narrow bank's assets, and there would be no fear of a run. Deposit insurance would be unnecessary.

Second, it may be possible to address the potential for systemic bank panics by improving the effectiveness of the lender of last resort. In 1873, Walter Bagehot suggested that, in order to prevent the failure of solvent but illiquid financial institutions, the central bank should lend freely on good collateral at a penalty rate. By lending freely, he meant providing liquidity on demand to any bank that asked. Good collateral would ensure that the borrowing bank was in fact solvent, and a high interest rate would penalize the bank for failing to manage its assets sufficiently cautiously. While such a system could work to stem financial contagion, it has a critical flaw. For it to work, central bank officials who approve the loan applications must be able to distinguish an illiquid from an insolvent institution. But during times of crisis computing the market value of a bank's asset is almost impossible, since there are no operating financial markets and no prices for financial instruments. Because a bank will go to the central bank for a direct loan only after having exhausted all opportunities to sell its assets and borrow from other banks without collateral, its illiquidity and its need to seek a loan from the government draw its solvency into question. Officials anxious to keep the crisis from deepening are likely to be generous in evaluating the bank's assets, and to grant a loan even if they suspect the bank might be insolvent. And, knowing this, bank managers will tend to take too many risks.

Finally, we could require that banks issue subordinated debt. These are unsecured bonds, with the lender being paid only after all other bondholders are paid. Someone who buys a bank's subordinate debt has a very strong incentive to monitor the risk-taking behaviour of the bank. The price of these publicly traded bonds then provides the market's evaluation of the quality of the bank's balance sheet and serves to discipline its management.

By eliminating the accountability of bank managers to their depositors, deposit insurance encourages risky behaviour. So, while financial stability is clearly in the public interest, deposit insurance may not be.

STEPHEN G. CECCHETTI

See also **Bagehot, Walter; financial intermediation; financial structure and economic development; moral hazard; risk.**

Bibliography

Bagehot, Walter. 1873. *Lombard Street: A Description of the Money Market.* London: Henry S. Kin & Co.

Cecchetti, S. 2006. *Money, Banking, and Financial Markets.* Boston, MA: McGraw-Hill Irwin.

Cecchetti, S. and Krause, S. 2005. Deposit insurance and external finance. *Economic Inquiry* 43, 531–41.

Demirgüç-Kunt, A. and Kane, E. 2002. Deposit insurance around the globe: where does it work? *Journal of Economic Perspectives* 16(2), 175–95.

Diamond, D. and Dybvig, P. 1983. Bank runs, deposit insurance, and liquidity. *Journal of Political Economy* 91, 401–19.

Friedman, M. and Schwartz, A. 1963. *A Monetary History of the United States: 1867–1960.* Princeton, NJ: Princeton University Press.

Gorton, G. 1988. Banking panics and business cycles. *Oxford Economic Papers* 40, 751–88.

White, E. 1995. Deposit insurance. Policy Research Working Paper No. 1541. Washington, DC: World Bank.

depreciation

Depreciation estimates the decline in the value of capital as a result of ageing, its maximum value being near its age of manufacture and its minimum value when it is dismantled and sold as scrap. It is of great importance to capital accounting, for the rate of dividend is calculated as the ratio of the surplus to the current value of assets. The reduction in value of equipment comes about from two causes – firstly that its productivity may fall with age; and secondly that, as time advances, the expected remaining earning life of the plant is shorter. Hence, the capitalized value of the present value of expected future stream of quasi-rents from an old piece of equipment is smaller for any given rate of interest than for a younger machine.

'One-hoss shay' assumption

The influence of declining productivity over time may be eliminated by assuming a 'one-hoss shay' type of equipment, which keeps its efficiency constant over its service life and falls to pieces at the end. However, the product of a process is not only its current output but the stock of equipment which remains at the end of the production period – as stressed by von Neumann (1933) and by Sraffa (who in 1960 referred to Robert Torrens as having insisted in the years 1818 and 1821 on its being considered as a part of output).

On account of the shorter remaining service life of equipment at the end of a period (and the consequent smaller number of expected items of quasi-rent in its stream of earnings), there is lesser value of capital remaining at the end of a production period – a so-called 'year'. This reduction in value of a stock output affects adversely the productivity in value terms (even with 'one-hoss shay' equipment) and it measures the depreciation. There is, therefore, an aggravated tendency of the value of capital embodied to fall as the plant is older.

Shape of decline in valuation curve

In a straight line approximation, depreciation is taken as constant in absolute amount per year. In a formula using the exponential concept depreciation is at a constant rate; hence, the fall in value is more when machines are younger and higher priced, than when they are older – as in radioactive decay, that is, it indicates a curve convex to the origin. But depreciation is at higher rates for older capital in service – not as would be given by an exponentially falling value of equipment at a constant rate with

respect to time. When there is a rising rate of reduction of value, it makes the decline more than exponential as the machines are older, and yields a steeply falling value towards the end of the service life, that is, it yields a curve with respect to time which is concave to the origin. The straight line approximation of value of capital (with respect to its age) which is used in some calculations is thus wide of the mark; and even the exponentially falling value according to a constant rate of reduction does not make the value of old machines decline sufficiently markedly.

In a Sraffa or von Neumann valuation of capital (of different ages taken as different commodities) this decline is well brought out automatically, for differences between value of the commodity called 'equipment t years old' and the one called '$t + 1$ years old', increases as t becomes larger.

This aspect of the Sraffa system (1960) was not known to Joan Robinson or to Professor Richard Kahn and D.G. Champernowne in 1954 when the text of *Accumulation of Capital* (1956) was being finalized – especially its Mathematical Appendix (to a part of which the latter two had contributed as authors, their names appearing in the original printed text). It is all the more remarkable that it was discovered that, in the measurement of value of ageing equipment, one could strike upon another useful device – of balanced age composition of capital.

Balanced age composition of capital

In demographic studies as part of the subject of manpower, it is well known that, for a population of human beings growing at g per cent, there are higher numbers of children of age t in comparison to those a year older (of age $t + 1$) by the factor $(1 + g)$. The same principle can be applied to a population of plants, and we can derive a universe of plants ordered according to their ages in this particular manner. One can try to ascertain what the number of plants in a cohort of each age is and the value of capital embodied in each cohort.

The value of plant at the centre of gravity of the age-composition pyramid may then be used as the standard unit of measurement of the value of a plant of any particular age. The result would be in agreement with the well-known, but rather mystifying, Kahn–Champernowne formula of the reciprocal value of a new plant in terms of value of the plant of average age. This reciprocal will be called a K–C unit in honour of those two authors who worked out the said formula.

Kahn–Champernowne units of measurement

In a generalized version of this concept, as the set of pieces of capital of constant physical productivity and of balanced age composition growing exponentially at a steady rate, keep the composition in terms of relative sizes of cohorts constant; hence the value of the average plant does not change. This is the justification of the K–C units.

In terms of a balanced age composition of equipment (with T years expected service life since its manufacture), a piece of equipment t 'years' old is replaced at the end of t years by a piece which was $t-1$ years old in the beginning of the year. Except for this replacement by equipment which is now of the same age as the piece it substitutes, there is no depreciation visible in the physical system or its statistical depiction.

Redundancy of gross and net concepts

It is to be remembered that Joan Robinson had correctly realized that depreciation was not a physical phenomenon but a notional or value one. The implication of depreciation not being a physical phenomenon in terms of effect upon the concepts of gross and net investment had to wait until the von Neumann model was integrated (in 1960) with the Robinsonian golden-age system. In traditional analysis the system is depicted as z machines (newly produced and added) in a factory, and at the same time another z machine rendered inoperative (by completion of their natural life). But the net investment is not an act of accretion–depreciation in physical terms; for the machines added through current investment are new ones and the depletion is of old machines – and it makes no sense if value measurement were not resorted to for calculating the excess of accretion over depreciation.

The balanced age composition is a device by which one can realize that in a von Neumann system as a growing economy m machines of age t years exist and $m(1 + g)$ of age $t - 1$ years are automatically substituted a year later by $m(1 + g)$ machines – also now of t years age. The stock as well as each age cohort grows at rate g, and depreciation of value by ageing is exactly counterbalanced by that much capital of erstwhile younger age and erstwhile higher value (but now of the same age and the same value as the m plants at the beginning of the year) replacing it. In addition one has mg times more machine of age t. The total stock grows at a given rate of growth, and depreciation is also compensated for exactly, for $m(1 + g)$ is equal to m for replacement, and mg for accumulation for each age cohort.

In Sraffa–von Neumann analysis (as a simplified purposive model combining the two general constituents of those two models and integrating the resultant with the Robinsonian golden-age system), this fact was noticed in 1961, and it was discovered that in a state of steady growth and balanced age composition depreciation of a stock of inputs in terms of writing down of value of equipment (due to ageing) is a dispensable concept (Mathur, 1965). Each age cohort is replenished exactly by an age cohort from within the system in physical numbers and value, and there is nothing to be written down of any piece of equipment by a chartered accountant at the end of the year. The pieces of equipment of each age are higher by the rate g, and valuation is required for finding out cumulative accumulation of

equipment in each age cohort. In a body of equipment of balanced age composition as the value of capital of different ages differs by the amount of depreciation, the concept of depreciation is required for measuring aggregate accumulation due to ageing, not for decumulation due to ageing as was required in the traditional concept.

It is because of the total absence of writing down of value of stock of any age that it was realized that there is no concept of gross or net necessary in such a reckoning (Mathur, 1965), and depreciation is important not as the difference between gross and net investment, but as the difference of value of an older machine in relation to a younger one for purposes of measuring accumulation (of positive-age equipment from within the firm and of new equipment from the manufacturers). It is only when the age composition is grossly unbalanced – as for newly established firms – that it may be necessary to use depreciation in the traditional sense of writing down value of stocks. But in that case measurement of depreciation or of amount to be written off is itself a procedure not entirely free from logical doubts.

Depreciation and maintenance

In manpower-employment terms, total new employment is given by gross investment and not by net investment, because the amount spent on activities of maintaining capital intact (repairing, renovating) also creates employment, and not only the building of new capital. Hence, in national income statistics, it is gross investment which creates manpower employment and not net investment by itself. The difference between gross and net is taken to be depreciation, but in manpower terms it does not so follow – for employment created for maintenance of a machine (like a sealed unit) might be very low, and yet the reduction of its value year by year very high due to ageing. When viewing manpower statistics, the activity of operatives of a particular type ought to be supplemented by statistics of valuation (Mathur, 1983). While figures in terms of counting heads are important for a physical count, greater economic significance would be acquired if the productivity of each type of human equipment were determined and its true value calculated in K–C units with respect to a balanced age composition and age structure (Mathur, 1964). But depreciation in value terms alone without the physical counterpart of replacement (of equipment or manpower) also tells us an incomplete story, and only valuation and quantification (in physical terms) together give the full picture.

GAUTAM MATHUR

See also **amortization**.

Bibliography

Champernowne, D.G. and Kahn, R.F. 1956. The value of invested capital. Mathematical Note appended at the end of Robinson (1956).

Mathur, G. 1964. The valuation of human capital for manpower planning. *Applied Economic Papers* (Hyderabad) 4(2), 14–35.

Mathur, G. 1965. *Planning for Steady Growth*. Oxford: Basil Blackwell; New York: Augustus Kelly.

Mathur, G. 1983. Web of inequity. Presidential Address to the Silver Jubilee Session of the Indian Labour Economic Association, Lucknow, 1982. *Indian Journal of Labour Economics*, Sec. VIII.

Neumann, J. von. 1933. A model of general economic equilibrium. *Review of Economic Studies* 13(1945), 1–9.

Robinson, J. 1956. *The Accumulation of Capital*. London: Macmillan.

Sraffa, P. 1960. *Production of Commodities by Means of Commodities*. Cambridge: Cambridge University Press.

derived demand

The idea that the demand for intermediate goods is *derived* from the demand for the final goods they help produce is obvious and appealing. It was implied by Cournot (1838, pp. 99–116) and explicitly stated by Gossen (1854, pp. 31, 113) and Menger (1871, pp. 63–7). That the British classical school failed to make use of such a perspective – Mill's famous proposition that 'demand for commodities is not demand for labour' (1848, Book I, ch. 5) came close to denying it – was doubtless due to the strong emphasis placed on prior accumulation of capital as a prerequisite for production. But it was Alfred Marshall in his *Principles of Economics* (1890, pp. 381–93, 852–6) who introduced the term 'derived demand' and developed the concepts of the derived demand curve for an input and the elasticity of derived demand.

Marshall focused on a case in which a commodity is produced by the cooperation of several inputs, which are thus jointly demanded for the purpose, the demand for each being derived from the demand for the product. His formal analysis proceeded on the assumption that the inputs were all combined in fixed proportions (which might vary with the scale of output) although he suggested that the variable-proportions case would be similar.

A derived demand curve can be constructed for a selected input on the assumptions that production conditions, the demand curve for output, and the supply curves for all other inputs remain fixed, and that the competitive markets for output and all other inputs are always in equilibrium. The resulting derived demand curve can most easily be interpreted as the outcome of a hypothetical experiment. Make the selected input available, perfectly elastically, at an arbitrary price, y, per unit. Now ascertain, under the above conditions about the markets for output and other inputs, what quantity, x, of the selected input would be demanded. All other markets must be in equilibrium, and each seller or buyer must be optimally adjusted to the assumed terms of availability of the selected input. Repeating this experiment for different values of y would generate the inverse of the relationship between x and y, $y = f(x)$, whose graphical representation is Marshall's derived demand curve for the selected input. Bringing this demand curve into conjunction with the actual supply curve of the selected input will determine the actual equilibrium price and quantity for this input and thereby implicitly determine the actual equilibrium prices and quantities of output and all other inputs. But the point of obtaining the derived demand curve is not to permit such a two-stage determination of the actual equilibrium. It is rather to permit a simplified analysis of the effect of *changes* in the supply conditions of the selected input when supply conditions of other inputs, as well as technology and the demand conditions for output, remain unaltered.

Marshall invoked a simple example in which the final product, a knife, is obtained by joining costlessly a unit each of the two inputs, blades and handles. The derived demand curve for handles is then given by the rule that y, the derived demand price for x handles, is the demand price for x knives less the supply price for x blades.

Marshall analysed the conditions producing a low elasticity of derived demand for an input, a condition which would encourage supply restriction. The first condition, the lack of a good substitute, is already implied by the fixity of production coefficients. The second is that the demand for the final output be inelastic. The third, aptly described by Henderson (1922, p. 59) as 'the importance of being unimportant', is that expenditure on the input in question be only a small fraction of total production cost. The final condition is that cooperating inputs be in inelastic supply. These last three conditions ensure that a large rise in the price of the input will not raise product price much, that a rise in the product price will not reduce sales much, and that a reduction in sales and production will lower the cost of cooperating inputs substantially.

The next major contribution was that of Hicks (1932, pp. 241–6) who formally relaxed the assumption of fixed production coefficients. He analysed the consequences of input substitutability for a two input case with constant returns to scale in production, making use of his newly invented concept of the elasticity of substitution. His principal finding was that, to get a low elasticity of derived demand, 'It is "important to be unimportant" only when the consumer can substitute more easily than the entrepreneur', that is, only when the elasticity of demand for the product exceeds the elasticity of input substitution (1932, p. 246). This finding, which is not easily explained intuitively, has been the subject of intermittent controversy, aptly summarized and resolved in Maurice (1975). The extension of Hicks's analysis to the many-input case has been accomplished by Diewert (1971), using an elegant dual approach based on the cost function concept. However, modern theoretical work is more prone to work explicitly and symmetrically with complete systems of input demand equations for firm and industry.

More or less contemporaneously with Hicks, Joan Robinson (1933, chs. 23, 24) was studying the derived demand curve for an input in cases where the final product is sold by a monopolist, who might also acquire cooperating inputs monopsonistically. The question of when areas under a derived demand curve can be given a welfare interpretation, analogous to consumer surplus for a final demand curve, has been broached by Wisecarver (1974).

The concept of derived demand finds its main application in discussions of labour-market questions, and Marshall's tools still play a significant part in the teaching and writing in that area.

J.K. WHITAKER

See also **acceleration principle; Marshall, Alfred.**

Bibliography

Cournot, A.A. 1838. *Mathematical Principles of the Theory of Wealth*. Trans., New York: Macmillan, 1897.
Diewert, W.E. 1971. A note on the elasticity of derived demand in the n-factor case. *Economica* 38, 192–8.
Gossen, H.H. 1854. *The Laws of Human Relations*. Trans., Cambridge, MA: MIT Press, 1983.
Henderson, D.H. 1922. *Supply and Demand*. London: Nisbet.
Hicks, J.R. 1932. *The Theory of Wages*. London: Macmillan.
Marshall, A. 1890. *Principles of Economics*, vol. 1, 8th edn. London: Macmillan, 1920.
Maurice, S.C. 1975. On the importance of being unimportant: an analysis of the paradox in Marshall's third rule of derived demand. *Economica* 42, 385–93.
Menger, C. 1871. *Principles of Economics*. Trans., New York: New York University Press, 1981.
Mill, J.S. 1848. *Principles of Political Economy*. London: Parker.
Robinson, J. 1933. *The Economics of Imperfect Competition*. London: Macmillan.
Wisecarver, D. 1974. The social costs of input-market distortions. *American Economic Review* 64, 359–72.

Destutt De Tracy, Antoine Louis Claude (1754–1836)

French philosopher and economist, Tracy was born into a noble family of the *ancien régime* at Paris on 20 July 1754 and died in the same city on 10 March 1836. His life spanned the most tumultuous period of French history, from the twilight of the Old Regime to the dawn of capitalism, romanticism and socialism. One of the last *philosophes*, Tracy began as an 18th-century classical metaphysician, preoccupied with the sensationalist doctrine of Locke and Condillac, and ended up, in the words of Auguste Comte, as the philosopher 'who had come closest to the positive state'. In the interim he knelt at the feet of Voltaire; served alongside Lafayette in the Royal Cavalry, and as deputy to the French Estates General and the Constituent Assembly; was imprisoned during the

Reign of Terror; released after Thermidor (escaping the guillotine by a mere two days); subsequently helped to establish his country's first successful national programme of public education; led the opposition to Napoleon from his seat in the French Senate; regained his title under the Bourbon Restoration; counted among his associates the likes of Mirabeau, Condorcet, Cabanis, DuPont de Nemours, Jefferson, Franklin, Lavoisier, Ricardo and Mill; and retained his early sympathies for liberty throughout.

Long before it took on its pejorative sense at the hands of Marx, Tracy coined the term 'ideology' (by which he meant the science of ideas) to describe his philosophy, which embraced and intertwined psychological, moral, economic and social phenomena, but which gave primacy to economics because he thought that the purpose of society was to satisfy man's material needs and multiply his enjoyments. Tracy rejected the Physiocratic notion of value, substituting a labour theory that Ricardo subsequently endorsed in his *Principles*. Like Say, he denied Smith's distinction between productive and unproductive labour. But unlike Smith or Say, he reduced all wealth, including land, to labour. On numerous other topics (that is, wages, profits, rents, exchange, price variations, international trade) he was far less thorough and rigorous than either Smith or Say, but his exposition of the capitalization theory of taxation was superior to the rest. In the final analysis, his *Traité* was not properly a treatise on political economy so much as a part of a general study of the human will. Yet the resulting lack of depth did not impair his remarkable ability to allure great minds. Ricardo found him 'a very agreeable old gentleman', and Jefferson was influenced to the point of including 'ideology' among the ten projected departments in his plan for the University of Virginia.

Along with Say, Destutt de Tracy was one of the earliest members of the French liberal school. Patrician, philosopher and patriot, caught in the grips of major social and economic upheaval, he denounced the interests of his own class (the *rentiers*) and became the spokesman of a nascent capitalism in which he had neither role nor vested interest.

R.F. HÉBERT

Selected works

1804–18. *Eléments d'idéologie*, 5 vols. Paris: Courcier.
1811. *A Commentary and Review of Montesquieu's Spirit of Laws*. Trans, Philadelphia: William Duane.
1817. *A Treatise on Political Economy*. Ed. T. Jefferson. New York: Augustus M. Kelley, 1970.

Bibliography

Allix, E. 1912. Destutt de Tracy, économiste. *Revue d'Economie Politique* 26, 424–51.
Kennedy, E. 1978. *Destutt de Tracy and the Origins of 'Ideology'*. Philadelphia: American Philosophical Society.
Picavet, F.J. 1918. *Les Idéologues*, chs 5 and 6. Paris: F. Alcan.

determinacy and indeterminacy of equilibria

1 Introduction

The Arrow–Debreu model of competitive markets is one of the cornerstones of economics. Part of the explanatory power of this model stems from its flexibility in capturing price-taking behaviour in many different markets, and from the predictive power arising from the great generality under which equilibrium can be shown to exist. This predictive power is significantly enhanced when equilibria are determinate, meaning that equilibria are locally unique and local comparative statics can be precisely described. Instead, when equilibria are indeterminate, even arbitrarily precise local bounds on variables might not suffice to give a unique equilibrium prediction, the model might exhibit infinitely many equilibria, and each might be infinitely sensitive to arbitrarily small changes in parameters.

Simple exchange economies cast in an Edgeworth box with two agents and two goods illustrate the possibility of indeterminacy in equilibrium. One easy example arises when agents view the goods as perfect substitutes. In this case, every profile of initial endowments leads to a continuum of equilibria. Another example comes from the opposite extreme, in which each agent views the goods as perfect complements. Every profile of initial endowments dividing equal social endowments of the two goods leads to a continuum of equilibria. These examples may seem degenerate, since they involve individual demand behaviour either extremely responsive to prices, or extremely unresponsive to prices. Similar examples can be constructed using preferences that are less extreme, however, and that can be chosen to satisfy a number of regularity conditions including strict concavity, strict monotonicity, and smoothness. Problems from standard graduate texts illustrate this possibility. In fact, indeterminacy is unavoidable, at least for some endowment profiles, in almost any model that may exhibit multiple equilibria for some choices of endowments. The conditions leading to unique equilibria or unambiguous global comparative statics are well-known to be very restrictive, suggesting that equilibrium indeterminacy may be a widespread phenomenon.

In a deeper sense, however, these examples of indeterminacy remain knife-edge. Under fairly mild conditions on primitives, if an initial endowment profile leads to indeterminacy in equilibrium, arbitrarily small perturbations in endowment profiles must restore the determinacy of equilibrium. More powerfully, the set of endowment profiles for which equilibria are determinate is generic, that is, an open set of full Lebesgue measure. Explaining this remarkable result – originally postulated and established by Debreu (1970) – and its many extensions and generalizations is the focus of this article. Section 2 lays out the basic question of determinacy of equilibrium in finite exchange economies, and sketches the results. Section 3 describes the general underlying principles,

together with various applications and extensions. Section 4 concludes by examining recent work on determinacy in markets with infinitely many commodities.

2 Determinacy in finite exchange economies

Imagine a family of exchange economies, each with a fixed set of L commodities and a fixed set of m agents, $i = 1, \ldots, m$, with given preferences $\{\succeq_i\}_{i=1,\ldots,m}$, indexed by varying individual endowments $(e_1, \ldots, e_m) \in \mathbf{R}_{++}^{mL}$. Denote the social endowment $\bar{e} := \sum_i e_i$ and a particular profile of individual endowments by $e := (e_1, \ldots, e_m) \in \mathbf{R}_{++}^{mL}$. An economy $E(e)$ then refers to the exchange economy with preferences $\{\succeq_i\}_{i=1,\ldots,m}$ and endowment profile e. For simplicity this article focuses on exchange economies. Mas-Colell (1985) is a comprehensive reference that includes discussion of extensions allowing for production.

The crucial departure in Debreu (1970) is to view each economy as a member of this parameterized family, and to ask whether perhaps almost no economies exhibit indeterminacy or pathological comparative statics when indexed this way. To formalize this, Debreu (1970) summarizes an agent's choice behaviour by a C^1 demand function $x_i : \mathbf{R}_{++}^L \times \mathbf{R}_{++}^L \to \mathbf{R}_+^L$ satisfying basic properties such as homogeneity of degree 0 in prices, Walras's Law, and boundary conditions as prices converge to zero. This leads to the familiar characterization of equilibria as zeros of excess demand:

$$0 = z(p, e) := \sum_i x_i(p, e_i) - \bar{e}.$$

Two simplifying normalizations are then commonly adopted. Demand functions derived from optimal choices of price-taking agents are homogeneous of degree zero in prices, so normalize by setting $p_1 \equiv 1$. Normalized prices thus can be taken to range over $\mathbf{R}_{++}^{L-1}$. Next, Walras's Law ensures that excess demand functions are not independent across markets, as $p \cdot z(p, e) = 0$ for each $p \in \mathbf{R}_{++}^{L-1}$. This renders one market clearing equation redundant, and leads to the characterization of equilibria by normalized price vectors $p \in \mathbf{R}_{++}^{L-1}$ such that

$$z_{-L}(p, e) = 0$$

where, adopting common conventions, the subscript $-L$ refers to all goods except L, so $z_{-L}(p, e) = (z_1(p, e), \ldots, z_{L-1}(p, e))$. Using these normalizations, the equilibrium correspondence can be defined by

$$E(e) := \{(x, p) \in \mathbf{R}_+^{mL} \times \mathbf{R}_{++}^{L-1} : z_{-L}(p, e) = 0,$$
$$x_i = x_i(p, e) \text{ for } i = 1, \ldots, m\}.$$

Fix a particular equilibrium price vector p^* in the economy $E(e)$. One way to answer local comparative statics questions at this equilibrium is to apply the classical implicit function theorem. If $D_p z_{-L}(p^*, e)$ is invertible, then the implicit function theorem provides several

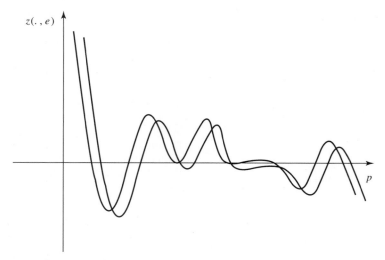

Figure 1   Generic determinacy for smooth excess demand

immediate predictions: the equilibrium price p^* is locally unique; locally, on neighbourhoods W of e and V of p^*, the equilibrium price set is described by the graph of a C^1 function $p : W \to \mathbf{R}_{++}^{L-1}$; and local comparative statics are given by the formula

$$Dp(e) = -[D_p z_{-L}(p^*, e)]^{-1} D_e z_{-L}(p^*, e).$$

If this analysis can be performed for each equilibrium, then there are only finitely many equilibria, because the equilibrium set is compact. Moreover, for each equilibrium $(x, p) \in E(e)$ there is a neighbourhoods U of (x, p) for which $E(\cdot) \cap U$ has a unique, C^1 selection on a neighbourhood W of e, with the comparative statics derived from the preceding formula. Call such a correspondence *locally C^1 at e*. The following definition offers a convenient way to summarize these properties.

Definition 1 The economy $E(e)$ is 'regular' if it has finitely many equilibria, and E is locally C^1 at e.

An alternative way to describe the problem uses the language of differential topology. For a C^1 function $f : \mathbf{R}^m \to \mathbf{R}^n, y \in \mathbf{R}^n$ is a *regular value* of f if $Df(x)$ has full rank for every $x \in f^{-1}(y)$. Notice that this is precisely the condition identified above, for the case of equilibrium prices, under which local uniqueness and local comparative statics could be derived from the implicit function theorem. Whenever 0 is a regular value of $z_{-L}(\cdot, e)$, the corresponding economy $E(e)$ is regular. For a fixed function f, a given value y may fail to be a regular value, but almost every other value is regular: this is the conclusion of Sard's theorem. Dually, the fixed value y may fail to be a regular value for a particular function f, but is a regular value for almost every other function. When the set of functions is limited to those drawn from a particular parameterized family, the

conclusion remains valid for almost all members of this family provided the parameterization is sufficiently rich. This idea of a rich parameterization can be expressed by requiring y to be a regular value of the parameterized family, and this parametric version of Sard's theorem is typically called the transversality theorem. Figure 1 depicts this idea for smooth excess demand functions.

These observations suggest that, while extremely restrictive assumptions might be required to ensure that *every* economy is regular, generic regularity might follow simply from the differentiability of demand functions once the problem is framed this way. Straightforward calculations verify that 0 is a regular value of the excess demand function (viewed as a function of both prices and initial endowment parameters). From the transversality theorem we conclude that there is a subset $R^* \subset \mathbf{R}^{mL}_{++}$ of full Lebesgue measure such that for all $e \in R^*, E(e)$ is regular. With the use of additional properties of excess demand and equilibria, it is similarly straightforward to show that the set of regular economies is also open, giving a strong genericity result for regular economies.

This discussion follows Debreu's original development closely. This approach takes demand functions as primitives, and gives conditions on individual demand functions under which regularity is a generic feature of exchange economies. To take a step back and start with preferences as primitives, we seek conditions on preferences sufficient to guarantee that individual demand is suitably differentiable. Debreu (1972) addresses this point by introducing a class of 'smooth preferences', depicted in Figure 2.

Definition 2 *The preference order $\succeq$ on $\mathbf{R}_+^L$ is 'smooth' if it is represented by a utility function U such that*

- *$U : \mathbf{R}_+^L \to \mathbf{R}$ is C^2 on $\mathbf{R}_{++}^L$*
- *for each $x \in \mathbf{R}_{++}^L$, $\{y \in \mathbf{R}_+^L : y \sim x\} \subset \mathbf{R}_{++}^L$*

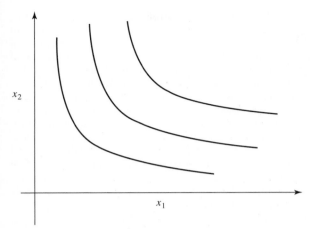

x_2

x_1

Figure 2   Smooth preferences

- *for each* $x \in \mathbf{R}^L_{++}$, $DU(x) \gg 0$
- *for each* $x \in \mathbf{R}^L_{++}$, $D^2U(x)$ *is negative definite on ker*
  $DU(x) := \{z \in \mathbf{R}^L : DU(x) \cdot z = 0\}$

Fairly straightforward arguments, again using the implicit function theorem, establish that individual demand functions derived from smooth preferences are C^1. Putting all of these results together yields:

Theorem 1 *Let* $\succeq_i$ *be a smooth preference order on* $\mathbf{R}^L_+$ *for each* $i = 1, \ldots, m$. *There exists an open set* $R^* \subset \mathbf{R}^{mL}_{++}$ *of full Lebesgue measure such that for all* $e \in R^*$, $E(e)$ *is regular.*

3 Determinacy and indeterminacy: a new approach to many problems

Behind this result for equilibria in finite exchange economies is a broad, powerful, and simple principle that has found many important and ingenious applications in the 35 years since Debreu's original 1970 paper. To cast the problem more generally, take a parameterized family of equations, captured by a function $f : \mathbf{R}^m \times \mathbf{R}^k \to \mathbf{R}^n$. This describes a problem with m variables and k parameters simultaneously entering n different equations. Imagine that for each parameter value $r \in \mathbf{R}^k$,

$$E(r) := \{x \in \mathbf{R}^m : f(x, r) = 0\}$$

gives the set of objects of interest. Moreover, imagine that the equations are sufficiently independent in determining the solutions, in the sense that 0 is a regular value of f. Counting the number of equations and unknowns produces three distinct cases, corresponding in turn to three different sorts of applications.

In the canonical case exemplified by the simple exchange economy described above, the number of relevant endogenous variables, m, is equal to the number of equations, n. In this case, 0 being a regular value of f

characterizes exactly the case in which the equations are sufficiently independent that the loose 'counting equations and unknowns' heuristic corresponds with the precise technical result of generic determinacy. One prominent illustration of this case is given by two-period incomplete markets models with real assets, that is, assets that pay off in bundles of commodities. In these models, there are as many distinct budget equations as there are states. If we let S denote the number of states, this means there are $S + 1$ distinct Walras's Law statements, leading to $S + 1$ redundant market clearing equations. Because asset payoffs are in real terms, all budget constraints are homogeneous of degree 0 in state prices. This generates $S + 1$ distinct normalizations of state prices, compensating exactly for the drop in independent market clearing equations determining equilibrium. Generic determinacy in this case is established by Geanakoplos and Polemarchakis (1987).

When $m < n$, there are fewer equations than unknowns, and the regularity of the system of equations means that it is generically overdetermined. In this case, generically it is impossible to satisfy the equations simultaneously, that is, generically $E(r)$ is empty. As a simple example of this argument, consider the prevalence of trade at equilibrium in an Edgeworth box economy. One market-clearing condition in one (normalized) price characterizes equilibria, and standard arguments show that this excess demand function has 0 as a regular value. In fact, varying the endowment of the first agent alone is enough. How often does equilibrium involve trade in some goods? With only two agents, trade occurs in equilibrium if and only if $x_2 \neq e_2$, so the additional two equations $x_2(p, e) - e_2 = 0$ characterize endowment and price combinations for which there is no trade in equilibrium. A simple calculation shows that 0 is a regular value of $f(p, e) := (z_{-2}(p, e), x_2(p, e) - e_2)$. Fixing the endowment profile e, however, this is a problem with three equations in a single variable, so there must be a set $R^{**} \subset \mathbf{R}^{mL}_{++}$ of full Lebesgue measure such that for every $e \in R^{**}$, there are no solutions to the equation $f(p, e) = 0$. For every endowment profile $e \in R^{**}$, every equilibrium then must involve trade, as every equilibrium price solves the first equation $z_{-2}(p^*, e) = 0$, so cannot also involve no trade, $x^2(p^*, e) \neq e_2$. Similar logic but more involved calculations show that equilibrium allocations are generically inefficient in incomplete markets models, and generically constrained inefficient in multi-good incomplete markets models. Geanakoplos and Polemarchakis (1987) pioneered this approach to efficiency with incomplete markets.

Finally, when $m > n$, generically indeterminacy arises, as generically the solution set $E(r)$ is an $(m - n)$-dimensional manifold. (A subset $M \subset \mathbf{R}^m$ is a d-dimensional C^ℓ manifold if for each $x \in M$ there exist open sets $V \subset \mathbf{R}^m$ and $W \subset \mathbf{R}^d$, where V is a neighbourhood of x, and a C^ℓ diffeomorphism $\phi : V \to W$ such that $\phi(V \cap M) = W$.) In this case, generically there is a continuum of solutions, and the set of solutions is locally, up to diffeomorphism, a set of dimension $m - n$. An important example of this case is provided by two-period incomplete financial markets models with nominal assets. Here, asset payoffs are in nominal terms, in some specified unit of account. As in the case of real assets described above, there are $S + 1$ independent budget constraints when there are S possible states of nature, so there are $S + 1$ redundant market clearing equations. Because asset payoffs are nominal, however, budget constraints are not all homogeneous of degree zero, and price levels matter. With only two homogeneity conditions, one for period one prices and one relating all commodity and asset prices, this leaves $S - 1$ dimensions of indeterminacy in equilibria generically. The detailed result is established by Geanakoplos and Mas-Colell (1989).

These three cases, and the generic properties of solution sets that follow, are collected below.

Theorem 2 *Let $f : \mathbf{R}^m \times \mathbf{R}^k \to \mathbf{R}^n$ be a C^ℓ function, where $\ell > \max\{m - n, 0\}$, and suppose 0 is a regular value of f.*

(a) *Suppose $m = n$. There exists a set $R^* \subset \mathbf{R}^k$ of full Lebesgue measure such that, for every $r \in R^*$, $E(r)$ contains only isolated points, $E(r)$ is finite when compact, and E is locally C^1 at r.*
(b) *Suppose $m < n$. There exists a set $R^* \subset \mathbf{R}^k$ of full Lebesgue measure such that for every $r \in R^*$, $E(r)$ is empty.*
(c) *Suppose $m > n$. There exists a set $R^* \subset \mathbf{R}^k$ of full Lebesgue measure such that for every $r \in R^*$, $E(r)$ is an $(m-n)$-dimensional C^ℓ manifold.*

The techniques pioneered by Debreu have found widespread applications, and have proven to be remarkably powerful. Nonetheless, the smoothness needed to study determinacy using the tools of differential topology does stem from assumptions that often carry real economic content. These assumptions restrict both the nature of admissible preferences and the nature of admissible constraints.

For example, to avoid problems arising when non-negativity constraints on consumption may become binding, these results rest on 'boundary' restrictions, both on endowments, because individual endowments are strictly positive, or on equilibrium consumption via boundary conditions on preferences that imply individual demands are strictly positive at all prices. Unless goods are aggregated extremely coarsely, neither pattern is supported by observations on consumer behaviour or characteristics. Relaxing the constraint on endowments turns out, perhaps surprisingly, to generate indeterminacy much more readily than relaxing the assumptions on positive consumptions, or incorporating other more general constraints on choices. Minehart (1997) shows by means of an example that for one natural case of

restricted endowments, in which each agent is constrained to hold a single, individual-specific, good, an open subset of such parameters leads to indeterminacy in equilibrium. Highlighting the fact that the choice of parameterization can be important, Mas-Colell (1985) shows that this conclusion is not robust to perturbations in preferences; generic determinacy, in a topological sense, is restored by considering variations in preferences as well as constrained endowments. If the assumption that individual endowments of every good are positive is maintained, the restriction to positive individual demand for every good can be relaxed. For example, Mas-Colell (1985) provides generic determinacy results for exchange economies allowing for boundary consumptions; Figure 3 depicts such preferences.

Smooth preferences, as defined by Definition 2 above, obviously rule out preferences with non-differentiabilities in level sets, a restriction that also has important behavioural content. Kinks have arisen as central manifestations of various behavioural phenomena, including loss aversion, ambiguity aversion, and reference dependence; examples include Kahneman and Tversky (1979), Tversky and Kahneman (1991), Koszegi and Rabin (2006), Sagi (2006), and Gilboa and Schmeidler (1989). Such kinks typically lead to excess demand functions that fail to be differentiable for some prices. Rader (1973), Pascoa and Werlang (1999), Shannon (1994), and Blume and Zame (1993) all develop methods to address such cases. With the exception of Blume and Zame (1993), these techniques can be roughly understood as expanding differential notions by adding to 'regularity' the condition that the function (for example, excess demand) is differentiable at every solution, and establishing that analogues of implicit function theorems, Sard's theorem or the transversality theorem remain valid for sufficiently nice non-smooth functions, such as Lipschitz continuous functions; in particular, see Shannon (1994; 2006). Blume and Zame (1993) instead use results that exploit the

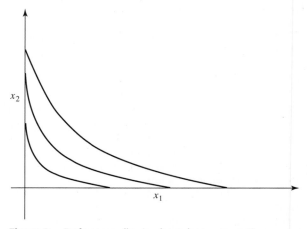

Figure 3   Preferences allowing boundary consumption

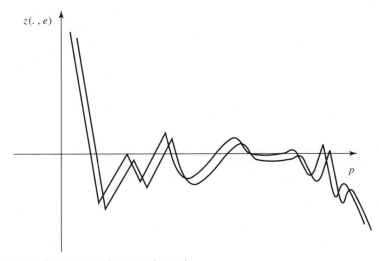

Figure 4   Generic determinacy for non-smooth excess demand

structure of algebraic sets to establish generic determinacy for utilities that are, roughly, finitely piecewise analytic, and need not be strictly concave. Examples in which determinacy has been studied using techniques along these various lines include asset market models with restricted participation (for example, see Cass, Siconolfi, and Villanacci, 2001) and models of ambiguity aversion (for example, see Rigotti and Shannon, 2006). Figure 4.

4 Determinacy in infinite-dimensional economies

Many economic models require an infinite number of marketed commodities. Important examples include dynamic infinite horizon economies, continuous-time trading in financial markets, and markets with differentiated commodities. Such infinite-dimensional models present big obstacles to studying determinacy, starting with the fact that individual demand is not defined for most prices, precluding any straightforward parallel of Debreu's arguments for finite economies. In addition, the positive cone in most infinite-dimensional spaces has empty interior in the relevant topologies, meaning individual consumption sets are 'all boundaries', and existence of equilibrium typically requires conditions, such as uniform properness or variants, that effectively bound marginal rates of substitution. Thus boundary conditions akin to those in Debreu's smooth preferences are likely either to be impossible to satisfy or to contradict equilibrium existence in many important applications.

Provided there are finitely many agents and no market distortions, using the welfare theorems and Negishi's argument provides an alternative characterization of equilibria, replacing excess demand with 'excess savings'. Some version of this characterization of equilibria provides the framework for much of the existing equilibrium analysis in economies with infinitely many commodities,

including the seminal work on existence of Mas-Colell (1986) and Aliprantis, Brown and Burkinshaw (1987), and the approach to determinacy for discrete-time infinite horizon models with time separability pioneered by Kehoe and Levine (1985). To explain this, let X denote the commodity space. The efficient allocations are the solutions to a social planner's problem of the following form: given $\lambda \in \Lambda := \{\lambda \in \mathbf{R}_+^m : \sum_{i=1}^m \lambda_i = 1\}$, choose a feasible allocation $x(\lambda)$ to solve:

$$\max \sum_{i=1}^m \lambda_i U_i(x_i)$$

$$s.t \sum_{i=1}^m x_i \leq \bar{e}.$$

Under standard assumptions, the solution $x(\lambda)$ to this problem is well-defined and unique for each $\lambda \in \Lambda$, and a unique price $p(\lambda)$ supporting $x(\lambda)$ can be characterized. Equilibria then correspond to the solutions λ to the budget equations

$$p(\lambda) \cdot (x_2(\lambda) - e_2) = 0$$
$$\vdots$$
$$p(\lambda) \cdot (x_m(\lambda) - e_m) = 0$$

where Walras's Law accounts for the missing equation. In parallel with excess demand, define the *excess savings map* $s : \Lambda \times X_+^m \to \mathbf{R}^{m-1}$

$$s(\lambda, e) := (p(\lambda) \cdot (x_2(\lambda) - e_2), \ldots,$$
$$p(\lambda) \cdot (x_m(\lambda) - e_m)).$$

Through this construction, the question of determinacy for infinite-dimensional economies can be cast in close parallel to finite economies, with the only change that the set of parameters is now infinite-dimensional. This raises several technical issues, most importantly the choice between topological and measure-theoretic notions of genericity due to the impossibility of defining a suitable analogue of Lebesgue measure in infinite-dimensional spaces (see Hunt, Sauer and Yorke, 1992, and Anderson and Zame, 2001, for a discussion of these issues). This construction also makes imperative the need to link conditions on excess savings used to imply determinacy with conditions on preferences since, in contrast with excess demand, excess savings depends on artificial and unobservable constructs. Somewhat surprisingly, Shannon (1999) and Shannon and Zame (2002) show that generic determinacy follows from conditions on preferences that closely resemble Debreu's (1972) smooth preferences, after suitable renormalization. As in the finite case, these conditions can roughly be understood as strengthened notions of concavity, requiring that near feasible bundles utility differs from a linear approximation by an amount quadratic in the distance to the given bundle. These notions of concavity thus rule out preferences displaying local or global substitutes. Shannon and Zame (2002) provide a simple geometric argument showing that the excess spending mapping is Lipschitz continuous. Generic determinacy then follows by arguments similar to those sketched above for other problems with non-differentiabilities, making use of Shannon (2006) on comparative statics and a version of the transversality theorem for this setting. The direct, geometric nature of these arguments render them applicable in a wide range of examples, including models of continuous-time trading, trading in differentiated commodities, and trading over an infinite horizon.

CHRIS SHANNON

Bibliography

Aliprantis, C., Brown, D.J. and Burkinshaw, O. 1987. Edgeworth equilibria. *Econometrica* 55, 1108–38.

Anderson, R. and Zame, W.R. 2001. Genericity with infinitely many parameters. *Advances in Theoretical Economics* 1.

Blume, L. and Zame, W.R. 1993. The algebraic geometry of competitive equilibrium. In *Essays in General Equilibrium and International Trade: In Memoriam Trout Rader*, ed. W. Neuefeind. New York: Springer-Verlag.

Cass, D., Siconolfi, P. and Villanacci, A. 2001. Generic regularity of competitive equilibrium with restricted participation on financial markets. *Journal of Mathematical Economics* 36, 61–76.

Debreu, G. 1970. Economies with a finite set of equilibria. *Econometrica* 38, 387–92.

Debreu, G. 1972. Smooth preferences. *Econometrica* 40, 603–15.

Dierker, E. 1972. Two remarks on the number of equilibria of an economy. *Econometrica* 40, 951–53.

Geanakoplos, J. and Mas-Colell, A. 1989. Real indeterminacy with financial assets. *Journal of Economic Theory* 47, 22–38.

Geanakoplos, J. and Polemarchakis, H. 1987. Existence, regularity and constrained suboptimality of competitive portfolio allocations when the asset market is incomplete. In *Essays in Honor of Kenneth J. Arrow*, vol. 3, ed. W.P. Heller, R.M. Starr and D.M. Starrett. Cambridge: Cambridge University Press.

Gilboa, I. and Schmeidler, D. 1989. Maxmin expected utility with non-unique prior. *Journal of Mathematical Economics* 18, 141–53.

Hunt, B.R., Sauer, T. and Yorke, J.A. 1992. Prevalence: a translation invariant 'almost every' on infinite dimensional spaces. *Bulletin (New Series) of the American Mathematical Society* 27, 217–38.

Kahneman, D. and Tversky, A. 1979. Prospect theory: an analysis of decision under risk. *Econometrica* 47, 263–91.

Kehoe, T. and Levine, D. 1985. Comparative statics and perfect foresight in infinite horizon economies. *Econometrica* 53, 433–52.

Koszegi, B. and Rabin, M. 2006. A model of reference-dependent preferences. *Quarterly Journal of Economics* 121, 1133–65.

Mas-Colell, A. 1985. *The Theory of General Economic Equilibrium: A Differentiable Approach*. Cambridge: Cambridge University Press.

Mas-Colell, A. 1986. The price equilibrium existence problem in topological vector lattices. *Econometrica* 54, 1039–54.

Minehart, D. 1997. A note on the generic finiteness of the set of equilibria in an exchange economy with constrained endowments. *Mathematical Social Sciences* 34, 75–80.

Pascoa, M.R. and Werlang, S. 1999. Determinacy of equilibrium in nonsmooth economies. *Journal of Mathematical Economics* 32, 289–302.

Rader, J.T. 1973. Nice demand functions. *Econometrica* 41, 913–35.

Rigotti, L. and Shannon, C. 2006. Sharing risk and ambiguity. Discussion paper, UC Berkeley.

Sagi, J. 2006. Anchored preference relations. *Journal of Economic Theory* 130, 283–95.

Shannon, C. 1994. Regular nonsmooth equations. *Journal of Mathematical Economics* 23, 147–66.

Shannon, C. 1999. Determinacy of competitive equilibria in economies with many commodities. *Economic Theory* 14, 29–87.

Shannon, C. 2006. A prevalent transversality theorem for Lipschitz functions. *Proceedings of the American Mathematical Society* 134, 2755–65.

Shannon, C. and Zame, W.R. 2002. Quadratic concavity and determinacy of equilibrium. *Econometrica* 70, 631–62.

Tversky, A. and Kahneman, D. 1991. Loss aversion in riskless choice: a reference dependent model. *Quarterly Journal of Economics* 106, 1039–61.

deterministic evolutionary dynamics

1 Introduction

Deterministic evolutionary dynamics for games first appeared in the mathematical biology literature, where Taylor and Jonker (1978) introduced the *replicator dynamic* to provide an explicitly dynamic foundation for the static evolutionary stability concept of Maynard Smith and Price (1973). But one can find precursors to this approach in the beginnings of game theory: Brown and von Neumann (1950) introduced differential equations as a tool for computing equilibria of zero-sum games. In fact, the replicator dynamic appeared in the mathematical biology literature long before game theory itself: while Maynard Smith and Price (1973) and Taylor and Jonker (1978) studied game theoretic models of animal conflict, the replicator equation is equivalent to much older models from population ecology and population genetics. These connections are explained by Schuster and Sigmund (1983), who also coined the name 'replicator dynamic', borrowing the word 'replicator' from Dawkins (1982).

In economics, the initial phase of research on deterministic evolutionary dynamics in the late 1980s and early 1990s focused on populations of agents who are randomly matched to play normal form games, with evolution described by the replicator dynamic or other closely related dynamics. The motivation behind the dynamics continued to be essentially biological: individual agents are preprogrammed to play specific strategies, and the dynamics themselves are driven by differences in birth and death rates. Since that time the purview of the literature has broadened considerably, allowing more general sorts of large population interactions, and admitting dynamics derived from explicit models of active myopic decision making.

This article provides a brief overview of deterministic evolutionary dynamics in game theory. More detailed treatments of topics introduced here can be found in the recent survey article by Hofbauer and Sigmund (2003), and in books by Maynard Smith (1982), Hofbauer and Sigmund (1988; 1998), Weibull (1995), Vega-Redondo (1996), Samuelson (1997), Fudenberg and Levine (1998), Cressman (2003), and Sandholm (2007).

2 Population games

Population games provide a general model of strategic interactions among large numbers of anonymous agents. For simplicity, we focus on games played by a single population, in which agents are not differentiated by roles; allowing for multiple populations is mostly a matter of introducing more elaborate notation.

In a single-population game, each agent from a unit-mass population chooses a strategy from the finite set $S = \{1, \ldots, n\}$, with typical elements i and j. The distribution of strategy choices at a given moment in time is described by a *population state* $x \in X = \{x \in \mathbf{R}_+^n : \sum_{i \in S} x_i = 1\}$. The *payoff* to strategy i, denoted $F_i : X \to \mathbf{R}$, is a continuous function of the population state; we use the notation $F : X \to \mathbf{R}^n$ to refer to all strategies' payoffs at once. By taking the set of strategies S as fixed, we can refer to F itself as a *population game*.

The simplest example of a population game is the most commonly studied one: random matching to play a symmetric normal form game $A \in \mathbf{R}^{n \times n}$, where A_{ij} is the payoff obtained by an agent choosing strategy i when his opponent chooses strategy j. When the population state is $x \in X$, the expected payoff to strategy i is simply the weighted average of the elements of the ith row of the payoff matrix: $F_i(x) = \sum_{j \in S} A_{ij} x_j = (Ax)_i$. Thus, the population game generated by random matching in A is the linear population game $F(x) = Ax$.

Many models of strategic interactions in large populations that arise in applications do not take this simple linear form. For example, in models of highway congestion, payoff functions are convex: increases in traffic when traffic levels are low have virtually no effect on delays, while increases in traffic when traffic levels are high increase delays substantially (see Beckmann, McGuire and Winsten, 1956; Sandholm, 2001). Happily, allowing non-linear payoffs extends the range of possible applications of population games without making evolutionary dynamics especially more difficult to analyse, since the dynamics themselves are nonlinear even when the underlying payoffs are not.

3 Foundations of evolutionary dynamics

Formally, an *evolutionary dynamic* is a map that assigns to each population game F a differential equation $\dot{x} = V^F(x)$ on the state space X. While one can define evolutionary dynamics directly, it is preferable to derive them from explicit models of myopic individual choice.

We can accomplish this by introducing the notion of a *revision protocol* $\rho : \mathbf{R}^n \times X \to \mathbf{R}_+^{n \times n}$. Given a payoff vector $F(x)$ and a population state x, a revision protocol specifies for each pair of strategies i and j a non-negative number $\rho_{ij}(F(x), x)$, representing the rate at which strategy i players who are considering switching strategies switch to strategy j. Revision protocols that are most consistent with the evolutionary paradigm require agents to possess only limited information: for example, a revising agent might know only the current payoffs of his own strategy i and his candidate strategy j.

A given revision protocol can admit a variety of interpretations. For one all-purpose interpretation, suppose each agent is equipped with an exponential alarm clock. When the clock belonging to an agent playing strategy i rings, he selects a strategy $j \in S$ at random, and then switches to this strategy with probability proportional to $\rho_{ij}(F(x), x)$. While this interpretation is always available, others may be simpler in certain instances. For example, if the revision protocol is of the imitative form $\rho_{ij} = x_j \times \hat{\rho}_{ij}$, we can incorporate the x_j term into our story by supposing that the revising agent selects his candidate strategy j not by drawing a strategy at random, but by drawing an opponent at random and observing this opponent's strategy.

A population game F and a revision protocol ρ together generate an ordinary differential equation $\dot{x} = V^F(x)$ on the state space X. This equation, which captures the population's *expected* motion under F and ρ, is known as the *mean dynamic* or *mean field* for F and ρ:

$$\dot{x} = V_i^F(x) = \sum_{j \in S} x_j \rho_{ji}(F(x), x) \\ - x_i \sum_{j \in S} \rho_{ij}(F(x), x). \tag{M}$$

The form of the mean dynamic is easy to explain. The first term describes the 'inflow' into strategy i from other strategies; it is obtained by multiplying the mass of agents playing each strategy j by the rate at which such agents switch to strategy i, and then summing over j. Similarly, the second term describes the 'outflow' from strategy i to other strategies. The difference between these terms is the net rate of change in the use of strategy i.

To obtain a formal link between the mean dynamic (M) and our model of individual choice, imagine that the population game F is played not by a continuous mass of agents but rather by a large, finite population with N members. Then the model described above defines a Markov process $\{X_t^N\}$ on a fine but discrete grid in the state space X. The foundations for deterministic evolutionary dynamics are provided by the following finite horizon deterministic approximation theorem: Fix a time horizon $T < \infty$. Then the behaviour of the stochastic process $\{X_t^N\}$ through time T is approximated by a solution of the mean dynamic (M); the approximation is uniformly good with probability close to 1 once the population size N is large enough. (For a formal statement of this result, see Benaïm and Weibull, 2003.)

In cases where one is interested in phenomena that occur over very long time horizons, it may be more appropriate to consider the infinite horizon behaviour of the stochastic process $\{X_t^N\}$. Over this infinite time horizon, the deterministic approximation fails, as a correct analysis must explicitly account for the stochastic nature of the evolutionary process. For more on the distinction between the two time scales, see Benaïm and Weibull (2003).

4 Examples and families of evolutionary dynamics

We now describe revision protocols that generate some of the most commonly studied evolutionary dynamics. In the table below, $\bar{F}(x) = \sum_{i \in S} x_i F_i(x)$ represents the population's average payoff at state x, and $B^F(x) = \arg\max_{y \in X} y' F(x)$ is the best response correspondence for the game F.

A common critique of evolutionary analysis of games is that the choice of a specific revision protocol, and hence the evolutionary analysis that follows, is necessarily arbitrary. There is surely some truth to this criticism: to the extent that one's analysis is sensitive to the fine details of the choice of protocol, the conclusions of the analysis are cast into doubt. But much of the force of this critique is dispelled by this important observation: *evolutionary dynamics based on qualitatively similar revision protocols lead to qualitatively similar aggregate behaviour.* We call a collection of dynamics generated by similar revision protocols a 'family' of evolutionary dynamics.

To take one example, many properties that hold for the replicator dynamic also hold for dynamics based on revision protocols of the form $\rho_{ij} = x_j \hat{\rho}_{ij}$ where $\hat{\rho}_{ij}$ satisfies

$$\text{sgn}\left((\hat{\rho}_{ki} - \hat{\rho}_{ik}) - (\hat{\rho}_{kj} - \hat{\rho}_{jk}) \right) \\ = \text{sgn}(F_i - F_j) \quad \text{for all } k \in S.$$

(In words: if i earns a higher payoff than j, then the net conditional switch rate from k to i is higher than that from k to j for all $k \in S$.) For reasons described in Section 3, dynamics generated in this way are called 'imitative dynamics'. (See Björnerstedt and Weibull, 1996, for a related formulation.) For another example, most properties of the pairwise difference dynamic remain true for dynamics based on protocols of the form $\rho_{ij} = \varphi(F_i - F_j)$, where $\varphi : \mathbf{R} \to \mathbf{R}_+$ satisfies sign-preservation:

$$\text{sgn}(\varphi(d)) = \text{sgn}([d]_+).$$

Dynamics in this family are called 'pairwise comparison dynamics'. For more on these and other families of dynamics, see Sandholm (2007, ch. 5).

5 Rest points and local stability

Having introduced families of evolutionary dynamics, we now turn to questions of prediction: if agents playing game F follow the revision protocol ρ (or, more broadly, a revision protocol from a given family), what predictions can we make about how they will play the game? To what extent do these predictions accord with those provided by traditional game theory?

A natural first question to ask concerns the relationship between the rest points of an evolutionary dynamic V^F and the Nash equilibria of the underlying game F. In fact, one can prove for a very wide range of evolutionary dynamics that if a state $x^* \in X$ is a Nash equilibrium (that is, if $x \in B(x)$), then x^* is a rest point as well.

One way to show that $NE(F) \subseteq RP(V^F)$ is to first establish a *monotonicity* property for V^F: that is, a property that relates strategies' growth rates under V^F with their payoffs in the underlying game (see, for example, Nachbar, 1990; Friedman, 1991; and Weibull, 1995). The most general such property, first studied by Friedman (1991) and Swinkels (1993), we call 'positive correlation':

$$\text{If } x \notin RP(V^F), \text{ then } F(x)'V^F(x) > 0.$$

$$(PC)$$

Property (PC) is equivalent to requiring a positive correlation between strategies' growth rates $V_i^F(x)$ and payoffs $F_i(x)$ (where the underlying probability measure is the uniform measure on the strategy set S). This property is satisfied by the first three dynamics in Table 1, and modifications of it hold for the remaining two as well. Moreover, it is not difficult to show that if V^F satisfies (PC), then all Nash equilibria of F are rest points of V^F: that is, $NE(F) \subseteq RP(V^F)$, as desired (see Sandholm, 2007, ch. 5).

In many cases, one can also prove that every rest point of V^F is a Nash equilibrium of F, and hence that $NE(F) = RP(V^F)$. In fact, versions of this statement are true for all of the dynamics introduced above, with the notable exception of the replicator dynamic and other imitative dynamics. The reason for this failure is easy to see: when revisions are based on imitation, unused strategies, even ones that are optimal, are never chosen. On the other hand, if we introduce a small number of agents playing an unused optimal strategy, then these agents will be imitated. Developing this logic, Bomze (1986) and Nachbar (1990) show that, under many imitative dynamics, every Lyapunov stable rest point is a Nash equilibrium.

As we noted at the onset, the original motivation for the replicator dynamic was to provide a foundation for Maynard Smith and Price's (1973) notion of an evolutionarily stable strategy (ESS). Hofbauer, Schuster and Sigmund (1979) and Zeeman (1980) show that an ESS is asymptotically stable under the replicator dynamic, but that an asymptotically state need not be an ESS.

More generally, when is a Nash equilibrium a dynamically stable rest point, and under which dynamics? Under differentiable dynamics, stability of isolated equilibria can often be determined by linearizing the dynamic around the equilibrium. In many cases, the question of the stability of the rest point x^* reduces to a question of the negativity of certain eigenvalues of the Jacobian matrix $DF(x^*)$ of the payoff vector field. In non-differentiable cases, and in cases where the equilibria in question form a connected component, stability can sometimes be established by using another standard approach: the construction of suitable Lyapunov functions. For an overview of work in these directions, see Sandholm (2007, ch. 6).

In the context of random matching in normal form games, it is natural to ask whether an equilibrium that is stable under an evolutionary dynamic also satisfies the restrictions proposed in the equilibrium refinements literature. Swinkels (1993) and Demichelis and Ritzberger (2003) show that this is true in great generality under even the most demanding refinements: in particular, any component of rest points that is asymptotically stable under a dynamic that respects condition (PC) contains a strategically stable set in the sense of Kohlberg and Mertens (1986). While proving this result is difficult, the idea behind the result is simple. If a component is asymptotically stable under an evolutionary dynamic, then this dynamic stability ought not to be affected by slight perturbations of the payoffs of the game. *A fortiori*, the *existence* of the component ought not to be affected by the payoff perturbations either. But this preservation of existence is precisely what strategic stability demands.

This argument also shows that asymptotic stability under evolutionary dynamics is a qualitatively stronger requirement than strategic stability: while strategic stability requires equilibria not to vanish after payoff perturbations, it does not demand that they be attracting

Table 1

Revision protocol	Evolutionary dynamic	Name	Origin
$\rho_{ij} = x_j(K - F_i)$, or $\rho_{ij} = x_j(K + F_j)$, or $\rho_{ij} = x_j[F_j - F_i]_+$	$\dot{x}_i = x_i(F_i(x) - \bar{F}(x))$	Replicator	Taylor and Jonker (1978)
$\rho_{ij} = [F_j - \bar{F}]_+$	$\dot{x}_i = [F_i(x) - \bar{F}(x)]_+ - x_i \sum_{j \in S} [F_j(x) - \bar{F}(x)]_+$	Brown–von Neumann–Nash (BNN)	Brown and von Neumann (1950)
$\rho_{ij} = [F_j - F_i]_+$	$\dot{x}_i = \sum_{j \in S} x_j[F_i(x) - F_j(x)]_+ - x_i \sum_{j \in S} [F_i(x) - F_j(x)]_+$	Pairwise difference (PD)	Smith (1984)
$\rho_{ij} = \dfrac{\exp(\eta^{-1}F_j)}{\sum_{k \in S}\exp(\eta^{-1}F_k)}$	$\dot{x}_i = \dfrac{\exp(\eta^{-1}F_i(x))}{\sum_{k \in S}\exp(\eta^{-1}F_k(x))} x_i$	Logit	Fudenberg and Levine (1998)
$\rho_{ij} = B_i^F(x)$	$\dot{x} = B_i^F(x) - x_i$	Best response	Gilboa and Matsui (1991)

under a disequilibrium adjustment process. For example, while all Nash equilibria of simple coordination games are strategically stable, only the pure Nash equilibria are stable under evolutionary dynamics.

Demichelis and Ritzberger (2003) establish their results using tools from index theory. Given an evolutionary dynamic V^F for a game F, one can assign each component of rest points an integer, called the *index*, that is determined by the behaviour of the dynamic in a neighbourhood of the rest point; for instance, regular, stable rest points are assigned an index of 1. The set of all indices for the dynamic V^F is constrained by the *Poincaré–Hopf theorem*, which tells us that the sum of the indices of the equilibrium components of V^F must equal 1. As a consequence of this deep topological result, one can sometimes determine the local stability of one component of rest points by evaluating the local stability of the others.

6 Global convergence: positive and negative results

To provide the most satisfying evolutionary justification for the prediction of Nash equilibrium play, it is not enough to link the rest points of a dynamic and the Nash equilibria of the underlying game, or to prove local stability results. Rather, one must establish convergence to Nash equilibrium from *arbitrary* initial conditions.

One way to proceed is to focus on a class of games defined by some noteworthy payoff structure, and then to ask whether global convergence can be established for games in this class under certain families of evolutionary dynamics. As it turns out, general global convergence results can be proved for a number of classes of games. Among these classes are *potential games*, which include common interest games, congestion games, and games generated by externality pricing schemes; *stable games*, which include zero-sum games, games with an interior ESS, and (perturbed) concave potential games; and *supermodular games*, which include models of Bertrand oligopoly, arms races, and macroeconomic search. A fundamental paper on global convergence of evolutionary dynamics is Hofbauer (2000); for a full treatment of these results, see Sandholm (2007).

Once we move beyond specific classes of games, global convergence to Nash equilibrium cannot be guaranteed; cycling and chaotic behaviour become possible. Indeed, Hofbauer and Swinkels (1996) and Hart and Mas-Colell (2003) construct examples of games in which all reasonable deterministic evolutionary dynamics fail to converge to Nash equilibrium from most initial conditions. These results tell us that general guarantees of convergence to Nash equilibrium are impossible to obtain.

In light of this fact, we might instead consider the extent to which solution concepts simpler than Nash equilibrium are supported by evolutionary dynamics. Cressman and Schlag (1998) and Cressman (2003) investigate whether imitative dynamics lead to subgame perfect equilibria in reduced normal forms of extensive form games – in particular, generic games of perfect information. In these games, interior solution trajectories do converge to Nash equilibrium components, and only subgame perfect components can be interior asymptotically stable. But even in very simple games interior asymptotically stable components need not exist, so the dynamic analysis may fail to select subgame perfect equilibria. For a full treatment of these issues, see Cressman (2003).

What about games with strictly dominated strategies? Early results on this question were positive: Akin (1980), Nachbar (1990), Samuelson and Zhang (1992), and Hofbauer and Weibull (1996) prove that dominated strategies are eliminated under certain classes of imitative dynamics. However, Berger and Hofbauer (2006) show that dominated strategies need not be eliminated under the BNN dynamic. Pushing this argument further, Hofbauer and Sandholm (2006) find that dominated strategies can survive under any continuous evolutionary dynamic that satisfies positive correlation and *innovation*; the latter condition requires that agents choose unused best responses with positive probability. Thus, whenever there is some probability that agents base their choices on direct evaluation of payoffs rather than imitation of successful opponents, evolutionary dynamics may violate even the mildest rationality criteria.

7 Conclusion

Because the literature on evolutionary dynamics came to prominence shortly after the literature on equilibrium refinements, it is tempting to view the former literature as a branch of the latter. But, while it is certainly true that evolutionary models have something to say about selection among multiple equilibria, viewing them simply as equilibrium selection devices can be misleading. As we have seen, evolutionary dynamics capture the behaviour of large numbers of myopic, imperfectly informed decision makers. Using evolutionary models to predict behaviour in interactions between, say, two well-informed players is daring at best.

The negative results described in Section 6 should be understood in this light. If we view evolutionary dynamics as an equilibrium selection device, the fact that they need not eliminate strictly dominated strategies might be viewed with disappointment. But, if we take the result at face value, it becomes far less surprising: if agents switch to strategies that perform reasonably well at the moment of choice, that a strategy is never optimal need not deter agents from choosing it.

A similar point can be made about failures of convergence to equilibrium. From a traditional point of view, persistence of disequilibrium behaviour might seem to undermine the very possibility of a satisfactory economic analysis. But the work described in this entry suggests that in large populations, this possibility is not only real but is also one that game theorists are well equipped to analyse.

WILLIAM H. SANDHOLM

See also **learning and evolution in games: adaptive heuristics; learning and evolution in games: an overview; learning and evolution in games: ESS; Nash equilibrium, refinements of; stochastic adaptive dynamics.**

I thank John Nachbar for helpful comments. Financial support from NSF Grant SES-0092145 is gratefully acknowledged.

Bibliography

Akin, E. 1980. Domination or equilibrium. *Mathematical Biosciences* 50, 239–50.

Beckmann, M., McGuire, C. and Winsten, C. 1956. *Studies in the Economics of Transportation*. New Haven, CT: Yale University Press.

Benaïm, M. and Weibull, J. 2003. Deterministic approximation of stochastic evolution in games. *Econometrica* 71, 873–903.

Berger, U. and Hofbauer, J. 2006. Irrational behavior in the Brown–von Neumann–Nash dynamics. *Games and Economic Behavior* 56, 1–6.

Björnerstedt, J. and Weibull, J. 1996. Nash equilibrium and evolution by imitation. In *The Rational Foundations of Economic Behavior*, ed. K. Arrow et al. New York: St Martin's Press.

Bomze, I. 1986. Non-cooperative two-person games in biology: a classification. *International Journal of Game Theory* 15, 31–57.

Brown, G. and von Neumann, J. 1950. Solutions of games by differential equations. In *Contributions to the Theory of Games I*, ed. H. Kuhn and A.Tucker. Annals of Mathematics Studies 24. Princeton: Princeton University Press.

Cressman, R. 2003. *Evolutionary Dynamics and Extensive Form Games*. Cambridge, MA: MIT Press.

Cressman, R. and Schlag, K. 1998. The dynamic (in)stability of backwards induction. *Journal of Economic Theory* 83, 260–85.

Dawkins, R. 1982. *The Extended Phenotype*. San Francisco: Freeman.

Demichelis, S. and Ritzberger, K. 2003. From evolutionary to strategic stability. *Journal of Economic Theory* 113, 51–75.

Friedman, D. 1991. Evolutionary games in economics. *Econometrica* 59, 637–66.

Fudenberg, D. and Levine, D. 1998. *Theory of Learning in Games*. Cambridge, MA: MIT Press.

Gilboa, I. and Matsui, A. 1991. Social stability and equilibrium. *Econometrica* 59, 859–67.

Hart, S. and Mas-Colell, A. 2003. Uncoupled dynamics do not lead to Nash equilibrium. *American Economic Review* 93, 1830–6.

Hofbauer, J. 2000. From Nash and Brown to Maynard Smith: equilibria, dynamics, and ESS. *Selection* 1, 81–8.

Hofbauer, J. and Sandholm, W. 2006. Survival of dominated strategies under evolutionary dynamics. Mimeo, University College London and University of Wisconsin.

Hofbauer, J., Schuster, P. and Sigmund, K. 1979. A note on evolutionarily stable strategies and game dynamics. *Journal of Theoretical Biology* 27, 537–48.

Hofbauer, J. and Sigmund, K. 1988. *Theory of Evolution and Dynamical Systems*. Cambridge: Cambridge University Press.

Hofbauer, J. and Sigmund, K. 1998. *Evolutionary Games and Population Dynamics*. Cambridge: Cambridge University Press.

Hofbauer, J. and Sigmund, K. 2003. Evolutionary game dynamics. *Bulletin of the American Mathematical Society N.S.* 40, 479–519.

Hofbauer, J. and Swinkels, J. 1996. A universal Shapley example. Mimeo, University of Vienna and Northwestern University.

Hofbauer, J. and Weibull, J. 1996. Evolutionary selection against dominated strategies. *Journal of Economic Theory* 71, 558–73.

Kohlberg, E. and Mertens, J.-F. 1986. On the strategic stability of equilibria. *Econometrica* 54, 1003–38.

Maynard Smith, J. 1982. *Evolution and the Theory of Games*. Cambridge: Cambridge University Press.

Maynard Smith, J. and Price, G. 1973. The logic of animal conflict. *Nature* 246, 15–18.

Nachbar, J. 1990. 'Evolutionary' selection dynamics in games: convergence and limit properties. *International Journal of Game Theory* 19, 59–89.

Samuelson, L. 1997. *Evolutionary Games and Equilibrium Selection*. Cambridge, MA: MIT Press.

Samuelson, L. and Zhang, J. 1992. Evolutionary stability in asymmetric games. *Journal of Economic Theory* 57, 363–91.

Sandholm, W. 2001. Potential games with continuous player sets. *Journal of Economic Theory* 97, 81–108.

Sandholm, W. 2007. *Population Games and Evolutionary Dynamics*. Cambridge, MA: MIT Press.

Schuster, P. and Sigmund, K. 1983. Replicator dynamics. *Journal of Theoretical Biology* 100, 533–8.

Smith, M. 1984. The stability of a dynamic model of traffic assignment – An application of a method of Lyapunov. *Transportation Science* 18, 245–52.

Swinkels, J. 1993. Adjustment dynamics and rational play in games. *Games and Economic Behavior* 5, 455–84.

Taylor, P. and Jonker, L. 1978. Evolutionarily stable strategies and game dynamics. *Mathematical Biosciences* 40, 145–56.

Vega-Redondo, F. 1996. *Evolution, Games, and Economic Behavior*. Oxford: Oxford University Press.

Weibull, J. 1995. *Evolutionary Game Theory*. Cambridge, MA: MIT Press.

Zeeman, E. 1980. Population dynamics from game theory. In *Global Theory of Dynamical Systems*, ed. Z. Nitecki and C. Robinson. Berlin: Springer.

deterrence (empirical), economic analyses of

Empirical economic analyses of deterrence seek to test the central prediction of the economic or rational-actor model of criminal behaviour that Becker (1968) pioneered. In the Beckerian model, a potential offender compares the expected costs and benefits of criminal activity, and when the expected utility of crime exceeds the expected utility loss of any punishment, the actor engages in the criminal activity. Economists have attempted to confirm or refute this model by relating geographic and temporal variation in punishment regimes, which proxy for the expected cost of offending, to aggregate crime rates, which measure the frequency of criminal activity. This approach poses two challenges. First, criminal justice policies are endogenous to crime rates, because jurisdictions often devote greater resources to crime control when the incidence of crime is higher. Second, even if the econometrician breaks the simultaneity of crime rates and crime-control policies, the estimates typically do not reveal whether deterrence or incapacitation is the operative mechanism.

Estimates of the causal effect of policing levels on crime rates

The criminal justice policies that have most often received empirical evaluation are the scale of policing and imprisonment, which in the economic model of crime correspond to the probability of apprehension and the magnitude of the sanction, respectively. Early studies tried to infer the causal effect of police levels on crime rates by drawing cross-sectional comparisons across cities or states, but Fisher and Nagin (1978) showed that cross-sectional estimates suffer from simultaneity bias because jurisdictions with higher crime rates respond by employing more police. In the 1990s a second wave of literature emerged.

The new studies employed more sophisticated econometric strategies to break the simultaneity problem. For example, Marvell and Moody (1996) used Granger causality to identify the impact of policing levels on crime rates. A variable 'Granger causes' another when changes in the first variable generally precede changes in the second, and thus Granger causality refers to a temporal relationship between two variables rather than actual causation (Granger, 1969). Marvell and Moody (1996) applied this technique to more than 20 years of state and city data and found that police Granger-caused lower crime, or that increases in police were associated with future declines in crime.

Levitt (1997) employed a different econometric strategy: an instrumental variables or 'natural experiment' approach. He argued that mayoral and gubernatorial elections were valid instruments, because they correlate with police but do not correlate with crime, except through the other explanatory variables in the crime equation. He showed that sizable increases in the police forces in major cities were concentrated in election years, perhaps because greater police generate electoral benefits for politicians. His estimate, that a ten per cent increase in

the police force produced at most a ten per cent reduction in crime rates, was comparable in magnitude to Marvell and Moody's (1996). McCrary (2002) argued that, when properly measured, electoral cycles induced insufficient variation in the size of police forces to measure the impact of crime. However, Levitt (2002) showed that an alternative instrumental variable, the number of firefighters, also produces negative and sizable estimates of the impact of police on crime. Recently, Evans and Owens (2005) demonstrated that the federal subsidies from the Clinton Crime Bill stimulated police hiring and produced similar reductions in crime rates.

Other authors used more finely disaggregated data to identify the effect of police on crime. In data with annual observations, any increase in crime and police occurring within a calendar year appears contemporaneous rather than sequential, and the short-term causal effect of police on crime is not observed. Corman and Mocan (2000) examined the short-term effect using nearly 30 years of monthly data from New York City and applying Granger causality techniques. They found that police hiring occurs approximately six months after a jump in crime and that the increase in police leads to reductions in crime as great as Levitt's (1997) largest estimate. Di Tella and Schargrodsky (2004) examined data decomposed to the level of city blocks. When the city of Buenos Aires reallocated police to temples and mosques in response to terrorist threats against them, Di Tella and Schargodsky observed that auto thefts immediately around those buildings declined abruptly but that the reduction in crime quickly decayed with distance.

Despite the use of different estimation procedures and different data-sets, the second wave of literature on policing and crime produced quite similar estimates of the crime-reducing effect of police levels. The marginal reduction in crime associated with hiring an additional police officer in large urban environments roughly equals the marginal cost.

Estimates of the causal effect of incarceration rates on crime rates

Empirical analyses of the crime-reducing effect of prisons evolved in a similar manner to studies of policing. Early efforts failed to recognize or address the simultaneity problem and prematurely concluded that imprisonment has neither deterrent nor incapacitating effects (see Zimring and Hawkins, 1991). In the 1990s researchers again applied more sophisticated empirical strategies that attempted to break the simultaneity problem. Marvell and Moody (1994) applied Granger causality techniques to a repeated cross-section of states and found that a ten per cent increase in the prison population produced nearly a two per cent fall in crime rates.

Levitt (1996) disentangled the simultaneity of crime and incarceration by using lawsuits challenging conditions in overcrowded prisons as instrumental variables. He showed that, when the suits produced court orders to reduce overcrowding, states typically complied by releasing prisoners who otherwise would have been incarcerated. His estimates implied that the reduction in crime from incarcerating an additional prisoner was two to three times larger than that predicted by Marvell and Moody (1994).

Although these studies indicate that imprisonment reduces crime, the relevance of their estimated parameters to social policy evaluation of present incarceration levels has already diminished. The prison population has grown so rapidly since the mid-1990s that its margin lies well outside the range in which the parameter estimates were generated. For most reasonable set of assumptions, the current scale of incarceration appears at or above the socially optimal level.

Estimates distinguishing deterrence from incapacitation

Although economists' understanding of the causal relationships among policing, incarceration, and crime has improved, they have made less progress on the question of whether the declines in crime are due to deterrence or incapacitation. Determining the operative effect is crucial for evaluating the economic model of crime and for designing crime-control policy.

A few empirical economic studies attempted to assess the relative importance of deterrence and incapacitation by exploring responses to increased punishments. Kessler and Levitt (1999) studied the effect of a California referendum that provided sentence enhancements for certain serious crimes. The sentence enhancement imposed an additional incapacitating effect only upon completion of the standard prison term, and any decline in crime before that date was arguably attributable to deterrence. Kessler and Levitt found that the rate of crimes covered by the referendum fell relative to other states and that the rate of crimes not covered by the referendum did not change. After the expiration of the standard prison terms, the rate of the affected crimes continued to fall, and this further decline indicated that the full impact of the sentence enhancements included both deterrent and incapacitating effects.

Another effort to distinguish deterrence from incapacitation proceeded from the observation that criminals do not specialize in particular types of offences, but instead are generalists who participate in potentially wide range of offences. Levitt (1998a) noted that, if deterrence is the operative mechanism, a longer sentence for a particular type of crime implies that generalist criminals should substitute to other kinds of crime. If the primary effect is instead incapacitation, then a longer sentence for a particular crime should lower the rate of alternative offences. Using arrest rate data, Levitt (1998a) found mixed evidence for deterrence.

Levitt (1998b) evaluated the responsiveness of criminal activity to the transition from the juvenile to the adult criminal justice system as another means of distinguishing deterrence from incapacitation. In states where the criminal justice system is substantially more punitive than the juvenile system, deterrence predicts that juveniles should reduce their criminal activity immediately upon reaching the age of majority (before there is time for incapacitation to become a factor). States where the adult system was especially punitive relative to the juvenile system experience sharp declines in crime at the age of majority relative to states where the transition to the adult system is most lenient, consistent with deterrence.

Other empirical analyses of deterrence

Capital punishment seemingly offers a direct test of the deterrence hypothesis, because the alternative sentence for a death-eligible offender is typically life imprisonment, and any crime-reducing effect of capital punishment is therefore arguably attributable to deterrence. Ehrlich (1975; 1977a; 1977b) produced some of the earliest and most contested claims of capital punishment's deterrent effect. Cameron (1994) reviews the large literature on the death penalty, and criticisms of Ehrlich's conclusions focus on the sensitivity of the estimates to the time period, the states, and the control variables included in the analysis. Recently, a number of studies examined the relationship between the death penalty and crime rates using repeated cross-sections of states in the period since the Supreme Court's 1976 reinstatement of capital punishment. These studies use data disaggregated at the monthly (Mocan and Gittings, 2003) or county-level (Dezhbakhsh, Rubin and Shepherd, 2003) and study the impact on different kinds of homicide (Shepherd, 2004). All claim deterrent effects at least as large as Ehrlich's original estimates, despite their continuing sensitivity to minor specification changes. In contrast, Katz, Levitt and Shustorovich (2003) used state-level panel data covering the period 1950–90 and detected no effect of the death penalty on crime rates. Unlike the literature on policing and incarceration, the use of higher frequency data and additional control variables has broadened, rather than narrowed, the range of estimated impacts of the death penalty.

Although most empirical economic analyses of deterrence evaluate the role of public law enforcement, a few studies consider the responsiveness of crime to the precautions taken by potential victims. A victim's precaution may have a general deterrent effect only if the prospective offender cannot observe it before deciding to commit the crime. Otherwise, the observation of a precaution may induce the offender to substitute to a more vulnerable victim but have no effect on the total rate of offending (see Clotfelter, 1978; Shavell, 1991). Ayres and Levitt (1998) analysed a particular kind of anti-theft device for automobiles as an unobservable precaution. The device contained a radio transmitter that allows police with special equipment to track the vehicle, but its lack of outward indications made it unobservable to potential offenders. Ayres and Levitt found that, when the device became available in a city, vehicle thefts fell sharply and that it did not induce car thieves to substitute to other types of crimes or to other geographic areas.

Another purported unobservable precaution that received extensive empirical analysis is surreptitious gun possession. Lott and Mustard (1997) and Lott (1998) claimed that laws relaxing the requirements for concealed weapons permits had a general deterrent effect on crime rates, but numerous researchers challenged the Lott findings. Ayres and Donohue (1999) found that in more recent years the law correlated either positively or not at all with crime rates, and Duggan (2000) showed that crime rates in states that adopted concealed-weapons laws began to decline before the passage of the laws. Other researchers argued that additional tests of the hypothesis failed to confirm it. Ludwig (1998) found that that passage of these laws was associated with large declines in the victimization of juveniles, a group not permitted to carry concealed weapons under these laws. Kovandzic and Marvell (2003) reported no relationship between the number of concealed weapons permits issued and violent crime rates in a single state.

STEVEN D. LEVITT AND THOMAS J. MILES

See also **Becker, Gary S.; causality in economics and econometrics; crime and the city; deterrence (theory), economics of; difference-in-difference estimators; Granger–Sims causality; law, public enforcement of; natural experiments and quasi-natural experiments.**

Bibliography

Ayres, I. and Donohue III, J. 1999. Nondiscretionary concealed weapons laws: a case study of statistics, standards of proof, and public policy. *American Law and Economics Review* 1, 436–70.

Ayres, I. and Levitt, S. 1998. measuring positive externalities from unobservable victim precautions: an empirical analysis of Lojack. *Quarterly Journal of Economics* 113, 43–77.

Becker, G. 1968. Crime and punishment: an economic approach. *Journal of Political Economy* 76, 169–217.

Cameron, S. 1994. A review of econometric evidence on the effects of capital punishment. *Journal of Socio-Economics* 23, 197–214.

Clotfelter, C. 1978. Private security and the public safety. *Journal of Urban Economics* 5, 388–402.

Corman, H. and Mocan, H. 2000. A time-series analysis of crime and drug use in New York City. *American Economic Review* 90, 584–604.

Dezhbakhsh, H., Rubin, P. and Shepherd, J. 2003. Does capital punishment have a deterrent effect? New evidence from postmoratorium panel data. *American Law and Economics Review* 5, 344–76.

Di Tella, R. and Schargrodsky, E. 2004. Do police reduce crime? Estimates using the allocation of police forces after a terrorist attack. *American Economic Review* 94, 115–33.

Duggan, M. 2000. More guns, more crime. *Journal of Political Economy* 109, 1086–114.

Ehrlich, I. 1975. The deterrent effect of capital punishment: a question of life and death. *American Economic Review* 65, 397–417.

Ehrlich, I. 1977a. The deterrent effect of capital punishment: reply. *American Economic Review* 67, 452–58.

Ehrlich, I. 1977b. Capital punishment: further thoughts and additional evidence. *Journal of Political Economy* 85, 741–88.

Evans, W. and Owens, E. 2005. Flypaper COPS. Mimeo. Department of Economics, University of Maryland. Online. Available at: http://www.bsos.umd.edu/econ/evans/wpapers/Flypaper%20COPS.pdf, accessed 20 October 2005.

Fisher, F. and Nagin, D. 1978. On the feasibility of identifying the crime function in a simultaneous equations model of crime and sanctions. In *Deterrence and Incapacitation: Estimating the Effects of Criminal Sanctions on Crime Rates*, ed. A. Blumstein, D. Nagin and J. Cohen. Washington, DC: National Academy of Sciences.

Granger, C. 1969. Investigating causal relations by econometric models and cross-spectral methods. *Econometrica* 37, 424–38.

Katz, L., Levitt, S. and Shustorovich, E. 2003. Prison conditions, capital punishment, and deterrence. *American Law and Economics Review* 5, 318–43.

Kessler, D. and Levitt, S. 1999. Using sentence enhancements to distinguish between deterrence and incapacitation. *Journal of Law and Economics* 17, 343–63.

Kovandzic, T. and Marvell, T. 2003. Right-to-carry concealed handguns and violent crime: crime control through gun decontrol? *Criminology and Public Policy* 2, 363–96.

Levitt, S. 1996. The effect of prison population size on crime rates: evidence from prison overcrowding litigation. *Quarterly Journal of Economics* 111, 319–25.

Levitt, S. 1997. Using electoral cycles in police hiring to estimate the effect of police on crime. *American Economic Review* 87, 270–90.

Levitt, S. 1998a. Why do increased arrest rates appear to reduce crime: deterrence, incapacitation, or measurement error? *Economic Inquiry* 36, 353–72.

Levitt, S. 1998b. Juvenile crime and punishment. *Journal of Political Economy* 106, 1156–85.

Levitt, S. 2002. Using electoral cycles in police hiring to estimate the effects of police on crime: reply. *American Economic Review* 92, 1244–50.

Lott, J., Jr. 1998. *More Guns, Less Crime.* Chicago: University of Chicago Press.

Lott, J., Jr. and Mustard, D. 1997. Crime, deterrence, and right-to-carry concealed handguns. *Journal of Legal Studies* 26, 1–68.

Ludwig, J. 1998. Concealed-gun-carrying laws and violent crime: evidence from state panel data. *International Review of Law and Economics* 18, 239–54.

Marvell, T. and Moody, C. 1994. Prison population growth and crime reduction. *Journal of Quantitative Criminology* 10, 109–40.

Marvell, T. and Moody, C. 1996. Specification problems, police levels, and crime rates. *Criminology* 34, 609–46.

McCrary, J. 2002. Using electoral cycles in police hiring to estimate the effect of police on crime: comment. *American Economic Review* 92, 1236–43.

Mocan, H. and Gittings, R. 2003. Getting off death row: commuted sentences and the deterrent effect of capital punishment. *Journal of Law and Economics* 46, 453–78.

Shavell, S. 1991. Individual precautions to prevent theft: private versus socially optimal behavior. *International Review of Law and Economics* 11, 123–32.

Shepherd, J. 2004. Murders of passion, execution delays, and the deterrence of capital punishment. *Journal of Legal Studies* 33, 283–322.

Zimring, F. and Hawkins, G. 1991. *The Scale of Imprisonment.* Chicago: University of Chicago Press.

deterrence (theory), economics of

The persistence of criminal activity throughout human history and the challenges it poses for determining optimal law-enforcement activity have already attracted the attention of utilitarian philosophers and early economists like Beccaria, Paley and Bentham. It was not until the late 1960s, however, especially following the seminal work by Becker (1968), that economists reconnected with the subject, using the modern tools of economic theory and econometrics.

In both its utilitarian and modern versions the economic approach to crime is predicated on what the new literature calls 'the deterrence hypothesis' – the assumption that potential and actual offenders respond to incentives, and that the volume of offences in the population is therefore influenced by law enforcement and other means of crime prevention. By its common connotation, deterrence generally refers to the threat of a criminal sanction, or any other form of punishment having some moderating effect on the willingness to engage in criminal activity. To interpret this hypothesis so narrowly misses, however, the basic idea on which it is founded (Ehrlich, 1979). The hypothesis relates to the role of both negative incentives (such as the prospect of apprehension, conviction and punishment) and positive incentives (such as opportunities for gainful employment in legitimate relative to illegitimate occupations) as deterrents to actual or would-be offenders. It follows that not just conventional law enforcement matters in influencing the flow of offences but external market and household conditions as well, to the extent that these affect prospective gains and losses from illegitimate activity. For this approach to provide a useful approximation of the complicated reality of crime, it is not necessary that all those who commit specific crimes respond to incentives, nor is the degree of individual responsiveness prejudged; it is sufficient that a significant number of potential offenders so behave on the margin. By the same token, the theory does not preclude a priori any category of crime, or any class of incentives, as non-conforming. Indeed, economists have applied the deterrence hypothesis to a myriad of illegal activities, from tax evasion, drug abuse and fraud to skyjacking, robbery and murder.

The economic approach

In Becker's analysis the equilibrium volume of crime reflects the interaction between offenders and the law-enforcement authority, and the focus is on optimal probability, severity, and type of criminal sanction – the implicit 'prices' society imposes on criminal behaviour – in view of their impact on offenders and the relative social costs associated with their imposition. Subsequent theoretical work has focused on more complete formulations of specific components of the system and their micro foundations – primarily the supply of offences, the production of specific law-enforcement activities, and alternative criteria for achieving socially optimal law enforcement. A later evolution has aimed to expand the analytical setting within which crime is analysed to address the interaction between potential offenders (supply), consumers and potential victims (private actual or indirect 'demand'), and deterrence and prevention by public authorities (government intervention). This 'market model of crime' (Ehrlich, 1981; 1996) has been further explored in recent years to include interactions between criminal activity and the general economy. For the specific articles on which the following discussion is based, see Ehrlich and Liu (2006, vols. 1 and 2).

Supply

The extent of participation in crime is generally modelled as an outcome of the allocation of time among competing legitimate and illegitimate activities by potential offenders. Since illegitimate activity carries the distinct risk of apprehension and punishment for illegitimate behaviour, individuals are assumed to act as expected-utility maximizers. This may generally lead many offenders to be multiple-job holders – being part-time offenders, or going in and out of criminal activity (see Ehrlich, 1973, and the empirical documentation in Reuter, MacCoun and Murphy, 1990). While the mix of pecuniary and non-pecuniary benefits varies across different crime categories, which attract persons of different earning opportunities or attitudes towards risk and moral values ('preferences'), the basic opportunities affecting choice are identified in all cases as the perceived probabilities of apprehension, conviction and punishment, the marginal penalties imposed, and the expected net return on illegal over legal activity. Net returns from crime rise with the level of community wealth, which enhances the potential loot from property crime, and fall with the probability of finding employment in the legitimate labour market and the prospective legitimate wages. Entry into criminal activity and the extent of involvement in crime are thus shown to be related inversely to deterrence variables and directly to the differential return it can provide over legitimate activity. Moreover, a one per cent increase in the probability of apprehension is shown to exert a larger deterrent effect than corresponding increases in the conditional probabilities of conviction and of any specific punishment if convicted (Ehrlich, 1975). Essentially due to conflicting income and substitution effects, however, sanction severity can have more ambiguous effects on active offenders: a strong preference for risk may weaken (Becker, 1968) or even reverse (Ehrlich, 1973) the deterrent effect of sanctions, and the results are even less conclusive if one assumes that the length of time spent in crime, not just the moral obstacle to entering it, generates disutility (Block and Heineke, 1975).

The results become less ambiguous at the *aggregate* level, however, as one allows for heterogeneity of potential offenders due to differences in legitimate employment opportunities or preferences for risk and crime: a more severe sanction can reduce the crime rate by deterring the entry of potential offenders even if it has little effect on actual ones (Ehrlich, 1973). In addition to heterogeneity across individuals in personal opportunities and preferences, the literature has also addressed the role of heterogeneity in individuals' perceptions about probabilities of apprehension, as affected by learning from past experience (Sah, 1991). As a result, current crime rates may react, in part, to past deterrence measures. A different type of heterogeneity that can affect variability in crime rates across different crime categories and geographical units is identified by Glaeser, Sacerdote and Scheinkman (1996) and Glaeser and Sacerdote (1999) as stemming from the degree of social interaction: that is, the extent to which potential offenders are influenced by the behaviour of their neighbours.

Private 'demand'

The incentives operating on offenders often originate from, and are partially controlled by, consumers and potential victims. Transactions in illicit drugs or stolen goods, for example, are patronized by consumers who generate a direct demand for the underlying offence. But even crimes that inflict pure harm on victims are affected by an indirect (negative) demand, which is derived from a positive demand for safety (Ehrlich, 1981). By their choice of optimal self-protection (lowering the risk of becoming a victim) or self-insurance (reducing the potential loss if victimized) through use of locks, guards, safes, and alarms, or selective avoidance of crime-prone

areas (Bartel, 1975; Shavell, 1991; Cullen and Levitt, 1999), potential victims influence the marginal costs to offenders, and thus the implicit return on crime. And since optimal self-protection generally increases with the perceived risk of victimization (the crime rate), private protection and public enforcement will be interdependent. The interaction between the two and its impact on possible fluctuations in the equilibrium volume of offences is explored in Clotfelter (1977), and Phillipson and Posner (1996).

Public intervention

Since crime, by definition, causes a net social loss, and crime control measures are largely a public good, collective action is needed to augment individual self-protection. Public intervention typically aims to 'tax' illegal returns through the threat of punishment, or to 'regulate' offenders via incapacitation and rehabilitation programmes. All control measures are costly. Therefore, the 'optimum' volume of offences as determined by the law-enforcement authority acting as a social planner cannot be nil, but must be set at a level where the marginal cost of each measure of enforcement or prevention equals its marginal benefit.

To assess the relevant net social loss, however, one must adopt a criterion for public choice. Becker (1968) and Stigler (1970) have chosen variants of aggregate income measures as the relevant social welfare function to be maximized, requiring the minimization of the sum of social damages from offences and the social cost of law-enforcement activities. This approach can lead to powerful propositions regarding the optimal magnitudes of probability and severity of punishments for different crimes and different offenders, or, alternatively, the optimal level and mix of expenditures on police, courts and corrections. The analysis reaffirms the classical utilitarian proposition that the optimal severity of punishment should 'fit the crime', and thus be assessed essentially by its deterrent value, as the marginal social cost is higher for offences causing greater marginal social damage. Moreover, it makes a strong case for the desirability of monetary fines as a deterring sanction, since fines involve pure transfer payments between offenders and the rest of society. Different criteria for public choice, however, yield different implications regarding the optimal mix of probability and severity of punishment, as is the case when the social welfare function is expanded to include concerns for the 'distributional consequences' of law enforcement on offenders and victims in addition to their aggregate income, in which case even fines can become socially costly. These considerations can be ascribed to aversion to risk (as in Polinsky and Shavell, 1979), or to aversion towards *ex post* inequality under the law as a result of the 'lottery' nature of law enforcement, by which only offenders caught and convicted for crime pay for the damage caused by all offenders, including the luckier ones who escape apprehension and conviction (as in Ehrlich, 1982). A positive analysis of enforcement must also address the behaviour of the separate agencies constituting the criminal justice system: police, courts, and prison authorities. For example, Landes's analysis (1971) of the courts, which focuses on the interplay between prosecutors and defence teams, explains why settling cases out of court may be an efficient outcome of many court proceedings.

The optimal enforcement policy arising from the income-maximizing criterion can be questioned from yet another angle: a public-choice perspective. The optimization rule invoked in the aforementioned papers assumes that enforcement is carried out by a social planner. In practice, public law enforcement can facilitate the interests of rent-seeking enforcers who are amenable to malfeasance and bribes. Optimal social policy needs to control malfeasance by properly remunerating public enforcers (Becker and Stigler, 1974) or, where appropriate, setting milder penalties (Friedman, 1999).

Market equilibrium

In Ehrlich's (1981; 1996) 'market model', the equilibrium flow of offences results from the interaction between aggregate supply of offences, direct or derived demand for offences (through self-protection), and optimal public enforcement, which operates like a tax on criminal activity. The model derives the equilibrium volume of offences as well as the equilibrium net return, or premium, to offenders from illegitimate over legitimate activity as a result of the interaction between the relevant aggregate supply, 'demand', and government net taxation of crime in a competitive setting. One important application concerns a comparison of deterrence, incapacitation and rehabilitation as instruments of crime control. This is because the efficacy of deterring sanctions cannot be assessed merely by the elasticity of the aggregate supply of offences schedule, as it depends on the elasticity of the private demand schedule as well. Likewise, the efficacy of rehabilitation and incapacitation programmes cannot be inferred solely from knowledge of their impact on individual offenders (see Cook, 1975). It depends crucially on the elasticities of the market supply and demand schedules, as these determine the extent to which successfully rehabilitated offenders will be replaced by others responding to the prospect of higher net returns. This market setting has also been applied by Viscusi (1986), who links observed net returns on specific crimes to underlying parameters of the model, and in works by Schelling (1967), Buchanan (1973), and Garoupa (2000), who analyse various aspects of organized crime by viewing it in a monopolistic rather than a competitive setting.

Crime and the economy

The 'market model' has been developed largely in a static, partial-equilibrium setting in which the general economy affects the illegal sector of the economy, but not vice versa. More recently, the model has been extended to deal with the interaction between the two under static and dynamic conditions as well. For example, Ehrlich (1973) argues that income inequality, serving as a proxy for relative earning opportunities in illegal versus legal activities, induces time allocation in favour of illegal activity. A number of subsequent studies interpreted this relation to imply that the volume of offences can be lowered through subsidies to legitimate employment by workers with low legitimate earning capacity. Using a general-equilibrium setting, Imrohoroglu, Merlo and Rupert (2000) show, however, that, to the extent that subsidies must be paid for by raising taxes on legitimate production, such income distribution policies have an ambiguous effect on crime. The subsidy raises the opportunity cost of crime to apprehended offenders, but it also works as a disincentive to legitimate production because of an increased tax rate, which lowers the tax revenue available for crime detection.

The choice between legitimate and illegitimate activity may have not just static effects on the economy's level of output but dynamic growth effects as well if it affects productive human capital formation, which serves as an engine of productivity growth. Bureaucratic corruption is a case in point. As Ehrlich and Lui (1999) argue, this is because, whenever government intervenes in private economic activity, bureaucrats have an opportunity to engage in rent seeking by collecting explicit or implicit bribes, which rise with their bureaucratic status. The return on corruption is thus higher the greater is one's investment in becoming a bureaucrat or attaining higher bureaucratic status, which competes with investment in productive human capital. The analysis explains why corruption is a barrier to growth especially in less developed countries, and why under benevolent autocratic regimes the rate of economic growth can be as high as under democratic regimes.

Investigating and implementing alternative versions of a comprehensive model of crime based on micro foundations remains an intriguing challenge for future research.

ISAAC EHRLICH

See also **Bentham, Jeremy; deterrence (empirical), economic analyses of; econometrics; equality; expected utility hypothesis; general equilibrium; labour economics; law, public enforcement of; microfoundations; rent seeking; risk; uncertainty; unemployment.**

Bibliography

Bartel, A. 1975. An analysis of firm demand for protection against crime. *Journal of Legal Studies* 4, 433–78.

Becker, G. 1968. Crime and punishment: an economic approach. *Journal of Political Economy* 76, 169–217.

Becker, G. and Stigler, G. 1974. Law enforcement, malfeasance, and compensation of enforcers. *Journal of Legal Studies* 3, 1–18.

Becker, G. and Landes, W. 1974. *The Economics of Crime and Punishment*. New York: Columbia University Press.

Block, M. and Heineke, J. 1975. A labor theoretic analysis of the criminal choice. *American Economic Review* 65, 314–25.

Buchanan, J. 1973. A defense of organized crime? In *The Economics of Crime and Punishment: A Conference Sponsored by American Enterprise Institute for Public Policy Research*. Washington, DC: American Enterprise Institute for Public Policy Research.

Clotfelter, C. 1977. Public services, private substitutes, and the demand for protection against crime. *American Economic Review* 67, 867–77.

Cook, Philip J. 1975. The correctional carrot: better jobs for parolees. *Policy Analysis* 1, 11–55.

Cullen, J. and Levitt, S. 1999. Crime, urban flight, and the consequences for cities. *Review of Economics and Statistics* 81, 159–69.

Ehrlich, I. 1973. Participation in illegitimate activities: theoretical and empirical investigation. *Journal of Political Economy* 81, 521–65. Reprinted with supplements as 'Participation in Illegitimate Activities: An Economic Analysis', in Becker and Landes (1974).

Ehrlich, I. 1975. The deterrent effect of capital punishment: a question of life and death. *American Economic Review* 65, 397–417.

Ehrlich, I. 1979. The economic approach to crime. In *Criminology Review Yearbook*, ed. S. Messingerand and E. Bittner. Beverly Hills: Sage Publications.

Ehrlich, I. 1981. On the usefulness of controlling individuals: an economic analysis of rehabilitation, incapacitation and deterrence. *American Economic Review* 71, 307–22.

Ehrlich, I. 1982. The optimum enforcement of laws and the concept of justice: a positive analysis. *International Review of Law and Economics* 2, 3–27.

Ehrlich, I. 1996. Crime, punishment, and the market for offenses. *Journal of Economic Perspectives* 10(1), 43–67.

Ehrlich, I. and Brower, G. 1987. On the issue of causality in the economic model of crime and law enforcement: some theoretical considerations and experimental evidence. *American Economic Review, Papers and Proceedings* 77(2), 99–106.

Ehrlich, I. and Liu, Z. 2006. *The Economics of Crime*. International Library of Critical Writings in Economics. Cheltenham, UK and Northampton, MA: Edward Elgar.

Ehrlich, I. and Lui, F. 1999. Bureaucratic corruption and endogenous economic growth. *Journal of Political Economy* 107, S270–S293.

Friedman, D. 1999. Why not hang them all: the virtues of inefficient punishment. *Journal of Political Economy* 107 (6, Part 2), S259–S269.

Garoupa, N. 2000. The economics of organized crime and optimal law enforcement. *Economic Inquiry* 38, 278–88.

Glaeser, E., Sacerdote, B. and Scheinkman, J. 1996. Crime and social interactions. *Quarterly Journal of Economics* 111, 507–48.

Glaeser, E. and Sacerdote, B. 1999. Why is there more crime in cities? *Journal of Political Economy* 107(6, Part 2), S225–S258.

Imrohoroglu, A., Merlo, A. and Rupert, P. 2000. On the political economy of income redistribution and crime. *International Economic Review* 41, 1–25.

Karpoff, J. and Lott, J., Jr. 1993. The reputational penalty firms bear from committing criminal fraud. *Journal of Law and Economics* 36, 757–802.

Landes, W. 1971. An economic analysis of the courts. *Journal of Law and Economics* 14, 61–107.

Philipson, T. and Posner, R. 1996. The economic epidemiology of crime. *Journal of Law and Economics* 39, 405–33.

Polinsky, A. and Shavell, S. 1979. The optimal trade-off between the probability and magnitude of fines. *American Economic Review* 69, 880–91.

Reuter, P., MacCoun, R. and Murphy, P. 1990. *Money from Crime: A Study of the Economics of Drug Dealing in Washington, D.C.* The RAND Corporation, R-3894-RF.

Sah, R. 1991. Social osmosis and patterns of crime. *Journal of Political Economy* 99, 1272–95.

Schelling, T. 1967. Economic analysis of organized crime. Appendix D in *Task Force on Organized Crime: The President's Commission on Law Enforcement and Administration of Justice*. Washington, DC: US Government Printing Office.

Shavell, S. 1991. Individual precautions to prevent theft: private versus socially optimal behavior. *International Review of Law and Economics* 11, 123–32.

Stigler, G. 1970. The optimum enforcement of laws. *Journal of Political Economy* 78, 526–35.

Viscusi, W. 1986. The risks and rewards of criminal activity: a comprehensive test of criminal deterrence. *Journal of Labor Economics* 4, 317–40.

development accounting. *See* **level accounting**

development economics

What we know as the 'developing world' is approximately the group of countries classified by the World Bank as having 'low' and 'middle' incomes. An exact description is unnecessary and not too revealing; suffice it to observe that these countries make up over five billion of the world's population, leaving out the approximately one billion who are part of the 'high' income 'developed world'. Together, the low- and middle-income countries generate approximately six trillion (2001) dollars of national income, to be contrasted with the 25 trillion generated by high-income countries. An index of income that controls for purchasing power would place these latter numbers far closer together (approximately 20 trillion and 26 trillion, according to the *World Development*

Report, World Bank, 2003), but the per capita disparities are large and obvious, and to those encountering them for the first time, still extraordinary.

Development economics, a subject that studies the economics of the developing world, has made excellent use of economic theory, econometric methods, sociology, anthropology, political science, biology and demography, and has burgeoned into one of the liveliest areas of research in all the social sciences. My limited approach in this brief article is one of deliberate selection of a few conceptual points that I consider to be central to our thinking about the subject. The reader interested in a more comprehensive overview is advised to look elsewhere (for example, at Dasgupta, 1993; Hoff, Braverman and Stiglitz, 1993; Ray 1998; Bardhan and Udry, 1999; Mookherjee and Ray, 2001; Sen, 1999).

I begin with a traditional framework of development, one defined by conventional growth theory. This approach develops the hypothesis that given certain parameters, say savings or fertility rates, economies inevitably move towards a steady state. If these parameters are the same across economies, then in the long run all economies converge to one another. If in reality we see the utter lack of such convergence – which we do (see, for example, Quah, 1996; Pritchett, 1997) – then such an absence must be traced to a presumption that the parameters in question are *not* the same. To the extent that history plays any role at all in this view, it does so by affecting these parameters – savings, demographics, government interventionism, 'corruption' or 'culture'.

This view is problematic for reasons that I attempt to clarify below. Indeed, the bulk of this article is organized around the opposite presumption: that two societies with the same fundamentals can evolve along very different lines – going forward – depending on past expectations, aspirations or actual history.

To some extent, the distinction between evolution and parameter is a semantic one. By throwing enough state variables ('parameters') into the mix, one might argue that there is no difference at all between the two approaches. Formally, that would be correct, but then 'parameters' would have to be interpreted so broadly as to be of little explanatory value. Ahistorical convergence and historically conditioned divergence express two fundamentally different world views, and there is little that semantic jugglery can do to bring them together.

1 Development from the viewpoint of convergence

Why are some countries poor while others are rich? What explains the success stories of economic development, and how can we learn from the failures? How do we make sense of the enormous inequalities that we see, both within and across questions? These, among others, are the 'big questions' of economic development.

It is fair to say that the model of economic growth pioneered by Robert Solow (1956) has had a fundamental impact on 'big-question' development economics. An entire literature, including theory, calibration and empirical exercises, emanates from this starting point (see, for example, Lucas, 1990; Mankiw, Romer and Weil, 1992; Barro, 1991; Parente and Prescott, 2000; Banerjee and Duflo, 2005). Solow's path-breaking work introduced the notion of *convergence*: countries with a low endowment of capital in relation to labour will have a high rate of return to capital (by the 'law' of diminishing returns). Consequently, a given addition to the capital stock will have a larger impact on per capita income. It follows that, if we suitably control for parameters such as savings rates and population growth rates, poorer countries will tend to grow faster and hence will catch up or *converge* to the levels of well-being enjoyed by their richer counterparts. According to this view, development is largely a matter of getting some economic and demographic parameters right and then settling down to wait.

It is true that savings and demography are not the only factors that qualify the argument. Anything that systematically affects the marginal addition to per capita income must be controlled for, including variables such as investment in 'human capital' or harder-to-quantify factors such as 'political climate' or 'corruption'. A failure to observe convergence must be traced to one or another of these 'parameters'.

Convergence relies on diminishing returns to 'capital'. If this is our assumed starting point, the share of capital in national income gives us rough estimates of the concavity of production in capital. The problem is that the resulting concavity understates observed variation in cross-country income by orders of magnitude. For instance, Parente and Prescott (2000) calibrate a basic Cobb–Douglas production function by using reasonable estimates of the share of capital income (0.25), but then huge variations in the savings rate do not change world income by much. For instance, doubling the savings rate leads to a change in steady-state income by a factor of 1.25, which is inadequate to explain an observed range of around 20:1 (in purchasing-power-parity incomes). Indeed, as Lucas (1990) observes, the discrepancy actually appears in a more primitive way, at the level of the production function. For the same simple production function to fit the data on per capita income differences, a poor country would have to have enormously higher rates of return to capital; say, 60 times higher if it is one-fifteenth as rich. This is implausible. And so begins the hunt for other factors that might explain the difference. What did we not control for, but should have?

This describes the methodological approach. The convergence benchmark must be pitted against the empirical evidence on world income distributions, savings rates, or rates of return to capital. The two will usually fail to agree. Then we look for the parametric differences that will bridge the model to the data.

'Human capital' is often used as a first port of call: might differences here account for observed cross-country variation? The easiest way to slip differences in human capital into the Solow equations is to renormalize labour. Usually, this exercise does not take us very far. Depending on whether we conduct the Lucas exercise or the Prescott–Parente variant, we would still be predicting that the rate of return to capital is far higher in India than in the United States, or that per capita income differences are only around half as much (or less) as they truly are. The rest must be attributed to that familiar black box – 'technological differences'. That slot can be filled in a variety of ways: externalities arising from human capital, incomplete diffusion of technology, excessive government intervention, within-country misallocation of resources, and so on. All these – and more – are interesting candidates, but by now we have wandered far from the original convergence model; and if that model still continues to illuminate, it is by way of occasional return to the recalibration exercise, after choosing plausible specifications for each of these potential explanations.

This model serves as a quick and ready fix on the world, and it organizes a search for possible explanations. Taken with the appropriate quantity of salt, and viewed as a first pass, such an exercise can be immensely useful. Yet playing this game too seriously reveals a particular world view. It suggests a fundamental belief that the world economy is ultimately a great leveller, and that if the levelling is not taking place we must search for that explanation in parameters that are somehow structurally rooted in a society.

While the parameters identified in these calibration exercises go hand in hand with underdevelopment, so do bad nutrition, high mortality rates, or lack of access to sanitation, safe water and housing. Yet there is no ultimate causal chain: many of these features go hand in hand with low income in self-reinforcing interplay. By the same token, corruption, culture, procreation and politics are all up for serious cross-examination: just because 'cultural factors' (for instance) seems more weighty an 'explanation', that does not permit us to assign them the status of a truly exogenous variable. In other words, the convergence predicted by technologically diminishing returns to inputs should not blind us to the possibility of nonconvergent behaviour when all variables are treated as they should be – as variables that potentially make for underdevelopment, but also as variables that are profoundly affected by the development process.

2 Development from the viewpoint of nonconvergence

This leads to a different way of asking the big questions, one that is not grounded in any presumption of convergence. The starting point is that two economies with the same fundamentals can move apart along very different paths. Some of the best-known economists writing on development in the first half of the 20th century were instinctively drawn to this view: Young (1928); Nurkse

(1953); Leibenstein (1957); and Myrdal (1957) among them.

Historical legacies need not be limited to a nation's inheritance of capital stock or GDP from its ancestors. Factors as diverse as the distribution of economic or political power, legal structure, traditions, group reputations, colonial heritage and specific institutional settings may serve as initial conditions – with a long reach. Even the accumulated baggage of unfulfilled aspirations or depressed expectations may echo into the future. Factors that have received special attention in the literature include historical inequalities, the nature of colonial settlement, the character of early industry and agriculture, and early political institutions.

2.1 Expectations and development

Consider the role of expectations. Rosenstein-Rodan (1943) and Hirschman (1958) (and several others following them) argued that economic development could be thought of as a massive *coordination failure*, in which several investments do not occur simply because other complementary investments are similarly depressed in the same bootstrapped way. Thus one might conceive of two (or more) equilibria *under the very same fundamental conditions*, 'ranked' by different levels of investment.

Such 'ranked equilibria' rely on the presence of a *complementarity*, a particular form of externality in which the taking of an action by an agent increases the marginal benefit to other agents from taking a similar action. In the argument above, sector-specific investments lie at the heart of the complementarity: more investment in one sector raises the return to investment in some related sector.

Once complementarities – and their implications for equilibrium multiplicity – enter our way of thinking, they seem to pop up everywhere. Complementarities play a role in explaining how technological inefficiencies persist (David, 1985; Arthur, 1994), why financial depth is low (and growth volatile) in developing countries (Acemoglu and Zilibotti, 1997), how investments in physical and human capital may be depressed (Romer, 1986; Lucas, 1988), why corruption may be self-sustaining (Kingston, 2005; Emerson, 2006), the growth of cities (Henderson, 1988; Krugman, 1991), the suddenness of currency crises (Obstfeld, 1994), or the fertility transition (Munshi and Myaux, 2006); I could easily go on. Even the traditional Rosenstein-Rodan view of demand complementarities has been formally resurrected (Murphy, Shleifer and Vishny, 1989).

An important problem with theories of multiple equilibria is that they carry an unclear burden of history. Suppose, for instance, that an economy has been in a low-level investment trap for decades. Nothing in the theory prevents the very same economy from abruptly shooting into the high-level equilibrium today. There is a literature that studies how the past might weigh on the present when a multiple equilibria model is embedded in real time (see, for example, Adserà and Ray, 1998; Frankel and Pauzner, 2000). When we have a better knowledge of such models we will be able to make more sense of some classical issues, such as the debate on balanced versus unbalanced growth. Rosenstein-Rodan argued that a 'big push' – a large, balanced infusion of funds – is ideal for catapulting an economy away from a low-level equilibrium trap. Hirschman argued, in contrast, that certain 'leading sectors' should be given all the attention, the resulting imbalance in the economy provoking salubrious cycles of private investment in the complementary sectors. To my knowledge, we still lack good theories to examine such debates in a satisfactory way.

2.2 Aspirations, mindsets and development

The aspirations of a society are conditioned by its circumstances and history, but they also determine its future. There is scope, then, for a self-sustaining failure of aspirations and economic outcomes, just as there is for ever-progressive growth in them (Appadurai, 2004; Ray, 2006).

Typically, the aspirations of an individual are generated and conditioned by the experiences of others in her 'cognitive neighbourhood'. There may be several reasons for this: the use of role models, the importance of relative income, the transmission of information, or peer-determined setting of internal standards and goals. Such conditioning will affect numerous important socio-economic outcomes: the rate of savings, the decision to migrate, fertility choices, technology adoption, adherence to norms, the choice of ethnic or religious identity, the work ethic, or the strength of mutual insurance motives.

As an illustration, consider the notion of an *aspirations gap*. In a relatively narrow economic context (though there is no need to restrict oneself to this) such a gap is simply the difference between the standard of living that is aspired to and the standard of living that one already has. The former is not exogenous; it will depend on the ambient standards of living among peers or near-peers, or perhaps other communities.

The aspirations gap may be filled, or neglected, by deliberate action. Investments in education, health, or income-generating activities are obvious examples. Does history, via the creation of aspirations gaps, harden existing inequalities and generate poverty traps? Or does the existence of a gap spur individuals on ever harder to narrow the distance? As I have argued in Ray (1998, Sections 3.3.2 and 7.2.4) and Ray (2006), the effect could go either way. A small gap may encourage investments, a large gap stifle it. This leads not only to history-dependence, but also a potential theory of the connections between income inequality and the rate of growth.

These remarks are related to Duflo's (2006) more general (but less structured) hypothesis that 'being poor almost certainly affects the way people think and decide'. This 'mindset effect' can manifest itself in many ways (an aspirations gap being just one of them), and can lead to poverty traps. For instance, Duflo and Udry (2004) find

that certain within-family insurance opportunities seem to be inexplicably forgone. In broadly similar vein, Udry (1996) finds that men and women in the same household farm land in a way that is not Pareto-efficient (gains in efficiency are to be had by simply reallocating inputs to the women's plots). These observations suggest a theory of the poor household in which different sources of income are treated differently by members of the household, perhaps in the fear that this will affect threat points in some intra-household bargaining game. This in itself is perhaps not unusual, but the evidence suggests that poverty itself heightens the salience of such a framework.

2.3 Markets and history dependence

I now move on to other pathways for history dependence, beginning with the central role of inequality. According to this view, historic inequalities persist (or widen) because each individual entity – dynasty, region, country – is swept along in a self-perpetuating path of occupational choice, income, consumption and accumulation. The relatively poor may be limited in their ability to invest productively, both in themselves and in their children. Such investments might include both physical projects, such as starting a business, and 'human projects', such as nutrition, health and education. Or the poor may have ideas that they cannot profitably implement, because implementation requires start-up funds that they do not have. Yet, faced with a different level of initial inequality, or jolted by a one-time redistribution, the very same economy may perform very differently. The ability to make productive investments is now distributed more widely throughout the population, and a new outcome emerges with not just lower inequality, but higher aggregate income. These are different steady states, and they could well be driven by distant histories (see, for example, Dasgupta and Ray, 1986; Banerjee and Newman, 1993; Galor and Zeira, 1993; Ljungqvist 1993; Ray and Streufert, 1993; Piketty 1997; Matsuyama, 2000).

The intelligent layperson would be unimpressed by the originality of this argument. That the past systematically preys on the present is hardly rocket science. Yet theories based on convergence *would rule out such obvious arguments*. Under convergence, the very fact that the poor have limited capital in relation to labour allows them to grow faster and (ultimately) to catch up. Economists are so used to the convergence mechanism that they sometimes do not appreciate just how unintuitive it is.

That said, it is time now to cross-examine our intelligent layperson. For instance, if all individuals have access to a well-functioning capital market, they should be able to make an efficient economic choice with no heed to their starting position, and the shadows cast by past inequalities must disappear (or at least dramatically shrink). For past wealth to alter current investments, imperfections in capital or insurance markets must play a central role.

At the same time, such imperfections are not sufficient: the concavity of investment returns would still guarantee convergence. A first response is that 'production functions' are simply not concave. A variety of investment activities have substantial fixed costs: business start-ups, nutritional or health investments, educational choices, migration decisions, crop adoptions. Indeed, it is hard to see how the presence of such non-convexities could *not* be salient for the ultra poor. Coupled with missing capital markets, it is easy to see that steady state traps, in which poverty breeds poverty, are a natural outcome (see, for example, Majumdar and Mitra, 1982; Galor and Zeira, 1993). Surveys of the economic conditions of the poor (Fields, 1980; Banerjee and Duflo, 2007) are eminently consistent with this point of view.

A related source of non-convexity arises from limited liability. A highly indebted economic agent may have little incentive to invest. Similarly, poor agents may enter into contracts with explicit or implicit lower bounds on liability. These bounds can create poverty traps (Mookherjee and Ray, 2002a).

Investment activities that go past these minimal thresholds are potentially open to 'convexification'. There are various stopping points for human capital acquisition, and a household can hold financial assets which are, in the end, scaled-down claims on other businesses. According to this point of view, dynasties that make it past the ultra-poor thresholds will exhibit ergodic behaviour (as in Loury, 1981; Becker and Tomes, 1986) and so the prediction is roughly that of a two-class society: the ultra poor are caught in a poverty trap and the remainder enjoy the benefits of convergence. History would matter in determining the steady-state proportions of the ultra-poor.

But this sort of analysis ignores the endogenous non-convexities brought about by the price system. For instance, even if there are many different education levels, the wage payoff to each such level will generally be determined by the market. There is good reason to argue (see, for example, Ljungqvist, 1993; Freeman, 1996; Mookherjee and Ray, 2002b; 2003) that the price system will sort individuals into different occupational choices, and that there will be persistent inequality across dynasties located at each of these occupational slots. Thus an augmented theory of history dependence might predict a particular proportion of the ultra-poor trapped by physical non-convexities (low nutrition, ill-health, debt, lack of access to primary education), as well as a persistently unequal dispersion of dynasties across different occupational choices, induced by the pecuniary externalities of relative prices.

Note that it is precisely the high-inequality, high-poverty steady states that are correlated with low average incomes for society as a whole, and it is certainly possible to build a view of underdevelopment from this basic premise. The argument can be bolstered by consideration of economy-wide externalities; for instance, in physical and human capital (Romer, 1986; Lucas, 1988; Azariadis and Drazen, 1990).

2.4 History, aggregates and the interactive world

Theories such as these might yield a useful model for the interactive world economy. Take, for instance, the notion of aspirations. Just as domestic aspirations drive the dynamics of accumulation *within* countries, there is a role, too, for national aspirations that are driven by inter-country disparities in consumption and wealth, with implications for the international distribution of income. Even the simplest growth framework that exhibits the usual features of convexity in its technology and budget constraints could give rise in the end to a bipolar world distribution. Countries in the middle of that distribution would tend to accumulate faster, be more dynamic and take more risks as they see the possibility of full catch-up within a generation or less. One might expect the greatest degree of 'country mobility' in this range. In contrast, societies that are far away from the economic frontier may see economic growth as too limited and too long-term an instrument, leading to a failure, as it were, of 'international aspirations'. Groups within these societies may well resort to other methods of potential economic gain, such as rent-seeking or conflict. (The aggregate impact of such activities would reinforce the slide.)

Of course, an entirely mechanical transplantation of the aspirations model to an international context is not a good idea. Countries are not individual units: a more complete theory must take into account the aspirations of various groups in the different countries, and the domestic and international components that drive such aspirations.

Next, consider the role of markets. Once again, tentatively view each country as a single economic agent in the framework of Section 2.3. The non-convexities to be considered are at the level of the country as a whole – Young's increasing returns on a grand scale, or economy-wide externalities as in Lucas–Azariadis–Drazen. This reinterpretation is fairly standard, but, less obviously, the occupational choice story stands up to reinterpretation as well. To see this, note that the pattern of production and trade in the world economy will be driven by patterns of comparative advantage across countries. But in a dynamic framework, barring non-reproducible resources such as land or mineral endowments, *every* endowment is potentially accumulable, so that comparative advantage becomes endogenous. Thus we may view countries as settling into subsets of occupational slots (broadly conceived), producing an incomplete range of goods and services in relation to the world list, and engaging in trade.

For instance, suppose that country-level infrastructure can be tailored to either high-tech or low-tech production, but not both. If both high-tech and low-tech are important in world production and consumption, then *one* country has to focus on low-tech and *another* on high-tech. Initial history will constrain such choices, if for no other reason than the fact that existing infrastructure (and national wealth) determines the selection of future infrastructure. This is not to say that no country can break free of those shackles. For instance, as the whole world climbs up the income scale, natural non-homotheticities in demand will push commodity compositions increasingly in favour of high-quality goods. As this happens, more countries will be able to make the transition. But on the whole, if national infrastructure is more or less conducive to some (but not the full) range of goods, the non-convergence model that we discussed for the domestic economy must apply to the world economy as well.

This raises an obvious question: what is so specific about 'national infrastructure'? Why is it not possible for the world to ultimately rearrange itself so that every country produces the same or similar mix of goods, thus guaranteeing convergence? Do current national advantages somehow manifest themselves in future advantages as well, thus ensuring that the world economy settles into a permanent state of global inequality? Might economic underdevelopment across countries, at least in this relative sense, always stay with us?

To properly address such questions we have to drop the tentative assumption that each country can be viewed as an individual unit. In a more general setting, there are individuals within countries, and then there is cross-country interaction. The former are subject to the forces of occupational structure (and possible fixed costs), as discussed in Section 2.3. The latter are subject to the specificities, if any, of 'national infrastructure', determining whether countries as a whole have to specialize (at least to some degree). The relative importance of within-country versus cross-country inequalities will rest, in large part, on considerations such as these.

I have not brought in international political economy so far (though see below); yet, as frameworks go, this is not a bad one to start thinking about the effects of globalization. It is certainly preferable to a view of the world as a set of disconnected, autarkic growth models.

2.5 Institutions and history

In many developing countries, the early institutions of colonial rule were directly set up for the purposes of surplus extraction. There would be variation, of course, depending on whether the areas were sparsely or densely populated to begin with, or whether there was widespread availability of mineral deposits. Resource deposits certainly favoured large-scale extractive industry (as in parts of South America), while soil and weather conditions might encourage plantation agriculture, often with the use of slave labour (as in the Caribbean). On the other hand, a high pre-existing population density would favour extraction of a different hue: the setting up of institutional systems to acquire rents (the British colonial approach in large parts of India).

It has been argued, perhaps most eloquently by Sokoloff and Engerman (2000), that initial institutional modes of production and extraction in distant history had far-reaching effects on subsequent development. In their words, scholars 'have begun to explore the possibility that initial conditions, or factor endowments broadly conceived, could have had profound and enduring impacts on long-run paths of institutional and economic development' (2000, p. 220). The inequalities generated by such initial conditions may subsequently be inimical to development in a variety of ways (via the market-based pathways discussed earlier, for instance). In contrast, where initial settlements did not go hand in hand with systems of tribute, land grants, or large-scale extractive industries (as in several regions of North America), one might expect broad-based development to occur.

This is consistent with the market-based processes considered earlier. But a principal strand of the Sokoloff–Engerman argument, as also the lines of reasoning pursued in Robinson (1998), Acemoglu, Johnson and Robinson (2001; 2002) and Acemoglu (2006), emphasizes political economy. In the words of Sokoloff and Engerman,

> [I]nitial conditions had lingering effects ... because government policies and other institutions tended to reproduce them. Specifically, in those societies that began with extreme inequality, elites were better able to establish a legal framework that insured them disproportionate shares of political power, and to use that greater influence to establish rules, laws, and other government policies that advantaged members of the elite relative to nonmembers contributing to persistence over time of the high degree of inequality ... In societies that began with greater equality or homogeneity among the population, however, efforts by elites to institutionalize an unequal distribution of political power were relatively unsuccessful ... (Sokoloff and Engerman, 2000, p. 223–4)

The elite – erstwhile collectors of tribute, land-grant recipients, plantation owners and the like – may survive long after the initial institutions that spawned them are gone. Such survival may nevertheless be compatible with the maximization of aggregate surplus, provided that the elite are the most efficient of the economic citizenry in the generations to come. But there is absolutely no reason why this should be the case. A new generation of entrepreneurs, economic and political, may be waiting to take over in the wings. It is an open question as to what will happen next, but the elite may well engage in policy that has as its goal not economic efficiency but the crippling of political opposition. Some evidence of this reluctance to let go may be seen in literature that argues that more unequal societies redistribute less (see Perotti, 1994; 1996; the survey by Bénabou, 1996).

There are other routes. The elite may be unable to avoid an oppositional showdown. A theory of bad policy may then have to be replaced by a model of social unrest and conflict generated by initial inequality. While this mechanism is clearly different, the end result is the same. The channelling of resources to ongoing conflict will surely inhibit the accumulation of productive resources (Benhabib and Rustichini, 1996; González, 2007). There may also be effects running through legal systems (see, for example, La Porta et al., 1997; 1998) or the varying nature of different colonial systems (see, for example, Bertocchi and Canova, 2002). There may be effects running through the insecurity of property rights or fear of elite expropriation (see, for example, Binswanger, Deininger and Feder, 1995).

We do not yet have a systematic exploration of these mechanisms, nor an accounting of their relative importance. But there is some reduced-form evidence that historical institutions affect growth in the manner described by Sokoloff and Engerman. The problem in establishing an empirical assertion of this sort is fairly obvious: good institutions and good economic outcomes may simply be correlated via variables we fail to observe or measure, or any observed causality may simply run from outcomes to institutions. Acemoglu, Robinson and Johnson (2001) propose a novel instrument for (bad) institutions: the mortality rate among European settlers (bishops, sailors and soldiers to be exact). This is a clever idea that exploits the following theory: only areas that could be settled by the Europeans developed egalitarian, broad-based institutions. In the other areas, the same Europeans settled for slavery, dictatorship, highly unequal land grants and unbridled extraction instead. (The implied instrument is more convincing when the analysis is combined with controls for the general disease environment, which could have a direct effect on performance.)

The Acemoglu–Johnson–Robinson results, which show that early institutions have an effect on current performance, are provocative and interesting. It bears reiteration, though, that IV estimates are suggestive of an institutional impact on development, but one just cannot be sure of what the mechanism is. By relinquishing more immediate institutional effects on the grounds of, say, endogeneity, it becomes much harder to identify the structural pathways of influence. This appears to be an endemic problem with large, sweeping cross-country studies that attempt to detect an institutional effect. Good instruments are hard to find, and when they exist, their effect could be the echo of one or more of a diversity of underlying mechanisms.

Iyer (2005) and Banerjee and Iyer (2005) consider a somewhat different channel of influence. Both these papers study the differential impact of colonial rule within a single country, India. Iyer studies British annexations of parts of India, and the effect today on public goods provision across annexed and non-annexed parts. There is obvious endogeneity in the areas chosen for annexation (a similar observation applies, in passing, to

countries 'selected' for colonization). Iyer instruments annexation by exploiting the so-called Doctrine of Lapse, under which the British annexed states in which a native ruler died without a biological heir. Banerjee and Iyer study the effect of variations in the land revenue systems set up by the British, starting from the latter half of the 18th century. In particular, they distinguish between landlord-based institutions, in which large landlords were used to syphon surplus to the British, and other areas based on rent payments, either directly from the cultivator or via village bodies. While these institutions of extraction no longer exist (India has no agricultural income tax), the authors argue that divided, unequal areas in the past cannot come together for collective action. Dispossessed groups are more worried about insecurity of tenure and fear of expropriation than about the absence of public goods, investment (public or private) or development expenditure.

2.6 Institutions and the interactive world

In Section 2.4, we applied market-based theories of occupational choice and persistent inequality to the interactive world economy, (tentatively) treating each country as an economic agent. Recall the main assumption for such an interpretation to be sensible: that countries must face infrastructural constraints that limit full diversification. With these constraints in place, there will be persistent inequality in the world income distribution, with countries in 'occupational niches' that correspond to their infrastructural choices.

Bring to this story the role of institutional origins. Then a particular institutional history may be more suited to particular subsets of occupations, driving the country in question into a determinate slot in the world economy. From that point on, the persistent cross-country inequalities generated by the market-based theory will continue to link past institutions to subsequent growth. In short, initial institutional differences may be correlated with subsequent performance, but the *magnitude* of that under- or over-performance is not to be entirely traced to initial history. Distant history could simply have served as a marker for some countries to supply a particular range of occupations, goods and services. Today's inequality may well be driven, not by that far-away history but simply by the world equilibrium path that follows on those initial conditions. If all goods are needed, there must be banana producers, sugar manufacturers, coffee growers, and high-tech enclaves, but there cannot be too little or too many of any of them.

The 'inefficient political power' argument used in Section 2.5 can also be transplanted to international interactions. It may well be that a large part of such interactions – protection of international property rights, restrictions on technology transfer, or barriers to trade – is used to deter the entry of developing countries onto a level playing field in which they can successfully compete with their compatriots in developed countries. It would certainly be naive to disregard this point of view altogether.

Looked at this way, our view of history fits in well with the entire debate on globalization. One might view one side of this debate as emphasizing the convergence attributes of globalization: outsourcing, the establishment of international production standards, technology transfer, political accountability and responsible macroeconomic policies may all be invoked as foot soldiers in the service of convergence.

On the other side of the battle lines are equally formidable opponents. A skewed playing field can only keep tipping, so goes the argument. The protection of intellectual property is just a way of maintaining or widening existing gaps in knowledge. Technology transfers are inappropriate because the input mix is not right. Non-convexities and increasing returns are endemic.

My goal here is not to take sides on this debate (though like everyone, I do have an opinion) but to clarify it from a 'non-convergence perspective' that has so far received more attention within the closed economy. There is a strong parallel between globalization (and those contented or discontented with it, to borrow a phrase from Joseph Stiglitz, 2002) and the questions of convergence and divergence in closed economies.

3 Digging deeper: the microeconomics of development

There is no getting away from the big questions, even if they cannot be fully answered with the knowledge and tools we have to hand. The issues we have discussed (and our intuitive first-takes on them) determine our world view, the cognitive canvas on which we arrange our overall thoughts. But only the most hard-bitten macroeconomist would feel no trepidation about taking these models literally, and applying them without hesitation across countries, regions and cultures.

The microeconomics of development enables us to dig below the macro questions, unearthing insight and structure with far more confidence than we can hope to have at the world or cross-country level. From the viewpoint of economic theory, the assumptions made can be more carefully motivated and are open to careful testing. From the viewpoint of empirical analysis, it is far easier to find instruments or natural experiments, or, for that matter, to conduct one's own experiments. There is the philosophical problem of scaling up the results, of using a well-controlled finding to predict outcomes elsewhere. In the end, the choice between the fuzzy, imprecise big picture and the small yet carefully delineated canvas is perhaps a matter of taste.

I need hardly add that my selectivity continues unabated: there is a whole host of issues, and I can but touch on a fraction of them. I focus deliberately on four important topics that are relevant to my overall theme of history dependence, and that have been the subject of much recent attention.

3.1 The credit market

As we have seen, a failure of the credit market to function is at the heart of market-based arguments for divergence.

The fundamental reason for imperfect or missing credit markets is that individuals cannot be counted upon (for reasons of strategy or luck) to fully repay their loans. If borrowers do not have deep pockets, or if a well-defined system for enforcing repayment is missing, then it stands to reason that lenders would be reluctant to advance those loans in the first place. There is little point in asserting that a carefully chosen risk premium will deal with these risks: the premium itself affects the default probability. Therefore some borrowers will be shut out of the market, *no matter what rate of interest they are willing to pay*. Such a market will typically clear by rationing access to credit, and not by an adjustment of the rate of interest.

Three fundamental features characterize different theories of imperfect credit markets. There is classical adverse selection, in which borrower (or project) characteristics may systematically adjust with the terms of the loan contract on offer. Stiglitz and Weiss (1981) initiate this literature for credit markets, arguing that the higher the interest rate, the more likely it is that the borrower pool will be contaminated by riskier types. Then there is the moral hazard problem (see, for example, Aghion and Bolton, 1997), in which the borrower must expend effort *ex post* to increase the chances of project success. Moral hazard also ties into 'debt overhang', in which existing indebtedness makes it less credible that a borrower will put in sustained effort in the project. Finally, there is the enforcement problem (see, for example, Eaton and Gersovitz, 1981), in which a borrower may be tempted to engage in strategic default. Ghosh, Mookherjee and Ray (2001) survey some of the literature.

The poor are particularly affected, not because they are intrinsically less trustworthy, but because in the event of a project failure they will not have the deep pockets to pay up. The poor may well possess collateral – a small plot of land or their labour – but such collateral may be hard to adequately monetize: a formal sector bank may be unwilling to accept a small rural plot as collateral, much less bonded labour; but other lenders (a rural landlord, for instance) might. It is therefore not surprising to see interlinkages in credit transactions for the poor: a small farmer is likely to borrow from a trader who trades his crop, while a rural tenant is likely to borrow from his landlord. Even when the entire market looks competitive, these niches may create pockets of exploitative local monopoly (Ray and Sengupta, 1989; Floro and Yotopoulos, 1991; Floro and Ray, 1997; Mansuri, 1997; Genicot, 2002).

In short, the very fact of their limited wealth puts the relatively poor under additional constraints in the credit market. This is why imperfect capital markets serve as a starting point for many of the models that study market-based history dependence.

The direct empirical evidence on the existence of credit constraints is surprisingly sparse, which is obviously not to say that they do not exist, but only to point out that this is an area for future research. Existing literature in a development context largely uses the existence of (presumably undesirable) consumption fluctuations in households to infer the lack of perfect financial markets (see Morduch, 1994; Townsend, 1995; Deaton, 1997). A direct test for credit constraints yields positive results for Indian firms (Banerjee and Duflo, 2004), though it is unclear how general this finding is (see, for example, Hurst and Lusardi, 2004). There is a sizeable literature dealing with the impact of credit constraints on outcomes such as health (Foster, 1995), education (Jacoby and Skoufias, 1997) or the acquisition of production inputs such as bullocks (Rosenzweig and Wolpin, 1993).

Chiappori and Salanie (2000) and Karlan and Zinman (2006) are two examples of specific tests for different frictions, such as adverse selection and enforcement. Udry's seminal (1994) paper on credit and insurance markets in northern Nigeria may be viewed as singling out enforcement as perhaps the most important binding constraint. The importance of enforcement constraints is, of course, not peculiar to credit or insurance; Fafchamps (2004) develops the point for a variety of markets in sub-Saharan Africa. For more on insurance, see Townsend (1993; 1995); Ligon (1998); Fafchamps (2003); and Fafchamps and Lund (2003). Coate and Ravallion (1993), Ligon, Thomas and Worrall (2002), Kocherlakota (1996) and Genicot and Ray (2003) develop some of the associated theory with limited enforcement.

Finally, there is a literature on *micro-credit*, the lending of relatively small amounts to the very poor; Armendáriz and Morduch (2005) is a good starting point.

3.2 Collective action for public goods

There is a growing literature on the political economy of development. Unlike some mainstream approaches in political science and political economy, this literature appears to largely eschew voting models. In my view this is not a bad thing. Perhaps the most important criticism of voting models is that even in vigorous democracies, most policies are not subject to referenda among the citizenry at large. Certainly, there are periodic elections, and the sum total of enacted policies – and the package of future promises – are then up for voter scrutiny, but, nevertheless, there is a large and significant gap between voting and the enactment of a *particular* policy. Between that policy and the voter falls the shadow of collective action, lobbies, capture and influence, cynical trade-offs across special interests, and covert or open conflict. For countries with a non-democratic history, these considerations are expanded by orders of magnitude.

An important literature concerns the determinants of collective action for the provision of public goods, and how poverty or inequality affects the ability to engage in such action. The relationship here is complex. There are two potential reasons why inequality in a community may enhance collective action. First, the elite in a high-inequality community might largely internalize all the benefits from the resulting public good, and therefore pay for it (Olson, 1965). Good examples involve military alliances (Sandler and Forbes, 1980), technology adoption (Foster and Rosenzweig, 1995) or even 'top-down interventions' by local rulers or elites (Banerjee, Iyer and Somanathan, 2007). Second, the elite has a low opportunity cost of money, while the poor have a low opportunity cost of labour; in some situations, the two resources can be usefully combined for collective action (an alliance for violent conflict, as in Esteban and Ray, 2007a, is a good example). But there are many situations in which inequality can dampen effective collective action: when all agents supply similar inputs – say effort – but their impact or cost of provision is nonlinear (Khwaja, 2004; Ray, Baland and Dagnielie, 2007), when there are unequally distributed private endowments (Baland and Platteau, 1998; Bardhan, Ghatak and Karaivanov, 2006), when different individuals in the same community want different things by virtue of their social differences or inequality (Alesina, Baqir and Easterly, 1999; Banerjee et al., 2001, Miguel and Gugerty, 2005; Alesina and La Ferrara, 2005), or when inequalities in wealth erode the informational basis of collective action (Esteban and Ray, 2006).

The importance of this area of research cannot be overemphasized. Several of the fundamental accompaniments of development require state intervention at a basic level: health, education, social safety nets and infrastructure. This is especially so in poor countries, where privatized health and education are often ruled out by the sheer force of economic necessity. Yet states often are set upon by numerous claims that compete for their attention. How are these claims resolved? The theory and practice of collective action demands more research.

Moreover, while it can be argued (as above) that inequality within a community might go either way in affecting that community's ability to obtain public goods, there is no escaping the fact that at the level of the entire society, high inequality serves to fracture and divide. Simply put, the very rich want state policy that is different from what the very poor desire, and rare is the society that has them in the same camp, and demanding the same things of their government. In the world of the median voter, one might simply resolve these issues by looking at the median voter's ideal policy, but even in this rarefied scenario there are complex issues that deserve our consideration. Political alliances can often redefine the median voter (Levy, 2004) and even without alliances it is unclear just who the median voter is (Bénabou, 2000). When we return to the 'real world' of collective action, these issues are magnified considerably. In that world, each citizen does not have an endowment of one vote. The real endowments are labour and money. How these commodities combine (or compete) is fundamental to our understanding of political economy and – via this channel – our views on persistent history-dependence.

3.3 Conflict

A more sinister expression of collective action is conflict. In the second half of the 20th century and well into the first decade of the 21st the loss of human life from conflicts in developing countries was immense; the costs are beyond measurement. Even the narrow economic costs of conflict can be extremely large (Hess, 2003).

That conflict contributes to economic regress is not surprising. But given our focus on history dependence, it is of equal interest to consider the causal chain running from underdevelopment to conflict. That chain has a natural and simple foundation: poverty reduces the opportunity cost of engaging in conflict. The grabbing of resources, often in an organized way, is often a far more lucrative alternative to the steady process of wealth accumulation. It is certainly a quicker alternative. (One might argue that there is less to gain as well, but this effect is attenuated in unequal societies.)

This unfortunate observation has substantial empirical support. For instance, Miguel, Satyanath and Sergenti (2004) use rainfall as an instrument for economic growth in 41 African countries and derive a striking negative effect of growth on civil conflict: a negative growth shock of five percentage points raises the likelihood of civil conflict by 50 per cent; see also Dube and Vargas (2006) and Hidalgo et al. (2007), both of which also instrument for economic shocks to find significant effects on conflictual outcomes. Collier and Hoeffler (1998), Sambanis (2001), Fearon and Laitin (2003), and Do and Iyer (2006) all establish strong correlations between economic adversity and conflict, the last of these countries establishing this over regions in a single country (Nepal).

Yet conflict is demonstrably wasteful, and if warring parties could sit down at the negotiating table, why would societies engage in it? This is a classical question to which there are a number of possible answers. First, there may be a Prisoner's Dilemma-like quality to conflictual incidents, in the sense that one party can precipitate attacks while the other remains passive (Leventoglu and Slantchev, 2005). Second, while conflict generates waste, there is no reason to believe that *every* group is thereby made worse off by it. It is entirely possible that a group prefers conflict to a peaceful outcome: the former involves a smaller pie, but the group may obtain a larger share of it (Esteban and Ray, 2001). Third, while one should be able to find a system of taxes and transfers that Pareto-dominate the conflict outcome, for various reasons – lack of commitment, a sparse informational base

for the levying of taxes, dynamics with rapid power shifts – it may not be possible to implement that system (Fearon, 1995; Powell, 2004; 2006). Fourth, it is certainly possible that conflict is over indivisible resources such as political power or religious hegemony. It may then be absurd to imagine that side A compensates side B with suitable transfers in exchange for political power: the lack of credibility involved is only too apparent. Finally, conflict may be endemic because both parties to it have incomplete information regarding chances of success, though this view has come under increasing criticism from political scientists (see, for example, Fearon, 1995).

The next question of relevance concerns ethnic and social divisions. Might the presence of potentially divisive markers (caste, religion, geography, ethnicity in general) exacerbate conflictual situations? For instance, Esteban and Ray (2007a) argue that non-economic ('ethnic') markers may play a salient role in the outbreak of conflict even when society exhibits high economic inequality and may look prima facie more ripe for a class war.

A standard tool for measuring ethnic and social divisions is that of *fractionalization*, roughly defined as the probability that two individuals drawn at random will come from two distinct groups. While fractionalization seems to have a negative effect on economic outcomes such as per capita GDP (Alesina et al., 2003), growth (Easterly and Levine, 1997), or governance (Mauro, 1995), its effect on civil conflict appears to be insignificant (Collier and Hoeffler, 2004; Fearon and Laitin, 2003). Of course, as Horowitz (2000) and others have observed, it is the presence of large cleavages that is potentially conflictual, whereas fractionalization continues to increase with diversity. The solution is to drop fractionalization altogether. Montalvo and Reynal-Querol (2005) adapt Esteban and Ray's (1994) measure of polarization to show that measures of ethnic and religious polarization *do* indeed have a significant impact on conflict (see also Do and Iyer, 2006). Obviously, more research is called for on questions such as these. For instance, it is unclear how polarization should enter an empirical specification: Esteban and Ray (2007b) argue that highly polarized societies may actually avoid a showdown through deterrence, though *conditional* on the outbreak of conflict, polarization must vary positively with the intensity of conflict.

The continuing study of conflict in development demands our highest priority. Certainly, the social waste of conflict dominates the inefficiency of misallocated resources that so many mainstream economists prefer to emphasize. Indeed, it is entirely possible that the much-maligned (and much-studied) inefficiencies of incomplete information are also of a lower order of magnitude. But, most of all, it is the chain of cumulative causation that must ultimately drive our interest, from underdevelopment to conflict and back again to continuing underdevelopment. Conflict is one channel through which history matters.

3.4 Legal matters

Contract enforcement, property rights, and expropriation risks: these are a few instances of legal matters that are central to development. They bear closely on that much-used catchall phrase, 'institutional effects on development'. For instance, Acemoglu, Robinson and Johnson (2001) as well as the recent survey by Pande and Udry (2007) clearly have the security of property rights high on the list when discussing 'institutions'. La Porta et al. (1997; 1998; 2002) and Djankov et al. (2003) begin with the premise that common (English commercial) law and civil (French commercial) law afford different degrees of protection and support to investors, creditors and litigants, and argue that it has had dramatic effects on a variety of indicators across countries: corruption, stock-market participation, corporate valuation, government interventionism, judicial efficiency – and presumably, via these, to economic indicators.

It is little surprise that the security of property rights is generally conducive to investment, and that long-term investment is especially encouraged by such security (see, for example, Demsetz, 1967). Short-term efforts, in contrast, may well be enhanced by insecurity of tenure. Depending on the exact form that property rights assume, there may be further positive effects – for example, via access to credit – that arise from the ability to mortgage or sell property (Feder et al., 1988).

Empirical research into these matters is invariably assailed by questions of endogeneity and omitted variables. For instance, long-gestation investments may provoke – and permit – the establishment of property rights, and high-ability agents might use their ability to both invest and secure their rights. Nevertheless, the evidence on property rights is that by and large they are good for investment and production (Besley, 1995; Banerjee, Gertler and Ghatak, 2002; Do and Iyer, 2003; Goldstein and Udry, 2005), and even more obviously, property values where these are reasonably well-defined (Alston, Libecap and Schneider, 1996; Lanjouw and Levy, 2002). Instances in which property titling creates better access to credit are, intriguingly enough, somewhat harder to come by (Field and Torero, 2006, and Dower and Potamites, 2006, are two of the rarer examples that do document better access, but with some qualifications).

Indeed, economists have little trouble in finding numerous instances of changed (or changing) property rights regimes. This is because there is a plethora of situations in which the absence of well-defined rights is the rule rather than the exception. In rural societies the world over, land rights can be highly ambiguous, and land titles can be missing even when an unambiguous definition of property exists. If one adds to this the sizable proportion of land under tenancy, the effective security for cultivators becomes more tenuous still (and indeed this complicates matters, because their rights may be inversely related to those of the owner!). In non-rural settings, there are substantial uncertainties for those who operate in the informal sector (such as the periodic 'cleansing' of informal retailers from city pavements). If the above studies are to be taken seriously, there are substantial production losses from such states of insecurity.

If imperfections of the law are so inimical to the fortunes of cultivators and producers (and especially for the small and the poor among them), why do we see such institutional 'failures' in equilibrium? The Coase–Posner view would presumably have none of this: in their view, legal systems would invariably develop to maximize social surplus. But of course, there could be several reasons for the persistence of 'inefficient institutions'. When side payments are not feasible or credible, economic agents often prefer a larger share of a reduced pie to a smaller share of a more efficient pie. For instance, domestic businesses that can rely on a trusted network of kin or extended family might prefer an ambiguous legal system, which prevents entry. Or workers might prefer imperfect enforceability of a work norm, so that efficiency wages need to be paid. Borrowers might prefer that loan repayment cannot be fully enforced, so that incentives to repay must be built into the loan contract. And when tenancy is widespread in agriculture, the very design of overall property rights to maximize efficiency can be a highly complex problem.

The last three examples possess another feature that is worth some emphasis: ambiguous property rights often have equity effects that do not go the same way that efficiency-minded economists would like them to go (see Weitzman, 1974; Cohen and Weitzman, 1975; Baland and Platteau, 1996). The ambiguity of property rights can serve as insurance, buffer, or redistributive device. As examples, consider broad access to water resources or grazing land, or the efficiency-wage premia that may need to be paid to workers or borrowers.

Most importantly, the ambiguity of property rights slows down the emergence of an overt assetless class, and that has its own social value (it should not be forgotten that the flip side of unambiguous rights is exclusion). For example, Goldstein and Udry (2005) develop this point of view in the context of rural Ghana, arguing that the ambiguity in property rights prevented the outbreak of extreme poverty (and had an interesting efficiency effect in the bargain, as individuals were reluctant to leave the land fallow – an important investment – in the fear that this would signal a lack of need for land).

The political economy of rights is a messy business, but of central importance in development economics. Poverty in general enhances the social and political need for ambiguity, while to the extent that such ambiguity wears on efficiency, we have an extremely important instance of non-convergence. Sometimes such non-convergence assumes particularly dramatic form. In West Bengal (India) 'Operation Barga' provided widespread – and welcome – use rights to registered sharecroppers (see, for example, Banerjee, Gertler and Ghatak, 2001). Those very use rights now lie at the heart of recent difficulties in converting agricultural land in India for use in industry. In the world of the second best, few policies have unambiguously one-directional effects.

4 A concluding note: theory and empirics

While I have tried to provide a conceptual overview in this article, recent research in development economics has been almost entirely empirical. A veritable explosion in computing power, the expansion of institutional data-sets and their increased availability in electronic form, and the growing ease of collecting one's own data have bred a new generation of development economists. Their empirical sensibilities are of a high order; they are extremely sensitive to issues of endogeneity, omitted variables, measurement error and biases induced by selection. They are constantly on the search for good instruments or natural experiments, and, when these are hard to find, they are adept at creating experiments of their own.

There is little doubt that we know little enough about the world we live in that it is often worth finding out the simple things, rather than continuing to engage in what some would term flights of theoretical fantasy. Are people really credit-rationed? Does rising income automatically make for better nutrition and health? If we had the option to throw in more textbooks, or reduce class size, or add more teachers, or install monitoring devices to track teacher attendance, which policy should we implement? Do women leaders behave differently from men in the policies that they adopt? Do households behave as one frictionless unit? Or, if one is the big-picture sort, have countries indeed converged over the last 200, or 500, years? Are richer countries more democratic? How many excess female deaths have occurred in China or India because of gender bias? Are poorer countries more 'corrupt'? And so on. The list is practically endless.

The somewhat churlish theoretically minded economist might ask, why are well-trained statisticians unable to answer these questions? Why do we need economists, who are supposed, at the very least, to combine two observations to form a deduction? The answer, at one level, is very simple and not overly supportive of the churlish theorist's complaint. While the questions are straightforward, the answers are often extremely difficult to tease out from the data, and one needs a well-trained *economist*, not a statistician, to understand the difficulty and eliminate it. Because of the aforementioned econometric issues, not a single one of the questions asked above admits a straightforward answer. Development economists spend a lot of time thinking of inventive ways to get around these problems, and it is no small feat of creativity, dedication and extremely hard work to pull off a convincing solution.

It is true that the very desire to obtain a clean, unarguable answer – with its attendant desire to have control over the empirical environment – sometimes narrows the scope of the enquiry. There is often great reluctance to rely on theoretical structure (for such reliance would contaminate the near-lexicographic desire for an unambiguous result). This means that the question to be asked is often akin to that for a simple production function (for example, 'do students do better in exams if they are given more textbooks?') or is focused on the direct effect of some policy intervention ('does the provision of health check-ups improve health outcomes?'). So it is that a boring but well-identified empirical question will often be treated with a great deal more veneration (especially if a clever instrument or randomization device is involved) than a model that relies on intuitive but undocumented assumptions.

That said, it is also a fact that we know very little about the answers to some of the most basic questions, such as the ones we have listed above. The great contribution of empirical development microeconomics is that we are building up this knowledge, piece by piece. Whether the search for that knowledge is informed by theory or not, there will be enough theorists to attempt to put these observations together. There will be enough empirical researchers to keep generating the hard knowledge. Development economics is alive and well.

DEBRAJ RAY

See also **agriculture and economic development; dual economies; emerging markets; endogenous growth theory; growth and institutions; poverty traps.**

Bibliography

Acemoglu, D. and Zilibotti, F. 1997. Was Prometheus unbound by chance? Risk, diversification and growth. *Journal of Political Economy* 105, 709–51.

Acemoglu, D., Johnson, S. and Robinson, J. 2001. The colonial origins of comparative development: an empirical investigation. *American Economic Review* 91, 1369–401.

Acemoglu, D., Johnson, S. and Robinson, J. 2002. Reversal of fortune: geography and institutions in the making of the modern world income distribution. *Quarterly Journal of Economics* 118, 1231–94.

Adserà, A. and Ray, D. 1998. History and coordination failure. *Journal of Economic Growth* 3, 267–76.

Aghion, P. and Bolton, P. 1997. A theory of trickle-down growth and development. *Review of Economic Studies* 64, 151–72.

Alesina, A., Baqir, R. and Easterly, W. 1999. Public goods and ethnic divisions. *Quarterly Journal of Economics* 114, 1243–84.

Alesina, A., Devleeschauwer, A., Easterly, W., Kurlat, S. and Wacziarg, R. 2003. Fractionalization. *Journal of Economic Growth* 8, 155–94.

Alesina, A. and La Ferrara, E. 2005. Ethnic diversity and economic performance. *Journal of Economic Literature* 43, 762–800.

Alston, L., Libecap, G. and Schneider, R. 1996. The determinants and impact of property rights: land titles on the Brazilian frontier. *Journal of Law, Economics, and Organization* 12, 25–61.

Appadurai, A. 2004. The capacity to aspire. In *Culture and Public Action*, ed. V. Rao and M. Walton. Stanford, CA: Stanford University Press.

Armendáriz, B. and Morduch, J. 2005. *The Economics of Microfinance*. Cambridge, MA: MIT Press.

Arthur, W. 1994. *Increasing Returns and Path-Dependence in the Economy*. Ann Arbor: University of Michigan Press.

Azariadis, C. and Drazen, A. 1990. Threshold externalities in economic development. *Quarterly Journal of Economics* 105, 501–26.

Baland, J.-M. and Platteau, J.-Ph. 1996. *Halting the Degradation of Natural Resources: Is there a Role for Rural Communities?* Oxford: Clarendon Press.

Baland, J.-M. and Platteau, J.-Ph. 1998. Wealth inequality and efficiency on the commons. Part II: the regulated case. *Oxford Economic Papers* 50, 1–22.

Banerjee, A. and Duflo, E. 2005. Growth theory through the lens of development economics. In *Handbook of Economic Growth*, ed. P. Aghion and S. Durlauf. Amsterdam: North-Holland.

Banerjee, A. and Duflo, E. 2007. The economic lives of the poor. *Journal of Economic Perspectives* 21(1), 141–67.

Banerjee, A., Gertler, P. and Ghatak, M. 2002. Empowerment and efficiency: tenancy reform in West Bengal. *Journal of Political Economy* 110, 239–80.

Banerjee, A. and Iyer, L. 2005. History, institutions and economic performance: the legacy of colonial land tenure systems in India. *American Economic Review* 95, 1190–213.

Banerjee, A., Iyer, L. and Somanathan, R. 2007. Public action for public goods. Working Paper No. 12911. Cambridge, MA: NBER.

Banerjee, A., Mookherjee, D., Munshi, K. and Ray, D. 2001. Inequality, control rights and efficiency: a study of sugar cooperatives in western Maharashtra. *Journal of Political Economy* 109, 138–90.

Banerjee, A. and Newman, A. 1993. Occupational choice and the process of development. *Journal of Political Economy* 101, 274–98.

Bardhan, P. and Udry, C. 1999. *Development Microeconomics*. New York: Oxford University Press.

Bardhan, P., Ghatak, M. and Karaivanov, A. 2006. Wealth inequality and collective action. Mimeo, London School of Economics.

Barro, R. 1991. Economic growth in a cross-section of countries. *Quarterly Journal of Economics* 106, 407–44.

Becker, G. and Tomes, N. 1986. Human capital and the rise and fall of families. *Journal of Labor Economics* 4, S1–39.

Bénabou, R. 1996. Inequality and growth. In *NBER Macroeconomics Annual*, ed. B. Bernanke and J. Rotemberg. Cambridge, MA: MIT Press.

Bénabou, R. 2000. Unequal societies: income distribution and the social contract. *American Economic Review* 90, 96–129.

Benhabib, J. and Rustichini, A. 1996. Social conflict and growth. *Journal of Economic Growth* 1, 125–42.

Bertocchi, G. and Canova, F. 2002. Did colonization matter for growth? An empirical exploration into the historical causes of Africa's underdevelopment. *European Economic Review* 46, 1851–71.

Besley, T. 1995. Property rights and investment incentives: theory and evidence from Ghana. *Journal of Political Economy* 103, 903–37.

Binswanger, H., Deininger, K. and Feder, G. 1995. Power, distortions, revolt and reform in agricultural land relations. In *Handbook of Development Economics*, vol. 3B, ed. J. Behrman and T. Srinivasan. Amsterdam: North-Holland.

Chiappori, P.-A. and Salanie, B. 2000. Testing for asymmetric information in insurance markets. *Journal of Political Economy* 108, 56–78.

Coate, S. and Ravallion, M. 1993. Reciprocity without commitment: characterization and performance of informal insurance arrangements. *Journal of Development Economics* 40, 1–24.

Cohen, J. and Weitzman, M. 1975. A Marxian view of enclosures. *Journal of Development Economics* 1, 287–336.

Collier, P. and Hoeffler, A. 1998. On economic causes of civil war. *Oxford Economic Papers* 50, 563–73.

Collier, P. and Hoeffler, A. 2004. Greed and grievance in civil war. *Oxford Economic Papers* 56, 563–95.

Dasgupta, P. 1993. *An Inquiry into Well-Being and Destitution*. Oxford: Clarendon Press.

Dasgupta, P. and Ray, D. 1986. Inequality as a determinant of malnutrition and unemployment: theory. *Economic Journal* 96, 1011–34.

David, P. 1985. Clio and the economics of QWERTY. *American Economic Review* 75, 332–7.

Deaton, A. 1997. *The Analysis of Household Surveys: A Microeconometric Approach to Development Policy*. Baltimore, MD: Johns Hopkins Press, for the World Bank.

Demsetz, H. 1967. Towards a theory of property rights. *American Economic Review* 57, 347–59.

Djankov, S., La Porta, R., Lopez-de-Silanes, F. and Shleifer, A. 2003. Courts. *Quarterly Journal of Economics* 118, 453–517.

Do, Q.-T. and Iyer, L. 2006. Poverty, social divisions and conflict in Nepal. Mimeo, Harvard Business School.

Dorfman, R., Samuelson, P. and Solow, R. 1958. *Linear Programming and Economic Analysis*. Tokyo: McGraw-Hill Kogashuka.

Dower, P. and Potamites, E. 2006. Signaling credit-worthiness: land titles, banking practices and access to formal credit in Indonesia. Mimeo, Department of Economics, New York University.

Dube, O. and Vargas, J. 2006. Are all resources cursed? Coffee, oil and armed conflict in Colombia. Documentos de CERAC 002748. Bogota, Colombia.

Duflo, E. 2006. Poor but rational? In *Understanding Poverty*, ed. A. Banerjee, R. Bénabou and D. Mookherjee. New York: Oxford University Press.

Duflo, E. and Udry, C. 2004. Intrahousehold resource allocation in Cote d'Ivoire: social norms, separate accounts and consumption choices. Working Paper No. 10498. Cambridge, MA: NBER.

Easterly, W. and Levine, R. 1997. Africa's growth tragedy: policies and ethnic divisions. *Quarterly Journal of Economics* 111, 1203–50.

Eaton, J. and Gersovitz, M. 1981. Debt with potential repudiation: theoretical and empirical analysis. *Review of Economic Studies* 48, 289–309.

Emerson, P. 2006. Corruption, competition and democracy. *Journal of Development Economics* 81, 193–212.

Esteban, J. and Ray, D. 1994. On the measurement of polarization. *Econometrica* 62, 819–51.

Esteban, J. and Ray, D. 2001. Social rules are not immune to conflict. *Economics of Governance* 2, 59–67.

Esteban, J. and Ray, D. 2006. Inequality, lobbying and resource allocation. *American Economic Review* 96, 257–79.

Esteban, J. and Ray, D. 2007a. On the salience of ethnic conflict. Mimeo, Department of Economics, New York University.

Esteban, J. and Ray, D. 2007b. Polarization, fractionalization and conflict. *Journal of Peace Research*, forthcoming.

Fafchamps, M. 2003. *Rural Poverty, Risk, and Development.* Cheltenham: Edward Elgar.

Fafchamps, M. 2004. *Market Institutions and Sub-Saharan Africa: Theory and Evidence.* Cambridge, MA: MIT Press.

Fafchamps, M. and Lund, S. 2003. Risk sharing networks in rural Philippines. *Journal of Development Economics* 71, 261–87.

Fearon, J. 1995. Rationalist explanations for war. *International Organization* 49, 379–414.

Fearon, J. and Laitin, D. 2003. Ethnicity, insurgency, and civil war. *American Political Science Review* 97, 75–90.

Feder, G., Onchan, T., Chalamwong, Y. and Hongladarom, C. 1988. *Land Policies and Farm Productivity in Thailand.* Baltimore, MD: Johns Hopkins University Press.

Field, E. and Torero, M. 2006. Do property titles increase credit access among the urban poor? Evidence from a Nationwide Titling Program. Mimeo. Department of Economics, Harvard University.

Fields, G. 1980. *Poverty, Inequality and Development.* London: Cambridge University Press.

Floro, M. and Ray, D. 1997. Vertical links between formal and informal financial institutions. *Review of Development Economics* 1, 34–56.

Floro, M. and Yotopoulos, P. 1991. *Informal Credit Markets and the New Institutional Economics: The Case of Philippine Agriculture.* Boulder, CO: Westview.

Foster, A. 1995. Prices, credit constraints, and child growth in low-income rural areas. *Economic Journal* 105, 551–70.

Foster, A. and Rosenzweig, M. 1995. Learning by doing and learning from others: human capital and technical change in agriculture. *Journal of Political Economy* 103, 1176–209.

Frankel, D. and Pauzner, A. 2000. Resolving indeterminacy in dynamic settings: the role of shocks. *Quarterly Journal of Economics* 115, 283–304.

Freeman, S. 1996. Equilibrium income inequality among identical agents. *Journal of Political Economy* 104, 1047–64.

Galor, O. and Zeira, J. 1993. Income distribution and macroeconomics. *Review of Economic Studies* 60, 35–52.

Genicot, G. 2002. Bonded labor and serfdom: a paradox of voluntary choice. *Journal of Development Economics* 67, 101–27.

Genicot, G. and Ray, D. 2003. Group formation in risk-sharing arrangements. *Review of Economic Studies* 70, 87–113.

Ghosh, P., Mookherjee, D. and Ray, D. 2001. Credit rationing in developing countries: an overview of the theory. In *Readings in the Theory of Economic Development*, ed. D. Mookherjee and D. Ray. London: Basil Blackwell.

Goldstein, M. and Udry, C. 2005. The profits of power: land rights and agricultural investment in Ghana. Mimeo, Department of Economics, Yale University.

González, F. 2007. Effective property rights, conflict and growth. *Journal of Economic Theory*, forthcoming.

Henderson, J. 1988. *Urban Development: Theory, Fact, and Illusion.* Oxford: Oxford University Press.

Hess, G. 2003. The economic welfare cost of conflict: an empirical assessment. Working Paper Series No. 852, CESifo, Munich.

Hidalgo, F., Naidu, S., Nichter, S. and Richardson, N. 2007. *Occupational choices: economic determinants of land invasions. Mimeo, Department of Political Science.* Berkeley CA: University of California.

Hirschman, A. 1958. *The Strategy of Economic Development.* New Haven, CT: Yale University Press.

Hoff, K., Braverman, A. and Stiglitz, J. 1993. *The Economics of Rural Organization: Theory, Practice and Policy.* London: Oxford University Press.

Hurst, E. and Lusardi, A. 2004. Liquidity constraints, household wealth, and entrepreneurship. *Journal of Political Economy* 112, 319–47.

Iyer, L. 2005. The long-term impact of colonial rule: evidence from India. Working Paper No. 05-041, Harvard Business School.

Jacoby, H. and Skoufias, E. 1997. Risk, financial markets, and human capital in a developing country. *Review of Economic Studies* 64, 311–35.

Karlan, D. and Zinman, J. 2006. Observing unobservables: identifying information asymmetries with a consumer credit field experiment. Mimeo, Department of Economics, Yale University.

Khwaja, A. 2004. Is increasing community participation always a good thing? *Journal of the European Economic Association* 2, 427–36.

Kingston, C. 2005. Social structure and cultures of corruption. Mimeo, Department of Economics, Amherst College.

Kocherlakota, N. 1996. Implications of efficient risk sharing without commitment. *Review of Economic Studies* 63, 595–609.

Krugman, P. 1991. *Geography and Trade*. Cambridge, MA: MIT Press.

La Porta, R., Lopez-de-Silanes, F., Shleifer, A. and Vishny, R. 1997. Legal determinants of external finance. *Journal of Finance* 52, 1131–50.

La Porta, R., Lopez-de-Silanes, F., Shleifer, A. and Vishny, R. 1998. Law and finance. *Journal of Political Economy* 106, 1113–55.

La Porta, R., Lopez-de-Silanes, F., Shleifer, A. and Vishny, R. 2002. Investor protection and corporate valuation. *Journal of Finance* 57, 1147–70.

Lanjouw, J. and Levy, P. 2002. Untitled: a study of formal and informal property rights in urban Ecuador. *Economic Journal* 112, 986–1019.

Leibenstein, H. 1957. *Economic Backwardness and Economic Growth*. New York: Wiley.

Leventoglu, B. and Slantchev, B. 2005. *The armed peace: a punctuated equilibrium theory of war. Mimeo, Department of Political Science*. San Diego: University of California.

Levy, G. 2004. A model of political parties. *Journal of Economic Theory* 115, 250–77.

Ligon, E. 1998. Risk-sharing and information in village economies. *Review of Economic Studies* 65, 847–64.

Ligon, E., Thomas, J. and Worrall, T. 2002. Mutual insurance and limited commitment: theory and evidence in village economies. *Review of Economic Studies* 69, 209–44.

Ljungqvist, L. 1993. Economic underdevelopment: the case of missing market for human capital. *Journal of Development Economics* 40, 219–39.

Loury, G. 1981. Intergenerational transfers and the distribution of earnings. *Econometrica* 49, 843–67.

Lucas, R. 1988. On the mechanics of economic development. *Journal of Monetary Economics* 22, 3–42.

Lucas, R. 1990. Why doesn't capital flow from rich to poor countries? *American Economic Review* 80, 92–6.

Majumdar, M. and Mitra, T. 1982. Intertemporal allocation with a non-convex technology: the aggregative framework. *Journal of Economic Theory* 27, 101–36.

Mankiw, N., Romer, D. and Weil, D. 1992. A contribution to the empirics of economic growth. *Quarterly Journal of Economics* 107, 407–38.

Mansuri, G. 1997. Credit layering in rural financial markets: theory and evidence from Pakistan. Ph.D. thesis, Boston University.

Matsuyama, K. 2000. Endogenous inequality. *Review of Economic Studies* 67, 743–59.

Mauro, P. 1995. Corruption and growth. *Quarterly Journal of Economics* 110, 681–712.

Miguel, E. and Gugerty, M. 2005. Ethnic diversity, social sanctions, and public goods in Kenya. *Journal of Public Economics* 89, 2325–68.

Miguel, E., Satyanath, S. and Sergenti, E. 2004. Economic shocks and civil conflict: an instrumental variables approach. *Journal of Political Economy* 112, 725–53.

Montalvo, J. and Reynal-Querol, M. 2005. Ethnic polarization, potential conflict, and civil wars. *American Economic Review* 95, 796–813.

Mookherjee, D. and Ray, D. 2001. *Readings in the Theory of Economic Development*. London: Basil Blackwell.

Mookherjee, D. and Ray, D. 2002a. Contractual structure and wealth accumulation. *American Economic Review* 92, 818–49.

Mookherjee, D. and Ray, D. 2002b. Is equality stable? *American Economic Review* 92, 253–9.

Mookherjee, D. and Ray, D. 2003. Persistent inequality. *Review of Economic Studies* 70, 369–94.

Munshi, K. and Myaux, J. 2006. Social norms and the fertility transition. *Journal of Development Economics* 80, 1–38.

Murphy, K., Shleifer, A. and Vishny, R. 1989. Industrialization and the big push. *Journal of Political Economy* 97, 1003–26.

Myrdal, G. 1957. *Economic Theory and Underdeveloped Regions*. London: Duckworth.

Nurkse, R. 1953. *Problems of Capital Formation in Underdeveloped Countries*. New York: Oxford University Press.

Obstfeld, M. 1994. The logic of currency crises. *Cahiers Economiques et Monétaires (Banque de France)* 43, 189–213.

Olson, M. 1965. *The Logic of Collective Action: Public Goods and the Theory of Groups*. Cambridge, MA: Harvard University Press.

Pande, R. and Udry, C. 2007. Institutions and development: a view from below. In *Proceedings of the 9th World Congress of the Econometric Society*, ed. R. Blundell, W. Newey and T. Persson. Cambridge: Cambridge University Press.

Parente, S. and Prescott, E. 2000. *Barriers to Riches*. Cambridge, MA: MIT Press.

Piketty, T. 1997. The dynamics of the wealth distribution and the interest rate with credit rationing. *Review of Economic Studies* 64, 173–89.

Perotti, R. 1994. Income distribution and investment. *European Economic Review* 38, 827–35.

Perotti, R. 1996. Growth, income distribution, and democracy: what the data say. *Journal of Economic Growth* 1, 149–87.

Powell, R. 2004. The inefficient use of power: costly conflict with complete information. *American Political Science Review* 98, 231–41.

Powell, R. 2006. War as a commitment problem. *International Organization* 60, 169–203.

Pritchett, L. 1997. Divergence, big time. *Journal of Economic Perspectives* 11(3), 3–17.

Quah, D. 1996. Twin peaks: growth and convergence in models of distribution dynamics. *Economic Journal* 106, 1045–55.

Ray, D., Baland, J.-M. and Dagnielie, O. 2007. Inequality and inefficiency in joint projects. *Economic Journal* 117, 922–35.

Ray, D. and Sengupta, K. 1989. Interlinkages and the pattern of competition. In *The Economic Theory of Agrarian Institutions*, ed. P. Bardhan. Oxford: Clarendon Press.

Ray, D. and Streufert, P. 1993. Dynamic equilibria with unemployment due to undernourishment. *Economic Theory* 3, 61–85.

Ray, D. 1998. *Development Economics*. Princeton, NJ: Princeton University Press.

Ray, D. 2006. Aspirations, poverty and economic change. In *Understanding Poverty*, ed. A. Banerjee, R. Bénabou and D. Mookherjee. New York: Oxford University Press.

Romer, P. 1986. Increasing returns and long-run growth. *Journal of Political Economy* 92, 1002–37.

Robinson, J. 1998. Theories of 'bad policy'. *Policy Reform* 1, 1–46.

Rosenstein-Rodan, P. 1943. Problems of industrialization of eastern and southeastern Europe. *Economic Journal* 53, 202–11.

Rosenzweig, M. and Wolpin, K. 1993. Credit market constraints and the accumulation of durable production assets in low-income countries: investments in bullocks. *Journal of Political Economy* 101, 223–44.

Sambanis, N. 2001. Do ethnic and nonethnic civil wars have the same causes? A theoretical and empirical inquiry (part 1). *Journal of Conflict Resolution* 45, 259–82.

Sandler, T. and Forbes, J. 1980. Burden sharing, strategy and the design of NATO. *Economic Inquiry* 18, 425–44.

Sen, A. 1999. *Development as Freedom*. New York: Alfred A. Knopf.

Sokoloff, K. and Engerman, S. 2000. History lessons: institutions, factor endowments, and paths of development in the new world. *Journal of Economic Perspectives* 14(3), 217–32.

Solow, R. 1956. A contribution to the theory of economic growth. *Quarterly Journal of Economics* 70, 65–94.

Stiglitz, J. 2002. *Globalization and its Discontents*. New York: W.W. Norton.

Stiglitz, J. and Weiss, A. 1981. Credit rationing in markets with imperfect information. *American Economic Review* 71, 393–410.

Townsend, R. 1993. Risk and insurance in village India. *Econometrica* 62, 539–91.

Townsend, R. 1995. Consumption insurance: an evaluation of risk-bearing systems in low-income economies. *Journal of Economic Perspectives* 99(3), 83–102.

Udry, C. 1994. Risk and insurance in a rural credit market: an empirical investigation in northern Nigeria. *Review of Economic Studies* 61, 495–526.

Udry, C. 1996. Gender, agricultural productivity and the theory of the household. *Journal of Political Economy* 104, 1010–45.

Weitzman, M. 1974. Free access vs. private ownership as alternative systems for managing common property. *Journal of Economic Theory* 8, 225–34.

World Bank. 2003. *World Development Report*. London: Oxford University Press.

Young, A. 1928. Increasing returns and economic progress. *Economic Journal* 38, 527–42.

dialectical reasoning

This notoriously elusive and multifaceted notion assumed importance in the history of political economy because Marx's 'critique of political economy', *Capital*, and particularly its first draft, the *Grundrisse* of 1857–8, was presented in a dialectical form. Part of the difficulty of encapsulating the dialectic within any concise definition derives from the fact that it may be conceived as a method of thought, a set of laws governing the world, the immanent movement of history or any combination of the three. The dialectic originated in ancient Greek philosophy. The original meaning of '*dialogos*' was to reason by splitting in two. In one form of its development, dialectic was associated with reason. Starting with Zeno's paradoxes, dialectical forms of reasoning were found in most of the philosophies of the ancient world and continued into medieval forms of disputation. It was this form of reasoning that Kant attacked in his distinction between the logic of understanding which, applied to the data of sensation, yielded knowledge of the phenomenal world, and dialectic or the logic of reasoning, which proceeded independently of experience and purported to give knowledge of the transcendent order of things in themselves. In another form of dialectic, the focus was primarily upon process: either an ascending dialectic in which the existence of a higher reality is demonstrated, or a descending form in which this higher reality is shown to manifest itself in the phenomenal world. Such conceptions were particularly associated with Christian eschatology, neo-platonism and illuminism, and typically patterned themselves into conceptions of original unity, division or loss, and ultimate reunification.

For practical purposes, however, the form in which the dialectic was inherited and modified by Marx was that in which it had been elaborated by Hegel. 'Hegel's dialectics is the basic form of all dialectics, but only *after* it has been stripped of its mystified form, and it is precisely this which distinguishes my method' (Marx, letter to Kugelmann, 6 March 1868).

In Hegel, the dialectic is a self-generating and self-differentiating process of reason (reason being understood both to be the process of cognition and the process of the world). The Hegelian Absolute actualizes itself by alienating itself from itself and then by restoring its self-unity. This corresponds to the three basic divisions of the Hegelian system: the *Logic*, the *Philosophy of Nature* and the *Philosophy of Mind*. It is free because self-determined. Its freedom consists in recognizing that its alienation into its other (nature) is but a free expression of itself. The truth is the whole and it unfolds through a dialectical progression of categories, concepts and forms of consciousness from the most simple and empty to the most complex and concrete. Each category reveals itself to the observer to be incomplete, lacking and contradictory; it thus passes over into a more adequate category capable of resolving the one-sided and contradictory aspects of its predecessor, though throwing up new contradictions in its turn. Against Kant, this process of dialectical reason is not concerned with the transcendent, but is immanent in reality itself. Reflective understanding is not false, but partial. It abstracts from reality and decomposes objects into their elements. Analytic understanding represents a localized standpoint which sets up an unsurpassable barrier between subject and object and thus cannot grasp the systematic interconnection between things or the total process of which it is a part. The absolute subject contains both itself and its other (both being and thought) which is revealed to be identical with itself. Human history, human thought are vehicles through which the absolute achieves self-consciousness, but humanity as such is not the subject of the process. Thus the absolute spirit dwells in human activity without being reducible to it, just as the categories of the *Logic* precede their embodiment in nature and history.

The character of the Marxian dialectic is yet harder to pin down than that of Hegel. In some well-known lines in the Post-Face to the Second Edition of *Capital* in 1873, Marx stated,

> I criticised the mystificatory side of the Hegelian dialectic nearly thirty years ago … [but] the mystification which the dialectic suffers in Hegel's hands by no means prevents him from being the first to present its general form of motion in a comprehensive and conscious manner. With him it is standing on its head. It must be inverted in order to discover the rational kernel within the mystical shell. (Marx, 1873, pp. 102–3)

This statement has satisfied practically no one. How can a dialectic be inverted? How can a rational kernel be extracted from a mystical shell? To critics from empiricist, positivist or structuralist traditions, anxious to free Marx from the clutches of Hegelianism, the dialectic is intrinsically unworkable and must either be dropped or stated in quite other terms (for example, Bernstein, 1899; Della Volpe, 1950; Althusser, 1965; Cohen, 1978; Elster, 1985). To a second group, the dialectical understanding of capitalism is only a particular instance of more general dialectical laws which govern reality as a whole, both natural and social (Engels, dialectical materialism). To a third group, the Hegelian roots of Marx's thought are not sufficiently emphasized in this statement; Marxism is only Hegelianism taken to its logical revolutionary conclusions in the discovery of the proletariat as the subject–object of history and the 'totality' as the distinguishing feature of its world-outlook (Lukács, 1923 and much of 20th-century Western Marxism). This

Methodenstreit cannot be discussed here. All that can be attempted is to give some sense to Marx's statement and in particular to indicate how it informed his critique of political economy.

Marx specifically criticized 'the mystificatory side of the Hegelian dialectic' in his 1843 *Critique of Hegel's Philosophy of Right* and in the concluding section of the *1844 Manuscripts* (both of which were only published in the 20th century). In these texts, Marx followed Feuerbach in considering Hegelian philosophy to be the conceptual equivalent of Christian theology; both were forms of alienation of man's species attributes; Christianity transposed human emotion into a religious Godhead, while Hegel projected human thinking into a fictive subject, the Absolute Idea, which in turn then supposedly generated the empirical world. Employing Feuerbach's 'transformative method' (the origin of the inversion metaphor), subject and predicate were reversed and hence the correct starting point of philosophy was the finite, man. Nature similarly was not the alienated expression of Absolute Spirit, it was irreducibly distinct. Thus there could be no speculative identity of being and thought. Man, however, as a natural being, could interact harmoniously with nature, his inorganic body. Once the absolute spirit had been dismantled and the identity of being and thought eliminated, it could be argued that the barrier against the harmonious interpenetration of man and nature and the free expression of human nature, was not 'objectification', the division between subject and object constitutive of the finite human condition, but rather the inhuman alienation of man's species life activity in property, religion and the state. True Communism, humanism, meant the re-appropriation of man's essential powers, the generic use of his conscious life activity. In contrast to the predominant Young Hegelian position, therefore, which counterposed Hegel's revolutionary 'method' (the dialectic) to his 'conservative system', Marx argued that there was no incompatibility between the two. For while Hegel's dialectic ostensibly negated the empirical world, it covertly depended upon it. Not only was the moment of contradiction a prelude to the higher moment of reconciliation and the restoration of identity, but the ideas themselves were tacitly drawn from untheorized experience. The effect of the dialectical chain which embodied the world was not to subvert the existing state of affairs, but to sanctify it.

In the crucial period that followed, that of the *German Ideology* and the *Poverty of Philosophy*, in which the basic architecture of the 'materialist conception of history' was elaborated, the attack upon speculative idealism was made more radical. The generic notion of 'conscious life activity', 'praxis', was replaced by the more specific notion of production. Hegel and the Idealist tradition were given credit for emphasizing the active transformative side of human history, but castigated for recognizing this activity only in the form of thought. Thought itself was now made a wholly derivative activity. The fundamental activity was labour and what developed in history were the productive powers men employed in their interaction with nature, 'the productive forces'. Stages in the development of these productive forces were accompanied by successive 'forms of human intercourse', what became 'the relations of production'. Finally, 'man' as a generic being was dispersed into the struggle between different classes of men, between those who produced and those who owned and controlled the means of production.

In this new theorization of history, explicit references to Hegel were few and the dialectic scarcely mentioned. But Hegel re-entered the story as soon as Marx attempted to write up a systematic theory of the capitalist mode of production in 1857–8. To see why, we must briefly survey his economic writings up to that date.

Marx's 1843 critique of Hegel had led him to the conclusion that civil society was the foundation of the state and that the anatomy of civil society was to be found in political economy. However, if his preoccupation with political economy dated from this point, it was not that of an economist. In the 1844 Manuscripts what is to be found is a humanist critique of both political economy and civil society: not an alternative theory of the economy, but rather a juxtaposition between the 'economic' and the 'human', the former being judged in terms of the latter. No distinction is made between political economy and the economic reality it purports to address, the one is simply seen as the mirror of the other.

The first attempt to define capitalism as an economic phenomenon occurred in the *Poverty of Philosophy* (1847). However, whatever the significance of that work in other respects, it did not outline any specifically Marxian portrayal of the capitalist economy. As in 1844 there was no internal critique of classical political economy. The main difference was that, whereas in 1844 Marx saw that economy through the eyes of Adam Smith, he now saw it through the eyes of Ricardo. In particular, he adopted what he took to be Ricardo's theory of value and belaboured Proudhon for positing as an ideal – the equivalence of value and price – what he considered to be the actual situation under capitalism. The only critique of Ricardo to be found there was a purely external historicist one: that Ricardo was the scientific expression of the epoch of capitalist triumph, but that that epoch had already passed away, that its gravediggers had already appeared and that its collapse was already at hand.

When Marx resumed his economic studies after the 1848 revolutions, Proudhonism was still the main object of attack. It occupied a major part of his unfinished economic manuscripts of 1850–1 and the attack on the Proudhonist banking schemes of Darimon took up the first part of the written-up notebooks of 1857–8, the *Grundrisse*. Proudhonism was the main object of attack because it could be taken for the predominant form of socialist or radical reasoning about the economy. Ricardo could again be utilized to attack such reasoning

in order to argue that it represented a nostalgia for petty commodity production under conditions of equal exchange, a situation supposedly preceding modern capitalism rather than representing an emancipation from it. However, if the capitalist mode of production and its historical limits were to be grasped in theory, this would have to involve a critique of Ricardo himself.

The form this critique took, involved problematizing Ricardo's theory of value (or rather Marx's reading of it). Steedman (1979) has argued strongly that Marx misconstrued Ricardo's theory, though Ricardo's shifting of position between the three editions of the *Principles* and the fact that Marx only used the third edition makes his mistake an understandable one). On the one hand, it raised a question never posed by Ricardo: the source of profit in a system of equal exchange. On the other hand, it involved juxtaposing wealth in the form of productive forces, that is, as a collection of use values against the translation of all wealth into exchange values within capitalism. Ricardo, it was argued, possessed no criterion for distinguishing between the content – or the material elements – and the form of the economy, such as Marx possessed in the distinction between forces and relations of production. Ricardo never problematized the 'value form'; he linked the object of measurement with the measurement itself. For this reason, Ricardo was considered to possess no conception of the historicity of capitalism. Once the material could be distinguished from the social, the content from the form, the capitalist mode of production could be conceived as a dynamic system whose principle of movement could be located in the contradictory relationship between matter and form.

It is here that Hegel came in. We know that during the writing of the *Grundrisse* at the beginning of 1858, Marx re-read Hegel, in particular the *Science of Logic*. He wrote to Engels, 'I am getting some nice developments, e.g. I have overthrown the entire doctrine of profit as previously conceived. In the method of working, it was of great service to me that by mere accident I leafed through Hegel's *Logic* again' (Marx to Engels, 16 January 1858).

What Marx found so useful in his reading of Hegel's *Logic* at this time is not really mysterious. It suggested a way of elaborating the contradictory elements that Marx had discerned in the value form into a theorization of the trajectory of the capitalist mode of production as a whole. The point is emphasized by Marx in his Post-Face to *Capital*: the dialectic includes in its positive understanding of what exists a simultaneous recognition of its negation, its inevitable destruction; because it regards every historically developed form as being in a fluid state, in motion, and therefore grasps its transient aspect as well (1873, p. 103). The dialectic offered a means of grasping a structure in movement, a process – the subtitle of *Capital*, Volume 1, was 'the process of capitalist production'. If capitalism could be represented as a process and not just a structure, then concomitantly its building blocks were not factors, but, as in Hegel, 'moments'. As Marx put it in the *Grundrisse*:

> When we consider bourgeois society in the long view and as a whole, then the final result of the process of social production always appears as the society itself i.e. the human being itself in its social relations. Everything that has a fixed form, such as the product etc., appears as merely a moment, a vanishing moment in this movement. The conditions and objectifications of the process are themselves equally moments of it, and its only subjects are the individuals, but individuals in mutual relationships, which they equally reproduce and produce anew .... in which they renew themselves even as they renew the world of wealth they create. (Marx [1857–8], p. 712)

Marx's attempt to utilize the *Logic* can be seen most clearly in the *Grundrisse*. There one can see the genesis of particular concepts which in *Capital* appear in more polished form. What is clear is that the *Logic* is used as a first means of setting terms in relation to each other. The text is littered with Hegelian expressions and turns of phrase; indeed, sometimes it appears as if lumps of Hegelian ratiocination have simply been transposed, undigested, to sketch the more intractable links in the chain. Here, for instance, is money striving to become capital: '... already for that reason, value which insists on itself as value preserves itself through increase; and it preserves itself precisely only by constantly driving beyond its quantitative barrier, which contradicts its character as form, its inner generality' (p. 270). But at the same time we can see Marx remind himself to correct the 'idealist manner of presentation, which makes it seem as if it were merely a matter of conceptual determination and of the dialectic of these concepts' (p. 151).

But the interest of dialectical logic for Marx was not simply that it offered him a way of outlining a structure in movement; more fundamentally it enabled him to depict contradiction as the motor of this movement. This was why the dialectic was 'in its very essence critical and revolutionary' (Marx, 1873, p. 103), in that both in Hegel and in ancient Greek usage movement was contradiction. This appears closely in the dramatic relationship that Marx sets up between the circulation system and the production system in *Capital*. The system of exchange of the market is the public face of capitalism. It is 'in fact a very Eden of the innate rights of Man' (p. 280). Exchanges are equal. To look for the source of inequality in the exchange system, like the Proudhonists, is to look in the wrong place. Yet, if exchanges are equal, how does capital accumulation take place? Equal exchange implies the principle of identity, of non-contradiction. It is, in Hegel's sense, the sphere of 'simple immediacy', the world as it first appears to the senses. It cannot move or develop, because it apparently contains no contradictory relations.

But this surface of things is not self-sufficient. It is 'the phenomenon of a process taking place behind it'. As a surface it is not nothing, but rather a boundary or limit. Contradiction and therefore movement is located in production. Here there is non-identity, the extraction of surplus labour disguised by the surface value form and its tendency to limitless expansion.

Thus, there are two processes, on the one hand that of the surface, that of immediate identity lacking the motive power of its own regeneration; on the other hand, that beneath the surface, a process of contradiction. Thus in Hegelian terms, the whole could then be defined as 'the identity of identity and non-identity'. In this whole, contradiction is the overriding moment, but the surface places increasingly formidable obstacles to its development, for instance, so-called 'realization' crises. Values can only be realized in an act of exchange and the medium of this exchange is money. But there is no guarantee that these exchanges must take place. The 'anarchy' of the market place is such that overproduction or disproportionality between sectors of production can only be seen after the event. Hence trade crises and slumps (see M. Nicolaus, Introduction to Marx [1857–8]).

This is only one example of how Marx employed dialectical principles in his attempt to conceptualize the process or movement of a contradictory whole. Another would be the six books Marx originally planned to write in 1857–8, the original blueprint of *Capital*. Their order would have been: Capital, Wage Labour, Landed Property, State, World Market, Crises. This plan is reminiscent of Hegel's *Encyclopaedia*. It describes a circle in a Hegelian sense. The point of departure is not capital per se, but commercial exchange as appearance, then proceeding through the contradictory world of production and eventually returning to commercial exchange again as the world market, but this time enriched by the whole of the preceding analysis.

There has been much controversy about the proximity or distance between the Hegelian and Marxian dialectics. Those who like Althusser (1965) argue for their radical dissimilarity, are on their strongest ground when arguing that in Marx the terms of the dialectic have been radically transformed. The contradiction between forces and relations of production cannot be reduced to the ultimate simplicity of that between Hegel's master and slave or of that between proletariat and bourgeoisie in the Hegelianized Marxist account of Lukacs. But it is far more difficult to establish as unambiguously the difference in the relationship between the terms in their respective dialectics. On the one hand, the relation between matter and form in Hegel is only one of apparent exteriority. Matter relates to form as other only because form is not yet posited within it. Once the terms are related, they are declared to be identical. Marx, on the other hand, insists upon the irreducible difference between matter and form, between the material and the social (even if he is not wholly successful in keeping them apart). Not only are matter and form different, but the one determines the other: value is determined in relation to the material production of use value; the opposite is not true. Relations of determination would seem to exclude identity, and this is confirmed by Marx's avoidance of the Hegelian notion of 'sublation' (*Aufhebung*), the higher moment of synthesis. The dialectical clash between forces and relations of production in the capitalist mode of production does not of itself produce a higher unity (socialism); rather what crises do, is to make manifest the otherwise hidden determination of value by use value, of form by matter. Against this, however, must be set one or two passages, including a famous peroration in *Capital* Volume 1, where Marx does conceive the end of capitalism as a return to a higher but differentiated unity and does employ the notion of the negation of the negation (Marx, 1873, p. 929), and, despite the best efforts of some modern commentators, it is difficult honestly to deny the strongly teleological imagination which underpins the whole enterprise of *Capital*.

Finally, in two important respects, Hegelian dialectic, however surreal, is less vulnerable than that of Marx. Firstly, Hegel's *Science of Logic* takes place outside spatio-temporal constraints. It is a purely logical progression of concepts, even if the principles on which one ontological category is derived from another 'have resisted analysis to this day' (Elster, 1985, p. 37). Marx's effort to avoid giving any impression of the 'self-determination' of the concept, took the form of attempting to demonstrate that 'the ideal is nothing but the material world reflected in the mind of man and translated into forms of thought' (Marx, 1873, p. 102). In practical terms this implied that there was some systematic relationship between the logical sequence of concepts in the exposition of the argument and the chronological order of their appearance in historical time. But this turned out to impose insurmountable difficulties in terms of presentation (and it is significant that, having begun with the product in the *Grundrisse*, he began with the commodity in *Capital*). Thus Marx both stated his position and violated it, bequeathing insoluble ambiguities surrounding his interpretation of value, of the meaning of 'reflection' and of the relationship between history and logic which have plagued even his closest followers ever since. Secondly, when it came to applying his dialectic to history, Hegel was categorical in refusing to project his theory into the future. The philosopher could explain the rationality of what had happened; it was only then that it could be grasped in thought. Marx, despite all his strictures against the voluntarism of other Young Hegelians and some of his fellow revolutionaries, was unable by the very nature of his project, fully to abide by the Hegelian restriction. Thus, while Hegel's owl of Minerva flew at dusk, the Marxian owl, unfortunately, took flight at high noon.

GARETH STEDMAN JONES

Bibliography

Althusser, L. 1965. *For Marx*. London: Allen Lane, 1969.
Bernstein, E. 1899. *Evolutionary Socialism*. Stuttgart. English trans. E.C. Harvey, London: Independent Labour Party, 1909.
Bhaskar, R. 1983. *Dialectic, Materialism and Human Emancipation*. London: New Left Books.
Cohen, G. 1978. *Karl Marx's Theory of History: A Defence*. London: Oxford University Press.
Della Volpe, G. 1950. *Logica come scienza positiva*. Messina: G. d'Anna.
Elster, J. 1985. *Making Sense of Marx*. Cambridge: Cambridge University Press.
Hegel, G.W.F. 1812–16. *The Science of Logic*. London: Allen & Unwin, 1961.
Kolakowski, L. 1978. *Main Currents of Marxism* Vol. I: *The Founders*, Oxford: Oxford University Press.
Lukács, G. 1923. *History and Class Consciousness*. London: Merlin, 1971.
Marx, K. 1844. *Economic and Philosophical Manuscripts of 1844*. In K. Marx and F. Engels, *Collected Works*, vol. 3. London: Lawrence & Wishart, 1975.
Marx, K. 1847. *The Poverty of Philosophy*. In *Collected Works*, vol. 6, London: Lawrence & Wishart, 1976.
Marx, K. [1857–8]. *Grundrisse*. Harmondsworth: Penguin 1973.
Marx, K. 1873. *Capital*, vol. I. 2nd edn., Harmondsworth: Penguin, 1976.
Rosdolsky, R. 1968. *The Making of Marx's Capital*. Trans. P. Burgeis, London: Pluto Press, 1977.
Steedman, I. 1979. Marx on Ricardo. Discussion Paper No. 10, Department of Economics, University of Manchester.

Díaz-Alejandro, Carlos (1937–1985)

Carlos Díaz-Alejandro was born in Havana and died in New York one day short of his 48th birthday. At 32, he became Yale's youngest ever full professor of economics. In 1983 he moved to Columbia, and at the time of his sudden death he had just accepted a chair at Harvard.

During sabbatical leaves, he visited many Latin American and European universities. Among numerous other activities he was a (dissenting) member of the Kissinger Commission on Central America. He strongly criticized US support for the 'Contras' in Nicaragua (para-military groups associated with the Somoza dictatorship, opposed to the Sandinista government), and insisted that if the United States were serious about Central America it should tie economic assistance to human rights and allow Central American exports free access into its own market. Needless to say, such quixotic attempts to influence US foreign policy were never among his greatest successes!

From a personal point of view I admired his sense of humour and wit, his approachability and 'bridge-building' capacity, his aversion to positions of administrative power, his independence of mind and his common sense.

His work gave us powerful insights into Latin America's trade and development, and its economic and financial history. He was particularly fascinated by the region's many financial crises. His contributions were characterized by a rare capacity to weave together history and theory, abstract economic theory and complex Latin American socio-political life.

In his doctoral dissertation at MIT, Díaz-Alejandro revisited the controversy between the 'elasticity' and 'absorption' approaches to the balance of payments in the context of Argentina's experience of devaluation, concluding that on balance it supported the first approach (1965a). He further argued that one of the main mechanisms through which devaluation influences both the balance of payments and economic growth is through its effects on income distribution. The apparent paradox that many devaluations improve the trade balance but negatively affect the overall growth of output could be explained by the complex redistributive effects of devaluation. In fact, the effectiveness of a devaluation may depend more on the nature of its distributional outcome than on its capacity to change relative prices. Therefore, the exchange rate could be seen as yet another sphere in the struggle between different groups over their shares in national income.

Another peculiar feature of semi-industrialized economies is that '[i]n the long run, the success or failure of a stabilisation effort will depend more on the capacity of governments to obtain a national consensus over the objectives and policy instruments than on the approval or help that they could receive from foreign investors or governments and international agencies' (1985, p. 201).

Díaz-Alejandro also maintained a keen interest in Latin American economic history, writing first on Argentina (1970). Then, in an article on the 1930s crisis, he identified the causes of the dissimilar performances of Latin American economies in the fact that some countries pursued an 'active' approach to fighting recession, while others stuck to conventional 'passive' adjustment mechanisms (1982). The 'active' countries were mainly the large ones, but also included Chile and Uruguay. They performed much better by abandoning the gold standard and by adopting flexible monetary and fiscal policies, real devaluations, moratoria on their foreign debt, and spending massively on public works. This heterodox response of some countries was in part a reaction to the emergence after the 1929 crash of a protectionist, interventionist and nationalistic Centre.

Díaz-Alejandro's articles on trade and development also discussed the high import intensity of import substitution (1965b), and the transition from import-substituting industrialization to export-led growth (1974). Diaz-Alejandro was particularly sceptical about the idea that this transition would help achieve both faster and more equitable growth. He strongly supported export orientation, but did not believe that it could be achieved simply by 'getting the prices right'; he also feared that it could contribute to 'stop-go' macroeconomics. Moreover, he thought that most of the advice given to Third World countries for their trade policies '… suggest evangelical fervour rather than scientific analysis' (1980, p. 332). Díaz-Alejandro re-examined all these issues in his book on Colombia (1975).

He was also a critic of the intervention of the International Monetary Fund (IMF) in markets which were not within its competence:

> It is the business of the IMF to insist on balance of payments targets … It is not the business of the IMF to make loans conditional on … food subsidies, utility rates, or controls over foreign corporations… It was a brilliant administrative stroke for the IMF staff to develop the 'monetary approach to the balance of payments' during the 1950s, allowing the translation of balance of payments targets into those involving domestic credit, but for many LDCs [less developed countries] the assumptions needed to validate such translation, such as a stable demand for money, have become less and less convincing. (1984, p. 169)

He also strongly criticized the IMF intervention in the debt crisis of the 1980s: 'Since August 1982 the world has lived with … a peculiar semi-cartelization shakily managed by central banks and the IMF [which] imposes on countries like Brazil the costs of monopoly (for example, larger spreads and fees) without some of its benefits (the ability to plan ahead)' (1983, p. 32).

The economic reforms of the late 1970s and 1980s provided another major intellectual challenge. Not since the 1930s had Latin America witnessed such dramatic economic and political experiments. The new military regimes of the Southern Cone applied their Chicago-oriented policies with a degree of ferocity that rivalled their treatment of political dissent. As Velasco said, Díaz-Alejandro's wisdom was twice as useful because it was delivered in a timely fashion (1988, p. 5). His papers of the late 1970s contain the basic ideas which later became accepted wisdom regarding the policy mistakes of the pro-Chicago governments in Latin America and the irrational behaviour of borrowers and lenders in (highly liquid) national and international financial markets. He particularly questioned the feasibility of simultaneous current and capital account liberalization, the lack of capital controls on speculative inflows, and the use of exchange rate policy to fight inflation.

Among his many articles from this period, his 'Southern Cone Stabilisation Plans' (1981) stands out. Appearing just before the Mexican moratorium which triggered the debt crisis, his argument ran completely against the tide of dominant opinion. Finally, a detailed analysis of the dynamics of the 1982 crisis was the last – and probably best known – of Díaz-Alejandro's contributions (see Palma, 2003).

Díaz-Alejandro began his studies at MIT at the time when Fidel Castro landed clandestinely in Cuba in 1956, and graduated at the time of the Bay of Pigs invasion (an unsuccessful CIA-planned and funded invasion by Cuban exiles in south-west Cuba in 1961). He felt that the complexity of the situation was such that he opted for the Miltonian hope that 'they also serve who only stand and wait'.

He had a fascination with Latin American economics. His approach was firmly grounded in the real world, and his work on economic history was rooted in the idea that all history is always the history of the present. As Gustav Ranis remarked, he always 'respected history, used data carefully, and theory selectively' (1989, p. xiv). Like his mentors Hirschman, Kindleberger, Lewis and Prebisch he basically belonged to the 'markets are good servants but bad masters' Keynesian school of economic thought, and always studied economic problems in their historical context, thus avoiding the sterility of pure formalistic theory that characterized so much of the economics of his own generation and the next.

JOSÉ GABRIEL PALMA

See also **elasticities approach to the balance of payments; Furtado, Celso; Kindleberger, Charles P.; Lewis, W. Arthur; Prebisch, Raúl; structuralism; terms of trade; third world debt.**

Selected works

1965a. *Exchange Rate Devaluation in a Semi-industrialized Country: Argentina 1955–1961.* Cambridge, MA: MIT Press.
1965b. On the import intensity of import substitution. Repr. in Velasco (1988).
1970. *Essays on the Economic History of Argentine.* New Haven, CT: Yale University Press.
1974. Some characteristics of recent export expansion in Latin America. Repr. in Velasco (1988).
1975. *Foreign Trade Regimes and Economic Development: Colombia.* New York: NBER.
1980. Discussions. *American Economic Review* 70, 330–5.
1981. Southern cone stabilization plans. Repr. in Velasco (1988).
1982. Latin America in the 1930's. Repr. in Velasco (1988).
1983. Some aspects of the 1982–83 Brazilian payments crisis. *Brookings Papers in Economic Activity* 1983(2), 515–2.
1984. Some economic lessons of the early 1980s. Repr. in Velasco (1988).

Bibliography

Palma, J.G. 2003. The three routes to financial crises. In *Rethinking Development Economics*, ed. H.-J. Chang. London: Anthem.
Ranis, G. 1989. Carlos Díaz-Alejandro: an appreciation. In *Debt, Stabilization and Development: Essays in Memory of Carlos Díaz-Alejandro*, ed. G. Calvo et al. Oxford: Blackwell.
Velasco, A., ed. 1988. *Trade, Development and the World Economy: Selected Essays of Carlos Díaz-Alejandro.* Oxford: Blackwell.

Dickinson, Henry Douglas (1899–1969)

Dickinson went from the King's School, Wimbledon, to Emmanuel College, Cambridge, where he took the Part II Tripos in both Economics and History. He carried out research at the London School of Economics under Cannan, then went to teaching posts at Leeds and Bristol, where he held the chair of economics from 1951 to 1964. Although his *Institutional Revenue* (1932) is of interest for generalizing the concept of institutional rents, he is deservedly known for a series of writings which attempted to reconcile choice and individual freedom with socialist planning, in the tradition of market socialism. Together with Taylor, Lange and Lerner he provided a rebuttal (based on actual markets) of von Mises's view that rational allocation under socialism was impossible. He saw 'the beautiful systems of economic equilibrium' not as 'descriptions of society as it is but prophetic visions of a socialist economy of the future' (1933, p. 247). During the 1930s his writings were well known to intellectuals of the Left, including Cole, Dalton, Durbin and Laski. The best-known of his works is the *Economics of Socialism* (1939). His technical prowess was later exhibited in a *Review of Economic Studies* article of 1954–5 in which he formulated a constant elasticity of substitution production function (CES) for the first time and anticipated some of the neoclassical growth results of Solow and Swan. 'Dick', as he was universally known, was a much loved, unworldly, eccentric figure with a keen sense of fun and a most astute mind.

DAVID COLLARD

Selected works

1932. *Institutional Revenue: A Study of the Influence of Social Institutions on the Distribution of Wealth.* London: Williams & Norgate.
1933. Price formation in a socialist community. *Economic Journal* 43, 237–50.
1939. *Economics of Socialism.* London: Oxford University Press.
1955. A note on dynamic economics. *Review of Economics Studies* 22(3), 169–79.

Dietzel, Heinrich (1857–1935)

Born in Leipzig, Dietzel was appointed to a chair at the University of Dorpat in 1885 after studies in economics and law in Heidelberg, Göttingen and Berlin. In 1890 he accepted a chair in the philosophy faculty in Bonn. There he died in 1935.

Dietzel was a respected figure in circles of 19th-century German economists (such as Rau, von Thünen, von Hermann, von Mangoldt and Wagner) who were endeavouring to defend, pursue and modify classical methods and principles. He kept a sceptical distance from both the younger Historical School and the Austrian School, and was sharply opposed to popular Marxism. Nevertheless his excellent biography of Rodbertus and his writings on the early socialists are proof of his academic openness and liberal fairness. Enthusiastically though not successfully engaged in propagating free trade, Dietzel (in contrast to Manchester liberalism) was not dogmatic concerning the functions of the state in a concrete mixed economy.

His most important contribution to theory, the *Theoretische Sozialökonomie* (1895), unfortunately remained a torso. It is a pioneering analysis of the two main orders of an economy, namely, the individualistic system of competitive markets and the collective system of compulsion of the state. This concept of the two (centralized and decentralized) elementary forms replaced the unscientific notions of capitalism and socialism, with their ideological bias. It opened the way to the foundation of an order theory that his disciple in Bonn, Walter Eucken, and the Freiburg School further developed and later on applied in Germany.

Though Dietzel dealt with self-interest, methodological theory (1911) and value theory, he and his followers (as Smithians) did not attempt to unify Smith's three systems of ethics, economics and politics to an integrated order theory via reconstructing and developing his 'obvious and simple system of natural liberty'. They also failed to produce an analysis of state and collective failures while they originally stressed the state's responsibility for ensuring sufficient market competition.

Nevertheless they made a number of contributions to the field and pointed to the right road to be taken in the future.

H.C. RECKTENWALD

Selected works

1882. *Über das Verhältnis der Volkswirtschaftslehre zur Sozialwirtschaftslehre*. Berlin: Puttkammer und Mühlbrecht.
1886–8. *Karl Rodbertus: Darstellung seines Lebens und seiner Lehre*. 2 vols. Jena: G. Fischer.
1895. *Theoretische Sozialökonomie, I*. Leipzig: Winter.
1911. Selbstinteresse und Methodenstreit in der Wirtschaftstheorie. In *Handwörterbuch der Staatswissenschaften*, vol. 7. Jena: G. Fischer.
1921. *Vom Lehrwert der Wertlehre und vom Grundfehler der Marxschen Verteilungslehre*. Leipzig-Erlangen: Scholl.
1922. *Technischer Fortschritt und Freiheit der Wissenschaft*. Bonn–Leipzig: Schroeder.

Bibliography

Recktenwald, H.C. 1985. Über das Selbstverständnis der ökonomischen Wissenschaft. *Jahrbuch der Leibniz-Akademie der Wissenschaften und der Literatur*. Wiesbaden: Steiner.
Recktenwald, H.C. and Samuelson, P.A. 1986. Über Thünen's 'Der isolierte Staat'. *Wirtschaft und Finanzen*, Darmstadt–Düsseldorf.

difference-in-difference estimators

Motivation and definition

Difference-in-differences (DID) estimators are often used in empirical research in economics to evaluate the effects of public interventions and other treatments of interest in the absence of purely experimental data.

The usual goal of evaluation studies is to estimate the average effect of a treatment (for example, participation in a vocational training programme) on some outcome variable of interest (for example, earnings or employment). Often researchers concentrate on estimating the average effect of the treatment on the treated, that is, on those individuals exposed to the treatment or intervention (for example, the trainees). In the typical setting of an evaluation study, we observe an outcome variable, Y_i, for a sample of treated individuals and also for a sample of untreated individuals. The main challenge in evaluation research is to find an appropriate comparison group among the untreated individuals, in the sense that the distribution of the outcome variable for the untreated comparison group can be taken as an approximation to the counterfactual distribution that the outcome variable, Y_i, would have followed for the treated in the absence of the treatment.

Sometimes the sample of untreated individuals may not provide an appropriate comparison group, and therefore differences in the distribution of the outcome variable between treated and untreated reflect not only the effect of the treatment but also intrinsic differences between the two groups. To address this problem, the DID estimator uses the assumption that in the absence of the treatment the average difference in the outcome variable, Y_i, between treated and untreated would have stayed roughly constant. Then, the average difference in the outcome variable between treated and untreated before the treatment can be used to approximate the part of the difference in average outcomes after the treatment that is created by intrinsic differences between the two groups and not by the effect of the treatment.

Let $\overline{Y}_t^T$ and $\overline{Y}_t^C$ be the average outcomes in period t ($t = 1, 2$) in the treated and untreated samples, respectively. Period $t = 1$ takes place before the treatment and period $t = 2$ takes place after the treatment. The difference in average outcomes between treated and untreated after the treatment is $\overline{Y}_2^T - \overline{Y}_2^C$. The same difference for the pre-treatment period is $\overline{Y}_1^T - \overline{Y}_1^C$. Then, the DID estimator is defined as follows:

$$\hat{\alpha} = \left(\overline{Y}_2^T - \overline{Y}_2^C\right) - \left(\overline{Y}_1^T - \overline{Y}_1^C\right). \qquad (1)$$

Figure 1 provides a graphical interpretation of the DID estimator. The solid lines represent the evolution in average outcomes for the treated and the untreated comparison group between the pre-treatment period ($t = 1$) and the post-treatment period ($t = 2$). The dashed line approximates the counterfactual evolution that the average outcome would have experienced for the treated in the absence of the treatment. This line is constructed under the DID assumption that, in the absence of the treatment, the difference in average outcomes between treated and untreated would have stayed roughly constant in the two periods. As reflected in Figure 1, an equivalent formulation of the DID assumption is that, in the absence of the treatment, average outcomes for treated and untreated would have followed a common trend. As a result, the untreated comparison group can be used to infer the counterfactual evolution of the average outcome for the treated in the absence of the treatment.

Difference in differences estimators have been applied to the study of a variety of issues in economics. Card and Krueger (1994) evaluate the employment effects of an increase in the minimum wage in New Jersey using a contiguous state (Pennsylvania), which did not increase the minimum wage, to approximate how employment would have evolved in New Jersey in the absence of the raise. Card (1990) applies DID estimators to evaluate the employment effects of the massive flow of Cuban immigrants to Miami during the 1980 Mariel boatlift. To estimate the effects of the boatlift, Card uses a group of four comparison cities to approximate how employment would have evolved in Miami in the absence of the 1980 immigration shock. Other applications of the DID estimator include studies of the effects of disability benefits on time out of work (Meyer, Viscusi and Durbin, 1995), the effect of anti-takeover laws on firms' leverage (Garvey and Hanka, 1999), and the effect of tax subsidies for health insurance on health insurance purchases (Gruber and Poterba, 1994).

The DID estimator has a simple regression representation. Let Y_{it} be the outcome of interest (for example, earnings) for individual i at time t, with $i = 1, \ldots, N$ and $t = 1, 2$. Let D_i be an indicator of membership to the treatment group, so $D_i = 1$ for the treated and $D_i = 0$ for the untreated. Finally, let $\Delta Y_i = Y_{i2} - Y_{i1}$ be the change in the outcome variable between the pre-treatment and the post-treatment period for individual i. The regression representation of the DID estimator is:

$$\Delta Y_i = \mu + \alpha D_i + u_i, \qquad (2)$$

where u_i is a regression error, which is mean independent of D_i (that is, $E[u_i|D_i = 1] = E[u_i|D_i = 0]$). It can be easily seen that the ordinary least squares estimator of α in eq. (2) is numerical identical to the DID estimator, $\hat{\alpha}$, in eq. (1). Regression standard errors along with the point estimate, $\hat{\alpha}$, can be used to construct confidence intervals for α and perform statistical hypothesis tests. As

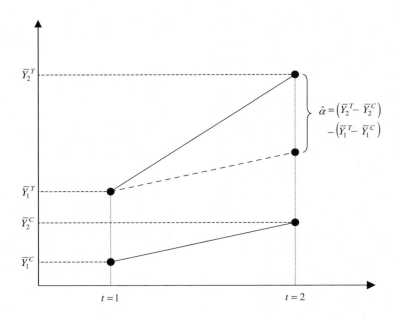

Figure 1

reflected in eq. (2) and emphasized in Blundell and MaCurdy (1999), the DID estimator is a particular case of fixed effects estimators for panel data, with only two time periods and a fraction of the sample exposed to the treatment in the second time period.

Extensions

In some instances, the common trend assumption adopted for DID is not plausible because treated and untreated differ in the distribution of some variables, X_i, that are thought to affect the trend of the outcome variable. In this situation, treated and untreated may exhibit different trends in the average of the outcome variable between $t = 1$ and $t = 2$, even if the treatment does not have any impact on the outcome of interest. The regression formulation of the DID estimator is useful to compute a conditional version of the DID estimator that corrects for the effect of X_i on the trend of Y_i:

$$\Delta Y_i = \mu + \alpha D_i + X_i'\beta + u_i.$$

Abadie (2005) and Heckman, Ichimura and Todd (1997) develop semiparametric and nonparametric versions of the conditional DID estimator.

Panel data are not always necessary to apply the DID estimator. A simple inspection of eq. (1) indicates that $\hat{\alpha}$ can be estimated from repeated cross sections, using a cross-section at time $t = 2$ to estimate $\overline{Y}_2^T - \overline{Y}_2^C$ and a cross section at time $t = 1$ to estimate $\overline{Y}_1^T - \overline{Y}_1^C$. A regression formulation of the DID estimator is also available for repeated cross sections (see, for example, Meyer, 1995; Abadie, 2005). When the DID estimator is constructed using repeated cross sections, it is important to check whether there exist compositional changes in the sample between the two periods. Compositional changes may constitute a threat to the assumption that the difference in the average outcome between treated and untreated would have stayed constant in the absence of the treatment.

In general, the DID assumption cannot be tested directly with data from $t = 1$ and $t = 2$ only. However, if the common trend assumption extends to more than one pretreatment period for which data are available, preexisting differences in the trends of the outcome variable between treated and untreated can be detected by applying the DID estimator to pretreatment data. This is done by constructing ΔY_i as the difference in the outcome variable for individual i between two pretreatment periods. Then, a test of the hypothesis $\alpha = 0$ in eq. (2) is a test of the common trend assumption. In addition, the DID assumption can sometimes be rejected when the dependent variable has bounded support (for example, when Y_i is a binary variable). If the dependent variable has bounded support the DID assumption may imply that, in the absence of the treatment, the average outcome for the treated would have lain outside the support of the dependent variable (see Athey and Imbens, 2006).

For a more detailed explanation of the theory behind DID estimators, see Abadie (2005), Angrist and Krueger (1999), Ashenfelter and Card (1985), Blundell and MaCurdy (1999), Heckman, Ichimura and Todd (1997), and Meyer (1995).

ALBERTO ABADIE

See also **econometric policy evaluation; fixed effects and random effects; treatment effect.**

Bibliography

Abadie, A. 2005. Semiparametric difference-in-differences estimators. *Review of Economic Studies* 72, 1–19.

Angrist, J.D. and Krueger, A.B. 1999. Empirical strategies in labor economics. In *Handbook of Labor Economics*, vol. 3A, ed. O. Ashenfelter and D. Card. Amsterdam: North-Holland.

Ashenfelter, O. and Card, D. 1985. Using the longitudinal structure of earnings to estimate the effects of training programs. *Review of Economics and Statistics* 67, 648–60.

Athey, S.C. and Imbens, G.W. 2006. Identification and inference in nonlinear difference-in-difference models. *Econometrica* 74, 431–98.

Blundell, R. and MaCurdy, T. 1999. Labor supply: a review of alternative approaches. In *Handbook of Labor Economics*, vol. 3A, ed. O. Ashenfelter and D. Card. Amsterdam: North-Holland.

Card, D. 1990. The impact of the Mariel Boatlift on the Miami labor market. *Industrial and Labor Relations Review* 44, 245–57.

Card, D. and Krueger, A.B. 1994. Minimum wages and employment: a case study of the fast-food industry in New Jersey and Pennsylvania. *American Economic Review* 84, 772–93.

Garvey, G.T. and Hanka, G. 1999. Capital structure and corporate control: the effect of antitakeover statutes on firm leverage. *Journal of Finance* 54, 519–46.

Gruber, J. and Poterba, J. 1994. Tax incentives and the decision to purchase health insurance: evidence from the self-employed. *Quarterly Journal of Economics* 109, 701–33.

Heckman, J.J., Ichimura, H. and Todd, P.E. 1997. Matching as an econometric evaluation estimator: evidence from evaluating a job training programme. *Review of Economic Studies* 64, 605–54.

Meyer, B.D. 1995. Natural and quasi-experiments in economics. *Journal of Business & Economic Statistics* 13, 151–61.

Meyer, B.D., Viscusi, W.K. and Durbin, D.L. 1995. Workers' compensation and injury duration: evidence from a natural experiment. *American Economic Review* 85, 322–40.

diffusion of agricultural technology

A high rate of technological change is a major feature of modern agriculture. New technologies are introduced gradually; diffusion is the process through which technologies spread throughout the farm sector over time. While adoption is the decision by an individual producer to use a new technology at a given moment, diffusion is the aggregate measure of adoption decisions. Early studies of diffusion were conducted by sociologists. Rogers (1962) measured technology usage as a fraction of farmers that had adopted a certain technology at a given point in time. Other studies measured diffusion by the fraction of land employed with the new technology. Rogers noticed that diffusion rates of hybrid corn in the United States fit very well as an S-shaped function of time:

$$S_t = \frac{K}{1 + exp^{-(a+bt)}},$$

where S_t is the level of diffusion at time t, K is the diffusion level at the limit and $K \leq 1$, a is a measure of initial diffusion, and b is a measure of the speed of diffusion. Rogers modelled diffusion as a process of imitation. In the early and late stages of diffusion, the level of diffusion is low because either the potential population of adopters or the population of users of the new technology to be imitated is small. During the middle period the diffusion rate takes off as there is a sufficient number of potential adopters, as well as a large population of established users to imitate. Rogers (1962) emphasized the role of distance from urban centres in explaining diffusion, finding that villages closer to urban centres had higher coefficients of diffusion.

Griliches (1957) argued that diffusion is an economic phenomenon and showed, using the diffusion data for hybrid corn in Iowa, that the three parameters K, a, and b are affected by profit. Other studies also found that the rate of diffusion tends to increase with farm size and the education of the farmer. However, as the review of Feder, Just and Zilberman (1985) suggests, the imitation model lacks a microeconomic foundation. An alternative model, the threshold model, suggests that the population of potential adopters is heterogeneous, and at every moment there is a critical variable that distinguishes between them. At every moment there is a threshold level of this variable that separates adopters from non-adopters.

Threshold models have three components: microeconomic behaviour, sources of heterogeneity, and a dynamic factor that drives the threshold level up or down. For example, adoption of mechanical innovation reflects the maximization of discounted net benefit. Farms vary in size, and at each moment there is a farm size threshold that distinguishes adopters from non-adopters. Over time, the cost of machinery may go down due to learning by doing, or the gain from adoption may go up because of learning by using, and that will reduce the adoption threshold. Empirical models, based on cross-sections of adopters, use discrete-choice estimation techniques to identify the key sources of heterogeneity. They found in many cases that size increases adoption of mechanical innovation, education explains adoption of more complex crops, and modern irrigation technologies that actually augment land quality are adopted earlier on lower-quality lands.

Much of the research has attempted to explain the diffusion of new 'Green Revolution' varieties in developing countries. In those cases, adoption was often partial (meaning farmers switched only a portion of their crops to the new technologies), and adoption rates were sometimes low, even given the significantly higher yields of Green-Revolution varieties. These facts emphasize the importance of risk considerations in explaining diffusion processes. Land allocation choices of risk-averse farmers were modelled as a portfolio, leading farmers to consider partial adoption of modern varieties because of their increased vulnerability to variable weather conditions. In addition to risk, wealth, human capital, and physical conditions, institutional forces have been identified as major determinants of diffusion rates. For example, renters are less likely to adopt new innovations than owners, especially when the rental contract is short. Lack of availability of credit is another deterrent to adoption. On the other hand, government policies, in the forms of output price subsidies and extension services that reduce the fixed costs of adoption, as well as technology and credit subsidies, can enhance the diffusion of modern agricultural technologies. For irrigation technologies, subsidies of water combined with restrictive trading regulations slow the diffusion of improved irrigation practices; water conservation can be enhanced by reducing constraints on water trading.

When demand for agricultural products is inelastic, the main beneficiary of the diffusion of more efficient technology is the consumer, while farmers are stuck on a 'technology treadmill'. Early adopters also benefit from the introduction of the new technology, but followers, who make up the majority of the farm population, may adopt only to stay competitive, while sometimes the laggards may go out of business. When the demand for agricultural products is elastic, then the gain from adoption of modern technologies contributes to enhanced land values, but the individual farm operators may not gain significantly because of the technology treadmill effect.

DAVID ZILBERMAN

See also **agricultural research; diffusion of technology.**

Bibliography

Feder, G., Just, R.E. and Zilberman, D. 1985. Adoption of agricultural innovations in developing countries: a survey. *Economic Development and Cultural Change* 32, 255–98.

Griliches, Z. 1957. Hybrid corn: an exploration in the economics of technological change. *Econometrica* 25, 501–22.

Rogers, E.M. 1962. *Diffusion of Innovation*. New York: Free Press.

diffusion of technology

The technology of a firm or country determines the efficiency with which inputs are mapped into outputs. Technological change may result in the ability to produce entirely new products, or it may allow an existing product to be produced with fewer inputs. This process has long been viewed as central to economic growth. The question of whether or not there is convergence across firms and countries raises issues related not only to the process of technical change but also to the diffusion of technology. Beginning in the late 1950s, economists have formalized their thinking as to how such technological knowledge diffuses from one economic entity to another. The early efforts were primarily directed to understanding firms' technology adoption decisions that often yield an S-shaped diffusion pattern over time. Since the 1990s, a vibrant literature has emerged in which the issues addressed are considerably broader, and where much more emphasis is placed on seeking high-quality empirical evidence.

A firm's technology and its productivity are closely related, and the two are identical if technology is identified with total factor productivity, an approach frequently adopted since the 1950s. The development of models of endogenous technical change in the early 1990s represented a step forward in that the R&D resources devoted to innovation were separated from the new technological knowledge itself. For example, consider the technology production function

$$\dot{N} = \eta N^{\lambda} H_N, \qquad (1)$$

where η and λ are parameters, η, $\lambda > 0$. The term H_N denotes the skilled-labour resources devoted the R&D, which according to eq. (1) lead to a flow of new technological knowledge of $\dot{N}$. A higher level of R&D produces a higher level of technology, N, and that in turn can be shown to result in higher productivity.

According to eq. (1), a higher stock of existing technological knowledge facilitates innovation. This stock of technological knowledge will rarely be entirely self-produced, so that (1) typically involves the diffusion of technology – diffusion between different persons, firms or countries. Technology is sometimes purchased or licensed in a market transaction, but, due to asymmetric information and other problems in the market for technology, non-market transactions in the form of technological externalities, called knowledge spillovers, are much more

important. What are the nature and the size of these knowledge spillovers? Since technological knowledge is non-rival, such externalities can in principle benefit many economic agents.

A useful benchmark is the complete diffusion of technology, which describes the case where technological knowledge created anywhere in the world is available worldwide immediately. This could underlie the assumptions of common-to-all and free technological knowledge of neoclassical growth theory. Clearly, this is not true in reality, where the diffusion of technology is gradual and uneven.

Why? First of all, acquiring technology involves making complementary investments, and the equilibrium choice for such investments often implies that not all technology diffuses. For instance, in Keller's (1996) model, international trade enables domestic producers to raise productivity by importing specialized foreign intermediate goods. Since these goods embody foreign R&D investments, this means the diffusion of technology from one country to another. For this imported technological knowledge to trigger domestic innovation, however, additional investments are necessary. According to Keller (1996), these investments mean additional training of workers so that they have the skills to manufacture products according to new blueprints. In addition, domestic innovators may have to invest resources in reverse engineering the foreign intermediate goods in order to fully comprehend the underlying foreign technological knowledge.

Second, another major determinant of the firm's decision to acquire the existing technology and innovate is the degree of product market competition. For example, in early Schumpeterian endogenous growth models, a higher degree of product market competition leads to lower monopoly profits and thus to a lower rate of innovation. More recent work by Aghion et al. (2001), for example, shows that, if technological laggards must first catch up with the leading-edge technology before battling for technological leadership in the future, the overall effect of more product market competition may be positive. The reason for this is that, even though more competition means lower monopoly profits, technological leaders now also have an incentive to innovate to avoid competition with technological laggards, and, if the latter effect is strong enough, product competition has a positive effect on technology diffusion and growth.

Third, there is no complete diffusion because it is simply not in the interest of the original creator of the technology, since his market for the technology would shrink if there were additional suppliers. In some cases, innovators obtain a patent that provides government-sanctioned protection of economic interests for a limited period of time in exchange for release of the technological information. Another strategy on the part of the original innovator is to use a varying amount of resources to keep the technological knowledge secret. At the same time, studies show that it often is no more than two years until new technology becomes publicly available.

Another, probably the most important, reason why knowledge spillovers are limited is that only the broad outlines of technology are codified – the remainder is the 'tacit' part of the knowledge. A person who is engaged in a problem-solving activity can often not fully define (and hence prescribe) what exactly he or she is doing. Along these lines, technology is only partially codified because it is impossible or at least very costly to fully codify it. For technology diffusion to occur completely, it may be necessary that the person who learns about the new technology can observe another person in the process of applying the technology. Even if this can be dispensed with, person-to-person contacts will generally be beneficial to the diffusion of technology.

Research has now turned to the essential task of assessing the importance of these processes empirically. As an intangible, technology is intrinsically difficult to measure, and economic data is hard to come by. This is even more the case for the non-market effects caused by technological knowledge. The main approach for quantifying technical change has been to study the relationship between R&D investments and productivity (Griliches, 1979). For example, Keller (2002a) estimates

$$tfp_{it} = \beta s_{it} + X'\gamma + \varepsilon_{it},$$
$$i = 1, \ldots I, \text{ and } t = 1, \ldots, T, \quad (2)$$

where tfp_{it} is log total factor productivity in industry i at time t, s_{it} are industry i's cumulative R&D investments (in logs) in period t, X is a vector of other observed determinants of productivity, and the error ε_{it} picks up unobserved effects. The parameter β, estimated in Keller (2002a) at $\beta = 0.15$, measures how R&D investments translate into higher productivity, thereby implicitly capturing the rate of technical change.

This approach is attractive since R&D spending is the main cause of technical change, and data on R&D expenditures is relatively easy to collect and compare across units (firms, industries and countries). A drawback is that measuring technical change this way requires an estimate of β. This can be complicated if productivity is badly measured, R&D is endogenous, or unobserved determinants on productivity are important, as in practice is often the case. Applications of instrumental-variable and control-function approaches have shown much promise in addressing the major estimation concerns (see Gong and Keller, 2003). Patents are an alternative measure of technology, with the advantage that patent data is available for a broader set of countries and a longer time horizon than is data on R&D (Jaffe and Trajtenberg, 2002). While patent counts are an imperfect measure of technology because the distribution of patent

values is extremely skewed, recent work using citations-weighted patent data has addressed this point since citations of a particular patent are a plausible indicator of its value. At the same time, patents cannot capture more than the codified part of technological knowledge, apart from the fact that across industries and firms the prevalence of patenting varies strongly for reasons that are difficult to fully ascertain.

Technology spillovers, as the major form of technology diffusion, are mainly analysed by extending eq. (2) above to estimate as well the effects of R&D investments conducted elsewhere. For example, in addition to the effects of own-industry R&D, Keller (2002a) estimates the effects of R&D in other domestic industries (s_{it}^{do}), as well as those of R&D in the same and other foreign industries (s_{it}^{f} and s_{it}^{fo}, respectively):

$$tfp_{it} = \beta_1 s_{it} + \beta_2 s_{it}^{do} + \beta_3 s_{it}^{f}$$
$$+ \beta_4 s_{it}^{fo} + \tilde{X}'\gamma + \varepsilon_{it}. \quad (3)$$

In this framework, the estimates of β_1, β_2, β_3, and β_4 determine the relative strength of intra- and inter-industry, and of domestic and international technology diffusion. For his sample of eight large Organisation for Economic Co-operation and Development (OECD) countries, Keller (2002a) finds that intra-industry effects dominate inter-industry spillovers, and that about 25 per cent of the total effect is due to international technology diffusion.

Other interesting approaches have employed multi-country extensions of recent models of endogenous technical change that include international technology diffusion (Eaton and Kortum, 1999). Because here the economic environment is fully specified, it is straightforward to simulate a model and perform interesting policy experiments. At the same time, typically there is little data on technology diffusion employed in the econometric estimation of these models. Consequently, the model's structure has a great influence on the results, while the implications for the diffusion of technology are not clear.

One major finding has been that the diffusion of technology is geographically localized, both domestically and internationally. For example, Keller (2002b) studies international technology diffusion between the G-5 countries (the United States, Japan, Germany, France, and England) and nine smaller OECD countries by estimating

$$tfp_{it} = \beta \left[s_{it} + \sum_{j \in G5} \exp(-\delta \, Dist_{ij}) s_{jt} \right]$$
$$+ X'\gamma + \varepsilon_{it},$$
$$i = 1, \ldots, I, \text{ and } t = 1, \ldots, T. \quad (4)$$

Here, $Dist_{ij}$ is the geographic distance between country i and G-5 country j. The parameter δ determines the extent of geographic localization: the higher is δ, the stronger is the degree of the localization of technological knowledge, while if $\delta = 0$, international technology diffusion is complete in the sense that geography has no impact whatsoever. The geographic reach of technology spillovers is a critical determinant of the cross-country income distribution, since global spillovers favour income convergence while local spillovers lead to income divergence. Keller's (2002b) results for the years 1970 to 1995 strongly reject the null hypothesis of complete diffusion. Instead, he estimates that with every additional 1,200 kilometres there is a 50 per cent drop in technology diffusion. The results imply that the benefits of being located next to major technology producers are substantial, highlighting the danger for isolated areas of being left behind.

While distance still shapes technology diffusion in a major way, there is also evidence that between 1970 and 1995 geography's grip on technology diffusion has weakened. Keller (2002b) estimates that the size of the δ parameter in eq. (4) fell substantially from the late 1970s to the 1990s, consistent with the idea that innovations in information and communication technologies have led to a major improvement in technology diffusion.

Such improvements in countries' abilities to draw on international innovations also imply that increasingly the ultimate sources of domestic productivity growth lie abroad. This is especially true for medium-sized and small countries, where the contribution of foreign technology to domestic productivity growth often exceeds 90 per cent. At the same time, because successful technology diffusion requires complementary investments in terms of adaptive R&D and/or human capital, domestic activities have a significant impact on the ease of technology diffusion.

WOLFGANG KELLER

See also **transfer of technology.**

Bibliography

Aghion, P., Harris, J., Howitt, P. and Vickers, J. 2001. Competition, imitation and growth with step-by-step innovation. *Review of Economic Studies* 68, 467–92.
Eaton, J. and Kortum, S. 1999. International technology diffusion: theory and measurement. *International Economic Review* 40, 537–70.
Gong, G. and Keller, W. 2003. Convergence and polarization in global income levels: a review of recent results on the role of international technology diffusion. *Research Policy* 32, 1055–79.
Griliches, Z. 1979. Issues in assessing the contribution of research and development to productivity growth. *Bell Journal of Economics* 10, 92–116.
Jaffe, A. and Trajtenberg, M. 2002. *Patents, Citations, and Innovations: A Window in the Knowledge Economy.* Cambridge, MA: MIT Press.
Keller, W. 1996. Absorptive capacity: on the creation and acquisition of technology in development. *Journal of Development Economics* 49, 199–227.
Keller, W. 2002a. Trade and the transmission of technology. *Journal of Economic Growth* 7, 5–24.
Keller, W. 2002b. Geographic localization of international technology diffusion. *American Economic Review* 92, 120–42.

directly unproductive profit-seeking (DUP) activities

Directly unproductive profit-seeking (DUP) activities are defined (Bhagwati, 1982a) as ways of making a profit (that is, income) by undertaking activities which are directly (that is, immediately, in their primary impact) unproductive, in the sense that they produce pecuniary returns but do not produce goods or services that enter a conventional utility function or inputs into such goods and services.

Typical examples of such DUP (pronounced appropriately as 'dupe') activities are (i) tariff-seeking lobbying which is aimed at earning pecuniary income by changing the tariff and therefore factor incomes; (ii) revenue-seeking lobbying which seeks to divert government revenues towards oneself as recipient; (iii) monopoly seeking lobbying whose objective is to create an artificial monopoly that generates rents; and (iv) tariff-evasion or smuggling which de facto reduces or eliminates the tariff (or quota) and generates returns by exploiting thereby the price differential between the tariff-inclusive legal and the tariff-free illegal imports.

While these are evidently profitable activities, their *output* is zero. Hence, they are wasteful in their primary impact, recalling Pareto's distinction between production and predation: they use real resources to produce profits but no output.

DUP activities of one kind or another have been analysed by several economic theorists, among them (i) the public-choice school's leading practitioners, their major work having been brought together in Buchanan, Tullock and Tollison (1980), (ii) Lindbeck (1976) who has worked on 'endogenous politicians', and (iii) the Chicago 'regulation' school, led by Stigler, Peltzman, Posner and also Becker (1983).

However, a central theoretical breakthrough has come from the work of trade theorists who have systematically incorporated the analysis of DUP activities in the main corpus of general equilibrium theory.

The early papers that defined this general-equilibrium-theoretic approach, and which were set in the context of the theory of trade and welfare, were: Bhagwati and

Hansen (1973) which analysed the question of illegal trade (that is, tariff-evasion), Krueger (1974) which analysed the question of rent-seeking for rents associated with import quotas specifically and quotas more generally, and (iii) Bhagwati and Srinivasan (1980) who analysed the phenomenon of revenue-seeking, the 'price' counterpart of Krueger's rent-seeking, where a tariff resulted in revenues which were then sought by lobbies.

The synthesis and generalization of these and other apparently unrelated contributions, showing that they all related to diversion of resources to zero-output activities, was provided in Bhagwati (1982a) where they were called DUP activities. The following significant aspects of the theoretical analysis of DUP activities are noteworthy.

First, they are generally related to policy interventions (but they need not be: plunder, for instance, pre-dates the organization of governments). In so far as policy interventions induce DUP activities, they are analytically divided into two appropriate categories (Bhagwati and Srinivasan, 1982):

Category I: Policy-triggered DUP activities. One class consists of *lobbying* activities. Examples include: rent-seeking analysis of the cost of protection *via* import licences (Krueger, 1974); revenue-seeking analysis of the cost of tariffs (Bhagwati and Srinivasan, 1980), of shadow prices in cost–benefit analysis (Foster, 1981), of price versus quantity interventions (Bhagwati and Srinivasan, 1982), of non-economic objectives (Anam, 1982), of rank-ordering of alternative distorting policies such as tariffs, production and consumption taxes (Bhagwati, Brecher and Srinivasan, 1984), of the optimal tariff (Dinopoulos, 1984), of the transfer problem (Bhagwati, Brecher and Hatta, 1985), and of voluntary export restrictions relative to import tariffs (Brecher and Bhagwati, 1987).

Another class consists of *policy-evading* activities. Examples include: analysis of smuggling (Bhagwati and Hansen, 1973), its implication for optimal tariffs (Johnson, 1974 and Bhagwati and Srinivasan, 1973), and alternative modelling by Kemp (1976), Sheikh (1974), Pitt (1981) and Martin and Panagariya (1984).

Category II: Policy-influencing DUP activities. The other generic class of DUP activities is not triggered by policies in place but is rather aimed at influencing the formulation of the policy itself. The most prominent DUP-theoretic contributions in this area relate to the analysis of tariff-seeking. Although Brock and Magee (1978; 1980) pioneered here, the general equilibrium analyses of endogenous tariffs began with Findlay and Wellisz (1982) and Feenstra and Bhagwati (1982), the two sets of authors modelling the government and the lobbying activities in contrasting ways. Notable among the later contributions are Mayer (1984), who extends the analysis formally to include factor income-distribution and therewith voting behaviour, and Wellisz and Wilson (1984). Magee (1984) has an excellent review of many of these contributions. The implication of endogenizing the

tariff for conventional measurement of the cost of protection has been analysed in Bhagwati (1980) and Tullock (1981).

The *choice* between alternative policy instruments when modelling the response of lobbies and governments to import competition has also been extensively analysed. The issue was raised by Bhagwati (1982b) and analysed further by Dinopoulos (1983) and Sapir (1983) in terms of how different agents (for example, 'capitalists' and 'labour') would profit from different policy responses such as increased immigration of cheap labour and tariffs when import competition intensified. It has subsequently been explored more fully by Rodrik (1986), who compares tariffs with production subsidies.

Second, Bhagwati (1982a) has noted, generalizing a result in Bhagwati and Srinivasan (1980), that DUP activities, while defined to be those that waste resources in their direct impact, cannot be taken as *ultimately* wasteful, that is, immiserizing, since they may be triggered by a suboptimal policy intervention. For, in that event, throwing away or wasting resources may be beneficial. The shadow price of a productive factor in such 'highly distorted' economies may be negative. This is the obverse of the possibility of immiserizing growth (Bhagwati, 1980). Thus, Buchanan (1980), who has addressed the issue of DUP activities and *defined* them as activities that (ultimately) cause waste, has been corrected in Bhagwati (1983): the definition of DUP activities cannot properly exclude the possibility that DUP activities are ultimately beneficial rather than wasteful. This central distinction between the direct and the ultimate welfare impacts of DUP activities is now universally accepted. DUP activities are therefore defined now, as in Bhagwati (1982b) and subsequent contributions, as wasteful only in the direct sense.

Third, Bhagwati, Brecher and Srinivasan (1984) have raised yet another fundamental issue concerning DUP activities. Thus, where DUP activities belong to Category II distinguished above, full endogeneity of policy can follow. If so, the conventional rank-ordering of policies is no longer possible. We have the *determinacy paradox:* policy is chosen in the solution to the full 'political-economy', DUP-theoretic solution and cannot be varied at will. These authors have therefore suggested that, where full endogeneity obtains, the appropriate way to theorize about policy is to take variations around the observed DUP-theoretic equilibrium. Thus, traditional economic parameters such as factor supply could be varied; similarly now the DUP-activity parameters such as, say, the cost of lobbying could be varied. The impact on actual welfare resulting from such variations can then be a proper focus of analysis, implying a wholly different way of looking at policy questions from that which economists have employed to date.

Finally, DUP activities are related to Krueger's (1974) important category of rent-seeking activities. The latter are a subset of the former, in so far as they relate to lobbying for quota-determined scarcity rents and are therefore part of DUP activities of Category II distinguished above (Bhagwati, 1983).

JAGDISH N. BHAGWATI

See also **bribery; rent seeking.**

Bibliography

Anam, M. 1982. Distortion-triggered lobbying and welfare: a contribution to the theory of directly-unproductive profit-seeking activities. *Journal of International Economics* 13(August), 15–32.

Becker, G.S. 1983. A theory of competition among pressure groups for political influence. *Quarterly Journal of Economics* 93, 371–400.

Bhagwati, J. 1980. Lobbying and welfare. *Journal of Public Economics* 14, 355–63.

Bhagwati, J. 1982a. Directly-unproductive profit-seeking (DUP) activities. *Journal of Political Economy* 90, 988–1002.

Bhagwati, J. 1982b. Shifting comparative advantage, protectionist demands, and policy response. In *Import Competition and Response*, ed. J. Bhagwati. Chicago: Chicago University Press.

Bhagwati, J. 1983. DUP activities and rent seeking. *Kyklos* 36, 634–7.

Bhagwati, J., Brecher, R. and Hatta, T. 1985. The generalized theory of transfers and welfare: exogenous (policy-imposed) and endogenous (transfer-induced) distortions. *Quarterly Journal of Economics* 100, 697–714.

Bhagwati, J., Brecher, R. and Srinivasan, T.N. 1984. DUP activities and economic theory. In *Neoclassical Political Economy: The Analysis of Rent-seeking and DUP Activities*, ed. D. Colander. Cambridge, MA: Ballinger & Co.

Bhagwati, J. and Hansen, B. 1973. Theoretical analysis of smuggling. *Quarterly Journal of Economics* 87, 172–87.

Bhagwati, J. and Srinivasan, T.N. 1973. Smuggling and trade policy. *Journal of Public Economics* 2, 377–89.

Bhagwati, J. and Srinivasan, T.N. 1980. Revenue-seeking: a generalization of the theory of tariffs. *Journal of Political Economy* 88, 1069–87.

Bhagwati, J. and Srinivasan, T.N. 1982. The welfare consequences of directly-unproductive profit-seeking (DUP) lobbying activities: price *versus* quantity distortions. *Journal of International Economics* 13, 33–44.

Brecher, R. and Bhagwati, J. 1987. Voluntary export restrictions and import restrictions: a welfare-theoretic comparison. In *Essays in Honour of W.M. Corden*, ed. H. Kierzkowski. Oxford: Basil Blackwell.

Brock, W. and Magee, S. 1978. The economics of special interest politics: the case of the tariff. *American Economic Review* 68, 246–50.

Brock, W. and Magee, S. 1980. Tariff formation in a democracy. In *Current Issues in International Commercial Policy and Diplomacy*, ed. J. Black and B. Hindley. New York: Macmillan.

Buchanan, J. 1980. Rent seeking and profit seeking. In *Towards a General Theory of the Rent-seeking Society*, ed. J. Buchanan, G. Tullock and R. Tollison. College Station: Texas A&M University Press.

Buchanan, J., Tullock, G. and Tollison, R., ed. 1980. *Towards a General Theory of the Rent-seeking Society*. College Station: Texas A&M University Press.

Dinopoulos, E. 1983. Import competition, international factor mobility and lobbying responses: the Schumpeterian industry cases. *Journal of International Economics* 14, 395–410.

Dinopoulos, E. 1984. The optimal tariff with revenue-seeking: a contribution to the theory of DUP activities. In *The Neoclassical Political Economy: The Analysis of Rent-seeking and DUP Activities*, ed. D. Colander. Cambridge, MA: Ballinger & Co.

Feenstra, R. and Bhagwati, J. 1982. Tariff seeking and the efficient tariff. In *Import Competition and Response*, ed. J. Bhagwati. Chicago: Chicago University Press.

Findlay, R. and Wellisz, S. 1982. Endogenous tariffs, the political economy of trade restrictions, and welfare. In *Import Competition and Response*, ed. J. Bhagwati. Chicago: Chicago University Press.

Foster, E. 1981. The treatment of rents in cost-benefit analysis. *American Economic Review* 71, 171–8.

Johnson, H.G. 1974. Notes on the economic theory of smuggling. In *Illegal Transactions in International Trade*, ed. J. Bhagwati. Series in International Economics. Amsterdam: North-Holland.

Kemp, M. 1976. Smuggling and optimal commercial policy. *Journal of Public Economics* 5, 381–4.

Krueger, A. 1974. The political economy of the rent-seeking society. *American Economic Review* 64, 291–303.

Lindbeck, A. 1976. Stabilization policies in open economies with endogenous politicians. Richard Ely Lecture. *American Economic Review* 66, 1–19.

Magee, S. 1984. Endogenous tariff theory: a survey. In *Neoclassical Political Economy: The Analysis of Rent-seeking and DUP Activities*, ed. D. Colander. Cambridge, MA: Ballinger & Co.

Martin, L. and Panagariya, A. 1984. Smuggling, trade, and price disparity: a crime-theoretic approach. *Journal of International Economics* 17, 201–18.

Mayer, W. 1984. Endogenous tariff formation. *American Economic Review* 74, 970–85.

Pitt, M. 1981. Smuggling and price disparity. *Journal of International Economics* 11, 447–58.

Rodrik, D. 1986. Tariffs, subsidies and welfare with endogenous policy. *Journal of International Economics* 21, 285–99.

Sapir, A. 1983. Foreign competition, immigration and structural adjustment. *Journal of International Economics* 14, 381–94.

Sheikh, M. 1974. Smuggling, production and welfare. *Journal of International Economics* 4, 355–64.

Tullock, G. 1981. Lobbying and welfare: a comment. *Journal of Public Economics* 16, 391–4.

Wellisz, S. and Wilson, J.D. 1984. Public sector inefficiency, a general equilibrium analysis. Discussion Paper No. 254, International Economics Research Center, Columbia University.

Director, Aaron (1901–2004)

Aaron Director's enduring contribution to economics came via his role in the development of the Chicago law and economics tradition. Director was born in Charterisk (in present-day Ukraine) in 1901 and emigrated to the United States with his family in 1913. He received his undergraduate degree from Yale University and his graduate training at the University of Chicago. Although he came to Chicago in 1927 to work with Paul Douglas on labour economics, it was Frank Knight and Jacob Viner who, via their price theory courses, had the greatest influence on him. Director remained at Chicago as a graduate student and part-time instructor until 1934. The 1930s were a heady period at Chicago, where the student body included George Stigler, Paul Samuelson (who credits Director's teaching with stimulating his interest in economics), and Milton Friedman – each of whom helped to reshape economic thinking in the middle third of the 20th century – as well as Rose Director (Aaron's sister and, eventually, Rose Friedman). Aaron Director was very much part of this milieu. He left the University of Chicago for the US Treasury Department in 1934 and, save for an aborted attempt to complete a dissertation on the history of the Bank of England, remained in Washington, DC, until 1946, when he returned to the University of Chicago to take up a position in the Law School, where he remained until his retirement in 1966.

Director's appointment in the Law School was a result of the efforts of Henry Simons, the first economist on the law faculty at Chicago, and Friedrich Hayek, whose *Road to Serfdom* was published in the United States largely because of Director's intervention with the University of Chicago Press. The plan, as laid out by Simons, was for Director to head up the 'Free Market Study', a Volker Fund-financed project, housed in the Law School and dedicated to undertaking 'a study of a suitable legal and institutional framework of an effective competitive system' (Coase, 1998, p. 246). However, Simons committed suicide in the summer of 1946, and Director was asked to take on Simons's basic Law School price theory course, 'Economic Analysis of Public Policy'. This provided Director with an initial forum for bringing the perspective he had learned from Knight and Viner into the Law School's teaching programme.

The transition from having an economist on the Law School faculty to the establishment of a law and economics tradition at Chicago began not long after this, when Edward Levi invited Director to collaborate in the teaching of the antitrust course. Levi would teach a traditional antitrust course for four days each week; Director would then come in on the fifth day and, using the tools of price theory, show that the traditional legal approach could not stand up to the rigours of economic analysis. The basic pattern was very simple: Director would ask whether the practice in question was, in general, consistent with monopolistic profit maximization. The answer was often negative, which meant that there had to be some sort of legitimate rationale for the supposedly anti-competitive practice in question. What Director's price theory showed was that the 'simple and obvious' answers were often wrong-headedly simplistic. This process had a profound impact on students and colleagues alike. Director's antitrust students – a group that included Robert H. Bork, Ward Bowman, Kenneth Dam, Edmund Kitch, Wesley J. Liebeler, John S. McGee, Henry Manne, and Bernard H. Siegan – have often spoken of the 'conversion' they experienced in this class, and even Levi himself became a partial convert (see Kitch, 1983; Director and Levi, 1951). What was perhaps Director's most significant contribution on the missionary front came after his retirement, when he and Richard Posner spent time together at Stanford in 1968 – Posner's first year on the Stanford Law School faculty. It was Director who taught Posner to think like a Chicago economist, introduced him to Stigler and Ronald Coase, and in this and other ways was instrumental in Posner's move to the Chicago Law School after only one year on the Stanford faculty. The rest, as they say, is history.

Although Director's published output was slight, his influence extended well beyond the classroom. His insights made their way into the antitrust literature – and, eventually, antitrust policy – through the writings of students and colleagues, as Sam Peltzman (2005) has detailed. Director's primary legacies are in the analysis of predatory pricing (via McGee, 1958), resale price maintenance (via Telser, 1960), and tie-in sales (see Director and Levi, 1951; Bowman, 1957; Burnstein, 1960), but his influence was also prominent in Stigler's view of oligopoly and antitrust policy, Posner's (1969) perspective on oligopoly and cartels, and Robert Bork's influential articles on antitrust (for example, Bork and Bowman, 1965; Bork, 1967). These contributions coalesced in a distinctive Chicago approach to antitrust analysis, an approach that Herbert Hovenkamp (1986, p. 1020) says 'has done more for antitrust policy than any other coherent economic theory since the New Deal', and whose influence is inescapable.

Director's impact at the Law School went far beyond antitrust: He was also the prime mover in the early professionalization of law and economics. Director formally established the nation's first law and economics programme, which maintained visiting fellowships for law and economics scholars, and, in 1958, founded the

Journal of Law and Economics. Within a few decades, Director's efforts at Chicago had been replicated in a set of thriving and well-funded law and economics programmes at major law schools around the country. One would be hard pressed to name an individual in our discipline who has had as much influence as Director without a much more extensive bibliography.

STEVEN G. MEDEMA

See also **antitrust enforcement; Chicago School; law, economic analysis of.**

Selected works

1933. *The Economics of Technocracy.* Public Policy Pamphlet No. 2. Chicago: University of Chicago Press.
1940. Does inflation change the economic effects of war? *American Economic Review* 35, 351–61.
1964. Parity and the economic marketplace. *Journal of Law and Economics* 7, 1–10.
1931. (With P. Douglas.) *The Problem of Unemployment.* New York: Macmillan.
1951. (With E. H. Levi.) Law and the future: trade regulation. *Northwestern Law Review* 51, 281–96.

Bibliography

Bork, R.H. 1967. Antitrust and monopoly: the goals of antitrust policy. *American Economic Review* 57, 242–53.
Bork, R.H. and Bowman, W.S., Jr. 1965. The crisis in antitrust. *Columbia Law Review* 65, 363–76.
Bowman, W.S., Jr. 1957. Tying arrangements and the leverage problem. *Yale Law Journal* 67, 19–36.
Burnstein, M.L. 1960. The economics of tie-in sales. *Review of Economics and Statistics* 42, 68–73.
Coase, R.H. 1998. Director, Aaron. In *The New Palgrave Dictionary of Economics and the Law.* London: Macmillan.
Hovenkamp, H. 1986. Chicago and its alternatives. *Duke Law Journal* 1986, 1014–29.
Kitch, E.W. 1983. The fire of truth: a remembrance of Law and Economics at Chicago, 1932–1970. *Journal of Law and Economics* 26, 163–233.
McGee, J.S. 1958. Predatory price cutting: the standard oil (NJ) case. *Journal of Law and Economics* 1, 137–69.
Peltzman, S. 2005. Aaron Director's influence on antitrust policy. *Journal of Law and Economics* 48, 313–30.
Posner, R.A. 1969. Oligopoly and the antitrust laws: a suggested approach. *Stanford Law Review* 21, 1562–1606.
Stigler, S.M. 2005. Aaron Director remembered. *Journal of Law and Economics* 48, 307–11.
Telser, L.G. 1960. Why should manufacturers want fair trade? *Journal of Law and Economics* 3, 86–105.

dispute resolution

Disputes may arise in a variety of settings, including labour negotiations, civil disputes and family conflict. If individuals fail to reach an agreement, there exist a variety of mechanisms for resolving the dispute. These include civil litigation, arbitration and, in labour relations, the strike. Resolving disputes in these ways is costly, thereby creating a contract zone within which both parties strictly prefer to settle. Given the high cost of disputes, considerable research has been devoted to understanding why settlement sometimes fails to occur and how different mechanisms affect the dispute rate. Here we focus on dispute resolution in the context of civil litigation and arbitration.

Why settlement fails

The dominant rational choice explanation for settlement failure is asymmetric information. An alternative explanation, not consistent with the assumption of full rationality, is optimism. If agents have symmetric information and beliefs about the expected outcome of a dispute, theory suggests a settlement will occur. However, if one party has private information about the expected outcome of the dispute, settlement failure can occur. Similarly, if one or both parties to the dispute are subject to optimism, a contract zone may fail to exist.

There are two basic models in the asymmetric information literature, which make different assumptions about the structure of information. When the uninformed party makes the offer, we have a screening model which was developed by Bebchuk (1984). When the informed party makes the offer we have a signalling model developed by Reinganum and Wilde (1986). Both models' predictions are consistent with the existence of costly disputes in equilibrium.

To explore the intuition behind these models, consider a civil dispute in which the failure of negotiations would result in a trial. Suppose a plaintiff known to be harmed has private information concerning the damages she has incurred and that this information would be revealed at trial. Further, suppose the plaintiff is one of two types: a weak type with a low expected payoff at trial or a strong type with a high expected payoff. A risk-neutral plaintiff will accept a settlement offer if and only if it equals or exceeds her expected net payoff from trial. The defendant knows the probability that he is facing a weak or strong plaintiff but not the plaintiff's exact type. In a screening model, the uninformed defendant makes an offer to the plaintiff. He will choose between a low (screening) offer that only weak plaintiffs would accept and a high (pooling) offer that both types would accept. If he makes the low offer, then a strong plaintiff would proceed to trial. The screening offer is more likely to be optimal for the defendant when there is a high prevalence of weak plaintiffs, when court costs are low, and when the difference in expected trial awards for the two plaintiff types is large.

If the informed plaintiff is allowed to make the offer, this is called the signalling game. While these games generally have multiple equilibria, the D1 refinement (Cho and Kreps, 1987) has been employed to focus on a separating equilibrium in which the weak plaintiff submits a low demand to the defendant, while the strong plaintiff submits a high demand. Under D1, it is assumed that an out of equilibrium offer is made by the plaintiff willing to make that offer for the largest set of acceptance probabilities. In equilibrium, the high demand must be rejected with a sufficiently high probability so as to discourage the weak plaintiff from also making this demand. These rejections lead to a positive probability of trial with the strong plaintiff.

While we used a two-type model to motivate the discussion, the Bebchuk and Reinganum and Wilde models employ a continuum of types whose distribution is known by the uninformed party. These models have been extended in numerous ways by allowing for two-sided information asymmetries (Schweizer, 1989; Daughety and Reinganum, 1994) and multiple offers (Spier, 1992) among other extensions. While the effects of policy variables (such as cost shifting at trial) are often sensitive to the modelling details, the prediction that asymmetric information can result in costly disputes is quite robust. Excellent surveys of the literature are provided by Spier (1998) and Daughety (1999).

The empirical studies by McConnell (1989), Conlin (1999) and Osborne (1999) support the model of asymmetric information.

The optimism or self-serving bias explanation for settlement failure relies on bargainers who have potentially inaccurate beliefs about the expected outcome at trial. For example, the plaintiff's belief about the probability she will prevail at trial may exceed the defendant's belief about this same probability. If these differences in beliefs are not based on differences in information, then we are in the realm of the optimism model. Versions of this model have been employed by Landes (1971), Posner (1973), Shavell (1982), and Priest and Klein (1984). Optimism violates rationality, but Bar-Gill (2002) finds that cautious optimism can allow the optimistic party to obtain a larger portion of the joint surplus from settlement. As a result, cautious optimism can persist in an evolutionary setting. Babcock and Loewenstein (1997) survey an experimental literature documenting the existence of a self-serving bias which leads players in the role of a plaintiff to expect a greater payoff at trial than the defendant, even though both are exposed to the same set of facts. When players are exposed to the facts of the case before being assigned their role as plaintiff or defendant, the self-serving bias tends to disappear.

Waldfogel (1998) and Farmer, Pecorino and Stango (2004) find empirical evidence that is consistent with the optimism model. Note that the optimism and asymmetric information explanations are not mutually exclusive.

It is possible that each factor is responsible for some proportion of observed disputes.

Mandatory discovery and voluntary disclosure

If asymmetric information causes disputes, it is logical to ask whether voluntary disclosures and mandatory discovery can eliminate these asymmetries. In a screening model where credible disclosure is costless, Shavell (1989) shows that plaintiffs with strong cases will reveal enough information to ensure that all cases settle. Plaintiffs who do not reveal their information (those with weak cases) receive a pooling offer that all accept. However, the work of Sobel (1989) shows that this result is not robust to the introduction of positive costs of disclosure. He also shows that a costless (to the plaintiff) discovery procedure will lead to greater settlement. Farmer and Pecorino (2005) consider costly discovery and disclosure in both the signalling and the screening games. Costly disclosures may be made in the signalling game but not the screening game, while costly discovery may be invoked in the screening game but not the signalling game. If the cost of these procedures is not too high, the combination of the two will lead to a great deal of information transmission and a large reduction in the dispute rate.

Why then do disputes persist? Perhaps, as Shavell (1989) suggests, private information has strategic value if withheld until trial. Hay (1995) develops a model in which an initial informational asymmetry on the merits of the case is resolved by mandatory discovery, but by the time this occurs a new asymmetry – namely, the extent of attorney preparation – has emerged. This second asymmetry leads to trials in the equilibrium of the model. Hay notes that the extent of attorney preparation is not subject to discovery. This is also true of preferences. Farmer and Pecorino (1994) show that asymmetric information on risk preferences can lead to trial, and that this information is neither subject to discovery nor easy to credibly transmit. As a result, this type of asymmetry may tend to persist in the face of mandatory discovery and opportunities for voluntary disclosure.

Other institutional features

There is a voluminous literature which examines how a variety of institutional features affect settlement in civil litigation. What follows is a much abbreviated discussion of a large and complex literature. One difficulty in addressing this question is that even a single institution is likely to have multiple effects on the litigation process. Thus, a single institution may have conflicting effects on the dispute rate and may also have important influences on other aspects of the litigation process.

A classic example of this difficulty is reflected in the analysis of the English rule under which the loser at trial pays the reasonable legal costs of the winner. If the probability of a finding for the plaintiff at trial is private information, then fee shifting at trial reduces settlement rates by, in effect, spreading out the distribution of player types (Bebchuk, 1984). If players are optimistic in their assessments of the probability that the plaintiff will prevail, then fee shifting will aggravate this optimism and reduce the probability that a contract zone will exist (Shavell, 1982).

This prediction – that fee shifting will increase the probability of trial – is made with expenditure at trial held constant. It is well established that the fee shifting at trial will increase expenditure (Braeutigam, Owen, and Panzar, 1984). If the expenditure effect is strong enough, it can result in fewer (but more costly) disputes (Hause, 1989). The English rule also affects the mix of cases which are filed. It discourages cases where there are large stakes but a low probability of success, and encourages low stakes cases with a high probability of success (Shavell, 1982).

Many of the theoretical predictions on fee shifting at trial appear to be borne out in the data (see Hughes and Snyder, 1998).

Under a contingency fee, the plaintiff's lawyer receives a percentage of the judgment at trial if the plaintiff wins the case and nothing if she loses. The effects of contingency fees on the litigation process are very complex and wide ranging (see Rubinfeld and Scotchmer, 1998, for a survey). However, one effect of contingency fees on settlement is clear: if the attorney controls the settlement decision, he will have an excessive incentive to settle the case relative to the interests of his client. The reason is that the attorney bears most of the costs of a trial but is paid only a fraction of the award. On the other hand, if the client controls the case, she may have an excessive incentive to reject a settlement offer and bring the case to trial. (This is particularly true if the contingency percentage is not lower for cases which settle early.)

When a single defendant faces multiple plaintiffs in sequence, some interesting issues regarding settlement arise. Spier (2003a; 2003b) and Daughety and Reinganum (2004) have analysed the use of most favoured nation (MFN) clauses in the context of repeat litigation. Suppose a plaintiff settles early under MFN. If another plaintiff later settles for more, the early settlement is adjusted upward. An MFN clause can be a mechanism whereby the defendant commits to not raising his offer to plaintiffs who settle later in the process. While there is some ambiguity of the effects of MFN on settlement rates and the overall dispute costs (see especially Spier, 2003a), the general thrust of these papers suggests that MFN clauses are efficiency enhancing in the sense that they will reduce the expected dispute costs associated with litigation.

Arbitration

Under conventional arbitration (CA), the arbitrator is free to impose her preferred settlement on the bargaining parties. Under final offer arbitration (FOA), each party to the dispute submits an offer to the arbitrator who must pick one of the submitted offers. While there is some evidence that submitted offers affect the outcome in CA (Farber and Bazerman, 1986), for the purpose of the following discussion we assume that they do not. From a modelling standpoint, this makes CA look exactly like a simple version of civil litigation. Under FOA, the submitted offers clearly affect the outcome, a feature which has important implications for dispute resolution.

Consider the two-type version of the screening model where the plaintiff can have a strong or a weak case. In CA, the defendant will either make an offer that only a weak plaintiff will accept or a pooling offer that both types will accept. If all negotiation takes place prior to the submission of offers to the arbitrator, then under FOA it is possible that the defendant will make an offer that neither type will accept, resulting in a 100 per cent dispute rate (Farmer and Pecorino, 2003). This can occur because the sequentially rational offer submitted to the arbitrator influences the acceptable settlement prior to arbitration. The lack of early settlement allows the defendant to commit to an offer which is optimal against the entire distribution of plaintiff types. Farmer and Pecorino (2003) also show (in contrast to CA) that costless voluntary disclosure never takes place when FOA is the dispute resolution mechanism. The reason is that information has strategic value in this game. Both of the impediments to settlement discussed above disappear if bargaining is permitted after offers are submitted to the arbitrator but prior to the arbitration hearing.

While not totally conclusive on this point, the results of Farmer and Pecorino (1998) also suggest that allowing for bargaining after offers are submitted to the arbitrator can increase settlement for reasons different from those discussed above. Because a submitted offer affects the outcome of arbitration, it will tend to reflect private information. This may in turn promote settlement. Taken together, the results on FOA suggest that the effects of this institution on settlement are sensitive to whether or not bargaining occurs in the face of offers submitted to the arbitrator. In major league baseball, a prominent use of FOA, a good deal of bargaining and settlement occurs after offers have been submitted to the arbitrator.

FOA was proposed by Stephens (1966) and has since become an important alternative to CA. Researchers continue to propose new arbitration mechanisms in the hope of improving the dispute resolution process. Combined arbitration (Brams and Merrill, 1986) is a mixture of FOA and CA. Other proposed mechanisms include tri-offer arbitration (Ashenfelter et al., 1992) and amended final offer arbitration (Zeng, 2003).

AMY FARMER AND PAUL PECORINO

See also **epistemic game theory: an overview; epistemic game theory: incomplete information.**

Bibliography

Ashenfelter, O., Currie, J., Farber, H. and Spiegel, M. 1992. An experimental comparison of dispute rates in alternative arbitration systems. *Econometrica* 60, 1407–33.

Babcock, L. and Loewenstein, G. 1997. Explaining bargaining impasse: the role of self-serving biases. *Journal of Economic Perspectives* 11, 109–26.

Bar-Gill, O. 2002. *The Success and Survival of Cautious Optimism: Legal Rule, and Endogenous Perceptions in Pre-Trial Settlement Negotiations*. Public Law Working Paper No. 35. Cambridge, MA: Harvard Law School.

Bebchuk, L. 1984. Litigation and settlement under imperfect information. *RAND Journal of Economics* 15, 404–15.

Braeutigam, R., Owen, B. and Panzar, J. 1984. An economic analysis of alternative fee shifting systems. *Law and Contemporary Problems* 47, 173–85.

Brams, S. and Merrill, S. 1986. Binding versus final-offer arbitration: a combination is best. *Management Science* 32, 1346–55.

Cho, I. and Kreps, D. 1987. Signaling games and stable equilibria. *Quarterly Journal of Economics* 102, 179–222.

Conlin, M. 1999. Empirical test of a separating equilibrium in National Football League contract negotiations. *RAND Journal of Economics* 30, 289–304.

Daughety, A. 1999. Settlement. In *Encyclopedia of Law and Economics*, Vol. 5, ed. B. Bouckaert and G. de Geest. Cheltenham: Edward Elgar.

Daughety, A. and Reinganum, J. 1994. Settlement negotiations with two-sided asymmetric information: model duality, information distribution, and efficiency. *International Review of Law and Economics* 14, 283–98.

Daughety, A. and Reinganum, J. 2004. Exploiting future settlements: a signaling model of most-favored-nation clauses in settlement bargaining. *RAND Journal of Economics* 35, 467–85.

Farber, H. and Bazerman, M. 1986. The general basis of arbitrator behavior: an empirical analysis of conventional and final offer arbitration. *Econometrica* 54, 819–54.

Farmer, A. and Pecorino, P. 1994. Pretrial negotiations with asymmetric information on risk preferences. *International Review of Law and Economics* 14, 273–81.

Farmer, A. and Pecorino, P. 1998. Bargaining with informative offers: an analysis of final offer arbitration. *Journal of Legal Studies* 27, 415–432.

Farmer, A. and Pecorino, P. 2003. Bargaining with voluntary transmission of private information: does the use of final offer arbitration impede settlement? *Journal of Law, Economics and Organization* 19, 64–82.

Farmer, A. and Pecorino, P. 2005. Civil litigation with mandatory discovery and voluntary transmission of private information. *Journal of Legal Studies* 34, 137–59.

Farmer, A., Pecorino, P. and Stango, V. 2004. The causes of bargaining failure: evidence from major league baseball. *Journal of Law and Economics* 47, 543–68.

Hause, J. 1989. Indemnity, settlement, and litigation, or I'll be suing you. *Journal of Legal Studies* 18, 157–79.

Hay, B. 1995. Effort, information, settlement, trial. *Journal of Legal Studies* 24, 29–62.

Hughes, J. W. and Snyder, E. A. 1998. Allocation of litigation costs: American and English rules. In *The New Palgrave Dictionary of Economics and the Law*, Vol. 1, ed. P. Newman. London: Macmillan.

Landes, W. 1971. An economic analysis of the courts. *Journal of Law and Economics* 14, 61–107.

McConnell, S. 1989. Strikes, wages and private information. *American Economic Review* 79, 801–15.

Osborne, E. 1999. Who should be worried about asymmetric information in litigation? *International Review of Law and Economics* 19, 399–409.

Posner, R. 1973. An economic approach to legal procedure and judicial administration. *Journal of Legal Studies* 2, 399–458.

Priest, G. and Klein, B. 1984. The selection of disputes for arbitration. *Journal of Legal Studies* 13, 215–43.

Reinganum, J. and Wilde, L. 1986. Settlement, litigation, and the allocation of litigation costs. *RAND Journal of Economics* 17, 557–66.

Rubinfeld, D. and Scotchmer, S. 1998. Contingent fees. In *The New Palgrave Dictionary of Economics and the Law*, vol. 1, ed. P. Newman. London: Macmillan.

Schweizer, U. 1989. Litigation and settlement under two-sided incomplete information. *Review of Economic Studies* 56, 163–78.

Shavell, S. 1982. Suit, settlement, and trial: a theoretical analysis under alternative methods for the allocation of legal costs. *Journal of Legal Studies* 11, 55–82.

Shavell, S. 1989. Sharing of information prior to settlement or litigation. *RAND Journal of Economics* 20, 183–195.

Sobel, J. 1989. An analysis of discovery rules. *Law and Contemporary Problems* 52, 133–59.

Spier, K. 1992. The dynamics of pretrial negotiation. *Review of Economic Studies* 59, 93–108.

Spier, K. 1998. Settlement of litigation. In *The New Palgrave Dictionary of Economics and the Law*, Vol. 3, ed. P. Newman. London: Macmillan.

Spier, K. 2003a. 'Tied to the mast': most-favored-nation clauses in settlement contracts. *Journal of Legal Studies* 32, 91–120.

Spier, K. 2003b. The use of 'most-favored-nation' clauses in settlement of litigation. *Rand Journal of Economics* 34, 78–95.

Stephens, C. 1966. Is compulsory arbitration compatible with bargaining? *Industrial Relations*, 5(1), 38–52.

Waldfogel, J. 1998. Reconciling asymmetric information and divergent expectations theories of litigation. *Journal of Law and Economics* 41, 451–76.

Zeng, D. 2003. An amendment to final offer arbitration. *Mathematical Social Sciences*, 46(1), 9–19.

distributed lags

Distributed lag models correlate a single dependent variable with its own lags and with current and lagged values of one or more explanatory variables. Examples concern the current and dynamic correlations between output and investment and between sales and advertising. Distributed lag models typically assume that the explanatory variable is exogenous. (In case of doubt, one usually resorts to vector autoregressive models where two or more variables can be endogenous; see Sims, 1980.)

This article highlights a few aspects of distributed lag models. The two main aspects are representation and interpretation. Useful extended surveys appear in Dhrymes (1971), Griliches (1967) and Hendry, Pagan and Sargan (1984).

Representation

Consider a dependent variable y_t and, for ease of notation, a single explanatory variable x_t. Indicator t runs from 1 to n and it can concern seconds, hours, days or even years. A general (autoregressive) distributed lag model is given by

$$y_t = \mu + \alpha_1 y_{t-1} + \ldots + \alpha_p y_{t-p}$$
$$+ \beta_0 x_t + \beta_1 x_{t-1} + \ldots + \beta_m x_{t-m} + \varepsilon_t,$$
$$(1)$$

where p and m can take any positive integer value, and where it is usually assumed that ε_t is an uncorrelated variable with mean zero and variance σ^2. (Part of the literature assumes the label distributed lags model for the case where $p = 0$ and $m = \infty$. Below we will see that such a model is often approximated by a model as in (1).)

As the model contains the lagged dependent variables, it is called an autoregressive distributed lag model with orders p and m, in short ADL(p, m). The model allows for delayed effects of x_t, as β_0 can be 0, and it also allows for time gaps in these effects when some β parameters are zero and others are not.

Reducing the number of parameters

Basically, given fixed and finite values of p and m, the parameters in (1) can be consistently estimated with ordinary least squares (OLS). (Typically one uses information criteria as those of Akaike or Schwarz to choose the relevant values of p and m in practice.) In practice, p and m can be large, and in theory even as large as ∞. This can be inconvenient, for two reasons. First, the variables y_t and x_t each can be strongly autocorrelated, and then the regression in (1) suffers from multicollinearity. Second, with many parameters in a model there might be many values to evaluate and interpret.

To reduce the number of parameters and to facilitate interpretation, one can impose restrictions. Early suggestions are the Almon and Shiller lag structures, where the parameters are made functions of i, $i = 0, 1, 2, \ldots, m$ (see Almon, 1965, and Shiller, 1973), and the so-called Koyck transformation (see Koyck, 1954).

Almon and Shiller transformations

Consider the version of (1) with $p = 0$ and $m = m$ and set μ at 0 for convenience, that is, consider

$$y_t = \beta_0 x_t + \beta_1 x_{t-1} + \ldots + \beta_m x_{t-m} + \varepsilon_t. \tag{2}$$

Almon (1965) proposes to reduce the number of parameters by assuming the approximation

$$\beta_i = \alpha_0 + \alpha_1 i + \alpha_2 i^2 + \ldots + \alpha_q i^q, \tag{3}$$

with $q > m$. This makes the sequence of β_i parameters a polynomial and hence a smooth function without possibly implausible spikes.

Working out the Almon lags, one can derive that the structure implies that

$$\beta_{i+1} - 2\beta_i + \beta_{i-1} = \gamma_i, \tag{4}$$

where γ_i is a function of α_i values. Shiller (1973) considers this as too restrictive and he proposes to assume that

$$\beta_{i+1} - 2\beta_i + \beta_{i-1} \sim N(0, \zeta^2), \tag{5}$$

for $i = 1, 2, \ldots, m - 1$.

Koyck transformation

The Koyck model can be interpreted as a model which includes adaptive expectations. Suppose that

$$y_t = \alpha + \beta x_t^* + \varepsilon_t, \tag{6}$$

where x_t^* denotes the expected value of x_t, an expectation formed at $t - 1$. When the adaptive expectations schedule is assumed, like

$$x_t^* = \lambda x_{t-1}^* + (1 - \lambda) x_t, \tag{7}$$

with again $|\lambda| < 1$, then substituting (7) into (6) gives

$$y_t = \alpha(1 - \lambda) + \lambda y_{t-1} + \beta(1 - \lambda) x_t$$
$$+ \varepsilon_t - \lambda \varepsilon_{t-1}. \tag{8}$$

The short-run effect of x_t on y_t is $\beta(1 - \lambda)$, while the long-run effect is $\frac{\beta(1-\lambda)}{1-\lambda} = \beta$, as could be expected given (6).

Consider the case were m equals ∞, and where all α parameters are set to zero. When it is further assumed that $\beta_j = \beta_0 \lambda^{j-1}$, with $|\lambda| < 1$ for j is $1, 2, \ldots$, then (1) becomes

$$
\begin{aligned}
y_t = {} & \mu + \beta_0 x_t + \beta_0 \lambda x_{t-1} \\
& + \beta_0 \lambda^2 x_{t-2} + \ldots + \varepsilon_t.
\end{aligned} \tag{9}
$$

Subtracting λy_{t-1} from this expression gives

$$
y_t = (1 - \lambda)\mu + \lambda y_{t-1} + \beta_0 x_t + \varepsilon_t - \lambda \varepsilon_{t-1}, \tag{10}
$$

which is again (8). This Koyck transformation leads to a rather simple model with a moving average (MA) error term. The appropriate estimation method is maximum likelihood, as it is described in, for example, Hamilton (1994, p. 132) for general ARMA models. Note that the parameter λ appears in the autoregressive part and in the MA part.

Restructuring the model

An alternative way to reduce the number of parameters, also in order to facilitate interpretation, is to restructure the model.

To overcome multicollinearity, one can rewrite model (1) in the so-called error correction format. This format combines levels and differences of levels, which is convenient as these are usually much less correlated than the levels themselves, and hence multicollinearity will be much less of a problem. An additional feature of the error correction format is that it provides an immediate look at key parameters such as the total effect, the current effect, and the speed at which the total effect is accomplished.

With Δ_j denoted as the j-th order differencing filter, that is, $\Delta_j y_t = y_t - y_{t-j}$, an error correction representation for (1) reads as

$$
\begin{aligned}
\Delta_1 y_t = {} & \mu + \left(\sum_{j=1}^{p} \alpha_j - 1 \right) \left[y_{t-1} - \frac{\sum_{i=0}^{m} \beta_i}{1 - \sum_{j=1}^{p} \alpha_j} x_{t-1} \right] \\
& + \beta_0 \Delta_1 x_t - \sum_{i=2}^{m} \beta_i \Delta_{i-1} x_{t-1} \\
& - \sum_{j=2}^{p} \alpha_j \Delta_{j-1} y_{t-1} + \varepsilon_t,
\end{aligned} \tag{11}
$$

where lagged levels are suitably combined into differenced variables such that at each lag a higher-order differenced variable appears. This representation even further reduces chances of having multicollinearity. Note that the model can also be written in terms of lagged levels and first differences only, that is as

$$
\begin{aligned}
\Delta_1 y_t = {} & \mu + \left(\sum_{j=1}^{p} \alpha_j - 1 \right) \left[y_{t-1} - \frac{\sum_{i=0}^{m} \beta_i}{1 - \sum_{j=1}^{p} \alpha_j} x_{t-1} \right] \\
& + \beta_0 \Delta_1 x_t + \sum_{i=1}^{m} \gamma_i \Delta_1 x_{t-i} \\
& + \sum_{j=1}^{p} \theta_j \Delta_1 y_{t-j} + \varepsilon_t
\end{aligned} \tag{12}
$$

With the use of (11), all but two parameters (that is, α_1 and β_1) can be directly estimated by using OLS, while $\hat{\alpha}_1$ and $\hat{\beta}_1$ straightforwardly follow from applying OLS to (1). Note that model (11) can also be written such that the levels (now at $t - 1$) enter at $t - 2$ or, say, $t - p$.

Interpretation

We now turn to the interpretation of distributed lag models.

Long-run and short-run effects

The error correction model in (11) provides immediate estimates of current and dynamic effects. (Fok et al., 2006, show that when the series y_t and x_t have a unit root and are cointegrated, as defined by Engle and Granger, 1987, one should speak of the long-run effect, while when the series are stationary there is a total or cumulative effect. For the latter, see also Hendry, Pagan and Sargan, 1984.) The current effect is β_0 and the long-run or total effect is

$$
\frac{\sum_{i=0}^{m} \beta_i}{1 - \sum_{j=1}^{p} \alpha_j}. \tag{13}
$$

Note that the long-run effect can be larger or smaller than the short-run effect, depending on the values of the parameters. The parameters in the error correction model, when written as

$$
\begin{aligned}
\Delta_1 y_t = {} & \mu + \rho[y_{t-1} - \gamma x_{t-1}] + \beta_0 \Delta_1 x_t \\
& - \sum_{i=2}^{m} \beta_i \Delta_{i-1} x_{t-1} \\
& - \sum_{j=2}^{p} \alpha_j \Delta_{j-1} y_{t-1} + \varepsilon_t,
\end{aligned} \tag{14}
$$

can be estimated using non-linear least squares. This method provides direct estimates of the long-run effect γ and its associated standard error.

Duration interval
As well as the long-run and short-run effects, one may also be interested in the speed with which the effect of x_t decays over time. To be able to compute this so-called duration interval, one needs explicit expressions for $\frac{\partial y_{t+k}}{\partial x_t}$ for all values of k running from 1 to, potentially, ∞. Given the expression in (1), these expressions are easily derived as

$$\frac{\partial y_t}{\partial x_t} = \beta_0$$

$$\frac{\partial y_{t+1}}{\partial x_t} = \beta_1 + \alpha_1 \frac{\partial y_t}{\partial x_t}$$

$$\frac{\partial y_{t+2}}{\partial x_t} = \beta_2 + \alpha_1 \frac{\partial y_{t+1}}{\partial x_t} + \alpha_2 \frac{\partial y_t}{\partial x_t}$$

$$\vdots$$

$$\frac{\partial y_{t+k}}{\partial x_t} = \beta_k + \sum_{j=1}^{k} \alpha_j \frac{\partial y_{t+(k-j)}}{\partial x_t}$$

where it should be noted that $\alpha_k = 0$ for $k > p$, and that $\beta_k = 0$ for $k > m$. Hence, the final form of a distributed lag model (see Harvey, 1990), is

$$y_t = \sum_{i=0}^{\infty} \delta_i x_{t-i} + error, \qquad (15)$$

where

$$\delta_i = \frac{\partial y_{t+i}}{\partial x_t}. \qquad (16)$$

With these δ_i, one can derive all kinds of summary effects (like mean and median, or half lives of shocks) of x_t on y_t.

When δ_i decays monotonically, it is useful to define the decay factor by

$$p_k = \frac{\frac{\partial y_t}{\partial x_t} - \frac{\partial y_{t+k}}{\partial x_t}}{\frac{\partial y_t}{\partial x_t}}. \qquad (17)$$

This can be computed only for discrete values of k as there are only discrete time intervals. This decay factor is a function of the model parameters. Through interpolation, one can decide on the time k it takes for the decay factor to be equal to some value of p, which typically is equal to 0.95 or 0.90. This estimated time k is then called

the p per cent duration interval. This measure is frequently used in advertising research (see Clarke, 1976; Leeflang et al., 2000; Franses and Vroomen, 2006).

Final issues
Distributed lag models continue to be a standard empirical approach. When the models are applied, there are at least two further issues that one needs to address, that is, next to selecting p and m and a useful transformation. The first concerns the statistical analysis of the model. For example, if y_t and x_t are not stationary, one needs to rely on cointegration techniques that involve non-standard asymptotic theory. The theory that is most relevant here is formulated in Boswijk (1995). Also, in the case of the Koyck model, one faces the so-called Davies (1987) problem. Under the null hypothesis that $\beta_0 = 0$, the model collapses to $y_t = \varepsilon_t$, and hence λ is not identified then.

The second issue concerns aggregation over time. It may be that y_t and x_t are not available at the same sampling frequency. For example, television commercials last for 30 seconds and recur each hour, say, while sales data are available only at the weekly level. Tellis and Franses (2006) have a few recent results, but more work is needed.

PHILIP HANS FRANSES

Bibliography
Almon, S. 1965. The distributed lag between capital appropriations and expenditures. *Econometrica* 33, 178–96.
Boswijk, H. 1995. Efficient Inference on cointegration parameters in structural error–correction models. *Journal of Econometrics* 69, 133–58.
Clarke, D. 1976. Econometric measurement of the duration of advertising effect on sales. *Journal of Marketing Research* 8, 345–57.
Davies, R. 1987. Hypothesis testing when a nuisance parameter is present only under the alternative. *Biometrika* 64, 247–54.
Dhrymes, P. 1971. *Distributed Lags: Problems of Estimation and Formulation*. San Francisco: Holden-Day.
Engle, R. and Granger, C. 1987. Cointegration and error correction: representation, estimation, and testing. *Econometrica* 55, 251–76.
Fok, D., Horvath, C., Paap, R. and Franses, P. 2006. A hierarchical Bayes error correction model to explain dynamic effects of price changes. *Journal of Marketing Research* 43, 443–61.
Franses, P. and Vroomen, B. 2006. Estimating confidence bounds for advertising effect duration intervals. *Journal of Advertising* 35(Summer), 33–37.
Griliches, Z. 1967. Distributed lags: a survey. *Econometrica* 35, 16–49.

Hamilton, J. 1994. *Time Series Analysis*. Princeton: Princeton University Press.

Hansen, B. 1996. Inference when a nuisance parameter is not identified under the null hypothesis. *Econometrica* 64, 413–30.

Harvey, A. 1990. *The Econometric Analysis of Time Series*. London: Philip Allan.

Hendry, D., Pagan, A. and Sargan, J. 1984. Dynamic specification. In *Handbook of Econometrics*, vol. 2, ed. Z. Griliches and M. Intriligator. Amsterdam: North-Holland.

Koyck, L. 1954. *Distributed Lags and Investment Analysis*. Amsterdam: North-Holland.

Leeflang, P., Wittink, D., Wedel, M. and Naert, P. 2000. *Building Models for Marketing Decisions*. Boston: Kluwer.

Shiller, R. 1973. A distributed lag estimator derived from smoothness priors. *Econometrica* 41, 775–88.

Sims, C. 1980. Macroeconomics and reality. *Econometrica* 48, 1–48.

Tellis, G. and Franses, P. 2006. The optimal data interval for econometric models of advertising. *Marketing Science* 23, 217–229.

distributive politics and targeted public spending

While conventional models of political economy, such as the median voter model, focus on the provision of national public goods, most federal spending programmes, such as the US interstate highway system, are more aptly characterized as local in nature. While in the United States the benefits of federal spending are concentrated in specific geographic units, such as states, counties, and Congressional districts, the associated tax costs are, by contrast, geographically dispersed. This common pool feature of federal spending – concentrated spending but dispersed financing – leads to a geographic tug-of-war in which jurisdictions attempt to increase own-jurisdiction spending but to reduce spending elsewhere due to the associated tax costs. This conflict between jurisdictions is reflected most intensely in the budget process within the US Congress, whose members are locally elected and thus naturally respond to these common pool incentives.

In this article, I first summarize evidence suggesting that Congressional representatives are responsive to the common pool incentives associated with concentrated spending but dispersed costs. Having established the empirical saliency of this common pool problem in Congress, I then summarize the literature examining how this conflict is resolved. In particular, I analyse the effects of Congressional delegation characteristics, such as size, ideology, seniority, and committee assignments, on the geographic allocation of federal funds. Finally, I review evidence on the effects of the geographic distribution of federal funds on electoral outcomes.

As described in Knight (2006), common pool problems underpin several theoretical models of the legislative process, such as the universalism model of Weingast, Shepsle, and Johnsen (1981) and the legislative bargaining model of Baron and Ferejohn (1989). Whether or not Congressional delegations respond to these incentives in practice, however, is primarily an empirical question. It may be the case, for example, that political parties, or related Congressional organizations, serve as collective mechanisms through which legislators internalize the tax costs in other jurisdictions associated with own-jurisdiction spending. One of the first papers to directly measure the responsiveness of representatives to common pool problems is by DelRossi and Inman (1999), who examine the geographic distribution of water projects authorized by the Water Resources Development Act of 1986. In particular, the authors compare the size of project requests before and after changes in local matching requirements, which significantly increased the fraction of project costs financed by local governments. As hypothesized, districts experiencing larger increases in matching rates requested significantly less funding for water projects. In a similar vein, Knight (2004b) examines Congressional voting in 1998 over whether to finance a set of transportation projects, which were earmarked for specific Congressional districts and were funded primarily via federal gasoline taxes. As predicted, support for funding was concentrated in those districts receiving more in funding and also in those districts with lower gasoline tax burdens.

How is this geographic battle between jurisdictions resolved? Which states and Congressional districts win and why? Regarding the mere size of delegations, an important feature of the US Congress is its bicameral structure in which each state has an equal number of delegates in the Senate but in which seats are apportioned between states according to population in the House of Representatives. This equality of delegation sizes in the US Senate provides small states with power disproportionate to their population; Senators from California, the largest state, currently have over 60 times as many constituents as do senators from Wyoming, the smallest state. In attempting to measure the magnitude of this small-state bias, Atlas et al. (1995) and Lee (1998) find that small states receive significantly more per capita in aggregate federal spending than do large states. While this finding is certainly provocative, it is difficult to distinguish between the role of Senate representation and other factors, such as population density, that make small states inherently different from larger states. In attempting to address this issue of unobserved differences between small and large states, Knight (2004a) demonstrates that small states receive considerably more per-capita funding in projects earmarked in Senate bills; in House bills, by contrast, small and large states receive similar project spending on a per-capita basis. Knight (2004b) also identifies two theoretical channels underlying this small-state bias in the

US Senate. Relative to their population, small states are disproportionately represented on key committees (the proposal power channel) but are also cheaper coalition partners (the vote cost channel) given that they pay a smaller share of federal taxes. Interestingly, both channels are shown to be empirically important and, taken together, explain over 90 per cent of the measured small-state bias. In a related study of the size of delegations, Falk (2006) studies discontinuities in the apportionment of seats in the US House arising from both timing (re-apportionment occurs once every ten years) and rounding issues (delegation sizes must be integers). Using this variation in delegation sizes, he finds that increases in seats per capita lead to statistically significant increases in federal spending per capita.

Delegations of similar sizes, however, may differ significantly in their composition. Key differences between delegations in the degree of political power include majority party affiliation, seniority, and representation on key committees. Regarding majority party affiliation, Levitt and Snyder (1995) find that the Democratic Party used its majority control of Congress to channel federal funds into Congressional districts with a high percentage of Democratic voters during the period 1984–90. However, they find no evidence that, conditional on the percentage of Democratic voters, districts represented by Democrats received higher federal spending. Levitt and Poterba (1999) report that states with very senior Democratic representatives experienced more rapid economic growth than did other states. However, they find no relationship between the partisan affiliation of delegations and the geography of federal spending, a key hypothesized channel of the measured differences in economic growth. Regarding the role of Congressional committees, Knight (2005) finds that Congressional districts represented on key committees received substantially more funding in projects earmarked in transportation bills authorized in 1991 and 1998. He interprets this result as evidence of the importance of proposal power associated with the committee's ability to set the legislative agenda. De Figueiredo and Silverman (2002) examine interactions between committee representation and lobbying in an empirical examination of earmarked projects for universities. In particular, they find a strong correlation between lobbying outlays by universities and the receipt of federal funding; this link between lobbying and spending, however, is found to be much stronger for those universities located in districts that are represented on key appropriations committees.

We have focused throughout this survey on the determinants of the geographic distribution of federal funds. Politicians, however, have an incentive to put forth the effort to secure project funding only if they perceive that the associated political gains are sufficiently high. While clearly important, measurement of the effects of federal spending on incumbent vote shares is plagued with endogeneity problems. For example, incumbents facing the strongest opposition have the strongest incentives to put forth effort in securing funds. Thus, there may be a downward bias in ordinary least squares (OLS) estimates of the effect of federal spending on incumbent vote shares. As an instrument for district-specific federal spending, Levitt and Snyder (1997) use federal spending outside of the district but within the state. The idea is that other actors, such as Senators or governors, also play a role in the geographic distribution of federal funds. Using this exogenous variation in federal spending, they conclude that an additional $100 per capita in spending translates into an additional two percentage points in incumbent vote shares.

We conclude that common pool problems associated with concentrated project benefits but dispersed costs are reflected not only in the behaviour of Congressional delegations but also in the resulting distribution of federal funds. Who wins and who loses in this geographic battle is determined in part by state size and the political power of delegations. Consistent with these results, evidence suggests that incumbent re-election prospects are significantly enhanced by increases in federal spending.

BRIAN G. KNIGHT

See also **campaign finance, economics of; fiscal federalism; intergovernmental grants; local public finance; political institutions, economic approaches to; public choice.**

Bibliography

Atlas, C.M., Gilligan, T.W., Hendershott, R.J. and Zupan, M.A. 1995. Slicing the federal government net spending pie: who wins, who loses, and why. *American Economic Review* 85, 624–29.

Baron, D.P. and Ferejohn, J.A. 1989. Bargaining in legislatures. *American Political Science Review* 83, 1881–207.

de Figueiredo, J.M. and Silverman, B.S. 2002. Academic earmarks and the returns to lobbying. Working Paper No. 9064. Cambridge, MA: NBER.

DelRossi, A.F. and Inman, R.P. 1999. Changing the price of pork: the impact of local cost sharing on legislators' demands for distributive public goods. *Journal of Public Economics* 71, 247–73.

Falk, J. 2006. The effects of Congressional district size and representative's tenure on the allocation of federal funds. Working paper, University of California, Berkeley.

Knight, B.G. 2004a. Legislative representation, bargaining power, and the distribution of federal funds: evidence from the U.S. Senate. Working Paper No. 10385. Cambridge, MA: NBER.

Knight, B.G. 2004b. Parochial interests and the centralized provision of local public goods: evidence from congressional voting on transportation projects. *Journal of Public Economics* 88, 845–66.

Knight, B.G. 2005. Estimating the value of proposal power. *American Economic Review* 95, 1639–52.

Knight, B.G. 2006. Common tax pool problems in federal systems. In *Democratic Constitutional Design and Public Policy Analysis and Evidence*, ed. R.D. Congleton and B. Swedenborg. Cambridge, MA: MIT Press.

Lee, F.E. 1998. Representation and public policy: the consequences of Senate apportionment for the geographic distribution of federal funds. *Journal of Politics* 60, 34–62.

Levitt, S.D. and Poterba, J.M. 1999. Congressional distributive politics and state economic performance. *Public Choice* 99, 185–216.

Levitt, S.D. and Snyder, J.M. 1995. Political parties and the distribution of federal outlays. *American Journal of Political Science* 39, 958–80.

Levitt, S.D. and Snyder, J.M. 1997. The impact of federal spending on House election outcomes. *Journal of Political Economy* 105, 30–53.

Weingast, B.R., Shepsle, K.A. and Johnsen, C. 1991. The political economy of benefits and costs: a neoclassical approach to distributive politics. *Journal of Political Economy* 89, 642–64.

dividend policy

There are two major ways in which a firm can distribute cash to its common stockholders. The firm can either declare a cash dividend which it pays to all its common stockholders or it can repurchase shares. Stock repurchases may take the form of registered tender offers, open market purchases, or negotiated repurchases from a large shareholder. In a share repurchase, shareholders may choose not to participate. In contrast, dividends are direct cash payments to shareholders and are distributed on a pro rata basis to all shareholders.

Most firms pay cash dividends on a quarterly basis. The dividend is declared by the firm's board of directors on a date known as the 'announcement date'. The board's announcement states that a cash payment will be made to stockholders who are registered owners on a given 'record date.' The dividend checks are mailed to stockholders on the 'payment date', which is usually about two weeks after the record date. Stock exchange rules generally dictate that the stock is bought or sold with the dividend until the 'ex-dividend date', which is a few business days before the record date. After the ex-dividend date, the stock is bought and sold without the dividend.

Dividends may be either labelled or unlabelled. Most dividends are not given labels by management. Unlabelled dividends are commonly referred to as 'regular dividends'. When managers label a dividend, the most common label is 'extra'.

A historical perspective

Prior to 1961, academic treatments of dividends were primarily descriptive, as, for example, in Dewing (1953).

To the extent that economists considered corporate dividend policy, the commonly held view was that investors preferred high dividend payouts to low payouts (see, for example, Graham and Dodd, 1951). The only question was how much value was attached to dividends relative to capital gains in valuing a security (Gordon, 1959). This view was concisely summarized with the saying that a dividend in the hand is worth two (or some multiple) of those in the bush. The only question was: what is the multiple?

In 1961, scientific inquiry into the motives and consequences of corporate dividend policy shifted dramatically with the publication of a classic paper by Miller and Modigliani. Perhaps the most significant contribution of the Miller and Modigliani paper was to spell out in careful detail the assumptions under which their analysis was to be conducted. The most important of these include the assumptions that the firm's investment policy is fixed and known by investors, that there are no taxes on dividends or capital gains, that individuals can costlessly buy and sell securities, that all investors have the same information, and that investors have the same information as the managers of the firm. With this set of assumptions, Miller and Modigliani demonstrate that a firm's stockholders are indifferent among the set of feasible dividend policies. That is, the value of the firm is independent of the dividend policy adopted by management.

Because investment policy is fixed in the Miller–Modigliani set-up, all feasible dividend policies involve the distribution of the full present value of the firm's free cash flow (that is, cash flow in excess of that required for investment) and are, therefore, equally valuable. If internally generated funds exceed required investment, the excess must be paid out as a dividend so as to hold investment constant. If internally generated funds are insufficient to fund the fixed level of investment, new shares must be sold. It is also possible for managers to finance a higher dividend with the sale of new shares.

The key insight from the Miller–Modigliani analysis is that investors will be indifferent among the feasible dividend choices because they can costlessly create their own dividend stream by buying and selling shares. If investors demand higher dividends than the amount paid by the firm, they can sell shares and consume the proceeds, leaving themselves in the same position as if the firm had paid a dividend. Alternatively, if shareholders prefer to reinvest rather than to consume, they can choose to purchase new shares with any dividends paid. In this instance, shareholders would be in the same position that that they would have been in had no dividends been paid. Thus, regardless of corporate dividend policy, investors can costlessly create their own dividend position. For this reason, stockholders are indifferent to corporate dividend policy, and, as a consequence, the value of the firm is independent of its dividend policy.

After a brief flurry of debate, the Miller–Modigliani irrelevance proposition was essentially universally accepted as correct under their set of assumptions. There nevertheless remained an underlying notion that dividend policy must 'matter' given that managers and security analysts spend time worrying about it. If so, and if the Miller–Modigliani proposition is accepted, it must be due to violation of one or more of the Miller–Modigliani assumptions in the real world.

Since the early 1960s, the dividend debate has been lively and interesting. Economists have analysed theoretically whether the relaxation of the various Miller–Modigliani assumptions alters their irrelevance proposition. In addition, economists have analysed the data from several perspectives. First, they have undertaken an array of analyses to determine the effect, if any, of dividend policy on stock value and firm performance. Second, they have sought to identify the characteristics associated with dividend payments (or the lack thereof) by individual firms. Third, they have attempted to characterize macroeconomic trends in the level and propensity of firms to pay dividends, and in the form of the payout. Our discussion of these issues focuses primarily (though not exclusively) on studies of US firms since these are the studies most accessible to us.

Relaxing the Miller–Modigliani assumptions

Taxes

Perhaps the obvious starting point for an investigation into the effect of relaxing the Miller–Modigliani assumptions is to introduce taxes. In the United States, dividend payments by a corporation do not affect that firm's taxes. However, at least historically, dividends have been taxed at a higher rate than capital gains at the personal level. Thus, superficially, the US tax code appears to favour a low dividend payout policy, with payouts occurring primarily through share repurchases.

Under the assumption that dividends and capital gains are taxed differentially, Brennan (1970) derives a model of stock valuation in which stocks with high payouts have higher required before-tax returns than stocks with low payouts. As a counterpoint to this proposition, Miller and Scholes (1978) argue that under the US tax code there exist sufficient loopholes so that investors can shelter dividend income from taxation, thereby driving the effective tax rate on dividends to zero. Early studies of the association between stock returns and dividend yield (for example, Black and Scholes, 1974; Litzenberger and Ramaswamy, 1979; Miller and Scholes, 1982) yielded mixed results using different definitions of dividend yield. Subsequent studies indicated that the correlation between dividend yield and stock returns (if any) appeared to be due to omitted risk factors that were correlated with dividend yield. For example, Chen, Grundy and Stambaugh (1990) report that dividend yield and risk measures are cross-sectionally correlated. Similarly, Fama and French (1993) show that, when a three-factor model for expected returns is used, there is

no significant relation between dividend yields and stock returns.

Other studies have analysed the potential effects of the differential taxation of dividends and capital gains by studying the behaviour of stock prices and trading volume around ex-dividend days. The logic of these studies is that, in order for investors to be indifferent between selling a stock just before it goes ex dividend and just after, stocks should be priced so that the marginal tax liability would be the same for each strategy. Thus, if dividends are taxed more heavily than are capital gains, stock prices should fall by less than the size of the dividend on the ex-dividend day. Evidence consistent with a tax effect in stock price behaviour around ex-dividend days is provided in Elton and Gruber (1970), Eades, Hess and Kim (1984), Green and Rydqvist (1999), Bell and Jenkinson (2002), and Elton, Gruber and Blake (2005). In addition, evidence of tax-motivated trading around ex-dividend days is provided in Lakonishok and Vermaelen (1986), Michaely and Vila (1995) and Green and Rydqvist (1999).

Collectively, the evidence in these studies indicates that the differential taxation of dividends and capital gains affects both ex-dividend day stock returns and trading activity. This conclusion has been reinforced in studies that examine changes in tax laws (for example, Poterba and Summers, 1984; Barclay, 1987; Michaely, 1991). Nonetheless, the fact that individual investors in high tax brackets receive large amounts of taxable dividends each year (Allen and Michaely, 2003) casts doubt on taxes being a first-order determinant of dividend policy.

Agency costs

A second real-world violation of the Miller–Modigliani assumptions is the existence of agency costs associated with stock ownership. In particular, managers of firms maximize their own utility, which is not necessarily the same as maximizing the market value of common stock. The costs associated with this potential conflict of interest include expenditures for structuring monitoring and bonding contracts between shareholders and managers, and residual losses due to imperfectly constructed contracts (Jensen and Meckling, 1976).

Several authors have argued that dividends may be important in helping to resolve manager–shareholder conflicts. If dividend payments reduce agency costs, firms may pay dividends even if these payments are taxed disadvantageously.

Easterbrook (1984) and Rozeff (1982) argue that establishing a policy of paying dividends enables managers to be evaluated periodically by the capital market. By paying dividends, managers are required to tap the capital market more frequently to obtain funds for investment projects. Periodic review by the market is one way in which agency costs are reduced, which in turn raises the value of the firm. Similarly, Jensen (1986) argues that establishing a policy of paying dividends reduces agency problems of overinvestment by reducing the amount of discretionary cash controlled by managers.

An implication of the agency models is that dividends will be more valuable in mature firms with substantial cash flow and poor investment opportunities. Early tests of this implication focused on the stock price reaction to dividend change announcements and produced mixed results. Lang and Litzenberger (1989) find that firms with less valuable growth opportunities exhibit a larger stock price reaction to dividend increase announcements than firms with more valuable growth opportunities. Although this finding is consistent with the agency cost hypothesis, Denis, Denis and Sarin (1994) find that when they control for other factors, particularly the change in dividend yield, they find no difference in the stock price reaction to dividend changes between firms with good growth opportunities and those with poor growth opportunities. Moreover, they find no evidence that increases in dividends reduce corporate investment.

More recent tests of the agency models have focused on the cross-sectional determinants of dividend policy. Fama and French (2001) find that the propensity to pay dividends is positively related to firm size and profitability, and negatively related to the value of future growth opportunities. DeAngelo, DeAngelo and Stulz (2006) find that the propensity to pay dividends is strongly associated with the proportion of the firm's equity that comes from retained earnings. These findings support the primary prediction of the agency models that dividends are more valuable for mature firms with high cash flow and poor growth opportunities.

La Porta et al. (2000) and Faccio and Lang (2002) provide further support for the agency models of dividend policy by analysing international evidence. La Porta et al. hypothesize that agency conflicts will differ across countries because of differences in the extent of investor protection. In a sample of 33 different countries, they find that dividend payments are higher in countries with better investor protection. This indicates that when investors are better able to monitor managers, they are able to force higher dividend payouts. Faccio and Lang (2002) show that in western Europe and in Asia dividend payments are higher when controlling shareholders have a higher ratio of voting rights to cash flow rights – that is, those situations in which minority shareholders are otherwise at greatest risk of expropriation by the controlling shareholder.

Asymmetric information
Contrary to the Miller–Modigliani assumption that investors have the same information as managers, a large number of studies assume that managers possess more information about the prospects of the firm than individuals outside the firm, and that dividend changes convey this information to outsiders. This idea was suggested by Miller and Modigliani and has roots in Lintner's (1956) classic study on dividend policy. Lintner

interviewed a sample of corporate managers. One of the primary findings of the interviews is that a high proportion of managers attempt to maintain a stable regular dividend. In Lintner's words, managers demonstrate a 'reluctance (common to all companies) to reduce regular rates once established and a consequent conservatism in raising regular rates' (1956, p. 84).

If managers change regular dividends only when the earnings potential of the firm has changed, changes in regular dividends are likely to provide some information to the market about the firm's prospects. More formal models in which dividends convey information to outsiders include Bhattacharya (1979; 1980), John and Williams (1985), and Miller and Rock (1985). The common assumption in these models is that managers have information not available to outside investors. Typically, the information has to do with the current or future earnings of the firm.

Empirical evidence on the information content of dividends has taken three forms. First, a large set of studies has analysed whether dividend changes are associated with abnormal stock returns of the same sign. Second, studies have analysed whether dividend changes are associated with subsequent earnings changes. Third, studies have analysed the association between dividend changes and changes in investor expectations regarding future earnings.

Studies have consistently documented that stock returns around the announcement of a dividend change are positively correlated with the change in the dividend (Aharony and Swary, 1980; Asquith and Mullins, 1983; Brickley, 1983; Healy and Palepu, 1988; Grullon, Michaely and Swaminathan, 2002; Michaely, Thaler and Womack, 1995; Pettit, 1972). These studies are robust over time and are robust to controls for contemporaneous earnings announcements. Moreover, in general, the studies indicate that the market reacts more strongly to a dividend decrease than to a dividend increase.

The findings described above indicate that dividend announcements provide information to the market. Subsequent studies have investigated whether this information is correlated with current or future earnings. On this issue, the evidence is more mixed. In a study of dividend initiations and omissions, Healy and Palepu (1988) find that the initiation of dividends follows a period of abnormal earnings growth and that earnings continue to grow in subsequent years. For omissions, however, earnings decline in the year of omission, then rebound in the following years. Using a comprehensive sample of dividend changes, Benartzi, Michaely and Thaler (1997) find no evidence that dividend changes are associated with subsequent earnings changes of the same sign. Miller's interpretation of the evidence (1987) is that dividends appear to be better described as lagging earnings than as leading earnings.

One difficulty in testing whether dividend changes 'signal' unexpected future earnings is that it is difficult to identify what level of earnings would be expected by the market if the dividend change did not take place. To address this issue, Ofer and Siegel (1987) study how analysts alter their estimates of current year earnings when firms announce dividend changes. They find that analysts revise their earnings estimates in the direction of the dividend change and that the size of the earnings revision is positively associated with the stock price reaction to the dividend change. Similarly, Fama and French (1998) report a positive association between dividends and firm value after controlling for past, current and future earnings, as well as investment and debt. They conclude that dividends contain information about value that is not contained in earnings, investment and debt.

The accumulated empirical evidence thus indicates that dividend announcements provide information to the market. Whether they convey information about future earnings is less clear. Moreover, other findings indicate that information signalling is unlikely to be a first-order determinant of dividend policy. For example, as noted earlier, dividends are paid primarily by larger, more mature firms with higher cash flow and poorer growth opportunities. These types of firm would seem to be least in need of signalling their true value to the market.

Firm value and the form of the payout

As with increases in regular cash dividends, specially labelled cash dividends and share repurchases have been shown to be accompanied by permanent increases in stock prices (Brickley, 1983; Dann, 1981; Vermaelen, 1981). However, there is little agreement on the factors that lead managers to choose one method over another.

Given the Miller–Modigliani assumptions, the choice of the payout mechanism, like the choice of dividend policy itself, does not affect the value of the firm. Therefore, if the form of the payout is to matter, it must be due to violation of one or more of the Miller–Modigliani assumptions. To develop a theory to explain the choice of payout mechanism, it must be that there are differential costs or benefits associated with the alternative payout methods. Furthermore, the relative benefits or costs must be especially significant because, in general, dividends have been tax-disadvantaged (at the personal level) relative to share repurchases.

Economists have explored several possible explanations as to why a particular form of payout is chosen, including adverse selection effects (Barclay and Smith, 1988; Miller and McConnell, 1995), the impact on equity ownership structure (Stulz, 1988; Denis, 1990), the signalling power of alternative payout mechanisms (Ofer and Thakor, 1987; Jagannathan, Stephens and Weisbach, 2000), and the impact of executive stock options (Fenn and Liang, 2001). The evidence indicates that share repurchases are more likely when recent earnings increases are temporary, when earnings are riskier, when firms make heavy use of stock options in executive compensation contracts and when firms seek to protect themselves from a hostile takeover.

As regards the choice between regular cash dividends and specially labelled cash dividends, reasonable explanations have been relatively scarce. Brickley (1983) does provide evidence that specially labelled dividends convey a less positive message about firm value than do increases in regular cash dividends. Nonetheless, it is unclear why this is so. Moreover, there has been little examination of the choice between special dividends and share repurchases.

What managers say

Lintner's (1956) classic empirical study began with a survey of corporate executives. The results of that survey and the accompanying evidence laid the foundation for much of the empirical and theoretical work that has followed over the succeeding half century. Brav et al. (2005) have conducted a new and more extensive survey of chief financial officers (CFOs) regarding their views of corporate payout policy. Their survey yields further insights into what managers think about dividend policy, and complements the existing empirical evidence.

Brav et al. report that CFOs view dividends as inflexible in that, once a dividend level has been established, any dividend cut is likely to have a significantly adverse impact on the company's stock price. Thus, consistent with Lintner's (1956) original observation, managers tend to be conservative when adjusting dividends upward in order to avoid having to cut the dividend at a later date. Rather than establishing a target payout ratio, managers set a per share payment that is downwardly inflexible. According to the survey, managers do not explicitly view dividends as a mechanism for signalling information that would distinguish their companies from competitors, and they consider tax effects only as an afterthought. These observations accord with the conclusions drawn from empirical studies in that both imply that taxes and signalling are not first-order determinants of dividend policy.

In contrast to dividends, repurchases are viewed by managers as a parallel but more flexible way to distribute cash to shareholders in that they can be initiated and discontinued as funds are available. This observation is consistent with the empirical evidence cited earlier that repurchases tend to be associated with temporary increases in earnings, while dividends are associated with earnings changes that are more permanent. Whether the modern survey of Brav et al. leads to the volume of additional empirical work that followed Lintner's study remains to be seen.

Summary and recent trends

Since the mid-1960s, rigorous consideration has added considerably to progress in what is known about dividend policy. We know that firms pay out to stock-holders substantial amounts of cash annually in the form of regular cash dividends, share repurchases and specially labelled dividends. We also know that stock prices increase permanently when regular dividends are increased, when special dividends are declared, and when shares are repurchased, and that stock prices decline when regular dividends are reduced. While these findings imply that dividend changes reflect information available to managers that is not otherwise available to outside investors, it is still not clear what information is being conveyed through the dividend payment. Moreover, although we now know a considerable amount about the empirical determinants of the size of payout and the form of payout, there is little agreement as to whether the level of cash payout affects the value of the firm or and whether the choice of the payout method matters.

We conclude by outlining several recent trends that pose additional challenges to our understanding of dividend policy. First, Fama and French (2001) document that the propensity to pay dividends has declined substantially since the late-1970s. Second, despite this decline in the propensity to pay dividends, aggregate dividends have not declined (DeAngelo, DeAngelo and Skinner, 2004). Rather, dividends and earnings have become increasingly concentrated among larger firms. Third, specially labelled dividends have nearly disappeared (DeAngelo, DeAngelo and Skinner, 2000). Fourth, share repurchases have increased substantially so that aggregate payouts through share repurchases now exceed those through regular dividends (Grullon and Michaely, 2002). These trends are difficult to explain given our current understanding of dividend policy. Undoubtedly, therefore, economists will continue to devote substantial effort to understanding the puzzles of dividend policy.

DAVID J. DENIS AND JOHN J. MCCONNELL

See also **corporate finance; finance (new developments); Modigliani–Miller theorem.**

Bibliography

Aharony, J. and Swary, I. 1980. Quarterly dividend and earnings announcements and stockholders' returns: an empirical analysis. *Journal of Finance* 35, 1–12.

Allen, F. and Michaely, R. 2003. Payout policy. In *Handbook of the Economics of Finance: Volume 1a*, ed. G. Constantinides, M. Harris and R. Stulz. Amsterdam: North-Holland.

Asquith, P. and Mullins, D. 1983. The impact of initiating dividend payments on shareholders' wealth. *Journal of Business* 56, 77–96.

Barclay, M. 1987. Dividends, taxes, and common stock prices: the ex-dividend day behavior of common stock prices before the income tax. *Journal of Financial Economics* 14, 31–44.

Barclay, M. and Smith, C. 1988. Corporate payout policy: cash dividends versus open-market repurchases. *Journal of Financial Economics* 22, 61–82.

Bell, L. and Jenkinson, T. 2002. New evidence of the impact of dividend taxation and on the identity of the marginal investor. *Journal of Finance* 57, 1321–46.

Benartzi, S., Michaely, R. and Thaler, R. 1997. Do changes in dividends signal the future or the past? *Journal of Finance* 52, 1007–43.

Bhattacharya, S. 1979. Imperfect information, dividend policy, and 'the bird in the hand' fallacy. *Bell Journal of Economics* 10, 259–70.

Bhattacharya, S. 1980. Nondissipative signaling structures and dividend policy. *Quarterly Journal of Economics* 95, 1–24.

Black, F. and Scholes, M. 1974. The effects of dividend yield and dividend policy on common stock prices and returns. *Journal of Financial Economics* 1, 1–22.

Brav, A., Graham, J., Michaely, R. and Harvey, C. 2005. Payout policy in the 21st century. *Journal of Financial Economics* 77, 483–527.

Brennan, M. 1970. Taxes, market valuation and financial policy. *National Tax Journal* 23, 417–29.

Brickley, J. 1983. Shareholders wealth, information signaling, and the specially designated dividend: an empirical study. *Journal of Financial Economics* 12, 187–209.

Chen, N., Grundy, B. and Stambaugh, R. 1990. Changing risk, changing risk premiums, and dividend yield effects. *Journal of Business* 63, S51–S70.

Dann, L. 1981. Common stock repurchases: an analysis of returns to bondholders and stockholders. *Journal of Financial Economics* 9, 113–38.

DeAngelo, H., DeAngelo, L. and Skinner, D. 2000. Special dividends and the evolution of dividend signaling. *Journal of Financial Economics* 57, 309–354.

DeAngelo, H., DeAngelo, L. and Skinner, D. 2004. Are dividends disappearing? Dividend concentration and the consolidation of earnings. *Journal of Financial Economics* 72, 425–56.

DeAngelo, H., DeAngelo, L. and Stulz, R. 2006. Dividend policy and the earned/contributed capital mix: a test of the life-cycle theory. *Journal of Financial Economics* 81, 227–54.

Denis, D. 1990. Defensive changes in corporate payout policy: share repurchases and special dividends. *Journal of Finance* 45, 1433–56.

Denis, D., Denis, D. and Sarin, A. 1994. The information content of dividend changes: cash flow signaling, overinvestment, and dividend clienteles. *Journal of Financial and Quantitative Analysis* 29, 567–87.

Dewing, A. 1953. *The Financial Policy of Corporations*, vol. 2, 5th edn. New York: Ronald Press Co.

Eades, K., Hess, P. and Kim, E. 1984. On interpreting security returns during the ex-dividend period. *Journal of Financial Economics* 13, 3–34.

Easterbrook, F. 1984. Two agency–cost explanations of dividends. *American Economic Review* 74, 650–9.

Elton, E. and Gruber, M. 1970. Marginal stockholders' tax rates and the clientele effect. *Review of Economics and Statistics* 52, 68–74.

Elton, E., Gruber, M. and Blake, C. 2005. Marginal stockholder tax effects and ex-dividend day behavior: evidence from taxable versus nontaxable closed-end funds *Review of Economics and Statistics* 87, 579–86.

Faccio, M. and Lang, L. 2002. The ultimate ownership of western European corporations. *Journal of Financial Economics* 65, 365–95.

Fama, E.F. and French, K. 1993. Common risk factors in the returns on stocks and bonds. *Journal of Financial Economics* 33, 3–56.

Fama, E.F. and French, K. 1998. Taxes, financing decisions, and firm value. *Journal of Finance* 53, 819–43.

Fama, E.F. and French, K. 2001. Disappearing dividends: changing firm characteristics or lower propensity to pay? *Journal of Financial Economics* 60, 3–43.

Fenn, G. and Liang, N. 2001. Corporate payout policy and managerial stock incentives. *Journal of Financial Economics* 60, 45–72.

Gordon, M. 1959. Dividends, earnings and stock prices. *Review of Economics and Statistics* 41, 99–105.

Graham, B. and Dodd, D. 1951. *Security Analysis: Principles and Technique*. New York: McGraw-Hill.

Green, R. and Rydqvist, K. 1999. Ex-day behavior with dividend preference and limitation to short-term arbitrage: the case of Swedish lottery bonds. *Journal of Financial Economics* 53, 145–87.

Grullon, G. and Michaely, R. 2002. Dividends, share repurchases and the substitution hypothesis. *Journal of Finance* 57, 1649–84.

Grullon, G., Michaely, R. and Swaminathan, B. 2002. Are dividend changes a sign of firm maturity? *Journal of Business* 75, 387–424.

Healy, P. and Palepu, K. 1988. Earnings information conveyed by dividend initiations and omissions. *Journal of Financial Economics* 21, 149–76.

Jagannathan, M., Stephens, C. and Weisbach, M. 2000. Financial flexibility and the choice between dividends and stock repurchases. *Journal of Financial Economics* 57, 355–84.

Jensen, M. 1986. Agency costs of free cash flow, corporate finance, and takeovers. *American Economic Review* 76, 323–9.

Jensen, M. and Meckling, W. 1976. Theory of the firm: managerial behavior, agency costs and ownership structure. *Journal of Financial Economics* 3, 305–60.

John, K. and Williams, J. 1985. Dividends, dilution, and taxes: a signaling equilibrium. *Journal of Finance* 40, 1053–70.

La Porta, R., Lopez-De Silanes, F., Shleifer, A. and Vishny, R. 2000. Agency problems and dividend policy around the world. *Journal of Finance* 55, 1–33.

Lakonishok, J. and Vermaelen, T. 1986. Tax induced trading around ex-dividend dates. *Journal of Financial Economics* 16, 287–319.

Lang, L. and Litzenberger, R. 1989. Dividend announcements: cash flow signaling vs. free cash flow hypothesis. *Journal of Financial Economics* 24, 181–92.

Lintner, J. 1956. Distribution of incomes of corporations among dividends, retained earnings, and taxes. *American Economic Review* 46, 97–113.

Litzenberger, R. and Ramaswamy, K. 1979. The effects of personal taxes and dividends on capital asset prices: theory and empirical evidence. *Journal of Financial Economics* 7, 163–95.

Michaely, R. 1991. Ex-dividend day stock price behavior: the case of the 1986 tax reform act. *Journal of Finance* 46, 845–60.

Michaely, R. and Vila, J. 1995. Investors' heterogeneity, prices and volume around the ex-dividend day. *Journal of Financial and Quantitative Analysis* 30, 171–98.

Michaely, R., Thaler, R. and Womack, K. 1995. Price reactions to dividend initiations and omissions: overreaction or drift? *Journal of Finance* 50, 573–608.

Miller, J. and McConnell, J. 1995. Open-market share repurchase programs and bid-ask spreads on the NYSE: implications for corporate payout policy. *Journal of Financial and Quantitative Analysis* 30, 365–82.

Miller, M. 1987. The information content of dividends. In *Macroeconomics: Essays in Honor of Franco Modigliani*, ed. J. Bossons, R. Dornbush and S. Fischer. Cambridge MA: MIT Press.

Miller, M. and Modigliani, F. 1961. Dividend policy, growth and the valuation of shares. *Journal of Business* 34, 411–33.

Miller, M. and Rock, K. 1985. Dividend policy under asymmetric information. *Journal of Finance* 40, 1031–51.

Miller, M. and Scholes, M. 1978. Dividends and taxes. *Journal of Financial Economics* 6, 333–64.

Miller, M. and Scholes, M. 1982. Dividends and taxes: empirical evidence. *Journal of Political Economy* 90, 1118–41.

Ofer, A. and Siegel, D. 1987. Corporate financial policy, information, and market expectations: an empirical investigation of dividends. *Journal of Finance* 42, 889–911.

Ofer, A. and Thakor, A. 1987. A theory of stock price responses to alternative corporate cash disbursement methods: stock repurchases and dividends. *Journal of Finance* 42, 365–94.

Pettit, R. 1972. Dividend announcements, security performance, and capital market efficiency. *Journal of Finance* 27, 993–1007.

Poterba, J. and Summers, L. 1984. New evidence that taxes affect the valuation of dividends. *Journal of Finance* 39, 1397–415.

Rozeff, M. 1982. Growth, beta and agency costs as determinants of dividend payout ratios. *Journal of Financial Research* 5, 249–59.

Stulz, R. 1988. Managerial control of voting rights: financing policies and the market for corporate control. *Journal of Financial Economics* 20, 25–54.

Vermaelen, T. 1981. Common stock repurchases and market signaling: an empirical study. *Journal of Financial Economics* 9, 139–83.

Divisia, François Jean Marie (1889–1964)

Divisia was born in Tizi-Ouzou, Algeria. He received baccalaureate degrees in mathematics and philosophy at Algiers. After two years in the Ecole Polytechnique he worked for the government as a civil engineer (Ponts et Chaussées). His graduate engineering work at the Ecole Nationale des Ponts et Chaussées was completed in 1919 after the interruption of the First World War. After nearly ten years as a government engineer he joined the ministry of national education to continue research and teaching economics. He became a professor of applied economics at the Ecole Nationale des Ponts et Chaussées (1932–50), the Conservatoire National des Arts et Métiers (1929–59), and the Ecole Polytechnic (1929–59). He was a founding member of the Econometric Society and its president in 1935. Subsequently he was also president of the Paris Statistics Society (1939) and of the International Econometric Society. He was a Fellow of the American Statistical Association and of the American Association for the Advancement of Science.

His major contributions to economics can be found centred in several books on economics and applied statistics. The Divisia Index, a variable-weight price index, was developed in *L'indice monétaire et la théorie de la monnaie* (1926). His *Economique rationnelle* (1928) was widely acclaimed in mathematical economics and was awarded prizes by the Academy of Sciences and by the Academy of Moral Sciences and Politics. Using a microeconomic perspective he cautioned against uncritical acceptance of macroeconomic research in *Traitement économétrique de la monnaie, l'intérêt, l'emploi* (1962).

DAVID E.R. GAY

Selected works

1926. *L'indice monétaire et la théorie de la monnaie.* Paris.

1928. *Economique rationnelle.* Paris.

1931. *L'epargne et la richesse collective.* Paris.

1951–65. *Exposés d'économique.* Paris.

1962. *Traitement économétrique de la monnaie, l'intérêt, l'emploi.* Paris.

Divisia index

The Divisia index is a continuous-time index number formula due to François Divisia (1925–6) that has been widely used in theoretical discussions of data aggregation and the measurement of technical change. It is defined with respect to the time paths of a set of prices $[P_1(t), \ldots, P_N(t)]$ and commodities $[X_1(t), \ldots, X_N(t)]$. Total expenditure on this group of commodities is given by:

$$Y(t) = P_t(t)X_1(t) + \ldots + P_N(t)X_N(t). \tag{1}$$

With dots over variables indicating derivatives with respect to time, total differentiation of (1) yields:

$$\frac{\dot{Y}(t)}{Y(t)} = \sum_{i=1}^{i=N} \frac{P_i(t)X_i(t)}{Y(t)} \frac{\dot{P}_i(t)}{P_i(t)}$$
$$+ \sum_{i=1}^{i=N} \frac{P_i(t)X_i(t)}{Y(t)} \frac{\dot{X}_i(t)}{X_i(t)}. \tag{2}$$

The growth rates of the Divisia price and quantity indexes are the respective weighted averages of the growth rates of the individual $P_i(t)$ and $X_i(t)$, where the weights are the components' shares in total expenditure. The levels of these indexes are obtained by line integration over the trajectory followed by the individual prices and quantities over the time interval $[0, T]$. For the quantity index, the line integral has the following form:

$$I_q(0, T) = \exp\left\{ \int \left[\sum_{i=1}^{N} \frac{P_i(t)X_i(t)}{\sum P_j(t)X_j(t)} \frac{\dot{X}_i(t)}{X_i(t)} \right] \right\}$$
$$= \exp\left\{ \int_r \phi(X)\mathrm{d}X \right\}, \tag{3}$$

where ϕ is a vector-valued function whose arguments are $P_i(t)/Y(t)$, prices are assumed to be a function of the X_i, and Γ is the curve described by X_i. A similar expression characterizes the Divisia price index (for a more extensive discussion of Divisia line integrals, see Richter, 1966; Hulten, 1973; Samuelson and Swamy, 1974).

The value of the index defined by (3) depends on the solution of the line integral. This can be obtained by identifying a 'potential function' Φ whose partial derivatives are the vector-valued function ϕ, that is, $\phi = \nabla\Phi$. Writing $\Phi = \log F$ function, the value of the index can be shown to equal $F[X(T)]/F[X(0)]$, implying that the index is unique only up to a scalar multiple.

In economic terms, the solution to (3) is associated with some underlying economic relationship among the variables being indexed. Assume, for example, there is a constant returns to scale production function $F(X)$ and $F_i = \lambda P_i$ (F_i denotes the partial derivative of F with respect to X_i and λ is a factor of proportionality). Then the function $\log F$ can serve as the requisite potential function for (3), and in this particular case, the Divisia index of inputs can be interpreted as the ratio of output at time T to output at time zero.

If the form of the potential function is known a priori, the value of the index could be computed directly from the function F. However, the rationale for the Divisia index is that it provides a way of obtaining the ratio $F(X(T))/F(X(0))$ by using data on prices and quantities alone, without direct knowledge of F. Intuitively, this is possible because, under sufficiently restrictive assumptions, information about the slope of the function F (as estimated by relative prices) over the path followed by the inputs is sufficient to characterize F up to a scalar multiple.

When the objective is to form an index of a subset of inputs – aggregate labour input, for example – the required potential function is a 'piece' of a production function. Specifically, if one wants to form a Divisia index of the first M inputs, the production function needs to be weakly separable into a function of these inputs, that is, $F\{G[X_1(t), \ldots, X_M(t)], X_{M+1}(t), \ldots, X_N(t)\}$. The function log G serves as the potential function for the line integration (see also Balk, 2005).

These considerations apply to Divisia price indexes as well. The relevant potential function is now the factor price frontier $\Psi[P_1(t), \ldots, P_N(t)]$. A basic result of duality theory shows that the partial derivatives of Ψ are proportional to the corresponding $X_i(t)$.

The discussion suggests that the existence of the Divisia index is closely linked to the conditions for consistent aggregation. Furthermore, the required existence of a potential function implies that aggregation cannot proceed with just any set of prices or quantities. There must be an a priori reason for supposing that the variables to be indexed are theoretically related. This is an important characteristic of the Divisia index, one which it shares with the broader class of economic index numbers (in contrast to the non-structural axiomatic approach associated with Irving Fisher, 1921; see also Balk, 2005). The potential function theorem establishes the conditions under which the Divisia index is an 'exact' index number (to use the terminology of Diewert, 1976) for some underlying economic structure.

Divisia indexes have the desirable property that they are invariant when the path of integration lies entirely in the same level set of the potential function. That is, if one input is substituted for another along a given isoquant, the value of the index will not change. However, there is no guarantee of invariance when the path of integration lies across several level sets. This reflects the mathematical property that line integrals are, in general, path dependent.

Path dependence means that the index (3) will generally have a different value for a path $\beta(t) \in \Gamma_1$ than path $\alpha(t) \in \Gamma$, even though the beginning and end points of Γ_1 and Γ are identical. This can lead to the following situation: the economy moves along Γ_1 from X from X' (which is on a different isoquant); the economy then returns along Γ to the original point X; because of path dependence, the vector of quantities represented by the vector X will have a different Divisia index value after the trip around the composite path, and subsequent circuits will produce still different values. The value of the Divisia index at any point X is thus arbitrary under path dependence. The uniqueness of the Divisia index thus involves path independence.

The condition for path independence is the existence of a homothetic potential function, log F, such that $\phi = \nabla \log F$, where ϕ is defined in (3). Given the existence of the potential function, the value of (3) is $F(X(T))/F(X(0))$, implying path independence since (3) depends only on the end points of the path, $X(0)$ and $X(T)$. Conversely, if (3) is path independent, there exists a potential function log F such that $\nabla \log F = \phi$. In some applications in productivity analysis, the homotheticity condition must be strengthened to linear homogeneity, but this can be weakened depending on data availability (Hulten, 2001, pp. 11–12).

We note, finally, that the Divisia index is defined using time as a continuous variable. Data on prices and quantities typically refer to discrete points in time, and the indexes constructed from them must therefore have a discrete-time form. The continuous-time Divisia index is nevertheless useful, both for informing the structure of these discrete-time indexes (for example, for the determining which variables are conceptually related), and for interpreting the results. The Divisia framework is also appropriate for the theoretical analysis of many economic problems, such as the use of Divisia indexes by Solow (1957) in growth accounting.

One approach to linking discrete and continuous index numbers is to approximate the continuous variables of (2) with their discrete time counterparts. Under the Törnqvist (1936) approach, the growth rates of prices and quantities are approximated by logarithmic differences, and the continuous weights by two period arithmetic averages. The Törnqvist approximation to the growth rate of the Divisia quantity index can then be written:

$$\sum_{i=1}^{i=T} \frac{1}{2} \left[\frac{P_{i,t}X_{i,t}}{Y_i} + \frac{P_{i,t-1}X_{i,t-1}}{Y_{t-1}} \right] [\log X_{i,t} \\ - \log X_{i,t-1}] \qquad (4)$$

A similar approximation applies to the growth rate of the Divisia index of prices.

While the Törnqvist index may be regarded as approximate, Diewert (1976) has shown that it is exact when the underlying potential function has the (continuous) translog form. This result is very important in its own right, but can also be regarded as an important conceptual link between the discrete and continuous-time families of index numbers, given the exact properties of the Divisia index in continuous time.

The continuous Divisia index can also be approximated by using chain indexing procedures (the Divisia index is sometimes regarded as a chain whose links are defined over infinitesimal time periods). Other numerical approximation techniques can also be employed.

CHARLES R. HULTEN

See also **Divisia, François Jean Marie; index numbers.**

Bibliography
Balk, B. 2005. Divisia price and quantity indices: 80 years after. *Statistica Neerlandica* 59, 119–58.
Diewert, W. 1976. Exact and superlative index numbers. *Journal of Econometrics* 4(2), May, 115–45.
Divisia, F. 1925–6. L'indice monétaire et la théorie de la monnaie. *Revue d'Economie Politique* 39(4), 842–64; (5), 980–1008; (6), 1121–51; 40(1), 49–81. Also separately: Paris: Société Anonyme du Recueil Sirey, 1926.
Fisher, I. 1921. *The Making of Index Numbers*. Boston, MA: Houghton Mifflin Co.
Hulten, C. 1973. Divisia index numbers. *Econometrica* 41, 1017–25.
Hulten, C. 2001. Total factor productivity: a short biography. In *New Developments in Productivity Analysis*, ed. C. Hulten, E. Dean and M. Harper. Studies in Income and Wealth, vol. 63. Chicago: University of Chicago Press for the NBER.
Richter, M. 1966. Invariance axioms and economic indexes. *Econometrica* 34, 739–55.
Samuelson, P. and Swamy, S. 1974. Invariant economic index numbers and canonical duality: survey and synthesis. *American Economic Review* 64, 566–93.
Solow, R. 1957. Technical change and the aggregate production function. *Review of Economics and Statistics* 39, 312–20.
Törnqvist, L. 1936. The Bank of Finland's consumption price index. *Bank of Finland Monthly Bulletin* 10, 1–8.

division of labour

Division of labour, or specialization, may be defined as the division of a process or employment into parts, each of which is carried out by a separate person, or any system of production in which tasks are separated to enable specialization to occur. This includes the separation of employments and professions within society at large or *social division of labour* as well as the division of labour which takes place within the walls of a factory building or within the limits of a of a single industry, the *manufacturing division of labour*. Division of labour as a form of specialization can also be practiced by small firms which all contribute to the production of parts (inputs) for the manufacturing of a complex output, as in the case of aircraft production or sophisticated electronic equipment. This form of business organization requires excellent coordination and communication between its various parts to ensure continuous supply of the necessary parts for the manufacturer of the final output. It is a geographical form of division of labour, developed from the notion of clustering related firms in a particular area or industrial district (for a survey, see Dosi, 1988).

Division of labour and its consequences for productivity were analysed as early as the time of the Greek philosophers, including Plato, Aristotle and Xenophon. Early analysis of the manufacturing division of labour had to await industrial developments of the 17th and 18th centuries and underwent further qualitative change in the 19th, 20th and 21st centuries. Hence manufacturing and more detailed division of labour should not be seen as a simple continuum of the social division of labour. By the end of the Middle Ages, social division of labour was extensively practiced; manufacturing division of labour, generally speaking, came with the Industrial Revolution. Under modern capitalism, social division of labour remains largely a market influenced phenomenon but manufacturing division of labour is enforced by those who plan and control the manufacturing process. Furthermore, the one divides society, the other human activity within the workshop, or within an industry: labour generally enhances 'the individual and the species, [a manufacturing division] of labour, when carried on without regard to human capabilities and needs, is a crime against the person and against humanity' (Braverman, 1974, p. 73). Division of labour was first practiced within the household, a *sexual division of labour* between women's activities in or near the house, and those of men further afield. When applied to local specialization of industries both nationally and internationally, it has produced a variety of conceptions of the *territorial* or *international (global) division of labour*.

Adam Smith (1776) placed the division of labour at the forefront of his discussion of economic growth and progress. Neither in its social nor in its manufacturing forms did the idea originate with him. It retained a varying, but often very prominent, place in 19th-century writings (particularly those of Senior, Babbage, John Stuart Mill, Marx and Marshall). 'About 1890, Schmoller, Semmel, Bücher, Durkheim and Maunier all wrote on religious and sociological aspects of specialization' (Salz, 1934, p. 284). For much of the 20th century, division of labour and specialization virtually disappeared as a major topic from economic texts. Reasons for this varied. Some economists believed such discussions were more appropriate to technical handbooks of production engineering and factory management. Other writers wished to confine analysis of its effects to sociological studies assessing the general impact of division of labour on society. The return of economic growth as an important part of the economist's research programme from the 1950s onwards, and earlier the work of Young (1928), brought renewed interest in the division of labour in its wake, as did growing dissatisfaction with the narrow view confining economics to studying 'the disposal of scarce commodities' (Robbins, 1932, p. 38). Global organization of manufacturing made possible by improvements in transport and communication implies modern adaptations of the division of labour which economists cannot ignore. An example is the formation of industrialized districts, first observed and analysed by Alfred Marshall (1890), to be rediscovered and adapted to the post-Second World War Italian situation by Becattini (for example, 1990; 2001) and his colleagues (for a survey, see Goodman and Pamford, 1989). The various dimensions of division of labour raised in these introductory paragraphs suggest that a broad-based treatment of the subject is warranted by featuring highlights within its continuous development.

The Greeks

Many of the major Greek philosophers discussed aspects of the division of labour in their writings. In Book 2 of the *Republic*, Plato stated the necessity for a division of labour or specialization in occupations for social well-being and the adequate satisfaction of primary wants linking the phenomenon with exchange, the requirements of 'a market, and a currency as a medium of exchange' (Plato, 380 BC, pp. 102–6). Aristotle, though very conscious of the social need for a division of labour, did not depart much from Plato's earlier discussion (see Bonar, 1893, p. 34). More importantly, Xenophon linked division of labour and specialization to great cities, because they provided a substantial demand for individual products while the subdivision of work raised the skill of individual workers. Extracts from the work of these Greek pioneers on the division of labour have been often reprinted (see, for example, Sun, 2005, chs 2–4). Knowledge of these Greek texts among Arabian Islamic scholars during the middle ages enabled them to produce sophisticated treatments of the division of labour. Examples are the writings of Islamic theologian, al-Ghazali (1058–1111) and, more importantly, the writings of fourteenth century Islamic philosopher and historian, Ibn Khaldun, whose *Muqaddima* contains a detailed account of the division of labour (Sun, 2005, pp. 7–8, ch. 5).

Subsequent pre-Smithian developments

Towards the end of the seventeenth century, English economic literature rediscovered the concept of the division of labour and began to analyse the more modern manufacturing forms, linking them to productivity growth, cost reduction, increased international competitiveness and associating its scope with the more extensive markets made possible through urbanization. For example, Petty's *Political Arithmetick* written in 1671 compared the benefits of division of labour in textile production with specialization in ship building:

> For as Cloth must be cheaper made, when one Cards, another Spins, another Weaves, another Draws, another Presses and Packs; than when all the Operations abovementioned, were clumsily performed by the same hand; so those who command the Trade of Shipping [need] to build…a particular sort of Vessels for each particular Trade. (Petty, 1671, pp. 260–1)

Ten years later, in *Another Essay on Political Arithmetick Concerning the Growth of the City of London* (1683, p. 473), Petty showed that a major gain from a vast city like London came from the improvement and growth of manufactures it encouraged:

> For in so vast a City *Manufacturers* will beget one another, and each *Manufacture* will be divided into as many parts as possible, whereby the Work of each *Artisan* will be simple and easy; As for example in the making of a *Watch*, if one Man shall make the *Wheels*, another the *Spring*, another shall Engrave the *Dial-plate*, and another shall make the *Cases*, then the *Watch* will be better and cheaper, than if the whole Work be put upon any one man.

In continuing this argument Petty also suggested that specialization benefits could be achieved from concentrating certain manufactures on a particular location, partly because of the savings in transport and communication costs such concentration entailed (Petty, 1683, pp. 471–2). The anonymous author of *Considerations on the East India Trade* (1701, pp. 590–2) illustrated productivity gains from the division of labour by examples drawn from cloth making, watch making and shipbuilding. He clearly indicated that sufficient demand and regular trade were a precondition for such improvements, which lowered manufacturing labour costs without the need to lower wages. During the 18th century, examples of authors aware of the benefits and preconditions for a division of labour become more common. Practical writers like Patrick Lindsay (1733), Richard Campbell (1747) and Joseph Harris (1757) tended to concentrate on manufacturing division of labour using examples from linen and pin production as well as from the familiar watch making. Those writing from the position of moral or political philosophy, like Mandeville (1729), Hutcheson (1755), Ferguson (1767) and Josiah Tucker (1755; 1774) concentrated more on aspects of the social division of labour.

Discussion of the division of labour was of course not confined to English economic literature. A treatise on wealth published in the 1720s by Ernst Ludwig Carl discussed the benefits of the division of labour, applying them also to demonstrate the gains from free trade through an international division of labour based on different climates, resource availability and locational advantages (cited in Hutchison 1988, pp. 161–2). Among the Physiocrats, Quesnay dealt briefly with the social aspects of the division of labour in his article 'Natural Right' (1765, p. 51). Turgot developed the subject more thoroughly, making it the starting point of his *Reflections*, subsequently associating it with the introduction of money, the extension of commerce and the accumulation of capital (1766, pp. 44–6, 64, 70). Earlier, Turgot (1751, pp. 242–3) had linked the spread of social division of labour to inequality, arguing that this particular consequence of inequality improved living standards for even the humblest members of society and made possible cultivation of the arts and sciences. Among the general principles with which Beccaria (1771, pp. 387–8) commenced the argument of his *Elementi*, the division of labour and its benefits in terms of increased skills and dexterity are clearly set out. Finally, it may be noted that the *Encyclopédie* of Diderot and d'Alembert in its article 'Art' discussed the essentials of the manufacturing division of labour, listing its consequences as improvements in skill, better quality products, saving of time and of materials, and 'of making the time or the labour go further, whether by the invention of a new machine or the discovery of a more suitable method'. In its article on pins ('Epingle') their manufacture is described as being generally subdivided into eighteen separate operations and thereby a prime example of the manufacturing division of labour (see Cannan, 1929, pp. 94–5).

Adam Smith's treatment of the division of labour

Adam Smith's discussion of the division of labour deserves separate treatment not because of its 'originality' or 'completeness of exposition' (Cannan, 1929, p. 96) but because 'nobody either before or after [him], ever thought of putting such a burden upon division of labour. With A. Smith, it is practically the only factor in economic progress' (Schumpeter, 1954, p. 187). The first three chapters of the *Wealth of Nations* were devoted to its analysis because it provided one of the two causes explaining increases in per capita output by which Smith defined the wealth of the nation. Although therefore only one of two causes, the other being 'the proportion between the number of those who are employed in useful labour, and that of those who are not so employed' (Smith, 1776, p. 10), it is the dominant one. Smith seems to have believed that scope for substantial increases in the proportion of the labour force to productive activities was limited. Using the equation, $g = (k \cdot p/w) - 1$, developed by Hicks (1965, p. 38) to summarize the Smithian growth progress, if a change in k, the proportion of productive labour in the labour force, is more or less ruled out, a substantial growth rate (g), given the real wage (w), depends exclusively on rising productivity (p) through extensions of the division of labour. Smith's emphasis on the division of labour as a factor in growth via its enormous influence on productivity makes his treatment of the subject so novel. Surprisingly, this aspect

of his contribution was taken up by few 19th-century writers and had to be largely rediscovered in the work of Young (1928) and Kaldor (1972) who reiterated dynamic aspects of the phenomenon Smith was analysing.

Even though it was the most frequently revised part of his economics (see Meek and Skinner, 1973), Smith's basic account of the division of labour contains a number of weaknesses. First, Smith failed to develop aspects of the manufacturing division of labour with which he ought to have been familiar. Marglin (1974) points out that Smith ignored organizational features from a division of labour taking place within the one building of relevance to some well-established industries like textiles and the manufacture of metal implements. These organizational features which Smith omitted were associated with growing labour discipline problems, wasting time and materials, inherent in the putting-out system, then the dominant form of manufacturing organization. In fact it can be suggested that if this aspect of the division of labour is more fully taken into account, its important role in explaining economic growth so much emphasized by Smith is more easily integrated as a major factor explaining the industrial revolution (see Groenewegen, 1977). Marglin (1974) also questioned the force of 'the three different circumstances' by which Smith (1776, p. 17) explained the productivity gains from the division of labour: increased dexterity, saving of time, and invention of machinery. Although increased dexterity is clearly a product of a division of labour in a manufacturing process, its scope there is rather limited when compared to that of the continual practice of surgeons, concert pianists and opera singers, to give some examples. Time saved in eliminating time lost in passing from job to job is trivial and not the 'very considerable' benefit Smith (1776, pp. 18–19) had suggested. Savings in materials and time through transforming a putting-out to a factory system, an organizational feature of the division of labour Smith had ignored, was more important, particularly through eliminating losses from pilfering. Rae (1834, pp. 164–5) saw savings in the use of tools as far more significant than time saved, and for him (pp. 352–7) this provided the basic reason for extending the division of labour. Other 19th century writers, particularly Babbage (1832), expanded further on this aspect of the matter. Smith's association of division of labour with inventions (1776, pp. 19–22) covered both 'on the job improvements' and scientific inventions by specialists originating from within a more sophisticated division of professions. It ignores, as Hegel (1821, p. 129) was one of the first to point out (cf. Stewart, 1858–75, vol. 8, pp. 318–19), that as division of labour makes 'work more and more mechanical,…man is able to step aside and install machines in his place'. This feature of the process was subsequently noted by Babbage (1832, pp. 173–4), Ure (1835, p. 21) and developed by Marx (1867). In short, the three circumstances Smith saw as explaining the productivity consequences from the division of labour derive their basic validity from reasons different to those Smith advanced. Further, Smith's remarks (1776, pp. 16–17) on the smaller benefits from applying the division of labour to agriculture than to manufacturing can be contrasted with his quite different and controversial analysis of the primacy of agricultural investment in terms of its employment of productive labour. Agriculture's more substantial contribution to gross revenue as Smith (1776, Book II, ch. 5) subsequently argued, was used by him to define the 'natural' course of economic development (Book II, ch. 1) and recommended as superior practice for newly settled regions like the American colonies. Perelman (1984, p. 185) explained this seeming contradiction in Smith by suggesting Smith was the 'first theorist of neo-imperialism' because his strategy of development forces developing regions to specialize in raw material production whose terms of trade with manufactures are invariably poor. More likely, Smith's views on the productivity of agriculture relative to manufacturing are posed in terms of different yardsticks: agricultural activity by the very nature of its processes is less amenable to division of labour, even though its ability to employ productive labour is greater than produced by equal investments in manufacture and trade. However, growing mechanization of agriculture, especially in the 20th century, together with the greater scope for exporting agricultural surplus with modern transport, encouraged specialization in agriculture and very large scale farming (Salz, 1934, p. 283).

A final controversial issue from Smith's treatment of the division of labour concerns its social consequences, an argument he placed in the context of public education. The 'few simple operations' which under a division of labour most ordinary labouring people are asked to perform, renders them 'as stupid and ignorant as it is possible for a human creature to become' and increased 'dexterity at his own particular trade' is purchased with a reduction in 'intellectual, social and martial virtues... unless government take some pains to prevent it' through providing general education (Smith, 1776, pp. 781–5). Smith was not alone in presenting this disadvantage in an extensive division of labour: similar views were put by Ferguson (1767, p. 280) and Kames (1774). Ferguson described 'ignorance as the mother of industry' and argued that prosperous manufactures arise 'where the mind is least consulted, and where the workshop may...be considered as an engine, the parts of which are men.' At the turn of the century, and after, German philosophers (for example, Schiller, 1793, Hegel, 1821 and the young Marx, 1844) developed this into a humanist critique of industrial society, suggesting like Smith that these detrimental consequences were removable by education, especially aesthetic education. Such sentiments were resurrected in mid-19th century England by Carlyle (1843) and Ruskin (1851–3, pp. 197–8). For others, Smith's remarks were an aberration, 'as unfounded [a statement] as can well be imagined'

(McCulloch, 1850, p. 350) or even a contradiction with the division of labour's ability to inspire inventive faculties in labourers (West, 1964).

Despite its deficiencies, Smith's account of the division of labour proved particularly hardy and was invariably praised in most general terms by major textbook writers of the 19th century and after, though few followed the emphasis Smith gave it as the key factor explaining growth. Cannan (1929, p. 97) ascribed this success to 'the popularity of its form'. It can also be attributed to the striking productivity increase inherent in the pin example (cf. Mill, 1821, p. 215) and the unambiguous connection Smith drew between increased division of labour, extending the market and human proclivities 'to truck and barter' (McCulloch, 1825, pp. 54–5). The account of the division of labour is undoubtedly one of Smith's best remembered performances in economics.

19th-century developments

With the growth of the factory system and more extensive use of increasingly sophisticated machinery, the manufacturing form of division of labour was considerably expanded. Consequently, some economic writers focused on a number of new aspects of the phenomenon, linking the division of labour with developments in the machine tool industry, large scale production and its advantages, and hence, on a more theoretical level, with increasing returns to scale and explicit recognition of a different pattern of productivity growth in manufacturing from that in agriculture.

Charles Babbage was in many respects the pioneer in presenting the division of labour as 'the most important principle on which the economy of a manufacture depends' (1832, p. 169). He therefore carefully revised the advantages of a division of labour as first expounded by Adam Smith. In this discussion, time (and cost) savings were also related to time saved in learning a skill and reduced waste of materials during the learning process (pp. 170–1), as well as economy in tool using (p. 172), while the association between division of labour, dexterity and the introduction of new machines was developed more precisely and rigorously. More significantly, Babbage pointed to a hitherto ignored additional advantage of the division of labour he had derived from observation. This had earlier been discussed by Gioja (1815–17) whose interesting contribution on this subject was analysed by Scazzieri (1981, ch. 3).

> By dividing the work to be executed into different processes of skill or of force,...the master manufacturer...can purchase exactly that precise quantity of both which is necessary for each process; whereas, if the whole work were executed by one workman, that person must possess sufficient skill to perform the most difficult, and sufficient strength to execute the most laborious, of the operations into which the art is divided. (Babbage, 1832, pp. 175–6; emphasis in original)

This economy of skill, Babbage demonstrated from a pin example, not only reinforced the cost advantages traditionally associated with division of labour, but was also a major cause of establishing large factories: 'When the number of processes into which it is most advantageous to divide it, and the number of individuals to be employed in it, are ascertained then all factories which do not employ a direct multiple of this number, will produce the article at a greater cost' (Babbage, 1832, p. 213). Detailed division of labour, Babbage also argued, as in its manufacturing form, can also be applied to mental labour (p. 191). An illustration of its application to mining highlights these control and information gathering features, two aspects of the division of labour to which Babbage paid particular attention. His analysis of the division of labour is even more important because the process as he described it is made interdependent with machine production, increased factory size, lower costs and prices from such concentration of industry and hence induces growth in demand and an extended market (see Corsi, 1984).

Ure's (1835) contribution must also be noted. It likewise linked development of the factory system to division of labour, summarizing 'the principle of the factory system...as substituting mechanical science for hand skills, and the partition of a process into its essential constituents' (1835, p. 20). Ure commented on two other consequences of the division of labour in modern factories: deskilling of the workforce when workers become 'mere overlookers of machines' and the development of mechanical engineering since the 'machine factory displayed the division of labour in manifold gradations' and facilitated the substitution of skilled hands by 'the planning, the key-groove cutting, and the drilling machines' (pp. 20–1).

Accounts of the division of labour by economists of the middle of the century were generally less innovative than those of Babbage and Ure, though they did occasionally provide some new points of departure. Senior (1836, pp. 74–5, 77), after classifying division of labour as one major advantage from the use of capital, concentrated on listing its benefits additional to those given by Smith. Illustrating from the post office, he argued that the fact that 'the same exertions which are necessary to produce a single given result are often sufficient to produce many hundreds or many thousands similar results' was one aspect of the division of labour omitted by Smith. The development of retailing as a separate profession was likewise something Smith had failed to consider adequately. More importantly, for a number of reasons, but particularly the division of labour, Senior suggested 'additional Labour when employed in Manufactures is MORE, when employed in Agriculture is LESS efficient in proportion', linking manufacturing activity implicitly to increasing returns to scale (1836, pp. 81–2). Mill (1848) treated division of labour as an important aspect of cooperation, arguing that irrespective of its well known productivity advantages, without this complex cooperation in the modern division of labour 'few things

would be produced at all' (Mill, 1848, p. 118) In discussing the productivity advantages, Mill cited the modification and additional advantages provided by Babbage (1832) and Rae (1834), adding little to their discussion. However, in Chapter 9 dealing with large scale and small scale production, he highlighted the point, so 'ably illustrated by Mr Babbage...[that] the larger the enterprise, the farther the division of labour may be carried...as one of the principal causes of large manufactories' (Mill, 1848, p. 131), thereby bringing the argument firmly into the corpus of economics. Mill's account was largely followed by Fawcett (1863) and in most of its essentials by Nicholson (1893).

Marx's account (1867, chs. 13–15) combines much of this discussion, endowing it in the process with sharper analytical insights derived from his study of both the technical literature and his appreciation of the significance of the qualitative changes underlying the evolution of the division of labour. To Marx is owed the important distinction between manufacturing and social division of labour, as well as the precise assessment of the organizational features of its application to modern manufacture, derived from his careful study of Babbage, Ure and many other sources. No wonder that Nicholson (1893, p. 105) described Marx's treatment as 'both learned and exhaustive and...well worth reading'. More recently, Rosenberg (1976) expressed regret that Marx's close study of 'both the history of technology, and its newly emerging forms' has had so few imitators among contemporary economists.

Marshall is another economist from the second half of the 19th century who fully appreciated the importance of the division of labour and revealed it in its more modern forms. In 1879, the *Economics of Industry*, written with his wife (Marshall and Marshall 1879), devoted Chapter 8 of Book I to the division of labour, immediately after its Chapter 7 on organization of industry. It distinguished the opportunity to apply a division of labour as inherent in the nature of the work, as dependent on direction and control by an entrepreneur as earlier indicated by Bagehot, and as applied to firms: 'If there are any producers, large and small, all engaged in the same process, *Subsidiary Industries* will grow up to meet their special wants.' These include special machine tool makers for the industry, improved transport to enhance communication between related firms, as well as auxiliary enterprises in banking and credit provision (Marshall and Marshall 1879, p. 52). Localization of industry also fosters 'education of skills and taste': and 'diffusion of linked knowledge', and encourages large firms. Hence division of labour is closely related to economies of scale, where size has enabled specialization to grow more and more. Marshall also devoted no less than three chapters to division of labour in his *Principles* (1890, Book IV, chs 9–11), not only covering points traditionally dealt with under this heading, but often introducing subtle modifications. For example, Marshall (1890, p. 263) discounted detrimental social consequences from monotonous work by pointing to the mental stimulus from the 'social surroundings of the factory' and the view that factory work was not inconsistent with 'considerable intelligence and mental resources'. Likewise, he extended Babbage's principle of 'economy of skill' to economy of machinery and materials (1890, p. 265), used it as a major explanatory factor for the localization of specialized industry (p. 271) and made it the chief advantage of large scale production in his famous discussion of economies of scale (p. 278). Later, Marshall applied these aspects of his work to his detailed study of industry and trade to explain such things as America's leadership in standardized production (seen by Marshall, 1919, p. 149, as an 'unprecedented' application of Babbage's 'great principle of economical production'), the successful specialization of plant during the First World War, and new issues concerning the growth of the firm. It is therefore paradoxical that Marshall's work in other respects induced the demise of the division of labour in theoretical literature. This arose from the incompatability of increasing returns to scale with stable demand and supply equilibrium (Marshall, 1890, Appendix H). Apart from this, modern equilibrium analysis found it difficult to come to grips with the dynamic features of the division of labour process, and it is presumably at least partly for this reason that division of labour was dropped as an important subject from the economic textbooks (see Kaldor, 1972). However, the locational aspects of the division of labour were further addressed by Becattini (for example, 1990; 2001) in his development of the notion of industrial districts as a concentration of related firms. Marshall had discovered this aspect of industrial organization through the factory tours in the British midlands and Scotland he engaged in from the late 1860s, on which he first reported in 1879. When division of labour for technical reasons could not take place within the same building, small firms spring up specializing in part of the manufacturing process, thereby generating a division of labour among firms concentrated in a particular geographical area (for a survey, see Goodman and Pamford, 1988).

International division of labour

Torrens (1808) appears to have been the first economist to distinguish the territorial division of labour from the mechanical division, suggesting that the former is inspired by 'different soils and climates [being] adapted to the growth of different production' thereby inducing regional specialization in those products which best suit 'the varieties of their soil' and climate. Taking advantage of territorial division of labour through regional and international trade enhances productivity and increases the wealth of nations as much as a manufacturing division of labour. Senior (1836, p. 76) also drew attention to this aspect of the division of labour, attributing its discovery to Torrens. Marshall (1890, pp. 267–77) covered

territorial division of labour under localization of industry while Taussig (1911, pp. 41–7) called it 'the geographical division of labour' with gains arising from 'the adaptation of different regions to specific articles' for climatic and resource endowment reasons as well as from the general increase in proficiency which all specialization brings. During the 1970s a new dimension of the international division of labour was analysed, concentrating on its direct foreign investment aspects. Its novel features were a tendency to 'undermine the traditional bisection of the world into a few industrialized countries on the one hand, and a great majority of developing countries integrated into the world economy solely as raw material producers on the other, and [secondly, to compel] the increasing subdivision of manufacturing processes into a number of partial operations at different industrial sites throughout the world' to take advantage of favourable labour market circumstances, relatively cheap transport opportunities, tax breaks and other government inducements for foreign investors (Fröbel, Heinrichs and Kreye, 1980, p. 45). This multinational dimension to application of the division of labour is a direct descendent from the concept as understood by Smith, Babbage, Ure and Marx.

The characteristics of the contemporary global division of labour have been well captured by Hobsbawm (2000, pp. 65–6):

> Thus, while the global division of labour was once confined to the exchange of products between particular regions, today it is possible to produce across the frontiers of states and continents. This is what the process is founded on. The abolition of trade barriers and liberalization of markets is, in my opinion, a secondary phenomenon. This is the real difference between the global economy before 1914 and today. Before the Great War, there was pan global movement of capital, goods and labor. But the emancipation of manufacturing and occasionally agricultural products from the territory in which they were produced was not yet possible. When people talked about Italian, British and American industry, they meant not only industries owned by citizens of these countries, but also something that took place almost entirely in Italy, Britain, or America, and was then traded with other countries. This is no longer the case. How can you say that a Ford is an American car, given that it is made of Japanese and European components, as well as parts manufactured in Detroit?

Sexual division of labour

The first explicit reference to a sexual division of labour in economic literature I could find is Hodgskin (1829, pp. 111–12). He argued that

> There is no state of society, probably, in which division of labour between the sexes does not take place. It is

and *must* be practiced the instant a family exists. Among even the most barbarous tribes, *war* is the exclusive business of the males; they are in general the principal hunters and fishers...the woman labours in and about the hut...In modern as well as in ancient times,...we find the men as the rule taking the out-door work to themselves, leaving the women most of the domestic occupations....The aptitude of the sexes for different employments, is only an example of the more general principle, that every human being...is better adapted than another to some particular occupation.

Marx and Engels (1845–6, pp. 42–3) ascribed beginnings of the division of labour 'originally [to] nothing but the division of labour in the sexual act' and only later to that 'spontaneously' or 'naturally' derived from predisposition, needs, accidents, and so on. Engels (1884, esp. p. 311) elaborated further on the matter presenting the sexual division of labour in the family as a barrier to the 'emancipation of women'. Such an emancipation, he argued, was 'possibly only as a result of modern large-scale industry [which] actually called for the participation of women in production and moreover, strives to convert private domestic work also into a public industry'. Both aspects of the sexual division of labour to which Engels referred in the context of women's emancipation have been taken up in more recent research. The role of domestic labour has been analysed by contemporary writers (see, for example, Himmelweit and Mohun, 1977; Gershuny, 1983) while attention has also been drawn to the shift in the provision of services from domestic production to production for the market (laundromats, take-away-food) as a result of the gradual break-down of the traditional sexual division of labour within the family (Gouverneur, 1978). Sexual division of labour issues have also been applied in segmented labour market analysis, thereby enriching this particular aspect of labour economics.

Becker (1985) has analysed the sexual division of labour in the context of human capital investment and allocating the work load of parties within the household. Thus both the allocation of effort within a household, and the advantages of investing in specific human capital are designed to enhance the social division of labour and its benefits without necessarily diminishing the exploitative aspects of such arrangements (Becker 1985, p. S41). Social factors are, however, equally important. Increasing returns by itself cannot explain the traditional division of labour within the household; a division of labour itself subject to change. The increased contribution to housework by men during the 1970s is one observed aspect of this social change (Becker 1985, p. S56). Furthermore, as Posner (1992, pp. 54, 129) has noted in particular, women were not fully brought into the work place on a large scale until the two world wars, and this only became a dominant pattern in employment from the 1950s onwards. Cigno (1991) discusses many of these issues as part of his economics of the family.

Decline and rehabilitation of division of labour in the 20th and 21st centuries

The association between division of labour and increasing returns, the consequent possibility of falling supply and cost curves, created problems for equilibrium analysis already noticed as a factor explaining decline in emphasis on the division of labour and induced its virtual elimination from much of the theoretical literature. Attempts to remove division of labour from economics were also based on other grounds. Robbins (1932, pp. 32–8) argued that study of the 'technical arts of production' belonged to engineering and not to economics or, in the case of 'motion study', to industrial psychology even if this meant removal of traditional topics like division of labour from economics. Robbins's approach followed Sidgwick's (1883, pp. 104–7) treatment, removing all technical aspects from the topic, leaving only what he called the pure economics side. Others suggested it was better to leave discussion of division of labour to sociologists because Durkheim, and before him Comte and Herbert Spencer, had absorbed it within this emerging discipline. However, some economists in the 20th century objected to removal of the division of labour from economics. In particular, this would reduce understanding of the dynamics of economic progress.

Allyn Young (1928) was one of these economists. He made Adam Smith's theorem that the division of labour is limited by the extent of the market the central theme of his address to section F of the British Association, arguing this was 'one of the most illuminating and fruitful generalizations which can be found in the whole literature of economics' (Young, 1928, p. 529). Rather than covering all aspects of the division of labour, Young concentrated on two interdependent matters: 'growth of indirect and roundabout methods of production and the division of labour [or increased specialization] among industries' (Young, 1928, p. 529) but the former, as Kaldor (1975, pp. 355–6) pointed out, was not to be confounded with the Austrian capital theoretic notion. From this he deduced division of labour as a cumulative, self-reinforcing process, because every re-organization of production, sometimes described as a new invention, involves fresh application of scientific progress to industry,

> alters the conditions of industrial activity and initiates responses elsewhere in the industrial structure which in turn have further unsettling effects....The apparatus of supply and demand in their relation to prices does not seem to be particularly helpful for the purpose of an inquiry into these broader aspects of increasing returns. (Young, 1928, p. 533)

However, apart from this damaging conclusion for competitive price theory, the 'possibility of economic progress' could not really be grasped by ignoring these factors of greater specialization, better combinations of advantages of location, and a consequent increased number of specialized producers between basic raw materials and final producers (Young, 1928, pp. 538–40).

Kaldor was a major economist who took up Young's challenge in both its critical (Kaldor, 1972; 1975) and more constructive aspects (Kaldor, 1966; 1967). The major thrust of Kaldor's positive argument proclaimed that faster growth is derived from faster growth in the manufacturing sector, partly from the cumulative features linking the growth of manufacturing to growth of labour productivity via static and dynamic economies of scale, or the notion of increasing returns as developed by Young from the division of labour. This strong and powerful interaction of productivity growth and manufacturing growth is also posited in Verdoorn's Law (1949) but its association with aspects of the division of labour is what is relevant here. Faster manufacturing growth draws labour from other sectors of the economy, inducing faster productivity growth, but as the scope of transferring such labour from lower productivity sectors like agriculture dries up, the growth process slows down (see Thirlwall, 1983). A key feature of the process, as Rowthorn (1975, p. 899, n. 1) noticed in one his skirmishes with Kaldor on the subject, is that it is an interdependent, cumulative historical process where 'higher productivity means more exports which means greater industrial output which via its effects on investment, innovation and *scale of production* reacts back on productivity growth'. The importance of such a process was given detailed empirical examination in a discussion of the Taiwan machine tools industry in the 1970s as an application of the division of labour, envisaged as increases in output increasing productivity, with 'technological change, broadly defined, sandwiched in between' (Amsden, 1985, p. 271). Writers in the new growth economics, who emphasized the impact of increasing returns from specialization on growth performance (Romer, 1987) drew in part for their inspiration on the literature of the division of labour, in Romer's case as represented by Marshall (1890) and possibly Young (1928).

Research from the 1990s has particularly stressed the importance of communication and co-ordination costs of the division of labour. Becker and Murphy (1992) portray these costs as setting limits on the division of labour more important than that exerted by the extent of the market so heavily emphasized by Adam Smith. Subsequent, Camacho (1996) has studied this aspect in more detail, drawing a clear and direct relationship between increases in the division of labour and rises in both communication and co-ordination costs, as an essential extension to the modern theory of the firm and the market. Perlin (1993) has treated inter-firm cooperation and its benefits from a similar angle, assessing the benefits for production from such cooperation as an economy of conventions and inter-firm agreements. This analysis thereby treats division of labour once again as part of the organizational theory of the firm or a production unit in which much emphasis is placed on the potential trade-offs between the economies reaped from specialization and the transaction costs it generates (Yang and Ng 1993). In this way, division of labour has also become an important part of the foundations for a new classical micro-economic analytic framework.

Conclusion

Viewed dynamically within the context of economic growth, as Smith (1776) and others had intended the division of labour to be contemplated, it continues to be a powerful tool for understanding the process of growth and development. On this ground alone it can therefore not be jettisoned from economics as unwanted baggage, as Robbins (1932) mistakenly suggested. When its importance for understanding aspects of the labour process, the labour market, the theory of production and the theory of the firm contemplated at the plant and the industry level are included, this argument is even stronger. As mentioned in the previous paragraph, on these grounds division of labour is making a definite come-back as part of the theory of a new classical micro-economics. Last, but not least, the importance of the division of labour for economics is underlined by the fact that some of the major economic minds from both past and present have invariably included it as an important part of their economic analysis.

PETER GROENEWEGEN

Bibliography

Amsden, A.H. 1985. The division of labour is limited by the rate of growth of the market; the Taiwan machine tool industry in the 1970s. *Cambridge Journal of Economics* 9, 271–84.

Anonymous. 1701. Considerations on the East-India trade. In *A Select Collection of Early English Tracts on Commerce*, ed. J. R. McCulloch. London, 1856; Cambridge: Cambridge University Press, 1954.

Babbage, C. 1832. *On the Economy of Machinery and Manufactures*. London. Fourth enlarged edition of 1835; New York: Augustus M. Kelley, 1963.

Becattini, G. 1990. The Marshallian industrial district as a socio-economic notion. In *Industrial Districts and Inter-firm Co-Operation in Italy*, ed. F. Pyke, G. Becattini and W. Sengenberger. Geneva: International Institute of Labour Studies.

Becattini, G. 2001. *The Caterpillar and the Butterfly*. Florence: Felice de Monier.

Beccaria, C. 1771. *Elementi di economia pubblica*. In *Opere*, ed. S. Romagnoli. Florence: Sansoni, 1958.

Becker, G.S. 1985. Human capital, effort, and the sexual division of labour. *Journal of Labour Economics* 3(1), S33–S58.

Becker, G.S. and Murphy, K.M. 1992. The division of labour, coordination costs, and knowledge. *Quarterly Journal of Economics* 107, 1137–60.

Bonar, J. 1893. *Philosophy and Political Economy.* 3rd edn. London: Allen and Unwin, 1967.

Braverman, H. 1974. *Labour and Monopoly Capital: The Degradation of Work in the Twentieth Century.* New York: Monthly Review Press.

Camacho, A. 1996. *Division of Labour, Variability, Coordination and the Theory of the Firm and Markets.* Dordrecht: Kluwer.

Campbell, R. 1747. *The London Tradesman.* London.

Cannan, E. 1929. *Review of Economic Theory.* London: P.S. King & Son.

Carlyle, T. 1843. *Past and Present.* London: Chapman & Hall. Another edn. London: G. Routledge & Sons, 1893.

Cigno, A. 1991. *Economics of the Family.* Oxford: Oxford University Press.

Corsi, M. 1984. Il sistema di fabbrica e la divisione del lavoro, il pensiero di Charles Babbage. *Quaderni di storia dell'economia politica* 3, 111–23.

Dosi, G. 1988. Sources, procedures, and micro-economic effects of innovation. *Journal of Economic Literature* 26, 1120–71.

Engels, F. 1884. *Origin of the Family, Private Property and the State.* In *Marx-Engels Selected Works,* vol. 2. Moscow: Progress Publishers.

Fawcett, H. 1863. *Manual of Political Economy.* London and Cambridge: Macmillan & Co.

Ferguson, A. 1767. *An Essay on the History of Civil Society.* Edinburgh.

Fröbel, F., Heinrichs, J. and Kreye, O. 1980. *The New International Division of Labour,* trans. P. Burgess. Cambridge: Cambridge University Press.

Gershuny, J. 1983. *Social Innovation and Division of Labour.* Oxford: Oxford University Press.

Gioja, M. 1815–17. *Nuove prospetto delle scienze economiche.* Milan: G. Pirotta.

Goodman, E. and Pamford, J., eds. 1989. *Small Firms and Industrial Districts in Italy.* London: Routledge.

Gouverneur, J. 1978. *Contemporary Capitalism and Marxist Economics.* Trans. R. le Farnu. Oxford: Robertson, 1983.

Groenewegen, P. 1977. Adam Smith and the division of labour: a bi-centenary estimate. *Australian Economic Papers* 16, 161–74.

Harris, J. 1757. *An Essay Upon Money and Coins.* Part I. London: G. Hawkins.

Hegel, G.W.F. 1821. *Philosophy of Right.* Trans. T.M. Knox, Oxford: Clarendon Press, 1982.

Hicks, J.R. 1965. *Capital and Growth.* Oxford: Clarendon Press.

Himmelweit, S. and Mohun, S. 1977. Domestic Labour and Capital. *Cambridge Journal of Economics* 1, 15–31.

Hobsbawm, E. 2000. *The New Century.* London: Abacus.

Hodgskin, T. 1827. *Popular Political Economy.* London: C. Tait. New York: A.M. Kelley, 1966.

Hutcheson, F. 1755. *A System of Moral Philosophy.* Glasgow.

Hutchison, T.W. 1988. *Before Adam Smith: The Emergence of Political Economy 1662–1776.* Oxford: Basil Blackwell.

Kaldor, N. 1966. *Causes of the Slow Rate of Economic Growth in the United Kingdom.* Cambridge: Cambridge University Press.

Kaldor, N. 1967. *Strategic Factors in Economic Development.* Ithaca: State School of Industrial and Labour Relations, Cornell University.

Kaldor, N. 1972. The irrelevance of equilibrium economics. *Economic Journal* 82, 1237–55.

Kaldor, N. 1975. What is wrong with economic theory? *Quarterly Journal of Economics* 89, 347–57.

Kames, H.H. 1774. *Sketches of the History of Man.* Edinburgh: W. Creech. London: W. Strahan and T. Cadell.

Lindsay, P. 1733. *The Interest of Scotland Considered.* Edinburgh: R. Fleming & Co.

McCulloch, J.R. 1825. *Principles of Political Economy.* London: Murray, 1870.

McCulloch, J.R. 1850. Introduction and notes to Adam Smith, *An Inquiry into the Nature and Causes of the Wealth of Nations.* 4th edn. Edinburgh: A. & C. Black; London: Longman.

Mandeville, B. 1729. *The Fable of the Bees.* London: A. Roberts.

Marglin, S. 1974. What do bosses do? The origins and functions of hierarchy in capitalist production, *Review of Radical Political Economics* 6, 60–112.

Marshall, A. 1890. *Principles of Economics.* 8th edn. London: Macmillan & Co., 1920.

Marshall, A. 1919. *Industry and Trade.* 4th edn. London: Macmillan & Co., 1923.

Marshall, A. and Marshall, M.P. 1879. *The Economics of Industry.* London: Macmillan and Company.

Marx, K. 1844. *Economic and Philosophic Manuscripts.* Moscow: Foreign Languages Publishing House, 1959.

Marx, K. 1867. *Capital,* vol. 1, Moscow: Foreign Languages Publishing House, 1959.

Marx, K. and Engels, F. 1845–6. *The German Ideology.* Moscow: Progress Publishers, 1964.

Meek, R.L. and Skinner, A.S. 1973. The development of Adam Smith's ideas on the division of labour. *Economic Journal* 83, 1094–116.

Mill, J.S. 1821. *Elements of Political Economy.* 3rd edn. In *The Selected Economic Writings of James Mill,* ed. D.N. Winch. Edinburgh: Oliver & Boyd for the Scottish Economic Society, 1966.

Mill, J.S. 1848. Principles of Political Economy. In *The Collected Works of John Stuart Mill,* vols. 2 and 3, ed. J.M. Robson. Toronto: Toronto University Press, 1965.

Nicholson, J.S. 1893. *Principles of Political Economy.* 2nd edn. London: A. & C. Black.

Perelman, M. 1984. *Classical Political Economy: Primitive Accumulation and the Social Division of Labour.* London: Rowman & Allenheld.

Pernin, J.L. 1993. La cooperation entre firmes: une approche par l'économie des conventions. *Economie appliqué* 46, 105–26.

Petty, W. 1671. *Political Arithmetick.* In *Economic Writings of Sir William Petty,* ed. C.H. Hull. New York: A.M. Kelley, 1963.

Petty, W. 1683. *Another Essay on Political Arithmetick concerning the Growth of the City of London*. In *Economic Writings of Sir William Petty*, ed. C.H. Hull. New York: A.M. Kelley, 1963.

Plato. 380 BC. *The Republic*. Harmondsworth: Penguin Classics, 1955.

Posner, R.A. 1992. *Sex and Reason*. Cambridge, MA: Harvard University Press.

Quesnay, F. 1765. Natural right. Extracts trans. R.L. Meek, *The Economics of Physiocracy*. London: Allen & Unwin, 1962.

Rae, J. 1834. *Statement of Some New Principles on the Subject of Political Economy*. New York: A.M. Kelley, 1964.

Robbins, L. 1932. *An Essay on the Nature and Significance of Economic Science*. London: Macmillan, 1935.

Romer, P.M. 1987. New theories of economic growth: growth based on increasing returns due to specialization. *American Economic Review* 77, 56–62.

Rosenberg, N. 1976. Marx as a student of technology. *Monthly Review* 28, 56–77. In N. Rosenberg, *Inside the Black Box: Technology and Economics*. Cambridge: Cambridge University Press, 1982.

Rowthorn, R.E. 1975. A reply to Lord Kaldor's comment. *Economic Journal* 85, 897–901.

Ruskin, J. 1851–3. *The Stones of Venice*. In *The Complete Works of John Ruskin*, ed. E.T. Cook and A. Wedderburn. London: George Allen, 1904.

Salz, A. 1934. Specialisation. In *International Encyclopaedia of the Social Sciences*, vol. 13, ed. E.R.A. Seligman. New York: Macmillan Company, 1948.

Scazzieri, R. 1981. *Efficienza, produttività e livelli di attitività*. Bologna: Il Mulino.

Schiller, F. 1793. *On the Aesthetic Education of Man*. Trans. R. Snell, New York: Ungar, 1980.

Schumpeter, J.A. 1954. *History of Economic Analysis*. London: Oxford University Press.

Senior, N. 1836. *An Outline of the Science of Political Economy*. London: George Allen & Unwin, 1938, 1951.

Sidgwick, H. 1883. *Principles of Political Economy*, 2nd edn. London: Macmillan & Co.

Smith, A. 1776. *An Inquiry into the Nature and Causes of the Wealth of Nations*, ed. R.H. Campbell and A.S. Skinner. Oxford: Clarendon Press, 1976.

Stewart, D. 1858–75. *Collected Works of Sir Dugald Stewart*, ed. Sir William Hamilton, Edinburgh.

Sun, G.Z. 2005. *Readings in the Economics of the Division of Labour*. Singapore: World Scientific Publishing Company.

Taussig, F.W. 1911. *Principles of Economics*. 3rd edn. New York: Macmillan, 1936.

Thirlwall, A.P. 1983. A plain man's guide to Kaldor's growth law. *Journal of Post Keynesian Economics* 5, 345–58.

Torrens, R. 1808. *The Economists Refuted*. Sydney: Department of Economics, University of Sydney, Reprints of Economic Classics, 1984.

Tucker, J. 1755. *The Elements of Commerce and the Theory of Taxes*. London.

Tucker, J. 1774. *Four Tracts on Political and Economic Subjects*, 2nd edn. Gloucester: R. Raikes.

Turgot, A.R.J. 1751. Lettre à Madame de Graffigny sur les letters d'un Péruvienne. In *Oeuvres de Turgot et documents le concernant*, vol. 1, ed. G. Schelle. Paris: F. Alcan, 1913.

Turgot, A.R.J. 1766. Reflections on the Production and Distribution of Wealth. In *The Economics of A.R.J. Turgot*, ed. P.D. Groenewegen. The Hague: Nijhoff, 1977.

Ure, A. 1835. *The Philosophy of Manufactures*. London: C. Knight; London: Frank Cass, 1967.

Verdoorn, P.J. 1949. Fattori che regolano lo sviluppo della produttività del lavoro. *L'industria* 1, 45–53.

West, E.G. 1964. Adam Smith's two views on the division of labour. *Economica* 3, 23–32.

Yang, X.-K. and Ng, Y.-K. 1993. *Specialisation and Economic Integration: A New Classical Micro-economic Framework*. New York: Elsevier Science.

Young, A. 1928. Increasing returns and economic progress. *Economic Journal* 38, 327–42.

Dobb, Maurice Herbert (1900–1976)

Maurice Dobb was undoubtedly one of the outstanding political economists of this century. He was a Marxist, and was one of the most creative contributors to Marxian economics. As Ronald Meek put it, in his obituary of Dobb for the British Academy, 'over a period of fifty years [Dobb] established and maintained his position as one of the most eminent Marxist economists in the world'. Dobb's *Political Economy and Capitalism* (1937) and *Studies in the Development of Capitalism* (1946) are his two most outstanding contributions to Marxian economics. The former is primarily concerned with economic theory (including such subjects as value theory, economic crises, imperialism, socialist economies), and the latter with economic history (particularly the emergence of capitalism from feudalism). These two fields – economic theory and economic history – were intimately connected in Dobb's approach to economics. He also wrote an influential book on Soviet economic development. This was first published under the title *Russian Economic Development since the Revolution* (1928), and later in a revised edition as *Soviet Economic Development since 1917* (1948).

Maurice Dobb was born on 24 July 1900 in London. His father Walter Herbert Dobb had a draper's retail business and his mother Elsie Annie Moir came from a Scottish merchant's family. He was educated at Charterhouse, and then at Pembroke College, Cambridge, where he studied economics. This was followed by two postgraduate years at the London School of Economics, where he did his Ph.D. on 'The Entrepreneur'. The thesis formed the basis of his book *Capitalist Enterprise and Social Progress* (1925). Dobb returned to Cambridge at the end of 1924 on being appointed as a lecturer in economics. He taught in Cambridge until his retirement in 1967. He was a Fellow of Trinity College, and was elected to a University Readership in 1959. He received honorary degrees from the Charles University of Prague, the University of Budapest, and Leicester University, and was elected a Fellow of the British Academy. After retirement he and his wife, Barbara, stayed on in the neighbouring village of Fulbourn. He died on 17 August 1976.

Dobb was a theorist of great originality and reach. He was also, throughout his life, deeply concerned with economic policy and planning. His foundational critique of 'market socialism' as developed by Oscar Lange and Abba Lerner, appeared in the *Economic Journal* of 1933, later reproduced along with a number of related contributions in his *On Economic Theory and Socialism* (1955). His relatively elementary book *Wages* (1928) presented not merely a simple introduction to labour economics, but also an alternative outlook on these questions, including their policy implications, leading to interesting disputations with John Hicks, among others. In later years Dobb was much concerned with planning for economic development. In three lectures delivered at the Delhi School of Economics, later published as *Some Aspects of Economic Development* (1951), Dobb discussed some of the central issues of development planning for an economy with unemployed or underutilized labour, and his ideas were more extensively developed in his later book, *An Essay on Economic Growth and Planning* (1960).

Maurice Dobb also published a number of papers on more traditional fields in economic theory, including welfare economics, and some of these papers were collected together in his *Welfare Economics and the Economics of Socialism* (1969). In his *Theories of Value and Distribution since Adam Smith: Ideology and Economic Theory* (1973), he responded *inter alia* to the new developments in Cambridge political economy, including the influential 'Prelude to a Critique of Economic Theory' by Piero Sraffa (1960). Maurice Dobb's association with Piero Sraffa extended over a long period, both as a colleague at Trinity College, and also as a collaborator in editing *Works and Correspondence of David Ricardo*, published in 11 volumes between 1951 and 1973 (on the latter, see Pollitt, 1990).

In addition to academic writings, Maurice Dobb also did a good deal of popular writing, both for workers' education and for general public discussion. He wrote a number of pamphlets, including *The Development of Modern Capitalism* (1922), *Money and Prices* (1924), *An Outline of European History* (1926), *Modern Capitalism* (1927), *On Marxism Today* (1932), *Planning and Capitalism* (1937), *Soviet Planning and Labour in Peace and War* (1942), *Marx as an Economist: An Essay* (1943), *Capitalism Yesterday and Today* (1958), and *Economic Growth and Underdeveloped Countries* (1963), and many others. Dobb was a superb communicator, and the nature of his own research was much influenced by policy

debates and public discussions. Dobb the economist was not only close to Dobb the historian, but also in constant company of Dobb the member of the public. It would be difficult to find another economist who could match Dobb in his extraordinary combination of genuinely 'high-brow' theory, on the one hand, and popular writing on the other. The author of *Political Economy and Capitalism* (from the appearance of which – as Ronald Meek (1978) rightly notes – 'that future historians of economic thought will probably date the emergence of Marxist economics as a really serious economic discipline') was also spending a good deal of effort writing pamphlets and material for labour education, and doing straightforward journalism. It is not possible to appreciate fully Maurice Dobb's contributions to economics without taking note of his views of the role of economics in public discussions and debates.

Another interesting issue in understanding Dobb's approach to economics concerns his adherence to the labour theory of value. The labour theory has been under attack not only from neoclassical economists, but also from such anti-neoclassical political economists as Joan Robinson and, indirectly, even Piero Sraffa. In his last major work, *Theories of Value and Distribution since Adam Smith* (1973), Maurice Dobb speaks much in support of the relevance of Sraffa's (1960) major contribution, which eschews the use of labour values (on this see Steedman, 1977), but without abandoning his insistence on the importance of the labour theory of value. It is easy to think that there is some inconsistency here, and it is tempting to trace the origin of this alleged inconsistency to Dobb's earlier writings, which made Abram Bergson remark that 'in Dobb's analysis the labour theory is not so much an analytic tool as excess baggage' (Bergson, 1949, p. 445).

The key to understanding Dobb's attitude to the labour theory of value is to recognize that he did not see it just as an intermediate product in explaining relative prices and distributions. He took 'the labour-principle' as 'making an important qualitative statement about the nature of the economic problem' (Dobb, 1937, p. 21). He rejected seeing the labour theory of value as simply a 'first approximation' containing 'nothing essential that cannot be expressed equally well and easily in other terms' (Dobb, 1973, pp. 148–9). The description of the production process in terms of labour involvement has an interest that extends far beyond the role of the labour value magnitudes in providing a 'first approximation' for relative prices. As Dobb (1973, pp. 148–9) put it,

> there is something in the first approximation that is lacking in later approximations or cannot be expressed so easily in those terms (e.g., the first approximation may be a device for emphasising and throwing into relief something of greater generality and less particularity).

Any description of reality involves some selection of facts to emphasize certain features and to underplay others, and the labour theory of value was seen by Dobb as emphasizing the role of those who are involved in 'personal participation in the process of production *per se*' in contrast with those who do not have such personal involvement.

> As such 'exploitation' is neither something 'metaphysical' nor simply an ethical judgement (still less 'just a noise') as has sometimes been depicted: it is a factual description of a socio-economic relationship, as much as is Marc Bloch's apt characterisation of Feudalism as a system where feudal Lords 'lived on the labour of other men'. (Dobb, 1973, p. 145)

The possibility of calculating prices without going through value magnitudes, and the greater efficiency of doing that (on this see Steedman, 1977), does not affect this descriptive relevance of the labour theory of value in any way. Maurice Dobb also outlined the relationship of this primarily descriptive interpretation of labour theory of value with evaluative questions, for example, assessing the 'right of ownership' (see especially Dobb, 1937).

The importance for Dobb of descriptive relevance is brought out also by his complex attitude to the utility theory of value. While he rejected the view that the utility picture is the best way of seeing relative values ('by taking as its foundation a fact of individual consciousness'), he lamented the descriptive impoverishment that is brought about by replacing the subjective utility theory by the 'revealed preference' approach.

> If all that is postulated is simply that men *choose*, without anything being stated even as to how they choose or what governs their choice, it would seem impossible for economics to provide us with any more than a sort of algebra of human choice. (Dobb, 1937, p. 171)

Indeed, as early as 1929, a long time before the 'revealed preference theory' was formally inaugurated by Paul Samuelson, Dobb (1929, p. 32) had warned:

> Actually the whole tendency of modern theory is to abandon such psychological conceptions: to make utility and disutility coincident with observed offers in the market; to abandon a 'theory of value' in pursuit of a 'theory of price'. But this is to surrender, not to solve the problem.

Maurice Dobb's open-minded attitude to non-Marxian traditions in economics added strength and reach to his own Marxist theorizing. He could combine Marxist reasoning and methodology with other traditions, and he was eager to be able to communicate with economists belonging to other schools. Dobb's honesty and lack of dogmatism were important for the development of the Marxist economic tradition in the English-speaking world, because he occupied a unique position in Marxist thinking in Britain. As Eric Hobsbawm (1967, p. 1) has noted,

> for several generations (as these are measured in the brief lives of students) he was not just the only Marxist

economist in a British university of whom most people had heard, but virtually the only don known as a communist to the wider world.

The Marxist economic tradition was well served by Maurice Dobb's willingness to engage in spirited but courteous debates with economists of other schools. Dobb achieved this without compromising the integrity of his position. The distinctly Marxist quality of his economic writings was as important as his willingness to listen and dispassionately analyse the claims of other schools of thought with which he engaged in systematic disputation. The gentleness of Dobb's style of disputation arose from strength rather than from weakness.

Dobb's willingness to appreciate positive elements in other economic traditions while retaining the distinctive qualities of his own approach is brought out very clearly also in his truly far-reaching critique of the theory of socialist pricing as presented by Lange, Lerner, Dickinson and others in the 1930s. Dobb noted the efficiency advantages of a price mechanism, especially in a static context. He was, however, one of the first economists to analyse clearly the conflict between the demands of efficiency expressed in the equilibrium conditions of the Langer–Lerner price mechanism (and also of course in a perfectly competitive market equilibrium), and the demands that would be imposed by the requirements of equality, given the initial conditions. In his paper called 'Economic Theory and the Problems of a Socialist Economy' published in 1933, Maurice Dobb argued thus:

> If carpenters are scarcer or more costly to train than scavengers, the market will place a higher value upon their services, and carpenters will derive a higher income and have greater 'voting power' as consumers. On the side of supply the extra 'costliness' of carpenters will receive expression, but only at the expense of giving carpenters a differential 'pull' as consumers, and hence vitiating the index of demand. On the other hand, if carpenters and scavengers are to be given equal weight as consumers by assuring them equal incomes, then the extra costliness of carpenters will find no expression in costs of production. Here is the central dilemma. Precisely because consumers are also producers, both costs and needs are precluded from receiving simultaneous expression in the same system of market valuations. Precisely to the extent that market valuations are rendered adequate in one direction they lose significance in the other. (1933, p. 37)

The fact that given an initial distribution of resources the demands of efficiency and those of equity may – and typically will – conflict is, of course, one of the major issues in the theory of resource allocation, with implications for market socialism as well as for competitive markets in a private ownership economy. As a matter of fact, Marx had *inter alia* noted this conflict in his *Critique of Gotha Programme,* but in the discussion centring around Langer–Lerner systems, this deep conflict had attracted relatively little attention, except in the arguments presented by Maurice Dobb. The fact that even a socialist economy has to cope with inequalities of initial resource distribution (arising from, among other things, differences in inherited talents and acquired skills) makes it a relevant question for a socialist economy as well as for competitive market economies, and Dobb's was one of the first clear analyses of this central question of resource allocation.

The second respect in which Maurice Dobb found the literature on market socialism inadequate concerns allocation over time. In discussing the achievements and failures of the market mechanism, Maurice Dobb argued that the planning of investment decisions

> may contribute much more to human welfare than could the most perfect micro-economic adjustment, of which the market (if it worked like the textbooks, at least, and there were no income-inequalities) is admittedly more fitted in most cases to take care. (Dobb, 1960, p. 76)

In his book *An Essay in Economic Growth and Planning* (1960), Dobb provided a major investigation of the basis of planned investment decisions, covering overall investment rates, sectoral divisions, choice of techniques, and pricing policies related to allocation (including that over time).

This contribution of Dobb relates closely to his analysis of the problems of economic development. In his earlier book *Some Aspects of Economic Development* (1951), Dobb had already presented a pioneering analysis of the problem of economic development in a surplus-labour economy, with shortage of capital and of many skills. While, on the one hand, he anticipated W.A. Lewis's (1954) more well-known investigation of economic growth with 'unlimited supplies of labour', he also went on to demonstrate the far-reaching implications of the over-all savings rates being socially sub-optimal and inadequate. Briefly, he showed that this requires not only policies directly aimed at raising the rates of saving and investment, but it also has implications for the choice of techniques, sectoral balances, and price fixation.

In such a brief note, it is not possible to do justice to the enormous range of Maurice Dobb's contributions to economic theory, applied economics and economic history. Different authors influenced by Maurice Dobb have emphasized different aspects of his many-sided works (see, for example, Feinstein, 1967, and the *Cambridge Journal of Economics*' Maurice Dobb Memorial Issue (1978)). He has also had influence even outside professional economics, particularly in history, especially through his analysis of the development of capitalism.

Dobb argued that the decline of feudalism was caused primarily by 'the inefficiency of Feudalism as a system of production, coupled with the growing needs of the ruling class for revenue' (1946, p. 42). This view of feudal decline, with its emphasis on *internal* pressures, became the subject of a lively debate in the early 1950s. An alternative position, forcefully presented by Paul Sweezy in particular, emphasized some *external* developments, especially the growth of trade, operating through the relations between the feudal countryside and the towns that developed on its periphery. No matter what view is taken as to 'who won' the debates on the transition from feudalism to capitalism, Dobb's creative role in opening up a central question in economic history as well as a major issue in Marxist political economy can scarcely be disputed. Indeed, *Studies in the Development of Capitalism* (1946) has been a prime mover in the emergence of the powerful Marxian tradition of economic history in the English-speaking world, which has produced scholars of the eminence of Christopher Hill, Rodney Hilton, Eric Hobsbawm, Edward Thompson and others.

It is worth emphasizing that aside from the explicit contributions made by Maurice Dobb to economic history, he also did use a historical approach to economic analysis in general. Maurice Dobb's deep involvement in descriptive richness (as exemplified by his analysis of 'the requirements of a theory of value'), his insistence on not neglecting the long-run features of resource allocation (influencing his work on planning as well as development), his concern with observed phenomena in slumps and depressions in examining theories of 'crises', and so on, all relate to the historian's perspective. Dobb's works in the apparently divergent areas of economic theory, applied economics and economic history are, in fact, quite closely related to each other.

Maurice Dobb was not only a major bridge-builder between Marxist and non-Marxist economic traditions (aside from pioneering the development of Marxist economics in Britain and to some extent in the entire English-speaking world): he also built many bridges between the different pursuits of economic theorists, applied economists and economic historians. Dobb's political economy involved the rejection of the narrowly economic as well as the narrowly doctrinaire. He was a great economist in the best of the broad tradition of classical political economy.

AMARTYA SEN

Selected works

1925. *Capitalist Enterprise and Social Progress*. London: Routledge.

1928. *Russian Economic Development since the Revolution*. London: Routledge.

1928. *Wages*. London: Nisbet; Cambridge: Cambridge University Press.

1929. A sceptical view of the theory of wages. *Economic Journal* 39, 506–19.

1933. Economic theory and the problems of a socialist economy. *Economic Journal* 43, 588–98.

1937. *Political Economy and Capitalism: Some Essays in Economic Tradition*. London: Routledge.

1946. *Studies in the Development of Capitalism*. London: Routledge.
1948. *Soviet Economic Development since 1917*. London: Routledge.
1950. Reply (to Paul Sweezy's article on the transition from feudalism to capitalism). *Science and Society* 14(2), 157–67.
1951. *Some Aspects of Economic Development: Three Lectures*. Delhi: Ranjit Publishers, for the Delhi School of Economics.
1955. *On Economic Theory and Socialism*. London: Routledge.
1960. *An Essay on Economic Growth and Planning*. London: Routledge.
1969. *Welfare Economics and the Economics of Socialism*. Cambridge: Cambridge University Press.
1973. *Theories of Value and Distribution since Adam Smith: Ideology and Economic Theory*. Cambridge: Cambridge University Press.

Bibliography

Bergson, A. 1949. Socialist economics. In *A Survey of Contemporary Economics*, ed. H.S. Ellis. Philadelphia: Blakiston.
Cambridge Journal of Economics. 1978. Maurice Dobb memorial issue. Vol. 2(2), June.
Hobsbawm, E.J. 1967. Maurice Dobb. In *Socialism, Capitalism and Economic Growth: Essays Presented to Maurice Dobb*, ed. C. Feinstein. Cambridge: Cambridge University Press.
Lewis, W.A. 1954. Economic development with unlimited supplies of labour. *Manchester School* 20(2), 139–91.
Meek, R. 1978. Obituary of Maurice Herbert Dobb. *Proceedings of the British Academy 1977* 53, 333–44.
Pollitt, B.H. 1990. Clearing the path for 'Production of Commodities by Means of Commodities': notes on the collaboration of Maurice Dobb in Piero Sraffa's edition of 'The Works and Correspondence of David Ricardo'. In *Essays on Piero Sraffa: Critical Perspectives on the Revival of Classical Theory*, ed. K. Bharadwaj and B. Schefold. London: Unwin Hyman.
Sraffa, P. 1960. *Production of Commodities by Means of Commodities: Prelude to a Critique of Economic Theory*. Cambridge: Cambridge University Press.
Sraffa, P., with the collaboration of M.H. Dobb. 1951–73. *Works and Correspondence of David Ricardo*. 11 vols, Cambridge: Cambridge University Press.
Steedman, I. 1977. *Marx after Sraffa*. London: New Left Books.

dollarization

Dollarization is a situation in which a foreign currency (often the US dollar) replaces a country's currency in performing one or more of the basic functions of money.

Thus in Ortiz (1983) the term 'dollarization' refers to the widespread usage of US dollars for transaction purposes in Mexico. More recently, Ize and Levy-Yeyati (2003) use 'financial dollarization' for episodes in which domestic financial contracts are denominated in dollars or another foreign currency.

In some countries, dollarization has been the outcome of official government policy. Examples include Ecuador in 2000 and El Salvador in 2001, where the domestic currency was retired from circulation and the US dollar became the official currency. An immediate implication of such 'official dollarization' is that domestic prices of tradable goods are tied to world prices, so domestic inflation is closely related to US inflation. Hence official dollarization has been advocated for countries suffering from chronic, high, and volatile inflation.

On the other side of the ledger, official dollarization implies the surrender of independent monetary policy, leaving only fiscal policy available as a stabilization tool. In addition, the domestic government gives up seigniorage, or the revenue from money creation, which accrues to the US Federal Reserve. While both effects are widely regarded as costly for the domestic economy, their welfare implications depend on details about the policy-making process and, in particular, on whether the monetary authorities can credibly commit to implement optimal policy (see Chang and Velasco, 2002, for a discussion).

Finally, official dollarization implies that the domestic central bank is no longer available as a lender of last resort, which may be conducive to financial fragility and crises. Calvo (2005) argues, however, that last resort lending can be provided by alternative arrangements.

Impetus for official dollarization as a policy alternative was greatest at the turn of the millennium, as emerging economies had to cope with a sequence of financial and exchange rate crises while several European countries were abandoning their national currencies in favour of the newly created euro. Support for official dollarization appears to have subsided since, however.

More frequently, dollarization has emerged as a spontaneous response of domestic agents to inflation. The special case in which such a process has resulted in the dollar becoming a widespread medium of exchange is known as 'currency substitution'. Currency substitution has been the subject of a large literature, much of it focused on the determinants of the relative demand for domestic vis-à-vis foreign currencies and on implications for monetary management. Early research followed Girton and Roper (1981) in postulating ad hoc aggregate demand functions for domestic and foreign currency, in the portfolio balance tradition. Somewhat later, Calvo (1985) derived similar demand functions from an optimizing model in which domestic and foreign currencies entered the representative household's utility function. Those approaches emphasized the possibility that increasing substitutability between the domestic and the foreign currencies would lead to monetary and exchange rate instability. However, they did not identify the basic determinants of substitutability, which was buried in the specification of the postulated demand function for foreign currency or the properties of the representative agent's utility function. Hence the early studies were of little use in understanding how to cure the ills associated with dollarization, and, in particular, they failed to trace the consequences of common policies designed to deal directly with currency substitution, such as outright prohibitions on the holdings of foreign currency.

Subsequent studies have attempted to address these shortcomings by modelling more explicitly the fundamental frictions underlying currency substitution. Thus Guidotti and Rodriguez (1992) developed a cash-in-advance model of currency substitution on the assumption that using foreign currency entailed fixed transaction costs, while Chang (1994) studied the implications of a similar assumption in an overlapping generations setting. These models still left unexplained where the assumed transaction costs were coming from. Therefore, recent work on this area models currency substitution entirely from first principles, in the search theoretic tradition (see, for instance, Craig and Waller, 2004).

Another focus of recent literature has been the increased use of the dollar as the currency of denomination of the debts of domestic residents in emerging economies, a problem that Calvo (2005) terms 'liability dollarization'. A substantial degree of liability dollarization places an economy in a vulnerable situation, since presumably many of the agents with dollar debts have assets denominated in domestic currency. Such a currency mismatch situation means that a depreciation of the domestic currency reduces the net worth of domestic agents. If, in turn, aggregate demand depends on net worth (as would be the case in the presence of financial imperfections), a currency depreciation may lead to a reduction in income and employment. In other words, liability dollarization may render depreciations contractionary, not expansionary as assumed by conventional analysis (Aghion, Bachetta and Banerjee, 2001; Cespedes, Chang and Velasco, 2004). The combination of liability dollarization and net worth effects has been blamed for the severity of the income and output contractions in recent emerging markets crises.

At this point, no consensus exists as to the causes of liability and financial dollarization, although research on this question is rather active. Ize and Levy-Yeyati (2003), in particular, have examined the choice of currency denomination of assets and liabilities from a capital asset pricing model (CAPM) perspective, while Jeanne (2005) models liability dollarization as the private sector response to the lack of credibility in monetary policy. Finally, several studies estimate how measures of financial dollarization depend empirically on other characteristics of an economy. For example, Arteta (2005) has found

that the dollarization of bank deposits is empirically more frequent in countries with a higher degree of exchange rate flexibility.

ROBERTO CHANG

See also **currency unions; money.**

Bibliography

Aghion, P., Bachetta, P. and Banerjee, A. 2001. Currency crises and monetary policy in an economy with credit constraints. *European Economic Review* 45, 1121–50.

Arteta, C. 2005. Exchange rate regimes and financial dollarization: does flexibility reduce currency mismatches in bank intermediation? *Topics in Macroeconomics* 5(1).

Calvo, G.A. 1985. Currency substitution and the real exchange rate: the utility maximization approach. *Journal of International Money and Finance* 4, 175–88.

Calvo, G.A. 2005. Capital markets and the exchange rate with special reference to the dollarization debate in Latin America. In *Emerging Capital Markets in Turmoil*, ed. G. Calvo. Cambridge, MA: MIT Press.

Cespedes, L., Chang, R. and Velasco, A. 2004. Balance sheets and exchange rate policy. *American Economic Review* 94, 1183–93.

Chang, R. 1994. Endogenous currency substitution, inflationary finance, and welfare. *Journal of Money, Credit, and Banking* 26, 903–16.

Chang, R. and Velasco, A. 2002. Dollarization: analytical issues. In *Dollarization*, ed. E. Levy-Yeyati and F. Sturzenegger. Cambridge, MA: MIT Press.

Craig, B. and Waller, C. 2004. Dollarization and currency exchange. *Journal of Monetary Economics* 51, 671–89.

Girton, L. and Roper, D. 1981. Theory and implications of currency substitution. *Journal of Money, Credit, and Banking* 13, 12–30.

Guidotti, P.E. and Rodriguez, C.A. 1992. Dollarization in Latin America: Gresham's law in reverse? *IMF Staff Papers* 39, 518–44.

Ize, A. and Levy-Yeyati, E. 2003. Financial dollarization. *Journal of International Economics* 59, 323–47.

Jeanne, O. 2005. Why do emerging economies borrow in foreign currency? In *Other People's Money*, ed. B. Eichengreen and R. Hausmann. Chicago: University of Chicago Press.

Ortiz, G. 1983. Currency substitution in Mexico: the dollarization problem. *Journal of Money, Credit and Banking* 15, 174–85.

Domar, Evsey David (1914–1997)

Domar (Domashevitsky) was born in 1914 in Lodz, Russia (now Poland), spent most of his early life in Harbin, Manchuria, and moved permanently to the United States in 1936. His undergraduate degree in economics (1939) was from the University of California (Los Angeles); his graduate work was at the Universities of Michigan (MA, Mathematical Statistics) and Harvard (Ph.D., 1947), where he studied with Alvin Hansen, the leading American Keynesian and most important single intellectual influence on Domar. Domar is best known for his leadership role, along with Roy Harrod, in the initiation of modern growth theory.

His first position was with the research staff of the Board of Governors of the Federal Reserve System, where he worked on fiscal problems from 1943 to 1946. His subsequent academic career took him briefly to the Carnegie Institute of Technology, the Cowles Foundation and the University of Chicago, the Johns Hopkins University in 1948 for ten years, and the Massachusetts Institute of Technology in 1958, from which he retired in 1984. An avid traveller, he held more than a dozen visiting professorships in universities at home and abroad.

While the claim to the earliest statement of the famous Harrod–Domar growth model was clearly Harrod's (1939), Domar arrived independently at a structurally similar model but from a different point of view (1946; 1947). By incorporating into static Keynesian analysis the capacity changes associated with investment, he found that steady-state capacity growth required investment to grow at a rate equal to the savings rate multiplied by the capital–output ratio. From this simple beginning, growth theory took off to become a major focus, one might almost say obsession, of the profession in the 1950s and 1960s. Domar also made important contributions to some of its conceptual and measurement problems, such as the proper treatment of depreciation (1953) and the measurement of technological change (1961), and he coined the term 'residual' for the fraction of expanding output unexplained by the contribution of factors of production.

In fiscal theory, his early investigation, with Richard Musgrave (1944a), of the effect of a proportional income tax, with and without loss offsets, on portfolio choice was very similar in style and approach to portfolio theory of a decade later. Given individual preferences, the portfolio decision was modelled as a choice between alternative portfolios weighing their expected net returns against their risks (expected losses). The unconventional conclusion was reached that, given risk aversion, the imposition of a proportional income tax with symmetrical treatment of gains and losses would induce individuals to adjust their portfolios towards riskier assets. The reminder that expected risks and yields are both reduced by an income tax was an important correction to a simplistic focus on yields alone.

As an applied theorist, Domar had the knack of getting important results with simple theory. At a time when deficit finance was harshly criticized for increasing the debt burden and tax rate, Domar showed (1944b) that in a growing economy even continuous deficit finance resulted in only limited debt–income ratios and tax rates.

Second, he made a fertile historical hypothesis (1970) – that the economic basis for the introduction of serfdom (or slavery) was a low land-to-labour cost. Third, he ingeniously modified the administrative rules that guided the behaviour of collective farms (1966) or that determined the compensation of socialist managers (1974) to induce them towards more efficient price–output decisions.

Domar's work was informed by a rare combination of historical, empirical and theoretical breadth. His profound scholarship, in several languages, periods, and areas, often resurrected important findings of earlier writers previously overlooked.

E. CARY BROWN

Selected works

1944a. (With R.A. Musgrave.) Proportional income taxation and risk-taking. *Quarterly Journal of Economics* 58, 388–422.
1944b. The burden of the debt and the national income. *American Economic Review* 34, 798–827. Reprinted in Domar (1957).
1946. Capital expansion, rate of growth, and employment. *Econometrica* 14, 137–47. Reprinted in Domar (1957).
1947. Expansion and employment. *American Economic Review* 37, 34–55. Reprinted in Domar (1957).
1948. The problem of capital accumulation. *American Economic Review* 38, 777–94. Reprinted in Domar (1957).
1952. Economic growth: an econometric approach. *American Economic Review, Papers and Proceedings* 42, 479–95. Reprinted in Domar (1957).
1953. Depreciation, replacement and growth. *Economic Journal* 63, 1–32. Reprinted in Domar (1957).
1957. *Essays in the Theory of Economic Growth.* New York: Oxford University Press.
1961. On the measurement of technological change. *Economic Journal* 71, December, 709–29.
1966. The Soviet collective farm as a producer cooperative. *American Economic Review* 56, 734–57.
1970. The causes of slavery or serfdom: a hypothesis. *Journal of Economic History* 30, 18–32.
1974. On the optimal compensation of a socialist manager. *Quarterly Journal of Economics* 88, 1–18.

Bibliography

Harrod, R.F. 1939. An essay in dynamic theory. *Economic Journal* 49, March, 14–33, Errata, June 1939, 377.
Harrod, R.F. 1948. *Towards a Dynamic Economics. Some Recent Developments of Economic Theory and their Applications to Policy.* London: Macmillan.

Dorfman, Joseph (1904–1991)

Historian of American economic thought, Dorfman was born in Russia in 1904 and educated at Reed College and at Columbia University, where he earned a Ph.D. degree in 1935 and taught from 1931 until his retirement 40 years later. Dorfman was a student of Clarence Ayres at Reed, and of Wesley C. Mitchell and John Maurice Clark at Columbia. Mitchell in turn had been a student of Thorstein Veblen. These four economists, all with institutional leanings, stand out among the formative influences that affected Dorfman's early career. He made Veblen the subject of his doctoral dissertation, which was published under the title *Thorstein Veblen and His America* in 1934. This was at the time the only book-length appraisal of a modern economist that gave close attention not only to the subject's writings but also to biographical detail, the contemporary climate of opinion, and the general social and cultural setting of the work.

This type of holistic approach is characteristic also of Dorfman's monumental *The Economic Mind in American Civilization,* a five-volume work that he published from 1946 to 1959. It is dedicated 'To the pioneering spirit of Thorstein Veblen and the first-born of his intellectual heirs, Wesley C. Mitchell'. The work is a detailed history of American economic thought from colonial times to 1933, the first of its kind and not likely to be replaced for many years. It is based on extensive research and in many instances provides the first comprehensive account of a writer's life and work. Dorfman sees a break of emphasis in the history of American economic thought at the time of the Civil War: it was commerce before, and industry later. He notes with respect the achievements of the past, and is a critical but tactful chronicler of past foibles. He was a pioneer in exploring not only the printed page but also archival material made up of 'papers', 'letters', and similarly elusive sources of information, the first writer to do so on a large and systematic scale in the history of economic thought.

HENRY W. SPIEGEL

Selected works

1934. *Thorstein Veblen and His America.* New York: Viking Press.

1935. (With R.G. Tugwell.) *William Beach Lawrence: Apostle of Ricardo.* New York, reprinted from *Columbia University Quarterly,* September, 1935.

1940. *The Economic Philosophy of Thomas Jefferson.* New York: Academy of Political Science.

1946–59. *The Economic Mind in American Civilization.* 5 vols. New York: Viking Press.

1954. Introduction to Adams, H.C., *Relation of the State to Industrial Action, and Economics and Jurisprudence. Two Essays,* ed. J. Dorfman. New York: Columbia University Press.

1960. (With R.G. Tugwell.) *Early American Policy: Six Columbian Contributors.* New York: Columbia University Press.

1967–9. Introduction to W.C. Mitchell, *Types of Economic Theory: From Mercantilism to Institutionalism,* ed. J. Dorfman. New York: A.M. Kelley.

Dornbusch, Rudiger (1942–2002)

Rudiger Dornbusch was born in Germany on 8 June 1942. He received his Licence es Sciences Politiques from the University of Geneva in 1966, and his Ph.D. in Economics from the University of Chicago in 1971. He was an assistant professor at the Department of Economics at the University of Rochester from 1972 to 1974, an associate professor at the Graduate School of Business at Chicago University from 1974 to 1975, and a member of the MIT Department of Economics from 1975 to 1978. He became a Professor of Economics at MIT in 1978. From 1984 until his death from cancer on 25 July 2002, he was Ford International Professor of Economics at MIT.

Dornbusch was, by any measure, one of the giants of late 20th century international macroeconomics. His celebrated *Journal of Political Economy* paper 'Expectations and exchange rate dynamics' (1976), which introduced the concept of exchange rate 'overshooting', became the workhorse of international macroeconomics over the ensuing two decades. His *American Economic Review* paper (with Stanley Fischer and Paul Samuelson) 'Comparative advantage, trade and payments in a Ricardian model with a continuum of goods' (1977) introduced a simple tractable framework that became similarly influential in the study of international trade.

This entry begins by reviewing Dornbusch's two most important scientific contributions, and goes on to give a brief sketch of his broader influence on the profession through students (he served as an advisor on over 125 doctoral dissertations), through his leading intermediate textbook *Macroeconomics* (written with Stanley Fischer), and through his role as an important voice in the public policy debate.

Exchange rate overshooting

Dornbusch's overshooting model of exchange rates (1976) captured the imagination of policymakers and academics alike during the early years of floating exchange rates. The model attracted enormous attention because, after the break-up of the Bretton Woods system of fixed exchange rates in the early 1970s, exchange rates seemed far too volatile relative to the underlying fundamentals. Although subsequent empirical work has undermined the model's original bold claim to explain floating exchange rates (see Meese and Rogoff, 1983), the model is still viewed as relevant, especially during episodes of major shifts in monetary policy. In fact, an informal survey conducted by Alan Deardorff of eight top economics departments found that, as late as 1990, Dornbusch's overshooting model was the only paper taught in every one of their graduate international finance courses.

The idea of overshooting is so simple and elegant that the small-country version can be illustrated with just a couple of equations (the analysis here draws on Rogoff, 2002). The assumption of 'uncovered interest parity' relates the home nominal interest rate to the exogenous foreign nominal interest rate and the expected rate of depreciation of the exchange rate:

$$i_t = i_t^* + E(e_{t+1} - e_t) \qquad (1)$$

where i_t is the level home nominal interest rate and e_t is the logarithm of the exchange rate (the home currency price of foreign currency), so that $E(e_{t+1} - e_t)$ is the expected rate of change in the exchange rate. The second key relationship is a money demand equation that relates the real balances to the nominal interest rate.

$$m_t - p_t = -\lambda i_t + \eta y_t \qquad (2)$$

where y denotes the log of output, m is the nominal money supply and p is the price level. Higher interest rates lower the demand for real balances, and an increase in output raises it. Dornbusch posed the question of what would happen if there were a one-time permanent increase in the money supply, m. If prices were fully flexible, it would be possible to maintain equilibrium in the above two equations by having prices and exchange rates all rise permanently in proportion to the increase in the money supply. In this case, money would be neutral and have no real effects.

In reality, however, while asset markets (including the exchange rate) adjust very quickly, goods markets adjust more slowly partly due to temporary price rigidities. Therefore, in this set-up money is neutral only in the long run (in which the price level rises proportionately to the money supply). But with goods markets clearing only slowly, what is the impact of a money shock on exchange rates and interest rates? Assume that output, y, is also fixed. If domestic prices are constant, then a rise in the money supply implies a rise in real balances, $m - p$. But this means that the home nominal interest rate i must fall, so there is a corresponding rise in the demand for real balances. Then, however, the uncovered interest parity equation (eq. (1) above) implies that e_t *must fall, or depreciate, relative to expectations of* e_{t+1}. That is, after any initial movement of the exchange rate in response to an unexpected shock, the currency must subsequently be expected to appreciate. But recall that in the long run, even with sticky prices, money is still neutral, so the exchange rate has to depreciate by the same amount as the rise in the domestic price level, thus producing no real effect.

How is all this possible? The answer, Dornbusch deduced, is that the initial money shock must cause the exchange rate to depreciate by more in the short run than it does in the long run. It 'overshoots'. Therefore, Dornbusch's model offered a highly plausible explanation of why exchange rates seem to be so volatile relative to fundamentals. At one level of abstraction, of course, 'overshooting' is an application of Paul Samuelson's 'Le Chatelier's principle' theorem: when prices in some markets are inflexible in the short run, prices in others may overreact in the short run. But Dornbusch's model did much more than innovatively contrast the fast adjustment of asset markets with the slow adjustment of goods markets (an insight that any realistic short-run dynamic macroeconomic model should take into account). It offered a concrete and coherent analysis of an extremely important practical phenomenon. Over the decades since Dornbusch's article appeared, the term 'overshooting' has become deeply woven into the popular economic lexicon.

Modern research has advanced considerably beyond the overshooting model, of course, and the Mundell-Fleming-Dornbusch model has largely been supplanted by 'new open economy macroeconomics' (see Obstfeld and Rogoff, 1996). And the notion of looking at money shocks via a money demand equation has increasingly been supplanted by frameworks which view the overnight interest rate as the key instrument of monetary policy. Nevertheless, these newer frameworks typically include sticky prices – perhaps the most fundamental, and controversial, element of Dornbusch's model – and hence can all replicate a similar phenomenon to 'overshooting.'

Although Dornbusch's overshooting paper was his best-known work, with over 900 citations in refereed journals, he published numerous other very well-known articles, including his 1973 *American Economic Review* paper that was among the first to incorporate non-traded goods in a monetary model (see also his elegant 1974 contribution to the collection edited by Robert Aliber), his 1983 *Journal of Political Economy* paper that illustrated how changes in the real interest rate could affect exchange rates and current accounts, and his 1987 *American Economic Review* paper that demonstrated a link between market structure and the adjustment of relative prices to exchange rate movements. Without doubt, however, his other extremely influential paper was not in international finance but in trade.

Ricardian model of trade

Dornbusch's 1977 *American Economic Review* paper with MIT colleagues Stanley Fischer and Paul Samuelson almost single-handedly revived the analysis of Ricardian trade; a 'Ricardian' model of trade is one with only one factor of production (usually taken to be labour). Trade is driven by differences in technology. The Ricardian model is contrasted with the Hecksher–Ohlin framework, where countries have identical technologies but different relative endowments of the factors of production (labour and capital, in the simplest canonical case). Prior to Dornbusch–Fischer–Samuelson (DFS), the Ricardian approach had been dormant for years, having been largely supplanted by the Hecksher–Ohlin framework. The Ricardian model had lost out not so much because of poor empirical results but because it had come be viewed as intractable for all but illustrative purposes. By introducing a continuum of goods (rather than a discrete number), DFS were able to analyse elegantly a broad range of comparative static questions that had previously seemed unapproachable. DFS showed, for example, how to mobilize the combination of comparative advantage and trade costs to endogenize the dividing line between 'traded' and 'non-traded' goods, and how to analyse the classic 'transfer' problem where one country owes debt to another. Although at first only a trickle of papers followed DFS, the power of their continuum specification has led to a recent explosion of related research. DFS have become the starting point for a number of applied papers (see, for example, Copeland and Taylor, 1994). In addition, DFS form the basis for a broad range of empirical papers (see, for example, Eaton and Kortum, 2002; Kehoe and Ruhl, 2002; Kraay and Ventura, 2002; Kei-Mu Yi, 2003; Ghironi and Melitz, 2005; see also Feenstra and Hanson, 1996). As the empirical work following DFS deepens, it is fair to say that trade economists have increasing faith in the fundamental underpinnings of the model.

Broader contributions

Aside from his path-breaking research, Dornbusch made important contributions to economics in a number of other dimensions. His intermediate undergraduate textbook with Stanley Fischer, *Macroeconomics*, written in the mid-1970s, became a worldwide best-seller. The book was really the first to integrate modern supply-side economics into the standard demand-driven framework of the day. As such, students were able to gain a far deeper understanding of problems such as the effects of oil price shocks.

Dornbusch was enormously influential as a graduate teacher at MIT. At his regular early-morning international economics 'breakfasts', Dornbusch would dissect recent models and serve up provocative questions in a fast-paced freewheeling style; many students remember these unique seminars as their most influential experiences as Ph.D. students. Dornbusch served as thesis advisor to scores of economists (as noted earlier, more than 125 in all), including Jeffrey Frankel, Paul Krugman, Maurice Obstfeld and Kenneth Rogoff. His dynamic, Socratic lecturing style also attracted students from outside MIT to his advanced graduate classes, including the likes of Jeffrey Sachs and Lawrence Summers. Many Dornbusch students went on to become finance ministers and heads of central banks throughout the world.

Through clear and incisive policy analysis embodied in editorials, speeches, and private meetings, Dornbusch exercised an enormous influence on global macroeconomic policy. He was a frequent guest of leading government officials throughout the world, who greatly valued and respected his advice. Arguably, no other recent economist has had so great an impact on the global macroeconomic policy debate, especially in emerging markets such as Brazil, Korea and Mexico, but also in more advanced countries such as Italy and Germany. Notably, in his later writing he succeeded in drawing ever more concrete insights from contemporary academic research, displaying a magnificent ability to translate complex theoretical models into ideas of immediate practical relevance. For example, his 1994 Brookings paper (with Alejandro Werner) argued that Mexico's pegged exchange rate had become overvalued to an extent that was unsustainable. Dornbusch's comments on markets prior to the currency collapse at the end of 1994 were highly influential. He also advanced a number of innovative ideas for dealing with international debt problems. His policy analysis was notable in that he managed to adopt strong views while continuing to be perceived as an independent and objective thinker. Over the last ten years of his life, Dornbusch became especially well-known for his monthly 'Economic Perspectives' newsletter, which covered with panache a broad range of topical global economic problems. One innovative idea, first developed in the newsletter and then formally published in his 'Primer on Emerging Market Crises' (2002) was to apply 'value at risk' analysis to the balance sheet of a country. In his primer, he wrote:

> …the right answer to crisis avoidance is controlling risk. The appropriate conceptual framework is *value at risk (VAR)* – a model-driven estimate of the maximum risk for a particular balance sheet situation over a specified horizon. There are surely genuine issues with the specifics of VAR surrounding modelling as has been widely discussed with respect to bank risk models used for meeting BIS requirements. But just as surely there is no issue whatsoever in recognizing that this general approach is the right one. If authorities everywhere enforced a culture of risk-oriented evaluation of balance sheets, extreme situations such as those of Asia in 1997 would disappear or, at the least, become a rare species. (2002, pp. 743–54)

In this short space it has not been possible to do full justice to the range and breadth of Dornbusch's contributions. But I hope the reader has gained some perspective on why he will have a lasting influence.

KENNETH ROGOFF

See also **comparative advantage; exchange rate volatility; extremal quantiles and value-at-risk; neo-Ricardian economics.**

Selected works

1973. Devaluation, money and nontraded goods. *American Economic Review* 63, 871–80.

1974. Real and monetary aspects of the effects of exchange rate regime changes. In *National Monetary Policies and the International Financial System*, ed. R. Aliber. Chicago: University of Chicago Press.

1974. Tariffs and nontraded goods. *Journal of International Economics* 4, 177–85.

1976. Expectations and exchange rate dynamics. *Journal of Political Economy* 84, 1161–76.

1977. (With S. Fischer and P. Samuelson.) Comparative advantage, trade and payments in a Ricardian model with a continuum of goods. *American Economic Review* 67, 823–39.

1980. *Open Economy Macroeconomics*. New York: Basic Books.

1983. Real interest rates, home goods and optimal external borrowing. *Journal of Political Economy* 91, 141–53.

1987. Exchange rates and prices. *American Economic Review* 77, 93–106.

1990. (With S. Fischer.) *Macroeconomics*, 5th edn. New York: McGraw-Hill.

1994. (With A. Werner.) Mexico: stabilization, reform and no growth. *Brookings Papers on Economic Activity*, 1994: 1, 253–315.

2002. A primer on emerging market crises. In *Preventing Currency Crises in Emerging Markets*, ed. S. Edwards and J.A. Frankel. Chicago: The University of Chicago Press.

Bibliography

Copeland, B. and Taylor, M. Scott. 1994. North–South trade and the environment. *Quarterly Journal of Economics* 109, 755–87.

Eaton, J. and Kortum, S. 2002. Technology, geography, and trade. *Econometrica* 70, 1741–79.

Feenstra, R. and Hanson, G. 1996. Globalization, outsourcing, and wage inequality. *American Economic Review* 86, 240–5.

Ghironi, F. and Melitz, M. 2005. International trade and macroeconomic dynamics with heterogeneous firms. Mimeo. Harvard University.

Kehoe, T. J. and Ruhl, K. J. 2002. How important is the new goods margin in international trade? Mimeo. University of Minnesota.

Kei-Mu, Yi. 2003. Can vertical specialization explain the growth of world trade? *Journal of Political Economy* 111, 52–102.

Kraay, A. and Ventura, J. 2002. Trade integration and risk sharing. *European Economic Review* 46, 1023–48.

Meese, R. and Rogoff, K. 1983. Empirical exchange rate models of the seventies: do they fit out of sample? *Journal of International Economics* 14, 3–24.

Obstfeld, M. and Rogoff, K. 1996. *Foundations of International Macroeconomics*. Cambridge, MA: MIT Press.

Rogoff, K. 2002. Dornbusch's overshooting model after 25 years: IMF Mundell-Fleming Lecture. *International Monetary Fund Staff Papers* 49 (Special Issue), 1–35 (including remarks by Rudiger Dornbusch).

double-entry bookkeeping

Firms of all kinds need, in different degrees, to maintain records of their transactions with other firms and persons, of the debts they owe or are owed, and of their assets. The records they keep for this purpose constitute their accounting records. Traditionally they have consisted of account-books of various kinds, but they can take the form also of magnetic tapes and so on. If the records are kept on a systematic basis, one can speak of an accounting system. From the accounting records one can prepare a variety of accounting statements in which the detailed accounting information is rearranged, regrouped and presented in summary form. The balance sheet and the profit-and-loss (or income) account or statement are important examples of such accounting statements.

Double-entry bookkeeping is a system or method for the arrangement and classification of accounting information. It developed in Italy, possibly in the second half of the 13th century. A description of the system was first published in Venice in 1494 as one part of a famous compendium of mathematical and commercial information: Luca Pacioli's *Summa de Arithmetica Geometria Proportioni et Proportionalità*. Knowledge of the double-entry system spread gradually from Italy to the rest of Europe by way of commercial contacts, schools and published treatises. It is not possible to establish how widely the system was used by merchants and others, say, in the 18th century. But by the late 19th century it had become the standard system for accounting records. Today it is used by virtually all corporate enterprises and many other firms as well as non-profit-making organizations in the West and also elsewhere. It has also proved suitable to serve as a useful scaffolding for the construction of the national income and related accounts for countries or regions.

Double-entry bookkeeping is no more than a system for arranging and organizing accounting information. It does not itself define the scope and detail of that information. Thus, for example, the double-entry system does not require that all transactions with third parties should be recorded, although it is the convention now to record all of them. What is more important, it does not prescribe which occurrences or changes that do not involve external transactions should be recorded in the accounts. Thus it does not prescribe whether changes in the value of the firm's assets should be recorded, how they should be determined, or how they should be recorded. Double entry neither generates nor requires any particular set of valuation rules or profit concepts. Different valuation bases or conventions, and different treatments for changes in the value of money, are all compatible with the use of the double-entry system. The system itself is highly adaptable, since it is concerned with arrangement and organization rather than with scope and content. Its adaptability has made it possible for it to serve as the basis for arranging the records needed by the relatively small-scale merchants in the early modern period of economic expansion as well as for those of the largest corporate enterprises operating today. But this does not mean that asset values were recorded and profits calculated in the same way by 17th-century merchants as they are by today's corporate enterprises. In fact, 17th-century merchants used several alternative bases for recording changes in asset values. And some of these would not be used by companies today.

Moreover, although all the companies within the same jurisdiction are subject to the same laws and the same institutional constraints (for example, those imposed by the stock-market authorities and those reflecting professional accounting standards), there is still scope for considerable variation in the determination and statement of accounting profits and asset values. However, because of developments in legislation and in the other constraining forces operating on corporate enterprises, it is no longer the case that a company chairman in the United Kingdom would be able to say (as Arthur Chamberlain, chairman of Tube Investments said in 1935) that he 'would almost undertake to draw up two balance-sheets for the same company, both coming within an auditor's statutory certificate, in which practically the only recognizable items would be the name and the capital authorised and issued'.

Double entry requires that each transaction (or other event) recorded in the accounting system must be recorded twice, and for the same money amount, once in debit form and once in credit form. In double entry, as Pacioli expressed it, 'all the entries placed in the ledger must be double, that is if you make a creditor (entry) you must make a debtor (entry)'. The debit and credit entries are made in the ledger, on the basis of the information entered in preliminary records. The ledger, which may for convenience be subdivided into a series of specialized ledgers, consists of a number of ledger accounts, pertaining, for example, to particular debtors or creditors, particular assets or particular categories of expenditure. It is the convention that the debit entry is made on the left-hand (debit) side of the appropriate ledger account, and the corresponding off-setting credit entry on the right-hand (credit) side of the other appropriate ledger account.

The duality of entries for each transaction (or other recorded event) ties together the ledger accounts into an interlocking system of recorded information. Moreover, as each transaction gives rise to two equal but opposite entries, the system of accounts (if properly kept) is always in balance or equilibrium. The total of debit entries must be equal to the total of credit entries. Similarly, the total of the balances on all ledger accounts that have debit balances must be equal to the total of the balances on all the remaining ledger accounts that have credit balances. (If debit balances are taken as positive amounts and credit balances as negative amounts, the algebraic sum of the balances on all ledger accounts is zero.) The equality of debits and credits is the basis for the trial balance. This is a list of the balances on all open (that is, unbalanced) accounts in the ledger, distinguishing between debit and credit balances. If the trial balance does not balance, there

is some error in the ledger. Postlethwayt in his Dictionary (1751) wrote of the 'agreeable satisfaction' of getting a trial balance to balance, and said that the trial balance will 'shew you that this [double entry], of all methods, is the most excellent'. The fact that a trial balance does not balance is proof that the ledger does contain some error. The converse is, of course, not correct.

Roger North, son of the prominent Turkey merchant Sir Dudley North, wrote in 1714 as follows: 'The making true Drs. (debtors) and Crs. (creditors) is the greatest Difficulty of Accompting, and perpetually exerciseth the Judgment; being an Act of the Mind, intent upon the Nature and Truth of Things.' Writers of instructional books on bookkeeping and accounts through the centuries have devised various lists, rules or approaches to help the accountant decide which debit and credit entries he should make for the various categories of transaction.

An early rule, widely used, was as follows (taken from a verse, 'Rules to be Observed', in a book of 1553 by James Peele):

> To make the thinges Received, or the receiver,
> Debter to the thinges delivered, or to the deliverer.

This rule is obviously readily applicable to many categories of transaction. If cash is received from a debtor, debit the cash account; and credit the debtor's account. If office furniture is bought on credit, debit the furniture account; and credit the supplier's account. If the owner withdraws cash from the business, debit the capital (that is, owner's) account; and credit the cash account. But it is evidently a straining of the language to say, when an amount is written off the book value of, say, a ship, in order to reflect diminution of value due to wear and tear, that the profit-and-loss account, which is to be debited, 'receives' something that has been 'delivered' to it by the ship account. Teachers and textbook writers not surprisingly looked for a rule that is robust enough to cover comfortably all transactions and events to be recorded, and to indicate unambiguously in each case where the debit and where the credit are to be placed.

The most common rule or approach adopted today in transaction analysis in the double-entry system derives from the so-called balance-sheet equation. The earliest formulation of this approach can be traced to the work of a Dutchman, Willem van Gezel, published in 1681.

> The basic balance-sheet equation is:
> Owner's Equity
> (or the firm's net worth) = Assets − Liabilities = Net Assets; or Owner's Equity + Liabilities = Assets.

The ledger contains accounts for the various assets and liabilities; and there are accounts in it for the capital contributed or withdrawn by the owner(s) and for any increases (decreases) in 'net worth' resulting from the activities of the firm. In the double-entry system, increases in assets are indicated by debits to an asset account – the extent to which assets are subdivided into separate ledger accounts is for each firm to decide. Conversely, decreases in assets are recorded as credits to asset accounts. The total of a firm's assets is represented by the total of claims on those assets; namely, its liabilities (that is, its debts to third parties) and its owner's equity. The total of these claims must be a credit amount that equals the debit amount representing the assets. An increase (decrease) in a claim is therefore represented by a credit (debit) in a liability account or an equity account. (Again, the extent to which claims are subdivided into various ledger accounts is a matter for each firm to decide. As regards the equity element, it is common for a ledger to contain separate accounts for each major category of business expenditure and income, a trading account, perhaps subdivided by type of activity, for showing the gross profit, and a profit-and-loss account to bring together the results from all the subordinate ledger accounts.)

Transaction analysis follows readily. The payment of salaries reduces the asset 'cash' and reduces the owner's equity, since the payment, taken by itself, represents a loss to the firm: hence, debit the salaries (eventually, profit-and-loss) account; and credit the cash account. The depreciation of an asset likewise reduces an asset and reduces the equity: debit the depreciation account (eventually profit-and-loss) account; and credit the ship account.

As has already been emphasized, the double-entry system does not itself dictate whether or in what circumstances increases or decreases in assets are to be recognized in the accounts. Neither does the system dictate the basis on which, or the circumstances in which, assets are to be revalued in the accounts. Decisions of these kinds are accounting decisions; and whenever such decisions are taken, the double-entry system of recording will accommodate them in accordance with its own logical structure. It follows from this that, although the value of the owner's equity in the ledger will always be equal to the value of the firm's net assets (that is, assets minus liabilities to those outside the firm) as stated in the accounts, those two values depend on the bases on which the values of assets are stated in the accounts.

Subject to this crucial qualification, it follows from the equilibrium feature of the double-entry system that the change (increase or decrease) in the value of the net assets of a firm over a period will be reflected as entries in the various ledger accounts that represent the owner's equity. Those entries in the various equity accounts that relate to the firm's operations, when they are brought together in the profit-and-loss account, yield a balance that is equal to the change in the value of the net assets over the period. It is the profit (loss) for the period. This profit is equal to the change in the value of the net assets over the period (allowance being made for any contributions or withdrawals of assets by the owner). It may be noted that the same profit figure would be established if one took the difference between the totals of two inventories of the firm's net assets taken, respectively, at the beginning and at the end of the period, provided that the same valuations were used and the same allowance made for the owner's contributions and withdrawals. The method of profit calculation by means of successive inventories of assets and liabilities was widely used in the past. The surviving 16th-century records of the large-scale commercial, financial and mining enterprise of the Fugger family of Augsburg provide examples of this procedure.

The equality – Profit (Loss) = Change in Net Assets – evidently holds only if all the changes recorded in asset and liability accounts (other than the owner's contributions or withdrawals) are also recorded in equity accounts that, in turn, are closed into the profit-and-loss account. In contemporary corporate financial accounting it is permissible to allow the counter-entries representing certain changes in asset values, depending upon the circumstances, to bypass the profit-and-loss account (for example, by recording these changes as debits or credits to one or other reserve account). This practice breaks the nexus between changes in net asset values and profits. It does, however, allow more 'realistic' values to be used in asset accounts where, otherwise, their use might produce 'distortions' in the profit figures that could mislead users such as investors and investment advisers. Both 'realistic' and 'distortions' are words that give rise to much debate in accounting circles. The double-entry recording system can accommodate the practice of bypassing the profit-and-loss account as comfortably as it can the alternative. The system itself imposes no discipline or constraint upon accountant or management – except the constraint that for each transaction or change recorded in the firm's accounting system, equal but offsetting debit and credit entries have to be made in accounts in the ledger.

The German economic historian, Werner Sombart, claimed that 'capitalism without double-entry bookkeeping is simply inconceivable', and that double-entry was one of the most significant inventions or creations of the human spirit. In similar vein, Oswald Spengler asserted that the creator of double-entry bookkeeping could take his place worthily beside his contemporaries Columbus and Copernicus. These scholars evidently attributed to the double-entry system a role that goes well beyond what one might think appropriate to ascribe to a system of organizing and arranging accounting data. In a nutshell, Sombart argued that, historically, the double-entry system opened up possibilities and provided stimuli that enabled capitalism to develop fully. It clarified the acquisitive ends of commerce and provided the rational basis on which this acquisition could be carried on. It provided the basis for the continued rational pursuit of profits, and virtually compelled its users to pursue the acquisition of wealth. It also enabled the firm or enterprise to be separated from its owners, thus facilitating the development of corporate enterprises.

These views are in their details either untenable or grossly exaggerated. To note only a few points: the profits of an enterprise and its capital employed can be calculated without double-entry bookkeeping; joint-stock companies, such as the Dutch East India Company, have existed and flourished without double-entry bookkeeping; 16th- and 17th-century merchants, like the Fugger, who did not use the system do not seem to have been any less acquisitive, rational and successful than those who did use the system; and the adoption of the double-entry system could not have changed, or even have reinforced, the temperament, commercial acumen, motivation or goals of those who adopted it for organizing their accounting records.

To reject grandiose claims made for double-entry bookkeeping is not to deny the more workaday usefulness of the system. A method or system for recording and classifying accounting data that has been used increasingly over a period of six centuries must indeed have substantial practical merit. Double entry is a useful and versatile method for organizing accounting data, its value increasing with the volume and complexity of the data to be organized. In turn, the efficient organization of data helps management at various levels in many ways, more notably in large organizations. But its contribution to efficiency does not proceed along the lines emphasized by Sombart.

BASIL S. YAMEY

See also **accounting and economics; assets and liabilities; Sombart, Werner.**

Bibliography

Yamey, B.S. 1964. Accounting and the rise of capitalism. *Journal of Accounting Research* 2, 117–36 (for a discussion of Sombart's views on double-entry bookkeeping and capitalism).

Douglas, Paul Howard (1892–1976)

Born in 1892 in Salem, Massachusetts, Paul Douglas attended Bowdoin College in Maine (BA, 1913) and Columbia University (Ph.D., 1921). After holding a number of teaching posts between 1916 and 1920, he joined the faculty of the University of Chicago where he remained (apart from service in the Second World War) until 1948, when he became a United States Senator from Illinois. After his retirement from the Senate in 1966, he taught at the New School for Social Research for two years (1967–9).

Paul Douglas first became well known for his massive theoretical and factual studies (for example, 1930) of all the available information on wages in the United States from 1890. This work required laborious following up of old, obscure records, and repairing gaps in the available

knowledge, such as domestic service wages. Douglas also collected information on prices so as to make an estimate of the movement of real wages.

In Britain there was almost complete cessation of the growth of real wages between 1896 and 1914. Understandably, it was a period of growing social tension. Sir Henry Phelps Brown called it the 'climacteric'. We still do not really understand its cause; there was some sociological evidence about the deterioration of the quality of businessmen. D.H. Robertson found at least a partial explanation in economic causes, namely, that, of the two leading British export industries, cotton was produced under constant returns and coal under diminishing returns.

This problem remains of primary interest to economic historians, and naturally they enquire whether there is any evidence of a similar 'climacteric' in other countries. In Germany there was a slowing down of the rate of rise in real wages, but not very marked. Douglas's American data likewise do not show such a 'climacteric'. Recent research, however, has thrown some doubt not on Douglas's wage data, but on his price data; and perhaps there was some slowing down of the rate of growth of real wages.

Douglas became famous to the whole economic world through the 'Cobb–Douglas function' (for example, 1934). Working in conjunction with Charles W. Cobb, a mathematician from Amherst College, and using Massachusetts State annual factory returns, Douglas in 1928 established the following relation: Let product be P, labour input L, capital input C, and k a constant. Then $P = kL^aC^b$. (The same formula, with land in place of capital, had already been used by Wicksell – for example, 1900 – but he gave it neither theoretical nor empirical development.)

We may, if we wish, constrain a and b to add up to 1; but we get much the same results unconstrained. If a and b add up to more than 1 this is an indication of economies of scale (increasing returns) – a uniform increase in the quantities of inputs giving a more than proportionate increase in product.

Annual data, which many economists have been using, give results mainly dependent on fluctuations in the short-period business cycle – which is not what we want at all. It is only when we have data for such a long period as to make it possible to average out the business cycle that we can draw conclusions about productivity. This has been done by Solow in the United States, Aukrust in Norway, and Niitamo in Finland. In each case it was found, in the long run, that the product was rising much more rapidly than expected from inputs and their exponents. This difference is generally held to be due to technical advance, though some look for economies of scale. Some difficult but promising work by Denison further analyses the labour input by numerous categories, male and female, adult and juvenile, and various levels of education. These methods reduce the unknown factor – but it does not disappear.

Differentiating the Cobb–Douglas formula to obtain marginal productivities, then aggregate earnings of the factors should be proportional to a:b – assuming that each factor is remunerated according to its marginal productivity. When he first made this calculation (so he told me), Douglas fully expected the aggregate income of labour to be below that indicated by its marginal productivity. He was surprised, however, to find that it was almost exactly what was to be expected – about 75 per cent of the product.

The Cobb–Douglas formula has had abundant application in agricultural economics, especially for cross-section studies, where each farm may be considered an independent piece of evidence. Land is introduced as a factor, and also data for other inputs – fertilizers, insecticides, and so on – even (in one study in Sweden) the age of the farmer – a negative factor.

Douglas was very much a political economist. Organized labour in the United States did not attempt to form a political party of its own as in Britain, but instead played the two existing parties off against each other in demanding concessions. But in the 1920s this was not fully agreed. The other element in the population with a grievance against the current state of affairs was the farmers, and an attempt was made to form a Farmer–Labour political party. Douglas took an active part in these negotiations, and was national treasurer of the organization. But with the Roosevelt reforms of the 1930s the prospects of a Farmer–Labour party died away.

Chicago had acquired a worldwide reputation for corruption and crime; and the ruling Democratic Party considered that its 'image' would be improved by an upright professor of economics on the city council. Douglas assured me that some improvement had taken place, though less than was hoped for. Later, the despotic Mayor Daley achieved a real reduction in crime. But once I asked Douglas whether, if I wished to set up a milk distribution business in Chicago, he could guarantee my safety. He replied that, 'regrettably', he could not.

Douglas was a Quaker, and in the First World War applied for exemption from military service on religious grounds. But in the Second World War he felt very differently. In spite of his age, he obtained a commission in the marines through President Roosevelt's personal intervention, and took part in the bloody landing on Iwojima. He sustained an injury to his hand which was with him for the rest of his life.

From city councillor he advanced to become Senator for Illinois. On the very day that he arrived in Washington he found a vanload of furniture which had been offered to him as a gift. He sent it back. This episode prompted him to write a little book, *Ethics in Government* (1952). He saw no harm in the small presents customarily exchanged among businessmen and politicians – calendars, cigars, and so on – but instructed his staff to return any present valued at over four dollars.

COLIN G. CLARK

See also **Cobb–Douglas functions.**

Selected works

1928. (With C.W. Cobb.) A theory of production. *American Economic Review*, Supplement, 18, 139–65.

1930. *Real Wages in the United States, 1890–1926.* Boston and New York: Houghton Mifflin Company.

1934. *The Theory of Wages.* New York: Macmillan.

1936. *Social Security in the United States: An Analysis and Appraisal of the Federal Social Security Act.* New York, London: Whittlesey House, McGraw-Hill.

1947. (With E.H. Schoenberg.) Studies in the supply curve of labour; the relation in 1929 between average earnings in American cities and the proportions seeking employment. *Journal of Political Economy* 45, February, 45–79.

1939a. The effect of wage increases upon employment. *American Economic Review* Supplement 29, 138–57.

1939b. (With H.G. Lewis.) Some problems in the measurement of income elasticities. *Econometrica* 7, 208–20.

1939c. (With M. Bronfenbrenner.) Cross-section studies in the Cobb–Douglas function. *Journal of Political Economy* 47, 761–85.

1948. Are there laws of production? *American Economic Review* 38, 1–41.

1952. *Ethics in Government.* The Godkin Lectures at Harvard University, 1951. Cambridge, MA: Harvard University Press.

1972. *In the Fullness of Time: The Memoirs of Paul H. Douglas.* New York: Harcourt Brace Jovanovich.

1976. The Cobb–Douglas production function once again: its history, its testing and some new empirical values. *Journal of Political Economy* 84, 903–15.

Bibliography

Wicksell, K. 1900. Marginal productivity as the basis of distribution in economics. *Ekonomisk Tidskrift.* English trans. in *K. Wicksell: Selected Papers in Economic Theory*, ed. E. Lindahl. London: Allen & Unwin, 1958.

dual economies

Dual economies have asymmetric sectors, the interaction between which influences the path of development. W. Arthur Lewis introduced this idea in his paper, 'Economic Development with Unlimited Supplies of Labour' (Lewis, 1954), which earned him the Nobel Prize for Economics in 1979. That paper contains two theoretical models, both designed to explain the intrinsic problems of underdevelopment. When the prize was awarded, Ronald Findlay wrote that 'a large part of ... development economics ... can be seen as an extended commentary on the meaning and ramifications [of this article]' (Findlay, 1980, p. 64). Here we focus primarily on the first of Lewis's two models of dualism – that of a single underdeveloped economy. We describe that model, trace the evolution of the ideas which grew from it, and discuss the continuing importance of these ideas in the study of economic development.

Long before Lewis wrote his article, there had been much thinking about 'dual' economies, conceived of as economies with both an industrial sector and an agricultural sector. Adam Smith and David Ricardo both focused on the interaction between these sectors during the Industrial Revolution; for Ricardo the outlook for industrial growth was 'dismal' because of diminishing returns in agriculture (see Hicks, 1965; Pasinetti, 1974). In the early 20th century, there was an extended discussion in the Soviet Union of the 'scissors problem', concerning the determination of the terms of trade between these two sectors. Evgeny Preobrazhensky (1924) argued that a decrease in the relative price of agricultural goods could be used to stimulate industrial investment; others replied that sufficient agricultural goods would not be available at lower relative prices and that these goods would need to be seized by force, something which the collectivization of agriculture made possible (see Sah and Stiglitz, 1984). And during the Great Leap Forward in China in the 1950s, Chairman Mao attempted to confiscate an increasing quantity of primary goods from the Chinese countryside in order to facilitate the development of urban manufacturing. These policies led to famine and to the deaths of approximately 30 million people. Thus both theorists and policymakers have long recognized that, in an economy with two very different sectors, growth prospects hinge on how these sectors interact.

In his Nobel Prize autobiography, Lewis (1979) writes that his interest was in the 'fundamental forces determining the rate of economic growth'. But he was not satisfied with the neoclassical model of growth that was emerging at the time (Solow, 1956; Swan, 1956), out of the work of Roy F. Harrod (1939) and Evsey D. Domar (1945). That neoclassical framework aimed to provide a *general* theory of growth. But to Lewis it seemed inadequate because it did not deal with interactions between the industrial and the agricultural sectors: in Lewis's words, this model contained no discussion 'of what determines the relative price of steel and coffee [namely, of industrial goods and agricultural goods]. The approach through marginal utility made no sense to me. And the Heckscher–Ohlin framework could not be used, since that assumes that trading partners have the same production functions, whereas coffee cannot be grown in most of the steel-producing countries.' Furthermore, the neoclassical theory seemed inadequate to him for historical reasons: '[a]pparently, during the first fifty years of the industrial revolution, real wages in Britain remained more or less constant while profits and savings soared. This could [also] not be squared with the neoclassical framework, in which a rise in investment

should raise wages and depress the rate of return on capital' (Lewis, 1979).

Then, Lewis continues:

> One day in August, 1952, walking down the road in Bangkok, it came to me suddenly that both problems have the same solution. Throw away the neoclassical assumption that the quantity of labour is fixed. An 'unlimited supply of labour' will keep wages down, producing cheap coffee in the first case and high profits in the second case. The result is a dual (national or world) economy, where one part is a reservoir of cheap labour for the other. The unlimited supply of labour derives ultimately from population pressure, so it is a phase in the demographic cycle. (Lewis, 1979, p. 397)

This key insight launched Lewis on the journey that led to his famous article. Spelling out the implications of his insight led him to use the term 'dualism' to describe economies in which there are differences between industrial and agricultural sectors that cannot be adequately explained by differences in production technologies or in factor endowments, in the manner normally used by economists.

The Lewis model

Lewis identified three such differences between industry and agriculture, which we term 'asymmetries' in this article (following Kanbur and McIntosh, 1987).

First, there are technological differences between the sectors. Labour is used in each sector. In agriculture it is combined with land in production, whereas industrial goods are produced by combining labour with reproducible capital. Moreover, industrial goods can be consumed or invested, whereas agricultural goods can only be consumed.

Second, there are organizational differences between the sectors. The large, rural agricultural sector functions on traditional lines and is primarily based on subsistence; industrial production happens in a modern, market-oriented sector, located in towns and cities. There is 'an unlimited supply of labour, available at [a] subsistence wage' (Lewis 1954, p. 139) to both sectors. Lewis interprets the word 'subsistence' broadly. The level of the wage is determined in some way by conventions in the underdeveloped agricultural sector. Lewis is non-committal as to whether wages in this sector are set according to actual subsistence needs, or living standards, or workers' average product. The central idea is that workers are paid above their *marginal* product. Labour can be transferred from agricultural sector to the industrial sector by the migration of workers to towns and cities. The overall stock of labour in the economy is normally fixed in supply (though Lewis, like Ricardo, did sometimes allow for Malthusian features). Workers in the cities are paid not much more than the subsistence wage, although there may be a gap, as discussed below.

Third, and finally, there are differences in the behaviour of the actors in the two sectors. Capitalists in the industrial sector save all their profits, because they are ambitious. Workers save nothing, in either sector, because they are poor (Lewis describes them as not belonging to the 'the saving class' – 1954, p. 157). And landlords in agriculture are assumed to consume all their income, which comes to them to the extent that agricultural workers receive a wage below their average product.

The general story is this: the profits in the modern, capitalist, sector create a growing supply of savings. This finances the formation of an increasing stock of capital, which is used to employ more and more labour in the urban workforce.

We can explain the story in detail, using a simplified version of the model. To do this we make four sets of extreme assumptions. (a) There is 'pure' surplus labour, by which we mean that the marginal product of workers withdrawn from agriculture is *zero*. Wages initially consist only of agricultural goods, the level of wages per worker is exogenous, and workers are indifferent between working in industry and in agriculture at the same wage. (b) When one individual worker leaves agriculture and no longer needs to be rewarded there, then all the increase in the agricultural surplus (that is, all the increase in the total of food produced minus the total of wages paid to agricultural workers) accrues to landlords and is spent by them on consumption of industrial goods. (c) Industrial capitalists employ labour up to the point at which the marginal physical product of labour is equal to the cost of the wage, measured in industrial goods. (d) All industrial profits are saved and then invested in industrial production.

Given these assumptions, there are two steps to the argument. First, given assumptions (a), (c) and (d), the rate of growth depends negatively on the relative price of agricultural goods in terms of industrial goods. This is because an increase in food prices raises the cost of the wage per worker in terms of steel, causing less labour-intensive methods of production to be adopted, that is, causing production to become more capital intensive. As a result of this, any given amount of savings, and the accumulation of capital that it causes, will 'go less far' in employing labour in industry, and, as a result, industrial output will grow less rapidly. Second, assumptions (a) and (b) determine the relative price of agricultural goods, in the following way: the accumulation of capital in industry increases the demand for industrial workers, who must be transferred from agriculture. The relative price of agricultural goods will need to be high enough to induce the workers' landlords to offer up those agricultural goods that they would have paid to the transferred workers but now receive as surplus, so as to receive industrial goods for consumption in exchange. Such trade enables workers to be paid in industry, where they now work. As Lewis (1954, p. 188) says, 'the capitalists

need the peasants' food, and … the demand for food is inelastic'.

Clearly, the relative prices of industrial and agricultural goods, and the growth rate of the economy, are *jointly* determined in this process – as Lewis's intuition had suggested to him. And it will clearly be true that the relative price of agricultural goods will need to be less high – and so the rate of growth will be higher – the lower is the price of agricultural goods required for landlords to release their surplus in exchange for consumable industrial goods.

Note that the share of income that accrues to industrial capitalists will increase during the growth process, as the capitalist sector grows in size. This suggested to Lewis (1954, p. 155) that a growth process of this kind might help to solve what he called the 'central problem' of development: the need to raise the savings rate enough to enable rapid growth to take place. In this model it is necessary to transfer labour into industry, in order to increase the overall savings rate of the economy. This is due to the behavioural assumption that agricultural income is not saved; we revisit this assumption below. Interestingly – from today's point of view – Lewis thought that a savings rate of ten to twelve per cent might be sufficient to achieve the 'rapid capital accumulation' that he believed integral to the process of development (Lewis, 1954, p. 155). Note also that increasing inequality is a frequent, if not necessary, correlate to this rising savings share, at least in the early stages of development (see, for example, Fei et al., 1979). This story thus also provides an explanation of the 'Kuznets curve'.

Generalizations

Lewis does sometimes enlist the extreme simplifications made above. They correspond most closely to those made by Gustav Ranis and John C. H. Fei (1961), who used them to explain, more formally than Lewis did, what they call the 'first phase' of economic development – a phase in which there is 'pure' surplus labour. But Lewis also hints at many ways in which these assumptions could be relaxed. Ranis and Fei, along with Dale W. Jorgenson (and many others), went on to consider the implications of dualism when there are sectoral asymmetries different from those outlined above. In what follows, we consider a number of these extensions.

The first, and most fundamental, generalization of Lewis's model was made by Ranis and Fei (1961), who demonstrated that the dualistic framework continued to give insight into the process of economic growth even when the condition of pure surplus labour does not hold. They initiated a large body work on this question by examining the microeconomic foundations of surplus labour and exploring what occurs when these conditions come to an end. This occurs when a sufficient number of workers have been removed from agriculture for the marginal productivity of the remaining agricultural workers to become positive. As a result, agricultural output declines as further workers leave. (This may happen even if there is technological progress in agriculture, providing that this progress is not sufficient to fully compensate for lost labour.) Consequently, the marginal agricultural surplus per worker, which accrues to landlords as each worker leaves – and which is traded by landlords for industrial goods – begins to decline, *even if* the wage per worker (measured in terms of agricultural goods) is exogenous. This means that the cost of labour to industry, measured in terms of industrial goods, will begin to rise above the level described in the sketch above – thereby constraining the rate of growth. This is the 'first turning point' identified by Ranis and Fei. It corresponds to the onset of Ricardo's 'dismal' diminishing returns. Ranis and Fei label what happens beyond this point as the 'second phase' of economic development. In that phase the economy is characterized by 'disguised unemployment', since labour in agriculture is still paid more than its marginal product.

Lewis himself was accused of not allowing for this possibility, even though he had written that the existence of zero marginal product is 'not … of fundamental importance to our analysis' (Lewis, 1954, p. 142). This accusation led to what Lewis later called an 'irrelevant and intemperate controversy' about the existence, or not, of 'pure' surplus labour (Lewis, 1972, p. 77). Ranis (2003, p. 8) agrees with Lewis's self-defence: in a retrospective assessment, he describes the postulation of a 'pure' labour surplus as a red herring. Amartya Sen (1966) helpfully clarifies the debate about this issue.

Growth becomes more difficult in this second stage of development. Recall that Lewis argues that the real wages per worker, and the level of welfare per worker, do not fall as growth proceeds. But growth is driven by the transfer of labour from agriculture to industry, which, in this second phase, causes agricultural output to fall. As a consequence of this the relative price of agricultural goods rises, and real wages can remain constant only if workers are able to substitute towards industrial goods in such a way as to avoid any damage to their welfare.

The agricultural sector as a constraint on growth
To highlight the essential role of such substitution, Mukesh Eswaran and Ashok Kotwal (1993) assume an extreme version of Engel's law. Consumers are assumed to spend *all* their income on food until they reach a particular threshold level of consumption, when they become sated with food. Beyond this point all further increases in consumption are devoted to industrial goods. At the same time they assume that labour always has a positive marginal product in agriculture. Under these assumptions, if workers remain so poor that they are not sated with food, then the transfer of labour across sectors – and therefore accumulation of industrial capital – becomes impossible. The inability of the poor to 'eat

shirts' – an extreme version of what Ranis (2003) describes as the 'product' dimension of dualism – becomes a constraint on whether savings can lead to development. (And this constraint will bind quite independently of how high the marginal *physical* product of labour is in industry.) Any attempt to increase savings rates, in the manner desired by Lewis, so as to draw labour out of agriculture, would fail in these circumstances. The withdrawal of labour would lead to a reduction in the supply of food per worker – the only thing that matters for workers' real wages – and so to a shortage of food. That shortage would turn the terms of trade against industry, depressing industrial profits and savings until the downward pressure on the supply of food had been removed, or until growth has ceased. As a result, all the gains from any increase in industrial production would accrue, in the form of lower prices, to those who consume industrial goods, rather than enabling growth, as in the Lewis model. It is thus clear that an important influence on whether development can proceed under dualism is the ability to shift workers' demands away from agricultural goods.

Of course, in a small, open, economy, the relative prices of tradables will be tied down, and the economy can respond to any developing shortage of food simply by exporting manufactures and importing food. That was Ricardo's insight, over 100 years earlier, about the gains to Britain from the abolition of the Corn Laws; Lewis's model of dualism in the world economy also incorporates such trade. But Lewis (1972, p. 94) cautions that there may be limits to this if export prices are not really exogenous, and if, instead, the county needs to cheapen its exports to pay for the imports of food – and other goods – that it will need as it grows. Perhaps partly because of this, Lewis (1954, p. 176) argues that a country which exhausts its surplus labour supply might instead export its savings, investing in industrial development in countries where the surplus labour condition continues to hold, and so enabling the output of manufactured goods to grow without driving down the rate of profit. In addition, the country might import labour from these countries. In this way Lewis's early contributions anticipated, and fed into, debates about the roles of outsourcing and immigration in contemporary globalization.

Jorgenson (1961) further develops the study of the dynamics of a dualistic economy in this second phase of development – when there is a positive marginal product of labour in agriculture and disguised unemployment. He incorporates a Malthusian perspective, by supposing that population growth is increasing in the amount of food consumed per capita, up to a biological ceiling that corresponds to the food-consumption threshold of Eswaran and Kotwal. This has the consequence that too rapid a rate of growth of population can cause a Malthusian trap by preventing the emergence of *any* significant agricultural surplus. Growth of manufacturing activity, such as that analysed by Lewis, can then be sustained only if technological progress in agriculture enables food production to outstrip population growth. (Capital accumulation in agriculture could have a similar effect in a model more general than that used by Jorgenson.) Only then can an agricultural surplus emerge, and grow, and so only then can labour progressively move away from agriculture. If this does not happen, then any increases in profits, savings and capital accumulation in industry become self-defeating, since they turn the terms of trade against industry and so bring down profits and savings, and bring growth to an end, in the way described two paragraphs above.

As stressed by Avinash Dixit (1973, p. 346), such a model focuses on 'the constraint on growth imposed by the rate of release of labour from agriculture', whereas in Lewis's model the focus had been on the ability of capital accumulation in industry to soak up the surplus labour force in agriculture. Nevertheless, as Dixit notes, growth paths in the two models will produce similar outcomes. In particular, in both models one would observe an endogenous rise in the savings rate as development proceeds. And in both models, it *may* be the case that any attempt to foster growth in industry, by a 'big push' to save more, is self-defeating. (This can be true in Jorgenson's model, and as we saw above, it can also be true beyond the 'first stage' of growth in the Lewis model, if it is not possible to induce workers to substitute away from agricultural goods.) This is why Jorgenson thought of increases in savings rates as an *outcome* of development, not as a policy tool which can be used to *promote* development (Jorgenson, 1961, p. 328).

It is worth contrasting this view of potential 'development traps' with that which had been put forward in the 1940s by Paul Rosenstein-Rodan (1943), who built on his experience of eastern Europe. Rosenstein-Rodan's viewpoint also came from thinking about the interaction between agriculture and industry; like Lewis, he argued that development could only come to an agricultural economy through a process of industrialization. This, he argued, is because only industrial capitalists could afford to pay for the large fixed costs that are necessary to enable them to produce goods in a modern way, with low marginal costs. But if most people live in an impoverished agricultural sector then this would constrain their incomes, and so would limit their demand for modern industrial goods. That might make it unprofitable to make the required investment, and so might thwart the process of development. Here, just as for Lewis, a shortage of savings can be *the* problem of development. But by contrast with Lewis, a big push might fix it, since, roughly speaking, if all capitalists invested at once and paid their workers higher wages, then the demand for industrial goods would grow, making the investment worthwhile. This insight gave birth to the other great analytical engine of development economics, subsequently formalized by Kevin M. Murphy, Andrei Shleifer and Robert W. Vishny (1989) and Kiminori Matsuyama

(1991), and well explained by Paul Krugman (1993). Since the pecuniary externalities that allow an economy to escape from a development trap are accessible only in the 'modern' sector, asymmetries between sectors are also central to this view.

Further aspects of labour transfer

The Lewis model was also generalized to explain the gap between the wage paid in the rural sector and that paid in the urban sector and to explore the consequences of such a gap. Lewis himself (1954, p. 150) acknowledged the existence of a wage gap, and suggested that it may result from the psychological costs of lifestyle changes, from the need to reward skills accumulated in the urban sector, or from the ability of workers in cities to bargain for higher wages. (This is particularly relevant when we recognize that the urban sector includes government employment and some services.) Subsequent authors took up this question, arguing, for example, that wage premia may arise because they lead to greater productivity through effects on health or employee motivation (for example, Dasgupta and Ray, 1986; Shapiro and Stiglitz, 1984).

The consequences of such a gap, for the process of labour transfer from agriculture to industry, were set out in the celebrated work of John Harris and Michael P. Todaro (1970). It may be that a wage floor in the urban formal sector prevents the market from clearing there. If a wage floor operates, then workers who choose to leave the rural sector face the prospect of receiving an urban wage which is above that of the rural sector, if they get employed, but also face some probability of becoming unemployed. In the simplest version of this model, equilibrium occurs when labour migration equalizes expected income across sectors – an outcome in which the rural wage equals a weighted average of the incomes received by employed and unemployed urban workers, weighted according to the probability of unemployment in the urban sector. Even without this extreme outcome, there are important policy implications in such a model. The more elastic is labour supply to the urban sector with respect to expected income there, the greater the amount of urban unemployment that will be induced by any policies that increase urban wages. This incorporation of urban unemployment into the model also enables one to begin to discuss the growth of a third sector: the production of services in cities (see Fields, 1975). Roughly speaking, we can say that services get produced by (some of) those who migrate to cites, but do not get a job in manufacturing.

It is clear that the expansion of the industrial sector will ultimately take the economy beyond the second phase of economic development, in which there is disguised unemployment. This is because withdrawal of labour from agriculture will eventually reach the point at which the marginal product of the remaining labour rises to equality with the subsistence wage. Ranis and Fei call this a 'second turning point'. At this point the marginal worker, offered a subsistence wage, can now instead offer his or her labour to a higher bidder. From then on the wage (measured in agricultural goods) will begin to rise in both sectors as growth continues. We can say that the 'dualistic' structure of the economy then comes to an end, in that the rural economy becomes 'commercialized'. (Something similar, too, will happen in any services sector.) That leads one back to a labour-scarce economy, the analysis of which is better suited to neoclassical theory. A two-sector neoclassical growth model – something like the model of Hirofumi Uzawa (1961; 1963) – may be a better way to think about growth in these circumstances.

One key strand of the story of dualism that we have been telling is the assumption that capitalists save, but workers (and landlords) do not. Lewis's explanation of this asymmetry is largely behavioural. But such differences in savings rates between the traditional and the modern sectors might also be explained *institutionally*, by means of credit-market imperfections. If a technological asymmetry precludes investment in rural areas, and if limited financial development means that rural residents lack access to investment opportunities in manufacturing, then the agricultural surplus will not be used directly to finance investment. Moreover, typical characterizations of credit-market imperfections highlight the moral hazard problems that persist in rural areas because the poor there are unable to provide the kind of collateral required for formal-sector loans. (Small rural landholdings are of limited use as collateral.) Such lack of collateral stands as a barrier to borrowing, even though loans might be used to facilitate growth by promoting education, or capital accumulation, or technical progress in agriculture. (See Ray, 1998, for a summary of these arguments.) By contrast, Abhijit Banerjee and Andrew Newman (1998) provide an alternative perspective, emphasizing a sectoral asymmetry in the *informational* dimensions of credit-market imperfections, and showing how this can affect the willingness of individual workers to migrate in a dualistic economy. They present a model in which there is access to credit for consumption in rural areas. Given that workers have limited collateral wherever they live, a crucial determinant of their access to credit is the amount of information that lenders have about prospective borrowers. In contrast with the relative anonymity of urban life, small communities of the rural sector may provide superior information about borrowers, and thus foster lending. Banerjee and Newman show that dualism, characterized in terms of this differential severity of information asymmetries, might lead to a suboptimal allocation of labour across sectors. By financing consumption in the rural sector, rural credit might actually provide an incentive for labour to remain there; this incentive could offset the relatively high wages of the modern sector and could thereby impede the development process. Their paper suggests – at the least – that the lens of asymmetric information can shed useful light on the development of such economies.

Defining characteristics of economic dualism

We conclude by noting that we have described a number of reasons for differences between the industrial and agricultural sectors of a developing economy. Just as in Lewis's original article, all these differences go beyond mere asymmetries in production technologies or factor endowments between the sectors. This is why, following Ravi Kanbur and James McIntosh (1987), we would not normally describe the two-sector growth models of Uzawa (1961; 1963) as models of dualism, even though in those models the two sectors have different factor intensities. Nor would we say that that the two-sector Heckscher–Ohlin model of international trade is a model of a dualistic economy – even when its two sectors have different factor intensities, and even when the two sectors are labelled 'agriculture' and 'industry'. Furthermore, although the specificity of factors to sectors appears central to Lewis's set-up (with land specific to agriculture and capital specific to industry), this feature does not seem to be sufficient to merit the label of 'dualism'. Thus, for example, we would not regard the short-run version of the Heckscher–Ohlin trade model presented by J. Peter Neary (1978), with factors specific in each of the two sectors, as portraying a dualistic economy.

Instead, we would argue that the defining characteristic of modern theories of economic dualism lies – just as it did in Lewis's article – in a focus on sectoral asymmetries that are not simply technological. For Lewis, and for Ranis and Fei, there were *organizational* differences between sectors – in that wages were assumed to be determined by institutional factors in the agricultural sector – and *behavioural* differences between sectors – in that those in the rural sector were assumed to be unwilling to save, while capitalists were assumed to save everything. A focus on these features might imply that 'pull' factors drive labour transfer, and hence economic growth, in a dualistic economy. But since Lewis, economists studying economic development have explored alternative asymmetries between sectors and have reached different conclusions. The model of Eswaran and Kotwal, in which the defining asymmetries are *product* asymmetries – an assumption that all income is spent on agricultural goods until some threshold – highlights the need for labour productivity increases in agriculture to avoid stagnation of real wages. This is a need that persists even in the presence of rising productivity in industry. Jorgenson, who coupled such a view with a demonstration that Malthusian pressures can prevent income from ever rising above this threshold, showed clearly that growth can be constrained unless the 'push' factor of growth in agricultural technology is strong enough. Banerjee and Newman, by contrast, have emphasized that *informational* asymmetries between traditional and modern sectors can constrain the growth process.

We thus believe that, in the study of any particular economy, it is important to understand which asymmetries impose binding constraints on growth. Different constraints imply the need for different policies. But identifying the relevant asymmetries is even more important if we wish to remove these underlying constraints themselves. Joseph Stiglitz has proposed that we do just this, advocating what he calls 'growth strategies based on duality's elimination' (Stiglitz, 1999, p. 56). Much empirical work is necessary if we are to understand what such strategies might require.

DAVID VINES AND ANDREW ZEITLIN

See also **labour surplus economies; Lewis, W. Arthur.**

Bibliography

Banerjee, A.V. and Newman, A.F. 1998. Information, the dual economy, and development. *Review of Economic Studies* 65, 631–5.

Dasgupta, P. and Ray, D. 1986. Inequality as a determinant of malnutrition and unemployment: theory. *Economic Journal* 96, 1011–34.

Dixit, A. 1973. Models of dual economies. In *Models of Economic Growth: Proceedings of a Conference held by the International Economic Association at Jerusalem*, ed. J.A. Mirrlees and N.H. Stern. London: Macmillan.

Domar, E.D. 1946. Capital expansion, rate of growth, and employment. *Econometrica* 14, 137–47.

Eckaus, R.S. 1955. The factor proportions problem in underdeveloped areas. *American Economic Review* 45, 539–65.

Eswaran, M. and Kotwal, A. 1993. A theory of real wage growth in LDCs. *Journal of Development Economics* 42, 243–69.

Fei, J.C.H., Ranis, G. and Kuo, S.W.Y. 1979. *Growth with Equity: The Taiwan Case*. Oxford: Oxford University Press.

Fields, G.S. 1975. Rural-urban migration, urban unemployment and underemployment, and job-search activity in LDCs. *Journal of Development Economics* 2, 165–87.

Findlay, R. 1980. On W. Arthur Lewis' contributions to economics. *Scandinavian Journal of Economics* 82, 62–79.

Harris, J.R. and Todaro, M.P. 1970. Migration, unemployment and development: a two-sector analysis. *American Economic Review* 60, 126–42.

Harrod, R.F. 1939. An essay in dynamic theory. *Economic Journal* 49, 13–33.

Hicks, J. 1965. *Capital and Growth*. Oxford: Oxford University Press.

Higgins, B. 1956. The 'dualistic theory' of underdeveloped areas. *Economic Development and Cultural Change* 4, 99–115.

Jorgenson, D.W. 1961. The development of a dual economy. *Economic Journal* 71, 309–34.

Kanbur, R. and McIntosh, J. 1987. Dual economies. In *The New Palgrave: A Dictionary of Economics*, vol. 1, ed. J. Eatwell, M. Milgate and P.K. Newman. London: Macmillan.

Krugman, P. 1993. Toward a counter-counterrevolution in development theory. In *Proceedings of the World Bank Annual Conference on Development Economics, 1992: Supplement to The World Bank Economic Review and The World Bank Research Observer*, ed. L.H. Summers and S. Shah. Washington, DC: World Bank.

Lewis, W.A. 1953. *Report on Industrialisation and the Gold Coast*. Accra, Gold Coast: Government Printing Department.

Lewis, W.A. 1954. Economic development with unlimited supplies of labour. *Manchester School* 28, 139–91.

Lewis, W.A. 1955. *The Theory of Economic Growth*. London: Allen & Unwin.

Lewis, W.A. 1972. Reflections on unlimited labour. In *International Economics and Development: Essays in Honour of Raul Prebisch*, ed. L.E. di Marco. London: Academic Press.

Lewis, W.A. 1978. *Growth and Fluctuations, 1870–1913*. London: Allen & Unwin.

Lewis, W.A. 1979. Autobiography. In Lindbeck (1992). Also online. Available at http://nobelprize.org/nobel_prizes/economics/laureates/1979/lewis-autobio.html, accessed 9 January 2007.

Lindbeck, A., ed. 1992. *Nobel Lectures: Economic Sciences 1969–1980*. Singapore: World Scientific Publishing.

Matsuyama, K. 1991. Increasing returns, industrialization, and the indeterminacy of equilibrium. *Quarterly Journal of Economics* 106, 617–50.

Murphy, K.M., Shleifer, A. and Vishny, R.W. 1989. Industrialization and the big push. *Journal of Political Economy* 97, 1003–26.

Neary, J.P. 1978. Short-run capacity specificity and the pure theory of international trade. *Economic Journal* 88, 488–510.

Pasinetti, L. 1974. *Growth and Income Distribution—Essays in Economic Theory*. Cambridge: Cambridge University Press.

Preobrazhensky, E. 1924. *The New Economics*. Trans. B. Pearce, Oxford: Clarendon Press, 1965.

Ranis, G. 2003. Is dualism worth revisiting? Discussion Paper No. 870. Economic Growth Center, Yale University.

Ranis, G. and Fei, J.C.H. 1961. A theory of economic development. *American Economic Review* 51, 533–65.

Ray, D. 1998. *Development Economics*. Princeton: Princeton University Press.

Rosenstein-Rodan, P. 1943. Problems of industrialisation of eastern and south-eastern Europe. *Economic Journal* 53, 202–11.

Sah, R.K. and Stiglitz, J.E. 1984. The economics of price scissors. *American Economic Review* 74, 125–38.

Sen, A. 1966. Peasants and dualism with or without surplus labour. *Journal of Political Economy* 74, 425–50.

Shapiro, C. and Stiglitz, J.E. 1984. Equilibrium unemployment as a worker discipline device. *American Economic Review* 74, 433–44.

Solow, R.M. 1956. A contribution to the theory of economic growth. *Quarterly Journal of Economics* 70, 65–94.

Stiglitz, J.E. 1999. Duality and development: some reflections on economic policy. In *Development, Duality, and the International Economic Regime: Essays in Honor of Gustav Ranis*, ed. G. Saxonhouse and T.N. Srinivasan. Ann Arbor: University of Michigan Press.

Swan, T.W. 1956. Economic growth and capital accumulation. *Economic Record* 32, 334–61.

Uzawa, H. 1961. On a two-sector model of economic growth. *Review of Economic Studies* 29, 40–47.

Uzawa, H. 1963. On a two-sector model of economic growth II. *Review of Economic Studies* 30, 105–118.

dual track liberalization

Dual track liberalization is a reform strategy of market liberalization in which a market track is introduced while the plan track is maintained at the same time. Under the plan track, economic agents are assigned rights to and obligations for a fixed quantity of goods and services at fixed planned prices as specified in the pre-existing plan. Under the market track, economic agents can participate in the market at free market prices, provided that they fulfil their obligations under the pre-existing plan. The essential feature of the dual track strategy to market liberalization is that prices are liberalized at the margin while inframarginal plan prices and quotas are maintained for some time before being phased out. Although the dual track reform strategy is widely adopted in China during its transition from plan to market, it is also used in other countries. For example, when introducing new legislation, a 'grandfathering' clause is often adopted to protect existing interests, which is a form of the dual track approach to reform.

Analysis of dual track liberalization follows two lines of approach. The first focuses on its Pareto-improvement property, that is, dual track liberation makes nobody worse off while it makes somebody better off – and therefore it has a political advantage in implementing reforms. Most efficiency-improving market liberalization reforms potentially create winners and losers, despite the fact that, in theory, efficiency gains should be large enough to allow the potential losers to be compensated. For example, the single track approach to liberalization (that is, where all the prices are freed at once) in general cannot guarantee an outcome without losers. Dual track liberalization means that planned quantity continues to be delivered at plan price but any additional quantity can be sold freely in the market. With the dual track, the surpluses of the rationed users and the planned suppliers

remain exactly the same. The purpose of maintaining the plan track is to provide implicit transfers to compensate potential losers from market liberalization by protecting status quo rents under the pre-existing plan. On the one hand, the introduction of the market track provides the opportunity for economic agents who participate in it to be better off. At the same time, the new users and suppliers outside the plan are also better off. Therefore, the intuitive appeal of dual track liberalization for reformers lies precisely in the fact that it represents a mechanism of the implementation of a reform without creating losers (Lau, Qian and Roland, 2000).

The second approach focuses on the efficiency property of dual track liberalization. Pareto-improvement property implies that it always improves efficiency. This is independent of other assumptions, for example, as to whether the market is competitive or not. In contrast, the single track approach to liberalization may improve efficiency under perfect competition, but may not improve efficiency if the market is monopolistic (Li, 1999). The more subtle and deeper point is that the dual track approach to liberalization may achieve allocative efficiency, despite the fact that it appears inefficient, by maintaining the inefficient planned track. The fundamental reason is that the compensatory transfers, which are implicitly embodied in the planned track, are inframarginal, and thus the distortion can be avoided.

To see this we look at the special case where the pre-reform status quo features efficient rationing and efficient planned supply in the sense that the planned output is allocated to users with the highest willingness to pay and the planned supply is delivered by suppliers with the lowest marginal costs. Nevertheless, the price of the good is fixed at an artificially low level and the production quota is fixed below market equilibrium (Figure 1). When the market track is introduced into this setting, it is clear that the market equilibrium quantity and price would be identical to the case without the planned price and quota to start with. Therefore, dual track liberalization achieves efficiency. Notice that efficiency is achieved without making anyone worse off. Indeed, the rents enjoyed by the buyers under rationing (area A in Figure 1) are preserved under dual track liberalization, but would be lost under single track liberalization.

In a more general case of inefficient rationing and/or inefficient planned supply, efficiency can still be achieved provided market liberalization is full, in the sense that market resales of plan-allocated goods and market purchases by planned suppliers for fulfilling planned delivery quotas are permitted after the fulfilment of the obligations of planned suppliers and rationed users under the plan. This removes any inefficiency associated with the original planned prices and quotas and makes imputed rents under planning inframarginal. This type of transaction takes many forms in practice, for example, subcontracting by inefficient planned suppliers to more efficient non-planned suppliers, and labour reallocation

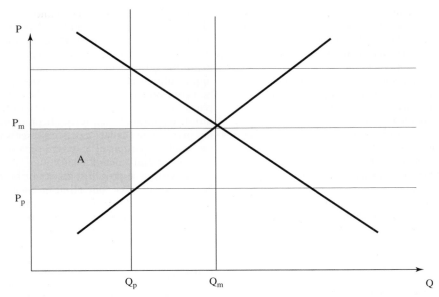

Figure 1   The case of efficient supply and efficient rationing

when workers in inefficient enterprises keep the housing while taking a new job in more efficient firms. In both examples, after fulfilling the obligations under the plan (planned delivery of supply and welfare support through housing subsidies), the market track functions to undo the inefficiency of the plan track.

This above partial equilibrium analysis can be generalized to a general equilibrium mode (Lau, Qian, and Roland, 1997). Efficiency requires full market liberalization under which market resales, subcontracting, and market purchases for redelivery are all allowed. Indeed, the distinction between limited and full market liberalization is a major difference between Lau, Qian, and Roland (1997) and Byrd (1991), and others who have studied the dual track approach.

If such resales and purchases are not allowed or cannot be achieved, then dual track liberalization is limited and efficiency in general cannot be achieved, although it can be improved. Of course, in the special case discussed above with efficient supply and efficient rationing, dual track with limited market liberalization is the same as dual track with full market liberalization. In general, dual track limited market liberalization need not be the same as dual track full market liberalization.

Sometimes dual track liberalization of the market takes the following sequential form: in a first stage, limited market liberalization is implemented, and then in a second stage full market liberalization is implemented. In the first stage, going from a centrally planned economy to limited market liberalization, Pareto improvement is clearly attained, but efficiency cannot be guaranteed. Specifically, limited market liberalization generally leads to inefficient overproduction due to market entry. In the

second stage, when full liberalization is introduced, efficiency is attained but Pareto improvement may not be. This is because the second-stage full market liberalization implies efficiency, and thus there must be a production contraction and some people have to reduce production and are made worse off. Therefore, the sequential dual track liberalization may result in some opposition to further reforms after the first and before the second stage, while the dual track full market liberalization that is implemented in one stroke will not. Nevertheless, it is also clear that, even under the sequential dual track liberalization, there are no losers at the end of the second stage compared with the status quo before the reform.

The dual track approach to market liberalization is an example of reform making the best use of existing information and institutions. First, it utilizes efficiently the existing information embedded in the original plan (that is, existing rents distribution) and its implementation does not require additional information. Second, it also enforces the plan through the existing plan institutions and does not need additional institutions. Enforcement of the plan track is crucial for preserving pre-existing rents. However, contrary to common understanding of the relationship between state power and reform, state enforcement power is needed here not to implement an unpopular reform, but to carry out one that creates only winners, without losers.

Economists sometimes find dual track liberalization puzzling and counter-intuitive, for several reasons. First, economists are used to the law of one price: in a competitive setting, multiple prices entail inefficiency. However, in dual price liberalization, the planned price comes together with planned quantity, when they are fixed, they do not entail inefficiency, at least not additional inefficiency. Second, dual track resembles price control, which is associated with inefficiency and rent seeking. But dual track is not price control; on the contrary, it is a move towards price liberalization. An important difference between the plan track under dual track and price control is that the plan track embodies both fixed prices and fixed quantities; it is a package of price and quantity control, not just price control. Under pure price control, the government fixes only prices, but not quantities. Third, to reformers, dual track seems a partial reform and not a complete reform. This is true under dual track with limited market liberalization, but not true with full market liberalization. Although dual track with limited market liberalization does not achieve efficiency, it improves efficiency and makes nobody worse off.

Dual track liberalization requires enforcement of the rights and obligations under the plan track. In fact, enforcement of the plan track alone would prevent any decline in aggregate output. Can the plan track be enforced? With a collapsing government, it cannot. But enforcing the pre-existing plan is informationally much less demanding for the government than drawing up a new plan. Under central planning, the information requirement for drawing up a plan is huge. Enforcing a pre-existing plan is different. In fact, the dual track approach uses minimal additional information as compared with other possible compensation schemes that may be used with other approaches to reform. Compliance with the plan by economic agents depends on their expectations of the credibility of state enforcement. If state enforcement is not credible, then the economic agents will have no incentive to fulfil their plan obligations. If people think that they are not going to receive the plan-mandated deliveries at plan prices, they will not make the plan-mandated sales at the fixed plan prices. In that case, dual track liberalization degenerates to single track liberalization.

Lack of enforcement of the plan track may result in supply diversion as analysed by Murphy, Shleifer and Vishny (1992). These authors studied a partial reform model with the following two crucial assumptions: (i) suppliers are free to sell to all users, and (ii) buyers who are not covered by the plan can freely purchase inputs at any price, but buyers who are covered by the plan are not allowed to purchase inputs above the plan price. This partial reform model differs from the dual track liberalization model in an important respect: there is no plan delivery quota enforced on the suppliers.

In their model, partial reform may lead to inefficient supply diversion to such an extent that the outcome can be worse than that without reform. Therefore, the partial reform is not only not Pareto improving, but also total welfare reducing. Consider the case where the initial condition is also characterized by efficient rationing and efficient supply as shown in Figure 1, where the planned price P_p is below the market clearing level P_m. Then, after

the partial liberalization as defined above, suppliers can sell the good freely to the highest bidders. While the firms under the plan are forced to buy the good at price P_p, the firms outside the plan are free to buy the good at any price. Then they will bid the good for price $P_p + \varepsilon$ where ε is a positive but small number. Because the firms under the plan are constrained to pay P_p, an amount will be diverted from them to those not covered by the plan. Because the willingness to pay from those not covered by the plan is lower than those covered by the plan (by the assumption of efficient rationing), this kind of partial reform induces a net efficiency loss. While the sector not covered by the plan gains, the sector covered by the plan loses, and the total welfare effect is unambiguously negative. Although the assumption of efficient rationing and efficient supply under central planning is too strong, the result of inefficient supply diversion under partial reform remains valid with weaker conditions about initial rationing and supply.

So which model is more relevant? It depends on the quota enforcement capability of the government. A good enforcement capability makes the dual track liberalization model of Lau, Qian and Roland more relevant, while a poor enforcement would make the partial reform model of Murphy, Shleifer, and Vishny more relevant. The dual track liberalization model is motivated mainly by the practice in China, where enforcement has been reasonably good, while the partial reform model is mainly motivated by the experiences of the last years of the Soviet Union, when the state enforcement power diminished quickly.

In China's context, lack of quota enforcement sometimes takes the following form. The government may be unable to freeze the plan by creating new quotas with (below market equilibrium) planned price and giving windfall rents to some people who are politically connected. This may lead to corruption: firms find it easier to make profits by lobbying the government for allocating more input goods delivery at low planned prices, without the corresponding obligations to deliver low price outputs as under central planning. They then sell the goods at the market price to receive the windfall gains. This type of corruption is often attributed to the dual track approach to liberalization. Indeed, without the coexistence of the planned prices and market prices, the above form of corruption is not possible. By eliminating the two prices, such form of corruption would disappear. However, the essence of the problem is the failure in the enforcement of the original planned track. If the planned track is strictly enforced, no new quotas should be created. (On the other hand, full market liberalization allows for market arbitrage, which may increase the welfare of those who were allocated with goods at below-market prices. This is essential for achieving efficiency. The difference is that the potential rents are inherited from the previous regime in this case, not from a new creation.)

Dual track liberalization in practice

Studies of dual track liberalization focus mostly on China, although other cases, such as that of Mauritius, are also mentioned. The origin of the dual track can be traced to the 1950s when China had two prices for grain, the official price and negotiated price. However, dual track approach to market liberalization as a reform strategy was used only after 1979, first in the agricultural goods markets, and then in other markets (Byrd, 1991; Naughton, 1995; Lau, Qian and Roland, 2000).

Agriculture goods. The agricultural reform in China started with a dual track approach to market liberalization. Under that reform, the commune (and later the household) was assigned the responsibility to sell a fixed quantity of output to the state procurement agency as previously mandated under the plan at predetermined plan prices and to pay a fixed tax (often in kind) to the government. It also had the right (and obligation) to receive a fixed quantity of inputs, principally chemical fertilizers, from state-owned suppliers at predetermined plan prices. Subject to fulfilling these conditions, the commune was free to produce and sell whatever it considered profitable, and retain any profit. Moreover, the commune could purchase from the market grain (or other) output for resale to the state in fulfilment of its responsibility. There was thus a full market liberalization.

Between 1978 and 1988 state procurement of domestically produced grain remained essentially fixed, with 47.8 million tons in 1978 and 50.5 million tons in 1988. During that same period, total grain output increased by almost one-third. But the dual track approach to liberalization applied to agricultural products other than grain: between 1978 and 1990, the share of transactions at plan prices in all agricultural goods fell from 94 per cent to 31 per cent, when the agricultural output in China doubled. There was a huge supply response to the introduction of the market track.

Industrial goods. The most noticeable and often cited application of the dual track approach to liberalization is to industrial goods (Byrd, 1991; McMillan and Naughton, 1992). The Chinese government issued a document in May 1984 stipulating that there would be two forms of production in state-owned enterprises: planned and non-planned. Correspondingly, there were two types of material supplies for enterprises, namely, state allocation and free purchase. Prices of goods in the former were fixed by the state and prices of goods above quota quantity could be sold in the market at price within a range up to 20 per cent higher or lower than of the planned price. In February 1985, the 20 per cent price cap was removed and the dual track for industrial goods was formally in place (Wu and Zhao, 1987). As a result, the share of transactions at plan prices, in terms of output value, fell from 100 per cent before the reform to 45 per cent in 1990.

Coal and steel are the two important industrial commodities most tightly controlled under central planning, and both coal and steel markets were liberalized through the dual track approach. For coal, China's principal energy source, the planned delivery led to some slight increases in absolute terms during the 1980s, but the market track increased dramatically from 293 million tons to 628 million tons over the same period – the supply came mainly from small rural mines run by Township–Village Enterprises. As a result, the share of the plan allocation declined from 53 per cent in 1981 to 42 per cent in 1990. For steel, the plan track was quite stable in absolute terms during the 1980s, but the share of plan allocation fell from 52 per cent in 1981 to 30 per cent in 1990. In the cases of both coal and steel, because the plan track was essentially frozen, the economy was able to 'grow out of the plan' on the basis of the expansion of the market track (Naughton, 1995).

Consumer goods. Prior to the economic reform of 1979, most essential consumer goods and services for urban residents, such as grain, cooking oil, meat, electricity, housing, and the monthly transport pass, were rationed with coupons at values lower than corresponding free market prices. With dual track liberalization, urban residents continued to have the right to purchase grain, meat, electricity and housing at the same pre-reform prices and within the limits of the pre-reform rationed quantities, but, at the same time, they were also free to buy consumer goods from the free market at generally higher prices. The proportion of transactions at plan prices declined from 97 per cent in 1978 to only 30 per cent in 1990.

Foreign exchange. Under central planning, foreign exchange transactions were strictly controlled by the government at the official exchange rate. Exporters were required to surrender to the state all foreign exchange they earned at the official exchange rate, and importers were allocated with planned quotas of foreign exchange, also at the official exchange rate. Foreign visitors to China were required to use 'foreign exchange certificates', which were available at the official exchange rate. Starting from May 1988, China allowed trading of foreign exchange at Foreign Exchange Adjustment Centres (more commonly referred to as 'swap centres') at the rate determined by market supply and demand, called 'swap rate'. This was the beginning of the dual track in the foreign exchange market. The swap rate was, not surprisingly, significantly higher than the official rate. The supply of foreign exchange in the swap markets was provided by exporters through the foreign exchange they were allowed to retain from net increases in their export earnings in relation to the base period. By the end of 1993, transactions at official exchange rates accounted only for about 20 per cent of the total; the rest were at the market rate.

Labour. As in many other centrally planned economies, the labour market in China was also distorted: most labour was allocated to unproductive, state-owned enterprises and few to the non-state sector. Dual track liberalization in the labour market takes two forms. In the first, the non-state sector (the liberalized sector) pays market wages and decides on hiring and firing. Between 1978 and 1994, employment in the non-state sector increased by 318.8 per cent, while employment in the state sector (including civil servants in government agencies and non-profit organizations) increased by only 50.5 per cent. Second, even within the state sector there are also two tracks. Beginning in 1980, while pre-existing employees maintained their permanent employment status, most new hires in the state sector were made under the more flexible contract system and often at lower effective wage rates. Employment in the plan track was virtually stationary – it declined from 87.14 million in 1983, on the eve of the introduction of economic reform in industry, to 83.61 million in 1994.

Special economic zones. Dual track liberalization can also have a geographical dimension: special economic zones are such examples. Although similar zones for processing exports can be seen in other Asian economies, special economic zones had a more profound effect in China because the whole country was still under central planning when they were created. Therefore, the purpose of special economic zones was more than for exporting; it was a strategy for market reform.

In 1980, China established four 'special economic zones', Shenzhen, Zhuhai and Shantou in Guangdong province and Xiamen in Fujian province. Most transactions relating to activities inside the zones were on the market track, including prices of input and output goods and wages of labour – at a time when the rest of the economy was still operating under central planning. The special economic zones were insulated from the rest of the economy to minimize the impact on and interaction with the rest of the economic system. Initially, firms inside the special economic zones had to import all their inputs and export all their outputs – thus creating no disruption to the domestic aggregate supply and demand. The principal purpose of this approach was to minimize the impact of new economic activities on the old-style domestic state-owned enterprises. Thus, once again, there were two tracks and the reform was Pareto improving.

In order for the special economic zones to work, merely creating them was not enough. One of the crucial conditions was the insulation of the non-liberalized sector from the liberalized sector so that the latter's existing rents could be maintained while the other sector was liberalized. Therefore, creation of special economic zones is a type of limited market liberalization. It is Pareto improving and efficiency enhancing, but cannot be fully efficient.

Phasing out the plan track

With rapid growth, the plan track will become a matter of little consequence to most potential losers, which in turn reduces the cost required for compensating them. In China, the plan track in product markets was largely phased out during the 1990s. By 1996, the plan track was reduced to 16.6 per cent in agricultural goods, 14.7 per cent in industrial producer goods, and only 7.2 per cent in total retail sales of consumer goods. However, this phasing-out of the plan track was generally accompanied by compensation. For example, urban food coupons (grain, meat, oil, and so on) were removed in the early 1990s with lump-sum compensation. But the cost of compensation was much smaller in relative terms as compared to the potential cost of compensation in the early 1980s. The dual track exchange rate ended on 1 January 1994, when the two exchange rates – the official rate and the swap rate – were merged into a single, market rate. In this last step of foreign exchange reform, those organizations that used to receive cheap foreign exchange were provided with annual lump-sum subsidies for a period of three years, which was sufficient for them to purchase the pre-reform allocation of foreign exchange. Because at that time the share of centrally allocated foreign exchange had already fallen to less than 20 per cent of the total, the cost of compensation was not too large.

YINGYI QIAN

See also **China, economics in.**

Bibliography

Byrd, W.A. 1991. *The Market Mechanism and Economic Reforms in China*. Armonk, NY: M.E. Sharpe.

Lau, L.J., Qian, Y. and Roland, G. 1997. Pareto-improving economic reforms through dual track liberalization. *Economics Letters* 55, 285–92.

Lau, L.J., Qian, Y. and Roland, G. 2000. Reform without losers: an interpretation of China's dual track approach to transition. *Journal of Political Economy* 108, 120–43.

Li, W. 1999. A tale of two reforms. *RAND Journal of Economics* 30, 120–36.

McMillan, J. and Naughton, B. 1992. How to reform a planned economy: lessons from China. *Oxford Review of Economic Policy* 8, 130–43.

Murphy, K.M., Shleifer, A. and Vishny, R.W. 1992. The transition to a market economy: pitfalls of partial reform. *Quarterly Journal of Economics* 107, 887–906.

Naughton, B. 1995. *Growing Out of the Plan*. Cambridge: Cambridge University Press.

Wu, J. and Zhao, R. 1987. The dual pricing system in China's industry. *Journal of Comparative Economics* 11, 309–18.

duality

1 Introduction

The word 'duality' is often used to invoke a contrast between two related concepts, as when the informal, peasant, or agricultural sector of an economy is labelled as dual to the formal, or profit-maximizing, sector. In microeconomic analysis, however, 'duality' refers to connections between quantities and prices which arise as a consequence of the hypotheses of optimization and convexity. Connected to this duality are the relationship between utility and expenditure functions (and profit and production functions), primal and dual linear programs, shadow prices, and a variety of other economic concepts. In most textbooks, the duality between, say, utility and expenditure functions arises from a sleight of hand with the first-order conditions for optimization. These dual relationships, however, are not naturally a product of the calculus; they are rooted in convex analysis and, in particular, in different ways of describing a convex set. This article will lay out some basic duality theory from the point of view of convex analysis, as a remedy for the microeconomic theory textbooks the reader may have suffered.

2 Mathematical background

Duality in microeconomics is properly understood as a consequence of convexity assumptions, such as laws of diminishing marginal returns. In microeconomic models, many sets of interest are closed convex sets. The mathematics here is surveyed in CONVEX PROGRAMMING. The urtext for this material is Rockafellar (1970).

Closed convex sets can be described in two ways: by listing their elements, the 'primal' description of the set, and by listing the closed half-spaces that contain it. A closed (upper) half-space in $\mathbf{R}^n$ is a set of the form $h_{pa} = \{x : p \cdot x \geq a\}$, where p is another n-dimensional vector, a is a number and $p \cdot x$ is the inner product. The vector p is the *normal vector* to the half-spaces h_{pa}. Geometrically speaking, this is the set of points lying on or above the line $p \cdot x = a$. The famous separation theorem for convex sets implies that every closed convex set is the intersection of the half-spaces containing it.

Suppose that C is a closed convex set, and that p is a vector in $\mathbf{R}^n$. How do we find all the numbers a such that $C \subset h_{pa}$? If there is an $x \in C$ such that $p \cdot x < a$, then a is too big. So the natural candidate is $w = \inf_{x \in C} p \cdot x$. If $a > w$ there will be an $x \in C$ such that $p \cdot x < a$ on the other hand, if $a < w$, then $p \cdot x > a$ for all $x \in C$. So the half-spaces h_{pa} for $a \leq w$ are the closed half-spaces containing C.

This construction can be applied to functions. A concave function on $\mathbf{R}^n$ is an $[-\infty, \infty)$ valued function f such that the *hypograph* of f, the set hypo $f = \{(x, a) \in \mathbf{R}^{n+1} : a \leq f(x)\}$, is convex. If hypo f is closed, f is said to be *upper semi-continuous* (usc). The *domain* dom f of concave f is the set of vectors in $\mathbf{R}^n$ for which f is finite-valued. Concave (and convex) functions are very well-behaved on the *relative interiors* of their effective domains. The relative interior ri C of a convex set C is the interior relative to the smallest affine set containing C (see CONVEX PROGRAMMING), and on ri dom f, f (concave or convex) is continuous.

Suppose that f is usc. The minimal level a such that $h_{(p,-1)a}$, the hyperplane in $\mathbf{R}^{n+1}$ with normal vector $(p, -1)$, contains hypo f is $f^*(p) \equiv \inf_x p \cdot x - f(x)$. Why the normal vector $(p, -1)$? Because the graph of the affine function $x \mapsto f^*(p) + px$ is a tangent line to f, the graph of f lies just beneath it, and no other line with the same slope and a smaller intercept has this property. The function $f^*(p)$ is the (concave) *Fenchel transform* or *conjugate* of f, and is traditionally denoted f^*. The construction of the preceding paragraph can be done just this way: the *concave indicator function* of a convex set C is the function $\delta_C(x)$ which is 0 on C and $-\infty$ otherwise, and $\delta_C^*(p) = \inf_{x \in C} p \cdot x$. For any function f, not necessarily usc or concave, the Fenchel transform f^* is usc and concave. If f is in fact both usc and concave, then $f^{**} = f$. This fact is known as the conjugate duality theorem. Convex functions with range $(-\infty, \infty]$ are treated identically. The function f is convex if and only if $-f$ is concave, but the definitions are handled slightly differently in order to preserve the intuition just described. The set epi $f = \{(x, a) : a > f(x)\}$, and the convex Fenchel transform is defined differently: $f^*(p) = \sup_x p \cdot x - f(x)$. The convex indicator function of a convex set C is the function $\delta^C(x)$ which is 0 on C and +otherwise; its (convex) conjugate is $\delta^{C*}(p) = \sup_x p \cdot x$. These facts are discussed in CONVEX PROGRAMMING.

If concave functions have tangent lines, then they must have something like gradients. A vector p is a sub*gradient* of f at x if $f(x) + p \cdot (y - x) \leq f(y)$. If f has a unique subgradient at x, then f is differentiable at x and $p = \nabla f(x)$, and conversely. But the subgradient need not be unique: the set $\partial f(x)$ of subgradients at x is the *subdifferential* of f at x. The *domain* of f, dom f, is the set of x such that $f(x) > -\infty$. The subdifferential is nonempty for all x in its *relative interior*. It follows from the definition of concavity (and is proved in CONVEX PROGRAMMING that the subdifferential correspondence is *monotonic*: if $p \in \partial f(x)$ and $q \in \partial f(y)$, then $(p-q) \cdot (x-y) \leq 0$. If f is convex, then the inequality is reversed, and $(p-q) \cdot (x-y) \geq 0$. Finally, suppose f is usc and concave. Then so is its conjugate f^*, and their subdifferentials have an inverse relationship: $p \in \partial f(x)$ if and only if $x \in \partial f^*(p)$.

3 Cost, profit and production

In the theory of the firm, profit functions and cost functions are alternative ways of describing the firms' technology choices. A technology is described by a set of vectors F in $\mathbf{R}^N$. Each vector $Z \in F$ is an input–output vector. We adopt the convention that negative coefficients correspond to input quantities and positive quantities correspond to outputs. Suppose that the first L goods are inputs and the last $M = N - L$ are outputs, so that $F \subset \mathbf{R}^L_- \times \mathbf{R}^M_+$. It is convenient to assume free disposal, so that if $(x, y) \in F$, and both $x' \leq x$ and $y' \leq y$ (more input and less output), then $(x', y') \in F$. Two important dual representations of the technology are the cost and profit functions. The profit function is $\pi(p, w) = \sup_{(x,y) \in F} p \cdot y + w \cdot x$ for $p \in$ and $w \in \mathbf{R}^L$, which is the conjugate of the convex indicator function of F. The cost function too can be obtained through conjugacy. The set $F(y) = \{x : (x, y) \in F\}$ is the set of all input bundles that produce y. Then $C(y, w) = -\sup_{x \in F(y)} w \cdot x$, that is, $C(y, \cdot) = -\delta^{F(y)*}$.

Immediately the properties of the Fenchel transform imply that $\pi(p, w)$ is convex in its arguments and $C(y, w)$ is concave in w, the profit function is lsc and the cost function is usc. (This implies that both functions are continuous on the relative interior of their effective domains.) Cost and profit functions are also linear homogeneous. Doubling all prices doubles both costs and revenues. Cost is also monotonic. If $w'_l < w_l$ for every input l, then $C(y, w') \leq C(y, w)$ and if $w'_l < w_l$ for all l, then $C(y, w') < C(y, w)$.

The point of duality is that, if the technology is closed and convex, then cost profit functions each characterize the technology F. The conjugate duality theorem (see CONVEX PROGRAMMING) implies that $\pi^*(x, y) \equiv \delta^{F**}(x, y) = \delta^F(x, y)$, the convex indicator function of F:

$$\sup_{(p,w) \in \mathbf{R}^N} p \cdot x + w \cdot y - \pi(p, w)$$
$$= \begin{cases} 0 & \text{if } (x, y) \in F, \\ +\infty & \text{otherwise.} \end{cases}$$

If F is closed and convex, then each $F(y)$ is convex. If F is closed then $F(y)$ will also be closed. Then $\delta^{F(y)}$ is concave and usc, so

$$\sup_{w \in \mathbf{R}_+^L} w \cdot x + C(y, w) = \sup_{w \in \mathbf{R}_+^L} w \cdot x \delta^{F(y)*}(w)$$
$$= \delta^{F(y)}(x).$$

Hotelling's lemma is a famous result of duality theory. It says that the net supply function of good i is the derivative of the profit function with respect to the price of good i. The usual proof is via the envelope theorem: the marginal change in profits from a change in price p is the quantity of good i times the change in the price plus the price of all goods times the changes in their respective quantities. But the quantity changes are second-order because the quantities solve the profit maximization first-order conditions, that price times the marginal change in quantities in technologically feasible directions is 0. Every advanced microeconomics text proves this. A result like this is true whenever the technology is convex, even if the technology is not smooth.

The convex version of Hotelling's lemma is a consequence of the inversion property of subdifferentials for concave and convex f; that $p \in \partial f(x)$ if and only if $x \in \partial f^*(p)$. See CONVEX PROGRAMMING for a brief discussion.

Hotelling's lemma $(x, y) \in \partial \pi(p, w)$ if and only if (x, y) is profit-maximizing at prices (p, w).

Hotelling's lemma is quickly argued. If $(x, y) \in \partial \pi(p, w) = \partial \delta^{F*}(p, w)$, then $(p, w) \in \partial \delta^{F**}(x, y) = \partial \delta^F(x, y)$. Then $\delta^F(x, y) + (p, w) \cdot ((x', y') - (x, y)) \leq \delta^F(x', y')$ for all (x', y'). This implies that $x \in F$ and furthermore that $(p, w) \cdot ((x', y') - (x, y)) \leq 0$ for all $(x, y) \in F$, in other words, that (x, y) is profit-maximizing at prices (p, w). Conversely, suppose that (x, y) is profit maximizing at prices (p, w). Then (p, w) satisfies the subgradient inequality of δ^F at (x, y), and so $(p, w) \in \partial \delta^F$. Consequently,

$$(x, y) \in \partial \delta^{F*}(p, w) \equiv \partial \pi(p, w).$$

The textbook treatment of duality observes that, if net supply is the first derivative of the profit function, then the own-price derivative of net supply must be the second own-partial derivative of profit with respect to price, and convexity of the profit function implies that this partial derivative should be positive, so net supply is increasing in price. The same fact follows in the convex framework from the monotonicity properties of the subgradients. Suppose that (w, p) and (w', p') are two price vectors, and suppose that (x, y) and (x', y') are two profit-maximizing production plans corresponding to the two price vectors. Then $(w - w', p - p')(x - x', y - y') \geq 0$. If the two price vectors are identical for all prices but, say, $p_k \neq p'_k$, then $(p_k - p'_k)(y_k - y'_k) \geq 0$, and net supply is non-decreasing in price. As with net supplies, some comparative statics of conditional factor demand with respect to input price changes follows from the monotonicity property of subgradients.

Another implication of profit function convexity and (twice continuous) differentiability is symmetry of the derivatives of net supply:

$$\frac{\partial y_k}{\partial p_l} = \frac{\partial^2 \pi}{\partial p_k \partial p_l} = \frac{\partial^2 \pi}{\partial p_l \partial p_k} = \frac{\partial y_l}{\partial p_k}.$$

The convex analysis version of this is that for any finite sequences of goods $i, \ldots, l$,

$$p_i \cdot (y_j - y_i) + p_j \cdot (y_k - y_j) + \cdots$$
$$+ p_l \cdot (y_i - y_l) \leq 0.$$

This requirement, which has a corresponding expression in terms of differences in prices, is called *cyclic monotonicity*. All subdifferential correspondences are cyclicly monotone. The connection with symmetry is not obvious, but it helps to know that Rockafellar (1974) leaves as an exercise (and so do we) that cyclic monotonicity is a property of a linear transformation corresponding to an $n \times n$ matrix M if and only if M is symmetric and positive semi-definite. Monotonicity is cyclic monotonicity for sequences of length 2.

The other famous result in duality theory for production is Shephard's lemma, which does for cost functions what Hotelling's lemma does for profit functions: conditional input demands are the derivatives of the cost functions. This is demonstrated in the same way, since the cost function and the indicator function for the set of inputs from which y is produceable are both convex and have closed hypographs.

4 Utility and expenditure functions

A quasi-concave utility function U defined on the commodity space $\mathbf{R}_+^n$ has upper contour sets, the sets R_u of consumptions bundles which have utility at least u, which are convex. If u is usc, these sets are closed as well.

The *expenditure function* gives for each utility level u and price vector p the minimum cost of realizing utility u at prices $p : e(p, u) = \inf\{p \cdot x : u(x) \geq u\}$. If the infimum is actually realized at a consumption bundle x, then x is the *Hicksian* or *compensated real income demand*.

In terms of convex analysis, $e(p, u)$ is the conjugate of the concave indicator function $\phi_u(x)$ of the set $R(u) = \{x : U(x) \geq u\}$, that is, $e(p, u) \equiv \phi_u^*(p)$. Thus $e(p, u)$ will be usc and concave in p for each u. The expenditure function is also linearly homogeneous in prices. If prices double, then the least cost of achieving u will double as well.

The duality of utility and expenditure functions is that each can be derived from the other; they are alternative characterizations of preference. Since the concave indicator function $\phi_u(x)$ is closed and convex, $e(\cdot, u)^* = \phi_u(x)$. For fixed u, the Fenchel transform of the expenditure function is the concave indicator function of $R(u)$; $\inf_p p \cdot x - e(p, u)$ is 0 if $U(x) \geq u$ and $-\infty$ otherwise. If $x \in R(u)$, then the cost of x at any price p can be no less than the minimum cost necessary to achieve utility u. The gap between the cost of x and the cost of utility level u is made by taking ever smaller prices, and so its minimum is 0. Suppose that x is not in $R(u)$. The separation theorem for convex sets says there is a price p such that $p \cdot x < \inf_{y \in R(u)} p \cdot y$; there is a price at which x is cheaper than the cost of u. Now, by taking ever larger multiples of p, the magnitude of the gap can be made arbitrarily large, and so the value of the conjugate is $-\infty$. Thus the conjugate is the concave indicator function of $R(u)$.

Among the most useful consequence of the duality between utility and expenditure functions is the relationship between derivatives of the expenditure function and the Hicksian, or compensated, demand. Hicksian demand. The compensated demand at prices p and utility u are those consumption bundles in $R(u)$ which minimize expenditure at prices p. This result is just Shephard's lemma for expenditure functions:

Hicks' compensated demand: Consumption bundle x is a Hick's compensated consumption bundle at prices p if and only if $x \in \partial_p e(p, u)$. Furthermore, if x is demanded at prices p and utility u, and y is demanded at prices q and the same utility u, then $(p-q) \cdot (x-y) \leq 0$.

The downward-sloping property just restates the monotonicity property of the subdifferential correspondence. For the special case of changes in a single price, the statement is that demand is non-increasing in its own price.

5 Equilibrium and optimality

The equivalence between Pareto optima and competitive equilibria can also be viewed as an expression of duality. When preferences have concave utility representations, *quasi-equilibrium* emerges from Lagrangean duality. Quasi-equilibrium entails feasibility, profit maximization, and expenditure minimization rather than utility maximization. That is, each trader's consumption allocation is expenditure minimizing for the level of utility it achieves. The now traditional route of Arrow (1952) and Debreu (1951) to the Second Welfare Theorem first demonstrates that a Pareto-optimal allocation can be regarded as a quasi-equilibrium for an appropriate set of prices. Under some additional conditions, the quasi-equilibrium is in fact a competitive equilibrium, wherein utility maximization on an appropriate budget set replaces expenditure minimization. Our concern here is with the first step on this path.

Suppose that each of I individuals has preferences represented by a concave utility function on $\mathbf{R}_+^N$, and that production is represented, as in Section 3, by a closed and convex set F of feasible production plans. Suppose that $0 \in F$ (it is possible to produce nothing) and that the aggregate endowment e is strictly positive. Assume, too, that there is free disposal in production. Every Pareto optimum is the maximum of a Bergson–Samuelson social welfare function of the form $\sum_i \lambda_i u_i$ defined on the set of all consumption allocations. An allocation is a vector (x, y) where $x \in \mathbf{R}_+^{NI}$ is a consumption allocation, a consumption bundle for each individual, and y is a production plan. The allocation is *feasible* if $y \in F$ and $y + e - \sum_i x_i \geq 0$. A Lagrangean for this convex program is

$$L(x, y, p) = \begin{cases} \sum_i u_i(x_i) + p \cdot (y + e - \sum_i x_i) & \text{if } x \in \mathbf{R}_+^{NI}, \ y \in F \text{ and } p \in \mathbf{R}_+^L, \\ +\infty & \text{if } x \in \mathbf{R}_+^{NI}, \ y \in F \text{ and } p \notin \mathbf{R}_+^L, \\ -\infty & \text{otherwise,} \end{cases}$$

where p is the vector of Lagrange multipliers for the L goods constraints.

The possibility of 0 production and the strict positivity of the aggregate endowment guarantee that the set of feasible solutions satisfies Slater's condition, and so a saddlepoint (x^*, y^*, p^*) exists; that is, $\sup_{x,y} L(x, y, p^*) \leq L(x^*, y^*, p^*) \leq L(x^*, y^*, p)$ for all $x \in \mathbf{R}^{NI}$, $y \in F$ and $p \in \mathbf{R}^L$. Then (x^*, y^*) is Pareto optimal and p^* solves the dual problem $\min_p \sup_{xy} L(x, y, p)$. The interpretation of (x^*, y^*, p^*) as a quasi-equilibrium comes from examining the dual problem. The dual problem can be rewritten as

$$
\begin{aligned}
&\inf_{p \in \mathbf{R}^L_+} \sup_{x \in \mathbf{R}^{NI}_+, y \in F} L(x, y, p) \\
&= \inf_{p \in \mathbf{R}^L_+} \sup_{x \in \mathbf{R}^{NI}_+, y \in F} \sum_i u_i(x_i) \\
&\quad + p \cdot \left(y + e - \sum_i x_i \right) \\
&= \inf_{p \in \mathbf{R}^L_+} \sum_i \sup_{x_i \in \mathbf{R}^L_+} \{ \lambda_i u_i(x_i) - p \cdot x_i \} \\
&\quad + \sup_{y \in F} p \cdot y.
\end{aligned}
\tag{1}
$$

In the dual problem, the Lagrange multipliers can be thought of as goods prices. The Second Welfare Theorem interprets the optimal allocation as an equilibrium allocation using the Lagrange multipliers as equilibrium prices. To see this, look at the second line of (1). At prices p, a production plan is chosen from y to maximize profits $p \cdot y$, so the value of this term is $\pi(p)$. Each consumer is asked to solve

$$
\begin{aligned}
\max_i \lambda_i u_i(x_0) - p \cdot x &= -\min p \cdot x - \lambda_i u_i(x_i) \\
&= \lambda_i u_i^* - \min p \cdot x \\
&\quad - \lambda_i(u_i(x_i) - u_i^*)
\end{aligned}
$$

where $u_i^* = u_i(x_i^*)$. The term being minimized is the Lagrangean for the problem of expenditure minimization, and so x_i^* is the Hicksian demand for consumer i at prices p and utility level $u_i^* = u_i(x_i^*)$. Finally, the optimal allocation is feasible, and so (x^*, y^*, p^*) is a quasi-equilibrium.

Given the observation about expenditure minimization, the saddle value of the Lagrangean is

$$
\sum_i \lambda_i u_i^* - e_i(p^*, u_i^*) + \pi(p^*)
$$

The planner chooses prices to minimize net surplus, which is the sum of profits from production and the excess of total Bergson–Samuelson welfare less the cost of the consumption allocation.

6 Historical notes

Duality ideas appeared very early in the marginal revolution. Antonelli, for instance, introduced the indirect utility function in 1886. The modern literature begins with Hotelling (1932), who provided us with Hotelling's lemma and cyclic monotonicity. Shephard (1953) was the first modern treatment of duality, making use of notions such as the support function and the separating hyperplane theorem.

The results on consumer and producer theory are surveyed more extensively in Diewert (1981), who also provides a guide to the early literature. In its focus on Fenchel duality, this review has not even touched on the duality between direct and indirect aggregators, such as utility and indirect utility, and topics that would naturally accompany this subject such as Roy's identity. Again, this is admirably surveyed in Diewert (1981).

LAWRENCE E. BLUME

See also **convex programming; convexity; duality; Lagrange multipliers; Pareto efficiency; quasi-concavity.**

Bibliography

Arrow, K.J. 1952. An extension of the basic theorems of classical welfare economics. In *Proceedings of the Second Berkeley Symposium on Mathematical Statistics and Probability*, ed. J. Neyman. Berkeley: University of California Press.

Debreu, G. 1951. The coefficient of resource utilization. *Econometrica* 19, 273–92.

Diewert, W.E. 1981. The measurement of deadweight loss revisited. *Econometrica* 49, 1225–44.

Hotelling, H. 1932. Edgeworth's taxation paradox and the nature of demand and supply. *Journal of Political Economy* 40, 577–616.

Rockafellar, R.T. 1970. *Convex Analysis*. Princeton, NJ: Princeton University Press.

Rockafellar, R.T. 1974. *Conjugate Duality and Optimization*. Philadelphia: SIAM.

Shephard, R.W. 1953. *Cost and Production Functions*. Princeton, NJ: Princeton University Press.

Dühring, Eugen Karl (1833–1921)

Dühring was born on 12 January 1833 in Berlin and died on 21 September 1921 at Nowawes bei Potsdam. The son of a Prussian state official, Dühring studied law, philosophy and economics at the University of Berlin and practised law until blindness obliged him to abandon this career. He then became a *Privatdozent* at the University of Berlin, where he taught philosophy and economics from 1863 to 1877, and began to write voluminously on a wide range of subjects, from the natural sciences to philosophy, social theory and socialism, his aim being to construct a system of social reform based upon positive science. His system was expounded in a series of books on capital and labour (1865), the principles of political economy (1866), a critical history of philosophy (1869), a critical history of political economy and socialism (1871), and courses in political economy and philosophy (1873; 1875). Dühring was an adherent of positivism, concerned in his philosophical works to expound a 'strictly scientific world outlook', in opposition particularly to the Hegelian dialectic. His economic writings emphasize the role of political factors in the development of capitalism, and he argued that social injustice is not caused primarily by the economic system, but by social and political circumstances, the remedy being to control the misuse of private property and capital (not abolish them) through workers' organizations and state intervention.

Schumpeter (1954, pp. 509–10), praised Dühring's history of mechanics (1873), which was awarded an academic prize, suggested that he would retain a prominent place in the history of anti-metaphysical and positivist currents of thought, and noted that he made an important criticism of Marxist theory in his argument that political causes had played a major part in constituting the property relations of capitalist society. In other respects, however, Schumpeter considered that Dühring had made no significant contribution to economic theory.

Engels, in his well-known book (originally published as a series of articles), *Herr Eugen Dühring's Revolution in Science* [*Anti-Dühring*] (1877–8), which has done more than anything else to keep Dühring's name alive, took a much more critical view, deriding his work as a prime example of the 'higher nonsense' which infected German academic life. His philosophical views were dismissed by Engels as 'vulgar materialism' and compared unfavourably with the 'revolutionary side' of Hegel's dialectics; and in the chapter of *Anti-Dühring* devoted to the history of political economy (largely written by Marx, but not published in full until the third edition of the book in 1894), Dühring was castigated for his superficiality and theoretical misconceptions. It was, however, the concern with Dühring's programme of social reform, and its possible baleful effect on the developing labour movement (Eduard Bernstein, for example, was initially impressed by Dühring's *Cursus* of 1873, though soon repelled by his anti-Semitism) that originally provoked Engels's articles, and was countered in the final section of the book (frequently reprinted later as a separate text under the title *Socialism, Utopian and Scientific*) by an exposition of Marxist socialism which became enormously influential.

It seems doubtful that Dühring occupies more than a minor place in the history of economic and social thought, except for this encounter with Marx and Engels, though Schumpeter (1954, p. 509) called him a 'significant thinker' and the entry in the *Encyclopedia of the Social Sciences* (1931, vol. 5, p. 273) described his writings

as 'among the important intellectual achievements of the nineteenth century'.

TOM BOTTOMORE

Selected works

1871. *Kritische Geschichte der Nationalökomie und des Sozialismus*. Berlin: T. Grieben.
1873. *Cursus der National- und Sozialökonomie einschliesslich der Hauptpunkte der Finanzpolitik*. Berlin: T. Grieben.
1875. *Cursus der Philosophie als streng wissenschaftlicher Weltanschauung und Lebensgestaltung*. Leipzig: E. Koschny.

Bibliography

Albrecht, G. 1927. *Eugen Dühring: ein Beitrag zur Geschichte der Sozialwissenschaften*. Jena: G. Fischer.
Engels, F. 1877–8. *Anti-Dühring. Herr Eugen Dühring's Revolution in Science*. Moscow: Progress Publishers, 1947.
Schumpeter, J.A. 1954. *A History of Economic Analysis*. London: Allen & Unwin.

dummy variables

In economics, as well as in other disciplines, qualitative factors often play an important role. For instance, the achievement of a student in school may be determined, among other factors, by his father's profession, which is a qualitative variable having as many attributes (characteristics) as there are professions. In medicine, to take another example, the response of a patient to a drug may be influenced by the patient's sex and the patient's smoking habits, which may be represented by two qualitative variables, each one having two attributes. The dummy-variable method is a simple and useful device for introducing, into a regression analysis, information contained in qualitative or categorical variables; that is, in variables that are not conventionally measured on a numerical scale. Such qualitative variables may include race, sex, marital status, occupation, level of education, region, seasonal effects, and so on. In some applications, the dummy-variable procedure may also be fruitfully applied to a quantitative variable such as age, the influence of which is frequently U-shaped. A system of dummy variables defined by age classes conforms to any curvature and consequently may lead to more significant results.

The working of the dummy-variable method is best illustrated by an example. Suppose we wish to fit an Engel curve for travel expenditure, based on a sample of *n* individuals. For each individual *i*, we have quantitative information on his travel expenditures (y_i) and on his disposable income (x_i), both variables being expressed in logarithms. A natural specification of the Engel curve is:

$$y_i = a + bx_i + u_i$$

where a and b are unknown regression parameters and u_i is a non-observable random term. Under the usual classical assumptions (which we shall adopt throughout this presentation), ordinary least-squares produce the best estimates for a and b.

Suppose now that we have additional information concerning the education level of each individual in the sample (presence or absence of college education). If we believe that the education level affects the travel habits of individuals, we should explicitly account for such an effect in the regression equation. Here, the education level is a qualitative variable with two attributes: college education; no college education. To each attribute, we can associate a dummy variable which takes the following form:

$$d_{1i} = \begin{cases} 1 & \text{if college education} \\ 0 & \text{if no college education} \end{cases}$$

$$d_{2i} = \begin{cases} 1 & \text{if no college education} \\ 0 & \text{if college education} \end{cases}$$

Inserting these two dummy variables in the Engel curve, we obtain the following expanded regression:

Specification I

$$y_i = a_1 d_{1i} + a_2 d_{2i} + bx_i + u_i$$

which may be estimated by ordinary least-squares. Alternatively, noting that $d_{1i} + d_{2i} = 1$ for all i, we can write:

Specification II

$$y_i = a_2 + (a_1 - a_2)d_{1i} + bx_i + u_i$$

which, again, may be estimated by ordinary least-squares.

It is easy to see how the procedure can be extended to take care of a finer classification of education levels. Suppose, for instance, that we actually have s education levels (s attributes). All we require is that the attributes be exhaustive and mutually exclusive. We then have the two following equivalent specifications:

Specification I

$$y_i = a_1 d_{1i} + a_2 d_{2i} + \ldots + a_s d_{si} + bx_i + u_i$$

Specification II

$$y_i = a_s + (a_1 - a_s)d_{1i} + \ldots + (a_{s-1} - a_s)d_{s-1,i} + bx_i + u_i.$$

Obviously, the two specifications produce the same results but give rise to different interpretations. Specification I includes all the s dummy variables but no constant term. In this case, the coefficient of d_{ji} gives the specific effect of attribute j. Specification II includes $s - 1$ dummy variables and an overall constant term. The constant term represents the specific effect of the omitted attribute, and the coefficients of the different d_{ji} represent the contrast (difference) of the effect of the jth attribute with respect to the effect of the omitted attribute. (Note that it is not possible to include all dummy variables plus an overall constant term, because of perfect collinearity.)

It is important to stress that by the introduction of additive dummy variables, it is implicitly assumed that the qualitative variable affects only the intercept but not the slope of the regression equation. In our example, the elasticity parameter, b, is the same for all individuals; only the intercepts differ from individual to individual depending on their education level. If we are interested in individual variation in slope, we can apply the same technique, as long as at least one explanatory variable has a constant coefficient over all individuals. Take the initial case of only two attributes. If the elasticity parameter varies according to the level of education, we have the following specification:

$$y_i = a_1 d_{1i} + a_2 d_{2i} + b_1 d_{1i}x_i + b_2 d_{2i}x_i + u_i.$$

Simple algebra shows that ordinary least-squares estimation of this model amounts to performing two separate regressions, one for each class of individuals. If, however, the model contained an additional explanatory variable, say z_i, with constant coefficient c, by simply adding the term cz_i to the above equation, we would simultaneously allow for variation in the intercept and variation in the slope (for x).

The dummy variable model also provides a conceptual framework for testing the significance of the qualitative variable in an easy way. Suppose we wish to test the hypothesis of no influence of the level of education on travel expenditures. The hypothesis is true if the s coefficients a_i are all equal; that is, if the $s - 1$ differences $a_j - a_s$, $j = 1, \ldots, s - 1$, are all zero. The test therefore boils down to a simple test of significance of the $s - 1$ coefficients of the dummy variables in Specification II. If $s = 2$, the t-test applied to the single coefficient of d_{1i} is appropriate. If $s > 2$, we may conveniently compute the following quantity:

$$\frac{(SS_c - SS)/(s - 1)}{SS/(n - s - 1)}$$

which is distributed as an F-variable with $s - 1$ and $n - s - 1$ degrees of freedom. In the above expression, SS is the sum of squared residuals for the model with the dummy variables (either Specification I or II), and SS_c is the sum of squared residuals for the model with no dummy variables but with an overall constant term.

In some economic applications the main parameter of interest is the slope parameter, the coefficients of the dummy variables being nuisance parameters. When, as in the present context, only one qualitative variable (with s attributes) appears in the regression equation, an easy computational device is available which eliminates the problem of estimating the coefficients of the dummy variables. To this end, it suffices to estimate, by ordinary least-squares, the simple regression equation:

Specification III

$$y_i^* = bx_i^* + u_i^*.$$

where the quantitative variables (both explained and explanatory) for each individual are expressed as deviations from the mean over all individuals possessing the same attribute. For the dichotomous case presented in the beginning, for an individual with college education, we subtract the mean over all individuals with college education and likewise for an individual with no college education. Note, however, that the true number of degrees of freedom is not $n - 1$ but $n - 1 - s$. The same procedure also applies when the model contains other quantitative explanatory variables. The interested reader may consult Balestra (1982) for the conditions under which this simple transformation is valid in the context of generalized regression.

The case of multiple qualitative variables (of the explanatory type) can be handled in a similar fashion. However, some precaution must be taken to avoid perfect collinearity of the dummy variables. The easiest and most informative way to do this is to include, in the regression equation, an overall constant term and to add for each qualitative variable as many dummy variables as there are attributes minus one. Take the case of our Engel curve and suppose that, in addition to the education level (only two levels for simplicity), the place of residence also plays a role. Let us distinguish two types of place of residence: urban and rural. Again, we associate to these two attributes two dummy variables, say e_{1i} and e_{2i}. A correct specification of the model which allows for both qualitative effects is:

$$y_i = a_1 + a_2 d_{1i} + a_3 e_{1i} + bx_1 + u_i.$$

Given the individual's characteristics, the measure of the qualitative effects is straightforward, as shown in the following table:

	Urban	Rural
College education	$a_1 + a_2 + a_3$	$a_1 + a_2$
No college education	$a_1 + a_3$	a_1

The specification given above for the multiple qualitative variable model corresponds to Specification II of the single qualitative variable model. Unfortunately, when there are two or more qualitative variables there is no easy transformation analogous to the one incorporated in Specification III, except under certain extraordinary circumstances (Balestra, 1982).

One such circumstance arises in connection with cross-section time-series models. Suppose that we have n individuals observed over t periods of time. If we believe in the presence of both an individual effect and a time effect, we may add to our model two sets of dummy variables, one corresponding to the individual effects and the other corresponding to the time effects. This is the so-called covariance model. The number of parameters to be estimated is possibly quite large when n or t or both are big. To avoid this, we may estimate a transformed model (with no dummies and no constant term) in which each quantitative variable (both explained and explanatory) for individual i and time period j is transformed by subtracting from it both the mean of the ith individual and the mean of the jth time period and by adding to it the overall mean. Note that, by this transformation, we lose $n + t - 1$ degrees of freedom.

To conclude, the purpose of the preceding expository presentation has been to show that the dummy-variable method is a powerful and, at the same time, simple tool for the introduction of qualitative effects in regression analysis. It has found and will undoubtedly find numerous applications in empirical economic research. Broadly speaking, it may be viewed as a means for considering a specific scheme of parameter variation, in which the variability of the coefficients is linked to the causal effect of some precisely identified qualitative variable. But it is not, by any means, the only scheme available. For instance, when the qualitative effects are generic, as in the cross-section time-series model, one may question the validity of representing such effects by fixed parameters. An interpretation in terms of random effects may seem more appealing. This type of consideration has led to the development of other schemes of parameter variation such as the error component model and the random coefficient model.

A final remark is in order. In the present discussion, qualitative variables of the explanatory type only have been considered. When the qualitative variable is the explained (or dependent) variable, the problem of these *limited* dependent variables is far more complex, both conceptually and computationally.

PIETRO BALESTRA

Bibliography

Balestra, P. 1982. Dummy variables in regression analysis. In *Advances in Economic Theory*, ed. Mauro Baranzini. Oxford: Blackwell.

Goldberger, A.S. 1960. *Econometric Theory*, pp. 218–27. New York: John Wiley.

Maddala, G.S. 1977. *Econometrics*, ch. 9. New York: McGraw-Hill.
Suits, D.B. 1957. Use of dummy variables in regression equations. *Journal of the American Statistical Association* 52, 548–51.

Dunlop, John Thomas (1914–2003)

John Dunlop was an extraordinary labour economist, Professor and Dean of the Faculty at Harvard University, Secretary of Labor of the United States, and mentor to students and practitioners in the world of labour. He was extraordinary because he was more than an economist and because he was driven by a moral vision of what economists and academics should do to make the world better. Labour economists and policymakers paid close attention to Dunlop's thoughts because he combined academic research with unparalleled practical experience in solving problems and building institutions. His academic writings, which include several classic articles as well as major books, reflect Dunlop's participation in events and direct observations of social behaviour.

Dunlop first attracted academic attention with his 1938 *Economic Journal* article on the movement of real and money wages over the business cycle, which forced Keynes to admit that the *General Theory* was wrong on this issue: real wages fall in recessions not in booms, contrary to simple marginal productivity analysis. Quite an achievement for a 24-year-old economist. Dunlop followed this with *Wage Determination Under Trade Unions* (1944), in which he modelled unions as optimizing organizations; with analyses of the cyclic variation of labour's share, with the concept of 'wage contours' that captured the notion that product markets influenced wages, and with numerous analysis of wage determination, labour relations, mediation and dispute resolution. Dunlop's book *Industrial Relations Systems* (1958) sought to develop a broader perspective on how labour relations fit into economics.

In the 1980s, concerned that labour economists were limited in their conceptual vision by narrow optimizing models and in their empirical analysis by extant government data-sets, Dunlop carped at them for failing to see what he could see in the labour market. Dunlop saw the labour market as pre-eminently a social institution to resolve labour problems, which should be analysed as such rather than as a bourse. His mode of analysis was that of a naturalist, who looks at the world with his own eyes and experience, with direct knowledge of the institutions and practitioners, without trying to force observation into a narrow conceptual framework.

Dunlop's career spanned a wide variety of activities. Earning his AB (1935) and Ph.D. (1939) from Berkeley, he rose to become professor of economics at Harvard and Dean of the University (1970–3), when he helped stabilize the university during a period of student disorders, and Lamont University Professor (1970–2003). He worked for the National War Labor Board (1943–54); served as member or chair on various national panels with responsibility for resolving labour disputes; led labour-management committees in areas ranging from missile sites to apparel, the public sector, and health; served as Director of the Cost of Living Council (1973–4), and as Secretary of Labor of the United States (1975–6). From 1993 to 1994 he chaired the Commission on the Future of Worker–Management Relations, popularly known as the Dunlop Commission, which was given the charge 'to recommend ways to improve labor–management cooperation and productivity'. The politics and economics of the time were not right, however, for bringing management and labour to a consensus on modernizing labour relations, so that much of the Commission's recommendations went unheeded.

Dunlop approached his work –advising presidents and cabinet officials and telling academics about the real world and practitioners about academic theory – with one goal: to help solve problems. The moral principle that guided him – that academics should use their knowledge and skill to help solve problems faced by real people, by workers and firms, and governments – represents social science at its best.

RICHARD B. FREEMAN

Selected works

1938. The movement of real and money wage rates. *Economic Journal* 48, 413–34.
1944. *Wage Determination under Trade Unions*. New York: Macmillan.
1948. Productivity and the wage structure. *In Income, Employment and Public Policy: Essays in Honor of Alvin H. Hansen*. New York: W.W. Norton.
1948. The development of labor organizations: a theoretical framework. In *Insights into Labor Issues*, ed. R. Lester and J. Shister. New York: Macmillan.
1957. The task of contemporary theory. In *New Concepts in Wage Determination*, ed. G. Taylor and F. Pierson. New York: McGraw-Hill.
1958. *Industrial Relation Systems*. New York: Henry Holt & Co.
1984. *Dispute Resolution*. Dover, MA: Auburn House Publishing Co.

Bibliography

Segal, M. 1986. Post-institutionalism in labor economics: the forties and fifties revisited. *Industrial and Labor Relations Review* 39, 388–403.

Du Pont de Nemours, Pierre Samuel (1739–1817)

Economic writer and editor. Born in Paris, he trained for various occupations including medicine and watch-making. A pamphlet on taxation (1763) brought him in contact with Mirabeau and Quesnay, under whose guidance he wrote a work on the grain trade (1764). He also befriended Turgot, with whom he diligently corresponded until Turgot's death. From 1766 to late 1768 he edited the *Journal de l'Agriculture* in the Physiocratic cause, then the *Ephémérides* until 1772. During this period he also published Quesnay's economics under the title *Physiocratie* (Du Pont, 1767) and summarized Mercier (1767), adding material on the history of the new science (Du Pont, 1768). From the early 1770s he developed a career as economic adviser through correspondence with the King of Sweden and the Margrave of Baden; the correspondence with the latter was subsequently published (Knies, 1892). In 1774 he was appointed tutor to the Polish royal family. On becoming *contrôleur-général*, Turgot required his friend's assistance and Du Pont was back in Paris by early 1775. Financial compensation for loss of his royal tutorship enabled him to purchase landed property near Nemours. Turgot's dismissal from office in 1776 did not end Du Pont's career in giving official economic advice; a highlight of which is his influence on the 1786 Anglo–French Commercial Treaty. Du Pont was politically active in the French Revolution, serving from 1789 as Deputy for Nemours in the National Assembly and becoming its President during 1790; in 1794 to 1797 he was imprisoned for short periods. He migrated to the United States in 1799 but returned to Paris in 1802. From 1803 to 1810 he served in the Paris Chamber of Commerce, and in addition edited Turgot's works (Du Pont, 1808–11). In 1815 he returned to the United States and settled in Delaware, the town where his son Irenée had started the gunpowder factory from which the Du Pont chemical conglomerate developed, and where he died in 1817. Du Pont is now mainly remembered as a major propagator of Physiocracy, an early historian of economics, a pioneer in the use of diagrams in economic argument and, most importantly, as the editor of Quesnay and Turgot, whose works he helped to preserve. An assessment of his work as economist needs to take all facets of his career into account, as the one full-length attempt at this (McLain, 1977) has in fact done.

Virtually all Du Pont's economic work is characterized by dogmatic adherence to the Physiocracy developed by Quesnay and codified by Mercier de la Rivière. Turgot criticized this 'servitude to the ideas of the master' as totally inappropriate in matters of science (Schelle, 1913–23, vol. 2, p. 677). Despite such criticism Du Pont allowed his dogmatism to colour excursions into the history of economics (Du Pont, 1769) and, more importantly, his preparation of Turgot's works for the press (see Groenewegen, 1977), particularly his editions of the *Reflections* (Turgot, 1766). Two examples of his more novel contributions to economics can be given. One is his use of diagrams in explaining economic policy, which Theocharis (1961, p. 60) described as the first use of a diagram by a professional economist for

'illustrating an economic argument set out in essentially dynamic time', thereby making Du Pont (1774) 'the earliest French contribution of importance in mathematical economics'. The problem analysed is the price effects of an excise reduction, the benefits of which are argued to accrue ultimately to the landowning class. The excise reduction's initial income effect on manufacturers and merchants allows them either to reduce their own prices or to pay higher prices for raw materials. By assuming this increased competition for raw materials to raise their price in each period by three-fourths of the increase in the preceding period, Du Pont shows how a new equilibrium price will be reached which transfers the benefits from excise reduction to the rural sector. His proof relies on the properties of diminishing geometrical progressions which also formed the basis for much of the analysis of the *Tableau économique*. Du Pont's analysis of the inflationary consequences from issuing assignats is a second example. Although much of this is similar to Turgot's (1749) analysis, some of it is of interest in explaining Smith's version of the specie mechanism to which Du Pont (1790, p. 28) explicitly refers. Issuing paper money by assignats makes silver superfluous as a circulating medium; this drives the metal out of the country because its only other use is to be sold abroad (Du Pont, 1790, p. 42), a specie mechanism like Smith's (1776, pp. 293–4) that is independent of relative price movements. Both examples of his more original economics relate to matters of economic policy and add force to the claim by McLain (1977, p. 255) that Du Pont represents 'the first important case of a professional economist turned government policy-maker, a tradition in which he would be followed by [many] others...'.

PETER GROENEWEGEN

Selected works

1763. *Réflexions sur l'écrit intitulé: Richesses de l'état*. Paris.
1764. *De l'exportation et de l'importation des grains*. Soissons and Paris.
1767. *Physiocratie, ou Constitution Naturelle du Gouvernement le plus avantageux au genre humain*. Leyden and Paris.
1768. *De l'origine et des progrès d'une science nouvelle*. In *Physiocrates*, ed. E. Daire. Paris, 1846.
1769. *Notice abrégée des différents écrits modernes qui ont concours en France à former la science de l'économie politique*. In *Oeuvres Oeconomiques et Philosophiques de François Quesnay*, ed. A. Oncken, Frankfurt and Paris, 1888.
1774. *On Economic Curves*, ed. H.W. Spiegel, Baltimore: Johns Hopkins Reprints of Economic Tracts, 1955.
1790. *The Dangers of Inflation*. Trans. E. E. Lincoln, Boston: Kress Library Publications, 1950.
1808–11. *Oeuvres de Turgot*. 9 vols, Paris.

Bibliography

Groenewegen, P.D. 1977. *The Economics of A.R.J. Turgot*. The Hague: Nijhoff.

Knies, K. 1892. *Carl Friedrichs von Baden Brieflicher Verkehr mit Mirabeau und Du Pont*. Heidelberg.

McLain, J.J. 1977. *The Economic Writings of Du Pont de Nemours*. Newark and London: University of Delaware Press.

Mercier de la Rivière, P.P. 1767. *L'Ordre Naturel et Essentiel des Sociétiés politiques*. Paris.

Schelle, G. 1913–23. *Oeuvres de Turgot et documents le concernant*. 5 vols. Paris: F Alcan.

Smith, A. 1776. *An Inquiry into the Nature and Causes of the Wealth of Nations*, ed. R.H. Campbell and A.S. Skinner, Oxford: Clarendon, 1976.

Theocharis, R.D. 1961. *Early Developments in Mathematical Economics*. London: Macmillan.

Turgot, A.R.J. 1749. *Letter on Paper Money*. In Groenewegen (1977).

Turgot, A.R.J. 1766. *Reflections on the Production and Distribution of Wealth*. In Groenewegen (1977).

Dupuit, Arsene-Jules-Emile Juvenal (1804–1866)

French engineer and economic theorist, born at Fossano, Piedmont, Italy on 18 May 1804, when this region was part of the French empire; died 5 September 1866 in Paris. After his parents returned to Paris in 1814, Dupuit continued his education in the secondary schools at Versailles, at Louis-le-Grand and at Saint-Louis, where he finished brilliantly by winning a physics prize in a large group of competitors. Accepted to the Ecole des Ponts et Chaussées in 1824, Dupuit soon distinguished himself as an engineer and, in 1827, was put in charge of an engineering district in the department of Sarthe, where he concentrated on roadway and navigation work. Dupuit's numerous and trenchant engineering studies on such topics as friction and highway deterioration, floods and hydraulics, and municipal water systems made him one of the most creative civil engineers of his day. Decorated for such contributions by the Legion of Honour in 1843, Dupuit ultimately became director-chief engineer in Paris in 1850 and Inspector-General of the Corps of Civil Engineers in 1855.

No less profound were Dupuit's contributions to general economic analysis and to the economic evaluation of public works (cost–benefit analysis). In fact, Dupuit was the most illustrious contributor in the long French tradition of study, teaching and writing on economic topics at the Ecole des Ponts et Chaussées, whose professors and students included Isnard, Henri Navier, Charles Minard, Emile Cheysson and Charles Ellet.

Led by a desire to evaluate the economic or *net* benefits of public provision, Dupuit directed his considerable analytical gifts to the utility foundation of demand and to its relevance to the welfare benefits of public works. In three substantial papers appearing in the *Annales des Ponts et Chaussées* (1844; 1849) and the *Journal des économistes* (1853), Dupuit became the first non-adventitious expositor of the theory of *marginal* utility, of (a variant of) marginal cost pricing, of simple and discriminating monopoly theory, and of pricing principles of the firm where location is a factor in expressing demand.

The font of Dupuit's contribution is the construction of a marginal utility curve and the identification of it with the demand curve or *courbe de consommation* (see Figure 1).

Arguing in the manner of Carl Menger, who later elaborated on the point, Dupuit showed that the marginal utility that an individual obtained from a homogeneous stock of goods is determined by the use to which the last units of the stock are put. In doing so, he clearly pointed out that the marginal utility of a stock or some particular good diminishes with increases in quantity and that each consumer attaches a different marginal utility to the same good according to the quantity consumed. The importance of Dupuit's invention rests in the fact that the psychological concept of diminishing marginal utility, and its ramifications, were carried over to the law of demand. With some, but not all, of the reservations and qualifications of Alfred Marshall, Dupuit *identified* the marginal utility curve with the demand curve, adding up the utility curves of individuals to obtain the market demand curve. Dupuit (1844, p. 106) described his construction (see Figure 1), which applied to all goods, public and private, as follows:

> If … along a line *Op* the lengths *Op*, *Op'*, *Op''* … represent various prices for an article, and that … *pn*, *p'n'*, *p''n''* … represent the number of articles consumed corresponding to these prices, then it is possible to construct a curve *Nn'n''P* which we shall call the curve of consumption. *ON* represents the quantity consumed when the price is zero, and *OP* the price at which consumption falls to zero.

The identification of marginal utility and demand, of course, sets up the demand curve as a welfare tool and Dupuit made specific calculations. A measure of the welfare produced by the good (*utilité absolue*) at quantity *Or* is the definite integral of the demand curve between *O* and *r*. Given that *Op* is the (average) cost of producing quantity *Or*, consumers earn a surplus (*utilité relative*) equal to absolute utility (*OrnP*) less costs of production (*Ornp*). (Relative utility (*pnP*) is none other than Marshall's consumers' surplus without all the reservations that Marshall attached to the concept.) Importantly, Dupuit identified area *rNn* as lost utility (*utilité perdue*). Under competitive conditions this loss was inevitable due to the opportunity cost of resources. Under a monopoly structure, for example, if, in Figure 1, *Op* were a monopoly price with zero production costs assumed, *utilité*

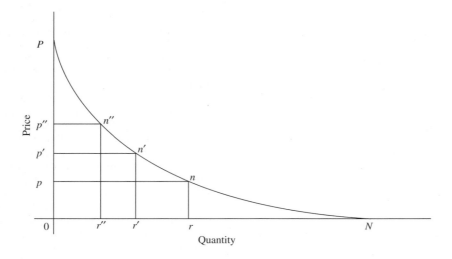

Figure 1

perdue would be a loss to society – the 'deadweight' loss associated with excise taxes, tariffs or monopoly. Further, Dupuit advanced the theorem that the loss in utility was proportional to the *square* of the tax of price above marginal cost. This theorem, with attendant analysis, formed the base for large areas of neoclassical welfare economics, including the taxation studies of F.Y. Edgeworth and the marginal cost pricing argument of Harold Hotelling.

From this theoretical base, Dupuit investigated an impressive number of pricing systems and market models (1849). While Dupuit was an ardent and stubborn defender of laissez faire in most markets (1861), he was equally concerned that public works, provided or regulated by government as a *last* resort, should produce the maximum amount of utility possible. Thus tools such as marginal cost pricing find their theoretical foundations in the writings of Dupuit. Although Dupuit did not provide an explicit formulation of the principle, one of his bridge pricing examples and other statements strongly suggest the possibilities of such a technique to maximize welfare, but as a *long-run* proposition.

Dupuit analysed, independently of Cournot, who was apparently unknown to him, the profit-maximizing behaviour of the simple monopolist. He saw monopoly at the apex of a range of problems regarding the production of total welfare, being unconcerned about the 'distribution' of welfare between producers and consumers. His point was that the amount of 'absolute utility' (or what could be called net benefit) was lessened by monopoly profit maximization. This led him to defend the private practice of price discrimination and to produce an economic theory of discrimination. Price discrimination could exist, in Dupuit's view, with differences in 'buyer estimates', with the ability to segment markets either naturally or artificially, and with some degree of monopoly power. The motive was profit

maximization, and although Dupuit discussed the effects of discrimination on price and revenue, he was primarily interested in the fact, as was Joan Robinson later, that discrimination could affect the size of the welfare benefit. This view was expanded to include the impact of price discrimination of welfare when buyers were spatially distributed (1849; 1854).

In the matter of policy, Dupuit recommended that tools be carefully fit to specific problems. If industries were to be collectivized or regulated by government, Dupuit proposed the maximization of net benefit under the constraint of covering total costs of production. The recovery of total cost might be achieved through regulated or constrained price discrimination or through a cost-based single price technique. However, Dupuit can hardly be credited with espousing an enlarged role for government or government intervention. A firm adherent of Smith's dictums concerning minimal government, Dupuit believed that free and open competition, along with vigorous antitrust or anticartel enforcement, would ensure optimal provisions in most cases, including transportation. Indeed, in the process of analysing the welfare principles of public works pricing, Dupuit discovered (in an uncommonly complete manner) some of the critical welfare-maximizing properties of a generalized competitive system.

ROBERT B. EKELUND, JR.

See also **consumer surplus; public utility pricing and finance.**

Selected works

1844. On the measurement of the utility of public works. Trans. R.H. Barback from the *Annales des Ponts et Chaussées*, in *International Economic Papers*, No. 2, London: Macmillan, 1952.

1849. On tolls and transport charges. Trans. E. Henderson from the *Annales des Ponts et Chaussées,* in *International Economic Papers,* No. 11, London: Macmillan, 1962.

1853. On utility and its measure – on public utility. *Journal des économistes* 36, 1–27.

1854. Péages. In *Dictionnaire de l'Economie Politique,* vol. 2. Paris: Guillaumin.

1861. *La Liberté Commerciale.* Paris: Guillaumin.

1934. *De l'Utilité et sa Mésure: écrits choisis et republiés,* ed. M. de Bernardi. Turin: La Riforma Sociale.

Bibliography

Ekelund, R.B., Jr. 1968. Jules Dupuit and the early theory of marginal cost pricing. *Journal of Political Economy* 76, 462–71.

Ekelund, R.B., Jr. 1970. Price discrimination and product differentiation in economic theory: an early analysis. *Quarterly Journal of Economics* 84, 268–78.

Ekelund, R.B., Jr. and Yeung-Nan Shieh. 1986. Dupuit, spatial economics, and optimal resource allocation: a French tradition. *Economica* 53, 483–96.

durable goods markets and aftermarkets

Durable goods are goods whose useful lifetime spans multiple periods.

This article surveys the extensive literature on durable-goods markets and aftermarkets. I begin with the main theoretical ideas, then turn to specific real-world issues such as durability choice and leasing, discuss aftermarket monopolization and then end with a brief conclusion. (A more in-depth survey appears in Waldman, 2003.)

Three theoretical building blocks

Much of our understanding of durable-goods markets derives from three theoretical contributions. The first is Coase's (1972) insight concerning time inconsistency. To see the basic logic, consider Bulow's (1982) formalization: a durable-goods monopolist sells its output in each of two periods and cannot commit in the first period to second-period actions. Bulow shows that, because in the second period the firm does not internalize how its actions affect the value of used units, its output is higher than under commitment. First-period purchasers anticipate this, pay less for new units and thus lower overall monopoly profitability.

Coase's insight has spawned a large literature. One branch of this literature focuses on the Coase conjecture, that is, the idea that in an infinite-period setting time inconsistency causes price to drop immediately to marginal cost. A second branch identifies tactics such as leasing that firms can employ to reduce or possibly avoid

time inconsistency. Finally, a third branch applies time inconsistency to other issues, including new-product introductions and repurchase prices.

The second major theoretical contribution is Akerlof's (1970) adverse-selection argument. This paper helped start the asymmetric-information revolution, but was not initially thought of as an important contribution to durable-goods theory. However, the paper's main example concerns second-hand markets. In Akerlof's model buyers have higher valuations than sellers, so efficiency requires that all units be traded. Further, each seller is privately informed of his own unit's quality. The result is a single price that reflects average quality, and sellers with high-quality units keep them because prices do not reflect actual quality, that is, trade is below the efficient level. (In Akerlof's analysis there is no trade, but this result is not robust.)

A small empirical literature looks for evidence of adverse selection in durable-goods markets. Most of these papers find some support. For example, Bond (1982) considers the used pickup truck market and finds support for adverse selection for older trucks, while Genesove (1993) finds some supporting evidence in used-car dealer auctions. More recently, Gilligan (2004) finds supporting evidence in business aircraft.

In terms of durable-goods theory, Akerlof's contribution was ignored for almost 30 years. Starting with Hendel and Lizzeri (1999a), however, a number of papers have extended Akerlof's analysis. There are three basic findings. First, Akerlof's main results continue to hold when new units are incorporated into the analysis. Second, because adverse selection in the used-unit market reduces the willingness to pay of new-unit buyers, firms will market new units in a manner that reduces adverse selection. Third, as discussed in detail later, new-unit leasing can be important for reducing adverse selection.

The third major theoretical contribution is that there is a close analogy between the product-line pricing problem and the durable-goods monopoly problem. This analogy is described in Waldman (1996). Consider Mussa and Rosen (1978), which analyses the product-line pricing problem of a non-durable-goods monopolist. The monopolist sells units of varying qualities to consumers who have heterogeneous valuations on quality. Because the substitutability between units links the various prices, the monopolist lowers below efficient levels the quality level sold to all but the highest-valuation group.

Now consider a durable-goods monopolist who controls the quality of a unit at every age. Further, assume heterogeneity in consumers' valuations for quality and a frictionless second-hand market. Then, if the firm can commit, quality choices are as above. That is, new-unit quality is efficient. But, because of the linkages between the various prices, all used-unit qualities are below efficient levels. As discussed later, a number of recent papers use this result to analyse various real-world issues concerning durable goods.

Three real-world issues
Optimal durability choice

A much debated issue is whether a durable-goods monopolist chooses socially optimal durability. Swan (1970; 1971) considers models that satisfy the once standard assumption that a unit is a bundle of 'service units', so some number of used units is a perfect substitute for a new unit. Swan's steady-state analysis shows durability choice to be socially optimal because the firm produces the steady-state flow of service units at minimum cost. (Swan's analysis corrected the conclusions of earlier papers that had concluded that in such settings the monopolist would choose inefficiently low durability levels.)

A large literature investigates the robustness of Swan's conclusions. There are two major findings. The first employs time inconsistency. Bulow (1986) moves away from Swan's assumption of steady-state behaviour by considering a model similar to his earlier one, but now allows endogenous durability choice. He shows that time inconsistency provides a rationale for a durable-goods monopolist to choose less than the socially-optimal durability level. The logic is that durability is what leads to time inconsistency, so reducing durability below the efficient level reduces time inconsistency and thus increases profitability.

The second major finding appears in Waldman (1996) and Hendel and Lizzeri (1999b), which drop the service units assumption and instead assume that new and used units vary in quality and that durability choice controls the speed of quality deterioration. The earlier discussion immediately translates into an incentive for the firm to choose less than the socially optimal durability level. That is, in this setting the incentive for the monopolist to sell output whose used-unit quality is below the efficient level translates into durability below the efficient level. (In Hendel and Lizzeri's analysis durability choice can be above, below, or equal to the first-best level, but it is always below the second-best level defined by actual outputs.)

Eliminating second-hand markets

Do durable-goods producers with market power have incentives to eliminate second-hand markets? For example, do textbook publishers introduce new editions in order to kill off the market for used books? Until recently, the standard argument, found, for example, in Swan (1980), was that, since the new-unit price reflects prices the product will sell for on the second-hand market in subsequent periods, the producer has no such incentive.

Two recent arguments show that this result is, in fact, quite limited. The first, which builds on the discussion above, appears in Waldman (1996; 1997) and Hendel and Lizzeri (1999b). The idea is that, because substitutability between new and used units means the price of a used unit on the second-hand market limits the amount the firm can charge for new units, the firm sometimes eliminates the second-hand market or similarly reduces used-unit

availability in order to raise the new-unit price. In particular, this is more likely when consumers of used units have low valuations for the firm's product. This is both because little revenue is lost by not serving such consumers and because serving them means a low used-unit price and thus a lower new-unit price. (A number of earlier papers find similar results starting with demand functions rather than utility maximization.)

The second argument, found initially in Waldman (1993), employs time inconsistency. As discussed, the early literature on time inconsistency focused on output choice. My 1993 paper shows time inconsistency also applies to actions such as new-product introductions that make used units unavailable because they become obsolete. The difference between this argument and the one above concerns commitment. Above it is assumed the firm can commit, so the firm eliminates the second-hand market only when it is profitable to do so. In contrast, here commitment is not assumed, so the firm may eliminate the second-hand market even though this lowers overall profitability.

A related empirical analysis appears in Iizuka (2004), which shows that the market share of used textbooks is an important determinant of whether or not a publisher introduces a new edition. This is consistent with new editions being used at least partly to eliminate second-hand markets, although Iizuka does not distinguish between the two possibilities described above for why a firm might want to do this. In future research, it might be possible to identify which argument is at work by focusing on how the decision to introduce new editions affects overall profitability.

Reasons for leasing

A number of reasons have been identified for why durable-goods producers frequently lease. (A reason I do not discuss is that there are sometimes tax advantages associated with leasing.) One reason, initially discussed in Coase (1972) and Bulow (1982), is that time inconsistency lowers profitability when a firm sells output because it chooses actions in later periods that inefficiently lower the value of used units. When the firm leases, however, it retains ownership of those units so the incentive to take inefficient actions disappears.

A second reason is also related to a previous discussion. As discussed in Waldman (1997) and Hendel and Lizzeri (1999b), when used-unit prices serve as important constraints on the new-unit price, leasing can be used to eliminate second-hand markets or at least reduce used-unit availability. The logic is that leasing allows a firm to eliminate the second-hand market by allowing the firm to retire returned used units. My 1997 paper shows this formally and argues that it is consistent with classic cases concerning the use of a lease-only policy such as United Shoe in the shoe machinery market, IBM in the computer market, and Xerox in the copier market. (One might argue that leasing is not needed because a firm can sell and then use high repurchase prices to purchase and retire used units. My 1997 paper shows this strategy is inferior to leasing because of time inconsistency.)

Finally, leasing is a response to adverse selection. This argument appears in Hendel and Lizzeri (2002) and Johnson and Waldman (2003). These papers show that, whether the new-unit market is monopolistic or competitive, in a world of asymmetric information leasing in the new-unit market can arise because it means used units are returned to the seller(s), which, in turn, avoids or at least reduces adverse selection in the used-unit market. The two papers develop different variants of the argument and show it is consistent with various empirical findings concerning the automobile market.

Aftermarket monopolization

Aftermarket monopolization is behaviour that stops alternative producers from selling aftermarket products to the firm's customers. The focus on this subject started after the US Supreme Court's 1992 decision in the case *Eastman Kodak Company v. Image Technical Services*. Aftermarkets are common with durable goods, where aftermarkets refer to markets for complementary products such as maintenance and upgrades. I consider three possibilities: (*a*) hold-up rationales; (*b*) price discrimination and efficiency rationales; and (*c*) other strategic rationales.

Hold-up

There are two distinct hold-up arguments, each of which focuses on aftermarket monopolization by competitive producers. In both, the firm prohibits other firms from selling the aftermarket product – for example, maintenance – and then exploits the locked-in positions of its customers in pricing the product. The result is a standard deadweight loss due to the high aftermarket price, although no transfer between the consumers and the firm since competition in the primary market means firms earn zero profits overall. In the 'costly-information' version, consumers ignore the aftermarket price when purchasing the primary product. In the 'lack-of-commitment' version, developed in Borenstein, Mackie-Mason and Netz (1995), consumers correctly anticipate the aftermarket price but, because firms cannot commit, time inconsistency causes firms to monopolize the aftermarket and inefficiently raise the aftermarket price after consumers are locked in. (A third hold-up theory is the 'surprise' theory. In this argument consumers are surprised by the aftermarket monopolization. Some discussions of this theory describe a transfer between the consumers and the firm, but it is unclear why competition does not result in zero profits, in which case the surprise and costly-information theories are equivalent.)

Price discrimination and efficiency rationales

For various reasons, such as that many buyers in the relevant industries are sophisticated firms for which the costly-information argument is implausible, attention has shifted towards other arguments many of which have either neutral or positive social-welfare implications.

One such argument is the price discrimination argument that appears in Chen and Ross (1993) and Klein (1993). Suppose the primary-good producer has market power. Then the firm may monopolize the aftermarket in order to raise the aftermarket price and in this way price discriminate by charging a high aggregate price to the high-volume/high-valuation consumers. From a social-welfare standpoint, this argument has neutral implications since an improved ability to price discriminate can either raise or lower social welfare. (Klein argues that this argument applies even when firms are competitive, although not perfectly competitive.)

A plausible efficiency rationale follows from Schmalensee's (1974) argument that, given a durable-goods monopolist (which means new units priced above marginal cost) and a competitive maintenance market (which means maintenance is priced at cost), consumers will sometimes inefficiently maintain rather than replace used units. Tirole (1988) shows this can lead to aftermarket monopolization in a durable-goods monopoly setting because having a monopoly in both markets allows the firm to avoid the inefficiency and thus increases its profits.

More recently, Morita and Waldman (2005) and Carlton and Waldman (2006) show that the argument extends to aftermarkets other than maintenance, and to competitive durable-goods markets given switching costs. In the latter case the inefficient substitution problem arises even with competition because switching costs create market power at the time of the maintenance/replacement decision. Interestingly, because competitive sellers earn zero profits in equilibrium, when aftermarket monopolization eliminates the distortion, both social welfare and consumer welfare increase.

Strategic rationales

There is an extensive literature on strategic rationales for the tying of complementary products. Since the tying of primary and aftermarket products is one potential way to achieve aftermarket monopolization, much of this literature is relevant to aftermarket monopolization.

Whinston (1990) shows that, if the primary good is not essential, tying may force the exit of an alternative producer of the complementary good and in this way increase the firm's profits by monopolizing the segment of the complementary-good market for which the primary good is not required.

In contrast, in Carlton and Waldman (2002) tying is sometimes used to preserve a monopoly in the primary-good market. They consider two-period settings in which a single potential entrant can enter the complementary market in either period but the primary market only in the second. In the presence of fixed costs of entry or network externalities, the primary-good monopolist sometimes ties in order to preserve its primary-good monopoly in the second period. For example, with entry costs tying stops the alternative producer from entering the complementary market in the first period. In turn, because of a possible inability to cover entry costs, the outcome can be no entry in either market in either period.

A third argument appears in Carlton and Waldman (2005). Whinston shows that in one-period settings there is never an incentive to tie if the monopolist's primary product is essential. Carlton and I show that in durable-goods settings, given the presence of complementary-good upgrades and switching costs, tying can be optimal even when the primary product is essential. The basic logic is that some profits are realized in later periods in the sale or lease of the upgraded complementary good, and the only way the monopolist can ensure it captures those profits is by tying and becoming the sole producer of the complementary good.

Conclusion

Starting in the early 1970s with the work of Akerlof, Coase and Swan, significant progress has been made in our understanding of durable-goods markets. In this entry I have surveyed this literature as well as the literature on the related issue of aftermarkets. Although I have referred throughout to various empirical papers, durable-goods markets is a topic for which theory is far ahead of empirical investigation. In the future I expect to see work that extends the theory in various important ways, but also empirical work that tests the validity of the various theoretical approaches that have been explored since the early 1970s.

MICHAEL WALDMAN

See also **adverse selection; Akerlof, George Arthur; bundling and tying; Coase, Ronald Harry; resale markets.**

Bibliography

Akerlof, G. 1970. The market for 'lemons': quality uncertainty and the market mechanism. *Quarterly Journal of Economics* 84, 488–500.

Bond, E. 1982. A direct test of the 'lemons' model: the market for used pick-up trucks. *American Economic Review* 72, 836–40.

Borenstein, S., Mackie-Mason, J. and Netz, J. 1995. Antitrust policy in aftermarkets. *Antitrust Law Journal* 63, 455–82.

Bulow, J. 1982. Durable goods monopolists. *Journal of Political Economy* 90, 314–32.

Bulow, J. 1986. An economic theory of planned obsolescence. *Quarterly Journal of Economics* 101, 729–49.

Carlton, D. and Waldman, M. 2002. The strategic use of tying to preserve and create market power in evolving industries. *RAND Journal of Economics* 33, 194–220.

Carlton, D. and Waldman, M. 2005. Tying, upgrades, and switching costs in durable goods markets. Mimeo, Cornell University.

Carlton, D. and Waldman, M. 2006. Competition, monopoly, and aftermarkets. Mimeo, Cornell University.

Chen, Z. and Ross, T. 1993. Refusals to deal, price discrimination and independent service organizations. *Journal of Economics and Management Strategy* 2, 593–614.

Coase, R. 1972. Durability and monopoly. *Journal of Law and Economics* 15, 143–9.

Genesove, D. 1993. Adverse selection in the wholesale used car market. *Journal of Political Economy* 101, 644–65.

Gilligan, T. 2004. Lemons and leases in the used business aircraft market. *Journal of Political Economy* 112, 1157–80.

Hendel, I. and Lizzeri, A. 1999a. Adverse selection in durable goods markets. *American Economic Review* 89, 1097–115.

Hendel, I. and Lizzeri, A. 1999b. Interfering with secondary markets. *RAND Journal of Economics* 30, 1–21.

Hendel, I. and Lizzeri, A. 2002. The role of leasing under adverse selection. *Journal of Political Economy* 110, 113–43.

Iizuka, T. 2004. An empirical analysis of planned obsolescence. Mimeo, Vanderbilt University.

Johnson, J. and Waldman, M. 2003. Leasing, lemons, and buy-backs. *RAND Journal of Economics* 34, 247–65.

Klein, B. 1993. Market power in antitrust: economic analysis after *Kodak*. *Supreme Court Economic Review* 3, 43–92.

Morita, H. and Waldman, M. 2005. Competition, monopoly maintenance, and consumer switching costs. Mimeo, Cornell University.

Mussa, M. and Rosen, S. 1978. Monopoly and product quality. *Journal of Economic Theory* 18, 301–17.

Schmalensee, R. 1974. Market structure, durability, and maintenance effort. *Review of Economic Studies* 41, 277–87.

Swan, P. 1970. Durability of consumption goods. *American Economic Review* 60, 884–94.

Swan, P. 1971. The durability of goods and regulation of monopoly. *Bell Journal of Economics* 2, 347–57.

Swan, P. 1980. Alcoa: the influence of recycling on monopoly power. *Journal of Political Economy* 88, 76–99.

Tirole, J. 1988. *The Theory of Industrial Organization.* Cambridge, MA: MIT Press.

Waldman, M. 1993. A new perspective on planned obsolescence. *Quarterly Journal of Economics* 108, 273–83.

Waldman, M. 1996. Durable goods pricing when quality matters. *Journal of Business* 69, 489–510.

Waldman, M. 1997. Eliminating the market for secondhand goods. *Journal of Law and Economics* 40, 61–92.

Waldman, M. 2003. Durable goods theory for real world markets. *Journal of Economic Perspectives* 17(1), 131–54.

Whinston, M. 1990. Tying, foreclosure, and exclusion. *American Economic Review* 80, 837–59.

Durbin-Watson statistic

The well-known Durbin–Watson, or DW, statistic, which was proposed by Durbin and Watson (1950; 1951), is used for testing the null hypothesis that the error terms of a linear regression model are serially independent.

Consider the linear regression model with AR(1) errors,

$$y_t = X_t b + u_t, \quad u_t = \rho u_{t-1} + \varepsilon_t,$$
$$\varepsilon_t \sim \text{IID}(0, \sigma^2). \tag{1}$$

Here the scalar y_t is an observation on a dependent variable, X_t is a $1 \times k$ vector of observations on independent variables that may be treated as fixed, and β is a k-vector of parameters to be estimated. There are n observations, and we wish to test the null hypothesis that $\rho = 0$, under which the model (1) reduces to

$$y_t = X_t b + u_t, \quad u_t \sim \text{IID}(0, \sigma^2), \tag{2}$$

for which the ordinary least squares (OLS) estimator is efficient. This estimator is usually written as $\hat{b} = (X'X)^{-1}X'y$, where the $n \times k$ matrix X has t^{th} row X_t and the n-vector y has t^{th} element y_t.

The Durbin–Watson d statistic for testing (2) against (1) is solely a function of the OLS residuals $\hat{u}_t = y_t - X_t\hat{b}$. It is defined as

$$d = \frac{\sum_{t=2}^{n}(\hat{u}_t - \hat{u}_{t-1})^2}{\sum_{t=1}^{n}\hat{u}_t^2}. \tag{3}$$

It is easy to see that d is approximately equal to $2 - 2\hat{\rho}$, where $\hat{\rho}$ is the OLS estimate of ρ in a regression of $\hat{u}_t$ on $\hat{u}_{t-1}$. Thus d will be approximately equal to 2 if the residuals do not display any serial correlation, and it will be less (greater) than 2 whenever $\hat{\rho}$ is more than a little bit greater (less) than 0.

The exact distribution of the DW statistic

The DW statistic can be written as a ratio of quadratic forms in the n-vector $\hat{u}$ of OLS residuals, the tth element of which is $\hat{u}_t$. Specifically,

$$d = \frac{\hat{u}'A\hat{u}}{\hat{u}'\hat{u}}, \tag{4}$$

where A is the $n \times n$ matrix

$$\frac{1}{2}\begin{bmatrix} 1 & -1 & 0 & 0 & \cdots & 0 & 0 & 0 \\ -1 & 2 & -1 & 0 & \cdots & 0 & 0 & 0 \\ 0 & -1 & 2 & -1 & \cdots & 0 & 0 & 0 \\ \vdots & \vdots & \vdots & \vdots & & \vdots & \vdots & \vdots \\ 0 & 0 & 0 & 0 & \cdots & -1 & 2 & -1 \\ 0 & 0 & 0 & 0 & \cdots & 0 & -1 & 1 \end{bmatrix}$$

Durbin and Watson (1950) actually considered a number of statistics that can be written in the form of (4) for different choices of the matrix A and chose to focus on d for reasons of computational and theoretical convenience. Because both the numerator and the denominator are proportional to σ^2, d is invariant to σ.

The exact distribution of d depends on X and the distribution of the u_t. When the error terms are i.i.d. normal, Durbin and Watson (1951) tabulated bounds on the critical values for tests based on d against the one-sided alternative that $\rho > 0$. These bounds, denoted d_L and d_U, depend on the sample size and the number of regressors. We can reject the null hypothesis when $d < d_L$, cannot reject it when $d < d_U$, and can draw no firm conclusion when $d_L < d < d_U$. To test against the alternative that $\rho < 0$, we would replace d by $4 - d$ and use the same procedure.

The original Durbin–Watson tables have been extended by various authors, notably Savin and White (1977). However, since $d_U - d_L$ can be quite large, tests based on the bounds often have indeterminate outcomes. It is much better to perform exact tests conditional on X, and this is easy to do with modern computing technology. There are two approaches.

The first approach is to calculate an exact P value for d using one of several methods for calculating the distribution of a ratio of quadratic forms in normal random variables. The method of Imhof (1961) is probably the best known of these, but the more recent method of Ansley, Kohn and Shively (1992) is faster. If a suitable computer program is readily available, this approach is the best one.

An alternative approach is to perform a Monte Carlo test. As can be seen from (4), the statistic d depends only on the vector u and the matrix X, since $\hat{u} = M_X u$, where $M_X = I - X(X'X)^{-1}X'$. Because of its invariance to σ, d does not depend on any unknown parameters. This implies that a Monte Carlo test will be exact.

To perform a Monte Carlo test at level α, we first choose B such that $\alpha(B + 1)$ is an integer (999 is often a reasonable choice) and generate B vectors u_j^*, each of which is multivariate standard normal. Each of the u_j^* is regressed on X to calculate a vector of residuals $M_X u_j^*$, which is then used to compute a simulated test statistic d_j^* according to (3). We can then calculate simulated P values for a one-tailed test against either $\rho > 0$ or $\rho < 0$ or for a two-tailed test. For example, the simulated P value for a one-tailed test against $\rho > 0$ is

$$P^*(d) = \frac{1}{B} \sum_{j=1}^{B} I(d_j^* < d),$$

where $I(\cdot)$ is the indicator function that is equal to 1 when its argument is true and equal to 0 otherwise. We reject the null hypothesis whenever $P^*(d) < \alpha$. For more on the calculation of P values for bootstrap and Monte Carlo tests, see Davidson and MacKinnon (2006).

Limitations of the DW statistic

The Durbin–Watson statistic is valid only when all the regressors can be treated as fixed. It is not valid, even asymptotically, when X_t includes a lagged dependent variable or any variable that depends on lagged values of y_t. Because $\hat{\rho}$ is biased towards 0 when X_t includes a lagged dependent variable, d is biased towards 2 in this case. Thus, a test based on the DW statistic will tend to under-reject when the null hypothesis is false.

Numerous procedures have been proposed for testing for serial correlation in models that include lagged dependent variables. The simplest is to rerun regression (2), with the addition of the lagged residuals from that regression. The test statistic is then the t statistic on the lagged residuals. This procedure, which is due to Durbin (1970) and Godfrey (1978), does not yield an exact test and should be bootstrapped when the sample size is small.

Of course, since the finite-sample distribution of the DW statistic depends on the distribution of the u_t, we cannot expect to obtain an exact test even when the X_t are exogenous if the normality assumption is not a good one. In principle, we could bootstrap d by using re-sampled residuals instead of multivariate standard normal vectors for the u_j^*. This would probably work very well in most cases, but it would not actually yield an exact test.

JAMES G. MACKINNON

See also **artificial regressions; serial correlation and serial dependence; testing.**

Bibliography

Ansley, C., Kohn, R. and Shively, T. 1992. Computing *p*-values for the generalized Durbin–Watson statistic and other invariant test statistics. *Journal of Econometrics* 54, 277–300.
Davidson, R. and MacKinnon, J. 2006. Bootstrap methods in econometrics. In *Palgrave Handbooks of Econometrics: Volume 1 Econometric Theory*, ed. T. Mills and K. Patterson. Basingstoke: Palgrave Macmillan.
Durbin, J. 1970. Testing for serial correlation in least-squares regression when some of the regressors are lagged dependent variables. *Econometrica* 38, 410–21.
Durbin, J. and Watson, G. 1950. Testing for serial correlation in least squares regression I. *Biometrika* 37, 409–28.
Durbin, J. and Watson, G. 1951. Testing for serial correlation in least squares regression II. *Biometrika* 38, 159–77.
Godfrey, L. 1978. Testing against general autoregressive and moving average error models when the regressors include lagged dependent variables. *Econometrica* 46, 1293–301.
Imhof, J. 1961. Computing the distribution of quadratic forms in normal variables. *Biometrika* 48, 419–26.
Savin, N. and White, K. 1977. The Durbin–Watson test for serial correlation with extreme sample sizes or many regressors. *Econometrica* 45, 1989–96.

Dutch Disease. *See* **de-industrialization, 'premature' de-industrialization and the Dutch Disease.**

dynamic models with non-clearing markets

This article studies a new class of models which synthesize the two traditions of general equilibrium with non-clearing markets and imperfect competition on the one hand, and dynamic stochastic general equilibrium (DSGE) models on the other hand. Although this line of models is still recent, it has clearly become in a short time a central paradigm of modern macroeconomics. The reasons are at least threefold.

The first is that it displays solid microeconomic foundations. This is quite natural since from the two constituent fields above this one inherited a strong general equilibrium framework where all agents (households or firms) maximize their respective objectives subject to well defined constraints.

The second is that it is a highly synthetic theory, which combines in a unified framework general equilibrium, non-clearing markets, imperfect competition, growth theory and rational expectations, so that it can appeal to macroeconomists with very different backgrounds.

The third reason is empirical. A key motivation for DSGE models is to compare the 'statistics' generated by these models with the real-world ones. In that respect the addition of non-clearing markets and imperfect competition has led to substantial progress in matching these statistics, and this has certainly been an important factor in the success of these models.

Now such a wide synthesis did not come all at once. So we begin by recalling briefly a little bit of history and some of the antecedents of the field.

We then present a series of models with explicit solutions. These will demonstrate analytically how the introduction of non-clearing markets allows us to substantially improve the ability of DSGE to reproduce a number of macroeconomic facts.

History
Early times
At the time when many of the developments leading to these models were initiated, there was a profound split between microeconomics and macroeconomics. On the one hand microeconomics, in its general equilibrium version, was dominated by Walras's (1874) model, as developed by Arrow and Debreu (1954), Arrow (1963), and Debreu (1959). In these models all adjustments are carried out via fully flexible prices, and agents never experience any quantity constraint. On the other hand in the standard macroeconomic model in the Keynes (1936) and Hicks (1937) tradition, as exemplified by the IS–LM model, there are price and wage rigidities, unemployment is present and most adjustments are carried out through variations in real income, a quantity, not a price.

Confronted with this inconsistency, the strategies of macroeconomists turned out to be quite diverse and they took two different routes.

General equilibrium with non-clearing markets
On the one hand, a first set of authors aimed at achieving a synthesis between the then existing microeconomics and macroeconomics. This was achieved by generalizing the traditional general equilibrium model, by introducing non-clearing markets, introducing quantity signals into demand and supply functions, and endogenizing prices in a framework of imperfect competition.

Patinkin (1956) and Clower (1965) showed that the presence of quantity constraints in non-clearing markets would drastically modify the demands for labour and goods, an insight further emphasized by Leijonhufvud (1968). Barro and Grossman (1971; 1976) combined these insights into a fixprice macromodel. Drèze (1975) and Bénassy (1975; 1982) constructed full general equilibrium concepts with price rigidities, where price movements are partially replaced by endogenous quantity constraints. Bénassy (1976) linked these concepts with general equilibrium under imperfect competition à la Negishi (1961). This link was furthered with the construction of a full general equilibrium concept of objective demand curve based on quantity constraints (Bénassy, 1988; see also Gabszewicz and Vial, 1972, for a Cournotian view). All these developments are reviewed in the dictionary entry 'non clearing markets in general equilibrium'.

Dynamic market clearing macroeconomics
A second set of authors achieved consistency between microeconomics and macroeconomics by importing into macroeconomics the basic assumption of the then dominant general equilibrium microeconomic models, market clearing. At the same time they paid strong attention to the issues of dynamics and expectations. A central part of these developments was the use of 'rational expectations' in the sense of Muth (1961). This was an important addition, as in the Keynesian system it was sometimes difficult to disentangle the results due to price or wage rigidity from those due to incorrect expectations. Rational expectations allowed the suppression of the second type of results. It appeared also that, even with rational expectations and market clearing, it was possible to build rigorous models displaying fluctuations (Lucas, 1972; Kydland and Prescott, 1982; Long and Plosser, 1983).

Non-Walrasian cycles
Starting in the mid-1980s authors began combining elements of the two paradigms described above, achieving the synthesis that is the subject of this article. Svensson (1986) studies a dynamic stochastic general equilibrium monetary economy subject to supply and demand shocks. Prices are preset one period in advance by monopolistically competitive firms, so we have both imperfect competition and sticky prices. Because of price presetting the model has multiple regimes.

Various types of rigidities have been then introduced in dynamic models, leading to different patterns of cycles. Andersen (1994) reviews various causes and consequences of price and wage rigidities.

A first type of rigidities is 'real' rigidities, which create an endogenous non-competitive wedge between various prices. As an example, monopolistic competition à la Dixit and Stiglitz (1977) introduces a markup between marginal cost and price. In this class Danthine and Donaldson (1990) introduce efficiency wages, Danthine and Donaldson (1991; 1992) introduce implicit contracts in the vein of Azariadis (1975), Baily (1974) and Gordon (1974). Rotemberg and Woodford (1992; 1995) study imperfect competition.

Models with nominal rigidities study situations where the nominal prices themselves (and not relative prices) are sluggish. Several devices have been used. The first, following the early works on wage and price contracts by Gray (1976), Fischer (1977), Phelps and Taylor (1977), Taylor (1979; 1980) and Calvo (1983), assumes that there is a system of contracts expiring at deterministic or stochastic dates. For that reason they are called 'time dependent'. Such contracts have been integrated in DSGE models by Cho (1993), Cho and Cooley (1995), Bénassy (1995; 2002; 2003a; 2003b), Yun (1996), Cho, Cooley and Phaneuf (1997), Andersen (1998), Jeanne (1998), Ascari (2000), Chari, Kehoe and McGrattan (2000), Collard and Ertz (2000), Ascari and Rankin (2002), Huang and Liu (2002), Smets and Wouters (2003) and Christiano, Eichenbaum and Evans (2005), to name only a few.

Another type of price rigidity, called 'state dependent', is based on costs of changing prices. Two specifications are favourite in the literature: quadratic costs of changing prices (Rotemberg, 1982a; 1982b), which have been implemented, for example, in Hairault and Portier (1993), and fixed costs of changing prices (Barro, 1972), often renamed 'menu costs'. Clearly these costs should be interpreted as surrogates for other unspecified causes, and identifying these causes is a challenge that faces this line of research.

Now most of the contributions of this field are based on numerical evaluations of various models. So we present next a number of models with explicit solutions which will make clear why this line of models has been successful in solving problems that were difficult to solve in market-clearing models.

An analytical illustration

We shall now show in this section in a series of explicitly solved models how the introduction of nominal rigidities in DSGE models allows to considerably improve the capacities of these models to reproduce the dynamic evolutions of actual economies.

We first present a basic model and compute as a reference its Walrasian equilibrium and dynamics. Then we introduce a first nominal rigidity, one-period wage contracts. This improves some correlations, but cannot create strong persistence as in reality. We next introduce multi-periodic wage contracts, and show that this allows us to obtain a persistent response of output to demand shocks. Finally, simultaneous rigidities of wages and prices are considered, and we show that one can obtain in this way with fairly realistic values of the parameters a persistent and hump-shaped response of both output and inflation.

The basic model

We study a dynamic monetary economy à la Sidrauski (1967) and Brock (1975), where goods are exchanged against money at the (average) price P_t and work against money at the (average) wage W_t. There are two types of agents: households and firms. Firms have a simple technology:

$$Y_t = Z_t N_t^\alpha \qquad (1)$$

where N_t is the quantity of labour used by firms and Z_t a technological shock common to all firms. Note that we do not introduce capital in this model. Because its rate of depreciation is low, it would not add much to our argument, and would substantially complicate the results and exposition.

The representative household works N_t, consumes C_t, and ends period t with a quantity of money M_t. It maximizes the expectation of its discounted utility:

$$U = E_0 \sum_{t=0}^{\infty} \beta^t \left[\log C_t + \omega \log \frac{M_t}{P_t} - \xi \frac{N_t^v}{v} \right].$$

$$(2)$$

At the beginning of period t the household faces a monetary shock à la Lucas (1972), whereby the quantity of money M_{t-1} coming from $t-1$ is multiplied by μ_t, so that its budget constraint for period t is:

$$C_t + \frac{M_t}{P_t} = \frac{W_t}{P_t} N_t + \frac{\mu_t M_{t-1}}{P_t}. \qquad (3)$$

There are thus two shocks in this economy, the technology shock Z_t and the monetary shock $\mu_t = M_t/M_{t-1}$. As an illustration we shall use below the following traditional processes (in all that follows lower-case letters represent the logarithm of the variable represented by the corresponding uppercase letter):

$$m_t - m_{t-1} = \frac{\varepsilon_{mt}}{1 - \rho L} \qquad z_t = \frac{\varepsilon_{zt}}{1 - \varphi L}$$

$$(4)$$

where ε_{zt} and ε_{m_t}, the innovations in z_t and m_t, are uncorrelated white noises with:

$$\text{var}(\varepsilon_{zt}) = \sigma_z^2 \qquad \text{var}(\varepsilon_{mt}) = \sigma_m^2. \qquad (5)$$

Walrasian dynamics

As a benchmark we shall study here the case where both labour and goods markets are in Walrasian equilibrium in each period, as in the first traditional real business cycle (RBC) models, and we shall see how this economy reacts to technological and monetary shocks. Solving the model we find that money holdings are a multiple of consumption:

$$\frac{M_t}{P_t C_t} = \frac{\omega}{1 - \beta} \qquad (6)$$

and that employment N_t is constant:

$$N_t = N = (\alpha/\xi)^{1/v}. \qquad (7)$$

Using (1) and (7) we find (we eliminate some irrelevant constant terms):

$$n_t = n \qquad y_t = z_t + \alpha n$$
$$w_t - p_t = y_t - n. \qquad (8)$$

Although we will not do any real calibration in this article, we can note at this stage a few issues that posed a problem to researchers in the RBC domain.

First, real wages are much too pro-cyclical in this Walrasian model. From (8) we see that the real wage–output correlation is equal to 1. Even though this correlation is lower than 1 in calibrated models where N_t varies, it is always quite above what is observed in real economies.

A second problem concerns the inflation–output correlation, a problem related to the literature on the Phillips curve. Whereas it is generally considered that this correlation is positive, the above Walrasian model yields a negative correlation:

$$\text{cov}(\Delta p_t, y_t) = -\frac{\sigma_z^2}{1 + \varphi} < 0. \qquad (9)$$

Finally, an important and recurrent critique of RBC-type models has been that they do not generate any internal propagation mechanism, and that the only persistence in output movements is that already present in the exogenous process of technological shocks z_t (see, for example, Cogley and Nason, 1993; 1995). This appears here in eq. (8), where the dynamics of output y_t is exactly the same as that of the technological shock z_t.

We shall now introduce wage contracts, first lasting one period, and then multi-period overlapping contracts, and we shall see that the above problems find a natural solution in this framework.

Single-period wage contracts

Let us thus assume (Bénassy, 1995, and Bénassy, 2002, for microfoundations), that the wages are predetermined at the beginning of each period at the expected value of the Walrasian wage (in logarithms), and that at this contractual wage the households supply the quantity of work demanded by firms (this type of contract was introduced by Gray, 1976).

Combining (6) and $C_t = Y_t$ we find that the Walrasian wage w_t^* is, up to an unimportant constant, equal to m_t, so that the preset wage w_t is given by:

$$w_t = E_{t-1}w_t^* = E_{t-1}m_t \qquad (10)$$

where $E_{t-1}m_t$ is the expectation of m_t formed at the beginning of period t, before shocks are known.

The difference with the Walrasian case is that employment N_t is now variable and demand determined. Equations (8) become:

$$y_t = z_t + \alpha n_t \qquad w_t - p_t = y_t - n_t \quad (11)$$

while $n_t = n$ is replaced by (10). So we first obtain the level of employment in period t:

$$n_t = n + m_t - E_{t-1}m_t = n + \varepsilon_{mt} \qquad (12)$$

since $m_t - E_{t-1}m_t = \varepsilon_{mt}$. Contrarily to what happened in the Walrasian version of the model, unanticipated monetary shocks now have an impact on the level of employment, and therefore output. We shall now use the preceding formulas to show that the hypothesis of preset wages allows to substantially improve some correlations relative to the Walrasian model.

Let us start with the real wage which, in the Walrasian model, has a much too high positive correlation with output. Let us combine (11) and (12), to obtain the values of output and real wage:

$$\begin{aligned} y_t &= z_t + \alpha \varepsilon_{mt} \\ w_t - p_t &= z_t - (1-\alpha)\varepsilon_{mt}. \end{aligned} \qquad (13)$$

We see that supply shocks create a positive correlation between the real wage and output. However, monetary shocks create a negative correlation. Our model thus allows us to combine this last characteristic, typical of traditional Keynesians models, with the usual results of

RBC models. If one considers the technological and monetary shocks (4), one obtains the following correlation:

$$\begin{aligned} &\text{corr}(w_t - p_t, y_t) \\ &= \frac{\sigma_z^2 - (1-\varphi^2)\alpha(1-\alpha)\sigma_m^2}{\left[\left(\sigma_z^2 + (1-\varphi^2)\,\alpha^2\sigma_m^2\right)\right]^{1/2}\left[\sigma_z^2 + (1-\varphi^2)\,(1-\alpha)^2\sigma_m^2\right]^{1/2}}. \end{aligned}$$
$$(14)$$

We see that the real-wage–output correlation is equal to 1 if there are *only* technological shocks. But this correlation diminishes as soon as there are monetary shocks, and it can even become negative. One can thus reproduce the correlations observed in reality by adequate combinations of technological and monetary shocks.

Let us now consider the relation between inflation and output, which are generally considered to be positively correlated, at least in Keynesian tradition. If we assume again the monetary and technological shocks (4), we find:

$$\text{Covariance } (\Delta p_t, y_t) = \alpha(1-\alpha)\,\sigma_m^2 - \frac{\sigma_z^2}{1+\varphi}.$$
$$(15)$$

Formula (15) shows us that the positive covariance (and thus correlation) between inflation and output is linked to the presence of demand shocks, and that the sign of this correlation may change if there are sufficiently strong technological shocks.

So we just saw that one-period contracts allow us to improve some important correlations. We now naturally ask a question already posed for the standard RBC model: is the response to shocks, and in particular to demand shocks, sufficiently persistent? Let us recall eq. (13):

$$y_t = z_t + \alpha \varepsilon_{mt}. \qquad (16)$$

We see that monetary shocks now have an immediate effect on output (and employment), but that, starting with the second period, the effect of these shocks is completely dampened. One-period contracts allow us to solve the puzzle raised by some correlations, but certainly not the persistence problem. We shall see in the next two sections that multi-periodic contracts allow us to solve that problem.

Multi-periodic wage contracts

The models that we have examined so far share with traditional RBC models the defect of having an extremely weak internal propagation mechanism. In particular, the response of output to monetary demand shocks is almost entirely transitory. But several empirical studies (see, for example, Christiano, Eichenbaum and Evans, 1999; 2005) have pointed out that in reality the response to monetary shocks not only was persistent but also had a hump-shaped response function. We shall now introduce multi-periodic wage contracts in rigorous stochastic dynamic models, and show that they allow us to reproduce these features. Models with such multi-periodic wage or price contracts have been studied notably by Yun (1996), Andersen (1998), Jeanne (1998), Ascari (2000), Chari, Kehoe and McGrattan (2000), Collard and Ertz (2000), and Bénassy (2002; 2003a; 2003b).

In order to make our demonstration analytically, we use a contract, inspired by Calvo (1983) and developed in Bénassy (2002; 2003a), which has three advantages: (a) the average duration of contracts can take any value from zero to infinity, (b) an analytical solution can be found with both wage and price contracts, and (c) it has explicit microfoundations.

In this framework in each period s a contract is made for wages at period $t \geq s$. As in the Gray contract, the contract wage is the expectation of the market-clearing wage in period t. So if we denote as x_{st} the contract wage made in s for period t:

$$x_{st} = E_s(w_t^*). \tag{17}$$

Now, as in Calvo (1983), each wage contract has a probability γ to stay unchanged, and a probability $1 - \gamma$ to be broken. If the contract is broken, a new contract is immediately renegotiated on the basis of current period information. So for $\gamma = 0$, wages are totally flexible, for $\gamma = 1$ they are totally rigid.

It is easy to compute the average duration of these contracts. The probability for a contract to be still valid j periods after the date it was concluded is equal to $(1 - \gamma)\gamma^j$. The expected duration of the contract is thus:

$$\sum_{j=0}^{\infty}(1 - \gamma)j\gamma^j = \frac{\gamma}{1 - \gamma}. \tag{18}$$

We thus see that varying γ from 0 to 1 the average duration of the contract varies from zero to infinity.

The average wage w_t is the mean of past x_{st}'s weighted by the probability for the corresponding contract to be still in effect. Because of the law of large numbers, and since the probability of survival of wage contracts is γ, the proportion of contracts coming from period $s \leq t$ is $(1 - \gamma)\gamma^{t-s}$. Therefore, the average wage in the economy is given by:

$$w_t = (1 - \gamma)\sum_{s=-\infty}^{t} \gamma^{t-s}x_{st}. \tag{19}$$

If we now solve the model with the shocks (4) we find that the dynamics of employment is characterized by (Bénassy, 2002; 2003a):

$$n_t = n + \frac{\gamma\varepsilon_{mt}}{(1 - \gamma L)(1 - \gamma\rho L)} \tag{20}$$

where L is the lag operator: $L^j X_t = X_{t-j}$. The response of output is deduced from that of employment through:

$$y_t = \alpha n_t + z_t. \tag{21}$$

Formula (20) shows clearly that, contrarily to the case of one-period contracts, the response to a monetary shock can be quite persistent. We can have an idea of the temporal profile of this response by computing the response function of output and employment to a monetary shock. The value of ρ most often found in the literature is $\rho = 0.5$. As for γ, we saw above (formula 18) that the average duration of wage contracts is equal to $\gamma/(1 - \gamma)$. One considers generally that the average duration of wage contracts is about one year (see, for example, Taylor, 1999), which corresponds to $\gamma = 4/5$. Figure 1 shows the response of employment (output is derived via 21) to a monetary shock for $\gamma = 4/5$.

We see that the response function displays persistence in the effects of monetary shocks, and has even a hump-shaped response. If we plot, however, the response function of inflation, we find that it is steadily decreasing after the initial jump, whereas it seems to have a delayed hump-shaped response in reality.

Wage and price multi-periodic contracts

We shall now enlarge our model by considering simultaneously wage and price multi-periodic contracts (see Bénassy, 2003b, for such a model with explicit microfoundations). Numerically solved models with both wage and price multi-periodic contracts are found in Christiano, Eichenbaum and Evans (2005), Huang and Liu (2002), Smets and Wouters (2003).

Wage contracts are exactly the same as in the preceding section: each contract is maintained with probability γ, or renegotiated with probability $1 - \gamma$. Symmetrically, price contracts are maintained with probability ϕ, or break down and are renegotiated with probability $1 - \phi$. The average price p_t is given by:

$$p_t = (1 - \phi)\sum_{s=-\infty}^{t} \phi^{t-s}q_{st} \tag{22}$$

where q_{st} is the price contract negotiated in period s for period t. Using again the shock processes (4), and taking $v = 1$, we find the following dynamics for output and inflation:

$$y_t = z_t - \frac{\phi\varepsilon_{zt}}{1 - \phi\varphi L} + \frac{\alpha\gamma\varepsilon_{mt}}{(1 - \gamma L)(1 - \gamma\rho L)}$$
$$+ \frac{\phi\varepsilon_{mt}}{(1 - \phi L)(1 - \phi\rho L)} - \frac{\alpha\gamma\phi\varepsilon_{mt}}{(1 - \gamma\phi L)(1 - \gamma\phi\rho L)}$$
$$\tag{23}$$

$$\pi_t = (1 - L)p_t = (1 - L)(m_t - y_t). \tag{24}$$

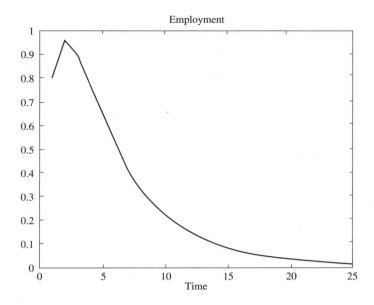

Figure 1

As in the preceding section we take as an illustration $\alpha = 2/3$, $\rho = 1/2$ and $\gamma = 4/5$ (one-year wage contracts). As for prices, we want to take a rather low duration of contracts, so we shall take $\phi = 1/2$ (one quarter). Simulations show that in that case we obtain a persistent and hump-shaped response for both output and inflation.

So we see that with only reasonable nominal rigidities we obtain some realistic response functions. Clearly the adjunction of 'real' rigidities would allow to reproduce even better the actual dynamic macroeconomic patterns.

JEAN-PASCAL BÉNASSY

See also **non-clearing markets in general equilibrium; real business cycles.**

Bibliography

Andersen, T. 1994. *Price Rigidity: Causes and Macroeconomic Implications.* Oxford: Oxford University Press.

Andersen, T. 1998. Persistency in sticky price models. *European Economic Review* 42, 593–603.

Arrow, K. 1963. The role of securities in the optimal allocation of risk–bearing. *Review of Economic Studies* 31, 91–6.

Arrow, K. and Debreu, G. 1954. Existence of an equilibrium for a competitive economy. *Econometrica* 22, 265–90.

Ascari, G. 2000. Optimising agents, staggered wages and persistence in the real effects of money shocks. *Economic Journal* 110, 664–86.

Ascari, G. and Rankin, N. 2002. Staggered wages and output dynamics under disinflation. *Journal of Economic Dynamics and Control* 26, 653–80.

Azariadis, C. 1975. Implicit contracts and underemployment equilibria. *Journal of Political Economy* 83, 1183–202.

Baily, M. 1974. Wages and employment under uncertain demand. *Review of Economic Studies* 41, 37–50.

Barro, R. 1972. A theory of monopolistic price adjustment. *Review of Economic Studies* 39, 17–26.

Barro, R. and Grossman, H. 1971. A general disequilibrium model of income and employment. *American Economic Review* 61, 82–93.

Barro, R. and Grossman, H. 1976. *Money, Employment and Inflation.* Cambridge: Cambridge University Press.

Bénassy, J.-P. 1975. Neo-Keynesian disequilibrium theory in a monetary economy. *Review of Economic Studies* 42, 503–23.

Bénassy, J.-P. 1976. The disequilibrium approach to monopolistic price setting and general monopolistic equilibrium. *Review of Economic Studies* 43, 69–81.

Bénassy, J.-P. 1977. A neoKeynesian model of price and quantity determination in disequilibrium. In *Equilibrium and Disequilibrium in Economic Theory,* ed. G. Schwödiauer. Boston: Reidel Publishing Company.

Bénassy, J.-P. 1982. *The Economics of Market Disequilibrium.* New York: Academic Press.

Bénassy, J.-P. 1988. The objective demand curve in general equilibrium with price makers. *Economic Journal* 98(supplement), 37–49.

Bénassy, J.-P. 1990. Non-Walrasian equilibria, money and macroeconomics. In *Handbook of Monetary Economics,* ed. B. Friedman and F. Hahn. Amsterdam: North-Holland.

Bénassy, J.-P. 1995. Money and wage contracts in an optimizing model of the business cycle. *Journal of Monetary Economics* 35, 303–15.

Bénassy, J.-P. 2002. *The Macroeconomics of Imperfect Competition and Nonclearing Markets: A Dynamic General Equilibrium Approach.* Cambridge, MA: M.I.T. Press.

Bénassy, J.-P. 2003a. Staggered contracts and persistence: microeconomic foundations and macroeconomic dynamics. *Recherches Economiques de Louvain* 69, 125–44.

Bénassy, J.-P. 2003b. Output and inflation persistence under price and wage staggering: analytical results. *Annales d'Economie et de Statistique,* n° 69, 1–30.

Brock, W. 1975. A simple perfect foresight monetary model. *Journal of Monetary Economics* 1, 133–50.

Calvo, G. 1983. Staggered prices in a utility-maximizing framework. *Journal of Monetary Economics* 12, 383–98.

Chari, V., Kehoe, P. and McGrattan, El. 2000. Sticky price models of the business cycle: can the contract multiplier solve the persistence problem? *Econometrica* 68, 1151–79.

Cho, J.-O. 1993. Money and the business cycle with one-period nominal contracts. *Canadian Journal of Economics* 26, 638–59.

Cho, J.-O. and Cooley, T. 1995. Business cycles with nominal contracts. *Economic Theory* 6, 13–34.

Cho, J.-O., Cooley, T. and Phaneuf, L. 1997. The welfare cost of nominal wage contracting. *Review of Economic Studies* 64, 465–84.

Christiano, L., Eichenbaum, M. and Evans, C. 1999. Monetary policy shocks: what have we learned and to what end? In *Handbook of Macroeconomics*, vol. 1A, ed. J. Taylor and M. Woodford. Amsterdam: North-Holland.

Christiano, L., Eichenbaum, M. and Evans, C. 2005. Nominal rigidities and the dynamic effects of a shock to monetary policy. *Journal of Political Economy* 113, 1–46.

Clower, R. 1965. The Keynesian counterrevolution: a theoretical appraisal. In *The Theory of Interest Rates*, ed. F. Hahn and F. Brechling. London: Macmillan.

Cogley, T. and Nason, J. 1993. Impulse dynamics and propagation mechanisms in a real business cycle model. *Economics Letters* 43, 77–81.

Cogley, T. and Nason, J. 1995. Output dynamics in real-business-cycle models. *American Economic Review* 85, 492–511.

Collard, F. and Ertz, G. 2000. Stochastic nominal wage contracts in a cash-in-advance model. *Recherches Economiques de Louvain* 66, 281–301.

Danthine, J.-P. and Donaldson, J. 1990. Efficiency wages and the business cycle puzzle. *European Economic Review* 34, 1275–301.

Danthine, J.-P. and Donaldson, J. 1991. Risk sharing, the minimum wage and the business cycle. In *Equilibrium Theory and Applications: A Conference in Honor of Jacques Drèze*, ed. W. Barnett et al. Cambridge: Cambridge University Press.

Danthine, J.-P. and Donaldson, J. 1992. Risk sharing in the business cycle. *European Economic Review* 36, 468–75.

Debreu, G. 1959. *Theory of value.* New York: Wiley.

Dixit, A. and Stiglitz, J. 1977. Monopolistic competition and optimum product diversity. *American Economic Review* 67, 297–308.

Drèze, J. 1975. Existence of an exchange equilibrium under price rigidities. *International Economic Review* 16, 301–20.

Fischer, S. 1977. Long-term contracts, rational expectations, and the optimal money supply rule. *Journal of Political Economy* 85, 191–205.

Gabszewicz, J. and Vial, J.-P. 1972. Oligopoly 'A la Cournot' in a general equilibrium analysis. *Journal of Economic Theory* 42, 381–400.

Gordon, D. 1974. A neo-classical theory of Keynesian unemployment. *Economic Inquiry* 12, 431–59.

Gray, J.-A. 1976. Wage indexation: a macroeconomic approach. *Journal of Monetary Economics* 2, 221–35.

Hairault, J.-O. and Portier, F. 1993. Money, new-Keynesian macroeconomics and the business cycle. *European Economic Review* 37, 1533–68.

Hicks, J. 1937. Mr. Keynes and the 'classics': a suggested interpretation. *Econometrica* 5, 147–59.

Huang, K. and Liu, Z. 2002. Staggered price-setting, staggered wage-setting, and business cycle persistence. *Journal of Monetary Economics* 49, 405–33.

Jeanne, O. 1998. Generating real persistent effects of monetary shocks: how much nominal rigidity do we really need? *European Economic Review* 42, 1009–32.

Keynes, J.M. 1936. *The General Theory of Employment, Interest and Money.* New York: Harcourt Brace.

Kydland, F. and Prescott, E. 1982. Time to build and aggregate fluctuations. *Econometrica* 50, 1345–70.

Leijonhufvud, A. 1968. *On Keynesian Economics and the Economics of Keynes.* Oxford: Oxford University Press.

Long, J. and Plosser, C. 1983. Real business cycles. *Journal of Political Economy* 91, 39–69.

Lucas, R., Jr. 1972. Expectations and the neutrality of money. *Journal of Economic Theory* 4, 103–24.

Muth, J. 1961. Rational expectations and the theory of price movements. *Econometrica* 29, 315–35.

Negishi, T. 1961. Monopolistic competition and general equilibrium. *Review of Economic Studies* 28, 196–201.

Patinkin, D. 1956. *Money, Interest and Prices,* 2nd edn. New York: Harper and Row, 1965.

Phelps, E. and Taylor, J. 1977. Stabilizing powers of monetary policy under rational expectations. *Journal of Political Economy* 85, 163–90.

Rotemberg, J. 1982a. Monopolistic price adjustment and aggregate output. *Review of Economic Studies* 44, 517–31.

Rotemberg, J. 1982b. Sticky prices in the United States. *Journal of Political Economy* 90, 1187–211.

Rotemberg, J. and Woodford, M. 1992. Oligopolistic pricing and the effects of aggregate demand on economic activity. *Journal of Political Economy* 100, 1153–207.

Rotemberg, J. and Woodford, M. 1995. Dynamic general equilibrium models with imperfectly competitive product markets. In *Frontiers of Business Cycle Research*, ed. T. Cooley. Princeton: Princeton University Press.

Sidrauski, M. 1967. Rational choice and patterns of growth in a monetary economy. *American Economic Review* 57(supplement), 534–44.

Smets, F. and Wouters, R. 2003. An estimated dynamic stochastic general equilibrium model of the euro area. *Journal of the European Economic Association* 1, 1123–75.

Svensson, L. 1986. Sticky goods prices, flexible asset prices, monopolistic competition and monetary policy. *Review of Economic Studies* 53, 385–405.

Taylor, J. 1979. Staggered wage setting in a macro model. *American Economic Review* 69, 108–13.

Taylor, J. 1980. Aggregate dynamics and staggered contracts. *Journal of Political Economy* 88, 1–23.

Taylor, J. 1999. Staggered price and wage setting in macroeconomics. In *Handbook of Macroeconomics*, vol. 1, ed. J. Taylor and M. Woodford. Amsterdam: North-Holland.

Walras, L. 1874. *Eléments d'économie politique pure.* Lausanne: Corbaz. Definitive edition trans. W. Jaffe as *Elements of Pure Economics*. London: Allen and Unwin, 1954.

Yun, T. 1996. Nominal price rigidity, money supply endogeneity, and business cycles. *Journal of Monetary Economics* 37, 345–70.

dynamic programming

1 Introduction

Dynamic programming is a recursive method for solving *sequential decision problems* (hereafter abbreviated as SDP). Also known as *backward induction*, it is used to find *optimal decision rules* in 'games against nature' and *subgame perfect equilibria* of dynamic multi-agent games, and competitive equilibria in dynamic economic models. Dynamic programming has enabled economists to formulate and solve a huge variety of problems involving sequential decision-making under uncertainty, and as a result it is now widely regarded as the single most important tool in economics. Section 2 provides a brief history of dynamic programming. Section 3 discusses some of the main theoretical results underlying dynamic programming, and its relation to game theory and optimal control theory. Section 4 provides a brief survey of numerical dynamic programming. Section 5 surveys the experimental and econometric literature that uses dynamic programming to construct empirical models economic behaviour.

2 History

The earliest reference to the use of the method of backward induction to solve decision problems appears to be Arthur Cayley's 1875 solution to the secretary problem (I am grateful to Arthur F. Veinott Jr. for alerting me to

this). In the mid-1940s a number of different researchers in economics and statistics appear to have independently discovered backward induction as a way to solve SDPs involving risk or uncertainty. Von Neumann and Morgenstern, in their seminal work on game theory (1944), used backward induction to find what we now call *subgame perfect equilibria of extensive form games*. ('We proceed to discuss the game Γ by starting with the last move $\mathscr{M}_\nu$ and then going backward from there through the moves $\mathscr{M}_{\nu-1}, \mathscr{M}_{\nu-2} \cdots$': 1944, p. 126.) Abraham Wald, who is credited with the invention of *statistical decision theory*, extended this theory to sequential decision-making in his 1947 book *Sequential Analysis*. Wald generalized the problem of gambler's ruin from probability theory and introduced the *sequential probability ratio test* that minimizes the expected number of observations in a sequential generalization of the classical hypothesis test. However, the role of backward induction is less obvious in Wald's work. It was more clearly elucidated in the 1949 paper by Arrow, Blackwell and Girshick. They studied a generalized version of the statistical decision problem and formulated and solved it in a way that is a readily recognizable application of modern dynamic programming. Following Wald, they characterized the optimal rule for making a statistical decision (for example, accept or reject a hypothesis), accounting for the costs of collecting additional observations. In the section 'The Best Truncated Procedure' they show how the optimal rule can be approximated 'Among all sequential procedures not requiring more than N observations …' and solve for the optimal truncated sampling procedure 'by induction backwards' (1949, p. 217).

Other early applications of backward induction include the work of Pierre Massé (1945, p. 196) on statistical hydrology and the management of reservoirs, and Dvoretsky, Kiefer and Wolfowitz's (1952) analysis of optimal inventory policy. Richard Bellman is widely credited with recognizing the common structure underlying SDPs, and showing how backward induction can be applied to solve a huge class of SDPs under uncertainty. Most of Bellman's work in this area was done at the RAND Corporation, starting in 1949. It was there that he invented the term 'dynamic programming' that is now the generally accepted synonym for backward induction. Bellman (1984, p. 159) explained that he invented the name 'dynamic programming' to hide the fact that he was doing mathematical research at RAND under a Secretary of Defense who 'had a pathological fear and hatred of the term, research'. He settled on 'dynamic programming' because it would be difficult give it a 'pejorative meaning' and because 'It was something not even a Congressman could object to'.

3 Theory

Dynamic programming can be used to solve for optimal strategies and equilibria of a wide class of SDPs and multiplayer games. The method can be applied both in discrete time and continuous time settings. The value of dynamic programming is that it is a 'practical' (that is, *constructive*) method for finding solutions to extremely complicated problems. However, continuous time problems involve technicalities that I wish to avoid in this survey. If a continuous time problem does not admit a closed-form solution, the most commonly used numerical approach is to solve an approximate discrete time version of the problem or game, since under very general conditions one can find a sequence of discrete time DP problems whose solutions converge to the continuous time solution the time interval between successive decisions tends to zero (Kushner, 1990). I start by describing how dynamic programming is used to solve single agent 'games against nature'. The approach can be extended to solve multiplayer games, dynamic contracts, principal–agent problems, and competitive equilibria of dynamic economic models. See RECURSIVE COMPETITIVE EQUILIBRIUM.

3.1 Sequential decision problems

There are two key variables in any dynamic programming problem: a *state variable* s_t, and a *decision variable* d_t (the decision is often called a 'control variable' in the engineering literature). These variables can be vectors in R^n, but in some cases they might be *infinite-dimensional* objects. For example, in Bayesian decision problems, one of the state variables might be a *posterior distribution* for some unknown quantity θ. In general, this posterior distribution lives in an infinite dimensional space of all probability distributions on θ. In heterogeneous agent equilibrium problems state variables can also be distributions. The state variable evolves randomly over time, but the agent's decisions can affect its evolution. The agent has a *utility* or *payoff function* $U(s_1, d_1, \ldots, s_T, d_T)$ that depends on the realized states and decisions from period $t = 1$ to the *horizon T*. In some cases $T = \infty$, and we say the problem is *infinite horizon*. In other cases, such as a life-cycle decision problem, T might be a random variable, representing a consumer's date of death. As we will see, dynamic programming can be adapted to handle either of these possibilities. Most economic applications presume a *discounted, time-separable* objective function, that is, U has the form

$$U(s_1, d_1, \ldots, s_T, d_T) = \sum_{t=1}^{T} \beta^t u_t(s_t, d_t)$$

(1)

where β is known as a *discount factor* that is typically presumed to be in the $(0, 1)$ interval, and $u_t(s_t, d_t)$ is the agent's *period t utility* (*payoff*) function. Discounted utility and profits are typical examples of time separable payoff functions studied in economics. However, the method of dynamic programming does not require time separability, and so I will describe it without imposing this restriction.

We model the uncertainty underlying the decision problem via a family of history and decision-dependent conditional probabilities $\{p_t(s_t|H_{t-1})\}$ where $H_{t-1} = (s_1, d_1, \ldots, s_{t-1}, d_{t-1})$ denotes the *history*, that is, the realized states and decisions from the initial date $t = 1$ to date $t = T$. Note that this includes all deterministic SDPs as a special case where the transition probabilities p_t are degenerate. In this case we can represent the 'law of motion' for the state variables by deterministic functions $s_{t+1} = f_t(s_t, d_t)$. This implies that in the most general case, $\{s_t, d_t\}$, evolves as a history dependent stochastic process. Continuing the 'game against nature' analogy, it will be helpful to think of $\{p_t(s_t|H_{t-1})\}$ as constituting a 'mixed strategy' played by 'nature' and the agent's optimal strategy as a 'best response' to nature's strategy.

The final item we need to specify is the *timing of decisions*. Assume that the agent selects d_t *after* observing s_t, which is 'drawn' from the distribution $p_t(s_t|H_{t-1})$. The alternative case where d_t is chosen before s_t is realized can also be handled, but requires a small change in the formulation of the problem. The agent's choice of d_t is restricted to a *state-dependent constraint (choice) set* $D_t(H_{t-1}, s_t)$. We can think of D_t as the generalization of a 'budget set' in standard static consumer theory. The choice set could be a finite set, in which case we refer to the problem as *discrete choice*, or D_t could be a subset of R^k with non-empty interior, then we have a *continuous choice* problem. In many cases, there is a mixture of types of choices, which we refer to as *discrete-continuous choice problems*. An example is commodity price speculation; see for example Hall and Rust (2006), where a speculator has a discrete choice of whether or not to order to replenish his inventory and a continuous decision of how much of the commodity to order. Another example is retirement: a person has a discrete decision of whether to retire and a continuous decision of how much to consume.

Definition A (single agent) *sequential decision problem* (SDP) consists of (1) a *utility function U*, (2) a sequence of *choice sets* $\{D_t\}$, and (3) a sequence of *transition probabilities* $\{pt(s_t|H_{t-1})\}$ where we assume that the process is initialized at some given initial state s_1.

In order to solve this problem, we have to make assumptions about how the decision-maker evaluates alternative risky strategies. The standard assumption is that the decision-maker maximizes *expected utility*. Backward induction does not necessarily result in optimal strategies for *non-expected utility maximizers*, except for certain classes of *recursive preferences*.

As the name implies, an expected utility maximizer makes decisions that maximize their *ex ante* expected utility. However, since information unfolds over time, it is generally not optimal to *pre-commit* to any fixed sequence of actions $(d_1, \ldots, d_T)$. Instead, the decision-maker can generally obtain higher expected utility by adopting a *history-dependent strategy* or *decision rule* $(\delta_1, \ldots, \delta_T)$. This is a sequence of *functions* such that for each time t the realized decision is a function of all available information. In the engineering literature, a decision rule that does not depend on evolving information is referred to as an *open-loop* strategy, whereas one that does is referred to as a *closed-loop* strategy. In deterministic control problems, the closed-loop and open-loop strategies are the same since both are simple functions of time. However in stochastic control problems, open-loop strategies are a strict subset of closed-loop strategies. Under our timing assumptions the information available at time t is (H_{t-1}, s_t), so we can write $d_t = \delta_t(H_{t-1}, s_t)$. By convention we set $H_0 = \emptyset$ so that the available information for making the initial decision is just s_1. A decision rule is *feasible* if it also satisfies $\delta_t(H_{t-1}, s_t) \in D_t(H_{t-1}, s_t)$ for all (s_t, H_{t-1}). Each feasible decision rule can be regarded as a 'lottery' whose payoffs are utilities, the expected value of which corresponds to expected utility associated with the decision rule. An *optimal decision rule* $\delta^* \equiv (\delta_1^*, \ldots, \delta_T^*)$ is simply a feasible decision rule that maximizes the decision-maker's expected utility

$$\delta^* = \underset{\delta \in \mathscr{F}}{\operatorname{argmax}} \, E\left\{ U\left(\{\tilde{s}_t, \tilde{d}_t\}_\delta \right) \right\}, \qquad (2)$$

where $\mathscr{F}$ denotes the class of feasible history-dependent decision rules, and $\{\tilde{s}_t, \tilde{d}_t\}_\delta$ denotes the stochastic process induced by the decision rule $\delta \equiv (\delta_1, \ldots, \delta_T)$. Problem (2) can be regarded as a static, *ex ante* version of the agent's problem. In game theory, (2) is referred to as the *normal form* or the *strategic form* of a dynamic game, since the dynamics are suppressed and the problem has the superficial appearance of a static optimization problem or game in which an agent's problem is to choose a best response, either to nature (in the case of single agent decision problems) or to other rational opponents (in the case of games). The strategic formulation of the agent's problem is quite difficult to solve since the solution is a *sequence of history-dependent functions* $\delta^* = (\delta_1^*, \ldots, \delta_T^*)$ for which standard finite dimensional constrained optimization techniques (for example, the Kuhn–Tucker theorem) are inapplicable. (If we consider problems where all states can assume only a finite number of values, it is possible to apply standard finite dimensional Kuhn–Tucker constrained optimization methods, but if the state variables can assume a continuum of possible values, the programming problem becomes an infinite dimensional programming problem for which optimal control and dynamic programming methods are more appropriate. See Luenberger, 1969, for a more thorough discussion of how Lagrange multipliers and Kuhn–Tucker methods

can be extended to problems where decisions are infinite-dimensional objects. These methods are usually applied in deterministic context, and there is a specialized literature on optimal control for solving such problems.) See PONTRYAGIN'S PRINCIPLE OF OPTIMALITY.

3.2 Solving sequential decision problems by backward induction

To carry out backward induction, we start at the last period, T, and for each possible combination (H_{T-1}, s_T) we calculate the time T *value function* and *decision rule* (we will discuss how backward induction can be extended to cases where T is random or where $T = \infty$ shortly).

$$V_T(H_{T-1}, s_T) = \max_{d_T \in D_T(H_{T-1}, s_T)} U(H_{T-1}, s_T, d_T)$$

$$\delta_T(H_{T-1}, s_T) = \underset{d_T \in D_T(H_{T-1}, s_T)}{\operatorname{argmax}} U(H_{T-1}, s_T, d_T),$$

$$(3)$$

where we have written $U(H_{T-1}, s_T, d_T)$ instead of $U(s_1, d_1, \ldots, s_T, d_T)$ since $H_{T-1} = (s_1, d_1, \ldots, s_{T-1}, d_{T-1})$. Next we move backward one time period to time $T-1$ and compute

$$V_{T-1}(H_{T-2}, s_{T-1})$$
$$= \max_{d_{T-1} \in D_{T-1}(H_{T-2}, s_{T-1})}$$
$$E\{V_T(H_{T-2}, s_{T-1}, d_{T-1}, \tilde{s}_T) | H_{T-2}, s_{T-1}, d_{T-1}\}$$
$$= \max_{d_{T-1} \in D_{T-1}(H_{T-2}, s_{T-1})}$$
$$\int V_T(H_{T-2}, s_{T-1}, d_{T-1}, s_T)$$
$$p_T(s_T | H_{T-2}, s_{T-1}, d_{T-1})$$

$$\delta_{T-1}(H_{T-2}, s_{T-1})$$
$$= \underset{d_{T-1} \in D_{T-1}(H_{T-2}, s_{T-1})}{\operatorname{argmax}}$$
$$E\{V_T(H_{T-2}, s_{T-1}, d_{T-1}, \tilde{s}_T) | H_{T-2}, s_{T-1}, d_{T-1}\},$$
$$(4)$$

where the integral in eq. (4) is the formula for the conditional expectation of V_T, where the expectation is taken with respect to the random variable $\tilde{s}_T$ whose value is not known as of time $T-1$. We continue the backward induction recursively for time periods $T-2$, $T-3$, ... until we reach time period $t = 1$. The equation for the value function V_t in an arbitrary period t is defined recursively by an equation that is now commonly called the *Bellman equation*

$$V_t(H_{t-t}, s_t) = \max_{d_t \in D_t(H_{t-1}, s_t)}$$

$$E\{V_{t+1}(H_{t-1}, s_t, d_t, \tilde{s}_{t+1}) | H_{t-1}, s_t, d_t\}$$

$$= \max_{d_t \in D_t(H_{t-1}, s_t)} \int V_{t+1}(H_{t-1}, s_t, d_t, s_{t+1})$$

$$p_{t+1}(s_{t+1} | H_{t-1}, s_t, d_t). \quad (5)$$

The decision rule δ_t is defined by the value of d_t that attains the maximum in the Bellman equation for each possible value of (H_{t-1}, s_t)

$$\delta_t(H_{t-t}, s_t) = \underset{d_t \in D_t(H_{t-1}, s_t)}{\text{argmax}}$$

$$E\{V_{t+1}(H_{t-1}, s_t, d_t, \tilde{s}_{t+1}) | H_{t-1}, s_t, d_t\}. \quad (6)$$

Backward induction ends when we reach the first period, in which case, as we will now show, the function $V_1(s_1)$ provides the expected value of an optimal policy, starting in state s_1 implied by the recursively constructed sequence of decision rules $\delta = (\delta_1, \ldots, \delta_T)$.

3.3 The principle of optimality
The key idea underlying why backward induction produces an optimal decision rule is called

The principle of optimality: *an optimal decision rule* $\delta^* = (\delta_1^*, \ldots, \delta_T^*)$ *has the property that given any* $t \in \{1, \ldots, T\}$ *and any history* H_{t-1} *in the support of the controlled process* $\{s_t, d_t\}_{\delta^*}$, δ^* *remains optimal for the 'subgame' starting at time t and history* H_{t-1}. *That is,* δ^* *maximizes the "continuation payoff" given by the conditional expectation of utility from period t to T, given history* H_{t-1}:

$$\delta^* = \underset{\delta}{\text{argmax}} \ E\{U(\{s_t, d_t\}_\delta) | H_{t-1}\}. \quad (7)$$

In game theory, the principle of optimality is equivalent to the concept of a *subgame perfect equilibrium* in an *extensive form game*. When all actions and states are discrete, the stochastic decision problem can be diagrammed as a *game tree*. The principle of optimality, which in game theory is equivalent to the concept of a *subgame perfect equilibrium*, guarantees that if δ^* is an optimal strategy (or equilibrium strategy) for the overall game tree, then it must also be an optimal strategy for every subgame, or, more precisely, *all subgames that are reached with positive probability from the initial node*.

It should now be evident why there is a need for the qualification 'for all H_{t-1} in the support of $\{s_t, d_t\}_{\delta^*}$' in the statement of the principle of optimality. There are some subgames that are never reached with positive

probability under an optimal strategy. Thus, it is easy to construct alternative optimal decision rules that do not satisfy the principle of optimality because they involve taking suboptimal decisions on 'zero probability subgames'. Since these subgames are never reached, such modifications do not jeopardize *ex ante* optimality. However we cannot be sure *ex ante* which subgames will be irrelevant *ex post* unless we carry out the full backward induction process. Dynamic programming results in strategies that are optimal in *every possible subgame*, even those which will never be reached when the strategy is executed. Since backward induction results in a decision rule δ that is optimal for all possible subgames, it is intuitively clear that δ is optimal for the game as a whole, that is, it is a solution to the *ex ante* strategic form of the optimization problem (2).

For a formal proof of this result for games against nature (with appropriate care taken to ensure measurability and existence of solutions), see Gihman and Skorohod (1979). If in addition to 'nature' we extend the game tree by adding another rational expected utility maximizing player, then backward induction can be applied in the same way to solve this alternating move dynamic game. Assume that player 1 moves first, then player 2, then nature, and so on. Dynamic programming results in a *pair of strategies* for both players. Nature still plays a 'mixed strategy' that could depend on the entire previous history of the game, including all the previous moves of both players. The backward induction process ensures that each player can predict the future choices of their opponent, not only in the succeeding move but in all future stages of the game. The pair of strategies (δ^1, δ^2) produced by dynamic programming are mutual best responses, as well as being best responses to nature's moves. Thus, these strategies constitute a *Nash equilibrium*. They actually satisfy a stronger condition: they are Nash equilibrium strategies in every possible subgame of the original game, and thus are *subgame-perfect* (Selten, 1975). Subgame-perfect equilibria exclude 'implausible equilibria' based on *incredible threats*. A standard example is an incumbent's threat to engage in a price war if a potential entrant enters the market. This threat is incredible if the incumbent would not really find it advantageous to engage in a price war (resulting in losses for both firms) if the entrant called its bluff and entered the market. Thus the set of all Nash equilibria to dynamic multiplayer games is strictly larger than the subset of subgame-perfect equilibria, a generalization of the fact that, in single agent decision problems, the set of optimal decision rules includes ones which take suboptimal decisions on subgames that have zero chance of being reached for a given optimal decision rule. Dynamic programming ensures that the decision-maker would never mistakenly reach any such subgame, similar to the way subgame perfection ensures that a rational player would not be fooled by an incredible threat.

3.4 Dynamic programming for stationary, Markovian, infinite-horizon problems

The *complexity* of dynamic programming arises from the exponential growth in the number of possible histories as the number of possible values for the state variables, decision variables, and/or number of time periods T increases. For example, in a problem with N possible values for s_t and D possible values for d_t in each time period t, there are $[ND]^T$ possible histories, and thus the required number of calculations to solve a general T period, history-dependent dynamic programming problem is $O([ND]^T)$. Bellman and Dreyfus (1962) referred to this exponential growth in the number of calculations as the *curse of dimensionality*. In the next section, I will describe various strategies for dealing with this problem, but an immediate solution is to restrict attention to *time separable Markovian decision problems*. These are problems where the payoff function U is additively separable as in eq. (1), and where both the choice sets $\{D_t\}$ and transition probabilities $\{p_t\}$ depend only on the contemporaneous state variable s_t and not on the entire previous history H_{t-1}. We say a conditional distribution p_t satisfies the *Markov property* if it depends on the previous history only via the most recent values, that is, if $p_t(s_t|H_{t-1}) = p_t(s_t|s_{t-1}, d_{t-1})$. In this case backward induction becomes substantially easier. For example, in this case the dynamic programming optimizations have to be performed only at each of the N possible values of the state variable at each time t, so only $O(NDT)$ calculations are required to solve a time T period time separable Markovian problem instead of $O([ND]^T)$ calculations when histories matter. This is part of the reason why, even though time non-separable utilities and non-Markovian forms of uncertainty may be more general, most dynamic programming problems that are solved in practical applications are both time separable and Markovian.

SDPs with random horizons $\tilde{T}$ can be solved by backward induction provided there is some finite time $\bar{T}$ satisfying $Pr\{\tilde{T} \leq \bar{T}\} = 1$. In this case, backward induction proceeds from the maximum possible value $\bar{T}$ and the *survival probability* $\rho_t = Pr\{\tilde{T} > t|\tilde{T} \geq t\}$ is used as to capture the probability that the problem will continue for at least one more period. The Bellman equation for the discounted, time-separable utility with uncertain lifetime is

$$V_t(s_t) = \max_{d \in D_t(s_t)} [u_t(s_t, d) + \rho_t \beta EV_{t+1}(s_t, d)]$$

$$\delta_t(s_t) = \underset{d \in D_t(s_t)}{\mathrm{argmax}}[u_t(s_t, d) + \rho_t \beta EV_{t+1}(s_t, d)],$$

$$\tag{8}$$

where

$$EV_{t+1}(s, d) = \int_{s'} V_{t+1}(s')p_{t+1}(s'|s, d). \tag{9}$$

In many problems there is no finite upper bound $\bar{T}$ on the horizon. These are called *infinite horizon problems* and they occur frequently in economics. For example, SDPs used to model decisions by firms are typically treated as infinite horizon problems. It is also typical in infinite horizon problems to assume *stationarity*. That is, the utility function $u(s, d)$, the constraint set $D(s)$, the survival probability ρ, and the transition probability $p(s'|s, d)$ do not explicitly depend on time t. In such cases, it is not hard to show the value function and the optimal decision rules are also stationary, and satisfy the following version of Bellman's equation

$$V(s) = \max_{d \in D(s)} [u(s, d) + \rho \beta EV(s, d)]$$

$$\delta(s) = \underset{d \in D(s)}{\mathrm{argmax}}[u(s, d) + \rho \beta EV(s, d)], \tag{10}$$

where

$$EV(s, d) = \int_{s'} V(s')p(s'|s, d). \tag{11}$$

This is a fully recursive definition of V, and as such there is an issue of existence and uniqueness of a solution. In addition, it is not obvious how to carry out backward induction, since there is no 'last' period from which to begin the backward induction process. However, under relatively weak assumptions one can show there is a unique V satisfying the Bellman equation, and the implied decision rule in eq. (10) is an optimal decision rule for the problem. Further, this decision rule can be approximated by solving an approximate finite horizon version of the problem by backward induction.

For example, suppose that $u(s, d)$ is a continuous function of (s, d), the state space S is compact, the constraint sets $D(s)$ are compact for each $s \in S$, and the transition probability $p(s'|s, d)$ is weakly continuous in (s, d) (that is, $EV(s, d) \equiv \int_{s'} W(s')p(s'|s, d)$ is a continuous function of (s, d) for each continuous function $W:S \to R$). Blackwell (1965a; 1965b), Denardo (1967) and others have proved that, under these sorts of assumptions, V is the unique fixed point to the *Bellman operator* $\Gamma: B \to B$, where B is the Banach space of continuous functions on S under the supremum norm, and Γ is given by

$$\Gamma(W)(s) = \max_{d \in D(s)}$$
$$\left[u(s, d) + \rho \beta \int_{s'} W(s')p(s'|s, d) \right].$$

$$\tag{12}$$

The existence and uniqueness of V is a consequence of the *contraction mapping theorem,* since Γ can be shown to satisfy the contraction property,

$$\|\Gamma W - \Gamma V\| \leq \alpha \|W - V\|, \tag{13}$$

where $\alpha \in (0,1)$ and $\|W\| = \sup_{s \in S}|W(s)|$. In this case, $\alpha = \rho\beta$, so the Bellman operator will be a contraction mapping if $\rho\beta \in (0,1)$.

The proof of the optimality of the decision rule δ in eq. (10) is somewhat more involved. Using the Bellman equation (10), we will show that (see eq. (34) in section 4),

$$V(s) = u(s, \delta(s)) + \rho\beta \int_{s'} V(s')p(s'|s, \delta(s))$$

$$= E\left\{ \sum_{t=0}^{\infty} [\rho\beta]^t u(s_t, \delta(s_t)) \middle| s_0 = s \right\},$$

$$(14)$$

that is, V is the value function implied by the decision rule δ. Intuitively, the boundedness of the utility function, combined with discounting of future utilities, $\rho\beta \in (0,1)$, implies that if we truncate the infinite horizon problem to a T period problem, the error in doing so would be arbitrarily small when T is sufficiently large. Indeed, this is the key to understanding how to find approximately optimal decision rules to infinite horizon SDPs: *we approximate the infinite horizon decision rule δ by solving an approximate finite horizon version of the problem by dynamic programming*. The validity of this approach can be formalized using a well-known property of contraction mappings, namely, that the *method of successive approximations* starting from any initial guess W converges to the fixed point of Γ, that is

$$\lim_{t \to \infty} V_t = \Gamma^t(W) = V \quad \forall W \in B, \qquad (15)$$

where $\Gamma^t W$ denotes t successive iterations of the Bellman operator Γ,

$$V_0 = \Gamma^0(W) = W$$
$$V_1 = \Gamma^1(W)$$
$$\cdots$$
$$V_t = \Gamma^t(W) = \Gamma(\Gamma^{t-1}W) = \Gamma(V_{t-1}).$$

$$(16)$$

If $W = 0$ (that is the zero function in B), then $V_T = \Gamma^T(0)$ is simply the period $t = 1$ value function resulting from the solution of a T period dynamic programming problem. Thus, this result implies that the optimal value function V_T for a T-period approximation to the infinite horizon problem converges to V as $T \to \infty$. Moreover, the difference in the two functions satisfies the bound

$$\|V_T - V\| \leq \frac{[\rho\beta]^T \|u\|}{1 - \rho\beta}. \qquad (17)$$

Let $\delta_T = \delta_{1,T}, \delta_{2,T} \ldots, \delta_{T,T}$ be the optimal decision rule to the T period problem. It can be shown that, if we follow this decision rule up to period T and then use $\delta_{1,T}$ in

every period after T, the resulting decision rule is approximately optimal in the sense that the value function for this infinite horizon problem also satisfies inequality (17), and thus can be made arbitrarily small as T increases.

In many cases in economics the state space S has no natural upper bound. An example might be where s_t denotes an individual's wealth at time t, or the capital stock of the firm. If the unboundedness of the state space results in unbounded payoffs, the contraction mapping argument must be modified since the Banach space structure under the *supremum* norm no longer applies to unbounded functions. Various alternative approaches have been used to prove existence of optimal decision rules for unbounded problems. One is to use an alternative norm (for example, a *weighted norm*) and demonstrate that the Banach space/contraction mapping argument still applies. However, there are cases where there are no natural weighted norms, and the contraction mapping property cannot hold since the Bellman equation can be shown to have multiple solutions. The most general conditions under which the existence and uniqueness of the solution V to the Bellman equation and the optimality of the implied stationary decision rule δ has been established is in Bhattacharya and Majumdar (1989). However, as I discuss in the next section, considerable care must be taken in solving unbounded problems numerically.

4 Numerical dynamic programming and the curse of dimensionality

The previous section showed that dynamic programming is a powerful tool that has enabled us to formulate and solve a wide range of economic models involving sequential decision-making under uncertainty – at least 'in theory'. Unfortunately, the cases where dynamic programming results in *analytical, closed-form solutions* are rare and often rather fragile in the sense that small changes in the formulation of a problem can destroy the ability to obtain an analytical solution. However even though most problems do not have analytical solutions, the theorems in the previous section guarantee the *existence* of solutions, and these solutions can be calculated (or approximated) by numerical methods. Since the 1980s, faster computers and better numerical methods have made dynamic programming a tool of substantial practical value by significantly expanding the range of problems that can be solved. In particular, it has led to the development of a large and rapidly growing literature on econometric estimation and testing of 'dynamic structural models' that I will discuss in the next section.

However, there are still many difficult challenges that prevent us from formulating and solving models that are as detailed and realistic as we might like, a problem that is especially acute in empirical applications. The principal challenge is what Bellman and Dreyfus (1962) called *the curse of dimensionality.* We have already illustrated this problem in Section 3.4: for history-dependent SDPs with a finite horizon T and a finite number of states N and actions D, dynamic programming requires $O([ND]^T)$ operations to find a solution. Thus it appears that the time required to compute a solution via dynamic programming increases *exponentially fast* with the number of possible decisions or states in a dynamic programming problem.

Fortunately, computer power (for example, operations per second) has also been growing exponentially fast, a consequence of *Moore's Law* and other developments in information technology, such as improved communications and massive parallel processing. Bellman and Dreyfus (1962) carried out calculations on RAND's 'Johnniac' computer (named in honour of Jon von Neumann, whose work contributed to the development of the first electronic computers) and reported that this machine could do 12,500 additions per second. Nowadays, in 2007, a typical laptop computer can do over a billion operations per second and we now have supercomputers that are approaching a *thousand trillion operations per second* – a level known as a 'petaflop'. In addition to faster 'hardware', research on numerical methods has resulted in significantly better 'software' that has had a huge impact on the spread of numerical dynamic programming and on the range of problems we can solve. In particular, algorithms have been developed that succeed in 'breaking' the curse of dimensionality, enabling us to solve in polynomial time classes of problems that were previously believed to be solvable only in exponential time. The key to breaking the curse of dimensionality is the ability to recognize and exploit *special structure* in an SDP problem. We have already illustrated an example of this in Section 3.4: if the SDP is Markovian and utility is time separable, a finite horizon, finite state SDP can be solved by dynamic programming in only $O(NDT)$ operations, compared to the $O([ND]^T)$ operations that are required in the general history-dependent case. There is only enough space here to discuss several of the most commonly used and most effective numerical methods for solving different types of SDPs by dynamic programming. I refer the reader to Puterman (1994), Rust (1996) and Judd (1998) for more in-depth surveys on the literature on numerical dynamic programming. See COMPUTATIONAL METHODS IN ECONOMETRICS.

Naturally, the numerical method that is appropriate or 'best' depends on the type of problem being solved. Different methods are applicable depending on whether the problem has (*a*) finite versus infinite horizon, (*b*) finite versus continuous-valued state and decision variables, and (*c*) single versus multiple players. In finite horizon problems, backward induction is the essentially the only approach, although as we will see there are many different choices about how to most implement it most efficiently – especially in discrete problems where the number of possible values for the state variables is huge

(for example, chess) or in problems with continuous state variables. In the latter case, it is clearly not possible to carry out backward induction for every possible history (or value of the state variable at stage t if the problem is Markovian and time separable), since there are infinitely many (indeed a continuum) of them. In these cases, it is necessary to *interpolate* the value function, whose values are only explicitly computed at a finite number of points in the state space. I use the term 'grid' to refer to the finite number of points in the state space where the backward induction calculations are actually performed. Grids might be *lattices* (that is, regularly spaced sets of points formed as Cartesian products of unidimensional grids for each of the continuous state variables), or they may be *quasi-random grids* formed by randomly sampling the state space from some probability distribution, or by generating deterministic sequences of points such as *low discrepancy sequences*. The reason why one might choose a random or low-discrepancy grid instead of regularly spaced lattice is to break the curse of dimensionality, as I discuss shortly. Also, in many cases it is advantageous to refine the grid over the course of the backward induction process, starting out with an initial 'coarse' grid with relatively few points and subsequently increasing the number of points in the grid as the backward induction progresses. I will have more to say about such *multigrid* and *adaptive grid* methods when I discuss solution of infinite horizon problems below.

Once a particular grid is chosen, the backward induction process is carried out in the way it would be normally be done in a finite state problem. On the assumption that the problem is Markovian and the utility is time separable and there are n grid points $\{s_1, \ldots, s_n\}$, this involves the following calculation at each grid point s_i, $i = 1, \ldots, n$

$$V_t(s_i) = \max_{d \in D_t(s_i)} [u_t(s_i, d) + \rho\beta\hat{E}V_{t+1}(s_i, d)],$$

$$(18)$$

where $\hat{E}V_{t+1}(s_i, d)$ is a numerical estimate of the conditional expectation of next period's value function. I will be more specific below about which numerical integration methods are appropriate, but at this point it suffices to note that they are all simple weighted sums of values of the value function at $t+1$, $V_{t+1}(s)$. We can now see that, even if the actual backward induction calculations are carried out only at the n grid points $\{s_1, \ldots, s_n\}$, we will still have to do numerical integration to compute $\hat{E}V_{t+1}(s_i, d)$ and the latter calculation may require values of $V_{t+1}(s)$ at points s off the grid, that is at points $s \notin \{s_1, \ldots, s_n\}$. This is why some form or interpolation (or in some cases *extrapolation*) is typically required. Almost all methods of interpolation can be represented as weighted sums of the value function at its known values $\{V_{t+1}(s_1), \ldots, V_{t+1}(s_n)\}$ at the n grid points, which were calculated by backward induction at the previous stage. Thus, we have

$$\hat{V}_{t+1}(s) = \sum_{j=1}^{n} w_i(s) V_{t+1}(s_i), \qquad (19)$$

where $w_i(s)$ is a weight assigned to the ith grid point that depends on the point s in question. These weights are typically positive and sum to 1. For example in *multi-linear interpolation* or *simplicial interpolation* the $w_i(s)$ weights are those that allow s to be represented as a convex combination of the vertices of the smallest lattice hypercube containing s. Thus, the weights $w_i(s)$ will be zero for all i except the immediate neighbours of the point s. In other cases, such as *kernel density* and *local linear regression*, the weights $w_i(s)$ are generally non-zero for all i, but the weights will be highest for the grid points $\{s_1, \ldots, s_n\}$ which are the *nearest neighbours* of s. An alternative approach can be described as *curve fitting*. Instead of attempting to interpolate the calculated values of the value function at the grid points, this approach treats these values as a *data-set* and estimates parameters θ of a flexible functional form approximation to $V_{t+1}(s)$ by *nonlinear regression*. Using the estimated $\hat{\theta}_{t+1}$ from this nonlinear regression, we can 'predict' the value of $V_{t+1}(s)$ at any $s \in S$

$$\hat{V}_{t+1}(s) = f(s, \hat{\theta}_{t+1}). \qquad (20)$$

A frequently used example of this approach is to approximate $V_{t+1}(s)$ as a linear combination of K 'basis functions' $\{b_1(s), \ldots, b_K(s)\}$. This implies that $f(s, \theta)$ takes the form of a *linear regression* function

$$f(s, \theta) = \sum_{k=1}^{K} \theta_k b_k(s), \qquad (21)$$

and $\hat{\theta}_{t+1}$ can be estimated by *ordinary least squares*. Neural networks are an example where f depends on θ in a nonlinear fashion. Partition θ into subvectors $\theta = (\gamma, \lambda, \alpha)$, where γ and λ are vectors in R^J, and $\alpha = (\alpha_1, \ldots, \alpha_J)$, where each α_j has the same dimension as the state vector s. Then the neural network f is given by

$$f(s, \theta) = f(s, \gamma, \lambda, \alpha) = \sum_{j=1}^{J} \gamma_j \phi(\lambda_j + \langle s, \alpha_i \rangle) \qquad (22)$$

where $\langle s, \alpha_j \rangle$ is the inner product between s and the conformable vector α_j, and ϕ is a 'squashing function' such as the logistic function $\phi(x) = \exp\{x\}/(1 + \exp\{x\})$. Neural networks are known to be 'universal approximators' and require relatively few parameters to provide good approximations to nonlinear functions of many variables. For further details on how neural networks are applied, see the book by Bertsekas and Tsitsiklis (1996) on *Neuro-Dynamic Programming*.

All these methods require extreme care for problems with *unbounded state spaces*. By definition, any finite grid can cover only a small subset of the state space in this case, and thus any of the methods discussed above would require *extrapolation* of the value function to predict its values in regions where there are no grid points, and thus 'data' on what its proper values should be. Not only may mistakes that lead to incorrect extrapolations in these regions lead to errors in the regions where there are no grid points, but the errors can 'unravel' and also lead to considerable errors in approximating the value function in regions where we do have grid points. Attempts to 'compactify' an unbounded problem by arbitrarily truncating the state space may also lead to inaccurate solutions, since the truncation is itself an implicit form of extrapolation (for example, some assumption needs to be made what to do when state variables approach the 'boundary' of the state space: do we assume a 'reflecting boundary', an 'absorbing boundary', and so on?). For example in life-cycle optimization problems, there is no natural upper bound on wealth, even if it is true that there is only a finite amount of wealth in the entire economy. We can always ask the question, if a person had wealth near the 'upper bound', what would happen to next period wealth if he invested some of it? Here we can see that, if we extrapolate the value function by assuming that the value function is bounded in wealth, this means that by definition there is no incremental return to saving as we approach the upper bound. This leads to lower saving, and this generally leads to errors in the calculated value function and decision rule far below the assumed upper bound. There is no good general solution to this problem except to solve the problem on a much bigger (bounded) state space than one would expect to encounter in practice, in the hope that extrapolation-induced errors in approximating the value function die out the further one is from the boundary. This property should hold for problems where the probability that the next period state will hit or exceed the 'truncation boundary' gets small the farther the current state is from this boundary.

When a method for interpolating/extrapolating the value function has been determined, a second choice must be made about the appropriate method for *numerical integration* in order to approximate the conditional expectation of the value function $EV_{t+1}(s, d)$ given by

$$EV_{t+1}(s, d) = \int_{s'} V_{t+1}(s') p_{t+1}(s'|s, d). \qquad (23)$$

There are two main choices here: (1) deterministic quadrature rules or (2) (quasi-) Monte Carlo methods. Both methods can be written as weighted averages of form

$$\hat{EV}_{t+1}(s, d) = \sum_{i=1}^{N} w_i(s, d) V_{t+1}(a_i), \qquad (24)$$

where $\{w_i(s, d)\}$ are *weights*, and $\{a_i\}$ are *quadrature abscissae*. Deterministic quadrature methods are highly accurate (for example, an N-point Gaussian quadrature rule is constructed to exactly integrate all polynomials of degree $2N-1$ or less), but become unwieldy in multivariate integration problems when *product rules* (tensor products of unidimensional quadrature) are used. *Any sort of deterministic quadrature method can be shown to be subject to the curse of dimensionality in terms of worst-case computational complexity* (see Traub and Werschulz, 1998). For example, if $N = O(1/\varepsilon)$ quadrature points are necessary to approximate a univariate integral within ε, then in a d-dimensional integration problem $N^d = O(1/\varepsilon^d)$ quadrature points would be necessary to approximate the integral with an error of ε, which implies that computational effort to find an ε-approximation increases exponentially fast in the problem dimension d. Using the theory of computational complexity, one can prove that *any* deterministic integration procedure is subject to the curse of dimensionality, at least in terms of a 'worst case' measure of complexity. The curse of dimensionality can disappear if one is willing to adopt a Bayesian perspective and place a 'prior distribution' over the space of possible integrands and consider an 'average case' instead of a 'worst case' notion of computational complexity.

Since multivariate integration is a 'sub-problem' that must be solved in order to carry out dynamic programming when there are continuous state variables (indeed, dynamic programming in principle involves infinitely many integrals in order to calculate $EV_{t+1}(s, d)$, one for each possible value of (s, d)), if there is a curse of dimensionality associated with numerical integration of a single multivariate integral, then it should also not be surprising that dynamic programming is also subject to the same curse. There is also a curse of dimensionality associated with global optimization of nonconvex objective functions of continuous variables. Since optimization is also a sub-problem of the overall dynamic programming problem, this constitutes another reason why dynamic programming is subject to a curse of dimensionality. Under the standard worst case definition of computational complexity, Chow and Tsitsiklis (1989) proved that *no* deterministic algorithm can succeed in breaking the curse of dimensionality associated with a sufficiently broad class of dynamic programming problems with continuous state and decision variables. This negative result dashes the hopes of researchers dating back to Bellman and Dreyfus (1962), who conjectured that there might be sufficiently clever deterministic algorithms that can overcome the curse of dimensionality.

However, there are examples of *random algorithms* that can circumvent the curse of dimensionality. Monte Carlo integration is a classic example. Consider approximating the (multidimensional) integral in eq. (23) by using *random* quadrature abscissae $\{\tilde{a}_i\}$ that are N independent and identically distributed (*IID*) draws from the distribution $p_{t+1}(s'|s, d)$ and uniform quadrature weights equal to $w_i(s, d) = 1/N$. Then the law of large numbers and the central limit theorem imply that the Monte Carlo integral $\hat{E}V_{t+1}(s, d)$ converges to the true conditional expectation $EV_{t+1}(s, d)$ at rate $1/\sqrt{N}$ *regardless of the dimension of the state space* d. Thus a random algorithm, Monte Carlo integration, succeeds in breaking the curse of dimensionality of multivariate integration. Unfortunately, randomization does *not* succeed in breaking curse of dimensionality associated with general nonconvex optimization problems with continuous multidimensional decision variables d (see Nemirovsky and Yudin, 1983).

However, naive application of Monte Carlo integration will not necessarily break the curse of dimensionality of the dynamic programming problem. The reason is that a form of *uniform convergence* (as opposed to pointwise) convergence of the conditional expectations $\hat{E}V_{t+1}(s, d)$ to $EV_{t+1}(s, d)$ is required in order to guarantee that the overall backward induction process converges to the true solution as the number of Monte Carlo draws, N, gets large. To get an intuition why, note that if separate *IID* sets of quadrature abscissae $\{\tilde{a}_i\}$ where drawn for each (s, d) point that we wish to evaluate the Monte Carlo integral $\hat{E}V_{t+1}(s, d)$ at, the resulting function would be an extremely 'choppy' and irregular function of (s, d) as a result of all the random variation in the various sets of quadrature abscissae. Extending an idea introduced by Tauchen and Hussey (1991) to solve rational expectations models, Rust (1997) proved that it is possible to break the curse of dimensionality in a class of SDPs where the choice sets $D_t(s)$ are finite, a class he calls *discrete decision processes*. The restriction to finite choice sets is necessary, since, as noted above, randomization does not succeed in breaking the curse of dimensionality of nonconvex optimization problems with continuous decision variables. The key idea is to choose, as a *random grid*, the same set of random points that are used quadrature abscissae for Monte Carlo integration. That is, suppose $p_{t+1}(s'|s, d)$ is a transition *density* and the state space (perhaps after translation and normalization) is identified with the d-dimensional *hypercube* $S = [0,1]^d$. Apply Monte Carlo integration by drawing N IID points $\{\tilde{s}_1, \ldots, \tilde{s}_N\}$ from the this hypercube (this can be accomplished by drawing each component of s_i from the uniform distribution on the $[0,1]$ interval). We have

$$\hat{E}V_{t+1}(s, d) = \frac{1}{N} \sum_{i=1}^{N} V_{t+1}(\tilde{s}_i) p_{t+1}(\tilde{s}_i|s, d).$$

$$(25)$$

Applying results from the theory of *empirical processes* (Pollard, 1989), Rust showed that this form of the Monte Carlo integral does result in uniform convergence (that is, $\mathrm{P}\hat{E}V_{t+1}(s, d) - EV_{t+1}(s, d)\mathrm{P} = O_p(1/\sqrt{N})$), and, using this, he showed that this randomized version of backward induction succeeds in breaking the curse of dimensionality of the dynamic programming problem. The intuition of why this works is, instead of trying to approximate the conditional expectation in (23) by computing *many independent Monte Carlo integrals* (that is, drawing separate sets of random abscissae $\{\tilde{a}_i\}$ from $p_{t+1}(s'|s,d)$ for each possible value of (s,d)), the approach in eq. (25) is to compute a *single Monte Carlo integral* where the random quadrature points $\{\tilde{s}_i\}$ are drawn from the uniform distribution on $[0,1]^d$, and the integrand is treated as the *function* $V_{t+1}(s')p_{t+1}(s'|s,d)$ instead of $V_{t+1}(s')$. The second important feature is that eq. (25) has a *self-approximating* property: that is, since the quadrature abscissae are the same as the grid points at which we compute the value function, no auxiliary interpolation or function approximation is necessary in order to evaluate $\hat{E}V_{t+1}(s,d)$. In particular, if $p_{t+1}(s'|s,d)$ is a smooth function of s, then $\hat{E}V_{t+1}(s,d)$ will also be a smooth function of s. Thus, backward induction using this algorithm is extremely simple. Before starting backward induction we choose a value for N and draw N IID random vectors $\{\tilde{s}_1,\ldots,\tilde{s}_N\}$ from the uniform distribution on the d-dimensional hypercube. This constitutes a random grid that remains fixed for the duration of the backward induction. Then we begin ordinary backward induction calculations, at each stage t computing $V_t(\tilde{s}_i)$ at each of the N random grid points, and using the self-approximating formula (25) to calculate the conditional expectation of the period $t+1$ value function using only the N stored values $(V_{t+1}(\tilde{s}_1),\ldots,V_{t+1}(\tilde{s}_N))$ from the previous stage of the backward induction. See Keane and Wolpin (1994) for an alternative approach, which combines Monte Carlo integration with the curve-fitting approaches discussed above. Note that the Keane and Wolpin approach will not generally succeed in breaking the curse of dimensionality since it requires approximation of functions of d variables which is also subject to a curse of dimensionality, as is well known from the literature on *nonparametric regression*.

There are other subclasses of SDPs for which it is possible to break the curse of dimensionality. For example, the family of *linear quadratic/Gaussian* (LQG) can be solved in polynomial time using highly efficient matrix methods, including efficient methods for solving the *matrix Ricatti equation* which is used to compute the *Kalman filter* for Bayesian LQG problems (for example, problems where agents only receive a noisy signal of a state variable of interest, and they update their beliefs about the unknown underlying state variable via Bayes rule).

Now consider stationary, infinite horizon Markovian decision problems. As noted in Section 3.4, there is no 'last' period from which to begin the backward induction process. However, if the utility function is time separable and discounted, then, under fairly general conditions, it will be possible to approximate the solution arbitrarily closely by solving a finite horizon version of the problem, where the horizon T is chosen sufficiently large. As we noted in Section 3.4, this is equivalent to solving for V, the fixed point to the contraction mapping $V=\Gamma(V)$ by the method of *successive approximations*, where Γ is the *Bellman operator* defined in eq. (12) of Section 3.4.

$$V_{t+1} = \Gamma(V_t). \tag{26}$$

Since successive approximations converges at a geometric rate, with errors satisfying the upper bound in eq. (17), this method can converge at an unacceptably slow rate when the discount factor is close to 1. A more effective algorithm in such cases is *Newton's method* whose iterates are given by

$$V_{t+1} = V_t - [I - \Gamma'(V_t)]^{-1}[V_t - \Gamma(V_t)], \tag{27}$$

where Γ' is the *Gateaux* or *directional derivative* of Γ, that is, it is the linear operator given by

$$\Gamma'(V)(W) = \lim_{t\to 0} \frac{\Gamma(V+tW) - \Gamma(V)}{t}. \tag{28}$$

Newton's method converges *quadratically* independent of the value of the discount factor, as long as it is less than 1 (to guarantee the contraction property and the existence of a fixed point). In fact, Newton's method turns out to be equivalent to the method of *policy iteration* introduced by Howard (1960). Let δ be any stationary decision rule, that is, a candidate *policy*. Define the policy-specific conditional expectation operator E_δ by

$$E_\delta V(s) = \int_{s'} V(s')p(s'|s,\delta(s)). \tag{29}$$

Given a value function V_t, let δ_{t+1} be the decision rule implied by V_t, that is

$$\delta_{t+1}(s) = \underset{d\in D(s)}{\text{argmax}}$$

$$\left[u(s,d) + \rho\beta\int_{s'} V_t(s')p(s'|s,d)\right]. \tag{30}$$

It is not hard to see that the value of policy δ_{t+1} must be at least as high as V_t, and for this reason, eq. (30) is called the *policy improvement step* of the policy iteration algorithm. It is also not hard to show that

$$\Gamma'(V_t)(W)(s) = \rho\beta E_{\delta_{t+1}}W(s), \tag{31}$$

and this implies that the Newton iteration, eq. (27), is numerically identical to *policy iteration*

$$V_{t+1}(s) = [I - \rho\beta E_{\delta_{t+1}}]^{-1}u(s,\delta_{t+1}(s)), \tag{32}$$

where δ_{t+1} is given in eq. (30). Equation (32) is called the *policy valuation step* of the policy iteration algorithm since it calculates the value function implied by the policy δ_{t+1}. Note that, since E_δ is an expectation operator, it is linear and satisfies $\|E_\delta\| \leq 1$, and this implies that the operator $[I - \rho\beta E_\delta]$ is invertible and has the following geometric series expansion

$$[I - \rho\beta E_\delta]^{-1} = \sum_{j=0}^{\infty} [\rho\beta]^j E_\delta^j, \qquad (33)$$

where E_δ^j is the j step ahead expectations operator. Thus, we see that

$$[I - \rho\beta E_\delta]^{-1} u(s, \delta(s))$$
$$= \sum_{j=0}^{\infty} [\rho\beta]^j E_\delta^j u(s, \delta(s))$$
$$= E\left\{ \sum_{t=0}^{\infty} [\rho\beta]^t u(s_t, \delta(s_t)) \,\middle|\, s_0 = s \right\}, \qquad (34)$$

so that value function V_t from the policy iteration (32) corresponds to the expected value implied by policy (decision rule) δ_t.

If there are an infinite number of states, the expectations operator E_δ is an infinite-dimensional linear operator, so it is not feasible to compute an exact solution to the policy iteration eq. (32). However if there are a finite number of states (or an infinite state space is discretized to a finite set of points, as per the discussion above), then E_δ is an $N \times N$ transition probability matrix, and policy iteration is feasible using ordinary matrix algebra, requiring at most $O(N^3)$ operations to solve a system of linear equations for V_t at each policy valuation step. Further, when there are a finite number of possible actions as well as states, there are only a finite number of possible policies $|D|^{|S|}$, where $|D|$ is the number of possible actions and $|S|$ is the number of states, and policy iteration can be shown to converge in a finite number of steps, since the method produces an improving sequences of decision rules, that is $V_t \leq V_{t+1}$. Thus, since there is an upper bound on the number of possible policies and policy iteration cannot cycle, it must converge in a finite number of steps. The number of steps is typically quite small, far fewer than the total number of possible policies. Santos and Rust (2004) show that the number of iterations can be bounded independent of the number of elements in the state space $|S|$. Thus, policy iteration is the method of choice for infinite horizon problems for which the discount factor is sufficiently close to 1. However, if the discount factor is far enough below 1, then successive approximations can be faster since policy iteration requires $O(N^3)$ operations per iteration whereas successive approximations requires $O(N^2)$ operations per

iteration. At most $T(\varepsilon, \beta)$ successive approximation iterations are required to compute an ε-approximation to an infinite horizon Markovian decision problem with discount factor β, where $T(\varepsilon, \beta) = \log((1 - \beta)\varepsilon) / \log(\beta)$. Roughly speaking, if $T(\varepsilon, \beta) < N$, then successive approximations are faster than policy iteration.

Successive approximations can be accelerated by a number of means discussed in Puterman (1994) and Rust (1996). *Multigrid algorithms* are also effective: these methods begin backward induction with a coarse grid with relatively few grid points N, and then as iterations proceed, the number of grid points is successively increased, leading to finer and finer grids as the backward induction starts to converge. Thus, computational time is not wasted early on in the backward induction iterations when the value function is far from the true solution. *Adaptive grid* methods are also highly effective in many problems: these methods can automatically detect regions in the state space where there is higher curvature in the value function, and in these regions more grid points are added in order to ensure that the value function is accurately approximated, whereas in regions where the value function is 'flatter' grid points can be removed, so as to direct computational resources to the regions of the state space where there is the highest payoff in terms of accurately approximating the value function. See Grüne and Semmler (2004) for more details and an interesting application of adaptive grid algorithms.

I conclude this section with a discussion of several other alternative approaches to solving stationary infinite horizon problems that can be extremely effective relative to 'discretization' methods when the number of grid points N required to obtain a good approximation becomes very large. Recall the curve-fitting approach discussed above in finite horizon SDPs: we approximate the value function V by a parametric function as $V_\theta(s) \equiv f(s, \theta)$ for some flexible functional form f, where θ are treated as unknown parameters to be 'estimated'. For infinite horizon SDPs, our goal is to find parameter values $\hat{\theta}$ so that the implied value function satisfies the Bellman equation as well as possible. One approach to doing this, known as the *minimum residual method*, is a direct analogue of nonlinear least squares: if θ is a vector with K components, we select $N \geq K$ points in the state space (potentially at random) and find $\hat{\theta}$ that minimizes the squared deviations or *residuals* in the Bellman equation

$$\hat{\theta} = \underset{\theta \in R^K}{\text{argmin}} \sum_{i=1}^{N} [\hat{\Gamma}(V_\theta)(s_i) - V_\theta(s_i)]^2, \qquad (35)$$

where $\hat{\Gamma}$ denotes an approximation to the Bellman operator, where some numerical integration and optimization algorithm are used to approximate the true expectation operator and maximization in the Bellman equation (12). Another approach, called the *collocation method*, finds $\hat{\theta}$

by choosing K grid points in the state space and setting the residuals at those K points to zero:

$$V_{\hat{\theta}}(s_1) = \hat{\Gamma}(V_{\hat{\theta}})(s_1)$$
$$V_{\hat{\theta}}(s_2) = \hat{\Gamma}(V_{\hat{\theta}})(s_2)$$
$$\cdots \qquad (36)$$
$$V_{\hat{\theta}}(s_K) = \hat{\Gamma}(V_{\hat{\theta}})(s_K).$$

Another approach, called *parametric policy iteration*, carries out the policy iteration algorithm in eq. (32) above, but, instead of solving the linear system (32) for the value function V_t at each policy valuation step, they approximately solve this system by finding θ_t that solves the regression problem

$$\theta_t = \underset{\theta \in R^K}{\mathrm{argmin}} \sum_{i=1}^{N} [V_{\hat{\theta}_t}(s_i)$$
$$- u(s_i, \delta_t(s_i)) - \rho\beta E_{\delta_t} V_{\theta_t}(s_i)]^2. \qquad (37)$$

Other than this, policy iteration proceeds exactly as discussed above. Note that, due to the linearity of the expectations operator, the regression problem above reduces to an ordinary linear regression problem when V_θ is approximated as a linear combination of basis functions as in (21) above.

There are variants of the minimum residual and collocation methods that involve parameterizing the *decision rule* rather than the value function. These methods are frequently used in problems where the control variable is continuous, and construct residuals from the *Euler equation* – a functional equation for the decision rule that can in certain classes of problems be derived from the first-order necessary condition for the optimal decision rule. These approaches then try to find $\hat{\theta}$ so that the Euler equation (as opposed to the Bellman equation) is approximately satisfied, in the sense of minimizing the squared residuals (minimum residual approach) or setting the residuals to zero at K specified points in the state space (collocation method). See Judd (1998) for further discussion of these methods and a discussion of strategies for choosing the grid points necessary to implement the collocation or minimum residual method.

There is a variety of other *iterative stochastic algorithms* for approximating solutions to dynamic programming problems that have been developed in the computer science and 'artificial intelligence' literatures on *reinforcement learning*. These methods include *Q-learning*, *temporal difference learning*, and *real time dynamic programming*. The general approach in all these methods is to iteratively update an estimate of the value function, and recursive versions of Monte Carlo integration methods are employed in order to avoid doing numerical integrations to calculate conditional expectations. Using methods adapted from the literature on *stochastic approximation*, it is possible to prove that these methods converge to the true value function in the limit as the number of iterations tends to infinity. A key assumption underlying the convergence proofs is that there is sufficient stochastic noise to ensure that all possible decisions and decision nodes are visited 'infinitely often'. The intuition of why such an assumption is necessary follows from the discussion in Section 3: suppose that at some state s an initial estimate of the value function for decision that is actually optimal happens to be so low that the action is deemed to be 'nonoptimal' relative to the initial estimate. If the agent does not 'experiment' sufficiently, and thus fails to choose suboptimal decisions infinitely often, the agent may fail to learn that the initial estimated value was an underestimate of the true value, and therefore the agent might never learn that the corresponding action really is optimal. There is a trade-off between learning and experimentation, of course. The literature on 'multi-armed bandits' (Gittins, 1979) shows that a fully rational Bayesian decision-maker will generally not find it optimal to experiment infinitely often. As a result such an agent can fail to discover actions that are optimal in an *ex post* sense. However, this does not contradict the fact that their behaviour is optimal in an *ex ante* sense: rather, it is a reflection that learning and experimentation is a costly activity, and thus it can be optimal to be incompletely informed, a result that has been known as early as Wald (1947). A nice feature of many of these methods, particularly the real time dynamic programming developed in Barto, Bradtke and Singh (1995), is that these methods can be used in 'real time', that is, we do not have to 'precalculate' the optimal decision rule in 'offline' mode. All these algorithms result in steady improvement in performance with experience. Methods similar to these have been used to produce highly effective strategies in extremely complicated problems. An example is IBM's 'Deep Blue' computer chess strategy, which has succeeded in beating the world's top human chess player, Garry Kasparov. However, the level of computation and repetition necessary to 'train' effective strategies is hugely time consuming, and it is not clear that any of these methods succeed in breaking the curse of dimensionality. For further details on this literature, see Bertsekas and Tsitsiklis (1996). Pakes (2001) applies these methods to approximate Markov perfect equilibria in games with many players. All types of stochastic algorithms have the disadvantage that the approximate solutions can be 'jagged' and there is always at least a small probability that the converged solution can be far from the true solution. However, they may be the only feasible option in many complex, high-dimensional problems where deterministic algorithms (for example, the Pakes and McGuire, 1994, algorithm for Markov perfect equilibrium) quickly become intractable due to the curse of dimensionality.

5 Empirical dynamic programming and the identification problem

The developments in numerical dynamic programming described in the previous section paved the way for a new, rapidly growing literature on empirical estimation and testing of SDPs and dynamic games. This literature began to take shape in the late 1970s, with contributions by Sargent (1978) on estimation of dynamic labour demand schedules in a linear quadratic framework, and Hansen and Singleton (1982), who developed a generalized method of moment estimation strategy for a class of continuous choice SDPs using the *Euler equation* as an *orthogonality condition*. About the same time, a number of papers appeared that provided different strategies for estimation and inference in *dynamic discrete choice models* including Gotz and McCall's (1980) model of retirements of air force pilots, Wolpin's (1984) model of a family's decision whether or not to have a child, Pakes's (1986) model of whether or not to renew a patent, and Rust's (1987) model of whether or not to replace a bus engine. Since 1987, hundreds of different empirical applications of dynamic programming models have been published. For surveys of this literature see Eckstein and Wolpin (1989), Rust (1994), and the very readable book by Adda and Cooper (2003) – which also provides accessible introductions to the theory and numerical methods for dynamic programming. The remainder of this section will provide a brief overview of estimation methods and a discussion of the identification problem.

In econometrics, the term *structural estimation* refers to a class of methods that tries to go beyond simply summarizing the behaviour of economic agents by attempting to infer their underlying *preferences* and *beliefs*. This is closely related to the distinction between the *reduced-form* of an economic model and the underlying *structure* that 'generates' it. (Structural estimation methods were first developed at the Cowles Commission at Yale University, starting with attempts to structurally estimate the linear simultaneous equations model, and models of investment by firms. Frisch, Haavelmo, Koopmans, Marschak, and Tinbergen were among the earliest contributors to this literature.) The reason why one would want to do structural estimation, which is typically far more difficult (for example, computationally intensive) than reduced-form estimation, is having knowledge of underlying structure enables us to conduct *hypothetical/counterfactual policy experiments*. Reduced-form estimation methods can be quite useful and yield significant insights into behaviour, but they are limited to summarizing behaviour under the status quo. However, they are inherently limited in their ability to forecast how individuals change their behaviour in response to various changes in the environment, or in *policies* (for example, tax rates, government benefits, regulations, laws, and so on) that *change the underlying structure* of agents' decision problems. As long as it is possible to predict how different policies change the underlying structure, we can use dynamic programming to re-solve agents' SDPs under the alternative structure, resulting in corresponding decision rules that represent predictions of how their behaviour (and welfare) will change in response to the policy change.

The rationale for structural estimation was recognized as early as Marschak (1953); however, his message appears to have been forgotten until the issue was revived in Lucas's (1976) critique of the limitations of reduced-form methods for policy evaluation. An alternative way to do policy evaluation is via *randomized experiments* in which subjects are randomly assigned to the *treatment group* (where the 'treatment' is some alternative policy of interest) and the *control group* (who continue with the policy under the status quo). By comparing the outcomes in the treatment and control groups, we can assess the behavioural and welfare impacts of the policy change. However, human experiments can be very time consuming and expensive to carry out, whereas 'computational experiments' using a structural model are very cheap and can be conducted extremely rapidly. The drawback of the structural approach, though, is the issue of *credibility* of the structural model. If the structural model is *misspecified*, it can generate incorrect forecasts of the impact of a policy change. There are numerous examples of how structural models can be used to make policy predictions: see Todd and Wolpin (2005) for an example that compares the prediction of a structural model with the results of a randomized experiment, where the structural model is estimated using subjects from the control group, and *out-of-sample predictions* are made to predict the behavioural response by subjects in the treatment group. They show that the structural model results in accurate predictions of how the treatment group subjects responded to the policy change.

I illustrate the main econometric methods for structural estimation of SDPs in the case of a stationary infinite horizon Markovian decision problem, although all the concepts extend in a straightforward fashion to finite horizon, nonstationary and non-Markovian problems. Estimation requires a specification of the *data generating process*. Assume we observe N agents, and we observe agent i from time period T_{-i} to $\bar{T}_i$ (or via appropriate re-indexing, from $t = 1, \ldots, T_i$). Assume observations of each individual are independently distributed realizations from the controlled process $\{s_t, d_t\}$. However, while we assume that we can observe the decisions made by each agent, it is more realistic to assume that we only observe a subset of the agent's state s_t. If we partition $s_t = (x_t, \varepsilon_t)$, assume that the econometrician observes x_t but not ε_t, so this latter component of the state vector constitutes an *unobserved state variable*. Then the reduced-form of the SDP is the decision rule δ

$$d = \delta(x, \varepsilon), \tag{38}$$

since the decision rule embodies all the behavioural content of the SDP model. The *structure* Λ consists of the objects $\Lambda = \{\beta, \rho, u(s,d), p(s'|s,d)\}$. Equation (10) specifies the mapping from the structure Λ into the reduced form, δ. The data-set consists of $\{(x_{i,t}, d_{i,t}), \ t = 1, \ldots, T_i, \ i = 1, \ldots, N\}$. The econometric problem is to infer the underlying structure Λ from our data on the observed states and decisions by a set of individuals. Although the decision rule is potentially a complicated nonlinear function of unobserved state variables in the reduced-form eq. (38), it is often possible to consistently estimate the decision rule under weak assumptions as $N \to \infty$, or as $T_i \to \infty$ if the data consists only of a single agent or a small number of agents i who are observed over long intervals. Thus, the decision rule δ can be treated as a *known function* for purposes of a theoretical analysis of identification. The *identification problem* is the question, *under what conditions is the mapping from the underlying structure Λ to the reduced form 1 to 1 (that is invertible)*? If this mapping is 1 to 1, we say that the structure is *identified* since in principle it can be inverted to uniquely determine the underlying structure Λ. In practice, we construct an *estimator* $\hat{\Lambda}$ based on the available data and show that $\hat{\Lambda}$ converges to the true underlying structure Λ as $N \to \infty$ and/or $T_i \to \infty$ for each i.

Unfortunately, rather strong a priori assumptions on the form of agents' preferences and beliefs are required in order to guarantee identification of the structural model. Rust (1994) and Magnac and Thesmar (2002) have shown that an important subclass of SDPs, *discrete decision processes* (DDPs), are *nonparametrically unidentified*. That is, if we are unwilling to make any *parametric* functional form assumptions about preferences or beliefs, then in general there are infinitely many different structures Λ consistent with any reduced form δ. In more direct terms, there are many different ways to *rationalize* any observed pattern of behaviour as being 'optimal' for different configurations of preferences and beliefs. It is likely that these results extend to continuous choice problems, since it is possible to approximate a continuous decision process (CDP) by a sequence of DDPs with expanding numbers of elements in their choice sets. Further, for dynamic games, Ledyard (1986) has shown that *any* undominated strategy profile can be a Bayesian equilibrium for some set of preferences and beliefs. Thus, the hypothesis of optimality or equilibrium per se does not have testable empirical content: further a priori assumptions must be imposed in order for SDPs models to be identified and result in empirically testable restrictions on behaviour.

There are two main types of identifying assumptions that have been made in the literature to date: (*a*) parametric functional form assumptions on preferences $u(s,d)$ and components of agents' beliefs $p(s'|s,d)$ that involve unobserved state variables ε and (*b*) *rational expectations*. Rational expectations states that an agent's

subjective beliefs $p(s'|s,d)$ coincide with *objective probabilities* that can be estimated from data. Of course, this restriction is useful only for those components of s, x, that the econometrician can actually observe. In addition, there are other more general *functional restrictions* that can be imposed to help identify the model. One example is monotonicity and shape restrictions on preferences (for example, concavity and monotonicity of the utility function), and another example is independence or *conditional independence* assumptions about variables entering agents' beliefs. I will provide specific examples below; however, it should be immediately clear why these additional assumptions are necessary.

For example, consider the two parameters ρ (the agent's subjective survival probability) and β (the agent's subjective discount factor). We have seen in Section 3 that *only the product of ρ and β enter the SDP model, and not ρ and β separately*. Thus, at most the product $\rho\beta$ can be identified, but without further assumptions it is impossible to separately identify the subjective survival probability ρ from the subjective discount factor β since both affect an agent's behaviour in a symmetrical fashion. However, we can separately identify ρ and β if we assume that an individual has *rational survival expectations*, that is, that their subjective survival probability ρ coincides with the 'objective' survival probability. Then we can estimate ρ 'outside' the SDP model, using data on the lifetime distributions of similar types of agents, and then β can be identified if other restrictions are imposed to guarantee that the product $\rho\beta$ is identified. However, it can be very difficult to make precise inferences about agents' discount factors in many problems, and it is easy to think of models where there is heterogeneity in survival probabilities and discount factors, and unobserved variables affecting one's beliefs about them (for example, family characteristics such as a predisposition for cancer, and so on, that are observed by an agent but not by the econometrician) where identification is problematic.

There are two main approaches for conducting inference in SDPs: (*a*) maximum likelihood and (*b*) 'simulation estimation'. The latter category includes a variety of similar methods such as *indirect inference* (Gourieroux and Monfort, 1997), *simulated method of moments* (McFadden, 1989; Gallant and Tauchen, 1996), *simulated maximum likelihood and method of simulated scores* (see SIMULATION-BASED ESTIMATION), and *simulated minimum distance* (Hall and Rust, 2006). To simplify the discussion I will define these initially for single agent SDPs and at the end discuss how these concepts naturally extend to dynamic games. I will illustrate maximum likelihood and show how a likelihood can be derived for a class of DDPs; however, for CDPs, it is typically much more difficult to derive a likelihood function, especially when there are issues of *censoring*, or problems involving mixed discrete and continuous choice. In such cases simulation estimation is often the only feasible way to do inference.

For discrete decision processes, assume that the utility function has the following parametric, *additively separable* representation

$$u(x, \varepsilon, d) = u(x, d, \theta_1) + \varepsilon(d) \quad (AS).$$
(39)

where $\varepsilon = \{\varepsilon(d) | d \in D(x)\}$, and $\varepsilon(d)$ is interpreted as an unobserved component of utility associated with choice of alternative $d \in D(x)$. Further, suppose that the transition density $p(x', \varepsilon' | x, \varepsilon, d)$ satisfies the following *conditional independence assumption*

$$p(x', \varepsilon' | x, \varepsilon, d) = p(x' | x, d, \theta_2)q(\varepsilon', \theta_3) \quad (CI).$$
(40)

The CI assumption implies that $\{\varepsilon_t\}$ is an *IID* 'noise' process that is independent of $\{x_t, d_t\}$. Thus all of the serially correlated dynamics in the state variables are captured by the observed component of the state vector x_t. If, in addition, $q(\varepsilon_t, \theta_3)$ is a distribution with unbounded support with finite absolute first moments, one can show that the following *conditional choice probabilities* exist

$$P(d | x, \theta) = \int I_\varepsilon \{d = \delta(x, \varepsilon, \theta)\}q(\varepsilon)d\varepsilon,$$
(41)

where $\theta = (\rho, \beta, \theta_1, \theta_2, \theta_3)$ constitute the vector of unknown parameters to be estimated. (Identification of fully parametric models is a 'generic' property, that is, if there are two different parameters θ that produce the same conditional choice probability $P(d | x, \theta)$ for all x and $d \in D(x)$ – and thus led to the same limiting expected log-likelihood – small perturbations in the parameterization will 'almost always' result in a nearby model for which θ is uniquely identified.) In general, the parametric functional form assumptions, combined with the assumption of rational expectations and the AS and CI assumptions, are sufficient to identify the unknown parameter vector θ^*. θ^* can be estimated by maximum likelihood, using the *full information* likelihood function L_f given by

$$\mathscr{L}_f(\theta | \{x_{i,t}, d_{i,t}\}, t = 1, \ldots, T_i, \ i = 1, \ldots, N)$$
$$= \prod_{i=1}^{N} \prod_{t=2}^{T_i} P(d_{i,t} | x_{i,t}, \theta)$$
$$\times p(x_{i,t} | x_{i,t-1}, d_{i,t-1}, \theta_2).$$
(42)

A particularly tractable special case is where $q(\varepsilon, \theta_3)$ has a *multivariate extreme value distribution* where θ_3 is a common scale parameter (linearly related to the standard deviation) for each variable in this distribution (see MCFADDEN, DANIEL; LOGIT MODELS OF INDIVIDUAL CHOICE for the exact formula for this density). This specification leads to a dynamic generalization of the *multinomial logit model*

$$P(d | x, \theta) = \frac{\exp\{v(x, d, \theta)/\theta_3\}}{\sum_{d' \in D(x)} \exp\{v(x, d', \theta)/\theta_3\}},$$
(43)

where $v(x, d, \theta)$ is the expected, discounted utility from taking action d in observed state x given by the unique fixed point to the following *smoothed Bellman equation*

$$v(x, d, \theta) = u(x, d, \theta_1) + \rho\beta \int_{x'} \theta_3 \log$$
$$\times \left(\sum_{d' \in D(x')} \exp\{v(x', d', \theta)/\theta_3\} \right)$$
$$\times p(x' | x, d, \theta_2)dx'.$$
(44)

Define Γ_θ by

$$\Gamma_\theta(W)(x, d) = u(x, d, \theta_1) + \rho\beta \int_{x'} \theta_3 \log$$
$$\times \left(\sum_{d' \in D(x')} \exp\{\exp\{W(x', d', \theta)/\theta_3\} \right)$$
$$\times p(x' | x, d, \theta_2)dx'.$$
(45)

It is not hard to show that under weak assumptions Γ_θ is a contraction mapping, so that $v(x, d, \theta)$ exists and is unique. Maximum likelihood estimation can be carried out using a nested *fixed point* maximum likelihood algorithm consisting of an 'outer' optimization algorithm to search for a value of θ that maximizes $\mathscr{L}_f(\theta)$, and an 'inner' fixed point algorithm that computes $v_\theta = \Gamma_\theta(v_\theta)$ each time the outer optimization algorithm generates a new trial guess for θ. The implicit function theorem guarantees that v_θ is a smooth function of θ. See Aguirregabiria and Mira (2004) for an ingenious alternative that 'swaps' the order of the inner and outer algorithms of the nested fixed-point algorithm resulting in significant computational speedups. See also Rust (1988) for further details on the nested fixed-point algorithm and the properties of the maximum likelihood estimator, and Rust (1994) for a survey of alternative less efficient but computationally simpler estimation strategies.

As noted above, econometric methods for CDPs, that is, problems where the decision variable is continuous (such as firm investment decisions, price settings, or consumption/savings decisions) are harder, since there is no tractable, general specification for the way unobservable state variables to enter the decision rule that result in a *nondegenerate* likelihood function (that is, where the likelihood $\mathscr{L}(\theta)$ is non-zero for any data-set and any value of θ). For this reason, maximum likelihood estimation of CDPs is rare, outside certain special subclasses, such at linear quadratic CDPs (Hansen and Sargent, 1980; Sargent, 1981). However, simulation-based methods of inference can be used in a huge variety of situations where a likelihood is difficult or impossible to derive. These methods have a great deal of flexibility, a high degree of generality, and often permit substantial computational savings. In particular, generalizations of McFadden's (1989) *method of simulated moments* (MSM) have enabled estimation of a wide range of CDPs. The MSM estimator minimizes a quadratic form between a set of moments constructed from the data, h_N and a vector of *simulated moments* $h_{N,S}(\theta)$, that is

$$h_N = \frac{1}{N} \sum_{i=1}^{N} h(\{x_{it}, d_{it}\})$$

$$h_{N,S}(\theta) = \frac{1}{S} \sum_{j=1}^{S} \frac{1}{N} \sum_{i=1}^{N} h(\{\tilde{x}_{it}^j(\theta), \tilde{d}_{it}^j(\theta)\})$$

$$(46)$$

where h is a vector of $J \geq K$ 'moments' (that is, functionals of the data that the econometrician is trying to 'match'), where K is the dimension of θ, $\{x_{it}, d_{it}\}$ are the data, and $\{\tilde{x}_{it}^j(\theta), \tilde{d}_{it}^j(\theta)\}$, $j=1,\ldots,S$ are S IID realizations of the controlled process. The estimate $\hat{\theta}$ is given by

$$\hat{\theta} = \underset{\theta \in R^K}{\operatorname{argmin}} [h_N - h_{N,S}(\theta)]' W_N [h_N - h_{N,S}(\theta)],$$

$$(47)$$

where W_N is a $J \times J$ positive–definite weighting matrix. The most efficient choice for W_N is $W_N = [\hat{\Omega}_N]^{-1}$ where $\hat{\Omega}_N$ is the variance-covariance matrix formed from the vector of sample moments h_N. Simulation estimators require a nested fixed-point algorithm since each time the outer minimization algorithm tries a new trial value for θ, the inner fixed point problem must be called to solve the CDP problem, using the optimal decision rule $d_{it}^j(\theta) = \delta(x_{it}^j, \varepsilon_{it}^j, \theta)$ to generate the simulated decisions, and the transition density $p(x_{i,t+1}^j, \varepsilon_{i,t+1}^j | x_{i,t}^j, \varepsilon_{i,t}^j, d_{it}^j, \theta_2)$ to generate $j=1,\ldots,S$ IID realizations for a simulated panel each potential value of θ. (It is important to simulate using 'common random numbers' that remain fixed as θ varies over the course of the estimation, in order to satisfy the *stochastic equicontinuity conditions* necessary to establish consistency and asymptotic normality of the simulation estimator.)

Simulation methods are extremely flexible for dealing with a number of data issues such as attrition, missing data, censoring and so forth. The idea is that, if we are willing to build a stochastic model of the data 'problem', we can account for it in the process of simulating the behavioural model. For example, Hall and Rust (2006) develop a dynamic model of commodity price speculation

in the steel market. An object of interest is to estimate the stochastic process governing wholesale steel prices; however, there is no public commodity market where steel is traded and prices are recorded on a daily basis. Instead, Hall and Rust observe only the actual wholesale prices of a particular steel trader, who records wholesale prices only on the days he actually buys steel in the wholesale market. Since the speculator makes money by 'buying low and selling high', the set of observed wholesale prices are *endogenously sampled*, and failure to account for this can lead to incorrect inferences about wholesale prices – a dynamic analogue of *sample selection bias*. However, in a simulation model it is easy to censor the simulated data in the same way it is censored in the actual data, that is, by discarding simulated wholesale prices on days where no simulated purchases are made. Hall and Rust show that even though moments based on the observed (censored) data are 'biased' estimates, the simulated moments are biased in exactly the same fashion, so minimizing the distance between actual and simulated biased moments nevertheless results in consistent and asymptotically normal estimates of the parameters of the wholesale price process and other parameters entering the speculator's objective function.

Simulation methods have also enabled the use of Bayesian methods, resulting in methods of inference that do not require asymptotic approximations, although they generally use Markov chain Monte Carlo methods to generate simulated draws from a distribution that approximates the exact finite sample posterior distribution for the parameters of interest (see for example, Lancaster, 1997; Imai, Jain and Ching, 2005; Nourets, 2006).

The most recent literature has extended the methods for estimation of single-agent SDPs to multi-agent dynamic games. For example, Rust (1994) described applications of dynamic discrete choice models to multiple-agent *discrete dynamic games*. The unobserved state variables ε_t entering any particular agent's payoff function are assumed to be unobserved both by the econometrician and by the other players in the game. The *Bayesian–Nash equilibria* of this game can be represented as a vector of conditional choice probabilities $(P_1(d_1|x), \ldots, P_n(d_n|x))$, one for each player, where $P_i(d_i|x)$ represents the econometrician's and the other players' beliefs about the probability player i will take action d_i, 'integrating out' over the unobservable states variable $\varepsilon_{i,t}$ affecting player i's decision at time t similar to eq. (41) for single-agent problems. If one adapts the numerical methods for Markov-perfect equilibrium described in Section 4, it is possible to compute Bayesian–Nash equilibria of discrete dynamic games using nested fixed-point algorithms. While it is relatively straightforward to write down the likelihood function for the game, actual estimation via a straightforward application of full information maximum likelihood is extremely computationally demanding since it requires a *doubly nested fixed point algorithm* (that is, an 'outer' algorithm to search over θ to maximize the likelihood, and then an inner algorithm to solve the dynamic game for each value of θ, but this inner algorithm is itself a nested fixed-point algorithm). Alternative, less computationally demanding estimation methods have been proposed by Aguirregabiria and Mira (2007), Bajari and Hong (2006), Bajari, Benkard and Levin (2007), and Pesendorfer and Schmidt-Dengler (2003). This research is at the current frontier of development in numerical and empirical applications of dynamic programming.

Besides econometric methods, which are applied for structural estimation for actual agents in their 'natural' settings, an alternative approach is to try to make inferences about agents' preferences and beliefs (and even their 'mode of reasoning') for artificial SDPs in a laboratory setting. The advantage of a laboratory experiment is *experimental control over preferences and beliefs*. The ability to control these aspects of decision-making can enable much tighter tests of theories of decision-making. For example, Binmore et al. (2002) structured a laboratory experiment to determine whether individuals do backward induction in one- and two-stage alternating offer games, and 'find systematic violations of backward induction that cannot be explained by payoff-interdependent preferences' (2002, p. 49).

6 Comments

There has been tremendous growth in research related to dynamic programming since the 1940s. The method has evolved into the main tool for solving sequential decision problems, and research related to dynamic programming has led to fundamental advances in theory, numerical methods and econometrics. As we have seen, while dynamic programming embodies the notion of rational decision-making under uncertainty, there is mixed evidence as to whether it provides a good literal description of how human beings actually behave in comparable situations. Although human reasoning and decision-making is undoubtedly both more complex and more 'frail' and subject to foibles and limitations than the idealized notion of 'full rationality' that dynamic programming embodies, the discussion of the identification problem shows that, if we are given sufficient flexibility about how to model individual preferences and beliefs, there exist SDPs whose decision rules provide arbitrarily good approximations to individual behaviour.

Thus, dynamic programming can be seen as a useful 'first approximation' to human decision-making, but it will undoubtedly be superseded by more descriptively accurate psychological models. Indeed, in the future one can imagine behavioural models that are not derived from some a priori axiomatization of preferences, but will result from empirical research that will ultimately deduce human behaviour from yet even deeper 'structure', that is the very underlying neuroanatomy of the human brain.

Even if dynamic programming is unlikely to be a descriptively accurate model of *human* decision-making, it will probably still remain highly relevant for the foreseeable future as the embodiment of *rational* decision-making. There are well-defined problems, for example, profit maximization or cost minimization, where there is agreement on the objective function to be maximized or minimized, and where there will be a demand for dynamic programming methods to find the optimal profit- or cost-minimizing strategies. There are many examples of this in the operations research literature. Practical applications include optimal inventory management (Hall and Rust, 2006) and optimal harvesting of timber (Paarsch and Rust, 2007).

Some observers such as Kurzweil (2005) predict that in the not too distant future (for example, approximately 2050) a *singularity* will occur, 'during which the pace of technological change will be so rapid, its impact so deep, that human life will be irreversibly transformed' (2005, p. 7). The singularity is a complex of accelerating improvements in computer hardware and software, and a merger of machine- and biological-based intelligence that will blur the distinction between 'artificial intelligence' and human intelligence, that will overcome many of current limitations of the human brain and human reasoning: 'By the end of this century, the nonbiological portion of our intelligence will be trillions and trillions of times more powerful than unaided human intelligence' (2005, p. 9). Dynamic programming will undoubtedly continue to be a critical tool in this brave new world.

Whether this prognosis will ever come to pass, or come to pass as soon as Kurzweil forecast, is debatable; but it does suggest that there will be continued interest in and research on dynamic programming. However, the fact that reasonably broad classes of dynamic programming problems are subject to a curse of dimensionality suggests that it may be too optimistic to think that human rationality will soon be superseded by 'artificial rationality'. While there are many complicated problems that we would like to solve by dynamic programming in order to understand what 'fully rational' behaviour actually looks like in specific situations, the curse of dimensionality still limits us to very simple 'toy models' that only very partially and simplistically capture the myriad of details and complexities we face in the real world. Although we now have a number of examples where artificial intelligence based on principles from dynamic programming outstrips human intelligence, for example computerized chess, all these cases are for very specific problems in very narrow domains. I believe that it will be a long time before technological progress in computation and algorithms produce truly general-purpose 'intelligent behaviour' that can compete successfully with human intelligence in widely varying domains and in the immensely complicated situations that we operate in every day. Despite all our psychological frailties and limitations, there is an important unanswered question of 'how do we do it?', and more research is required to determine if human behaviour is simply suboptimal, or whether the human brain uses some powerful implicit 'algorithm' to circumvent the curse of dimensionality that digital computers appear to be subject to for solving problems such as SDPs by dynamic programming. For a provocative theory that deep principles of quantum mechanics can enable human intelligence to transcend computational limitations of digital computers, see Penrose (1989).

JOHN RUST

See also **Bellman equation; game theory; logit models of individual choice; McFadden, Daniel; recursive competitive equilibrium; recursive preferences; sequential analysis; simulation-based estimation.**

This article has benefited from helpful feedback from Kenneth Arrow, Daniel Benjamin, Larry Blume, Moshe Buchinsky, Larry Epstein, Chris Phelan and Arthur F. Veinott, Jr.

Bibliography

Adda, J. and Cooper, R. 2003. *Dynamic Economics Quantitative Methods and Applications.* Cambridge, MA: MIT Press.

Aguirregabiria, V. and Mira, P. 2004. Swapping the nested fixed point algorithm: a class of estimators for discrete Markov decision models. *Econometrica* 70, 1519–43.

Aguirregabiria, V. and Mira, P. 2007. Sequential estimation of dynamic discrete games. *Econometrica* 75, 1–53.

Arrow, K.J., Blackwell, D. and Girshik, M.A. 1949. Bayes and minimax solutions of sequential decision problems. *Econometrica* 17, 213–44.

Bajari, P. and Hong, H. 2006. Semiparametric estimation of a dynamic game of incomplete information. Technical Working Paper No. 320. Cambridge, MA: NBER.

Bajari, P., Benkard, L. and Levin, J. 2007. Estimating dynamic models of imperfect competition. *Econometrica* 75, 1331–70.

Barto, A.G., Bradtke, S.J. and Singh, S.P. 1995. Learning to act using real-time dynamic programming. *Artificial Intelligence* 72, 81–138.

Bellman, R. 1957. *Dynamic Programming.* Princeton, NJ: Princeton University Press.

Bellman, R. and Dreyfus, S. 1962. *Applied Dynamic Programming.* Princeton, NJ: Princeton University Press.

Bellman, R. 1984. *Eye of the Hurricane.* Singapore: World Scientific.

Bertsekas, D.P. 1995. *Dynamic Programming and Optimal Control,* vols 1 and 2. Belmont, MA: Athena Scientific.

Bertsekas, D.P. and Tsitsiklis, J. 1996. *Neuro-Dynamic Programming.* Belmont, MA: Athena Scientific.

Bhattacharya, R.N. and Majumdar, M. 1989. Controlled semi-Markov models – the discounted case. *Journal of Statistical Planning and Inference* 21, 365–81.

Binmore, K., McCarthy, J., Ponti, G., Samuelson, L. and Shaked, A. 2002. A backward induction experiment. *Journal of Economic Theory* 104, 48–88.

Blackwell, D. 1962. Discrete dynamic programming. *Annals of Mathematical Statistics* 33, 719–26.

Blackwell, D. 1965a. Positive dynamic programming. *Proceedings of the 5th Berkeley Symposium* 3, 415–28.

Blackwell, D. 1965b. Discounted dynamic programming. *Annals of Mathematical Statistics* 36, 226–35.

Cayley, A. 1875. Mathematical questions and their solutions. Problem No. 4528. *Educational Times* 27, 237.

Chow, C.S. and Tsitsiklis, J.N. 1989. The complexity of dynamic programming. *Journal of Complexity* 5, 466–88.

Denardo, E. 1967. Contraction mappings underlying the theory of dynamic programming. *SIAM Review* 9, 165–77.

Dvoretzky, A., Kiefer, J. and Wolfowitz, J. 1952. The inventory problem: I. Case of known distributions of demand. *Econometrica* 20, 187–222.

Eckstein, Z. and Wolpin, K.I. 1989. The specification and estimation of dynamic stochastic discrete choice models: a survey. *Journal of Human Resources* 24, 562–98.

Gallant, A.R. and Tauchen, G.E. 1996. Which moments to match? *Econometric Theory* 12, 657–81.

Gihman, I.I. and Skorohod, A.V. 1979. *Controlled Stochastic Processes.* New York: Springer.

Gittins, J.C. 1979. Bandit processes and dynamic allocation indices. *Journal of the Royal Statistical Society* B 41, 148–64.

Gotz, G.A. and McCall, J.J. 1980. Estimation in sequential decision-making models: a methodological note. *Economics Letters* 6, 131–6.

Gourieroux, C. and Monfort, A. 1997. *Simulation-Based Methods of Inference.* Oxford: Oxford University Press.

Grüne, L. and Semmler, W. 2004. Using dynamic programming with adaptive grid scheme for optimal control problems in economics. *Journal of Economic Dynamics and Control* 28, 2427–56.

Hall, G. and Rust, J. 2006. Econometric methods for endogenously sampled time series: the case of commodity price speculation in the steel market. Manuscript, Yale University.

Hansen, L.P. and Sargent, T.J. 1980. Formulating and estimating dynamic linear rational expectations models. *Journal of Economic Dynamics and Control* 2, 7–46.

Hansen, L.P. and Singleton, K. 1982. Generalized instrumental variables estimation of nonlinear rational expectations models. *Econometrica* 50, 1269–81.

Howard, R.A. 1960. *Dynamic Programming and Markov Processes.* New York: Wiley.

Imai, S., Jain, N. and Ching, A. 2005. Bayesian estimation of dynamic discrete choice models. Manuscript, University of Illinois.

Judd, K. 1998. *Numerical Methods in Economics.* Cambridge, MA: MIT Press.

Keane, M. and Wolpin, K.I. 1994. The solution and estimation of discrete choice dynamic programming models by simulation: Monte Carlo evidence. *Review of Economics and Statistics* 76, 648–72.

Kurzweil, R. 2005. *The Singularity is Near when Humans Transcend Biology.* New York: Viking Press.

Kushner, H.J. 1990. Numerical methods for stochastic control problems in continuous time. *SIAM Journal on Control and Optimization* 28, 999–1048.

Lancaster, A. 1997. Exact structural inference in optimal job search models. *Journal of Business Economics and Statistics* 15, 165–79.

Ledyard, J. 1986. The scope of the hypothesis of Bayesian equilibrium. *Journal of Economic Theory* 39, 59–82.

Lucas, R.E. Jr. 1976. Econometric policy evaluation: a critique. In *The Phillips Curve and Labour Markets*, ed. K. Brunner and A.K. Meltzer. Carnegie-Rochester Conference on Public Policy. Amsterdam: North-Holland.

Lucas, R.E. Jr. 1978. Asset prices in an exchange economy. *Econometrica* 46, 1426–45.

Luenberger, D.G. 1969. *Optimization by Vector Space Methods*. New York: Wiley.

Magnac, T. and Thesmar, D. 2002. Identifying dynamic discrete decision processes. *Econometrica* 70, 801–16.

Marschak, T. 1953. Economic measurements for policy and prediction. In *Studies in Econometric Method*, ed. W.C. Hood and T.J. Koopmans. New York: Wiley.

Massé, P. 1945. *Application des probabilités en chaîne á l'hydrologie statistique et au jeu des réservoirs*. Report to the Statistical Society of Paris. Paris: Berger-Levrault.

Massé, P. 1946. *Les réserves et la régulation de l'avenir*. Paris: Hermann.

McFadden, D. 1989. A method of simulated moments for estimation of discrete response models without numerical integration. *Econometrica* 57, 995–1026.

Nemirovsky, A.S. and Yudin, D.B. 1983. *Problem Complexity and Method Efficiency in Optimization*. New York: Wiley.

Nourets, A. 2006. Inference in dynamic discrete choice models with serially correlated unobserved state variables. Manuscript, University of Iowa.

Paarsch, H.J. and Rust, J. 2007. Stochastic dynamic programming in space: an application to British Columbia forestry. Working paper.

Pakes, A. 1986. Patents as options: some estimates of the values of holding European patent stocks. *Econometrica* 54, 755–84.

Pakes, A. 2001. Stochastic algorithms, symmetric Markov perfect equilibria and the 'curse' of dimensionality. *Econometrica* 69, 1261–81.

Pakes, A. and McGuire, P. 1994. Computing Markov perfect Nash equilibrium: numerical implications of a dynamic differentiated product model. *RAND Journal of Economics* 25, 555–89.

Penrose, R. 1989. *The Emperor's New Mind*. New York: Penguin.

Pesendorfer, M. and Schmidt-Dengler, P. 2003. Identification and estimation of dynamic games. Manuscript, University College London.

Pollard, D. 1989. Asymptotics via empirical processes. *Statistical Science* 4, 341–86.

Puterman, M.L. 1994. *Markovian Decision Problems*. New York: Wiley.

Rust, J. 1985. Stationary equilibrium in a market for durable goods. *Econometrica* 53, 783–805.

Rust, J. 1987. Optimal replacement of GMC bus engines: an empirical model of Harold Zurcher. *Econometrica* 55, 999–1033.

Rust, J. 1988. Maximum likelihood estimation of discrete control processes. *SIAM Journal on Control and Optimization* 26, 1006–24.

Rust, J. 1994. Structural estimation of Markov decision processes. In *Handbook of Econometrics*, vol. 4, ed. R.F. Engle and D.L. McFadden. Amsterdam: North-Holland.

Rust, J. 1996. Numerical dynamic programming in economics. In *Handbook of Computational Economics*, ed. H. Amman, D. Kendrick and J. Rust. Amsterdam: North-Holland.

Rust, J. 1997. Using randomization to break the curse of dimensionality. *Econometrica* 65, 487–516.

Rust, J. and Hall, G.J. 2007. The (S, s) rule is an optimal trading strategy in a class of commodity price speculation problems. *Economic Theory* 30, 515–38.

Rust, J. and Phelan, C. 1997. How Social Security and Medicare affect retirement behavior in a world with incomplete markets. *Econometrica* 65, 781–832.

Rust, J., Traub, J.F. and Woźniakowski, H. 2002. Is there a curse of dimensionality for contraction fixed points in the worst case? *Econometrica* 70, 285–329.

Santos, M. and Rust, J. 2004. Convergence properties of policy iteration. *SIAM Journal on Control and Optimization* 42, 2094–115.

Sargent, T.J. 1978. Estimation of dynamic labor demand schedules under rational expectations. *Journal of Political Economy* 86, 1009–44.

Sargent, T.J. 1981. Interpreting economic time series. *Journal of Political Economy* 89, 213–48.

Selten, R. 1975. Reexamination of the perfectness concept for equilibrium points in extensive games. *International Journal of Game Theory* 4, 25–55.

Tauchen, G. and Hussey, R. 1991. Quadrature-based methods for obtaining approximate solutions to nonlinear asset pricing models. *Econometrica* 59, 371–96.

Todd, P. and Wolpin, K.I. 2005. Ex ante evaluation of social programs. Manuscript, University of Pennsylvania.

Traub, J.F. and Werschulz, A.G. 1998. *Complexity and Information*. Cambridge: Cambridge University Press.

Von Neumann, J. and Morgenstern, O. 1944. *Theory of Games and Economic Behavior*. Princeton, NJ: Princeton University Press. 3rd edn, 1953.

Wald, A. 1947. Foundations of a general theory of sequential decision functions. *Econometrica* 15, 279–313.

Wald, A. 1947. *Sequential Analysis*. New York: Dover Publications.

Wald, A. and Wolfowitz, J. 1948. Optimum character of the sequential probability ratio test. *Annals of Mathematical Statistics* 19, 326–39.

Wolpin, K. 1984. An estimable dynamic stochastic model of fertility and child mortality. *Journal of Political Economy* 92, 852–74.

E

Easterlin hypothesis

The Easterlin, or 'relative cohort size', hypothesis as originally formulated posits that, other things constant, the economic and social fortunes of a cohort (those born in a given year) tend to vary as a function of its relative size, approximated by the crude birth rate surrounding the cohort's birth (Easterlin, 1987). This hypothesis has since been extended to suggest a wider range of effects on the economy as a whole (Macunovich, 2002).

Although cohort size effects were originally expected to be symmetrical around the peak of the baby boom, which in the United States entered the labour market around 1980, it is now thought that they are tempered by aggregate demand effects and by feedback effects from adjustments made by young adults on the 'leading edge' of a baby boom. As a result, cohorts – and the economy generally – on the 'leading edge of a baby boom fare much better than those on the 'trailing edge', when all else is equal.

The ultimate effects of changing relative cohort size are hypothesized to fall into these three categories:

1. Direct or first-order effects of relative cohort size on male relative income (the earnings of young men relative to their aspirations); male unemployment and hours worked; men's and women's college wage premium (the extra earnings of a college graduate relative to those of a secondary school graduate); and levels of income inequality generally.
2. Second-order effects operating through male relative income, especially the demographic adjustments people make in response to changing relative income, such as changes in women's labour force participation and their occupational choices; men's and women's college enrolment rates; marriage and divorce; fertility; crime, drug use, and suicide rates; out-of-wedlock childbearing and the incidence of female-headed families; and living arrangements.
3. Third-order effects on the economy of changing relative cohort size and the resulting demographic adjustments, such as changes in average wage growth; the overall demand for goods and services in the economy and hence the growth rate of the economy; inflation, interest rates, and savings rates; stock market performance; industrial structure; measures of gross domestic product (GDP); and productivity measures.

The three categories of effect are discussed first in this article, followed by a consideration of feedback effects and a discussion of empirical evidence.

First-order effects

The linkage between higher birth rates and adverse social and economic effects arises from 'crowding mechanisms'

operating within three major social institutions, the family, school and the labour market. Within the family, a sustained upsurge in the birth rate is likely to entail an increase in the average number of siblings, higher average birth order, and a shorter average birth interval, and there is a substantial literature in psychology, sociology and economics linking child development negatively to one or more of these magnitudes (Ernst and Angst, 1983; Heer, 1985). The negative effects that have been investigated range over a wide variety of phenomena. With regard to mental health, for example, there is evidence that problem behaviours such as fighting, breaking rules, and delinquency are associated with increased family size. Adverse effects on morbidity and mortality of children have been found to be associated with increased family size and shorter birth spacing. A negative association between IQ and number of siblings has been found in a number of studies, and, with IQ controlled for, between educational attainment and family size. The principal mechanism underlying such developments is likely to be the dilution of parental time and energy per child and family economic resources per child, associated with increased family size.

The family mechanisms just discussed imply that, on average, a larger cohort is likely to perform less well in school. But even in the absence of any adverse effects within the family, a large cohort is likely to experience crowding in schools, which reduces average educational performance (Freeman, 1976). At any given time the human and physical capital stock comprising the school system tends to be either fixed in amount or to expand at a fairly constant rate, so that a surge in entrants into the school system tends to be accompanied by a reduction in physical facilities and teachers per student. In the United States, school planning decisions are divided among numerous local governments and private institutions, and expansion has tended to occur in reaction to, rather than in anticipation of, a large cohort's entry. Moreover, even when expansion occurs it is usually not accompanied by maintenance of curriculum standards, partly because of the diminishing pool of qualified teachers available to supply the needs of educational expansion.

The experience of a large cohort both in the family and in school is likely, in turn, to leave the cohort less well prepared, on reaching adulthood, for success in the labour market. But even if there were no prior effects, the entry of a large proportion of young and relatively inexperienced workers into the labour market creates a new set of crowding phenomena, because the expansion of complementary factor inputs is unlikely to be commensurate with that of the youth labour force. Additions to physical capital stock tend to be dominated by considerations other than the relative supply of younger workers, and the growth in older, experienced, workers is largely governed by prior demographic conditions. Growth in the relative supply of younger workers results, in consequence, in a deterioration of their relative wage rates, unemployment conditions and upward job mobility (Welch, 1979). The adverse effects of labour market crowding tend to reinforce those of crowding within the school and family. For example, the deterioration in relative wage rates of the young translates into lower returns to education and consequent adverse impact on school drop-out rates and college enrolment (Freeman, 1976). Also, problems encountered in finding a good job may reinforce feelings of inadequacy or frustration already stirred up by some prior experiences at home or in school, and lead to lower labour force participation among young men.

Second-order effects

The relative economic standing of successive generations at a given point in time may be altered systematically by fluctuations in relative cohort size. If parents' living levels play an important role in setting their children's material aspirations, as socialization theory leads one to believe, then an increase in the shortfall of children's wage rates relative to parents will cause the children to feel relatively deprived and under greater pressure to keep up. The importance of relative status influences of this type in affecting attitudes or behaviour has been widely recognized in social science theory (Duesenberry, 1949).

Confronted with the prospect of a deterioration in its living level relative to that of its parents, a large young adult cohort may make a number of adaptations in an attempt to preserve its comparative standing. Foremost among these are changes in behaviour related to family formation and family life (Macunovich and Easterlin, 1990; McNown and Rajbhandary, 2003). To avoid the financial pressures associated with family responsibilities, marriage may be deferred. If marriage occurs, wives are more likely to work and to put off childbearing. If a wife bears children, she is more likely to couple labour force participation with childrearing, and to have a smaller number of children more widely spaced (Macunovich, 2002; Jeon and Shields 2005).

The process of demographic adjustment to changing relative income can best be thought of in terms of *ex ante* and *ex post* income; that is, the disposable per capita income of individuals prior to and then following the adjustments. Analyses of baby boom cohorts in the United States have found that a cohort's male relative income – individual earning potential of baby boomers relative to that of their parents – was significantly lower than the individual earning potential of *pre-boom* cohorts relative to *their* parents. But after making the type of demographic adjustments indicated above, the boomers managed to bring their per capita disposable income on a par with that of their parents (Easterlin, Macdonald and Manucnovich, 1990).

Other reactions to the psychological stresses induced by large cohort size may be viewed as socially dysfunctional. Feelings of inadequacy and frustration, for example, may lead to disproportionate consumption of alcohol and drugs, to mental depression, and, at the extreme, to a higher rate of suicide (Pampel, 2001; Stockard and O'Brien, 2002). Feelings of bitterness, disappointment and rage may induce a higher incidence of crime (O'Brien, Stockard and Isaacson, 1999). Within marriage, the stresses of conflicting work and motherhood roles for women, and feelings of inadequacy as a breadwinner for men, are likely to result in a higher incidence of divorce (Macunovich, 2002). In the political sphere, the disaffection felt by a large cohort because of its lack of success may make it more responsive to the appeals of those who are politically alienated (O'Brien and Gwartney-Gibbs, 1989).

Third-order effects

The second-order effects described in the previous section will, through reduced marriage rates and increased divorce and female labour force participation rates, reduce the proportion of households with stay-at-home spouses, which increases the tendency to purchase market replacements for the goods and services traditionally produced by women in the home. The result is a 'commoditization' of many goods and services that used to be produced in the home. They are now exchanged in the market – and thus counted in official measures of GDP and productivity – whereas previously they were part of the excluded 'non-market' economy.

This commoditization of goods and services causes measures of industrial structure to skew strongly toward services and retail, away from agriculture and manufacturing, creating low-wage service jobs. In addition, the influx of inexperienced young workers as members of a large birth cohort – both men and women – into the labour market exacerbates any decline in productivity growth by changing the composition of the workforce to one dominated by inexperienced and therefore lower-productivity workers. This decline in relative wages of younger workers resulting from their oversupply would lead employers to substitute cheaper labour for more expensive capital, thus lowering the young workers' productivity still further by providing those low-wage workers with less productivity-enhancing machinery and technology.

Although some analysts maintain that the potential age structure effect of the baby boomers on personal savings is not large enough to explain the full drop in US national savings rates since the 1980s, studies of this phenomenon to date have focused only on the behaviour of the baby boomers themselves. However, one might argue that the baby boomers have affected the propensity to save in age groups other than their own. For example, because boomers' earnings were depressed and they experienced an inflated housing market when they went to buy homes (both the effects of their own large cohort size), many parents of baby boomers drew on their own savings in order to help with down payments.

When the age structure of children is permitted to affect consumption and savings, a very strong age-related pattern of expenditures and saving can be identified. Children induce savings on the part of their parents between the ages of five and 16, possibly in anticipation of later educational expenses. When the relationships identified in this way are combined with the changing age distribution in the US population during the 20th century, they produce a savings rate that fluctuates by plus or minus 25 per cent around the mean, simply as a result of changing age structure (Macunovich, 2002).

Similarly, a strong effect has been identified of changing age structure (measured simply as the proportion of young to old in the population) on real interest rates and inflation, because of differential patterns of savings and consumption with age (McMillan and Baesel, 1990). A higher proportion of young adults in a population will produce lower aggregate savings levels – and hence higher interest rates. In this model, today's lower interest and inflation rates are the result of the ageing of the baby boomers, as they begin to acquire assets for their retirement years. The converse of this phenomenon – the potential 'meltdown' effect of a retiring baby boom on financial markets, asset values and interest rates – has been described as well (Schieber and Shoven, 1994).

Some research has estimated a strong effect of age structure on housing prices in the United States, with the entry of the baby boom into the housing market causing the severe house price inflation of the 1970s and 1980s, and the entry of the baby bust causing house price deflation (Mankiw and Weil, 1989). Although some have disputed the magnitude of the effect estimated there, most researchers have confirmed its existence. A later study, for example, found significant effects of detailed (single year) age structure in the adult population on all forms of consumption, including housing demand, and on money demand (Fair and Dominguez, 1991).

These potential effects on aggregate demand, savings rates, interest rates and inflation suggest that there might have been a connection between changing age structure and macroeconomic fluctuations in the United States and elsewhere during the 20th century. When the population of young adults is expanding, the resultant growth in demand for durable goods creates confidence in investors, while an unexpected slowdown in the growth rate of young adults could cause cutbacks in production and investment in response to inventory buildups, with a snowball effect throughout the economy. There was a close correspondence in the United States in the 20th century between 'turnaround points' of growth in the key age group of 15–24, and significant economic dislocations in 1908, 1929, 1938 and 1974. Similarly, there was a correlation between age structure and economic performance in industrialized nations in the 1930s, and in both industrialized and developing nations since the 1980s, with the 'Asian Tigers' some of the most recent examples (Macunovich, 2002).

Feedback effects on the relationship between relative cohort size and relative income

Easterlin's original statements recognized the potential effects of outside influences on the relative cohort size mechanism (Easterlin, 1987). However, the dynamic nature of the mechanism – the fact that many of these other factors would, in fact, be secondary and tertiary results of changing relative cohort size, and thus endogenous in any empirical application – has not been fully appreciated in most analyses to date. As a result, it is often concluded that the hypothesis may have been relevant in the post-Second World War period up to about 1980, but that it fails to extend beyond one full cycle to apply to the period since 1980.

The aggregate demand effect of changing relative cohort size, discussed in the previous section, is hypothesized to contribute significantly to the observed asymmetry in relative cohort size effects on male relative income. Although cohorts on the leading edge of a baby boom experience declining wages relative to those of older workers, they do so in an economy experiencing strong growth in aggregate demand resulting from the increasing relative cohort size among young adults. Cohorts on the lagging edge of a baby boom, however, enter a labour market weakened by the economic slump resulting from a transition from expanding to contracting relative cohort size.

Similarly, as one of the secondary effects of changing relative cohort size discussed earlier, female labour force participation is hypothesized to have increased in response to declining male relative income as the leading edge of the baby boom entered the labour market. If, as hypothesized, these young women also increased their levels of educational attainment in anticipation of future labour market participation, they would have in many cases competed directly with the male members of their cohort and exacerbated the effects of relative cohort size on male relative income. This effect would have been greatest for cohorts on the lagging edge of the boom – those who should have benefited from declining relative cohort size. It is important in empirical analyses to recognize the potential endogeneity of these other factors, rather than treat relative cohort size effects as 'contingent' on exogenous changes in female labour force participation, educational attainment and wages. Wage analyses based on relative cohort size which control for a cohort's position in the US baby boom – and thus allow for aggregate demand and female labour force changes – can explain most of the observed change in young men's entry level wages and in their returns to experience and education (Macunovich, 2002).

Empirical analyses

Empirically, the most important application of the hypothesis has been to explain the varying experience of young adults in the United States since the Second World War. There is, however, some evidence of its relevance to the experience of developed countries more generally in this period (Korenman and Neumark, 2000; Pampel, 2001; Stockard and O'Brien, 2002; Jeon and Shields, 2005), and perhaps as a mechanism leading to fertility decline during the demographic transition in developing countries (Macunovich, 2002).

Overall, however, empirical analyses testing various aspects of the Easterlin hypothesis have produced fairly mixed results. By 2007 there have been two comprehensive analyses of the literature on the Easterlin hypothesis, and one meta-analysis of 19 studies completed between 1976 and 2002. The meta-analysis (Waldorf and Byun, 2005) focused on the age structure–fertility link, and concluded that analytical problems contribute to an apparent lack of empirical support for the Easterlin hypothesis. Most significant among these were the failure to recognize the endogeneity of an income variable when combined with a relative cohort size variable, and the use of very broad age groups in defining relative cohort size.

The first of the literature reviews considered a broad range of topics, including labour market experience and education; marriage, fertility and divorce; and crime, suicide and alienation. It concluded:

> [T]he evidence for the Easterlin effect proves mixed at best and plain wrong at worst... Aggregate data support the hypothesis more than individual level data, period-specific or time-series data support the hypothesis more than cohort-specific data, experiences from 1945–1980 support the hypothesis more than the years since 1980, and trends in the United States support the hypothesis more than trends in European nations. (Pampel and Peters, 1995, p. 189)

The second literature review evaluated 76 published analyses focused solely on fertility, and concluded:

> With an equal number of micro- and macro-level analyses using North American data (twenty-two), the 'track record' of the hypothesis is the same in both venues, with fifteen providing significant support in each case. The literature suggests unequivocal support for the relativity of the income concept in fertility but is less clear regarding the source(s) of differences in material aspirations, and suggests that the observed relationship between fertility and cohort size has varied across countries and time periods due to the effects of additional factors not included in most models. (Macunovich, 1998, p. 53)

This review suggests that, because of data limitations and idiosyncratic interpretations of the hypothesis by individual researchers, many of the studies with unfavourable findings have been only peripherally related to the Easterlin hypothesis.

Conclusion

Since the early 1980s, demographic concepts have encroached modestly on economic theory, as evidenced by the appearance of life cycle, overlapping generations and vintage models. The cohort size hypothesis might be viewed as another in this sequence. Its roots, however, extend beyond economics, reaching out into sociology, demography and psychology, and it seeks to encompass a wider range of attitudinal and behavioural phenomena than is traditionally considered economic.

DIANE J. MACUNOVICH AND RICHARD A. EASTERLIN

See also **demographic transition; economic demography.**

Bibliography

Duesenberry, J.S. 1949. *Income, Saving, and the Theory of Consumer Behaviour.* Cambridge, MA: Harvard University Press.

Easterlin, R.A. 1980. *Birth and Fortune*, 1st edn. New York: Basic Books.

Easterlin, R.A. 1987. *Birth and Fortune*, 2nd edn. Chicago: University of Chicago Press.

Easterlin, R.A., Macdonald, C. and Macunovich, D.J. 1990. How have the American baby boomers fared? Earnings and well-being of young adults 1964–1987. *Journal of Population Economics* 3, 277–90.

Ernst, C. and Angst, J. 1983. *Birth Order: Its Influence on Personality.* Berlin: Springer.

Fair, R.C. and Dominguez, K. 1991. Effects of the changing U.S. age distribution on macroeconomic equations. *American Economic Review* 81, 1276–94.

Freeman, R.B. 1976. *The Overeducated American.* New York: Academic Press.

Heer, D.M. 1985. Effects of sibling number on child outcome. *Annual Review of Sociology* 11, 27–47.

Jeon, Y. and Shields, M.P. 2005. The Easterlin hypothesis in the recent experience of higher-income OECD countries: a panel-data approach. *Journal of Population Economics* 18, 1–13.

Korenman, S. and Neumark, D. 2000. Cohort crowding and the youth labour market: a cross-national analysis. In *Youth Employment and Joblessness in Advanced Countries*, ed. D.G. Blanchflower and R.B. Freeman. NBER Comparative Labour Market Series. Chicago: University of Chicago Press.

Macunovich, D.J. 1998. Fertility and the Easterlin hypothesis: an assessment of the literature. *Journal of Population Economics* 11, 1–59.

Macunovich, D.J. 2002. *Birth Quake: The Baby Boom and Its After Shocks.* Chicago: University of Chicago Press.

Macunovich, D.J. and Easterlin, R.A. 1990. How parents have coped: the effect of life cycle decisions on the economic status of pre-school age children, 1964–1987. *Population and Development Review* 16, 301–25.

Mankiw, N.G. and Weil, N.D. 1989. The baby boom, the baby bust and the housing market. *Regional Science and Economics* 19, 235–58.

McMillan, H.M. and Baesel, J.B. 1990. The macroeconomic impact of the baby boom generation. *Journal of Macroeconomics* 12, 167–95.

McNown, R. and Rajbhandary, S. 2003. Time series analysis of fertility and female labour market behaviour. *Journal of Population Economics* 16, 501–23.

O'Brien, R.M. and Gwartney-Gibbs, P.A. 1989. Relative cohort size and political alienation: three methodological issues and a replication supporting the Easterlin hypothesis. *American Sociological Review* 54, 476–80.

O'Brien, R.M., Stockard, J. and Isaacson, L. 1999. The enduring effects of cohort characteristics on age-specific homicide rates 1960-1995. *American Journal of Sociology* 104, 1061–95.

Pampel, F.C. 2001. *The Institutional Context of Population Change: Patterns of Fertility and Mortality across High-Income Nations*. Chicago: University of Chicago Press.

Pampel, F.C. and Peters, H.E. 1995. The Easterlin effect. *Annual Review of Sociology* 21, 163.

Schieber, S.J. and Shoven, J.B. 1994. The consequences of population aging on private pension fund saving and asset markets. Working Paper No. 4665. Cambridge, MA: NBER.

Stockard, J. and O'Brien, R.M. 2002. Cohort effects on suicide rates: international variations. *American Sociological Review* 67, 854–72.

Waldorf, B. and Byun, P. 2005. Meta-analysis of the impact of age structure on fertility. *Journal of Population Economics* 18, 14–40.

Welch, F. 1979. Effects of cohort size on earnings: the baby boom babies' financial bust. *Journal of Political Economy* 87, 65–97.

Eckstein, Otto (1927–1984)

Eckstein was an entrepreneur who moved a whole technology from the research community into the marketplace. Until he founded Data Resources, Inc., macro-econometric models were research vehicles and not vehicles for aiding business decision making. Under his direction Data Resources came to dominate the marketplace for this type of information, but more importantly it changed the nature of the game. To be taken seriously after his innovation, all economic forecasts had to be buttressed with econometric equations and no large firm would attempt to begin its decision-making processes without an understanding of the national and international economic forecasts emanating from such models.

Born in Ulm, Germany, in 1927, Dr Eckstein fled to England in 1938 and came to the United States in 1939.

He graduated from Stuyvesant High School in New York City and served in the United States Army Signal Corps from 1946 to 1947. He received an AB degree from Princeton University in 1951 and a Ph.D. from Harvard University in 1955.

In 1968, he and Donald B. Marron founded Data Resources, Inc., which has grown into the largest economic information company in the world. The firm became a subsidiary of McGraw-Hill, Inc. in 1979. He directed the development of the Data Resources Model of the US economy, and was responsible for its forecasting operations.

As an immigrant to the United States from Nazi Germany, Otto Eckstein wanted to contribute something to America's future success. Better economic policies that would lead to a higher American standard of living were not an abstraction to him. They were the centre of his professional life.

His professional career began with the analysis of large scale multi-year water resources projects and how one might better allocate national resources in such projects. In the late 1950s he was the principal intellectual director of a Joint Economic Committee study on how the United States might break out of what was then seen as the stagnation of the mid-1950s. His study on growth, full employment and price stability laid the basis for the successful economic policies that were followed in the first two-thirds of the 1960s. But he went on to implement those intellectual foundations as a member of the President's Council of Economic Advisers under President Johnson.

No one who knew the enthusiasm of Otto Eckstein for studying, teaching, and practising economics could thereafter think of economics as the dismal science.

LESTER C. THUROW

Selected works

1958a. *Water Resources Development: The Economics of Project Evaluation*. Cambridge, MA: Harvard University Press.

1958b. (With J.V. Krutilla.) *Multiple Purpose River Development*. Baltimore: Johns Hopkins Press.

1964a. *Public Finance*. New York: Prentice-Hall. 4th edn, 1979.

1964b. (With E.S. Kirschen and others.) *Economic Policy in Our Time*. Amsterdam: North-Holland.

1967. (ed.) *Studies in the Economics of Income Maintenance*. Washington, DC: Brookings.

1970. (ed.) *The Econometrics of Price Determination*. Washington, DC: Board of Governors of the Federal Reserve System and Social Science Research Council.

1976. (ed.) *Parameters and Policies in the U.S. Economy*. Amsterdam: North-Holland.

1978. *The Great Recession*. Amsterdam: North-Holland.

1981. *Core inflation*. New York: Prentice-Hall.

1983. *The DRI Model of the U.S. Economy*. New York: McGraw-Hill.

1984. (With C. Caton, R. Brinner and P. Duprey.) *The DRI Report on U.S. Manufacturing Industries.* New York: McGraw-Hill.

ecological economics

Ecology can be regarded as the study of living species such as animals, plants and microorganisms, and the relations among them and their natural environment. In this context, an ecosystem includes these species and their non-living environment, their interactions, and their evolution in time and space (see, for example, Roughgarden, May and Levin, 1989). Economics, meanwhile, is the study of how human societies use scarce resources to produce commodities and to distribute them among their members.

The need for an interdisciplinary approach – 'ecological economics' – stems from the fact that natural ecosystems and human economies are closely linked. In the process of production and consumption, human beings use ecosystems and their services, influence their evolution, and are the recipients of feedbacks originating from their actions upon ecosystems. As Kenneth Boulding (1965) notes in his classic paper 'Earth as a space ship', which can be regarded as a landmark in the emergence of ecological economics, 'Man is finally going to have to face the fact that he is a biological system living in an ecological system, and that his survival power is going to depend on his developing symbiotic relationships of a closed-cycle character with all the other elements and populations of the world of ecological systems.'

Thus, ecological economics can be regarded as the study of the interactions and co-evolution in time and space of human economies and the ecosystems in which human economies are embedded. This implies that the task of ecological economics is to bridge the gap between economy and ecology by uncovering the links and the feedbacks between human economies and ecosystems, and by using these links and feedbacks to provide a unified picture of ecology and economy and their interactions and co-evolution. In a sense, ecological economics aims at linking ecological models and economic models in order to provide insights into complex and inter-related phenomena stemming from and affecting both ecosystems and human economies.

The natural link between ecology and human economies has been manifested in the traditional development of resource management or bio-economic models (for example, Clark, 1990), in which the main focus has been on fishery or forestry management where the impact of humans on ecosystems is realized through harvesting. More close links have been developed, however, as both disciplines evolve.

Common methodological approaches may also be encountered in ecology and economics. Optimality behaviour, which is fundamental in economics, has also been used to provide insights into the structure of ecological systems, in the context of optimal foraging behaviour, species competition, or net energy maximization by organisms (for example, Tschirhart, 2000; Tilman, Polasky and Lehman, 2005) with the purpose of founding macro-behaviours in ecosystems – such as those emerging from population dynamics – on micro-foundations.

In the same context, the classical phenomenological-descriptive approach to species competition based on Lotka–Volterra systems has recently been complemented by mechanistic resource-based models of species competition for limiting resources (Tilman, 1982; 1988). This approach has obvious links to competition among economic agents for limited resources. Furthermore, by linking the functioning of natural ecosystems with the provision of useful services to humans, or by using concepts such as ecosystems productivity, insurance from the genetic diversity of ecological systems against catastrophic events, or development of new products using genetic resources existing in natural ecosystems (Heal, 2000), new insights into the fundamental issues of the valuation of ecosystems or the valuation of biodiversity have been derived. (Examples of useful services to humans include provisioning services, such as food, water, fuel, genetic material; regulation services, such as climate regulation, disease regulation; and cultural services and supporting services, such as soil formation, nutrient cycling; see Millennium Ecosystem Assessment, 2005.)

Ecological models

The traditional bio-economic models (Clark, 1990), which describe the evolution of the population or the biomass of species when harvesting takes place, have formed the building blocks of ecological-economic modelling. These models can be extended along various lines to provide a more realistic picture of ecosystems (for a detailed analysis, see Murray, 2003) and help build meaningful ecological-economic models. To start with, let $x(t)$ denote the biomass of a certain species at time t. Then evolution of the biomass is described by an ordinary differential equation

$$\frac{dx(t)}{dt} = \text{birth} - \text{natural death} \\ + \text{migration} - \text{harvesting.} \tag{1}$$

In the analysis of population models it is common, unless it is a specific case, to set the migration rate at zero, and to represent the natural rate of population growth (birth-natural death) by a function $F(x)$. The most common specification of this function is the famous *logistic function*, which is $F(x) = rx(1 - x/K)$. In this function r is a positive constant called *intrinsic growth rate* and K is the *carrying capacity* of the environment which depends on factors such as resource availability or environmental pollution. If we denote by $h(t)$ the rate of harvesting of the species biomass by humans, the population model becomes:

$$\frac{dx(t)}{dt} = F(x) - h(t), \quad x(0) = x_0. \quad (2)$$

If $h(t) \equiv F(x)$, the population remains constant and the harvesting rate corresponds to *sustainable yield*. Harvesting rate is usually modelled as population dependent or $h = qEx$, where q is a positive constant, referred to as a *catchability coefficient* in fishery models, and E is *harvesting effort*. Human activities can affect the species population, in addition to harvesting, by affecting parameters such as the intrinsic growth rates or the carrying capacity. For example, if the stock of environmental pollution of a certain pollutant (such as phosphorus in a lake) in a natural ecosystem is denoted by P, with dynamics described by

$$\frac{dP(t)}{dt} = g(s(t), P(t)), \quad P(0) = P_0, \quad (3)$$

where $s(t)$ is the rate of emissions (such as phosphorus loadings), and the pollutant affects parameters of the population model, then the combined model will be (3) along with

$$\frac{dx(t)}{dt} = r(P)x \left[1 - \frac{x}{K(P)} \right] \quad (4)$$
$$-qEx, r'(P) < 0, \quad K'(P) < 0.$$

If the catchability coefficient is affected by technical change, then it can be expressed by a function of time as $q(t)$. In this case (4) is not autonomous. Alternatively q can be a function of technological variables like R&D evolving in the economic module.

The population model (2) can be generalized to age-structured populations and multi-species populations. In multi-species populations the Lotka–Volterra predator–prey models are classic. If we denote the prey population by $x(t)$ and the predator population by $y(t)$ and ignore harvesting for the moment to simplify things, the model can be written as

$$\frac{dx(t)}{dt} = x \left[r \left(1 - \frac{x}{K} \right) - yR(x) \right], \quad x(0) = x_0 \quad (5)$$

$$\frac{dy(t)}{dt} = ym \left(1 - \frac{ny}{x} \right), \quad y(0) = y_0 \quad m, n > 0 \quad (6)$$

where $R(x)$ is a function called the *predation term*, which can be specified as $\gamma x / (x^2 + \delta^2)$, $\gamma, \delta > 0$. A more general multi-species model with J prey and J predators can

be written, for $i = 1, \ldots, J$, as

$$\frac{dx_i(t)}{dt} = x_i \left[a_i - \sum_{j=1}^{J} \beta_{ij} y_j \right], \quad x_i(0) = x_{i0} \quad (7)$$

$$\frac{dy_i(t)}{dt} = y_i \left[\sum_{j=1}^{J} \gamma_{ij} x_j - \delta_i \right], \quad y_i(0) = y_{i0} \quad (8)$$

where all parameters are positive constants. An even more general model of interacting populations can be obtained by the generalized Kolmogorov model where the evolution of each species biomass is described by:

$$\frac{dx_i(t)}{dt} = x_i F_i(x_1, x_2, x_3, \ldots) \quad i = 1, 2, 3, \ldots \quad (9)$$

In the mechanistic resource-based models of species competition emerging from the work of Tilman (for example, Tilman, 1982; 1988; species compete for limiting resources. (For the use of this model in ecological-economic modelling, see Brock and Xepapadeas, 2002; Tilman, Polasky and Lehman, 2005.) In these models the growth of a species depends on the limiting resource, and interactions among species take place through the species' effects on the limiting resource. Let $\mathbf{x} = (x_1, \ldots, x_n)$ be the vector of species biomasses and R the amount of the available limiting resource. Then a mechanistic resource-based model with a single limiting factor in a given area and $i = 1, \ldots, n$ species can be described by the following equations:

$$\frac{\dot{x}_i}{x_i} = g_i(R) - d_i, \quad x_i(0) = x_{i0} \quad (10)$$

$$\dot{R} = S - aR - \sum_{i=1}^{n} w_i x_i g_i(R) \quad (11)$$

where $g_i(R)$ is resource-related growth, d_i is the species' natural death rate, S is the amount of resource supplied, a is the natural resource removal rate (leaching rate), and w_i is specific resource consumption by species i. The main result in this framework relates to an exclusion principle stating that, in a landscape free of disturbances, the species with the lowest resource requirement in equilibrium will competitively displace all other species, driving the system to a monoculture. Species coexistence and polycultures in equilibrium can be supported in a system with more than one limiting resource, or even in single resource systems if there is temperature-dependent growth and temperature variation in the ecosystem, spatial or temporal variations in resource ratios, differences in local palatabilities and local abundance of herbivores.

In addition to the temporal variation captured by the models described above an important characteristic of ecosystems is that of spatial variation. Biological resources tend to disperse in space under forces promoting 'spreading' or 'concentrating' (Okubo, 2001); these processes, along with intra- and inter-species interactions, induce the formation of spatial patterns for species. A central concept in modelling the dispersal of biological resources is that of *diffusion*. Diffusion is defined as a process whereby the microscopic irregular movement of particles such as cells, bacteria, chemicals, or animals results in some macroscopic regular motion of the group. Biological diffusion is based on random walk models which, when coupled with population growth equations, lead to general reaction-diffusion systems (see, for example, Okubo and Levin, 2001; Murray, 2003). When only one species is examined, the coupling of classical diffusion with a logistic growth function leads to the so-called Fisher–Kolmogorov equation, which can be written as

$$\frac{\partial x(z,t)}{\partial t} = F(x(z,t)) + D_x \frac{\partial^2 x(z,t)}{\partial z^2} \qquad (12)$$

where $x(z,t)$ denotes the concentration of the biomass at spatial point z at time t. The biomass grows according to a standard growth function $F(x)$ which determines the resource's kinetics but also disperses in space with a constant diffusion coefficient D_x. (Nonlinear reaction diffusion equations are associated with propagating wave solutions.) In general, a diffusion process in an ecosystem tends to produce a uniform population density, that is, spatial homogeneity. Thus it might be expected that diffusion would 'stabilize' ecosystems where species disperse and humans intervene through harvesting.

There, is however, one exception, known as 'diffusion induced instability' or 'diffusive instability'. It was Alan Turing (1952) who suggested that under certain conditions reaction-diffusion systems can generate spatially heterogeneous patterns. This is the so-called 'Turing mechanism' for generating diffusion instability. With two interacting species evolving according to

$$\frac{\partial x(z,t)}{\partial t} = F(x,y) + D_x \frac{\partial^2 x(z,t)}{\partial z^2} \qquad (13)$$

$$\frac{\partial y(z,t)}{\partial t} = G(x,y) + D_y \frac{\partial^2 y(z,t)}{\partial z^2}, \qquad (14)$$

if in the absence of diffusion $(D_x = D_y = 0)$ the system tends to a spatially uniform stable steady state, then under certain conditions, depending on the relationship D_x/D_y, spatially heterogeneous patterns can emerge due to diffusion-induced instability.

Spatial variations in ecological systems can also be analysed in terms of meta-population models. A meta-population is a set of local populations occupying isolated patches which are connected by migrating individuals. Meta-population dynamics can be developed for single or many species (Levin, 1974). For the single species case the dynamics become

$$\frac{d\mathbf{x}}{dt} = \mathbf{F}(\mathbf{x})\mathbf{x} + \mathbf{D}\mathbf{x} \qquad (15)$$

where $\mathbf{x} = (x_1, \ldots, x_J)$ is a column vector of species densities, $\mathbf{F}$ has its ith row depending on the ith row of $\mathbf{x}$, and $\mathbf{D} = [d_{ij}]$ is a connectivity matrix, where d_{ij} is the rate of movement from patch j to patch i $(j \neq i)$. Thus dynamics are local with the exception of movements from one patch to the other.

A more general model encompassing $i = 1, \ldots, n$ species competing for $j = 1, \ldots, J$ limiting resources, with density-dependent growth and interactions across patches $c = 1, \ldots, C$ in a given landscape, can be written as

$$\frac{\dot{x}_{ci}}{x_{ci}} = F_{ic}(\mathbf{x}_c, \mathbf{x}_{-c}) g_{ic}(\mathbf{R}_c, d_{ic}), \quad \forall i, c \qquad (16)$$

$$\dot{R}_{jc} = S_{jc}(\mathbf{R}_c, \mathbf{R}_{-c}) - D_{jc}(\mathbf{x}_c, \mathbf{x}_{-c}, \mathbf{R}_c, \mathbf{R}_{-c}), \quad \forall j, c \qquad (17)$$

where $\mathbf{R}_{-c}, \mathbf{x}_{-c}$ are respectively vectors of resources and species outside patch c.

(For a detailed analysis, see Brock and Xepapadeas, 2002.) A more general set-up can be obtained in the context of co-evolutionary models which describe the interactions between population (or biomass) dynamics and mutation (or trait dynamics). Antagonistic co-evolution of species on the one hand and pests or parasites or the other can be described by the so-called Red Queen hypothesis (see, for example, Van Valen, 1973, and Kawecki, 1998). According to this hypothesis, parasites evolve ceaselessly in response to perpetual evolution of species' (or hosts') resistance. The co-evolution of the parasites' ability to attack (virulence) and the hosts' resistance is expected to indicate persistent fluctuations of resistance and virulence. In this context the Red Queen hypothesis generates a continuous need for variation, resulting in a limit cycle or other non-point attractor in trait space dynamics, which are called Red Queen races. Red Queen cycles are observed in a slow time scale, since trait dynamics are assumed to evolve slowly, in contrast to the population, host-parasite, dynamics which are assumed to evolve fast (see Dieckmann and Law, 1996).

A simple co-evolutionary model can be developed in a system with one harvested ('useful') species or host species whose biomass is denoted by x and a parasite denoted by y, where the abundance of x and y depends on the evolution of two characteristics or traits denoted by d

and γ (see, for example, the Red Queen dynamic models developed by Krakauer and Jansen, 2002), where d affects the fitness of x and γ affects the fitness of y.

Let the growth rates of x and the pathogen y be given by

$$g_x = \frac{\dot{x}}{x} = (s - rx - yQ(d,\gamma)) \tag{18}$$

$$g_y = \frac{\dot{y}}{y} = (xQ(d,\gamma) - \delta). \tag{19}$$

If we measure fitness by growth rates, then $\frac{\partial Q(d,\gamma)}{\partial d} < 0$, so that an increase in d increases fitness of x. In the same way, $\frac{\partial Q(d,\gamma)}{\partial \gamma} > 0$ for an increase in γ to increase fitness of y. In equilibrium of the fast population system where $\dot{x} = \dot{y} = 0$, it holds that

$$\hat{x} = \frac{\delta}{Q(d,\gamma)}, \quad \hat{y} = \frac{s - r\hat{x}}{Q(d,\gamma)}, \quad s \geq r\hat{x}. \tag{20}$$

On the assumption of constant mutation rates μ_d and μ_γ, the evolutionary dynamics for the traits d and γ, when population dynamics have reached the asymptotically stable steady state, are given by

$$\dot{d} = -\mu_d \hat{x}\hat{y}\frac{\partial Q(d,\gamma)}{\partial d} \tag{21}$$

$$\dot{\gamma} = \mu_\gamma \hat{x}\hat{y}\frac{\partial Q(d,\gamma)}{\partial \gamma}. \tag{22}$$

See Krakauer and Jansen (2002) who, by considering the slow time scale trait dynamics, show that the equilibrium point $(d^*, \gamma^*) : \dot{d} = \dot{\gamma} = 0$ is not attracting; the dynamics spiral away from this point. This behaviour is the oscillatory, Red Queen dynamics.

Ecological-economic modelling

The ecological models developed above are the cornerstones of the development of meaningful ecological-economic models. The impact of humans on the population of species can be realized through direct harvesting h as described in (1) and (2). This type of impact can be easily incorporated into the more general population dynamic models by selecting the harvested species. Human influence can also be realized in an indirect way by having the environmental carrying capacity affected by environmental pollution generated in the non-harvesting sector of the economy, as in (3) and (4), or by having technological considerations affecting catchability coefficients. It is also possible that external environmental conditions which are anthropogenic, such as global warming, can make some parameters associated

with population dynamics or mutation dynamics change slowly. This can be modelled in (21) and (22) by considering μ_d and μ_γ as slow varying parameters, defined as $\mu_d(\varepsilon t)$ and $\mu_\gamma(\varepsilon t)$, where $0 < \varepsilon \ll 1$ is the *adiabatic parameter*. This slowly varying system could be used to model slow anthropogenic impacts on ecosystem structure.

However, the size and the severity of the impact of human economies in ecosystems depend on the way in which variables, such as harvesting or other variables which can be chosen by humans (such as emissions, investment in harvesting capacity) and which influence the evolution of ecosystems, are actually chosen. These variables can be regarded as *control variables*, and the way in which they are chosen affects the evolution of ecological variables, such as species biomasses or traits, which can be considered as the *state variables* of the problem.

The typical approach in economics is to associate the choice of the control variables with *optimizing behaviour*. Thus, the control variables are chosen so that a criterion function is optimized, and the economic problem of ecosystem management – where management means choice of control variables – is defined as a formal optimal control problem. In this problem the objective is the optimization of the criterion function subject to the constraints imposed by the structure of the ecosystem. These constraints, which provide the transition equations of the optimal control problem, are the dynamic equations of the ecological models described in the previous section.

The solution of the ecological-economic model, provided it exists, will determine the paths of the state and the control variables and the steady state of the system, which will determine the long-run equilibrium values of the ecological populations as well as the approach dynamics to the steady state. In this context, managed ecological systems which are predominantly nonlinear could exhibit dynamic behaviour characterized by multiple, locally stable and unstable steady states, limit cycles, or the emergence of hysteresis, bifurcations or irreversibilities.

The way in which the objective function is set up and the ecological constraints which are taken into account determine the solution of the ecological-economic model. In principle, a *socially optimal solution* can be distinguished from a *privately optimal solution*. The socially optimal solution corresponds to the so-called problem of the *social planner*, where the objective function takes into account not only benefits from harvesting certain resources of the ecological system, which corresponds to harvesting commercially valuable biomass, but in addition a wide spectrum of flows of services generated by the whole ecosystem. These include, as described above, regulation, cultural or supporting services, existence values, or benefits associated with productivity or insurance gains. If $V(\mathbf{h}(t))$ denotes harvesting benefits at time t associated with harvesting vector $\mathbf{h}$, and $U(\mathbf{x}(t))$

denotes the flow of benefits associated with ecosystem service generated by species biomasses existing in the ecosystem and not removed by harvesting, then the total flow of benefit is $V(\mathbf{h}(t)) + U(\mathbf{x}(t))$. In this formulation, the $V(\cdot)$ and $U(\cdot)$ functions are usually assumed to be monotonically increasing and concave. In a more general setup, the total benefit function can be non-separable, defined as $u(\mathbf{h}(t), \mathbf{x}(t))$.

The objective can then be written as:

$$\max_{\{\mathbf{h}(t)\}} \int_0^\infty e^{-\rho t}[V(\mathbf{h}(t)) + U(\mathbf{x}(t))]dt$$

$$(23)$$

where $\rho \geq 0$ is a discount rate. It should be noted that in principle benefits associated with $V(\mathbf{h}(t))$ can be estimated using market data, while benefits associated with $U(\mathbf{x}(t))$ are hard to estimate because markets for the larger part of the spectrum of ecosystem services are missing. (Valuation of ecosystem services is an open question. For details, see, for example, Bingham et al., 1995.) The social optimum corresponds to the maximization of (23), subject to the constraints imposed by the ecological system. For example, if we use the generalized model of resource competition, the constraints are:

$$\frac{\dot{x}_{ci}}{x_{ci}} = F_{ic}(\mathbf{x}_c, \mathbf{x}_{-c})g_{ic}(\mathbf{R}_c, d_{ic}) - h_{ic}, \quad \forall i, c$$

$$(24)$$

$$\dot{R}_{jc} = S_{jc}(\mathbf{R}_c, \mathbf{R}_{-c}) - D_{jc}(\mathbf{x}_c, \mathbf{x}_{-c}, \mathbf{R}_c, \mathbf{R}_{-c}), \quad \forall j, c.$$

$$(25)$$

A solution $(\mathbf{h}^*(t), \mathbf{x}^*(t))$ is regarded as the socially optimal solution.

The privately optimal solution is distinguished from the socially optimal by the fact that only harvesting benefits enter the objective function. The assumption is that management is carried out by a 'small' profit-maximizing private agent that ignores the general flows of ecosystem services. In this case, the private agents do not take into account *externalities* associated with their management practices on ecosystem service flow and $U(\mathbf{x}(t)) \equiv 0$. Market externalities associated with the definition of $V(\mathbf{h})$ could relate to imperfections in the markets for the harvested commodities, or to property rights-related externalities, as the well known 'tragedy of the commons' emerging in the harvesting of open access resources.

In general the privately optimal solution $(\mathbf{h}^0(t), \mathbf{x}^0(t))$ will deviate from the socially optimal solution. Another type of externality can be associated with strategic behaviour in resource harvesting if more than one private agent harvests the resource. If $l = 1, \ldots, L$

harvesters are present, then the biomass equation (24) for patch c becomes

$$\frac{\dot{x}_{ci}}{x_{ci}} = F_{ic}(\mathbf{x}_c, \mathbf{x}_{-c})g_{ic}(\mathbf{R}_c, d_{ic}) - \sum_{l=1}^{L} h_{ic}, \quad \forall i, c.$$

$$(26)$$

In this case the privately optimal solution can be obtained as an *open loop or feedback* Nash equilibrium.

Privately optimal solutions can also be distinguished from the socially optimal by the extent to which the ecological constraints are taken into account. For example, if resource dynamics or trait dynamics are not taken into account in the optimization problem, the management rule will deviate from the social optimum. Furthermore, since *all* the ecological constraints are operating, there will be discrepancies between the perceived evolution of ecosystems under management that ignores certain constraints, and the actual evolution of the ecosystem. Brock and Xepapadeas (2003), show that, by ignoring genetic constraints associated with the development of resistance to genetically modified organisms, the actual system loses any productivity advantage because of resistance development.

These discrepancies might be a cause for *surprises* in ecosystem management. For example, with reference to the co-evolutionary model (18) – (22), profit-maximizing decisions which ignore evolution might steer the system to a certain steady state on a fast time scale, but then the underlying trait dynamics might move the system in slow time to another attractor.

The deviations between the private solution and the social optimum provide a basis for regulation which is similar to the rationale behind the regulation of environmental externalities. Regulation could take the form, in general spatial models of ecosystem management, of species-specific and site-specific taxes on harvesting, or equivalent quota and zoning systems.

ANASTASIOS XEPAPADEAS

See also **approximate solutions to dynamic models (linear methods); common property resources; consumption externalities; dynamic programming; environmental economics; spatial economics; spatial econometrics.**

Bibliography

Bingham, G. et al. 1995. Issues inecosystem valuation: improving information for decision making. *Ecological Economics* 14(2), 73–90.

Boulding, K. 1965. Earth as a space ship. Washington State University Committee on Space Sciences. Kenneth E. Boulding Papers, Archives (Box # 38), University of Colorado at Boulder Libraries. Online. Available at http://www.colorado.edu/econ/Kenneth.Boulding/spaceship-earth.html, accessed 13 July 2005.

Brock, W. and Xepapadeas, A. 2002. Optimal ecosystem management when species compete for limiting resources. *Journal of Environmental Economics and Management* 44, 189–230.

Brock, W. and Xepapadeas, A. 2003. Valuing biodiversity from an economic perspective: a unified economic, ecological and genetic approach. *American Economic Review* 93, 1597–614.

Clark, C. 1990. *Mathematical Bioeconomics: The Optimal Management Of Renewable Resources.* 2nd edn. New York: Wiley.

Dieckmann, U. and Law, R. 1996. The dynamical theory of coevolution: a derivation from stochastic ecological processes. *Journal of Mathematical Biology* 34, 579–612.

Kawecki, T. 1998. Red Queen meets Santa Rosalia: arms races and the evolution of host specialization in organisms with parasitic lifestyles. *American Naturalist* 152(4), 635–51.

Krakauer, D. and Jansen, V. 2002. Red Queen dynamics in protein translation. *Journal of Theoretical Biology* 218, 97–109.

Heal, G. 2000. *Nature and the Marketplace: Capturing the Value of Ecosystem Services.* Washington, DC: Island Press.

Levin, S. 1974. Dispersion and population interactions. *American Naturalist* 108, 207–28.

Millennium Ecosystem Assessment. 2005. *Ecosystems and Human Well-being. Volume 3: Policy Responses.* Washington, DC: Island Press.

Murray, J. 2003. *Mathematical Biology*, 3rd edn. Berlin, Springer.

Okubo, A. 2001. Introduction: the mathematics of ecological diffusion. In *Diffusion and Ecological Problems: Modern Perspectives*, 2nd edn., ed. A. Okubo and S. Levin. Berlin: Springer.

Okubo, A. and Levin, S. 2001. The basics of diffusion. In *Diffusion and Ecological Problems: Modern Perspectives*, 2nd edn., ed. A. Okubo and S. Levin. Berlin: Springer.

Roughgarden, J., May, R. and Levin, S. 1989. *Perspectives in Ecological Theory.* Princeton, NJ: Princeton University Press.

Tilman, D. 1982. *Resource Competition and Community Structure.* Princeton, NJ: Princeton University Press.

Tilman, D. 1988. *Plant Strategies and the Dynamics and Structure of Plant Communities.* Princeton, NJ: Princeton University Press.

Tilman, D., Polasky, S. and Lehman, C. 2005. Diversity, productivity and temporal stability in the economies of humans and nature. *Journal of Environmental Economics and Management* 49, 405–26.

Tschirhart, J. 2000. General equilibrium of an ecosystem. *Journal of Theoretical Biology* 203, 13–32.

Turing, A. 1952. The chemical basis of morphogenesis. *Philosophical Transactions of the Royal Society of London. Series B: Biological Sciences* 237(641), 37–72.

Van Valen, L. 1973. A new evolutionary law. *Evolutionary Theory* 1, 1–30.

ecological inference

1 The ecological inference problem

For expository purposes, we discuss only an important but simple special case of ecological inference, and adopt the running example and notation from King (1997: ch. 2). The basic problem has two observed variables (T_i and X_i) and two unobserved quantities of interest (β_i^b and β_i^w) for each of p observations. Observations represent aggregate units, such as geographic areas, and each individual-level variable within these units is dichotomous.

To be more specific, in Figure 1 we observe for each electoral precinct $i(i = 1, \ldots, p)$ the fraction of voting age people who turnout to vote (T_i) and who are black (X_i), along with the number of voting age people (N_i). The quantities of interest, which remain unobserved because of the secret ballot, are the proportions of blacks who vote (β_i^b) and whites who vote (β_i^w). The proportions β_i^b and β_i^w are not observed because T_i and X_i are from different data sources (electoral results and census data, respectively) and record linkage is impossible (and illegal), and so the cross-tabulation cannot be computed.

Also of interest are the district-wide fractions of blacks and whites who vote, which are respectively

$$B^b = \frac{\sum_{i=1}^{p} N_i X_i \beta_i^b}{\sum_{i=1}^{p} N_i X_i}, \qquad \text{and} \qquad (1)$$

$$B^w = \frac{\sum_{i=1}^{p} N_i (1 - X_i) \beta_i^w}{\sum_{i=1}^{p} N_i (1 - X_i)}. \qquad (2)$$

These are weighted averages of the corresponding precinct-level quantities. Some methods aim to estimate only B^b and B^w without giving estimates of β_i^b and β_i^w for all i.

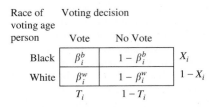

Race of voting age person	Voting decision		
	Vote	No Vote	
Black	β_i^b	$1 - \beta_i^b$	X_i
White	β_i^w	$1 - \beta_i^w$	$1 - X_i$
	T_i	$1 - T_i$	

Figure 1   Notation for Precinct i. *Note*: The goal is to estimate the quantities of interest, β_i^b (the fraction of blacks who vote) and β_i^w (the fraction of whites who vote), from the aggregate variables X_i (the fraction of voting age people who are black) and T_i (the fraction of people who vote), along with N_i (the known number of voting age people).

2 Deterministic and statistical approaches

The ecological inference literature before King (1997) was bifurcated between supporters of the method of bounds, originally proposed by Duncan and Davis (1953), and supporters of statistical approaches, proposed even before Ogburn and Goltra (1919) but first formalized into a coherent statistical model by Goodman (1953; 1959). (For the historians of science among us: although these two monumental articles were written by two colleagues and friends in the same year and in the same department and university – the Department of Sociology at the University of Chicago – the principal did not discuss their work prior to completion. Even by today's standards, nearly a half century after their publication, the articles are models of clarity and creativity.) Although Goodman and Duncan and Davis moved on to other interests following their seminal contributions, most of the ecological inference literature in the five decades since 1953 was an ongoing war between supporters of these two key approaches, and often without the usual academic decorum.

2.1 Extracting deterministic information: the method of bounds

The purpose of the method of bounds and its generalizations is to extract deterministic information, known with certainty, about the quantities of interest.

The intuition behind these quantities is simple. For example, if a precinct contained 150 African-Americans and 87 people in the precinct voted, then how many of the 150 African-American actually cast their ballot? We do not know exactly, but bounds on the answer are easy to obtain: in this case, the answer must lie between 0 and 87. Indeed, conditional only on the data being correct, [0,87] is a 100 per cent confidence interval. Intervals like this are sometimes narrow enough to draw meaningful inferences, and sometimes they are too wide, but the ability to provide (non-trivial) 100 per cent confidence intervals in even some situations is quite rare in any statistical field.

In general, before any data are seen, the unknown parameters β_i^b and β_i^w are each bounded on the unit interval. Once we observe T_i and X_i they are bounded more narrowly, as:

$$\beta_i^b \in \left[\max\left(0, \tfrac{T_i - (1 - X_i)}{X_i}\right), \ \min\left(\tfrac{T_i}{X_i}, 1\right) \right]$$
$$\beta_i^w \in \left[\max\left(0, \tfrac{T_i - X_i}{1 - X_i}\right), \ \min\left(\tfrac{T_i}{1 - X_i}, 1\right) \right].$$

$$(3)$$

Deterministic bounds on the district-level quantities B^b and B^w are weighted averages of these precinct-level bounds.

The bounds then indicate that the parameters in each case fall within these deterministic bounds with certainty, and in practice they are almost always narrower than

[0,1]. Whether they are narrow enough in any one application depends on the nature of the data.

2.2 Extracting statistical information: Goodman's regression

Leo Goodman's (1953; 1959) approach is very different from, but just as important as, Duncan and Davis's. He looked at the same data and focused on the statistical information. His approach examines variation in the marginals (X_i and T_i) over the precincts to attempt to reason back to the district-wide fractions of blacks and whites who vote, B^b and B^w. The outlines of this approach and the problems with it have been known at least since Ogburn and Goltra (1919). For example, if in precincts with large proportions of black citizens we observe that many people do not vote, then it may seem reasonable to infer that blacks turn out at lower rates than whites. Indeed, it often is reasonable, but not always. The problem is that it could instead be the case that the whites who happen to live in heavily black precincts are the ones who vote less frequently, yielding the opposite ecological inference to the individual-level truth.

What Goodman accomplished was to formalize the logic of the approach in a simple regression model, and to give the conditions under which estimates from such a model are unbiased. To see this, note first that the accounting identity

$$T_i = X_i \beta_i^b + (1 - X_i)\beta_i^w \qquad (4)$$

holds exactly. Then he showed that a regression of T_i on X_i and $(1 - X_i)$ with no constant term could be used to estimate B^b and B^w, respectively. The key assumption necessary for unbiasedness that Goodman identified is that the parameters and X_i be uncorrelated: $\mathrm{Cov}(\beta_i^b, X_i) = \mathrm{Cov}(\beta_i^w, X_i) = 0$. In the example, the assumption is that blacks vote in the same proportions in homogeneously black areas as in more integrated areas. Obviously, this is true sometimes and it is false other times. (King, 1997: ch. 3, showed that Goodman's assumption was necessary but not sufficient. To have unbiasedness, it must also be true that the parameters and N_i are uncorrelated.)

As Goodman recognized, when this key assumption does not hold, estimates from the model will be biased. Indeed, they can be very biased, outside the deterministic bounds, and even outside the unit interval. This technique has been used extensively since the 1950s, and impossible estimates occur with considerable frequency (some estimates range to a majority of real applications; Achen and Shively, 1995).

3 Extracting both deterministic and statistical information: King's EI approach

From 1953 until 1997, the only two approaches used widely in practice were the method of bounds and

Goodman's regression. King's (1997) idea was that the insights from these two conflicting literatures in fact do not conflict with each other; the sources of information are largely distinct and can be combined to improve inference overall and synergistically. The idea is to combine the information from the bounds, applied to both quantities of interest for each and every precinct, with a statistical approach for extracting information within the bounds. The amount of information in the bounds depends on the data-set, but for many data-sets it can be considerable. For example, if precincts are spread uniformly over a scatterplot of X_i by T_i, the average bounds on β_i^b and β_i^w are narrowed from [0,1] to less than half of that range – hence eliminating half of the ecological inference problem with certainty. This additional information also helps make the statistical portion of the model far less sensitive to assumptions than previous statistical methods which exclude the information from the bounds.

To illustrate these points, we first present all the information available without making any assumptions, thus extending the bounds approach as far as possible. As a starting point, the left graph in Figure 2 provides a scatterplot of a sample data set as observed, X_i horizontally by T_i vertically. Each point in this figure corresponds to one precinct, for which we would like to estimate the two unknowns. We display the unknowns in the right graph of the same figure; any point in the right graph portrays values of the two unknowns, β_i^b which is plotted horizontally, and β_i^w which is plotted vertically. Ecological inference involves locating, for each precinct, the one point in this unit square corresponding to the true values of β_i^b and β_i^w, since values outside the square are logically impossible.

To map the knowns onto the unknowns, King begins Goodman's accounting identity from eq. (4). From this equation, which holds exactly, King solves for one unknown in terms of the other:

$$\beta_i^w = \left(\frac{T_i}{1 - X_i} \right) - \left(\frac{X_i}{1 - X_i} \right) \beta_i^b, \qquad (5)$$

which shows that β_i^w is a *linear* function of β_i^b, where the intercept and slope are known (since they are functions of the data, X_i and T_i).

King then maps the knowns from the left graph onto the right graph by using the linear relationship in eq. (5). A key point is that each dot on the left graph can be expressed, without assumptions or loss of information, as what King called a 'tomography' line within the unit square in the right graph. It is precisely the information lost due to aggregation that causes us to have to plot an entire line (on which the true point must fall) rather than the goal of one point for each precinct on the right graph. In fact, the information lost is equivalent to having a graph of the β_i^b by β_i^w points but having the ink smear, making the points into lines and partly but not entirely obscuring the correct positions of the (β_i^b, β_i^w) points. (King also showed that the ecological inference problem is mathematically equivalent to the ill-posed 'tomography' problem of many medical imaging procedures, such as CAT and PET scans, where one attempts to reconstruct the inside of an object by passing X-rays through it and gathering information only from the outside. Because the line sketched out by an X-ray is closely analogous to eq. (5), King labels the latter a *tomography line* and the corresponding graph a *tomography graph*.)

What does a tomography line tell us? Before we know anything, we know that the true (β_i^b, β_i^w) point must lie somewhere within the unit square. After X_i and

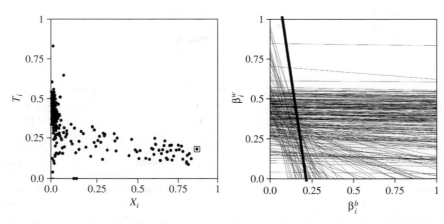

Figure 2   Two views of the same data. *Note*: The left graph is a scatterplot of the observables, X_i by T_i. The right graph displays this same information as a tomography plot of the quantities of interest, β_i^b by β_i^w. Each precinct i that appears as a point in the left graph is a line (rather than a point because of information lost due to aggregation) in the right graph. For example, precinct 52 appears as the dot with a little square around it in the left graph and the dark line in the right graph. *Source*: The data are from King (1997: Figures 5.1 and 5.5).

T_i are observed for a precinct, we also know that the true point must fall on a specific line represented by eq. (5) and appearing in the tomography plot in Figure 2. In many cases narrowing the region to be searched for the true point from the entire square to the one line in the square can provide a significant amount of information. To see this, consider the point enclosed in a box in the left graph, and the corresponding dark line in the right graph. This precinct, number 52, has observed values of $X_{52} = 0.88$ and $T_{52} = 0.19$. As a result, substituting into eq. (5) gives $\beta_i^w = 1.58 - 7.33\beta_i^b$, which when plotted appears as the dark line on the right graph. This particular line tells us that, in our search for the true $\beta_{52}^b, \beta_{52}^w$ point on the right graph, we can eliminate with certainty all area in the unit square except that on the line, which is clearly an advance over not having the data. Translated into the quantities of interest, this line tells us (by projecting the line downward to the horizontal axis) that, wherever the true point falls on the line, β_{52}^b must fall in the relatively narrow bounds of $[0.07, 0.21]$. Unfortunately, in this case, β_i^w can only be bounded (by projecting to the left) to somewhere within the entire unit interval. More generally, lines that are relatively steep, like this one, tell us a great deal about β_i^b and little about β_i^w. Tomography lines that are relatively flat give narrow bounds on β_i^w and wide bounds on β_i^b. Lines that cut off the bottom left (or top right) of the figure give narrow bounds on both quantities of interest.

If the only information available to learn about the unknowns in precinct i is X_i and T_i, a tomography line like that in Figure 2 exhausts all this available information. This line immediately tells us the known bounds on each of the parameters, along with the precise relationship between the two unknowns, but it is not sufficient to narrow in on the right answer any further. Fortunately, additional information exists in the other observations in the same data set (X_j and T_j for all $i \neq j$) which, under the right assumptions, can be used to learn more about β_i^b and β_i^w in our precinct of interest.

In order to borrow statistical strength from all the precincts to learn about β_i^b and β_i^w in precinct i, some assumptions are necessary. The simplest version of King's model (that is, the one most useful for expository purposes) requires three assumptions, each of which can be relaxed in different ways.

First, the set of (β_i^b, β_i^w) points must fall in a single cluster within the unit square. The cluster can fall anywhere within the square; it can be widely or narrowly dispersed or highly variable in one unknown and narrow in the other; and the two unknowns can be positively, negatively, or not at all correlated over i. An example that would violate this assumption would be two or more distinct clusters of (β_i^b, β_i^w) points, as might result from subsets of observations with fundamentally different data generation processes (such as from markedly different regions). The specific mathematical version of this one-cluster assumption is that β_i^b and β_i^w follow a truncated bivariate normal density

$$\text{TN}(\beta_i^b, \beta_i^w | \breve{\mathfrak{B}}, \breve{\Sigma}) = \text{N}(\beta_i^b, \beta_i^w | \breve{\mathfrak{B}}, \breve{\Sigma}) \frac{1(\beta_i^b, \beta_i^w)}{R(\breve{\mathfrak{B}}, \breve{\Sigma})},$$

$$(6)$$

where the kernel is the untruncated bivariate normal,

$$\text{N}(\beta_i^b, \beta_i^w | \breve{\mathfrak{B}}, \breve{\Sigma}) = (2\pi)^{-1} |\breve{\Sigma}|^{-1/2}$$
$$\times \exp\left[-\frac{1}{2}(\beta_i - \breve{\mathfrak{B}})'\breve{\Sigma}^{-1}(\beta_i - \breve{\mathfrak{B}})\right], \quad (7)$$

and $1(\beta_i^b, \beta_i^w)$ is an indicator function that equals 1 if $\beta_i^b \in [0, 1]$ and $\beta_i^w \in [0, 1]$ and zero otherwise. The normalization factor in the denominator, $R(\breve{\mathfrak{B}}, \breve{\Sigma})$, is the volume under the untruncated normal distribution above the unit square:

$$R(\breve{\mathfrak{B}}, \breve{\Sigma}) = \int_0^1 \int_0^1 \text{N}(\beta^b, \beta^w | \breve{\mathfrak{B}}, \breve{\Sigma}) d\beta^b d\beta^w$$

$$(8)$$

When divided into the untruncated normal, this factor keeps the volume under the truncated distribution equal to 1. The parameters of the truncated density, which we summarize as

$$\breve{\psi} = \{\breve{\mathfrak{B}}^b, \breve{\mathfrak{B}}^w, \breve{\sigma}_b, \breve{\sigma}_w, \breve{\rho}\} = \{\breve{\mathfrak{B}}, \breve{\Sigma}\}, \quad (9)$$

are on the scale of the untruncated normal (and so, for example, $\breve{\mathfrak{B}}^b$ and $\breve{\mathfrak{B}}^w$ need not be constrained to the unit interval even though β_i^b and β_i^w are constrained by this density).

The second assumption, which is necessary to form the likelihood function, is the absence of spatial autocorrelation: conditional on X_i, T_i and T_j are mean independent. Violations of this assumption in empirically reasonable (and even some unreasonable) ways do not seem to induce much if any bias.

The final, and by far the most critical, assumption is that X_i is independent of β_i^b and β_i^w. The three assumptions together produce what has come to be known as King's 'basic' EI model. (The use of EI to name this method comes from the name of his software, available at http://GKing.Harvard.edu.) King also generalizes this assumption, in what has come to be known as the 'extended' EI model, by allowing the truncated normal parameters to vary as functions of measured covariates, Z_i^b and Z_i^w, giving:

$$\breve{\mathfrak{B}}_i^b = \left[\phi_1(\breve{\sigma}_b^2 + 0.25) + 0.5\right] + (Z_i^b - \bar{Z}^b)\alpha^b$$
$$\breve{\mathfrak{B}}_i^w = \left[\phi_2(\breve{\sigma}_w^2 + 0.25) + 0.5\right] + (Z_i^w - \bar{Z}^w)\alpha^w$$

$$(10)$$

where α^b and α^w are parameter vectors to be estimated along with the original model parameters and that have as many elements as Z_i^b and Z_i^w have columns. This relaxes the mean independence assumptions to:

$$E(\beta_i^b|X_i, Z_i) = E(\beta_i^b|Z_i)$$
$$E(\beta_i^w|X_i, Z_i) = E(\beta_i^w|Z_i).$$

Note that this extended model also relaxes the assumptions of truncated bivariate normality, since there is now a separate density being assumed for each observation. Because the bounds, which differ in width and information content for each i, generally provide substantial information, even X_i can be used as a covariate in Z_i. (The recommended default setting in EI includes X_i as a covariate with a prior on its coefficient.) In contrast, under Goodman's regression, which does not include information in the bounds, including X_i leads to an unidentified model (King, 1997: sec. 3.2).

These three assumptions – one cluster, no spatial autocorrelation, and mean independence between the regressor and the unknowns conditional on X_i and Z_i – enable one to compute a posterior (or sampling) distribution of the two unknowns in each precinct. A fundamentally important component of EI is that the quantities of interest are not the parameters of the likelihood but instead come from conditioning on T_i and producing a posterior for β_i^b and β_i^w in each precinct. Failing to condition on T_i and examining the parameters of the truncated bivariate normal only makes sense if the model holds exactly and so is much more model-dependent than King's approach. Since the most important problem in ecological inference modelling is precisely model misspecification, failing to condition on T assumes away the problem without justification. This point is widely regarded as a critical step in applying the EI model (Adolph and King, with Herron and Shotts, 2003).

When bounds are narrow, EI model assumptions do not matter much. But, for precincts with wide bounds on a quantity of interest, inferences can become model dependent. This is especially the case with ecological inference problems precisely because of the loss of information due to aggregation. In fact, this loss of information can be expressed by noting that the joint distribution of β_i^b and β_i^w cannot be fully identified from the data without some untestable assumptions. To be precise, distributions with positive mass over *any* curve or combination of curves that connects the bottom left point ($\beta_i^b = 0, \beta_i^w = 0$) to the top right point ($\beta_i^b = 1$, $\beta_i^w = 1$) of a tomography plot cannot be rejected by the data (King, 1997: 191). Other features of the distribution are estimable. This fundamental indeterminacy is, of course, a problem because it prevents pinning down the quantities of interest with certainty, but it can also be something of an opportunity since different distributional assumptions can lead to the same estimates, especially since only those pieces of the distributions above the tomography lines are used in the final analysis.

4 Alternative approaches to ecological inference

In the continuing search for more information to bring to bear on ecological inferences, King, Rosen and Tanner (1999) extend King's (1997) model another step. They incorporate King's main advance of combining deterministic and statistical information but begin modelling a step earlier at the individuals who make up the counts. They also build a hierarchical Bayesian model, using easily generalizable Markov chain Monte Carlo (MCMC) technology (Tanner, 1996).

To define the model formally, let T_i' denote the *number* of voting age people who turn out to vote. At the top level of the hierarchy they assume that T_i' follows a binomial distribution with probability equal to $\theta_i = X_i\beta_i^b + (1 - X_i)\beta_i^w$ and count N_i. Note that at this level it is assumed that the *expectation* of T_i', rather than T_i', is equal to $X_i\beta_i^b + (1 - X_i)\beta_i^w$. In other words, King (1997) models T_i as a continuous proportion, whereas King, Rosen, and Tanner (1999) recognize the inherently discrete nature of the counts of voters that go into computing this proportion. The two models are connected, of course, since T_i/N_i approaches T_i as N_i gets large.

The connection to King's tomography line can be seen in the contribution of the data from precinct i to the likelihood, which is

$$(X_i\beta_i^b + (1 - X_i)\beta_i^w)^{T_i'}(1 - X_i\beta_i^b - (1 - X_i)\beta_i^w)^{(N_i - T_i')}.$$
$$(11)$$

By taking the logarithm of this contribution to the likelihood and differentiating with respect to β_i^b and β_i^w, King, Rosen and Tanner show that the maximum of (11) is not a unique point, but rather a line whose equation is given by the tomography line in eq. (5). Thus, the log-likelihood for precinct i looks like two playing cards leaning against each other. As long as T_i is fixed and bounded away from 0.5 (and X_i is a fixed known value between 0 and 1), the derivative at this point is seen to increase with N_i, that is, the pitch of the playing cards increases with the sample size. In other words, for large N_i, the log-likelihood for precinct i degenerates from a surface defined over the unit square into a single playing card standing perpendicular to the unit square and oriented along the corresponding tomography line.

At the second level of the hierarchical model, β_i^b is distributed as a beta density with parameters c_b and d_b and β_i^w follows an independent beta with parameters c_w and d_w. While β_i^b and β_i^w are assumed *a priori* independent, they are *a posteriori* dependent. At the third and final level of the hierarchical model, the unknown parameters c_b, d_b, c_w and d_w follow an exponential distribution with a large mean.

A key advantage of this model is that it generalizes immediately to arbitrarily large $R \times C$ tables. This approach was pursued by Rosen et al. (2001), who also provided a much faster method of moment-based estimator. For an application, see King et al. (2003).

Wakefield (2004) presents an alternative approach based on the Bayesian paradigm using a Markov chain Monte Carlo inference scheme. King, Rosen and Tanner (2004) survey the latest strategies for solving ecological inference problems in various fields, many of which do not fit the textbook case of a 2 × 2 table with known marginals and unknown cell entries. Staniswalis (2005) proposes a nonparametric model for ecological inference with an application to renal failure data.

GARY KING, ORI ROSEN AND MARTIN TANNER

Bibliography

Achen, C. and Shively, W. 1995. *Cross-Level Inference.* Chicago: University of Chicago Press.

Adolph, C., King, G., Herron, M. and Shotts, K. 2003. A consensus position on second stage ecological inference models. *Political Analysis* 11, 86–94.

Duncan, O. and Davis, B. 1953. An alternative to ecological correlation. *American Sociological Review* 18, 665–6.

Goodman, L. 1953. Ecological regressions and the behavior of individuals. *American Sociological Review* 18, 663–6.

Goodman, L. 1959. Some alternatives to ecological correlation. *American Journal of Sociology* 64, 610–24.

King, G. 1997. *A Solution to the Ecological Inference Problem: Reconstructing Individual Behavior from Aggregate Data.* Princeton: Princeton University Press.

King, G., Rosen, O. and Tanner, M. 1999. Binomial-beta hierarchical models for ecological inference. *Sociological Methods and Research* 28, 61–90.

King, G., Rosen, O. and Tanner, M., eds. 2004. *Ecological Inference: New Methodological Strategies.* Cambridge: Cambridge University Press.

King, G., Rosen, O., Tanner, M. and Wagner, A. 2003. The ordinary election of Adolf Hitler: a modern voting behavior approach. Online. Available at http://gking.harvard.edu/files/naziV.pdf, accessed 16 August 2006.

Ogburn, W. and Goltra, I. 1919. How women vote: a study of an election in Portland, Oregon. *Political Science Quarterly* 34, 413–33.

Rosen, O., Jiang, W., King, G. and Tanner, M. 2001. Bayesian and frequentist inference for ecological inference: the $R \times C$ case. *Statistica Neerlandica* 55(2), 134–56.

Staniswalis, J. 2005. On fitting generalized non-linear models with varying coefficients. *Computational Statistics and Data Analysis* 50, 893–902.

Tanner, M. 1996. *Tools for Statistical Inference: Methods for the Exploration of Posterior Distributions and Likelihood Functions*, 3rd edn. New York: Springer-Verlag.

Wakefield, J. 2004. Prior and likelihood choices in the analysis of ecological data. In *Ecological Inference: New Methodological Strategies*, ed. G. King, O. Rosen and M. Tanner. Cambridge: Cambridge University Press.

econometrics

1 What is econometrics?

Broadly speaking, econometrics aims to give empirical content to economic relations for testing economic theories, forecasting, decision making, and for *ex post* decision/policy evaluation. The term 'econometrics' appears to have been first used by Pawel Ciompa as early as 1910, although it is Ragnar Frisch who takes the credit for coining the term, and for establishing it as a subject in the sense in which it is known today (see Frisch, 1936, p. 95, and Bjerkholt, 1995). By emphasizing the quantitative aspects of economic relationships, econometrics calls for a 'unification' of measurement and theory in economics. Theory without measurement can have only limited relevance for the analysis of actual economic problems; while measurement without theory, being devoid of a framework necessary for the interpretation of the statistical observations, is unlikely to result in a satisfactory explanation of the way economic forces interact with each other. Neither 'theory' nor 'measurement' on its own is sufficient to further our understanding of economic phenomena.

As a unified discipline, econometrics is still relatively young and has been transforming and expanding very rapidly since an earlier version of this article was published in the first edition of *The New Palgrave: A Dictionary of Economics* in 1987 (Pesaran, 1987a). Major advances have taken place in the analysis of cross-sectional data by means of semiparametric and nonparametric techniques. Heterogeneity of economic relations across individuals, firms and industries is increasingly acknowledged, and attempts have been made to take them into account either by integrating out their effects or by modelling the sources of heterogeneity when suitable panel data exists. The counterfactual considerations that underlie policy analysis and treatment evaluation have been given a more satisfactory foundation. New time series econometric techniques have been developed and employed extensively in the areas of macroeconometrics and finance. Nonlinear econometric techniques are used increasingly in the analysis of cross-section and time-series observations. Applications of Bayesian techniques to econometric problems have been given new impetus largely thanks to advances in computer power and computational techniques. The use of Bayesian techniques has in turn provided the investigators with a unifying framework where the tasks of forecasting, decision making, model evaluation and learning can be considered as parts of the same interactive and iterative process; thus paving the way for establishing the foundation of 'real time econometrics'. See Pesaran and Timmermann (2005a).

This article attempts to provide an overview of some of these developments. But to give an idea of the extent to which econometrics has been transformed over the past decades we begin with a brief account of the literature that pre-dates econometrics, and discuss the birth of econometrics and its subsequent developments to the present. Inevitably, our accounts will be brief and non-technical. Readers interested in more details are advised to consultant the specific entries provided in the *New Palgrave* and the excellent general texts by Maddala (2001), Greene (2003), Davidson and MacKinnon (2004), and Wooldridge (2006), as well as texts on specific topics such as Cameron and Trivedi (2005) on microeconometrics, Maddala (1983) on econometric models involving limited-dependent and qualitative variables, Arellano (2003), Baltagi (2005), Hsiao (2003), and Wooldridge (2002) on panel data econometrics, Johansen (1995) on cointegration analysis, Hall (2005) on generalized method of moments, Bauwens, Lubrano and Richard (2001), Koop (2003), Lancaster (2004), and Geweke (2005) on Bayesian econometrics, Bosq (1996), Fan and Gijbels (1996), Horowitz (1998), Härdle (1990), Härdle and Linton (1994) and Pagan and Ullah (1999) on nonparametric and semiparametric econometrics, Campbell, Lo and MacKinlay (1997) and Gourieroux and Jasiak (2001) on financial econometrics, Granger and Newbold (1986), Lütkepohl (1991) and Hamilton (1994) on time series analysis.

2 Quantitative research in economics: historical backgrounds

Empirical analysis in economics has had a long and fertile history, the origins of which can be traced at least as far back as the work of the 16th-century political arithmeticians such as William Petty, Gregory King and Charles Davenant. The political arithmeticians, led by Sir William Petty, were the first group to make systematic use of facts and figures in their studies. They were primarily interested in the practical issues of their time, ranging from problems of taxation and money to those of international trade and finance. The hallmark of their approach was undoubtedly quantitative, and it was this which distinguished them from their contemporaries. Although the political arithmeticians were primarily and understandably preoccupied with statistical measurement of economic phenomena, the work of Petty, and that of King in particular, represented perhaps the first examples of a unified quantitative–theoretical approach to economics. Indeed Schumpeter in his *History of Economic Analysis* (1954, p. 209) goes as far as to say that the works of the political arithmeticians 'illustrate to perfection, what Econometrics is and what Econometricians are trying to do'.

The first attempt at quantitative economic analysis is attributed to Gregory King, who was the first to fit a linear function of changes in corn prices on deficiencies in the corn harvest, as reported in Charles Davenant (1698). One important consideration in the empirical work of King and others in this early period seems to have been the discovery of 'laws' in economics, very much like those in physics and other natural sciences.

This quest for economic laws was, and to a lesser extent still is, rooted in the desire to give economics the status that Newton had achieved for physics. This was in turn reflected in the conscious adoption of the method of the physical sciences as the dominant mode of empirical enquiry in economics. The Newtonian revolution in physics, and the philosophy of 'physical determinism' that came to be generally accepted in its aftermath, had far-reaching consequences for the method as well as the objectives of research in economics. The uncertain nature of economic relations began to be fully appreciated only with the birth of modern statistics in the late 19th century and as more statistical observations on economic variables started to become available.

The development of statistical theory in the hands of Galton, Edgeworth and Pearson was taken up in economics with speed and diligence. The earliest applications of simple correlation analysis in economics appear to have been carried out by Yule (1895; 1896) on the relationship between pauperism and the method of providing relief, and by Hooker (1901) on the relationship between the marriage rate and the general level of prosperity in the United Kingdom, measured by a variety of economic indicators such as imports, exports, and the movement in corn prices.

Benini (1907), the Italian statistician was the first to make use of the method of multiple regression in economics. But Henry Moore (1914; 1917) was the first to place the statistical estimation of economic relations at the centre of quantitative analysis in economics. Through his relentless efforts, and those of his disciples and followers Paul Douglas, Henry Schultz, Holbrook Working, Fred Waugh and others, Moore in effect laid the foundations of 'statistical economics', the precursor of econometrics. The monumental work of Schultz, *The Theory and the Measurement of Demand* (1938), in the United States and that of Allen and Bowley, *Family Expenditure* (1935), in the United Kingdom, and the pioneering works of Lenoir (1913), Wright (1915; 1928), Working (1927), Tinbergen (1929–30) and Frisch (1933) on the problem of 'identification' represented major steps towards this objective. The work of Schultz was exemplary in the way it attempted a unification of theory and measurement in demand analysis; while the work on identification highlighted the importance of 'structural estimation' in econometrics and was a crucial factor in the subsequent developments of econometric methods under the auspices of the Cowles Commission for Research in Economics.

Early empirical research in economics was by no means confined to demand analysis. Louis Bachelier (1900), using time-series data on French equity prices, recognized the random walk character of equity prices, which proved to be the precursor to the vast empirical literature on market efficiency hypothesis that has evolved since the early 1960s. Another important area was research on business cycles, which provided the basis of the later development in time-series analysis and macroeconometric model building and forecasting. Although, through the work of Sir William Petty and other early writers, economists had been aware of the existence of cycles in economic time series, it was not until the early 19th century that the phenomenon of business cycles began to attract the attention that it deserved. Clement Juglar (1819–1905), the French physician turned economist, was the first to make systematic use of time-series data to study business cycles, and is credited with the discovery of an investment cycle of about 7–11 years duration, commonly known as the Juglar cycle. Other economists such as Kitchin, Kuznets and Kondratieff followed Juglar's lead and discovered the inventory cycle (3–5 years duration), the building cycle (15–25 years duration) and the long wave (45–60 years duration), respectively. The emphasis of this early research was on the morphology of cycles and the identification of periodicities. Little attention was paid to the quantification of the relationships that may have underlain the cycles. Indeed, economists working in the National Bureau of Economic Research under the direction of Wesley Mitchell regarded each business cycle as a unique phenomenon and were therefore reluctant to use statistical methods except in a non-parametric manner and for purely descriptive purposes (see, for example, Mitchell, 1928; Burns and Mitchell, 1947). This view of business cycle research stood in sharp contrast to the econometric approach of Frisch and Tinbergen and culminated in the famous methodological interchange between Tjalling Koopmans and Rutledge Vining about the roles of theory and measurement in applied economics in general and business cycle research in particular. (This interchange appeared in the August 1947 and May 1949 issues of the *Review of Economics and Statistics*.)

3 The birth of econometrics

Although, quantitative economic analysis is a good three centuries old, econometrics as a recognized branch of economics began to emerge only in the 1930s and the 1940s with the foundation of the Econometric Society, the Cowles Commission in the United States, and the Department of Applied Economics (DAE) in Cambridge, England. (An account of the founding of the first two organizations can be found in Christ, 1952; 1983, while the history of the DAE is covered in Stone, 1978.) This was largely due to the multidisciplinary nature of econometrics, comprising of economic theory, data, econometric methods and computing techniques. Progress in empirical economic analysis often requires synchronous developments in all these four components.

Initially, the emphasis was on the development of econometric methods. The first major debate over econometric method concerned the applicability of the probability calculus and the newly developed sampling theory of R.A. Fisher to the analysis of economic data. Frisch (1934) was highly sceptical of the value of sampling theory and significance tests in econometrics. His objection was not, however, based on the epistemological reasons that lay behind Robbins's and Keynes's criticisms of econometrics. He was more concerned with the problems of multicollinearity and measurement errors which he believed were pervasive in economics; and to deal with the measurement error problem he developed his confluence analysis and the method of 'bunch maps'. Although used by some econometricians, notably Tinbergen (1939) and Stone (1945), the bunch map analysis did not find much favour with the profession at large. Instead, it was the probabilistic rationalizations of regression analysis, advanced by Koopmans (1937) and Haavelmo (1944), that formed the basis of modern econometrics.

Koopmans did not, however, emphasize the wider issue of the use of stochastic models in econometrics. It was Haavelmo who exploited the idea to the full, and argued for an explicit probability approach to the estimation and testing of economic relations. In his classic paper published as a supplement to *Econometrica* in 1944, Haavelmo defended the probability approach on two grounds. First, he argued that the use of statistical measures such as means, standard errors and correlation coefficients for inferential purposes is justified only if the process generating the data can be cast in terms of a probability model. Second, he argued that the probability approach, far from being limited in its application to economic data, because of its generality is in fact particularly suited for the analysis of 'dependent' and 'non-homogeneous' observations often encountered in economic research.

The probability model is seen by Haavelmo as a convenient abstraction for the purpose of understanding, or explaining or predicting, events in the real world. But it is not claimed that the model represents reality in all its details. To proceed with quantitative research in any subject, economics included, some degree of formalization is inevitable, and the probability model is one such formalization. The attraction of the probability model as a method of abstraction derives from its generality and flexibility, and the fact that no viable alternative seems to be available. Haavelmo's contribution was also important as it constituted the first systematic defence against Keynes's (1939) influential criticisms of Tinbergen's pioneering research on business cycles and macroeconometric modelling. The objective of Tinbergen's research was twofold: first, to show how a macroeconometric model may be constructed and then used for simulation and policy analysis (Tinbergen, 1937); second, 'to submit to statistical test some of the theories which have been put forward regarding the character and causes of cyclical fluctuations in business activity' (Tinbergen, 1939, p. 11). Tinbergen assumed a rather limited role for the econometrician in the process of testing economic theories, and argued that it was the responsibility of the 'economist' to specify the theories to be tested. He saw the role of the econometrician as a passive one of estimating the parameters of an economic relation already specified on a priori grounds by an economist. As far as statistical methods were concerned, he employed the regression method and Frisch's method of confluence analysis in a complementary fashion. Although Tinbergen discussed the problems of the determination of time lags, trends, structural stability and the choice of functional forms, he did not propose any systematic methodology for dealing with them. In short, Tinbergen approached the problem of testing theories from a rather weak methodological position. Keynes saw these weaknesses and attacked them with characteristic insight (Keynes, 1939). A large part of Keynes's review was in fact concerned with technical difficulties associated with the application of statistical methods to economic data. Apart from the problems of the 'dependent' and 'non-homogeneous' observations mentioned above, Keynes also emphasized the problems of misspecification, multicollinearity, functional form, dynamic specification, structural stability, and the difficulties associated with the measurement of theoretical variables. By focusing his attack on Tinbergen's attempt at testing economic theories of business cycles, Keynes almost totally ignored the practical significance of Tinbergen's work for econometric model building and policy analysis (for more details, see Pesaran and Smith, 1985a).

In his own review of Tinbergen's work, Haavelmo (1943) recognized the main burden of the criticisms of Tinbergen's work by Keynes and others, and argued the need for a general statistical framework to deal with these criticisms. As we have seen, Haavelmo's response, despite the views expressed by Keynes and others, was to rely more, rather than less, on the probability model as the basis of econometric methodology. The technical problems raised by Keynes and others could now be dealt with in a systematic manner by means of formal probabilistic models. Once the probability model was specified, a solution to the problems of estimation and inference could be obtained by means of either classical or of Bayesian methods. There was little that could now stand in the way of a rapid development of econometric methods.

4 Early advances in econometric methods

Haavelmo's contribution marked the beginning of a new era in econometrics, and paved the way for the rapid development of econometrics, with the likelihood method gaining importance as a tool for identification, estimation and inference in econometrics.

4.1 Identification of structural parameters

The first important breakthrough came with a formal solution to the identification problem which had been formulated earlier by Working (1927). By defining the concept of 'structure' in terms of the joint probability distribution of observations, Haavelmo (1944) presented a very general concept of identification and derived the necessary and sufficient conditions for identification of the entire system of equations, including the parameters of the probability distribution of the disturbances. His solution, although general, was rather difficult to apply in practice. Koopmans, Rubin and Leipnik (1950) used the term 'identification' for the first time in econometrics, and gave the now familiar rank and order conditions for the identification of a single equation in a system of simultaneous *linear* equations. The solution of the identification problem by Koopmans (1949) and Koopmans, Rubin and Leipnik (1950) was obtained in the case where there are a priori linear restrictions on the structural parameters. They derived rank and order conditions for identifiability of a single equation from a complete system of equations without reference to how the variables of the model are classified as endogenous or exogenous. Other solutions to the identification problem, also allowing for restrictions on the elements of the variance–covariance matrix of the structural disturbances, were later offered by Wegge (1965) and Fisher (1966).

Broadly speaking, a model is said to be identified if all its structural parameters can be obtained from the knowledge of its implied joint probability distribution for the observed variables. In the case of simultaneous equations models prevalent in econometrics, the solution to the identification problem depends on whether there exists a sufficient number of a priori restrictions for the derivation of the structural parameters from the reduced-form parameters. Although the purpose of the model and the focus of the analysis on explaining the variations of some variables in terms of the unexplained variations of other variables is an important consideration, in the final analysis the specification of a minimum number of identifying restrictions was seen by researchers at the Cowles Commission to be the function and the responsibility of 'economic theory'. This attitude was very much reminiscent of the approach adopted earlier by Tinbergen in his business cycle research: the function of economic theory was to provide the specification of the econometric model, and that of econometrics to furnish statistically optimal methods of estimation and inference. More specifically, at the Cowles Commission the primary task of econometrics was seen to be the development of statistically efficient methods for the estimation of structural parameters of an a priori specified system of simultaneous stochastic equations.

More recent developments in identification of structural parameters in context of semiparametric models is discussed below in Section 12. See also Manski (1995).

4.2 Estimation and inference in simultaneous equation models

Initially, under the influence of Haavelmo's contribution, the maximum likelihood (ML) estimation method was emphasized as it yielded consistent estimates. Anderson and Rubin (1949) developed the limited information maximum likelihood (LIML) method, and Koopmans, Rubin and Leipnik (1950) proposed the full information maximum likelihood (FIML). Both methods are based on the joint probability distribution of the endogenous variables conditional on the exogenous variables and yield consistent estimates, with the former utilizing all the available a priori restrictions and the latter only those which related to the equation being estimated. Soon, other computationally less demanding estimation methods followed, both for a fully efficient estimation of an entire system of equations and for a consistent estimation of a single equation from a system of equations.

The two-stage least squares (2SLS) procedure was independently proposed by Theil (1954; 1958) and Basmann (1957). At about the same time the instrumental variable (IV) method, which had been developed over a decade earlier by Reiersol (1941; 1945), and Geary (1949) for the estimation of errors-in-variables models, was generalized and applied by Sargan (1958) to the estimation of simultaneous equation models. Sargan's generalized IV estimator (GIVE) provided an asymptotically efficient technique for using surplus instruments in the application of the IV method to econometric problems, and formed the basis of subsequent developments of the generalized method of moments (GMM) estimators introduced subsequently by Hansen (1982). A related class of estimators, known as k-class estimators, was also proposed by Theil (1958). Methods of estimating the entire system of equations which were computationally less demanding than the FIML method were also advanced. These methods also had the advantage that, unlike the FIML, they did not require the full specification of the entire system. These included the three-stage least squares method due to Zellner and Theil (1962), the iterated instrumental variables method based on the work of Lyttkens (1970), Brundy and Jorgenson (1971), and Dhrymes (1971) and the system k-class estimators due to Srivastava (1971) and Savin (1973). Important contributions have also been made in the areas of estimation of simultaneous nonlinear equations (Amemiya, 1983), the seemingly unrelated regression equations (SURE) approach proposed by Zellner (1962), and the simultaneous rational expectations models (see Section 7.1 below).

Interest in estimation of simultaneous equation models coincided with the rise of Keynesian economics in early 1960s, and started to wane with the advent of the rational expectations revolution and its emphasis on the GMM estimation of the structural parameters from the Euler equations (first-order optimization conditions). See Section 7 below. But, with the rise of the dynamic stochastic general equilibrium models in macro-econometrics, a revival of interest in identification and estimation of nonlinear simultaneous equation models seems quite likely. The recent contribution of Fernandez-Villaverde and Rubio-Ramirez (2005) represents a start in this direction.

4.3 Developments in time series econometrics

While the initiative taken at the Cowles Commission led to a rapid expansion of econometric techniques, the application of these techniques to economic problems was rather slow. This was partly due to a lack of adequate computing facilities at the time. A more fundamental reason was the emphasis of the research at the Cowles Commission on the simultaneity problem almost to the exclusion of other econometric problems. Since the early applications of the correlation analysis to economic data by Yule and Hooker, the serial dependence of economic time series and the problem of nonsense or spurious correlation that it could give rise to had been the single most important factor explaining the profession's scepticism concerning the value of regression analysis in economics. A satisfactory solution to the spurious correlation problem was therefore needed before regression analysis of economic time series could be taken seriously. Research on this topic began in the mid-1940s at the Department of Applied Economics (DAE) in Cambridge, England, as a part of a major investigation into the measurement and analysis of consumers' expenditure in the United Kingdom (see Stone et al., 1954). Although the first steps towards the resolution of the spurious correlation problem had been taken by Aitken (1934–5) and Champernowne (1948), the research in the DAE introduced the problem and its possible solution to the attention of applied economists. Orcutt (1948) studied the autocorrelation pattern of economic time series and showed that most economic time series can be represented by simple autoregressive processes with similar autoregressive coefficients. Subsequently, Cochrane and Orcutt (1949) made the important point that the major consideration in the analysis of stationary time series was the autocorrelation of the error term in the regression equation and not the autocorrelation of the economic time series themselves. In this way they shifted the focus of attention to the autocorrelation of disturbances as the main source of concern. Although, as it turns out, this is a valid conclusion in the case of regression equations with strictly exogenous regressors, in more realistic set-ups where the regressors are weakly exogenous the serial correlation of the regressors is also likely to be of concern in practice. See, for example, Stambaugh (1999).

Another important and related development was the work of Durbin and Watson (1950; 1951) on the method of testing for residual autocorrelation in the classical regression model. The inferential breakthrough for testing serial correlation in the case of observed time-series data had already been achieved by von Neumann (1941; 1942), and by Hart and von Neumann (1942). The contribution of Durbin and Watson was, however, important from a practical viewpoint as it led to a bounds test for residual autocorrelation which could be applied irrespective of the actual values of the regressors. The independence of the critical bounds of the Durbin–Watson statistic from the matrix of the regressors allowed the application of the statistic as a general diagnostic test, the first of its type in econometrics. The contributions of Cochrane and Orcutt and of Durbin and Watson marked the beginning of a new era in the analysis of economic time-series data and laid down the basis of what is now known as the 'time-series econometrics' approach.

5 Consolidation and applications

The work at the Cowles Commission on identification and estimation of the simultaneous equation model and the development of time series techniques paved the way for widespread application of econometric methods to economic and financial problems. This was helped significantly by the rapid expansion of computing facilities, advances in financial and macroeconomic modelling, and the increased availability of economic data-sets, cross section as well as time series.

5.1 Macroeconometric modelling

Inspired by the pioneering work of Tinbergen, Klein (1947; 1950) was the first to construct a macroeconometric model in the tradition of the Cowles Commission. Soon others followed Klein's lead. Over a short space of time macroeconometric models were built for almost every industrialized country, and even for some developing and centrally planned economies. Macroeconometric models became an important tool of *ex ante* forecasting and economic policy analysis, and started to grow in both size and sophistication. The relatively stable economic environment of the 1950s and 1960s was an important factor in the initial success enjoyed by macroeconometric models. The construction and use of large-scale models presented a number of important computational problems, the solution of which was of fundamental significance, not only for the development of macroeconometric modelling but also for econometric practice in general. In this respect advances in computer technology were clearly instrumental, and without them it is difficult to imagine how the complicated computational problems involved in the estimation and simulation of large-scale models could have been solved. The increasing availability of better and faster computers was also instrumental as far as the types of problems studied and the types of solutions offered in the literature were concerned. For example, recent developments in the area of microeconometrics (see Section 10 below) could hardly have been possible if it were not for the very important recent advances in computing facilities.

5.2 Dynamic specification

Other areas where econometrics witnessed significant developments included dynamic specification, latent variables, expectations formation, limited dependent variables, discrete choice models, random coefficient models, disequilibrium models, nonlinear estimation, and the analysis of panel data models. Important advances were also made in the area of Bayesian econometrics, largely thanks to the publication of Zellner's textbook (1971), which built on his earlier work including important papers with George Tiao. The Seminar on Bayesian Inference in Econometrics and Statistics (SBIES) was founded shortly after the publication of the book, and was key in the development and diffusion of Bayesian ideas in econometrics. It was, however, the problem of dynamic specification that initially received the greatest attention. In an important paper, T. Brown (1952) modelled the hypothesis of habit persistence in consumer behaviour by introducing lagged values of consumption expenditures into an otherwise static Keynesian consumption function. This was a significant step towards the incorporation of dynamics in applied econometric research, and allowed the important distinction to be made between the short-run and the long-run impacts of changes in income on consumption. Soon other researchers followed Brown's lead and employed his autoregressive specification in their empirical work.

The next notable development in the area of dynamic specification was the distributed lag model. Although the idea of distributed lags had been familiar to economists through the pioneering work of Irving Fisher (1930) on the relationship between the nominal interest rate and the expected inflation rate, its application in econometrics was not seriously considered until the mid-1950s. The geometric distributed lag model was used for the first time by Koyck (1954) in a study of investment. Koyck arrived at the geometric distributed lag model via the adaptive expectations hypothesis. This same hypothesis was employed later by Cagan (1956) in a study of demand for money in conditions of hyperinflation, by Friedman (1957) in a study of consumption behaviour and by Nerlove (1958a) in a study of the cobweb phenomenon. The geometric distributed lag model was subsequently generalized by Solow (1960), Jorgenson (1966) and others, and was extensively applied in empirical studies of investment and consumption behaviour. At about the same time Almon (1965) provided a polynomial generalization of I. Fisher's (1937) arithmetic lag distribution which was later extended further by Shiller (1973). Other forms of dynamic specification considered in the literature included the partial adjustment model (Nerlove, 1958b; Eisner and Strotz, 1963) and the multivariate flexible accelerator model (Treadway, 1971) and Sargan's (1964) work on econometric time series analysis which formed the basis of error correction and cointegration analysis that followed next. Following the contributions of Champernowne

(1960), Granger and Newbold (1974) and Phillips (1986) the spurious regression problem was better understood, and paved the way for the development of the theory of cointegration. For further details see Section 8.3 below.

5.3 Techniques for short-term forecasting

Concurrent with the development of dynamic modelling in econometrics there was also a resurgence of interest in time-series methods, used primarily in short-term business forecasting. The dominant work in this field was that of Box and Jenkins (1970), who, building on the pioneering works of Yule (1921; 1926), Slutsky (1927), Wold (1938), Whittle (1963) and others, proposed computationally manageable and asymptotically efficient methods for the estimation and forecasting of univariate autoregressive-moving average (ARMA) processes. Time-series models provided an important and relatively simple benchmark for the evaluation of the forecasting accuracy of econometric models, and further highlighted the significance of dynamic specification in the construction of time-series econometric models. Initially univariate time-series models were viewed as mechanical 'black box' models with little or no basis in economic theory. Their use was seen primarily to be in short-term forecasting. The potential value of modern time-series methods in econometric research was, however, underlined in the work of Cooper (1972) and Nelson (1972) who demonstrated the good forecasting performance of univariate Box–Jenkins models relative to that of large econometric models. These results raised an important question about the adequacy of large econometric models for forecasting as well as for policy analysis. It was argued that a properly specified structural econometric model should, at least in theory, yield more accurate forecasts than a univariate time-series model. Theoretical justification for this view was provided by Zellner and Palm (1974), followed by Trivedi (1975), Prothero and Wallis (1976), Wallis (1977) and others. These studies showed that Box–Jenkins models could in fact be derived as univariate final form solutions of linear structural econometric models. In theory, the pure time-series model could always be embodied within the structure of an econometric model and in this sense it did not present a 'rival' alternative to econometric modelling. This literature further highlighted the importance of dynamic specification in econometric models and in particular showed that econometric models that are outperformed by simple univariate time-series models most probably suffer from specification errors.

The papers in Elliott, Granger and Timmermann (2006) provide excellent reviews of recent developments in economic forecasting techniques.

6 A new phase in the development of econometrics

With the significant changes taking place in the world economic environment in the 1970s, arising largely from the breakdown of the Bretton Woods system and the quadrupling of oil prices, econometrics entered a new phase of its development. Mainstream macroeconometric models built during the 1950s and 1960s, in an era of relative economic stability with stable energy prices and fixed exchange rates, were no longer capable of adequately capturing the economic realities of the 1970s. As a result, not surprisingly, macroeconometric models and the Keynesian theory that underlay them came under severe attack from theoretical as well as from practical viewpoints. While criticisms of Tinbergen's pioneering attempt at macroeconometric modelling were received with great optimism and led to the development of new and sophisticated estimation techniques and larger and more complicated models, the disenchantment with macroeconometric models in 1970s prompted a much more fundamental reappraisal of quantitative modelling as a tool of forecasting and policy analysis.

At a theoretical level it was argued that econometric relations invariably lack the necessary 'microfoundations', in the sense that they cannot be consistently derived from the optimizing behaviour of economic agents. At a practical level the Cowles Commission approach to the identification and estimation of simultaneous macroeconometric models was questioned by Lucas and Sargent and by Sims, although from different viewpoints (Lucas, 1976; Lucas and Sargent, 1981; Sims, 1980). There was also a move away from macroeconometric models and towards microeconometric research with greater emphasis on matching of econometrics with individual decisions.

It also became increasingly clear that Tinbergen's paradigm where economic relations were taken as given and provided by 'economic theorist' was not adequate. It was rarely the case that economic theory could be relied on for a full specification of the econometric model (Leamer, 1978). The emphasis gradually shifted from estimation and inference based on a given tightly parameterized specification to diagnostic testing, specification searches, model uncertainty, model validation, parameter variations, structural breaks, and semiparametric and nonparametric estimation. The choice of approach often governed by the purpose of the investigation, the nature of the economic application, data availability, computing and software technology.

What follows is a brief overview of some of the important developments. Given space limitations there are inevitably significant gaps. These include the important contributions of Granger (1969), Sims (1972) and Engle, Hendry and Richard (1983) on different concepts of 'causality' and 'exogeneity', the literature on disequilibrium models (Quandt, 1982; Maddala, 1983; 1986), random coefficient models (Swamy, 1970; Hsiao and Pesaran, 2008, unobserved time series models (Harvey, 1989), count regression models (Cameron and Trivedi, 1986; 1998), the weak instrument problem (Stock, Wright and Yogo, 2002), small sample theory

(Phillips, 1983; Rothenberg, 1984), econometric models of auction pricing (Hendricks and Porter, 1988; Laffont, Ossard and Vuong, 1995).

7 Rational expectations and the Lucas critique

Although the rational expectations hypothesis (REH) was advanced by Muth in 1961, it was not until the early 1970s that it started to have a significant impact on time-series econometrics and on dynamic economic theory in general. What brought the REH into prominence was the work of Lucas (1972; 1973), Sargent (1973), Sargent and Wallace (1975) and others on the new classical explanation of the apparent breakdown of the Phillips curve. The message of the REH for econometrics was clear. By postulating that economic agents form their expectations *endogenously* on the basis of the true model of the economy, and a *correct* understanding of the processes generating exogenous variables of the model, including government policy, the REH raised serious doubts about the invariance of the structural parameters of the mainstream macroeconometric models in the face of changes in government policy. This was highlighted in Lucas's critique of macroeconometric policy evaluation. By means of simple examples Lucas (1976) showed that in models with rational expectations the parameters of the decision rules of economic agents, such as consumption or investment functions, are usually a mixture of the parameters of the agents' objective functions and of the stochastic processes they face as historically given. Therefore, Lucas argued, there is no reason to believe that the 'structure' of the decision rules (or economic relations) would remain invariant under a policy intervention. The implication of the Lucas critique for econometric research was not, however, that policy evaluation could not be done, but rather than the traditional econometric models and methods were not suitable for this purpose. What was required was a separation of the parameters of the policy rule from those of the economic model. Only when these parameters could be identified separately given the knowledge of the joint probability distribution of the variables (both policy and non-policy variables) would it be possible to carry out an econometric analysis of alternative policy options.

There have been a number of reactions to the advent of the rational expectations hypothesis and the Lucas critique that accompanied it.

7.1 Model consistent expectations
The least controversial reaction has been the adoption of the REH as one of several possible expectations formation hypotheses in an otherwise conventional macro-econometric model containing expectational variables. In this context the REH, by imposing the appropriate cross-equation parametric restrictions, ensures that 'expectations' and 'forecasts' generated by the model are consistent. In this approach the REH is regarded

as a convenient and effective method of imposing cross-equation parametric restrictions on time series econometric models, and is best viewed as the 'model-consistent' expectations hypothesis. There is now a sizeable literature on solution, identification, and estimation of linear RE models. The canonical form of RE models with forward and backward components is given by

$$\mathbf{y}_t = \mathbf{A}\mathbf{y}_{t-1} + \mathbf{B}E(\mathbf{y}_{t+1}|F_t) + \mathbf{w}_t,$$

where $\mathbf{y}_t$ is a vector of endogenous variables, $E(.|F_t)$ is the expectations operator, F_t the publicly available information at time t, and $\mathbf{w}_t$ is a vector of forcing variables. For example, log-linearized version of dynamic general equilibrium models (to be discussed) can all be written as a special case of this equation with plenty of restrictions on the coefficient matrices A and B. In the typical case where $\mathbf{w}_t$ are serially uncorrelated and the solution of the RE model can be assumed to be unique, the RE solution reduces to the vector autoregression (VAR)

$$\mathbf{y}_t = \Phi\mathbf{y}_{t-1} + \mathbf{G}\mathbf{w}_t,$$

where Φ and $\mathbf{G}$ are given in terms of the structural parameters:

$$\mathbf{B}\Phi^2 - \Phi + \mathbf{A} = 0, \quad \text{and} \quad \mathbf{G} = (\mathbf{I} - \mathbf{B}\Phi)^{-1}.$$

The solution of the RE model can, therefore, be viewed as a restricted form of VAR popularized in econometrics by Sims (1980) as a response in macroeconometric modelling to the rational expectations revolution. The nature of restrictions is determined by the particular dependence of A and B on a few 'deep' or structural parameters. For general discussion of solution of RE models see, for example, Broze, Gouriéroux and Szafarz (1985) and Binder and Pesaran (1995). For studies of identification and estimation of linear RE models see, for example, Hansen and Sargent (1980), Wallis (1980), Wickens (1982) and Pesaran (1981; 1987b). These studies show how the standard econometric methods can in principle be adapted to the econometric analysis of rational expectations models.

7.2 Detection and modelling of structural breaks
Another reaction to the Lucas critique has been to treat the problem of 'structural change' emphasized by Lucas as one more potential econometric 'problem'. Clements and Hendry (1998; 1999) provide a taxonomy of factors behind structural breaks and forecast failures. Stock and Watson (1996) provide extensive evidence of structural break in macroeconomic time series. It is argued that structural change can result from many factors and need not be associated solely with intended or expected changes in policy. The econometric lesson has been to pay attention to possible breaks in economic relations. There now exists a large body of work on testing for

structural change, detection of breaks (single as well as multiple), and modelling of break processes by means of piece-wise linear or non-linear dynamic models (Chow, 1960; Brown, Durbin and Evans, 1975; Nyblom, 1989; Andrews, 1993; Andrews and Ploberger, 1994; Bai and Perron, 1998; Pesaran and Timmermann, 2005b; 2007. See also the surveys by Stock, 1994; Clements and Hendry, 2006). The implications of breaks for short-term and long-term forecasting have also begun to be addressed (McCulloch, and Tsay, 1993; Koop and Potter, 2004a; 2004b; Pesaran, Pettenuzzo and Timmermann, 2006).

8 VAR macroeconometrics

8.1 Unrestricted VARs

The Lucas critique of mainstream macroeconometric modelling also led some econometricians, notably Sims (1980; 1982), to doubt the validity of the Cowles Commission style of achieving identification in econometric models. Sims focused his critique on macroeconometric models with a vector autoregressive (VAR) specification, which was relatively simple to estimate; and its use soon became prevalent in macroeconometric analysis. The view that economic theory cannot be relied on to yield identification of structural models was not new and had been emphasized in the past, for example, by Liu (1960). Sims took this viewpoint a step further and argued that in presence of rational expectations a priori knowledge of lag lengths is indispensable for identification, even when we have distinct strictly exogenous variables shifting supply and demand schedules (Sims, 1980, p. 7). While it is true that the REH complicates the necessary conditions for the identification of structural models, the basic issue in the debate over identification still centres on the validity of the classical dichotomy between exogenous and endogenous variables (Pesaran, 1981). In the context of closed-economy macroeconometric models where all variables are treated as endogenous, other forms of identification of the structure will be required. Initially, Sims suggested a recursive identification approach where the matrix of contemporaneous effects was assumed to be lower (upper) triangular and the structural shocks orthogonal. Other non-recursive identification schemes soon followed.

8.2 Structural VARs

One prominent example was the identification scheme developed in Blanchard and Quah (1989), who distinguished between permanent and transitory shocks and attempted to identify the structural models through long-run restrictions. For example, Blanchard and Quah argued that the effect of a demand shock on real output should be temporary (that is, it should have a zero long-run impact), while a supply shock should have a permanent effect. This approach is known as 'structural VAR' (SVAR) and has been used extensively in the literature. It continues to assume that structural shocks are orthogonal, but uses a mixture of short-run and long-run restrictions to identify the structural model. In their work Blanchard and Quah considered a bivariate VAR model in real output and unemployment. They assumed real output to be integrated of order 1, or $I(1)$, and viewed unemployment as an $I(0)$, or a stationary variable. This allowed them to associate the shock to one of the equations as permanent, and the shock to the other equation as transitory. In more general settings, such as the one analysed by Gali (1992) and Wickens and Motto (2001), where there are m endogenous variables and r long-run or cointegrating relations, the SVAR approach provides $m(m - r)$ restrictions which are not sufficient to fully identify the model, unless $m = 2$ and $r = 1$ which is the simple bivariate model considered by Blanchard and Quah (Pagan and Pesaran, 2007). In most applications additional short-term restrictions are required. More recently, attempts have also been made to identify structural shocks by means of qualitative restrictions, such as sign restrictions. Notable examples include Canova and de Nicolo (2002), Uhlig (2005) and Peersman (2005).

The focus of the SVAR literature has been on impulse response analysis and forecast error variance decomposition, with the aim of estimating the time profile of the effects of monetary policy, oil price or technology shocks on output and inflation, and deriving the relative importance of these shocks as possible explanations of forecast error variances at different horizons. Typically such analysis is carried out with respect to a single model specification, and at most only parameter uncertainty is taken into account (Kilian, 1998). More recently the problem of model uncertainty and its implications for impulse response analysis and forecasting have been recognized. Bayesian and classical approaches to model and parameter uncertainty have been considered. Initially, Bayesian VAR models were developed for use in forecasting as an effective shrinkage procedure in the case of high-dimensional VAR models (Doan, Litterman and Sims, 1984; Litterman, 1985). The problem of model uncertainty in cointegrating VARs has been addressed in Garratt et al. (2003b; 2006), and Strachan and van Dijk (2006).

8.3 Structural cointegrating VARs

This approach provides the SVAR with the decomposition of shocks into permanent and transitory and gives economic content to the long-run or cointegrating relations that underlie the transitory components. In the simple example of Blanchard and Quah this task is trivially achieved by assuming real output to be $I(1)$ and the unemployment rate to be an $I(0)$ variable. To have shocks with permanent effects some of the variables in the VAR must be non-stationary. This provides a natural link between the SVAR and the unit root and cointegration literature. Identification of the cointegrating relations can be achieved by recourse to economic theory, solvency or arbitrage conditions (Garratt et al., 2003a). Also there are often long-run over-identifying restrictions that can be tested. Once identified and empirically validated, the long-run relations can be embodied within a VAR structure, and the resultant structural vector error correction model identified using theory-based short-run restrictions. The structural shocks can be decomposed into permanent and temporary components using either the multivariate version of the Beveridge and Nelson (1981) decompositions, or the one more recently proposed by Garratt, Robertson and Wright (2006).

Two or more variables are said to be cointegrated if they are individually integrated (or have a random walk component), but there exists a linear combination of them which is stationary. The concept of cointegration was first introduced by Granger (1986) and more formally developed in Engle and Granger (1987). Rigorous statistical treatments followed in the papers by Johansen (1988; 1991) and Phillips (1991). Many further developments and extensions have taken place with reviews provided in Johansen (1995), Juselius (2006) and Garratt et al. (2006). The related unit root literature is reviewed by Stock (1994) and Phillips and Xiao (1998).

8.4 Macroeconometric models with microeconomic foundations

For policy analysis macroeconometric models need to be based on decisions by individual households, firms and governments. This is a daunting undertaking and can be achieved only by gross simplification of the complex economic interconnections that exists across millions of decision-makers worldwide. The dynamic stochastic general equilibrium (DSGE) modelling approach attempts to implement this task by focusing on optimal decisions of a few representative agents operating with rational expectations under complete learning. Initially, DSGE models were small and assumed complete markets with instantaneous price adjustments, and as a result did not fit the macroeconomic time series (Kim and Pagan, 1995). More recently, Smets and Wouters (2003) have shown that DSGE models with sticky prices and wages along the lines developed by Christiano, Eichenbaum and Evans (2005) are sufficiently rich to match most of the statistical features of the main macroeconomic time series. Moreover, by applying Bayesian estimation techniques, these authors have shown that even relatively large models can be estimated as a system. Bayesian DSGE models have also shown to perform reasonably well in forecasting as compared with standard and Bayesian vector autoregressions. It is also possible to incorporate long-run cointegrating relations within Bayesian DSGE models. The problems of parameter and model uncertainty can also be readily accommodated using data-coherent DSGE models. Other extensions of the DSGE models to allow for learning, regime switches, time variations in shock variances, asset prices, and

multi-country interactions are likely to enhance their policy relevance (Del Negro and Schorfheide, 2004; Del Negro et al., 2005; An and Schorfheide, 2007; Pesaran and Smith, 2006). Further progress will also be welcome in the area of macroeconomic policy analysis under model uncertainty, and robust policymaking (Brock and Durlauf, 2006; Hansen and Sargent, 2007).

9 Model and forecast evaluation

While in the 1950s and 1960s research in econometrics was primarily concerned with the identification and estimation of econometric models, the dissatisfaction with econometrics during the 1970s caused a shift of focus from problems of estimation to those of model evaluation and testing. This shift has been part of a concerted effort to restore confidence in econometrics, and has received attention from Bayesian as well as classical viewpoints. Both these views reject the 'axiom of correct specification' which lies at the basis of most traditional econometric practices, but they differ markedly as how best to proceed.

It is generally agreed, by Bayesians as well as by non-Bayesians, that model evaluation involves considerations other than the examination of the statistical properties of the models, and personal judgements inevitably enter the evaluation process. Models must meet multiple criteria which are often in conflict. They should be relevant in the sense that they ought to be capable of answering the questions for which they are constructed. They should be consistent with the accounting and/or theoretical structure within which they operate. Finally, they should provide adequate representations of the aspects of reality with which they are concerned. These criteria and their interaction are discussed in Pesaran and Smith (1985b). More detailed breakdowns of the criteria of model evaluation can be found in Hendry and Richard (1982) and McAleer, Pagan, and Volker (1985). In econometrics it is, however, the criterion of 'adequacy' which is emphasized, often at the expense of relevance and consistency.

The issue of model adequacy in mainstream econometrics is approached either as a model selection problem or as a problem in statistical inference whereby the hypothesis of interest is tested against general or specific alternatives. The use of absolute criteria such as measures of fit/parsimony or formal Bayesian analysis based on posterior odds are notable examples of model selection procedures, while likelihood ratio, Wald and Lagrange multiplier tests of nested hypotheses and Cox's centred log-likelihood ratio tests of non-nested hypotheses are examples of the latter approach. The distinction between these two general approaches basically stems from the way alternative models are treated. In the case of model selection (or model discrimination) all the models under consideration enjoy the same status and the investigator is not committed a priori to any one of the alternatives. The aim is to choose the model which is likely to perform best with respect to a particular loss function. By contrast, in the hypothesis-testing framework the null hypothesis (or the maintained model) is treated differently from the remaining hypotheses (or models). One important feature of the model-selection strategy is that its application always leads to one model being chosen in preference to other models. But, in the case of hypothesis testing, rejection of all the models under consideration is not ruled out when the models are non-nested. A more detailed discussion of this point is given in Pesaran and Deaton (1978).

Broadly speaking, classical approaches to the problem of model adequacy can be classified depending on how specific the alternative hypotheses are. These are the *general specification tests, the diagnostic tests*, and the *non-nested tests*. The first of these, pioneered by Durbin (1954) and introduced in econometrics by Ramsey (1969), Wu (1973), Hausman (1978), and subsequently developed further by White (1981; 1982) and Hansen (1982), are designed for circumstances where the nature of the alternative hypothesis is kept (sometimes intentionally) rather vague, the purpose being to test the null against a *broad* class of alternatives. (The pioneering contribution of Durbin, 1954, in this area has been documented by Nakamura and Nakamura, 1981.) Important examples of general specification tests are Ramsey's regression specification error test (RESET) for omitted variables and/or misspecified functional forms, and the Durbin–Hausman–Wu test of misspecification in the context of measurement error models and/or simultaneous equation models. Such general specification tests are particularly useful in the preliminary stages of the modelling exercise.

In the case of diagnostic tests, the model under consideration (viewed as the null hypothesis) is tested against more specific alternatives by embedding it within a general model. Diagnostic tests can then be constructed using the likelihood ratio, Wald or Lagrange multiplier (LM) principles to test for parametric restrictions imposed on the general model. The application of the LM principle to econometric problems is reviewed in the papers by Breusch and Pagan (1980), Godfrey and Wickens (1982), and Engle (1984). An excellent review is provided in Godfrey (1988). Examples of the restrictions that may be of interest as diagnostic checks of model adequacy include zero restrictions, parameter stability, serial correlation, heteroskedasticity, functional forms, and normality of errors. The distinction made here between diagnostic tests and general specification tests is more apparent than real. In practice some diagnostic tests such as tests for serial correlation can also be viewed as a general test of specification. Nevertheless, the distinction helps to focus attention on the purpose behind the tests and the direction along which high power is sought.

The need for non-nested tests arises when the models under consideration belong to separate parametric families in the sense that no single model can be obtained from the others by means of a suitable limiting process. This situation, which is particularly prevalent in econometric research, may arise when models differ with respect to their theoretical underpinnings and/or their auxiliary assumptions. Unlike the general specification tests and diagnostic tests, the application of non-nested tests is appropriate when specific but rival hypotheses for the explanation of the same economic phenomenon have been advanced. Although non-nested tests can also be used as general specification tests, they are designed primarily to have high power against specific models that are seriously entertained in the literature. Building on the pioneering work of Cox (1961; 1962), a number of such tests for single equation models and systems of simultaneous equations have been proposed (Pesaran and Weeks, 2001).

The use of statistical tests in econometrics, however, is not a straightforward matter and in most applications does not admit of a clear-cut interpretation. This is especially so in circumstances where test statistics are used not only for checking the adequacy of a *given* model but also as guides to model construction. Such a process of model construction involves specification searches of the type emphasized by Leamer (1978) and presents insurmountable pre-test problems which in general tend to produce econometric models whose 'adequacy' is more apparent than real. As a result, in evaluating econometric models less reliance should be placed on those indices of model adequacy that are used as guides to model construction, and more emphasis should be given to the performance of models over other data-sets and against rival models.

A closer link between model evaluation and the underlying decision problem is also needed. Granger and Pesaran (2000a; 2000b) discuss this problem in the context of forecast evaluation. A recent survey of forecast evaluation literature can be found in West (2006). Pesaran and Skouras (2002) provide a review from a decision-theoretic perspective.

The subjective Bayesian approach to the treatment of several models begins by assigning a prior probability to each model, with the prior probabilities summing to 1. Since each model is already endowed with a prior probability distribution for its parameters and for the probability distribution of observable data conditional on its parameters, there is then a complete probability distribution over the space of models, parameters, and observable data. (No particular problems arise from non-nesting of models in this framework.) This probability space can then be augmented with the distribution of an object or vector of objects of interest. For example, in a macroeconomic policy setting the models could include VARs, DSGEs and traditional large-scale macroeconomic models, and the vector of interest might include future output growth, interest rates, inflation and unemployment, whose distribution is implied by each of the models considered. Implicit in this formulation is the conditional distribution of the vector of interest conditional on the observed data. Technically, this requires the integration (or marginalization) of parameters in each model as well as the models themselves. As a practical matter this usually proceeds by first computing the probability of each model conditional on the data, and then using these probabilities as weights in averaging the posterior distribution of the vector of interest in each model. It is not necessary to choose one particular model, and indeed to do so would be suboptimal. The ability to actually carry out this simultaneous consideration of multiple models has been enhanced greatly by recent developments in simulation methods, surveyed in Section 16 below; recent texts by Koop (2003), Lancaster (2004) and Geweke (2005) provide technical details. Geweke and Whiteman (2006) specifically outline these methods in the context of economic forecasting.

10 Microeconometrics: an overview

Partly as a response to the dissatisfaction with macroeconometric time-series research and partly in view of the increasing availability of micro data and computing facilities, since the mid-1980s significant advances have been made in the analysis of micro data. Important micro data-sets have become available on households and firms especially in the United States in such areas as housing, transportation, labour markets and energy. These data sets include various longitudinal surveys (for example, University of Michigan Panel Study of Income Dynamics, and Ohio State National Longitudinal Study Surveys), cross-sectional surveys of family expenditures, population and labour force surveys. This increasing availability of micro-data, while opening up new possibilities for analysis, has also raised a number of new and interesting econometric issues primarily originating from the nature of the data. The errors of measurement are likely to be important in the case of some micro data-sets. The problem of the heterogeneity of economic agents at the micro level cannot be assumed away as readily as is usually done in the case of macro data by appealing to the idea of a 'representative' firm or a 'representative' household.

The nature of micro data, often being qualitative or limited to a particular range of variations, has also called for new econometric models and techniques. Examples include categorical survey responses ('up', 'same' or 'down'), and censored or truncated observations. The models and issues considered in the microeconometric literature are wide ranging and include fixed and random effect panel data models (for example, Mundlak, 1961; 1978), logit and probit models and their multinominal extensions, discrete choice or quantal response models (Manski and McFadden, 1981), continuous time duration models (Heckman and Singer, 1984),

and microeconometric models of count data (Hausman, Hall and Griliches, 1984; Cameron and Trivedi, 1986).

The fixed or random effect models provide the basic statistical framework and will be discussed in more detailed below. Discrete choice models are based on an explicit characterization of the choice process and arise when individual decision makers are faced with a finite number of alternatives to choose from. Examples of discrete choice models include transportation mode choice (Domenich and McFadden, 1975), labour force participation (Heckman and Willis, 1977), occupation choice (Boskin, 1974), job or firm location (Duncan 1980), and models with neighbourhood effects (Brock and Durlauf, 2002). Limited dependent variables models are commonly encountered in the analysis of survey data and are usually categorized into truncated regression models and censored regression models. If all observations on the dependent as well as on the exogenous variables are lost when the dependent variable falls outside a specified range, the model is called *truncated*, and, if only observations on the dependent variable are lost, it is called *censored*. The literature on censored and truncated regression models is vast and overlaps with developments in other disciplines, particularly in biometrics and engineering. Maddala (1983, ch. 6) provides a survey.

The censored regression model was first introduced into economics by Tobin (1958) in his pioneering study of household expenditure on durable goods, where he explicitly allowed for the fact that the dependent variable, namely, the expenditure on durables, cannot be negative. The model suggested by Tobin and its various generalizations are known in economics as Tobit models and are surveyed in detail by Amemiya (1984), and more recently in Cameron and Trivedi (2005, ch. 14). Continuous time duration models, also known as survival models, have been used in analysis of unemployment duration, the period of time spent between jobs, durability of marriage, and so on. Application of survival models to analyse economic data raises a number of important issues resulting primarily from the non-controlled experimental nature of economic observations, limited sample sizes (that is, time periods), and the heterogeneous nature of the economic environment within which agents operate. These issues are clearly not confined to duration models and are also present in the case of other microeconometric investigations that are based on time series or cross-section or panel data.

Partly in response to the uncertainties inherent in econometric results based on non-experimental data, there has also been a significant move towards social experimentation, and experimental economics in general. A social experiment aims at isolating the effects of a policy change (or a treatment effect) by comparing the consequences of an exogenous variation in the economic environment of a set of experimental subjects known as the 'treatment' group with those of a 'control' group that have not been subject to the change. The basic idea goes back to the early work of R.A. Fisher (1928) on randomized trials, and has been applied extensively in agricultural and biomedical research. The case for social experimentation in economics is discussed in Burtless (1995). Hausman and Wise (1985) and Heckman and Smith (1995) consider a number of actual social experiments carried out in the United States, and discuss their scope and limitations.

Experimental economics tries to avoid some of the limitations of working with observations obtained from natural or social experiments by using data from laboratory experiments to test economic theories by fixing some of the factors and identifying the effects of other factors in a way that allows *ceteris paribus* comparisons. A wide range of topics and issues are covered in this literature, such as individual choice behaviour, bargaining, provision of public goods, theories of learning, auction markets, and behavioural finance. A comprehensive review of major areas of experimental research in economics is provided in Kagel and Roth (1995).

These developments have posed new problems and challenges in the areas of experimental design, statistical methods and policy analysis. Another important aspect of recent developments in microeconometric literature relates to the use of microanalytic simulation models for policy analysis and evaluation to reform packages in areas such as health care, taxation, social security systems, and transportation networks. Cameron and Trivedi (2005) review the recent developments in methods and application of microeconometrics. Some of these topics will be discussed in more detail below.

11 Econometrics of panel data

Panel data models are used in many areas of econometrics, although initially they were developed primarily for the analysis of micro behaviour, and focused on panels formed from cross-section of N individual households or firms surveyed for T successive time periods. These types of panels are often refereed to as 'micropanels'. In social and behavioural sciences they are also known as longitudinal data or panels. The literature on micro-panels typically takes N to be quite large (in hundreds) and T rather small, often less than ten. But more recently, with the increasing availability of financial and macroeconomic data, analyses of panels where both N and T are relatively large have also been considered. Examples of such data-sets include time series of company data from Datastream, country data from International Financial Statistics or the Penn World Table, and county and state data from national statistical offices. There are also pseudo panels of firms and consumers composed of repeated cross sections that cover cross-section units that are not necessarily identical but are observed over relatively long time periods. Since the available cross-section observations do not (necessarily) relate to the same individual unit, some form of grouping of the cross-section

units is needed. Once the grouping criteria are set, the estimation can proceed using fixed effects estimation applied to group averages if the number of observations per group is sufficiently large; otherwise possible measurement errors of the group averages also need to be taken into account. Deaton (1985) pioneered the econometric analysis of pseudo panels. Verbeek (2008) provides a recent review.

Use of panels can enhance the power of empirical analysis and allows estimation of parameters that might not have been identified using the time or the cross-section dimensions alone. These benefits come at a cost. In the case of linear panel data models with a short time span the increased power is usually achieved under assumptions of parameter homogeneity and error cross-section independence. Short panels with autocorrelated disturbances also pose a new identification problem, namely, how to distinguished between dynamics and state dependence (Arellano, 2003, ch. 5). In panels with fixed effects the homogeneity assumption is relaxed somewhat by allowing the intercepts in the panel regressions to vary freely over the cross-section units, but continues to maintain the error cross-section independence assumption. The random coefficient specification of Swamy (1970) further relaxes the slope homogeneity assumption, and represents an important generalization of the random effects model (Hsiao and Pesaran, 2007). In micro-panels where T is small cross-section dependence can be dealt with if it can be attributed to spatial (economic or geographic) effects. Anselin (1988) and Anselin, Le Gallo and Jayet (2007) provide surveys of the literature on spatial econometrics. A number of studies have also used measures such as trade or capital flows to capture economic distance, as in Conley and Topa (2002), Conley and Dupor (2003), and Pesaran, Schuermann and Weiner (2004).

Allowing for dynamics in panels with fixed effects also presents additional difficulties; for example, the standard within-group estimator will be inconsistent unless $T \to \infty$ (Nickell, 1981). In linear dynamic panels the incidental parameter problem (the unobserved heterogeneity) can be resolved by first differencing the model and then estimating the resultant first-differenced specification by instrumental variables or by the method of transformed likelihood (Anderson and Hsiao, 1981; 1982; Holtz-Eakin, Newey and Rosen, 1988; Arellano and Bond, 1991; Hsiao, Pesaran and Tahmiscioglu, 2002). A similar procedure can also be followed in the case of short T panel VARs (Binder, Hsiao and Pesaran, 2005). But other approaches are needed for nonlinear panel data models. See, for example, Honoré and Kyriazidou (2000) and review of the literature on nonlinear panels in Arellano and Honoré (2001). Relaxing the assumption of slope homogeneity in dynamic panels is also problematic, and neglecting to take account of slope heterogeneity will lead to inconsistent estimators. In the presence of slope heterogeneity Pesaran and Smith (1995) show that the within-group estimator remains inconsistent even if both N and $T \rightarrow \infty$. A Bayesian approach to estimation of micro dynamic panels with random slope coefficients is proposed in Hsiao, Pesaran and Tahmiscioglu (1999).

To deal with general dynamic specifications, possible slope heterogeneity and error cross-section dependence, large T and N panels are required. In the case of such large panels it is possible to allow for richer dynamics and parameter heterogeneity. Cross-section dependence of errors can also be dealt with using residual common factor structures. These extensions are particularly relevant to the analysis of purchasing power parity hypothesis (O'Connell, 1998; Imbs et al., 2005; Pedroni, 2001; Smith et al., 2004), output convergence (Durlauf, Johnson and Temple, 2005; Pesaran, 2007b), the Fisher effect (Westerlund, 2005), house price convergence (Holly, Pesaran and Yamagata, 2006), regional migration (Fachin, 2006), and uncovered interest parity (Moon and Perron, 2007). The econometric methods developed for large panels has to take into account the relationship between the increasing number of time periods and cross-section units (Phillips and Moon, 1999). The relative expansion rates of N and T could have important consequences for the asymptotic and small sample properties of the panel estimators and tests. This is because fixed T estimation bias tend to magnify with increases in the cross-section dimension, and it is important that any bias in the T dimension is corrected in such a way that its overall impact disappears as both N and $T \rightarrow \infty$, jointly.

The first generation panel unit root tests proposed, for example, by Levin, Lin and Chu (2002) and Im, Pesaran and Shin (2003) allowed for parameter heterogeneity but assumed errors were cross-sectionally independent. More recently, panel unit root tests that allow for error cross-section dependence have been proposed by Bai and Ng (2004), Moon and Perron (2004) and Pesaran (2007a). As compared with panel unit root tests, the analysis of cointegration in panels is still at an early stage of its development. So far the focus of the panel cointegration literature has been on residual-based approaches, although there has been a number of attempts at the development of system approaches as well (Pedroni, 2004). But once cointegration is established the long-run parameters can be estimated efficiently using techniques similar to the ones proposed in the case of single time-series models. These estimation techniques can also be modified to allow for error cross-section dependence (Pesaran, 2007a). Surveys of the panel unit root and cointegration literature are provided by Banerjee (1999), Baltagi and Kao (2000), Choi (2006) and Breitung and Pesaran (2008).

The micro and macro panel literature is vast and growing. For the analysis of many economic problems, further progress is needed in the analysis of nonlinear panels, testing and modelling of error cross-section dependence, dynamics, and neglected heterogeneity. For general reviews of panel data econometrics, see Arellano (2003), Baltagi (2005), Hsiao (2003) and Wooldridge (2002).

12 Nonparametric and semiparametric estimation

Much empirical research is concerned with estimating conditional mean, median, or hazard functions. For example, a wage equation gives the mean, median or, possibly, some other quantile of wages of employed individuals conditional on characteristics such as years of work experience and education. A hedonic price function gives the mean price of a good conditional on its characteristics. The function of interest is rarely known a priori and must be estimated from data on the relevant variables. For example, a wage equation is estimated from data on the wages, experience, education and, possibly, other characteristics of individuals. Economic theory rarely gives useful guidance on the form (or shape) of a conditional mean, median, or hazard function. Consequently, the form of the function must either be assumed or inferred through the estimation procedure.

The most frequently used estimation methods assume that the function of interest is known up to a set of constant parameters that can be estimated from data. Models in which the only unknown quantities are a finite set of constant parameters are called 'parametric'. A linear model that is estimated by ordinary least squares is a familiar and frequently used example of a parametric model. Indeed, linear models and ordinary least squares have been the workhorses of applied econometrics since its inception. It is not difficult to see why. Linear models and ordinary least squares are easy to work with both analytically and computationally, and the estimation results are easy to interpret. Other examples of widely used parametric models are binary logit and probit models if the dependent variable is binary (for example, an indicator of whether an individual is employed or whether a commuter uses automobile or public transit for a trip to work) and the Weibull hazard model if the dependent variable is a duration (for example, the duration of a spell of employment or unemployment).

Although parametric models are easy to work with, they are rarely justified by theoretical or other a priori considerations and often fit the available data badly. Horowitz (2001), Horowitz and Savin (2001), Horowitz and Lee (2002), and Pagan and Ullah (1999) provide examples. The examples also show that conclusions drawn from a convenient but incorrectly specified model can be very misleading. Of course, applied econometricians are aware of the problem of specification error. Many investigators attempt to deal with it by carrying out a specification search in which several different models are estimated and conclusions are based on the one that appears to fit the data best. Specification searches may be unavoidable in some applications, but they have many undesirable properties. There is no guarantee that a specification search will include the correct model or a good approximation to it. If the search includes the correct model, there is no guarantee that it will be selected by the investigator's model selection criteria. Moreover, the search process invalidates the statistical theory on which inference is based.

Given this situation, it is reasonable to ask whether conditional mean and other functions of interest in applications can be estimated nonparametrically, that is, without making a priori assumptions about their functional forms. The answer is clearly 'yes' in a model whose explanatory variables are all discrete. If the explanatory variables are discrete, then each set of values of these variables defines a data cell. One can estimate the conditional mean of the dependent variable by averaging its values within each cell. Similarly, one can estimate the conditional median cell by cell.

If the explanatory variables are continuous, they cannot be grouped into cells. Nonetheless, it is possible to estimate conditional mean and median functions that satisfy mild smoothness conditions without making a priori assumptions about their shapes. Techniques for doing this have been developed mainly in statistics, beginning with Nadaraya's (1964) and Watson's (1964) nonparametric estimator of a conditional mean function. The Nadaraya–Watson estimator, which is also called a kernel estimator, is a weighted average of the observed values of the dependent variable. More specifically, suppose that the dependent variable is Y, the explanatory variable is X, and the data consist of observations $\{Y_i, X_i : i = 1, \ldots, n\}$. Then the Nadaraya–Watson estimator of the mean of Y at $X = x$ is a weighted average of the Y_i's. Y_i's corresponding to X_i's that are close to x get more weight than do Y_i's corresponding to X_i's that are far from x. The statistical properties of the Nadaraya–Watson estimator have been extensively investigated for both cross-sectional and time-series data, and the estimator has been widely used in applications. For example, Blundell, Browning and Crawford (2003) used kernel estimates of Engel curves in an investigation of the consistency of household-level data and revealed preference theory. Hausman and Newey (1995) used kernel estimates of demand functions to estimate the equivalent variation for changes in gasoline prices and the deadweight losses associated with increases in gasoline taxes. Kernel-based methods have also been developed for estimating conditional quantile and hazard functions.

There are other important nonparametric methods for estimating conditional mean functions. Local linear estimation and series or sieve estimation are especially useful in applications. Local linear estimation consists of estimating the mean of Y at $X = x$ by using a form of weighted least squares to fit a linear model to the data. The weights are such that observations (Y_i, X_i) for which X_i is close to x receive more weight than do observations for which X_i is far from x. In comparison with the Nadaraya–Watson estimator, local linear estimation has important advantages relating to bias and behaviour near the boundaries of the data. These are discussed in the book by Fan and Gijbels (1996), among other places.

A series estimator begins by expressing the true conditional mean (or quantile) function as an infinite series expansion using basis functions such as sines and cosines, orthogonal polynomials, or splines. The coefficients of a truncated version of the series are estimated by ordinary least squares. The statistical properties of series estimators are described by Newey (1997). Hausman and Newey (1995) give an example of their use in an economic application.

Nonparametric models and estimates essentially eliminate the possibility of misspecification of a conditional mean or quantile function (that is, they consistently estimate the true function), but they have important disadvantages that limit their usefulness in applied econometrics. One important problem is that the precision of a nonparametric estimator decreases rapidly as the dimension of the explanatory variable X increases. This phenomenon is called the 'curse of dimensionality'. It can be understood most easily by considering the case in which the explanatory variables are all discrete. Suppose the data contain 500 observations of Y and X. Suppose, further, that X is a K-component vector and that each component can take five different values. Then the values of X generate 5^k cells. If $K = 4$, which is not unusual in applied econometrics, then there are 625 cells, or more cells than observations. Thus, estimates of the conditional mean function are likely to be very imprecise for most cells because they will contain few observations. Moreover, there will be at least 125 cells that contain no data and, consequently, for which the conditional mean function cannot be estimated at all. It has been proved that the curse of dimensionality is unavoidable in nonparametric estimation. As a result of it, impractically large samples are usually needed to obtain acceptable estimation precision if X is multidimensional.

Another problem is that nonparametric estimates can be difficult to display, communicate, and interpret when X is multidimensional. Nonparametric estimates do not have simple analytic forms. If X is one- or two-dimensional, then the estimate of the function of interest can be displayed graphically, but only reduced-dimension projections can be displayed when X has three or more components. Many such displays and much skill in interpreting them can be needed to fully convey and comprehend the shape of an estimate.

A further problem with nonparametric estimation is that it does not permit extrapolation. For example, in the case of a conditional mean function it does not provide predictions of the mean of Y at values of x that are outside of the range of the data on X. This is a serious drawback in policy analysis and forecasting, where it is often important to predict what might happen under conditions that do not exist in the available data.

Finally, in nonparametric estimation it can be difficult to impose restrictions suggested by economic or other theory. Matzkin (1994) discusses this issue.

The problems of nonparametric estimation have led to the development of so-called semiparametric methods that offer a compromise between parametric and non-parametric estimation. Semiparametric methods make assumptions about functional form that are stronger than those of a nonparametric model but less restrictive than the assumptions of a parametric model, thereby reducing (though not eliminating) the possibility of specification error. Semiparametric methods permit greater estimation precision than do nonparametric methods when X is multidimensional. Semiparametric estimation results are usually easier to display and interpret than are nonparametric ones, and provide limited capabilities for extrapolation.

In econometrics, semiparametric estimation began with Manski's (1975; 1985) and Cosslett's (1983) work on estimating discrete-choice random-utility models. McFadden had introduced multinomial logit random utility models. These models assume that the random components of the utility function are independently and identically distributed with the Type I extreme value distribution. (The Type I extreme value distribution and density functions are defined, for example, in eqs (3.1) and (3.2) Maddala, 1983, p. 60.) The resulting choice model is analytically simple but has properties that are undesirable in many applications (for example, the well-known independence-of-irrelevant-alternatives property). Moreover, estimators based on logit models are inconsistent if the distribution of the random components of utility is not Type I extreme value. Manski (1975; 1985) and Cosslett (1983) proposed estimators that do not require a priori knowledge of this distribution. Powell's (1984; 1986) least absolute deviations estimator for censored regression models is another early contribution to econometric research on semiparametric estimation. This estimator was motivated by the observation that estimators of (parametric) Tobit models are inconsistent if the underlying normality assumption is incorrect. Powell's estimator is consistent under very weak distributional assumptions.

Semiparametric estimation has continued to be an active area of econometric research. Semiparametric estimators have been developed for a wide variety of additive, index, partially linear, and hazard models, among others. These estimators all reduce the effective dimension of the estimation problem and overcome the curse of dimensionality by making assumptions that are stronger than those of fully nonparametric estimation but weaker than those of a parametric model. The stronger assumptions also give the models limited extrapolation capabilities. Of course, these benefits come at the price of increased risk of specification error, but the risk is smaller than with simple parametric models. This is because semiparametric models make weaker assumptions than do parametric models, and contain simple parametric models as special cases.

Semiparametric estimation is also an important research field in statistics, and it has led to much interaction between statisticians and econometricians. The early statistics and biostatistics research that is relevant to econometrics was focused on survival (duration) models. Cox's (1972) proportional hazards model and the Buckley and James (1979) estimator for censored regression models are two early examples of this line of research. Somewhat later, C. Stone (1985) showed that a nonparametric additive model can overcome the curse of dimensionality. Since then, statisticians have contributed actively to research on the same classes of semiparametric models that econometricians have worked on.

13 Theory-based empirical models

Many econometric models are connected to economic theory only loosely or through essentially arbitrary parametric assumptions about, say, the shapes of utility functions. For example, a logit model of discrete choice assumes that the random components of utility are independently and identically distributed with the Type I extreme value distribution. In addition, it is frequently assumed that the indirect utility function is linear in prices and other characteristics of the alternatives. Because economic theory rarely, if ever, yields a parametric specification of a probability model, it is worth asking whether theory provides useful restrictions on the specification of econometric models, and whether models that are consistent with economic theory can be estimated without making non-theoretical parametric assumptions. The answers to these questions depend on the details of the setting being modelled.

In the case of discrete-choice, random-utility models, the inferential problem is to estimate the distribution of (direct or indirect) utility conditional on observed characteristics of individuals and the alternatives among which they choose. More specifically, in applied research one usually is interested in estimating the systematic component of utility (that is, the function that gives the mean of utility conditional on the explanatory variables) and the distribution of the random component of utility. Discrete choice is present in a wide range of applications, so it is important to know whether the systematic component of utility and the distribution of the random component can be estimated nonparametrically, thereby avoiding the non-theoretical distributional and functional form assumptions that are required by parametric models. The systematic component and distribution of the random component cannot be estimated unless they are identified. However, economic theory places only weak restrictions on utility functions (for example, shape restrictions such as monotonicity, convexity, and homogeneity), so the classes of conditional mean and utility functions that satisfy the restrictions are large. Indeed, it is not difficult to show that observations of individuals' choices and the values of the explanatory variables, by themselves, do not identify the systematic component of utility and the distribution of the random component without making assumptions that shrink the class of allowed functions.

This issue has been addressed in a series of papers by Matzkin that are summarized in Matzkin (1994). Matzkin gives conditions under which the systematic component of utility and the distribution of the random component are identified without restricting either to a finite-dimensional parametric family. Matzkin also shows how these functions can be estimated consistently when they are identified. Some of the assumptions required for identification may be undesirable in applications. Moreover, Manski (1988) and Horowitz (1998) have given examples in which infinitely many combinations of the systematic component of utility and distribution of the random component are consistent with a binary logit specification of choice probabilities. Thus, discrete-choice, random-utility models can be estimated under assumptions that are considerably weaker than those of, say, logit and probit models, but the systematic component of utility and the distribution of the random component cannot be identified using the restrictions of economic theory alone. It is necessary to make additional assumptions that are not required by economic theory and, because they are required for identification, cannot be tested empirically.

Models of market-entry decisions by oligopolistic firms present identification issues that are closely related to those in discrete-choice, random utility models. Berry and Tamer (2006) explain the identification problems and approaches to resolving them.

The situation is different when the economic setting provides more information about the relation between observables and preferences than is the case in discrete-choice models. This happens in models of certain kinds of auctions, thereby permitting nonparametric estimation of the distribution of values for the auctioned object. An example is a first-price, sealed bid auction within the independent private values paradigm. Here, the problem is to infer the distribution of bidders' values for the auctioned object from observed bids. A game-theory model of bidders' behaviour provides a characterization of the relation between bids and the distribution of private values. Guerre, Perrigne and Vuong (2000) show that this relation nonparametrically identifies the distribution of values if the analyst observes all bids and certain other mild conditions are satisfied. Guerre, Perrigne and Vuong (2000) also show how to carry out nonparametric estimation of the value distribution.

Dynamic decision models and equilibrium job-search models are other examples of empirical models that are closely connected to economic theory, though they also rely on non-theoretical parametric assumptions. In a dynamic decision model, an agent makes a certain decision repeatedly over time. For example, an individual may decide each year whether to retire or not. The optimal decision depends on uncertain future events (for example, the state of one's future health) whose probabilities may change over time (for example, the probability of poor health increases as one ages) and depend on the decision. In each period, the decision of an agent who maximizes expected utility is the solution to a stochastic, dynamic programming problem. A large body of research, much of which is reviewed by Rust (1994), shows how to specify and estimate econometric models of the utility function (or, depending on the application, cost function), probabilities of relevant future events, and the decision process.

An equilibrium search model determines the distributions of job durations and wages endogenously. In such a model, a stochastic process generates wage offers. An unemployed worker accepts an offer if it exceeds his reservation wage. An employed worker accepts an offer if it exceeds his current wage. Employers choose offers to maximize expected profits. Among other things, an equilibrium search model provides an explanation for why seemingly identical workers receive different wages. The theory of equilibrium search models is described in Albrecht and Axell (1984), Mortensen (1990), and Burdett and Mortensen (1998). There is a large body of literature on the estimation of these models. Bowlus, Kiefer and Neumann (2001) provide a recent example with many references.

14 The bootstrap

The exact, finite-sample distributions of econometric estimators and test statistics can rarely be calculated in applications. This is because, except in special cases and under restrictive assumptions (for example, the normal linear model), finite sample distributions depend on the unknown distribution of the population from which the data were sampled. This problem is usually dealt with by making use of large-sample (asymptotic) approximations. A wide variety of econometric estimators and test statistics have distributions that are approximately normal or chi-square when the sample size is large, regardless of the population distribution of the data. The approximation error decreases to zero as the sample size increases. Thus, asymptotic approximations can to be used to obtain confidence intervals for parameters and critical values for tests when the sample size is large.

It has long been known, however, that the asymptotic normal and chi-square approximations can be very inaccurate with the sample sizes encountered in applications. Consequently, there can be large differences between the true and nominal coverage probabilities of confidence intervals and between the true and nominal probabilities with which a test rejects a correct null hypothesis. One approach to dealing with this problem is to use higher-order asymptotic approximations such as Edgeworth or saddlepoint expansions. These received much research attention during 1970s and 1980s, but analytic higher-order expansions are rarely used in applications because of their algebraic complexity.

The bootstrap, which is due to Efron (1979), provides a way to obtain sometimes spectacular improvements in the accuracy of asymptotic approximations while avoiding algebraic complexity. The bootstrap amounts to treating the data as if they were the population. In other words, it creates a pseudo-population whose distribution is the empirical distribution of the data. Under sampling from the pseudo-population, the exact finite sample distribution of any statistic can be estimated with arbitrary accuracy by carrying out a Monte Carlo simulation in which samples are drawn repeatedly from the empirical distribution of the data. That is, the data are repeatedly sampled randomly with replacement. Since the empirical distribution is close to the population distribution when the sample size is large, the bootstrap consistently estimates the asymptotic distribution of a wide range of important statistics. Thus, the bootstrap provides a way to replace analytic calculations with computation. This is useful when the asymptotic distribution is difficult to work with analytically.

More importantly, the bootstrap provides a low-order Edgeworth approximation to the distribution of a wide variety of asymptotically standard normal and chi-square statistics that are used in applied research. Consequently, the bootstrap provides an approximation to the finite-sample distributions of such statistics that is more accurate than the asymptotic normal or chi-square approximation. The theoretical research leading to this conclusion was carried out by statisticians, but the bootstrap's importance has been recognized in econometrics and there is now an important body of econometric research on the topic. In many settings that are important in applications, the bootstrap essentially eliminates errors in the coverage probabilities of confidence intervals and the rejection probabilities of tests. Thus, the bootstrap is a very important tool for applied econometricians.

There are, however, situations in which the bootstrap does not estimate a statistic's asymptotic distribution consistently. Manski's (1975; 1985) maximum score estimator of the parameters of a binary response model is an example. All known cases of bootstrap inconsistency can be overcome through the use of subsampling methods. In subsampling, the distribution of a statistic is estimated by carrying out a Monte Carlo simulation in which the subsamples of the data are drawn repeatedly. The subsamples are smaller than the original data-set, and they can be drawn randomly with or without replacement. Subsampling provides estimates of asymptotic distributions that are consistent under very weak assumptions, though it is usually less accurate than the bootstrap when the bootstrap is consistent.

15 Programme evaluation and treatment effects

Programme evaluation is concerned with estimating the causal effect of a treatment or policy intervention on some population. The problem arises in many disciplines, including biomedical research (for example, the effects of a new medical treatment) and economics (for example, the effects of job training or education on earnings). The most obvious way to learn the effects of treatment on a group of individuals by observing each individual's outcome in both the treated and the untreated states. This is not possible in practice, however, because one virtually always observes any given individual in either the treated state or the untreated state but not both. This does not matter if the individuals who receive treatment are identical to those who do not, but that rarely happens. For example, individuals who choose to take a certain drug or whose physicians prescribe it for them may be sicker than individuals who do not receive the drug. Similarly, people who choose to obtain high levels of education may be different from others in ways that affect future earnings.

This problem has been recognized since at least the time of R.A. Fisher. In principle, it can be overcome by assigning individuals randomly to treatment and control groups. One can then estimate the average effect of treatment by the difference between the average outcomes of treated and untreated individuals. This random assignment procedure has become something of a gold standard in the treatment effects literature. Clinical trials use random assignment, and there have been important economic and social experiments based on this procedure. But there are also serious practical problems. First, random assignment may not be possible. For example, one cannot assign high-school students randomly to receive a university education or not. Second, even if random assignment is possible, post-randomization events may disrupt the effects of randomization. For example, individuals may drop out of the experiment or take treatments other than the one to which they are assigned. Both of these things may happen for reasons that are related to the outcome of interest. For example, very ill members of a control group may figure out that they are not receiving treatment and find a way to obtain the drug being tested. In addition, real-world programmes may not operate the way that experimental ones do, so real-world outcomes may not mimic those found in an experiment, even if nothing has disrupted the randomization.

Much research in econometrics, statistics, and biostatistics has been aimed at developing methods for inferring treatment effects when randomization is not possible or is disrupted by post-randomization events. In econometrics, this research dates back at least to Gronau (1974) and Heckman (1974). The fundamental problem is to identify the effects of treatment or, in less formal terms, to separate the effects of treatment from those of other sources of differences between the treated and untreated groups. Manski (1995), among many others, discusses this problem. Large literatures in statistics, biostatistics, and econometrics are concerned with developing identifying assumptions that are reasonable in applied settings. However, identifying assumptions are not testable empirically and can be controversial. One widely accepted way of dealing with this problem is to conduct a sensitivity analysis in which the sensitivity of the estimated treatment effect to alternative identifying assumptions is assessed. Another possibility is to forgo controversial identifying assumptions and to find the entire set of outcomes that are consistent with the joint distribution of the observed variables. This approach, which has been pioneered by Manski and several co-investigators, is discussed in Manski (1995; 2003), among other places. Hotz, Mullin and Sanders (1997) provide an interesting application of bounding methods to measuring the effects of teenage pregnancy on the labour market outcomes of young women.

16 Integration and simulation methods in econometrics

The integration problem is endemic in economic modelling, arising whenever economic agents do not observe random variables and the behaviour paradigm is the maximization of expected utility. The econometrician inherits this problem in the expression of the corresponding econometric model, even before taking up inference and estimation. The issue is most familiar in dynamic optimization contexts, where it can be addressed by a variety of methods. Taylor and Uhlig (1990) present a comprehensive review of these methods; for later innovations see Keane and Wolpin (1994), Rust (1997) and Santos and Vigo-Aguiar (1998).

The problem is more pervasive in econometrics than in economic modelling, because it arises, in addition, whenever economic agents observe random variables that the econometrician does not. For example, the economic agent may form expectations conditional on an information set not entirely accessible to the econometrician, such as personal characteristics or confidential information. Another example arises in discrete choice settings, where utilities of alternatives are never observed and the prices of alternatives often are not. In these situations the economic model provides a probability distribution of outcomes conditional on three classes of objects: observed variables, available to the econometrician; latent variables, unobserved by the econometrician; and parameters or functions describing the preferences and decision-making environment of the economic agent. The econometrician typically seeks to learn about the parameters or functions given the observed variables.

There are several ways of dealing with this task. Two approaches that are closely related and widely used in the econometrics literature generate integration problems. The first is to maintain a distribution of the latent

variables conditional on observed variables, the parameters in the model, and additional parameters required for completing this distribution. (This is the approach taken in maximum likelihood and Bayesian inference.) Combined with the model, this leads to the joint distribution of outcomes and latent variables conditional on observed variables and parameters. Since the marginal distribution of outcomes is the one relevant for the econometrician in this conditional distribution, there is an integration problem for the latent variables. The second approach is weaker: it restricts to zero the values of certain population moments involving the latent and observable variables. (This is the approach taken in generalized method of moments, which can be implemented with both parametric and nonparametric methods.) These moments depend upon the parameters (which is why the method works) and the econometrician must therefore be able to evaluate the moments for any given set of parameter values. This again requires integration over the latent variables.

Ideally, this integral would be evaluated analytically. Often – indeed, typically – this is not possible. The alternative is to use numerical methods. Some of these are deterministic, but the rapid growth in the solution of these problems since (roughly) 1990 has been driven more by simulation methods employing pseudo-random numbers generated by computer hardware and software. This section reviews the most important these methods and describes their most significant use in non-Bayesian econometrics, namely, simulated method of moments. In Bayesian econometrics the integration problem is inescapable, the structure of the economic model notwithstanding, because parameters are treated explicitly as unobservable random variables. Consequently simulation methods have been central to Bayesian inference in econometrics.

16.1 Deterministic approximation of integrals

The evaluation of an integral is a problem as old as the calculus itself. In well-catalogued but limited instances analytical solutions are available: Gradshteyn and Ryzhik (1965) is a useful classic reference. For integration in one dimension there are several methods of deterministic approximation, including Newton-Coates (Press et al., 1986, ch. 4; Davis and Rabinowitz, 1984, ch. 2), and Gaussian quadrature (Golub and Welsch, 1969; Judd, 1998, s. 7.2). Gaussian quadrature approximates a smooth function as the product a polynomial of modest order and a smooth basis function, and then uses iterative refinements to compute the approximation. It is incorporated in most mathematical applications software and is used routinely to approximate integrals in one dimension to many significant figures of accuracy.

Integration in several dimensions by means of deterministic approximation is more difficult. Practical generic adaptations of Gaussian quadrature are limited to situations in which the integrand is approximately the product of functions of single variables (Davis and Rabinowitz, 1984, pp. 354–9). Even here the logarithm of computation time is approximately linear in the number of variables, a phenomenon sometimes dubbed 'the curse of dimensionality.' Successful extensions of quadrature beyond dimensions of four or five are rare, and these extensions typically require substantial analytical work before they can be applied successfully.

Low discrepancy methods provide an alternative generic approach to deterministic approximation of integrals in higher dimensions. The approximation is the average value of the integrand computed over a well-chosen sequence of points whose configuration amounts to a sophisticated lattice. Different sequences lead to variants on the approach, the best known being the Halton (1960) sequence and the Hammersley (1960) sequence. Niederreiter (1992) reviews these and other variants.

A key property of any method of integral approximation, deterministic or non-deterministic, is that it should provide as a by-product some indicator of the accuracy of the approximation. Deterministic methods typically provide upper bounds on the approximation error, based on worst-case situations. In many situations the actual error is orders of magnitude less than the upper bound, and as a consequence attaining desired error tolerances may appear to be impractical, whereas in fact these tolerances can easily be attained. Geweke (1996, s. 2.3) provides an example.

16.2 Simulation approximation of integrals

The structure of integration problems encountered in econometrics makes them often more amenable to attack by simulation methods than by non-deterministic methods. Two characteristics are key. First, integrals in many dimensions are required. In some situations the number is proportional to the size of the sample, and, while the structure of the problem may lead to decomposition in terms of many integrals of smaller dimension, the resulting structure and dimension are still unsuitable for deterministic methods. The second characteristic is that the integration problem usually arises as the need to compute the expected value of a function of a random vector with a given probability distribution P:

$$I = \int_S g(\mathbf{x})p(\mathbf{x})d\mathbf{x}, \qquad (1)$$

where p is the density corresponding to P, g is the function, $\mathbf{x}$ is the random vector, and I is the number to be approximated. The probability distribution P is then the point of departure for the simulation.

For many distributions there are reliable algorithms, implemented in widely available mathematical applications software, for simulation of random vectors x. This yields a sample $\{g(x^{(m)})\}(m = 1, \dots, M)$ whose arithmetic mean provides an approximation of I, and for which a central limit theorem provides an assessment of the accuracy of the approximation in the usual way. (This requires the existence of the first two moments of g, which must be shown analytically.) This approach is most useful when p is simple (so that direct simulation of $\mathbf{x}$ is possible) but the structure of g precludes analytical evaluation of I.

This simple approach does not suffice for the integration problem as it typically arises in econometrics. A leading example is the multinomial probit (MNP) model with J discrete choices. For each individual i the utility of the last choice u_{iJ} is normalized to be zero, and the utilities of the first $J - 1$ choices are given by the vector

$$\mathbf{u}_i \sim N(\mathbf{X}_i\beta, \Sigma), \qquad (2)$$

where $\mathbf{X}$ is a matrix of characteristics of individual i, including the prices and other properties of the choices presented to that individual, and β and Σ are structural parameters of the model. If the j'th element of $\mathbf{u}_i$ is positive and larger than all the other elements of $\mathbf{u}_i$ the individual makes choice j, and if all elements of $\mathbf{u}$ are negative the individual makes choice J. The probability that individual i makes choice j is the integral of the $(n - 1)$-variate normal distribution (1) taken over the subspace $\{\mathbf{u}_i : u_{ik} \leq u_{ij} \forall k = 1, \dots, n\}$. This computation is essential in evaluating the likelihood function, and it has no analytical solution. (For discussion and review, see Sandor and Andras, 2004.)

Several generic simulation methods have been used for the problem (1) in econometrics. One of the oldest is acceptance sampling, a simple variant of which is described in von Neumann (1951) and Hammersley and Handscomb (1964). Suppose it is possible to draw from the distribution Q with density q, and the ratio $p(\mathbf{x})/q(\mathbf{x})$ is bounded above by the known constant a. If $\mathbf{x}$ is simulated successively from Q but accepted and taken into the sample with probability $p(\mathbf{x})/[aq(\mathbf{x})]$, then the resulting sample is independently distributed with the identical distribution P. Proofs and further discussion are widely available; for example, Press et al. (1992, s. 7.4), Bratley, Fox and Schrage (1987, s. 5.2.5), and Geweke (2005, s. 4.2.1). The unconditional probability of accepting draws from Q is $1/a$. If a is too large the method is impractical, but when acceptance sampling is practical it provides draws directly from P. This is an important component of many of the algorithms underlying the 'black box' generation of random variables in mathematical applications software.

Alternatively, in the same situation all of the draws from Q are retained and taken into a stratified sample in which the weight $w(\mathbf{x}^{(m)}) = p(\mathbf{x}^{(m)})/q(\mathbf{x}^{(m)})$ is associated with the m'th draw. The approximation of I in (1) is then the weighted average of the terms $g(\mathbf{x}^{(m)})$. This approach dates at least to Hammersley and Handscomb (1964, s. 5.4), and was introduced to econometrics by Kloek and van Dijk (1978). The procedure is more general than acceptance sampling in that a known upper bound of w is not required, but if in fact a is large then the weights will display large variation and the approximation will be poor. This is clear in the central limit theorem for the accuracy of approximation provided in Geweke (1989a), which as a practical matter requires that a finite upper bound on w be established analytically. This is a key limitation of acceptance sampling and importance sampling.

Markov chain Monte Carlo (MCMC) methods provide an entirely different approach to the solution of the integration problem (1). These procedures construct a Markov process of the form

$$\mathbf{x}^{(m)} \sim p(\mathbf{x}|\mathbf{x}^{(m-1)}) \qquad (3)$$

in such a way that

$$M^{-1} \sum_{m=1}^{M} g(x^{(m)})$$

converges (almost surely) to I. These methods have a history in mathematical physics dating back to the algorithm of Metropolis et al. (1953). Hastings (1970) focused on statistical problems and extended the method to its present form known as the Hastings–Metropolis (HM) algorithm. HM draws a candidate $\mathbf{x}^*$ from a convenient distribution indexed by $\mathbf{x}^{(m-1)}$. It sets $\mathbf{x}^{(m)} = \mathbf{x}$ with probability $\alpha(\mathbf{x}^{(m-1)}, \mathbf{x}^{(m)})$ and sets $\mathbf{x}^{(m)} = \mathbf{x}^{(m)-1}$ otherwise, the function α being chosen so that the process (3) defined in this way has the desired convergence property. Chib and Greenberg (1995) provide a detailed introduction to HM and its application in econometrics. Tierney (1994) provides a succinct summary of the relevant continuous state space Markov chain theory bearing on the convergence of MCMC.

A version of the HM algorithm particularly suited to image reconstruction and problems in spatial statistics, known as the Gibbs sampling (GS) algorithm, was introduced by Geman and Geman (1984). This was subsequently shown to have great potential for Bayesian computation by Gelfand and Smith (1990). In GS the vector $\mathbf{x}$ is subdivided into component vectors, $\mathbf{x}' = (\mathbf{x}'_1, \ldots, \mathbf{x}'_B)$, in such a way that simulation from the conditional distribution of each $\mathbf{x}_j$ implied by $p(\mathbf{x})$ in (1) is feasible. This method has proven very advantageous in econometrics generally, and it revolutionized Bayesian approaches in particular beginning about 1990.

By the turn of the century HM and GS algorithms were standard tools for likelihood-based econometrics. Their structure and strategic importance for Bayesian econometrics were conveyed in surveys by Geweke (1999) and Chib (2001), as well as in a number of textbooks, including Koop (2003), Lancaster (2004), Geweke (2005) and Rossi, Allenby and McCulloch (2005). Central limit theorems can be used to assess the quality

of approximations as described in Tierney (1994) and Geweke (2005).

16.3 Simulation methods in non-Bayesian econometrics
Generalized method of moments estimation has been a staple of non-Bayesian econometrics since its introduction by Hansen (1982). In an econometric model with $k \times 1$ parameter vector $\boldsymbol{\theta}$ economic theory provides the set of sample moment restrictions

$$\mathbf{h}(\boldsymbol{\theta}) = \int_S \mathbf{g}(\mathbf{x})p(\mathbf{x}|\boldsymbol{\theta}, \mathbf{y})d\mathbf{x} = 0, \qquad (4)$$

where $\mathbf{g}(\mathbf{x})$ is a $p \times 1$ vector and $\mathbf{y}$ denotes the data including instrumental variables. An example is the MNP model (2). If the observed choices are coded by the variables $d_{ij} = 1$ if individual i makes choice j and $d_{ij} = 0$ otherwise, then the expected value of d_{ij} is the probability that individual i makes choice j, leading to restrictions of the form (4).

The generalized method of moments estimator minimizes the criterion function $\mathbf{h}(\boldsymbol{\theta})'\mathbf{W}\mathbf{h}(\boldsymbol{\theta})$ given a suitably chosen weighting matrix $\mathbf{W}$. If the requisite integrals can be evaluated analytically, $p \geq k$, and other conditions provided in Hansen (1982) are satisfied, then there is a well-developed asymptotic theory of inference for the parameters that by 1990 was a staple of graduate econometrics textbooks. If for one or more elements of $\mathbf{h}$ the integral cannot be evaluated analytically, then for alternative values of it is often possible to approximate the integral appearing in (4) by simulation. This is the situation in the MNP model.

The substitution of a simulation approximation

$$M^{-1} \sum_{m=1}^{M} \mathbf{g}(\mathbf{x}^{(m)})$$

for the integral in (4) defines the method of simulated moments (MSM) introduced by McFadden (1989) and Pakes and Pollard (1989), who were concerned with the MNP model (2) in particular and the estimation of discrete response models using cross-section data in general. Later the method was extended to time series models by Lee and Ingram (1991) and Duffie and Singleton (1993). The asymptotic distribution theory established in this literature requires that the number of simulations M increase at least as rapidly as the square of the number of observations. The practical import of this apparently severe requirement is that applied econometric work must establish that changes in M must have little impact on the results; Geweke, Keane and Runkle (1994; 1997) provide examples for MNP. This literature also shows that in general the impact of using direct simulation, as opposed to analytical evaluation of the integral, is to increase the asymptotic variance of the GMM estimator of $\boldsymbol{\theta}$ by the factor M^{-1}, typically trivial in view of the number of simulations required. Substantial surveys of the details of MSM and leading applications of the method can be found in Gourieroux and Monfort (1993; 1996), Stern (1997) and Liesenfeld and Breitung (1999).

The simulation approximation, unlike the (unavailable) analytical evaluation of the integral in (4), can lead to a criterion function that is discontinuous in θ. This happens in the MNP model using the obvious simulation scheme in which the choice probabilities are replaced by their proportions in the M simulations, as proposed by Lerman and Manski (1981). The asymptotic theory developed by McFadden (1989) and Pakes and Pollard (1989) copes with this possibility, and led McFadden (1989) to used kernel weighting to smooth the probabilities. The most widely used method for smoothing probabilities in the MNP model is the Geweke–Hajivassiliou–Keane (GHK) simulator of Geweke (1989b), Hajivassiliou, McFadden and Ruud (1991) and Keane (1990); a full description is provided in Geweke and Keane (2001), and comparisons of alternative methods are given in Hajivassiliou, McFadden and Ruud (1996) and Sandor and Andras (2004).

Maximum likelihood estimation of θ can lead to first-order conditions of the form (4), and thus becomes a special case of MSM. This context highlights some of the complications introduced by simulation. While the simulation approximation of (1) is unbiased, the corresponding expression enters the log likelihood function and its derivatives nonlinearly. Thus for any finite number of simulations M, the evaluation of the first-order conditions is biased in general. Increasing M at a rate faster than the square of the number of observations eliminates the squared bias relative to the variance of the estimator; Lee (1995) provides further details.

16.4 Simulation methods in Bayesian econometrics
Bayesian econometrics places a common probability distribution on random variables that can be observed (data) and unobservable parameters and latent variables. Inference proceeds using the distribution of these unobservable entities conditional on the data – the posterior distribution. Results are typically expressed in terms of the expectations of parameters or functions of parameters, expectations taken with respect to the posterior distribution. Thus, whereas integration problems are application-specific in non-Bayesian econometrics, they are endemic in Bayesian econometrics.

The development of modern simulation methods had a correspondingly greater impact in Bayesian than in non-Bayesian econometrics. Since 1990 simulation-based Bayesian methods have become practical in the context of most econometric models. The availability of this tool has been influential in the modelling approach taken in addressing applied econometric problems.

The MNP model (2) illustrates the interaction in latent variable models. Given a sample of n individuals, the $(J-1) \times 1$ latent utility vectors $\mathbf{u}_1,\ldots,\mathbf{u}_n$ are regarded explicitly as $n(J-1)$ unknowns to be inferred along with the unknown parameters β and Σ. Conditional on these parameters and the data, the vectors $\mathbf{u}_1,\ldots,\mathbf{u}_n$ are independently distributed. The distribution of $\mathbf{u}_i$ is (2) truncated to an orthant that depends on the observed choice j: if $j<J$ then $u_{ik}<u_{ij}$ for all $k \neq j$ and $u_{ij}>0$, whereas for choice J:, $u_{ik}<0$ for all k. The distribution of each u_{ik}, conditional on all of the other elements of $\mathbf{u}_i$, is truncated univariate normal, and it is relatively straightforward to simulate from this distribution. (Geweke, 1991, provides details on sampling from a multivariate normal distribution subject to linear restrictions.) Consequently GS provides a practical algorithm for drawing from the distribution of the latent utility vectors conditional on the parameters.

Conditional on the latent utility vectors – that is, regarding them as observed – the MNP model is a seemingly unrelated regressions model, and the approach taken by Percy (1992) applies. Given conjugate priors the posterior distribution of β, conditional on Σ and utilities, is Gaussian, and the conditional distribution of Σ, conditional on β and utilities, is inverted Wishart. Since GS provides the joint distribution of parameters and latent utilities, the posterior mean of any function of these can be approximated as the sample mean. This approach and the suitability of GS for latent variable models were first recognized by Chib (1992). Similar approaches in other latent variable models in include McCulloch and Tsay (1994), Chib and Greenberg (1998), McCulloch, Polson and Rossi (2000) and Geweke and Keane (2001).

The Bayesian approach with GS sidesteps the evaluation of the likelihood function and, of any moments in which the approximation is biased given a finite number of simulations, two technical issues that are prominent in MSM. On the other hand, as in all MCMC algorithms, there may be sensitivity to the initial values of parameters and latent variables in the Markov chain, and substantial serial correlation in the chain will reduce the accuracy of the simulation approximation. Geweke (1992; 2005) and Tierney (1994) discuss these issues.

17 Financial econometrics
Attempts at testing of the efficient market hypothesis (EMH) provided the impetus for the application of time series econometric methods in finance. The EMH was built on the pioneering work of Bachelier (1900) and evolved in the 1960s from the random walk theory of asset prices advanced by Samuelson (1965). By the early 1970s a consensus had emerged among financial economists suggesting that stock prices could be well approximated by a random walk model and that changes in stock returns were basically unpredictable. Fama (1970) provides an early, definitive statement of this position. He distinguished between different forms of the EMH: the 'weak' form that asserts all price information is fully reflected in asset prices; the 'semi-strong' form that requires asset price changes to fully reflect all publicly

available information and not only past prices; and the 'strong' form that postulates that prices fully reflect information even if some investor or group of investors have monopolistic access to some information. Fama regarded the strong form version of the EMH as a benchmark against which the other forms of market efficiencies are to be judged. With respect to the weak form version he concluded that the test results strongly support the hypothesis, and considered the various departures documented as economically unimportant. He reached a similar conclusion with respect to the semi-strong version of the hypothesis. Evidence on the semi-strong form of the EMH was revisited by Fama (1991). By then it was clear that the distinction between the weak and the semi-strong forms of the EMH was redundant. The random walk model could not be maintained either, in view of more recent studies, in particular that of Lo and MacKinlay (1988).

This observation led to a series of empirical studies of stock return predictability over different horizons. It was shown that stock returns can be predicted to some degree by means of interest rates, dividend yields and a variety of macroeconomic variables exhibiting clear business cycle variations. See, for example, Fama and French (1989), Kandel and Stambaugh (1996), and Pesaran and Timmermann (1995) on predictability of equity returns in the United States; and Clare, Thomas and Wickens (1994), and Pesaran and Timmermann (2000) on equity return predictability in the UK.

Although it is now generally acknowledged that stock returns could be predictable, there are serious difficulties in interpreting the outcomes of market efficiency tests. Predictability could be due to a number of different factors such as incomplete learning, expectations heterogeneity, time variations in risk premia, transaction costs, or specification searches often carried out in pursuit of predictability. In general, it is not possible to distinguish between the different factors that might lie behind observed predictability of asset returns. As noted by Fama (1991) the test of the EMH involves a joint hypothesis, and can be tested only jointly with an assumed model of market equilibrium. This is not, however, a problem that is unique to financial econometrics; almost all areas of empirical economics are subject to the joint hypotheses problem. The concept of market efficiency is still deemed to be useful as it provides a benchmark and its use in finance has led to significant insights.

Important advances have been made in the development of equilibrium asset pricing models, econometric modelling of asset return volatility (Engle, 1982; Bollerslev, 1986), analysis of high frequency intraday data, and market microstructures. Some of these developments are reviewed in Campbell, Lo and MacKinlay (1997), Cochrane (2005), Shephard (2005), and McAleer and Medeiros (2007). Future advances in financial econometrics are likely to focus on heterogeneity, learning and model uncertainty, real time analysis, and

further integration with macroeconometrics. Finance is particularly suited to the application of techniques developed for real time econometrics (Pesaran and Timmermann, 2005a).

18 Appraisals and future prospects

Econometrics has come a long way over a relatively short period. Important advances have been made in the compilation of economic data and in the development of concepts, theories and tools for the construction and evaluation of a wide variety of econometric models. Applications of econometric methods can be found in almost every field of economics. Econometric models have been used extensively by government agencies, international organizations and commercial enterprises. Macroeconometric models of differing complexity and size have been constructed for almost every country in the world. In both theory and practice, econometrics has already gone well beyond what its founders envisaged. Time and experience, however, have brought out a number of difficulties that were not apparent at the start.

Econometrics emerged in the 1930s and 1940s in a climate of optimism, in the belief that economic theory could be relied on to identify most, if not all, of the important factors involved in modelling economic reality, and that methods of classical statistical inference could be adapted readily for the purpose of giving empirical content to the received economic theory. This early view of the interaction of theory and measurement in econometrics, however, proved rather illusory. Economic theory is invariably formulated with *ceteris paribus* clauses, and involves unobservable latent variables and general functional forms; it has little to say about adjustment processes, lag lengths and other factors mediating the relationship between the theoretical specification (even if correct) and observables. Even in the choice of variables to be included in econometric relations, the role of economic theory is far more limited than was at first recognized. In a Walrasian general equilibrium model, for example, where everything depends on everything else, there is very little scope for a priori exclusion of variables from equations in an econometric model. There are also institutional features and accounting conventions that have to be allowed for in econometric models but which are either ignored or are only partially dealt with at the theoretical level. All this means that the specification of econometric models inevitably involves important auxiliary assumptions about functional forms, dynamic specifications, latent variables, and so on, with respect to which economic theory is silent or gives only an incomplete guide.

The recognition that economic theory on its own cannot be expected to provide a complete model specification has important consequences for testing and evaluation of economic theories, for forecasting and real time decision making. The incompleteness of economic theories makes the task of testing them a formidable undertaking. In general it will not be possible to say whether the results of the statistical tests have a bearing on the economic theory or the auxiliary assumptions. This ambiguity in testing theories, known as the Duhem–Quine thesis, is not confined to econometrics and arises whenever theories are conjunctions of hypotheses (on this, see for example Cross, 1982). The problem is, however, especially serious in econometrics because theory is far less developed in economics than it is in the natural sciences. There are, of course, other difficulties that surround the use of econometric methods for the purpose of testing economic theories. As a rule economic statistics are not the results of designed experiments, but are obtained as by-products of business and government activities often with legal rather than economic considerations in mind. The statistical methods available are generally suitable for large samples while the economic data typically have a rather limited coverage. There are also problems of aggregation over time, commodities and individuals that further complicate the testing of economic theories that are micro-based.

Econometric theory and practice seek to provide information required for informed decision-making in public and private economic policy. This process is limited not only by the adequacy of econometrics but also by the development of economic theory and the adequacy of data and other information. Effective progress, in the future as in the past, will come from simultaneous improvements in econometrics, economic theory and data. Research that specifically addresses the effectiveness of the interface between any two of these three in improving policy – to say nothing of all of them – necessarily transcends traditional sub-disciplinary boundaries within economics. But it is precisely these combinations that hold the greatest promise for the social contribution of academic economics.

JOHN GEWEKE, JOEL HOROWITZ AND HASHEM PESARAN

Bibliography

Aitken, A.C. 1934–5. On least squares and linear combinations of observations. *Proceedings of the Royal Society of Edinburgh* 55, 42–8.
Albrecht, J.W. and Axell, B. 1984. An equilibrium model of search unemployment. *Journal of Political Economy* 92, 824–40.
Allen, R.G.D. and Bowley, A.L. 1935. *Family Expenditure*. London: P.S. King.
Almon, S. 1965. The distributed lag between capital appropriations and net expenditures. *Econometrica* 33, 178–96.
Amemiya, T. 1983. Nonlinear regression models. In *Handbook of Econometrics*, vol. 1, ed. Z. Griliches and M.D. Intriligator. Amsterdam: North-Holland.
Amemiya, T. 1984. Tobit models: a survey. *Journal of Econometrics* 24, 3–61.

An, S. and Schorfheide, F. 2007. Bayesian analysis of DSGE models. *Econometric Reviews* 26(2–4), 113–72.
Anderson, T.W. and Hsiao, C. 1981. Estimation of dynamic models with error components. *Journal of the American Statistical Society* 76, 598–606.
Anderson, T.W. and Hsiao, C. 1982. Formulation and estimation of dynamic models using panel data. *Journal of Econometrics* 18, 47–82.
Anderson, T.W. and Rubin, H. 1949. Estimation of the parameters of a single equation in a complete system of stochastic equations. *Annals of Mathematical Statistics* 20, 46–63.
Andrews, D.W.K. 1993. Tests for parameter instability and structural change with unknown change point. *Econometrica* 61, 821–56.
Andrews, D.W.K. and Ploberger, W. 1994. Optimal tests when a nuisance parameter is present only under the alternative. *Econometrica* 62, 1383–414.
Anselin, L. 1988. *Spatial Econometrics: Methods and Models*. Boston: Kluwer Academic Publishers.
Anselin, L., Le Gallo, J. and Jayet, H. 2007. Spatial panel econometrics. In *The Econometrics of Panel Data: Fundamentals and Recent Developments in Theory and Practice*, 3rd edn. ed. L. Matyas and P. Sevestre. Dordrecht: Kluwer (forthcoming).
Arellano, M. 2003. *Panel Data Econometrics*. Oxford: Oxford University Press.
Arellano, M. and Bond, S.R. 1991. Some tests of specification for panel data: Monte Carlo evidence and an application to employment equations. *Review of Economic Studies* 58, 277–97.
Arellano, M. and Honoré, B. 2001. Panel data models: some recent developments. In *Handbook of Econometrics*, vol. 5, ed. J.J. Heckman and E. Leamer. Amsterdam: North-Holland.
Bachelier, L.J.B.A. 1900. *Théorie de la Speculation*. Paris: Gauthier-Villars. Reprinted in *The Random Character of Stock Market Prices*, ed. P. H. Cootner. Cambridge, MA: MIT Press, 1964.
Bai, J. and Ng, S. 2004. A panic attack on unit roots and cointegration. *Econometrica* 72, 1127–77.
Bai, J. and Perron, P. 1998. Estimating and testing linear models with multiple structural changes. *Econometrica* 66, 47–78.
Baltagi, B. 2005. *Econometric Analysis of Panel Data*, 2nd edn. New York: Wiley.
Baltagi, B.H. and Kao, C. 2000. Nonstationary panels, cointegration in panels and dynamic panels: a survey. In *Nonstationary Panels, Panel Cointegration, and Dynamic Panels*, ed. B. Baltagi. Advances in Econometrics, vol. 15. Amsterdam: JAI Press.
Banerjee, A. 1999. Panel data unit roots and cointegration: an overview. *Oxford Bulletin of Economics and Statistics* 61, 607–29.
Basmann, R.L. 1957. A generalized classical method of linear estimation of coefficients in a structural equation. *Econometrica* 25, 77–83.

Bauwens, L., Lubrano, M. and Richard, J.F. 2001. *Bayesian Inference in Dynamic Econometric Models.* Oxford: Oxford University Press.

Benini, R. 1907. Sull'uso delle formole empiriche a nell'economia applicata. *Giornale degli economisti*, 2nd series 35, 1053–63.

Berry, S. and Tamer, E. 2006. Identification in models of oligopoly entry. In *Advances in Economics and Econometrics: Theory and Applications, Ninth World Congress*, ed. R. Blundell, W.K. Newey and T. Persson. Cambridge: Cambridge University Press.

Beveridge, S. and Nelson, C.R. 1981. A new approach to the decomposition of economic time series into permanent and transitory components with particular attention to measurement of the 'business cycle'. *Journal of Monetary Economics* 7, 151–74.

Binder, M., Hsiao, C. and Pesaran, M.H. 2005. Estimation and inference in short panel vector autoregressions with unit roots and cointegration. *Econometric Theory* 21, 795–837.

Binder, M. and Pesaran, M.H. 1995. Multivariate rational expectations models and macroeconometric modelling: a review and some new results. In *Handbook of Applied Econometrics, Volume 1 – Macroeconomics*, ed. M.H. Pesaran and M.R. Wickens. Oxford: Basil Blackwell.

Bjerkholt, O. 1995. Ragnar Frisch, Editor of *Econometrica*. *Econometrica* 63, 755–65.

Blanchard, O.J. and Quah, D. 1989. The dynamic effects of aggregate demand and supply disturbances. *American Economic Review* 79, 1146–64.

Blundell, R.W., Browning, M. and Crawford, I.A. 2003. Nonparametric Engel curves and revealed preference. *Econometrica* 71, 205–40.

Bollerslev, T. 1986. Generalised autoregressive conditional heteroskedasticity. *Journal of Econometrics* 51, 307–27.

Boskin, M.J. 1974. A conditional logit model of occupational choice. *Journal of Political Economy* 82, 389–98.

Bosq, D. 1996. *Nonparametric Statistics for Stochastic Processes.* New York: Springer.

Bowlus, A.J., Kiefer, N.M. and Neumann, G.R. 2001. Equilibrium search models and the transition from school to work. *International Economic Review* 42, 317–43.

Box, G.E.P. and Jenkins, G.M. 1970. *Time Series Analysis: Forecasting and Control.* San Francisco: Holden-Day.

Bratley, P., Fox, B.L. and Schrage, L.E. 1987. *A Guide to Simulation.* New York: Springer-Verlag.

Breitung, J. and Pesaran, M.H. 2008. Unit roots and cointegration in panels. In *The Econometrics of Panel Data: Fundamentals and Recent Developments in Theory and Practice*, 3rd edn. ed. L. Matyas and P. Sevestre. Dordrecht: Kluwer.

Breusch, T.S. and Pagan, A.R. 1980. The Lagrange multiplier test and its applications to model specification in econometrics. *Review of Economic Studies* 47, 239–53.

Brock, W. and Durlauf, S. 2002. A multinomial choice model with neighborhood effects. *American Economic Review* 92, 298–303.

Brock, W. and Durlauf, S. 2006. Macroeconomics and model uncertainty. In *Post-Walrasian Macroeconomics: Beyond the Dynamic Stochastic General Equilibrium Model*, ed. D. Colander. New York: Cambridge University Press.

Brown, R.L., Durbin, J. and Evans, J.M. 1975. Techniques for testing the constancy of regression relationships over time (with discussion). *Journal of the Royal Statistical Society, Series B* 37, 149–92.

Brown, T.M. 1952. Habit persistence and lags in consumer behaviour. *Econometrica* 20, 355–71.

Broze, L., Gouriéroux, C. and Szafarz, A. 1985. Solutions of dynamic linear rational expectations models. *Econometric Theory* 1, 341–68.

Brundy, J.M. and Jorgenson, D.N. 1971. Efficient estimation of simultaneous equations by instrumental variables. *Review of Economics and Statistics* 53, 207–24.

Buckley, J. and James, I. 1979. Linear regression with censored data. *Biometrika* 66, 429–36.

Burdett, K. and Mortensen, D.T. 1998. Wage differentials, employer size, and unemployment. *International Economic Review* 39, 257–73.

Burns, A.F. and Mitchell, W.C. 1947. *Measuring Business Cycles.* New York: Columbia University Press for the NBER.

Burtless, G. 1995. The case for randomized field trials in economic and policy research. *Journal of Economic Perspectives* 9(2), 63–84.

Cagan, P. 1956. The monetary dynamics of hyperinflation. In *Studies in the Quantity Theory of Money*, ed. M. Friedman. Chicago: University of Chicago Press.

Cameron, A.C. and Trivedi, P.K. 1986. Econometric models based on count data: comparisons and applications of some estimators and tests. *Journal of Applied Econometrics* 1, 29–53.

Cameron, A.C. and Trivedi, P.K. 1998. *Regression Analysis for Count Data.* Econometric Society Monograph No. 30.Cambridge: Cambridge University Press.

Cameron, A.C. and Trivedi, P.K. 2005. *Microeconometrics: Methods and Applications.* Cambridge: Cambridge University Press.

Campbell, J.Y., Lo, A.W. and MacKinlay, A.C. 1997. *The Econometrics of Financial Markets*, Princeton: Princeton University Press.

Canova, F. and de Nicolo, G. 2002. Monetary disturbances matter for business fluctuations in the G7. *Journal of Monetary Economics* 49, 1131–59.

Champernowne, D.G. 1948. Sampling theory applied to autoregressive sequences. *Journal of the Royal Statistical Society, Series B* 10, 204–31.

Champernowne, D.G. 1960. An experimental investigation of the robustness of certain procedures for estimating means and regressions coefficients. *Journal of the Royal Statistical Society* 123, 398–412.

Chib, S. 1992. Bayes inference in the tobit censored regression model. *Journal of Econometrics* 51, 79–99.

Chib, S. 2001. Markov chain Monte Carlo methods: computation and inference. In *Handbook of*

Econometrics, vol. 5, ed. J. J. Heckman and E. Leamer. Amsterdam: North-Holland.

Chib, S. and Greenberg, E. 1995. Understanding the Metropolis–Hastings algorithm. *The American Statistician* 49, 327–35.

Chib, S. and Greenberg, E. 1998. Analysis of multivariate probit models. *Biometrika* 85, 347–61.

Choi, I. 2006. Nonstationary panels. In *Palgrave Handbooks of Econometrics*, vol. 1, ed. T.C. Mills and K. Patterson. Basingstoke: Palgrave Macmillan.

Chow, G.C. 1960. Tests of equality between sets of coefficients in two linear regressions. *Econometrica* 28, 591–605.

Christ, C.F. 1952. *Economic Theory and Measurement: A Twenty-Year Research Report. 1932–52.* Chicago: Cowles Commission for Research in Economics.

Christ, C.F. 1983. The founding of the Econometric Society and Econometrica. *Econometrica* 51, 3–6.

Christiano, L.J., Eichenbaum, M. and Evans, C. 2005. Nominal rigidities and the dynamic effects of a shock to monetary policy. *Journal of Political Economy* 113, 1–45.

Clare, A.D., Thomas, S.H. and Wickens, M.R. 1994. Is the gilt–equity yield ratio useful for predicting UK stock return? *Economic Journal* 104, 303–15.

Clements, M.P. and Hendry, D.F. 1998. *Forecasting Economic Time Series.* Cambridge: Cambridge University Press.

Clements, M.P. and Hendry, D.F. 1999. *Forecasting Non-stationary Economic Time Series.* Cambridge, MA: MIT Press.

Clements, M.P. and Hendry, D.F. 2006. Forecasting with breaks. In *Handbook of Economic Forecasting*, vol. 1, ed. G. Elliott, C.W.J. Granger and A. Timmermann. Amsterdam: North-Holland.

Cochrane, J. 2005. *Asset Pricing*, rev. edn. Princeton: Princeton University Press.

Cochrane, P. and Orcutt, G.H. 1949. Application of least squares regression to relationships containing autocorrelated error terms. *Journal of the American Statistical Association* 44, 32–61.

Conley, T.G. and Dupor, B. 2003. A spatial analysis of sectoral complementarity. *Journal of Political Economy* 111, 311–52.

Conley, T.G. and Topa, G. 2002. Socio-economic distance and spatial patterns in unemployment. *Journal of Applied Econometrics* 17, 303–27.

Cooper, R.L. 1972. The predictive performance of quarterly econometric models of the United States. In *Econometric Models of Cyclical Behavior*, ed. B.G. Hickman. New York: NBER.

Cosslett, S.R. 1983. Distribution free maximum likelihood estimation of the binary choice model. *Econometrica* 51, 765–82.

Cox, D.R. 1961. Tests of separate families of hypotheses. *Proceedings of the Fourth Berkeley Symposium on Mathematical Statistics and Probability*, vol. 1. Berkeley: University of California Press.

Cox, D.R. 1962. Further results of tests of separate families of hypotheses. *Journal of the Royal Statistical Society*, Series B 24, 406–24.

Cox, D.R. 1972. Regression models and life tables. *Journal of the Royal Statistical Society*, Series B 34, 187–220.

Cross, R. 1982. The Duhem–Quine thesis, Lakatos and the appraisal of theories in macroeconomics. *Economic Journal* 92, 320–40.

Davenant, C. 1698. *Discourses on the Publick Revenues and on the Trade of England*, Vol. 1, London.

Davidson, R. and MacKinnon, J.G. 2004. *Econometric Theory and Methods.* Oxford: Oxford University Press.

Davis, P.J. and Rabinowitz, P. 1984. *Methods of Numerical Integration.* Orlando, FL: Academic Press.

Deaton, A. 1985. Panel data from time series of cross-sections. *Journal of Econometrics* 30, 109–26.

Del Negro, M. and Schorfheide, F. 2004. Priors from equilibrium models for VAR's. *International Economic Review* 45, 643–73.

Del Negro, M., Schorfheide, F., Smets, F. and Wouters, R. 2005. On the fit and forecasting performance of new Keynesian models. Working Paper No. 491. Frankfurt: European Central Bank.

Dhrymes, P. 1971. A simplified estimator for large-scale econometric models. *Australian Journal of Statistics* 13, 168–75.

Doan, T., Litterman, R. and Sims, C.A. 1984. Forecasting and conditional projections using realistic prior distributions. *Econometric Reviews* 3, 1–100.

Domenich, T. and McFadden, D. 1975. *Urban Travel Demand: A Behavioral Analysis.* Amsterdam: North-Holland.

Duffie, D. and Singleton, K. 1993. Simulated moments estimation of Markov models of asset prices. *Econometrica* 61, 929–52.

Duncan, G. 1980. Formulation and statistical analysis of the mixed continuous/discrete variable model in classical production theory. *Econometrica* 48, 839–52.

Durbin, J. 1954. . Errors in variables. *Review of the International Statistical Institute* 22, 23–32.

Durbin, J. and Watson, G.S. 1950. Testing for serial correlation in least squares regression I. *Biometrika* 37, 409–28.

Durbin, J. and Watson, G.S. 1951. Testing for serial correlation in least squares regression II. *Biometrika* 38, 159–78.

Durlauf, S.N., Johnson, P.A. and Temple, J.R.W. 2005. Growth econometrics. In *Handbook of Economic Growth*, vol. 1A, ed. P. Aghion and S. N. Durlauf. Amsterdam: North-Holland.

Efron, B. 1979. Bootstrap methods: another look at the jackknife. *Annals of Statistics* 7, 1–26.

Eisner, R. and Strotz, R.H. 1963. Determinants of business investment. In *Impacts of Monetary Policy*. Englewood Cliffs, NJ: Prentice-Hall, for the Commission on Money and Credit.

Elliott, G., Granger, C.W.J. and Timmermann, A. 2006. *Handbook of Economic Forecasting*, vol. 1, Amsterdam: North-Holland.

Engle, R.F. 1982. Autoregressive conditional heteroscedasticity, with estimates of the variance of United Kingdom inflation. *Econometrica* 50, 987–1007.

Engle, R.F. 1984. Wald likelihood ratio and Lagrange multiplier tests in econometrics. In *Handbook of Econometrics*, vol. 2, ed. Z. Griliches and M.D. Intriligator. Amsterdam: North-Holland.

Engle, R.F. and Granger, G. 1987. Cointegration and error-correction: representation, estimation and testing. *Econometrica* 55, 251–76.

Engle, R.F., Hendry, D.F. and Richard, J.-F. 1983. Exogeneity. *Econometrica* 51, 277–304.

Fachin, S. 2006. Long-run trends in internal migrations in Italy: a study in panel cointegration with dependent units. *Journal of Applied Econometrics* (forthcoming).

Fama, E.F. 1970. Efficient capital markets: a review of theory and empirical work. *Journal of Finance* 25, 383–417.

Fama, E.F. 1991. Efficient capital markets: II. *Journal of Finance* 46, 1575–617.

Fama, E.F. and French, K.R. 1989. Business conditions and expected returns on stocks and bonds. *Journal of Financial Economics* 25, 23–49.

Fan, J. and Gijbels, I. 1996. *Local Polynomial Modelling and Its Applications*. London: Chapman & Hall.

Fernandez-Villaverde, J. and Rubio-Ramirez, J. 2005. Estimating dynamic equilibrium economies: linear versus nonlinear likelihood. *Journal of Applied Econometrics* 20, 891–910.

Fisher, F.M. 1966. *The Identification Problem in Econometrics*. New York: McGraw-Hill.

Fisher, I. 1930. *The Theory of Interest*. New York: Macmillan. Reprinted, Philadelphia: Porcupine Press, 1977.

Fisher, I. 1937. Note on a short-cut method for calculating distributed lags. *Bulletin de l'Institut International de Statistique* 29, 323–7.

Fisher, R.A. 1928. *Statistical Methods for Research Workers*, 2nd edn. London: Oliver and Boyd.

Friedman, M. 1957. *A Theory of the Consumption Function*. Princeton: Princeton University Press.

Frisch, R. 1933. *Pitfalls in the Statistical Construction of Demand and Supply Curves*. Leipzig: Hans Buske Verlag.

Frisch, R. 1934. *Statistical Confluence Analysis by Means of Complete Regression Systems*. Oslo: University Institute of Economics.

Frisch, R. 1936. A note on the term 'econometrics'. *Econometrica* 4, 95.

Gali, J. 1992. How well does the IS–LM model fit postwar US data? *Quarterly Journal of Economics* 107, 709–38.

Garratt, A., Robertson, D. and Wright, S. 2006. Permanent vs transitory components and economic fundamentals. *Journal of Applied Econometrics* 21, 521–42.

Garratt, A., Lee, K., Pesaran, M.H. and Shin, Y. 2003a. A long run structural macroeconometric model of the UK. *Economic Journal* 113(487), 412–55.

Garratt, A., Lee, K., Pesaran, M.H. and Shin, Y. 2003b. Forecast uncertainty in macroeconometric modelling: an application to the UK economy. *Journal of the American Statistical Association* 98(464), 829–38.

Garratt, A., Lee, K., Pesaran, M.H. and Shin, Y. 2006. *Global and National Macroeconometric Modelling: A Long-Run Structural Approach*. Oxford: Oxford University Press.

Geary, R.C. 1949. Studies in relations between economic time series. *Journal of the Royal Statistical Society*, Series B 10, 140–58.

Gelfand, A.E. and Smith, A.F.M. 1990. Sampling based approaches to calculating marginal densities. *Journal of the American Statistical Association* 85, 398–409.

Geman, S. and Geman, D. 1984. Stochastic relaxation, Gibbs distributions and the Bayesian restoration of images. *IEEE Transactions on Pattern Analysis and Machine Intelligence* 6, 721–41.

Geweke, J. 1989a. Bayesian inference in econometric models using Monte Carlo integration. *Econometrica* 57, 1317–0.

Geweke, J. 1989b. Efficient simulation from the multivariate normal distribution subject to linear inequality constraints and the evaluation of constraint probabilities. Discussion paper, Duke University.

Geweke, J. 1991. Efficient simulation from the multivariate normal and student-t distributions subject to linear constraints. In *Computing Science and Statistics: Proceedings of the Twenty-Third Symposium on the Interface*, ed. E.M. Keramidas. Fairfax: Interface Foundation of North America, Inc.

Geweke, J. 1992. Evaluating the accuracy of sampling-based approaches to the calculation of posterior moments. In *Bayesian Statistics 4*, ed. J.M. Bernardo et al. Oxford: Clarendon Press.

Geweke, J. 1996. Monte Carlo simulation and numerical integration. In *Handbook of Computational Economics*, ed. H.M. Amman, D.A. Kendrick and J. Rust. Amsterdam: North-Holland.

Geweke, J. 1999. Using simulation methods for Bayesian econometric models: Inference, development and communication (with discussion and rejoinder). *Econometric Reviews* 18, 1–126.

Geweke, J. 2005. *Contemporary Bayesian Econometrics and Statistics*. New York: Wiley.

Geweke, J. and Keane, M. 2001. Computationally intensive methods for integration in econometrics. In *Handbook of Econometrics*, vol.5, ed. J. Heckman and E.E. Leamer. Amsterdam: North-Holland.

Geweke, J., Keane, M. and Runkle, D. 1994. Alternative computational approaches to statistical inference in the multinomial probit model. *Review of Economics and Statistics* 76, 609–32.

Geweke, J., Keane, M. and Runkle, D. 1997. Statistical inference in the multinomial multiperiod probit model. *Journal of Econometrics* 80, 125–65.

Geweke, J. and Whiteman, C. 2006. Bayesian forecasting. In *Handbook of Economic Forecasting*, ed. G. Elliott, C.W.J.

Granger and A. Timmermann. Amsterdam: North-Holland.

Godfrey, L.G. 1988. *Misspecification Tests in Econometrics: The LM principle and Other Approaches*. Cambridge: Cambridge University Press.

Godfrey, L.G. and Wickens, M.R. 1982. Tests of mis-specification using locally equivalent alternative models. In *Evaluation and Reliability of Macro-economic Models*, ed. G.C. Chow and P. Corsi. New York: John Wiley.

Golub, G.H. and Welsch, J.H. 1969. Calculation of Gaussian quadrature rules. *Mathematics of Computation* 23, 221–30.

Gourieroux, C. and Jasiak, J. 2001. *Financial Econometrics: Problems, Models, and Methods*. Oxford: Oxford University Press.

Gourieroux, C. and Monfort, A. 1993. Simulation based inference: a survey with special reference to panel data models. *Journal of Econometrics* 59, 5–33.

Gourieroux, C. and Monfort, A. 1996. *Simulation-Based Econometric Methods*. New York: Oxford University Press.

Gradshteyn, I.S. and Ryzhik, I.M. 1965. *Tables of Integrals, Series and Products*. New York: Academic Press.

Granger, C.W.J. 1969. Investigating causal relations by econometric models and cross-spectral methods. *Econometrica* 37, 424–38.

Granger, C.W.J. 1986. Developments in the study of co-integrated economic variables. *Oxford Bulletin of Economics and Statistics* 48, 213–28.

Granger, C.W.J. and Newbold, P. 1974. Spurious regressions in econometrics. *Journal of Econometrics* 2, 111–20.

Granger, C.W.J. and Newbold, P. 1986. *Forecasting Economic Time Series*, 2nd edn. San Diego: Academic Press.

Granger, C.W.J. and Pesaran, M.H. 2000a. A decision theoretic approach to forecast evaluation. In *Statistics and Finance: An Interface*, ed. W.S. Chan, W.K. Li and H. Tong. London: Imperial College Press.

Granger, C.W.J. and Pesaran, M.H. 2000b. Economic and statistical measures of forecast accuracy. *Journal of Forecasting* 19, 537–60.

Greene, W.H. 2003. *Econometric Analysis*, 5th edn. New Jersey: Prentice Hall.

Gronau, R. 1974. Wage comparisons – a selectivity bias. *Journal of Political Economy* 82, 1119–43.

Guerre, E., Perrigne, I. and Vuong, Q. 2000. Optimal nonparametric estimation of first-price auctions. *Econometrica* 68, 525–74.

Haavelmo, T. 1943. Statistical testing of business cycle theories. *Review of Economics and Statistics* 25, 13–18.

Haavelmo, T. 1944. The probability approach in econometrics. *Econometrica* 12(Supplement), 1–118.

Hajivassiliou, V., McFadden, D. and Ruud, P. 1991. Simulation of multivariate normal rectangle probabilities. Methods and programs mimeo, University of California, Berkeley.

Hajivassiliou, V., McFadden, D. and Ruud, P. 1996. Simulation of multivariate normal rectangle probabilities

and their derivatives: theoretical and computational results. *Journal of Econometrics* 72, 85–134.

Hall, A.R. 2005. *Generalized Method of Moments*. Oxford: Oxford University Press.

Halton, J.M. 1960. On the efficiency of evaluating certain quasi-random sequences of points in evaluating multi-dimensional integrals. *Numerische Mathematik* 2, 84–90.

Hamilton, J.D. 1994. *Time Series Analysis*. Princeton: Princeton University Press.

Hammersley, J.M. 1960. Monte Carlo methods for solving multivariate problems. *Annals of the New York Academy of Sciences* 86, 844–74.

Hammersley, J.M. and Handscomb, D.C. 1964. *Monte Carlo Methods*. London: Methuen.

Hansen, L.P. 1982. Large sample properties of generalized method of moments. *Econometrica* 50, 1029–54.

Hansen, L.P. and Sargent, T.J. 1980. Formulating and estimating dynamic linear rational expectations models. *Journal of Economic Dynamics and Control* 2, 7–46.

Hansen, L.P. and Sargent, T.J. 2007. *Robustness*. Princeton: Princeton University Press.

Hart, B.S. and von Neumann, J. 1942. Tabulation of the probabilities for the ratio of mean square successive difference to the variance. *Annals of Mathematical Statistics* 13, 207–14.

Härdle, W. 1990. *Applied Nonparametric Estimation*. Cambridge: Cambridge University Press.

Härdle, W. and Linton, O. 1994. Applied nonparametric methods. In *Handbook of Econometrics*, vol. 4, ed. R.F. Engle and D. McFadden. Amsterdam: North-Holland.

Harvey, A. 1989. *Forecasting, Structural Time Series Models and Kalman Filter*. Cambridge: Cambridge University Press.

Hastings, W.K. 1970. Monte Carlo sampling methods using Markov chains and their applications. *Biometrika* 57, 97–109.

Hausman, J.A. 1978. Specification tests in econometrics. *Econometrica* 46, 1251–72.

Hausman, J.A., Hall, B.H. and Griliches, Z. 1984. Econometric models for count data with application to the patents–R&D relationship. *Econometrica* 52, 909–1038.

Hausman, J.A. and Newey, W.K. 1995. Nonparametric estimation of exact consumers surplus and deadweight loss. *Econometrica* 63, 1445–76.

Hausman, J.A. and Wise, D.A., eds. 1985. *Social Experimentation*. NBER Conference Report.Chicago: University of Chicago Press.

Heckman, J.J. 1974. Shadow prices, market wages, and labor supply. *Econometrica* 42, 679–94.

Heckman, J.J. and Singer, B. 1984. Econometric duration analysis. *Journal of Econometrics* 24, 63–132.

Heckman, J.J. and Smith, A.J. 1995. Assessing the case for social experimentation. *Journal of Economic Perspectives* 9(2), 85–110.

Heckman, J.J. and Willis, R. 1977. A beta-logistic model for the analysis of sequential labour force participation by married women. *Journal of Political Economy* 85, 27–58.

Hendricks, K. and Porter, R.H. 1988. An empirical study of an auction with asymmetric information. *American Economic Review* 78, 865–83.

Hendry, D.F. and Richard, J.-F. 1982. On the formulation of empirical models in dynamic econometrics. *Journal of Econometrics* 20, 3–33.

Holly, S., Pesaran, M.H. and Yamagata, T. 2006. A spatio-temporal model of house prices in the US. Mimeo, University of Cambridge.

Holtz-Eakin, D., Newey, W.K. and Rosen, H.S. 1988. Estimating vector autoregressions with panel data. *Econometrica* 56, 1371–95.

Honoré, B. and Kyriazidou, E. 2000. Panel data discrete choice models with lagged dependent variables. *Econometrica* 68, 839–74.

Hooker, R.H. 1901. Correlation of the marriage rate with trade. *Journal of the Royal Statistical Society* 44, 485–92.

Horowitz, J.L. 1998. *Semiparametric Methods in Econometrics*. New York: Springer-Verlag.

Horowitz, J.L. 2001. Semiparametric models. In *International Encyclopedia of Behavioral and Social Sciences*, ed. N.J. Smelser and P.B. Baltes. Elsevier. Amsterdam: Elsevier.

Horowitz, J.L. and Lee, S. 2002. Semiparametric methods in applied econometrics: do the models fit the data? *Statistical Modelling* 2, 3–22.

Horowitz, J.L. and Savin, N.E. 2001. Binary response models: logits, probits, and semiparametrics. *Journal of Economic Perspectives* 15(4), 43–56.

Hotz, V.J., Mullin, C.H. and Sanders, S.G. 1997. Bounding causal effects using data from a contaminated natural experiment: analyzing the effects of teenage childbearing. *Review of Economic Studies* 64, 575–603.

Hsiao, C. 2003. *Analysis of Panel Data*, 2nd edn. Cambridge: Cambridge University Press.

Hsiao, C. and Pesaran, M.H. 2007. Random coefficient panel data models. In *The Econometrics of Panel Data: Fundamentals and Recent Developments in Theory and Practice*, 3rd edn. ed. L. Matyas and P. Sevestre. Dordrecht: Kluwer.

Hsiao, C., Pesaran, M.H. and Tahmiscioglu, A.K. 1999. Bayes estimation of short-run coefficients in dynamic panel data models. In *Analysis of Panels and Limited Dependent Variables Models*, ed. C. Hsiao et al. Cambridge: Cambridge University press.

Hsiao, C., Pesaran, M.H. and Tahmiscioglu, A.K. 2002. Maximum likelihood estimation of fixed effects dynamic panel data models covering short time periods. *Journal of Econometrics* 109, 107–50.

Im, K.S., Pesaran, M.H. and Shin, Y. 2003. Testing for unit roots in heterogenous panels. *Journal of Econometrics* 115, 53–74.

Imbs, J., Mumtaz, H., Ravn, M.O. and Rey, H. 2005. PPP strikes back, aggregation and the real exchange rate. *Quarterly Journal of Economics* 120, 1–43.

Johansen, S. 1988. Statistical analysis of cointegration vectors. *Journal of Economic Dynamics and Control* 12, 231–54. Reprinted in *Long-run Economic Relationships*, ed. R.F. Engle and C.W.J. Granger. Oxford: Oxford University Press, 1991.

Johansen, S. 1991. Estimation and hypothesis testing of cointegrating vectors in Gaussian vector autoregressive models. *Econometrica* 59, 1551–80.

Johansen, S. 1995. *Likelihood-based Inference in Cointegrated Vector Autoregressive Models*. Oxford: Oxford University Press.

Jorgenson, D.W. 1966. Rational distributed lag functions. *Econometrica* 34, 135–49.

Judd, K.L. 1998. *Numerical Methods in Economics*. Cambridge, MA: MIT Press.

Juselius, K. 2006. *The Cointegrated VAR Model: Econometric Methodology and Macroeconomic Applications*. Oxford: Oxford University Press.

Kagel, J. and Roth, A.E., eds. 1995. *The Handbook of Experimental Economics*. Princeton: Princeton University Press.

Kandel, S. and Stambaugh, R.F. 1996. On the predictability of stock returns: an asset-allocation perspective. *Journal of Finance* 51, 385–424.

Keane, M.P. 1990. A computationally practical simulation estimator for panel data, with applications to estimating temporal dependence in employment and wages. Discussion paper, University of Minnesota.

Keane, M. and Wolpin, K.I. 1994. The solution and estimation of discrete choice dynamic programming models by simulation: Monte Carlo evidence. *Review of Economics and Statistics* 76, 648–72.

Keynes, J.M. 1939. The statistical testing of business cycle theories. *Economic Journal* 49, 558–68.

Kilian, L. 1998. Small-sample confidence intervals for impulse response functions. *Review of Economics and Statistics* 80, 218–29.

Kim, K. and Pagan, A.R. 1995. The econometric analysis of calibrated macroeconomic models. In *Handbook of Applied Econometrics: Macroeconomics*, ed. M.H. Pesaran and M. Wickens. Oxford: Basil Blackwell.

Klein, L.R. 1947. The use of econometric models as a guide to economic policy. *Econometrica* 15, 111–51.

Klein, L.R. 1950. *Economic Fluctuations in the United States 1921–1941*. Cowles Commission Monograph No. 11. New York: John Wiley.

Kloek, T. and van Dijk, H.K. 1978. Bayesian estimates of equation system parameters: an application of integration by Monte Carlo. *Econometrica* 46, 1–20.

Koop, G. 2003. *Bayesian Econometrics*. Chichester: Wiley.

Koop, G., Pesaran, M.H. and Potter, S.M. 1996. Impulse response analysis in nonlinear multivariate models. *Journal of Econometrics* 74, 119–47.

Koop, G. and Potter, S. 2004a. Forecasting and estimating multiple change-point models with an unknown number of change-points. Mimeo, University of Leicester and Federal Reserve Bank of New York.

Koop, G. and Potter, S. 2004b. Prior elicitation in multiple change-point models. Mimeo, University of Leicester and Federal Reserve Bank of New York.

Koopmans, T.C. 1937. *Linear Regression Analysis of Economic Time Series*. Haarlem: De Erven F. Bohn for the Netherlands Economic Institute.

Koopmans, T.C. 1949. Identification problems in economic model construction. *Econometrica* 17, 125–44.

Koopmans, T.C., Rubin, H. and Leipnik, R.B. 1950. Measuring the equation systems of dynamic economics. In *Statistical Inference in Dynamic Economic Models*, ed. T.C. Koopmans. Cowles Commission Monograph No. 10. New York: John Wiley.

Koyck, L.M. 1954. *Distributed Lags and Investment Analysis*. Amsterdam: North-Holland.

Laffont, J-J., Ossard, H. and Vuong, Q. 1995. Econometrics of first-price auctions. *Econometrica* 63, 953–80.

Lancaster, T. 2004. *An Introduction to Modern Bayesian Econometrics*. Malden MA: Blackwell.

Leamer, E.E. 1978. *Specification Searches: Ad Hoc Inference with Non-experimental Data*. New York: John Wiley.

Lee, B.S. and Ingram, B. 1991. Simulation estimation of time-series models. *Journal of Econometrics* 47, 197–205.

Lee, L.F. 1995. Asymptotic bias in simulated maximum likelihood estimation of discrete choice models. *Economic Theory* 11, 437–83.

Lenoir, M. 1913. *Etudes sur la formation et le mouvement des prix*. Paris: Giard et Brière.

Lerman, S. and Manski, C.S. 1981. On the use of simulated frequencies to approximate choice probabilities. In *Structural Analysis of Discrete Data with Econometric Applications*, ed. C.F. Manski and D. McFadden. Cambridge, MA: MIT Press.

Levin, A., Lin, C. and Chu, C.J. 2002. Unit root tests in panel data: asymptotic and finite-sample properties. *Journal of Econometrics* 108, 1–24.

Liesenfeld, R. and Breitung, J. 1999. Simulation based method of moments. In *Generalized Method of Moment Estimation*, ed. L. Tatyas. Cambridge: Cambridge University Press.

Litterman, R.B. 1985. Forecasting with Bayesian vector autoregressions: five years of experience. *Journal of Business and Economic Statistics* 4, 25–38.

Liu, T.C. 1960. Underidentification, structural estimation and forecasting. *Econometrica* 28, 855–65.

Lo, A. and MacKinlay, C. 1988. Stock market prices do not follow random walks: evidence from a simple specification test. *Review of Financial Studies* 1, 41–66.

Lucas, R.E. 1972. Expectations and the neutrality of money. *Journal of Economic Theory* 4, 103–24.

Lucas, R.E. 1973. Some international evidence on output-inflation tradeoffs. *American Economic Review* 63, 326–34.

Lucas, R.E. 1976. Econometric policy evaluation: a critique. In *The Phillips Curve and Labor Markets*, ed. K. Brunner and A. M. Meltzer. Amsterdam: North-Holland.

Lucas, R.E. and Sargent, T. 1981. Rational expectations and econometric practice. Introduction to *Rational Expectations and Econometric Practice*. Minneapolis: University of Minnesota Press.

Lütkepohl, H. 1991. *Introduction to Multiple Time Series Analysis*. New York: Springer-Verlag.

Lyttkens, E. 1970. Symmetric and asymmetric estimation methods. In *Interdependent Systems*, ed. E. Mosback and H. Wold. Amsterdam: North-Holland.

Maddala, G.S. 1983. *Limited Dependent and Qualitative Variables in Econometrics*. Cambridge: Cambridge University Press.

Maddala, G.S. 1986. Disequilibrium, self-selection, and switching models. In *Handbook of Econometrics*, vol. 3, ed. Z. Griliches and M.D. Intriligator. Amsterdam: North-Holland.

Maddala, G.S. 2001. *Introduction to Econometrics*, 3rd edn. New York: John Wiley and Sons.

Manski, C.F. 1975. Maximum score estimation of the stochastic utility model of choice. *Journal of Econometrics* 3, 205–28.

Manski, C.F. 1985. Semiparametric analysis of discrete response: asymptotic properties of the maximum score estimator. *Journal of Econometrics* 27, 313–34.

Manski, C.F. 1988. Identification of binary response models. *Journal of the American Statistical Association* 83, 729–38.

Manski, C.F. 1995. *Identification Problems in the Social Sciences*. Cambridge, MA: Harvard University Press.

Manski, C.F. 2003. *Partial Identification of Probability Distributions*. New York: Springer-Verlag.

Manski, C.F. and McFadden, D. 1981. *Structural Analysis of Discrete Data with Econometric Applications*. Cambridge, MA: MIT Press.

Matzkin, R.L. 1994. Restrictions of economic theory in nonparametric methods. In *Handbook of Econometrics*, vol. 4, ed. R. F. Engle and D. L. McFadden. Amsterdam: North-Holland.

McAleer, M., Pagan, A.R. and Volker, P.A. 1985. What will take the con out of econometrics? *American Economic Review* 75, 293–307.

McAleer, M. and Medeiros, M.C. 2007. Realized volatility: a review. *Econometric Reviews*.

McCulloch, R.E. and Rossi, P.E. 1994. An exact likelihood analysis of the multinomial probit model. *Journal of Econometrics* 64, 207–40.

McCulloch, R.E. and Tsay, R.S. 1993. Bayesian inference and prediction for mean and variance shifts in autoregressive time series. *Journal of the American Statistical Association* 88, 965–78.

McCulloch, R.E. and Tsay, R.S. 1994. Bayesian analysis of autoregressive time series via the Gibbs sampler. *Journal of Time Series Analysis* 15, 235–50.

McCulloch, R.E., Polson, N.G. and Rossi, P.E. 2000. A Bayesian analysis of the multinomial probit model with fully identified parameters. *Journal of Econometrics* 99, 173–93.

McFadden, D. 1989. A method of simulated moments for estimation of multinomial probits without numerical integration. *Econometrica* 57, 995–1026.

Metropolis, N., Rosenbluth, A.W., Rosenbluth, M.N., Teller, A.H. and Teller, E. 1953. Equation of state calculations by fast computing machines. *Journal of Chemical Physics* 21, 1087–92.

Mitchell, W.C. 1928. *Business Cycles: The Problem in its Setting*. New York: NBER.

Moon, R. and Perron, B. 2004. Testing for unit root in panels with dynamic factors. *Journal of Econometrics* 122, 81–126.

Moon, R. and Perron, B. 2007. An empirical analysis of nonstationarity in a panel of interest rates with factors. *Journal of Applied Econometrics* 22, 383–400.

Moore, H.L. 1914. *Economic Cycles: Their Law and Cause*. New York: Macmillan.

Moore, H.L. 1917. *Forecasting the Yield and the Price of Cotton*. New York: Macmillan Press.

Mortensen, D.T. 1990. Equilibrium wage distributions: a synthesis. In *Panel Data and Labor Market Studies*, ed. J. Hartog, G. Ridder and J. Theeuwes. New York: North Holland.

Mundlak, Y. 1961. Empirical production function free of management bias. *Journal of Farm Economics* 43, 44–56.

Mundlak, Y. 1978. On the pooling of time series and cross section data. *Econometrica* 46, 69–85.

Muth, J.F. 1961. Rational expectations and the theory of price movements. *Econometrica* 29, 315–35.

Nadaraya, E.A. 1964. On estimating regression. *Theory of Probability and Its Applications* 10, 141–2.

Nakamura, A. and Nakamura, M. 1981. On the relationships among several specification error tests presented by Durbin, Wu, and Hausman. *Econometrica* 49, 1583–88.

Nelson, C.R. 1972. The prediction performance of the FRB-MIT-Penn model of the US economy. *American Economic Review* 62, 902–17.

Nerlove, M. 1958a. Adaptive expectations and the cobweb phenomena. *Quarterly Journal of Economics* 72, 227–40.

Nerlove, M. 1958b. *Distributed Lags and Demand Analysis*. Washington, DC: USDA.

Newey, W.K. 1997. Convergence rates and asymptotic normality for series estimators. *Journal of Econometrics* 79, 147–68.

Nickell, S. 1981. Biases in dynamic models with fixed effects. *Econometrica* 49, 1399–416.

Niederreiter, H. 1992. *Random Number Generation and Quasi-Monte Carol Methods*. Philadelphia: SIAM.

Nyblom, J. 1989. Testing for the constancy of parameters over time. *Journal of the American Statistical Association* 84, 223–30.

O'Connell, P. 1998. The overvaluation of purchasing power parity. *Journal of International Economics* 44, 1–19.

Orcutt, G.H. 1948. A study of the autoregressive nature of the time series used for Tinbergen's model of the economic system of the United States, 1919–1932 (with discussion). *Journal of the Royal Statistical Society* Series B 10, 1–53.

Pagan, A.R. and Pesaran, M.H. 2007. On econometric analysis of structural systems with permanent and transitory shocks and exogenous variables. Unpublished manuscript.

Pagan, A. and Ullah, A. 1999. *Nonparametric Econometrics*. Cambridge: Cambridge University Press.

Pakes, A. and Pollard, D. 1989. Simulation and the asymptotics of optimization estimators. *Econometrica* 57, 1027–58.

Peersman, G. 2005. What caused the early millennium slowdown? Evidence based on autoregressions. *Journal of Applied Econometrics* 20, 185–207.

Pedroni, P. 2001. Purchasing power parity tests in cointegrated panels. *Review of Economics and Statistics* 83, 727–31.

Pedroni, P. 2004. Panel cointegration: asymptotic and finite sample properties of pooled time series tests with an application to the PPP hypothesis. *Econometric Theory* 20, 597–625.

Percy, D.F. 1992. Prediction for seemingly unrelated regressions. *Journal of the Royal Statistical Society*, Series B 54, 243–52.

Pesaran, M.H. 1981. Identification of rational expectations models. *Journal of Econometrics* 16, 375–98.

Pesaran, M.H. 1987a. Econometrics. In *The New Palgrave: A Dictionary of Economics*, vol. 2, ed. J. Eatwell, M. Milgate and P. Newman. London: Macmillan.

Pesaran, M.H. 1987b. *The Limits to Rational Expectations*. Oxford: Basil Blackwell.

Pesaran, M.H. 2006. Estimation and inference in large heterogeneous panels with cross section dependence. *Econometrica* 74, 967–1012.

Pesaran, M.H. 2007a. A simple panel unit root test in the presence of cross section dependence. *Journal of Applied Econometrics* 22, 265–312.

Pesaran, M.H. 2007b. A pair-wise approach to testing for output and growth convergence. *Journal of Econometrics* 138, 312–55.

Pesaran, M.H. and Deaton, A.S. 1978. Testing non-nested nonlinear regression models. *Econometrica* 46, 677–94.

Pesaran, M.H., Pettenuzzo, D. and Timmermann, A. 2006. Forecasting time series subject to multiple structural breaks. *Review of Economic Studies* 73, 1057–84.

Pesaran, M.H., Schuermann, T. and Weiner, S.M. 2004. Modelling regional interdependencies using a global error-correcting macroeconometric model (with discussion). *Journal of Business and Economic Statistics* 22, 129–62, 175–81.

Pesaran, M.H. and Skouras, S. 2002. Decision-based methods for forecast evaluation. In *A Companion to Economic Forecasting*, ed. M.P. Cements and D.F. Hendry. Oxford: Blackwell Publishing.

Pesaran, M.H. and Smith, R.P. 1985a. Keynes on econometrics. In *Keynes' Economics: Methodological Issues*, ed. T. Lawson and M.H. Pesaran. London: Croom Helm.

Pesaran, M.H. and Smith, R.P. 1985b. Evaluation of macroeconometric models. *Economic Modelling* 2, 125–34.

Pesaran, M.H. and Smith, R. 1995. Estimating long-run relationships from dynamic heterogeneous panels. *Journal of Econometrics* 68, 79–113.

Pesaran, M.H. and Smith, R.P. 2006. Macroeconometric modelling with a global perspective. *Manchester School* 74, 24–49.

Pesaran, M.H. and Timmermann, A. 1995. The robustness and economic significance of predictability of stock returns. *Journal of Finance* 50, 1201–28.

Pesaran, M.H. and Timmermann, A. 2000. A recursive modelling approach to predicting UK stock returns. *Economic Journal* 110, 159–91.

Pesaran, M.H. and Timmermann, A. 2005a. Real time econometrics. *Econometric Theory* 21, 212–31.

Pesaran, M.H. and Timmermann, A. 2005b. Small sample properties of forecasts from autoregressive models under structural breaks. *Journal of Econometrics* 129, 183–217.

Pesaran, M.H. and Timmermann, A. 2007. Selection of estimation window in the presence of breaks. *Journal of Econometrics* 137, 134–61.

Pesaran, M.H. and Weeks, M. 2001. Non-nested hypothesis testing: an overview. In *Companion to Theoretical Econometrics*, ed. B.H. Baltagi. Oxford: Basil Blackwell.

Phillips, P.C.B. 1983. Exact small sample theory in the simultaneous equations model. In *Handbook of Econometrics*, vol. 1, ed. Z. Griliches and M. D. Intrilgator. Amsterdam: North-Holland.

Phillips, P.C.B. 1986. Understanding spurious regressions in econometrics. *Journal of Econometrics* 33, 311–40.

Phillips, P.C.B. 1991. Optimal inference in cointegrated systems. *Econometrica* 59, 283–306.

Phillips, P.C.B. and Moon, H.R. 1999. Linear regression limit theory for nonstationary panel data. *Econometrica* 67, 1057–111.

Phillips, P.C.B. and Xiao, Z. 1998. A primer on unit root testing. *Journal of Economic Surveys* 12, 423–69.

Powell, J.L. 1984. Least absolute deviations estimation for the censored regression model. *Journal of Econometrics* 25, 303–25.

Powell, J.L. 1986. Censored regression quantiles. *Journal of Econometrics* 32, 143–55.

Press, W.H., Flannery, B.P., Teukolsky, S.A. and Vetterling, W.T. 1986. *Numerical Recipes: The Art of Scientific Computing*, 1st edn. Cambridge: Cambridge University Press.

Press, W.H., Flannery, B.P., Teukolsky, S.A. and Vetterling, W.T. 1992. *Numerical Recipes: The Art of Scientific Computing*, 2nd edn. Cambridge: Cambridge University Press.

Prothero, D.L. and Wallis, K.F. 1976. Modelling macroeconomic time series. *Journal of the Royal Statistical Society*, Series A 139, 468–86.

Quandt, R.E. 1982. Econometric disequilibrium models. *Econometric Reviews* 1, 1–63.

Ramsey, J.B. 1969. Tests for specification errors in classical linear least squares regression analysis. *Journal of the Royal Statistical Society*, Series B 31, 350–71.

Reiersol, O. 1941. Confluence analysis by means of lag moments and other methods of confluence analysis. *Econometrica* 9, 1–24.

Reiersol, O. 1945. Confluence analysis by means of instrumental sets of variables. *Arkiv for Mathematik Astronomi och Fysik* 32, 1–119.

Rossi, P.E., Allenby, G.M. and McCulloch, R. 2005. *Bayesian Statistics and Marketing*. Chichester: Wiley.

Rothenberg, T.J. 1984. Approximating the distributions of econometric estimators and test statistics. In *Handbook of Econometrics*, vol. 2, ed. Z. Griliches and M.D. Intriligator. Amsterdam: North-Holland.

Rust, J. 1994. Structural estimation of Markov decision processes. In *Handbook of Econometrics*, vol. 4, ed. R.F. Engle and D.L. McFadden. Amsterdam: North-Holland.

Rust, J. 1997. Using randomization to break the curse of dimensionality. *Econometrica* 65, 487–516.

Samuelson, P. 1965. Proof that properly anticipated prices fluctuate randomly. *Industrial Management Review* 6, 41–9.

Sandor, Z. and Andras, P. 2004. Alternative sampling methods for estimating multivariate normal probabilities. *Journal of Econometrics* 120, 207–34.

Santos, M.S. and Vigo-Aguiar, J. 1998. Analysis of a numerical dynamic programming algorithm applied to economic models. *Econometrica* 66, 409–26.

Sargan, J.D. 1958. The estimation of economic relationships using instrumental variables. *Econometrica* 26, 393–415.

Sargan, J.D. 1964. Wages and prices in the United Kingdom: a study in econometric methodology. In *Econometric Analysis for National Economic Planning*, ed. P.E. Hart, G. Mills and J.K. Whitaker. London: Butterworths.

Sargent, T.J. 1973. Rational expectations, the real rate of interest and the natural rate of unemployment. *Brookings Papers on Economic Activity* 1973(2), 429–72.

Sargent, T.J. and Wallace, N. 1975. Rational expectations and the theory of economic policy. *Journal of Monetary Economics* 2, 169–84.

Savin, N.E. 1973. Systems k-class estimators. *Econometrica* 41, 1125–36.

Schultz, M. 1938. *The Theory and Measurement of Demand*. Chicago: University of Chicago Press.

Schumpeter, J.A. 1954. *History of Economic Analysis*. London: George Allen & Unwin.

Shephard, N., ed. 2005. *Stochastic Volatility: Selected Readings*. Oxford: Oxford University Press.

Shiller, R.J. 1973. A distributed lag estimator derived from smoothness priors. *Econometrica* 41, 775–88.

Sims, C.A. 1972. Money, income and causality. *American Economic Review* 62, 540–52.

Sims, C.A. 1980. Macroeconomics and reality. *Econometrica* 48, 1–48.

Sims, C.A. 1982. Policy analysis with econometric models. *Brookings Papers on Economic Activity* 1982(1), 107–64.

Slutsky, E. 1927. The summation of random causes as the source of cyclic processes. In *Problems of Economic Conditions*, vol. 3. Moscow. English trans. in *Econometrica* 5 (1937), 105–46.

Smets, F. and Wouters, R. 2003. An estimated stochastic dynamic general equilibrium model of the euro area. *Journal of the European Economic Association* 1, 1123–75.

Smith, V., Leybourne, S., Kim, T.-H. and Newbold, P. 2004. More powerful panel data unit root tests with an application to mean reversion in real exchange rates. *Journal of Applied Econometrics* 19, 147–70.

Solow, R.M. 1960. On a family of lag distributions. *Econometrica* 28, 393–406.

Srivastava, V.K. 1971. Three-stage least-squares and generalized double k-class estimators: a mathematical relationship. *International Economic Review* 12, 312–16.

Stambaugh, R.F. 1999. Predictive regressions. *Journal of Financial Economics* 54, 375–421.

Stern, S. 1997. Simulation-based estimation. *Journal of Economic Literature* 35, 2006–39.

Stock, J.H. 1994. Unit roots, structural breaks and trends. In *Handbook of Econometrics*, ed. R.F. Engle and D.L. McFadden. Amsterdam: North-Holland.

Stock, J.H. and Watson, M.W. 1996. Evidence on structural instability in macroeconomic time series relations. *Journal of Business and Economic Statistics* 14, 11–30.

Stock, J.H., Wright, J.H. and Yogo, M. 2002. A survey of weak instruments and weak identification in generalized method of moments. *Journal of Business and Economic Statistics* 20, 518–29.

Stone, C.J. 1985. Additive regression and other nonparametric models. *Annals of Statistics* 13, 689–705.

Stone, J.R.N. 1945. The analysis of market demand. *Journal of the Royal Statistical Society*, Series A 108, 286–382.

Stone, J.R.N. 1978. Keynes, political arithmetic and econometrics. Seventh Keynes Lecture in Economics, British Academy.

Stone, J.R.N. et al. 1954. *Measurement of Consumers' Expenditures and Behavior in the United Kingdom. 1920– 38*, 2 vols. London: Cambridge University Press.

Strachan, R.W. and van Dijk, H.K. 2006. Model uncertainty and Bayesian model averaging in vector autoregressive processes. Discussion Papers in Economics 06/5, Department of Economics, University of Leicester.

Swamy, P.A.V.B. 1970. Efficient inference in a random coefficient regression model. *Econometrica* 38, 311–23.

Taylor, J.B. and Uhlig, H. 1990. Solving nonlinear stochastic growth models: a comparison of alternative solution methods. *Journal of Business and Economic Statistics* 8, 1–18.

Theil, H. 1954. Estimation of parameters of econometric models. *Bulletin of International Statistics Institute* 34, 122–8.

Theil, H. 1958. *Economic Forecasts and Policy*. Amsterdam: North-Holland; 2nd edn, 1961.

Tierney, L. 1994. Markov chains for exploring posterior distributions with discussion and rejoinder. *Annals of Statistics* 22, 1701–62.

Tinbergen, J. 1929–30. Bestimmung und Deutung von Angebotskurven: ein Beispiel. *Zeitschrift für Nationalökonomie* 1, 669–79.

Tinbergen, J. 1937. *An Econometric Approach to Business Cycle Problems*. Paris: Herman & Cie Editeurs.

Tinbergen, J. 1939. *Statistical Testing of Business Cycle Theories. Vol. 1: A Method and its Application to Investment activity; Vol. 2: Business Cycles in the United States of America. 1919– 1932*. Geneva: League of Nations.

Tobin, J. 1958. Estimation of relationships for limited dependent variables. *Econometrica* 26, 24–36.

Treadway, A.B. 1971. On the multivariate flexible accelerator. *Econometrica* 39.

Trivedi, P.K. 1975. Time series analysis versus structural models: a case study of Canadian manufacturing behaviour. *International Economic Review* 16, 587–608.

Uhlig, H. 2005. What are the effects of monetary policy: results from an agnostic identification approach. *Journal of Monetary Economics* 52, 381–419.

Verbeek, M. 2008. Pseudo panels and repeated cross-sections. In *The Econometrics of Panel Data: Fundamentals and Recent Developments in Theory and Practice*, 3rd edn. ed. L. Matyas and P. Sevestre. Dordrecht: Kluwer.

von Neumann, J. 1941. Distribution of the ratio of the mean square successive difference to the variance. *Annals of Mathematical Statistics* 12, 367–95.

von Neumann, J. 1942. A further remark on the distribution of the ratio of the mean square successive difference to the variance. *Annals of Mathematical Statistics* 13, 86–8.

von Neumann, J. 1951. Various techniques used in connection with random digits. *Applied Mathematics Series* 12, 36–8. US National Bureau of Standards.

Wallis, K.F. 1977. Multiple time series analysis and the final form of econometric models. *Econometrica* 45, 1481–97.

Wallis, K. 1980. Econometric implications of the Rational Expectations Hypothesis. *Econometrica* 48, 49–73.

Watson, G.M. 1964. Smooth regression analysis. *Sankhyâ*, Series A 26, 359–72.

Watson, M.W. 1994. Vector autoregressions and cointegration. In *Handbook of Econometrics*, vol. 4, ed. R.F. Engle and D.L. McFadden. Amsterdam: North-Holland.

Wegge, L.L. 1965. Identifiability criteria for a system of equations as a whole. *Australian Journal of Statistics* 7, 67–77.

West, K.D. 2006. Forecast evaluation. In *Handbook of Economic Forecasting*, vol. 1, ed. G. Elliott, C. Granger and A. Timmermann. Amsterdam: North-Holland.

Westerlund, J. 2005. Panel cointegration tests of the Fisher effect. Working Papers 2005:10, Lund University, Department of Economics.

White, H. 1981. Consequences and detection of misspecified nonlinear regression models. *Journal of the American Statistical Association* 76, 419–33.

White, H. 1982. Maximum likelihood estimation of misspecified models. *Econometrica* 50, 1–26.

Whittle, P. 1963. *Prediction and Regulation by Linear Least-squares Methods*. London: English Universities Press.

Wickens, M. 1982. The efficient estimation of econometric models with rational expectations. *Review of Economic Studies* 49, 55–68.

Wickens, M.R. and Motto, R. 2001. Estimating shocks and impulse response functions. *Journal of Applied Econometrics* 16, 371–87.

Wold, H. 1938. *A Study in the Analysis of Stationary Time Series*. Stockholm: Almqvist and Wiksell.

Wooldridge, J.M. 2002. *Econometric Analysis of Cross Section and Panel Data*. Cambridge, MA: MIT Press.

Wooldridge, J.M. 2006. *Introductory Econometrics: A Modern Approach*, 3rd edn. Stamford: Thomson-South-Western.

Working, E.J. 1927. What do statistical 'demand curves' show? *Quarterly Journal of Economics* 41, 212–35.

Wright, P.G. 1915. Review of economic cycles by Henry Moore. *Quarterly Journal of Economics* 29, 631–41.

Wright, P.G. 1928. *The Tariff on Animal and Vegetable Oils*. London: Macmillan for the Institute of Economics.

Wu, D. 1973. Alternatives tests of independence between stochastic regressor and disturbances. *Econometrica* 41, 733–50.

Yule, G.U. 1895, 1896. On the correlation of total pauperism with proportion of out-relief. *Economic Journal* 5, 603–11; 6, 613–23.

Yule, G.U. 1921. On the time-correlation problem, with special reference to the variate-difference correlation method. *Journal of the Royal Statistical Society* 84, 497–526.

Yule, G.U. 1926. Why do we sometimes get nonsense correlations between time-series? A study in sampling and the nature of time-series. *Journal of the Royal Statistical Society* 89, 1–64.

Zellner, A. 1962. An efficient method of estimating seemingly unrelated regressions and tests for aggregation bias. *Journal of the American Statistical Association* 57, 348–68.

Zellner, A. 1971. *An Introduction to Bayesian Inference in Econometrics*. New York: John Wiley and Sons.

Zellner, A. 1984. *Basic Issues in Econometrics*. Chicago: University of Chicago Press.

Zellner, A. 1985. Bayesian econometrics. *Econometrica* 53, 253–70.

Zellner, A. and Palm, F. 1974. Time series analysis and simultaneous equation econometric models. *Journal of Econometrics* 2, 17–54.

Zellner, A. and Theil, H. 1962. Three-stage least squares: simultaneous estimation of simultaneous equations. *Econometrica* 30, 54–78.

economic anthropology

Economic anthropology is an empirical science that seeks to describe how production, exchange and consumption operate outside the West (compare Hunt, 1997). The second edition (1952) of Herskovits's (1940) text, titled *Economic Anthropology*, labelled this sub-discipline in anthropology. The broader mission of anthropology has been to make sense of the diversity in the human experience, which became apparent to Europeans during progressive stages of exploration, colonialization and globalization. Underlying anthropological research is the premise that human societies have developed parallel institutions of aesthetics, religion, kinship, politics, and of course economics. All societies have economies, and the economic patterns observed in non-Western economies both comfort and confront theories developed by Western scholars.

Common economic processes, such as rational decision-making, law of competitive advantage, and institutional economics help explain many patterns across human economies based on variable conditions of cost, demand and availability. Additionally, however, human economies appear often to be structured quite differently from Western models, and these differences in institutional structure and motivation are of theoretical significance. From the beginning, economic anthropology has contained, and more or less successfully resolved, a tension between the desire to find cross-culturally general theories and to recognize the uniqueness of each individual case. In economic anthropology this tension has been represented in the formalist–substantivist debate.

Few anthropologists identify themselves primarily as economic anthropologists, but study economic matters as part of a broadly integrative approach to human societies. Founded in 1980, the Society of Economic Anthropology is the primary organization for anthropologists with such interests. Members include ethnographers, applied development anthropologists, archaeologists and ethnohistorians, suggesting that economic studies bridge the diversity of the discipline. The society sponsors annual meetings on themes that range across topics including key institutions of labour, property, markets and consumption, and special topics from the gift to slow foods. *Research Series in Economic Anthropology* and *Society for Economic Anthropology Monographs* offer edited volumes on the sub-discipline.

History of economic anthropology

From early in the 20th century, anthropologists have questioned whether theories developed to understand Western market economies apply only to those Western societies for which they were generated. To answer this question, anthropologists have described traditional economies, which survived into the 20th century, which existed in the past, and which have been transformed by engagement with the West. Largely empirical, the work is of substantial theoretical significance for understanding economies cross-culturally. Gudeman (1998) has compiled many of the most highly referenced articles.

Economic anthropology's beginning traces to the landmark ethnography *Argonauts of the Western Pacific*, in which Malinowski (1922) described the circulation of shell valuables among the islands of the Kula Ring. Malinowski used the Trobriand Islanders' obsession with certain shell valuables to challenge simplistic notions of 'economic man', and he argued that a non-Western economy could be fundamentally different from modern market economies in values and socialized exchange relationships. Anthropological studies of traditional economies thrived during the first half of the 20th century. As part of British functionalism, Malinowski and his students developed the approach; in French structuralism, Mauss (1925) focused on the gift as a social phenomenon; and, within American anthropology, Herskovits (1940) defined the sub-field. Much of the work was descriptive, emphasizing how traditional people meet basic needs and how the exchange of primitive valuables fashioned and maintained social relationships.

By mid-century, however, studies of traditional economy were increasingly adopting the terms and concepts of Western economic theory. Both Herskovits (1940) and Firth (1939) revised their original books on traditional economies so as to clarify underlying similarities across world economies. They each took concepts, like scarcity and specialization, and generalized them to show that they apply well to societies in which market penetration is not great. They were making the essential point that traditional economies were not simply driven by the food quest. Although most anthropologists took pains to emphasize the differences between traditional economies and market-integrated systems, some seemed to homogenize the human experience, and a sharp reaction followed.

In the tradition of Max Weber, economic historian Karl Polanyi (1944) wrote his famous treatise *The Great Transformation* to argue that the integrating structure of modern markets, for which prices are set by supply and demand, are a very recent creation of industrialism and capitalism. Theories based on scarcity, rationality, equilibrating price mechanisms operated, he argued, only in the special case of Western capitalism. Modern market conditions should not be taken as inherent in the human experience, but as a recent social artifact malleable in future societies.

Polanyi's impact on economic anthropology was profound and created the debate between substantivists and formalists that raged in the sub-discipline for a generation. *Trade and Markets in the Early Empires* (Polanyi, Arensberg and Pearson, 1957), the seminal edited book, came out of a discussion group which Polanyi led at Columbia University and which included anthropologists who would be influential in the field. Polanyi's chapter 'The Economy as Instituted Process' characterized the substantivist approach. He defined three forms of distribution found in societies with different structured relationships: reciprocity in egalitarian relationships; redistribution in hierarchical relationships; and market exchange in the anonymous relationships of the market. Because economic relations were so deeply embedded in social structure, variation in social organization was thought to explain the differences in the economies. Substantivists recognized that markets were found widely in traditional economies, but argued that those markets were peripheral to most economic activities, which were deeply embedded in social relationships (Bohannan

and Dalton, 1962). A compendium collected by Dalton (1967) provided empirical cases that illustrate the embedded nature of traditional economies.

In his critique of those using economic theory in non-Western contexts, Polanyi labelled them as 'formalist', meaning that they focused on 'formal' (mathematical) maximizing models to predict how individuals choose among alternative possibilities to allocate limited time, money and other resources. The substantivists, in contrast, focused on how economies were embedded within cultural institutions to meet the material desires that particular culture might have. The debate raged between the two factions through the 1960s and 1970s. Much of the argument became focused on how extensive markets were in traditional societies. In a classic cross-cultural study, Pryor (1977) showed that markets were very broadly distributed, sometimes moving primitive valuables, tools and food. They certainly did not originate with modern capitalism. In his famously acerbic article, Cook (1969) criticized substantivists for being romantic and naive; after all, even if they had useful points to make, the penetration of market economies, he argued, was so pervasive that formalist theories were *now* effectively universal.

Articles representing the two sides were collected in a reader by LeClair and Schneider (1968) that has been used to teach the debate ever since. Articulating the substantivist position, Sahlins (1972) then argued that many concepts of Western economic theory were inapplicable to traditional economy. He discussed the affluence of hunter-gatherers, underproduction in household economies, and the social determinants of reciprocal exchanges. Schneider (1974) countered with the fully articulated formalist position, summarizing how Western economics can be applied cautiously to a wide range of non-Western transactions and decisions, including marriage payments, primitive money, the prestige economy and household production. The debate came to focus on definitions of rationality, scarcity and institutional constraints, but those reading the papers increasingly saw that the participants were talking past each other.

The formalist and substantivist factions represented the inherent tension within anthropology: on the one hand, to seek cross-cultural regularities that reflect shared social process; on the other, to recognize the cultural relativity and uniqueness of each culture. The two sides of the debate fought to exhaustion, as both presented compelling approaches that could be seen as more complementary than alternative. In 1980, Schneider helped organize the Society for Economic Anthropology in order to resolve the debate by bringing the full spectrum of economic anthropologists together. The first meeting, published as *Economic Anthropology: Topics and Theories* (Ortiz, 1983) gathered an eclectic group of scholars to bridge the theoretical divides within the sub-discipline, with broad interests in marketing, institutions, Marxism, ecology, and economic development. An edited text,

Economic Anthropology (Plattner, 1989), provided a new generation of students with the breadth of economies and economic conditions that anthropologists were trying to make sense of.

Important to the new harmony has been respect for the different objectives of economic anthropologists, including ethnographic work on traditional economies, applied work on developing economies, and archaeological and historical studies of economies. The field has recognized diversity in both the theoretical and historical nature of human economies. To maintain a proper balance between substantivists and formalists (relativists and universalists) in economic anthropology, the role of archaeological and historical studies has been especially important. As ethnographers increasingly study variants of a single modern system, historical and archaeological studies continue to study the true variation in how human economies are organized and operate. Earle (2002), for example, looks at the alternative means by which political economies have emerged to finance the evolution of chiefdoms and states, showing that the development of market systems is quite rare and specific in that process. Although no careful comparative study exists, the extent of exchange in prehistory appears to have been highly variable.

During the 1980s and 1990s, as economic anthropology matured as a sub-discipline, it became marginalized within anthropology. As in many of the social sciences and humanities, postmodernism became popular, and its anti-materialist, anti-scientific critiques were antithetical to much of what the sub-discipline advocated. As the excesses of postmodernism have receded, however, economic anthropology has regained some of its former popularity, and its potential significance for anthropology and economics seems promising. Perhaps the greatest challenge now is that economics and economic anthropology have remained far apart because of the strongly formal (theoretical) basis of the former and empirical basis of the later. The two approaches would, however, seem complementary.

Economic anthropology and its perspective on world economies

Economists should consider the empirical value of economic anthropology, and a good place to begin is the compendium *Theory in Economic Anthropology* (Ensminger, 2002a). Economic anthropologists are committed to models of reality. The empirical observations and theoretical inferences of anthropology should help recognize the specific frames of applicability for grand theories. In essence, anthropology makes clear that all things are never equal. In this section, I summarize a few conclusions derived from economic anthropology that make a difference to studies of economies. These involve human rationality, consumer behaviour, commodity chains, and the multi-sectored quality of human economies. This list is not meant to be exhaustive, but only to illustrate the importance of cross-cultural evaluations for the models that economists develop. As economics begins to look at such concepts as behavioural economics and personalized networks, the relevance of anthropology's research on these topics becomes particularly significant.

Human decision-making is to a degree rational, and empirical anthropological work significantly improves an understanding of decision-making processes from a cross-cultural and evolutionary perspective. Although rationality underpins much economic theorizing, human cognitive abilities and goals have been under theorized. Recent trends to rectify this within behavioural economics emphasize that individuals do not always act rationally with primary economic objectives and it would appear that economic anthropology could provide valuable cross-cultural validation of these new ideas. Humans prove to be fairly poor decision makers; they appear rather to use simplified proximate measures to estimate such considerations as value and cost (Henrich, 2002). Anthropologists have experimented with various economic games given under controlled conditions in non-Western societies, and their results are often counter-intuitive (Ensminger, 2002b). In a sample of societies representing different levels of economic development, for example, as market integration increases cooperation can be shown in such game-playing experiments to become more highly prized.

To understand the evolutionary roots of human rationality, anthropological research has looked at decision-making in small-scale hunting and gathering societies (see for example, Cashdan, 1989). As seen by the rapid expansion in brain size deep in history, humans must have been under strong selective pressure for expanded cognitive abilities, and this selective pressure took place when humans were low-density hunter-gatherers. Such hunter-gatherers make daily a wide range of decisions about what foods to eat, where to camp, what groups to join, and the like, and the relative scarcity and abundance of food and their different nutritional qualities appear to be considered. Human cognitive skill determines the ability of hunter-gatherers to adjust rapidly to changing conditions of food availability, to occupy diverse habitats from the Arctic to the tropical forests, and to intensify food procurement as required by population growth. In short, cognitive abilities in the food quest, in movement through the landscape, and in deciding which groups to join must have provided a strong selective advantage that resulted in the moulding of human rationality.

As illustrated by economic anthropology, human decisions often have little direct relationship to economic factors of cost and financial gain. Although of more interest recently to economists, with the notable exception of Thorstein Veblen, economic theory has not attempted systematically to explain how potential consumer outcomes are ranked. Rather, within the West, consumer behaviour has been studied with a rather eclectic and under-theorized set of assumptions. Anthropologists, however, have tried to understand consumption cross-culturally as a social process involving issues of identity and association (Rutz and Orlove, 1989). From the anthropological literature, we know how valued objects signify social relationships. The giving and receiving of gifts impart form and meaning to social relationships, and materialize the social distance between actors (Sahlins, 1972).

Economic anthropologists frequently study the movements of objects around the globe. These commodity chains describe how goods are produced, distributed and transformed as they move through a sequence of markets (Hansen, 2002; Obukhova and Guyer, 2002). Commodity chains illustrate how goods, like used clothing, are transformed in value, form and meaning as they pass through a sequence of social worlds and economic sectors. Social considerations of prestige and personal worth are always of great concern in this highly creative process of economic decision-making.

Economic anthropologists have emphasized that economies are multilayered and that the specific character of an economy has historical routes. Although economists often refer to 'dual economy', implying a vestigial survival of traditional practices, they have been reluctant to accept that economies are always multilayered mosaics with spheres of exchange that only partially articulate the different sectors. Economic theory thus radically simplifies reality by focusing on decision-making and outcomes under market conditions, and this simplification makes very different economies appear superficial similar. In the emergence and development of capitalism, since wealth was made in the markets, the primary concern of economists became directed there. As anthropologists seek to understand the different motives and dynamics of economies as articulated in specific social contexts, they have, however, realized that human economies are highly variable, combining subsistence, social, political, and market sectors, each with distinct logics and historical traditions.

The subsistence sector is family-based and involves the daily struggles to meet basic needs. It is universal and represents the economic world of survival in which humans evolved as a species. The primary motivation of humans has probably always been the satisfaction of a family's basic needs. The construction of a general theory of human economies should thus start with how households and communities make a living. Until recently, household requirements were handled largely by family production. Although markets have a long history in human societies, they were typically quite marginal to subsistence needs. Theorized as the domestic mode of production (DMP; Sahlins, 1972), households were oriented to meeting their subsistence needs, and distribution involved sharing between family members with different tasks appropriate to an elementary division of labour by age and gender. In the model, the household is economically self-sufficient, and the economy is not inherently growth-oriented. The amazing conclusion of considerable anthropological research is that the DMP is often at least the model of what the economy should be, and the amounts of goods consumed by households that are produced outside the family have often been but a fraction of the households' overall consumption budget. Prior to the development of full-scale markets, households probably produced 75 per cent or more of everything that they consumed.

The social sector is community-based and involves the lifetime strategies of individuals to define identity and relationships within a broader social group. The social sector is probably universal, finding its roots among early hunter-gatherers and their need to form networks of support, cooperation and exclusion. In cross-cultural perspective, much of the social sector involves reciprocal exchanges within highly social worlds that can be manipulated to emphasize personal prestige. In traditional societies, such competitive exchanges commonly produce social ranking in what has been called a 'prestige economy'. The social sector was elaborated following the Neolithic revolution, as the creation of local corporate groups must have placed a premium on group identity and status. With deep and enduring roots in human history, the social sector would seem to provide a cross-cultural understanding of consumer behaviour as part of processes much broader than capitalism.

Economics now questions assumptions about anonymous markets organized independently of other social institutions. Goods and services are seen as flowing through personalized networks that create the institutions for expanding economic transactions. Greif (2006), for example, argues that the social networks of medieval Europe provided the frameworks for an emergent modern economy. Almost self-evident to anthropologists, such conclusions suggest how economic theory can gain from insights from comparative empirical studies of non-industrial political forms.

Political sectors mobilize and allocate goods to finance regional and interregional institutions of domination and stratification (Earle, 2002). Importantly, political economies are not universal. From the fourth millennium bc, the political sector of the economy developed along with chiefdoms and then states. Goods became mobilized as a tax or tribute and then 'redistributed' by dominant political organizations as means to finance their activities. Recent archaeology has studied how political sectors were developed and functioned. An inherent contrast is between staple finance and wealth finance. In staple finance, food goods are mobilized and stored centrally as a means to support craftsmen, warriors and labourers working for the state. Many of these systems, especially in chiefdoms but also some states, functioned with few or no markets. Subsistence and social sectors continued largely unchanged, but new patterns of land ownership and domination required the production of a surplus for ruling institutions. Wealth finance worked similarly, but the local surplus was used to support the production of wealth for tribute payments.

And what about the market sector, so fundamental to most economic theorizing? Archaeological evidence documents that exchange and markets were not universal. From case to case, the amount and types of goods exchanged varied greatly according to specific conditions of availability and production costs and to specific objects of value. Based on ethnographic analogies, until quite recently most of the goods traded were probably handled by down-the-line exchanges between social partners. Goods moving any distance were primarily primitive valuables, items of display and tribute. The extent of exchange in Neolithic and later Bronze Age communities, for example, has been discussed for Europe, where the comparative advantage of one region over another would have been based on the availability of special materials (Sherratt, 1997). Subsistence and technological items were rarely exchanged over long distances until the end of the medieval age. Earlier, some market exchange certainly existed, but their extent and elaboration were apparently quite small.

This empirical record from economic anthropology contests economic theories based on asserted long-term trends in the emergence of marketing. A common assumption among economists from Adam Smith onwards has been that the creation of wealth is an outcome of the development of efficiencies associated with specialization and trade. For example, in his analysis of institutional economics, North (1990) argues that states developed to lower transaction costs between locally specialized but political independent regions. To simplify the logic, technological development and specialization should have created increasing productive economies that, with the emergence of integrating political systems to guarantee the peace of the market, would generate the surplus used to support the growth of civilizations.

The development of markets, however, was quite late and episodic. Following North, economics might suggest that such failure of markets to develop was an outcome of high transaction costs that made exchange unprofitable. Empirically such a conclusion, however, can be shown to be wrong. As political superstructures were developed and imposed broad regional peace that would have radically lowered transactions costs, markets surprisingly did not emerge. The reason appears to be linked to the nature of finance. When finance was based on staples, markets were only rudimentary and peripheral. The complex Hawaiian chiefdoms, for example, conquered and integrated several islands with local specialties in food, stone and other materials, but trade remained very small-scale and local despite the regional peace. Archaeology has documented only minor trade in basaltic adzes and obsidian in Hawaiian prehistory, and these exchanges did not increase with the formation of the large-scale chiefdoms. As a dramatic example, the Inka empire conquered a massive territory that extended 3,000 km up the spine of the Andes, imposed an effective regional peace across that territory, and constructed nearly 30,000 km of roads to integrate it. Although these actions would certainly have lowered transaction costs, the regional and distant movements of goods, like metal, ceramics and foods, remained very limited and completely unchanged from the pre-imperial period (Earle, 2002).

Both markets and currencies seem to have expanded in other circumstances where they were linked with wealth finance of states. In the Aztec empire, tribute to the state was in wealth objects like textiles that could be easily transported long distances, centrally stored, and then used as payment to those working for the state. But the use of wealth objects in payment required that the objects be convertible into the staple goods and other consumables desired by state personnel. The Aztec market system provided the mechanism for conversion and was apparently developed by the state (Brumfiel, 1980). Afterwards, markets appear to have escaped from state sponsorship and control to take on many of the characterizations commonly associated with market systems.

What are the possibilities for a grand theory of economies? The relatively low status of historical and comparative studies within economics is not promising, but economics would do well to test theories claimed for generally applicability by looking closely at the anthropological literature. To the degree that economic models are used to design economic development in non-Western societies, the general relevance of the economic models must demonstrated. Using a uniform method of analysis, the economist Pryor (2005) has compared industrial economies and traditional (hunter-gatherer and agricultural) economies. His primary conclusions are startling, suggesting the advantages of such comparative analyses. All economies appear to consist of a small number of component parts, probably reflecting the processes and constraints involved in the production and movement of material goods. Economies are thus comparable. Furthermore, the factors that affect such variables as gross productivity or volume of exchange appear not to be determined by social structure but by the particular internal characteristics of the economy. Thus, Polanyi would appear to be wrong; economies are rather independent engines of essential processes. As recent work in economics has relaxed simplifying assumptions about information, frictionless trade and anonymity of markets, the potential links between economics and economic anthropology take on reciprocal value.

TIMOTHY EARLE

See also **behavioural economics and game theory; hunting and gathering economies; 'political economy'; property rights; stratification.**

Bibliography

Bohannan, P. and Dalton, G., eds. 1962. *Markets in Africa*. Evanston, IL: Northwestern University Press.

Brumfiel, E. 1980. Specialization, market exchange, and the Aztec state: a view from Huexotla. *Current Anthropology* 21, 459–78.

Cashdan, E. 1989. Hunters and gatherers: economic behavior in bands. In *Economic Anthropology*, ed. S. Plattner. Stanford: Stanford University Press.

Cook, S. 1969. The 'anti-market' mentality: a critique of the substantive approach to economic anthropology. *Southwestern Journal of Anthropology* 25, 378–406.

Dalton, G., ed. 1967. *Tribal and Peasant Economies*. Garden City, NY: Natural History Press.

Earle, T. 2002. *Bronze Age Economics*. Boulder, CO: Westview.

Ensminger, J., ed. 2002a. *Theory in Economic Anthropology*. Walnut Creek, CA: AltaMira.

Ensminger, J. 2002b. Experimental economics: a powerful new method for theory testing in anthropology. In Ensminger (2002a).

Firth, R. 1939. *Primitive Polynesian Economy*. London: Routledge & Kegan Paul, 1965.

Greif, A. 2006. *Institutions and the Path to the Modern Economy: Lessons from Medieval Trade*. Cambridge: Cambridge University Press.

Gudeman, S. 1998. *Economic Anthropology*. Cheltenham: Edward Elgar.

Hansen, K.T. 2002. Commodity chains and the international secondhand clothing trade: *Salaula* and the work of consumption in Zambia. In Ensminger (2002a).

Henrich, J. 2002. Decision-making, cultural transmission and adaptation in economic anthropology. In Ensminger (2002a).

Herskovits, M. 1940. *Economic Anthropology*. New York: Knopf, 1952.

Hunt, R. 1997. Economic anthropology. In *The Dictionary of Anthropology*, ed. T. Barfield. Oxford: Blackwell.

Johnson, A. and Earle, T. 2000. *The Evolution of Human Societies*, 2nd edn. Stanford: Stanford University Press.

LeClair, E.E. and Schneider, H.K., eds. 1968. *Economic Anthropology*. New York: Holt, Rinehart.

Malinowski, B. 1922. *Argonauts of the Western Pacific*. London: Routledge.

Mauss, M. 1925. *The Gift: Forms and Functions of Exchange in Archaic Societies*. London: Cohen and West, 1969.

North, D. 1990. *Institutions, Institutional Change, and Economic Performance*. Cambridge: Cambridge University Press.

Obukhova, E. and Guyer, J.I. 2002. Transcending the formal/informal distinction: commercial relations in Africa and Russia in the post-1989 world. In Ensminger (2002a).

Ortiz, S., ed. 1983. *Economic Anthropology: Topics and Theories*. Lanham, MD: University Press of America.

Plattner, S., ed. 1989. *Economic Anthropology*. Stanford: Stanford University Press.

Polanyi, K. 1944. *The Great Transformation*. New York: Rinehart.

Polanyi, K., Arensberg, C. and Pearson, H., eds. 1957. *Trade and Market in the Early Empires*. New York: Free Press.

Pryor, F.L. 1977. *The Origins of the Economy*. New York: Academic Press.

Pryor, F.L. 2005. *Economic Systems of Foraging, Agricultural, and Industrial Societies*. Cambridge: Cambridge University Press.

Rutz, H., Orlove, B., eds. 1989. *The Social Economy of Consumption*. Lanham, MD: University Press of America.

Sahlins, M. 1972. *Stone Age Economics*. Chicago: Aldine.

Schneider, H.K. 1974. *Economic Man: The Anthropology of Economics*. New York: Free Press.

Sherratt, A. 1997. *Economy and Society in Prehistoric Europe*. Edinburgh: Edinburgh University Press.

economic calculation in socialist countries

Economic calculation and political decisions

An important result of the archival revolution of the 1990s (that is, the access to former Soviet archives made possible by the collapse of the USSR) was the additional knowledge it provided about economic decision-making in the USSR in the Stalin period. This made it clear that in the 1930s, when the socialist economic system emerged, economic decisions were based not on detailed and precise economic methods of calculation but on rough and ready political methods. Interesting light has been thrown on the significance of this for macroeconomic, mesoeconomic and microeconomic decision making.

Macroeconomic policy in the Stalin era aimed to maximize investment subject to the need to provide sufficient consumer goods (mainly food) to maintain labour productivity. The consumer goods were obtained from agriculture by force and allocated by the state in a way which it was hoped would enable investment to be maximized. A schematic representation of short-term macroeconomic calculation under these circumstances is set out in Figure 1.

Figure 1 shows an output curve OQ which depends on the effort the workers provide, and an effort curve E_oE_{max} which depends on the real wage and the level of coercion. If the state chooses too low a level of wages, output will decline and the intended investment level will be impossible to meet. If wages are set at the fair wage level, output will be maximized but investment less than desired. At the wage level W^*, investment will be maximized. Hence, macroeconomic calculation involved gathering information about worker attitudes (via the state security organizations), allocating the available food to crucial groups of workers, and using coercive or ideological methods to reduce the food–output ratio.

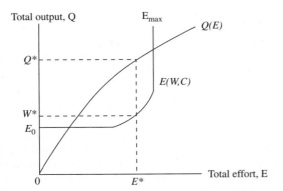

Figure 1   Maximizing the investible surplus. *Source*: adapted from Gregory and Harrison (2005, p. 732).

Mesopolicy aimed at developing heavy industry and the defence sector. An important result was what has been termed the 'structural militarization' of the Soviet economy. This resulted from the Soviet view of international relations, the stress on mobilization planning, the lessons of 1941, and the use by the general staff of absurdly inflated estimates of the mobilization capacity of the USA and other countries. An example is the USSR's capacity at the end of the 1980s to produce about four million tons of aluminium annually. This was greatly in excess of the peacetime economy's need for aluminium. However, in the event of mobilization it would have enabled the country to produce huge numbers of military airplanes. This situation arose as a result of using as a method of economic calculation the attainment of Western levels and of these levels in the military sphere being systematically exaggerated.

On the microeconomic level, Lazarev and Gregory (2003) have studied the allocation of motor vehicles (cars/autos and lorries/trucks) from the central reserve fund in 1932 and 1933. This showed that an economic planning model was unable to explain their allocation (in the regressions the economic variables were insignificant and frequently had the wrong signs). But a political model, in which their allocation was explained as part of a gift-exchange process, explained the data quite well.

Incrementalism

A basic method of economic calculation used in the state socialist countries – particularly in the post-Stalin period – was that of incrementalism, or, as it was known in the USSR, 'planning from the achieved level'. The starting point of all economic plans was the actual or expected outcome of the previous period. The planners adjusted this by reference to anticipated growth rates, current economic policy, shortages and technical progress. For nearly all products, the planned output for next year was the anticipated output for this year plus a few per cent added on. The advantages of incrementalism as a method

of economic calculation were its simplicity, realism and compatibility with the functioning of a hierarchical bureaucracy. Its disadvantages were that it provided no method for making technically efficient or consistent decisions, nor did it ensure that the population derived maximum satisfaction from the resources available.

Planning and counter-planning

A widely used method of economic calculation was that of planning and counter-planning. If the plan were simply handed down to the enterprises from above, in accordance with the planners' view of national economic requirements but in ignorance of the real possibilities of each enterprise, then it would be unfeasible (if it was too high) or wasteful (if it was too low) or both at the same time (that is, unfeasible for some products and wasteful for others). Conversely, if plans were simply drawn up by each enterprise, they might have failed to use resources in accordance with national economic requirements. The process of planning and counter-planning involved a mutual submission and discussion of planning suggestions, designed to lead to the adoption of a plan which was feasible for the enterprise and ensured that the resources of each enterprise were used in accordance with national requirements.

Unfortunately, the bureaucratic complexity of this procedure militated against both efficiency and consistency.

Input norms

The main method of economic calculation used to ensure efficiency was that of input norms. An input norm is simply a number assumed to describe an efficient process of transformation of inputs into outputs. For example, suppose that the norm for the utilization of coal in the production of one ton of steel is x tons. Then the efficient production of z tons of steel is assumed to require zx tons of coal.

The method of norms was widely used in Soviet planning, and considerable effort was devoted to updating them. Very detailed norm fixing took place for expenditures of fuel and energy. Much attention was devoted to the development of norms for the expenditure of metal, cement, and timber in construction. All this work was directed by the department of norms and normatives of Gosplan (the State Planning Commission). Responsibility for elaborating and improving the norms lay with Gosplan's Scientific Research Institute of Planning and Norms.

Nevertheless, the method of norms was incapable of ensuring efficiency. The norms used in planning calculations were simply averages of input requirements, weighted somewhat in favour of efficient producers. Actual technologies showed a wide dispersion in input–output relations. Furthermore, given norms took no account of the possibilities of substitution of inputs for one another in the production process, non-constant returns to scale, and the results of technical progress. Thus in general, the method of norms did not make it possible to calculate efficient input requirements, and plans calculated in this way were always inefficient.

The method of norms was used not only in inter-industry planning but also in consumption planning. In calculating the volume of particular consumer goods and services required, the planners used two main methods. One was forecasts of consumer behaviour, based on extrapolation, expenditure patterns of higher-income groups, income and price elasticities of demand, and consumer behaviour in the more advanced countries. The other method was that of consumption norms. The former method attempted to foresee consumer demand, the latter to shape it.

An example of the method of norms, and its policy implications, is set out in Table 1.

Table 1 makes clear the logic of the Soviet policy in the Brezhnev era (1964–82) of expanding the livestock sector, and also importing fodder and livestock products. Since the consumption of livestock products was below the norm level, the government sought to make possible an increase in their consumption.

The method of consumption norms was an alternative to the price mechanism for the determination of output. It has also been used, however, in Western countries. It is used there in those cases where distribution on the basis of purchasing power has been replaced by distribution on the basis of need. Examples include the provision of housing, hospitals, schools and parks. Calculations of the desirable number of rooms, hospital beds and school places per person are a familiar tool of planning in welfare states.

There are two main problems with the norm method of consumption planning. The first is that of substitution between products. Although consumers may well have a medically necessary need for x grams of protein per day,

they can obtain these proteins from a wide variety of foods. Second, consumers may choose to spend their money 'irrationally', for example, to buy spirits instead of children's shoes.

Material balances

A material balance is a balance sheet for a particular commodity showing, on the one hand, the economy's resources and potential output, and, on the other, the economy's need for a particular product. Material (and labour) balances were the main methods used in calculating production and distribution plans for goods, supply plans and labour plans. Soviet planners took great pride in the balance method and considered it one of the greatest achievements of planning theory and practice. Material balances were drawn up for different periods (for example, for annual or five year periods), by different organizations (for example, Gosplan, Gossnab – the body responsible for allocating supplies of inputs – and the ministries) and at different levels (for example, national and republican). The material balances were also drawn up with different degrees of aggregation. Highly aggregated balances were drawn up for the Five Year Plans, and highly disaggregated balances by the chief administrations of Gossnab for annual supply planning. The aim of the material balance method was to ensure the consistency of the plans.

Normally, at the start of the planning work, the anticipated availability of a commodity was not sufficient to meet anticipated requirements. To balance the two, the planners sought possibilities of economizing on scarce products and substituting for scarce materials; they investigated the possibilities of increasing production or importing raw materials or equipment, or in the last resort they determined the priority needs to be fulfilled by the scarce commodity. Even with great efforts, achieving a balance was difficult. The complexity of an economy in which a great variety of goods are produced by different processes, all of which are subject to continuous technological change, was often too great for anything more than a balance that balanced only on paper. Hence it was normal, during the 'planned' period, for the plan to be altered, often repeatedly, as imbalances came to light. Particularly important problems with the use of material balances were the highly aggregated nature of the balances and their interrelated nature.

Input–output

A wide variety of input–output tables were regularly constructed in socialist countries. *Ex post* national tables in value terms, planning national tables in value and physical terms, regional tables, and capital stock matrices were widely constructed and used. An interesting and important use concerned variant calculations of the structure of production in medium-term planning.

Table 1   *The Soviet diet*

	Norm (kgs/head/year)	Per capita consumption in 1976 as % of norm
Bread and bread products	120	128
Potatoes	97	123
Vegetables and melons	37	59
Vegetable oil and margarine	7	85
Meat and meat products	82	68
Fish and fish products	18	101
Milk and milk products	434	78
Eggs	17	72

Sources: Weitzman (1974); Agababyan and Yakovleva (1979, p. 142).

Because an input–output table can be represented by a simple mathematical model, and because of the assumption of constant coefficients, an input–output table can be utilized for variant calculations.

$$X = (I - A)^{-1} Y$$

On the assumption that A is given, X can be calculated for varying values of Y. Variant calculations of the structure of production were not undertaken with material balances because of their great labour intensity. Variant calculations played a useful role in medium-term planning because they enabled the planners to experiment with a wide range of possibilities. The first major use of variant calculations of the structure of production in Soviet national economic planning was in connection with the 1966–70 Five Year Plan. Gosplan's economic research institute analysed the results of various possible shares of investment in the national income for 1966–70. It became clear that stepping up the share of investment in the national income would increase the rate of growth of the national income, but that this would have very little effect on the rate of growth of consumption (because almost all of the increased output would be producer goods). The results of the calculations are set out in Tables 2 and 3.

The five variants are for the share of investment in the national income, I being the lowest and V the highest. A sharp increase in the share of investment in the national income in the Five Year Plan 1966–70 would have led to a sharp fall in the share of consumption in the national income, and only a small increase in the rate of growth of consumption (within a Five Year Plan period). What is

Table 2   *Output of steel on various assumptions*

	Variants				
	I	II	III	IV	V
Production of steel in 1970 (millions of tonnes)	109	115	121	128	136

Table 3   *Average annual growth rates of selected industries, 1966–1970 (%)*

	Variants				
	I	II	III	IV	V
Engineering and metal working	7.1	8.2	9.3	10.4	11.4
Light industry	6.3	6.6	6.8	7.0	7.2
Food industry	7.1	7.3	7.4	7.5	7.6

Source: Ellman (1973, p. 71).

very sensitive to the share of investment in the national income is the output of the producer goods industries, as Tables 2 and 3 show.

These results are along the lines of what one would expect on the basis of Fel'dman's model, but the input–output technique improves on Fel'dman's model since it enables the effect of different strategies to be seen at the industry level rather than merely in terms of macroeconomic aggregates.

Another example of the use of input–output for economic calculations concerns the statistical data about the relations between industries contained in the national *ex post* tables in value terms. In his controversial 1968 book *Mezhotraslevye svyazi sel'skogo khozyaistva*, M. Lemeshev, then deputy head of the sector for forecasting the development of agriculture of the USSR Gosplan's Economic Research Institute, used the Soviet input–output table for 1959 as the basis for a powerful plea for more industrial inputs to be made available to agriculture.

He began by observing that from the 1959 input–output table it was clear that of the current material inputs into agriculture in that year only 23.4 per cent came from industry, while 54.7 per cent came from agriculture itself (feed, seed and so on). He argued that this was most unsatisfactory. In the section on the relationship between agriculture and engineering Lemeshev argued that the supply to agriculture of agricultural machinery was inadequate, in the section on the relationship between agriculture and the chemical industry he argued that the supply of fertilizers was inadequate, and in the section on agriculture and electricity he argued that the supply of electricity to the villages for both productive and unproductive needs was inadequate. In addition, in the section on the relationship between agriculture and the processing industry he argued that the latter was not helping agriculture as it should do; for example, it was sometimes impossible to accept vegetables (although the consumption of these in the towns was below the norms) because of inadequate processing and distribution facilities. Furthermore, he argued that the supply of concentrated feed was inadequate and the processing of milk wasteful. In view of the inadequate development of the food processing industry, he argued for the development of processing enterprises by the farms themselves.

The chapter on the productive relations between agriculture and the building industry was an extensive critique of the practice of productive, and of housing and communal, building in the villages. Lemeshev argued that the state should take on responsibility for building on the collective farms. The chapter on the relationship between agriculture and transport was critical of the shortage of river freight boats. The chapter on investment argued that investment in agriculture was inadequate, and that in the period 1959–65 there was an unwarranted increase in the proportion of investment in the collective farms which they had to finance themselves. He also argued that a greater proportion of agricultural investment should be financed by bank loans, and that as a criterion of investment efficiency the recoupment period was satisfactory. The concluding chapter was concerned with improving the productive relations between agriculture and the rest of the economy. The author argued for improving central planning by the use of input–output, for replacing procurement plans by free contracts between farms and the procurement organs (if a shortage of a particular product threatened then its price could be raised), and for the elimination of the supply system (that is, the rationing of producer goods) which hindered farms from receiving the goods they wanted and sometimes supplied them with goods that they did not want. Lemeshev also argued for higher pay in agriculture and for the reorganization of the labour process within state and collective farms on the basis of small groups which were paid by results.

This book was a good example of the use of input–output to provide statistical data which could be used, alongside other information, to provide a description of important economic relations and to support a case for important institutional and policy changes.

Project evaluation

In the USSR of the 1930s, it was officially considered that there was no problem of project evaluation to which economists could contribute. The sectoral allocation of investment was a matter for the central political leadership to decide. It was they who decided in which sectors and at which locations production should be expanded. These decisions were based on the experience of the more advanced countries, the traditions of the Russian state (for example, stress on railway building) and of the Bolshevik movement (for example, stress on electrification and on the metal-using industries) and on the needs of defence. As far as decisions within sectors were concerned, here the main idea was to fulfil the plan by using the world's most advanced technology.

The practical study of methods for choosing between variants within sectors was begun by engineers in the electricity and railway industries. The problem analysed was that of comparing the cost of alternative ways of meeting particular plan targets. A classic example of the type of problem considered was the choice between producing electricity by a hydro station and by a thermal station.

During Stalin's lifetime, the elaboration by orthodox economists and the adoption by the planners of economic criteria for project evaluation were impossible because they were outside Stalin's conception of the proper role of economists (apologetics). When economists did make a contribution in this area, as was done by Novozhilov, it was ignored. After Stalin's death, however, it became possible for Soviet economists to contribute to the elaboration of methods of economic calculation for use in the decision-making process. An

early and important example was in the field of project evaluation. An official method for project evaluation was adopted in 1960, and revised versions in 1964, 1966, 1969 and 1981. In a very abbreviated and summary form, the 1981 version was as follows.

In evaluating investment projects, a wide variety of factors have to be taken into account, for example, the effect of the investment on labour productivity, capital productivity, consumption of current material inputs (such as metals and fuel), costs of production, environmental effects, technical progress, the location of economic activity and so on. Two indices which give useful synthetic information about economic efficiency (but are not necessarily decisive in choosing between investment projects) are the coefficient of absolute economic effectiveness and the coefficient of relative economic effectiveness.

At the national level, the coefficient of absolute effectiveness is defined as the incremental output–capital ratio.

$$E_p = \frac{\Delta Y}{I}$$

where E_p is the coefficient of absolute effectiveness for a particular project, ΔY is the increase in national income generated by the project, and I is the investment cost. The value of E_p calculated in this way for a particular investment has to be compared with E_a, the normative coefficient of absolute effectiveness, which is fixed for each Five Year Plan and varies between sectors. In the 11th Five Year Plan (1981–85) it was 0.16 in industry, 0.07 in agriculture, 0.05 in transport and communications, 0.22 in construction and 0.25 in trade.

If $E_p > E_a$

then the project is considered efficient.

For calculating the criterion of absolute effectiveness at the level of individual industries, net output is used in the numerator instead of national income. At the level of individual enterprises and associations, in particular when a firm's own money or bank loans are the source of finance, profit is used instead of national income.

The coefficient of relative effectiveness is used in the comparison of alternative ways of producing particular products. In the two products case

$$E = \frac{C_1 - C_2}{K_2 - K_1}$$

where E is the coefficient of relative effectiveness, C_i is the current cost of the ith variant, and K_i is the capital cost of the ith variant.

If $E > En$, where En is the officially established normative coefficient of relative economic efficiency, then the more capital intensive variant is economically justified. In the 11th Five Year Plan, En was in general

0.12, but exceptions were officially permitted in the range 0.08/0.10–0.20/0.25.

In the more than two variants case, they should be compared according to the formula

$$C_i + E_n K_i \rightarrow \text{minimum}$$

that is, choose that variant which minimizes the sum of current and capital costs.

At one time a rationalist misinterpretation of socialist planning was widespread. According to this view, a planned economy was one in which rational decisions were made after a dispassionate analysis by omniscient and all-powerful planners of all the alternative possibilities. In such a system, the adoption of rational criteria for project evaluation would have been of enormous importance. Socialist planning, however, was just one part of the social relations between individuals and groups in the course of which decisions were taken, all of which were imperfect and many of which produced results quite at variance with the intentions of the top economic and political leadership.

A good example of the factors actually influencing investment decisions under state socialism was the commencement of the construction of the Baoshan steel plant near Shanghai. The site was apparently chosen because of the political influence of a high-ranking Shanghai party official. The location decision ignored the fact that, because of the swampy nature of the site, necessitating large expenditures on the foundations, this was in fact the most expensive of the sites considered. Very expensive, dogged with cost overruns, involving major pollution problems, the whole project was kept alive for some time by a powerful steel lobby. In due course, as a result of a national policy reversal in Beijing, the second phase was deferred and those involved publicly criticized. To judge from its initial costs of production, it produced gold rather than steel.

In general, the choice of projects owed more to inter-organization bargaining in an environment characterized by investment hunger than it did to the detached choice of a cost-minimizing variant. The development of new and better criteria for project evaluation turned out to be no guarantee that project evaluation would improve since the criteria were often not in fact used to evaluate projects. Their main function was to provide an acceptable common language in which various bureaucratic agencies conducted their struggles. Agencies adopted projects on normal bureaucratic grounds and then tried to get them adopted by higher agencies, or defended them against attack, by presenting efficiency calculations using the official methodology but relying on carefully selected data.

Linear programming and extensions

Linear programming was discovered by the Soviet mathematician Kantorovich in the late 1930s. Its relevance for Soviet planning was widely discussed in the USSR in the 1960s and extensive efforts were made actually to use it in Soviet planning in the 1970s. Three examples of its use follow.

Production scheduling in the steel industry

Linear programming was discovered by Kantorovich in the course of solving the problem, presented to him by the Laboratory of the all-Union Plywood Trust, of allocating productive tasks between machines in such a way as to maximize output given the assortment plan. From a mathematical point of view, the problem of optimal production scheduling for tube mills and rolling mills in the steel industry, which was tackled by Kantorovich in the 1960s, is very similar to the Plywood Trust problem, the difference being its huge dimensions.

The problem arose in the following way. As part of the planning of supply, Soyuzglavmetal (the department of Gossnab concerned with the metal industries), after the quotas had been specified, had to work out production schedules and attachment plans in such a way that all the orders were satisfied and none of the producers received an impossible plan. In the 1960s an extensive research programme was initiated by the department of mathematical economics (which was headed by Academician Kantorovich) of the Institute of Mathematics of the Siberian branch of the Academy of Sciences, to apply optimizing methods to this problem. The chief difficulties were the huge dimensions of the problem and the lack of the necessary data. About 1,000,000 orders, involving 60,000 users, more than 500 producers and tens of thousands of products, were issued each year for rolled metal. Formulated as a linear programming problem it had more than a million unknowns and 30,000 constraints. Collecting the necessary data took about six years. Optimal production scheduling was first applied to the tube mills producing tubes for gas pipelines (these were a scarce commodity in the USSR). In 1970 this made possible an output of tubes 108,000 tons greater than it would otherwise have been, and a substantial reduction in transport costs was also achieved.

The introduction of optimal production scheduling into the work of Soyuzglavmetal was only part of the work initiated in the late 1960s on creating a management information and control system in the steel industry. This was intended to be an integrated computer system which would embrace the determination of requirements, production scheduling, stock control, the distribution of output and accounting. Such systems were widely introduced in Western steel firms in the late 1960s. Work on the introduction of management information and control systems in the Soviet economy was widespread in the 1970s, but by the 1980s there was widespread scepticism in the USSR about their usefulness. This largely resulted from the failure to fulfil the earlier exaggerated hopes about the returns to be obtained from their introduction in the economy.

Industry investment plans

In the state socialist countries investment plans were worked out for the country as a whole, and also for industries, ministries, departments, associations, enterprises, republics, economic regions and cities. An important level of investment planning was the industry. Industry investment planning is concerned with such problems as the choice of products, of plants to be expanded, location of new plants, technology to be used, and sources of raw materials.

The main method used in the 1970s and 1980s in the Council for Mutual Economic Assistance (CMEA, known in the West as Comecon) countries for processing the data relating to possible investment plans into actual investment plans was mathematical programming. After extensive experience in this field, in 1977 a Standard Methodology for doing such calculations was adopted by the Presidium of the USSR Academy of Sciences.

The Soviet Standard Methodology presented models for three standard problems. They were: a static multi-product production problem with discrete variables, a multi-product dynamic production problem with discrete variables, and a multi-product static problem of the production-transport type with discrete variables. The former can be set out as follows:

Let $i = 1, \ldots, n$ be the finished goods or resources, $j = 1, \ldots, m$ be the production units, $r = 1, \ldots, R_j$ be the production technique in a unit, a_{ij}^r be the output of good $I = 1, \ldots, n'$ or input of resource $i = n' + 1, \ldots, n$, using technique r of production in unit j; C_j^r are the costs of production using technique r in unit j; D_i is the given level of output of good i, $I = 1, \ldots, n'$; P_i is the total use of resource i, $i = n' + 1, \ldots, n$ allocated to the industry; Z_j^r is the unknown intensity of use of technique r at unit j.

The problem is to find values of the variables Z_j^r that minimize the objective function

$$\sum_{j=1}^{m} \sum_{j=1}^{R_j} C_j^r Z_j^r \tag{1}$$

that is, minimize costs of production subject to

$$\sum_{j=1}^{m} \sum_{r=1}^{R_j} a_{ij}^r Z_j^r \geq D_i, \qquad i = 1, \ldots, n' \tag{2}$$

that is, each output must be produced in at least the required quantities

$$\sum_{j=1}^{m} \sum_{r=1}^{R_j} a_{ij}^r Z_j^r \leq P_i, \qquad i = n' + 1, \ldots, n \tag{3}$$

that is, the total use of resources cannot exceed the level allocated to the branch

$$\sum_{r=1}^{R_j} Z_j^r \leq 1, \qquad j = 1, \ldots, m \tag{4}$$

$$Z_j^r = 0 \text{ or } 1, \quad j = 1, \ldots, m, \quad r = 1, \ldots, R_j \tag{5}$$

that is, either a single technique of production for unit j is included in the plan or unit j is not included in the plan.

In order to illustrate the method, an example will be given which is taken from the Hungarian experience of the 1950s in working out an investment plan for the cotton weaving industry for the 1961–65 Five Year Plan. The method of working out the plan can be presented schematically by looking at the decision problems, the constraints, the objective function and the results.

The decision problems to be resolved were:

(a) How should the output of fabrics be increased, by modernizing the existing weaving mills or by building new ones?
(b) For part of the existing machinery, there were three possibilities. It could be operated in its existing form, modernized by way of alterations or supplementary investments, or else scrapped. Which should be chosen?
(c) For the other part of the existing machinery, it could be either retained or scrapped. What should be done?
(d) If new machines are purchased, a choice has to be made between many types. Which types should be chosen, and how many of a particular type should be purchased?

The constraints consisted of the output plan for cloth, the investment fund, the hard currency quota, the building quota and the material balances for various kinds of yarn. The objective function was to meet the given plan at minimum cost.

The results provided answers to all the decision problems. An important feature of the results was the conclusion that it was cheaper to increase production by modernizing and expanding existing mills than by building new ones.

It would clearly be unsatisfactory to optimize the investment plan of each industry taken in isolation. If the calculations show that it is possible to reduce the inputs into a particular industry below those originally envisaged, then it is desirable to reduce planned outputs in other industries, or increase the planned output of the industry in question, or adopt some combination of these strategies. Accordingly, the experiments in working out optimal industry investment plans, begun in Hungary in the 1950s, led to the construction of multi-level plans linking the optimal plans of the separate industries to each other and to the macroeconomic plan variables. Multi-level planning of this type was first developed in Hungary, but subsequently spread to the other CMEA countries. Extensive work on the multi-level optimization of investment planning was undertaken in the USSR in connection with the 1976–90 long-term plan. (The 1976–90 plan, like all previous Soviet attempts to compile a long-term plan, was soon overtaken by events. The plan itself seems never to have been finished and was replaced by ten-year guidelines for 1981–90.)

The determination of costs in the resource sector
In view of the wide dispersion of production costs in the resource sector, the use of average costs (and of prices based on average costs) in allocation decisions is likely to lead to serious waste. An important outcome of the work of Kantorovich and his school for practical policy was (after a long lag) official acceptance of this proposition and of linear programming as a way of calculating the relevant marginal costs. For example, in 1979 in the USSR the State Committee for Science and Technology and the State Committee for Prices jointly approved an official method for the economic evaluation of raw material deposits. This was a prescribed method for the economic evaluation of exploration and development of raw material deposits. What was new in principle about this document was that it permitted the output derived from the deposits to be evaluated either in actual (or forecast) wholesale prices or in marginal costs. For the fuel-energy sector, a lot of work was done to calculate actual (and forecast) marginal costs for each fuel at different locations throughout the country and for different periods. These figures were regularly calculated on optimizing models (they were the dual variables to the output maximizing primal) and were widely used in planning practice for many years.

Comparison with the West

An important method of economic calculation in socialist countries was comparison with the West. If a particular product or method of production had already been introduced (or phased out) in the West, this was generally considered a good argument to introduce it (or phase it out) in the socialist countries, subject to national priorities and economic feasibility. Obtaining advanced technology from abroad (by purchase, Lend-Lease, reparations, espionage, direct investment) was an integral part of socialist planning, the importance of the different elements varying over time. Comparisons with the West were particularly important in an economic system which lagged behind the leading countries, lacked institutions which automatically introduced innovations into production (that is, profit-seeking business firms), and found it difficult (because of the ignorance of the planners, stable cost-plus prices and the self-interest of

rival bureaucratic agencies) to notice, appraise realistically when noticed, and adopt, innovations.

Economic calculation and economic results

It is important not to exaggerate the influence of methods of economic calculation on the performance of an economy. The performance of an economy is largely determined by external factors (such as the world market), economic policy (for example, the decision to import foreign capital or to declare a moratorium), economic institutions (like collective farms) and the behaviour of the actors within the system (for example, underestimation of investment costs by initiators of investment projects). It is entirely possible for an improvement in the methods of economic calculation to coincide with a worsening of economic performance (as happened in the USSR in the Brezhnev period). Realization of these facts led in the 1970s to a shift from the traditional normative approach (which concentrates on the methods of economic calculation and which regards their improvement as the main key to improved economic performance and the main role of the economist) in the study of planned economies, to the systems and behavioural approaches.

Economic calculation and economic intuition

In view of bounded rationality, and the huge volume, and distorted nature, of the information available to the central leadership, really existing decision-making relied heavily on rules of thumb and the 'feel' for reality of the top decision-makers (sometimes known as 'planning by feel'). This could quickly lead to an equilibrium, but an inefficient one.

MICHAEL ELLMAN

See also **behavioural public economics; Kantorovich, Leonid Vitalievich; Leontief, Wassily; Soviet Union, economics in; Stalinism, political economy of.**

Bibliography

Agababyan, E. and Yakovleva, Ye., eds. 1979. *Problemy raspredeleniya i rost narodnogo blagosostoyaniya*. Moscow: Nauka.
Birman, I. 1978. From the achieved level. *Soviet Studies* 30(2), 153–72.
Birman, I. 1996. Otraslevoe optimal'noe. Ch. 9 of *Ya – ekonomist*. Novosibirsk: EKOR.
Boltho, A. 1971. *Foreign Trade Criteria in Socialist Economies*. Cambridge: Cambridge University Press.
Ellman, M. 1973. *Planning Problems in the USSR: The Contribution of Mathematical Economics to their Solution 1960– 1971*. Cambridge: Cambridge University Press.
Ellman, M. 1983. Changing views on central planning: 1958–1983. *ACES Bulletin* [now *Comparative Economic Studies*] 25(1), 11–34.

Gács, J. and Lackó, M. 1973. A study of planning behaviour on the national-economic level. *Economics of Planning* 13, 91–119.

Giffen, J. 1981. The allocation of investment in the Soviet Union. *Soviet Studies* 33(4), 593–609.

Granick, D. 1990. Planning as coordination. Ch. 3 of *Chinese State Enterprises*. Chicago: University of Chicago Press.

Gregory, P. and Harrison, M. 2005. Allocation under dictatorship: research in Stalin's archives. *Journal of Economic Literature* 43, 721–61.

Kornai, J. 1967. *Mathematical Planning of Structural Decisions*. Amsterdam: North-Holland.

Kornai, J. 1980. *Economics of Shortage*, 2 vols. Amsterdam: North-Holland.

Kornai, J. 1992. Planning and direct bureaucratic control. Ch. 7 of *The Socialist System*. Oxford: Oxford University Press.

Kueh, Y. 1985. *Economic Planning and Local Mobilization in Post-Mao China*. London: Contemporary China Institute.

Kushnirsky, F. 1982. *Soviet Economic Planning, 1965–1980*. Boulder, CO: Westview, ch. 4.

Lazarev, V. and Gregory, P. 2003. Commissars and cars: a case study in the political economy of dictatorship. *Journal of Comparative Economics* 31, 1–19.

Lemeshev, M. 1968. *Mezhotraslevye svyazi sel'skogo khozyaistva*. Moscow: Ekonomika.

Levine, H. 1959. The centralized planning of supply in Soviet industry. In *Comparisons of the United States and Soviet Economies*. Washington, DC: Joint Economic Committee, US Congress.

Malinovskii, B. 1995. *Istoriya vychislitel'noi tekhniki v litsakh*. Kyiv: KIT/A.S.K.

Matekon. 1978. Standard methodology for calculations to optimize the development and location of production in the long run. Vol. 15(1), 75–96.

Qian, Y., Roland, G. and Xu, C. 2000. Coordinating activities under alternative organizational forms. In *Planning, Shortage, and Transformation*, ed. E. Maskin and A. Simonovits. Cambridge, MA: MIT Press.

Shlykov, V. 2004. The economics of defense in Russia and the legacy of structural militarization. In *The Russian military*, ed. S. Miller and D. Trenin. Cambridge, MA: MIT Press.

Stalin, J. 1952. Concerning the errors of comrade L.D. Yaroshenko. In *Economic Problems of Socialism in the USSR*. Moscow: Foreign Languages Publishing House.

Tretyakova, A. and Birman, I. 1976. Input–output analysis in the USSR. *Soviet Studies* 28(2), 157–86.

Weitzman, P. 1974. Soviet long term consumption planning: distribution according to rational need. *Soviet Studies* 26(3), 305–21.

World Bank. 1992. *China: Reform and the Role of the Plan in the 1990s*. Washington, DC: World Bank.

economic demography

Economic demography is an area of study that examines the determinants and consequences of demographic change, including fertility, mortality, marriage, divorce, location (urbanization, migration, density), age, gender, ethnicity, population size, and population growth. An applied area of research, economic demography draws upon the theoretical and applied fields of economics. For example, the determinants of fertility or migration primarily draw upon microeconomic theory and labour economics, while the consequences of population growth or ageing primarily draw upon macroeconomic theory and development economics.

The field has had a long tradition of controversy, beginning with the publication in 1798 of *An Essay on the Principle of Population* by the Reverend Thomas Malthus. The basic Malthusian model is founded on two propositions: (*a*) population, *when unchecked*, increases at a geometric rate (for example, 1, 2, 4, 8 …) and (*b*) food, in contrast, expands at an arithmetic rate (for example, 1, 2, 3, 4 …). The result is a population trapped at a meagre standard of living. Short of 'preventive checks' (birth control), population is constrained to live at subsistence by 'positive checks' (deaths, war, famines and pestilence). In later writings Malthus admitted the possibility of 'moral restraint' that could deter births, primarily through the postponement of marriage. However, he held little hope for a notable attenuation of the 'natural passions' of the working class.

While much of the controversy relating to Malthusianism has focused on the determinants of population growth, a second premise of his model relates to its economic underpinnings: the determinants of agricultural growth. Here Malthus appealed to the historical law of diminishing returns in agriculture. While this proposition engendered relatively little dispute at the time, history has since documented widespread and sometimes notable improvements in agricultural technology. Indeed, food production has represented an engine of growth in many of the areas that Malthus investigated. In some areas today, governments worry about 'excess' food production that depresses prices and farmers' living standards. Unfortunately, the pessimistic food-production predictions, when confronted by rapid population growth, caused economics to be dubbed the 'dismal science'.

The enormous popularity of the Malthusian ideas was the result of several factors: the model's simplicity and its explanation of poverty (the poor failed to exercise moral restraint, ending up with large families); the appeal of the message that subsidizing the poor is of questionable efficacy; and the plausibility of the Malthusian argument given the unexpected 'population explosion' revealed by the 1801 census. These and other elements of the 'Malthusian debate' provide a useful taxonomy for organizing the present article.

Specifically, we highlight the macroeconomic dimensions of the economic consequences of population growth since 1950. As with the early Malthusian debates, an assessment of the macroeconomic impacts of demographic change on economic production has resulted in an outpouring of research, which has spawned further debate. There are periods when vigorous Malthusian-like alarmism has carried the day; there are periods of counter-challenges; and, since the mid-1980s, there has been a productive 'revisionist' movement. In short, the simplistic Malthusian notion of diminishing returns in production has given way to more informed modelling of economic–demographic interactions. An assessment of the historical evolution of this literature will constitute the bulk of this review and appropriately delimits the scope of our essay since a wide range of important microeconomic themes are taken up in other articles in this dictionary (see FERTILITY IN DEVELOPING COUNTRIES, FAMILY DECISION MAKING, MARRIAGE AND DIVORCE, RETIREMENT, and multiple articles dealing with the topics of gender, ageing and mortality).

We begin by examining population impacts in one-sector growth models. This leads nicely into a more detailed assessment of factor accumulation, and in particular, the impacts of demography on saving, investment and technological change. This is in turn followed by an analytical description of the evolution of economic–demographic thinking since 1950. Such a perspective exposes many of the key analytical and empirical linkages of interest. The article concludes with an examination of 'convergence modelling', a useful paradigm that exposes the roles of changing demographic structures that take place over the demographic transition.

1. Theory: modelling economic–demographic change

1.1. One-sector growth models

The aggregate production function constitutes the primary organizing device for delineating the impacts of demographic change on economic growth. Within this model, labour productivity depends on the availability of complementary factors of production (land, natural resources, human and physical capital) and technology. If we assume, for convenience, that labour is a constant fraction of population, then population size directly affects aggregate output.

In a production function with constant returns to scale, an increase in population growth will lower the average availability of other factors of production – a 'resource-shallowing' effect, and, through diminishing returns, reduce the growth of worker productivity. Such an adverse demographic impact can be magnified (or attenuated) if population growth diminishes (raises) the growth rate of complementary factors.

In a standard growth model with factor inputs of labour and capital, and a saving rate and pace of technological change that are exogenous with respect to population growth, demography affects the long-run

level but not the long-run *growth rate* of output per capita. This is because the capital-shallowing effect of increased population will eventually reduce the capital per worker ratio to a level sufficient to be maintained by a fixed rate of saving. In this case, long-run growth is determined by the pace of technological change. The determinants of the 'fixed' saving rate and pace of technology growth, both considered in more detail below, are central to the analysis.

If one relaxes some of the assumptions of this model, the impact of population growth on per capita output growth can be ambiguous. Negative impacts can arise through diminishing returns, diseconomies of scale, and perhaps savings, while positive impacts can arise through induced technological change, economies of scale, and possibly savings. Most economists believe that adverse capital-shallowing impacts will dominate positive feedback effects, although the magnitude of the demographic impacts may not be all that large.

1.2. Saving

Possibly the most investigated linkage of population growth to economic growth has been the impact of demographic change on saving. Two perspectives dominate.

Adult equivalency. Rapid (slow) rates of population growth result in a disproportionate number of children (elderly adults) who consume, but contribute relatively little to, household income. In recognizing that these 'dependents' consume less than a working-age adult, the notion of an 'adult equivalent' consumer was born. The financing of an additional child's 'adult-equivalent' consumption has been hypothesized to be out of saving. Such a view, however, has been challenged by consideration of several offsetting alternatives. Specifically, children may (a) substitute for other forms of consumption, (b) contribute directly to household market and non-market income, (c) encourage parents to work more (or less), (d) stimulate the amassing (or reduction) of estates, and (e) encourage (or discourage) the accumulation of certain types of assets (for example, education or farm implements). The net impact of changing dependency rates on saving is therefore theoretically ambiguous. This is particularly the case if one views human capital as an investment financed in part by households and governments. At any rate, empirical evidence showing negative impacts of youth dependency on saving are found in several studies.

The life-cycle. A second population-saving linkage is based on a life-cycle formulation incorporated into a lifetime household utility function. Specifically, households attempt to even out their lifetime consumption by setting aside earnings during working years to finance consumption by their children as well as for their own retirement. This formulation can yield positive or negative impacts on aggregate saving depending on the relative sizes of the dissaving youth and elderly cohorts. While empirical evidence from life-cycle modelling is mixed, those studies do tend to show linkages between age structure and saving. However, the direction and magnitude of that impact depends upon time and place. (See, for example, Mason, 1987; Higgins, 1998; and Lee, Mason and Miller, 2001.)

1.3. Population-sensitive government spending

Government spending on population-sensitive activities such as schooling (youth) and health (elderly) has been alleged both to reduce saving and to crowd out spending on relatively growth-oriented investments. These two hypotheses constitute the core of Ansley J. Coale and Edgar M. Hoover's (1958) path-breaking study of India. While these premises are appealing, they require quali-fication. Governments have many options to accommo-date population pressures. Indeed, limited empirical evidence (for example, Schultz, 1987) has shown that education financing can be met all or in part by (a) trade-offs within the public sector, (b) reductions in per pupil expenditures, and (c) efficiency gains. While the second approach can be expected to reduce the quality of education (and therefore future productivity), the importance of population pressures on government spending or educational quality is uncertain.

1.4. Technological change: density, size and endogenous growth

While development economists have for decades hark-ened the pace of technological change as a (the?) major source of economic growth, most standard growth theory models take the rate of technological change as exogenous. With technological change independent of demographic change, population growth per se will have no impact on the pace of economic growth in long-run equilibrium. By contrast, if technological change is all or in part *embodied* in new investment, then a vintage specification is appropriate whereby new capital is rela-tively more productive than old. In this set-up, popula-tion growth can be economic-growth enhancing by expanding the rate at which technology is incorporated into production. In yet another specification, population growth can directly affect the rate of technological change and/or its form (factor bias). Kenneth J. Arrow (1962) has hypothesized that learning by doing is quickened in an environment of rapid employment growth.

A fourth linkage between technology and demography is found in 'endogenous growth' models that relate the pace of technology directly to population *size*. In parti-cular, the benefits of R&D are assumed to be available to all firms without cost; that is, an R&D industry generates a non-rival stock of knowledge. As a result, if we hold constant the share of resources used for research, an increase in population size advances technological change without limit. This somewhat controversial prediction has been qualified by models that incorporate various firm- or industry-specific constraints on R&D produc-tion. Such models typically reduce, but do not eliminate,

the positive impacts of population size which, as in the embodiment models above, are manifested largely during the 'transition' to long-run equilibrium.

Evidence on the roles of demographic-technology linkages and growth has been fragmentary and sparse. A pioneering study by Hollis Chenery and Moises Syrquin (1975) draws upon the experience of 101 countries across the income spectrum over the period 1950–70. They find that the structure of development reveals strong and pervasive scale effects (measured by population size) that vary by stage of development. Basically, small countries develop a modern productive industrial structure more slowly and later, while large countries have higher levels of accumulation and (presumably) higher rates of tech-nological change. Although these roles for demography may have been important historically, the impacts plausibly have waned somewhat: (a) economies in infra-structure are judged to be substantially exhausted in cities of moderate size; (b) specialization through inter-national trade provides a means of garnering some or many of the benefits of size; and (c) scale effects are most prevalent in industries with relatively high capital–labour ratios and such industries are inappropriate to the factor proportions of developing countries.

It is in agriculture where the positive benefits of popu-lation size have been most discussed. Higher population densities can lower per unit costs and increase the effi-ciency of transport, irrigation, extension services, markets and communications (Glover and Simon, 1975). Possibly the most cited work is that by Ester Boserup (1965; 1981), who observes that increasingly productive agri-cultural technologies are made economically attractive in response to higher land densities. While this is prob-ably true, the issue becomes one of identifying the quantitative magnitude of such effects over varying population sizes and in differing institutional settings. One must be cautious in attributing causation. For example, while high population densities may have accounted for a portion of expanded agricultural output in recent decades, in several important Asian countries these densities were sufficiently high decades ago to jus-tify the investments associated with the new technologies. Boserup in more recent writing has been less sanguine about the benefits of population size because densities appropriate to modern technologies in Asia are three to four times the average for Africa and Latin America.

In short, a wide-ranging review of the literature does not provide a strong consensus on the quantitative linkages between the size and growth of population, on the one hand, and the pace of technological change and economic growth, on the other hand.

1.5. The bottom line

An evaluation of population growth on economic growth through the filter of formal economic-growth modelling yields limited results: population growth affects the level but not the growth of per capita income in long-run equilibrium. Moreover, the key determinants of long-run growth are saving and technology. Only if these factors depend on demographic change does population matter. This somewhat constraining limitation of growth theory has caused researchers to branch out and explore a host of economic–demographic interactions using less formal paradigms. This blossoming literature has been extensive, lively and sometimes contentious.

2. Evolution of population-impacts thinking: 1950–90

Four major studies, two by the United Nations (1953; 1973) and two by the National Academy of Sciences (1971; 1986), reveal well the evolution of thinking on population matters over the period 1950–90. Three individual scholars, Coale and Hoover and Simon, also played prominent and important roles. (This section draws on Kelley, 2001.)

2.1. United Nations, 1953

The 1953 United Nations report, *Determinants and Consequences of Population Trends*, easily represents the most important contribution to population thinking since the writings of Malthus. Unlike Malthus, however, the UN study was balanced and exhaustive both in detail and in coverage. Some 21 linkages between population and the economy were taken up. For example, the impacts of population on the economy can be: (*a*) positive due to economies of scale and organization; (*b*) negative due to diminishing returns; or (*c*) neutral due to technology and social progress. An evaluation of these and other linkages led to a mildly negative overall assessment that was both cautious and qualified.

The most notable feature of this report was its methodology. More than any major study on population to that time, the UN Report embraced a methodology that would ultimately represent elements of modern-day 'revisionism'. Specifically, the report (*a*) downgraded the importance of population growth's impact on economic growth by placing it on a par with several other determinants of equal or greater impact; (*b*) assessed the consequences of population over a long period of time; and (*c*) emphasized the importance of feedbacks within and between the economic and political systems.

2.2. Coale and Hoover, 1958

The next major contribution to the population-impacts literature was provided by Ansley J. Coale and Edgar M. Hoover in their 1958 book *Population Growth and Economic Development in Low-Income Countries*. Based on simulations of a mathematical model calibrated with Indian data, they concluded that India's development would be enhanced by lower population growth. This was due to the hypothesized adverse impacts of population on household saving. It was also proffered that 'unproductive' investments in human capital (such as

health and education) would partially displace investments in 'relatively productive' forms (such as machines and factories). Economic growth would diminish in response.

Empirically, the above hypotheses have not been convincingly established. While several studies have exposed negative dependency-rate impacts on saving, there are others that show little or no impact. Overall, the findings are mixed, with a tilt toward supporting the Coale and Hoover formulation. (See Section 1.2 above for a discussion of the trade-offs that households can make to maintain saving in response to expanding family size.)

Similarly, there are alternative ways for governments to organize and finance schooling in response to population pressures. Unfortunately, studies of this are limited, although one by T. Paul Schultz (1987) finds no support for the Coale and Hoover (1958) formulation.

2.3. National Academy of Sciences, 1971

Arguably the most pessimistic assessment of the consequences of population growth was a study compiled by the National Academy of Sciences (NAS). The panel's final submission, *Rapid Population Growth: Consequences and Policy Implications*, issued in 1971, appeared in two volumes: Volume 1, *Summary and Recommendations*, and Volume 2, *Research Papers*. Unfortunately, the *Summary* volume appeared to be more political than academic in goal and orientation, and was not faithful to many of the underlying research reports assembled by the panel. Indeed, the *Summary* volume highlighted some 25 alleged negative consequences of population growth, whereas it downplayed or eliminated impacts that could be considered as 'neutral' or 'favourable'. As a result, the *Summary* represents an upper bound on the negative consequences of population growth. (A detailed documentation exposing the somewhat controversial way in which the *Summary* was compiled is provided by Kelley, 2001.)

What can be learned from the NAS study? First, given its apparent bias and the lack of a systematic vetting of Volume 1 by members of the panel, it is difficult to use that volume, either in full or in part. However, the individual papers are available and they, in total, offer a more balanced treatment. Second, by its own acknowledgment, the study focused on the short run when negative impacts of population change are most likely to prevail. ('We have limited ourselves to relatively short term issues'; 1971, p. vi.) By contrast, 'direct' (short-run) impacts of demographic change are almost always attenuated (and sometimes offset) by 'indirect feedbacks' that occur over longer periods of time. Thus the decision by the NAS panel to focus only on the short-run direct impacts resulted in an overly negative assessment of the consequences of population growth.

Third, economists were underrepresented on both the panel and in providing background reports. This is relevant since economists have substantial faith in the capacity of markets, individuals and institutions to adjust in the face of population pressures. Such adjustments, of course, take time and they are not without cost. Finally, this NAS Report provides a striking example of the difficulty of maintaining objectivity when social science research enters the public policy domain.

2.4. United Nations, 1973

In 1973 the United Nations weighed in with an update of its previous seminal work (United Nations, 1953). In contrast to the broadly eclectic stance in the earlier report, the new one ended with a mild to moderate negative overall assessment of rapid population growth. The authors were concerned with the ability of agriculture to feed expanding populations (à la Malthus) and the difficulty of offsetting capital shallowing (à la Coale and Hoover). Still, the 1973 Report, whose conclusions are highly qualified, is not alarmist, nor is it all that pessimistic. The reason for this moderate stance was the exceptionally influential empirical finding of Simon Kuznets (1960, pp. 19–20, 63) that notable negative correlations between population growth and per capita output growth were largely absent in the data. Given the strong priors of some contributors to the UN study, a failure to find a negative association in the aggregate data by a scholar with impeccable credentials had a profound impact. Indeed, this singular finding arguably kept the population debate alive for yet another round of assessments in the 1980s.

2.5. Revisionism, 1980s and beyond

The 1980s represented a decade when many of the underlying assumptions and conclusions of earlier studies of population–development interactions were subjected to critical scrutiny. The result was a revisionist rendering that was both surprising and controversial. Specifically, the revisionists downgraded the prominence of population growth as either a major source of, or a constraint on, economic prosperity in the Third World. The basis of this somewhat startling conclusion was the revisionists' methodology that (a) assessed the consequences of demographic change over longer periods of time and (b) expanded the analysis to take into account *indirect feedbacks* within economic and political systems. In general, empirical assessments of population growth will be smaller (less negative or less positive) when using the revisionist's methodology than when focusing on the short run and ignoring feedbacks. On net, most revisionists conclude that many, if not most, Third World countries would benefit from slower population growth.

2.6. Julian L. Simon, 1981

No one was more important in stimulating the new round of debates in the 1980s than Julian L. Simon, author of *The Ultimate Resource* (1981). This book attracted enormous attention, substantially because of two factors. First, it concluded that population growth would likely provide a *positive* impact on economic development of many developed, and some less developed, countries. Second, the book was accessible, well written, and organized in a 'debating', confrontational style. This included goading and prodding, the setting up and knocking down of straw men, and an examination of albeit popular, but somewhat extreme, anti-natalist positions. Simon's powerful book helped spawn a group of survey articles in the 1980s.

What accounts for Simon's positive assessments? Simon was an early advocate of evaluating the full effects of population over the intermediate to long run. He argued that the negative 'direct' impacts in the short run will probably be moderated, or sometimes overturned, when households, businesses, and/or governments react to changing prices which signal problems of resource scarcity. Two important examples of responses to population pressures can be cited: those relating to technological change and those relating to natural resource scarcity, both highlighted by Simon.

Technological change. Simon hypothesized and attempted to document that the pace of technological change, and its bias, can be stimulated by population pressures. Technological change, in turn, plays a central role in economic growth theory and has been shown in sources-of-growth studies to be a (the?) key to economic growth. Additionally, with respect to population size impacts in general, Simon observes that major social overhead projects (for example, roads, communications and irrigation) have benefited from expanded populations and scale. (For more detail, see Section 1.4 above.)

Resource depletion. Consider next the impacts of population growth on natural resource depletion. Theoretically an exhaustion of non-renewable resources (for example, coal and minerals) would appear to be inevitable in the long run. However, such a period may be in the indeterminably distant future. By contrast, Simon argued that the most relevant measure of resource scarcity is its price. He prepared many graphs of US non-renewable resource prices (deflated by price indexes in order to focus on 'real' resource trends).

Surprisingly, virtually every resource has experienced a *declining* real price over lengthy periods of time. This means, à la Simon, that resources are becoming *more* abundant over time. It seems that the more resources are used, the more abundant they become! How can this happen? Simple. A rising resource price, due in part to population pressures, triggers several reactions that reduce or even eliminate the apparent resource scarcity. Specifically, in the short run, rising prices encourage an economizing of the resource at every level of production and consumption. In the longer run, rising prices stimulate exploration, new methods of extraction and process, and the search for substitutes.

Nevertheless, Simon recognized that market failures, institutional failures, and political factors can all result in less-than-complete adjustments when population and economic development press against resource availabilities.

This is particularly the case with renewable resources (such as rain forests, fisheries, the environment, and so forth) where market or institutional failures are pervasive. Without mechanisms to assign and maintain property rights, internalize externalities, and address free rider problems of public and quasi-public goods, government regulation may be required to safeguard renewable resources over time.

2.7. National Academy of Sciences, 1986

Some 15 years after the 1971 National Academy Report that highlighted 25 negative consequences of population growth, a new National Academy Report was released. In contrast to the previous study, the new report was balanced, eclectic and non-alarmist. A careful examination of its bottom line is instructive.

'*On balance*, we reach the *qualitative* conclusion that *slower* population growth would be beneficial to economic development of *most* developing countries.' (1986, p. 90; emphasis added)

This qualified assessment reveals key features found in most population assessments in the 1980s. Specifically: (*a*) there are both positive and negative impacts of demographic change (thus 'on balance'); (*b*) the magnitude of the net impacts cannot be determined given current evidence (thus 'qualitative'); (*c*) only the direction of the impact from high to low growth rates can be ascertained (thus 'slower' rather than 'slow'); and (*d*) the net impact varies from country to country. In most cases it will be negative; in some positive; and in others of little impact (thus, 'most developing countries').

What accounts for the dramatic turnaround in the two National Academy assessments? Several factors can be advanced. First, the 1986 report extends the short-run time horizon of the 1971 report to examine individual and institutional responses to the initial impacts of population change: conservation in response to scarcity, substitution of abundant for scarce factors of production, innovation and adoption of technologies to exploit profitable opportunities, and the like. These responses are considered to be pervasive and they are judged to be important. According to the report writers: 'the key [is the] mediating role that human behavior and human institutions play in the relation between population growth and economic processes' (1986, p. 4).

Second, the 1986 study was assembled almost entirely by economists whose understanding of and faith in markets to induce responses that modify initial direct impacts of population change is far greater than that of other social and biological scientists.

Third, research accumulating over the 15 years between the two reports revealed a need to downgrade: (1) the concern about non-renewable resource exhaustion; (2) the adverse impact of children on the capacity to save, and in turn to undertake productive investments; and (3) the inability to invest in schooling and health facilities.

Finally, the 1986 Report upgrades the concern about population impacts on *renewable* natural resources (such as fishing areas and rain forests) where property rights are difficult to assign and maintain. Overuse can result. It is recognized that the problems of overuse are not solely due to population growth per se, but rather institutional failure. Cutting population growth by one half, or even to zero, would not solve the problem. Rather it would slow the process and postpone the date of resource exhaustion. Government policies are needed to account for negative externalities and market failure. Slowing population growth provides time for institutional response.

3. New paradigms for modelling demography's role in economic growth: 1990 and beyond

As noted previously, Kuznets's empirical finding of an absence of notable negative correlations between population growth and per capita output growth influenced the population debate throughout the 1970s and 1980s. Simple correlations stimulated research during the 1990s as well. This time, however, statistically significant negative correlations during the 1980s drove the discussion. Interestingly, economic–demographic modelling continued in the 'revisionist' vein, incorporating positive and negative as well as short- and long-run influences into an economic growth model. The modelling challenge remains one of accommodating correlations that can be negative, positive or insignificant depending upon time and place.

3.1. Convergence growth models: a framework for assessing demography's impact

Renewed interest in modelling the impacts of demographic change on economic growth coincided with the emergence in the economic growth literature of the 'technology gap' or 'convergence' model. This model, formulated initially by Barro and Sala-i-Martin (1991), has been used widely to explore many hypothesized influences on economic growth, including openness to trade, form of government, and the rule of law. Since this type of modelling highlights the dynamics of the adjustment process, it is particularly relevant to examining the impacts of major shifts in the population's age distribution associated with birth and death rates that change systematically over the demographic transition. As a result, economic demographers have employed convergence paradigms to explore demographic–economic interactions.

Briefly stated, convergence models focus on the pace at which countries move from their current level of labour productivity to their long-run or steady-state level of labour productivity. The model assumes that all countries converge at the same rate from their current to their long-run levels (which can vary across countries and over time). The greater the productivity gap, the greater are the gaps of physical capital, human capital and technical efficiency from their long-run levels. Large gaps allow for 'catching up' through (physical and human) capital accumulation, and technology creation and diffusion across countries and over time. Indeed, many empirical studies indicate that growth rates do slow down as a country approaches its long-run productivity level, especially those studies that provide for country- and period-specific conditions that influence the long-run level of labour productivity.

Since long-run labour productivity is unobservable, empirical implementations of the model substitute a vector of 'conditioning' variables thought to influence long-run labour productivity. The actual specification of these conditioning variables varies notably. Consider two of their many representations. The first, by Barro (1997), highlights inflation, government consumption ratios, the rule of law, the form of the political system, terms of trade, human capital, the total fertility rate, and life expectancy at birth (a proxy for health). The second formulation, by Bloom and Williamson (1998), highlights two categories of growth-rate determinants: economic structure variables (natural resources, schooling, access to ports, location in the tropics, whether landlocked, and extent of coastline); and economic and political policies (openness to trade, quality of institutions, and government savings share of GDP). Clearly there are many defensible perspectives on variable choice, and much is yet to be learned about the appropriate configuration of conditioning variables that influence long-run productivity levels.

3.2. Alternative demographic renderings within a convergence framework

The 1990s witnessed attempts by various researchers to model demography in a manner that accommodates both the insignificant correlations of the 1960s and 1970s as well as the significant negative correlations of the 1980s and 1990s. Three different approaches are described here. All three employ a convergence-type growth model and all employ a broad set of countries spanning the income spectrum.

Modelling through aggregate measures of fertility and mortality. Barro (1997) includes two demographic aggregate measures among his list of conditioning variables, the total fertility rate (TFR) and life expectancy. Barro's formulation thus has demography impacting the long-run equilibrium level of per capita income. The TFR captures, for example, the adverse capital-shallowing impact of more rapid population growth as well as the resource opportunity costs of bringing up children. Furthermore, while Barro treats life expectancy as a human capital proxy for health, demographers consider it to be a demographic variable. Both are statistically significant, with a higher TFR inhibiting, and longer life expectancy enhancing economic growth.

Modelling through population growth components. Kelley and Schmidt (1995) decompose population growth by examining two components (births and deaths) and by modelling their contemporaneous and lagged impacts. This approach allows for disparate impacts of fertility and mortality as well as negative short-run effects (costs of high birth and death rates) and positive long-run effects (favourable impacts of past births on current labour force growth and declining mortality). Consistent with Kuznets's earlier work, they found an absence of a net demographic impact on economic growth in the 1960s and 1970 – the separate impacts of births and deaths are notable but offsetting. Consistent with empirical work of the early 1990s, they found negative impacts throughout the 1980s. These negative correlations were in part the result of (*a*) rising short-run costs of high birth rates, (*b*) declining benefits of mortality reduction, and (*c*) insufficient labour force entry from past births to offset these increased costs.

Modelling through differential age-structure growth. In a series of papers beginning in the late 1990s, several Harvard economists argued for a demographic rendering that incorporates not only population growth but also labour growth (see, for example, Bloom and Williamson, 1998; and Bloom, Canning and Malaney, 2000). They note that, while theorists conceptualize the economic growth process in labour productivity terms, empirical growth models are generally specified in per capita terms. This makes no difference when population and labour grow at the same rate, but does when they grow at different rates.

The authors argue that the post-war period was exactly such a time since during that period demographic transitions took place in different countries at different times and at different paces. At various stages of the demographic transition, the population and working ages (used within this framework as a proxy for labour) can grow at very different rates. In a predictable pattern, the population initially grows faster, then slower, and then faster than the working-aged population during the transition from a high-fertility, high-mortality to a low-fertility, low-mortality demographic steady-state equilibrium. (For an historical evolution of economic, sociological, and biological factors during the demographic transition, see R.A. Easterlin, 1978.)

Without allowing for differential growth rates of the population and working ages, demographic coefficient estimates (mainly population growth) will be biased. In that case the population–growth coefficient captures net demographic impacts that can be positive, negative, or neutral, depending upon time and place. Bloom and Williamson (1998) demonstrate this point for a broad cross-section of countries over the period 1965–90 in a convergence model that also includes life expectancy as a human capital variable. Consistent with some studies, their simple demographic rendering results in a positive but insignificant coefficient for the population growth rate. When supplemented by the working-age growth rate, however, that coefficient turns negative and the coefficient for the working-age growth rate is positive, both statistically significant.

Effectively, the Harvard economists append an accounting structure to translate labour productivity impacts into per capita terms. The resulting demographic specification is elegant in its simplicity, incorporating only two demographic variables that have unambiguous predicted coefficient values of −1 (for population rate of growth, Ngr) and +1 (for working-age population rate of growth, WAgr) when used to expose demography's impact on income growth per capita relative to income growth per working-age population. In that context, demography exerts its primary impact on the pace at which the long-run equilibrium is reached (Bloom and Williamson, 1998, p. 419) rather than on the long-run equilibrium level of productivity.

This is an intriguing specification. The interpretation is clear: if labour force growth exceeds population growth, then the rate of per capita income growth is boosted by demography. The Harvard economists label this phenomenon the 'demographic gift' that may be reaped for several decades after the onset of fertility decline as new labour force entrants from earlier large birth cohorts outpace fertility. The 'gift' was large throughout the 1965–90 period for Japan and other Asian Tigers because of the early and rapid pace of their demographic transition. Of course, the converse of the 'gift' began to be felt in the 1990s as new labour force entry from smaller birth cohorts was outpaced by labour force exit of the aging population. The model predicts productivity outpacing per capita income growth over several decades into the future in these Asian (and other) countries.

Note that the qualitative predictions are based on theoretically determined coefficients on WAgr and Ngr of +1 and −1, respectively. To the extent that estimated coefficients deviate from +1 and −1, WAgr and Ngr play an additional role in the determination of the long-run productivity level. The Harvard studies provide some guidance in this area. In their earlier study, Bloom and Williamson (1998) estimate coefficients that differ significantly from +1 and −1. However, in a later study that further elucidates the accounting, Bloom, Canning and Malaney (2000) find no significant difference from those values. If that is the case, then the model at once makes an important contribution and is somewhat narrower than many in the literature which admit both short-run and long-run impacts of demographic change as a part of the theoretical structure. Yet modelling demography in growth equations tends to be both imprecise and ad hoc. In contrast, the Bloom and Williamson model is relatively clear in interpretation, and it targets the shorter-run impacts that are of primary interest to policymakers.

3.3 The bottom line

Bloom and Williamson (1998) estimated that as much as one-third of the average per capita income growth rate in East Asian countries over the period 1965–90 is explained by population dynamics. Kelley and Schmidt (2001) evaluated eight distinct demographic renderings within a convergence model using a consistent set of conditioning variables – those described above for Barro's variant. Among others, these renderings included Barro's TFR; a 'naive' variant predating the 1990s work that simply includes Ngr; a 'components' model (contemporaneous and lagged birth rates and the death rate: Kelley and Schmidt, 2001); two variants of the Harvard transitions framework; and demographic extensions to several variants.

Kelley and Schmidt (2001) find that on average, across all eight demographic formulations and over their full 86-country sample (covering the full income spectrum), approximately 21 per cent of the combined impacts on change in the per capita income growth rate is accounted for by changes in the demographic variables in the various models. What is striking about this result is that the 21 per cent is fairly stable across all eight demographic renderings, from one that is quite simplistic (Ngr only) to those that incorporate short-, intermediate- and long-term population effects. On the one hand, this should not be terribly surprising because of the interconnectedness of all of the demographic measures. On the other hand, while population matters, it is still important to determine why.

Although there is an emerging consensus that the magnitude of the impacts of population growth have been sizeable (for example, 21 per cent globally and as much as 33 per cent in East Asia), the reasons why this is the case are still both contestable and not well understood. Are the demographic determinants primarily longer-run impacts, or are they mainly shorter-run transitional dynamics that are diminishing? Will the so-called 'demographic gift' of these dynamics in the past reveal themselves as a 'demographic drag' in the future, deriving from reduced fertility, slow population growth and ageing? Or will a new mechanism reveal itself? For example, (a) will future modelling better expose the components of labour force change (for example, utilization rates, age- and/or gender-specific participation rates); and (b) will fertility and mortality be endogenously specified to better reveal the dynamics of the demographic transition about which the field of economic demography has much to say? Whatever the outcome, the stage is set for another round of research, pinning down the results of the past with the goal of understanding the future.

ALLEN C. KELLEY AND ROBERT M. SCHMIDT

See also **family decision making; fertility in developing countries; marriage and divorce; retirement.**

Bibliography

Arrow, K.J. 1962. The economic implications of learning by doing. *Review of Economic Studies* 29, 155–73.

Barro, R.J. 1997. *Determinants of Economic Growth: A Cross-Country Empirical Study.* Cambridge, MA: MIT Press.

Barro, R.J. and Sala-i-Martin, X. 1991. Convergence. *Journal of Political Economy* 100, 223–51.

Birdsall, N., Kelley, A.C. and Sinding, S., eds. 2001. *Demography Matters: Population Change, Economic Growth and Poverty in the Developing World.* Oxford: Oxford University Press.

Bloom, D.E. and Williamson, J.G. 1998. Demographic transitions and economic miracles in emerging Asia. *World Bank Economic Review* 12, 419–55.

Bloom, D.E., Canning, D. and Malaney, P. 2000. Demographic change and economic growth in Asia. *Population and Development Review* 26, 257–1990.

Boserup, E. 1965. *Conditions of Agricultural Growth.* Chicago: Aldine.

Boserup, E. 1981. *Population and Technological Change.* Chicago: University of Chicago Press.

Chenery, H. and Syrquin, M. 1975. *Patterns of Development: 1950– 1970.* Oxford: Oxford University Press.

Coale, A.J. and Hoover, E.M. 1958. *Population Growth and Economic Development in Low-Income Countries.* Princeton: Princeton University Press.

Easterlin, R.A. 1978. The economics and sociology of fertility: a synthesis. In *Historical Studies of Changing Fertility*, ed. C. Tilly. Princeton: Princeton University Press.

Glover, D.R. and Simon, J.L. 1975. The effect of population density on infrastructure: the case of road building. *Economic Development and Cultural Change* 23, 453–68.

Higgins, M. 1998. Demography, national savings, and international capital flows. *International Economic Review* 39, 343–69.

Kelley, A.C. 1988. Economic consequences of population change in the Third World. *Journal of Economic Literature* 26, 1685–728.

Kelley, A.C. 2001. The population debate in historical perspective: revisionism revisited. In Birdsall, Kelley and Sinding (2001).

Kelley, A.C. and Schmidt, R.M. 1995. Aggregate population and economic growth correlations: the role of the components of demographic change. *Demography* 32, 543–55.

Kelley, A.C. and Schmidt, R.M. 2001. Economic and demographic change: a synthesis of models, findings and perspectives. In Birdsall, Kelley and Sinding (2001).

Kuznets, S. 1960. Population change and aggregate output. In *Demographic and Economic Change in Developed Countries.* National Bureau of Economic Research. Princeton: Princeton University Press.

Lee, R.D., Mason, A. and Miller, T. 2001. Saving, wealth and population. In Birdsall, Kelley and Sinding (2001).

Malthus, T.R. 1798. *An Essay on the Principle of Population.* Harmondsworth: Penguin, 1970.

Mason, A. 1987. National saving rates and population growth: a new model and new evidence. In *Population Growth and Economic Development: Issues and Evidence*, eds. D.G. Johnson and R.D. Lee. Madison: University of Wisconsin Press.

National Academy of Sciences. 1971. *Rapid Population Growth: Consequences and Policy Implications*, vol. 1: *Summary and Recommendations;* vol. 2: *Research Papers.* Baltimore: Johns Hopkins University Press.

National Academy of Sciences. 1986. *Population Growth and Economic Development: Policy Questions.* Washington, DC: National Research Council.

Schultz, T.P. 1987. Schooling Expenditures and Enrollments 1960–1980: The Effects on Income, Prices and Population Growth. In *Population Growth and Economic Development Issues and Evidence*, eds. D. Gale Johnson and R.D. Lee. Madison: University of Wisconsin Press.

Simon, J.L. 1981. *The Ultimate Resource.* Princeton: Princeton University Press.

Simon, J.L. 1996. *The Ultimate Resource 2.* Princeton: Princeton University Press.

Srinivasan, T.N. 1988. Modeling growth and economic development. *Journal of Policy Modeling* 10, 7–28.

United Nations. 1953. *The Determinants and Consequences of Population Trends.* New York: United Nations.

United Nations. 1973. *The Determinants and Consequences of Population Trends.* New York: United Nations.

economic development and the environment

Economic development depends on sustained per capita income growth and entails dramatic changes in production structure. In low-income economies, growth typically stimulates markets and promotes the evolution of institutions that constrain behaviour according to social norms. The expansion of trade in relation to GDP is another common accompaniment to growth. Each of these has effects on 'the environment', which in a developing-country setting refers not only to phenomena such as water and air quality but also, importantly, to natural resource stocks such as forests, fisheries and soils.

Conversely, changes in environmental quality, including resource stock drawdowns, may affect economic development in a dynamic interaction. This feedback is hard to quantify; however, the World Bank's *World Development Indicators* series now includes 'adjusted' national accounts data reporting GDP and savings net of the implied value of resource depletion and environmental damage (Bolt, Matete and Clemens, 2002). These indicate that environmental damage can reduce GDP growth by as much as one to two per cent per year. On a broader scale, growth of large low-income economies like China and India is beginning to have ramifications not only for their own environmental conditions, but also for the global environment through transboundary pollution spillovers and greenhouse gas (GHG) emissions.

The welfare of the poor in low-income countries is intimately linked to their access to environmental assets, and especially to the natural resource base. Despite this, the central concerns of environmental and resource economics – the economic costs of pollution and natural resource depletion – have only recently begun to be linked to models of economic development. Publication of the so-called Brundtland Report (WCED, 1987) was a watershed event; since then, 'no account of economic development would be regarded as adequate if the environmental-resource base were absent from it' (Dasgupta and Mäler, 1995, p. 2734).

Growth in low-income economies is inevitably associated with higher resource demands and increased pollution intensity per unit of income generated. Other things equal, more economic activity generates more environmental damage monotonically through a scale effect. The relationship may be nonlinear, however. As income grows, environmental damage per unit of additional income may initially rise, then decline. This conjecture, known as the environmental Kuznets curve (EKC), posits that scale effects dominate all other influences on the growth–environment relationship at low income levels, but that, as incomes rise, changes in the composition of production, technological improvements, and income-elastic preferences for conservation and a cleaner environment become more influential (Grossman and Krueger, 1993). Institutional and legal constraints on pollution and resource depletion, initially so weak as to create a form of open access for polluters and resource depleters, may also evolve or be applied with greater vigour as incomes increase, whether due to income effects or to increased recognition of limits to growth imposed by pollution and resource scarcity (Stokey, 1998). Despite the heuristic value of EKC, however, empirical tests in low-income economies are plagued by data and measurement problems. Most notably, there is no robust evidence of an EKC for resource-depleting activities such as deforestation.

Changes in production structure and factor demands are also inherent to development. The most prominent manifestation of structural change in low-income countries is the relative decline of agricultural and resource sectors as contributors to GDP and employment. This has clear environmental implications when the majority of the population is initially dependent on the natural resource base. In capital-scarce economies, forest and land conversion for agriculture and the exploitation of fisheries and other resource stocks are standard strategies for increasing labour productivity and generating surpluses. Accordingly, early stages of development are characterized by rapid resource depletion – most visibly in the form of tropical deforestation. Such processes are abetted by conditions of open access (Barbier, 2005).

Whether the depletion rate eventually slows – a prerequisite for sustainable development – depends largely on the extent to which surpluses are used to build capacity in secondary and tertiary industries making more intensive use of reproducible resources such as labour, technology and human capital. In this way, the central story of structural change in low-income economies is intimately linked to the evolution of demands on the environmental and natural resource base. Sustained growth leads to a relative reduction in dependence on natural resources, and thus makes it easier for society to agree to promote conservation, biodiversity retention and non-use amenities. Conversely, macroeconomic failures, often in combination with rapid population growth, high transactions costs and market failures, can lead low-income economies into unsustainable cycles of poverty, resource over-exploitation, and institutional failure.

Trade is another influential source of structural change. Early development policies stressing import substitution and de-emphasizing trade have, in most countries, been supplanted by greater outward orientation. Trade-to-GDP ratios have risen and domestic prices have tended to converge on world market prices, thus altering domestic production and investment incentives. With the exception of resource-poor East Asian countries like Korea and Taiwan, the pursuit of comparative advantage in low-income countries initially means expanded exports of tropical agriculture, forestry and fisheries and of resource-based semi-manufactures such as sawnwood. Both the growth of global demand and the pro-trade effects of policy reforms encourage accelerated resource drawdowns; unless property rights and externalities are adequately dealt with, these are likely to occur at socially excessive rates (Coxhead and Jayasuriya, 2003). A related idea known as the pollution haven hypothesis posits that weak environmental laws and unresolved externalities may lead developing countries to specialize in pollution-intensive industrial activities (Copeland and Taylor, 1994).

Whereas early policy advice to developing countries typically stressed the desirability of exploiting resource wealth to create jobs and earn foreign exchange, contemporary concerns about exhaustibility and the integrity of ecological systems have led to more cautious counsel and an emphasis on sustainable development. Such advice, however, is often difficult to implement as policy in the face of pressures to promote growth and alleviate poverty in the current generation.

New issues in the development–environment relationship continue to emerge as economies grow and become more globalized. Traditionally, trade-environment analyses used Ricardian or Heckscher–Ohlin models of North–South interactions in which welfare growth in resource-abundant South is contingent on trade with industrialized North and on domestic externalities or market failures (for example, Chichilnisky 1994).

However, South–South trade – or, in the case of China's emergence as a major market for resource exports from Asia, Africa, and Latin America, 'East–South' trade – is now growing much faster than trade of the North–South type. South–South trade is a form of internationally fragmented production in which primary products or semi-manufactures are exported from one low-income country to another to be used in production of final goods. The latter low-income economy thus moves to 'clean' growth based on labour-intensive manufactures, while growth in the former becomes more resource-intensive. Countries in the South may have comparative advantage in either clean or dirty goods – or both. Conventional models and measures for evaluating environmental costs of growth must be adapted to such new modalities.

Other new trends reflect the growing global influence of large developing economies. In poor countries, about 50 per cent of carbon dioxide emissions (the primary sources of GHGs) comes from land conversion. But total emissions increase rapidly with energy demands driven by growth, urbanization and industrialization. According to the International Energy Agency, China accounted for 13 per cent of global energy-related CO_2 emissions in 2006, and is expected to overtake the USA as the largest CO_2 source by 2009; India is now following a similar path (IEA, 2006). Under the 1997 Kyoto Protocol, these economies are not required to limit GHG emissions. But, even if they do take major steps to limit pollution intensity, scale effects of their growth will ensure that global pollution externalities will continue to expand for the foreseeable future. In turn, concerns over the global environmental consequences of growth in low-income countries will find increasingly forceful expression in international negotiations not only on the environment but also on trade and other forms of international integration.

IAN COXHEAD

See also **climate change, economics of; environmental economics; environmental Kuznets curve; poverty alleviation programmes; sustainability.**

Bibliography

Barbier, E.B. 2005. *Natural Resources and Economic Development*. Cambridge: Cambridge University Press.
Bolt, K., Matete, M. and Clemens, M. 2002. Manual for calculating adjusted net savings. Mimeo, Environment Department, World Bank.
Chichilnisky, G. 1994. North–South Trade and the global environment. *American Economic Review* 84, 851–74.
Copeland, B.R. and Taylor, M.S. 1994. North–South trade and the environment. *Quarterly Journal of Economics* 109, 755–87.
Coxhead, I. and Jayasuriya, S.K. 2003. *The Open Economy and the Environment: Development, Trade and Resources in Asia*. Cheltenham, UK and Northampton, MA: Edward Elgar.
Dasgupta, P. and Mäler, K.-G. 1995. Poverty, institutions and the natural resource base. In *Handbook of Development Economics*, vol. 3A, ed. J. Behrman and T.N. Srinivasan. Amsterdam: North-Holland.
Grossman, G.M. and Krueger, A.B. 1993. The environmental impacts of a North American Free Trade Agreement. In *The US-Mexico Free Trade Agreement*, ed. P. Garber. Cambridge, MA: MIT Press.
IEA (International Energy Agency). 2006. *World Energy Outlook 2006*. Paris: IEA.
Stokey, N. 1998. Are there limits to growth? *International Economic Review* 39, 1–31.
WCED (World Commission on Environment and Development). 1987. *Our Common Future*. Oxford and New York: Oxford University Press.

economic epidemiology

The fast-growing literature on the economic analysis of epidemiological issues (see Philipson, 2000, for a review) delivers very different implications about disease occurrence and its optimal control from those of traditional analysis of the same issues in the field of public health. At the risk of vastly oversimplifying the positive component of the public health approach, the traditional analysis comprises empirical methods and analysis aimed at identifying and quantifying the effects of 'risk factors' on health outcomes. These factors are typically defined as covariates that negatively affect the measured health outcomes – for example, the effects of smoking on lung cancer or the effects of obesity on heart disease. Thereafter, the normative component of the public health approach is concerned with attempts to reduce the measured risk factors, whether through private or public intervention, and to thereby improve health outcomes.

This approach drastically differs from that of economic epidemiology, which attempts to explain undesirable disease occurrence as the result of self-interested behaviour in the presence of constraints. The effects and desirability of disease-reducing public interventions are then evaluated in terms of how they improve the private behaviour essential to controlling disease in the first place. In some sense, the public health approach aims to improve health, whereas the economic approach aims to improve economic efficiency, even if that does not necessarily improve health. Just as closing highways would improve health but impair economic efficiency, the two approaches often clash in desired interventions. The public health approach, therefore, more often favours public intervention, and sometimes simply assumes that the existence of a health problem is sufficient cause for intervention, potentially because it lacks a theory about how private incentives affect the observed level of disease across time and populations.

Economic epidemiology and infectious disease

Infectious diseases cause roughly one-third of all deaths worldwide and represent the primary cause of mortality in the world. Historically, the share of worldwide mortality due to infectious diseases has been even greater, although data tend to be less reliable for earlier periods. Morbidity and mortality from infectious diseases such as tuberculosis, malaria and acute respiratory infection have always been at the forefront of public policy in developing countries, where infectious diseases accounted for nearly one-half of mortality in the 1990s.

Worldwide concern about infectious disease has received renewed interest in public policy discussions given the disastrous impacts of HIV/AIDS and the potential threat of bird flu. Like most communicable diseases, especially those that are potentially fatal, HIV has incited an extensive governmental response, consisting of regulatory measures, subsidies for research, education, treatment, testing and counselling. Here we review the main contributions of economic epidemiology in predicting both the short- and the long-run behaviour of infectious disease, as well as the effects and desirability of public health interventions that attempt to reduce such disease.

Philipson and Posner (1993) provide the first systematic analysis of rational infectious disease epidemics in the context of AIDS. Kremer (1996) analyses the effects of a reduction in the number of one's sexual partners on the growth of disease. The predictions of such models rely crucially on the prevalence elasticity of private demand for prevention against disease, that is, the degree to which prevention increases in response to disease occurrence. Prevalence-elastic behaviour has different implications for the susceptibility to infection than standard epidemiological models of disease occurrence as discussed in Philipson (1995). Evidence of the degree of prevalence-elastic demand is discussed in Ahituv, Hotz and Philipson (1996) and Auld (2003; 2006). Oster (2006) attempts to explain the lack of prevalence-elastic demand in Africa by the competing risks that lower the demand for prevention in that part of the world. Lakdawalla, Sood and Goldman (2006) provide evidence that demand is sensitive to overall risk, both in terms of prevalence and the cost of infection as when reduced by new medical technologies.

This type of prevalence-elastic behaviour has two major implications. First, growth of infectious disease is self-limiting because it induces preventive behaviour. Second, since the decline of a disease discourages prevention, initially successful public health efforts actually make it progressively harder to eradicate infectious diseases. Geoffard and Philipson (1996) discuss a very general result concerning the inability of private markets to eradicate disease when demand is prevalence-elastic because a disappearing disease implies less prevention. Barrett (2003; 2004) also analyses the implications of economic efficiency for optimal eradication. See also Gersovitz and Hammer (2003; 2004; 2005).

Regarding the value of public health interventions, Mechoulan (2004) analyses the prevalence and efficiency implications of HIV testing. Geoffard and Philipson (1996) argue that eradication is never Pareto optimal when only the current generation is considered. However, the missing market is dynamic: future generations cannot pay vaccine producers for the benefit they derive from the producers' product. Brito, Sheshinski and Intriligator (1991) analyse the non-standard efficiency implications of mandatory vaccinations.

Moreover, the prevalence elasticity of demand lowers the price elasticity of demand, which implies that Pigouvian-style subsidies to stimulate prevention may have only limited success. This occurs because demand rises among those who are subsidized and falls among those who are not – in the extreme case, total demand is inelastic to subsidies. In addition, prevalence competes with public interventions in inducing protective activity, which makes the timing of the public intervention a crucial factor in determining its economic efficiency. If the subsidy is not prompt enough, the growth in prevalence will have already induced protection.

A growing literature examines the optimal control of infectious diseases in the presence of antibiotic resistance (see, for example, Laxminarayan and Brown, 2001; Laxminarayan and Weitzman, 2002; Laxminarayan, 2002; and Horowitz and Moehring, 2004). The standard, positive external effect of treating more individuals with an infectious disease is partly or fully offset by the negative external effect induced by increased antibiotic resistance. The R&D problem induced by external consumption effects such as antibiotic resistance is discussed in Philipson, Mechoulan and Jena (2006).

Economic epidemiology has also considered the welfare losses induced by disease, the welfare effects of R&D in developing new methods of prevention and treatment (Philipson, 1995), and how these contrast with cost-of-illness studies of disease burden.

Spread of economic epidemiology to other fields

Several other topics have grown out of this more systematic analysis of infectious disease by economists. One strand is the analysis of public health-related issues such as obesity (Philipson and Posner, 2003; Lakdawalla, Philipson and Bhattacharya, 2005). The addictive aspect of obesity is analysed by Cawley (1999). Empirical studies explaining the observed growth in obesity, whether it includes a rise in caloric intake or fall in caloric expenditure, include Cutler, Glaeser and Shapiro (2003). Chou, Grossman and Saffer (2004) and Rashad and Grossman (2004) analyse the co-variation between the growth of obesity and smoking and fast-food establishments in the United States. The important and rich set of issues raised by growth in obesity promises a useful role for economic analysis.

Another area in which economic analysis of epidemiological issues has emerged is the economic analysis of clinical trials (see, for example, Philipson and DeSimone, 1997; Philipson and Hedges, 1998; Malani, 2006). This literature deals with the non-traditional aspects of programme evaluation that are unique to clinical trials – for example, the blinding of subjects. Economic analysis of clinical trials differs from bio-statistical analysis in that subjects are assumed to act in their best interest rather than be passively observed.

The stark difference between economic explanations of disease occurrence on the one hand and the evaluation of public interventions aimed at limiting disease on the other implies that economics may have a very useful role to play in understanding these issues.

TOMAS J. PHILIPSON

See also **health economics.**

Bibliography

Ahituv, A., Hotz, J. and Philipson, T. 1996. Is AIDS self-limiting? Evidence on the prevalence elasticity of the demand for condoms. *Journal of Human Resources* 31, 869–98.

Auld, M.C. 2003. Choices, beliefs, and infectious disease dynamics. *Journal of Health Economics* 22, 361–77.

Auld, M.C. 2006. Estimating behavioral response to the AIDS epidemic. *Contributions to Economic Analysis and Policy* 5(1), Article 1.

Barrett, S. 2003. Global disease eradication. *Journal of the European Economic Association* 1, 591–600.

Barrett, S. and Hoel, M. 2004. Optimal disease eradication. Working Paper No. 50.04, FEEM.

Brito, D., Sheshinski, E. and Intriligator, M. 1991. Externalities and compulsory vaccinations. *Journal of Public Economics* 45, 69–90.

Cawley, J. 1999. Obesity and addiction. Ph.D. dissertation, Department of Economics, University of Chicago.

Chou, S.-Y., Grossman, M. and Saffer, H. 2004. An economic analysis of adult obesity, results from the Behavioral Risk Factor Surveillance System. *Journal of health Economics* 23, 565–87.

Cutler, D.M., Glaeser, E.L. and Shapiro, J.M. 2003. Why have Americans become more obese? *Journal of Economic Perspectives* 17(3), 93–118.

Geoffard, P.-Y. and Philipson, T. 1996. Rational epidemics and their public control. *International Economic Review* 37, 603–24.

Gersovitz, M. and Hammer, J.S. 2003. Infectious diseases, public policy, and the marriage of economics and epidemiology. *The World Bank Research Observer* 18, 129–57.

Gersovitz, M. and Hammer, J.S. 2004. The economical control of infectious diseases. *Economic Journal* 114, 1–27.

Gersovitz, M. and Hammer, J.S. 2005. Tax/subsidy policies toward vector-borne infectious diseases. *Journal of Public Economics* 89, 647–74.

Grossman, M. and Rashad, I. 2004. The economics of obesity. *Public Interest* (156), 104–12.

Horowitz, B.J. and Moehring, B.H. 2004. How property rights and patents affect antibiotic resistance. *Health Economics* 13, 575–83.

Kremer, M. 1996. Integrating behavioral choice into epidemiological models of the AIDS epidemic. *Quarterly Journal of Economics* 111, 549–73.

Lakdawalla, D., Philipson, T. and Bhattacharya, J. 2005. Welfare enhancing technological change and the growth of obesity. *American Economic Review* 95, 253–8.

Lakdawalla, D., Sood, N. and Goldman, D. 2006. HIV breakthroughs and risky sexual behavior. *Quarterly Journal of Economics* 121, 1063–102.

Laxminarayan, R. 2002. How broad should the scope of antibiotics patents be? *American Journal of Agricultural Economics* 84, 1287–92.

Laxminarayan, R. and Brown, G.M. 2001. Economics of antibiotics resistance: a theory of optimal use. *Journal of Environmental Economics and Management* 42, 183–206.

Laxminarayan, R. and Weitzman, M.L. 2002. On the implications of endogenous resistance to medications. *Journal of Health Economics* 21, 709–18.

Malani, A. 2006. Identifying placebo effects with data from clinical trials. *Journal of Political Economy* 114, 236–56.

Mechoulan, S. 2004. HIV testing, a Trojan horse? *Topics in Economic Analysis and Policy* 4(1), article 18.

Oster, E. 2006. HIV and sexual behavior change: why not Africa? Working paper, Graduate School of Business, University of Chicago.

Philipson, T. 1995. The welfare loss of disease and the theory of taxation. *Journal of Health Economics* 14, 387–96.

Philipson, T. 2000. Economic epidemiology and infectious disease. In *Handbook of Health Economics*, ed. T. Culyer and J. Newhouse. Amsterdam: North-Holland.

Philipson, T. and DeSimone, J. 1997. Experiments and subject sampling. *Biometrika* 84, 618–32.

Philipson, T. and Hedges, L. 1998. Subject evaluation in social experiments. *Econometrica* 66, 381–409.

Philipson, T., Mechoulan, S. and Jena, A.B. 2006. IP and external consumption effects: generalizations from health care markets. Working Paper No. 11930. Cambridge, MA: NBER.

Philipson, T. and Posner, R.A. 1993. *Private Choices and Public Health, An Economic Interpretation of the AIDS Epidemic*. Cambridge, MA: Harvard University Press.

Philipson, T. and Posner, R.A. 2003. The long run growth of obesity as a function of technological change. *Perspectives in Biology and Medicine* 46(3), 87–108.

Rashad, I. and Grossman, M. 2004. The economics of obesity. *Public Interest* (156), 104–12.

economic governance

Formal and informal institutions arise and evolve to underpin economic activity and exchange by protecting property rights, enforcing contracts, and collectively providing physical and organizational infrastructure.

The field of economic governance studies and compares these institutions: state politico-legal institutions, private ordering within the law (credible contracting, arbitration), for-profit governance (credit-rating agencies, organized crime), and social networks and norms. Private institutions can outperform the state's legal system in obtaining and interpreting relevant information, and imposing social sanctions on the violators of norms. But private institutions are often limited in size; as economic activity expands, a transition towards more formal institutions is usually observed.

Concepts and taxonomies

The term 'governance' has exploded from obscurity to ubiquity in economics since the 1970s. A search of the EconLit database shows clear evidence of this explosion. In the relevant categories (title, keywords and abstracts), there are just five occurrences of the word from 1970 to 1979. The number jumps to 112 for the 1980s and 3,825 for the 1990s. Since 2000 to the time of this writing (December 2005), there are already 7,948.

The Oxford English Dictionary gives several definitions of the word 'governance': (*a*) the action or manner of governing; controlling, directing, or regulating influence; control, sway, mastery; the state of being governed; good order; (*b*) the office, function, or power of governing; authority or permission to govern; that which governs; (*c*) the manner in which something is governed or regulated; method of management, system of regulations; a rule of practice, a discipline; and (*d*) the conduct of life or business; mode of living; behaviour, demeanour; discreet or virtuous behaviour; wise self-command. These diverse meanings allow the word to be used (and sometimes misused) for almost any context of economic decision-making or policy.

Two areas of application merit special mention. One is *corporate governance*. This analyses the internal management of a corporation – organizational structure and the design of incentives for managers and workers – and the rules and procedures by which the corporation deals with its shareholders and other stakeholders.

The second is *economic governance*; Williamson (2005) expresses its theme as the 'study of good order and workable arrangements'. This includes the institutions and organizations that underpin economic transactions by protecting property rights, enforcing contracts, and organizing collective action to provide the infrastructure of rules, regulations, and information that are needed to lend feasibility or workability to the interactions among different economic actors, individual and corporate. Different economies at different times have used different institutions to perform these functions, with different degrees of success. The field of economic governance studies and compares these different institutions. It includes theoretical models and empirical and case studies of the performance of different institutions under

different circumstances, of how they relate to each other, of how they evolve over time, and of whether and how transitions from one to another occur as the nature and scope of economic activity and its institutional requirements change.

Corporate governance and economic governance are connected because the boundary of a corporation is itself endogenous, determined by the same considerations of information and commitment costs that raise problems of internal organization as well as those of property and contract (Coase, 1937). Specifically, the nature of transaction costs may make it more efficient to handle some problems of governance by merging the two parties, for example by vertical integration (Williamson, 1975; 1995). But it is analytically convenient to separate the two. This article concerns economic governance. To avoid constant repetition, I will simply call it 'governance' here unless some explicit reference to corporate governance is relevant.

Governance was neglected by economists for a long time, perhaps because they expected the government to provide it efficiently. However, experience with less developed and reforming economies, and observations from economic history, have led economists to study non-governmental institutions of governance.

Governance is not a field per se; it is an organizing or encompassing concept that bears on issues in many fields, including institutions and organizational behaviour, economic development and growth, industrial organization, law and economics, political economy, comparative economic systems, and various subfields of these.

We can organize the subject by classifying institutions along different dimensions. As is usual with such taxonomies, these are conceptual categories to help organize our thinking and analysis. In reality, there are significant differences within each category and overlaps across categories.

The first dimension concerns the purpose of the institution. The categories are: (*a*) protection of property rights against theft by other individuals and usurpation by the state itself or its agents, (*b*) enforcement of voluntary contracts among individuals, and (*c*) provision of the physical and regulatory infrastructure to facilitate economic activity and the functioning of the first two categories of institutions. We might also consider a fourth category, namely, the deep institutions that are essential to avoid serious cleavages or alienation that threaten the cohesion of the society itself. But this has not been studied in this context so far.

The second dimension concerns the nature of the institution. The categories are: (*a*) the formal state institutions that enact and enforce the laws, including the legislature, police, judiciary and regulatory agencies, (*b*) institutions of private ordering that function under the umbrella of state law, for example various forums for arbitration, (*c*) private for-profit institutions that provide information and enforcement, and (*d*) self-enforcement within social or ethnic groups and network. My discussion is organized in sections along this dimension.

A third dimension distinguishes institutions that arise and evolve organically from those that are designed purposively; self-enforcing groups are often organic while the first three categories in the second dimension usually require some measure of design. This matters for the evolution of institutions of governance (see Greif, 2006, especially ch. 6; Williamson 2005 p. 1).

Formal institutions of the state

There is broad agreement that the quality of institutions of governance significantly affects economic outcomes. The importance of protecting property rights, both from other individuals and from predation by the state itself, is generally recognized and documented (for example, De Soto, 2000). But serious disputes about the precise measures of quality of institutions, and about many details of the causal mechanisms by which they affect economic outcomes, remain.

At the broadest level, the distinction is between democracy and authoritarianism, each of which comes in many different varieties. Democracy has many normative virtues, but its worth in governance is less clear. Barro (1999, p. 61) finds an inverse U-shaped relationship between economic growth and a continuous measure of democracy – 'more democracy raises growth when political freedoms are weak, but depresses growth when a moderate amount of freedom is already established' – but the fit is relatively poor. Persson (2005), using cross-sectional as well as panel data, finds that the crude distinction between democratic and non-democratic forms of government is not enough. The precise form of democracy matters for policy design and economic outcomes: 'parliamentary, proportional, and permanent democracies seem to foster the adoption of more growth-promoting structural policies, whereas ... presidential, majoritarian, and temporary democracy do not' (Persson, 2005, p. 22). However, Keefer (2004, p. 10), after surveying a wide-ranging literature on electoral rules and legislative organizations, concludes that they affect policies but are not a crucial determinant of success: 'electoral rules ... almost surely do not explain why some countries grow and others do not', and 'the mere fact that developing countries are more likely to have presidential forms of government is unlikely to be a key factor to explain slow development.'

Democracy can be important for governance because its reliance on rules and procedures provides citizens with protection against predation by the state or its agents. Indeed, the elite, which might otherwise prefer to rule unconstrained, may find it in its own interest to make a credible commitment not to steal from the population by creating and fostering democracy (Acemoglu, 2003; Acemoglu and Robinson, 2005). Greif, Milgrom and Weingast (1994) discuss how groups of traders (guilds) in late medieval Europe took collective action to counter rulers' incentives to violate their members' property rights.

Even in a democracy, agents of the state may pursue their private interests using corruption, complex regulations to extract rent, and favouritism. In fact, an emerging literature argues that economic growth, at least in its early stages, is better promoted under suitably authoritarian regimes. Glaeser et al. (2004) argue that less developed countries that achieve economic success do so by pursuing good policies, often under dictatorships, and only then do they democratize. While these conclusions are controversial, these authors' criticisms of the measures of institutions used in the research that argues for the primacy of institutions in general, and of democracy in particular, are telling. Giavazzi and Tabellini (2005) find a positive feedback between economic and political reform, but they also find that the sequence of reforms matters, and countries that implement economic liberalization first and then democratize do much better in most dimensions than those that follow the opposite route. In practice, of course, it is difficult to ensure *ex ante* that an authoritarian ruler will implement good governance.

Many different measures of institutional quality exist. World Bank researchers Kaufman, Kraay and Mastruzzi (2005, which contains citations to their earlier work) have constructed six: (*a*) Voice and Accountability – measuring political, civil and human rights; (*b*) Political Instability and Violence – measuring the likelihood of violent threats to, or changes in, government, including terrorism; (*c*) Government Effectiveness – measuring the competence of the bureaucracy and the quality of public service delivery; (*d*) Regulatory Burden – measuring the incidence of market-unfriendly policies; (*e*) Rule of Law – measuring the quality of contract enforcement, the police, and the courts, as well as the likelihood of crime and violence; and (*f*) Control of Corruption – measuring the exercise of public power for private gain, including both petty and grand corruption and state capture. Of these, (*e*), (*f*) and also (*b*) concern the most basic institutions for protection of property rights and enforcement of contracts, (*a*) relates to governance because voice and accountability can reduce the severity of the agency problem between the citizens and the agencies of the state, and (*c*) and (*d*) pertain to what I called provision of the infrastructure of governance. Conceptually they are a mixed bag; the quality of some of them can itself depend on the quality of other more basic ones, and some are closer to being measures of effects than of causes. Their method of construction relies on subjective perceptions, and is subject to error. But when used with caution, they have proved significant as explanatory variables in empirical studies of economic growth, and for observing changes in governance quality over time in specific countries. Corruption and regulatory burdens are major themes of the World Bank's research on governance in many countries (see World Bank Institute, website).

Empirical estimations of the level or growth of GDP on various measures of institutional quality confront many conceptual and econometric problems. Researchers have tackled the issue of reverse causation by using various instruments, such as the nationality of colonizers (Hall and Jones, 1999), mortality among colonizers (Acemoglu, Johnson and Robinson, 2001), and whether a colony had rich mineral resources or climatic and soil conditions conducive to plantation agriculture and a large or dense native population, or was sparsely populated and poor in the 1500s (Engerman and Sokoloff, 2002; Acemoglu, Johnson and Robinson, 2002). The general idea is that in the former circumstances the European colonizers established institutions of slavery and inequality to facilitate the exploitation of labour on a large scale, whereas in the latter conditions, where the colonizers had to exert their own effort, their institutions provided the correct incentives and became conducive to longer-term economic success. The debate on the factual and econometric validity and the economic interpretation of these findings is fierce and continuing; Hoff (2003) surveys and discusses this literature in detail.

La Porta et al. (1998; 1999) contrast different legal traditions for protecting the rights of small shareholders. If such protection is poor, that will inhibit the flows of capital to its most efficient uses. They find that systems based on common law are better in this regard than those based on civil law. But Rajan and Zingales (2003) and Lamoreaux and Rosenthal (2005) argue that in practice there was little difference between the systems during critical periods of industrialization.

These debates are sure to continue, and this section will get out of date very quickly.

At the international level, formal governance works through bodies like the World Trade Organization. Their members are sovereign countries; therefore their procedures must be subject to self-enforcement in repeated interactions, whether through bilateral or multilateral sanctions. These institutions are therefore basically similar to the social networks discussed below. See Maggi (1999) and Bagwell and Staiger (2003) for detailed analyses.

Private institutions

The policing functions for property right protection supplied by the state are often supplemented by private security systems that serve specific clients and purposes – firms employ or hire security personnel, gated communities and neighbourhoods have private (hired or volunteer) patrols. These generally merely supplement the functions of the police for their specific context and work cooperatively with the police, but the two may clash if the private security system goes beyond its permissible functions.

Private institutions of contract enforcement similarly coexist with formal law, and become essential when the latter is weak or nonexistent. Explicit or implicit private contractual arrangements are also important for assignment of property rights as a part of Coasean contracting for efficient outcomes. Therefore, analyses of private institutions often focus on the governance of contracts.

The basic problem of contract enforcement is control of opportunism. If one or both parties have to make transaction-specific investments, the other can attempt to secure a greater part of the benefit by reneging or demanding renegotiation. The prospect of this can jeopardize the potentially mutually beneficial deal in the first place. Williamson (1975; 1995) pioneered the analysis of this issue under the title of transaction cost economics.

Information constitutes a major source of advantage for private ordering over formal law. Enforcement of a contract in a court requires offering proof of misconduct by the other party in the dispute; the relevant information must be verifiable to outsiders. Therefore, formal contracts can stipulate actions by the parties conditional only on verifiable information. Other or more detailed information may be observable to the parties themselves, or can be inferred by specialist insiders to the industry, but cannot be verified to non-specialist judges or juries of the state's legal system, or can be verified only at excessive cost.

The informational advantage of private ordering may be offset by a disadvantage in enforcement. Informal arrangements must be made to overcome each participant's temptation to behave opportunistically at the others' expense. Different methods of this kind underlie the various institutions of informal governance, and achieve different degrees of success. Some are able to exert coercion for immediate punishment of misbehaviour. Others create long-run costs, typically in the form of exclusion from future participation or worse future opportunities, to offset the short-run advantages of opportunism. This is the standard theory of self-enforcing cooperation in repeated Prisoner's Dilemmas. The following sections discuss some of these alternatives.

Private ordering with formal law in the background

Perhaps the most remarkable thing about formal legal institutions and mechanisms for the enforcement of commercial contracts is how rarely they are actually used. Business transactions often do have underlying formal contracts, but when disputes arise recourse to the law is often the last resort. Other private alternatives are tried first; these include bilateral negotiation, arbitration by industry experts, and so on. Filing a suit in a formal court of law often signals the end of a business relationship. Most actual practice in business contracting is therefore better characterized as 'private ordering under the shadow of the law' (Macaulay, 1963; Williamson, 1995, pp. 95–100, 121–2).

If one of the parties to an ongoing informal relationship behaves opportunistically, the most common alternative is to fall back on a formal contract based on verifiable contingencies alone. Suppose an outcome based on a tacit understanding of what each party should do in any one exchange (including good-faith negotiation to adapt to changing circumstances) yields both of them higher payoffs than does a formal contract. Consider the implicit arrangement where, if one party deviates from the agreed course of action to its own advantage and to the detriment of the other, their future exchanges will be governed by the formal contract. This yields a subgame-perfect (credible) equilibrium of the repeated game if each party's one-time gain from opportunism does not exceed the capitalized value of the future difference of payoffs between the tacit and the formal contracts. Williamson (2005, p. 2) expresses this well: 'continuity can be put in jeopardy by defecting from the spirit of cooperation and reverting to the letter.'

When such relationship-based implicit contracting prevails, partial improvement in the formal system can worsen the outcome, due to a problem of the second-best. The partial improvement raises the payoffs the two parties could get from the fallback formal contract. This in turn reduces the future cost of a current deviation from the implicit contract or spirit of cooperation. It tightens the incentive-compatibility constraints, and therefore worsens what can be achieved by relational contracting (Baker, Gibbons and Murphy, 1994; Dixit, 2004, ch. 2).

Arbitration comes in two prominent forms. One is industry-specific, based on expert knowledge of insiders. More information is verifiable in such settings; therefore richer contracts specifying actions for more detailed contingencies become feasible. In many industries there is a large common-knowledge basis of custom and practice, which may even make it unnecessary to write down a contingent contract in great detail. Arbitration can also provide an opportunity for the parties to communicate and renegotiate adaptations to new circumstances. Formal legal systems often recognize these advantages of expert arbitration, and courts stand ready to enforce the decisions of arbitrators if the losing party tries to evade the sanctions. However, industry arbitrators often have severe sanctions at their own disposal; they can essentially drive the miscreant out of business, and even ostracize him or her from the social group of that business community. Examples of arbitration institutions include Bernstein's (1992) classic study of the diamond industry. For further discussion and modelling, see Dixit (2004, ch. 2) and Williamson (2005, p. 14).

The other prominent forums of arbitration deal with international contracts (Dezalay and Garth, 1996; Mattli, 2001). There are several of these, specializing in different legal traditions. They lack direct power to enforce their decisions, but are backed by treaties that ensure enforcement by national courts. These forums do not have

industry-specific knowledge, their processes can be slow and costly, and their decisions can be somewhat arbitrary. But parties in transnational transactions may prefer them to either country's courts, suspecting that these will be biased in favour of their own nationals.

For-profit private institutions

If the state is unwilling to protect certain kinds of property or enforce certain kinds of contracts (for example in illegal activities), or is unable to do so (for example in weak and failing states), or is itself predatory, then private institutions can emerge to perform these functions for a profit. Organized crime often fills the niches uncovered by the state. Gambetta (1993), Bandiera (2003) and others argue that the Mafia emerged in just such a situation to fill the vacuum of protection in late 19th-century Sicily. Landowners began to hire guards of former feudal lords, and even the toughest among bandits, to protect their property. Gambetta describes how the Mafia's role expanded to providing contract enforcement in illegal or grey markets. Similarly, the Japanese Yakuza was instrumental in organizing markets at the end of the Second World War in August and September 1945 when the Japanese state had collapsed (Dower, 1999, pp. 140–8), and mafias grew in Russia after the collapse of the Soviet regime (Varese, 2001).

Gambetta (1993, p. 19) argues that this 'business of protection' is the core business of the Mafia. It may engage in other activities using in-house protection, but that is just downstream vertical integration – the opposite of upstream integration where an ordinary business firm has its in-house security department. A transaction-cost analysis of the internal organization of mafias, and of their vertical integration decisions, may provide an interesting link between economic governance and corporate governance. Another dimension in which the protection business can expand is extortion; although private protectors may be welcome when state protection has collapsed, 'protectors, once enlisted, invariably overstay their welcome' (Gambetta, 1993, p. 198).

The Mafia can provide contract enforcement because, even though two traders may not have sufficiently frequent dealings with each other to achieve good outcomes in an ongoing bilateral relationship, each trader can be a regular customer of the enforcer. This converts multiple one-shot Prisoner's Dilemma games among the whole group of traders into several bilateral repeated games of each trader with the enforcer. The intermediary can provide information (keeping track of previous contract violations and informing a customer of the history of a potential trading partner) and/or actual punishment if a customer's trading partner violates their contract. The information role of the Mafioso is similar to that of credit rating agencies and Better Business Bureaus in the United States. Dixit (2004, ch. 4) constructs a model of such for-profit governance, and establishes the conditions for an equilibrium with for-profit private enforcement. These are lower bounds on the shares of the surplus that the customer and the Mafioso must have, so as to overcome the trader's temptation to cheat and the Mafioso' temptation to double-cross the customer. Milgrom, North and Weingast (1990) have a related and complementary model of private judges at medieval European trade fairs. They specify the game of each trade, and investigation in the event of cheating, in greater detail, but do not examine the issue of the judges' honesty.

Group enforcement through social networks and norms

Any institution of contract enforcement must solve three key problems: (*a*) detection of opportunistic deviations from the contractually stipulated behaviour, (*b*) preservation and dissemination of information about the histories of the participants' behaviour, and (*c*) inflicting appropriate punishments to reduce future payoffs of any deviators. The first is often constrained by the available technology of monitoring, although institutions and regulations such as reporting requirements and auditing can improve the technology. The second and third problems are best resolved in bilateral ongoing relationships: each party has a natural incentive to detect and remember the other's cheating, and can punish the other by breaking off the relationship. However, governance is often needed in groups each of whose members interacts frequently with someone else in the group, but not necessarily bilaterally with the same person every time. Now remembering and transmitting information about your current partner's behaviour to others, and refusing a potentially beneficial deal because the counter-party has cheated someone else in the past, are privately costly activities and therefore require their own governance mechanisms.

Formal state institutions of governance can solve these problems by fiat; the legal system compels the whole group of traders to commit to good behaviour by subjecting themselves to detection and punishment if they cheat. A third-party supplier of information or enforcement serves similar functions. In the case of a Mafia enforcer, anyone who trades with a customer of the Mafioso subjects himself to the grim punishment if he cheats. In the case of a Better Business Bureau, a firm that joins the organization thereby gives hostage to its own good behaviour: if it misbehaves it will get a poor rating or blacklisting. Transactions vary in their characteristics; therefore we should expect the effectiveness of such reputation mechanisms to vary also, and should not expect universal success from any one.

An institution of social networks and norms can solve the problems of information and punishment in a decentralized manner. Each participant can transmit information about his or her current trading partner's behaviour to others in the group to whom he or she is linked. And each can play his or her assigned part in

punishment, typically by refusing to trade, if he or she gets matched with a potential partner who is known to have misbehaved in past dealings with others in the group. Incentives to transmit information or refuse potentially good trades can be established by a norm that regards refusal to do so as itself a punishable offence, as in Abreu's (1986) penal codes for repeated games; see Calvert (1995a; 1995b). Extrinsic incentives may even be unnecessary if people have sufficiently strong natural instincts to punish social cheaters, as found by Fehr and Gächter (2000).

Numerous empirical and case studies of governance based on social relations have been conducted; space constraints allow mention of only a few. Greif's (1993) historical analysis of Maghribi traders' system of communication and collective punishment is well known. So is Ostrom's (1990) synthesis of the evidence on common-pool resource management; she emphasizes the importance of local knowledge and communication, of appropriately designed (generally graduated) punishments, and of incentives for individuals to perform their assigned roles and actions in the system. Fafchamps (2004) studies and compares many different market institutions in Africa; his work highlights the importance of designing systems appropriate to the conditions of each country or group. Ensminger (1992) describes a similarly rich complex of arrangements for trade and employment relationships among the Orma tribe of Kenya, and examines how formal institutions of property right enforcement including title registration can interact dysfunctionally with traditional arrangements based on family and tribal connections. Johnson, McMillan and Woodruff (2002) present and analyse findings from survey research in former socialist economies. Of particular interest are the links between evolving formal and informal governance. Even without a backup of courts, trust in bilateral relationships can build quickly in response to good experiences. New or transient customers are more likely to be offered credit if courts work better, but the effectiveness of courts becomes largely irrelevant for the functioning of established relationships. Casella and Rauch (2002) study the role of ethnic networks in international trade.

Li (2003) points out a key difference between the costs of operating such a system and those of formal governance. A relation-based system of networks and norms has low fixed costs, but high and rising marginal costs. Trading on a small scale naturally starts among the most closely connected people who have sufficiently good communication and common understanding to sustain honesty. No fixed costs need be incurred to establish any formal rules or mechanisms of enforcement. But as trade expands, potential partners added at the margin are almost by definition less well-connected, making it harder to communicate information with them and to ensure their participation in any punishments. By contrast, formal or rule-based governance has high fixed costs of setting up the legal system and the information mechanism, but once these are incurred, marginal costs of dealing with strangers are low. Therefore, relation-based governance is better for small groups and rule-based governance better for large groups. Greif's (1994) comparison between the relation-based system of Maghribi traders and the formal institutions of Genoese traders supports this theory. Dixit (2004, ch. 3) constructs a formal model that compares relation-based and rule-based systems. This characterizes the maximum size of a self-enforcing group, and finds that, when the group exceeds this critical size, the maximum scope of sustainable honesty shrinks absolutely. The intuition is as follows. At the critical size, each trader is indifferent between honesty and cheating when dealing with the most distant person. When more traders are added, this weakens the communication between the previously marginal person and other almost equally distant ones, tipping the balance toward cheating.

Kranton (1996) models individuals who can either choose bilateral long-lived self-enforcing trading relationships or search for one-time trading partners in an anonymous market with external enforcement. The market thus provides the outside opportunity in the repeated game of bilateral trade. If more people trade in the anonymous market, it becomes thicker and offers better prospects for successful search. Then parties in bilateral relationships have better outside opportunities, which makes it harder to sustain tacit cooperation there, further increasing the relative attraction of the market. Therefore the system can have multiple equilibria – no one uses the market because no one else uses it, or everyone uses the market because everyone else does – and can get locked into a Pareto-inferior equilibrium.

Evolution and transformation of governance institutions

A persistent theme in this survey has been that different governance institutions are optimal for different societies, for different kinds of economic activity, and at different times. Changes in underlying technologies of production, exchange and communication change the relative merits of different methods of governance. As the volume and scope of trade expand, formal institutions generally become superior to informal ones, but informal ones serve useful roles under the shadow of formal ones even in the most advanced economies and sectors. All this raises the question of whether we should expect institutions to adapt and evolve optimally.

Williamson's famous 'discriminating alignment hypothesis' says that transactions, with their different attributes, align with institutions, with their different costs and competencies; see his recent exposition (2005, p. 6). This gives ground for optimism for synergistic evolution of the need for governance and the institutions that supply it. Others are less sanguine. North (1990) and others argue that institutional change is subject to long

delays due to resistance by organized interests favouring the status quo, problems of coordinating collective action to bring about a discrete change in equilibrium, and so on. Dixit (2004, pp. 79–85) discusses some of these problems for transition from relation-based to rule-based contract enforcement. Eggertson (2005) gives a dramatic example of how institutions restricting fishing and requiring costly mutual insurance persisted in Iceland for centuries after they had become obstacles to good economic performance.

I believe that a balanced approach is needed, recognizing the tendency towards synergistic alignment but also the obstacles to its realization. The net outcome will depend on many specifics of each context. Understanding and predicting the process requires a combination of approaches: case-based and analytical, inductive and deductive. Greif (2006) discusses, develops and applies such methodologies using historical studies of trade in medieval Europe.

AVINASH K. DIXIT

See also **cooperation; corporations; growth and institutions; hold-up problem; law, economic analysis of; law, public enforcement of; market institutions; property rights; social norms; spontaneous order; transition and institutions.**

I thank Tore Ellingsen, Diego Gambetta, Karla Hoff, Eva Meyersson-Milgrom, Dani Rodrik, Oliver Williamson, and the editors for comments on previous drafts, and the National Science Foundation for research support.

Bibliography

Abreu, D. 1986. Extremal equilibria of oligopolistic supergames. *Journal of Economic Theory* 39, 191–225.
Acemoglu, D. 2003. Why not a political Coase Theorem? Social conflict, commitment and politics. *Journal of Comparative Economics* 31, 620–52.
Acemoglu, D., Johnson, S. and Robinson, J. 2001. The colonial origins of comparative development: an empirical investigation. *American Economic Review* 91, 1369–401.
Acemoglu, D., Johnson, S. and Robinson, J. 2002. Reversal of fortune: geography and institutions in the making of the modern world income distribution. *Quarterly Journal of Economics* 117, 1231–94.
Acemoglu, D. and Robinson, J. 2005. *Economic Origins of Dictatorship and Democracy.* New York: Cambridge University Press.
Bagwell, K. and Staiger, R. 2003. *The Economics of the World Trading System.* Cambridge, MA: MIT Press.
Baker, G., Gibbons, R. and Murphy, K. 1994. Subjective performance measures in optimal incentive contracts. *Quarterly Journal of Economics* 109, 1125–56.
Bandiera, O. 2003. Land reform, the market for protection and the origins of the Sicilian Mafia: theory and

evidence. *Journal of Law, Economics and Organization* 19, 218–44.

Barro, R. 1999. *Determinants of Economic Growth: A Cross-Country Empirical Study.* Cambridge, MA: MIT Press.

Bernstein, L. 1992. Opting out of the legal system: extralegal contractual relations in the diamond industry. *Journal of Legal Studies* 21, 115–57.

Calvert, R. 1995a. The rational choice theory of social institutions: cooperation, communication, and coordination. In *Modern Political Economy: Old Topics, New Directions*, ed. J. Banks and E. Hanushek. Cambridge: Cambridge University Press.

Calvert, R. 1995b. Rational actors, equilibrium, and social institutions. In *Explaining Social Institutions*, ed. J. Knight and I. Sened. Ann Arbor: University of Michigan Press.

Casella, A. and Rauch, J. 2002. Anonymous market and group ties in international trade. *Journal of International Economics* 58, 19–47.

Coase, R. 1937. The nature of the firm. *Economica* 4, 386–406.

De Soto, H. 2000. *Mystery of Capital: Why Capitalism Triumphs in the West and Fails Everywhere Else.* New York: Basic Books.

Dezalay, Y. and Garth, B. 1996. *Dealing in Virtue: International Commercial Arbitration and the Construction of a Transnational Order.* Chicago, IL and London: University of Chicago Press.

Dixit, A. 2004. *Lawlessness and Economics: Alternative Modes of Governance.* Princeton, NJ: Princeton University Press.

Dower, J. 1999. *Embracing Defeat: Japan in the Wake of World War II.* New York: W.W. Norton.

Eggertson, T. 2005. *Imperfect Institutions: Possibilities and Limits of Reform.* Ann Arbor: University of Michigan Press.

Engerman, S. and Sokoloff, K. 2002. Factor endowments, inequality, and paths of development among New World economies. *Economia* 3, 41–109.

Ensminger, J. 1992. *Making a Market: The Institutional Transformation of an African Society.* New York: Cambridge University Press.

Fafchamps, M. 2004. *Market Institutions in Sub-Saharan Africa: Theory and Evidence.* Cambridge, MA: MIT Press.

Fehr, E. and Gächter, S. 2000. Cooperation and punishment in public goods experiments. *American Economic Review* 90, 980–94.

Gambetta, D. 1993. *The Sicilian Mafia: The Business of Private Protection.* Cambridge, MA: Harvard University Press.

Giavazzi, F. and Tabellini, G. 2005. Economic and political liberalizations. *Journal of Monetary Economics* 57, 1297–330.

Glaeser, E., La Porta, R., Lopez-de-Silanes, F. and Shleifer, A. 2004. Do institutions cause growth? *Journal of Economic Growth* 9, 271–303.

Greif, A. 1993. Contract enforceability and economic institutions in early trade: the Maghribi traders' coalition. *American Economic Review* 83, 525–48.

Greif, A. 1994. Cultural beliefs and the organization of society: a historical and theoretical reflection on collectivist and individualist societies. *Journal of Political Economy* 102, 912–50.

Greif, A. 2006. *Institutions and the Path to the Modern Economy: Lessons from Medieval Trade.* New York: Cambridge University Press.

Greif, A., Milgrom, P. and Weingast, B. 1994. Coordination, commitment, and enforcement: the case of the merchant guild. *Journal of Political Economy* 102, 745–76.

Hall, R. and Jones, C. 1999. Why do some countries produce so much more output than others? *Quarterly Journal of Economics* 114, 83–116.

Hoff, K. 2003. Paths of institutional development: a view from economic history. *World Bank Research Observer* 18, 205–26.

Johnson, S., McMillan, J. and Woodruff, C. 2002. Courts and relational contracts. *Journal of Law, Economics and Organization* 18, 221–77.

Kaufman, D., Kraay, A. and Mastruzzi, M. 2005. Governance matters IV: Updated governance indicators 1996–2004. Washington, DC: World Bank research paper. Online. Available at http://www.worldbank.org/wbi/governance/pubs/govmatters4.html, accessed 20 April 2006.

Keefer, P. 2004. What does political economy tell us about economic development – and vice versa? *Annual Review of Political Science* 7, 247–72.

Kranton, R. 1996. Reciprocal exchange: a self–sustaining system. *American Economic Review* 86, 830–51.

La Porta, R., Lopez-de-Silanes, F., Shleifer, A. and Vishny, R. 1998. Law and finance. *Journal of Political Economy* 106, 1113–55.

La Porta, R., Lopez-de-Silanes, F., Shleifer, A. and Vishny, R. 1999. The quality of government. *Journal of Law, Economics and Organization* 15, 222–79.

Lamoreaux, N. and Rosenthal, J.-L. 2005. Legal regime and contractual flexibility: a comparison of business's organizational choices in France and the United States during the era of industrialization. *American Law and Economics Review* 7, 28–61.

Li, J.S. 2003. Relation-based versus rule-based governance: an explanation of the East Asian miracle and Asian crisis. *Review of International Economics* 11, 651–73.

Macaulay, S. 1963. Non-contractual relationships in business: a preliminary study. *American Sociological Review* 28, 55–70.

Maggi, G. 1999. The role of multilateral institutions in international trade cooperation. *American Economic Review* 89, 190–214.

Mattli, W. 2001. Private justice in a global economy: from litigation to arbitration. *International Organization* 55, 919–47.

Milgrom, P., North, D. and Weingast, B. 1990. The role of institutions in the revival of trade: the law merchant, private judges, and the Champagne fairs. *Economics and Politics* 2, 1–23.

North, D. 1990. *Institutions, Institutional Change, and Economic Performance*. Cambridge: Cambridge University Press.

Ostrom, E. 1990. *Governing the Commons: The Evolution of Institutions for Collective Action*. Cambridge, UK, and New York: Cambridge University Press.

Persson, T. 2005. Forms of democracy, policy, and economic development. Working Paper No. 11171. Cambridge, MA: NBER.

Rajan, R. and Zingales, L. 2003. The great reversals: the politics of financial development in the twentieth century. *Journal of Financial Economics* 69, 5–50.

Varese, F. 2001. *The Russian Mafia: Private Protection in a New Market Economy*. Oxford: Oxford University Press.

Williamson, O. 1975. *Markets and Hierarchies: Analysis and Antitrust Implications*. New York: Free Press.

Williamson, O. 1995. *The Mechanisms of Governance*. New York: Oxford University Press.

Williamson, O. 2005. The economics of governance. *American Economic Review* 95, 1–18.

World Bank Institute. Governance and Anti-Corruption. Online. Available at http://www.worldbank.org/wbi/governance, accessed 20 April 2006.

economic growth

Economic growth is typically measured as the change in per capita gross domestic product (GDP). Sustained long-term economic growth at a positive rate is a fairly recent phenomenon in human history, most of it having occurred in the last 200 years. According to Maddison's (2001) estimates, per capita GDP in the world economy was no higher in the year 1000 than in the year 1, and only 53 per cent higher in 1820 than in 1000, implying an average annual growth rate of only one-nineteenth of one per cent over the latter 820-year period. Some time around 1820, the world growth rate started to rise, averaging just over one-half of one per cent per year from 1820 to 1870, and peaking during what Maddison calls the 'golden age', the period from 1950 to 1973, when it averaged 2.93 per cent per year. By 2000, world per capita GDP had risen to more than 8.5 times its 1820 value.

Growth has been uneven not only across time but also across countries. Since 1820, living standards in Western Europe and its offshoots in North America and the Antipodes have raced ahead of the rest of the world, with the exception of Japan, in what is often referred to as the 'Great Divergence'. As shown in Figure 1 below, the proportional gap in per capita GDP between the richest group of countries and the poorest group (as classified by Maddison) grew from three in 1820 to 19 in 1998. Pritchett (1997) tells a similar story, estimating that the proportional gap between the richest and poorest countries grew more than fivefold from 1870 to 1990.

This widening of the cross-country income distribution seems to have slowed during the second half of the 20th century, at least among a large group of nations. Indeed, Figure 1, which is drawn on a proportional scale, shows that with the acceleration of growth in Asia there has been a narrowing of the spread between the richest and the second poorest group since 1950. Evans (1996) shows a narrowing of the top end of the distribution (that is, among Organisation for Economic Co-operation and Development, OECD, countries) over the period. However, not all countries have taken part in this convergence process, as the gap between the leading countries as a whole and the very poorest countries has

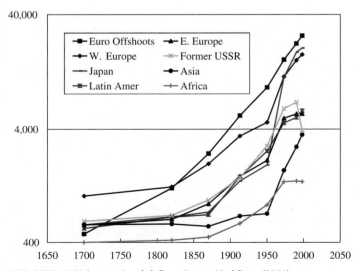

Figure 1   Per capita GDP, 1650–2000, 1990 international dollars. *Source*: Maddison (2001).

continued to widen. In Figure 1 the gap between the Western Offshoots and Africa grew by a factor of 1.75 between 1950 and 1998. Likewise, the proportional income gap between Mayer-Foulkes's (2002) richest and poorest convergence groups grew by a factor of 2.6 between 1960 and 1995.

Jones (1997) argues that continuing divergence of the poorest countries from the rest of the world does not imply rising income inequality among the world's population, mainly because China and India, which contain about 40 per cent of that population, are rising rapidly from near the bottom of the distribution. Indeed, Sala-i-Martin (2006) shows, using data on within-country income distributions, that the cross-individual distribution of world income narrowed considerably between 1970 and 2000, even as the cross-country distribution continued to widen somewhat. But between-country inequality is still extremely important; in 1992 it explained 60 per cent of overall world inequality (Bourguignon and Morrison, 2002). Another reason that growth economists are typically more concerned with the cross-country than the cross-individual distribution is that many of the determinants of economic growth vary across countries but not across individuals within countries.

The production function approach

The main task of growth theory is to explain this variation of living standards across time and countries. One way to organize one's thinking about the sources of growth is in terms of an aggregate production function, which indicates how a country's output per worker y depends on the (per worker) stocks of physical, human, and natural capital, represented by the vector k, according to

$$y = f(k, A),$$

where A is a productivity parameter. Economic growth, as measured by the growth rate of y, depends therefore on the rate of capital accumulation and the rate of productivity growth. Similarly, countries can differ in their levels of GDP per capita either because of differences in capital or because of differences in productivity. Much recent work on the economics of growth has focused on trying to identify the relative contributions of these two fundamental factors to differences in growth rates or income levels among countries.

Modern growth theory started with the neoclassical model of Solow (1956) and Swan (1956), who showed that in the long run growth cannot be sustained by capital accumulation alone. In their formulation, the diminishing marginal product of capital (augmented by an Inada condition that makes the marginal product asymptote to zero as capital grows) will always terminate any temporary burst of growth in excess of the growth rate of labour-augmenting productivity. But this perspective has been challenged by more recent endogenous

growth theory. In the *AK* theory of Frankel (1962) and Romer (1986), growth in productivity is functionally dependent on growth in capital, through learning by doing and technology spillovers, so that an increase in investment rates in physical capital can also sustain a permanent increase in productivity growth and hence in the rate of economic growth. In the innovation-based theory that followed *AK* theory, the Solow model has been combined with a Schumpeterian theory of productivity growth, in which capital accumulation is one of the factors that can lead to a permanently higher rate of productivity growth (Howitt and Aghion, 1998).

Capital

Having introduced the production function in a general sense, we now examine the accumulation of different types of capital in more detail, and then turn to an assessment of the relative importance of factor accumulation and productivity in explaining income differences among countries and growth over time.

Physical capital

Physical capital is made up of tools, machines, buildings, and infrastructure such as roads and ports. Its key characteristics are, first, that it is produced (via investment), and second that it is in turn used in producing output. Physical capital differs importantly from technology (which, as is discussed below, is also both produced and productive) in that physical capital is rival in its use: only a limited number of workers can use a single piece of physical capital at a time.

Differences in physical capital between rich are poor countries are very large. In the year 2000, for example, physical capital per worker was 148,091 dollars in the United States, 42,991 dollars in Mexico, and 6,270 dollars in India. These large differences in physical capital are clearly contributors to income differences among countries in a proximate sense. That is, if the United States had India's level of capital it would be a poorer country. The magnitude of this proximate effect can be calculated by using the production function. For example, using a value for capital's share of national income of 1/3 (which is consistent with the findings of Gollin (2002) for a cross-section of countries), the ratio of capital per worker in the United States to that in India would by itself explain a ratio of income per capita in the two countries of 7.9 ($=(148,091/6,270)^{1/3}$).

Differences in physical capital among countries can result from several factors. First, countries may differ in their levels of investment in physical capital relative to output. In an economy closed to external capital flows, the investment rate will equal the national saving rate. Saving rates can differ among countries because of differences in the security of property rights, due to the availability of a financial system to bring together savers and investors, because of government policies like budget deficits or pay-as-you-go old age pensions, differences in cultural attitudes towards present versus future consumption, or simply because deferring consumption to the future is a luxury that very poor people cannot afford.

A second factor that drives differences in investment rates among countries is the relative price of capital. The price of investment goods in relation to consumption goods is two to three times as high in poor countries as in rich countries. If one measures both output and investment at international prices, investment as a fraction of GDP is strongly correlated with GDP per capita (correlation of 0.50), and poor countries have on average between one half and one quarter of the investment rate of rich countries. When investment rates are expressed in domestic prices, the correlation between investment rates and GDP per capita falls to 0.05 (Hsieh and Klenow, 2007).

But levels of capital can also differ among countries for reasons that have nothing to do with the rate of accumulation. Differences in productivity (the A term in equation 1) will produce different levels of capital even in countries with the same rates of physical capital investment. Similarly, differences in the accumulation of other factors of production will produce differences in the level of physical capital per worker.

Human capital
Human capital refers to qualities such as education and health that allow a worker to produce more output and which themselves are the result of past investment. Like physical capital, human capital can earn an economic return for its owner. However, the two types of capital differ in several important respects. Most significantly, human capital is 'installed' in a person. This makes it very difficult for one person to own human capital that is used by someone else. Human capital investment is a significant expense. In the United States in the year 2000, spending by governments and families on education amounted to 6.2 per cent of GDP; forgone wages by students were of a similar magnitude.

Information on the productivity of human capital can be derived from comparing wages of workers with different levels of education. So called 'Mincer regressions' of log wage on years of education, controlling by various means for bias due to the endogeneity of schooling, yield estimated returns to schooling of about ten per cent per year. In the year 2000, the average schooling of workers in advanced countries was 9.8 years and among workers in developing countries 5.1 years. Applying a rate of return of ten per cent implies that the average worker in the advanced countries supplied 56 per cent more labour input because of this education difference. If labour's share in a Cobb–Douglas production function is two-thirds, this would imply that education differences would explain a factor of 1.35 difference in income between the advanced and developing countries, which is very small

relative to the observed gap in income. Allowing for differences in school quality increases somewhat the income differences explained by human capital in the form of schooling.

A second form of human capital is health. The importance of health as an input into production can be estimated by looking at microeconomic data on how health affects individual wages. Health differences between rich and poor countries are large, and in wealthy countries worker health has improved significantly over the last 200 years (Fogel, 1997). Weil (2007), using the adult survival rate as a proxy for worker health, estimates that eliminating gaps in worker health among countries would reduce the log variance of GDP per worker by 9.9 per cent.

Natural capital
Natural capital is the value of a country's agricultural and pasture lands, forests and subsoil resources. Like physical and human capital, natural capital is an input into production of goods and services. Unlike other forms of capital, however, it is not itself produced.

Natural capital per worker and GDP per worker are positively correlated, but the link is much weaker than for the other measures of capital discussed above. The poor performance of many resource-rich countries has led many observers to identify a 'resource curse' by which the availability of natural capital undermines other forms of capital accumulation or reduces productivity. Among the suggested channels by which this happens are that resource booms lead countries to raise consumption to unsustainable levels, thus depressing saving and investment (Rodriguez and Sachs, 1999); that exploitation of natural resources suppresses the development of a local manufacturing sector, which holds back growth because manufacturing is inherently more technologically dynamic than other parts of the economy (this is the so called Dutch disease); and that economic inefficiencies are associated with political competition or even civil war to appropriate the rents generated by natural resources.

Population and economic growth
Population affects the accumulation of all three forms of capital discussed above, and through them the level of output per worker. Rapid population growth dilutes the quantities of physical and human capital per worker, raising the rates of investment and school expenditure required to maintain output per worker. The interaction of natural capital with population growth is at the centre of the model of Malthus (1798). For a fixed stock of natural capital, higher population lowers output per capita. Combined with a positive feedback from the level of income to population growth, this resource constraint produces a stable steady state level of output per capita and, with technology fixed, a stable level of population as well. This Malthusian feedback is the explanation for the

long period of nearly constant living standards that pre-ceded the Industrial Revolution (Galor and Weil, 2000). Because of resource-saving technological progress, as well as expansion of international trade, which allows coun-tries to evade resource constraints, the interaction of population and natural capital is much less important today than in the past, with the exception of very poor countries that are reliant on subsistence agriculture.

In addition to its effect on the level of factors of production per worker, population also matters for economic growth because demographic change produces important changes in the age structure of the population. A reduction in fertility, for example, will produce a long period of reduced dependency, in which the ratio of children and the elderly, on the one hand, to working age adults, on the other, is temporarily below its sustainable steady state level. This is the so-called 'demographic dividend' (see POPULATION AGEING).

In addition to these effects of population on the level of income per capital, there is also causality that runs from the economic to the demographic. Over the course of economic development, countries generally move through a demographic transition in which mortality rates fall first, followed by fertility rates. While the decline in mortality is easily explained as a consequence of higher income and technological progress, the decline in fertility is not fully understood. Among the factors thought to contribute to the decline in fertility are falling mortality, a shift along a quality–quantity trade-off due to rising returns to human capital, the rise of women's relative wages, the reduced importance of children as a means of old age support, and improvements in the availability of contraception.

Growth accounting and development accounting

The discussion above makes clear that stocks of different forms of capital are positively correlated with GDP per capita. Similarly, as countries grow, levels of capital per worker grow as well. It is natural to ask whether these variations in capital are sufficiently large to explain the matching variations in growth. The techniques of growth accounting (Solow, 1957) and development accounting (Klenow and Rodriguez-Clare, 1997; Hall and Jones, 1999) attempt to give quantitative answers to this ques-tion. Using a parameterized production function and measures of the quantities of human and physical capital, one can back out relative levels of productivity among countries and rates of productivity growth within a country.

Caselli (2005) presents a review of development accounting along with his own thorough estimates. His finding is that if human and physical capital per worker were equalized across countries, the variance of log GDP per worker would fall by only 39 per cent. In other words, the majority of variation in income is due to differences in productivity, not factor accumulation. Differences in

productivity growth, rather than differences in the growth of physical and human capital, are also the domi-nant determinants of differences in income growth rates among countries (Weil, 2005, ch. 7; Klenow and Rodriguez-Clare, 1997); differences in productivity levels among countries are striking. For example, comparing the countries at the 90th and 10th percentiles of the income distribution (which differ in income by a factor of 21), the former would produce seven times as much output as the latter with equal quantities of human and physical capital.

Productivity, technology and efficiency

Development accounting shows that productivity differences among countries are the dominant explana-tion for income differences. Similarly, differences in productivity growth are the most important explanation for differences in income growth rates among countries. And as a theoretical matter, the Solow model shows that as long as there are decreasing returns to capital per worker, productivity growth can be the only source of long-term growth. The question is: what explains these changes over time and differences in the level of pro-ductivity? Over the long term it is natural to associate productivity growth with technological change. However, especially as an explanation for differences in productiv-ity at a given point in time, a second possibility is that productivity differences reflect differences not in tech-nology, in the sense of inventions, blueprints, and so on, but rather differences in how economies are organized and use available technology and inputs. We label this second contributor to productivity as 'efficiency'.

Technology

Technology consists of the knowledge of how to transform basic inputs into final utility. This knowledge can be thought of as another form of capital, an intan-gible intellectual capital. What distinguishes technology from human or physical capital is its non-rival character. For example, the knowledge that a particular kind of corn will be immune to caterpillars, or the knowledge of how to produce a 3 GHz CPU for a portable computer, can be used any number of times by any number of people without diminishing anyone's ability to use it again. By contrast, if you drive a lorry for an hour, or if you employ the skills of a doctor for an hour, then that lorry or those skills are not available to anyone else during that hour.

Different growth theories have different approaches to modelling the accumulation of technology – that is, technical progress. According to neoclassical theory, for example, the relationship between technology and the economy is a one-way street, with all of the causation running from technology to the economy. It portrays technical progress as emanating from a scientific progress that operates outside the realm of economics, and thus takes the rate of technical progress as being given exogenously.

This neoclassical view has never been accepted universally. Specialists in economic history and the economics of technology have generally believed that technical progress comes in the form of new products, new techniques and new markets, which do not spring directly from the scientific laboratory; instead they come from discoveries made by private business enterprises, operating in competitive markets, and motivated by the search for profits. For example, the transistor, which underlies so much recent technological progress, was discovered by scientists working for the AT&T telephone company on the practical problem of how to improve the performance of switch boxes that were using vacuum tubes. Rosenberg (1981) describes many other examples of scientific and technological breakthroughs that originated in profit-oriented economic activity.

What kept this view of endogenous technology from entering the mainstream of economics until recently was the difficulty of incorporating increasing returns to scale into dynamic general equilibrium theory. Increasing returns arise once one considers technology as a kind of capital that can be accumulated, because of its non-rival nature; that is, the cost of developing a technology for producing a particular product is a fixed set-up cost, which does not have to be repeated when more of the product is produced. Once the technology has been developed then there should be at least constant returns to scale in the factors that use that technology, on the grounds that if you can do something once then you can do it twice. But this means that there are increasing returns in the broad set of factors that includes the technology itself. Increasing returns creates a problem because it generally implies that a competitive equilibrium will not exist, at least not without externalities.

These technical difficulties were overcome by the new 'endogenous growth theory' introduced by Romer (1986) and Lucas (1988), which incorporated techniques that had been developed for dealing with increasing returns in the theories of industrial organization and international trade. The first generation of endogenous growth theory to enter the mainstream was the 'AAK theory', according to which technological progress takes place as a result of externalities in learning to produce capital goods more efficiently. The second generation was the innovation-based theory of Romer (1990) and Aghion and Howitt (1992), which emphasizes the distinction between technological knowledge and other forms of capital, and analyses technological innovation as a separate activity from saving and schooling.

Historically, technical progress has engendered much social conflict, because it involves what Schumpeter (1942) called 'creative destruction'; that is, new technologies render old technologies obsolete. As a result, technical progress is a game with losers as well as winners. From the handloom weavers of early 19th century Britain to the former giants of mainframe computing in the late 20th century, many people's skills, capital equipment and technological knowledge have been devalued and their livelihoods imperilled by the same innovations that have created fortunes for others.

The destructive side of technical progress shows up most clearly during periods when a new 'general purpose technology' (GPT) is being introduced. A GPT is a basic enabling technology that is used in many sectors of the economy, such as the steam engine, the electric dynamo, the laser or the computer. As Lipsey, Carlaw and Bekar (2005) have emphasized, a GPT typically arrives only partially formed, creates technological complementarities and opens a window on new technological possibilities. Thus it is typically associated with a wave of new innovations. Moreover, the period in which the new GPT is diffusing through the economy is typically a period of rapid obsolescence, costly learning and wrenching adjustment. Greenwood and Yorukoglu (1997) argue that the productivity slowdown of the 1970s is attributable to the arrival of the computer, and Howitt (1998) argues that the rapid obsolescence generated by a new GPT can cause per capita income to fall for many years before eventually paying off in a much higher standard of living.

New technologies are often opposed by those who would lose from their introduction. Some of this opposition takes place within the economic sphere, where workers threaten action against firms that adopt labour-saving technologies and firms try to pre-empt innovations by rivals. But much of it also takes place within the political sphere, where governments protect favoured firms from more technically advanced foreign competitors, and where people sometimes vote for politicians promising to preserve traditional ways of life by blocking the adoption of new technologies.

The leading industrial nations of the world spend large amounts on R&D for generating innovations. In the United States, for example, R&D expenditures constituted between 2.2 and 2.9 per cent of GDP every year from 1957 to 2004. But not much cutting-edge R&D takes place outside a small group of countries. In 1996, for example, 73 per cent of the world's R&D expenditure, as measured by UNESCO, was accounted for by just five countries (in decreasing order of R&D expenditure they are the United States, Japan, Germany, France and United Kingdom). In the majority of countries that undertake very little measured R&D, technology advances not so much by making frontier innovations as by implementing technologies that have already been developed elsewhere. But the process of implementation is not costless, because technologies tend to be context-dependent and technological knowledge tends to be tacit. So implementation requires an up-front investment to adapt the technology to a new environment (see, for example, Evenson and Westphal, 1995). This investment plays the same role analytically in the implementing country as R&D does in the original innovating country.

Implementation is important in accounting for the patterns of cross-country convergence and divergence noted above. This is because a country in which firms are induced to spend on implementation have what Gerschenkron (1952) called an 'advantage of backwardness'. That is, the further they fall behind the world's technology frontier, the faster they will grow with any given level of implementation expenditures, because the bigger is the improvement in productivity when they implement any given foreign technology. In the long run, as Howitt (2000) has shown, this force can cause all countries that engage in R&D or implementation to grow at the same rate, while countries in which firms are not induced to make such investments will stagnate. But technology transfer through implementation expenditures is no guarantee of convergence, because the technologies that are being developed in the rich R&D-performing countries are not necessarily appropriate for conditions in poor implementing countries (Basu and Weil, 1998; Acemoglu and Zilibotti, 2001) and because financial constraints may prevent poor countries from spending at a level needed to keep pace with the frontier (Aghion, Howitt and Mayer-Foulkes, 2005).

Efficiency
The efficiency with which a technology is used is not likely to play a major role in accounting for long-run growth rates, because there is a finite limit to how high you can raise living standards simply by using the same technologies more efficiently. But there is good reason to believe that differences in efficiency account for much of the cross-country variation in the level of productivity.

Inefficiencies take several different forms. Economic resources are sometimes allocated to unproductive uses, or even unused, as when union featherbedding agreements kick in. Resources can be misallocated as the result of taxes, subsidies and imperfect competition, all of which create discrepancies between marginal rates of substitution. Technologies can be blocked by those who would lose from their implementation and have more market power or political influence than those who would win.

The distinction between differences in technology and differences in efficiency is often unclear. Suppose firms in country A are using the same machinery and the same number of workers per machine as in country B, but output per worker is higher in A than B. This may appear to be an obvious case of inefficiency, since the technology embodied in the machines used by workers in the two countries is the same. But maybe it is just that people in country B lack the knowledge of how best to use the machines, in which case it may actually be a case of differences in technology. As an example, General Motors has had little success in their attempts to emulate the manufacturing methods that Toyota has deployed successfully for many years even in their US operations.

Moreover, identical technologies will have different effects in different countries, because of differences in language, raw materials, consumer preferences, workers expectations and the like. Euro Disney, for example, was plagued initially with labour disputes when it first opened its park in the outskirts of Paris in 1987. It took the American managers several years to realize that the problem was not recalcitrant workers but rather that French workers consider it an intolerable indignity to be forced to wear items such as mouse ears when serving the public. A minor adjustment in amusement park technology was needed to make it as productive in France as it had been in the United States.

Deeper determinants of growth
Even if we knew how much of the cross-country variation in growth rates or income levels to attribute to different kinds of capital or to technology or efficiency, we would still be faced with the deeper question of why these differences in capital and productivity arise. A large number of candidate explanations have been offered in the literature. These candidates can be classified into four broad categories: geography, institutions, policy and culture.

Geographical differences are perhaps the most obvious. As Sachs (2003) has emphasized, countries that are landlocked, that suffer from a hazardous disease environment and that have difficult obstacles in the way of internal transport, will almost certainly produce at a lower level than countries without these problems, even if they use the same technology and the same array of capital. In addition, the lower productivity of these countries will serve to reduce the rate of return to accumulating capital and to generating new technologies.

Institutions matter because of the way they affect private contracts and also because of the way they affect the extent to which the returns to different kinds of investments can be appropriated by the government. The origin of a country's legal system has been shown by La Porta et al. (1998) to have an important effect on private contracts. In particular, these authors show that countries with British legal origins tend to offer greater protection of investor and creditor rights, which in turn is likely to affect both capital accumulation and investment in technology by making outside finance more easily available.

Because long-term productivity growth requires technical progress, it depends on political, institutional and regulatory factors that affect the way the conflict between the winners and losers of technical progress will be resolved, and hence affect the incentives to create and adopt new technologies. For example, the way intellectual property is protected will affect the incentive to innovate, because on the one hand no one will want to spend resources creating new technologies that his or her rivals can easily copy, while on the other hand a firm that is protected from competition by patent laws that make it difficult for rivals to innovate in the same product lines will be under less pressure to innovate. Likewise, a populist political regime may erect barriers to labour-saving innovation, resulting in slower technical progress.

Economic policies matter not only because of the way they affect the return to investing in capital and technology but also because of the inefficiencies that can be created by taxes and subsidies. But how these policies affect economic growth can vary from one country to another. In particular, Aghion and Howitt (2006) have argued that growth-promoting policies in technologically advanced countries are not necessarily growth-promoting in poorer countries, because innovation and implementation are affected differently by the same variables. For example, tighter competition policy in a relatively backward country might retard technology development by local firms that will be discouraged by the threat of foreign entry, whereas in more advanced countries firms will be spurred on to make even greater R&D investments when threatened by competition.

As this example suggests, international trade is one of the policy domains most likely to matter for growth and income differences, because of the huge productivity advantage that is squandered by policies that run counter to comparative advantage, because protected firms tend to become technologically backward firms, and because for many countries international trade is the only way for firms to gain a market large enough to cover the expense of developing leading-edge technologies. So it is probably no accident that export promotion has been a prominent feature of all the East Asian countries that began escaping from the lower end of the world income distribution towards the end of the 20th century, whereas import substitution was a prominent feature of several Latin American countries that fell from the upper end of the distribution early in the 20th century.

Culture is a difficult factor to measure. In principle, however, it is capable of explaining a great deal of cross-country variation in growth, because a society in which people are socialized to trust each other, to work hard, to value technical expertise and to respect law and order is certainly going to be thriftier and more productive than a society in which these traits do not apply. Recent work has begun to quantify the role of culture using measures of social capital, social capability, ethno-linguistic fractionalization, religious belief, the spread of Anglo-Saxon culture and many other variables.

PETER HOWITT AND DAVID WEIL

See also **economic growth, empirical regularities in; economic growth in the very long run; endogenous growth theory; growth accounting; growth and institutions; growth take-offs; level accounting; population ageing.**

Bibliography

Acemoglu, D. and Zilibotti, F. 2001. Productivity differences. *Quarterly Journal of Economics* 116, 563–606.

Aghion, P. and Howitt, P. 1992. A model of growth through creative destruction. *Econometrica* 60, 323–51.

Aghion, P. and Howitt, P. 2006. Appropriate growth policy: an integrating framework. *Journal of the European Economic Association* 4, 269–314.

Aghion, P., Howitt, P. and Mayer-Foulkes, D. 2005. The effect of financial development on convergence: theory and evidence. *Quarterly Journal of Economics* 120, 173–222.

Basu, S. and Weil, D.N. 1998. Appropriate technology and growth. *Quarterly Journal of Economics* 113, 1025–54.

Bourguignon, F. and Morrison, C. 2002. Inequality among world citizens: 1820–1992. *American Economic Review* 92, 727–44.

Caselli, F. 2005. Accounting for cross-country income differences. In *Handbook of Economic Growth*, vol. 1, ed. P. Aghion and S.N. Durlauf. Amsterdam: North-Holland.

Evans, P. 1996. Using cross-country variances to evaluate growth theories. *Journal of Economic Dynamics and Control* 20, 1027–49.

Evenson, R.E. and Westphal, L.E. 1995. Technological change and technology strategy. In *Handbook of Development Economics*, vol. 3A, ed. T.N. Srinivasan and J. Behrman. Amsterdam: Elsevier.

Fogel, R. 1997. New findings on secular trends in nutrition and mortality: some implications for population theory. In *Handbook of Population and Family Economics*, vol. 1A, ed. M.R. Rosenzweig and O. Stark. Amsterdam: North-Holland.

Frankel, M. 1962. The production function in allocation and growth: a synthesis. *American Economic Review* 52, 995–1022.

Galor, O. and Weil, D.N. 2000. Population, technology, and growth: from Malthusian stagnation to the demographic transition and beyond. *American Economic Review* 90, 806–28.

Gerschenkron, A. 1952. Economic backwardness in historical perspective. In *The Progress of Underdeveloped Areas*, ed. B.F. Hoselitz. Chicago: University of Chicago Press.

Gollin, D. 2002. Getting income shares right. *Journal of Political Economy* 110, 458–74.

Greenwood, J. and Yorukoglu, M. 1997. *1974 Carnegie-Rochester Conference Series on Public Policy* 46, 49–95.

Hall, R. and Jones, C. 1999. Why do some countries produce so much more output per worker than others? *Quarterly Journal of Economics* 114, 83–116.

Howitt, P. 1998. Measurement, obsolescence, and general purpose technologies. In *General Purpose Technologies and Economic Growth*, ed. E. Helpman. Cambridge, MA: MIT Press.

Howitt, P. 2000. Endogenous growth and cross-country income differences. *American Economic Review* 90, 829–46.

Howitt, P. and Aghion, P. 1998. Capital accumulation and innovation as complementary factors in long-run growth. *Journal of Economic Growth* 3, 111–30.

Hsieh, C.-T. and Klenow, P. 2007. Relative prices and relative prosperity. *American Economic Review* 97.

Jones, C.I. 1997. On the evolution of the world income distribution. *Journal of Economic Perspectives* 11(3), 19–36.

Klenow, P. and Rodriguez-Clare, A. 1997. The neoclassical revival in growth economics: has it gone too far? In *NBER Macro Annual*, ed. B. Bernanke and J. Rotemberg. Cambridge, MA: MIT Press.

La Porta, R., Lopez-de-Silanes, F., Shleifer, A. and Vishny, R.W. 1998. Law and finance. *Journal of Political Economy* 106, 1113–55.

Lipsey, R.G., Carlaw, K.I. and Bekar, C.T. 2005. *Economic Transformations: General Purpose Technologies and Long Term Economic Growth.* New York: Oxford University Press.

Lucas, R.E. Jr. 1988. On the mechanics of economic development. *Journal of Monetary Economics* 22, 3–42.

Maddison, A. 2001. *The World Economy: A Millennial Perspective.* Development Centre Studies. Paris: OECD.

Malthus, T.R. 1798. *An Essay on the Principle of Population, as it Affects the Future Improvement of Society with Remarks on the Speculations of Mr. Godwin, M. Condorcet, and Other Writers.* London: printed for J. Johnson in St Paul's Churchyard.

Mayer-Foulkes, D. 2002. Global divergence. Documento de Trabajo del CIDE, SDTE 250, División de Economía. Mexico: CIDE.

Pritchett, L. 1997. Divergence, big-time. *Journal of Economic Perspectives* 11(3), 3–17.

Rodriguez, F. and Sachs, J.D. 1999. Why do resource-abundant economies grow more slowly? *Journal of Economic Growth* 4, 277–303.

Romer, P.M. 1986. Increasing returns and long-run growth. *Journal of Political Economy* 94, 1002–37.

Romer, P.M. 1990. Endogenous technological change. *Journal of Political Economy* 98, S71–102.

Rosenberg, N. 1981. How exogenous is science? In *Inside the Black Box: Technology and Economics*, ed. N. Rosenberg. New York: Cambridge University Press.

Sachs, J.D. 2003. Institutions don't rule: direct effects of geography on per capita income. Working Paper No. 9490. Cambridge, MA: NBER.

Sala-i-Martin, X. 2006. The world distribution of income: falling poverty and … convergence, period. *Quarterly Journal of Economics* 121, 351–97.

Schumpeter, J.A. 1942. *Capitalism, Socialism and Democracy.* New York: Harper.

Solow, R.M. 1956. A contribution to the theory of economic growth. *Quarterly Journal of Economics* 70, 65–94.

Solow, R.M. 1957. Technical change and the aggregate production function. *Review of Economics and Statistics* 39, 312–20.

Swan, T.W. 1956. Economic growth and capital accumulation. *Economic Record* 32, 334–61.

Weil, D.N. 2005. *Economic Growth.* Boston: Addison-Wesley.

Weil, D.N. 2007. Accounting for the effect of health on economic growth. *Quarterly Journal of Economics* 122.

economic growth, empirical regularities in

The evolution of economic growth theory throughout the post-war period has been deeply influenced by the effort to explain broad patterns in cross-country behaviour. In this entry, we discuss some of the salient empirical regularities associated with neoclassical and new growth economics and consider the shift in focus that has occurred. We first describe the role of empirical regularities in neoclassical growth theory as it emerged in the 1950s. Next, we consider how a switch in focus to a different class of regularities is associated with the new growth economics that developed in the 1980s and continues to dominate contemporary research. Finally, we assess this shift. Durlauf, Johnson and Temple (2005) contains details of the data and methods used to substantiate the claims made here.

Empirical regularities and neoclassical growth

Neoclassical growth theory is commonly associated with Kaldor's (1961) well-known 'stylized facts' of long-run economic behaviour, which primarily focused on the invariance of long run behaviour for advanced economies. Four of his six facts – (1) the constancy of the growth rate of output per worker over long time horizons, (2) the constancy of the growth rate of capital which is lower than the growth rate of the labour supply, (3) the absence of any systematic trends in the capital–output ratio and (4) the constancy of the rate of profit (and, by implication with the other facts, factor shares in national income) – emphasize common behaviour across countries. Only the fifth and sixth facts – the presence of substantial differences in output per worker across countries, and the positive relationship between the rate of profit and the investment–output ratio – focus on heterogeneity. Kaldor (1957) cites the prediction of constant factor shares as an important test of alternative growth models. An important empirical study at the time was Klein and Kosobud (1961) who investigated constancy by testing for a trend in labour's share, finding none using US data from 1900 to 1953.

While these facts are generally cited as a motivation of neoclassical growth models, their actual relationship to the theory is in fact more complicated. In Solow (1956), for example, the objective is the explanation of long-run economic growth and the constancy of factor shares is only mentioned in passing as an implication of the Cobb–Douglas technology. Indeed Solow (1958) criticizes the literature studying the constancy of factor shares for lacking a precise notion of constancy given that exact constancy cannot reasonably be expected. Bronfenbrenner (1960) argues that, for a wide range of values of the elasticity of substitution between capital and labour, and for reasonable variation in the capital–labour ratio, the theoretical variation in factor shares is consistent with that observed. He concludes that the constancy or otherwise of factor shares is not useful in the assessment of (distribution) theories. Put differently, the first three of Kaldor's stylized facts seem most important to understanding the motivation of the neoclassical program; Solow (2000, p. 4) (in a discussion originally published in 1970) remarks that growth theory is largely

> devoted to analyzing the properties of steady states and to finding out whether an economy not initially in a steady state will evolve into one …

How do Kaldor's stylized facts appear from the vantage point of modern empirical growth research? Barro and Sala-i-Martin (2004, pp. 12–16) assess the concordance of Kaldor's stylized facts with the data and conclude that, with the exception of the constant rate of profit, each of the first five holds 'reasonably well' for developed economies. They cite evidence suggesting some tendency for the real rate of return to decline in some economies. The evidence they present, and that which we discuss below, shows that, at least as far as it concerns the rate of growth of labour productivity, the sixth of Kaldor's facts also fits well with the data.

Kaldor's stylized facts are therefore of contemporary use in understanding long-run output behaviour. That said, the facts are no longer central to the research efforts in growth economics as other regularities (or the lack thereof) have become the primary focus of research. We therefore turn to those regularities that have become the focus of contemporary work.

Empirical regularities and the new growth economics

The renaissance in growth theory associated with the rise of endogenous growth models was influenced by interest in the determinants of heterogeneity in growth experiences. While not usually called stylized facts, there is a set of general propositions about heterogeneity that have been very important in influencing research. The most prominent global features evident in the data are the divergence in living standards over the past three centuries and the large disparities in living standards at the end of the 20th century. By modern standards, all countries were poor in 1700 but since then sustained growth, first in the United Kingdom and parts of Western Europe, and more recently in the United States and parts of the Asia–Pacific region, has resulted in large cross-country differences in living standards. In 2000 average GDP per worker in some countries was about one-fiftieth that in the United States while more than 40 per cent of the

world's population lived in countries with average levels of GDP per worker of no more than ten per cent of that in the United States.

Divergence in living standards over the 1960–2000 period is also evident in the large group of countries covered by the Penn World Tables (PWT) (Heston, Summers and Aten, 2002). While a substantial group of countries has exhibited prolonged growth over this period, there remains a large mass of countries at the bottom of the distribution. One result was a hollowing out of the middle of the distribution – a phenomenon labelled 'twin peaks' by Quah (1996; 1997). Moreover, there is strong persistence within the cross-country income distribution with a Spearman rank correlation of 0.84 between GDP per worker in 1960 and that in 2000. This degree of correlation is not peculiar to the PWT data. Easterly et al. (1993) report a rank correlation of 0.82 between GDP per capita in 1988 and that in 1870 for the 28 countries in Maddison (1989). This sense of a lack of mobility is reinforced by Bianchi (1997), who found that very few of the possible crossings from one end of the distribution to the other actually occurred between 1970 and 1989.

The persistence in levels of GDP per worker contrasts sharply with the wide cross-country variation in the growth rates of GDP per worker especially for those countries with relatively low levels of GDP per worker in 1960. The data show scant support for the proposition that the countries of the world are converging to a common level of income per person or for the belief that poor countries have always grown slowly. Both growth 'miracles' – countries exhibited consistently strong growth over the 1960–2000 period – and growth 'disasters' – countries that did poorly, often having negative average growth rates – are present in the data. East and South East Asian countries are well represented among the former group while the later is dominated by countries in sub-Saharan Africa. Taiwan, for example, grew at an average annual rate of over six per cent during this 40-year period and increased GDP per worker by a factor of 11 in the process. Hong Kong, Korea and Singapore were not far behind in either respect. By contrast, Mauritania, Senegal, Chad, Mozambique, Madagascar, Zambia, Mali, Niger, Nigeria, the Central African Republic, Angola and the Democratic Republic of the Congo all had negative average growth over this period.

For most countries, the average growth rate from 1980 to 2000 was lower than that from 1960 to 1980. The notable exceptions to this observation are China and India. Moreover, past growth does not seem to be a good predictor of future growth as, for example, the correlation between growth in 1960–80 and that in 1980–2000 is just 0.40. Easterly et al. (1993) suggest that the lack of persistence in growth rates indicates the importance of good luck in economic development. Nevertheless, the cross-decade correlations in growth rates have tended to increase during the 1960–80 period, indicating a sorting of countries into distinct groups of winners and losers.

There seems to be little relationship between the 1960 level of GDP per worker and subsequent average growth rates. The cross-country dispersion of growth rates tends to fall as initial income rises largely due to the rarity of poor performance among the countries with relatively high levels of GDP per worker in 1960. There is, however, a close relationship between geographical group membership and economic growth between 1960 and 2000. As alluded to above, the countries of sub-Saharan Africa performed poorly over this period, with three-quarters of them growing at an average annual rate of less than just 1.3 per cent. The countries in South and Central America did somewhat better with three-quarters of them having grown at an average of less than 1.5 per cent. Among the East and South East Asian countries, three-quarters grew at an average rate of over 3.8 per cent, and a similar fraction of the South Asian countries grew at over 1.9 per cent.

Many of the poor countries of the world were unable to break out of stagnation between 1960 and 2000. A country growing at two per cent per year for 40 years would enjoy a 120 per cent increase in income per person over that period. Yet, between 1960 and 2000, about a quarter of countries never exceeded their 1960 income level by more than 60 per cent, and about ten per cent of countries never exceeded their 1960 level by more than 30 per cent. One reason for this stagnation is the disposition of some economies to large, abrupt output collapses. About half of countries experienced a three-year output collapse of 15 per cent or more between 1960 and 2000. Over the same period, the largest three-year output collapse in the United States was 5.4 per cent, and in the United Kingdom 3.6 per cent, both in 1979–82.

In sum, there are large cross-countries disparities in GDP per worker and hence in living standards. These disparities have grown wider since 1960 and the middle of cross-country income distribution has thinned since 1960. There is substantial immobility in a country's position in the distribution. Growth rates are much less persistent and have tended to fall since 1980. In general, the countries of sub-Saharan Africa performed poorly over the 1960–2000 period. The countries in South and Central America did somewhat better while the South Asian countries did better still. The East and South East Asian countries did best of all.

The changing empirical focus of growth economics

The two sets of empirical regularities we have described, while appearing to differ greatly in terms of their implications for understanding the determinants of the growth process, may in fact be reconciled. A key difference between neoclassical and modern growth economics is its domain of explanation: whereas neoclassical theory attempted to understand the long-run behaviour of advanced industrialized economies, the new growth economics attempts to understand worldwide growth patterns. As a result, the differences between the advanced industrialized economies and the rest of the world take on primary importance. Lucas (2002, pp. 2–3) describes his motivation as

> to see whether modern growth theory could also be adapted for use as a theory of economic development. Adaptation of some kind was evidently necessary: The balanced path of growth theory, with constant income growth, and the assumed absence of population pressures, obviously did not fit all of economic history or even all the behavior that can be seen in today's world. The theory is, and was designed to be, a model of the recent past of a subset of countries.

Thus, as the domain of inquiry in growth economics has evolved, the stylized facts of interest have shifted to identifying features of international divergence rather than international convergence.

Further, the effort to identify patterns that characterize the differences in cross-country growth experiences has led to empirical research that focuses on the identification of particular factors in generating the divergence. Theoretical work in growth economics moved away from the traditional emphasis on factor accumulation and towards the analysis of a wide range of social, historical, geographic, and political factors as sources of cross-country heterogeneity. For example, a major strand of contemporary research focuses on the ways that institutional quality affects growth and development; see Acemoglu, Johnson and Robinson (2005) for a detailed survey. The richness of the modern growth literature has led to the widespread use of regression methods to allow for the simultaneous consideration of multiple growth determinants, with a focus on identifying which determinants in fact matter.

The move towards regression methods as the basis for empirical growth research has altered the nature of the sorts of regularities that link data and theory. It is still the case that theoretical analyses are often motivated by the identification of a bivariate relationship between some factor of interest and growth rates. However, relationships of this type do not represent basic growth regularities in the way that Kaldor's stylized facts did. The reason for this transition is that the different growth factors that have expanded the domain of growth economics are typically mutually consistent (Brock and Durlauf, 2001) and so the empirical significance of one factor can only be assessed when others are considered as well. Put differently, the finding of a bivariate relationship, or lack thereof, can always be rationalized as reflecting a failure to control for other factors.

As a result, the empirical regularities that matter for contemporary research, such as the coefficient relating a measure of institutional quality to growth, are derivative from statistical analyses of the entire growth process.

But statistical models of growth are subject to many forms of model uncertainty, ranging from uncertainty about the appropriate theories to employ to uncertainty about the empirical measurement of the qualitative factors identified by a theory to uncertainty about the details of the statistical specification of a model; see Durlauf and Quah (1999) and Durlauf, Johnson and Temple (2005) for a delineation of these issues. Model uncertainty has meant that there is relatively little consensus on the empirically salient determinants of growth and so little consensus on which regularities should be of primary interest. Thus current growth economics has been handicapped as different papers identify different salient empirical regularities, with inadequate attention to the robustness of such claims. The development of sturdy inferences about the growth process thus represents a very active area of current work.

STEVEN N. DURLAUF AND PAUL A. JOHNSON

See also **economic growth; endogenous growth theory; growth accounting; level accounting.**

Bibliography

Acemoglu, D., Johnson, S. and Robinson, J. 2005. Institutions as the fundamental cause of long-run growth. In *Handbook of Economic Growth*, ed. P. Aghion and S. Durlauf. Amsterdam: North-Holland.

Barro, R. and Sala-i-Martin, X. 2004. *Economic Growth*, 2nd edn. Cambridge, MA, and London: MIT Press.

Bianchi, M. 1997. Testing for convergence: evidence from nonparametric multimodality tests. *Journal of Applied Econometrics* 12, 393–409.

Brock, W. and Durlauf, S. 2001. Growth empirics and reality. *World Bank Economic Review* 15, 229–72.

Bronfenbrenner, M. 1960. A note on relative shares and the elasticity of substitution. *Journal of Political Economy* 68, 284–7.

Durlauf, S., Johnson, P. and Temple, J. 2005. Growth econometrics. In *Handbook of Economic Growth*, ed. P. Aghion and S. Durlauf. Amsterdam: North-Holland.

Durlauf, S. and Quah, D. 1999. The new empirics of economic growth. In *Handbook of Macroeconomics*, ed. J. Taylor and M. Woodford. Amsterdam: North-Holland.

Easterly, W., Kremer, M., Pritchett, L. and Summers, L. 1993. Good policy or good luck? Country growth performance and temporary shocks. *Journal of Monetary Economics* 32, 459–83.

Heston, A., Summers, R. and Aten, B. 2002. *Penn World Table Version 6.1*. Philadelphia: Center for International Comparisons at the University of Pennsylvania (CICUP).

Kaldor, N. 1957. A model of economic growth. *Economic Journal* 67, 591–624.

Kaldor, N. 1961. Capital accumulation and economic growth. In *The Theory of Capital, Proceedings of a Conference held by the International Economic Association*, ed. F. Lutz and D. Hague. London: Macmillan.

Klein, L. and Kosobud, R. 1961. Some econometrics of growth: great ratios of economics. *Quarterly Journal of Economics* 125, 173–98.

Lucas, R. 2002. *Lectures on Economic Growth*. Cambridge, MA: Harvard University Press.

Maddison, A. 1989. *The World Economy in the 20th Century*. Paris: OECD.

Quah, D. 1996. Twin peaks: growth and convergence in models of distribution dynamics. *Economic Journal* 106, 1045–55.

Quah, D. 1997. Empirics for growth and distribution: stratification, polarization, and convergence clubs. *Journal of Economic Growth* 2(1), 27–59.

Solow, R. 1956. A contribution to the theory of economic growth. *Quarterly Journal of Economics* 70, 65–94.

Solow, R. 1958. A skeptical note on the constancy of relative shares. *American Economic Review* 48, 618–31.

Solow, R. 2000. *Growth Theory: An Exposition*. Oxford: Oxford University Press.

economic growth in the very long run

The evolution of economies during the major portion of human history was marked by Malthusian stagnation. Technological progress and population growth were minuscule by modern standards, and the average growth rates of income per capita in various regions of the world were even slower due to the offsetting effect of population growth on the expansion of resources per capita.

In the past two centuries the pace of technological progress increased significantly in association with the process of industrialization. Various regions of the world departed from the Malthusian trap and experienced a considerable rise in the growth rates of income per capita and population. Unlike episodes of technological progress in the pre-Industrial Revolution era that failed to generate sustained economic growth, the increasing role of human capital in the production process in the second phase of industrialization ultimately prompted a demographic transition, liberating the gains in productivity from the counterbalancing effects of population growth. The decline in the growth rate of population and the enhancement of human capital formation and technological progress paved the way for the emergence of the modern state of sustained economic growth. Variations in the timing of the transitions from a Malthusian epoch to a state of sustained economic growth across countries lead to a considerable rise in the ratio of GDP per capita between the richest and the poorest regions of the world from 3:1 in 1820 to 18:1 in 2000 (see Figure 1).

The transition from stagnation to growth and the associated phenomenon of the great divergence have been the subject of intensive research in the growth literature in recent years (Galor and Weil, 1999; 2000; Galor and Moav, 2002; Lucas, 2002; Hansen and Prescott, 2002; Jones, 2001; Hazan and Berdugo, 2002; Doepke, 2004;

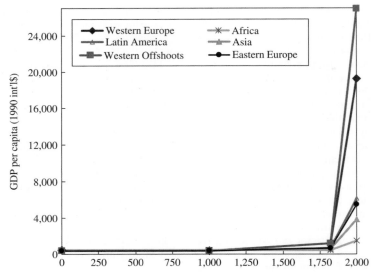

Figure 1   The evolution of regional income per capita, 1–2000. *Source*: Maddison (2001).

Lagerlof, 2003; 2006; Galor and Mountford, 2003; 2006). The inconsistency of exogenous and endogenous growth models with some of the most fundamental features of the process of development has led to a search for a unified theory that would unveil the underlying micro-foundations of the growth process in its entirety, and would capture in a single framework the epoch of Malthusian stagnation that characterized most of human history, the contemporary era of modern economic growth, and the driving forces that triggered the recent transition between these regimes.

The advance of unified growth theory was fuelled by the conviction that the understanding of the contemporary growth process would be fragile and incomplete unless growth theory were based on proper microfoundations that reflect the various qualitative aspects of the growth process and their central driving forces. Moreover, it has become apparent that a comprehensive understanding of the hurdles faced by less developed economies in reaching a state of sustained economic growth would remain obscure unless the factors that prompted the transition of the currently developed economies into a state of sustained economic growth could be identified and modified to account for the differences in the growth structure of less developed economies in an interdependent world.

Unified growth theory explores the fundamental factors that generated the remarkable escape from the Malthusian epoch and their significance in understanding the contemporary growth process of developed and less developed economies. Moreover, it sheds light on the perplexing phenomenon of the great divergence in income per capita across regions of the world in the past two centuries. It suggests that the transition from stagnation to growth is an inevitable outcome of the process of development. The inherent Malthusian interaction between the level of technology and the size and the composition of the population accelerated the pace of technological progress and ultimately raised the importance of human capital in the production process. The rise in the demand for human capital in the second phase of industrialization and its impact on the formation of human capital as well as on the onset of the demographic transition brought about significant technological advances along with a reduction in fertility rates and population growth, enabling economies to convert a larger share of the fruits of factor accumulation and technological progress into growth of income per capita, and paving the way for the emergence of sustained economic growth.

Differences in the timing of the take-off from stagnation to growth across countries (for example, England's earlier industrialization in comparison with China) contributed significantly to the great divergence and to the emergence of convergence clubs. These variations reflect initial differences in geographical factors and historical accidents and their manifestation in diversity in institutional, demographic, and cultural factors, trade patterns, colonial status, and public policy. In particular, once a technologically driven demand for human capital emerged in the second phase of industrialization, the prevalence of human capital-promoting institutions determined the extensiveness of human capital formation, the timing of the demographic transition, and the pace of the transition from stagnation to growth. Thus, unified growth theory provides the natural framework of analysis in which variations in the economic performance across countries and regions could be examined based on the effect of variations in educational, institutional, geographical, and cultural factors on the pace of the transition from stagnation to growth.

The process of development

The process of economic development has been characterized by of three fundamental regimes: the Malthusian epoch, the post-Malthusian regime, and the sustained growth regime.

The Malthusian epoch

During the Malthusian epoch that characterized most of human history, humans were subjected to a persistent struggle for existence. Resources generated by technological progress and land expansion were channelled primarily towards an increase in the size of the population, with a minor long-run effect on income per capita. Improvements in the technological environment or in the availability of land generated temporary gains in income per capita, leading eventually to a larger but not richer population. Technologically superior countries ultimately had denser populations but their standard of living did not reflect the degree of their technological advancement.

During the Malthusian epoch the average growth rate of output per capita was negligible and the standard of living did not differ greatly across countries. The average level of income per capita in the world during the first millennium fluctuated around $450 per year (in 1990 international dollars) and the average growth rate of output per capita was nearly zero (Maddison, 2001). This state of Malthusian stagnation persisted until the end of the 18th century. In the years 1000–1820, the average level of income per capita in the world economy was below $670 per year, and the average growth rate of the world income per capita was minuscule, creeping at a rate of about 0.05 per cent per year. Nevertheless, income per capita fluctuated significantly within regions, deviating from their sluggish long-run trend over decades and sometimes centuries.

Population growth over this era followed the Malthusian pattern as well. The gradual increase in income per capita during the Malthusian epoch was associated with a monotonic increase in the average rate of growth of world population. The slow pace of resource expansion in the first millennium was reflected in a modest increase in the population of the world from 231 million people in 1 CE to 268 million in 1000 CE: a minuscule average growth rate of 0.02 per cent per year. The more rapid (but still very slow) expansion of resources in the period 1000–1500 permitted the world population to increase by 63 per cent, from 268 million in 1000 to 438 million in 1500; a slow 0.1 per cent average growth rate per year. Resource expansion over the period 1500–1820 had a more significant impact on the world population, which grew 138 per cent from 438 million in 1500 to 1,041

million in 1820: an average pace of 0.27 per cent per year.

Variations in population density across countries during the Malthusian epoch reflected primarily cross-country differences in technology and land productivity. Due to the positive adjustment of the population to an increase in income per capita, differences in technology or in land productivity across countries resulted in variations in population density rather than in the standard of living. For instance, China's technological advancement in the period 1500–1820 permitted its share of world population to increase from 23.5 per cent to 36.6 per cent, while its income per capita in the beginning and the end of this time interval remained approximately $600 per year.

The post-Malthusian regime

During the post-Malthusian regime, the pace of technological progress markedly increased in association with the process of industrialization, triggering a take-off from the Malthusian trap. The growth rate of income per capita increased significantly but the positive Malthusian effect of income per capita on population growth was still maintained, generating a sizeable increase in population growth that offset some of the potential gains in income per capita.

The take-off of developed regions from the Malthusian regime occurred at the beginning of the 19th century and was associated with the Industrial Revolution, whereas the take-off of less developed regions occurred towards the beginning of the 20th century and was delayed in some countries well into the 20th century. During the post-Malthusian regime the average growth rate of output per capita increased significantly and the standard of living began to differ considerably across countries. The average growth rate of output per capita in the world soared from 0.05 per cent per year during the period 1500–1820 to 0.53 per cent per year in the years 1820–70, and 1.3 per cent per year during the period 1870–1913. The timing of the take-off and its magnitude differed across regions. The take-off from the Malthusian epoch and the transition to the post-Malthusian regime occurred in western Europe, the Western offshoots (that is, the United States, Canada, Australia and New Zealand), and eastern Europe at the beginning of the 19th century, whereas in Latin America, Asia and Africa it occurred towards the beginning of the 20th century.

The rapid increase in income per capita in the post-Malthusian regime was channelled partly towards an increase in the size of the population. During this period, the Malthusian mechanism linking higher income to higher population growth continued to function. However, the effect of higher population on the dilution of resources per capita was counteracted by accelerated technological progress and capital accumulation, allowing income per capita to rise despite the offsetting effects of population growth.

The western European take-off along with that of the Western offshoots brought about a sharp increase in population growth in these regions and consequently a modest rise in population growth in the world as a whole. The subsequent take-off of less developed regions, and the associated increase in their rates of population growth, brought about a significant rise in population growth in the world. The rate of population growth in the world increased from an average rate of 0.27 per cent per year in the period 1500–1820 to 0.4 per cent per year in the years 1820–70, and to 0.8 per cent per year in the time interval 1870–1913. Despite the decline in population growth in western Europe and the Western offshoots towards the end of the 19th century and the beginning of the 20th century, the delayed take-off of less developed regions, and the significant increase in their income per capita prior to their demographic transitions, generated a further increase in the rate of population growth in the world to 0.93 per cent per year in the period 1913–50, and 1.92 per cent per year in the period 1950–73. Ultimately, the onset of the demographic transition in less developed economies during the second half of the 20th century reduced population growth rates to 1.66 per cent per year in the 1973–98 period (Maddison, 2001).

It appears that the significant rise in income per capita in the post-Malthusian regime increased the desired number of surviving offspring and thus, despite the decline in mortality rates, fertility increased significantly so as to enable households to reach this higher desired level of surviving offspring. Fertility was controlled during this period, despite the absence of modern contraceptive methods, partly via adjustment in marriage rates. Increased fertility was achieved by earlier female age of marriage, and a decline in fertility by a delay in the marriage age.

The take-off in the developed regions was accompanied by a rapid process of industrialization. Per-capita level of industrialization increased significantly in the United Kingdom, rising 50 per cent over the 1750–1800 period, quadrupling in the years 1800–60, and nearly doubling in the time period 1860–1913. Similarly, per capita level of industrialization accelerated in the United States, doubling in the 1750–1800 as well as 1800–60 periods, and increasing sixfold in the years 1860–1913. A similar pattern was experienced in Germany, France, Sweden, Switzerland, Belgium and Canada. The take-off of less developed economies in the 20th century was associated with increased industrialization as well. However, during the 19th century these economies experienced a decline in per capita industrialization, reflecting the adverse effect of the sizeable increase in population on the level of industrial production per capita as well as the forces of globalization and colonialism, which induced less developed economies to specialize in the production of raw materials (Galor and Mountford, 2003; 2006).

The acceleration in technological progress during the post-Malthusian regime and the associated increase in income per capita stimulated the accumulation of human capital in the form of literacy rates, schooling, and health. The increase in the investment in human capital was induced by the rise in income per capita, as well as by qualitative changes in the economic environment that increased the demand for human capital and induced households to invest in the education of their offspring.

In the first phase of the Industrial Revolution, human capital had a limited role in the production process. Education was motivated by a variety of reasons, such as religion, enlightenment, social control, moral conformity, socio-political stability, social and national cohesion, and military efficiency. The extensiveness of public education was therefore not necessarily correlated with industrial development, and it differed across countries due to political, cultural, social, historical and institutional factors. In the second phase of the Industrial Revolution, however, the demand for education increased, reflecting the increasing skill requirements in the process of industrialization. The economic interests of capitalists were a significant driving force behind the implementation of educational reforms (Galor and Moav, 2006). The process of industrialization has been characterized by a gradual increase in the relative importance of human capital in less developed economies as well and educational attainment increased significantly across all less developed regions in the post-Malthusian regime.

The sustained growth regime

The acceleration in the rate of technological progress in the second phase of industrialization, and its interaction with human capital formation, triggered a demographic transition, paving the way to a transition to an era of sustained economic growth. In the post demographic-transition period, the rise in aggregate income due to technological progress and factors accumulation was no longer counterbalanced by population growth, permitting sustained growth in income per capita in regions that experienced sustained technological progress and accumulation of physical and human capital.

The transition of the developed regions of western Europe and the Western offshoots to the state of sustained economic growth occurred towards the end of the 19th century, and their income per capita in the 20th century has advanced at a stable rate of about two per cent per year. The transition of some less developed countries in Asia and Latin America occurred towards the end of the 20th century. Africa, in contrast, is still struggling to make this transition.

The transition to a state of sustained economic growth was characterized by a gradual increase in the importance of the accumulation of human capital relative to physical capital as well as with a sharp decline in fertility rates. In the first phase of the Industrial Revolution (1760–1830), capital accumulation as a fraction of GDP significantly increased whereas literacy rates remained largely unchanged. Skills and literacy requirements were minimal, the state devoted virtually no resources to raise the level of literacy of the masses, and workers developed skills primarily through on-the-job training (Green, 1990; Mokyr, 1993). Consequently, literacy rates did not increase during the period 1750–1830 (Sanderson, 1995).

In the second phase of the Industrial Revolution, however, the pace of capital accumulation subsided, skills became necessary for production and the education of the labour force markedly increased. The investment ratio in the UK, which increased from six per cent in 1760 to 11.7 per cent in 1831, remained at around 11 per cent on average in the years 1856–1913 (Crafts, 1985). In contrast, the average years of schooling of males in the labour force that did not change significantly until the 1830s tripled by the beginning of the 20th century. The drastic rise in the level of income per capita in England as of 1865 was associated with an increase in school enrolment of ten-year-old children from 40 per cent in 1870 to 100 per cent in 1900. Moreover, total fertility rate in England sharply declined over this period from about five in 1875, to nearly two in 1925.

The demographic transition swept the world in the course of the 20th century. The unprecedented increase in population growth during the post-Malthusian regime was reversed and the demographic transition brought about a significant reduction in fertility rates and population growth in various regions of the world, enabling economies to convert a larger share of the fruits of factor accumulation and technological progress into growth of income per capita. The demographic transition enhanced the growth process via three channels: (a) reductions in the dilution of the stocks of capital and natural resources, (b) enhancements in human capital formation, and (c) changes in the age distribution of the population, temporarily increasing the size of the labour force relative to the population as a whole.

The timing of the demographic transition differed significantly across regions. The reduction in population growth occurred in Western Europe, the Western offshoots, and eastern Europe towards the end of the 19th century and in the beginning of the 20th century, whereas Latin America and Asia experienced a decline in the rate of population growth only in the last decades of the 20th century. Africa's population growth, in contrast, has been rising steadily.

The process of industrialization was characterized by a gradual increase in the relative importance of human capital in the production process. The acceleration in the rate of technological progress gradually increased the demand for human capital, inducing individuals to invest in education, and stimulating further technological advancement. Moreover, in developed as well as less developed regions, the onset of the process of human capital accumulation preceded the onset of the demographic transition, suggesting that the rise in the demand for human capital in the process of industrialization and the subsequent accumulation of human capital played a significant role in the demographic transition and the shift to a state of sustained economic growth.

Notably, the reversal of the Malthusian relation between income and population growth during the demographic transition corresponded to an increase in the level of resources invested in each child. For example, literacy rate among men in England was stable at around 65 per cent in the first phase of the Industrial Revolution and increased significantly during the second phase, reaching nearly 100 per cent at the end of the 19th century. In addition, the proportion of children aged 5 to 14 in primary schools increased from 11 per cent in 1855 to 74 per cent in 1900. A similar pattern is observed in other European societies (Flora, Kraus and Pfenning, 1983).

The process of industrialization was characterized by a gradual increase in the relative importance of human capital in less developed economies as well. Educational attainment increased significantly across all less developed regions. Moreover, in line with the pattern that emerged among developed economies in the 19th century, the increase in educational attainment preceded or occurred simultaneously with the decline in total fertility rates.

The great divergence

The differential timing of the take-off from stagnation to growth across countries and the corresponding variations in the timing of the demographic transition led to a great divergence in income per capita as well as population growth. Inequality in the world economy was negligible till the 19th century. The ratio of GDP per capita between the richest region and the poorest region in the world was only 1.1:1 in 1000, 2:1 in 1500 and 3:1 in 1820. In the past two centuries, however, the ratio of GDP per capita between the richest group (Western offshoots) and the poorest region (Africa) has widened considerably from a modest 3:1 ratio in 1820, to 5:1 ratio in 1870, 9:1 ratio in 1913, 15:1 ratio in 1950, and 18:1 ratio in 2001.

An equally momentous transformation occurred in the distribution of world population across regions. The earlier take-off of western European countries increased the amount of resources that could be devoted for the increase in family size, permitting a 16 per cent increase in the share of their population in the world from 12.8 per cent in 1820 to 14.8 per cent in 1870. However, the early onset in the western European demographic transition and the long delay in the demographic transition of less developed regions, well into the second half of the 20th century, led to a decline in the share of western European population in the world, from 14.8 per cent in 1870 to 6.6 per cent in 1998. In contrast, the prolongation of the post-Malthusian period among less developed regions, in association with the delay in their demographic transition well into the second half of 20th century, channelled their increased resources towards a significant increase in their population. Africa's share of world population increased from seven per cent in 1913 to 12.9 per cent in 1998, Asia's share of world population increased from 51.7 per cent in 1913 to 57.4 per cent in 1998, and Latin American countries increased their share in world population from two per cent in 1820 to 8.6 per cent in 1998.

Unified growth theory

Galor and Weil (2000) advanced a unified growth theory that captures the three regimes that have characterized the process of development as well as the fundamental driving forces that generated the transition from an epoch of Malthusian stagnation to a state of sustained economic growth. The theory replicates the observed time paths of population, income per capita, and human capital, generating: (a) the Malthusian oscillations in population and output per capita during the Malthusian epoch, (b) an endogenous take-off from Malthusian stagnation that is associated with an acceleration in technological progress and is accompanied initially by a rapid increase in population growth, and (c) a rise in the demand for human capital, followed by a demographic transition and sustained economic growth. These qualitative patterns are confirmed in the calibration of the theory by Lagerlof (2006).

The theory proposes that in early stages of development economies were in the proximity of a stable Malthusian equilibrium. Technology advanced rather slowly, and generated proportional increases in output and population. The inherent positive interaction between population and technology in this epoch, however, gradually increased the pace of technological progress, and due to the delayed adjustment of population, output per capita advanced at a minuscule rate. The slow pace of technological progress in the Malthusian epoch provided a limited scope for human capital in the production process and parents, therefore, had no incentive to reallocate resources towards human capital formation in their offspring.

The Malthusian interaction between technology and population accelerated the pace of technological progress and permitted a take-off to the post-Malthusian regime. The expansion of resources was partially counterbalanced by the enlargement of population, and the economy was characterized by rapid growth rates of income per capita and population. The acceleration in technological progress eventually increased the demand for human capital, generating two opposing effects on population growth. On the one hand, it eased households' budget constraints, allowing the allocation of more resources for raising children. On the other hand, it induced a reallocation of resources towards child quality. In the post-Malthusian regime, due to the modest demand for human capital, the first effect dominated, and the rise in

real income permitted households to increase the number as well the quality of their children.

As investment in human capital took place, the Malthusian steady-state equilibrium vanished and the economy started to be attracted by the gravitational forces of the modern growth regime. The interaction between investment in human capital and technological progress generated a virtuous circle: human capital generated faster technological progress, which in turn further raised the demand for human capital, inducing further investment in child quality, and eventually triggering the onset of the demographic transition and the emergence of a state of sustained economic growth.

The theory suggests that the transition from stagnation to growth is an inevitable outcome of the process of development. The inherent Malthusian interaction between the level of technology and the size of the population accelerated the pace of technological progress, and ultimately raised the importance of human capital in the production process. The rise in the demand for human capital in the second phase of the Industrial Revolution and its impact on the formation of human capital as well as on the onset of the demographic transition brought about significant technological advancements along with a reduction in fertility rates and population growth, enabling economies to convert a larger share of the fruits of factor accumulation and technological progress into growth of income per capita, and paving the way for the emergence of sustained economic growth. Quantitative analysis of unified growth theories (Doepke, 2004); Lagerlof, 2006) indeed suggest that the rise in the demand for human capital was a significant force behind the demographic transition and the emergence of a state of sustained economic growth.

Variations in the timing of the transition from stagnation to growth and thus in economic performance across countries reflect initial differences in geographical factors and historical accidents and their manifestation in diversity in institutional, demographic, and cultural factors, trade patterns, colonial status, and public policy. In particular, once a technologically driven demand for human capital emerged in the second phase of industrialization, the prevalence of human capital-promoting institutions determined the extensiveness of human capital formation, the timing of the demographic transition, and the pace of the transition from stagnation to growth.

The theory proposes that the growth process is characterized by stages of development and it evolves nonlinearly. Technological leaders experienced a monotonic increase in the growth rates of their income per capita. Their growth was rather slow in early stages of development, increased rapidly during the take-off from the Malthusian epoch, and continued to rise, often stabilizing at higher levels. In contrast, technological followers that made the transition to sustained economic growth experienced a non-monotonic increase in the growth rates of their income per capita. Their growth rate was rather slow in early stages of development, but increased rapidly in the early stages of the take-off from the Malthusian epoch, boosted by the adoption of technologies from the existing technological frontier. However, once these economies reached the technological frontier, their growth rates dropped to the level of the technological leaders. Hence, consistently with contemporary evidence about the existence of multiple growth regimes (Durlauf and Quah, 1999), the differential timing of the take-off from stagnation to growth across economies generated convergence clubs characterized by a group of poor countries in the vicinity of the Malthusian equilibrium, a group of rich countries in the vicinity of the sustained growth equilibrium, and a third group in the transition from one club to another.

ODED GALOR

See also **growth take-offs; human capital, fertility and growth.**

Bibliography

Crafts, N.F.R. 1985. *British Economic Growth during the Industrial Revolution*. Oxford: Oxford University Press.

Doepke, M. 2004. Accounting for fertility decline during the transition to growth. *Journal of Economic Growth* 9, 347–83.

Durlauf, S.N. and Quah, D. 1999. The new empirics of economic growth. In *Handbook of Macroeconomics*, ed. J. B. Taylor and M. Woodford. Amsterdam: North-Holland.

Flora, P., Kraus, F. and Pfenning, W. 1983. *State, Economy and Society in Western Europe 1815–1975*. Chicago: St. James Press.

Galor, O. 2005. From stagnation to growth: unified growth theory. In *Handbook of Economic Growth*, ed. P. Aghion and S.N. Duraluf. Amsterdam: North-Holland.

Galor, O. and Moav, O. 2002. Natural selection and the origin of economic growth. *Quarterly Journal of Economics* 117, 1133–92.

Galor, O. and Moav, O. 2006. Das human kapital: a theory of the demise of the class structure. *Review of Economic Studies* 73, 85–117.

Galor, O. and Mountford, A. 2003. Trading population for productivity. Working paper, Brown University.

Galor, O. and Mountford, A. 2006. Trade and the great divergence: the family connection. *American Economic Review* 96, 299–303.

Galor, O. and Weil, D.N. 1999. From Malthusian stagnation to modern growth. *American Economic Review* 89, 150–4.

Galor, O. and Weil, D.N. 2000. Population, technology and growth: from the Malthusian regime to the demographic transition and beyond. *American Economic Review* 110, 806–28.

Green, A. 1990. *Education and State Formation*. New York: St. Martin's Press.

Hansen, G. and Prescott, E. 2002. Malthus to Solow. *American Economic Review* 92, 1205–17.

Hazan, M. and Berdugo, B. 2002. Child labor, fertility and economic growth. *Economic Journal* 112, 810–28.

Jones, C.I. 2001. Was an industrial revolution inevitable? Economic growth over the very long run. *Advances in Macroeconomics* 1, 1–43.

Lagerlof, N. 2003. From Malthus to modern growth: the three regimes revisited. *International Economic Review* 44, 755–77.

Lagerlof, N. 2006. The Galor–Weil model revisited: a quantitative exploration. *Review of Economic Dynamics* 9, 116–42.

Lucas, R.E. 2002. *The Industrial Revolution: Past and Future*. Cambridge, MA: Harvard University Press.

Maddison, A. 2001. *The World Economy: A Millennia Perspective*. Paris: OECD.

Mokyr, J. 1993. The new economic history and the industrial revolution. In *The British Industrial Revolution: An Economic Perspective*, ed. J. Mokyr. Boulder, CO: Westview Press.

Sanderson, M. 1995. *Education, Economic Change and Society in England 1780–1870*. Cambridge: Cambridge University Press.

economic growth nonlinearities

Nonlinear growth models

Nonlinear growth models are characterized by a country's subsequent performance being critically dependent upon its initial conditions. In particular, these models tend to imply that countries which have unfavourable initial conditions may either experience substantial periods of time in low-growth/low-income poverty traps or be altogether caught in one. In some cases, it has been explicitly suggested that active (exogenous) policy interventions may be necessary in order to kick-start a country into a more favourable equilibrium. Nonlinear growth models can be broadly classified into two classes: structural change (or 'stages of development') models, and models that emphasize endogenous technological development and cross-country interactions in terms of technological diffusion.

Structural change models focus on the (internal) transformations of an economy as it transits through critical phases or 'stages' (see Lewis, 1956; Rostow, 1960) leading to industrialization. The aim of this work is to clarify the conditions for such transitions to occur. Early work in the economic development literature (see Rosenstein-Rodan, 1943; Nurkse, 1953; Scitovsky, 1954; Fleming, 1955; formalized by Murphy, Shleifer, and Vishy, 1989) emphasized the importance of increasing returns and the size of the market in industrialization. The key idea behind this view is that countries could be locked in a no-industrialization trap because of the small size of the market for each sector of the economy.

No single sector can achieve growth on its own. However, the growth of one sector results in the enlargement of markets for other sectors. The enlargement of markets then encourages investment and growth in the corresponding sectors. These spillover effects and strategic complementarities imply that a 'big push' – that is, coordinated investments (or 'balanced growth') across sectors – may be sufficient to push the economy out of the trap and into a 'take-off' towards industrialization. Other models are explicitly informed by the analysis of historical data (see Maddison, 2004), and emphasize the importance of explaining simultaneously both historical patterns of other state variables associated with growth and growth itself. An important recent work that models the demographic transition in growth take-offs is Galor and Weil (2000). Because these models require that certain conditions be met before countries are able to achieve take-off, those who do not meet these requirements could find themselves trapped in a phase of economic stagnation for extended periods of time.

The second class of models focuses on the role of technological progress in growth. In particular, the emphasis of these models is on the diffusion of technology from countries which are technological leaders to less developed countries. Lucas (2000) is a seminal work in this area (see also Basu and Weil, 1998; Parente and Prescott, 1994; Howitt and Mayer-Foulkes, 2005). Particular attention has been paid to exploring the channels through which less advanced countries imitate or adopt technologies in leader countries. If there are no barriers to technological diffusion across countries, then these models typically predict that rich and poor countries would gradually converge in per capita income. However, if such barriers exist, then countries may differ in their ability to adopt technologies leading to the creation of 'clusters' of countries defined by a set of common barriers to technological adoption. Countries within each of these clusters or 'convergence clubs' converge to common levels of mean per capita income. Nevertheless, the per capita incomes across convergence clubs need never converge and the polarization of per capita incomes across countries may be permanent.

Growth empirics
In both classes of models, therefore, the primary concern is that countries may become separated – perhaps permanently – into multiple growth regimes corresponding to different levels of long-run per capita income. The fact that nonlinear growth models imply that global inequality may be persistent has sparked major advances in the area of cross-country growth empirics. Driven by such concerns, the central preoccupation of growth empirics has been to evaluate the conditions under which poor countries catch up with rich ones or fail to do so. Initial work along these lines focused on the concept of 'conditional convergence'. Conditional convergence is said to occur if permanent per capita income differences between countries can be accounted for solely by structural differences (and not initial conditions). Researchers initially argued that because conditional convergence was predicted by the canonical neoclassical growth model (see Ramsey, 1928; Solow, 1956; Swan, 1956; Cass, 1965; Koopmans, 1965) whereas nonlinear growth models potentially predict dependence on initial conditions, tests for conditional convergence could be used to discriminate between these classes of theories.

Following Mankiw, Romer and Weil (1992) and Barro and Sala-i-Martin (1992), the canonical way such tests were conducted was to first construct a linearized version of the neoclassical growth model about the (unique) steady state with average growth rates across a time period as the dependent variable, and measures of physical and human capital, population growth rates, and initial per capita income as covariates. Researchers then applied the linearized neoclassical model to cross-country data with the aim of testing to see whether the data supported a negative coefficient on initial per capita income. A finding of a negative coefficient on initial per capita income was taken to imply that, conditional on countries having similar structural characteristics (as defined by the set of covariates), poorer countries would close the income gap with the rich – that is, conditional convergence.

An important outcome of the, oftentimes heated (see Sala-i-Martin, 1996), convergence debates of the 1990s was precisely to weaken the idea that such tests of convergence could be interpreted as model selection tests. In a highly influential work, Bernard and Durlauf (1996) strongly disputed the interpretation of such 'conditional convergence' tests by pointing out that these tests were not able to discriminate against a class of nonlinear growth theories that have dramatically different ergodic implications from the neoclassical model. The class of models they were referring to was developed by Azariadis and Drazen (1990). Azariadis and Drazen extended the spillover models of Lucas (1988) and Romer (1986) and showed that, if (local) nonconvexities in the production function were sufficiently strong, then countries that are similar in all aspects except for initial conditions may nevertheless be organized into multiple growth regimes, each of which corresponds to a different steady state for long-run per capita income.

Bernard and Durlauf showed that the multiple-regimes Azariadis–Drazen model was theoretically consistent with a finding of conditional convergence in the data. Therefore, even in the narrowly restricted sense of countries being structurally similar, the finding of a negative coefficient to initial income in the data was no guarantee that countries would converge to a common steady state. Galor (1997) lent further support to the relevance of the Azariadis–Drazen model by arguing that standard ways of augmenting the traditional Solow model increased the likelihood that the true data-generating process followed a multiple-regimes rather than a single steady-state model. Clearly, evidence of multiple regimes and nonlinearities in growth raises questions about misspecification in empirical studies that assume that all countries follow the same growth process, and casts doubt on inferences and policy recommendations that are drawn from these studies.

The work by Bernard and Durlauf has spurred a large quantity of research searching for the existence of multiple-growth regimes. One direction of this new research has been to argue that the finding of parameter heterogeneity in the neoclassical model may be suggestive of the existence of multiple growth regimes. In a seminal work, Durlauf and Johnson (1995), employing a classification and regression tree methodology, implemented a version of Azariadis and Drazen's model and showed that there was evidence in the data to suggest that countries grouped according to initial per capita income and literacy rates correspond to four different growth regimes. Their work has inspired a long list of confirmatory works using a wide variety of econometric approaches (for example, Bloom, Canning and Sevilla, 2003; Canova, 2004; Durlauf, Kourtellos and Minkin, 2001; Kourtellos, 2005; Liu and Stengos, 1999; and Tan, 2005).

While there now is a strong consensus in the literature that there exists substantial heterogeneity across countries, it should be emphasized that this finding is only suggestive of multiple-growth regimes and is not conclusive evidence of it. These heterogeneities could arise because of small deviations in the specification of the production function (see Masanjala and Papageorgiou, 2004) which need not correspond to multiple-growth regimes. Further, even within the context of Azariadis–Drazen model, if non-convexities in the production function are not strong enough, the finding of parameter heterogeneity would not imply the existence of multiple regimes (see Durlauf and Johnson, 1995, Figure 2).

An alternative approach to investigating the existence of multiple regimes or convergence clubs has focused on the evolution of the world distribution of per capita income. The aim of this research has been to look for evidence of emerging multimodality (typically, bimodality) in the world income distribution. A secondary aim has been to evaluate the degree of churning within the multimodal distribution. If the world income distribution is characterized by emerging multimodality with little evidence of countries moving freely within the distribution (that is, churning), then this finding would suggest, in a manner analogous with the finding of multiple-growth regimes, that global income inequality is real, intensifying and persistent in nature. In fact, these are the precise findings by Quah (1993). By estimating transition probabilities for the cross-country per capita income distribution, Quah finds emerging 'twin peaks' in the world income distribution as well as substantial persistence within the distribution. Quah's seminal work has

been confirmed by subsequent work (for example, Bianchi, 1997; Fiaschi and Lavezzi, 2003; and Paap and van Dijk, 1998) even though there had been questions about the robustness of his initial methodology (see Kremer, Onatski and Stock, 2001).

While the findings of the 'twin peaks' literature have been suggestive of growth nonlinearities and multiple equilibria, it is not definitive. It is quite possible, for instance, that the aggregate production functions across countries actually exhibit decreasing marginal productivity of capital, so that there is only one steady state. However, other growth factors are sufficiently strong to overcome the convergence effect of diminishing marginal returns to produce divergence and bimodality in cross-country incomes nevertheless. Without an explicit theory to explain the observed income divergence, there is also the question of whether the bimodality in the cross-country income distribution is a transitional or permanent feature of growth (see Galor, 1997; Lucas, 2000).

Conclusion

Nonlinearities in growth have been highly influential in shaping the thinking of both growth theorists and empiricists in recent years. The work on multiple-growth regimes and the world income distribution suggests that there may exist growth factors strong enough to overcome the decreasing marginal productivity of the neoclassical production function, thereby producing increasing inequality across countries. Nevertheless, while an increasingly large body of work finds evidence that is suggestive of growth nonlinearities, many questions remain open and are the subject of current research. What are the factors that are responsible for generating multiple growth regimes or convergence clubs? Are the effects of these factors transient or permanent? If the former, what are the applicable timescales? This area of research continues to be promising and fruitful.

CHIH MING TAN

See also **balanced growth; diffusion of technology; economic growth, empirical regularities in; economic growth in the very long run; endogenous growth theory; growth take-offs; Rosenstein-Rodan, Paul Narcyz; structural change.**

Bibliography

Azariadis, C. and Drazen, A. 1990. Threshold externalities in economic development. *Quarterly Journal of Economics* 105, 501–26.
Barro, R.J. and Sala-i-Martin, X. 1992. Convergence. *Journal of Political Economy* 100, 223–51.
Basu, S. and Weil, D.N. 1998. Appropriate technology and growth. *Quarterly Journal of Economics*, 1025–54.
Bernard, A. and Durlauf, S. 1996. Interpreting tests of the convergence hypothesis. *Journal of Econometrics* 71, 161–73.

Bianchi, M. 1997. Testing for convergence: evidence from nonparametric multimodality tests. *Journal of Applied Econometrics* 12, 393–409.

Bloom, D., Canning, D. and Sevilla, J. 2003. Geography and poverty traps. *Journal of Economic Growth* 8, 355–78.

Canova, F. 2004. Testing for convergence clubs in income per capita: a predictive density approach. *International Economic Review* 45, 49–77.

Cass, D. 1965. Optimum growth in an aggregative model of capital accumulation. *Review of Economic Studies* 32, 233–40.

Durlauf, S. and Johnson, P. 1995. Multiple regimes and cross-country growth behavior. *Journal of Applied Econometrics* 10, 363–84.

Durlauf, S., Kourtellos, A. and Minkin, A. 2001. The local Solow growth model. *European Economic Review* 45, 928–40.

Fiaschi, D. and Lavezzi, M. 2003. Distribution dynamics and nonlinear growth. *Journal of Economic Growth* 8, 355–78.

Fleming, J.M. 1955. External economies and the doctrine of balanced growth. *Economic Journal* 65, 241–56.

Galor, O. 1997. Convergence? Inferences from theoretical models. *Economic Journal* 106, 1056–69.

Galor, O. and Weil, D.N. 2000. Population, technology, and growth: from the Malthusian regime to the demographic transtition and beyond. *American Economic Review* 90, 806–28.

Howitt, P. and Mayer-Foulkes, D. 2005. R&D, implementation, and stagnation: a Schumpeterian theory of convergence clubs. *Journal of Money, Credit and Banking* 37, 147–77.

Koopmans, T.C. 1965. On the concept of optimal growth. In *The Econometric Approach to Development Planning.* Amsterdam: North Holland.

Kremer, M., Onatski, A. and Stock, J. 2001. Searching for prosperity. *Carnegie-Rochester Conference Series on Public Policy* 55, 275–303.

Kourtellos, A. 2005. Modeling parameter heterogeneity in cross-country growth regression models. Working paper, Department of University of Cyprus.

Lewis, A. 1956. *The Theory of Economic Growth.* London: Allen & Unwin.

Liu, Z. and Stengos, T. 1999. Non-linearities in cross-country growth regressions: a semiparametric approach. *Journal of Applied Econometrics* 14, 527–38.

Lucas, R. 1988. On the mechanics of economic development. *Journal of Monetary Economics* 22, 3–42.

Lucas, R. 2000. Some macroeconomics for the 21st century. *Journal of Economic Perspectives* 11, 159–68.

Maddison, A. 2004. *The World Economy: Historical Statistics.* Paris: Development Studies Centre, OECD.

Mankiw, N.G., Romer, D. and Weil, D. 1992. A contribution to the empirics of economic growth. *Quarterly Journal of Economics* 107, 407–37.

Masanjala, W. and Papageorgiou, C. 2004. The Solow model with CES technology: nonlinearities and parameter heterogeneity. *Journal of Applied Econometrics* 19, 171–202.

Murphy, K., Shleifer, A. and Vishny, R. 1989. Industrialization and the big push. *Journal of Political Economy* 97, 1003–26.

Nurkse, R. 1953. *Problems of Capital Formation in Underdeveloped Countries.* New York: Oxford University Press.

Paap, R. and van Dijk, H. 1998. Distribution and mobility of wealth of nations. *European Economic Review* 42, 1269–93.

Parente, S.L. and Prescott, E.C. 1994. Barriers to technology adoption and development. *Journal of Political Economy* 102, 298–321.

Quah, D.T. 1993. Empirical cross-section dynamics for economic growth. *European Economic Review* 37, 426–34.

Ramsey, F. 1928. A mathematical theory of saving. *Economic Journal* 38, 543–59.

Romer, P.M. 1986. Increasing returns and long-run growth. *Journal of Political Economy* 94, 1002–37.

Rosenstein-Rodan, P. 1943. Problems of industrialization of Eastern and South-Eastern Europe. *Economic Journal* 53, 202–11.

Rostow, W.W. 1960. *The Stages of Economic Growth.* Oxford: Oxford University Press.

Sala-i-Martin, X. 1996. The classical approach to convergence analysis. *Economic Journal* 106, 1019–36.

Scitovsky, T. 1954. Two concepts of external economies. *Journal of Political Economy* 62, 143–51.

Solow, R. 1956. A contribution to the theory of economic growth. *Quarterly Journal of Economics* 70(1), 65–94.

Swan, T.W. 1956. Economic growth and capital accumulation. *Economic Record* 32, 334–61.

Tan, C.M. 2005. No one true path to development: uncovering the interplay between geography, institutions, and ethnic fractionalization in economic development. Mimeo, Tufts University.

economic history

Economic history is a sub-discipline within economics and, to a lesser degree, within history, whose main focus is the study of economic growth and development over time. It is to be distinguished from the history of economic thought, a branch of intellectual history.

Studies in economic growth, whether historical or contemporary, develop and analyse quantitative measures of increases in output and output per capita, emphasizing in particular changes in saving rates and rates of technological innovation and their consequences. Economic development is a larger and more encompassing rubric, also including consideration of the role of cultural changes and changes in formal institutions.

Economic history has its origins in two main traditions. The first is the German historical school, a group of scholars in the 19th and early 20th centuries, including

Gustav Schmoller and Max Weber, who ranged widely over human history with special emphasis on the consequences of institutional variation for economic as well as political performance. The second tradition stems from the efforts of a group of writers who viewed the complex of innovations in steam power, iron manufacture, and textiles in late 18th-century Britain as an epochal event – an industrial revolution – equivalent in its significance for human welfare to the Neolithic revolution which gave birth to agriculture around ten millennia earlier. The study of the causes, dimensions and consequences of the emergence of sustained increases in per capita incomes – what Simon Kuznets (1966) called modern economic growth – along with a focus on the consequences of institutional variation, continues to define much of what economic historians do.

Although historians have practised their craft at least since the time of Herodotus (the fifth century BC), economics emerged as a separate social science with the work of Adam Smith or, perhaps, as some have argued, that of the Mercantilists and the Physiocrats. Classical economists, with the notable exception of Ricardo, were almost all also historical economists. The reader of Smith, Mill, Marx, or even Marshall ploughs through thick volumes in which propositions in economic theory are embedded in often lengthy descriptions of historical events or the course of economic history. Throughout most of the 19th century, the divide between economists and economic historians was weak.

With the professionalization of economics that picked up speed in the 20th century (the American Economics Association was founded in 1885; the Royal Economic Society in 1890), economic history began to emerge as a distinct and to some degree separate sub-discipline. The Economic History Society was founded in Britain in 1926; the Economic History Association in the United States in 1941. The trend towards a separate identity accelerated in the third quarter of the 20th century, with the increasing emphasis within economics on formal mathematical modelling and the weakening within the general profession of ties to historical traditions. In the economic history societies, in contrast, those trained as historians as well as economists remained active; in Britain, distinctiveness was accentuated by the establishment of separate university departments of economic history.

As the intellectual paths taken by economics and economic history seemed increasingly to diverge, a countervailing intellectual movement known variously as 'cliometrics' or the 'new economic history' emerged. Its pioneers knew their history, but emphasized by argument and example that, if economic history was to remain influential within economics, it had to make more use of formal models as well as place increased emphasis on quantitative (rather than just qualitative) data and more advanced statistical techniques (econometrics) to analyse them.

The use of mathematical models was anathema within historical traditions, but by the 1960s widely accepted in economics. Thus, the new economic history represented something of a gauntlet thrown down to those trained in history or allied with its traditions. The push for quantitative data analysis, in contrast, was more cross-cutting in the challenges it implied. Many traditional economic historians had in fact examined such data, although the statistical techniques they used were often quite rudimentary. Within some economic circles, on the other hand, an emphasis on data was becoming suspect. Here, some scholars were comfortable with the evolution of economic theory as a branch of applied mathematics, constrained and judged by the rules of logic and consistency, but governed in its realism, if at all, by intuition rather than systematic empirical inquiry.

The effort to force formal theory upon traditional economic history often lacked acknowledgement that the relation between economic history and formal theory might usefully be a two-way street. The emphasis on data analysis, in contrast, offered a bridge between economics and economic history. It helped reaffirm within economics the importance of empirical inquiry, and encouraged those historically trained to become more sophisticated in their statistical analyses.

Nevertheless, the stress on quantitative data could not help but draw attention away from economic history's traditional concern with legal and institutional variation, where the source documents were almost uniformly qualitative. How would this theme, one of the defining features of economic history since its inception, survive the new economic history? The initial 'solution' was to try to make institutions endogenous. Blending a mix of influences from technologically deterministic Marxism to the emerging law and economics and public choice literatures, a number of scholars suggested that institutions could be understood as epiphenomenal: reflective of more fundamental givens. The high point of such efforts was probably the short book by North and Thomas (1973).

These efforts, however, gradually disintegrated under the force of the ad hoc twists required to make the framework consistent with known historical evidence (Field, 1981), and even proponents such as North eventually backed away from this agenda. Formal rules often vary where technologies and endowments are similar, and are often similar when more fundamental givens differ, and such variation has consequences for economic performance. Had the endogenization initiative been successful, it would have eliminated from economic history one of the most important perspectives it offers to general economics.

The old economic historians had taken it as obvious that, at critical historical junctures, changes in formal institutions such as laws or constitutions had powerful influences on the course of a country or region's economic development, and that these changes were not always predictable *ex ante*. The breakdown of the former Soviet Empire, and the opportunities afforded to Western scholars actually to influence the design of formal institutions, gave a powerful impetus to returning to thinking about such designs as consequential, and increasingly this perspective came to be reflected in research by scholars who did not necessarily think of themselves as economic historians.

If the main subject of economic history continues to be the history of economic growth and development, the influence of variations in formal and informal institutions in both the private and public arenas will remain an important theme. These institutions and a broader economic culture help structure the environment in which individuals pursue their interests. But the success of an economy in raising output and output per person also depends on available technologies, on the size, composition, and characteristics of the labour force, on natural resources, and on the accumulation of physical capital. The study of the evolution of these inputs suggests some of the other themes around which economic historians organize their work. In particular, there is a rich tradition, particularly in the United States, examining issues in and applying methods from modern labour economics within an historical context.

The basic agenda of economic history has not changed since the first edition of *The New Palgrave*. Interest in the causes and dimensions of the Industrial Revolution, for example, remains strong, particularly in Britain. But the field has evolved in new directions, with several discernable trends. First, scholars have concerned themselves with a broadening range of topics under the umbrella of growth and development. In the 1960s and 1970s, especially in the United States, railroads and slavery dominated much of the discussion. In recent decades, it is not possible to point to one or two issues around which research and discussion has coalesced to the same degree. Instead, there have been a number of new initiatives; one example would be the growing exploitation of anthropometric data to make inferences about variation in standards of living.

Associated with this has been a broadening of the scope of the discipline, both in terms of the countries in which economic history research is conducted and in the geographical range of topics, which extends, somewhat more so than in earlier decades, beyond Western Europe and North America to Asia, Latin America, Australia, and Africa. One illustration of this has been a range of crossnational studies, exploring such issues as economic convergence.

A third trend has been a growing willingness to think of the 20th century as an historical epoch in its own right. When the new economic history began, the Second World War had barely ended and the Great Depression was recent history. The main focus of research was the 18th and 19th centuries. Treating the 20th century as an historical period promises to reduce the gap between economics and economic history. The Great Depression, of course, continues to attract attention, but interest in the 20th century is beginning to expand beyond this. The data and events of recent decades can now more easily be seen in an historical context. The result can be a smoother continuum between topics understood as economic history and the analysis of contemporary data.

Placing more recent developments within a longer-run perspective has already begun to pay important dividends. Many trends that economists and economic historians expected at mid-century would characterize the 20th century as a whole moderated, became erratic, or in some cases reversed themselves in the last quarter of the century (Field, 2001). In 1950, for example, it looked as if the United States (and other countries) would continue to experience decreases in wealth and income inequality, robust and perhaps rising shares of union membership in the labour force, a growing role for government, and a continuing high contribution of total factor productivity (TFP) growth to growth in output per hour. In fact, inequality has generally increased, union membership has fallen, and TFP growth basically disappeared in much of the developed world between 1973 and the 1990s. The size and role of government, which many predicted would continue to expand, has in fact displayed a more complex dynamic.

A fourth and related trend has been a reinvigoration within mainstream economics of interest in what has always been a primary subject of economic history: economic growth. Much of economic theory in the 1950s and 1960s modelled production and allocation within a static economy. The revived interest in the study of growth, combined with the growing willingness of economists to adopt traditional institutional approaches, reflects the persisting influence of the original concerns and approaches of economic history within the larger profession.

Whatever the labels people apply to themselves and others, if we want better understanding of the processes of growth and development, we will continue to need scholars familiar with how to work with data and interpret the influences on economic outcomes of institutional, political, and cultural variation. Doctoral training with a specialization in economic history is well suited to imparting such knowledge and the skills for acquiring it, capabilities that will remain essential in developing improved theory and policy in the area.

ALEXANDER J. FIELD

See also **cliometrics; growth and cycles; growth and institutions; Historical School, German; institutionalism, old; technical change.**

Bibliography

Cairncross, A. 1989. In praise of economic history. *Economic History Review* 42, 173–85.

Field, A. 1981. The problem with neoclassical institutional economics. *Explorations in Economic History* 18, 174–98.

Field, A. 1987. *The Future of Economic History*. Boston: Kluwer-Nijhoff.

Field, A. 2001. Not what it used to be: the Cambridge Economic History of The United States, vols. II and III. *Journal of Economic History* 61, 806–18.

Kuznets, S. 1966. *Modern Economic Growth: Rate, Structure, and Spread*. New Haven: Yale University Press.

McCloskey, D. 1987. *Econometric History*. Basingstoke: Macmillan Education.

North, D. and Thomas, R. 1973. *The Rise of the Western World*. Cambridge: Cambridge University Press.

Solow, R. 1985. Economic history and economics. *American Economic Review* 75, 328–31.

economic laws

The social sciences, and economics in particular, separated from moral and political philosophy in the second half of the 18th century when the results of the myriad of intentional actions of people were perceived to produce regularities resembling the laws of a system. Both Physiocratic thought and Smith's *Wealth of Nations* reflect this extraordinary discovery: scientific laws thought to be found only in nature could also be found in society. This extension poses several problems. A serious one refers to the tension of combining individuals' freedom of action with the scientists' desire to discover the systematic aspects of the unintended and quite often unpredictable consequences of human action, that is, the desire to arrive at laws characterized by a certain degree of generality and permanence.

In the history of economic thought this fundamental tension has been solved in different ways. In the 18th century, the mechanistic ideal of the natural sciences, combined with the natural law idea of a harmonious order of nature, determined the way social phenomena were treated. There was a desire to discover the 'natural laws' of economic life and to formulate the natural precepts which rule human conduct. The classical economists upheld the notion that natural laws are embedded in the economic process as beneficial laws, along with the belief in the existence of rules of nature capable of being discovered. Thus the belief that things could follow the beneficial 'natural course' only in a rationally organized society which it was a duty to create according to the precepts of nature. The economic system is the mechanism by which the individual is driven to fostering the prosperity of society while pursuing his private interest. Hence the automatic operation of the economic system may be combined with freedom of individual action. This is the core of the doctrine of economic harmony. Besides being causal laws of a mechanical type, the laws of nature are providentially imposed norms of conduct. In such a setting it would have been pointless to separate

means and ends, since the implementation of natural laws is both an end and a means, and even more pointless to think of a tension between 'explaining' and 'understanding' economic behaviour. Causal and teleological, positive and normative, theoretical and practical started being seen as separate categories only when the economic discourse freed itself from the philosophy of natural law and all its implications.

Post-classical economics set out to be a science of the laws regulating the economic order and of the conditions allowing these laws to operate. It became the basis of a theory that, in Jevons's own terms, proposed to construct a 'social physics'. The view of a social world ordered according to transcendent ends was abandoned in favour of an ideal of objective knowledge of economic phenomena gained through a 'positive' study of the laws that regulate market activities. In so doing, neoclassical 'positive' economics solves the aforementioned tension by extrapolating the theoretical model of natural sciences to economics: economics is to produce the laws of motion similar to those of physics, chemistry, astronomy.

But what is a scientific law and which role do laws play within the logical positivist's perspective adopted by neoclassical economics? Laws provide the foundation of a deductive scientific method of inquiry. According to the deductive–nomological conception of explanation, due to C. Hempel, laws are universal statements not requiring reference to any one particular object or spatio–temporal location. To be valid, laws are constrained neither to finite populations nor to particular times and places; they are, in effect, expressions of natural stationarities. This interpretation of the notion of law provides the so-called covering-law model of explanation with an unquestionably firm inferential foundation. Deductive logic is employed to ensure the truth status of propositions and, since the deductions are (by hypothesis) predicated on true universal statements (laws), the empirical validity of these statements may be ascertained. However, what sort of constraints on economic discourse are imposed by this positivistic structure? On the one hand this structure constitutes its object; on the other hand it generates specific economic questions together with their method of solution. Following the model of natural sciences and its success in controlling a natural world made up of objects and unvarying relations among them expressed in the form of laws, the neoclassical approach arrives at a study of regularities conceived of as specifying the nature of its objects.

To capture the different interpretations of the notion of law by classical and neoclassical economists let us refer to one of the most famous of economic laws: the law of diminishing returns, also known as the law of variable proportions. Studying agricultural production, Ricardo had noted that different quantities of labour, assisted by certain quantities of other inputs (farm tools, fertilizers, and so on), could be employed on a given piece of land, that is, it was possible to vary the proportions in which land and complex labour (labour assisted by other inputs) are employed. He accordingly arrived at the law which states that production increases resulting from equal increments in the employment of complex labour, while the quantity of land farmed remains constant, will initially be increasing and then decreasing. (To be sure, the first statement of the law is due to the Physiocratic economist Turgot.)

Three points deserve attention. First, Ricardo and classical authors in general offer no formal demonstration of this law. To them, it is basically an empirical law, on which no functional association between output and variable inputs can be built. Second, the classics' use of the law refers to their theories of distribution and development: as the supply of land in the whole system is fixed, sooner or later a point will be reached at which economic growth will come to a halt, notwithstanding any countervailing effects due to technical progress. Finally, the law presupposes a comparative statics framework: the pattern of the marginal products of complex labour refers to different observable equilibrium positions and not to hypothetical or virtual variations.

With the advent of the marginalist revolution, two subtle changes in the interpretation of the law took place. (a) The de facto elimination of the distinction between the extensive case (the case of the simultaneous cultivation of pieces of land of different fertility) and the intensive case (the application of successive doses of capital and labour to the same piece of land) with an over-evaluation of the latter. Classical economists, being interested in the explanation of rent, concentrated on the extensive case; they took also the intensive case into consideration but with many qualifications. Indeed, whereas the various levels of productivity of different qualities of land is a circumstance which may be directly observed in a given situation, the marginal productivity of a given input is related to a virtual increment in output and therefore to a virtual change in the situation. (b) The change in the method of analysis – it was preferred to reason in terms of hypothetical rather than observable changes – brought about by the shift of interest towards the intensive margin, supported the thesis of the symmetrical nature of land and other inputs. This in turn favoured the extension of the substitutability between land and complex labour from agricultural production to all kinds of production, including those in which land does not figure as a direct input. It so happened that whereas in classical economics the substitutability between land and complex labour presupposes that simple labour and equipment are strictly complementary, in neoclassical economics this substitutability is applied to all inputs indiscriminately.

However, the neoclassical interpretation of the law poses serious problems. In the first place, there is the problem of justifying, on empirical grounds, the general applicability of the substitution principle. Secondly, and more importantly, in order to allow the substitution of inputs to take place, a certain lapse of time is required during which the required modifications to the productive structure can be made. (It is certainly true that coal can replace oil to provide heating, but before this can happen it will be necessary to change the heating system.) The well-known distinction between the short run and the long run is a partial and indirect way to take the temporal element into consideration. In the short run the plant is fixed by definition. It is therefore the fixed input which, in the neoclassical interpretation of the law, plays the same role as land in the classical interpretation. Now, neoclassical theory correctly states the law of diminishing returns with respect to the short run; however it is in the long run that the substitutability of inputs becomes actually feasible. One is therefore confronted with a dilemma: the neoclassical interpretation of the law seems to be more plausible in a long-run framework when there exists the necessary time to accommodate input adjustments; on the other hand, fixed inputs cannot, by definition, exist in the long run so that the law of variable proportions cannot be stated in such a context.

This dilemma is the price neoclassical theory has to pay for its interpretation of the law in accordance with the positivistic statute. Indeed, the power of deductive, truth-preserving rules of scientific inference is not purchased without a cost. A school of economic thought which is not prepared to sustain such a cost is the neo-Austrian. The neo-Austrian economists solve what has been called the fundamental tension by arguing economics cannot and should not provide general laws since, by its very nature, it is an idiographic and not a nomothetical discipline. The general target of economics is 'understanding' grounded in Verstehen doctrine: by introspection and empathy, the study of the economic process should aim at explaining individual occurrences, not abstract classes of phenomena. It follows that if by a scientific law one should mean a universal conditional statement of type 'for all x, if x is A, then x is B', statements regarding unique events cannot by definition express any regularity for the simple reason that any regularity presupposes the recurrence of what is defined as regular. In the words of L. von Mises, who shares with F. von Hayek the paternity of the neo-Austrian school, what assigns economics its peculiar and unique position in the orbit of pure knowledge '... is the fact that its particular theorems are not open to any verification or falsification on the ground of experience ... the ultimate yardstick of an economic theorem's correctness or incorrectness is solely reason unaided by experience' (von Mises, 1949, p. 858).

There is indeed a place for economic 'laws' in the framework of Austrian economics. The familiar 'laws' of economics (diminishing marginal utility, supply and demand, diminishing returns to factors, Say's Law and so on) are seen as 'necessary truths' which explain the essential structure of the economic world but with no predictive worth. In other words, economic laws are not generalizations from experience, as it is the case within the positivistic paradigm, but are theorems which enable us to understand the economic world. It is ironic that Mises' position of radical apriorism joined to Hayek's attack on scientism and methodological monism are completely at variance with the position taken by the father of the Austrian school, Carl Menger (1883), who announced that in economic theories exact laws are defined which are just as rigorous as in fact are the laws of nature.

Between the extreme positions of neoclassical positive economic and neo-Austrian economics are those who, without denying that economics is in search for laws in the same sense in which natural sciences are and that laws perform an explanatory as well as a predictive function, underline that the explicative structure of economics, albeit nomothetical, substantially differs from that of natural sciences. This intermediate position can be traced back to Keynes's (1973) methodology which considers the conditions of truth and universality of the positivistic conception of scientific laws as far too rigid for a discipline such as economics. Two main reasons account for the different epistemological status of laws in natural sciences and in economics. First, the knowledge of economic phenomena is itself an economic variable, that is, it changes, along with the process of its own acquisition, the economic situation to which it refers. The formulation of a new physical law does not change the course of physical processes; it does not influence the truth or falsity of the prognosis. This is not the case in economics where the prognosis, say, that in two years time there will be a boom can cause overproduction and a resulting recession. In turn, this specific aspect is strictly connected to the fact that the object of study of economics possesses an historical dimension. Economics is in time in a way that natural sciences are not. The ensuing mutability of observed regularities is well expressed by Keynes when he writes, 'As against Robbins, economics is essentially a moral science and not a natural science. That is to say it employs introspection and judgements of value' (1973, p. 297) to which he adds, 'It deals with motives, expectations, psychological uncertainties. One has to be constantly on guard against treating the material as constant and homogenous' (p. 300).

Second, the role played by *ceteris paribus* clauses in natural sciences and in economics is substantially different. The modern economists appeal to the 'other things being equal' clause – which according to Marshall is invariably attached to any economic law – in all those cases where the classical economists were talking of 'disturbing causes'. J.S. Mill's (1836) discussion of inexact sciences is suggestive here:

When the principles of Political Economy are to be applied to a particular case then it is necessary to take into account all the individual circumstances of that case ... These circumstances have been called *disturbing*

causes. This constitutes the only uncertainty of Political Economy. (1836, p. 300)

Also in natural sciences we find *ceteris paribus* clauses. Indeed, a scientific theory that could dispense with them would in effect achieve perfect closure, which is a rarity. So where lies the difference? The example of the science of tides used by Mill is revealing. Physicists know the laws of the greater causes (the gravitational pull of the moon) but do not know the laws of the minor causes (the configuration of the sea bottom). The 'other things' which scientists hold equal are the lesser causes. So could we conclude that just about all generalizations in both natural sciences and economics express in fact *tendency laws*, in the sense that these 'laws' truly capture only the functioning of 'greater causes' within some domain? Certainly not, since there is a world of difference between the two cases. Galileo's law of falling bodies certainly presupposes a *ceteris paribus* clause, so much so that he had to employ the idealization of a 'perfect vacuum' to get rid of the resistance of air. However, he was able to give estimates of the magnitudes of the amount of distortion that friction and the other 'accidents' would determine and which the law ignored. In other words, whereas in natural sciences the 'disturbing causes' have their own laws, this is not the case in economics where we find tendency statements with unspecified *ceteris paribus* clauses or, if specified, specified only in qualitative terms. In economics it is generally impossible to list all the conceivable inferences implied in a lawlike statement and to replace the *ceteris paribus* clause with precise conditions. So, for example, the law that 'less will be bought at a higher price' is not refuted by panic buying, nor is it confirmed by organized consumer boycotts. No test is decisive unless *ceteris* are really *paribus*.

These remarks help to understand the role acknowledged by Keynes to laws in economic inquiry. Besides general laws, there are also rules and norms which are significant in the explanation of economic behaviour. To Keynes, it makes no sense to reduce all forms of explanation in economics to that of the covering-law model. Indeed, whereas to justify a law one has to show that it is logically derivable from some other more general statements, often called principles or postulates, the justification of rules occurs through the reference to goals and the justification of norms through the reference to values which are not general sentences, but rather intended singular patterns or even ideal entities. Since no scientific law, in the natural scientific sense, has been established in economics, on which economists can base predictions, what are used and have to be used to explain or to predict are tendencies or patterns expressed in empirical or historical generalizations of less than universal validity, restricted by local and temporal limits. Recently, Arrow has amazed orthodox economists when raising doubts about the mechanistically inspired understanding of economic processes: 'Is economics a subject like physics, true for all time or are its laws historically conditioned?' (Arrow, 1985, p. 322).

The list of generally accepted economic laws seems to be shrinking. The term itself has come to acquire a somewhat old-fashioned ring and economists now prefer to present their most cherished general statements as theorems or propositions rather than laws. This is no doubt a healthy reaction: for too long economists have been under the nomological prejudice, of positivistic origin, that the only route towards explanation and prediction is the one paved with laws, and laws as forceful as Newton's laws. Images in science are never innocent: wrong images can have disastrous effects.

STEFANO ZAMAGNI

Bibliography

Arrow, K. 1985. Economic history: a necessary though not sufficient condition for an economist. *American Economic Review, Papers and Proceedings* 75 , 320–3.

Keynes, J.M. 1973. *The General Theory and After.* Part II: *Defence and Development.* In *The Collected Writings of John Maynard Keynes*, vol. 14, London: Macmillan.

Menger, C. 1883. *Unterschungen über die Methode der Sozialwissenschaften.* Leipzig: Duncker & Humblot.

Mill, J.S. 1836. On the definition of political economy and the method of investigation proper to it. Repr. in *Collected Works of John Stuart Mill*, vol. 4, *Essays on Economy and Society*, ed. J.M. Robson, Toronto: University of Toronto Press, 1967.

Mises, L. von. 1949. *Human Action: A Treatise on Economics.* London: William Hodge.

economic man

Among the many different portrayals of economic agents, the title of *homo economicus* is usually reserved for those who are rational in an instrumental sense. For example, this is how agency is defined in neoclassical economics. In its ideal type case the agent has complete, fully ordered preferences (defined over the domain of the consequences of his or her feasible actions), perfect information and all the necessary computing power. After deliberation, he or she chooses the action that satisfies their preferences better (or at least no worse) than any other. No questions are raised about the source or worth of preferences, reason focuses on the efficient selection of the means to given ends.

This basic model is then made more sophisticated. The theory of risk allows for the point that an action may have several possible consequences. When preferences are represented via the device of a utility function, the agent assesses his or her expected utility by discounting the utility of each consequence by how likely it is to be the actual one. That requires the agent to have a probability distribution for the consequences, even if only a

subjective one. Other refinements include allowance for costs of acquiring information, of processing it and of action. Then there are complexities, illustrated by game theory, when actions of other agents form part of the environment in which the person acts. The basic vision remains, however, one of agents who are rational in the sense that they maximize an objective function subject to constraints (or act 'as if' this were the case).

This vision is not unique to neoclassical economics. For example, Marx's profit-maximizing capitalist fits the same instrumental model of rationality. Institutionalist accounts of, for instance, banks or trade unions often conceive economic bodies as similar unitary rational agents. Nor is the vision confined to any specific motivating desire in agents, like a selfish pleasure-maximizing drive. There is scope for allowing ethical preferences alongside the symptomatic textbook desires for apples and oranges. Agents are, however, regarded as self-interested, in the looser sense that they are moved to satisfy whatever preferences they happen to have. Furthermore, granted that *de gustibus non est disputandum*, this modest base is enough to ground a full-blown social theory on a model of agency which can be exported to other social sciences.

Such a social theory is individualist and contractarian, with a pedigree that includes Hobbes's *Leviathan* and Benthamite utilitarianism. The satisfaction of individual preference, aided by felicific calculation, is what makes the social world go round. Social relations become instrumental, in the sense that they embody exchanges in the service of individual preferences (see Becker, 1976). For instance, marriage has been analysed in this spirit as an arrangement to secure the mutual benefit of exchange between two agents with different endowments. Crime has been claimed to occur because calculation of costs and benefits proves it to be the action that maximizes expected utility. Meanwhile, institutions, which feature in elementary microeconomics as constraints on individual choice, become deposits left by earlier transactions, often deliberately so as devices to prevent preferences being frustrated by situations of the Prisoner's Dilemma type. Government policies are explained on the hypothesis that the political arena is also peopled by individuals maximizing expected utility, who form coalitions in support of policies that will secure re-election (see Downs, 1957). In short, *homo economicus* morphs into a universal *homo sapiens*.

Such a full-blown social theory may be too ambitious because assumptions that are plausible for simple market transactions become suspect when scaled up. For example, the ideal-type case makes agents, so to speak, transparent to themselves, and does not allow for history occurring behind their backs. Freudians would object to transparency of preferences and Marxians would invoke theories of false consciousness. (Although Marx's capitalists are instrumentally rational, their desire to maximize profit is an alienated one, 'forced' on them by a competitive capitalist system.) Many other social theorists would object to the treatment of norms and social relations as instrumental, on the grounds that norms are prior to preferences. For instance, cultural forms like the rules of orchestral composition are a source of musical preferences rather than a solution to a priori problems of maximizing musical enjoyment. Or, to put this differently, game theory yields too many instances of indeterminacy for an ambitious programme of reducing all social practices to the exercise of instrumental reason by the individual participating agents.

Such objections, of course, need not affect the more modest enterprise of explaining economic transactions within the parameters of social institutions like the market. But even here *homo economicus* has critics. Philosophically, it is not plain that preferences can be taken as given in a sense which makes them impervious to the agent's beliefs about the moral quality of his or her actions. In supposing that only desires can motivate agents, the economist is taking sides in a continuing philosophical dispute between Humeans, who regard reason as the slave of the passions, and Kantians, who make place for the rational monitoring of desire. This dispute surfaces plainly in welfare economics, when it is asked whether all preferences should count equally or whether 'capabilities' are more appropriate for the evaluation of social states than degrees of preference satisfaction, but bears on the elementary model of action too (see Sen, 1999).

There are also methodological doubts about the empirical standing of the model. What would falsify the claim that economic agents seek the most effective means to satisfy their preferences? Apparent counter-examples can always be dealt with by treating them as evidence that preferences have changed or been dismissed through a careful individuation of outcomes. Indeed, since preferences are unobservable, they can be identified only if the correctness of the model is presupposed. In other words, there is room for deeper dispute about the foundations of orthodox microeconomics than is always realized.

Even within economics there are critics. The most substantial attack comes from those who think that perfect information is not a useful limiting case of imperfect information. Granted that there is often no way of calculating the likely marginal costs and benefits of acquiring extra information (short of actually acquiring it), how shall the agent decide rationally when to stop? Simon (1976) uses the question to argue for 'satisficing' models, in place of maximizing ones, and for 'procedural' or 'bounded' rationality. Rationality, he suggests, is a matter of following a procedure that halts with a good solution, and should not be defined in terms of best solutions. While this is a tempting thought, it is not obvious that searching for a 'good' solution is any easier than the best one if 'good' is some kind of second-best version of the 'best'. As a result, 'behavioural economists' have been drawn to the large experimental literature in psychology on how people actually behave and have

produced economic models of decision-making that incorporate a variety of psychological processes such as 'self-serving biases', the 'law of small numbers' and 'reference dependence' (see Kahneman, 2003). In this way, *homo economicus* has become more psychologically complex and more of an institutional or organizational person than an abstract maximizer.

The rational expectations hypothesis offers a different approach to the information issue. A rational agent who is short of information should not use an information-generating mechanism that gives rise to systematic errors. If errors are systematic, the agent should be able to learn how to eliminate them by amending the mechanism. There is an incentive to do so, because improved estimates of future variables will be profitable. On the face of it this makes rational expectations the natural ally of the pure economic-man models. Economic Man can proceed much as before, in the assurance that inadequate information involves nothing more systematic than 'white noise' and with the benefit of fresh analytic results that flow from a rational expectations hypothesis.

But this is to sidestep the informational problem set earlier, unless one sees how rational agents will learn to remove systematic errors. When there are costs to learning then it may not be rational to expend the effort that achieves a rational expectation. If we set such costs aside, in some simple learning situations a Bayesian updating procedure turns a rational expectations-generating process into an approximation of adaptive expectations, which could be construed as a procedural rule of thumb. But no general rapprochement between maximizing and procedural models of rationality follows. In more general learning situations the rational agent is trying to learn the rational expectations equilibrium relationship between variables – the one which, if used by agents to form their expectations, would reproduce itself in experience (white noise apart). This sounds easy, in that repeated experience of a particular relationship should lead to convergence on accurate parameter estimates. However, ignorance of the rational expectations equilibrium values produces behaviour that departs from those values. So observed values of variables embody a distortion which agents cannot correct without knowing the dimensions of their own ignorance. To know this, however, they would have to know the rational expectations equilibrium values already. To put it as the procedural critics might, learning would be feasible only if there were nothing to learn. The information question has been begged; and the door again opens on to psychology and its rich literature on what people actually do.

Nevertheless, the ideal-type Economic Man remains a powerful model of action not only in neoclassical theories, where insights in comparative statics have been especially notable, but elsewhere too. How powerful it finally is depends, within economics, on what becomes of the informational difficulties and on whether procedural or bounded models can come up with rival results of equal scope and elegance. For the wider social sciences, it offers a tempting analysis of social behaviour at large both for transactions in other social arenas and for the emergence of the institutions that govern those arenas. But the greater its ambitions, the more serious become the unresolved doubts about the origin of preferences and their relation to norms and institutions.

SHAUN HARGREAVES HEAP

See also **altruism, history of the concept; rational behaviour; rationality, history of the concept; utilitarianism and economic theory.**

Bibliography

Becker, G. 1976. *The Economic Approach to Human Behaviour.* Chicago: Chicago University Press.
Downs, A. 1957. *An Economic Theory of Democracy.* New York: Harper Row.
Kahneman, D. 2003. Maps of bounded rationality: psychology for behavioural economics. *American Economic Review* 93, 1449–75.
Sen, A. 1999. *Commodities and Capabilities.* Oxford: Oxford University Press.
Simon, H.A. 1976. From substantive to procedural rationality. In *Method and Appraisal in Economics,* ed. S. Latsis. Cambridge: Cambridge University Press.

economic sanctions

Economic sanctions are tools of statecraft used to influence the behaviour of foreign countries by the threat or actual withdrawal of trade and sources of finance. Traditional means of coercion include trade embargoes, withholding development assistance, and asset freezes. The objective is to confront a foreign country with a choice: either bear the cost of lost trade and finance, or change policies to comply with the demands of those imposing the sanctions (the sender countries). Projecting power through economic coercion is deemed more forceful than diplomatic reproach yet less drastic than military intervention. In practice, economic measures generally are deployed as part of a broader programme of foreign policy responses encompassing diplomatic entreaties, covert or quasi-military intrusions, and threat of or preparation for military action.

Countries impose sanctions in pursuit of a variety of foreign policy goals. Historically, economic sanctions have preceded and then accompanied military conflict. The oil embargo of Japan was a prelude to the Second World War in the Pacific; so, too, were the United Nations' sanctions against Iraq following its invasion of Kuwait in 1990. Obviously, sanctions are part and parcel of 'hot wars' that sever economic ties between the combatants; but they are also prevalent in 'cold war' episodes, where the goal is to impair military capabilities through denial of weapons and dual-use technologies (for example, post-war sanctions against the Soviet Union and its satellites under the auspices of the Consultative Group and Coordinating Committee for Multilateral Export Controls, or CoCom, and efforts to blunt the development of nuclear weapons in Iran and North Korea). In addition, sanctions have sought to impede or reverse military incursions across borders (for example, the League of Nations effort to get Italy to withdraw from Abyssinia in 1936) and between warring factions within a country (the sad recent history of several West African states).

Not all sanctions episodes respond to or presage military actions. Many post-war cases have been advanced to counter other types of aberrant behaviour such as state sponsored terrorism, proliferation of weapons of mass destruction, or human rights abuses. In these cases, sender countries impose sanctions in an effort to redress foreign outrages, to deter emulation by others (the rationale in most anti-proliferation cases), and to punish the target regime for its misdeeds (for example, the US grain embargo after the Soviet invasion of Afghanistan in 1989). In a number of cases, sanctions pursue the goal of regime change *sotto voce* – whether the target is Moammar Gaddafi in Libya, Kim Jong-il in North Korea, or the Afrikaaners in South Africa. Sanctions that portend regime change obviously meet stauncher resistance than those that seek narrow changes in governance by the target government.

Do sanctions 'work'?

Foreign policy ventures seldom yield unambiguous results. Gauging the effectiveness of sanctions involves a combination of quantitative method and intuition, and often requires subjective evaluation of incomplete results. Sanctions alone seldom are sufficient to change foreign practices, but they can *contribute* to the achievement of policy goals in conjunction with other instruments of statecraft, if properly designed and implemented. That is easier said than done.

Sanctions are blunt policy instruments; they are better at impairing economic performance over time than at inflicting surgical strikes on target countries. Senders that expect immediate gratification often tire of the effort, especially if the sanctions impose significant costs on their own firms and workers. Moreover, when sanctions are hard hitting, it is difficult to avoid innocent victims within the target country and in neighbouring states; in such cases, the debilitating effect of sanctions often results in substantial suffering among the civilian population. Humanitarian exemptions from the sanctions designed to soften the blow to the general public invariably weaken the economic impact of the sanctions and muddy the policy signal to the target regime. To be sure, such loopholes in the sanctions net are important both on moral grounds and to maintain the cohesion of the coalition of sender countries, but the loopholes are prone to abuse (witness the scandalous operation of the United Nations' oil-for-food programme, which was supposed to channel Iraqi oil export revenues to humanitarian assistance) and reduce the economic pressure to comply with the sender's demands.

Almost all sanctions leak; targeted countries can evade the full thrust of the economic restrictions by redirecting trade and finance to non-sanctioning states or by engaging in clandestine operations. Countries seeking economic or political influence with the target regime often conspire to evade the sanctions; the Cold War period was replete with examples of 'Black Knight' countries coming to the rescue of targeted regimes with aid to offset the impact of sanctions imposed by the United States or the Soviet Union. Smugglers still outwit even the most comprehensive embargoes – witness the billions of dollars earned by Saddam Hussein, the former Iraqi president, from illicit oil exports during the period of 'comprehensive' UN sanctions against Iraq. For a price, targeted regimes can still procure goods, services and technologies; the profit motive seems to be an irresistible force regardless of region or culture!

That said, sanctions have contributed to a few notable successes in the post-war era, including the collapse of the apartheid regime in South Africa and the renunciation of terrorism by President Gaddafi in Libya. Hufbauer et al. (2007) found success – measured by the partial fulfillment or better of policy goals – in more than a quarter of the almost 200 sanctions episodes documented in the 20th century. (The third edition of this comprehensive study of economic sanctions contains updated policy analysis and case studies, and an extensive bibliography. See also Baldwin, 1985, for an examination of the tools of economic statecraft, and Martin, 1992, for analysis of the use of multilateral economic sanctions.) Most of these cases, however, involved relatively modest demands on the target country. When the stakes are high, resistance by the target regime stiffens. Accordingly, most high-profile sanctions cases – like those seeking to oust President Castro in Cuba or to deter support for terrorism and the development of nuclear weapons by the ayatollahs in Iran – have been abject failures.

Can sanctions be effective in an era of rampant globalization?

Economic sanctions traditionally have been the domain of big powers, acting unilaterally or as part of a broader international coalition. Until recently, the big powers controlled the trade lanes and purse strings of international commerce, and held a near monopoly on advanced technologies. Since the mid-1980s, however, the success of post-war economic development, spurred in part by the spread of technological innovation, has eroded the franchise of the big powers and created alternative sources of goods, technology and capital for countries targeted by economic sanctions. Simply put, globalization has made it much harder to design an effective sanctions policy.

In addition, global politics are now more complex than in the period of East–West rivalry. Former allies differ regarding strategies and priorities for using sanctions to deal with regional trouble spots. For example, Europe is more vulnerable than the United States to an interruption of energy supplies from the Middle East, and thus is less willing to constrain oilfield development and to take actions that risk political retaliation. Similarly, China and Japan are highly dependent on imported energy and thus sensitive to sanctions against Iran and other oil-producing states.

Globalization also has contributed to the decentralization of power, allowing smaller countries – especially those rich in energy resources – to provide offsetting assistance to blunt the economic impact of sanctions. But the influence of globalization goes beyond the realm of state-to-state intervention; terrorism, for example, now operates in a stateless domain of sleeper cells and territories outside of governmental control linked through informal financial and telecommunications networks. For that reason, sanctions policies increasingly seek to target individuals and corporations as well as governmental bodies, and to favour financial measures to interdict inter-bank electronic transfers in addition to the more traditional controls on trade, investment and development assistance.

In sum, economic sanctions continue to play a major role in international relations. However, the familiar goals of economic coercion now must be pursued through measures adapted to the changing conditions in global markets. The use of economic sanctions needs to be reconsidered and revamped, but not abandoned.

JEFFREY J. SCHOTT

See also **foreign aid; trade policy, political economy of; transfer of technology.**

Bibliography

Baldwin, D.A. 1985. *Economic Statecraft*. Princeton, NJ: Princeton University Press.

Hufbauer, G.C., Elliott, K., Schott, J.J. and Oegg, B. 2007. *Economic Sanctions Reconsidered*, 3rd edn. Washington, DC: Peterson Institute for International Economics.

Martin, L.L. 1992. *Coercive Cooperation: Explaining Multilateral Economic Sanctions*. Princeton, NJ: Princeton University Press.

economic sociology

The first recorded use of the term 'economic sociology' is in a 1879 work by Stanley W. Jevons; and it is clear from the context that Jevons viewed economic sociology as part of the overall enterprise of economics rather than as an area belonging to another social science, such as sociology. Today, in contrast, the term 'economic sociology' is used primarily by sociologists, and they define it as *the application of sociological concepts and methods of analysis to economic phenomena*. While it is definitely possible to treat the great concern with institutions in New Institutional Economics, for example, as a kind of economic sociology, the reader is referred to the entry for this topic for this type of analysis. Similarly, while Gary Becker at times has referred to his extension of the economic model to non-economic topics as 'economic sociology', the reader is similarly referred to the entry for his work.

Here, the first section, on classical economic sociology, is followed by sections on more recent economic sociology. This way of proceeding not only follows the general development of the field of economic sociology but is often how economic sociology is taught today, since the classics play a somewhat different role in economic sociology (as in sociology itself) to that in economics. In brief, while sociologists are trained through work with the classics as well as modern material, today's economists read the classics primarily when they study the history of their discipline.

Classical economic sociology

The work of Karl Marx (1818–83) can be seen as a type of economic sociology, in the sense just mentioned. More generally, Marx closely linked classical economic categories, such as value, price and capital, to distinctly social categories, such as class, work and relations of production. Nevertheless, Marx has played a marginal role in economic sociology as an academic enterprise – except as a catalyst and inspiration for a number of scholars, including Max Weber and Joseph Schumpeter.

Modern academic sociology is generally regarded as having three founders – Max Weber, Emile Durkheim and Georg Simmel – all of whom were interested in the economy. Georg Simmel (1858–1918), who pioneered sociology in Germany, wrote on the sociological role of money, competition and trust in the economy (Simmel, 1900; 1908). He closely linked different types of money to different types of social authority, and also attempted to show how money is linked to the element of relativism in modern society. Competition, he argued, releases the energy of all participants to the benefit of the public, whereas in a conflict combatants are pitted against each other and block each other's efforts. Trust, finally, is central to the economy as well as society at large; without trust, the economy as well as society would collapse.

Emile Durkheim (1858–1917), unlike Simmel, attempted to institutionalize economic sociology, partly by encouraging some of his students to specialize in this field. Durkheim's own most important contribution to economic sociology can be found in his doctoral study of the division of labour, which contains a sharp critique of the argument in Adam Smith's *The Wealth of Nations* (Durkheim, 1893). According to Durkheim, while Adam Smith had seen the significance of division of labour exclusively from the perspective of the creation of wealth, he had neglected its importance for the cohesion of society. More precisely, Smith had failed to realize that the primary function of the division of labour in modern society is to tie people together: people who do very different things need each other, and this is also what gives cohesion to modern society.

The most sustained effort to lay a solid theoretical foundation for economic sociology and also to carry out empirical studies can be found in the work of Max Weber (1864–1920) (Swedberg, 1998). While Weber is famous for *The Protestant Ethic and the Spirit of Capitalism* (1905), it is less well known that his work is part of a more general attempt to develop a new academic field that would complement economic history and economic theory, namely, economic sociology.

At first Weber carried out empirical and historical studies with this goal in mind, and of these *The Protestant Ethic* is by far the best known (but see also Weber, 1909; 1895). Weber's thesis, which holds that a certain type of religion ('ascetic Protestantism') had helped to create the mentality of modern capitalism in the 16th and 17th centuries ('rational capitalism'; Weber, 1905), has led to a heated debate. Most commentators have found Weber's thesis unconvincing, but it should be emphasized that the debate is still going on with as much fervour as in the early 20th century (see, for example, Marshall, 1982).

The heart of Weber's economic sociology is to be found in *Economy and Society*, a work that was incomplete when Weber died. It is here, for example, that Weber set out his well-known typology of capitalism: political capitalism, traditional capitalism and rational capitalism. While the former two have existed for thousands of years, rational capitalism has emerged only in modern times and in the West. While traditional capitalism is non-dynamic and centred around small enterprises involving trade and the exchange of money, political capitalism is profit-making that either takes place through the state or under its direct protection, as in imperialism. Rational capitalism, in contrast, gets its name from the strong element of conscious and methodical calculation: the activities of the firm are carried out with the help of accountants and a trained staff; similarly, the activities of the state bureaucracy (including in the legal system) are predictable and rational. All of this makes possible a truly dynamic and revolutionary form of capitalism, according to Weber.

Economy and Society also contains a serious attempt by Weber to develop the central theoretical categories of economic sociology (Weber, 1914, pp. 63–211). The basic unit of analysis is 'economic social action', which differs from economic action in economic theory by partly being determined by its social dimension.

Economic social action is defined by Weber as behaviour that is (*a*) invested with meaning, (*b*) aimed at utility and (*c*) *oriented to another actor*. Utility is what makes the action 'economic'; and Weber's definition of 'social' is to be found in the formula 'orientation to another actor'. The emphasis on meaning explains why Weber's sociology is called an interpretive sociology; his economic sociology was to be a form of *interpretive economic sociology*.

Weber then proceeds to economic relationships in which two actors orient their actions to one another. These relationships can be either open or closed; and there is a general tendency for open economic relationships to become closed when there are not enough resources to go round. Economic organizations are defined as closed social relationships of a certain type; there also has to be a staff. Economic systems, finally, can be oriented either to profit-making (as in capitalism) or to the provision for a household (as in socialism or earlier non-market economies). Weber also discusses a host of other topics, including trade, money, division of labour and different ways of appropriation.

After the classics

While the founding fathers of sociology were all interested in economic sociology and promoted it, the topic did not become popular among sociologists until the mid-1980s with the emergence of so-called 'new economic sociology'. The reason for this is not clear, but may well have been a strong sense among sociologists that the economists were better equipped to deal with economic topics. In any case, very little work on economic sociology was produced between 1920 and the mid-1980s.

There were, however, a few exceptions. For one thing, sociologists did discuss topics relating to the economy, even if they did so under labels other than 'economic sociology'. One example is industrial sociology, which saw as its main task to analyse situations when people work in groups, in the factory as well as the office. An important research result is that workers develop norms in a number of areas, including what is seen as the maximum effort. Those who breach these norms are punished (for example, Whyte, 1955).

Three individuals who all made important contributions to economic sociology also appeared during the period after the classics: Joseph Schumpeter, Karl Polanyi and Talcott Parsons. According to Schumpeter (1885–1950), economics should be a broad science ('social economics') and encompass four areas: economic theory, economic history, economic statistics and economic sociology (Schumpeter, 1954, pp. 12–24). Schumpeter did work in each of these fields, including economic sociology. According to Schumpeter, economic sociology deals with institutions, while economic theory deals with economic mechanisms. Schumpeter's three most famous essays in economic sociology deal with the issues of social class in economic life, the role of taxation ('fiscal sociology') and imperialism (Schumpeter, 1991). Schumpeter thought highly of these essays and they are all considered minor classics today.

But one can also find elements of economic sociology in some of Schumpeter's non-sociological writings. This goes for the famous analysis of entrepreneurship in *Theory of Economic Development*, not least the element of resistance from the environment that the entrepreneur usually confronts (Schumpeter, 1934). Similarly in *Capitalism, Socialism and Democracy*, we find a sociological portrait of contemporary capitalism. The US economy was doing very well, according to Schumpeter, but its institutions were decaying (Schumpeter, 1942).

Like Schumpeter, Karl Polanyi (1886–1964) came from the Austro-Hungarian Empire and ended his life on the American continent. Like Schumpeter, he wrote a famous book on capitalism – *The Great Transformation* – and contributed to the economic sociology of his days (Polanyi, 1957). It is to Polanyi that we owe the term 'embeddedness', even if he used it in his own, very political sense: all economies had been embedded in politics and religion before the advent of capitalism, and were disembedded by the traumatic 'great transformation'. The political task of the day, in other words, was to re-embed the economy into political and human values.

Polanyi covered historical distances with great ease and was as much at home in ancient Babylonia as in 19th-century Britain or 20th-century United States. The scope of his knowledge about the economy is also reflected in one of his most useful sets of categories: the concepts of reciprocity, redistribution and exchange (for example, Polanyi, 1971). In a kinship situation, for example, reciprocity may be used as a way of distributing resources. A political centre, like the state, would in contrast redistribute resources; and a market distributes resources through exchange. Most economic systems draw on each of these three ways of distributing resources, with their corporate sectors ('exchange'), state sectors ('redistribution') and household sectors ('reciprocity').

Talcott Parsons (1902–1979) had begun his career as an economist, only to switch to sociology, since he thought that utilitarian thought was unable to properly capture the structure of modern society. Parsons argued for a general systems perspective in social theory, and suggested in *Economy and Society* (together with Neil Smelser) that the economy should be conceptualized as a sub-system of the general system of society (Parsons and Smelser, 1956). Just as each society has to have a distinct goal ('Polity') and a value-system ('Latent-Pattern-Maintenance'), it also has to adapt to nature and reality ('Economy'). While it is part of society, the economy is also its own society, with a 'polity', 'latent-pattern-maintenance', and so on.

New economic sociology

Around the mid-1980s American sociologists suddenly started to become interested in economic sociology, and it is this development that is generally known as 'new economic sociology'. One article in particular operated as a catalyst in this process, and that is Mark Granovetter's 'Economic action and social structure: the problem of embeddedness' (1985). Its central argument is that all economic actions are embedded in personal networks, and it is this quality that brings them into the sociologist's domain. While this message was important enough in itself, the article's implicit or subliminal message that sociology had neglected a whole area of social life which lent itself to sociological analysis, namely, the economy, also explains its great impact. Since sociological skills had not been applied to economic problems, sociologists might also be able to solve a number of important puzzles that the economists had failed to do, according to Granovetter.

Since the mid-1980s economic sociology has advanced steadily, and it is now fully institutionalized in the United States. It is routinely taught in sociology departments in all the major universities and also has a strong presence among the major journals of the profession. The American Sociological Association has a special section for economic sociology; a number of readers have been published as well as a huge handbook (Smelser and Swedberg, 1994; 2005).

Economic sociology is becoming increasingly popular and accepted in Europe as well, though in a somewhat different form than in the United States, which is only natural given the various national traditions in sociology. While interesting contributions can be found in many European countries, it is especially in France that one can find highly original contributions that stand up well to international competition (for England, see for example Dodd, 1994; for Scotland, MacKenzie, 2003; for Germany, Beckert, 2004; for Italy, Trigilia, 2002; and for Sweden, Aspers, 2001).

The three key figures in French economic sociology are Pierre Bourdieu, Luc Boltanski and Michel Callon (see also the works of Lebaron, 2000, and Steiner, 2005). Bourdieu (1930–2002) has, among other things, analysed consumption in an innovative manner in his celebrated study Distinction (1986); he has also sketched a whole programme for economic sociology, drawing on his three key concepts of habitus, field, and different types of capitals (Bourdieu, 1979; 2005). Luc Boltanski has contributed to the discussion of modern capitalism through an important study of class formation and also co-authored a provocative volume on 'the new spirit of capitalism' (Boltanski, 1987; Boltanski and Chiapello, 1999). And Michel Callon (1998) has introduced the so-called theory of performativity or the idea that economic theory may be as successful as an explanatory approach for the simple reason that it analyses phenomena that it has helped to create in the first place.

The number of studies in economic sociology (books and articles) amounts to several thousand by now, which makes it hard to summarize its achievements. One way to convey a sense of this literature, however, would be to discuss the methods that are being used to gather and analyse data as well as some of the most important topics. That economic sociology indeed has a distinctive profile that sets it off from mainstream economics emerges very clearly from a discussion of these two themes.

The data that is being used in economic sociology has often been put together by the analyst, and it is considerably less common than in mainstream economics to draw on official data of the type that is produced by government agencies. One example is historical studies in economic sociology, as illustrated by Bruce Carruther's City of Capital (1996). The focus in this work is the emergence of one of the world's first financial markets, and the author draws heavily on various primary and secondary sources. In particular, Carruthers succeeds in showing that early trade in shares often followed party lines; that is, sellers were reluctant to trade with political opponents.

Comparative studies are long-standing in economic sociology and have also been popular in new economic sociology. In one of these, Forging Industrial Policy, Frank Dobbin (1994) compares the ways in which the railroad industry developed in the 19th century in the United States, Britain and France. The author shows that industrial policy has largely mirrored the general political culture in its approach to solving problems in each of these three countries. In the United States, there has been scepticism towards the state and reliance on the corporations; in France, the state has been the central actor; and in Britain there has been an attempt to protect the individual firm from competition as well as from interventions from the state. Dobbin claims to have found that there is no one best way of doing things. Rather, people generalize from how they themselves do things and proclaim this to be the universally rational way to proceed.

Economic sociologists also draw on ethnography and participant observation, two methods that allow the researcher to handle huge amounts of empirical detail and to approach things from the perspective of the actors. Michael Burawoy (1979), for example, worked as a shop steward in order to better understand how workers interact and deal with the demands of their work (especially boredom); and Mitchel Abolafia (1996; 1998) passed an examination as a stockbroker in order to better understand what goes on in various stock and bond exchanges.

By far the most significant single method used by economic sociologists today, however, is that of networks. This is a very flexible tool, which allows for quantification and therefore goes well with a large number of research tasks. It has been used, for example, to analyse the links that exist between corporations by virtue of having the same individual on their boards (so-called interlocks). Through the resultant system of communication, various ways of doing things may be diffused. The so-called poison pill (a measure against hostile takeovers) has, for example, been shown to diffuse quickly among corporations linked by common board members (Davis, 1991). That links between corporations are not to be understood exclusively in terms of instrumental actions may be exemplified by the fact that, when a board member resigns or dies, he or she is only replaced in something like half of the cases (Palmer, 1983).

Using networks is also a popular way in economic sociology to approach collaboration between corporations as well as the relationship between firms and their customers and suppliers (see, for example, Gulati and Gargiulo, 1999). The area where it has been most successful, however, may well be the labour market; and here the classic study is Mark Granovetter's *Getting a Job* (Granovetter, 1974). While one may have thought that the most important source of assistance for a person seeking a job is that's person's closest friends and family ('strong ties'), in fact it is his or her more casual contacts ('weak ties'), whose number depends on how many jobs a person has had. The reason for this 'strength of weak ties' is simply that, whereas one's 'strong ties' all share the same information, 'weak ties' can provide access to new and varied information, including information about job opportunities.

In European economic sociology an attempt has also been made to expand the notion of networks to include not only people and organizations in the category of actors but also objects (so-called actor-network-theory; see, for example, Law and Hassard, 1999). That objects can be actors in the conventional sense of this term is no doubt wrong; the weaker claim that objects can be part of networks is, however, more interesting. One may, for example, see a machine as a link between people, some objects may be used for communication between people, and so on – and all this can affect the structure of the network. More generally, the advocates of actor-network-theory also argue that the traditional approach of economists and sociologists tends totally to ignore the role that objects play in the economy and to focus exclusively on actions, social relations and the like. The perspective that argues for including objects in the analysis is usually referred to as 'materiality'.

When it comes to the topics that are often analysed, new economic sociologists have first and foremost tried to focus on economic institutions as opposed to phenomena situated at the boundary of, say, religion and the economy or politics and the economy. The reason for this has been a desire to take on truly 'economic' topics and go beyond the old division of labour between economics and sociology, when the former dealt with the economy and the latter with society minus the economy. As examples of this is the interest among contemporary economic

sociologists in markets and corporations, which have attracted a large number of studies.

One type of study has attempted to develop a general model for markets that differs sharply from the standard economic model of the perfect market. The most prominent example of this is the work of Harrison White (1981; 2002) on so-called production markets, by which he roughly means industrial markets. Production markets, it is argued, differ from so-called exchange markets primarily because their participants have permanent roles as either sellers or buyers and do not switch between these two roles as is common in financial markets.

According to White, the typical production market holds about a dozen actors who closely follow what the other actors are up to. Markets come into being, White argues, precisely because economic actors position themselves in relation to the products of other actors. Prices are not set through demand and supply but by producers relating the revenue of their goods to the volume that is being sold. Individual markets, finally, are connected to each other in giant networks, either 'upstreams' (suppliers) or 'downstreams' (customers).

A number of studies of financial markets have also been carried out, and here the work of Donald MacKenzie is outstanding (for example, MacKenzie, 2003; MacKenzie and Millo, 2003). MacKenzie has picked up from Callon the theme of performativity, and he uses it, for example, in his analysis of trade in options. The pricing of options was very difficult, the argument goes, until Black, Scholes and Merton suggested a solution for which the latter two would win the Nobel Prize in 1997. While this formula covers most cases with much precision, according to MacKenzie it does not cover all – and this was to have important consequences. Since this fact was not well understood, however, and since economic reality was mistaken for how it was portrayed in finance theory (performativity), there have been cases in which people were unprepared for what was happening (as in the case of Long-Term Capital Management). MacKenzie traces this development and also shows how actors have tried to protect themselves against exceptional cases by keeping a margin against the price predicted according to the Black–Scholes–Merton formula.

Economic sociologists have suggested several new ways to approach consumer markets. Viviana Zelizer (1979), for example, has analysed the growth of the market in life insurance in the United States and shown how the idea of putting a price on a human life initially attracted hostility, for religious reasons. But as people moved into the cities and religion had to adjust to new circumstances, a different view of life insurance emerged. Zelizer has recently also started to look at consumption among children, both how children are socialized into becoming consumers and the ways in which they themselves relate to objects and goods in their environment (Zelizer, 2005).

DiMaggio and Louch (1998) have attempted to use networks to analyse consumption. While it is well known that people will turn to others in their surroundings to find out where to buy something, and which merchants, traders and so on are reliable ('search embeddedness'), DiMaggio and Louch examine situations in which people approach someone in their personal network in order to buy something ('within-networks exchange'). As it happens, this is quite common, especially infrequent purchases of the type that involve legal services, home repair maintenance and the buying of a car or a home.

The number of studies in economic sociology that deal with corporations is very great, but a few studies nonetheless stand out. One of these is Mark Granovetter's pioneering 1994 article on business groups. Against R.H. Coase, Granovetter argues that it is not so much the existence of the individual firm that needs to be explained but the common phenomenon of groups of firms. In many countries, such as India, South Korea and Japan, these business groups control large parts of the economy, but have not received the scholarly attention that they deserve. The impact of business groups in the United States is not clear from Granovetter's work, except that US antitrust legislation has ruled out some common forms of this phenomenon.

The business groups that Granovetter studies lend themselves to a networks approach, and so do the corporations that Ronald Burt (1983) has analysed in his study of US industrial markets. Each firm, according to Burt, can be conceptualized as situated at the centre of a network in which there are a number of competitors, suppliers and customers. The fewer competitors there are, the more suppliers, and the more customers, the more the corporation is characterized by 'structural independence'. And with more structural independence comes more profit, as Burt shows.

The emphasis on corporations in interaction, as opposed to the single corporation, is also obvious in another landmark study in economic sociology, *Regional Advantage* by AnnaLee Saxenian (1994). Following Alfred Marshall in analysing industrial districts, Saxenian carries out a comparative study of the computer industry during the post-war period in Silicon Valley and the area around Route 128 in Boston. Silicon Valley has clearly overtaken Route 128 during recent decades, and the reason for this, according to Saxenian, has to do with the nature of the interaction in the two regions. While in Route 128 the corporations are loath to cooperate, rely on banks for finance, and prosecute employees who switch to competitors, in Silicon Valley there is plenty of cooperation, finance comes from venture capital firms, and employees are free to switch as they like. A much more decentralized and flexible form of entrepreneurship, in brief, has emerged in Silicon Valley.

Saxenian's fascination with entrepreneurship is shared by many economic sociologists. While she argues that a radical decentralized industrial region represents the best conditions for entrepreneurship, there exist other perspectives as well. Granovetter, for example, argues that entrepreneurs often come from those parts of the social system which are far away from the controlling centre (for example, Granovetter, 2005). While this may be termed a theory of peripheral entrepreneurship, Granovetter suggests several other situations that are favorable to entrepreneurship. An entrepreneur may, for example, be someone who crosses a social boundary in society and thereby becomes the first to unite resources from two otherwise separated regions (for example, Granovetter, 1995). On immigration, Granovetter also points out that some ethnic groups that are not entrepreneurial in their country of origin may be highly entrepreneurial in their new country because they often leave parts of the extended family behind (Granovetter, 1995). This means that they do not have to provide jobs for their relations or share their wealth with relatives.

Economic sociologists have been very active in studying ethnic entrepreneurship, (for example, Light, 2005). Ethnic entrepreneurs, for example, often have to overcome the fact that their initial market consists of their countrymen ('the ethnic market'), and that they will have to go beyond this market if they are to expand. In many cases they have become entrepreneurs simply because they have no other way of making a living ('forced entrepreneurship').

Economic sociologists have also emphasized the collective nature of entrepreneurship and attempted to explode the myth of the creative Schumpeterian individual. One important example of this can be found in the research by Rosabeth Moss Kanter (1983) on entrepreneurship within the corporation, so-called intrapreneurship. Through a combination of ethnographic studies and survey research, Kanter has attempted to show the conditions under which it is possible to put together creative and entrepreneurial groups in modern corporations. Someone has to suggest the creation of such groups and provide them with resources and legitimacy. The group also has to be defended from outside intervention while it operates, internal conflicts have to be solved, and so on. According to Kanter, this type of group is common among modern corporations.

While economic sociologists have been unable to present a general theory of entrepreneurship, it is nonetheless clear that a number of insights have been accumulated. Economic sociologists are also expanding their work into such topics as social entrepreneurship and the diffusion of courses among business schools (for example, Swedberg, 2000).

Concluding remarks

Economic sociology is currently in a very active phase of its development, and all signs indicate that this trend will continue. Economic sociologists are also gradually expanding their range of topics of study. There has recently, for example, been an attempt to introduce law into the analysis, and some economic sociologists are trying to formulate a position on the relationship between the economy and technology. Some economic sociologists are also in the process of investigating the role of emotions in the economy; and there is a growing number of studies of gender and the economy. What all of this adds up to, again, is a steady growth of studies in economic sociology and a confirmation that economic sociology is established as a distinct and accepted area of sociology. But it remains to be seen whether economic sociology will be able to make inroads into economics itself and gain respect from economists, along the lines of, say, behavioural economics.

RICHARD SWEDBERG

See also **Akerlof, George Arthur; cartels; entrepreneurship.**

Bibliography

Abolafia, M. 1996. *Making Markets: Opportunism and Restraint on Wall Street.* Cambridge, MA: Harvard University Press.

Abolafia, M. 1998. Markets as culture: an ethnographic approach. In *The Laws of the Markets*, ed. M. Callon. Oxford: Blackwell.

Aspers, P. 2001. *A Market in Vogue: A Study of Fashion Photography in Sweden.* Stockholm: City University Press.

Beckert, J. 2004. *Unverdients Vermögen. Soziologie des Erbrechtes.* Frankfurt: Campus Verlag.

Boltanski, L. 1987. *The Making of a Class: Cadres in French Society.* Cambridge: Cambridge University Press.

Boltanski, L. and Chiapello, E. 1999. *Le Nouvel Esprit du Capitalisme.* Paris: Gallimard.

Bourdieu, P. 1979. *Algeria 1960.* Cambridge: Cambridge University Press.

Bourdieu, P. 1986. *Distinction: A Social Critique of the Judgment of Taste.* London: Routledge.

Bourdieu, P. 2005. Principles of an economic anthropology. In *The Handbook of Economic Sociology*, 2nd edn., ed. N. Smelser and R. Swedberg. Princeton: Princeton University Press.

Burawoy, M. 1979. *Manufacturing Consent: Changes in the Labor Process under Monopoly Capitalism.* Chicago: University of Chicago Press.

Burt, R. 1983. *Corporate Profits and Cooptation: Networks of Market Constraints and Directorate Ties in the American Economy.* New York: Academic Press.

Burt, R. 1992. *Structural Holes: The Social Structure of Competition.* Cambridge, MA: Harvard University Press.

Callon, M., ed. 1998. *The Laws of the Markets.* Oxford: Blackwell.

Carruthers, B. 1996. *City of Capital: Politics and Markets in the English Financial Revolution.* Princeton, NJ: Princeton University Press.

Davis, G. 1991. Agents without principles? The spread of the poison pill throughout the intercorporate network. *Administrative Science Quarterly* 36, 583–613.

DiMaggio, P. and Louch, H. 1998. Socially embedded consumer transactions: for what kind of purchases do people most often use networks? *American Sociological Review* 63, 619–37.

Dobbin, F. 1994. *Forging Industrial Policy: The United States, Britain, and France in the Railway Age.* New York: Cambridge University Press.

Dodd, N. 1994. *The Sociology of Money: Economics, Reason and Contemporary Society.* Cambridge: Polity Press.

Durkheim, E. 1893. *The Division of Labor in Society.* New York: Free Press, 1984.

Granovetter, M. 1974. *Getting a Job: A Study of Contacts and Careers.* Cambridge. MA: Harvard University Press.

Granovetter, M. 1985. Economic action and social structure: the problem of embeddedness. *American Journal of Sociology* 91, 481–510.

Granovetter, M. 1994. Business groups. In *The Handbook of Economic Sociology*, ed. N. Smelser and R. Swedberg. Princeton, NJ: Princeton University Press.

Granovetter, M. 1995. The economic sociology of firms and entrepreneurship. In *The Economic Sociology of Immigration*, ed. A. Portes. New York: Russell Sage Foundation.

Granovetter, M. 2005. The impact of social structure on economic outcomes. *Journal of Economic Perspectives* 19(1), 33–50.

Gulati, R. and Gargiulo, M. 1999. Where do interorganizational networks come from? *American Journal of Sociology* 104, 1439–93.

Jevons, S. 1879. *The Theory of Political Economy*, 5th edn. New York: Augustus M. Kelley, 1965.

Kanter, R. 1983. *The Change Masters: Innovation and Entrepreneurship in America.* New York: Simon and Schuster.

Law, J. and Hassard, J., eds. 1999. *Actor Network Theory and After.* Oxford: Blackwell.

Lebaron, F. 2000. *La Croyance Economique: Les Economistes entre Science et Politique.* Paris: Seuil.

Light, I. 2005. The ethnic economy. In *The Handbook of Economic Sociology*, 2nd edn., ed. N. Smelser and R. Swedberg. Princeton, NJ: Princeton University Press.

MacKenzie, D. 2003. Long-term capital management and the sociology of arbitrage. *Economy and Society* 32, 349–80.

MacKenzie, D. and Millo, Y. 2003. Constructing a market, performing theory: the historical sociology of a financial derivatives exchange. *American Journal of Sociology* 109, 107–45.

Marshall, G. 1982. *In Search of the Spirit of Capitalism: An Essay on Max Weber's Protestant Ethic Thesis.* London: Hutchinson.

Palmer, D. 1983. Broken ties: interlocking directorates and intercorporate coordination. *Administrative Science Quarterly* 28, 40–55.

Parsons, T. and Smelser, N. 1956. *Economy and Society: A Study in the Integration of Economic and Social Theory.* New York: The Free Press.

Polanyi, K. 1957. *The Great Transformation*. Boston: Beacon Hill.

Polanyi, K. 1971. The economy as instituted process. In *Trade and Market in the Early Empires*, ed. K. Polanyi, C. Arensberg and H. Pearson. Chicago, IL: Henry Regnery.

Saxenian, A. 1994. *Regional Advantage: Culture and Competition in Silicon Valley and Route 128*. Cambridge, MA: Harvard University Press.

Schumpeter, J. 1934. *The Theory of Economic Development*. Cambridge, MA: Harvard University Press.

Schumpeter, J. 1954. *History of Economic Analysis*. London: Allen & Unwin.

Schumpeter, J. 1991. *The Economics and Sociology of Capitalism*, ed. R. Swedberg. Princeton, NJ: Princeton University Press.

Schumpeter, J. 1942. *Capitalism, Socialism and Democracy*. London: Routledge, 1994.

Simmel, G. 1900. *The Philosophy of Money*. London: Routledge, 1978.

Simmel, G. 1908. Competition. In *Conflict and the Web of Group-Affiliation*. New York: The Free Press, 1955.

Smelser, N. and Swedberg, R., eds. 1994. *The Handbook of Economic Sociology*. Princeton, NJ: Princeton University Press.

Smelser, N. and Swedberg, R., eds. 2005. *The Handbook of Economic Sociology*, 2nd edn. Princeton, NJ: Princeton University Press.

Steiner, P. 2005. *L'Ecole Durkheimienne et l'Economie*. Geneva: Droz.

Swedberg, R. 1998. *Max Weber and the Idea of Economic Sociology*. Princeton, NJ: Princeton University Press.

Swedberg, R. 2000. The social science view of entrepreneurship. In *Entrepreneurship: The Social Science View*, ed. R. Swedberg. Oxford: Oxford University Press.

Trigilia, C. 2002. *Economic Sociology: State, Market and Society in Modern Capitalism*. Oxford: Blackwell.

Weber, M. 1895. The national state and economic policy (Freiburg Address). *Economy and Society* 9, 428–49 (1980).

Weber, M. 1905. *The Protestant Ethic and the Spirit of Capitalism*. New York: Charles Scribner's Sons, 1958.

Weber, M. 1909. *The Agrarian Sociology of Ancient Civilizations*. London: New Left Books, 1976.

Weber, M. 1914. *Economy and Society: An Outline of Interpretive Sociology*. Berkeley: University of California Press, 1978.

White, H. 1981. Where do markets come from? *American Journal of Sociology* 87, 517–47.

White, H. 2002. *Markets from Networks: Socioeconomic Models of Production*. Princeton, NJ: Princeton University Press.

Whyte, W. 1955. *Money and Motivation*. New York: Harper.

Zelizer, V. 1979. *Morals and Markets: The Development of Life Insurance in the United States*. New York: Columbia University Press.

Zelizer, V. 2005. Culture and consumption. In *The Handbook of Economic Sociology*, 2nd edn., ed. N. Smelser and R. Swedberg. Princeton, NJ: Princeton University Press.

economic surplus and the equimarginal principle

Marginal analysis is actually only a particular case of a more general theory, the theory of surpluses and the economy of markets, which, if considered first, facilitates the discussion of the equimarginal principle.

The general theory of surpluses and the economy of markets – fundamental concepts and theorems

To simplify the exposition, it is assumed that one good (U), enters all preference and production functions, and that its quantity can vary continuously. Except for the hypothesis of continuity with respect to this good (U), the discussion in this first part is free of any restrictive hypothesis of continuity, differentiability or convexity for the goods $(V),\ldots,(W)$ considered, and the preference indexes and production functions. (For an exposition of the following theory in the case where no one good plays a particular role, see Allais, 1985, Section II, pp. 139–41.)

Structural conditions

The needs of every unit of consumption, individual or collective, can be entirely defined by considering a preference index

$$I_i = f_i(U_i, V_i, \ldots, W_i) \qquad (1)$$

increasing as it passes from a given situation to one it finds preferable. Every quantity V_i is counted positively if it refers to a consumption, negatively if it refers to a service supplied.

The set of feasible techniques for a unit of production j can be represented by a condition of the form

$$f_j(U_j, V_j, \ldots, W_j) \geq 0$$

where every quantity V_j is considered as representing a consumption or an output depending on whether it is positive or negative. The extreme points corresponding to the boundary between possible and impossible situations represent states of maximum efficiency for the production unit considered. They may be represented by the condition

$$f_j(U_j, V_j, \ldots, W_j) = 0. \qquad (2)$$

The function f_j may be called the production function. It is defined up to any transformation which leaves its sign unchanged.

From a technical point of view, maximum efficiency implies quite specific conditions. If, for instance, one considers a production technique $A = A(X, Y, \ldots, Z)$ and if n production units are technically preferable to a single one, we should have (Allais, 1943, pp. 187–8; 1981, pp. 319–22)

$$\sum_j A(X_j, Y_j, \ldots, Z_j) > A\left[\sum_j X_j, \sum_j Y_j, \ldots, \sum_j Z_j\right].$$

(3)

In the opposite case we have

$$A\left[\sum_j X_j, \sum_j Y_j, \ldots, \sum_j Z_j\right] > \sum_j A(X_j, Y_j, \ldots, Z_j).$$

(3*)

An industry is referred to as differentiated if the use of distinct production units is technically more advantageous than the concentration of all production operations into a single production unit. It is called non-differentiated in the opposite case. Conditions (3) and (3*) are two particular illustrations of differentiation (Allais, 1943, p. 637).

From inequality (3) it is possible to show that the whole production function of a differentiated industry is asymptotically homogeneous. In this case ($n \gg 1$) there is quasihomogeneity (Allais, 1943, pp. 201–6; 1974b).

Distributable Surplus Corresponding to a Given Modification of the Economy

The distributable surplus σ_u relative to a good (U) and to a realizable modification of the economy which leaves all preference indexes unchanged is defined as the quantity of that good which can be released following this shift (Allais, 1943, pp. 610–16). The surplus considered here differs essentially from the concepts of consumer surplus as normally considered in the literature (for example, Samuelson, 1947, pp. 195–202; Blaug, 1985, pp. 355–70; Allais, 1981, pp. 297–8, and 1985, nn. 12–13).

Let us consider an initial state $(\mathscr{E}_1)$ characterized by consumption values $U_i, V_I, \ldots, W_i$ and $U_j, V_j, \ldots, W_j$ (positive or negative) of the different units of consumption and production. We have

$$\sum_i U_i + \sum_j U_j = U_0;$$

$$\sum_i V_i + \sum_j V_j = V_0; \ldots; \sum_i W_i + \sum_j W_j = W_0$$

(4)

where $U_0, V_0, \ldots, W_0$ designate available resources. Let $(\delta\mathscr{E}_1)$ be a feasible modification of $(\mathscr{E}_1)$ characterized by finite variations $\delta U_i, \delta V_i, \ldots, \delta W_i, \delta U_j, \delta V_j, \ldots, \delta W_j$, and let

$$(\mathscr{E}_2) = (\mathscr{E}_1) + \delta(\mathscr{E}_1)$$

represent the new state.

According to (4) we naturally have

$$\sum_i \delta V_i + \sum_j \delta V_j = 0$$

for every good $(U), (V), \ldots, (W)$. From (2) we also have for every unit of production j

$$f_j(U_j + \delta U_j, V_j + \delta V_j, \ldots, W_j + \delta W_j) = 0.$$

According to (1) the preference indexes become

$$I_i + \delta I_i = f_i(U_i + \delta U_i, V_i + \delta V_i, \ldots, W_i + \delta W_i).$$

The δI_i can be positive, zero, or negative.

Let us now define a third state $(\mathscr{E}_3)$ by the condition that by the modification $-\delta\sigma_{ui}$ of just the quantities $U_i + \delta U_i$ all the preference indexes return to their initial values.

We then have the conditions

$$f_i(U_i + \delta U_i - \delta\sigma_{ui}, V_i + \delta V_i, \ldots, W_i + \delta W_i)$$
$$= f_i(U_i, V_i, \ldots, W_i).$$

(5)

The state $(\mathscr{E}_3)$ can be termed 'isohedonous' with the state $(\mathscr{E}_1)$. In passing from $(\mathscr{E}_1)$ to $(\mathscr{E}_3)$ the quantity

$$\delta\sigma_u = \sum_i \delta\sigma_{ui}$$

(6)

of the good (U) is released, as all the units of consumption find themselves again in situations which they consider equivalent, since their preference indexes return to the same values (Allais, 1943, pp. 637–8).

The surplus $\delta\sigma_u$ has been released during the passage from $(\mathscr{E}_1)$ to $(\mathscr{E}_3)$. It may then be considered that in the situation $(\mathscr{E}_1)$ this surplus was both realizable and distributable. It may further be considered that in passing from $(\mathscr{E}_1)$ to $(\mathscr{E}_2)$, it has in effect been distributed.

The distributable surplus thus defined covers the whole economy, but this definition can be used for any group of agents. It is necessary only to consider the functions f_i and f_j and the resources relating to this group in the preceding relations.

Any exchange system, with the corresponding production operations it implies, is deemed 'advantageous' when a distributable surplus is achieved and distributed, so that the preference index of any consumption unit concerned increases. If an exchange and production system is advantageous, there must be at least one system of prices which allows it, the prices used by each pair of agents being specific to them. The distribution of the realized surplus between agents is determined by the system of prices used in the exchanges between them.

Conditions of equilibrium and maximum efficiency
In essence all economic operations of whatever type may be considered as reducing to the search for, the

achievement of, and the distribution of surpluses. Thus stable general economic equilibrium exists if, and only if, in the situation under consideration, there is no realizable surplus, which means

$$\delta\sigma_u \leq 0 \qquad (7)$$

for all feasible modifications of the economy (Allais, 1943, pp. 606–12).

In such a situation the distributable surplus is zero or negative for all possible modifications of the economy compatible with its structural relations, and it is impossible to find any set of prices that would permit effective bilateral or multilateral exchanges (accompanied by the implied production operations) which are advantageous to all the agents concerned.

A situation of maximum efficiency can be defined as a situation in which it is impossible to improve the situation of some people without undermining that of others, i.e. to increase certain preference indices without decreasing others. The set of states of maximum efficiency represents the boundary between the possible and the impossible (Figure 1).

From those definitions of the situations of maximum efficiency and stable general economic equilibrium, it follows, with the greatest generality and without any restrictive hypothesis of continuity, differentiability or convexity, except for the common good (U), that:

> Any state of stable general economic equilibrium is one of maximum efficiency (*First theorem of equivalence*).
> Any state of maximum efficiency is one of stable general economic equilibrium (*Second theorem of equivalence*).

Since there can be no stable general economic equilibrium if there is any distributable surplus, every state of stable general economic equilibrium is a state of maximum efficiency. Conversely, if there is maximum efficiency, there is no realizable surplus which could be used to increase at least one preference index without decreasing the others, and consequently, every state of maximum efficiency is a state of stable general economic equilibrium.

Because of the theorems of equivalence, the terms 'conditions of stable general economic equilibrium' and 'conditions of maximum efficiency' are used interchangeably below.

The dynamic process of the economy: decentralized search for surpluses

In their essence all economic operations, whatever they may be, can be thought of as boiling down to the pursuit, realization and allocation of distributable surpluses. The corresponding model is the Allais model of the economy of markets (1967), defined by the fundamental rule that every agent tries to find one or several other agents ready to accept at specific prices a bilateral or multilateral exchange (accompanied by corresponding production decisions) which will release a positive surplus that can be shared out, and which is realized and distributed once discovered. Thus the evolution of the market's economy is characterized by the condition

$$\delta I_i \geq 0$$

for every consumption unit.

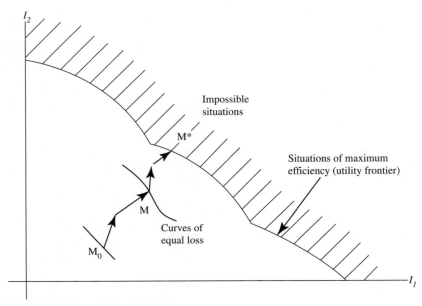

Figure 1   Process of dynamic evolution. Illustrative diagram

Since in the evolution of an economy of markets surpluses are constantly being realized and allocated, the preference indexes of the consumption units are never decreasing, at the same time as some are increasing. This means that for a given structure, that is to say, for given preferences, resources, and technical know-how, the working of an economy of markets tends to bring it nearer and nearer to a state of stable general economic equilibrium, hence a state of maximum efficiency (Figure 1), which is the third fundamental theorem.

Naturally such evolution takes place only if sufficient information exists about the actual possibilities of realizing surpluses.

To any given initial situation whatsoever, assumed not to be a situation of equilibrium, there corresponds an infinite number of possible equilibrium situations, each corresponding to a particular path and each satisfying the general condition that no index of preference should take on a lower value than in the initial situation (Figure 1).

Economic loss

The loss σ_u^* which is associated with a given situation is defined as the greatest quantity of the good (U) which can be released in a transformation of the economy for which all the preference indexes remain unchanged (Figure 1) (Allais, 1943, pp. 638–49).

It is a well determined function

$$\sigma_u^* = F[I_1, I_2, \ldots, I_n, U_0, V_0, \ldots, W_0]$$

(8)

of the preference indexes I_i and of the resources V_0 which characterize this situation. The loss σ_u^* is an indicator of inefficiency, and $-\sigma_u^*$ an indicator of the efficiency of the economy as a whole.

The loss is minimum and nil in every state of maximum efficiency, and positive in every feasible situation which is not a state of maximum efficiency. It decreases in any modification of the economy, whereby some preference indexes increase, others remaining unchanged, or whereby some surpluses are released with no decline in some preference indexes.

Paths to states of economic equilibrium and maximum efficiency

Since the preference indices I_i are continuous functions of the quantities U_i of the common good (U), the boundary between the possible and the impossible situations in the hyperspace of preference indexes is constituted by a continuous surface. On this surface the loss σ_u^* is nil. This representation allows an immediate demonstration by simple topological considerations of propositions whose proof would otherwise be very difficult. (The paternity of this representation has been unduly attributed to P. Samuelson, 1950, but it was in fact published for the first time in Allais, 1943, and systematically used by Allais

in later years especially 1945 and 1947; see Allais, 1971, n.11, p. 385; and 1974a, n.18, pp. 176–7.)

For every feasible situation which is not a state of maximum efficiency, represented by a point such as M_0, there are an infinity of realizable displacements M_0M enabling a situation of maximum efficiency M^* to be approached, such that all the preference indexes have greater values than in the initial situation M_0.

Figure 1 presents an illustration of the process of dynamic evolution by releasing and sharing out of surpluses during which the loss σ_u^* is constantly decreasing (Allais, 1943; 1974b; and 1981, p. 121).

The changing structure of the economy

As psychological patterns vary, as techniques are improved, or as new resources are discovered (or existing resources depleted), the set of situations of maximum efficiency relative to the indexes of preference constantly undergoes change over time. Consequently, situations of equilibrium and maximum efficiency are never reached, and what is really important is to determine the rules of the game which must be applied to come constantly closer to them as rapidly as possible. At a given time t, if information is sufficient and if the adjustments are sufficiently rapid, the point representing the economy will never be very far from the maximum efficiency surface of that time t.

General comment

An economy of markets can be defined as one in which the agents – consumption, production, and arbitrage units – coexist and are free to undertake any exchange transaction or production operation which can result in rendering some distributable surplus available. The principle of the market economy is that any surplus realized is shared among the operators involved. How the surpluses achieved are shared out depends on the specific systems of prices used in the exchanges between the agents concerned. The prices used are always specific to the exchange and production operations considered and there is never a unique system of prices used in common by all the agents.

Diagrammatic representation like that of Figure 1 reveals clearly three basic facts:

1. There is an infinity of situations of maximum efficiency corresponding to a given initial situation characterized by some distribution of property.
2. To each situation of maximum efficiency there corresponds a final distribution of property.
3. This final distribution depends on the initial situation and the distribution of surpluses in the course of the transition.

Thus there is a very strong interdependence between the point of view of efficiency corresponding to the discovery and realization of surpluses and the ethical point of view corresponding to their sharing.

In any event, since only what is produced can be shared, the incentive stemming from the partial or total appropriation of the surpluses by the various agents appears as a fundamental factor for the functioning of the economy of markets.

On the general theory of surpluses and the economy of markets in the general case, and on the fundamental theorems see Allais (1943, pp. 112–77; 181–211; 604–56), (1967, § 8–65), (1968a, vol. 2), (1968b), (1971), (1974a), (1981, pp. 27–48), (1985).

The equimarginal principle

Continuity and differentiability

The preceding definitions and theorems are very general and do not make any hypothesis of continuity, derivability or convexity, except the hypothesis of continuity for the common good (U).

We now assume in addition only that all the quantities and functions considered are continuous and that all functions have first and second order derivatives, the following developments being totally independent of any hypothesis of general convexity.

From the sign conventions adopted earlier it follows that for any i, j and V

$$f'_{iv} = \partial f_i / \partial V_i \geq 0, \quad f'_{jv} = \partial f_j / \partial V_j \geq 0.$$

The second partial derivatives are written

$$f''_{ivw} = \partial^2 f_i / \partial V_i \delta W_i, \quad f''_{jvw} = \partial^2 f_j / \partial V_j \delta W_j.$$

In the following, the symbol $\overline{d^2 g}$ represents the second differential

$$\overline{d^2 g} = \sum_U^W g''_{v^2} \, dV^2 + 2 \sum_{U,V} g''_{vw} dV \, dW$$

of a function $g(U, V, \ldots, W)$ when all parameters in that function are taken as independent, while the symbol $\overline{d^2 g_u}$ represents what this second differential becomes after du has been replaced by its expression derived from

$$dg = \sum_U^W g'_v dV = 0$$

(Allais, 1968a, vol. 2, pp. 77–8; 1973b, pp. 151–5; 1981, pp. 688–9).

Convexity and concavity

The local properties of diminishing or increasing marginal returns are related to local conditions of convexity or concavity. Convexity is defined as follows:

Ordinal fields of preference A field of choice is said to be convex in the whole space (postulate of general convexity) if, at all points of the field, the condition

$$I(M_0) \leq I(M_1)$$

entails

$$I(M_0) \leq I(M)$$

with

$$M = \lambda M_0 + (1 - \lambda)M_1 \quad 0 < \lambda < 1.$$

There is local convexity at M_0 if this condition is satisfied only for

$$|M_0 M_1| < \varepsilon$$

where ε is a given positive number.

When differentiability is assumed local convexity implies

$$\overline{d^2 f_{iu}} \leq 0 \quad \text{for } df_1 = 0.$$

Fields of production A field of production is said to be convex over the whole space (postulate of general convexity) if, for any two possible points M_0 and M_1, the centre of gravity defined by the relation

$$M = \lambda M_0 + (1 - \lambda)M_1$$

is likewise a possible point for

$$0 < \lambda < 1.$$

Local convexity obtains at M_0 if the preceding condition is satisfied only for

$$|M_0 M_1| < \varepsilon$$

where ε is a given positive number.

When differentiability is assumed, local convexity implies

$$\overline{d^2 f_{iu}} \leq 0 \quad \text{for } df_1 = 0.$$

In fact there is no production operation that does not begin by providing increasing marginal returns, and it is only beyond a certain threshold that diminishing marginal returns are observed. That is a general physical law of nature (Allais, 1943, pp. 193–5; 1968a, vol. 2, pp. 68–96; 1971, pp. 362–4; 1974a, pp. 153–7). Similarly it can be considered as an introspective datum that psychological returns begin by increasing but in the end always decrease beyond certain threshold values. That is a general psychological law (Allais, 1968a, vol. 2, pp. 109–38; 1971, pp. 360–2; 1974a, pp. 153–5). These are two fundamental properties of fields of choice and production. They rule out the postulate of general convexity which is generally accepted in the contemporary literature.

Generation of distributable surplus

Consider any economic state $(\mathscr{E})$ and a realizable modification $(\delta\mathscr{E})$ such that all the preference indexes I_i remain constant (isohedonous modification). Let the conditions of constancy of these indexes and the conditions corresponding to the production functions be written in the same general form

$$g_k(U_k, V_k, \ldots, W_k) = 0 \qquad (9)$$

where $U_k, V_k, \ldots, W_k$ represent the consumption of both consumption and production units. By convention, any quantity V_k, if positive, represents consumptions, either by a consumption or a production unit. For any production or consumption unit, any parameter V_k, if negative, represents production of a good or a service.

Let $dU_k, dV_k, \ldots, dW_k$, be the first order differentials of the variations $\delta U_k, \delta V_k, \ldots, \delta W_k$ of consumptions U_k, $V_k, \ldots, W_k$ in the displacement $(\delta\mathscr{E})$. From (9), we have

$$g'_{ku}dU_k + g'_{kv}dV_k + \cdots + g'_{kw}dW_k = 0. \qquad (10)$$

Let δV_{kl} be the quantity of (V) received by the consumption or the production unit k from the consumption or production unit l. By definition, we have

$$\delta V_k = \sum_{k \neq 1} \delta V_{kl} \qquad (11)$$

$$\delta V_{lk} = -\delta V_{kl}. \qquad (12)$$

Assuming that the displacement $(\delta\mathscr{E})$ is such that

$$\sum_k \delta V_k = 0, \ldots, \sum_k \delta W_k = 0. \qquad (13)$$

Let

$$\varepsilon^k_{v,u} = g'_{kv}/g'_{ku}. \qquad (14)$$

The ratio $\varepsilon^k_{v,u}$ is the coefficient of marginal equivalence (or marginal rate of substitution) of goods (V) and (U) for agent k (Allais, 1943, pp. 609–10, and 617–21).

From (10) and (14) we have the relation

$$dU_k = -\left[E^k_{vu}dV_k + \cdots + E^k_{wu}dW_k\right] \qquad (15)$$

between the first order differential $dU_k, dV_k, \ldots, dW_k$.

If dU_k is positive, agent k receives a quantity dU_k to within the second order. If dU_k is negative, agent k supplies a quantity $-dU_k$ to within the second order.

From the condition (13), it follows that the displacement considered releases a global distributable surplus

$$\delta\sigma_u = -\sum_k \delta U_k$$

representing the excess of the quantities supplied over the quantities received of good (U) whose first order differential is

$$\delta\sigma_u = -\sum_k dU_k.$$

From (11) and (15)

$$dU_k = -\sum_v^w \left[\varepsilon^k_{vu} \sum_{\substack{k,l \\ k<l}} dV_{kl}\right]$$

and from (12), we have (Allais, 1952c, p. 31; 1968a, vol. 2, p. 174; 1981, p. 88)

$$d\sigma_u = \sum_v^w \sum_{\substack{k,l \\ k<l}} (\varepsilon^k_{vu} - \varepsilon^l_{kl})dV_{kl}. \qquad (16)$$

According to definitions (5) and (6) $d\sigma_u$ is the first differential of the global distributable surplus $\delta\sigma_u$ released in the displacement considered. For all economic agents the unit of value is defined by condition $u_k = u = 1$. The marginal values $v_k, \ldots, w_k$ of goods $(V), \ldots, (W)$ for unit k are defined with respect to the u_k by the relations

$$\frac{g'_{ku}}{u_k} = \frac{g'_{kv}}{v_k} = \cdots = \frac{g'_{kw}}{w_k} \qquad (17)$$

$$u_k = u = 1. \qquad (18)$$

Under the adopted sign convention, all the v_k are positive. We have from (14) and (18)

$$\varepsilon^k_{vu} = v_k \qquad (19)$$

and relation (16) is written

$$d\sigma_u = \sum_v^w \sum_{\substack{k,l \\ k<l}} (v_k - v_l)dV_{kl} \qquad (20)$$

where v_k and v_l are the marginal values of good (V) for units k and l. This summation covers all agents, both consumption and production units. It can thus be seen that all the differences between the marginal values in the situation $\mathscr{E}$ can give rise to the release of potential surpluses which can be released and distributed.

The meaning of relation (20) is immediate. Thus if $v_k > v_l$ the relative value of good (V) is higher for agent k than for agent l. The transfer of a positive quantity dV_{kl} of good (V) from agent l to agent k therefore creates an additional positive value

$$d\sigma_{ukl} = (v_k - v_l)dV_{kl}.$$

If in this 'isohedone' transformation surpluses are released, all positive, they can be distributed in such a way as to increase all preference indexes. In such a modification of the economy, the maximum distributable surplus diminishes, and the point representing the economic situation considered moves closer to the surface of maximum efficiency in the hyperspace of preference indexes. Naturally, for this condition to obtain, the corresponding exchanges and the changes of the consumptions and productions they imply in the production system, must effectively occur.

Psychological values and marginal psychological values
Naturally, the v_k are only marginal values for the agents. The psychological values v_i^* of the consumption V_i of a subject i is defined by the relation

$$f_i(U_i + v_i^* V_i, 0, \dots, W_i) = f(U_i, V_i, \dots, W_i)$$

where $v_i^* V_i$ is the sum he would accept to receive to offset the drop in his consumption V_i to zero. The unit value v_i^* is generally much higher than the marginal value v_i corresponding to relations (17), (18) and (19).

In any event, a consumption is only advantageous when its psychological value is higher than its marginal value, because, if this were not so, it would be in the subject's interest to reduce his consumption V_i.

Conditions of Stable General Economic Equilibrium and Maximum Efficiency of the Economy
From condition (7) it follows that the necessary and sufficient condition for a situation ($\mathscr{E}$) to be of stable equilibrium and maximum efficiency is that the distributable surplus $\delta\sigma_u$ defined by (5) and (6) be negative or zero for every feasible modification ($\delta\mathscr{E}$), that is every modification that is compatible with the constraint conditions, that is, the structural relations of the economy (2) and (4) above.

Condition (7) implies the two conditions (Allais, 1943, p. 612)

$$d\sigma_u = 0 \quad \text{(first order condition)}$$
$$(21)$$

$$d^2\sigma_u \leq 0 \quad \text{(second order condition)}$$
$$(22)$$

for any realizable and reversible modification ($\delta\mathscr{E}$) in which the expressions of $d\sigma_u$ and $d^2\sigma_u$ represent the first and second differential of $\delta\sigma_u$.

Thus we have according to (21) and (22) using the above notations

$$d\sigma_u = \sum_i d\sigma_{ui} = \sum_i dI_i / I'_{iu} = 0 \qquad (23)$$

$$d^2\sigma_u = \sum_i \overline{\frac{d^2 f_{iu}}{f'_{iu}}} + \sum_j \overline{\frac{d^2 f_{ju}}{f'_{ju}}} \leq 0 \quad \text{for} \quad d\sigma_u = 0.$$
$$(24)$$

Actually, and according to relation (20), the first order condition (23) implies that when the quantities V_k are not nil, all the marginal values v_k are equal to a same value v and a same system of prices $u, v, \dots, w$ then exists for all the agents k concerned, such that

$$\frac{g'_{ku}}{u} = \frac{g'_{kv}}{v} = \dots = \frac{g'_{kw}}{w}. \qquad (25)$$

These equalities condense the general equimarginal principle into a single formulation. They express the fact that in a situation of equilibrium and maximum efficiency, the psychological (or objective) value v_k of the last dollar is the same, for any agent (consumption or production unit), whatever use it is put to.

For the quantities V_k which are nil (terminal equilibria), we necessarily have

$$v_k \leq v$$

since, if this were not true, the operator's interest would be to increase V_k from the value $V_k = 0$; he could indeed do this because of the existence of other operators who are in a situation of tangential equilibrium for good (V).

The second order condition (24) holds whether or not the df_i are equal to zero. It is only subject to the constraint (21). If we consider only the modifications of the economy involving units k and l, condition (24) is written

$$d^2\sigma_u = \overline{\frac{d^2 f_{ku}}{f'_{ku}}} + \overline{\frac{d^2 f_{lu}}{f'_{lu}}} \leq 0 \quad \text{for} \quad d\sigma_{uk} + d\sigma_{ul} = 0$$

shows that when in a situation of maximum efficiency consumption or production units consume (or produce) the same goods, one unit at most is in a situation of local concavity, that is, in a situation of marginal increasing returns (Allais, 1968a, pp. 196–9; 1974a, n.125, p. 184; 1981, p. 65).

Consequently, when maximum efficiency obtains, most operators are in a situation of local convexity and marginal decreasing returns. However, this condition cannot be interpreted as meaning that all fields of choice and production are convex everywhere, this hypothesis being totally contradicted by observed data.

When local convexity obtains for a consumption unit, its index of preference is effectively at a maximum, subject to the budgetary constraint, equilibrium prices being taken as given. Similarly, if local convexity obtains for a production unit, the unit's income is effectively at a maximum, equilibrium prices again being taken as given. However, these two principles, which in any case could be valid only for a situation of maximum efficiency, cannot be considered as corresponding in all cases to optimum behaviour, and they cannot be taken to be of general value. As a matter of fact and for instance, if, in a situation of maximum efficiency, a production unit is in a situation of local concavity, its income is minimum, the equilibrium prices being considered as given.

Conditions (25) and (24) show the total symmetry of the implications of the psychological and technical structures of the economy.

Approximate value of the economic loss corresponding to the non-equality of marginal values in the neighbourhood of a situation of maximum efficiency
The integration of eq. (20) along a path leading to a state of maximum efficiency leads to the following approximate estimate to within third order accuracy of the global loss involved in the initial situation (relation 8)

$$\sigma_u^* \sim \frac{1}{2u} \sum_v \sum_{\substack{k,l \\ k<l}} (v_k - v_l) \delta V_{kl}^*. \qquad (26)$$

In this relation, the quantities $v_k - v_l$ represent the differences of marginal values in the initial state considered, and the δV_{kl}^* are the quantities of the good (V) received by operator k from operator l in the transition from the initial to the final state. Relation (26) is of the broadest generality, and holds whatever the initial state (Allais, 1952a, pp. 31–2, n. 8; 1968a, vol. 2, p. 207; 1981, p. 110).

Its simplicity is really extraordinary in view of the complexity of the concept it represents, namely the maximum of the distributable surplus for all the modifications which the economy can undergo while leaving the preference indexes unchanged.

In the neighbourhood of a situation of maximum efficiency, the $(v_k - v_l)$ and δV_{kl}^* are of the first order quantities, where as the loss σ_u^* is only of the second order. However, since the δV_{kl}^* are of the first order, the variations δI_i of the preference indexes are also of the first order. As a result, and for instance, in the neighbourhood of a situation of maximum efficiency, taxes have major first order effects on the distribution of income but only second order effects on the efficiency of the economy.

On the theoretical foundations of the equimarginal principle, see Allais (1943, pp. 604–56), (1945), (1952a, pp. 28–32), (1967), (1968a, vol. 2), (1971), (1973a), (1973b), (1974a), (1974b), (1981) and (1986). Illustrative models: Allais (1943, Annexe I, pp. 4–24), (1945, pp. 57–69). On its extension see: cases of perfect and imperfect foresight: Allais (1943, pp. 343–84), (1947, pp. 23–228), (1964), (1967), (1968a, vol. 2). Illustrative models: (1947, pp. 631–771). Capitalistic optimum theory: Allais (1947, pp. 179–228), (1962), (1963). Demographic optimum theory: Allais (1943, pp. 749–85). Case of risk: Allais (1952b). Application of marginal analysis to

transport: Allais (1964) and (1987). For a general overview on the meaning, limits, generalizations, and history of the equimarginal analysis see Allais (1987).

General overview
Theory of surpluses and marginal analysis
As a matter of fact a single relation, the relation (20) (or the equivalent relation (16)) condenses the whole marginal approach as it has developed for over a century. Subject only to the hypotheses of continuity and derivability implied by any marginal theory, it applies in all cases, and its simplicity is really extraordinary.

It also shows that equilibrium and maximum efficiency can obtain only when all marginal values are equal, which is the equimarginal principle.

The equimarginal principle was discovered first by Gossen (1854), and rediscovered, broadened and introduced independently into economics by Jevons (1871), Menger (1871) and Walras (1874–7). In the following years numerous new developments of the principle have been presented by their immediate successors, especially by Edgeworth (1881), Irving Fisher (1892) and Vilfredo Pareto (1896–1911). Particularly striking illustrations of the role of differences in marginal equivalences are Ricardo's theory of comparative costs (1817) and Dupuit's theory of economic losses (1844–53).

This principle corresponds to the outcome of the dynamic process of the economy induced by differences in marginal equivalences. According to Irving Fisher (1892), with whose judgement I agree fully, 'No idea has been more fruitful in the history of economic science.' Its applications and generalizations dominate all economic analysis in real terms.

From the foregoing a double conclusion emerges: the classical theory of marginal equivalences is irreplaceable to make understandable the underlying nature of all economic phenomena; the general theory of surplus, of which classical marginal theory is only a special case, allows one to extend the propositions of marginal analysis to the most general case of discrete variations and indivisibilities.

As important as the analysis of the conditions of general equilibrium and maximum efficiency may be, the analysis of the dynamic processes which enable surpluses to be generated from a given situation is much more important. From this point of view the analyses by Dupuit, Jevons, Edgeworth, Pareto, and the marginal school and its predecessors in general, appear much more realistic than the contributions which rest only upon the consideration of Walras's general model of equilibrium.

In fact, what is really important is not so much the knowledge of the properties of a state of maximum efficiency as the rules of the game which have been applied to the economy effectively to move nearer to a state of maximum efficiency.

The decentralized search for surpluses is truly the dynamic principle from which a thorough and yet very simple conception of the operation of the whole economy can be derived. Whereas in the market economy model the search for efficiency is essentially focused on the determination of a certain set of prices, the analysis of the model of the economy of markets is based on the search for potential surpluses and their realization. Not only is the economy of markets model much more realistic than the market economy model while lending itself to much simpler proofs, but also these proofs are not subordinated to any restrictive assumptions relating to continuity, differentiability of functions, or convexity. All of economic dynamics is reduced to a single principle: the search for and realization of potential surpluses, which leads to the minimization of loss for the economy as a whole.

On all these points see especially: Allais (1971) and (1974a).

The tendencies of the contemporary literature

From Walras on, the literature became progressively – and unduly – concentrated on equilibrium analysis which, however interesting it could be, is less so than the analysis of the processes by which the economy tends at any time towards situations of equilibrium which in fact are never reached.

Today there is a tendency to neglect the dynamic marginal approach based on the consideration of differences in marginal equivalences; and in the name of a so-called rigour it has been replaced by new theories. A fortiori, the general theory of surpluses which generalizes marginal analysis is simply ignored. This development, which in reality, and despite the too-widely held belief to the contrary, represents an immense step backward, basically stems from the unquestioning acceptance of 'established truths' taught by the dominant 'establishments', whose only real basis is their incessant repetition.

As a matter of fact the guiding principles of the contemporary theories descending from Walras: the adoption of the market economy model; the hypothesis that a common price system applicable to all operators prevails at each instant; the assumption of general convexity; and the exaltation of mathematical formalism of the theory of sets to the detriment of conformity with actual facts, constitute an impediment to any genuine progress in analysis of the economy in real terms.

The essential difference between the market economy model and the model of the economy of markets is that, in the latter, the exchanges leading to equilibrium take place successively at different prices, and that, at any given moment, the price sets used by different operators are not necessarily the same. Whereas in the first model the final situation is determined totally by the initial situation, which correspondingly plays a privileged role without any real justification, in the second the final situation depends both on the initial situation and the path taken from it to the final situation (Figure 1).

Whereas the market economy model postulates perfect competition and a large number, if not an infinity, of operators, the model of the economy of markets applies just as well to the cases of monopoly as to the cases of competition.

Not only is the market economy model unrealistic, but it also gives rise to considerable mathematical difficulties when an attempt is made to demonstrate the above three fundamental theorems. Whether differential calculus or set theory is used, the theorems can only be demonstrated under extremely restrictive conditions, and the difficulties they imply are, from an economic standpoint, completely artificial, for they arise solely from the unrealistic nature of the model used. Paradoxically, whereas these restrictive assumptions are totally unrealistic, most of the theoretical difficulties encountered disappear, as shown above, once they are discarded.

The market economy approach leads to imposing on any economic model, for it to be considered satisfactory, conditions which actually apply to a particular model, which are generally not fulfilled in reality, and for which, at all events, no rigorous justification can be found.

By departing from the great tradition of marginal theory and by adopting an unrealistic model and unrealistic assumptions, the contemporary theories, purely mathematical, have doomed themselves to sterility as regards the understanding of reality.

On the contemporary theories see especially: Samuelson (1947); Arrow (1968); Debreu (1959) and (1985); Blaug (1979 and 1985); Arrow and Hahn (1971); Hutchison (1977, pp. 62–97 and 161–70); Woo (1985); and Allais (1952b; 1968b; 1968e; 1971; 1974a; and 1981).

MAURICE ALLAIS

See also **Allais, Maurice; efficient allocation; general equilibrium; optimality and efficiency; surplus.**

Bibliography

Allais, M. 1943. *A la recherche d'une discipline économique. Première partie. l'économie pure.* 2 vols. Paris: Ateliers Industria. 2nd edn, *Traité d'économie pure*, 5 vols, Paris: Imprimerie Nationale, Paris, 1952. (The second edition is identical to the first, except for a new Introduction.)

Allais, M. 1945. *Economie pure et rendement social.* Paris: Sirey.

Allais, M. 1947. *Economie et intérêt.* 2 vols. Paris: Imprimerie Nationale and Librairie des Publications Officielles.

Allais, M. 1952a. *Introduction to the 2nd edn of Allais (1943).* Paris: Imprimerie Nationale.

Allais, M. 1952b. L'éxtension des théories de l'équilibre économique général et du rendement social au cas du risque. *Colloques Internationaux du Centre National de la Recherche Scientifique* 40, Econométrie, 81–120. A summarized version was published under the same title in *Econometrica*, (1953), 269–90.

Allais, M. 1962. The influence of the capital output ratio on real national income. *Econometrica* 30, 700–28. Republished in American Economic Association, *Readings in Welfare Economics*, vol. 12, with an additional Note, 1969.

Allais, M. 1963. The role of capital in economic development. In *Study Work on the Econometric Approach to Development Planning*, Pontificiae Academiae Scientiarum Scripta Varia 28, Pontifica Academia Scientiarum, Amsterdam: North-Holland, 1963 and Chicago: Rand McNally, 1965.

Allais, M. 1964. La theorie economique et la tarification optimum de l'usage des infrastructures de transport. *La Jaune et la Rouge* (publication of the Société Amicale des Anciens Eléves de l'Ecole Polytechnique), special issue *Les Transports*, Paris, 1964.

Allais, M. 1967. Les conditions de l'éfficacite dans l'économie. Fourth International Seminar, Centro Studi e Ricerche su Problemi Economico-Sociali, Milan. Italian translation: 'Le condizioni dell' efficienza nell' economia' in *Programmazione E Progresso Economico*, Milan: Franco Angeli, 1969. Original French text in M. Allais, *Les Fondements du Calcul Economique*, vol. 1, Paris: Ecole Nationale Supérieure des Mines de Paris, 1967.

Allais, M. 1968a. *Les fondements du calcul économique*. 3 vols. Paris: Ecole Nationale Supérieure des Mines, Paris, vol. 1, 1967, and vols 2 and 3, 1968.

Allais, M. 1968b. The conditions of efficiency in the economy. *Economia Internazionale* 21(3), 399–420.

Allais, M. 1968c. Pareto, Vilfredo: contributions to economics. In *International Encyclopedia of the Social Sciences*, vol. 2 New York: Macmillan and Free Press.

Allais, M. 1968d. Fisher, Irving. In *International Encyclopedia of the Social Sciences*, vol. 5 New York: Macmillan and Free Press.

Allais, M. 1968e. L'économie en tant que science. *Revue d'Economie Politique*, January-February, 5–30. Trans. as 'Economics as a Science', *Cahiers Vilfredo Pareto*, (1968), 5–24.

Allais, M. 1971. Les théories de l'équilibre economique général et de l'éfficacité maximale – impasse récentes et nouvelles perspectives. Congrés des Economistes de Langue Francçaise, 2–6 June. *Revue d'Economie Politique* 3, 331–409. Spanish translation: 'Las theorias del equilibrio economico general y de la eficacia maxima – recientes callejones sin salida y nuevas perspectivas'. *El Trimestre Economico* (Mexico), 39 (1972), 557–633; English translation: see Allais (1974a).

Allais, M. 1973a. La théorie générale des surplus et l'apport fondamental de Vilfredo Pareto. *Revue d'Economie Politique* 6, 1044–97.

Allais, M. 1973b. The general theory of surplus and Pareto's fundamental contribution. *Convegno Internazionale Vilfredo Pareto*. Roma, 25–27 October, Rome: Accademia Nazionale dei Lincei, 1975 (English trans. of Allais, 1973a.)

Allais, M. 1974a. Theories of general economic equilibrium and maximum efficiency. Vienna Institute for Advanced Studies. In *Equilibrium and Disequilibrium in Economic Theory*, ed. G. Schwödiauer. Dordrecht: Reidel, 1977. (English version of Allais, 1971, with some additions.)

Allais, M. 1974b. Les implications de rendements croissants et décroissants sur les conditions de l'équilibre economique général et d'uned éfficacité maximale. In *Hommage à François Perroux*. Grenoble: Presses Universitaires de Grenoble, 1978.

Allais, M. 1981. *La théorie générale des surplus. Economies et Sociétés*, 2 vols. Montrouge institut de sciences mathématique et économies appliquées.

Allais, M. 1985. The concepts of surplus and loss and the reformulation of the theories of stable general economic equilibrium and maximum efficiency. In *Foundations and Dynamics of Economic Knowledge*, ed. M. Baranzini and R. Scazzieri. Oxford: Basil Blackwell.

Allais, M. 1987. *The Equimarginal Principle, Meaning, Limits, and Generalisations*. Centre d'Analyse Economique. *Revista internazionale di scienze economiste e commerciale*.

Arrow, K.J. 1968. Economic equilibrium. In *International Encyclopedia of the Social Sciences*, vol. 4, New York: Macmillan and Free Press.

Arrow, K.J. and Hahn, F.H. 1971. *General Competitive Analysis*. San Francisco: Holden-Day; Edinburgh: Oliver.

Blaug, M. 1979. *Economic Theory in Retrospect*. 4th edn, London: Heinemann Educational Books, 1985.

Debreu, G. 1959. *Theory of Value*. New York: Wiley.

Debreu, G. 1985. *Theoretic Models: Mathematical Form and Economic Content*. Frisch Memorial Lecture, Fifth World Congress of the Econometric Society, MIT, 17–24 August.

Dupuit, J. 1844. De la mesure de l'utilité des travaux publics. *Annales des Ponts et Chaussées*, 2nd series, Mémoires et Documents No. 116, vol. 8.

Dupuit, J. 1849. De l'influence des péages sur l'utilité des voies de communication. *Annales des Ponts et Chaussées*, 2nd series.

Dupuit, J. 1853. De l'utilite et de sa mesure. *Journal des Economistes* 36(147), 1–28.

Edgeworth, F.Y. 1881. *Mathematical Psychics: An Essay on the Application of Mathematics to the Moral Sciences*. London: Kegan Paul. Reprinted, New York: Kelley, 1953.

Fisher, I. 1892. *Mathematical Investigations in the Theory of Value and Prices*. New Haven: Yale University Press, 1925.

Gossen, H.H. von. 1854. *Entwickelung der Gesetze des menschlichen Verkehrs und der daraus fliessender Regeln für menschliches Handeln*. 3rd edn, introduction by Friedrich Hayek, Berlin: Präger, 1927.

Hutchison, T.W. 1977. *Knowledge and Ignorance in Economics*. Oxford: Blackwell.

Jevons, W.S. 1871. *The Theory of Political Economy*. London: Macmillan. 5th edn trans. as *La théorie de l'économie politique*. Paris: Giard, 1909.

Menger, C. 1871. *Grundsätze der Volkswirtschaftslehre*. Vienna: Braumneller.

Pareto, V. 1896–7. *Cours d'économie politique*. 2 vols. Lausanne: Rougé. Reprinted, Geneva: Droz, 1964.

Pareto, V. 1901. Anwendungen der Mathematik auf Nationalökonomie. *Encyklopädie der Mathematichen Wissenschaften*, vol. 1, Leipzig.

Pareto, V. 1906. *Manuale d'economia politica*. Milan. Trans. as *Manuel d'économie politique*. Paris: Giard et Brière, 1909, and Geneva: Droz, 1966, and as *Manual of Political Economy*. Reprinted, New York: Kelley, 1971.

Pareto, V. 1911. Economie mathématique. *Encyclopedie des Sciences Mathématiques*, Paris: Gauthier–Villars, 1911, also in *Statistique et économie mathématique*, Geneva: Droz, 1966, published in English as 'Mathematical economics', *International Economic Papers* No. 5. New York, 1955.

Ricardo, D. 1817. On the Principles of Political Economy and Taxation. Vol. I of *The Works and Correspondence of David Ricardo*, ed. P. Sraffa. Cambridge: Cambridge University Press, 1951–1955.

Samuelson, P.A. 1947. *Foundations of Economic Analysis*. 2nd edn, Cambridge, MA: Harvard University Press, 1948.

Samuelson, P.A. 1950. Evaluation of real national income. *Oxford Economic Papers* NS, 2, 1–29.

Walras, L. 1874–7. *Eléments d'économie politique pure – théorie de la richesse sociale*. 6th edn. Paris: Guillaumin; reprinted, Paris: Pichon et Durand-Auzias, 1952. English trans. of the 6th edn. as *Elements of Pure Economics*, ed. W. Jaffé, London: Allen & Unwin, 1954.

Woo, H.K.H. 1985. *What's Wrong with Formalization in Economics? – An Epistemological Critique*. Hong Kong: Hong Kong Institute of Economic Science.

economics, definition of

The definition of economics has evolved significantly over time, influenced by and influencing the focus of economic study. The definition often attributed to Jacob Viner, 'economics is what economists do', reflects the difficulty of providing an unambiguous definition. The problem, of course, is that definitions of the field are proposed *ex post* in an attempt to impose order upon a body of work that has grown up as economists have sought to tackle diverse practical and intellectual problems. Viner's statement suggests that there is no need for a tight, specific definition of the subject, which may explain the tendency of economists blithely to ignore definitions, and hence to not analyse them in detail, except sporadically. However, definitions of the subject do have effects through influencing what economists choose to study and the methods they think legitimate for analysing them.

The root of the word 'economics' lies in the Greek οἰκονομία, meaning the management of a household, as in Xenophon's Οἰκονομικος, written around 400 BC. In the 18th century, the idea of efficiently providing for the wants of a household was extended to the nation as a whole, under the heading 'political economy', the term first used for the discipline that later became economics. The first systematic English-language book on the subject was James Steuart's *An Inquiry into the Principles of Political Oeconomy* (1767, p. 16). Though Steuart made an analogy between 'providing for all the wants of a family, with prudence and frugality' and doing the same for the state, there was a difference, for the ruler of the state could not direct people in the way that the head of a household was able to do. This had the consequence that,

> The great art therefore of political oeconomy is, first to adapt the different operations of it [the state] to the spirit, manners, habits and customs of the people; and afterwards to model these circumstances so, as to be able to introduce a set of new and more useful institutions. (Steuart, 1767, p. 16)

No doubt influenced by German Cameralism, Steuart saw institutional design as lying at the heart of political economy. This usage was followed by Adam Smith, who saw political economy as 'a branch of the science of a statesman or legislator' with two objects: providing the people with 'plentiful revenue or subsistence' and providing the state with enough revenue to provide public services (Smith, 1776, p. 428).

Many of the classical economists, however, disagreed with the focus on policy, arguing that political economy was concerned with the laws that govern the production, distribution and consumption of wealth, the clearest example of this being Jean Baptiste Say, whose major work (1803) is *Traité d'économie politique, ou simple exposition de la manière dont se forment, se distribuent et se consomment les richesses* (A treatise on political economy, or a simple account of the way in which wealth is formed, distributed and consumed). This definition formed the basis for Nassau Senior's *Outline of the Science of Political Economy* (1836) in which he argued that the science was based on four propositions, the first and most important of which was 'That every man desires to obtain additional Wealth with as little sacrifice as possible' (Senior, 1836, p. 26).

Neither of these definitions was acceptable to John Stuart Mill, whose 'On the Definition of Political Economy; and the Method of Investigation Proper to It', first published in 1836, was the last of his *Essays on Some Unsettled Questions of Political Economy* (1844). To define political economy as the rules for making a nation rich was to confuse 'art' and 'science'. However, it was not enough to define it as the laws relating to the production and use of wealth, for these included many physical laws that lay outside its remit. He thus favoured a more limited definition: 'The science which treats of the production and distribution of wealth, so far as they depend upon the laws of human nature' or 'The science relating to the moral or psychological laws of the production and distribution of wealth' (Mill, 1844, p. 318). Mill went on to argue that even this definition was too broad, for political economy related only to man in society.

The most significant challenge to this definition of political economy as, loosely, the science of wealth, came from Alfred Marshall, who offered the well-known definition:

> Political Economy or Economics is a study of mankind in the ordinary business of life; it examines that part of individual and social action which is most closely connected with the attainment and with the use of the material requisites of wellbeing. (Marshall, 1890, p. 1)

This definition is significant not so much for changing the name of the discipline to economics as for its focus on the study of mankind. For Marshall, as for many of his generation, the evolution of human character was of crucial importance: it was important to study actual human behaviour, but it was important, especially in the longer run, to consider how activities and consumption served to influence character and hence behaviour. Wants could not be taken as given but depended on activities.

In these discussions there was, as Neville Keynes pointed out, an ambiguity in the use of the word 'economic'. On the one hand it referred to attaining an end 'with the least possible expenditure of money, time and effort' (Keynes, 1891, pp. 1–2) whilst on the other hand it was used as an adjective corresponding to the noun, wealth. The economists who laid most emphasis on the first of these were the Austrians – Carl Menger and his successors – who focused on economizing behaviour. It was his familiarity with this literature that led Lionel Robbins to deny originality for his much-quoted definition, 'Economics is the science which studies human behaviour as a relationship between ends and scarce means which have alternative uses' (Robbins, 1932, p. 16). Robbins's definition put scarcity and choice at the centre of economic analysis. He emphasized that 'any kind of human behaviour' that demonstrates the scarcity aspect falls within the scope of economics, and that there are 'no limitations on the subject-matter of Economic Science' beyond involving 'the relinquishment of other desired alternatives' (choice) (1932, p. 17). The significance of this definition lies in its analytical nature: instead of defining economics in terms of its subject matter, it defines it as an aspect of behaviour.

In spite of Robbins's claim that he was simply describing professional practice, the initial reaction of the profession to his definition of economics, at least as it surfaced in academic journal articles and introductory textbooks (where the definition of economics was primarily discussed), was negative (for a detailed discussion, see Backhouse and Medema, 2007). Throughout the

1930s and 1940s, textbook writers continued to define economics in terms more reminiscent of Mill and Marshall than Robbins, in that, even where reference was made to scarcity, this was frequently qualified: economics was described as a social science concerned with the study of wealth, of earning a living or a study of the system of free enterprise. Robbins's choice-based definition was seen as too wide, and needed to be restricted so as to rule out matters that did not come within the 'traditional' boundaries of economics. The acceptance of the Robbins definition came piecemeal. First, scarcity came to be stressed as important to the subject. The first edition of Paul Samuelson's *Economics* (1948), undoubtedly the leading textbook in the post-war period, captures well the qualified attitude with which the Robbins definition was approached. Samuelson explained that economics was about scarcity, for 'the American way of life' required more resources than were available, but he chose to define the subject in terms of 'what', 'how' and 'for whom' – that is, as concerning the production and consumption of goods and services. There is nothing here that is inconsistent with Robbins, but this approach was equally consistent with a more traditional approach. Books such as George Stigler's *Theory of Price* (1946), which adopted the Robbins definition, laid great stress on both scarcity and choice, but others carefully refrained from doing so.

It was only in the late 1950s and 1960s that the use of Robbins's definition became widespread. By the late 1960s, Samuelson's *Economics* was claiming that economists agreed on 'a general definition something like the following':

> Economics is the study of how men and society *choose*, with or without the use of money, to employ *scarce* productive resources, which could have alternative uses, to produce various commodities over time and distribute them for consumption, now and in the future, among various people and groups in society. (Samuelson, 1967, p. 5).

However, support for this was still not universal. For Richard Lipsey, whose *Introduction to Positive Economics* was one of the most successful rivals to Samuelson's *Economics*, scarcity was '*one of* the basic problems encountered in *most* aspects of economics', not the entire subject (Lipsey, 1963/71, p. 50). Economics also dealt with questions related to failure to achieve a point on the production possibility frontier, such as explaining unemployment, which could not be reduced to problems of scarcity.

The move by Robbins to define economics as an aspect of behaviour made it just a short step to defining economics in terms of a method – that of rational choice – which could be applied not simply to production and consumption choices, but to all of human behaviour. This move was encouraged by the tendency, in the aftermath of the Second World War, to see economics though the lens of operations research, as social engineering, in which optimization techniques were central and game theory played a significant role. It has also been argued that this move towards emphasising rational choice had ideological attractions during the Cold War. During the 1960s, economics became increasingly conceived as the 'science of choice', without reference to a particular social domain, even, at times, without reference to scarcity: the subject could encompass non-market as well as market activities. The work of Theodore Schulz and Gary Becker on human capital, James Buchanan, Anthony Downs and Gordon Tullock on political processes, and Becker on discrimination and on crime and punishment laid foundation for what came to be called 'economics imperialism', the application of economics to fields including politics, law, history, and sociology. These theoretical moves were reinforced by advances on the empirical side, where the techniques developed by, for example, James Heckman and Daniel McFadden for analysing cross-section data sets on individuals and households were used to investigate phenomena, such as non-marital fertility, that lie outside the traditional domain of economics as concerned with market behaviour.

Robbins's definition of economics in terms of the allocation of scarce resources remains the most widely cited definition of the subject, but it has never commanded universal assent. Though scarcity can be defined in such a way as to make it true, there have always been significant numbers of economists who have considered that it does not encompass all aspects of their discipline and that qualifications or extensions are required. These result in definitions closer to those found in the 19th-century literature, focusing on phenomena such as the production and distribution of wealth. At the other end of the spectrum, there are economists for whom rational choice is more fundamental than scarcity. To this extent, then, there is no universally agreed upon definition of the subject.

The reason this does not present a problem is that economists can proceed with their work irrespective of how their subject is defined. Definitions of fields generally come only after the field is established; as fields change, so definitions change. Despite this, however, definitions can matter. As Mill recognized, questions of method and definition are linked. The clearest example of this is Robbins, who sought to derive all the main propositions of economics from the premise of scarcity. His definition, therefore, was the basis for claiming that economic theory was central to economics – that it was far more important than Marshall had believed it to be. Also significant was his reference to economic science, for the word science is far from neutral. Robbins had argued that value judgements, including those necessary to make interpersonal welfare comparisons, did not come within the scope of economic science, but belonged instead to the realm of 'political economy'. In claiming this, he was arguably attempting to clarify the status

of economists' arguments, for, as he later made very clear, offering any advice on economic policy requires such value judgements. Thus if economics includes policy advice it must encompass more than economic science as Robbins defines it. However, such is the prestige of 'science' that Robbins's definition caused many economists to try to dispense with value judgements altogether, even in welfare economics. An exercise in clarification (and no doubt a critique of certain views of the subject) thus had the effect of significantly narrowing the subject. Attempting to define economics thus was not and is not simply a descriptive exercise; it has consequences for what economists do, and how they go about doing it.

ROGER BACKHOUSE AND STEVEN MEDEMA

See also **altruism, history of the concept; Mill, John Stuart; rationality, history of the concept; Robbins, Lionel Charles; Samuelson, Paul Anthony; United States, economics in (1945 to present).**

Bibliography

Backhouse, R.E. and Medema, S.G. 2007. Defining economics: Robbins's definition in theory and practice. SSRN working paper. Abstract online. Available at http://ssrn.com/abstract=969994, accessed 19 May 2007.
Keynes, J.N. 1891. *The Scope and Method of Political Economy*. London: Macmillan.
Lipsey, R.G. 1963. *An Introduction to Positive Economics*, 3rd edn. London: Weidenfeld & Nicolson, 1971.
Marshall, A. 1890. *Principles of Economics*. London: Macmillan, 1949.
Mill, J.S. 1844. *Essays on Some Unsettled Questions of Political Economy*. In *Collected Works of John Stuart Mill*, ed. J.M. Robson. Toronto: University of Toronto Press, 1967.
Robbins, L.C. 1932. *An Essay on the Nature and Significance of Economic Science*, 2nd edn. London: Macmillan, 1935.
Samuelson, P.A. 1948. *Economics*. New York: McGraw Hill.
Samuelson, P.A. 1967. *Economics*, 7th edn. New York: McGraw Hill.
Say, J.-B. 1803. *Traité d'économie politique, ou simple exposition de la manière dont se forment, se distribuent et se consomment les richesses*. Paris: Deterville.
Senior, N. 1836. *An Outline of the Science of Political Economy*. New York: Augustus Kelley, 1965.
Smith, A. 1776. *An Inquiry into the Nature and Causes of the Wealth of Nations*, 2 vols. Indianapolis: Liberty Press, 1976.
Steuart, J. 1767. *An Inquiry into the Principles of Political Oeconomy*, 2 vols. Edinburgh: Oliver & Boyd, 1966.
Stigler, G.J. 1946. *The Theory of Price*. New York: Macmillan.

economy as a complex system

Introduction

The term 'complex system' has been widely used in science and many different definitions have been given. Frequently, rather than give a definition of such a system, scientists have fallen back on certain characteristics that these systems exhibit. For example, emergence, self-organization, synergetics, collective behaviour, and non-equilibrium have all been cited in this regard. It is useful at the outset to make the distinction between 'complexity' and 'complex system'. The former involves a number of ideas which are important in economics and which are inherited from computer science but which will not be dealt with here. In particular, the notion of computational complexity, as it applies to decision-making or to the computation of equilibria or of dynamic programming problems is central to certain aspects of economic theory. However, here the discussion will turn on the idea of the economy as a complex, adaptive, evolving system. For economists the first real incarnation of this approach was with the introduction of deterministic chaos. The idea of complex dynamic behaviour, which would not explode or cycle or converge to a steady state, was fascinating for a science long dominated by the ideas of convergence to a static equilibrium or to a steady state. Jean-Michel Grandmont (1985) developed a simple model of 'business cycles' involving the 'tent map', which gave rise to such chaotic behaviour. Apart from the idea of the complicated dynamics involved, it was clear that the fact that a small perturbation in the initial conditions governing such a process could produce radically different trajectories was also of great intellectual interest. Two important innovations were involved. Firstly, there was the idea that the economy should be thought of as a truly dynamic system and that the initial conditions of such a system might play a key role. Secondly, there was the idea that there might be no continuity in the dependence on those initial conditions and that small changes might radically influence the trajectory of the system; hence the famous allusion to the influence of the fluttering of a butterfly's wing on the world's weather. These two aspects led economists to focus their attention on deterministic chaos. Yet in making such a close link between complexity and chaos, economists may have lost sight of the broader implications of complexity for the analysis of economic systems.

To see why this is so, consider what sort of systems are referred to as 'complex' in other disciplines. Typically they have some, or all, of the following characteristics:

- The agents are heterogeneous and interact directly with each other.
- The interaction and the information of agents are 'local'.
- The agents' behaviour is governed by simple 'rules of thumb'.
- The aggregate behaviour of the system is not that of an 'average' or representative agent.
- This aggregate behaviour 'emerges' from the complicated interaction between the individuals.

To someone who has not studied theoretical economics, all of these characteristics might seem rather intuitive as features of an economy. Yet they are very different from the traditional view. In that view, the economy is a system in which the only interaction is through the market. By this it is meant that agents react to signals from some central authority such as an auctioneer. In some way the central prices adjust so as to coordinate the activities of the agents. The system adjusts in this way until the activities are coordinated – for example, in a market economy, until aggregate demand for all products is equal to the aggregate supply of those products. Once this is achieved, the signals will not change and no agent has an incentive to modify his behaviour and to deviate from this 'equilibrium state'.

This description reveals another important feature of the collective model. No agent takes account of any influence that he might have on the outcome of the system. Many economists will react to this description by arguing that models of 'imperfect competition' abound, and in these models agents take into account the impact of their actions on the state of the system and know that other agents do the same.

This brings us to a second view of the economy, that based on game theory. Here all agents take account of the reciprocal impacts of their actions and know that all the other agents do the same. This view is very different from the basic model of the economy. However, it is also very different from that of a complex system, since it attributes unlimited calculating capacity and depth of reasoning to the agents.

The vision of the economy as a complex system falls between these two approaches. It requires neither the central coordinating mechanism of the competitive market, nor the analytically sophisticated players of game theory.

A good comparison might be between, on the one hand, an economy organized as a set of markets that are open simultaneously, each with an auctioneer and, on the other, an ants' nest. In the former there are structured central price-giving mechanisms, and the actors gradually reveal their willingness or non-willingness to pay until the goods are allocated efficiently. In the latter, the individuals pursue their own different activities and react to each other and to outside stimuli. The system organizes itself but there is no central mechanism for achieving such organization. No one would think of trying to describe the activity of an ants' nest by examining the behaviour of the 'representative ant' yet many would describe the allocation of effort and resources as 'efficient'.

The sort of system that could be described as complex in the sense outlined above can be physical or biological or social. A typical reaction to the use of physical or biological analogies in economics or other social sciences is that social systems are populated by individuals that have intentions and undertake purposeful activity, while the other systems are composed of purposeless molecules or particles. It is therefore argued that the sort of analysis that can be applied to the other systems is not pertinent to the analysis of economic systems. This reasoning does not stand up to close inspection. If individuals follow well-defined rules and their interaction is well specified, the simple models that are used in physical and biological models can be applied. Precisely why the individuals should follow these rules is a different question.

Why is the complex systems approach of particular interest currently to economists? Economic theory has recently been attacked on two fronts. The first is the problem of aggregation: how is the behaviour of the economic system related to that of the individuals that make it up? The second is the question of why individuals behave as they do. The answer to the first question is simple but undermines much of modern macroeconomics that is based on the idea that the behaviour of the aggregate can be treated as the behaviour of an individual. Yet what is known is that the standard model of a system composed of isolated individuals each solving his own maximizing problem does not allow one to treat the system as an individual. (This is not the place to enter into the details of this assertion but the basic argument is given in Kirman, 1992, and stems from the results of Sonnenschein, Mantel and Debreu.) The second question is that posed by behavioural economics that questions the idea of the isolated maximizing individual. Ideas from Simon (1957) onwards have suggested that individuals reason in a limited and local way. Experiments, observation, and examination of the neural processes utilized in making decisions all suggest that *homo economicus* is not an accurate or adequate description of human decision making. (For a good survey of the relevant literature, see Rabin, 1998.)

All of this suggests that one might want to take a very different view of how the economy functions. In particular, the notion of a complex system as used in many parts of science seems to correspond well to an intuitive vision of the economy. Just as in an ants' nest individuals perform tasks without having any idea of the behaviour of the system, individuals in an economy go about their business and achieve a remarkable degree of coordination. Take a simple example that of bees in a hive. The tasks for house bees are varied but temperature control is one of the important duties. When the temperature is low, bees cluster to generate heat for themselves, but when it is high some of them fan their wings to circulate air throughout the hive. The general hive temperature required is between 33° and 36°C, while the brood chamber requires a constant heat of 35°. Honey has to be cured in order to ripen, and this also requires the help of circulating air. According to Crane (1999), 12 fanning

bees positioned across a hive entrance 25 cm wide can produce an air flow amounting to 50–60 litres per minute. This fanning can go on day and night during the honey-flow season. Honeybees' wings beat 11,400 times per minute, thus making their distinctive buzz.

What is the lesson here for us? The typical economist's response to this phenomenon would be to consider a representative bee and then study how its behaviour responds to the ambient temperature. This would be a smooth function of temperature, wing beats going up or down with the temperature. Yet this is not what happens at all. Bees have different threshold temperatures and they are either on (beating at 11,400 beats per minute) or off. As the temperature rises more bees join in. Thus collectively with very simple 1, 0 rules the bees produce a smooth response. This sort of coordination, with each agent doing something simple, can only be explained by having a distribution of temperature thresholds across bees. Aggregation of individuals with specific local and differentiated behaviour produces smooth and sophisticated aggregate behaviour.

Nobody would argue that, in social systems, all coordination is achieved by simple interaction. Markets make a powerful contribution to economic coordination. Yet the important question is not whether such mechanisms exist, but how they come into being and develop and modify their rules. As already explained, the idea that the existence of such markets facilitates the allocation of resources is clear and generally accepted. What is not so clear is that the abstract idea of a market governed by centralized prices which are adjusted to equilibrate the market has any descriptive value. The idea of markets and networks of communication and transactions as emergent and changing phenomena is much more persuasive.

Considering the economy in this light is far from a new idea. When Adam Smith discusses the 'invisible hand' some of these notions are apparent, Pareto's work contains some of these ideas and Hayek is perhaps he who was closest to this vision. Schelling in his *Micromotives and Macrobehavior* (1978) clearly foresaw the role of self-organization. A recent development of these ideas had an introduction on the formal level by Foellmer (1974), who adopted the basic Ising model. He posited a system in which individuals were situated in space and whose preferences were dependent on those around them. This dependence was stochastic, that is, the probability of having certain preferences depended on the preferences of an individual's neighbours. If all the preferences are independently drawn, then one can determine the expected values of the equilibrium prices. However, if the interdependence of the individuals is too strong, this is no longer true. The 'law of large numbers' no longer applies. There is no easy transition from the micro to the macro level by simple averaging.

Foellmer's contribution was left to one side for a long time. However, the complexity approach to economics took on new life with the work at the Santa Fe Institute of a number of economists, physicists and other scientists such as Arthur, Bak, Blume, Durlauf, Geanakoplos, and Holland. A good picture of this sort of work can be found in *The Economy as an Evolving Complex System* (Anderson, Arrow and Pines, 1988) and the two additional volumes that followed it (Arthur, Durlauf and Lane, 1997; Blume and Durlauf, 2006).

The emphasis on the increasing 'socialization' of economics, which is intrinsic to models of interacting agents, permits one to introduce the influence of neighbours and groups on individual behaviour. Such an approach is standard in sociology and anthropology but has remained a very thinly populated field in economics. A good survey of this work is to be found in Durlauf and Young (2001).

One important part of the research on complex systems in economics has been that on *agent-based* models. Here the idea is to look at a set of linked individuals whose behaviour is influenced by and which influences their neighbours, and to simulate the dynamics of that interaction. Perhaps the best-known early example of this was Axelrod's work on the Prisoner's Dilemma, which is summarized in Axelrod (1997). He started from a series of tournaments. The strategies used for these were those that individuals proposed for a repeated Prisoner's Dilemma game. These strategies were then played against each other in a series of tournaments and the winning strategy turned out to be 'tit for tat', which is basically cooperative.

Axelrod was concerned that those who had entered his tournament had already anticipated the strategies that would be proposed by others. To overcome this he ran simulations in which new strategies were introduced into the pool of existing strategies. To do this he assigned existing strategies randomly to his artificial agents and then modified them using a 'genetic algorithm'. (For an introduction to the theory and use of genetic algorithms see Mitchell, 1996.) The set of strategies thus evolved in two ways. After the strategies had played against each other a new generation with more of the successful strategies was created. To these were also added new strategies generated by mutations and crossovers from the current population. After a while reciprocating strategies – that is, strategies which respond to cooperation with cooperation but which defect in the face of defection – took over, giving high payoffs. Here we have a selection process working on strategies that evolved rather than were consciously chosen. The behaviour of this basic but complex system – indeed, Axelrod refers to himself as a complexity scientist – led to the evolution of interesting aggregate characteristics. In this context it is also interesting to look at the work of Lindgren (1991), who also allowed the evolution of the strategy pool and generated periods of stability in which one strategy dominated, followed by periods of instability as the population was invaded by another strategy. This corresponds to the idea of 'punctuated equilibria'

introduced into evolutionary theory by Eldredge and Gould (1972).

The notion of evolution, which can also be interpreted in the human or social context as adaptive learning, is important here. We can think of selection among a population of automata endowed with single strategies or of the idea that individuals learn to use more successful strategies.

Phase transitions

Recalling the characterization of complex systems given above, it is worth considering a few examples.

In complex systems governed by local interactions, it may be the case that as a result of some perturbation there is a major change in aggregate behaviour. This is an important idea which is central to *statistical mechanics*. The idea here is that local interaction can generate a rapid transition from one 'phase' to another of an economic system and, more importantly, that one cannot simply apply the 'law of large numbers' to evaluate the impact of stochastic shocks. An example of this is provided by Bak et al. (1993), who consider a model of 'self-organized criticality' to describe an economy composed of a large number of productive units, each supplying a limited number of customers and, in turn, each supplied by a limited number of suppliers; both customers and suppliers are located near the productive unit.

The graph outlining the location of productive units is a cylindrical lattice. In other words, each production unit is supplied by the firms above it on a vertical line and supplies the customers next to it on a horizontal line. The demand for each final good producer is characterized by stochastic fluctuations, which affects the variability of orders received by the suppliers. Such orders (and shocks) are locally and vertically correlated, as every final producer is supplied by the two upstream firms situated a line up along the network representing the productive system. In such a context, characterized by local interaction, Bak et al. (1993) prove that, if individual costs are non-convex, the aggregation of small independent individual shocks may lead to large aggregate fluctuations in the productive system, breaking therefore the law of large numbers. These small shocks do not cancel each other out but are amplified by their interaction. Thus fluctuations at the aggregate level cannot be explained by reducing the whole model to one of an individual.

Coordination: the Schelling model

Now let us pursue the discussion of the relationship between aggregate and individual behaviour. One of the important features of complex systems is that the system can coordinate on a solution which could not be predicted from a careful analysis of the average or typical individual. In other words, patterns at the aggregate level can emerge as the individuals in an economy or market interact with each other. The emergence of such aggregate patterns cannot be forecast from the specification of the individual characteristics. A good example of this was provided by Tom Schelling at the end of the 1960s (for a summary see Schelling, 1978). He introduced a model of segregation involving local interaction, in the sense that peoples' utility depends on the race of their neighbours. He showed that, even if people have only a very mild preference for living with neighbours of their own colour, as they move to satisfy their preferences complete segregation will occur.

The basic model is very simple. Take a large chess board, and place a certain number of black and white counters on the board, leaving some free places. A counter prefers to be on a square where half or more of the counters in his Moore neighbourhood, (the eight squares around him) are of its own colour (utility 1) to the opposite situation (utility 0). From the counters with utility zero, one is chosen at random and moves to a preferred location. This model, when simulated, yields complete segregation even though people's preferences for being with their own colour are not strong. Indeed, the result holds when individuals are happy even when more than half of their neighbours are of a colour different from their own. This result was greeted with surprise and has generated a large literature.

In fact, this result is not surprising and some simple physical theory (see Vinkovic and Kirman, 2006), can explain the segregation phenomenon. Numerous variants on Schelling's original model have been developed. In particular, the form of the utility function used by Schelling, the size of neighbourhoods, the rules for moving, and the amount of unoccupied space have all been studied (see Pancs and Vriend, 2007, for a survey). The physical model encompasses all of these variants.

An attempt to provide a formal structure has been made by Pollicot and Weiss (2001). They however, examine the limit of a Laplacian process in which individuals' preferences are strictly increasing in the number of like neighbours. In this situation it is intuitively clear that there is a strong tendency to segregation. Yet Schelling's result has become famous because the preferences of individuals for segregation were not particularly strong. The model is of interest because it illustrates the emergence of an aggregate phenomenon which is not directly foreseen from individual behaviour and because it concerns an important economic problem, that of segregation.

The physical analogue to Schelling's model, developed in Vinkovic and Kirman (2006), exhibits three features of the resultant segregation. The first is the organization of the system into 'regions' or clusters, each containing individuals of only one colour. Second, it explains the shape of the frontier between the regions. Lastly, in the case where several clusters of one colour may form it allows one to analyse the size distribution of the clusters.

The basic idea is simple. Think of utility as the negative of energy. Particles with high energy in the physical system correspond to individuals with low utility in the social system. Where are the unhappy or high-energy individuals to be found? Clearly they are individuals on the frontiers of clusters. Those within clusters of their own colour are happy and have no possibility of increasing their utility by moving. Those on the frontier, on the other hand, are in contact with those of the other colour and there may be too many of the latter. In this case these individuals correspond to particles with high energy. A physical system with these characteristics will seek to minimize its energy. The energy is highest on the frontier between clusters. Thus the way for the system to minimize its energy is to reduce the length of these frontiers. It will achieve this by organizing itself into clusters, and the shape and size of these clusters will depend on the precise variant of the model. In the original model the system will organize itself into two giant clusters, each composed of individuals of one colour. If we only allow people to move to currently free places, then the number of these will be important for the outcome. If there are not enough, the system will 'freeze' with many small clusters. If, on the other hand, individuals can swap places the system will segregate, but there will be perpetual movement within it. Thus, a simple physical model generates the result obtained by Schelling and, furthermore, shows how the form of the segregation depends on the exact version of the model. (For a discussion of the emergent properties of the Schelling model see EMERGENCE.)

The 'El Farol Bar'

Another interesting example of emergent coordination is that provided by Brian Arthur (1994) in his 'El Farol Bar' problem. The simple model that he develops and which has been taken up by many physicists under the name of 'the minority game' shows how individuals using rules of thumb can come to coordinate in a way which yields a satisfactory social outcome even though no individual had any such intention. The idea is that the bar can hold 100 people. Being at the bar with fewer than 60 people is, by common consent, better than staying at home. However once attendance goes over 60 the bar becomes too crowded and home is the preferable alternative. The question then is how people will decide whether to go to the bar. Suppose that they all reason strategically. In this case they must decide in function of what their neighbours will decide. Thus, to anticipate whether there will be more than 60 people at the bar they must reflect on the strategies employed by the others. However, they must also take into account that the others are doing the same and know that the others know that they know that they are behaving in this way. This leads to an infinite regress that poses logical problems for the foundations of such game-theoretic reasoning. Rather than attribute such calculating capacities to his agents, Brian Arthur imagined that each was endowed with a set of forecasting rules based on previous attendance at the bar. Given his set of rules the individual chooses that rule which has forecast best up to the present, 'best' meaning the forecast that has the smallest sum of squared prediction errors, for example. Now, each agent uses, as information, just the attendance observed at the bar, and updates in consequence. There is no coordinating mechanism, yet the model quickly settles to the 'equilibrium' solution with 60 people at the bar with occasional small deviations. Furthermore, each agent receives a fixed number of forecasting rules, some of which may be rather stupid. Nevertheless, coordination is achieved at the aggregate level.

Some things about this model are worth noting. It is not guaranteed that all agents will learn to forecast correctly; some may persist in erroneous forecasts. The way in which the model is set up means that whenever attendance goes to 61 many people are unhappy, which is not the case when it goes to 59. This asymmetry does not prevent the achievement of collective coordination, however. Thus, the relation between satisfactory performance at the aggregate level and satisfaction at the individual level is tenuous. While many may find this example intriguing, one might enquire as to how it can be directly applied to economic problems. An interesting answer is to be found in a book by some Oxford physicists who specialize in complex systems and who apply the model to financial markets (see Johnson, Jeffries and Hui, 2003, pp. 81–136).

Financial markets

This brings us to another important example, that of financial markets. Models of economies with interacting agents in the spirit of complex systems may, as we have just seen, be able to show how certain aggregate coordination may emerge. They may also help us to analyse some of the observed features of markets which normal economic analysis has difficulty explaining. For example, one of the major problems with the standard model of financial markets is that they do not reproduce certain well-established stylized facts about empirical price series. In standard models, where there is uncertainty about the evolution of prices, the usual way of achieving consistency is to assume that agents have common and 'rational' expectations. Yet, if agents have such common expectations, how can there be trade? Indeed there are many 'no trade' theorems for such markets. How, then, do we deal with the fact that the volume of trade on financial markets is very important and that agents do, in fact, differ in their opinions and forecasts and that this is one of the main sources of such trade? There is also an old problem of 'excess volatility,' that is,

prices have a higher variance than the returns on the assets on which they are based. One answer is to allow for direct interaction between agents other than through the market mechanism. Models reminiscent of the Ising model from physics have been used to doing this. For example, one might suggest that individuals may change their opinions or forecasts as a function of those of other agents. In simple models of financial markets such changes may be self-reinforcing. If agents forecast an increase in the price of an asset and others are persuaded by their view, the resultant demand will drive the price up, thereby confirming the prediction. However, the market will not necessarily 'lock on' to one view for ever. Indeed, under certain rather reasonable assumptions, if agents make stochastic rather than deterministic choices, then it is certain that the system will swing back to a situation in which another opinion dominates. The stochastic choices are not irrational, however. The better the results obtained when following one opinion, the higher is the probability of continuing to hold that opinion.

Such models will generate swings in opinions, regime changes and 'long memory', all of which are hard to explain with standard analysis. An essential feature of these models is that agents are wrong for some of the time, but whenever they are in the majority they are essentially right. Thus they are not systematically irrational. (For examples of this sort of model see, Lux and Marchesi, 1999; Brock and Hommes, 1997; and Kirman and Teyssiere, 2005, and for a recent survey, De Grauwe and Grimaldi, 2006.) Thus the behaviour of the agents in the market cannot correctly be described as 'irrational exuberance', in the well-known words of Alan Greenspan, Chairman of the Board of Governors of the Federal Reserve from 1987 to 2006.

Economists faced with this sort of model are often troubled by the lack of any equilibrium notion. The process is always moving; agents are neither fully rational nor systematically mistaken. Worse, the process never settles down to a particular price even without exogenous shocks. Suppose that we accept this kind of model: can we say anything analytic about the time series that result? If we consider some of these models, for certain configurations of parameters they could become explosive. There are two possible reactions to this. Since we will never observe more than a finite sample, it could well be that the underlying stochastic process is actually explosive, but this will not prevent us from trying to infer something about the data that we observe. Suppose, however, that we are interested in being able, from a theoretical point of view, to characterize the long-run behaviour of the system. In particular, if we treat the process as being stochastic and do not make a deterministic approximation, then we have to decide what, if anything, constitutes an appropriate long-run equilibrium notion. Such a concept provides an answer to those who consider that complex systems, by their nature, are not amenable to formal analysis. Foellmer, Horst and Kirman (2005), examined the sort of price process discussed here and produced some analytical results characterizing the process. Furthermore, they provided a long-run equilibrium notion that is not the convergence to a particular price vector.

If prices change all the time, as they will do in an evolving complex system, how may one speak of 'equilibrium'? The idea is to look at the evolving distribution of prices and to try to characterize its long-run behaviour. Foellmer, Horst and Kirman (2005) examined the process governing the evolution of asset prices and the profits made by traders, and gave conditions under which it is ergodic, that is, the proportion of time that the price takes on each possible value converges over time and that the *limit distribution* is unique. (For a discussion of the mathematical background, see ERGODICITY AND NONERGODICITY IN ECONOMICS.) This means that, unlike the 'anything can happen' often associated with deterministic chaos, in the long run the price and profits process does have a well-defined structure.

Conclusion

To view the economy as a complex system implies a fundamental rethinking of theoretical economics. The basic idea is that of a decentralized system with no central source of signals, whose aggregate behaviour cannot be reduced to that of an individual. Furthermore, the individuals are endowed with local information and interact directly with each other, and their behaviour can be characterized by simple rules. Such a vision is far from new in economics. Its origins can be traced back at least to Adam Smith and a long chain of economists leads from him to Hayek and Simon, who preceded the developments described here. The most recent contributions borrow heavily from other disciplines such as statistical physics and the appearance of 'econophysics' represents a shift from the path that led from classical mechanics to axiomatic mathematical models as the basic paradigm of economic theory. This sort of approach has already allowed economists to analyse problems such as contagion, neighbourhood effects, financial bubbles, and herding behaviour, none of which fits well into the standard economic framework. In addition, many of the features that are imposed on standard models emerge as a result of the interaction between agents (see EMERGENCE).

Perhaps, most importantly, looking at economies in this way provides a very different and more intuitive vision of the economy as a vast interactive system whose aggregate properties reflect the self-organization of the system and its continual adaptation. However, entrenched ideas die hard and it remains to be seen whether Steven Hawking's prediction that the 21st century will be the 'age of complexity' will hold true for economics.

ALAN KIRMAN

See also **emergence; ergodicity and nonergodicity in economics; interacting agents in finance; social interactions (empirics); social interactions (theory); social networks, economic relevance of; statistical mechanics.**

Bibliography

Anderson, P.W., Arrow, K.J. and Pines, D. 1988. *The Economy as an Evolving Complex System*. Redwood City, CA: Addison-Wesley.

Arthur, W.B. 1994. Inductive reasoning and bounded rationality. *American Economic Review* 84, 406–11.

Arthur, W.B., Durlauf, S.N. and Lane, D.A., eds. 1997. *The Economy as an Evolving Complex System II*. Reading, MA: Addison-Wesley.

Axelrod, R. 1997. *The Complexity of Cooperation: Agent-Based Models of Competition and Collaboration*. Princeton, NJ: Princeton University Press.

Bak, P., Chen, K., Scheinkman, J. and Woodford, M. 1993. Aggregate fluctuations from independent sectoral shocks: self-organized criticality in a model of production and inventory dynamics. *Ricerche Economiche* 47, 3–30.

Blume, L. and Durlauf, S.N. 2006. *The Economy as an Evolving Complex System III*. Oxford: Oxford University Press.

Brock, W.A. and Hommes, C. 1997. A rational route to randomness. *Econometrica* 65, 1059–95.

Crane, E. 1999. *The World History of Beekeeping and Honey Hunting*. London: Duckworth.

De Grauwe, P. and Grimaldi, M. 2006. *The Exchange Rate in a Behavioral Finance Framework*. Princeton, NJ: Princeton University Press.

Durlauf, S.N. and Young, H.P. 2001. *Social Dynamics: Economic Learning and Social Evolution*. London: MIT Press.

Eldredge, N. and Gould, S.J. 1972. Punctuated equilibria: an alternative to phyletic gradualism. In *Models in Paleobiology*, ed. T.J.M. Schopf. San Francisco: Freeman, Cooper and Co.

Foellmer, H. 1974. Random economies with many interacting agents. *Journal of Mathematical Economics* 1, 51–62.

Foellmer, H., Horst, U. and Kirman, A. 2005. Equilibrium in financial markets with heterogeneous agents: a new perspective. *Journal of Mathematical Economics* 41, 123–55.

Grandmont, J.-M. 1985. On endogenous competitive business cycles. *Econometrica* 53, 995–1046.

Johnson, N., Jeffries, P. and Hui, P.M. 2003. *Financial Market Complexity*. Oxford: Oxford University Press.

Kirman, A. 1992. What or whom does the representative individual represent? *Journal of Economic Perspectives* 6(2), 117–36.

Kirman, A. and Teyssiere, G. 2005. Testing for bubbles and change points. *Journal of Economic Dynamics and Control* 29, 765–99.

Lindgren, K. 1991. Evolutionary phenomena in simple dynamics. In *Artificial Life II*, ed. C.G. Langton, C.

Taylor, J.D. Farmer and S. Rasmussen. Redwood City, CA: AddisonWesley.

Lux, T. and Marchesi, M. 1999. Scaling and criticality in a stochastic multi-agent model of a financial market. *Nature* 397, 498–50.

Mitchell, M. 1996. *An Introduction to Genetic Algorithms*. Cambridge, MA: MIT Press.

Pancs, R. and Vriend, N.J. 2007. Schelling's spatial proximity model of segregation revisited. *Journal of Public Economics* 91, 1–24.

Pollicott, M. and Weiss, H. 2001. The dynamics of Schelling-type segregation models and a nonlinear graph Laplacian variational problem. *Advances in Applied Mathematics* 27, 17–40.

Rabin, M. 1998. Psychology and economics. *Journal of Economic Literature* 36, 11–46.

Schelling, T. 1978. *Micromotives and Macrobehavior*. New York: W.W. Norton and Co.

Simon, H.A. 1957. *Models of Man: Social and Rational*. New York: Wiley.

Vinkovic, D. and Kirman, A. 2006. A physical analogue of the Schelling model. *Proceedings of the National Academy of Sciences* 103, 19261–5.

econophysics

According to Bikas Chakrabarti (2005, p. 225), the term 'econophysics' was neologized in 1995 at the second Statphys-Kolkata conference in Kolkata (formerly Calcutta), India, by the physicist H. Eugene Stanley, who was also the first to use it in print (Stanley, 1996). Mantegna and Stanley (2000, pp. viii–ix) define 'the multidisciplinary field of econophysics' as 'a neologism that denotes the activities of physicists who are working on economics problems to test a variety of new conceptual approaches deriving from the physical sciences'.

The list of such problems has included distributions of returns in financial markets (Mantegna, 1991; Levy and Solomon, 1997; Bouchaud and Cont, 1998; Gopakrishnan et al., 1999; Sornette and Johansen, 2001; Farmer and Joshi, 2002), the distribution of income and wealth (Drăgulescu and Yakovenko, 2001; Bouchaud and Mézard, 2000; Chatterjee, Yarlagadda and Charkrabarti, 2005), the distribution of economic shocks and growth rate variations (Bak et al., 1993; Canning et al., 1998), the distribution of firm sizes and growth rates (Stanley et al., 1996; Takayasu and Okuyama, 1998; Botazzi and Secchi, 2003), the distribution of city sizes (Rosser, 1994; Gabaix, 1999), and the distribution of scientific discoveries (Plerou et al., 1999; Sornette and Zajdenweber, 1999), among other problems, all of which are seen at times not to follow normal or Gaussian patterns that can be described fully by mean and variance. The main sources of conceptual approaches from physics used by the econophysicists have been from models of statistical mechanics (Spitzer, 1971), geophysical models of earthquakes (Sornette, 2003), and 'sandpile' models of avalanches, the latter involving self-organized criticality (Bak, 1996). An early physicist to assert the essential identity of statistical methods used in physics and the social sciences was Majorana (1942).

A common theme among those who identify themselves as econophysicists is that standard economic theory has been inadequate or insufficient to explain the non-Gaussian distributions empirically observed for various of these phenomena, such as 'excessive' skewness and leptokurtotic 'fat tails' (McCauley, 2004). With their sense of creating and developing a new science based on physics that is superior to the older conventional economics, many of the econophysicists have focused their publishing efforts in physics journals, notably *Physica A*, *Physical Review E*, and *European Physical Journal B*, to name some of the most frequently used ones, along with the general science journal *Nature* and some more clearly multidisciplinary journals such as *Quantitative Finance*. However, increasingly some of the econophysicists have begun to publish jointly with economists, with some of these papers appearing in economics journals as well. This should not be surprising in that the emergence of econophysics followed fairly shortly after the influential interactions and discussions that occurred between groups of physicists and economists at the Santa Fe Institute (Anderson, Arrow and Pines, 1988; Arthur, Durlauf and Lane, 1997), with some of the physicists involved in these discussions also becoming involved in the econophysics movement.

Now we come to a great curiosity and irony in this matter: some of the main techniques used by econophysicists were initially developed by economists (with many others developed by mathematicians), and some of the ideas associated with economists were developed by physicists. Thus, in a sense, these efforts by physicists resemble carrying coals to Newcastle, except that it must be admitted that many economists either forgot or never knew of these issues or methods. This is true of the most canonical of such models, the Pareto distribution.

The empirical focus on scaling laws (power laws)

If there is a single issue that unites the econophysicists it is the insistence that many economic phenomena occur according to distributions that obey scaling laws rather than Gaussian normality. Whether symmetric or skewed, the tails are fatter or longer than they would be if Gaussian, and they appear to be linear in figures with the logarithm of a variable plotted against its cumulative probability distribution. They search for physics processes, most frequently from statistical mechanics, that can generate these non-Gaussian distributions that obey scaling laws.

The canonical (and original) version of such a distribution was discovered by the mathematical economist and sociologist, Vilfredo Pareto, in 1897. Let N be the number of observations of a variable that exceed a value x with A and α positive constants. Then

$$N = Ax^{-\alpha}. \qquad (1)$$

This exhibits the scaling property in that

$$\ln(N) = \ln A - \alpha \ln(x). \qquad (2)$$

This can be generalized to a more clearly stochastic form by replacing N with the probability that an observation will exceed x. Pareto formulated this to explain the distribution of income and wealth, and believed that there was a universally true value for α that equalled about 1.5. More recent studies (Clementi and Gallegati, 2005) suggest that it is only the upper end of income and wealth distributions that follow such a scaling property, with the lower ends following the lognormal form of the Gaussian distribution that is associated with the random walk, originally argued for the whole of the income distribution by Gibrat (1931).

The random walk and its associated lognormal distribution is the great rival to the Pareto distribution and its relatives in explaining stochastic economic phenomena. It was only a few years after Pareto did his work that the random walk was discovered in a Ph.D. thesis about speculative markets by the mathematician Louis Bachelier (1900), five years prior to Einstein using it to model Brownian motion, its first use in physics (Einstein, 1905). Although the Paretian distribution would have its advocates for explaining stochastic price dynamics (Mandelbrot, 1963), the random walk would become the standard model for explaining asset price dynamics for many decades, although it would be asset returns that would be so modelled rather than asset prices themselves directly as Bachelier did originally. As a further irony, it was a physicist, M. F. M. Osborne (1959), who was among the influential advocates of using the random walk to model asset returns. It was the Gaussian random walk that would be assumed to underlie asset price dynamics when such basic financial economics concepts as the Black–Scholes formula would be developed (Black and Scholes, 1973). If we let p be price, R be the return due to a price increase, B be debt, and σ be the standard deviation of the Gaussian distribution, then Osborne characterized the dynamic price process by

$$dp = Rpdt + \sigma p dB. \qquad (3)$$

Meanwhile, a variety of efforts were made over a long time by physicists, mathematicians and economists to model a variety of phenomena using either the Pareto distribution or one its relatives or generalizations, such as the stable Lévy (1925) distribution, prior to the clear emergence of econophysics. Alfred Lotka (1926) saw scientific discoveries as following this pattern. George Zipf (1941) would see city sizes as doing so. Benoit

Mandelbrot (1963) saw cotton prices doing so and was inspired to discover fractal geometry from studying the mathematics of the scaling property (Mandelbrot, 1983; 1997). Ijiri and Simon (1977) saw firm sizes also following this pattern, a result more recently confirmed by Axtell (2001).

Economists doing econophysics?

Also, economists would move to use statistical mechanics models to study a broader variety of economic dynamics prior to the emergence of econophysics as such. Those doing so included Hans Föllmer (1974), Lawrence Blume (1993), Steven Durlauf (1993), William Brock (1993), Duncan Foley (1994) and Michael Stutzer (1994), with Durlauf (1997) providing an overview of an even broader set of applications. However, by 1993 the econophysicists were fully active even if they had not yet identified themselves by this term.

While little of this work explicitly focuses on generating outcomes consistent with scaling laws, it is certainly reasonable to expect that many of them could. It is true that the more traditional view of efficient markets with all agents possessing full information rational expectations about a single stable equilibrium is not maintained in these models, and therefore the econophysics critique carries some weight. However, many of these models do make assumptions of at least forms of bounded rationality and learning, with the possibility that some agents may even conform to the more traditional assumptions. Stutzer's (1994) reconciles the maximum entropy formulation of Gibbsian statistical mechanics with a relatively conventional financial economics formulation of the Black–Scholes options formula, based on Arrow–Debreu contingent claims (Arrow, 1974). Brock and Durlauf (2001) formalize heterogeneous agents socially interacting within a utility maximizing, discrete choice framework. Neither of these specifically generates scaling law outcomes, but there is nothing preventing them from doing so potentially.

While some econophysicists seek to integrate their findings with economic theory, as noted above many seek to replace conventional economic theory, seeing it as useless and limited. An irony in this effort is that it has been argued that conventional neoclassical economic theory itself was substantially a result of importing 19th-century physics conceptions into economics, with not all observers approving of this (Mirowski, 1989). The culmination of this effort is seen by many as being Paul Samuelson's *Foundations of Economic Analysis* (1947), whose undergraduate degree was in physics at the University of Chicago. Samuelson himself noted approvingly that Irving Fisher's 1892 dissertation (1926) was partly supervised by the pioneer of statistical mechanics, J. Willard Gibbs (1902), and as far back as 1801 Nicholas-François Canard conceived of supply and demand ontologically being contradicting 'forces' in a physics sense. So the interplay between economics and physics has been going on for far longer and is considerably more complicated than is usually conceived.

Related trans-disciplinary movements

Curiously but unsurprisingly given the tremendous attention given to the new econophysics movement, it has spawned imitators since 2000 in the form of *econo-chemistry* and *econobiology*, although these have not had nearly the same degree of development. The former term is the title of a course of study established at the University of Ulm by Barbara Mez-Starke, and was used to describe the work of Hartmann and Rössler (1998) at a conference in 2002 in Urbino, Italy (see also Padgett, Lee and Collier, 2003, for a more recent effort). The latter term first appeared in Hens (2002), although McCauley (2004, pp. 196–9) dismisses it as not a worthy competitor for econophysics. Nevertheless, there has long been a tradition among economists of advocating drawing more from biology for inspiration than from physics (Hodgson, 1993), going back at least as far as Alfred Marshall's famous declaration that economics is 'a branch of biology broadly interpreted' (Marshall, 1920, p. 637), even as Marshall's actual analytical apparatus arguably drew more from physics than from biology.

In any case, one trend we can expect for some time is an increase in coauthoring between economists and physicists within the area of econophysics (Lux and Marchesi, 1999; Li and Rosser, 2004). Very likely we shall eventually see the more useful ideas of econophysics coming to be absorbed into economics proper. As that comes to pass, it may also come to pass that the separate and distinct movement we now know as econophysics will cease to exist and will be forgotten, just as most economists do not think about the physics roots of standard neoclassical economic theory today.

J. BARKLEY ROSSER, JR.

See also **economy as a complex system; evolutionary economics; Gibrat, Robert Pierre Louis; Gibrat's Law; inequality (global); inequality (measurement); lognormal distribution; nonlinear time series analysis; Pareto distribution; Pareto, Vilfredo; power laws; redistribution of income and wealth; science, economics of; systems of cities; transfer of technology; urban growth.**

Bibliography

Anderson, P., Arrow, K. and Pines, D., eds. 1988. *The Economy as an Evolving Complex System*. Redwood City, CA: Addison-Wesley.

Arrow, K. 1974. *Essays in the Theory of Risk Bearing*. Amsterdam: North-Holland.

Arthur, W., Durlauf, S. and Lane, D., eds. 1997. *The Economy as an Evolving Complex System II*. Redwood City, CA: Addison-Wesley.

Axtell, R. 2001. Zipf distribution of firm sizes. *Science* 293, 1818–20.

Bachelier, L. 1900. Théorie de la spéculation. *Annales Scientifique de l'École Normale Supérieure* III-17, 21–86. (English translation in P. Cootner, ed. *The Random Character of Stock Market Prices*. Cambridge, MA: MIT Press, 1964.)

Bak, Per. 1996. *How Nature Works: The Science of Self-Organized Criticality*. New York: Copernicus Press for Springer-Verlag.

Bak, P., Chen, K., Scheinkman, J. and Woodford, M. 1993. Aggregate fluctuations from independent sectoral shocks: self-organized criticality in a model of production and inventory dynamics. *Ricerche Economiche* 47, 3–30.

Black, F. and Scholes, M. 1973. The pricing of options and corporate liabilities. *Journal of Political Economy* 81, 637–54.

Blume, L. 1993. The statistical mechanics of strategic interaction. *Games and Economic Behavior* 5, 387–424.

Botazzi, G. and Secchi, A. 2003. A stochastic model of firm growth. *Physica A* 324, 213–19.

Bouchaud, J.-P. and Cont, R. 1998. A Langevin approach to stock market fluctuations and crashes. *European Physical Journal B* 6, 543–50.

Bouchaud, J.-P. and Mézard, M. 2000. Wealth condensation in a simple model of economy. *Physica A* 282, 536–45.

Brock, W. 1993. Pathways to randomness in the economy: emergent nonlinearity and chaos in economics and finance. *Estudios Económicos* 8, 3–55.

Brock, W. and Durlauf, S. 2001. Discrete choice with social interactions. *Review of Economic Studies* 68, 235–60.

Canard, N.-F. 1801. *Principes d'Économie Politique*. Rome: Edizioni Bizzarri, 1969.

Canning, D., Amaral, L., Lee, Y., Meyer, M. and Stanley, H. 1998. A power law for scaling the volatility of GDP growth rates with country size. *Economics Letters* 60, 335–41.

Chakrabarti, B. 2005. Econphys-Kolkata: a short story. In *Econophysics of Wealth Distributions*, ed. A. Chatterjee, S. Yarlagadda and B. Charkrabarti. Milan: Springer.

Chatterjee, A., Yarlagadda, S. and Charkrabarti, B., eds. 2005. *Econophysics of Wealth Distributions*. Milan: Springer.

Clementi, F. and Gallegati, M. 2005. Power law tails in the Italian Personal income distribution. *Physica A* 350, 427–38.

Drăgulescu, A. and Yakovenko, V. 2001. Exponential and power-law probability distributions of wealth and income in the United Kingdom and the United States. *Physica A* 299, 213–21.

Durlauf, S. 1993. Nonergodic economic growth. *Review of Economic Studies* 60, 349–66.

Durlauf, S. 1997. Statistical Mechanics approaches to socioeconomic behavior. In *The Economy as a Complex Evolving System II*, ed. W. Arthur, S. Durlauf and D. Lane. Redwood City, CA: Addison-Wesley.

Einstein, A. 1905. Über die von der molekularkinetischen Theorie der Wärme geforderte Bewegung von der ruhenden Flüssigkeiten suspendierten Teichen. *Annalen der Physik* 17, 549–60.

Farmer, J. and Joshi, S. 2002. The price dynamics of common trading strategies. *Journal of Economic Behavior and Organization* 49, 149–71.

Fisher, I. 1926. *Mathematical Investigations into the Theory of Value and Prices*. New Haven: Yale University Press.

Foley, D. 1994. A statistical equilibrium theory of markets. *Journal of Economic Theory* 62, 321–45.

Föllmer, H. 1974. Random economies with many interacting agents. *Journal of Mathematical Economics* 1, 51–62.

Gabaix, X. 1999. Zipf's law for cities: an explanation. *Quarterly Journal of Economics* 114, 739–67.

Gibbs, J. 1902. *Elementary Principles in Statistical Mechanics*. New Haven: Yale University Press.

Gibrat, R. 1931. *Les Inégalités Économiques*. Paris: Sirey.

Gopakrishnan, P., Plerou, V., Amaral, L., Meyer, M. and Stanley, H. 1999. Scaling of the distributions of fluctuations of financial market indices. *Physical Review E* 60, 5305–16.

Hartmann, G. and Rössler, O. 1998. Coupled flare attractors – a discrete prototype for economic modelling. *Discrete Dynamics in Nature and Society* 2, 153–9.

Hens, T. 2002. Evolutionary portfolio theory. *Asset allocation Almanac: Special Report #4*. Merrill Lynch. Online. Available at http://www.evolutionaryfinance.ch/uploads/media/MerrillLynch.pdf, accessed 22 May 2006.

Hodgson, G. 1993. *Economics and Evolution: Bringing Life Back into Economics*. Ann Arbor: University of Michigan Press.

Ijiri, Y. and Simon, H. 1977. *Skew Distributions and the Sizes of Business Firms*. Amsterdam: North-Holland.

Levy, M. and Solomon, S. 1997. New evidence for the power-law distribution of wealth. *Physica A* 242, 90–4.

Lévy, P. 1925. *Calcul des Probabilités*. Paris: Gauthier-Villars.

Li, H. and Rosser, J., Jr. 2004. Market dynamics and stock price volatility. *European Physical Journal B* 39, 409–13.

Lotka, A. 1926. The frequency distribution of scientific productivity. *Journal of the Washington Academy of Sciences* 12, 317–23.

Lux, T. and Marchesi, M. 1999. Scaling and criticality in a stochastic multi-agent model of a financial market. *Nature* 397, 498–500.

Majorana, E. 1942. Il valore delle leggi statistiche nelle fisica e nelle scienze sociali. *Scientia* 36, 58–66.

Mandelbrot, B. 1963. The variation of certain speculative prices. *Journal of Business* 36, 394–419.

Mandelbrot, B. 1983. *The Fractal Geometry of Nature*. San Francisco: W.H. Freeman.

Mandelbrot, B. 1997. *Fractals and Scaling in Finance*. New York: Springer-Verlag.

Mantegna, R. 1991. Lévy walks and enhanced diffusion in Milan stock exchange. *Physica A* 179, 232–42.

Mantegna, R. and Stanley, H. 2000. *An Introduction to Econophysics: Correlations and Complexity in Finance*. Cambridge: Cambridge University Press.

Marshall, A. 1920. *Principles of Economics*, 8th edn. London: Macmillan.

McCauley, J. 2004. *Dynamics of Markets: Econophysics and Finance*. Cambridge: Cambridge University Press.

Mirowski, P. 1989. *More Heat than Light: Economics as Social Physics, Physics as Nature's Economics*. Cambridge: Cambridge University Press.

Osborne, M. 1959. Brownian motion in stock markets. *Operations Research* 7, 145–73.

Padgett, J., Lee, D. and Collier, N. 2003. Economic production as chemistry. *Industrial and Corporate Change* 12, 843–77.

Pareto, V. 1897. *Cours d'Économie Politique*. Paris and Lausanne. Trans. A. Schwier, *Manual of Political Economy*. New York: Kelly, 1971.

Plerou, V., Amaral, L., Gopakrishnan, P., Meyer, M. and Stanley, H. 1999. Similarities between the growth dynamics of university research and competitive economic activities. *Nature* 400, 433–37.

Rosser, J., Jr. 1994. Dynamics of emergent urban hierarchy. *Chaos, Solitons & Fractals* 4, 553–62.

Samuelson, P. 1947. *Foundations of Economic Analysis*. Cambridge, MA: Harvard University Press.

Sornette, D. 2003. *Why Stock Markets Crash: Critical Events in Complex Financial Systems*. Princeton: Princeton University Press.

Sornette, D. and Johansen, A. 2001. Significance of log-periodic precursors to financial crashes. *Quantitative Finance* 1, 452–71.

Sornette, D. and Zajdenweber, D. 1999. Economic returns of research: the Pareto law and its implications. *European Physical Journal B* 8, 653–64.

Spitzer, F. 1971. *Random Fields and Interacting Particle Systems*. Providence: American Mathematical Society.

Stanley, H., Afanasyev, V., Aamaral, L., Buldyrev, S., Goldberger, A., Havlin, S., Leschhorn, H., Maass, P., Mantegna, R., Peng, C.-K., Prince, P., Salinger, M., Stanley, M. and Viswanathan, G. 1996. Anomalous fluctuations in the dynamics of complex systems: from DNA and physiology to econophysics. *Physica A* 224, 302–21.

Stanley, M., Amaral, L., Buldyrev, S., Havlin, S., Leschhorn, H., Maass, P., Salinger, M. and Stanley, H. 1996. Scaling behavior in the growth of companies. *Nature* 379, 804–6.

Stutzer, M. 1994. The statistical mechanics of asset prices. In *Differential Equations, Dynamical Systems, and Control Science: A Festschrift in Honor of Lawrence Markus*, vol. 152, ed. K. Elworthy, W. Everitt and E. Lee. New York: Marcel Dekker.

Takayasu, H. and Okuyama, K. 1998. Country dependence on company size distributions and a numerical model based on competition and cooperation. *Fractals* 6, 67–79.

Zipf, G. 1941. *National Unity and Disunity*. Bloomington, IN: Principia Press.

Eden, Frederick Morton (1766–1809)

The son of Sir Robert Eden, F.M. Eden was educated at Oxford, gaining a Master's degree in 1789. A co-founder of the Globe Insurance Company, he published in 1797 the three volumes of his investigation into the conditions of the labouring poor, *The State of the Poor*. This work was perhaps the most detailed appraisal of social legislation and its actual workings that had appeared, and the findings provided ample material for ensuing debate on the best form of dealing with poverty and pauperism. In the years that followed Eden wrote a number of pamphlets on related issues.

The greater part of *The State of the Poor* records Eden's findings relating to the actual conditions prevailing in the parishes of England. Stimulated by the high prices prevailing in 1794–5, Eden initially set out to study the condition of the poor, but later extended this to the labouring classes. He encountered at times great resistance from local parish authorities, but despite this he was able to gather a considerable amount of information on wage levels, diet and prices. This was linked to an appraisal of the nutritional value of available foodstuffs, such that it was possible to arrive at some kind of comparative assessment of levels of poverty and want. It emerged from his empirical findings that the actual conditions and treatment of the poor varied greatly from parish to parish, this in part reflecting the patchwork of legislation that had grown up over the years in relation to the pauper and the workless. He argued however that existing legislation implied a policy of support for the indigent, and that in general a civilized society had an obligation to make such provision.

K. TRIBE

Selected work

1797. F. M. Eden, *The State of the Poor: Or an History of the Labouring Classes in England*, 3 vols. London.

Edgeworth, Francis Ysidro (1845–1926)

Biographia

Francis Ysidro Edgeworth (1845–1926) was born in Edgeworthstown in County Longford, Ireland. The background into which he was born was dominated by the 'larger than life' figure of his grandfather Richard Lovell Edgeworth (1744–1817), whose life was documented in a two-volume memoir (1820) by his oldest daughter, the famous novelist Maria Edgeworth (1767–1849). Richard Lovell's many scientific and mechanical experiments were helped by his strong association with the Lunar Society of Birmingham, whose members included Watt, Bolton, Wedgwood, Priestley, Darwin and Galton. In addition, Maria's scientific acquaintances included Davy, Humboldt, Herschel, Babbage, Hooker and Faraday. The

marriage of F. Y. Edgeworth's cousin Harriet Jessie Edgeworth (daughter of Richard Lovell's seventh and youngest son Michael Pakenham, 1812–81) to Arthur Gray Butler provided links with another large and eminent academic family. These connections extend even further since A. G. Butler's sister, Louisa Butler, married Francis Galton, a cousin of Charles Darwin.

Richard Lovell's sixth son, and 17th surviving child, was Francis Beaufort Edgeworth (1809–46), who met his wife, Rosa Florentina Eroles, the daughter of a Spanish refugee from Catalonia and then aged 16, while on the way to Germany to study philosophy; they married within three weeks in 1831. F. Y. Edgeworth was their fifth son. With his family background and his knowledge of French, German, Spanish and Italian, Edgeworth had wide international sympathies. On the family background, see Butler and Butler (1927) and for a full-length treatment of Edgeworth's work, see Creedy (1986).

Edgeworth was educated by tutors in Edgeworthstown until the age of 17, when in 1862 he entered Trinity College Dublin to study languages. In 1867 Edgeworth entered Exeter College, Oxford, but after one term transferred to Magdalen Hall. He transferred to Balliol in 1868, where in Michaelmas 1869 he obtained a first in Literae Humaniares. He was called to the bar in 1877, the same year in which his first book, *New and Old Methods of Ethics*, was published. Edgeworth applied unsuccessfully for a professorship of Greek at Bedford College, London, in 1875, but later lectured there on English language and literature for a brief period from late 1877 to mid-1878. He had earlier lectured on logic, mental and moral sciences and metaphysics to prospective Indian civil servants, at a private institution run by a Mr Walter Wren. In 1880 he applied for a chair of philosophy, also unsuccessfully, but began lecturing on logic to evening classes at King's College London. Soon after the publication of his second book, *Mathematical Psychics*, in 1881, he applied for a professorship of logic, mental and moral philosophy and political economy at Liverpool. Testimonials for two of Edgeworth's applications were given by Jevons (see Black, 1977, v, pp. 98, 145) and Marshall.

Edgeworth had to wait until 1890 until he obtained a professorial appointment: this was at King's College London, where he succeeded Thorold Rogers in the Tooke Chair of Economic Science and Statistics. In the next year, 1891, he again succeeded Rogers, this time to become Drummond Professor and Fellow of All Souls' College, Oxford, a position he held until his retirement in 1922. Edgeworth therefore finally settled in Oxford at the age of 46 in what was to become one of the most illustrious British chairs in economics. At the same time he became the first editor of the *Economic Journal*. He was editor or co-editor from its first issue until his death. He was supported by Henry Higgs from 1892 to 1905, when the latter became the Prime Minister's Private

Secretary, with further assistance provided at a later stage by Alfred Hoare. Keynes was a co-editor for 15 years. After a tremendously creative period of the late 1870s and 1880s, Edgeworth had become firmly established as the leading economist, after Marshall, in Britain.

In addition to his work in economics, Edgeworth began a series of statistical papers in 1883. He was President of section F of the British Association in 1889, a position he held again in 1922. Edgeworth's work on mathematical statistics played an increasingly important role. Indeed, of about 170 papers which he published, approximately three-quarters were concerned with statistical theory. He became a Guy Medalist (Gold) of the Royal Statistical Society in 1907 and was President of the Society during 1912–14. His main contributions to statistics concern work on inference and the law of error, the correlation coefficient, transformations (what he called 'methods of translation'), and the 'Edgeworth expansion'. The latter, a series expansion which provides an alternative to the Pearson family of distributions, has been widely used (particularly since the work of Sargan, 1976) to improve on the central limit theorem in approximating sampling distributions. It has also been used to provide support for the bootstrap in providing an Edgeworth correction. Edgeworth's work in probability and statistics has been collected by McCann (1996). His third and final book was *Metretike: or the Method of Measuring Probability and Utility* (1887). These contributions are not examined here; see Bowley (1928) and Stigler (1978).

Approach to economics

A dominant characteristic of Edgeworth's approach to economics is that it is mathematical, characterized by an original use of techniques, although he does not appear to have received a formal training in mathematics. However, he came to economics from moral philosophy. The central question of distributive justice, rather than simply the application of mathematics, dominated his attitude towards economics. His main argument was that mathematics provided powerful assistance to 'unaided' reason, and could check the conclusions reached by other methods. Thus:

> He that will not verify his conclusions as far as possible by mathematics, as it were bringing the ingots of common sense to be assayed and coined at the mint of the sovereign science, will hardly realise the full value of what he holds, will want a measure of what it will be worth in however slightly altered circumstances, a means of conveying and making it current. (1881, p. 3)

Edgeworth's approach contrasts sharply with that of Marshall. The contrast between Edgeworth and Marshall was neatly summarized by Pigou as follows:

> During some thirty years until their recent deaths in honoured age, the two outstanding names in English economics were Marshall ... and Edgeworth ... Edgeworth, the tool-maker, gloried in his tools ... Marshall, on the other hand, had what almost amounted to an obsession for hiding his tools away. (Pigou and Robertson, 1931, p. 3)

Although both men turned to economics from mathematics and moral philosophy, Marshall generally used biological analogies, and was concerned with developing maxims. In contrast, Edgeworth generally used mechanical analogies, and was more concerned with developing theorems.

In the 1880s and 1890s the deductive method encountered a great deal of criticism, especially from the 'Historical School' of economists. Edgeworth's defence of the deductive method often involved showing how other economists had advocated its use. His interest in the natural sciences often led him to make comparisons with scientific laws, and especially to show that the physical sciences also relied on abstraction and approximation.

Edgeworth argued carefully that the assumptions used in economics are often untestable, and he therefore took precautions against the accusation of 'plucking assumptions from the air'. He was conscious of the fact that the difficulty is in making the crucial abstractions which make the particular problem under consideration tractable, but which are not question begging. His attitude to many a priori assumptions was directly related to his approach to statistical inference. In *Mathematical Psychics*, for example, he referred to 'the first principle of probabilities, according to which cases about which we are equally undecided ... count as equal' (1881, p. 99). This was then transferred to economics. The appropriate assumption was that all feasible values, say, of elasticities, were equally likely, until evidence is obtained. Hence, 'There is required, I think ... in order to override the a priori probability, either very definite specific evidence, or the consensus of high authorities' (1925, ii, pp. 390-391). This also illustrates Edgeworth's attitude to authority and his many allusions to the views of other leading economists. Price (1946, p. 38) referred to his frequent 'reference to authority for ... support of tentative opinion waveringly advanced'.

Edgeworth was also prone to stress negative results. For example, in discussing taxation, where the criterion of minimum sacrifice does not alone provide a simple tax formula, he stated:

> Yet the premises, however inadequate to the deduction of a definite formula, may suffice for a certain negative conclusion. The ground which will not serve as the foundation of the elaborate edifice designed may yet be solid enough to support a battering-ram capable of being directed against simpler edifices in the neighbourhood. (1925, p. 261)

Edgeworth's position as editor of the *Economic Journal* enabled him to combine both his critical attitude and his appetite for a wide range of reading. He contributed 32 book reviews, and in sending books to other reviewers he would include 'apposite remarks on particular points in the text' (Bowley, 1934, p. 123). These reviews should also be placed beside his 17 reviews in the Academy, and 131 articles in the original *Palgrave's Dictionary of Political Economy*. Furthermore, Edgeworth's later articles in the *Economic Journal*, such as those on international trade and on taxation, took the form of extended commentaries on contemporary work.

Early work in moral philosophy

Before turning to economics, Edgeworth published a brief note in *Mind* in 1876, and his first (privately printed) book on *New and Old Methods of Ethics* in 1877. The description by Keynes of Edgeworth's first book could just as well be applied to his other two books:

> Edgeworth's peculiarities of style, his brilliance of phrasing, his obscurity of connection, his inconclusiveness of aim, his restlessness of direction, his courtesy, his caution, his shrewdness, his wit, his subtlety, his learning, his reserve – all are there full-grown. Quotations from the Greek tread on the heels of the differential calculus. (Keynes, 1972, p. 257)

The main focus of this early work, strongly influenced by the great Cambridge philosopher Henry Sidgwick (1838–1900), was to examine in detail the implications of utilitarianism for the optimal distribution of resources. Edgeworth's special and original contribution was to apply advanced mathematics to this problem. Edgeworth's approach was dominated by his utilitarianism, but the influence of contemporary psychological research and the impact of evolutionary ideas can also be traced. Both aspects led to explicit consideration of differences between individuals and changes which take place over time.

Edgeworth was also influenced by the major fierce debates in the last half of the 19th century between egoism, evolutionism, idealism, intuitionism, and of course utilitarianism. His brand of utilitarianism became extremely eclectic, and embraced the majority of the above principles (except for those of the Hegelian idealists) while regarding utilitarianism as the 'sovereign principle'. His note in *Mind* discussed Matthew Arnold's views of Joseph Butler, who had examined egoism at great length. Arnold had argued that Butler's term 'self love' should be interpreted to mean 'the pursuit of our temporal good'. However, Edgeworth argued that egoism and utilitarianism could be subsumed under the same principle. He believed Butler to be saying, 'duty and interest are perfectly coincident; for the most part in this world, but entirely and in every instance, if we take in the future and the whole' (1876, p. 571).

Edgeworth generally distinguished between 'impure' and 'pure' utilitarianism. In the latter case individuals are assumed to be concerned with the welfare of society as a whole. The former case in fact corresponds more closely with a 'short term' version of egoism. Economic exchange can usefully be analysed in terms of 'jostling egoists', but he believed that ultimately individuals would evolve to become pure utilitarians. A reason for believing that individuals would make such a transition was later to be developed by Edgeworth in the form of his contractarian justification of utilitarianism as the appropriate principle of distributive justice.

Edgeworth's early utilitarianism was influenced by his wide knowledge of work in experimental psychology. In his books of 1877 and 1881 there are many references to the work of Delboeuf, Fechner, Helmholtz, Weber and Wundt. These references occur in the context of discussing the nature of utility functions and, although Edgeworth at this time was not aware of the earlier work of Jevons, the same range of psychological work was also important to Jevons. Edgeworth in 1877 explicitly suggested, in connection with Fechner, that an additive form would not be appropriate.

A further aspect of Edgeworth's utilitarianism is his attitude towards authority. An important issue for early utilitarians involved the nature of inductive evidence about the consequences of acts. Most people cannot know the full consequences of their acts, so that rules of moral conduct must be followed (in contrast with intuitionism where individuals are assumed to have immediate consciousness of moral rules). In arriving at such rules, the opinions of highly regarded individuals are taken to be credible though it may not be possible to show conclusively that they are 'correct'. Edgeworth argued, for example, that 'we ought to defer even to the undemonstrated dicta and opinions of the wise, who have a power of mental vision acquired by experience' (1925, ii, p. 149).

Edgeworth defined the problem of determining the optimal utilitarian distribution as follows: 'given a certain quantity of stimulus to be distributed among a given set of sentients … to find the law of distribution productive of the greatest quantity of pleasure' (1877, p. 43). In treating this problem mathematically Edgeworth used Lagrange multipliers, without any explanation, and concluded that, 'unto him that hath greater capacity for pleasure shall be added more of the means of pleasure' (1877, p. 43). In using Lagrange multipliers Edgeworth was also careful to discuss possible complications, referring to the possibility of multiple solutions and explicitly discussing corner solutions and inequality constraints.

Further complexities were then examined, where Edgeworth emphasized that utilitarianism implies equality of the 'means of pleasure' only under a special set of assumptions, and in the general case the prescribed solution will be some form of inequality. In dealing with the distribution of effort, he argued not surprisingly that most work should be provided by those most capable of providing it. In a yet more general treatment of the problem, Edgeworth used the calculus of variations, but again provided the reader with virtually no help in following his mathematical argument. Edgeworth's analysis of the utilitarian optimal distribution was continued in his paper on 'The Hedonical Calculus' (1879), which was later reprinted as the third part of *Mathematical Psychics*.

Early work in economics

The turning point in Edgeworth's work was his introduction to Jevons in 1879 by a mutual friend James Sully, who in 1878 moved to Hampstead, where Edgeworth had lodgings in Mount Vernon and where Jevons also lived; see Sully (1918, pp. 180, 223). His first knowledge of Marshall came from Jevons, who 'highly praised the then recently published Economics of Industry' (in Pigou, 1925, p. 66). Edgeworth became interested in the problem of the indeterminacy of the rate of exchange, arising from the existence of only a small number of transactors. This led rapidly to Edgeworth's second and most important book *Mathematical Psychics: An Essay on the Application of Mathematics to the Moral Sciences* (1881), which was clearly written in a state of considerable enthusiasm for his new subject. This slim volume of 150 pages was known only to a small group of experts. Marshall's review began, 'this book shows clear signs of genius, and is a promise of great things to come' (Whitaker, 1975, p. 265). Jevons began by stating that 'whatever else readers of this book may think about it, they would probably all agree that it is a very remarkable one' (1881, p. 581). It was not until the middle of the 20th century that many of its central ideas began to be more fully appreciated.

Part 1 of *Mathematical Psychics* (1881, pp. 1–15) was devoted mainly to a justification of the use of mathematics in economics where precise data are not available. There is probably no other 'apology' in the whole of economic literature which compares with Edgeworth's plea for the application of mathematics. For example, when considering individual utility maximization:

> Atoms of pleasure are not easy to distinguish and discern; more continuous than sand, more discrete than liquid; as it were nuclei of the just-perceivable, embedded in circumambient semi-consciousness. We cannot count the golden sands of life; we cannot number the 'innumerable smile' of seas of love; but we seem to be capable of observing that there is here a greater, there a less, multitude of pleasure-units; mass of happiness; and that is enough. (1881, pp. 8–9)

Great stress was placed on comparison with Lagrange's 'principle of least action' in examining the overall effects produced by the interactions among many particles. The connection with Edgeworth's analysis of competition, involving interaction among a large number of competitors to produce a determinate rate of exchange, is central here. The fact that in the natural sciences so much could

be derived from a single principle was important for both Jevons and Edgeworth. But Edgeworth took this to its ultimate limit in arguing that the comparable single principle in social sciences, that of maximum utility, would produce results of comparable value. Referring to Laplace's massive work, *Mécanique Céleste*, he suggested that:

> 'Mécanique Sociale' may one day take her place along with 'Mécanique Celeste' [sic], throned each upon the double-sided height of one maximum principle, the supreme pinnacle of moral as of physical science … the movements of each soul, whether selfishly isolated or linked sympathetically, may continually be realising the maximum energy of pleasure, the Divine love of the universe. (1881, p. 12)

Jevons's work in the *Theory of Political Economy* involved the application of very basic mathematics and of psychological research to the analysis of exchange in competitive markets. In addition to this direct stimulus, Edgeworth was also influenced by an anonymous review of Jevons's book in the *Saturday Review* (1871).

The crucial development following Edgeworth's contact with Jevons was not simply the realization that mathematics could be used to examine equilibrium in exchange. Rather, it was that in his analysis Jevons explicitly assumed, through his 'law of indifference', that all individuals take the equilibrium prices as given, that is, outside their control. In using this law as 'one of the central pivots of the theory', Jevons stated that, 'there can only be one ratio of exchange of one uniform commodity at any moment' (1871, p. 87). His theory was explicitly limited to the static equilibrium conditions. He deliberately excluded the role of the number of competitors from his analysis via the awkward notion of the 'trading body', following correspondence with Fleeming Jenkin (1833–85), who raised the question of indeterminacy with just two traders; see Black (1977, iii, pp. 166–78). Jenkin could not see why two isolated individuals should accept the price-taking equilibrium, whereas Jevons wished to consider the behaviour of two typical individuals in a large market.

In a section on 'Failure of the Laws of Exchange', Jevons discussed cases in which some indeterminacy would result. His most notable example was of house sales, where it was suggested that indeterminacy would result from the discrete nature of the good being exchanged. The *Saturday Review* article took exception to this, suggesting that indeterminacy 'is really owing in our opinion to the assumed absence of competition' (see Black, 1981, p. 157). The stress on indeterminacy was also influenced by Marshall's discussion of wage bargaining: Edgeworth (1881, p. 48 n.1) referred to Thornton's comparison of the determination of prices in Dutch and English auctions, and cited Alfred and Mary Paley Marshall's joint book on the *Economics of Industry* (1879).

It was this gap in Jevons's analysis that Edgeworth set out to fill. His achievement was to show the conditions under which competition between buyers and sellers, through a barter process, leads to a 'final settlement' which is equivalent to one in which all individuals act independently as price takers. As he later stated (1925, p. 453), 'the existence of a uniform rate of exchange between any two commodities is perhaps not so much axiomatic as deducible from the process of competition in a perfect market'.

Exchange and contract

Having argued that 'the conception of Man as a pleasure machine may justify and facilitate the employment of mechanical terms and Mathematical reasoning in social science' (1881, p. 15), Edgeworth moved on to the analysis of the 'economical calculus', the starting point of which was the assumption that 'every agent is actuated only by self-interest' (1881, p. 16).

In modern economic analysis the analytical tools invented by Edgeworth in 1881, such as the indifference map and the contract curve, are now used in a vast range of contexts. They were introduced by Edgeworth to examine the nature of barter among individuals. He wanted to see if a determinate rate of exchange would be likely to result in barter situations where it is assumed only that individuals wish to maximize their own utility, considered solely as a function of their own consumption. With full knowledge of individuals' utility functions, and their initial endowments of goods, would it be possible to work out a 'determinate' rate of exchange at which trade would take place? Edgeworth's direct statement of the problem is as follows:

> The PROBLEM to which attention is specially directed in this introductory summary is: How far contract is indeterminate – an inquiry of more than theoretical importance, if it show not only that indeterminateness tends to [be present] widely, but also in what direction an escape from its evils is to be sought. (1881, p. 20)

Edgeworth began his analysis of this problem by taking the simplest case of two individuals exchanging fixed quantities of two goods. The basic framework is that described by Jevons, where the first individual holds all of the initial stocks of the first good, and the second individual holds all the stocks of the second good. He wrote the utility functions of each individual in terms of the amounts exchanged rather than consumed, using the general utility function ('utility is regarded as a function of the two variables, not the sum of two functions of each', 1881, p. 104). He then immediately defined the contract curve and indifference curves, in that order.

In the sentence which follows Edgeworth's introduction of the general utility function, he raised the question of the equilibrium which may be reached with 'one or both refusing to move further'. In barter the conditions of

exchange must be reached by voluntary agreement, or contract, between the two parties, and of course it is fundamental that no egoist would agree to a contract which would make him worse off than before the exchange. The question thus concerns the nature of the settlement reached by two contracting parties. He immediately answered that contract supplies only part of the answer so that 'supplementary conditions ... supplied by competition or ethical motives' are required, and then wrote the equation of his famous contract curve (1881, pp. 20–1).

The problem of obtaining the equilibrium values of x and y which, 'cannot be varied without the consent of the parties to it' was stated as follows: 'It is required to find a point (x, y) such that, in whatever direction we take an infinitely small step, $[U_A]$ and $[U_B]$ do not increase together, but that, while one increases, the other decreases' (1881, p. 21). The locus of such points 'it is here proposed to call the contract-curve'. Edgeworth's alternative derivations of the contract curve involved the movement, from an arbitrary position, along one person's indifference curve; 'motion is possible so long as, one party not losing, the other gains' (1881, p. 23). He thus used the Lagrange multiplier method of maximizing one person's utility subject to the condition that the other person's utility remains constant.

In the diagram drawn by Edgeworth (1881, p. 28) he did not use a box construction. Furthermore the only indifference curves shown fully were those which each individual is able to reach in isolation, and which therefore specify the limits beyond which each is not prepared to move. Also part of the offer or reciprocal demand curves of each individual were drawn on the same diagram, although they were not defined until ten pages later.

After presenting the results for the two-person two-good case, Edgeworth (1881, p. 26) examined the contract curve in the case where three individuals exchange three goods, stated that it is given by the 'eliminant', and then gave three lines of three sets of partial derivatives. In fact, the contract curve in this context is defined by $\left| \frac{\partial U_i}{\partial x_j} \right| = 0$, where $\frac{\partial U_i}{\partial x_j}$, is the marginal utility of person i with respect to good j, but Edgeworth did not use the modern notation for determinants and did not set the Jacobian equal to zero. This early use of determinants in economics would probably have confused many of his readers.

The problem of indeterminacy

The concepts of indifference curves and the contract curve therefore help to specify a range of 'efficient exchanges' of goods between individuals. The essential feature of the analysis from Edgeworth's point of view is precisely that there is a range rather than a unique point: 'the settlements are represented by an indefinite number of points' (1881, p. 29). At any particular settlement, the rate of exchange is expressed simply in terms of the amount of one good which is given up in order to obtain a specified amount of the other good. Hence the existence of a range of efficient contracts means that the rate of exchange is 'indeterminate'. The rate of exchange achieved in practice will thus depend to a large extent on bargaining strength. It was this result which led Edgeworth to make his often quoted remark that 'an accessory evil of indeterminate contract is the tendency, greater than in a full market, towards dissimulation and objectionable arts of higgling' (1881, p. 30).

Edgeworth argued that his analysis of indeterminacy in contract between two traders could be applied to a very wide variety of contexts. In particular, the tendency of large groups to form 'combinations', as in the case of trade unions and employers' associations, would serve to increase the extent of indeterminacy. The general applicability of his analysis of contract and indeterminacy was summarized by Edgeworth as follows:

> What it has been sought to bring clearly into view is the essential identity (in the midst of diversity of fields and articles) of contract; a sort of unification likely to be distasteful to those excellent persons who are always dividing the One into the Many, but do not appear very ready to subsume the Many under the One. (1881, p. 146; Plato's expression 'the one in the many' was later used by Marshall as the motto for his 1919 book on *Industry and Trade*.)

Having shown the possibilities of indeterminacy, Edgeworth then went on to show how 'the escape from its evils' requires either competition or arbitration.

Competition and the number of traders

The central question which Edgeworth was trying to resolve in the second part of *Mathematical Psychics* was that of the conditions necessary to remove the indeterminacy which exists in the case of barter between two traders. The question naturally arises as to the extent to which this indeterminacy is the result of the absence of competition in the simple two-person market. Edgeworth thus quickly moved on to the introduction of further traders.

In Edgeworth's earlier problem of two traders exchanging two goods, the definition of a range of efficient exchanges (along the contract curve) is of course analytically separate from the question of whether or not two isolated traders would actually reach a settlement on the contract curve. However, these two aspects were not clearly separated by Edgeworth because at the beginning of his analysis he introduced his stylized description of the process of barter: this is the famous 'recontracting' process. Edgeworth did not wish to assume that individuals initially have perfect knowledge. Instead, he supposed that, 'There is free communication throughout a normal competitive field. You might suppose the

constituent individuals collected at a point, or connected by telephones – an ideal supposition, but sufficiently approximate to existence or tendency for the purposes of abstract science' (1881, p. 18). The knowledge of the other traders' dispositions and resources could be obtained by the formation of tentative contracts which are not assumed to involve actual transfers, and can be broken when further information is obtained. Edgeworth introduced this in typical style:

> 'Is it peace or war?' asks the lover of 'Maud', of economic competition, and answers hastily: it is both, pax or pact between contractors during contract, war, when some of the contractors without the consent of others recontract. (1881, p. 17; the allusion here is to Alfred Tennyson's poem *Maud: A Monodrama*, part 1, verse VII.)

An important role of the recontracting process is thus to disseminate information among traders. It allows individuals who initially agree to a contract, which is not on the contract curve, to discover that an opportunity exists for making an improved contract according to which at least one person gains without another suffering.

However, the real importance of the recontracting process lies in the fact that it allows for Edgeworth's analysis of the role of the number of individuals in a market. With numerous individuals, the recontracting process makes it possible to analyse the use of collusion among some of the traders. Individuals are allowed to form coalitions in order to improve bargaining strength. Recontracting enables the coalitions to be broken up by outsiders who may attract members of a group away with more favourable terms of exchange.

Edgeworth's analysis was extremely terse and the following discussion does not therefore follow his own presentation. The analysis begins by introducing a second person A and a second person B. The new traders are assumed to be exact replicas of the initial pair, with the same tastes and endowments. This simplification is useful because the dimensions of the Edgeworth box and the utility curves are identical for each pair of traders. Hence, it enables the same diagram to be used as in the case when only two traders are considered in isolation. Two basic points can be stated immediately. First, in the final settlement all individuals will be at a common point in the Edgeworth box. Second, the settlement must be on the contract curve. The first point arises because if two individuals have identical tastes then their total utility is maximized by sharing their resources equally. It is useful to consider other types of contract which will eventually be broken, in order to illustrate the way in which the introduction of additional traders provides a role for some kind of competitive process.

The major question at issue is whether the range of indeterminacy along the contract curve is reduced by the addition of these traders. Consider Figure 1 and suppose that when A_1 and B_1 are trading independently of A_2 and

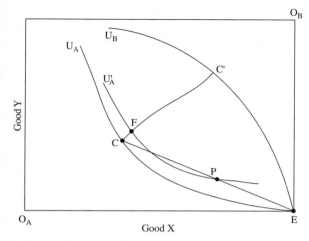

Figure 1   Two pairs of traders

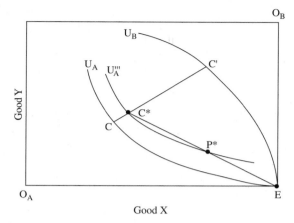

Figure 2   The new limit to the contract curve

B_2, trader B_1 has all the bargaining power and is able to appropriate all the gains from trade by pushing A_1 to the limit of the contract curve at point C. Suppose also that the same applies to A_2 and B_2. If the two pairs of traders are then able to communicate with each other, A_2 can now simply refuse to trade with B_2 at C. With no transaction costs, A_2 was previously indifferent between trading at C and consuming at the endowment point, E. This endowment position is effectively the 'threat point' of the As: it is the position in which they would find themselves if the bargaining process were to break down. But A_2 no longer needs to remain in isolation after refusing to trade with B_2, and instead can trade with A_1, after A_1 has traded with B_1 at C and has therefore obtained some of good Y. The two As can share their stocks of X and Y equally, arriving at point P; such an equal division maximizes their total utility.

By reaching point P, halfway between C and E, the convexity of the indifference curves implies that they are both better off than anywhere on the no-trade indifference curve. The two As would be on a higher common indifference curve, and thus better-off, if they could consume at a point along the CPE which is to the north-west of point P. However, they do not have enough resources to move beyond the halfway point P.

Trader B_2, who has been isolated, cannot prevent such a bargain. Thus B_1 is at C, both As are at P and B_2 is at the initial endowment point E. In this situation B_1 has no incentive to change, but B_2 has a strong incentive to offer a better deal to one of the As than the one offered by trader B_1. So long as B_2 offers one of the As, say A_2, a trade on the contract curve which allows A_2 to reach a higher indifference curve than U'_A, the initial agreement with B_1 will be broken and recontracting will take place.

The implication is that the ability of the As to turn to someone else, rather than deal with a single trader, means that the Bs now compete against each other. However,

trader B_1, who cannot prevent the recontracting, has an incentive to make yet a better offer. Hence, the recontracting process continues. The stylized process of recontracting with the two Bs competing against each other will produce a final settlement at the point C^* in Figure 2. This has the property that the indifference curve U'''_A passes through C^* and P^*, where P^* is halfway between C^* and E. This means that the two As are indifferent between C^* and P^*, and since they cannot both reach any point between C^* and P^* along the line C^*E, they are unable to improve on C^*. Hence there is no need to leave one of the Bs in isolation and the two Bs will trade with the two As at point C^*.

This argument has shown that at the final settlement all traders are at a common point on the contract curve and the limit has moved inwards along the old contract curve. The analysis can be repeated by starting with an alternative situation whereby the As are initially assumed to be able to appropriate all the gains from trade. The point C' would then no longer qualify as a point on the new contract curve. The introduction of the additional pair of traders means that the contract curve shrinks, and the range of indeterminacy involved in barter is correspondingly reduced.

The extent to which the contract curve shrinks when the additional pair of traders is introduced is influenced by the fact that the As cannot get further than halfway along a ray from a point on the contract curve to the endowment position. However, if there are three pairs of As and three pairs of Bs, the repetition of the above analysis involves two of the As dealing with two of the Bs at a point on the contract curve. The two As then share their resources equally with the remaining A while the third B is isolated. The As are able to consume together at a point which is two-thirds of the way along the ray from the initial endowment position to the point on the contract curve where the trade involving the two As and two Bs takes place.

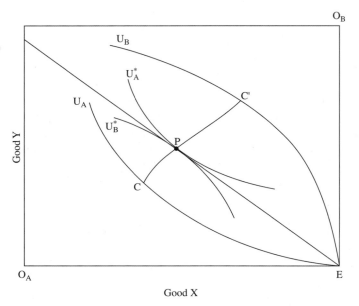

Figure 3   Final settlement with many traders

With N pairs, the As can reach a proportion $\frac{(N-1)}{N}$ of the way from the endowment point to the contract curve. Thus as N increases, the values of k approaches unity. This means that the As can reach all the way from E to the contract curve, so that the final settlement must be such that the indifference curve is tangential to the ray from the origin. A final settlement with many traders is therefore shown in Figure 3 as point P on the contract curve. The effect of working in from the point C' would lead to an equivalent result for an indifference curve of the Bs, shown as U_B^*.

The result is that the final settlement looks just like a price-taking equilibrium. The figure illustrates the case where there is a single price-taking equilibrium. If there are multiple equilibria, the recontracting process causes the number of final settlements, with sufficiently large N, to shrink to the number of price-taking equilibria. (For discussion of utility functions involving multiple equilibria, and comparison of bargaining, competitive and utilitarian solutions, see Creedy, 1994a.) This argument relating to the shrinking contract curve, first established by Edgeworth, is often referred to as the *limit theorem*.

After Edgeworth's terse discussion, he stated:

> If this reasoning does not seem satisfactory, it would be possible to give a more formal proof; bringing out the important result that the common tangent to both indifference curves … is the vector from the origin. (1881, p. 38)

The price-taking solution is necessarily on the contract curve. This gives rise to what is now referred to as the 'first fundamental theorem' of welfare economics – that a

price-taking equilibrium is Pareto efficient. Furthermore, the use of price-taking provides a considerable reduction in the amount of information required by traders when compared with the recontracting process. Given an equilibrium set, individuals need to know only the prices of goods, whereas in the recontracting process they have to learn a considerable amount of information about other individuals' preferences and endowments. But Edgeworth placed more stress on the equivalence of the competitive price-taking solution with a recontracting barter process involving large numbers.

Given that coalitions among traders are allowed in the recontracting process, a price-taking equilibrium cannot be blocked by a coalition of traders. In this sense the competitive equilibrium is robust. The argument that a complex process of bargaining among a large number of individuals produces a result which replicates a price-taking equilibrium, allowing for the free flow of information using recontracting and enabling coalitions of traders to form and break up, is an important result that is far from intuitively obvious. The recontracting process can be said to represent a competitive process, and the contract curve shrinks essentially because of the competition between suppliers of the same good, although it is carried out in a barter framework in which explicit prices are not used (although rates of exchange are equivalent to price ratios).

The price-taking equilibrium, in contrast, does not actually involve a competitive process. Individuals simply believe that they must take market prices as given and outside their control. They respond to those prices without any reference to other individuals. But the result is that the price-taking equilibrium looks just like a situation in which all activity is perfectly coordinated.

Edgeworth suggested that similar results apply when some of the assumptions are relaxed. Thus, 'when we suppose plurality of natures as well as persons, we have to suppose a plurality of contract-curves … Then, by considerations analogous to those already employed, it may appear that the quantity of final settlements is diminished as the number of competitors is increased' (1881, p. 40). He then briefly considered different numbers of As and Bs, concluding that 'the theorem admits of being extended to the general case of unequal numbers and natures' (1881, p. 43). However, some of the results do not hold in the general case; for example, equality within the group of As no longer holds when there are unequal numbers of As and Bs. A considerable number of articles have been written, since the late 1950s, examining various aspects of the Edgeworth recontract model under different assumptions.

Reciprocal demand curves

It has been mentioned that Edgeworth included in his diagram (1881, p. 28) the reciprocal demand curve, or offer curve, of each individual, although such curves were then called 'demand-and-supply curves'. Edgeworth mentioned them only briefly in the text (1881, p. 39), but the lack of emphasis is understandable since in imperfect competition they are not relevant. Edgeworth's contribution was to provide the basic 'analytics' of the offer curve in terms of indifference curves, whereby it is 'the locus of the point where lines from the origin touch curves of indifference' (1881, p. 113).

When there is a lack of competition, giving rise to indeterminacy, there is nothing to ensure that individuals will trade on their offer curves and, as Edgeworth argued, 'the conceptions of demand and supply at a price are no longer appropriate' (1881, p. 31). It is this general preference, in favour of the analysis of barter in non-competitive situations, to which Marshall objected and which led to the controversy discussed below.

The utilitarian calculus

Having shown how indeterminacy can be removed by increasing the number of traders, Edgeworth turned to consider the role of arbitration in resolving the conflict between traders, in a 'world weary of strife' (1881, p. 51). The principle of arbitration examined was, not surprisingly, the utilitarian principle, which Edgeworth had earlier used to examine the optimal distribution. However, the new context of indeterminacy led him to a deeper justification of utilitarianism as a principle of distributive justice. Having arrived at this new link between 'impure' (egoistic) and 'pure' utilitarianism, Edgeworth had only to reorientate his earlier analysis of optimal distribution, contained in his paper in *Mind* of 1879.

The need for arbitration with indeterminacy had been stated by Jevons as follows:

> The dispositions and force of character of the parties … will influence the decision. These are motives more or less extraneous to a theory of economics, and yet they appear necessary considerations in this problem. It may be that indeterminate bargains of this kind are best arranged by an arbitrator or third party. (1871, pp. 124–5)

Edgeworth's statement of the same point was as usual rather less prosaic: 'The whole creation groans and yearns, desiderating a principle of arbitration, and end of strifes' (1881, p. 51). Edgeworth argument involved two steps. First, he showed that the principle of utility maximization places individuals on the contract curve, because the first-order conditions are equivalent to the tangency of indifference curves.

> It is a circumstance of momentous interest that one of the in general indefinitely numerous settlements between contractors is the utilitarian arrangement … the contract tending to the greatest possible total utility of the contractors. (1881, p. 53)

Edgeworth recognized that this result was not sufficient to justify the use of utilitarianism as a principle of arbitration. It is only a necessary condition of a principle of arbitration that it should place the parties somewhere on the contract curve. Edgeworth's justification for utilitarianism as a principle of justice, comparing points along the contract curve, was as follows:

> Now these positions lie in a reverse order of desirability for each party; and it may seem to each that as he cannot have his own way, in the absence of any definite principle of selection, he has about as good a chance of one of the arrangements as another … both parties may agree to commute their chance of any of the arrangements for … the utilitarian arrangement. (1881, p. 55)

The important point to stress about this statement is that Edgeworth clearly viewed distributive justice in terms of choice under uncertainty. He argued that the contractors, faced with uncertainty about their prospects, would choose to accept an arrangement along utilitarian lines. A crucial component of this argument, also clearly stated by Edgeworth in this quotation, is the use of equal a priori probabilities.

The importance to him of this new justification of utilitarianism cannot be exaggerated. Indeed the whole of *Mathematical Psychics* seems to be imbued with a feeling of excitement generated by his discovery of a justification based on a 'social contract'. This provided the crucial link between 'impure' and 'pure' utilitarianism in a more satisfactory way than his earlier appeal to evolutionary forces.

Edgeworth believed that he had provided an answer to an age-old question, stating 'by what mechanism the force of self-love can be applied so as to support the structure of utilitarian politics, neither Helvetius, nor Bentham, nor any deductive egoist has made clear' (1881, p. 128). Nevertheless this argument was neglected until restatements along similar lines were made by Harsanyi (1953; 1955) and Vickrey (1960). The maximization of expected utility, with each individual taking the a priori view that any outcome is equally likely, was shown to lead to the use of a social welfare function which maximizes the sum of individual utilities. This approach is now usually described as 'contractarian neo-utilitarianism'.

In discussing the utilitarian solution as a principle of arbitration in indeterminate contract, Edgeworth did not clearly indicate in 1881 that the utilitarian solution of maximum total utility could specify a position which makes one of the parties worse off than in the no-trade situation. This was nevertheless later made explicit when, after proposing arbitration along utilitarian lines, he added 'subject to the condition that neither should lose by the contract' (1925, ii, p. 102). This possibility of course depends largely on the initial endowments of the individuals.

Later work in economics

After the publication of *Mathematical Psychics*, Edgeworth concentrated increasingly on mathematical statistics, in particular on the problem of statistical inference, but, following his appointment to the Drummond Chair at Oxford, Edgeworth again made important contributions to economics, although this work mainly involved reactions to, and discussions arising from, the later work of other authors.

Demand and exchange

In the *Principles of Economics* (1890, Appendix F) Marshall included a brief discussion of Edgeworth's analysis of barter, and produced a figure showing the contract curve. During the following year, in the course of a review written in Italian (translated in Edgeworth, 1925, ii, pp. 315–19), Edgeworth criticized Marshall for not having dealt sufficiently with the problem of indeterminacy. The basic problem was that Marshall, using a model in which a series of trades are allowed to take place at disequilibrium prices, believed he had shown that prices will eventually settle at the price-taking equilibrium. However, the argument was not transparent. The adjustment process involves moving from the initial endowment point in a series of trades, where trading at 'false' prices is allowed at each step. The process must conclude with both individuals at a point on the contract curve. A feature of the process is the assumption that each stage or iteration of the sequence involves Pareto improvements: individuals trade only if it makes them better off. Furthermore, it involves trading at the 'short end' of the market, that is, the minimum of supply and demand. This arises from the impossibility of forcing any

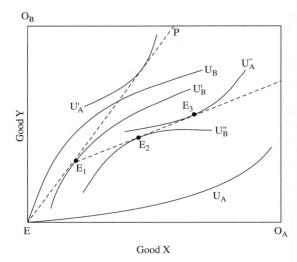

Figure 4 Disequilibrium trades

individual either to buy or sell more than desired at any price.

An example of two disequilibrium trades is shown in Figure 4, where the endowment moves from E to E_1, and then to E_2. With a price line represented by EP, there is an excess supply of good X as person A tries to reach the indifference curve U'_A and person B wishes to reach U'_B. Trade takes place at E_1, the short end of the market. Point E_1 then becomes the new endowment point. At the second trading stage, the price of X must be lowered to induce person B to purchase more. At a price represented by the line E_1P_1 through the new endowment point, the excess supply is lower than formerly and trade takes place at E_2. Comparing U'_A and U'_B with U''_A and U''_B respectively, it can be seen that E_2 is a Pareto improvement relative to E_1. It is also clear that person A is better off the slower the fall is in the price of X relative to Y at each stage.

The combination of Pareto-efficient moves at each stage and an adjustment process such that an excess supply leads to a price reduction, and vice versa, produces a stable process that converges to an equilibrium somewhere on the contract curve. (This type of sequence of disequilibrium trades was later used by Launhardt; see Creedy, 1994b.)

The basic problem was that Marshall believed that his assumption of an additive utility function, combined with the assumption that the marginal utility of one good is constant for both individuals, guaranteed a determinate price, if the good having constant marginal utility was money. Indeed, this case was mentioned by Edgeworth (see 1925, ii, p. 317 n.1). The contract curve is a straight line parallel to the y axis (where this good is the one with constant marginal utility), along which the rate of exchange is constant. So the equilibrium price does not depend on the sequence of trades. However,

Edgeworth's point was that the total amount spent on good x remains indeterminate.

There was a later, though much milder, disagreement between Marshall and Edgeworth over the so-called Giffen good. In a book review, Edgeworth argued that, 'even the milder statement that the elasticity of demand for wheat may be positive, though I know it is countenanced by high authority, appears to me so contrary to a priori probability as to require very strong evidence (1909, p. 104). The 'authority' was of course Marshall (1890, p. 132), who replied directly to Edgeworth that, 'I don't want to argue … But … the matter has not been taken quite at random' (Pigou, 1925, p. 438). Marshall gave a numerical example involving a journey travelled by two methods, where the distance travelled by the cheaper and slower method must increase when its price increases. For further details, see Creedy (1990).

It has been mentioned that Edgeworth introduced the generalized utility function. An implication is that it allows for complementarity, although Edgeworth did not explicitly consider this in 1881. The first formal definition of complementarity is attributed to Auspitz and Lieben, and it was used by Edgeworth in his paper on the pure theory of monopoly, and also by Pareto: this amounts to what is now called 'gross' complementarity, defined in terms of cross-price elasticities. It is also sometimes referred to, using the initials of the four people mentioned above, as ALEP complementarity.

The first major criticism came from Johnson (1913), who pointed out that the criterion was not invariant with respect to monotonic transformations of the utility function. His treatment was extended by Hicks and Allen (1934), so that the modern definition involves 'net' complements in terms of compensated price changes. There is no symmetry between gross substitutes and complements as only the matrix of (compensated) substitution elasticities is assumed to be symmetric.

Monopoly and oligopoly

In a paper first published in Italian in 1897, and not translated until the collected *Papers* (1925), Edgeworth examined several problems relating to monopoly. He began his discussion with Cournot's (1838) example of the 'source minérale' in which there are 'two monopolists' (that is, duopolists), each owning a spring of mineral water. It would be natural for Edgeworth to expect an indeterminate price in this 'small numbers' context. Cournot had arrived at a determinate solution for price and output, but Edgeworth showed that 'when two or more monopolists are dealing with competitive groups, economic equilibrium is indeterminate' (1925, p. 116). The daily output from each spring was assumed to be limited to identical fixed amounts, delivery costs were zero and all consumers had the same demand curve (purchasing one unit only of output). Hence demand is $n(1 - p)$ where n is the number of customers and p is the

price. Cournot's solution was that the price would be $p = 1/4$, but Edgeworth argued that one of the 'monopolists' had an incentive to raise the price back to $p = 1/2$, which is the revenue maximizing price, so that there is not a determinate price. He argued that:

> at every stage … it is competent to each monopolist to deliberate whether it will pay him better to lower his price against his rival as already described, or rather to raise it to a higher … for that remainder of customers of which he cannot be deprived by his rival. … Long before the lowest point has been reached, that alternative will have become more advantageous than the course first described' (1925, p. 120)

Edgeworth went on to say 'the matter may be put in a clearer light', and he then defined what are now called the reaction curve and isoprofit lines (in that order) for variations in prices. However, it was not until Bowley's (1924) discussion that these matters began to be presented in a more transparent manner.

Edgeworth then considered the case of complementary demand within the context of 'bilateral monopoly', where the two goods are demanded in fixed proportions for use in the production of a further article. An interesting feature is that he wrote the equations of the reaction curves and explicitly dealt with what are now called conjectural variations, reflecting the extent to which one duopolist is expected to change price in response to changes made by the second duopolist. In discussing this problem Edgeworth also introduced the further important concept of the 'saddle point', which he called the 'hog's back', clearly indicating its importance for stability.

The no-profit entrepreneur

Walras (1874, p. 225) had introduced the concept of the entrepreneur who neither gains nor loses. This result applied only to the competitive equilibrium, where there are no incentives for entrepreneurs to enter any industry. This does not of course mean that there are no profits, in the accounting sense, since the returns to homogeneous units of inputs of organization and management services are subsumed in the costs of the firm.

Edgeworth's criticisms of this concept of the no-profit entrepreneur, reproduced in his *Papers* (1925), recognized that with Walras's assumptions there was nothing illogical about the argument. The theory simply means that nothing remains 'after the entrepreneur has paid a normal salary to himself ' (1925, pp. 26, 30). Furthermore, 'if [the general expenses] are taken into account, the argument becomes a fortiori. For why should not a substantial remuneration for the entrepreneur be included in the general expenses of the business' (1925, ii, p. 469). Edgeworth's difference with Walras was to some extent 'only verbal', but he was also unhappy with the idea that entrepreneurship is homogeneous and divisible.

The theory of taxation

In the 1890s Edgeworth produced two surveys of considerable importance. These surveys, of the pure theory of taxation and of the pure theory of international values, were both published in the *Economic Journal* and subsequently reproduced (with alterations) in his *Papers* (1925, vol. ii). Each survey consisted of three separate parts, and displayed a staggering breadth of knowledge and command of the subject. They represent his most serious attempts to produce any kind of synthesis of a branch of economic literature. Edgeworth began his survey with the rather strong statement that 'the science of taxation comprises two subjects to which the character of pure theory may be ascribed; the laws of incidence, and the principle of equal sacrifice' (1925, p.64). He then considered a variety of special cases and contexts of tax incidence. The basic framework for incidence analysis was the simple partial equilibrium approach, still used in many basic textbooks, in which the incidence depends on the relative values of supply and demand elasticities.

The basic approach to incidence analysis actually stemmed from the important paper by Jenkin (1871). It suggests that in general the price of the taxed good will either remain constant (in the extreme case of inelastic supply) or will increase. However, this result ignores interrelationships among commodities. Edgeworth showed that, when such interrelationships are explicitly allowed, there are some circumstances in which the price of the taxed good will actually fall. When discussing this 'paradox', Edgeworth reproduced his argument which had in fact been explored in more detail in his paper on monopoly, published in Italian in the same year (translated in Edgeworth, 1925, i, pp. 111–42). Edgeworth first stated his 'tax paradox' in the following terms:

> when the supply of two or more correlated commodities – such as the carriage of passengers by rail first class or third class – is in the hands of a single monopolist, a tax on one of the articles – e.g. a percentage of first class fares – may prove advantageous to the consumers as a whole. ... The fares for all the classes might be reduced. (1925, p. 139)

Edgeworth regarded this result as an example of a situation where, 'the abstract reasoning serves as a corrective to what has been called the "metaphysical incumbus" of dogmatic laisser faire' (1925, i, p. 139; see also 1925, ii, pp. 93–4). Essentially the two commodities must be substitutes in consumption and production, and the result is partly brought about by the fact that the monopolist has an incentive to increase the supply of the untaxed commodity. Edgeworth also recognized that the result could occur in competitive markets (see 1925, p. 63). As with many of Edgeworth's original results, this tax paradox was not a subject of continuous development. Its main practical importance perhaps arises from the fact that in the early 1930s it attracted the attention of Hotelling

(1932). For further discussion of the paradox, see Creedy (1988).

The section of the taxation survey which attracted most immediate attention was Edgeworth's discussion of the various 'sacrifice' theories of the distribution of the tax burden, and his qualified support for progressive taxation. Edgeworth's attitude to taxation was similar to that of the major classical economists in that he rejected a benefit approach, on the argument that taxation is not an economic bargain governed by competition. Thus in his view the problem was to determine 'the distribution of those taxes which are applied to common purposes, the benefits whereof cannot be allocated to particular classes of citizens' (1925, p. 103). A principle of justice is thus required. His approach can be seen as marking a crucial stage in the transition towards a 'welfare economics' view of public finance, rather than using a special set of 'tax maxims' such as the famous criteria laid down by Adam Smith.

Not surprisingly, Edgeworth (1925, p. 102) argued along neo-contractarian lines set down in *Mathematical Psychics* that the utilitarian arrangement would be accepted by individuals uncertain of their own prospects and taking an equal a priori view of the probabilities. He suggested that

> each party may reflect that, in the long run of various cases ... of all the principles of distribution which would afford him now a greater, now a smaller proportion of the sum-total utility obtainable ... the principle that the collective utility should be on each occasion a maximum is most likely to afford the greatest utility in the long run to him individually

Having established the use of utilitarianism as a principle of distributive justice, Edgeworth then succinctly stated the main argument:

> The condition that the total net utility procured by taxation should be a maximum then reduces to the condition that the total disutility should be a minimum ... it follows in general that the marginal disutility incurred by each taxpayer should be the same. (1925, p. 103)

The implication is that, if all individuals have the same cardinal utility function, after-tax incomes would be equalized. Edgeworth also clearly recognized that, if there is considerable dispersion of pre-tax incomes relative to the total amount of tax to be raised, where there is 'not enough tax to go around' (1925, ii, p. 103), the equimarginal condition cannot be fully satisfied unless there is a 'negative income tax' which raises the incomes of the poorest individuals to a common level. Thus, 'the acme of socialism is for a moment sighted' (1925, p. 104). But Edgeworth immediately considered the practical limitations to such high progressive taxation. The following quotation illustrates one of Edgeworth's favourite metaphors, his respect for Sidgwick, his attitude to authority, his views on utilitarianism and the applicability of pure theory, and of course his unmistakable style:

> In this misty and precipitous region let us take Professor Sidgwick as our chief guide. He best has contemplated the crowning height of the utilitarian first principle, from which the steps of a sublime deduction lead to the high tableland of equality; but he also discerns the enormous interposing chasms which deter practical wisdom from moving directly towards that ideal. (1925, p. 104)

Among the various limitations, Edgeworth noted differences in individual utility functions, population effects, the disincentives to work, growth of culture and knowledge, savings, and of course the problem of evasion.

International trade

Edgeworth's survey of the pure theory of international values was in some ways responsible for a change of emphasis in the approach to trade theory, despite the fact that it contained few original analytical contributions. Indeed, he said that, 'Mill's exposition of the general theory is still unsurpassed' (1925, p. 20), and acknowledged further that, 'what is written ... after a perusal of [Marshall's] privately circulated chapters ... can make no claim to originality' (1925, p. 46). Edgeworth saw trade theory as an application of the general theory of exchange:

> The fundamental principle of international trade is that general theory ... the Theory of Exchange ... which ... constitutes the 'kernel' of most of the chief problems in economics. It is a corollary of the general theory that all the parties to a bargain look to gain by it ... This is the generalised statement of the theory of comparative cost. (1925, p. 6)

Thus the gains from trade are analogous to the gains from exchange in simple barter and 'It is useful ... to contemplate the theory of distribution as analogous to that of international trade proper' (1925, p. 19). Hence trade theory is to Edgeworth simply one more application of the general method of *Mathematical Psychics*. In directly applying the theory of exchange to that of trade, Edgeworth was quite content to use community indifference curves without clearly specifying how aggregation might be carried out. He said only that 'by combining properly the utility curves for all the individuals, we obtain what may be called a collective utility curve' (1925, p. 293).

One of Edgeworth's criticisms of Mill (1848) was that the latter took as his measure of the gain from trade the change in the ratio of exchange of exports against imports. Thus Mill in this case 'confounds "final" with integral utility' (1925, p. 22). The same point had in fact been made by Jevons (1871, pp. 154–6). However, Edgeworth, while preferring total utility, admitted that

Mill was not otherwise led to serious error in using his own measure.

Edgeworth's survey was, as always, extremely wide-ranging, though for later developments the most interesting parts are concerned with his elucidation of Mill's 'recognition of the case in which an impediment may be beneficial – or an improvement prejudicial – to one of the countries' (1925, p. 9). These cases would now be discussed under the headings of the 'optimal tariff' and 'immiserizing growth'. In the case of an optimal tariff, a country acts as monopolist and imposes a price which enables that country to attain its highest indifference curve, subject to the other country's offer curve. However, this position is not on the contract curve. The detailed specification of the optimum tariff in terms of elasticities had to wait until Bickerdike (1906), Pigou (1908) and the later revivals of interest in the 1940s. Edgeworth's judgement of Bickerdike was that he had 'accomplished a wonderful feat. He has said something new about protection' (1925, ii, p. 344).

Edgeworth could not of course be expected to support the use of such tariffs in practice. He acknowledged the possibility of retaliation, but also:

> For one nation to benefit itself at the expense of ... others is contrary to the highest morality ... But in an abstract study upon the motion of projectiles in vacuo, I do not think it necessary to enlarge upon the horrors of war. (1925, p. 17 n. 5)

The 'highest morality' was, of course, the principal of utilitarianism.

Conclusions

It has been seen that Edgeworth did not begin working and writing in economics until his mid-30s, but in common with the majority of neoclassical economists he soon pursued an academic career as a professor of economics. Indeed, in a period which saw the rapid and widespread professionalization of the subject Edgeworth held an academic position in England that was regarded as second only to that of Alfred Marshall. In spite of his wide range of reading and sympathies, Edgeworth's work was characterized by the fact that it was virtually all addressed to his fellow professional economists. So uncompromising was he in his view that economics is a very difficult subject offering only remote and nearly always negative policy advice that it may fairly be said that his work was addressed to just a small number of 'fellow travellers' in the rarefied atmosphere of the 'higher regions' of pure theory. However, Edgeworth imposed no geographical limitations, and with his considerable linguistic skills and international sympathies was in contact with the majority of leading economists around the world.

The distinguishing feature of the neoclassical 'revolution' was its emphasis on exchange as the central economic problem. The success of this shift of focus from production and distribution to exchange was closely associated with the fact that it had as its foundation a model based on utility maximization. This allowed for a deeper treatment of the gains from exchange and the wider considerations of economic welfare. Schumpeter summarized the point by stating that utility analysis must be understood in terms of exchange as the central 'pivot' and 'the whole of the organism of pure economics thus finds itself unified in the light of a single principle' (1954, p. 913). This is indeed the context in which Edgeworth's work in economics must be seen. Schumpeter's remark is merely a more prosaic expression of Edgeworth's view quoted above that '"Méchanique Sociale" may one day take her place along with "Méchanique Celeste" [sic], throned each upon the double-sided height of one maximum principle'. The central theme of Edgeworth's work is also clear in his revealing statement, taken from his presidential address to Section F of the Royal Society, that:

> It may be said that in pure economics there is only one fundamental theorem, but that is a very difficult one: the theory of bargain in a wide sense. (1925, ii, p. 288)

This perspective helps the major thread which runs through all Edgeworth's work in economics to be seen. His earlier mathematical analysis of the implications of utilitarianism for the optimal distribution, written before he turned to economics, was not only highly original (and esoteric) but laid the foundation for his work in economics. Thus, the transition from *New and Old Methods of Ethics* to *Mathematical Psychics* was not a shift in major preoccupations but rather a change of emphasis. Distribution was then seen as an important concomitant of exchange, so that the analysis of contract became central for Edgeworth. Edgeworth's emphasis on the indeterminacy (the inability of utility maximization alone to determine the rate of exchange, only a range of efficient exchanges) which results from the existence of a small number of traders led him to his path-breaking analysis of the role of numbers in competition, along with the efficiency properties of competitive equilibria.

The analysis of the utilitarian objective as an arbitration rule led Edgeworth directly to his new 'social contract' argument in explaining the acceptance of utilitarianism as a principle of social justice. It was the realization of this new justification of utilitarianism, using his newly developed analytical tools, which generated the excitement that is clearly evident in his first work in economics. While *Mathematical Psychics* developed the techniques of indifference curves and the contract curve within the 'Edgeworth box' – tools which are now ubiquitous in economic analysis – Edgeworth himself was clearly driven mainly by his ability to link the analysis of private contracts in markets to that of a social contract in which utilitarianism is the 'sovereign principle'. The integration of his analysis of barter, and the effects of the introduction of additional traders into the market, with

the demonstration that the utilitarian arrangement prescribes a point on the contract curve of efficient exchanges and is acceptable to risk-averse traders, was to Edgeworth nothing short of 'momentous'.

The results are of course highly abstract. In discussing their ultimate value suggested that:

> Considerations so abstract it would of course be ridiculous to fling upon the flood-tide of practical politics … it is at a height of abstraction in the rarefied atmosphere of speculation that the secret springs of action take their rise, and a direction is imparted to the pure foundation of youthful enthusiasm whose influence will ultimately affect the broad current of events. (1881, p. 128)

The intellectual pleasure derived from being able to draw together so many different subjects of analysis, and strands of his enormous range of learning, is clearly evident. However, it is precisely this wide field of vision, combined with the technical level and idiosyncratic style of writing, which made *Mathematical Psychics* so difficult for his contemporaries, and which continue to make the book seem so strange and yet so rewarding to the modern reader.

JOHN CREEDY

I am grateful to Denis O'Brien and Steven Durlauf for comments on an earlier draft of this article.

Selected works

1876. Mr. Matthew Arnold on Bishop Butler's doctrine of self love. *Mind* 1, 570–1.
1877. *New and Old Methods of Ethics: or 'Physical Ethics' and 'Methods of Ethics'*. Oxford: Parker.
1879. The hedonical calculus. *Mind* 4, 394–408.
1881. *Mathematical Psychics: An Essay on the Application of Mathematics to the Moral Sciences*. London: Kegan Paul.
1887. *Metretike, or the Method of Measuring Probability and Utility*. London: Temple.
1909. Review of Free Trade in Being. *Economic Journal* 19, 104–5.
1925. *Papers Relating to Political Economy*, 3 vols. London: Macmillan, for the Royal Economic Society.

Bibliography

Bickerdike, C. 1906. The theory of incipient taxes. *Economic Journal* 16, 529–35.
Black, R., ed. 1977. *Papers and Correspondence of William Stanley Jevons*. London: Macmillan, for the Royal Economic Society.
Black, R., ed. 1981. *Papers and Correspondence of William Stanley Jevons. Vol. VII, Papers on Political Economy*. London: Macmillan, for the Royal Economic Society.
Bowley, A. 1924. *The Mathematical Groundwork of Economics*. Oxford: Clarendon Press.
Bowley, A. 1928. *Edgeworth's Contribution to Mathematical Statistics*. London: Royal Statistical Society.

Bowley, A. 1934. Francis Ysidro Edgeworth. *Econometrica* 1, 113–24.

Butler, J. and Butler, H. 1927. *The Black Book of Edgeworthtown and other Edgeworth Memories 1585–1817*. London: Faber and Gwyer.

Cournot, A. 1838. *Researches into the Mathematical Principles of the Theory of Wealth*, trans. N. Bacon, ed. I. Fisher. London: Stechert-Hafner, 1927.

Creedy, J. 1986. *Edgeworth and the Development of Neoclassical Economics*. Oxford: Basil Blackwell.

Creedy, J. 1988. Wicksell on Edgeworth's tax paradox. *Scandinavian Journal of Economics* 90, 101–12.

Creedy, J. 1990. Marshall and Edgeworth. *Scottish Journal of Political Economy* 37, 18–39.

Creedy, J. 1994a. Exchange equilibria: bargaining, utilitarian and competitive solutions. *Australian Economic Papers* 33, 34–52.

Creedy, J. 1994b. Launhardt's model of exchange. *Journal of the History of Economic Thought* 16, 40–60.

Edgeworth, M. 1820. *Memories of Richard Lovell Edgeworth Esq., Begun by Himself and Concluded by his Daughter, Maria Edgeworth*, 2 vols. London: Hunter.

Harsanyi, J. 1953. Cardinal utility in welfare economics and in the theory of risk taking. *Journal of Political Economy* 61, 434–5.

Harsanyi, J. 1955. Cardinal welfare, individualistic ethics, and interpersonal comparisons of utility. *Journal of Political Economy* 63, 309–21.

Hicks, J. and Allen, R. 1934. A reconsideration of the theory of value. *Economica* 14, 52–76, 196–219.

Hotelling, H. 1932. Edgeworth's taxation paradox and the nature of demand and supply functions. *Journal of Political Economy* 40, 577–616.

Jenkin, F. 1871. On the principles which regulate the incidence of taxes. Reproduced in *Readings in the Economics of Taxation*, ed. R. Musgrave and C. Shoup. London: Allen and Unwin, 1959.

Jevons, W. 1881. Review of *Mathematical Psychics*. *Mind* 6, 581–3.

Jevons, W. 1871. *The Theory of Political Economy*, 5th edn, ed. H. Jevons. New York: Augustus Kelly, 1957.

Johnson, W. 1913. The pure theory of utility curves. *Economic Journal* 23, 483–513.

Keynes, J.M. 1972. *Essays in Biography. Vol. X of the Collected Writings of Keynes*. London: Macmillan, for the Royal Economic Society.

McCann, C., ed. 1996. *F.Y. Edgeworth: Writings in Probability, Statistics and Economics*, 3 vols. Cheltenham: Edward Elgar.

Marshall, A. 1890. *Principles of Economics*, 2 vols, ed. C. Guillebaud (variorum edn). London: Macmillan. 1961.

Marshall, A. and Marshall, M. 1879. *Economics of Industry*. London: Macmillan.

Mill, J.S. 1848. *Principles of Political Economy*. Reprinted with editorial material by W. Ashley. London: Longmans, Green, 1920.

Pigou, A. 1908. *Protective and Preferential Import Duties*. London: Macmillan.

Pigou, A. (ed.) 1925. *Memorials of Alfred Marshall 1842–1924*. London: Macmillan.

Pigou, A. and Robertson, D. 1931. *Economic Essays and Addresses*. London: King.

Price, L. 1946. Memoirs and notes on British economists 1881–1946. MSS, Brotherton Library, University of Leeds.

Sargan, J. 1976. Econometric estimators and the Edgeworth expansion. *Econometrica* 44, 421–48.

Schumpeter, J. 1954. *History of Economic Analysis*. London: Allen and Unwin.

Stigler, S. 1978. Francis Ysidro Edgeworth, statistician. *Journal of the Royal Statistical Society, A* 141, 287–322.

Sully, J. 1918. *My Life and Friends*. London: Fisher Unwin.

Vickrey, W. 1960. Utility, strategy and social decision rules. *Quarterly Journal of Economics* 74, 507–35.

Walras, L. 1874. *Elements of Pure Economics*, trans. W. Jaffe. London: Allen and Unwin, 1954.

Whitaker, J. (ed.) 1975. *The Early Economic Writings of Alfred Marshall 1867–1890*. London: Macmillan.

education in developing countries

Most economists who study economic growth agree that an educated citizenry is necessary for sustained economic growth, and virtually all international development organizations concur (UNDP, 1990; World Bank, 2001), and so those organizations provide substantial financial resources and policy advice to promote education in developing countries. Yet in many developing countries, especially the poorest, many children leave school at a young age and learn little during the time they spend in school. These problems have led many economists and other social scientists to turn their attention to education in developing countries.

This article summarizes recent research on the factors that affect the amount of time that children spend in school and the factors that determine how much they learn during their time in school. Thus, it focuses on the factors that shape education outcomes as opposed to the impact of education on income, economic growth and other phenomena (for a recent assessment of the impact of education on other socio-economic outcomes, see Glewwe, 2002). This article also omits, due to space constraints, a discussion of estimation issues (see Glewwe, 2002, and Glewwe and Kremer, 2006, for thorough discussions of estimation problems and possible solutions).

Factors that determine years of schooling

In developing countries, parents usually decide how many years their children will attend school. Each year, parents consider the costs and expected benefits of an additional year of schooling and then enrol their children for another year if the expected benefits outweigh the estimated costs. The main costs are school fees and other payments required by schools, transportation and (occasionally) meals and housing, and the opportunity cost of the children's time. There may also be an additional, 'psychic' cost; some parents may dislike particular values that schools attempt to instil in students. For many parents, the largest of these costs is the value of their children's time; in developing countries, especially in rural areas, children's time is valuable because they can help in household farming activities.

The main benefits of schooling are the skills learned (which usually reap substantial monetary returns in the labour market), increased employment opportunities that come with educational credentials, and the direct satisfaction and social approval that parents receive from having educated children. While the decision rule to continue schooling when the benefits outweigh the costs would seem to hold as a tautology, there are circumstances in which children are not enrolled in school even when the economic benefits outweigh the costs. This could occur because the costs are incurred today while the benefits accrue over many years in the future. In particular, parents who have low incomes and cannot obtain credit may not send their children to school even though the present discounted value at prevailing interest rates is positive.

Given this type of decision making by parents, policies to increase school enrolment must focus on reducing the costs of schooling, increasing the benefits of education, or providing access to credit. Reductions in fees are easy to implement, and in some countries (such as Mexico) parents with low incomes receive monthly payments if their children are enrolled in school. Of course, this entails potentially large budgetary costs, so some governments try to limit fee exemptions and outright subsidies to households or communities that are particularly needy. Evidence from many developing countries indicates that reducing fees or providing payments conditional on school enrolment can lead to large increases in enrolment; studies in Honduras, Kenya, Mexico and Nicaragua document these impacts (see Glewwe and Kremer, 2006, for further details and references).

The main alternative policy for increasing school enrolment is to increase the expected returns. These returns will increase if the relative price of skilled labour increases, and if schools become more effective at providing academic skills. While some economists have shown that increased returns to education does raise school enrollment (Foster and Rosenzweig, 1996), most policy research has focused on what makes schools more efficient at raising students' skills. This research is discussed in the next section.

Three additional points regarding policies to increase years of schooling deserve attention. First, improvements in the health and nutritional status of both very young and school-age children are another potentially important route to increase the time that children spend in school (see Glewwe and Miguel, 2006, for a review of this literature). Second, many policy discussions presume that the main reason children are not in school is that no school is available, yet in most countries schools are available but parents opt not to enrol their children because they judge that the costs outweigh the benefits (see Glewwe and Zhao, 2005). Third, the role of credit constraints in determining years in school is an under-researched topic, in terms of both the impact of credit constraints and policies that could loosen those constraints.

Factors that determine student learning

In principle, student learning can be depicted as a production process in which student, household, teacher and school characteristics combine to produce students' academic skills. While the existence of an academic skills production function is true almost by definition, there are serious problems that confound attempts to estimate this process. The main problem is omitted variables bias: students, households, teachers and schools can vary in hundreds of ways, and no data-set contains all variables that are potentially important. Indeed, important factors such as student innate ability, teacher effort and parental encouragement are almost impossible to measure and likely to be correlated with the observed variables. This problem applies to virtually all studies based on retrospective (non-experimental) data; indeed, it is probably the main reason that different studies find very different results (the main alternative explanation is that educational production functions are very different in different countries). A second serious estimation problem is attenuation bias. Much of the data on students, households, teachers and schools has a substantial amount of measurement error. This typically leads to underestimation of the true impacts of variables, which may explain, at least in part, why many variables in estimates of the determinants of student learning are statistically insignificant.

In recent years economists and other social scientists have turned to natural experiments and randomized trials to estimate the impacts of particular school characteristics, policies and programmes on student academic achievement. Natural experiments result from institutions and policies that cause random variation in school or student characteristics, which can be used to analyse the impact of those characteristics on student learning (and on time spent in school). Randomized trials are controlled experiments designed by researchers and school officials that generate random variation in a school characteristic or policy, which again allows one to estimate the impact of the characteristic or policy on learning. Natural experiments are relatively rare, but in recent years randomized trials have been implemented in many countries in Africa, Asia and Latin America.

One of the first randomized trials was conducted in Nicaragua in the late 1970s. The results indicated that workbooks and radio instruction had significant impacts on pupils' math scores. In the Philippines in the early 1980s, provision of textbooks raised students' performance on academic tests, but in Kenya in the late 1990s the only effect of textbooks was among the better students, perhaps because the textbooks provided were too difficult for most students. Other randomized trials conducted in Kenya suggest little impact on test scores from reductions in class size, provision of flip charts, and provision of deworming medicine. On a more positive note, school meals in Kenya raised test scores in schools that had well-trained teachers, but not in schools with poorly trained teachers. In public schools in an urban area of India, a remedial education programme increased test scores at a relatively low cost. Finally, a computer-assisted learning programme in India also appears to have increased test scores. The positive impacts of radio education in Nicaragua and computer instruction in India suggest that using modern technologies may be particularly helpful in schools with weak teachers. (For citations and more detailed discussion, see Glewwe and Kremer, 2006.)

While natural experiments and especially randomized trials may seem to avoid the estimation problems that plague retrospective studies, more randomized studies are needed before general conclusions can be drawn that can guide policy in countries that have not yet had such studies. Moreover, randomized trials can also suffer from estimation problems. One problem is that parents of students in the control schools (or schools excluded from the evaluation) may try to enrol their children in the treatment schools. This may affect the results by increasing class size (if class size affects learning). This would not occur if the policy were implemented nation-wide. In addition, children who transfer into treatment schools may not be a random sample of the general student population. A related problem is that marginal students in the treatment schools are less likely to drop out (if the intervention raises student achievement), which leads to underestimation of the impact of the policy on learning if comparisons are made based on all students currently enrolled in school. A final problem with randomized trials is that the evaluation itself may lead the treatment group to change its behaviour, or the control group to change its behaviour, because both groups know that their results are being used in an evaluation.

In summary, recent research on education in developing countries has provided fairly convincing evidence of the impact on time in school and on learning for particular policies in particular countries. Many additional studies are currently under way, and as these results accumulate it is likely that general conclusions can be drawn. This should lead to better education policies, which will contribute to higher economic growth and, ultimately, a higher quality of life in developing countries.

PAUL GLEWWE

See also **development economics; education production functions; human capital; returns to schooling.**

Bibliography

Foster, A. and Rosenzweig, M. 1996. Technical change and human capital returns and investments: evidence from the Green Revolution. *American Economic Review* 86, 931–53.

Glewwe, P. 2002. Schools and skills in developing countries: education policies and socioeconomic outcomes. *Journal of Economic Literature* 40, 436–82.

Glewwe, P. and Kremer, M. 2006. Schools, teachers and education outcomes in developing countries. In *Handbook on the Economics of Education*, ed. E. Hanushek and F. Welch. Amsterdam: North-Holland.

Glewwe, P. and Miguel, E. 2006. The impact of child health and nutrition on education in less developed countries. In *Handbook of Agricultural Economics*, vol. 4, ed. R. Evenson and T. Schultz. Amsterdam: North-Holland.

Glewwe, P. and Zhao, M. 2005. Attaining universal primary completion by 2015: how much will it cost? Department of Applied Economics, University of Minnesota.

UNDP (United Nations Development Programme). 1990. *Human Development Report.* New York: UNDP.

World Bank. 2001. *World Development Report 2000/2001: Attacking Poverty.* Washington, DC: World Bank.

education production functions

A simple production model lies behind much of the analysis in the economics of education. The common inputs are things like school resources, teacher quality, and family attributes; and the outcome is student achievement. Knowledge of the production function for schools can be used to assess policy alternatives and to judge the effectiveness and efficiency of public provided services. This area is, however, distinguished from many because the results of analyses enter quite directly into the policy process.

Historically, the most frequently employed measure of schooling has been attainment, or simply years of schooling completed. The value of school attainment as a rough measure of individual skill has been verified by a wide variety of studies of labour market outcomes (for example, Mincer, 1970; Psacharopoulos and Patrinos, 2004). However, the difficulty with this common measure of outcomes is that it assumes a year of schooling produces the same amount of student achievement, or skills, over time and in every country. This measure simply counts the time spent in schools without judging what happens in schools – thus, it does not provide a complete or accurate picture of outcomes.

Recent direct investigations of cognitive achievement find significant labour market returns to individual differences in cognitive achievement (for example, Lazear, 2003; Mulligan, 1999; Murnane et al., 2000). Similarly, society appears to gain in terms of productivity; Hanushek and Kimko (2000) demonstrate that quality differences in schools have a dramatic impact on productivity and national growth rates. (A parallel line of research has employed school inputs to measure quality but has not been as successful. Specifically, school input measures have not proved to be good predictors of wages or growth.)

Because outcomes cannot be changed by fiat, much attention has been directed at inputs – particularly those perceived to be relevant for policy such as school resources or aspects of teachers.

Analysis of the role of school resources in determining achievement begins with the Coleman Report, the US government's monumental study on educational opportunity released in 1966 (Coleman et al., 1966). That study's greatest contribution was directing attention to the distribution of student performance – the outputs as opposed to the inputs.

The underlying model that has evolved as a result of this research is very straightforward. The output of the educational process – the achievement of individual students – is directly related to inputs that both are directly controlled by policymakers (for example, the characteristics of schools, teachers, and curricula) and are not so controlled (such as families and friends and the innate endowments or learning capacities of the students). Further, while achievement may be measured at discrete points in time, the educational process is cumulative; inputs applied sometime in the past affect students' current levels of achievement.

Family background is usually characterized by such socio-demographic characteristics as parental education, income, and family size. Peer inputs, when included, are typically aggregates of student socio-demographic characteristics or achievement for a school or classroom. School inputs typically include teacher background (education level, experience, sex, race, and so forth), school organization (class sizes, facilities, administrative expenditures, and so forth), and district or community factors (for example, average expenditure levels). Except for the original Coleman Report, most empirical work has relied on data constructed for other purposes, such as a school's standard administrative records. Based upon this, statistical analysis (typically some form of regression analysis) is employed to infer what specifically determines achievement and what is the importance of the various inputs into student performance.

Measured school inputs

The state of knowledge about the impacts of resources is best summarized by reviewing available empirical studies. Most analyses of education production functions have directed their attention at a relatively small set of resource measures, and this makes it easy to summarize the results (Hanushek, 2003). The 90 individual publications that appeared before 1995 contain 377 separate production function estimates. For classroom resources, only nine per cent of estimates for teacher education and 14 per cent for teacher–pupil ratios yielded a positive and statistically significant relationship between these factors and student performance. Moreover, these studies were offset by another set of studies that found a similarly negative correlation between those inputs and student achievement. Twenty-nine per cent of the studies found a positive correlation between teacher experience and student performance; however, 71 per cent still provided no support for increasing teacher experience (being either negative or statistically insignificant). Studies on the effect of financial resources provide a similar picture. These indicate that there is very weak support for the notion that simply providing higher teacher salaries or greater overall spending will lead to improved student performance. Per pupil expenditure has received the most attention, but only 27 per cent of studies showed a positive and significant effect. In fact, seven per cent even suggested that adding resources would harm student achievement. It is also important to note that studies involving pupil spending have tended to be the lowest-quality studies as defined below, and thus there is substantial reason to believe that even the 27 per cent figure overstates the true effect of added expenditure.

These studies make a clear case that resource usage in schools is subject to considerable inefficiency, because schools systematically pay for inputs that are not consistently related to outputs.

Study quality

The previous discussions do not distinguish among studies on the basis of any quality differences. The available estimates can be categorized by a few objective components of quality. First, while education is cumulative, frequently only current input measures are available, which results in analytical errors. Second, schools operate within a policy environment set almost always at higher levels of government. In the United States, state governments establish curricula, provide sources of funding, govern labour laws, determine rules for the certification and hiring of teachers, and the like. In other parts of the world, similar policy setting, frequently at the national level, affects the operations of schools. If these attributes are important – as much policy debate would suggest – they must be incorporated into any analysis of performance. The adequacy of dealing with these problems is a simple index of study quality.

The details of these quality issues and approaches for dealing with them are discussed in detail elsewhere (Hanushek, 2003) and only summarized here. The first problem is ameliorated if one uses the 'value added' versus 'level' form in estimation. That is, if the achievement relationship holds at different points in time, it is possible to concentrate on the growth in achievement and on exactly what happens educationally between those points when outcomes are measured. This approach ameliorates problems of omitting prior inputs of schools and families, because they will be incorporated in the initial achievement levels that are measured (Hanushek, 1979). The latter problem of imprecise measurement of the policy environment can frequently be ameliorated by studying performance of schools operating within a consistent set of policies – for example, within individual states in the USA or similar decision-making spheres elsewhere. Because all schools within a state operate within the same basic policy environment, comparisons of their performance are not strongly affected by unmeasured policies (Hanushek, Rivkin and Taylor, 1996).

If the available studies are classified by whether or not they deal with these major quality issues, the prior conclusions about research usage are unchanged (Hanushek, 2003). The best quality studies indicate no consistent relationship between resources and student outcomes.

An additional issue, which is particularly important for policy purposes, concerns whether this analytical approach accurately assesses the causal relationship between resources and performance. If, for example, school decision-makers provide more resources to those they judge as most needy, higher resources could simply signal students known for having lower achievement. Ways of dealing with this include various regression discontinuity or panel data approaches. When done in the case of class sizes, the evidence has been mixed (Angrist and Lavy, 1999; Rivkin, Hanushek and Kain, 2005).

An alternative involves the use of random assignment experimentation rather than statistical analysis to break the influence of sample selection and other possible omitted factors. With one major exception, this approach nonetheless has not been applied to understand the impact of schools on student performance. The exception is Project STAR, an experimental reduction in class sizes that was conducted in the US state of Tennessee in the mid-1980s (Word et al., 1990). To date, it has not had much impact on research or our state of knowledge. While Project STAR has entered into a number of policy debates, the interpretation of the results remains controversial (Krueger, 1999; Hanushek, 1999).

Magnitude of effects

Throughout most consideration of the impact of school resources, attention has focused almost exclusively on whether a factor has an effect on outcomes that is statistically different from zero. Of course, any policy consideration would also consider the magnitude of the impacts and where policies are most effective. Here, even the most refined estimates of, say, class size impacts does

not give very clear guidance. The experimental effects from Project STAR indicate that average achievement from a reduction of eight students in a classroom would increase by about 0.2 standard deviations, but only in the first grade of attendance in smaller classes (kindergarten or first grade) (see Word et al., 1990; Krueger, 1999). Angrist and Lavy (1999), with their regression discontinuity estimation, find slightly smaller effects in grade five and approximately half the effect size in grade four. Rivkin, Hanushek and Kain (2005), with their fixed effects estimation, find effects half of Project STAR in grade four and declining to insignificance by grade seven. Thus, from a policy perspective the alternative estimates are both small in economic terms when contrasted with the costs of such large class size reductions and inconsistent across studies.

Do teachers and schools matter?

Because of the Coleman Report and subsequent studies discussed above, many have argued that schools do not matter and that only families and peers affect performance. Unfortunately, these interpretations have confused measurability with true effects.

Extensive research since the Coleman Report has made it clear that teachers do indeed matter when assessed in terms of student performance instead of the more typical input measures based on characteristics of the teacher and school. When fixed effect estimators that compare student gains across teachers are used, dramatic differences in teacher quality are seen.

These results can also be reconciled with the prior ones. These differences among teachers are simply not closely correlated with commonly measured teacher characteristics (Hanushek, 1992; Rivkin, Hanushek and Kain, 2005). Moreover, teacher credentials and teacher training do not make a consistent difference when assessed against student achievement gains (Boyd et al., 2006; Kane, Rockoff and Staiger, 2006). Finally, teacher quality does not appear to be closely related to salaries or to market decisions. In particular, teachers exiting for other schools or for jobs outside of teaching do not appear to be of higher quality than those who stay (Hanushek et al., 2005).

Some conclusions and implications

The existing research suggests inefficiency in the provision of schooling. It does not indicate that schools do not matter. Nor does it indicate that money and resources never impact achievement. The accumulated research surrounding estimation of education production functions simply says there currently is no clear, systematic relationship between resources and student outcomes.

ERIC A. HANUSHEK

See also **human capital; local public finance; returns to schooling.**

Bibliography

Angrist, J.D. and Lavy, V. 1999. Using Maimondides' rule to estimate the effect of class size on scholastic achievement. *Quarterly Journal of Economics* 114, 533–75.

Boyd, D., Grossman, P., Lankford, H., Loeb, S. and Wyckoff, J. 2006. How changes in entry requirements alter the teacher workforce and affect student achievement. *Education Finance and Policy* 1, 176–216.

Coleman, J.S., Campbell, E.Q., Hobson, C.J., McPartland, J., Mood, A.M., Weinfeld, F.D. and York, R.L. 1966. *Equality of Educational Opportunity*. Washington, DC: US Government Printing Office.

Hanushek, E.A. 1979. Conceptual and empirical issues in the estimation of educational production functions. *Journal of Human Resources* 14, 351–88.

Hanushek, E.A. 1992. The trade-off between child quantity and quality. *Journal of Political Economy* 100, 84–117.

Hanushek, E.A. 1999. Some findings from an independent investigation of the Tennessee STAR experiment and from other investigations of class size effects. *Educational Evaluation and Policy Analysis* 21, 143–63.

Hanushek, E.A. 2003. The failure of input-based schooling policies. *Economic Journal* 113, F64–F98.

Hanushek, E.A., Kain, J.F., O'Brien, D.M. and Rivkin, S.G. 2005. The market for teacher quality. Working Paper No. 11154. Cambridge, MA: NBER.

Hanushek, E.A. and Kimko, D.D. 2000. Schooling, labor force quality, and the growth of nations. *American Economic Review* 90, 1184–208.

Hanushek, E.A., Rivkin, S.G. and Taylor, L.L. 1996. Aggregation and the estimated effects of school resources. *Review of Economics and Statistics* 78, 611–27.

Kane, T.J., Rockoff, J.E. and Staiger, D.O. 2006. What does certification tell us about teacher effectiveness? Evidence from New York City. Working Paper No. 12155. Cambridge, MA: NBER.

Krueger, A.B. 1999. Experimental estimates of education production functions. *Quarterly Journal of Economics* 114, 497–532.

Lazear, E.P. 2003. Teacher incentives. *Swedish Economic Policy Review* 10(3), 179–214.

Mincer, J. 1970. The distribution of labor incomes: a survey with special reference to the human capital approach. *Journal of Economic Literature* 8, 1–26.

Mulligan, C.B. 1999. Galton versus the human capital approach to inheritance. *Journal of Political Economy* 107(pt. 2), S184–S224.

Murnane, R.J., Willett, J.B., Duhaldeborde, Y. and Tyler, J.H. 2000. How important are the cognitive skills of teenagers in predicting subsequent earnings? *Journal of Policy Analysis and Management* 19, 547–68.

Psacharopoulos, G. and Patrinos, H.A. 2004. Returns to investment in education: a further update. *Education Economics* 12, 111–34.

Rivkin, S.G., Hanushek, E.A. and Kain, J.F. 2005. Teachers, schools, and academic achievement. *Econometrica* 73, 417–58.

Word, E., Johnston, J., Bain, H.P., DeWayne Fulton, B., Zaharies, J.B., Lintz, M.N., Achilles, C.M., Folger, J. and Breda, C. 1990. *Student/Teacher Achievement Ratio (STAR), Tennessee's K-3 Class Size Study: Final Summary Report, 1985– 1990*. Nashville: Tennessee State Department of Education.

educational finance

This article deals with the government-financed system of education in the United States, which is referred to as 'public' education. Educational finance in the United States is different from that of other nations, which typically fund education from national taxes. Within each American state, a substantial portion of education is financed by local governments, although the proportion financed locally has declined from 83.2 per cent in 1920 to 43.2 per cent in 2000.

The state–local system of finance stems from the history and geography of the United States and the federal nature of its government. The 50 states are, in the eyes of the national government, primarily responsible for education. In most states, implementation of this responsibility is delegated to local municipal corporations called 'school districts'. The school district is more than a local administrative agency of the state. It is a distinct political entity that usually has some correspondence with the geographic area of a municipality. The district, however, has a separate board of directors, which is locally elected. The board then selects a superintendent of schools to manage the district's education. Boards have the authority to levy taxes, which are almost always on property within their district, and spend the revenue they derive from them. The state government may prescribe curricular standards for public schools, but the method of achieving these standards is the responsibility of the local district.

School districts and school boards were once the most common form of local government in the United States, numbering about 200,000 in 1900. The number of school districts declined steadily throughout the 20th century, which can largely be accounted for by the consolidation of rural one-room school districts into larger units. By 1970, one-room schools were essentially extinct, and since 1970 the total number of school districts has declined only slightly, numbering about 16,000 at the beginning of the 21st century.

Despite their numerical decline in rural areas, there are many school districts in most metropolitan areas. Urban households that are already on the move for job-related reasons have the luxury of choosing a home within one of several school districts in most regions of the nation. Choosing among school districts and the resulting competition among districts to obtain residents is consistent with the model proposed by Tiebout (1956). Numerous tests of the Tiebout model indicate that the quality of schooling is important to most home buyers (Oates,

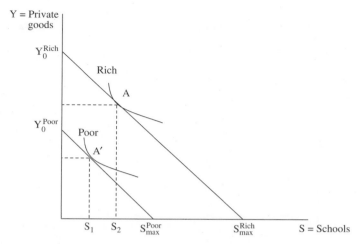

Figure 1 School spending in rich and poor districts

1969; Bradbury, Case and Mayer, 2001). There is also evidence that spatial competition makes school districts more efficient in delivering education services (Hoxby, 2000).

One-room schools of the 19th century were usually 'ungraded'. Students were instead divided into skill-specific recitation groups, formed without regard for chronological age. In this system, uniformity of education was not critical. New pupils could be placed according to what they knew in particular subjects rather than by age. But when almost all schools were age-graded, it paid for each district to offer an age-specific curriculum that allowed both teachers and pupils to be interchangeable among schools and districts (Fischel, 2006a).

Standardization of age-graded curricula became widespread by about 1940 and was brought about by two forces, one local and the other statewide. Property-owning voters in a given district would find that potential homebuyers would shun them if they did not offer a standard, public-school education. Voters would thus support taxes necessary to fund standardized schools. However, differences in the economic make-up and tax-bases of local districts sometimes made this difficult to do.

Figure 1 illustrates the problem for attempts to fund schools from local sources. It depicts a trade-off between local school spending and other goods for the median voter (the voter with the median income, assumed always to be in the majority in local elections) in two separate communities, a rich district and a poor district. The decisive voter chooses the mix of school spending and private goods that achieves the highest indifference curve that his private–public budget line allows (Bergstrom and Goodman, 1973). Because at the local level education is essentially a private good, the slope of the budget lines is the 'tax price' of school spending for the median voter in each community.

The tax price is not a tax rate. A school district composed exclusively of mansions will have, for a given level of spending, a much lower property tax *rate* than a district composed of modest-sized homes. But if the second moment of the distribution of wealth is the same in both communities, the *tax price* faced by the median voter in each will be the same. A 1,000 dollar increase in per-pupil spending will cost the median voter the same amount of money in both cases, if one assumes that the number of public-school children per household is the same in both.

The other generalization that Figure 1 illustrates is that average income of a district accounts for much of the differences in spending per pupil. Even though the tax prices are the same, the positive income elasticity of demand for education (estimated at somewhere between 0.5 and 1.0) causes the richer community to choose a higher level of school inputs (Bergstrom, Rubinfeld and Shapiro, 1982). While much of the criticism of these differences is based on equity concerns, there are efficiency reasons to promote a relatively uniform system of education (Benabou, 1996).

The way most states have attempted to equalize education opportunities is to reduce the tax price of spending in poorer districts. State funds (from statewide taxes) are offered to the poorer community in proportion to the district's own tax effort. The poorer median voter thus perceives, as indicated by the dotted budget line in Figure 2, that for every dollar raised locally, the state will send it another dollar. The tax price has been cut in half in the graphical example, so that the poorer community will choose to spend an amount closer to that of the richer district.

By manipulating the local tax price, state governments can in principle induce a substantial equality of school spending in nominally independent districts, though

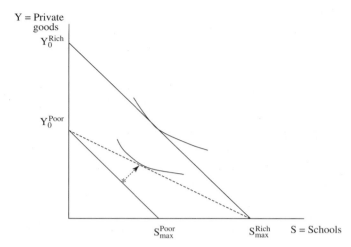

Figure 2   Subsidies to poor districts

state officials still seem surprised that there is an income effect as well as a substitution effect from lowering the tax price. They seem to expect that the arrow in Figure 2 should point horizontally to the right. Instead, local voters use the subsidy (the reduced tax price) to both increase local spending on schools, which is the desired substitution effect, and to reduce their own local taxes (nudging the arrow's direction upwards), which is the income effect.

Another factor can also account for differences in local tax prices. The poorer district may have a substantial amount of non-residential property to tax. Commercial and industrial uses do not come with children attached (at least in metropolitan areas, where workers can live in other communities), and so their tax revenues amount to a subsidy to their school district. The effect of this is the same as a matching-grant subsidy by the state. And the effect is not trivial. Nationally, almost one-half of all property taxes are paid by non-residential property owners, which puts them on the same order of magnitude as state funds for public education.

Although both state subsidies and a large non-residential tax base reduce the tax price, they have been treated differently in recent years. The school finance litigation movement began with *Serrano v. Priest* in California in 1971 (Brunner and Sonstelie, 2006). Its objective was to use state constitutional directives (equal protection and school funding clauses) to improve schools in poor districts. For strategic reasons, the movement focused its remedial efforts on differences in tax base per pupil rather than differences in spending per pupil or on educational outcomes. Many state courts thus ruled that unequal tax bases, not unequal spending, were constitutionally suspect and ordered legislatures to transfer funds from the 'property rich' to the 'property poor'.

What this remedy overlooked is that low-income communities are as likely to be 'property rich' (on the

widely used 'tax base per-pupil' standard) as high-income communities. This is because many urban districts have a large non-residential property tax base that offsets the lower valued residential tax base. (The poor may have migrated there for jobs or rezoned land to attract industry, something most affluent suburbs are reluctant to do.) Besides this, poorer cities often have relatively few children in public schools because of an aged population or because low-quality public schools encourage the use of private schools. In any case, many of the court-induced 'equalization' remedies have actually caused state funds to be removed from low-income (but 'property rich') districts to higher-income districts that are 'property poor' because of their modest nonresidential tax base and large school-age population.

An alternative response to the difficulties of distributing state funds to school districts is simply to have the state government run the schools without the intermediation of local school boards and districts. Another is a voucher system, in which the state gives public funds to parents and allows them to select whatever school they want. Both are certainly viable means of school finance, and it is worth asking why they have not been embraced.

Full state funding forgoes the local monitoring of school performance by voters. Capitalization of school quality in local home values creates a feedback mechanism for local governance. The median voter in most jurisdictions is a homeowner, and voters therefore care about the consequences of school governance. School superintendents who waste local taxpayers' money will find that their tenure is short as voters become dissatisfied. Even if they keep their jobs, the declines in taxable property value due to inefficient policies will leave them with less revenue to spend in the future (Hoxby, 1999). Neither of these desirable feedback effects is likely to occur under a state-managed system.

The drawback of school vouchers appears to be that voters are reluctant to embrace them as a general practice. American voters appear to perceive benefits from local public schools that go beyond educational qualities. One benefit I have advanced is that public schools create location-specific social capital among adults (Fischel, 2006b). Adults with children are more likely to know the parents of their children's schoolmates. This creates a network of adult social capital that lowers the transaction costs of public participation in municipal affairs. A voucher system disperses children to various schools and thus does not create the same location-specific social capital that public schools do. In any case, America's continuing embrace of locally run and locally financed public education reflects the school's central role in facilitating local self-governance.

WILLIAM A. FISCHEL

See also **exit and voice; fiscal federalism; local public finance; property taxation; public choice; school choice and competition; Tiebout hypothesis.**

Bibliography

Benabou, R. 1996. Heterogeneity, stratification, and growth: macroeconomic implications of community structure and school finance. *American Economic Review* 86, 584–609.

Bergstrom, T.C. and Goodman, R.P. 1973. Private demand for public goods. *American Economic Review* 63, 280–96.

Bergstrom, T.C., Rubinfeld, D.L. and Shapiro, P. 1982. Micro-based estimates of demand functions for local school expenditures. *Econometrica* 50, 1183–205.

Bradbury, K.L., Case, K.E. and Mayer, C. 2001. Property tax limits, local fiscal behavior, and property values: evidence from Massachusetts under proposition 2 1/2. *Journal of Public Economics* 80, 287–311.

Brunner, E.J. and Sonstelie, J. 2006. California's school finance reform: an experiment in fiscal federalism. In *The Tiebout Model at Fifty*, ed. W.A. Fischel. Cambridge, MA: Lincoln Institute of Land Policy.

Fischel, W.A. 2006a. Will I see you in September? An economic explanation for the standard school calendar. *Journal of Urban Economics* 59, 236–51.

Fischel, W.A. 2006b. Why voters veto vouchers: public schools and community-specific social capital. *Economics of Governance* 7, 109–32.

Hoxby, C.M. 1999. The productivity of schools and other local public goods producers. *Journal of Public Economics* 74, 1–30.

Hoxby, C.M. 2000. Does competition among public schools benefit students and taxpayers? *American Economic Review* 90, 1209–38.

Oates, W.E. 1969. The effects of property taxes and local public spending on property values: an empirical study of tax capitalization and the Tiebout hypothesis. *Journal of Political Economy* 77, 957–71.

Tiebout, C.M. 1956. A pure theory of local expenditures. *Journal of Political Economy* 64, 416–24.

effective demand

'Effective demand' is the term used by Keynes in his *General Theory* (1936a) to represent the forces determining changes in the scale of output and employment as a whole. Keynes attributed the first discussions of the determinants of the supply and demand for output as a whole to the classical economists, in particular the debate between Ricardo and Malthus concerning the possibility of 'general gluts' of commodities, or what has come to be known as Say's Law of Markets. Indeed, Keynes's theory was intended to replace Say's Law, although the emergence of effective demand from his *Treatise on Money* (1930) critique of the quantity theory of money, and his insistence on its application in what he originally called a 'monetary production economy', suggests that it should also be seen in antithesis to classical monetary theory. For Adam Smith (1776, p. 285), 'A man must be perfectly crazy who ... does not employ all the stock which he commands, whether it be his own or other peoples' on consumption or investment. As long as there was what Smith called 'tolerable security', economic rationality implied that it was impossible for demand for output as a whole to diverge from aggregate supply. Although Smith (1776, p. 73) did call the demand 'sufficient to effectuate the bringing of the commodity to the market', the 'effectual demand' 'of those who are willing to pay the natural price' of the commodity, the idea referred to divergence of market from natural price of particular commodities and the process of gravitation of prices to their natural values. J.B. Say's discussion of the problem of the 'disposal of commodities' adopted Smith's position. Against those who held that 'products would always be abundant, if there were but a ready demand, or market for them,' Say's 'law of markets' argued 'that it is production which opens a demand for products' (1855, pp. 132–3); if production determined ability to buy, then demand could not be deficient. While excesses in particular markets were admitted, they would always be offset by deficiencies in others. Ricardo used similar arguments against Malthus, who responded by suggesting that:

> from the want of a proper distribution of the actual produce, adequate motives are not furnished to continued production, ... the grand question is whether it [actual produce] is distributed in such a manner between the different parties concerned as to occasion the most effective demand for future produce ... (Malthus, 1821)

Malthus argues that the composition of output affects its quantity by producing doubts in the minds of Smith's rational entrepreneurs concerning the 'security' of their future profit.

The final word in the classical debate was J.S. Mill's 'On the Influence of Consumption on Production', which sought exceptions to the proposition that 'All of which is produced is already consumed, either for the purpose of reproduction or enjoyment' so that 'There will never, therefore, be a greater quantity produced, of commodities in general, than there are customers for' (1874, pp. 48–9). Mill accused those who argued that demand limits output of a fallacy of composition, for the individual shopkeeper's failure to sell is due to a disproportion of demand which cancels out for the nation as a whole. Mill also notes that the argument that every purchaser must be a seller presumes barter, for money enables exchange 'to be divided into two separate acts' so one 'need not buy at the same moment when he sells' (p. 70). To avoid this problem 'money must itself be considered as a commodity', for 'there cannot be an excess of all other commodities, and an excess of money at the same time' (p. 71). Mill admits that if money were 'collected in masses', there might be an excess of all commodities, but this would mean only a temporary fall in the value of all commodities relative to money. Similarly to Smith's 'tolerable security', Mill explains an excess of commodities in general by 'a want of commercial confidence', which he denies may be caused by an overproduction of commodities (p. 74).

Mill's defence of Say's Law highlights the importance of the classical quantity theory, which was originally formulated to oppose the undue emphasis given to precious metals as components of national wealth by the mercantilists. Hume noted that labour, not gold, produced the commodities which composed national wealth; that gold was only as good as the labour it commanded to produce output. Thus the classical position that the velocity of circulation of money was independent of its quantity was built on the view that money would only be held to be spent. Money could at best cause temporary general gluts; in the long term, 'rational' men would not choose to hold money rather than spend it.

On the eve of the marginal revolution, classical theory thus admitted the temporary occurrence of general gluts explained by cyclical disproportions in demand for money and commodities due to crises of confidence. It is paradoxical that, while the marginal revolution was motivated by the failure of classical theory to give sufficient attention to the role of demand in value theory, it failed to extend its analysis of demand to output as a whole in either the long or the short period. Indeed, the emphasis on individual equilibrium produced by the subjective theory of value which replaced the classical theory, made separate discussion of aggregate supply and demand redundant. Thus Keynes's reference to 'the disappearance of the theory of demand and supply for output as a whole, that is the theory of employment *after* it has been for a quarter of a century the most discussed thing in economics' (Keynes, 1936c).

But it was discussion, not Say's Law, which disappeared from neoclassical economics. Thus Keynes classed economists from Smith and Ricardo to Marshall and Pigou as 'Classical', for, despite antagonistic theories of value and distribution, they all held a similar theory of supply and demand for output as a whole.

Keynes suggests that this was due more to the failure of neoclassical economists to heed Mill's warning concerning the extension of the conditions faced by the individual to the economy as a whole, than to positive analysis. If consumers (producers) maximize utility (profit) subject to an income (cost) constraint, reaching the maximum by substituting in consumption (production) goods (inputs) which were cheaper per unit of utility (output), then excess supply of any good (resource) is due to its price exceeding its marginal utility (productivity). Market competition would lead to relative price adjustments which eliminate excess supply. Since it was impossible for any single good (resource) to be unsold (unemployed), it was natural to extend this analysis to the aggregate level to deny the possibility of general gluts without further analysis.

Any divergence from this position was explained, not by reference to hoarding money due to crises of confidence, but by temporary impediments to the automatic adjustment of relative prices in competitive markets. Thus, despite their new marginal theory of value, Keynes's contemporaries reached a similar result that divergence of employment from its full employment level would be determined by temporary non-persistent causes eliminated in the long run.

From 1921 to 1939 the unemployment rate in the United Kingdom never fell below ten per cent, peaking in 1932 at 22.5 per cent (over 2.7 million). This exceeded the limits that most economists attributed to short-period frictions. The self-adjusting nature of the neoclassical version of Say's Law that Keynes chose to criticize was thus contradicted by reference to economic events as well as by Keynes's conception of effective demand.

Keynes was not concerned with impediments to the equality of the supply and demand, but with the

> problem of the equilibrium of supply and demand for output as a whole, in short, of effective demand ... When one is trying to discover the volume of output and employment, it must be this point of equilibrium for which one is searching.

While the Classics solved the problem by assuming the identity of savings and expenditure on investment goods, neoclassical theory presumed Say's Law 'without giving the matter the slightest discussion' (1936b, p. 215).

Keynes's theory of effective demand thus had to replace Say's Law. To do this Keynes departed from the Classical position on two points. The first was to assume that wages exceed subsistence so that expenditure on consumption goods does not exhaust factor incomes. As expressed in Keynes's psychological law of consumption, this implied that as output increased, the gap between aggregate expenditure and factor costs increased, so that unless investment expenditure expanded to fill the gap, entrepreneurs would experience losses.

The second departure was from the assumption that rationality dictated that entrepreneurs' savings represented productive investment expenditure. If investment could produce losses, or changes in interest rates change capital values, then greater future enjoyment might be assured by not investing; holding money might be 'rational' in such conditions. Further, in a monetary economy, nothing guarantees that maximization of returns in money will maximize either productive capacity or the demand for labour.

In Keynes's theory the propensity to consume and the multiplier produce the proposition that it is the level of output which adjusts saving to investment, rather than the rate of interest, while the explanation of the decisions over the level of investment in a monetary economy requires an explanation of rates of interest in money terms. The two factors are closely related.

In a 1934 letter to Kahn, Keynes gives a 'precise definition of what is meant by effective demand' (1934a, p. 422). If O is the level of output, W the marginal prime cost of production for that output, and P the expected selling price, 'Then OP is effective demand'. The classical theory that 'supply creates its own demand' assumes that OP equals OW, irrespective of the value of O, 'so that effective demand is incapable of setting a limit to employment which consequently depends on the relation between marginal product in wage-goods industries and marginal disutility of employment'. Thus, what Keynes later called (1936a, ch. 2) the two 'classical' postulates limit O at full employment. In contrast,

> On my theory $OW \neq OP$ for *all* values of O, and entrepreneurs have to choose a value of O for which it is equal – otherwise the equality of price and marginal prime cost is infringed. This is the real starting point of everything.

The key point was thus the impact of different levels of O on the difference between costs and prices, that is on entrepreneurs' profits. Keynes took up this question, in an undated exchange with Sraffa of about the same time (1934b, pp. 157ff). Keynes notes that a non-unitary marginal propensity to consume implies $OP \neq OW$ for any O, and generates

> the general principle that *any* expansion of output gluts the market unless there is a *pari passu* increase of investment appropriate to the community's marginal propensity to consume; and any contraction leads to windfall profits to producers unless there is an appropriate *pari passu* contraction of investment.

The level of O at which $OP = OW$ will be determined by the level of investment and the propensity to consume. Changes in the rate of investment, based on entrepreneurs' expectations of their future profits, will determine O.

In an early draft of the *General Theory* Keynes (1973a, p. 439) put it this way:

> Effective demand is made up of the sum of two factors based respectively on the expectation of what is going to be consumed and on the expectation of what is going to be invested.

Thus the theory of effective demand required, in addition to explanation of consumption based on the propensity to consume, an explanation of variations in the level of investment. Since neoclassical theory resolved this problem by presuming that investment was brought into balance with full employment saving by means of the rate of interest, Keynes located the 'flaw being largely due to the failure of the Classical doctrine to develop a satisfactory theory of the rate of interest' (1934c, p. 489).

Keynes concentrated his efforts to produce a theory of interest compatible within this theory of effective demand within what he called a monetary production economy. The *Treatise on Money* (1930) had explained changes in prices in terms of households' consumption decisions relative to entrepreneurs' production decisions. If these decisions were incompatible, investment diverged from saving and prices of consumption goods adjusted producing windfall profits or losses. The prices of investment goods were determined separately from this process, by means of the interaction of the bearishness of the public reflecting their decisions to hold bank deposits or securities on the one hand, and the monetary policy of the banking system on the other.

Investment goods are held because their present costs or supply prices are lower than the present value of their anticipated future earnings or demand prices; the larger this difference, the higher the expected rate of return. Since any change in the price of a durable capital asset will influence its rate of return, a theory that explains the price of capital assets also explains rates of return (which Keynes called marginal efficiency). With the demand price of an asset based on the value of expected future earnings discounted by the rate of interest, it is clear why a satisfactory theory of interest is crucial to the explanation of effective demand.

But money was a durable asset like any other, and as such it has a spot or demand price and a supply price or forward price, which determine the money rate of interest. Keynes thus transformed his concept of bearishness into liquidity preference which, together with banking policy, would determine the rate of interest. For Keynes, 'the money rate of interest ... is nothing more than the percentage excess of a sum of money contracted for forward delivery ... over what we may call the "spot" or cash price of the sum thus contracted for forward delivery' (1936a, p. 222), it is:

> the premium obtainable on current cash over deferred cash ... No one would pay this premium unless the possession of cash served some purpose, that is had some efficiency. Thus we may conveniently say that interest on money measures the marginal efficiency of money measured in terms of itself as a unit. (1937a, p. 101)

Since both money and capital assets had marginal efficiencies representing their rates of return, profit-maximizing individuals in a monetary economy would demand money and capital assets in proportions which equated their respective returns. The equilibrium level of output chosen by entrepreneurs would then be represented by equality of the marginal efficiency of capital and the rate of interest (the marginal efficiency of money). The question of the effect of an increase in output on profit raised by a propensity to consume less than unity can now be seen as the effect of an increase in investment on the marginal efficiency of money relative to the marginal efficiencies of capital assets. Since these marginal efficiencies reflect pairs of spot and forward asset prices, the question can also be put as the effect of an increase in investment on relative money prices. Thus Keynes's independent variables, the propensity to consume, the efficiency of capital and liquidity preference, given expectations and monetary policy, interact to determine effective demand.

Since this equilibrium could be described by S = I, or equality between the rate of interest and the marginal efficiency of capital, the level of output which equates aggregate demand and supply also equates marginal efficiency with the rate of interest. To complete his theory of effective demand, Keynes faced the question first raised by Wicksell of the causal relation between the natural and the money rate of interest. Just as Keynes rejected the determination of the level of O at which OP = OW by the equality of the marginal productivity and disutility of labour, he rejected marginal productivity as the determinant of marginal efficiency and the real rate of interest determining the money rate because it was based on 'circular reasoning' (1937b, p. 212).

Keynes argues instead that it is the marginal efficiency of capital assets which adapts to the money rate of interest rather than vice versa. These two points of departure are discussed in Chapters 16 and 17 of the *General Theory*, where Keynes points out that the money rate of return to be expected from a capital asset depends on the relation of anticipated money receipts relative to expected money costs, and that there is no reason to believe that these will be related in any predictable way to the asset's physical productivity. Wicksell's natural rate, derived from physical relations of production and exchange, has no application in a monetary economy; Keynes thus substitutes the concept of marginal efficiency.

Keynes also notes that increased investment in particular capital assets increases supply prices and reduces demand prices, causing a decline in marginal efficiencies; an increase in output thus leads to investment in assets with lower rates of return. At some point the marginal efficiency of money will make investment in money as profitable as the purchase of capital assets. At this point the rate of interest equals the marginal efficiency of capital, and any further increase in output would confirm Keynes's

'general principle' that any further expansion in output gluts the market, for increased income is not spent but held in the form of money which becomes a 'generalised sink for purchasing power'.

The question that distinguishes Keynes's theory is thus why money's liquidity premium does not fall as output expands, for this is what prevents investment from rising by just the amount to fill the gap created by the propensity to consume being less than one. To describe these 'essential properties of interest and money', Keynes departs from Mill's position that money is just another commodity. When money is the debt of the banking system its price and quantity behaviour will differ from physical commodities, for it has no real costs of production nor real substitutes. Thus an asset which has a negligible elasticity of production and substitution with respect to a change in effective demand, will have a rate of return which responds less rapidly to an expansion in demand. As long as the rate of interest falls less rapidly than the marginal efficiencies of capital assets, its rate will be the one which sets the point at which further expansion creates losses.

Thus the propensity to consume shows that investment will have to increase by the amount of the gap between incomes and expenditures as incomes rise if entrepreneurs are not to make losses, while the marginal efficiency of capital and liquidity preference in a monetary production economy explain why the behaviour of the rate of interest relative to the marginal efficiency of capital makes it unlikely that the rate of investment should adjust by just that amount. Since entrepreneurs maximize monetary returns, not employment or physical output, there is no reason why their investment decisions should lead to an equilibrium at full employment. Keynes's explanation of the limit to the level of employment permits any level as a stable equilibrium, including full employment; it is thus more general than the classical Say's Law position, in which the only stable equilibrium was the limit set by full employment as given in the labour market.

J. A. KREGEL

See also **Say's Law.**

Bibliography

Keynes, J.M. 1930. *A Treatise on Money*. Reprinted in Keynes (1971).

Keynes, J.M. 1934a. Letter to R.F. Kahn, 13 April. Reprinted in Keynes (1973b).

Keynes, J.M. 1934b. Letter to P. Sraffa, undated. Reprinted in Keynes (1979).

Keynes, J.M. 1934c. Poverty in plenty: is the economic system self-adjusting? Reprinted in Keynes (1973b).

Keynes, J.M. 1936a. *The General Theory of Employment, Interest and Money*. Reprinted in Keynes (1973a).

Keynes, J.M. 1936b. Letter to A. Lerner, 16 June. Reprinted in Keynes (1979).

Keynes, J.M. 1936c. Letter to R.F. Harrod, 30 August. Reprinted in Keynes (1973c).

Keynes, J.M. 1937a. The theory of the rate of interest. Reprinted in Keynes (1973c).

Keynes, J.M. 1937b. Alternative theories of the rate of interest. Reprinted in Keynes (1973c).

Keynes, J.M. 1971–83. *The Collected Writings of John Maynard Keynes*, ed. D. Moggridge. London: Macmillan for the Royal Economic Society: 1971. Vols. 5 and 6. *A Treatise on Money* (1930). 1973a. Vol. 7. *The General Theory of Employment, Interest and Money* (1936). 1973b. Vol. 13. *The General Theory and After: Part I – Preparation*. 1973c. Vol. 14 *The General Theory and After: Part II – Defence and Development*. 1979. Vol. 29. *The General Theory and After – A Supplement*.

Malthus, T.M. 1821. Letter from Malthus to Ricardo, 7 July. Reprinted in Ricardo (1952), 9–10.

Mill, J.S. 1874. On the influence of consumption on production. In J.S. Mill, *Essays on Some Unsettled Questions of Political Economy*, 2nd edn, reprinted Clifton, NJ: A.M. Kelley, 1974.

Ricardo, D. 1952. *Works and Correspondence of David Ricardo*, vol. 9, ed. P. Sraffa with the collaboration of M. Dobb. Cambridge: Cambridge University Press.

Say, J.B. 1855. *A Treatise on Political Economy*. 6th American edn, Philadelphia: J.B. Lippincott.

Smith, A. 1776. *An Inquiry into the Nature and Causes of the Wealth of Nations*. Oxford: Oxford University Press, 1976.

efficiency bounds

Oftentimes we want to compare estimators. For a given parameter in which we are interested, there are typically many estimators that can estimate it consistently. We need to choose the best estimator, or the estimator that is the closest to the true parameter value. The mean square error (MSE), $E(\hat{\theta} - \theta)^2$, is frequently used as a measure of closeness. However, there can be many other various measures of closeness, and often they do not agree with each other. See, for example, Amemiya (1994, pp. 116–24).

Even with a given measurement of closeness, such as the MSE, it is typically not possible to rank two estimators. For two estimators X and Y of θ, X is better than Y only if $E(X - \theta)^2 \leq E(Y - \theta)^2$ for all $\theta \in \Theta$. An estimator that is not dominated by another estimator in the above sense is called admissible.

A uniformly 'most' efficient estimator does not exist. To find an efficient estimator, one needs to confine the analysis to a limited class of estimators, such as unbiased estimators or equivariant estimators. Alternatively, one can rely on a subjective strategy such as average risk optimality which requires a prior distribution over the parameter space, or use a pessimistic and risk-averse approach such as minimax optimality.

In large sample analysis, the performance measures of estimators can often be approximated by their asymptotic distribution. Under suitable regularity conditions, many estimators are consistent and converge to the true parameter values at $\sqrt{n}$ rate. These estimators can be compared based on their asymptotic variance. The notation of efficiency bound usually refers to the largest lower bound for the variances that can be achieved by $\sqrt{n}$ consistent and asymptotically normal estimators under suitable regularity conditions.

Asymptotic efficiency in parametric models

In parametric models, the variance of an unbiased estimator has to be larger than the Cramer–Rao lower bound, which is defined as the inverse of the information matrix:

$$V(\hat{\theta}) \geq -\left(E\frac{\partial^2 \log L}{\partial \theta^2}\right)^{-1},$$

where L is the likelihood function. Proofs of this result can be found, for example, in Amemiya (1994, pp. 138–39; 1985, pp. 14–17). A consistent estimator is said to be asymptotically efficient if its asymptotic variance achieves the Cramer–Lao lower bound. Under suitable regularity assumptions such as those given in Theorem 4.1.3 in Amemiya (1985), the maximum likelihood estimator is asymptotically efficient.

There exist super-efficient estimators whose asymptotic variances are smaller than the Cramer–Rao lower bound on a set of parameter θ with Lebesgue measure zero, such as Hodges's estimator defined as

$$w_T = \begin{cases} 0 & \text{if } |\hat{\theta}| < T^{1/4} \\ \hat{\theta} & \text{if } |\hat{\theta}| \geq T^{-1/4}. \end{cases}$$

where $\sqrt{T}(\hat{\theta} - \theta) \xrightarrow{d} N(0, v(\theta))$. One can show that $\sqrt{T}(w_T - \theta) \xrightarrow{d} N(0, v(\theta))$ if $\theta \neq 0$ and $\sqrt{T}(w_T) \xrightarrow{d} 0$ if $\theta = 0$. However, the better behaviour of w_T at $\theta = 0$ comes at the expense of erratic behaviour when θ is close to 0. See, for example, van der Vaart (1999, p. 110).

A common alternative to maximum likelihood is generalized method of moment estimators (GMM). Its asymptotic efficiency is extensively discussed in Newey and McFadden (1994). While GMM estimators are less efficient than maximum likelihood (see, for example, the proof in Newey and McFadden (1994, p. 2163), oftentimes they are easy to compute, especially when maximum likelihood is computationally infeasible. For a given set of unconditional moment conditions, a proper choice of the weighting matrix or the linear combination matrix minimizes the asymptotic variance. For a given set of conditional moment conditions, a proper choice of instruments can also minimize the asymptotic variance.

A GMM estimator can be formed from the over-identified moment conditions $Em(z; \theta) \equiv 0$ by minimizing a quadratic form based on a weighting matrix W:

$$\frac{1}{T}\sum_{t=1}^{T} m(z_t; \hat{\theta}) W \frac{1}{T}\sum_{t=1}^{T} m(z_t; \hat{\theta}).$$

The resulting estimator has asymptotic variance $(G'WG)^{-1}$ $(G'W\Omega WG)(G'WG)^{-1}$, where $G = E\frac{\partial}{\partial\theta}m(z;\theta)$ and $\Omega = Var(m(z;\theta))$. Hansen (1982) showed that the optimal choice of $W = \Omega^{-1}$, which equates $G'WG = G'W\Omega WG$. In this case the asymptotic variance is reduced to $(G'\Omega^{-1}G)^{-1}$.

Alternatively, a set of over-identified moment conditions $Em(z;\theta) \equiv 0$ can be translated into a set of exactly identified moment conditions by a linear combination matrix $AEm(z;\theta) \equiv 0$. Given A, the resulting method of moment estimator that equates $A\sum_{t=1}^{T}m(z_t;\hat{\theta})$ to zero has asymptotic variance $(AG)^{-1}(A\Omega A')(G'A')^{-1}$. As a rule of thumb, the optimal choice of A should simplify this asymptotic variance, by equating $AG = A\Omega A' = G'A'$. The resulting optimal $A = G'\Omega^{-1}$ gives rise to the same asymptotic distribution as the above optimally weighted GMM estimator of Hansen (1982), which minimizes

$$\frac{1}{T}\sum_{t=1}^{T} m(z_t; \hat{\theta})\ \Omega^{-1} \frac{1}{T}\sum_{t=1}^{T} m(z_t; \hat{\theta}).$$

Many economic models, such as those based on Euler equations, are stated in terms of conditional moment conditions of the form $E(m(z;\beta)|x) = 0$ for almost all x. These conditional moment conditions can be translated into exactly identified unconditional moment conditions using an instrument matrix $A(x) : EA(x)m(z;\beta) = 0$. The question arises as to what is the optimal instrument matrix $A(x)$. For a given choice of $A(x)$, the resulting method of moment estimator that equates $\frac{1}{T}\sum_{t=1}^{T}(x_t)$ $m(z_t;\beta) = 0$ has asymptotic variance $(EA(x)G(x))^{-1}$ $EA(x)\Omega(x)A(x)'$ $(EG(x)'A(x)')^{-1}$, where $G(x) = E(\frac{\partial}{\partial\theta}m(z;\theta)|x)$ and $\Omega(x) = Var(m(z;\beta)|x)$. We can then equate

$$EA(x)G(x) = EA(x)\Omega(x)A(x)'$$

to obtain the optimal instrument matrix $A(x) = G(x)'\Omega(x)^{-1}$. The resulting efficient asymptotic variance is therefore $(EG(x)'\Omega(x)^{-1}G(x))^{-1}$.

Formal proofs of these derivations can be found in, for example, Newey and McFadden (1994). Estimators that achieve these efficiency bounds typically involve two-step or multi-step procedures and possibly nonparametric methods, such as Newey and Powell (1990).

Asymptotic efficiency in semiparametric models

Semiparametric models are extensions of parametric models where some components are specified nonparametrically with unknown functional forms. Generalized method of moment models are semiparametric models if the data-generating process is not fully specified.

A partial linear model is another example. Other popular semiparametric models are surveyed in Powell (1994).

Intuitively, the variance of an estimator for a semiparametric model should be larger than the Cramer–Rao lower bound for any parametric sub-model that satisfies the semiparametric restrictions. The semiparametric efficiency bound is therefore defined to be the supremum of the Cramer–Rao bounds for all parametric models that satisfy the semiparametric restrictions. Extensive results for semiparametric efficiency bounds are developed in, among others, Bickel et al. (1993) and Newey (1990). In this section we give a brief summary of some of the results presented in Newey (1990). The next section will apply these results to a particular estimation problem.

Because of pathological cases such as the super-efficient estimator, the semiparametric efficiency bound is used to provide a lower bound only for *regular* estimators. Consider a parameter of interest that is a smooth function of the underlying parametric path: $\beta(\theta)$. A regular estimator $\hat{\beta}$ is one where for each θ_0 the limiting distribution of $\sqrt{T}(\hat{\beta} - \beta(\theta_T))$ does not depend on θ_T as long as $\sqrt{T}(\theta_T - \theta_0)$ is bounded. The super-efficient estimator is not regular.

Most estimators in econometrics are asymptotically linear, in the sense that they have an influence function representation as

$$\sqrt{T}(\hat{\beta} - \beta_0) = \frac{1}{\sqrt{T}}\sum_{t=1}^{T}\psi(z_t) + o_p(1).$$

In particular, almost all econometric estimators asymptotically solve some moment conditions $\frac{1}{\sqrt{T}}\sum_{t=1}^{T}m(z_t;\hat{\beta}) = o_p(1)$, in which case the linear influence function is given by $\psi(z_t) = -G^{-1}m(z_t;\beta)$ for $G = E\frac{\partial}{\partial\beta}m(z_t;\beta)$.

Asymptotically linear estimators are regular if and only if for all parametric sub-models $\frac{\partial}{\partial\theta}\beta(\theta) = E\psi S_\theta'$. When $\psi(z_t) = -G^{-1}m(z_t;\beta)$, this follows from differentiating $E_\theta m(z;\beta(\theta)) = 0$ with respect to θ. The asymptotic variance of an asymptotically linear estimator is $E\psi\psi'$, which is apparently larger than that of the maximum likelihood estimator $\beta(\hat{\theta})$ of any parametric sub-model, which is given through information matrix and the delta method as

$$\left(\frac{\partial}{\partial\theta}\beta(\theta)\right)(E(S_\theta S_\theta'))^{-1}\left(\frac{\partial}{\partial\theta}\beta(\theta)\right)'$$
$$= E[\psi S_\theta'](ES_\theta S_\theta')^{-1}E[S_\theta\psi'].$$

A starting point for calculating the semiparametric efficiency bound is to restrict attention to differentiable parameters $\beta(\theta)$ which satisfies $\frac{\partial\beta(\theta)}{\partial\theta} = E(dS_\theta')$ for some d and all parametric sub-models. Such d are not unique. Adding a random vector that is orthogonal to S_θ preserves the validity of d. In fact, any linear influence function ψ can serve as a d. For differentiable parameters, if we use the invariance principle and the delta method, the Cramer–Rao lower bound for estimating $\beta(\theta)$ is

$$
\left(\frac{\partial}{\partial \theta}\beta(\theta)\right)(E(S_\theta S_\theta'))^{-1}\left(\frac{\partial}{\partial \theta}\beta(\theta)\right)'
$$
$$
= E[dS_\theta'](ES_\theta S_\theta')^{-1}E[S_\theta d'].
$$

Obviously, this is the variance of $d_\theta = E[dS_\theta]$ $(E[S_\theta S_\theta])^{-1}S_\theta$, which is the projection of d onto the linear space spanned by the score functions S_θ.

As the class of parametric sub-models expands, the linear space it spans also increases and the variance of d_θ also increases. The semiparametric efficiency bound should be the limit of this progress of increments. Formally, the tangent space is defined to be the mean square closure of all linear combinations of scores S_θ for smooth parametric sub-models, and the efficiency bound is given by the variance of the projection of d onto the tangent space T. In other words, the efficiency bound is given by $V = E[\delta\delta']$ where $\delta \in T$ and $E[(d-\delta)'\iota] = 0$ for all $\iota \in T$.

Application

In this section we illustrate the computation of semiparametric efficiency bound using a model of non-classical measurement errors, studied in Chen, Hong and Tamer (2005) and Chen, Hong and Tarozzi (2004), where information from a primary data-set and from an auxiliary data-set need to be efficiently combined. Their models extend the results in the treatment effect literature on the mean parameter (see Hahn, 1998, Hirano, Imbens and Ridder, 2003 and Imbens, Newey and Ridder, 2005), to measurement error models where parameters are generically defined through nonlinear moment conditions.

Consider the following model. The researcher is interested in a parameter β defined by the moment condition $Em(Y;\beta) = 0$ if and only if $\beta = \beta_0$. The researcher has access to a primary data-set which is a random sample from the population of interest. However, the true variable Y is not always observed in the primary data-set. Instead, a proxy variable X is observed throughout the primary data. For a subset of the primary data-set, which we will call the auxiliary data-set, X is validated so that both Y and X are observed. We will use the random variable $D = 0$ to denote observations in the auxiliary data-set where both X and Y are observed, and will use $D = 1$ to denote the rest of the primary data-set where only X is observed. Chen, Hong and Tarozzi (2004) call this the 'verify-in-sample' case. They make the following conditional independence assumption:

Assumption 4.1 $Y \perp D|X$.
Under this assumption, we follow the framework of Newey (1990) to show that the efficiency bound for

estimating β is given by $\left(J_\beta \Omega_\beta^{-1} J_\beta\right)^{-1}$, where for $p(X) = p(D = 1|X)$:

$$
\mathscr{J}_\beta = \frac{\partial}{\partial \beta}E[m(Y;\beta)] \quad \text{and}
$$

$$
\Omega_\beta = E\left[\frac{1}{1-p(X)}V[m(Y;\beta)|X] + \mathscr{E}(X;\beta)\mathscr{E}(X;\beta)'\right].
$$

To demonstrate this result, we follow the steps in the efficiency framework of Newey (1990). First we characterize the properties of the tangent space under assumption 4.1. Next we write the parameter of interest in its differential form and therefore find a linear influence function d. Finally, we conjecture and verify the projection of d onto the tangent space and the variance of this projection gives rise to the efficiency bound. We first go through these three steps under the assumption that the moment conditions exactly identify β. Finally, the results are extended to over-identified moment conditions by considering their optimal linear combinations.

First we assume that the moment conditions exactly identify β.

Step 1. Consider a parametric path θ of the joint distribution of Y, X and D. Define $p_\theta(x) = P_\theta(D = 1|x)$. Under assumption 1, the joint density function for Y, D and X can be factorized into

$$
f_\theta(y,x,d) = f_\theta(x)p_\theta(x)^d[1-p_\theta(x)]^{1-d}f_\theta(y|x)^{1-d}.
$$
(1)

The resulting score function is then given by

$$
S_\theta(d,y,x) = (1-d)s_\theta(y|x)
$$
$$
+ \frac{d - p_\theta(x)}{p_\theta(x)(1-p_\theta(x))}\dot{p}_\theta(x) + t_\theta(x),
$$

where

$$
s_\theta(y|x) = \frac{\partial}{\partial \theta}\log f_\theta(y|x), \quad \dot{p}_\theta(x) = \frac{\partial}{\partial \theta}p_\theta(x),
$$
$$
t_\theta(x) = \frac{\partial}{\partial \theta}\log f_\theta(x).
$$

The tangent space of this model is therefore given by:

$$
\mathscr{T} = \{(1-d)s_\theta(y|x) + a(x)(d - p_\theta(x)) + t_\theta(x)\}
$$
(2)

where $\int s_\theta(y|x)f_\theta(y|x)dy = 0$, $\int t_\theta(x)f_\theta(x)dx = 0$, and $a(x)$ is any square integrable function.

Step 2. As in the method of moment model in Newey (1990), the differential form of the parameter β

can be written as

$$
\begin{aligned}
\frac{\partial \beta(\theta)}{\partial \theta} &= -(\mathscr{I}_\beta)^{-1} E\left[m(Y;\beta) \frac{\partial \log f_\theta(Y,X)}{\partial \theta'} \right] \\
&= -(\mathscr{I}_\beta)^{-1} \{ E[m(Y;\beta)(s_\theta(Y|X)' \\
&\quad + t_\theta(X)')]\} \\
&= -(\mathscr{I}_\beta)^{-1} \{ E[m(Y;\beta)s_\theta(Y|X)'] \\
&\quad + E[\mathscr{E}(X)t_\theta(X)']\}.
\end{aligned}
\tag{3}
$$

Therefore $d = -\mathscr{I}_\beta^{-1} m(Y;\beta)$. Since $\mathscr{I}_\beta$ is only a constant matrix of nonsingular transformation. The projection of d onto the tangent space will be $-\mathscr{I}_\beta$ multiplied by the projection of $m(Y;\beta)$ onto the tangent space. Therefore we only need to consider the projection of $m(Y;\beta)$ onto the tangent space.

Step 3. We conjecture that this projection takes the form of

$$
\tau(Y,X,D) = \frac{1-D}{1-p(X)}[m(Y;\beta) - \mathscr{E}(X)] + \mathscr{E}(X)
$$

To verify that this is the efficient influence function we need to check that $\tau(Y,X,D)$ lies in the tangent space and that

$$
E[(m(Y;\beta) - \tau(Y,X,D))s_\theta(Y,X)] = 0.
$$

or that

$$
E[m(Y;\beta)s_\theta(Y,X)] = E[\tau(Y,X,D)s_\theta(Y,X)].
\tag{4}
$$

To see that $\tau(Y, X, D)$ lies in the tangent space, note that the first term in $\tau(Y, X, D)$ has mean zero conditional on X, and corresponds to the first term of $(1-d)s_\theta(y|x)$ in the tangent space. The second term in $\tau(Y,X,D)$, $\mathscr{E}(x)$, has unconditional mean zero and obviously corresponds to the $t_\theta(x)$ in the tangent space.

To verify (4), one can make use of the representation of $E[m(Y;\beta)s_\theta(Y, X)]$ in (3), by verifying the two terms in $\tau(Y, X, D)$ separately. The second term is obvious and tautological. The first part,

$$
\begin{aligned}
E&\left[\frac{1-D}{1-p(X)}[m(Y;\beta) - E(X)]s_\theta(Y,X) \right] \\
&= E[m(Y;\beta)s_\theta(Y,X)],
\end{aligned}
$$

follows from the conditional independence assumption 4.1 and the score function property $E[s_\theta(Y,X)|X] = 0$. Therefore we have verified that $\tau(Y, X, D)$ is the efficient

projection and that the efficiency bound is given by

$$
\begin{aligned}
V &= (\mathscr{I}_\beta)^{-1} E[\tau(Y,X,D)\tau(Y,X,D)'](J_\beta)'^{-1} \\
&= (\mathscr{I}_\beta)^{-1} E\left[\frac{1}{1-p(X)} Var(m(Y;\beta)|X) \right. \\
&\quad \left. + \mathscr{E}(X)\mathscr{E}(X)' \right] (\mathscr{I}_\beta)'^{-1}.
\end{aligned}
$$

Finally, consider the extensions of these results to the over-identified case. When $d_m > d_\beta$, the moment condition is equivalent to the requirement that for any matrix A of dimension $d_\beta \times d_m$ the following exactly identified system of moment conditions holds

$$
\mathscr{A} E[m(Y;\beta)] = 0.
$$

Differentiating under the integral again, we have

$$
\begin{aligned}
\frac{\partial \beta(\theta)}{\partial \theta} &= -\left(\mathscr{A} E\left[\frac{\partial m(Y;\beta)}{\partial \beta} \right] \right)^{-1} \\
& E\left[\mathscr{A} m(Y;\beta) \frac{\partial \log f_\theta(Y,X|D=1)}{\partial \theta'} \right].
\end{aligned}
$$

Therefore, any regular estimator for β will be asymptotically linear with influence function of the form

$$
-\left(\mathscr{A} E\left[\frac{\partial m(Y;\beta)}{\partial \beta} \right] \right)^{-1} \mathscr{A} m(Y;\beta).
$$

For a given matrix $\mathscr{A}$, the projection of the above influence function onto the tangent set follows from the previous calculations, and is given by

$$
-[\mathscr{A} J_\beta]^{-1} \mathscr{A} \tau(y,x,d).
$$

The asymptotic variance corresponding to this efficient influence function for fixed $\mathscr{A}$ is therefore

$$
[\mathscr{A} J_\beta]^{-1} \mathscr{A} \Omega \mathscr{A}'[J_\beta \mathscr{A}']^{-1}
\tag{5}
$$

where

$$
\Omega = E[\tau(Y,X,D)\tau(Y,X,D)']
$$

as calculated above. Therefore, the efficient influence function is obtained when $\mathscr{A}$ is chosen to minimize this efficient variance. It is easy to show that the optimal choice of $\mathscr{A}$ is equal to $J_\beta' \Omega^{-1}$, so that the asymptotic variance becomes

$$
V = \left(\mathscr{I}_\beta' \Omega^{-1} \mathscr{I}_\beta \right)^{-1}.
$$

Different estimation methods can be used to achieve this semiparametric efficiency bound. In particular, Chen, Hong and Tarozzi (2004) showed that both a semiparametric conditional expectation projection estimator and a semiparametric propensity score estimator based on a sieve nonparametric first-stage regression achieve this efficiency bound.

Conclusion

As discussed in Newey (1990), while the calculation of the tangent space and the efficient projection is easy in several important examples, including the one above, it can be difficult in general. A variety of techniques are available to characterize the tangent space and the efficient projection. Some of these are discussed in details in Newey (1990) and Bickel et al. (1993).

Even in parametric models, the notion of asymptotic efficiency is more complex when one compares estimators that do not converge at $\sqrt{n}$ rate or are not asymptotically distributed. Comparing these estimators requires the choice of a loss function, and different loss functions can lead to different efficiency rankings (see Ibragimov and Has'minskii, 1981). In econometrics, these estimators sometimes arise in structural models in labour economics and in industrial organization. The efficiency properties of these estimators are analysed in Hirano and Porter (2003) and Chernozhukov and Hong (2004).

HAN HONG

See also **generalized method of moments estimation; maximum likelihood; measurement error models; nonparametric structural models; semiparametric estimation; stratification.**

Bibliography

Amemiya, T. 1985. *Advanced Econometrics*. Cambridge, MA: Harvard University Press.
Amemiya, T. 1994. *Introduction to Statistics and Econometrics*. Cambridge, MA: Harvard University Press.
Bickel, P., Klaassen, C.A., Ritov, Y. and Wellner, J. 1993. *Efficient and Adaptive Estimation for Semiparametric Models*. New York: Springer-Verlag.
Chen, X., Hong, H. and Tamer, E. 2005. Measurement error models with auxiliary data. *Review of Economic Studies* 72, 343–66.
Chen, X., Hong, H. and Tarozzi, A. 2004. Semiparametric efficiency in GMM models of nonclassical measurement errors. Working paper, Duke University and New York University.
Chernozhukov, V. and Hong, H. 2004. Likelihood inference for a class of nonregular econometric models. *Econometrica* 72, 1445–80.
Hahn, J. 1998. On the Role of propensity score in efficient semiparametric estimation of average treatment effects. *Econometrica* 66, 315–32.
Hansen, L. 1982. Large sample properties of generalized method of moments estimators. *Econometrica* 50, 1029–54.
Hirano, K., Imbens, G. and Ridder, G. 2003. Efficient estimation of average treatment effects using the estimated propensity score. *Econometrica* 71, 1161–89.
Hirano, K. and Porter, J. 2003. Asymptotic efficiency in parametric structural models with parameter-dependent support. *Econometrica* 71, 1307–38.
Ibragimov, I. and Has'minskii, R. 1981. *Statistical Estimation: Asymptotic Theory*. New York: Springer-Verlag.
Imbens, G., Newey, W. and Ridder, G. 2005. Mean-squared-error calculations for average treatment effects. Working paper.
Newey, W. 1990. Semiparametric efficiency bounds. *Journal of Applied Econometrics* 5(2), 99–135.
Newey, W. and McFadden, D. 1994. Large sample estimation and hypothesis testing in *Handbook of Econometrics*, vol. 4, ed. R. Engle and D. McFadden. Amsterdam: North-Holland.
Newey, W. and Powell, J. 1990. Efficient estimation of linear and type in censored regression models under conditional quantile restrictions. *Econometric Theory* 6, 295–317.
Powell, J. 1994. Estimation of semiparametric models. In *Handbook of Econometrics*, vol. 4, ed. R. Engle and D. McFadden. Amsterdam: North-Holland.
van der Vaart, A. 1999. *Asymptotic Statistics*. Cambridge: Cambridge University Press.

efficiency wages

'Efficiency wages' is a term used to express the idea that labour costs can be described in terms of efficiency units of labour rather than in terms of hours worked, and that wages affect the performance of workers. In this respect, labour differs from most other inputs (with the notable exception of credit), in which inputs are well defined independently of prices. Models of efficiency wages explore the implications of the interconnections between compensation and productivity. On the macroeconomic level, efficiency wages can explain persistent unemployment without relying on either structural imperfections such as search costs or fixed-length contracts or irrational behaviour such as money illusion, which would cause real wages to fail to adjust to market conditions. (For some of the earliest such models, see Futia, 1977; Salop, 1979; Solow, 1979; Shapiro and Stiglitz, 1984; Weiss, 1981.) At the level of the firm, efficiency wages can result in job queues (excess supply of labour) and can explain why seemingly identical workers may receive different wages at different firms, and why these observed wage differentials are positively correlated with firm characteristics such as profitability, high capital–labour ratios, and establishment size (Brown and Medoff, 1989). These market imperfections arise because employers cannot costlessly observe the ability and productivity of workers or because of capital market imperfections that prevent workers from 'buying' the high-wage jobs.

Efficiency wage models have one or more of the following characteristics:

1. Compensation levels and rules affect the types of workers who are attracted to, and retained by, the firm – this is normally referred to as the sorting effect of wages.
2. Compensation rules create incentives for workers to behave in ways that increase firm profits.
3. Wages affect the nutrition and health of workers and thus higher wages directly increase productivity (these 'nutrition' models are most applicable in poor countries).

Consequences of the use of efficiency wages are:

1. Compensation levels within a firm may not be proportionate to relative productivity.
2. Compensation could be a function of characteristics of the establishment employing the worker.
3. Wages could rise more steeply with tenure than does productivity.
4. Some firms could have an excess supply of workers.
5. A frictionless economy could be in a long-run equilibrium with unemployment.

The sorting effects of wages enable a firm to benefit from private information that the employee knows about himself and that is either not available to the firm or would be costly for the firm to acquire. High compensation enables the firm to draw from a larger and better pool of workers. Firms that test job applicants will also find that, by offering a higher wage, the expected quality of the worker hired, conditional on the applicants test score, will also be higher.

The test could be in the form of a low-wage probation period for new hires. Using a low-wage probation period, followed by a significant wage increase, followed by high wages for workers who perform well during the probation period, the firm can attract job applicants with positive private information about their ability. If the test is imperfect, the use of a low-wage probation period will also discourage applications from risk-averse applicants as well as applicants with a higher cost of capital. Wages that increase steeply with tenure will attract workers who have low quit propensities (aside from their incentive effect of deterring quits). Groshen and Loh (1993) have found that much of the return to tenure takes place at the end of low-wage probationary periods.

Sorting effects of efficiency wages may also explain why firms do not cut wages in response to a fall in demand. If a firm were to cut the wages, it may find that its better workers are most likely to quit. Thus, a profit maximizing firm could find that its best response to a fall in demand for its product would be to fire workers rather than to cut wages.

Most of the efficiency wage models have focused on the ways in which compensation affects the behaviour of workers.

The incentive effects of wages stem from the effect of the level of compensation on the cost to the worker of being fired. Thus, wages above the market clearing level will increase effort, decrease employee theft, decrease absenteeism, and decrease quits. See, for example, Salop and Salop (1976), Klein, Spady and Weiss (1991), and Weiss (1984) on quits; Shapiro and Stiglitz (1984) on effort; Lazear and Rosen (1981), Weiss (1985) on absenteeism.

Levels of compensation also affect the attitude of the employee towards the firm. Thus, paying wages above the market clearing level may have multiple beneficial effects for the firm including: reducing employee theft, increasing unobserved effort, and inducing higher levels of care, which will decrease costs incurred from damage to the firm's property. Greater loyalty to the firm will also encourage workers to acquire firm-specific human capital, to report theft of firm property, and to allocate the worker's effort in ways that benefit the firm. See Akerlof (1984) on gift exchange.

Higher levels of compensation will also reduce the time needed to fill vacancies (Lang, 1991). In this case the behaviour being affected is the application process.

Wages directly affect the productivity of workers through their effect on the nutrition of workers as well as their access to clean water and medical care and other goods and services that directly improve their productivity. These 'nutrition' effects are strongest in poor countries and could also possibly explain poverty traps for particularly poor workers who do not have access to firms that are offering efficiency wages.

The importance of these effects will vary across firms. For instance, we would expect that capital-intensive firms will derive the greatest benefit from reductions in absenteeism and quits, and from increased productivity of their employees. Capital-intensive firms will also tend to be most vulnerable to careless behaviour by workers that would damage the valuable property. Larger firms have more difficulty monitoring individual effort and directing the effort in ways that fit the needs of the firm. Consequently, the efficiency wage models would predict that compensation would be correlated with firm size. The direct effects of wages through better nutrition and health take some time to affect productivity, so we would expect that firms with lower costs of capital will offer higher wages – in poor countries these tend to be foreign firms. (In poor countries, in which the nutrition effects are strongest, we might see that wages would be correlated with a firm's cost of capital as well as with the ability of the firm to retain workers after their productivity has been enhanced by the higher wages. The nutrition effects of wages may take some time to affect productivity.) Finally, if high wages are used to attract better workers, then we would expect that when workers are laid off from firms in high-wage industries they will tend to get jobs in other high-wage industries (see Gibbons and Katz, 1992).

All of these implications of the efficiency wage model have been confirmed by empirical studies of the relationship between firm characteristics and wages. (In cases in which wages directly affects productivity we would expect that firms that are likely to be able to retain their workers will also pay higher wages. However, since wages directly affects turnover, and prices vary according to the presence of competitive firms, this implication of the nutrition version of the efficiency wage model is more difficult to verify.) Of course, many if not all of these empirical findings can be explained by other models. For example, the relationship between prior and posterior industry wages for laid-off workers can be explained by competitive models in which workers are being selected based on attributes, such as pulchritude, that are directly observed by the firm but not by the researchers.

Thus, efficiency wages can explain why empirical studies of the relationship between wage and characteristics of establishments find that large, capital-intensive establishments are most likely to pay wages that are above market clearing levels – and in the case of poor countries why foreign firms tend to pay higher wages. The efficiency wage models also can explain why firms fire workers rather than cutting wages, offer wages that attract an excess supply of workers, and pay some of their workers to take early retirement or seek to impose mandatory retirement. See, for instance, Brown and Medoff (1989). Finally, efficiency wage theory can explain the persistence of involuntary unemployment in a free market economy.

ANDREW WEISS

Bibliography

Akerlof, G.A. 1982. Labor contracts as partial gift exchange. *Quarterly Journal of Economics* 97, 543–69.

Akerlof, G.A. 1984. Gift exchange and efficiency-wage theory: four views. *American Economic Review* 74(2), 79–83.

Brown, C. and Medoff, J. 1989. The employer size-wage effect. *Journal of Political Economy* 97, 1027–59.

Futia, C. 1977. Excess supply equilibria. *Journal of Economic Theory* 14, 200–20.

Gibbons, R. and Katz, L. 1992. Does unmeasured ability explain inter-industry wage differentials? *Review of Economic Studies* 59, 515–35.

Groshen, E. and Loh, E.S. 1993. What do we know about probationary periods? Proceedings of the 45th Annual Meeting of the Industrial Relations Research Association, Madison, WI.

Klein, R., Spady, R. and Weiss, A. 1991. Factors affecting the output and quit propensities of production workers. *Review of Economic Studies* 58, 929–54.

Landau, H. and Weiss, A. 1984. Wages, hiring standards and firm size. *Journal of Labor Economics* 2, 477–99.

Lang, K. 1991. Persistent wage dispersion and involuntary unemployment. *Quarterly Journal of Economics* 106, 181–202.

Lazear, E. and Rosen, S. 1981. Rank-order tournaments as optimum labor contracts. *Journal of Political Economy* 89, 841–64.

Salop, J. and Salop, S. 1976. Self-selection and turnover in the labor market. *Quarterly Journal of Economics* 90, 619–27.

Salop, S. 1979. A model of the natural rate of unemployment. *American Economic Review* 69, 117–25.

Shapiro, C. and Stiglitz, J. 1984. Equilibrium unemployment as a worker discipline device. *American Economic Review* 74, 433–44.

Solow, R. 1979. Another possible source of wage stickiness. *Journal of Macroeconomics* 1, 79–82.

Weiss, A. 1981. Job queues and layoffs in labor markets with flexible wages. *Journal of Political Economy* 88, 526–38.

Weiss, A. 1984. Determinants of quit behavior. *Journal of Labor Economics* 2, 371–87.

Weiss, A. 1985. Absenteeism and wages. *Economics Letters* 19, 277–9.

Weiss, A. and Wang, R. 1998. Probation, layoffs, and wage-tenure profiles: a sorting explanation. *Labour Economics* 5, 359–83.

efficient allocation

Analysis of efficiency in the context of resource allocation has been a central concern of economic theory from ancient times, and is an essential element of modern microeconomic theory. The ends of economic action are seen to be the satisfaction of human wants through the provision of goods and services. These are supplied by production and exchange and limited by scarcity of resources and technology. In this context efficiency means going as far as possible in the satisfaction of wants within resource and technological constraints. This is expressed by the concept of Pareto optimality, which can be stated informally as follows: a state of affairs is Pareto optimal if it is within the given constraints and it is not the case that everyone can be made better off in his own view by changing to another state of affairs that satisfies the applicable constraints.

Because knowledge about wants, resources and technology is dispersed, efficient outcomes can be achieved only by coordination of economic activity. Hayek (1945) pointed out the role of knowledge or information, particularly in the context of prices and markets, in coordinating economic activity. Acquiring, processing and transmitting information are costly activities themselves subject to constraints imposed by technological and resource limitations. Hayek pointed out that the institutions of markets and prices function to communicate information dispersed among economic agents so as to bring about coordinated economic action. He also drew attention to motivational properties of those institutions, or incentives. In this context, the concept of efficiency takes account of the organizational constraints on information processing and transmission in addition to those on production of ordinary goods and services. The magnitude of resources devoted to business or governmental bureaucracies, and to some of the functions performed by industrial salesmen, attests to the importance of these constraints. Economic analysis of efficient allocation has formally imposed only the constraints on production and exchange, and until recently recognized organizational constraints only in an informal way. But it is these constraints that motivate the pervasive and enduring interest in decentralized modes of economic organization, particularly the competitive mechanism.

It is necessary to limit the scope of this essay so that it is not coextensive with microeconomic theory. The main limitation imposed here is to confine attention to models in which either the role of information is ignored, or in which agents do not behave strategically on the basis of private information. In so doing, a large and important class of models involving problems of efficient allocation in the presence of incentive constraints is excluded.

The main ideas of efficient resource allocation are present in their simplest form in the linear activity analysis model of production. We begin with that model.

Efficiency of Production: Linear Activity Analysis

The analysis of production can to some extent be separated from that of other economic activity. The concept of efficiency appropriate to this analysis descends from that of Pareto optimality, which refers to both productive and allocative efficiency in the full economy in which production is embedded. It is useful to begin with a model in which technological possibilities afford constant returns to scale, that is, with the (linear) activity analysis model of production pioneered by Koopmans (1951a, 1951b, 1957), and closely related to the development of linear programming associated with Dantzig (1951a, 1951b) and independently with the Russian mathematician Kantorovitch (1939, 1942) and Kantorovitch and Gavurin (1949).

The two primitive concepts of the model are *commodity* and *activity*. A list of n commodities is postulated; a commodity *bundle* is given by specifying a sequence of n numbers $a_1, a_2, \ldots, a_n$. Technological possibilities are thought of as knowledge of how to transform commodities. Such knowledge may be described in terms of collections of activities called *processes*, much as knowledge of how to prepare food is described by recipes. A recipe commonly has two parts, a list of ingredients or inputs and of the output(s) of the recipe, and a description of how the ingredients are to be combined to produce the output(s). In the activity analysis model the description of productive activity is suppressed. Only the specification of inputs and outputs is retained; this defines the production process.

Commodities are classified into 'desired', 'primary' and 'intermediate' commodities. Desired commodities are those whose consumption or availability is the recognized goal of production; they satisfy wants. Primary commodities are those available from nature. (A primary commodity that is also desired is listed separately among the desired commodities and must be transformed by an act of production into its desired form.) Intermediate commodities are those that merely pass from one stage of production to another. Each commodity can exist in any non-negative amount (*divisibility*). Addition and subtraction of the numbers measuring the amount of a commodity represent joining and separating corresponding amounts of the commodity.

An activity is characterized by a *net output number* for each commodity, which is positive if the commodity is a net output, negative if it is a net input and zero if it is neither. The term *input-output vector* is also used for this ordered array of numbers. Activity analysis postulates a finite number of basic activities from which all technologically possible activities can be generated by suitable combination. Allowable combinations are as follows. If two activities are known to be possible, then the activity given by their algebraic sum is also possible, i.e. if $a = (a_1, a_2, \ldots, a_n)$ and $b = (b_1, b_2, \ldots, b_n)$, then $a + b = (a_1 + b_1, a_2 + b_2, \ldots, a_n + b_n)$ is also possible. Thus, additivity embodies an assumption of non-interaction between productive activities, at least at the level of knowledge. Furthermore, if an activity is possible, then so is every non-negative multiple of it (*proportionality*), i.e. if $a = (a_1, a_2, \ldots, a_n)$ is possible, then so is $\mu a = (\mu a_1, \mu a_2, \ldots, \mu a_n)$ for any non-negative real number μ. This expresses the assumption of constant returns to scale. The family of activities consisting of all non-negative multiples of a given one forms a process. Since there is a finite number of basic activities, there is also a finite number of basic processes, each intended to describe a basic method of production capable of being carried out at different levels, or intensities.

The assumptions of additivity and proportionality determine a linear model of technology that can be given the following form. Let A be an n by k matrix whose jth column is the input-output vector representing the basic activity that defines the jth basic process, and let $x = (x_1, x_2, \ldots, x_n)$ be the vector whose jth component x_j is the scale (level or intensity) of the jth basic process. Let $y = (y_1, y_2, \ldots, y_n)$ be the vector of commodities. Technology is represented by a linear transformation mapping the space of activity levels into the commodity space, i.e.

$$y = Ax \quad x \geq 0.$$

With the properties assumed, a process can be represented geometrically in the commodity space by a halfline from the origin including all non-negative multiples of some activity in that process. The finite number of halflines representing basic processes generate a convex polyhedral cone consisting of all activities that can be expressed as sums of activities in the basic processes, or equivalently, as non-negative linear combinations of the basic activities, sometimes called a *bundle of basic activities*. This cone is called the *production set*, or set of *possible productions*.

Two other assumptions are made about the production set itself, rather than just the individual activities. First, there is no activity, whether basic or derived, in the production set with a positive net output of some commodity and non-negative net outputs of all commodities. This excludes the possibility of producing something from nothing, whether directly or indirectly. Second, it is assumed that the production set contains at least one activity with a positive net output of some commodity.

If the availability of primary commodities is subject to a bound, the technologically possible productions described by the production set are subject to another restriction; only those possible productions that do not require primary inputs in amounts exceeding the given bounds can be produced. Furthermore, because intermediate commodities are not desired in themselves, their net output is required to be zero. (Strictly speaking, the technological constraint on intermediate commodities is that their net output be non-negative. The requirement that they be zero can be viewed as one of elementary efficiency, excluding accumulation or necessity to dispose of unwanted goods.) With these restrictions the model can be written

$$y = Ax, \quad x \geq 0, \quad y_i = 0$$

if i is an intermediate commodity, and

$$y_i \geq r_i$$

if i is a primary commodity, where r_i is the (non-positive) limit on the availability of primary commodity i. This leads to the concept of an *attainable* activity.

A bundle of basic activities is *attainable* if the resulting net outputs are non-negative for all desired commodities, zero for intermediate commodities and non-positive for primary commodities, and if the total inputs of primary commodities do not exceed (in absolute amount) the prescribed bounds of availability of those commodities. The set of activities satisfying these conditions is a truncated convex polyhedral cone in the commodity space called the *set of attainable productions*.

The concept of productive efficiency in this model is as follows. An activity (a bundle of basic activities) is *efficient* if it is attainable and if every activity that provides more of some desired commodity and no less of any other is not attainable.

This concept can be seen to be a specialization of Pareto optimality. If for each desired commodity there is at least one consumer who is not satiated in that commodity, at least in the range of production attainable within the given resource limitations, then increasing the

amount of any desired commodity without decreasing any other can improve the state of some non-satiated consumer without worsening that of any other.

Characterizing efficient production in terms of prices

Efficient production can be characterized in terms of *implicit prices*, also called *shadow prices*, or in the context of linear programming, *dual variables*. Efficient activities are precisely those that maximize profit for suitably chosen prices. The profit returned by a process carried out at the level x is

$$x \sum_i p_i a_i,$$

where the prices are $p = (p_1, \ldots, p_n)$, and $a = (a_1, \ldots, a_n)$ is the basic activity defining the process; the profit on the bundle of activities Ax at prices p is given by the inner product $py = pAx$.

This characterization is the economic expression of an important mathematical fact about convex sets in $n-1$ dimensional Euclidean space, namely that through every point of the space not interior to the convex set in question there passes a hyperplane that contains the set in one of its two halfspaces (Fenchel, 1950; Nikaido, 1969, 1970). (A hyperplane in n dimensional space is a level set of a linear function of n variables, and thus is a translate of an $n-1$ dimensional linear subspace. A hyperplane is given by an equation of the form $c_1x_1 + c_2x_2 + \cdots + c_nx_n = k$, where the x's are variables, the c's are coefficients defining the linear function and k is a constant identifying the level set. A hyperplane divides the space into two halfspaces corresponding to the two inequalities $c_1x_1 + c_2x_2 + \cdots + c_nx_n \gtrless k$ respectively.) It can also be seen that a point of a convex set is a boundary point if and only if it maximizes a linear function on the (closure of the) set. These facts can be used to characterize efficient production because the attainable production set is convex and efficient activities are boundary points of it. Because the efficient points are those, roughly speaking, on the 'north-east' frontier of the set, the linear functions associated with them have non-negative coefficients, interpreted as prices. On the other hand, if a point of the attainable set maximizes a linear function with strictly positive coefficients (prices), then it is on the 'north-east' frontier of the set.

In Figure 1 the set enclosed by the broken line and the axes is the projection of the attainable set on the output coordinates; inputs are not shown. The point y' in the figure is efficient; the point y' is not; both y' and y'' maximize a linear function with non-negative coefficients (the level set containing y' is labelled a and also contains y''). However, y' maximizes a linear function with positive coefficients (one such, whose level set through y' is labelled b, is shown), while y'' does not.

These implicit, or efficiency prices arise from the logic of efficiency or maximization when the relevant sets are

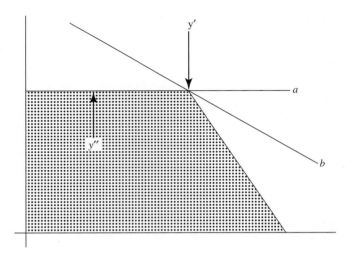

Figure 1

convex, not from any institutions such as markets or exchange. An important reason for interest in them is the possibility of achieving efficient performance by decentralized methods. As described above, under the assumptions of additivity and constant returns to scale the production set can be seen to be generated by a finite number of basic processes, each of which consists of the activities that are non-negative multiples of a basic activity, the multiple being the scale (level, or intensity) at which the process is operated. Following the presentation of Koopmans (1957), each basic process is controlled by a manager, who decides on its level. The manager of a process is assumed to know only the input-output coefficients of his process. Each primary resource is in the charge of a resource holder, who knows the limit of its availability. Efficiency prices are used to guide the choices of managers and resource holders. (Under constant returns to scale, if an activity yields positive profit at a given system of prices, then increasing the scale of the process containing that activity increases the profit. Since the scale can be increased without bound, if the profitability of a process is not zero or negative, then, in the eyes of its manager, who does not know the aggregate resource constraints, it can be made infinite. Therefore, the systems of prices that can be considered for the role of efficiency prices must be restricted to those *compatible with the given technology*, namely prices such that no process is profitable and at least one process breaks even.) Two propositions characterize efficient production by prices and provide the basis for an interpretation in terms of decentralized control of production.

In a given linear activity analysis model, if there is a given system of prices compatible with the technology, in which the prices of all desired commodities are positive, then any attainable bundle of basic activities

selected only from processes that break even and which utilizes all positively priced primary commodities to the limit of their availability and does not use negatively priced primary commodities at all, is an efficient bundle of activities.

In a given linear activity analysis model, each efficient bundle of activities has associated with it at least one system of prices compatible with the technology such that every activity in that bundle breaks even and such that prices of desired commodities are positive, and the price of a primary commodity is non-negative, zero or non-positive, according as its available supply is full, partly, or not used at all (Koopmans, 1957).

These propositions are stated in a static form. There is no reference to managers raising or lowering the levels of the processes they control, or to resource holders adjusting prices. A dynamic counterpart of these propositions would be of interest, but because of the linearity of the model such dynamic adjustments are unstable (Samuelson, 1949).

It should also be noted that the concept of decentralization is not explicitly defined in this literature; the interpretation is by analogy with the competitive mechanism. Nevertheless, the interest in characterizing efficiency by prices and their interpretation in terms of decentralization is an important theme in the study of efficient resource allocation.

The linear activity analysis model has been generalized in several directions. These include dropping the assumption of proportionality, dropping the restriction to a finite number of basic activities, dropping the restriction to a finite number of commodities and dropping the restriction to a finite number of agents. Perhaps the most directly related generalization is to the nonlinear activity analysis, or nonlinear programming, model.

Efficiency of production: nonlinear programming

In the nonlinear programming model there is, as in the linear model, a finite number of basic processes. Their levels are represented by a vector $x = (x_1, x_2, \ldots, x_k)$, where k is the number of basic processes. Technology is represented by a nonlinear transformation from the space of process levels to the commodity space (still assumed to be finite dimensional), written

$$y = F(x), \quad x \geq 0.$$

The production set in this model is the image in the commodity space of the non-negative orthant of the space of process levels. Under the assumptions usually made about F, the production set is convex, though, of course, not a polyhedral cone.

In this model as in the linear activity analysis model a central result is the characterization of efficient production in terms of prices. The simplest case to begin with is that of one desired commodity, say, one output, with perhaps several inputs. In this case the (vector-valued) function F can be written

$$F(x) = [f(x), g_1(x), g_2(x), \ldots, g_m(x)],$$

where the value of f is the output, and $g_1, \ldots, g_m$ correspond to the various inputs. Resource constraints are expressed by the conditions

$$g_j(x) \geq 0, \quad for \ j = 1, 2, \ldots, m,$$

and non-negativity of process levels by the condition, $x \geq 0$. (Here the resource constraints $r_j \leq h_j(x) \leq 0$ are written more compactly as $h_j(x) - r_j = g_j(x) \geq 0$.)

In this model the definition of efficient production given in the linear model amounts to maximizing the value of f subject to the resource and non-negativity constraints just mentioned.

Problems of constrained maximization are intimately related to saddle-point problems. Let L be a real valued function defined on the set $X \times Y$ in R^n. A point (x^*, y^*) in $X \times Y$ is a *saddle point* of L if

$$L(x, y^*) \leq L(x^*, y^*) \leq L(x^*, y),$$

for all x in X and all y in Y. The concept of a concave function is also needed. A real valued function f defined on a convex set X in R^n is a *concave function* if for all x and y in X and all real numbers $0 \leq a \leq 1$

$$f(ax + (1 - a)y) \geq af(x) + (1 - a)f(y).$$

The following mathematical theorem is fundamental.

Theorem (Kuhn and Tucker, 1951; Uzawa, 1958): Let f and $g_1, g_2, \ldots, g_m$ be real valued concave functions defined on a convex set X in R^n. If f achieves a maximum on X subject to $g_j(x) \geq 0$, $j = 1, 2, \ldots, m$ at the point x^* in X, then there exist non-negative numbers $p_0^*, p_0^*, \ldots, p_m^*$, not

all zero, such that $p_0^* f(x) + p^* g(x) \leq p_0^* f(x^*)$ for all x in X, and furthermore, $p^* g(x^*) = 0$. (Here the vectors $p^* = (p_1^*, p_2^*, \ldots, p_m^*)$, and $g(x) = [g_1(x), g_2(x), \ldots, g_m(x)]$) The vector p^* may be chosen so that

$$\sum_0^m p_j^* = 1.$$

An additional condition (Slater, 1950) is important. (It ensures that the coefficient p_0 of f is not zero.)

Slater's Condition: There is a point x' in X at which $g_j(x') > 0$ for all $j = 1, 2, \ldots, m$.

If attention is restricted to concave functions, as in the Kuhn-Tucker-Uzawa Theorem, the relation between constrained maxima and saddle points can be summarized in the following theorem.

Theorem: If f and g_j, $j = 1, 2, \ldots, m$ are concave functions defined on a convex subset X in R^n, and if Slater's Condition is satisfied, then x^* in X maximizes f subject to $g_j(x) \geq 0$, $j = 1, 2, \ldots, m$, if and only if there exists $\lambda^* = (\lambda_1^*, \lambda_2^*, \ldots, \lambda_m^*)$, $\lambda_j^* \geq 0$ for $j = 1, 2, \ldots, m$, such that (x^*, λ^*) is a saddle point of $L(x, \lambda) = f(x) + \lambda g(x)$ on $X \times R_+^n$.

This theorem is easily seen to cover the case where some constraints are equalities, as in the case of intermediate commodities. The sufficiency half of this theorem holds for functions that are not concave.

The auxiliary variables $\lambda_1, \lambda_2, \ldots, \lambda_m$, called *Lagrange multipliers*, play the role of efficiency prices, or shadow prices; they evaluate the resources constrained by the condition $g(x) \geq 0$. The maximum characterized by the theorem is a global one, as in the case of linear activity analysis.

If the functions involved are differentiable, a saddle point of the Lagrangean can be studied in terms of first-order conditions. The first-order conditions are necessary conditions for a saddle point of L. If the functions f and the g's are concave on a convex set X, then the first-order conditions at a point (x^*, λ^*) are also sufficient; that is, they imply that (x^*, λ^*) is a saddle point of L. Thus,

Theorem: If $f, g_1, g_2, \ldots, g_m$ are concave and differentiable on an open convex set X in R^n, and if Slater's Condition is satisfied, then x^* maximizes f subject to $g_j(x) \geq 0$ for $j = 1, 2, \ldots, m$ if and only if there exists numbers $\lambda_1^*, \lambda_2^*, \ldots \lambda_m^*$ such that the first-order conditions for a saddle point of $L(x, \lambda) = f(x) + \lambda g(x)$ are satisfied at (x^*, λ^*).

If there are non-negativity conditions on the x's,

$$g_j(x) \geq 0, x \geq 0, \quad x \ in \ R^n$$

and the first-order conditions can be written

$$f_x^* + \lambda^* g_x^* \leq 0, \qquad (f_x^* + \lambda^* g_x^*)x^* = 0,$$
$$\lambda^* g(x^*) = 0, \quad g(x^*) \geq 0, \quad g(x^*) \geq 0,$$
$$\lambda^* \geq 0 \qquad \text{and} \quad \lambda^* g(x^*) = 0,$$

where f_x^* denotes the derivative of f evaluated at x^*. In more explicit notation, the conditions $f_x^* + \lambda^* g_x^* = 0$ can be written as

$$\partial f/\partial x_i + \sum_{j=1}^{m} \lambda_j^* \partial g_j/\partial x_i = 0, \qquad i = 1, 2, \ldots, n$$

When the assumption of concavity is dropped, it is no longer possible to ensure that the local maximum is also a global one. However, it is still possible to analyse local constrained maxima in terms of local saddle-point conditions. In this case a condition is needed to ensure that the first-order conditions for a saddle point are indeed necessary conditions. The Kuhn-Tucker Constraint Qualification is such a condition. Arrow, Hurwicz and Uzawa (1961) have found a number of conditions, more useful in application to economic models, that imply the Constraint Qualification.

The case of more than one desired commodity leads to what is called the *vector maximum problem*, Kuhn and Tucker (1951). This may be defined as follows. Let $f_1, f_2, \ldots, f_k$ and $g_1, g_2, \ldots, g_m$ be real valued functions defined on a set X in R^n. We say x^* in X achieves a (global) *vector maximum* of $f = (f_1, f_2, \ldots, f_k)$ subject to $g_j(x) \geq 0, j = 1, 2, \ldots, m$ if,

(I) $g_j(x^*) \geq 0, j = 1, 2, \ldots, m,$
(II) there does not exist x' in X satisfying $f_i(x') \geq f_i(x^*)$ for $i = 1, 2, \ldots, k$ with $f_i(x') > f_i(x^*)$ for some value of i, and $g_j(x') \geq 0$ for $j = 1, 2, \ldots, m$.

This is just the concept of an efficient point expressed in the present notation.

A vector maximum has a saddle-point characterization similar to that for a scalar valued function.

Theorem: Let $f_1, f_2, \ldots, f_k$ and $g_1, g_2, \ldots, g_m$ be real valued concave functions defined on a convex X set in R^n. Suppose there is x^0 in X such that $g_j(x^0) > 0, j = 1, 2, \ldots, m$ (Slater's Condition). If x^* achieves a vector maximum of f subject to $g(x) \geq 0$ then there exist $a = (a_1, a_2, \ldots, a_k)$ and $\lambda^* = (\lambda_1^*, \lambda_2^*, \ldots, \lambda_m^*)$ with $a_j \geq 0$ for all j, $a \neq 0$ and $\lambda \geq 0$ such that (x^*, λ^*) is a saddle point of the Lagrangean $L(x, \lambda) = af(x) + \lambda g(x)$.

Several different 'converses', to this theorem are known. One states that if x^* maximizes $L(x, \lambda^*)$ for some strictly positive vector a and non-negative λ^*, and if $\lambda^* g(x^*) = 0$ and $g(x^*) \geq 0$, then x^* gives a vector maximum of f subject to $g(x) \geq 0$, and x in X. Another, parallel

to the result for the case of one desired commodity, is the following.

Theorem: Let f and g be functions as in the theorem above. If there are positive real numbers $a_1, a_2, \ldots, a_k$ and if (z^*, λ^*) is a saddle point of the Lagrangean L (defined as above) then (I) x^* achieves a maximum of f subject to $g(x) \geq 0$ on X, and (II) $\lambda^* g(x^*) = 0$.

The positive numbers $a_1, \ldots, a_k$ are interpreted as prices of desired commodities, and the non-negative numbers λ_j^* are prices of the remaining commodities. The condition $\lambda^* g(x^*) = 0$ which arises in these theorems states that the value of unused resources at the efficiency prices λ^* is zero; that is, resources not fully utilized at a vector maximum have a zero price.

The connection between vector maxima and Pareto optima is as follows. Because a vector maximum is an efficient point (for the vectorial ordering of the commodity space), it is a Pareto optimum for appropriately specified (non-satiated) utility functions, as was already pointed out in the case of the linear activity analysis model. Furthermore, if the functions $f_1, \ldots, f_k$ are themselves utility functions, and the variable x denotes allocations, with the constraints g defining feasibility, then a vector maximum of f subject to the constraints $g(x) \geq 0$ and x in X is a Pareto optimum, and vice versa. Hence the saddle-point theorems give a characterization of Pareto optima by prices. The interpretation of prices in terms of decentralized resource allocation described in the linear activity analysis model also applies in this nonlinear model. The proofs of these theorems reveal an important logical role played by the principle of marginal cost pricing.

The basic theorems of nonlinear programming, especially the Kuhn-Tucker-Uzawa Theorem in the setting of the vector maximum problem, have been extended to the case of infinitely many commodities. (Hurwicz, 1958, first obtained the basic results in this field.) Technicalities aside, the theorems carry over to certain infinite dimensional spaces, namely linear topological spaces, or in the case of first-order conditions, Banach spaces.

Dropping the restriction to a finite number of basic processes leads to classical production or transformation function models of production, whose properties depend on the detailed specifications made.

Samuelson (1947) used Lagrangean methods to analyse interior maxima subject to equality constraints in the context of production function models, as well as that of optimization by consumers. He also gave the interpretation of Lagrange multipliers as shadow prices.

Efficient allocation in an economy with consumers and producers

In an economy with both consumption and production decisions, efficiency is concerned with distribution as well as production. Data about restrictions on consumption and the wants of consumers must be specified in addition to the data about production. The elements of the models are as follows.

The commodity space is denoted X; it might be l-dimensional Euclidean space, or a more abstract space such as an additive group in which, for example, some coordinates are restricted to have integer values. There is a (finite) list of consumers, $1, 2, \ldots, n$, and a similar list of producers, $1, 2, \ldots, m$. A *state* of the economy is an array consisting of a commodity bundle for each agent in the economy, consumer or producer. This may be written $(\langle x^i \rangle, \langle y^j \rangle)$, where $\langle x^i \rangle = (x^1, x^2, \ldots, x^n)$ and $\langle y^j \rangle = (y^1, y^2, \ldots, y^m)$ and x^i and y^j are commodity bundles. Absolute constraints on consumption are expressed by requiring that the allocation $\langle x^i \rangle$ belong to a specified subset X of the space X^n of allocations. Examples of such constraints are:

1. The requirement that the quantity of a certain commodity be non-negative.
2. The requirement that a consumer requires certain minimum quantities of commodities in order to survive.

Each consumer i has a preference relation, denoted $\succsim_i$, defined on X. This formulation admits externalities in consumption, including physical externalities and externalities in preferences; for example, preferences that depend on the consumption of other agents, termed non-selfish preferences. The consumption set of the ith consumer is the projection X^i of X onto the space of commodity bundles whose coordinates refer to the holdings of the ith consumer.

Technology is specified by a production set Y, a subset of X^m, consisting of those arrays $\langle y^j \rangle$ of input-output vectors that are jointly feasible for all producers. The production set of the jth producer, denoted Y^j, is the projection of Y onto the subspace of X^m whose coordinates refer to the jth producer.

The (aggregate) initial endowment of the economy is denoted by w, a commodity bundle in X.

These specifications define an *environment*, a term introduced by Hurwicz (1960) in this usage and according to him suggested by Jacob Marschak. This term refers to the primitive or given data from which analysis begins. Each environment determines a set of *feasible* states. These are the states $(\langle x^i \rangle, \langle y^j \rangle)$ such that $\langle x^i \rangle$ is in X, $\langle y^j \rangle$ is in Y and

$$\sum x^i - \sum y^j \le w.$$

An environment determines the set of states that are Pareto optimal for that environment. Explicitly, they are the states $(\langle x^{*i} \rangle, \langle y^{*j} \rangle)$ that are feasible in the given environment, and such that if any other state $(\langle x^i \rangle, \langle y^j \rangle)$ has the property that $\langle x^i \rangle \succsim_i \langle x^{*i} \rangle$ for all i with $\langle x^i \rangle \succ_i \langle x^{*i} \rangle$ for some i', then $(\langle x^i \rangle, \langle y^j \rangle)$ is not feasible in the given environment.

It is important to note that the set of feasible states and the set of Pareto optimal states are completely determined by the environment; specification of economic organization is not involved.

At this level of generality, where externalities in consumption and production are admitted as possibilities, and where commodities may be indivisible, no general characterization of Pareto optima in terms of prices is possible. (Indeed, Pareto optima may not exist. Conditions that make the set of feasible allocations non-empty and compact and preferences continuous suffice to ensure the existence of Pareto optima.) In environments with externalities, or other non-neoclassical features, Pareto optima are generally not attainable by decentralized processes.

If the class of environments under consideration is restricted to the neoclassical environments, the fundamental theorems of welfare economics provide a characterization of Pareto optimal states via efficiency prices. That characterization has a natural interpretation in terms of a decentralized mechanism for allocation of resources.

The framework for these results is obtained by restricting the class of environments specified above as follows. The commodity space is to be Euclidean space of l dimensions, i.e. $X = R^l$. The consumption set for the economy is to be the product of its projections, i.e. $X = X^1 \times X^2 \times \cdots \times X^n$. This expresses the fact that if each agent's consumption is feasible for him, the total array is jointly feasible. Furthermore, each agent is restricted to having selfish preferences; that is, agent i's preference relation depends only on the coordinates of the allocation that refer to his holdings. In that case the preference relation $\succsim_i$ may be defined only on X^i, for each i. Similarly, externalities are ruled out in production, i.e. $Y = Y^1 \times Y^2 \times \cdots \times Y^m$.

The concept of an *equilibrium relative to a price system* (Debreu, 1959) serves to characterize Pareto optima by prices. A price system, denoted p, is an element of R^l; the environment $e = [(X^i), (\succsim_i), (Y^j), w]$ is of the restricted type specified above (free of externalities and indivisibilities).

A state $[(x^{*i}), (y^{*j})]$ of e is an *equilibrium relative to price system p* if:

1. For every consumer i, x^{*i} maximizes preference $\succsim_i$ on the set of consumption bundles whose value at the prices p does not exceed the value of x^{*i} at those prices, i.e. if x^i is in $\{x^i$ in $x^i : px^i \le px^{*i}\}$ then $x^i \precsim_i x^{*i}$.
2. For every producer j, y^{*j} maximizes profit py^j on Y^j.
3. Aggregate supply and demand balance, i.e.

$$\sum_i x^{*i} - \sum_i y^{*j} = w.$$

An equilibrium relative to a price system differs from a competitive equilibrium (see below) in that the former does not involve the budget constraints applying to consumers in the latter concept. In an equilibrium relative to a price system the distribution of initial endowment and of the profits of firms among consumers need not be specified.

The first theorem of neoclassical welfare economics states, subject only to the exclusion of externalities and a mild condition that excludes preferences with thick indifference sets, that a state of an environment e that is an equilibrium relative to a price system p is a Pareto optimum of e (Koopmans, 1957).

The second welfare theorem is deeper and holds only on a smaller class of environments, sometimes referred to in the literature as the *classical environments* (called neoclassical above). One version of this theorem is as follows. Let $e = [(X^i), (\succsim_i), (Y^j), \ w]$ be an environment such that for each i

1. X^i is convex.
2. The preference relation $\succsim_i$ is continuous.
3. The preference relation $\succsim_i$ is convex.
4. The set $\Sigma_j Y_j$ is convex.

Let $[(x^{*i}), (y^{*j})]$ be a Pareto optimum of e such that there is at least one consumer who is not satiated at x^{*i}. Then there is a price system p, with not all components equal to 0, such that – except for Arrow's (1951) 'exceptional case', where p is such that for some i the expenditure px^{*i} is a minimum on the consumption set X^i – the state $[(x^{*i}), (y^{*j})]$ is an equilibrium relative to p.

(The condition that preferences are convex and not satiated is sufficient to exclude 'thick' indifference sets. A preference relation on X^i is convex if whenever x' and x'' are points of X^i with x' strictly preferred to x'' then the line segment connecting them (not including the point x'') is strictly preferred to x'. The consumption set X^i must be convex for this property to make sense. A preference relation is not satiated if there is no consumption preferred to all others.)

Hurwicz (1960) has given an alternative formalization of the competitive mechanism in which Arrow's exceptional case presents no difficulties.

If the exceptional case is not excluded, then it can still be said that:

1. x^{*i} minimizes expenditure at prices p on the upper contour set of x^{*i}, for every i, and
2. y^{*j} maximizes 'profit' py^j on the production set Y^j, for every j.

The state (x^*, y^*) together with the prices p, constitute a *valuation equilibrium* (Debreu, 1954).

As in the case of efficiency prices in pure production models, these prices have in themselves no institutional significance. They are, however, in the same way as other efficiency prices, suggestive of an interpretation in terms of decentralization.

If, in addition to the restriction to classical environments, the economic organization is specified to be that of a system of markets in a private ownership economy, and if agents are assumed to take prices as given, then the welfare theorems can translate into the assertion that the set of Pareto optima of an environment e and the set of competitive equilibria for e (subject to the possible redistribution of initial endowment and ownership shares) are identical. More precisely, the specification of the environment given above is augmented by giving each consumer a bundle of commodities, his initial endowment, denoted w^i. The total endowment is $w = \Sigma_i w^i$. Furthermore, each consumer has a claim to a share of the profits of each firm; the claims for the profit of each firm are assumed to add up to the entire profit. When prices and the production decisions of the firms are given, the profits of the firms are determined and so is the value of each consumer's initial endowment. Therefore, the income of each consumer is determined. Hence, the set of commodity bundles a consumer can afford to buy at the given prices, called his *budget set*, is determined; this consists of all bundles in his consumption set whose value at the given prices does not exceed his income at the given prices. Competitive behaviour of consumers means that each consumer treats the prices as given constants and chooses a bundle in his budget set that maximizes his preference: that is, a bundle x^i that is in X^i and such that if any other bundle $x'^{\ i}$ is preferred to it, then $x'^{\ i}$ is not in his budget set.

Competitive behaviour of firms is to maximize profits computed at the given prices p, regarded by the firms as constants; that is, a firm chooses a production vector y^j in its production set with the property that any other vector affording higher profits than py^j is not in the production set of firm j.

A *competitive equilibrium* is a specification of a commodity bundle for each consumer, a production vector for each firm, and a price system, together denoted $[(x^{*i}), (y^{*j}), p^*]$, where p^* has no negative components, satisfying the following conditions:

1. For each consumer i the bundle x^{*i} maximizes preference on the budget set of i.
2. For each firm j the production vector y^{*j} maximizes profit $p^* y^j$ on the production set Y^j.
3. For each commodity, the total consumption does not exceed the net total output of all firms plus the total initial endowment, i.e. $\Sigma_i x^{*i} - \Sigma_j y^{*j} \leq w = \Sigma_i w^i$;
4. For those commodities k for which the inequality in 3 is strict; that is, the total consumption is less than initial endowment plus net output, the price p_k^* is zero.

The welfare theorems stated in terms of equilibrium relative to a price system translate directly into theorems stated in terms of competitive equilibrium. Briefly, every competitive equilibrium allocation in a given classical environment is Pareto optimal in that environment, and every Pareto optimal allocation in a given classical environment can be made a competitive equilibrium allocation of an environment that differs from the given one only in the distribution of the initial endowment. (Arrow (1951), Koopmans (1957), Debreu (1959) and Arrow and Hahn (1971) give modern and definitive treatment of the classical welfare theorems.)

It should be noted that the equilibria involved must exist for these theorems to have content. Sufficient conditions for existence of competitive equilibrium, which, since a competitive equilibrium is automatically an equilibrium relative to a price system, are also sufficient for existence of an equilibrium relative to a price system, include convexity and continuity of consumption sets and preferences and of production sets, as well as some assumptions which apply to the environment as a whole, restricting the ways in which individual agents may fit together to form an environment (Arrow and Debreu, 1954; Debreu, 1959; McKenzie, 1959).

The second welfare theorem involves redistribution of initial endowment. This is essential because the set of competitive equilibria from a given initial endowment is small (essentially finite) (Debreu, 1970), while the set of Pareto optima is generally a continuum. The set of Pareto optima cannot in general be generated as competitive allocations without varying the initial point. If redistribution is done by an economic mechanism, then it should be a decentralized one to support the interpretation given of the second welfare theorem. No such mechanism has been put forward as yet. Redistribution of initial endowment by lump-sum taxes and transfers has been discussed. A customary interpretation views these as brought about by a process outside economics, perhaps by a political process; no claim is made that such processes are decentralized. Some economists consider dependence on redistribution unsatisfactory because information about initial endowment is private; only the individual agent knows his own endowment. Consequently the expression of that information through political or other action can be expected to be strategic. The theory of second-best allocations has been proposed in this context. Redistribution of endowment is excluded, and the mechanism is restricted to be a price mechanism, but the price system faced by consumers is allowed to be different from that faced by producers; all agents behave according to the rules of the (static) competitive mechanism. The allocations that satisfy these conditions, when the price systems are variable, are maximal allocations in the sense that they are Pareto optimal within the restricted class just defined. These are so-called *second-best* allocations. This analysis was pioneered by Lipsey and Lancaster (1956) and Diamond and Mirrlees (1971).

Efficient allocation in non(neo)classical environments

The term *nonclassical* refers to those environments that fail to have the properties of classical ones; there may be indivisible commodities, nonconvexities in consumption sets, preferences or production sets, or externalities in production or consumption. An example of non-convex preference would arise if a consumer preferred living in either Los Angeles or New York to living half the time in each city, or living halfway between them, depending on the way the commodity involved is specified. A production set representing a process that affords increasing returns to scale is an example of nonconvexity in production. A large investment project such as a road system is an example of a significant indivisibility. Phenomena of air or water pollution provide many examples of externalities in consumption and production.

The characterization of optimal allocation in terms of prices provided by the classical welfare theorems does not extend to nonclassical environments. If there are indivisibilities, equilibrium prices may fail to exist. Lerner (1934, 1947) has proposed a way of optimally allocating resources in the presence of indivisibilities. It would typically require adding up consumers' and producers' surplus.

Increasing returns to scale in production generally results in non-existence of competitive equilibrium, because of unbounded profit when prices are treated as given. Nash equilibrium, a concept from the theory of games, can exist even in cases of increasing returns. The difficulty is that such equilibria need not be optimal. Similar difficulties occur in cases of externalities.

Failure of the competitive price mechanism to extend the properties summarized in the classical welfare theorems to nonclassical environments has led economists to look for alternative ways of achieving optimal allocation in such cases. Such attempts have for the most part sought institutional arrangements that can be shown to result in optimal allocation. Ledyard (1968, 1971) analysed a mechanism for achieving Pareto optimal performance in environments with externalities. The use of taxes and subsidies advocated by Pigou (1932) to achieve Pareto optimal outcomes in cases of externalities is such an example. In a similar spirit Davis and Whinston (1962) distinguish externalities in production that leave marginal costs unaffected from those that do change marginal costs. In the former case they propose a pricing scheme, but one that involves lump-sum transfers. Marginal cost pricing, including lump-sum transfers to compensate for losses, which was extensively discussed as a device to achieve optimal allocation in the presence of increasing returns (Lerner, 1944; Hotelling, 1938; and many others) is another example of a scheme to realize optimal outcomes in nonclassical environments in a way that seeks to capture the benefits associated with decentralized resource allocation. In the case of production under conditions of increasing returns, the use of nonlinear prices has been suggested in an effort to achieve optimality with at least some of the benefits of decentralization. (See Arrow and Hurwicz,

1960; Heal, 1971; Brown and Heal, 1982; Brown et al., 1986; Jennergren, 1971; Guesnerie, 1975.)

In the case of indivisibilities, and in the context of productive efficiency, integer programming algorithms exist for finding optima in specific problems, but a general characterization in terms of prices such as exists for the classical environments is not available. A decentralized process, involving the use of randomization, whose equilibria coincide with the set of Pareto optima has been put forward by Hurwicz, Radner and Reiter (1975). This process has the property that the counterparts of the classical welfare theorems hold for environments in which all commodities are indivisible, and the set of feasible allocations is finite, or in which there are no indivisible commodities, or externalities, but there may be nonconvexities in production or consumption sets, or in preferences. This, of course, includes the possibility of increasing returns to scale in production.

The schemes and processes that have been proposed, including many not described here, are quite different from one another. If attention is confined to pricing schemes without additional elements, such as lump-sum transfers, it may be satisfactory to proceed on the basis of an informal intuitive notion of decentralization. This amounts in effect to identifying decentralization with the competitive mechanism, or more generally with price or market mechanisms. If a broader class of processes is to be considered, including some already mentioned in this discussion, then a formal concept of decentralized resource allocation process is needed.

Efficient allocation through informationally decentralized processes

A formal definition of a concept of *allocation process* was first given by Hurwicz (1960). He also gave a definition of *informational decentralization* applying to a broad class of allocation mechanisms, based in part on a discussion by Hayek (1945) of the advantages of the competitive market mechanism for communicating knowledge initially dispersed among economic agents so that it can be brought to bear on the decisions that determine the allocation of resources. Hurwicz's formulation is as follows.

There is an initial dispersion of information about the environment; each agent is assumed to observe directly his own characteristic, e^i, but to know nothing directly about the characteristics of any other agent. In the absence of externalities, specifying the array of individual characteristics specifies the environment, i.e. $e = (e^1, \ldots, e^n)$. When there are externalities, an array of individual characteristics, each component of which corresponds to a possible environment, may not together constitute a possible environment. In more technical language, when there are externalities the set of environments is not the Cartesian product of its projections onto the sets of individual characteristics.

The goal of economic activity, whether efficiency, Pareto optimality or some other desideratum such as fairness, can be represented by a relation between the set of environments and the set of allocations, or outcomes. This relation assigns to each environment the set of allocations that meet the criterion of desirability. In the case of the Pareto criterion, the set of allocations that are Pareto optimal in a given environment is assigned to that environment. Formally, this relation is a correspondence (a set-valued function) from the set of environments to the set of allocations.

An allocation process, or mechanism, is modelled as an explicitly dynamic process of communication, leading to the determination of an outcome. In formal organizations standardized forms are frequently used for communication; in organized markets like the Stock Exchange, these include such things as order forms; in a business, forms on which weekly sales are reported; in the case of the Internal Revenue Service, income tax forms. A form consists of entries or blanks to be filled in a specified way. Thus, a form can be regarded as an ordered array of variables whose values come from specified sets. In the Hurwicz model, each agent is assumed to have a *language*, denoted M^i for the ith agent, from which his (possibly multi-dimensional) *message*, m^i, is chosen. The *joint message* of all the agents, $m = (m^1, \ldots, m^n)$ is in the *message space* $M = M^1 \times \cdots \times M^n$. Communication takes place in time, which is discrete; the message $m_t = (m_t^1, \ldots, m_t^n)$ denotes the message at time t. The message an agent emits at time t can depend on anything he knows at that time. This consists of what the agent knows about the environment by direct observation, by assumption, (*privacy*) his own characteristics, e^i for agent i, and what he has learned from others via the messages received from them. The agents' behaviour is represented by *response functions*, which show how the current message depends on the information at hand. Agent i's message at time t is

$$m_t^i = f^i(m_{t-1}, m_{t-2}, \ldots, e^i), \quad i = 1, \ldots, n,$$
$$t = 0, 1, 2, \ldots$$

If it is assumed that memory is finite, and bounded, it is possible without loss of generality to take the number of past periods remembered to be one. (If memory is unbounded, taking the number of periods remembered to be one excludes the possibility of a finite dimensional message space.) In that case the response equations become a system of first order temporally homogeneous difference equations in the messages. Thus:

$$m_t^i = f_i(m_{t-1}; e^i) \quad i = 1, \ldots, n, \quad t = 0, \ldots,$$

which can be written more compactly as

$$(*) \quad m_t = f(m_{t-1}; e).$$

(This formulation can accommodate the case of directed communication, in which some agents do not receive some messages; if agent i is not to receive the message of j, then f^i is independent of m^j, although m^j appears formally as an argument.) Analysis of informational properties of mechanisms is to begin with separated from that of incentives. When the focus is on communication and complexity questions, the response functions are not regarded as chosen by the agent, but rather by the designer of the mechanism.

The iterative interchange of messages modelled by the difference equation system $(^*)$ eventually comes to an end, by converging to a stationary message. (It is also possible to have some stopping rule, such as to stop after a specified number of iterations.) The stationary message, which will be referred to as an *equilibrium message*, is then translated into an outcome, by means of the *outcome function*:

$$h : M \to Z,$$

where Z is the space of outcomes, usually allocations or trades. An allocation mechanism so modelled is called an *adjustment process*; it consists of the triple (M, f, h). Since no production or consumption takes place until all communication is completed, these processes are *tâtonnement* processes.

A more compact and general formulation was given by Mount and Reiter (1974) by looking only at message equilibria when attention is restricted to static properties. A correspondence is defined, called the *equilibrium message correspondence*. It associates to each environment the set of equilibrium messages for that environment. In order to satisfy the requirement of privacy, namely that each agent's message depend on the environment only through the agent's characteristic, the equilibrium message correspondence must be the intersection of individual message correspondences, each associating a set of message acceptable to the individual agent as equilibria in the light of his own characteristic. Thus the equilibrium message correspondence

$$\mu : E \to M,$$

is given by

$$\mu(e) = \bigcap_i \mu^i(e^i),$$

where $\mu^i : E^i \to M$, is the individual message correspondence of agent i. Note that here the message space M need not be the Cartesian product of individual languages. In the case of an adjustment process, the equilibrium message correspondence is defined by the conditions

$$\mu^i(e^i) = \{m \text{ in } M | f^i(m; e^i) = m^i\}, \quad i = 1, \dots, n$$

together with the condition that μ is the intersection of the μ^i. Specification of the outcome function $h : M \to Z$ completes the model, (M, μ, h).

The performance of a mechanism of this kind can be characterized by the mapping defined by the composition of the equilibrium message correspondence μ and the outcome function h. The mapping $h\mu; E \to Z$, possibly a correspondence, specifies the outcomes that the mechanism (M, μ, h) generates in each environment in E. A mechanism, whether in the form of an adjustment process, or in the equilibrium form, is called *Pareto-satisfactory* (Hurwicz, 1960) if for each environment in the class under consideration, the set of outcomes generated by the mechanism coincides with the set of Pareto optimal outcomes for that environment. Allowance must be made for redistribution of initial endowment, as in the case of the second welfare theorem. (A formulation in the framework of mechanisms is given in Mount and Reiter, 1977.)

The competitive mechanism formalized as a static mechanism is as follows. (Hurwicz, 1960, has given a different formulation, and Sonnenschein, 1974, has given an axiomatic characterization of the competitive mechanism from a somewhat different point of view.) The message space M is the space of prices and quantities of commodities going to each agent (it has dimension $n(l-1)$ when there are n agents and l commodities, taking account of budget constraints and Walras' Law), the individual message correspondence μ^i maps agent i's characteristic e^i to the graph of his excess demand function. The equilibrium message is the intersection of the individual ones, and is therefore the price-quantity combinations that solve the system of excess demand equations. The outcome function h is the projection of the equilibrium message onto the quantity components of M. Thus $h\mu(e)$ is a competitive equilibrium allocation (or trade) when the environment is e. The classical welfare theorems state that for each e in E_c, $h[\mu(e)] = p(e)$, where E_c denotes the set of classical environments and P is the Pareto correspondence. (Allowance must be made for redistribution of initial endowment in connection with the second welfare theorem. Explicit treatment of this is omitted to avoid notational complexity. The decentralized redistribution of initial endowment is, as in the case of the second welfare theorem, not addressed.) The welfare theorems can be summarized in the Mount–Reiter diagram (Figure 2) (Reiter, 1977).

The welfare theorems state that this diagram *commutes* in the sense that starting from any environment e in E_c one reaches the same allocations via the mechanism, that is, via $h\mu$, as via the Pareto correspondence P.

With welfare theorems as a guide, the class of environments E_c can be replaced by some other class E, and the Pareto correspondence can be replaced by a correspondence, P, embodying another criterion of optimality, and one can ask whether there is a mechanism, (M, μ, h) that makes the diagram commute, or, in other words, *realizes P*? Without further restrictions on the mechanism, this is a triviality, because one agent can act as a central agent to whom all others communicate their

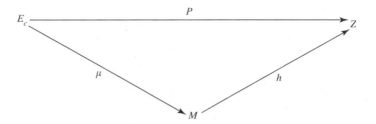

Figure 2

environmental characteristics; the central agent then has the information required to evaluate P.

The concept of an *informationally decentralized mechanism* defined by Hurwicz (1960) makes explicit intuitive notions underlying the view that the price mechanism is decentralized.

Informationally decentralized processes are a subclass of so-called *concrete processes*, introduced by Hurwicz (1960). These are processes that use a language and response rules that allow production and distribution plans to be specified explicitly. The informationally decentralized processes are those whose response rules permit agents to transmit information only about their own actions, and which in effect require each agent to treat the rest of the economy either as one aggregate, or in a symmetrical way that, like the aggregate, gives anonymity to the other agents.

In the case of static mechanisms, the requirements for informational decentralization boil down to the condition that the message space have no more than a certain finite dimension, and in some cases only that it be of finite dimension. In the case of classical environments this can be seen to include the competitive mechanism, and to exclude the obviously centralized one mentioned above.

Without going deeply into the matter, an objective of this line of research is to analyse explicitly the consequences of constraints on economic organization that come from limitations on the capacity of economic agents to observe, communicate and process information. One important result in this field is that there is no mechanism (M, μ, h) where μ preserves privacy, that uses messages smaller (in dimension) than those of the competitive mechanism (Hurwicz, 1972b; Mount and Reiter, 1974; Walker, 1977; Osana, 1978). Similar results have been obtained for environments with public goods, showing that the Lindahl mechanism uses the minimal message space (Sato, 1981). Another objective is to analyse effects on incentives arising from private motivations in the presence of private information; that is, information held by one agent that is not observable by others, except perhaps at a cost. (There is a large literature on this subject under the rubric 'incentive compatibility', or 'strategic implementation' (Dasgupta, Hammond and Maskin, 1979; Hurwicz, 1971, 1972a).

The informational requirements of achieving a specified performance taking some aspects of incentive compatibility into account have been studied by Hurwicz (1976), Reichelstein (1984a, 1984b) and by Reichelstein and Reiter (1985).

Some important results for non-neoclassical environments can be mentioned. Hurwicz (1960, 1972a) has shown that there can be no informationally decentralized mechanism that realizes Pareto optimal performance on a class of environments that includes those with externalities. Calsamiglia (1977, 1982) has shown in a model of production that if the set of environments includes a sufficiently rich class of those with increasing returns to scale in production, then the dimension of the message space of any mechanism that realizes efficient production cannot be bounded.

Efficient allocation with infinitely many commodities

An infinite dimensional commodity space is needed when it is necessary to make infinitely many distinctions among goods and services. This is the case when commodities are distinguished according to time of availability and the time horizon in the model is not bounded or when time is continuous, or according to location when there is more than a finite number of possible locations; differentiated commodities provide other examples, and so does the case of uncertainty with infinitely many states. The bulk of the literature deals with the infinite horizon model of allocation over time, though recently more attention is given to models of product differentiation. Ramsey (1928) studied the problem of saving in a continuous time infinite horizon model with one consumption good and an infinitely lived consumer. He used as the criterion of optimality the infinite sum (integral) of undiscounted utility. Ramsey's contribution was largely ignored, and rediscovered when attention returned to problems of economic growth. A model of maximal sustainable growth based on a linear technology with no unproduced inputs was formulated by von Neumann (1937 in German; English translation, 1945–6). This contribution was unknown among English-speaking economists until after World War II. Study of intertemporal allocation by Anglo-American economists effectively began with the contributions of Harrod (1939) and Domar (1946). These models were concerned with stationary growth at a constant sustainable rate (stationary growth paths) rather than full intertemporal efficiency. Malinvaud (1953) first addressed this problem in a pioneering model of intertemporal allocation with an infinite horizon.

Efficient allocation over (discrete) time would be covered by the finite dimensional models described above if the time horizon were finite. It might be thought that a model with a sufficiently large but still finite horizon would for all practical purposes be equivalent to one with an infinite horizon, while avoiding the difficulties of infinity, but this is not the case, because of the dependence of efficient or optimal allocations on the value given to final stocks, a value that must depend on their uses beyond the horizon.

Malinvaud (1953) formulated an important infinite horizon model, which is the infinite dimensional counterpart of the linear activity analysis model of Koopmans. In Malinvaud's model time is discrete. The time horizon consists of an infinite sequence of time periods. At each date there are finitely many commodities. All commodities are desired in each time period, and no distinction is made between desired, intermediate and primary commodities. As in the activity analysis model, there is no explicit reference to preferences of consumers. Productive efficiency over time is analysed in terms of the output available for consumption, rather than the resulting utility levels.

Technology is represented by a production set X^t for each time period $t = 1, 2,\ldots$, an element of X^t being an ordered pair (a^t, b^{t+1}) of commodity bundles where a^t represents inputs to a production process in period t, and b^{t+1} represents the outputs of that process available at the beginning of period $t+1$. Here both a^t and b^{t+1} are non-negative. The set X^t is the aggregate production set for the economy during period t. The net outputs available for consumption are given by

$$y^t = b^t - a^t, \quad \text{for } t \geq 1,$$

where b^1 is the initial endowment of resources available at the beginning of period 1. A *programme* is an infinite sequence $\langle (a^t, b^{t+1}) \rangle$; it is a *feasible programme* if (a^t, b^{t+1}); is in X^t, and $b^t - a^t \geq 0$ for each $t \geq 1$, given b^1. The sequence $y = \langle y^t \rangle$ is called the *net output programme* associated with the given programme; it is a *feasible net output programme* if it is the net output programme of a feasible programme. A programme is *efficient* if it is (1) feasible and (2) there is no other programme that is feasible, from the same initial resources b^1, and provides at least as much net output in every period and a larger net output in some period. This is the concept of efficient production, already seen in the linear activity analysis model, now extended to an infinite horizon model. The main aim of this research is to extend to the

infinite horizon model the characterization of efficient production by prices seen in the finite model. This goal is not quite reached, as is seen in what follows.

The main difficulties presented by the infinite horizon are already present in a special case of the Malinvaud model with one good and no consumers. Let Y be the set of all non-negative sequences $y = (y_t)$ that satisfy $0 \leq y_t = f(a_{t-1}) - a_t$ for $t \geq 1$, and $0 \leq y^0 = b^1 - a^0$, $b^1 > 0$, where f is a real-valued continuous concave function on the non-negative real numbers (the production function), $f(0) = 0$, and b^1 is the given initial stock. The set Y is the set of all feasible programmes. A programme $y' - y > 0$. A price system is an infinite sequence $p = (p^t)$ of non-negative numbers. Denote by P the set of all price systems.

Malinvalud recognized the possibility that an efficient net output programme (y^t) need not have an associated system of non-zero prices (p^t) relative to which the production programme generating y satisfies the condition of intertemporal profit maximization, namely that

$$p^{t+1} f(a^t) - p^t a^t \geq p^{t+1} f(a) - p^t a$$

for all t and every $a \geq 0$. (Here (a^t) is the sequence of inputs producing y.) A condition introduced by Malinvaud, called *nontightness*, is sufficient for the existence of such non-zero prices. Alternative proofs of Malinvaud's existence theorem were given by Radner (1967) and Peleg and Yaari (1970). (An example showing the possibility of non-existence given by Peleg and Yaari (1970) is as follows. Suppose f is as shown in Figure 3.

At an interior efficient, and therefore value maximizing, programme the first-order necessary conditions for a maximum imply $p^{t+1} f'(a^t) = p^t$. If there is a time at which $a^t = a^*$, in an efficient programme, then, since $f'(a^*) = 0$, it follows that prices at all prior and future times are 0. (Nontightness rules out such examples.)

On the side of sufficiency, Malinvaud showed that intertemporal profit maximization relative to a strictly positive price system p is not enough to ensure that a feasible programme is efficient. An additional (transversality) condition is needed. In the present model the

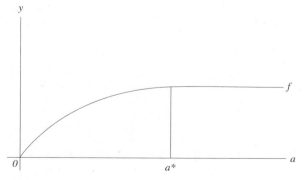

Figure 3

following is such a condition:

$$\lim_{t \to \infty} p^t y^t = 0.$$

Cass (1972) has given a criterion that completely characterizes the set of efficient programmes in a one-good model with strictly concave and smooth production technology that satisfies endpoint conditions $0 \leq f'(\infty) < 1 < f'(x) < \infty$ for some $x > 0$. Cass's criterion, states that a programme is *inefficient* if and only if the associated competitive prices – that is, satisfying $p^{t+1}f'(a^t) = p^t$ – also satisfy $\Sigma_{t=1}^{\infty}(1/p^t) < \infty$. This criterion may be interpreted as requiring the terms of trade between present and future to deteriorate sufficiently fast. Other similar conditions have been presented (Benveniste and Gale, 1975; Benveniste, 1976; Majumdar, 1974; Mitra, 1979). It is hard to see how any transversality condition can be interpreted in terms of decentralized resource allocation.

An alternate approach to characterizing efficient programmes was taken by Radner (1967), based on value functions as introduced in connection with valuation equilibrium by Debreu (1954). (Valuation equilibrium was discussed in connection with Arrow's exceptional case, above.) The value function approach was followed up by Majumdar (1970, 1972) and by Peleg and Yaari (1970). A price system defines a continuous linear functional, (a real-valued linear function) on the commodity space. This function assigns to a programme its present value. The present value may not be well-defined, because the infinite sequence that gives it diverges. This creates certain technical problems passed over here. A more important difficulty is that linear functionals exist that are not defined by price systems. Radner's approach was to characterize efficient programmes in terms of maximization of present value relative to a linear functional on the commodity space. Radner showed, technical matters aside, that:

1. If a feasible programme maximizes the value of net output (consumption) relative to a strictly positive continuous linear functional, then it is efficient.
2. If a given programme is efficient, then there is a non-zero non-negative continuous linear functional such that the given programme maximizes the value of net output relative to that functional on the set of feasible programmes.

These propositions seem to be the precise counterparts of the ones characterizing efficiency in the finite horizon model. Unfortunately, a linear functional may not have a representation in the form of the inner product of a price sequence with a net output sequence. (The production function $f(a) = a^{\beta}$, with $0 < \beta < 1$ provides an example. It is known that the programme with constant input sequence $x_t = (1/\beta)^{\beta/\beta-1}$ and output sequence $y_t = (1/\beta)^{\beta/\beta-1} - (1/\beta)^{1/\beta-1}t = 1, 2, \ldots$, is efficient, and

therefore there is a continuous linear functional relative to which it is value maximizing. But there is no price sequence (p^t) that represents that linear functional.) This presents a serious problem, because in the absence of such a representation it is unclear whether this characterization has an interpretation in terms of decentralized allocation processes; profit in any one period can depend on 'prices at infinity'.

This approach has the advantage that it is applicable not only to infinite horizon models, but to a broader class in which the commodity space is infinite dimensional. Bewley (1972), Mas-Colell (1977) and Jones (1984) among others discuss Pareto optimality and competitive equilibrium in economies with infinitely many commodities. Hurwicz (1958) and others analysed optimal allocation in terms of nonlinear programming in infinite dimensional spaces. Theorems of programming in infinite dimensional spaces are also used in some of the models mentioned in this discussion.

The basic difficulties encountered in the one-good model, apart from the numerous technical problems that tend to make the literature large and diverse as different technical structures are investigated, are on the one hand the fact that transversality conditions are indispensable, and on the other the possibility that linear functionals, even when they exist, may not be representable in terms of price sequences. These problems raise strong doubt about the possibility of achieving efficient intertemporal resource allocation by decentralized means, though they leave open the possibility that some other decentralized mechanism, not using prices, might work. Analysis of this possibility has just begun, and is discussed below.

The difficulties seen in the one-good production model persist in more elaborate ones, including multi-sectoral models with efficiency as the criterion, and models with consumers in which Pareto optimality is the criterion. McFadden, Mitra and Majumdar (1980) studied a model in which there are firms, and overlapping generations of consumers, as in the model first investigated by Samuelson (1958). Each consumer lives for a finite time and has a consumption set and preferences like the consumers in a finite horizon model. A model with overlapping generations of consumers presents the fundamental difficulty that consumers cannot trade with future consumers as yet unborn. This difficulty can appear even in a finite horizon model if there are too few markets. The economy is closed in the sense that there are no nonproduced resources; the von Neumann growth model is an example of such a model. Building on the results of an earlier investigation (Majumdar, Mitra and McFadden, 1976), these authors introduced several notions of price systems, of competitive equilibrium, efficiency and optimality, and sought to establish counterparts of the classical welfare theorems. To summarize, in the 1976 paper they strengthen an earlier result of Bose (1974) to the effect that the problem of proper distribution of goods in essentially a short-run problem, and that the only long-run problem, one created by the infinite horizon, is that of inefficiency through overaccumulation of capital. In the 1980 paper the focus is on the relationships among various notions of equilibrium and Pareto optimality. The force of their results is, as might be expected, that the difficulties already seen in one-good model without consumers persist in this model. A transversality condition is made part of the definition of competitive equilibrium in order to obtain the result that an equilibrium is optimal. A partial converse requires some additional assumptions on the technology (reachability) and on the way the economy fits together (nondecomposability). These results certainly illuminate the infinite horizon model with overlapping generations of consumers and producers, but the possibility of efficient or optimal resource allocation by decentralized means is not different from that in the one-good Malinvaud model.

Hurwicz and Majumdar in an unpublished manuscript dated 1983, and later Hurwicz and Weinberger (1984), have addressed this issue directly, building on the approach of mechanism theory.

Hurwicz and Majumdar have studied the problem of efficiency in a model with an infinite number of periods. In each period there are finitely many commodities, one producer who is alive for just one period, and no consumers' choices. The criterion is the maximization of the discounted value of the programme (well-defined in this model). The producer alive in any period knows only the technology in that period. The question is whether there is a (static) privacy preserving mechanism using a finite dimensional message space whose equilibria coincide with the set of efficient programmes. The question can be put as follows. In each period a message is posted. The producer alive in that period responds 'Yes' or 'No'. If every producer over the entire infinite horizon answers 'Yes', the programme is an outcome corresponding to the equilibrium consisting of the infinite succession of posted messages. Since each producer knows only the technology prevailing in the period when he is alive, the process preserves privacy. If in addition the message posted in each period is finite dimensional, the process is informationally decentralized. Period-by-period profit maximization using period-by-period prices is a mechanism of this type; the message posted in each period consists of the vector of prices for that period, and the production plan for that period, both finite dimensional. The object is to characterize all efficient programmes as equilibria of such a mechanism. This would be an analogue of the classical welfare theorems, but without the restriction to mechanisms that use prices in their messages.

The main result is in the nature of an impossibility theorem. If the technology is constant over time, and that fact is common knowledge at the beginning, the problem is trivial since knowledge of the technology in the first period automatically means knowledge of it in every period. On the other hand, if there is some period whose technology is not known in the first period, then there is no finite dimensional message that can characterize efficient programmes, and in that sense, production cannot be satisfactorily decentralized over time.

Hurwicz and Weinberger (1984) have studied a model with both producers and consumers. As with producers, there is a consumer in each period, who lives for one period. The consumer in each period has a one-period utility function, which is not known by the producer; similarly the consumer does not know the production function. The criterion of optimality is the maximization of the sum of discounted utilities over the infinite horizon. Hurwicz and Weinberger show that there is no privacy preserving mechanism of the type just described whose equilibria correspond to the set of optimal programmes. It should be noted that their mechanism requires that the first-period actions (production, consumption and investment decisions) be made in the first period, and not be subject to revision after the infinite process of verification is completed. (On the other hand, under tâtonnement assumptions it may be possible to decentralize. In this model tâtonnement entails reconsideration 'at infinity'.)

If attention is widened to efficient programmes, and if technology is constant over time, there is an efficient programme with a fixed ratio of consumption to investment. This programme can be obtained as the equilibrium outcome of a mechanism of the specified type. However, this corresponds to only one side of the classical welfare theorems. It says that the outcome of such a mechanism is efficient; but it does not ensure that every efficient programme can be realized as the outcome of such a mechanism. The latter property fails in this model.

STANLEY REITER

See also **incentive compatibility; linear programming; welfare economics.**

Bibliography

Arrow, K. 1951. An extension of the basic theorems of classical welfare economics. In *Proceedings of the Second Berkeley Symposium on Mathematical Statistics and Probability*, ed. J. Neyman. Berkeley: University of California Press.

Arrow, K. and Debreu, G. 1954. Existence of an equilibrium for a competitive economy. *Econometrica* 22(July), 265–90.

Arrow, K. and Hahn, F. 1971. *General Competitive Analysis*. San Francisco: Holden-Day.

Arrow, K. and Hurwicz, L. 1960. Decentralization and computation in resource allocation. In *Essays in Economics and Econometrics*, ed. R.W. Pfouts. Chapel Hill: University of North Carolina Press, 34–104.

Arrow, K., Hurwicz, L. and Uzawa, H. 1961. Constraint qualifications in maximization problems. *Naval Research Logistics Quarterly* 8(2), June, 175–91.

Benveniste, L. 1976. Two notes on the Malinvaud condition for efficiency of infinite horizon programs. *Journal of Economic Theory* 12, 338–46.

Benveniste, L. and Gale, D. 1975. An extension of Cass' characterization of infinite efficient production programs. *Journal of Economic Theory* 10, 229–38.

Bewley, T. 1972. Existence of equilibria in economies with infinitely many commodities. *Journal of Economic Theory* 4, 514–40.

Bose, A. 1974. Pareto optimality and efficient capital accumulation. Discussion Paper No. 74–4. Department of Economics, University of Rochester.

Brown, D. and Heal, G. 1982. Existence, local-uniqueness and optimality of a marginal cost pricing equilibrium in an economy with increasing returns. Cal. Tech. Social Science Working Paper No. 415.

Brown, D., Heal, G., Ali Khan, M. and Vohra, R. 1986. On a general existence theorem for marginal cost pricing equilibria. *Journal of Economic Theory* 38, 371–9.

Calsamiglia, X. 1977. Decentralized resource allocation and increasing returns. *Journal of Economic Theory* 14, 263–83.

Calsamiglia, X. 1982. On the size of the message space under non-convexities. *Journal of Mathematical Economics* 10, 197–203.

Cass, D. 1972. On capital over-accumulation in the aggregative neoclassical model of economic growth: a complete characterization. *Journal of Economic Theory* 4(2), April, 200–23.

Dantzig, G.B. 1951a. The programming of interdependent activities. In *Activity Analysis of Production and Allocation*, ed. T. Koopmans, Cowles Commission Monograph No. 13, New York: Wiley, ch. 2, 19–32.

Dantzig, G.B. 1951b. Maximization of a linear function of variables subject to linear inequalities. In *Activity Analysis of Production and Allocation*, ed. T. Koopmans, Cowles Commission Monograph No. 13, New York: Wiley, ch. 21, 339–47.

Dasgupta, P., Hammond, P. and Maskin, E. 1979. The implementation of social choice rules: some general results on incentive compatibility. *Review of Economic Studies* 46, 185–216.

Davis, O.A. and Whinston, A.B. 1962. Externalities welfare and the theory of games. *Journal of Political Economy* 70, 214–62.

Debreu, G. 1954. Valuation equilibrium and Pareto optimum. In *Proceedings of the National Academy of Sciences of the USA* 40(7), 588–92.

Debreu, G. 1959. *Theory of Value*. New York: Wiley.

Debreu, G. 1970. Economies with a finite set of equilibria. *Econometrica* 38(3), May, 387–92.

Diamond, P. and Mirrlees, J. 1971. Optimal taxation and public production. I: Production efficiency; II: Tax rules. *American Economic Review* 61, 8–27; 261–78.

Domar, E. 1946. Capital expansion, rate of growth, and employment. *Econometrica* 14(April), 137–47.

Fenchel, W. 1950. Convex cones, sets, and functions. Princeton University (hectographed).

Guesnerie, R. 1975. Pareto optimality in non-convex economies. *Econometrica* 43, 1–29.

Harrod, R.F. 1939. An essay in dynamic theory. *Economic Journal* 49, 14–33.

Hayek, F. von. 1945. The use of knowledge in society. *American Economic Review* 35, 519–53. Reprinted in F. von Hayek, *Individualism and Economic Order*, Chicago: University of Chicago Press, 1949, 77–92.

Heal, G. 1971. Planning, prices and increasing returns. *Review of Economic Studies* 38, 281–94.

Hotelling, H. 1938. The general welfare in relation to problems of taxation and of railway and utility rates. *Econometrica* 6, 242–69.

Hurwicz, L. 1958. Programming in linear spaces. In *Studies in Linear and Non-Linear Programming*, ed. K. Arrow, L. Hurwicz and H. Uzawa. Stanford: Stanford University Press.

Hurwicz, L. 1960. Optimality and informational efficiency in resource allocation processes. In *Mathematical Methods in the Social Sciences, 1959*, ed. K.J. Arrow, S. Karlin and P. Suppes. Stanford: Stanford University Press.

Hurwicz, L. 1971. Centralization and decentralization in economic processes. In *Comparison of Economic Systems: Theoretical and Methodological Approaches*, ed. A. Eckstein. Berkeley: University of California Press, ch. 3.

Hurwicz, L. 1972a. On informationally decentralized systems. In *Decision and Organization*, ed. C. McGuire and R. Radner. Amsterdam, London: North-Holland, ch. 14, 297–336.

Hurwicz, L. 1972b. On the dimensional requirements of in formationally decentralized Pareto-satisfactory processes. Presented at the Conference Seminar in Decentralization North-western University. In *Studies in Resource Allocation Processes*, ed. K.J. Arrow and L. Hurwicz. Cambridge: Cambridge University Press, 1977.

Hurwicz, L. 1976. On informational requirements for nonwasteful resource allocation systems. In *Mathematical Models in Economics: Papers and Proceedings of a US-USSR Seminar, Moscow*, ed. S. Shulman. New York: National Bureau of Economic Research.

Hurwicz, L., Radner, R. and Reiter, S. 1975. A stochastic decentralized resource allocation process. *Econometrica* 43 : Part I, 187–221; Part II, 363–93.

Hurwicz, L. and Weinberger, H. 1984. Paper presented at IMA seminar in Minneapolis.

Jennergren, L. 1971. Studies in the mathematical theory of decentralized resource-allocation. PhD dissertation, Stanford University.

Jones, L. 1984. A competitive model of commodity differentiation. *Econometrica* 52, 507–30.

Kantorovitch, L. 1939. *Matematicheskie metody organizatii i planirovania proizvodstva* (Mathematical methods in the organization and planning of production). Izdanie

Leningradskogo Gosudarstvennogo Universiteta, Leningrad. Trans. in *Management Science* 6(4), July 1960, 363–422.

Kantorovitch, L. 1942. On the translocation of masses. (In English.) *Comptes Rendus (Doklady) de l'Academie des Sciences d l'URSS* 37(7–8).

Kantorovitch, L. and Gavurin, M. 1949. Primenenie matematicheskikh metodov v voprosakh analyza grusopotokov (The application of mathematical methods to problems of freight flow analysis). In *Problemy Povysheniia Effektivnosty Raboty Transporta* (Problems of raising the efficiency of transportation), ed. V. Zvonkov. Moscow and Leningrad: Izdatel'stvo Akademii Nauk SSSR.

Koopmans, T.C. 1951a. Analysis of production as an efficient combination of activities. In *Activity Analysis of Production and Allocation*, ed. T. Koopmans, Cowles Commission Monograph No. 13, New York: Wiley, ch. 3, 33–97.

Koopmans, T.C. 1951b. Efficient allocation of resources. *Econometrica* 19, 455–65.

Koopmans, T.C. 1957. *Three Essays on the State of Economic Science*. New York: McGraw-Hill, 66–104.

Kuhn, H. and Tucker, A. 1951. Nonlinear programming. In *Proceedings of the Second Berkeley Symposium on Mathematical Statistics and Probability*, ed. J. Neyman. Berkeley: University of California Press, 481–92.

Ledyard, J. 1968. Resource allocation in unselfish environments. *American Economic Review* 58, 227–37.

Ledyard, J. 1971. A convergent Pareto-satisfactory non-tâtonnement adjustment process for a class of unselfish exchange environments. *Econometrica* 39, 467–99.

Lerner, A. 1934. The concept of monopoly and measurement of monopoly power. *Review of Economic Studies* 1(3), June, 157–75.

Lerner, A. 1944. *The Economics of Control*. New York: Macmillan.

Lipsey, R. and Lancaster, K. 1956. The general theory of second best. *Review of Economic Studies* 24, 11–32.

McFadden, D., Mitra, T. and Majumdar, M. 1980. Pareto optimality and competitive equilibrium in infinite horizon economies. *Journal of Mathematical Economics* 7, 1–26.

McKenzie, L. 1959. On the existence of general equilibrium for a competitive market. *Econometrica* 27(1), January, 54–71.

Majumdar, M. 1970. Some approximation theorems on efficiency prices for infinite programs. *Journal of Economic Theory* 2, 399–410.

Majumdar, M. 1972. Some general theorems of efficiency prices with an infinite dimensional commodity space. *Journal of Economic Theory* 5, 1–13.

Majumdar, M. 1974. Efficient programs in infinite dimensional spaces: a complete characterization. *Journal of Economic Theory* 7, 355–69.

Majumdar, M., Mitra, T. and McFadden, D. 1976. On efficiency and Pareto optimality of competitive programs in closed multisector models. *Journal of Economic Theory* 13, 26–46.

Malinvaud, E. 1953. Capital accumulation and efficient allocation of resources. *Econometrica* 21, 233–68.

Mas-Colell, A. 1977. Regular nonconvex economies. *Econometrica* 45, 1387–407.

Mitra, T. 1979. On optimal economic growth with variable discount rates: existence and stability results. *International Economic Review* 20, 133–45.

Mount, K. and Reiter, S. 1974. The informational size of message spaces. *Journal of Economic Theory* 8, 161–92.

Mount, K. and Reiter, S. 1977. Economic environments for which there are Pareto satisfactory mechanisms. *Econometrica* 45, 821–42.

Neumann, J. von. 1937. A model of general economic equilibrium. *Ergebnisse eines mathematischen Kolloquiums*, No. 8. Trans. from German, *Review of Economic Studies* 13(1), (1945–6), 1–9.

Nikaido, H. 1969. *Convex Structures and Economic Theory*. New York: Academic Press.

Nikaido, H. 1970. *Introduction to Sets and Mappings in Modern Economics*. Trans. K. Sato, Amsterdam: North-Holland (Japanese original, Tokyo, 1960).

Osana, H. 1978. On the informational size of message spaces for resource allocation processes. *Journal of Economic Theory* 17, 66–78.

Peleg, B. and Yaari, M. 1970. Efficiency prices in an infinite dimensional commodity space. *Journal of Economic Theory* 2, 41–85.

Pigou, A. 1932. *The Economics of Welfare*. 4th edn, London: Macmillan.

Radner, R. 1967. Efficiency prices for infinite horizon production programs. *Review of Economic Studies* 34, 51–66.

Ramsey, F. 1928. A mathematical theory of saving. *Economic Journal* 38, 543–59.

Reichelstein, S. 1984a. Dominant strategy implementation, incentive compatibility and informational requirements. *Journal of Economic Theory* 34(1), October, 32–51.

Reichelstein, S. 1984b. Information and incentives in economic organizations. PhD dissertation, Northwestern University.

Reichelstein, S. and Reiter, S. 1985. Game forms with minimal strategy spaces. Discussion Paper No. 663, The Center for Mathematical Studies in Economics and Management Science, Northwestern University, Evanston, Ill.

Reiter, S. 1977. Information and performance in the (new)[2] welfare economics. *American Economic Review* 67, 226–34.

Samuelson, P. 1947. *Foundations of Economic Analysis*. Cambridge, Mass.: Harvard University Press.

Samuelson, P. 1949. Market mechanisms and maximization, I, II, III. Hectographed memoranda, The RAND Corporation, Santa Monica.

Samuelson, P. 1958. An exact consumption-loan model of interest with or without the social contrivance of money. *Journal of Political Economy* 66(December), 467–82.

Sato, F. 1981. On the informational size of message spaces for resource allocation processes in economies with public goods. *Journal of Economic Theory* 24, 48–69.

Slater, M. 1950. Lagrange multipliers revisited: a contribution to non-linear programming. *Cowles Commission Discussion Paper*, Math. 403, also RM–676, 1951.

Sonnenschein, H. 1974. An axiomatic characterization of the price mechanism. *Econometrica* 42, 425–34.

Uzawa, H. 1958. The Kuhn-Tucker Theorem in concave programming. In *Studies in Linear and Non-Linear Programming*, ed. K. Arrow, L. Hurwicz and H. Uzawa. Stanford: Stanford University Press.

Walker, M. 1977. On the informational size of message spaces. *Journal of Economic Theory* 15, 366–75.

efficient markets hypothesis

There is an old joke, widely told among economists, about an economist strolling down the street with a companion. They come upon a $100 bill lying on the ground, and as the companion reaches down to pick it up, the economist says, 'Don't bother – if it were a genuine $100 bill, someone would have already picked it up'. This humorous example of economic logic gone awry is a fairly accurate rendition of the efficient markets hypothesis (EMH), one of the most hotly contested propositions in all the social sciences. It is disarmingly simple to state, has far-reaching consequences for academic theories and business practice, and yet is surprisingly resilient to empirical proof or refutation. Even after several decades of research and literally thousands of published studies, economists have not yet reached a consensus about whether markets – particularly financial markets – are, in fact, efficient.

The origins of the EMH can be traced back to the work of two individuals in the 1960s: Eugene F. Fama and Paul A. Samuelson. Remarkably, they independently developed the same basic notion of market efficiency from two rather different research agendas. These differences would propel the them along two distinct trajectories leading to several other breakthroughs and milestones, all originating from their point of intersection, the EMH.

Like so many ideas of modern economics, the EMH was first given form by Paul Samuelson (1965), whose contribution is neatly summarized by the title of his article: 'Proof that Properly Anticipated Prices Fluctuate Randomly'. In an informationally efficient market, price changes must be unforecastable if they are properly anticipated, that is, if they fully incorporate the information and expectations of all market participants. Having developed a series of linear-programming solutions to spatial pricing models with no uncertainty, Samuelson came upon the idea of efficient markets through his interest in temporal pricing models of storable commodities that are harvested

and subject to decay. Samuelson's abiding interest in the mechanics and kinematics of prices, with and without uncertainty, led him and his students to several fruitful research agendas including solutions for the dynamic asset-allocation and consumption-savings problem, the fallacy of time diversification and log-optimal investment policies, warrant and option-pricing analysis and, ultimately, the Black and Scholes (1973) and Merton (1973) option-pricing models.

In contrast to Samuelson's path to the EMH, Fama's (1963; 1965a; 1965b, 1970) seminal papers were based on his interest in measuring the statistical properties of stock prices, and in resolving the debate between technical analysis (the use of geometric patterns in price and volume charts to forecast future price movements of a security) and fundamental analysis (the use of accounting and economic data to determine a security's fair value). Among the first to employ modern digital computers to conduct empirical research in finance, and the first to use the term 'efficient markets' (Fama, 1965b), Fama operationalized the EMH hypothesis – summarized compactly in the epigram 'prices fully reflect all available information' – by placing structure on various information sets available to market participants. Fama's fascination with empirical analysis led him and his students down a very different path from Samuelson's, yielding significant methodological and empirical contributions such as the event study, numerous econometric tests of single- and multi-factor linear asset-pricing models, and a host of empirical regularities and anomalies in stock, bond, currency and commodity markets.

The EMH's concept of informational efficiency has a Zen-like, counter-intuitive flavour to it: the more efficient the market, the more random the sequence of price changes generated by such a market, and the most efficient market of all is one in which price changes are completely random and unpredictable. This is not an accident of nature, but is in fact the direct result of many active market participants attempting to profit from their information. Driven by profit opportunities, an army of investors pounce on even the smallest informational advantages at their disposal, and in doing so they incorporate their information into market prices and quickly eliminate the profit opportunities that first motivated their trades. If this occurs instantaneously, which it must in an idealized world of 'frictionless' markets and costless trading, then prices must always fully reflect all available information. Therefore, no profits can be garnered from information-based trading because such profits must have already been captured (recall the $100 bill on the ground). In mathematical terms, prices follow martingales.

Such compelling motivation for randomness is unique among the social sciences and is reminiscent of the role that uncertainty plays in quantum mechanics. Just as Heisenberg's uncertainty principle places a limit on what we can know about an electron's position and momentum if quantum mechanics holds, this version of the EMH places a limit on what we can know about future price changes if the forces of economic self-interest hold.

A decade after Samuelson's (1965) and Fama's (1965a; 1965b; 1970) landmark papers, many others extended their framework to allow for risk-averse investors, yielding a 'neoclassical' version of the EMH where price changes, properly weighted by aggregate marginal utilities, must be unforecastable (see, for example, LeRoy, 1973; M. Rubinstein, 1976; and Lucas, 1978). In markets where, according to Lucas (1978), all investors have 'rational expectations', prices do fully reflect all available information and marginal-utility-weighted prices follow martingales. The EMH has been extended in many other directions, including the incorporation of non-traded assets such as human capital, state-dependent preferences, heterogeneous investors, asymmetric information, and transactions costs. But the general thrust is the same: individual investors form expectations rationally, markets aggregate information efficiently, and equilibrium prices incorporate all available information instantaneously.

The random walk hypothesis

The importance of the EMH stems primarily from its sharp empirical implications many of which have been tested over the years. Much of the EMH literature before LeRoy (1973) and Lucas (1978) revolved around the random walk hypothesis (RWH) and the martingale model, two statistical descriptions of unforecastable price changes that were initially taken to be implications of the EMH. One of the first tests of the RWH was developed by Cowles and Jones (1937), who compared the frequency of *sequences* and *reversals* in historical stock returns, where the former are pairs of consecutive returns with the same sign, and the latter are pairs of consecutive returns with opposite signs. Cootner (1962; 1964), Fama (1963; 1965a), Fama and Blume (1966), and Osborne (1959) perform related tests of the RWH and, with the exception of Cowles and Jones (who subsequently acknowledged an error in their analysis – Cowles, 1960), all of these articles indicate support for the RWH using historical stock price data.

More recently, Lo and MacKinlay (1988) exploit the fact that return variances scale linearly under the RWH – the variance of a two-week return is twice the variance of a one-week return if the RWH holds – and construct a variance ratio test which rejects the RWH for weekly US stock returns indexes from 1962 to 1985. In particular, they find that variances grow faster than linearly as the holding period increases, implying positive serial correlation in weekly returns. Oddly enough, Lo and MacKinlay also show that individual stocks generally do satisfy the RWH, a fact that we shall return to below.

French and Roll (1986) document a related phenomenon: stock return variances over weekends and exchange holidays are considerably lower than return variances over the same number of days when markets are open. This difference suggests that the very act of trading creates volatility, which may well be a symptom of Black's (1986) noise traders.

For holding periods much longer than one week – for example, three to five years – Fama and French (1988) and Poterba and Summers (1988) find negative serial correlation in US stock returns indexes using data from 1926 to 1986. Although their estimates of serial correlation coefficients seem large in magnitude, there is insufficient data to reject the RWH at the usual levels of significance. Moreover, a number of statistical artifacts documented by Kim, Nelson and Startz (1991) and Richardson (1993) cast serious doubt on the reliability of these longer-horizon inferences.

Finally, Lo (1991) considers another aspect of stock market prices long thought to have been a departure from the RWH: long-term memory. Time series with long-term memory exhibit an unusually high degree of persistence, so that observations in the remote past are non-trivially correlated with observations in the distant future, even as the time span between the two observations increases. Nature's predilection towards long-term memory has been well-documented in the natural sciences such as hydrology, meteorology, and geophysics, and some have argued that economic time series must therefore also have this property.

However, using recently developed statistical techniques, Lo (1991) constructs a test for long-term memory that is robust to short-term correlations of the sort uncovered by Lo and MacKinlay (1988), and concludes that, despite earlier evidence to the contrary, there is little support for long-term memory in stock market prices. Departures from the RWH can be fully explained by conventional models of short-term dependence.

Variance bounds tests

Another set of empirical tests of the EMH starts with the observation that in a world without uncertainty the market price of a share of common stock must equal the present value of all future dividends, discounted at the appropriate cost of capital. In an uncertain world, one can generalize this *dividend-discount model* or *present-value relation* in the natural way: the market price equals the conditional expectation of the present value of all future dividends, discounted at the appropriate risk-adjusted cost of capital, and conditional on all available information. This generalization is explicitly developed by Grossman and Shiller (1981).

LeRoy and Porter (1981) and Shiller (1981) take this as their starting point in comparing the variance of stock market prices to the variance of *ex post* present values of future dividends. If the market price is the conditional expectation of present values, then the difference between the two, that is, the forecast error, must be uncorrelated with the conditional expectation by construction. But this implies that the variance of the *ex post* present value is the sum of the variance of the market price (the conditional expectation) and the variance of the forecast error. Since volatilities are always non-negative, this variance decomposition implies that the variance of stock prices cannot exceed the variance of *ex post* present values. Using annual US stock market data from various sample periods, LeRoy and Porter (1981) and Shiller (1981) find that the variance bound is violated dramatically. Although LeRoy and Porter are more circumspect about the implications of such violations, Shiller concludes that stock market prices are too volatile and the EMH must be false.

These two papers ignited a flurry of responses which challenged Shiller's controversial conclusion on a number of fronts. For example, Flavin (1983), Kleidon (1986), and Marsh and Merton (1986) show that statistical inference is rather delicate for these variance bounds, and that, even if they hold in theory, for the kind of sample sizes Shiller uses and under plausible data-generating processes the sample variance bound is often violated purely due to sampling variation. These issues are well summarized in Gilles and LeRoy (1991) and Merton (1987).

More importantly, on purely theoretical grounds Marsh and Merton (1986) and Michener (1982) provide two explanations for violations of variance bounds that are perfectly consistent with the EMH. Marsh and Merton (1986) show that if managers smooth dividends – a well-known empirical phenomenon documented in several studies of dividend policy – and if earnings follow a geometric random walk, then the variance bound is violated in theory, in which case the empirical violations may be interpreted as *support* for this version of the EMH.

Alternatively, Michener constructs a simple dynamic equilibrium model along the lines of Lucas (1978) in which prices do fully reflect all available information at all times but where individuals are risk averse, and this risk aversion is enough to cause the variance bound to be violated in theory as well.

These findings highlight an important aspect of the EMH that had not been emphasized in earlier studies: tests of the EMH are always tests of joint hypotheses. In particular, the phrase 'prices fully reflect all available information' is a statement about two distinct aspects of prices: the information content and the price formation mechanism. Therefore, any test of this proposition must concern the *kind* of information reflected in prices, and *how* this information comes to be reflected in prices.

Apart from issues regarding statistical inference, the empirical violation of variance bounds may be interpreted in many ways. It may be a violation of EMH, or a sign that investors are risk averse, or a symptom of dividend smoothing. To choose among these alternatives, more evidence is required.

Overreaction and underreaction

A common explanation for departures from the EMH is that investors do not always react in proper proportion to new information. For example, in some cases investors may overreact to performance, selling stocks that have experienced recent losses or buying stocks that have enjoyed recent gains. Such overreaction tends to push prices beyond their 'fair' or 'rational' market value, only to have rational investors take the other side of the trades and bring prices back in line eventually. An implication of this phenomenon is price reversals: what goes up must come down, and vice versa. Another implication is that *contrarian* investment strategies – strategies in which 'losers' are purchased and 'winners' are sold – will earn superior returns.

Both of these implications were tested and confirmed using recent US stock market data. For example, using monthly returns of New York Stock Exchange (NYSE) stocks from 1926 to 1982, DeBondt and Thaler (1985) document the fact that the winners and losers in one 36-month period tend to reverse their performance over the next 36-month period. Curiously, many of these reversals occur in January (see the discussion below on the 'January effect'). Chopra, Lakonishok and Ritter (1992) reconfirm these findings after correcting for market risk and the size effect. And Lehmann (1990) shows that a zero-net-investment strategy in which long positions in losers are financed by short positions in winners almost always yields positive returns for monthly NYSE/AMEX stock returns data from 1962 to 1985.

However, Chan (1988) argues that the profitability of contrarian investment strategies cannot be taken as conclusive evidence against the EMH because there is typically no accounting for risk in these profitability calculations (although Chopra, Lakonishok and Ritter, 1992 do provide risk adjustments, their focus was not on specific trading strategies). By risk-adjusting the returns of a contrarian trading strategy according to the capital asset pricing model, Chan (1988) shows that the expected returns are consistent with the EMH.

Moreover, Lo and MacKinlay (1990c) show that at least half of the profits reported by Lehmann (1990) are not due to overreaction but rather the result of positive cross-autocorrelations between stocks. For example, suppose the returns of two stocks A and B are both serially uncorrelated but are positively cross-autocorrelated. The lack of serial correlation implies no overreaction (which is characterized by negative serial correlation), but positive cross-autocorrelations yields positive expected returns to contrarian trading strategies. The existence of several economic rationales for positive cross-autocorrelation that are consistent with EMH suggests that the profitability of contrarian trading strategies is not sufficient evidence to conclude that investors overreact.

The reaction of market participants to information contained in earnings announcements also has implications for the EMH. In one of the earliest studies of the information content of earnings, Ball and Brown (1968) show that up to 80 per cent of the information contained in the earnings 'surprises' is anticipated by market prices.

However, the more recent article by Bernard and Thomas (1990) argues that investors sometimes under-react to information about future earnings contained in current earnings. This is related to the 'post-earnings announcement drift' puzzle first documented by Ball and Brown (1968), in which the information contained in earnings announcement takes several days to become fully impounded into market prices. Although such effects are indeed troubling for the EMH, their economic significance is often questionable – while they may violate the EMH in frictionless markets, very often even the smallest frictions – for example, positive trading costs, taxes – can eliminate the profits from trading strategies designed to exploit them.

Anomalies

Perhaps the most common challenge to the EMH is the anomaly, a regular pattern in an asset's returns which is reliable, widely known, and inexplicable. The fact that the pattern is regular and reliable implies a degree of predictability, and the fact that the regularity is widely known implies that many investors can take advantage of it.

For example, one of the most enduring anomalies is the 'size effect', the apparent excess expected returns that accrue to stocks of small-capitalization companies – in excess of their risks – which was first discovered by Banz (1981). Keim (1983), Roll (1983), and Rozeff and Kinney (1976) document a related anomaly: small capitalization stocks tend to outperform large capitalization stocks by a wide margin over the turn of the calendar year. This so-called 'January effect' seems robust to sample period, and is difficult to reconcile with the EMH because of its regularity and publicity. Other well-known anomalies include the Value Line enigma (Copeland and Mayers, 1982), the profitability of short-term return-reversal strategies in US equities (Rosenberg, Reid and Lanstein,1985; Chan, 1988; Lehmann, 1990; and Lo and MacKinlay, 1990c), the profitability of medium-term momentum strategies in US equities (Jegadeesh, 1990; Chan, Jegadeesh and Lakonishok, 1996; and Jegadeesh and Titman, 2001), the relation between price/earnings ratios and expected returns (Basu, 1977), the volatility of orange juice futures prices (Roll, 1984), and calendar effects such as holiday, weekend, and turn-of-the-month seasonalities (Lakonishok and Smidt, 1988).

What are we to make of these anomalies? On the one hand, their persistence in the face of public scrutiny seems to be a clear violation of the EMH. After all, most of these anomalies can be exploited by relatively simple trading strategies, and, while the resulting profits may not be riskless, they seem unusually profitable relative to their risks (see, especially, Lehmann, 1990).

On the other hand, EMH supporters might argue that such persistence is in fact evidence in favour of EMH or, more to the point, that these anomalies cannot be exploited to any significant degree because of factors such as risk or transactions costs. Moreover, although some anomalies are currently inexplicable, this may be due to a lack of imagination on the part of academics, not necessarily a violation of the EMH. For example, recent evidence suggests that the January effect is largely due to 'bid–ask bounce', that is, closing prices for the last trading day of December tend to be at the bid price and closing prices for the first trading day of January tend to be at the ask price. Since small-capitalization stocks are also often low-price stocks, the effects of bid–ask bounce in percentage terms are much more pronounced for these stocks – a movement from bid to ask for a $5.00 stock on the NYSE (where the minimum bid-ask spread was $0.125 prior to decimalization in 2000) represents a 2.5 per cent return.

Whether or not one can profit from anomalies is a question unlikely to be settled in an academic setting. While calculations of 'paper' profits of various trading strategies come easily to academics, it is virtually impossible to incorporate in a realistic manner important features of the trading process such as transactions costs (including price impact), liquidity, rare events, institutional rigidities and non-stationarities. The economic value of anomalies must be decided in the laboratory of actual markets by investment professionals, over long periods of time, and even in these cases superior performance and simple luck are easily confused.

In fact, luck can play another role in the interpretation of anomalies: it can account for anomalies that are not anomalous. Regular patterns in historical data can be found even if no regularities exist, purely by chance. Although the likelihood of finding such spurious regularities is usually small (especially if the regularity is a very complex pattern), it increases dramatically with the number of 'searches' conducted on the same set of data. Such *data-snooping* biases are illustrated in Brown et al. (1992) and Lo and MacKinlay (1990b) – even the smallest biases can translate into substantial anomalies such as superior investment returns or the size effect.

Behavioural critiques

The most enduring critiques of the EMH revolve around the preferences and behaviour of market participants. The standard approach to modelling preferences is to assert that investors optimize additive time-separable expected utility functions from certain parametric families – for example, constant relative risk aversion. However, psychologists and experimental economists have documented a number of departures from this paradigm, in the form of specific behavioural biases that are ubiquitous to human decision-making under uncertainty, several of which lead to undesirable outcomes for an individual's economic welfare – for example, overconfidence (Fischoff and Slovic, 1980; Barber and Odean, 2001; Gervais and Odean, 2001), overreaction (DeBondt and Thaler, 1985), loss aversion (Kahneman and Tversky, 1979; Shefrin and Statman, 1985; Odean, 1998), herding (Huberman and Regev, 2001), psychological accounting (Tversky and Kahneman, 1981), miscalibration of probabilities (Lichtenstein, Fischoff and Phillips, 1982), hyperbolic discounting (Laibson, 1997), and regret (Bell, 1982). These critics of the EMH argue that investors are often – if not always – irrational, exhibiting predictable and financially ruinous behaviour.

To see just how pervasive such behavioural biases can be, consider the following example which is a slightly modified version of an experiment conducted by two psychologists, Kahneman and Tversky (1979). Suppose you are offered two investment opportunities, A and B: A yields a sure profit of $240,000, and B is a lottery ticket yielding $1 million with a 25 per cent probability and $0 with 75 per cent probability. If you had to choose between A and B, which would you prefer? Investment B has an expected value of $250,000, which is higher than A's payoff, but this may not be all that meaningful to you because you will receive either $1 million or zero. Clearly, there is no right or wrong choice here; it is simply a matter of personal preferences. Faced with this choice, most subjects prefer A, the sure profit, to B, despite the fact that B offers a significant probability of winning considerably more. This behaviour is often characterized as 'risk aversion' for obvious reasons. Now suppose you are faced with another two choices, C and D: C yields a sure loss of $750,000, and D is a lottery ticket yielding $0 with 25 per cent probability and a loss of $1 million with 75 per cent probability. Which would you prefer? This situation is not as absurd as it might seem at first glance; many financial decisions involve choosing between the lesser of two evils. In this case, most subjects choose D, despite the fact that D is more risky than C. When faced with two choices that both involve losses, individuals seem to be 'risk seeking', not risk averse as in the case of A versus B.

The fact that individuals tend to be risk averse in the face of gains and risk seeking in the face of losses can lead to some very poor financial decisions. To see why, observe that the combination of choices A and D is equivalent to a single lottery ticket yielding $240,000 with 25 per cent probability and −$760,000 with 75 per cent probability, whereas the combination of choices B and C is equivalent to a single lottery ticket yielding $250,000 with 25 per cent probability and −$750,000 with 75 per cent probability. The B and C combination has the same probabilities of gains and losses, but the gain is $10,000 higher and the loss is $10,000 lower. In other words, B and C is formally equivalent to A and D plus a sure profit of $10,000. In light of this analysis, would you still prefer A and D?

A common response to this example is that it is contrived because the two pairs of investment opportunities were presented sequentially, not simultaneously. However, in a typical global financial institution the London office may be faced with choices A and B and the Tokyo office may be faced with choices C and D. Locally, it may seem as if there is no right or wrong answer – the choice between A and B or C and D seems to be simply a matter of personal risk preferences – but the globally consolidated financial statement for the entire institution will tell a very different story. From that perspective, there *is* a right and wrong answer, and the empirical and experimental evidence suggests that most individuals tend to select the wrong answer. Therefore, according to the behaviouralists, quantitative models of efficient markets – all of which are predicated on rational choice – are likely to be wrong as well.

Impossibility of efficient markets

Grossman and Stiglitz (1980) go even farther – they argue that perfectly informationally efficient markets are an *impossibility* for, if markets are perfectly efficient, there is no profit to gathering information, in which case there would be little reason to trade and markets would eventually collapse. Alternatively, the degree of market *inefficiency* determines the effort investors are willing to expend to gather and trade on information, hence a nondegenerate market equilibrium will arise only when there are sufficient profit opportunities, that is, inefficiencies, to compensate investors for the costs of trading and information gathering. The profits earned by these attentive investors may be viewed as 'economic rents' that accrue to those willing to engage in such activities. Who are the providers of these rents? Black (1986) gave us a provocative answer: 'noise traders', individuals who trade on what they consider to be information but which is, in fact, merely noise.

The supporters of the EMH have responded to these challenges by arguing that, while behavioural biases and corresponding inefficiencies do exist from time to time, there is a limit to their prevalence and impact because of opposing forces dedicated to exploiting such opportunities. A simple example of such a limit is the so-called 'Dutch book', in which irrational probability beliefs give rise to guaranteed profits for the savvy investor. Consider, for example, an event E, defined as 'the S&P 500 index drops by five per cent or more next Monday', and suppose an individual has the following irrational beliefs: there is a 50 per cent probability that E will occur, and a 75 per cent probability that E will *not* occur. This is clearly a violation of one of the basic axioms of probability theory – the probabilities of two mutually exclusive and exhaustive events must sum to 1 – but many experimental studies have documented such violations among an overwhelming majority of human subjects.

These inconsistent subjective probability beliefs imply that the individual would be willing to take both of the following bets B_1 and B_2:

$$B_1 = \begin{cases} \$1 & \text{if } E \\ -\$1 & \text{otherwise} \end{cases}, \quad B_2 = \begin{cases} \$1 & \text{if } E^c \\ -\$1 & \text{otherwise} \end{cases}$$

where E^c denotes the event 'not E'. Now suppose we take the opposite side of both bets, placing \$50 on B_1 and \$25 on B_2. If E occurs, we lose \$50 on B_1 but gain \$75 on B_2, yielding a profit of \$25. If E^c occurs, we gain \$50 on B_1 and lose \$25 on B_2, also yielding a profit of \$25. Regardless of the outcome, we have secured a profit of \$25, an 'arbitrage' that comes at the expense of the individual with inconsistent probability beliefs. Such beliefs are not sustainable, and market forces – namely, arbitrageurs such as hedge funds and proprietary trading groups – will take advantage of these opportunities until they no longer exist, that is, until the odds are in line with the axioms of probability theory. (Only when these axioms are satisfied is arbitrage ruled out. This was conjectured by Ramsey, 1926, and proved rigorously by de Finetti, 1937, and Savage, 1954.) Therefore, proponents of the classical EMH argue that there are limits to the degree and persistence of behavioural biases such as inconsistent probability beliefs, and substantial incentives for those who can identify and exploit such occurrences. While all of us are subject to certain behavioural biases from time to time, according to EMH supporters market forces will always act to bring prices back to rational levels, implying that the impact of irrational behaviour on financial markets is generally negligible and, therefore, irrelevant.

But this last conclusion relies on the assumption that market forces are sufficiently powerful to overcome any type of behavioural bias, or equivalently that irrational beliefs are not so pervasive as to overwhelm the capacity of arbitrage capital dedicated to taking advantage of such irrationalities. This is an empirical issue that cannot be settled theoretically, but must be tested through careful measurement and statistical analysis. The classic reference by Kindleberger (1989) – where a number of speculative bubbles, financial panics, manias, and market crashes are described in detail – suggests that the forces of irrationality can overwhelm the forces of arbitrage capital for months and, in several well-known cases, years.

So what does this imply for the EMH?

The current state of the EMH

Given all of the theoretical and empirical evidence for and against the EMH, what can we conclude? Amazingly, there is still no consensus among economists. Despite the many advances in the statistical analysis, databases, and theoretical models surrounding the EMH, the main result of all of these studies is to harden the resolve of the proponents of each side of the debate.

One of the reasons for this state of affairs is the fact that the EMH, by itself, is not a well-defined and empirically refutable hypothesis. To make it operational, one must specify additional structure, for example, investors' preferences or information structure. But then a test of the EMH becomes a test of several auxiliary hypotheses as well, and a rejection of such a joint hypothesis tells us little about which aspect of the joint hypothesis is inconsistent with the data. Are stock prices too volatile because markets are inefficient, or due to risk aversion, or dividend smoothing? All three inferences are consistent with the data. Moreover, new statistical tests designed to distinguish among them will no doubt require auxiliary hypotheses of their own which, in turn, may be questioned.

More importantly, tests of the EMH may not be the most informative means of gauging the efficiency of a given market. What is often of more consequence is the efficiency of a particular market *relative* to other markets – for example, futures vs. spot markets, auction vs. dealer markets. The advantages of the concept of relative efficiency, as opposed to the all-or-nothing notion of absolute efficiency, are easy to spot by way of an analogy. Physical systems are often given an efficiency rating based on the relative proportion of energy or fuel converted to useful work. Therefore, a piston engine may be rated at 60 per cent efficiency, meaning that on average 60 per cent of the energy contained in the engine's fuel is used to turn the crankshaft, with the remaining 40 per cent lost to other forms of work, such as heat, light or noise.

Few engineers would ever consider performing a statistical test to determine whether or not a given engine is perfectly efficient – such an engine exists only in the idealized frictionless world of the imagination. But measuring relative efficiency – relative, that is, to the frictionless ideal – is commonplace. Indeed, we have come to expect such measurements for many household products: air conditioners, hot water heaters, refrigerators, and so on. Therefore, from a practical point of view, and in light of Grossman and Stiglitz (1980), the EMH is an idealization that is economically unrealizable, but which serves as a useful benchmark for measuring relative efficiency.

The desire to build financial theories based on more realistic assumptions has led to several new strands of literature, including psychological approaches to risk-taking behaviour (Kahneman and Tversky, 1979; Thaler, 1993; Lo, 1999), evolutionary game theory (Friedman, 1991), agent-based modelling of financial markets (Arthur et al., 1997; Chan et al., 1998), and direct applications of the principles of evolutionary psychology to economics and finance (Lo, 1999; 2002; 2004; 2005; Lo and Repin, 2002). Although substantially different in methods and style, these emerging sub-fields are all directed at new interpretations of the EMH. In particular, psychological models of financial markets focus on the

the manner in which human psychology influences the economic decision-making process as an explanation of apparent departures from rationality. Evolutionary game theory studies the evolution and steady-state equilibria of populations of competing strategies in highly idealized settings. Agent-based models are meant to capture complex learning behaviour and dynamics in financial markets using more realistic markets, strategies, and information structures. And applications of evolutionary psychology provide a reconciliation of rational expectations with the behavioural findings that often seem inconsistent with rationality.

For example, in one agent-based model of financial markets (Farmer, 2002), the market is modelled using a non-equilibrium market mechanism, whose simplicity makes it possible to obtain analytic results while maintaining a plausible degree of realism. Market participants are treated as computational entities that employ strategies based on limited information. Through their (sometimes suboptimal) actions they make profits or losses. Profitable strategies accumulate capital with the passage of time, and unprofitable strategies lose money and may eventually disappear. A financial market can thus be viewed as a co-evolving ecology of trading strategies. The strategy is analogous to a biological species, and the total capital deployed by agents following a given strategy is analogous to the population of that species. The creation of new strategies may alter the profitability of pre-existing strategies, in some cases replacing them or driving them extinct.

Although agent-based models are still in their infancy, the simulations and related theory have already demonstrated an ability to understand many aspects of financial markets. Several studies indicate that, as the population of strategies evolves, the market tends to become more efficient, but this is far from the perfect efficiency of the classical EMH. Prices fluctuate in time with internal dynamics caused by the interaction of diverse trading strategies. Prices do not necessarily reflect 'true values'; if we view the market as a machine whose job is to set prices properly, the inefficiency of this machine can be substantial. Patterns in the price tend to disappear as agents evolve profitable strategies to exploit them, but this occurs only over an extended period of time, during which substantial profits may be accumulated and new patterns may appear.

The adaptive markets hypothesis

The methodological differences between mainstream and behavioural economics suggest that an alternative to the traditional deductive approach of neoclassical economics may be necessary to reconcile the EMH with its behavioural critics. One particularly promising direction is to view financial markets from a biological perspective and, specifically, within an evolutionary framework in which markets, instruments, institutions and investors interact and evolve dynamically according to the 'law' of economic selection. Under this view, financial agents compete and adapt, but they do not necessarily do so in an optimal fashion (see Farmer and Lo, 1999; Farmer, 2002; Lo, 2002; 2004; 2005).

This evolutionary approach is heavily influenced by recent advances in the emerging discipline of 'evolutionary psychology', which builds on the seminal research of E.O. Wilson (1975) in applying the principles of competition, reproduction, and natural selection to social interactions, yielding surprisingly compelling explanations for certain kinds of human behaviour, such as altruism, fairness, kin selection, language, mate selection, religion, morality, ethics and abstract thought (see, for example, Barkow, Cosmides and Tooby, 1992; Gigerenzer, 2000). 'Sociobiology' is the rubric that Wilson (1975) gave to these powerful ideas, which generated a considerable degree of controversy in their own right, and the same principles can be applied to economic and financial contexts. In doing so, we can fully reconcile the EMH with all of its behavioural alternatives, leading to a new synthesis: the adaptive markets hypothesis (AMH).

Students of the history of economic thought will no doubt recall that Thomas Malthus used biological arguments – the fact that populations increase at geometric rates whereas natural resources increase at only arithmetic rates – to arrive at rather dire economic consequences, and that both Darwin and Wallace were influenced by these arguments (see Hirshleifer, 1977, for further details). Also, Joseph Schumpeter's view of business cycles, entrepreneurs and capitalism have an unmistakable evolutionary flavour to them; in fact, his notions of 'creative destruction' and 'bursts' of entrepreneurial activity are similar in spirit to natural selection and Eldredge and Gould's (1972) notion of 'punctuated equilibrium'. More recently, economists and biologists have begun to explore these connections in several veins: direct extensions of sociobiology to economics (Becker, 1976; Hirshleifer, 1977); evolutionary game theory (Maynard Smith, 1982); evolutionary economics (Nelson and Winter, 1982); and economics as a complex system (Anderson, Arrow and Pines, 1988). And publications like the *Journal of Evolutionary Economics* and the *Electronic Journal of Evolutionary Modeling and Economic Dynamics* now provide a home for research at the intersection of economics and biology.

Evolutionary concepts have also appeared in a number of financial contexts. For example, Luo (1995) explores the implications of natural selection for futures markets, and Hirshleifer and Luo (2001) consider the long-run prospects of overconfident traders in a competitive securities market. The literature on agent-based modelling pioneered by Arthur et al. (1997), in which interactions among software agents programmed with simple heuristics are simulated, relies heavily on evolutionary dynamics. And at least two prominent practitioners have proposed Darwinian alternatives to the EMH. In a chapter titled 'The Ecology of Markets', Niederhoffer (1997, ch. 15) likens financial markets to an ecosystem with dealers as 'herbivores', speculators as 'carnivores', and floor traders and distressed investors as 'decomposers'. And Bernstein (1998) makes a compelling case for active management by pointing out that the notion of equilibrium, which is central to the EMH, is rarely realized in practice and that market dynamics are better explained by evolutionary processes.

Clearly the time is now ripe for an evolutionary alternative to market efficiency.

To that end, in the current context of the EMH we begin, as Samuelson (1947) did, with the theory of the individual consumer. Contrary to the neoclassical postulate that individuals maximize expected utility and have rational expectations, an evolutionary perspective makes considerably more modest claims, viewing individuals as organisms that have been honed, through generations of natural selection, to maximize the survival of their genetic material (see, for example, Dawkins, 1976). While such a reductionist approach can quickly degenerate into useless generalities – for example, the molecular biology of economic behaviour – nevertheless, there are valuable insights to be gained from the broader biological perspective. Specifically, this perspective implies that behaviour is not necessarily intrinsic and exogenous, but evolves by natural selection and depends on the particular environment through which selection occurs. That is, natural selection operates not only upon genetic material but also upon social and cultural norms in *homo sapiens*; hence Wilson's term 'sociobiology'.

To operationalize this perspective within an economic context, consider the idea of 'bounded rationality' first espoused by Nobel-prize-winning economist Herbert Simon. Simon (1955) suggested that individuals are hardly capable of the kind of optimization that neoclassical economics calls for in the standard theory of consumer choice. Instead, he argued that, because optimization is costly and humans are naturally limited in their computational abilities, they engage in something he called 'satisficing', an alternative to optimization in which individuals make choices that are merely satisfactory, not necessarily optimal. In other words, individuals are bounded in their degree of rationality, which is in sharp contrast to the current orthodoxy – rational expectations – where individuals have unbounded rationality (the term 'hyper-rational expectations' might be more descriptive). Unfortunately, although this idea garnered a Nobel Prize for Simon, it had relatively little impact on the economics profession. (However, his work is now receiving greater attention, thanks in part to the growing behavioural literature in economics and finance. See, for example, Simon, 1982; Sargent, 1993; A. Rubinstein, 1998; Gigerenzer and Selten, 2001.) Apart from the sociological factors discussed above, Simon's framework was commonly dismissed because of one specific criticism: what determines the point at which an individual stops optimizing and reaches a satisfactory solution? If such a point is determined by the usual cost–benefit calculation underlying much of microeconomics (that is, optimize until the marginal benefits of the optimum equals the marginal cost of getting there), this assumes the optimal solution is known, which would eliminate the need for satisficing. As a result, the idea of bounded rationality fell by the wayside, and rational expectations has become the de facto standard for modelling economic behaviour under uncertainty.

An evolutionary perspective provides the missing ingredient in Simon's framework. The proper response to the question of how individuals determine the point at which their optimizing behaviour is satisfactory is this: such points are determined not analytically but through trial and error and, of course, natural selection. Individuals make choices based on past experience and their 'best guess' as to what might be optimal, and they learn by receiving positive or negative reinforcement from the outcomes. If they receive no such reinforcement, they do not learn. In this fashion, individuals develop heuristics to solve various economic challenges, and, as long as those challenges remain stable, the heuristics will eventually adapt to yield approximately optimal solutions to them.

If, on the other hand, the environment changes, then it should come as no surprise that the heuristics of the old environment are not necessarily suited to the new. In such cases, we observe 'behavioural biases' – actions that are apparently ill-advised in the context in which we observe them. But rather than labelling such behaviour 'irrational', it should be recognized that suboptimal behaviour is not unlikely when we take heuristics out of their evolutionary context. A more accurate term for such behaviour might be 'maladaptive'. The flopping of a fish on dry land may seem strange and unproductive, but under water the same motions are capable of propelling the fish away from its predators.

By coupling Simon's notion of bounded rationality and satisficing with evolutionary dynamics, many other aspects of economic behaviour can also be derived. Competition, cooperation, market-making behaviour, general equilibrium, and disequilibrium dynamics are all adaptations designed to address certain environmental challenges for the human species, and by viewing them through the lens of evolutionary biology we can better understand the apparent contradictions between the EMH and the presence and persistence of behavioural biases.

Specifically, the adaptive markets hypothesis can be viewed as a new version of the EMH, derived from evolutionary principles. Prices reflect as much information as dictated by the combination of environmental conditions and the number and nature of 'species' in the economy or, to use the appropriate biological term, the *ecology*. By 'species' I mean distinct groups of market participants, each behaving in a common manner. For example, pension funds may be considered one species; retail investors, another; market-makers, a third; and hedge-fund managers, a fourth. If multiple species (or the members of a single highly populous species) are competing for rather scarce resources within a single market, that market is likely to be highly efficient – for example, the market for 10-Year US Treasury Notes reflects most relevant information very quickly indeed. If, on the other hand, a small number of species are competing for rather abundant resources in a given market, that market will be less efficient – for example, the market for oil paintings from the Italian Renaissance. Market efficiency cannot be evaluated in a vacuum, but is highly context-dependent and dynamic, just as insect populations advance and decline as a function of the seasons, the number of predators and prey they face, and their abilities to adapt to an ever-changing environment.

The profit opportunities in any given market are akin to the amount of food and water in a particular local ecology – the more resources present, the less fierce the competition. As competition increases, either because of dwindling food supplies or an increase in the animal population, resources are depleted which, in turn, causes a population decline eventually, decreasing the level of competition and starting the cycle again. In some cases cycles converge to corner solutions, that is, certain species become extinct, food sources are permanently exhausted, or environmental conditions shift dramatically. By viewing economic profits as the ultimate food source on which market participants depend for their survival, the dynamics of market interactions and financial innovation can be readily derived.

Under the AMH, behavioural biases abound. The origins of such biases are heuristics that are adapted to non-financial contexts, and their impact is determined by the size of the population with such biases versus the size of competing populations with more effective heuristics. During the autumn of 1998, the desire for liquidity and safety by a certain population of investors overwhelmed the population of hedge funds attempting to arbitrage such preferences, causing those arbitrage relations to break down. However, in the years prior to August 1998 fixed-income relative-value traders profited handsomely from these activities, presumably at the expense of individuals with seemingly 'irrational' preferences (in fact, such preferences were shaped by a certain set of evolutionary forces, and might be quite rational in other contexts). Therefore, under the AMH, investment strategies undergo cycles of profitability and loss in response to changing business conditions, the number of competitors entering and exiting the industry, and the type and magnitude of profit opportunities available. As opportunities shift, so too will the affected populations. For example, after 1998 the number of fixed-income relative-value hedge funds declined dramatically – because of outright

failures, investor redemptions, and fewer start-ups in this sector – but many have reappeared in recent years as performance for this type of investment strategy has improved.

Even fear and greed – the two most common culprits in the downfall of rational thinking according to most behaviouralists – are the product of evolutionary forces, adaptive traits that enhance the probability of survival. Recent research in the cognitive neurosciences and economics, now coalescing into the discipline known as 'neuroeconomics', suggests an important link between rationality in decision-making and emotion (Grossberg and Gutowski, 1987; Damasio, 1994; Elster, 1998; Lo and Repin, 2002; and Loewenstein, 2000), implying that the two are not antithetical but in fact complementary. For example, contrary to the common belief that emotions have no place in rational financial decision-making processes, Lo and Repin (2002) present preliminary evidence that physiological variables associated with the autonomic nervous system are highly correlated with market events even for highly experienced professional securities traders. They argue that emotional responses are a significant factor in the real-time processing of financial risks, and that an important component of a professional trader's skills lies in his or her ability to channel emotion, consciously or unconsciously, in specific ways during certain market conditions.

This argument often surprises economists because of the link between emotion and behavioural biases, but a more sophisticated view of the role of emotions in human cognition shows that they are central to rationality (see, for example, Damasio, 1994; Rolls, 1999). In particular, emotions are the basis for a reward-and-punishment system that facilitates the selection of advantageous behaviour, providing a numeraire for animals to engage in a 'cost–benefit analysis' of the various actions open to them (Rolls, 1999, ch. 10.3). From an evolutionary perspective, emotion is a powerful adaptation that dramatically improves the efficiency with which animals learn from their environment and their past (see Damasio, 1994). These evolutionary underpinnings are more than simple speculation in the context of financial market participants. The extraordinary degree of competitiveness of global financial markets and the outsize rewards that accrue to the 'fittest' traders suggest that Darwinian selection – 'survival of the richest', to be precise – is at work in determining the typical profile of the successful trader. After all, unsuccessful traders are eventually eliminated from the population after suffering a certain level of losses.

The new paradigm of the AMH is still under development, and certainly requires a great deal more research to render it 'operationally meaningful' in Samuelson's sense. However, even at this early stage it is clear that an evolutionary framework is able to reconcile many of the apparent contradictions between efficient markets and behavioural exceptions. The former may be viewed as the steady-state limit of a population with constant environmental conditions, and the latter involves specific adaptations of certain groups that may or may not persist, depending on the particular evolutionary paths that the economy experiences. More specific implications may be derived through a combination of deductive and inductive inference – for example, theoretical analysis of evolutionary dynamics, empirical analysis of evolutionary forces in financial markets, and experimental analysis of decision-making at the individual and group level.

For example, one implication is that, to the extent that a relation between risk and reward exists, it is unlikely to be stable over time. Such a relation is determined by the relative sizes and preferences of various populations in the market ecology, as well as institutional aspects such as the regulatory environment and tax laws. As these factors shift over time, any risk–reward relation is likely to be affected. A corollary of this implication is that the equity risk premium is also time-varying and path-dependent. This is not so revolutionary an idea as it might first appear – even in the context of a rational expectations equilibrium model, if risk preferences change over time, then the equity risk premium must vary too. The incremental insight of the AMH is that aggregate risk preferences are not immutable constants, but are shaped by the forces of natural selection. For example, until recently US markets were populated by a significant group of investors who had never experienced a genuine bear market – this fact has undoubtedly shaped the aggregate risk preferences of the US economy, just as the experience since the bursting of the technology bubble in the early 2000s has affected the risk preferences of the current population of investors. In this context, natural selection determines who participates in market interactions; those investors who experienced substantial losses in the technology bubble are more likely to have exited the market, leaving a markedly different population of investors. Through the forces of natural selection, history matters. Irrespective of whether prices fully reflect all available information, the particular path that market prices have taken over the past few years influences current aggregate risk preferences. Among the three fundamental components of any market equilibrium – prices, probabilities, and preferences – preferences is clearly the most fundamental and least understood. Several large bodies of research have developed around these issues – in economics and finance, psychology, operations research (also called 'decision sciences') and, more recently, brain and cognitive sciences – and many new insights are likely to flow from synthesizing these different strands of research into a more complete understanding of how individuals make decisions (see Starmer, 2000, for an excellent review of this literature). Simon's (1982) seminal contributions to this literature are still remarkably timely and their implications have yet to be fully explored.

Conclusions

Many other practical insights and potential breakthroughs can be derived from shifting our mode of thinking in financial economics from the physical to the biological sciences. Although evolutionary ideas are not yet part of the financial mainstream, the hope is that they will become more commonplace as they demonstrate their worth – ideas are also subject to 'survival of the fittest'. No one has illustrated this principal so well as Harry Markowitz, the father of modern portfolio theory and a Nobel laureate in economics in 1990. In describing his experience as a Ph.D. student on the eve of his graduation, he wrote in his Nobel address (Markowitz, 1991, p. 476):

> ... [W]hen I defended my dissertation as a student in the Economics Department of the University of Chicago, Professor Milton Friedman argued that portfolio theory was not Economics, and that they could not award me a Ph.D. degree in Economics for a dissertation which was not Economics. I assume that he was only half serious, since they did award me the degree without long debate. As to the merits of his arguments, at this point I am quite willing to concede: at the time I defended my dissertation, portfolio theory was not part of Economics. But now it is.

In light of the sociology of the EMH controversy (see, for example, Lo, 2004), the debate is likely to continue. However, despite the lack of consensus in academia and industry, the ongoing dialogue has given us many new insights into the economic structure of financial markets. If, as Paul Samuelson has suggested, financial economics is the crown jewel of the social sciences, then the EMH must account for half the facets.

ANDREW W. LO

See also **financial market anomalies; rational expectations; rationality, bounded**.

I thank John Cox, Gene Fama, Bob Merton, and Paul Samuelson for helpful discussions.

Bibliography

Anderson, P., Arrow, K. and Pines, D., eds. 1988. *The Economy as an Evolving Complex System*. Reading, MA: Addison-Wesley Publishing Company.

Arthur, B., Holland, J., LeBaron, B., Palmer, R. and Tayler, P. 1997. Asset pricing under endogenous expectations in an artificial stock market. In *The Economy as an Evolving Complex System II*, ed. B. Arthur, S. Durlauf, and D. Lane. Reading, MA: Addison Wesley.

Ball, R. and Brown, P. 1968. An empirical evaluation of accounting income numbers. *Journal of Accounting Research* 6, 159–78.

Banz, R. 1981. The relationship between return and market value of common stock. *Journal of Financial Economics* 9, 3–18.

Barber, B. and Odean, T. 2001. Boys will be boys: gender, overconfidence, and common stock investment. *Quarterly Journal of Economics* 116, 261–29.

Barkow, J., Cosmides, L. and Tooby, J. 1992. *The Adapted Mind: Evolutionary Psychology and the Generation of Culture*. Oxford: Oxford University Press.

Basu, S. 1977. The investment performance of common stocks in relation to their price–earnings ratios: a test of the efficient market hypothesis. *Journal of Finance* 32, 663–82.

Becker, G. 1976. Altruism, egoism, and genetic fitness: economics and sociobiology. *Journal of Economic Literature* 14, 817–26.

Bell, D. 1982. Risk premiums for decision regret. *Management Science* 29, 1156–66.

Bernard, V. and Thomas, J. 1990. Evidence that stock prices do not fully reflect the implications of current earnings for future earnings. *Journal of Accounting and Economics* 13, 305–40.

Bernstein, P. 1998. Why the efficient market offers hope to active management. In *Economics and Portfolio Strategy*, 1 October. New York: Peter Bernstein, Inc.

Black, F. 1986. Noise. *Journal of Finance* 41, 529–44.

Black, F. and Scholes, M. 1973. Pricing of options and corporate liabilities. *Journal of Political Economy* 81, 637–54.

Brown, S., Goetzmann, W., Ibbotson, R. and Ross, S. 1992. Survivorship bias in performance studies. *Review of Financial Studies* 5, 553–80.

Campbell, J. and Shiller, R. 1988. The dividend–price ratio and expectations of future dividends and discount factors. *Review of Financial Studies* 1, 195–228.

Chan, K. 1988. On the contrarian investment strategy. *Journal of Business* 61, 147–64.

Chan, L., Jegadeesh, N. and Lakonishok, J. 1996. Momentum strategies. *Journal of Finance* 51, 1681–713.

Chan, N., LeBaron, B., Lo, A. and Poggio, T. 1998. Information Dissemination and Aggregation in Asset Markets with Simple Intelligent Traders. Laboratory Technical Memorandum No. 1646. Cambridge, MA: MIT Artificial Intelligence.

Chopra, N., Lakonishok, J. and Ritter, J. 1992. Measuring Abnormal Performance: Do Stocks Overreact? *Journal of Financial Economics* 31, 235–86.

Cootner, P. 1962. Stock prices: random vs. systematic changes. *Industrial Management Review* 3, 24–45.

Cootner, P. 1964. *The Random Character of Stock Market Prices*. London: Risk Publications.

Copeland, T. and Mayers, D. 1982. The Value Line enigma (1965–1978): a case study of performance evaluation issues. *Journal of Financial Economics* 10, 289–322.

Cowles, A. 1960. A revision of previous conclusions regarding stock price behavior. *Econometrica* 28, 909–15.

Cowles, A. and Jones, H. 1937. Some a posteriori probabilities in stock market action. *Econometrica* 5, 280–294.

Damasio, A. 1994. *Descartes' Error: Emotion, Reason, and the Human Brain*. New York: Avon Books.

Dawkins, R. 1976. *The Selfish Gene*. Oxford: Oxford University Press.

de Finetti, B. 1937. La Prévision: Ses Lois Logiques, Ses Sources Subjectives. *Annales de l'Institut Henri Poincaré* 7, 1–68. English translation in *Studies in Subjective Probability*, ed. H. Kyburg and H. Smokler. New York: John Wiley & Sons, 1964.

DeBondt, W. and Thaler, R. 1985. Does the stock market overreact? *Journal of Finance* 40, 793–807.

Eldredge, N. and Gould, S. 1972. Punctuated equilibria: an alternative to phyletic gradualism. In *Models in Paleobiology*, ed. T. Schopf. San Francisco: Freeman, Cooper.

Elster, J. 1998. Emotions and economic theory. *Journal of Economic Literature* 36, 47–74.

Fama, E. 1963. Mandelbrot and the stable Paretian hypothesis. *Journal of Business* 36, 420–29.

Fama, E. 1965a. The behavior of stock market prices. *Journal of Business* 38, 34–105.

Fama, E. 1965b. Random walks in stock market prices. *Financial Analysts Journal* 21, 55–9.

Fama, E. 1970. Efficient capital markets: a review of theory and empirical work. *Journal of Finance* 25, 383–417.

Fama, E. and Blume, M. 1966. Filter rules and stock market trading profits. *Journal of Business* 39, 226–41.

Fama, E. and French, K. 1988. Permanent and temporary components of stock prices. *Journal of Political Economy* 96, 246–73.

Farmer, D. 2002. Market force, ecology and evolution. *Industrial and Corporate Change* 11, 895–953.

Farmer, D. and Lo, A. 1999. Frontiers of finance: evolution and efficient markets. *Proceedings of the National Academy of Sciences* 96, 9991–2.

Fischoff, B. and Slovic, P. 1980. A little learning…: confidence in multicue judgment tasks. In *Attention and Performance, VIII*, ed. R. Nickerson. Hillsdale, NJ: Erlbaum.

Flavin, M. 1983. Excess volatility in the financial markets: a reassessment of the empirical evidence. *Journal of Political Economy* 91, 929–56.

French, K. and Roll, R. 1986. Stock return variances: the arrival of information and the reaction of traders. *Journal of Financial Economics* 17, 5–26.

Friedman, D. 1991. Evolutionary games in economics. *Econometrica* 59, 637–66.

Gervais, S. and Odean, T. 2001. Learning to be overconfident. *Review of Financial Studies* 14, 1–27.

Gigerenzer, G. 2000. *Adaptive Thinking: Rationality in the Real World*. Oxford: Oxford University Press.

Gigerenzer, G. and Selten, R. 2001. *Bounded Rational: The Adaptive Toolbox*. Cambridge, MA: MIT Press.

Gilles, C. and LeRoy, S. 1991. Econometric aspects of the variance-bounds tests: a survey. *Review of Financial Studies* 4, 753–92.

Grossberg, S. and Gutowski, W. 1987. Neural dynamics of decision making under risk: affective balance and cognitive-emotional interactions. *Psychological Review* 94, 300–18.

Grossman, S. and Shiller, R. 1981. The determinants of the variability of stock market prices. *American Economic Review* 71, 222–7.

Grossman, S. and Stiglitz, J. 1980. On the impossibility of informationally efficient markets. *American Economic Review* 70, 393–408.

Hirshleifer, J. 1977. Economics from a biological viewpoint. *Journal of Law and Economics* 20, 1–52.

Hirshleifer, D. and Luo, G. 2001. On the survival of overconfident traders in a competitive securities market. *Journal of Financial Markets* 4, 73–84.

Huberman, G. and Regev, T. 2001. Contagious speculation and a cure for cancer: a nonevent that made stock prices soar. *Journal of Finance* 56, 387–96.

Jegadeesh, N. 1990. Evidence of predictable behavior of security returns. *Journal of Finance* 45, 881–98.

Jegadeesh, N. and Titman, S. 2001. Profitability of momentum strategies: an evaluation of alternative explanations. *Journal of Finance* 56, 699–720.

Kahneman, D. and Tversky, A. 1979. Prospect theory: an analysis of decision under risk. *Econometrica* 47, 263–91.

Keim, D. 1983. Size-related anomalies and stock return seasonality: further empirical evidence. *Journal of Financial Economics* 12, 13–32.

Kim, M., Nelson, C. and Startz, R. 1991. Mean reversion in stock prices? a reappraisal of the empirical evidence. *Review of Economic Studies* 58, 515–28.

Kindleberger, C. 1989. *Manias, Panics, and Crashes: A History of Financial Crises*. New York: Basic Books.

Kleidon, A. 1986. Variance bounds tests and stock price valuation models. *Journal of Political Economy* 94, 953–1001.

Laibson, D. 1997. Golden eggs and hyperbolic discounting. *Quarterly Journal of Economics* 62, 443–77.

Lakonishok, J. and Smidt, S. 1988. Are seasonal anomalies real? A ninety-year perspective. *Review of Financial Studies* 1, 403–25.

Lehmann, B. 1990. Fads, martingales, and market efficiency. *Quarterly Journal of Economics* 105, 1–28.

Leroy, S. 1973. Risk aversion and the martingale property of stock returns. *International Economic Review* 14, 436–46.

LeRoy, S. and Porter, R. 1981. The present value relation: tests based on variance bounds. *Econometrica* 49, 555–74.

Lichtenstein, S., Fischoff, B. and Phillips, L. 1982. Calibration of probabilities: the state of the art to 1980. In *Judgment Under Uncertainty: Heuristics and Biases*, ed. D. Kahneman, P. Slovic and A. Tversky. Cambridge: Cambridge University Press.

Lo, A. 1991. Long-term memory in stock market prices. *Econometrica* 59, 1279–313.

Lo, A. (ed.) 1997. *Market Efficiency: Stock Market Behavior in Theory and Practice*, 2 vols. Cheltenham: Edward Elgar Publishing Company.

Lo, A. 1999. The three P's of total risk management. *Financial Analysts Journal* 55, 87–129.

Lo, A. 2001. Risk management for hedge funds: introduction and overview. to appear in *Financial Analysts Journal* 57, 16–33.

Lo, A. 2002. Bubble, rubble, finance in trouble? *Journal of Psychology and Financial Markets* 3, 76–86.

Lo, A. 2004. The adaptive markets hypothesis: market efficiency from an evolutionary perspective. *Journal of Portfolio Management* 30, 15–29.

Lo, A. 2005. Reconciling efficient markets with behavioral finance: the adaptive markets hypothesis. *Journal of Investment Consulting* 7, 21–44.

Lo, A. and MacKinlay, C. 1988. Stock market prices do not follow random walks: evidence from a simple specification test. *Review of Financial Studies* 1, 41–66.

Lo, A. and MacKinlay, C. 1990a. An econometric analysis of nonsynchronous trading. *Journal of Econometrics* 45, 181–212.

Lo, A. and MacKinlay, C. 1990b. Data snooping biases in tests of financial asset pricing models. *Review of Financial Studies* 3, 431–68.

Lo, A. and MacKinlay, C. 1990c. When are contrarian profits due to stock market overreaction? *Review of Financial Studies* 3, 175–206.

Lo, A. and MacKinlay, C. 1999. *A Non-Random Walk Down Wall Street*. Princeton, NJ: Princeton University Press.

Lo, A. and Repin, D. 2002. The psychophysiology of real-time financial risk processing. *Journal of Cognitive Neuroscience* 14, 323–39.

Loewenstein, G. 2000. Emotions in economic theory and economic behavior. *American Economic Review* 90, 426–32.

Lucas, R. 1978. Asset prices in an exchange economy. *Econometrica* 46, 1429–46.

Luo, G. 1995. Evolution and market competition. *Journal of Economic Theory* 67, 223–50.

Markowitz, H. 1991. Foundations of portfolio theory. *Journal of Finance* 46, 469–77.

Marsh, T. and Merton, R. 1986. Dividend variability and variance bounds tests for the rationality of stock market prices. *American Economic Review* 76, 483–98.

Maynard Smith, J. 1982. *Evolution and the Theory of Games*. Cambridge: Cambridge University Press.

Merton, R. 1973. Theory of rational option pricing. *Bell Journal of Economics and Management Science* 4, 141–83.

Merton, R. 1987. On the current state of the stock market rationality hypothesis. In *Macroeconomics and Finance: Essays in Honor of Franco Modigliani*, ed. R. Dornbusch, S. Fischer and J. Bossons. Cambridge, MA: MIT Press.

Michener, R. 1982. Variance bounds in a simple model of asset pricing. *Journal of Political Economy* 90, 166–75.

Nelson, R. and Winter, S. 1982. *An Evolutionary Theory of Economic Change*. Cambridge, MA: Belknap Press of Harvard University Press.

Niederhoffer, V. 1997. *Education of a Speculator*. New York: John Wiley & Sons.

Odean, T. 1998. Are investors reluctant to realize their losses? *Journal of Finance* 53, 1775–98.

Osborne, M. 1959. Brownian motion in the stock market. *Operations Research* 7, 145–73.

Poterba, J. and Summers, L. 1988. Mean reversion in stock returns: evidence and implications. *Journal of Financial Economics* 22, 27–60.

Ramsey, F. 1926. Truth and probability. In *Foundations of Mathematics and Other Logical Essays*, ed. R. Braithwaite. New York: Harcourt Brace & Co.

Richardson, M. 1993. Temporary components of stock prices: a skeptic's view. *Journal of Business and Economics Statistics* 11, 199–207.

Roberts, H. 1959. Stock-market 'patterns' and financial analysis: methodological suggestions. *Journal of Finance* 14, 1–10.

Roberts, H. 1967. Statistical versus clinical prediction of the stock market. Unpublished manuscript, Center for Research in Security Prices, University of Chicago.

Roll, R. 1983. Vas is das? The turn-of-the-year effect and the return premia of small firms. *Journal of Portfolio Management* 9, 18–28.

Roll, R. 1984. Orange juice and weather. *American Economic Review* 74, 861–80.

Rolls, E. 1999. *The Brain and Emotion*. Oxford: Oxford University Press.

Rosenberg, B., Reid, K. and Lanstein, R. 1985. Persuasive evidence of market inefficiency. *Journal of Portfolio Management* 11, 9–17.

Rozeff, M. and Kinney, W., Jr. 1976. Capital market seasonality: the case of stock returns. *Journal of Financial Economics* 3, 379–402.

Rubinstein, A. 1998. *Modeling Bounded Rationality*. Cambridge, MA: MIT Press.

Rubinstein, M. 1976. The valuation of uncertain income streams and the pricing of options. *Bell Journal of Economics* 7, 407–25.

Samuelson, P. 1947. *Foundations of Economics Analysis*. Cambridge, MA: Harvard University Press.

Samuelson, P. 1965. Proof that properly anticipated prices fluctuate randomly. *Industrial Management Review* 6, 41–9.

Sargent, T. 1993. *Bounded Rationality in Macroeconomics*. Oxford: Clarendon Press.

Savage, L. 1954. *Foundations of Statistics*. New York: John Wiley & Sons.

Schumpeter, J. 1939. *Business Cycles: A Theoretical, Historical, And Statistical Analysis of the Capitalist Process*. New York: McGraw-Hill.

Shefrin, M. and Statman, M. 1985. The disposition to sell winners too early and ride losers too long: theory and evidence. *Journal of Finance* 40, 777–90.

Shiller, R. 1981. Do stock prices move too much to be justified by subsequent changes in dividends? *American Economic Review* 71, 421–36.

Simon, H. 1955. A behavioral model of rational choice. *Quarterly Journal of Economics* 69, 99–118.

Simon, H. 1982. *Models of Bounded Rationality*, 2 vols. Cambridge, MA: MIT Press.

Starmer, C. 2000. Developments in non-expected utility theory: the hunt for a descriptive theory of choice under risk. *Journal of Economic Literature* 38, 332–82.

Thaler, R. (ed.) 1993. *Advances in Behavioral Finance*. New York: Russell Sage Foundation.

Tversky, A. and Kahneman, D. 1981. The framing of decisions and the psychology of choice. *Science* 211, 453–8.

Wilson, E. 1975. *Sociobiology: The New Synthesis*. Cambridge, MA: Belknap Press of Harvard University Press.

egalitarianism

All modern political theories assume that persons are in some relevant sense moral equals, entitled to equal concern, respect or treatment, and that a theory of justice must interpret and reflect that moral equality. This commitment is sometimes dubbed the 'egalitarian plateau', and it has been a common foundational moral assumption since Locke. Contemporary theories differ in how they *interpret* the egalitarian plateau. Two kinds of theory of justice are usually counted as egalitarian. Theories of distributive equality concern themselves with the relative standing of individuals in the distribution of benefits and burdens; theories of relational equality concern themselves with the relative standing of individuals when they face each other in the public sphere.

The metric

One key question concerns the metric of equality: what, precisely, is it that egalitarians should seek to equalize? The literature falls into three main camps. Resourcists argue that people should be equal in the space of resources, meaning that they should have equal opportunity for achieving holdings of alienable goods. How are holdings priced? Ronald Dworkin imagines a hypothetical auction in which persons with equal holdings of some currency bid for available goods until markets clear (Dworkin, 2000). The distribution after the auction is equal if no one prefers anyone else's bundle of goods to her own; the distribution is then said to pass the 'envy test'. The intuitive idea is that the price of some good is set by the opportunity cost to others of that good. We have to tailor our preferences to our resources; equality is achieved when all face the same budget constraint, not when all achieve equal satisfaction.

Equality of resources has difficulty with the intuition that those with less socially valued talent, and in particular those with serious impairments, should receive compensation. Two strategies are available. One is to adopt a view that talent is socially constructed, so that much of the disadvantage faced by the less talented and the impaired is a consequence not of their lack of talent but of the fact that social institutions are maladapted to their natural endowments (Pogge, 2003). This view allows resourcists to call for the reform of social institutions in the name of equality, without demanding compensation for impairments. The problem with this strategy is that some mental and physical impairments *intrinsically* cause disadvantage; there is no feasible set of social arrangements that would not make it more difficult for people with the impairments to derive satisfaction from resources. So an alternative strategy is to make the cut between persons and resources in a different place, regarding talents as resources and disabilities as resource-deficits. Dworkin's own version of this strategy proposes compensating the less talented with additional income, the amount calculated by looking at the insurance that talented individuals would have bought against a lack of talents if they had no knowledge of their probability of having the talents.

An alternative metric is welfare; egalitarians of welfare would seek to equalize levels of welfare (understood sometimes as idealized preference satisfaction, sometimes in terms of internal states such as happiness). This view handles talent-inequality in a straightforward manner; the less talented and the disabled should be compensated up to the level where they enjoy as much welfare as anyone else. But it faces the problem that there is no reason for people to moderate their preferences; since welfare is a direct target, those with expensive tastes receive more resources than those with inexpensive tastes, which is widely regarded as intuitively unfair. An alternative view – equality of opportunity for welfare – deals with this problem by seeking equality of welfare except when inequalities are the result of voluntary well-informed choices rather than bad luck or circumstances outside the agent's control (Arneson, 1989). Again, the less talented are straightforwardly compensated for the way in which they find it harder than others to derive satisfaction, but those who cultivate expensive tastes are not. However, those with non-cultivated expensive preferences are also compensated, even if they could easily be overcome; this view does not see lack of talent, and disability in particular, as morally more urgent than expensive preferences. (See Roemer, 1986, for an argument that equality of resources implies equality of welfare.)

All of the views deploying an 'opportunity' metric, including Dworkin's resourcist view, presume the desirability of holding people accountable for their voluntary choices, but compensating them for deficits that are beyond their control. Views of this kind are sometimes referred to as varieties of 'luck egalitarianism'. Inequalities resulting from voluntary choice are acceptable because they reflect a deeper sense in which we are equal as moral agents; choice legitimizes inequality, brute luck does not. (For an elegant attempt both to conceptualize and operationalize equality of opportunity *tout court*, see Roemer, 1998.)

The main rival account – namely, the capabilities approach developed by Amartya Sen and Martha Nussbaum – focuses on the *preconditions* of agency (Sen, 1999; Nussbaum, 2000). Equality of capabilities demands that people be equal in the space of the functionings or livings that they are substantively able to achieve. Walking is a functioning, so are eating, reading, mountain climbing, and chatting. 'The concept of functionings … reflects the various things a person may value doing or being – varying from the basic (being adequately nourished) to the very complex (being able to take part in the life of the community)' (Sen, 1999, p. 75). But when we make interpersonal comparisons of well-being we should find a measure that incorporates references to functionings but also reflects the intuition that what matters is not merely achieving the functioning but being free to achieve it. So we should look at 'the freedom to achieve actual livings that one can have a reason to value' (Sen, 1999, p. 73) or, to put it another way, 'substantive freedoms – the capabilities – to choose a life one has reason to value'. The idea is that people should be equal in this space.

The capabilities approach avoids the problems of the standard welfarist approaches by focusing on *choice* (thus treating inequalities arising from voluntary choices differently from those arising from circumstances). It avoids the difficulty resourcist accounts have with unequal talent by focusing on *functionings*; talent deficits are compensated for by looking not at what others would pay to avoid them but at the valuable activities the deficits deprive people of access to. Some theorists place the capabilities account in the welfarist camp (Williams, 2002) but it is not implausible to think of it as a variant of resourcism, distinguished by its approach to the valuation of talents.

A major recent development in the debates about egalitarianism has involved criticisms of luck egalitarianism. Each of the luck egalitarian principles, taken alone, imposes heavy costs on those who endure misfortunes for which they can be held responsible, even if those costs place the agent below the threshold for full participation in social affairs. An alternative has developed which is best described as 'relational egalitarianism'. Relational egalitarianism is not directly concerned with equality in terms of the distribution of any particular currency, but endorses the idea that individuals should have equal standing in the public sphere. This vague idea has several instantiations. Elizabeth Anderson (1999, p. 304) talks of seeking 'a social order in which persons stand in relations of equality'; Nancy Fraser (1998, p. 30) says that 'Justice requires social arrangements that permit all (adult) members of society to interact with one another as peers'. Both fill out their theories with more details. According to Fraser (1998, p. 24), 'It is unjust that some individuals and groups are denied the status of full partners in social interaction, simply as a consequence of institutionalized patterns of interpretation and evaluation in whose construction they have not equally participated and that disparage their distinctive characteristics or the distinctive characteristics assigned to them'. A third variant of relational egalitarianism spells it out specifically in terms of political equality, the idea being that it is particularly important that people enjoy equal availability of or opportunity for political power or influence (Christiano, 1995). This variant is typically less hostile than other variants to luck egalitarianism.

Each of the views reviewed in this section allows inequality along some dimensions. Relational egalitarianisms allow such inequalities of income, wealth, welfare or capabilities as are compatible with equal political influence, or interaction as peers, or 'equal opportunity for participation as a peer'. These permitted inequalities may be great or very small, and how great or small may vary by social context. Principles demanding equality of opportunity are consistent with great inequalities in outcome, and consistent also with some being very badly off in absolute terms. While equality of opportunity conceptions place no limit on how badly off someone may be as a result of her own imprudent choices, equality of social standing demands that no one fall below the threshold needed for equal participation, even if she makes numerous imprudent choices.

The distributive rules

Do egalitarians even care about *equality*? Principles demanding equality of X seem vulnerable to an obvious objection. In some dynamic situations it is possible to produce more of X by distributing X unequally, and to ensure that even those with least have more than under an equal distribution. For example, we can sometimes produce more wealth by judiciously attaching higher income to more productive positions in the economy, and to longer work hours; the higher income acts both as a signal and as an incentive to produce more. That greater production can be turned to the benefit of those with least. But, the objection goes, it would be perverse to prefer an equal situation in which everyone has less to one in which everyone has more, even if we have to sacrifice equality for the sake of that additional product.

This is known as the 'levelling down' objection to equality. Egalitarians make two distinct responses. The first is to concede the argument, abandoning 'equality' and replacing it with 'giving priority to the interests of the least advantaged'. John Rawls's difference principle, which states that 'social and economic inequalities are to be arranged to the maximum benefit of the least advantaged', embodies one variant of this response, a variant that gives *absolute* priority to the prospects of the least advantaged (Rawls, 1971; 2001). A weaker variant in this family of views, usually known as 'prioritarianism', simply says that it is more urgent to provide benefits to those with less advantage than to those with more (Parfit, 2000).

An alternative response is to assert value pluralism. This response acknowledges that priority to the least advantaged is an important value and perhaps more important than equality, so that when it comes to policy or action prioritarian principles should govern. But it says that equality nevertheless matters some; there is one way in which an unequal distribution is worse than an equal distribution, even if, all things considered, it is better; the way in which it is worse is that it is unequal and for that reason unfair (Temkin, 2002). This response is bolstered by the observation that there is nothing eccentric about endorsing a principle that values distributions that benefit nobody; the retributive principle of proportionality between punishment and crime, for example, calls for harming the criminal even when there is no gain to anyone else in harming him.

Some reject principles of equality and priority on the grounds that all that matters for the purposes of justice is that all have enough. Sufficientarian theories are not usually counted as within the egalitarian family, because they eschew any fundamental concern with relativities. Relativities may matter in determining what is enough for people to live a decent life in any given social environment, but ultimately what matters is not where someone ranks in the distribution of resources (or anything else) but whether she has enough. However, as suggested above, sufficientarian principles also have a place in some variants of egalitarianism. While relational egalitarianism places no principled limits on the level of material or welfare inequality, and gives no general priority to the least advantaged, it does set a floor – all must have sufficient resources to be full participants in social interaction. Equality of political influence demands that all have sufficient resources, personal and financial, to play an equal role in political life, but, as long as it is possible to insulate politics from residual inequalities of wealth, it is not concerned with equalizing or prioritizing benefit to the least advantaged.

Many theories of justice that do not fit the above characterizations of egalitarianism nevertheless incorporate some elements of egalitarian thinking. John Rawls's theory of justice, for example, prioritizes the principle that certain basic liberties (not including strong property rights) be equally distributed, then demands that within that constraint fair equality of opportunity should be implemented, and then that social and economic inequalities be arranged to the greatest benefit of the least advantaged in so far as that is possible without jeopardizing the equal liberty and fair equality of opportunity principle (Rawls, 1971; 2001). Michael Walzer's (1983) theory of 'complex equality' takes seriously widely shared intuitions that different goods are subject to different distributive rules. For example, while income should be distributed according to productive contribution, as will tend to result from market interactions, the inequalities this norm generates should be prevented from translating into unequal access to certain key goods like health care and educational opportunities, the distribution of which should be governed by need and the requirements of equal opportunity respectively. It is unclear in what sense Walzer's 'complex equality' is genuinely an egalitarian position, since it is in principle consistent with unequal and coinciding distributions of all goods that are not themselves governed by egalitarian norms.

Priority and equality coincide in practice for one class of goods: positional goods. These have the property that the contribution an individual's share of the good makes to her absolute position is determined by how much of the good she has relative to others. The credentialing aspect of education is a paradigm case; how useful a degree is in landing a job (as opposed to the learning one achieved in the process of getting the degree) depends entirely on the credentials of one's competitors for that job. (Other cases are detailed in Hirsch, 1976.) Those who give priority to the worst-off will countenance inequalities in positional goods only in so far as they are required by or result in the least advantaged benefiting overall (Brighouse and Swift, 2006).

The scope of equality

Whatever the right distribuendum, and whatever the appropriate distributive principle, it is a further question who should be equal to whom. Some limit the application of their egalitarianism to members of the same society or system of cooperation, or to those subject to the same coercive structure (Nagel, 2005), or hold that it is states that owe their citizens a particular duty to treat them with equal concern and respect (Dworkin, 2000). Others believe that egalitarian principles should apply to all human beings, irrespective of the relations that obtain between them. If we restrict the application of egalitarian principles to schemes of cooperation, that does not exclude the possibility of a global egalitarianism, since most now accept that in the modern world social cooperation extends well beyond national boundaries (Julius, 2006). But consider this version of Derek Parfit's divided world case. All the people in A are half as well off as all the people in B, but A and B have no knowledge of or contact with each other (Parfit, 2000). Is there anything regrettable from the perspective of injustice about this inequality? If so, then the scope of justice is cosmic, not simply social. In the stated version of the divided world case this difference is motivationally inert, since the people in B do not have the relevant knowledge. But, if they did, cosmic egalitarianism would give them a reason to try to find a way to contact and interact with the people on A, while intra-societal egalitarianism would provide them with no such reason.

The divided world case brings out another difference in orientation. Where members of A and B have no interaction, or even knowledge of each other, equality can be valued only intrinsically rather than because of its effects on members of A or B. Often, however, inequality with respect to some goods is devalued, and equality valued, instrumentally, because of its absolute effects on those subject to the unequal distribution – usually its effects on the relatively disadvantaged. Thus, for example, economic inequalities are thought to undermine the fairness of legal or political processes, or occupational or other status hierarchies are claimed to harm the health of those on the lower rungs. Those who value equality intrinsically would hold that there is a reason to level down for the sake of equality or fairness, whereas instrumental egalitarians might seek the more equal distribution of some goods, not for egalitarian reasons *stricto sensu*, but to eliminate the bad effects of certain kinds of inequality.

The subject of justice

A further dividing line between egalitarians concerns the subject of justice. Rawls stipulates that the subject is the 'basic structure of society', which consists of some of the central, interaction-shaping institutions of a society: for example, the constitution, the legally recognized forms of property, the structure of the economy, the design of the legislature, and the judiciary. The idea is that these institutions govern the division of the advantages that accrue from social cooperation, and they assign the basic rights and responsibilities to citizens. So a society is just when those institutions are arranged according to the correct principles.

Rawls officially exempts individual actions and motives from evaluation from the perspective of egalitarian justice, as long as individuals obey the rules set by a just basic structure. But this has the consequence that a society in which talented individuals take advantage of the prerogatives not to serve the least advantaged that are built into the principles that he thinks justice requires of coercive institutions is no less just than one in which they are much more strongly motivated by the desire to benefit the least advantaged through their choices regarding work. A society with an egalitarian governing ethos, on this view, is no more just than one without, even when the least advantaged are much better off. But the motivations and actions of talented individuals affect the prospects and status of others in ways that have 'profound and pervasive influence on persons' (Rawls, 2001, p. 55), which is Rawls's central reason for focusing on the basic structure. So some egalitarians regard justice as commenting not only on the broad coercive outline of society, but also on less officially coercive institutions such as a society's ethos (Cohen, 1997). For a powerful defence of an account intermediate between Cohen's and Rawls's, see Julius, 2003).

Other values

Most egalitarian theorists are value pluralists; they believe that equality (or priority) of their preferred metric matters, but so do other principles. Observing that equality or priority is sometimes in conflict with liberty or privacy or efficiency does not require us to reject one of the conflicting values. It requires us, instead, to evaluate reasons for considering one of the values more morally important than the others, and, in the light of that evaluation, to establish which should give way in different conflicts. Unless the relationship between values is one of lexical priority (in which case the prior value always trumps subordinate values, which can be pursued only when there is no conflict), different trade-offs between values will be mandated in different conflicts. But lexical priority is unlikely to hold between genuine values. If a value matters *at all*, it is hard to believe it could never be the case that a very large amount of it was greater than a very small amount of a conflicting value *however great that conflicting value is*.

HARRY BRIGHOUSE AND ADAM SWIFT

See also **equality of opportunity; ethics and economics; liberalism and economics; libertarianism; Pareto efficiency; satisficing; Sen, Amartya.**

Bibliography

Anderson, E. 1999. What is the point of equality? *Ethics* 109, 287–337.

Arneson, R. 1989. Equality and equal opportunity for welfare. *Philosophical Studies* 56, 77–93.

Arneson, R. 2002. Egalitarianism. *Stanford Encyclopedia of Philosophy*. Online. Available at http://plato.stanford.edu/entries/egalitarianism, accessed 14 October 2006.

Brighouse, H. and Swift, A. 2006. Equality, priority and positional goods. *Ethics* 116, 471–97.

Christiano, T. 1995. *The Rule of the Many*. Boulder, CO: Westview Press.

Cohen, G.A. 1997. Where the action is: on the site of distributive justice. *Philosophy and Public Affairs* 26, 3–30.

Dworkin, R. 1981. What is equality? Part 2: equality of resources. *Philosophy and Public Affairs* 10, 283–345.

Dworkin, R. 2000. *Sovereign Virtue*. Cambridge, MA: Harvard University Press.

Fraser, N. 1998. Social justice in the age of identity politics. *The Tanner Lectures on Human Values*. Stanford University, 29 April–2 May. Online. Available at http://www.tannerlectures.utah.edu/lectures/Fraser98.pdf, accessed 14 October 2006.

Hirsch, F. 1976. *Social Limits to Growth*. London: Routledge & Kegan Paul.

Julius, A.J. 2003. Basic structure and the value of equality. *Philosophy and Public Affairs* 31, 321–55.

Julius, A.J. 2006. Nagel's atlas. *Philosophy and Public Affairs* 34, 176–92.

Nagel, T. 2005. The problem of global justice. *Philosophy and Public Affairs* 33, 113–47.

Nussbaum, M. 2000. *Women and Human Development*. Cambridge: Cambridge University Press.

Parfit, D. 2000. Equality or priority? In *The Ideal of Equality*, ed. M. Clayton and A. Williams. London and New York: Palgrave Macmillan and St Martin's Press.

Pogge, T. 2003. Can the capability approach be justified? *Philosophical Topics* 30(2), 167–228.

Rawls, J. 1971. *A Theory of Justice*. Cambridge, MA: Harvard University Press.

Rawls, J. 2001. *Justice as Fairness*. Cambridge, MA: Harvard University Press.

Roemer, J.E. 1986. Equality of resources implies equality of welfare. *Quarterly Journal of Economics* 101, 751–84.

Roemer, J.E. 1998. *Equality of Opportunity*. Cambridge, MA: Harvard University Press.

Sen, A. 1999. *Development as Freedom*. New York: Knopf.

Temkin, L. 2002. Equality, priority, and the levelling down objection. In *The Ideal of Equality*, ed. M. Clayton and A. Williams. London and New York: Palgrave Macmillan and St Martin's Press.

Walzer, M. 1983. *Spheres of Justice*. New York: Basic Books.

Williams, A. 2002. Dworkin on capability. *Ethics* 113, 23–39.

Einaudi, Luigi (1874–1961)

An outstanding Italian economist and influential figure on the broader political and cultural scene, Einaudi was born in Carru (Piedmont) on 24 March 1874 and died in Rome on 30 October 1961. He graduated in law from Turin in 1895 and then, while continuing with this studies, embarked on a career in journalism. The success he achieved in both fields underlined his rare talent and his endless capacity for work. In fact, his academic progress was so rapid that in 1907 he was appointed as professor of public finance at the University of Turin. Meanwhile, he wrote articles for the most influential Italian daily newspaper of the period, the *Corriere delle Serra*, which not only brought him national recognition but also earned him the reputation of 'educator' of the entire country. He became a member of the Senate in 1919, but retired from all political and public activity with the advent of fascism. Towards the end of the First World War he went into exile in Switzerland. On his return, he was appointed Governor of the Bank of Italy (1945), Vice-President of the Cabinet and Minister in charge of the Budget (1947), and was finally elected President of the Republic of Italy (1948–1955). At the end of his seven-year presidential term of office, he was made a life member of the Senate.

The most important aspect of Einaudi's achievements is the use he made of his academic and journalistic ability, as foundations for his activity as a statesman and politician. In addition, close study of his strictly scientific works reveals the extent to which he drew on the wealth of knowledge and experience which he had gained also in other fields. The 3,800 recorded items of Einaudi's works cover such a wide range of interests that it is necessary here to concentrate on his contributions to the study of public finance and his ideas on economic policy. Einaudi's main contributions to the study of public finance were investigations, based on the classical ideas of John Stuart Mill, which gave a solid logical basis to the principle of the exclusion of savings from taxable income; his research into the theory of capitalization of taxation; his critical and constructive contributions on the effects of certainty and stability of fiscal principles; his important analysis of the concept of taxable income which he identified with normal income, or, in other words, with the average income potentiality of the person subject to taxation.

Einaudi's position vis-à-vis public intervention in the economy was not hostile in principle, though he undoubtedly took a limited view of state interference in economic life. Since, for Einaudi, 'All liberties were jointly liable', autonomous sources of income were a necessity to prevent people from being subjected to a single centralizing order of the state. He asserted this during the 20 years of fascism, when he continued to teach with the same independence of mind and without compromising his fidelity to economic liberalism. Even though Einaudi had been stressing the usefulness of productive public expenditure since 1919, he showed a singular lack of comprehension of the Keynesian contribution, in the belief that it would be an inevitable cause of inflation.

F. CAFFÈ

Selected works

On Luigi Einaudi himself there is a *Bibliografia degli scritti* edited by Luigi Firpo under the auspices of the Bank of Italy, Turin, 1971. It is useful to divide his work into the three main areas which he outlined: theory, politics and history. Representative works of the three sections are as follows:

1912. *Intorno al concetto di reddito imponibile e di un sistema di imposte sul reddito consumato*. Turin: V. Bona.
1919. *Osservazioni critiche intorno alla teoria dell'ammortamento dell'imposta e teoria delle variazioni nei redditi e nei valori capitali sussequenti all'imposta*. Turin: Fratelli Bocca.
1929. *Contributo all ricerca della 'ottima imposta'*. Milan: Bocconi.
1938. *Miti e paradossi delli giustizia tributaria*. Turin: Luigi Einaudi.
The following handbooks are available:
1914. *Corso di scienza delle finanze*. Turin: Tip. e Bono.
1932–66. *Principi di scienza delle finanze*. Turin: La Riforma Sociale.
1932. *Il sistema tributario italiano*. Turin: La Riforma Sociale.
With reference to the history of finance and the history of ideas see:
1908. *La finanza sabauda all'aprirsi del secolo XVIII e durante la guerra di successione spagnola*. Turin: Società Tip. Editrice Nazionale.
1927. *La guerra e il sistema tributario italiano*. Bari: Laterza.

1953. *Saggi bibliografici e storici intorno alle dottrine economiche*. Rome: Ediz. Storia e Litteratura.

Einaudi's journalistic work has been largely collected in eight volumes comprising the *Cronache economiche e politiche di un trentennio* (1893–1925), Turin: Ed. Einaudi, 1959–65, and in *Lo scrittoio del Presidente 1948–1955*, Turin: Ed. Einaudi, 1956. For many years Einaudi was Italian correspondent for the *Economist*.

Eisner, Robert (1922–1998)

Robert Eisner, a leading American macroeconomist and theorist of the investment function, graduated in history from College of the City of New York in 1940, took an MA in sociology from Columbia University in 1942 and, following service in the army and the Office of Price Administration, a Ph.D. in economics under Fritz Machlup at Johns Hopkins University in 1951. He joined the faculty of Northwestern University in 1952, rising to hold the William R. Kenan Professorship of Economics from 1974 until his retirement in 1994. He served as President of the American Economic Association in 1988.

Eisner was an architect of the Keynesian ascendancy in post-war America. Much of his work was devoted to technical developments in that tradition; his singular distinction lay in taking the accounting foundations of Keynesian macroeconomics seriously and in developing their implications with utmost rigour. This thread runs through his writing from his earliest papers on the 'Invariant Multiplier', the permanent income hypothesis, liquidity preference and the liquidity trap. It reaches its apogee in his work on a Total Income System of Accounts (TISA). It suffuses his later, policy-oriented writings on the meaning and implications of deficits in the budget, current account, and Social Security system.

No shrinking violet, Eisner liked to call his shots. Thus, H. S. Houthaker 'has not performed [a] test correctly'; 'Bronfenbrenner and Mayer... confound... issues of elasticity with those of slope'; 'Re-estimation with Pifer's data and application of appropriate statistical tests contradict Pifer's conclusions' (1998a, pp. 8, 27, 48). The tone is ever tactful, the intent always the pursuit of truth, the subtext a certain delight in finding the exact, fatal weakness of an opposing view. Late in his life, this author heard Eisner speak to a room of senior officials in China on the error and futility of the one-child policy, a delicate issue which he raised in the same spirit and with deeply impressive effect.

Underpinning his technical precision lay an unflagging commitment to larger social goals, especially full employment, peace, and justice. Eisner actively advocated all three throughout his career, but especially in the later years when he appeared frequently on the opinion pages of the *Wall Street Journal*, as a leading director of Economists Allied for Arms Reduction, and in causes devoted to the advancement of women in the economics profession.

For instance, in a 1952 paper in the *American Economic Review* (1998a, 106–17) Eisner analysed the relationship of replacement costs to depreciation allowances in a growing economy. In doing so he called attention to the fact that growth in the latter usually exceeded that in the former, resulting in reported profits that were understated for purposes of both taxation and collective bargaining. Pointedly, he suggested the work ought to interest both revenue officers and trade unionists.

Yet Eisner's views were often unfashionable and politically inconvenient. In important papers in the 1980s, at a time when Democrats had taken the veil of fiscal virtue, he undertook with Paul Pieper to show that (among numerous other difficulties with budget accounting) inflation had rendered the deficit meaningless, introducing vast inconsistencies between the nominal budget deficit and the change in the real public debt. Thus, the Reagan deficits were far smaller than normally supposed, while those of Carter were surpluses in real terms – likely to produce fiscal drag and so to bear partial responsibility for the stagnation of those years. Correctly accounting for inflation, Eisner argued, might have forestalled the new classical critique that led many in those years to abandon Keynesian principles.

A closely related cause was the misunderstanding of 'national saving' and the fallacious popular argument that to reduce deficits would lead to increased capital formation. In 1995, Eisner argued that to take the accounting relation between public and private saving

> as evidence that reducing the federal deficit must raise national saving should be recognized, on even the slightest reflection, as patently absurd. It is startlingly akin to the assumption, more than half a century ago, that saving and investment would be increased if we all undertook to save more by consuming less. Perhaps! But that is exactly the proposition to be proved, or supported by empirical evidence, not assumed. (1998a, p. 322)

Second only to correct reasoning, evidence mattered. In the 1990s Eisner took up arms against the 'governing myth' of economic policy, the natural rate of unemployment introduced by Friedman and Phelps in 1968. From this strangely self-damaging justification for perpetually high unemployment, Eisner hoped for a 'NAIRU escape'. His method was largely econometric, and in what may have been his final paper, published in 1998 (1998a, pp. 454–87), he argued that a separate analysis of low-unemployment cases showed no relationship between full employment and rising inflation. This position was to be vindicated dramatically in the two years following his death.

Eisner embraced capital budgeting, so that the liabilities acquired by the government might be properly offset against corresponding assets. This position helped underpin a strong advocacy of liberal expenditure on infrastructure, education, and research and development. It also provides one bridge between the Keynesian Eisner and his counterpart, the theorist of investment, public finance, and peace economics and stalwart defender of Social Security, all of which he was.

Eisner's investigations of investment involved pioneering use of corporate records. They permitted cross-section analysis of firm decisions, showing that the concepts of macro models, such as the accelerator, operated differently on firms from different industries or with differing recent growth histories. In numerous studies, Eisner criticized neoclassical investment theories. Rejecting the notions of a desired capital stock and unit relative price elasticity, he adhered to a Keynesian relation of investment to expected profitability and of expected profits to the rate of growth. An important theme in this work concerns the appropriate level of aggregation at which to take measurements. Eisner found that firms appropriately assess the growth of their own industry to be the most relevant to profit prospects, not the inherently variable growth of individual firms or the potentially irrelevant growth of generalized aggregate demand.

Eisner's Total Income System of Accounts marked the peak of his campaign to rationalize economic measurement and theory. The importance of changing household relations appears vividly in his initial motivation for this work: 'What happens to income, output, and productivity when clotheswashing moves from the washtub and the professional laundry to the laundromat and to the automatic washer and dryer...?' (1998b, p. 188). Particularly noteworthy is capital accumulation by households in a country where transportation is provided mainly and increasingly by private car. The challenge of TISA remains to be taken up by most economists and national income statisticians.

Finally, midway through the Vietnam War Eisner deflated the view that President Johnson might have forestalled inflation by raising taxes; the only sure way to that end, he showed, would have been to avoid the war. This insight led to papers on the 'staggering cost' of the Vietnam war, much in the spirit of total accounts, and on post-cold war disarmament. Equally, to the end of his life Robert Eisner defended Social Security from all those who would cut it. Spurious and persistent allegations of financial 'crisis' notwithstanding, he believed that a rich and civilized society can, and should, provide decent incomes and care for its old.

JAMES K. GALBRAITH

See also **government budget constraint; labour supply; national accounting, history of; Social Security in the United States; war and economics.**

Selected works

1986. *How Real is the Federal Deficit?* New York: Free Press.

1994. *The Misunderstood Economy: What Counts and How to Count It.* Cambridge, MA: Harvard Business School Press.

1997. *The Great Deficit Scares: The Federal Budget, Trade, and Social Security.* New York: The Century Foundation.

1998a. *The Keynesian Revolution, Then and Now: The Selected Essays of Robert Eisner,* vol. 1. Cheltenham: Edward Elgar.

1998b. *Investment, National Income and Economic Policy: The Selected Essays of Robert Eisner,* vol. 2. Cheltenham: Edward Elgar.

elasticities approach to the balance of payments

The substance of a theory is independent of the manner in which it is dressed. In particular, it is a matter of style only whether or not formulae are expressed in terms of elasticities of demand and supply, or in terms of ordinary derivatives. To speak of an 'elasticities approach' to the balance of payments is therefore to speak no sense at all.

However, behind the nonsensical label there hides a coherent and distinctive theory of what determines the response of a country's balance of payments to parametric changes in its rate of exchange, that is, to changes in the terms on which its currency exchanges for other currencies. The theory goes back to a paper published by Charles Bickerdike (1920).

Consider a simplified world containing just two countries (the 'home' country and the 'foreign') and producing and trading just two commodities. Let R be the price of foreign currency in terms of home currency, let p_i be the home price of the ith commodity in terms of home currency (so that, in arbitrage equilibrium, $p_i^* \equiv p_i/R$ is the foreign price of the commodity in terms of foreign currency), and let B be the home balance of trade in terms of foreign currency. Then, writing $z_i(p_i)$ and $z_i^*(p_i^*)$ as the home and foreign excess demands for the ith commodity, Bickerdike's model of the balance of payments reduces to the system of three equations

$$z_i(p_i) + z_i^*(p_i/R) = 0 \quad (i = 1, 2)$$
$$B = -(1/R)[p_1 z_1(p_1) + p_2 z_2(p_2)] \quad (1)$$

In this system the rate of exchange R is treated as a parameter and p_1, p_2 and B as variables to be determined. Differentiating (1) with respect to R, solving for dB and the dp_i, and converting to elasticities, we obtain

$$dB = \left\{ -p_2^* z_2^* \left[\frac{\eta_1^*(1 + \eta_1)}{\eta_1^* - \eta_1} - \frac{\eta_2^*(1 + \eta_2)}{\eta_2^* - \eta_2} \right] - B \right\} \frac{dR}{R} \tag{2}$$

and

$$\frac{dp_i}{p_i} = \frac{\eta_i^*}{\eta_i^* - \eta_i} \frac{dR}{R}, \quad i = 1, 2 \tag{3}$$

where $\eta_i \equiv (dz_i/dp_i)(p_i/z_i)$ and $\eta_i^* \equiv (dz_i^*/dp_i^*)(p_i^*/z_i^*)$. In the special case in which B is initially zero, (2) takes the simpler form

$$dB = -p_2^* z_2^* \left[\frac{\eta_1^*(1 + \eta_1)}{\eta_1^* - \eta_1} - \frac{\eta_2^*(1 + \eta_2)}{\eta_2^* - \eta_2} \right] \frac{dR}{R}. \tag{2'}$$

Equation (2) is often referred to as the Bickerdike–Robinson–Metzler formula: however, the role of Robinson (1947) and of Metzler (1949) was that of expositor only.

Suppose for concreteness that the home country exports the first commodity and imports the second, so that η_1 and η_2^* are export-supply elasticities and η_2 and η_1^* import-demand elasticities. Suppose further that all marginal propensities to buy are positive, so that η_1 and η_2^* are positive, η_2 and η_1^* negative. Then for the balance of payments to improve in response to devaluation it suffices that the sum of the two import demand elasticities exceed 1 in magnitude, that is, that the Marshall–Lerner condition be satisfied. Thus eq. (2') can be rewritten as

$$dB = -p_2^* z_2^* \left[\frac{\eta_1 \eta_2^*(1 + \eta_1^* + \eta_2) - \eta_1^* \eta_2(1 + \eta_1 + \eta_2^*)}{(\eta_1 - \eta_1^*)(\eta_2 - \eta_2^*)} \right] \frac{dR}{R}$$

with all terms of known sign except $(1 + \eta_1^* + \eta_2)$. For a positive response of the balance of payments to devaluation it suffices also that the terms of trade improve, or at least that they not worsen. For changes in the terms of trade are indicated by changes in p_1/p_2 and, from eq. (3),

$$\frac{d(p_1/p_2)}{p_1/p_2} = \frac{dp_1}{p_1} - \frac{dp_2}{p_2} = \left(\frac{\eta_1^*}{\eta_1^* - \eta_1} - \frac{\eta_2^*}{\eta_2^* - \eta_2} \right) \frac{dR}{R}$$

If this expression is non-negative then, from (2'), dB must be positive.

Bickerdike's theory is very special in that the excess demand for each commodity depends on the money price of that commodity only. Implicitly, all 'cross' price elasticities are set equal to zero. For more general theories and, in particular, more general versions of (2'), the reader is referred to Negishi (1968), Kemp (1970), Dornbusch (1975) and Kyle (1978).

MURRAY C. KEMP

Bibliography

Bickerdike, C.F. 1920. The instability of foreign exchange. *Economic Journal* 30(March), 118–22.

Dornbusch, R. 1975. Exchange rates and fiscal policy in a popular model of international trade. *American Economic Review* 65, 859–71.

Kemp, M.C. 1970. The balance of payments and the terms of trade in relation to financial controls. *Review of Economic Studies* 37, 25–31.

Kyle, J.F. 1978. Financial assets, non-traded goods and devaluation. *Review of Economic Studies* 45, 155–63.

Metzler, L.A. 1949. The theory of international trade. In *A Survey of Contemporary Economics*, ed. H.S. Ellis. Philadelphia: Blakiston.

Negishi, T. 1968. Approaches to the analysis of devaluation. *International Economic Review* 9, 218–27.

Robinson, J. 1947. *Essays in the Theory of Employment*. 2nd edn. Oxford: Basil Blackwell.

elasticity

One day in the winter of 1881–2 Alfred Marshall came down from the sunny rooftop of his hotel in Palermo 'highly delighted', for he had just invented elasticity of demand (Keynes, 1925, pp. 39 n. 3, 45 n. 2). So delighted was he that within a mere four years he had introduced the word *elasticity* into the technical literature of economics (Marshall, 1885), which by his own standards was rushing pell-mell into print. But if the speed of its introduction was uncharacteristic the manner of it was not, tucked away as it was at the end of a lecture dull even for its time, and giving no hint that elasticity was new and exciting (1885, p. 187).

The notion that demand varies less or more than price can of course be found rather often in classical economics, especially in John Stuart Mill (Edgeworth, 1894, p. 691). But to turn that trite idea into something useful requires a firm grip on the prior idea of quantity demanded *at a price*. So it is not surprising that the only ancient who came close to Marshall's idea was Cournot himself, the inventor of (among much else) the demand function.

In fact Cournot came so close that it is hard to understand, first, why he did not go all the way, and second, why Marshall gave him no credit for showing that way. Such lack of generosity is the more puzzling since we know that between the time when (according to Mrs Marshall) he invented elasticity, and the late spring of 1882 when he first drafted the chapter on Elasticity for the *Principles*, Marshall reread Cournot (Whitaker, 1975, vol. 1, p. 85).

Starting with the demand function $D = F(p)$, Cournot pointed out that $pF(p)$ is total revenue, so that for maximum revenue the price p must be such that $F(p) + pF'(p) = 0$ (1838, p. 56). Thus total revenue will increase or decrease with increase in price according as $\Delta D / \Delta p$ is larger or smaller than D/p, where ΔD is the absolute value of the change in quantity demanded.

Commercial statistics should therefore be required to separate articles of high economic importance into two categories, according as their current prices are above or below the value which makes a maximum of $pF(p)$. We shall see that many economic problems have different solutions, according as the article in question belongs to one or other of these two categories. (Bacon's translation, 1897, p. 54)

Let f be a real-valued nonzero differentiable function whose domain is some open interval I of the real line. In conformity with Marshall's Mathematical Appendix (1890, Note IV, pp. 738–40), the *elasticity of f at the point x*, denoted by $\eta_f(x)$, is defined here to be the *number* $xf'(x)/f(x)$. The *function* η_f defined by this formula is called the *elasticity of f*. To define the elasticity of *demand*, some authors prefer to follow the convention $f(x) = -xf'(x)/f(x)$, which is not used here. Unfortunately there is no standard notation for elasticity, since the obvious candidates are already taken, e for e and E for the expectations operator.

Cournot's critical value of p, his criterion for sorting out commodities, is simply that p^* for which $\eta_f(p^*) = -1$; he was close indeed. However, unlike Marshall (who is crystal clear on the point) there is no trace in Cournot of the crucial property that the elasticity measure is *invariant* to changes in units of measurement of quantities and prices, and it is this property alone that makes it so important in pure and applied economics.

A little calculus will prove such invariance, but is more enlightening to apply the dimensional analysis of Jevons and Wicksteed. Let the dimension of x be X and that of $f(x) = y$ be Y, so that $f'(x)$ has dimension YX^{-1}. The dimension of $\eta_f(x)$ is then $X \cdot YX^{-1} \cdot Y^{-1}$ and everything cancels. The elasticity of f at x is a pure number, unaffected by change in the units of either x or y. (This application is so obvious that the most plausible explanation of why it was not included in Wicksteed, 1894, is that his entry was actually written before Marshall's *Principles* appeared.) Although invariance to transformation of units is the key property of elasticities, partly as a consequence the measure has a number of other agreeable properties. For example, it is easily seen that $\eta_f(x) = d \log f(x)/d \log x$, which paves the way for a whole calculus of elasticities in terms of logarithmic derivatives (Champernowne, 1935; Allen, 1938, pp. 251–4). One simple application of this calculus is the formula $\eta_{fg}(x) = \eta_f(x) + \eta_g(x)$, where fg is the product of f and g (with a corresponding formula for the quotient function f/g), while another is the characterization of constant elasticity functions as those which are linear in logarithms, that is, of Wicksell–Cobb–Douglas type. Incidentally, Douglas's paper of 1927 was apparently intended to introduce elasticity of supply, which is odd since it had already appeared 20 years before (and rather late at that) in the fifth edition of the *Principles* (see Marshall, 1961, vol. 2, p. 521).

The extension of elasticity to functions of more than one variable is easy – one simply uses the partial derivatives f_i rather than the derivative f' – and is staple fare in textbooks (see for example Allen, 1938, pp. 310–12). However, many of those textbooks underplay another useful property of elasticities of strictly monotonic functions (such as the usual demand and supply curves) which follows from the inverse function theorem. Considering just functions of one variable, if we write $\Phi = f^{-1}$ then from that theorem $\Phi' = f^{-1}$, so from this and the definition of elasticity,

$$\eta_\Phi(y) = y\Phi'(y)/\Phi(y) = f(x)/xf'(x) = (\eta_f(x)^{-1}),$$

that is, the elasticity of the inverse function is the inverse of the elasticity. Two obvious applications of this to the elementary theory of the firm are:
(i) Since the revenue function is $R(q) = pq = q\Phi(q)$,

$$\begin{aligned}
\text{marginal revenue } (mr) &= \Phi(q) + q\Phi'(q) \\
&= \Phi(q)[1 + (q\Phi'(q)/\Phi(q)] \\
&= \Phi(q)(1 + \eta_\Phi(q)),
\end{aligned}$$

from which one can derive the more usual but less intuitive formula $mr = p[1 + (1/\eta_f(p))]$; and (ii) since at the firm's profit maximizing output marginal cost $mc = mr$, the Lerner (1934) measure of monopoly power $(p-mc)/p$ may be written $[\Phi(q) - mr]/\Phi(q) = 1 - [\Phi(q)(1 + \eta_\Phi(q)]/\Phi(q) = -\eta_\Phi(q)$.

Arc elasticity, which is really ordinary elasticity with the index number problem thrown in, was introduced quite early by Dalton (1920, pp. 192–7). But the heyday of elasticities of all kinds came later, in the 1930s, so much so that it is small wonder that in the immediate post-war period Samuelson (1947, pp. 4–5) used elasticity statements to exemplify what he meant both by 'meaningful theorems' and by non-meaningful theorems in economics. A peculiar aspect of some of the elasticity measures introduced then was their definition not in terms of the properties of a given *function f* (as here), but rather as the ratio of proportionate change in one variable to proportionate change in another, allegedly causative, variable, without any explicit functional relationship intervening. Thus with Hicks's 'elasticity of expectations' (1939, p. 205) there *is* no 'expectation function' of which it is an elasticity, as that term is defined above. Similarly, although the elasticity of substitution (σ) invented by Hicks (1932) and Robinson (1933) immediately provoked many articles in response (for example, Lerner, 1933), at no time was a 'substitution function' introduced whose elasticity it was. The lack of a generating function for σ might help to explain why its use often occasions technical difficulty.

It is of some interest to apply duality theory to the problem of deriving simple formulas for entities like σ

(cf. Woodland, 1982, p. 31). Consider the elasticity of substitution σ between two consumer's goods x and y, with no restriction being placed on preferences apart from the smoothness conditions implicit at this level of analysis. First, take advantage of homogeneity in both the ordinary and compensated demand functions to write the former function as $f(p, m)$ and the latter as $h(p, t)$, where p is the price of x in terms of y, m is the consumer's income in terms of y, and t is the *maximized* level of utility for the price-income situation (p, m). Put $x^* = f(p, m)$. Finally, observe that σ is wholly determined by the price slope corresponding to p together with the indifference curve corresponding to t, so that we may write $\sigma = \sigma(p, t)$.

From a modern version of the fundamental equation of value theory (Hicks, 1939, p. 309),

$$f_p(p, m) = h_p(p, t) - x^* f_m(p, m) \qquad (1)$$

where f_p, h_p and f_m are, in sequence, the partial derivatives of f and h with respect to p, and of f with respect to m. Multiplying (1) by $p/f(p, m)$ and writing η_{fp}, η_{fm} for the two partial elasticities of f, we obtain

$$\begin{aligned}
\eta_{fp}(p, m) &= ph_p(p, t)/x^* - px^*mf_m(p, m)/(mf(p, m) \\
&= ph_p(p, t)/x^* - k\eta_{fm}(p, m) \quad (2)
\end{aligned}$$

where $k = px^*/m$, that is, the fraction of m spent on x. Now since t is the maximized level of utility, given local non-satiation $x^* = h(p, t)$. Hence, the first term on the right-hand side of (2) is $\eta_{hp}(p, t)$, the partial elasticity of h with respect to p, and (2) becomes

$$\eta_{fp}(p, m) = \eta_{hp}(p, t) - k\eta_{fm}(p, m). \qquad (3)$$

A standard result of Hicks and Allen (1934; see Hicks, 1981, p. 20) for the two-good case can be written in the present notation as

$$-\eta_{fp}(p, m) = k\eta_{fh}(p, m) + (1 - k)\sigma(p, t) \qquad (4)$$

so from (3) and (4),

$$(k - 1)\sigma(p, t) = \eta_{hp}. \qquad (5)$$

Let the cost (expenditure) function for this problem be $c(p, t)$, and denote its partial derivative with respect to p by c_p. Then, writing η_{cpp} for the partial elasticity of c_p with respect to p, since Shephard's Lemma implies $c_p = h$ we have

$$\eta_{hp}(p, t) = \eta_{cpp}(p, t). \qquad (6)$$

Now $k = px^*/m = ph(p, m)/m = pc_p(p, t)/m$. Because t is the maximized level of utility $m = c(p, t)$, so $k = pc_p(p, t)/c(p, t) = \eta_{cp}(p, t)$, where η_{cp} is the partial elasticity of c with respect to p. Substituting from this and (6) into (5),

$$\sigma(p,t) = \eta_{cpp}(p,t)/(\eta_{cp}(p,t)-1). \qquad (7)$$

Thus the elasticity of substitution in this two-good case can be expressed entirely in terms of the cost function.

PETER NEWMAN

See also **Marshall, Alfred.**

Bibliography

Allen, R.G.D. 1938. *Mathematical Analysis for Economists.* London: Macmillan.
Champernowne, D.G. 1935. A mathematical note on substitution. *Economic Journal* 15, 246–58.
Cournot, A.A. 1838. *Recherches sur les principes mathématiques de la théorie des richesses.* Paris: Hachette. New edn, ed. G. Lutfalla, Paris: Riviére, 1938. English trans. by N.T. Bacon, 1897. Reprinted, New York: A.M. Kelley, 1960.
Dalton, H. 1920. *Some Aspects of the Inequality of Incomes in Modern Communities.* London: Routledge.
Douglas, P.H. 1927. Elasticity of supply as a determinant of distribution. In *Economic Essays Contributed in Honor of John Bates Clark*, ed. J.H. Hollander. New York: Macmillan.
Edgeworth, F.Y. 1894. Elasticity. In *Dictionary of Political Economy*, vol. 1, ed. R.H.I. Palgrave. London: Macmillan.
Hicks, J.R. 1932. *The Theory of Wages.* London: Macmillan.
Hicks, J.R. 1939. *Value and Capital.* Oxford: Clarendon Press.
Hicks, J.R. 1981. *Collected Essays on Economic Theory.* Vol. 1: Wealth and Welfare. Cambridge, MA: Harvard University Press.
Hicks, J.R. and Allen, R.G.D. 1934. A reconsideration of the theory of value. *Economica* 1, 52–76, 196–219. Reprinted in Hicks (1981, vol. 1).
Keynes, J.M. 1925. Alfred Marshall, 1842–1924. In Pigou (1925).
Lerner, A.P. 1933. The diagrammatical representation of elasticity of substitution. *Review of Economic Studies.* Reprinted in Lerner (1953).
Lerner, A.P. 1934. The concept of monopoly and the measurement of monopoly power. *Review of Economic Studies* 1(June), 157–75. Reprinted in Lerner (1953).
Lerner, A.P. 1953. *Essays in Economic Analysis.* London: Macmillan.
Marshall, A. 1885. The graphic method of statistics. *Journal of the Royal Statistical Society.* Reprinted in Pigou (1925).
Marshall, A. 1890. *Principles of Economics*, vol. 1, London: Macmillan.
Marshall, A. 1961. *Principles of Economics.* 9th (variorum) edn, with annotations by C.W. Guillebaud. 2 vols. London: Macmillan.
Pigou, A.C. (ed.) 1925. *Memorials of Alfred Marshall.* London: Macmillan. Reprinted, New York: A.M. Kelley, 1966.
Robinson, J.V. 1933. *The Economics of Imperfect Competition.* London: Macmillan.
Samuelson, P.A. 1947. *Foundations of Economics Analysis.* Cambridge, MA: Harvard University Press.
Whitaker, J.K. (ed.) 1975. *The Early Economic Writings of Alfred Marshall, 1867– 1890.* 2 vols. New York: Free Press.
Wicksteed, P.H. 1894. Dimensions of economic quantities. In *Dictionary of Political Economy*, vol. I, ed. R.H.I. Palgrave. London: Macmillan.
Woodland, A.D. 1982. *International Trade and Resource Allocation.* Amsterdam: North-Holland.

elasticity of intertemporal substitution

The EIS and consumption theory

The elasticity of intertemporal substitution (EIS) is an important number in macroeconomic theory. It measures the willingness on the part of the consumer to substitute future consumption for present consumption. This parameter plays a key role in the theory of consumption and saving, in particular in the life-cycle version of that theory. For a start we examine the role of the EIS in a basic life-cycle model. In that model there is complete certainty concerning prices, future income, and preferences present and future. The consumer can lend and borrow at will at a single invariant rate of interest, subject only to a lifetime budget constraint. Preferences are additively separable. The consumer chooses present and future consumption to maximize:

$$\sum_{t=1}^{T} \delta^{t-1} U[c_t] \qquad (1)$$

where c_t is consumption in period t, and $0 < \delta < 1$, and where δ is the rate at which utility is discounted. Lifetime utility (1) is maximized subject to the lifetime budget constraint:

$$\sum_{t=1}^{T} c_t \left(\frac{1}{1+r}\right)^{t-1} \le \sum_{t=1}^{T} y_t \left(\frac{1}{1+r}\right)^{t-1} \qquad (2)$$

where the y values are incomes in the various periods, and r is the real rate of interest. Assuming positive consumptions in all periods, the maximization of (1) requires:

$$\frac{dU[c_t]}{dc_t} \delta^{t-1} - \lambda \left(\frac{1}{1+r}\right)^{t-1} = 0 \qquad (3)$$

where λ is the Lagrange multiplier. From (3), taking logs:

$$\ln\frac{dU[c_{t+1}]}{dc_{t+1}} - \ln\frac{dU[c_t]}{dc_t} = \ln\left(\frac{1}{1+r}\right) - \ln\ \delta$$

$$(4)$$

Differentiating (4) with respect to r and holding c_t constant gives:

$$\frac{\frac{d^2U[c_{t+1}]}{dc_{t+1}^2}}{\frac{dU[c_{t+1}]}{dc_{t+1}}}\frac{dc_{t+1}}{dr} = -\frac{1}{(1+r)}$$

$$(5)$$

A useful way of writing (5) is:

$$\frac{1}{c_{t+1}}\frac{dc_{t+1}}{dr} = -\frac{1}{(1+r)}\frac{\frac{dU[c_{t+1}]}{dc_{t+1}}}{c_{t+1}\frac{d^2U[c_{t+1}]}{dc_{t+1}}}$$

$$(6)$$

Or equivalently:

$$\frac{1}{c_{t+1}}\frac{dc_{t+1}}{dr} = \frac{\sigma(c_{t+1})}{1+r}$$

$$(7)$$

where σ is the EIS, defined as:

$$\sigma(c) = -\frac{\frac{dU[c]}{dc}}{c\frac{d^2U[c]}{dc^2}}$$

$$(8)$$

Equation (7) indicates that the size of the EIS will be a crucial determinant of how far consumption levels will respond to changes in the interest rate.

The effect of a small change in r analysed above is a standard partial equilibrium result, in which enough is held constant to obtain a definite result. The calculation shows how two solution paths compare with regard to c_{t+1}, as r is varied slightly, when for each of these paths c_t takes the same optimal value. For that special case, (7) says that c_{t+1} increases with r, which is to say that c_{t+1} increases relative to c_t. In that particular sense a small increase in r encourages saving. Even for the two-period model popular for classroom exposition, it cannot be shown that a rise in r encourages saving. However in the two-period model it is true for any separable lifetime utility function, as (1), that c_1 declines as r increases, provided that $c_2 > y_2$, the usual case. When r increases the substitution effect always favours lower early consumption. When $y_2 > c_2$, however, the income effect opposes the substitution effect, and the outcome is uncertain.

Equation (8) shows that the second derivative of the utility function, how curvy it is if one likes, is crucial in giving a specific value to the EIS. If:

$$U[c] = c^{\frac{\sigma-1}{\sigma}}$$

$$(9)$$

then the EIS is constant, independent of c, and equal to σ.

Consumption smoothing and risk aversion

The EIS as defined in (8) is the same as the Arrow–Pratt measure of relative risk aversion. It is no accident that consumption substitution through time, with no uncertainty whatsoever, and risk aversion, where uncertainty is necessarily involved, should involve the same parameter. Absolute risk aversion is related to the willingness of a consumer to accept a lottery ticket in preference to a sum of money available for certain, the certain sum being lower than the expected value of the lottery. One can think of the extra expected value in the better-than-fair lottery as a premium needed to entice the agent to accept the risk. The higher is relative risk aversion, the larger must be the expected-value premium in the lottery. Arrow (1971, ch. 3) provides a detailed discussion, and references the parallel and independent work of Pratt.

Now consider the life-cycle maximization of (1) subject to (2). To make the explanation as simple as possible let δ and r both be zero. The consumer maximizes:

$$\sum_{t=1}^{T} U\ [c_t]$$

$$(10)$$

subject to the lifetime budget constraint:

$$\sum_{t=1}^{T} c_t \leq \sum_{t=1}^{T} y_t$$

$$(11)$$

With U [ ] a concave function, it is evident that the consumer will consume at the same level in each period:

$$c_t = \frac{\sum_{t=1}^{T} y_t}{T}$$

$$(12)$$

In the particular sense defined by this special case, the consumer is averse to consumption variability over time. It is the same as the risk-averse consumer disliking variations in wealth when different states of the world are realized. That each period of time will certainly arrive, whereas only one state of the world will be realized, is irrelevant in the *ex ante* view of the consumer facing uncertainty. A risk-averse agent can be induced to accept a gamble if the odds are sufficiently favourable, that is, if the expected-value premium is sufficiently large. Similarly, a life-cycle planner will opt for a non-constant consumption plan if it provides a larger total consumption sufficient to compensate for the unattractive variability. A positive rate of interest plays the same role as an expected-value premium. It is the sweetener that persuades the consumer to accept variability. For this reason it is no surprise to find that the extent to which the consumer will respond to the sweetener, in either case, is governed by precisely how much the consumer dislikes variability. And the EIS, or the coefficient of relative risk aversion, as the case may be, measures that dislike of variability.

The argument just completed ignores the part played by δ, the utility discount rate. The presence of a positive δ means that, were r zero, the consumer would choose a plan with consumption falling through time. Then a positive r, and especially an r greater than δ, persuades the consumer to select a consumption plan with consumption falling less rapidly or rising through time. How far an optimal plan responds to a given change in r is governed again by the EIS.

A constant or a variable coefficient?

The EIS has been compared above to the coefficient of relative risk aversion. In the theory of risk aversion the emphasis is on the variability of the coefficient. On this turns the issue of whether the wealthy will be more or less willing to undertake risk than the poor. With the EIS the most common assumption is that it is a constant. A popular special case of (1) is:

$$U[c_1, c_2, \ldots, c_n] = c_1^{\frac{\sigma-1}{\sigma}} + \delta c_2^{\frac{\sigma-1}{\sigma}} + \ldots + \delta^{n-1} c_n^{\frac{\sigma-1}{\sigma}}$$

$$(13)$$

This is the love-of-variety utility function of Dixit and Stiglitz (1977), with discounting added. The EIS measured at any of the consumptions above is σ.

The elegance and convenience of forms such as (13) has made them appealing. Thus Barro and Sala-i-Martin (1995), in their influential study of economic growth, assume that different countries or regions solve independent Ramsey optimal model problems. This leads to the condition:

$$\frac{1}{c}\frac{dc}{dt} = \sigma[AF_1\{k, 1\} - \delta]$$

$$(14)$$

where F_1 is the marginal product of capital, c is consumption, k is capital, δ is the utility discount rate, A measures total factor productivity as it is affected by policy, culture, corruption, and so on, and σ is the EIS. The lower is k the larger is F_1. If this effect is not offset by poor countries having lower total factor productivities, and if all countries share the same values of δ and σ, then conditional β-convergence follows from (14), meaning that poor countries grow faster.

The poor will be reluctant to save if their value of σ is low. And this is a most plausible specification. When all the meals that one eats are small, it is rationally more difficult to postpone eating now for a larger meal later. This point has been recognized in the literature. For example, King and Rebelo (1993) allow for a utility function of the Stone–Geary form, where the consumer gives priority to a fixed basket of essentials until that basket has reaches a critical scale. With those preferences, the poorest consumers will not save at all, and there is the possibility of a poverty trap. The Stone–Geary utility function implies a

zero value for the EIS at low consumptions, and positive values for higher consumptions.

The EIS in consumption studies

Many applied economists used to take the view that the value of σ is close to zero (see Hall, 1988; Mankiw, Rotenberg and Summers, 1985). This reflects the failure of consumption studies to find a significant effect of the rate of interest on saving. Such estimates are seriously biased if the consumer is constrained from borrowing freely (a feature ignored in the computations above) or if, as in Deaton (1992), most consumers save only to replenish precautionary balances following negative shocks. Then the optimizing substitution-based theory does not apply. Blundell, Browning and Meghir (1994) and Attanasio and Browning (1995) show that representative consumer models give seriously misleading results when applied to aggregate consumption data. They use UK household expenditure data to model consumption at the individual level and obtain a greatly improved fit when they allow the rich to have a higher EIS than the poor. Does that mean that as economies grow richer over time, the average EIS will increase? This remains an unanswered question.

VEIS functions

Let the utility function be chosen from a class of which the simplest case is:

$$U[c] = \int_0^c \exp\left\{\frac{1}{\beta x}\right\} dx$$

$$(15)$$

where β is a positive constant and c is the level of consumption. This is a VEIS utility function, where VEIS stands for *variable elasticity of intertemporal substitution*. Then:

$$\frac{dU[c]}{dc} = \exp\left\{\frac{1}{\beta x}\right\} > 0$$

$$(16)$$

and:

$$\frac{d^2U[c]}{dc^2} = -\exp\left\{\frac{1}{\beta c}\right\}\frac{1}{\beta c^2} < 0$$

$$(17)$$

$U[\bullet]$ is an increasing concave function. Now the EIS may be computed as:

$$-\frac{\frac{dU[c]}{dc}}{c\frac{d^2U[c]}{dc^2}} = \beta c$$

$$(18)$$

This increases linearly with consumption at rate β. The poor have a lower EIS and β-convergence will not necessarily prevail.

A variable EIS in the Diamond capital model

In their deep study of the Diamond overlapping generations model with capital, De La Croix and Michel (2002) more or less dismiss the importance of multiple stable equilibria. To summarize, it is possible to obtain multiple stable steady-state solutions with simple functional forms, but these cases are unsatisfactory at best. If the production function is Cobb–Douglas and with a simple separable utility function, there are no cases of multiple stable steady states. With a logarithmic utility function and the constant elasticity of substitution in production $\rho > 0$, there can be two positive steady-states, but it may be that only the corner degenerate outcome is stable.

Rather than using given simple functional forms and looking for a few steady-state solutions, try for a continuum of solutions as follows. Assume:

$$-\frac{\frac{dU[c]}{dc}}{c\frac{d^2U[c]}{dc^2}} = \sigma(c) \qquad (19)$$

where $\sigma(c)$ is an arbitrary positive increasing function of c. Then:

$$\frac{\frac{d^2U[c]}{dc^2}}{\frac{dU[c]}{dc}} = -\frac{1}{\sigma(c)c} \qquad (20)$$

Integrating (20) gives:

$$\ln\frac{dU[c]}{dc} = -\int_a^c \frac{1}{\sigma(x)x}dx + \ln D \qquad (21)$$

where a is a positive constant, and D is a constant of integration.

In a steady state solution to the Diamond model we must have:

$$\ln\frac{dU[c_1]}{dc} - \ln\frac{dU[c_2]}{dc} = \ln\delta - \ln R \qquad (22)$$

where c_1 and c_2 are consumption in respectively the first and second period of a life, R is the gross rate of return to saving, and δ is the discount factor. From (21) and (22):

$$\int_{c_1}^{c_2} \frac{1}{\sigma(x)x}dx = \ln\delta - \ln R \qquad (23)$$

Now in steady state c_1, c_2 and R all depend upon capital per head k. If over some range of values of k every value gives a steady state, then (23) will be an identity in k. Let the per capita production function be Cobb–Douglas with coefficient α. Then (23) takes the form:

$$\int_{(1-\alpha)k^\alpha - k}^{k+\alpha k^\alpha} \frac{1}{\sigma(x)x}dx = \ln\delta - \ln(1 + \alpha k^{\alpha-1})$$

$$(24)$$

When (24) is an identity in k, over an interval at least, then differentiating both sides of (24) gives:

$$\frac{1}{\sigma(k+\alpha k^\alpha)}\frac{1}{k+\alpha k^\alpha} - \frac{1}{\sigma((1-\alpha)k^\alpha - k)}\frac{1}{(1-\alpha)k^\alpha - k}$$
$$= \frac{\alpha(1-\alpha)k^{\alpha-2}}{1+\alpha k^{\alpha-1}} \qquad (25)$$

Take a given a value of k, and let $\sigma(c_1)$ values be known for the c_1 value implied by that k all the way up to the c_2 defined by the same k. Then $\sigma(c_2)$ values are determined by (25), which rolls out a solution for σ such that all values of k on a connected interval are steady-state equilibrium levels. The contrast to the case advanced by De La Croix and Michel is striking.

Concluding remarks

The EIS is an important value, just as is its cousin, the coefficient of relative risk aversion. The use of a simple functional form has too often frozen the EIS as a constant. When it is allowed to vary, the β-convergence of growth theory is no longer secure; cross-section consumption studies perform better; and multiple equilibrium in the Diamond capital model is seen to be far more probable than previous studies indicate.

CHRISTOPHER BLISS

See also **consumer expenditure; consumer expenditure (new developments and the state of research).**

Bibliography

Arrow, K.J. 1971. *Essays in the Theory of Risk Bearing.* Amsterdam: North-Holland.

Attanasio, O.P. and Browning, M. 1995. Consumption over the life-cycle and over the business cycle. *American Economic Review* 85, 1118–37.

Barro, R.J. and Sala-i-Martin, X. 1995. *Economic Growth.* New York: McGraw-Hill.

Blundell, R., Browning, M. and Meghir, C. 1994. Consumer demand and the life-cycle allocation for household expenditures. *Review of Economic Studies* 61, 57–80.

Deaton, A. 1992. *Understanding Consumption.* Clarendon Lectures in Economics, Oxford: Clarendon Press.

De La Croix, D. and Michel, P. 2002. *A Theory of Economic Growth: Dynamics and Policy in Overlapping Generations.* Cambridge: Cambridge University Press.

Dixit, A.K. and Stiglitz, J.E. 1977. Monopolistic competition and optimum product diversity. *American Economic Review* 67, 297–308.

Hall, R.E. 1988. Intertemporal substitution in consumption. *Journal of Political Economy* 96, 339–57.

King, R.T. and Rebelo, S.T. 1993. Transitional dynamics and economic growth in the neoclassical model. *American Economic Review* 83, 908–31.

Mankiw, N.G., Rotenberg, J.J. and Summers, L. 1985. Intertemporal substitution in macroeconomics. *Quarterly Journal of Economics* 100, 225–81.

elasticity of substitution

The concept of the elasticity of substitution, developed by Joan Robinson and John Hicks separately in the 1930s, represented an important addition to the marginal theory of the 1870s, in the tradition of Marshall, Edgeworth and Pareto. It brought together two concepts which were already well established in the literature – the ideas of elasticities (which derive from Mill) and those of substitution (which go back to Smith). The relationship defined by the concept is a mathematical one relating to utility and production functions, with considerable economic implications. It has two applications: to the theory of production, and in particular the isoquant relationship between factor inputs, and to consumer behaviour and the indifference curve. Let us look at each in turn.

The two inventors of the concept – Joan Robinson, in her *Economics of Imperfect Competition* (1933), and John Hicks in his *Theory of Wages* (1932) – each developed Marshall's formula for the elasticity of derived demand. Each defined the concept somewhat differently. For Hicks, the definition was the percentage change in the relative amount of the factors employed resulting from a given percentage change in the relative marginal products or relative prices, that is (following Samuelson, 1968):

$$\sigma = \sigma_{12} = (F_1 F_2 / F F_{12}) = \sigma_{21},$$

where $F(V_1, V_2)$ is a standard neoclassical production function, and the subscripts are the partial derivatives. This is sometimes called the direct elasticity of substitution. For Joan Robinson, on the other hand, concerned with relative shares and hence distributional issues, the elasticity of substitution was defined as 'the proportionate change in the ratio of the amounts of the factors employed divided by the proportionate change in the ratio of their prices' (1933, p. 256):

$$\sigma = - \frac{\partial (V_1 / V_2)/(V_1 / V_2)}{\partial (W_1 / W_2)/(W_1 / W_2)}$$

where W_1 is the price of the V_1 factor.

These two definitions of the concept gave rise to a considerable debate in the early issues of the *Review of Economic Studies*, with in particular a notable contribution from Kahn (1933) concerned to identify how these concepts related to each other. It turns out that these two original definitions are identical when the production function is confined to two factors of production, where the partial derivatives of the production function are the marginal productivities of the factor inputs and yield the relevant factor prices. In addition, the contributors to the debate attempted to identify the implications of these somewhat abstract concepts. Amongst these were the joint determination by the elasticity of substitution and the factor supplies of the relative shares of the factor reward (wages and profits), and implications for the

definition of imperfect competition with increasing returns to scale.

It is not surprising that it is with the cases where the restrictive neoclassical assumptions for the production function are not met that most interest arises. Two important developments are where production function involves three or more factors and in extending from Cobb–Douglas to constant elasticity of substitution (CES) production functions. But although considerable emphasis has been placed on the elasticity of substitution in production, it remains a technical concept concerning factor substitutability. It has no direct allocation consequence. Diminishing elasticity of substitution does not imply diminishing returns to scale, since for returns we must have prices. Thus it is restricted to describing the technical conditions of production. But, being a technical concept, it can be generalized to all forms of transformation. Thus, as we noted above, along with a number of other concepts, these tools developed for production were taken over to consumer theory. Because of the implications the concept had for the development of consumer behaviour, and because of the insight which the resulting difficulties threw up concerning the concept more generally, this application is of special interest.

It was Hicks (and Allen) who made that step. While Joan Robinson's development of the concept was closely related to her extension of Marshall's theory of the industry, Hicks was familiar with a very different approach to value theory, that of Edgeworth, Pareto and Walras. While Joan Robinson had focused on production substitutions, and hence isoquants, Hicks took the idea developed in that domain, and translated it across to consumer theory, and to the indifference curves which he had got from Edgeworth. In the two goods case, price elasticity could be represented in terms of his fundamental formula, according to which:

$$\text{Price elasticity} = k(\text{income elasticity}) + (1 - k) \text{ (e.s.)}$$

where k is the total expenditure that is spent on the commodity. Thus, with income elasticity, consumer theory led into a representation of the effect of a price change in terms of the income and substitution effects, with elasticity being thus of prime importance in classifying goods by their demand characteristics.

But whereas the elasticity concept in production theory naturally led on to the possibility of measurement, that step in consumer theory was more contentious. For although this technical concept represented one important step in the development of the marginalist approach to the theory of value, the theory of demand behaviour requires a behavioural theory of choice. The elasticity of substitution with respect to the indifference curve is one technical component. But, as with production theory, prices, and in this case the budget line, are also required.

Technical concepts thus aided the formulation of modern consumer theory as outlined in Hicks and

Allen's 'A Reconsideration of the Theory of Value' (1934) and the opening chapters of *Value and Capital* (1939), a path from which it has scarcely deviated. But, despite the mathematical elegance of this construction, it may be argued that it disguised many of the important underlying questions. The increased power of the indifference curve analysis begged the question of whether consumer preferences could in reality be represented in this abstract way. Ultimately, whether consumer behaviour is well described by concepts like the elasticity of substitution, depends upon whether preferences can be represented by complete, transitive, utility functions. Much recent evidence from psychologists and decision theorists suggests otherwise. Likewise for production theory, the concepts of capital and labour may be themselves ambiguous.

D.R. HELM

See also **CES production function; Cobb–Douglas functions; production functions.**

Bibliography

Hicks, J.R. 1932. *The Theory of Wages.* London: Macmillan.
Hicks, J.R. 1939. *Value and Capital.* Oxford: Clarendon Press.
Hicks, J.R. 1970. Elasticity of substitution again: substitutes and complements. *Oxford Economic Papers* 22, 289–96.
Hicks, J.R. and Allen, R.G.D. 1934. A reconsideration of the theory of value I–II. *Economica* 1, Pt 1, February, 52–76; Pt II, May, 196–219.
Kahn, R.F. 1933. The elasticity of substitution and the relative share of a factor. *Review of Economic Studies* 1(October), 72–8.
Robinson, J. 1933. *The Economics of Imperfect Competition.* London: Macmillan.
Samuelson, P.A. 1968. Two generalizations of the elasticity of substitution. In *Value, Capital and Growth*, ed. J.N. Wolfe. Edinburgh: Edinburgh University Press.

electricity markets

In many parts of the world buyers and sellers now trade electrical energy in liberalized markets. These markets have partially replaced cost-based regulation and government ownership.

Since the 1980s, governments in many countries have privatized and restructured their electricity industries. Liberalized electricity markets now operate in much of Europe, North and South America, New Zealand and Australia. These changes were primarily motivated by the perception that the previous regimes of either state ownership or cost-of-service regulation yielded inefficient operations and poor investment decisions. Liberalization of the electricity industry also reflected the progression of a deregulation movement that had already transformed

infrastructure industries, including water, communications and transportation, in many countries. Although electricity shares many characteristics with other deregulated industries, the differences have proven to be more important than the similarities. Electricity has been one of the most challenging industries to liberalize and in most places new layers of regulations have replaced the old.

Historically, electricity was viewed as a natural monopoly. Typically, a single utility company generated, transmitted and distributed all electricity in its service territory. In much of the world, the monopoly was a state-owned utility. Within the United States, private investor-owned companies supplied the majority of customers, although federally and municipally owned companies played an important minority role. These companies operated under multiple layers of local, state and federal regulation.

Restructured electricity markets share a common basic organization. The three segments – generation, transmission, and distribution – have been unbundled. Wholesale generation, no longer viewed as a natural monopoly, is priced through a market process. Transmission and distribution remain regulated, although in many cases some form of incentive regulation has replaced cost-of-service regulation or state ownership.

Most wholesale electricity is traded through long-term (a week or longer) forward contracts. Many markets also feature day-ahead auction-based exchanges. Because supply and demand must be continually balanced to preserve transmission stability, transmission system operators run real-time balancing markets. Prices in these high-frequency markets can be highly volatile since electricity is non-storable and real-time demand fluctuates dramatically. To meet unforeseen contingencies, transmission system operators also contract for and occasionally use standby or reserve generation services. Many markets reflect price differences across geographical locations when parts of the transmission grid are congested (Schweppe et al., 1988; Chao and Peck, 1992). Game theorists and experimental economists are involved in the ongoing process of designing electricity markets (Wilson, 2002), while empirical researchers have used detailed auction data to estimate how well predictions from theoretical models describe firm behaviour (Wolak, 2000; Hortascu and Puller, 2004).

At the retail level, the vision of liberalization was to provide customers a choice among competing retailers who would operate as either resellers or integrated providers with access to customers through a regulated common-carriage distribution network. In most restructured US markets, retail competition for residential customers is very weak (Joskow, 2005). Retail competition is more advanced in the United Kingdom, although evidence suggests that customers have been slow to take advantage of the ability to switch to a lower-priced retailer (Waddams, 2004). Several authors have noted the economic benefits of allowing retail prices to vary to reflect real-time changes in the wholesale prices, although this sort of real-time retail pricing has been slow to take hold in practice (Borenstein and Holland, 2005; Joskow and Tirole, 2004).

Oligopoly simulation analysis indicates the potential for serious market power problems because suppliers face extremely inelastic demand and entry requires long lead times (Green and Newbery, 1992). Empirical work has indicated that market power has indeed been present, although to varying degrees in different markets. Wolfram (1999) found that prices in England and Wales were lower than static oligopoly models would suggest. By contrast, extreme levels of market power in California contributed to record high prices in 2000–1 (Borenstein, Bushnell and Wolak, 2002). The explanations for these differences have focused on variations in the threat of future regulation and in the extent of long-term fixed price contracts (Bushnell, Mansur and Saravia, 2005).

Although the main motivation for market liberalization was to improve economic efficiency, there have been few attempts to measure efficiency changes. Newbery and Pollitt (1997) and Fabrizio, Rose and Wolfram (2004) find modest positive effects of market liberalization on, respectively, industry efficiency in the United Kingdom and plant-level efficiency in the United States.

As electricity industry restructuring moves forward, the major unresolved question is the degree to which public policy will influence investment decisions. Electric generating plants are long-lived, so while operating efficiency gains appear to be real, the potential gains from improved investment stand to be larger. Also, policies to limit the environmental impact of electricity generation could affect the types of technologies in which we invest.

JAMES BUSHNELL AND CATHERINE WOLFRAM

See also **competition; energy economics; privatization impacts in transition economies.**

Bibliography

Borenstein, S., Bushnell, J. and Wolak, F. 2002. Measuring market inefficiencies in California's deregulated wholesale electricity market. *American Economic Review* 92, 1376–405.
Borenstein, S. and Holland, S. 2005. On the efficiency of competitive electricity markets with time-invariant retail prices. *RAND Journal of Economics* 36, 469–93.
Bushnell, J., Mansur, E. and Saravia, C. 2005. Vertical arrangements, market structure, and competition: an analysis of restructured US electricity markets. CSEM Working Paper WP-126. Berkeley: University of California Energy Institute.
Chao, H. P. and Peck, S. 1992. A market mechanism for electric power transmission. *Journal of Regulatory Economics* 10, 25–60.
Fabrizio, K., Rose, N. and Wolfram, C. 2004. Does competition reduce costs? Assessing the impact of

regulatory restructuring on US electric generation efficiency. CSEM Working Paper WP-135. Berkeley: University of California Energy Institute.
Green, R. and Newbery, D. 1992. Competition in the British electricity spot market. *Journal of Political Economy* 100, 929–53.
Hortascu, A. and Puller, S. 2004. Understanding strategic bidding in restructured electricity markets: a case study of ERCOT. Working paper. Chicago: University of Chicago. Online. Available at http://econweb.tamu.edu/puller/AcadDocs/Hortacsu_Puller.pdf, accessed 20 July 2005.
Joskow, P. 2005. The difficult transition to competitive electricity markets in the US. In *Electricity Restructuring: Choices and Challenges*, ed. J. Griffen and S. Puller. Chicago: University of Chicago Press.
Joskow, P. and Tirole, J. 2004. Retail electricity competition. CMI Working Paper 44. Cambridge, MA: MIT.
Newbery, D. and Pollitt, M. 1997. The restructuring and privatization of Britain's CEGB – was it worth it? *Journal of Industrial Economics* 45, 269–303.
Schweppe, F., Caramanis, M., Tabors, R. and Bohn, R. 1988. *Spot Pricing of Electricity*. Norwell, MA: Kluwer Academic Publishers.
Waddams, C. 2004. Spoilt for choice? The costs and benefits of opening UK residential energy markets. Working Paper 04-1. Norwich: Centre for Competition and Regulation, University of East Anglia.
Wilson, R. 2002. Architecture of power markets. *Econometrica* 70, 1299–340.
Wolak, F. 2000. An empirical analysis of the impact of hedge contracts on bidding behavior in a competitive electricity market. *International Economic Journal* 14, 1–39.
Wolfram, C. 1999. Measuring duopoly power in the British electricity spot market. *American Economic Review* 89, 805–26.

electronic commerce

In this article, electronic commerce is defined as the exchange, distribution, or marketing of goods or services over the Internet.

There is, unfortunately, no standard definition used in the academic literature or the popular press. A broader definition would include all business facilitated by telephones, fax machines, televisions, and other technologies that are 'electronic'. This broad definition, however, becomes so large that it encompasses a substantial fraction of all economic activity since the 1950s. A narrower definition would focus only on items sold over the World Wide Web, the browser-enabled portion of the Internet. This definition omits much of the important business-to-business segment of electronic commerce and the numerous advertising-supported websites.

The definition used here encompasses a variety of ways in which businesses have used the Internet. The Internet is a worldwide network of computers that connect to each other using the communication protocols defined by TCP/IP. Electronic commerce includes businesses that have used the Internet to reach other businesses and to reach consumers directly. It includes businesses that sell products directly to their customers and businesses that function as intermediaries. This definition also includes businesses that operate only online, the online business of those that operate online and offline, and businesses that use the Internet but not as their primary business function.

Adoption of electronic commerce by industry

While most attention has focused on those few businesses where the Internet is a fundamental part of their strategy, electronic commerce is just one aspect of business processes for most businesses. As of 2000, nearly 90 per cent of large US establishments used the Internet (Forman, Goldfarb and Greenstein, 2002). Nearly all industries and cities had adoption rates well over 70 per cent. For the vast majority of these establishments, the Internet was used to send and receive email, to help automate some basic processes like inventory management, and/or for web browsing. This basic level of use was particularly important to establishments in rural areas (Forman, Goldfarb and Greenstein, 2005). Overall, the impact on most industries, from nursing homes to construction to furniture manufacturing to petrol stations, has been limited. The Internet is used in day-to-day business activities, but it is a small piece in a much larger puzzle. Even in retail, the US Census reported that Internet sales (totalling $26.3 billion) were just 2.7 per cent of total US retail sales in the second quarter of 2006 (U.S. Census Bureau, 2006b).

Still, a small portion of businesses have used the Internet to enhance business processes at a deep level. While little research has examined why some industries adopted quickly and others did not, it is the businesses that adopted quickly that get the majority of the attention. The Internet has had a profound effect on publishing, securities trading, some wholesaling, and some retailing (for example, books and computers). In particular, businesses that rely heavily on electronic commerce can be divided into four (not necessarily mutually exclusive) groups: retail, media, business-to-business (B2B), and other intermediaries.

Retail

Electronic commerce represents the introduction of a new sales channel. While the size of the online channel is still small relative to the entire retail sector, electronic commerce has had a large effect on some retail markets. According to the U.S. Census, Internet sales made up over ten per cent of 2004 retail sales in two broad categories if online-only stores are included: electronics and appliance stores (that is, NAICS 443) and sporting goods,

hobby, book, and music stores (that is, NAICS 451) (U.S. Census Bureau, 2006a). Much of the literature on electronic commerce has focused on these categories, as well as motor vehicles and travel.

A new channel has the potential to create channel conflict. There is considerable evidence that consumers compare prices and options across channels (Prince, 2006; Ellison and Ellison, 2006). Forman, Ghose and Goldfarb (2006) show that use of the online channel depends on local offline retail options. Also, Hendershott and Jie Zhang (2006) argue that manufacturers may face resistance from their retailers to setting up a direct online channel. They show that the benefits of selling directly to consumers (rather than though a retailer) depend on the relative online–offline search costs. The benefits of the online channel are largest for goods that are not widely available in retail stores (that is, high offline search costs) and for goods that do not need to be touched to assess quality (that is, low online search costs).

Media websites

In addition to a new retail channel, the Internet has provided a new media outlet. This outlet has developed a market structure similar to the magazine industry (Goldfarb, 2004). Media websites provide information to visitors and earn money (mostly) through advertising. In particular, entry is easy but distribution is difficult to achieve; concentration is largely determined by market size and distribution costs; large media conglomerates coexist with small niche players; and there is a high mortality rate. Online media appear to be particularly important to overcome local isolation (Sinai and Waldfogel, 2004). The two-sided nature of the media market and the digital nature of the product mean that competition between media websites is different in nature from competition between online retailers.

Intermediaries

According to Alexa.com, six of the top seven most popular websites in October 2006 had roles as intermediaries: Yahoo, MSN, Google, MySpace, YouTube, and eBay. While these intermediaries may share features of media websites (Google) or retailers (eBay), their primary business is to facilitate online interactions. Without physical storefronts or displays, intermediaries help individuals (and firms) find each other online. Intermediaries allow people with heterogeneous tastes to find better matches in terms of media, products, and people (Scott Morton, 2006).

Business to business

Business-to-business (B2B) electronic commerce is a relatively under-researched area, perhaps because of the difficulties in obtaining data. Still, B2B transactions are many times the size of business-to-consumer transactions. Lucking-Reiley and Spulber (2001) summarize many of the key questions and opportunities in B2B electronic commerce including B2B exchanges, automatic ordering, and outsourcing. Some aspects of the Internet, such as asynchronous communication, may be particularly important for international B2B interactions. Many B2B applications can also be done on electronic data interchange (EDI) rather than the Internet.

Key features of electronic commerce for general economic research

In addition to its widespread usage across industries and its profound impact on a small set of them, electronic commerce has a number of features that make it a particularly interesting area of study for economists.

Fewer economic frictions

The Internet reduces a number of economic frictions that are often cited as key contributors to observed imperfections in markets. To the consumer, search and switching costs are reduced substantially. To the firm, menu and distribution costs may fall.

For consumers, the Internet makes it relatively easy to search through several retail options. Instead of having to walk from store to store, consumers can simply click from one company to another without leaving their desks. Furthermore, a number of intermediaries exist that reduce search costs even further. These 'shopbots' allows consumers to compare prices and features from several websites during a single keyword search. In addition to lower search costs, switching costs are also lower online than offline. It is not difficult to switch from one competitor to another. Much of the earliest research examining electronic commerce focused on why price dispersion persisted in this environment. Broadly speaking, this literature concluded that, all else equal, search and switching costs are lower online; however, firms created search and switching costs to overcome this challenge (Ellison and Ellison, 2004). Consequently, there is still substantial price dispersion online. Still, low search costs do not mean zero search costs. Visibility matters to the long-term prospects of any business-to-consumer company. Many early Internet companies struggled because they misinterpreted low search costs as zero search costs, mistakenly assuming customers would arrive once they set up the website.

Firms also benefit from fewer frictions online. In particular, the menu costs of changing prices and updating product offerings are much lower online than offline. In addition to the reduction in menu costs, some firms benefit from lower distribution costs: for digital goods (namely, music, news and images) online distribution costs are near zero. Low menu costs combined with the digital nature of many online products allow for mass customization of products (Murthi and Sarkar, 2003) and creative bundling, licensing, versioning and pricing strategies. Shapiro and Varian (1999) and Bakos and Brynjolfsson (1999) provide examples of a number of

situations in which online firms are better able to match customers needs and therefore are better able to price discriminate.

Lower communication costs

The Internet reduces communication costs considerably. It provides an additional means of communication that creates new potential to interact with customers, suppliers and with other branches of the same firm. Internet communication differs from telephone communication in two primary ways. First, the marginal cost of communication is effectively zero, even over long distances. While establishing a connection is costly, each additional e-mail, web page viewed, and instant messaging interaction has no monetary cost to the communicator. Second, Internet communication is often asynchronous. Unlike telephone communications, the people communicating do not necessarily have to be available at the same time. This has many important applications. For example, it facilitates communication across time zones. Together, these features of Internet communication mean that geography may be less important online. Given access, people can communicate with any other person who has access, irrespective of location. Still, despite the substantial fall in long-distance communications costs, most online communication is local because social networks are local (Wellman, 2001).

Lower marginal costs

Many goods sold over the Internet are digital in nature (for example, newspaper content, music, information). The marginal cost of replication for digital goods is near zero. Depending on the particular good, fixed costs may be high (software) or low (blogs). Shapiro and Varian (1999) discuss in detail the economics of goods with high fixed and low marginal costs. If fixed costs are high enough, this cost structure allows monopolists with broad flexibility in pricing, versioning and bundling policies. It also leads to substantial economies of scale and incentives to sell a broad scope of products. In markets with more than one player, this cost structure can lead to fierce competition and little profit. If fixed costs are low and entry is easy then prices should approach zero.

One misunderstood aspect of electronic commerce is that many Internet business models have not benefited from low marginal costs, and therefore have no cost advantage over offline competition. Low marginal costs apply only to digital goods and services. In the late 1990s, many companies failed because their business models shipped heavy items to consumers. For example, taking orders for pet food and shipping it to customers involves very high marginal costs per item sold.

Rich data

By definition, all online activity is digital. This means that it is relatively easy to record and store information on the behaviour of consumers and firms online. In contrast, it is extremely expensive to track all a shopper's activity in a typical offline store. Online, however, every item browsed and the time spent looking is easily recorded. This presents an opportunity for both firms and researchers. Firms can use this data to better understand their customers, which leads to more effective customization. Researchers can use this data to answer many questions that previously could not be answered due to data constraints. Online data has greatly enhanced of our understanding of a number of economic concepts including auctions (for example, Bajari and Hortacsu, 2003), the economics of information (for example, Jin and Kato, 2005), and social interactions (for example, Mayzlin and Chevalier, 2006).

In summary, this article has identified some important features of electronic commerce and the some of the main areas of related economic research. Useful surveys of electronic commerce and related subjects include Scott Morton (2006), Hendershott (2007), and Ellison and Ellison (2005).

AVI GOLDFARB

See also **computer industry; information technology and the world economy; Internet, economics of the; price dispersion.**

Bibliography
Bajari, P. and Hortacsu, A. 2003. The winner's curse, reserve prices, and endogenous entry: empirical insights from eBay auctions. *RAND Journal of Economics* 34, 329–55.
Bakos, Y. and Brynjolfsson, E. 1999. Bundling information goods: price, profits, and efficiency. *Management Science* 45, 1613–30.
Ellison, G. and Ellison, S.F. 2004. Search, obfuscation, and price elasticities on the Internet. Working Paper, No. 10570. Cambridge, MA: NBER.
Ellison, G. and Ellison, S.F. 2005. Lessons about markets from the Internet. *Journal of Economic Perspectives* 19(2), 139–58.
Ellison, G. and Ellison, S.F. 2006. Internet retail demand: taxes, geography, and online-offline competition. Working paper No. 12242. Cambridge, MA: NBER.
Forman, C., Ghose, A. and Goldfarb, A. 2006. Geography and electronic commerce: measuring convenience, selection, and price. Working Paper No. 06-15. New York: NET Institute.
Forman, C., Goldfarb, A. and Greenstein, S. 2002. Digital dispersion: an industrial and geographic census of commercial Internet use. Working Paper No. 9287. Cambridge, MA: NBER.
Forman, C., Goldfarb, A. and Greenstein, S. 2005. How did location affect adoption of the commercial Internet? Global village vs. urban leadership. *Journal of Urban Economics* 58, 389–420.
Goldfarb, A. 2004. Concentration in advertising-supported online markets: an empirical approach. *Economics of Innovation and New Technology* 13, 581–94.
Hendershott, T., ed. 2007. *Handbook of Economics and Information Systems.* Amsterdam: North-Holland.
Hendershott, T. and Jie Zhang. 2006. A model of direct and intermediated sales. *Journal of Economics & Management Strategy* 15, 279–316.
Jin, G.Z. and Kato, A. 2005. Price, quality, and reputation: evidence from an online field experiment. *RAND Journal of Economics* (forthcoming).
Lucking-Reiley, D. and Spulber, D.F. 2001. Business-to-business electronic commerce. *Journal of Economic Perspectives* 15(1), 55–68.
Mayzlin, D. and Chevalier, J.A. 2006. The effect of word-of-mouth on sales: online book reviews. *Journal of Marketing Research* 43, 345–54.
Murthi, B.P.S. and Sarkar, S. 2003. The role of the management sciences in research on personalization. *Management Science* 49, 1344–62.
Prince, J. 2006. The beginning of online/retail competition and its origins: an application to personal computers. in the *International Journal of Industrial Organization* (forthcoming).
Scott Morton, F. 2006. Consumer benefit from use of the Internet. In *Innovation Policy and the Economy*, vol. 6, ed. A.B. Jaffe, L. Lerner and S. Stern. Cambridge, MA: MIT Press.
Shapiro, C. and Varian, H.R. 1999. *Information Rules: A Strategic Guide to the Network Economy.* Boston: Harvard Business School Press.
Sinai, T. and Waldfogel, J. 2004. Geography and the Internet: is the Internet a substitute or a complement for cities? *Journal of Urban Economics* 56, 1–24.
U.S. Census Bureau. 2006a. E-Stats, 25 May. Online. Available at http://www.census.gov/eos/www/papers/2004/2004reportfinal.pdf, accessed 13 January 2007.
U.S. Census Bureau. 2006b. Quarterly retail e-commerce sales, 2nd quarter 2006. U.S. Census Bureau News, 17 August. Online. Available at http://www.census.gov/mrts/www/data/html/06Q2.html, accessed 13 January 2007.
Wellman, B. 2001. Computer networks as social networks. *Science* 29, 2031–4.

elites and economic outcomes

A ruling elite (from the Latin *eligere*, 'to elect') is a small, dominant group that enjoys the power of decision in the various sectors of the economic and social organization of a state. It includes the bureaucrats and civil servants who rule the macro-environment; the political elite that governs and operates the executive, legislative and judicial structures; and the business elite. Non-ruling elites include the members of the media, academia and the intelligentsia.

Even in a democratic regime in which the power is meant to reside in the *demos* ('the people'), power is really concentrated in the hands of a few. All political organizations, even democracies, tend towards domination by an oligarchy, which Mills (1956) called the *power elite*. This is the *iron law of oligarchy* as stated by Michels (1915). This stratification of society based on the accumulation of decision-making power therefore differs from the familiar stratification based on income and economic means, or on ownership of the factors of production as emphasized by Marx.

The effects of elite actions on the economy operate through several channels: economic growth and development; social mobility; inequality; and the political system, which in turn affects the economy. The characteristics that affect these economic realms are (a) the extent of the intertwining and inter-connections of elites; and (b) the stability and recruitment of the elite.

Elites' interconnections

The ruling elite can display unity and collusion, acting as a monolithic group, or it can be fragmented and characterized by dissociation and diversification of power, a 'polyarchy' that permits competition among its members.

The elite in non-democratic polities displays unity, has unlimited political and economic power, and typically acts on behalf of its own interests. But democracy should a priori impose some control on the power of the ruling elite. Indeed, Schumpeter (1954) claimed that the democratic process permits 'free competition among would-be leaders for the vote of the electorate' and that the masses can choose between various elites. In contrast, classical elite theorists such as Mosca (1939), Pareto (1935), Michels and Mills emphasized that there can be collusion even in democracies. Numerous elites may not be mutually competitive and may not control and balance each other; instead, they may be intertwined as a unanimous, cohesive power elite.

Economic consequences of the extent of interconnection

Inequality

The elite's plurality and competition ensures its responsiveness to the demands of the public, while a consensual elite might use its power for its own interests. Etzioni-Halevi (1997) claims that a unified elite does not use its power to reduce inequality and promote the development of a more egalitarian society, due to common recruitment and common interests. It is the plurality and differentiation of the members of the elite that enables them to countervail each others' power and to increase their responsiveness to the will of public. In consequence, elite homogeneity might actually increase the gap between the elite and the masses.

When the political elite controls wealth and the main factors of production, then elite and class stratifications coincide, and consequently power and wealth are in the hands of the same happy few. Engerman and Sokoloff (1997) showed that members of the elite who have power and wealth establish institutions that serve their own interests and exclude the masses from benefits. In consequence, inequality persists through institutional development in the elite's own favour. Justman and Gradstein (1999) added that elite unity leads to greater inequality through regressive redistribution policy. A power elite that controls wealth may refrain from investment in human capital of the majority because education would increase the latter's political voice and weaken the elite's hold on power (Easterly, 2001); yet in some cases, the elite deliberately decides to forfeit power by investing in human capital as a consequence of a cost–benefit analysis (Bourguignon and Verdier, 2000).

The extent of elite unity can be endogenously determined (Sokoloff and Engerman, 2000), and elite unity can also be affected by revolutions, wars and economic growth. Justman and Gradstein (1999) argue that economic growth dilutes the power of the elite by broadening political participation and reducing inequality.

Economic growth

A strong interconnection among elites has the consequence that all sectors of the economy are ruled by a group that thinks in a monolithic way. Two lines of thoughts have related a monolithic group to economic growth. The first one underlines that a monolithic group leads to the stagnation of ideas and attitudes, which in turn may prevent the adoption of major technological breakthroughs (Bourdieu, 1977). The lack of competition in a monolithic powerful group also generates corruption, with harmful consequences for growth.

The second line of thought argues that wealthy elites with enough political power to block changes will not accept adopting institutions that would enhance growth, since they might hurt them. Acemoglu, Johnson and Robinson (2001) developed this line of thought in relation to colonial impacts, showing that, wherever colonial governments were composed of few elite members, economic progress was reduced.

Following the same line of reasoning, Acemoglu and Robinson (2000) and Gradstein (2007) stressed that elite plurality, in which the political and economic elites are separate, explains the adoption of political franchise and industrialization in western Europe; while 19th-century eastern Europe, where elite unity was strong, did not adopt growth-enhancing institutions, since its elites held on to their wealth and power.

Paradoxically, in countries in which the elite was united and consensual, with common aims, the transition to capitalist production in the 1990s took place without violence, as in Poland and the Czech Republic. In contrast, wherever the elite was divided and fragmented, there were conflicts, especially on the ethno-nationalist level, as in Yugoslavia and Romania (Pakulski, 1999).

Recruitment and training of elites

Plato claimed that government should be in the hands of the most able members of society, that is, the *aristocracy*

(Greek for 'rule by the best'), a term that became pejorative and was later changed to *meritocracy* (coined by Young, 1958). Pareto argued that a stable economic system needs a *circulation of elites*, so that the most capable and talented are in the governing class. He stressed that the quality of the ruling class can be maintained only if social mobility is allowed, so that the non-elite has the possibility of entering the elite: 'History is a cemetery of aristocracies' (Pareto, 1935). His theory may be viewed as a sort of social Darwinism in which mobility is needed, just as evolution relied upon competition and selection.

For millennia, recruitment of the Western elite was based on social inheritance and was carried out via heredity, nepotism and violence. Hereditary monarchy was considered the most legitimate means of recruitment for rulers, and the upper elite was made up of wealthy large landowners, an *état de fait* considered normal in agrarian societies. Nevertheless, there were some channels of entrance into the elite, such as military prowess and exploits or involvement in government finance (Brezis and Crouzet, 2004).

In democracies, the political elite came to be recruited mainly by election. Yet for a long time, the franchise was not for all. Big landowners and members of the upper middle class were the overwhelming majority in parliaments and cabinets, even though some prominent business people entered the political elite. Only in the late 19th century did members of the lower middle class and working class enter the political elite.

From the 19th century onwards, the circulation of the business elite took two differing yet concurrent paths. The first was that economic growth led to spurts of new firms and the decline of others, allowing a new business elite to emerge (Schumpeter, 1961). The second path was the rise of the professions, with competitive and meritocratic exams that led to circulation of elites (Perkin, 1978). After the Second World War, the elite was mainly recruited through education into elite universities to which admission started to be conferred following success at meritocratic exams.

Economic consequences of the recruitment of elites

Social mobility in the economy

Prior to recruitment through meritocracy, social mobility, and in particular the potential for non-elite members to enter the elite, was low. Temin (1999a; 1999b) showed that today, as in the 1900s, and despite meritocracy, the American economic elite is composed almost entirely of white Protestant males who have been educated for the most part in Ivy League colleges. Although in 1900 the political elite was quite similar to the business elite, today the former is more diversified; the political elite has changed in its recruitment, while the economic elite has not. In other words, minorities have not penetrated the economic elite in the United States (see also Friedman

and Tedlow, 2003, which summarizes studies on US elite mobility, and Foreman-Peck and Smith, 2004 on British elites).

Recruitment to a university through meritocratic entrance exams, does not, indeed, lead to enrolment from all classes of society according to distribution or ability, nor does it necessarily lead to the admission of the most talented. Recruitment by entrance exam still encompasses a bias in favour of elite candidates because this type of exam requires a pattern of aptitude and thinking that favours candidates from an elite background. All elite positions may be open to all applicants with the right qualifications, but they are more accessible to those with specific social, cultural and symbolic capital (Arrow, Bowles and Durlauf, 2000). Thus the power elite maintains its status and power by a *strategy of distinction*, or a cultural bias that is necessary for accessing it (Bourdieu, 1977). A small difference in culture and education leads to narrow recruitment, and in turn to class-based stratification in the recruitment of the elite, despite meritocratic selection for universities (Brezis and Crouzet, 2006).

The relationship between mobility and the political system, as emphasized by Pareto, has been analysed by sociologists. For instance, Lengyel (1999) showed that circulation in the elite occurs at times of political upheavals and revolutions: the existing elite is eliminated and replaced by a new one. The first-generation members of the elite following a political change have neither specific training and education nor specific origin; they are the trailblazers, the entrepreneurs who seized power on the strength of their competence. In the next generation, the elite becomes narrowly recruited from the best educated, and members are selected mostly by training and education. The elite returns to an occupational specialization, similar to the meritocratic profession criterion of earlier industrialization (Perkin, 1978).

Economic growth

A crucial element of economic growth is that the recruited elite be of the highest quality. Countries in which elites are recruited in a non-meritocratic way face the problem of the quality of their elites. However, the prevalence of meritocratic recruitment does not necessarily lead to the selection of the best ruling elites. Brezis and Crouzet (2006) argue that, when a country faces only mild technological and structural changes, the narrow recruitment, due to meritocracy, optimally fulfils its purpose, since the cultural bias of the elites is an advantage in the given type of technology. However, at times of major changes in technology, elites recruited this way are not the best for adopting new technologies.

Moreover, the homogeneity of the recruitment of elites through similar curricula leads to convergence of views; this, in turn, leads to a monolithic elite, which, as we have claimed above, may have negative consequences for economic growth.

Conclusion

In this short article, we have summarized the modern research that has examined recruitment schemes and incentives for elites to discover how they can be used to promote, rather than impede, economic growth. There is also an entire economic history literature that has enriched us with a wealth of knowledge on the business elite. The main works in this literature are by Cassis (1997), Crouzet (1999) and Lachmann (2000).

The literature cited herein seems to show that the structure of this small group called the elite has numerous effects on the world economy. In the opposite direction, globalization will also affect the elite, as we are now facing a globalization of education of the elite.

In its first wave, globalization of education will probably create a new collection of elites and elicit some changes, yet the unity and uniformity of the elite will be even greater, not only at the national level but also at the global level. National elites will be replaced by a worldwide elite, along with uniformity in culture and education. We will face an international technocratic elite with its own norms, ethos, and identity, as well as its private clubs like the Davos World Economic Forum – a transnational oligarchy.

ELISE S. BREZIS AND PETER TEMIN

See also **economic growth, empirical regularities in; income mobility; Pareto, Vilfredo; social status, economics and.**

Bibliography

Acemoglu, D., Johnson, S. and Robinson, J.A. 2001. The colonial origins of comparative development: an empirical investigation. *American Economic Review* 91, 1369–401.

Acemoglu, D. and Robinson, J. 2000. Political losers as a barrier to economic development. *American Economic Review* 90, 126–44.

Arrow, K., Bowles, S. and Durlauf, S. 2000. *Meritocracy and Economic Inequality*. Princeton: Princeton University Press.

Bourdieu, P. 1977. *Reproduction in Education, Society and Culture*. London: Sage.

Bourguignon, F. and Verdier, T. 2000. Oligarchy, democracy, inequality and growth. *Journal of Development Economics* 62, 285–313.

Brezis, E.S. and Crouzet, F. 2004. Changes in the training of the power elites in Western Europe. *Journal of European Economic History* 33, 33–58.

Brezis, E.S. and Crouzet, F. 2006. The role of higher education institutions: recruitment of elites and economic growth. In *Institutions and Economic Growth*, ed. T. Eicher and Garcia-Penalosa. Cambridge, MA: MIT Press.

Brezis, E.S. and Temin, P., eds. 1999. *Elites, Minorities, and Economic Growth*. Amsterdam: Elsevier.

Cassis, Y. 1997. *Big Business. The European Experience in the Twentieth Century*. Oxford: Oxford University Press.

Crouzet, F. 1999. Business dynasties in Britain and France. In Brezis and Temin (1999).

Easterly, W. 2001. The middle class consensus and economic development. *Journal of Economic Growth* 6, 317–35.

Engerman, S.L. and Sokoloff, K.L. 1997. Factor endowments, institutions, and differential paths of growth among New World economies: a view from economic historians of the United States. In *How Latin America Fell Behind: Essays on the Economic History of Brazil and Mexico, 1800–1914*, ed. S. Haber. Stanford: Stanford University Press.

Etzioni-Halevy, E. 1997. *Classes and Elites in Democracy and Democratization*. New York: Garland.

Foreman-Peck, J. and Smith, J. 2004. Business and social mobility into the British elite 1870–1914. *Journal of European Economic History* 55, 485–518.

Friedman, W.A. and Tedlow, R.S. 2003. Statistical portraits of American business elites: a review essay. *Business History* 45, 89–113.

Gradstein, M. 2007. Inequality, democracy and the protection of property rights. *Economic Journal* 117, 252–69.

Higley, J. and Burton, M. 1989. The elite variable in democratic transitions and breakdowns. *American Sociological Review* 54, 17–32.

Justman, M. and Gradstein, M. 1999. Industrial Revolution, political transition and the subsequent decline in inequality in nineteenth-century Britain. *Explorations in Economic History* 36, 109–127.

Lachmann, R. 2000. *Capitalist in Spite of Themselves: Elite Conflict and Economic Transition*. Oxford: Oxford University Press.

Lengyel, G. 1999. Two waves of professionalization of the Hungarian economic elite. In Brezis and Temin (1999).

Michels, R. 1915. *Political Parties*. New York: Dover, 1959.

Mills, C.W. 1956. *The Power Elite*. New York: Oxford University Press.

Mosca, G. 1939. *The Ruling Class*. New York: McGraw.

Pakulski, J. 1999. Elites, ethnic mobilization and democracy in postcommunist Europe. In Brezis and Temin (1999).

Pareto, V. 1935. *The Mind and Society*. New York: Harcourt Brace.

Perkin, H. 1978. The recruitment of elites in British society since 1800. *Journal of Social History* 12, 222–34.

Sokoloff, K. and Engerman, S. 2000. Institutions, factor endowment and paths of development in the New World. *Journal of Economic Perspectives* 14(3), 217–32.

Schumpeter, J.A. 1954. *Capitalism, Socialism and Democracy*. London: Routledge.

Schumpeter, J.A. 1961. *Theory of Economic Development*. New York: Oxford University Press.

Temin, P. 1999a. The American business elite in historical perspective. In Brezis and Temin (1999).

Temin, P. 1999b. The stability of the American business elite. *Industrial and Corporate Change* 8, 189–210.

Young, M. 1958. *The Rise of Meritocracy, 1870–2033*. London: Thames & Hudson.

Ellet, Charles, Jr. (1810–1862)

American engineer and economic theorist, Ellet was born on 1 January 1810 at Penn's Manor, Pennsylvania, and died on 21 June 1862, a victim of the Civil War. Ellet grew up on a family farm but showed little inclination for agriculture: at age 17 he joined a surveying crew. With no formal education or training, he soon became an assistant engineer to Benjamin Wright, chief engineer of the Chesapeake and Ohio Canal. With ability and hard work Ellet taught himself mathematics and French, earning the respect of influential engineers. Letters of introduction to Lafayette and the American ambassador helped secure Ellet a place at the Ecole des Ponts et Chaussées, Dupuit's alma mater, in 1830. On his return to America in 1832 Ellet became the premier suspension bridge designer in America, building in 1849 the (then) longest suspension bridge in the world across the Ohio River at Wheeling. Colonel Ellet designed, constructed and commanded the ram fleet of the Union forces at the naval battle at Memphis, Tennessee. He died as a result of a wound received in the heat of that battle.

Ellet spent most of his professional life as an engineer, but, in one major work and in a number of contributions to the *Journal of the Franklin Institute* between 1840 and 1844, he significantly advanced the economic theory of monopoly, input selection, spatial economics, benefit–cost theory and econometric estimation. All Ellet's contributions were facilitated by the use of the differential calculus, which permitted him to express the simple theory of the firm, and some of its extensions, in mathematical terms. In his *Essay on the Laws of Trade* (1839) Ellet established the demand curve for a monopoly railroad with distance as a variable. Utilizing first-order conditions and solving for the gross toll on passenger traffic, Ellet demonstrated that the profit-maximizing toll would be equal to one-half the costs of transportation added to a constant quantity, a well-known result.

Ellet considered not one monopoly model but a multiplicity of them, including those dealing with freight transport, duopoly conditions and the principles of monopoly price discrimination. Further, Ellet's particular insights into simple and discriminatory pricing systems led him to provide, with distance as a variable, an amazingly complete mathematical and graphical analysis of the impact of changes in the pricing system upon the market area served by a profit-maximizing railroad (1840a). In this important contribution to market area analysis Ellet argued that a set of (constrained) discriminatory tolls inverse to distance, in contrast to tolls proportional to distance, could be devised whereby all interested parties (management, shippers, the state) could be made better off. In a series of papers (1842–4) Ellet extended his theoretical analysis of inputs and input selection (1839) to one of the earliest attempts to develop, empirically specify and test a theoretical cost function. Utilizing a 'law' of costs which included his selected determinants of annual total railway costs, Ellet estimated the empirical dimensions from data collected from the mid-1830s. He then reaffirmed the power of his initial equation with new and supplementary data.

In all, the calibre and completeness of Ellet's theoretical and empirical inventions would not compare unfavourably with those of von Thünen, Cournot, Dupuit or Lardner. Ellet, who was primarily an engineer, was America's best representative among the pioneer contributors to scientifically oriented economics in the 19th century.

ROBERT B. EKELUND, JR.

Selected works

1839. *An Essay on the Laws of Trade in Reference to the Works of Internal Improvement in the United States.* Richmond. Reprint from the 1st edn, New York: Augustus Kelley, 1966.
1840a. The laws of trade applied to the determination of the most advantageous fare for passengers on railroads. *Journal of the Franklin Institute* 30, 369–79.
1840b. A popular exposition of the incorrectness of the tariffs of tolls in use on the public improvements of the United States. *Journal of the Franklin Institute* 29, 225–32.
1842–4. Cost of transportation on railways. *Journal of the Franklin Institute*, various issues.

Bibliography

Baumol, W. and Goldfeld, S.M., eds. 1968. *Precursors in Mathematical Economics: An Anthology.* London: London School of Economics and Political Science.
Calsoyas, C.D. 1950. The mathematical theory of monopoly in 1839: Charles Ellet, Jr. *Journal of Political Economy* 58(April), 162–70.
Ekelund, R.B., Jr. and Hooks, D. 1972. Joint demand, discriminating two-part tariffs and location theory: an early American contribution. *Western Economic Journal* 10, 84–94.
Viner, J. 1958. *The Long View and the Short: Studies in Economic Theory and Policy.* New York: Glencoe.

Ely, Richard Theodore (1854–1943)

Ely was born in Ripley, New York, on 13 April 1854 and died at Old Lyme, Connecticut, on 4 October 1943.

Ely's long and vigorous career epitomizes the general proposition that an economist can exert a major constructive influence on his subject and profession even though his original contribution to economic theory is negligible. A highly effective teacher and maker of careers for his former students; prolific author of popular articles, scholarly volumes, and publications series; organizer and fund-raiser for major research projects; founder of various academic institutes and associations; leader or participant in numerous reform societies; and centre of innumerable controversies, Ely was the most widely known, even notorious, economist in the USA around the turn of the 20th century.

After a brief spell as a country schoolteacher and a preliminary year at Dartmouth College, Ely graduated from Columbia College in 1876 and was awarded a three-year fellowship to study philosophy in Germany. He soon switched to political economy, came under the influence of Karl Knies at Heidelberg, where he obtained a Ph.D., summa cum laude, in 1878, and later attended Adolph Wagner's lectures in Berlin. Returning to the USA he was unemployed for more than a year before his appointment, initially on a half-time basis, at Johns Hopkins, where he taught from 1881 to 1892. He then moved to Wisconsin, founding an outstanding school of Economics, Political Science and History including such luminaries as F.J. Turner, E.A. Ross, and J.R. Commons. A unique collaboration developed between the social scientists and the state legislators, especially under the La Follette governorship, which pioneered major social and economic reform legislation. In 1925 Ely took his Institute for Research in Land Economics and Public Utilities, founded in 1920, from Madison to Northwestern University, and remained there until 1932, when he launched a new, but impoverished Institute for Economic Research in New York City. Eventually hit by the depression, Ely was forced to depend on the support of friends and former students as he completed his autobiography and failed to complete a massive history of American economic thought initiated 50 years earlier.

An ardent Christian Socialist and outspoken critic of laissez-faire individualism and 'old school' English classical economics, Ely delighted social reformers and outraged conservatives by his writings on such controversial current topics as socialism and the American labour movement. Prone to emotional overstatement and careless in exposition, his public pronouncements and reputation frequently embarrassed the aspiring young professional economists with whom he founded the American Economic Association, in 1885, and for a time discouraged some moderate and conservative economists from joining. Although Ely's original draft prospectus had been rejected, and the association's original constitution was toned down, and then dropped, the organization hovered uneasily between missionary evangelism and scholarly objectivity until he was obliged to relinquish his secretaryship in 1892.

Two years later, at Wisconsin, Ely's fellow professionals rallied around him when he was denounced for preaching socialism and encouraging strikes, and, although he was completely exonerated in a 'trial' that attracted national attention, Ely gradually became more conservative. Ironically, in the 1920s his institute was attacked, no doubt unfairly, as a tool of the public utilities, and was referred to disparagingly in a report on professional ethics by a committee of the American Association of University Professors, in 1930.

During his long lifetime Ely wrote extensively on an extraordinarily wide variety of topics, often in a popular and journalistic fashion. Nevertheless, he repeatedly opened up new research topics that were developed by his colleagues and former students – for example, in labour history, state taxation, land economics, and natural resources – and his various textbooks, especially the multi-edition *Outlines of Economics* which sold 350,000 copies, were both widely used and highly regarded.

At Wisconsin he helped to launch the American Association for Labor Legislation, of which he became President, and raised private resources to finance John R. Commons's massive *Documentary History of American Society* (11 vols, 1910–11). He served as President of the American Economic Association in 1900–1901.

Ely was a stimulating teacher whose ideas formed a direct link between the doctrines of the German Historical School and American institutionalism, a link most clearly evident in his neglected two-volume study of *Property and Contract in their Relations to the Distribution of Wealth* (1914). Many of his students went on to distinguished careers in academic and/or public life. He was undoubtedly an outstanding academic entrepreneur, and his contribution to the American Economic Association is recognized in its annual invited Richard T. Ely lecture, which was inaugurated in 1963.

A.W. COATS

Selected works

1883. *French and German Socialism in Modern Times*. New York: Harper & Brothers. Reprinted, 1911.

1884a. *The Past and the Present of Political Economy*. Baltimore: N. Murray for Johns Hopkins.

1884b. *Recent American Socialism*. Baltimore: N. Murray for Johns Hopkins. Reprinted, 1885.

1886. *The Labour Movement in America*. New York: T.Y. Crowell Co. New edn, revised and enlarged, New York: Macmillan Co, 1905.

1888a. *Taxation in American States and Cities*. New York: T.Y. Crowell Co.

1888b. *Social Aspects of Christianity*. Boston: W.L. Greene and Co.

1889a. *An Introduction to Political Economy*. New York: Chautauqua Press. New and revised edn, New York: Eaton and Mains; Cincinnati: Jennings and Pye, 1901.

1889b. *Social Aspects of Christianity and other Essays*. New York: T.Y. Crowell Co. Reprinted, 1895.

1893. *Outlines of Economics*. Meadville, Pennsylvania and New York: Flood and Vincent. 6th edn with Ralph Hess, New York: Macmillan Co., 1938.

1894. *Socialism: An Examination of its Nature, its Strength, its Weakness. With Suggestions for Social Reform*. London: S. Sonnenschein Co.; New York and Boston: T.Y. Crowell Co.

1900. *Monopolies and Trusts*. New York: Macmillan Co. Reprinted, 1912.

1903. *Studies in the Evolution of Industrial Society*. New York and London: Macmillan Co. Reprinted, 1918.

1914. *Property and Contract in their Relations to the Distribution of Wealth*. 2 vols. New York: Macmillan Co. Reprinted, 1922.

1924. (With E.W. Morehouse.) *Elements of Land Economics*. New York: Macmillan Co. Reprinted, 1932.

1928. (With G.S. Wehrwein.) *Land Economics*. Ann Arbor, Michigan: Edwards Bros. Revised edn, Madison: University of Wisconsin Press, 1964.

1938. *Ground Under Our Feet: An Autobiography*. New York: Macmillan Co.

Bibliography

Rader, B.G. 1966. *The Academic Mind and Reform: The Influence of Richard T. Ely in American Life*. Lexington: University of Kentucky Press.

emergence

Having acquired widespread use among life scientists and science writers since the early 1990s, the term 'emergence' in economics is more evocative than precise, reflects influence from physics and biology, and has come to be associated with phenomena involving evolution of economic structures into qualitatively different forms. These phenomena exhibit properties that are emergent in the sense that they are novel and apply at an aggregate more 'complex' level but lack individual analogues and therefore are not describable at, or reducible to, the individual level. A good case in point is the statement that consciousness is an emergent property of the brain. The notion of emergence originates in the philosophy of science, with John Stuart Mill being an important precursor (see Stanford Encyclopedia of Philosophy, 2002).

This article reviews, albeit selectively, the recent usage of the term by emphasizing applications with predominantly economic phenomena where emergence of macroscopic properties may be elucidated by means of economic arguments. These range from neighbourhood tipping and evolution of patterns in international trade to emergence of urban structure and the establishment of norms and institutions and of a common currency, among many others.

More generally, emergent properties or behaviours have been studied in a variety of circumstances in nature, such as emergence of differentiated behaviour in colonies of animals, of herding behaviour in organizations and markets, of specialization of individuals into occupations and of cities and of regions and countries in specific products, of groups of biological cells in multicellular biological organisms and even of groups of processors in computer simulations involving cellular automata (see Holland, 1998). The World Wide Web is an example of a decentralized engineering system that is continuously being modified by human initiatives in the form of actions by individuals and firms. The web has not been deliberately designed and no central organization administers how different sites are linked to others. Some of the properties of the graph topology of the web may be termed as *emergent*, such as that the number of links pointing to each page follows approximately a power law, with a few pages being pointed to by many others and most others seldom, and the fact that any pair of pages can be connected to each other through a relatively short chain of links in the average.

The presence of 'emergence' within the vocabulary of economists does suggest some interplay with multidisciplinary research by scientists who have been associated with the Santa Fe Institute (http://www.santafe.edu). To quote from Kauffman (1995, p. 24), an alternative definition of emergence is that '[t]he whole is greater than the sum of its parts'. And 'life itself is an emergent phenomenon ... arising as the molecular diversity of a prebiotic chemical system increases beyond a threshold of complexity. If true, then life is not located in the property of any single molecule – in the details – but is a collective property of systems of interacting molecules.' The entirety of complex molecules together is able to reproduce and evolve, a 'stunning property'.

Blume and Durlauf (2001) argue that emergence plays an important role even within the body of neoclassical economics proper. For example, the extent to which macroeconomics is a distinct discipline from microeconomics would be explained by emergent properties as alluded to by the statement 'aggregation is not summation' (see Kirman, 1992). Consider, within microeconomics and general equilibrium theory, the metaphor of the invisible hand of the market (which goes back to Adam Smith), whereby individuals' pursuit of their own selfish aims leads to social outcomes that obey important social properties. Under certain conditions, after markets have brought about an equilibrium, it is impossible to make anyone better off without making someone worse off. Thus, the first fundamental theorem of welfare economics is an emergent property of social outcomes. However, the more modern work on emergence in economics has emphasized emergence of patterns. Similarly, Hayek's concept of spontaneous order may be considered an instance of emergence.

There are numerous other contexts where emergence has been alleged to occur. This article explores a number of examples of emergence that are limited to social and economic settings. They underscore the scope of the concept of emergence in such settings. As discussed earlier, there are many other contexts in socioeconomic settings and beyond, ranging from computation to the life sciences.

Emergent social interconnections

Suppose that a society consists of I individuals, where I is large, where any two individuals may be linked in a way that allows for communication, social relations, or social interactions. Let p_k denote the probability that each individual is connected with exactly k other individuals. A literature going back to Erdös and Renyi (1960) and continuing at the time of writing up to Newman, Strogatz and Watts (2001) has studied the topological properties of the (random) graph formed by the agents as nodes and connections between agents as edges when each agent's connections with other follows a given distribution p_k and the number of agents is large. According to Newman, Strogatz, and Watts (2001), depending upon whether the quantity $E[k^2] - 2E[k]$ is greater than or equal to 0, or falls below 0, there emerges, as I tends to infinity, a *proportion* of all individuals being interconnected, or, alternatively, the economy consists of different groups of finite sizes. In other words, the social structure undergoes a *phase transition* when this quantity exceeds 0: a giant interconnected component emerges. Intuitively, starting from a connected component of the graph, consider adding a new edge that connects with a previously isolated node of degree k. Doing so will change the number of nodes on the boundary of the connected component by $-1 + (k - 1) = k - 2$. The likelihood that a node is on the boundary of the connected component is proportional to k. The expected change in the number of nodes on the boundary when an additional node is connected is given by $\sum_i k_i(k_i - 2)/\sum_i k_i$. If this quantity is negative, then the number of nodes on the boundary decreases and therefore the connected component will stop growing. If it is positive, on the other hand, then the number of boundary nodes will grow and the connected component will grow, limited only by the size of the network.

In the simple case of the Erdös and Renyi random graph, where the number of connections is proportional to the number of individuals, the phase transition occurs when the factor of proportionality is equal to $\frac{1}{2}$ and the corresponding average number of connections per person is equal to 1. Below this value, there are too few edges

and the components of the random graph are small; above that value, a proportion of the entire graph belongs to a single, *giant* component. In this case, emergence of a qualitatively different social structure depends on the value of a single parameter (Kirman, 1983; Ioannides, 1990; Durlauf, 1997). Individual behaviour that leads to a law for the number of individuals' connections does not necessarily imply the same macroscopic outcome in all circumstances. Similarly, social outcomes are not described by means of mere summation of individual actions; aggregation is not summation (Kirman, 1992). Kauffman (1995, p. 57) invokes this in the context of autocatalytic reactions and goes as far as seeing this 'as a toy version of phase transition that I believe led to the origin of life'.

Patterns of residential segregation

Now we turn to a description of neighbourhood tipping, which is originally due to Thomas C. Schelling (1978) and has been adapted here from recent works. Suppose that individual i is white and would live in a neighbourhood provided that the percentage of whites among her neighbours, $\omega \in [0, 1]$, is at least w_i, $\omega \geq w_i$. She moves out otherwise. Individuals differ in terms of preference characteristic w_i, which is assumed to be distributed in a typical neighbourhood according to $F(w)$, when the analysis starts. For any neighbourhood with a share of white residents equal to ω, the percentage of white individuals who would find living there acceptable are those with $w < \omega$. Their share is given by the value of the cumulative distribution function at ω, $F = F(\omega)$.

In Figure 1, let the horizontal axis e_1 denote ω and w_i, the vertical axis e_2 the cumulative distribution F, and $(O, \bar{O})$ the 45-degree line. As long as $\omega > F(\omega)$, whites

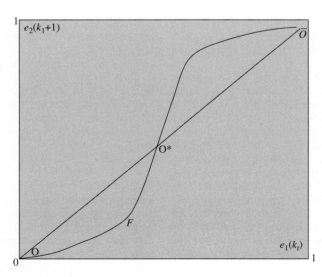

Figure 1   Neighbourhood tipping, poverty traps

have an incentive to exit the neighbourhood, causing a reduction of ω, and this process continues until there are no whites left; $\omega = 0$. If, on the other hand, $\omega \leq F(\omega)$, additional whites have an incentive to enter, and this process continues until $\omega = 1$. Thus, the process has three equilibria, $(O, O^*, \bar{O})$, of which the two extreme ones, either only blacks or no blacks in the neighbourhood, are stable, and the mixed one, with ω^* whites in the neighbourhood, where $\omega^* = F(\omega^*)$, unstable. The mixed equilibrium defines the *tipping point*. Individuals' preferences differ widely, but only extreme outcomes *emerge* at the social equilibrium. Schelling (1978) underscores how outcomes that persist may not be what individuals had intended.

Could such a stark outcome be due to the fact that the respective populations of individuals are not being replenished? It turns out that, if one goes deeper and allows for turnover and stochastic shocks, persistence of stable states may be rigorously characterized by means of the tools of stochastic stability theory (Blume and Durlauf, 2003; Young, 1998). Multiplicity of equilibria allows, of course, for accidents of history to become reinforced over time.

Emergence of urbanization

The concentrated economic activity that we associate with the emergence of cities punctuates the physical and economic landscape throughout the world. How did it emerge? While small-scale agriculture and home production could be reasonably accurately referred to as spatially uniform distribution of economic activity, the world population is increasingly concentrated in cities. Also, urbanization has been closely associated with economic development.

Let us consider a simple setting where utility U depends on individual productivity, itself an increasing function $f(n_\ell)$, of the total number of others in the same location, n_ℓ, and on the share of a fixed resource, $\frac{R}{n_\ell}$. Even when utility is assumed to be increasing and concave in both arguments, it is initially increasing, as a function of n_ℓ, may reach a peak at n^*, and then may start decreasing. In other words, a larger population initially means more innovation and mutually beneficial interaction until congestion offsets them. Consider then two alternative locations, $\ell = 1, 2$, that do not interact spatially, and a total of N individuals who wish to locate so as to maximize utility. At a locational equilibrium, individuals must be indifferent as to where they locate. If $N < 2n^*$, the symmetric equilibrium, where $n_1 = n_2 = \frac{1}{2}N$, is unstable and agglomeration – that is, either site occupied by the entire population – is stable. Therefore, the trade-off between the value of agglomeration and the cost of congestion moves the economy away from the symmetric outcome (Anas, 1992).

Consider next a setting where interactions do explicitly depend on distance to others, as with accessibility to

others being valued and congestion disliked. If individuals are allowed to relocate, with probabilities that depend on expected utilities in each site relative to all other sites, then a dynamic model may be formulated that describes locational outcomes for an entire population. The economy may attain steady states that are either uniform (populations are equal across all sites) or uneven (with some sites having large and others small populations). Such a stylized reduced-form model of spatial patterns of human settlements (see Papageorgiou and Smith, 1983) yields spatially uniform outcomes that are either stable or unstable. Agglomeration is determined by the interplay between the value of agglomeration and the cost of congestion. If the former dominates, spatially uniform steady states are unstable. Fujita, Krugman and Venables (1999, chs 6 and 17) develop a model with ingredients from economic geography that incorporates trading costs and also allows for uniform distributions of economic activity to exhibit different stability properties. Again, conditions under which agglomerations prevail possess intuitive economic appeal.

Emergence of poverty traps

In a standard neoclassical growth model that extends over discrete time, with a demographic structure consisting of two overlapping generations and individuals living for two periods, working only in the first and retiring in the second, individual savings would be proportional to the wage rate under Cobb–Douglas preferences. Let the aggregate production function expressing output Y_t as a function of capital, labour and total factor productivity, K_t, L_t, A, respectively, be of the constant elasticity of substitution form,

$$Y_t = A \left(\delta K_t^{1 - \frac{1}{\sigma}} + (1 - \delta) L_t^{1 - \frac{1}{\sigma}} \right)^{\frac{\sigma}{\sigma - 1}}.$$

If the elasticity of substitution is sufficiently small – that is, complementarity between capital and labour is high – and total factor productivity sufficiently large, the time map of the economy – that is, the amount of capital per person next period (axis e_2) as a function of the amount of capital per person in the present period (axis e_1) – may be loosely graphed, as in Figure 1. Therefore, depending upon the economy's starting point, it may end up at a steady state either with high or with low capital per person at a steady state. The mid-range ('symmetric') steady state is unstable. Therefore, conditions of productive complementarities, (even small) initial differences in capital per person, and possibly historical accidents as well across countries in terms of characteristics and endowments when growth starts, mitigate in favour of an explanation for inequalities in incomes per person across different countries. The same mechanism worldwide produces sharply different outcomes (see Azariadis and Stachurski, 2006, for an in-depth treatment).

Similar arguments may be developed in order to understand persistence in the inequality of the distribution of wealth within an economy. Matsuyama (2006) presents a model of emergent class structure, in which a society inhabited by inherently identical households may, depending upon parameter values, be endogenously split into the rich bourgeoisie and the poor proletariat. For some parameter values, the model has no steady state where all households remain equally wealthy. The model predicts emergent class structure or the rise of class societies. Even if every household starts with the same amount of wealth, the society will experience 'symmetry breaking' and will be polarized into two classes in steady state, where the rich maintain a high level of wealth partly due to the presence of the poor, who have no choice but to work for the rich at a wage rate strictly lower than the 'fair' value of labour.

It is worth noting that similar modelling tools may be used to express Adam Smith's famous dictum that 'the division of labour is limited by the extent of the market' and thus endogenize specialization (Weitzman, 1994). The division of labour emerges as individuals in an economy acquire specialized roles.

Emergent structures in international economics: autarky, specialization, and international currencies

Krugman (1995) and Matsuyama (1995) discuss how a world economy where all countries are initially identical and live in autarky (a 'symmetric' outcome) leads to a world that is separated into rich and poor regions, once countries engage in international trade. International trade *causes* specialization and agglomeration of different economic activities in different regions of the world to emerge, with some countries being rich and others poor. In several similarly motivated papers, Matsuyama (in particular, 2004; 2006) shows the effects of financial market globalization on the cross-country pattern of development in the world economy. In the absence of the international financial market, the world economy converges to the symmetric steady state, and the cross-country difference disappears in the long run. Financial market globalization causes the instability of the symmetric steady state and generates stable asymmetric steady states, in which the world economy is polarized into the rich and the poor. The world output is smaller, the rich are richer and the poor are poorer in these asymmetric steady states than in the (unstable) symmetric steady state. The model thus demonstrates the possibility that financial market globalization may cause, or at least magnify, inequality among nations, and that the international financial market is a mechanism through which some countries become rich at the expense of others. Furthermore, the poor countries cannot jointly escape from the poverty trap by merely cutting their links to the rich. Nor would foreign aid from the rich to the poor eliminate inequality; as in a game of musical chairs, some countries must be excluded from being rich.

Especially at times of political and economic upheavals, many different national currencies may circulate simultaneously within and across countries. From a modelling viewpoint, such circumstances fit neatly multiplicity of equilibria. Emergence of a particular currency as an international currency, which in turn depends on the degree of economic and financial integration, may be more of a decentralized phenomenon then the emergence and establishment of a national currency (Matsuyama, Kiyotaki and Matsui, 1993). To start with, a national currency is typically fiat money, whose use is decreed although not necessarily ensured. World monetary history suggests that a bewildering variety of commodities have served as medium of exchange, unit of account and store of value, and may have coexisted at times of financial uncertainties. It has been known at least since Menger (1892) that fiat money comes to dominate other options, thus leading to establishment of monetary equilibria, because individuals accept fiat money in trade when it is convenient and they trust that others will do the same. Such an outcome may be fragile, when trust in the currency is weakened, especially in time of war and other upheavals. Howitt and Clower (2000) employ 'rules' concerning transactor behaviour (instead of relying on a priori principles of equilibrium and rationality) to show computationally commodity 'money' as a possible emergent property of interactions between gain-seeking transactors who are unaware of any system-wide consequences of their own actions. Similar is the emergence of standards in new industries described by many writers.

Concluding remarks

The scientific literature, along with popular science literature, on emergence has sought to explain the emergence of persistent patterns as outcomes of dynamic interactions between individuals, groups of individuals and other entities. Such emergence is typically intrinsic to specific nonlinear dynamic processes and represents international currency. Not all possible outcomes may be sustained at equilibrium, and economic and political structures emerge as a result of self-organization. Future research needs to go beyond evolutionary thinking and also deal with emergence in the context of purposeful action by forward-looking agents, as opposed to social outcomes of decentralized interactions of many agents.

YANNIS M. IOANNIDES

See also **poverty traps; spontaneous order.**

Bibliography

Anas, A. 1992. On the birth and growth of cities: laissez-faire and planning compared. *Regional Science and Urban Economics* 22, 243–58.

Azariadis, C. and Stachurski, J. 2006. Poverty traps. In *Handbook of Economic Growth*, ed. P. Aghion and S.N. Durlauf. Amsterdam: North-Holland.

Blume, L.E. and Durlauf, S.N. 2001. The interactions-based approach to socioeconomic behavior. In *Social Dynamics*, ed. S.N. Durlauf and H. Peyton Young. Princeton, NJ: Princeton University Press.

Blume, L.E. and Durlauf, S.N. 2003. Equilibrium concepts for social interaction models. *International Game Theory Review* 5, 193–209.

Durlauf, S.N. 1997. Statistical mechanics approaches to socioeconomic behavior. In *The Economy as an Evolving complex System II*, ed. W.B. Arthur, S.N. Durlauf and D. Lane. Redwood City, CA: Addison-Wesley.

Erdös, P. and Renyi, A. 1960. On the evolution of random graphs. *Publications of the Mathematical Institute of the Hungarian Academy of Sciences* 5, 17–61.

Fujita, M., Krugman, P.R. and Venables, A.J. 1999. *The Spatial Economy*. Cambridge, MA: MIT Press.

Holland, J.H. 1998. *Emergence: From Chaos to Order*. Reading, MA: Addison-Wesley.

Howitt, P. and Clower, R. 2000. The emergence of economic organization. *Journal of Economic Behavior and Organization* 41, 55–84.

Ioannides, Y.M. 1990. Trading uncertainty and market form. *International Economic Review* 31, 619–38.

Kauffman, S.A. 1995. *At Home in the Universe: The Search for the Laws of Self-Organization and Complexity*. Oxford: Oxford University Press.

Kirman, A.P. 1983. Communication in markets: a suggested approach. *Economic Letters* 12, 1–5.

Kirman, A.P. 1992. Whom or what does the representative individual represent? *Journal of Economic Perspectives* 6(2), 117–36.

Krugman, P.R. 1995. Complexity and emergent structure in the international economy. In *New Directions in Trade Theory*, ed. A.V. Deardorff, J. Levinsohn and R.M. Stern. Ann Arbor: University of Michigan Press.

Matsuyama, K. 1995. Comment on P. Krugman, 'Complexity and Emergent Structure in the International Economy'. In *New Directions in Trade Theory*, ed. A.V. Deardorff, J. Levinsohn and R.M. Stern. Ann Arbor: University of Michigan Press.

Matsuyama, K. 2004. Financial market globalization, symmetry breaking and endogenous and endogenous inequality of nations. *Econometrica* 72, 853–84.

Matsuyama, K. 2006. The 2005 Lawrence R. Klein lecture: emergent class structure. *International Economic Review* 47, 327–60.

Matsuyama, K., Kiyotaki, N. and Matsui, A. 1993. Toward a theory of international currency. *Review of Economic Studies* 60, 283–307.

Menger, C. 1892. On the origins of money. *Economic Journal* 2, 239–255.

Newman, M.E.J., Strogatz, S.H. and Watts, D.J. 2001. Random graphs with arbitrary degree distribution and their applications. *Physical Review E* 64, 026118.

Papageorgiou, Y.Y. and Smith, R.S. 1983. Agglomeration as a local instability of spatially uniform steady-states. *Econometrica* 51, 1109–19.

Schelling, T.C. 1978. *Micromotives and Macrobehavior*. New York: Norton.

Stanford Encyclopedia of Philosophy. 2002. Emergent properties. Online. Available at http://plato.stanford.edu/entries/properties-emergent, accessed 17 November 2006.

Weitzman, M.L. 1994. Monopolistic competition with endogenous specialization. *Review of Economic Studies* 61, 45–56.

Young, H.P. 1998. *Individual Strategy and Social Structure: An Evolutionary Theory of Institutions*. Princeton: Princeton University Press.

emerging markets

'Emerging markets' are countries or markets that are not well established economically and financially, but are making progress in that direction.

The growing focus on emerging markets follows exciting developments during the second half of the 20th century – the emergence of a growing class of (formerly) poor countries that took off, and managed to close half of their income gap with the OECD countries within a generation or two. Remarkably, from 1960 to 1989 seven high-performing Asian economies (HPAEs) experienced unprecedented growth rates of the real GDP per capita in the range of four to seven per cent. This phenomenon has been the focus of a notable research report by the World Bank (1992), whose title *The East Asian Miracle* suggests a possible, though controversial, interpretation. The big story of recent years has been that the two most populous countries, China and India, joined the HPAE club. With few exceptions (such as Chile and Botswana), the club of high-performing emerging markets is fairly concentrated in East Asia. The HPAEs' remarkable growth rates during recent decades imply a sizable drop in global poverty rates, also entailing greater concentration of the incidence of extreme poverty, mostly in Africa (see Fischer, 2003). Yet the emerging markets phenomenon goes well beyond Asia, encompassing a growing share of developing countries that are closing, though at a lower rate than the HPAEs, their income gap with the OECD countries.

These developments were in sharp contrast to the pessimistic predictions made in the 1950–60s by several influential economic growth models (for a review, see Easterly, 1999). The HPAE experience dispelled most of these fears. The superior performance of the HPAEs illustrated that the fast growth option is viable, raising pertinent questions, and stirring a lively debate. While the World Bank (1992) dubbed the experience of the HPAEs a 'miracle', Young (1995) questioned this 'miraculous' interpretation, arguing that it is in line with Solow's growth model. Specifically, he reasoned that most of the growth has been the outcome of very high rates of investment in tangible and human capital, and a sizable increase in labour market participation. Controlling for these factors, Young found that the HPAEs' total factor productivity growth is in line with the historical experience of other countries. The debate about the role of accumulation in accounting for the HPAE experience is not over, yet the large drop of the growth rate of Japan in the 1990s, and the East Asian financial crisis of 1997, somehow deflated the 'East Asian miracle' hypothesis, suggesting the onset of Solow's growth convergence. Even if Young's thesis is correct, the speed and relative smoothness of the convergence of the HPAEs to the OECD's development level are without precedent. It raises questions about the obstacles preventing other countries from accomplishing this task, and about the ways to facilitate the take-off process in other regions.

The HPAE take-offs have been associated with fast growth of exports climbing, over time, the technology ladder of trade. This led to a lively debate about the importance of exports as the engine of growth: is the dominant causal association from exports to growth or vice versa? Earlier studies inferred that trade liberalization enhances growth (Ben-David, 1993; Edwards, 1998), a point disputed by Rodríguez and Rodrik (2001). Several authors revisited this issue, applying better controls, inferring strong growth effects of trade openness. Frankel and Romer (1999) applied measures of the geographic component of countries' trade to obtain instrumental variables estimates of the effect of trade on income. They inferred that ordinary least square (OLS) estimates understate the effects of trade, and that trade has a significant large positive effect on income. The contrast between the economic performance of the Soviet Union and that of China in the second part of the 20th century suggests another advantage of export orientation: it imposes a powerful market test on domestic output. Since exports must meet the quality and pricing tests of the global market, export-led growth limits potential distortions induced by 'growth promoting' domestic policies. Specifically, it prevents Soviet Union-type superficial economic growth induced by forced investment, growth that may result in inferior products that would be wiped out in the absence of protection. Export-oriented growth also forces countries to move faster towards the technological frontier in order to survive competitive global pressures.

Some of the obstacles preventing countries from taking off arise from political economy factors. Specifically, as growth is frequently associated with the emergence of new sectors and new elites, incumbent policymakers opt to block development in an attempt to preserve their rents and their grip on power. This phenomenon was vividly illustrated at the micro level by De Soto

(1989), and was shown to be a major impediment to growth (see Parente and Prescott, 2005). As the burden of the low growth would mostly affect future generations, the low growth equilibrium may persist with limited opposition. Proponents of this view point out that free commerce, both internal (between provinces or states in a union) and international, provides a powerful constraint on an incumbent's ability to block development.

The importance of external financing and financial integration in the development process remains a hotly debated topic. Advocates of financial liberalization in the early 1990s argued that external financing would alleviate the scarcity of saving in developing countries, inducing higher investments and thus higher growth rates. In contrast, Rodrik (1998) and Stiglitz (2002) questioned the gains from financial liberalization. Indeed, the 1990s experience with financial liberalization suggests that the gains from external financing are overrated – the bottleneck inhibiting economic growth is less the scarcity of saving and more the scarcity of good governance. This can be illustrated by tracing the patterns of self-financing ratios, measuring the share of tangible capital financed by past national saving (see Aizenman, Pinto and Radziwill, 2004). Higher self-financing rates of the nation's stock of capital are associated with a significant *increase* in growth rates. Remarkably, the wave of financial reforms in the 1990s led to deeper diversification, where greater inflows from the OECD financed comparable outflows from developing countries, with little effect on the availability of resources to finance tangible investment.

These findings are consistent with several interpretations. The first deals with risk: agents in various countries may react to exposure to financial risk differently. The desire to diversify these risks may lead to two-way capital flows, with little change in net positions (see Dooley, 1988). The ultimate obstacles limiting external financing may be related to acute moral hazard and agency problems – sovereign states, decision makers and corporate insiders pursue their own interests at the expense of outside investors (see Gertler and Rogoff, 1990; Stulz, 2005). An alternative interpretation follows Caroll and Weil (1994), who found that statistical causality runs from higher growth rates to higher saving rates. They conjectured that the growth-saving causality may be explained by habit formation, where consumers' utility depends on both present and past consumption. 'Habit formation', however, may be observationally equivalent to adaptive learning in the presence of uncertainty – in countries where private savings are taxed in arbitrary and unpredictable ways, credibility must be acquired as an outcome of a time-consuming learning process. In these circumstances, a higher growth rate provides a positive signal about the competence and the intentions of the administration, increasing saving and investment over time. Consequently, agents in countries characterized by greater political instability and polarization would be more cautious in increasing their saving and investment rates following a reform. Hence, accomplishing take-offs in Latin America may be much harder than in Asia, explaining Latin America's relatively low growth rate. (Various studies pointed out that policy uncertainty and political instability reduce private investment and growth; see Ramey and Ramey, 1995; Aizenman and Marion, 1999).

I close this review with an outline of open issues. The positive association between the equality of institutions and growth is well documented, yet the precise role of institutions in the development process remains debatable. Acemoglu et al. (2003) inquired how the colonial history of a developing country affects the quality of institutions, concluding that distortionary macroeconomic policies are more likely to be symptoms of underlying institutional problems rather than the main causes of economic volatility. Yet this interpretation does not satisfactorily explain the role of institutions in the growth process. The remarkable take-offs of China and India in recent decades, episodes directly affecting about a third of the global population, cannot obviously be explained by reference to institutional changes. This suggests that there is no simple correspondence or causality between growth and institutions. A tentative answer is provided by Rodrik (1999), who identifies a nonlinear interaction between shocks, polarization of a society and the quality of institutions. This argument suggests the key importance of the capacity of societies to adjust policies to shocks. A deeper understanding of the interaction between history, geography, polarization and institutions remains a challenge awaiting future research.

The exciting developments associated with the emergence of a growing class of (formerly) poor countries that took off implies that the rewards for adopting the proper growth incentives are high. A remaining challenge is how to facilitate the widening of the emerging market club, and how to minimize the prospects of new conflicts associated with the emergence of new economic powers like China and India.

JOSHUA AIZENMAN

See also **development economics; growth and institutions; growth and international trade; Solow, Robert.**

Bibliography

Acemoglu, D., Johnson, S., Robinson, J. and Thaicharoen, Y. 2003. Institutional causes, macroeconomic symptoms: volatility, crises and growth. *Journal of Monetary Economics* 50, 49–123.

Aizenman, J. and Marion, N. 1999. Volatility and investment: interpreting evidence from developing countries. *Economica* 66, 157–79.

Aizenman, J., Pinto, B. and Radziwill, A. 2004. Sources for financing domestic capital – is foreign saving a viable option for developing countries? Working Paper No. 1007. Department of Economics, University of California, Santa Cruz.

Ben-David, D. 1993. Equalizing exchange: trade liberalization and income convergence. *Quarterly Journal of Economics* 108, 653–79.

Caroll, C. and Weil, D. 1994. Saving and growth: a reinterpretation. *Carnegie Rochester Conference Series* 40, 133–92.

De Soto, H. 1989. *The Other Path.* New York: Harper and Row.

Dooley, M. 1988. Capital flight: a response to differences in financial risks. *IMF Staff Papers* 35, 422–36.

Easterly, W. 1999. The ghost of financing gap: testing the growth model used in the international financial institutions. *Journal of Development Economics* 60, 423–38.

Edwards, S. 1998. Openness, productivity and growth: what do we really know? *Economic Journal* 108, 383–98.

Fischer, S. 2003. Globalization and its challenges. *American Economic Review* 93, 1–30.

Frankel, J. and Romer, D. 1999. Does trade cause growth? *American Economic Review* 89, 379–99.

Gertler, M. and Rogoff, K. 1990. North-South lending and endogenous domestic capital market inefficiencies. *Journal of Monetary Economics* 26, 245–66.

Parente, S. and Prescott, E. 2005. A unified theory of the evolution of international income levels. In *Handbook of Economic Growth*, vol. 1, ed. P. Aghion and S. Durlauf. Amsterdam: North-Holland.

Ramey, G. and Ramey, V. 1995. Cross-country evidence on the link between volatility and growth. *American Economic Review* 85, 1138–51.

Rodríguez, F. and Rodrik, D. 2001. Trade policy and economic growth: a skeptic's guide to the cross-national evidence, *Macroeconomics Annual 2000*, ed. B Bernanke and K. Rogoff. Cambridge, MA: MIT Press for NBER.

Rodrik, D. 1998. Who needs capital-account convertibility? In *Should the IMF Pursue Capital Account Convertibility? Essays in International Finance*, No. 207, ed. P. Kenen. Princeton, NJ: Princeton University Press.

Rodrik, D. 1999. Where did all the growth go? External shocks, social conflict, and growth collapses. *Journal of Economic Growth* 4, 358–412.

Stiglitz, J. 2002. *Globalization and Its Discontents.* New York: W. W. Norton.

Stulz, R. 2005. The limits of financial globalization. *Journal of Finance* 60, 1595–638.

World Bank. 1992. *The East Asian Miracle.* Washington, DC: World Bank.

Young, A. 1995. The tyranny of numbers: confronting the statistical realities of the East Asian growth experience. *Quarterly Journal of Economics* 110, 641–80.

empirical likelihood

1 Introduction

Empirical likelihood (EL) is a method for estimation and inference without making distributional assumptions. The main feature of EL is the use of a discrete distribution to approximate the unknown distribution function nonparametrically, where the approximating discrete distribution is typically supported by empirical observations. Owen (1988) and subsequent papers considered applications of this approach to moment condition models. Their important discovery is that EL, which can be interpreted as a nonparametric maximum likelihood estimation (NPMLE) method, possesses many desirable asymptotic properties that are analogous to those of parametric likelihood procedures. To describe more details of empirical likelihood, consider i.i.d. data $\{z_i\}_{i=1}^n$, where each z_i is distributed according to an unknown probability distribution F_0. Suppose the expectation of an $\mathbb{R}^q$-valued function $g(z, \theta_0)$, which is known up to the finite-dimensional parameter θ_0 in $\Theta \subset \mathbb{R}^k$, is restricted to be zero:

$$E[g(z, \theta_0)] = \int g(z, \theta_0) \mathrm{d}F_0(z) = 0. \quad (1.1)$$

Let Δ denote the simplex $\{(p_1, \ldots, p_n) : \sum_{i=1}^n p_i = 1, 0 \le p_i, \ i = 1, \ldots, n\}$. Each vector $(p_1, \ldots, p_n) \in \Delta$ 'parametrizes' the unknown distribution F_0 by $F_n(z) = \sum_{i=1}^n p_i 1\{z_i \le z\}, z \in \mathbb{R}$ ($1\{\cdot\}$ signifies the usual indicator function). This is the approximating discrete distribution mentioned above. The nonparametric log-likelihood function to be maximized is

$$\ell_{\mathrm{NP}} = \sum_{i=1}^n \log p_i, \quad \sum_{i=1}^n g(z_i, \theta) p_i = 0,$$

$$(p_1, \ldots, p_n) \in \Delta, \theta \in \Theta.$$

Let $(\hat{\theta}_{\mathrm{EL}}, \hat{p}_{\mathrm{EL}1}, \ldots, \hat{p}_{\mathrm{EL}n})$ denote the value of $(\theta, p_1, \ldots, p_n) \in \Theta \times \Delta$ that maximizes ℓ_{NP}. This is called the (maximum) empirical likelihood estimator. The NPMLE for θ and F are $\hat{\theta}_{\mathrm{EL}}$ and $\hat{F}_{\mathrm{EL}} = \sum_{i=1}^n \hat{p}_{\mathrm{EL}i} 1\{z_i \le z\}$. One might expect that the high dimensionality of the parameter space $\Theta \times \Delta$ makes the above maximization problem intractable for any practical application. Fortunately, that is not the case, if one uses the following nested procedure. First, fix θ at a value in Θ and consider the log-likelihood with the parameters $(p_1, \ldots, p_n)$ 'profiled out':

$$\ell(\theta) = \max \ell_{\mathrm{NP}}(p_1, \ldots, p_n)$$

$$\text{subject to } \sum_{i=1}^n p_i = 1, \sum_{i=1}^n p_i g(z_i, \theta) = 0.$$

$$(1.2)$$

A straightforward application of the Lagrange multiplier method shows that $\ell(\theta)$ is represented by

$$\ell(\theta) = \min_{\gamma \in \mathbb{R}^q} - \sum_{i=1}^n \log(1 + \gamma' g(z_i, \theta))) - n \log n$$

$$(1.3)$$

(see, for example, Kitamura, 2006). The numerical evaluation of the function $\ell(\cdot)$ is easy, because (1.3) is a low-dimensional convex maximization problem, for which a simple Newton algorithm works. Second, obtain the empirical likelihood estimator $\hat{\theta}_{\mathrm{EL}}$ as the maximizer of (1.3). The maximization of $\ell(\theta)$ with respect to θ is typically carried our using a nonlinear optimization algorithm.

Basic properties of the empirical likelihood procedure are now well-understood. The EL estimator $\hat{\theta}_{\mathrm{EL}}$ is $n^{1/2}$-consistent and asymptotically normal. Let D and S denote $E[\nabla_\theta g(z, \theta_0)]$ and $E[g(z, \theta_0)g(z, \theta_0)']$, then its asymptotic distribution is given by $N(0, (D'SD)^{-1})$. Also, suppose R is a known $\mathbb{R}^s$-valued function of θ, and the econometrician poses a hypothesis that θ_0 is restricted as $R(\theta_0) = 0$, where the s restrictions are independent. This can be tested by forming a nonparametric analogue of the parametric likelihood ratio statistic. Let $r = -2(\sup_{\theta:R(\theta)=0} \ell(\theta) - \sup_{\theta \in \Theta} \ell)$, then this obeys the chi-square distribution with s degrees of freedom asymptotically under the null. The factor r is called the empirical likelihood ratio (ELR) statistic. ELR also applies to testing overidentifying restrictions: see Section 2. These properties and other basics of EL and related methods have been studied extensively in the literature (see Qin and Lawless, 1994; Imbens, 1997; Kitamura, 1997; Kitamura and Stutzer, 1997; Smith, 1997; Imbens, Spady and Johnson, 1998; Newey and Smith, 2004).

An alternative way to motivate EL is to use a minimum divergence estimation framework. Let f and g denote the density functions or the probability functions of distribution functions F and G. Define a 'divergence measure' between F and G to be

$$D(F, G) = \int \phi\left(\frac{f(z)}{g(z)}\right) g(z) \mathrm{d}z, \quad (1.4)$$

for a convex function ϕ. It is easy to see that $D(\cdot, G)$ is minimized at G. Let

$$\mathscr{F}(\theta) = \left\{F : \int g(z, \theta) \mathrm{d}F = 0, \quad F \text{ is a CDF}\right\}.$$

Then $\mathscr{F} = \cup_{\theta \in \Theta} \mathscr{F}(\theta)$ is the set of all probability distributions that are compatible with the moment restriction (1.1). Now consider the problem of minimizing the divergence $D(F, F_0)$ with respect to $F \in \mathscr{F}$. In other words, a distribution that is 'closest' to the true distribution F_0 in the class of distributions $\mathscr{F}$ is sought. Pick a value $\theta \in \Theta$ and define

$$\nu(\theta) = \inf_F D(F, F_0) \quad \text{subject to}$$
$$\int g(z, \theta) f \mathrm{d}z = 0, \int f \mathrm{d}z = 1. \tag{P}$$

The value $\nu(\theta)$ is regarded as the minimum divergence between F_0 and the set of distributions that satisfy the moment restriction with respect to $g(z, \theta)$. The non-negativity of f is maintained if ϕ is modified so that $\phi(z) = \infty$ for $z < 0$ (see Borwein and Lewis, 1991). The primal problem (P) has a dual problem

$$\nu^*(\theta) = \max_{\lambda \in \mathbb{R}, \gamma \in \mathbb{R}^q} \left[\lambda - \int \phi^*(\lambda + \gamma' g(z, \theta)) \mathrm{d}F_0(z) \right], \tag{DP}$$

where ϕ^* is the convex conjugate (or the Legendre transformation) of ϕ, that is $\phi^*(y) = \sup_x[xy - \phi(x)]$. (DP) is a finite-dimensional unconstrained convex maximization problem. The Fenchel duality theorem implies that $\nu(\theta) = \nu^*(\theta)$. Since the true value θ_0 minimizes $\nu(\theta)$ over Θ, it follows that

$$\theta_0 = \operatorname{argmin}_{\theta \in \Theta} \nu^*(\theta). \tag{1.5}$$

Note that the integral in the definition of ν^* is the expected value of $\phi^*(\lambda + \gamma' g(z, \theta))$ with respect the true distribution F_0, which is unknown in practice. A feasible procedure is obtained by replacing the expectation with the sample average, that is

$$\hat{\nu}^*(\theta) = \max_{\lambda \in \mathbb{R}, \gamma \in \mathbb{R}^q} \left[\lambda - \frac{1}{n} \sum_{i=1}^n \phi^*(\lambda + \gamma' g(z_i, \theta)) \right]. \tag{1.6}$$

Corresponding to (1.5), an appropriate minimum distance estimator takes the form

$$\hat{\theta} = \operatorname{argmin}_{\theta \in \Theta} \hat{\nu}^*(\theta).$$

This minimum divergence framework yields empirical likelihood as a special case with $\phi(x) = -\log(x)$ (or equivalently, $\phi^*(x) = -1 - \log(-y)$). Other choices for ϕ are, of course, possible. For example, $\phi(x) = x \log(x)$ yields the 'exponential tilt' estimator (Kitamura and Stutzer, 1997), while $\phi(z) = \frac{1}{2}(x^2 - 1)$ corresponds to the continuous updating GMM estimator (CUE) (Hansen, Heaton and Yaron, 1996). A convenient parametric family of convex functions known as the Cressie–Read family (Read and Cressie, 1988) subsumes these three important cases. If ϕ belongs to the Cressie–Read family, one can show that the minimum divergence estimator can

be written as

$$\hat{\theta} = \operatorname{argmin}_{\theta \in \Theta} \max_{\gamma \in \mathbb{R}^q} \left[\frac{1}{n} \sum_{i=1}^n \kappa(\gamma' g(z_i, \theta)) \right] \tag{1.7}$$

where $\kappa(y) = -\phi^*(y + 1)$. This is essentially equivalent to the generalized empirical likelihood (GEL) estimator by Smith (1997). Smith (2004) provides a detailed account for GEL.

2 EL and the large deviation principle

Like the conventional asymptotic method, the large deviation principle (LDP) offers first order approximations for various estimators and tests. Unlike the conventional theory, which produces local linear approximations, the LDP provides global nonlinear approximations. It is the latter feature that enables the LDP to yield results not obtained by the conventional linear approximations. For example, the LDP shows that EL enjoys many optimality properties that are not shared by, for example, the conventional GMM estimator.

To introduce the concept of the LDP in the context of moment condition models, suppose the econometrician observes i.i.d. data $(z_1, \ldots, z_n)$, where z_i satisfies the restriction (1.1). Let A_n be an event as a result of estimation or testing: for example, if one uses an estimator θ_n to estimate θ_0, one may consider $A_n = 1\{\|\theta_n - \theta_0\| > c\}$ for a constant c. Then $\Pr\{A_n\}$ is the probability of the estimator missing the true value by a margin larger than c. Or, in testing a null hypothesis H_0, A_n can represent the event that H_0 is accepted. If the null is incorrect, $\Pr\{A_n\}$ is the probability of type II errors. In either way, $\lim_{n\to\infty} \Pr\{A_n\} = 0$ if the estimator or the test is consistent. The LDP also deals with asymptotic properties, but it is concerned with the limit of the form $\lim_{n\to\infty} \frac{1}{n} \log \Pr\{A_n\}$. (If the limit does not exist, one needs to consider lim inf or lim sup, depending on the purpose of analysis.) Let $-d \leq 0$ denote the above limit so that $\Pr\{A_n\} \approx e^{-nd}$, which characterizes how fast $\Pr\{A_n\}$ decays. The goal is to obtain a procedure that maximizes the speed of decay d.

Kitamura and Otsu (2005) study the estimation of models of the form (1.1) using the LDP. One complication in the application of the LDP to an estimation problem in general is that an estimator that maximizes the limiting decay rate d with $A_n = 1\{\|\theta_n - \theta_0\| > c\}$ uniformly in unknown parameters does not exist in general, unless the model belongs to the exponential family. A possible way around this issue is to pursue minimax optimality, rather uniform optimality. See Puhalskii and Spokoiny (1998) for a general discussion on such a minimax framework. Note that the probability of the event $A_n = 1\{\|\hat{\theta} - \theta_0\| > c\}$ depends on θ_0 and F_0, therefore the worst case scenario is given by the pair (allowed in the model (1.1)) that maximizes $\Pr\{A_n\}$. Suppose an estimator θ_n minimizes this worst-case probability, thereby achieving minimaxity. The limit inferior of the minimax probability provides an asymptotic minimax criterion. Kitamura and Otsu (2005) show that an estimator that attains the lower bound of the asymptotic minimax criterion can be obtained from the EL objective function $\ell(\theta)$ in (1.2) as follows:

$$\hat{\theta}_{\text{ld}} = \operatorname*{argmin}_{\theta \in \Theta} Q_n(\theta),\ Q_n(\theta) = \sup_{\theta^* \in \Theta: \|\theta^* - \theta\| > c} \ell(\theta^*).$$

Calculating $\hat{\theta}_{\text{ld}}$ in practice is straightforward. If the dimension of θ is high, it is also possible to focus on a low-dimensional sub-vector of θ and obtain a large deviation minimax estimator for it, treating the rest as nuisance parameters.

Kitamura (2001) shows that empirical likelihood dominates other methods in terms of the LDP when applied to overidentifying restrictions testing. Researchers routinely test overidentifying restrictions of the form

$$\int g(z, \theta)\mathrm{d}F = 0 \quad \text{for some } \theta \in \Theta \text{ and}$$

$$\text{for some distribution function } F,$$

$$\text{(O)}$$

with $\dim(\Theta) = k$ and $g \in \mathbb{R}^q$, $q > k$. The log empirical likelihood under the restriction (O) is $\sup_{\theta \in \Theta} \ell(\theta)$; without the restriction, it is $-n \log n$. The ELR test statistic for (O) is the difference of the two multiplied by -2. It is asymptotically distributed according to the χ^2 distribution with $q-k$ degrees of freedom under (O) (Qin and Lawless, 1994). Using the notation in the previous section, rewrite the above null in an equivalent form: $(\text{O})' : F_0 \in \mathscr{F}$. It turns out that ELR for $(\text{O})'$ has a property of being uniformly most powerful in an LDP criterion. To state this optimality property of ELR formally, let $\mathscr{F}$ denote the set of all probability distribution functions. Practically all reasonable tests for (O) (or $(\text{O})'$) can be represented by a partition $\Omega = (\Omega_1, \Omega_2)$ of $\mathscr{F}$, such that if the empirical distribution function F_n falls into Ω_1 (Ω_2) one rejects (accepts) (O). It is a straightforward exercise to show that the ELR test rejects the null if the Kullback–Leibler divergence $K(F_n, G)$ between F_n and G, minimized over $G \in \mathscr{F}$, is too large. Therefore ELR is represented by the following partition of $\mathscr{F}$: $\Lambda = (\Lambda_1, \Lambda_2)$, $\Lambda_1 = \{F : \inf_{G \in \mathscr{F}} K(F, G) < \eta\}$, $\Lambda_2 = \Lambda_1^c$ for a positive number η. Following Owen (2001), for an event A_n that involves observations $z_1, \ldots, z_n$ that are randomly sampled from F, let $\Pr\{A_n; F\}$ denote the probability of the event. By applying a mathematical result called Sanov's theorem, it can be shown that

$$\sup_{F^* \in \mathscr{F}} \limsup_{n \to \infty} \frac{1}{n} \log \Pr\{F_n \in \Lambda_2; F^*\} \leq -\eta.$$

Kitamura (2001) also shows that if the following inequality holds for a test $\Omega = (\Omega_1, \Omega_2)$ that satisfies some regularity conditions (see Kitamura, 2001, for the regularity conditions):

$$\sup_{F^* \in \mathscr{F}} \limsup_{n \to \infty} \frac{1}{n} \log \Pr\{F_n \in \Omega_2; F^*\} \leq -\eta,$$

then it must be that

$$\limsup_{n \to \infty} \frac{1}{n} \log \Pr\{F_n \in \Omega_1; F^{**}\}$$

$$\geq \limsup_{n \to \infty} \frac{1}{n} \log \Pr\{F_n \in \Lambda_1; F^{**}\}$$

for every $F^{**} \notin \mathscr{F}$. The first two of the above three inequalities mean that the ELR test Λ and the arbitrary regular test Ω are comparable in terms of its LDP property of type I error probabilities. But the third inequality implies that the ELR test is no less powerful than the arbitrary test if the LDP of type II error probabilities are used to measure the asymptotic powers of the tests. Note that the third inequality holds for every $F^{**} \notin \mathscr{F}$: that is, it holds uniformly over alternatives. Since the test (Ω_1, Ω_2) is arbitrary, this shows that ELR is uniformly most powerful in an LDP sense. Such a property is sometimes referred to as the Generalized Neyman–Pearson (GNP) optimality.

3 Higher-order asymptotics

An alternative way to see why EL works well is to analyse it using higher-order asymptotics. Newey and Smith (2004) investigate higher-order properties of the GEL family of estimators. To illustrate their findings, it is instructive to look at the first-order condition that the EL estimator satisfies, that is $\nabla_\theta \ell(\hat{\theta}_{\text{EL}}) = 0$. A straightforward calculation shows that this condition, using the notation $\hat{D}(\theta) = \sum_{i=1}^n \hat{p}_{\text{EL}i} \nabla_\theta g(z_i, \theta)$ and $\hat{S}(\theta) = \sum_{i=1}^n \hat{p}_{\text{EL}i} g(z_i, \theta) g(z_i, \theta)'$, can be written as

$$\hat{D}(\hat{\theta}_{\text{EL}})' \hat{S}^{-1}(\hat{\theta}_{\text{EL}}) \bar{g}(\hat{\theta}_{\text{EL}}) = 0; \quad (3.1)$$

see Theorem 2.3 of Newey and Smith (2004). The factor $\hat{D}(\hat{\theta}_{\text{EL}})' \hat{S}^{-1}(\hat{\theta}_{\text{EL}})$ can be interpreted as a feasible version of the optimal weight for the sample moment $\bar{g}(\theta) = \frac{1}{n}\sum_{i=1}^n g(z_i, \theta)$. Equation (3.1) is similar to the first-order condition for GMM, though there are important differences. Notice that the Jacobian term D and the variance term S are estimated by $\hat{D}(\hat{\theta}_{\text{EL}})$ and $\hat{S}(\hat{\theta}_{\text{EL}})$ in (3.1). It can be shown that these are semiparametrically efficient estimators of D and S under the moment restriction (1.1). This means that they are asymptotically uncorrelated with $\bar{g}(\theta_0)$, removing the important source of the second-order bias of GMM. Moreover, the EL estimator does not involve a preliminary estimator, thereby eliminating another source of the second-order bias in GMM. Newey and Smith (2004) formalize this intuition and obtain an important conclusion that the second-order bias of the EL estimator is equal to that of the infeasible method-of-moments estimator that optimally weights $\bar{g}$ by the unknown factor $D'S^{-1}$. In contrast, the first-order condition of GMM takes a similar form, but the terms that correspond to D and S are inefficiently estimated, causing bias. Newey and Smith (2004) note that the first-order conditions of GEL estimators have a form where D is efficiently estimated but S is not, leaving a source of bias that is not present for EL.

Higher-order properties of ELR tests have been studied in the literature as well. One of the significant findings in the early literature of empirical likelihood is the Bartlett correctability of the empirical likelihood ratio test, discovered by DiCiccio, Hall and Romano (1991). Consider the ELR test statistic for $H_0: \theta = \theta_0$ in the model (1.1) with $q = k$. DiCiccio, Hall and Romano (1991) show that the accuracy of the χ^2 asymptotic approximation for the distribution of the ELR statistic can be improved from the rate n^{-1} to the much faster rate n^{-2} by multiplying it by a factor called the Bartlett coefficient.

4 Some variations of EL

EL is applicable to many problems other than (1.1), but they sometimes require extending and modifying the standard EL method described so far. For example, suppose economic theory implies that the conditional mean of $g(z, \theta_0)$ given a vector of covariates x is zero:

$$E[g(z, \theta_0)|x] = 0. \qquad (4.1)$$

This restriction is stronger than (1.1). Though one can choose an arbitrary function $a(x)$ of x as an instrument, this can be problematic since (a) choosing an instrument that delivers strong identification may be a difficult task, and (b) an arbitrary instrument does not achieve efficiency in general. Kitamura, Tripathi and Ahn (2004) use the kernel regression technique to incorporate the information in the conditional moment restriction into empirical likelihood. Their estimator achieves the semiparametric efficiency bound of the model (4.1) under weak regularity conditions. While there exist estimators that achieve efficiency in the model, the EL-based estimator has an advantage that finding a preliminary estimator that is consistent is not necessary. A simulation study in Kitamura, Tripathi and Ahn (2004) indicates that the conditional EL estimator and tests based on it work remarkably well in finite samples. Donald, Imbens and Newey (2003) propose an alternative estimator for (4.1). Their idea is to use a sequence functions of x as a vector of instruments, then apply EL to the resulting unconditional moment restriction model. By letting the dimension of the instrument vector grow with the sample size in such a way that it spans the 'optimal

instrument' asymptotically, their procedure also achieves the semiparametric efficiency bound.

A topic that is closely related to the above is non-parametric specification testing. Suppose, for example, one is interested in testing the specification of a parametric regression model $E[y|x] = m(x, \theta_0)$, where m is parametrized by a vector $\theta_0 \in \Theta$. The null hypothesis of correct specification can be written in terms of a conditional moment restriction for the function $g(z, \theta) = y - m(x, \theta), z = (x', y)'$:

$$E[g(z, \theta)|x] = 0 \text{ for some } \theta \in \Theta. \qquad (C)$$

Tripathi and Kitamura (2003) shows that a conditional version of the ELR test applies to the above problem. They propose a simple procedure: reject (C) if the maximized value of the conditional empirical likelihood function, which is essentially the one used in Kitamura, Tripathi and Ahn (2004), is too small. They also calculate the asymptotic power of their test. Their analysis shows that the EL-based testing procedure has an asymptotic optimality property in terms of an average power criterion.

Another example in which EL needs an appropriate modification is a time series model. Suppose the researcher observes a strictly stationary and weakly dependent time series $\{z_1, \ldots, z_t\}$, and each z_t satisfies the moment condition $E[g(z_t, \theta_0)] = 0, \theta_0 \in \Theta$. Applying EL to this model ignoring dependence is inappropriate; it leads to efficiency loss, and the chi-square asymptotics of the ELR test break down.

There are at least three alternative ways to deal with the problem caused by dependence. The first approach is to parametrize the dynamics using a reduced form time series model such as a vector autoregression (VAR) model (Kitamura, 2006). While straightforward, this approach involves the risk of mis-specifying the dynamics, and reduces the appeal of EL as nonparametric likelihood. The second approach is the blocking method proposed by Kitamura and Stutzer (1997) and Kitamura (1997). The idea is to form data blocks by taking consecutive observations, and apply EL to them. This is termed blockwise empirical likelihood (BEL). BEL preserves the dependence information in the data, in a fully nonparametric manner. The third approach is a hybrid of the first and the second approaches (Kitamura, 2006). That is, one applies a low order parametric filter to lessen the degree of dependence in the data, then applies BEL to the filtered data. While this does not change the desirable asymptotic property of BEL, it appears to have advantages in finite samples when applied to a time series that is highly persistent.

YUICHI KITAMURA

See also **generalized method of moments estimation; semiparametric estimation; vector autoregressions.**

Borwein, J.M. and Lewis, A.S. 1991. Duality relationships for
entropy-type minimization problems. *SIAM Journal of
Control and Optimization* 29, 325–38.

DiCiccio, T., Hall, P. and Romano, J. 1991. Empirical
likelihood is Bartlett-correctable. *Annals of Statistics* 19,
1053–61.

Donald, S.G., Imbens, G.W. and Newey, W.K. 2003.
Empirical likelihood estimation and consistent tests with
conditional moment restrictions. *Journal of Econometrics*
117, 55–93.

Hansen, L.P., Heaton, J. and Yaron, A. 1996. Finite-sample
properties of some alternative GMM estimators. *Journal
of Business and Economic Statistics* 14, 262–80.

Imbens, G.W. 1997. One-step estimators for over-identified
generalized method of moments models. *Review of
Economic Studies* 64, 359–83.

Imbens, G.W., Spady, R.H. and Johnson, P. 1998.
Information theoretic approaches to inference in
moment condition models. *Econometrica* 66, 333–57.

Kitamura, Y. 1997. Empirical likelihood methods with
weakly dependent processes. *Annals of Statistics* 25,
2084–102.

Kitamura, Y. 2001. Asymptotic optimality of empirical
likelihood for testing moment restrictions. *Econometrica*
69, 1661–72.

Kitamura, Y. 2006. Empirical likelihood methods in
econometrics: theory and practice. In *Advances in
Economics and Econometrics: Theory and Applications,
Ninth World Congress*, ed. R. Blundell, W.K. Newey and
T. Persson. Cambridge: Cambridge University Press.

Kitamura, Y. and Otsu, T. 2005. Minimax estimation and
testing for moment condition models via large
deviations. Manuscript, Department of Economics, Yale
University.

Kitamura, Y. and Stutzer, M. 1997. An information theoretic
alternative to generalized method of moments
estimation. *Econometrica* 65, 861–74.

Kitamura, Y., Tripathi, G. and Ahn, H. 2004. Empirical
likelihood based inference in conditional moment
restriction models. *Econometrica* 72, 1667–714.

Newey, W.K. and Smith, R.J. 2004. Higher order properties
of GMM and generalized empirical likelihood estimators.
Econometrica 72, 219–55.

Owen, A. 1988. Empirical likelihood ratio confidence
intervals for a single functional. *Biometrika* 75, 237–49.

Owen, A. 2001. *Empirical Likelihood*. New York: Chapman
and Hall/CRC.

Puhalskii, A. and Spokoiny, V. 1998. On large-deviation
efficiency in statistical inference. *Bernoulli* 4, 203–72.

Qin, J. and Lawless, J. 1994. Empirical likelihood and general
estimating equations. *Annals of Statistics* 22, 300–25.

Read, T.R.C. and Cressie, N.A.C. 1988. *Goodness-of-Fit
Statistics for Discrete Multivariate Data*. Berlin: Springer.

Smith, R.J. 1997. Alternative semi-parametric likelihood
approaches to generalized method of moments
estimation. *Economic Journal* 107, 503–19.

Smith, R.J. 2004. GEL criteria for moment condition
models. Working paper, University of Warwick.

Tripathi, G. and Kitamura, Y. 2003. Testing conditional
moment restrictions. *Annals of Statistics* 31, 2059–95.

encompassing

Introduction and motivation

Imaginative and productive disciplines like economics
generate many new theories, partly to extend the range
of phenomena that they embrace but also to improve on
existing theories. New theories require rigorous evaluation to establish their worth if they are to be relevant, reliable, and robust. In addition to checking their
logical consistency and relevance it is important to assess
their coherence with observation. The latter usually
involves the development of a model that embodies the
essential characteristics of the theory and has observable
implications.

The analysis presented here concentrates on the
evaluation of empirical models. Numerous criteria have
been proposed for assessing the coherence of an empirical model with observation. Measures of goodness of fit
and selection criteria based on likelihood functions (usually degrees of freedom adjusted) are common (Schwarz,
1978), and are often used both to assess coherence with
observation and to select the preferred model. Probably
the most comprehensive and demanding criterion for
data coherence is that of congruence (Hendry, 1995;
Bontemps and Mizon, 2003), which requires a model to
be a valid reduction of whatever process actually generates the observed data – the data generation process
(DGP). When $\mathbf{x}_t$ contains the full set of variables involved
in an investigation, let the DGP be denoted by the joint
density $D_x(\mathbf{x}_t|\mathbf{X}_{t-1}, f)$ for $\mathbf{x}_t$ conditional on its history
$\mathbf{X}_{t-1}$ with parameters $\boldsymbol{\phi}$. Knowledge of the DGP endows
one with omniscience and in particular the ability to
derive the properties of all models involving the same
variables such as $f_x(\mathbf{x}_t|\mathbf{X}_{t-1}, x)$, but, alas, for practical
purposes it is unattainable. In empirical modelling,
therefore, congruence means that, given the available
information, the model is indistinguishable from the
DGP for the chosen variables, that is, no evidence has
been evinced that the model is not the DGP. Testing the
latter requires that extensive, not limited, searching is
done for evidence of non-congruence. This leads to the
adoption of statistical tests of model mis-specification
(for example, wrong functional form, heteroskedastic or
serially correlated residuals) as indirect but practical tests
of congruence (Hendry, 1995; Mizon, 1995). Since in
practice a congruent model will not be the DGP, it will
not necessarily be able to explain the properties of
other models, and in particular those that constitute the

current best knowledge and practice. Thus, a valuable part of the evaluation of a model is an assessment of whether it represents an advance on existing knowledge. 'The encompassing principle is concerned with the ability of a model to account for the behaviour of others, or less ambitiously, to explain the behaviour of relevant characteristics of other models' (Mizon, 1984, p. 136). A well-known illustration in physics, discussed by Okasha (2002), for example, is provided by Newton's laws of motion and gravitation that encompassed Kepler's laws of motion and gravitation as well as Galileo's law of free-fall, and as a result the same laws explained the motion of bodies in both the terrestrial and the celestial domains. This added credence to Newton's laws, as it does for all models that encompass their rivals. It was widely believed for a long time that Newton's theory revealed the workings of nature and had the ability to explain everything in principle. However, Newton's laws have been superseded or encompassed by Einstein's relativity theory and quantum mechanics. This illustrates the fact that modelling, like discovery, is not a once-for-all event, but a continuous process of development. Progress in science, however, is achieved in many ways, with confidence and persistence playing a role in some instances as a consequence of rejection not being accepted as final or corroboration of models that are subsequently superseded not being taken as definitive.

Background

The idea underlying the encompassing principle has a long pedigree; for example, the comparison of competing theories has been long recognized as a basic ingredient of a scientific research strategy (Nagel, 1961). The implementation via a statistical contrast equally has a long history; Cox (1961; 1962) are the most significant early examples. These papers introduced statistical tests for separate families of hypotheses, and discussed several examples to illustrate their practical relevance. The tests were later developed in the literature on non-nested hypothesis testing (Pesaran, 1974; Davidson and MacKinnon, 1981), and encompassing (Mizon, 1984). The latter paper contains a general presentation of the concept of encompassing and discussion of numerous applications, and Mizon and Richard (1986) provides a theoretical framework for encompassing, on which other theoretical papers have built extensions. Davidson et al. (1978) is one of the first attempts to develop a framework for a scientific comparison of alternative economic theories and econometric models implementing them. Different econometric models for the series of UK consumption, which rely on different economic hypotheses about consumption behaviour, were embedded in a general model and shown to imply different testable restrictions on its coefficients.

Distinguished natural scientists have expressed surprise that social scientists are able to learn anything from empirical observation when they rarely have experimental evidence. However, the encompassing principle provides precisely the analogue of the physical experiment. Experiments enable physicists and chemists to sift through alternative theories by evaluating the veracity of their implications or predictions in controlled conditions, and thus to eliminate those theories whose predictions perform badly. Congruence is the analogue of setting up controlled experimental conditions. The need to distinguish between alternative theories that each appear to be coherent with outcomes, experimental or non-experimental, leads to the search for dominant theories. For disciplines that are largely non-experimental, having a principle such as encompassing is essential for discriminating between alternative models. Typically, alternative empirical models use different information sets and possibly different functional forms, and are thus separate or non-nested. This non-nested feature enables more than one model to be congruent with respect to sample information – each can be congruent with respect to its own information set – and so it is important to assess their relative merits. Using the encompassing principle, Ericsson and Hendry (1999) analyse this issue and show that the corroboration of more than one model can imply the inadequacy of each, and Mizon (1989) provides an illustration by comparing a Keynesian and a monetarist model of inflation. Hence, congruence and encompassing are inextricably linked; in particular, encompassing comparisons of non-congruent models can be misleading. For example, general models will not always encompass simplifications of themselves even though that might seem to be an obvious characteristic of a general model, but a congruent general model will always encompass simpler models (Hendry, 1995; Gouriéroux and Monfort, 1995; Bontemps and Mizon, 2003).

Principle

Underlying all empirical econometric analyses is an information set (collection of variables or their sigma field), and a corresponding probability space. This information set has to be sufficiently general to include all the variables thought to be relevant to the empirical implementation of theoretical models in the form of statistical models. It is also important that this information set include the variables needed for all competing models that are to be compared. When these variables are $\mathbf{x}_t$ the DGP for the observed sample is the joint density $D_x(\mathbf{x}_t|\mathbf{X}_{t-1}, f)$ at the particular parameter value $\phi = \phi_0$. Let a parametric statistical model of the joint distribution be $M_f = \{f_x(\mathbf{x}_t|\mathbf{X}_{t-1}, x) \quad x \in \Xi \subset \mathrm{R}^k\}$. Let $\hat{x}$ be the maximum likelihood estimator of ξ so that $\hat{x} \xrightarrow[M_f]{P} x$ and $\hat{x} \xrightarrow[DGP]{P} x(f_0) = x_0$ which is the pseudo-true value of $\hat{x}$. Note that the parameters of a model are not arbitrary in that M_f and its parameterization ξ are chosen to correspond to phenomena of interest such as elasticities and partial responses within the chosen probability space. For the two alternative models $M_1 = \{f_1(\mathbf{x}_t|\mathbf{X}_{t-1}, q_1), q_1 \in \Theta_1 \subset R^{p_1}\}$ and $M_2 = \{f_2(\mathbf{x}_t|\mathbf{X}_{t-1}, q_2), q_2 \in \Theta_2 \subset R^{p_2}\}$ the concept of parametric encompassing, in accordance with the approach in Mizon (1984), Mizon and Richard (1986), and Hendry and Richard (1989), can be defined as follows. M_1 encompasses M_2 (denoted $M_1 \mathscr{E} M_2$) if and only if $q_{20} = \mathbf{h}_{21}(q_{10})$ when θ_{i0} is the pseudo-true value of the maximum likelihood estimator $\hat{q}_i$ of $q_i \quad i =_p 1, 2$, and $\mathbf{h}_{21}(q_{10})$ is the binding function given by $\hat{q}_2 \xrightarrow{M} \mathbf{h}_{21}(q_{10})$ (Mizon and Richard, 1986; Hendry and Richard, 1989; Gouriéroux and Monfort, 1995). Note that this definition of encompassing applies when M_1 and M_2 are non-nested as well as nested. However, Hendry and Richard (1989) showed that when M_1 and M_2 are non-nested $M_1 \mathscr{E} M_2$ is equivalent to M_1 being a valid reduction of the minimum completing model $M_c = M_1 \cup M_2^{\perp}$ (so that $M_1, M_2 \subset M_c$) when $M_2^{\perp}$ is the model which represents all aspects of M_2 that are not contained in M_1. When this condition is satisfied, M_1 is said to parsimoniously encompass M_c (denoted $M_1 \mathscr{E}_p M_c$). Parsimonious encompassing is the property that a model is a valid reduction of a more general model. When a general-to-simple modelling strategy is adopted, the general unrestricted model (GUM) will have been chosen to embed the different econometric models implementing rival economic theories for the phenomenon of interest. Hence searching for the model that parsimoniously encompasses the congruent GUM is an efficient way to find congruent and encompassing models in practice. Hendry and Krolzig (2003) describe and illustrate the performance of a computer program that implements a general-to-specific modelling strategy.

The comparison of Gaussian linear regression models provides a simple and convenient framework to illustrate the main ideas. Consider the two models M_1 and M_2 defined in:

$$
\begin{aligned}
M_1 &\curvearrowleft \mathbf{y} = \mathbf{Z}_1 b + \mathbf{u}_1, & \mathbf{u}_1 &\curvearrowleft N\left(0, s_1^2 \mathbf{I}_n\right) \\
M_2 &\curvearrowleft \mathbf{y} = \mathbf{Z}_2 g + \mathbf{u}_2, & \mathbf{u}_2 &\curvearrowleft N\left(0, s_2^2 \mathbf{I}_n\right) \\
M_c &\curvearrowleft \mathbf{y} = \mathbf{Z}_1 \mathbf{b} + \mathbf{Z}_2 \mathbf{c} + e & e &\curvearrowleft N\left(0, s_c^2 \mathbf{I}_n\right)
\end{aligned}
$$
$$(1)$$

when $\mathbf{y}$ is $n \times 1$, and $\mathbf{Z}_i$ is $n \times k_i (i = 1, 2)$ containing n observations on the independent and two sets of explanatory variables respectively with no variables in common. The explanatory variables are distributed independently of the error vectors $\mathbf{u}$, $\mathbf{v}$, and $\boldsymbol{\varepsilon}$. When M_1, M_2 and M_c are each hypotheses about the distribution of $y|z$, the models M_1 and M_2 are non-nested in that neither is a special case of the other, whereas both M_1 and M_2 are nested within M_c. A test of the hypothesis that M_1 encompasses M_2 (denoted $M_1 \mathscr{E} M_2$) is possible using the contrast $\hat{\psi}_\gamma = \hat{\gamma} - \hat{\gamma}_1 = (\mathbf{Z}_2' \mathbf{Z}_2)^{-1} \mathbf{Z}_2' \mathbf{Q}_1 \mathbf{y}$ with $\mathbf{Q}_1 = (\mathbf{I}_n - \mathbf{Z}_1 (\mathbf{Z}_1' \mathbf{Z}_1)^{-1} \mathbf{Z}_1')$ between the maximum likelihood estimator of γ,

$\hat{g} = (\mathbf{Z}_1' \mathbf{Z}_2)^{-1} \mathbf{Z}_2' \mathbf{y}$, and an estimate $\hat{g}_1 = (\mathbf{Z}_2' \mathbf{Z}_2)^{-1} \mathbf{Z}_2' (\mathbf{Z}_1' \mathbf{Z}_1)^{-1} \mathbf{Z}_1' \mathbf{y}$ of the pseudo-true value of $\hat{g}$ under M_1 given by $g_1 = p \lim_{n \to \infty} (\hat{g})$. The sample complete parametric encompassing^{M_1} test statistic is given by $\eta_c = \hat{\psi}_\gamma'(\mathbf{Z}_2' \mathbf{Z}_2)(\mathbf{Z}_2' \mathbf{Q}_1 \mathbf{Z}_2)^{-1}(\mathbf{Z}_2' \mathbf{Z}_2)\hat{\psi}_\gamma/k_2 \hat{\sigma}_c^2$ when $\hat{\sigma}_c^2 = \mathbf{y}'(\mathbf{I}_n - \mathbf{Z}(\mathbf{Z}'\mathbf{Z})^{-1}\mathbf{Z}')\mathbf{y}/(n - k_1 - k_2)$ is the unbiased estimator of σ_c^2 with $\mathbf{Z} = (\mathbf{Z}_1, \mathbf{Z}_2)$. Under the complete parametric encompassing hypothesis $H_c : \psi_\gamma = \gamma - \gamma_1 = 0$ the statistic η_c is distributed as $F(k_2, n - k_1 - k_2)$. Mizon and Richard (1986) showed that this is precisely the same statistic as that for testing the hypothesis $\mathbf{c} = 0$ in (1), that is, the test statistic for $M_1 \mathscr{E} M_2$ is exactly the same as that for $M_1 \mathscr{E}_p M_c$ in this case. Variance encompassing is based on the contrast $\hat{\psi}_{\sigma_2^2} = \hat{\sigma}_2^2 - \hat{\sigma}_{21}^2$ between $\hat{\sigma}_2^2$ and an estimator of $\sigma_{21}^2 = \sigma_2^2 + (\sigma_1^2/n)b_1'(\mathbf{Z}_1' \mathbf{Q}_2 \mathbf{Z}_1)b_1$ the pseudo-true value $\hat{\sigma}_2^2$ under M_1 when $\mathbf{Q}_2 = (\mathbf{I}_n - \mathbf{Z}_2(\mathbf{Z}_2' \mathbf{Z}_2)^{-1} \mathbf{Z}_2')$. Mizon and Richard (1986) showed that the resulting variance encompassing test statistic is asymptotically equivalent to each of the one degree of freedom non-nested test statistics developed by Cox (1961; 1962), Perasan (1974), and Davidson and MacKinnon (1981), among others. The fact that variance dominance is a necessary but not a sufficient condition for variance encompassing highlights a serious limitation of choosing models on the basis of goodness-of-fit selection criteria rather than comparing the alternative models using encompassing test statistics.

Further developments

This analysis illustrates the fact that the choice of statistic for the encompassing contrast is very important, and may depend very much on the purpose of the analysis or the nature of the models being investigated. For example, when the GUM is not easily available or the calculation of pseudo-true values for other encompassing test statistics is difficult, comparison of the forecasting abilities provides an alternative basis for an encompassing test. Although selecting models on the basis of forecast performance can be very misleading for some purposes in a non-stationary environment with regime shifts (Hendry and Mizon, 2005), the concept of forecast encompassing is a valuable method of model comparison. Forecast encompassing statistics were presented by Chong and Hendry (1990), and Ericsson (1993) and Lu and Mizon (1991) extend this analysis in several directions, including multi-step ahead forecasts from nonlinear dynamic models with estimated coefficients. Similarly, when the analytic calculation of pseudo-true values is intractable simulation methods may be used to estimate the pseudo-true values and hence compute the non-nested test statistics (Hendry and Richard, 1989; Pesaran and Pesaran, 1993). Gouriéroux, Monfort and Renault (1993) developed a comprehensive framework for such simulation known as indirect inference, which allows choice of auxiliary functions as the basis for parameter estimation. A consistent estimator of the parameters involved in the encompassing contrast can be obtained when a correction based on the simulated pseudo-true values of the testing statistics is applied. This approach has the potential to extend the application of the encompassing principle enormously. The relationship between encompassing and conditional moment or m-tests (Newey, 1985) is discussed in White (1994) and Lu and Mizon (1996). The possibility that the encompassing principle be used as a generator of test statistics is discussed in Mizon and Richard (1986). Govaerts, Hendry and Richard (1994) consider the application of encompassing in dynamic models, and Hendry and Mizon (1993) apply it to the comparison of alternative dynamic simultaneous equations models containing integrated and cointegrated variables. A Bayesian approach to encompassing is presented in Florens, Hendry and Richard (1996) and, as a result of using statistical procedures rather than pseudo-true values as in Mizon and Richard (1986), argues that encompassing can be interpreted as a property of model specificity analogous to that of sufficiency for statistics. The encompassing relationship between nonparametric models is considered in Bontemps, Florens and Richard (2006). Finally, Marcellino and Mizon (2006) contains a comprehensive statement and analysis of encompassing as well as many applications of the principle.

GRAYHAM E. MIZON

See also **artificial regressions; forecasting; model selection; models; testing.**

Bibliography

Bontemps, C., Florens, J. and Richard, J. 2006. Encompassing in regression models: parametric and non-parametric procedures. In *Progressive Modelling: Non-nested Testing and Encompassing*, ed. M. Marcellino and G. Mizon. Oxford: Oxford University Press.

Bontemps, C. and Mizon, G. 2003. Congruence and encompassing. In *Econometrics and the Philosophy of Economics*, ed. B. Stigum. Princeton: Princeton University Press.

Chong, Y. and Hendry, D. 1990. Econometric evaluation of linear macro-economic models. In *Modelling Economic Series*, ed. C. Granger. Oxford: Clarendon Press.

Cox, D. 1961. Tests of separate families of hypotheses. *Proceedings of the Fourth Berkeley Symposium on Mathematical Statistics and Probability* 1, 105–23. Berkeley: University of California Press.

Cox, D. 1962. Further results on tests of separate families of hypotheses. *Journal of the Royal Statistical Society B* 24, 406–24.

Davidson, J., Hendry, D., Srba, F. and Yeo, J. 1978. Econometric modelling of the aggregate time-series relationship between consumers' expenditure and income in the United Kingdom. *Economic Journal* 88, 661–92.

Davidson, R. and MacKinnon, J. 1981. Several tests for model specification in the presence of alternative hypotheses. *Econometrica* 49, 781–93.

Ericsson, N. 1993. Comment on 'On the limitations of comparing mean squared forecast errors', by M. P. Clements and D. F. Hendry. *Journal of Forecasting* 12, 644–51.

Ericsson, N. and Hendry, D. 1999. Encompassing and rational expectations: how sequential corroboration can imply refutation. *Empirical Economics* 24, 1–21.

Florens, J.-P., Hendry, D. and Richard, J.-F. 1996. Encompassing and specificity. *Econometric Theory* 12, 620–56.

Gouriéroux, C. and Monfort, A. 1995. Testing, encompassing, and simulating dynamic econometric models. *Econometric Theory* 11, 195–228.

Gouriéroux, C., Monfort, A. and Renault, E. 1993. Indirect inference. *Journal of Applied Econometrics* 8, 85–118.

Govaerts, B., Hendry, D. and Richard, J.-F. 1994. Encompassing in stationary linear dynamic models. *Journal of Econometrics* 63, 245–70.

Hendry, D. 1995. *Dynamic Econometrics*. Oxford: Oxford University Press.

Hendry, D. and Krolzig, H.-M. 2003. New developments in automatic general-to-specific modelling. In *Econometrics and the Philosophy of Economics*, ed. B. Stigum. Princeton: Princeton University Press.

Hendry, D. and Mizon, G. 1993. Evaluating dynamic econometric models by encompassing the VAR. In *Models, Methods and Applications of Econometrics*, ed. P. Phillips. Oxford: Basil Blackwell.

Hendry, D. and Mizon, G. 2005. Forecasting in the presence of structural breaks and policy regime shifts. In *Identification and Inference for Econometric Models: Festschrift in Honor of Tom Rothenberg*, ed. D. Andrews and J. Stock. Cambridge: Cambridge University Press.

Hendry, D. and Richard, J.-F. 1989. Recent developments in the theory of encompassing. In *Contributions to Operations Research and Economics. The XXth Anniversary of CORE*, ed. B. Cornet and H. Tulkens. Cambridge, MA: MIT Press.

Lu, M. and Mizon, G. 1991. Forecast encompassing and model evaluation. In *Economic Structural Change, Analysis and Forecasting*, ed. P. Hackl and A. Westlund. Berlin: Springer-Verlag.

Lu, M. and Mizon, G. 1996. The encompassing principle and hypothesis testing. *Econometric Theory* 12, 845–58.

Marcellino, M. and Mizon, G., eds. 2006. *Progressive Modelling: Non-nested Testing and Encompassing*. Oxford: Oxford University Press.

Mizon, G. 1984. The encompassing approach in econometrics. In *Econometrics and Quantitative Economics*, ed. D. Hendry and K. Wallis. Oxford: Blackwell.

Mizon, G. 1989. The role of econometric modelling in economic analysis. *Revista Espanola de Economia* 6, 167–91.

Mizon, G. 1995. Progressive modelling of macroeconomic time series: the LSE methodology. In *Macroeconometrics: Developments, Tensions and Prospects*, ed. K. Hoover. Dordrecht: Kluwer Academic Press.

Mizon, G. and Richard, J.-F. 1986. The encompassing principle and its application to non-nested hypothesis tests. *Econometrica* 54, 657–78.

Nagel, E. 1961. *The Structure of Science*. New York: Harcourt Brace.

Newey, W. 1985. Maximum likelihood specification testing and conditional moment tests. *Econometrica* 53, 1047–70.

Okasha, S. 2002. *Philosophy of Science: A Very Short Introduction*. Oxford: Oxford University Press.

Pesaran, M. 1974. On the general problem of model selection. *Review of Economic Studies* 41, 153–71.

Pesaran, M. and Pesaran, B. 1993. A simulation approach to the problem of computing Cox's statistic for testing non-nested models. *Journal of Econometrics* 57, 377–92.

Schwarz, G. 1978. Estimating the dimension of a model. *Annals of Statistics* 6, 461–64.

White, H. 1994. *Estimation, Inference and Specification Analysis*. Cambridge: Cambridge University Press.

endogeneity and exogeneity

Endogeneity and exogeneity are properties of variables in economic or econometric models. The specification of these properties for respective variables is an essential component of the entire process of model specification. The words have an ambiguous meaning, for they have been applied in closely related but conceptually distinct ways, particularly in the specification of stochastic models. We consider in turn the case of deterministic and stochastic models, concentrating mainly on the latter.

A deterministic economic model typically specifies restrictions to be satisfied by a vector of variables **y**. These restrictions often incorporate a second vector of variables **x**, and the restrictions themselves may hold only if **x** itself satisfies certain restrictions. The model asserts

$$\forall \mathbf{x} \in R, \ \ G(\mathbf{x}, \mathbf{y}) = 0.$$

The variables **x** are exogenous and the variables **y** are endogenous. The defining distinction between **x** and **y** is that **y** may be (and generally is) restricted by **x**, but not conversely. This distinction is an essential part of the specification of the functioning of the model, as may be seen from the trivial model,

$$\forall \mathbf{x} \in \mathbf{R}^1, \ \ x + y = 0.$$

The condition $x + y = 0$ is symmetric in x and y; the further stipulation that x is exogenous and y is endogenous specifies that in the model x restricts y and not conversely, a property that cannot be derived from $x + y = 0$. In many instances the restrictions on **y** may *determine* **y**, at least for $\mathbf{x} \in R^* \subset R$, but the existence of a *unique* solution has no bearing on the endogeneity and exogeneity of the variables.

The formal distinction between endogeneity and exogeneity in econometric models was emphasized by the Cowles Commission in its path-breaking work on the estimation of simultaneous economic relationships. The class of models it considered is contained in the specification

$$\mathbf{B}(L)\mathbf{y}(t) + \mathbf{\Gamma}(L)\mathbf{x}(t) = \mathbf{u}(t);$$
$$\mathbf{A}(L)\mathbf{u}(t) = \varepsilon(t);$$
$$\mathrm{cov}[\varepsilon(t), \mathbf{y}(t - s)] = \mathbf{O}, \ \ s > 0;$$
$$\mathrm{cov}[\varepsilon(t), \mathbf{x}(t - s)] = \mathbf{O}, \ \ \text{all } s;$$
$$\varepsilon(t) \sim \mathbf{IIDN}(\mathbf{O}, \Sigma).$$

The vectors $\mathbf{x}(t)$ and $\mathbf{y}(t)$ are observed, whereas $\mathbf{u}(t)$ and $\varepsilon(t)$ are underlying disturbances not observed but affecting **y**(t). The lag operator L is defined by $L\mathbf{x}(t) = \mathbf{x}(t-1)$; the roots of $|\mathbf{B}(L)|$ and $|\mathbf{A}(L)|$ are assumed to have modulus greater than 1, a stability condition guaranteeing the non-explosive behaviour of **y** given any stable path for **x**. The Cowles Commission definition of exogeneity in this model (Koopmans and Hood, 1953, pp. 117–20) as set forth in Christ (1966, p. 156) is as follows:

> An exogenous variable in a stochastic model is a variable whose value in each period is statistically independent of the values of all the random disturbances in the model in all periods.

All other variables are endogenous. In the prototypical model set forth above **x** is exogenous and **y** is endogenous.

The Cowles Commission distinction between endogeneity and exogeneity applied to a specific class of models, with linear relationships and normally distributed disturbances. The exogenous variables **x** in the prototypical model have two important but quite distinct properties. First, the model may be solved to yield an expression for $\mathbf{y}(t)$ in terms of current and past values of **x** and **ε**,

$$y(t) = \mathbf{B}(L)^{-1}\mathbf{\Gamma}(L)\mathbf{x}(t) + \mathbf{B}(L)^{-1}\mathbf{A}(L)\varepsilon(\mathrm{t}).$$

Given suitably restricted $\mathbf{x}(t)$ (for example, all **x** uniformly bounded, or being realizations of a stationary stochastic process with finite variance) it is natural to complete the model by specifying that it is valid for all **x** meeting the restrictions, and this is often done. The variables **x** are therefore exogenous here as **x** is exogenous in a deterministic economic model. A second, distinct property of these variables is that in estimation $\mathbf{x}(t)$ $(-\infty < t < \infty)$ may be regarded as fixed, thus extending to the environment of simultaneous equation models methods of statistical inference initially designed for experimental settings. It was generally recognized that exogeneity in the prototypical model was a sufficient but not a necessary condition to justify treating variables as fixed for purposes of inference. If $\mathbf{u}(t)$ in the model is serially independent (that is, $A(L)=I$) then lagged values of $\mathbf{y}$ may also be treated as fixed for purposes of the model; this leads to the definition of 'predetermined variables' (Christ, 1966, p. 227) following Koopmans and Hood (1953, pp. 117–21):

> A variable is predetermined at time t if all its current and past values are independent of the vector of current disturbances in the model, and these disturbances are serially independent.

These two properties were not explicitly distinguished in the prototypical model (Koopmans, 1950; Koopmans and Hood, 1953) and tended to remain merged in the literature over the next quarter-century (for example, Christ, 1966; Theil, 1971; Geweke, 1978). By the late 1970s there had developed a tension between the two, due to the increasing sophistication of estimation procedures in nonlinear models, treatment of rational expectations, and the explicit consideration of the respective dynamic properties of endogenous and exogenous variables (Sims, 1972; 1977; Geweke, 1982). Engle, Hendry and Richard (1983), drawing on this literature and discussions at the 1979 Warwick Summer Workshop, formalized the distinction of the two properties we have discussed. Drawing on their definitions 2.3 and 2.5 and the discussions in Sims (1977) and Geweke (1982), $\mathbf{x}$ is *model exogenous* if given $\{\mathbf{x}(t), t \leq T\} \in R(T)$ the model may restrict $\{\mathbf{y}(t), t \leq T\}$, but given

$$\{\mathbf{x}(t), t \leq T + J\} \in R(T + J)$$

there are no further restrictions on $\{\mathbf{y}(t), t \leq T\}$, for any $J > 0$. If the model in fact does restrict $\{\mathbf{y}(t), t \leq T\}$, then $\mathbf{y}$ is model endogenous. As examples consider

Model 1:

$$y(t) = ay(t - 1) + bx(t) + u(t),$$
$$x(t) = cx(t - 1) + v(t);$$

Model 2:

$$y(t) = ay(t - 1) + bx(t) + u(t),$$
$$x(t) = cx(t - 1) + dy(t) + v(t);$$

Model 3:

$$y(t) = \quad\quad ay(t - 1) + b\{x(t)$$
$$+ E[x(t)|x(t - s), s > 0]\} + u(t),$$
$$x(t) = cx(t - 1) + v(t).$$

In each case $u(t)$ and $v(t)$ are mutually and serially independent, and normally distributed. The parameters are

assumed to satisfy the usual stability restrictions guaranteeing that x and y have normal distributions with finite variances. In all three models y is model endogenous, and x is model exogenous in Models 1 and 3 but not 2. For estimation the situation is different. In Model 1, treating $x(t)$, $x(t-1)$ and $y(t-1)$ as fixed simplifies inference at no cost; $y(t-1)$ is a classic predetermined variable in the sense of Koopmans and Hood (1953) and Christ (1966). Similarly in Model 2, $x(t-1)$ and $y(t-1)$ may be regarded as fixed for purposes of inference despite the fact that x and y are both model endogenous. When Model 3 is re-expressed

$$y(t) = ay(t - 1) + bx(t) + bcx(t - 1) + u(t),$$
$$x(t) = cx(t - 1) + v(t),$$

it is clear that $x(t)$ cannot be treated as fixed if the parameters are to be estimated efficiently since there are cross-equation restrictions involving the parameter c. Model exogeneity of a variable is thus neither a necessary nor a sufficient condition for treating that variable as fixed for purposes of inference.

The condition that a set of variables can be regarded as fixed for inference can be formalized, following Engle, Hendry and Richard (1983) along the lines given in Geweke (1984). Let

$$\mathbf{X} \equiv [\mathbf{x}(1), \dots, \mathbf{x}(n)] \quad \text{and} \quad \mathbf{Y} \equiv [\mathbf{y}(1), \dots, \mathbf{y}(n)]$$

be matrices of n observations on the variables $\mathbf{x}$ and $\mathbf{y}$ respectively. Suppose the likelihood function $L(\mathbf{X}, \mathbf{Y}|\boldsymbol{\Theta})$ can be reparameterized by $\boldsymbol{\lambda} = F(\boldsymbol{\Theta})$ where F is a one-to-one transformation; $\lambda' = (\lambda_1, \lambda_2)'$, $(\lambda_1, \lambda_2) \in \Lambda_1 X \Lambda_2$; and the investigator's loss function depends on parameters of interest λ_1 but not nuisance parameters λ_2. Then $\mathbf{x}$ is *weakly exogenous* if

$$L(\mathbf{X}, \mathbf{Y}|\lambda_1, \lambda_2) = L_1(\mathbf{Y}|\mathbf{X}, \lambda, \lambda_1) \cdot L_2(\mathbf{X}|\lambda_2),$$

and in this case $\mathbf{y}$ is *weakly endogenous*. When this condition is met the expected loss function may be expressed using only $L_1(\mathbf{Y}|\mathbf{X}, \lambda_1)$, that is, $\mathbf{x}$ may be regarded as fixed for purposes of inference.

The concepts of model exogeneity and weak exogeneity play important but distinct roles in the construction, estimation, and evaluation of econometric models. The dichotomy between variables that are model exogenous and model endogenous is a global property of a model, drawing in effect a logical distinction between the inputs of the model $\{x(t), t \leq T\} \in R(T)$ and the set of variables restricted by the model $\{\mathbf{y}(t), t \leq T\}$. Since model exogeneity stipulates that $\{\mathbf{x}(t), t \leq T + J\}$ places no more restrictions on $\{\mathbf{y}(t), t \leq T\}$ than does $\{\mathbf{x}(t), t \leq T\}$, the global property of model exogeneity is in principle testable, either in the presence or absence of other restrictions imposed by the model. When conducted in the absence of most other restrictions this test is often termed a 'causality test', and its use as a test of specification was introduced by Sims (1972). The distinction between weakly exogenous and weakly endogenous variables permits a simplification of the likelihood function that depends on the subset of the model's parameters that are of interest to the investigator. It is a logical property of the model: the same results would be obtained using $L(\mathbf{X}, \mathbf{Y}|\lambda_1, \lambda_2)$ as using $L(\mathbf{Y}|\mathbf{X}, \lambda_1)$. The stipulation of weak exogeneity is therefore not, by itself, testable.

JOHN GEWEKE

See also **causality in economics and econometrics; identification; simultaneous equations models.**

Bibliography

Christ, C.F. 1966. *Econometric Models and Methods.* New York: Wiley.
Engle, R.F., Hendry, D.F. and Richard, J.-F. 1983. Exogeneity. *Econometrica* 51, 277–304.
Geweke, J. 1978. Testing the exogeneity specification in the complete dynamic simultaneous equation model. *Journal of Econometrics* 7, 163–85.
Geweke, J. 1982. Causality, exogeneity, and inference. In *Advances in Econometrics*, ed. W. Hildenbrand. Cambridge: Cambridge University Press.
Geweke, J. 1984. Inference and causality. In *Handbook of Econometrics*, vol. 2, ed. Z. Griliches and M.D. Intriligator. Amsterdam: North-Holland.
Koopmans, T.C. 1950. When is an equation system complete for statistical purposes? In *Statistical Inference in Dynamic Economic Models*, ed. T.C. Koopmans. New York: Wiley.
Koopmans, T.C. and Hood, W.C. 1953. The estimation of simultaneous economic relationships. In *Studies in Econometric Method*, ed. W.C. Hood and T.C. Koopmans. New York: Wiley.
Sims, C.A. 1972. Money, income, and causality. *American Economic Review* 62, 540–52.
Sims, C.A. 1977. Exogeneity and causal ordering in macroeconomic models. In *New Methods in Business Cycle Research*, ed. C.A. Sims. Minneapolis: Federal Reserve Bank of Minneapolis.
Theil, H. 1971. *Principles of Econometrics.* New York: Wiley.

endogenous growth theory

Endogenous growth is long-run economic growth at a rate determined by forces that are internal to the economic system, particularly those forces governing the opportunities and incentives to create technological knowledge.

In the long run the rate of economic growth, as measured by the growth rate of output per person, depends on the growth rate of total factor productivity (TFP), which is determined in turn by the rate of technological progress. The neoclassical growth theory of Solow (1956) and Swan (1956) assumes the rate of technological progress to be determined by a scientific process that is separate from, and independent of, economic forces. Neoclassical theory thus implies that economists can take the long-run growth rate as given exogenously from outside the economic system.

Endogenous growth theory challenges this neoclassical view by proposing channels through which the rate of technological progress, and hence the long-run rate of economic growth, can be influenced by economic factors. It starts from the observation that technological progress takes place through innovations, in the form of new products, processes and markets, many of which are the result of economic activities. For example, because firms learn from experience how to produce more efficiently, a higher pace of economic activity can raise the pace of process innovation by giving firms more production experience. Also, because many innovations result from R&D expenditures undertaken by profit-seeking firms, economic policies with respect to trade, competition, education, taxes and intellectual property can influence the rate of innovation by affecting the private costs and benefits of doing R&D.

AK theory

The first version of endogenous growth theory was AK theory, which did not make an explicit distinction between capital accumulation and technological progress. In effect it lumped together the physical and human capital whose accumulation is studied by neoclassical theory with the intellectual capital that is accumulated when innovations occur. An early version of AK theory was produced by Frankel (1962), who argued that the aggregate production function can exhibit a constant or even increasing marginal product of capital. This is because, when firms accumulate more capital, some of that increased capital will be the intellectual capital that creates technological progress, and this technological progress will offset the tendency for the marginal product of capital to diminish.

In the special case where the marginal product of capital is exactly constant, aggregate output Y is proportional to the aggregate stock of capital K:

$$Y = AK \qquad (1)$$

where A is a positive constant. Hence the term 'AK theory'.

According to AK theory, an economy's long-run growth rate depends on its saving rate. For example, if a fixed fraction s of output is saved and there is a fixed rate of depreciation δ, the rate of aggregate net investment is:

$$\frac{dK}{dt} = sY - \delta K$$

which along with (1) implies that the growth rate is given by:

$$g \equiv \frac{1}{Y}\frac{dY}{dt} = \frac{1}{K}\frac{dK}{dt} = sA - \delta.$$

Hence an increase in the saving rate s will lead to a permanently higher growth rate.

Romer (1986) produced a similar analysis with a more general production structure, under the assumption that saving is generated by intertemporal utility maximization instead of the fixed saving rate of Frankel. Lucas (1988) also produced a similar analysis focusing on human capital rather than physical capital; following Uzawa (1965) he explicitly assumed that human capital and technological knowledge were one and the same.

Innovation-based theory

AK theory was followed by a second wave of endogenous growth theory, generally known as 'innovation-based' growth theory, which recognizes that intellectual capital, the source of technological progress, is distinct from physical and human capital. Physical and human capital are accumulated through saving and schooling, but intellectual capital grows through innovation.

One version of innovation-based theory was initiated by Romer (1990), who assumed that aggregate productivity is an increasing function of the degree of product variety. In this theory, innovation causes productivity growth by creating new, but not necessarily improved, varieties of products. It makes use of the Dixit–Stiglitz–Ethier production function, in which final output is produced by labour and a continuum of intermediate products:

$$Y = L^{1-\alpha} \int_0^A x(i)^\alpha \ di, \quad 0<\alpha<1 \qquad (2)$$

where L is the aggregate supply of labour (assumed to be constant), $x(i)$ is the flow input of intermediate product i, and A is the measure of different intermediate products that are available for use. Intuitively, an increase in product variety, as measured by A, raises productivity by allowing society to spread its intermediate production more thinly across a larger number of activities, each of which is subject to diminishing returns and hence exhibits a higher average product when operated at a lower intensity.

The other version of innovation-based growth theory is the 'Schumpeterian' theory developed by Aghion and Howitt (1992) and Grossman and Helpman (1991). (Early models were produced by Segerstrom, Anant and Dinopoulos, 1990, and Corriveau, 1991). Schumpeterian theory focuses on quality-improving innovations that render old products obsolete, through the process that Schumpeter (1942) called 'creative destruction.'

In Schumpeterian theory aggregate output is again produced by a continuum of intermediate products, this time according to:

$$Y = L^{1-\alpha} \int_0^1 A(i)^{1-\alpha} x(i)^\alpha di, \qquad (3)$$

where now there is a fixed measure of product variety, normalized to unity, and each intermediate product i has a separate productivity parameter $A(i)$. Each sector is monopolized and produces its intermediate product with a constant marginal cost of unity. The monopolist in sector i faces a demand curve given by the marginal product: $\alpha \cdot (A(i)L/x(i))^{1-\alpha}$ of that intermediate input in the final sector. Equating marginal revenue (α time this marginal product) to the marginal cost of unity yields the monopolist's profit-maximizing intermediate output:

$$x(i) = \xi L A(i)$$

where $\xi = \alpha^{2/(1-\alpha)}$. Using this to substitute for each $x(i)$ in the production function (3) yields the aggregate production function:

$$Y = \theta A L \qquad (4)$$

where $\theta = \xi^\alpha$, and where A is the average productivity parameter:

$$A \equiv \int_0^1 A(i) \ di.$$

Innovations in Schumpeterian theory create improved versions of old products. An innovation in sector i consists of a new version whose productivity parameter $A(i)$ exceeds that of the previous version by the fixed factor $\gamma > 1$. Suppose that the probability of an innovation arriving in sector i over any short interval of length dt is $\mu \cdot dt$. Then the growth rate of $A(i)$ is

$$\frac{dA(i)}{A(i)} \cdot \frac{1}{dt} = \begin{cases} (\gamma - 1) \cdot \frac{1}{dt} & \text{with probability } \mu \cdot dt \\ 0 & \text{with probability } 1 - \mu \cdot dt \end{cases}.$$

Therefore the expected growth rate of $A(i)$ is:

$$E(g) = \mu(\gamma - 1). \qquad (5)$$

The flow probability μ of an innovation in any sector is proportional to the current flow of productivity-adjusted R&D expenditures:

$$\mu = \lambda R/A \qquad (6)$$

where R is the amount of final output spent on R&D, and where the division by A takes into account the force of increasing complexity. That is, as technology advances it becomes more complex, and hence society must make an ever-increasing expenditure on research and development just to keep innovating at the same rate as before.

It follows from (4) that the growth rate g of aggregate output is the growth rate of the average productivity parameter A. The law of large numbers guarantees that g equals the expected growth rate (5) of each individual productivity parameter. From this and (6) we have:

$$g = (\gamma - 1)\, \lambda R/A.$$

From this and (4) it follows that the growth rate depends on the fraction of GDP spent on research and development, $n = R/Y$, according to:

$$g = (\gamma - 1)\, \lambda \theta L n. \qquad (7)$$

Thus, innovation-based theory implies that the way to grow rapidly is not to save a large fraction of output but to devote a large fraction of output to research and development. The theory is explicit about how R&D activities are influenced by various policies, who gains from technological progress, who loses, how the gains and losses depend on social arrangements, and how such arrangements affect society's willingness and ability to create and cope with technological change, the ultimate source of economic growth.

Empirical challenges

Endogenous growth theory has been challenged on empirical grounds, but its proponents have replied with modifications of the theory that make it consistent with the critics' evidence. For example, Mankiw, Romer and Weil (1992), Barro and Sala-i-Martin (1992) and Evans (1996) showed, using data from the second half of the 20th century, that most countries seem to be converging to roughly similar long-run growth rates, whereas endogenous growth theory seems to imply that, because many countries have different policies and institutions, they should have different long-run growth rates. But the Schumpeterian model of Howitt (2000), which incorporates the force of technology transfer, whereby the productivity of R&D in one country is enhanced by innovations in other countries, implies that all countries that perform R&D at a positive level should converge to parallel long-run growth paths.

The key to this convergence result is what Gerschenkron (1952) called the 'advantage of backwardness'; that is, the further a country falls behind the technology frontier, the larger is the average size of innovations, because the larger is the gap between the frontier ideas incorporated in the country's innovations and the ideas incorporated in the old technologies being replaced by innovations. This increase in the size of innovations keeps raising the laggard country's growth rate until the gap separating it from the frontier finally stabilizes.

Likewise, Jones (1995) has argued that the evidence of the United States and other OECD countries since 1950 refutes the 'scale effect' of Schumpeterian endogenous growth theory. That is, according to the growth equation (7) an increase in the size of population should raise long-run growth by increasing the size of the workforce L, thus providing a larger market for a successful innovator and inducing a higher rate of innovation. But in fact productivity growth has remained stationary during a period when population, and in particular the number of people engaged in R&D, has risen dramatically. The models of Dinopoulos and Thompson (1998), Peretto (1998) and Howitt (1999) counter this criticism by incorporating Young's (1998) insight that, as an economy grows, proliferation of product varieties reduces the effectiveness of R&D aimed at quality improvement by causing it to be spread more thinly over a larger number of different sectors. When modified this way the theory is consistent with the observed coexistence of stationary TFP growth and rising population, because in a steady state the growth-enhancing scale effect is just offset by the growth-reducing effect of product proliferation.

As a final example, early versions of innovation-based growth theory implied, counter to much evidence, that growth would be adversely affected by stronger competition laws, which by reducing the profits that imperfectly competitive firms can earn ought to reduce the incentive to innovate. However, Aghion and Howitt (1998, ch. 7) describe a variety of channels through which competition might in fact spur economic growth. One such channel is provided by the work of Aghion et al. (2001), who show that, although an increase in the intensity of competition will tend to reduce the absolute level of profits realized by a successful innovator, it will nevertheless tend to reduce the profits of an unsuccessful innovator by even more. In this variant of Schumpeterian theory, more intense competition can have a positive effect on the rate of innovation because firms will want to escape the competition that they would face if they lost whatever technological advantage they have over their rivals.

Much more work needs to be done before we can claim to have a reliable explanation for why economic growth is faster in some countries and in some time periods than in others. But the fact that much of the cross-country variation in growth rates is attributable to differences in productivity growth rather than differences in rates of capital accumulation suggests that endogenous growth theory, which aims to provide an economic explanation of these differences in productivity growth, will continue to attract economists' attention for years to come.

PETER HOWITT

See also **Schumpeterian growth and growth policy design.**

Bibliography

Aghion, P. and Howitt, P. 1992. A model of growth through creative destruction. *Econometrica* 60, 323–51.

Aghion, P. and Howitt, P. 1998. *Endogenous Growth Theory*. Cambridge, MA: MIT Press.

Aghion, P., Harris, C., Howitt, P. and Vickers, J. 2001. Competition, imitation and growth with step-by-step innovation. *Review of Economic Studies* 68, 467–92.

Barro, R.J. and Sala-i-Martin, X. 1992. Convergence. *Journal of Political Economy* 100, 223–51.

Corriveau, L. 1991. Entrepreneurs, growth, and cycles. Doctoral dissertation, University of Western Ontario.

Dinopoulos, E. and Thompson, P. 1998. Schumpeterian growth without scale effects. *Journal of Economic Growth* 3, 313–35.

Evans, P. 1996. Using cross-country variances to evaluate growth theories. *Journal of Economic Dynamics and Control* 20, 1027–49.

Frankel, M. 1962. The production function in allocation and growth: a synthesis. *American Economic Review* 52, 995–1022.

Gerschenkron, A. 1952. Economic backwardness in historical perspective. In *The Progress of Underdeveloped Areas*, ed. B.F. Hoselitz. Chicago: University of Chicago Press.

Grossman, G.M. and Helpman, E. 1991. *Innovation and Growth in the Global Economy*. Cambridge, MA: MIT Press.

Howitt, P. 1999. Steady endogenous growth with population and R&D inputs growing. *Journal of Political Economy* 107, 715–30.

Howitt, P. 2000. Endogenous growth and cross-country income differences. *American Economic Review* 90, 829–46.

Jones, C.I. 1995. R&D-based models of economic growth. *Journal of Political Economy* 103, 759–84.

Lucas, R.E., Jr. 1988. On the mechanics of economic development. *Journal of Monetary Economics* 22, 3–42.

Mankiw, N.G., Romer, D. and Weil, D.N. 1992. A contribution to the empirics of economic growth. *Quarterly Journal of Economics* 107, 407–37.

Peretto, P.F. 1998. Technological change and population growth. *Journal of Economic Growth* 3, 283–311.

Romer, P.M. 1986. Increasing returns and long-run growth. *Journal of Political Economy* 94, 1002–37.

Romer, P.M. 1990. Endogenous technological change. *Journal of Political Economy* 98, S71–S102.

Schumpeter, J.A. 1942. *Capitalism, Socialism and Democracy*. New York: Harper.

Segerstrom, P.S., Anant, T.C.A. and Dinopoulos, E. 1990. A Schumpeterian model of the product life cycle. *American Economic Review* 80, 1077–91.

Solow, R.M. 1956. A contribution to the theory of economic growth. *Quarterly Journal of Economics* 70, 65–94.

Swan, T.W. 1956. Economic growth and capital accumulation. *Economic Record* 32, 334–61.

Uzawa, H. 1965. Optimal technical change in an aggregative model of economic growth. *International Economic Review* 6, 18–31.

Young, A. 1998. Growth without scale effects. *Journal of Political Economy* 106, 41–63.

endogenous market incompleteness

An asset trading arrangement is incomplete if it is too restrictive to ensure a fully Pareto-optimal allocation of risk. Endogenously incomplete models derive such trading arrangements from primitive frictions. They are to be contrasted with models that *assume* a particular incomplete asset markets structure.

Recent contributions to the endogenous incompleteness literature have emphasized imperfections in the enforcement and monitoring technologies available to societies. They derive endogenous market structures, sometimes supplemented with a tax system, as decentralizations of planning problems in which the planner faces one or both of these imperfections. These market-tax structures ensure that agents are provided with incentives to honour promises that cannot be costlessly enforced or that are contingent on states that cannot be costlessly observed. By construction they admit equilibria that are constrained efficient.

Models with endogenous incompleteness have received a variety of applications in macroeconomics. They have been used to enhance understanding of risk sharing, asset pricing and business cycles; on the normative side they have been applied to analyses of optimal fiscal policy. Here I review some of these applications and the models that underpin them.

Limited enforcement

The canonical example of a limited enforcement model is the bilateral insurance game of Kocherlakota (1996). In this game, two risk-averse agents are endowed with random and imperfectly correlated income processes. Neither agent can be compelled to deliver resources to the other, even if they have promised to do so in the past.

Equilibrium allocations in this setting can be implemented with strategies that revert to autarky following an agent defection. Agents with high-income shocks can be induced to share some of their resources by the threat of such reversion and, when this is insufficient, by promises of extra resources in the future. Such promises introduce additional dynamics into optimal equilibrium allocations; shocks that cannot be smoothed over states are smoothed over time instead, ensuring that individual consumption is persistent even when aggregate consumption is not.

Constrained-efficient allocations in limited enforcement economies can be decentralized using a complete set of Arrow security markets coupled with endogenous debt limits (see Alvarez and Jermann, 2000). Intuitively, agents can borrow only up to the amount that they are willing to pay back in the future given that the penalty for default is consignment to autarky. Thus, the limited enforcement friction provides a micro-foundation for the often-made assumption of a debt limit tighter than that implied by an agent's intertemporal budget constraint. In the limited enforcement case, however, the debt limit is state-contingent; it depends upon the value of autarky to the agent. Since this value is a function of individual and aggregate shocks, the parameters of the shock process and, in richer models, the agent's opportunities for self- or public insurance after exclusion from markets, so too is the debt limit.

When agents' endogenous debt limits periodically bind, risk sharing is disrupted; individual consumption, conditional on the aggregate state, is positively correlated with current and past individual income. Qualitatively, such departures from full risk-sharing cohere well with evidence on individual consumption. In Alvarez and Jermann's (2001) quantitative analysis of a calibrated limited enforcement model, the endogenous debt limits bind fairly often and permit relatively little risk sharing. This is consistent with evidence on the sharing of low-frequency risks. Alvarez and Jermann's analysis also has implications for asset pricing. They obtain a volatile asset pricing kernel and risk premia that are large and time varying. These implications are consistent with asset pricing data, but contrast with those of the benchmark representative agent asset pricing model.

Cross-country consumption data also exhibit apparent departures from full risk sharing. Standard models (with complete markets) imply co-movements in consumption that exceed those in output, yet the data suggests the reverse. Kehoe and Perri (2002) show that a limited enforcement model augmented with production and physical capital accumulation can go some way to explaining this anomaly.

Recent papers have considered alternative penalties for default including the confiscation of an endogenously valued collateral asset (see, Lustig, 2005) or the payment of a fixed default cost (Cooley, Marimon and Quadrini, 2004). These contributions illustrate the scope of limited enforcement models: Lustig explores the implications of endogenously valued collateral for asset pricing and obtains a large and time-varying price of risk; Cooley, Marimon and Quadrini examine the role of limited enforcement frictions in propagating business cycle shocks. Cordoba (2005) and Arpad and Cárceles-Poveda (2005), however, sound cautionary notes. They provide calibrated models in which the introduction of collateral relaxes endogenous debt limits so much that agents can fully diversify risk.

Private information

An alternative line of research has analysed environments in which risk-averse agents privately observe shocks to their endowments, tastes or productivity (see, for example, Atkeson and Lucas, 1992). In this setting, agents must be provided with incentives to reveal information. The socially efficient provision of incentives requires the conditioning of current consumption on an agent's history of shock reports. Intuitively, agents are rewarded for reporting a low current need for resources with the promise of more consumption in the future. Thus, intertemporal consumption smoothing is enhanced and interstate smoothing disrupted.

Albanesi and Sleet (2006) and Kocherlakota (2005) show that optimal information-constrained allocations can be implemented with a mixture of non-contingent debt markets and taxes. Thus, these authors derive joint restrictions on the market structure *and* the tax system from primitive informational frictions. Central to their analyses is an 'inverted Euler equation'. If $\{c_t^*\}_{t=0}^{\infty}$ denotes the optimal consumption allocation, this equation is given by:

$$\frac{1}{u'(c_t^*(z^t, \theta^t))} = \beta\lambda_{t+1}(z^{t+1})E_t\left[\frac{1}{u'(c_{t+1}^*)}|z^{t+1}, \theta^t\right].$$
(1)

Here θ^t denotes an agent's period t history of privately observed shocks, z^t and z^{t+1} denote t and $t+1$ histories of observable aggregate shocks, β is the agent's discount factor and u' her marginal utility of consumption. λ_{t+1} is a social stochastic discount factor (SSDF) that 'prices' resources delivered after each history z^{t+1}. Golosov, Kocherlakota and Tsyvinski (2003) show that such equations hold in a large class of dynamic moral hazard models. They imply a wedge between an agent's conditional expected intertemporal marginal rate of substitution (IMRS) and the SSDF. This wedge provides a rationale for asset taxation; intuitively, agents must be discouraged from saving at date t since greater wealth at $t+1$ undermines incentives at that date. However, the implications for asset taxation are subtle. The optimal allocation cannot be implemented with an asset tax that merely 'matches the wedge' and equates the conditional expectation of an agent's IMRS to the SSDF. Instead, marginal asset taxes at $t+1$ are used to generate a positive covariance between the after-tax asset return and the agent's consumption that deters savings. In some cases, the expected asset tax is zero and the wedge is entirely generated by this covariance effect.

Positive analyses of dynamic moral hazard are relatively scarce. Green and Oh (1991) contrast the empirical implications of various incomplete market models, including those with moral hazard. Kocherlakota and Pistaferri (2005) identify λ_{t+1} with the market discount factor, assume that utility has the constant relative risk aversion property and use (1) to derive expressions for λ_{t+1} in terms of cross-sectional moments of the consumption distribution. They then investigate the implications of this dynamic moral hazard model for asset pricing and, in particular, the equity premium and risk-free rate. They find that plausible values of the coefficient of relative risk aversion set the equity premium pricing error to zero.

In all of the dynamic moral hazard models described so far, the consumption of agents is observable. An alternative assumption is that agents can undertake asset trades that are hidden from society. Agents must now be given incentives to reveal information *and* save an appropriate amount. This places additional constraints on risk sharing. When agents can control their publicly observable histories and can save at the prices implied by an exogenously given sequence of SSDFs, these constraints are severe. In this case, the optimal allocation is identical to that in an economy with riskless debt (see Cole and Kocherlakota, 2001). This result is important as it provides a micro-foundation for models that exogenously restrict agents to the trading of such debt.

Government incentive problems

Governments or mechanism designers may also have difficulty keeping their promises. There is a long tradition of considering commitment problems in Ramsey models. In these, a socially benevolent government typically has access to a restricted set of linear tax mechanisms and an asset market in which it can trade claims to resources. *Ex ante* optimal policy entails implicit promises over future allocations and, in particular, the expected value of the government's future stream of primary surpluses that it is rarely in the government's interests to keep. For example, if the government can default on its debt it will, since in this way it can avoid the distortionary taxes necessary for debt repayment. As in the limited-enforcement models described above, reversion to autarky after a default can sustain some equilibrium borrowing by the government, though typically it implies a tight endogenous debt limit (Chari and Kehoe, 1993). Sleet (2004) and Sleet and Yeltekin (2006a) consider models in which the government's true spending needs are not publicly observable. Although the government has access to a complete set of contingent claims markets, in equilibrium it is required to adopt a debt-trading policy consistent with truthful revelation of its spending needs. This limits its ability to buy claims against high spending-needs states and sell them against low spending-needs ones. The outcome is enhanced intertemporal, as opposed to inter-state, smoothing of taxes.

The optimal allocations and market-tax implementations implied by dynamic moral hazard models also involve promises from a planner (or government) to an agent. These allocations often entail the absorption of almost all agents by a minimal utility immiserating state; they thus place strong demands on the planner's ability to commit. Sleet and Yeltekin (2006b) remove this ability. They show that optimal allocations without planner commitment solve the problems of *committed* planners who discount the future *less heavily* than agents. Coupling this result with the work of Farhi and Werning (2005), who directly assume a planner discount factor in excess of the agents, suggests that constrained optimal allocations can be implemented with non-contingent debt, an income tax and a progressive estate tax. Analysis of dynamic moral hazard models without societal commitment is, however, still in its infancy and much remains to be done.

CHRISTOPHER SLEET

See also **default and enforcement constraints; optimal fiscal and monetary policy (without commitment); social insurance.**

Bibliography

Arpad, A. and Cárceles-Poveda, E. 2005. Endogenous trading constraints with incomplete markets. Working paper, University of Rochester.

Albanesi, S. and Sleet, C. 2006. Dynamic optimal taxation with private information. *Review of Economic Studies* 73, 1–30.

Alvarez, F. and Jermann, U. 2000. Efficiency, equilibrium and asset pricing with risk of default. *Econometrica* 68, 775–97.

Alvarez, F. and Jermann, U. 2001. Quantitative asset pricing implications of endogenous solvency constraints. *Review of Financial Studies* 14, 1117–51.

Atkeson, A. and Lucas, R. 1992. On efficient distribution with private information. *Review of Economic Studies* 59, 427–53.

Chari, V. and Kehoe, P. 1993. Sustainable plans and debt. *Journal of Economic Theory* 60, 175–95.

Cole, H. and Kocherlakota, N. 2001. Efficient allocations with hidden income and storage. *Review of Economic Studies* 68, 523–42.

Cooley, T., Marimon, R. and Quadrini, V. 2004. Aggregate consequences of limited contract enforceability. *Journal of Political Economy* 112, 817–47.

Cordoba, J.-C. 2005. US inequality: debt constraints or incomplete markets? Working paper, Rice University.

Farhi, E. and Werning, I. 2005. Inequality, social discounting and estate taxation. Working Paper No. 11408. Cambridge, MA: NBER.

Golosov, M., Kocherlakota, N. and Tsyvinski, A. 2003. Optimal indirect and capital taxation. *Review of Economic Studies* 70, 569–87.

Green, E. and Oh, S. 1991. Contracts, constraints and consumption. *Review of Economic Studies* 58, 883–99.

Kehoe, P. and Perri, F. 2002. International business cycles with endogenously incomplete markets. *Econometrica* 70, 907–28.

Kocherlakota, N. 1996. Implications of efficient risk sharing without commitment. *Review of Economic Studies* 63, 595–609.

Kocherlakota, N. 2005. Zero expected wealth taxes: a
    Mirrleesian approach to dynamic optimal taxation.
    Econometrica 73, 1587–622.
Kocherlakota, N. and Pistaferri, L. 2005. Asset pricing
    implications of Pareto optimality with private
    information. Discussion Paper No. 4930. London: CEPR.
Lustig, H. 2005. The market price of aggregate risk and the
    wealth distribution. Working Paper No. 11132.
    Cambridge, MA: NBER.
Sleet, C. 2004. Optimal taxation with private government
    information. *Review of Economic Studies* 71, 1217–39.
Sleet, C. and Yeltekin, S. 2006a. Optimal taxation with
    endogenously incomplete markets. *Journal of Economic
    Theory* 127, 36–73.
Sleet, C. and Yeltekin, S. 2006b. Credibility and endogenous
    societal discounting. *Review of Economic Dynamics* 9,
    410–37.

energy economics

Energy is crucial to the economic progress and social development of nations. Energy can be neither created nor destroyed but its form can be changed. Energy comes from the physical environment and ultimately returns there. The demand for energy is a derived demand. The value of energy is assessed by its ability to provide a set of desired services in both industry and in the household.

Energy commodities are economic substitutes. Energy resources are depletable or renewable and storable or non-storable. On a global scale the 20th century was dominated by the use of fossil fuels. According to the US Department of Energy, in the year 2000 global commercial energy consumption consisted of petroleum (39 per cent), coal (24 per cent), natural gas (23 per cent), hydro (6 per cent), nuclear (7 per cent) and others (1 per cent). In 1999, of the total sources of energy consumed in the United States, 92 per cent were from depletable resources and only 8 per cent from renewable resources (EIA, 2001). No one doubts that fossil fuels are subject to depletion, and that depletion leads to scarcity, which in turn leads to higher prices. Resources are defined as 'non-conventional' when they cannot be produced economically at today's prices and with today's technology. With higher prices, however, the gap between conventional and non-conventional oil resources narrows. Ultimately, a combination of escalating prices and technological enhancements can transform the non-conventional into the conventional. Much of the pessimism about oil resources has been focused entirely on conventional resources.

Demand for energy

Bohi and Toman (1996) suggest a link between energy and economy. An abundance of empirical research suggests a strong correlation between increases in oil prices and decreases in macroeconomic performance for oil-importing industrialized countries. Higher import costs may lead to higher price levels and inflation. Industrial energy demand increases most rapidly at the initial stages of development, but growth slows steadily throughout the industrialization process (Medlock and Soligo, 2001). Energy demand for transportation rises steadily, and takes the major share of total energy use at the latter stages of developments.

Elasticity of energy demand

Is energy an essential good? In economics, an essential good is one for which the demand remains positive no matter how high its price. Energy is often described as an essential good because human activity would be impossible absent use of energy. Although energy is essential to humans, neither particular energy commodities nor any purchased energy commodities are essential goods because consumers can convert one form of energy into another.

The income elasticity of energy demand is defined as the percentage change in energy demand given a one per cent change in income holding all else constant, or

$$\varepsilon_y = \frac{\%\Delta e}{\%\Delta y} = \frac{de}{dy} \cdot \frac{y}{e}$$

where e denotes energy demand and y denotes income. 'The household sector's share of aggregate energy consumption tends to fall with income, the share of transportation tends to rise, and the share of industry follows an inverse-U pattern' (Judson, Schmalensee and Stoker, 1999).

The price elasticity of energy demand is defined as the percentage change in energy demand given a one per cent change in price, with all else held constant, or

$$\varepsilon_p = \frac{\%\Delta e}{\%\Delta p} = \frac{de}{dp} \cdot \frac{p}{e}$$

where p denotes the price of energy.

Cooper (2003) uses a multiple regression model derived from an adaptation of Nerlove's (1958) partial adjustment model to estimate both the short-run and the long-run elasticity of demand for crude oil in 23 countries over a 30-year period from 1971 to 2000. The estimates so obtained confirm that the demand for crude oil internationally is highly insensitive to changes in price.

Demand substitution between energy commodities and others

Denny, Fuss, and Waverman (1981) used time-series data for 18 US manufacturing two-digit industries (1948–71) and 18 Canadian manufacturing industry groups (1962–75). Their results were also mixed: for both the United States and Canada, energy and capital were substitutes in the food industry, but they were complements in the tobacco industry.

Energy consumption can be modelled either as providing utility to households or as an input in the production process for firms. To express the former problem mathematically, a representative consumer maximizes utility, $U(z, e)$, which is function of energy consumption, e, and all other consumption, z, subject to the constraint that expenditures cannot exceed income, y. Let the energy variable be a vector of n energy products, $e = (e_1, e_2, \ldots, e_n)$; we could examine the substitution possibilities across energy products. Allowing the price of good j to be represented as p_j, the consumer is assumed to

$$\max_{z, e_1, \ldots e_n} U(z, e_1, \ldots e_n)$$

$$\text{subject to: } y \geq p_z Z + p_{e_1} e_1 + \ldots + p_{e_n} e_n$$

The first order necessary conditions for a maximum for this problem can be solved to yield demand equations for each of the energy products and for all other consumption. With some adjustments, the above method can be applied to a representative firm.

Recent research focuses mainly on dynamic models. Dynamic models allow for a more complete analysis of the energy demand because they are capable of capturing factors that generate the asymmetries. In addition, dynamic models incorporate the intertemporal choices that a consumer/firm must make when maximizing utilities or profits over some time horizon. Medlock and Soligo (2002) developed a useful framework. Let z_t be multiple types of capital and e_t be multiple types of energy consumption. Denoting time using the subscript t, the consumer will maximize the discounted sum of lifetime utility, $\sum_{t=0}^{T} \beta^t U(z_t, e_t)$, subject to the constraint whereby capital goods purchases (i_t), purchases of other goods (z_t), purchases of energy (e_t), and savings (s_t) in each period cannot exceed this period's income (y_t), plus the return of last period's saving ($(1 + r)s_{t-1}$). It is assumed that capital goods depreciate at a rate δ, savings earn a rate return r, the discount rate is $0 < \beta < 1$, and all initial conditions are given.

Consumers will

$$\max_{z, e, s} \sum_{t=0}^{T} \beta^t U(z_t, e_e) \text{ subject to}$$

$$p_{zt} z_t + p_{et} e_t + p_{kt} i_t + s_t \leq y_t + (1 + r)s_{t-1}$$

$$i_t = k_t - (1 - \delta)k_{t-1}$$

$$z_t, u_t, k_t \geq 0 \text{ for } t = 1, \ldots, T$$

Medlock and Soligo (2002) indicate that the income elasticity of passenger vehicle demand is decreasing as the real GDP per capita increases, no matter in the long run or in the short run. For example, with 1988 purchasing power parity dollar, if the real GDP per capita is $500, the

short-run elasticity is 0.74 and the long-run elasticity is 3.61; if the real GDP per capita is $20,000, the short-run and the long-run elasticity are 0.02 and 0.09, respectively.

Energy supply

OPEC

The Organization of the Petroleum Exporting Countries (OPEC) comprises countries that have organized for the purpose of negotiating with oil companies on matters of petroleum production, prices, and future concession rights. Founded on 14 September 1960 at a Baghdad conference, OPEC originally consisted of only five countries – Iran, Iraq, Kuwait, Saudi Arabia and Venezuela – but has since expanded to include several others: Algeria, Indonesia, Libya, Nigeria, Qatar and United Arab Emirates. The members of OPEC, which constitute a cartel, agree on the quantity and the prices of the oil exported. OPEC seeks to regulate oil production, and thereby manage oil prices, primarily by setting quotas for its members. Member countries hold about 75 per cent of the world's oil reserves, and supply 40 per cent of the world's oil. Loury (1990) is an excellent clarification; it studies a dynamic, quantity-setting duopoly game. The author considers a model of competition between two independent firms, A and B, facing indivisibility in production, with given limitations on their cumulative capacities to produce. At date t the flow rates of production of firms A and B are denoted by q_t^a and q_t^b respectively. The demand side of the market is passively modelled; buyers do not behave strategically. There is an inverse demand function, $P(\cdot)$, which is time invariant and dependent only on the total rate of flow of output of the two firms. Define the discount factor δ^t, a dollar received on date t is worth dollars at date zero. Then their respective payoffs are V_A and V_B where:

$$V_A = \beta \sum_t \delta^t [q_t^a P(Q_t)]; \text{ and}$$

$$V_B = \beta \sum_t \delta^t [q_t^b P(Q_t)]$$

for β, the lump sum equivalent of the flow of one dollar. It is shown that the ability to precommit can be disadvantageous. Loury (1990) also formalizes the intuition that, when indivisibilities are important, tacit coordination of plans so as to avoid destructive competition is facilitated by establishing a convention of 'taking turns', that is, a self-enforcing norm of mutual, alternate forbearance. Since worldwide oil sales are denominated in US dollars, changes in the value of the dollar against other world currencies affect OPEC's decisions on how much oil to produce. After the introduction of the euro, Iraq unilaterally decided it wanted to be paid for its oil in euros instead of US dollars.

OPEC decisions have a strong influence on international oil prices. A good example is the 1973 energy crisis, in which OPEC refused to ship oil to Western countries that had supported Israel in its conflict with Egypt, the Yom Kippur War. This refusal caused a fourfold increase in oil prices, which lasted five months, starting on 17 October 1973 and ending on 18 March 1974. OPEC nations then agreed, on 7 January 1975, to raise crude oil prices by ten per cent. The high and rising price of oil burdens industrial oil-importing countries in two ways. First, it renders the standard of living lower than otherwise. Second, it affects the economy in ways that are difficult for policymakers to manage: on the one hand, the rising oil price spurs general inflation; on the other hand, it depresses domestic demand and employment. Unlike many other cartels, OPEC has been successful at increasing the price of oil for extended periods. Much of OPEC's success can be attributed to Saudi Arabia's flexibility. It has tolerated cheating on the part of other cartel members, and cut its own production to compensate for other members exceeding their production quotas. This actually gives them good leverage because, with most members at full production, Saudi Arabia is the only member with spare capacity and the ability to increase supply, if needed. The policy has been successful. However, OPEC's ability to raise prices does have some limits. An increase in oil price decreases consumption, and could cause a net decrease in revenue. Furthermore, an extended rise in price could encourage systematic behaviour change, such as alternative energy utilization, or increased conservation. As of August 2004, OPEC has been communicating that its members have little excess pumping capacity, indicating that the cartel is losing influence over crude oil prices.

The six major non-OPEC oil-producing nations are Norway, Russia, Canada, Mexico, the United States and Oman. Russian production increases dominated non-OPEC production growth from 2000 onward and was responsible for most of the non-OPEC increases since the turn of the century. In 2001, a weakening US economy and increases in non-OPEC production put downward pressure on prices. In response OPEC once again entered into a series of reductions in member quotas, cutting production by 3.5 million barrels per day by 1 September 2001. In the absence of the September 11, 2001 terrorist attack this would have been sufficient to moderate or even reverse the trend.

In the wake of that attack the crude oil price plummeted. Under normal circumstances a drop in price of this magnitude would have resulted in another round of quota reductions, but, given the political climate, OPEC delayed additional cuts until January 2002, when it reduced its quota by 1.5 million barrels per day and was joined by several non-OPEC producers, including Russia, which promised combined daily production cuts of an additional 462,500 barrels. This had the desired effect, with oil prices moving into the $25 per barrel range by March 2002. By mid-year the non-OPEC members were restoring their production cuts, but prices continued to rise and US inventories reached a 20-year low later in the year. By year's end oversupply was not a problem. Problems in Venezuela led to a strike at Petroleos de Venezuela (PDVSA) causing Venezuelan production to plummet. In the wake of the strike Venezuela was never able to restore capacity to its previous levels. On 19 March 2003, just as some Venezuelan production was beginning to return, military action began in Iraq. Meanwhile, inventories remained low in the United States and other OECD countries. With an improving economy US demand was increasing, and Asian demand for crude oil was growing at a rapid pace. The loss of production capacity in Iraq and Venezuela, combined with increased production to meet growing international demand, led to the erosion of excess oil production capacity. During much of 2004 and 2005 the spare capacity to produce oil has been less than one million barrels per day. A million barrels per day is not enough spare capacity to cover an interruption of supply from almost any OPEC producer. In a world that consumes over 80 million barrels of petroleum products per day, that adds a significant risk premium to crude oil price and is largely responsible for prices in excess of $40 per barrel. For further information, see Energy Information Administration (EIA).

Future energy supply

Undoubtedly, depletable resource use cannot dominate forever. Therefore, a future transition from depletable resources, particularly from fossil fuels, is inevitable. However, which renewable energy sources will dominate future consumption is unclear. And there is great uncertainty about the timing of a shift to renewable energy resources. Although this is a formidable question, Wiser et al. (2004) introduce green pricing programmes, which represent one way whereby consumers can voluntarily support renewable energy. Their analysis yields several interesting results. Programme duration affects customer response. The longer a programme has been operating, the more likely it is that its message has spread and the higher the probability of strong programme success. Initial customer participants in green pricing programmes may not be highly sensitive to cost, and may be willing to purchase higher quantities of renewable energy, which makes the case for utilities focusing on maximizing renewable energy sales, not customer participation rates. Price premiums and minimum monthly costs are not the primary determinants of programme success. Price may become a more important determinant as green pricing programmes expand beyond the early innovator customers. And smaller utilities appear to have a greater likelihood of achieving success.

The prospect of producing clean, sustainable power in substantial quantities from renewable energy sources is arousing renewed interest worldwide. Hydroelectricity is the only renewable energy source today that makes a large contribution to world energy production. Its long-term technical potential is believed to be 9 to 12 times current production, but increasingly environmental concerns block new dams. The large areas affected may have a negative environmental impact. Hydroelectricity dams, like the Aswan Dam, have adverse consequences both upstream and downstream.

Wind power is one of the most cost-competitive renewable sources today. Its long-term technical potential is believed to be five times current global energy consumption. But this requires 12.7 per cent of all land area and the facilities have to be built at certain height. Geothermal power and tidal power are the only renewable sources not dependent on the sun, but are today limited to special locations. Most renewable sources are diffuse and require large land areas and great quantities of construction material for significant energy production. There is some doubt that they can be built rapidly enough to replace fossil fuels. The large and sometimes remote areas may also increase energy loss and cost from distribution. On the other hand, some forms allow small-scale production and may be placed very close to or directly at consumer households, businesses, and industries. We may forecast the future coexistence of multi-renewable energy sources. Boyle (1996) provides a comprehensive overview of the principal renewable energy sources: solar thermal, biomass, tidal, wave, photovoltaic, hydro, wind and geothermal.

Forecasts of the energy markets

According to Energy Information Administration (EIA, 2005b), based on its expectations for world energy prices, world energy consumption is projected to increase by 57 per cent from 2002 to 2025. World oil use is expected to grow from 78 million barrels per day in 2002 to 103 million barrels per day in 2015 and 119 million barrels per day in 2025. The projected increment in worldwide oil use would require an increment in world oil production capacity of 42 million barrels per day above 2002 levels.

Members of OPEC are expected to be the major suppliers of the increased production that will be required to meet demand, accounting for 60 per cent of the projected increase in world capacity. In addition, non-OPEC suppliers are expected to add nearly 17 million barrels per day of oil production capacity between 2002 and 2025. Substantial increments in new non-OPEC supply are expected to come from the Caspian Basin, Western Africa, and Central and South America.

Natural gas is projected to be the fastest-growing component of world primary energy consumption. Consumption of natural gas worldwide increases in the forecast by an average of 2.3 per cent annually from 2002 to 2025, compared with projected annual growth rates of 1.9 per cent for oil consumption and 2.0 per cent for coal consumption. From 2002 to 2025, consumption of natural gas is projected to increase by 69 per cent, and its share of total energy consumption is projected to grow from 23 to 25 per cent.

Natural gas is seen as a desirable alternative to electricity generation in many parts of the world, given its relatively efficiency in comparison with other energy sources, as well as the fact that it burns more cleanly than either coal or oil and thus is an attractive alternative for countries pursuing reductions in greenhouse gas emission.

World coal consumption is projected to increase at an average rate of 2.5 per cent per year. From 2015 to 2025, the projected rate of increase in world coal consumption slows to 1.3 per cent annually. Coal is expected to maintain its importance as an energy source in both the electric power and industrial sectors.

Hydroelectricity and other renewable energy sources are expected to maintain their 8 per cent share of total energy use worldwide throughout the projection period. Much of the projected growth in renewable electricity generation is expected to result from the completion of large hydroelectric facilities in emerging economies, particularly in Asia.

Energy policies

The study of depletable resource economics began with articles by H. Hotelling (1931), which examined economically intertemporal optimal extraction from a perfectly known stock of the resource, with perfectly predictable future prices of the extracted commodity. Sweeney (1977) and Stiglitz (1976) both clarified the Hotelling rule in the presence of monopoly, and Gilbert and Richard (1978) and Salant (1976) extended this to the case of a dominant producer with a competitive fringe and several dominant producers, analogous to the case of OPEC. Pindyck (1982) and Kolstad (1994) extended the model to several imperfectly substitutable exhaustible resources.

Energy security refers to loss of economic welfare that may occur as a result of a change in the price of availability of energy. In the years following the 1973 oil price rise, US energy policy could be characterized as generally suspicious of the market. Supply augmentation was a major strategy pursued by the US government in addressing the 'energy crisis'. The security dimensions of energy supply have always been viewed as appropriate concerns of the government. One could argue that the Gulf War in the early 1990s was simply a form of energy policy, protecting Western oil supplies originating in the Middle East. Countries other than the United States (such as Japan and China) have tried to diversify their sources of energy to reduce the risk of disruption. Security was also viewed as threatened by sudden fluctuations in the price of oil, hence the establishment in the United States of the Strategic Petroleum Reserve (SPR): petroleum.

Stocks are maintained by the federal government for use during periods of major supply interruption. The idea is that, if the price of oil were to rise rapidly due to disruption in supply, then the SPR could be called upon to provide supplies, thus reducing the price shock.

Nuclear power was declared dead in the United States because it is too expensive and unacceptably risky. Around the world, nuclear plant ended up achieving less than ten per cent of the new capacity and one per cent of the new orders (all from countries with centrally planned energy systems) forecast in the early 1980. The industry has suffered the greatest collapse of any enterprise in industrial history. Scientists still have not developed reliable ways to handle nuclear wastes and decommissioned plants, which remain dangerously radioactive for far longer than societies last or geological foresight extends.

Strong economic growths across the globe and new global demands for more energy have meant the end of sustained surplus capacity in hydrocarbon fuels and the beginning of capacity limitations. In fact, the world is currently precariously close to utilizing all of its available oil-production capacity, raising the chances of an oil-supply crisis with more substantial consequences than seen since the early 1970. These limits mean that the United States can no longer assume that oil-producing states will provide more oil. Nor is it strategically and politically desirable for the United States to remedy its present tenuous situation by simply increasing its dependence on a few foreign sources. As a result, expanding demand for energy will change US policy towards the Middle East, Russia and China. A recent example is that, in 2005, the state-owned Chinese company CNOOC eventually abandoned its bid for Unocal due to strong political opposition in the United States.

Effects of energy demand

Energy and macroeconomics
In fact, almost every recession since the Second World War in the United States, as well as many other energy-importing nations, has been preceded by a spike in the price of energy (Hamilton, 1983; Ferderer, 1996; Mork, Mysen and Olsen, 1994). The oil price movement affects certain sectors: oil-dependent manufacturing such as paper and packaging, consumer-related sectors such as autos, refiners' margins, the energy-intensive utility sector, and of course exploration companies and the big oil majors themselves.

Energy, economy and environment
Many important environment damages stem from the production, conversion, and consumption of energy. The costs of these environmental damages generally are not incorporated into prices for energy commodities and resources; this omission leads to overuse of energy. It has been shown that estimates of damage costs resulting from combustion of fossil fuels, if internalized into the price of the resulting output of electricity, could clearly

lead to a number of renewable technologies being financially competitive with generation from coal plants. Environmental impacts currently receiving most attention are associated with the release of greenhouse gases in the atmosphere, primarily carbon dioxide, from the combustion of fossil fuels. During combustion, carbon combines with oxygen to produce carbon dioxide, the primary greenhouse gas. Carbon dioxide accumulates in the atmosphere and is expected to result in significant detrimental impacts on the world's climate, including global warming, rises in the ocean levels, increased intensity of tropical storms, and losses in biodiversity. Concern about this issue is common to energy economics, environmental economics, and ecological economics. Cropper and Oates (1992) suggest measuring benefits and costs with a review of cases where benefit–cost analyses have actually been used in the setting of environmental standards. Owen (2004) suggests that penalizing high pollutant-emitting technologies not only creates incentives for 'new' technologies but also encourages the adoption of energy-efficiency measures with existing technologies and consequently lower pollutants per unit of output.

World carbon dioxide emissions are expected to increase by 1.9 per cent annually between 2001 and 2025. Much of this increase is expected to occur in developing countries. The United States produces about 25 per cent of global carbon dioxide emissions from burning fossil fuels, primarily because of it has the largest economy in the world and meets 85 per cent of its energy needs through burning fossil fuels. The United States is projected to lower its carbon intensity by 25 per cent from 2001 to 2025. There are numerous proposals aimed at reducing the carbon dioxide emissions, of which the Kyoto Protocol is a well-known and influential one. During 1–11 December 1997, more than 160 nations met in Kyoto, Japan, to negotiate binding limitations on greenhouse gases for the developed nations, pursuant to the objectives of the Framework Convention on Climate Change of 1992. The outcome of the meeting was the Kyoto Protocol, in which the developed nations agreed to limit their greenhouse gas emissions relative to the levels emitted in 1990. The United States agreed to reduce emissions from 1990 levels by seven per cent during the period 2008 to 2012.

Sickles and Jeon (2004) evaluate the role that undesirable outputs of the economy, such as carbon dioxide and other greenhouse gases, play on the frontier production process. This paper also explores implications for growth of total factor productivity in the OECD and Asian economies.

Natural disasters shock the energy market, too. According to the Minerals Management Service (MMS, 2005), Gulf of Mexico daily oil production was reduced by 89 per cent as a result of Hurricane Katrina in 2005. The MMS also reports that 72 per cent of daily Gulf of Mexico natural gas production was shut in. In 2004,

Hurricane Ivan caused lasting damage to the energy infrastructure in the Gulf of Mexico and interrupted oil supplies to the United States. US Secretary of Energy Spencer Abraham agreed to release 1.7 million barrels of oil in the form of a loan from the Strategic Petroleum Reserve.

A concluding comment

The world runs on energy, primarily energy generated from coal and petroleum. The current war against terrorism and the tensions in the Middle East have raised new questions about the reliability of America's oil supply from that region. Concerns about global climate change have also focused increased attention on the search for cleaner fuels and energy-generating methods. Russia's determination to become a major petroleum supplier, OPEC's periodic moves to restrict oil production and the rising energy needs in China and other developing countries are all important issues forming the future world energy market.

I would like to thank Robert Thomure, Rice University, for his research assistance.

ROBIN SICKLES

See also **environmental economics; oil and the macroeconomy; Organization of the Petroleum Exporting Countries (OPEC).**

Bibliography

Bohi, D. and Toman, M. 1996. *The Economics of Energy Security.* Boston: Kluwer Academic Publishers.

Boyle, G. 1996. *Renewable Energy: Power for a Sustainable Future.* Oxford: Oxford University Press.

Cooper, J. 2003. Price elasticity of demand for crude oil: estimates for 23 countries. *OPEC Review: Energy Economics & Related Issues* 27(1), 1–8.

Cropper, M. and Oates, W. 1992. Environmental economics: a survey. *Journal of Economic Literature* 30, 675–740.

Denny, M., Fuss, M. and Waverman, L. 1981. Substitution possibilities for energy: evidence from U.S. and Canadian manufacturing. In *Modeling and Measuring Natural Resource Substitution*, ed. E. Berndt and B. Field. Cambridge, MA: MIT Press.

EIA (Energy Information Administration). 2001. *Annual Energy Review 2000.* Washington, DC: EIA.

EIA. 2004. *Annual Energy Review 2004.* Washington, DC: EIA.

EIA. 2005a. *Annual Energy Review 2005.* Washington, DC: EIA.

EIA. 2005b. *Annual Energy Outlook 2005 with Projections to 2025.* Washington, DC: EIA.

Fang, F. and Sickles, R. 2004. The role of environmental factors in growth accounting. *Journal of Applied Econometrics* 19, 567–91.

Ferderer, P. 1996. Oil price volatility and macroeconomy. *Journal of Macroeconomics* 18, 1–26.

Gilbert, R. and Richard, J. 1978. Dominant firm pricing policy in a market for an exhaustible resource. *Bell Journal of Economics* 9, 385–95.

Hamilton, J. 1983. Oil and the macroeconomy since World War II. *Journal of Political Economy* 91, 228–48.

Hotelling, H. 1931. The economics of exhaustible resources. *Journal of Political Economy* 39, 137–75.

Judson, R., Schmalensee, R. and Stoker, T. 1999. Economic development and the structure of the demand for commercial energy. *Energy Journal* 20(2), 29–7.

Kolstad, C. 1994. Hotelling rents in hotelling space: product differentiation in exhaustible resource markets. *Journal of Environmental Economics and Management* 26, 163–80.

Loury, G. 1990. Tacit collusion in a dynamic duopoly with indivisible production and cumulative capacity constraints. Working Paper No. 557. Department of Economics, MIT.

Medlock, K. and Soligo, R. 2001. Economic development and end-use energy demand. *Energy Journal* 22(2), 77–105.

Medlock, K. and Soligo, R. 2002. Car ownership and economic development with forecasts to 2015. *Journal of Transport Economics and Policy* 36, 163–88.

MMS (Minerals Management Service). 2005. *Hurricane Information.* Washington, DC: MMS.

Mork, K., Mysen, H. and Olsen, O. 1994. Macroeconomic responses to oil price increases and decreases in seven OECD countries. *Energy Journal* 15(4), 19–35.

Nerlove, M. 1958. Distributed lags and demand analysis for agricultural and other commodities. In *Agriculture Handbook* No. 141. Washington, DC: US Department of Agriculture.

Owen, A. 2004. Environmental externalities, market distortions and the economics of renewable energy technologies. *Energy Journal* 25(3), 127–56.

Pindyck, R. 1982. Jointly produced exhaustible resources. *Journal of Environmental Economics and Management* 9, 291–303.

Salant, S. 1976. Exhaustible resources and industrial structure – Nash–Cournot approach to the world oil market. *Journal of Political Economy* 84, 1079–93.

Sickles, R. B. and Jeon, B. M. 2004. The role of environmental factors in growth accounting. *Journal of Applied Econometrics* 19, 567–91.

Stiglitz, J. 1976. Monopoly and rate of extraction of exhaustible resources. *American Economic Review* 66, 655–61.

Sweeney, J. 1977. Economics of depletable resources – market forces and intertemporal bias. *Review of Economic Studies* 44, 125–41.

Wiser, R., Olson, S., Bird, L. and Swezey, B. 2004. *Utility Green Pricing Programs: A Statistical Analysis of Program Effectiveness.* Berkeley, CA and Golden, CO: Ernest Orlando Lawrence Berkeley National Laboratory and National Renewable Energy Laboratory. Online. Available at http://www.eere.energy.gov/greenpower/resources/pdfs/lbnl_54437.pdf, accessed 1 August 2006.

Engel, Ernst (1821–1896)

Born in Dresden, Engel was a German statistician best known for the discovery of the Engel curve and of Engel's Law. In his early years he was associated with the French sociologist Frédéric Le Play, whose interest in the family led him to conduct household surveys. The expenditure data collected in these surveys convinced Engel that there was a relation between a household's income and the allocation of its expenditures between food and other items. This was one of the first functional relations ever established quantitatively in economics. Furthermore, he observed that households with higher incomes tended to spend more on food than poorer households, but that the share of food expenditures in the total budget tended to vary inversely with income. From this empirical regularity he went on to infer that in the course of economic development agriculture would decline relative to other sectors of the economy (Engel, 1857). From 1860 to 1882 Engel was director of the Prussian statistical bureau in Berlin, in which capacity he did much to expand and strengthen official statistics. His resignation resulted from his opposition to Bismarck's protectionist policies. In his own research he dealt particularly with the value of human life (Engel, 1877), which he approached from the cost side. He also investigated the influence of price on demand. His influence on official statistics extended well beyond Germany, and in 1885 he was among the founders of the International Statistical Institute. He died in Radebeul in 1896.

H.S. HOUTHAKKER

Selected works

1857. Die Productions- und Consumptionsverhaeltnisse des Koenigsreichs Sachsen. Reprinted with Engel (1895), *Anlage* I, 1–54.

1877. *Der Kostenwerth des Menschen*. Berlin.

1895. Die Lebenskosten Belgisher Arbeiter-Familien fruether und jetzt. Reprinted in *International Statistical Institute Bulletin* 9, 1–124.

Engel curve

An Engel curve is the function describing how a consumer's expenditures on some good or service relate to the consumer's total resources, with prices fixed, so $q_i = g_i(y, z)$, where q_i is the quantity consumed of good i, y is income, wealth, or total expenditures on goods and services, and z is a vector of other characteristics of the consumer, such as age and household composition. Usually y is taken to be total expenditures, to separate the problem of allocating total consumption to various goods from the decision of how much to save or dissave out of current income. Engel curves are frequently expressed in the budget share form $w_i = h_i[\log(y), z]$ where w_i is the fraction of y that is spent buying good i. The goods are typically aggregate commodities such as

total food, clothing or transportation, consumed over some weeks or months, rather than discrete purchases. Engel curves can be defined as Marshallian demand functions, with the prices of all goods fixed.

The term 'Engel curve' is also used to describe the empirical dependence of q_i on y, z in a population of consumers sampled in one time and place. This empirical or statistical Engel curve coincides with the above theoretical Engel curve definition if the law of one price holds (all sampled consumers paying the same prices for all goods), and if all consumers have the same preferences after conditioning on z and possibly on some well-behaved error terms. Since these conditions rarely hold, it is important in practice to distinguish between these two definitions.

Using data from Belgian surveys of working class families, Ernst Engel (1857; 1895) studied how households' expenditures on food vary with income. He found that food expenditures are an increasing function of income and of family size, but that food budget shares decrease with income. This relationship of food consumption to income, known as Engel's law, has since been found to hold in most economies and time periods, often with the function h_i for food i close to linear in $\log(y)$.

Engel curves can be used to calculate a good's income elasticity, which is roughly the percentage change in q_i that results from a one per cent change in y, or formally $\partial \log g_i(y, z)/\partial \log(y)$. Goods with income elasticities below zero, between zero and 1, and above 1 are called inferior goods, necessities and luxuries respectively, so by these definitions what Engel found is that food is a necessity. Elasticities can themselves vary with income, so a good that is a necessity for the rich can be a luxury for the poor.

Some empirical studies followed Engel (1895), such as Ogburn (1919), but Allen and Bowley (1935) firmly connected their work to utility theory. They estimated linear Engel curves $q_i = a_i + b_i y$ on data-sets from a range of countries, and found that the resulting errors in these models were sometimes quite large, which they interpreted as indicating considerable heterogeneity in tastes across consumers. Working (1943) proposed the linear budget share specification $w_i = a_i + b_i \log(y)$, which is known as the Working–Leser model, since Leser (1963) found this functional form to fit better than some alternatives. However, Leser obtained still better fits with what would now be called a rank-three model, namely, $w_i = a_i + b_i \log(y) + c_i y^{-1}$, and in a similar, earlier, comparative statistical analysis Prais and Houthakker (1955) found $q_i = a_i + b_i \log(y)$ to fit best. More recent work documents sometimes considerable nonlinearity in Engel curves. Motivated by this nonlinearity, one of the earlier empirical applications of nonparametric regression methods in econometrics was kernel estimation of Engel curves. Examples include Bierens and Pott-Buter (1990), Lewbel (1991), and Härdle and Jerison (1991). More recent studies that control for complications like measurement error and other covariates z, including Hausman, Newey and Powell (1995) and Banks, Blundell and Lewbel (1997), find Engel curves for some goods are close to Working–Leser, while others display considerable curvature, including quadratics or S shapes. Even Allen and Bowley (1935, p. 123) noted 'there is a good fit, allowance being made for observation and sampling errors,…, to a linear expenditure relation and occasionally to a parabolic relation'.

Other variables z also help explain cross-section variation in demand. Commonly used covariates include the number, ages and gender of family members, location measures, race and ethnicity, seasonal effects, and labour market status. Variables indicating ownership of a home, a car or other large durables can also have considerable explanatory power, though these are themselves consumption decisions.

Engel's original work showed the relevance of family size, and later studies confirm that larger families typically have larger budget shares of necessities than smaller families at the same income level. Adult equivalence scales model the dependence of utility functions on family size, and use this dependence to compare welfare across households, assuming that a large family with a high income is as well off as a smaller family with a lower income if both families have demands that are similar in some way, such as equal food budget shares or equal expenditures on adult goods such as alcohol. The ratio of total expenditures needed to equate food budget shares across households are known as Engel equivalence scales, while the ratio that equates expenditures on adult goods are called Rothbarth scales (Rothbarth, 1943).

Shape invariance assumes that budget share Engel curves for one type of consumer, such as a household with children, is a linear transformation of the budget-share Engel curves for other types of consumers, such as households without children. Shape invariance is necessary for constructing what are known as exact or independent of base equivalence scales, and has been found to at least approximately hold in some data-sets. See Lewbel (1989), Blackorby and Donaldson (1991), Gozalo (1997), Pendakur (1999), and Blundell, Browning and Crawford (2003).

The level of aggregation across goods affects Engel curve estimates. Demand for a narrowly defined good like apples varies erratically across consumers and over time, while Engel curves based on broad aggregates like food are affected by variation in the mix of goods purchased. The aggregate necessity food could include inferior goods like cabbage and luxuries like caviar, which may have very different Engel curve shapes.

Other empirical Engel curve complications include unobserved variations in the quality of goods purchased, and violations of the law of one price. When price or quality variation is unobserved, their effects may correlate with, and so be erroneously attributed to, y or z. Examples of such correlations could include the wealthy

systematically favouring higher quality goods, and the poor facing higher prices than other consumers because they cannot afford to travel to discount stores.

Assume a consumer (household) h determines demands q_{hi} facing prices p_i for each good i by maximizing a well-behaved utility function over goods (which could depend on z_h), subject to a budget constraint $\sum_i p_i q_{hi} \leq y_h$. This yields Marshallian demand functions $q_{hi} = G_{hi}(p, y_h, z_h)$, with Engel curves given by these functions with the price vector p fixed. Utility functions that yield Engel curves of the form $q_{hi} = b_i(z)y_h$ are called homothetic, and $q_{hi} = a_i(z) + b_i(z)y_h$ are quasihomothetic. Many theoretical results regarding two-stage budgeting and aggregation across goods require homotheticity or quasihomotheticity, most notably Gorman (1953).

The shape of Engel curves plays an important role in the determination of macroeconomic demand relationships. For example, if we ignore z for now, suppose individual consumers h each have Engel curves of the quasihomothetic form $q_{hi} = a_{hi} + b_i y_h$. Then, letting Q_i and Y be aggregate per capita quantities and total expenditures in the population, we get $Q_i = A_i + b_i Y$ by averaging q_{hi} across consumers h. This is a representative consumer model, in the sense that the distribution of y affects aggregate demand Q_i only through its mean $E(y) = Y$. Gorman (1953) showed that only linear Engel curves have this property, though linear Engel curve aggregation dates back at least to Antonelli (1886). Gorman's linearity requirement, which does not usually hold empirically, can be relaxed given restrictions on the distribution of y; for example, Lewbel (1991) shows that $E(y \log y)/Y - \log(Y)$ is very close to constant in US data, and if it is constant then Working–Leser household Engel curves yield Working–Leser aggregate, representative consumer demands.

Exactly aggregable demands are defined by $q_i = \sum_{j=1}^{J} A_{ji}(p)c_j(y,z)$, and so have Engel curves $q_i = \sum_{j=1}^{J} a_{ji}c_j(y,z)$ that are linear in the functions $c_j(y,z)$. These models have the property that aggregate demands Q_i depend only on the means of $c_j(y,z)$. Utility theory imposes constraints on the functional forms of $c_j(y,z)$. Properties of exactly aggregable demands and associated Engel curves are derived in Muellbauer (1975), Jorgenson, Lau and Stoker (1982), and Lewbel (1990), but primarily by Gorman (1981), who proved the surprising result that utility maximization forces the matrix of Engel curve coefficients a_{ji} to have rank three or less.

Lewbel (1991) extends Gorman's rank idea to arbitrary demands, not just those in the exactly aggregable class, by defining the rank of a demand system as the dimension of the space spanned by its Engel curves. Engel curve rank can be nonparametrically tested, and has implications for utility function separability, welfare comparisons, and for aggregation across goods and across consumers. Many empirical studies find demands have rank three.

One area of current research concerns the observable implications of collective models, that is, households that determine expenditures based on bargaining among members. For example, the Engel curves of such households could violate Gorman's rank theorem, even if each member had exactly aggregable preferences. Another topic attracting current attention is the role of errors in demand models, particularly their interpretation as unobserved preference heterogeneity, random utility model parameters. This matters in part because another of Allen and Bowley's (1935) findings remains true today, namely, Engel curve and demand function models still fail to explain most of the observed variation in individual consumption behaviour.

ARTHUR LEWBEL

See also **aggregation (theory); consumer expenditure; demand theory; Engel, Ernst; Engel's Law; equivalence scales; Gorman, W.M. (Terence); utility.**

Bibliography

Allen, R. and Bowley, A. 1935. *Family Expenditure: A Study of its Variation*. London: P.S. King and Son.

Antonelli, G. 1886. Sulla teoria metematica della economia politica. Pisa: Nella Tipografia del Fochetto. Translated as 'On the mathematical theory of political economy', in *Preferences, Utility and Demand*, ed. J. Chipman et al. New York: Harcourt Brace Jovanovich, 1971.

Banks, J., Blundell, R. and Lewbel, A. 1997. Quadratic Engel curves and consumer demand. *Review of Economics and Statistics* 79, 527–39.

Bierens, H. and Pott-Buter, H. 1990. Specification of household expenditure functions and equivalence scales by nonparametric regression. *Econometric Reviews* 9, 123–210.

Blackorby, C. and Donaldson, D. 1991. Adult-equivalence scales, interpersonal comparisons of well-being, and applied welfare economics. In *Interpersonal Comparisons of Well-Being*, ed. J. Elster and J. Roemer. Cambridge: Cambridge University Press.

Blundell, R., Browning, M. and Crawford, I. 2003. Nonparametric Engel curves and revealed preference. *Econometrica* 71, 205–40.

Engel, E. 1857. Die Productions- und Consumptionsverhaeltnisse des Koeniglreichs Sachsen. *Zeitschrift des Statistischen Bureaus des Koniglich Sachsischen Ministeriums des Inneren, No. 8 und 9.* Reprinted in the Appendix of Engel (1895).

Engel, E. 1895. Die Lebenskosten Belgischer Arbeiter-Familien Fruher und jetzt. *International Statistical Institute Bulletin* 9, 1–74.

Gorman, W. 1953. Community preference fields. *Econometrica* 21, 63–80.

Gorman, W. 1981. Some Engel curves. In *Essays in the Theory and Measurement of Consumer Behaviour in Honor of Sir Richard Stone*, ed. A. Deaton. Cambridge: Cambridge University Press.

Gozalo, P. 1997. Nonparametric bootstrap analysis with applications to demographic effects in demand functions. *Journal of Econometrics* 81, 357–93.

Härdle, W. and Jerison, M. 1991. Cross section Engel curves over time. *Recherches Economiques de Louvain* 57, 391–431.

Hausman, J., Newey, W. and Powell, J. 1995. Nonlinear errors in variables: estimation of some Engel curves. *Journal of Econometrics* 65, 205–53.

Jorgenson, D., Lau, L. and Stoker, T. 1982. The transcendental logarithmic model of aggregate consumer behavior. In *Advances in Econometrics*, ed. R. Basman and G. Rhodes. Greenwich: JAI Press.

Leser, C. 1963. Forms of Engel functions. *Econometrica* 31, 694–703.

Lewbel, A. 1989. Household equivalence scales and welfare comparisons. *Journal of Public Economics* 39, 377–91.

Lewbel, A. 1990. Full rank demand systems. *International Economic Review* 31, 289–300.

Lewbel, A. 1991. The rank of demand systems: theory and nonparametric estimation. *Econometrica* 59, 711–30.

Muellbauer, J. 1975. Aggregation, income distribution, and consumer demand. *Review of Economic Studies* 62, 269–83.

Ogburn, W. 1919. Analysis of the standard of living in the District of Columbia in 1916. *Journal of the American Statistical Association* 16, 374–89.

Pendakur, K. 1999. Estimates and tests of base-independent equivalence scales. *Journal of Econometrics* 88, 1–40.

Prais, S. and Houthakker, H. 1955. *The Analysis of Family Budgets*. Cambridge: Cambridge University Press (2nd edn. 1971).

Rothbarth, E. 1943. Note on a method of determining equivalent income for families of different composition. Appendix 4 in *War-time Pattern of Saving and Spending*, ed. C. Madge. Cambridge: Cambridge University Press.

Working, H. 1943. Statistical laws of family expenditures. *Journal of the American Statistical Association* 38, 43–56.

Engel's Law

Engel's law states that food is not a luxury. This is one of the earliest empirical regularities in economics and also one of the most robust. The widespread finding is that regressions of food expenditures, quantities or budget shares on income or total expenditure and other variables such as prices, demographics and regional dummies uniformly imply that the income elasticity of food is less than 1 (and greater than zero). For example, time series from individual countries, cross-sections within countries and cross-country analyses all find the same qualitative empirical finding.

This correlation seems to have been highlighted for a number of reasons. First, food is an important component of household budgets everywhere so that it is intrinsically of interest. Second, the finding suggests that over the long run countries experiencing significant growth will find that agriculture provides an increasingly unimportant part of national income. This argues against balanced growth in long-run development. Third, we do not observe such a consistent pattern for any other wide commodity grouping such as clothing or durables. Finally, the fact that the food budget share is a decreasing function of the material standard of living (if other factors are held constant) suggested at one time that it can be used as an indicator of the latter. In particular, iso-prop ('same proportion') methods have been used to compute adult equivalence scales by finding the level of income that would equate the food budget share across different demographic groups. The conditions under which the iso-prop method is valid are very strong – essentially, extra people in the household have to make the household behave as though it is poorer and should not cause any change in the structure of demands above this – and such methods have fallen out of favour (see Deaton and Muellbauer, 1986, for discussion and references).

Despite the venerability of the literature on Engel's law, the inferences that can be drawn from it are limited. For example, the cross-section finding is consistent with all households having a decreasing relationship so that increasing the income of a household will lead to a decrease in the food budget share. On the other hand, the correlation might be completely spurious if it is due to poorer households having a higher 'taste' for food. In this case the apparent dependence is simply due to heterogeneity in tastes, which is correlated with income. The fact that studies using aggregate time series-data find different elasticities from those found in cross-section data from the same country and time period suggests that the empirical finding is a combination of both causes. The paucity of panel data with full expenditure information makes any inference hazardous. Thus Engel's law remains what it has always been: a very robust but unsurprising partial correlation with many alternative interpretations.

MARTIN BROWNING

See also **Engel, Ernst; Engel curve; equivalence scales.**

Bibliography

Deaton, A. and Muellbauer, J. 1986. On measuring child costs: with applications to poor countries. *Journal of Political Economy* 94, 720–44.

Engels, Friedrich (1820–1895)

Born in Barmen, the eldest son of a textile manufacturer in Westphalia, Engels was trained for a merchant's profession. From school onwards, however, he developed radical literary ambitions which eventually brought him into contact with the Young Hegelian circle in Berlin in 1841. In 1842, Engels left for England to work in his father's Manchester firm. Already converted by Moses Hess to a belief in 'communism' and the imminence of an English social revolution, he used his two-year stay to study the conditions which would bring it about. From this visit came two works which were to make an important contribution to the formation of Marxian socialism: *Outlines of a Critique of Political Economy* (generally called the *Umrisse*) published in 1844, and *The Condition of the Working Class in England*, published in Leipzig in 1845.

Returning home via Paris in 1844, Engels had his first serious meeting with Marx. Their lifelong collaboration dated from this point with an agreement to produce a joint work (*The Holy Family*), setting out their positions against other tendencies within Young Hegelianism. This was followed by a second unfinished joint enterprise (*The German Ideology*, 1845–6), where their materialist conception of history was expounded systematically for the first time.

Between 1845 and 1848, Engels was engaged in political work among German communist groups in Paris and Brussels. In the 1848 revolution itself, he took a full part, first as a collaborator of Marx on the *Neue Rheinische Zeitung* and subsequently in the last phase of armed resistance to counter-revolution in the summer of 1849.

In 1850, Engels returned once more to Manchester to work for his father's firm and remained there until he retired in 1870. During this period, in addition to numerous journalistic contributions, including attempts to publicize Marx's *Critique of Political Economy* (1859) and *Capital*, Volume 1 (1867, second edition 1873), he first developed his interest in the relationship between historical materialism and the natural sciences. These writings were posthumously published as *The Dialectics of Nature* (1925). In 1870 Engels moved to London.

As Marx's health declined, Engels took over most of his political work in the last years of the First International (1864–72) and took increasing responsibility for corresponding with the newly founded German Social Democratic Party and other infant socialist parties. Engels's most important work during this period was his polemic against the positivist German socialist, Eugen Dühring. The *Anti-Dühring* (1877) was the first comprehensive exposition of a Marxian socialism in the realms of philosophy, history and political economy. The success of this work, and in particular of extracts from it like *Socialism, Utopian and Scientific*, represented the decisive turning point in the international diffusion of Marxism and shaped its understanding as a theory in the period before 1914.

In his last years after Marx's death in 1883, Engels devoted most of his time to the editing and publishing of the remaining volumes of *Capital* from Marx's manuscripts. Volume 2 appeared in 1885, Volume 3 in 1894, a year before his death. Engels had also hoped to prepare the final volume dealing with the history of political economy. But the difficulty of deciphering Marx's handwriting, his own failing eyesight and the formidable editorial problems encountered in constructing Volumes 2 and 3, induced him to hand over this task to Karl Kautsky, who subsequently published it under the title *Theories of Surplus Value*.

Engels's work was of importance, both in the construction and interpretation of Marxian economic theory and in the laying down of important guidelines in the subsequent development of Marxist economic policy.

In the realm of theory, his contribution is of particular significance in three respects.

First, and of real importance in the formation of a distinctively Marxian stance towards political economy was Engels's *Outlines of a Critique of Political Economy* (the *Umrisse*), published in 1844. In 1859 in his own *Critique of Political Economy*, Marx acknowledged this sketch as 'brilliant', and its impact is discernible in Marx's 1844 writings. The *Umrisse* represented the first systematic confrontation between the 'communist' strand of Young Hegelianism and political economy. The communist aspiration was expressed in Feuerbachian language, while the mode of analysis was Hegelian. But, as has recently been demonstrated (Claeys, 1984), the content of Engels's critique was first and foremost a product of his early stay in Manchester. For, apart from some indebtedness to Proudhon's *What is Property?* (1841), the main source of Engels's essay was John Watts, *The Facts and Fictions of Political Economy* (1842), a resumé of the Owenite case against the propositions of political economy. At this stage, Engels's own acquaintance with the work of political economists seems to have been mainly at second-hand.

The *Umrisse* was an attempt to demonstrate that all the categories of political economy presupposed competition which in turn presupposed private property. He began with an analysis of value, which juxtaposed a 'subjective' conception of value as utility ascribed to Say with an 'objective' conception as cost of production attributed to Ricardo and McCulloch. Reconciling these two definitions in Hegelian fashion, Engels defined value as the relation of production costs to utility. This was the equitable basis of exchange, but one impossible to implement on the basis of competition which was responsive to market demand rather than social need. (Engels still adhered to this definition of value 30 years later in the *Anti-Dühring*. Discussing the disappearance of the 'law of value' with the end of commodity production, he wrote:

> As long ago as 1844, I stated that the above mentioned balancing of useful effects and expenditure of labour would be all that would be left, in a communist society, of the concept of value as it appears in political economy … The scientific justification for this statement, however,… was only made possible by Marx's *Capital*. (Engels, 1877, pp. 367–8)

This shows how much greater continuity of thought there was between the young and the old Engels than is normally imagined.)

He next analysed rent, counterposing a Ricardian notion of differential productivity to one attributed to Smith and T.P. Thompson based upon competition. Interestingly, in this analysis Engels differed both from Watts and Proudhon, in denying the radical form of the labour theory – the right to the whole product of labour – both by citing the case of the need to support children and in querying the possibility of calculating the share of labour in the product.

Finally, after an attack on the Malthusian population theory, which closely followed Alison and Watts, Engels attacked competition itself, both because it provided no mechanism of reconciling general and individual interest, and because it was argued to be self-contradictory. Competition based on self-interest bred monopoly. Competition as an immanent law of private property led to polarization and the centralization of property. Thus private property under competition is self-consuming.

What particularly impressed Marx was the argument that all the categories of political economy were tied to the assumption of competition based on private property. This, for him, represented an important advance over Proudhon whose notion of equal wage would lead to a society conceived as 'abstract capitalist' and whose conception of labour right presupposed private property. Proudhon had not seen that labour was the essence of private property. His critique was of 'political economy from the standpoint of political economy'. He had not 'considered the further creations of private property, e.g. wages, trade, value, price, money etc. as forms of private property in themselves' (Engels and Marx, 1844b, p. 312). The *Umrisse* suggested a new means of underpinning the Marxian ambition to transcend the categorical world of political economy and private property altogether. Moreover, by representing competition as a law which would produce its opposite, monopoly, the elimination of private property and revolution, Engels preceded Marx in positing the 'free trade system' as a process moving towards self-destruction through the operation of laws immanent within it.

These conclusions were amplified in Engels's other major work of this period, *The Condition of the Working Class in England*. Here, the law of competition by engendering 'the industrial revolution' had created a revolutionary new force, the working class. The single thread underlying the development of the working class movement had been the attempt to overcome competition. Such an analysis prefigured the famous statement in the *Communist Manifesto* that the capitalists were begetting their own gravediggers (Stedman Jones, 1977).

Between the mid-1840s and the mid-1870s, Engels played no discernible part in the elaboration of *Capital* beyond supplying Marx with practical business information. His vital contributions to the prehistory of the theory were forgotten and it was only in his better-known role as interpreter and publicist of Marx's work that his writings received widespread attention. During the Second International period, these writings attained almost canonical status, but in the 20th century they generally provided a polemical target for all those attempting to re-theorize Marx in the light of the publication of his early writings.

In the realm of political economy more narrowly conceived, Engels helped to set up the 'transformation' debate by his dramatization of Marx's switch from value to production price in his introductions to Volumes 2 and 3 of *Capital*. Engels's own contribution to this debate in his last published article in *Neue Zeit* in 1895 (now published as 'Supplement and Addendum' to Volume 3 of *Capital*) was to argue that the shift from value to production price was not merely a logical development entailed by the enlargement of the scope of investigation to include circulation and the 'process of capitalist production as a whole', but also reflected a real historical transition from the stage of simple commodity production to that of capitalism proper. 'The Marxian law of value has a universal economic validity for an era lasting from the beginning of the exchange that transforms products into commodities down to the fifteenth century of our epoch' (Marx, 1894, p. 1037).

Leaving aside the empirical question whether during the pre-capitalist era commodities were exchanged in accordance with the amount of labour embodied in them, commentators as diverse as Bernstein and Rubin have objected that this makes no sense in terms of Marx's theory, since during this epoch there existed 'no mechanism of the general equalisation of different individual labour expenditures in separate economic units on the market' and that consequently it was not appropriate to speak of 'abstract and socially necessary labour which is the basis of the theory of value' (Rubin, 1928, p. 254). They have further objected, appealing to Marx's 1857 'Introduction to the Critique of Political Economy', that there is no necessary connection between the logical and historical sequence of concepts, and that the order of appearance of concepts in *Capital* is determined simply by the logical place they occupy in an exposition of the theory of the capitalist mode of production.

Engels could certainly claim explicit textual support from Volume 3 for his historical interpretation of value ('It is also quite apposite to view the value of commodities not only as theoretically prior to the prices of production, but also as historically prior to them. This applies to those conditions in which the means of production belong to the worker...'; Marx, 1894, p. 277). It should also be stressed that there was nothing new in Engels's representation of the character of Marx's theory. Back in 1859, in a review of Marx's *Critique of Political Economy*, Engels stated, 'Marx was, and is, the only one who could undertake the work of extracting from the Hegelian Logic the kernel which comprised Hegel's real

discoveries... and to construct the dialectical method divested of its idealistic trappings'; and in characterizing that method as a form of identity between logical and historical progression, he continued, 'the chain of thought must begin with the same thing that this history begins with, and its further course will be nothing but the mirror image of the historical course in abstract and theoretically consistent form...' (Engels, 1859). It is implausible to suppose that Marx at this time should have sanctioned a fundamental distortion of his method and it is suggestive that he himself, describing his relationship to Hegel, should have endorsed the metaphor of discovering 'the rational kernel in the mystical shell' in his 1873 Postface to the second edition of *Capital*, Volume 1 (Marx, 1873, p. 103). Perhaps the real difficulty lies not in Engels but in Marx himself. It may be, as Louis Althusser has claimed, that Marx did not find a suitable language in which to characterize the distinctiveness of his approach, or it may be more simply that Marx remained ambivalent about how to characterize the theory. In any event, it is not difficult to establish disjunctions between the way he proceeds and the descriptions he gives of his procedures. Engels stuck fairly closely to Marx's descriptions of his procedures and can hardly be reproached for taking Marx at his word.

The problem of Engels's role as an interpreter of Marx's theory debouches onto a third and potentially yet more contentious aspect of Engels's legacy, his role as editor of *Capital*, Volumes 2 and 3. Engels's work was not confined to the transcription of Marx's illegible handwriting. He had to make active editorial choices. The published versions of these volumes contain over 1,300 pages, but the original manuscripts amount to almost twice as many. For Volume 2, for instance, Marx had composed eight versions of his treatment of the process of circulation, from which Engels made a collation. In the absence of an independent transcription and publication of the manuscripts, from which Engels worked, it is impossible to assess whether the emphasis and meaning of the published volumes differ in any significant way from the original. What seems clear, is that in his cautious desire to reproduce as much of the original material as possible, Engels produced a much bulkier and more repetitive version than Marx originally intended. Marx, it seems, always hoped that *Capital* should consist of two volumes and a further volume on the history of political economy (Rubel, 1968; Levine, 1984). From a detailed comparison of Volume 2, Part 1, with the original manuscripts, it appears that Engels also occasionally committed inaccuracies in the citation of the manuscripts he had used (Levine, 1984). Much more doubtful, given all we know of Engels's caution as an editor, is the further suggestion that Engels's editing procedures may have shifted the meaning of the text in ways that lent support to a 'collapse theory' of capitalism (*Zusammenbruchstheorie*) (Levine, 1984). Apart from the smallness of the sample and Engels's own reservations about such a theory, the fact is that proponents of such a position already had sufficient ammunition from *Capital*, Volume 1. Moreover, it simply begs the question whether Marx's attitude to the collapse of capitalism was any more or less apocalyptic than that of Engels.

This discussion by no means exhausts Engels's importance in the history of economic theory or policy. A fuller treatment would have to discuss his analysis of the 'peasant question' which included the important prescription that collectivization must be by example rather than force, his definition of political economy in the *Anti-Dühring*, his interpolations in *Capital*, Volume 3, on banks, the stock exchange and cartels which set the agenda for the early 20th-century discussion of finance capital, his various writings on the relationship between the state and economic forces and his later surveys of English developments since 1844 which prepared the way for later Marxist theories of labour aristocracy. These are only some of the more salient examples.

Finally, at a time when it seems that the technical debate on value seems to have reached a moment of exhaustion, it is perhaps worth going back to Engels if only to remind us of the anti-economic purpose underlying Marx's attempt to construct a theory of value in the first place.

GARETH STEDMAN JONES

Selected works

1844a. *Outlines of a Critique of Political Economy*. In Karl Marx and Frederick Engels, *Collected Works* [MECW], vol. 3. London: Lawrence & Wishart, 1975.
1844b. (With K. Marx.) *The Holy Family*. In MECW, vol. 4.
1845. *The Condition of the Working Class in England*. In MECW, vol. 4.
1845–6. (With K. Marx.) *The German Ideology*. London: Lawrence & Wishart, 1987.
1859. Karl Marx, 'A Contribution to the Critique of Political Economy'. *Das Volk*, Nos. 14 and 16, 6 and 20 August.
1877. *Anti-Dühring*. Moscow: Foreign Languages Publishing House, 1954.
1894. *The Peasant Question in France and Germany*. In Karl Marx and Frederick Engels, *Selected Works*, vol. 3, Moscow: Progress Publishers, 1970.
1925. *The Dialectics of Nature*. Moscow: Foreign Languages Publishing House, 1954.
1938. *Engels on Capital*. London: Lawrence & Wishart.

Bibliography

Claeys, G. 1984. Engels' *Outlines of a Critique of Political Economy* (1843) and the origins of the Marxist critique of capitalism. *History of Political Economy* 16, 207–32.
Levine, N. 1984. *Dialogue within Dialectics*. London: Allen & Unwin.
Marx, K. 1859. *A Contribution to the Critique of Political Economy*. In MECW, vol. 16.

Marx, K. 1873. *Capital*, vol. 1, 2nd edn. Harmondsworth: Penguin, 1976.
Marx, K. 1894. *Capital*, vol. 3. Harmondsworth: Penguin, 1981.
Rubel, M. ed. 1968. *Karl Marx, Oeuvres*, vol. 2. Paris: Gallimard.
Rubin, I. 1928. *Essays on Marx's Theory of Value*. Detroit: Black & Red, 1972.
Stedman Jones, G. 1977. Engels and the history of Marxism. In *The History of Marxism*, ed. E.J. Hobsbawm. Hassocks: Harvester, 1983.

Bibliographic addendum

T. Carver, *Friedrich Engels: His Life and Thought*, London, Palgrave Macmillan, 1990 and J.D. Hunley, *The Life and Thought of Friedrich Engels: A Reinterpretation of His Life and Thought*, New Haven: Yale University Press, 1991, are useful in understanding Engels as an original thinker in his own right.

Engle, Robert F. (born 1942)

Robert F. Engle was born in 'Syracuse, upstate New York,' on 10 November 1942. Shortly thereafter his family moved to Philadelphia, and Engle graduated from high school there in 1960. He majored in physics as an undergraduate at Williams College, and went on to enrol as a Ph.D. student in physics at Cornell University. However, after one year he decided to switch to the Ph.D. programme in economics, where he wrote his thesis on temporal aggregation and the relationship between macroeconomic models estimated at different frequencies, under the direction of T.C. Liu. After graduating from Cornell in 1969, Engle was hired as an assistant professor at MIT. He moved on to University College at San Diego (UCSD) in 1975, where he was promoted to full professor in 1977 and a Chancellors' Associates Chair in 1993. He also chaired the UCSD Economics Department from 1990 to 1994. In 2000 his growing interest in financial markets prompted him to accept the Michael Armellino Professorship in Finance at the Stern School of Business at New York University, and he now lives on Manhattan with his wife of many years, Marianne, for most of the year. Together they have two grown children.

Engle has written and published extensively on a wide array of topics, ranging from urban economics to band spectrum regression, electricity demand, state-space modelling, testing, exogeneity, seasonality, option pricing, and market microstructure finance. However, he is particularly well-known for his contributions to time series econometrics and his path-breaking work on cointegration and AutoRegressive Conditional Heteroskedasticity (ARCH). The 2003 Bank of Sweden Prize in Economic Sciences in Memory of Alfred Nobel was explicitly awarded to Engle for 'methods of analyzing economic time series with time-varying volatility

(ARCH)', a prize he shared with Clive W. J. Granger for his seminal contributions to the theory of cointegration. It is hardly an exaggeration to say that since the 1980s the concepts of cointegration and ARCH have completely revolutionized the field of time series econometrics and the practice of empirical macroeconomics and asset pricing finance, respectively. The blossoming new research field of financial econometrics and corresponding developments in practical risk management and measurement may also in large part be attributed to the insights afforded by the ARCH class of models and some of Engle's many other pioneering research contributions.

Encouraged by his senior colleagues Franklin M. Fisher, Robert Solow and Jerome Rothenberg, much of Engle's work as an assistant professor at MIT was in the area of urban economics. In fact, Engle was hired by UCSD as an urban economist, and he continued to teach, and occasionally publish in, urban economics almost up until he left San Diego in 2000. It was Clive Granger, whom Engle had first met at the 1970 World Congress of the Econometric Society in Cambridge, who persuaded Engle to move to the West Coast. Granger had himself just accepted a permanent position at UCSD in 1974 and, only a few years after Engle's arrival in 1975, Halbert White also joined the department. The ensuing two decades may rightfully be referred to as the golden age of modern time series econometrics, and UCSD, along with Yale, home of the group led by Peter Phillips, was *the* place to be. The list of visitors to the UCSD Economics Department over this period reads like a who's who in time series econometrics. Engle's hospitality and generosity with his time, as well as the many successful conferences he organized in San Diego, played a crucial role in fostering this nexus. The group was further strengthened by the arrival of James Hamilton, Graham Elliott and Allan Timmermann as additional faculty members in the early 1990s, and the Engle–Granger UCSD econometrics tradition continues to this day. Many of Engle's former Ph.D. students from that period have also gone on to successful academic careers, continuing the UCSD legacy.

Albert Einstein's famous maxim 'Everything should be made as simple as possible, but not simpler' succinctly characterizes Engle's approach to econometric modelling. Consider his early research on band spectrum regression. The static OLS regression approach routinely employed throughout economics implicitly assumes that the identical linear relationship holds across all frequencies. Yet in many situations this is obviously a gross oversimplification. For instance, the relation between interest rates and housing starts arguably differs between the short run and the long run. Similarly, the Phillips-curve trade-off between unemployment and inflation may be primarily a business cycle phenomenon. Rather than building a fully fledged complicated dynamic model for analysing these types of temporal dependencies, the band spectrum regression approach offers a simple way of estimating separate regression coefficients, and therefore different relationships, for different frequencies. The idea of estimating different short-run and long-run regressions may also be seen as a precursor to Engle's later work on cointegration and error correction models.

The original idea of cointegration came from Granger. Nonetheless, it was the seminal joint paper by Engle and Granger (1987a) that devised the first empirical test for cointegration and formally established the link between cointegration and the error-correction type models popularized by Denis Sargan and David Hendry at the LSE during the 1960s and 1970s. More specifically, suppose that the two univariate time series y_t and x_t are both non-stationary, or $I(1)$, so that their first differences, $\Delta y_t \equiv y_t - y_{t-1}$ and $\Delta x_t \equiv x_t - x_{t-1}$, are stationary, or $I(0)$. Most nominal macroeconomic and financial time series may be characterized in this way. Any linear combination of the two series, say $z_t = y_t - \beta x_t$, will then generally also be non-stationary. However, it is possible that z_t may actually be stationary, or $I(0)$, in which case y_t and x_t are said to be cointegrated, with cointegrating vector $(1, -\beta)$. Indeed, many of the 'classical ratios' in macroeconomics and finance (such as consumption/income and dividends/prices) are naturally thought of as cointegrating relationships when expressed in logs. Engle and Granger showed that in this situation a satisfactory vector autoregression for the stationary bivariate process of first differences, $\{\Delta y_t, \Delta x_t\}$, must necessarily include the z_t 'error-correction' term in at least one of the two equations, the so-called Granger Representation Theorem. Intuitively, while both y_t and x_t are stochastically trending, they trend together, so that in the long run they do not stray too far apart. The inclusion of the stationary z_t term as an additional explanatory variable ensures this condition. On the other hand, if the two variables are not cointegrated z_t will be non-stationary, resulting in an unbalanced regression. Hence, empirically the null hypothesis of no cointegration may be assessed on the basis of the popular Engle–Granger cointegration test for a unit root in z_t or, if β is not known, a least-squares estimate thereof. The cointegration concept has had a profound impact on practical macroeconomic time series modelling in government and private institutions around the world. The academic literature also abounds with hundreds, if not thousands, of papers expanding upon the basic testing and modelling approach first developed by Engle and Granger. Engle's subsequent work on common features may also be seen as a natural extension of the cointegration concept.

Another more technical theme brought to the fore by Engle's research entails the powerful use of one-step-ahead prediction error decompositions and conditional Gaussian likelihoods. For instance, the beauty of his influential work on testing, including the simple-to-implement Lagrange Multiplier (LM) chi-square type test statistics constructed by multiplying the number of time series observations with the R^2 from an auxiliary regression of either unity on the vector of scores evaluated under the null hypothesis, or, alternatively, a regression of the squared residuals on the derivatives of the conditional mean, hinges directly on recursively expressing the likelihood function in terms of conditional one-step-ahead densities. Engle's pioneering contributions on dynamic factor models and Kalman filtering are similarly based on the powerful idea of representing the likelihood function in terms of successive conditional densities. Most important, however, the seminal ARCH class of models is also formulated directly in terms of one-step-ahead conditional expectations and densities.

The ARCH model (aptly named so by David Hendry) was conceived during Engle's sabbatical visit to the LSE in 1979. Engle's interest in modelling variance dynamics was spurred by the assertion in Milton Friedman's 1976 Nobel Lecture on a trade-off between unemployment and inflationary uncertainty rather than a trade-off between unemployment and the level of inflation as stipulated by the conventional Phillips curve. The actual formulation of the first ARCH model was also influenced by Granger's ongoing work on bilinear models. At the time Granger had noted that in a non-Gaussian setting white noise series need not necessarily be unpredictable, and, in particular, when the squared residuals from otherwise well-specified linear models were regressed on their own lagged squared values, the regression coefficients often turned out to be highly significant. Engle realized that this was not actually a test for bilinearity but rather the optimal LM test for some other nonlinear model. Putting this together, Engle brought forth the ARCH model.

The particular ARCH(p) model first analysed and estimated by Engle (1982) may be succinctly expressed as

$$y_t = m_t + \varepsilon_t, \quad \varepsilon_t | I_{t-1} \sim N(0, h_t),$$
$$\text{and} \quad h_t = \alpha_0 + \alpha_1 \varepsilon_{t-1}^2 + \ldots + \alpha_p \varepsilon_{t-p}^2,$$

where I_{t-1} refers to the set of information available at time $t-1$, m_t denotes the conditional mean of the y_t time series, and all of the $\alpha_0, \ldots, \alpha_p$ parameters are restricted to be non-negative. The first equation for the conditional mean is, of course, completely standard (in his original application to UK consumer prices Engle used an error correction model for the mean). However, the key difference – Engle's brilliant new insight – comes from recognizing that even though the residuals, ε_t, must be serially uncorrelated, their conditional variance, and therefore the conditional variance of y_t, need not be constant but may in fact be predictable. Moreover, by explicit parameterizing h_t as a function of the past squared residuals and by assuming conditional normality, the joint density for all of the observations, say y_t, $t = 1, 2, \ldots, T$, may easily be evaluated through a prediction error decomposition type argument, and the log likelihood function maximized with respect to all of the

model parameters, in turn resulting in a time series of positively serially correlated conditional variance estimates, $\hat{h}_t$, $t = 1, 2, \ldots, T$ (that is, estimates of inflationary uncertainty in Engle's original application).

While Engle's initial work and empirical applications of the ARCH model were rooted in macroeconomics, the model has shone most brightly in the area of finance. Since Mandelbrot's work in the early 1960s on the behaviour of speculative prices, it had been recognized that, even though most returns are approximately serially uncorrelated (at least over shorter daily or weekly horizons), 'large changes tend to be followed by large changes – of either sign – and small changes tend to be followed by small changes' (Mandelbrot, 1963). However, the empirical finance literature up until the mid-1980s had largely ignored this fact, focusing instead on best characterizing the unconditional return distributions. Meanwhile, Engle soon realized that the ARCH model was ideally suited to this type of data: little, or no, serial correlation in the mean, but strong serial correlation in the second moments. Moreover, the ability to directly quantify the risk through a parametric model for the conditional variance, or more generally the conditional covariance matrix, for the returns strikes directly at the heart of the risk-return trade-off central to asset pricing finance. Consequently, Engle quickly shifted the focus of his research agenda to finance. Over the next 20 years, along with his many students and other collaborators, he developed numerous refinements to the basic ARCH model described above designed to account better for specific features of the data and/or questions of economic import: richer ARMA-type representations for the variance, including unit-root and long-memory type dependencies, models in which the variance directly influences the conditional mean, asymmetries or leverage effects in the variance, alternative parametric and non-parametric conditional distributions in place of the normal, multivariate factor models and cointegration in variance, to mention but a few. The corresponding long list of new acronyms is also legendary: ARCH-M, GARCH, IGARCH, EGARCH, TARCH, GJR-GARCH, NARCH, QARCH, STARCH, VGARCH, SWARCH, FIGARCH – the list goes on. Empirical applications of these models have in turn resulted in many important new insights into the pricing and hedging of financial instruments and functioning of financial markets, and it is no exaggeration to say that the day-to-day risk management and monitoring in financial institutions have been completely altered by the advent of the ARCH class of models.

Not one to rest on his laurels, Engle continues to push forward the research frontier in financial econometrics. Most recently he has worked extensively on new methods for analysing ultra high-frequency, or tick-by-tick, financial data. In particular, whereas most procedures in time series econometrics, including most of Engle's own earlier work, are explicitly designed for modelling discretely sampled equidistant observations, high-frequency financial data are typically not observed at fixed time intervals. Engle's recent Autoregressive Conditional Duration (ACD) model, which derives many of its statistical properties from the ARCH class of models, provides a particularly convenient way of accommodating this feature by explicitly modelling the times between observations as a serially correlated process. His Dynamic Conditional Correlation (DCC) model, which allows for the estimation of large-scale dynamic covariance matrices, represents another recent noteworthy advance. In keeping with his trademark, this latest research represents the perfect blend between sophisticated yet simple-to-implement econometric techniques explicitly designed for answering genuinely interesting economic questions. Like most of his research since the 1970s, his latest work has already found widespread use both inside and outside academia, and spurred a number of ongoing new developments by other researchers in the field.

In addition to the much-deserved recognition bestowed on him by the Nobel Prize Committee, Engle is a long-standing fellow of the Econometric Society, of the American Statistical Association, and of the American Academy of Arts and Sciences. He is also an excellent speaker, and he has a long list of invited talks and keynote addresses to his name, including the prestigious A.W. Philips and Fisher-Schultz lectures sponsored by the Econometric Society. (For a more in-depth discussion of Engle's work along with some personal reflections, see Diebold, 2004; 2003.)

In conclusion, it is simply impossible to imagine what the field of time series econometrics, let alone the new field of financial econometrics, would have looked like today had it not been for Engle's seminal contributions, both direct and indirect, through the substantial subsequent research programmes his work has helped stimulate. But Engle isn't merely one of the greatest econometricians of his time. He has a wide range of other interests and talents. For example, he is an outstanding ice skater, having competed at the US national level, finishing second in the 1996 and 1999 ice dancing championship competition.

TIM BOLLERSLEV

See also **ARCH models; cointegration; econometrics; extremal quantiles and value-at-risk; forecasting; Granger, Clive W. J.; measurement error models; risk.**

Selected works

1972. (With F. Fisher, J. Harris and J. Rothenberg.) An econometric simulation model of intra-metropolitan housing location: housing, business, transportation and local government. *American Economic Review* 62, 87–97.

1973. Band spectrum regression. *International Economic Review* 15, 1–11.

1976. Interpreting spectral analysis in terms of time domain models. *Annals of Economic and Social Measurement* 5, 89–109.

1979. (With C. Granger, R. Ramanathan and A. Andersen.) Residential load curves and time-of-day pricing: an econometric analysis. *Journal of Econometrics* 9, 13–32.

1981. (With M. Watson.) A one-factor multivariate time series model of metropolitan wage rates. *Journal of the American Statistical Association* 76, 774–81.

1982. Autoregressive conditional heteroskedasticity with estimates of the variance of UK inflation. *Econometrica* 50, 987–1008.

1983a. (With D. Hendry and J. Richard.) Exogeneity. *Econometrica* 51, 277–304.

1983b. (With M. Watson). Alternative algorithms for the estimation of dynamic factor, MIMIC, and varying coefficient regression models. *Journal of Econometrics* 23, 385–400.

1984. Wald, likelihood ratio, and Lagrange multiplier tests in econometrics. In *Handbook of Econometrics*, vol. 3, ed. Z. Griliches and M. Intrilligator. Amsterdam: North-Holland.

1986a. (With C. Granger, J. Rice and A. Weiss.) Semi-parametric estimation of the relation between weather and electricity demand. *Journal of the American Statistical Association* 81, 310–20.

1986b. (With T. Bollerslev.) Modeling the persistence in conditional variances. *Econometric Reviews* 5, 1–50.

1987a. (With C. Granger.) Co-integration and error correction: representation estimation and testing. *Econometrica* 55, 251–76.

1987b. (With D. Lilien and R. Robins.) Estimation of time varying risk premia in the term structure: the ARCH-M model. *Econometrica* 55, 391–407.

1987c. (With S. Yoo.) Forecasting and testing in co-integrated systems. *Journal of Econometrics* 35, 143–59.

1988. (With T. Bollerslev and J. Wooldridge.) A capital asset pricing model with time varying covariances. *Journal of Political Economy* 96, 116–31.

1990a. (With T. Ito and W. Lin.) Meteor showers or heat waves? Heteroskedastic intra-daily volatility in the foreign exchange market. *Econometrica* 58, 525–42.

1990b. Asset pricing with a factor ARCH covariance structure: empirical estimates for Treasury bills (with V. Ng and M. Rothschild). *Journal of Econometrics* 45, 213–37.

1990c. Seasonal integration and cointegration (with S. Hylleberg, C.W.J. Granger and B.S. Yoo). *Journal of Econometrics* 40, 45–62.

1991a. (With C. Granger.) *Long Run Economic Relations: Readings in Cointegration*. Oxford: Oxford University Press.

1991b. (With G. Gonzales.) Semi-parametric ARCH models. *Journal of Business and Economic Statistics* 9, 345–59.

1992. (With C. Mustafa.) Implied ARCH models from options prices. *Journal of Econometrics* 52, 289–311.

1993a. (With T. Bollerslev.) Common persistence in conditional variances. *Econometrica* 61, 167–86.

1993b. (With S. Kozicki). Testing for common features. *Journal of Business and Economic Statistics* 11, 369–80.

1994a. *Handbook of Econometrics*, vol. 4, ed. with D. McFadden. Amsterdam: North-Holland.

1994b. (With T. Bollerslev and D. Nelson.) ARCH models. In *Handbook of Econometrics*, vol. 4, ed. R. Engle and D. McFadden. Amsterdam: North-Holland.

1995a. *ARCH: Selected Readings*. Oxford: Oxford University Press.

1995b. (With K. Kroner.) Multivariate simultaneous ARCH. *Econometric Theory* 11, 122–50.

1998. (With J. Russell). Autoregressive conditional duration: a new model for irregularly spaced transaction data. *Econometrica* 66, 1127–62.

1999. (With G. Lee). A permanent and transitory component model of stock return volatility. In *Cointegration, Causality, and Forecasting: A Festschrift in Honor of Clive W. J. Granger*, ed. R. Engle and H. White. Oxford: Oxford University Press, 475–97.

2000. The econometrics of ultra high frequency data. *Econometrica* 68, 1–22.

2002a. Dynamic conditional correlation: a simple class of multivariate GARCH models. *Journal of Business and Economic Statistics* 20, 339–50.

2002b. (With J. Rosenberg.) Empirical pricing kernels. *Journal of Financial Economics* 64, 341–72.

2004. Risk and volatility: econometric models and financial practice. *American Economic Review* 94, 405–20.

Bibliography

Diebold, F. 2003. The ET interview: Professor Robert F. Engle. *Econometric Theory* 19, 1159–93.

Diebold, F. 2004. The Nobel memorial for Robert F. Engle. *Scandinavian Journal of Economics* 106, 165–85.

Mandelbrot, B. 1963. The variation in certain speculative prices. *Journal of Business* 36, 394–419.

English School of political economy

The 'English School' of political economy comprises all major British economists of the 19th century, together with J.-B. Say and perhaps Karl Marx.

'Important changes have taken place in the meaning of the term "political economy," as used by leading writers, since it was first employed', wrote Henry Sidgwick in Palgrave's original *Dictionary of Political Economy* (Palgrave, 1899, pp. 128–9). As first used by Mayerne-Turquet and Montchrétien, '*œconomie politique*' signified an attempt to extend the art of estate management to the entire kingdom of Louis XIII and his successors (Waterman, 2004, p. 225). This usage, generalized

to mean a 'system' of policy designed to 'increase the riches and power' of a country (Smith, 1776, I.xi.n.1; II.5.31; IV.1.3) remained current until the end of the 18th century and was so employed by Steuart.

Adam Smith disliked the usage because of its implicit mercantilism. He recognized it, but proposed a better definition. 'What is properly called Political Œconomy' is 'a branch of the science of a statesman or legislator': namely '*an inquiry*', which is in principle disinterested and open-ended, into '*the nature and causes of the wealth of nations*' (Smith, 1776, IV. intro; IV.ix.38; emphasis added). The prestige that *Wealth of Nations* quickly acquired, amplified by Dugald Stewart's widely influential Edinburgh lectures in the new science, redefined 'political economy' as a 'part of the science of human society' (Palgrave,1899, p. 129; cf. Winch, 1983, who appears to disagree with this interpretation) and created a circle of younger thinkers committed both to criticizing and refining Smith's ideas and to propagating them among the governing classes. Though the *Edinburgh Review*, founded in 1802, was at first the principal means of propagation, most of the prime movers soon migrated to London, which from the second or third decade of the 19th century became the home of what was soon called the 'English School'.

It is important to recognize that to describe the small community of anglophone political economists in the 1820s and after as a 'school' is to imply neither a quasi-apostolic succession of doctrine in some leading university nor a closed shop of experts defined by their adherence to any orthodoxy. It is rather the fact, as T.S. Eliot observed of all such intellectual circles in general, that 'they are driven to each other's company by their common dissimilarity from everybody else, and by the fact that they find each other the most profitable people to disagree with' (Kojecky, 1971, p. 244). Members of the English School, like all subsequent economists, were notorious for their disagreements, both with Adam Smith and with each other. But they did not find it profitable to disagree with hostile critics of their enterprise, such as the Lake Poets (from whom they were all indeed markedly 'dissimilar'), because the latter chose not to acquire the viewpoint and vocabulary of the new, political-economy conversation, but resorted rather to the idioms of a very different conversation: that of Romantic aesthetics and non-utilitarian ethics.

In attending to the conversation of the English School it is necessary first to establish its identity, secondly to consider its members and its literature, and thirdly to distinguish its chief analytical features, especially as these differ both from the economic thought that preceded it and from economics of the present day. Finally, since the boundary between the political economy of the English School and what is generally thought of as 'modern' economics is vague and permeable, some attention should be paid to continuity and 'revolution', if any.

Identity of the English School of political economy

Writing of 'the English School of Political Economy' in the original Palgrave dictionary, James Bonar (Palgrave, 1894, p. 730) observed that 'The English writers on political economy before Adam Smith do not at any time present the marks of a "school" properly so called.' What Bonar called 'Modern Economics' – meaning 'political economy' in the new, Smithian sense – he then divided into four periods headed respectively: Adam Smith, Malthus and Ricardo, John Stuart Mill, and W.S. Jevons; with all other authors subsumed under these canonical names.

Adam Smith was not an Englishman and he died before Malthus and Ricardo had begun to write. Though the *Edinburgh Review* (Anon., 1837, p. 73) referred to the English School as 'the school of which Adam Smith was the founder', this is Caledonian hyperbole. Smith founded no 'school'. His most influential disciple, Dugald Stewart, was the intermediary between Smith and those the *Edinburgh Review* more accurately described later in the article as the 'followers of Dr Smith' practising 'Political Economy, using the word in the sense of Ricardo and Malthus' (Anon., 1837, pp. 77, 79). Subject to this important qualification, Bonar's chronology is helpful. Roughly speaking the English School lasted for about three generations. The first generation, from 1798 to the 1830s is that of which Malthus, Henry Thornton, Chalmers, James Mill, Torrens, West, Ricardo, and Thomas Tooke are now the best remembered. A second generation, whose members were active in some cases before 1830, but who flourished for the most part until the 1860s or even later, included Whately, Senior, McCulloch and J.S. Mill. Political economy of the English School never really died out. It changed, very gradually and almost imperceptibly, into the international, professionalized 'economics' of the mid-20th century. Yet a third and last generation can be detected – and was in fact detected in the 1890s – which included W.T. Thornton, the Fawcetts, Cairnes, Jevons, Bagehot, Foxwell, Sidgwick, J.N. Keynes and Nicholson. The positions of Marshall, Edgeworth and Wicksteed are problematic and will be considered below.

The English School was recognized by its difference from 'the foreign school' (Anon., 1837, p. 77) which included Sismondi, Cherbuliez and Villeneuve, but not J.-B. Say who from the first was deemed an honorary Englishman. The English writers distinguished 'the art of government' from the 'science' of political economy. With respect to the former, the latter is 'only one of many subservient sciences; which involves the consideration only of motives, of which the desire for wealth is only one among many, and aims at objects to which the possession of wealth is only a subordinate means' (Senior, 1836, pp. 129–30). The foreign writers rejected this minimalist construal, labelled it 'chrematistics' or 'chrysology', and continued to maintain that political economy embraces both the art and the science of government.

The incipient distinction between 'art' and 'science' seemed to imply that any practitioner of the latter must abstain – qua political economist – from political judgements. His analytical conclusions, being strictly positive and abstracted from ethical considerations, 'do not authorize him in adding a single syllable of advice'. His business is 'neither to recommend nor to dissuade, but to state general principles which it is fatal to neglect' (Senior, 1836). McCulloch for one strongly disagreed with Senior on this point: the general principles, he thought, had already been completely enounced by Ricardo. What remained was 'to exhibit some of their more important applications' (McCulloch, 1843, p. vi). Though Senior's view of the scope of political economy was tidied up and assimilated by the end of the 19th century (J.N. Keynes, 1891), all members of the 'school' were agreed at the outset on at least one most important 'application'. The 'great principles of free exchange and natural distribution' that Smith had developed from 'the philosophers of the Continent' (that is, Quesnay, Turgot and so on) showed it to be economically unprofitable 'for the legislature to intermeddle' with trade and income distribution (Anon., 1837, pp. 80, 78). Though Cairnes later averred that 'political economy has nothing to do with laisser faire', Sidgwick thought this 'too daring a paradox'.

> There can be no doubt that the interest of Adam Smith's book for ordinary readers is largely due to the decisiveness with which he offers to statesmen the kind of practical counsels which, according to Senior and Cairnes, he ought carefully to have abstained from giving. (Palgrave, 1899, pp. 130–1)

Rightly or wrongly, the political economy of the English School was associated in the popular mind with free trade and attacks on corporate privilege, and was denounced for these disturbing ideas by a wide variety of hostile critics.

Both the methodological tendencies of the new science and its 'more important applications' owed much to Dugald Stewart: the former to his influential *Philosophy of the Human Mind* (1792, 1814, 1827) the latter to his annual public lectures at the University of Edinburgh, beginning in the winter of 1800/1.

Though preferring a broader definition of 'political economy' than that of either Smith or his English followers, and emphasizing the historical character of economic knowledge, Stewart argued in *Human Mind* that the hypothetical and a priori reasoning so characteristic of what he called the 'new science' – and which became one of the hallmarks of the English School – was perfectly legitimate, and compatible with the testing of theories against experience (Fontana, 1985, pp. 99–102; Waterman, 2004, ch. 8).

Stewart's Edinburgh lectures were crucial in what a recent author has aptly called 'the process of Anglicisation of Scottish thought after 1790' (Fontana, 1985, p. 9). Not only were they attended by Jeffrey, Horner,

Brougham, Chalmers and the newly arrived Englishman, Sydney Smith, all of whom were influential in propagating political economy; Pryme (1823, p. vii) records that they 'attracted so much attention that several members of our own university [namely, Cambridge] went from the South of England to pass the Winter at Edinburgh, for the purpose of attending them': one of these seems to have been John Bird Sumner (Waterman, 1991a, pp. 159–60). According to a later account, 'a wave of young Englishmen … went North in lieu of the grand tour made impossible by the renewal of war' (Checkland, 1951, p. 43). Though the lectures were diffuse and circumspect, their underlying message was that contained in an early paper that Adam Smith had entrusted to Stewart before his death:

> Little else is required to carry a state to the highest degree of opulence from the lowest barbarism, but peace, easy taxes and a tolerable administration of justice; all the rest being brought about by the natural course of things (Smith, 1755). (Winch, 1996, p. 90)

Leading members of Stewart's circle – Jeffrey, Horner, Brougham, and Sydney Smith – founded the *Edinburgh Review* to urge this message upon the Holland House Whigs from whom they hoped to receive patronage. First Smith, then Jeffrey, served as editor until 1829, when replaced by McVey Napier. By that date its contributors on political economy had included all the leading members of the English School save Ricardo (who declined out of modesty, and who died in 1823): Malthus, James Mill, Chalmers, Torrens and McCulloch (Fontana, 1985, p. 8).

Of these authors, all save Chalmers were members of the Political Economy Club, a London dining club founded in 1821 which, in addition to Malthus, Mill and Torrens, included from the outset Ricardo, George Warde Norman and Thomas Tooke. J.-B. Say was elected as an Honorary Foreign Member in 1822, the only such member until 1919. McCulloch was elected in 1829, shortly after his migration from Scotland; Senior (in 1823), Pryme (1828) and Whately (1831) were elected as Honorary Members by virtue of their professorships in political economy. Cairnes (1862), Cliffe Leslie (1862), Fawcett (1862), Jevons (1873), Foxwell (1882), Marshall (1886), Nicholson (1888) and Edgeworth (1891) were all subsequently elected under this rule. Among those political economists now remembered as influential authors of the English School, only Henry Thornton, Sir Edward West, Archbishop J. B. Sumner, Thomas Chalmers, Poulett Scrope, and Richard Jones were never members of the Club: Thornton because he died in 1815, West because he went to India, Chalmers because he stayed in Scotland, and Sumner because he announced in 1818 – to Ricardo's regret – that he intended to give up political economy for the study of theology (Waterman, 1991a, p. 157). Scrope and Jones were on the outer edge of the 'School'.

It has been suggested that the English School was a 'scientific community' of which the Political Economy Club was a 'vital hub' (O'Brien, 2004, pp. 12–13). There is merit in this suggestion, but it should be recognized that the original purpose of the club, though including the 'mutual instruction' of members, was chiefly propagandist: 'the diffusion amongst others of the just principles of Political Economy' and

> to watch carefully the proceedings of the Press, and to ascertain if any doctrines hostile to sound views on Political Economy have been propagated … to refute such erroneous doctrines, and counteract their influence … and to limit the influence of hurtful publications. (Political Economy Club, 1921, p. 375)

Many members were Whig or liberal statesmen who knew a 'hurtful publication' when they saw one: 52 of the 115 elected between 1821 and 1870 sat in either the upper or lower House of Parliament; and included Lord Althorp, the Marquis of Landsdown (a descendent of Sir William Petty), Earl Grey and W. E. Gladstone. Fetter (1980) has documented the activities of 'the economists in Parliament'.

Almost from the first there was a desire by the Club to recognize and foster the academic study of political economy. Though there had been high-level economic analysis at British universities before the end of the 18th century, it was but a small ingredient of 'moral and political philosophy' (for example, see Waterman, 1995) and never known as 'political economy'. But in the decade of the 1820s chairs in political economy were established in Oxford, London and Cambridge and their incumbents immediately co-opted (Checkland, 1951).

We may therefore identify the English School roughly speaking as that subset of Political Economy Club members in the 19th century who published and disputed with each other on the subject, together with half a dozen or so other major authors who at some time or other were part of their conversation. Despite its name, several leading members were Scotch immigrants, and it included one Frenchman. Though Karl Marx lived in London from 1848 and thoroughly digested the literature of anglophone political economy over the next two decades, he was not known or recognized by the Club. But in the 'Afterword' to the second German edition of *Capital* (Marx, 1873, vol. 1, p. 26) he explicitly identified his own work, in method at least, with that of the English School.

Literature of the English School

Literature of the English School begins with Malthus's first *Essay on Population* (1798). For as an unintended consequence of his Whiggish polemic against Godwin's (1793) romantic anarchism, Malthus analysed the effect of population growth under land scarcity to show what was later called 'diminishing returns' (Stigler, 1952). Though diminishing returns in agriculture had been identified by Steuart (1767) and Turgot (1768), and had actually been used by Anderson (1777) to adumbrate the 'Ricardian' theory of rent, the concept was not integrated into 18th-century economic thought. Notwithstanding Samuelson's influential interpretation, land scarcity plays little or no analytical part in *Wealth of Nations* (Samuelson, 1978; cf. Hollander, 1998; Waterman, 1999). When Malthus (1815a), West (1815), Torrens (1815) and Ricardo (1815) worked out the implications of Malthus (1798) they believed that they were correcting Smith and saying something new and important (McCulloch, 1845, p. 68). Diminishing returns immediately became part of the hard core of the so-called classical political economy of the English School.

Ricardo made diminishing returns in agriculture the cornerstone of his *Principles* (1817), combined it with 'Malthusian' population theory, Smith's account of accumulation and growth, and an ad hoc '93% Labor Theory of Value' (Stigler, 1958) to produce a complete account of value, distribution and growth in a two-sector market economy. The labour theory of value (LTV) was also the key concept in Ricardo's rigorous and elegant analysis of comparative advantage in international trade. Looking back 30 years later, McCulloch (1845, p. 16) called the LTV 'the fundamental theorem of the science of value'. An authoritative and exhaustive account of Ricardo's contribution – which it treats, à la McCulloch, as virtually identical with 'classical economics' – appeared in the first edition of *The New Palgrave Dictionary of Economics* (Blaug, 1987).

In addition to the above works, the 'English' literature that already existed by the time the Political Economy Club was founded in 1821 included Malthus's (1800) *High Price of Provisions*, which formally specified a demand function of price and inaugurated the supply-and-demand value theory that eventually 'won out' over the Ricardo–Marx LTV (Smith, 1956; Schumpeter, 1954, p. 48) which it generalizes, Thornton's (1802) *Paper Currency*, which analysed the macroeconomic relations between monetary and real variables in a manner reinvented by Wicksell a century later, and numerous pamphlets by many authors on monetary questions provoked by the Parliamentary Bullion Committee of 1810. It was this controversy that brought Malthus and Ricardo together, and which seems to have been a catalyst for the nascent 'scientific community'. The pre-1821 literature also includes J.-B. Say's (1803) *Traité d'économie politique*; Lauderdale's (1804) *Inquiry*, dismissed by McCulloch (1845, p. 15) as without value; Chalmers's (1808) strikingly original but completely neglected *Nature and Stability of National Resources* (see Waterman, 1991b); Malthus's (1815b) heretical pamphlet, 'Restricting the Importation of Foreign Corn', which led to his excommunication by the *Edinburgh Review* (Fontana, 1985, p. 75); J. B. Sumner's (1816) *Records of the Creation* that Ricardo (1951–73, vol. 7, pp. 247–8) deemed a 'clever book' and which McCulloch (1845, p. 261 described as 'an

excellent work'; the fifth edition of Malthus's Essay on Population (1817) substantially modified as a result of Sumner's arguments; Mrs Marcet's (1817) influential work of popularization, *Conversations on Political Economy*; and Copleston's (1819a; 1819b) two brilliant and penetrating *Letters to Peel* that grasped more clearly than Malthus himself the connection between population and poverty, and between the latter and inflation of the currency – and which Ricardo so admired that he made a detailed paragraph-by-paragraph summary (Waterman, 1991a, pp. 186–95; Hollander, 1932, p. 135–45). Finally, shortly before or just after the first meeting of the Club there appeared important monographs by three of the founding fathers: Malthus's (1820) *Principles*, which quarrelled with Ricardo over value theory and put forward a heterodox macroeconomics of 'general gluts' that Keynes was later to find so appealing, James Mill's (1821) *Elements of Political Economy*, and Torrens's (1821) long undervalued *Essay on the Production of Wealth*.

It is apparent that during the first two decades of the 19th century, and for a further ten years or more, Malthus was at the centre of the political-economy conversation of the English School. This fact has been obscured by the excessive attention paid to Ricardo by those eager to praise or blame him for present-day economics, and by textbook authors wanting a handle on which to hang a student-friendly chapter on 'classical economics'. A long process of reappraisal, beginning with J. M. Keynes's (1972, vol. 10) biographical essay of 1933, has gradually restored the true picture (Waterman, 1998). Donald Winch's (1996) *Riches and Poverty* is the latest and most authoritative intellectual history of political economy, covering the period 1750–1834. Nearly half his book is concerned with Malthus. Ricardo, 'treated largely as a foil to Malthus' (Winch, 1996, p. 15) gets a few scattered references. Samuel Hollander's (1997) magisterial *Economics of Thomas Robert Malthus* shows that the analytical differences between Malthus and Ricardo have been exaggerated, and that the former was a theoretician of the same order, and of at least as much historical importance as the latter.

Malthus was central because the first *Essay* began a century-long transformation of 'political economy' (the science of wealth) into 'economics' (the science of scarcity). The theological implications of this, totally ignored by most historians, are a vital part of the intellectual context of the English School. Economic thought of the 18th century was believed by all to be wholly compatible with Christianity. But the seeming inevitability of 'misery' or 'vice' produced by human fecundity and resource scarcity challenges the goodness of God; and the political economy of Malthus and Ricardo was therefore condemned as 'hostile to religion'. For most of the 19th century, England was both officially and actually a Christian society. In such a society it is part of the duty of a scientist – essential if his work is to receive serious attention – to reconcile his findings with Christian theology. Malthus attempted this in 1798 and failed. His failure stimulated an important branch of the literature of the English School now known as 'Christian Political Economy' (Waterman, 1991a). Works by William Paley (1802), by Malthus himself (1803; 1817), and by J. B. Sumner (1816) who eventually became Archbishop of Canterbury, demonstrated that the new science could be co-opted as theodicy; and even better, be used to demonstrate the benevolent 'design' of the Creator. The approval that Ricardo and McCulloch evinced for Sumner's 'clever book' had less to do with their own religious convictions than with their relief that political economy had been convincingly defended against the damaging charge of irreligion.

Quite different circumstances in the 1820s revived the need to defend political economy against religion, and created a new need: to defend religion against political economy. Jeremy Bentham, James Mill and other Benthamites, who were later called the 'Philosophic Radicals', founded the *Westminster Review* in 1824 to propagate a 'radical' reformism as against the Whiggish reformism of the *Edinburgh*. Anti-clerical and at times anti-religious, the radicals hijacked political economy to mount a strictly utilitarian attack on the Establishment in Church and State. Animated by James Mill's puritanical hatred of the Arts, the *Westminster* compounded the injury by gratuitous attacks on the Lake Poets and other romantic authors. Influential Tories at the two universities (then exclusively Anglican) were alarmed, and opposition was made to the teaching of, and the establishment of chairs in, political economy. In this crisis, both political economy and Christian theology were authenticated and insulated against mutual encroachment by two Oxford men, Richard Whately, a former pupil and friend of Copleston, and Nassau Senior, Whately's former pupil and friend (Waterman, 1991a, pp. 196–215).

Whately engineered the election of Senior as first Drummond Professor of Political Economy in 1826, and accepted the chair himself when it fell vacant in 1830. His seminal Introductory Lectures (1831) argued for an epistemological demarcation between 'religious and 'scientific' knowledge; and explained how, like all scientific knowledge, political economy depends upon both a priori deduction and the possibility of falsification. Whately thus established the methodological tradition of the English School that runs through Senior, J.S. Mill, J.N. Keynes and Lionel Robbins. Pietro Corsi (1987) has shown that Whately's philosophical apparatus was based on Dugald Stewart's *Philosophy of the Human Mind*, transmitted to Oxford through the friendship between Stewart and Copleston created by the migration from Edinburgh to Oxford in 1799 of J.W. Ward, 1st Earl of Dudley.

Whately's decisive intervention healed a potentially disastrous schism in the young 'scientific community' between Benthamite radicals and Malthusian Whigs.

Elections to the Political Economy Club in the 1820s and 1830s included both Whigs and radicals and even the liberal Tory, Lord Althorp. When McVey Napier edited the 1824 *Supplement to the Encyclopaedia Britannica* he commissioned articles on political economy from Malthus and Sumner on the one hand, and from Mill and McCulloch on the other. (Ricardo's contribution, on the Funding System, was posthumous.) The Royal Commission on the Poor Laws (1832) which included Sumner, then Bishop of Chester, united all in the common cause once again. Malthus was the most important witness. The report, which led to the Poor Law Amendment Act (1834), was jointly written by the Benthamite Chadwick and the Whatelian Senior, and was based on Copleston's (1819b, p. 28) crucial distinction between 'propagation' and 'preservation' of human life.

One of the most interesting, certainly the most revealing, contributions to literature of the English School is McCulloch's compendious *Literature of Political Economy* (1845) which appeared about halfway through the life of the 'school'. The usual English and Scotch authors from Mun and Petty are listed, and many of their works praised or censured in light of McCulloch's doctrinal preconceptions. Malthus is predictably belittled. All the leading French authors of the 18th and early 19th centuries appear save Boisguilbert and Cournot. Condillac's path-breaking *Le Commerce et le gouvernement* (1776) is dismissed with a patronizing comment of J.-B. Say (McCulloch, 1845, p. 63; cf. Eltis and Eltis, 1997, pp. 30–4). Considerable respect is paid to Italian authors (McCulloch, 1845, pp. 28–31, 86), but the Spanish are written off as intellectually impotent until Napoleon's invasion (1845, pp. 31–2, 326). McCulloch seems never to have heard of Thünen, and no other German author is mentioned. Omissions of anglophone authors are equally telling. Whewell's pioneering mathematical economics is ignored, presumably for the same reason as the omission of Cournot and Thünen. Dugald Stewart is cited merely as a biographer of Adam Smith and Robertson (1845, pp. 8, 104, 162). McCulloch seems not have read or understood either Chalmers (1808) or Copleston (1818), nor to have grasped the analytical significance of Malthus (1800). Everything is viewed through the powerful but slightly distorting lenses of Adam Smith and Ricardo.

Three years later there appeared the single most important production of the School: J.S. Mill's *Principles of Political Economy* (1848), perceptively reviewed by Bagehot (1848) among many others. Mill's *Principles* is the definitive statement of the English School of political economy. It went through seven editions in the author's lifetime; the 1909 scholarly edition by Ashley was based on the seventh (1871), and may be taken as the terminus ad quem of the English School. For though Mill continued to be the principal textbook in political economy until the 1930s at many universities throughout the English-speaking world, Anglophone economic literature of the 20th century gradually became less insular

(Palgrave, 1894, p. 735) and was formed in the cautiously new idiom of Marshall and Pigou, with at least some peripheral awareness of Jevons and Edgeworth, Walras and Pareto, Weiser and Böhm-Bawerk, Cassel and Wicksell, J.B. Clark and Fisher.

Though Mill dominated, there were many other significant contributions to the literature in the last third of the 19th century. Henry Fawcett's *Manual of Political Economy* (1863) encapsulated Mill's *Principles* for faint-hearted undergraduates; his wife's even more elementary *Political Economy for Beginners* (1870) went through ten editions over the next 41 years. W.T. Thornton's *On Labour* (1869) introduced the concept of multiple equilibria, as Mill (1869, p. 637) admitted. Cairnes's *Leading Principles* first appeared in 1874, Cliffe Leslie's *Essays* in 1879, Bagehot's posthumous *Economic Studies* in 1880, and Henry Sidgwick's *Political Economy* in 1883. Sidgwick's importance in the incipient 'Cambridge' mutation of the English School has lately been documented (Backhouse, 2006). J.N Keynes's classic *Scope and Method* first appeared in 1891. Perhaps the last major production of the English School was J. Shield Nicholson's three-volume *Principles of Political Economy* (1893–1901), a basically Millian exposition with the occasional bow to Marshall, used as a textbook in many parts of the British Empire in the early 20th century. Nicholson's appears to be the last widely read work of political economy to consider explicitly the relation between that science and Christian theology (1893–1901, vol. 3, ch. 20).

Stanley Jevons (1871, p. 275) went out of his way to challenge 'the noxious influence of authority' in the English School, above that of Mill. Though elected to the Political Economy Club as a professor in 1873 and as an Ordinary Member in 1882 (the year of his death), he was therefore handled with caution by his fellow economists – including the powerfully influential Marshall. Whilst crediting him with the intellectual defeat of Ricardian and Marxian value theory, Bonar (Palgrave, 1894, p. 735) thought that 'the ideas of Jevons have had greater power since his death than during his life'. Jevons and his two most creative English followers, Edgeworth and Wicksteed, were 'often spoken of as a school by itself, the mathematical school' (Palgrave, 1894). The original Palgrave article on 'Recent Developments of Political Economy' (Palgrave, 1894, p. 148) alludes to Jevons's *State in Relation to Labour* (1882) but ignores his *Theory of Political Economy* (1871).

Literature of the English School was augmented and popularized by *The Economist* newspaper, founded in 1843 and edited by Walter Bagehot from 1860 to 1877, which, like the Political Economy Club, sought to relate economic analysis to public policy in the spirit of Adam Smith. That literature may be said to have culminated in the three-volume *Dictionary of Political Economy* (1894–1899) edited by R.H. Inglis Palgrave.

Some analytical features of the English School

Political economists of the English School inherited much of their economic analysis from their 18th-century predecessors, especially Cantillon, Hume, Quesnay, Smith and Turgot. However, some features of their analysis were as 'novel' as any idea ever is in the social sciences. And despite loose talk about a 'marginal revolution', much of their analysis, both what they inherited and what they originated, has become part of the stock-in-trade of present-day economics. The standard account by D.P. O'Brien (2004) should be supplemented by S.J. Peart's and D. Levy's (2003) review of the period 1830–1870, which considers catallactics, methodological egalitarianism and the new ideological alliance – a mutation of the old Whig-Liberal orthodoxy – between political economists and reformist Evangelicals in the Church of England.

The central conception of 18th-century economic thought was that of a surplus of production in one period over and above what is necessary (as inputs into production) to sustain that level of production in the next. The agricultural sector is an obvious source of the surplus since land normally produces more than the (food) cost of necessary labour and capital inputs. But Smith generalized the concept to include all produced goods capable of use as inputs. Masters incur production costs in advance, hence control the entire output at the end of the process. Some of this they consume either directly, or in the employment of unproductive labour. The remainder is used to feed and equip productive labour. This unconsumed portion of output is the (circulating) capital stock of a master, firm or community, the growth, stationarity or decay of which depends on a psychological propensity of masters: the extent of their 'frugality' or parsimony (Eltis, 2000, pp. 75–100). These ideas, and the necessarily dynamic analytical framework they imply, were taken for granted by most the English School despite its seeming incompatibility with such other conceptions as comparative advantage in trade (Blaug, 1987, vol. 1, pp. 439–42). Other characteristically 18th-century ideas accepted by 'the followers of Dr Smith' included that of a labour supply perfectly elastic in the (Malthusian) long period at a socially determined zero-population-growth real wage; enough factor mobility to produce uniform rates of wages and profit throughout the economy; a negative relation between the real wage and the rate of profit; a positive relation between the general price level and the stock of money, and the Cantillon–Hume price-specie-flow mechanism of international monetary adjustment which follows from that relation. Most accepted Smith's account of natural prices that correspond, more or less, to Marshall's long-period equilibrium prices, but O'Brien (2004, ch. 4) has shown in detail how much variation there was in this matter. Perhaps the most important 18th-century idea, certainly that which gave the English School its ideological momentum, was Boisguilbert's vision – derived from the Jansenist theology of Pierre Nicole and Jean Domat – of a self-regulating market economy driven by 'self-love' and producing some kind of social optimum at competitive equilibrium (Faccarello, 1999). This powerful conception was transmitted by Mandeville, Cantillon and Quesnay and canonized by Smith in *Wealth of Nations*.

As we have seen, the English School made at least one sharp analytical break with 18th-century thought. The explicit incorporation of diminishing returns (though as yet in agricultural production only) created a fundamentally different view of the economic universe. Though all recognized increasing returns to scale (IRS) resulting from the division of labour, IRS plays a small or negligible part in the implicit growth models of Malthus and his successors (Eltis, 2000). The salient feature of the new growth theory was rather a tendency for the rate of profit to fall: either because of rising costs in agriculture as in Malthus and Ricardo, or because of increasing capital intensity in manufactures as in Marx. In the former case, falling real factor payments retarded the growth of capital and labour, leading to a stationary state in the absence of technical progress. Samuelson (1978) has shown that the variable factor in agriculture was conceived as a single 'labor-cum-capital' unit, and though all 'classical' economists recognized the possibility of factor substitution especially in manufacturing, the capital–labour ratio was generally taken as a parameter. The same was true of technique. Improvements were seen to occur from time to time, and their effect upon wages, profits and employment analysed. Malthus, and perhaps some others, recognized that technical progress could become endogenous (Eltis, 2000, pp. 150 ff.) and few if any of the English School regarded it, as some do today, as 'manna from Heaven'. Two other new, or somewhat new, analytical features of the English School deserve note. The first is the LTV theory of comparative advantage, later improved by Mill's analysis of reciprocal demand. The second is Say's Law of Markets, which in its strong form (Say's identity) implies the neutrality of money (Blaug, 1996, pp. 143–60). Whether Samuel Hollander (for example, 1987, pp. 6–7) is correct in maintaining that Ricardo and his contemporaries and successors, including Marx, recognized 'a fundamentally important core of general-equilibrium economics accounting for resource allocation in terms of the rationing function of relative prices' is still a matter of debate (Blaug, 1987, vol. 1, pp. 442–3).

It is evident that most of these analytical characteristics, both those inherited from the 18th century and those that were new, have been transmitted to present-day economic thought. The obvious exception is the concept of a surplus with its concomitant distinction between 'productive' and 'unproductive' labour; though in the spirit of Feyerabend's (1988) methodological anarchism this venerable doctrine has lately been brought back to useful life (Bacon and Eltis, 1976). For the most part however,

present-day economists prefer to rely on a putatively constant-returns-to-scale (CRS) general equilibrium model that abstracts from time, and in which each factor-owner is paid the value of his factor's marginal product. The surplus is therefore regarded as a museum piece and left to heterodox Marxists and Sraffians (Walsh and Gram, 1980; cf. Blaug, 1987, vol. 1, pp. 440–2). It is important to recognize, however, that the eventual disappearance of the surplus in a neoclassical theory of distribution was brought about by an ever wider application of the marginal analysis originally applied by Steuart, Turgot, and Anderson, and then by Malthus, Ricardo and their contemporaries to agricultural production costs alone (Blaug, 1987, vol. 1, p. 441). Authors of the next generation such as Longfield and Lloyd began the analysis of marginal utility (O'Brien, 2004, pp. 119–22). Replacement of the dynamic surplus macroeconomics by a static general-equilibrium microeconomics dependent on universal CRS created perhaps the most significant analytical difference between political economists of the English School and the new professionalized economists of the early 20th century: an almost complete lack of interest among the latter in macroeconomics and growth theory. Not until Keynes's rediscovery of Malthus (Kates, 1994) and Harrod's (1939) critique of Keynesian 'equilibrium' did these return to the theoretical agenda. As for Adam Smith's IRS, quietly forgotten by most of the English School – save Marshall – for most of the time and ignored by their successors, its reintroduction by Sraffa (1926) and Young (1928) has remained a thorn in the flesh for general equilibrium theorists.

Revolution and continuity

Present-day 'economics' looks quite different from 'political economy' of the English School. Yet despite Samuelson's remarks about Marshall in *Foundations* (1947, pp. 6, 142, 311–12) and despite his focus on Walrasian general equilibrium in that work, the microeconomic part of his immensely influential *Economics* (1948) is unmistakeably Marshallian, at any rate as mediated by Chamberlin (1933) and Joan Robinson (1933). And though Marshall had digested Thünen and Cournot, knew the work of Menger and the Austrian School, and admitted that 'there are few writers of modern times who have approached as near to the brilliant originality of Ricardo as Jevons has done' (Marshall, 1920, p. 673), yet he 'consistently discounted the "Jevonian revolution"' (Schumpeter, 1954, p. 826) and used all his influence, which was great, to insist that in science, as in the world it contemplates, *Natura non facit saltum*. There are few references to Jevons in his famous *Principles*, and in the most extended of these (Appendix I) Marshall went out of his way to counter the former's 'antagonism to Ricardo and Mill' and to defend their value theory against his intemperate exaggerations' (Marshall, 1920, pp. 673–6; see also O'Brien, 1994, vol. 2, pp. 325–61).

Upon the evidence of Palgrave's original dictionary it appears that by the last decade of the 19th century the effect of Marshall's efforts had been to co-opt Jevons and his 'marginalist' followers into the mainstream of English political economy with a minimum of fuss, and with a minimum of attention to the continental marginalists. Jevons's 'final utility' became 'marginal utility' in Marshall's *Principles* (1920, pp. 78–85), and there was used with deceptive innocence (see Blaug, 1996, p. 322–37) to generate a market demand function of price. Though Edgeworth himself contributed 17 articles to the dictionary, including 'Cournot', 'Curves' and 'Demand Curve' in volume 1, 'Mathematical Methods' in volume 2 and 'Pareto', 'Pareto's Law', 'Supply Curve' and 'Utility' in volume 3, his own work was ignored in the general surveys of 'Political Economy' and 'The English School' and his name omitted from the index of volume 1, along with those of Menger and J.B. Clark. Walras received three short references in that volume. Not until volume 3 (1899, pp. 652–5) was his work recognized, and then only for its use of marginal utility. There is no awareness of general equilibrium in that article, and the term appears nowhere else in the original *Dictionary*.

It would appear from the foregoing that if there really was any such thing as a marginal revolution in Anglophone political economy, it began as early as 1767 with Steuart's *Political Œconomy* and still had some way to go by the time volume 3 of the Palgrave dictionary appeared in 1899. Thünen's (1826) generalization of diminishing returns to all factors of production remained unnoticed by any save Marshall. Though Wicksteed (1894) and Flux (1894) reinvented this wheel, Wicksteed's (Palgrave, 1899, pp. 140–2) own contribution to the Palgrave article on 'Political Economy' only hints at what later became known as the neoclassical theory of distribution. In 1895 Edgeworth rejected Barone's submission to the *Economic Journal* showing that product exhaustion is implied by Walras's (1894) cost-minimization equations. A companion article to Wicksteed's baldly states that 'the law of DIMINISHING RETURNS points to an increase in the cost of agricultural produce accompanying increase of population' (Palgrave, 1899, p. 140). For that author at any rate, nothing had changed since Malthus.

In summary, it would appear that the English School was alive and well in the first decade of the 20th century. Elections to the Political Economy Club included Pigou (in 1906) and J.M. Keynes (1912), along with the Bishop of Stepney (1904), the Rt Hon. Herbert Samuel MP, the Viscount Ridley (1907) and John Buchan (1909). Mill's *Principles* was still perhaps the most widely used textbook. Questions on Adam Smith still appeared in university examinations in political economy (for example, at Edinburgh, 21 November 1898, 17 March 1899). Mathematics was still an unwelcome eccentricity. Jevons (1871, p. vii) had asserted that economics 'must be a mathematical science in matter if not in language'. Marshall (for example, 1890, p. ix) threw all his influence against this

doctrine and locked up his own sophisticated mathematics in well-guarded appendices (Keynes, 1972, pp. 182–8). Despite his dependence upon mathematical reasoning and his prominence in the emerging profession of economics, Edgeworth's deference for Marshall deterred him from challenging a Cambridge, anti-mathematical orthodoxy that persisted until the 1950s.

Edgeworth was unusual, too, in his ability and willingness to read foreign authors and to recognize their contributions (Keynes, 1972, pp. 263–5). In general, the insularity of the English School persisted until well into the 20th century. When Harrod was about to begin his studies in economics, Keynes advised him not to waste his time on the Continent 'where they knew nothing at all of economics' (Harrod, 1952, pp. 317–19). The 'market socialists' of the 1930s, none of whom was English, were the first to specify the complete set of marginal conditions required for a welfare optimum in general competitive equilibrium. J.R. Hicks (1939, p. 6) believed himself to be the first English author to 'free the Lausanne School from the reproach of sterility brought against it by the Marshallians'.

It might have been expected that political economists in the United States, at any rate, would have identified with the English School. In the early 19th century authors such as Wayland (1837) had assimilated Malthus and Ricardo, and as late as 1888 Amasa Walker regarded Jevons and Marshall as 'an extension of the English School' (Goodwin, 1972, p. 562). But throughout much of the century protectionist sentiment in the USA was at variance with the ideology of free trade promoted by the English School. And towards the end of that century there was 'an estrangement from British scholarly life' created by a 'growing attachment to German thought' (Goodwin, 1972, p. 563). The American Economic Association was originally formed to promote the Liberal-Protestant 'social gospel', very different in spirit and substance from the aristocratic Whiggery of the Political Economy Club.

A.M.C. WATERMAN

Bibliography

Anderson, J. 1777. *Observations on the Means off Exciting a Spirit of National Industry; Chiefly Intended to Promote the Agriculture, Commerce, Manufactures, and Fisheries, of Scotland*. Dublin: Price.

Anon. 1837. 1. *An Outline of the Science of Political Economy.* By Nassau W. Senior. London: 1836. 2. *Principes Fondamentaux de l'Économie Politiques, tirés de leçons édites et inédites, de M. N. W. Senior.* Par le Compte Jean Arrivabe. Paris: 1836. *Edinburgh Review* (October), 73–102.

Backhouse, R.E. 2006. Sidgwick, Marshall and the Cambridge school of economics. *History of Political Economy* 38, 15–44.

Bacon, R. and Eltis, W. 1976. *Britain's Economic Problem: Too Few Producers*. London: Macmillan.

Bagehot, W. 1848. Review of J.S. Mill's Principles of Political Economy. *Prospective Review* 4, 460–502.

Bagehot, W. 1880. *Economic Studies*, ed. R.H. Hutton. London: Longmans, Green.

Blaug, M. 1987. Classical economics. In *The New Palgrave: A Dictionary of Economics*, vol. 1, ed. J. Eatwell, M. Milgate and P. Newman. London: Macmillan.

Blaug, M. 1996. *Economic Theory in Retrospect*, 5th edn. Cambridge: Cambridge University Press.

Cairnes, J.E. 1874. *Some Leading Principles of Political Economy Newly Expounded*. London: Macmillan.

Chalmers, T. 1808. *An Enquiry into the Nature and Stability of National Resources*. Edinburgh: Moir.

Chamberlin, E.H. 1933. *The Theory of Monopolistic Competition: a Reorientation of the Theory of Value*. Cambridge, MA: Harvard University Press.

Checkland, S.G. 1951. The advent of academic economics in England. *The Manchester School* 19, 43–70.

Cliffe Leslie, T.E. 1879. *Essays in Political and Moral Philosophy*. London: Longmans, Green.

Copleston, E. 1819a. *A Letter to the Right Hon. Robert Peel, MP for the University of Oxford, on the Pernicious Effect of a Variable Standard of Value, especially as it regards the Condition of the Lower Orders and the Poor Laws ...* Oxford: Murray.

Copleston, E. 1819b. *A Second Letter to the Right Hon. Robert Peel, MP for the University of Oxford, on the Causes of the Increase in Pauperism, and on the Poor Laws ...* Oxford: Murray.

Corsi, P. 1987. The heritage of Dugald Stewart: Oxford philosophy and the method of political economy. *Nuncius* 2, 89–143.

Eltis, S. and Eltis, W. 1997. The life and contribution to economics of the Abbé de Condillac. In E.B. Abbé de Condillac, *Commerce and Government*, trans. S. Eltis. Cheltenham: Elgar.

Eltis, W. 2000. *The Classical Theory of Economic Growth*, 2nd edn. Basingstoke: Palgrave.

Faccarello, G. 1999. *The Foundations of Laissez-Faire: The Economics of Pierre de Boisguilbert*. London: Routledge.

Fawcett, H. 1863. *Manual of Political Economy*. London: Macmillan.

Fawcett, M.G. 1870. *Political Economy for Beginners*. London: Macmillan.

Fetter, F. 1980. *The Economist in Parliament, 1780–1868*. Durham, NC: Duke University Press.

Feyerabend, P. 1988. *Against Method*. London: Verso.

Flux, A.W. 1894. Review of Wicksteed. Repr. in *Precursors in Mathematical Economics*, ed. W. Baumol and S.M. Goldfield. London School of Economics, 1968.

Fontana, B. 1985. *Rethinking the Politics of a Commercial Society: The Edinburgh Review 1802–1832*. Cambridge: Cambridge University Press.

Godwin, W. 1793. *Enquiry Concerning Political Justice and its Influence on Morals and Happiness*. London: Robinson.

Goodwin, C.G.W. 1972. Marginalism moves to the New World. *History of Political Economy* 4, 551–70.

Harrod, R.F. 1939. An essay in dynamic theory. *Economic Journal* 49, 14–33.

Harrod, R.F. 1952. *The Life of John Maynard Keynes*. London: Macmillan.

Hicks, J.R. 1939. *Value and Capital*. Oxford: Clarendon Press.

Hollander, J.H. 1932. *Minor Papers on the Currency Question, 1809–1823 by David Ricardo*. Baltimore: Johns Hopkins Press.

Hollander, S. 1987. *Classical Economics*. Oxford: Blackwell.

Hollander, S. 1997. *The Economics of Thomas Robert Malthus*. 2 vols. Toronto: University of Toronto Press.

Hollander, S. 1998. The canonical classical growth model: content, adherence and priority. *Journal of the History of Economic Thought* 20, 253–77.

Jevons, W.S. 1871. *Theory of Political Economy*. London: Macmillan.

Jevons, W.S. 1882. *The State in Relation to Labour*. London: Macmillan.

Kates, S. 1994. The Malthusian origins of the General Theory or how Keynes came to write a book about Say's Law and effective demand. *History of Economics Review* 21, 10–20.

Keynes, J.M. 1972. Essays in biography. In *The Collected Writings of John Maynard Keynes*, vol. 10, ed. E. Johnson and D. Moggridge. London: Macmillan.

Keynes, J.N. 1891. *The Scope and Method of Political Economy*. London: Macmillan.

Kojecky, R. 1971. *Eliot's Social Criticism*. London: Faber.

Lauderdale, J. and Maitland, E. 1804. *An Inquiry into the Nature and Origin of Public Wealth: and into the Means and Causes of its Increase*. Edinburgh: Constable.

Malthus, T.R. 1798. *An Essay on the Principle of Population as It Affects the Future Improvement of Society, with Remarks upon the Speculations of Mr Godwin, M. Condorcet, and Other Writers*. London: Johnson.

Malthus, T.R. 1800. *An Investigation of the Cause of the Present High Price of Provisions. By the Author of the Essay on the Principle of Population*. London: Johnson.

Malthus, T.R. 1803. *An Essay on the Principle of Population, or, A View of its Past and Present Effects on Human Happiness, with an Inquiry into our Prospects Respecting the Future Removal or Mitigation of the Evils which it Occasions*. London: Johnson.

Malthus, T.R. 1815a. *An Inquiry into the Nature and Progress of Rent, and the Principles by which is Regulated*. London: Murray.

Malthus, T.R. 1815b. *The Grounds of an Opinion on the Policy of Restricting the Importation of Foreign Corn ...* London: Murray.

Malthus, T.R. 1817. *An Essay on the Principle of Population ... 4th edn of Malthus (1803), described as 5th edn*. London: Hunter.

Malthus, T.R. 1820. *Principles of Political Economy, Considered with a View to their Practical Application*. London: Pickering.

[Mrs Marcet (Jane Haldimand)] 1817. *Conversations on Political Economy: in which the Elements of that Science are Familiarly Explained*. London: Longman et al.

Marshall, A. 1890. *Principles of Economics*. London: Macmillan.

Marshall, A. 1920. *Principles of Economics*, 8th edn. London: Macmillan1952.

Marx, K. 1873. *Capital: A Critique of Political Economy*, ed. F. Engels, trans. S. Moore and E. Aveling. 3 vols. Moscow: Progress Publishers, 1954.

McCulloch, J.R. 1843. *Principles of Political Economy*. Edinburgh: Tait.

McCulloch, J.R. 1845. *Literature of Political Economy: A Classified Catalogue*. London: Longmans.

Mill, J. 1821. *Elements of Political Economy*. London: Baldwin et al.

Mill, J.S. 1869. Thornton on labour and its claims. In *Collected Works of John Stuart Mill*. vol. 5, ed. J.M. Robson. Toronto: University of Toronto Press, 1967.

Mill, J.S. 1871. *Principles of Political Economy*, 7th edn. ed. W.J. Ashley. London: Longmans, Green, 1909.

Nicholson, J.S. 1893–1901. *Principles of Political Economy*. 3 vols. London: Macmillan.

O'Brien, D.P. 1994. *Methodology, Money and the Firm*. 2 vols. Aldershot: Elgar.

O'Brien, D.P. 2004. *The Classical Economists Revisited*. Princeton, NJ: Princeton University Press.

Paley, W. 1802. *Natural Theology*. London: Wilkes and Taylor.

Palgrave, R.H. 1894–1899. *Dictionary of Political Economy*. 1899. Vol. 1, 1894; vol. 2, 1896; vol. 3, London: Macmillan.

Peart, S.J. and Levy, D.M. 2003. 1830–1870: Post-Ricardian British economics. In *A Companion to the History of Economic Thought*, ed. W.J. Samuels, J.E. Biddle and J.B. Davis. Oxford: Blackwell.

Political Economy Club. 1921. *Minutes of Proceedings, 1899– 1920, Roll of Members and Questions Discussed, 1821– 1920, with Documents bearing on the History of the Club*. London: Macmillan.

Pryme, G. 1823. *Introductory Lecture and Syllabus*. Cambridge: Cambridge University Press.

Ricardo, D. 1815. *An Essay on the Influence of a Low Price of Corn on the Profits of Stock*. London: Murray.

Ricardo, D. 1817. *On the Principles of Political Economy and Taxation*. London: Murray.

Ricardo, D. 1951–73. *The Works and Correspondence of David Ricardo*, 11 vols, ed. P. Sraffa. Cambridge: Cambridge University Press.

Robinson, J.V. 1933. *The Economics of Imperfect Competition*. London: Macmillan.

Samuelson, P.A. 1947. *Foundations of Economic Analysis*. Cambridge MA: Harvard University Press.

Samuelson, P.A. 1948. *Economics: an Introductory Analysis*. New York: McGraw Hill.

Samuelson, P.A. 1978. The canonical classical model of political economy. *Journal of Economic Literature* 16, 1415–34.

Say, J.-B. 1803. *Traitè d'economie politique: ou, Simple exposition de la manière dont se forment, se distribuent et se consomment les richesses*. Paris: Dèterville.

Schumpeter, J.A. 1954. *History of Economic Analysis*. London: Allen & Unwin.

Senior, N.W. 1836. *An Outline of the Science of Political Economy*. London: W. Clowes.

Sidgwick, H. 1883. *The Principles of Political Economy*. London: Macmillan.

Smith, A. 1776. *An Inquiry into the Nature and Causes of the Wealth of Nations*, 2 vols. ed. R.H. Campbell, A.S. Skinner and W.B. Todd. Oxford: Oxford University Press, 1976.

Smith, V.E. 1956. Malthus's theory of demand and its influence on value theory. *Scottish Journal of Political Economy* 3, 205–20.

Sraffa, P. 1926. The laws of return under competitive conditions. *Economic Journal* 35, 535–50.

Steuart, J. 1767. *An Inquiry into the Principles of Political Œconomy: being an Essay on the Science of domestic Policy in free Nations. In which are particularly considered Population, Agriculture, Trade, Industry, Money, Coin, Interest, Circulation, Banks, Exchange, Public Credit, and Taxes*. 2 vols. London: Millar and Cadell.

Stewart, D. 1792, 1814, 1827. *Elements of the Philosophy of the Human Mind*, 3 vols. Reprinted in *The Collected Works of Dugald Stewart*, 11 vols. ed. W. Hamilton. Edinburgh: Constable, 1854–60.

Stigler, G. 1952. The Ricardian theory of value and distribution. *Journal of Political Economy* 60, 187–207.

Stigler, G. 1958. Ricardo and the 93% labor theory of value. *American Economic Review* 48, 357–67.

Sumner, J.B. 1816. *A Treatise on the Records of the Creation: with Particular Reference to Jewish History, and the Consistency of the Principle of Population with the Wisdom and Goodness of the Deity*. 2 vols. London: Hatchard.

Thornton, H. 1802. *An Enquiry into the Nature and Effects of the Paper Credit of Great Britain*. London: Hatchard.

Thornton, R.W.T. 1869. *On Labour: its Wrongful Claims and Rightful Dues, its Actual Present and Possible Future*. London: Macmillan.

Thünen, J. H. von. 1826. *Der isolirte Staat in Beziehung auf Landwirthschaft und Nationalökonomie*, part 1. Hamburg: Pethes.

Torrens, R. 1815. *An Essay on the External Corn Trade*. London: Longman et al.

Torrens, R. 1821. *An Essay on the production of Wealth: with an Appendix, in which the Principles of Political Economy are applied to the Actual Circumstances of this Country*. London: Longman et al.

Turgot, A.R.J. 1768. Observations sur le mémoire de M. de Saint-Péravy en faveur de l'impôt indirect. In *Écrits Économiques*, ed. B. Cazes. Paris: Calman-Lévey, 1970.

Walras, L. 1894. *Élèments d'èconomie politique pure: ou, Thèorie de la richesse sociale*. Lausanne: Rouge.

Walsh, V. and Gram, H. 1980. *Classical and Neoclassical Theories of General Equilibrium*. New York: Oxford University Press.

Waterman, A.M.C. 1991a. *Revolution, Economics and Religion: Christian Political Economy, 1798– 1833*. Cambridge: Cambridge University Press.

Waterman, A.M.C. 1991b. The 'canonical classical model' in 1808 as viewed from 1825: Thomas Chalmers on the national resources. *History of Political Economy* 23, 221–41.

Waterman, A.M.C. 1995. Why William Paley was 'The first of the Cambridge Economists'. *Cambridge Journal of Economics* 20, 673–86.

Waterman, A.M.C. 1998. Reappraisal of 'Malthus the economist', 1933–97. *History of Political Economy* 30, 293–334.

Waterman, A.M.C. 1999. Hollander on the 'canonical classical growth model': a comment. *Journal of the History of Economic Thought* 21, 311–13.

Waterman, A.M.C. 2004. *Political Economy and Christian Theology since the Enlightenment.* Basingstoke: Palgrave Macmillan.

Wayland, Francis. 1837. *The Elements of Political Economy.* Boston: Gould and Lincoln.

West, E. 1815. *Essay on the Application of Capital to Land; with Observations Shewing the Impolicy of Any Great Restriction on the Importation of Corn … by a Fellow of University College, Oxford.* London: Underwood.

Whately, R. 1831. *Introductory Lectures in Political Economy.* London: Fellowes.

Wicksteed, P.H. 1894. *Essay on the Coordination of the Laws of Distribution.* London: Macmillan.

Winch, D.N. 1983. Science and the legislator: Adam Smith and after. *Economic Journal* 93, 501–20.

Winch, D.N. 1996. *Riches and Poverty. An Intellectual History of Political Economy in Britain, 1750– 1834.* Cambridge: Cambridge University Press.

Young, A.A. 1928. Increasing returns and economic progress. *Economic Journal* 38, 527–42.

Enlightenment, Scottish

Between 1740 and 1790 Scotland provided one of the most distinguished branches of the European Enlightenment. David Hume and Adam Smith were the pre-eminent figures in this burst of intellectual activity; and around them clustered a galaxy of major thinkers, including Francis Hutcheson, Lord Kames, Adam Ferguson, William Robertson, Thomas Reid, Sir James Steuart and John Millar. The interests of individual thinkers ranged from metaphysics to the natural sciences; but the distinctive achievements of the Scottish Enlightenment as a whole lay in those fields associated with the enquiry into 'the progress of society' – history, moral and political philosophy and, not least, political economy.

'Enlightenment' and 'Scottish Enlightenment' were usages unknown in the 18th century: the term 'Scottish Enlightenment' was first coined in the early 20th century, and began to be generally used by historians in the 1960s. (*Lumières* and *Aufklärung* were in 18th-century use, but not to denote a European Enlightenment as a whole.) As a historian's construction, however, the term 'Scottish Enlightenment' is supported by the consciousness of

those named above that they shared common intellectual interests (which did not preclude disagreement between them) and a common standing as men of letters in 18th-century Scottish society. This awareness of belonging to a broad intellectual movement extended to the continent of Europe: led by Hume, the Scottish thinkers cultivated connections with Paris, the Enlightenment's acknowledged metropolitan centre. But the Scottish Enlightenment is perhaps best understood when it is compared with the Enlightenment in Italy or in Germany. The concern with economic improvement and its moral and political conditions and consequences was as urgent, for instance, in the distant Kingdom of Naples as in Scotland; and political economy was equally absorbing to the Neapolitan philosophers Antonio Genovesi and Ferdinando Galiani.

At the same time, the experience of Scotland in the 18th century was distinctive in a number of respects, which offered a particular stimulus to Scottish thinkers. First of all, there was the actual achievement of economic growth. The late 17th-century Scottish economy supported an uneasy balance between population and food supply; bad harvests, which occurred in a sequence in the 1690s, could cause severe shortages and even localized famine. Overseas trade was likewise vulnerable. Nevertheless the elites, both landed and urban, were committed to economic development, and showed a marked propensity to invest. Agriculture gradually became commercialized, and land-owners joined merchants to invest in manufactures, and, most spectacularly, in the 'Darien venture', intended to establish a Scottish trading colony in Panama. The failure of the latter persuaded many of the elite that economic development could only come through closer union with England. In the event, the economic fruits of the Union were disappointingly slow in coming; but by the third quarter of the 18th century it was clear to contemporaries that agriculture, trade and manufactures were all on an upward curve. The thinkers of the Scottish Enlightenment thus enjoyed an unusually direct acquaintance with the phenomena of economic development.

Scotland's political position was also unusual. Many of Europe's monarchies sought to bring their constituent kingdoms into closer union over the 18th century, for economic as well as administrative reasons. But none did so as successfully as the British monarchy. The Union of 1707 with England was in no simple or direct sense the cause of Scotland's economic growth (or of its Enlightenment). But it secured a common framework of law and a common market, and it also established that the Scottish Presbyterian and the English Anglican Churches should coexist in peace. These gains were important to the great majority of the Scottish elites, and it was never in their interest to back the Jacobite challenge to the Hanoverian monarchy.

Culturally and intellectually, the position of Scotland looked unpropitious before 1700. There were pockets of interest in the new science, Newton having a group of

Scottish adherents; but the latest developments in French philosophy were shunned for their Epicurean, materialist and sceptical tendencies. After the Revolution of 1688, however, change gradually got under way in the institutions most important for intellectual life, making possible the infiltration of new ideas. The fierce, covenanting Presbyterianism of the 17th century was dissipated, as the 'Moderate' group of clergy rose to power in the Kirk. The universities of Edinburgh, Glasgow, Aberdeen and St Andrews were reformed, allowing professorial specialization; and around the universities there developed a vigorous informal culture of voluntary clubs, most famous of which was the Select Society of Edinburgh, founded by David Hume and his friends in 1754. Together these changes secured for Scottish thinkers unprecedented intellectual freedom and social support; and they provided an object lesson in the importance of the moral and cultural as well as the material dimensions of progress.

The intellectual interests which distinguished the Scottish Enlightenment had two more specific sources. One was the explicit preoccupation with the conditions and means of economic development which was fostered by the debate which preceded the Union of 1707. The preoccupation was by no means unique to the Scots, but the contributions of John Law (the future author of the French Mississippi Scheme) and others ensured a high quality of discussion. The other, two decades later, was the initiative taken by two very different philosophers, Francis Hutcheson and David Hume, to transform the agenda by which philosophy was taught and discussed in Scotland. Drawing on the moral philosophy of Shaftesbury and the natural jurisprudence of Pufendorf, Hutcheson taught his Glasgow students, who included Adam Smith, a moderate, benevolent, providential Stoicism. More disturbingly, Hume drew on the scepticism of Pierre Bayle and the Epicurean morals of Bernard Mandeville to offer in his *Treatise of Human Nature* (1739–40) and his two later *Enquiries* (1748; 1751) an account of justice and morals which had no need of divine support. Most of those now associated with the Scottish Enlightenment found Hutcheson's philosophy more congenial; but it was Hume's challenge which galvanized them. It was Hume, moreover, who turned their attention back to economic matters. Recognizing that philosophy alone would never make the Scots into virtuous atheists, Hume decided instead to educate them in political economy, the subject of the leading essays in his *Political Discourses* of 1752.

For Hume as for all the Scottish thinkers, political economy was not a science apart. It belonged within a wider enquiry into the 'progress of society'. There were three principal dimensions to this enquiry: the historical, the moral and the political.

The historical theory of the Scottish Enlightenment developed a line of argument from later 17th-century natural jurisprudence, a tradition made familiar to the Scots by its incorporation in the moral philosophy curriculum of the reformed universities. Discarding the older jurisprudential thesis of the contractual foundations of society and government, the Scots focused on the new insights of Pufendorf and Locke into the origin and development of property. According to Pufendorf, there had never been an original state of common ownership of land and goods; from the first, property was the result of individual appropriation. As increasing numbers made goods scarce, individual property became the norm, and systems of justice and government were established to secure it. What the Scots added to this argument was a scheme of specific stages of social development, the hunting, the pastoral, the agricultural and the commercial. At each of the four stages the extent of property ownership was related to the society's means of subsistence, and these shaped the nature and sophistication of the society's government. Different versions of the theory were offered by Adam Ferguson in his *Essay on the History of Civil Society* (1767) and by John Millar in his *Origin of the Distinction of Ranks* (1770), and it underlay both Lord Kames's investigations into legal history and William Robertson's historical narratives. The locus classicus of the theory, however, was Adam Smith's Lectures on Jurisprudence, delivered to his students in Glasgow in the early 1760s.

As Smith's exposition makes particularly clear, the stages theory of social development provided the historical premises for political economy. An explicitly conjectural theory – a model of society's 'natural' progress – it provided a framework for a comparably theoretical treatment of economic development as 'the natural progress of opulence'. By positing the systematic interrelation of economic activity, property and government, with consequences which could be neither foreseen nor controlled by individuals, the theory also underlined the limits of effective government action. 'Reason of state', the standby of rulers and their advisers for over two centuries, still had the capacity to distort and obstruct the economic activity of subjects and those with whom they would trade; but the Scots' historical perspective showed it to be a doctrine inadequate to the complexity of a modern commercial economy.

The moral thought of the Scottish Enlightenment was closely related to the historical, sharing a common origin in 17th-century natural jurisprudence. Here the inspiration was the jurisprudential thinkers' increasingly sophisticated treatment of needs. These, it was recognized, could no longer be thought of primarily in relation to subsistence; with the progress of society, needs must be understood to cover a much wider range of scarce goods, luxuries as well as necessities. The potential of this insight was seen by every Scottish moral philosopher, but again it was Smith who exploited it to the full, in the *Theory of Moral Sentiments* (1759). Beyond the most basic necessities, Smith acknowledged, men's needs were always relative, a matter of status and emulation, of bettering one's individual condition. But it was precisely the vain desires of the rich and the envy of others which served, by 'an invisible hand', to stimulate men's industry and hence to increase the stock of goods available for all ranks.

Such an argument, however, had to overcome two of the most deeply entrenched convictions of European moral thought: the Aristotelian view that the distribution of goods was a matter for justice, and the classical or civic humanist view that luxury led to corruption and the loss of moral virtue. The Scots answered the first more confidently (but perhaps less satisfactorily) than the second. Following Grotius, Hobbes and Pufendorf, they defined justice in exclusively corrective terms, setting aside questions of distribution. On the issue of corruption, they were divided. Hume, who ridiculed fears of luxury, was the most confident; Ferguson, who defiantly reasserted the ancient ideal of virtue, was the most pessimistic. Smith was closer to Hume in preferring propriety to virtue, at least for the great majority; but he showed that he shared Ferguson's doubts when he added, at the end of his life, that the disposition to admire the rich and the great did tend to corrupt moral sentiments. At a fundamental level, however, there was general agreement. As a consequence of the progress of society, the multiplication of needs was not only irreversible; it was the essential characteristic of a 'cultivated' or 'civilized' as distinct from a 'barbarian' society. And civilization, however morally ambiguous, was preferable to barbarism. With consensus on this, the moral premises of political economy were secure.

The definition of justice in simple corrective terms provided the starting-point for the political dimension of the Scottish enquiry. The priority of any government, the Scots believed, must be the security of life and property, ensuring every individual liberty under the law. This, as Smith put it, was freedom 'in our present sense of the word'; and there was a general confidence that it was tolerably secure under the governments of modern Europe, including the absolute monarchies. In principle, individual liberty was a condition of a fully commercial society: its provision, therefore, was the institutional premise of political economy.

Few of the Scots took their analysis beyond this relatively simple, if vital, point; the theory of the modern commercial state was not a Scottish achievement. Both Hume and Smith were more concerned to limit the opportunities for enlarging government at the expense of 'productive' society, by confining the former to the minimum necessary provision of justice, defence and public works. But they also recognized that the proliferation of interests in a commercial society would require more sophisticated institutional mechanisms to ensure their adequate representation within the political system. Smith's analysis in Book IV of the *Wealth of Nations* of the growing alienation of the colonial elites in North America from parliamentary authority was an object lesson in the need for such representation – and a strong

hint that it was incompatible with maintaining an extended empire.

A large part of the originality of the Scottish Enlightenment's conception of political economy lay in this exploration of the historical, moral and institutional framework of economic activity. But of course the Scots also engaged directly in economic analysis; and one such work of analysis, Adam Smith's *Wealth of Nations* (1776), would so outshine all others that it came to be regarded as having established political economy as a science in its own right.

The Scots' attention focused on growth in a context of international rivalry. In contemporary terms, Hont has shown, the issue was the means by which poor countries (of which Scotland might be regarded as one) could best hope to catch up on rich countries (such as England certainly was). What is striking is the hard-headedness with which Hume and Smith tackled the issue. Responding to French economists – Hume to Jean-François Melon, Smith to the Physiocrats – who argued that agriculturally endowed countries should follow a different path from purely commercial nations, the Scots insisted that one analysis applied to all. Protection for agricultural economies and their manufactures, a policy supported by the former Jacobite exile Sir James Steuart in his *Principles of Political Economy* (1767), was futile and damaging. But theirs was no naive optimism in the equalizing powers of commerce. The ideal of *doux commerce*, by which trade would be the agent of global peace and prosperity, was as much of a panacea as the belief that commercial success would be self-cancelling, because the advantage of low labour costs would always pass on to others. Instead, Hume and Smith suggested that rich countries could expect to maintain their advantage over poorer ones, whether by flexible specialization and product innovation (Hume) or by constantly increasing industrial productivity through the division of labour (Smith). What distinguished commercial superiority from military conquest was that it was achieved 'without malice'; poor countries would also develop if they followed the same route, even if they might never catch up on the rich.

Brilliant as Hume's economic essays were, it was Adam Smith's *Wealth of Nations* (1776) which set the standard of Enlightenment political economy. To be systematic and comprehensive had earlier been the ambition, at least, of Quesnay's *Tableau Economique* (1758–9), Genovesi's *Lezioni di Commercio* (1765) and Steuart's *Principles;* but the *Wealth of Nations* eclipsed them all. Its success, moreover, was such as to suggest that political economy had an identity all of its own. Smith himself did not admit such an implication, continuing to insist that political economy was but 'a branch of the science of a statesman or legislator': his own engagement with both jurisprudence and moral philosophy left him disinclined to drop the wider intellectual framework in which political economy had been conceived. But a work at once as extensive and as self-contained as the *Wealth of Nations* made it at least plausible to suppose that what it presented was a distinct, autonomous science of political economy.

Smith's death in 1790 coincided with the end of the Scottish Enlightenment. In Scotland as throughout Europe, the French Revolution transformed the conditions and assumptions of intellectual life, while political economy had to come to terms with the increasingly obvious impact of machinery. Within Scotland Dugald Stewart set himself to adapt the Enlightenment conception of political economy to these new circumstances; but while he had French admirers, his expansive, didactic approach had few followers in Britain. Another Scot, Thomas Chalmers, took the lead alongside Malthus in attaching political economy to newly urgent theological concerns, while Ricardo and his followers simply took a narrower view of the subject. Even so, it would be a mistake to see 19th-century classical political economy as a new departure. As the philosophical analysis of Hegel (who learnt much from Steuart) and the radical critiques of Marx and the early socialists pointed out, the historical, moral and institutional premises on which political economy rested were still those elucidated by the Scots.

JOHN ROBERTSON

See also **Hume, David; Hutcheson, Francis; Mandeville, Bernard; Pufendorf, Samuel von; Smith, Adam; Steuart, Sir James; Stewart, Dugald.**

Bibliography

Berry, C.J. 1997. *Social Theory and the Scottish Enlightenment*. Edinburgh: Edinburgh University Press.
Broadie, A. ed. 2003. *The Cambridge Companion to the Scottish Enlightenment*. Cambridge: Cambridge University Press.
Devine, T.M. 1994. *The Transformation of Rural Scotland: Social change and the Agrarian Economy 1660–1815*. Edinburgh: Edinburgh University Press.
Hont, I. and Ignatieff, M., eds. 1983. *Wealth and Virtue: The Shaping of Political Economy in the Scottish Enlightenment*. Cambridge: Cambridge University Press.
Hont, I. 2005. *Jealousy of Trade: International Competition and the Nation-State in Historical Perspective*. Cambridge, MA: Harvard University Press.
Phillipson, N.T. 1981. The Scottish Enlightenment. In *The Enlightenment in National Context*, ed. R. Porter and M. Teich. Cambridge: Cambridge University Press.
Robertson, J.C. 2005. *The Case for the Enlightenment. Scotland and Naples 1680–1760*. Cambridge: Cambridge University Press.
Sakamoto, T. and Tanaka, H., eds. 2003. *The Rise of Political Economy in the Scottish Enlightenment*. London: Routledge.
Sher, R.B. 1985. *Church and University in the Scottish Enlightenment*. Princeton and Edinburgh: Princeton:

Princeton University Press; Edinburgh: Edinburgh University Press.
Sher, R.B. 2006. *The Enlightenment and the Book. Scottish Authors and their Publishers in Eighteenth-Century Britain, Ireland and America*. Chicago and London: Chicago University Press.
Winch, D. 1996. *Riches and Poverty: An Intellectual History of Political Economy in Britain 1750–1834*. Cambridge: Cambridge University Press.

enterprise zones

Enterprise zone programmes are geographically targeted tax, expenditure, and regulatory inducements used by US state and local governments since the early 1980s and by the federal government since 1993. While they differ in their specifics, all the programmes provide development incentives, including tax preferences to capital and/or labour, in an attempt to induce private investment location or expansion to depressed areas and to enhance employment opportunities for zone residents. Most enterprise zones are designated in urban areas, but there are some rural zones. Typically, state and local zone programmes provide larger tax credits for business investment than for employment incentives. Investment incentives include the exemption of business-related purchases from state sales and use taxes, investment tax credits and corporate income or unemployment tax rebates. Labour subsidies include employer tax credits for all new hires or zone-resident new hires, employee income tax credits and job-training tax credits. Some programmes assist firms financially with investment funds or industrial development bonds.

Enterprise zones have been criticized as ineffective and inefficient in stimulating new economic activity. This criticism is part of a long-standing debate on the effects of intersite tax differentials on the location of capital investment. It is argued that if tax-induced investment represents only relocation from another state, then tax competition is a zero-sum game for the country as a whole. In addition, the preferential treatment of certain types of investment or employment within enterprise zones may induce decisions that would not be economically sound in the absence of the tax incentives. Often, however, redistribution of economic activity within a state may be a desirable goal. If investment is relocated from local labour markets with low unemployment to local labour markets with higher unemployment, the incentives may generate efficiency gains for the economy as underutilized resources are tapped (Bartik, 1991). Efficiency gains may also result if reductions in unemployment produce positive externalities, such as reductions in social unrest.

A partial equilibrium model predicts that a labour subsidy or an equal-cost subsidy to both zone capital and zone resident labour will raise zone wages. A capital subsidy alone may actually reduce zone wages – yet many of the subsidies are for capital investment in the zone (Gravelle, 1992; Papke, 1994).

Empirical evaluations of zone programmes typically measure the amount of investment undertaken after the designation, for example, or the increase in the number of firms in the zone, and the change in zone employment. Two key methodological issues in empirical evaluations are (a) to separate the effects of zone designation from jobs and investments arising from other factors – for example, general upswings in the economy; (b) to account for the depressed economic characteristics that led to the initial zone designation. If zone sites are better randomly selected, the effect of the programme can be measured by comparing the performance of the experimental and control groups. But zone designation in the 43 state and local programmes in the United States depends on comparative unemployment rates, population levels and trends, poverty status, median incomes, and percentage of welfare recipients, so the data are non-experimental. This sample selection problem can be addressed with a variety of econometric techniques.

Econometric analysis of a zone's success faces a practical difficulty in that conventional economic data are not available by zone. In most states, zones do not coincide with census tracts or taxing jurisdictions. As a result, zone areas cannot be pinpointed in standard data collections. Zip code level data is available from the Census, but outcome measures are ten years apart.

Econometric evaluations of the Indiana and New Jersey programmes find mixed effects on investment and employment. Indiana zones are estimated to have greater inventory growth and fewer unemployment claims than they would have in the absence of the zone designation (from 1983 to 2006, an inventory tax credit was the most lucrative incentive). However, in the 1980s, inventory investment came at the cost of a drop in the value of depreciable property (Papke, 1994). Moreover, despite the reduction in unemployment rates in the zones, a comparison of incomes from the 1980 and 1990 Censuses suggests that zone residents are not appreciably better off after the first decade of the Indiana zone programme (Papke, 1993) and there is no discernable increase in capital investment or land values (Papke, 2001). Similar econometric analysis of the New Jersey enterprise zone programme finds no positive effects on either business investment or employment (Boarnet and Bogart, 1996). Multi-state econometric analyses that combine data from many states – thereby assuming zone programmes have similar effects in every state – typically find no positive zone effects on business activity or employment (Bondonio, 2003; Bondonio and Engberg, 2000). Peters and Fisher (2002) survey state evaluations.

Cost-per-job estimates from zone programmes are rare. The literature also lacks a discussion of the distribution of the cost of the zone programme between state and local governments. For example, local governments may bear the brunt of the cost of a state enterprise zone programme if tax incentives are provided against local taxes without state reimbursement.

Congress established the Empowerment Zone and Enterprise Community (EZ/EC) programme in 1993 and the Renewal Community (RC) programme in 2000 to provide assistance to the nation's distressed communities. By 2007, there had been three rounds of EZs, two rounds of ECs, and one round of RCs leading to a total of 40 empowerment zones (30 urban and 10 rural), 95 enterprise communities (65 urban, 30 rural) and 40 renewal communities.

Empowerment zone incentives include a 20 per cent employer wage credit for the first 15,000 dollars of wages for zone residents who work in the zone, additional expensing of equipment investments of qualified zone businesses, and expanded tax exempt financing for certain zone facilities. Each zone is eligible for 100 million dollars in Social Services Block Grant funds. Selected areas needed to demonstrate pervasive poverty, unemployment and general distress, and applicants had to outline a plan of action that included local business and community interests. The residence-based approach of the income tax credit differs significantly from another federal programme designed to increase employment of the disadvantaged. The Targeted Jobs Tax Credit provides firms with a similar-sized subsidy for wages paid to targeted individuals – primarily welfare recipients and poor youth. Providing a subsidy based on individual characteristics may create a stigma that actually reduces the probability of being hired. Residence-based eligibility may eliminate this problem and encourage individuals who become employed to continue to live in the zone.

Features of the programmes have changed over time. Round I and II EZs and ECs received different combinations of grant funding and tax benefits. By round III, EZs and the RCs received mainly tax benefits. The GAO (1991; 2004; 2006) reports that Round I and II EZs and ECs are continuing to access their grant funds and Internal Revenue Service (IRS) data show that businesses are claiming some tax benefits (Brashares, 2000). However, the IRS does not collect data on other tax benefits and cannot always identify the communities in which they were used. The lack of tax benefit data limits evaluation of the programmes.

Evaluation of the federal programme is also confounded by its hybrid structure. The federal EZ/EC programme is based on the idea that effective community revitalization results when the strategy is tailored to the local site. The diverse nature of the Round I EZ/ECs – each may differ in terms of objective, size of targeted area, type of designation, governance structure, projects used, grant money, and strategies for implementation – has made it difficult to generate general conclusions about even the early stages of Round I implementation (GAO, 2004; 2006). Further, the tax incentives changed over the three rounds of the federal programme. Third, no easy method of data collection was included in the tax forms so even usage is hard to measure.

Using Census data, Hanson (2007) finds no effect of the first round zone programme on local employment or poverty rates in the targeted areas, but instead finds capitalization into property values. Busso and Kline (2006) find modest improvements in labour market conditions, but sizable increases in owner-occupied housing values and rents along with small changes in the demographic composition of neighbourhoods. Taken together, these two papers suggest that improvements for residents have been limited at best, but that property owners have benefited from the federal programme.

LESLIE E. PAPKE

See also **economic development and the environment; fiscal federalism; local public finance; public finance; regional development, geography of; taxation and poverty.**

Bibliography

Bartik, T.J. 1991. *Who Benefits from State and Local Economic Development Policies?* Kalamazoo, MI: W.E. Upjohn Institute for Employment Research.

Boarnet, M.G. and Bogart, W.T. 1996. Enterprise zones and employment: evidence from New Jersey. *Journal of Urban Economics* 40, 198–215.

Bondonio, D. and Engberg, J. 2000. Enterprise zones and local employment: evidence from the states' programs. *Regional Science and Urban Economics* 30, 519–49.

Brashares, E. 2000. Empowerment zone tax incentive use: what the 1996 data indicate. *Statistics of Income Bulletin.*

Busso, M. and Kline, P. 2006. Do local economic development programs work? Evidence from the federal empowerment zone program. Mimeo, University of Michigan.

Engberg, J. and Greenbaum, R. 1999. State enterprise zones and local housing markets. *Journal of Housing Research* 10, 163–87.

GAO (General Accounting Office). 1991. Businesses' use of empowerment zone incentives. RCED-99-253. US Government Accounting Office, Washington, DC.

GAO. 2004. Community development: federal revitalization programs are being implemented, but data on the use of tax programs are limited. RCED 04-306, US Government Accounting Office, Washington, DC.

GAO. 2006. Empowerment zone and enterprise community program: improvements occurred in communities but the effect of the program is unclear. RCED-06-727, US Government Accounting Office, Washington, DC.

Gravelle, J.G. 1992. Enterprise zones: the design of tax incentives. CRS Report for Congress 92-476 S. Congressional Research Service, Library of Congress, Washington, DC.

Hanson, A. 2007. Poverty reduction and local employment effects of geographically targeted tax incentives: an instrumental variables approach. Mimeo, Syracuse University.

HUD (U.S. Department of Housing and Urban Development). 1992. *State Enterprise Zone Update: Summaries of the State Enterprise Zone Programs.* Washington, DC: U.S. Department of Housing and Urban Development.

Papke, L.E. 1993. What do we know about enterprise zones? In *Tax Policy and the Economy*, vol. 7, ed. J.M. Poterba. Cambridge, MA: MIT Press.

Papke, L.E. 1994. Tax policy and urban development: evidence from the Indiana enterprise zone program. *Journal of Public Economics* 54, 37–49.

Papke, L.E. 2001. The Indiana enterprise zone revisited: effects on capital investment and land values. National Tax Association Proceedings of the Ninety-Third Annual Conference, National Tax Association, Washington DC.

Peters, A.H. and Fisher, P.S. 2002. *State Enterprise Zone Programs: Have They Worked?* Kalamazoo, MI: W.E. Upjohn Institute for Employment Research.

entitlements in laboratory experiments

Entitlements are rights granted by contract, law or practice. Under the assumption of pure self-interest, modelling games with entitlements is fairly straightforward; however, work in behavioural economics has consistently demonstrated the existence of other-regarding preferences, with strong effects of perceptions of what is fair. In the laboratory, behaviour is affected not only by the entitlement per se but also by the procedure by which entitlements come about. One form of laboratory entitlement is a more advantageous position in an economic game, where the advantage arises from a larger endowment, favourable exchange rules or greater decision-making authority. A second type of entitlement is a guaranteed payoff or a payoff floor. Experimental results show that the means by which entitlements are acquired is one cue that influences the nature of other-regarding behaviour. This is important both for understanding behaviour and the design of experiments.

In early experimental work on entitlements, Hoffman and Spitzer (1985) demonstrate that both the existence of an entitlement and its source determine economic outcomes. They study bilateral bargaining problems where one of the two subjects, called the 'controller', has unilateral authority to decide the outcome of a negotiation game in the event of disagreement. Authority is assigned based on either the outcome of a coin flip or the result of a simple test of a skill that is irrelevant to the experimental task. They find that controllers are most willing to exploit their power when they are assigned their role based on the skill test and are told that they 'earned' the right to be the controller – that is, that they have moral authority. These results are consistent with Burrows and Loomes (1994).

The subjects' behaviour illustrates Rawls's (1971) notion of 'desert', which requires that people deserve the conditions underlying their actions as well as the fruits of their actions. Thus subjects divided an endowment equally when the controller was chosen according to the flip of a coin and had low moral authority. On the other hand, both earning the right to be controller and higher moral authority triggered changes in observed allocations, so that outcomes favoured the controller. Entitlements that were earned or that involved 'morally unequal' agents were sufficient to trigger unequal outcomes. Equity theory developed by social psychologists is similar in spirit to this theory of justice.

Ideas of procedural fairness also affect perceptions of government entitlements. Fong (2001) looks at poll data on perceptions of poverty and opportunity, and finds that beliefs about others' effort, luck and opportunity play the largest role in determining support for government entitlement programmes. In particular these beliefs outweigh concerns about tax costs in supporting these programmes. These results are consistent with the experimental results discussed above, where low payoffs are acceptable if one displays low effort. If one's situation is determined by poor luck, however, one will give up some of one's earnings to increase the earnings of others.

A number of experimental studies on income redistribution examine Rawls's claim that individuals prefer an income redistribution rule that maximizes the position of the poorest member of society (Frohlich and Oppenheimer, 1990). Studies where subjects must choose a principle of distributive justice and a tax system in addition to participating in a production task find that people choose rules that maximize the productivity of society while maintaining a minimum floor for the worst off members. Subjects generate greater output in experiments where they are able to determine the entitlements for the worst off individual in their group, again demonstrating that the source of entitlements matters.

These results show that researchers need to pay attention to how entitlements are determined. This is a complication for theories of behavioural economics or psychological games. People do not have a pure taste for fair allocations; they are more self-interested, altruistic or fair according to circumstances that depend on how advantage arises. This behaviour is closely related to reciprocity, but that is often modelled as 'if you are nice to me I'll be nice to you' (Bowles and Gintis, 2001). In contrast, this collection of results can be interpreted as, 'I will respect your entitlement if you deserve it'.

A preference for procedural factors also complicates experimental design, since subjects behave in a more self-interested manner when entitlements are earned than when they are randomly assigned. Researchers must be careful to consider how subjects will interpret the rules by which advantages are assigned or they may risk introducing nuisance variables. Future work might deliberately award entitlements in a manner that subjects view

as unjust to see whether that produces yet another pattern of behaviour.

SHERYL BALL

See also **behavioural game theory; Coase theorem; experimental economics; fair allocation; justice; psychological games.**

Bibliography

Burrows, P. and Loomes, G. 1994. The impact of fairness on bargaining behavior. *Empirical Economics* 19, 201–21.
Bowles, S. and Gintis, H. 2001. The inheritance of economic status: education, class and genetics. In *Genetics, Behavior and Society*, ed. M. Feldman. In *International Encyclopedia of the Social and Behavioral Sciences*, ed. N. Smelser and P. Baltes. Oxford: Elsevier.
Fong, C. 2001. Social preferences, self-interest, and the demand for redistribution. *Journal of Public Economics* 82, 225–46.
Frohlich, N. and Oppenheimer, J.A. 1990. Choosing justice in experimental democracies with production. *American Political Science Review* 84, 461–77.
Hoffman, E. and Spitzer, M.L. 1985. Entitlements, rights and fairness: an experimental examination of subjects' concepts of distributive justice. *Journal of Legal Studies* 14, 259–97.
Lissowski, G., Tyszka, T. and Okrasa, W. 1991. Principles of distributive justice: experiments in Poland and America. *Journal of Conflict Resolution* 35, 98–119.
Rawls, J. 1971. *A Theory of Justice.* Cambridge, MA: Belknap, Harvard University Press.

entrepreneurship

An entrepreneur is an individual who organizes, operates, and assumes the risk of creating new businesses. There are two types. A replicative entrepreneur organizes a new business firm that is like other firms already in existence. An innovating entrepreneur provides something new – a new product or process, or a new type of business structure, a new approach to marketing, and so on. These innovations need not be productive or beneficial. For example, Richard Cantillon (one of the first great economic theorists) spoke of thieves who are entrepreneurs (Cantillon, 1730, pp. 54–5). And Joseph A. Schumpeter, arguably the contributor of the most important analysis of entrepreneurship, included as an entrepreneurial act '…the creation of a monopoly position (for example, through trustification)….' (Schumpeter, 1911, p. 66). Entrepreneurs (interpreted as the self-employed) are estimated to constitute about seven percent of the labour force in the United States (U.S. Bureau of the Census, 2004). Most of them are probably replicative, not innovative, entrepreneurs.

It is widely agreed that the entrepreneur plays an important role in economic growth. But the evidence shows little correlation between an economy's number of replicative entrepreneurs and its growth rate. Innovative entrepreneurs do make a substantial difference to a nation's growth rate, having introduced many breakthrough innovations like the telephone and the airplane. The primary social contribution of replicative entrepreneurship is as a means for individuals to escape poverty, because such undertakings require little capital, education or experience. Still, the data show that entrepreneurs, on average, earn less than employees with similar education and experience (Freeman, 1978; Astebro, 2003; Benz and Frey, 2004).

Although economists have recently exhibited a resurgence of interest in entrepreneurship, the entrepreneur nevertheless rarely shows up in contemporary mainstream economic theory.

Early writings and the origin of the term

Until the 20th century, writings in English referred to entrepreneurs as 'adventurers' or 'undertakers' (see, for example, Marshall, 1923, p. 172). Apparently, the term 'entrepreneur' was introduced by Cantillon in the French translation of his great work, *Essai Sur la Nature de Commerce en Général* (1730, p. 54), but what is apparently *his* English text uses the word 'undertaker'. The early writings on entrepreneurship were descriptive rather than theoretical. Cantillon's discussion (1730, ch. 11) is brief, focusing on replicative entrepreneurs: '…wholesalers in Wool and Corn, Bakers, Butchers, Manufacturers and Merchants of all kinds….' (1730, p. 51). Cantillon's main point, like that of Frank H. Knight (1921), was the task's riskiness: 'These Undertakers can never know how great will be the demand in their City, nor how long their customers will buy of them since their rivals will try all sorts of means to attract customers from them. All this causes so much uncertainty among these Undertakers that every day one sees some of them become bankrupt' (1730, p. 51).

Nearly a century later, Jean-Baptiste Say's (1819) discussion is still brief, but richer. Say seems interested primarily in innovating entrepreneurs dealing with three types of 'producers': scientists, entrepreneurs and labourers. Using mechanical locks as an example, the scientist investigates '…the properties of iron, the method of extracting from the mine and refining the ore…' The entrepreneurs deal with '…application of this knowledge to a useful purpose…,' while the third group – the workers – actually make the product (1819, p. 80). And any successful economy needs all three: 'Nor can [industry] approximate to perfection in any nation, till that nation excel in all three branches' (1819, p. 80).

Thus, Say blames poverty in Africa on the absence of scientists and entrepreneurs. Lack of entrepreneurs alone can undercut prosperity, even with scientific knowledge abundant, for without the entrepreneur, '…that knowledge might possibly have lain dormant in the memory of one or two persons, or in the pages of literature' (1819, p. 81). This is precisely the explanation that one of the present authors proposed for the failure of medieval China and the Soviet Union to translate an abundance of non-military inventions into viable consumer products (Baumol, 2002, chs. 5, 14). Say also foreshadows some of Schumpeter's analysis (see below): 'In manufacture…if success [in innovation] ensue, the adventurer is rewarded by a longer period of exclusive advantage, because his process is less open to observation' (1819, p. 84).

Finally, Say mentions the spillovers of innovation and their justification for governmental financing: 'The charges of experiment, when defrayed by the government… [are] hardly felt at all, because the burthen is divided among innumerable contributors; and the advantages resulting from success being a common benefit to all, it is by no means inequitable that the sacrifices, by which they are obtained, should fall on the community at large' (1819, p. 85).

Before Schumpeter's breakthrough (see below), the subject was touched upon by economists like J.S. Mill, Alfred Marshall and (a bit later) Knight. Generally, their focus was not on innovative entrepreneurship, and they emphasized management's directing of going concerns rather than establishment of new firms. (But Marshall, 1923, p. 172, does digress briefly to mention Matthew Boulton's significant role as an entrepreneur dealing with James Watt's inventions.) Today, however, these discussions would hardly be considered theory. Rather, they are usually narratives containing illuminating observations. They assert that the entrepreneur's payment is a residual after other inputs are compensated, and that compensation is determined by the entrepreneur's ability and the supply of entrepreneurship in the market. They note that entrepreneurs employ themselves, so that unlike other inputs there is no demand function, as for other inputs.

Disappearance of the entrepreneur from modern mainstream economics

Given the acknowledged importance of the entrepreneur's role, it could be hoped that modern theoretical economics, with its powerful analytic tools, would have produced an extensive entrepreneurship analysis. Instead, the opposite happened – the entrepreneur became the 'invisible man' in mainstream theory. There are at least two reasons for this. First, the most advanced and powerful microeconomic models predominately study timeless static equilibria. But, for the entrepreneur, the transition process is the heart of the story. Schumpeter (1911) shows the entrepreneur as a destroyer of equilibria by constant innovation, while Israel Kirzner (1979) tells how the alert entrepreneur seeks out the arbitrage opportunities presented by disequilibria, thereby moving the economy back toward equilibrium. Such a relentless attack upon both equilibria and disequilibria does not fit a stationary model from which firm creation and invention are excluded.

The second reason for the entrepreneur's disappearance from mainstream theory is that, by definition, an invention is something never available before. So invention is the ultimate heterogeneous product. This impedes the optimality analysis underlying most microeconomic theory. Explicitly or implicitly, an optimality calculation entails a comparison among possible substitute choices, while the innovating entrepreneur normally deals with no well-defined substitutes with quantifiable attributes. In contrast, the standard theory of the firm analyses repetitive decisions of management in fully operational enterprises where the entrepreneur has already completed his job and left to create other firms.

Thus, neoclassical theory is justified in excluding the entrepreneur, because it deals with subjects for which the entrepreneur is irrelevant. That does not mean that no theory of entrepreneurship is needed, or that such a theory is lacking, but it means that a theory of entrepreneurship must be sought elsewhere, and that is what Schumpeter succeeded in doing.

Brief summary: Schumpeter's model – the supply and earnings of entrepreneurial activity

The basic Schumpeterian model (1911) notes that the successful innovative entrepreneur's reward is profit temporarily exceeding that of perfect competition. This attracts rivals who seek to share those profits by imitating the innovation, and thereby erode its super-competitive earnings. To prevent termination of these rewards, the entrepreneur can never desist from further innovation and cannot rest on his laurels.

Perhaps most important, the Schumpeterian analysis shows how the entrepreneur is driven to work without let-up for economic growth. Thus, it clearly reveals the tight association between innovative entrepreneurship and growth.

Allocation between productive and unproductive entrepreneurship

Some work of one of the present authors (Baumol, 2002, ch. 14) tells much of the rest of the story about the supply and allocation of productive entrepreneurship and the key role of evolving institutions. In the economic growth literature, it has often been asserted that an expanded supply of entrepreneurs effectively stimulates growth, while shrinkage in the supply undermines growth. But the standard explanation of the entrepreneurs' appearance and disappearance is shrouded in mystery, with hints about cultural developments and vague psychological and sociological changes. The historical evidence suggests a more mundane explanation: that entrepreneurs are always present but, as the structure of rewards in the economy changes, entrepreneurs switch their activities, moving to where payoffs become more attractive. In doing so, they move in and out of the activities usually recognized as entrepreneurial, exchanging them for other activities that also require enterprising talent but are often distant from production of goods and services. The generals of ancient Rome, the Mandarins of the Tang, Sung, and Ming Chinese empires, the captains of late medieval private and mercenary armies, the rent-seeking contemporary lawyers, and the Mafia Dons – all are clearly enterprising and often successful. And when institutions have changed so as to modify profoundly the relative payoffs offered by the different enterprising activities, the supply of entrepreneurs has shifted accordingly. Here, it is helpful to distinguish two categories of entrepreneurs, the productive and the unproductive entrepreneurs, with the latter, in turn, divided into subgroups such as rent-seeking entrepreneurs and destructive entrepreneurs, including the organizers of private armies or criminal groups. Once there is a pertinent change in the institutions that govern the relative rewards, the entrepreneurs will shift their activities between productive and unproductive occupations, so the set of productive entrepreneurs will appear to expand or contract autonomously. For example, when institutions change to prohibit private armies, entrepreneurs are led to look elsewhere to realize their financial ambitions. If, simultaneously, rules against confiscation of private property and for patent protection of inventions are adopted, entrepreneurial talent will shift into productive, innovative directions.

Recent studies: other disciplines and empirical approaches

Outside mainstream economic analysis, research on entrepreneurship has expanded rapidly since the 1980s, particularly that by specialists in management, psychology, and sociology. We focus here on three streams of work that have attracted the most scholarly attention: (*a*) how differences among individuals influence entry into (and success in) entrepreneurship, (*b*) how environment influences entrepreneurship, and (*c*) the strategies and forms of organization used by entrepreneurs.

Differences among individuals

There are numerous studies investigating how differences among individuals (in attributes such as education, age, experience, social position and psychology) are associated with a propensity to become self-employed and the likelihood of success at entrepreneurship. A wide variety of studies have indicated that individuals with higher education than the general population are more likely to become entrepreneurs (Shane, 2003). Robinson and Sexton (1994) and others have found that number of years of education is significantly related to likelihood of becoming self-employed, and Bates (1995) found that individuals with a graduate education were significantly more likely to become self-employed. Age appears to have an inverted U-shaped relationship with likelihood of forming a new venture. Entrepreneurship first increases with age because of experience, and then decreases with age because of opportunity costs and uncertainty premiums (Bates, 1995; Shane, 2003).

A number of studies that look at how experience influences likelihood of starting a business and the success of the new venture have found that general business experience (Evans and Leighton, 1989; Robinson and Sexton, 1994), experience specific to the industry in which the entrepreneur later founds a business (Aldrich, 1999), and prior self-employment (Carroll and Mosakowski, 1987) all increase the likelihood that an individual will found a new business. Furthermore, such experience tends to improve new venture performance and survival rates (Gimeno et al., 1997).

Studies have revealed that, in general, social status increases the likelihood of forming a new venture (for example, Stuart, Huang and Hybels, 1999). The number and diversity of an individual's social ties also increase the likelihood of founding a company (Aldrich, Rosen and Woodward, 1987), as well as the success of the venture (Hansen, 1995). Psychological factors also influence an individual's likelihood of becoming an entrepreneur (Shane, 2003). In particular, extraversion (Babb and Babb, 1992), need for achievement (Hornaday and Aboud, 1973), risk-taking propensity (Astebro, 2003), self-efficacy (Zietsma, 1999), overconfidence (Arabsheibani et al., 2000), and creativity (Ames and Runco, 2005) have all been shown to be significantly related to an individual's likelihood of becoming an entrepreneur.

Environmental factors

A number of industry characteristics influence new venture formation. Market size (Pennings, 1982) and growth (Dean and Meyer, 1992) increase the likelihood of new firm formation, while uncertainty from technological change decreases the rate of business start-ups (Audretsch and Acs, 1994). Capital intensity also reduces new firm formation by raising entry costs (Dean and Meyer, 1992). The density of firms has an inverted U-shaped relationship with new firm formation (Carroll and Wade, 1991). Too few firms in an industry may signal that there is no opportunity worth pursuing, or scarcity of market information. Thus, initial increases in the density of firms in the industry encourage business start-ups (Shane, 2003), although high density can increase competition for resources and create an entry barrier.

Not surprisingly, the institutional environment of an industry or region also affects new firm formation. Capital availability (for example, low-cost debt or venture capital) enhances firm formation (McMillan and Woodruff, 2002). Higher marginal federal income tax rates decrease self-employment (Gentry and Hubbard,

2000) and business tax concessions increase business start-ups (Dana, 1987). Stronger property rights encourage entrepreneurship, presumably because they assure entrepreneurs that they can appropriate the fruits of their efforts (McMillan and Woodruff, 2002). Researchers have also investigated the role of university technology-transfer offices on entrepreneurship, with most research indicating that such offices increase rates of new venture formation, particularly when technology-transfer offices are structured to profit from the transfers (Markman et al., 2005). Finally, socio-cultural norms about the desirability of self-employment or the risks of failure are significantly related to rates of business start-ups in a nation or ethnic group (Butler and Herring, 1991).

Strategy and organization

The area of entrepreneurial strategy that has received most research attention is method of financing. Consistent with Knight's (1921) argument that self-financing is needed to overcome moral hazard problems, most entrepreneurs finance their ventures primarily with their own capital (Aldrich, 1999; Shane, 2003). However, funds provided by 'angel' investors (wealthy individuals who invest in entrepreneurial companies, usually at an early stage) and venture capitalists are also important. The research on angel investment is sparse, but there is more research on venture capitalst investment. A number of researchers have investigated how venture capitalists choose their investments, mitigate risk, and influence new venture survival and growth (Bygrave and Timmons, 1992). Some studies have also examined how entrepreneurs identify opportunities (Shane, 2003), their degree of reliance on patent protection (Shane, 2001), the effect of entrepreneurs' new product development strategies (Zahra and Bogner, 2000), and their breadth of market focus (Bhide, 2000; Gimeno et al., 1997).

Finally, there also has been some research on the organization of new ventures – how they are formed as legal entities, the performance implication of this choice (Delmar and Shane, 2004), and the effect of venture team size and background (Eisenhardt and Schoonhoven, 1990). In general, formation as a legal entity and a large, diverse venture team appear to improve new venture performance.

On the state of the theory of entrepreneurship

Our discussion demonstrates that the beginnings of a significant theory of entrepreneurship already exist. The analysis uses little mathematics to derive any formal theorems, and its results are primarily qualitative. But this nascent theory of entrepreneurship does tell us about its supply and earnings, its role in the pricing of its products and the role of the price mechanism in its allocation among alternative activities. The Schumpeterian model tells us about the determination of entrepreneurs' profits and the prices of their products, as well as their influence on the supply of their activity. The model of productive and unproductive entrepreneurship tells us more about supply, as well as about the allocation of this resource. The empirical research adds further insight into the factors that increase the likelihood of individuals engaging in, and being successful at, entrepreneurship.

Beyond the stationary analysis of standard micro-economic theory, we see that the entrepreneurship models enable us to deal with such important questions as what features of the structure of the free market economy have caused it to outperform by an order of magnitude the innovation and growth of any alternative economic system. The institutional changes that reallocated much of entrepreneurship from redistributive to productive activities are, according to the model, the key to the answer. And this has profound policy implications both for developing countries seeking desperately to escape their poverty and for developed economies seeking to keep up the pace of their growth.

WILLIAM J. BAUMOL AND MELISSA A. SCHILLING

See also **Cantillon, Richard; growth and institutions; intellectual property; Knight, Frank Hyneman; Schumpeterian growth and growth policy design.**

Bibliography

Aldrich, H. 1999. *Organizations Evolving*. London: Sage.

Aldrich, H., Rosen, B. and Woodward, W. 1987. The impact of social networks on business foundings and profit: a longitudinal study. In *Frontiers of Entrepreneurship Research*, ed. N. Churchill et al. Babson Park, MA: Babson College.

Ames, M. and Runco, M. 2005. Predicting entrepreneurship from ideation and divergent thinking. *Creativity and Innovation Management* 14, 311–15.

Arabsheibani, G., De Meza, D., Maloney, J. and Pearson, B. 2000. And a vision appeared unto them of a great profit: evidence of self-deception among the self-employed. *Economics Letters* 67, 35–41.

Astebro, T. 2003. The return to independent invention: evidence of unrealistic optimism, risk seeking or skewness loving. *Economic Journal* 113, 226–38.

Audretsch, D. and Acs, Z. 1994. New firm startups, technology, and macroeconomic fluctuations. *Small Business Economics* 6, 439–49.

Babb, E. and Babb, S. 1992. Psychological traits of rural entrepreneurs. *Journal of Socio-Economics* 21, 353–62.

Bates, T. 1995. Self-employment entry across industry groups. *Journal of Business Venturing* 10, 143–56.

Baumol, W. 2002. *The Free-Market Innovation Machine: Analyzing the Growth Miracle of Capitalism*. Princeton, NJ: Princeton University Press.

Benz, M. and Frey, B. 2004. Being independent raises happiness at work. *Swedish Economic Policy Review* 11, 95–134.

Bhide, A. 2000. *The Origin and Evolution of New Businesses*. New York: Oxford University Press.

Butler, J. and Herring, C. 1991. Ethnicity and entrepreneurship in America: toward an explanation of racial and ethnic group variations in self-employment. *Sociological Perspectives* 34, 79–95.

Bygrave, W. and Timmons, J. 1992. *Venture Capital at the Crossroads*. Boston, MA: Harvard Business School Press.

Cantillon, R. 1730. *Essai Sur la Nature de Commerce en Général*, trans. H. Higgs. London: Macmillan, 1931.

Carroll, G. and Mosakowski, E. 1987. The career dynamics of self employment. *Administrative Science Quarterly* 32, 570–89.

Carroll, G. and Wade, J. 1991. Density dependence in organizational evolution of the American brewing industry across different levels of analysis. *Social Science Research* 20, 271–302.

Dana, L. 1987. Entrepreneurship and value creation – an international comparison of five commonwealth nations. In *Frontiers of Entrepreneurship Research*, ed. N. Churchill et al. Babson Park, MA: Babson College.

Dean, T. and Meyer, G. 1992. New venture formation in manufacturing industries: a conceptual and empirical analysis. In *Frontiers of Entrepreneurship Research*, ed. N. Churchill et al. Babson Park, MA: Babson College.

Delmar, F. and Shane, S. 2004. Legitimating first: organizing activities and the survival of new ventures. *Journal of Business Venturing* 19, 385–410.

Eisenhardt, K. and Schoonhoven, K. 1990. Organizational growth: linking founding team, strategy, environment, and growth among U.S. semiconductor ventures, 1978–1988. *Administrative Science Quarterly* 35, 504–29.

Evans, D. and Leighton, L. 1989. Some empirical aspects of entrepreneurship. *American Economic Review* 79, 519–35.

Freeman, R. 1978. Job satisfaction as an economic variable. *American Economic Review* 68, 135–41.

Gentry, W. and Hubbard, R. 2000. Tax policy and entrepreneurial entry. *American Economic Review Papers and Proceedings* 90, 283–92.

Gimeno, J., Folta, T., Cooper, A. and Woo, C. 1997. Survival of the fittest? Entrepreneurial human capital and the persistence of underperforming firms. *Administrative Science Quarterly* 42, 750–83.

Hansen, E. 1995. Entrepreneurial networks and new organization growth. *Entrepreneurship Theory and Practice* 19(4), 7–19.

Hornaday, J. and Aboud, J. 1973. Characteristics of successful entrepreneurs. *Personnel Psychology* 24, 141–53.

Kirzner, I. 1979. *Perception, Opportunity and Profit*. Chicago: University of Chicago Press.

Knight, F. 1921. *Risk, Uncertainty and Profit*. Boston and New York: Houghton Mifflin Company.

Markman, G., Phan, P., Balkin, D. and Gianiodis, P. 2005. Entrepreneurship and university-based technology transfer. *Journal of Business Venturing* 20, 241–63.

Marshall, A. 1923. *Industry and Trade*. London: Macmillan.

McMillan, J. and Woodruff, C. 2002. The central role of entrepreneurs in transition economies. *Journal of Economic Perspectives* 16(3), 153–70.

Pennings, J. 1982. Organizational birth frequencies: an empirical investigation. *Administrative Science Quarterly* 27, 120–44.

Robinson, T. and Sexton, E. 1994. The effect of education and experience on self-employment success. *Journal of Business Venturing* 9, 141–56.

Say, J.-B. 1819. *Traite d'économie politique.* 4th edn, trans. C. Prinsep. Boston: Wells and Lilly, 1821.

Schumpeter, J. 1911. *The Theory of Economic Development*, trans. R. Opie. Cambridge, MA: Harvard University Press, 1934.

Shane, S. 2001. Technology opportunities and new firm creation. *Management Science* 47, 205–20.

Shane, S. 2003. *A General Theory of Entrepreneurship: The Individual-Opportunity Nexus.* Northampton, MA: Edward Elgar.

Stuart, T., Huang, H. and Hybels, R. 1999. Interorganizational endorsements and the performance of entrepreneurial ventures. *Administrative Science Quarterly* 44, 315–49.

U.S. Bureau of the Census. 2004. *Statistical Abstract of the United States: 2004–2005.* Washington, DC: Bureau of the Census.

Zahra, S. and Bogner, W. 2000. Technology strategy and software new ventures' performance: exploring the moderating effect of the competitive environment. *Journal of Business Venturing* 15, 135–73.

Zietsma, C. 1999. Opportunity knocks – or does it hide? An examination of the role of opportunity recognition in entrepreneurship. In *Frontiers of Entrepreneurship Research*, ed. P. Reynolds et al. Babson Park, MA: Babson College.

envelope theorem

The origin of this famous theorem is the discussion between Jacob Viner (1931) and his draughtsman Y.K. Wong concerning the relationship between short- and long-run average cost curves. Viner had apparently reasoned that since in the long run average costs should be at a minimum, the long-run average cost (LRAC) curve should not only always be below the short-run average cost (SRAC) curves, but should also pass through the minimum points of each short-run curve. Wong pointed out the impossibility of this joint occurrence, and Viner opted to draw the long-run curve through the minimum points, thereby necessarily passing above sections of the short run curves. It was also puzzling (in the now corrected diagram) that at the point of tangency between the LRAC and a SRAC, the rate of change of average cost with respect to output was the same when capital was fixed as when it was allowed to vary. The puzzle was solved by Samuelson (1947), who showed in a general

way why the long-run curve would be the 'envelope' curve to the set of short-run curves. Perhaps the most surprising result of all was that this seeming mathematical curiosity turned out to be the fundamental basis for the development of refutable comparative statics implications in economics.

Unconstrained maximization models

The most general comparative statics model with explicit maximizing behaviour is *maximize* $y = f(x, \alpha)$ subject to $g(x, \alpha) = 0$, where $x = (x_1, \dots, x_n)$ is a vector of decision variables, $\alpha = (\alpha_1, \dots, \alpha_m)$ is a vector of parameters (though for simplicity, we treat α as a scalar in the discussion below), and $g(\cdot)$ represents one or more constraints. Models at this level of generality, however, imply no refutable implications and are hence largely uninteresting. In particular, there are never refutable implications for parameters that enter the constraint (see, for example, Silberberg and Suen, 2000). We therefore initially restrict the analysis to models of unconstrained maximization:

$$\text{maximize } y = f(x, \alpha) \qquad (1)$$

The necessary first-order conditions (NFOC) are

$$f_i(x, \alpha) = 0 \quad i = 1, \dots, n \qquad (2)$$

The sufficient second-order conditions (SSOC) are that the *Hessian* matrix $\mathbf{H} = (f_{ij})$ is negative definite. Alternatively, the principal minors of order (size) k of the Hessian determinant $H = |f_{ij}|$ have sign $(-1)^k$. Assuming the sufficient second-order conditions hold, we can in principle 'solve' for the n explicit choice functions $x = x^*(\alpha)$. Of course, since these choice functions are the result of solving the NFOC simultaneously, each individual x_i is a function of *all* the parameters, not just ones which might appears in some f_i.

Substituting the x_i^*'s into the objective function yields the *indirect objective function* $\varphi(\alpha) = f(x^*(\alpha), \alpha)$, the maximum value of f for given α. Since $\varphi(\alpha)$ is by definition a maximum value, $\varphi(\alpha) \geq f(x, \alpha)$, but $\varphi(\alpha) = f(x, \alpha)$ when $x = x^*$. In Figure 1, a typical $\varphi(\alpha)$ is plotted. For an arbitrary α^0, an $x^0 = x^*(\alpha^0)$ is implied. Consider the behaviour of $f(x, \alpha)$ when the x_i's are held fixed at x^0 as opposed to when they are variable. When $\alpha = \alpha^0$, the 'correct' x_i's are chosen, and therefore $\varphi(\alpha) = f(x^0, \alpha)$ at that one point. However, both to the left and to the right of α^0, the 'wrong' (that is non-maximizing) x_i's are chosen, and, since $\varphi(\alpha)$ is the *maximum* value of f for given α, $f(x^0, \alpha) \leq \varphi(\alpha)$ in any neighbourhood around α^0. This implies that φ and f must be tangent at α^0 (assuming differentiability), and, moreover, f must be either more concave or less convex than φ there. Since this must happen for arbitrary α, similar tangencies occur at other values of α. It is apparent from Figure 1 that $\varphi(\alpha)$ is the *envelope* of the $f(x_1, x_2, \alpha)$'s for each α. What surprised most researchers was the discovery that all comparative

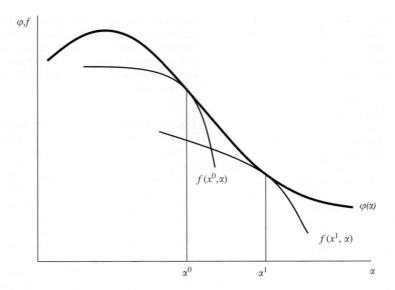

Figure 1

statics theorems in maximization models are in fact consequences of the relative curvatures of φ and f.

From the above discussion, the function $F(x, \alpha) = f(x, \alpha) - \varphi(\alpha)$ has a maximum of zero, with respect to both x and α. Thus we consider the *primal–dual* model

$$maximize\ F(x, \alpha) = f(x, \alpha) - \varphi(\alpha) \qquad (3)$$

where the maximization runs over x and also α. (In the latter instance, we ask, for given x_i's, what values of the parameters would make these x_i's the maximizing values?) The NFOC with respect to x are the same as in the original model. With respect to α, the NFOC yield the famous 'envelope theorem' which is the tangency of f and φ in Figure 1:

$$F_\alpha = f_\alpha - \varphi_\alpha = 0 \qquad (4)$$

In the α dimensions, the second-order conditions are simply

$$F_{\alpha\alpha} = f_{\alpha\alpha} - \varphi_{\alpha\alpha} \leq 0. \qquad (5)$$

This inequality says that in the α dimensions, f is relatively more concave than φ. (When α is a vector, this second-order condition is that the Hessian matrix $(F_{\alpha\alpha})$ is negative semi-definite.)

This is the fundamental geometrical property that underlies all comparative statics relationships. The NFOC (4) are identities when $x = x^*$. That is,

$$\varphi_\alpha(\alpha) \equiv f_\alpha(x^*(\alpha), \alpha) \qquad (6)$$

Differentiating with respect to α,

$$\varphi_{\alpha\alpha} \equiv \sum_1^n f_{\alpha i} \frac{\partial x_i^*}{\partial \alpha} + f_{\alpha\alpha} \qquad (7)$$

Rearranging terms, using (5) and invariance to the order of differentiation,

$$\varphi_{\alpha\alpha} - f_{\alpha\alpha} \equiv \sum_1^n f_{i\alpha} \frac{\partial x_i^*}{\partial \alpha} \geq 0 \qquad (8)$$

This is the fundamental relation of comparative statics. From it, we can derive Samuelson's famous 'conjugate pairs' theorem that refutable implications occur in maximization models when and only when a parameter enters one and only one first-order condition. For in that case, where say α enters only $f_i = 0$, $f_{j\alpha} \equiv 0$, $j \neq i$, and so (8) reduces to one term:

$$f_{i\alpha} \frac{\partial x_i^*}{\partial \alpha} \geq 0 \qquad (9)$$

In this case we can say that the response of x_i is in the same direction as the disturbance to the equilibrium (or, in the case of minimization models, in the opposite direction). For example, consider the profit-maximization model

$$maximize\ \pi = f(x, w, p) = p\,\theta(x_1 \ldots, x_n) - \sum w_i x_i$$

Each parameter w_i enters only the ith NFOC, and $f_{x_i w_i} = -1$, so that (9) yields the slope property $\partial x_i^* / \partial w_i \leq 0$; the factor demand functions are downward sloping in their own price.

The envelope theorem also yields the non-intuitive 'reciprocity' conditions. Suppose there are two parameters α and β. Then from invariance of second partial derivatives to the order of differentiation (Young's theorem), $\varphi_{\alpha\beta} = \varphi_{\beta\alpha}$. Using equation (6) above,

$$\sum f_{i\alpha} \frac{\partial x_i^*}{\partial \beta} = \sum f_{i\beta} \frac{\partial x_i^*}{\partial \alpha} \qquad (10)$$

When the objective function contains a linear expression such as in the profit maximization model, that is, $w_1 x_1 + \cdots + w_n x_n$, we have $f_{x_i w_i} = -1$ and $f_{x_i w_j} = 0$, $i \neq j$. In that case, (11) reduces to the simple expression $\frac{\partial x_i^*}{\partial w_j} = \frac{\partial x_j^*}{\partial w_i}$. This result also occurs in consumer theory for the Hicksian demands.

Constrained maximization models

Consider now the general comparative statics model with constraints, *maximize* $y = f(x, \alpha)$ subject to $g(x, \alpha) = 0$, where $g(\cdot)$ represents one or more constraints. Assuming just one constraint for the moment, the Lagrangian for this model is $L = f(x, \alpha) + \lambda g(x)$, producing the NFOC

$$L_i = f_i(x, \alpha) + \lambda g_i(x, \alpha) = 0 \quad i = 1, \ldots, n \qquad (11)$$

$$L_\lambda = g(x, \alpha) = 0 \qquad (12)$$

Assuming the SSOC, we can in principle 'solve' for the $n+1$ explicit choice functions $x = x^*(\alpha)$ and $\lambda = \lambda^*(\alpha)$. We derive the indirect objective function as before by substituting the x_i^*'s into the objective function producing $\varphi(\alpha) = f(x^*(\alpha), \alpha)$, the maximum value of f for given α, now also subject to the constraint. Proceeding as above, since $\varphi(\alpha)$ is by definition a maximum value, $\varphi(\alpha) \geq f(x, \alpha)$, but $\varphi(\alpha) = f(x, \alpha)$ when $x = x^*$. Thus the function $F(x, \alpha) = f(x, \alpha) - \varphi(\alpha)$ has a (constrained) maximum of zero, with respect to both x and α. Thus we consider the *primal–dual* model

$$\textit{maximize } F(x, \alpha) = f(x, \alpha) - \varphi(\alpha) \qquad (13)$$

$$\textit{subject to } g(x, \alpha) = 0 \qquad (14)$$

where the maximization runs over x and also α. The Lagrangian for this model is

$$L = f(x, \alpha) - \varphi(\alpha) + \lambda g(x, \alpha) \qquad (15)$$

The first-order conditions with respect to x are the same as in the original model. With respect to α, we get the envelope theorem in its most general form,

$$\varphi_\alpha = L_\alpha = f_\alpha + \lambda g_\alpha \qquad (16)$$

At this level of generality, it is not possible to generate any useful curvature properties of $\varphi(\alpha)$. However, consider the case where α does not enter any constraint. In that case, $g_\alpha \equiv 0$ and the NFOC reduce to (4) above, that is, $F_\alpha = f_\alpha - \varphi_\alpha = 0$. Moreover, when α does not enter the constraint, the primal–dual model is *an unconstrained maximization in* α. Hence in the α dimensions, the second-order conditions are as before:

$$F_{\alpha\alpha} = f_{\alpha\alpha} - \varphi_{\alpha\alpha} \leq 0. \qquad (17)$$

Thus in this important class of models, the comparative statics are identical to the models with no constraints. We obtain the inequalities (8) and (9) in the same manner as above.

Consider now an important class of models having the structure maximize $f(x)$ subject to $g(x) = k$, where we suppress all parameters except k, which is the focus of this analysis. The Lagrangian for this model is $L = f(x) + \lambda(k - g(x))$; assuming the NFOC and SSOC are valid, we solve for the explicit choice functions $x = x^*(k)$ and $\lambda^*(k)$. The indirect objective function is $\varphi(k) = f(x^*(k))$, the maximum value of f for given k. The envelope theorem (16) yields

$$\varphi_k = \lambda^*(k) \qquad (18)$$

Suppose the function f represents the value of output, and the constraint describes a limitation on that value due to the scarcity of some resource, measured by the value of k. Then the Lagrange multiplier imputes a 'shadow price', a marginal evaluation of that resource, since $\lambda^*(k)$ is the rate of change of the maximum value of output with respect to a change in the availability of that resource. This is a very widespread use of Lagrangian analysis in economics. For example, the fundamental model from which we derive the cost curves for a firm is, *minimize* $C = \sum w_i x_i$ subject to $f(x) = y$, where y is a parameter. Using (17), the Lagrange multiplier in this model is the marginal cost function $\partial C^* / \partial y = \lambda^*(w, y)$.

To further show the powerful nature of this analysis, consider the two-factor, two-goods model that plays an important part of international trade theory:

$$\textit{maximize } NNP = p_1 y_1 + p_2 y_2$$

subject to :

$$y_1 = f^1(L_1, K_1) \quad y_2 = f^2(L_2, K_2) \quad L_1 + L_2 = L$$

$$K_1 + K_2 = K$$

where f^1 and f^2 are production functions using labour (L) and capital (K) in each of two industries with outputs y_1 and y_2; output prices p_1 and p_2 and labour and capital endowments L and K are parametric. We can enumerate the salient properties of this model just by inspection, using the above results. The Lagrangian for this model is

$$L = p_1 y_1 + p_2 y_2 + \lambda_1 (f^1(L_1, K_1) - y_1)$$
$$+ \lambda_2 (f^2(L_2, K_2) - y_2)$$
$$+ \lambda_L (L - L_1 - L_2) + \lambda_K (K - K_1 - K_2).$$

Assuming the NFOC and SSOC hold, we solve the NFOC for the output supply functions $y_1^*(p_1, p_2, L, K)$ and $y_2^*(p_1, p_2, L, K)$, and the Lagrange multipliers, particularly $\lambda_L^*(p_1, p_2, L, K)$ and $\lambda_K^*(p_1, p_2, L, K)$. Substituting $y_1^*(\cdot)$ and $y_2^*(\cdot)$ into the objective function, we get the maximum value of NNP for given prices and resource constraints, $\varphi(p_1, p_2, L, K)$. Since prices enter the objective function only, and in the classic linear form, (9) immediately yields the envelope relations $\varphi_{p_i} = y_i^*(\cdot)$. We also note $\varphi_L = \lambda_L^*(\cdot)$ *and* $\varphi_K = \lambda_K^*(\cdot)$. The primal–dual model is, *maximize* $F = p_1 y_1 + p_2 y_2 - \varphi(p_1, p_2, L, K)$ subject to the same constraints above. Since p_1 and p_2 do not enter the constraints, F is concave in p_1 and p_2. Since the first two terms are linear and φ enters negatively, φ is convex in p_1 and p_2, and thus $\varphi_{p_i p_i} = \partial y_i^* / p_i > 0$; the supply curves are upward sloping. Furthermore, from (17), the Lagrange multipliers λ_L^* and λ_K^* are the imputed values of labour and capital. If an additional increment of labour, say, became available, λ_L^* would represent its marginal value product, and hence its implied wage in a competitive economy. Without further assumptions (for example, concavity of the production functions), we cannot determine a sign for how these imputed values change when the resource endowment changes: $\partial \lambda_L^* / \partial L / 0$. The reciprocity relationships are straightforward: $\varphi_{p_1 p_2} = \partial y_1^* / \partial p_2 = \partial y_2^* / \partial p_1 = \varphi_{p_2 p_1}$, and similarly, $\varphi_{LK} = \partial \lambda_L^* / \partial K = \partial \lambda_K^* / \partial L = \varphi_{KL}$. We also find $\varphi_{p_1 L} = \partial y_1^* / \partial L = \partial \lambda_L / \partial p_1 = \varphi_{Lp_1}$, and so on. It seems unlikely that Jacob Viner could have imagined what the corrected version of his diagram would eventually lead to!

EUGENE SILBERBERG

See also **cost functions; duality; Hicksian and Marshallian demands; Le Chatelier principle.**

Bibliography

Samuelson, P.A. 1947. *Foundations of Economic Analysis*. Cambridge, MA: Harvard University Press.

Silberberg, E. and Suen, W. 2000. *The Structure of Economics*, 3rd edn. New York: McGraw-Hill.

Viner, J. 1931. Cost curves and supply curves. *Zeitschrift für Nationalökonomie* 3, 23–46. Repr. in American Economic Association, *Readings in Price Theory*, Homewood, IL: Irwin, 1952.

environmental economics

The fundamental theoretical argument for government activity in the environmental realm is that pollution is an externality – an unintended consequence of market decisions which affect individuals other than the decision maker. Providing incentives for private actors to internalize the full costs of their actions was long thought to be the theoretical solution to the externality problem. The primary advocate of this view was Arthur Pigou, who in *The Economics of Welfare* (1920) proposed that the government should impose a tax on emissions equal to the cost of the related damages at the efficient level of control.

A response to the Pigouvian perspective was provided by Ronald Coase in 'The problem of social cost' (1960). Coase demonstrated that, in a bilateral bargaining environment with no transaction costs, wealth or income effects, or third-party impacts, two negotiating parties will reach socially desirable agreements, and the overall amount of pollution will be independent of the assignment of property rights. At least some of the specified conditions are unlikely to hold for most environmental problems. Hence, private negotiation will not – in general – fully internalize environmental externalities.

Criteria for environmental policy evaluation

More than 100 years ago Vilfredo Pareto (1896) enunciated the well-known normative criterion for judging whether a social change makes the world better off: a change is *Pareto efficient* if at least one person is made better off and no one is made worse off. This criterion has considerable normative appeal, but virtually no public policies meet the test. Nearly 50 years later Nicholas Kaldor (1939) and John Hicks (1939) postulated a more pragmatic criterion that seeks to identify 'potential Pareto improvements': a change is welfare-improving if those who gain from the change could – in principle – fully compensate the losers, with (at least) one gainer still being better off.

The Kaldor–Hicks criterion – a test of whether total social benefits exceed total social costs – is the theoretical foundation for the use of the analytical device known as benefit–cost (or net present value) analysis. If the objective is to maximize the difference between benefits and costs (net benefits), then the related level of environmental protection (pollution abatement) is defined as the efficient level of protection:

$$\max_{\{q_i\}} \sum_{i=1}^{N} [B_i(q_i) - C_i(q_i)] \rightarrow q_i^* \qquad (1)$$

where q_i is abatement by source i ($i = 1$ to N), $B_i(\cdot)$ is the benefit function for source i, $C_i(\cdot)$ is the cost function for the source, and q_i^* is the efficient level of protection (pollution abatement). The key necessary condition that emerges from the maximization problem of equation (1) is that marginal benefits be equated with marginal costs (on the assumption of convexity of the respective functions).

The Kaldor–Hicks criterion is clearly more practical than the strict Pareto criterion, but its normative standing is less solid. Some have argued that other factors should be considered in a measure of social well-being, and that criteria such as distributional equity should trump efficiency considerations in some collective decisions (Sagoff, 1993). Many economists would agree with this assertion, and some have noted that the Kaldor–Hicks criterion should be considered neither a necessary nor a sufficient condition for public policy (Arrow et al., 1996).

Benefit–cost analysis of environmental regulations

The soundness of empirical benefit–cost analysis rests upon the availability of reliable estimates of social benefits and costs, including estimates of the social discount rate. The present value of net benefits (PVNB) is defined as:

$$PVNB = \sum_{t=0}^{T}\{(B_t - C_t) \cdot (1 + r)^{-t}\} \qquad (2)$$

where B_t are benefits at time t, C_t are costs at time t, r is the discount rate, and T is the terminal year of the analysis. A positive PVNB means that the policy or project has the potential to yield a Pareto improvement (meets the Kaldor–Hicks criterion). Thus, carrying out benefit–cost or 'net present value' (NPV) analysis requires discounting to translate future impacts into equivalent values that can be compared. In essence, the Kaldor–Hicks criterion provides the rationale both for benefit–cost analysis and for discounting (Goulder and Stavins, 2002).

Choosing the discount rate to be employed in an analysis can be difficult, particularly where impacts are spread across a large number of years involving more than a single generation. In theory, the social discount rate could be derived by aggregating the individual time preference rates of all parties affected by a policy. Evidence from market behaviour and from experimental economics indicates that individuals may employ lower discount rates for impacts of larger magnitude, higher discount rates for gains than for losses, and rates that decline with the time span being considered (Cropper, Aydede and Portney, 1994; Cropper and Laibson, 1999). In particular, there has been support for the use of hyperbolic discounting and similar approaches with declining discount rates over time (Ainslie, 1991; Weitzman, 1994; 1998), but most of these approaches are subject to time inconsistency.

The costs of environmental regulations

In the environment context, the economist's notion of cost (or, more precisely, opportunity cost) is a measure of

the value of whatever must be sacrificed to prevent or reduce the risk of an environmental impact. A full taxonomy of environmental costs ranges from the most obvious to the least direct (Jaffe et al., 1995).

Methods of direct compliance cost estimation, which measure the costs to firms of purchasing and maintaining pollution-abatement equipment plus costs to government of administering a policy, are acceptable when behavioural responses, transitional costs, and indirect costs are small. Partial and general equilibrium analysis allows for the incorporation of behavioural responses to changes in public policy. Partial equilibrium analysis of compliance costs incorporates behavioural responses by modelling supply and/or demand in major affected markets, but assumes that the effects of a regulation are confined to one or a few markets. This may be satisfactory if the markets affected by the policy are small in relation to the overall economy; but, if an environmental policy is expected to have large consequences for the economy, general equilibrium analysis is required, such as through the use of computable general equilibrium models (Hazilla and Kopp, 1990; Conrad, 2002). The potential interaction of abatement costs with pre-existing taxes indicates the importance of employing general equilibrium models for comprehensive cost analysis. Revenue recycling (using emission tax or auctioned permit revenues to reduce distortionary taxes) can make the costs of pollution control significantly less than they would otherwise be (Goulder, 1995).

In a retrospective examination of 28 environmental and occupational safety regulations, Harrington, Morgenstern, and Nelson (2000) found that 14 cost estimation analyses had produced *ex ante* cost estimates that exceeded actual *ex post* costs, apparently due to technological innovation stimulated by market-based instruments (see below).

The benefits of environmental regulations

Protecting the environment usually involves active employment of capital, labour, and other scarce resources. The benefits of an environmental policy are defined as the sum of individuals' aggregate willingness to pay (WTP) for the reduction or prevention of environmental damages or individuals' willingness to accept (WTA) compensation to tolerate such environmental damages. In theory, which measure of value is appropriate for assessing a particular policy depends upon the related assignment of property rights, the nature of the status quo, and whether the change being measured is a gain or a loss; but under a variety of conditions the difference between the two measures may be expected to be relatively small (Willig, 1976). Empirical evidence suggests larger than expected differences between willingness to pay and willingness to accept (Fisher, McClelland and Schulze, 1988). Theoretical explanations include psychological aversion to loss and poor substitutes for environmental amenities (Hanemann, 1991).

The benefits people derive from environmental protection can be categorized as (a) related to human health (mortality and morbidity), (b) ecological (both market and non-market), or (c) materials damage. The distinction between use value and non-use value is critical. In addition to the direct benefits (use value) people receive through protection of their health or through use of a natural resource, they derive passive or non-use value from environmental quality, particularly in the ecological domain. For example, an individual may value a change in an environmental good because she wants to preserve the good for her heirs (bequest value). Still other people may envision no current or future use by themselves or their heirs, but still wish to protect the good because they believe it should be protected or because they derive satisfaction from simply knowing it exists (existence value).

How much would individuals sacrifice to achieve a small reduction in the probability of death during a given period of time? How much compensation would individuals require to accept a small increase in that probability? These are reasonable economic questions because most environmental regulations result in very small changes in individuals' mortality risks. Hedonic wage studies, averted behaviour, and contingent valuation (all discussed below) can provide estimates of marginal willingness to pay or willingness to accept related to small changes in mortality risk, and such estimates can be normalized as the 'value of a statistical life' (VSL).

The VSL is *not* the value of an individual life, whether in ethical or technical, economic terms. Rather it is simply a convention:

$$VSL = \frac{MWTP \text{ or } MWTA \text{ (from hedonic wage or CV)}}{\text{Small risk change}}$$

(3)

where $MWTP$ and $MWTA$, respectively, refer to marginal willingness to pay and marginal willingness to accept. For example, if people are willing, on average, to pay $12 for a risk reduction from 5 in 500,000 to 4 in 500,000, equation (3) would yield:

$$VSL = \frac{\$12}{0.000002} = \$6,000,000$$

(4)

Thus, VSL quantifies the aggregate amount that a group of individuals are willing to pay for small reductions in risk, standardized (extrapolated) for a risk change of 1.0. It is not the economic value of an individual life because the VSL calculation does not signify that an individual would pay $6 million to avoid (certain) death this year, or accept (certain) death this year in exchange for $6 million.

Revealed preference methods of environmental benefit estimation

The *averting behaviour method*, in which values of willingness to pay are inferred from observations of people's behavioural responses to changes in environmental quality, is grounded in the household production function framework (Bockstael and McConnell, 1983). People sometimes take actions to reduce the risk (averting behaviour) or lessen the impacts (mitigating behaviour) of environmental damages, for example by purchasing water filters or bottled water. In theory, people's perceptions of the cost of averting behaviour and its effectiveness should be measured (Cropper and Freeman, 1991), but in practice actual expenditures on averting and mitigating behaviours are typically employed. An additional challenge is posed by the necessity of disentangling attributes of the market good or service.

Recreational activities represent a potentially large class of benefits that are important in assessing policies affecting the use of public lands. The models used to estimate recreation demand fall within the class of household production models. *Travel cost models* (or Hotelling–Clawson–Knetsch models) use information about time and money spent visiting a site to infer the value of that recreational resource (Bockstael, 1996). The simplest version of the method involves one site and uses data from surveys of users from various geographic origins, together with estimates of the cost of travel and opportunity cost of time, to infer a demand function relating the number of trips to the site to a function of people's willingness to pay for the experience. *Random utility models* explicitly model the consumer's decision to choose a particular site from among recreation locations, assessing the probability of visiting each location. Such models can be used to value changes in environmental quality by comparing decisions to visit alternative sites (Phaneuf and Smith, 2004).

All recreation demand models share limitations. First, the valuation of costs depends on estimates of the opportunity cost of (leisure) time, which is notoriously difficult to estimate. Also, most trips to a recreation site are part of a multi-purpose experience. In addition, random utility models rely on people's perceptions of environmental quality changes. Finally, like all revealed-preference approaches, recreation demand models can be used to estimate use value only; non-use value cannot be examined.

An alternative approach to assessing people's willingness to pay for recreational experiences is to draw on evidence from *private options to use public goods*. This approach also fits within the household production framework, and is based upon the notion of estimating the derived demand for a privately traded option to utilize a freely available public good. In particular, the demand for state fishing licences has been used to infer the benefits of recreational fishing. Using panel data on fishing license sales and prices, combined with data on substitute prices and demographic variables, Bennear, Stavins and Wagner (2005) estimated a licence demand function from which the expected benefits of a recreational fishing day were derived.

Hedonic pricing methods are founded on the proposition that people value goods in terms of the bundles of attributes that constitute those goods. *Hedonic property value methods* employ data on residential property values and home characteristics, including structural, neighbourhood, and environmental quality attributes (Palmquist, 2003). By regressing the property value on key attributes, the hedonic price function is estimated:

$$P = f(\underset{\sim}{x}, \underset{\sim}{z}, e) \qquad (5)$$

where P = housing price (includes land); $\underset{\sim}{x}$ = vector of structural attributes; $\underset{\sim}{z}$ = vector of neighbourhood attributes; and e = environmental attribute of concern.

From the estimated hedonic price function of equation (5), the marginal implicit price of any attribute, including environmental quality, can be calculated as the partial derivative of the housing price with respect to the given attribute:

$$\frac{\partial P}{\partial e} = \frac{\partial f(\cdot)}{\partial e} = P_e \qquad (6)$$

This marginal implicit price, P_e, measures the aggregate marginal willingness to pay for the attribute in question. For purposes of benefit estimation, the demand function for the attribute is required, and so it is necessary to examine how the marginal implicit price of the environmental attribute varies with changes in the quantity of the attribute and other relevant variables. If the hedonic price equation (5) is nonlinear, then fitted values of P_e can be calculated as e is varied, and a second-stage equation can be estimated:

$$\hat{P}_e = g(e, \underset{\sim}{y}) \qquad (7)$$

where $\hat{P}_e$ = the fitted value of the marginal implicit price of e from the first-stage equation; and y = a vector of factors that affect marginal willingness tõ pay for e, including buyer characteristics.

Equation (7), above, has been interpreted as the demand function for the environmental attribute, from which benefits (consumers surplus) can be estimated in the usual way; but there are problems. Most important among these is the question of whether a demand function has actually been estimated, since environmental quality may affect both the demand for housing and its supply, raising the classic identification problem. In addition, informational asymmetries may distort the analysis. Also, because the hedonic property method is based on analysis of marginal changes, it should not be applied to analysis of policies with large anticipated effects.

A related benefit-estimation technique is the *hedonic wage method*, based on the reality that individuals in well-functioning labour markets make trade-offs between wages and risk of on-the-job injuries (or death). A job is a bundle of characteristics, including its wage, responsibilities and risk, among others factors. Two jobs that require the same skill level but have different risks of on-the-job mortality will pay different wages. On the labour supply side, employees tend to require extra compensation to accept jobs with greater risks; and on the labour demand side, employers are willing to offer higher wages to attract workers to riskier jobs. Hence, labour market data on wages and job characteristics can be used to estimate people's marginal implicit price of risk, that is, their valuation of risk. By regressing the wage on key attributes, the hedonic price function is estimated:

$$W = h(\underset{\sim}{x}, r) \qquad (8)$$

where W = wage (in annual terms); $\underset{\sim}{x}$ = vector of worker and job characteristics; and r = mortality risk of job.

The marginal implicit price of risk is calculated as the partial derivative of the annual wage with respect to the measured mortality risk:

$$\frac{\partial W}{\partial r} = \frac{\partial h(\cdot)}{\partial r} = W_r \qquad (9)$$

This marginal implicit price of risk is the average annual income necessary to compensate a worker for a marginal change in risk throughout the year, and it varies with the level of risk.

Many of the issues that arise with the hedonic property value method have parallels here. First, there is the possibility of simultaneity: causality between risk and wages can run in both directions. Also, if individuals' perceptions of risk do not correspond with actual risks, then the marginal implicit price of risk calculated from a hedonic wage study will be biased, and imperfections in labour markets (less than perfect mobility) can cause further problems.

Direct application of the method in the environmental realm is limited to occupational (as opposed to environmental) exposures and risks. Yet hedonic wage methods are of considerable importance in the environmental policy realm, because the results from hedonic wage studies have frequently been used through 'benefit transfer' to infer the VSL. In such applications, the hedonic wage method brings with it possible bias, because studies typically focus on risky occupations, which may attract workers who are systematically less risk-averse.

Standard economic theory would suggest that younger people would have higher values for risk reduction because they have a longer expected life remaining before them and thus a higher expected lifetime utility (Moore and Viscusi, 1988; Cropper and Sussman, 1990). In contrast, some models and empirical evidence suggest that older people may in fact have a higher demand for reducing mortality risks than younger people, and that the value of a life may follow an 'inverted-U' shape over the life cycle, with its peak during mid-life (Shepard and Zeckhauser, 1982; Mrozek and Taylor, 2002; Viscusi and Aldy, 2003; Alberini et al., 2004).

Stated preference methods of environmental benefit estimation

In the best known stated preference method, *contingent valuation* (CV), survey respondents are presented with scenarios that require them to trade off, hypothetically, something for a change in an environmental good or service (Mitchell and Carson, 1989; Boyle, 2003). The simplest approach is to ask people for their maximum willingness to pay, but as there are few real markets in which individuals are actually asked to generate their reservation prices, this method is considered unreliable. In a bidding game, the researcher begins by stating a willingness-to-pay number, asks for a yes–no response, and then increases or decreases the amount until indifference is achieved. The problem with this approach is starting-point bias. A related approach is the use of a payment card shown to the respondent, but the range of WTP on the card may introduce bias, and the approach cannot be used with telephone surveys. Finally, the referendum (discrete choice) approach is favoured by researchers. Each respondent is offered a different WTP number, to which a simple yes–no response is solicited.

The primary advantage of contingent valuation is that it can be applied to a wide range of situations, including use as well as non-use value; but potential problems remain. Respondents may not understand what they are being asked to value. This may introduce greater variance, if not bias, in responses. Likewise, respondents may not take the hypothetical market seriously because no budget constraint is imposed. This can increase variance and bias. Yet if the scenario is 'too realistic,' strategic bias may be expected to show up in responses. Finally, the 'warm glow effect' may plague some stated preference surveys: people may purchase moral satisfaction with large but unreal statements of their willingness-to-pay (Andreoni, 1995).

The 1989 Exxon Valdez oil spill off the coast of Alaska led to massive litigation, and resulted in the most prominent use ever of the concept of non-use value and the method of contingent valuation for its estimation. The result was a symposium sponsored by the Exxon Corporation attacking the CV method (Hausman, 1993), and the subsequent creation of a government panel – established by the National Oceanic and Atmospheric Administration (NOAA) and chaired by two Nobel laureates in economics – to assess the scientific validity of the CV method. The NOAA panel concluded that 'CV studies can produce estimates reliable enough to be the starting point of a judicial process of damage assessment, including lost passive (non-use) values' (Arrow et al., 1993, p. 4610). The panel offered its approval of CV methods subject to a set of best-practice guidelines.

It is important to distinguish between legitimate methods of benefit estimation and approaches sometimes encountered in the policy process that do not measure willingness-to-pay or willingness-to-accept. Frequently misused techniques include: (*a*) employing, as proxies for the benefits of a policy, estimates of the 'cost avoided' by not using the next most costly means of achieving the policy's goals; (*b*) 'societal revealed preference' models, which seek to infer the benefits of a proposed policy from the costs of previous regulatory actions; and (*c*) cost-of-illness or human-capital measures which estimate explicit market costs resulting from changes in morbidity or mortality. Because none of these approaches provides estimates of WTP or WTA, these techniques do not provide valid measures of economic benefits.

Choosing instruments: the means of environmental policy

Even if the goals of environmental policies are given, economic analysis can bring insights to the assessment and design of environmental policies. One important criterion is *cost-effectiveness*, defined as the allocation of control among sources that results in the aggregate target being achieved at the lowest possible cost, that is, the allocation which satisfies the following cost-minimization problem:

$$\min_{\{r_i\}} \; C = \sum_{i=1}^{N} c_i(r_i) \qquad (10)$$

$$s.t. \quad \sum_{i=1}^{N} [u_i - r_i] \leq \bar{E} \qquad (11)$$

$$and \quad 0 \leq r_i \leq u_i \qquad (12)$$

where r_i = reductions in emissions (abatement or control) by source i ($i = 1$ to N); $c_i(r_i)$ = cost function for source i; C = aggregate cost of control; u_i = uncontrolled emissions by source i; and $\bar{E}$ = the aggregate emissions target imposed by the regulatory authority.

If the cost functions are convex, then necessary and sufficient conditions for satisfaction of the constrained optimization problem posed by equations (10) to (12) are the following (among others) (Kuhn and Tucker, 1951):

$$\frac{\partial c_i(r_i)}{\partial r_i} - \lambda \geq 0 \qquad (13)$$

$$r_i \cdot \left[\frac{\partial c_i(r_i)}{\partial r_i} - \lambda \right] = 0 \qquad (14)$$

Equations (13) and (14) together imply the crucial condition for cost-effectiveness that all sources (that exercise some degree of control) experience the same marginal abatement costs (Baumol and Oates, 1988). Thus, when one examines environmental policy instruments, a key question is whether marginal abatement costs are likely to be being equated across sources.

Command-and-control versus market-based instruments

Conventional approaches to regulating the environment – frequently characterized as command-and-control – allow relatively little flexibility in the means of achieving goals. Such policy instruments tend to force firms to take on equal shares of the pollution-control burden, regardless of the cost. The most prevalent form of uniform command-and-control standards is technology standards that specify the adoption of specific pollution-control technologies, and performance standards that specify uniform limits on the amount of pollution a facility can generate. In theory, non-uniform performance standards could be made to be cost-effective, but the government typically lacks the requisite information (on marginal costs of individual sources).

Market-based instruments encourage behaviour through market signals rather than through explicit directives regarding pollution-control levels or methods. Market-based instruments fall within four categories: pollution charges, tradable permits, market-friction reductions, and government subsidy reductions. Liability rules may also be thought of as a market-based instrument, because they provide incentives for firms to take into account the potential environmental damages of their decisions.

Where there is significant heterogeneity of abatement costs, command-and-control methods will not be cost-effective. In reality, costs can vary enormously due to production design, physical configuration, age of assets, and other factors. For example, the marginal costs of controlling lead emissions have been estimated to range from \$13 to \$56,000 per ton (Hartman, Wheeler and Singh, 1994; Morgenstern, 2000). But where costs are similar among sources, command-and-control instruments may perform as well as (or better than) market-based instruments, depending on transactions costs, administrative costs, possibilities for strategic behaviour, political costs, and the nature of the pollutants (Newell and Stavins, 2003).

In theory, market-based instruments allow any desired level of pollution clean-up to be realized at the lowest overall cost by providing incentives for the greatest reductions in pollution by those firms that can achieve the reductions most cheaply. Rather than equalizing pollution levels among firms, market-based instruments equalize their marginal abatement costs (Montgomery, 1972). In addition, market-based instruments have the potential to bring down abatement costs over time by providing incentives for companies to adopt cheaper and better pollution-control technologies. This is because, with market-based instruments, most clearly with emission taxes, it pays firms to clean up a bit more if a sufficiently low-cost method (technology or process) of doing so can be identified and adopted (Downing and White, 1986; Maleug, 1989; Milliman and Prince, 1989; Jaffe and Stavins, 1995). However, the ranking among policy instruments in terms of their respective impacts on technology innovation and diffusion is ambiguous (Jaffe, Newell and Stavins, 2003).

Closely related to the effects of instrument choice on technological change are the effects of vintage-differentiated regulation on the rate of capital turnover, and thereby on pollution abatement costs and environmental performance. Vintage-differentiated regulation is a common feature of many environmental policies, whereby the standard for regulated units is fixed in terms of their date of entry, with later vintages facing more stringent regulation. Such vintage-differentiated regulations can be expected to retard turnover in the capital stock, and thereby to reduce the cost-effectiveness of regulation. Under some conditions the result can be higher levels of pollutant emissions than would occur in the absence of regulation. Such economic and environmental consequences are not only predictions from theory (Maloney and Brady, 1988); both types of consequences have been validated empirically (Gruenspecht, 1982; Nelson, Tietenberg and Donihue, 1993).

Pollution charges

Pollution charge systems assess a fee or tax on the amount of pollution that firms or sources generate (Pigou, 1920). By definition, actual emissions are equal to unconstrained emissions minus emissions reductions, that is, $e_i = u_i - r_i$. A source's cost minimization problem in the presence of an emissions tax, t, is given by:

$$\min_{\{r_i\}} \; [c_i(r_i) + t \cdot (u_i - r_i)] \tag{15}$$

$$s.t. \quad r_i \geq 0 \tag{16}$$

The result for each source is:

$$\frac{\partial c_i(r_i)}{\partial r_i} - t \geq 0 \tag{17}$$

$$r_i \cdot \left[\frac{\partial c_i(r_i)}{\partial r_i} - t \right] = 0 \tag{18}$$

Equations (17) and (18) imply that each source (that exercises a positive level of control) will carry out abatement up to the point where its marginal control costs are equal to the tax rate. Hence, marginal abatement costs are equated across sources, satisfying the condition for cost-effectiveness specified by equations (13) and (14), at least in the simplest case of a uniformly mixed pollutant. In the non-uniformly mixed pollutant case, where 'hot spots' can be an issue, the respective cost-effective instrument is an 'ambient charge'.

A challenge with charge systems is identifying the appropriate tax rate. For social efficiency, it should be set equal to the marginal benefits of clean-up at the efficient level of clean-up (Pigou, 1920); but policymakers are more likely to think in terms of a desired level of clean-up, and they do not know beforehand how firms will respond to a given level of taxation. An additional problem is that, although such systems minimize aggregate social costs, these systems may be *more* costly than comparable command-and-control instruments *for regulated firms*, because firms pay both their abatement costs *and* taxes on their residual emissions.

If charges are broadly defined, many applications can be identified (Stavins, 2003). Coming closest to true Pigouvian taxes are the increasingly common *unit-charge* systems for financing municipal solid waste collection, where households and businesses are charged the incremental costs of collection and disposal. Another important set of charge systems has been *deposit refund systems*, whereby consumers pay a surcharge when purchasing potentially polluting products, and receive a refund when returning the product to an approved centre for recycling or disposal. A number of countries and states have implemented this approach to control litter from beverage containers and to reduce the flow of solid waste to landfills (Bohm, 1981; Menell, 1990), and the concept has also been applied to lead-acid batteries. There has also been considerable use of *environmental user charges*, through which specific environmentally related services are funded. Examples include *insurance premium taxes* (Barthold, 1994). Another set of environmental charges are *sales taxes* on motor fuels, ozone-depleting chemicals, agricultural inputs, and low-mileage motor vehicles. Finally, *tax differentiation* has been used to encourage the use of renewable energy sources.

Tradable permit systems
Tradable permits can achieve the same cost-minimizing allocation as a charge system, while avoiding the problems of uncertain firm responses and the distributional consequences of taxes. Under a tradable permit system, an allowed overall level of pollution, $\bar{E}$, is established and allocated among sources in the form of permits. Firms that keep emission levels below allotted levels may sell surplus permits to other firms or use them to offset excess emissions in other parts of their operations. Let q_{0i} be the initial allocation of emission permits to source i, such that:

$$\sum_{i=1}^{N} q_{0i} = \bar{E} \qquad (19)$$

Then, if p is the market-determined price of tradable permits, a single firm's cost minimization problem is given by:

$$\min_{\{r_i\}} \; [c_i(r_i) + p \cdot (u_i - r_i - q_{0i})] \qquad (20)$$

$$s.t. \quad r_i \geq 0 \qquad (21)$$

The result for each source is:

$$\frac{\partial c_i(r_i)}{\partial r_i} - p \geq 0 \qquad (22)$$

$$r_i \cdot \left[\frac{\partial c_i(r_i)}{\partial r_i} - p \right] = 0 \qquad (23)$$

Equations (22) and (23) together imply that each source (that exercises a positive level of control) will carry out abatement up to the point where its marginal control costs are equal to the market-determined permit price. Hence, the environmental constraint, $\bar{E}$, is satisfied, and marginal abatement costs are equated across sources, satisfying the condition of cost-effectiveness. The unique cost-effective equilibrium is achieved independently of the initial allocation of permits (Montgomery, 1972), which is of great political significance.

The performance of a tradable permit system can be adversely affected by: concentration in the permit market (Hahn, 1984; Misolek and Elder, 1989); concentration in the product market (Maleug, 1990); transaction costs (Stavins, 1995); non-profit maximizing behaviour, such as sales or staff maximization (Tschirhart, 1984); the pre-existing regulatory environment (Bohi and Burtraw, 1992); and the degree of monitoring and enforcement (Montero, 2003).

Tradable permits have been the most frequently used market-based system (US Environmental Protection Agency, 2000). Significant applications include: the emissions trading programme (Tietenberg, 1985; Hahn, 1989); the leaded gasoline phase-down; water quality permit trading (Hahn, 1989; Stephenson, Norris and Shabman, 1998); CFC trading (Hahn and McGartland, 1989); the sulphur dioxide (SO_2) allowance trading system for acid rain control (Schmalensee et al., 1998; Stavins, 1998; Carlson et al., 2000; Ellerman et al., 2000); the RECLAIM programme in the Los Angeles metropolitan region (Harrison, 1999); tradable development rights for land use; and the European Union's greenhouse gas emission trading scheme.

Market friction reduction
Market friction reduction can serve as a policy instrument for environmental protection. *Market creation* establishes markets for inputs or outputs associated with environmental quality. Examples of market creation include

measures that facilitate the voluntary exchange of water rights and thus promote more efficient allocation and use of scarce water supplies (Howe, 1997), and policies that facilitate the restructuring of electricity generation and transmission. Since well-functioning markets depend, in part, on the existence of well-informed producers and consumers, *information programmes* can help foster market-oriented solutions to environmental problems. These programmes have been of two types. *Product labelling requirements* have been implemented to improve information sets available to consumers, while other programmes have involved *reporting requirements* (Hamilton, 1995; Konar and Cohen, 1997; Khanna, Quimio and Bojilova, 1998).

Government subsidy reduction

Government subsidy reduction constitutes another category of market-based instruments. Subsidies are the mirror image of taxes and, in theory, can provide incentives to address environmental problems. Although subsidies can advance environmental quality (see, for example, Jaffe and Stavins, 1995), it is also true that subsidies, in general, have important disadvantages relatives to taxes (Dewees and Sims, 1976; Baumol and Oates, 1988). Because subsidies increase profits in an industry, they encourage entry, and can thereby increase industry size and pollution output (Mestelman, 1982; Kohn, 1985). In practice, rather than internalizing externalities, many subsidies promote economically inefficient and environmentally unsound practices. In such cases, reducing subsidies can increase efficiency and improve environmental quality. For example, because of concerns about global climate change, increased attention has been given to cutting inefficient subsidies that promote the use of fossil fuels.

Implications of uncertainty for instrument choice

The dual task facing policymakers of choosing environmental goals and selecting policy instruments to achieve those goals must be carried out in the presence of the significant uncertainty that affects the benefits and the costs of environmental protection. Since Weitzman's (1974) classic paper on 'Prices vs. quantities', it has been widely acknowledged that benefit uncertainty on its own has no effect on the identity of the efficient control instrument, but that cost uncertainty can have significant effects, depending upon the relative slopes of the marginal benefit (damage) and marginal cost functions. In particular, if uncertainty about marginal abatement costs is significant, and if marginal abatement costs are flat relative to marginal benefits, then a quantity instrument is more efficient than a price instrument.

In the environmental realm, benefit uncertainty and cost uncertainty are usually both present, with benefit uncertainty of greater magnitude. When marginal benefits are positively correlated with marginal costs (which, it turns out, is not uncommon), then there is an additional argument in favour of the relative efficiency of quantity instruments (Stavins, 1996). Nevertheless, the regulation of stock pollutants will often favour price instruments, because the marginal benefit function – linked with the stock of pollution – will tend to be flatter than the marginal cost function – linked with the flow of pollution (Newell and Pizer, 2003). In theory, there would be considerable efficiency advantages in the presence of uncertainty of hybrid systems – for example, quotas combined with taxes – or nonlinear taxes (Roberts and Spence, 1976; Weitzman, 1978; Kaplow and Shavell, 2002; Pizer, 2002), but such systems have not been adopted.

Conclusion

The growing use of economic analysis to inform environmental decision-making marks greater acceptance of the usefulness of these tools in improving regulation. But debates about the normative standing of the Kaldor–Hicks criterion and the challenges inherent in making benefit–cost analysis operational will continue. Nevertheless, economic analysis has assumed a significant position in the regulatory state. At the same time, despite the arguments made for decades by economists, there is only limited political support for broader use of benefit–cost analysis to assess proposed or existing environmental regulations. These analytical methods remain on the periphery of policy formulation. In a growing literature (not reviewed here), economists have examined the processes through which political decisions regarding environmental regulation are made (Stavins, 2004).

The significant changes that have taken place over the past 20 years with regard to the means of environmental policy – that is, acceptance of market-based environmental instruments – may provide a model for progress with analysis of the ends – the targets and goals – of public policies in this domain. The change in the former realm has been dramatic. Market-based instruments have moved centre stage, and policy debates today look very different from those of 20 years ago, when these ideas were routinely characterized as 'licences to pollute' or dismissed as completely impractical. Market-based instruments are now considered seriously for nearly every environmental problem that is tackled, ranging from endangered species preservation to regional smog and global climate change. Of course, no individual policy instrument – whether market-based or conventional – is appropriate for all environmental problems. Which instrument is best in any given situation depends upon a variety of characteristics of the environmental problem, and the social, political, and economic context in which it is regulated.

ROBERT N. STAVINS

See also **climate change, economics of; common property resources; contingent valuation; ecological economics;**

energy economics; environmental Kuznets curve; hedonic prices; household production and public goods; pollution haven hypothesis; pollution permits; social discount rate; value of life.

Bibliography

Ainslie, G. 1991. Derivation of rational economic behavior from hyperbolic discount curves. *American Economic Review* 81, 334–40.

Alberini, A., Cropper, M., Krupnick, A. and Simon, N. 2004. Does the value of a statistical life vary with age and health status? Evidence from the US and Canada. *Journal of Environmental Economics and Management* 48, 769–92.

Andreoni, J. 1995. Warm-glow versus cold-prickle: the effects of positive and negative framing on cooperation in experiments. *Quarterly Journal of Economics* 110, 1–21.

Arrow, K., Cropper, M., Eads, G., Hahn, R., Lave, L., Noll, R., Portney, P., Russell, M., Schmalensee, R., Smith, K. and Stavins, R. 1996. Is there a role for benefit–cost analysis in environmental, health, and safety regulation? *Science* 272, 221–2.

Arrow, K., Solow, R., Portney, P., Leamer, E., Radner, R. and Schuman, H. 1993. Report of the NOAA Panel on Contingent Valuation. *Federal Register* 58, 4601–14.

Barthold, T. 1994. Issues in the design of environmental excise taxes. *Journal of Economic Perspectives* 8(1), 133–51.

Baumol, W. and Oates, W. 1988. *The Theory of Environmental Policy*. Cambridge: Cambridge University Press.

Bennear, L., Stavins, R. and Wagner, A. 2005. Using revealed preferences to infer environmental benefits: evidence from recreational fishing. *Journal of Regulatory Economics* 28, 157–79.

Bockstael, N. 1996. Travel cost methods. In *The Handbook of Environmental Economics*, ed. D. Bromley. Oxford: Blackwell Publishers.

Bockstael, N. and McConnell, K. 1983. Welfare measurement in the household production framework. *American Economic Review* 73, 806–14.

Bohi, D. and Burtraw, D. 1992. Utility investment behavior and the emission trading market. *Resources and Energy* 14, 129–53.

Bohm, P. 1981. *Deposit-Refund Systems: Theory and Applications to Environmental, Conservation, and Consumer Policy*. Baltimore, MD: Resources for the Future/Johns Hopkins University Press.

Boyle, K. 2003. Contingent valuation in practice. In *A Primer on Nonmarket Valuation*, ed. P. Champ, K. Boyle and T. Brown. Dordrecht: Kluwer Academic Publishers.

Carlson, C., Burtraw, D., Cropper, M. and Palmer, K. 2000. Sulfur dioxide control by electric utilities: what are the gains from trade? *Journal of Political Economy* 108, 1292–326.

Coase, R. 1960. The problem of social cost. *Journal of Law and Economics* 3, 1–44.

Conrad, K. 2002. Computable general equilibrium models in environmental and resource economics. In *The International Yearbook of Environmental and Resource Economics 2002/2003: A Survey of Current Issues*, ed. T. Tietenberg and H. Folmer. Northampton: Edward Elgar.

Cropper, M., Aydede, S. and Portney, P. 1994. Preferences for life saving programs: how the public discounts time and age. *Journal of Risk and Uncertainty* 8, 243–65.

Cropper, M. and Freeman, A. 1991. Environmental health effects. In *Measuring the Demand for Environmental Quality*, ed. B. Braden and C. Kolstad. Amsterdam: Elsevier Science Publications.

Cropper, M. and Laibson, D. 1999. The implications of hyperbolic discounting for project evaluation. In *Discounting and Intergenerational Equity*, ed. P. Portney and J. Weyant. Washington, DC: Resources for the Future.

Cropper, M. and Sussman, F. 1990. Valuing future risks to life. *Journal of Environmental Economics and Management* 19, 160–74.

Dewees, D. and Sims, W. 1976. The symmetry of effluent charges and subsidies for pollution control. *Canadian Journal of Economics* 9, 323–31.

Downing, P. and White, L. 1986. Innovation in pollution control. *Journal of Environmental Economics and Management* 13, 18–27.

Ellerman, D., Joskow, P., Schmalensee, R., Montero, J. and Bailey, E. 2000. *Markets for Clean Air: The US Acid Rain Program*. New York: Cambridge University Press.

Fisher, A., McClelland, G. and Schulze, W. 1988. Measures of willingness to pay versus willingness to accept: evidence, explanations, and potential reconciliation. In *Amenity Resource Valuation: Integrating Economics with Other Disciplines*, ed. G. Peterson, B. Driver and R. Gregory. State College, PA: Venture.

Goulder, L. 1995. Environmental taxation and the double dividend: a reader's guide. *International Tax and Public Finance* 2, 157–83.

Goulder, L. and Stavins, R. 2002. An eye on the future: how economists' controversial practice of discounting really affects the evaluation of environmental policies. *Nature* 419, 673–4.

Gruenspecht, H. 1982. Differentiated regulation: the case of auto emissions standards. *American Economic Review Papers and Proceedings* 72, 328–31.

Hahn, R. 1984. Market power and transferable property rights. *Quarterly Journal of Economics* 99, 753–65.

Hahn, R. 1989. Economic prescriptions for environmental problems: how the patient followed the doctor's orders. *Journal of Economic Perspectives* 3, 95–114.

Hahn, R. and McGartland, A. 1989. Political economy of instrument choice: an examination of the U.S. role in implementing the Montreal Protocol. *Northwestern University Law Review* 83, 592–611.

Hamilton, J.T. 1995. Pollution as news: media and stock market reactions to the toxic release inventory data. *Journal of Environmental Economics and Management* 28, 98–113.

Hanemann, M. 1991. Willingness to pay and willingness to accept: how much can they differ? *American Economic Review* 81, 635–47.

Harrington, W., Morgenstern, R. and Nelson, P. 2000. On the accuracy of regulatory cost estimates. *Journal of Policy Analysis and Management* 19, 297–322.

Harrison, D., Jr. 1999. Turning theory into practice for emissions trading in the Los Angeles air basin. In *Pollution for Sale: Emissions Trading and Joint Implementation*, ed. S. Sorrell and J. Skea. Edward Elgar.

Hartman, R., Wheeler, D. and Singh, M. 1994. The cost of air pollution abatement. Policy Research Working Paper No. 1398. Washington, DC: World Bank.

Hausman, J., ed. 1993. *Contingent Valuation: A Critical Assessment*. Amsterdam: North-Holland.

Hazilla, M. and Kopp, R. 1990. Social cost of environmental quality regulations: a general equilibrium analysis. *Journal of Political Economy* 98, 853–73.

Hicks, J. 1939. The foundations of welfare economics. *Economic Journal* 49, 696–712.

Howe, C. 1997. Increasing efficiency in water markets: examples from the western United States. In *Water Marketing – The Next Generation*, ed. T. Anderson and P. Hill. Lanham, MD: Rowman and Littlefield Publishers.

Jaffe, A., Newell, R. and Stavins, R. 2003. Technological change and the environment. In *The Handbook of Environmental Economics*, ed. K. Mäler and J. Vincent. Amsterdam: North-Holland/Elsevier Science.

Jaffe, A. and Stavins, R. 1995. Dynamic incentives of environmental regulation: the effects of alternative policy instruments on technological diffusion. *Journal of Environmental Economics and Management* 29, S43–S63.

Jaffe, A., Peterson, S., Portney, P. and Stavins, R. 1995. Environmental regulation and the competitiveness of U.S. manufacturing: what does the evidence tell us? *Journal of Economic Literature* 33, 132–65.

Kaldor, N. 1939. Welfare propositions of economics and interpersonal comparisons of utility. *Economic Journal* 49, 549–52.

Kaplow, L. and Shavell, S. 2002. On the superiority of corrective taxes to quantity regulation. *American Law and Economics Review* 4, 1–17.

Khanna, M., Quimio, W. and Bojilova, D. 1998. Toxic release information: a policy tool for environmental protection. *Journal of Environmental Economics and Management* 36, 243–66.

Kohn, R. 1985. A general equilibrium analysis of the optimal number of firms in a polluting industry. *Canadian Journal of Economics* 18, 347–54.

Konar, S. and Cohen, M. 1997. Information as regulation: the effect of community right to know laws on toxic emissions. *Journal of Environmental Economics and Management* 32, 109–24.

Kuhn, H. and Tucker, A. 1951. Nonlinear programming. In *Proceedings of the Second Berkeley Symposium on Mathematical Statistics and Probability*, ed. J. Neyman. Berkeley: University of California Press.

Maleug, D. 1989. Emission credit trading and the incentive to adopt new pollution abatement technology. *Journal of Environmental Economics and Management* 16, 52–7.

Maleug, D. 1990. Welfare consequences of emission credit trading programs. *Journal of Environmental Economics and Management* 18, 66–77.

Maloney, M. and Brady, G. 1988. Capital turnover and marketable property rights. *Journal of Law and Economics* 31, 203–26.

Menell, P. 1990. Beyond the throwaway society: an incentive approach to regulating municipal solid waste. *Ecology Law Quarterly* 17, 655–739.

Mestelman, S. 1982. Production externalities and corrective subsidies: a general equilibrium analysis. *Journal of Environmental Economics and Management* 9, 186–93.

Milliman, S. and Prince, R. 1989. Firm incentives to promote technological change in pollution control. *Journal of Environmental Economics and Management* 17, 247–65.

Misolek, W. and Elder, H. 1989. Exclusionary manipulation of markets for pollution rights. *Journal of Environmental Economics and Management* 16, 156–66.

Mitchell, R. and Carson, R. 1989. *Using Surveys to Value Public Goods: The Contingent Valuation Method*. Washington, DC: Resources for the Future.

Montero, J. 2003. Tradeable permits with imperfect monitoring: theory and evidence. Working paper. Cambridge, MA: Center for Energy and Environmental Policy, MIT.

Montgomery, D. 1972. Markets in licenses and efficient pollution control programs. *Journal of Economic Theory* 5, 395–418.

Moore, M. and Viscusi, W. 1988. The quantity-adjusted value of life. *Economic Inquiry* 26, 369–88.

Morgenstern, R. 2000. Decision making at EPA: economics, incentives and efficiency. Draft paper for a conference: EPA at thirty: evaluating and improving the environmental protection agency, Duke University, 7–8 December.

Mrozek, J. and Taylor, L. 2002. What determines the value of life? A meta analysis. *Journal of Policy Analysis and Management* 21, 253–70.

Nelson, R., Tietenberg, T. and Donihue, M. 1993. Differential environmental regulation: effects on electric utility capital turnover and emissions. *Review of Economics and Statistics* 75, 368–73.

Newell, R. and Pizer, W. 2003. Regulating stock externalities under uncertainty. *Journal of Environmental Economics and Management* 45, 416–32.

Newell, R. and Stavins, R. 2003. Cost heterogeneity and the potential savings from market-based policies. *Journal of Regulatory Economics* 23, 43–59.

Palmquist, R. 2003. Property value models. In *Handbook of Environmental Economics*, vol. 2, ed. K. Mäler and J. Vincent. Amsterdam: North-Holland/Elsevier Science.

Pareto, V. 1896. *Cours d'Economie Politique*, vol. 2. Lausanne: F. Rouge.

Phaneuf, D. and Smith, V. 2004. Recreation demand models. In *Handbook of Environmental Economics*, ed. K. Mäler and J. Vincent. Amsterdam: North-Holland/Elsevier Science.

Pigou, A. 1920. *The Economics of Welfare*. London: Macmillan.

Pizer, W. 2002. Combining price and quantity controls to mitigate global climate change. *Journal of Public Economics* 85, 409–34.

Porter, M. and van der Linde, C. 1995. Toward a new conception of the environment-competitiveness relationship. *Journal of Economic Perspectives* 9(4), 97–118.

Revesz, R. and Stavins, R. 2005. Environmental law and policy. In *The Handbook of Law and Economics*, ed. A. Polinsky and S. Shavell. Amsterdam: North-Holland/Elsevier Science.

Roberts, M. and Spence, M. 1976. Effluent charges and licenses under uncertainty. *Journal of Public Economics* 5, 193–97.

Sagoff, M. 1993. Environmental economics: an epitaph. *Resources* (111), 2–7.

Schmalensee, R., Joskow, P., Ellerman, A., Montero, J. and Bailey, E. 1998. An interim evaluation of sulfur dioxide emissions trading. *Journal of Economic Perspectives* 12(3), 53–68.

Shepard, D. and Zeckhauser, R. 1982. Life-cycle consumption and willingness to pay for increased survival. In *The Value of Life and Safety*, ed. M. Jones-Lee. Amsterdam: North-Holland.

Stavins, R. 1995. Transaction costs and tradeable permits. *Journal of Environmental Economics and Management* 29, 133–46.

Stavins, R. 1996. Correlated uncertainty and policy instrument choice. *Journal of Environmental Economics and Management* 30, 218–25.

Stavins, R. 1998. What have we learned from the grand policy experiment: lessons from SO_2 allowance trading. *Journal of Economic Perspectives* 12(3), 69–88.

Stavins, R. 2003. Experience with market-based environmental policy instruments. In *The Handbook of Environmental Economics*, ed. K. Mäler and J. Vincent. Amsterdam: North-Holland/Elsevier Science.

Stavins, R. 2004. *The Political Economy of Environmental Regulation*. Northampton, MA: Edward Elgar.

Stephenson, K., Norris, P. and Shabman, L. 1998. Watershed-based effluent trading: the nonpoint source challenge. *Contemporary Economic Policy* 16, 412–21.

Tietenberg, T. 1985. *Emissions Trading: An Exercise in Reforming Pollution Policy*. Washington, DC: Resources for the Future.

Tschirhart, J. 1984. Transferable discharge permits and the control of stationary source air pollution: a survey and synthesis. In *Economic Perspectives on Acid Deposition Control*, ed. T. Crocker. Boston: Butterworth.

US Environmental Protection Agency. 2000. *Guidelines for Preparing Economic Analyses*. Washington, DC: US EPA.

Viscusi, W. and Aldy, J. 2003. The value of a statistical life: a critical review of market estimates throughout the world. *Journal of Risk and Uncertainty* 27(1), 5–76.

Weitzman, M. 1974. Prices vs. quantities. *Review of Economic Studies* 41, 477–91.

Weitzman, M. 1978. Optimal rewards for economic regulation. *American Economic Review* 68, 683–91.

Weitzman, M. 1994. On the environmental discount rate. *Journal of Environmental Economics and Management* 26, 200–9.

Weitzman, M. 1998. Why the far-distant future should be discounted at its lowest possible rate. *Journal of Environmental Economics and Management* 36, 201–8.

Willig, R. 1976. Consumer's surplus without apology. *American Economic Review* 66, 589–97.

environmental Kuznets curve

Some forms of pollution appear first to worsen and later to improve as countries' incomes grow. The world's poorest and richest countries have relatively clean environments, while middle-income countries are the most polluted. Because of its resemblance to the pattern of inequality and income described by Simon Kuznets (1955), this pattern of pollution and income has been labelled an 'environmental Kuznets curve' (EKC).

Grossman and Krueger (1995) and the World Bank (1992) first popularized this idea, using a simple empirical approach. They regress data on ambient air and water quality in cities worldwide on a polynomial in GDP per capita and other city and country characteristics. They then plot the fitted values of pollution levels as a function of GDP per capita, and demonstrate that many of the plots appear inverse-U-shaped, first rising and then falling. The peaks of these predicted pollution-income paths vary across pollutants, but 'in most cases they come before a country reaches a per capita income of $8000' in 1985 dollars (Grossman and Krueger, 1995, p. 353).

In the years since these original observations were made, researchers have examined a wide variety of pollutants for evidence of the EKC pattern, including automotive lead emissions, deforestation, greenhouse gas emissions, toxic waste and indoor air pollution. Some investigators have experimented with different econometric approaches, including higher-order polynomials, fixed and random effects, splines, semi- and nonparametric techniques, and different patterns of interactions and exponents. Others have studied different groups of jurisdictions and different time periods, and have added control variables, including measures of corruption, democratic freedoms, international trade openness, and even income inequality (bringing the subject full circle back to Kuznets's original idea).

Some generalizations across these approaches emerge. Roughly speaking, pollution involving local externalities begins improving at the lowest income levels. Fecal coliform in water and indoor household air pollution are examples. For some of these local externalities, pollution appears to decrease steadily with economic growth, and we observe no turning point at all. This is not a rejection of the EKC; pollution must have increased at some point in order to decline with income eventually, and there simply are no data from the earlier period. By contrast, pollutants involving very dispersed externalities tend to have their turning points at the highest incomes, or even no turning points at all, as pollution appears to increase steadily with income. Carbon emissions provide one such example. This, too, is not necessarily a rejection of the EKC; the turning points for these pollutants may come at levels of income per capita higher than in today's wealthiest economies.

Another general empirical result is that the turning points for individual pollutants differ across countries. This difference shows up as instability in empirical approaches that estimate one fixed turning point for any given pollutant. Countries that are the first to deal with a pollutant do so at higher income levels than following countries, perhaps because the following countries benefit from the science and engineering lessons of the early movers.

Most researchers have been careful to avoid interpreting these reduced-form empirical correlations structurally, and to recognize that economic growth does not automatically cause environmental improvements. All of the studies omit country characteristics correlated with both income and pollution levels, the most important being environmental regulatory stringency. The EKC pattern does not provide evidence of market failures or efficient policies in rich or poor countries. Rather, there are multiple underlying mechanisms, some of which have begun to be modelled theoretically.

In theory, the EKC relationship can be divided into three parts: scale, composition, and technique (see Brock and Taylor, 2005). If as an economy grows the *scale* of all activities increases proportionally, pollution will increase with economic growth. If growth is not proportional but is accompanied by a change in the *composition* of goods produced, then pollution may decline or increase with income. If richer economies produce proportionally fewer pollution-intensive products, because of changing tastes or patterns of trade, this composition effect can lead to a decline in pollution associated with economic growth. Finally, if richer countries use less pollution-intensive production *techniques*, perhaps because environmental quality is a normal good, growth can lead to falling pollution.

The EKC summarizes the interaction of these three processes.

Beyond this aggregate decomposition of the EKC, some attempts have been made to formalize structural models that lead to inverse-U-shaped pollution-income patterns. Many describe economies at some type of corner solution initially, where residents of poor countries are willing to trade environmental quality for income at a faster rate than possible using available technologies or resources. As the model economies become wealthier and their environments dirtier, eventually the marginal utility of income falls and the marginal disutility from pollution rises, to the point where people choose costly abatement mechanisms. After that point, the economies are at interior solutions, marginal abatement costs equal marginal rates of substitution between environmental quality and income, and pollution declines with income (see Stokey, 1998). In frameworks of this type, there is typically zero pollution abatement until some threshold income level is crossed, after which abatement begins and pollution starts declining with income.

To date, the practical lessons from this theoretical literature are limited. Most of the models are designed to yield inverse-U-shaped pollution-income paths, and succeed using a variety of assumptions and mechanisms. Hence, any number of forces may be behind the empirical observation that pollution increases and then decreases with income. Moreover, that pattern cannot be interpreted causally, and is consistent with either efficient or inefficient growth paths. Perhaps the most important insight is in Grossman and Krueger's original paper: 'We find no evidence that economic growth does unavoidable harm to the natural habitat' (1995, p. 370). Economists have long argued that environmental degradation is not an inevitable consequence of economic growth. The EKC literature provides empirical support for that claim.

ARIK LEVINSON

See also **environmental economics; growth and international trade; pollution haven hypothesis.**

Bibliography

Brock, W. and Taylor, M. 2005. Economic growth and the environment: a review of theory and empirics. In *The Handbook of Economic Growth*, vol. 1, ed. S. Durlauf and P. Aghion. Amsterdam: North-Holland.

Grossman, G. and Krueger, A. 1995. Economic growth and the environment. *Quarterly Journal of Economics* 110, 353–77.

Kuznets, S. 1955. Economic growth and income inequality. *American Economic Review* 45, 1–28.

Stokey, N. 1998. Are there limits to growth? *International Economic Review* 39, 1–31.

World Bank. 1992. *World Development Report 1992*. New York: Oxford University Press.

Ephémérides du citoyen ou chronique de l'esprit national

French economic periodical issued in three series under different names from 1766 to 1772, 1774 to 1776 and in 1788. Published first as a bimonthly by its founder and first editor, l'Abbé Baudeau, it became a monthly as from January 1767 after Baudeau's conversion to Physiocracy by Mirabeau and Le Trosne. Its contents included contributed articles on economic and political subjects, book reviews, comments and letters to the editor, together with a chronicle of public events of interest to its readership. This provided its format from January 1769, when Du Pont de Nemours took over the editorship. Although censorship problems troubled the journal persistently (as disclosed in the Turgot–Du Pont correspondence, for this reason many issues appeared well after the ostensible month of publication) the first series was terminated by l'Abbé Terray in November 1772, presumably because it contained much vigorous criticism of his abolition of domestic free trade in grain. The first series produced therefore six issues in 1766 as a bi-monthly and 63 monthly issues from January 1767 to March 1772 inclusive. Under the title *Nouvelles Ephémérides ou Bibliothèque raisonnée de l'histoire, de la morale et de la politique*, it was revived by Baudeau after Turgot became Contrôleur-général in 1774, publishing 18 issues in all from January 1775 to June 1776, that is, the month after Turgot's dismissal from the ministry. A third series, *Nouvelles Ephémérides économiques* published three issues from January to March 1788, again under Baudeau's editorship, but his failing mental powers were presumably the reason why this final series ended so quickly.

Although initially set up by Baudeau in imitation of the English *Spectator*, within a year of its inception economics began to dominate its contents and many of the leading Physiocrats, in particular Mirabeau, Baudeau and Du Pont de Nemours, contributed most of the articles. A detailed discussion of its contents is given in Bauer (1894) and in Coquelin and Guillaumin (1854, pp. 710–12). Perhaps the most important piece it contained is Turgot's *Réflexions sur la formation et distribution des richesses* in serial form (*Ephémérides*, 1769, No. 11, pp. 12–56; No. 12, pp. 31–98; and 1770, No. I, pp. 113–73), although with considerable unauthorized alterations and notes by Du Pont (see Groenewegen, 1977, pp. xix–xxi). It also published foreign contributions in French translation, including Beccaria's inaugural lecture with copious notes and comments by Du Pont (*Ephémérides*, 1769, No. 6, pp. 57–152) and a contribution by Franklin on the increasing troubles between England and her American colonies (*Ephémérides*, 1768, No. 8, pp. 159–92). As an early, if not the first, economic journal, the *Ephémérides* remains an important part of economic literature and an indispensable source for those interested in the study of Physiocracy.

PETER GROENEWEGEN

See also **physiocracy.**

Bibliography

Bauer, S. 1894. Ephémérides. In *Dictionary of Political Economy*, vol. 1, ed. R.H.I. Palgrave. London: Macmillan.

Coquelin, C. and Guillaumin, H., ed. 1854. Ephémérides. In *Dictionnaire de l'économie politique*, vol. 1. Paris: Guillaumin, Hachette.

Groenewegen, P.D. 1977. *The Economics of A.R.J. Turgot*. The Hague: Martinus Nijhoff.

epistemic game theory: an overview

The following three articles survey some aspects of the foundations of non-cooperative game theory. The goal of work in foundations is to examine in detail the basic ingredients of game analysis.

The starting point for most of game theory is a 'solution concept' – such as Nash equilibrium or one of its many variants, backward induction, or iterated dominance of various kinds. These are usually thought of as the embodiment of 'rational behaviour' in some way and used to analyse game situations.

One could say that the starting point for most game theory is more of an endpoint of work in foundations. Here, the primitives are more basic. The very idea of rational – or irrational – behaviour needs to be formalized. So does what each player might know or believe about the game – including about the rationality or irrationality of other players. Foundational work shows that even what each player knows or believes about what other players know or believe, and so on, can matter.

Investigating the basis of existing solution concepts is one part of work in foundations. Other work in foundations has uncovered new solution concepts with useful properties. Still other work considers changes even to the basic model of decision making by players – such as departures from the expected utility model or reasoning in various formal logics.

The first article, EPISTEMIC GAME THEORY: BELIEFS AND TYPES, by Marciano Siniscalchi, describes the formalism used in most work on foundations. This is the 'types' formalism going back to Harsanyi (1967–8). Originally proposed to describe the players' beliefs about the structure of the game (such as the payoff functions), the types approach is equally suited to describing beliefs about the play of the game or beliefs about both what the game is and how it will be played. Indeed, in its most general form, the formalism is simply a way to describe any multi-person uncertainty. Harsanyi's conception of a 'type' was a crucial breakthrough in game theory. Still, his work left many fundamental questions about multi-person uncertainty unanswered. Siniscalchi's article surveys these later developments.

The second and third articles apply these tools to the two kinds of uncertainty mentioned. The second article, EPISTEMIC GAME THEORY: COMPLETE INFORMATION, concerns the case where the matrix or tree itself is 'transparent' to the players, and what is uncertain are the actual strategies chosen by the players. The third article, EPISTEMIC GAME

THEORY: INCOMPLETE INFORMATION, by Aviad Heifetz, has the opposite focus: it covers the case of uncertainty about the game itself. (Following Harsanyi, the third article focuses on uncertainty about the payoffs, in particular.)

Both cases are important to the foundations programme. Because Nash equilibrium is 'as if' each player is certain (and correct) about the strategies chosen by the other players (Aumann and Brandenburger, 1995, Section 7h), uncertainty of the first kind has played a small role in game theory to date. Uncertainty of the second kind is the topic of the large literatures on information asymmetries, incentives, and so on.

Interestingly, though, von Neumann and Morgenstern (1944) already appreciated the significance of both complete and incomplete information environments. Indeed, they asserted that phenomena often thought to be characteristic of incomplete-information settings could, in fact, arise in complete-information settings (1944, p. 31):

> Actually, we think that our investigations – although they assume 'complete information' without any further discussion – do make a contribution to the study of this subject. It will be seen that many economic and social phenomena which are usually ascribed to the individual's state of 'incomplete information' make their appearance in our theory and can be satisfactorily interpreted with its help.

This is indeed true, as work in the modern foundations programme shows. (Some instances are mentioned in what follows.) Overall, the foundations programme aims at a 'neutral' and comprehensive treatment of all ingredients of a game.

ADAM BRANDENBURGER

See also **epistemic game theory: beliefs and types; epistemic game theory: complete information; epistemic game theory: incomplete information; game theory; Nash equilibrium, refinements of.**

My thanks to Rena Henderson and Michael James. The Stern School of Business provided financial support.

Bibliography

Aumann, R. and Brandenburger, A. 1995. Epistemic conditions for Nash equilibrium. *Econometrica* 63, 1161–80.

Harsanyi, J. 1967–8. Games with incomplete information played by 'Bayesian' players, I–III. *Management Science* 14, 159–82, 320–34, 486–502.

Von Neumann, J. and Morgenstern, O. 1944. *Theory of Games and Economic Behavior*. Princeton: Princeton University Press. 60th anniversary edn, 2004.

epistemic game theory: beliefs and types

John Harsanyi (1967–8) introduced the formalism of type spaces to provide a simple and parsimonious representation of belief hierarchies. He explicitly noted that his formalism was not limited to modelling a player's beliefs about payoff-relevant variables: rather, its strength was precisely the ease with which Ann's beliefs about Bob's beliefs about payoff variables, Ann's beliefs about Bob's beliefs about Ann's beliefs about payoff variables, and so on, could be represented.

This feature plays a prominent role in the epistemic analysis of solution concepts (see EPISTEMIC GAME THEORY: COMPLETE INFORMATION), as well as in the literature on global games (Morris and Shin, 2003) and on robust mechanism design (Bergemann and Morris, 2005). All these applications place particular emphasis on the expressiveness of the type-space formalism. Thus, a natural question arises: just how expressive is Harsanyi's approach?

For instance, solution concepts such as Nash equilibrium or rationalizability can be characterized by means of restrictions on the players' mutual beliefs. In principle, these assumptions could be formulated directly as restrictions on players' hierarchies of beliefs; but in practice the analysis is mostly carried out in the context of a type space à la Harsanyi. This is without loss of generality only if Harsanyi type spaces do not themselves impose restrictions on the belief hierarchies that can be represented. Similar considerations apply in the context of robust mechanism design.

A rich literature addresses this issue from different angles, and for a variety of basic representations of beliefs. This article focuses on hierarchies of probabilistic beliefs; however, some extensions are also mentioned. For simplicity, attention is restricted to two players, denoted '1' and '2' or 'i' and '$-i$.'

Probabilistic type spaces and belief hierarchies

Begin with some mathematical preliminaries. A topology on a space X is deemed Polish if it is separable and completely metrizable; in this case, X is itself deemed a Polish space. Examples include finite sets, Euclidean space $\mathbb{R}^n$ and closed subsets thereof. A countable product of Polish spaces, endowed with the product topology, is itself Polish. For any topological space X, the notation $\Delta(X)$ indicates the set of Borel probability measures on X. If the topology on X is Polish, then the weak* topology on $\Delta(X)$ is also Polish (for example, Aliprantis and Border, 1999, Theorem 14.15). A sequence $\{\mu^k\}_{k \geq 1}$ in $\Delta(X)$ converges in the weak* sense to a measure $\mu \in \Delta(X)$, written $\mu^k \xrightarrow{w^*} \mu$, if and only if, for every bounded, continuous function $\psi : X \to \mathbb{R}$, $\int_X \psi d\mu^k \to \int_X \psi d\mu$. The weak* topology on $\Delta(X)$ is especially meaningful and convenient when X is a Polish space: see Aliprantis and Border (1999, ch. 14) for an overview of its properties. Finally, if μ is a measure on some product space $X \times Y$, the marginal of μ on X is denoted $\text{marg}_X \mu$.

The basic ingredient of the players' hierarchical beliefs is a description of payoff-relevant or fundamental uncertainty. Fix two sets S_1 and S_2, hereinafter called the *uncertainty domains*; the intended interpretation is that S_{-i} describes aspects of the strategic situation that Player i is uncertain about. For example, in an independent private-values auction, each set S_i could represent bidder i's possible valuations of the object being sold, which is not known to bidder $-i$. In the context of interactive epistemology, S_i is usually taken to be Player i's strategy space. It is sometimes convenient to let $S_1 = S_2 \equiv S$; in this case, the formalism introduced below enables one to formalize the assumption that each player observes different aspects of the common uncertainty domain S (for instance, different signals correlated with the common, unknown value of an object offered for sale).

An (S_1, S_2)-*based type space* is a tuple $\mathcal{T} = (T_i, g_i)_{i=1,2}$ such that, for each $i = 1, 2$, T_i is a Polish space and $g_i : T_i \rightarrow \Delta(S_{-i} \times T_{-i})$ is continuous. As noted above, type spaces can represent hierarchies of beliefs; it is useful to begin with an example. Let $S_1 = S_2 = \{a, b\}$ and consider the type space defined in Table 1. To interpret, for every $i = 1, 2$, the entry in the row corresponding to t_i and (s_{-i}, t_{-i}) is $g_i(t_i)(\{(s_{-i}, t_{-i})\})$. Thus, for instance, $g_1(t_1)(\{(a, t_2')\}) = 0$; $g_2(t_2)(\{b\} \times T_1) = 0.5$.

Consider type t_1 of Player 1. She is certain that $s_2 = a$; furthermore, she is certain that Player 2 believes that $s_1 = a$ and $s_1 = b$ are equally likely. Taking this one step further, type t_1 is certain that Player 2 assigns probability 0.5 to the event that Player 1 believes that $s_2 = b$ with probability 0.7.

These intuitive calculations can be formalized as follows. Fix an (S_1, S_2)-based type space $\mathcal{T} = (T_i, g_i)_{i=1,2}$; for every $i = 1, 2$ define the set X_{-i}^0 and the function $h_i^1 : T_i \rightarrow \Delta(X_0^{-i})$ by

$$X_{-i}^0 = S_{-i} \quad \text{and} \quad \forall t_i \in T_i, \; h_i^1(t_i) = \text{marg}_{S_{-i}} g_i(t_i).$$
$$(1)$$

Thus, $h_i^1(t_i)$ represents the *first-order beliefs* of type t_i in type space T – her beliefs about the uncertainty domain S_{-i}. Note that each $X_{-i}^0 = S_{-i}$ is Polish. Proceeding inductively, assuming that $X_{-i}^0, \ldots, X_{-i}^{k-1}$ and $h_i^1, \ldots, h_i^k$ have been defined up to some $k > 0$ for $i = 1, 2$, and that all sets X_{-i}^ℓ, $\ell = 0, \ldots, k - 1$ are Polish, define the set X_{-i}^k and the functions $h_i^{k+1} : T_i \rightarrow \Delta(X_{-i}^k)$ for $i = 1, 2$ by

$$X_{-i}^k = X_{-i}^{k-1} \times \Delta(X_{-i}^{k-1}) \quad \text{and}$$

$$\forall t_i \in T_i, \; h_i^{k+1}(t_i)(E) = g_i(t_i)(\{(s_{-i}, t_{-i})$$
$$\in S_{-i} \times T_{-i} : (s_{-i}, h_{-i}^k(t_{-i})) \in E\}) \quad (2)$$

Table 1   *A type space*

T_1	a, t_2		a, t_2'	b, t_2		b, t_2'
t_1	1		0	0		0
t_1'	0		0.3	0		0.7
T_2	a, t_1		a, t_1'	b, t_1		b, t_1'
t_2	0		0.5	0.5		0
t_2'	0		0	0		1

for every Borel subset E of X_{-i}^k. Thus, $h_1^2(t_1)$ represents the *second-order beliefs* of type t_1 – her beliefs about *both* the uncertainty domain $S_2 = X_2^0$ and Player 2's beliefs about S_1, which by definition belong to the set $\Delta(X_1^0) = \Delta(S_1)$. Similarly, $h_i^{k+1}(t_i)$ represents type t_i's $(k+1)$-th order beliefs.

Observe that type t_i's second-order beliefs are defined over $X_2^0 \times \Delta(X_1^0) = S_2 \times \Delta(S_1)$, rather than just over $\Delta(X_1^0) = \Delta(S_1)$; a similar statement holds for her $(k+1)$-th order beliefs. This is crucial in many applications. For instance, a typical assumption in the literature on epistemic foundations of solution concepts is that Player 1 believes that Player 2 is rational. Letting S_i be the set of actions or strategies of Player i in the game under consideration, this can be modelled by assuming that the support of $h_1^2(t_1)$ consists of pairs $(s_2, \mu_1) \in S_2 \times \Delta(S_1)$ wherein s_2 is a best response to μ_1. Clearly, such an assumption could not be formalized if $h_1^2(t_1)$ only conveyed information about type t_1's beliefs on Player 2's first-order beliefs: even though type t_i's beliefs about the action played by Player 2 could be retrieved from $h_1^1(t_1)$, it would be impossible to tell whether each action that type t_1 expects to be played is matched with a belief that rationalizes it.

Note that, since X_i^{k-1} and X_{-i}^{k-1} are assumed Polish, so are $\Delta(X_i^{k-1})$ and X_{-i}^k. Also, each function h_i^k is continuous.

Finally, it is convenient to define a function that associates to each type $t_i \in T_i$ an entire *belief hierarchy*: to do so, define the set H_i and, for $i = 1, 2$, the function $h_i : T_i \rightarrow H_i$ by

$$H_i = \prod_{k \geq 0} \Delta(X_{-i}^k) \quad \text{and} \quad \forall t_i \in T_i, \; h_i(t_i)$$
$$= (h_i^1(t_i), \ldots, h_i^{k+1}(t_i), \ldots).$$
$$(3)$$

Thus, H_i is the set of all hierarchies of beliefs; notice that, since each X_{-i}^k is Polish, so is H_i.

Rich type spaces

The preceding construction suggests a rather direct way to ask how expressive Harsanyi's notion of a type space is: can one construct a type space that generates *all* hierarchies in H_i?

A moment's reflection shows that this question must be refined. Fix a type space $(T_i, g_i)_{i=1,2}$ and a type $t_i \in T_i$; recall that, for reasons described above, the first- and second-order beliefs of type t_i satisfy $h_i^1(t_i) \in \Delta(S_{-i})$ and $h_i^2(t_i) \in \Delta(X_{-i}^0 \times \Delta(X_i^0)) = \Delta(S_{-i} \times \Delta(S_i))$ respectively. This, however, creates the potential for redundancy or even contradiction, because both $h_i^1(t_i)$ and $\text{marg}_{s_{-i}} h_i^2(t_i)$ can be viewed as 'type t_i's beliefs about S_{-i}'. A similar observation applies to higher-order beliefs. Fortunately, it is easy to verify that, for every type space $(T_i, g_i)_{i=1,2}$ and type $t_i \in T_i$, the following *coherency* condition holds:

$$\forall k > 1, \quad \text{marg}_{X_{-i}^{k-2}} h_i^k(t_i) = h_i^{k-1}(t_i); \quad (4)$$

To interpret, recall that $h_i^k(t_i) \in \Delta(X_{-i}^{k-1}) = \Delta(X_{-i}^{k-2} \times \Delta(X_{-i}^{k-2}))$. Thus, in particular, $\text{marg}_{s_{-i}} h_i^2(t_i) = h_i^1(t_i)$.

Since H_i is defined as the set of *all* hierarchies of beliefs for Player i, some (in fact, 'most') of its elements are not coherent. As noted above, no type space can generate incoherent hierarchies; more importantly, coherency can be viewed as an integral part of the interpretation of interactive beliefs. How could an individual simultaneously hold (infinitely) many distinct first-order beliefs? Which of these should be used, say, to verify whether she is rational? This motivates restricting attention to coherent hierarchies, defined as follows:

$$H_i^c = \left\{ (\mu_i^1, \mu_i^2, \ldots) \in H_i : \forall k > 1, \text{marg}_{X_{-i}^{k-2}} \mu_i^k = \mu_i^{k-1} \right\}.$$

(5)

Since $\text{marg}_{X_{-i}^{k-2}} : \Delta(X_{-i}^{k-1}) \to \Delta(X_{-i}^{k-2})$ is continuous, H_i^c is a closed, hence Polish subspace of H_i.

Brandenburger and Dekel (1993, Proposition 1) show that there exist homeomorphisms $g_i^c : H_i^c \to \Delta(S_{-i} \times H_{-i})$: that is, every coherent hierarchy corresponds to a distinct belief over the uncertainty domain and the hierarchies of the opponent, and conversely. Furthermore, this homeomorphism is canonical, in the following sense. Note that $S_{-i} \times H_{-i} = S_{-i} \times \prod_{k \geq 0} \Delta(X_i^k) = X_{-i}^k \times \prod_{\ell > k} \Delta(X_i^\ell)$. Then it can be shown that, if $\mu_i = (\mu_i^1, \mu_i^2, \ldots) \in H_i^c$, then $\text{marg}_{X^k} g_i^c(\mu_i) = \mu_i^{k+1}$. Intuitively, the marginal belief associated with μ_i over the first k orders of the opponent's beliefs is precisely what it should be, namely μ_i^{k+1}. The proof of these results builds upon Kolmogorov's extension theorem, as may be suggested by the similarity of the coherency condition in eq. (5) with the notion of Kolmogorov consistency: cf. for example Aliprantis and Border (1999, theorem 14.26).

This result does not quite imply that all coherent hierarchies can be generated in a suitable type space; however, it suggests a way to obtain this result. Notice that the belief on $S_{-i} \times H_{-i}$ associated by the homeomorphism g_i^c to a coherent hierarchy μ_i may include *in*coherent hierarchies $\nu_{-i} \in H_{-i} \backslash H_{-i}^c$ in its support. This can be interpreted in the following terms: if Player i's hierarchical beliefs are given by μ_i, then she is coherent, but she is not certain that her opponent is. On the other hand, consider a type space $(T_i, g_i)_{i=1,2}$; as noted above, for every player i, each type $t_i \in T_i$ generates a coherent hierarchy $h_i(t_i) \in H_i^c$. So, for instance, if (s_1, t_1) is in the support of $g_2(t_2)$ then t_1 also generates a coherent hierarchy. Thus, not only is type t_2 of Player 2 coherent: he is also certain (believes with probability one) that Player 1 is coherent. Iterating this argument suggests that *hierarchies of beliefs generated by type spaces display common certainty of coherency.*

Motivated by these considerations, let

$$H_i^0 = H_i^c \quad \text{and} \quad \forall k > 0,$$
$$H_i^k = \{ \mu_i \in H_i^{k-1} : g_i^c(\mu_i)(S_{-i} \times H_{-i}^{k-1}) = 1 \}.$$

(6)

Thus, H_i^0 is the set of coherent hierarchies for Player i; H_i^1 is the set of hierarchies that are coherent and correspond to beliefs that display certainty of the opponent's coherency; and so on. Finally, let $H_i^* = \cap_{k \geq 0} H_i^k$. Each element of H_i^* is intuitively consistent with coherency and common certainty of coherency.

Brandenburger and Dekel (1993, Proposition 2) show that the restriction g_i^* of g_i^c to H_i^* is a homeomorphism between H_i^* and $\Delta(S_{-i} \times H_{-i}^*)$; furthermore, it is canonical in the sense described above. This implies that the tuple $(H_i^*, g_i^*)_{i=1,2}$ is a type space in its own right – the (S_1, S_2)-based *universal type space*.

The existence of a universal type space fully addresses the issue of richness. Since the homeomorphism g_i^* is canonical, it is easy to see that the hierarchy generated as per eqs (1) and (2) by any 'type' $t_i = (\mu^1, \mu^2, s) \in H_i^*$ in the universal type space $(H_i^*, g_i^*)_{i=1,2}$ is t_i itself; thus, since H_i^* consists of all hierarchies that are coherent and display common certainty of consistency, the universal type space also *generates* all such hierarchies.

The type space $(H_i^*, g_i^*)_{i=1,2}$ is rich in two additional, related senses. First, as may be expected, every belief hierarchy for Player i generated by an arbitrary type space is an element of H_i^*; this implies that every type space $(T_i, g_i)_{i=1,2}$ can be uniquely embedded in $(H_i^*, g_i^*)_{i=1,2}$ as a 'belief-closed' subset: see Battigalli and Siniscalchi (1999, Proposition 8.8). Call a type space *terminal* if, like $(H_i^*, g_i^*)_{i=1,2}$, it embeds all other type spaces as belief-closed subsets.

Second, since each function g_i^* is a homeomorphism, in particular it is a surjection (that is, onto). Call a type space $(T_i, g_i)_{i=1,2}$ *complete* if every map g_i is onto. (This should not be confused with the topological notion of completeness.) Thus, the universal type space $(H_i^*, g_i^*)_{i=1,2}$ is complete. It is often the case that, when a universal type space is employed in the epistemic analysis of solution concepts, the objective is precisely to exploit its completeness. Furthermore, for certain representations of beliefs, it is not known whether universal type spaces can be constructed; however, the existence of complete type spaces can be established, and is sufficient for the purposes of epistemic analysis. The next section provides examples.

Alternative constructions and extensions

The preceding discussion adopts the approach proposed by Brandenburger and Dekel (1993), which has the virtue of relying on familiar ideas from the theory of stochastic processes. However, the first constructions of universal type spaces consisting of hierarchies of beliefs are due to Armbruster and Böge (1979), Böge and Eisele (1979) and Mertens and Zamir (1985).

From a technical point of view, Mertens and Zamir (1985) assume that the state space S is compact Hausdorff and beliefs are regular probability measures. Heifetz and Samet (1998b) instead drop topological assumptions altogether: in their approach, both the underlying set of states and the sets of types of each player are modelled as measurable spaces. They show that a terminal type space can be explicitly constructed in this environment.

In all the contributions mentioned so far, beliefs are modelled as countably additive probabilities. The literature has also examined other representations of beliefs, broadly defined.

A *partitional structure* (Aumann, 1976) is a tuple $(\Omega,(\sigma_i, P_i)_{i=1,2})$, where Ω is a (typically finite) space of 'possible worlds', every $\sigma_i : \Omega \to S_i$ indicates the realization of the basic uncertainty corresponding to each element of Ω, and every P_i is a partition of Ω. The interpretation is that, at any world $\omega \in \Omega$, Player i is only informed that the true world lies in the cell of the partition P_i containing ω, denoted $P_i(\omega)$. The *knowledge operator* for Player i can then be defined as

$$\forall E \subset \Omega, \quad K_i(E) = \{\omega \in \Omega : P_i(\omega) \subseteq E\}.$$

Notice that no probabilistic information is provided in this environment (although it can be easily added).

Heifetz and Samet (1998a) show that a terminal partitional structure does not exist. This result was extended to more general 'possibility' structures by Meier (2005). Brandenburger and Keisler (2006) establish related non-existence results for complete structures. However, recent contributions show that topological assumptions, which play a key role in the constructions of Mertens and Zamir (1985) and Brandenburger and Dekel (1993), can also deliver existence results in non-probabilistic settings. For instance, Mariotti, Meier and Piccione (2005) construct a structure that is universal, complete and terminal for possibility structures.

Other authors investigate richer probabilistic representations of beliefs. Battigalli and Siniscalchi (1999) construct a universal, terminal, and complete type space for *conditional probability system*, or collections of probability measures indexed by relevant conditioning events (such as histories in an extensive game) and related by a version of Bayes's rule. This type space is used in (2002) to provide an epistemic analysis of forward induction. Brandenburger, Friedenberg and Keisler (2006) construct a complete type space for *lexicographic sequences*, which may be thought of as an extension of lexicographic probability systems (Blume, Brandenburger and Dekel, 1991) for infinite domains. They then use it to provide an epistemic characterization of iterated admissibility.

Non-probabilistic representations of beliefs that reflect a concern for ambiguity (Ellsberg, 1961) have also been considered. Heifetz and Samet (1998b) observe that their measure-theoretic construction extends to beliefs represented by continuous *capacities*, that is non-additive set functions that preserve monotonicity with respect to set inclusion. Motivated by the multiple-priors model of Gilboa and Schmeidler (1989), Ahn (2006) constructs a universal type space for sets of probabilities.

Epstein and Wang (1996) approach the richness issue taking *preferences*, rather than beliefs, as primitive objects. In their setting, an S-based type space is a tuple $(T_i, g_i)_{i=1,2}$, where, for every type t_i, $g_i(t_i)$ is a suitably regular preference over *acts* defined on the set $S \times T_{-i}$. The analysis in the preceding section can be viewed as a special case of Epstein and Wang (1996), where preferences conform to expected-utility theory. Epstein and Wang construct a universal type space in this framework (see also Di Tillio, 2006).

Finally, constructions analogous to that of a universal type space appear in other, unrelated contexts. For instance, Epstein and Zin (1989) develop a class of recursive preferences over infinite-horizon temporal lotteries; to construct the domain of such preferences, they employ arguments related to Mertens and Zamir's. Gul and Pesendorfer (2004) employ analogous techniques to analyse self-control preferences over infinite-horizon consumption problems.

MARCIANO SINISCALCHI

See also **epistemic game theory: an overview; epistemic game theory: complete information; epistemic game theory: incomplete information.**

Bibliography

Ahn, D. 2007. Hierarchies of ambiguous beliefs. *Journal of Economic Theory* 136, 286–301.

Aliprantis, C. and Border, K. 1999. *Infinite Dimensional Analysis*, 2nd edn. Berlin: Springer.

Armbruster, W. and Böge, W. 1979. Bayesian game theory. In *Game Theory and Related Topics*, ed. O. Moeschlin and D. Pallaschke. Amsterdam: North-Holland.

Aumann, R. 1976. Agreeing to disagree. *Annals of Statistics* 4, 1236–9.

Battigalli, P. and Siniscalchi, M. 1999. Hierarchies of conditional beliefs and interactive epistemology in dynamic games. *Journal of Economic Theory* 88, 188–230.

Battigalli, P. and Siniscalchi, M. 2002. Strong belief and forward induction reasoning. *Journal of Economic Theory* 106, 356–91.

Bergemann, D. and Morris, S. 2005. Robust mechanism design. *Econometrica* 73, 1521–34.

Blume, L., Brandenburger, A. and Dekel, E. 1991. Lexicographic probabilities and choice under uncertainty. *Econometrica* 59, 61–79.

Böge, W. and Eisele, T. 1979. On solutions of Bayesian games. *International Journal of Game Theory* 8, 193–215.

Brandenburger, A. and Dekel, E. 1993. Hierarchies of beliefs and common knowledge. *Journal of Economic Theory* 59, 189–98.

Brandenburger, A., Friedenberg, A. and Keisler, H.J. 2006. Admissibility in games. Unpublished, Stern School of Business, New York University.

Brandenburger, A. and Keisler, J. 2006. An impossibility theorem on beliefs in games. *Studia Logica* 84, 211–40.

Di Tillio, A. 2006. Subjective expected utility in games. Working Paper No. 311, IGIER, Università Bocconi.

Ellsberg, D. 1961. Risk, ambiguity, and the Savage axioms. *Quarterly Journal of Economics* 75, 643–69.

Epstein, L. and Wang, T. 1996. Beliefs about beliefs without probabilities. *Econometrica* 64, 1343–73.

Epstein, L. and Zin, S. 1989. Substitution, risk aversion, and the temporal behavior of consumption and asset returns: a theoretical framework. *Econometrica* 57, 937–69.

Gilboa, I. and Schmeidler, D. 1989. Maxmin-expected utility with a non-unique prior. *Journal of Mathematical Economics* 18, 141–53.

Gul, F. and Pesendorfer, W. 2004. Self-control and the theory of consumption. *Econometrica* 72, 119–58.

Harsanyi, J. 1967–8. Games of incomplete information played by Bayesian players. Parts I, II, III. *Management Science* 14, 159–82, 320–34, 486–502.

Heifetz, A. and Samet, D. 1998a. Knowledge spaces with arbitrarily high rank. *Games and Economic Behavior* 22, 260–73.

Heifetz, A. and Samet, D. 1998b. Topology-free typology of beliefs. *Journal of Economic Theory* 82, 324–81.

Mariotti, T., Meier, M. and Piccione, M. 2005. Hierarchies of beliefs for compact possibility models. *Journal of Mathematical Economics* 41, 303–24.

Meier, M. 2005. On the nonexistence of universal information structures. *Journal of Economic Theory* 122, 132–9.

Mertens, J.F. and Zamir, S. 1985. Formulation of Bayesian analysis for games with incomplete information. *International Journal of Game Theory* 14, 1–29.

Morris, S. and Shin, H. 2003. Global games: theory and applications. In *Advances in Economics and Econometrics (Proceedings of the Eight World Congress of the Econometric Society)*, ed. M. Dewatripont, L. Hansen and S. Turnovsky. Cambridge: Cambridge University Press.

epistemic game theory: complete information

1 Epistemic analysis

Under the epistemic approach, the traditional description of a game is augmented by a mathematical framework for talking about the rationality or irrationality of the players, their beliefs and knowledge, and related ideas.

The first step is to add sets of *types* for each of the players. The apparatus of types goes back to Harsanyi (1967–8), who introduced it as a way to talk formally about the players' beliefs about the payoffs in a game, their beliefs about other players' beliefs about the payoffs, and so on. (See EPISTEMIC GAME THEORY: INCOMPLETE INFORMATION.) But the technique is equally useful for talking about uncertainty about the actual play of the game – that is, about the players' beliefs about the strategies chosen in the game, their beliefs about other players' beliefs about the strategies, and so on. This survey focuses on this second source of uncertainty. It is also possible to treat both kinds of uncertainty together, using the same technique.

We give a definition of a type structure as commonly used in the epistemic literature, and an example of its use.

Fix an *n*-player finite strategic-form game $\langle S^1, \ldots, S^n, \pi^1, \ldots, \pi^n \rangle$. Some notation: given sets $X^1, \ldots, X^n$, let $X = \times_{i=1}^n X^i$ and $X^{-i} = \times_{j \neq i} X^j$. Also, given a finite set Ω, write $\mathcal{M}(\Omega)$ for set of all probability measures on Ω.

Definition 1.1 *An* $(S^1, \ldots, S^n)$**-based** **(finite) type structure** *is a structure*

$$\langle S^1, \ldots, S^n; T^1, \ldots, T^n; \lambda^1, \ldots, \lambda^n \rangle,$$

where each T^i *is a finite set, and each* $\lambda^i : T^i \to \mathcal{M} (S^{-i} \times T^{-i})$. *Members of* T^i *are called* **types** *for player i. Members of* $S \times T$ *are called* **states** *(of the world).*

For some purposes – see, for example, Sections 4 and 6 – it is important to consider infinite type structures. Topological assumptions are then made on the type spaces T_i.

A particular state $(s^1, t^1, \ldots, s^n, t^n)$ describes the strategy chosen by each player, and also each player's type. Moreover, a type t^i for player i induces, via a natural induction, an entire hierarchy of beliefs – about the strategies chosen by the players $j \neq i$, about the beliefs of the players $j \neq i$, and so on. (See EPISTEMIC GAME THEORY: BELIEFS AND TYPES.)

The following example is similar to one in Aumann and Brandenburger (1995, pp. 1166–7).

Example 1.1 (**A coordination game**). *Consider the coordination game in* Figure 1.1 *(where Ann chooses the row and Bob the column), and the associated type structure in* Figure 1.2.

There are two types t^a, u^a for Ann, and two types t^b, u^b for Bob. The measure associated with each type is as shown. Fix the state (D, t^a, R, t^b). At this state, Ann plays D and Bob plays R. Ann is 'correct' about Bob's strategy. (Her type t^a assigns probability 1 to Bob's playing R.) Likewise, Bob is correct about Ann's strategy. Ann, though, thinks it possible Bob is wrong about her strategy. (Her type assigns probability 1/2 to type u^b for Bob, which assigns probability 1/2 to Ann's playing U, not D.) Again, likewise with Bob.

What about the rationality or irrationality of the players? At state (D, t^a, R, t^b), Ann is rational. Her strategy

	L	R
U	2, 2	0, 0
D	0, 0	1, 1

Figure 1.1

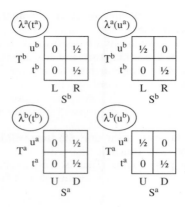

Figure 1.2

maximizes her expected payoff, given her first-order belief (which assigns probability 1 to R). Likewise, Bob is rational. Ann, though, thinks it possible Bob is irrational. (She assigns probability 1/2 to (R, u^b). With type u^b, Bob gets a higher expected payoff from L than R.) The situation with Bob is again symmetric.

Summing up, the example is just a description of a game situation, not a prediction. A type structure is a descriptive tool. Note, too, that the example includes both rationality and irrationality, and also allows for incorrect as well as correct beliefs (for example, Ann thinks it possible Bob is irrational, though in fact he isn't). These are typical features of the epistemic approach.

Two comments on type structures. First, we can ask whether Definition 1.1 above is to be taken as primitive or derived. Arguably, hierarchies of beliefs are the primitive, and types are simply a convenient tool for the analyst. See EPISTEMIC GAME THEORY: BELIEFS AND TYPES for further discussion.

Second, note that Definition 1.1 applies to finite games. These will be the focus of this survey. There is nothing yet approaching a developed literature on epistemic analysis of infinite games.

2 Early results

A major use of type structures is to identify conditions on the players' rationality, beliefs, and so on, that yield various solution concepts.

A very basic solution concept is iterated dominance. This involves deleting from the matrix all strongly dominated strategies, then deleting all strategies that become strongly dominated in the resulting submatrix, and so on until no further deletion is possible. (It is easy to check that in finite games – as considered in this survey – the residual set will always be non-empty.) Call the remaining strategies the *iteratively undominated* (IU) strategies. There is a basic equivalence: a strategy is not strongly dominated

if and only if there is a probability measure on the product of the other players' strategy sets under which it is optimal. Using this, IU can also be defined as follows: delete from the matrix any strategy that isn't optimal under some measure on the product of the other players' strategy sets. Consider the resulting sub-matrix and delete strategies that don't pass this test on the sub-matrix, and so on.

The second definition suggests what a formal epistemic treatment of IU should look like. A rational player will choose a strategy which is optimal under some measure. This is the first round of deletion. A player who is rational and believes the other players are rational will choose a strategy which is optimal under a measure that assigns probability 1 to the strategies remaining after the first round of deletion. This gives the second round of deletion. And so on.

Type structures allow a formal treatment of this idea. First the formal definition of rationality. This is a property of strategy-type pairs. Say (s^i, t^i) is **rational** if s^i maximizes player i's expected payoff under the marginal on S^{-i} of the measure $\lambda^i(t^i)$.

Say type t^i of player i **believes** an event $E \subseteq S^{-i} \times T^{-i}$ if $\lambda^i(t^i)(E) = 1$, and write

$$B^i(E) = \{t^i \in T^i : t^i \text{ believes } E\}.$$

Now, for each player i, let R_1^i be the set of all rational pairs (s^i, t^i), and for $m > 0$ define R_m^i inductively by

$$R_{m+1}^i = R_m^i \cap [S^i \times B^i(R_m^{-i})].$$

Definition 2.1 *If* $(s^1, t^1, \ldots, s^n, t^n) \in R_{m+1}$, *say there is* ***rationality and mth-order belief of rationality (RmBR)*** *at this state. If* $(s^1, t^1, \ldots, s^n, t^n) \in \bigcap_{m=1}^{\infty} R_m$, *say there is* ***rationality and common belief of rationality (RCBR)*** *at this state.*

These definitions yield an epistemic characterization of IU: *Fix a type structure and a state* $(s^1, t^1, \ldots, s^n, t^n)$ *at which there is RCBR. Then the strategy profile* $(s^1, \ldots, s^n)$ *is IU. Conversely, fix an IU profile* $(s^1, \ldots, s^n)$. *There is a type structure and a state* $(s^1, t^1, \ldots, s^n, t^n)$ *at which there is RCBR.* Results like this can be found in the early literature – see, among others, Brandenburger and Dekel (1987) and Tan and Werlang (1988).

An important stimulus to the early literature was the pair of papers by Bernheim (1984) and Pearce (1984), which introduced the solution concept of *rationalizability*. This differs from IU by requiring on each round that a player's probability measure on the product of the other players' (remaining) strategy sets be a product measure – that is, be independent. Thus the set of rationalizable strategy profiles is contained in the IU set. It is well known that there are games (with three or more players) in which inclusion is strict.

The argument for the independence assumption is that in non-cooperative game theory it is supposed that players do not coordinate their strategy choices. Interestingly though, correlation is consistent with the non-cooperative approach. This view is put forward in Aumann (1987). (Aumann, 1974, introduced the study of correlation into non-cooperative theory.) Consider an analogy to coin tossing. A correlated assessment over coin tosses is possible, if there is uncertainty over the coin's parameter or 'bias'. (The assessment is usually required to be conditionally i.i.d., given the parameter.) Likewise, in a game, Charlie might have a correlated assessment over Ann's and Bob's strategy choices, because, say, he thinks Ann and Bob have observed similar signals before the game (but is uncertain what the signal was).

The same epistemic tools used to understand IU can be used to characterize other solution concepts on the matrix. Aumann and Brandenburger (1995, Preliminary Observation) point out that pure-strategy Nash equilibrium is characterized by the simple condition that each player is rational and assigns probability 1 to the actual strategies chosen by the other players. (Thus, in Example 1.1 above, these conditions hold at the state (D, t^a, R, t^b), and (D, R) is indeed a Nash equilibrium.) As far as mixed strategies are concerned, in the epistemic approach to games these don't play the central role that they do under equilibrium analysis. Built into the set-up of Section 1 is that each player makes a definite choice of (pure) strategy. (If a player does have the option of making a randomized choice, this can be added to the – pure – strategy set. Indeed, in a finite game, a finite number of such choices can be added.) It is the other players who are uncertain about this choice. Harsanyi (1973) originally proposed this shift in thinking about randomization. Aumann and Brandenburger (1995) give an epistemic treatment of mixed-strategy Nash equilibrium along these lines.

Aumann (1987) asks a question about an outside observer of a game. He provides conditions under which the observer's assessment of the strategies chosen will be the distribution of a correlated equilibrium (as defined in his 1974 paper). The distinctive condition in (1987) is the so-called Common Prior Assumption, which says that the probability assessment associated with each player's type is the same as the observer's assessment, except for being conditioned on what the type in question knows. A number of papers have investigated foundations for this assumption – see, among others, Morris (1994), Samet (1998), Bonanno and Nehring (1999), Feinberg (2000), Halpern (2002), and also the exchange between Gul (1998) and Aumann (1998).

3 Next steps: the tree

An important next step in the epistemic programme was extending the analysis to game trees. A big motivation for this was to understand the logical foundation of *backward induction* (*BI*). At first sight, BI is one of the easiest ideas in game theory. If Ann, the last player to move, is

rational, she will make the BI choice. If Bob, the second-to-last player to move, is rational and thinks Ann is rational, he will make the choice that is maximal given that Ann makes the BI choice – that is, he too will make the BI choice. And so on back in the tree, until the BI path is identified (Aumann, 1995).

For example, Figure 3.1 is a three-legged centipede (Rosenthal, 1981). (The top payoffs are Ann's, and the bottom payoffs are Bob's.) BI says Ann plays *Out* at her first node. But what if she doesn't? How will Bob react? Perhaps Bob will conclude that Ann is an irrational player, who plays *Across*. That is, Bob might play *In*, hoping to get a payoff of 6 (better than 4 from *Out*). Perhaps, anticipating this, Ann will in fact play *Down*, hoping to get 4 (better than 2 from playing *Out*).

Many papers have examined this conceptual puzzle with BI – see, among others, Binmore (1987), Bicchieri (1988, 1989), Basu (1990), Bonanno (1991), and Reny (1992).

A key step in resolving the puzzle is extending the epistemic tools of Section 1, to be able to talk formally about rationality, beliefs and so on in the tree.

Example 3.1 (three-Legged centipede). Figure 3.2 *is a type structure for three-legged Centipede.*

There are two types t^a, u^a *for Ann. Type* t^a *for Ann has the measure shown in the top-left matrix. It assigns probability 1 to* (In, t^b) *for Bob. Type* u^a *has two associated measures – shown in the top-right matrix. The first measure (the numbers without parentheses) assigns probability 1 to* (Out, u^b) *for Bob. In this case, we also specify a second measure for Ann, because we want to specify what Ann thinks at her second node, too. Reaching this node is assigned positive probability (in fact, probability 1) under Ann's type* t^a, *but probability 0 under her type* u^a. *So, for type* u^a, *there isn't a well-defined conditional probability measure at Ann's second node. This is why we (separately) specify a second measure for Ann's type* u^a: *it is the measure in square brackets. If type* u^a, *Ann assigns probability 1 to* (In, t^b) *at her second node.*

There are also two types t^b, u^b *for Bob. Both types initially assign probability 1 to Ann's playing Out. For both of Bob's types, there isn't a well-defined conditional probability measure at his node. At his node, Bob's type* t^b *assigns probability 1 to* {($Across$, t^a)}, *while his type* u^b *assigns probability 1 to* {($Down$, t^a)}.

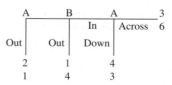

Figure 3.1

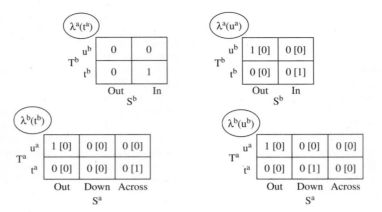

Figure 3.2

This is a simple illustration of the concept of a *conditional probability system* (*CPS*), due to Rényi (1955). A CPS specifies a family of conditioning events E and a measure p_E for each such event, together with certain restrictions on these measures. The interpretation is that p_E is what the player believes, after observing E. Even if $p_\Omega(E) = 0$ (where Ω is the entire space), the measure p_E is still specified. That is, even if E is 'unexpected', the player has a measure if E nevertheless happens. This is why CPS's are well-suited to epistemic analysis of game trees – where we need to be able to describe how players react to the unexpected.

Myerson (1991, ch. 1) provided a preference-based axiomatization of a class of CPS's. Battigalli and Siniscalchi (1999; 2002) further developed both the pure theory and the game-theoretic application of CPS's (see below).

Suppose the true state in Figure 3.2 is (*Down*, t^a, *In*, t^b). In particular, Ann plays *Down*, expecting Bob to play *In*. Bob plays *In*, expecting (at his node) Ann to play *Across*. Ann expects a payoff of 4 (and gets this). Bob expects a payoff of 6 (but gets only 3). In everyday language, we can say that Ann successfully bluffs Bob. (At the state (*Down*, t^a, *In*, t^b), the bluff works. By contrast, at the state (*Down*, t^a, *Out*, u^b), Ann attempts the bluff and it fails.)

But what about epistemic conditions? Are the players rational in this situation? Does each think the other is rational? And so on.

To answer, we need a definition of rationality with CPS's. Fix a strategy-type pair (s^i, t^i), where t^i is associated with a CPS. Call this pair **rational (in the tree)** if the following holds: fix any information set H for i allowed by s^i, and look at the measure on the other players' strategies, given H. (This means given the event that the other players' strategies allow H.) Require that s^i maximizes i's expected payoff under this measure, among all strategies r^i of i that allow H.

With this definition, the rational strategy-type pairs in Figure 3.2 are (*Down*, t^a), (*Out*, u^a), (*In*, t^b), and (*Out*, u^b).

Next, what does Ann think about Bob's rationality? To answer, we need a CPS-analogue to belief (as defined in Section 2). Ben Porath (1997) proposed the following (we have taken the liberty of changing terminology, for consistency with 'strong belief' below): Say player i **initially believes** event E if, under i's CPS, E gets probability 1 at the root of the tree. (Formally, the conditioning event consists of all strategy profiles of the other players.) Battigalli and Siniscalchi (2002) strengthened this definition to: Say player i **strongly believes** event E if, under i's CPS, E gets probability 1 at every information set at which E is possible. Under initial belief, E also gets probability 1 at any information set H that gets positive probability under i's initial measure (that is, i's measure given the root). This is just standard conditioning on non-null events. But under strong belief, this conclusion holds for any information set H which has a non-empty intersection with E – even if H is null under i's initial measure. This is why strong belief is stronger than initial belief.

Let us apply these definitions to Figure 3.2. Does Ann initially believe that Bob is rational? Yes. Both of Ann's types initially believe Bob is rational. Type t^a initially assigns probability 1 to the rational pair (*In*, t^b). Type u^a initially assigns probability 1 to the rational pair (*Out*, u^b). In fact, both types strongly believe Bob is rational. Since, under type t^a, Ann's second node gets positive probability (in fact, probability 1) under her initial measure, we need only check this for type u^a. But at Ann's second node, type u^a assigns probability 1 to the rational pair (*In*, t^b).

Turning to Bob, both of his types initially believe that Ann is rational. Type u^b even strongly believes Ann is rational; but type t^b doesn't. This is because, at Bob's node, type t^b assigns positive probability (in fact, probability 1) to the irrational pair (*Across*, t^a).

Staying with initial belief (we come back to strong belief below), we can parallel Definition 2.1 and define inductively **rationality and *m*th-order initial belief of** rationality (R*m*IBR) at a state of a type structure, and **rationality and common initial belief of rationality (RCIBR)** (see Ben Porath, 1997). In Figure 3.2, since all four types initially believe the other player is rational, a simple induction gives that at the state (*Down, t^a, In, t^b*) for instance, RCIBR holds.

In words, Ann plays across at her first node, believing (initially) that Bob will play *In*, so she can get a payoff of 4. Why would Bob play In? Because he initially believes that Ann plays *Out*. But in the probability-0 event that Ann plays across at her first node, Bob then assigns probability 1 to Ann's playing across at her second node – that is, to Ann's being irrational. He therefore (rationally) plays *In*. All this is consistent with RCIBR.

4 Conditions for backward induction

Interestingly, this is exactly the line of reasoning which, as we said, was the original stimulus for investigating the foundations of BI. So, there is no difficulty with it – we've just seen a formal set-up in which it holds. The resolution of the BI puzzle is simply to accept that the BI path may not result.

But one can also argue that RCIBR is not the right condition: it is too weak. In the above example, Bob realizes that he might be 'surprised' in the play of the game – that's why he has a CPS, not just an ordinary probability measure. If he realizes he might be surprised, should he abandon his (initial) belief that Ann is rational when he is surprised? Bob's type t^b does so. This is the step taken by Battigalli and Siniscalchi (2002) with their concept of strong belief. The argument says that we want t^b to strongly believe, not just initially believe, that Ann is rational. Type t^b will strongly believe Ann is rational if we move the probability-1 weight (in square brackets) on (*Across, t^a*) to (*Down, t^a*). But now (*In, t^b*) isn't rational for Bob, so Ann doesn't (even initially) believe Bob is rational. It looks as if the example unravels.

We can again parallel Definition 2.1 and define inductively **rationality and mth-order strong belief of rationality (R*m*SBR)**, and **rationality and common strong belief of rationality (RCSBR)** (see Battigalli and Siniscalchi, 2002). The question is then: does RCSBR yield BI?

The answer is yes. *Fix a CPS-based type structure for n-legged Centipede* (Figure 4.1), *and a state at which there is RCSBR. Then Ann plays Out.* The result follows from Friedenberg (2002), who shows that in a PI game

(satisfying certain payoff restrictions), RCSBR yields a Nash-equilibrium outcome. In Centipede, there is a unique Nash path and it coincides with the BI path. Of course, this isn't true in general.

Example 4.1 (A second coordination game) *Consider the coordination game in* Figure 4.2 *and the associated CPS-based type structure in* Figure 4.3.

The rational strategy-type pairs are (Out, t^a) and (Out, t^b) for Ann and Bob respectively. Ann's type t^a strongly believes {(Out, t^b)}, and Bob's type t^b strongly believes {(Out, t^a)}. By induction, RCSBR holds at the state (Out, t^a, Out, t^b).

Here, the BI path need not be played under RCSBR. The key is to see that both (*Down, t^a*) and (*Across, t^a*) are irrational for Ann, since she (strongly) believes Bob plays *Out*. So at his node, Bob can't believe Ann is rational. If he considers it sufficiently more likely Ann will play *Down* rather than *Across*, he will rationally play *Out* (as happens). In short, if Ann doesn't play *Out*, she is irrational and so 'all bets are off' as to what she will do. She could play *Down*.

This situation may be surprising, at least at first blush, but there does not appear to be anything conceptually wrong with it. Indeed, it points to an interesting way in which the players in a game can literally be trapped by their beliefs – which here prevent them from getting their mutually preferred (3, 3) outcome.

But one can also argue differently. If Ann forgoes the payoff of 2 she can get by playing *Out* at the first node, then surely she must be playing *Across* to get 3. Playing *Down* to get 0 makes little sense since this is lower than the payoff she gave up at the first node. (This is forward-induction reasoning à la Kohlberg and Mertens, 1986, Section 2.3, introduced in the context of non-PI games. Interestingly, epistemic analysis makes clear that the issue already arises in PI games, such as Figure 4.2.) But if Bob considers *Across* (sufficiently) more likely than

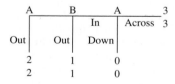

Figure 4.2

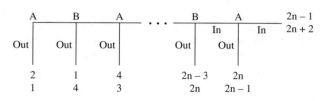

Figure 4.1

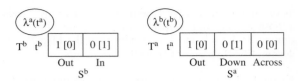

Figure 4.3

Down, he will play *In*. Presumably then, Ann will indeed play *Across*, and the BI path results.

There is no contradiction with the previous analysis because in Figure 4.3 Ann is irrational once she doesn't play *Out*, so we can't say Ann should then rationally play *Across* not *Down*. To make *Across* rational for Ann, we have to add more types to the structure – specifically, we would want to add a second type for Ann that assigns (initial) probability 1 to Bob's playing *In* not *Out*. This key insight is due to Stalnaker (1998) and Battigalli and Siniscalchi (2002).

Battigalli and Siniscalchi formulate a general result of this kind. They consider a **complete** CPS-based type structure, which contains, in a certain sense, every possible type for each player (a complete type structure will be uncountably infinite), and prove: *Fix a complete CPS-based type structure. If there is RCSBR at the state $(s^1, t^1, \ldots, s^n, t^n)$, then the strategy profile $(s^1, \ldots, s^n)$ is extensive-form rationalizable. Conversely, if the profile $(s^1, \ldots, s^n)$ is extensive-form rationalizable, then there is a state $(s^1, t^1, \ldots, s^n, t^n)$ at which there is RCBR.*

The extensive-form rationalizability strategies (Pearce, 1984) yield the BI outcome in a PI game (under an assumption ruling out certain payoff ties; Battigalli, 1997), so the Battigalli and Siniscalchi analysis gives epistemic conditions for BI.

There are other routes to getting BI in PI games. Asheim (2001) develops an epistemic analysis using the properness concept (Myerson, 1978). Go back to Example 4.1. The properness idea says that Bob's type t^b should view (*Across*, t^a) as infinitely more likely than (*Down*, t^a) since *Across* is the less costly 'mistake' for Ann, given her type t^a. Unlike the completeness route taken above, the irrationality of both *Down* and *Across* (given Ann's type t^a) is accepted. But the relative ranking of these 'mistakes' must be in the right order. With this ranking, Bob is irrational to play *Out* rather than *In*. Ann presumably will play *Across*, and we get BI again. Asheim (2001) formulates a general such result.

Another strand of the literature on BI employs knowledge models rather than belief models. As pointed out in Example 1.1, players' beliefs don't have to be correct in any sense. For example, a type might even assign probability 1 to a strategy-type pair for another player different from the actual one. Knowledge as usually formalized is different, in that if a player knows an event E, then E indeed happens.

Aumann (1995) formulates a knowledge-based epistemic model for PI trees. In his set-up, the condition of common knowledge of rationality implies that the players choose their BI strategies. Stalnaker (1996) finds that non-BI outcomes are possible, under a different formulation of the same condition. The explanation lies in differences in how counterfactuals are treated. These play an important role in a knowledge-based analysis, when we talk about what a player thinks at an information set that cannot be reached given what he knows. Halpern (2001) provides a synthesis in which these differences can be understood. See also the exchange between Binmore (1996) and Aumann (1996), and the analyses by Samet (1996), Balkenborg and Winter (1997), and Halpern (1999).

Aumann (1998) provides knowledge-based epistemic conditions under which Ann plays *Out* in Centipede. The conditions are weaker than in his (1995) paper, and the conclusion weaker (about outcomes not strategies). There is an obvious parallel between this result and the belief-based result on Centipede we stated above (also about outcomes). More generally, there may be an analogy between counterfactuals in knowledge models and extended probabilities in belief models. But, for one thing, completeness is crucial to the belief-based approach, as we have seen, and an analogous concept does not appear to be present in the knowledge-based approach. As yet, there does not appear to be any formal treatment of the relationship between the two approaches.

5 Next steps: weak dominance

Extending the epistemic analysis of games from the matrix to the tree has been the focus of much recent work in the literature. Another area has been extending the analysis on the matrix from strong dominance (described in Section 2) to weak dominance.

Weak dominance (admissibility) says that a player considers as possible (even if unlikely) any of the strategies for the other players. In the game context, we are naturally led to consider *iterated admissibility* (*IA*) – the weak-dominance analogue to IU. This is an old concept in game theory, going back at least to Gale (1953). Like BI, it is a powerful solution concept, delivering sharp answers in many games – Bertrand, auctions, voting games, and others. (Mertens, 1989, p. 582, and Marx and Swinkels, 1997, pp. 224–5, list various games involving weak dominance.)

But, also like BI, there is a conceptual puzzle. Suppose Ann conforms to the admissibility requirement, so that she considers possible any of Bob's strategies. Suppose Bob also conforms to the requirement, and this leads him not to play a strategy, say *L*. If Ann thinks Bob adheres to the requirement (as he does), then she can rule out Bob's playing *L*. But this conflicts with the requirement that she not rule anything out (see Samuelson, 1992).

Can a sound argument be made for IA? To investigate this, the epistemic tools of Section 1 have to be extended again.

Example 5.1 (**Bertrand**) Figure 5.1 *is a Bertrand pricing game, where each firm chooses a price in {0, 1, 2, 3}.* (*Ken Corts kindly provided this example.*) *The left payoff is to A, the right payoff to B. Each firm has capacity of two units and zero cost. Two units are demanded. If the firms charge the same price, they each sell one unit.* Figure 5.2 *is an associated type structure* (*with one type for each player*).

The rational strategy-type pairs are $R_1^a = \{0, 1, 2, 3\} \times \{t^a\}$ and $R_1^b = \{0, 1, 2, 3\} \times \{t^b\}$. Since both types assign positive probability only to a rational strategy-type pair for the other player, we get $R_m^a = R_1^a$ and $R_m^b = R_1^b$ for all m. In particular, there is RCBR at the state $(3, t^a, 3, t^b)$.

But a price of 3 is inadmissible (as is a price of 0). The IA set is just $\{(1,1)\}$, where each firm charges the lowest price above cost. (This is a plausible scenario: while pricing at cost is inadmissible, competition forces price down to the first price above cost.)

A tool to incorporate admissibility is *lexicographic probability systems* (*LPS's*), introduced and axiomatized by Blume, Brandenburger and Dekel (1991a; 1991b). An LPS specifies a sequence of probability measures. The interpretation is that the first measure is the player's primary hypothesis about the true state. But the player recognizes that his primary hypothesis might be mistaken, and so also forms a secondary hypothesis. This is his second measure. Then his tertiary hypothesis, and so on. The primary states can be thought of as infinitely more likely than the secondary states, which are infinitely more likely than the tertiary states, and so on. Stahl (1995), Stalnaker (1998), Asheim (2001), Brandenburger,

Friedenberg and Keisler (2006), and Asheim and Perea (2005), among other papers, use LPS's.

Example 5.2 (**Bertrand contd.**) Figure 5.3 *is a type structure for Bertrand (Figure 5.1) that now specifies LPS's.*

Each player has a primary hypothesis which assigns probability 1 to the other player's charging a price of 0. But each player also has a secondary hypothesis that assigns equal probability to each of the three remaining choices for the other player. This measure is shown in parentheses. Note that every state (that is, strategy-type pair) gets positive probability under some measure. But states can also be ruled out, in the sense that they can be give infinitely less weight than other states.

What about epistemic conditions? Are the players rational in this situation? Does each think the other is rational? And so on.

To answer, we need a definition of rationality with LPS's. Fix strategy-type pairs (s^i, t^i) and (r^i, t^i) for player i, where t^i is now associated with an LPS. Calculate the tuple of expected payoffs to i from s^i, using first the primary measure associated with t^i, then the secondary measure associated with t^i, and so on. Calculate the corresponding tuple for r^i. If the first tuple lexicographically exceeds the second, then s^i is preferred to r^i. (If $x = (x_1, \ldots, x_n)$ and $y = (y_1, \ldots, y_n)$, then x lexicographically exceeds y if $y_j > x_j$ implies $x_k > y_k$ for some $k < j$.) A strategy-type pair (s^i, t^i) is **rational** (**in the lexicographic sense**) if s^i is maximal under this ranking.

So $(3, t^a)$ and $(3, t^b)$ are irrational. All choices give each player an expected payoff of 0 under the primary measure. But a price of 2 gives each player an expected payoff of 2 under the secondary measure, as opposed to an expected payoff of 1 from a price of 3. Conceptually, we want $(3, t^a)$ and $(3, t^b)$ to be irrational (because a price of 3 is inadmissible).

What does each player think about the other's rationality? For this, we again need an LPS-based definition. An early candidate in the literature was: Say player i **believes** event E **at the 1st level** if E gets primary probability 1 under i's LPS (Börgers, 1994; Brandenburger, 1992). A stronger concept is: Say i **assumes** E if all states not in E are infinitely less likely than all states in E, under i's LPS (Brandenburger, Friedenberg and Keisler, 2006). In other words, a player who assumes E recognizes E may not happen, but is prepared to 'count on' E versus not-E.

In Figure 5.3, type t^a doesn't 1st-level believe (so certainly doesn't assume) the other player is rational. Likewise with t^b. Again, this is right conceptually.

	B			
	3	2	1	0
3	3, 3	0, 4	0, 2	0, 0
2	4, 0	2, 2	0, 2	0, 0
1	2, 0	2, 0	1, 1	0, 0
0	0, 0	0, 0	0, 0	0, 0

(A labels the rows)

Figure 5.1

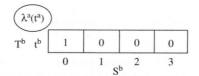

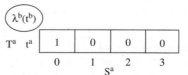

Figure 5.2

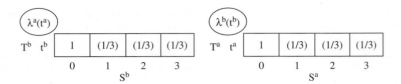

Figure 5.3

6 Conditions for iterated admissibility

Once again we can parallel Definition 2.1 and define inductively **rationality and *m*th-order 1st-level belief of rationality (R*m*1BR)** at a state of a type structure, and **rationality and common 1st-level belief of rationality (RC1BR)**. Likewise, one can define **rationality and *m*th-order assumption of rationality (R*m*AR)**, and **rationality and common assumption of rationality (RCAR)**. What do these conditions yield?

In fact, just as we saw in Sections 3 and 4 that neither RCIBR not RCSBR yields BI, so neither RC1BR nor RCAR yields IA. RC1BR is characterized by the $S^\infty W$ concept (Dekel and Fudenberg, 1990), that is , the set of strategies that remain after one round of deletion of inadmissible strategies followed by iterated deletion of strongly dominated strategies. RCAR is characterized by the self-admissible set concept (Brandenburger, Friedenberg and Keisler, 2006). Self-admissible sets may be viewed as the weak-dominance analogue to Pearce (1984) best-response sets.

But while the IA set is one self-admissible set in a game, there may well be others. To select the IA set, a completeness assumption is needed, similar to Section 4: *Fix a complete LPS-based type structure. If there is RmAR at the state $(s^1, t^1, ..., s^n, t^n)$, then the strategy profile $(s^1, ..., s^n)$ survives $(m+1)$ rounds of iterated admissibility. Conversely, if the profile $(s^1, ..., s^n)$ survives $(m+1)$ rounds of iterated admissibility, then there is a state $(s^1, t^1, ..., s^n, t^n)$ at which there is RmAR* (Brandenburger, Friedenberg and Keisler, 2006).

This result is stated for R*m*AR and not RCAR. See the next section for the reason. Of course, for a given game, there is an *m* such that IA stabilizes after *m* rounds.

IA yields the BI outcome in a PI game (again ruling out certain payoff ties; Marx and Swinkels, 1997), so, understanding IA gives, in particular, another analysis of BI.

Related analyses of IA include Stahl (1995) and Ewerhart (2002). Stahl uses LPS's and directly assumes that Ann considers one of Bob's strategies infinitely less likely than another if the first is eliminated on an earlier round of IA than the second. Ewerhart gives an analysis of IA couched in terms of provability (from mathematical logic).

7 Strategic versus extensive analysis

Kohlberg and Mertens (1986, Section 2.4) argued that a 'fully rational' analysis of games should be invariant – that is , should depend only on the fully reduced strategic form of a game. (This is the strategic form after elimination of any – pure – strategies that are duplicates or convex combinations of other strategies.) In this, they appealed to early results in game theory (Dalkey, 1953; Thompson, 1952) which established that two trees sharing the same reduced strategic form differ from each other by a (finite) sequence of elementary transformations of the tree, each of which can be argued to be 'strategically inessential'. Kohlberg and Mertens added a fourth transformation involving convex combinations, to get to the fully reduced strategic form.

In decision theory, invariance is implied by (and implies) admissibility. (Kohlberg and Mertens, 1986, Section 2.7, gave the essential idea. See Brandenburger, 2007, for the decision-theory argument.) If we build up our game analysis using a decision theory that satisfies admissibility, we can hope to get invariance at this level too. LPS-based decision theory satisfies admissibility. Indeed, IA, and also the $S^\infty W$ and self-admissible set concepts, are invariant in the Kohlberg–Mertens sense. The extensive-form rationalizability concept (Section 4) is not.

There does appear to be a price paid for invariance, however. The extensive-form conditions of RCSBR and (CPS-based) completeness are consistent (in any tree). That is, for any tree, we can build a complete type structure and find a state at which RCSBR holds. But Brandenburger, Friedenberg and Keisler (2006) show the strategic-form conditions of RCAR and (LPS-based) completeness are inconsistent (in any matrix satisfying a non-triviality condition).

A possible interpretation is that rationality, even as a theoretical concept, appears to be inherently limited. There are purely theoretical limits to the Kohlberg-Mertens notion of a 'fully rational' analysis of games.

The epistemic programme has uncovered a number of impossibility results (see EPISTEMIC GAME THEORY: BELIEFS AND TYPES for some others). We don't see this as a deficiency of the programme, but rather as a sign it has reached a certain depth and maturity. Also, central to the programme is the analysis of scenarios (we have seen several in this survey) that are 'a long way from' these theoretical limits. Under the epistemic approach to game theory there is not one right set of assumptions to make about a game.

ADAM BRANDENBURGER

See also **epistemic game theory: an overview; epistemic game theory: beliefs and types; epistemic game theory: incomplete information; game theory; Nash equilibrium, refinements of.**

This survey is based on Brandenburger (2007). I am grateful to Springer for permission to use this material. I owe a great deal to joint work and many conversations with Robert Aumann, Eddie Dekel, Amanda Friedenberg, Jerry Keisler and Harborne Stuart. My thanks to Konrad Grabiszewski for important input, John Nachbar for very important editorial advice, and Michael James for valuable assistance. The Stern School of Business provided financial support.

Bibliography

Asheim, G. 2001. Proper rationalizability in lexicographic beliefs. *International Journal of Game Theory* 30, 453–78.

Asheim, G. and Perea, A. 2005. Sequential and quasi-perfect rationalizability in extensive games. *Games and Economic Behavior* 53, 15–42.

Aumann, R. 1974. Subjectivity and correlation in randomized strategies. *Journal of Mathematical Economics* 1, 67–96.

Aumann, R. 1987. Correlated equilibrium as an expression of Bayesian rationality. *Econometrica* 55, 1–18.

Aumann, R. 1995. Backward induction and common knowledge of rationality. *Games and Economic Behavior* 8, 6–19.

Aumann, R. 1996. Reply to Binmore. *Games and Economic Behavior* 17, 138–46.

Aumann, R. 1998. On the centipede game. *Games and Economic Behavior* 23, 97–105.

Aumann, R. 1998. Common priors: a reply to Gul. *Econometrica* 66, 929–38.

Aumann, R. and Brandenburger, A. 1995. Epistemic conditions for Nash equilibrium. *Econometrica* 63, 1161–80.

Balkenborg, D. and Winter, E. 1997. A necessary and sufficient epistemic condition for playing backward induction. *Journal of Mathematical Economics* 27, 325–45.

Basu, K. 1990. On the existence of a rationality definition for extensive games. *International Journal of Game Theory* 19, 33–44.

Battigalli, P. 1997. On rationalizability in extensive games. *Journal of Economic Theory* 74, 40–61.

Battigalli, P. and Siniscalchi, M. 1999. Hierarchies of conditional beliefs and interactive epistemology in dynamic games. *Journal of Economic Theory* 88, 188–230.

Battigalli, P. and Siniscalchi, M. 2002. Strong belief and forward-induction reasoning. *Journal of Economic Theory* 106, 356–91.

Ben Porath, E. 1997. Rationality, Nash equilibrium, and backward induction in perfect information games. *Review of Economic Studies* 64, 23–46.

Bernheim, D. 1984. Rationalizable strategic behavior. *Econometrica* 52, 1007–28.

Bicchieri, C. 1988. Strategic behavior and counterfactuals. *Synthese* 76, 135–69.

Bicchieri, C. 1989. Self-refuting theories of strategic interaction: a paradox of common knowledge. *Erkenntnis* 30, 69–85.

Binmore, K. 1987. Modelling rational players I. *Economics and Philosophy* 3, 179–214.

Binmore, K. 1996. A note on backward induction. *Games and Economic Behavior* 17, 135–7.

Blume, L., Brandenburger, A. and Dekel, E. 1991a. Lexicographic probabilities and choice under uncertainty. *Econometrica* 59, 61–79.

Blume, L., Brandenburger, A. and Dekel, E. 1991b. Lexicographic probabilities and equilibrium refinements. *Econometrica* 59, 81–98.

Bonanno, G. 1991. The logic of rational play in games of perfect information. *Economics and Philosophy* 7, 37–65.

Bonanno, G. and Nehring, K. 1999. How to make sense of the common prior assumption under incomplete information. *International Journal of Game Theory* 28, 409–34.

Börgers, T. 1994. Weak dominance and approximate common knowledge. *Journal of Economic Theory* 64, 265–76.

Brandenburger, A. 1992. Lexicographic probabilities and iterated admissibility. In *Economic Analysis of Markets and Games*, ed. P. Dasgupta, D. Gale, O. Hart and E. Maskin. Cambridge, MA: MIT Press.

Brandenburger, A. 2007. The power of paradox: some recent results in interactive epistemology. *International Journal of Game Theory* 35, 465–92.

Brandenburger, A. and Dekel, E. 1987. Rationalizability and correlated equilibria. *Econometrica* 55, 1391–402.

Brandenburger, A., Friedenberg, A. and Keisler, H.J. 2006. Admissibility in games. Unpublished, Stern School of Business, New York University.

Dalkey, N. 1953. Equivalence of information patterns and essentially determinate games. In *Contributions to the Theory of Games*, vol. 2, ed. H. Kuhn and A. Tucker. Princeton: Princeton University Press.

Dekel, E. and Fudenberg, D. 1990. Rational behavior with payoff uncertainty. *Journal of Economic Theory* 52, 243–67.

Ewerhart, C. 2002. Ex-ante justifiable behavior, common knowledge, and iterated admissibility. Unpublished, Department of Economics, University of Bonn.

Feinberg, Y. 2000. Characterizing common priors in terms of posteriors. *Journal of Economic Theory* 91, 127–79.

Friedenberg, A. 2002. When common belief is correct belief. Unpublished, Olin School of Business, Washington University.

Gale, D. 1953. A theory of *n*-person games with perfect information. *Proceedings of the National Academy of Sciences* 39, 496–501.

Gul, F. 1998. A comment on Aumann's Bayesian view. *Econometrica* 66, 923–7.

Halpern, J. 1999. Hypothetical knowledge and counterfactual reasoning. *International Journal of Game Theory* 28, 315–30.

Halpern, J. 2001. Substantive rationality and backward induction. *Games and Economic Behavior* 37, 425–35.

Halpern, J. 2002. Characterizing the common prior assumption. *Journal of Economic Theory* 106, 316–55.

Harsanyi, J. 1967–8. Games with incomplete information played by 'Bayesian' players, I–III. *Management Science* 14, 159–82, 320–34, 486–502.

Harsanyi, J. 1973. Games with randomly disturbed payoffs: a new rationale for mixed strategy equilibrium points. *International Journal of Game Theory* 2, 1–23.

Kohlberg, E. and Mertens, J.-F. 1986. On the strategic stability of equilibria. *Econometrica* 54, 1003–37.

Marx, L. and Swinkels, J. 1997. Order independence for iterated weak dominance. *Games and Economic Behavior* 18, 219–45.

Mertens, J.-F. 1989. Stable equilibria – a reformulation. *Mathematics of Operations Research* 14, 575–625.

Morris, S. 1994. Trade with heterogeneous prior beliefs and asymmetric information. *Econometrica* 62, 1327–47.

Myerson, R. 1978. Refinements of the Nash equilibrium concept. *International Journal of Game Theory* 1, 73–80.

Myerson, R. 1991. *Game Theory*. Cambridge, MA: Harvard University Press.

Pearce, D. 1984. Rational strategic behavior and the problem of perfection. *Econometrica* 52, 1029–50.

Reny, P. 1992. Rationality in extensive form games. *Journal of Economic Perspectives* 6(4), 103–18.

Rényi, A. 1955. On a new axiomatic theory of probability. *Acta Mathematica Academiae Scientiarum Hungaricae* 6, 285–335.

Rosenthal, R. 1981. Games of perfect information, predatory pricing and the chain-store paradox. *Journal of Economic Theory* 25, 92–100.

Samet, D. 1996. Hypothetical knowledge and games with perfect information. *Games and Economic Behavior* 17, 230–51.

Samet, D. 1998. Common priors and the separation of convex sets. *Games and Economic Behavior* 24, 172–4.

Samet, D. 1998. Iterated expectations and common priors. *Games and Economic Behavior* 24, 131–41.

Samuelson, L. 1992. Dominated strategies and common knowledge. *Games and Economic Behavior* 4, 284–313.

Stahl, D. 1995. Lexicographic rationalizability and iterated admissibility. *Economic Letters* 47, 155–9.

Stalnaker, R. 1996. Knowledge, belief and counterfactual reasoning in games. *Economics and Philosophy* 12, 133–63.

Stalnaker, R. 1998. Belief revision in games: forward and backward induction. *Mathematical Social Sciences* 36, 31–56.

Tan, T. and Werlang, S. 1988. The Bayesian foundations of solution concepts of games. *Journal of Economic Theory* 45, 370–91.

Thompson, F. 1952. Equivalence of games in extensive form. Research Memorandum RM-759. The RAND Corporation.

epistemic game theory: incomplete information

A game of incomplete information is a game in which at least some of the players possess private information which may be relevant to the strategic interaction. The private information of a player may be about the payoff functions in the game, as well as about some exogenous, payoff-irrelevant events. The player may also form beliefs about other players' beliefs about payoffs and exogenous events, about their beliefs about the beliefs of others, and so forth.

Harsanyi (1967–8) introduced the idea that such a state of affairs can be succinctly described by a *type space*. With this formulation, T_i denotes the set of player i's *types*. Each type $t_i \in T_i$ is associated with a belief $\lambda_i(t_i) \in \Delta(K \times T_{-i})$ about some basic space of uncertainty, K, and the combination T_{-i} of the other players' types. The basic space of uncertainty K is called the space of *states of nature*, and $\Omega = K \times \Pi_{i \in I} T_i$, where I is the set of players, is called the space of *states of the world*.

A type space models a game of incomplete information once each state of nature $k \in K$ is associated with a payoff matrix of the game, or, more generally, with a payoff function u_i^k for each player $i \in I$. This payoff function specifies the player's payoff $u_i^k(s)$ for each combination of strategies $s = (s_i)_{i \in I} \in S = \Pi_{i \in I} S_i$ of the players. (In the particular case in which k is associated with a payoff matrix, that is, the game is such that each player has finitely many strategies, the payoffs $u_i^k(s)$ to the players $i \in I$ appear in the entry of the matrix corresponding to the combination of strategies $s = (s_i)_{i \in I}$.) As usual, the set of strategies S_i of player $i \in I$ may be a complex object by itself. For instance, it may be the set of mixed strategies over some set of pure strategies S_i^0. The payoff function of player i in the state of nature k is $u_i^k : S \to \mathbb{R}$.

Obviously, different types of a player may want to choose different strategies. Thus, a *Bayesian strategy* of player i in a game of incomplete information specifies the strategy $\sigma_i(t_i) \in S_i$ that the player chooses given each one of her types $t_i \in T_i$.

Given a profile of Bayesian strategies $\sigma = (\sigma_j : T_j \to S_j)_{j \in I}$ of the players, the expected payoff of player i of type t_i is

$$U_i(\sigma, t_i) = \sum_{(k,t_{-i}) \in K \times T_{-i}} u_i^k(\sigma_i(t_i), \sigma_{-i}(t_{-i})) \times \lambda_i(t_i)(k, t_{-i})$$

where $\sigma_{-i}(t_{-i}) = (\sigma_j(t_j))_{j \neq i}$. If there is a continuum of states of nature and types, the sum becomes an integral:

$$U_i(\sigma, t_i) = \int_{K \times T_{-i}} u_i^k(\sigma_i(t_i), \sigma_{-i}(t_{-i})) d\lambda_i(t_i)(k, t_{-i})$$

(In this case, the expected payoff function $U_i(\sigma, t_i)$ is well defined if the Bayesian strategies $\sigma_j : T_j \to S_j$ are measurable functions and if the payoff function $u_i : K \times S \to R$ is measurable as well; we omit the details of this technical requirement.)

We assume that the players are expected payoff maximizers. Thus, player i prefers the Bayesian strategy σ over σ' if and only if $U_i(\sigma, t_i) \geq U_i(\sigma', t_i)$ for each of her types $t_i \in T_i$. It follows that given a Bayesian strategy profile σ_{-i} of the other players, the Bayesian strategy σ_i is a *best reply* of player i if for any other strategy σ'_i of hers, $U_i((\sigma_i, \sigma_{-i}), t_i) \geq U_i((\sigma', \sigma_{-i}), t_i)$ for each of her types $t_i \in T_i$. A *Bayes–Nash equilibrium* or a *Bayesian equilibrium* is a profile of Bayesian strategies $\sigma^* = (\sigma^*_i)_{i \in I}$ such that σ^*_i is a best reply against σ^*_{-i} for every player $i \in I$.

A simple, discrete variant of an example by Gale (1996) may clarify these abstract definitions. There are two investors $i = 1, 2$ and three possible states of nature $k \in K = \{-1, 0, 1\}$. Each investor i only knows her own type

$$t_i \in T_i = \{-10, -6, -2, 2, 6, 10\}.$$

Every type t_i of investor i believes that all of the other investor's types $t_j \in T_j$, $j \neq i$, are equally likely, so that each of them has probability $\frac{1}{6}$. Moreover, every type t_i believes that the state of nature is $k = 1$ when $t_i + t_j > 0$; that the state of nature is $k = 0$ when $t_i + t_j = 0$; and that the state of nature is $k = -1$ when $t_i + t_j < 0$. Formally, the belief $\lambda_i(t_i)$ of type $t_i \in T_i$ is defined by

$$\lambda_i(t_i)(k, t_j) = \begin{cases} \frac{1}{6} & k \text{ has the same sign as } t_i + t_j \\ 0 & \text{otherwise.} \end{cases}$$

The investors cannot communicate their types to one another. They can invest in at most one of two available investment periods. Each investor has three relevant strategies: invest *immediately*, in the first period; *wait* to the second period and invest only if the other investor has invested in the first period; or *never* invest. The payoff of each of the investors depends on the state of nature $k \in K = \{-1, 0, 1\}$ and on her own investment strategy, but not on the investment strategy of the other investor. The payoffs are as follows:

- Investing *immediately* when the state of nature is k yields investor i a payoff of k

  $$u^k_i(immediately', \cdot) = k$$

  (The $\cdot$ stands for the investment decision of the other investor $j \neq i$, which, as we said, does not effect the payoff of investor i.)
- If investor i chooses to *wait* to the second period and invest only if the other investor has invested in the first

period, investor i's payoff in the state of nature k is

$$u^k_i(wait', \cdot) = \frac{3}{4}k.$$

- If the investor *never* invests, her payoff is 0 irrespective of the state of nature:

  $$u^k_i(never', \cdot) = 0.$$

How will the different types behave at a Bayesian equilibrium?

The type $t_i = 10$ assesses that by investing immediately her expected payoff is

$$U_i(immediately', 10) = \frac{1}{6} \times 0 + \frac{5}{6} \times 1 = \frac{5}{6}$$

(immediate investment yields 0 in case $t_j = -10$, and yields 1 in case $t_j = -6, -2, 2, 6, 10$). This is higher than $\frac{3}{4}$, the maximum payoff she could possibly get by waiting for the second period, and higher than the payoff 0 of never investing. So at a Bayesian equilibrium

$$\sigma^*_i(10) = immediately', \quad i = 1, 2.$$

Next, the expected payoff to the type $t_i = 6$ from immediate investment is

$$U_i(immediately', 6) = \frac{1}{6} \times (-1) + \frac{1}{6} \times 0 + \frac{4}{6} \times 1 = \frac{1}{2}$$

(immediate investment yields 1 unless $t_j = -10$, in which case the payoff is -1, or $t_j = -6$, in which case the payoff is 0). So investing immediately is preferred for her over never investing. But how about waiting until the second period? That's an inferior option as well, since the types $t_j = -10, -6, -2$ will never invest in the first period (this would yield them a negative expected payoff). So only the positive types $t_j = 2, 6, 10$ could *conceivably* invest immediately, with overall probability reaching at most $\frac{3}{6}$. So waiting to see if they invest yields to the type $t_i = 6$ an expected payoff not higher than $\frac{3}{6} \times \frac{3}{4} = \frac{3}{8}$, which is smaller than $\frac{1}{2}$. We conclude that the preferable strategy of $t_i = 6$ at equilibrium is

$$\sigma^*_i(6) = immediately', \quad i = 1, 2.$$

What about $t_i = 2$? Immediate investment yields her

$$U_i(immediately', 2) = \frac{2}{6} \times (-1) + \frac{1}{6} \times 0 + \frac{3}{6} \times 1 = \frac{1}{6}$$

(-1 is the payoff when $t_j = -10, -6$; 0 is the payoff when $t_j = -2$; the payoff is 1 otherwise). However, given that the types $t_j = 6, 10$ invest immediately at equilibrium, and that the negative types $t_j = -10, -6, -2$ do not invest immediately, the type $t_i = 2$ figures out that by waiting and investing only if the other investor has invested first would yield her an expected payoff

$$U_i(wait', 2) \geq \frac{2}{6} \times \frac{3}{4} = \frac{1}{4} > \frac{1}{6}$$

($\frac{2}{6}$ is the probability assigned by $t_i = 2$ to the event that $t_j \in \{6, 10\}$ and hence j invests immediately, and $\frac{3}{4}$ is the payoff from the second period investment). The preferred strategy of $t_i = 2$ at equilibrium is therefore

$$\sigma_i^*(2) = wait', \quad i = 1, 2.$$

We can now compute inductively, in a similar way, that also

$$\sigma_i^*(-2) = wait', \quad i = 1, 2$$
$$\sigma_i^*(-6) = wait', \quad i = 1, 2$$

and that

$$\sigma_i^*(-10) = never', \quad i = 1, 2.$$

Notice that the equilibrium in the example is inefficient. For instance, when the pair of types is $(t_1, t_2) = (2, 2)$ the investment is profitable, but both investors wait to see if the other one invests, and thus end up not investing at all. In this case, behaviour would become efficient if the investors could communicate their types to each other. Indeed, they would have been happy to do so, because their interests are aligned.

Obviously, there are other strategic situations with incomplete information in which the interests of the players are not completely aligned. For example, a potential seller of an object would like to strike a deal with a potential buyer at a price which is as high as possible, while the potential buyer would like the price to be as low as possible. That's why the traders might not volunteer to communicate honestly their private valuations of the object, even if they are technically able to do so. Still, in case the buyer values the object more than the seller, they would both prefer to trade at some price in-between their valuations rather than forgoing trade altogether. Therefore, the traders would nevertheless like to avoid a complete lack of communication. Myerson and Satterthwaite (1983) phrase general conditions under which no Bayesian equilibrium of any trade mechanism is ever fully efficient due to this tension between interests alignment and interests mismatch. Under these conditions, even if the traders are able to communicate their private information, at no Bayesian equilibrium does trade take place in all instances in which there exist gains from trade.

In the above variant of Gale's example we were able to find the unique Bayesian equilibrium using iterative dominance arguments. We have iteratively crossed out strategies that are inferior for some types, which enabled us to eliminate inferior strategies for other types, and so

forth. As in games of complete information, this technique is not applicable in general, and there are games with incomplete information in which a Bayesian equilibrium is not the outcome of any process of iterative elimination of dominated strategies (Battigalli and Siniscalchi, 2003; Dekel, Fudenberg and Morris, 2007).

Games with incomplete information are discussed in many game theory textbooks (for example, Dutta, 1999; Gibbons, 1992; Myerson, 1991; Osborne, 2003; Rasmusen, 1989; Watson, 2002). Aumann and Heifetz (2002), Battigalli and Bonanno (1999) and Dekel and Gul (1997) are advanced surveys.

AVIAD HEIFETZ

See also **epistemic game theory: an overview; epistemic game theory: beliefs and types; epistemic game theory: complete information; game theory.**

Bibliography

Aumann, R.J. and Heifetz, A. 2002. Incomplete information. In *Handbook of Game Theory*, vol. 3, ed. R.J. Aumann and S. Hart. Amsterdam: North-Holland.

Battigalli, P. and Bonanno, G. 1999. Recent results on belief, knowledge and the epistemic foundations of game theory. *Research in Economics* 53, 149–225.

Battigalli, P. and Siniscalchi, M. 2003. Rationalization with incomplete information. *Advances in Theoretical Economics* 3(1), article 3. Online. Available at http://www.bepress.com/bejte/advances/vol3/iss1/art3, accessed 25 April 2007.

Dekel, E., Fudenberg, D. and Morris, S. 2007. Interim correlated rationalizability. *Theoretical Economics* 2, 15–40.

Dekel, E. and Gul, F. 1997. Rationality and knowledge in game theory. In *Advances in Economics and Econometrics*, ed. D. Kreps and K. Wallis. Cambridge, UK: Cambridge University Press.

Dutta, P.K. 1999. *Strategies and Games: Theory and Practice*. Cambridge, MA: MIT Press.

Gale, D. 1996. What have we learned from social learning? *European Economic Review* 40, 617–28.

Gibbons, R. 1992. *Game Theory for Applied Economists*. Princeton: Princeton University Press.

Harsanyi, J.C. 1967–8. Games with incomplete information played by Bayesian players, parts I–III. *Management Science* 14, 159–82, 320–34, 486–502.

Myerson, R. 1991. *Game Theory: Analysis of Conflict*. Cambridge, MA: Harvard University Press.

Myerson, R. and Satterthwaite, M. 1983. Efficient mechanisms for bilateral trading. *Journal of Economic Theory* 29, 265–81.

Osborne, M. 2003. *Introduction to Game Theory*. Oxford: Oxford University Press.

Rasmusen, E. 1989. *Games and Information: An Introduction to Game Theory*. Oxford: Basil Blackwell.

Watson, J. 2002. *Strategy: An Introduction to Game Theory*. New York: W.W. Norton.